Handbook of Human Symbol

This book would not have been possible were it not for three remarkable women: Bonnie O'Brien, Val Service, and one whose name we do not know, but who may have lived in Africa between 100 000 and 200 000 years ago, and is often referred to as 'Eve' (see Campbell, this volume, Chapter 2, and Waddell and Penny, this volume, Chapter 3). This book is for them.

Handbook of Human Symbolic Evolution

Edited by
Andrew Lock
and Charles R. Peters

BLACKWELL
Publishers

This edition copyright © Blackwell Publishers Ltd, 1999

First published in hardback 1996 by Oxford University Press. Hardback edition copyright © Andrew Lock, Charles R. Peters and contributors listed on pp. xxviii to xxix, 1996

Paperback edition published 1999 by Blackwell Publishers Ltd

2 4 6 8 10 9 7 5 3 1

Blackwell Publishers Ltd
108 Cowley Road
Oxford OX4 1JF
UK

Blackwell Publishers Inc.
350 Main Street
Malden, Massachusetts 02148
USA

All rights reserved. Except for the quotation of short passages for the purposes of criticism and review, no part of this publication may be reproduced, stored in a retrieval system, or transmitted, in any form or by any means, electronic, mechanical, photocopying, recording or otherwise, without prior permission of the publisher.

Except in the United States of America, this book is sold subject to the condition that it shall not, by way of trade or otherwise, be lent, resold, hired out, or otherwise circulated without the publisher's prior consent in any form of binding or cover other than that in which it is published and without a similar condition including this condition being imposed on the subsequent purchaser.

British Library Cataloguing in Publication Data

A CIP catalogue record for this book is available from the
British Library.

Library of Congress Cataloging-in-Publication Data

Handbook of human symbolic evolution / edited by Andrew Lock and Charles R. Peters.
   p.   cm.
Originally published: Oxford: Clarendon Press: New York: Oxford
University Press, 1996, in series: Oxford science publications.
Includes bibliographical references and index.
ISBN 0-631-21690-1 (pbk.: alk. paper)
1. Symbolism (Psychology) 2. Symbolism--History 3.Psycholinguistics. 4. Biolinguistics
I. Lock. Andrew. II. Peters, Charles R.
[BF458. H25 1999]
302.2--dc21   99-37853
                                                                                      CIP

ISBN 0-198-52153-7 (hbk)
ISBN 0-631-21690-1 (pbk)

Typeset by Pure Tech Corporation, Pondicherry, India
Printed in Great Britain by TJ International Ltd., Padstow, Cornwall

This book is printed on acid-free paper

Preface to the paperback edition

We 'prefaced the preface' of the hardback edition of this volume with a reproduction of the 1866 Statutes of the Societé de Linguistique de Paris that banned contributions on the origins of language at its meetings. While that ban may never be repealed, we also noted that there has remained a large interest in the issue. The current level of interest in questions of human origins may be dated back to the popular books by Robert Ardrey in the late 1950s and early 1960s, and Jacob Bronowski's 1970s TV serial *The Ascent of Man*. Both media continue to address the question, and new media, such as the World Wide Web, add further resources and repositories of information and speculation.

For example, in May 1999 a search of the catalogue of Amazon Books using the key phrase 'human evolution' returns a listing of 1,197 available titles, of which 265 have been published since the beginning of 1996. These figures are clearly an underestimation of what is available, since that phrase does not turn up this volume, although it is in the catalogue (with a '5 star rating', no less). Similarly, a search of web sites on the Hotbot search engine returns 7,750 sites that deal with various aspects of 'human evolution', and that leaves out topics concerned with the origins of language, paleolithic art, and so on.

The aim of this book is to create a reference work that sets out and evaluates the basic knowledge and theory relevant to the evolutionary origin of human symbolic behaviour that has accumulated in scientific literature, especially over the last few decades. It is a compendium, and a guide, to general topics that might help in understanding human symbolic evolution.

Fortunately, science never rests. New findings are made, new interpretations are offered. Thus you will not find mention here of recent finds, such as a *Homo* jaw bone from Hadar dated at 2.33 million years ago; *Homo antecessor* from the Atapuerca hills of Northern Spain, dated around 800,000 years ago, and hypothesized as a common ancestor of both *H. sapiens* and *H. neanderthalensis;* early stone tools at Gona in Ethiopia, dated at 2.5 million years old; or the Chauvet-Pont-d'Arc Cave in France, nor discussions of their possible significance. This *Handbook* does, however, provide an overview of issues against which such new discoveries can be evaluated.

To keep the material it surveys updated is a task beyond the possibilities of traditional publication technologies. Any multi-authoured work that spans such a range of material has a production time-span that guarantees it will lag behind the most recent newsworthy finds. New electronic media allow this time-span to be reduced. Consequently, we will attempt to update the material periodically using the World Wide Web. Abstracts to the chapters and links to emerging resources can be accessed via Blackwell's site at http://www.blackwellpublishers.co.uk and also, at the time of going to press, at http://www.massey.ac.nz/ ALock/hbook/start.htm.

We are grateful to Philip Carpenter and Martin Davies at Blackwell's for their support in making this volume available at an affordable price. We are grateful to Oxford University Press for their efforts in enabling the original publication of the volume in hardback, and for assigning their rights to Blackwell Publishers so that this paperback edition could be published. We are also grateful to Massey University in New Zealand for server space that makes the Web revisions possible.

Andrew Lock
Pohangina, New Zealand

Charles Peters
Athens, Georgia, USA

June 1999

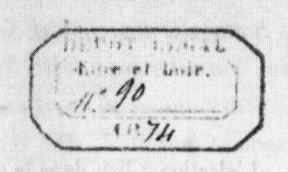

SOCIÉTÉ DE LINGUISTIQUE DE PARIS.

STATUTS

Approuvés par décision ministérielle du 8 Mars 1866.

Article premier.
La Société de linguistique a pour but l'étude des langues, celle des légendes, traditions, coutumes, documents, pouvant éclairer la science ethnographique. Tout autre objet d'études est rigoureusement interdit.

Article 2.
La Société n'admet aucune communication concernant, soit l'origine du langage, soit la création d'une langue universelle.

Article 3.
La Société publie chaque année au moins un volume de mémoires.

Article 4.
Elle peut y insérer des travaux de savants étrangers.

Article 5.
La Société correspond avec les Sociétés savantes de la France et de l'étranger.

Article 6.
Le nombre des membres de la Société ne peut être supérieur à cinq cents.

Article 7.
Tout candidat est présenté par deux membres qui font connaître son nom, sa demeure, et, s'il y a lieu, ses titres à l'admission.

Preface

When the Société de Linguistique de Paris was established in 1865, its statutes expressly forbade papers on the origins of language (see facing page). The first two of its dozen statutes precluded as acceptable inquiry much of what appears in this *Handbook*. They have been translated by Stam (1976, p. 255) as follows.

Article I: The Society of Linguistics has as its object the study of languages, and of legends, traditions, customs, and documents which could clarify ethnographic science. All other objects of study are rigorously forbidden.

Article II: The Society will accept no communication dealing with either the origin of language or the creation of a universal language.[1]

Much has transpired in the one hundred and thirty or so years since that perhaps laudable SLP ban. Explanations and accounts of our own origins have become one of the most popular of all the areas in science that are now regularly brought into the public arena via television, lavishly illustrated books, and even cartoons. The discovery of fossils and artefacts has its own intrinsic interest, but it is the origin of our characteristically human abilities—to be able to speak, create images, read and write, etc.—that holds the imagination. For it is in these areas that science bears directly on a demand that many hold to be the first step in making sense of this life of ours: 'Know thyself.'

Our aim in this book is to create a reference work that sets out and evaluates the basic knowledge and theory relevant to these origins that has accumulated in the scientific literature, especially over the last few decades. We have titled it a *Handbook* in the sense of a compendium and a guide, and not in the literal sense that it may easily be held in the hand. We offer no apologies for its size, except in so far as we have not been able to include every topic that might help in understanding human symbolic evolution.

There are several meanings of 'evolution' and 'symbolism' that fall within the scope of this inquiry. The appropriate senses of 'evolution' include: the origin of species by a process of development from earlier forms; the process of unrolling, unfolding, or opening out, as in an orderly succession, of a long train of events; the process of developing from a rudimentary to a mature state; the working out in detail of what is implicitly or potentially contained in a principle or idea; and the development or growth, according to its inherent tendencies, of anything that may be compared to a living organism—for example, that of a political constitution, a science, a language, and so forth.

The appropriate senses of 'symbol' include: something that stands for, represents, or denotes something else, concrete, abstract, or immaterial, not by exact resemblance, but by vague suggestion, or some conventional or accidental relation, for example, a gesture or a word; or a written character or mark used to represent something, such as a figure or sign conventionally standing for some object, process, quality, or condition.

Our concern, in part, is with the establishment of the symbolic, specifying those capacities that are necessary to allow symbols to be created, and how those capacities may themselves have been established. Secondly, we are concerned with the elaboration and use of the symbols themselves. We have, then, to distinguish between the abilities that make symbols possible, and the capabilities that are made possible by the use of symbols.

Our primary purpose is to provide the reference materials that would help in developing a clearer picture of *what* has occurred over time in the performance and elaboration of

---

[1] The Philological Society of London did not make its ban official, but the intellectual climate there was no less disapproving. See Stam (1976, pp. 255–66) for the 1873 remarks of that Society's president, Alexander J. Ellis.

human symbolic abilities. We see this as a necessary pre-requisite to any theorizing as to *how* that elaboration is to be explained. The history of the scientific investigation of human symbolic evolution is briefly summarized by Hewes (this volume, Chapter 21). There are a number of points in his outline that should be kept clearly in mind, lest anyone's expectations of what we can accomplish here exceed all reasonable possibility.

First, it is a short history, essentially having its scientific origins in the mid-twentieth century. Since then the field has developed as new fossil specimens and artefacts have been recovered, as new dating techniques have been developed, and as evidence has accrued in allied fields such as primatology, linguistics, and developmental psychology that allows wider and more informed hypotheses about, and interpretations of, the significance of the material record. But the amount of time within which this development has occurred is not sufficient for a mature science to have been established. There is little consensus as yet; no long-established and widely accepted core to the field.

Thus, secondly, there is a sense in which this area of inquiry has not yet, so to speak, left the nineteenth century, that stage in proto-scientific time when ruling theories and single-minded ideas held sway, before Chamberlin's (1890) reform placed the method of multiple working hypotheses at centre stage. Earlier abuses, like current successes, may be exaggerated; but it was characteristic of that period for theories to be advocated and proliferate without any hope or means of being tested. This justifiably led to the ban by the Paris Linguistic Society, and to Whitney's observation (1870) that: 'No theme in linguistic science is more often and more voluminously treated than this, by scholars of every grade and tendency; nor any, it may be added, with less profitable result in proportion to the labor expended.'

Advocacy; the omission of contrary evidence; and in some cases a predilection for untestable theories: all can still be found in the field today. But the immense intrinsic interest in the topic remains unabated, and this *Handbook* marks a new attempt to bring us closer to a science of symbolic origins and evolution. It is indicative of the field's having reached that healthy period in which it has articulated a sufficiently rich framework of concepts and knowledge to begin to be thoroughly self-critical: to take stock and establish the *limits* of what it can claim. Progress has been made since the unearthing of Neanderthal Man in 1856 marked the turning-point that forced humans to realize the antiquity of their heritage, and we should not seek to deny that it has. But approach the field with a healthy scepticism; beware of interpretative errors and overattribution: progress is more likely to come from that frame of mind than any other.

There is, moreover, a third point to be made here. The number of disciplines that contribute, either directly or indirectly, to this area of inquiry is such that no single investigator can reasonably hope to judge all the sources of evidence that bear on the topic. This is one of the motivating factors behind this volume: to have evaluated and make accessible as much of this material as possible. We hope, then, that *ex post* this volume, a palaeoanthropologist, for example, will be better able to invoke the role of, say, linguistic factors as important in hominid evolution, and to do so in a more informed way.

A glance at the Contents will show that we have cast our net widely. We have tried to include authoritative reviews on the many topic areas that contribute fundamentally to our goal. Inevitably we have missed some, through various combinations of ill luck, limitations of time and talent, and oversight. Perhaps some of the shortcomings can be put right in a second edition, some time in the twenty-first century. Our most notable omission is the anthropological literature dealing with culture as a set of symbols, typically represented by a group of publications by Geertz (1971, 1973), Turner (1974), Sperber (1975), Douglas (1966, 1970), Firth (1973), and Lévi-Strauss (1963, 1969, 1971, etc.).

The idea of this *Handbook* came out of our meeting in Paris in August 1981, appropriately at the first transatlantic conference on the evolution of language. The reality of it has grown from inviting numerous experts to take responsibility for various chapters. In the course of that growth we have filled two filing cabinets with a correspondence that would constitute a goldmine for a historian of science or psychoanalyst were it not for our destroying it to save all our embarrassments. We have also received help during this period from all over the world.

Over the years, the manuscript has travelled about a good deal. We are exceedingly grateful to the Universities of Adelaide, Georgia, Lancaster, Massey, Murcia, and Otago for the space and facilities they have offered us. The library staff at Lancaster and Georgia have been especially helpful in tracking down material for us. On the administration and support side Linda Adams at UGA, and Hazel Dale, Tracy Newsham, and Sheila Whalley at Lancaster have managed wonders. We are most indebted to Tracy for her amazing skills on both typewriters and word processors.

We have been helped in the overall shaping of the book, and in refereeing the chapters, by an editorial board consisting of Ben Blount, Peter Bryant, Kevin Connolly, Robin Dennell, and Henry Plotkin, whose wide experience and knowledge have assisted us in many ways. Most contributors of chapters have also acted as referees; but of these we owe a special debt to Len Rolfe's enquiring mind and editorial skills, which we discovered quite early on, and exploited to the depths of his good nature.

It is difficult to convey all our individual 'thank you's', for were we to acknowledge all our sources of assistance and advice we would fill another book. We have reduced our list here to those who gave us 'significant' help—sometimes by an extra close reading of and copious comments on draft chapters, yet sometimes only by a crucial line in a letter which pointed us in a new direction. Our thanks go to Diana Adlam, Chris Anderson, Elaine Anderson, Paul Atkinson, Paul Bahn, Elizabeth Bates, Basil Bernstein, Steven Brandt, Eve Clark, Cliff Davies, Leb Gold, Jack Goody, Charles Hudson, Jared Klein, Scott Kleiner, Stephen Kowalewski, Robert Layton, Greg McLennan, Doug Mcleod, Heidi Marcos, Alexander Marshack, Harald Martinsen, Paul Mellars, Lyn Miles, Salikoko Mufwene, John Napier, Netman@uk.ac.lancs.cent1, Jenny Pinkus, Colin Renfrew, Sue Savage-Rumbaugh, W. John Smith, John Stanton, Chris Stringer, Philip Tobias, Geoff White, and Jan Wind.

Finally, we owe an enormous debt to a number of people at Oxford University Press. In the same way as correspondence with contributors might have constituted the 'text' for a study in the history of science, so that between the us and Oxford University Press could constitute a paradigmatic exemplification of the vicissitudes of book production. Our original editor at the Press, Bruce Wilcock, maintained this project over a number of years with his belief that it was possible: his support was invaluable. On his retirement, Julia Maidment carried the project to the edge of production, and extended us an opportunity to make more last-minute revisions than is usual. Finally, Stuart McRobbie came up with an injection of realism that brought the process to a conclusion with the help of John Grandidge. Throughout this time John Harrison has overseen the details of production and retained a standard of efficiency and patience that is rare. Lastly, David Phelps took on the task of copy editor. His contribution to the volume cannot be overestimated. We thought copy editors just dotted 'i's and crossed 't's. David certainly did, and with a thoroughness that shamed our belief we had submitted a 'clean' manuscript. But in addition, he brought an incredible wealth of scholarship to the text, raising intellectual and conceptual points over a whole range of issues. Anyone who finds anything of use to their work in the following pages should raise their hat to him: we cannot thank him enough.

Palmerston North, New Zealand A. L.
Athens, Georgia, USA C. R. P.

References

Chamberlin, T. C. (1890). The method of multiple working hypotheses. *Science* (OS), **15**, 92–6. [Reprinted in *Science* (NS), **148**, 754–9 (1965)].
Douglas, M. (1966). *Purity and danger*. Praeger, New York.
Douglas, M. (1970). *Natural symbols*. Pantheon, New York.
Firth, R. (1973). *Symbols, public and private*. Cornell University Press, Ithaca, NY.
Geertz, C. (ed.) (1971). *Myth, symbol, and culture*. Norton, New York.
Geertz, C. (1973). *The interpretation of cultures*. Basic Books, New York.
Lévi-Strauss, C. (1963). *Structural anthropology*. Basic Books, New York.

Lévi-Strauss, C. (1969). *The raw and the cooked.* Harper & Row, New York.
Lévi-Strauss, C. (1971). *Mythologies IV: l'homme nu.* Plon, Paris.
Sperber, D. (1975). *Rethinking symbolism.* Cambridge University Press.
Stam, J. H. (1976). *Inquiries into the origins of language.* Harper & Row, New York.
Turner, V. (1974). *Dramas, fields, and metaphors.* Cornell University Press, Ithaca, NY.
Whitney, W. D. (1870). On the present state of the question as to the origin of language. *Transactions of the American Philological Association.* (Cited by Marshall, J. C. (1985). Speechless in Java. *Nature,* **316**, 685–6, and reprinted in Whitney's (1973) *Oriental and linguistic studies.* Scribner, New York.)

Contents

List of contributors — xxviii

Part I: Palaeoanthropology

1. Photogallery of fossil skulls — 3

2. An outline of human phylogeny: *Bernard G. Campbell* — 31
- 2.1 The nature of the evidence — 31
- 2.2 The earliest apes — 32
- 2.3 The earliest Hominidae: *Australopithecus* — 36
- 2.4 The earliest humans: *Homo habilis* — 41
- 2.5 *Homo erectus* — 43
- 2.6 Modern humans: *Homo sapiens* — 46
- 2.7 Discussion — 49
- Editorial notes — 50
- References — 51

3. Evolutionary trees of apes and humans from DNA sequences: *Peter J. Waddell and David Penny* — 53
- 3.1 Introduction — 54
- 3.2 Reconstructing relationships: from DNA sequences to evolutionary history — 54
  - 3.2.1 Basic steps in obtaining a tree for a selected stretch of DNA — 54
  - 3.2.2 Putting dates on a tree — 56
  - 3.2.3 Results from other molecular data — 58
  - 3.2.4 Polymorphisms and population variability — 59
  - 3.2.5 Total error on estimated divergence times — 59
- 3.3 Human genetic data: mtDNA sequences — 60
  - 3.3.1 Out-of-Africa, or mitochondrial Eve — 60
  - 3.3.2 Problems with trees from large numbers of sequences — 62
  - 3.3.3 Results from re-analysing the data — 62
  - 3.3.4 When, where, who, and how — 64
    - 3.3.4.1 When and where — 64
    - 3.3.4.2 Dating trees with archaeological evidence — 65
    - 3.3.4.3 Who and how — 66
- 3.4 Trees of human relationships from nuclear genetic data — 66
  - 3.4.1 Alleles and polymorphisms — 66
  - 3.4.2 Ingroup dating of the tree — 67
- 3.5 Conclusions and prospects — 69

	Epilogue	70
	Notes	71
	References	72

4. Evolution of the human brain: *Ralph Holloway* — 74

4.1 Introduction — 74
4.2 The human brain — 76
4.3 Lines of evidence regarding human brain evolution — 85
4.4 Palaeoneurological evidence — 87
    4.4.1 Brain size — 87
        4.4.1.1 Absolute brain size — 87
        4.4.1.2 Encephalization quotients — 89
    4.4.2 Organization of the brain — 90
        4.4.2.1 Relative increase in parietal lobe association cortex — 90
        4.4.2.2 A more human-like third inferior frontal convolution — 91
        4.4.2.3 Asymmetries of the brain and laterality — 93
        4.4.2.4 Towards a synthesis — 95
4.5 Conclusion — 97
Appendix: Sexual dimorphism and the brain — 98
Notes — 100
References — 102
Editorial appendix I: Endocranial volumes — 108
Editorial appendix II: Evolution of the human vocal apparatus — 116

5. Evolution of the hand and bipedality: *Mary W. Marzke* — 126

5.1 Introduction — 126
5.2 Non-human primate hands — 127
    5.2.1 Hand morphology shared by primates — 127
    5.2.2 The hand of the Great Apes — 128
5.3 Human hands — 130
    5.3.1 The morphological basis of tool-use in humans — 131
    5.3.2 Contribution of bipedality to tool-use — 134
    5.3.3 The brain, the hand, and tool-use — 135
5.4 Fossil hominoid hands and locomotor apparatus — 135
    5.4.1 Miocene fossil Hominoidea — 135
    5.4.2 Pliocene and Early Pleistocene Hominidae — 137
    5.4.3 Middle Pleistocene — 141
    5.4.4 Late Pleistocene — 141
5.5 Origin and evolution of the hominid hand and bipedality — 142
    5.5.1 The evidence from the comparative morphology of extant primates — 142
    5.5.2 The fossil evidence — 143
    5.5.3 The hand and the origin of bipedality — 144
    5.5.4 New perspectives on the role of tool-use and tool-making in the evolution of the hominid hand and bipedality — 144

5.5.5 The hand in symbolic behaviour	145
5.6 Conclusions	145
Notes	146
References	147

Part II: Social and socio-cultural systems

Editorial introduction to Part II: Social and socio-cultural systems

	157
References	163

Part II (a): Comparative perspectives

6. Primate communication, lies, and ideas:
Alison Jolly

	167
6.1 Introduction	167
6.2 Overview of the primates: taxonomy, social structure, intelligence	168
6.3 Communication and privacy	170
6.4 Deception	172
6.5 Symbolic behaviour in apes	174
References	176

7. Social relations, human ecology, and the evolution of culture: an exploration of concepts and definitions:
Tim Ingold

	178
7.1 Introduction	178
7.2 Animal society and human society	179
7.2.1 The superorganic	180
7.2.2 Society versus culture	181
7.2.3 Interactions and relationships	182
7.2.4 The constitution of persons	183
7.2.5 Sociality and consciousness	184
7.3 Social and ecological relations	185
7.3.1 The environment of an organism	186
7.3.2 Making an artificial environment	187
7.3.3 The ecological approach to sociality	188
7.3.4 The separation of social and ecological domains	189
7.3.5 The essence of production	190
7.4 Culture: its transmission and evolution	191
7.4.1 Reason and tradition	191
7.4.2 Individual and social learning	193
7.4.3 Learning and teaching	195
7.4.4 The analogy between organic and cultural evolution	196
7.4.5 Variation and invention	197

	7.4.6 Natural and artificial selection	197
7.5	Conclusion	199
	Notes	200
	References	200

8. Social relations, communication, and cognition:
Andrew Lock and Kim Symes — 204

8.1 Introduction	205
8.2 Non-verbal communication in relation to cultural complexity	205
8.2.1 The unadorned body	205
8.2.1.1 Innate versus learned repertoires	205
8.2.1.2 Innate versus learned emotions	206
8.2.1.3 Empirical research	206
8.2.2 The adorned body	206
8.2.2.1 Body adornment in simple societies	206
8.2.2.2 Body adornment in complex societies	207
8.2.3 Summary	208
8.3 Linguistic communication	208
8.3.1 Sociolinguistics: origins and perspectives	208
8.3.2 Sex differences in speech	209
8.3.3 Sex differences, conservatism, and prestige	210
8.3.4 Language styles in socially-stratified language communities	211
8.3.4.1 Rules of address	211
8.3.4.2 Differences between speakers from different social groups	211
8.3.4.3 Differences between speakers of different ages	212
8.3.4.4 Differences within an individual's speech depending on social context	212
8.3.4.5 Institutionalized language contexts	213
8.3.5 Summary	213
8.4 The significance of linguistic and non-verbal variation	213
8.5 Ecological determinants of cultural patterns	214
8.5.1 Social differentiation and its measurement	214
8.5.2 Social organization and 'ecological factors'	215
8.5.3 Socialization practices and exploitive patterns	216
8.5.4 Summary	217
8.6 Language, thought, and social structure in a cross-cultural perspective	217
8.6.1 Fischer's work on Truk and Ponape language and social structure	217
8.6.1.1 Fischer's theoretical perspective	217
8.6.1.2 Fischer's data	218
8.6.2 Greenberg's analytic–synthetic ordination	219
8.6.3 Colour-term differentiation and social structure	219
8.6.4 Chinese and English grammatical structures and problem-solving	220
8.6.4.1 Counter-factuality	221

8.6.4.2 Generics	221
8.6.4.3 Entification	221
8.6.4.4 Experimental work	221
8.7 Self, culture, and communication	222
8.8 Bernstein's theorizing	223
8.9 The reproduction of culture	225
8.9.1 Parent–child interaction and the transmission of cognitive skills	225
8.9.2 Cognitive style and social practices across cultures	227
8.9.2.1 Socialization practices	228
8.9.2.2 Social tightness (social conformity)	228
8.9.2.3 Ecological adaptation	228
8.9.3 Are Bernstein's and Witkin's approaches comparable?	228
8.9.4 Cognitive style and brain function	229
8.10 Overview	229
Notes	231
References	232

Part II (b): Palaeoanthropological perspectives

9. On the evolution of human socio-cultural systems: Randall K. White — 239

9.1 Introduction	239
9.2 Archaeological methods of social analysis	240
9.2.1 Social demography	240
9.2.2 Degree and nature of material exchange	240
9.2.3 Degree of internal differentiation	240
9.2.4 Degree of formal boundaries	240
9.2.5 Degree of local integration	241
9.3 Archaeology's refinement of the social evolutionary framework	241
9.4 The evolution of human social patterns	241
9.4.1 Social life in the Lower Palaeolithic (2 m.y.a. to 125 000 years ago)	241
9.4.2 Social life and interassemblage variation in the Middle Palaeolithic (125 000–35 000 years ago)	242
9.4.3 Social life and culture-as-we-know-it: the Upper Palaeolithic/Late Stone Age (35 000–12 000 years ago)	243
9.4.4 Social life at the end of the Pleistocene (12 000 to $c.$8 000 years ago)	244
9.4.5 Stratification and the State ($c.$5000–$c.$3000 years ago)	246
9.5 Some general observations	246
Editorial appendix: The development of complex societies	247
Notes	251
References	255

10. The evolution of tools and symbolic behaviour: *Thomas G. Wynn* — 263

    10.1 Introduction — 263
    10.2 Non-human tools — 264
    10.3 Hominid tools at 1.8 million years ago — 264
    10.4 Hominid tools at 1 million years ago — 268
    10.5 Tools at 300 000 years ago — 271
    10.6 Tools at 85 000 years ago — 275
    10.7 Tools at 15 000 years ago — 278
    10.8 Recent developments — 282
    10.9 Summary and conclusions — 284
    Notes — 285
    References — 285

11. A history of the interpretation of European 'palaeolithic art': magic, mythogram, and metaphors for modernity: *Margaret W. Conkey* — 288

    11.1 Introduction — 289
    11.2 Some intellectual context — 289
    11.3 The materials that we call 'Palaeolithic art' — 292
        11.3.1 Concerns about the evidence: diversity and preservation — 292
        11.3.2 Concerns about the evidence: chronology — 293
        11.3.3 Techniques, conventions, and media — 294
        11.3.4 The 'subject-matter' of the imagery — 296
    11.4 The major interpretations — 299
        11.4.1 Art as hunting magic: the foundation interpretation — 299
        11.4.2 The structuralist 'break-out' — 300
    11.5 Alternative interpretative perspectives — 302
        11.5.1 Art as adaptive: attempts at context and process — 303
        11.5.2 Alexander Marshack: from calendars to 'post-structuralist'? — 306
    11.6 Assessing the interpretative terrain: into the 1990s — 307
    11.7 What becomes of the study of origins? — 309
    11.8 Some closing thoughts — 311
    Notes — 311
    References — 344

12. Photogallery of contemporary hunter–gatherer rock art — 350

    12.1 Australian Aboriginal art: Andrew Lock and Margaret Nobbs — 351
    12.2 Southern African Bushmen rock art: J. David Lewis-Williams, Thomas Dowson, Andrew Lock, and Charles R. Peters — 360

12.3 Editorial note on the significance of contemporary hunter–gatherer art for understanding the palaeolithic materials	363
Notes	367
References	367

Part III: *Ontogeny and symbolism*

Editorial introduction to Part III: Ontogeny: symbolic development and symbolic evolution — 371

1 Ontogeny and phylogeny	371
1.1 Mosaic evolution	372
1.2 Embryological mosaics, or dissociations	372
1.3 Heterochrony and evolution	373
2 A role for heterochronic processes in human evolution	374
3 Relating human behavioural ontogeny and phylogeny: issues	377
3.1 Dissociable systems in cognition and language	377
3.1.1 Parallels between action and language: analogy or homology?	379
3.1.2 Dissociable developmental mechanisms in early language development	382
3.2 Implications of cognitive ontogeny for phylogeny	383
3.2.1 Specific ontogenetic adaptations	383
3.2.2 Maturational schedules	383
3.2.3 The existence of 'end products' as ontogenetic substrates	384
3.2.4 The 'unit' problem and the co-evolutionary system	385
3.2.5 The 'attribution' problem and the socio-cultural system	385
4 Relating human behavioural ontogeny and phylogeny: strategies	386
4.1 The suite of separable capacities	387
4.1.1 Gestural, referential communication	387
4.1.2 Imitation	387
4.1.3 Means–end analysis	388
4.1.4 Comprehension	388
4.1.5 Additional pre-requisites	388
4.2 The provision of a scaffolding system	389
5 Concluding remarks	394
Notes	394
References	395

13. The role of ontogenesis in human evolution and development: *Christopher G. Sinha* — 400

13.1 Adaptation and representation	400
Note	405
References	405

14. The ontogeny and evolution of the brain, cognition, and language: *Kathleen Gibson* — 407

- 14.1 Neoteny and recapitulation — 407
- 14.2 Quantity versus reorganization in the evolution of the brain — 409
- 14.3 Comparative brain maturation patterns — 410
- 14.4 Cognitive maturation and neocortical development — 412
- 14.5 Can the neo-Piagetian ontogenetic framework be placed in an evolutionary perspective? — 418
- 14.6 What can be 'reconstructed' phylogenetically? — 421
- Notes — 425
- References — 426

15. Early interaction and cognitive skills: implications for the acquisition of culture: *David Messer and Glyn Collis* — 432

- 15.1 Introduction — 433
- 15.2 Developmental theory and the origins of social behaviour — 434
  - 15.2.1 Trevarthen — 434
  - 15.2.2 Piagetian approaches — 436
  - 15.2.3 Vygotsky — 437
  - 15.2.4 Kaye — 437
  - 15.2.5 Summary — 438
- 15.3 The beginnings of cultural understanding: infant capacities and social interaction — 439
  - 15.3.1 Social responsiveness and infant capacities — 439
  - 15.3.2 Social interaction and infant capacities — 440
- 15.4 Social interaction and the development of relationships — 442
  - 15.4.1 Attachment — 442
- 15.5 Relationships and social cognition in non-human primates — 446
- 15.6 Social interaction and the development of cognitive skills — 447
  - 15.6.1 Imitation — 447
  - 15.6.2 The co-ordination of attention — 449
  - 15.6.3 Social interaction and the acquisition of language — 452
    - 15.6.3.1 Simplified speech to children — 452
    - 15.6.3.2 Speech and non-linguistic context — 453
    - 15.6.3.3 Pre-linguistic and linguistic communication — 453
    - 15.6.3.4 Social interaction and linguistic innovation — 454
    - 15.6.3.5 Summary — 455
- 15.7 The influence of social processes — 455
  - 15.7.1 Cross-cultural comparisons of social interaction and language-acquisition — 455
  - 15.7.2 Social interaction and cognitive development — 456
- 15.8 Communication with language and the extension of cognitive skills — 457
- 15.9 Summary — 459
- References — 460

16. The origins of language and thought in early childhood: *George Butterworth* — 469

   16.1 Introduction — 470
   16.2 Classical theories of the relation between language and thought in young children: Piaget and Vygotsky — 470
      16.2.1 Piaget's theory — 470
      16.2.2 Vygotsky's theory — 471
   16.3 Perceptual, intellectual, and social precursors of thought and language in young children — 472
      16.3.1 Intellectual development and the pre-requisites for linguistic communication — 472
      16.3.2 Linguistic aspects of thought in young children — 475
         16.3.2.1 Effects of language on categorization: the Whorfian hypothesis — 475
         16.3.2.2 The verbal regulation of behaviour — 475
   16.4 Ontogeny and phylogeny: stages of development and the recapitulation hypothesis — 476
      16.4.1 'Terminal addition' — 477
      16.4.2 Neoteny — 477
      16.4.3 Common constraint — 478
   16.5 Conclusion — 479
   Editorial appendix: Recent studies of the relation between cognition and language development — 479
   References — 480

17. Theories of symbolization and development: *Christopher G. Sinha* — 483

   17.1 Sign and object — 483
      17.1.1 Signification, representation, and language — 483
      17.1.2 Theories of the sign: Peirce, Frege, Saussure — 484
         17.1.2.1 Charles Sanders Peirce (1839–1914) — 484
         17.1.2.2 Gottlob Frege (1848–1925) — 485
         17.1.2.3 Ferdinand de Saussure (1857–1913) — 486
      17.1.3 Social life as semiosis: Mead, Bakhtin, Barthes — 487
         17.1.3.1 George Herbert Mead (1863–1931) — 487
         17.1.3.2 Mikhail Bakhtin (1895–1975) — 488
         17.1.3.3 Roland Barthes (1915–1980) — 489
   17.2 Language, thought, and symbol in human development: Freud, Piaget, Vygotsky — 489
      17.2.1 Darwin and the mind of the child — 489
      17.2.2 Freud: *Ursprache* and the Unconscious — 490
         17.2.2.1 Freud as evolutionist — 490
         17.2.2.2 Freud as semiotician — 491
         17.2.2.3 Freud as cognitive theorist — 492
      17.2.3 Piaget: from solipsism to structure — 492
      17.2.4 Vygotsky: language and the sociogenesis of reasoning — 495

17.3	Conclusion	497
	Notes	498
	References	498

18. Children's drawings and the evolution of art: *J. Gavin Bremner* — 501

18.1	Introduction	502
18.2	Children's spontaneous drawings	502
	18.2.1 Eng (1931)	502
	18.2.2 Luquet (1927)	503
	18.2.3 Kellogg (1970)	503
18.3	The development of drawing within the Piagetian framework	505
18.4	Some recent experimental approaches to children's drawings	506
	18.4.1 Drawing as problem-solving	506
	18.4.2 Drawing as a spatial skill	507
	18.4.3 Representation of the third dimension in drawings	508
	18.4.4 Intellectual realism re-assessed	510
18.5	Summary	512
18.6	Two accounts of the evolution of art	512
	18.6.1 Gablik (1976)	513
	18.6.2 Gombrich (1960)	514
	18.6.3 Pointers from developmental psychology	516
	Note	518
	References	518

Part IV: Language systems

Editorial introduction to Part IV: Language systems in an evolutionary perspective — 523

1	Contruction vs. discontinuity: modularity and grammar again	523
2	Language reconstruction	525
	Notes	529
	References	530

19. Map gallery of the distribution and classification of extant human languages — 532

20. Spoken language and sign language: *Margaret Deuchar* — 553

20.1	Introduction	553
20.2	Spoken and signed languages	554
	20.2.1 Spoken languages	554
	20.2.2 Signed languages	554
20.3	Definition of language	554
	20.3.1 Focus on structure	557

	20.3.2 Focus on function	557
20.4	Arbitrariness	558
	20.4.1 Iconicity versus arbitrariness	558
	20.4.2 Iconicity in spoken language	558
	20.4.3 Iconicity versus arbitrariness in sign languages	559
	20.4.4 Arbitrariness and evolution	562
	20.4.5 Summary	562
20.5	Structure in spoken and signed language	562
	20.5.1 Structure below the word/sign level	562
	20.5.2 Grammar	564
	20.5.3 Summary	565
20.6	Language acquisition	565
	20.6.1 Summary	566
20.7	Sign languages and creoles	566
20.8	Language use and language evolution	568
20.9	Spoken and signed languages: a summary	568
	References	568

21. A history of the study of language origins and the gestural primacy hypothesis: *Gordon W. Hewes*   571

21.1	Introduction	571
21.2	A brief history of speculations about language origins	572
	21.2.1 Classical and pre-Renaissance times	572
	21.2.2 The seventeenth to mid-nineteenth centuries	572
	21.2.3 The mid-nineteenth century and the Darwinian revolution	576
	21.2.4 The early twentieth century	576
	21.2.5 The 1920s to the 1950s	576
	21.2.5.1 Primate field studies	579
	21.2.5.2 The Chomskyan revolution in linguistics	579
	21.2.6 Renewed interest in gestural languages	582
	21.2.7 1970 to the present	582
21.3	Specific evidence for the gestural origin of language	582
	21.3.1 The questionable suitability of sound as the original basis for language	583
	21.3.2 Weakness of some alternative models for the emergence of language	583
	21.3.3 Manual motor control and the cerebral lateralization of language	584
	21.3.4 Some advantages of gestural communication	584
	21.3.5 Distribution of gestural communication	585
	21.3.6 Volar depigmentation as evidence for the primacy of gestural communication	586
	21.3.7 Phonemes and language	586
	21.3.8 Fetal eavesdropping	587
	21.3.9 Possible cognitive and other advantages of speech	587
21.4	A gestural-origin model for language	587

	21.4.1	Biological and cultural pressures for continuing language evolution	588
	21.4.2	Cerebral lateralization	588
	21.4.3	Deixis as a starting-point for the early hominids	588
	21.4.4	Hypothetical onset of gestural communication	589
	21.4.5	Beginnings of vocal language, hypothetically as an accompaniment of gestural communication	589
	21.4.6	Attainment of habitual but still non-phonemic vocal language	590
	21.4.7	World-wide phonemicization of speech	590
21.5	Conclusions		591
	Notes		591
	References		592

22. Cognitive abilities in a comparative perspective:
Andrew Lock and Michael Colombo   596

22.1	Animal cognition	597
22.2	Memory	597
	22.2.1 Spatial memory	597
	22.2.2 Delayed matching to sample paradigm (DMTS)	598
	22.2.3 Serial position and clustering effects in recall	598
22.3	Other cognitive systems	599
22.4	Some caveats	600
22.5	The uses of cognition: what can animals do or 'learn'?	601
	22.5.1 Simple conditioning and discrimination learning	604
	22.5.2 Complex discrimination learning	604
	22.5.2.1 Discrimination learning set (DLS)	604
	22.5.2.2 Reversal learning	606
	22.5.3 Time-based, response-patterned, and match-to-sample learning	607
	22.5.4 Probability learning	609
	22.5.5 Cross-modal recognition and transfer of learning	610
	22.5.6 Representation: categorization and concepts	610
22.6	Same/different judgements by animals	611
	22.6.1 Non-primates	612
	22.6.2 Monkeys	613
	22.6.3 Apes	614
22.7	Symmetry and transitivity of conditional relations	615
	22.7.1 Non-primates and monkeys	615
	22.7.2 Apes	616
22.8	Representations of lists in pigeons and monkeys	616
	22.8.1 Interim conclusion	617
22.9	Observational learning	617
22.10	Piagetian studies of animal cognition	617
	22.10.1 Sensorimotor intelligence	619
	22.10.2 Object permanence	619
	22.10.3 Other sensorimotor acquisitions	620

	22.10.4 Representational intelligence	620
22.11	Tool-use, insight, self-recognition, and mirror-use in primates	621
	22.11.1 Tool-use and insight	621
	22.11.2 Mirror-use and self-recognition	622
	22.11.3 'Theory of mind': understanding others	624
	22.11.3.1 Phylogenetic origins	627
	22.11.3.2 Ontogenetic origins	629
22.12	Imitation, problem-solving, and cognition	630
22.13	Evolutionary implications of comparative studies of 'cognition'	631
	22.13.1 The ecology of cognition	631
	22.13.2 The testing of cognition	631
	22.13.3 The attribution of cognition	632
	22.13.4 Accounting for human cognition from a comparative perspective	632
	Notes	633
	References	634

23. Animal language and cognition projects:
Carolyn A. Ristau   644

23.1 Introduction	645
23.2 A brief history of the ape-language projects	645
23.3 Provisos when interpreting the results of the ape-language and cognition studies	646
23.4 Theoretical issues	647
23.4.1 Definition of language: the meaning of the word and the utterance	647
23.4.2 Definitions of language and the nature of American Sign Language	648
23.4.3 Production versus comprehension	649
23.5 The findings with regard to symbolic abilities	650
23.5.1 Linguistic-like capacities	650
23.5.1.1 Grammar	650
23.5.1.2 Meaning	653
23.5.1.2.1 Procedures to establish reference	653
23.5.1.2.2 The nature of the internal representation	655
23.5.1.2.3 Displacement testing	657
23.5.1.3 Other instances of possible symbolism by human-reared apes, signing apes, and apes in the wild	658
23.5.2 Cognitive and symbolic capacities	658
23.5.2.1 Quantitative abilities	658
23.5.2.2 Reasoning	661
23.6 Methodological problems	662
23.6.1 The 'Clever Hans' phenomenon	662

		23.6.2	Strict experimental control or casual interaction between ape and communicator—which is more appropriate?	663
		23.6.3	The significance of the novel or few-time event	664
		23.6.4	The importance of ethological indices of behaviour	664
	23.7	The changing interpretations of linguistic development and of the ape language projects		665
	23.8	Evolutionary significance		667
	23.9	Conclusions		667
		Note		668
		References		680

24. Symbols and structures in language-acquisition:
Carolyn Johnson, Henry Davis, and Marlys Macken   686

	24.1	Introduction		687
	24.2	Language as a symbol system		687
	24.3	Child-language study in a historical perspective		688
	24.4	Theories of language-acquisition		689
		24.4.1	The interactive approach	691
		24.4.2	The cognitive approach	696
			24.4.2.1 Correlations between linguistic and cognitive developments	696
			24.4.2.2 General cognitive mechanisms	697
		24.4.3	The autonomous approach	697
	24.5	Acquisition of the phonological system		698
		24.5.1	The learner's task	698
		24.5.2	Three developmental issues	699
			24.5.2.1 The initial state	699
			24.5.2.2 The transition from babbling to language	701
			24.5.2.3 Learning the phonological system	701
		24.5.3	Theories of phonological acquisition	702
			24.5.3.1 Jakobson's structuralist theory	703
			24.5.3.2 Stampe's natural phonology	703
			24.5.3.3 Generative phonology	703
			24.5.3.4 Prosodic theory	703
			24.5.3.5 Cognitive theory	703
	24.6	Acquisition of the lexical semantic system		705
		24.6.1	The learner's task	705
		24.6.2	Early lexical development	705
			24.6.2.1 Children's errors in word-use	706
		24.6.3	Theories of the acquisition of word-meaning	707
			24.6.3.1 The semantic feature hypothesis	707
			24.6.3.2 The functional core hypothesis	708
			24.6.3.3 The contrastive hypothesis	709
			24.6.3.4 Prototype theories	709
			24.6.3.5 The state of the art	710
		24.6.4	The lexicon and human symbolic capacity: developmental questions	711

	24.6.4.1 When is a word a symbol?	711
	24.6.4.2 Is the developing lexical system part of a general symbolic capacity?	712
	24.6.4.2.1 Gesture and words	712
	24.6.4.2.2 Words and concepts	713
24.7	Acquisition of the syntactic system	716
	24.7.1 The learner's task	716
	24.7.2 Overview of syntactic acquisition	718
	24.7.2.1 The presyntactic stage	718
	24.7.2.2 The syntactic stage	720
	24.7.2.3 The postsyntactic stage	722
24.8	Acquisition of speech acts	724
	24.8.1 Primitive speech acts	724
	24.8.2 Communicative functions in early childhood	726
	24.8.3 The relation between communicative functions and linguistic forms in childhood	726
24.9	Language-acquisition and the evolution of language: some closing thoughts	727
	24.9.1 Theories of language-acquisition and symbolic evolution	728
24.10	Conclusion	729
	Editorial appendix: Gestures and words: the early stages of communicative development	729
	Notes	734
	References	737

25. The reconstruction of the evolution of human spoken language: *Mary LeCron Foster*   747

25.1	Introduction	748
25.2	Language as classification by analogy	748
25.3	Analogy and language as system	749
25.4	Reconstructive methodology	751
	25.4.1 The comparative method	751
	25.4.2 Systems in sound-change	752
	25.4.3 Family trees and wave theory	752
	25.4.4 Linguistic groupings	753
	25.4.5 Typology	754
	25.4.6 Internal reconstruction	755
25.5	Reconstructed languages	756
25.6	Hypotheses suggesting further groupings	759
25.7	Guidelines for remote reconstruction	760
25.8	Towards primordial language: the monogenetic hypothesis	762
	25.8.1 Background	762
	25.8.2 The phememic system	762
	25.8.3 Comparison, internal reconstruction, and system for phememes	763
	25.8.4 Toward a theory of the PL stem	767

	25.8.5 Sorting the isoglosses	768
25.9	Wider implications of monogenetic reconstruction	768
	25.9.1 The role of analogy	768
	25.9.2 Internal and external analogy	769
	25.9.3 The complexity of modern 'roots'	769
	25.9.4 Advantages of inclusion of dissimilar languages	770
25.10	Systematic correlations in human evolution	770
	25.10.1 The beginning	770
	25.10.2 Hypothetical evolutionary stages	771
25.11	Language in culture	772
	Notes	772
	References	772

26. Theoretical stages in the prehistory of grammar: *Leonard Rolfe* — 776

26.1	Introduction	777
26.2	Patterns of grammar	778
26.3	The frame of dialogue	779
26.4	Ostension and deixis	780
26.5	Naming and words	781
26.6	Thematization—the phonemic principle	782
26.7	Topic–comment and nominals	784
26.8	Topic–comment and case-relations	785
26.9	Narrative	786
26.10	The epistemic pattern: objectivity in discourse	787
26.11	Conclusion	789
	Notes	789
	References	791

27. Social and cognitive factors in the historical elaboration of writing: *David Barton and Mary Hamilton* — 793

27.1	What is writing?	793
	27.1.1 Introduction	793
	27.1.2 Defining writing	794
	27.1.2.1 Definitions of writing-systems	796
	27.1.3 Writing and earlier forms of symbolic representation	797
	27.1.4 Apparent functions and processes in early societies	799
	27.1.4.1 Functions of writing	799
	27.1.4.2 Materials and techniques for writing	799
27.2	The development of writing	800
	27.2.1 Types of writing-system	800
	27.2.1.1 Logographic script	800
	27.2.1.2 Syllabic script	801
	27.2.1.3 Alphabetic script	801
	27.2.2 Development from one system to another	801

	27.2.3	Advantages and disadvantages of particular types of writing-system	802
		27.2.3.1 The efficiency of scripts	803
		27.2.3.2 Two principles of writing-systems	804
27.3	Literacy and cognition		804
	27.3.1	Cognitive 'effects'	804
	27.3.2	Speech as 'deficit'	805
27.4	Social influences		806
	27.4.1	Characterizing literacy	806
	27.4.2	Greek literacy	807
	27.4.3	Social and economic correlates of literacy	808
		27.4.3.1 Franchise and political participation	809
		27.4.3.2 Employment and economic development	809
		27.4.3.3 Modernity	810
	27.4.4	The technology of printing	810
	27.4.5	Restricted access to written language	811
27.5	Conclusions		813
Editorial appendices	I: Script evolution		813
	II: Ehlich's developmental account of writing		814
	III: Numbers and mathematics		818
	IV: Music		824
	V: Dance and choreography		826
	VI: Cartography		826
	VII: Printing		834
	Notes		854
	References		854

Part V: Epilogue

28. Tempo and mode of change in the evolution of symbolism: *Charles R. Peters* — 861

28.1	Introduction	861
28.2	Chart I	862
28.3	Charts II and III	864
28.4	Chart IV	868
28.5	Modernity's complexity	872
	Notes	873
	References	874

Time Charts — 878

Index — 887

Contributors

David Barton
Department of Linguistics and
Modern English Language
University of Lancaster
Lancaster LA1 4YT
UK

J. Gavin Bremner
Department of Psychology
University of Lancaster
Fylde College
Lancaster LA1 4YF
UK

George Butterworth
Division of Psychology
University of Sussex
Falmer
Brighton BN1 9QU
UK

Bernard G. Campbell
Sedgeford Hall
Sedgeford
Hunstanton
Norfolk PE36 5LT
UK

Glyn Collis
Department of Psychology
University of Warwick
Coventry
CV4 7AL
UK

Michael Colombo
Department of Psychology
University of Otago
Dunedin
New Zealand

Margaret W. Conkey
Department of Anthropology
University of California
Berkeley
California 94720
USA

Henry Davis
Department of Linguistics
University of British Columbia
Vancouver, BC
Canada V6T 1W5

Margaret Deuchar
Department of Linguistics
School of English and Linguistics
University of Wales, Bangor
Bangor
Gwynedd LL57 2DG
UK

Thomas Dowson
Archaeology Department
University of Southampton
Highfield
Southampton SO17 1BJ
UK

Mary LeCron Foster
Department of Anthropology
University of California
Berkeley
California 94720
USA

Kathleen R. Gibson
University of Texas
Dental Branch
6516 John Freeman Avenue
PO Box 20068
Houston
Texas 77225
USA

Mary Hamilton
Department of Educational
Research
University of Lancaster
Lancaster LA1 4YJ
UK

Gordon W. Hewes
University of Colorado
Department of Anthropology
Campus Box 233

Boulder
Colorado 80309
USA

Ralph L. Holloway
Department of Anthropology
Columbia University
New York
New York 10027
USA

Tim Ingold
Dept of Social Anthropology
University of Manchester
Manchester
M13 9PL
UK

Carolyn Johnson
School of Audiology and Speech
Sciences
Faculty of Medicine
University of British Columbia
5804 Fairview
Vancouver, BC
Canada V6T 1Z3

Alison Jolly
Department of Biology
Princeton University
Princeton
New Jersey 08544
USA

J. David Lewis-Williams
Rock Art Research Unit
Department of Archaeology
University of the Witwatersrand
1 Jan Smuts Avenue
Johannesburg 2050
South Africa

Andrew Lock
Department of Psychology
Massey University
Palmerston North
New Zealand

Marlys Macken
Department of Linguistics
University of Wisconsin
Madison
Wisconsin 53706
USA

Mary W. Marzke
Department of Anthropology
Arizona State University
Tempe
Arizona 85287
USA

David Messer
Psychology Department
School of Health and Social Sciences
University of Hertfordshire
Hatfield Campus
College Lane
Hatfield
Herts AL10 9AB
UK

Margaret Nobbs
8 Hazelwood Avenue
Hazelwood Park 5066
South Australia

David Penny
Department of Plant Biology and Biotechnology
Massey University
Palmerston North
New Zealand

Charles R. Peters
Department of Anthropology
Badwin Hall
University of Georgia
Athens
Georgia 30602
USA

Carolyn A. Ristau
Department of Psychology
Barnard College of Columbia University
3009 Broadway
New York, NY 10023
USA

Leonard Rolfe
101 Lockington Crescent
Stowmarket
Suffolk
IP14 1DA
UK

Christopher G. Sinha
Institute of Psychology
Aarhus University
Asylvej 4
DK-8240 Risskov
Denmark

Kim Symes
8 Crummock Street
Carlisle
Cumbria CA2 5PT
UK

Peter Waddell
Department of Mathematics
Massey University
Palmerston North
New Zealand

Randall K. White
Department of Anthropology
New York University
25 Waverly Place
New York
New York 10003
USA

Thomas G. Wynn
Department of Anthropology
University of Colorado
Colorado Springs
Colorado 80933–7056
USA

Part I
Palaeoanthropology

1
Photogallery of fossil skulls

This photogallery of fossil hominid skulls provides views of some of the key specimens and morphological changes discussed in the *Handbook*. Two other excellent sources for photos are Michael Day's (1986 edition) *Guide to fossil man* (University of Chicago Press) and Eric Delson's (1995) *Ancestors—the hard evidence* (Alan R. Liss, New York).

Hominid taxonomy is seemingly in a perpetual state of flux. Some notes alerting the reader to major classifactory changes are provided in the introductory chapter by Campbell, the editorial appendices to Holloway's chapter, and the epilogue chapter by Peters. Both the stability and the change characteristic of palaeoanthropology over the past few decades are testimony to the field's maturation.

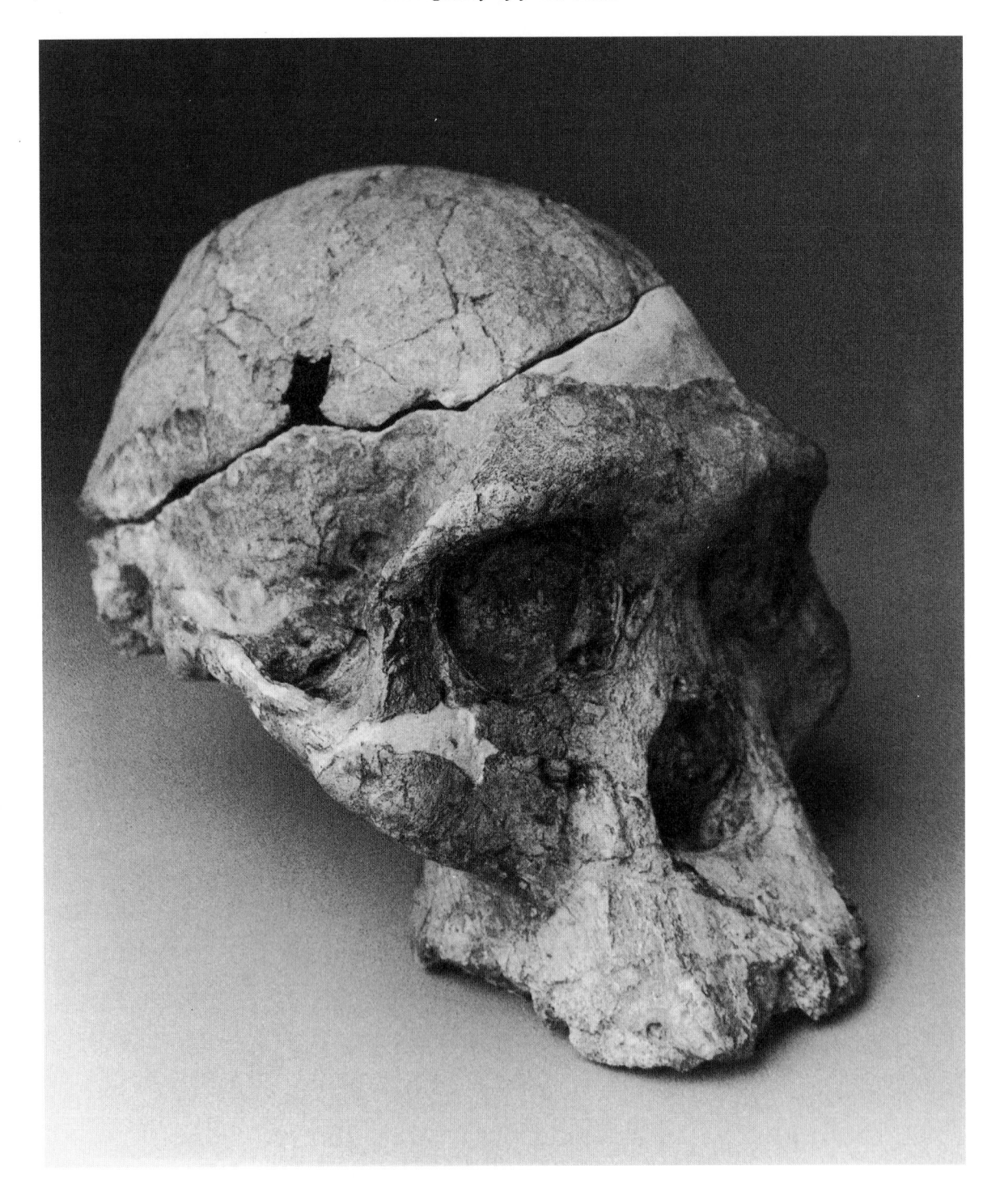

Plate I. Sts 5: *Australopithecus africanus*, Sterkfontein, South Africa, c.2.4–2.5 million years old. Three-quarter, lateral, and basal views. Courtesy of P. V. Tobias. Photos by A. R. Hughes.

Photogallery of fossil skulls

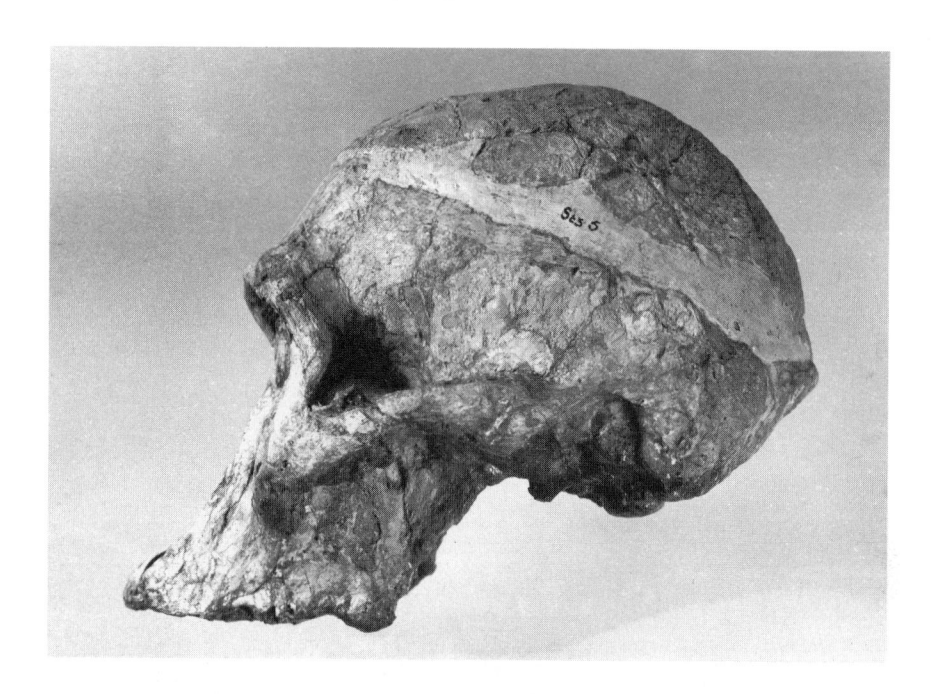

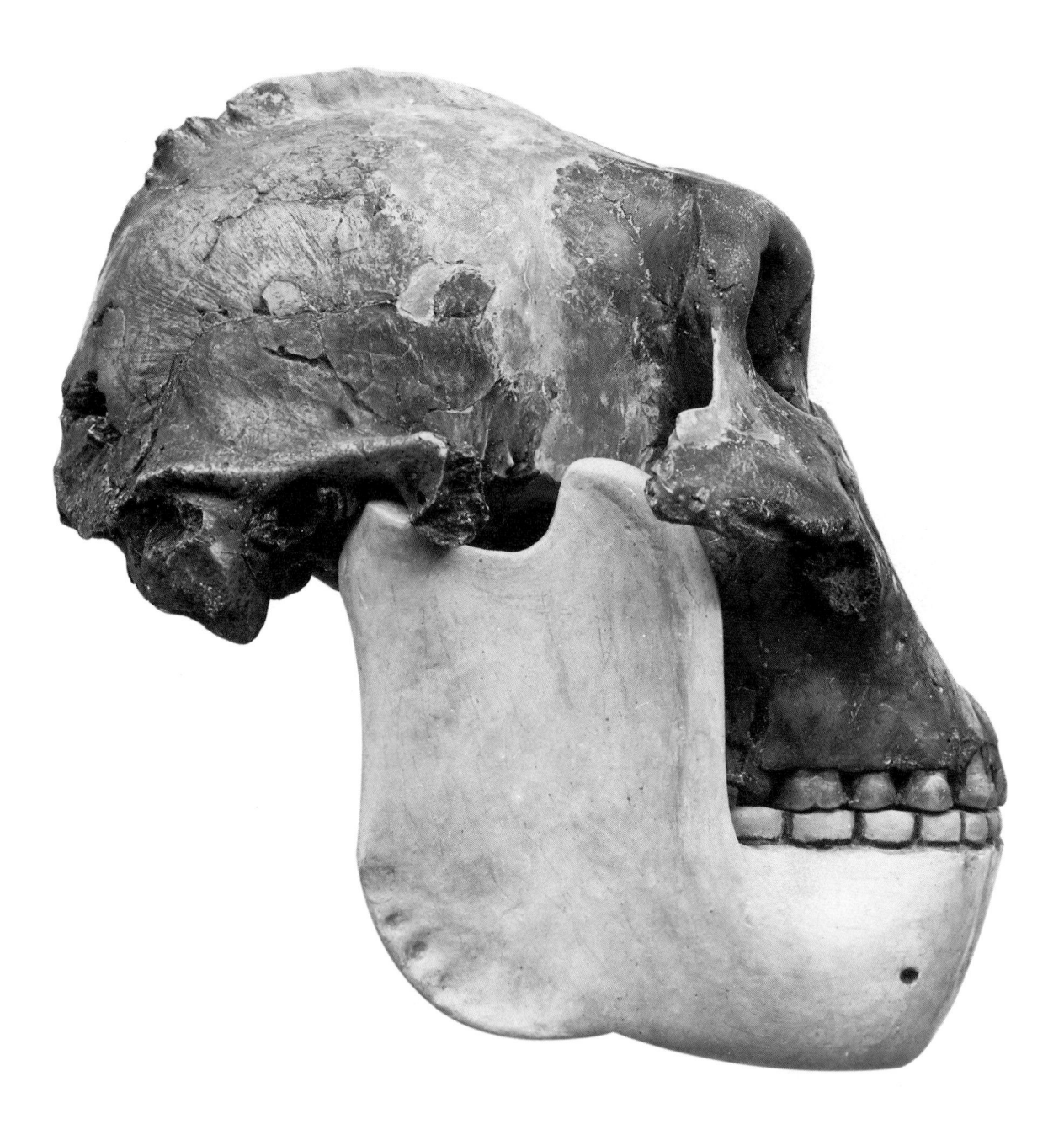

Plate II. OH5: *Australopithecus boisei*, Olduvai Gorge, Tanzania, c.1.75–1.8 million years old. Lateral and frontal views (with a model of the mandible), plus basal view. Courtesy of P. V. Tobias. Lateral and frontal photos by Bob Campbell.

Photogallery of fossil skulls 7

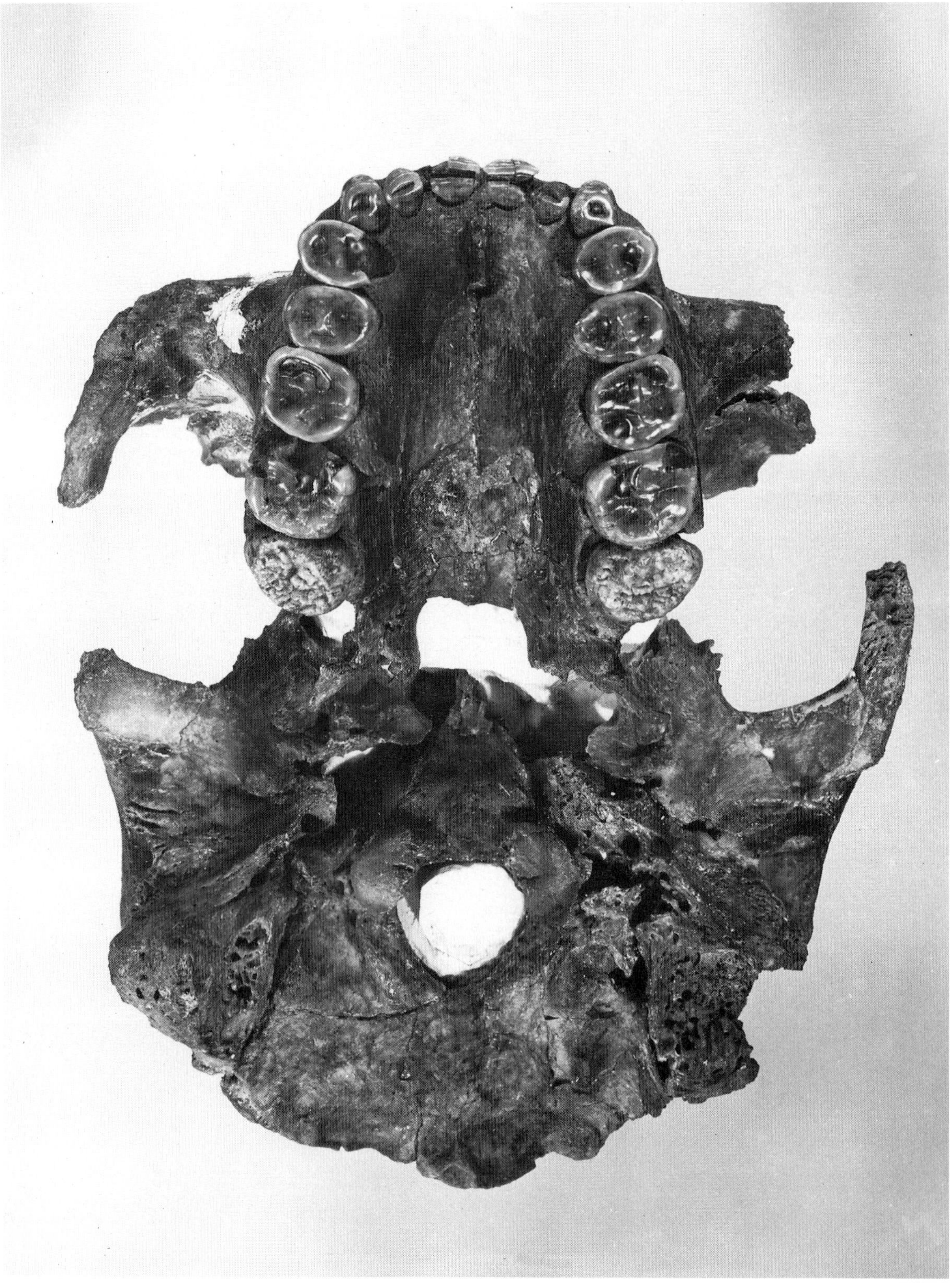

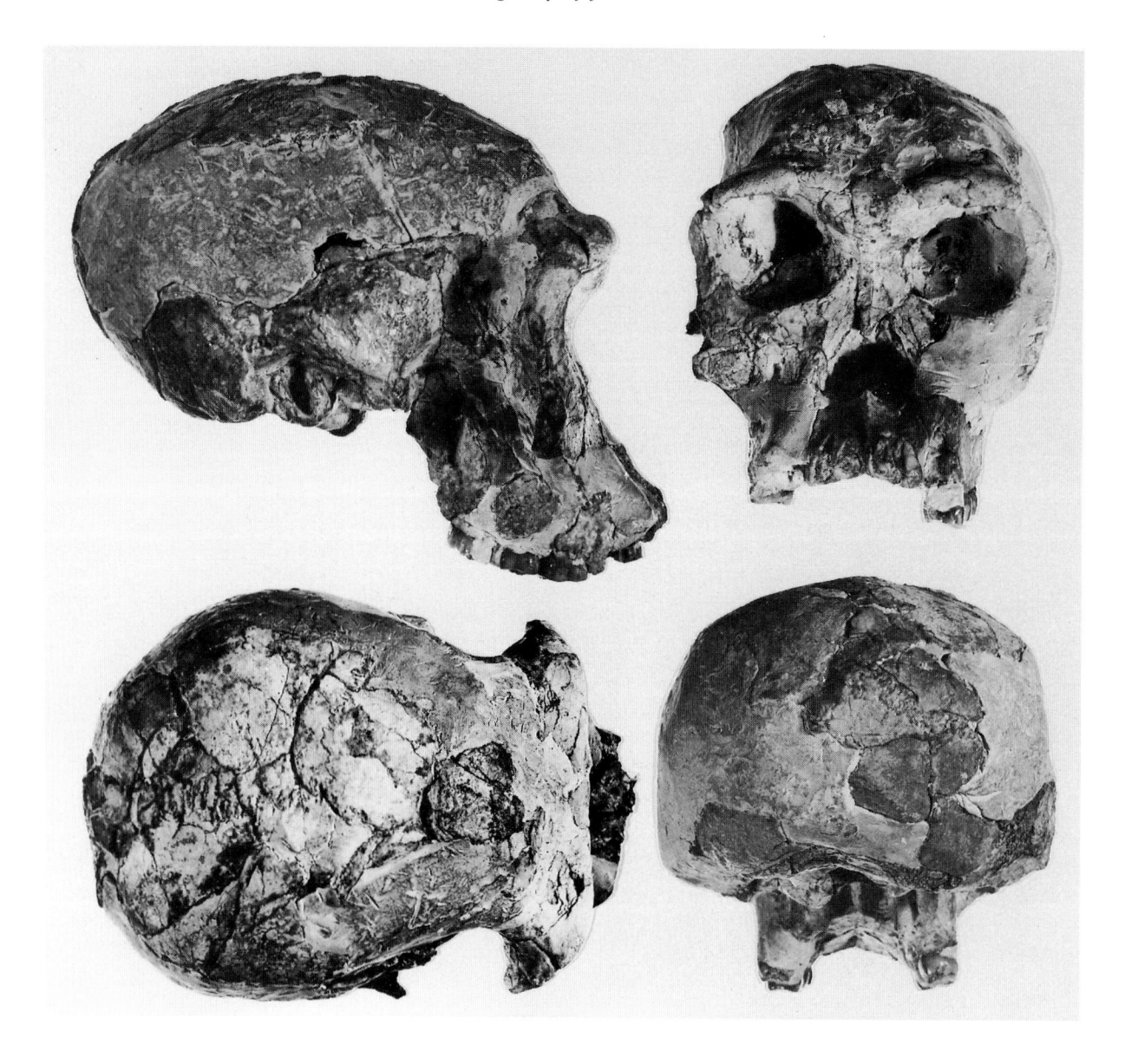

Plate III. OH24: *Homo habilis*, Olduvai Gorge, Tanzania, *c*.1.8–1.9 million years old. Lateral, frontal, superior, occipital, and basal views. Courtesy of P. V. Tobias. Cranium reconstructed and restored by R. J. Clarke.

Photogallery of fossil skulls

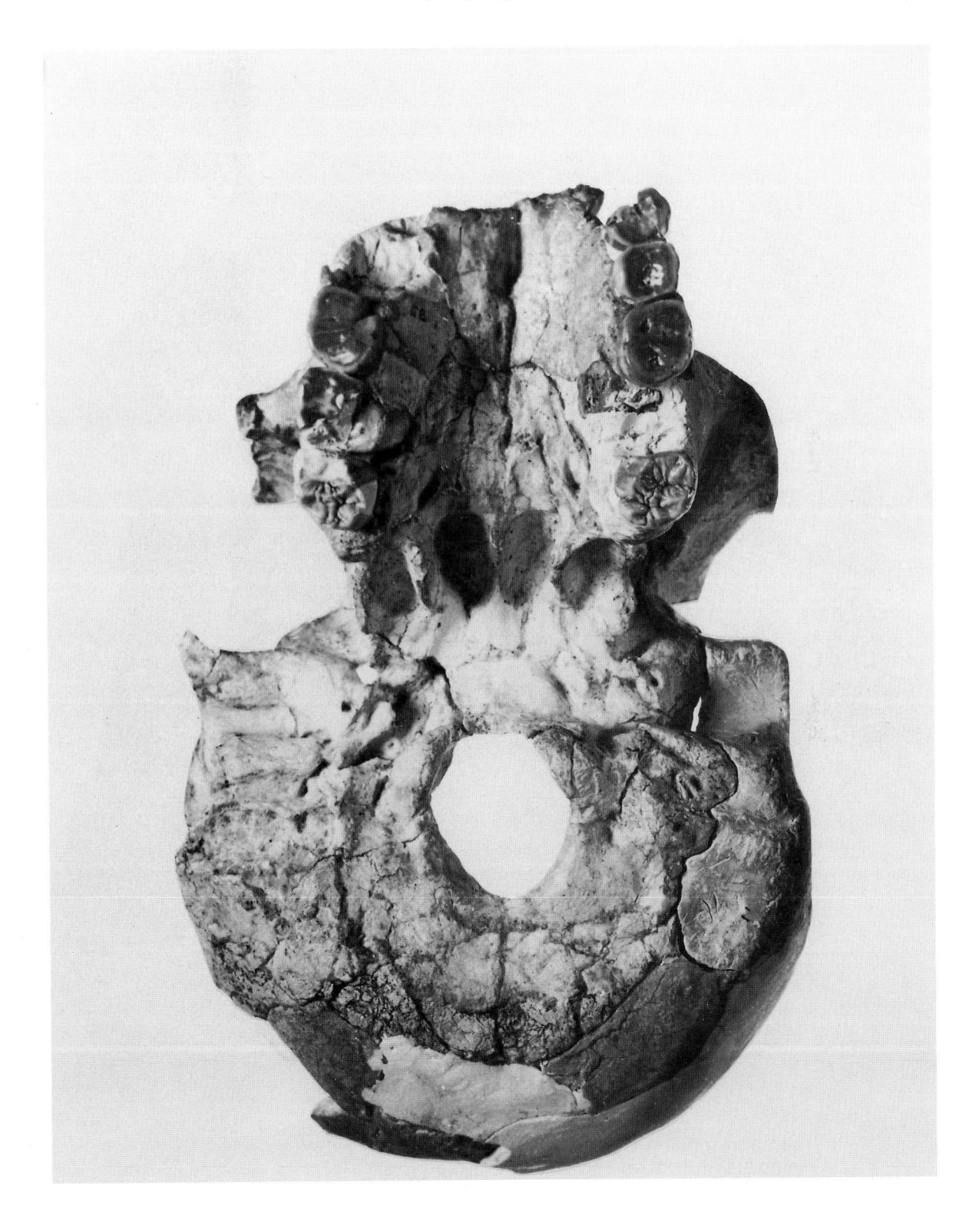

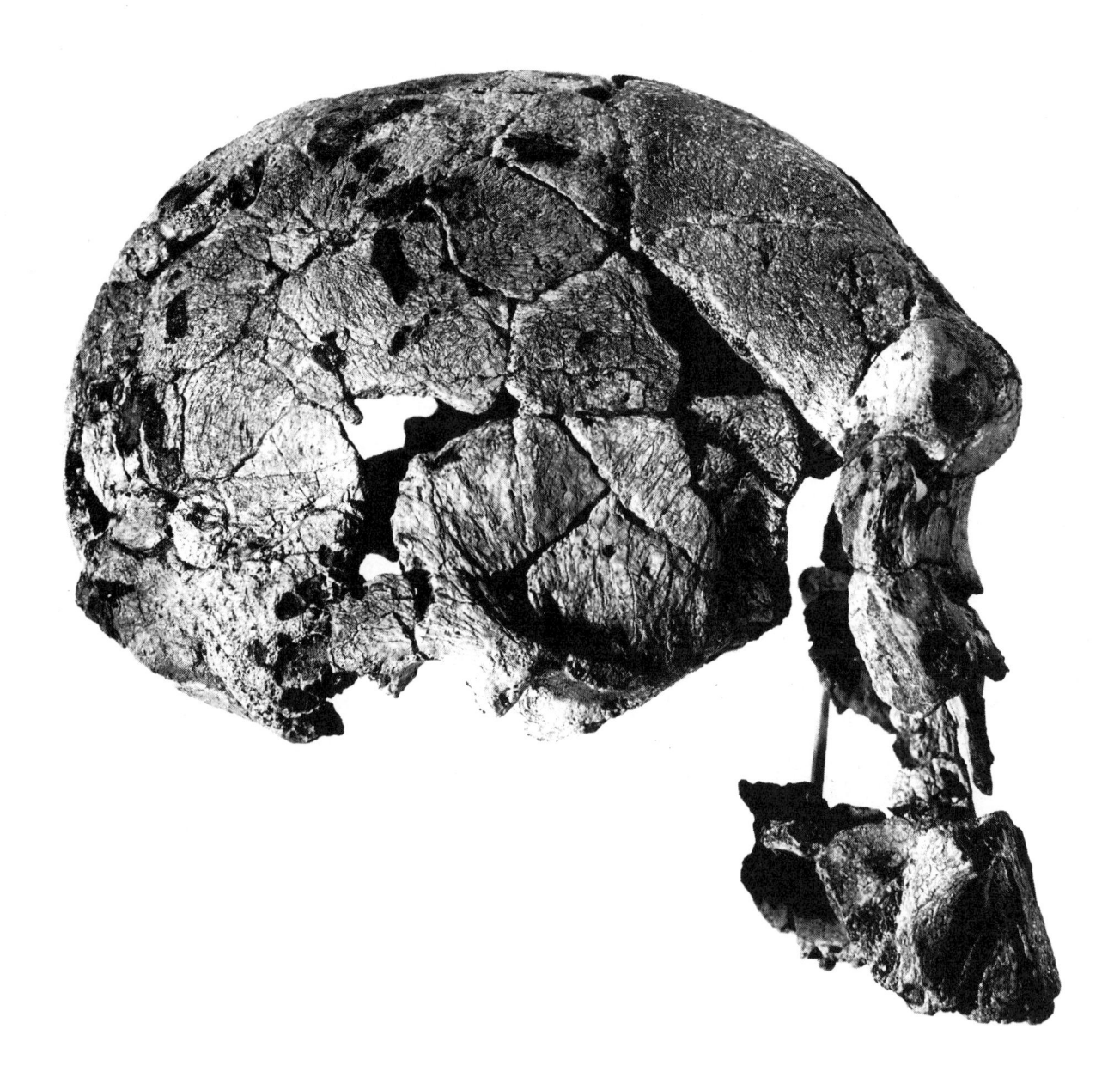

Plate IV. KNM-ER 1470: *Homo habilis*, East Turkana, Kenya, *c*.1.9 million years old. Lateral and frontal views. Courtesy of R. E. Leakey and the National Museums of Kenya. Specimen recently reclassified as *Homo rudolfensis*.

Photogallery of fossil skulls

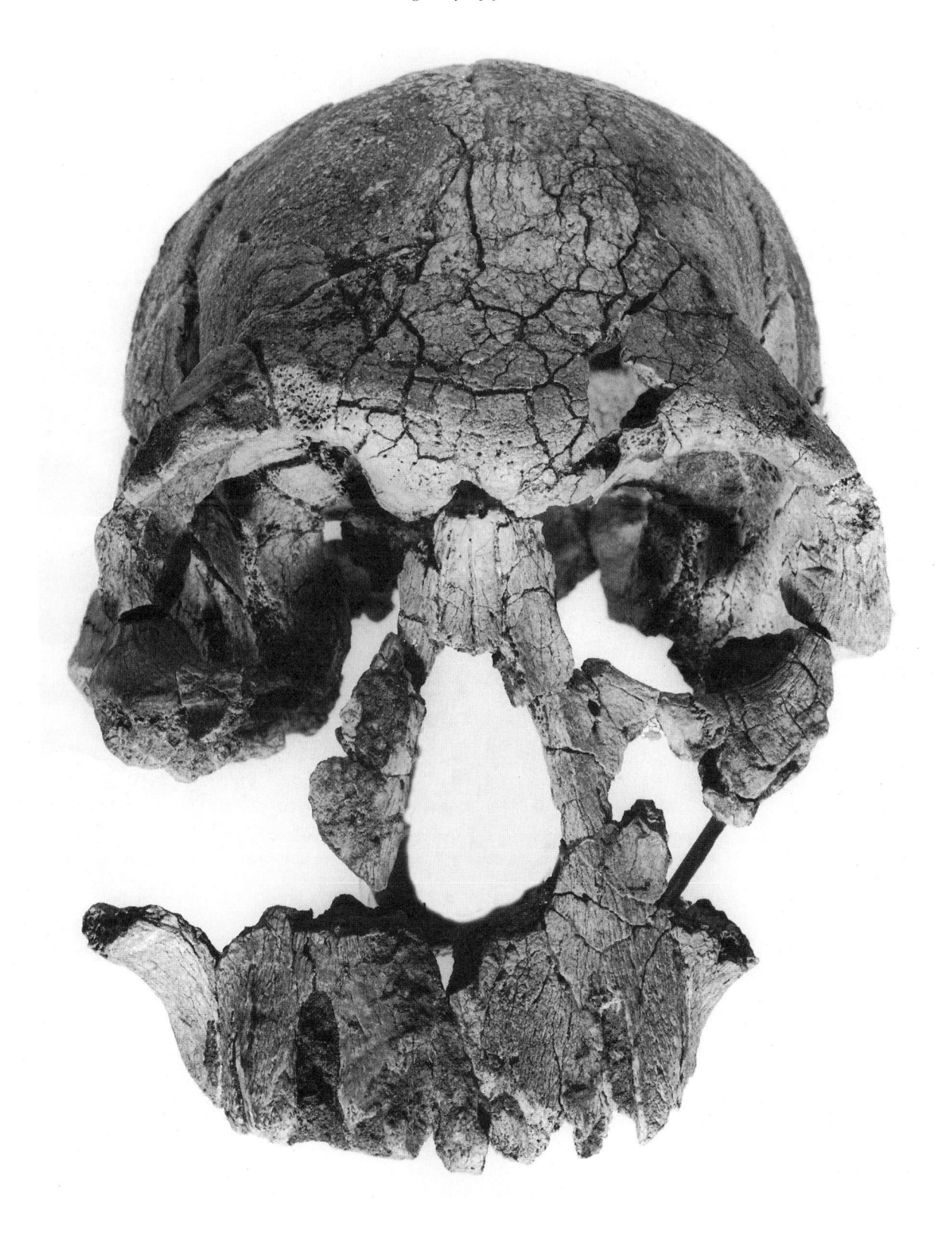

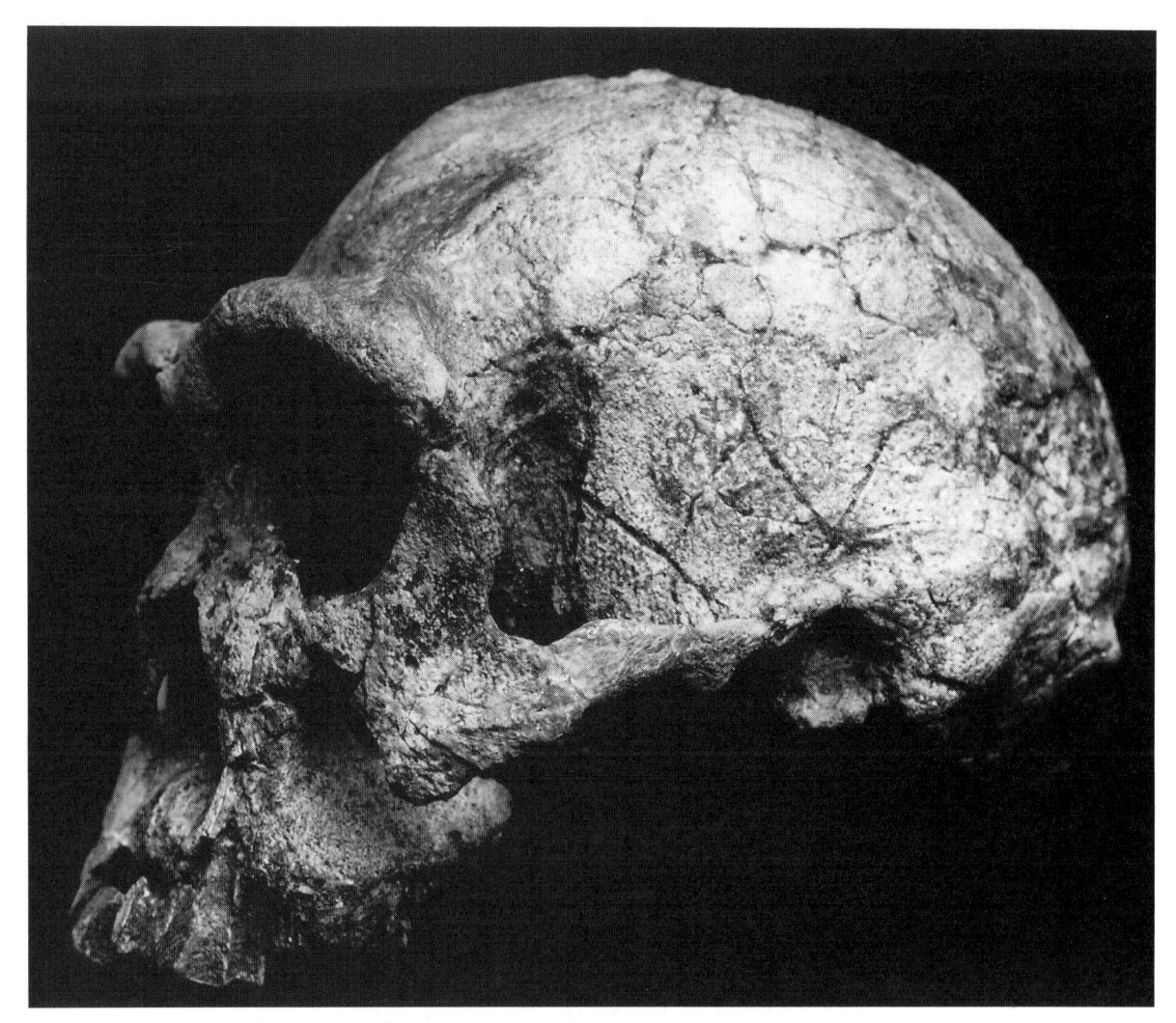

Plate V. KNM-ER 3733: *Homo erectus*, East Turkana, Kenya, c.1.7–1.8 million years old. Near-lateral and basal views. Courtesy of M. Crabtree/AAAS, R. E. Leakey, and the National Museums of Kenya. Specimen reclassified by some authorities as *Homo ergaster*.

Photogallery of fossil skulls

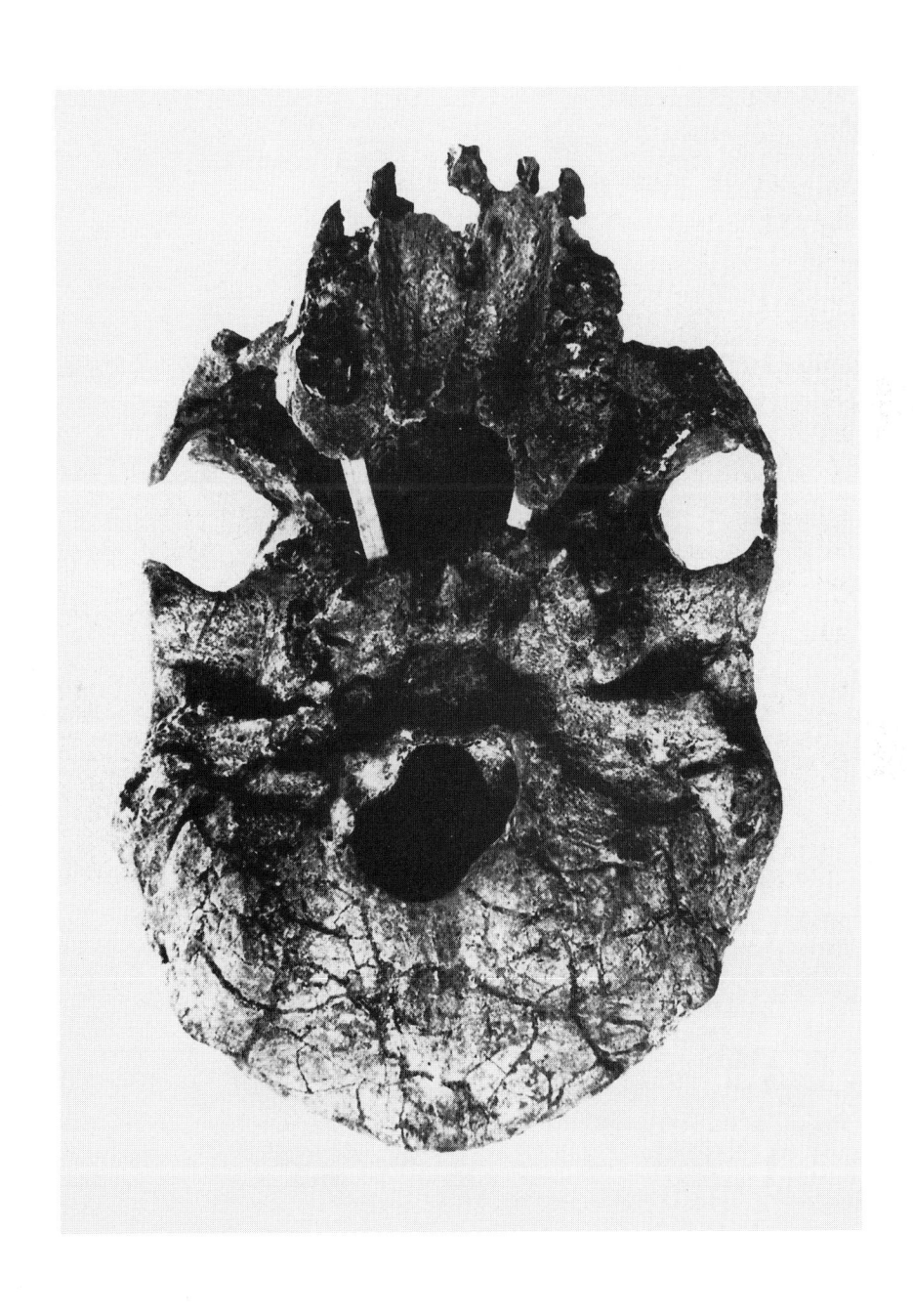

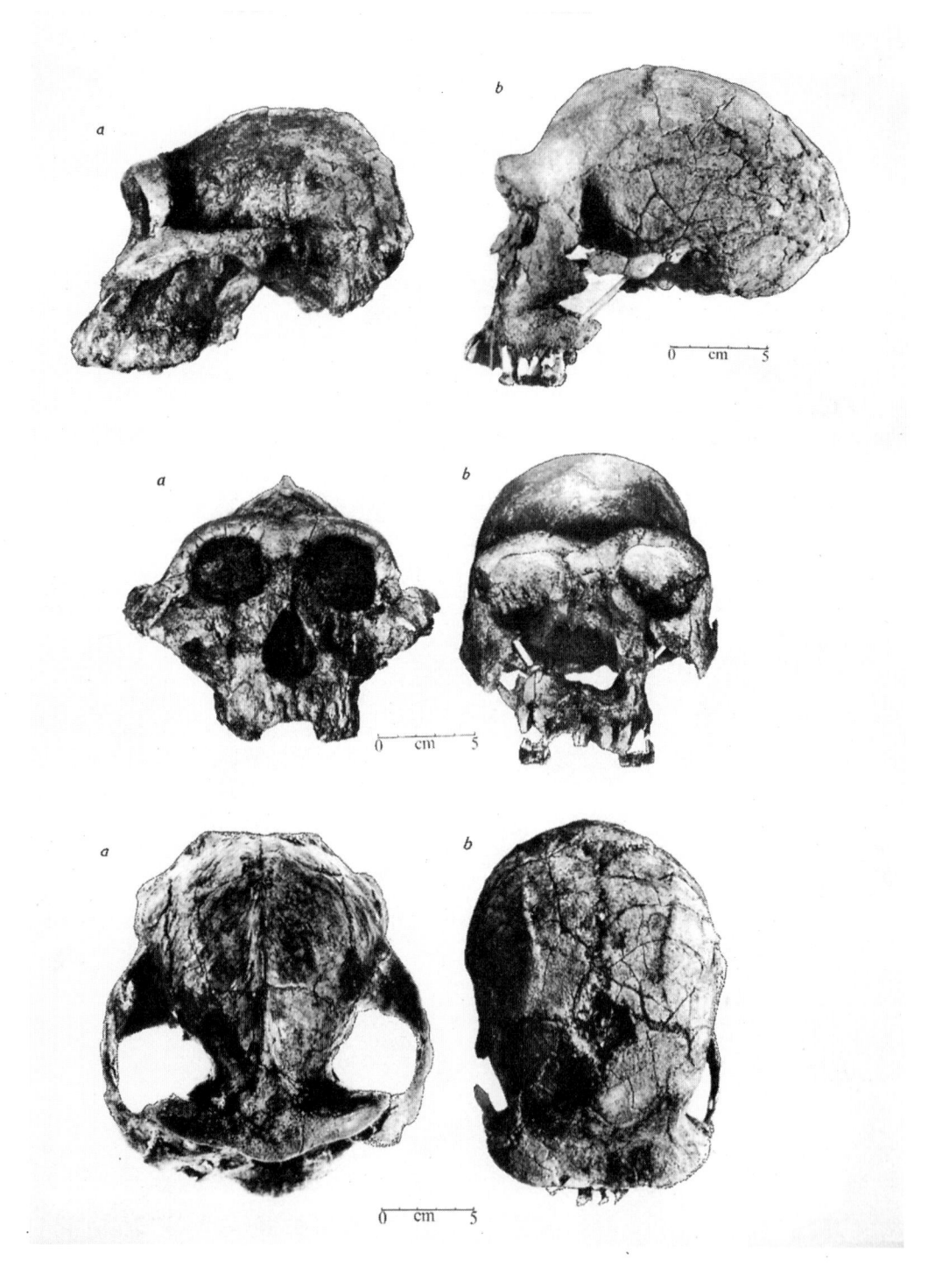

Plate VI. KNM-ER 3733 and KNM-ER 406: Geologically contemporaneous *Homo erectus* and *Australopithecus boisei*, East Turkana, Kenya, c.1.7–1.8 million years old. Lateral, frontal, and superior views with outlines enhanced. After Leakey, R. E. and Walker, A. C. (1976). *Nature* (London) **261**, 572–4.

Plate VII. Petralona: Archaic *Homo sapiens*, Petralona, Greece, *c*.300 000–400 000 years old. Lateral and basal views. Courtesy of R. Murrill and C. B. Stringer.

Photogallery of fossil skulls

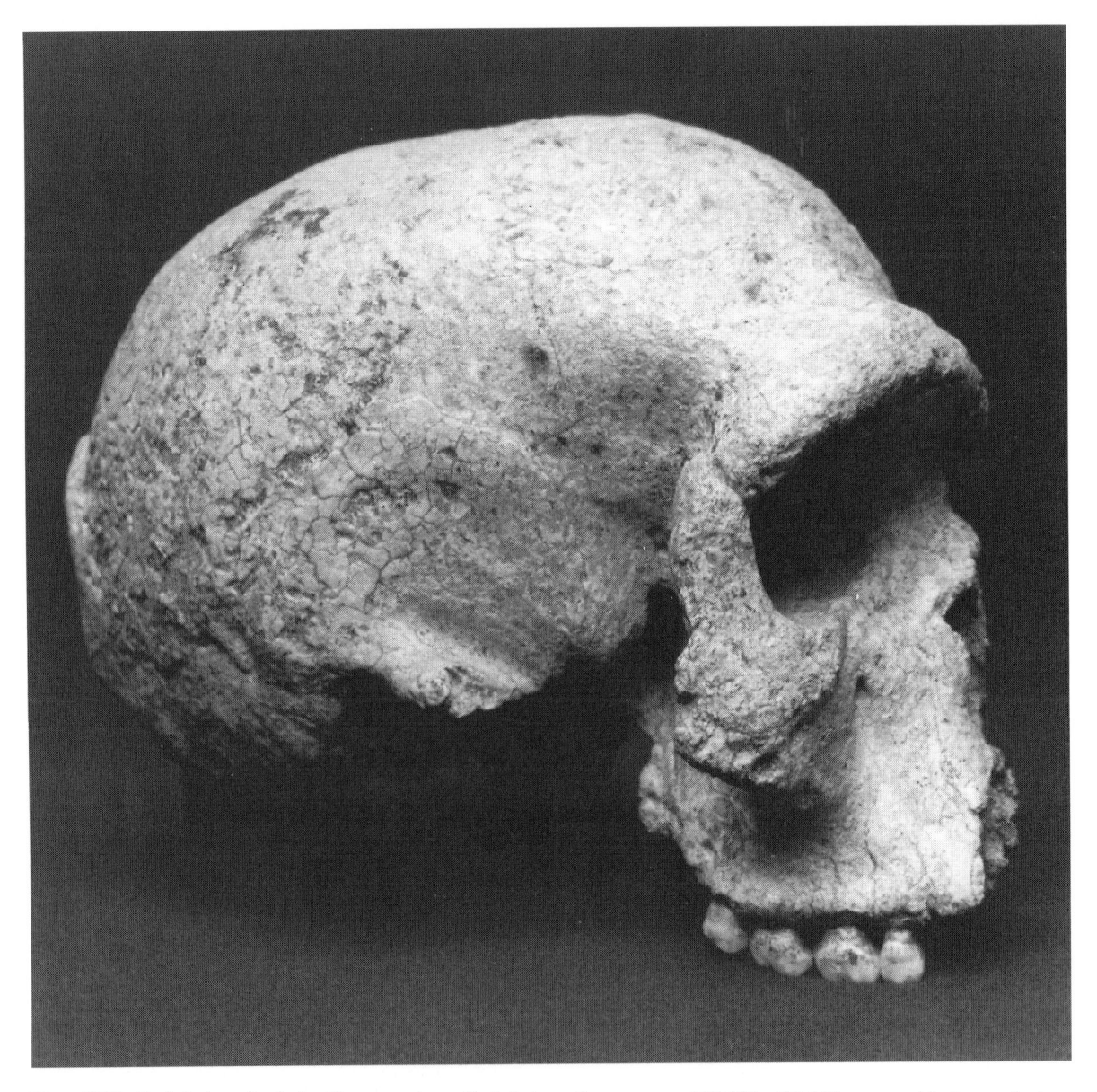

Plate VIII. Steinheim: Archaic *Homo sapiens*, Steinheim, Germany. *c.*300 000–400 000 years old. Lateral view. Courtesy of M. Crabtree/AAAS and the Staatliches Museum für Naturkunde.

Photogallery of fossil skulls

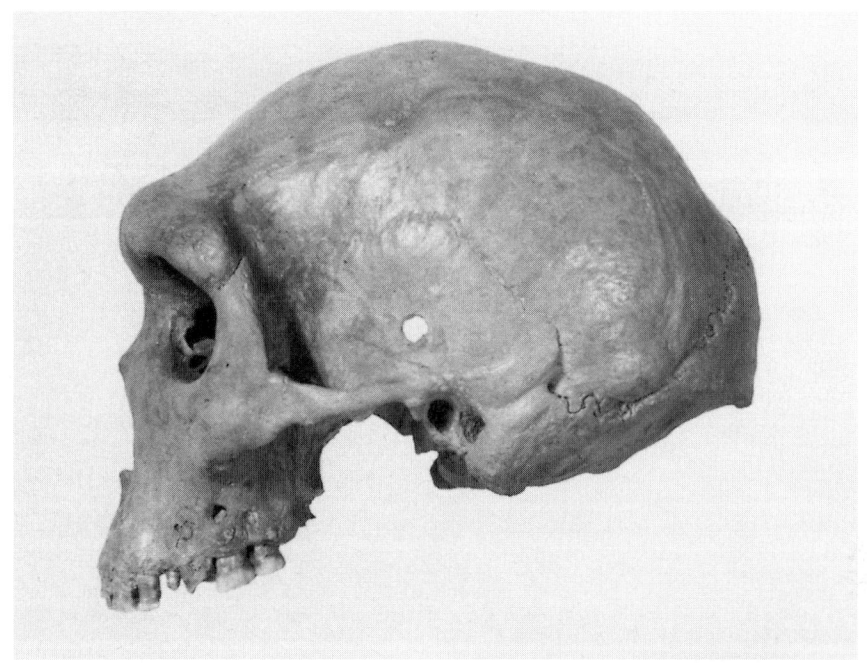

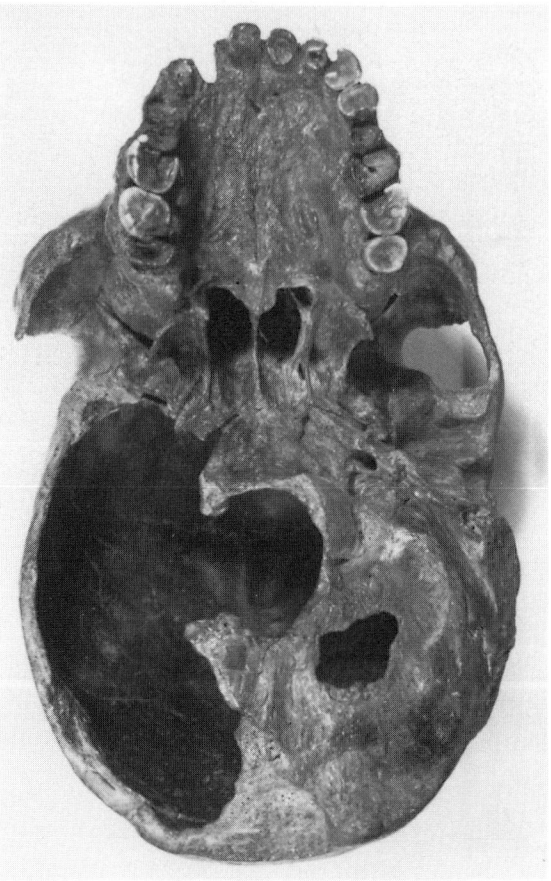

Plate IX. Broken Hill: Archaic *Homo sapiens*, Kabwe, Zambia, c.150 000–300 000 years old. Lateral and basal views. Courtesy of C. B. Stringer and the Trustees of the British Museum (Natural History).

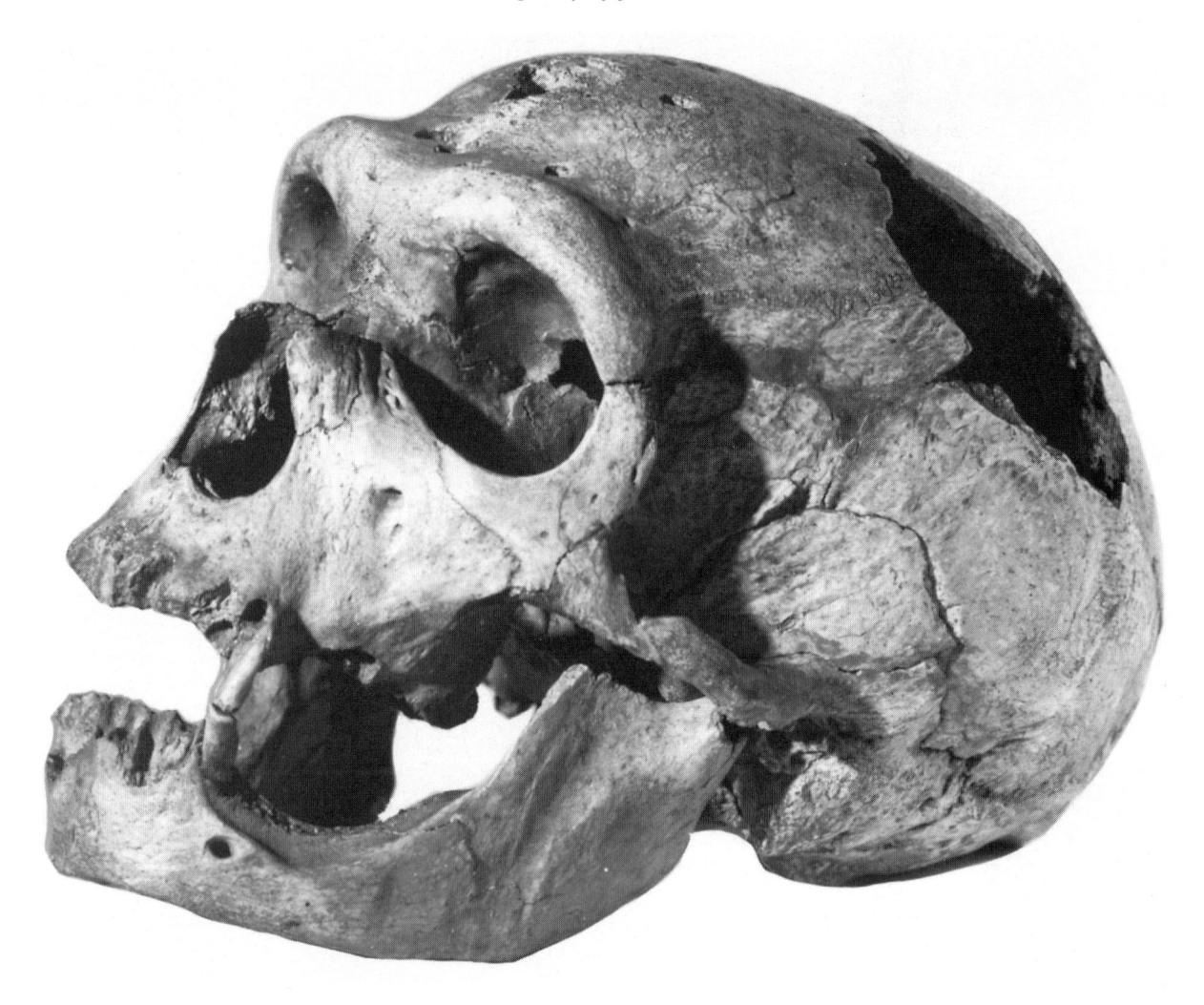

Plate X. La Chapelle-aux-Saints: *Homo sapiens neanderthalensis*, La Chapelle-aux-Saints, France, *c*.40 000–60 000 years old. Three-quarter and basal views. Courtesy of Y. Coppens and the Musée de l'Homme.

Photogallery of fossil skulls

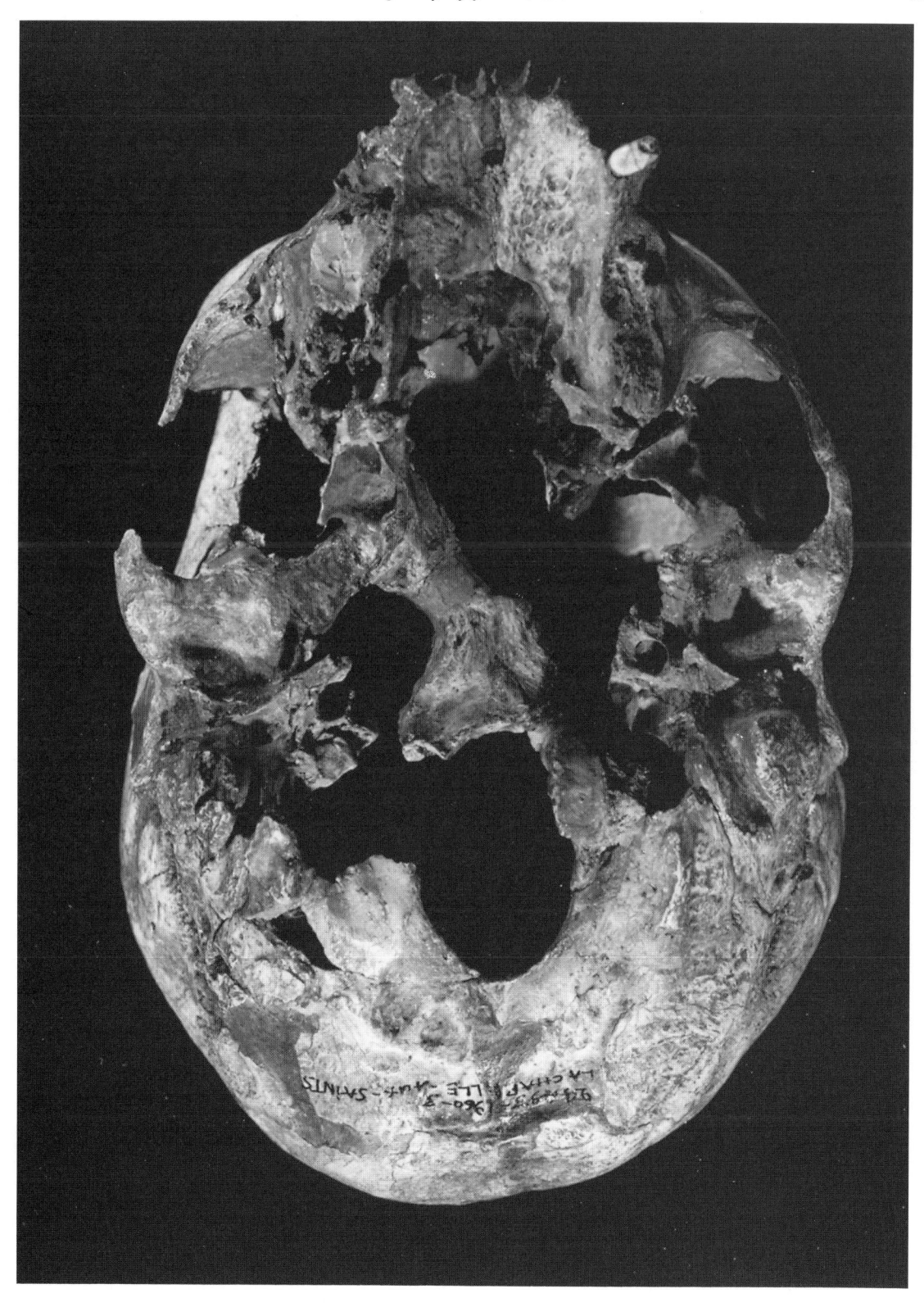

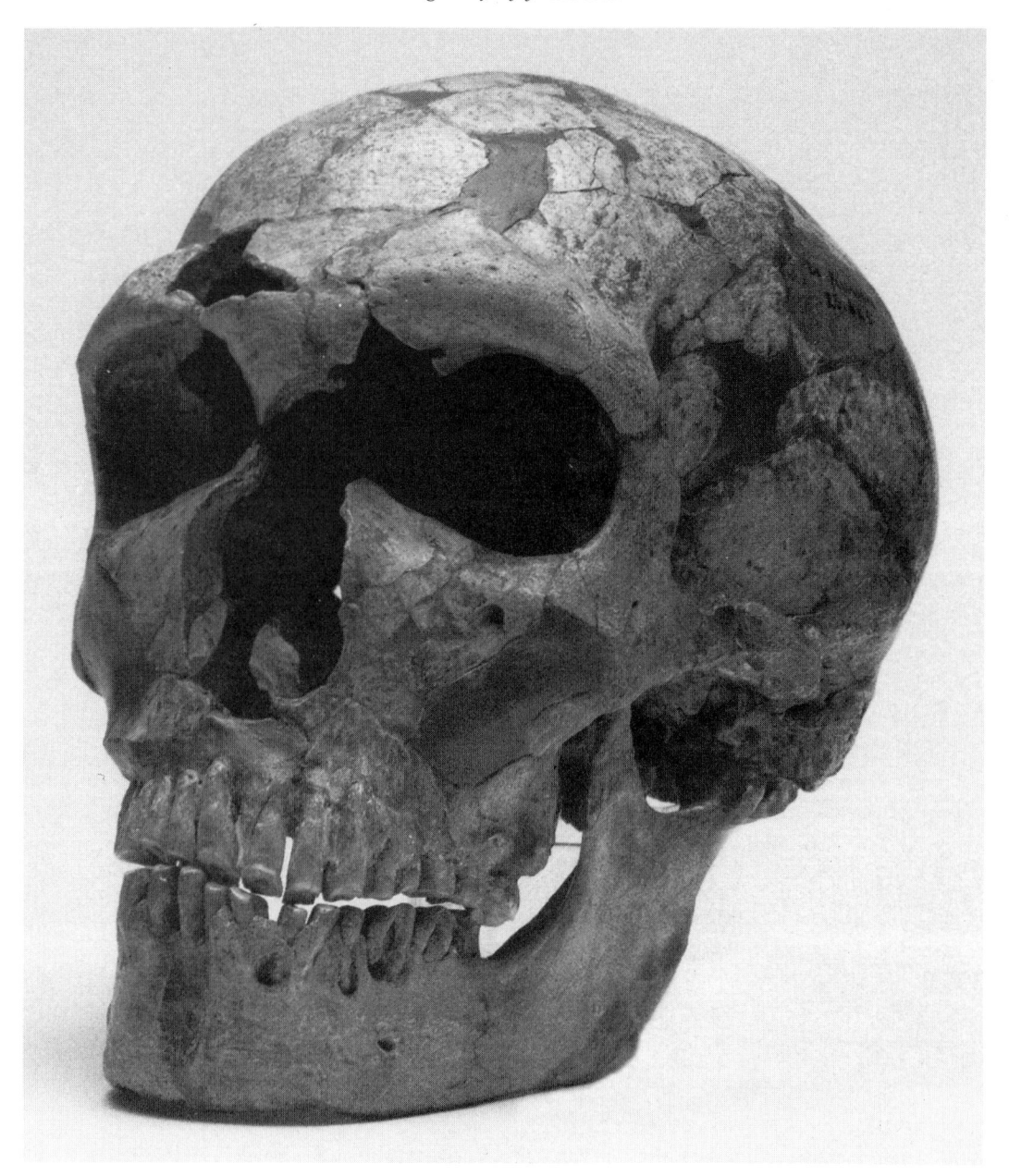

Plate XI. La Ferrassie 1: *Homo sapiens neanderthalensis*, La Ferrassie, France, *c*.40 000–70 000 years old. Three-quarter and basal views. Courtesy of M. Crabtree/AAAS, Y. Coppens, and the Musée de l'Homme.

Photogallery of fossil skulls

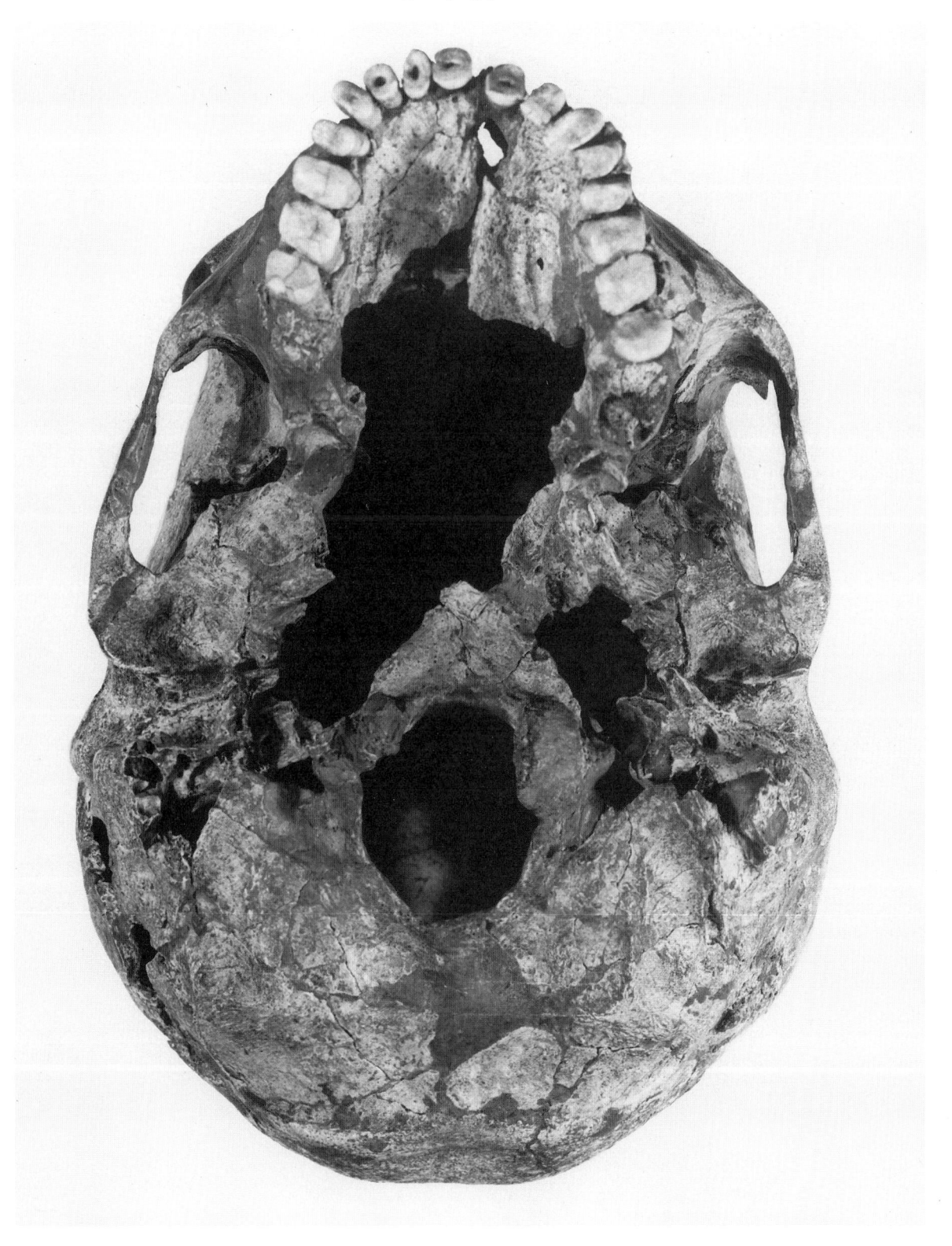

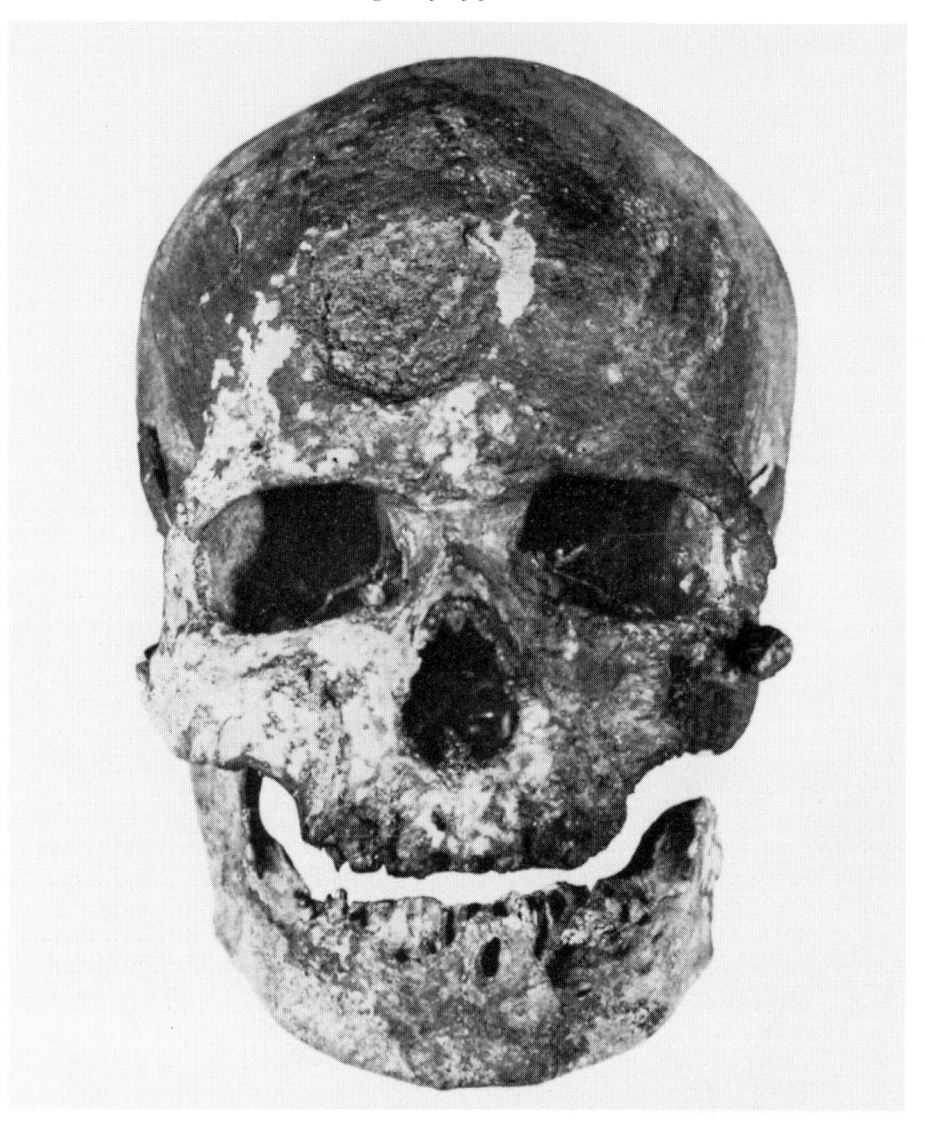

Plate XII. Cro-Magnon 1: *Homo sapiens sapiens*, Cro-Magnon, France, c.28 000–34 000 years old. Frontal, lateral, three-quarter, and basal views. Courtesy of Y. Coppens and the Musée de l'Homme.

Photogallery of fossil skulls

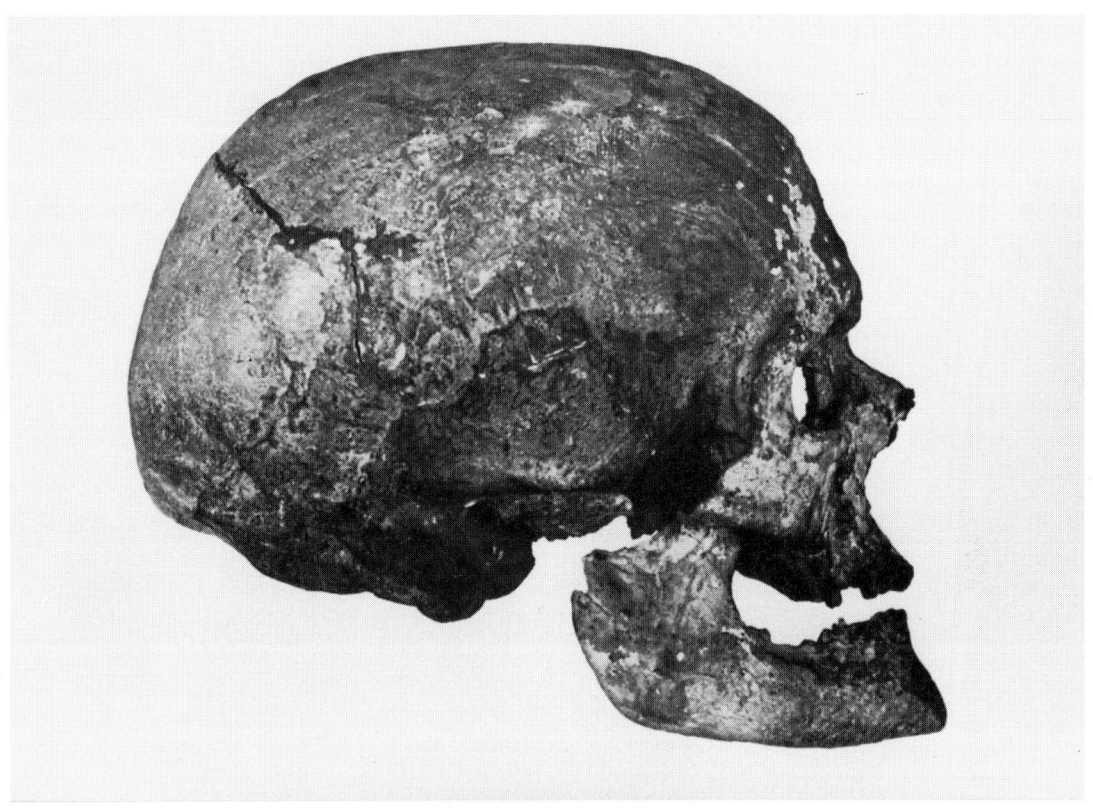

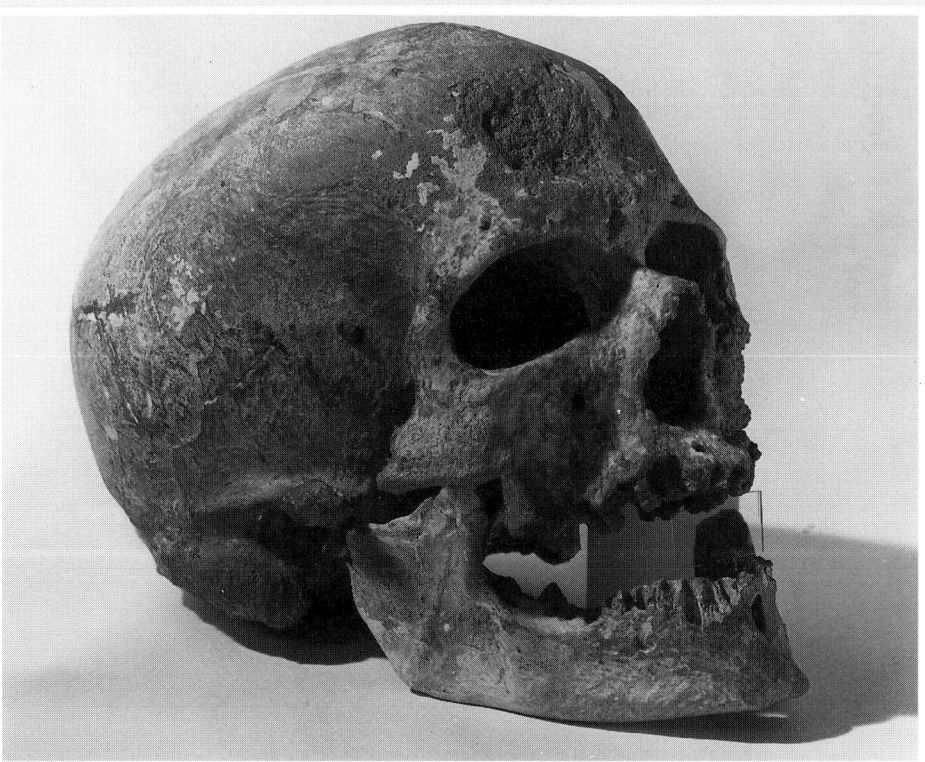

Photogallery of fossil skulls

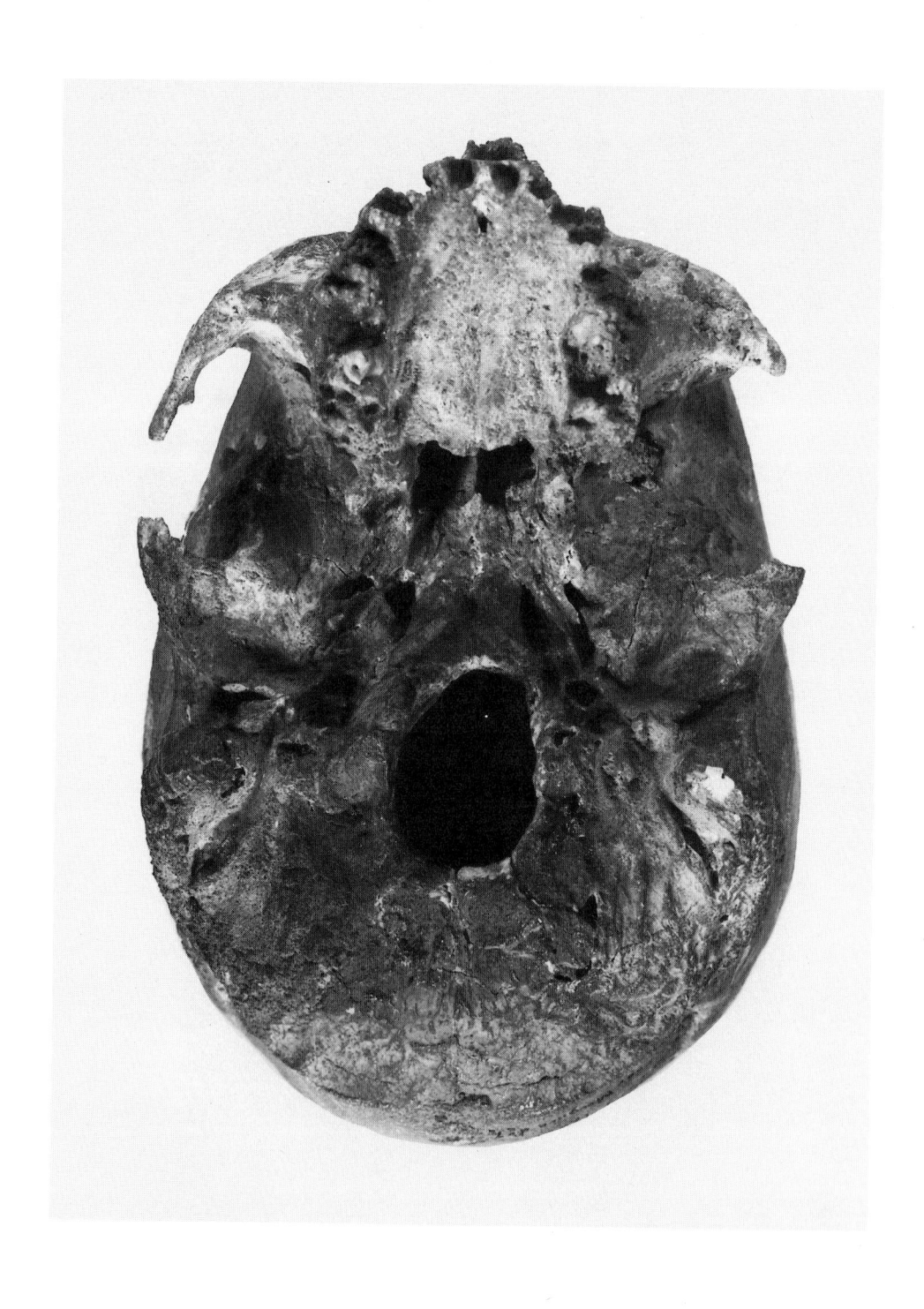

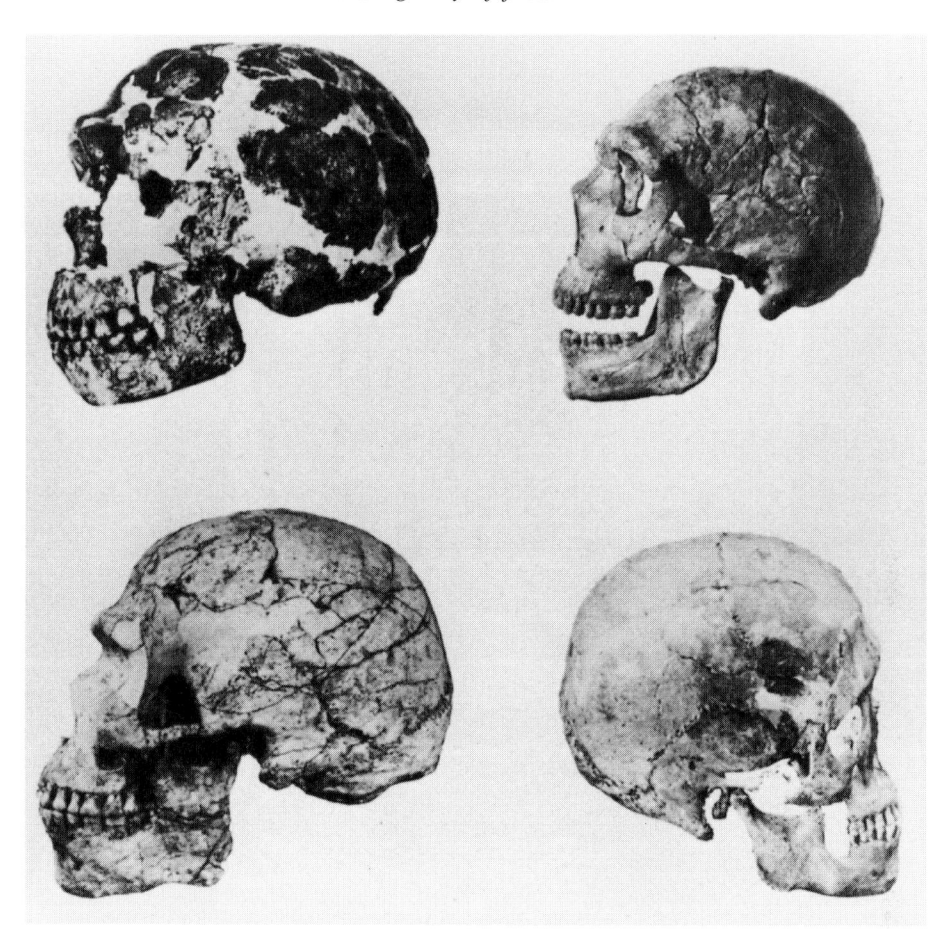

Plate XIII. A sequence of skulls from Israel showing some of the changes that occurred in the transition from archaic to anatomically modern *Homo sapiens* within the last 100 000 years. From left to right, they are (above) from et-Tabūn and Amud, and (below) from es-Skhūl and an Upper Palaeolithic Natufian site in Israel. Courtesy of B. G. Campbell.

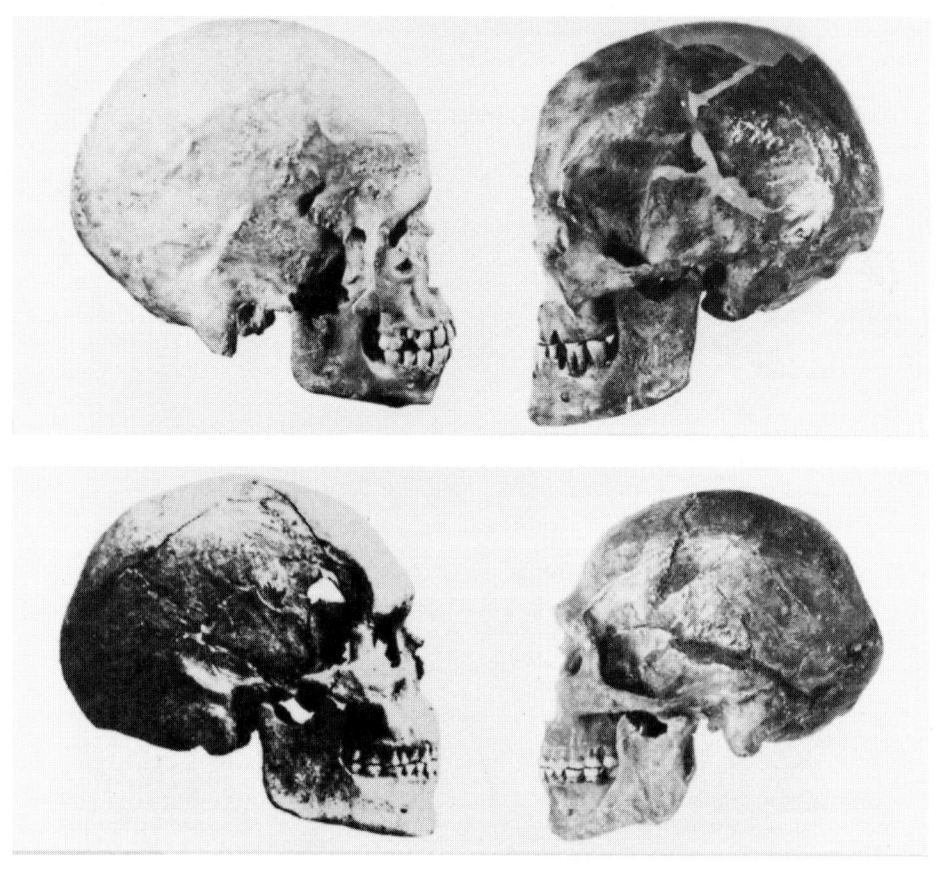

Plate XIV. Four European Cro-Magnon-type skulls showing some of the variation in European Upper Palaeolithic *Homo sapiens sapiens*. These are from Grimaldi in Italy, near the border with France on the Mediterranean coast; from Chancelade and Combe Capelle in France; and from Předmostí in Czechoslovakia. All are dated to *c.*20 000–25 000 years ago. Courtesy of B. G. Campbell.

2

An outline of human phylogeny

Bernard G. Campbell

Abstract

In 1871 Charles Darwin was able to propose that we were most probably of African origin and most closely related to the Great Apes of Africa. Biochemical evidence now reinforces this conclusion and indicates that the divergence of our lineage, the Hominidae, from the African apes took place between 5 and 8 million years ago (m.y.a.).

There are no fossils now believed to lie within our hominid lineage before $c.6.0$ m.y.a. The earliest group of well-known undoubted hominid fossils comes from Laetoli in Tanzania, and dates from $c.3.7$ m.y.a. though earlier fossils are known. These belong to the genus *Australopithecus*, which is considered to range in time from $c.5.5$ m.y.a. to 1 m.y.a., and appears to have been confined to the continent of Africa. *Australopithecus* was a bipedal, small-brained hominid, which later diversified into 2–3 more robustly built species, as well as probably giving rise to members of our own genus, *Homo*.

The earliest fossil remains classified as *Homo*, and thought to be our direct ancestors, come from Ethiopia and Kenya. They are dated to $c.2$ m.y.a. This species, named *Homo habilis*, possessed a somewhat larger brain than *Australopithecus*, and appears at approximately the same time as the earliest stone tools. The successor to *Homo habilis* was the much more modern-looking *Homo erectus*. The earliest specimens are from Asia and Africa, and date to $c.1.9$–1.7 m.y.a. Archaic forms of *Homo sapiens* are variously recognized from Afro-Eurasian specimens dated to $c.300\,000$ years ago. Recent biochemical data suggest that modern humans, *Homo sapiens sapiens*, arose in Africa $c.200\,000$ years ago. This fits in well with the available fossil evidence from Africa and the Near East, where human skeletal material with completely modern features is known from an earlier date than elsewhere.

2.1 The nature of the evidence

The human species belongs to the zoological family Hominidae. This group includes all those species and populations that share a common ancestor with the Great Apes, but have evolved as an independent evolutionary lineage distinct from that leading to the African apes (Fig. 2.1).

We shall consider human phylogeny under seven headings: (1) The nature of the evidence; (2) The earliest apes; (3) The earliest Hominidae: *Australopithecus*; (4) The first of the genus *Homo*: *Homo habilis*; (5) *Homo erectus*; (6) Modern humans: *Homo sapiens*; and (7) Discussion.

The evidence for human origins comes from two sources: living species and fossils. Already in 1758 Linnaeus had classified humans in the order Primates; and in 1871 Charles Darwin (1871) was able to propose that we were most probably of African origin and most closely related to the Great Apes of central Africa (Tables 2.1, 2.2, 2.3). This conclusion was reached without any evidence from fossils. Huxley reinforced the relationship by his careful analysis (Huxley 1863), and they were both correct. Although since that time many other theories have been proposed, such as an origin among the prosimians (Kurten 1972) or from an Asian stock of apes (Campbell and Bernor 1976; Schwartz 1987), the evidence

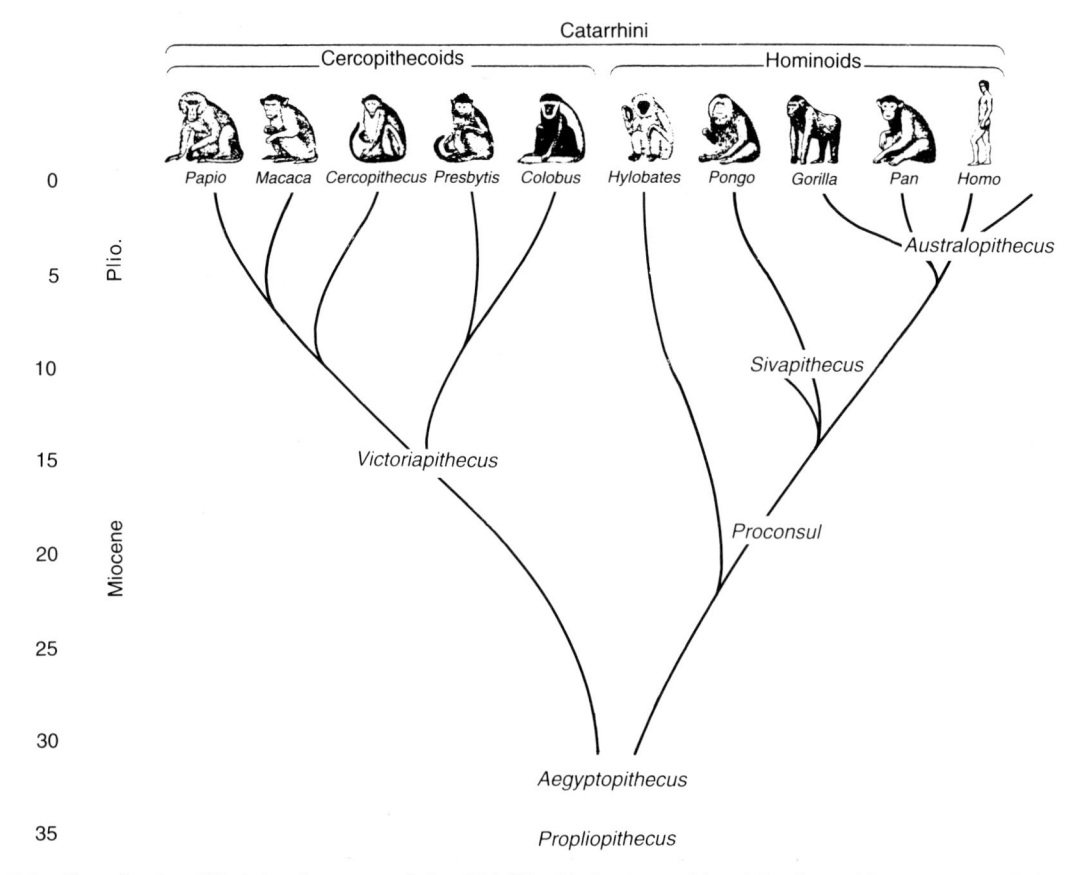

Fig. 2.1 Greatly simplified dendrogram of the Old World Anthropoidea. The branching sequence is based on biochemical estimates of relationships, and the branch timing is based on both biochemical age calculations and on geological evidence of the age of the fossils. The reality this diagram represents is very complex, and all the drawn lines are hypothetical (after Pilbeam 1984).

for an African origin is now compelling. As subdivisions of this evidence we can consider both comparative anatomy and comparative biochemistry.

Thanks to over 100 years of active research since Darwin's day, we now possess a priceless collection of fossil specimens which go some way to indicate the route of human evolution from our remote primate ancestors (Delson 1985). The fossil record is necessarily incomplete, since we are considering a period approaching 35 million years since the first creatures with ape-like traits are recorded. Nevertheless, the fossil evidence is compelling, especially during the last 4 million years, and the story of our evolution is now being gradually unfolded.[1]

No lineage of fossil hominids can be reconstructed without good evidence of the age of the fossils, both in relation to each other and in absolute terms. This is not the place to review dating methods; but it is a fortunate fact that most of the fossil evidence for human phylogeny comes from the area of the Great Rift valley in Africa (which stretches from Ethiopia to Lake Nyasa). Because of its volcanic origin, this geological formation contains precisely the types of deposit which are most suitable for fossil preservation and for geochemical dating techniques. We are, therefore, in a position to place the phylogeny in a reasonably well-based time-scale.

2.2 The earliest apes

Our present knowledge suggests that the primates themselves may have evolved during the Late Cretaceous period in North America and were certainly present in North Africa (Morocco) during the Palaeocene. Others were present in Asia at this time and by the Eocene, fossils are found in Africa, Eurasia, and N. America. The early history of the group is

Table 2.1 A simplified and traditional classification of living primates showing the relationships of the different taxonomic groups

ORDER: PRIMATES	
Suborder: Prosimii	
Infraorder: Lemuriformes	Lemurs
	Indris
	Sifakas
Lorisiformes	Lorises
	Bush-babies
Tarsiiformes	Tarsiers
Suborder: Anthropoidea	
Infraorder: Platyrrhini	New World monkeys
Catarrhini	Old World monkeys, apes, and humans
Superfamily: Cercopithecoidea	Old World monkeys
Hominoidea	Apes and humans
Family: Hylobatidae	Lesser apes
Pongidae	Orang-utans
Panidae	African apes
Hominidae	Humans

still poorly understood, but the primate characteristics can be broadly understood as a process of adaptation to insect predation in an arboreal environment. In the following period, the Oligocene, we have an extensive fossil sample from the Fayum region of N.Egypt. This group includes the earliest known specimens of any higher primate; that is to say, the fossils are the first known members of the Anthropoidea—the group containing the monkeys, apes, and humans (see Fig. 2.1).[2] Human phylogeny, then, really begins here, where, among fossils discovered by the Yale expeditions to Egypt over the last fifty years, we have examples of kinds of primitive ape-like creatures which could well have given rise both to the later apes and humans, and probably to the monkeys as well. The Fayum fossils contain two genera which require consideration in a study of human phylogeny:

Propliopithecus. A few mandibles have been attributed to this genus and dated between 35 and 33 m.y.a. Their dentition carries certain characteristics which suggest they may have given rise to the later Hominoidea, but the evidence is still very limited. At the higher level in the deposits, numerous remains of a second genus have been found which carries more strikingly ape-like characters in its dentition.

Aegyptopithecus, found in strata dated to *c.*33 m.y.a. (Kappelman 1992) bears a dentition which suggests that it may well stand at the base of the lineage of all the apes, leading eventually to humans. Although it was no bigger than a spaniel, had a head shaped somewhat like that of a present-day monkey, and carried a long tail, the evidence of jaws and teeth suggests it was already an ancestral ape (Fleagle and Kay 1983). Some authors believe that it may also have been ancestral to monkeys, because both apes and monkeys share some derived characters not present in *Aegyptopithecus* (Kay et al. 1981). Simons, who led the expeditions which made the discoveries of this important fossil, believes that it was directly ancestral to the later Miocene ape, *Proconsul.*

Proconsul. First found in Kenya in the early 1930s, and dating from 20 to 18 m.y.a., this fossil genus is now well represented and generally accepted as having a remarkably close resemblance to the later Great Apes of Central Africa—the chimpanzee and gorilla. It has been found in three sizes, usually classified as three species, and most parts of the skeleton of the small species, *P. africanus*, are known. It weighed 9–11 kg, and stood 45 cm high at the shoulder when standing on all four limbs. It was a robust unspecialized ape—probably a powerful quadrupedal climber, not yet adapted to either brachiation, knuckle-walking, or fast terrestrial quadrupedalism. Its ape-like character was still expressed most clearly in the dentition. Although intermediate fossils are lacking, there seems to be general agreement that the genus *Proconsul* is very likely to be ancestral to the living African apes and humans.

Figure 2.1 summarizes the phylogeny outlined. Although the dendrogram is drawn with a single line, that line represents a complex succession of diverging and converging populations, with numerous speciation events and extinctions. What does seem almost certain is that this early phase of Hominoid evolution took place in Africa, probably in the central forested

Table 2.2 A list of living primate genera placed in their different families with their common names and geographical ranges (after Campbell and Loy 1966)

Suborder	Superfamily	Family	Genus	Common name	Location
Prosimii	Lemuroidea		9 genera	Lemurs	Madagascar
	Daubentonioidea		1 genus	Aye-ayes	Madagascar
	Lorisioidea		5 genera	Lorises	Africa/Asia
	Tarsioidea		1 genus	Tarsiers	South-east Asia
Anthropoidea	Ceboidea			New World monkeys	Central/South America
		Callithricidae	*Callithrix*	Marmosets	South America
			Cebuella	Pygmy marmosets	South America
			Saguinus	Tamarins	Central/South America
			Leontideus	Golden lion tamarins	South America
			Callimico	Goeldi's marmosets	South America
		Cebidae	*Pithecia*	Sakis	South America
			Chiropotes	Bearded sakis	South America
			Cacajao	Uakaris	Central/South America
			Aotus	Douroucoulis	Central/South America
			Callicebus	Titis	Central/South America
			Saimiri	Squirrel monkeys	Central/South America
			Cebus	Capuchins	Central/South America
			Alouatta	Howler monkeys	Central/South America
			Ateles	Spider monkeys	Central/South America
			Lagothrix	Woolly monkeys	South America
			Brachyteles	Woolly spider monkeys	South America
	Cercopithecoidea			Old World monkeys	Africa/Asia
		Cercopithecidae	*Cercopithecus*	Guenons	Africa
			Erythrocebus	Patas monkey	Africa
			Cercocebus	Mangabeys	Africa
			Mandrillus	Mandrills	Africa
			Papio	Baboons	Africa
			Theropithecus	Geladas	Africa
			Macaca	Macaques	Asia/Africa
			Cynopithecus	Celebes black ape	Asia
		Colobidae	*Colobus*	Guerezas	Africa
			Presbytis	Langurs	Asia
			Pygathrix	Douc langurs	Asia
			Rhinopithecus	Snub-nosed langurs	Asia
			Nasalis	Proboscis monkeys	Asia
			Simias	Pagai Island langurs	Asia
	Hominoidea			Apes and humans	World-wide
		Hylobatidae	*Hylobates*	Gibbons	South-east Asia
			Symphalangus	Siamangs	South-east Asia
		Pongidae	*Pongo*	Orang-utans	South-east Asia
		Panidae	*Pan*	Chimpanzees	Africa
			Gorilla	Gorilla	Africa
		Hominidae	*Homo*	Humans	World-wide

Note: Authors differ in details of Primate classification. This table presents a classification that is widely accepted.

region. The fossils we have were preserved because the populations they represent lived near the then developing Rift Valley, where changes in land levels resulted in considerable alluvial deposits interleaved with volcanic ash and lavas. Thus we are able to give approximate dates to the fossils, and get an idea of the chronological extension of human phylogeny.

Particular lines of biochemical evidence vary in their interpretation, but all point to the general conclusion that humans are more closely related to the

Table 2.3 Revised classification of the Primates: this recent classification shows some changes in taxonomy and nomenclature at present in progress but not yet widely recognized

Suborder	Superfamily	Family	Subfamily	Genus	Common name
Strepsirhini	Lemuroidea	Cheirogaleidae		*Allocebus*	Hairy-eared dwarf lemur
				Cheirogaleus	Dwarf lemur
				Microcebus	Mouse lemur
				Mirza	Coquerel's dwarf lemur
				Phaner	Fork-marked lemur
		Daubentoniidae		*Daubentonia*	Aye-aye
		Indriidae		*Avahi*	Woolly lemur
				Indri	Babakoto
				Propithecus	Sifaka
		Lemuridae		*Lemur*	Lemur
				Varecia	Ruffed lemur
		Lepilemuridae	Hapalemurinae	*Hapalemur*	Gentle lemur
			Lepilemurinae	*Lepilemur*	Sportive lemur
	Lorisoidea	Lorisidae	Galaginae	*Euoticus*	Needle-clawed galago
				Galago	Bush-baby
			Lorisinae	*Arctocebus*	Angwantibo
				Loris	Slender loris
				Nycticebus	Slow loris
				Perodicticus	Potto
Haplorhini	Tarsii	Tarsioidea	Tarsiidae		*Tarsius*	Tarsier
	Platyrrhini	Ceboidea	Callimiconidae		*Callimico*	Goeldi's marmoset
			Callitrichidae		*Callithrix*	Marmoset
					Cebuella	Pygmy marmoset
					Leontopithecus	Golden lion tamarin
					Saguinus	Tamarin
			Cebidae	Alouattinae	*Alouatta*	Howler monkey
				Aotinae	*Aotus*	Owl or night monkey
				Atelinae	*Ateles*	Spider monkey
					Brachyteles	Woolly spider monkey
					Lagothrix	Woolly monkey
				Callicebinae	*Callicebus*	Titi monkey
				Cebinae	*Cebus*	Capuchin monkey
				Pitheciinae	*Cacajao*	Uakari
					Chiropotes	Bearded saki
					Pithecia	Saki
				Saimiriinae	*Saimiri*	Squirrel monkey
	Catarrhini	Cercopithecoidea	Cercopithecidae	Cercopithecinae	*Allenopithecus*	Swamp monkey
					Cercocebus	Mangabey
					Cercopithecus	Guenon
					Erythrocebus	Patas
					Macaca	Macaque
					Miopithecus	Talapoin
					Papio	Savanna baboon
					Theropithecus	Gelada baboon
					Colobus	Colobus monkey
					Nasalis	Proboscis monkey

Table 2.3 (*Contd.*)

Suborder	Superfamily	Family	Subfamily	Genus	Common name
			Colobinae	*Presbytis*	Langur
				Pygathrix	Douc langur
				Rhinopithecus	Golden monkey
	Hominoidea*	Hominidae	Gorillinae	*Gorilla*	Gorilla
				Pan	Chimpanzee
			Homininae	*Homo*	Human
		Pongidae		*Pongo*	Orang-utan
		Hylobatidae		*Hylobates*	Gibbon

* This classification is based on recent molecular evidence. Note that the family Hominidae here includes the African Apes. Traditional classifications put *Gorilla* and *Pan* in a separate family from *Homo*.
From: Richard (1985).

African apes than to other primates (see Waddell and Penny, this volume, Chapter 3). In 1969 Wilson and Sarich published data on the biochemical differences between the albumens of a number of living primate species, using immunological techniques (Wilson and Sarich 1969). Phylogenetic relationships based on only one protein, such as albumen, cannot be regarded as being wholly definitive. However, the albumen studies suggested that the living African apes and humans are much more closely related than their morphological differences might suggest.

Wilson and Sarich proceded to calculate the rate of immunological change by calibrating 'immunological distance' against a known and dated event in the evolutionary calendar (viz the divergence of Anthropoidea and Prosimians at 70 m.y.a.). This produced the result that the African apes and Hominids diverged between 5 and 8 m.y.a. Within these limits, this result has proved to be congruent with the palaeontological evidence. The technique, and others related to it, are of great interest to palaeontologists. By combining the biochemical data of a group of similar measures (derived from electrophoretic analysis, protein sequencing, nucleic acid hybridization, and restriction endonuclease analysis of mitochondrial DNA), a branching sequence can be generated (Fig. 2.2) which is widely accepted, with the branching points dated as follows:

Table 2.4 Estimated branching points in human evolution (Andrews and Cronin 1982)

Human–gorilla–chimpanzee	5 ± 1.5 m.y.a.
Human–orang-utan	10 ± 3.0 m.y.a.
Human–gibbon	12 ± 3.0 m.y.a.

2.3 The earliest Hominidae: *Australopithecus*

Following *Proconsul*, there are few fossils that are now believed to lie on the lineage leading to humans until we come to about 5 m.y.a. (Andrews 1992). Fossils of *Ramapithecus* and *Dryopithecus* which at previous times were thought possibly to lie on the human lineage now seem most likely to form part of lineages which, respectively, lead to the orang-utan or become extinct.

Most of the earliest hominid specimens have been found in East Africa. Recent dicoveries from Aramis in Ethiopia dated to *c*.4.4 m.y.a have been named *Australopithecus ramidus* (alternatively *Ardipithecus ramidus*) (White *et al.* 1994).[3] Somewhat younger and better known early hominid specimens from Tanzania and Ethiopia have been named *Australopithecus afarensis*. These are finds from Laetoli in Tanzania dated *c*.3.8–3.5 m.y.a.[4] and from Hadar in Ethiopia (not far from Aramis) dated to 3.0–3.4 m.y.a.[5] *Australopithecus* is extremely well known, considering its

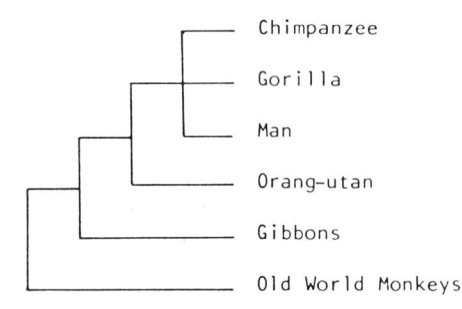

Fig. 2.2 A study of shared and derived characters will generate a branching sequence such as this. Here, biochemical data are used to demonstrate the relationships between humans and the various apes. The branching points are dated in Table 2.4 (from Andrews and Cronin 1982).

An outline of human phylogeny

Table 2.5 Four widely recognized species to which *Australopithecus* fossils have been assigned

Name	First found	Geographic range	Age (m.y.a.)	Type
A. afarensis	1973	Northeast Tanzania, Ethiopia	3.7–3.0	gracile
A. africanus	1924	Transvaal, South Africa	3.0–2.3	gracile
A. robustus	1937	Transvaal, South Africa	1.9–1.6	robust
A. boisei	1959	Tanzania, Kenya, Ethiopia	2.5–1.2	hyper-robust

Table 2.6 Mean crown areas of maxillary (cheek) teeth in mm^2 (Tobias 1967)

Tooth	A. africanus	A. boisei	Modern humans (USA)
PM 1	110	185	66
PM 2	117	212	60
M 1	173	269	126
M 2	213	361	106
M 3	206	336	91

PM = premolar; M = molar.
Measurements for *A. boisei* are from the first complete cranium, discovered at Olduvai Gorge.

antiquity. It can be considered to range from *c*.5 to 1 m.y.a. One skeleton is 40 per cent complete (which means about 70 per cent if we take a mirror image), and the number of specimens (i.e. informative fragments) known lies in the hundreds. The fossil sample is therefore relatively large, and is highly variable, and thus difficult to interpret with assurance; it is generally agreed, however, that at least four and possibly six species are represented. The genus appears to have been confined to the continent of Africa.

Australopithecus was a bipedal, small-brained hominid, and most of the various fossils have been assigned to one of two main types, one heavily built and robust, the other comparatively slender and delicate, or gracile. Many investigators, however, have divided the genus into four well defined species, and these are shown in Table 2.5.

Morphologically we can put these four species in an approximate size order: *A. afarensis* – *A. africanus* – *A. robustus* – *A. boisei*, passing from the most gracile to the most robust, although *A. afarensis* and *A. africanus* are very close in this respect (Fig. 2.3). However, the series does show another trend which is of the greatest interest: an increase in the importance of the molar teeth and a decrease in importance of the front teeth—especially the canines, but, in the most robust groups, also the incisors.

The increase in relative and absolute molar size, especially in the robust species, is particularly striking; and this development has also affected the premolar teeth. These have become greatly enlarged, and have evolved more cusps, so that they look increasingly like molar teeth. This process has been called 'molarization', and it is a product of the adaptive increase in dental crushing and grinding surface. The crown areas of these teeth in the upper jaw are shown in Table 2.6, together with figures for modern humans.

This development makes it clear that a remarkable adaptation was under way towards a very powerful crushing and grinding dentition which was associated with a reduction in the anterior teeth. Nothing remotely comparable with this is seen in any other primate lineage, dead or living. Bearing in mind that the dating of South African finds is less reliable than that of the East African fossils, we can nevertheless identify what may be an evolutionary sequence, as shown in Table 2.7.

Table 2.7 Time-range of *Australopithecus* species

Approximate age; m.y.a.	Species
2.5–1.2	A. boisei and A. robustus
3.0–2.3	A. africanus
3.7–3.0	A. afarensis
4.4	A. ramidus

We will now consider the four major species in turn. *Australopithecus afarensis* is represented by the Hadar and Laetoli specimens, and is characterized by having

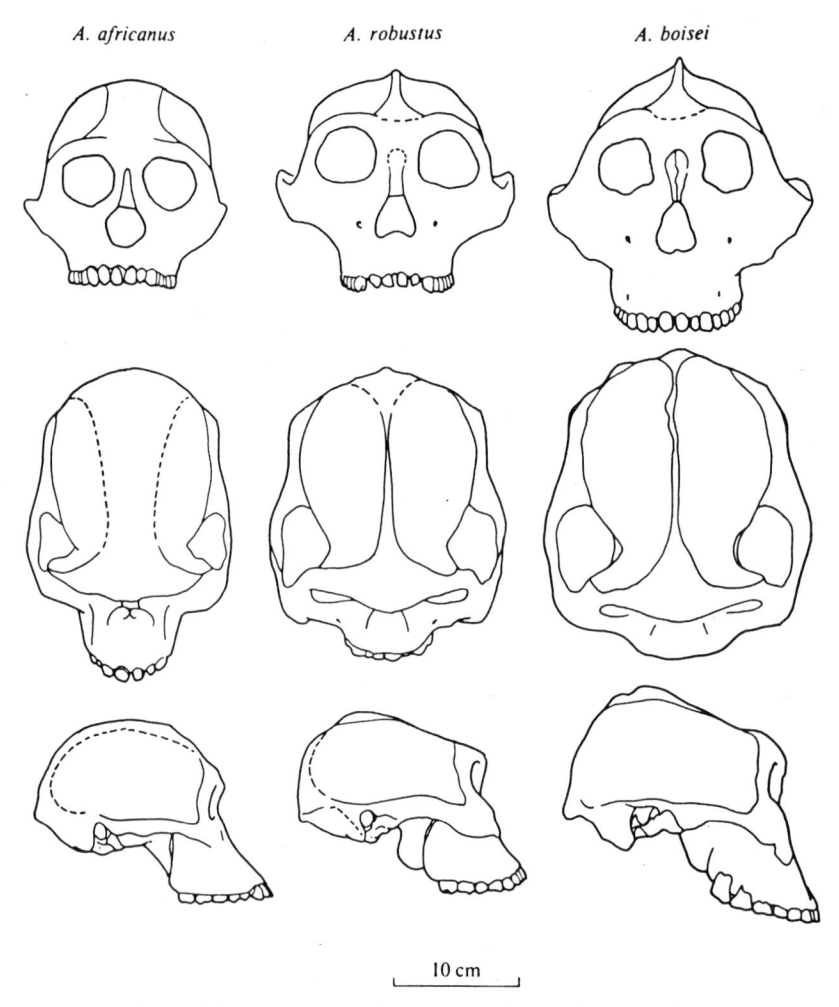

Fig. 2.3 Frontal (top), superior (middle), and lateral (bottom) views of the skulls of *Australopithecus africanus*, *A. robustus*, and *A. boisei* (from Tobias 1967).

a skull somewhat reminiscent of that of a chimpanzee, but with a relatively large brain (343–485 cc) (cf. Holloway, in Chapter 4 of this volume, Table 4.1), a large masticatory apparatus, and extensive pneumatization (development of air cells) within the thicker skull bones (Kimbel *et al.* 1984). The dentition, however, shows characteristics that clearly distinguish it from that of the apes, indicating that evolution toward the human condition was already well under way (Fig. 2.4). It is clear from the structure of the base of the skull that it was balanced on the vertebral column as in humans, and is fundamentally different in this respect from that of the apes (which are quadrupedal). Evidence of the limbs also shows that the creatures were fully bipedal, but in this species this was accompanied by some apelike features of the hand and foot. The stature of the 40 per cent complete Hadar skeleton (Lucy) (see Fig. 2.5) was estimated to have been approximately 1.1 m (Table 2.8).

The skull of *A. africanus* is again superficially apelike in appearance. Its estimated cranial capacity ranges from 428 to 485 cc. The brain-case is slightly larger and more rounded than that of *A. afarensis*, especially in the larger-brained individuals. The face is again relatively massive; the jaws are large, and the supraorbital torus well developed. The big jaws have no chin; but a dental arcade and dentition quite unlike that of any ape, and closer to that of modern humans, are present. The canine teeth are totally distinct from those of the apes and similar to our own, being nearly spatulate like the incisors; and all the front teeth are much reduced compared with those of apes. The molar and premolar teeth are large, but humanlike in general appearance.

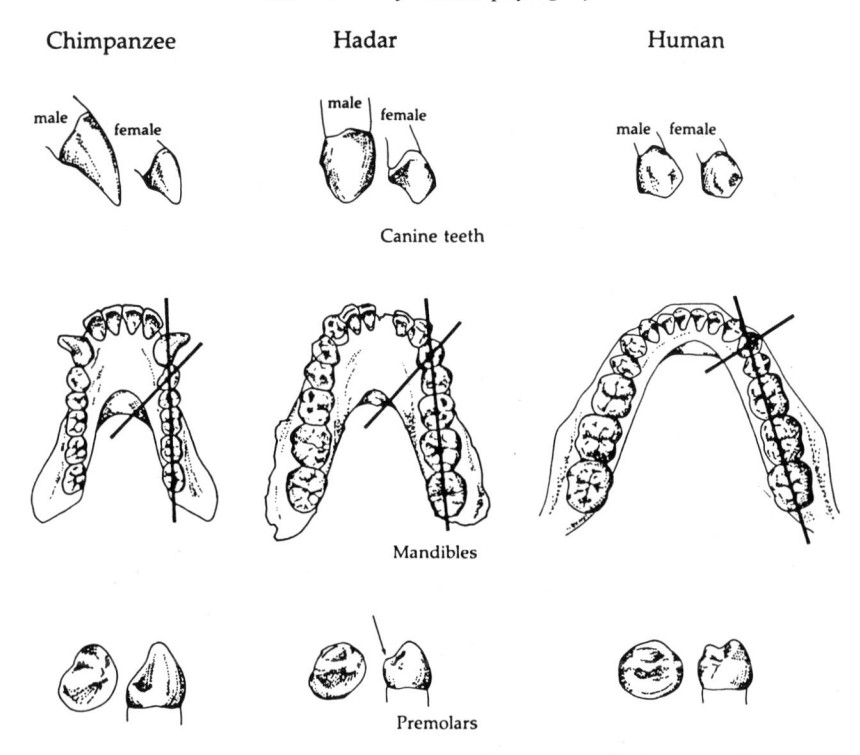

Fig. 2.4 Comparison of the dentition of *Australopithecus afarensis* (Hadar), ape (chimpanzee), and human. At the top, the male and female canines of each species. Note the pointed ape canines and the blunt human teeth, together with the sexual dimorphism in the ape and *Australopithecus* teeth. The three mandibles show the *Australopithecus* mandible to be intermediate; the lines show the alignment of the lower PM1 in relation to the molar series as a whole. Below, the first lower molar is illustrated. The ape premolar is unicuspid, the human premolar bicuspid. The *Australopithecus* premolar is intermediate; it has a small lingual cusp (Johanson and Edey 1981).

The best-preserved skeleton belonging to *A. africanus* comes from the Sterkfontein cave site near Johannesburg. There are fragments of the scapula and humerus, some rib fragments, nine vertebrae from the thoracic and lumbar regions, the sacrum, pelvis, and a poorly preserved femur. The stature of this skeleton was probably about 1.3 m. The skull and other parts of the skeleton of this species are known from many fragments of other individuals from this and other sites.

The anatomy of the postcranial skeleton is more obviously humanlike than that of the skull (see Fig. 2.6), though there are minor differences from modern humans in most features.[6] It is quite different in almost every feature from that of living monkeys and apes. The implications of a detailed analysis of the skeletal material are that both *Australopithecus afarensis* and *A. africanus* were efficiently adapted bipedal animals. The brain, however, was still small relative to body size, compared with a modern human brain, though relatively somewhat larger than that of an ape; and the large jaws gave the head an apelike appearance.

In summary, these species carry an extraordinary combination of human and apelike characteristics, but the human characteristics predominate, particularly in *A. africanus*.

Because the features of *A. africanus* are broadly similar to those of *A. afarensis*, some authors (for instance, Tobias 1980) consider *A. afarensis* no more than a northern subspecies of *A. africanus*. (The latter name has priority, as it dates from 1925; *A. afarensis* dates from 1978.) White *et al.* (1981), however, have carefully reviewed the characteristics of both groups, and have identified distinctive features that they believe justify recognition of the groups as two distinct, but successive, species. Clearly, if they were successive, their features might blend in a continuous series, even though their mean characteristics were distinct. For descriptive purposes herein it is simpler to use the terminology of species than subspecies, and the evidence seems to weigh in favour of this approach. What

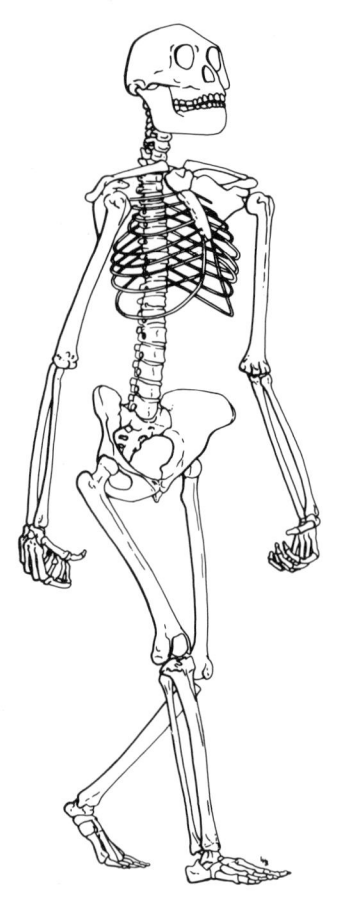

Fig. 2.5 Reconstruction of *Australopithecus afarensis* by C. O. Lovejoy and his students. Based upon one partial skeleton (Lucy) and other specimens from Hader. Drawn from a photograph in *National Geographic* (Nov. 1985), with permission. Copyright © by Charles R. Peters.

is important is the anatomy and age of these fossils, and the part they have played in human evolution.

The species *A. robustus* comes from sites in the Transvaal that are broadly similar to but later than those that have yielded *A. africanus*. Again, there has been controversy as to whether these fossils are really specifically distinct from the older *A. africanus*. They could prove to represent a descendant population that may be either subspecifically distinct or specifically distinct from its ancestors. For our purposes it seems most convenient to recognize the group as a distinct species, even though the matter is not finally resolved.

Australopithecus robustus differs from *A. africanus* in having a more robust skull (Fig. 2.3, p. 38) and possibly a slightly larger and more robust skeleton. The differences in the skull are related to the dentition, in the form of the molar and premolar teeth, and the heavier bone and muscle apparatus needed to carry them. This means that the important distinctions lie in the teeth and the skull. Only one fairly complete skull, with a cranial capacity of about 530 cc, is known. Little is known of the postcranial skeleton, but it appears to indicate the same kind of bipedalism as that found in *A. africanus*.

The species *Australopithecus boisei* stands out clearly from the other three species of the genus. This hominid, of which fossils have been found only in eastern Africa, responded to environmental change, not as did the gracile species, whose descendants were destined to become adventurous tool-using scavengers and eventually predators, but more in the manner of the other higher primates with terrestrial adaptations, in the direction of becoming a more orthodox herbivore. The teeth of *A. boisei* underwent exactly the kind of evolutionary changes that we associate with a vegetable diet; that is, a vast increase in the crushing and grinding surface of the molars and a reduction of the front teeth—features also characteristic of, but less pronounced in, *A. robustus*. That the species is an *Australopithecus* is obvious at a glance (Fig. 2.3, p. 38); but it is also clear from the dental anatomy and microscopic examination of tooth wear that the species's adaptation was somewhat different, and involved a full commitment to the mastication and digestion of tough and/or coarse, abrasive vegetable food, including seeds, nuts, and tubers. The molars are vast (Table 2.6), bigger than those of any living primate; and the jaws are the most efficient known in terms of the power brought to bear on these teeth. Like many other herbivores, this is a robust animal, far more heavily built than the other species of the genus.

Beyond this we know very little about *A. boisei*. The few bones of its skeleton that we possess do not tell us for sure if the species was as fully bipedal as *A. afarensis* and *A. africanus*. But *A. boisei* had a brain of 500–530 cc, and brought all the advantages of advanced primate nature to its new environment. The fossil record of the species may begin over 2.5 million years ago (Walker *et al.* 1986); and the last remains date from a little over one million years ago.[7]

Probably the most important single characteristic of the genus *Australopithecus* is that it was bipedal. The best evidence for bipedalism comes from Sterkfontein in the Transvaal and Hadar in Ethiopia. From Sterkfontein there is a fairly well preserved pelvis (Fig. 2.6) plus partial knee-joints and foot bones, and from Hadar some pelvic fragments and a complete knee-joint. The evidence for bipedalism by 3.0 m.y.a. is clear; detailed comparison of *Australopithecus* pelves and femora with those of apes, monkeys, and humans leaves no room for doubt.

An outline of human phylogeny

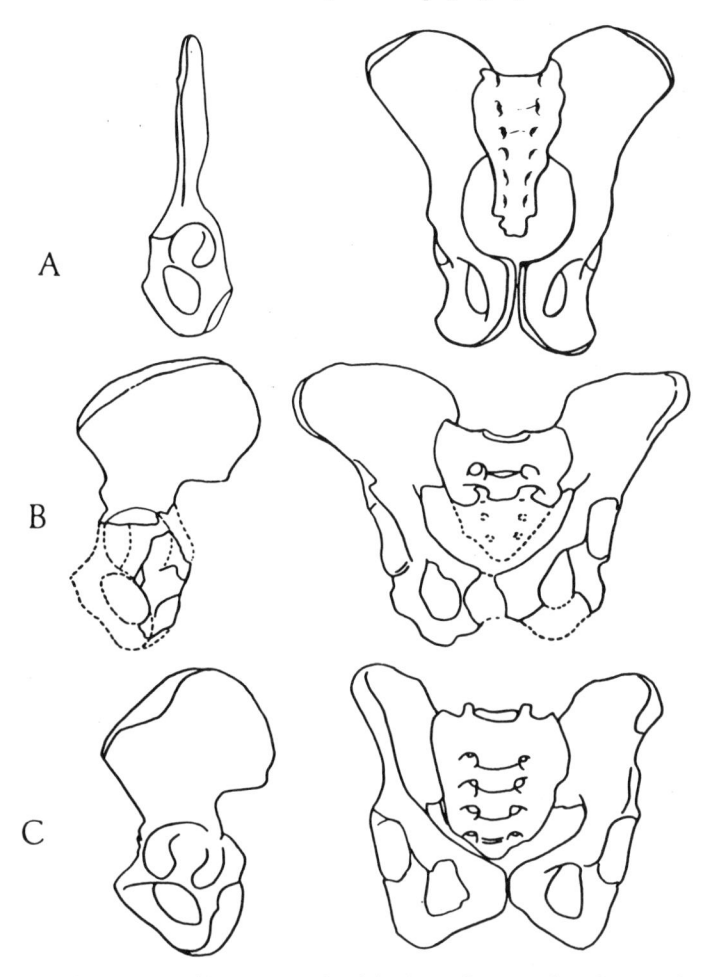

Fig. 2.6 The pelvis of a chimpanzee (A) compared with that of *Australopithecus africanus* from Sterkfontein (B) and a modern human (San Bushman) (C). Left: left lateral view; right: the entire pelvis from the front. Note the striking similarity between the *Australopithecus* and the human pelvis (Le Gros Clark and Campbell 1978).

There is no fossil evidence of *Australopithecus* after 1.2 m.y.a., by which time the very robust forms had evidently become extinct. What is more important, however, is that although its members were bipedal hominids, and clearly related to *Homo*, the *Australopithecus* series could not have given rise to *Homo* at its termination, as it had become morphologically too specialized.

For many years it has been clear (on anatomical and chronological grounds) that the last member of the *Australopithecus* series which could possibly have given rise to *Homo* was *A. africanus*. The first creatures assigned to the genus *Homo* are known from the fossil record of about 2.0 m.y.a., and were much more like modern humans than the robust australopithecines which were contemporary or later in age.

The well-documented *Australopithecus* lineage is now thought to represent a side branch of hominids, which differentiated from the line leading to *Homo* c.3 m.y.a. and became extinct 2 million years later. The number of specimens of the lineage that we have in our museums is a witness to its successful adaptation.

2.4 The earliest humans: *Homo habilis*

The earliest fossil remains classified as *Homo*, and thought to represent our direct ancestors, are teeth found in South Ethiopia close to the Omo river at a place called Shungura, and are dated to a little over 2 m.y.a. They can be considered as representing the species that was named *Homo habilis*.[8]

The first specimens designated *Homo habilis* were skull fragments found at Olduvai Gorge in Tanzania by Louis and Mary Leakey between 1960 and 1964.[9] At first publication (Leakey and Leakey 1964) many people rejected the fragments as being in no way significantly different from *A. africanus* from South Africa. The authors claimed, however, that not only was *H. habilis* often 'associated' (i.e. found on the same limited area of ancient land-surface) with stone tools (while *Australopithecus* was more rarely if ever so associated), but also the specimens demonstrated a mean cranial capacity larger than that of *A. africanus* (Table 2.9, p. 45). The data, though based on very small samples, do suggest that *Homo habilis* had evolved a significantly larger cranial capacity than *Australopithecus africanus* (cf. Holloway, this volume, Table 4.1).

A difference in overall body size could account for these differences in cranial capacity; and such size differences would not imply a major taxonomic distinction. The evidence for the body size of these samples suggests, however, that they were small. Although at present based on a single individual, an estimated stature of only *c.*1 metre adds to the significance of the enhanced cranial capacity. The only long bones which can be associated with the cranial material suggest that the arms were significantly longer than those of modern humans in relation to leg length. Predictably, the intermembral index probably lies somewhere between that of human and chimpanzee (Johanson *et al.* 1987).

Much discussion was centred on the premolar teeth of *H. habilis* when the fossils were first discovered. It was observed that though they were molarized, as are those of *Australopithecus africanus*, they had a narrower shape. This lends support to the view that *Homo* branched off from the rest of the australopithecines at an early stage, before those with very large premolar teeth had evolved.

The 'association' of the Olduvai *H. habilis* with stone tools is important. At Hadar, stone tools have been found with a date of perhaps 2.5 m.y.a. (Harris 1983), and at Omo-Shungura there are well-dated tools from over 2 m.y.a. (but see Isaac 1984). It is usually assumed that these tools were made by *Homo habilis*, although they and the bones do not occur in the same deposits. However, the discovery of tools dated 1.7–1.8 m.y.a. and *H. habilis* remains in the same deposits at Olduvai tends to confirm the hypothesis that it was *Homo habilis* who chipped the small chunks of quartzite and flaked the cobble-stones that are the earliest recognizable stone tools.

On a gradualist model of hominid evolution we could expect that, if *Australopithecus* gave rise to *Homo habilis*, they would intergrade; and this would mean that no clear line of demarcation could be drawn between the two sequent species. In practice, the line is drawn according to convenience and convention. The convention used is that *Homo* was a stone-tool-maker, and *Australopithecus* was not; many workers therefore recognize the transition at the first appearance of recognizable lithic artefacts.

Given that there are species of *Homo* and *Australopithecus* which are known to be contemporary at Olduvai and Omo, and given that each is associated somewhere with tools on different 'living floors' (activity areas found on excavated ancient land-surfaces), it is theoretically difficult to know which was the tool-maker. The hominid bones are found on the 'living floors' in a state just as fragmentary as that of the other bone fragments which are believed to be food remains. The individuals represented may therefore be the victims of the tool-maker rather than the tool-makers themselves. In practice, the distinction between the tool-makers and their dinner cannot be reliably drawn in the archaeological record. In attempts to make this distinction, the following considerations are taken into account.

1. Where remains of two species are found together, the more advanced species (in terms of relative brain size as well as of morphological similarity to modern humans) is more likely to be the tool-maker.

2. There is no way of showing that the less advanced form was not also a tool-maker.

3. Either of these species may have been an occasional cannibal. (The descendant species, *Homo sapiens*, practised at least some cannibalism.)

4. From our knowledge of animal behaviour and from the archaeological record we know that these creatures were scavenging and possibly killing small game, and bringing it to their 'living floor'.

5. We do not know how they dealt with their dead or how their skeletal remains would have found their way on to the ancient land surfaces that have been excavated.

When the activity area FLK I at Olduvai Gorge (Bed I) was first discovered it was assumed that the bones of *Australopithecus boisei* found there by Louis and Mary Leakey were those of the tool-maker. Later discoveries of a 'more advanced' hominid (*H. habilis*) also associated with tools in five contemporary sites suggested to them that *A. boisei* was more likely to be a prey species, and therefore was probably not a tool-maker, though it might have been a tool-user.

On theoretical grounds, however, it seems probable that *Australopithecus* was a tool-user and a very primitive tool-maker, characteristics that cannot easily be detected from the palaeoanthropological record.

In summary, the available evidence suggests that *Homo habilis*, the first species of *Homo*, and the first maker of recognizable stone tools, evolved from a gracile population of *Australopithecus*, and appeared in East Africa somewhat over 2 m.y.a. Whether there was more than one species of *Homo* at this period and at this level of cultural development is questionable. Some authors recognize 2 species—a larger and a smaller one; but the data are still very limited.[10]

2.5 *Homo erectus*

The successor to *Homo habilis* was the much more modern-looking species *Homo erectus*. Fossils of this species have been found through much of the Old World as far north as latitude 50°, and it is remarkable for the variety of environments to which it became adapted (Fig. 2.7a). The finest specimens, dating from 1.5–1.8 m.y.a., come from tropical regions of Africa; but by this time we find that populations have spread into Eurasia.

The specimens from East Africa include a remarkably complete skeleton of a young male (though lacking hands and feet), estimated (by American standards) to be 12 years of age and 1.6 m tall (Brown *et al.* 1985). Not only was this a notable find, but it shows us that, at this relatively early date, *H. erectus* was significantly taller than any earlier hominids (Walker and Leakey 1993). This skeleton comes from deposits west of Lake Turkana; other remains come from the east of the lake. Skulls and long bones suggest tall, strong people, with brains larger than those of *Homo habilis* (Tables 2.8, 2.9, pp. 44–5).

In terms of age, these Kenyan specimens are equalled by some from Asia (Swisher *et al.* 1994). Again, the skulls are heavily built; and at many sites the material culture is more developed. We do not have good examples of transitional forms between *H. habilis* and *H. erectus*, but we can hypothesize that, if *H. erectus* arose in Africa, which it apparently did, it must have spread from there to the more easterly tropical and subtropical regions of the Old World; and that as it occupied these different environments it maintained an increased stature and heavy build. In Africa, at Koobi Fora and at Olduvai, *H. erectus* appears to succeed *H. habilis*, and may be derived from the earlier form, although intermediate skulls are not known.

Skulls of *H. erectus* typically are heavily built (Fig. 2.8). The cranial bones are thick, and the facial bones well buttressed. The jaws are still more powerful than our own, but less prognathous than in *Homo*

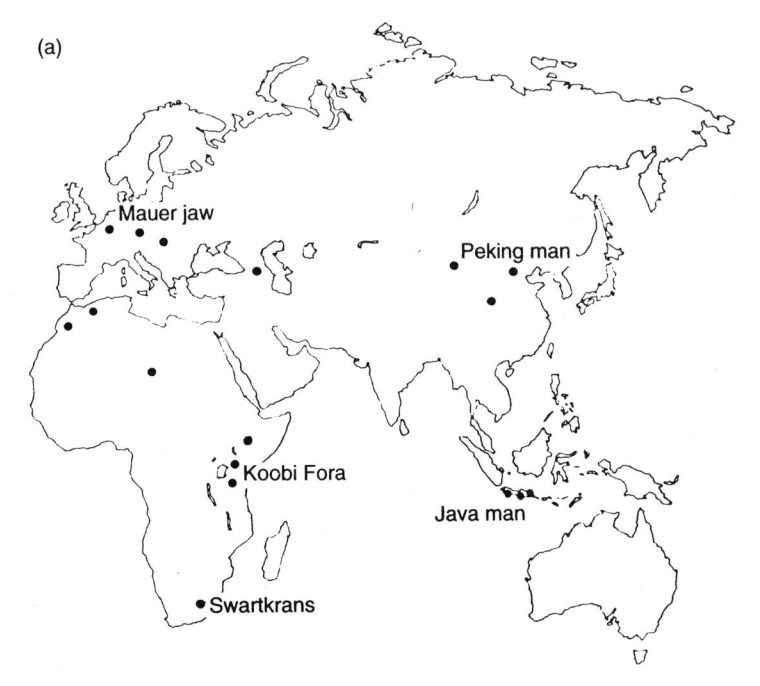

Fig. 2.7 (a) and (b) Maps of the Old World showing sites at which fossil remains of *Homo erectus* and Neanderthals have been found. Fig. 2.7a (above) shows the distribution of *Homo erectus* fossils. Some researchers prefer to assign the North African and European specimens to archaic *H. sapiens* (from Campbell 1985).

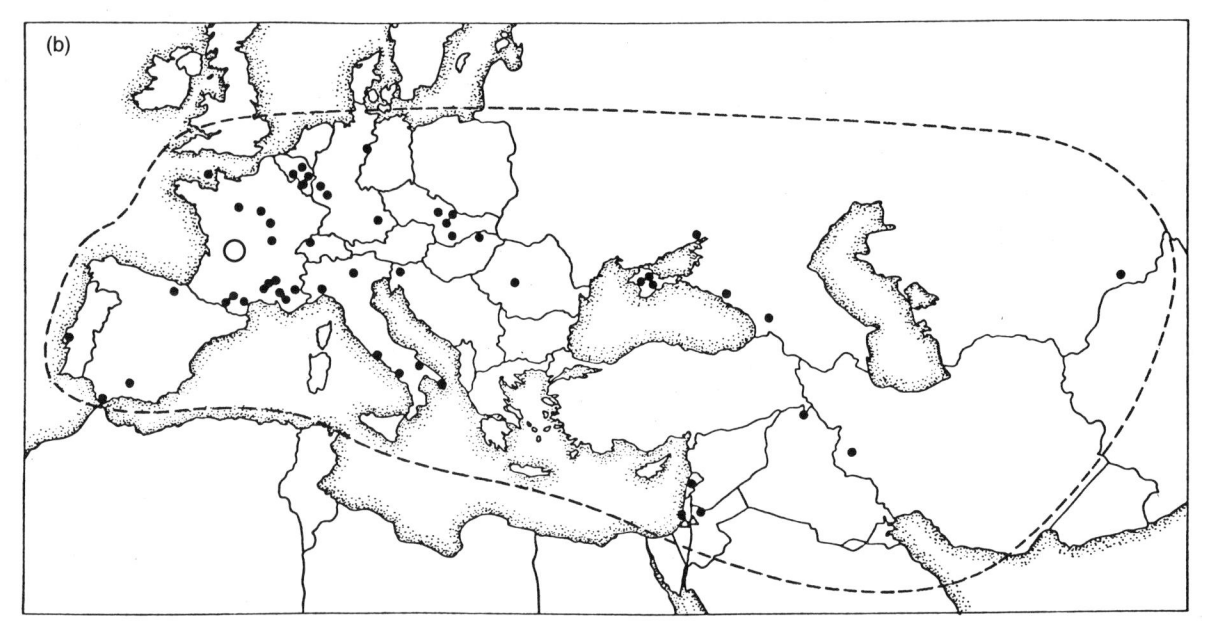

Fig. 2.7b Neanderthals (from Trinkaus and Howells 1979. Copyright © 1979 by Scientific American, Inc. All rights reserved.) The circle encloses a further ten Neanderthal sites.

habilis, because the brain-case extends further forwards above the eyes than it does in that species. Most striking, perhaps, is the heavy supraorbital torus, which makes the whole face unmistakable. The best preserved and most typical specimens come from Zhoukoudian, near Beijing, where a cave site contained extensive deposits with remains of about fifty individuals, loosely associated with stone tools, dating from *c*.500 000 years ago (see references in Binford and Ho 1986; Wolpoff 1986).

Although the skull is strikingly primitive in appearance, the dentition is becoming modern; and although the teeth are still a little larger than our own, they are similar to those of modern humans in shape and cusp pattern. The jaw is also heavier than ours, and there is no chin. The post-cranial skeleton is modern in form, and except for their thickness individual bones cannot easily be distinguished from those of modern humans. Taken as a whole the skeleton is robust in comparison to our own; the limb bones are slightly curved, and have well-developed areas for the insertion of the powerful musculature. The stature of these people was probably significantly greater than that of their ancestral species, and may have averaged *c*.1.7 m (Table 2.8). Only a few hand and foot bones have been discovered (see Marzke, Chapter 5, this volume).

Apart from its size, probably the most striking feature of *H. erectus*, in comparison with earlier forms, is its larger brain. Mean cranial capacity had more than doubled between 2.0 and 0.5 m.y.a., from 500 to 1200 cc. By 0.5 m.y.a. the cranial capacity falls within the lower end of the range for modern humans (Table 2.9, p. 45).

In discussing brain size as a feature that changes during evolution it should be noted that brain size is allometrically related to body size. Among mammals in general, there is a positive correlation between body weight and brain weight. All species of anthropoid primates deviate positively from this overall mammal relationship, i.e. they have heavier brains in relation to their size than 'average' mammals of the same body weight. The species that deviates most is *Homo sapiens*, whose brain is 3.1 times larger than that predicted from a non-human primate of similar body size (Passingham 1982). Among the Hominidae, the incompleteness of the fossil evidence does not enable us to calculate values for body weights. Instead, we calculate stature by extrapolation from the length of fossil long bones. However, the data are rather meagre (Table 2.8).

Table 2.8 The stature of three fossil hominid species and of modern humans (McHenry 1991; see also McHenry 1992)

Australopithecus afarensis ($n = 1$)	1.1–1.5 m
A. africanus ($n = 4$)	1.2–1.4 m
Homo erectus ($n = 2$)	1.6–1.8 m
H. sapiens ($n = >100$)	1.3–2.0 m

Table 2.9 Cranial capacities of primates, especially Hominidae (cf. Editorial appendix I to Chapter 4 of this volume)

Species	Sample size	Range of cranial capacity: cm^3	Average cranial capacity: cm^3
Lemur	—	10–70	—
Chimpanzee	—	282–500	383
Gorilla	—	340–752	505
A. afarensis	2	343–485	414
A. africanus	5	428–485	444
A. robustus	1	—	530
A. boisei	5	500–530	516
H. habilis	5	590–752	661
H. erectus	15	780–1225	953
Modern adult human	> 1000	900–2000	1345

Note: The brain itself is always considerably smaller in size than the cranial cavity, since the latter also contains other structures and fluid. The figures in the Table are approximations only, as they are obtained from small samples.

Within *H. sapiens*, brain size is positively correlated with stature. However, the extent of the evolution of the hominid brain is indicated by the fact that Andamanese and other pygmy people are not any taller (about 1.35 m) than the *Australopithecus* species mentioned above, yet have mean cranial capacities rather over two and a half times as large as them, that is, about 1100–1200 cm^3 (Flower 1889; Martin 1959).

This increase in brain size was surely adaptive, and is most probably linked to the following factors:

(1) developing technology and associated skills: food gathering, scavenging, and hunting;

(2) the variety of environments occupied, including northern temperate habitats with their more extreme seasonal temperature variations;

(3) the development of more complex social organization;

(4) inter-group competition; and

(5) the development of symbolic cognition and communication.

The relationship between these new behavioural developments and the brain must have been a reciprocal, closely interrelated one. That is to say, while the evolution of a larger brain seemingly would have facilitated the evolution of these five behavioural complexes, they themselves apparently created circumstances in which yet larger brain size was adaptive. The factor that apparently triggered this process was the early use of tools, and the adaptive advantage that tools gave to a social primate in extracting food from the environment and providing a means of self-defence. Once the process began, the most important factor was almost certainly that of the increasingly complex social environment.

The fact that a single species could occupy such a range of habitats, from savannah in East Africa to tropical forest in Java to temperate woodland in Europe and China, suggests that culture was becoming important in enabling humans to exploit their environment.

Humans were the first species to modify their environment in a substantial manner. Two factors were especially important in this respect: the invention of *energy constructs*, and the control of *fire*. Both are forms of energy control and transformation. While the evidence for both prior to the appearance of *Homo sapiens* is very poor, they may both have found their beginning during *Homo erectus* times.

While food resources are the primary energy source for every species, early humans began to tackle energy problems in a more direct way. Energy constructs (sometimes called 'facilities': Wagner 1960) may be defined as being objects that restrict or prevent motion or energy exchanges, or amplify them. Simple examples are dams and insulation, as well as tents, houses, or clothing, which retain heat. Containers of various sorts (skins, baskets, pots, boats, fences, traps, or even cords) also fall into this category, since they restrict motion. Slings are examples of energy constructs which restrict motion while amplifying it. Temperate environments require far more energy (perhaps one to two orders of magnitude more) than tropical environments; and energy constructs made it possible for humans that originally evolved in the tropics to overcome a number of critical limiting factors in a northern climate.

Secondly, fire is equally important. Its capture, handling, and control may have been difficult and unreliable at first, for natural fires are rare; but its use was probably essential for survival through harsh north temperate winters. The control of fire is one of *H. erectus*'s or early *H. sapiens*'s most remarkable achievements, and was a necessary precursor of highly advanced lithic and metal technology. The use of energy constructs and fire was our first adventure into the direct control and production of energy. Thus would we change the face of the earth.

Thus there are a number of reasons why north temperate regions presented a great challenge to *Homo erectus*. A number of very important new environmental factors are present in these regions when compared with tropical and subtropical zones:

1. Large variations in annual temperatures impose direct stress on animals and plants. *Homo erectus* probably wore skins as clothing at least during the winter months.

2. Low winter temperatures either restrict or inhibit plant growth. We believe that tropical hominids would have consumed normally no more than about 30 per cent of meat in an otherwise vegetarian diet, and probably less. During the late winter months, when seeds and fruits had been eaten, there would be no vegetable foods available in the colder temperate regions. Without ample storage, *Homo erectus* would therefore have become totally dependent on scavenging (and probably hunting) during this period for survival, and effective hunting techniques would surely have become vital.

3. There would be great advantage in the use of a permanently sheltered home base, such as a cave.

4. Food-storage techniques would have become very important.

5. Fire would have become of great value, especially in the more northerly regions and in cold cave dwellings.

There are many other factors that could be listed. The important feature of colonization of northern latitudes is that some degree of culture must have been a necessary pre-condition for the colonization to occur. Furthermore, colonization of relatively hostile habitats would have favoured further cultural evolution. Thus cultural evolution must have had a reciprocal effect on other aspects of human biology besides brain size, such as diet, social organization, food-finding behaviour, and the general exploitation of the habitat. The same environmental factors would apply in extreme southern latitudes (for example, Cape Province, R.S.A.), though here there was little scope for migration because of the geography of the continent.

2.6 Modern humans: *Homo sapiens*

The appearance of *Homo sapiens* represents a culmination of the trend towards increased cranial capacity. In association with the rapidly increasing brain size of *H. sapiens* the cranium expanded over the jaws, to produce the vertical forehead and facial structure of modern humans. This development was accentuated by some reduction in the size of the jaws and the dentition as a whole (Fig. 2.8). However, the extent of this reduction varies in different races. Some populations among the Aborigines of Australia, for example, show little reduction of the dentition from that of *Homo erectus*. With the further increase of

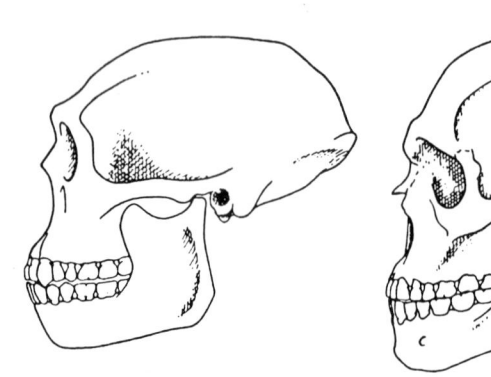

Fig. 2.8 Comparison of the reconstructed skull of *Homo erectus* from China and that of a modern (extremely robust) Australian aborigine. Note that the modern skull has a larger brain-case and lacks the protruding jaws of *H. erectus* (Le Gros Clark and Campbell 1978).

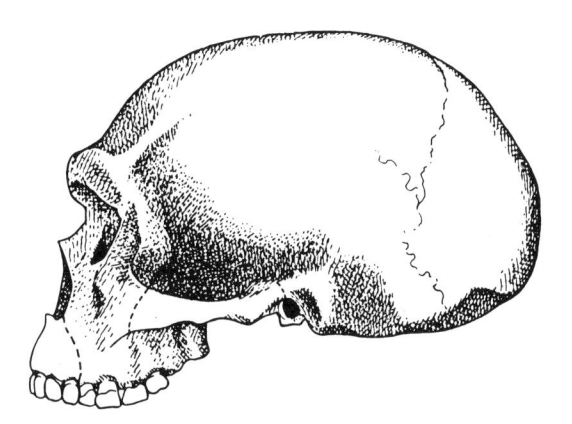

Fig. 2.9 The Steinheim skull (partly restored) (Le Gros Clark and Campbell 1978).

cranial capacity, there came a reduction in the thickness of the skull bones, and a general lightening of the skeleton. Today, different races of modern humans vary considerably in height and robustness; we can reasonably assume that such variation is usual and probably ancient, so that both widespread species of *Homo* would always have been highly variable in these characters.

The earliest members of *Homo sapiens*, the archaic members of our species, dating from about 300 000 years ago, are theoretically indistinguishable from the later members of *H. erectus* from which they evolved. Consequently, the status of specimens from Vértesszöllös (Hungary), Heidelberg (W. Germany), and Petralona (Greece) which fall near the boundary is controversial; and the problem is similar to that found at other lineage boundaries. By 300 000 years ago, however, most fossil skulls have cranial capacities well within the modern range (1000–2000 cm^3; average 1330 cm^3) and a relatively small modern dentition. The earliest human remains from this time-period that are perhaps not borderline are those from Swanscombe in Kent, from Steinheim in Germany (Fig. 2.9), and from Arago in south-west France. They vary from a robust and *erectus*-like appearance (Arago) to a more modern form (Swanscombe); and at the site of Arago we find much variability in the two jaws preserved, which can be attributed not only to normal individual variation but also to a fairly high level of sexual dimorphism.

The last million years at least have been characterized in northern temperate zones by striking climatic oscillations, the glacial and interglacial periods (Covey 1979) (Fig. 2.10). During the glacial periods, little of Europe was habitable; only Spain, Italy, and Greece had climates suitable for a temperate fauna and flora. During interglacial periods, the climate was probably not very different from what it is today (for we are presently enjoying an interglacial). Because of the environmental problems posed by glacial conditions, *Homo erectus* and early archaic *H. sapiens* would have entered the central latitudes of Eurasia only during interglacial times, and would then have retreated before each glacial advance. Slowly humans learned to cope with colder conditions until, about 70 000 years ago, we have evidence of people living in north-west Europe and north-western Asia (see Fig. 2.7b, p. 44) under conditions as rigorous as the Eskimos experience today in North America. These were the Neanderthal people, much maligned in fiction, but tough, intelligent, and technologically advanced in fact.

The neanderthalers of western Europe are the first hominids associated with a periglacial fauna, and were possibly the first to cope with what we would call near-arctic conditions. To do so, they had developed a technology far ahead of anything known before that time. The fact that they ultimately died out in western Europe should not obscure their re-

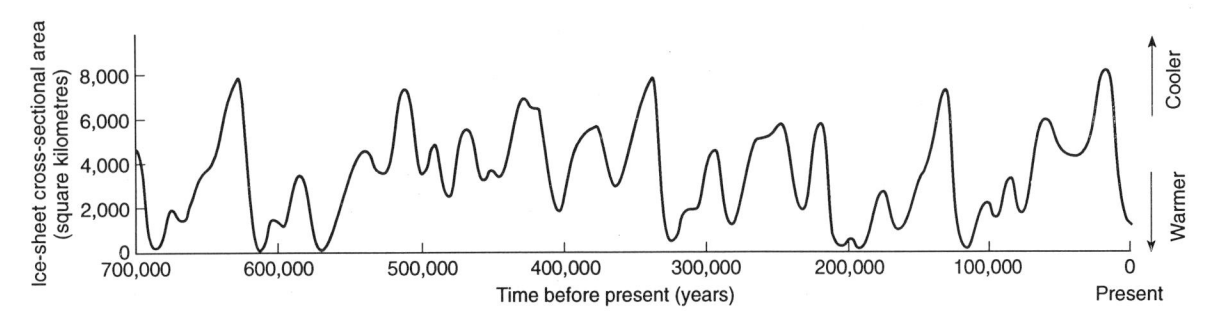

Fig. 2.10 Pleistocene ice ages were caused by periodic variations in the geometry of the Earth's orbit; and such variations are believed to concur with the evidence of variations in the total amount of frozen sea-water that can be derived from measurements of the oxygen-isotope content of the sea obtainable from deep-sea cores. The main peaks of cold temperature have occurred at 100 000-year intervals, but many other oscillations occur between these peaks (after Covey 1979).

markable achievement. They were physically characterized by considerable strength, which was expressed in their faces as protruding brow ridges and prognathous jaws (Fig. 2.11). An examination of the muscle attachment areas on their bones shows them to have been very muscular in comparison with ourselves. But these tough and ingenious people disappeared from their hunting grounds in Europe and were replaced (by about 30 000 years ago) by people of modern, paedomorphic, gracile appearance, who were anatomically indistinguishable from ourselves. What happened to the neanderthalers of western Europe is a mystery: they may have been extinguished by an extreme climatic oscillation (see Fig. 2.10, p. 47), or by genetic introgression as a result of interbreeding with invaders from the east, or by a combination of both. Their fate may have been similar to the virtual genocide that befell the Native Americans and the Australian Aborigines. Many of these people were killed, but others interbred with early settlers. Today, millions of Americans have Indian ancestors; but Native American cultures are now extinct in most states.

At about the same time as the disappearance of the Neanderthal people in western Europe (about 40 000–30 000 years ago) advanced modern populations adapted to arctic conditions crossed the Bering Straits (which were at that time intermittently dry land) into North America and dispersed south and east. The earliest evidence of human activity from South America may be dated at over 30 000 years ago (Guidon and Delibrias 1986). In south-east Asia, meanwhile, people had crossed the Timor Sea to Australia at least 40 000 years ago and occupied much of that vast continent (Wolpoff *et al.* 1984).

Through a combination of anatomical specializations, coupled with adaptability, intelligence, and cultural developments in technology and language, the genus *Homo* had come to occupy almost every part of the earth.

Modern humans fulfil the biological definition of a species: a group of actually or potentially interbreeding natural populations, reproductively isolated from other species. *Homo sapiens* is a polytypic species which can be subdivided into geographical races, although there is little agreement on the number which can be recognized. The commonest classifications list the following five or six major geographical races: Africans, Native Americans, Asiatics, Australians, Caucasians, and Oceanics (Garn 1971; Brues 1990). Further subdivisions have been made; and some authors recognize more than sixty racial groups. Classification at this level is essentially subjective and impossible to justify genetically.

A. C. Wilson and colleagues (Cann *et al.* 1987) have published important biochemical data on human mitochondrial DNA (mtDNA). Because this molecule is relatively small, simple, and well understood, and is inherited only from the female parent (being found not in the nucleus but in the cytoplasm of the egg cell), knowledge of mtDNA differences (assumed, with reason, to result from neutral mutations) enables us to assess relationships between individuals, populations, and races. By measuring mtDNA differences between individuals from a wide selection of different human races, it has proved possible to estimate the relationships between races; and by calculating mutation rates (in a similar way to those calculated for albumen differences) it has been possible to estimate within a wide margin the time of the common

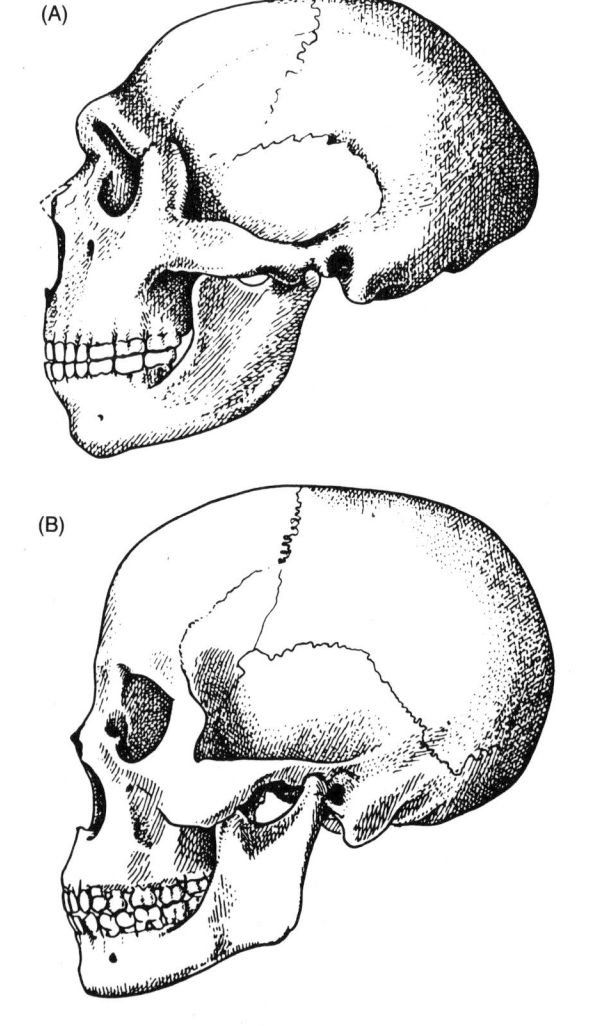

Fig. 2.11 The skull of a Neanderthal male (from Monte Circeo, Italy) (A) and a modern European (B) (Le Gros Clark and Campbell 1978).

ancestor of all living humans. The results of this research suggest convincingly that all modern humans have evolved from one or a relatively small number of closely related African women who lived between 140 000 and 290 000 years ago, presumably in Africa.

This hypothesis is not accepted by all workers (see Gibbons 1992) who claim to see morphological continuity in human crania in many parts of the world, especially in central and eastern Europe and in East Asia. However, recent work on nuclear DNA has indicated an African origin for modern humans, and a number of other recent publications support this hypothesis (for example, Wainscoat *et al.* 1986; Stringer and Andrews 1988; and other references in Cann *et al.* 1987).

If this interpretation of Wilson's results is accepted, it leads to a number of important hypotheses about the more recent stages of human phylogeny.

1. It now seems that at all the stages of human evolution of which we have fossil documentation, Africa has generated the most rapidly evolving populations. *Australopithecus*, *Homo habilis*, *Homo erectus*, and *Homo sapiens* all originated in Africa, and the latter two expanded from this continent into Eurasia.

2. If early modern humans first appeared in Africa at about 200 000 years ago, Eurasian *H. erectus* and even archaic Eurasian *H. sapiens* probably made very little, if any, contribution to our modern gene pool. In particular, theories that East Asian *H. erectus* evolved into modern Australian Aborigines would become untenable (Wolpoff *et al.* 1984).

3. Later Eurasian *H. sapiens* (for example, the Neanderthal people and their immediate ancestors) also probably made very little, if any, contribution to the present human gene pool.

4. The origin of modern humans is now seen to have been in Africa. This ties in well with the available fossil evidence from Africa, where human skeletal material with modern features is known from an earlier date than elsewhere, for example, at Border Cave, Natal, *c.*50 000 years before the present (BP); Klasies River, Cape Province, *c.*100 000 BP; Omo, S. Ethiopia, 100 000 BP; Singa, Sudan, *c.*100 000 BP (Bräuer 1984).

5. All differentiation of the living races has occurred within the last 200 000 years, and some races may be much younger than this.

These hypotheses are dealt with in detail by Waddell and Penny in the following chapter.

2.7 Discussion

The meaning of species in palaeontology has caused problems in the interpretation of human phylogeny. So long as the fossil record is poor, different groups of fossils from different strata and different places have often been recognized as belonging to distinct species and genera, and there is no real basis for confusion between them. Type specimens are selected and taxa described and published in the usual way. These *chronospecies* are to all intents treated taxonomically as if they were biospecies (defined by biological criteria). They are not, however. Whether evolution is a process of gradual change in morphology or proceeds in jumps, it is clear that chronospecies are quite different from biospecies; they consist of a succession of populations of a biospecies over a finite period of time. They also have an extra dimension, their longevity. What is more, specimens assigned to chronospecies cannot, as we have seen, be demarcated clearly from their ancestor and descendant chronospecies. (Palaeontologists follow the convention, therefore, that the variability in the time dimension of a chronospecies should be approximately of the same order of magnitude as the variability in the morphological dimensions of a typical biospecies.)

The problems arising from species taxonomy are especially pressing in the hominid fossil record because the lineage is better documented than many other similar mammalian lineages. The question of how to classify intermediate fossils (that is, those on the *Australopithecus*–*Homo* boundary or the *H. erectus* – *H. sapiens* boundary) is always liable to attract discussion. The time-lines given in Table 2.10, can by convention be used to divide the successive taxa of the Hominidae we have discussed above.

Table 2.10 Taxonomy of the Hominid lineage, with temporal divisions drawn by convention

Species	Time-line (m.y.a.)
Australopithecus – *Homo*	2.50 or 2.25 or 2.00
Homo habilis – *H. erectus*	1.75 or 1.50
H. erectus – *H. sapiens*	0.30 or 0.25

Preferred dates underlined.

However, by no means all palaeoanthropologists accept such an approach. In the first place it is not always the case that there is universal agreement that, for any two fossil populations, one is unequivocally the ancestor and the other the descendant. Secondly, dating techniques are not sufficiently precise to enable us to be sure that all fossils are accurately dated. Finally, this convention assumes that hominid

Table 2.11 Two classifications of the Hominidae: (a) a lumper's classification; (b) a splitter's classification*

(a) Genus	Species	(b) Genus	Species
Australopithecus	*africanus*	*Ardipithecus*	*ramidus*
Homo	*erectus*	*Australopithecus*	*afarensis*
Homo	*sapiens*		*africanus*
			aethiopicus
		Paranthropus	*robustus*
		Zinjanthropus	*boisei*
		Pithecanthropus	*erectus*
		Sinanthropus	*pekinensis*
		Atlanthropus	*mauritanicus*
		Homo	*habilis*
			rudolfensis
			heidelbergensis
			neanderthalensis
			sapiens

* For recent views of this type see Groves (1989) and Wood (1992b) (eds).

evolution was a gradual process. Some palaeontologists subscribe to the view that hominid evolution may have involved a series of quantum leaps (Simpson 1944) or punctuated equilibria (Stanley 1979). In this case lines may be drawn at precisely those points at which the leaps have occurred, once we know when these events happened. Clearly there have been periods of fast evolution, and these we can hope will one day form the basis for a taxonomy (Campbell 1963) in which morphology is the key dimension.

Taxonomists, who are concerned with these matters, are also sometimes divided into 'splitters' and 'lumpers'. Splitters divide their samples into taxa, each with small variability, so that many species and genera are recognized. Lumpers accept wide ranges of variability within taxa, and so recognize only a few species and genera. Table 2.11 shows two extreme classifications of the family Hominidae.

Different theories of taxonomic practice and interpretations of speciation events have bedevilled the construction of a human phylogeny, yet today most workers would agree that the overall pattern is now clear: that the human lineage found its origin among the apes of Africa, between 8 and 5 m.y.a., and that the first anatomical changes were those involving bipedalism and some modification of the hands, to be followed later by the expansion of the brain. The earliest bipeds were of the genus *Australopithecus*, and a species of this genus, more lightly built than the others, made the transition from being a bipedal hominid to becoming the earliest human species, more skilled in tool-making, with a more human-like brain, and later developing language. The further evolution of the brain and culture made possible the expansion of successive waves of migrants from the African continent into Eurasia. The expansion of which we ourselves are a product may have occurred somewhat less than 200 000 years ago; and these, our ancestors, displaced the archaic inhabitants of the Old World, and entered Australasia and America some 50 000–20 000 years ago.

Home sapiens is a Mammal, a Higher Primate, with an exceptionally large brain; the maker of a complex material culture and possessor of a rapidly evolving social culture. In one characteristic humans are unique among animals; they communicate with language. It is our linguistic ability above all other species-specific characters which has given us, animals as we are, both unprecedented power and knowledge. (Campbell 1985) The outline of our evolution is now clear, but the details still remain to be filled in. Considering the knowledge that has been gathered within a hundred years, the development of palaeoanthropology has been a striking achievement.

Editorial notes

1. Day (1986) provides a useful guide to most of the important fossil sites spanning the nearly four million years of well-documented hominid evolution.

2. See Martin (1993) for references to the recent redating of the Fayum deposits (now thought to straddle the Eocene–Oligocene boundary), indicating that the anthropoids must have emerged by the late

Eocene, if not earlier. They are represented by perhaps as many as eleven extinct genera in these deposits.

3. Further hominid remains nearly as old have recently been discovered in strata dated 4.2.–3.9 m.y.a. at Kanapoi and Allia Bay, Kenya by Leakey *et al.* (1995). These finds have been named *Australopithecus anamensis*.

4. The Leakey and Harris (1987) volume provides photographs and references to the descriptions of this material, plus a chapter on the Laetoli hominid footprints. Fleagle *et al.* (1991) report that hominid dental remains of a similar geological age, if not somewhat older, have been found in southern Ethiopia.

5. See Kimbel *et al.* (1994) for new discoveries at Hadar and current dating of this material.

6. See Aiello and Dean (1990) for an evolutionarily oriented introduction to hominid skeletal anatomy.

7. See Grine (1988) for a variety of views on the evolution of the robust and hyperrobust australopithecines.

8. Claims of earlier dates for fossil evidence of *Homo* are briefly discussed by Wood (1992*a*).

9. See Tobias's (1991) monographic treatment of these specimens.

10. Groves (1989) and Wood (1991, 1992*b*) provide the key analyses that recognize the probability of two species of *Homo* within the specimens commonly subsumed under *Homo habilis*: these are *H. habilis sensu stricto* and *H. rudolfensis*.

References

Aiello, L. and Dean, C. (1990). *An introduction to human evolutionary anatomy*. Academic Press, London.
Andrews, P. (1992). Evolution and environment in the Hominoidea. *Nature*, **360**, 641–46.
Andrews, P. and Cronin, J. E. (1982). The relationships of *Sivapithecus* and *Ramapithecus* and the evolution of the Orang-utan. *Nature*, **297**, 541–6.
Binford, L. R. and Ho, C. K. (1985). Taphonomy at a distance. Zhoukoudian, "The cave of Beijing Man"? *Current Anthropology*, **26**, 413–42.
Bräuer, G. (1984). A craniological approach to the origin of anatomically modern *Homo sapiens* in Africa and implications for the appearance of modern europeans. In *The origin of modern humans: a world survey of the fossil evidence* (ed. F. H. Smith and F. Spencer), pp. 327–410. A. R. Liss, New York.
Brown, F., Harris, J., Leakey, R. E. F., and Walker, A. (1985). Early *Homo erectus* skeleton from west Lake Taskana, Kenya. *Nature*, **316**, 788–92.
Brues, A. M. (1990 [1977]). *People and races*. Waveland Press, Prospect Heights, Illinois.
Campbell, B. G. (1963). Quantitative taxonomy and human evolution. In *Classification and human evolution* (ed. S. L. Washburn), pp. 50–74. Aldine, New York.
Campbell, B. G. (1985). *Human evolution* (3rd edn). Aldine, New York.
Campbell, B. G., and Loy, J. D. (1996). *Humankind emerging* (7th edn). Harper Collins, New York.
Campbell, B. G. and Bernor, R. L. (1976). The origin of the Hominidae: Africa or Asia? *Journal of Human Evolution*, **5**, 445–58.
Campbell, B. G., and Loy, J. D. (1996). *Humankind emerging* (7th edn). Harper Collins, New York.
Cann, R. L., Stoneking, M., and Wilson, A. C. (1987). Mitochondrial DNA and human evolution. *Nature*, **325**, 31–6.
Covey, C. (1979). The Earth's orbit and the Ice Ages. *Scientific American*, **241**(2), 42–50.
Darwin, C. (1871). *The descent of man, and selection in relation to sex*. J. Murray, London.
Day, M. H. (1986). *Guide to fossil man* (4th edn). University of Chicago Press.
Delson, E. (1985). *Ancestors: the hard evidence*. A. R. Liss, New York.
Fleagle, J. G. and Kay, R. F. (1983). New interpretations of the phyletic position of Oligocene hominoids. In *New interpretations of ape and human ancestry* (ed. R. L. Ciochon and R. S. Corruccini), pp. 181–210. Plenum, New York and London.
Fleagle, J. G. Rasmussen, D. T., Yirga, S., Bown T. M., and Grine, F. E. (1991). New hominid fossils from Fejej, southern Ethoipia. *Journal of Human Evolution*, **21**, 145–52.
Flower, W. H. (1889). Description of two skeletons of Akkas, a pygmy race from Central Africa. *Journal of the Royal Anthropological Institute*, **18**, 3–19.
Garn, S. M. (1971). *Human races*, (3rd edn). Thomas, Springfield, Ill.
Gibbons, A. (1992). Mitochondrial eve: wounded but not yet dead. *Science*, **257**, 873–5.
Grine, F. E. (ed.) (1988). *Evolutionary history of the "robust" australopithecines*. Aldine de Gruyter, New York.
Groves C. P. (1989). *A theory of human and primate evolution*. Clarendon Press, Oxford.
Guidon, N. and Delibrias, G. (1986). Carbon-14 dates point to man in the Americas 32,000 years ago. *Nature*, **321**, 769–71.
Harris, J. W. K. (1983). Cultural beginnings: Plio-Pleistocene archaeological occurrences from the Afar, Ethiopia. *African Archaeological Review*, **1**, 3–31.
Huxley, T. H. (1863). *Evidence as to man's place in nature*. Williams and Norgate, London.
Isaac, G. L. (1984). The archaeology of human origins. *Advances in World Archaeology*, **3**, 1–87.
Johanson, D. C. and Edey, M. A. (1981). *Lucy: the beginnings of humankind*. Simon & Schuster, New York.
Johanson, D. C., Masao, F. T., Eck, G. G., White, T. D., Walter, R. C., Kimbel, W. H., *et al.* (1987). New partial skeleton of *Homo habilis* from Olduvai Gorge, Tanzania. *Nature*, **327**, 205–9.

Kappelman, J. (1992). The age of the Fayum primates as determined by paleomagnetic reversal stratigraphy. *Journal of Human Evolution*, **22**, 495–503.

Kay, R. F., Fleagle, J. G., and Simons, E. L. (1981). A revision of the Oligocene apes from the Fayum province, Egypt. *American Journal of Physical Anthropology*, **55**, 293–322.

Kimbel, W. H., White, T. D., and Johanson, D. C. (1984). Cranial morphology of *Australopithecus afarensis*: a comparative study based on a composite reconstruction of the adult skull. *American Journal of Physical Anthropology*, **64**, 337–88.

Kimbel, W. H., Johanson, D. C., and Rak, Y. (1994). The First skull and other new discoveries of *Australopithecus afarensis* at Hadar, Ethiopia. *Nature*, **368**, 449–51.

Kurten, B. (1972). *Not from the apes*. Pantheon, New York.

Leakey, M. D. and Harris, J. M. (eds) (1987). *Laetoli: a Pliocene site in northern Tanzania*. Clarendon Press, Oxford.

Leakey, M. G., Feibel, C. S., McDougall, I., and Walker, A. (1995). New four-million-year-old hominid species from Kanapoi and Allia Bay, Kenya. *Nature*, **376**, 565–71.

Leakey, L. S. B. and Leakey, M. D. (1964). Recent discoveries of fossil hominids in Tanganyika at Olduvai and near Lake Natron. *Nature*, **202**, 5–7.

Le Gros Clark, W. E. and Campbell, B. G. (1978). *Fossil evidence for human evolution* (3rd edn). Chicago University Press.

McHenry, H. M. (1991). Femoral lengths and stature in Plio-Pleistocene hominids. *American Journal of Physical Anthropology*, **85**, 149–58.

McHenry, H. M. (1992). Body size and proportions in early hominids. *American Journal of Physical Anthropology*, **87**, 407–31.

Martin, R. (1959). *Lehrbuch der Anthropologie in Systematischer Darstellung*, (3rd edn), Vol. 2. G. Fischer, Stuttgart.

Martin, R. D. (1993). Primate origins: plugging in the gaps. *Nature*, **363**, 223–34.

Passingham, R. E. (1982). *The human primate*. Freeman, San Francisco.

Pilbeam, D. R. (1984). The descent of Hominoids and Hominids. *Scientific American*, **250**(3), 60–9.

Richard, A. (1985). *Primates in nature*. Freeman, New York.

Schwartz, H. H. (1987). *The red ape*. Elm Tree Books/Hamish Hamilton, London.

Simpson, G. G. (1944). *Tempo and mode in evolution*. Columbia University Press, New York.

Stanley, S. M. (1979). *Macroevolution, pattern and progress*. Freeman, San Francisco.

Stringer, C. B. and Andrews, P. (1988). Genetic and fossil evidence for the origin of modern humans. *Science*, **239**, 1263–8.

Swisher, C. C., Curtis, G. H., Jacob, T., Getty A. C., and Sprijo, A. (1994). The age of the earliest known hominids from Java, Indonesia. *Science*, **263**, 1118–21.

Tobias, P. V. (1967). *The cranium of Australopithecus (Zinjanthropus) boisei*. Cambridge University Press.

Tobias, P. V. (1980). *Australopithecus* and early *Homo*. In *Proceedings of the 8th Pan-African Congress of Prehistory and Quartenary Studies* (ed. R. E. F. Leakey and B. A. Ogot), pp. 161–8. National Museums of Kenya, Nairobi.

Tobias, P. V. (1991). *The skulls, endocasts and teeth of Homo habilis* (Olduvai Gorge, Vol. 4). 2 vols. Cambridge University Press.

Wagner, P. (1960). *The human use of the earth*. Free Press, Glencoe.

Wainscoat, J. S., Hill, A. V. S., Boyce, A. L., Flint, J., Hernandez, M., Thein, S. L., et al. (1986). Evolutionary relationships of human populations from an analysis of nuclear DNA polymorphisms. *Nature*, **319**, 491–3.

Walker, A. and Leakey, R. (1993). *The Nariokotome Homo erectus skeleton*. Harvard University Press, Cambridge, Mass.

Walker, A., Leakey, R. E., Harris, J. M., and Brown, F. H. (1986). 2.5-Myr *Australopithecus boisei* from west of Lake Turkana, Kenya. *Nature*, **322**, 517–22.

White, T. D., Johanson, D. C., and Kimbel, W. H. (1981). *Australopithecus africanus*: its phyletic position reconsidered. *South African Journal of Science*, **77**, 445–70.

White, T. D., Suwa, G., and Asfaw, B. (1994). *Australopithecus ramidus*, a new species of early hominid from Aramis, Ethiopia. *Nature*, **371**, 306–12; (1995 Corrigendum, **375**, 88).

Wilson, A. C. and Sarich, V. M. (1969). A molecular time scale for human evolution. *Proceedings of the National Academy of Sciences USA*, **63**, 1088–93.

Wolpoff, M. H. (1986). More on Zhoukoudian. *Current Anthropology*, **27**, 45–7.

Wolpoff, M. H., Wu, X. Z., and Thorne, A. G. (1984). Modern *Homo sapiens* origins: a general theory of Hominid evolution involving the fossil evidence from East Asia and Australasia. In *The origins of modern humans: a world survey of the fossil evidence* (ed. F. H. Smith and F. Spencer), pp. 411–83. A. R. Liss, New York.

Wood, B. (1991). *Hominid cranial remains: Koobi Fora research project*, Vol. 4. Clarendon Press, Oxford.

Wood, B. (1992a). Old bones match old stones. *Nature*, **355**, 678–9.

Wood, B. (1992b). Origin and evolution of the genus *Homo*. *Nature*, **355**, 783–90.

3
Evolutionary trees of apes and humans from DNA sequences

Peter J. Waddell and David Penny

Abstract

Developments over the past decade have made DNA sequences the primary source of information for inferring relationships between organisms. Originally sequences were used for studying relationships between species, but increasingly they are now used to study relationships between individuals and between populations. In this chapter we show how sequences have changed, and continue to change, our views of human origins and evolution. Techniques used to go from DNA sequences to evolutionary inference are outlined, because they are crucial in evaluating this vast new source of data. In addition to a review we report some of the latest research findings, and where necessary have developed appropriate statistical methods. The main points of this chapter are:

1. There is consistently strong support for the human and chimpanzee lineages' being the closest relatives to each other, and the next closest the gorilla lineage, with the orang-utan being the closest non-African relative of these African hominoids.

2. A calibration of these evolutionary trees is given, with estimated dates of divergence for the living hominoids, together with estimates of the expected errors—an important consideration to those interested in assessing the compatibility or otherwise of fossil (or palaeoanthropological) data with molecular inferences. We estimate that the divergence of human and chimpanzee lineages took place approximately 6.5 million years ago, while the standard error of such dating methods is at present about 1 million years.

3. Our evaluation of the 'Out of Africa hypotheses' (mitochondrial 'Eve') leads to the conclusion that this set of four hypotheses (pertaining to the when, where, who, and how of modern humans' origins) does indeed stand up to scrutiny; a point reinforced by our reanalysis of specific features of the data. No single data-set gives overwhelming support to all four aspects of the Out-of-Africa scenario; but it is consistent with several data-sets, while overall the data contradict the 'multiregion' hypothesis of human origins.

4. A re-evaluation of the molecular evidence confirms that the 'when' was almost certainly less than 200 000 years ago, as inferred from both mitochondrial and nuclear DNA data calibrated using both biological and palaeoanthropological data. Africa is most consistently inferred as the 'where'. The mitochondrial DNA sequences give us a glimpse of 'who' founded populations outside Africa and 'how', as populations appear to have expanded rapidly at some point after their arrival into new lands.

Novel maximum likelihood methods were developed to estimate trees with other statistical techniques to infer the reliability of branching points and species divergence dates.

3.1 Introduction

DNA sequences are now used to study two important aspects of human evolution, relationships between humans and higher primates, and relationships among modern humans (*Homo sapiens sapiens*). This chapter illustrates both aspects. For reconstructing evolutionary relationships (phylogeny), the rationale for each step, from sequence data to statistically-justifiable inferences of evolutionary events, is explained as simply as possible. We illustrate these steps with recent data which allow us to address questions such as whether common (*Pan troglodytes*) and pygmy (*Pan paniscus*) chimpanzees are the closest living relatives of modern humans.

For elucidating relationships within the group of modern humans, we introduce and extend the analyses of the data used to support the 'Eve' hypothesis (Wilson and Cann 1992)—that of a recent, African origin for all modern groups. We conclude that, despite recent controversies, these and other molecular data are consistent with the hypotheses that *Homo sapiens sapiens*:

(1) is a very recent species (less than 200 000 years old);

(2) originated in a localized region of Africa; and

(3) close to 100 000 years ago spread out of Africa to replace all other hominids living in Europe (Neanderthals) and Asia (for example, the Solo specimens).

There is no evidence as yet of any interbreeding between modern humans and these other species, which in some areas (for example, Western Europe) appear to have become extinct shortly (perhaps less than 2000 years) after they came into contact with modern humans (Stringer 1990). This does not necessarily imply any direct interaction between species, such as warfare, but could result from indirect competition (Zubrow 1989).

The use of DNA sequences is now standard for inferring evolutionary relationships between species, though it certainly was controversial two decades ago. The use of sequences for studies of relationships within populations and species is newer, and still controversial in some quarters. Nevertheless, the scope and power of such studies is increasing rapidly, and we expect them to quickly become routine. A common theme we use is to build trees from different data-sets and compare the results to see if there is more agreement than would be expected to come about by chance (Penny *et al.* 1982). We conclude the chapter with a discussion of the advances in our understanding of human evolution that we might expect to achieve over the next ten years through the use of molecular data.

3.2 Reconstructing relationships: from DNA sequences to evolutionary history

We will outline the general biochemical approach to establishing evolutionary relations in the phylogeny of apes and humans (that is, hominoids). Using DNA sequences, the three steps in analysing the data are:

(1) estimating the separation (branching) pattern of all species of living hominoids, thus establishing an evolutionary tree;

(2) calibrating this tree, so that we can infer how many millions of years ago (m.y.a.) different lineages diverged; and

(3) placing statistical confidence limits on the estimates of divergence times.

3.2.1 *Basic steps in obtaining a tree for a selected stretch of DNA*

DNA is made up of ordered sequences of the four nucleotide bases that are abbreviated as a, c, t, and g. The chromosomes of living hominoids contain approximately 3 billion bases, arranged linearly on 44 to 48 chromosomes (the number varies in different species owing to chromosome fusions and/or splitting). The chromosomes are in the nucleus of each cell, and contain the nuclear DNA. Chromosomes are inherited equally, but randomly, from both parents (excepting the male Y chromosome). In addition, there are just under 17 000 base pairs (b.p.) of DNA in the mitochondria (mtDNA). Mitochondria are organelles in the cytoplasm of cells, and are inherited solely from the mother via the egg cell. In the present context we are only interested in what DNA sequences can tell about evolutionary history. For our purpose a DNA locus (plural loci) is a contiguous stretch of DNA.

Our interest in DNA is not its function, but how we can use the changes (mutations) at particular loci to trace the evolution of the DNA, and hence to gain insight into the evolution of the species. Figure 3.1 shows a short piece of DNA sequence from a human mitochondrion lined up with the equivalent sequences from apes. During each cell division the total DNA of an individual is copied with great precision, although very occasionally (about once per billion nucleotide replications) a mistake is made.[1] This mistake, if it is either advantageous or effectively neutral (neither helping nor hindering an individual) may persist and spread in later generations. Almost all the substitutions occurring in the DNA are neutral (Nei 1987; Penny 1994).

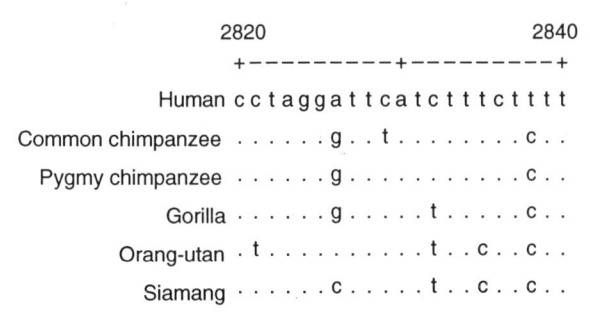

```
                        2820              2840
                        +---------+---------+
              Human  c c t a g g a t t c a t c t t t c t t t t
Common chimpanzee  . . . . . . g . . t . . . . . . . . . c . .
 Pygmy chimpanzee  . . . . . . g . . . . . . . . . . . . c . .
           Gorilla  . . . . . . g . . . . t . . . . . c . .
        Orang-utan  . t . . . . . . . . . . . t . . c . . c . .
           Siamang  . . . . . . c . . . . . t . . c . . c . .
```

Fig. 3.1 An example of 21 nucleotide sites (base pairs) of aligned DNA sequences from hominoids (sites 2820–2840 from Horai *et al.* 1992). The convention of having a dot when a species has the same nucleotide as the first species allows patterns in the data to be seen more readily. The pattern of changes at site 2832 groups humans and chimpanzees together, while site 2835 groups humans, chimpanzees, and gorillas. There are also sites where only one species differs from the rest (for example, sites 2821, 2829, 2838); these provide evidence for the length of time since the divergence of that species. Site 2826 is an example of a site that must have changed at least twice. Maximum likelihood takes all these patterns into account when working out which tree best fits the data, and also allows a test of which, if any, tree model fits the data adequately.

Since neutral changes are essentially invisible to the processes of natural selection their occurrences can be modelled accurately using mathematical and statistical theory. A neutral mutation will usually disappear by chance (it is initially present in only one individual); but occasionally it may—also by chance—spread throughout a species.[2] Between the different hominoid species there has been sufficient time to ensure that most neutral mutations have by chance either become lost or become predominant within a particular species; when predominant, they are called substitutions.

It can scarcely be emphasized highly enough that treating changes as neutral removes one of the most important difficulties that beset earlier generations of researchers, namely, the difficulty involved in treating differences between humans as evidence for inferiority or superiority. The traditional European view of Natural Theology was that everything, including differences, must have a 'purpose'. This, combined with the idea of a Great Chain of Being comprising a hierarchy of living forms led naturally to explanations involving value judgements about human differences. Nothing of which we are aware limits the capacity to coin such explanations to European cultures. Last century Darwinian evolutionary theory introduced probabilistic reasoning as a major concept in science; and this, and other aspects of the theory (Penny 1994) removed the need for assuming everything must have a purpose. The application of probabilistic thinking to evolutionary studies has increased, particularly over the past few decades, with the development of Kimura's theory of neutral evolution. Treating the vast majority of DNA changes as neutral therefore has the double advantage both of allowing more detailed mathematical modelling and of removing value judgements from the study of human variation.

When substitutions in DNA are relatively rare it is likely that sequences sharing a substitution are more closely related to each other than to others with a different base. Thus, for example, site 2832 of mtDNA as shown in Fig. 3.1 suggests that humans and chimpanzees are a group separate from the other apes, i.e. the site has a 'pattern' supporting a chimpanzee–human grouping. Occasionally, however, two or more individuals might share the same innovatory base at a site as a result of independent mutations. Such events can lead to erroneous conclusions about relationships if considered just by themselves, since some evidence can be found for almost any hypothesis! This problem of parallel mutation is much less likely to occur if the total amount of change in the DNA being examined is reasonably low; and, additionally, with longer sequences we get more accurate estimates of the true frequencies.

Evolution is a stochastic (probabilistic) process, and so the same change will occur on different lineages just by chance. Because of these multiple changes it can be difficult to get the correct tree directly from DNA sequences (see Fig. 3.2 for commonly used terms). A useful statistical criterion for deciding which weighted tree best fits the data is 'maximum likelihood' (Swofford and Olsen 1990). This assesses the likelihood of the observed nucleotide patterns given a weighted tree (Fig. 3.2 (II)) and the relative rates of substitutions (such as a → c, a → g). The 'weight' of an edge is the average number of substitutions per site expected on that edge. Computer programs are available that search for both the weighted tree and the mechanism of change that gives the best overall fit between the model and the data. The maximum likelihood criterion has the advantage of allowing confidence estimates on all parameters in the model. An important test is whether a tree model even fits the data adequately. A model may fail such a test for a number of reasons (including selection for certain changes); and in such cases we must be careful in placing confidence in the results. These tests may indicate that we need to consider more complicated models that include factors such as hybridization between different lineages.

Fig. 3.2 Types of evolutionary tree:

(I) An unrooted tree (H = human, C = common chimpanzee, G = gorilla, O = orang-utan).

(II) An unrooted weighted tree, where edge (= internode) lengths are proportional to the amount of change (here DNA substitutions) on that edge. The X marks the position where the siamang joins this tree.

(IIIa) A rooted weighted tree; the tree in II rooted by the outgroup method (i.e. the point at which the sequence of a more distantly related species joins the tree). Here the siamang is used as the outgroup. **(IIIb)** The same rooted tree as in IIIa. **(IIIc)** The same tree again, but drawn so that the edge lengths are represented by only the vertical component. This is not intended to indicate anything different about the process of evolution, but rather, is useful in comparing edge lengths in a rooted weighted tree.

(IV) The maximum likelihood tree of 5 kilobases of mtDNA sequence, with the pygmy chimpanzee also included, calibrated by dating the origin of the edge leading to the orang-utan at 16 million years ago. CC is the common chimpanzee, while PC is the pygmy chimpanzee *Pan paniscus*. This tree is the optimal tree for the best-fitting maximum likelihood model (Kimura's 3ST mechanism of nucleotide change, with the rate of change across sites falling into two classes), with the constraint that the edge tips all meet at the same level (i.e. follow a molecular clock). The dates of the earliest known fossils attributed to specific lineages are shown, as well as the dates of less certain, more fragmentary, but similar remains (Groves 1989; Martin 1990; Campbell, this volume, Chapter 2).

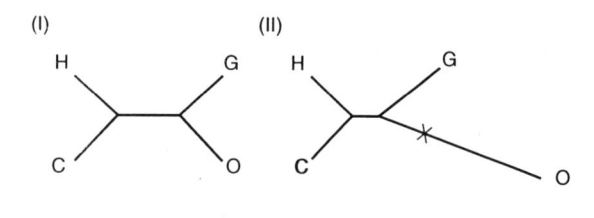

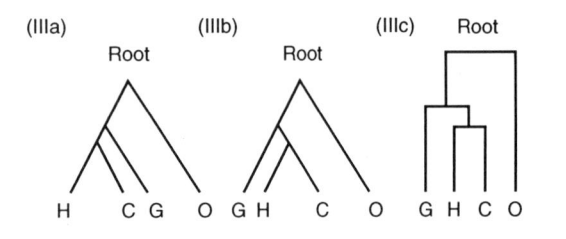

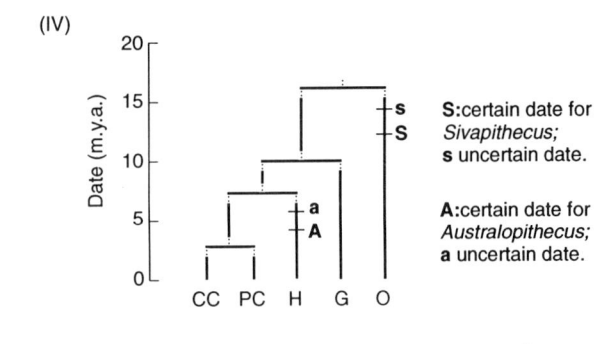

Figure 3.2 (IV) shows the tree which best fits 5000 base pairs of mtDNA from apes and humans (Horai *et al.* 1992). One of the interesting things about this tree is that it groups humans and chimpanzees together to the exclusion of other apes—a conclusion that is consistent with other data-sets discussed below. The sampling error resulting from a finite sequence length is shown for each edge, and the error on even the shortest edge (human–chimpanzee) is clearly less than the support for that edge.[3] The weighted tree we have illustrated is the maximum likelihood tree, with the constraint of a molecular clock (see below) so that all tips meet at the same level (Fig. 3.2(IV)). This constraint results in a slightly worse fit of model to data (as is to be expected, because it has fewer adjustable parameters); but it is the type of model that is most useful when calibrating the tree. Further details of the maximum likelihood models used to analyse the data are given in a footnote to the caption of Table 3.1.

3.2.2 Putting dates on a tree

There are two ways in which we can add a temporal dimension to these trees: with respect to other species; and using the 'molecular clock' hypothesis. The orientation of the hominoid tree with respect to time (called 'rooting' the tree, Fig. 3.2(III)) has been inferred using the knowledge that the Great Apes are more closely related to each other than to the gibbons (lesser apes), a view confirmed when trees are built from larger sets of species, including Old and New World monkeys, tarsiers, and lemurs. We will refer to this technique as the 'outgroup' method of tree-rooting.

The conclusion that a sequence of DNA will evolve at the same average rate in different species is called the 'molecular clock' hypothesis. The average rate is characteristic for each DNA sequence, and depends on the mutation rate and the number of constraints on the sequence. The 'clock' is a natural outcome of

Table 3.1 Estimated divergence dates of DNA sequences for humans, chimps, and gorillas, calibrated assuming an orang-utan divergence date of 16 m.y.a. Because of polymorphism in the ancestral populations these dates will tend on average to be greater than the actual dates of species divergence (see text). The loci are mtDNA (Horai *et al.* 1992), and ψη and γ-globin nuclear DNA loci (Bailey *et al.* 1992). The 'difference' column is the estimated time between the gorilla separation and the subsequent splitting of the lineages to humans and chimpanzees. These sequences were analysed using a variety of maximum likelihood models based around the Kimura 3ST model, allowing the rate of substitution to vary between sites (Steel *et al.* 1993; plus see technical notes at the bottom of this table). The values for the DNA hybridization experiments designated 'DNA hybrid.-1' are maximum and minimum values from trees shown in Sibley *et al.* (1990) and constrained to fit a molecular clock. The values in 'DNA hybrid.-2' are for the clock-constrained tree estimated using a least-squares fit on the DNA hybridization data of Caccone and Powell (1989). Note the discrepancy of the inferred time of the lineages leading to humans and gorillas for these two data-sets, a feature which must make us cautious in over-interpreting the DNA hybridization results, at least until the reasons for it are understood. Values are in millions of years (values in parentheses are relative divergence dates, i.e. proportion of time back to orang-utan divergence). 'Chimp–chimp' values are for the divergence dates found for common vs Pygmy chimpanzees (*Pan troglodytes* vs *Pan paniscus*).

	human–chimp	human–gorilla	difference	chimp–chimp
mtDNA	7.2(0.45)	9.8(0.61)	2.6(0.16)	2.7(0.17)
gamma	6.5(0.41)	7.8(0.49)	1.3(0.08)	—
psi–eta	7.6(0.47)	8.2(0.51)	0.6(0.04)	—
DNA hybrid.-1	5.9(0.37) to 8.1(0.51)	8.2(0.52) to 9.1(0.57)	0.6(0.04) to 2.3(0.14)	3.0(0.19)
DNA hybrid.-2	7.2(0.45)	11.4(0.71)	4.2(0.26)	3.6(0.22)

Technical notes: Our evaluations with maximum likelihood included novel models assuming that substitution rates of sites followed a gamma distribution (similar in shape to the lognormal). The general model used here is the extended Kimura 3ST model of Steel *et al.* (1993). All free parameters in all models were optimized using a Newton method. The 'logarithm of the likelihood of the data' is the sum over all site patterns of {observed frequency of the i-th site pattern multiplied by the natural logarithm of its probability under the model}. The fit of model to data gives strong evidence that rates of substitution in the mtDNA sites vary considerably (due, no doubt, to stabilizing selection, i.e. many sites do not accept substitutions because they are functionally constrained). Here, the log likelihood ratio fit statistic (lnLR or $G^2/2$) decreased from 391.6 to 151.6 when the gamma distribution was allowed (with the optimized shape parameter equalling 0.351). An 'invariant sites' model gave an even better fit, suggesting a demarcation between those sites which can *vs* cannot change (lnLR 139.7 with 59.2% of sites assumed unable to change; the variable sites are mostly 'third' position sites, see Nei 1987). Interestingly, a mixed gamma-invariant sites model did not further improve the fit. We found no evidence to suggest unequal site rates in the nuclear 'non-coding regions'. For the mtDNA alone, the divergence times in this table are the average of 21 different submodels of the generalized Kimura 3ST. This allowed inference of errors due to choice of evolutionary model. These fluctuations in divergence times, relative to orang-utan, had range: human–chimp (0.38 to 0.53), human-gorilla (0.56 to 0.66), and chimp-chimp (0.14 to 0.21). Further details of these analyses are available from PJW (email: farside@ massey.ac.nz).

a probabilistic process where most changes are neutral (Penny 1994). Because changes are stochastic the 'clock' can only be an average rate; but it gives another way of rooting a tree, by estimating the midpoint in the tree (that farthest from the living species) and taking this as the earliest ancester. The molecular clock hypothesis can be difficult to test without reliable outgroup rooting; but it is useful, as we will see in the later section on relationships within the human species. For the mtDNA data, and two other hominoid nuclear DNA loci that we analyse below, both outgroup and molecular clock methods of rooting are in agreement, thus providing further reassurance that the root is reasonably placed.

There are two reliable methods used to calibrate the branching points on a tree. The first is a well-dated fossil reliably associated with a particular lineage in the tree, preferably close to the origin of that lineage. In the case of hominoids the *Sivapithecus* fossils known from Asia, with good fossils from Pakistan (Pilbeam 1984), seem to fit this role, as may *Dryopithecus* from Europe (Solà and Köhler 1993). In the last twenty years better examples of these fossil apes, and a revision of systematics, have shown that they share a number of unique skeletal features with orang-utans (*Pongo pygmaeus*). This suggests that *Sivapithecus* was somewhere along the lineage leading to the orang-utan. Since these fossils are known to date back 12 million years, the point where the orang-utan edge joins the rest of the tree must indicate an event at least 12 million years old (Pilbeam 1984; Groves 1989).

Speculation about the time required for *Sivapithecus* to acquire its unique features, and the trend of the past twenty years of finding somewhat older *Sivapithecus* fossils, suggests that the true date of the origin of the edge leading to the orang-utan was probably 2 to 6 million years older again. Thus fossil evidence suggests that the node where the orang-utan last shared an ancestor with the other Great Apes would be 16 m.y.a. ± myr (million years), consistent with Pilbeam's (1984) expectations. The most crucial assumption here is that *Sivapithecus* is most closely related to orang-utans amongst the living apes. Estimates of the relationships of fossil hominoids have changed much over the past twenty years, and it would be reassuring if the discovery of post-cranial bones of *Sivapithecus* supported its present placement.

Other fossils which can help calibrate the hominoid tree are australopithecine fossils on the edge of the tree leading to humans. At present these are dated back with certainty to 4 million years, and possibly even to 5.5 m.y.a. (Campbell, this volume, Chapter 2; Groves 1989). Figure 3.2(IV) shows these dates plotted on to the tree calibrated with the expected divergence time of the orang-utan. At present there are no other fossils that we can confidently associate with any other edge of the tree. Groves (1989) gives a useful overview of the status of fossils which, with better evidence, may eventually be assigned to particular parts of the hominoid tree.

A second way of estimating the date of a split in a tree is by a dated biogeographic event. An often-cited example is the separation of the lineages of the ratite birds ostrich (Africa) and rhea (South America) due to the continental rifting which caused the Atlantic Ocean to appear approximately 80 m.y.a. The opposite biogeographic effect occurred approximately 18 m.y.a., when the Arabian peninsula collided with Eurasia and allowed the biota of both areas to mix (Thomas 1985; Pilbeam 1984). Since the earliest fossil apes come from Africa it seems plausible that the ancestors of living and fossil Asian apes emigrated from Africa not earlier than 20 m.y.a. The evidence of other fossil groups emigrating from Africa to Asia at this time is consistent with the fossil dating given above for the orang-utan.

There are other methods for dating the branching points of trees, but these are less reliable than those just noted. Rates estimated by the above techniques in one group are extrapolated to those of another group. An example is to estimate an average rate of nucleotide substitution for different mammalian orders, using a diversification time of approximately 60–80 million years ago. This average can then be applied to any group of mammals that has a poor fossil record. Unfortunately, it is suggested that even within the mammals DNA substitutions may be as much as three times higher than the average in some groups, such as certain rodents (Nei 1987). Accordingly this approach is good for a ballpark figure, but has additional uncertainties.

The calibration of hominoid divergences has been attempted using DNA from a range of older primate divergences, such as: the divergence of Old World monkeys and hominoids; the earlier split of the ancestors of these two groups from the New World monkeys; and earlier events, going back to the supposed origin of primates (as estimated from fragmentary fossils). Such an approach has two major drawbacks:

(i) The suggested divergence times for the above events vary, because the earliest known fossils for a group may occur well after the origin of that group (Martin 1990); and

(ii) there is evidence that the rate of DNA substitution in hominoids, especially the larger ones, has slowed down in relation to that in other primates (Bailey *et al.* 1992).

Both of these effects are expected to cause an underestimation of the divergence dates of hominoids. Together with sampling errors they largely explain why some published dates for the divergence of humans and chimpanzees from DNA sequences are too recent (as little as 3.3 m.y.a. in one case (Hasegawa *et al.* 1985)).

3.2.3 *Results from other molecular data*

In addition to the 5 kilobases of mtDNA sequence used above, there are sequences approximately 10 000 base pairs long for two regions of nuclear DNA, the ψη (psi–eta) and γ (gamma) globin loci (Bailey *et al.* 1992). These two loci are contiguous, and form part of a region of about 100 kilobases known as the β globin gene cluster. Trees for these loci have been estimated using maximum likelihood methods based upon Kimura's 3ST (Nei 1987) model of nucleotide change, with the option of allowing the relative probability of substitution to vary at different nucleotide sites (Steel *et al.* 1993; Waddell, in preparation).

Statistical tests indicate that in all cases the fit of the tree model to the data is acceptable. Using a likelihood ratio statistic none of the other trees (including any tree grouping chimpanzee and gorilla, *Gorilla gorilla*) provide an adequate fit for the data. Thus three loci clearly favour humans and chimpanzees as closest relatives, with the gorilla being the next closest living relative. Results such as these have overturned the prevailing view of the last hundred years that the Great Apes are all more closely related to each other than any of them is to humans. Even though it is accepted that the African hominoids are our closest relatives, there is still some reluctance among morphologists to separate the knuckle-walking apes (chimpanzees and gorillas). As we show below, the molecular evidence is consistently in favour of the human–chimpanzee grouping.

Another set of data often referred to in studies of hominoids comes from the method of DNA hybridization. This estimates the overall nucleotide change between two species, but without determining the actual sequences. There are four published data-sets of DNA hybridization distances that include at least four living hominoids (Sibley *et al.* 1990). All favour the human–chimpanzee tree over the chimpanzee-gorilla tree, with one data-set (that of Caccone and Powell 1989) giving this tree over 99.7 per cent of the time in a statistical resampling procedure (Marshall 1991). Note, however, that different experimental procedures (hybrid-1 *vs* hybrid-2 in Table 3.1) can yield different results, sounding a note of caution in interpreting this type of data. In general, the DNA hybridization results are reassuringly consistent with results from sequence data, and, experimental errors aside, are expected to be indicative of the tree for the majority of DNA loci in these species.

Since we have reasonable dates on at least one node in our tree we can estimate dates for others. Table 3.1 shows these dates from three DNA loci on the basis of models assuming a molecular clock. In general they agree quite well. A notable exception is the estimated time from the divergence of the gorilla lineage to the separation of humans and chimpanzes (Table 3.1, column 4). This is a point of some interest to researchers; for example, if the combined human–chimpanzee lineage was relatively long humans may still share some ancestral characteristics with chimpanzees (perhaps language abilities or behaviour) that are not shared with gorillas.

3.2.4 *Polymorphisms and population variability*

A probable reason for the differences in these estimates of the divergence time from gorillas, apart from sampling error, is molecular polymorphism. There are different, but related, sequences (alleles) in a population at any one time.[4] The degree of DNA polymorphism in a population is directly proportional to the long-term size of its population (also known as effective population size: Nei 1987). Consequently, at any time a population will have alleles that originated well back in the past, and these times will vary for different alleles. When a population subdivides, leading eventually to two species, there is a random component as to which of the ancestral alleles become prevalent in each population. Consequently we require many DNA sequences before we can confidently predict the exact separation time of species, and not just the earlier divergence times of alleles.[5]

The trees for the three loci discussed are also consistent with a number of shorter sequences. These loci, such as 28S rRNA and the associated spacer region (Gonzalez *et al.* 1991), favour the human–chimpanzee tree, though individually they do not statistically reject the alternatives (in the mentioned example a claimed significant result is doubtful after we found four possible alignment errors in the original data). Less decisive data-sets, such as chromosome structure and allozyme frequencies, are also consistent to the limit of their resolution with the human–chimpanzee grouping. Given the molecular results, palaeontologists are reappraising the fossil data, which some now contend (for example, Begun 1992) are really most consistent with the human–chimpanzee grouping.

3.2.5 *Total error on estimated divergence times*

So far we have identified four independent sources of error on divergence times: fossil calibration; sequence length (sampling error); ancestral polymorphism; and the variety in methods and models that can be used to infer trees, edge lengths, and node times. If these errors are independent and additive we may estimate the total error on divergence times, since the overall variance from independent sources is then the sum of the individual variances. In this example all errors are independent, but some of them are multiplicative, and since they are such the total error we derive here will be an underestimate. (Later, in note 10, we show how to calculate some of these multiplicative errors in the context of dating the origin of modern humans.) We will describe the exact statistical error structure of molecular divergence times elsewhere (Waddell in preparation). We now illustrate these calculations with the divergence time of human and chimpanzee mtDNA, plus the additional step of inferring the divergence times of the actual populations. The standard error is the square root of the variance estimated from the sample.

1. The fossil calibration for the origin of the orang-utan edge has a standard deviation of about 1 million years (myr), so the variance of this is $1^2 = 1$. But, since the human edge is only about half as long as the orang-utan edge, the relative error becomes ½ myr, giving a variance of $0.5^2 = 0.25$.

2. The standard deviation (due to sampling error) of the ratio of the height of the human–chimpanzee node to the divergence of the orang-utan lineage is approximately 0.05, which translates to ¾ myr (variance = $0.75^2 = 0.56$).

3. Different models of sequence evolution will also give slightly different ratios of edge lengths, and it is hard to be sure which will give the best estimate. In addition there are alternative ways of estimating divergence times without imposing a molecular clock. For these mitochondrial data the observed standard deviation of edge-length ratios due to these two causes together is equivalent to about ½ myr (variance = 0.25) (Waddell, unpublished).

4. While the polymorphism effect is still an unknown quantity, in our samples it may have introduced an average error of about 1 myr if the effective population size of the last human–chimpanzee ancestor was about the same as that of chimpanzees before human impact. If there was a similar amount of polymorphism at the origin of the orang-utan lineage then this effect is reduced by about a half, to about ½ myr (if the effective population size was constant for long enough, then the distribution will be approximately exponential, which implies the mean is equal to the s.d., so the variance = ¼. This is an approximation.

Adding up all these variances we have an overall variance of $0.25 + 0.56 + 0.25 + 0.25 = 1.31$. The overall standard deviation of our estimates is then $\sqrt{1.31}$ or about 1.15, which is expected to be close to normally distributed, and consequently the 95 per cent confidence interval is ± 2 standard errors. Thus we estimate the time of human–chimpanzee mtDNA divergence at 7.2 m.y.a., with a 95 per cent confidence interval of ± 2(1.14), this is approximately 4.9 to 9.5 m.y.a. There is an expected upward bias of 0.5 to 1.0 myr due to ancestral DNA polymorphism, making about 6.2–6.7 m.y.a.[6] the most likely time for the divergence of the actual populations, but still with a standard deviation of just over 1 myr.

The results shown in Table 3.1 still have a rather large uncertainty in estimating divergence times (even if we could fix the divergence time of the orang-utan lineage exactly). While a refined estimate of divergence times will require the sequencing and analysis of more DNA loci, it appears most likely that the lineages leading to humans and chimpanzees diverged about 6 to 7 m.y.a. The population leading to gorillas probably diverged somewhere between 0.5 to 2.5 myr earlier again. Similarly, the evidence points to the divergence of the two chimpanzee species about 2.0 to 2.5 m.y.a.[7]

In conclusion, almost all the data collected so far are consistent with humans and chimpanzees being the closest relatives, and from the diverse data amassed it would be surprising if this view were overturned. The example illustrates the usefulness of molecular data when resolving what was probably a fairly closely-spaced series of population divergences. Even with molecular data quite long sequences are needed to have confidence in the results. The relative duration of an ancestral population leading eventually to humans and chimpanzees should become clear with additional sequences, and these are becoming available at an increasing rate. This will allow reliable inferences about some of the population dynamics of these ancestral species. Having established a reliable phylogeny we consider next what can be learned from molecular data regarding the origin and expansion of one particular species, *Homo sapiens sapiens*.

3.3 Human genetic data: mtDNA sequences

Here we look at the genetic evidence of the origin and interrelationships of humans. We consider the controversial findings from human mtDNA, and then how these results compare with evidence of genetic relationships from nuclear DNA.

3.3.1 *Out-of-Africa, or mitochondrial Eve*

The 'Out-of-Africa' (or mitochondrial 'Eve') hypothesis was the result of Cann, Stoneking, and Wilson's (1987) study using sequence markers on mtDNA to trace the maternal ancestry of 147 people from widely-dispersed indigenous populations (see Wilson and Cann 1992 for a general review). The Eve hypothesis is a set of hypotheses (see below) proposing that all modern humans have a common maternal ancestor who lived in Africa around 200 000 years ago. Similar ideas had been developed earlier for nuclear-coded protein polymorphisms (Nei and Roychoudhury 1982). An opposing hypothesis, that modern human races evolved *in situ* from interbreeding populations of *Homo erectus* and its descendant populations (for example Neanderthals), is referred to as the 'multi-regional hypothesis' (Wilson and Cann 1992). Perhaps the most common misunderstanding of the Eve hypothesis is that there was just a single female in the population. However, calculations referred to below suggest a population of 1000 to 10 000 females. It is probabilistic processes, referred to earlier under the neutral theory, that eventually lead to all mitochondria being derived from just a single female.

Because sequences are expected to be more informative than sequence markers for building trees, we will use the data of Vigilant *et al.* (1991). They used 630 base pairs from the fastest-evolving region of mitochondrial DNA (the origin of DNA replication, also known as the D-loop), and 135 different sequences were obtained. We will use this data-set both to explore the 'Out-of-Africa hypothesis' and to outline the techniques developed, and which need to be developed, to analyse such data fully.

The 'Out-of-Africa' hypothesis is really a set of hypotheses, each one of which predicts a different characteristic of the tree of human maternal relationships. The hypotheses are:

1. All human populations can trace their maternal ancestry back to a common ancestor, who was surprisingly recent (possibly less than 200 000 years ago). Further, because of the effect of molecular polymorphism (described above), this date may be an overestimation of the actual age of the species *H. s. sapiens*.

2. The most probable ancestral region of modern humans was Africa, because the earliest divergences in the rooted tree seemed to have purely African descendants.

3. Most major populations (for example Asians, Europeans) show a number of distinct maternal lineages, suggesting that they were founded by populations of diverse individuals, and not by small, closely-related groups.

A more recently identified feature of human mtDNA trees that adds to the original hypothesis is:

4. Some time after the deepest divergences in the tree there appear many lineage separations, consistent with a rapid expansion in the size of the human population (Di Rienzo and Wilson 1991). This feature is noted especially amongst the non-African sequences; it is a feature predictable from 'Out-of-Africa', but not specifically predicted by the 'multi-regional hypothesis'.

The first two parts, the 'when' and the 'where', are the most critical in deciding which hypothesis (Out-of-Africa or the multiregional) is better. The second two parts, the 'who' and the 'how' of founding populations, add detail. All four of these parts are logically independent (a subset of them could be true and the others false), and each requires a specific type of test.

In conjunction with fossil evidence, the maternal mtDNA data implies some startling features in human evolution. There is good fossil evidence that *Homo erectus* occupied much of the Old World (Africa and Eurasia) by 1 m.y.a. and after that differentiated into regional forms (such as Neanderthals in Europe). An age of only 200 000 years for the last common ancestor of all modern humans implies that only one of these *Homo erectus* populations has left any maternal descendants. In other words, a single geographically localized lineage of *Homo erectus* must have evolved into modern humans and then spread out to replace other living hominids. Molecular population geneticists are generally comfortable with such a hypothesis, because it is consistent with known processes of mutation and replacement. They do not have to appeal to unknown mechanisms or to mechanisms 'special' to humans.

Africa has been suggested as the ancestral area of humans by two criteria: (1) the location of the root of the human mtDNA tree, with its first lineages leading to large African branches; and (2) the fact that present-day African populations include the most divergent human mtDNA sequences (Vigilant *et al.* 1991). The mere presence of such large blocks of purely African sequences argues strongly against the multiregional hypothesis, which requires a large amount of interbreeding between all descendant populations of *Homo erectus* (Wilson and Cann 1992). Further, the very recent nature of all human mtDNA diversity makes even the whole of Africa look unlikely as the place where *Homo sapiens* first evolved, and suggests that the earliest origin of modern humans was in a part of Africa perhaps supporting a population of the order of 10 000 or even less (from the expected polymorphism of different population sizes under the neutral model). As yet there is no direct genetic evidence for the exact location.

Such major claims regarding the evolution of modern humans have not gone uncriticized (for example: Templeton 1993). There are criticisms directed against the reliability of trees obtained from such data, and also of hypotheses related to the location of the root (for example: Maddison, *et al.* 1992) and the date of the root (for example: Nei 1992) of the human mtDNA tree.

We will proceed as follows. First, we shall note some of the problems that are involved in trying to infer evolutionary hypotheses from the available data. Second, we shall briefly note the approach we are currently pursuing in a re-analysis of Vigilant *et al.*'s 1991 data-set. Third, we shall draw on our earlier discussion of tree-building and dating techniques (developed above with respect to the hominoids) to consider the status of the related hypotheses of when, where, who, and how. And finally, we shall consider other sources of molecular data, independent of the mtDNA sets, and attempt to link them to fossil and other data. This enables us to estimate the accumulated sources of error, and to establish the 'ballpark' within which the evolution of the hominid lineage occurred. We attempt, in the midst of a great number of sources of uncertainty, to indicate what seems to us, at this point, the best interpretation of the data available. For an opposing view, see Templeton 1993; but even here the critics of the Out-of-Africa model do not find support in DNA sequences for a multiregion model.

For this section, we shall introduce a new term: *branch*. A branch is a collection of edges in a tree emanating from one node, and, just as with a real tree, when you detach a branch you take with it all the edges further out from the root. As such, the term 'branch' is not interchangeable with the term 'edge' (internode).

3.3.2 *Problems with trees from large numbers of sequences*

The problems of determining the branching order and the rooting of human mtDNA trees are, for a number of reasons, extreme: the tree has many edges; the sequences are relatively short; only ⅓ of the sites have patterns directly useful in estimating the branching pattern; and the distance to the closest human outgroup (chimpanzees) is approximately 20 times as long as the maximum distance between human sequences. Maximum likelihood methods of tree reconstruction slow down[8] as the number of sequences increases, and are as yet impractical for the types of study these data require.

Instead we have chosen to use the method of parsimony (Swofford and Olsen 1990), which searches for the tree requiring the smallest number of nucleotide substitutions (mutations). When the overall rate of change is small parsimony is expected to perform in a similar manner to maximum likelihood. A complication is that, with a similar number of sequences and sites informative to parsimony, there can be many trees with equivalent support. This is especially true with so many sequences (there are over 10^{260} possible trees for 135 sequences—compared with about 10^{70} elementary particles in the universe).

In such cases the fundamental questions the researcher needs answered are: do the optimal trees seem to be converging towards a common answer; and, if so, what are the general features of the best-fitting trees; are these trees consistent with other relevant but independent data?

3.3.3 *Results from re-analysing the data*

To study the reliability of the analysis of Vigilant *et al.* (1991) we began searches for optimal trees from many random starting-points, using a recently developed search method (the Great Deluge algorithm, Penny *et al.* 1994) that has been shown to work well on other problems of similar complexity. The results of over 400 separate runs have been studied (Penny *et al.* 1995). By using measures of distances between trees and basic geometry, the locally optimal trees can be viewed as forming a single peak, implying that shorter trees are indeed converging to a relatively small subset of similar trees that we expect to be good estimators of the main features of the underlying tree.

Here we have chosen to infer the form of the true tree by finding the median tree (Fig. 3.3) that is, in a geometrical sense, the middle of all the best trees found. It turns out that this median tree is also one of the shortest trees on this data-set (one step longer than the shortest). Many of the edges in this tree have estimated lengths of zero, and for robustness we only show those edges which are supported by one or more changes. The tree has been rooted by the midpoint method, which depends on a molecular clock. This tree is generally similar to those found by other tree-building methods (for example, the neighbour-joining method: Swofford and Olsen 1990). It is different from the original tree of Vigilant *et al.* (1991) in that it unites the !Kung.

Other interesting features are found in the tree. Some African sequences tend to form clusters according to ethnic origin, for example the !Kung and the Pygmies. Such features (assuming a random sample) probably indicate moderate population sizes and a degree of isolation from other groups, rather than founding by a few closely-related individuals. The depth of these clusters indicates that these populations have existed for a relatively long time with respect to the depth of the root. Sequences from other African groups (the Yoruban, the Herero, and the Hadza) also form distinct clusters within larger assemblages of African sequences, but are also found intermixed with the Asian, European, and New Guinean sequences. Such a pattern is indicative of either a sudden population expansion or else a period of exponential population growth (Di Rienzo and Wilson 1991; Rogers and Harpending 1992; Harpending *et al.* 1993). The relationships between African and non-African sequences is revealing; the Africans clearly form a superset, with only some of the African clusters being

Evolutionary trees of apes and humans

found amongst non-Africans—which is in agreement with an African origin of *H. s. sapiens*, followed by expansion out of Africa by a subset of the total African diversity.

The positions in the tree of the African-American sequences are consistent with their being members of the African groups (excepting the Pygmies and the !Kung) recently displaced by the slave trade. Are the two Asian sequences in the midst of these otherwise African clusters similarly due to a slave trade or a mixed Afro-American/Chinese ancestry of Californian subjects.

---

the Great Deluge search algorithm, starting from random trees). It is rooted by the midpoint method, that is the root is located in the middle of the longest path through the weighted tree, in this case between WP43 and Eu102. The most noticeable differences between this tree and that of Vigilant *et al.* (1991) are:

(1) the African !Kung (San Bushmen) sequences form a single group, rather than indicating the preservation of many independent ancient lineages of mtDNA (also found by others); and

(2) the location of the root has shifted from being on the branch leading to the !Kung plus the sequences near WP1, a position suggested by outgroup rooting, to the present position, which appears more consistent with the molecular clock hypothesis. Notice how predominant African sequences are amongst the deepest branches of this tree—just what would be expected if only a subset of the variation originally in Africa had spread with a migration of people out of Africa.

Another noticeable feature is that most European, Asian, and New Guinean sequences trace back to a closely packed series of branching events, which are largely unresolved. This is suggested to be the result of a rapid population expansion that followed the migration of groups of *Homo sapiens sapiens* out of Africa. Notice the variability in edge lengths caused by sampling error when there are relatively few DNA substitutions between sequences. This feature makes the development of methods which take the contribution of all edges leading back to a branch point very desirable for both rooting and dating such large trees. The sequence numbers are the same as those used by Vigilant *et al.*, while the labels used are: AA = African American, Ai = Asian (the * indicates sequences hypothesized as possibly belonging to the descendants of slaves traded between Africa and Asia), Au = Australian Aborigine, EP = Eastern Pygmy (Central African Republic), Eu = European, Ha = Hadza (East Africa), He = Herero (Southern Africa), Ku = !Kung (Southern Africa), Na = Naron (Southern Africa), NG = Papua New Guinea highlander, WP = Western Pygmy (Zaïre), Yo = Yoruban (West Africa).

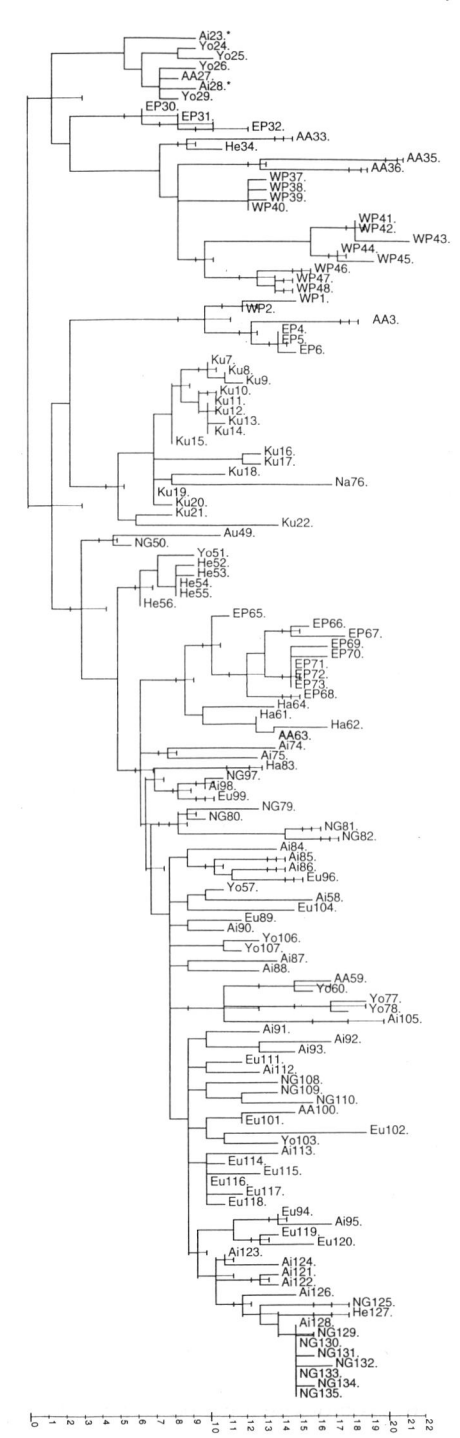

Fig. 3.3 Median tree from the re-analysis of the 135 human mtDNA sequences of Vigilant *et al.* (1991). This is the median tree of over 400 independent runs with the maximum parsimony criterion (as found by

If the time between the origin of *Homo sapiens sapiens* and major expansions in its range were reasonably close (say expansion was less than one-third the time after origin) then this will add to the difficulty in locating the root. In addition, populations such as the !Kung have possibly lost some of their original diversity, and this loss increases the difficulty of finding evidence for an early rapid expansion among African populations prior to the expansion of *Homo sapiens sapiens* out of Africa.

In general, we suggest that the above pattern is most consistent with an African origin, followed relatively soon by migrations out of Africa by people who possessed only a subset of the mtDNA diversity, resulting in the superset–subset relationship between the African sequences and those from the rest of the world. Some of the African groups, such as the !Kung, were possibly separate by this time, and may have had a relatively small effective population size since then. Alternatively, if the tree and its root are substantially correct, and the population size of the !Kung has not been so small as to lose mtDNA diversity over the millennia, then we have the suggestion that the !Kung were an early migration into Southern Africa from elsewhere. That our tree alone, of all those yet published, places the Naron (another group of San people) with the !Kung reinforces this possibility (also relevant to this hypothesis is Deacon (1992), who argues that humans in Southern Africa were quiet isolated from 125 000 to 10 000 years ago). This in turn suggests that Southern Africa was *not* the place of origin of modern humans. If this conjecture is correct then little by little we are whittling down the area in Africa which provides genetic evidence of being the place of origin of modern humans. Interestingly, since this analysis an associate has found very high mtDNA diversity in a small region near the border of Kenya and Ethiopia (E. Watson, in preparation), and this has fired these expectations further.

3.3.4 *When, where, who, and how*

We now look in more detail at these four aspects of the Out-of-Africa hypothesis.

3.3.4.1 *When and where*

It is difficult to date the trees of human sequences using the same methods for the human–ape trees, because the sequences of our closest living relatives, chimpanzees, are over twenty times more divergent than the greatest differences within human mtDNA. The short sequences presently available make it difficult to determine accurately where these very different sequences should join to the human tree. A recent critique of the evidence for the Out-of-Africa hypothesis pointed out that the chimpanzee sequences can sometimes join the group of human sequences at positions which do not support an Out-of-Africa scenario, with minimal change to the fit of tree to data (Maddison *et al.* 1992). With the present data the chimpanzee sequences are too divergent to indicate the root accurately by themselves. However, in that study the main alternative contender for the location of the root was along an Australian or a New Guinean sequence, areas which by other criteria (for example archaeology) are considered very unlikely to be the ancestral region of *Homo sapiens*, leaving Africa as the front-runner of the favoured regions by using an outgroup to root the tree. (It was interesting that by giving a slight down-weighting to the fastest evolving sites, estimating the tree of human sequences first, and then locating the chimpanzee sequences onto the human tree, we obtained more consistent rooting than previous analyses. This approach has theoretical justification, and in our analysis always implied an African root.)

An alternative to outgroup rooting is using the molecular clock. This assumption is probably well founded within humans, because of our close relatedness and the fact that there is no evidence of differential mutation rates. As mentioned earlier, we can then use the midpoint to identify the root. Preliminary results again are generally consistent with the root being among the Africans. This is one of the most simple forms of clock-compatible rooting, and is not dependent on the number of sequences in a cluster. It may be susceptible to sequences that are more divergent than others, either from chance or from sequencing errors (for example the 5 errors found in an earlier African-American sample resequenced by Kocher and Wilson 1991). Taking averages of sets of edges should result in more robust rooting, and this is currently being studied, with similar results. Perhaps the most reliable method would be similar to that used for the apes, which allocates edge lengths, consistent with a clock, to the predetermined unrooted tree in order to maximize the likelihood. Such a method may be available in the next few years for application to large sets of sequences.

In order to date the time of divergence of the human sequences it is necessary to estimate the relative length of the branch leading from chimpanzees to humans, using the longest D loop sequences.[9] With data for 1000 nucleotide positions under the ideal model (all changes equally likely, all sites equally likely to change) the standard error of the distance measured will be at least 4 per cent. In reality, measuring the true distance from human to chimpanzee D loop sequences is more difficult; and, after taking into account factors such as variation of rates at different sites and different rates of substitutions, the relative error climbs to about 33 per cent! (Tamura and Nei 1993). We have to add to this the uncertainty of the exact time of divergence of chimpanzee from human mtDNA, which we estimated above to be about 5 to 9 million years ago once the different sources of error were taken into account.

The same statistical reasoning used earlier to estimate the total error of the human–chimpanzee divergence date can be applied to Kocher and Wilson's data[9] in order to estimate the date of the deepest root in the human mtDNA tree by Tamura and Nei's (1993) method of measuring relative distances. We estimate the age of the human mtDNA ancestor from the D-loop sequences to be 240 000 years ago, with a standard error of about 220 000 years.[10] This is a substantially larger standard error than that calculated by Tamura and Nei (1993) for the human mtDNA ancestor.[11]

Recently, Hasegawa et al. (1993) have used approximately 300 of the fastest-evolving (third position) sites of a protein-coding region of the mtDNA (from Kocher and Wilson 1991), and have estimated the human mtDNA root using a maximum likelihood method to be 100 000 years ago, with a standard deviation of 50 000 years. However, they took the divergence of human and chimpanzee mtDNA as 4 m.y.a., which seems too recent given the australopithecine fossils and the expected polymorphisms in the ancestral population. Their analysis of these data calculates the ratio of the root of human mtDNA sequences to the divergence of human and chimpanzee mtDNA as 1/40, with a standard error of 1/80. The expected divergence date, using a figure of 7.2 m.y.a. for human–chimpanzee divergence (with variance = 1.31), then becomes $1/40 \times 7.2$ m.y.a. = 0.18 m.y.a., with a standard error of 1/80 (so variance = $(1/80)^2$). The variance of this product estimated by the formula given in note 10 is $(1/40 \times 1.31 + 7.2 \times (1/80)^2 + (1/80)^2 \times (1.31))$ which gives a standard error of 0.185 m.y.a. or 185 000 years, which is close to that of the D-loop region.[12]

Because both of the above estimates are for the origin of the same thing (human mtDNA), we can further improve our estimate of the date of the root of the human mtDNA tree as their weighted average (here for simplicity we ignore the weights since they are nearly equal). This gives $(180\,000 + 240\,000)/2 = 210\,000$ years ago, and, because the standard errors of the two estimates are about equal, the variance of the average is approximately halved, reducing the standard error by about 30 per cent, in this case about 150 000 years. This date is a useful calibration point for other studies of human mitochondrial DNA, such as Harpending et al. (1993). Estimates of average mammalian rates of mtDNA evolution also give estimated divergence dates that are close to those obtained by the above two methods (Wilson and Cann 1992), increasing our confidence that we are not wrong in the basic tenet that the genetic evidence indicates that our species comprised a single small population less than 200 000 years ago.

Even though the variance of the estimated date of the deepest root in the human mtDNA tree is large, its answer to 'When?' gives strong evidence that *Homo sapiens* is a recent species derived from the descendants of *Homo erectus* on only one of the continents of Africa, Asia, or Europe. If *Homo sapiens* were derived from a mixture of intercontinental populations then some human mtDNA types should date back to before the time that *Homo erectus* colonized Europe and Asia, a time generally taken to be close to 1 m.y.a. This is a highly unlikely date given the present data, even with their large variance. The evidence from mtDNA so far has tended to rebut the multiregional hypothesis of human evolution as an adequate explanation.[13]

3.3.4.2 Dating trees with archaeological evidence

Archaeological evidence can also help give more precision in dating the human mtDNA tree. At least two major events promise to help here. The first is the arrival of people into Australia (which, owing to the lower sea levels of the last Ice Age, was then connected to New Guinea and some of Melanesia). Occupation of this region has been pushed back over 10 000 years in the past decade, to approximately 50 000 BP, by archaeological finds (Jones 1989). Given that the archaeological sites almost certainly represent minimum ages, the first colonization of the Australian region could be about 60 000 years ago, with an approximate 95 per cent confidence interval being 50 000 to 70 000 years ago. This period also coincides with a period of minimum sea levels that would have reduced the largest sea crossing between South-East Asia and Australia–New Guinea to about 100 kilometres—still a major feat, however, with no evidence of its having been accomplished by *Homo erectus*.

A concern in using such an 'ingroup' dating technique is sampling enough sequences to be reasonably accurate in estimating the closest relatives among populations. Unfortunately, there are still few mtDNA sequences of Australians (including New Guineans) and South-East Asians. Accuracy would be enhanced if we could identify the oldest exclusively Australian assemblages of types, so as not to risk biasing the results towards a greater degree of Australian divergence due to the effect of polymorphism. While the present sample of Australian–New Guinean sequences is small (about 15 sequences) they tend to branch quite deeply, and often most closely with Asian sequences. A crude estimate of the depth of these branchings relative to the root is 2/5 that of the root. A similar picture emerges from the data of Cann *et al.* (1987), who had more Australians in their study, and estimated sequence divergence across the mitochondria using the observed changes in genetic markers.

A second event is the migration of people into the Ameircas; but there is still uncertainty on this dating, though it is generally accepted to have occurred 15 to 40 thousand years ago. Recently larger samples (72 and 63: Horai *et al.* 1992) of American Indian mtDNA have been sequenced, allowing more confidence in the true depth of indigenous mtDNA groups. These groups appear to be approximately 7 per cent of the depth of the root of human mtDNA, which is consistent with the first colonization of the America's 14 to 20 thousand years ago (though there could still be older immigrants not represented in the study). The results of these two studies, particularly that of the Australians, lend support to the hypothesis of a last common ancestor for human mtDNA 200 000–250 000 years ago.

3.3.4.3 *Who and how*

The diversity and distribution of mtDNA from people outside Africa (as shown, for example, in Fig. 3.3 indicates that the colonization of each of the main continents involved genetically diverse individuals, or possibly more than one wave of colonists. The mixture of Yoruban, Herero, and Hadza mtDNA lineages with those from outside Africa suggests that some of the ancestors of all these groups may have been among those who left Africa (possibly in contrast to more specialized peoples such as the Pygmy and the !Kung, who may already have been ethnically distinct at that time). Questions of 'Who' left Africa should become much clearer with further sampling of other African ethnic groups.

The final feature of the mtDNA tree we shall discuss pertains to the 'How' question of human population expansion. Nearly all the deepest lineages connecting Asians, Europeans, New Guineans, and some African sequences arose in a very short time. This feature has also been studied, using pairwise distances between sequences, by a number of authors (for example Di Rienzo and Wilson 1991; Rogers and Harpending 1992; Harpending *et al.* 1993). While their methods do not yet have formal statistical tests, simulations performed by Rogers and Harpending show that the real data fit well with a rapid population explosion, but fit very poorly with a constant population size. The most rapid branching occurs at approximately half the height to the root, which, with present estimates of the time to the root, makes this diversification about 100 000 years ago (and may coincide with modern people settling the Middle East and adjacent lands). Such a feature is expected under the Out-of-Africa model. Archaeological evidence fits this picture quite well, with some interesting punctuations. For example, modern humans did not colonize the bulk of Europe until about 35 000 years ago (a glacial period), although they may have been in the Middle East from 100 000 years ago (Stringer 1990). It will be interesting to see if genetic evidence can help explain such mysteries. In addition, their analysis (Rogers and Harpending 1992; Harpending *et al.* 1993) also suggests that *Homo sapiens sapiens* probably evolved from a population with a breeding population of 1000 to 10 000 females, suggesting a total population of at most a few tens of thousands of humans. Such analyses help show the tremendous potential power of DNA sequences.

3.4 Trees of human relationships from nuclear genetic data

3.4.1 *Alleles and polymorphisms*

Nuclear DNA evolves more slowly than mtDNA (which has a higher mutation rate) and consequently nuclear gene sequences would need to be long (perhaps 10 000 base pairs each) to provide a similar amount of resolution of the branching patterns within humans. However, over the past thirty years a large data-base has been built up of the frequencies of different alleles of proteins for many human populations. The majority of differences in these protein alleles are due to the neutral evolution of the protein (that is, the stochastic replacement of one amino acid with another equally suitable amino acid, so that the protein continues to function quite adequately). More recently, additional alleles at different DNA loci have been detected by using enzymes which cut the DNA only at specific short sequences (4 to 8 base pairs) (Bowccock *et al.* 1991).

A general feature of these data is that there are significantly more allelic variants in Africa than in any other region of the world, and that the non-African populations appear as subsets of the diversity in Africa. These data do not rely upon tree-building, but parallel the situation found for human mtDNA: the most divergent forms of mtDNA are all African, with the non-African forms being derived from a subset of the deepest branches. To a non-specialist the concentration on amounts of diversity may not seem significant; but under population genetics models the amount of diversity increases as a function of time after a population has expanded in numbers, and so the amount and type of genetic diversity is a powerful indicator of population history.

After a population divides the relative frequencies of the various alleles change (become less alike), and we can thus measure the genetic distance between two populations. The rate at which the frequencies of neutral alleles diverge is also a function of breeding population size; but if population size remains fairly constant then the degree of divergence in allele frequencies is expected to be proportional to time. To keep the variance of such measures of genetic distance reasonable it is necessary to measure many loci (100 or more if possible).

Data on the frequency of alleles are also useful for estimating the phylogenetic relationships of human populations. If a sufficient number of genetic loci are mapped in large samples in different populations this allows polymorphism in the ancestral populations to be taken into account. Early results with allelic frequencies (including blood-group data, Nei and Roychoudhury 1982) argued for an African origin about 100 000 years ago. Figure 3.4 reproduces the results of a later study undertaken by Bowcock *et al.* (1991). Their data-set is large, and the model used to estimate the phylogeny is one of the most detailed yet developed. Initial analysis of these data showed that a tree did not fit the data well (Bowcock *et al.* 1991); but they then modelled the possibility that each population was founded by a mixing of two others. Of the different possibilities only that of Europeans' being founded by a mixing of people with both Asian and African origins allowed the data and model to agree within statistical limits. The data are also consistent with a constant rate of evolution in all lineages in the phylogeny (not a tree, since it has rejoining or reticulate lineages), and these can be rooted by assuming a molecular clock. With trees based on allele frequencies the alternative of rooting or dating trees using chimpanzees as the outgroup is even less certain than in the case of mtDNA sequences, because most allelic variants become fixed (either becoming predominant or else disappearing altogether) during the time of separation of humans and chimpanzees.

3.4.2 Ingroup dating of the tree

Bowcock *et al.* (1991) dated the root of the tree in Fig. 3.4 by assuming that human populations first moved out of Africa 100 000 years ago, on the basis of fossil evidence (which has been disputed) of apparent *H. s. sapiens* in Israel at about that time. We have recalibrated this tree by using an estimated time of divergence of Melanesians and Asians. This time should be nearly coincidental with the people crossing the sea channels to reach the greater Australian continent, an estimated 60 000 ± 10 000 years ago (see above). To this date we will add 5000 years to allow for the separation between South-East Asian people (assuming they are the closest relatives of the Australian–New Guinean–Melanesian peoples) and the more northerly Asians who constitute part of the Chinese sample used here (see the tree figure in Cavalli-Sforza 1991). We will also raise the standard deviation to 7000 years to take into account some uncertainty as to exactly how much difference should

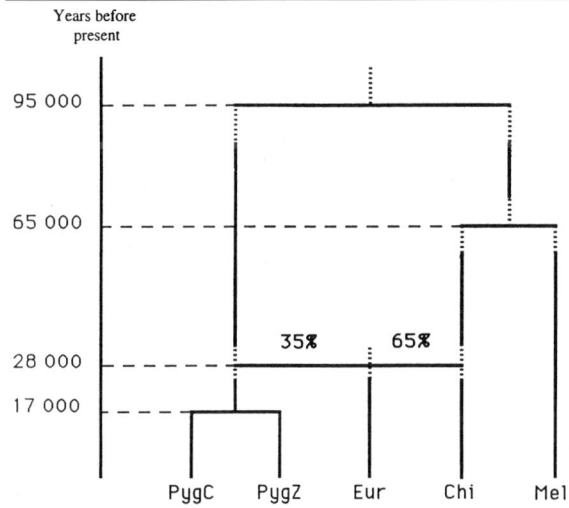

Fig. 3.4 A redrawing of the phylogeny of selected human populations as estimated by Bowcock *et al.* (1991) from the frequencies of DNA variants at 100 loci (PygC and PygZ = Pygmy populations in the Central African Republic and Zaire respectively, Eur = Europeans, Chi = Chinese, Mel = Melanesians). We have calibrated the phylogeny with an assumed date for the divergence between Melanesians and Asians of 65 000 years ago. The dots show the standard error on each edge resulting from sampling a finite number of DNA loci. The standard error for the estimated divergence time between African and non-African peoples (the root of this tree), after taking into account other known sources of error, is close to 20 000 years (see text). The percentages on the edges leading to Europeans (which they treat as a hybrid population) are the maximum likelihood estimates of the amount of genetic material contributed from each ancestral lineage.

be allowed for in using the more northerly Chinese population.

When we recalibrate the tree this way two other dates in the tree agree with known fossil evidence of the spread of humans. The first appearance of fossils of modern human form in the Middle East is about 100 000 years ago; and the first appearance of modern humans in Europe (for example the Cro-Magnon specimen) is about 28 000–34 000 years ago (Stringer 1990). Note that this last date excludes the possibility that Neanderthals contributed substantially to the genetics of modern Europeans, as they are a unique lineage that evolved in Europe over at least 120 000 years, and are now favoured as being in a line of descent back to the earliest European *Homo erectus*-like fossils of 500 000 or more years ago (Stringer 1990). That is, if they had contributed even 1/8 of the genetic material of modern Europeans, then modern Europeans should form a noticeably deeper edge in the phylogeny of Fig. 3.4. This is strong evidence that Neanderthals were a species distinct from *Homo sapiens sapiens*, but of uncertain biological status with respect to other lineages descended from *Homo erectus*. Thus an independent dating point not relying upon a cententious assignment of fossils to fully modern people, gives a similar result to Bowcock *et al.* (1991), and clearly supports the recent origin of modern humans.

There is also good evidence that, on the whole, modern Europeans (including Basque people) are closely related to peoples of the Middle East (for example, Iraqi, Iranian: see Cavalli-Sforza 1991) which could well have been a mixing place of the peoples moving between the South and East (Africa and Asia respectively). Such a hypothesis is also testable with mtDNA data. Notably, it does seem that more European sequences of mtDNA associate with either an African or an Asian sequence then African and Asian sequences are inferred as direct relatives (Fig. 3.3), although there are as yet relatively few sequences and it is not certain what effect sampling errors may be having. Many of the published trees derived from the genetic distances of protein alleles are in good agreement with these findings (Cavalli-Sforza 1991) despite not taking into account the mixing of populations (such as is implicated in the origin of Europeans and Polynesians, for example).

We can now take a further step, and attempt to estimate the date for the movement of modern humans out of Africa. The date we have assigned for the divergence of Melanesian and Chinese populations is 65 000 years ago, with a standard deviation of 7000 years (variance = 7000^2). The ratio of the root of this tree (Figure 1, Bowcock *et al.* 1991) to the separation of Melanesians from Chinese is 100/68 = 1.47. Making the approximation that the errors in estimating the relative times of the root and the separation of Chinese and Melanesians are independent (with variances of 10.0^2 and 7.5^2 respectively; see Bowcock *et al.* 1991 and Fig. 3.4), then the variance of this ratio = $(100/65)^2 \times (10^2/100^2 + 7.5^2/68^2) = 0.05$ (as given by the formula already used in footnote 10).

Thus the ratio of the divergence of African from non-African populations relative to the separation of Chinese and Melanesians is 1.47, with a standard error of $\sqrt{0.05} = 0.23$. The two numbers that make up this ratio are in fact positively correlated, which makes this a slight overestimate of the true variance. Our estimate of the Out-of-Africa event from this data-set is $1.47 \times 65\,000$ year = 95 500. As was described above, the variance of this last number is given by the formula for the variance of a product of two independent numbers, and equals $(65\,000^2 \times 0.05 + 1.47^2 \times 49\,000\,000 + 0.05 \times 49\,000\,000 = 319\,600\,000$, giving a standard error of about 18 000 years. While we have not taken into account variability due to choice of model, this estimate is probably reasonably accurate, given the overestimate we made of the variance of the ratio we calculated.

This estimate and its variance (which is noticeably smaller than for the mtDNA estimates) clearly reject the idea of *Homo sapiens sapiens* being anything like one million years old, while the expansion out of Africa was almost certainly less than 140 000 years ago. Most importantly, this estimate comes from a random sample of over 100 of our DNA loci, making it highly improbable that it is atypical, something which is never so certain when studying a few loci (one, in the case of the mtDNA data).

The dates we have produced here from both the mtDNA and the nuclear data suggest that the Out-of-Africa event most probably occurred about 100 000 years ago. Such a date is very informative in the light of known fossil evidence. The first skulls with a distinctly modern aspect appear in Africa about 100 000 to 120 000 years ago, and are preceded over the previous 200 000 years by skulls from throughout Africa sharing some unique features with modern humans (Groves 1989). About 100 000 years ago quite modern-looking human skeletons are known from caves in the Middle East, where they were apparently contemporaneous with Neanderthal forms (Stringer 1990). While the exact nature of these modern-looking skulls is still in dispute (for instance, whether they are fully modern), the dating of the genetic evidence, so far, is consistent with the notion that they were amongst the first modern humans to have migrated out of Africa. There is no clear fossil evidence as yet that these two lineages interbred—another finding consistent with the genetic evidence.

Finally, there are now sufficient nuclear sequences to begin to make some statements about longer-term aspects of our genetic structure. Applying aspects of genetic drift theory to these sequences, they give a hint of the long-term population size of the lineage of hominids that led to modern humans. Recent analyses such as that of Takahata (1993) suggest that the long-term effective population size of the hominid lineage leading to modern humans never fell below 10 000 for any noticeable period. While this is just the beginning of a most interesting area of research, where more data and theoretical work is eagerly awaited, it highlights just how quickly genetic information is uncovering new sources of knowledge about the pattern and demographics of human evolution.

3.5 Conclusions and prospects

Molecular data have over the last thirty years elucidated many points of the evolutionary history of hominoids, though each problem considered was initially controversial. Sequence data have confirmed the findings of the early immunological explorations of the relationships of apes and humans: that the African apes and humans are the closest relatives; and that their divergence was much more recent than had previously been believed (5 to 10 million years, vs 20 to 30 million years). Such results have forced palaeontologists to reappraise their own assumptions about fossil relationships and to reconsider their methodologies. The most recent data and analyses most strongly support the grouping of humans and chimpanzees as the closest relatives, contradicting the apparent morphological similarities of gorillas and chimpanzees.

As we noted, the exact dating of the divergences of hominoids is still somewhat general, but importantly does not exclude any of the australopithecine fossils as hominid, and frames the origin of human ancestors in a 4.5 to 8.5 million-year period, despite the lack of any decisive fossil evidence. Claims of australopithecine fossils dating back to 5.5 m.y.a. challenge those molecular biologists who confidently estimate 4 m.y.a. as the human–chimpanzee divergence time, and highlight the importance of considering the compound uncertainties in calculating evolutionary dates.

It is only recently that relevant molecular data have been available for a large number of humans, and these have led to an argument for a very recent African origin of modern humans (Cann *et al.* 1987). First critiques of these data and the conclusions drawn from them were often relatively easy: biologists simply had not, and in many ways still have not, developed the techniques with which to analyse such large data-sets adequately. It was easy to criticize the original results for inadequate analyses without considering any alternative hypotheses or trying to integrate all the lines of evidence available. Reanalyses with appropriate techniques are now supporting the original claims when all the evidence and alternatives are considered. Balancing this there is also a need to consider the myriad possible sources of error in making quantitative estimates from molecular data. Molecular data do not stand on their own in the larger field of biological knowledge. Yet it remains true that the most powerful way to test the evolutionary ideas we have discussed here will be with the adequate sampling, sequencing, and analysis of other DNA loci.

Using a statistical framework we have outlined the major sources of error in reconstructing an accurate chronological phylogeny of ape and human evolution. We expect that within ten years the uncertainty from each of these sources will be more than halved. Given the trend of the past ten years we expect new fossil and archaeological finds that will improve the absolute time calibration of human–ape evolutionary trees, although the overall problem remains of assigning fossils accurately to lineages. There will also be a much greater number (20+) of independent DNA loci with which to build trees, while the methods expected to be available will allow more refined statistical estimates. By that time we should be getting a clear picture of the long-term breeding population size of our distant ancestors, answering questions such as: 'How big was the population of our ancestors before it diverged into the lineages leading to chimpanzees and humans?' or 'Was the long-term effective population size of the hominids leading to humans really as small as 10 000?'; 'Was it perhaps even smaller?'

Another area of interest is obtaining DNA sequences from (sub)fossil bones over 30 000 years old. Much effort is being put into this, especially to obtain verified sequences from Neanderthal bones that are not contaminated from handling by modern humans. This would allow a direct test of the hypothesis that the deepest divergence in the mtDNA of modern humans significantly postdated the divergence of non-African descendants of *Homo erectus*. It may eventually be possible to use DNA sequences from ancient bones to determine relationships amongst the populations of *Homo erectus*. Unfortunately there is evidence that the DNA in such old bones (unless frozen or else preserved in exceptionally dry conditions) is much degraded, making sequencing impractical with current techniques. The recovery of DNA from specimens in amber (now of *Jurassic Park* fame) is a different matter; but humans have yet to turn up in this predicament.

We expect an even more profound understanding of human evolution to be exposed by molecular genetics over the next few decades. The human genome project will supply a huge amount of detailed information on the genetic structure of our own species and those of our closest living relatives. Combining this with advances in our understanding of developmental biology we may finally be able to identify which sets of genes regulate such features as body and brain development. Phylogenetic analysis of such sequences should allow us to estimate when such genes changed their function, and hence when, for example, areas of the brain associated with language evolved. It is no exaggeration to say that we will have previously unimagined insights into how we ended up being, well, human.

Epilogue

We take the opportunity to update the latest developments since the completion of the main manuscript in September 1993. There have been some exciting new developments, most of which reinforce our main conclusions. The whole mtDNA genome of all the Great Apes has been sequenced (Horai et al. 1995) and this again verifies both the closer relationship of human and chimpanzee sequences and that the gap back to the gorilla sequence is about 2 myr. The non-synonymous substitutions in this data set show very few multiple substitutions and seem to allow inference of the divergence dates without need of a specific model (although we would like the full outgroup sequences of a gibbon to confirm that the orang-utan lineage is not evolving faster than the African hominoids). When we use the methods in this paper to estimate divergence times (plus total standard error) for mtDNA using just the non-synonymous substitutions, we arrive at the following divergence dates: human–gorilla, 7.6 myr (1 s.e. = 0.71), human–chimp 5.6 (s.e. = 0.60), and chimp–pygmy chimp 3.0 (s.e. = 0.44). Taking into account unknown ancestral polymorphism (using the assumptions already made in the text), the species divergence date is expected to be about 1/2 myr more recent in each case (and the overall s.e. will rise by approximately 0.25 myr in each case). The full sequence of a divergent African mtDNA by Horai et al. (1995) strongly supports the hypothesis that the root of the human mt DNA was less than 200 000 years ago (although exact dating is still contingent upon reducing the other 3 main sources of error in making this calibration). Recent fossil finds have also been claimed to support a *Dryopithicus* (*Sivapithecus*, orang-utan) group (Solà, and Köhler 1993). This hypothesis also looks reasonable on biogeographic grounds, since dryopithecines are found in Europe, *Sivapithecus* near the Indian subcontinent, and orangutan in S.E. Asia. This reinforces our anticipation that orang-utan divergence was in the period 14 to 18 m.y.a., especially since dryopithecine fossils are known to date back to approximately 14 m.y.a.

There have been more papers showing evidence of the greater genetic diversity in Africa (e.g. alu elements, Batzer et al. 1994; mtDNA; E. E. Watson, in preparation). There are also analyses of nuclear sequence variation which are beginning to rival the lineage resolving power of mtDNA (Bowcock et al. 1994), and these are supporting Out-of-Africa. Recalibration of the age of some Javan fossils at nearly 2 myr old (Swisher et al. 1994), and new fossil finds in the near east (Gabunia and A. Vekua 1995) suggest *Homo erectus* spread out of Africa even earlier than assumed, making the multi-regional explanation of human origins even harder to defend. Overdue analyses are revealing that each chimp species has much more genetic variation than all humans (Morin et al. 1994), and similar results are coming to light for gorillas. This further bolsters the argument for humans evolving from a relatively small local population in the recent past, fully consistent with Out-of-Africa. On the theoretical front, there are signs that a variety of computationally feasible maximum likelihood models for estimating population histories (e.g. expected ancestral size, evidence of population expansion or migration) will be available in the next few years (Mary Kuhner and Joe Felsenstein, Bob Griffiths and Simon Tavare, pers. comm.). These should greatly help in the quantitative interpretation of past population events.

Substantial finds of fossils 4 to 4.5 myr old from Ethiopia have been assigned to a new species *Australopithecus ramidus* (White et al. 1994). Overall they appear chimp-like, and so far no characters amongst are conclusive in assigning this taxon to either human, chimp, or human–chimp ancestor lineages (although the base of the skull particularly appears to have some special features in common with at least one australopithecine species). More material is required and is rumoured likely to be reported soon. If this does turn out to be an early hominid we strongly resist it being assigned to a new genus. Most mammalian genera are at least 5 million years old. To have 3 or 4 named genera within the human lineage goes against the very truth revealed by the genetic studies of hominoid relationships—that humans are a recent group surprisingly closely related to the African apes.

Taken all together the evidence now seems to be favouring the period 4.5 to 6.5 m.y.a. more than the period 6.5 to 8.5 m.y.a. as the divergence time of human and chimp lineages. A conclusive resolution of the question exactly when, will require more sequences, better fossil calibrations of gene trees, and techniques to reliably infer the genetic variability of ancestral populations. Overall, results relating to human origins are, as expected, accumulating at an ever increasing rate, with the human genome project yet to make its presence felt. Perhaps it is just as well that we learn more about our own past history, before we answer the next millennium's issues relating to modifying our future evolution.

Notes

1. A good analogy is the form of a surname, which can have related forms (e.g. Davey, Davis, Davies, etc.), each usually the result of a change that is passed on to direct descendants.

2. A good analogy is again given by surnames. If a family has only daughters (who to make the analogy strictly correct must marry to have children who inherit the husband's surname) that branch of the family name will die out. Conversely, a family may have all sons, and the family surname is then more likely to increase in frequency. These effects are often noted in small villages, where, after a period of time, many people end up with the same surname.

3. When a Chi-square goodness-of-fit test was performed on the observed and predicted nucleotide patterns the optimal tree model was not rejected. However the fit of all other possible trees was very poor (including having chimpanzees and gorilla (*Gorilla gorilla*) together), and accordingly we reject all alternative trees.

4. A good analogy for this effect is again the inheritance of family names (surnames). Envisage a situation where all the surnames in a town have evolved from one form; they are slightly different, but clearly related. The town was then divided into two parts, which were isolated from each other (by a dragon, a spell, or nationalistic armies). In each town one of the forms of the ancient name became predominant. A linguist then came along who knew how quickly names change in form, and deduced accurately how long it had taken for the two names to have changed from the ancestral form into their present forms. However she realized it was still not possible to estimate exactly when the two towns were separated, because the initial differences between the names probably predated the division of the old town.

5. In compensation there is the bonus that the distribution of the divergence dates from many different alleles will allow us to estimate the effective population size of all the nodes of the tree, including our distant ancestors. With the human genome project, which is also committed to sequence large stretches of DNA from apes for comparison with humans, data relevant to doing just this should be flowing in at an increased rate over the next decade.

6. These apparent divergence dates need not be minimal population divergence dates, because a large degree of polymorphism existing at the time the organ-utan lineage diverged could bias downwards the estimated times of later divergences.

7. Allowing 1 myr as the mean increase in divergence expected from DNA polymorphism in their ancestral population.

8. That is, the numbers of calculations required to estimate them are non-trivial, and could occupy even modern computers for long periods of time, and searching across the tree space could take years.

9. The sequences used here are from Kocher and Wilson (1991), who sequenced complete Dloop sequences (1135 base pairs) for about 20 individuals. They sequenced the DNA going in both directions, and we expect that the sequencing error rate was about 1 in 2000. This can be expected to inflate by about 5% our estimate (below) of the age of the root of human mtDNA.

10. (A) Estimated distance between the most divergent human lineages is 0.024, with variance $= 0.006^2 = 0.000\,036$.

(B) Estimated distance between human and chimp mtDNA $= 0.752$, with variance $= 0.224^2 = 0.0502$.

(C) Estimated divergence time of human and chimpanzee mtDNA is 7.2 m.y.a., with variance $1.15^2 = 1.32$.

The ratio of the deepest human distances relative to human–chimpanzee distances is $0.024 / 0.722 = 0.033$. The variance of this ratio is given by the formula

$(x_1 / x_2)^2 (\text{var}(x_1) / x_1^2 + \text{var}(x_2)/x_2^2)$ where x is the average and var(x) its variance (Stuart and Ord 1987, p. 325), p. 325, which in our case is 0.000467, with the standard error being 0.021. Taking the time of divergence of human and chimpanzee mtDNA to be 7.2 myr (standard deviation = 1 m.y.a.), we have an estimated age for the human mtDNA ancestor of 7.2 myr $\times$ 0.033 = 0.24 m.y.a.. The variance of the product of these two independent numbers $= x_1 \times \text{var}(x_2) + x_2 \times \text{var}(x_1) + \text{var}(x_1) \times \text{var}(x_2) = (7.2 \times 0.000467 + 0.033 \times 1.31 + 1.31 \times 0.000467)^{0.5} =$, 0.047 which gives a standard error of 0.22 (Stuart and Ord 1987, p. 325), which equates to 220 000 years.

11. This is due to omitting the interaction of errors on the two distances, in taking first the ratio, and

then the errors on the product calculated above. We have not included the errors expected from the choice of model used for the distance measure (including the estimate of the variation of rates across sites) nor the difficulty in locating the exact root of the tree of human sequences.

12. Notice that when 300 sites were used in the above study the standard error was very similar to the standard error for the 630 sites from the D-loop, which shows the gain that can be made using sites that are evolving in a more predictable way. If the sequencing error rate is approximately 1 in 2000 this last figure is probably biased less than 4% upwards (expected error rate (1/2000) × sequence length (300) divided by average difference between human sequences in this region (4), times 100).

13. The sequencing of 5 kilobase stretches of the mtDNA from different humans could well reduce the standard error (due to finite sequence length) of the age of the last known human mtDNA ancestor to within 25 000 years. While such data may be available in the next couple of years, any further significant reduction in the region of error inherent in the above method of dating will require a significant improvement in the dating of the divergence time of human and chimpanzee mtDNA.

References

Bailey, W. J., Hayasaka, K., Skinner, C. G., Kehoe, S., Sieu, L. C., Slightom, J. L., and Goodman, M. (1992). Reexamination of the African Hominoid trichotomy with additional sequences from the primate β-globin gene cluster. *Molecular Phylogeny and Evolution*, **1**, 97–135.

Batzer, M. A., Stoneking, M., Algeria-Hartman, M., Bazan, H., Kass, D. H., Shaikh, T. H., *et al.* (1994). African origin of human specific polymorphic Alu insertions. *Proceedings of the National Academy of Sciences (USA)*, **91**, 12288–92.

Begun, D. R. (1992). Miocene fossil Hominoids and the chimp–human clade. *Science*, **257**, 1929–33.

Bowcock, A. M., Kidd, J. R., Mountain, J. L., Hebert, J. M., Carotenuto, L., Kidd, K. K., and Cavalli-Sforza, L. L. (1991). Drift, admixture, and selection in human evolution: A study with DNA polymorphisms. *Proceedings of the National Academy of Sciences (USA)*, **88**, 839–43.

Bowcock, A. M., Ruiz-Linares, A., Tomfohrde, J., Minch, E., Kidd, J. R., and Cavalli-Sforza, L. L. (1994). High resolution of human evolutionary trees with polymorphic microsatellites. *Nature*, **368**, 455–7.

Caccone, A. and Powell, J. R. (1989). DNA divergence among hominoids. *Evolution*, **43**, 925–42.

Cann, R. L., Stoneking, M., and Wilson, A. C. (1987). Mitochondrial DNA and human evolution. *Nature*, **325**, 31–6.

Cavalli-Sforza, L. L. (1991). Genes, peoples and languages. *Scientific American*, **265**, 72–8.

Deacon, H. L. (1992). Southern Africa and modern human origins. *Phil. Trans. R. Soc. Lond. B.*, **337**, 177–83.

Di Rienzo, A. and Wilson, A. C. (1991). Branching pattern in the evolutionary tree for human mitochondrial DNA. *Proceedings of the National Academy of Sciences (USA)*, **88**, 1597–1601.

Gabunia, L. and Vekua, A. (1995). A Plio-Pleistocene hominid from Dmanisi, East Georgia, Caucasus. *Nature*, **373**, 509–12.

Gonzalez, I. L., Sylvester, J. E., Smith, T. F., Stambolian, D., and Schmickel, R. D. (1990). Ribosomal RNA gene sequences and Hominoid phylogeny. *Molecular Biology and Evolution*, **7**, 203–19.

Groves, C. P. (1989). *A theory of human and primate evolution*. Clarendon Press, Oxford.

Harpending, H. C., Sherry, S. T., Rogers, A. R., and Stoneking, M. (1993). The genetic structure of ancient human populations. *Current Anthropology*, **34**, 483–96.

Hasegawa, M., Kishino, H., and Yano, T. (1985). Dating of the human–ape splitting by a molecular clock of mitochondrial DNA. *Journal of Molecular Evolution*, **22**, 160–74.

Hasegawa, M., Di Rienzo, A., Kocher, T. D., and Wilson, A. C. (1993). Toward a more accurate time scale for the human mitochondrial DNA tree. *Journal of Molecular Evolution*, **37**, 347–54.

Horai, S., Satta, Y., Hayasaka, K., Kondo, R., Inoue, T., Ishida, T., *et al.* (1992). Man's place in the Hominoidea revealed by mitochondrial DNA genealogy. *Journal of Molecular Evolution*, **35**, 32–43.

Horai, S., Kondo, R., Nakagawa-Hattori, Y., Hayashi, S., Sonoda, S., Tajima, K. (1993). Peopling of the Americas, founded by four major lineages of mitochondrial DNA. *Molecular Biology and Evolution*, **10**, 23–47.

Horai, S., Hayasaka, K., Kondo, R., Tsugane, K., and Takahata, N. (1995). Recent African origin of modern humans revealed by complete sequences of hominoid mitochondrial DNAs. *Proceedings of the National Academy of Sciences (USA)*, **92**, 532–6.

Jones, R. (1989). East of Wallace's line: issues and problems in the colonisation of the Australian continent. In *The human revolution* (ed. P. Mellars and C. Stringer), pp. 743–82. Princeton University Press.

Kocher, T. D. and Wilson, A. C. (1991). Sequence evolution of mitochondrial DNA in humans and chimpanzees: control region and a protein-coding region. In *Evolution of life* (ed. S. Osawa and T. Honjo), pp. 391–413. Springer-Verlag, Tokyo.

Maddison, D. R., Ruvolo, M., and Swofford, D. L. (1992). Geographic origins of human mitochondrial DNA: phylogenetic evidence from control region sequences. *Systematic Biology*, **41**, 111–24.

Marshall, C. R. (1991). Statistical tests and bootstrapping: assessing the reliability of phylogenies based on distance data. *Molecular Biology and Evolution*, **8**, 386–91.

Martin, R. D. (1990). *Primate origins and evolution*. Chapman and Hall, London.

Morin, P. A., Moore, J. J., Chakraborty, R., Jin, L., Goodall, J., and Woodruff, D. S. (1994). Kin selection, social structure, gene flow, and the evolution of chimpanzees. *Science* **265**, 1193–201.

Nei, M. (1987). *Molecular evolutionary genetics*. Columbia University Press, New York.

Nei, M. (1992). Age of the common ancestor of human mitochondrial DNA. *Molecular Biology and Evolution*, **9**, 1176–8.

Nei, M., and Roychoudhury, A. K. (1982). Genetic relationship and evolution of human races. *Evolutionary Biology*, **14**, 1–59.

Penny, D. (1994). Darwinian evolution; molecular evolution; molecular phylogeny. In *Encyclopedia of molecular biology* (ed. J. Kendrew). Blackwell, Oxford.

Penny, D., Foulds, L. R., and Hendy, M. D. (1982). Testing the theory of evolution by comparing phylogenetic trees constructed from five different protein sequences. *Nature*, **297**, 197–200.

Penny, D., Steel, M. A., Waddell, P. J., and Hendy M. D. (1995). Improved analyses of human mtDNA sequence support a recent African origin for *Homo sapiens*. *Molecular Biology and Evolution*, **12(5)** 863–82

Pilbeam, D. (1984). The descent of hominoids and hominids. *Scientific American*, **250**, 60–9.

Rogers, A. R. and Harpending, H. (1992). Population growth makes waves in the distribution of pairwise genetic differences. *Molecular Biology and Evolution*, **9**, 552–69.

Sibley, C. G., Comstock, J. A. and Ahlquist, J. E. (1990). DNA hybridization evidence of Hominoid phylogeny: a reanalysis of the data. *Journal of Molecular Evolution*, **30**, 202–36.

Solà, S. M. and Köhler, M. (1993). Recent discoveries of *Dryopithecus* shed new light on evolution of great apes. *Nature*, **365**, 543–5.

Steel, M. A., Székely, L., Erdös, P. L., and Waddell, P. J. (1993). A complete family of phylogenetic invariants for any number of taxa under Kimura's 3ST model. *New Zealand Journal of Botany* (Conference Issue). 31 289–96.

Stringer, C. B. (1990). The emergence of modern humans. *Scientific American*, **263**, 68–74.

Stuart, A. and Ord, J. K. (1987). *Kendall's advanced theory of statistics*. Charles Griffin and Co., London.

Swisher, C. C., Curtis, G. H., Jacob, T., Getty, A. G., Suprijo, A., and Widiasmoro (1994). Age of the earliest known hominoids in Java, Indonesia. *Science*, **263**, 1118–21.

Swofford, D. L. and Olsen, G. J. (1990). Phylogeny reconstruction. In *Molecular systematics* (ed. D. M. Hillis and C. Moritz), pp. 411–501. Sinauer Associates, Sunderland, Mass.

Takahata, N. (1993). Allelic genealogy and human evolution. *Molecular Biology and Evolution*, **10**, 2–22.

Tamura, K. and Nei, M. (1993). Estimation of the number of nucleotide substitutions in the control region of mitochondrial DNA in humans and chimpanzees. *Molecular Biology and Evolution*, **10**, 513–26.

Templeton, A. R. (1993). The "Eve" hypothesis: a genetic critique and reanalysis. *American Anthropologist*, **95**, 51–72.

Thomas, H. (1985). The early and middle Miocene land connection of the Afro-Arabian plate and Asia: a major event for Hominoid dispersal? In *Ancestors: the hard evidence* (ed. E. Delson), pp. 42–50. Alan R. Liss, New York.

Vigilant, L., Stoneking, M., Harpending, H., Hawkes, K., and Wilson, A. C. (1991). African populations and the evolution of human mitochondrial DNA. *Science*, **253**, 1503–7.

Waddell, P. J., Penny, D., Hendy, M. D., and Arnold, G. C. (1994). Variance–covariance matrices for evolutionary trees using Hadamard transforms. *Molecular Biology and Evolution*, **11**, 630–42.

White, T. D., Suwa, G., and Asfaw, B. (1994). *Australopithecus ramidus*, a new species of early hominid from Aramis, Ethiopia. *Nature*, **371**, 306–12.

Wilson, A. C. and Cann R. L. (1992). The recent African genesis of humans. *Scientific American*, **266**, 22–7.

Zubrow, E. (1989). The demographic modelling of Neanderthal extinction. In *The human revolution* (ed. P. Mellars and C. Stringer), pp. 212–31. Princeton University Press.

4
Evolution of the human brain

Ralph Holloway

Abstract

Direct palaeoneurological evidence about the evolution of the hominid brain comes from study of the size and surface features of the endocasts of once-living brains. Because of intervening tissues, the detailed surface features of the brain are seldom clearly expressed on the inside surface of the skull. Therefore convolutional details of the brain's surface are the least reliably preserved features. Cerebral asymmetries are more reliably preserved. Overall size, despite its questionable significance, is the most reliable evidence of evolutionary change.

In the last 3–4 million years brain volume within the hominid lineage has increased from less than 400 ml to roughly 1400 ml. The first clear increase in hominid brain size is seen in early *Homo*, at *c*. 2 m.y.a. in East Africa (most reliably in cranial specimen KNM-ER 1470). This is an evolutionarily significant change that cannot be simply accounted for in terms of increased body size alone. From the appearance of *H. erectus* at *c*.1.7 m.y.a. to the present, the brain increases nearly twofold: from *c*.800 ml to 1500 ml in Late Pleistocene *H. sapiens*, without any apparent change in body size.

With regard to brain reorganization, left–right cerebral hemispheric asymmetries exist in extant pongids and the australopithecines, but neither the pattern nor direction is as strongly developed as in modern or fossil *Homo*. KNM-ER 1470 shows a strong pattern that may be related to handedness and tool-use/manufacture. The degree of asymmetry appears to increase in later hominids.

The appearance of a more human-like third inferior frontal convolution provides another line of evidence about evolutionary reorganization of the brain. None of the australopithecine endocasts show this region preserved satisfactorily. There is a consensus among palaeoneurologists that the endocast of the specimen KNM-ER 1470 does show, however, a somewhat more complex and modern-human-like third inferior frontal convolution compared with those of pongids. This region contains Broca's area, which in humans is related to the motor control of speech. Unfortunately, later hominid endocasts, including *H. habilis* and *H. erectus* through archaic *H. sapiens* to the present, seldom show the sulcal and gyral patterns faithfully. Thus nothing palaeoneurological can be said with confidence about possible changes with the emergence of anatomically modern *H. sapiens*. On the other hand, there is nothing striking about Neanderthal brain casts in comparison to more recent *H. sapiens*, except their slightly larger size, suggesting no significant evolutionary change thereon [eds].

4.1 Introduction

The evolution of the brain from some primitive *Australopithecus* stage to our present condition has taken some three million years to achieve. At the least, this has certainly involved an increase in brain size of roughly 3+ times. The expression 'at the least' is used here because how one views human brain evolution is often dependent on how one views the product of brain function, i.e., human behaviour. For most of us, this involves the concept of culture, and whether or not we perceive this phenomenon as unique to humankind. Is culture species-specific? Do any other animals 'have it'? Is there a discontinuity between human and other animal behaviour? Are human beings simply more clever than their nearest relatives, the chimpanzees and gorillas, or do human beings possess brains that provide both continuity and emergent properties when behaviour is compared? Different accounts of human brain evolution will often reflect how these questions are answered (cf. Ingold, this volume; Chapter 7; Gibson and Ingold 1993). Conversely, views of the similarities and differences in human and other primates' behaviour can effect how the brain is viewed in structural, functional, and evolutionary terms. For those preferring an approach of total continuity between ourselves and other primates, the size of the brain is a sufficient neural variable. After some Rubicon is reached (for example, 750 ml) human behaviour suddenly cuts in. For others preferring to believe in some discontinuity, the size of the brain is important, but insufficient as an explanatory variable: it is also necessary to consider how the brains of different animals are organized.

Since many animal species overlap with regard to their brain weights, yet demonstrate species-specific behavioural repertoires, it is difficult to understand how brain size alone can account for the behavioural differences in sensorimotor function, sexual, and agonistic behaviour, special sensory adaptations (for example, vision, auditory, and olfactory modes), and the integration of these with both the general and specific cognitive orientations to ecological diversity and specialization. Even between and within genera as closely related as *Papio* and *Macaca* monkeys, there are only minor differences in brain size, yet clear behavioural differences do exist that cannot be explained at the neural level.

In my view, human beings are unique in their ability to maintain a behavioural system based on culture, using both extrinsic arbitrary and iconic symbol systems to depict reality and unreality (Holloway 1967, 1969*a*, 1976*a*; 1981*a*, cf. Mundinger 1980). However clever other primates may appear, whether in their natural settings or within human manipulated laboratories, only humans have the temerity to study themselves and other species, and share their findings and hypotheses.

There are many difficulties in the task of understanding how our brain evolved. Firstly, there are no brains to study except those of the living. Comparative neuroanatomy can study only the present terminal products of separate evolutionary developments. Thus, in a strict empirical sense, we have no evidence for human brain evolution beyond its size and other critical morphological features to allow us access to the forces of natural selection that worked on past behaviour patterns. These patterns are the important but missing interfaces between evolving brain structure and function (Holloway 1970, 1979), some of which (for example tool-making, hunting and/or scavenging, food-sharing) may exist in the archaeological record, but require interpretation, and are always difficult to interpret without evoking controversy.

Secondly, the relationships between neural variables (for example, brain size, neocortical size, types of nuclei) and behaviour are not thoroughly understood, and only recently are relatively non-invasive techniques such as MRI (magnetic resonance imaging) or PET (positron emission tomography) scanning beginning to suggest how different parts of the brain interact and relate to complex cognitive behaviours. Thirdly, the actual evidence from brain evolution in any animal lineage can only be related to surface features of the brain, which in turn relate only to a limited subset of all behavioural repertoires. This problem is particularly severe in hominid brain endocasts.

Finally, there is a vast hiatus in our knowledge regarding variation in species-specific behaviour and its relationship to neuroanatomical variation of the neural substrate. This problem is compounded by the lack of such knowledge for within-species variability, which in the human case is almost always attributed to cultural factors alone. The appendix to this chapter, p. 98, on sexual dimorphism of the corpus callosum, is one such example.

What follows in this chapter is a preliminary examination of our knowledge of how the human brain evolved based on several lines of evidence, written explicitly from the viewpoint that while size is important, other phenotypic characters must be given consideration, as size, taken alone as a neural variable or parameter, cannot explain species-specific behaviour beyond general formulations relating to intelligence, however defined.

For accounts written from other perspectives, see Jerison (1973), for example, who focuses on the relationships between overall brain size and information-processing, i.e., intelligence—a term that it is very difficult to define without controversy and to compare across different taxonomic units. Tobias's (1971) book on hominid brain evolution is similarly oriented, and in particular adopts Jerison's (op. cit.) 'extra neuron numbers' approach, and is directed toward a positive-feedback interaction between behavioural complexity (culture) and brain size. In earlier versions of my own work (Holloway 1964, 1966, 1967, 1968, 1969*b*, 1970, 1979, 1981*a*) I tried to explain the evolution of brain size as an outcome of positive feedback between behavioural complexity and the neural components (nerve cells) that make up the brain, as well as of interactions between its components. Many anatomists concerned with human evolution, aside from Dart and his mentor G. E. Smith, have thought and do think of brain size as *the* most important ingredient of hominid evolution, and most appear to have great faith in 'cerebral Rubicon' models that have been around since Darwin's time. The value of 750 cc stated by Keith (1948) is the more or less implicitly assumed value at which 'true' hominid behaviour (culture) emerges. Indeed, by focusing only on brain size, hominid evolution is most often viewed as a process in which the brain was the last organ to undergo any evolutionary change (see for example, Washburn 1960). My own perspective is that the brain was always undergoing evolutionary change, from pre-*Australopithecus* to the Upper Pleistocene. I find Rubicon models too confining, as they rely only on brain size and do not consider the interaction of neural variables or the organization of the brain as important substrates for biobehavioural evolution. Additionally, there is something suspect about a parameter that is continuous but will evince qualitative functional changes with a simple increase in quantity alone (Holloway 1964, 1967). Finally, brain size is a variable over which one can all too easily find oneself hoisted by one's own petard. Too literal a reliance on a close causal relationship between size and function leads to all kinds of interpretative problems within species, i.e., as between subspecies, sexes, etc.

Other workers have focused on energy models, particularly metabolism, in attempts to understand the unique size of the human brain, both in relative and absolute terms (see for example Martin 1981, 1982, 1983; Little 1989; Parker 1990). Longevity and prenatal and postnatal developmental durations have been studied in depth by Sacher (for example Sacher 1982; Sacher and Staffeldt 1974) the better to understand the comparative situation among living animals. Others, such as Blumenberg (1983), have proposed complex feedback schemes between hunting behaviour, diet, neuropeptides, and the enlarged hominid brain. Passingham (1982) (see also Sawaguchi and Kudo 1990) has focused on the role of the cerebral cortex in human evolution, relying heavily on quantitative data on the brain structures of living primates obtained through the study of allometry. This region of the brain has most recently been hypothetically associated with language as a form of social grooming, with primate brain evolution viewed as simply an ever-increasing capacity for social grooming (Dunbar 1992; Aiello and Dunbar 1993). For a critique, see Holloway (1993). In the human animal, as group sizes became too large for physical social grooming, language evolved as a cheap substitute for manual grooming. Others, such as Milton (1981, 1993), believe brain size is essentially related to the food quest. Parker and Gibson (1979, 1990; cf. Gibson 1990, and this volume, Chapter 14) appear to believe that the cognitive stages elaborated by Piaget can be correlated with evolutionary developments in primate cognition, and directly related to both brain size and ontogeny.

All these writers and many others not mentioned here ignore to one degree or another the organization of the brain as an integral part of primate neurological evolution that must also be integrated with size variables. Indeed, most of the writings of the 1970s and 1980s have tended to focus on brain–body size relationships, in which the brain, treated as a dependent variable, enlarges mainly through selection pressures operating on body size. Radinsky's many articles (for example, 1972, 1975, 1977, 1979) have stressed the allometric approach within the palaeoneurological context. Still others, for example Armstrong (1983, 1985, 1990), have looked more carefully at certain neural structures other than the cortex (for example, the thalamus), and have proposed models that emphasize the *quantitative organization* of primate brains in relationship to social behaviour. For recent reviews of these models and their histories, see Blumenberg (1983, 1986), Falk (1980a, 1982, 1987), and Armstrong (1990).

This chapter will focus mainly on integrative work that attempts to synthesize comparative and palaeoneurological approaches. The 1970s and 1980s have witnessed a virtual explosion in the neurosciences generally (with the early 1990s particularly spectacular), and the evolution of the brain taken as an integrated topic has shifted enormously from explanations of a single neural variable (for example, brain size) and a single selection pressure (for example, for 'intelligence'), to a complex interweaving of many neural variables and a multifaceted view of probable selection pressures involving multiple behavioural levels. Steklis and Erwin's (1988) volume, *Neurosciences* (Comparative Primate Biology, Vol. 4) is an invaluable compendium of recent advances in our knowledge of the Primate Brain, particularly as regards newer knowledge about cortical cytoarchitectonics in a growing list of primate species. (In particular, see the papers there by Allman and McGuinness, Yin and Medjbeur, Pandya *et al.*, Kaas and Pons, and Kaas and Huerta. Elsewhere, see Allman 1990 and Pandya and Yeterian 1990.)

4.2 The human brain

In overview, the human brain is the largest among the primates, but certainly not the largest in either absolute or relative terms among the mammals.

Whales, dolphins, and elephants have larger absolute brain sizes, while some small mammals, including some primates, have relatively larger brains. The human brain, averaging approximately 1330 grams[1] (Tobias 1971), represents some 2 per cent of our body weight, yet continuously uses 15 per cent of our cardiac output, and consumes about 20 per cent of our metabolic resources (see Chien (1981) and Martin (1983) for examples and further references). There are no 'new' evolutionarily-derived structures in the human brain as compared to that of other mammals and, in particular, to that of other primates. Nuclear masses and the fibre systems interconnecting them appear to be the same, that is they exist and are homologous structures; they need not be structurally 'identical'. Deacon's (1988a, b, and 1990a, b) writings are an important reminder of the close homologies of human and macaque cortical fibre systems in those regions classically regarded as language 'centres'. What seem to vary are the quantitative relationships between and among these nuclei and fibre tracts, and the different ways in which the cerebral cortex becomes structurally and functionally subdivided and ultimately integrated. I am referring here to cytoarchitectonic differences in the cerebral mantle, as commonly illustrated by the famous 'maps' of Brodmann (1909) (see Fig. 4.1 [eds]). (Recent discussions of some of these differences may be found in Kaas and Pons 1988, Kaas and Huerta 1988, and Armstrong and Falk 1982.)

We must assume that species-specific behaviour depends on the size and underlying organization of each species's brain, its ontogenetic development, and how that occurs within varying environments, both material and social. When this chapter refers to *reorganization* of the brain during evolution, it means that natural selection has worked upon *quantitative shifts* in the relative sizes of brain *components*, and that such changes have had important consequences for behaviour (see, for example, Holloway 1964, 1968, 1970, 1979, in press, a). Such reorganizational changes have come about largely through heterochrony (Gould 1977; Shea 1983; Deacon 1990a, b), that is, changes in the timing (initiation, duration, and termination) of mitotic divisions and selective death of cell populations, leading to species-specific differences in both hyperplasia and hypertrophy of nerve-cells. (Hyperplasia refers to the number of cells produced, while hypertrophy refers to the size of the cells, both of which determine, at least quantitatively, synaptic connectivity.) Thus far, no significant differences among primates have been discovered at the neurochemical or molecular neurobiological levels.

The brain is an extremely complex set of organs, containing billions of parts if one is referring to nerve-cells alone. These cells are in one of two states: firing, or not. The effects of their firing can be either excitatory or inhibitory, thus leading to a dual set of 'digital' states. However, whether or not a nerve-cell fires will depend on a process of summation of many thousands of inhibitory or excitatory connections with other nerve-cells. Estimates of one nerve-cell in the visual cortex's having as many as 10 000 connections are common. This might be considered the 'analogue' condition of the nerve-cells. The complexity increases vastly when one adds to this picture the fact the brain has both 'serial' and 'parallel' organization among its many components, such that information about the environment can be evaluated both directly and indirectly, in present and in future perspectives depending on how experience becomes organized in both short- and long-term memory, how it is stored, how it is retrieved, and how it is transformed. These functions involve other brain structures as well as the cerebral cortex (for example, the thalamic nuclei, the hippocampus, the septum, the reticular formation, etc.).

The brain is also organized hierarchically (see Fig. 4.2 [eds]). This refers to the relationships between the cerebral cortex, the underlying basal ganglia, the limbic system, and the olfactory bulbs (the *telencephalon* or forebrain), which surround the *diencephalon*, including the thalamus, epithalamus, hypothalamus, and pineal gland. Next, moving 'downward', there is the brainstem, which contains the superior and inferior colliculi, which are visual and auditory in function (the *mesencephalon* or midbrain). Lastly, there are the more 'primitive' structures, which consist of the cerebellum, the pons, the medulla, and the third and fourth ventricles, which are integrated with the spinal chord (the *metencephalon* and *myelencephalon*).

This kind of structural similarity is found in almost all vertebrate brains, suggesting an extraordinary degree of genetic conservatism underlying the ontogenetic development of the brain. Additionally, the cerebral cortex is organized into vertical columnar units (Mountcastle 1978; Szentagothai 1978) containing very similar numbers of neuronal cells (both neurons and their metabolically-supporting neuroglial cells) in very similar ways in almost all mammals, indicating another structural and possibly functional level of great genetic conservatism. This suggests that it is the interconnections between neurons, and their growth and development, that are partly responsible for species-specific differences in behaviour.

Obviously, given the enormous differences between humans and other animals in the size of our brain, and in particular our cerebral cortex (this accounts for 76 per cent of our brain volume, as calculated

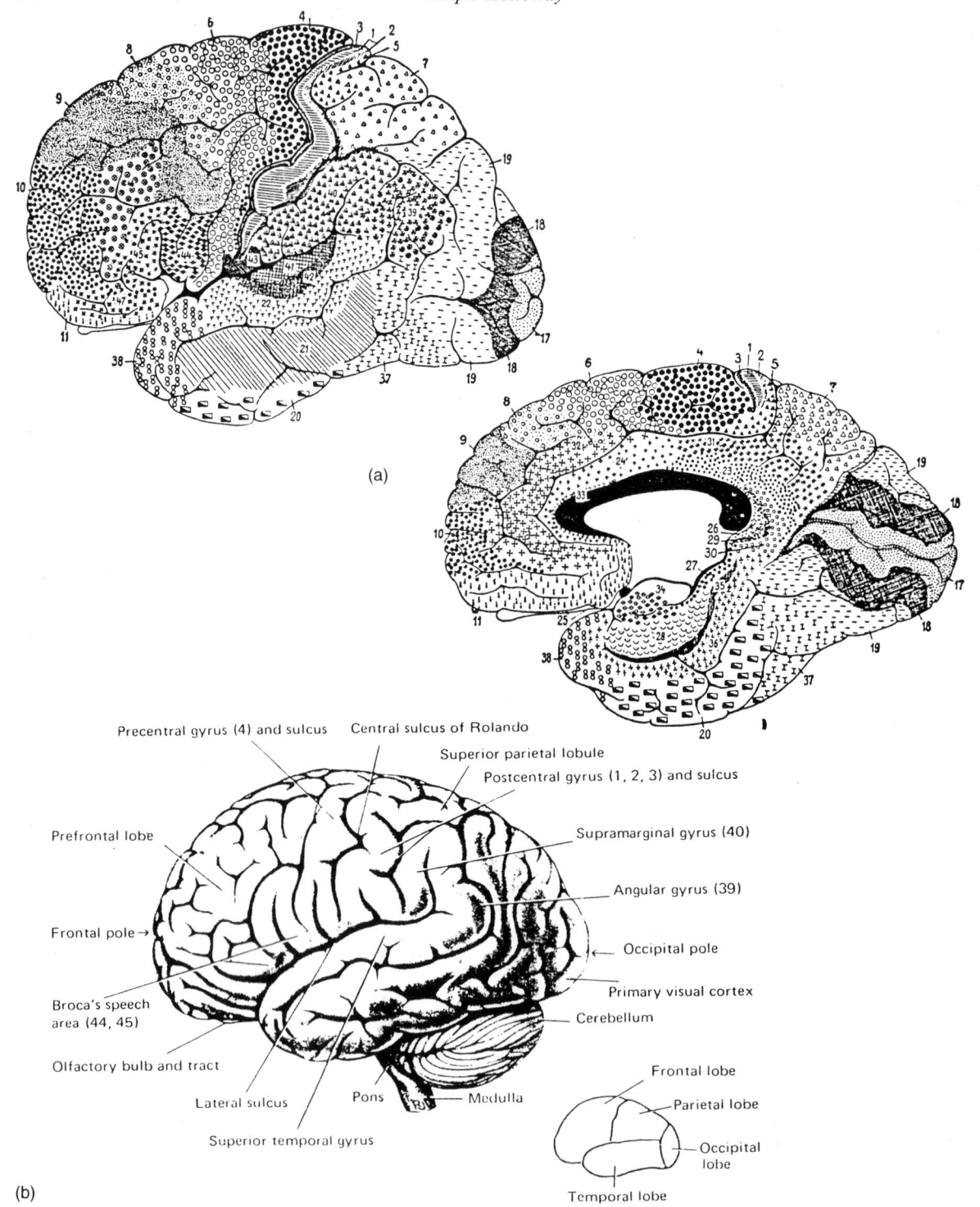

Fig. 4.1 (a) Maps from Brodmann (1909) of areas of the human cortex, each of which possesses a distinctive cytoarchitectonic structure. Top figure: lateral view; bottom figure: medial view. [eds].
(b) Lateral surface anatomy of human brain [eds].

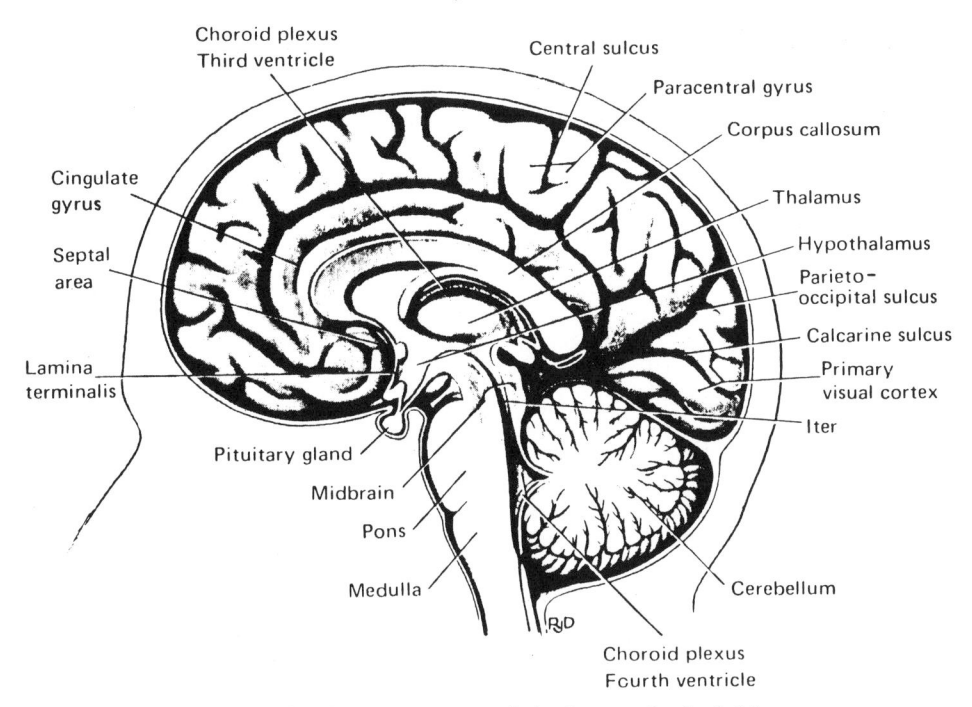

Fig. 4.2 Gross structure of the human brain [eds].

from the data of Stephan *et al.* 1981), genetic changes controlling both the rates and duration of mitotic division of certain neural masses have occurred during evolution. Both hyperplasia (the number of cells), hypertrophy (the size of neural units), and cortical columnar interconnections have been key evolutionary events in reorganizing brains and species-specific behaviour patterns.

While no one is certain how many 'genes' control the formation of the brain ontogenetically, it is estimated that perhaps as many as 40 000 genes may be involved. Obviously, an enormous amount of potential genetic variability exists for natural selection to work upon now, as it has in the past. Thus, one of the most formidable challenges facing any attempt to understand brain evolution is how to account for the complex mixture of both conservative and new genetic expression relating to all parts of the brain, and how these relate to behaviour, adaptation, and evolution within the primates, or, for that matter, any animal group. Deacon's (1990*a*, *b*) articles are unique in his appreciation of and attempts to clarify this complexity.

We know next to nothing about within-species neural variability and behaviour (see for example Holloway 1968, 1969*a*, 1976*a*, *b*, 1979, 1980, 1983*a*; Holloway *et al.*, in press, *b*). Even between-species neural differences cannot be directly related to different species' behaviour. We do not know which genes exist or control the ontogenetic unfolding of particular brain regions or nuclei or fibre tracts. Neurochemistry does not at present provide any convincing relationships between brains and behaviour except, say, between neurotransmitters and psychopathological states. The data for a neuroscientific explanation of readily observable behavioural differences between different breeds of mice, rats, cats, or dogs do not exist. Simply ask anyone what are the neural differences that might explain the behavioural differences between orang-utans, chimpanzees, and gorillas, or various species of *Macaca* and *Papio*. As for between species behaviour and quantitative variations in the brain, the best known mammalian examples are the *Chiroptera* (bats) as published by Pirlot and his colleagues (see Jolicoeur *et al.* (1984) for a similar perspective on these hiatuses and references to his own and his colleagues' works on bats and other animals).

Considering how much is known about animal behaviour under naturalistic field conditions, it is disappointing that so little synthesis can be made directly with neuroanatomical data. The Jolicoeur *et al.* 1984 studies on quantitative structures of the bat brain and their relation to feeding behaviours (herbivory, fructivory, predation) stand almost alone as one promising direction. If we consider for a moment the very wide range of behavioural differences among the living primates—i.e. lemurs, tarsiers, New and Old

World monkeys, chimpanzees, gorillas, orang-utans, and gibbons—we find an embarrassing lack of reliable neurological synthesis. None of the behavioural differences can as yet be linked with the animals' respective brain sizes or organizations. Brain size is simply insufficient for such a task of synthesis, although it is an essential starting-point, given that it comprises most of our reliable data-bases.

Considerable knowledge regarding brain size as a correlate with behavioural and other anatomical variables has been gained through allometric studies. In these, brain size is usually considered a dependent variable, and relationships are made to body weight, gestational duration, growth stages, longevity, metabolism, precocial or altricial development, and broad ecological areas relating to subsistence (for example folivory, frugivory, omnivory, and predation) (see Passingham 1982; Martin 1981, 1982, 1983; Milton 1981, 1993; Clutton-Brock and Harvey 1980; Harvey and Clutton-Brock 1985; Dunbar 1992; Aiello and Dunbar 1992; Armstrong 1983, 1990; Hofman 1982, 1983; Leutenegger 1982, 1987; Sawaguchi 1988, 1990; Sawaguchi and Kudo 1990; Sacher 1982; Shea 1983). All these analyses treat brain size as if the brain were an organ in its own right, seldom with any realization that many aspects of the behaviour being examined cannot be related to brain size in any causal manner. Brain size is good as a starting-point; but it seems to become a reified end in itself.

Most allometric studies plot the size of an organ (in this case the brain, or a part of the brain) against a large variable, such as body weight or total brain weight. There is inherent in such studies the 'mouse–whale' phenomenon, in which the values, once transformed into log (base 10) values, cannot do other than appear as a straight line, as the transformation is an often-used technique in reducing statistical variances in raw variables. (We can substitute *Microcebus*—the dwarf lemur—and *Gorilla* as examples of 'mouse' and 'whale' for the primates: see R. J. Smith 1980.) When such log–log plots are done, it is generally the slope of the regression line that is of most interest, as it specifies how one organ is scaling against another, or a total weight. Figure 4.3 shows a log–log plot of brain weight against body weight for some 85 species of primates, based on data kindly given to the author by Dr Heinz Stephan. The value for our own species is in the extreme upper right-hand corner of the figure. The closest three squares are the pongids, the gorilla, chimpanzee, and orang-utan. The correlation coefficient is about 0.97, without the *Homo sapiens* value, which is about three times higher than its predicted value based on body weight. The slope of the regression line without the *Homo sapiens* value is about 0.78 (and not 0.66, as was earlier declared by Jerison (1973) and many others; see Martin 1983, for references and further discussion). This number of about 0.78 (approximately 3/4) for the

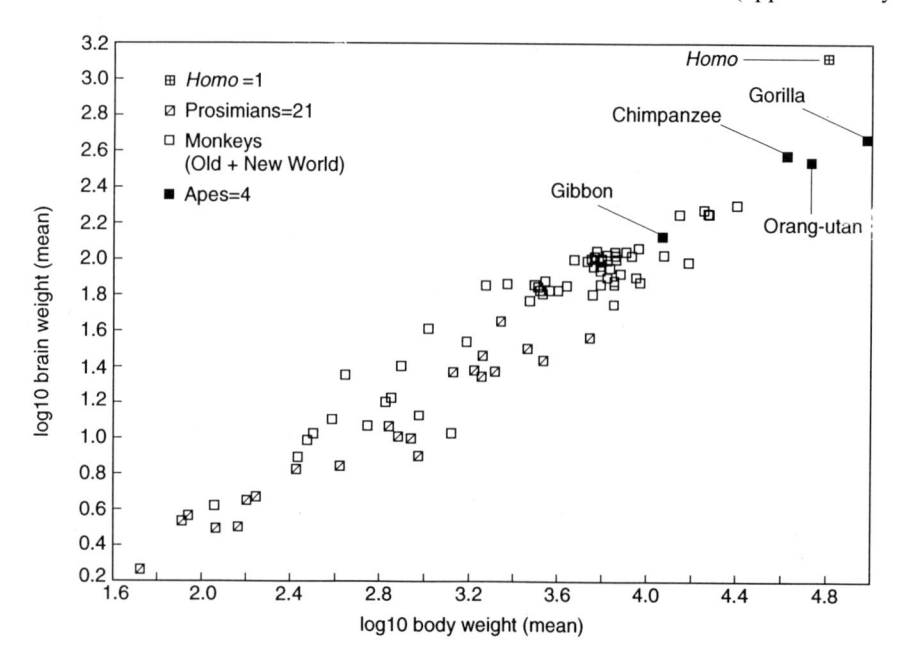

Fig. 4.3 A log–log (base 10) plot of the mean brain and body weights for 85 species of primates, including *Homo sapiens sapiens* (top right).

order as a whole is suggestive of a metabolic constraint between body weight and the weight of the brain, although no precise formulation has proved adequate as yet. If the points are plotted *within* different taxonomic categories—i.e., prosimians alone, New World cebids alone, Old World monkeys, etc.—each group scales somewhat differently. Within families, the slope is close to roughly 0.66. This latter exponent is suggestive of a geometric relationship between surface area and volume, i.e., the ratio 2:3. Lower-level taxa scale at lower exponents, such as roughly 0.3 between species of the same genus, or around 0.1 to 0.2 within a species (see Holloway and Post 1982; Holloway 1980 for further discussion). It is for this reason that encephalization quotients (see Section 4.4.1.2. below) are 'relative', as each species value depends on the allometric equation used.

An important point is that the slopes, whether 0.76 or 0.66 (or whatever value), reflect not a *law*, but *constraints* around which different species *vary*. While it is possible that some of the discrepancies between predicted and observed values may be purely statistical in nature—i.e., may arise from sampling phenomena—it is also possible that some of these departures may contain interesting and provocative insights into the neural biologies of particular primate species. The human case is simply the most obvious among primates.

The picture becomes very much more complex when components of the brain are log–log regressed against each other, or against brain weight. For example, the human animal shows enormous departures (in terms of percentages) of actual from predicted volumes in a number of brain structures. I have, elsewhere, mentioned the primary visual striate cortex, which, in a sample of 45 primate species, falls 121 per cent below expected volume (calculated from the data of Stephan *et al.* 1981, and discussed earlier in Holloway 1976a, 1979, and again in 1988a, 1992, yet ignored in Passingham *et al.* 1986 and Armstrong *et al.* 1991). The lateral geniculate body of the thalamus is similarly 'off target' (i.e., about 146 per cent below the value expected for *Homo sapiens*), as would be expected from its close relationship to the visual cortex. Indeed, I have found differences of up to 7000 per cent for some of the smaller structures in the human brain. It is interesting that the volume of the ventricles, which in the fetal brain provide the neuroblasts that eventually become the ten billion or so neurons in the adult cerebral cortex, is roughly 52 per cent greater than expected, which correlates with the fact that the human brain has the highest percentage of cerebral cortex among the primates.

Herein lies a rich treasure-trove barely explored (see also Deacon 1988a, b), which could have interesting potential for understanding, at least quantitatively, the differences in organization between our brains and those of other animals. I explicitly wish to suggest that it might at least be an interesting place to begin. As is always the case, it is tempting to interpret such differences in size of the neural components as direct evolutionary statements, assuming reasonably (at first glance) that what is big is important, and what is smaller has become that way through natural selection operating directly on behaviour, and thus on the genes controlling the ontogenetic development of particular neural masses. We really need many more quantitative data before such explanatory leaps can even legitimately be made, let alone reliably evaluated.

The cerebral cortex is an interesting and provocative example in this discussion. After all, we do pride ourselves as being that species that has so much of it . . . Yet, as Passingham and Ettlinger (1973) showed, and as has been discussed more recently by Passingham (1982), log–log regressions of cortical volume against brain weight show that the human animal has about as much cerebral cortex as would be expected for a primate of its brain weight. The human brain is 76 per cent neocortex. This is the highest primate ratio, followed next by the chimpanzee, with about 72 per cent; average values for most primates are about 50–60 per cent. (These figures are based on my analyses of the data of Stephan *et al.* 1981.) The actual value for human cerebral cortex volume is less than 1 per cent different from its allometrically expected value. The cerebellum for the human species is about 6 per cent greater than expected. Given statistical variation, and small sample sizes of such neural data within species, differences of 10 to 25 per cent are probably not significant. However, in the visual system, i.e., the primary visual striate cortex and the lateral geniculate nuclei, differences (−121 per cent and −146 per cent respectively) *are* significant, *and* these reductions signal a *relative increase* in the volume of parietal 'association' cortex, which is usually related to complex cognitive activities such as visuo-spatial integration, etc. Here is a prime example of reorganization.

However, one must consider carefully these regression operations. In the case of the cerebral cortex, one is regressing the cortex against brain volume, *of which over 50 per cent is represented by the cerebral cortex alone* in primates generally, and 76 per cent in *Homo* specifically. It is thus hardly unexpected that the correlations are so tight, and the differences between observed and expected values so close. Perhaps some other measurement or set of ratio data would be more useful in underlining the unique relative volume of the human cerebral cortex (see Passingham 1982,

Passingham et al. 1986, Deacon 1990a, b, and Dunbar 1992 for other examples, and different emphases).

Quantitative studies on primate brains that go beyond brain weight or volume are still in their infancy (see for example Deacon 1988a, b; Frahm et al. 1982, 1984; Stephen et al. 1981; Matano et al. 1985a, b; Armstrong and Falk 1982; Passingham 1982; Passingham et al. 1986). Much of the quantitative evidence is based on a sample size of one for most species. Not all neural structures have been measured, including functionally meaningful divisions of the cerebral cortex according to cytoarchitectonic patterns (i.e., whether sensorimotor or 'associative', cf. Brodmann's areas; (see Fig. 4.4 [eds]). These examples hardly diminish the possible list of true gaps which exist in our knowledge at present.

Another example concerns that most favoured part of the cerebral cortex, the frontal lobe, which is regarded by so many as the chief expanding unit during human evolution. In 1964 (see also also Holloway 1968) I tried to show that the quantitative evidence for a *unique* increase in this part of the human brain was suspect. This was before the days of any careful realization of the usefulness of allometrical analysis. More recently, Uylings and van Eden (1990) have claimed that the prefrontal cortex in humans has shown an allometric increase beyond that of other primates. While the amount of prefrontal cortex does scale positively to the total amount of cerebral cortex (the slope when plotted with isocortex volume is 1.069), the slope when the prefrontal cortex is plotted against total brain volume is 1.108, and this is reported as significantly different from 1.0. This positive allometry led the authors to suggest that the human prefrontal cortex was proportionately larger than in apes. The rat, marmoset, macaque, orangutan, and human all fall nicely on the same straight log–log regression lines. In other words, the 'positive' allometric increase from rat to marmoset, then to macaque, and then to orang-utan, is proportionately the same as that from orang-utans to humans. To me, this signals that human prefrontal cortex is exactly what would be expected for a primate with a human brain weight. Is this the same as saying that humans have relatively more prefrontal cortex than other primates? I don't believe so; but this does not diminish the importance of the frontal lobe, either with regard to its size or its organization.

Another very important topic is that of the surface convolutions of the brain. While the human brain is in actuality some 3–4 times heavier than the chimpanzee brain, there is considerable similarity between the two species with regard to the convolutional details (see Armstrong et al. 1991 in particular, as well

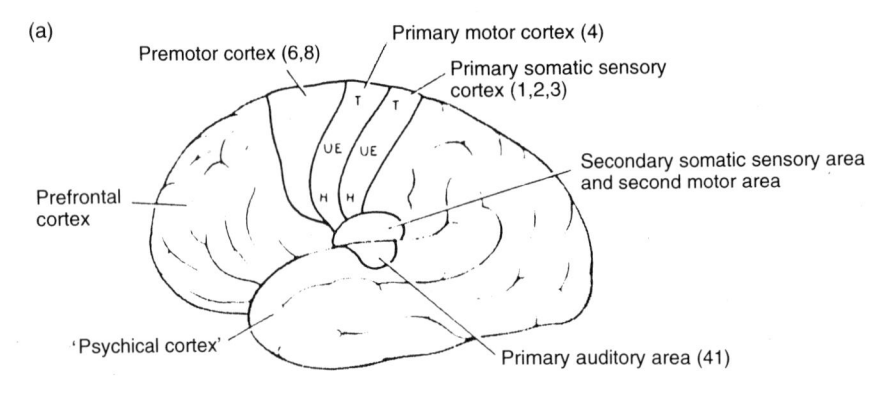

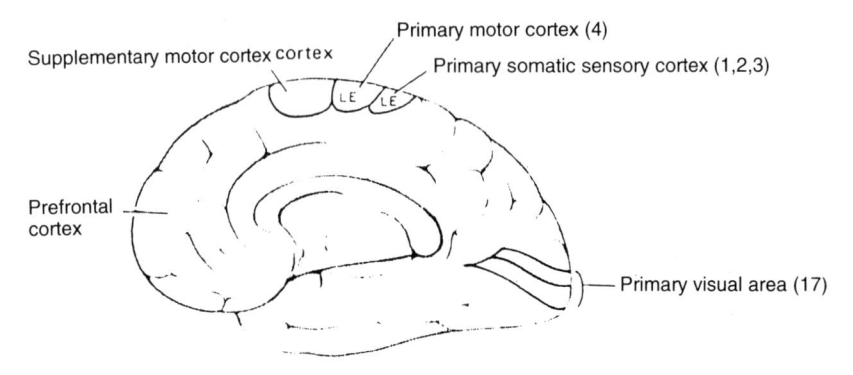

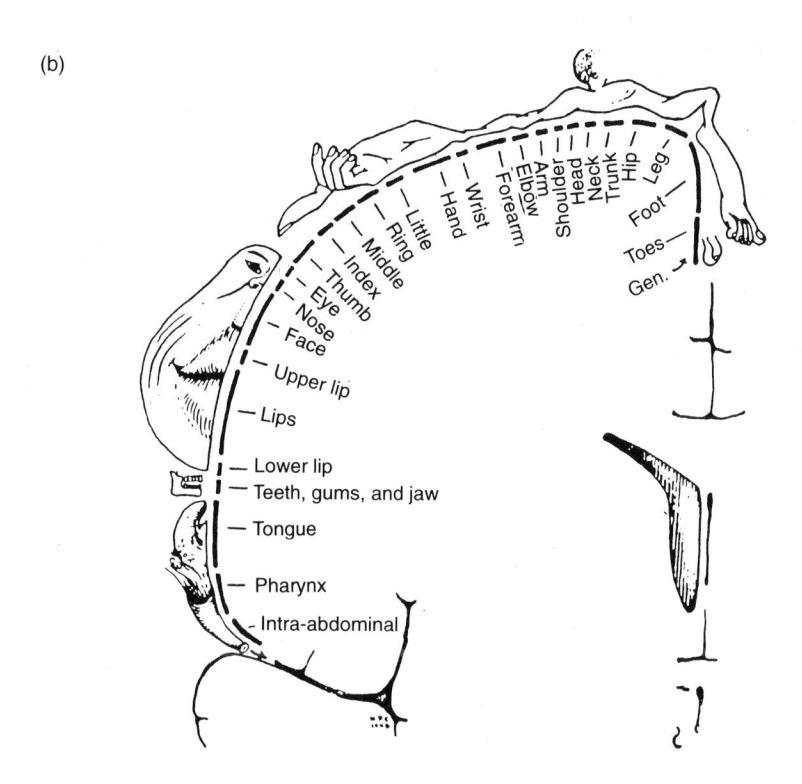

Fig. 4.4 Gross functional areas of the cerebral cortex [eds]. (a) The location of several functional areas: the representation of body parts on the primary motor and somatic sensory cortices includes the head (H), upper externity (UE), trunk (T), and lower extremity (LE), numbers represent areas of Brodmann. (b) Diagram showing the relative sizes of the parts of the central cortex from which sensations localized to distant parts of the body can be elicited on electrical stimulation in many (from Penfield and Rasmussen (1950)).

as my (Holloway 1992) critique regarding australopithecine brain endocasts). Although the human brain has more convolutions (a fact which is related to brain weight), and very considerable variation of its gyri and sulci, particularly in the parietal and frontal lobes, the primary and secondary gyri (the hills) and sulci (the valleys) are very similar and often the same between hominoid species. Of considerable interest to those studying the palaeoneurology of our fossil ancestors are the sulci labelled the lunate, the intraparietal, the Sylvian, and the lateral calcarine (see for example Holloway 1983b; see Fig. 4.5). In apes, such as the chimpanzee, the lunate sulcus is always present, and is the anterior boundary of primary visual striate cortex (area 17 of Brodmann), which subserves visual functions. Furthermore, in apes the intraparietal sulcus, in its posterior part, always terminates against the lunate sulcus, and divides the parietal portion of cerebral cortex into superior and inferior lobules. The calcarine fissure always runs medial from the occipital pole to a lateral position, but terminates before it reaches the lunate sulcus. Thus these sulci should not be confused with each other, but taken together represent an important neuroanatomical unit. The lunate sulcus of a human brain is in a very posterior position relative to where it can be found in apes (see Connolly 1950 and Holloway 1985a for a review of the history of this sulcus and its significance to human brain evolution). As the figures for the volume of visual striate cortex previously discussed indicate, the human brain has relatively less of this cortex making up its cerebrum than those of the apes. This means that the relative amount of parietal 'association' cortex has increased in the human species. The challenge is to document when such changes took place in hominid evolution. Unfortunately, endocasts seldom show the convolutions that existed in the brain.

The central sulcus divides the frontal from the parietal lobe, and functionally marks the separation between the mainly motoric anterior gyrus and the posterior sensory gyrus. Both the inferior third frontal

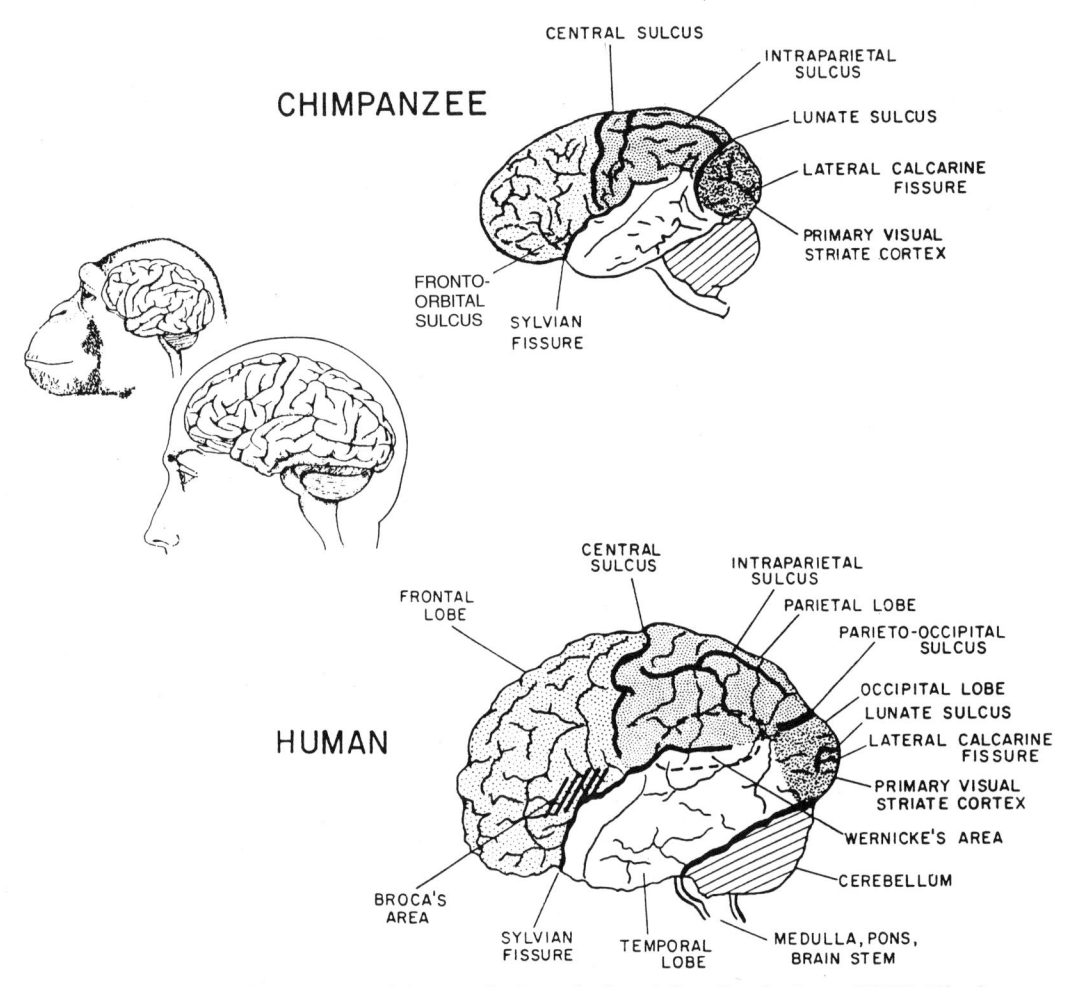

Fig. 4.5 The brains of chimpanzee and human in lateral view (after Passingham (1982) *The human primate*, Freeman, San Francisco) [eds].

convolution (with Broca's area) and the posterior temporal and middle parietal lobes (containing Wernicke's area) appear more convoluted in the human species, and have important relationships to both the motor and sensory (receptive) aspects of linguistic communication. These particular regions are seldom well preserved on fossil endocasts, and are quite variable with regard to tertiary convolutions (for the finest details see for example Connolly 1950), and are areas of considerable interpretative controversy among palaeoneurologists.

Figure 4.6 (cf. Holloway 1983a) attempts to depict a few possibilities that might help to explain human brain evolution (see Section 4.4), in which brain size, reorganization, differences in 'wiring' of components, and asymmetries can be seen as phenotypic manifestations of changes in unknown parts of the underlying genetic code. Figure 4.7 provides a model (Holloway 1979) that attempts to synthesize the different viewpoints between those who stress mass or size, and those who stress reorganization. Brain size is simply the most obvious and reliably measured of such possible phenotypic windows on evolutionary changes. As one can see, many important changes enhancing hominid behavioural adaptations (see for example Holloway 1970, 1983a) *could* have taken place without necessarily involving brain-size increase. This is meant as a distinct warning to those who would simply plot brain size against time and conclude the brain was showing evidence for 'stasis' between hominid populations (for example Eldredge and Gould 1972; Eldredge and Tattersall 1982; Cronin *et al.* 1981). Many other events could have taken place that were important factors in human

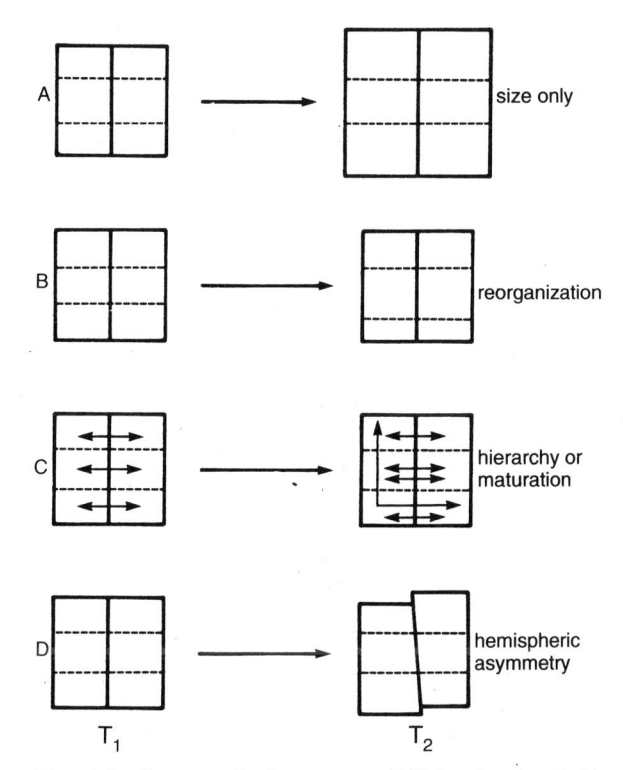

Fig. 4.6 Some evolutionary possibilities for hominid brains (adapted from Holloway 1983.) Four different (but not necessarily exclusive) possibilities of brain evolutionary change through time (T_1 to T_2). In A, the brain is represented with two hemispheres (left and right). The two dotted transverse lines are representations of the central or Rolandic sulcus (top) and the lunate sulcus (bottom). The change in time is simply an increase in absolute brain size, with or without concomitant body-size increase, and *without any changes in the size of the cerebral components or the connections between them*. This change could have occurred isometrically or allometrically. An example might be the change from *Homo erectus* in Indonesia to the later forms of the same species in China.

In B, the change from T_1 to T_2 does not necessarily, involve any change in brain size. Instead, *there is a change in the relative size of the components*. In this case, the lunate sulcus has moved back posteriorly, increasing the relative size of the parietal association cortex. This is a *reorganizational* model. An example could be the change from a pongid pre-australopithecine precursor to *Australopithecus afarensis*, or to *A. africanus*. (Combining A and B might be an example of the evolutionary change from a primitive *Australopithecus* to early *Homo*.)

Model C shows changes in the development of interconnections between cerebral components (hierarchical development, see Holloway 1979), *without any necessary change in absolute or relative brain size*. The arrows represent different fibre systems maturing at different rates and/or increasing in number between different cortical regions through the corpus callosum.

In model D, the absolute brain size remains constant from T_1 to T_2, but a more human type of hemispheric asymmetrical pattern develops (that is, a left-occipital right-frontal torsion pattern). For example, the change in brains from *Homo erectus* to *Homo sapiens* might have involved minimal increase in size, but changes in both hierarchy or maturation rates *and* hemispheric asymmetry.

It is important to note that these four models hardly exhaust the possibilities of different brain changes through time, and all of these changes and particular combinations of them may have been realized in human brain evolution. In addition, model B could be a true case of 'punctuated equilibrium' (as could A, C, or D), and thus be overlooked in hominid evolution. The change in A, rather than being a case of 'punctuated equilibrium', could be a simple matter of allometry, i.e., due to an increase in body size, without any substantial behavioural differences between T_1 and T_2.

brain and behavioural evolution that did not effect brain size *per se*. Punctuated equilibria and stasis models based on brain size alone never take these possibilities into account.

This does not mean that brain size should be ignored, or was unimportant during hominid evolution. It simply means that we should cast our nets wider for other phenotypic measures that pertain to the brain and how it works, both within and between species.

4.3 Lines of evidence regarding human brain evolution

There are three lines of evidence available to us for studying the evolution of the human brain: (1) palaeoneurology; (2) comparative neuroanatomy; and (3) both the products of behaviour resulting from past hominid activity and the fossil remains of cranial and postcranial anatomy. The first is a direct line of evidence; the others provide indirect evidence.

Palaeoneurology involves the study of endocasts, which are limited to only the surface features of once-living brains. The data available from such studies of casts made of the interiors of fossil crania include endocranial volume, convolutional details of the cerebral cortex, traces of meningeal vessels which may have some taxonomic if not functional significance, and the shape and asymmetries of the cerebral cortex.

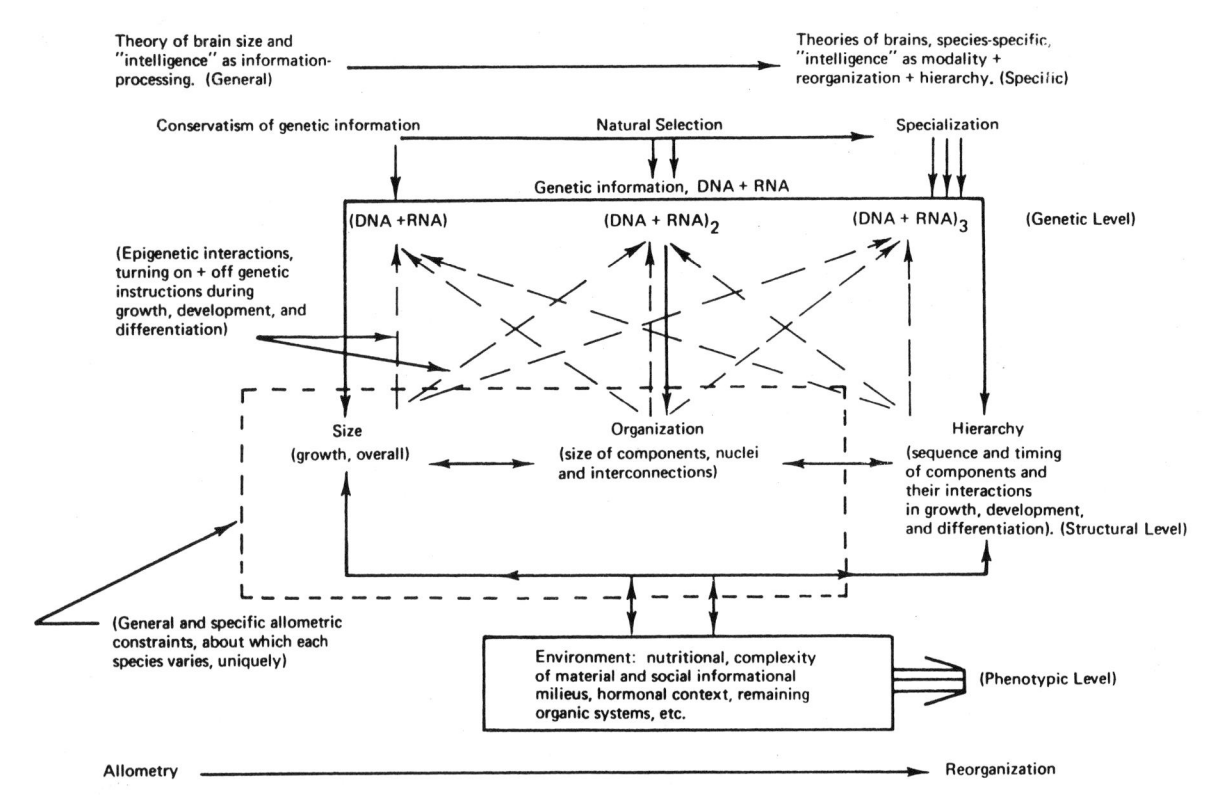

Fig. 4.7 A model of how brain size (absolute), reorganization (differential sizes of components), and hierarchy might be conceived. The 'phenotypic level' toward the bottom right portion of the diagram is almost exclusively regarded as brain size by most authors, but in this model is meant to include more than absolute size. For allometrists, only the left side appears to be of interest, the rest being 'trivial'. For anyone concerned about species-specific brain-behavioural evolution, i.e., *Homo sapiens*, the left portion cannot explain the totality that is the human brain (or any other animal's brain), as allometry is only the constraint around which other species vary, and brain size alone cannot be related to species-specific repertoires of behaviour, or unique evolutionary histories. This model explicitly regards the final phenotypic level as a complex orchestration between the neural events which unfold through the interaction of structural and regulatory genes, with natural selection operating upon at least three realms of genetic information (Holloway 1983a: adapted from Holloway 1979, where a fuller discussion can be found).

Unfortunately, in life the brain is covered with three meningeal tissues—dura, arachnoid, and pia mater—which tend to obscure the imprinting of underlying cortical gyri and sulci on the internal table of cortical bone, to say nothing of the effects of the cerebrospinal fluid. This means that it is extraordinarily rare for all cortical convolutions to be preserved on endocasts. Interpretations of what *is* preserved are thus often controversial; and hence it is the size of the brain and possible asymmetries of the cortex that provide the most reliable evidence of evolutionary changes.[2]

The study of *comparative neuroanatomy* provides evidence regarding the size and organization of the brains of living animals, each of which is a terminal end-product of its own evolutionary development, and not a living stage, as it were, of human evolution. This first line of indirect evidence is absolutely essential for understanding how neural nuclei, fibre tracts, and cortical cytoarchitecture vary in different animals, and how these variations relate to species-specific behaviour. While the present living chimpanzee (*Pan troglodytes* or *Pan paniscus*) is not a true stage in human evolution, most neuroscientists seem to believe that modern *Pan* is probably similar in many ways to the line from which the earliest hominids diverged 5 to 8 million years ago. Thus *Pan* is a useful comparative foil for providing quantitative data regarding neural organization, against which we can compare human brains. To a lesser extent the same applies to *Macaca* monkeys, although these are *the* favoured primates for comparisons with the human brain, and the species on which much of modern neuroscience is done. For example, Heilbroner and Holloway (1988, 1989) have shown that there is Sylvian fissure asymmetry favouring the left side, and thus the temporal lobe as a whole, in *Macaca*, *Saimiri*, and *Callicebus*—results consistent with those reported by Falk *et al.* (1986). Our second study (1989) found less compelling evidence in other regions of the cortex and limbic system. It is also within this type of comparative evidential base that the study of allometry has become a valuable tool of studies of the brain (see Section 4.2 above).

The second line of indirect evidence relies upon the *products* of behaviour *resulting from past hominid activity* (see especially Chapters 9, 10, and 11 by White, Wynn, and Conkey respectively, of the present volume). This includes stone tools made to both non-standardized and standardized forms, and the palaeoarchaeological sites which have preserved past patterns of hominid activity, such as gathering, stone tool-making, scavenging, butchering, and hunting. Another aspect of this line of evidence relates to the actual *skeletal remains* of the hominids themselves (see Marzke, this volume, Chapter 5), which provide information regarding their locomotory patterns (bipedalism), their manipulatory capabilities (hand bones), and their other anatomical characteristics, including details of accidents and pathology. While all these provide only the most indirect of clues, major patterns of locomotory adaptations and complicated cognitive processes such as tool-making cannot evolve in a neural vacuum. The central nervous system has an intimate, if not a controlling, relationship with musculo-skeletal organization. Certainly, hominid evolution has been mosaic (McHenry 1975, 1982, 1988; White 1980), but there have been mosaics within the mosaic (Armstrong 1983; Holloway 1970, 1983a; Holloway and Post 1982). Empirically, the brain did become significantly larger in size *after* bipedalism developed; but size refers to but one phenotypic manifestation of the brain, and it is surely premature to ascribe a terminal role to brain evolution, i.e., to treat the brain as the last organ to evolve. Thus, Toth's suggestion (1985) that early stone tools show signs of having been made by right-handers is an important piece of possible evidence that can be related to the asymmetrical organization of the cerebral cortex in early *Homo* (see Section 4.4.2.3 below), and thus to handedness and other possible cognitive specializations.

Given the scarcity of truly empirical evidence regarding brain evolution, the challenge is to use all three of the above lines of evidence judiciously, synthesizing the best, and hopefully framing hypotheses that are capable of being refuted later. This is no easy task. While there is convincing evidence from the neurosciences in general relating to asymmetries of the human brain, and the relationships between handedness and language, or the hippocampus and memory, or the organization of the cerebral cortex into both serial and parallel processing devices (with unequal representations of cortical regions such as Broca's or Wernicke's areas), it is very difficult to relate this knowledge directly to poorly preserved, fragmentary endocast portions of fossil hominid crania.

The above discussion places the problems of attempting to understand human brain evolution in a fuller light. It is now time to approach these evidential bases more thoroughly.

4.4. Palaeoneurological evidence

4.4.1 *Brain size*

4.4.1.1. *Absolute brain size*

Table 4.1 (and see also Editorial appendix I—[eds]) provides a listing of the brain volumes (endocranial volumes) presently known for the major fossil hominid finds. Starting more than three million years ago with *Australopithecus afarensis*, there has been an increase in brain volume within the Hominidae from about 400 ml to roughly 1400 ml. That evolutionary gain in size, 1000 ml, is coincidentally approximately equivalent to the total known variation in normal human beings, based on large population sizes (see for example Dekaban and Sadowsky 1978; and Table 4.8b of Appendix I to this chapter—[eds]). Specific individual cases would include such old anthropological 'chestnuts' as Anatole de France, with roughly 1000 ml, and Jonathan Swift, with more than 2000 ml. Until recently, no reliable evidence has ever been presented that this variation in size has any meaningful relationship to behaviour in humans, either of a quantitative or qualitative nature, aside from known pathological conditions such as microcephaly, hydrocephaly, etc.: there are, for example, no published accounts of 'geniuses' with brain sizes in the range of pongids, i.e., 275 ml (lower limit of female pygony chimpanzee) to 752 ml (upper limit of male gorilla). (Values taken from Tobias (1971b) = Table 4.8b of Editorial appendix I to this chapter.) Recently, using MRI scans, Andrensen *et al.* (1993) have shown that there are strong correlations between the Wechsler Adult Intelligence Scale and various volume estimates of parts of the brain, with correlation coefficients of between 0.3 and 0.5. These are relatively high for a within-species sample of 67 volunteers.

Realizing that human pathologies have their limitations in discussing evolutionary trends, it is still intriguing to reflect on microcephaly. There are recorded

Table 4.1. Endocranial brain volumes of reconstructed hominids

Specimen		Region	Endocranial volume (ml)	Method	Eval.
Taung	*A. africanus*	S.A.	440*	A	1
STS60	"	"	428	A	1
STS71	"	"	428	C	2–3
STS19/58	"	"	436	B	2
STS5	"	"	485	A	1
MLD37/38	"	"	435	D	1
MLD1	"	"	500 ± 20	B	3
SK1585	*A. robustus*	"	530	A	1
OH5	*A. boisei*	E.A.	530	A	1
ER406	"	"	525	D	2
ER407	"	"	510	A	1
ER732	"	"	500	A	1
ER1805	*H.?*	"	582	A	1
ER1813	"	"	510	A	1
ER1470	*H. habilis*	"	752	A	1
OH7	"	"	687	B	2
OH13	"	"	650	C	2
OH24	"	"	590	A	2–3
OH9	*H. erectus(?)*	"	1067	A	1
ER3733	"	"	848	A	1
ER3883	"	"	804	A	1
HE1 (1892)	"	Indonesia	953	A	1
HE2 (1937)	"	"	815	A	1
HE4 (1938)	"	"	900	C	2–3
HE6 (1963)	"	"	855	A	2
HE7 (1965)	"	"	1059	C	1–2
HE8 (1969)	"	"	1004	A	1
SOLO I	"	"	1172	A	1
SOLO V	"	"	1250	A	1
SOLO VI	"	"	1013	A	1
SOLO X	"	"	1231	A	1
SOLO XI	"	"	1090	A	1
SALÉ	"	Morocco	880	A	1
SPY I	*H. sapiens (N)*	Europe"	1305	A	1
SPY II	"	"	1553	A	1
La Chapelle	"	"	1625	X	
La Ferassie I	"	"	1640	X	
Neandertal	"	"	1525	X	2
La Quina	"	"	1350	X	1
Jebel Irhoud1	"	Morocco	1305	A	1
AL 333–45	*A. afarensis*	Ethiopia	485**	C	2
AL 162–28	"	"	375–400**	est.	2
AL 333–105	"	"	310–320**	C	2

Some selected cranial capacities for different hominids that the author has examined. Method A, direct water displacement of either a full or a hemi-endocast with minimal distortion and plasticene reconstruction; B, partial endocast determination as described by Tobias (1971); C, extensive plasticene reconstruction amounting to half of the total endocast; D, determination from regression formulae. X refers to previously published values now confirmed by the author. An evaluation of 1 indicates the highest reliability; 3 the lowest, depending on the completeness of the specimen, distortion, and the author's techniques. An asterisk * refers to estimated adult volume from a juvenile or a child's endocast. The double asterisked items **, confined to the Hadar, Ethiopian *Australopithecus afarensis* materials, are provisional estimates based on current research of the author. The AL 333–105 endocast is severely distorted, mostly incomplete, and that of a young child.

cases of humans born with this condition who nevertheless have been claimed to develop and use 'language' via 'arbitrary symbol systems' though their brain sizes are less than those of some chimpanzees and gorillas (see the references below). True, their intelligence is subnormal, and many do not speak. Nevertheless, this pathology indicates that it may be possible for a brain with less volume than that of some apes to be organized in some distinctly human fashion. (Discussions of this condition have been more extensively treated by Seckel 1960; Lenneberg 1967; Yakovlev 1960; and Holloway 1964, 1966, 1968.)

Given the above, brain size taken alone presents something of a conundrum for interpreting the significance of the large increase in brain size from the fossil record. Intuitively, the more than trebling of brain size (even taking body-size increases into account through allometry) over three million years must have been important; but *where* is the evidence that brain-size increase in small increments is useful in some adaptational sense? Does this mean that one *Homo erectus* with a brain size of 749 ml was at a behavioural disadvantage compared to a sibling with 751 ml? Does it mean that the latter could 'talk' but the former couldn't (assuming that the famed 'Rubicon' *à la* Keith for symbolic language is 750 ml)? One thousand ml divided by three million years is an increase of 0.0003 ml per year, or with a generation span of 20 years, an increase of 0.0066 ml per generation. Thus it takes about 1000 generations to aggregate only about 6 ml of brain-size increase. It can be argued that this example mistakenly mixes analyses of population variability with evolutionary changes; but the point remains that in order to describe evolutionary changes in organs sensibly we must have some knowledge relating structural to functional variability. This is very much lacking for the brain. This is one reason, for example, why Dunbar's (1993) ideas regarding the relationships between group size, social grooming, and communication and neocortical size are difficult to accept. In fact, modern Europeans probably have less neocortex than did large-brained Neanderthals, and Australian Aborigines, famous for social complexity that has challenged many a social anthropologist, have demonstrably less neocortex than caucasians (Klekamp *et al.* 1987) but there is scant sign of any rush to conclude on the basis of this evidence that the Neanderthals must therefore have had a more sophisticated social structure and language than do modern humans.

Nevertheless, as Table 4.1 indicates, it would appear that the australopithecines had ape-sized brains, and retained them for perhaps two million years; and it is not until the first fossil evidence for the genus *Homo* (*H. habilis*, about 1.8 million years ago, using KNM-ER-1470, with about 752 ml in brain volume, for data) that we have evidence for a dramatic increase in brain size.[3]

In my view it is difficult to explain how and why hominid brains evolved after *Homo habilis*: that is, from *Homo erectus* at roughly 1.7 million years ago to the present. In this matter, the brain increases roughly twofold: from about 750 ml in late *Homo habilis* to 1500 ml in late Pleistocene *Homo sapiens sapiens* (for example, Cro-Magnon), while body size increases (or decreases?) only a small amount, if at all. (See Walker and Leakey (1993) for estimates of body size in early *H. erectus*—[eds]) This could appear to rule out simple allometry as a causal explanation.[4] Indeed, both Neanderthals and late Pleistocene specimens have, on the average, larger cranial capacities than the mean for recent *Homo sapiens*. Neanderthal brain volumes on average are slightly larger than those of modern *Homo*.

4.4.1.2 *Encephalization quotients*

Encephalization once meant only that the cerebral cortex had taken on more functions during the course of evolution, and that coritical organization is more specialized in advanced than in primitive animals. A more recently developed meaning given to encephalization is purely quantitative—the encephalization quotient, or EQ. In this latter sense, one is referring to a ratio, where an animal's brain weight is divided by an allometric equation derived from a particular taxon.

The equation below:

$$EQ = \frac{\text{brain weight (grams)}}{0.09908 \times (\text{body weight, grams})^{0.76237}}$$

is an example where the denominator is derived from the allometric equation for 88 species of primates, where the log (base 10) values for brain weights and body weights were regressed together (Martin 1983). In this example, using an average brain weight for *Homo sapiens* of 1330 grams, and a body weight of 65 000 grams, the EQ is 2.87. The chimpanzee and gorilla values, respectively, would be 1.14, and 0.75. If the allometric equation for basal insectivores is used (*per* Stephan *et al.* 1970), the human, chimpanzee, and gorilla values would be 28.8, 11.3, and 6.67 respectively.[5]

Two important points emerge from EQ studies: first, the human animal always has the highest EQ whatever denominator value (or allometric equation) is used. Secondly, the EQ values and their relative values for different species can vary by as much as 20 per cent, and Spearman rank-ordering between species can vary, i.e., the correlations are less than perfect. In other words, there is a definite 'relativity' to relative brain sizes, depending on the taxonomic groups used to generate the denominator (see Holloway and Post (1982) for other examples of this phenomenon).

It is important to underline the fact that when this quantitative approach is used with fossil hominids, the relative closeness to modern *Homo* or other primates (such as the chimpanzee or gorilla) will vary considerably depending on the equation chosen. There is no consensus at present regarding which equations to use—all vertebrates (Jerison 1973), all mammals, all primates, all basal insectivores (Stephan *et al.* 1970, 1981), the Anthropoidea, or the Pongidae (see also Martin 1983 for discussion).

One way around this dilemma is simply to take a '*Homo*centric' approach (since we humans have the highest EQ) and measure other animals' EQs to a standard in which the EQ for modern *Homo sapiens* is regarded as 1.0 or 100 per cent. This is achieved by using the average brain and body weight for the species as one point, and the 'origin' (i.e., 0 brain weight and 0 body weight) as the other point. The denominator then becomes body weight raised to the 0.64906 power, with a constant of 1.0. The advantage of this equation is that all other animal EQs are expressed directly as a percentage of *Homo*. In this case, the chimpanzee is 0.39, and the gorilla, 0.23. The disadvantage, of course, is that this approach is blatantly '*Homo*centric', and thus something of a matter of taste; but then again, so are all other EQ equations, and the purpose is almost always a comparison between ourselves and other animals.

Holloway and Post (1982) provided ten empirical allometric equations and some 20 examples of australopithecine brain and body-weight combinations (from low to heavy) and the resulting EQs, and compared these to modern *Homo* and *Pan*. High brain and low body weights for fossil hominids naturally provide higher EQs as a percentage of modern *Homo*, as compared to *Pan*, but the percentage can vary from about 40 per cent to 50–65 per cent of the modern *Homo* value *depending on the basal equation used.* Obviously, the evolutionary implications for early hominid brain evolution, and the interpretations of these, will vary considerably depending upon the provisional EQ established. All of this assumes, moreover, that we possess accurate estimates of the body weights of our fossil ancestors. This, again, is dubious and highly controversial (see for example McHenry 1975, 1982, 1988, who, however, continues to ignore the relative nature of these EQs). More solid data regarding fossil hominid brain and body weights (see McHenry 1992 for more recent estimates of body weights) and some consensus regarding the most appropriate allometric equations to be used in generating EQ scores are needed before any clear conclusions can be drawn.

4.4.2 *Organization of the brain*

As a palaeoneurologist, I look for *three indications* of brain reorganization. First, can one find a reduction in the relative amount of primary visual striate cortex, usually delimited anteriorally by the *lunate sulcus*? Secondly, can one 'make a case' for the third inferior frontal convolution's being human rather than apelike? Thirdly, does the endocast show evidence of cortical asymmetries that follow the well-known right-frontal, left-occipital torsion petalial patterns (to be discussed below) described by LeMay and her colleagues (LeMay 1976, 1985; LeMay *et al.* 1982)? As the literature and following discussion will attest, the first two questions are highly controversial.

4.4.2.1 *Relative increase in parietal lobe association cortex*

The position of the lunate sulcus is one feature that is very different between ape and human brains (see Fig. 4.5, p. 84). All apes show the lunate sulcus in a relatively forward position; in humans, the position is much more posterior, if it occurs at all (see G. E. Smith 1904*a,b*; Levin 1937; Connolly 1950; Holloway 1983*b, c*). The importance of this fact is as follows: the lunate sulcus in apes is the anterior boundary of primary visual striate cortex, and the posterior boundary of parietal lobe 'association cortex', where major cross-modality integration is performed (see for example Geschwind 1965). (I am referring here to the role of posterior parietal cortex in mediating perceptions of spatial relationships among objects and places, as well as the cross-modal aspects made clear in Geschwind's famous 1965 papers on 'disconnection syndromes', which he claimed were an important foundation for human language). In humans, part of the posterior parietal cortex includes Wernicke's area, an important receptive and integrative zone for decoding and understanding speech. To state the fact that the amount of primary visual striate cortex is relatively reduced in humans is tantamount to saying that the relative amount of parietal lobe association cortex is larger. The comparative evidence for this is indisputable.

The human animal has roughly 121 per cent *less* primary visual striate cortex than would be expected for a primate of its brain size (as calculated from the data of Stephan *et al.* 1981).[6] Thus the correct identification of the lunate sulcus as an anatomical landmark on early hominid endocasts is an essential first step toward discerning any cortical reorganization in early hominid evolution. Consequently, if a hominid endocast (such as that of the Taung or Hadar A.L. 162-28 specimens) were to show the lunate sulcus to be in a posterior position, it would be indicative of some cortical reorganization toward a more human-like pattern as having taken place early in hominid evolution, whatever the size of the brain. Conversely, if the lunate sulcus appears in a relatively anterior pongid-like position, one can infer that cortical reorganization of this area was not present during early hominid evolution: that is, early hominids, such as australopithecines, would be seen as having retained a primitive pongid-like cortical organization of this area of the brain. Obviously, these two alternative possibilities will have an important bearing on how one interprets the behavioural capacities of early hominids and their possible mosaic evolution.

There are two endocast portions that play a predominant role in the discussion of these questions. These are, first, that from the famous Taung specimen, named *Australopithecus africans* and described by Dart in 1925 and 1953 (and by Broom and Schepers 1946; cf. Connolly 1950); and, second, that from the more recently discovered Hadar (Ethiopia) A.L. 162–28 specimen of *Australopithecus afarensis*, the latter appearing to date from between 3.0 and 3.3 million years ago.[7]

My own interpretation (Holloway 1969a, 1972a, 1975a,b, 1981b,c, 1983c, 1984, 1985a, 1992, in press a; Holloway and Kimbel 1986; Holloway and Shapiro 1992) of these is that while early hominid brains were small, and within ape-sized limits, the organization of the cerebral cortex had already altered toward a more human-like pattern, in that there was a relative increase in posterior parietal 'association' cortex, and a relative reduction of primary visual sensory cortex. Falk's (1980b, 1983a,b, 1985a,b, 1986) position is that this reorganizational change is not present in these australopithecines. Tobias (1987, 1991) does not take a position, except to agree with me that one cannot demonstrate where the lunate sulcus is, but only show where it is not—a position I have long made public in my articles (in particular, see Holloway 1985a for a historical review of this matter).

The Hadar AL 162–28 *Australopithecus afarensis* endocast supports further comment, since the posterior end of the intraparietal sulcus is in a posterior position when the endocast is correctly oriented (cf. Falk 1986). Most recently, Holloway and Shapiro (1992) have shown that the squamous suture on all hominids has a relatively high arching pattern compared with that of all pongids. The remnant of the squamous suture on the AL 162–28 cranial fragment shows an arched configuration (Kimbel *et al.* 1982), well above the cranial landmark, the asterion, when the cranial fragment is approximated to a *norma lateralis* orientation. In addition, and most importantly, the distance from the occipital pole to the posterior end of the intraparietal sulcus (Falk and I apparently agree on this landmark) is one half ($c.0.5$) the distance on chimpanzee brain casts, which are usually *smaller* than the endocranial capacity of AL 162–28 (Holloway 1983b; Holloway 1988b; Holloway 1995; Holloway and Kimbel 1986). If there was a lunate sulcus at the position where the posterior part of the intraparietal sulcus was, then it was in a very un-pongid-like position, suggesting that by this early phase in hominid evolution there had already been significant cerebral reorganization, providing an expansion of the posterior parietal cerebral cortex beyond that of apes *prior* to any major expansion of the brain.

Unquestionably, this controversy is far from finished (in particular see Armstrong *et al.* 1991, who have not used the above measurements, and Holloway 1992); but the goals of testable hypotheses still remain well worth the effort, despite the sometimes acrimonious discussions. Needless to say, only more discoveries of better-preserved and more complete crania of early hominids are likely to settle this issue. Finally, in this palaeoneurological context, it is important that more investigators study these specimens (both fossil and extant) and come forth with new suggestions and hypotheses capable of quantitative testing to take us out of the present situation, where there are so few scientists studying these problems within the context of palaeoneurology.

Finally, it is necessary to point out that hominid endocasts subsequent to those known as australopithecine, such as those of *Homo habilis* and *Homo erectus* and late Neanderthal specimens, are unhappily devoid of clear convolutional details, thus making it impossible to be certain about possible reorganizational changes within our own genus.

4.4.2.2 *A more human-like third inferior frontal convolution*

As for the frontal lobe and a Broca's area, none of the australopithecine endocasts show enough morphological convolutional relief to be satisfactory, despite Falk's (1980b, 1983a) claims of a pongid-like fronto-orbital sulcus in Taung. The specimens in question are either damaged in these regions, retain matrix, or have the area missing (for example Taung, STS 60, STS 5). None of the adult Hadar *Australopithecus afarensis* specimens have the anterior portions available.

The exception to this state of affairs is the famous KNM-ER 1470 habiline (see Fig. 4.8), with a cranial capacity of 752 ml and dating at $c.1.9$ million years ago.

The ER 1470 endocast does indicate a somewhat more complex and modern-human-like third inferior frontal convolution containing Broca's area related to motor control of vocalization (Holloway 1976a,

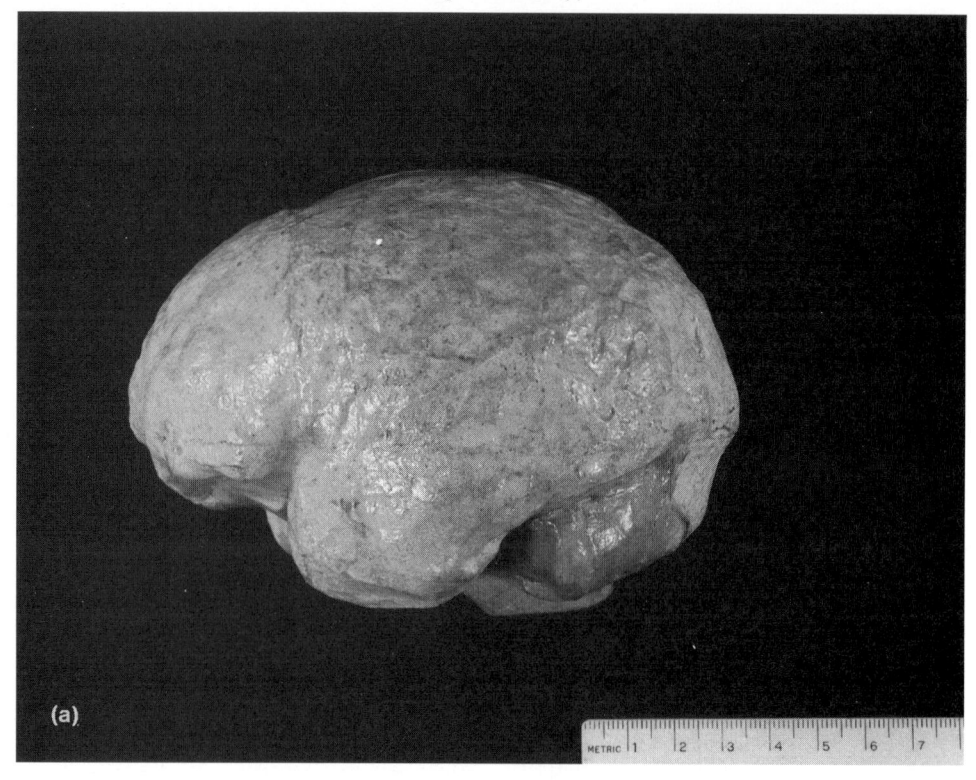

Fig. 4.8 Endocast of KNM-ER 1470.
(a) left lateral view. The third inferior frontal convolution, or Broca's area, appears *Homo*-like.

(b) dorsal view, showing typical *Homo* right-handed petalial pattern of a protruding left occipital lobe and a wider right frontal lobe.

1983b; Falk 1983b). Unfortunately, the posterior parietal region, the anterior occipital zone, and the superior portion of the temporal lobe of fossil endocasts do not provide enough cortical details to prove that a human-like receptive and associative Wernicke's region was present at that time, nor can one assess the position of the lunate sulcus, if indeed there was one.

In the chimpanzee the inferior prefrontal portion of the frontal lobe is structurally homologous to Broca's region in living *humans*, at least cytoarchitectonically, although the degree of macroscopic convolutedness is not as great, and one cannot really parcellate the *Pan* region into strict *Homo* homologues. The same applies to Wernicke's area, although in this case it is not really clear whether the cytoarchitectonic evidence proves homology. As for functions, fortunately no ablation research has been (or is) being published which can bear on the problems of functional homology between *Pan* and *Homo*, and one can only hope that this situation remains the same. Radical neurosurgery on chimpanzees, simply to see if their motoric vocalization (Broca's) suffers or their receptive understanding of sounds (Wernicke's) diminishes, as it does in aphasic humans, would be criminal. It seems safest to conclude that while the human precursors for such specialized tissue may be nascent in *Pan*, they are fully evolved in living *Homo*, and would appear to be present in *Homo habilis*, at about 2.0 million years ago, judging by ER 1470.

Recently, Tobias (1987) has claimed that language began with *Homo habilis*, an opinion based mainly on the ER 1470 endocast that I originally prepared in Nairobi in the mid-1970s. As Leakey (1981) and Leakey and Lewin (1977) indicated, I was then of the opinion that the endocast showed a human third inferior frontal convolution, with a good example of Broca's area (see also Holloway 1976a). I believe Tobias is correct, but this is hardly a new position.[8]

Unfortunately, later hominid endocasts from *H. habilis* to the present seldom show the sulcal and gyral patterns faithfully. It thus becomes impossible to test whether or not there had been further cortical reorganization from *H. habilis* times to the present, for instance in either Broca's or Wernicke's areas.[9]

Broca's and Wernicke's areas are not discrete structures, and are defined better by function than as precise anatomical locations with particular sulcal boundaries. In general, Broca's area in humans includes the third inferior frontal gyrus, including the *pars orbitalis*, *pars triangularis*, and *pars opercularis*. From numerous neurophysiological studies, it is known that this region of the frontal lobe in humans has a close relationship to the motor control of vocalization, and one presumes, to the serialization of motor sequencing. Lesions in this region, particularly on the dominant side (which is the left cerebral cortex in right-handers) often lead to motor aphasia, an inability to articulate language through speech. Wernicke's area is even more difficult to localize in any exact neuroanatomical sense. In general it is in the posterior half of the human brain, behind the sensorimotor cortex, and involves the inferior parietal and superior temporal lobes in the region of the Sylvian fissure. This region appears to be strongly related to receptive functions of speech and language, and lesions in this region (both sides, but it appears more lateralized in males to the left side) produce receptive *aphasias*, an inability to understand spoken and sometimes written language. (These descriptions are oversimplified, and the reader is encouraged to consult a good neuropsychology text—for example Kolb and Whishaw 1985—and Ojemann's (1991) article on the cortical organization of language.)

These two regions can seldom be recognized *unambiguously* on endocasts. Broca's area is somewhat easier to demonstrate than Wernicke's area on endocasts because of the more distinctive human morphology of the third inferior frontal convolution, particularly with its *pars triangularis*. Indeed, the ER 1470 endocast of *Homo habilis* shows a human-like shape and morphology in this region, in contrast to monkeys and apes, which show very little differentiation of this region, as I demonstrated in 1974 to Richard Leakey. Lesions in these zones, however, in both monkeys and apes, can disrupt cognitive processes related to communication. There is some indication that even in macaques (see Dewson 1977) there is some auditory lateralization in what might be roughly analogous to Wernicke's area in humans. Deacon's (1988a) paper on language circuits and homologies in the fibre systems in the brain challenges us all to understand what differs neurologically between ourselves and our pongid cousins.

4.4.2.3 Asymmetries of the brain and laterality

Observations regarding cerebral hemispheric asymmetries extend back into the nineteenth century, but have only been fully corroborated and certified within the last decade.[10] The topic has an enormous literature and is very complicated, being wholly relevant to handedness, hemispheric specialization, left–right brain differences, sexual dimorphism in the corpus callosum and behaviour, symbolic abilities, linguistics, comparative behaviour, and certainly human brain evolution. The neurological literature is very rich, but the comparative and evolutionary records are not, and, as with any neurologically-based subject, are still controversial.

Basically, there is considerable evidence that differing degrees of cognitive competence exist between the cerebral hemispheres with regard to symbol comprehension and manipulation (left hemisphere in the overwhelming majority of right-handers) and visuospatial integration and emotional appreciation of context (right hemisphere) (see especially Witelson 1977, 1982, for further commentary). Of particular interest is Scheibel's (1990) finding that neurons in the left Broca's region tend to have more dendritic branching than on the right side.

Asymmetries in brain structures are *not* limited to the human animal. Asymmetries have been found in chimpanzees (Yeni-Komshian and Benson 1976), monkeys (Dewson 1977; Falk 1978; Falk *et al.* 1986; Heilbroner and Holloway 1988, 1989), and rats, fish, and amphibians (Denneberg 1981; Diamond *et al.* 1981), and have been beautifully demonstrated in certain bird species (see for example Nottebohm 1977). Other anthropologically-oriented investigators such as Marshack (1976) and Jaynes (1976) have relied upon some of the above neurological literature to discuss aspects of human brain and cultural evolution, taking the point of view that true human symbolic behaviour did not emerge until hemispheric specialization took place, or until one has examples of explicit symbolic depictions, such as in cave art, or inscribed artefacts. The fossil hominid endocasts clearly show the kind of asymmetry associated with hemispheric specialization, but are not archeologically associated with any art.

While cerebral (and subcortical) asymmetries appear widespread throughout the animal kingdom, it remains a possibility that the patterns of asymmetries and their quantitative extent could differ in important ways between different species or taxonomic levels. Certainly, handedness, the ability to throw objects with force and accuracy (Holloway 1976*a*, *b*, 1983*a*; see W. H. Hall 1983 and Calvin 1983 for a much more expanded analysis), and systematic tool-making in standardized stylistic forms, as well as the use of arbitrary symbol systems, do appear confined to the human species. Further comparative work with different primate species could well upset this notion. At present, however, there is no reliable evidence for any of these behavioural patterns in other species.

It was really the work of LeMay (1976; see also LeMay *et al.* 1982 and LeMay 1985 for more extensive citations and recent findings) on petalias (asymmetrical projections of occipital and frontal cortex, anterior and posterior, as well as laterally) that opened up the possibility of establishing relationships between external cortical morphology and handedness, and trying to find such patterns in different species and in the fossil record. Basically, right-handers show a left-occipital petalia combined with a torsional right-frontal petalia, while mixed- and left-handers tend to show the opposite configuration. Such asymmetries were shown by LeMay and her colleagues to be strongly correlated with handedness. (The correlations are strong but not 100 per cent, which is why LeMay and her colleagues have emphasized that such petalial configurations are *not totally obligatory*.)

Brain endocasts of apes have been studied for petalial configurations and asymmetries. LeMay *et al.* (1982) and LeMay (1985) continue to find petalial asymmetries in pongids based on fairly small sample sizes, with the gorilla showing the strongest degree of petalial differences between right and left occipital lobes. Holloway and de Lacoste-Lareymondie (1982) studied some 190 endocasts, with chimpanzee and gorilla species numbering roughly 40 each. In general, we found, on the basis of *ordinal* observations, that (1) asymmetries were clearly present in pongids, most strongly in the gorilla; but (2) the combination of left-occipital/right-frontal petalias was fairly rare in pongids, including the gorilla; yet (3) strong in modern humans, and certainly present in early *Homo*. LeMay *et al.* (1982), LeMay (1985), and Geschwind (1984) conclude that pongids show asymmetries in the same direction as *Homo*. Our study suggests that while asymmetries certainly exist in pongids, neither the pattern nor direction is anywhere near as strong as in *Homo*.

It remains, therefore, an intriguing possibility that cerebral cortical asymmetries have evolved independently in several primate species, or are part of a common evolutionary heritage, but are more pronounced in hominids, and particularly modern humans, although weaker asymmetries appear presaged in the earlier and later australopithecines (Holloway 1976*a*, 1983*b*; Holloway and de Lacoste-Lareymondie 1982). The KNM-ER 1470 early *Homo* specimen shows a very strong left-occipital and right-frontal petalial pattern (Holloway and de Lacoste-Lareymondie 1982). In the light of Toth's (1985) marginal evidence for right-handedness based on stone-tool analyses, this correlation provides a tantalizing glimpse offering both a structural and a functional synthesis between some of the palaeoneurological, modern neurobiological, and archaeological lines of evidence. Obviously, such interpretations, which try to link handedness, stone-tool-making, and endocasts, not to say language, together, are highly speculative.[11] Once again, correlations are *not proof* of causal connections, and it would be healthy to remain sceptical of such interpretations. Nevertheless, this is one set of findings, which, with further research, might prove promising in coming to a better understanding of how and when the human brain and its associated behavioural elements evolved.

Thus another morphological aspect, asymmetry between left and right hemispheres, becomes an important focus in studying the human brain. While this question is still being studied, Holloway and de Lacoste-Lareymondie (1982) have suggested that the degree of asymmetry may increase with later hominids—i.e., those later than *H. habilis*, although it is not clear whether the greater amount of asymmetry, which favours the left hemisphere and current models of right-handedness, could be explained allometrically. Other descriptions of some of the fossil hominid endocasts may be found in Holloway 1983c.

In the early Pleistocene East African *Homo erectus* specimens, KNM-ER 3733 and ER 3883, the internal table of bone has been damaged, but each shows a pattern of cerebral asymmetry suggestive of right-handedness, i.e., with left-occipital and right-frontal endocast petalias. The degree of these asymmetries is somewhat less strong than in ER 1470, but certainly within the modern *Homo* range. Both *Homo erectus* and fossil *H. sapiens* have hemispheric asymmetries well within the range of variation for living *Homo sapiens*. Indeed, some of the Upper Palaeolithic specimens of *H. sapiens*, such as Predmostí, show petalial asymmetries at the stronger end of the human distribution, while *Homo erectus* would be at the weaker end. This suggests some possible evolutionary change of *degree* in brain organization between *Homo erectus* and *Homo sapiens sapiens*, although the very small sample sizes of such specimens preclude proof of this assertion.

The australopithecine evidence is very scanty given its fragmentary preservation, seldom of both sides in the same individual. The SK 1585 endocast (Holloway 1972b) does show considerable asymmetry of the occipital petalias, suggesting a right-handed pattern. Earlier australopithecine endocasts that possibly led to the robust line show very small asymmetries, if any (Holloway 1988b).

4.4.2.4 *Towards a synthesis*

Tables 4.2, 4.3, and 4.4 provide a synthesis between the direct evidence of brain endocasts and the reorganizational and size changes that occurred during the evolution of the hominid brain.

There are four major reorganizational changes that have occurred during hominid brain evolution, viz.: (1) reduction of the relative volume of primary visual striate cortex area, with a concomitant relative increase in the volume of posterior parietal cortex, which in humans contains Wernicke's area; (2) reorganization of the frontal lobe, mainly involving the third inferior frontal convolution, which in humans contains Broca's area; (3) the development of strong cerebral asymmetries of a torsional pattern consistent with human right-handedness (left-occipital and right-frontal in conjunction); and (4) refinements in cortical organization to a modern human pattern, most probably involving tertiary convolutions. (This last 'reorganization' is inferred; in fact, there is no direct palaeoneurological evidence for it.)

Integrated with the first three reorganizations are no less than five episodes of brain-size change, all positive except for the last (present?) episode. These changes in size have been regarded as either allometric or not, the criterion being whether or not there has been a significant increase in body size. This is a judgemental process, and these judgements could prove erroneous in reality. For example, the brain-size increase from *A. afarensis* to *A. africanus* is judged to be small, on the basis of only three fragmentary *A. afarensis* endocasts, all of which have required considerable reconstruction, and on little in the way of postcranial remains, in a situation in which sexual dimorphism in body size is likely to have been very high in *A. afarensis*. The increase could of course be a statistical artefact; but it could also be a combination of allometric and non-allometric increments. The fossil record simply does not allow any finer-grained analysis as yet.

Similarly, the increase from *A. africanus* to *Homo habilis* opens the thorny issue of *which* specimens are habiline! Are OH 24, OH 13, OH 16, and OH 7 habilines, or a more advanced species of *Australopithecus*? I frankly do not know; but in this proposed scheme I mean by *Homo habilis* something akin to ER 1470 or OH 7. Depending on which one chooses, the brain-size increase will be differently proportioned between allometric and non-allometric increases.

These details will only emerge with more specimens of a less fragmentary nature. Still, despite these and other problems, I suggest that the evolution of the hominid brain has been a reticulated process of reorganizational and phasic allometric and non-allometric increases in brain size, and that it is highly unlikely that these processes were independent. At least two forms of dependence are likely: (1) reorganization and size-increase could involve each other—i.e., as reorganization of the brain occurs, so does an overall size-increase (see Armstrong *et al*. 1991 and my critique, Holloway 1992); and (2) an increase in size could alter succeeding selection pressures for reorganization and vice versa. I think it is foolhardy to suggest that all selection pressures were of a constant (for example, brain-size and intelligence) nature. Why should the brain-size increase between *A. africanus* and early *Homo* be for the same reasons or involve exactly the same selection pressures as between *Homo*

Table 4.2 Reorganizational changes in the evolution of the human brain (Holloway 1995)

Brain changes (*Reorganization*)	Taxa	Time (m.y.a.)	Evidence
(1) Reduction of primary visual striate cortex (area 17); and relative increase in posterior parietal cortex	*A. afarensis*	3.5 to 3.0	AL 162–28 brain endocast
(2) Reorganization of frontal lobe (third inferior frontal convolution, Broca's area)	*Homo habilis*	2.0 to 1.8	KNM-ER 1470 endocast
(3) Cerebral asymmetries, left occipital, right frontal petalias	? *Homo habilis*	2.0 to 1.8	KNM-ER 1470 endocast
(4) Refinements in cortical organization to a modern *Homo* pattern	? *Homo erectus* to Present ?	1.5 to 0.10	*Homo* endocasts (*erectus, neanderthalensis, sapiens*)

Table 4.3 Brain-size change in human evolution (Holloway 1995)

Brain changes (brain-size related)	Taxa	Time (m.y.a.)	Evidence
(1) Small increase, allometric*	*A. afarensis* to *A. africanus*	3.0 to 2.5	Brain endocasts increase from 400 ml to 450 ml.
(2) Major increase, rapid, both allometric and non-allometric	*A. africanus* to *Homo habilis*	2.5 to 1.8	KNM-1470, 752 ml (*c.*300 ml)
(3) Modest allometric increase in brain size to 800 ml–1000 ml (assumes *habilis* is KNM-ER 1470-like)	*Homo habilis* to *Homo erectus*	1.8 to 0.5	*Homo erectus* brain endocasts and postcranial bones, e.g., KNM-ER 17000
(4) Gradual and modest size increase to archaic *Homo sapiens*, non-allometric	*Homo erectus* to *Homo sapiens neanderthalensis*	0.5 to 0.075	Archaic *Homo* and Neanderthal endocasts 1200 to 1700+ ml
(5) Small reduction in brain size among modern *Homo sapiens*	*Homo s. sapiens*	0.015 to the present	Modern endocranial capacities

* *Note*: Allometric means related to body-size increase.

Table 4.4 Major cortical regions in early hominid evolution

Cortical regions	Brodmann areas	Functions
posterior occipital striate cortex	17	primary visual
posterior parietal and anterior occipital (peri-and parastriate cortex)	18, 19	secondary and tertiary visual; integration with area 17
posterior parietal, superior lobule	5, 7	secondary somatosensory
posterior parietal, inferior lobule (mostly right-side. Left-side processes symbolic–analytical)	39	angular gyrus: perception of spatial relations among objects
posterior parietal, inferior lobule (mostly right-side. See above)	40	supramarginal gyrus: spatial ability
posterior superior temporal cortex	22	Wernicke's area, posterior superior temporal gyrus: comprehension of language
posterior inferior temporal	37	polymodal integration, visual, auditory; perception and memory of objects' qualities
anterior prefrontal cortex	44, 45 (also 8, 9, 10, 46)	Broca's area: motor control of vocalization, language

habilis and *Homo erectus*? In my 1980 paper on within-species variation I questioned the need for a homogeneous explanation for all brain-size increases, and suggested instead that the brain could have increased in size for different reasons at different times. I see no reason to withdraw that suggestion.

Thus this reticulated process is hypothesized not to have involved constant selection pressures for some general 'intelligence' except at particular times; and it is doubtful that these could ever be reconstructed from the fossil and archaeological records.

Finally, as Table 4.4 indicates, a major portion of the reorganizational changes involved posterior parietal and anterior occipital cortex. In humans these regions (Brodmann's 18, 19, 5, 7, 39, 40) are mostly involved with visuo-spatial relationships, while the adjacent area 22, mostly Wernicke's area, is involved with comprehension of communication (language), whereas area 37, the inferior temporal cortex, is involved with polymodal integration and the perception of, and memory of, an object's qualities. In particular, see Holloway (1995).

Each of these are rather specific kinds of intelligence that integrate to become a major adaptive mode for humans: expanding the world in which we live and the materials we utilize to secure an existence in which to reproduce and evolve. If there is one common denominator in all of these complexities, it is (for me, at least) that all relate to social behaviour; hence my insistence over the years that our brains are the product of varying selection pressures for different aspects of social behaviour, of which, incidentally, tool-making is but one small portion, albeit the greatest from the standpoint of the archaeological evidence (Holloway 1981*a*). At this point, perhaps it will now become evident why there is an Appendix I on the sexual dimorphism of the corpus callosum. This structure, after all, particularly in its posterior or splenial portion, is the structure that unites the two posterior parietal cerebral cortices, left and right. I believe that the sexual dimorphism to be described is the result of natural selection's operating to maximize a complemental strategy of male and female behaviours supporting offspring that remain immature for longer periods of time, and thus dependent on a complemental social order requiring some gender based differentiation of behavioural and economic roles. I am not so sure that this differentiation is particularly adaptive now ...

4.5 Conclusion

It is difficult not to admire that Gallic opinion that once relegated such discussions as that on the origins of human language to non-scientific audiences as totally untestable, and proscribed them from open discussion in scholarly gatherings during the nineteenth century. Thus I prefer to take the position that while there is no palaeoneurological evidence that can prove the presence or absence of speech and the use of a symbol system, certain combinations of evidence do increase the probability that either signed or spoken proto-language was an early hominid invention, based on the evolution of the brain from some more primitive hominoid precursor. I doubt that language was present in the australopithecines, but I do believe their brains were organized differently from like-sized ape brains in important ways that relate to visuo-spatial integration and communication, and that they were more social-behaviourally adapted toward a human direction than are the present living apes. I certainly believe that some form of primitive language was present in early *Homo*, and that stone tools made to standardized patterns are the best chances we have of learning about early hominid cognitive behaviour.

To me social behaviour and its evolution was of considerable importance in the evolution of the human brain, and vice versa. For me the two cannot be dissociated from each other, as I have tried to explain in previous publications (for example 1964, 1967, 1970, 1972*a*, 1975*a*, *b*, 1981*a*, 1983*b*, in press *a*). This is, of course, a perspective which weights social behaviours as prime interactive agents in human brain evolution somewhat more heavily than other explanations, such as bipedalism, hunting and/or gathering *per se*, or tool-using and making.[12] I specifically *do not* mean that these were unimportant factors contributing to the totality of pressures which eventually shaped human brains. Rather, our humanness resides mostly in our brains, endowed with symbolic abilities, which have permitted the human animal extraordinary degrees of control over its adaptive powers in both the social and material realms. Conceivably, these powers may eventually themselves become powerless to undo many of the human-generated conditions that terrify us all. Given the resurgence of 'ethnic cleansing' (Bosnia), the killing fields of Cambodia, the omnipresent anti-Semitism and xenophobia (to mention but a few horrors), and all the disastrous environmental waste and destruction still manifested by human-kind, despite the grim lessons of the Second World War and more than ample evidence regarding ecology, I feel rather pessimistic about the future of human evolution. These things are products of human brains and social structures *in concert*. More optimistically perhaps, it is also conceivable that a better understanding of how we became what we are, and how our brains and behaviour were shaped during millions of years of evolution, would allow our socially-derived brains to control those very inventions (conceived and designed through the use of symbols) which so threaten ourselves and all other forms of life on this planet. First, we must cope with ourselves...

Appendix: Sexual dimorphism and the brain

Associated with the cerebral asymmetries discussed above in Section 4.4.2.3 is growing evidence that the brains of many animals are sexually dimorphic, and that structures other than those related to reproductive functions (for example, the preoptic nucleus of the hypothalamus) *might* also be sexually dimorphic. Sexual dimorphism in brain structures related to reproduction is not surprising when one considers the vast amount of variation in the sexual behaviour and courtship patterns that exists among sexually reproducing animals. Nevertheless, the possibility that the human species *might* evidence inherent sexual dimorphism in their brains that is unrelated to reproduction (see Swaab and Fliers (1985) on the preoptic nucleus), with associated cognitive differences (at least in *degree*) between females and males, is regarded as a heresy in current anthropological (and social-science) circles.

While some nineteenth-century and early twentieth-century neuroanatomists were deeply concerned with the topic, and claimed many differences in cerebral morphology and behavioural functioning (see both Mall 1909 and Papez 1927 for excellent reviews of some of these early works), the topic is extraordinarily controversial. There has been a discernible shift in the last decade that recognizes some dimorphism between male and female brains (beyond brain weight, which is well known, but functionally empty—see for example Holloway 1980), and it has become an almost-respected line of inquiry today, in contrast to the situation just a decade ago, as described for example in McGlone's fine 1980 review. Indeed, an entire volume of *Progress in Brain Research*, edited by DeVries *et al.* (1984) is devoted to the topic.[13]

We have found significant differences in the midline area of the corpus callosum and particularly in the posterior splenial portion of this structure (de Lacoste-Utamsing and Holloway 1982; Holloway and de Lacoste 1986, Holloway *et al.*, 1993.). It is the corpus callosum which interconnects the two cerebral hemispheres, and the splenial portion carries fibres which interconnect the posterior portions of the cerebral cortex, particularly the parietal lobes. Our research indicates that in females the average total area of the corpus callosum is roughly the same or slightly smaller than in males, but not significantly so. Relatively, however, i.e., correcting for brain weights (significantly higher in males), the corpus callosal area is statistically significantly larger in females. The same applies, but more strongly so, to the posterior portion, the splenium, as measured by its dorsal–ventral width. The two early papers (1982, 1986) were on small, but independent samples ($n=16$ in the latter reference), and we did find that the *absolute* area of the corpus callosum was larger in females, a finding not replicated in a recent study (Holloway *et al.* 1993). These results, on a sample of over 100 brains, roughly equally divided between males and females, show statistically strong differences in the *relative* size of the corpus callosum and the splenium in favour of females. Baack *et al.* (1982) and, recently, de Lacoste *et al.* (1986) have shown that these differences are apparent by the age of 26 weeks prenatal. Papez (1927), in his classic work on the brain of Helen Gardener, found differences in a small sample of males and females; but unfortunately the necessary statistical methods were not used for in-depth analyses of those differences. It must be remembered that these are average relative differences (differences of sample means), and while statistically significant, are based on small sample sizes. There is certainly overlap in the values, both absolute and relative. Only larger samples will be adequate for accurately assessing the degree of overlap. Earlier, Witelson (1985) had argued that her data did not show sex differences; rather, she claimed the difference was related to handedness. Our analysis of her data, however, leaves her finding exciting but doubtful (Holloway and de Lacoste 1986). More recently Witelson suggests that there is a sex difference in the so-called isthmus of the splenium (Witelson 1989).

Since we wrote our replication study (Holloway and de Lacoste 1986) a number of papers have appeared attempting to prove that there is no significant sexual dimorphism in the human corpus callosum (see below). Reviews of this literature may be found in Holloway (1990), Holloway *et al.* (1993), Peters (1988), and Clarke *et al.* (1989). These last authors, while finding partial support for sexual dimorphism in human corpora callosa, have failed to consider brain weight correctly in either their post-mortem (autopsied) samples or those from MRI (Magnetic Resonance Imaging). Unfortunately, at the present time it is not feasible to obtain accurate brain volumes from MRI materials, although de Lacoste (pers. comm.) has found some linear measurements that correlated with brain size to roughly $r = 0.94$.

More recently, some additional support for our original findings has been published by Allen *et al.* (1991) and Steinmetz *et al.* (1992, 1995), particularly with regard to the splenial portion of the corpus callosum's being larger and more bulbous in females.

It is interesting to consider some of the reports claiming no sexual dimorphism in the corpus callosum. For example, Weber and Weis (1986) used a sample with an average age of 74.7 years, and a male average brain weight of 1029 grams and 890 grams for females. This is a rather high average age, and the brain weights are obviously on the low side, reflecting the known fact that brain weight decreases with age. The standard deviations for the callosal measures were roughly 1/4 the value of the mean, which is very high, and yet the average values for the corpus callosal area and splenium were roughly equal between the two sexes. No attempt was made to correct for brain size, despite the large average difference between male and female brain weights. MRI studies by Bleier et al. (1986), Byne et al. (1988), Kertesz et al. (1987), Oppenheim et al. (1987), Weis et al. (1988), and Yoshii et al. (1986), do *not* provide brain-size data, but find that the values between male and female measurements on the corpus callosum are roughly equal. Demeter et al. (1988) claim from autopsied material that there is no significant sexual dimorphism, even though splenial area is identical between males and females while male brain sizes go as high as 1700 cc, and their Fig. 4 (p. 222) shows great dimorphism in brain size, with practically no overlap between 22 males and 12 females. Yet they adamantly dismiss brain size as relevant in their studies!

All of the above papers have ignored the point that my colleague and I attempted to make, which is that *it is the relative size of the corpus callosum and its splenial portion that is dimorphic between human females and males*. In fact, many of these reports suggest that if brain size were correctly taken into account, the relative size of the corpus callosum would be larger in females. (In particular, see Appendix I in Holloway et al. 1993) Unfortunately, most of the studies were done using MRI techniques, which do not allow for easy assessment of brain size.

The dimorphism of the corpus callosum *contrasts* remarkably with all other regions of the brain (Holloway et al. 1993). For example, if one takes the cerebellum, hippocampus, pulvinar, putamen, pallidum, ventricles, or amygdala and compares their volumes between human males and females, one finds that there is a significant size difference favouring males. If one looks at brain size, there is also a significant difference favouring males. If one then corrects for brain size, the ratios of these structures divided by brain volumes *show no significant differences between males and females*. This pattern is in complete contrast to the corpus callosum, where the female sizes either equal or are larger than in males, and the relative sizes are statistically significantly larger in females.

It is interesting to recall the monumental task that Maccoby and Jacklin (1974) undertook in exposing so much of the current and past psychological literature on sex differences in behaviour related to symbolic and language skills and visuo-spatial integration, exposing sexual biases and poor experimental design. Even so, they were left with an essential core of differences that they believed could not simply be explained away as cultural artefacts. Those 'residua' were repeated findings about 'rough and tumble play', mathematical and visuo-spatial integration, and language (symbolic competence) abilities, in which males on average consistently scored higher in the first two components, and females in the third. Hall's (1984) book, reviewing sex differences in non-verbal behaviour, also found consistent average differences in the same directions. It would appear from the above-mentioned studies on the corpus callosum that a neuroanatomical basis might exist, in so far as these data suggest that males are perhaps on average more lateralized and asymmetrical than females with regard to the posterior part of the cerebral cortex. With regard to the frontal lobe, however, it is possible that females may be on average more lateralized than males for motor control of vocalization, according to recent research published by Kimura (1980, 1983, 1992) and Kimura and Harshman (1984).

Using an experimental technique of measuring sex differences in test responses to visual images, either words or faces, Pollack (1987) has demonstrated that human females score on average fewer errors and with shorter reaction times than males. Her models and tests were based on a callosal model to test our previous anatomical findings (de Lacoste-Utamsing and Holloway 1982). As her samples' sizes were 25 of each gender, and many of the results were statistically significant between the two, male–female differences should be consistently tested cross-culturally.

If such results prove replicable, we must eventually face the questions of whether there is some true causal relationship between brain structure and functioning, and then of how these dimorphic brain differences came about. As these differences in the corpus callosum appear by the age of 26 weeks prenatal, this suggests that purely cultural influences can be ruled out. Furthermore, common sense, as well as everyday experience (not to mention many scientific studies) indicates that *even if* such differences exist, they do not necessarily become biologically-determined fates for individual members of the two genders. With training and equal access to opportunities, whatever inherent differences might exist would appear to be highly malleable. My interest in this issue, however, is not the present or future, but rather the past evolution of hominid brains and behaviour.

It is my own conviction (Holloway 1983*c*, Holloway 1990, Holloway *et al.* 1993) that these differences are evolutionarily-derived residua, based on past selection pressures for female precocity in neural and behavioural development, and upon a complemental social and cognitive structuring of intellectual and sensorimotor tasks that were related to a stronger degree of sexual division of labour than currently exists. Obviously, *palaeoneurology cannot provide any evidence relating to these speculations*. Given that at some time in past hominid evolution there necessarily occurred an increased period of social and material nurturance of longer-growing offspring with increased dependency times, I believe a complemental behavioural adaptation between males and females was necessary to support such a change successfully. Perhaps the palaeoanthropological record could help in this regard, if it were more complete.[14] In other words, understanding human evolution really requires an analysis that includes the differential cognitive abilities of males and females that complement each other and together enabled the species to be more successful in its social-behavioural adaptations, particularly those associated with the nurturance and education of offspring with prolonged learning periods gained through a delay in maturation. In this model males are regarded as having (on average) superior visuo-spatial integrative abilities (including orientation in space relative to distant food and water resources), and these would have complemented the female's superior abilities at social communication and the nurturance of children.

I also believe that the sexual dimorphism in the human corpus callosum may be species-specific, as there appears to be no solid evidence from prosimians and monkeys (either Old or New World) of any dimorphism in this structure (Heilbronet and Holloway 1989). An earlier report by de Lacoste and Woodward (1988) regarding a dimorphism in the Great Apes is based on averaging small numbers of each sex, and needs replication, particularly on a large sample of chimpanzees. These data are, for the time being, almost impossible to get.

Notes

1. Most neurobiologists speak of endocranial volume and brain size as the same things. Brain size, as measured by its volume, is very close to endocranial volume, as the specific gravity of brain tissue is essentially 1.0. Brain *weight*, however, is always somewhat less than brain *volume*, but the difference is seldom more than 10 per cent, which is roughly the amount of the meninges which surround the brain, and which are often included in brain volume. In this chapter, I am taking the liberty, except where noted, of using these terms interchangeably.

2. A case in point is a recent publication by Falk *et al.* (1989), which claims to have traced the sulci on the Taung infant brain endocast, and, by comparing the lengths of the sulci with a formula established by Jerison (1982) based on comparative (adult animals) work by Elias and Schwartz (1969), to conclude that Taung definitely had an ape-like brain. Unfortunately, there are no agreements as to which sulci and gyri are actually on the Taung endocast (see for example Connolly 1950; Holloway 1981*b*, 1983*b*, 1984, 1985*a*, 1988*b*).

3. The majority of palaeoanthropologists would consider specimens KNM-ER 1813 and 1805 to be smaller-brained versions of *Homo habilis*. (Falk (1982) appears to be alone in considering ER 1805 an australopithecine.) But both of these specimens, while relatively complete, are very puzzling to those emphasizing the importance of brain size to taxonomy. It is far from clear whether these are *H. habilis* or perhaps an advanced version of *Australopithecus*, given their relatively small cranial capacities. See Wood (1992) for a recent discussion of these hominids and the possible diversity of species of early *Homo*. (Wood follows Groves (1989) in placing ER 1470 in the species *H. rudolfensis*—[eds])

4. Although it is possible to suggest that as the tool industries become more complex from *H. habilis* to the late Pleistocene, so does the brain, it remains conceptually difficult causally to connect brain-size increase with technological advances solely related to tool-making and use. In essence, the relationship between behavioural and brain complexity (as measured by size) is a correlational, rather than causal one. After all, stone-tool-making is but one subset of the totality of social behaviour.

5. The equations used for the Primates (or indeed, any other taxon) will depend on the number of species used and the values selected to represent brain and body weights. In the example provided above for 88 primate species drawn from the list in Stephan *et al.* (1970) list (see also Bauchot and Stephan 1969), *Homo* was excluded. If *Homo* is included the equation becomes EQ = brain weight (grams) ÷ (0.09 × body weight (grams), exp. 0.77531). The correlation coefficient for the first equation is 0.97062, while for the latter equation it is 0.96863. Whether *Homo* is included or not, the correlation between the $\log_{10}$ of brain weight and $\log_{10}$ body weight is very high. This

example is included to give some idea of how one point out of 88, i.e. *Homo*, can affect the equations. It should be noted that the exponent for the power equation between brain and body weights is of the order of 0.76–0.77, and not the value of 0.66 which Jerison (1973) claimed. Jerison's (1973) formula was not an empirical one based on the actual data for the primates, but rather a perceived 0.66 slope was put through an array of mammalian data (Jerison's 'polygon') to satisfy Jerison's (1973) preconceived notions regarding a 0.66 (2/3) scaling factor. However, some physical anthropologists, such as McHenry (1988) and Tobias (1987), continue to use Jerison's (1973) equation of EQ = brain weight (grams) × 0.12 body weight (grams), exp. 0.66 (see Holloway 1988*a* for additional comments). Given the incorrectness of Jerison's (1973) equation for mammals and for the primates in particular, it is erroneous that Tobias (1971, 1987, 1991) continues to calculate 'extra neurons' from Jerison's equations. As I have indicated elsewhere (Holloway 1966, 1968, 1974), these numbers are fictitious, whether they occur in Jerison (1973) or in Tobias's works. These numbers are calculated from a formula which doubles the exponent of 0.66, and is biased toward higher body weights.

6. This was shown by Holloway (1976*a*, 1979) on four independent samples, confirming Passingham's (1982) and Passingham and Ettlinger's (1973) observations. (Passingham *et al.* (1986) does not provide the percentage deviations of humans from a pongid base-line, and interprets *reorganization* somewhat differently.)

7. The Taung specimen includes a fossil endocast portion. Unfortunately, none of the other natural endocasts of *Australopithecus africanus*, such as STS 60, or the robust australopithecines, such as SK 1585, provide any confirming data, as these portions are either missing, as is the case with STS 60, or not visible, as in SK 1585 (see also Holloway 1972*a, b* [eds]). This area is also not visible on the constructed endocast of STS 5 (Ms. Ples.).

8. Tobias's claims regarding a *Homo*-like brain and language functions based on endocasts of *Homo habilis* fossil specimens differ from those of previous authors (for example Holloway or Falk) in being too adamant, and in including fossil specimens which are *extraordinarily incomplete* or *fragmented*, such as OH 16, OH 24, etc. These simply cannot be assigned to any taxon without controversy. Neither are they supporting evidence for strong *Homo*-like petalial asymmetries in purported habilines (*pace* Tobias 1987). OH 16 has been likened to a few bits of bone floating in sea of plaster. To describe petalias on such fragmentary specimens is probably quite erroneous. The OH 7 parietals were found flattened in Bed I at Olduvai Gorge. How one can claim a slightly larger parietal petalia on one side, as does Tobias, is beyond this writer. Additionally, OH 24 was found crushed in five layers before Dr Ron Clarke's excellent reconstructions of the crania. Any petalias found on this specimen are similarly suspect.

9. For example, the degree of replication of convolutions in Neanderthal specimens is simply horrible, making fine-grained statements about mental functioning in these hominids impossible. Still, there is nothing striking about Neanderthal brain casts in comparison to those of more recent *Homo sapiens* (for example, Cro-Magnon) suggesting any significant evolutionary change. (See also Holloway 1985*b*.)

10. For reviews of this literature, and more recent analyses, please refer to Corballis (1983); Damasio and Geschwind (1984); Helige (1983); Strauss *et al.* (1983); Witelson (1982); Young (1983); Kimura (1983); Kimura and Harshman (1984); and Geschwind and Galaburda (1984). More recent papers by some of the above authors may be found in Glick (1985). Kolb and Whishaw (1985) provide an excellent text for introducing the subject, and many of these topics can also be found in Kandel and Schwartz's (1981) textbook.

11. Also along these lines, Frost (1980) has argued that in hominid evolution lateralized representation was an evolutionary consequence of the requirement for asymmetrical employment of the forelimbs in the making and using of tools. The colateralization of language mechanisms was held to be a consequence of the coupling of these to the motoric mechanisms already lateralized at an earlier point in hominid evolution [eds].

12. In particular, the increased postnatal dependency period and the increased duration of growth appear to me as extraordinary evolutionary events, as they appear to imply an energetic shift toward nurturance that would require a more complemental set of behavioural patterns between the sexes. This implies a cohesive and co-operative adaptation within groups. To increase brain size threefold beyond pongid levels must surely have required socially cohesive adaptations.

13. For other reviews of this general topic, see Kimura 1980, 1983; Kimura and Harshman 1984; Kinsbourne 1978; McGlone 1980, with associated peer commentary; Witelson 1982; Arnold and Gorski 1984; J. A. Hall 1984; and Khan and Cataio 1984 for a wealth of references on human sexual dimorphism in general, but particularly the brain; for sexual differentiation of the brain, see Toran-Allerand 1986; Juraska 1986; and Kelly 1981 in Kandel and Schwartz. For a speculative account and an attempted

synthesis with the fossil record and hominid evolution, see Holloway 1983c, 1990, and Holloway et al., in press).

14. I think it is worth stressing that considerable caution must be exercised regarding these preliminary findings. No one knows exactly what these dimorphisms mean in the modern context, and there is still much vital ethnographic and anatomical evidence to be collected regarding sexual dimorphism in brain and behaviour. I frankly do not know what the practical significance of such findings is within Western societies. Furthermore, the presence of an anatomical difference need not necessarily prove a functional difference. In this case, however, there is an overwhelming amount of clinical evidence suggesting that these differences in the corpus callosum might form a possible anatomical substrate to help explain such behavioural differences. This, in part, should explain my preference for regarding these differences as evolutionarily derived, and as another example of reorganization of the brain, involving important internal changes without necessarily creating an increase in brain size.

References

Aiello, L. C. and Dunbar, R. I. M. (1993). Neocortex size, group size, and the evolution of language. *Current Anthropology*, **34**, 184–94.
Allen, L. S., Richey, M. F., Chai, Y. M., and Gorski, R. A. (1991). Sex differences in the corpus callosum of the living human being. *Journal of Neuroscience*, **11**, 933–42.
Allman, J. (1990). Evolution of neocortex. In *Cerebral cortex*, Vol. 8A, (ed. E. G. Jones and A. Peters,) pp. 269–83. Plenum, New York.
Allman, J. and McGuinness, E. (1988). Visual cortex in primates. In *Neurosciences, Comparative primate biology*, Vol. 4 (ed. H. D. Steklis and J. Erwin), pp. 279–336. A. R. Liss, New York.
Armstrong, E. (1983). Relative brain size and metabolism in mammals. *Science*, **220**, 1302–4.
Armstrong, E. (1985). Enlarged limbic structures in the human brain: the anterior thalamus and medial mamillary body. *Brain Research*, **362**, 394–7.
Armstrong, E. (1990). Evolution of the brain. In *The human nervous system* (ed. G. Paxinos), pp. 1–16. Academic Press, New York.
Armstrong, E. and Falk, D. (eds.) (1982). *Primate brain evolution: methods and concepts*. Plenum, New York.
Armstrong, E., Zilles, K., Kurtis, M., and Schleicher, A. (1991). Cortical folding, the lunate sulcus and the evolution of the human brain. *Journal of Human Evolution*, **20**, 341–8.
Andrensen, N. C., Flaum, M., Swayze II, V., O'Leary, D. S., Alliger, R., Cohen, G., Ehrhardt, N., and Yuh, W. T. C. (1993). Intelligence and brain structure in normal individuals. *American Journal of Psychiatry*, **150**, 130–4.
Arnold, A. P. and Gorski, R. A. (1984). Gonadal steroid induction of structural sex differences in the central nervous system. *Annual Review of Neuroscience*, **7**, 413–42.
Baack, J., de Lacoste-Utamsing, M. C. and Woodward, D. J. (1982). Sexual dimorphism in human fetal corpora callosa. *Neuroscience Abstract*, **8**, 18.
Bauchot, R. and Stephan, H. (1969). Encéphalisation et niveau évolutif chez les simiens. *Mammalia*, **33**, 225–75.
Bleier, R., Houston, L. and Byne, W. (1986). Can the corpus callosum predict gender, age, handedness, or cognitive differences? *Trends in Neurosciences*, **9**, 391–4.
Blumenberg, B. (1983). The evolution of the advanced hominid brain. *Current Anthropology*, **24**, 589–624.
Blumenberg, B. (1986). Population characteristics of extinct hominid endocranial volume. *American Journal of Physical Anthropology*, **68**, 269–79.
Brodmann, K. (1909). *Vergleichende Lokalizationzlehre der Grosshirnrinde*. J. A. Barth, Leipzig.
Broom, R. and Schepers, C. W. H. (1946). The South African fossil ape-men: the *Australopithecinae*. *Transvaal Museum Memoirs*, **2**, 1–272.
Byne, W., Bleier, R., and Houston, L. (1988). Variations in human corpus callosum do not predict gender: a study using magnetic resonance imaging. *Behavioral Neuroscience*, **102**, 222–7.
Calvin, W. (1983). A stone's throw and its launch window: timing precision and its implications for language and hominid brains. *Journal of Theoretical Biology*, **104**, 121–35.
Chien, S. (1981). Cerebral circulation and metabolism. In *Principles of neural science* (ed. E. R. Kandel and J. H. Schwartz), pp. 660–6. North-Holland Elsevier, Amsterdam.
Clarke, S., Kraftsik, R., Van der Loos, H., and Innocenti, G. M. (1989). Forms and measure of adult and developing human corpus callosum: is there sexual dimorphism? *Journal of Comparative Neurology*, **280**, 213–30.
Clutton-Brock, T. H. and Harvey, P. H. (1980). Primates, brains, and ecology. *Journal of Zoology* (London), **190**, 309–23.
Connolly, C. J. (1950). *The external morphology of the primate brain*. C. C. Thomas, Springfield, Illinois.
Corballis, M. C. (1983). *Human laterality*. Academic Press, New York.
Cronin, J. E., Boaz, N. T., Stringer, C. B., and Rak, Y. (1981). Tempo and mode in hominid evolution. *Nature*, **292**, 113–22.
Damasio, A. R. and Geschwind, N. (1984). The neural basis for language. *Annual Review of Neuroscience*, **7**, 127–47.
Dart, R. (1925). *Australopithecus africanus*: the man-ape of South Africa. *Nature*, **115**, 195–9.
Dart, R. (1953). The relationship of brain size and brain pattern to human status. *South African Journal of Medical Science*, **21**, 23–45.
Deacon, T. W. (1988a). Human brain evolution. 1. Evolution of language circuits. In *Intelligence and evolutionary*

biology, NATO series, ASI, Series G, Vol. 17 (ed. H. J. and I. Jerison), pp. 363–82. Springer-Verlag, Berlin.

Deacon, T. W. (1988*b*). Human brain evolution. 2. Embryology and brain allometry. In *Intelligence and evolutionary biology*, NATO series, ASI, Series G, Vol. 17 (ed. H. J. and I. Jerison), pp. 383–416. Springer-Verlag, Berlin.

Deacon, T. W. (1990*a*). Fallacies of progression in theories of brain-size evolution. *International Journal of Primatology*, **11**, 193–236.

Deacon, T. W. (1990*b*). Problems of ontogeny and phylogeny in brain-size evolution. *International Journal of Primatology*, **11**, 237–82.

Dekaban, A. S. and Sadowsky, D. (1978). Changes in brain weights during span of human life: relation of brain weights to body heights and body weights. *Annals of Neurology*, **4**, 345–56.

de Lacoste-Utamsing, M. C. and Holloway, R. L. (1982). Sexual dimorphism in the corpus callosum. *Science*, **216**, 1431–2.

de Lacoste, M. C., Holloway, R. L., and Woodward, D. J. (1986). Sex differences in the fetal human corpus callosum. *Human Neurobiology*, **5**, 93–6.

de Lacoste, M. C. and Woodward, D. (1988). The corpus callosum in nonhuman primates: determinates of size. *Brain, Behavior, Evolution*, **31**, 318–23.

Demeter, S., Ringo, J. L., and Doty, R. W. (1988). Morphometric analysis of the human corpus callosum and anterior commissure. *Human Neurobiology*, **6**, 219–26.

Denneberg, V. H. (1981). Hemispheric laterality in animals and the effects of early experience. *Behavioral and Brain Sciences*, **4**, 1–49.

DeVries, G. J., DeBruin, R. M., Uylings, H. M. B., and Corner, M. A. (eds) (1984). *Sex differences in the brain*, Progress in Brain Research, Vol. 61. Elsevier, Amsterdam.

Dewson, J. H. (1977). Preliminary evidence of hemispheric asymmetry of auditory function in monkeys. In *Lateralization in the nervous system* (ed. S. Harnad, R. W. Doty, L. Goldstein, J. Jaynes, and G. Kranthamer), pp. 63–71. Academic Press, New York.

Diamond, M. C., Dowling, G. A., and Johnson, R. E., (1981). Morphologic cerebral cortical asymmetry in male and female rats. *Experimental Neurology*, **71**, 261–8.

Dunbar, R. I. M. (1992). Neocortex size as a constraint on group size in primates. *Journal of Human Evolution*, **20**, 469–93.

Dunbar, R. I. M. (1993). Co-evolution of neocortex size, group size and language in humans. *Behavioral and Brain Sciences*, **16**, 681–734.

Eldredge, N. and Gould, S. J. (1972). Punctuated equilibria: an alternative to phyletic gradualism. In *Models in paleobiology*, (ed. T. J. M. Schopf), pp. 82–115. Freeman, Cooper, San Francisco.

Eldredge, N. and Tattersall, I. (1982). *The myths of evolution*. Columbia University Press, New York.

Elias, N. and Schwartz, D. (1969). Surface areas of the cerebral cortex of mammals determined by stereological methods. *Science*, **166**, 11–13.

Falk, D. (1978). Cerebral asymmetry in Old World monkeys. *Acta Anatomica*, **101**, 334–9.

Falk, D. (1980*a*). Hominid brain evolution: the approach from paleoneurology. *Yearbook of Physical Anthropology*, **23**, 93–107.

Falk, D. (1980*b*). A reanalysis of the South African Australopithecine natural endocasts. *American Journal of Physical Anthropology*, **53**, 525–39.

Falk, D. (1982). Primate neuroanatomy: an evolutionary perspective. In *A history of American physical anthropology, 1930–1980* (ed. F. Spencer), pp. 75–103. Academic Press, New York.

Falk, D. (1983*a*). The Taung endocast. A reply to Holloway. *American Journal of Physical Anthropology*, **60**, 479–90.

Falk, D. (1983*b*). Cerebral cortices of East African early hominids. *Science*, **222**, 1072–4.

Falk, D. (1985*a*). Hadar AL 162–28 endocast as evidence that brain enlargement preceeded cortical reorganization in hominid evolution. *Nature*, **313**, 45–7.

Falk, D. (1985*b*). Apples, oranges, and the lunate sulcus. *American Journal of Physical Anthropology*, **67**, 313–15.

Falk, D. (1986). Reply to Holloway and Kimbel. *Nature*, **321**, 536–7.

Falk, D. (1987). Brain lateralization in primates and its evolution in hominids. *Yearbook of Physical Anthropology*, **30**, 107–25.

Falk, D., Cheverud, J., Vannier, M. W., and Conroy, C. G. (1986). Advanced computer graphics technology reveals cortical asymmetry in endocasts of rhesus monkeys. *Folia Primatologica*, **46**, 98–103.

Falk, D., Hildebolt, C., and Vannier, M. W. (1989). Reassessment of the Taung early hominid from a neurological perspective. *Journal of Human Evolution*, **18**, 485–92.

Frahm, H. D., Stephan, H., and Stephan, M. (1982). Comparison of brain structure volumes in Insectivora and Primates: I. Neocortex. *Journal für Hirnforschung*, **23**, 375–89.

Frahm, H. D., Stephan, H., and Baron, G. (1984). Comparison of brain structure volumes in Insectivora and Primates. V. Area striata. *Journal für Hirnforschung*, **25**, 537–57.

Frost, G. T. (1980). Tool behavior and the origins of laterality. *Journal of Human Evolution*, **9**, 447–59.

Geschwind, N. (1965). "Disconnexion" syndromes in animals and man. *Brain*, **88**, 237–94, 585–644.

Geschwind, N. (1984). Historical introduction. In *Cerebral dominance: the biological foundations* (eds. N. Geschwind and A. M. Galaburda), pp. 1–8. Harvard University Press, Cambridge, Mass.

Geschwind, N. and Galaburda, A. M. (eds) (1984). *Cerebral dominance: the biological foundations*. Harvard University Press, Cambridge, Mass.

Gibson, K. R. (1990). New perspectives on instincts and intelligence: brain size and the emergence of hierarchical mental construction skills. In *Language and intelligence in monkeys and apes* (ed. S. T. Parker and K. R. Gibson), pp. 97–128. Cambridge University Press.

Gibson, K. R. and Ingold, T. (eds.) (1993). *Tools, language and cognition in human evolution.* Cambridge University Press.

Glick, S. D. (ed.) (1985). *Cerebral lateralization in non-human species.* Academic Press, New York.

Gould, S. J. (1977). *Ontogeny and phylogeny.* Bellknap Press, Cambridge, Mass.

Groves, C. P. (1989). *A theory of human and primate evolution.* Clarendon Press, Oxford.

Hall, J. A. (1984). *Nonverbal sex differences.* Johns Hopkins University Press, Baltimore.

Hall, W. H. (1983). Did throwing stones shape hominid brain evolution? *Ethology and Sociobiology*, **3**, 115–24.

Harvey, P. H. and Clutton-Brock, T. H. (1985). Life history variation in primates. *Evolution*, **39**, 559–81.

Heilbroner, P. and Holloway, R. L. (1988). Anatomical brain asymmetries in New World and Old World monkeys: stage of temporal lobe development in primate evolution. *American Journal of Physical Anthropology*, **76**, 39–48.

Heilbroner, P. and Holloway, R. L. (1989). Anatomical brain asymmetry in monkeys: frontal, parietal, and limbic cortex in *Macaca*. *American Journal of Physical Anthropology*, **80**, 203–11.

Helige, J. B. (ed.) (1983). *Cerebral hemisphere asymmetry.* Praeger, New York.

Hofman, I. (1982). Encephalization in mammals in relationship to the size of the cerebral cortex. *Brain, Behavior and Evolution*, **20**, 84–96.

Hofman, I. (1983). Evolution of brain size in neonatal and adult placental mammals: a theoretical approach. *Journal of Theoretical Biology*, **105**, 317–32.

Holloway, R. L. (1964). Some aspects of quantitative relations in the primate brain. Unpublished Ph.D. dissertation. University of California, Berkeley.

Holloway, R. L. (1966). Cranial capacity, neural reorganization and hominid evolution: a search for more suitable parameters. *American Anthropologist*, **68**, 103–21.

Holloway, R. L. (1967). The evolution of the human brain: some notes toward a synthesis between neural structure and the evolution of complex behavior. *General Systems*, **12**, 3–19.

Holloway, R. L. (1968). The evolution of the primate brain: some aspects of quantitative relationships. *Brain Research*, **7**, 121–72.

Holloway, R. L. (1969a). Culture: a human domain. *Current Anthropology*, **10**, 359–412.

Holloway, R. L. (1969b). Some questions on parameters of neural evolution in primates. *Annals of the New York Academy of Sciences*, **167**, 332–40.

Holloway, R. L. (1970). Neural parameters, hunting and the evolution of the human brain. In *Advances in primatology*: Vol. 1 (ed. C. R. Noback and W. Y. Montagna), pp. 299–309. Appleton-Century-Crofts, New York.

Holloway, R. L. (1972a). Australopithecine endocasts, brain evolution in the Hominoidea and a model of hominid evolution. In *The functional and evolutionary biology of primates* (ed. R. Tuttle), pp. 185–204. Aldine Press, Chicago.

Holloway, R. L. (1972b). New Australopithecine endocast, SK1585, from Swartkrans, S. Africa. *American Journal of Physical Anthropology*, **37**, 173–86.

Holloway, R. L. (1974). On the meaning of brain size. A review of H. J. Jerison's 1973 *Evolution of the brain and intelligence. Science*, **184**, 677–9.

Holloway, R. L. (1975a). *The role of human social behavior in the evolution of the brain*, 43rd James Arthur Lecture at American Museum of Natural History (1973). American Museum of Natural History, New York.

Holloway, R. L. (1975b). Early hominid endocasts: volumes, morphology and significance. In *Primate functional morphology and evolution* (ed. R. Tuttle), pp. 393–416. Mouton, The Hague.

Holloway, R. L. (1976a). Paleoneurological evidence for language origins. In *Origins and evolution of language and speech. Annals of the New York Academy of Sciences*, **280**, 330–48.

Holloway, R. L. (1976b). Some problems of hominid brain endocast reconstruction, allometry, and neural reorganization. In *Colloquium VI of the IX Congress of the USPP Nice, 1976 Congress.* Prétirage, pp. 69–119.

Holloway, R. L. (1979). Brain size, allometry, and reorganization: toward a synthesis. In *Development and evolution of brain size: behavioral implications* (ed. M. E. Hahn, C. Jensen, and B. C. Dudek), pp. 59–88. Academic Press, New York.

Holloway, R. L. (1980). Within-species brain–body weight variability: a re-examination of the Danish data and other primate species. *American Journal of Physical Anthropology*, **53**, 109–21.

Holloway, R. L. (1981a). Cultural symbols and brain evolution: a synthesis. *Dialectical Anthropology*, **5**, 287–303.

Holloway, R. L. (1981b). Revisiting the S. African Australopithecine endocasts: results of stereoplotting the lunate sulcus. *American Journal of Physical Anthropology*, **56**, 43–58.

Holloway, R. L. (1981c). Exploring the dorsal surface of hominid brain endocasts by stereoplotter and discriminant analysis. *Philosophical Transactions of the Royal Society of London*, **B 292**, 155–66.

Holloway, R. L. (1983a). Human brain evolution: a search for units, models and synthesis. *Canadian Journal of Anthropology*, **3**, 215–32.

Holloway, R. L. (1983b). Cerebral brain endocast pattern of *A. afarensis* hominid. *Nature*, **303**, 420–2.

Holloway, R. L. (1983c). Human paleontological evidence relevant to language behavior. *Human Neurobiology*, **2**, 105–14.

Holloway, R. L. (1984). The Taung endocast and the lunate sulcus: a rejection of the hypothesis of its anterior position. *American Journal of Physical Anthropology*, **64**, 285–8.

Holloway, R. L. (1985a). The past, present, and future significance of the lunate sulcus in early hominid evolution. In *Hominid evolution: past, present, and future* (ed. P. V. Tobias), pp. 47–62. A. R. Liss, New York.

Holloway, R. L. (1985b). The poor brain of *Homo sapiens neanderthalensis*: see what you please... In *Ancestors: the hard evidence* (ed. E. Delson), pp. 319–24. A. R. Liss, New York.

Holloway, R. L. (1988a). The brain. In *Encyclopedia of human evolution and prehistory* (ed. I. Tattersall, E. Delson, and J. Van Couvering), pp. 98–105. Garland, New York.

Holloway, R. L. (1988b). 'Robust' Australopithecine brain endocasts: some preliminary observations. In *Evolutionary history of the "robust" Australopithecines* (ed. F. E. Grine), pp. 97–105. Aldine–de Gruyter, New York.

Holloway, R. L. (1990). Sexual dimorphism in the human corpus callosum: its evolutionary and clinical implication. In *From apes to angels: essays in anthropology in honor of Philip V. Tobias* (ed. G. Sperber), pp. 221–8. Wiley–Liss, New York.

Holloway, R. L. (1992). The failure of the gyrification index (GI) to account for volumetric reorganization in the evolution of the human brain. *Journal of Human Evolution*, **22**, 163–70.

Holloway, R. L. (1993). Another primate brain fiction: brain (cortex) weight and homogeneity. *Behavioral and Brain Sciences*, **16**, 707–8.

Holloway, R. L. (1995). Toward a synthetic theory of human brain evolution. In *Origins of the human brain* (ed. J. P. Changeur and J. Chavaillon), pp. 42–55. Clarendon Press, Oxford.

Holloway, R. L. and de Lacoste-Lareymondie, M. C. (1982). Brain endocast asymmetry in pongids and hominids: some preliminary findings on the paleontology of cerebral dominance. *American Journal of Physical Anthropology*, **58**, 101–10.

Holloway, R. L. and de Lacoste, M. C. (1986). Sexual dimorphism in the human corpus callosum: an extension and replication study. *Human Neurobiology*, **5**, 87–91.

Holloway, R. L. and Heilbroner, P. (1992). Corpus callosum in sexually dimorphic and nondimorphic primates. *American Journal of Physical Anthropology*, **87**, 349–57.

Holloway, R. L. and Kimbel, W. H. (1986). Endocast morphology of Hadar hominid AL 162-28. *Nature*, **321**, 538.

Holloway, R. L., and Post, D. (1982). The relativity of relative brain measures and hominid evolution. In *Primate brain evolution: methods and concepts* (ed. E. Armstrong and D. Falk), pp. 57–76. Plenum, New York.

Holloway, R. L. and Shapiro, J. S. (1992). Relationship of squamosal suture to asterion in pongids (*Pan*): relevance to early hominid brain evolution. *American Journal of Physical Anthropology*, **89**, 275–82.

Holloway, R. L., Anderson, P. J., Defidini, R. and Harper, C. (1993). Sexual dimorphism of the human corpus callosum from three independent samples: relative size of the corpus callosum. *American Journal of Physical Anthropology*, **92**, 481–98.

Jaynes, J. (1976). The evolution of language in the late Pleistocene. *Annals of the New York Academy of Sciences*, **280**, 312–25.

Jerison, H. J. (1973). *Evolution of brain and intelligence*. Academic Press, New York.

Jerison, H. J. (1982). Allometry, brain size, cortical surface, and convolutedness. In *Primate brain evolution: methods and concepts* (ed. E. Armstrong and D. Falk), pp. 71–84. Plenum, New York.

Jolicoeur, P., Pirlot, P., Baron, G. and Stephan, H. (1984). Brain structure and correlation patterns in Insectivores, Chiroptera, and Primates. *Systematic Zoology*, **33**, 14–29.

Juraska, J. M. (1986). Sex differences in developmental plasticity of behavior and the brain. In *Developmental neuropsychobiology* (ed. W. T. Greenhough and J. M. Juraska), pp. 409–22. Academic Press, New York.

Kaas, J. and Huerta, M. F. (1988). The subcortical visual system of primates. In *Neurosciences, Comparative Primate Biology, Vol. 4* (ed. H. D. Steklis and J. Erwin), pp. 327–92. A. R. Liss, New York.

Kaas, J. and Pons, T. P. (1988). The somatosensory system of primates. In *Neurosciences, Comparative Primate Biology, Vol 4* (ed. H. D. Steklis and J. Erwin), pp. 421–68. A. R. Liss, New York.

Kandel, E. R. and Schwartz, J. H. (eds) (1981). *Principles of neural science*. North Holland–Elsevier, Amsterdam.

Keith, A. (1948). *A new theory of human evolution*. Watts, London.

Kelly, D. (1981). Sexual differentiation of the nervous system. In *Principles of neural science* (ed. E. R. Kandel and J. H. Schwartz), pp. 533–46. Elsevier, Amsterdam.

Kertesz, A., Polk, M., Howell, J., and Black, S. E. (1987). Cerebral dominance, sex, and callosal size on M.R.I. *Neurology*, **37**, 1385–8.

Khan, A. U. and Cataio, J. (1984). *Men and women in biological perspective: a review of the literature*. Praeger, New York.

Kimbel, W. H., Johannson, D. C., and Coppens, Y. (1982). Pliocene hominid cranial remains from the Hadar Formation, Ethiopia. *American Journal of Physical Anthropology*, **57**, 453–500.

Kimura, D. (1980). Sex differences in intra-hemispheric organization of speech. *Behavioral and Brain Sciences*, **3**, 240–1.

Kimura, D. (1983). Sex differences in cerebral oraganization of speech and praxic functions. *Canadian Journal of Psychology*, **37**, 19–35.

Kimura, D. (1992). Sex differences in the brain. *Scientific American*, **267**, 118–25.

Kimura, D. and Harshman, R. A. (1984). Sex differences in brain organization for verbal and non-verbal functions. *Progress in Brain Research*, **61**, 423–41.

Kinsbourne, M. (ed.) (1978). *Asymmetrical function of the brain*. Cambridge University Press.

Klekamp, G., Reidel, A., Harper, C., and Kretschmann, H.-J. (1987). A quantitative study of Australian Aborigine and Caucasian brains. *Journal of Anatomy*, 150–210.

Kolb, K. and Whishaw, I. Q. (1985). *Fundamentals of human neurophysiology*. Freeman, New York.

Leakey, R. E. (1981). *The making of mankind*. Dutton, New York.

Leakey, R. E. and Lewin, R. (1977). *Origins*. Dutton, New York.

LeMay, M. (1976). Morphological cerebral asymmetries of modern man, fossil man, and nonhuman primates. *Annals of the New York Academy of Sciences*, **280**, 349–66.

LeMay, M. (1985). Asymmetries of the brains and skulls of nonhuman primates. In *Cerebral lateralization in nonhuman species* (ed. S. D. Glick), pp. 233–45. Academic Press, Orlando.

LeMay, M., Billig, M. S. and Geschwind, N. (1982). In *Primate brain evolution: methods and concepts* (ed. E. Armstrong and D. Falk), pp. 263–78. Plenum Press, New York.

Lennenberg, E. H. (1967). *Biological foundations of language*. Wiley, New York.

Leutenegger, W. (1982). Encephalization and obstetrics in primates with particular reference to human evolution. In *Primate brain evolution: methods and concepts* (ed. E. Armstrong and D. Falk), pp. 85–95. Plenum, New York.

Leutenegger, W. (1987). Neonatal brain size and neurocranial dimensions in Pliocene hominids: implications for obstetrics. *Journal of Human Evolution*, **16**, 291–6.

Levin, G. (1937). Racial and "inferiority" characters in the human brain. *American Journal of Physical Anthropology*, **22**, 345–80.

Little, B. B. (1989). Gestation length, metabolic rate, and body and brain weights in primates: epigenetic effects. *American Journal of Physical Anthropology*, **80**, 213–18.

Maccoby, E. E. and Jacklin, C. N. (1974). *The psychology of sex differences*. Stanford University Press.

McGlone, J. (1980). Sex differences in human brain asymmetry: a critical review. *Behavioral and Brain Sciences*, **3**, 215–63.

McHenry, H. M. (1975). Fossils and the mosaic theory of human evolution. *Science*, **190**, 425–31.

McHenry, H. M. (1982). The pattern of human evolution: studies on bipedalism, mastication, and encephalization. *Annual Review of Anthropology*, **11**, 151–73.

McHenry, H. M. (1988). New estimates of body weight in early hominids and their significance to encephalization and megadontia in 'robust' australopithecines. In *Evolutionary history of the 'robust' austrolopithecines* (ed. F. E. Grine), pp. 133–48. Aldine–de Gruyter, New York.

McHenry, H. M. (1992). Body size and proportions in early hominids. *American Journal of Physical Anthropology*, **87**, 407–31.

Mall, F. P. (1909). On several anatomical characters of the numan brain, said to vary according to race and sex. *American Journal of Anatomy*, **9**, 1–32.

Marshack, A. (1976). Some implications of the Paleolithic symbolic evidence for the origin of language. *Annals of the New York Academy of Sciences*, **280**, 289–311.

Martin, R. D. (1981). Relative brain size and metabolic rate in terrestrial vertebrates. *Nature*, **293**, 57–60.

Martin, R. D. (1982). Allometric approaches to the evolution of the primate nervous system. In *Primate brain evolution: methods and concepts* (ed. E. Armstrong and D. Falk), pp. 39–56. Plenum, New York.

Martin, R. D. (1983). *Human evolution in an ecological context*, James Arthur Lecture (1982). American Museum of Natural History, New York.

Matano, S., Stephan, H., and Baron, G. (1985a). Volume comparisons in the cerebellar complex of primates. I. Ventral pons. *Folia Primatologica*, **44**, 177–81.

Matano, S., Baron, G., Stephan, H., and Frahm, H. D. (1985b). Volume comparisons in the cerebellar complex of primates. II. Cerebellar nuclei. *Folia Primatologica*, **44**, 183–203.

Milton, K. (1981). Distribution patterns of tropical plant foods as an evolutionary stimulus to primate mental development. *American Anthropologist*, **83**, 534–48.

Milton, K. (1993). Diet and primate evolution. *Scientific American*, **269**, 86–93.

Mountcastle, V. B. (1978). An organizing principle for cerebral function: the unit module and the distributed system. In *The mindful brain* (ed. G. M. Edelman and V. B. Mountcastle), pp. 7–50. MIT Press, Cambridge, Mass.

Mundinger, P. (1980). Animal cultures and a general theory of cultural evolution. *Ethology and Sociobiology*, **1**, 183–223.

Nottebohm, F. (1977). Asymmetries in neural control of vocalization in the canary. In *Lateralization of the nervous system* (ed. S. Harnad, R. Doty, L. Goldstein, J. Jaynes, and G. Krauthamer), pp. 23–44. Academic Press, New York.

Ojemann, G. A. (1991). Cortical organization of language. *Journal of Neuroscience*, **11**, 2281–7.

Oppenheim, J. S., Lee, B. C. P., Nass, R., and Gazzaniga, M. S. (1987). No sex-related differences in human corpus callosum based on magnetic resonance imagery. *Annals of Neurology*, **21**, 604–6.

Pandya, D. P. and Yeterian, E. H. (1990). Architecture and connections of cerebral cortex: implications for brain evolution and function. In *Neurobiology of cognitive function*, No. 29, UCLA Forum in Medical Science (ed. A. B. Scheibel and A. F. Wechsler), pp. 53–83. Guilford Press, New York.

Pandya, D. P., Seltzer, B., and Barbas, H. (1988). Input–output organization of the primate cerebral cortex. In *Neurosciences, Comparative Primate Biology, vol. 4* (ed. H. D. Steklis and J. Erwin), pp. 39–80. A. R. Liss, New York.

Papez, J. W. (1927). The brain of Helen H. Gardener. *American Journal of Physical Anthropology*, **11**, 29–88.

Parker, S. T. (1990). Why big brains are so rare: energy costs of intelligence and brain size in anthropoid primates. In *Language and intelligence in monkeys and apes* (ed. S. T. Parker and K. R. Gibson), pp. 129–54. Cambridge University Press.

Parker, S. T. and Gibson, K. R. (1979). A model of the evolution of language and intelligence in early hominids. *Behavioral and Brain Sciences*, **2**, 367–407.

Parker, S. T. and Gibson, K. R. (eds) (1990). *Language and intelligence in monkeys and apes*. Cambridge University Press.

Passingham, R. E. (1982). *The human primate*. Freeman, Oxford.
Passingham R. E. and Ettlinger, G. (1973). A comparison of cortical function in man and other primates. *International Review of Neurobiology*, **16**, 233–99.
Passingham, R. E., Heywood, C. A., and Nixon, P. D. (1986). Reorganization in the human brain as illustrated by the thalamus. *Brain, Behavior, and Evolution*, **29**, 68–76.
Peters, M. (1988). The size of the corpus callosum in males and females: implications of a lack of allometry. *Canadian Journal of Psychology*, **42**(3), 313–24.
Pollack, J. M. (1987). Sex differences in the visual functions of the human corpus callosum. Unpublished Ph.D. dissertation, Columbia University, New York.
Radinsky, L. B. (1972). Endocasts and studies of primate brain evolution. In *The functional and evolutionary biology of primates* (ed. R. Tuttle), pp. 175–84. Aldine, Chicago.
Radinsky, L. B. (1975). Primate brain evolution. *American Scientist*, **63**, 656–63.
Radinsky, L. B. (1977). Early primate brains: fact and fiction. *Journal of Human Evolution*, **6**, 79–86.
Radinsky, L. B. (1979). *The fossil record of primate brain evolution*, James Arthur Lecture. American Museum of Natural History, New York.
Sacher, G. A. (1982). The role of brain maturation in the evolution of the primates. In *Primate brain evolution: methods and concepts* (ed. E. Armstrong and D. Falk), pp. 97–112. Plenum, New York.
Sacher, G. A. and Staffeldt, E. F. (1974). Relation of gestation time to brain weight for placental mammals: implications for a theory of vertebrate growth. *American Naturalist*, **108**, 593–615.
Sawaguchi, T. (1988). Correlations of cerebral indices for "extra" cortical parts and ecological variables in Primates. *Brain, Behavior and Evolution*, **32**, 129–40.
Sawaguchi, T. (1990). Relative brain size, stratification and social structure in Anthropoids. *Primates*, **31**, 257–72.
Sawaguchi, T. and Kudo, H. (1990). Neocortical development and social structure in primates. *Primates*, **31**, 283–90.
Scheibel, A. B. (1990). Dendritic correlates of higher cognitive function. In *Neurobiology of cognitive function*, Number 29, UCLA Forum in Medical Science (ed. A. B. Scheibel and A. F. Wechsler), pp. 239–68. Guilford Press, New York.
Schultz, A. H. (1962). Die Schädelkapazität männlicher Gorillas und ihr Hochstwert. *Anthropologie Anzeiger*, **25**, 197–203.
Seckel, H. P. G. (1960). *Birdheaded dwarfs: studies in developmental anthropology*. Karger, New York.
Shea, B. T. (1983). Phyletic size change and brain/body allometry: a consideration based on the African pongids and other primates. *International Journal of Primatology*, **4**, 33–62.
Smith, G. E. (1904a). Studies in the morphology of the human brain with special reference to that of the Egyptians. No. 1. The occipital region. *Records. Egyptian Government School of Medicine*, **2**, 725.
Smith, G. E. (1904b). The morphology of the occipital region of the cerebral hemispheres in man and apes. *Anatomischer Anzeiger* **24**, 436.
Smith, R. J. (1980). Rethinking allometry. *Journal of Theoretical Biology*, **87**, 98–112.
Steinmetz, H., Janke, L., Kleinschmidt, A., Schlaug, M. D., Volkmann, J., and Huang, Y. (1992). Sex but no hand difference in the isthmus of the corpus callosum. *Neurology*, **42**, 749–52.
Steinmetz, H., Staiger, J. F., Schlaug, G., Huang, Y., and Jancke, J. (1995). Corpus callosum and brain volume in men and women. *Neuroreport*, **6**, 1002–4.
Steklis, H. D. and Erwin, J. (eds) (1988). *Neurosciences, Comparative Primate Biology, Vol. 4*. A. R. Liss, New York.
Stephan, H., Bauchot, R., and Andy, O. J. (1970). Data on size of the brain and various brain parts in insectivores and primates. In *The primate brain* (ed. C. R. Noback and W. Montagna), pp. 289–97. Appleton-Century-Crofts, New York.
Stephan, H., Frahm, H., and Baron, G. (1981). New and revised data on volumes of brain structures in Insectivores and Primates. *Folia Primatologica*, **35**, 1–29.
Strauss, E., Kosaka, B., and Wada, J. (1983). The neurobiological basis of lateralized cerebral function. A review. *Human Neurobiology*, **2**, 115–28.
Swaab, D. F. and Fliers, E. (1985). A sexually dimorphic nucleus in the human brain. *Science*, **228**, 1112–15.
Szentagothai, J. (1978). The neuron network of the cerebral cortex: a functional interpretation. The Ferrier Lecture, 1977. *Proceedings of the Royal Society of London*, Series B, **201**, 219–48.
Tobias, P. V. (1971). *The brain in hominid evolution*. Columbia University Press, New York.
Tobias, P. V. (1987). The brain of *Homo habilis*: a new level of organization in cerebral evolution. *Journal of Human Evolution*, **16**, 741–62.
Tobias, P. V. (1991). *Olduvai Gorge: The skulls, endocasts, and teeth of* Homo habilis, Vol. 4. Cambridge University Press.
Toran-Allerand, C. D. (1986). Sexual differentiation of the brain. In *Developmental neuropsychobiology*, (ed. W. T. Greenhough and J. M. Juraska), pp. 175–211. Academic Press, New York.
Toth, N. (1985). Archaeological evidence for preferential right-handedness in the lower and middle Pleistocene, and its possible implications. *Journal of Human Evolution*, **14**, 607–14.
Uylings, H. B. M. and van Eden, C. G. (1990). Qualitative and quantitative comparison of the prefrontal cortex in rats and primates, including humans. *Progress in Brain Research*, **85**, 31–62.
Walker, A. and Leakey, R. (1993). *The Nariokotome* Homo erectus *skeleton*. Harvard University Press, Cambridge, Mass.
Washburn, S. L. (1960). Tools and human evolution. *Scientific American*, **203**, 63–75.
Weber, G. and Weis, S. (1986). Morphometric analysis of the human corpus collosum fails to reveal sex-related differences. *Journal für Hirnforschung*, **27**, 237–40.

Weis, S., Weber, G., Weneger, E., and Kimbacher, M. (1988). The human corpus callosum and the controversy about sexual dimorphism. *Psychobiology*, **16**, (4), 411–15.

White, T. D. (1980). Evolutionary implications of Pliocene hominid footprints. *Science*, **208**, 175–6.

Witelson, S. F. (1977). Anatomic asymmetry in the temporal lobes: its documentation, phylogenesis, and relationship to functional asymmetry. *Annals of the New York Academy of Sciences*, **299**, 328–54.

Witelson, S. F. (1982). Bumps on the brain: right–left asymmetry in brain anatomy and function. In *Language functions and brain organization* (ed. S. Segalowitz), pp. 117–43. Academic Press, New York.

Witelson, S. F. (1985). The brain connection: the corpus callosum is larger in left-handers. *Science*, **229**, 665–8.

Witelson, S. F. (1989). Hand and sex differences in the isthmus and genu of the human corpus callosum: a postmortem morphological study. *Brain*, **112**, 799–835.

Wood, B. (1992). Origin and evolution of the genus *Homoe Nature*, **355**, 783–90.

Yakovlev, P. I. (1960). Anatomy of the human brain and the problem of mental retardation. In *Proceedings of the First International Medical Conference* (ed. P. W. Bownman and H. V. Mautner), pp. 1–43. Grune and Stratton, New York.

Yeni-Komshian, G. H. and Benson, D. A. (1976). Anatomical study of cerebral asymmetry in the temporal lobe of humans, chimpanzees, and rhesus monkeys. *Science*, **192**, 387–9.

Yin, T. C. T. and Medjbeur, S. (1988). Cortical association areas and visual attention. In *Neurosciences, comparative primate biology, Vol. 4* (ed. H. D. Steklis and J. Erwin), pp. 393–422. A. R. Liss, New York.

Yoshii, F., Barker, W. Apicella, J. Chang, J. Sheldon, J. and Duara, R. (1986). Measurements of the corpus callosum (CC) on magnetic resonance (MR) scans: effects of age, sex, handedness, and disease. *Neurology*, **36** (Suppl. 1), 133.

Young, A. W. (ed.) (1983). *Functions of the right cerebral hemisphere*. Academic Press, New York.

Editorial appendix I: Endocranial volumes

In addition to the figures given in Table 4.1, there have been numerous published determinations of the endocranial volumes of many of the fossil specimens by a number of workers. Table 4.5 is probably not exhaustive, but lists those we have located. Table 4.6 gives summary statistics of the measures in Table 4.5. Table 4.7 is from Tobias (1987), and samples a different, though occasionally overlapping, population of specimens. Table 4.8a and b, the former by Holloway, the latter from Tobias (1971b), give comparative figures for extant hominoids. Given the material and methods involved in determining these volumes, we can at least conclude that hominid brain capacities have increased in the human evolutionary line, while noting (1) brain–body ratios need to be taken into account (see the discussion of encephalization quotients Section 4.4.1.2 above); (2) extant *Homo sapiens sapiens* figures appear to show smaller capacities than Neanderthals; and (3) exact capacities determined by any one researcher are more informative when viewed in the context of figures generated by several investigators over a period of years [eds].

Table 4.5 Hominid endocranial volumes [eds]

Taxon	Specimen	Country/Region	Endocranial volume (cm^3)	Reference
A. afarensis	AL 162–28	Ethiopia	350–400	Falk (1985)
			c.400	Holloway (1983a, b)
			375–400	Holloway (this chapter)[1]
	AL 333–45	Ethiopia	485–500	Holloway (1983b)
			485	Holloway (this chapter)[1]
	AL 333–105[2]	Ethiopia	352[3]	Falk (1985)
			343[3]	Falk (1987)
			310–320	Holloway (this chapter)[1]
			310–320; c.400[3]	Holloway (1983b)
A. africanus	MLD 1	S.A.	c.500	Holloway (1973)
	MLD 37/38	S.A.	435	Holloway (1970, 1973)
			480	Tobias (1971a)[4]
	STS 5	S.A.	485	Holloway (1970, 1973, 1981a)
			480	Tobias (1971a)[4]
	STS 19/58	S.A.	436	Holloway (1970, 1973)
			530	Tobias (1971a)[4]
	STS 60	S.A.	428	Holloway (1970, 1973)
			435	Tobias (1971a)[4]

Table 4.5 (*Contd.*)

Taxon	Specimen	Country/Region	Endocranial volume (cm^3)	Reference
A. africanus (cont.)	STS 71	S.A.	428	Holloway (1970, 1973)
			480–520	Tobias (1971a)[4]
	TAUNG[2]	S.A.	412[3]	Falk (1987)
			405; 440[3]	Holloway (1970)
			500; 540[3]	Tobias (1971a)[4]
A. robustus	SK 1585	S.A.	475	C. K. Brain (cited by Tobias 1971a)
			530	Holloway (1970, 1973)
A. boisei	OH 5	Tanzania	530	Holloway (1970, 1973)
			522	Holloway (1975)
			530	Tobias (1971a)
	L388y-6[2]	Ethiopia	427; 448[3]	Holloway (1981d)[5]
	ER 406	Kenya	c.510	Holloway (1973)
			525	Holloway (1983b)
	ER 407	Kenya	506	Falk and Kasinga (1983)
			510	Holloway (1983b)
	ER 732	Kenya	500	Holloway (1973)
	ER 13750	Kenya	450–480	Holloway (1988)
			c.530	Leakey and Walker (1988)
	WT 17000	Kenya	410	Walker *et al.* (1986)[6]
	WT 17400[2]	Kenya	390–400	Holloway (1988)
Homo habilis?	ER 1805	Kenya	582	Holloway (1978)
	ER 1813	Kenya	509	Holloway (1978)
H. habilis	OH 7[2]	Tanzania	687	Holloway (1978)
			700–750	Holloway (1980a)
			657; 684[3]	Tobias (1971a)
			647; 674[3]	Tobias (1987, 1991)
			690	Vaisnys *et al.* (1984)
			580–600	Wolpoff (1981)[7]
	OH 13[2]	Tanzania	650	Holloway (1973)
			639; 652[3]	Tobias (1971a, 1975)
			673[3]	Tobias (1987, 1991)
	OH 16[2]	Tanzania	620; 633[3]	Tobias (1971a, 1975)
			625; 638[3]	Tobias (1987, 1991)
	OH 24	Tanzania	590	Holloway (1973)
			597	Tobias (1975)
	ER 1470	Kenya	752	Holloway (1978)
	ER 1590[2]	Kenya	810	Blumenberg (1985)[8]
H. erectus	ER 3733	Kenya	848	Holloway (1983b)
	ER 3883	Kenya	804	Holloway (1983b)
	WT 15000[2]	Kenya	880; 909[3]	Begun and Walker (1993)
	OH 9	Tanzania	1067	Holloway (1975)
			1000[9]	Tobias (1971a)
	OH 12	Tanzania	727	Holloway (1978)
			750	Holloway (1980a)
	TRINIL 2 (Pith.I)	Java	940	Holloway (1981c)
			953	Holloway (this chapter)
			850	Tobias (1967)
			935	Weinert (cited by Weidenreich 1943)
	SANGIRAN 2 (Pith.II)	Java	815	Boule and Vallois (1957)
			813	Holloway (1981c)
			815	Holloway (this chapter)
			775	Weidenreich (1943)
	SANGIRAN 4 (Pith.IV)	Java	908	Holloway (1981c)
			900	Holloway (this chapter)
			750	von Koenigswald (1962)
			c.880	Weidenreich (1943)

Table 4.5 (*Contd.*)

Taxon	Specimen	Country/Region	Endocranial volume (cm^3)	Reference
H. erectus (cont.)	SANGIRAN 10 (Pith.V, also Skull 6)	Java	855 975 975	Holloway (1981*c*) Jacob (1966) von Koenigswald (1962)
	SANGIRAN 12 (Pith.VII)	Java	1059 915	Holloway (1981*c*) Tobias (1971*a*)[4]
	SANGIRAN 17 (Pith.VIII)	Java	1004 1029	Holloway (1981*c*) Sartono (1971)
	SOLO I (Ngandong)	Java	1143 1172 1158 1035	Dubois (1937) Holloway (1980*b*) Oppenoorth (1937) Weidenreich (1943)
	SOLO V (Ngandong)	Java	1284 1251 1316[10] 1255	Dubois (1937) Holloway (1980*b*) Oppenoorth (1937) Weidenreich (1943)
	SOLO VI (Ngandong)	Java	1087 1013 1189[10] 1040 1035	Dubois (1937) Holloway (1980*b*) Oppenoorth (1937) Schaefer (1963) Weidenreich (1943)
	SOLO IX (Ngandong)	Java	1135	Weidenreich (1943)
	SOLO X (Ngandong)	Java	1231 1055[9]	Holloway (1980*b*) Weidenreich (1943)
	SOLO XI (Ngandong)	Java	1090 1095 1060	Holloway (1980*b*) Schaefer (1963) Weidenreich (1943)
	LANTIAN	China	780	Woo (1966)
	ZHOUKOUDIAN II	China	1030	Weidenreich (1943)
	ZHOUKOUDIAN III[2]	China	915	Weidenreich (1943)
	ZHOUKOUDIAN X	China	1225	Weidenreich (1943)
	ZHOUKOUDIAN XI	China	1015	Weidenreich (1943)
	ZHOUKOUDIAN XII	China	1030	Weidenreich (1943)
	SALÉ[11]	Morocco	880	Holloway (1981*b*)
Archaic *H. sapiens*	BROKEN HILL (Kabwe)	Zambia	1280[9] 1325	Day (1986) Weidenreich (1943)
	FLORISBAD	S.A.	>1280?	Singer (cited by Beaumont *et al.* 1978)
	SALDANHA	S.A.	1200[9]–1250	Drennan (1953)
	LAETOLI, LH18	Tanzania	1200	Day *et al.* (1980)
	NDUTU	Tanzania	1070–1120	Holloway (pers. comm., cited by Rightmire 1983)
	OMO II	Ethiopia	1430 1435± 20	Day (1972) Day (1986)
	ARAGO XXI–XLVII	Europe	1100–1200	Holloway (cited by Day 1986)
	PETRALONA	Europe	1220 1230 1190–1210	Poulianos (cited by Day 1986) Protsch (pers. comm. cited by Stringer 1984) Stringer *et al.* (1979)
	STEINHEIM	Europe	1150–1175 1070	Howell (1960) Weinert (cited by Day 1986)
	SWANSCOMBE	Europe	1250–1300 1325	Howell (1960) Swanscombe Committee (cited by Day 1986)
	VÉRTESSZÖLLÖS II	Europe	± 1300 1115–1437	Thoma (1981) Wolpoff (1977)
	DALI	China	1120 1200	Wu (1981) Wu (pers. comm., cited by Day 1986)

Table 4.5 (*Contd.*)

Taxon	Specimen	Country/Region	Endocranial volume (cm^3)	Reference
H. sapiens neanderthalensis	LA CHAPELLE	Europe	1626	Boule (cited by Holloway 1981*b*)[12]
	LA FERRASSIE I	Europe	1689* 1641	Heim (1976) Boule (cited by Holloway 1981*b*)[12]
	LA QUINA 5	Europe	1350	Boule (cited by Holloway 1981*b*)[12]
	LE MOUSTIER	Europe	1565	Olivier and Tissier (1975)
	MONTE CIRCEO I	Europe	1550	Olivier and Tissier (1975)
	NEANDERTAL	Europe	1525	Boule (cited by Holloway 1981*b*)[12]
	SACCOPASTORE I	Europe	1245	Olivier and Tissier (1975)
	SPY I	Europe	1305	Holloway (1981*b*)
	SPY II	Europe	1553	Holloway (1981*b*)
	AMUD I	Near East	1740	Ogawa *et al.* (1970)
	SHANIDAR I	Near East	1600	Stewart (1977)
	TABŪNIN I	Near East	1271	McCown and Keith (1939)
	JEBEL IRHOUD 1[13]	Morocco	1305	Holloway (1981*b*)
	JEBEL IRHOUD 2[13]	Morocco	1450	Olivier and Tissier (1975)
Early *H. sapiens sapiens*	JEBEL QAFZEH VI	Near East	1568[10]	Vallois and Vandermeersch (cited by Day 1986) Early *H. sapiens*
	SKHŪL I[2]	Near East	1150;1450[3]	McCown and Keith (1939)
	SKHŪL IV	Near East	1554[9]	McCown and Keith (1939)
	SKHŪL V	Near East	1450–1518	McCown and Keith (1939)
	SKHŪL IX	Near East	1587	McCown and Keith (1939)

[1] Provisional estimates from current research, see Table 4.1, in this chapter.

[2] Immature individual.

[3] Estimated adult volume from immature individual.

[4] These values from Tobias (1971*a*) are based on published figures of earlier authors.

[5] Holloway (1981*d*) has questioned the assignment of this specimen to *A. boisei*.

[6] WT 17000 has been reassigned by some authors to *A. aethiopicus* (e.g. Kimbel *et al.* 1988).

[7] See Tobias (1991) for critical notes on Wolpoff's methods.

[8] Blumenberg's (1985) value seems to be his own guesstimate. Specimen ER 1590, which is approximately the same stratigraphic age as ER 1470, is probably somewhat larger in cranial capacity, although given the incompleteness of the fragments a reliable endocranial capacity cannot be calculated (Holloway, pers. comm.). Also note that both ER 1470 and ER 1590 have been reclassified as *H. rudolfensis* by Groves (1989) and Wood (1992).

[9] Olivier and Tissier's (1975) morphometric analysis questions a quantity this low for this specimen.

[10] Olivier and Tissier's (1975) morphometric analysis questions a quantity this high for this specimen.

[11] The Salé specimen may be an early Archaic *Homo sapiens* (Hublin 1985).

[12] Volume estimate confirmed by Holloway, this chapter, Table 4.1.

[13] The Jebel Irhoud specimens should probably not be assigned to the Neanderthal clade (see references in Day 1986).

Table 4.6 Hominid Endocranial Volumes: Summary [eds]

Taxon	Specimen notes	n	Adult cranial volume estimates (range, cm^2)
Australopithecus afarensis	All are incomplete specimens requiring extensive reconstructions. Falk (1988) believes that the high value for AL 333–45 will decrease significantly once the frontal part of the endocast is reconstructed on the basis of new information gained from the endocast of WT 17000.	3	c.343 to 485
A. africanus	The range cited here is based on the most reliable estimates (see notes in Holloway 1975 and this chapter, Table 4.1). Falk (1987) gives a new adult estimate for Taung that would lower the range to 412.	3	428 to 485
A. aethiopicus sensu auctt.	Only one skull is available (WT 17000). The reconstructed endocast is pictured in Leakey and Walker (1988).	1	410
A. robustus	Only the estimated values for one specimen are available (SK 1585, a partial fossil endocast; see Tobias 1971a). The estimate cited here is considered to be reliable (see discussion in Holloway 1975).	1	530
A. boisei	There are three specimens with what are considered to be reliable estimates (see Holloway 1975, and this chapter, Table 4.1). Note that the value of 522 for OH 5 is taken from Holloway 1975.	3	500 to 522
Homo habilis sensu stricto	This range is based on the Olduvai specimens, all of which required extensive reconstructions (see notes in Holloway 1978, 1980a, and Tobias 1991). (If, following Wood 1992, we added ER 1813 to the *H. habilis* hypodigm then the observed range would be lowered to 509.)	4	590 to c.700 (?)
H. rudolfensis (cf. Groves 1989)	Only the value for one specimen is available (ER 1470), but that is considered reliable (see Holloway 1978). Another specimen assigned to this species (a calotte and partial fossil endocast, ER 3732) is similar in size and shape to ER 1470 (Wood 1991). The immature and fragmentary specimen ER 1590, also assigned to this species, probably had a somewhat larger endocranial volume (the conjoined parietal fragments fit over ER 1470: Holloway pers. comm.).	1(3)	752+
H. ergaster sensu Wood (1992)	This species contains ER 3733, ER 3883 and WT 15000, all of which provide reliable endocranial estimates.	3	804 to 909
H. erectus sensu stricto	This includes the Zhoukoudian specimens from China and the Trinil plus Sangiran specimens from Java. The range cited here is based on the most reliable estimates (see notes in Holloway 1975, and this chapter, Table 4.1, plus the notes in Weidenreich 1943, p. 113).	8	815 to 1225
H. sapiens neanderthalensis	For a broad range of reliable estimates we have those of Holloway, this chapter, Table 4.1. If we add Amud I (Ogawa *et al.* 1970) the upper value increases to 1740, and in so far as one can judge from text and photographs this seems to be a reliable estimate.	5(6)	1305 to 1640 (1740)
H. sapiens sapiens	The values cited here are based on the example of maximum normal variation provided by Hrdlička (1939), and may represent less than 1% of the non-pathological distribution in pre-industrial human populations. Tobias (1971b) gives a value of 800 for the lower part of the range.	1000s	(800) 910 to 2100

Table 4.7 Endocranial capacity values for various fossil hominid series (cm^3): means, standard deviations, coefficients of variation and 95 per cent population limits. From Tobias (1967), courtesy of P. V. Tobias.[a]

Taxon	n	Mean	SD	V per cent	95 per cent limits of population (rounded off to nearest cm^3)
A. afarensis	3	?413.5	—	—	352–?493[f]
A. africanus	6	441.2	19.60	4.44	391–492
A. robustus	1	530.0	—	—	
A. boisei	4	513.0	11.49	2.24	476–550
A. robustus/A. boisei	5	516.4	12.52	2.42	482–551
H. habilis	6	640.2	82.23	12.85	429–852
H. erectus erectus[b]	7	895.6	93.57	10.45	667–1125
H. erectus erectus[c]	6	929.8	91.67	9.86	694–1165
H. erectus pekinensis	5	1043.0	112.51	10.79	731–1355
H. erectus (Asia and Africa)	15	937.2	135.48	14.46	647–1228
H. sapiens soloensis[d]	6	1090.8	75.39	6.91	897–1285
H. sapiens soloensis[e]	5	1151.4	99.51	8.64	896–1407

[a] In this table no attempt has been made to separate the series into presumptive male and female sub-sets.

[b] Based on Tobias's (1975) estimate, but with the incorporation of that author's new value for Trinil 2, based on Holloway's (1975) new value for Sangiran 2.

[c] Based on Holloway's (1981b) new values for six Indonesian specimens.

[d] Based on Weidenreich (1943). (Ngandong specimens [eds])

[e] Based on Holloway (1980b). (Ngandong specimens [eds])

[f] Observed range.

Table 4.8a Hominoid cranial volumes—means, ranges, SDs

Species	Sex	Sample size	Volume (mean)	SD	Range
Gibbon (*Hylobates lar*)					
	male	44	106.3	7.23	92–125*
	female	37	104.2	7.01	90–116
Siamang (*Symphalangus syndactylus*)					
	male	8	127.7	8.15	99–140†
	female	12	125.9	12.71	102–143
Chimpanzee (*Pan troglodytes*)					
	male	159	397.2	39.4	322–503‡
	female	204	365.7	31.9	270–450
Chimpanzee (*Pan paniscus*)					
	male	28	351.8	30.6	295–440§
	female	30	349.0	37.7	265–420
Orang-utan (*Pongo pygmaeus*)					
	male	66	415.6	33.6	334–502¶
	female	63	343.1	33.6	276–431
Gorilla (*Gorilla gorilla*)					
	male	283	535.5	55.3	410–715[11]
	female	199	452.2	41.6	345–553
Human (*Homo sapiens sapiens*)					
	male	502	1457.2	119.8	1160–1850**
	female	165	1317.9	109.8	1040–1615

From Tobias (1971b), by permission.

* The individual cases on which these statistics are based came from Dr A. Schultz's collection and were kindly provided to Dr Holloway by Dr D. Passingham.

† These figures are again based on Schultz's collections.
‡ The specimens summarized here are drawn from various collections.
§ These statistics are based on the individual specimens from the collection at Terveuren, Belgium, which were kindly provided to Dr Holloway by Dr D. Kramer.
¶ The orang values include values from the Smithsonian collection, the Cleveland Museum of Natural History, the American Museum of Natural History, and Schultz's collection.
‖ These values do not include the famous 752 cc case published by Schultz (1962), as Dr Holloway was never able to locate the proposed specimen in Zurich. Two 715 cc values come from the Powell–Cotton collection. These figures include specimens from the Powell–Cotton, Todd, American Museum of Natural History, and Schultz collections. We are grateful to Dr Bernard Wood for the Powell–Cotton data, and Dr W. Kimbel for the Todd collection values in the Cleveland Museum. These data were analysed using the SPSS^X statistical package. The values presented in this table are quite similar to those in Tobias (1971b) (see table 4.8b) but herein include the standard deviations.
** These cases are from Holloway (1980c), and are based on a culled Danish sample previously published by Pakkenberg and Voigt (1964). All pathological cases were removed, including extreme low and high body weights and statures. In fact, world-wide, the range of normal variation for the species *Homo* is roughly 1000–2200, the SD being roughly 10 per cent of the mean. There are ethnic variations in brain weight, but these appear to be mainly related to body size.

Table 4.8b Hominoid cranial volumes—means and sample ranges

Species	Size of sample	Mean volume (cm³)	Range (cm³)
Gibbon (*Hylobates lar*)			
Males	95	104.0	89–125 (=36)
Females	85	100.9	82–116 (=34)
Combined males and females	180	102.5	82–125 (=43)
Gibbon (*Hylobates agilis*)			
Combined males and females	21	98.8	81–120 (=39)
Siamang (*Symphalangus syndactylus*)			
Males	23	125.8	100–150 (=50)
Females	17	122.8	105–152 (=47)
Combined males and females	40	124.5	100–152 (=52)
Chimpanzee (*Pan troglodytes*)			
Males	163	398.5	292–500 (=208)
Females	200	371.1	282–460 (=178)
Combined males and females	363	383.4	282–500 (=218)
Pygmy chimpanzee (*Pan paniscus*)			
Males	6	356.0	334–381 (=47)
Females	5	329.0	275–358 (=83)
Combined males and females	11	343.7	275–381 (=106)
Orang-utan (*Pongo pygmaeus*)			
Males	203	434.4	320–540 (=220)
Females	199	374.5	276–494 (=218)
Combined males and females	402	404.8	276–540 (=264)
Gorilla (*G. gorilla gorilla*)			
Males	414	534.6	412–752 (=340)
Females	254	455.6	340–595 (=255)
Combined males and females	668	504.6	340–752 (=412)
Modern Human (*Homo sapiens sapiens*)			
Males	1000s	1345.0	900–2000 (=1100) 800–2100 (=1300)

From Tobias (1971b) by permission.

References to Editorial appendix I

Beaumont, P. B., de Villiers, H., and Vogel, J. C. (1978). Modern man in sub-Saharan Africa prior to 49 000 years B.P.: a review and evaluation with particular reference to Border Cave. *South African Journal of Science*, **74**, 409–19.

Begun, D. and Walker, A. (1993). The endocast of the Nariokotome hominid. In *The Nariokotome* Homo erectus *skeleton* (ed. A. Walker and R. E. Leakey), pp. 326–58. Harvard University Press, Cambridge, Mass.

Blumenberg, B. (1985). Population characteristics of extinct hominid endocranial volume. *American Journal of Physical Anthropology*, **68**, 269–79.

Boule, M. and Vallois, H. V. (1957). *Fossil men*. The Dryden Press, New York.

Day, M. (1972). The Omo human skeletal remains. In *The origin of* Homo sapiens (ed. F. Bordes), pp. 31–5. UNESCO, Paris.

Day, M. (1986). *Guide to fossil man* (4th edn). University of Chicago Press.

Day, M. H., Leakey, M. D., and Magori, C. (1980). A new hominid fossil skull (L. H. 18) from the Ngaloba Beds, Laetoli, northern Tanzania. *Nature*, **284**, 55–6.

Drennan, M. R. (1953). A preliminary note on the Saldanha skull. *South African Journal of Science*, **50**, 7–11.

Dubois, E. (1937). Early man in Java and *Pithecanthropus erectus*. In *Early man* (ed. G. G. MacCurdy), pp. 315–22. Lippincott, Philadelphia.

Falk, D. (1985). Hadar AL 162-28 endocast as evidence that brain enlargement preceded cortical reorganization in hominid evolution. *Nature*, **313**, 45–7.

Falk, D. (1987). Hominid paleoneurology. *Annual Review of Anthropology*, **16**, 13–30.

Falk, D. (1988). Enlarged occipital/marginal sinuses and emissary foramina: their significance in hominid evolution. In *Evolutionary history of the "robust" australopithecines* (ed. F. E. Grine), pp. 85–96. Aldine de Gruyter, New York.

Falk, D. and Kasinga, S. (1983). Cranial capacity of a female robust australopithecine (KNM-ER 407) from Kenya. *Journal of Human Evolution*, **12**, 515–18.

Groves, C. P. (1989). *A theory of human and primate evolution*. Clarendon Press, Oxford.

Heim, J.-L. (1976). Les hommes fossiles de La Ferrassie. Tome I. *Archives de l'Institut de Paléontologie Humaine Mémoir*, **35**, 1–331. Masson et Cie, Paris.

Holloway, R. L. (1970). Australopithecine endocast (Taung Specimen, 1924): a new volume determination. *Science*, **168**, 966–8.

Holloway, R. L. (1973). Endocranial volumes of early African hominids and the role of the brain in human mosaic evolution. *Journal of Human Evolution*, **2**, 449–58.

Holloway, R. L. (1975). Early hominid endocasts: volumes, morphology and significance for hominid evolution. In *Primate functional morphology and evolution* (ed. R. H. Tuttle), pp. 393–415. Mouton, The Hague.

Holloway, R. L. (1978). Problems of brain endocast interpretation and African hominid evolution. In *Early hominids in Africa* (ed. C. Jolly), pp. 379–401. Duckworth, London.

Holloway, R. L. (1980a). The O.H.7 (Olduvai Gorge, Tanzania) hominid partial brain endocast revisited. *American Journal of Physical Anthropology*, **53**, 267–74.

Holloway, R. L. (1980b). Indonesian "Solo" (Ngandong) endocranial reconstructions: some preliminary observations and comparisons with Neanderthal and *Homo erectus* groups. *American Journal of Physical Anthropology*, **53**, 285–95.

Holloway, R. L. (1980c). Within-species brain–body weight variability: a re-examination of the Danish data and other primate species. *American Journal of Physical Anthropology*, **53**, 109–21.

Holloway, R. L. (1981a). Exploring the dorsal surface of hominid brain endocasts by stereoplotter and discriminant analysis. *Philosophical Transactions of the Royal Society of London B*, **292**, 155–66.

Holloway, R. L. (1981b). Volumetric and asymmetry determinations on recent hominid endocasts: Spy I and II, Djebel Irhoud I, and the Salé *Homo erectus* specimens, with some notes on Neanderthal brain size. *American Journal of Physical Anthropology*, **55**, 385–93.

Holloway, R. L. (1981c). The Indonesian *Homo erectus* brain endocasts revisited. *American Journal of Physical Anthropology*, **55**, 503–21.

Holloway, R. L. (1981d). The endocast of the Omo L338y-6 juvenile hominid: gracile or robust *Australopithecus*? *American Journal of Physical Anthropology*, **54**, 109–18.

Holloway, R. L. (1983a). Cerebral brain endocast pattern of *Australopithecus afarensis* hominid. *Nature*, **303**, 420–2.

Holloway, R. L. (1983b). Human paleontological evidence relevant to language behavior. *Human Neurobiology*, **2**, 105–14.

Holloway, R. L. (1988). "Robust" australopithecine brain endocasts: some preliminary observations. In *Evolutionary history of the "robust" australopithecines* (ed. F. E. Grine), pp. 97–105. Aldine de Gruyter, New York.

Holloway, R. L. and Post, D. G. (1982). The relativity of relative brain measures and hominid mosaic evolution. In *Primate brain evolution: methods and concepts* (ed. E. Armstrong and D. Falk), pp. 57–76. Plenum Press, New York.

Howell, F. C. (1960). European and northwest African Middle Pleistocene hominids. *Current Anthropology*, **1**, 195–232.

Hrdlička, A. (1939). Normal micro-and macrocephaly in America. *American Journal of Physical Anthropology*, **25**, 1–91.

Hublin, J. J. (1985). Human fossils from the north African Middle Pleistocene and the origin of *Homo sapiens*. In *Ancestors: the hard evidence* (ed. E. Delson), pp. 283–8. Alan R. Liss, New York.

Jacob, T. (1966). The sixth skull cap of *Pithecanthropus erectus*. *American Journal of Physical Anthropology*, **25**, 243–59.

Kimbel, W. H., White, T. D., and Johanson, D. C. (1988). Implications of KNM-WT 17000 for the evolution of "robust" *Australopithecus*. In *Evolutionary history of the "robust" australopithecines* (ed. F. E. Grine), pp. 259–68. Aldine de Gruyter, New York.

Leakey, R. E. F., and Walker, A. (1988). New *Australopithecus boisei* specimens from east and west Lake Turkana, Kenya. *American Journal of Physical Anthropology*, **76**, 1–24.

McCown, T. D., and Keith, A. (1939). *The Stone Age of Mount Carmel:* Vol. 2: *The fossil human remains from the Levalloiso-Mousterain*. Clarendon Press, Oxford.

Ogawa, T., Kamiya, T., Sakai, S., and Hosokawa, H. (1970). Some observations on the endocranial cast of the Amud Man. In *The Amud Man and his cave site* (ed. H. Suzuki and F. Takai), pp. 407–19. Academic Press of Japan, Tokyo.

Olivier, G. and Tissier, H. (1975). Determination of cranial capacity in fossil men. *American Journal of Physical Anthropology*, **43,**D 353–62.

Oppenoorth, W. F. F. (1937). The place of *Homo soloensis* among fossil men. In *Early man* (ed. G. G. MacCurdy), pp. 349–60. Lippincott, Philadelphia.

Pakkenberg, H. and Voigt, J. (1964). Brain weight of the Danes. *Acta Anatomica*, **56**, 297–307.

Rightmire, G. P. (1983). The Lake Ndutu cranium and early *Homo sapiens* in Africa. *American Journal of Physical Anthropology*, **61**, 245–54.

Sartono, S. (1971). Observations on a new skull of *Pithecanthropus erectus* (Pithecanthropus VIII) from Sangiran, Central Java. *Proceedings, Koninklijke Nederlandse Akadademie der Wetenschappen, Amsterdam, Ser. B*, **74**, 185–94.

Schaefer, U. (1963). Die Grösse der Hirnschadelkapazität und ihre Bestimmung bei rezenten und vorgeschichtlichen Menschen. *Zeitschrift für Morphologie und Anthropologie*, **53**, 165–70.

Schultz, A. H. (1962). Die Schädelkapazität männlicher Gorillas und ihr Hochstwert. *Anthropologie Anzeiqer*, **25**, 197–203.

Stewart, T. D. (1977). The Neanderthal skeletal remains from Shanidar cave, Iraq: a summary of the findings to date. *Proceedings of the American Philosophical Society*, **121**, 121–65.

Stringer, C. B. (1984). The definition of *Homo erectus* and the existence of the species in Africa and Europe. *Courier des Forschungs Instituts Senckenberg*, **69**, 131–43.

Stringer, C. B., Howell, F. C., and Melentis, J. K. (1979). The significance of the fossil hominid skull from Petralona, Greece. *Journal of Archaeological Science*, **6**, 235–53.

Thoma, A. (1981). The position of the Vértesszöllös find in relation to *Homo erectus*. In Homo erectus: *Papers in honour of Davidson Black* (ed. B. A. Sigmon and J. S. Cybulski), pp. 105–14. University of Toronto Press, Toronto.

Tobias, P. V. (1967). *Olduvai Gorge*, Vol. II: *The cranium and maxillary dentition of* Australopithecus (Zinjanthropus) boisei. Cambridge University Press.

Tobias, P. V. (1971*a*). *The brain in hominid evolution*. Columbia University Press, New York.

Tobias, P. V. (1971*b*). The distribution of cranial capacity values among living hominoids. *Proceedings of the Third International Congress of Primatology, Zurich 1970*, Vol. 1, pp. 18–35.

Tobias, P. V. (1975). Brain evolution in the Hominoidea. In *Primate functional morphology and evolution* (ed. R. H. Tuttle), pp. 353–92. Mouton, The Hague.

Tobias, P. V. (1987). The brain of *Homo habilis*: a new level of organization in cerebral evolution. *Journal of Human Evolution*, **16**, 741–61

Tobias, P. V. (1991). *The skulls, endocasts and teeth of* Homo habilis. 4 Vols. Cambridge University Press.

Vaisnys, J. R., Lieberman, D., and Pilbeam, D. (1984). An alternative method of estimating the cranial capacity of Olduvai Hominid 7. *American Journal of Physical Anthropology*, **65**, 71–81.

von Koenigswald, G. H. R. (1962). *The evolution of man*. University of Michigan Press, Ann Arbor.

Walker, A., Leakey, R. E., Harris, J. M., and Brown, F. H. (1986). 2.5-Myr *Australopithecus boisei* from west of Lake Turkana, Kenya. *Nature*, **322**, 517.

Weidenreich, F. (1943). The skulls of *Sinanthropus pekinensis*: a comparative study on a primitive hominid skull. *Palaeontologia Sinica*, ns D, **10**, 1–291.

Wolpoff, M. H. (1977). Some notes on the Vértesszöllös occipital. *American Journal of Physical Anthropology*, **47**, 357–63.

Wolpoff, M. H. (1981). Cranial capacity estimates for Olduvai Hominid 7. *American Journal of Physical Anthropology*, **56**, 297–304.

Woo, J.-K. (1966). The hominid skull of Lantian, Shensi. *Vertebrata Palasiatica*, **10**, 14–22.

Wood, B. (1991). *Hominid cranial remains: Koobi Fora research project*, Vol. 4. Clarendon Press, Oxford.

Wood, B. (1992). Origin and evolution of the genus *Homo*. *Nature*, **355**, 783–90.

Wu, X. (1981). A well preserved cranium of an archaic type of early *Homo sapiens* from Dali China. *Scientia Sinica*, **24**, 530–9.

Editorial appendix II: Evolution of the human vocal apparatus[1]

Introduction

The mammalian upper respiratory system is often informally referred to as the 'vocal tract'. It is composed of the larynx and pharynx, plus the nasal and oral cavities (see Fig. 4.9). In fact, this anatomical region is the crossroads of both our respiratory and our alimentary systems, as well as the site for the production of vocal sounds. In particular, the location of the larynx is very important in determining the way we breathe, swallow, and vocalize. Nineteenth-century anatomical studies noted that the larynx of mammals is placed high in the neck. Negus (1929, 1949, 1965), documented the high position of the larynx in many mammalian species, and

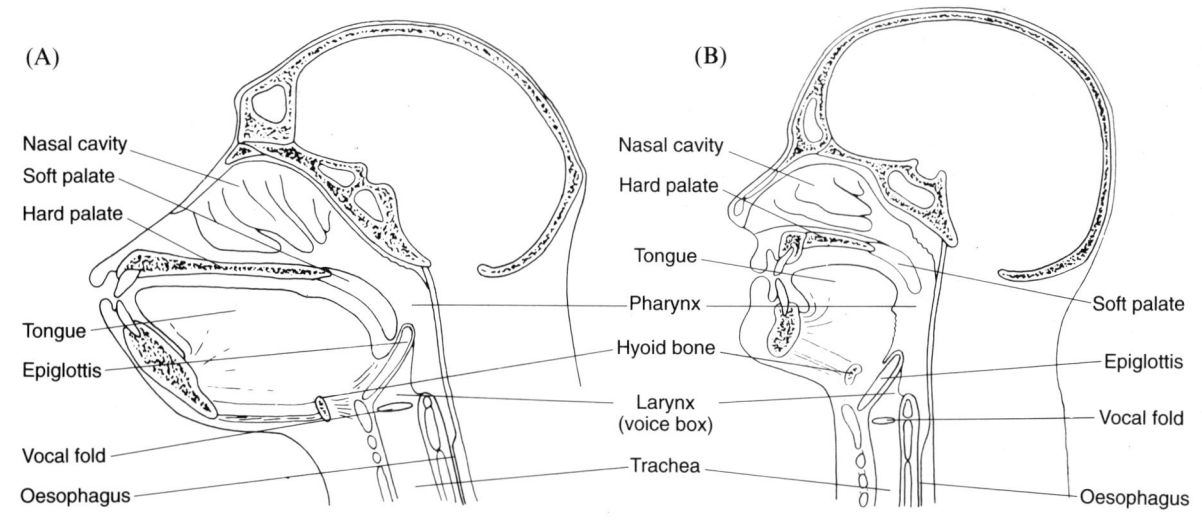

Fig. 4.9 Drawing of a midsagittal section through the head and neck of A, an adult chimpanzee and B, an adult human. Note the high position of the larynx in the neck of the chimpanzee, similar to that in a human infant. In the adult human the larynx is located markedly lower in the neck. The lower position of the larynx in humans increases the height of the pharynx; that is, of the space situated between the back of the nasal cavity and the larynx below. This extension to the pharynx allows us to utter the sounds typical of spoken language. In apes and monkeys the limited capacity of the pharyngeal space allows of only a slight ability to modulate the sounds produced by the vocal folds of the larynx (After Laitman 1986).

postulated that the high position of the larynx probably permitted the maintenance of a direct airway from the nasal cavity to the lungs, while the alimentary canal remained open so as to allow an animal to breathe and swallow simultaneously. Radiographic and cineradiographic studies of breathing, swallowing, and vocalizing in a range of mammals have confirmed what Negus and others had postulated (see Laitman 1977; Laitman *et al.* 1977; Sasaki *et al.* 1977; Laitman and Crelin 1980*a*; Reidenberg 1988; Reidenberg and Laitman 1990). This anatomical configuration permits the laryngeal airway to remain open while streams of liquid are transmitted around each side of the larynx during swallowing. This, in essence, creates two largely separate pathways: a respiratory tract from the nose to the lungs, and a digestive tract from the oral cavity to the oesophagus. Like other mammals, non-human primates also exhibit this arrangement of a larynx positioned high in the neck; and these radiographic studies have demonstrated that non-human primates are among the species possessing the ability to breathe and to swallow at least certain foods simultaneously.

Human newborns and young infants closely resemble the basic mammalian and primate pattern described above. In newborns and young infants up to 1½ to 2 years of age, the larynx is situated high in the neck. While the high position of the larynx in newborn humans or non-human primates enables them to breathe and swallow simultaneously, it severely limits the array of sounds that they can produce. The studies by Lieberman, Laitman, and co-workers, in particular, have shown that the high position of the larynx severely limits the supralaryngeal portion of the pharynx, which is responsible for modifying the initial, or fundamental, sounds produced at the vocal folds or 'vocal cords' (see Lieberman 1968, 1983, 1984; Lieberman and Crelin 1971; Lieberman *et al.* 1972, 1992; Laitman 1977, 1983, 1986; Laitman and Crelin 1976, 1980*b*; Laitman and Reidenberg 1988, 1993*a, b*). Hence a newborn baby or a monkey with a larynx situated high in the neck would have a more restricted range of vocalizations available to it than would an adult human with its larynx placed lower in the neck. The linguistic analyses by Lieberman, in particular, identified the quantal vowels [i], [u], and [a] as sounds that infants or non-human primates cannot produce. As these vowels are the limiting articulations of a vowel triangle that is language-universal (Lieberman 1984; Liljencrants and Lindblom 1972) their absence would considerably limit the human speech capabilities of an individual unable to produce them.

Until approximately 1½ to 2 years of age the larynx of the human infant remains positioned very high in the neck, approximating the condition found in non-human primates. By the third year the position of the larynx has been significantly lowered (Laitman 1977;

Table 4.9 Fossil hominids studied to reconstruct upper respiratory anatomy*

Specimen	Site	Possible taxon†	Approx. age‡
MLD 37/38	Makapansgat South Africa	*Australopithecus africanus*	3 m.y.a.
Sts 5	Sterkfontein South Africa	*A. africanus*	2.5 m.y.a.
Taung	Taung South Africa	*A. africanus*	2 m.y.a.+
SK 47	Swartkrans South Africa	*A. robustus*	1.5–2.0 m.y.a.
SK 48	Swartkrans South Africa	*A. robustus*	1.5–2.0 m.y.a.
SK 83	Swartkrans South Africa	*A. robustus*	1.5–2.0 m.y.a.
OH 5	Olduvai Gorge Tanzania	*A. boisei*	1.7–1.8 m.y.a.
OH 24	Olduvai Gorge Tanzania	*Homo habilis*	1.8–1.9 m.y.a.
KNM-ER 406	East Turkana Kenya	*A. boisei*	1.7–1.8 m.y.a.
KNM-ER 3733	East Turkana Kenya	*Homo erectus*	1.7–1.8 m.y.a.
KNM-ER 3883	East Turkana Kenya	*Homo erectus*	1.5–1.6 m.y.a.
Salé	Salé Morocco	*Homo erectus/Homo sapiens*	200 000–500 000 BP
Petralona	Petralona Greece	*Homo sapiens*	300 000–400 000 BP
Steinheim	Steinheim Germany	*Homo sapiens*	300 000–400 000 BP
Broken Hill ('Rhodesian Man')	Broken Hill (Kabwe) Zambia	*Homo sapiens*	150 000–300 000 BP
Saccopastore 2	Saccopastore Italy	*Homo sapiens*	125 000 BP
Tesik-Tash	Tesik-Tash Soviet Union	*Homo sapiens neanderthalensis*	40 000–100 000 BP
Monte Circeo (Circeo 1)	Monte Circeo Italy	*Homo sapiens neanderthalensis*	40 000–100 000 BP
Gibraltar 1	Forbes' Quarry Gibraltar	*Homo sapiens neanderthalensis*	40 000–100 000 BP
La Chapelle-aux-Saints	La Chapelle-aux-Saints France	*Homo sapiens neanderthalensis*	40 000–100 000 BP
La Ferrassie 1	La Ferrassie France	*Homo sapiens neanderthalensis*	40 000–100 000 BP
Cro-Magnon 1	Les Eyzies France	*Homo sapiens sapiens*	28 000–34 000 BP
Predmosti 4	Predmosti Czechoslovakia	*Homo sapiens sapiens*	26 000 BP
Afalou 5	Afalou-bou-Rhummel Algeria	*Homo sapiens sapiens*	Holocene
Taforalt 12	Taforalt Morocco	*Homo sapiens sapiens*	Holocene
Ain Dokhara	Ain Dokhara Algeria	*Homo sapiens sapiens*	Holocene

* Partial list of the best-preserved specimens studied by Laitman for upper respiratory tract reconstruction. Discussions of the specimens are in publications by the author. Many other fossil hominids have been examined, and are the subject of ongoing investigation. These include Plio-Pleistocene specimens often attributed to *H. habilis*, middle Pleistocene *H. erectus*, and early *H. sapiens*. Many partial specimens, with portions of the basicranium preserved, have also been the subject of study. These include hominids such as australopithecines Sts 19 and 71, and *H. erectus* material OH 9 and Sangiran 4.

† The taxonomic assignments are meant to illustrate groups to which the specimens have been frequently assigned. Arguments for including these hominids in different taxa have often been advanced (see discussions in Day (1986); or Campbell, this volume, Chapter 2).

‡ Ages listed are offered as only approximate guides.

Laitman and Crelin 1976, 1980*b*); and this process appears to continue into young adulthood. The lower position of the larynx alters dramatically the way humans, after the early years of life, breathe and swallow. The loss of the ability of the epiglottis to make contact with the soft palate means that the possibility of having two largely separate pathways, one for air and one for liquid, no longer exists. The respiratory and digestive tracts now cross each other in the area of the pharynx. Because of this crossing it is no longer possible to breathe and swallow at the same time. This new configuration can, and does, have unfortunate drawbacks. The major problem is that a bolus of food can become lodged in the entrance of the larynx. If this material cannot be expelled rapidly an individual may literally choke to death. This unfortunate event is often referred to as a 'café coronary', so named because it frequently occurs in restaurants and is mistaken for a heart attack (Haugen 1963). Another disadvantage of the crossed pathways is the relative ease with which vomit can be aspirated into the larynx and trachea, and thus pass into the lungs. The lower position of the larynx in the neck has, however, provided one major positive aspect: a greatly expanded supralaryngeal portion of the pharynx. Pharyngeal modification of sounds produced at the vocal folds is thus considerably greater in adults than that possible for newborns, early infants, or any non-human mammal. In essence, it is this expanded pharynx which gives us the anatomical ability to produce our species-typical speech sounds.

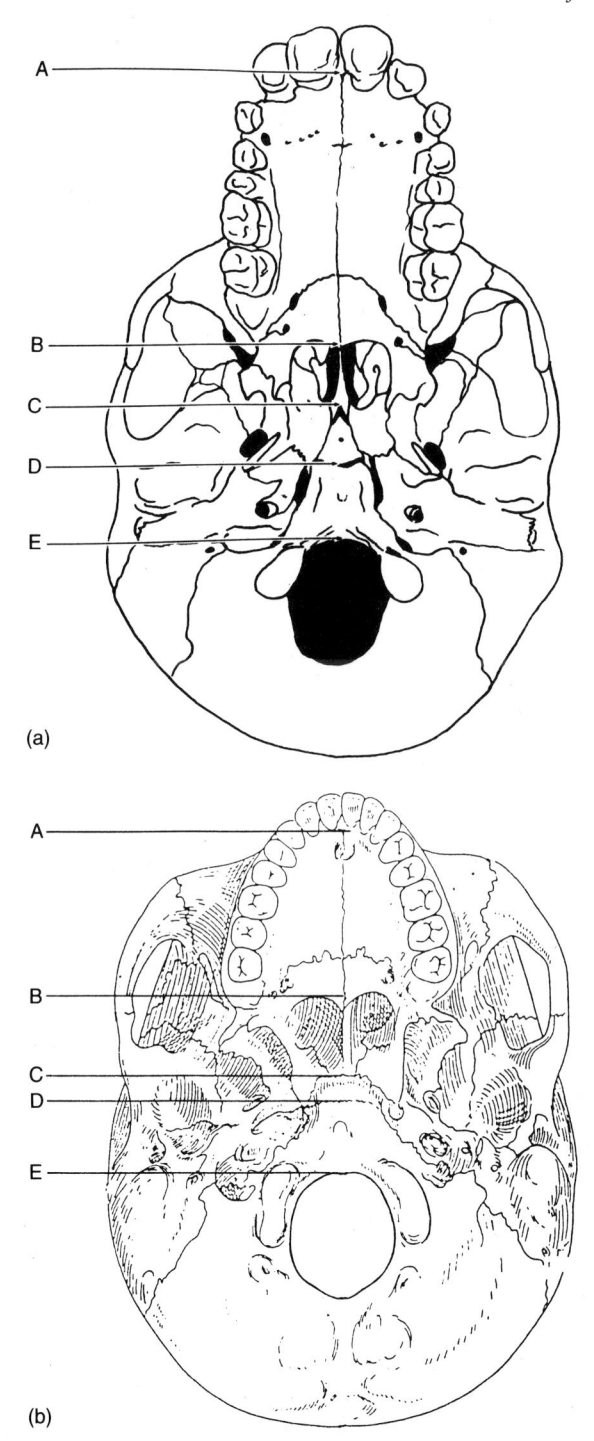

Fig. 4.10 Basicranial views of the craniometric points on the midlines of the basicrania of (a) an adult female chimpanzee (*Pan troglodytes*), and (b) an adult human (*Homo sapiens sapiens*) (not to scale), illustrating the anatomical points from which Laitman's measurements are taken. A, prosthion; B, staphylion; C, hormion; D, sphenobasion; E, endobasion (from Laitman *et al.* 1979; by permission of the author).

Palaeoanthropological inference

One method for making inferences about the upper respiratory anatomy of fossil hominids draws upon the anatomical relationship between the respiratory system and the base of the skull, or basicranium. The basicranium is the superior limit of the upper respiratory tract, literally the 'roof' of the upper respiratory region. Studies by Lieberman and Crelin (1971), Laitman and Crelin (1976), Laitman *et al.* (1978), Laitman (1983), Gibbons (1974), George (1978), Grossmangin (1979), Reidenberg (1988), and Reidenberg and Laitman (1991) have shown that the position of the larynx and pharynx is related to the orientation of the cranial base. In mammals such as cats, dogs, monkeys, and apes, in which the larynx is positioned high in the neck, the basicranium is relatively non-flexed. Only older humans show a larynx placed absolutely and relatively lower in the neck, and only these individuals exhibit a markedly flexed basicranium. This can be seen by comparisons of a basicranial line which describes exocranial flexion between the front of the hard palate and the anterior margin of the foramen magnum (Laitman *et al.* 1978; Laitman and Heimbuch 1984*a*), determined by a series of linear measurements taken along five craniometric points identified on the midline of the exocranial surface of the basicranium (see Figs. 4.10, 4.11, and 4.12). Thus, to judge from these studies, it appears that two basic patterns exist concerning the relationship of the basicranium and the upper respiratory tract. In the first, basicrania that are largely non-flexed correspond to larynges placed quite high in the neck, as in non-human primates. In the second, markedly flexed basicrania correspond to upper respiratory structures placed considerably lower in the neck, such as are found in modern adult humans.

One other point needs to be brought out here. Overall, the basicranium shows considerably less variability than do most regions of the face or the neurocranial vault. The basicranium may thus be a more stable area than other regions of the skull. As a result, basicranial alteration may indicate changes of a more substantial nature than the more frequent, and easily achieved, changes in the more plastic parts of the skull (see Laitman 1977, 1983, and Reidenberg 1988 for more detailed discussion). Hence, developmental and evolutionary alteration of the basicranium may directly effect the position of structures such as the larynx and pharynx, and this, in turn, upper respiratory function.

tric analysis of the basicranial line of fossil hominids whose basicrania have been sufficiently preserved in this anatomical region. The basicranial lines of those hominids have then been compared to those generated previously for extant non-human primates and humans. The upper respiratory tract profiles of the hominids are then 'reconstructed'. For example, if a skull of a fossil hominid were found to exhibit little or no basicranial flexion, and thus to resemble the condition observed in living monkeys and apes, the larynx and tongue would be inferred to have been positioned high in the neck. Conversely, a fossil hominid that exhibited marked flexion of the skull base would be inferred to have had a throat structured very much like our own. Unless fossil hominids exhibited different biomechanical or neuromuscular mechanisms from those found in extant primates, which seems unlikely, it is reasonable to assume that similarities in structures between extant and fossil hominids would produce similarities in function. In this way the consistent relationships observed in living mammals between the angulation of the basicranium and the position of the larynx have proved to be an important tool for reconstructing this region in fossil hominids.

The australopithecines

Analysis of a number of australopithecines—such as the well-preserved skull of Sts. 5 from Sterkfontein in South Africa and that of KNM-ER 406 from the eastern shore of Lake Turkana in Kenya—has shown that they exhibit basicrania that closely approximate to the general non-flexed condition of the extant Great Apes (Laitman and Heimbuch 1982) (see Fig. 4.13; see also the Fossil Hominid Photogallery for Sts. 5 and ER 406). The basicranial similarities between the australopithecines and the extant apes suggest that their upper respiratory tracts were also similar. As with the living apes, the high position of the larynx would greatly restrict the supralaryngeal portion of the pharynx. This would in turn markedly reduce the pharyngeal area available to modify the initial, or fundamental, sound generated at the vocal folds. Thus it is inferred that the range of sounds available to these hominids was probably not much greater than that shown by the living apes and monkeys.

Homo erectus

It is among some members of this species that we start to observe changes in the cranial base that may indicate corresponding change in the associated upper respiratory tract. Analysis of the largely complete African cranium KNM-ER 3733 from the eastern

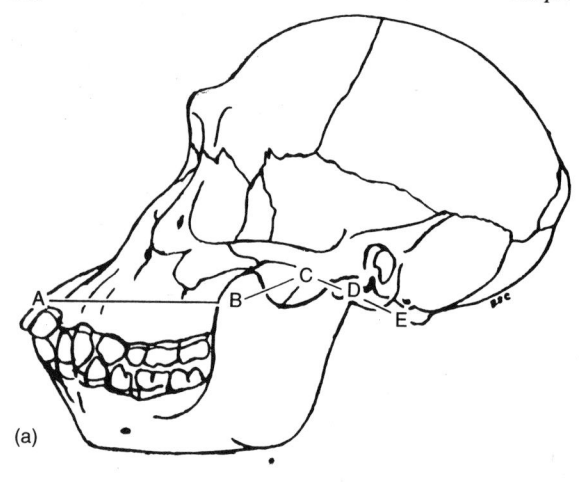

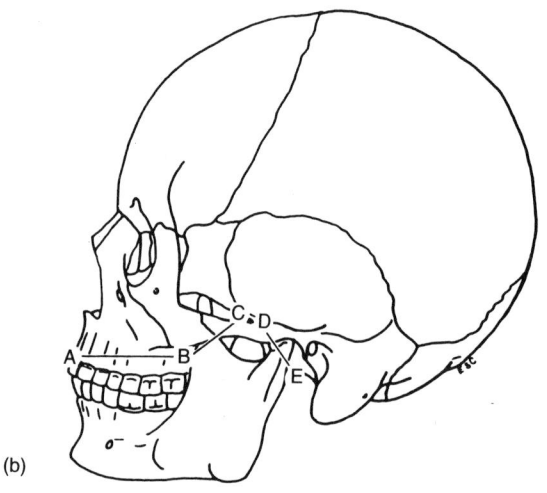

Fig. 4.11 Side-view sagittal cross-sections and topographic projections of the basicranial lines upon the skulls of (a) an adult chimpanzee and (b) an adult human. The skulls are positioned in the palatal plane (A–B). The letters represent the anatomical points between which measurements are taken: A, prosthion; B, staphylion; C, hormion; D, sphenobasion; E, endobasion. Note the relatively flat skull of the chimpanzee in comparison with the highly flexed (arched) skull of the human. (From Laitman 1983, by permission of the author.)

On the basis of these studies the relationship between the skull base and the upper respiratory tract of extant primates has been used as a guide for making inferences about the upper respiratory systems of fossil hominids (see for example Laitman et al. 1979; Laitman and Heimbuch 1982; Laitman and Reidenberg 1988). This has been done by craniome-

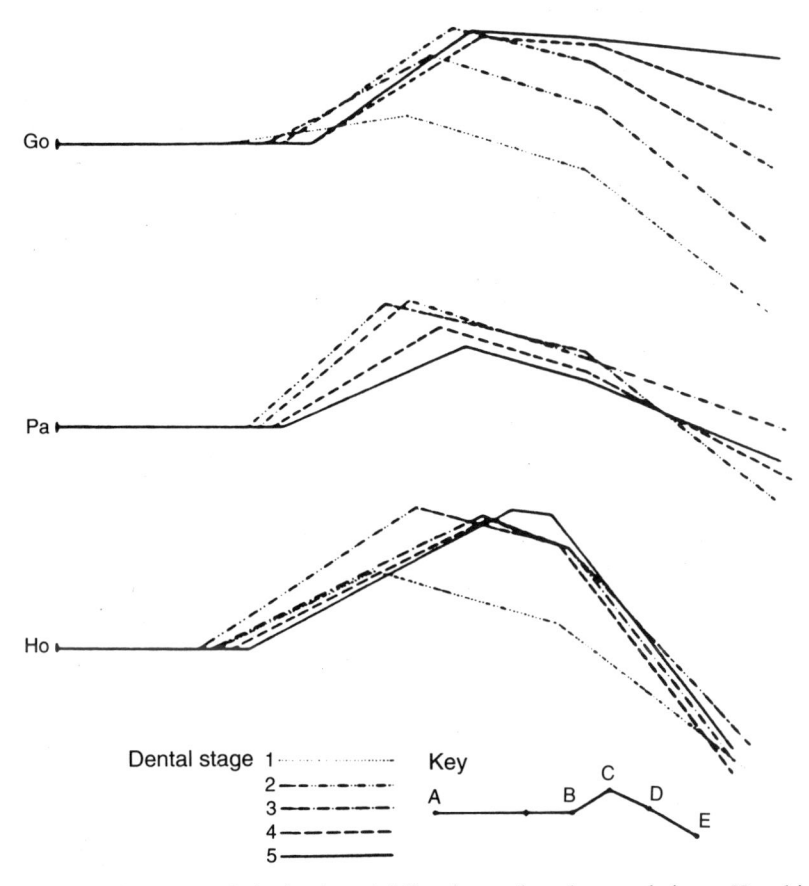

Fig. 4.12 Ontogenetic development of the basicranial line in modern human beings, *Ho*; chimpanzees, *Pa*; and gorillas, *Go* (from Laitman *et al.* (1978)). Stage 1 refers to samples of skulls from individuals who died before the eruption of the deciduous dentition; stage 2, from the eruption of the first central incisor to the completion of the deciduous dentition; stage 3, the eruption of the first permanent molar; stage 4, the eruption of the second permanent molar; and stage 5, eruption of the third permanent molar. Note that the human new-born (stage 1) basicranial line is similar to that of both the chimpanzee and the gorilla. While the chimpanzee and gorilla lines become progressively flatter, the human basicranium becomes more flexed, with the greatest change occurring between stages 1 and 2.

shore of Lake Turkana indicates what may be the first example of incipient basicranial flexion away from the non-flexed pattern of the australopithecines and towards that shown by modern humans (Laitman and Heimbuch 1984*b*; Laitman *et al.* 1992). The partial flexion in this specimen may well correspond to partial descent of the larynx. While the descent of the larynx was probably not comparable to the extreme descent found in modern *H. sapiens*, the inferred positional change would undoubtedly have altered the breathing and swallowing abilities of these hominids away from the basic mammalian pattern retained by their australopithecine ancestors.

The new lower position of the larynx in *H. erectus* would have increased this species' ability to make more human-like sounds. Owing to the descent of the larynx an increased area of the supralaryngeal portion of the pharynx could now be used to modify the fundamental frequencies produced at the vocal folds. As a result, the variety of sounds that *H. erectus* could have produced was probably considerably greater than the limited range of vocalizations available to the australopithecines, yet still less than that possible for modern humans.

Whether or not the changes described first began with *H. erectus*, or even earlier, with some hominids ascribed to the Plio-Pleistocene taxon of *H. habilis*, is still unclear. Examination of the Olduvai specimen OH 24, a hominid assigned to *H. habilis*, indicates that its basicranium, although somewhat distorted,

Australopithecus africanus

Sterkfontein 5

Taung

MLD 37/38

Australopithecus robustus

Swartkrans 47

Australopithecus boisei

KNM–ER 406

Swartkrans 48

Olduvai Hominid 5

Swartkrans 83

Homo habilis

Olduvai Hominid 24

Archaic *Homo sapiens*

Steinheim

Broken Hill

Homo sapiens neanderthalensis

Gibraltar 1

La Chapelle

Homo sapiens sapiens

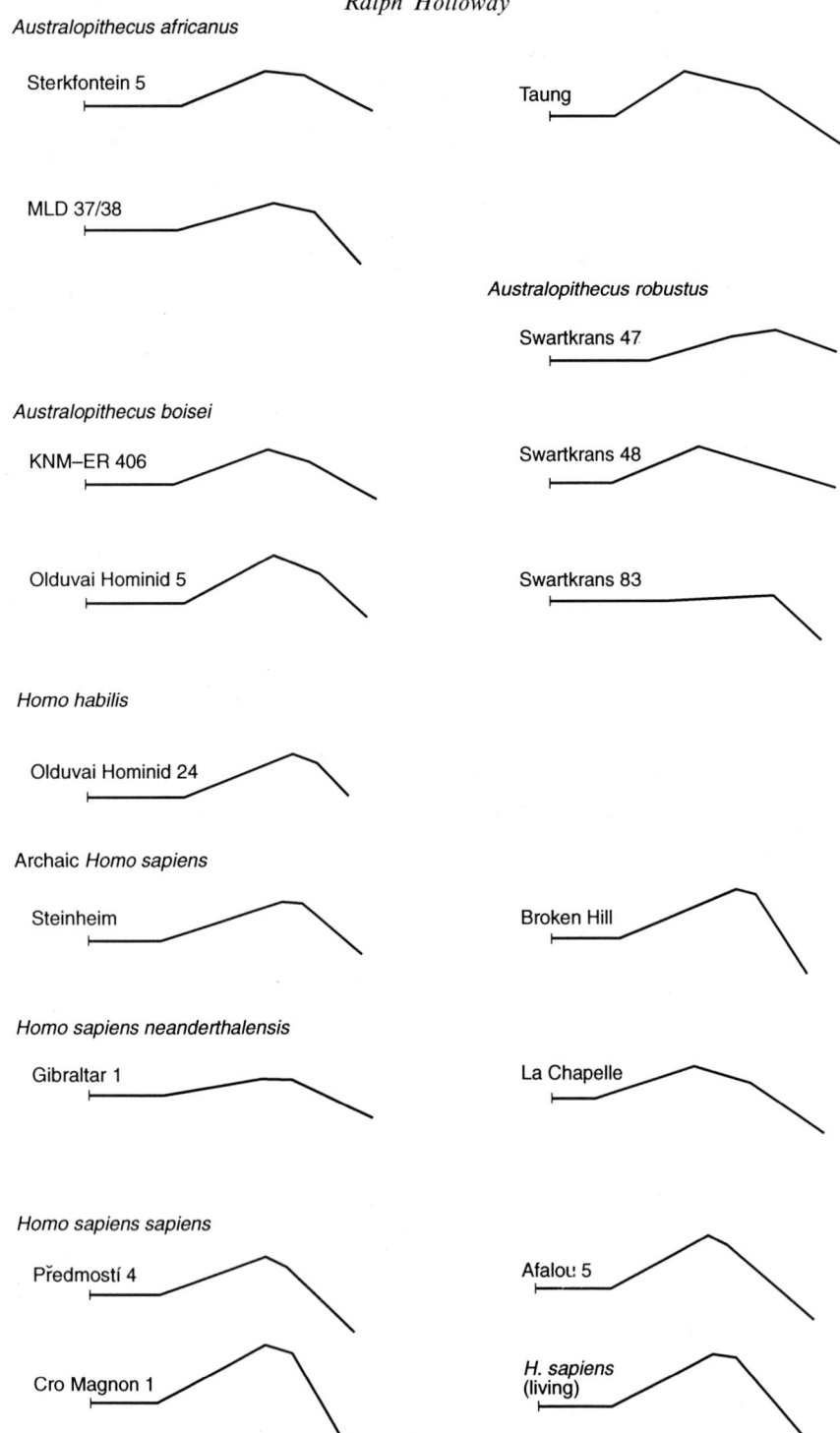

Předmostí 4

Afalou 5

Cro Magnon 1

H. sapiens (living)

Fig. 4.13 Basicranial lines of hominid fossils published by Laitman and his associates, arranged by current species assignment. The basicranial line of a contemporary adult human is shown for comparison. A standardized segment of the front part of the palate has been removed from all the basicranial lines in this figure; they have also been standardized for size (see Laitman and Heimbuch 1982 for details).

appeared more similar to that of the extant apes than to that of living humans (Laitman and Heimbuch 1982). Accordingly, its upper respiratory anatomy was deemed to be similar to that of the australopithecines. Unfortunately, other important specimens attributed to *H. habilis sensu lato*, such as KNM-ER 1470, lack the greater part of their cranial bases. As a result, assessment of their upper respiratory tracts has not as yet been possible.

Early Homo sapiens and Neanderthals

Analysis of a number of middle to late Pleistocene hominids—in particular, Petralona from Greece, Broken Hill (Kabwe) from Zambia,[2] Steinheim from Germany, and the partial cranium from Salé in Morocco[3]—indicate that their basicrania were largely like those of living humans. On the basis of these analyses it is inferred that their upper respiratory tracts were also similar to our own (Laitman 1983; Laitman *et al.* 1979, 1992). The advent of a human-like upper respiratory tract thus may well have occurred some 200 000 to 400 000 years before the present, with hominids usually described as early or archaic *H. sapiens*.

While an essentially modern upper respiratory tract was probably present in some early *H. sapiens*, one group of late Pleistocene hominids may have had an upper respiratory tract that differed from those of modern humans. This group is the Neanderthals.

Investigations of a number of late Pleistocene hominids (Lieberman and Crelin 1971; Lieberman *et al.* 1972; Laitman 1977; Laitman *et al.* 1979; Grossmangin 1979) have highlighted the fact that Neanderthals exhibit a number of basicranial characteristics that differ from those of both living humans and also certain other middle-to late-Pleistocene hominids such as those found at Petralona and Broken Hill. For example, some 'classic' Neanderthals, such as the specimens from La Chapelle and La Ferrassie, exhibit less basicranial flexion than that shown by either living populations or other early specimens of *H. sapiens*.

The original studies by Lieberman and Crelin emphasized the differences between Neanderthals and living humans, describing features Neanderthals may have had in common with newborn humans and champanzees. Laitman's analyses, while also finding differences, emphasize that Neanderthals were more like humans in a number of important parameters. For example, these studies noted that the larynx of Neanderthals would not have been positioned like that of an ape. Rather, it would have been positioned slightly higher in the neck than that found in living humans. This higher position would reduce the area available to them with which to modify certain sounds as compared with ourselves, yet would still allow overall breathing and swallowing patterns similar to those of living humans. Recently, Laitman and his co-workers have attributed some of the differences found in some 'classic' Neanderthals to respiratory-related specializations that evolved to accommodate the requirements of breathing in a cold and dry environment (Laitman *et al.* 1993).

While Neanderthals may not have had a vocal repertoire as extensive as our own, these large-brained hominids (see earlier sections of this chapter) could still be hypothesized to have had a highly complex form of language. For discussion of the controversies surrounding Neanderthals and language see most recently Lieberman *et al.* (1992) (and also Carlisle and Siegel 1974; Lieberman and Crelin 1974; Morris 1974; Falk 1975; LeMay 1975; Burr 1976*b*, DuBrul 1977; Laitman 1977; Grossmangin 1979; Lieberman 1984; Crelin 1987; Arensburg *et al.* 1989; and Lieberman *et al.* 1989).[4] Certainly, the final words on the linguistic capabilities of Neanderthals, in terms of both their vocal apparatus and their brain structure, have not yet been written.[5]

Notes to Editorial appendix II

1. We thank Jeffrey Laitman for his generous assistance in the preparation of this appendix. He was one of the original authorities contributing to the creation of the *Handbook*, but after numerous delays in its publication other responsibilities prevented him from preparing a revision of his chapter for this volume.

2. Burr (1976*a*) was the first to show that the Broken Hill hominid had features of the basicranium related to a modern vocal anatomy.

3. The Salé basicranium is incomplete, but the preserved portion (see Hublin 1985, and the colour plates in that volume) exhibits basicranial features indicating a highly flexed condition similar to that of modern *Homo sapiens* (Laitman and Crelin 1980*c*; Laitman 1985). Otherwise the partial cranium is largely archaic in appearance; and Day (1986) retains the assignment of the specimen within *Homo erectus*.

4. There are a minority of modern human languages—for example, Kabardian (see Spuhler 1977)—that would appear to be compatible with the supposed limitations of Neanderthal vocal production.

5. There are other additional aspects of cranial anatomy that may also contribute to a fuller understanding of the evolution of spoken language abilities. For example, first, Bauer (1986) has comparatively

analysed the sound production of adult chimpanzees and human infants, since these two groups have more mutually similar vocal tract anatomies—associated with similar basicranial configurations—than do human infants and human adults. He notes that 'vocalic sound durations, fundamental frequency contours and ranges for human infants about a year old appeared much more variable than those observed for adult chimpanzees' (ibid., p. 327); chimpanzee sound patterns 'lack consonants and thus the segmental contrastivity of varied consonants and vowels in rapid succession, which are characteristic of human infant sounds' (ibid., p. 342). He hypothesizes this difference as resulting from differing degrees of ability for functional closing of the nasal ports (openings from the nose to upper pharynx) in chimpanzees and infant humans. Thus there may be processes involved in establishing the speech capability of humans that are additional to the anatomical rearrangement of the components of the upper respiratory system.

Second, an evolving ability for speech production implies an equal ability for speech perception. Rosowski (1992) reports different frequency and threshold responses for the auditory responses of humans and chimpanzees. And Masali (1992) notes different anatomical structuring of the ear ossicles in humans, gorillas, and chimpanzees, and hypothesizes that these differences create different perceptual parameters with respect to the reception of speech sounds. And finally, Daniel (1989) notes that the labyrinthine structures of the inner ear are differently orientated in humans as against quadrupedal mammals, and that this reorientation also occurs ontogenetically in humans. This reorientation would appear to be related to maintaining balance during the attainment of bipedal posture. Thus there has probably been an integrated suite of structural and functional reorientations of ear, nose, and throat relations integral to the evolutionary underwriting of human speech abilities.

References to Editorial appendix II

Arensburg, B., Tillier, A. M., Vandermeersch, B., Duday, H, Schepartz, L., and Rak, Y. (1989). A Middle Palaeolithic human hyoid bone. *Nature*, **338**, 758–60.

Bauer, H. R. (1986). A comparative study of common chimpanzee and human infant sounds. In *Current perspectives in primate social dynamics* (ed. D. M. Taub, and F. A. King), pp. 327–45. Van Nostrand Reinhold, New York.

Burr, D. B. (1976a). Rhodesian man and the evolution of speech. *Current Anthropology*, **17**, 762–3.

Burr, D. (1976b). Further evidence concerning speech in Neandertal man. *Man*, **11**, 104–10.

Carlisle, R. C. and Siegel, M. I. (1974). Some problems in the interpretation of Neanderthal speech capabilities: a reply to Lieberman. *American Anthropologist*, **76**, 319–22.

Crelin, E. S. (1987). *The human vocal tract*. Vantage Press, New York.

Daniel, H. J. (1989). The vestibular system and language evolution. In *Studies in language origins*, Vol. 1 (ed. J. Wind, E. G. Pulleybank, E. de Grolier, and B. H. Bichakajian), pp. 257–71. John Benjamins, Amsterdam.

Day, M. H. (1986). *Guide to fossil man* (4th edn). University of Chicago Press.

DuBrul, E. L. (1977). Origin of the speech apparatus and its reconstruction in fossils. *Brain and Language*, **4**, 365–81.

Falk, D. (1975). Comparative anatomy of the larynx in man and chimpanzee: implications for language in Neanderthal. *American Journal of Physical Anthropology*, **43**, 123–32.

George, S. L. (1978). The relationship between cranial base angle morphology and infant vocalizations. Ph.D. dissertation, University of Connecticut, University Microfilms, Ann Arbor, Michigan.

Gibbons, M. F. (1974). Anatomical and quantitative approaches to the evolution of the speech generation potential. Ph.D. dissertation, Yale University. University Microfilms, Ann Arbor, Michigan.

Grossmangin, C. (1979). Base du crâne et pharynx dans leur rapports avec l'appareil du langage articulé. Ph.D. thesis, Paris: Mémoire du Laboratoire d'Anatomie de la Faculté de Médecine de Paris, No. 40–1979.

Haugen, R. K. (1963). The café coronary: sudden deaths in restaurants. *Journal of the American Medical Association*, **186**, 142–3.

Hublin, J.-J. (1985). Human fossils from the North African Middle Pleistocene and the origin of *Homo sapiens*. In *Ancestors: the hard evidence* (ed. E. Delson), pp. 283–8. Alan R. Liss, New York.

Laitman, J. T. (1977). The ontogenetic and phylogenetic development of the upper respiratory system and basicranium in man. Ph.D. dissertation, Yale University. University Microfilms, Ann Arbor, Michigan.

Laitman, J. T. (1983). The evolution of the hominid upper respiratory system and implications for the origins of speech. In *Glossogenetics: the origin and evolution of language* (ed. E. de Grolier), pp. 63–90. Harwood Academic Press, Paris.

Laitman, J. T. (1985). Latter Middle Pleistocene hominids. In *Ancestors: the hard evidence* (ed. E. Delson), pp. 265–7. Alan R. Liss, New York.

Laitman, J. T. (1986). L'origine du langage articulé. *La Recherche*, **17**, (181), 1164–73.

Laitman, J. T. and Crelin, E. S. (1976). Postnatal development of the basicranium and vocal tract region in man. In *Symposium on the development of the basicranium* (ed. J. F. Bosma), pp. 206–20. US Government Printing Office, Washington, DC.

Laitman, J. T. and Crelin, E. S. (1980a). Tantalum markers as an aid in identifying the upper respiratory structures of experimental animals. *Laboratory Animal Science*, **30**(2), 245–8.

Laitman, J. T. and Crelin, E. S. (1980b). Developmental change in the upper respiratory system of human infants. *Perinatology/Neonatology*, **4**, 15–22.

Laitman, J. T. and Crelin, E. S. (1980c). An analysis of the Salé cranium: a possible early indicator of a modern upper respiratory tract. *American Journal of Physical Anthropology*, **52**, 245–6 (abstract).

Laitman, J. T. and Heimbuch, R. C. (1982). The basicranium of Plio-Pleistocene hominids as an indicator of their upper respiratory systems. *American Journal of Physical Anthropology*, **59**, 323–43.

Laitman, J. T. and Heimbuch, R. C. (1984a). A measure of basicranial flexion in the pygmy chimpanzee, *Pan paniscus*. In *The pygmy chimpanzee: evolutionary biology and behavior* (ed. R. L. Susman), pp. 49–63. Plenum Press, New York.

Laitman, J. T. and Heimbuch, R. C. (1984b). The basicranium and upper respiratory system of African *Homo erectus* and early *H. sapiens*. *American Journal of Physical Anthropology*, **63**, 180 (abstract).

Laitman, J. T., and Reidenberg, J. S. (1988). Advances in understanding the relationship between the skull base and larynx with comments on the origins of speech. *Human Evolution*, **3**, 99–109.

Laitman, J. T. and Reidenberg, J. S. (1993a). Comparative and developmental anatomy of human laryngeal position. In *Head and neck surgery—Otolaryngology*, Vol. 1 (ed. B. Bailey), pp. 36–43. Lippincott, Philadelphia.

Laitman, J. T. and Reidenberg, J. S. (1993b). Specializations of the human upper respiratory and upper digestive systems as seen through comparative and developmental anatomy. *Dysphagia*, **8**, 318–25.

Laitman, J. T., Crelin, E. S., and Conlogue, G. J. (1977). The function of the epiglottis in monkey and man. *Yale Journal of Biology and Medicine*, **50**, 43–9.

Laitman, J. T., Heimbuch, R. C., and Crelin, E. S. (1978). Developmental change in a basicranial line and its relationship to the upper respiratory system in living primates. *American Journal of Anatomy*, **152**, 467–82.

Laitman, J. T., Heimbuch, R. C., and Crelin, E. S. (1979). The basicranium of fossil hominids as an indicator of their upper respiratory systems. *American Journal of Physical Anthropology*, **51**, 15–34.

Laitman, J. T., Reidenberg, J. S., and Gannon, P. J. (1992). Fossil skulls and hominid vocal tracts: new approaches to charting the evolution of human speech. In *Language origin: a multidisciplinary approach* (ed. J. Wind, B. Chiarelli, B. Bichakjian, A. Nocentini, and A. Jonker), pp. 395–407. Kluwer Academic Publishers, Dordrecht, The Netherlands.

Laitman, J. T., Reidenberg, J. S., Friedland, D. R., Reidenberg, B. E., and Gannon, P. J. (1993). Neandertal upper respiratory specializations and their effect upon respiration and speech. *American Journal of Physical Anthropology, Supplement*, **6**, 129 (abstract).

LeMay, M. (1975). The language capability of Neandertal man. *American Journal of Physical Anthropology*, **42**, 9–14.

Lieberman, P. (1968). Primate vocalizations and human linguistic ability. *Journal of the Acoustical Society of America*, **44**, 1574–84.

Lieberman, P. (1983). On the nature and evolution of the biological bases of language. In *Glossogenetics: the origin and evolution of language* (ed. E. de Grolier), pp. 91–114. Harwood Academic Press, Paris.

Lieberman, P. (1984). *The biology and evolution of language*. Harvard University Press, Cambridge, Mass.

Lieberman, P. and Crelin, E. S. (1971). On the speech of Neanderthal man. *Linguistic Inquiry*, **2**, 203–22.

Lieberman, P. and Crelin, E. S. (1974). Speech and Neanderthal man: a reply to Carlisle and Siegel. *American Anthropologist*, **76**, 323–5.

Lieberman, P., Crelin, E. S., and Klatt, A. H. (1972). Phonetic ability and related anatomy of the newborn, adult human, Neanderthal man, and the chimpanzee. *American Anthropologist*, **74**, 287–307.

Lieberman, P., Laitman, J. T., Reidenberg, J. S., Landahl, K., and Gannon, P. J. (1989). Folk physiology and talking hyoid bones. *Nature*, **342**, 486–7.

Lieberman, P., Laitman, J. T., Reidenberg, J. S., and Gannon, P. J. (1992). The anatomy, physiology, acoustics and perception of speech: essential elements in analysis of the evolution of human speech. *Journal of Human Evolution*, **23**, 447–67.

Liljencrants, J. and Lindblom, B. (1972). Numerical simulation of vowel quality systems. The role of perceptual contrast. *Language*, **48**, 839–62.

Masali, M. (1992). The ear ossicles and the evolution of the primate ear: a biomechanical approach. *Human Evolution*, **7**, 1–5.

Morris, D. H. (1974). Neanderthal speech. *Linguistic Inquiry*, **5**, 144–50.

Negus, W. E. (1929). *The mechanism of the larynx*. Mosby, St Louis.

Negus, V. E. (1949). *The comparative anatomy and physiology of the larynx*. Heinemann, London.

Negus, V. E. (1965). *The biology of respiration*. Livingstone, Edinburgh.

Reidenberg, J. S. (1988). Experimental alteration of the rat skull base and its effect upon the position of the larynx and hyoid bone. Ph.D. dissertation, City University of New York. University Microfilms, Ann Arbor, Michigan.

Reidenberg, J. S., and Laitman, J. T. (1990). A new method for radiographically locating upper respiratory and upper digestive tract structures in rats. *Laboratory Animal Science*, **40**(1), 72–6.

Reidenberg, J. S., and Laitman, J. T. (1991). Effect of basicranial flexion on larynx and hyoid position in rats: an experimental study of skull and soft tissue interactions. *Anatomical Record*, **203**, 557–69.

Rosowski, J. J. (1992). Hearing in transitional mammals. In *The evolutionary biology of hearing* (ed. D. B. Webster, R. R. Fay, and A. N. Popper), pp. 615–31. Springer-Verlag, New York.

Sasaki, C. T., Levine, P. A., Laitman, J. T., and Crelin, E. S. (1977). Postnatal descent of the epiglottis in man: a preliminary resport. *Archives of Otolaryngology*, **103**, 169–71.

Spuhler, J. N. (1977). Biology, speech and language. *Annual Review of Anthropology*, **6**, 509–61.

5
Evolution of the hand and bipedality

Mary W. Marzke

Abstract

Symbolic behaviour among humans and non-human primates incorporates the hands, and in human ancestors opportunities to use the hand for this purpose must have increased with the evolution of habitual bipedal posture and locomotion. In tracing the evolution of human symbolic behaviour it is therefore important to trace the origins of human bipedality, and to explore the progressive changes in hominid hand structure and functions that may have affected the use of the hands in communication.

A comparison of modern human hands with those of non-human primates reveals features unique to humans. Functional analyses of these unique features have shown that they are consistent with the stresses and requirements for joint movements associated with effective use of hand-held paleolithic stone tools. Hominid fossil hands from the Pliocene and Pleistocene provide some evidence of the sequence in which these features evolved. Structural adjustments to bipedal posture in the earliest hominids may have been an important correlate to developments in the hand, facilitating the use of the trunk as leverage in accelerating the hand during tool-use.

The evolutionary state of manipulatory potential of hominid hands has probably never been a limiting factor in gestural communication or in the manual creation of symbols.

5.1 Introduction

With the emergence of hominid bipedality there was a major shift in hand functions from the broad spectrum of support, locomotion, and manipulation to a narrower one, emphasizing manipulation. Changes in hand morphology occurred as hominid survival became increasingly dependent upon the use of tools for gathering food and repelling predators. Bipedality significantly enhanced the effectiveness of tool-use by adding the leverage of the trunk to movements of the hand.

Distinctively human features of the hand and locomotor apparatus compatible with tool-use appeared early in hominid evolution, at first incorporated into a morphological pattern adapted in our prehominid ancestors to positional as well as manipulative behaviour. This chapter will critically review the evidence bearing upon the ancestral pattern, and will then trace the evolution of derived elements in the hand and locomotor apparatus. It begins in Section 5.2 with a description of morphological features shared by all primates which lie at the root of our manipulative abilities. It then focuses on hand structure in a few species closely related to hominids. These species are selected because some features of their hands appear in fossil and living hominids, and their functional analysis thus throws light on hand functions in our ancestors. Section 5.3 examines the modern human hand and bipedality. Continuities with non-human primate hands are described first, followed by a functional analysis of distinctive patterns of hominid morphology. Section 5.4 identifies in fossil hominids additional features shared with non-human primates, and traces the sequence of appearance of derived human characteristics through the fossil record. Section 5.5 draws together the diverse sources of evidence in the previous sections, presenting a functional interpretation of the sequence of changes in the mosaic of primtive and derived characteristics during the evolution of the hominid hand and bipedality.

5.2 Non-human primate hands

Versatility is the hallmark of the primate hand, and, as one would expect, it requires a complex and subtle mechanism. With minor variations between species, thirty-five joints accommodate the palm and fingers to branches and objects of all sizes, shapes, and orientations. Six layers of muscles produce movements that propel the animals and effect gripping patterns used in maintaining feeding and resting positions, securing an infant's hold on its mother, removing parasites from the fur, catching insects, plucking fruits, extracting foods from their source, and positioning objects for tactile, olfactory, and visual scrutiny. Multiple structural constraints on mobility stabilize joints in regions that are habitually exposed to stresses during these positional and manipulatory activities. The locations and configurations of these constraints vary considerably among species, as do the relative proportions of hand segments, reflecting the diversity of their locomotor and feeding patterns (Fig. 5.1).

5.2.1 Hand morphology shared by primates

The key to the versatility of primate hands is to be found in the nature of the thumb and the fingertips. The thumb is structurally and functionally differen-

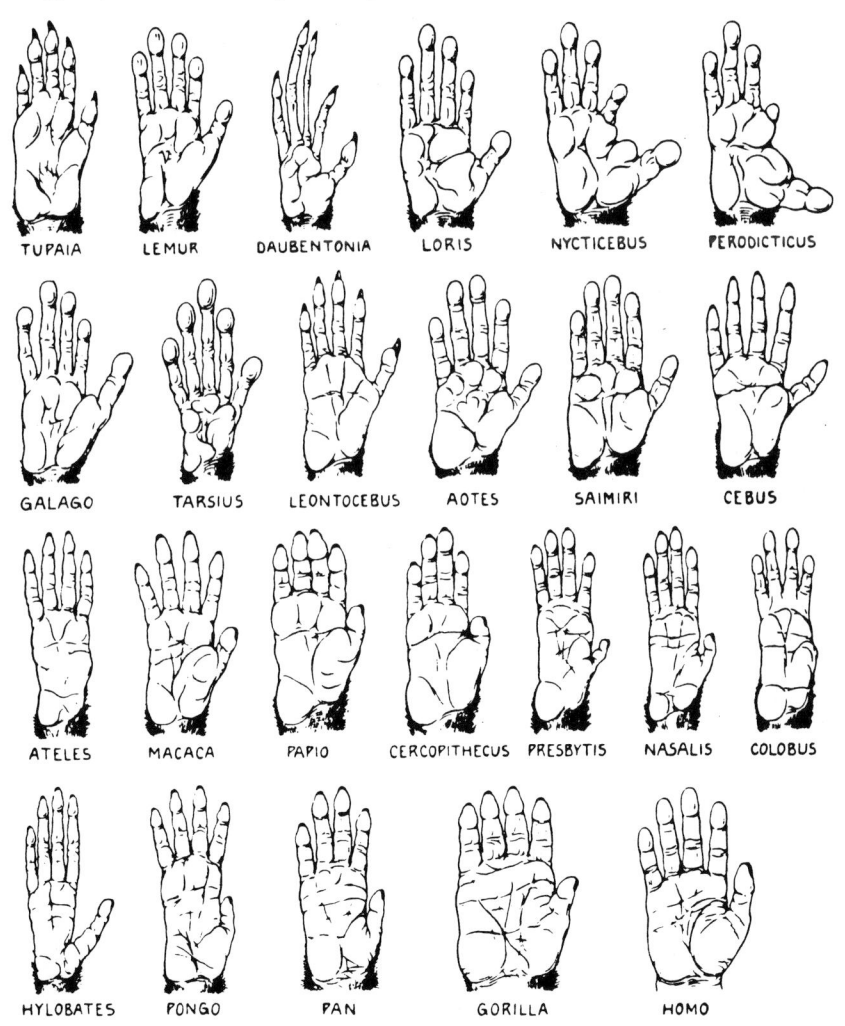

Fig. 5.1 Hands of several primate genera, reduced to the same length (from Schultz 1969). *Tupaia* are the common treeshrews, sometimes classified in the order Primates; *Leontocebus, Leontideus* and *Leontopithecus*, and *Aotes* = *Aotus* in Tables 2.2 and 2.3 of Chapter 2.

tiated from the rest of the fingers. The tips of all five digits are relatively broad, with moist, ridged, sensitive palmar pads that are supported by nails rather than claws (see Napier 1961 and Biegert 1963). This combination of features, which facilitates the grasping of branches and objects (Fig. 5.2), is present in all but a few species. The advantages of a grasping hand are most apparent in the levels of the forest where vines, bushes, and the slender upper and outer branches of the canopy offer the least purchase to a paw with claws on the fingertips (Cartmill 1974). It is possible, in fact, that the initial adaptation of primates was to feeding in the slender branches of the forest (Cartmill 1974; Rasmussen 1990; Sussman 1991). Differentiation of the thumb provides the ability for prehension of objects by one hand (Napier 1961). The variety and skill of prehensile activities depend upon the details of joint structure, the relative length of the thumb and fingers, the sensory nerve-supply to the distal digital pads, and the motor control of hand movements by the brain. Both sensory supply and motor control are advanced in Anthropoidea over Prosimii, and among the Anthropoidea studied by Bishop (1964) independent control of digits II–V has been found to be restricted to Hominoidea and some cercopithecoid species. There has been recent progress in the understanding of the neurophysiological basis of this control (Muir 1985).

The extent to which manual skill is expressed in Anthropoidea seems to be related to a certain degree to the demands of particular ecological niches for fine hand-control in the gathering of food (Welles 1975).

5.2.2 The hand of the Great Apes

Orang-utan. Characteristic features of the orang-utan hand are (1) proportionately long, curved fingers and a short thumb (Fig. 5.1), (2) strong muscles that flex the fingers, and (3) a mobile wrist. During arboreal feeding and locomotion the four fingers often function as a hook, suspending the body below the branch and pulling it upward, either alone or, during climbing, with the help of the feet (Fig. 5.3). The hook is maintained by powerful superficial and deep flexor muscles which originate on the arm and forearm, and terminate in long tendons that run along the palmar aspect of the fingers to their attachments on the middle and distal phalanges. (See Figs. 5.4 and 5.5 for illustrations of bones and muscles described in this and following sections.) Because of their large size and the strength of the muscle contraction, the tendons are contained in tunnels, formed on one side by an arched wrist skeleton and the phalanges, and on the other by tough ligaments. The ligaments run across the palmar aspect of the wrist (the flexor retinaculum) and across the fingers (annular ligaments), attaching to a projecting pisiform and projections on the other marginal carpal bones, and to ridges along the margins of the proximal and middle phalanges. While one hand maintains its hook-like grip of a branch, the body is able to pivot around it as the other hand reaches for food and supports,

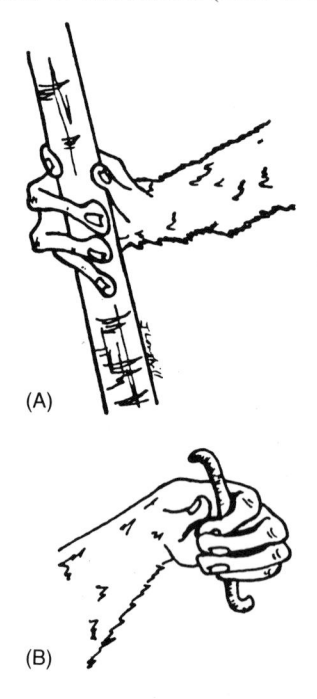

Fig. 5.2 Bush-baby hand (*Galago senegalensis*). A. Grasping a branch. B. Grasping an insect.

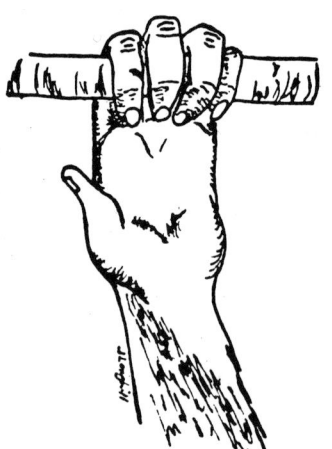

Fig. 5.3 Orang-utan hand (*Pongo pygmaeus*) with fingers flexed at the proximal interphalangeal joints.

Evolution of the hand and bipedality

in humans insert exclusively into the proximal phalanx and metacarpal. (As in all primates, the thumb has only two phalanges.) Adductor muscles (contrahentes) to digits II, IV, and V, which in some primates draw the fingers toward the midline of the hand, are generally absent (Forster 1917; Marzke, personal observation).

African apes (gorilla, common chimpanzee, and pygmy chimpanzee): African apes climb like orangutans, and use their similarly long, hook-like hands to hoist the body and to pull in vegetation. Their fingers are long relative to the thumb (Schultz 1956; Fig. 5.1), and the wrist (Napier 1961) and hand skeleton (Susman 1979) are modified for support of a strong flexor apparatus. The range of mobility at the wrist is smaller than in the orang-utan (Tuttle 1969*b*; Sarmiento 1988), and the specialized configuration of the central wrist for rotation during suspension by the hand is not present (Jenkins 1981). The ulna does not articulate with the wrist in the gorilla, but has been found to articulate with the triquetrum (but not the pisiform) in some chimpanzees (Lewis 1972, 1974).

On the ground, African apes walk on the back of the flexed middle phalanges, with the metcarpals and proximal phalanges held in line with the forearm (Tuttle 1967, 1969*a*, *b*; see Fig. 5.6). There are several interlocking and ligamentous restraining features at the wrist-joints, which together are unique to the African apes, and stabilize the hand in this knuckle-walking posture. Irregular contours of the joint surfaces between the carpals and metacarpals II–V, and a strong deep transverse metacarpal ligament connecting the metacarpal heads (through their attachments to the joint capsules) interfere with sliding of the carpals on the metacarpals and splaying of the metacarpals as the weight is transmitted across this narrow region of the forelimb (Marzke 1971, 1983).

Two grips are favoured in manipulatory behaviour (Napier 1960). Large objects are secured by hook-like flexion of digits II–V. Chimpanzees have been filmed and illustrated using this grip when they crack nuts with stones and wood clubs (Boesch and Boesch 1983*a*, 1993), and when they brandish sticks and throw large rocks (van Lawick-Goodall 1971, Appendix C). Smaller objects, such as termite-catching and ant-fishing sticks, are often held between the opposable thumb and the side of the index finger (van Lawick-Goodall 1968, 1971, 1973; Nishida 1972; McGrew 1974; McGrew *et al.* 1979; Jordan 1982; Christel 1993). The flexor pollicis longus muscle is frequently absent in African apes, but the adductor pollicis and flexor pollicis brevis muscles have attachments on the distal phalanx (Tuttle 1970), and are in a position to maintain stability of the thumb against the pressure of the side of the index finger.

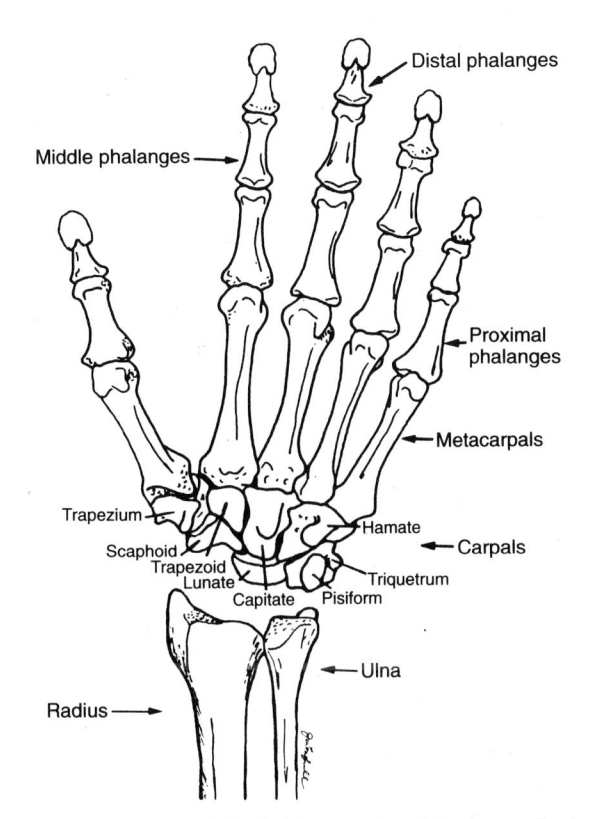

Fig. 5.4 Bones of the left human hand (palmar view).

exploiting a special configuration of the central wrist joints (Jenkins 1981). This configuration is shared with the gibbon and spider monkey, which also suspend the body by the hand during feeding (ibid.). A considerable range of movement is permitted at the joint between the forearm and wrist (Tuttle 1969*b*; Sarmiento 1988). Contributing to this range is the lack of a direct articulation between the ulna and the carpals (Lewis 1972, 1974). Articular surfaces at the joints between the carpals and metacarpals II–V are low in relief, offering neither extensive mobility nor marked bony constraints on sliding to their mutual contiguous surfaces.

Although the thumb is only about 43 per cent of total hand-length (Schultz 1956), it is used against the side of digit II in manipulating objects. A saddle joint between the first metacarpal and the trapezium permits opposition of the thumb to the palm and fingers, as it does in most other Anthropoidea (Lewis 1977). The long extrinsic flexor muscle to the distal phalanx of the thumb (flexor pollicis longus) is rarely present in the orang-utan. However, the phalanx is supplied by tendons from the intrinsic pollical muscles, including the adductor pollicis muscle (Tuttle 1969*b*), which

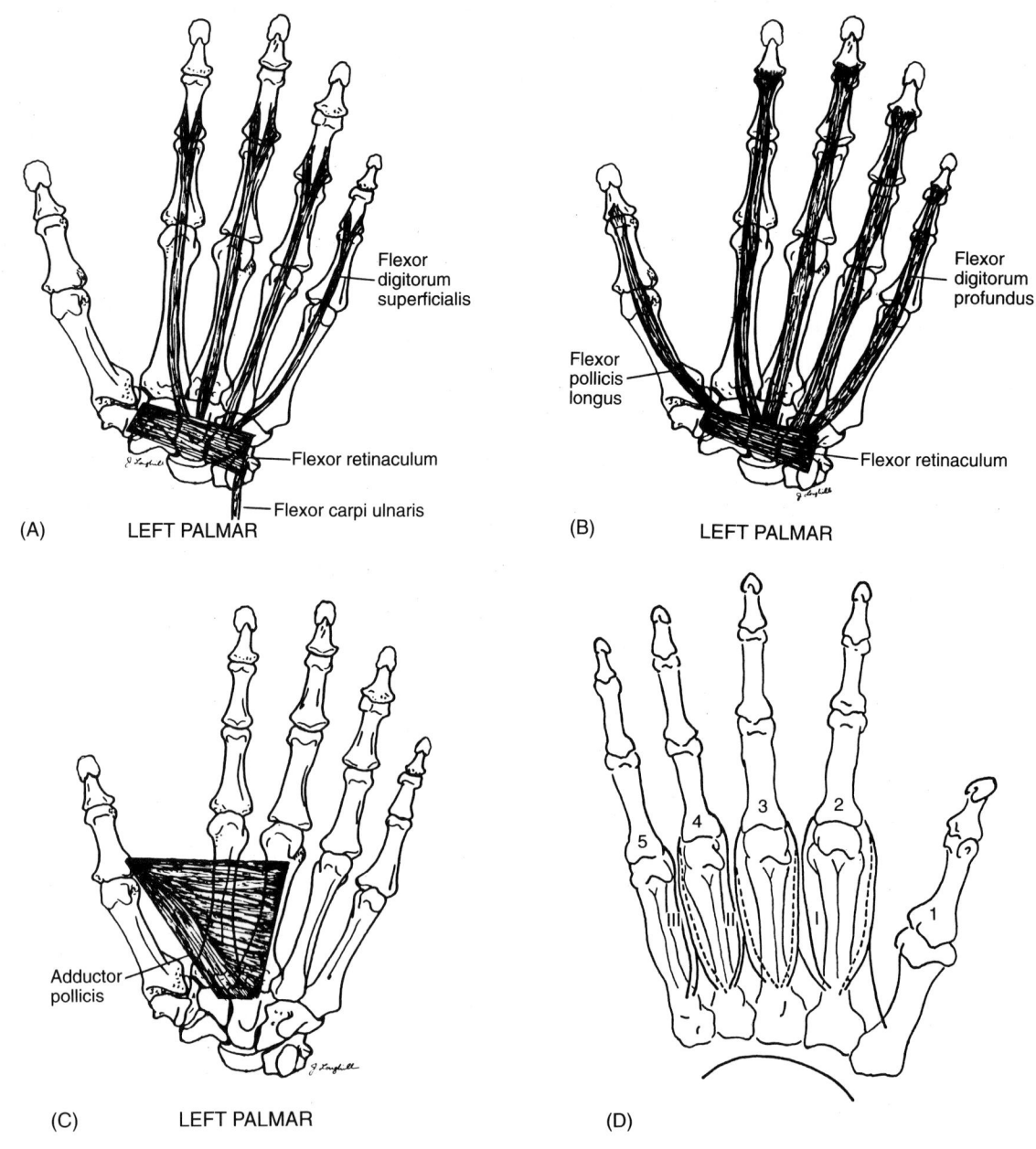

Fig. 5.5 Muscles of the human hand referred to in the text, in order from superficial to deep layers, palmar view. A. Flexor digitorum superficialis. B. Flexor digitorum profundus. C. Adductor pollicis. D. Interossei (from Marzke 1971). The three palmar interossei which adduct digits II, IV, and V are indicated by Roman numerals. The four dorsal interossei are represented by solid lines, without numbers. The interrupted lines represent the remaining palmar interossei, which are incorporated into the adjacent dorsal interossei in humans.

Contrahentes adductor muscles to digits II, IV and V are generally absent (Forster 1917 [common chimpanzee]; Marzke personal observation [common chimpanzee and gorilla]).

5.3 Human hands

Humans share with Great Apes two derived features which form part of the pongid pattern associated with

Fig. 5.6 Gorilla hand (*Gorilla gorilla gorilla*) in knuckle-walking posture.

use of the hand as a mobile hook-like unit in suspensory behaviour (Table 5.1, Nos. 1–2). One is a reduced articulation between the ulna and wrist. Neither the pisiform nor the triquetrum articulates directly with the ulna. The other is the general lack of contrahentes adductor muscles to digits II, IV, and V. Three muscles in the palm of the hand, the second, fifth, and seventh palmar interossei (Fig. 5.5 D), are uniquely positioned in humans to function as adductors of the fingers in place of the contrahentes (Lessertisseur 1958; Jouffroy and Lessertisseur 1959). Also, like the African apes, humans have a well-developed deep transverse intermetacarpal ligament connecting the metacarpophalangeal joint-capsules of digits II–V.

Humans lack other derived features of Asiatic and African apes related to use of the hand in suspensory and knuckle-walking behaviour (Table 5.1, Nos. 4–6). However, they have a distinctive morphological pattern which has been analysed in connection with the manipulative behaviour that characterizes hominids, namely tool-use and tool-making (Napier 1956, 1960, 1962*a*, 1980; Washburn 1959; Lewis 1977; Marzke 1983, 1986; Marzke and Shackley 1986; Marzke and Marzke 1987; Marzke *et al.* 1992). From the prehistoric record we know that hominids have been making hand-held stone tools for *c*.2.5 million years, and in some areas of the world stone tools continue to be made (Schick and Toth 1993). One would therefore expect to find constraints and buttressing at joints repeatedly stressed by percusive blows with hand-held stones, and contours and orientation of joint facets which facilitate the effective gripping of stone tools and their preforms. Analysis of gripping postures and hand movements used in the experimental replication and manipulation of palaeolithic tools has shown that most of the morphological features distinguishing hominid from pongid hands (Table 5.1, Nos. 7–19) are indeed consistent with the demands of habitual stone-tool-use and stone-tool-making (Marzke and Shackley 1986; Marzke and Marzke 1987). The results of this analysis are summarized below.

5.3.1 *The morphological basis of tool-use in humans*

Pounding with hand-held hammerstones has possibly been the tool-using and tool-making activity with the greatest frequency and antiquity in hominid evolution. It is an activity which directs large, repetitive forces toward the central region of the palm. Production of forceful and accurate blows by a hand-held stone requires control of the hammerstone by firm precision grips which assure both retention of the stone in the hand and fine adjustments in its orientation by the thumb and fingers. Stabilization of preforms that are held in the other hand and pounded by hammerstones in the production of tools also requires firm precision grips and the ability to vary the orientation of the stone.

The central region of the modern human palm is stabilized, buttressed, and protected against intrinsic and extrinsic forces associated with the grasp and manipulation of stones in pounding by robust bones, a styloid process on M III, a ligament from the pisiform to M III, and a fat-pad (Table 5.1, No. 7–11; Fig. 5.7). A secure grasp and controlled manœuvring of stones by the thumb, fingers, and palm are facilitated by a unique pattern of hand proportions and joint-and-muscle configurations that permit cupping of the hand and the formation of a wide variety of grips (Table 5.1, Nos. 6, 12–19). The proportionately long thumb and short fingers with broad fingertip pads are able to manœuvre the stones and to hold them firmly, exploiting the leverage of the fingers, or bracing the stones against the palm. The unique arrangement of intrinsic musculature and orientation of joints along the second, third, and fifth rays, favouring rotation of the fingers, allow optimal positioning of the thumb and fingers for grasping and orienting the stones.

Grips that were found through experimentation (see below) to accommodate and control the stones most comfortably and effectively involved primarily the thumb, index, and third fingers. These included the pad-to-side and three-jaw-chuck thumb/finger grips and extensions of these grips that incorporate the palm as a passive buttress. (The grips described here are illustrated in Figs. 5.8–5.10. A more extensive classification and definition of grips may be found in Marzke and Shackley (1986).) The three-jaw-chuck thumb/finger grip is most effective both for wielding hammerstones and for throwing stones. The stones (in the range of about 500 grams, comparable in size to tennis balls) are held by the thumb, index, and third fingers, frequently against the side of the flexed fourth finger (buttressed by the flexed fifth finger) as a support. The thumb-tip and index and third finger-tips control the orientation of the stone and keep it away from the palm, so that the leverage of these rays is exploited in propelling the stone. The pressure and leverage of these rays are important factors in controlling the rotation and speed of an object thrown by the hand (Alston and Weiskopf 1972; Watts 1973; Seaver with Lowenfish 1984).

The modern human structure of the joints along the fifth ray (Table 5.1, Nos. 18, 19) probably contributes to the effectiveness of the finger/active-palm squeeze grip, which employs all the fingers and active

Table 5.1 Comparison of Great Apes and humans in hand morphology

	Pongo	Pan and Gorilla	Homo (sapiens)
1. Ulnar articulation with carpals[1]	none	restricted or none	none
2. Mm. contrahentes II, IV, V[2]	rare	rare	rare
3. Deep transverse intermetacarpal ligament[2]	weak	well developed	well developed
4. Carpal and metacarpophalangeal constraints for knuckle-walking[3]	absent	present	absent
5. Flexor apparatus[4]	well developed	well developed	less well developed
6. Thumb length/hand length[5]	short	short	long
7. M III styloid process[6]	absent	absent	generally present
8. Pisometacarpal ligament to M III and associated groove on hamate[7]	absent	absent	present
9. Deep palmar fat pad[8]	thin	thin	thick
10. M II and PP II relative robustness[9]	less	less	more
11. Capitate shape[10]	'waisted'	'waisted'	bulbous
12. Flexor pollicis longus[11]	usually absent	weak or absent	present
13. Number of palmar interosseous muscles[11]	7	7	3
14. M II head asymmetry[12]	less	less	more
15. Plane of joints between M II and the capitate and trapezium[13]	sagittal	sagittal	away from sagittal
16. M III head orientation[9]	palmad	palmad	slightly radiad
17. Distal phalanges: shaft and tip[14]	narrow	narrow	broad
18. Carpometacarpal V joint[15]	relatively narrow	relatively narrow	relatively broad
19. M V head asymmetry[16]	less	less	more

[1] Lewis et al. (1970).
[2] Forster (1917); Marzke, personal observation.
[3] Tuttle (1967); Marzke (1983).
[4] Tuttle (1969b).
[5] Schultz (1956).
[6] See Marzke and Marzke (1987).
[7] Lewis (1977); Marzke (1986); Marzke and Marzke (1987).
[8] Spinner (1984); Marzke, personal observation.
[9] Susman (1979).
[10] Lewis (1985).
[11] See Marzke (1971).
[12] Dorsal bevelling and palmar protrusion of the radial side of the M II head cause the collateral ligament to stretch when the proximal phalanx flexes, resulting in rotation of the palmar surface of the phalanx toward the fifth finger (Lewis 1977). The orientation of the carpometacarpal joints permits rotation of M II in the same direction (ibid.).
[13] Marzke (1983).
[14] See Shrewsbury and Johnson (1983).
[15] Marzke et al. (1992).
[16] Dorsal bevelling and palmar protrusion of the ulnar side of the M V head permits rotation of the palmar surface of the proximal phalanx toward the thumb as it flexes (see note 12 above).

Evolution of the hand and bipedality

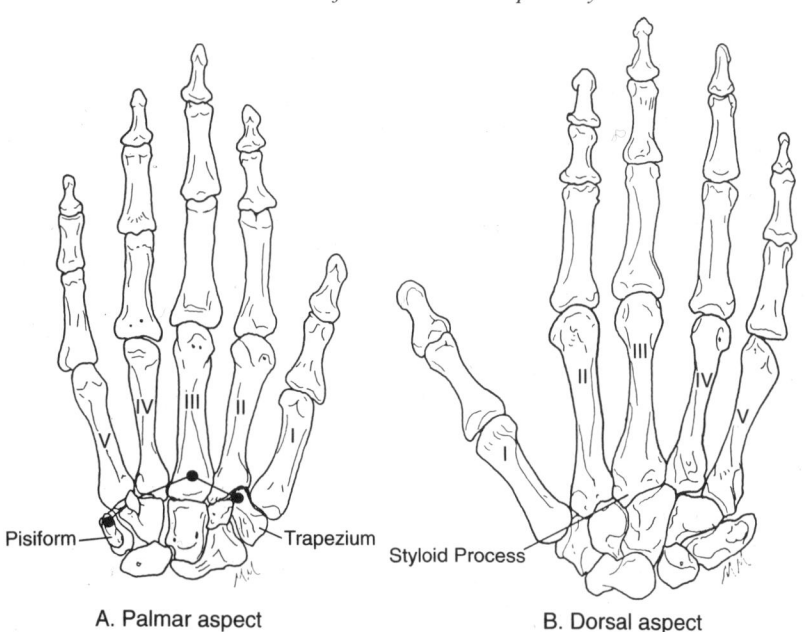

Fig. 5.7 Human hand-skeleton, with structures identified that stabilize the joints and bolster the hand against percussive forces associated with tool-use. A. Lines between the dots represent the pisometacarpal and trapeziometacarpal ligaments running to their attachment on the palmar aspect of the third metacarpal base. B. Styloid process on the dorsal aspect of the third metacarpal base.

convergence of the palm around a cylindrical tool (such as an antler hammer) to secure it, so that the tool functions as an extension of the hand and forearm.

The use of small modern tools such as needles and pencils involves the rotation and translation of objects by the pads of the fingertips opposed to the thumb-tip pad, exploiting a unique human compartmentalization of these pads described by Shrewsbury and Johnson (1983). There has been a tendency in the literature to apply the term 'precision grip' exclusively to this

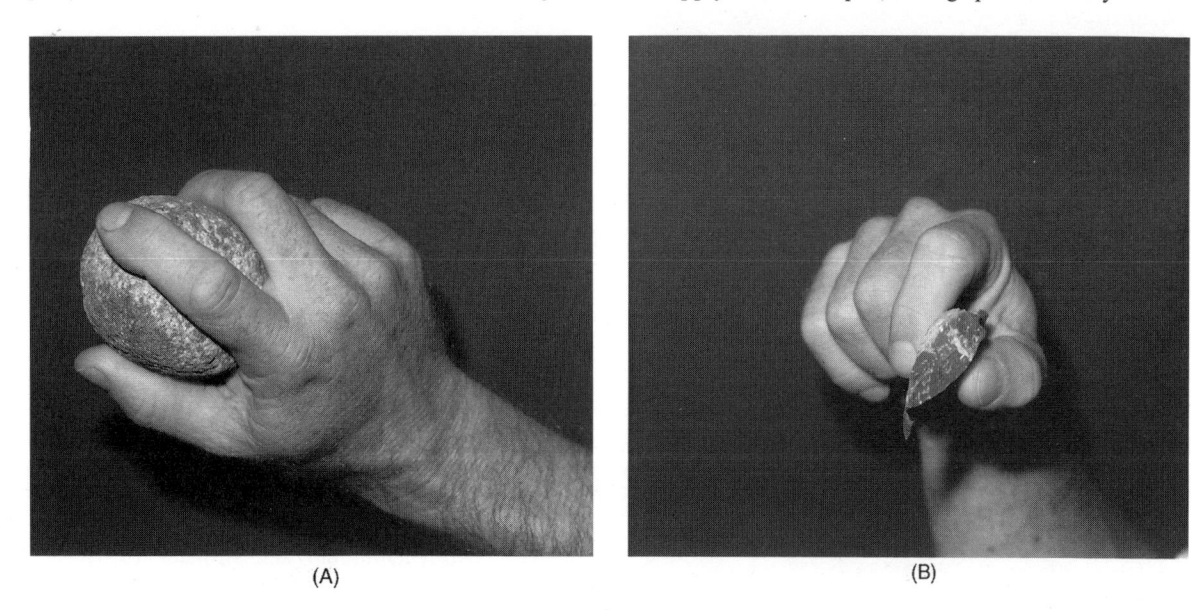

Fig. 5.8 Human finger grips. A. Three-jaw-chuck thumb–finger grip. B. Pad-to-side grip.

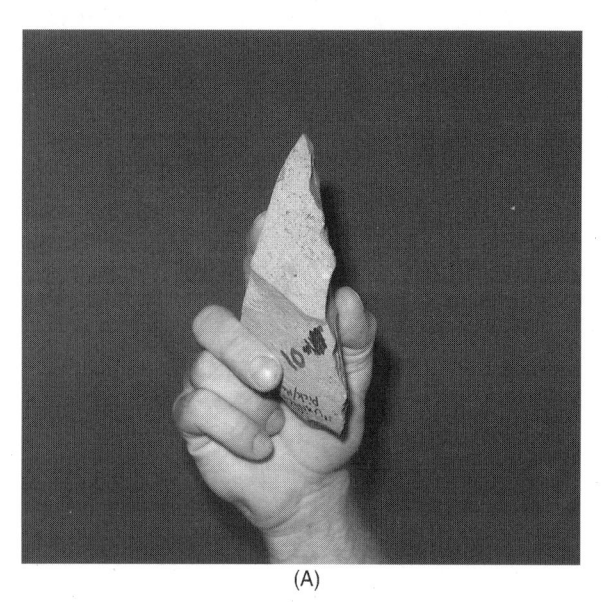

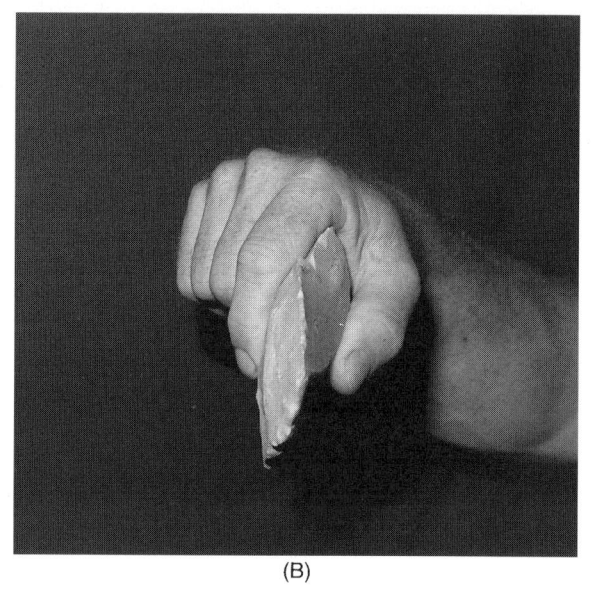

Fig. 5.9 Human finger–passive palm grips. A. Extended three-jaw-chuck grip. B. Buttressed pad-to-side grip.

Fig. 5.10 Human finger–active palm squeeze grip.

thumb-tip–fingertip prehension. It was thus surprising to find that these postures and movements rarely occurred in the experimental stone-tool-making sessions. Precision grips (defined as those exploiting leverage of the fingers without the palm) instead involved the full palmar surfaces of the thumb and fingers, or the thumb and the side of the index finger.

5.3.2 Contribution of bipedality to tool-use

Try throwing, digging, or pulling from a sitting position, and then again bipedally. It will be immediately apparent that with the leverage of the trunk these activities become more effective. Indeed, the trunk contributes almost 50 per cent of the speed to a ball thrown overhand (Toyoshima *et al.* 1974). In order to use the trunk effectively to accelerate a tool held in the hand, one must be able both to maintain balance of the trunk on the hindlimbs in bipedal posture, and to control the rotation and tilt of the trunk on the hindlimb as the forelimb wields the tool and as one hindlimb steps forward with the thrust of the arm. Humans are more steady on their hindlimbs than pongids because of their heavier hindlimbs in proportion to trunk and forelimbs, and because the lower limb-joints are aligned with the line of gravity in erect posture (Zihlman and Brunker 1979). The human ilium is short, with the sacroiliac and hip joints closely approximated, contributing to stability of the trunk on the hindlimbs (Straus 1962). Control of truncal movements on the hindlimbs is achieved by contraction of the gluteal muscles, which extend, abduct, and rotate the pelvis when the hindlimbs are fixed (Karlsson and Jonsson 1964), and by the extensor and rotator muscles of the back (see, for example, Floyd and Silver 1955). Monitoring of the gluteus maximus muscle by electromyography during throwing by the contralateral hand has revealed the key role of this muscle during throwing. It contracts consistently at a point when the trunk reaches the frontal plane and ceases to rotate laterally on the hindlimb. By reversing the rotation of the trunk, the muscle allows the angular momentum of the arm to increase as it moves ahead of the trunk. The uniquely human attachment of the gluteus maximus muscle to the ilium in the region of the posterior superior spine gives the muscle mechanical advantage in moving the trunk on the hindlimbs (Marzke et al. 1988).

5.3.3 The brain, the hand, and tool-use

Cortical mapping has revealed a striking contrast between non-human primates and humans in the proportionate size of cortical areas controlling the hand (Penfield and Rasmussen, and Woolsey and Settlage, cited by Washburn 1959). Calvin (1983) has advanced a theory to account for the initial increase in area of the cortex related to the hand and for cortical lateralization of rapid motor sequencing. He postulates that the rewards of rapid, one-handed throwing of small rocks may have been an important factor in selection for redundant timing circuits (increasing brain size) and for centralization of rapid motor sequencing on one side of the brain (lateralization). Both the multiplication and centralization of timing circuits would have contributed to the speed of the throw, and thereby to the distance from which a hominid might throw a stone at prey and to the 'stopping power' (p. 203) of the projectile. Calvin relates the emergence of language (which requires rapid sequencing) to these developments in the brain for rapid sequencing connected with tool-use. (Holloway (1976) has also noted the unique human ability to throw a missile over a distance with force and accuracy, and the spatiovisual integration required for this.)

Jones (1949), Oakley (1972a), and others have stressed the importance of the large human brain size to our dexterity, but with the assumption that the hand is unspecialized in structure because of its freedom from stresses encountered in locomotion. It is becoming clear now that changes in both hand structure and cortical control of its movments have been factors in the refinement of hominid dexterity. More specific suggestions have been made as to changes in cortical control of the hand which must have accompanied the evolution of tool-use and tool-making. Boesch and Boesch (1984) note the effect that a tool has on the nature of sensory information relayed to the brain by the hands as they strike objects with tools, and the adjustments which must be made in motor responses to this altered information. Marshack (1984a, b) calls attention to advances in interhemispheric and visual co-ordination of the two hands which must have occurred as the hands assumed increasingly differentiated functions in the course of hominid evolution.

Toth (1985) makes an interesting case for the existence of right-handedness (with concomitant marked lateralization of the brain) in early hominids by 1.9–1.4 m.y.a., based upon his analysis of flaking patterns on Early and Middle Pleistocene stone artefacts from Koobi Fora and Ambrona.

5.4 Fossil hominoid hands and locomotor apparatus

5.4.1 Miocene fossil Hominoidea (Table 5.2)

Among the known fossil species of the Miocene are two groups of genera which will be identified here as the ramamorphs and dryomorphs, following Ward and Pilbeam (1983). These two groups probably incorporated species that were in the line to modern pongids and hominids. The dryomorphs first appeared in Africa at least 20 million years ago (m.y.a.) and persisted in Africa and Europe until about 9 m.y.a. The ramamorphs appear in the African fossil record at least 16 m.y.a., and persisted in Africa and Eurasia until about 7 m.y.a. (with *Gigantopithecus* persisting until the Pleistocene (Conroy 1990)).

There is an abundance of dryomorph hand bones from the early Miocene genus *Proconsul*, some of which were found quite recently (Napier and Davis 1959; Walker and Teaford 1986; Walker et al. 1985; Beard et al. 1986; Begun et al. 1994). The hand lacks the curvature and cresting of the phalanges (Napier and Davis 1959) which in pongids are related to strong flexor musculature used in suspensory behaviour. Lewis (1971, 1972, 1985, 1989) has maintained that the wrist shares with living hominoid wrists derived features permitting a range of movement at the forearm/wrist and mid-carpal joints compatible with suspensory behaviour. However, most other functional and morphometric analyses have led to the conclusion that movement potential in this region is indicative of more cautious arboreal quadrupedal progression (see McHenry and Corruccini 1983; Rose 1983; Walker and Pickford 1983; Robertson 1984;

Table 5.2 Fossil hand-bones of dryomorphs and ramamorphs*

	Carpals	Metacarpals	Phalanges
Dryomorphs *Proconsul hesloni,* formerly *P. africanus* (Walker *et al.* 1993) (KNM-RU 2036) Kenya	L. hand: scaphoid, lunate, triquetrum (fragment), pisiform, trapezium, capitate, hamate	L. hand I–V, epiphyses I–V	L. hand proximal: I–IV; epiphyses III–V. Middle: II–V; epiphysis II. Distal: one (lost)
	R. hand: triquetrum	R. hand: I, IV, V; II epiphysis	R. hand middle: one; 2 epiphyses. Distal: I, two others
Additional *Proconsul* specimens	scaphoid (KNM-SO 999)† capitate (KNM-SO 1000) capitate (KNM-SO 1001) capitate (KNM-SO 1002) capitate (KNM-RU 1907)		
Ramamorphs *Sivapithecus?* (GSP 17119) Pakistan	capitate		
Sivapithecus parvada (GSP 6664, 19833, 17154, 19700; NG 940, 933)	hamate	L. hand: I	proximal I proximal (II?) proximal (III?) proximal (IV)

* Compiled from Pilbeam *et al.* (1980), Walker and Pickford (1983), Robertson (1984), Rose (1984a, 1986), Walker and Teaford (1986), Kelley (1988), Spoor *et al.* (1991). Additional recently-discovered hand specimens of *Proconsul hesloni* and *Proconsul nyanzae* are listed in Beard *et al.* (1986) and in Begun *et al.* (1994). Leakey *et al.* (1988a, b) list hand-bones found recently with other remains of *Turkanapithecus kalakolensis* and *Afropithecus turkanensis*. They are described as having general similarities with the hands of *Proconsul*, although there are several distinctive features. A comparative study of the bones is reported to be in progress. Rose (1992) has found shared derived morphology in the first carpometacarpal joint of *Afropithecus*, *Proconsul*, and modern hominoids.
† Kenya National Museum numbers.

and references therein). The newly-discovered bones are described and illustrated as exhibiting a forearm-wrist articulation quite unlike that of living Hominoidea, and a mosaic structure whose function was probably not closely comparable with that of any living monkeys or apes (Beard *et al.* 1986). However, the *Proconsul* functional pattern appears to be a good model for the ancestry of modern pongid patterns (Begun *et al.* 1994).

The structure of the hand in late Miocene ramamorphs is known only from a partial capitate, a hamate, a first metacarpal, and four proximal phalanges, all found in Pakistan. The capitate has some features indicative of suspensory activity and climbing, but there are also resemblances to the capitate of species that walk quadrupedally on the palm of the hand (Rose 1984a). The hamate similarly resembles that of less suspensory species (Spoor *et al.* 1991) Three proximal phalanges described by Rose (1986) similarly exhibit a mosaic of features, some found today in the chimpanzee, and others characteristic of living plantigrade quadrupeds.

The structure of the pollical proximal phalanx indicates the ability to oppose the pulp surface securely

Table 5.3 Fossil hominid hand bones from the Pliocene and Early and Middle Pleistocene[1]

	Carpals	Metacarpals	Phalanges
Laetoli (3.59–3.77 m.y.a.)[2]		II or III (1)	2 proximal 1 middle
Hadar (A.L. 288; A.L. 333/333w) (2.8–3.3 m.y.a.)[3]	pisiform, trapezium, 2 capitates, hamate	I (3), II (3), III (3), IV (3), V (6)	16 proximal 10 middle 2 distal
Omo (2.4–2.6 m.y.a.)[4]			1 middle
Sterkfontein (1.5–3.0 m.y.a.)[5,6]	capitate	I (1), II (1), III (3), IV (4), V (1)	5 proximal 1 middle 1 distal
Swartkrans (1.7–1.9 m.y.a.)[7]	triquetrum	I (2), III (1), IV (1), V (1)	6 proximal 8 middle 3 distal
Olduvai (O.H.7) (Bed I, 1.82 ± 0.13 m.y.a.)[8]	scaphoid, trapezium, capitate	II (base)	2 proximal 4 middle 3 distal
Koobi Fora (KNM-ER 164b, 1.6–1.9 m.y.a.)[9]		metacarpal head (1)	2 (3?) proximal
Koobi Fora (KNM-ER 803t, 1.6–1.9 m.y.a.)[9]		metacarpal fragment (1)	
Koobi Fora (KNM-ER3735, 1.88–1.91 m.y.a.)[10]			2 proximal
Olduvai (O.H.18) Bed II (1.15–1.71 m.y.a.)[11]			1 phalanx
Baringo (0.23–0.7 m.y.a.)[12]			2 proximal
Zhoukoudian (0.4–0.5 m.y.a.)[13]	lunate		
Jinniu Shan (0.28 m.y.a.)[14]	capitate, 2 lunates triquetrum hamate, scaphoid, trapezoid, pisiform, trapezium	II(1), III(1)	4 proximal 2 middle 1 distal
Nariokotome (KNM-WT 15 000)[15]		I(2)	1 proximal 1 middle

[1] Compiled from Black *et al.* (1933), Musgrave (1970), Leakey (1971), Howell and Coppens (1976), Day (1976, 1978), Howell (1978), Leakey and Leakey (1978), Leakey *et al.* (1976), Tobias (1978), White (1980), Bush *et al.* (1982), Ricklan (1986), and Lu (1990).
[2] Leakey *et al.* (1978).
[3] Brown (1982).
[4] Brown and Shuey (1976).
[5] Tobias (1980).
[6] Tobias (1978); Ricklan (1986, 1987, 1990).
[7] Howell (1978); Susman (1988, 1989).
[8] Curtis and Hay (1972).
[9] F. H. Brown (personal communication).
[10] Leakey *et al.* (1989).
[11] Hay (1976).
[12] Tallon (1978).
[13] Howells (1980).
[14] Lu (1990); but see Schick and Zhuan (1993).
[15] Walker and Leakey (1993).

against either the side of digit II or the pulp surfaces of the remaining digits (Pilbeam *et al.* 1980). The exact nature of the opposition is unknown, since there are no fossil remains of middle and distal phalanges of the other fingers.

5.4.2 *Pliocene and Early Pleistocene Hominidae*

The following sections will focus on fossil evidence of hand structure and on the morphology of the locomotor apparatus relevant to questions in this chapter regarding movements of the trunk on the hindlimb during activities involving the hands. For more complete reviews of the evolution of hominid bipedalism see McHenry and Temerin (1979), Zihlman and Brunker (1979), McHenry (1982), and Aiello and Dean (1990).

Bones of the hand which have been recovered from hominids through the Middle Pleistocene are listed by site in Table 5.3.

Hadar. One of the earliest, and the most extensive series, with excellent preservation of several joint-complexes, comes from the Hadar region of Ethiopia. The bones, attributed to the species *Australopithecus afarensis*, have been described by Bush *et al.* (1982) and Johanson *et al.* (1982*a*). Functional analyses of some regions have been made by Bush (1980), Tuttle (1981), Marzke (1983), McHenry (1983), Stern and Susman (1983), and Susman *et al.* (1984). The Hadar hands exhibit three pongid features (Table 5.4, Nos. 1–3) which together imply the existence of well-developed finger and wrist-flexor musculature. Apparently the hand of these hominids had the potential to maintain the fingers in a hook-like grasp of a branch by digits II–V and to flex the wrist strongly. There are other pongid traits, but evidence of knuckle-walking has not been found (Tuttle 1981; Marzke 1983).

There are interesting departures from the pongid pattern in the Hadar hands that approach the human pattern (Table 5.4, Nos. 9–10). These involve features that facilitate and stabilize grips by the thumb and the index and third fingers, and one that possibly indicates stabilization of the palm by a pisometacarpal III ligament (Marzke and Shackley 1986; see Fig. 5.7a, p. 133). Among the grips enhanced by these features are the three-jaw chuck thumb–finger, extended three-jaw chuck, pad-to-side, and buttressed pad-to-side grips.[1]

The fifth carpometacarpal joint is distinctive in the considerable depth of the anteroposterior concavity and the distal extent of the joint dorsally. It appears from the manner in which the fifth metacarpal was cradled by the hamate that it may not have moved over the same range that it does in modern humans, and thus may not have facilitated an effective squeeze grip.

The locomotor apparatus of the Hadar hominids is represented by a considerable portion of one skeleton (Lucy) and by bones from several other individuals. Descriptions of each region were published together in one issue of the *American Journal of Physical Anthropology* (Johanson *et al.* 1982*a, b*; Latimer *et al.* 1982; and Lovejoy *et al.* 1982). Several functional interpretations have been published of the segment proportions, the shape and orientation of the bones, and configurations at the hip, knee, ankle, and meta-

Table 5.4 Morphology of the Hadar hands

Pongid features	Human features	Unique features
1. Prominent marginal ridges on proximal phalanges for annular ligaments	8. Thumb length/hand length (closer to human than to apes)[2]	11. Deep hamate concavity for proximal M V
2. Longitudinally curved proximal phalanges	9. Orientation of carpometacarpal II joints away from sagittal plane[2]	
3. Long pisiform	10. Groove on hamate between M V facet and hook, possibly for pisometacarpal III ligament[3]	
4. Markedly concavo-convex carpometacarpal I joint, recalling *Pan*[1]		
5. Relatively narrow apical tufts on distal phalanges[1]		
6. M III styloid process absent		
7. Capitate 'waisting' at midline		

[1] Stern and Susman (1983).
[2] Marzke (1983).
[3] Marzke (1986).

tarsophalangeal joints. There is general agreement that the morphology of the locomotor apparatus reflects terrestrial bipedality, but there is considerable debate about the nature and energy requirements of these hominids' bipedal locomotion and about the extent to which they moved about in the trees. A review of the debate, focusing on evidence for arboreality and a unique form of bipedality, may be found in Susman et al. (1984).

The length of the humerus bore approximately the same relation to body size that it does in modern humans (Jungers 1982; Jungers and Stern 1983; Wolpoff 1983a, b), indicating that the trunk may have been better balanced on the hindlimbs than it is in pongids (Jungers and Stern 1983). The pelvis was short, with close approximation of the sacroiliac and hip joints, probably facilitating stable transmission of trunk weight to the hindlimbs (Johanson et al. 1982b; Berge et al. 1984). There is some question, however, about the precise orientation of the lumbo-sacral region (Abitbol 1995a). Unfortunately damage on the ilium in the region of the posterior gluteal line (Johanson et al. 1982a) precludes determination of whether the gluteus maximus muscle had an iliac attachment. Knee-joint structure in some specimens indicates possible differences from modern humans in the full range of movement capability (Stern and Susman 1983; Tardieu 1981, 1983; but see Bedford and Lovejoy (1985)). Ankle and foot morphology is consistent with bipedality, although there are pongid elements in the pattern which cause Stern and Susman (1983) and Susman et al. (1984) to infer a climbing component in the locomotor repertoire, and Gebo (1992) to suggest some functional differences from modern humans. In the view of Latimer (1983), Latimer and Lovejoy (1989, 1990), and Latimer et al. (1987) the structure of the Hadar foot skeleton indicates commitments to bipedality that would have constrained climbing. However, relative proportions of the foot and its components indicate possible biomechanical differences from modern humans in bipedal gait (Jungers 1988a). There is some evidence at the knee and ankle joints of dimorphism in the range of mobility, which has been interpreted as evidence of sexual (Stern and Susman 1983) or intergeneric (Tardieu 1983) difference in the ability to climb trees.

There are several indications that the adaptation of *A. afarensis* to bipedality differed from that of modern humans. The hindlimbs were short relative to estimated body weight—comparable in this ratio to pygmy chimpanzee hindlimbs— and the pelvic inlet is wider relative to body size than in modern humans (Rak 1991). Whether the ratio indicates a requirement for more energy consumption during bipedal walking has been a subject of debate (Jungers 1982; Jungers and Stern 1983; Wolpoff 1983a, b). Associated with the shorter hindlimb length in A.L. 288-1 was a relative hip-joint size smaller than in modern humans (Jungers 1988b). From the orientation of the ilium it appears that the muscles involved in balancing the trunk on a single hindlimb during bipedal progression may have functioned differently from those in humans (Jungers 1982; Jungers and Stern 1983). Relatively long toes may also have affected gait (Jungers 1982; Jungers and Stern 1983).

Sterkfontein. A capitate from Member 4 at Sterkfontein, in South Africa, dated to c.2.5 m.y.a. (Tobias 1980) and attributed to the species *Australopithecus africanus* (Broom and Schepers 1946), is similar to the Hadar capitates in its mosaic of pongid and hominid features (McHenry 1983; Marzke 1983). It is intermediate in size between the two Hadar specimens. Other hand bones from Members 4 and 5 at Sterkfontein (Tobias 1978; Ricklan 1986, 1987, 1990) have not yet been fully described. From a functional analysis of selected hand bones from Member 4, attributed to *Australopithecus africanus*, Ricklan (1988) has concluded that they exhibit a capacity for at least the range of modern human activities facilitated by hand structure in *Australopithecus afarensis*, with potential for a powerful grip and powerful hand and wrist movement. Attributes of the thumb metacarpal and distal phalanx are suggestive of a potential for thumb/finger tip-pad precision opposition, assuming human-like relative proportions of the metacarpals and phalanges (Ricklan 1990).

More is preserved of the *Australopithecus africanus* postcranial skeleton, which is very similar to that of *Australopithecus afarensis* (McHenry 1986; Abitbol 1995b). Remains of the pelvis of the same species have also been found at Makapansgat (Dart 1949), in a level whose probable age is c.3 m.y.a. (Cadman and Rayner 1989). There is evidence on one of a restricted but distinctively human attachment of the cranial fibres of the gluteus maximus muscle to the ilium (Dart 1949). Bipedality in this species is indicated by a suite of morphological features that are incorporated into a mosaic whose implications for locomotion are not fully understood (McHenry 1982 and references therein).

Olduvai. More information about hand function and potential grips in early hominids may be gleaned from the hand and wrist bones found at the Olduvai FLK NN site in Tanzania, on the same ancient land surface with Oldowan tools and skull fragments, a mandible, and a foot (Napier 1962a, b; Leakey 1971; Day 1976; Susman and Creel 1979). Napier inferred from the

Table 5.5 Morphology of the Olduvai hand

Pongid features	Human features
1. Scaphoid morphology (recalls African ape morphology)[1]	3. Broad apical tufts on distal phalanges, with rugosities[3] indicating specialized anchoring of pads[4]
2. Middle phalanges robust, curved, with marked insertion area for flexor digitorum superficialis[1,2]	4. Extensive, relatively shallow trapezial facet for M I[3,5,6]

[1] Susman and Stern (1979).
[2] The middle phalanx from the third finger of a left hand at Omo, Ethiopia (Coppens 1973), is said by Day (1978) to be similar in morphology to these phalanges.
[3] Susman and Creel (1979).
[4] Shrewsbury and Sonek (1986).
[5] Stern and Susman (1983).
[6] Trinkaus (1989).

morphology of the hands an ability to make the associated tools. Structurally the scaphoid and phalanges find their closest parallel among African apes (Susman and Creel 1979). Climbing and suspensory behaviour are suggested by the structure of the middle phalanges (Susman and Stern 1979). The modern human features of the thumb and fingertips (Table 5.5) are those which allow manœuvring of an object while still maintaining a firm grip (Marzke and Shackley 1986). A flatter trapeziometacarpal articulation than is characteristic of modern humans may reflect large forces associated with grips (Trinkaus 1989).

A tibia and fibula (OH 35) from the FLK site may belong to the same individual as the foot from FLK NN (Susman and Stern 1982). There is general agreement that bipedality is indicated by the morphology of the hindlimb remains, although debate continues as to evidence in the foot for climbing (Lewis 1980, 1989; Oxnard and Lisowski 1980; White and Suwa 1987; Susman and Brain 1988).

A partial skeleton (OH 62), found at a level dating to c.1.8 m.y.a. and attributed to *Homo habilis*, (Johanson *et al.* 1987) is more similar than *Australopithecus afarensis* to African apes in relative limb dimensions compared by Hartwig-Sherer and Martin (1991).

Swartkrans. One wrist bone and 25 hand-bones have been found at the Swartkrans site in South Africa, from levels dating to approximately the age of the FLK NN site with the Olduvai hand. Stone and bone tools also occur at these levels (Susman 1988, 1989). Susman has described all but two of the bones, SK 84 (a thumb metacarpal described by Napier in 1959) and SKW 14147 (a fifth metacarpal described by Day and Scheuer (1973)). Since two genera (*Paranthropus* and *Homo*) occur at all levels in which hand bones have been found, generic identification of these bones cannot be made with certainty (see Trinkaus and Long 1990). Susman's description of the new first metacarpal specimen unfortunately is not sufficient to distinguish it from SK 84 generically on the basis of distal articular morphology or to indicate whether either or both had a proximal articular surface consistent with the flat facet for the first metacarpal on the Olduvai trapezium. Broad apical tufts on the distal phalanges, and an extensive insertion area for the flexor pollicis longus muscle on the distal phalanx of the thumb, are both present in these fossils, and are both essential for achieving a firm grip of large stones and for manœuvring them with control. There is disagreement about the potential range of mobility at the joint between the fifth metacarpal and the hamate (Day and Scheuer 1973; Lewis 1977; Marzke 1983). Without a fossil hamate, and without adequate data on movements at this joint in humans and non-human primates, the functional implications of muscle markings and proximal facet contours on the fifth metacarpal are ambiguous.

Koobi Fora. Little can be determined about hand function from the two metacarpal fragments and phalanges from Koobi Fora. Two proximal phalanges (164b) are described by Howell (1978) as being long and straight. The other two phalanges exhibit evidence of extremely powerful flexion potential (Leakey *et al.* 1989).

Habitual bipedal locomotion is suggested by the morphology of a hip-bone from Koobi Fora (KNM-ER 3228), which is dated to about 1.9 m.y.a. (Rose 1984b). From the orientation of the ilium, it may be inferred that the position of the trunk on the hindlimb during bipedal locomotion was controlled by the gluteal muscles in the same way as it is in modern humans (ibid.). Femora of the same general age from Koobi Fora indicate bipedality, but there is some disagreement as to the taxonomic implications of features which distinguish these femora from those of earlier hominids and from modern humans (Kennedy 1983a, b; Trinkaus 1984a; Walker and Leakey 1993). The functional implications of these features are not fully understood. On the basis of similarities to the chimpanzee in a partial skeleton (KNM-ER 3735), Leakey et al. (1989) suggest the possibility that more than one non-robust species coexisted in the region at this time.

West Lake Turkana. A large portion of a skeleton from a juvenile male has recently been found at the Nariokotome III site on the west side of Lake Turkana and is dated to 1.53 ± 0.05 m.y.a. (Brown and McDougall 1993). The first metacarpal shafts and shafts of a first proximal phalanx and a middle phalanx have been described (Walker and Leaky 1993). There is some question as to whether the metacarpals may be attributed to the hominid.

The pelvis and hip region have been reconstructed from remains of the right and left hip bones and sacrum and right and left femora. Bipedality is clearly indicated, with evidence for the biomechanical advantages of a narrow pelvis and long femoral neck.

5.4.3 *Middle Pleistocene*

Until now remains of hands from the Middle Pleistocene have not been extensive enough to throw much light on the range of potential prehensile postures and manipulative behaviour of these hominids. The size and proportions of the Zhoukoudian lunate are described by Black et al. (1933) as being similar to those of humans. The complete proximal phalanx of the fifth finger from Baringo is curved, with well-developed marginal ridges for the annular ligament (Musgrave 1970). Considerably more information will be obtained from hand bones found recently at Jinniu Shan (Lu 1990) and Atapuerca (Aguirre et al. 1990)

A hip-bone from Olduvai Bed IV (O.H.28), dated to 0.8–0.6 m.y.a. (Hay 1976), and another which is possibly as recent as 0.25 m.y.a. from Arago in France (Arago XLIV) closely resemble the Koobi Fora KNM-ER 3228 os coxae (Rose 1984b). Femora are known from Middle Pleistocene levels at Olduvai and Zhoukoudian. While their morphology reflects bipedality, there is debate about taxonomic and functional implications of some structural details (see Day 1973; Kennedy 1983b; Ruff et al. 1993).

5.4.4 *Late Pleistocene*

There are several partial hand-skeletons and a large number of additional hand-bones of Neanderthals which together provide information about relative proportions of hand segments, the development of hand muscles, and movement potential at all the joints (Table 5.6). Functional analyses of the hand have considered primarily the proportions, muscular development, and contours of some of the joint surfaces (Musgrave 1971, 1973; Trinkaus 1983a, b;

Table 5.6 Morphology of Neanderthal hands*

General modern human features	Exceptional features
1. Hand size same relative to arm. 2. Relative proportions of hand segments same, with exception of thumb (see No. 4). 3. M III styloid process present.	4. Relatively long distal phalanx and short proximal phalanx of thumb 5. Consistently prominent ridges on metacarpals for attachments of interossei, especially first dorsal interosseus 6. Well-developed crests on M I and V for opponens muscles 7. Marked insertion areas on distal phalanges for flexor digitorum profundus 8. Well-developed tuberosities on carpals for flexor retinaculum attachment 9. Relatively broad metacarpal heads and proximal phalangeal bases 10. Very broad tubercles on distal phalanges

* Compiled from Musgrave (1970, 1971, 1973); Trinkaus (1983a, b); and Stoner and Trinkaus (1981).

Stoner and Trinkaus 1981; Trinkaus 1989), relating these in a general way to manipulative behaviour. Trinkaus (1989) has found that the Neanderthal trapezia, like the Olduvai trapezium, were flatter than in more recent humans, and the trapezial surface of the first metacarpal was dorsopalmarly straight or convex, as opposed to concave, in approximately three-quarters of the specimens. He attributes this morphology to large axial forces at the joint generated by well-developed extrinsic and intrinsic hand muscles that secured the grip by broad distal phalangeal apical tufts. Metrical comparisons have been made with hands of early anatomically modern *Homo sapiens* (Musgrave 1970, 1973; Trinkaus 1983a, 1989). Detailed comparisons with these hominids in joint morphology, considered in the light of requirements for the manipulation of Middle and Upper Palaeolithic tools, have not yet been published.

Relative proportions of the hand and its segments are similar to those of modern humans, save for the relatively long distal phalanx and short proximal phalanx of the thumb (Musgrave 1970, 1971; Trinkaus 1983b; Trinkaus and Villemeur 1991). There is a styloid process on the third metacarpal, indicating stability at the centre of the palm (see Fig. 5.7B, p. 133). In comparison to modern humans, larger muscle-markings on the bones, larger projecting tuberosities on the wrist-bones, and proportionately larger joint surfaces reflect the greater size and force of the extrinsic and intrinsic musculature used in gripping and manœuvring tools. Two possible explanations for the great breadth of the tubercles on the distal phalanges (implying very broad overlying pads) are (1) that the pads secured a powerful grip (Musgrave 1971; Trinkaus 1983b) and (2) that the digits probably were highly vascular, and therefore of advantage in cold environments (Musgrave 1971).

The structure and potential functions of the fifth carpometacarpal joint have not been thoroughly examined throughout the Neanderthal sample, but have been described as being similar to the pattern in modern humans (Trinkaus 1983b). (Account should be taken of variability in structure of this joint in both Neanderthal and modern human hands (Musgrave 1970; Marzke, personal observation)).

The morphology of the Neanderthal locomotor apparatus is well known from a number of skeletons, and has been the subject of functional analysis by Trinkaus (1984b and references therein), Rak and Arensburg (1987), and Ruff *et al.* (1993). Some departures from the mechanics of modern human bipedal locomotion may be indicated by the low crural index (Trinkaus 1984b) and by the externally rotated os coxae (Rak and Arensburg 1987).

5.5 Origin and evolution of the hominid hand and bipedality

5.5.1 *The evidence from the comparative morphology of extant primates*

The hand has been central in debates concerning the relation of humans to apes. Blumenbach (1791) classified humans in an order 'Bimana' ('two hands'), separate from other primates with four 'hands' ('Quadrumana'). Huxley (1871) found evidence in the hand that African apes were more similar to humans than to other primates. With the turn of the century, interest came to focus primarily on the relative size of the thumb, which in pongids is so unlike the human thumb in its length relative to the fingers. Jones (1948, 1949) postulated human ancestry among generalized, tarsier-like primates, with hand proportions similar to ours. Osborn (1927a, b, 1928, 1930) and Straus (1949) considered the hand proportions of Old World monkeys to be a suitable model for the ancestral human hand, and the fossil hand of *Proconsul* was also offered as a model (Clark 1962). Implicit in these models was the search in non-human primate hands for human manipulatory potential. Gregory (1927, 1928, 1930, 1934) and later Washburn (1959, 1967, 1968) followed another approach, identifying in hominid hands remnants of pongid features compatible with pongid postures and locomotion. According to their model, the human hand is a modified pongid one, which underwent changes in thumb–finger proportions with the evolution of bipedality and tool-use. Since 1960, functional analyses of hominoid hands (see references in Sections 5.2 and 5.3) have been directed toward identifying more specifically the morphological correlates of suspensory behaviour in pongid hands, of knuckle-walking in African ape hands, and of tool-use in hominid hands. It is possible now to examine with greater depth and breadth the question of human ancestral hand uses, and the relationship of tool-use to the evolution of the hominid hand.

A search in modern human hands for evidence of ancestral locomotor behaviour has turned up very little, although two features are strongly suggestive of suspensory behaviour. One is the lack of a direct articulation between the ulna and the wrist. A similar condition (partial or total elimination of direct articulation) is found in apes and in the Prosimian subfamily Lorisinae (Cartmill and Milton 1977). Articulation of the ulna with both the triquetrum and pisiform is characteristic of the remaining primates. A possible explanation for the parallel reduction of ulnar participation in the forearm–wrist joint in Hominoidea and Lorisinae may be found in their similar habits of bridging gaps between branches by careful grasping of the new support followed by release of the old, a behaviour which requires mobility of the wrist (Cartmill and Milton 1977).[3]

The second human feature suggestive of suspensory behaviour is the reduction of the contrahentes layer of muscles, shared with great apes. Separate contrahentes adductor muscles for the second, fourth, and fifth fingers may have been lost in our common ancestors with these apes, as their arboreal postures and locomotion increasingly exploited the unified flexion of the four long fingers to maintain a hook-like grasp of the branches (Marzke 1971).

A possible clue to use of the ancestral hand in terrestrial locomotion may be the presence in modern humans of a well-developed deep transverse intermetacarpal ligament. All non-human species known to have this ligament walk on the ground primarily on the fingers, with the metacarpals held perpendicular to the ground in line with the forearm. Weight is supported either on the back of the fingers (in the knuckle-walking of African apes) or on the palmar aspect of the fingers (in the digitigrade walking of baboons, some macaques, the patas monkey, and cats).[2] However, one cannot rule out the possibility that the ligament may have evolved separately and more recently in hominids in response to tool-using activities requiring stabilization of the palm against blows directed toward the metacarpal heads.

The source of modern human manipulative facility may be found in the opposable thumb, the relatively broad, sensitive touch-pads on the fingertips, and the well-differentiated intrinsic musculature of the thumb, which are shared with most Anthropoidea; to these our derived features were added as stresses on the hands came to be associated more with tool-use and tool-making than with locomotion. The comparison of living primate hands can presently lead us no further than this. It is the fossils which reveal the extent to which the hominid hand was committed to suspensory behaviour and perhaps limited in manipulatory behaviour by awkward thumb–finger proportions before our ancestors became bipedal. We now consider the most important evidence from this source.

5.5.2 The fossil evidence

Miocene hominoids and Pliocene and Lower Pleistocene hominids

Known fossils of hominoid hands from the early Miocene do not have the full suite of morphological features shared by hominids and the Great Apes. It cannot be determined from the available fossil record when the full suite was attained.

At the other extreme, the Pliocene Hadar hands have more in common with the hands of the Great Apes than do modern human hands. Some of their pongid features indicate a well-developed flexor apparatus that may have been used for climbing. However, built into this complex were a few distinctively human features that should have facilitated more firm and controlled grips of objects than is reported for pongid hands. These grips would have been effective for the use of stones in pounding nuts, digging, butchering small game, and cutting vegetation, and for throwing stones the size of tennis balls with speed and accuracy at small game. Perhaps the groove on the hamate, which in modern humans accommodates the pisometacarpal ligament (see Fig. 5.7A, p. 133) that stabilizes the third metacarpal base, announces an early stage in hominid adjustment to the percussive forces endured during repeated pounding with stones.

The Early Pleistocene Olduvai hand still retained a pongid-like flexor apparatus, but, compared with the Hadar hands, the Olduvai one was more modern in a few features that facilitate the retention of a firm grasp and the control of stones by the thumb and fingers during tool-use and tool-making (Marzke and Shackley 1986). This hand and the hand-bones from Swartkrans are consistent in their morphology with the occurrence of stone tools at these sites. However, the claim by Susman (1988, 1994) for evidence in these hands of tool-making capabilities cannot be sustained (Marzke, submitted; see also Aiello (1994), Hamrick and Inouye (1995), and Ohman *et al.* (1995)).

The stresses of tool-use and tool-making are reflected in the robustness and muscularity of the early Upper Pleistocene Neanderthal hands, while all traces of pongid modifications in the phalanges for strong flexion of the fingers are gone. We also see in these hominids evidence of structure in the carpometacarpal V region that may be necessary for effective use of the full palm and fourth and fifth fingers in controlling and securing cylindrical tools. Zvelebi (1984) cautions that Neanderthal dexterity may have been limited by mental and physical constraints. Hand morphology alone, however, does not seem to indicate any such physical limitations. The trend in the Upper Palaeolithic toward more gracile hands without the markedly enlarged joint surfaces (Trinkaus 1983*a*) may have been related to shifts in tool technology which reduced demands on the hand musculature.

If our functional interpretation of Neanderthal hand-structure is correct, one may expect to find similar morphology in the hands of Middle Pleistocene hominids, whose stone tools reflect the same requirements for strength and finger mobility.

5.5.3 The hand and the origin of bipedality

The origin of bipedality is often attributed to the advantage it gives by freeing the hands for use in foraging and feeding, tool-use, or the carrying of food and other objects.[4]

Our experiments with tool-use indicate that for early hominids it may have been not so much the passive freeing of the hands but the active exploitation of the trunk in tool-use that contributed to selection for morphological adjustments to bipedality. Bipedal posture enhances the effectiveness of the hands in wielding tools by providing balance during movement of the forelimbs and by adding the leverage of the trunk to that of the forelimb (Marzke 1983, 1986; Marzke et al. 1988). Although Dart (1959) and Robinson (1972) have previously noted these advantages of bipedality, their relevance to the origin of bipedal posture and locomotion seems to have been overlooked. The sequence of appearance of structural characteristics related to bipedality is interesting in this regard. *Australopithecus afarensis* had morphological features compatible with the balance of the trunk on the hindlimb in bipedal posture, but lacked the full complement of adjustments in morphology and hindlimb length to the modern human type of striding locomotion (Jungers and Stern 1983). The limb proportions of the OH 62 skeleton and the hip morphology of the KNM-WT 15 000 skeleton further record the disjunction of postural and locomotor evolution. Evolution of the unique attachments and the proportionately large size of the gluteus maximus muscle in hominids may be related as much to the requirements for control of the trunk on the hindlimbs during movements involving forelimb activities in tool-using as to those for maintaining the trunk above the hindlimbs in bipedal locomotion (Marzke 1986; Marzke et al. 1988). Additional unique features of the pelvis and hindlimb in later hominids appear to be related to loading and energetic factors associated with bipedal locomotion (see Rodman and McHenry 1980 and Jungers 1988b).

5.5.4 New perspectives on the role of tool-use and tool-making in the evolution of the hominid hand and bipedality

The functional analyses of derived features in the hominid hand and locomotor apparatus in this chapter have focused on tool-use, since this is a category of behaviour that distinguishes modern humans from non-human primates in the extent to which it places demands on morphological structure. The concurrent appearance, in the earliest fossil hominids, of morphological features in both the hand and the locomotor apparatus that are essential to the distinctively human facility in tool-use is consistent with the hypothesis that tool-use was a factor in the divergence of hominids from the pongid line. Stone tools manufactured by hominids appear much later in the pre-historic record than these morphological features, but consideration should be given to the obvious advantages that the controlled and forceful manipulation of unmodified sticks and stones (which tend not to be preserved or recognized in the record) might have given the earliest hominids in their exploitation of savannah and woodlands food sources (Peters and Maguire 1981; Peters and O'Brien 1984), predominantly in 'extractive foraging' (Parker and Gibson 1977, 1979). We consider briefly some of these advantages in connection with the following activities.

Digging. The potential advantages to early hominids of using a digging- or probing-stick to reach underground roots, water, and animals are widely recognized.[5] A precedent for the use of digging-sticks is seen in the use of the hand for digging by baboons (De Vore and Hall 1965; Hamilton et al. 1978; and Dunbar 1983), the mountain gorilla (Schaller 1963), and the pygmy chimpanzee (Kano 1979). Digging-sticks today are an important component of the tool kit among hunter–gatherers (Tanaka 1976; Hill 1982). The probability that foods requiring the use of a digging-stick would have been a component of the diet for early hominids is suggested by their inclusion in lists of foods determined to have been available to and edible by early hominids (Hatley and Kappelman 1980; Peters and O'Brien 1981; Stahl 1984). Wear of a type that appears today on digging-sticks used to extract bulbs has been found on some bones in Member I at Swartkrans, from 1.5–2.0 m.y.a. (Brain 1985). Wear on two early stone-flakes characteristic of wood-whittling (Keeley and Toth 1981) may also be indicative of the manufacture of digging-sticks (Isaac 1983). The morphology of the australopithecine trunk and hindlimbs, allowing use of the trunk as leverage, should have facilitated forceful wielding of these digging-tools.

Throwing. Throwing has not been widely discussed in connection with tool-use by early hominids, although brief reference to the likelihood of its having been within their repertoire has been made (Robinson 1963; Holloway 1976; Isaac 1979, 1984; and Isaac and Crader 1981). Isaac (1987), in an extensive review of historical accounts of throwing, presents a compelling argument for examining critically the potential role of throwing in hominid origins. Parker and Gibson

(1979) speculate that the hominid capacity for precision throwing developed in connection with hunting with the emergence of *Homo habilis*, and O'Brien (1981, 1984) and Calvin (1993) consider the Acheulean 'hand-axes' to have been projectile weapons. Spheroids and unmodified stones that could have been projectiles for use in hunting have been found in many assemblages. Chimpanzees throw objects during displays, sometimes aiming the objects, but not often hitting their targets, although accuracy has been observed to improve with increase in the frequency of throwing (van Lawick-Goodall 1970). Throwing of objects in displays has been suggested as a model of the initial stage (or of one factor) in the evolution of weapons (Hall 1963; Kortlandt and Kooij 1963; Washburn and Jay 1967; van Lawick-Goodall 1968; Lancaster 1968). That it may in fact have played a decisive role in food-acquisition and self-protection is suggested by the damage which a stone, spear, or club is known to inflict when a skilful person today throws it at another human or at prey, and the advantage this technique gives of killing at a distance (Darlington 1975; Calvin 1983, 1993). Recently it was suggested, on the basis of an apparent potential for using a three-jaw-chuck grip and for relatively steady bipedal posture and control of the trunk in *Australopithecus afarensis*, that throwing could have been an effective component of this species's strategies for food-acquisition and self-protection from predators (Marzke 1983, 1986). The advantage of the three-jaw-chuck thumb-finger grip is that it permits manœuvring of the stone by the thumb and fingertips and exploitation of finger leverage in imparting direction and speed to the missile. Relatively small, light stones may be thrown with greater velocity than heavier stones. Toyoshima and Miyashita (1973) have demonstrated this size–velocity relationship in the throwing of balls. The ability to throw stones weighing of the order of 500 grams using the three-jaw-chuck grip thus might have contributed to the precision and effectiveness with which animals could be struck by these early hominids. Controlled rotation of the trunk on the hindlimb increases the velocity still more by its contribution of trunk leverage to the leverage of the arm, forearm, wrist, and fingers. Marked rotation of the trunk does not seem to be a characteristic element of chimpanzee throwing motions (Kortlandt 1972).

Clubbing and pounding. Clubbing with a heavy stick would have been another potential means of securing small animals for meat (Bartholomew and Birdsell 1953; Dart 1959; Oakley 1961, 1972*b*; Robinson 1963). This technique has been noted among the contemporary Ache hunters (Hill 1982). Brandishing of sticks by chimpanzees during aggressive encounters suggests a behavioural forerunner of clubbing (Kortlandt 1965; Washburn 1963). Morphological adjustments to bipedal posture should have lent stability and the leverage of the trunk to the force and accuracy of this potential foraging technique in early hominids.

Many animal species pound foods with stones and wood to remove inedible coverings (Beck 1980; Boesch and Boesch 1981, 1983*a*, *b*; Sugiyama 1981; Sugiyama and Koman 1979). The relatively short fingers and long thumb should have facilitated a firm and adaptable grip of hammerstones by australopithecines. Controlled and forceful use of cylindrical wood or bone tools as stable extensions of the forelimb would have benefited from the ability to use a secure squeeze grip, which is facilitated by a pattern of morphological features that may have been acquired relatively recently (Marzke *et al.* 1992).

5.5.5 *The hand in symbolic behaviour*

The potential in chimpanzees as well as in humans for use of the hands in symbolic behaviour suggests the same potential among early hominids (see Hewes, this volume, Chapter 21). Modern human hand morphology is not essential to gesturing. If hand-gestures were used by early hominids, the ability for even the earliest known hominids to maintain good balance in upright posture while the hands were signalling might have been an advantage. The later use of stone tools for engraving and writing symbols should have benefited by some of the same developments in hominid hand morphology that we have found to enhance the manipulation of these tools for other purposes.

5.6 Conclusions

The neurological potential for using and making tools, and the ecological circumstances eliciting these activities, have been well documented in chimpanzees and other primates (see Beck 1980 and McGrew 1992, for reviews, and for surveys of literature demonstrating precedents for tool-using in early hominids). It is reasonable to assume that hominid ancestors, even if they lacked the hand proportions, sensitivity, and dexterity of modern humans, possessed tool-using capabilities at least comparable to those of chimpanzees, and applied them to the problems of food-acquisition unique to their niche. There can be little doubt that the gripping and wielding of tools put progressively more consistent and extreme demands on human hands between the time that hominids diverged from their common ancestor with the chimpanzee and the stage at which stone-tool technology was developed.

The possibility should therefore be considered that some morphological features of the hand and of the trunk and hindlimb enhancing the effectiveness of tool-use may have become established in the australopithecines in response to the demands of tool-use. The combination in australopithecines of potential for a partial human repertoire of gripping postures with the capacity for balance and possibly some trunk-control in bipedal posture equipped these early hominids for using tools to acquire foods such as roots and small animals that are accessible only in small quantities or in some cases are inaccessible to most competitors. Probably no single behaviour alone can account for the emergence of hominid bipedality (Napier 1964; Sigmon 1971; Robinson 1972; Teleki 1975; Rose 1976, 1982). But tool-use should be considered as one important parameter in the interplay of behaviours which must have given direction to the trend toward modern human hand and hindlimb structure in early hominids.

Between the Pliocene and the earliest Pleistocene, when the first tentative anatomical evidence associated with human manipulative capabilities appears in the Hadar, Sterkfontein, Swartkrans, and Olduvai hands, and the Late Pleistocene, when most of the modern human skeletal features are present in Neanderthal hands, there is a long period during the Early and Middle Pleistocene that is virtually undocumented by fossils bearing clues of manipulative behaviour. This is the period during which stone-tool-making technology emerged and became a fundamental component of the hominid terrestrial adaptation. Functional analysis of recently-discovered hominid hand-bones from this period will be of great interest in tracing stages in the evolution of hominid hand-structure and manipulative capabilities, and in relating morphological change to changes in the brain and to advances in hominid technology and symbolic behaviour. There have been suggestions in the past that the Swartkrans and Olduvai hominids, whose hand-bones have been found alongside stone tools, were advanced in their ability to use a 'precision grip' using the thumb and some of the fingertips in addition to a 'power grip' using the palm and fingers. Our present understanding of the gripping patterns which most comfortably and effectively facilitate tool-use and tool-making, and of the complexity and subtlety of the forces and movements accompanying the manipulation of tools, caution us to avoid a simple categorization of gripping patterns into 'power' and 'precision' grips, and the placing of these general categories into a temporal sequence. The advance at Swartkrans and Olduvai over earlier hominids seems to have been primarily in the ability to position the fingers and secure the grasp for a more forceful and finely-controlled manipulation of objects, though using some of the same grips that were within the potential of earlier hominids.

As new fossil hands are recovered, it is to be hoped that there will be a search for evidence of the full range of postures and movements that may have been within their capabilities, and that more experiments will be performed with tools to determine the degree of skill, strength, and manual mobility necessary to manipulate them.

Acknowledgements

The author gratefully acknowledges her indebtedness to Dr E. Brandt and Dr R. Marzke for critical reading of an early draft of the manuscript.

Notes

1. The term 'three-jawed-chuck' grip used by Shrewsbury and Sonek (1986) is comparable to only one of several grips grouped under this term by Marzke and Shackley (1986), the 'three-jaw pad-to-pad' grip, in which objects are held exclusively by the pads over the distal phalanges. Shrewsbury and Sonek conclude from their analysis of the Hadar hand-structure that it may have been incapable of using this grip. Their disagreement with the conclusion of Marzke (1983), that a three-jaw chuck grip was within Hadar hominid capabilities, appears to stem from a misinterpretation of Marzke's term 'three-jaw chuck', which in this context referred to a grip in which all the phalanges contact the surface of the object (the 'three-jaw chuck thumb/fingers' grip of Marzke and Shackley).

2. Forster (1917) describes transverse fibres in the position of the deep transverse intermetacarpal ligament in an orang-utan. Fibres in this position were found in the hands of two orangutan individuals dissected in our laboratory, but they were not well developed. We did find a stout ligament in several digitigrade species, including an olive baboon, a pigtail macaque, a patas monkey, and a cat. The ligament has not been found in our dissections of any of a large number of palmigrade species. These findings suggest that a stout ligament may be related specifically to digitigrade and knuckle-walking locomotion (Marzke and Marzke 1987).

3. Lewis (1985a) has refuted the claim that lorisine wrist joint specializations parallel those of hominoids.

4. See, for example, Darwin (1874); Bartholomew and Birdsell (1953); Washburn and Howell (1960); Washburn (1960, 1967); Hewes (1961, 1962); Brace (1962); Du Brul (1962); Gruber (1962); Hockett and Ascher (1964); Kortlandt (1967); Jolly (1970); Mann (1972); Rose (1974, 1976, 1982); Lancaster (1978); Peters (1979); Sugiyama and Koman (1979); Wrangham (1980); Lovejoy (1981); Hill (1982); and Hunt (1994). Some of these theories, and others explaining the origin of bipedality which do not involve the hand, have been reviewed by McHenry (1982) and by Rose (1982).

5. See Bartholomew and Birdsell (1953); Washburn and Avis (1958); Washburn (1960); Robinson (1963); Mann (1972); Coursey (1972, 1973); Hamilton (1973); Tanner and Zihlman (1976); Hatley and Kappelman (1980); Isaac and Crader (1981); Zihlman (1981); Toth (1982); Wolpoff (1982); Isaac (1984); and McGrew (1984).

References

Abitbol, M. M. (1995a). Lateral view of *Australopithecus afarensis*: primitive aspects of bipedal positional behaviour in the earliest hominids, *Journal of Human Evolution*, **28**, 211–29.

Abitbol, M. M. (1995b). Reconstruction of the STS 14 (*Australopithecus afarensis*) pelvis. *American Journal of physical Anthropology*, **96**, 143–53.

Aguirre, E., Arsuaga, J. L, Bermudez de Castro, J. M, Carbonell, E., Ceballos, M., Diez, C., et al. (1990). The Atapuerca sites and the Ibeas hominids. *Human Evolution*, **5**, 55–73.

Aiello, L. (1994). Thumbs up for our early ancestors. *Science*, **265**, 1540–1.

Aiello, L. and Dean, C. (1990). *An introduction to human evolutionary anatomy*. Academic Press, London.

Alston, W. and Weiskopf, D. (1972). *The complete baseball handbook*. Allyn and Bacon, Boston.

Bartholomew, G. A., jun. and Birdsell, J. B. (1953). Ecology and the protohominids. *American Anthropologist*, **55**, 481–98.

Beard, K. C., Teaford, M. F., and Walker, A. (1986). New wrist bones of *Proconsul africanus* and *P. nyanzae* from Rusinga Island, Kenya. *Folia Primatologica*, **47**, 97–118.

Beck, B. B. (1980). *Animal tool behavior: the use and manufacture of tools by animals*. Garland STPM Press, New York.

Bedford, M. E. and Lovejoy, C. O. (1985). Morphological correlates of differential knee joint reaction forces in primates. *American Journal of Physical Anthropology*, **66**, 143.

Begun, D. R., Teaford, M. F., and Walker, A. (1994). Comparative and functional anatomy of *Proconsul* phalanges from the Kasawanga Primate Site, Rusinga Island, Kenya. *Journal of Human Evolution* **26**, 89–165.

Berge, C., Orban-Segebarth, R., and Schmid, P. (1984). Obstetrical interpretation of the Australopithecine pelvic cavity. *Journal of Human Evolution*, **13**, 573–87.

Biegert, J. (1963). The evaluation of characteristics of the skull, hands and feet for primate taxonomy. In *Classification and human evolution* (ed. S. L. Washburn), pp. 116–45. Aldine, Chicago.

Bishop, A. (1964). Use of the hand in lower primates. In *Evolutionary and genetic biology of primates* (ed. J. Buettner-Janusch), pp. 133–225. Academic Press, New York.

Black, D., de Chardin, T., Young, C. C., and Pei, W. C. (1933). *Fossil man in China*. The Geological Survey of China and The Section of Geology of the National Academy of Peiping, Peiping [Peking].

Blumenbach, J. F. (1791). *Handbuch der Naturgeschichte*. Dieterich, Göttingen.

Boesch, C. and Boesch, H. (1981). Sex differences in the use of natural hammers by wild chimpanzees; a preliminary report. *Journal of Human Evolution*, **10**, 585–93.

Boesch, C. and Boesch, H. (1983a). Nut-cracking behavior of wild chimpanzees, Ivory Coast. Film shown at the Xth Congress of the International Primatological Society, Nairobi, Kenya, July 22–27, 1984.

Boesch, C. and Boesch, H. (1983b). Optimisation of nut-cracking with natural hammers by wild chimpanzees. *Behaviour*, **83**, 265–86.

Boesch, C. and Boesch, H. (1984). Possible causes of sex differences in the use of natural hammers by wild chimpanzees. *Journal of Human Evolution*, **13**, 415–40.

Boesch, C. and Boesch, H. (1993). Different hand postures for pounding nuts with natural hammers by wild chimpanzees. In *Hands of primates* (ed. H. Preuschoft and D. J. Chivers), pp. 31–43. Springer–Verlag, New York.

Brace, C. L. (1962). Comments on 'Food transport and the origin of hominid bipedalism'. *American Anthropologist*, **64**, 606–7.

Brain, C. K. (1985). Cultural and taphonomic comparisons of hominids from Swartkrans and Sterkfontein. In *Ancestors: the hard evidence* (ed. E. Delson), pp. 72–5. Alan R. Liss, New York.

Broom, R. and Schepers, G. W. H. (1946). *The South African fossil ape-men: the Australopithecinae*. Transvaal Museum Memoirs, **2**, 1–272.

Brown, F. H. (1982). Tulu Bor tuff at Koobi Fora correlated with Sidi Hakoma tuff at Hadar. *Nature*, **300**, 631–3.

Brown, F. H. and Shuey, R. T. (1976). Magnetostratigraphy of the Shungura and Usno Formations, Lower Omo Valley, Ethiopia. In *Earliest man and environments in the Lake Rudolf basin* (ed. Y. Coppens, F. C. Howell, G. L. Isaac, and R. E. F. Leakey), pp. 64–78. University of Chicago Press.

Brown, F. H. and McDougall, I. (1993). Geologic setting and age. In *The Nariokotome* Homo erectus *Skeleton* (ed. A. Walker and R. Leakey), pp. 9–20. Harvard University Press) Cambridge, M.A.

Bush, M. E. (1980). The thumb of *Australopithecus afarensis*. *American Journal of Physical Anthropology*, **52**, 210. (Abstract.)

Bush, M. E., Lovejoy, C. O., Johanson, D. C., and Coppens, Y. (1982). Hominid carpal, metacarpal, and phalangeal bones recovered from the Hadar Formation: 1974–1977 collections. *American Journal of Physical Anthropology*, **57**, 651–77.

Cadman, A. and Rayner, R. J. (1989). Climatic change and the appearance of *Australopithecus africanus* in the Makapansgat sediments. *Journal of Human Evolution*, **18**, 107–13.

Calvin, W. H. (1983). *The throwing madonna*. McGraw-Hill, New York.

Calvin, W. H. (1993). The unitary hypothesis: a common neural circuitry for novel manipulations, language, plan-ahead, and throwing? In *Tools, Language and Cogntion in Human evolution* (ed. K. R. Gibson and T. Ingold), pp. 230–50. Cambridge University Press.

Cartmill, M. (1974). Rethinking primate origins. *Science*, **184**, 436–43.

Cartmill, M. and Milton, K. (1977). The lorisiform wrist joint and the evolution of 'brachiating' adaptations in the Hominoidea. *American Journal of Physical Anthropology*, **47**, 249–72.

Christel, M. (1993). Grasping techniques and hard preferences in hominoidea In *Hands of primates* (ed. H. Preuschoft and D. J. Chivers), pp. 91–108. Springer-Verlag, New York.

Clark, W. E. Le Gros (1962). *The antecedents of man* (2nd edn). Edinburgh University Press.

Conroy, G. C. (1990). *Primate evolution*. W. W. Norton and Company, New York.

Coppens, M. Y. (1973). Les restes d'Hominidés des séries inférieures et moyennes des formations plio-villafranchiennes de l'Omo en Éthiopie (récoltes 1970, 1971 et 1972). *Comptes Rendus Hebdomadaires des Séances de l'Académie des Sciences*, 276, Série D, 1823–6.

Coursey, D. G. (1972). The civilizations of the yam: interrelationships of man and yams in Africa and the Indo-Pacific area. *Archaeology and Physical Anthropology in Oceania*, **7**, 215–33.

Coursey, D. G. (1973). Hominid evolution and hypogeous plant foods. *Man* (NS.), **8**, 634–5.

Curtis, G. H. and Hay, R. L. (1972). Further geological studies and potassium–argon dating at Olduvai Gorge and Ngorongoro Crater. In *Calibration of hominoid evolution* (ed. W. W. Bishop and J. A. Miller), pp. 289–301. Scottish Academic Press, Edinburgh.

Darlington, P. J., jun. (1975). Group selection, altruism, reinforcement, and throwing in human evolution. *Proceedings of the National Academy of Sciences USA*, **72**, 3748–52.

Dart, R. (1949). Innominate fragments of *Australopithecus prometheus*. *American Journal of Physical Anthropology*, **7**, 301–34.

Dart, R. (1959). *Adventures with the missing link*. Harper and Brothers, New York.

Darwin, C. (1874). *The descent of man and selection in relation to sex* (2nd edn). The Wheeler Publishing Company, New York.

Day, M. H. (1973). Locomotor features of the lower limb in hominids. *Symposia of the Zoological Society of London*, **33**, 29–51.

Day, M. H. (1976). Hominid postcranial material from Bed I, Olduvai Gorge. In *Human origins* (ed. G. L. Isaac and E. R. McCown), pp. 363–74. W. A. Benjamin, Menlo Park, California.

Day, M. H. (1978). Functional interpretation of the morphology of postcranial remains of early African hominids. In *Early hominids of Africa* (ed. C. J. Jolly), pp. 311–45. Duckworth, London.

Day, M. H. and Scheuer, J. L. (1973). SKW 14147: a new hominid metacarpal from Swartkrans. *Journal of Human Evolution*, **2**, 429–38.

De Vore, I. and Hall, K. R. L. (1965). Baboon ecology. In *Primate behaviour, field studies of monkeys and apes* (ed. I. De Vore), pp. 20–52. Holt, Rinehart, and Winston, New York.

Du Brul, E. L. (1962). The general phenomenon of bipedalism. *American Zoologist*, **2**, 205–8.

Dunbar, R. I. M. (1983). Theropithecines and hominids: contrasting solutions to the same ecological problem. *Journal of Human Evolution*, **12**, 647–58.

Floyd, W. F. and Silver, P. H. S. (1955). The function of the erectores spinae muscles in certain movements and postures in man. *Journal of Physiology*, **129**, 184–203.

Forster, A. (1917). Die Mm. contrahentes und interossei manus in der saugetierreihe und beim Menschen. *Archiv für Anatomie, Anatomische Abteilung des Archives für Anatomie und Physiologie*, pp. 101–378.

Gebo, D. L. (1992). Plantigrady and foot adaptation in African apes: implications for hominid origins. *American Journal of Physical Anthropology*, **89**, 29–58.

Gregory, W. K. (1927). How near is the relationship of man to the chimpanzee–gorilla stock? *Quarterly Review of Biology*, **2**, 549–60.

Gregory, W. K. (1928). Were the ancestors of man primitive brachiators? *Proceedings of the American Philosophical Society*, **67**, 129–50.

Gregory, W. K. (1930). A critique of Professor Osborn's theory of human origins. *American Journal of Physical Anthropology*, **14**, 133–61.

Gregory, W. K. (1934). *Man's place among the anthropoids*. Clarendon Press, Oxford.

Gruber, A. (1962). Comments and queries on the origin of hominid bipedalism. *American Anthropologist*, **64**, 605–6.

Hall, K. R. L. (1963). Tool-using performances as indicators of behavioral adaptability. *Current Anthropology*, **4**, 479–94.

Hamilton, W. J. III (1973). *Life's color code*. McGraw-Hill, New York.

Hamilton, W. J. III, Buskirk, R. E., and Buskirk, W. K. (1978). Environmental determinants of object manipulation by Chacma baboons (*Papio ursinus*) in two South African environments. *Journal of Human Evolution*, **7**, 205–16.

Hamrick, M. W. and Inouye, S. E. (1995). Thumbs, tools, and early humans. *Science*, **268**, 586–7.

Hartwig-Scherer, S. and Martin, R. D. (1991). Was "Lucy" more human than her "child"? Observations on early hominid postcranial skeletons. *Journal of Human Evolution*, **21**, 439–49.

Hatley, T. and Kappelman, J. (1980). Bears, pigs, and Plio-Pleistocene hominids: a case for the exploitation of belowground food resources. *Human Ecology*, **8**, 371–87.

Hay, R. L. (1976). *Geology of the Olduvai Gorge*. University of California Press, Berkeley.

Hewes, G. W. (1961). Food transport and the origin of hominid bipedalism. *American Anthropologist*, **63**, 687–710.

Hewes, G. W. (1962). Hominid bipedalism: independent evidence for the food-carrying theory. *Science*, **146**, 416–18.

Hill, K. (1982). Hunting and human evolution. *Journal of Human Evolution*, **11**, 521–44.

Hockett, C. F. and Ascher, R. (1964). The human revolution. *Current Anthropology*, **5**, 135–68.

Holloway, R. (1976). Some problems of hominid brain endocast reconstruction, allometry, and neural reorganization. In *Colloquium VI ('Les plus anciens hominides', ed. Philip V. Tobias and Y. Coppens) of the Ninth Congress of the Union Internationale des Sciences Préhistoriques et Protohistoriques*, pp. 69–119. Centre National de la Recherche Scientifique, Paris.

Howell, F. C. (1978). Hominidae. In *Evolution of African mammals* (ed. V. J. Maglio and H. B. S. Cooke), pp. 154–248. Harvard University Press, Cambridge, Mass.

Howell, F. C. and Coppens, Y. (1976). An overview of Hominidae from the Omo Succession, Ethiopia. In *Earliest man and environments in the Lake Rudolf basin* (ed. Y. Coppens, F. C. Howell, G. L. Isaac and R. E. F. Leakey), pp. 522–32. University of Chicago Press.

Howells, W. W. (1980). *Homo erectus*—who, when, and where: a survey. *Yearbook of Physical Anthropology*, **23**, 1–23.

Hunt, K. D. (1994). The evolution of human bipedality: ecology and functional morphology. *Journal of Human Evolution*, **26**, 183–202.

Huxley, T. (1871). *Evidence as to man's place in nature*. D. Appleton and Company, New York.

Isaac, B. (1987). Throwing and human evolution. *The African Archaeological Review*, **5**, 3–17.

Isaac, G. L. (1979). Comment. *The Behavioral and Brain Sciences*, **2**, 388.

Isaac, G. L. (1983). Early stages in the evolution of human behavior: the adaptive significance of stone tools. *Zesde Kroon-Voordracht Gehouden von de Stichting Nederlands Museum von Anthropologie en Praehistorie te Amsterdam op 22 April 1983*. Joh. Enschede en Zonen, Haarlem.

Isaac, G. L. (1984). The archaeology of human origins: studies of the Lower Pleistocene in East Africa 1971–1981. *Advances in World Archaeology*, **3**, 1–87.

Isaac, G. L. and Crader, D. C. (1981). To what extent were early hominids carnivorous? An archaeological perspective. In *Omnivorous primates* (ed. R. S. O. Harding and G. Teleki), pp. 37–103. Columbia University Press, New York.

Jenkins, F. A., jun. (1981). Wrist rotation in primates: a critical adaptation for brachiators. *Symposia of the Zoological Society of London*, **48**, 429–51.

Johanson, D. C., Lovejoy, C. O., Kimbel, W. K., White, T. D., Ward, S. C., Bush, M. E., et al. (1982a). Morphology of the Pliocene partial hominid skeleton (A. L. 288–1) from the Hadar Formation, Ethiopia. *American Journal of Physical Anthropology*, **57**, 403–51.

Johanson, D. C., Taieb, M., and Coppens, Y. (1982b). Pliocene hominids from the Hadar Formation, Ethiopia (1973–1977): stratigraphic, chronologic, and paleoenvironmental contexts, with notes on hominid morphology and systematics. *American Journal of Physical Anthropology*, **57**, 373–402.

Johanson, D. C., Masao, F. T., Eck, G. G., White, T. D., Walter, R. C., Kimbel, W. H., et al. (1987). New partial skeleton of *Homo habilis* from Olduvai Gorge, Tanzania. *Nature*, **327**, 205–9.

Jolly, C. J. (1970). The seed-eaters: a new model of hominid differentiation based on a baboon analogy. *Man* (NS), **5**, 1–26.

Jones, F. Wood (1948). *Hallmarks of mankind*. Williams and Wilkins, Baltimore.

Jones, F. Wood (1949). *The principles of anatomy as seen in the hand*. Baillière, Tindall, and Cox, London.

Jordan, C. (1982). Object manipulation and tool-use in captive pygmy chimpanzees (*Pan paniscus*). *Journal of Human Evolution*, **11**, 35–9.

Jouffroy, F. K. and Lessertisseur, J. (1959). Réflections sur les muscles contracteurs des doigts et des orteils (contrahentes digitorum) chez les primates. *Annales des Sciences Naturelles de Zoologie et Biologie Animale*, Série 12(1), 211–35.

Jungers, W. L. (1982). Lucy's limbs: skeletal allometry and locomotion in *Australopithecus afarensis*. *Nature*, **297**, 676–8.

Jungers, W. L. (1988a). Lucy's length: stature reconstruction in *Australopithecus afarensis* (A. L. 288–1) with implications for other small-bodied hominids. *American Journal of Physical Anthropology*, **76**, 227–31.

Jungers, W. L. (1988b). Relative joint size and hominoid locomotor adaptations with implications for the evolution of hominid bipedalism. *Journal of Human Evolution*, **17**, 247–65.

Jungers, W. L. and Stern, J. T., jun. (1983). Body proportions, skeletal allometry and locomotion in the Hadar hominids: a reply to Wolpoff. *Journal of Human Evolution*, **12**, 673–84.

Kano, T. (1979). A pilot study on the ecology of pygmy chimpanzees, *Pan paniscus*. In *The Great Apes* (ed. D. A. Hamburg and E. R. McCown), pp. 123–35. The Benjamin/Cummings Publishing Company, Menlo Park, California.

Karlsson, E. and Jonsson, B. (1964). Function of the gluteus maximus muscle. An electromyographic study. *Acta Morphologica Neerlando-Scandinavica*, **6**, 161–9.

Keeley, L. and Toth, N. (1981). Micro wear polishes in early stone tools from Koobi Fora, Kenya. *Nature*, **293**, 464–5.

Kelley, J. (1988). A new large species of *Sivapithecus* from the Siwaliks of Pakistan. *Journal of Human Evolution*, **17**, 305–24.

Kennedy, G. E. (1983a). A morphometric and taxonomic assessment of a hominine femur from the

Lower Member, Koobi Fora, Lake Turkana. *American Journal of Physical Anthropology*, **61**, 429–36.

Kennedy, G. E. (1983*b*). Some aspects of femoral morphology in *Homo erectus*. *Journal of Human Evolution*, **12**, 587–616.

Kortlandt, A. (1965). How do chimpanzees use weapons when fighting leopards? *The American Philosophical Society Year Book*, **1965**, pp. 327–32.

Kortlandt, A. (1967). Handgebrauch bei freilebenden Schimpansen. In *Handgebrauch und Verstandigung bei Affen und Fruhmenschen* (ed. B. Rensch), pp. 59–102. Huber, Bern and Stuttgart.

Kortlandt, A. (1972). *New perspectives on ape and human evolution*, First preliminary edn. Stichting voor Psychobiologie, Amsterdam.

Kortlandt, A. and Kooij, M. (1963). Protohominid behaviour in primates. *Symposia of the Zoological Society of London*, **10**, 61–88.

Lancaster, J. B. (1968). On the evolution of tool-using behavior. *American Anthropologist*, **70**, 56–66.

Lancaster, J. B. (1978). Carrying and sharing in human evolution. *Human Nature*, **1**, 82–9.

Latimer, B. (1983). The anterior foot skeleton of *Australopithecus afarensis*. *American Journal of Physical Anthropology*, **60**, 217.

Latimer, B. and Lovejoy, C. O. (1989). The calcaneus of *Australopithecus afarensis* and its implications for the evolution of bipedality. *American Journal of Physical Anthropology*, **78**, 369–86.

Latimer, B. and Lovejoy, C. O. (1990). Metatarsophalangeal joints of *Australopithecus afarensis*. *American Journal of Physical Anthropology*, **83**, 13–23.

Latimer, B. M., Lovejoy, C. O., Johanson, D. C., and Coppens, Y. (1982). Hominid tarsal, metatarsal, and phalangeal bones recovered from the Hadar Formation: 1974–1977 collections. *American Journal of Physical Anthropology*, **57**, 701–19.

Latimer, B., Ohman, J. C., and Lovejoy, C. O. (1987). Talocrural joint in African hominoids: implications for *Australopithecus afarensis*. *American Journal of Physical Anthropology*, **74**, 155–75.

Leakey, M. D. (1971). *Olduvai Gorge* Vol. 3. Cambridge University Press, London.

Leakey, M. D., Hay, R. L., Curtis, C. H., Drake, R. E., Jackes, M. K., and White, T. D. (1976). Fossil hominids from the Laetolil Beds, Tanzania. *Nature*, **262**, 460–5.

Leakey, R. E. and Leakey, M. G. (1978). *Koobi Fora Research Project* (Vol. 1). Clarendon Press, Oxford.

Leakey, R. E., Leakey, M. G., and Walker, A. C. (1988*a*). Morphology of *Afropithecus turkanensis* from Kenya. *American Journal of Physical Anthropology*, **76**, 289–307.

Leakey, R. E., Leakey, M. G., and Walker, A. C. (1988*b*). Morphology of *Turkanapithecus kalakolensis* from Kenya. *American Journal of Physical Anthropology*, **76**, 277–88.

Leakey, R. E., Walker, A., Ward, C. V., and Grausz, H. M. (1989). A partial skeleton of a gracile hominid from the Upper Burgi Member of the Koobi Fora Formation, East Lake Turkana, Kenya. In *Hominidae: Proceedings of the 2nd International Congress of Human Paleontology* (ed. G. Giacobini), pp. 167–73. Jaca Book, Milan.

Lessertisseur, J. (1958). Doit-on distinguer deux plans de muscles interosseux à la main et au pied des primates? *Annales des Sciences Naturelles de Zoologie et Biologie Animale*, **Série 11 (20)**, 77–104.

Lewis, O. J. (1971). Brachiation and the early evolution of the Hominoidea. *Nature*, **230**, 577–8.

Lewis, O. J. (1972). Osteological features characterizing the wrists of monkeys and apes, with a reconsideration of this region in *Dryopithecus (Proconsul) africanus*. *American Journal of Physical Anthropology*, **36**, 45–58.

Lewis, O. J. (1974). The wrist articulations of the Anthropoidea. In *Primate locomotion* (ed. F. A. Jenkins, jun.), pp. 143–69. Academic Press, New York.

Lewis, O. J. (1977). Joint remodelling and the evolution of the human hand. *Journal of Anatomy*, **123**, 157–201.

Lewis, O. J. (1980). The joints of the evolving foot. Part III. The fossil evidence. *Journal of Anatomy*, **131**, 275–98.

Lewis, O. J. (1985*a*). Derived morphology of the wrist articulation and theories of hominind evolution. Part I: the lorisine joints. Journal of Anatomy, **140**, 447–60.

Lewis, O. J. (1985*b*). Derived morphology of the wrist articulations and theories of hominoid evolution. *Journal of Anatomy*, **142**, 151–72.

Lewis, O. J. (1989). *Functional morphology of the evolving hand and foot*. Clarendon Press, Oxford.

Lewis, O. J., Hamshere, R. J., and Bucknill, T. M. (1970). The anatomy of the wrist joint. *Journal of Anatomy*, **106**, 539–52.

Lovejoy, C. O. (1981). The origin of man. *Science*, **211**, 341–50.

Lovejoy, C. O., Johanson, D. C., and Coppens, Y. (1982). Hominid lower limb bones recovered from the Hadar formation: 1974–1977 collections. *American Journal of Physical Anthropology*, **57**, 637–49.

Lu, Z. (1990). La découverte de l'homme fossile de Jing-Niu-Shan. Première étude. *L'Anthropologie*, **94**, 899–902.

McGrew, W. C. (1974). Tool use by wild chimpanzees in feeding upon driver ants. *Journal of Human Evolution*, **3**, 501–8.

McGrew, W. C. (1984). Comment. *Current Anthropology*, **25**, 160.

McGrew, W. C. (1992). *Chimpanzee material culture*. Cambridge University Press.

McGrew, W. C., Tutin, C. E. G., and Baldwin, P. J. (1979). Chimpanzees, tools, and termites: cross-cultural comparisons of Senegal, Tanzania, and Rio Muni. *Man*, **14**, 185–214.

McHenry, H. M. (1982). The pattern of human evolution: studies on bipedalism, mastication and encephalization. *Annual Review of Anthropology*, **11**, 151–73.

McHenry, H. M. (1983). The capitate of *Australopithecus afarensis* and *A. africanus*. *American Journal of Physical Anthropology*, **62**, 187–98.

McHenry, H. M. (1986). The first bipeds: a comparison of the *A. afarensis* and *A. africanus* postcranium and implications for the evolution of bipedalism. *Journal of Human Evolution*, **15**, 177–91.

McHenry, H. M. and Corruccini, R. S. (1983). The wrist of *Proconsul africanus* and the origin of hominid postcranial adaptations. In *New interpretations of ape and human ancestry* (ed. R. L. Ciochon and R. S. Corruccini), pp. 353–67. Plenum, New York.

McHenry, H. M. and Temerin, L. A. (1979). The evolution of hominid bipedalism: evidence from the fossil record. *Yearbook of Physical Anthropology*, **22**, 105–31.

Mann, A. (1972). Hominid and cultural origins. *Man*, **7**, 379–86.

Marshack, A. (1984a). The ecology and brain of two-handed bipedalism: an analytic, cognitive, and evolutionary assessment. In *Animal cognition* (ed. H. L. Roitblat, T. G. Bever, and H. S. Terrace), pp. 491–511. Lawrence Erlbaum, Hillsdale, New Jersey.

Marshack, A. (1984b). *Hierarchical evolution of the human capacity: the paleolithic evidence*, James Arthur Lecture. American Museum of Natural History, New York.

Marzke, M. W. (1971). Origin of the human hand. *American Journal of Physical Anthropology*, **34**, 61–84.

Marzke, M. W. (1983). Joint functions and grips of the *Australopithecus afarensis* hand, with special reference to the region of the capitate. *Journal of Human Evolution*, **12**, 197–211.

Marzke, M. W. (1986). Tool use and the evolution of hominid hands and bipedality. *Proceedings of the Tenth Congress of the International Primatological Society*, Volume 1 (ed. J. G. Else and P. C. Lee), pp. 203–9. Cambridge University Press, London.

Marzke, M. W. (submitted). Precision grips, hand morphology, and tools.

Marzke, M. W. and Marzke, R. F. (1987). The third metacarpal styloid process in humans: origin and functions. *American Journal of Physical Anthropology*, **73**, 415–31.

Marzke, M. W. and Shackley, M. S. (1986). Hand use by early hominids: evidence from experimental archeology and comparative morphology. *Journal of Human Evolution*, **15**, 439–60.

Marzke, M. W., Longhill, J. M., and Rasmussen, S. A. (1988). Gluteus maximus muscle function and the origin of hominid bipedality. *American Journal of Physical Anthropology*, **77**, 519–28.

Marzke, M. W., Wullstein, K. L., and Viegas, S. F. (1992). Evolution of the power ("squeeze") grip and its morphological correlates in hominids. *American Journal of Physical Anthropology*, **89**, 283–98.

Muir, R. B. (1985). Small hand muscles in percision [*sic*] grip: Á corticospinal prerogative?. *Experimental Brain Research*, **Suppl. 10**.

Musgrave, J. H. (1970). An anatomical study of the hands of Pleistocene and recent man. Unpublished Ph. D. thesis, University of Cambridge.

Musgrave, J. H. (1971). How dextrous was Neanderthal Man? *Nature*, **233**, 538–41.

Musgrave, J. H. (1973). The phalanges of Neanderthal and Upper Paleolithic hands. In *Human evolution* (ed. M. H. Day), pp. 59–85. Taylor and Francis, London.

Napier, J. R. (1956). The prehensile movements of the human hand. *The Journal of Bone and Joint Surgery*, **38-B**, 902–13.

Napier, J. R. (1959). Fossil metacarpals from Swartkrans. *Fossil Mammals of Africa* (British Museum Natural History), **17**, 1–18.

Napier, J. R. (1960). Studies of the hands of living primates. *Proceedings of the Zoological Society of London*, **134**, 647–57.

Napier, J. R. (1961). Prehensility and opposability in the hands of primates. *Symposia of the Zoological Society of London*, **5**, 115–32.

Napier, J. R. (1962a). The evolution of the hand. *Scientific American*, **207** (6), 56–62.

Napier, J. R. (1962b). Fossil hand bones from Olduvai Gorge. *Nature*, **196**, 409–11.

Napier, J. R. (1964). The evolution of bipedal walking in the hominids. *Archives de Biologie*, **75**, 673–708.

Napier, J. R. (1980). *Hands*. Pantheon Books, New York.

Napier, J. R. and Davis, P. R. (1959). The fore-limb skeleton and associated remains of *Proconsul africanus*. *Fossil Mammals of Africa* (British Museum Natural History), **16**, 1–69.

Nishida, T. (1972). The ant-gathering behaviour by the use of tools among wild chimpanzees of the Mahali Mountains. *Journal of Human Evolution*, **2**, 357–70.

Oakley, K. P. (1961). On man's use of fire, with comments on tool-making and hunting. In *Social life of early man* (ed. S. L. Washburn), pp. 176–93. Aldine, Chicago.

Oakley, K. P. (1972a). Skill as a human possession. In *Perspectives on human evolution* (ed. S. L. Washburn and P. Dolhinow), pp. 14–50. Rinehart and Winston, New York.

Oakley, K. P. (1972b). *Man the tool-maker*. University of Chicago Press.

O'Brien, E. M. (1981). The projectile capabilities of an Acheulean handaxe from Olorgesailie. *Current Anthropology*, **22**, 76–9.

O'Brien, E. M. (1984). What was the Acheulean hand ax? *Natural History*, **93**, 20–3.

Ohman, J. C., Slonina, M., Baker, G., and Mensforth, R. P. (1995). Thumbs, tools, and early humans. *Science*, **268**, 587–9.

Osborn, H. F. (1927a). Recent discoveries relating to the origin and antiquity of man. *Proceedings of the American Philosophical Society*, **66**, 373–89.

Osborn, H. F. (1927b). *Man rises to Parnassus: critical epochs in the prehistory of man*. Princeton University Press.

Osborn, H. F. (1928). The influence of bodily locomotion in separating man from the monkeys and apes. *Scientific Monthly*, **28**, 385–99.

Osborn, H. F. (1930). The discovery of tertiary man. *Science*, **71**, 1–7.

Oxnard, C. E. and Lisowski, F. P. (1980). Functional articulation of some hominoid foot bones: implications for the Olduvai (Hominid 8) foot. *American Journal of Physical Anthropology*, **52**, 107–17.

Parker, S. T. and Gibson, K. R. (1977). Object manipulation, tool use and sensorimotor intelligence as feeding adaptations in *Cebus* monkeys and great apes. *Journal of Human Evolution*, **6**, 623–41.

Parker, S. T. and Gibson, K. R. (1979). A developmental model for the evolution of language and intelligence in early hominids. *The Behavioral and Brain Sciences*, **2**, 367–408.

Peters, C. R. (1979). Toward an ecological model of African Plio-Pleistocene hominid adaptations. *American Anthropologist*, **81**, 261–78.

Peters, C. R. and Maguire, B. (1981). Wild plant foods of the Makapansgat area: a modern ecosystems analogue for *Australopithecus africanus* adaptations. *Journal of Human Evolution*, **10**, 565–83.

Peters, C. R. and O'Brien, E. M. (1981). The early hominid plant-food niche: insights from an analysis of plant exploitation by *Homo*, *Pan*, and *Papio* in eastern and southern Africa. *Current Anthropology*, **22**, 127–40.

Peters, C. R. and O'Brien, E. M. (1984). On hominid diet before fire. *Current Anthropology*, **25**, 358–60.

Pickford, M. (1983) Sequence and environments of the Lower and Middle Miocene hominoids of Western Kenya. In *New interpretations of ape and human ancestry* (ed. R. L. Ciochon and R. S. Corruccini), pp. 421–39. Plenum Press, New York.

Pilbeam, D. R., Rose, M. D., Badgley, C., and Lipschutz, B. (1980). Miocene hominoids from Pakistan. *Postilla*, **181**, 1–94.

Rak, Y. (1991). Lucy's pelvic anatomy: its role in bipedal gait. *Journal of Human Evolution*, **20**, 283–90.

Rak, Y. and Arensburg, B. (1987). Kebara 2 Neanderthal pelvis: first look at a complete inlet. *American Journal of Physical Anthropology*, **73**, 227–31.

Rasmussen, T. (1990). Primate origins: lessons from a neotropical marsupial. *American Journal of Primatology*, **22**, 263–77.

Ricklan, D. E. (1986). The differential frequency of preservation of early hominid hand and wrist bones. *Human Evolution*, **1**, 373–82.

Ricklan, D. E. (1987). Functional anatomy of the hand of *Australopithecus africanus*. *Journal of Human Evolution*, **16**, 643–64.

Ricklan, D. E. (1990). The precision grip in *Australopithecus africanus*: anatomical and behavioural correlates. In *From apes to angels* (ed. G. H. Sperber), pp. 171–83. Wiley-Liss, New York.

Robertson, M. L. (1984). *The carpus of Proconsul africanus: functional analysis and comparison with selected nonhuman primates*. Unpublished Ph.D. dissertation, The University of Michigan.

Robinson, J. T. (1963). Adaptive radiation in the Australopithecines and the origin of man. In *African ecology and human evolution*, Viking Fund Publications in Anthropology 36 (ed. F. C. Howell and F. Bourlière), pp. 385–416. Aldine, Chicago.

Robinson, J. T. (1972). *Early hominid posture and locomotion*. University of Chicago Press.

Rodman, P. S. and McHenry, H. M. (1980). Bioenergetics and the origin of hominid bipedalism. *American Journal of Physical Anthropology*, **52**, 103–6.

Rose, M. D. (1974). Postural adaptations in new and old world monkeys. In *Primate locomotion* (ed. F. A. Jenkins), pp. 201–22. Academic Press, New York.

Rose, M. D. (1976). Bipedal behavior of olive baboons (*Papio anubis*) and its relevance to an understanding of the evolution of human bipedalism. *American Journal of Physical Anthropology*, **44**, 247–62.

Rose, M. D. (1982). Food acquisition and the evolution of positional behavior: the case of bipedalism. In *Food acquisition and processing in primates* (ed. D. J. Chivers), pp. 509–24. Plenum, New York.

Rose, M. D. (1983). Miocene hominoid postcranial morphology: monkey-like, ape-like, neither, or both? In *New interpretations of ape and human ancestry* (ed. R. L. Ciochon and R. S. Corruccini), pp. 405–17. Plenum Press, New York.

Rose, M. D. (1984a). Hominoid postcranial specimens from the middle Miocene Chinji Formation, Pakistan. *Journal of Human Evolution*, **13**, 503–16.

Rose, M. D. (1984b). A hominine hip bone, KNM-ER 3228, from East Lake Turkana, Kenya. *American Journal of Physical Anthropology*, **63**, 371–78.

Rose, M. D. (1986). Further hominoid postcranial specimens from the Late Miocene Nagri Formation of Pakistan. *Journal of Human Evolution*, **15**, 333–67.

Rose, M. D. (1992). Kinematics of the trapezium–1st metacarpal joint in extant anthropoids and Miocene hominoids. *Journal of Human Evolution*, **22**, 255–66.

Ruff, C. B., Trinkaus, E., Walker, A., and Larson, C. S. (1993). Postcranial robusticity in *Homo*. I: Temporal trends and mechanical interpretation. *American Journal of Physical Anthropology*, **91**, 21–53.

Sarmiento, E. E. (1988). Anatomy of the hominoid wrist joint: its evolution and functional implications. *International Journal of Primatology*, **9**, 281–345.

Schaller, G. (1963). *The mountain gorilla*. University of Chicago Press.

Schick, K. D. and Toth, N. (1993). *Making silent stones speak*. Simon and Schuster Inc., New York.

Schick, K. D. and Zhuan, D. (1993). Early Paleolithic of China and Eastern Asia. *Evolutionary Anthropology*, **2**, 22–35.

Schultz, A. H. (1956). Postembryonic age changes. In *Primatologia* (ed. H. Hofer, A. H. Schultz, and D. Starck), Vol. 1, pp. 887–964. S. Karger, Basle.

Schultz, A. H. (1969). *The life of primates*. Weidenfeld and Nicolson, London.

Seaver, T. with Lowenfish, L. (1984). *The art of pitching*. Hearst Books, New York.

Shrewsbury, M. M. and Johnson, R. K. (1983). Form, function, and the evolution of the distal phalanx. *The Journal of Hand Surgery*, **8**, 475–9.

Shrewsbury, M. M. and Sonek, A. (1986). Precision holding in humans, non-human primates, and Plio-Pleistocene hominids. *Human Evolution*, **1**, 233–42.

Sigmon, B. A. (1971). Bipedal behavior and the emergence of erect posture in man. *American Journal of Physical Anthropology*, **34**, 55–60.

Spinner, Morton (1984). *Kaplan's functional and surgical anatomy of the hand* (3rd edn). Lippincott, Philadelphia.

Spoor, C. F., Sondaar, P. Y., and Hussain, S. T. (1991). A new hominid hamate and first metacarpal from the Late Miocene Nagri Formation of Pakistan. *Journal of Human Evolution*, **21**, 413–23.

Stahl, A. B. (1984). Hominid dietary selection before fire. *Current Anthropology*, **25**, 151–68.

Stern, J. T., jun. and Susman, R. L. (1983). The locomotor anatomy of *Australopithecus afarensis*. *American Journal of Physical Anthropology*, **60**, 279–317.

Stoner, B. P. and Trinkaus, E. (1981). Getting a grip on the Neandertals: were they all thumbs? *American Journal of Physical Anthropology*, **54**, 281–2.

Straus, W. L., jun. (1949). The riddle of man's ancestry. *Quarterly Review of Biology*, **24**, 200–23.

Straus, W. L., jun. (1962). Fossil evidence of the evolution of the erect, bipedal posture. *Clinical Orthopaedics*, **25**, 9–19.

Sugiyama, Y. (1981). Observations on the population dynamics and behavior of wild chimpanzees at Bossou, Guinea, in 1979–1980. *Primates*, **22**, 435–44.

Sugiyama, Y. and Koman, J. (1979). Tool-using and making behavior in wild chimpanzees at Bossou, Guinea. *Primates*, **20**, 513–24.

Susman, R. L. (1979). Comparative and functional morphology of hominoid fingers. *American Journal of Physical Anthropology*, **50**, 215–36.

Susman, R. L. (1988). New postcranial remains from Swartkrans and their bearing on the functional morphology and behavior of *Paranthropus robustus*. In *Evolutionary history of the 'robust' Australopithecines* (ed. F. E. Grine), pp. 149–72. Aldine de Gruyter, New York.

Susman, R. L. (1989). New hominid fossils from the Swartkrans formation (1979–1986 excavations): postcranial specimens. *American Journal of Physical Anthropology*, **79**, 451–74.

Susman, R. C. (1994). Fossil evidence for early hominid tool use. *Science*, **265**, 1570–3.

Susman, R. L. and Brain, T. M. (1988). New first metatarsal (SKX 5017) from Swartkrans and the gait of *Paranthropus robustus*. *American Journal of Physical Anthropology*, **77**, 7–15.

Susman, R. L. and Creel, N. (1979). Functional and morphological affinities of the subadult hand (O. H. 7) from Olduvai Gorge. *American Journal of Physical Anthropology*, **51**, 311–32.

Susman, R. L. and Stern, J. T., jun. (1979). Telemetered electromyography of flexor digitorum profundus and flexor digitorum superficialis in *Pan troglodytes* and implications for interpretation of the O. H. 7 hand. *American Journal of Physical Anthropology*, **50**, 565–74.

Susman, R. L. and Stern, J. T., jun. (1982). Functional morphology of *Homo habilis*. *Science*, **217**, 931–4.

Susman, R. L., Stern, J. T., jun., and Jungers, W. L. (1984). Arboreality and bipedality in the Hadar hominids. *Folia Primatologica*, **43**, 113–56.

Sussman, R. W. (1991). Primate origins and the evolution of angiosperms. *American Journal of Primatology*, **23**, 209–23.

Tallon, P. W. J. (1978). Geological setting of the hominid fossils and Acheulean artifacts from the Kapthurin Formation, Baringo District, Kenya. In *Geological background to fossil man* (ed. W. W. Bishop), pp. 361–73. Scottish Academic Press, Edinburgh.

Tanaka, J. (1976). Subsistence ecology of Central Kalahari San. In *Kalahari hunter-gatherers* (ed. R. B. Lee and I. De Vore), pp. 98–119. Harvard University Press, Cambridge, Mass.

Tanner, N. and Zihlman, A. L. (1976). Women in evolution. Part 1: Innovation and selection in human origins. *Signs: Journal of Women in Culture and Society*, **1** (3), Part 1, 585–608.

Tardieu, C. (1981). Morpho-functional analysis of the articular surfaces of the knee-joint in primates. In *Primate evolutionary biology* (ed. A. B. Chiarelli and R. S. Corruccini), pp. 68–80. Springer-Verlag, Berlin.

Tardieu, C. (1983). *L'Articulation du genou*. Éditions du Centre National de la Recherche Scientifique, Paris.

Teleki, G. (1975). Primate subsistence patterns: collector-predators and gatherer–hunters. *Journal of Human Evolution*, **4**, 125–84.

Tobias, P. V. (1978). The place of *Australopithecus africanus* in hominid evolution. In *Recent advances in primatology* (ed. D. J. Chivers and K. A. Josey), Vol. 3, pp. 373–94. Academic Press, New York.

Tobias, P. V. (1980). *Australopithecus afarensis* and *A. africanus*: a critique and an alternative hypothesis. *Palaeontologia Africana*, **23**, 1–17.

Toth, N. (1982). *The stone technologies of early hominids at Koobi Fora, Kenya: an experimental approach*. Unpublished Ph.D. dissertation, University of California, Berkeley.

Toth, N. (1985). Archaeological evidence for preferential right-handedness in the Lower and Middle Pleistocene, and its possible implications. *Journal of Human Evolution*, **14**, 607–14.

Toyoshima, S. and Miyashita, M. (1973). Force velocity relation in throwing. *Research Quarterly*, **44**, 86–95.

Toyoshima, S.. Hoshikawa, T., Miyashita, M., and Oguri, T. (1974). Contribution of the body parts to throwing performance. In *Biomechanics IV* (ed. R. C. Nelson and C. A. Morehouse), pp. 169–74. University Park Press, Baltimore.

Trinkaus, E. (1983a). Human biocultural change in the Upper Pleistocene. In *The Mousterian legacy*, British Archaeological Reports Int. Ser. 164, (ed. E. Trinkaus), pp. 165–200. BAR, Oxford.

Trinkaus, E. (1983b). *The Shanidar Neandertals*. Academic Press. New York.

Trinkaus, E. (1984a). Does KNM-ER 1481A establish *Homo erectus* at 2.0 myr BP? *American Journal of Physical Anthropology*, **64**, 137–9.

Trinkaus, E. (1984b). Western Asia. In *The origins of modern humans* (ed. F. H. Smith and F. Spencer), pp. 251–93. Alan R. Liss, New York.

Trinkaus, E. (1989). Olduvai Hominid 7 trapezial metacarpal 1 articular morphology: contrasts with recent humans. *American Journal of Physical Anthropology*, **80**, 411–16.

Trinkaus, E. and Long, J. (1990). Species attribution of the Swartkrans Member 1 first metacarpals, SK 84 and SKX 5020. *American Journal of Physical Anthropology*, **83**, 419–24.

Trinkaus, E. and Villemeur, I. (1991). Mechanical advantages of the Neandertal thumb in flexion. *American Journal of Physical Anthropology*, **84**, 249–60.

Tuttle, R. H. (1967). Knuckle-walking and the evolution of hominoid hands. *American Journal of Physical Anthropology*, **26**, 171–206.
Tuttle, R. H. (1969a). Knuckle-walking and the problem of human origins. *Science*, **166**, 953–61.
Tuttle, R. H. (1969b). Quantitative and functional studies on the hands of the Anthropoidea. I. The Hominoidea. *Journal of Morphology*, **128**, 309–63.
Tuttle, R. H. (1970). Postural, propulsive, and prehensile capabilities in the cheiridia of chimpanzees and other great apes. In *The chimpanzee* (ed. G. H. Bourne), Vol. 2, pp. 167–253. Karger, Basle.
Tuttle, R. H. (1981). Evolution of hominid bipedalism and prehensile capabilities. *Philosophical Transactions of the Royal Society of London*, **B 292**, 89–94.
van Lawick-Goodall, J. (1968). The behaviour of free-living chimpanzees in the Gombe Stream Reserve. *Animal Behaviour Monographs*, **1**, 161–311.
van Lawick-Goodall, J. (1970). Tool-using in primates and other vertebrates. In *Advances in the study of behaviour* (ed. D. Lehrman, R. Hinde, and E. Shaw), Vol. 3, pp. 195–249. Academic Press, New York.
van Lawick-Goodall, J. (1971). *In the shadow of man*. Houghton Mifflin, Boston.
van Lawick-Goodall, J. (1973). Cultural elements in a chimpanzee community. In *Precultural primate behavior* (ed. E. W. Menzel), Vol. 1, pp. 144–84. Karger, Basle.
Walker, A. C. and Leakey, R. E. (1986). *Homo erectus* skeleton from West Lake Turkana, Kenya. *American Journal of Physical Anthropology*, **69**, 275. (Abstract.)
Walker, A. and Leakey, R. (1993). The postcranial bones. In *The Nariokotome* Homo erectus *skeleton* (ed. A. Walker and R. Leakey), pp. 95–160. Harvard University Press, Cambridge, MA.
Walker, A. C. and Pickford, M. (1983). New postcranial fossils of *Proconsul africanus* and *Proconsul nyanzae*. In *New interpretations of ape and human ancestry* (ed. R. L. Ciochon and R. S. Corruccini), pp. 325–51. Plenum, New York.
Walker, A. and Teaford, M. F. (1986). New information concerning the R114 *Proconsul* site, Rusinga Island, Kenya. In *Proceedings of the Tenth Congress of the International Primatological Society*, Vol. 1 (ed. J. G. Else and P. C. Lee), pp. 143–9. Cambridge University Press, London.
Walker, A., Teaford, M. F., and Leakey, R. E. (1985). New *Proconsul* fossils from the Early Miocene of Kenya. *American Journal of Physical Anthropology*, **66**, 239–40.
Walker, A., Teaford, M. F., Martin, L., and Andrews, P. (1993). A new species of *Proconsul* from the early Miocene of Rusinga/Mfangano Islands, Kenya. *Journal of Human Evolution*, **25**, 43–56.
Ward, S. C. and Pilbeam, D. R. (1983). Maxillofacial morphology of Miocene Hominoidea from Africa and Indo-Pakistan. In *New interpretations of ape and human ancestry* (ed. R. L. Ciochon and R. S. Corruccini), pp. 211–38. Plenum, New York.

Washburn, S. L. (1959). Speculations on the interrelations of the history of tools and biological evolution. In *The evolution of man's capacity for culture* (ed. J. N. Spuhler), pp. 21–31. Wayne State University Press, Detroit.
Washburn, S. L. (1960). Tools and human evolution. *Scientific American*, **203**, (September), 3–15.
Washburn, S. L. (1963). Comment. *Current Anthropology*, **4**, 492.
Washburn, S. L. (1967). Behaviour and the evolution of man. *Proceedings of the Royal Anthropological Institute of Great Britain and Ireland*, **1967**, 21–7.
Washburn, S. L. (1968). *The Study of human evolution*, Condon Lectures. Oregon State System of Higher Education, Eugene, Oregon.
Washburn, S. L. and Avis, V. (1958). Evolution of human behavior. In *Behavior and evolution* (ed. A. Roe and G. G. Simpson), pp. 421–36. Yale University Press, New Haven.
Washburn, S. L. and Howell, F. C. (1960). Human evolution and culture. In *Evolution after Darwin*, Vol. 2 (ed. S. Tax), pp. 33–56. University of Chicago Press.
Washburn, S. L. and Jay, P. (1967). More on tool-use among primates. *Current Anthropology*, **8**, 253–4.
Watts, L. (1973). *The fine art of baseball* (2nd edn). Prentice-Hall, Englewood Cliffs, New Jersey.
Welles, J. F. (1975). The anthropoid hand. A comparative study of prehension. In *Contemporary primatology. Proceedings of the Fifth Congress of the International Primatological Society* (ed. S. Kondo, M. Kawai, and A. Ehara), pp. 30–3. Karger, Basle.
White, T. D. (1980). Additional fossil hominids from Laetoli, Tanzania: 1976–1979 specimens. *American Journal of Physical Anthropology*, **53**, 487–504.
White, T. D. and Suwa, G. (1987). Hominid footprints at Laetoli: facts and interpretations. *American Journal of Physical Anthropology*, **72**, 485–514.
Wolpoff, M. (1982). *Ramapithecus* and hominid origins. *Current Anthropology*, **23**, 501–22.
Wolpoff, M. (1983a). Lucy's little legs. *Journal of Human Evolution*, **12**, 443–53.
Wolpoff, M. H. (1983b). Lucy's lower limbs: long enough for Lucy to be fully bipedal? *Nature*, **304**, 59–61.
Wrangham, R. W. (1980). Bipedal locomotion as a feeding adaptation in gelada baboons, and its implications for hominid evolution. *Journal of Human Evolution*, **9**, 329–31.
Zihlman, A. L. (1981). Women as shapers of the human adaptation. In *Woman the gatherer* (ed. F. Dahlberg), pp. 75–120. Yale University Press, New Haven.
Zihlman, A. and Brunker, L. (1979). Hominid bipedalism: then and now. *Yearbook of Physical Anthropology*, **22**, 132–62.
Zvelebil, M. (1984). Clues to recent human evolution from specialized technologies? *Nature*, **307**, 314–15.

Part II
Social and socio-cultural systems

Editorial introduction to Part II: Social and socio-cultural systems

The anatomies reviewed in Part I provide, literally, the bare bones of the course of human evolution. The most recent anatomy supports a way of living that is situated in an 'imagined' as much as in 'the real' world. Members of all present human cultures share intersubjective beliefs that constitute for them a reflexively-meaningful, lived-in, symbolically-interpreted reality just as real as 'the real' world (for them), as well as an interactionist world in between these two, a world of culture constructed by the synergistic interactions of individuals wherein cognition is externalized, symbolically embodied, and manipulated and transformed in new ways more intricate and multi-layered than previously imaginable (cf. Gumperz and Levinson 1991). To continue to borrow from the phrasing of these authors, this domain of the socio-cultural is filled with objects and events of perplexing ontology, because their essence is both physical and ideational.

The symbolically-mediated and constituted systems that sustain present-day human lives represent the ultimate product of extra-somatic evolution, a transformation of behavioural systems, many of which have a deep phylogeny, elaborated from abilities that are characteristic of the Primate order. The challenge is to account for how this transformation and elaboration was accomplished; for how it is that modern humans do not just behave, but can *act*: that is, can be socio-cultural animals.

This, and the remaining Parts of this volume, attempt to marshal together some of the material that is self-evidently fundamental to meeting this challenge. Behind this material lie three main issues. *First*, what was the temporal course of this transformation; essentially, what happened, where, and when? *Second*, the factors that may be hypothesized as playing a causative role in this transformation, in establishing the abilities that underpin and sustain it. *Third*, how bio-behavioural, cultural, and historical systems interrelate with each other to enable and sustain this transformation.

A number of separate lines of evidence provide indices that point to the relatively recent evolutionary emergence of the characteristics of modern human forms of social life. Patterns of spatial organization and the geographical movement of raw materials and artefacts in human groups exhibit modern characteristics from around 35 000 years BP (White, this volume, Chapter 9); a temporal and spatial volatility of artefact form characteristic of modern patterns emerges around the same time (Wynn, this volume, Chapter 10); so too does image-making (Conkey, this volume, Chapter 11). The beginnings of social differentiation, as marked by 'jewellery' have been documented recently by White (for example, 1989*b*; see also White 1989*a*, and Lock and Symes, this volume, Chapter 8) as occurring at around the same time.

There is still controversy concerning these dates. There is little archaeological evidence for symbolism from any regions of the Old World before the Late Palaeolithic, for either Neanderthals or morphologically-modern humans (Chase and Dibble 1987; Lindly and Clark 1990). As Lindly and Clark (p. 234) observe: 'if the contexts surrounding... [early] morphologically modern human remains have little or no evidence of symbolic behavior, it will be clear that no correlation of modern behavior with [early] modern morphology can be proposed.' There is clear evidence in the late Upper Palaeolithic (*c*.20 000–15 000 years BP) for a 'symbolic explosion' (see White, this volume, and Conkey, this volume, Chapters 9 and 11), and that this use of symbols can be regarded as 'modern' in its behavioural implications.

Lindly and Clark argue (ibid., 239) that this characteristic modernity originated during this recent period: 'while we are not suggesting there is no evidence for symbolism in the early Upper Palaeolithic of Europe... the major shift in adaptation occurred late in the Upper Palaeolithic/Late Stone Age and was largely unrelated to the perceived transition from the Middle to the Upper Palaeolithic'. In commenting on this claim, Mellars (1990) cites Cabrera-Valdes and Bischoff 1989; Gamble 1983; Mellars 1989; Whallon 1989; and White 1985; 1989*a*, *b*, *c* to put the view that the later Upper Palaeolithic represents a cultural 'intensification' of 'far more radical innovations in behavior at the *start* of the Upper Palaeolithic':

the whole character of [early Upper Palaeolithic] culture shows a dramatic contrast with earlier 'Middle Palaeolithic' culture not only in the character of the lithic industries but in such features as complex personal ornaments, elaborately shaped bone, antler, and ivory artefacts, far-travelled marine shells, increased use of other 'exotic' materials, and the earliest well-documented (and remarkably complex) art (Mellars 1990, 264).

But, irrespective of the exact timing, some characteristically modern patterns of human social activity have been a common feature of human life for only a relatively short period. There is a decoupling of the cultural emergence of modern human behaviours from that of the biological emergence of modern human skeletal morphology: characteristically modern human behaviours and socio-cultural patterns were the achievement of already-anatomically-modern humans. The evidence points to these patterns being predicated on the exploitation of the possibilities afforded by symbolic systems.

Thus, one question for the Palaeoanthropological section of Part II concerns the nature of the socio-cultural revolution. In the Ice Age art, notational systems, and apparent mnemonic devices of the Upper Palaeolithic we see the first clear evidence of the development of technology in the service of cognition. Following the discussion provided by Gumperz and Levinson (1991, 614), we can point out that the enabling effects of these developments consisted not only in overcoming the limitations of time and space associated with acts of communicative gesture and speech, but also in the externalization of cognition that makes it possible to examine *post hoc* various aspects of thinking, thereby facilitating a new multidimensional manipulation and recontextualization of what is in the vocal–auditory and gestural channels a largely linear medium. The development of these conceptual tools provided new socio-cultural means for interactive discourse whose potentialiaties continued to unfold over the next few thousand years.

In the first section of Part II, Lock and Symes (this volume, Chapter 8) review some of the literature on social relations and the socio-cultural 'scaffolding' of cognitive abilities. The important point to note here is that the somatic substrate of cognition, the cognitive technologies instantiated by symbol systems, and the social organization of human groups are all clearly interrelated phenomena. How they are interrelated needs to be explored in future research before the above questions can be both framed and addressed in more appropriate ways than are currently possible. But, that they *are* interrelated is of clear importance to developing a better understanding of the evolution of modern human abilities.

Another point to address here concerns the meaning of 'evolution' as it might be applied to understanding and explaining the changes that have occurred in the sphere of human social organization. There is a large literature addressing the 'evolution of society', running the gamut from that which rejects the applicability of evolutionary concepts in their entirety to that in which society is regarded as essentially just another biological phenomenon. This range represents a continuing historical debate prompted by our awareness that the first essays on 'social evolution' were embedded in a Victorian ideology that equated 'evolution' with 'progress', and regarded anyone other than white, male Western Europeans, be they women, children, or 'savages', as 'primitives' (or even—as in the early encounters of the Spanish with Americans or the English with Tasmanians—as 'non-human'); see, for example, Runciman (1989, chapter 2); and Benton (1991, especially pp. 9–15).[1]

First, in their rejection of these Victorian ideological presuppositions, some social anthropologists effectively deny that there is any problem to be addressed:

It is virtually dogma among social anthropologists of my sort that cultural otherness does not carry with it any necessary hierarchy of superiority/inferiority which can be appropriately labelled by such terms as 'primitive', 'backward', 'underdeveloped', 'childish', 'ignorant', 'simple', 'primeval', 'pre-literate', or whatever. My interest in the others arises because they are other, not because they are inferior.... In my own view there is no significant dichotomy in terms of either structure or form between 'modern' societies and 'primitive' societies. The social anthropologist can find what he is looking for in either.... is it possible to formulate a useful stereotype of what this notional entity 'a primitive society' or 'a savage (wild) society' is like? The answer is: No! (Leach (1982) 123–4, 141; cited by Hallpike (1986) 12).

But, as Hallpike comments on these passages:

When the concepts of 'pre-literate', 'simple', and 'primitive' are equated with 'backward', 'childish', 'ignorant', and 'inferior', and an intellectual fascination with the origin of things is regarded as a debased desire to gloat over the underprivileged, there is a powerful inducement to take the final step and deny that there is any such thing as primitive society at all (ibid., 12).

With no difference to be apprehended, there is nothing to explain.

A variant of this extreme, adopted by sociological theorists such as Giddens (1984, for example, especially Chapter 5), is also informed by the implicit assumption of ubiquitous human equality. Here, that there are differences between human societies is apparently accepted; but these differences are cast as differences of degree rather than of kind, hence retaining an equality of humanity at a fundamental level related to human capacity, the differences being variants of 'similar' otherness. On this ground, Giddens rejects the applicability of evolutionary theories to practically all spheres of human action, claiming human social orders are never the result of functional exigencies, but rooted in human choice. For Giddens, *all* human life is a matter of history, not biology, and:

> Human history does not have an evolutionary "shape", and positive harm can be done by attempting to compress it into one.... An evolutionary "shape"—a trunk with branches, or a climbing vine, in which the elapsing of chronological time and the progression of the species are integrated—is an inappropriate metaphor by which to analyse human society (ibid., 236–7).

Second, other sociologically-oriented theorists (for example, Mann 1986, especially chapter 2) maintain a distinction between history and prehistory, and see each as subject to different organizational principles: thus,

> general evolutionary theory may be applied [up] to the Neolithic Revolution, but its relevance then diminishes.... *No general social evolution occurred beyond the rank societies of early, settled neolithic societies* (ibid., 39, 69–70).

Beyond this point, we must move to what he terms 'local history', in which the consequences of human choice, not deterministic evolution, assume the role of prime mover:

> general processes were '*devolutions*'—movement back toward rank and egalitarian societies—and a *cyclical* process of movement around these structures, failing to reach permanent stratification and state structures. In fact, human beings devoted a considerable part of their cultural and organizational capacities to ensure that further evolution did *not* occur. They seem not to have wanted to increase their collective powers, because of the distributive powers involved (ibid., 39).

This distinction is not drawn by theorists of the *third* kind, who adopt a contrary stance to Giddens (1984). Where he seems to reject the distinction between biology and history, subsuming everything under 'history', these theorists subsume everything under 'biology', especially neo-Darwinian evolutionary theory. Some sociobiologists (such as Alexander 1979; Durham 1976) have attempted to demonstrate that many fundamental social institutions, from the emergence of the state, kinship, marriage, and warfare to the celibacy of the priesthood, can be explained in terms of increases accruing to the inclusive fitness of the individuals pursuing the strategies that constitute these institutions. Social institutions are here viewed as 'adaptive' responses, and subject to the process of selection, and are claimed as analogous to genes. Indeed, 'social genes' of various forms have been proposed as direct counterparts to biological ones. Following Hallpike (1986) we can quote examples of this position:

> In social systems the gene is the image or the idea in the mind of man.... The parallel with the gene is exact (Boulding (1970) 20).

> ...culture is acquired in tiny, unrelated snippets, which are specific interneural instructions transmitted from generation to generation. These 'corpuscles of culture' are transmitted and acquired with fidelity and ease because the organisms in question are phylogenetically adapted for transmitting and acquiring cultural corpuscles (Cloak (1975) 167).

> Examples of memes are tunes, ideas,...ways of making pots.... Just as genes propagate themselves in the gene pool by leaping from body to body via sperms or eggs, so memes propagate themselves in the meme pool by leaping from brain to brain via... imitation (Dawkins (1976) 206).

Another genetic analogue, the 'culturgen', has been proposed by Lumsden and Wilson (1981, for example p. 27); but as Hallpike (1986, 45) notes, 'this idea is so vague and all-encompassing that a culturgen could be any discriminable aspect of thought or behaviour whatsoever. It is as though the "thing" were to be proposed as the basic unit of physics.' These are clearly metaphors of inanity almost sufficient to make one side with Giddens in his rejection of the relevance of evolution to social issues, for,

> obviously, if one regards societies as nothing but jumbles of bits and pieces brought together by contingencies of history and cultural diffusion, theories of social evolution are indeed a complete waste of time (Hallpike (1986) 7).

These neo-Darwinist positions have kindred variants in biological anthropology (for example, Chagnon and Irons 1979; Foley 1987); ecological anthropology (for example, Rappaport 1973, 1984; Vayda 1969); and evolutionary archaeology (for example, Dunnell 1980). These suffer from some of the same problems as the neo-Darwinian accounts: the difficulty of going beyond metaphor to the identification of units of transmission; the problem of adducing organizational principles whereby a random set of cultural elements could be organized into the semblance of structure human social organizations exhibit at any level; their reliance on concepts that are problematic in their own right—for example, adaptation, for which there exists an extensive critical literature; and a disregard for individual human agency to an extent that warrants some of the extreme reactions of the social-science fraternity (for example, theories of the first type discussed above).

One apparent advance in the neo-functional or adaptationist socio-cultural approach is seen in Fig. 1, Rappaport's (1971) graphic model of the ecology of the Maring-speaking people living in the Bismarck Range of New Guinea. The model gives prominence to *emic*[2] factors in an account that attempts to integrate Maring cosmology (the cognized model) with the specifics of demographic and food-production processes at the local level, and a long-term ritual cycle that determines social relations at the regional level. It represents an externalization of indigenous

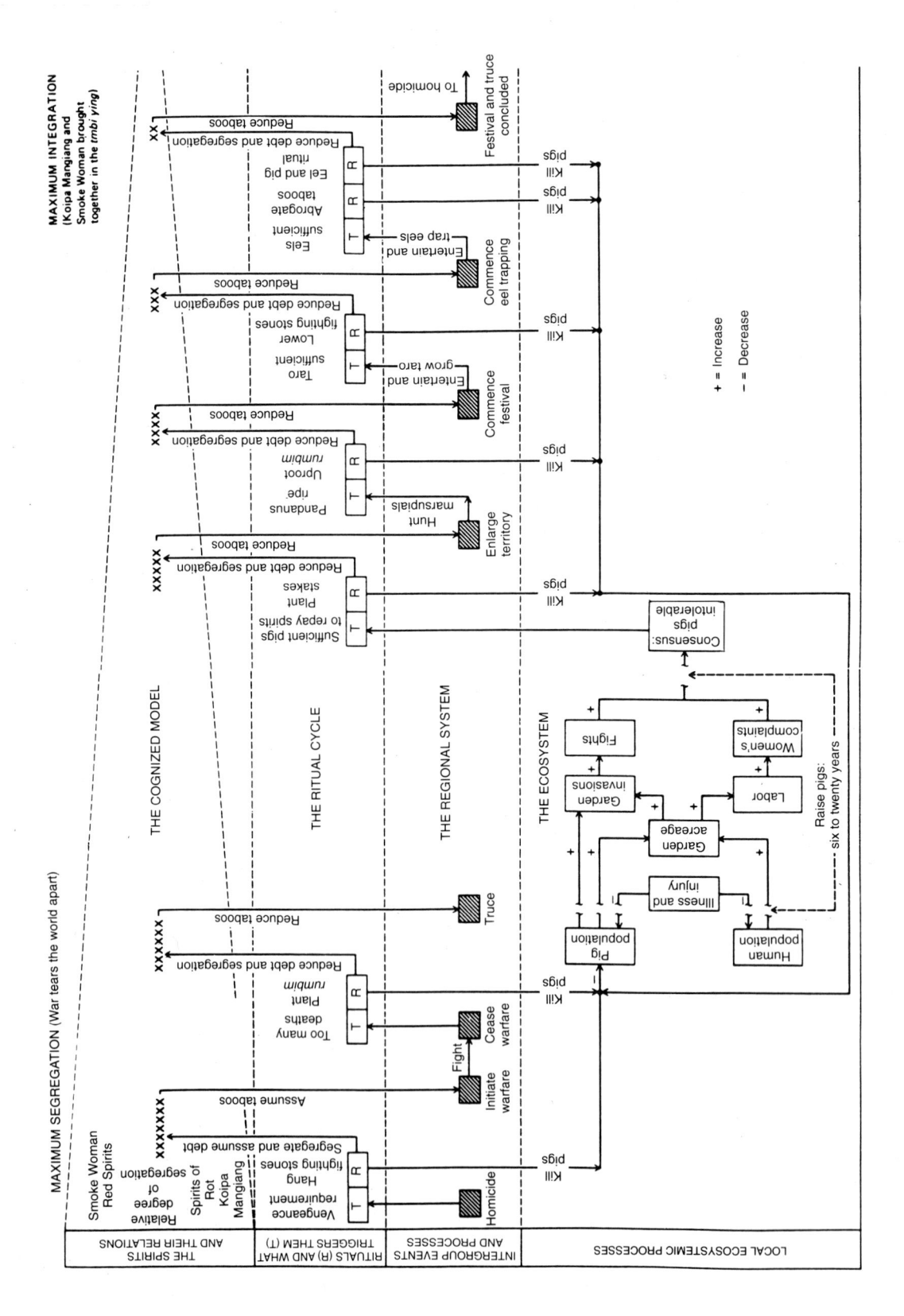

Fig. 1 Ritual regulation and ecosystem function among the Maning (Rappaport 1971).

cosmology, indigenous and Western knowledge of biological ecology, and Rappaport's hypothesis about how relations between local populations are regulated by a protracted ritual cycle that integrates a number of processes and functional levels. Rappaport acknowledges that no satisfactory answer can yet be offered for how such systems develop, but argues that they may be adaptive in particular environments. They can be seen as examples of non-unilineal sociocultural evolution. We should note once again that the kind of functionalism being assumed here is quite different from that argued by Hallpike (1986) to be characteristic of the changes associated with the emergence of the state.

Fourth, as Smith (1992, 130) notes: 'Grand theory is back'. The works of Giddens and Mann (above) are in this mould. But they essentially portray a discontinuist perspective, even though they do not neglect time as a significant aspect of human social organization. By contrast, there are evolutionist positions within the grand tradition. The most explicitly evolutionary recent 'grand theory' is that of Runciman (for example, 1989), who takes the concepts of 'evolution' and 'adaptation' as central to his project.[3] However, Runciman's notion that social evolution is a process whereby competitive practices change distributions of power would seem to be subject to Hallpike's (1986) criticisms of the adaptationist paradigm: that is, that it is a recent illusion that the same degree of 'functionality' exists in all societies; rather, functional and adaptive efficiency are emergent properties of society, associated in particular with the emergence of the state.

The *fifth* and final group to be noted here is the most heterogeneous: they may be called 'systems theories'. What unites them is a concern with the natural elaboration of complex patterns. They range from purely mathematical renderings and descriptions of change, from catastrophe theory to what is now termed the 'science of chaos', through ones that employ a less formal framework that emphasizes concepts such as 'emergence', to those that attempt to find a place for human activities themselves in their schemes. Dodgshon (1987, pp. 4–10) distinguishes three broad groupings of systems theories that we follow here.

(1) Regulative systems
Social systems (as a sub-category of systems in general) tend to be conservative, variations that arise being 'damped' by processes of self-regulative negative feedback to maintain a steady state. But, occasionally, 'the amplitude of variation becomes excessive, so much so that the entire system becomes relocated, via positive feedback and irreversible change, around an entirely new equilibrium' (ibid., 8).

Often an explicitly evolutionary terminology is adopted: Rappaport (1977), for example, considers excessive variation through the failure of regulators to cope with structural fluctuations as 'maladaptive', and hence the shift in equilibrium points to new forms of re-integrative organization as 'adaptive' responses. Such 'theories' face the criticism of being just metaphorical re-descriptions of change that are additionally pitched at such a level as to fail to illuminate the nature of the human activities giving rise to those changes.

(2) Dissipative systems
Drawing from work by Prigogine (for example Prigogine *et al.* 1977), this general approach dispenses with the notion of equilibrium, and considers the way that order can result from non-equilibrium systems, with dissipation as an agent of order. While standing equilibrium theories on their head, this approach is open to the same criticism.

(3) Interactive systems
Where the above two approaches see order and change as internal properties of systems, this third approach concerns itself with the effects of external factors on them, that is, they are couched 'in terms of how societal systems coped or reacted to what was going on around them' (Dodgshon (1987) 9). Examples of this approach are to be found in the appendix to White's chapter (Chapter 9); and for a general critique, see Mann (1986, Chapter 2). Perhaps the major criticism that may be advanced concerns the neglect of *emic* factors in *etic* accounts (cf. Note 2 to this Introduction).

One attempt to resolve this issue has been made by Hallpike (1986). He regards the evolution of human social organization as an unfolding, directional process:

without being deterministic, inevitable or purposive, . . . it can also be regarded, in certain ways, as a process by which latent potential is realized. It is also, quite clearly, a process involving increased elaboration of structural forms, and for all of these reasons we are therefore entitled to talk of social evolution, and not merely of social change (ibid., 15).[4]

There is an avowedly Piagetian flavour to Hallpike's position (stemming from his previous work, for example, 1979); and from this criticisms of Hallpike's specific approach can be put forward, in that it may be that other developmental metaphors, drawn from, say, a Vygotskyan perspective (see Sinha, this volume, Chapter 17), would be more appropriate in dealing with the shift in focus Hallpike's approach entails, foregrounding as it does the issue of the origin and trajectory of human belief systems. But there are clearly affinities here with theories noted earlier, for example by Giddens and Mann, that have their roots in sociology rather than, ultimately, biology. These affinities are ones that we may expect to see explored in the immediate future of 'grand theorizing', and to be crucial to the elaboration of the interactionist position needed to comprehend *human* evolution, since they force a consideration of the interdependence of the somatic and extra-somatic systems that are characteristic of anatomically-modern *Homo sapiens*.

The data at present point toward the conclusion that the modern human cognitive substrate was probably in place before the modern 'cognitive processes' it currently supports. Additionally, there is an emerging view across a number of disciplines that ' "cognitive processes" cannot be fully located within the individual' (Gumperz and Levinson 1991, 614; cf. Bruner 1990 for an overview; and Sinha, this volume, Chapter 17), but may be partially 'embodied' in cultural practices and the symbol systems that enable these. *Practices*, their origins, maintenance, and enabling properties, are thus the potential pivot around which a fuller understanding may be anticipated of human symbolic evolution, in terms of both its individual and its social components.

Notes

1. Thus, for example, Spencer (1961/1873, 64) conflates evolution and progress:

if there does exist an order among those structural and functional changes which societies pass through, knowledge of that order can scarcely fail to affect our judgements as to what is progressive and what is retrograde.

And Marx (1968, 183):

In broad outlines we can designate the Asiatic, the ancient, the feudal, and the modern bourgeois modes of production as epochs in the progress of the formation of society. Bourgeois relations are the last antagonistic form of the social process of production.

To which Runciman (1989, 47), from whom these quotes are taken, responds, summing up the problem, 'there are indeed qualitatively different epochs but their progressiveness is in the eye of the beholder . . .'.

It was in the decades surrounding the turn of the last century, in a first reaction to the above, that the still-central anti-naturalistic themes of the modern social sciences were set in place. Durkheim constituted the social world as a causal order in its own right. Thus the social world was disconnected from the biological, in that while biological factors may still determine individual behaviour, social processes became conceived as more than just the result of individuals' actions. Other central theoretical strands in sociology are similarly anti-naturalistic, for example, the Weberian, the Marxist, etc., as are influential traditions of social anthropology—those developed from the work of Boas and Kroeber, for example, which are inherently anti-evolutionist (see Stocking 1968; Freeman 1984; Horigan 1988; and Sanderson 1990 for fuller accounts). These disciplinary histories continue to exert their influence today.

2. The terms 'emic' and 'etic' originate from Pike (1967), and have been adopted by many anthropologists. Harris (1979, 32) distinguishes them thus:

Emic operations have as their hallmark the elevation of the native informant to the status of ultimate judge of the adequacy of the observer's descriptions and analyses. The test of the adequacy of *emic* analyses is their ability to generate statements the native accepts as real, meaningful or appropriate . . . *Etic* operations have as their hallmark the elevation of observers to the status of ultimate judges of the categories and concepts used in descriptions and analyses. The test of the adequacy of *etic* accounts is simply their ability to generate scientifically productive theories about the causes of socio-cultural differences and similarities. Rather than employ concepts that are necessarily real, meaningful, and appropriate from the native point of view, the observer is free to use alien categories and rules derived from the data language of science. Frequently, *etic* operations involve the measurement and juxtaposition of activities and events that native informants find inappropriate or meaningless.

3. Runciman's own summary of his position sets his question and answer thus:

it ought . . . to be possible . . . to say how it is that societies evolve—that is, to specify the process by which they change from one distinguishable type or mode to another, the units selected by that process, the functions which the units perform, and the direction which the process has taken (or perhaps I should say: has *happened* to take) thus far. Briefly, my suggested answer is that the process is one of competitive selection whereby certain roles and institutions come to replace or supersede others; that the units of selection are not roles or institutions but the practices of which classes, status-groups, orders, factions, sects, communities, age-sets, and so forth are the carriers; that their function lies in maintaining or augmenting the power which attaches to the roles and thereby institutions which they constitute and thus in preserving or changing the mode of the distribution of power in societies (or 'social aggregates' or 'social formations') taken as a whole; and that the direction which evolution has thus far taken has in consequence been one of both increasing and diminishing variation—increasing as mutant or recombinative practices create new roles and institutions, and decreasing as the competitive advantages which they confer on their carriers compel pre-existing ones to adapt to them (1989, 44).

4. However, while being an evolutionist, Hallpike's view is that

it is impossible to apply Darwinian principles to the evolution of human society because they are inherently of the wrong type to be applied to socio-cultural systems. . . . such basic Darwinian concepts as the unit of selection, fitness, adaptation, competition, and mutation are irrelevant to social evolution (1986, pp. 32, 36).

The essential orientation he adopts is clear in the following delineation he offers of his position in contradistinction to those we have listed above:

The faults of various brands of determinism (materialist, idealist, or structural functionalist) is to attribute powers of efficient causality to structures, instead of to individuals. The opposite error, of treating efficient causes as if they could of themselves generate social structures, is committed by naïve individualists, such as some psychologists and socio-biologists, who suppose that it is possible to deduce the structure of society from the simple aggregation of individual motives and actions alone.

Social structures [need to be] brought into relation with the world of thoughts and actions of individuals so that there is, on the one hand, a process of accommodation to reality, and on the other the assimilation of that reality to the particular structures of society and its belief system. While in one way social evolution can be regarded as a development of the structural potential of different societies, structures do not develop by themselves, but only by the ways in which the participating individuals interact with one another and with the physical world (ibid., 28).

References

Alexander, R. D. (1979). *Darwinism and human affairs*. University of Washington Press, Seattle and London.

Benton, T. (1991). Biology and social science: why the return of the repressed should be given a (cautious) welcome. *Sociology*, 25, 1–29.

Boulding, K. E. (1970). *A primer of social dynamics: history as dialectics and development*. Free Press, New York.

Bruner, J. S. (1990). *Acts of meaning*. Harvard University Press, Cambridge, Mass.

Cabrera-Valdes, V. and Bischoff, J. L. (1989). Accelerator 14C ages for basal Aurignacian at El Castillo (Spain). *Journal of Archaeological Science*, 16, 577–84.

Chagnon, N. A. and Irons, W. (1979). *Evolutionary biology and human social behavior: anthropological perspectives*. Duxbury Press, London.

Chase, P. and Dibble, H. (1987). Middle Paleolithic symbolism, a review of current evidence and interpretations. *Journal of Anthropological Archaeology*, 6, 263–96.

Cloak, F. T. (1975). Is a cultural ethology possible? *Human Ecology*, 3, 161–82.

Dawkins, R. (1976). *The selfish gene*. Oxford University Press.

Dodgshon, R. A. (1987). *The European past: social evolution and spatial order*. Macmillan, London.

Dunnell, R. (1980). Evolutionary theory and archaeology. In *Archaeological method and theory*, Vol. 3 (ed. M. B. Schiffer), pp. 35–99. Academic Press, London.

Durham, W. H. (1976). Resource competition and human aggression. Part 1: a review of primitive war. *Quarterly Review of Biology*, 51, 385–415.

Foley, R. (1987). *Another unique species*. Longman, London.

Freeman, D. (1984). *Margaret Mead and Samoa*. Penguin Books, Harmondsworth.

Gamble, C. (1983). Culture and society in the Upper Palaeolithic of Europe. In *Hunter–gatherer economy in prehistory* (ed. G. Bailey), pp. 201–11. Cambridge University Press.

Giddens, A. (1984). *The constitution of society: outline of the theory of structuration*. Polity Press, Cambridge.

Gumperz, J. J. and Levinson, S. C. (1991). Rethinking linguistic relativity. *Current Anthropology*, 32, 613–23.

Hallpike, C. R. (1979). *The foundations of primitive thought*. Clarendon Press, Oxford.

Hallpike, C. R. (1986). *The principles of social evolution*. Clarendon Press, Oxford.

Harris, M. (1979). *Cultural materialism: the struggle for a science of culture*. Random House, New York.

Horigan, S. (1988). *Nature and culture in Western discourses*. Routledge and Kegan Paul, London.

Leach, E. R. (1982). *Social anthropology*. Oxford University Press.

Lindly, J. M. and Clark, G. A. (1990). Symbolism and modern human origins. *Current Anthropology*, 31, 233–61.

Lumsden, C. J. and Wilson, E. O. (1981). *Genes, mind, and culture*. Harvard University Press, Cambridge, Mass.

Mann, M. (1986). *The sources of social power*: Vol. I, *A history of power from the beginning to AD 1760*. Cambridge University Press.

Marx, K. (1968). Preface to a contribution to the critique of political economy. In *Selected works* (K. Marks and F. Engels). Lawrence and Wishart–Progress Publishers–International Publishers Inc., London–Moscow–New York.

Mellars, P. (1989). Major issues in the emergence of modern humans. *Current Anthropology*, 30, 349–85.

Mellars, P. (1990). Comment on "Symbolism and modern human origins". *Current Anthropology*, 34, 245–6.

Pike, K. (1967). *Language in relation to a unified theory of the structure of human behavior* (2nd edn). Mouton, The Hague.

Prigogine, I., Allen, P. M., and Herman, R. (1977). Long-term trends and the evolution of complexity. In *Studies in the conceptual foundations of biology* (ed. E. Laszlo and J. Bierman), pp. 1–26. Pergamon Press, New York.

Rappaport, R. A. (1971). Nature, culture, and ecological anthropology. In *Man, culture, and society*, (rev. edn, ed. H. L. Shapiro), pp. 237–67. Oxford University Press.

Rappaport, R. A. (1973). The sacred in human evolution. *Annual Review of Ecology and Systematics*, 2, 23–44.

Rappaport, R. A. (1977). Maladaptation in social systems. In *The evolution of social systems* (ed. J. Friedman and M. J. Rowlands), pp. 49–71. Duckworth, London.

Rappaport, R. A. (1984). *Pigs for the ancestors: ritual in the ecology of a New Guinea people* (new, enlarged edn.) Yale University Press, New Haven.

Runciman, W. G. (1989). *Confessions of a reluctant theorist*. Routledge and Kegan Paul, London.

Sanderson, S. K. (1990). *Social evolutionism, a critical history*. Basil Blackwell, Oxford.

Smith, D. (1992). *The rise of historical sociology*. Polity Press, Cambridge.

Spencer, H. (1961/1873). *The study of sociology*. University of Michigan Press, Ann Arbor.

Stocking, G. W. (1968). *Race, culture and evolution*. Collier Macmillan, London.

Vayda, A. P. (1969). *Environment and cultural behavior*. Natural History Press, New York.

Whallon, R. (1989). Elements of cultural change in the Later Palaeolithic. In *The human revolution: behavioural and biological perspectives on the origins of modern humans* (ed. P. Mellars and C. Stringer), pp. 433–54. Edinburgh University Press.

White, R. (1985). Some thoughts on social relationships and language in hominid evolution. *Journal of Social and Personal Relationships*, **2**, 95–115.

White, R. (1989a). Visual thinking in the Ice Age. *Scientific American*, **261**, 92–9.

White, R. (1989b). Production complexity and standardization in early Aurignacian bead and pendant manufacture, evolutionary implications. In *The human revolution: behavioural and biological perspectives on the origins of modern humans* (ed. P. Mellars and C. Stringer), pp. 366–90. Edinburgh University Press.

White, R. (1989c). Toward a contextual understanding of the earliest body ornaments. In *The emergence of modern humans*, (ed. E. Trinkaus), pp. 211–31. Cambridge University Press.

Part II(a)
Comparative perspectives

6
Primate communication, lies, and ideas

Alison Jolly

Abstract

Communications which are interestingly symbolic involve a partial detachment from the referent: one criterion of a proto-linguistic mentality is how good it is at lying. Primate taxonomy does not correspond one-to-one with social structures. It correlates more closely with complexity of social intelligence. Our near relatives, the chimpanzees and bonobos, have male-bonded societies in which females migrate between troops, and individuals leave and rejoin the group. This means an individual potentially has private information it could share or withhold. Vocalizations of monkeys, and probably apes, contain semantic detail about social relations as well as external threats. Chimpanzees give food-calls in the wild which attract others; in captivity they can lead others to hidden food, and convey its quality. Apes, and occasionally monkeys, deliberately deceive others, concealing both food and sex, and even facial expressions or erections. Apes (but not monkeys) recognize themselves, removing marks from their faces in mirrors, and can take others' roles in shared experiments. The capacities to give or withhold information and to be aware of others' intentions may be pre-requisites for the capacity to manipulate signs detached from the immediate: in other words, to have an idea.

6.1 Introduction

As a Darwinian, I start from the assumption that language was useful to our evolving ancestor. Selection for rapid advances in linguistic capacity probably related to the advantages of efficient communication between members of social groups. It would seem that one approach to understanding the origins of language is to examine complex communication in the context of primate societies.

There is, however, a paradox. When we start to quote particular examples of primate behaviour that seem to prove intentional, cognitive, or even symbolic capacities, the examples are not smoothly functioning communications. Instead, they nearly always involve blocked communication. We turn to deception and to play for our best cases. In a formal sense, as Bateson (1973, pp. 150–166) pointed out, such cases involve negation: a tag, whether conscious or not, that says 'This is not real.'

A criterion of symbolic language is that it communicates what is not here, not now, not visible: 'There are wildebeest round that hill: if you go this way and I go that way, we'll catch one' (de Laguna 1963). An animal which can convey such information may also be able to communicate what is not even true. In other words, one criterion of language, and of the kind of mind that can use language, is how good it is at lying.

The evolution of lies is a topical concern in the study of animal behaviour (see for instance Krebs and Dawkins 1984; Harper 1991). It is perhaps a sidelight on the sociology of science that the Reagan–Thatcher years redefined what used to be called communication into the art of the deal. Even evolved signals may misinform. These include, for instance, camouflaged coloration, or displaying eyespots or erecting fur to 'look threatening' toward predators or rivals; also if possible *not* signalling during a fight 'I am weakening and close to surrender.' This paper looks at a more complex level: the communications which have as part of the act itself, 'This is both true and not true.'

6.2 Overview of the primates: taxonomy, social structure, intelligence

The Primate Order is divided into two suborders, the Strepsirhini and the Haplorhini (Table 6.1) (detailed references in Jolly 1985). Strepsirhini means 'turned nose'; the strepsirhines have a damp muzzle connected to their upper lip. They include the lemurs of Madagascar, the bush-babies and pottos of Africa, and the lorises of Asia. Most Strepsirhini are nocturnal. They usually forage alone, whatever their social links with other individuals. On Madagascar, however, where there are no monkeys, lemurs have evolved into monkeys' niches. Diurnal lemurs may live in cohesive troops just like some monkeys and apes, or in monogamous pairs with their maturing young. These large, diurnal lemurs feed on leaves and fruit, unlike the nocturnal forms, which primarily eat insects and tree-gums.

About a thousand years ago, thirteen known genera of lemurs became extinct following man's arrival on Madagascar. Their sizes ranged up to at least that of a chimpanzee. Two were apparently terrestrial, like African baboons and patas monkeys. Several were convergent anatomically with sloths; one with a huge

Table 6.1 Primate taxonomy

Strepsirhini:
Lemuroidea — Malagasy strepsirhines
  Lemuridae — Lemurs
    Cheirogaleinae — mouse and dwarf lemurs
    Lepilemurinae — lepilemurs
    Lemurinae — true lemurs
  Indriidae — indri, sifaka, avahi
  Daubentoniidae — aye-aye
Lorisoidea — African and Asian strepsirhines
  Lorisidae — lorises, pottos
    Lorisinae
    Galaginae — bush-babies

Haplorhini
Tarsioidea*
  Tarsiidae — tarsiers
Ceboidea (†) — New World monkeys
  Callitrichidae — marmosets and tamarins
    Callitrichinae
    Callimiconinae — Goeldi's monkey
  Cebidae
    Cebinae — capuchins
    Aotinae — owl monkeys
    Callicebinae — titi monkeys
    Saimirinae — squirrel-monkeys
    Pithecinae — sakis, uakaris
    Alouattinae — howlers
    Atelinae — spider monkeys, woolly monkeys
Cercopithecoidea — Old World monkeys
  Cercopithecidae — macaques, baboons, guenons
    Cercopithecinae — colobus, langurs
    Colobinae
Hominoidea — Apes and humans
  Hylobatidae — gibbons and siamang
  Pongidae — Great Apes
  Hominidae — humans

* Tarsiers are often classified with lemurs and lorises as Prosimii. The group of monkeys and apes then becomes Anthropoidea.
† New World monkeys may also be classed as Platyrrhini, as opposed to a grouping of Old World monkeys, apes, and humans as Catarrhini. Categories may change status: for example, Lepilemurinae and Cheirogaleinae may deserve family rank as Lepilemuridae and Cheirogaleidae, while gibbons may be merely Hylobatinae, a subfamily of the Pongidae.

koala bear. Perhaps these giant lemurs resembled higher primates in diverse, complex social structures: we do not know.

The Haplorhini, or the monkeys, the apes, and the tarsier, have simple nostrils, set off from the mouth by dry or furry skin. The important distinctions from strepsirhines are the greater development of the cerebellum and the visual areas of the brain, and the more intimate, and presumably more efficient, placental structure. The Haplorhini again can be divided into two great groups: the New World Monkeys as against the Old World Monkeys, apes, and humans. The tarsier of Borneo, Sulawesi, and the Philippines is a relict form, related to other Haplorhini.

The origin of New World Monkeys is much disputed. They may be a separate lineage, derived from a tarsier or a lemur-like ancestor, or they may have crossed directly from Africa to South America, descending from an Old World ancestor already at the anthropoid level of brain and reproductive development.

Among the New World monkeys, the marmosets and tamarins form one taxonomic family, the Callitrichidae. Female callitrichids in captivity fight and even kill each other. They were long thought to be strictly monogamous, a good example of a taxonomic group with just one social structure. Recent field studies reveal that several tamarin populations are, instead, polyandrous. There is only one breeding female in each group, but non-related males join as helpers, and more than one male may mate with the female during a single oestrus period. Larger juveniles also commonly stay for a time with the group, helping with the current infants. Both monogamous and polyandrous social structures reflect the peculiar callitrichid reproductive pattern. The female gives birth to twins twice a year, often becoming pregnant during a post-partum oestrus. In other primates lactation suppresses ovulation; but a callitrichid female is commonly lactating for twins while pregnant with the next set. It is the males and juvenile helpers who carry the growing twins, and who hunt insects for them, a vital contribution of parenting energy.

The rest of the New World monkeys, the Cebidae, also tend toward monogamy, with the father contributing both transport and care for the growing infant. Monogamy is characteristic of marmoset-like Goeldis' monkeys, the nocturnal owl monkey, titis, and sakis. However, many cebid monkeys instead live in groups with several males and several females. There the mother usually carries her young. These troops differ in internal structure and bonding. Some, the spider and woolly spider monkeys, are bands of male kin with immigrant females. Only a few New World monkeys, such as the red howler, the white-fronted capuchin, and the brown capuchin regularly form groups with several females but just one male, who jealously guards his prerogatives in mating.

In the Old World monkeys, the balance between monogamous and polygamous species is very different. Only three species apparently have some monogamous populations. The vast majority of Old World monkey females share their lives and range with other females and with one or more male associates.

Within the Old World monkeys, the two subfamilies have very different build and habits. Cercopithecinae are largely fruit-eaters. They are the arboreal African guenons and mangabeys, the partly terrestrial African baboons, and the Afro-Asian macaques. The Colobinae, in contrast, are leaf-eaters: the langurs of Asia and the colobus of Africa. Colobinae and guenons tend toward single-male groups; baboons, macaques, and mangabeys toward multi-male groups; but there are exceptions in each.

Finally, the apes, with a mere five surviving genera, display every structure but polyandry. Gibbons and siamang are monogamous; orang-utans solitary; gorillas single-male; chimpanzees multi-male, with fluid subgroups (Tables 6.2, 6.3, taken from Ghiglieri 1987, 1989).

The conclusion from this taxonomic survey is that within each major group there are general trends. However, many genera and even populations within a species can differ in grouping patterns. The social structure of a species cannot be predicted confidently from that of its relatives. Specialized ecological niches and local adaptations play as large a role as taxonomic inheritance in determining social structure.

Ecology seems fundamental to understanding the size of the social group that an animal will communicate with, and the distribution of ages, sex, kin, and competitors within the group.

Taxonomy, on the other hand, has much to do with intellectual capacity. Other contributions to this volume present the different gradations of brain size to body weight in primates (Holloway, Chapter 4, this volume). In the purely behavioural sphere, even social lemurs do not manipulate objects as monkeys do (Jolly 1985). In the more interesting social sphere, lemurs seem to have more fragmentary and unstable dominance hierarchies than monkeys (Pereira 1993, 7). Even short-term lemur interactions show little or no tripartite social support and no reconciliation with former enemies (Kappeler 1993). This means we finally have analysed and measured ways in which lemur social relations are less complex than those of at least Old World monkeys.

At the other taxonomic extreme, it seems that Great Apes have a 'theory of mind', while monkeys do not (Premack and Woodruff 1978). Many apes recognize themselves in mirrors, which is one firm test for a

Table 6.2 Summary of social structures of Great Apes (modified after Ghiglieri 1989)

Social characteristic	Mountain gorilla	Orang-utan	Chimpanzee	Bonobo (Pygmy chimpanzee)
1. Female exogamy	Common	Rare	Common	Common
2. Male exogamy	Less common, but frequent	Most common	Rare; none after infancy	Rare or None
3. Associations between females	Due to common attraction to silverback harem male	Mother and immature daughter only	Due to common attraction to males, sometimes matriarchal	unrelated 'friends'
4. Bonds between adult females	Weak; dissolve with loss of harem male.	Weak, but strong enough to cause matriarchal home-range cluster	Matriarchal may include mutual aid and rare adoption of orphans.	Strongest among Hominoidea: include plant-food sharing, genito-genital rubbing, and mutual grooming.
5. Type of social group: tendency for fusion–fission sociality	One-male harem, stable, closed. No tendency for fusion–fission sociality.	Solitary, independent adults. *Very* slight tendency for occasional meeting among females.	Multi-male kin-group: stable, closed community with multiple females. Fusion–fission sociality.	Multi-male, stable, closed community with multiple females. Fusion–fission sociality.
6. Mating system; mate-sharing	Polygynous, little male sharing of mates.	Polygynous, and polyandrous	Community polygyny, and polyandry	Community polygyny and polyandry.
7. Territoriality: defence by males; level of aggression	Neither sex territorial.	Males vary; most not territorial; a few fight for dominance.	Communal territoriality by male kin-group; expansion by killing unrelated males.	Communal territoriality by co-operating males; some fights serious.
8. Mating competition between males	Intense between individuals, severe fights.	Intense between individuals, fights.	Lethal between communities; also occurs within.	Intense between communities; much milder within.
9. Infanticide	By males against infants sired by rival males.	Not seen: may not occur.	By males against infants sired by rival males.	Has not been seen.
10. Parental investment by adult males	Slight; consists of protecting infant from rival males.	None.	Slight; mostly consists of protection of infant from rival males.	Slight, but greater than among chimps. Includes food-sharing and grooming.
11. Sexual dimorphism of adults	Extreme; males are 221% the weight of females.	Extreme; males are 237% the weight of females.	Moderate; males are 123% the weight of females.	Moderate; males are 136% the weight of females.

Note: See Ghiglieri (1987) for sources and discussion.

sense of self (Gallup 1970, 1979, 1982). So far no monkey has done so. Great Apes deceive, and play roles, in a way that suggests they credit other animals (or people) with differing points of view. The evolution of ape mind was a clear prerequisite for the eventual differentiation of humans within the apes.

6.3 Communication and privacy

Communication presumably tells other animals what they do not know. At a minimum it reinforces and emphasizes what they might know by other means, thus influencing their behaviour. The information

Table 6.3 Elements of social behaviour common to chimpanzees, bonobos, and humans. (Modified after Ghiglieri 1989; data from Table 6.2 and Ghiglieri 1987 modified.)

1. Female exogamy.
2. Bonds between females exist (contrast gorillas).
3. Social groups are closed.
4. Social groups are stable, multi-male, multi-female communities.
5. Males are active in territorial defence.
6. Males see, attack, and, in two species, may kill rival-group males.
7. Polygynous and polyandrous mating pattern.
8. Males sometimes travel alone.
9. Females sometimes travel alone.
10. Fusion–fission sociality common *within* communities.
11. Female associations due partly to attraction to same male(s), also sometimes to kin or friends.
12. Male-retention.
13. Communal territoriality typical.
14. Mating competition between males of same community mild in relation to that between communities.
15. Sexual dimorphism is moderate; males co-operate in alliance against rival-group males.

which is conveyed may be simple or complex, and vague or surprisingly 'semantically' detailed (see Smith 1977, 1981).

The long-distance, inter-troop calls and scents of primates are probably fairly simple. Any such message apparently conveys the identity of the sender. Marler and Hobbet (1975) showed this for the pant-hoots of chimpanzees, and Waser (1977) for the whoop-gobble of gray-cheeked mangabeys. At most, however, one imagines that such long-distance calls declare 'I am me, I am here at this direction and distance, I have a wife/six friends/etc., and keep out of my patch!' Female scent-marks in 'solitary' species might say 'I am me, I am here, and in three days I'll be receiving company.' However, loud alarm calls may be semantically precise. Seyfarth *et al.* (1980a, b) and Cheney and Seyfarth (1980, 1988, 1990) showed by elegant playback that vervet monkeys alarms indicate different dangers: hawk, ground predator, snake. Many other mammals and birds also have different air and ground alarms, which provoke different responses (running upward or else down under cover). Following the Cheney-Seyfarth work we now suppose these code for different external stimuli, not just different levels of alarm.

Short-distance messages, in contrast, seem exceedingly complex. Their subdivisions elude us, and like language, seem best understood by the reactions of the perceivers. The cooing of Japanese macaques, for instance, sounds like an indistinguishably graded set of sounds to human ears—or indeed to other species of macaques. However, the Japanese macaques hear subtle differences, and use each type of call in appropriate situations (Green and Marler 1979). Isolation peeps of infant squirrel-monkeys have similar specificities (Snowdon *et al.* 1982).

So far as we can tell, monkeys and apes integrate each message with previous and ongoing actions, so that not only the detailed structure of the call, but also its context, contributes, to the eventual meaning. Thus, some of the phonological and mental complexity which we associate with language is evident in the calls that monkeys and apes use for fine-tuning of social structures within the troop. It seems likely that the juggling of coalitions, friendships, kin support, and threats which emerges in a multi-male-multi-female troop would be most demanding of precise, subtle, voluntarily controlled communication.

Two studies analyse communications in Old World monkeys that reveal, and are interpreted by the monkeys in terms of, the social structure. Cheney and Seyfarth (1980) played back the distress screams of juvenile vervets in Amboseli Park, Kenya. Cheney and Seyfarth chose moments when the little vervets were behind a bush, out of sight of their mothers. Not surprisingly, each juvenile has an individual voice which the mother recognizes. More important, *other* females of the troop looked toward the mother after the playback. That is, each knew the *others*' kin relations.

In rhesus monkeys, Gouzoules *et al.* (1984) demonstrated that juveniles give five distinct types of scream. They scream to recruit aid from their mothers in dealing with threats from troop members. Rhesus have tightly nepotistic female hierarchies: all the members of one matriarchy are usually dominant to the next matriarchy, and so on down the line. The scream types do not just vary with the degree of fear, or of the severity of the threat. Noisy, blurred screams usually indicate that the juvenile is threatened by a monkey dominant to its mother. Arched screams, with rising and falling pitch, correlate with threat from a subordinate. Tonal and pulsed screams are given to attackers from the same matriarchy. Actual physical contact is most frequent with noisy and pulsed screams; that is, a monkey may be bitten by either relatives or dominants. However, mothers looked quickest and longest toward playback of noisy or arched screams. That is, mothers responded more if any non-kin monkey was apparently scaring their offspring, whether subordinate or dominant. They were more likely to ignore mere punishment by relatives.

Such specificity has its role in conveying what another animal had *not* seen. The rhesus scream is not simple alert to the mother, but a message defining the situation, and how urgently to take action. It is a major question how much information is so 'private' in a monkey troop that it merits this kind of specificity.

Dennett (1987) in an important paper called 'Out of the armchair and into the field' points out how little privacy there is in a close-knit monkey troop. Vervets and macaques are rarely out of sight of their troop-mates. One animal rarely has information about the environment which another cannot see for itself. Even more rarely does one monkey have information which it can be sure that the other lacks. Food-sources are widespread fruiting trees; predators are visible across the savannah. Almost the only everyday secret is who is mating behind a bush. Forest monkeys are more often out of sight of each other, so one might occasionally find a food-source the others do not know, or see a predator poised to pick off a rival. However, even this information is not likely to remain private more than a few minutes.

In contrast chimpanzees and organ-utans are far more likely to have knowledge gained in the course of their ranging which they might impart or else conceal from others whom they meet. Mackinnon (1978) made a case for the Miocene radiation of apes and hominids having far more solitary tendencies than the ancestral stock of monkeys. Of today's surviving apes, only gorillas live in coherent bands, and even in these the females are usually unrelated to each other, but bound to the harem male. Mackinnon argues that the ancestral apes, like modern chimpanzees, evolved good individual memories for ranging patterns and somewhat flexible social strategies to cope with shifting alliances when they did meet. This individualistic pattern of evolution thus set the stage for more detailed, informative communication among apes and humans because, in Dennett's terms, we had more potential secrets.

Wild male chimpanzees give pant-hoots and food-barks when they find a rich source of desirable food. Wrangham (1979) queried why they should thus attract competitors to eat the food. He concluded that the short-term loss is compensated by the long-term gain in friendship and bonding between males, who jointly defend their community territory. Chimpanzees also use much subtler signals to lead each other to food. Menzel (1971) in a famous series of experiments, concealed food in different spots in a field. One of his band of chimpanzees was carried out and shown the food; it was then returned to the group in their indoor house. All the chimpanzees were young, and dared not venture into the field alone. The chimpanzee who knew the location led the others out by striding off purposefully, or, if it was more subordinate in the group, by tugging on hands and fur and glancing in the direction to go. At times, when no one would follow, the 'leader' threw a tantrum.

The group soon became sophisticated enough to follow the one of two animals who had seen the bigger, more desirable food cache, or to follow a leader to desirable food right past an apple displayed on a stick. Then Menzel hid frightening objects such as rubber snakes. These were also approached, but with bristling fur and nervous slaps, or poking through the leaves with a stick, not the hands. Finally Menzel showed a hidden snake to a leader, but removed the rubber reptile before the group could return. They searched initially in the hiding-place, but then all round the area and up and down along their boundary fence. In short, these chimpanzees did indeed communicate the existence, location, and even quality of out-of-sight objects: whether desirable, more desirable, or frightening. (See also Lock and Colombo, Chapter 22, this volume.)

We have then, examples of precise semantic communication in the vocalizations of Old and New World monkeys. We have also examples of chimpanzees' conveying detailed information about the immediate environment. We have not yet analysed ape vocalizations for 'semantic' content, nor have we experiments on monkeys which parallel the indication of food caches by chimpanzees. However, the speculations of Mackinnon and Dennett suggest good reason for the chimpanzee communication to be more 'semantic'—in societies where individuals split and join, travelling either alone or in sociable groups, they should have both much private information and much incentive to share it.

6.4 Deception

Byrne and Whiten (1988) and Whiten and Byrne (1988) summarized all known accounts and then extended the literature on primate deception in their book, *Machiavellian intelligence*. They argued that social behaviour was the matrix which allowed the evolution of primate intelligence, and that deception reveals the workings of this matrix. They classified deceptive acts according to Dennett's (1987) levels of intentionality—whether an animal adjusts its acts to what another is potentially aware of, or even whether the other animal is aware of the first's awareness (see also Jolly 1988, 1991).

Goodall (1971) first reported chimpanzee deception. Figan, as an adolescent male in the Gombe Stream Reserve, realized that he could lead high-ranking males away from the banana pile. Figan would stride confidently into the woods; often the others followed him. Later he circled back to eat alone and unmolested. Once yet another high-ranking adult had appeared in the mean time; Figan stared at him, then threw a tantrum. Goodall (1986) later compiled a detailed comparison of such examples of chimpanzee mental activity, wild and captive, that can be listed according to Byrne and Whiten's categories. De Waal (1986) also explored chimpanzee deceit. A straightforward case may be 'not noticing'. Once the experimenters in the Antwerp Zoo buried three grapefruit in the sand of the chimpanzees' two-acre corral. Dandy, a young male, walked directly over the spot without breaking stride, so that even the humans supposed they had buried the fruit too deep to detect. Three hours later, when the older chimps were taking a siesta, Dandy returned directly to the spot, and without hesitation dug up the fruit, leaving no doubt that he had known its whereabouts all along.

Such 'not noticing' may be even more active. When one male is displaying, another may become intensely interested in something up in the sky, or in suddenly playing with the juveniles. Similarly, two males who have challenged each other may seem fascinated by something in the grass, and thus positioned to begin a bout of reconciliation grooming.

'Not noticing' appears in much cruder form in monkeys. Macaques may simply turn their heads away, avoiding looking at or responding to an overt threat (Altmann 1962; Chevalier-Skolnikoff 1982). De Waal (1986) speculates that macaques may even be fooled, in a sense, by such obvious nonresponse, which would not deceive either apes or humans.

Actual hiding happens in both monkeys and apes. The commonest case is, in fact, females mating with subordinate animals out of sight of the dominant male. Kummer (1982) reports that hamadryas baboon males, when groomed by bonded females out of sight of the harem leader, suppress the loud vocalization they usually give when groomed. He also describes one female who spent twenty minutes inching her forequarters behind a rock. Eventually her harem overlord could check her whereabouts by looking at her back and the top of her head—but her hidden hands were grooming a sub-adult male.

Chimpanzees do even more self-conscious hiding, such as the dangling hand and upraised knee that blocks a dominant's view of the erect penis, while revealing all to females on the other side. One young male, caught in flagrant exhibition, dropped both hands to cover his erection. De Waal's (1982) most cogent example is when Luit, the dominant male, was being challenged by Nikkie, another adult male. After Luit and Nikkie had displayed in each other's presence for over ten minutes, Nikkie was driven into a tree, but a little later he began to hoot at the leader again. Luit was sitting at the bottom of the tree with his back to his challenger. When he heard the renewed sounds of provocation he bared his teeth, but immediately put his hand to his mouth and pressed his lips together.

> I saw the nervous grin appear on his face again, and once more he used his fingers to press his lips together. The third time Luit finally succeeded in wiping the grin off his face; only then did he turn round. A little later he displayed at Nikkie as if nothing had happened, and with Mama's help chased him back into the tree. Nikkie watched his opponents walk away. All of a sudden he turned his back and, when the others could not see him, a grin appeared on his face and he began to yelp very softly. I could hear Nikkie because I was not very far away, but the sound was so suppressed that Luit probably did not notice that his opponent was also having trouble concealing his emotions.

Thus, chimpanzees seem to be aware of their own communications. Further, they seem aware of others' limitations. Dandy, seeing a third male mating with a female friend of Dandy's out of sight of the dominant, ran to the dominant barking loudly, and led him to chastise the couple. In a milder case, Premack and Premack (1983) have done a series of experiments with a blindfolded trainer. Young chimpanzees at first touched the trainer, then started off to lead him to unlock a food-box in a field, just as they would if he could see. Three of four young chimps learned after a few trials to keep hold and lead him by the hand or by his key-chain. The fourth simply pulled down the blindfold.

Finally, what is the clear evidence that apes envisage others' viewpoints or intentions? Premack and Woodruff (1978) allowed Sara to finish videos of a trainer attempting a task. Sara chose successful solutions for videos of a trainer she liked (for example, the trainer successfully stacking the boxes to reach a banana). She chose unsuccessful ones for a trainer she did not like (for example, the trainer on the floor underneath the boxes.) More simply, Povinelli *et al.*

(1992a,b) have shown that chimpanzees, but not monkeys, can instantly switch roles with the trainer in a two-part task. That is, the apes understood not only their own part of the task, but the complementary role from a different point of view.

Thus it seems clear that chimpanzees have at least two pre-requisites for symbolic language: first, they have some awareness of their own communications as communications, which may or may not be sent. That is, communications are not an inevitable part of the emotional or factual situation. Second, they have some awareness of others as receivers, who may or may not know what is going on, and who may thus sometimes be deceived or alerted to facts. At least chimpanzees, and perhaps other primates, can deliberately give information, tell lies, or keep secrets.

6.5 Symbolic behaviour in apes

Aside from taught 'language' there are a few cases of elaborate symbolic play in apes. Washoe, aged 1½ – 2 began to bathe her doll, including soaping it and rinsing it and drying it (Gardner and Gardner 1971). (This is much the same age that human girls begin such play.) Koko the gorilla has much symbolic play, such as a toy alligator with which she pretends to frighten people though, as Patterson and Linden (1981) remark, it is funny to think that a nearly full-grown gorilla needs a plastic alligator to frighten anyone! The most extensive account, never queried or refuted, is Catherine Hayes' (1951) description of the imaginary pull-toy. Viki, a home-raised chimpanzee who could pronounce four 'words', was just at the toddler stage when everything that is possible to drag along becomes a pull-toy. While playing in the bathroom she began to trail one arm behind her as though dragging a toy on a string that did not exist. She repeated the game over and over again. One day, the 'rope' apparently caught on the plumbing pipes; she fumbled, pulled, gave a jerk, and was off again. This was incorporated into her routine. Still later, she sat on the potty 'fishing' the object from the floor hand-over-hand. Then after several weeks, wrote Cathy Hayes, Viki's foster-mother:

> It was one of those days when Viki loves me to distraction. She had pattered along in my shadow [and] at every little crisis she called for 'Mama'. I was combing my hair before the bathroom mirror while Viki dragged the unseen pull-toy around the toilet. I was scarcely noticing what had become commonplace, until she stopped once more at the knob and struggled with the invisible tangled rope. But this time she gave up after exerting very little effort. She sat down abruptly with her hands extended as if holding a taut cord. She looked up at my face in the mirror and then she called loudly, 'Mama, Mama!'

Cathy Hayes, still half-unbelieving, went through an elaborate pantomime of untangling the rope and handed it back to the little chimp. 'Then I saw [her] expression—a look of sheer devotion—her whole face reflected the wonder in children's faces when they are astonished at a grownup's escape into make-believe. But perhaps Viki's look was just a good hard stare.'

A few days later Cathy Hayes decided to improve on the game. She invented a pull-toy of her own, which went clackety-clackety on the floor and squush-squush on the carpets. Viki stared at the point on the floor where the imaginary rope would have met the imaginary toy, uttered a terrified 'oo-oo-oo', leapt into Cathy's arms, and never played that game again.

On a less complex level, apes (and monkeys and pigeons) can recognize and categorize pictures. But apes, unlike monkeys, also recognize themselves in mirrors. Gallup's (1970, 1979; and also Povinelli *et al.*, in press) experiments, daubing paint on the ear or the brow-ridge of an anaesthetized ape, show clearly that the apes then groom the paint-spots while looking in the mirror. (Gallup found such recognition in chimpanzees and orang-utans, not gorillas. In view of Koko's interest in mirrors, and gorillas' known tendency not to play experimenters' games, this result is still open.) Kohler (1927) remarked that his chimpanzees observed not only themselves, but each other and distant objects, in puddles and stray bits of metal, glancing back and forth from the object to its reflection. Menzel *et al.* (1985) have analysed the use by Sherman and Austin (Savage-Rumbaugh's two language-trained chimpanzees) of closed-circuit television. They look at the television screen spontaneously, for instance, to peer inside their own mouths. In Menzel's experiments the chimpanzees reached behind a curtain to pick up small rewards, guiding their reach by watching the image of their own hand as seen on the TV screen. As with mirrors, this second-hand representation seems to present no problems to sophisticated chimpanzees (see also Ristau, this volume, Chapter 23; and Lock and Colombo, Chapter 22).

What connection has such symbolic play, and such ability to deal with pictorial representations, with the ability to communicate or deceive? I think that it is fairly clear that some mental framing, in Bateson's terms (1973, 159–63), is necessary to both. Although wild chimpanzees have never been seen to use unambiguous symbols, and possibly never will be, one relation I can see to their behaviour in the wild is the kind of complex communication and deceit employed by Figan and Luit in dealing with their fellow apes. The kind of mental detachment which allows one to deal with representations of objects seems somewhat like the imagination which allows one deliberately to lead a group to hidden food, and like the detachment involved in conveying a lie.

In my summary article in Byrne and Whiten (1988) I continued an earlier theme of Humphrey, to argue that the mind of a social primate functions in terms of what other animals want—that is, it presumes that other creatures are goal-oriented (Humphrey 1976; Jolly 1988). Such a mind might then also operate on the world as though inanimate objects and series of events were goal-oriented, because this device works for its social companions. The early part of this paper concentrated on the evolution of analytic symbols: that is, the partial detachment of symbols from their referents so that they can be used independently. The imbuing of a series of events with purpose in ape and early human social behaviour also may have contributed to the creation of synthetic symbols: that is, symbols which combine observed properties with imagined ones.

Over millennia of evolution, minds progressed from 'That monkey wants to bite me,' to 'That ape doesn't like me,' to 'That thunder wants to scare me,' to 'Zeus is angry with me.' The personification of tree-dryads and water-nymphs and thunder-gods may go back to ancient mnemonics for dealing with a primate's world. In this case, the importance of a symbol is not whether it is true, but that, in Bruner's phrase, it goes 'beyond the information given' (Humphrey 1976; Bruner 1973).

This ability again relates to Bateson's formulation, but in an affirmative, not a negative sense. Such synthesized symbols, or conclusions, are not marked by being true or untrue. They are marked by being imagined instead of seen. That is, they are ideas. When both aspects come together, such that the imaginer is aware that he, she, or it is having an idea, which can be in part detached from its referent, we arrive at the symbolic capacity which makes us human. We can delight in ideas, aware that they are not here, not now, and very possibly not true.

Editorial note

In conjectures about human social organization, appeals are made to primate data because of our close taxonomic relationship to primates. It has been claimed that primate societies show the highest level of social organization of all mammalian species (e.g. Wilson 1975). However, it is not clear how social complexity should be defined (but see Pearl and Schulman 1983); nor are there sufficient studies of mammalian social organization to allow such a conclusion. In addition, mamalian societies which exploit similar ecological niches to humans, rather than show close taxonomic relations, can provide information germane to human society.

The killing of prey that are larger than an individual predator requires a refined social coordination. Of the various carnivores that adopt this strategy, there are now a number of studies of African wild dogs (*Lycaeon pictus*) (Kuhme 1965; Estes and Goddard 1967; van Lawick and van Lawick-Goodall 1970; van Lawick 1974) and lions (*Panthera leo*) (Schaller 1972; Bertram 1975, 1976, 1978, 1979; Packer and Pusey 1982, 1983). The social organization of both these species provide many parallels with certain primates, but also contain unique features, such as close cooperation in food 'harvesting', or the division of labour between harvesting and infant care. Crook (1980, p. 111), drawing on Schaller (1972), maintains that:

Among the various carnivores that practice such social skills the following traits have emerged:
(1) Co-ordination of action within a group, often with some division of labour.
(2) Food-sharing and often transport to the den.
(3) Elaborate communication of affect and intention.
(4) Subtle quilibration of behaviour control in relation both to the behaviour of the prey and that of the collaborators.
(5) Hierarchical social structuring of the hunting group.

(See also Roeder (1985)). It may be, then, that the most plausible conjectures as to the evolutionary determinants of human social forms will come from our understanding of primate communicative repertoires and abilities, allied to the ecological demands made by an atypical primate food-harvesting strategy.

Regarding these ecological demands, it is only recently (see, for example, Cheney *et al.* 1986) that primatologists have begun to take into account the social ecology of primate society, and recognize that the very societies established by primates as adaptive strategies *constitute* a new set of pressures that must be adapted to. In a manner of speaking, primate societies thus select for the behavioural skills of the individuals that comprise them. Individuals within them that are socially adept will be at an 'advantage' over less adept members. There thus emerges

the intriguing hypothesis that primate intelligence—including our own—originally evolved to solve the challenges of interacting with one another (ibid., p. 1365).

References

Altmann, S. A. (1962). A field study of the sociobiology of rhesus monkeys, *Macaca mulatta*. *Annals of the New York Academy of Sciences*, **102**, 338–435.

Bateson, G. (1973). *Steps to an ecology of mind*. Paladin, St Albans.

Bertram, B. C. R. (1975). Social factors influencing reproduction in wild lions. *Journal of Zoology (London)* **177**, 463–82.

Bertram, B. C. R. (1976). Kin selection in lions and in evolution. In *Growing points in ethology* (ed. P. P. G. Bateson and R. A. Hinde), pp. 281–301. Cambridge University Press, Cambridge.

Bertram, B. C. R. (1978). Living in groups: predators and prey. In *Behavioural ecology* (ed. J. R. Krebs and N. B. Davies), pp. 64–96. Sinauer, Sunderland, Mass.

Bertram, B. C. R. (1979). Serengeti predators and their social systems. In *Serengeti: dynamics of an ecosystem* (ed. A. R. E. Sinclair and M. Norton-Griffiths), pp. 221–248. Chicago University Press, Chicago.

Bruner, J. S. (1973). *Beyond the information given*. Norton, New York.

Byrne, R. and Whiten, A. (eds) (1988). *Machiavellian intelligence: social expertise and the evolution of intellect in monkeys, apes, and humans*. Clarendon Press, Oxford.

Cheney, D. L. and Seyfarth, R. M. (1980). Vocal recognition in free-ranging vervet monkeys. *Animal Behavior*, **28**, 362–7.

Cheney, D. L. and Seyfarth, R. M. (1988). Social and non-social knowledge in vervet monkeys. In *Machiavellian intelligence: Social expertise and the evolution of intellect in monkeys, apes, and humans* (ed. R. Byrne and A. Whiten), pp. 255–70. Clarendon Press, Oxford.

Cheney, D. L. and Seyfarth, R. M. (1990). *How monkeys see the world*. Chicago University Press.

Cheney, D., Seyforth, R., and Smuts, B. (1986). Social relationships and social cognition in primates. *Science*, **234**, 1361–6.

Chevalier-Skolnikoff, S. (1982). A cognitive analysis of facial behavior in Old World monkeys, apes and humans. In *Primate communication* (ed. C. Snowdon, C. Brown, and M. Peterson), pp. 308–68. Cambridge University Press.

Crook, J. H. (1980). *The evolution of human consciousness*. Clarendon Press, Oxford.

de Laguna, G. A. (1963). *Speech, its function and development*. Indiana University Press, Bloomington. [Originally published 1927.]

Dennett, D. (1987). *The intentional stance*. Bradford Books, MIT Press, Cambridge, Mass.

de Waal, F. B. M. (1982). *Chimpanzee politics*. Harper and Row, New York.

de Waal, F. B. M. (1986). Deception in the natural communication of chimpanzees. In *Deception: perspectives on human and non-human deceit* (ed. R. Mitchell and N. Thompson), pp. 221–4. SUNY Press, Albany, NY.

Estes, R. D. and Goddard, J. (1967). Prey selection and hunting behaviour of the African wild dog. *Journal of Wildlife Management*, **31**, 52–70.

Gallup, G. (jun.) (1970). Chimpanzees: self-recognition. *Science*, **167**, 86–7.

Gallup, G. G. (1979). Self-awareness in primates. *American Scientist*, **67**, 417–21.

Gallup, G. G. (1982). Self-awareness and the emergence of mind in primates. *American Journal of Primatology*, **2**, 237–48.

Gardner, B. T. and Gardner, R. A. (1971). Two way communication with an infant chimpanzee. In *Behavior of nonhuman primates* (ed. A. M. Schrier and F. Stollnitz), pp. 117–85. Academic Press, New York.

Ghiglieri, M. P. (1987). Sociobiology of the great apes and the hominid ancestor. *Journal of Human Evolution*, **16**, 319–57.

Ghiglieri, M. P. (1989). Hominoid sociobiology and hominid social evolution. In *Understanding chimpanzees* (ed. P. G. Heltne and L. A. Marquardt), pp. 370–9. Harvard University Press, Cambridge, Mass.

Goodall, J. van Lawick (1971). *In the shadow of man*. Collins, London.

Goodall, J. (1986). *The chimpanzees of Gombe*. Harvard University Press, Cambridge, Mass.

Gouzoules, S., Gouzoules, H., and Marler, P. (1984). Rhesus monkey (*Macaca mulatta*) screams: representational signalling in the recruitment of agonistic aid. *Animal Behavior*, **32**, 182–93.

Green, S. and Marler, P. (1979). The analysis of animal communication. In *Handbook of behavioral neurology, Vol. 3: Social behavior and communication* (ed. P. Marler and J. G. Vandenbergh), pp. 73–158. Plenum, New York.

Harper, D. G. C. (1991). Communication. In *Behavioural ecology* (3rd edn, ed. J. R. Krebs and N. B. Davies), pp. 374–97. Blackwell, Oxford.

Hayes, C. (1951). *The ape in our house*. Harper, New York.

Humphrey, N. K. (1976). The social function of intellect. In *Growing points in ethology* (ed. P. P. G. Bateson and R. Hinde), pp. 303–17. Cambridge University Press.

Jolly, A. (1985). *The evolution of primate behavior* (2nd edn). Macmillan, New York.

Jolly, A. (1988). The evolution of purpose. In *Machiavellian intelligence: social expertise and the evolution of intellect in monkeys, apes, and humans* (ed. R. Byrne and A. Whiten), pp. 363–78. Clarendon Press, Oxford.

Jolly, A. (1991). Conscious chimpanzees? a review of recent literature. In *Cognitive ethology* (ed. C. Ristau), pp. 231–52. Erlbaum, Hillsdale, NJ.

Kappeler, P. H. (1993). Reconciliation and post-conflict behavior in ringtailed lemurs, *Lemur catta*, and red fronted lemurs, *Eulemur fulvus*. *Animal Behavior*, **40**, 774–6.

Kohler, W. (1927). *The mentality of apes* (2nd edn). Routledge and Kegan Paul, London.

Krebs, J. R. and Dawkins, R. (1991). Animal signals: mind-reading and manipulation. In *Behavioural ecology* (2nd edn), (ed. J. R. Krebs and N. B. Davies), pp. 380–402. Blackwell, Oxford.

Kuhme, W. (1965). Communal food distribution and division of labour in African hunting dogs. *Nature*, **205**, 443–4.

Kummer, H. (1982). Social knowledge in free-ranging primates. In *Animal mind, human mind*, Dahlem Konferenzen (ed. D. R. Griffin), pp. 113–30. Springer-Verlag, Berlin.

Mackinnon, J. (1978). *The ape within us*. Collins, London.

Marler, P. and Hobbet, P. (1975). Individuality in a long-range vocalization of wild chimpanzees. *Zeitschrift für Tierpsychologie*, **38**, 97–109.

Menzel, E. W. (jun.) (1971). Communication about the environment in a group of young chimpanzees. *Folia Primatologica*, **15**, 220–32.

Menzel, E. W., Savage-Rumbaugh, E. S., and Lawson, J. (1985). Chimpanzee (*Pan troglodytes*) spatial problem-solving with the use of mirrors and televised equivalents of mirrors. *Journal of Comparative Psychology*, **99**, 211–17.

Packer, C. and Pusey, A. E (1982). Cooperation and competition within coalitions of male lions: kin selection or game theory? *Nature*, **296** 750–52.

Packer, C. and Pusey, A. E. (1983). Adaptation of female lions to infanticide by incoming males. *American Naturalist*, **121**, 716–28.

Patterson, F. and Linden, E. (1981). *The education of Koko*. Holt, Rinehart, and Winston, New York.

Pearl, M. A. and Schulman, S. R. (1983). Techniques for the analysis of social structure in animal societies. *Advances in the Study of Behavioour*, **13**, 107–46.

Pereira, M. E. (1993). Agonistic interaction, dominance relations and ontogenetic trajectories in ringtailed lemurs. In *Juvenile primates* (ed. M. E. Pereira and L. A. Fairbanks), pp. 285–305. Oxford University Press.

Povinelli, D. J., Nelson, K. E., and Boysen, S. T. (1992*a*). Comprehension of social role reversal by chimpanzees: evidence of empathy? *Animal Behavior*, **43**, 633–40.

Povinelli, D. J., Parks, K. A., and Novak, M. A. (1992*b*). Role reversal by rhesus monkeys, but no evidence of empathy. *Animal Behavior*, **44**, 269–81.

Povinelli, D. J., Rulf, A. R., Landau, K. and Bierschwale, D. T. (1993). Self-recognition in chimpanzees (Pan troglodytes): distribution, ontogeny, and patterns of emergence. *Journal of Comparative Psychology* **107**, 347–72.

Premack, D. and Premack, A. J. (1983). *The mind of an ape*. Norton, New York.

Premack, D. and Woodruff, G. (1978). Does the chimpanzee have a theory of mind? *Behavioral and Brain Sciences*, **1**, 515–26.

Roeder, J.-J. (1985). *Évolution des systèmes de communication chez les carnivores et les primates: organisation social et modalités de communication*. Éditions du CRNS, Paris.

Schaller, G. B. (1972). *The Serengeti lion: a study of predator-prey relations*. University of Chicago Press, Chicago.

Seyfarth, R. M., Cheney, D. L., and Marler, P. (1980*a*). Vervet monkey alarm calls. *Animal Behavior*, **28**, 1070–94.

Seyfarth, R. M., Cheney, D. L., and Marler, P. (1980*b*). Monkey responses to three different alarm calls: evidence of predator classification and semantic communication. *Science*, **210**, 801–3.

Smith, W. J. (1977). *The behavior of communicating*. Harvard University Press, Cambridge, Mass.

Smith, W. J. (1981). Referents of animal communication. *Animal Behavior*, **29**, 1273–5.

Snowdon, C. T., Brown, C. H., and Peterson, M. R. (1982). *Primate communication*. Cambridge University Press.

van Lawick, H. (1974). *Solo: the story of an African wild dog*. Houghton-Mifflin, Boston.

van Lawick, H. and van Lawick-Goodall, J. (1970). *Innocent killers*. Collins, London.

Waser, P. M. (1977). Individual recognition, intragroup cohesion and intergroup spacing: evidence from sound playback to forest monkeys. *Behaviour*, **60**, 28–74.

Whiten, A. and Byrne, R. (1988). Tactical deception in primates. *Behavioral and Brain Sciences*, **11**, 233–73.

Wilson, E. O. (1975). *Sociobiology: the new synthesis* Belknap, Cambridge, Mass.

7

Social relations, human ecology, and the evolution of culture: an exploration of concepts and definitions[1]

Tim Ingold

Abstract

'Society' and 'culture' are among the most contentious concepts of the human sciences. Sometimes treated as virtually synonymous, sometimes radically distinguished, their study has been maintained as the particular preserve of social and cultural anthropology, at the same time as it has been opened up by biologists to embrace almost the entire field of animal behaviour. This chapter is an attempt to resolve some of the conceptual ambiguities surrounding these notions, through an exploration of both the continuities and the contrasts between the worlds of humanity and of non-human animals. The argument is presented in three main parts. The first examines alternative foundations of sociality, distinguishing its interactive, regulative, and constitutive forms, and establishes the connections between social life, consciousness, and culture. This leads, in the second part, to a discussion of the ways in which human beings and other animals construct their environments, and to a characterization of the connection, established in production, between social and ecological systems. In the third part, a contrast is set up between definitions of culture that emphasize its non-genetic mode of transmission, and those that rest on the symbolic organization of experience. This contrast is linked to the distinction between learning and teaching, and to alternative views of the possible analogies and contrasts between 'biological' and 'cultural' evolution. The chapter concludes with a word on the relation between learning, thinking, and consciousness.

7.1 Introduction

It is apparent, even from a cursory glance, that human society differs quite markedly in its organization from the societies of animals, including those of non-human primates. Yet the clarification of these differences has been continually impeded by conceptual ambiguities surrounding the very notions of 'society' and 'sociality'. Many biologists, working within a theoretical framework of modern evolutionary ecology, have posited a basic continuity from non-human animal to human sociality, taking social organization in both instances to comprise observable patterns of interaction and co-operation among individuals of the same species. At the same time, social anthropologists have jealously guarded the study of 'society' as their special preserve, even doubting whether the organization of non-human populations deserves to be called 'social' at all. Social relations, they say, presuppose the emergence of rules, embodied within a framework of institutions. Moreover such rules depend upon a distinctively human mode of objective self-awareness, which is also taken to be a precondition for 'culture' in its widest anthropological sense. Arguing thus, the domains of the social and of the cultural appear to merge into one domain of 'socio-cultural' phenomena, which have in common that they are composed and organized by symbolic meaning. It follows that the essence of sociality lies not in patterned interaction but in its constitution within a matrix of significant symbols (Sahlins 1976, p. 117).

From this initial contrast, it is obvious that we cannot construct a coherent account of the evolution of sociality without first drawing out the different senses of society and social relations, which are so frequently confused even in contemporary discussions of the subject, and without also clarifying the distinction between 'society' and 'culture'. This is what I attempt to do in the first part of this chapter. Beginning with a historical overview of the debates and misunderstandings that grew up around the notion of the superorganic, I distinguish between society as an aggregate of interacting individuals and as a system of normative regulations. I then go on to introduce a third way of regarding social relations, which I call 'constitutive'. Such relations bind persons conceived as intentional agents rather than as players of parts in a regulative structure. Constitutive sociality, I argue, presupposes consciousness, but not objective selfconsciousness; intentionality, but not the discursive articulation of prior intentions. While in these respects it goes beyond the kinds of interactive sociality that may be discovered in practically every branch of the animal kingdom, it underwrites the development of the symbolically mediated, regulative sociality that is the hallmark of humanity. Whereas the latter seems to establish a radical discontinuity between ourselves and other animals, by focusing on constitutive relations we can rediscover the underlying, evolutionary continuity from non-human to human forms of sociality.

In the second part of the chapter my aim is to correlate the different kinds of sociality already adduced with alternative characterizations of the 'environment' of human beings, and hence also of their ecological relations. For this purpose it is necessary to show how the human 'making' of the environment goes beyond the sense in which any organism, by the mere fact of its presence, may be said to 'construct' its niche. The man-made environment, just like the outward forms of 'culture' or 'society' (artefacts and institutions), seems uniquely constituted in relation to self-conscious human subjects through the imposition upon external reality of an arbitrary framework of symbolic meaning. But making, in this sense, does not always entail the physical modification of objects, although both aspects are commonly confused under the rubric of 'production', with its twin connotations of appropriation and transformation. I argue that the essence of production rather lies in intentional action initiated by socially constituted subjects which is directed upon the physical world, and that culture—including its component of regulative relations—serves as the mediator or conductor of this activity.

In the third part of the chapter, I proceed to focus on the concept of culture, surely one of the most contentious in the human sciences. Whereas in many anthropological definitions the world of culture is simply identified with what is man-made or artificial, thus presupposing the symbolic organization of experience, biologists have been more inclined to extend culture to include any learning-transmitted tradition, examples of which abound in the animal kingdom. The only condition for the transmission of behavioural instructions by observational learning is interactive sociality: hence culture in this broad sense does not necessarily entail symbolic representation. The latter, however, is fundamental to the transmission of culture through teaching. The consequences of the distinction between learning and teaching, or between the 'innate' and the 'artificial' in culture, are far-reaching. Teaching implies a priority of rule over execution, or of theory over practice, which makes possible the deliberate design of new forms, or innovation by invention, to meet novel environmental circumstances. As a result the evolution of culture comes to conform to a schema that, in a certain limited sense, is more 'Lamarckian' than 'Darwinian', involving the operation of a principle of artificial rather than natural selection. Yet this is not entirely so, and I conclude with the suggestion that in so far as culture-change is 'Darwinian', the kind of selection involved is not external—as in organic evolution—but internal.

7.2 Animal society and human society

It is generally agreed that human beings are pre-eminently social animals; indeed it is often claimed that no non-human species lives socially in quite the way that we do. What, then, is so distinctive about human social life? Is it a distinction of degree or of kind? These questions are vitally important in our efforts to comprehend the evolution of sociality among animals and humans, especially since ethologists and sociobiologists engaged in the study of animal behaviour commonly adopt quite different criteria from those invoked by anthropologists for identifying what is, or is not 'social'. A century ago, the same differences lay at the heart of certain misunderstandings surrounding the notion of the superorganic; today they frustrate our attempts to uphold a clear distinction between the social and the cultural. These issues form the point of departure for our discussion in this part of the chapter, from which I shall proceed to introduce the three alternative senses in which we might speak of 'social relations': namely interactive, regulative, and constitutive.

7.2.1 *The superorganic*

It was Herbert Spencer who, in 1876, coined the term 'superorganic' to denote the subject-matter of the nascent science of sociology. In introducing the subject he dwelt at some length on the 'societies' of insects, birds, and mammals (1876, I: 4–8), and included them—alongside human societies—in his class of superorganic entities. He did not imagine such entities to exist on a level of being transcending the organic, but meant rather to convey the idea of an extension of organization beyond the boundaries of the separate organisms making up the social aggregate. That is to say, Spencer's superorganism was superindividual, but not suprabiological (where the prefix 'super-' denotes 'greater than, but including', and 'supra-' a plane that lies above and beyond—see Fig. 7.1). Moreover, society had no emergent properties, being merely the resultant of the voluntary association of individuals acting in accordance with their own best interests, ultimately prompted by predispositions, needs, or desires that were essentially innate. Thus for Spencer there was nothing in society not already prefigured in the properties of its original constituents.

Spencer's principal adversary in the field of sociology was Emile Durkheim, who launched a sustained attack on his notion of society and the superorganic. For Durkheim, society was superorganic in the sense that social facts *'have a different substratum'*, in a domain that is not reducible to the organic (1982 [1895], p. 40). Formed not through the external contact and association of discrete individuals but through the interpenetration of minds, Durkheim's society was suprabiological because superindividual, corresponding to that increment by which the whole thus constituted exceeds the sum of its parts—see Fig. 7.1. This emergent component, moreover, was conceived to impose a determination of its own upon individual conduct, one liable to conflict with—or at least to curb—the spontaneous expression of innate dispositions. And whereas Spencer was forever drawing parallels between animal and human societies, for Durkheim these were phenomena of totally different kinds, essentially non-comparable. Writing in 1917, he put the contrast in the following terms:

The great difference between animal societies and human societies is that in the former, the individual creature is governed exclusively from *within itself*, by the instincts.... On the other hand human societies present a new phenomenon of a special nature, which consists in the fact that certain ways of acting are imposed, or at least suggested *from outside* the individual and are added on to his own nature: such is the character of [social] 'institutions'. (1982, p. 248)

Throughout his work, Durkheim identified society with that arrangement of normative institutions here reserved for humanity, with the implication that non-

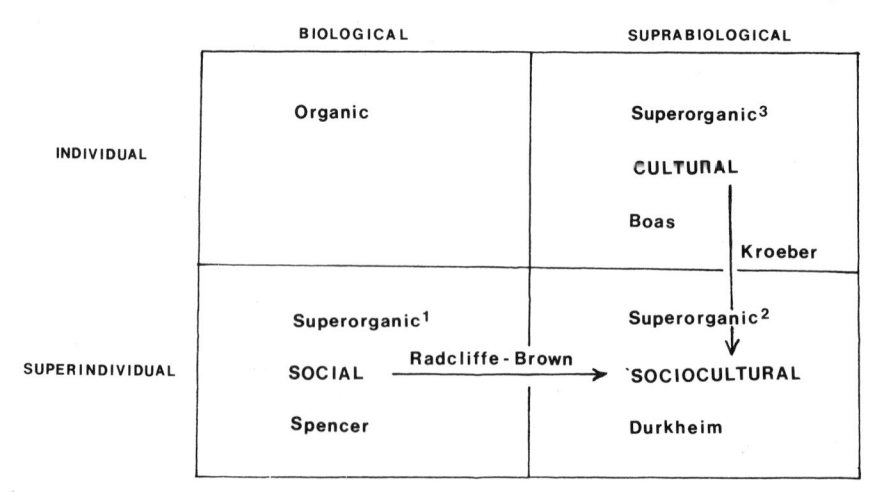

Fig. 7.1 Three senses of the superorganic: (1) Spencer: superindividual but not suprabiological; (2) Durkheim: suprabiological because superindividual; (3) Boas: suprabiological yet individual. Where Kroeber drew culture off from individuals and located it on the level of the collectivity, Radcliffe-Brown raised human society from the level of association between individual organisms to the suprabiological plane of social structure, consisting of relations between persons. The social and the cultural thus converge on the 'socio-cultural'.

human animals, however much they may associate in the execution of innate behavioural programmes, do not, strictly speaking, enjoy a social life at all.

The notion of the superorganic was first introduced into the mainstream of American cultural anthropology by A. L. Kroeber, in a celebrated paper dating from 1917 (1952, pp. 22–51). Kroeber's 'superorganic' is synonymous with 'culture' or 'civilization', and while acknowledging a debt to Spencer as the originator of the term, Kroeber recognized that his usage was quite different. His concern was to demonstrate the suprabiological character of culture, specifically in opposition to those (including Spencer himself) whose continued belief in the doctrine that acquired characteristics could be literally inherited, long after its empirical refutation by Weismann, perpetuated the unfortunate conflation of culture with race that was so common around the turn of the century (Stocking 1968, pp. 265–6).

In the course of his argument, Kroeber made claims for culture that almost exactly replicated what Durkheim had claimed for society. These particularly concerned its properties as a superindividual emergent. A number of Kroeber's contemporaries, all strongly influenced by the cultural anthropology of Franz Boas, felt that although culture was suprabiological in essence, it was nevertheless 'contained' as an internal property of discrete individual minds, rather than having its locus in a 'consciousness of the collectivity' formed through their interpenetration. Kroeber took a much more radical view of the autonomy of culture, drawing it off from individuals and situating it on a 'superpsychic' level, where it was supposed to enjoy a life of its own, revealed in culture-history, superimposed upon the residually organic biographies of its individual carriers. Where for Durkheim social life meant the life of society, for Kroeber cultural life meant the life of culture: for the one society, for the other culture, denoted a living superorganic entity (Fig. 7.1).

7.2.2 *Society versus culture*

Are there, then, any grounds for distinguishing between culture and society? Even in contemporary writing, the terms are often loosely interchangeable. Their conflation can be traced back to an elementary confusion, at the heart of Kroeber's 1917 paper, between the mode of *transmission* of culture and its mode of *existence*. Kroeber had imagined that to demonstrate the autonomy of tradition from hereditary constraint—to show that culture is socially transmitted, through a learning process—was *ipso facto* to prove that it is social in essence, having the character of a superindividual emergent. Many years later he renounced his original views, and insisted that the social and the cultural should, after all, be strictly distinguished.

One of the reasons for this change of heart lay in a reconsideration of the social life of non-human animals. Looking at a colony of ants or a hive of bees, Kroeber reasoned, one could scarcely deny the existence among them of societies of a scale and complexity approached only by those that appeared at an advanced stage in the social evolution of humanity. Yet if it is granted that the behaviour of insects is almost entirely pre-programmed by genetically inherited instructions, we would have no grounds for crediting them with culture. Since it thus appears that cultureless animals can enjoy a social life, society and culture must be distinguishable. However there can be no culture without social life, for the association and interaction of individuals is a pre-requisite for the transmission of tradition by learning (whether or not accompanied by formal teaching). Therefore, Kroeber concluded, 'developmentally, evolutionistically, society far antedates and thus underlies culture' (1963, p. 122; see also Stern 1929; Hallowell 1960, p. 329; Bonner 1980, p. 76).

Though the logic of this argument cannot be faulted, it rests on a conception of society—the resultant of associative interactions among a plurality of discrete individuals acting in accordance with innate predispositions—that is essentially Spencerian. Nor is this sense of the social altered if we allow what is considered to be 'innate' to include learning-transmitted or traditional as well as genetic instructions, so long as it is assumed that such instructions govern the behaviour of their carriers *from within*, and are not—as Durkheim and the early Kroeber supposed—imposed from an external, supraindividual source. Thus Kluckhohn, a committed advocate of the cultural determination of behaviour, was nevertheless compelled to warn against confusing culture with society: 'A "society" refers to a group of people who interact more with each other than they do with other individuals—who co-operate with each other for the attainment of certain ends. You can see and indeed count the individuals who make up a society. A "culture" refers to the distinctive ways of life of such a group of people' (1949, p. 24). Where Spencer's 'society' was an association of individual organisms, Kluckhohn's was an association of individual *culture-bearing* organisms, engaged in the pursuit of ends specific to their tradition rather than grounded in the universal attributes of human nature—see Fig. 7.2.

Turning from the American cultural anthropology represented by Kroeber and Kluckhohn to the British tradition of social anthropology, we find the same

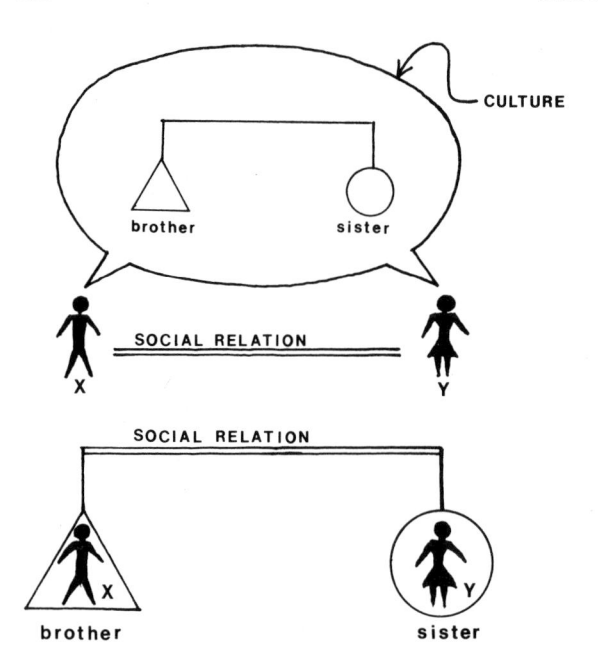

Fig. 7.2 The 'social relation' as an interaction between culture-bearing individuals (above) and as an arrangement of positions in an institutional order (below). In this example, two individuals, X and Y, relate as brother and sister. Adapted from Ingold (1986b, p. 240).

issues reappearing in the theoretical writings of its founder, A. R. Radcliffe-Brown. For his 'preliminary definition of social phenomena', Radcliffe-Brown appealed to Spencer: 'what we have to deal with are relations of association between individual organisms', manifested just as well in a hive of bees or a herd of deer as in a human community (1952, p. 189). Though all such associations are super*organic* in Spencer's sense, manifesting a superindividual level of organization, they are not *super*organic in the sense of having their source in a suprabiological, or cultural domain of being. The confusion of these two senses of the superorganic, connoting the social and the cultural respectively, would—as Radcliffe-Brown noted—lead to the absurd conclusion that by virtue of their association, the bees in his hive must have a 'culture' (1947, p. 79, n. 1; see Ingold 1986a, p. 347).

But once his attention turned to *human* society, Radcliffe-Brown slipped into quite another idiom, derived from Durkheim rather than Spencer (Fig. 7.1). Social phenomena among humans are not, he observed, simply the result of the interplay of individual natures; they are rather governed from above by an entity called 'social structure' (1952, pp. 190–1). Initially defined as a 'network of actually existing relations' between concrete individuals, social structure subsequently reappeared as an 'ordered arrangement of parts or components'. These components are *persons*, as distinct from (organic) individuals. Each person is a position in a structure, and the relations between these positions—embodied in institutions—regulate the conduct of their individual incumbents (see Fig. 7.2). That is to say, behaviour is governed by norms (1952, pp. 9–10, 193–4).

Once we conceive of society, in Radcliffe-Brown's terms, as a system of relations between persons, we have to admit that its existence depends upon the capacity of subjects to project and externalize their common experience on the level of ideas, that is, to form collective representations (Hallowell 1960, p. 352). In other words, social life presupposes a capacity for symbolic thought fundamental to *human* culture. Society, then, becomes conditional upon culture rather than vice versa; and by the same token, we would have to deny the sociality of non-human animals whose conduct is not bound by explicit rules and obligations (Fortes 1983, p. 34).

7.2.3 *Interactions and relationships*

Nowadays the distinction between the social and the cultural is commonly glossed over by vague references to human 'socioculture'. This notion once again conflates the two meanings of the superorganic that have caused so much confusion in the past. It indicates, moreover, that anthropologists (with the notable exception of Murdock 1972, p. 19) have not yet shaken off Kroeber's original error of imputing a superindividual (social) existence to that which is social in its mode of transmission. For the sake of clarity, it is essential that a distinction should be maintained between the contrasting senses of the social so far identified, corresponding on the one hand to the association of individual conspecifics, and on the other to the arrangement of positions within an imposed, institutional order. I shall refer to the relations implied by each sense as *interactive* and *regulative* respectively.

Since regulative relations are, apparently, uniquely human, biologists concerned with the social behaviour of non-human animals frequently adopt an interactive view of sociality, and like Spencer, they find the *raison d'être* for association in the mutual benefits or fitness-enhancing effects of co-operation. Dobzhansky, for example, defines society as 'a complex of individuals bound by co-operative interactions that serve to maintain a common life' (1962, p. 58). And for E. O. Wilson, society is 'a group of individuals belonging to the same species and organized in a co-operative manner. The principal criterion for applying the term "society" is the existence of reciprocal communication of a co-operative nature that extends beyond mere sexual activity' (1978, p. 222; for comparable formulations see Emerson 1958, p. 331; Altmann 1965, p. 519; Alexander 1974, p. 326).

Such definitions may be serviceable enough for the study of insects, but come nowhere near to capturing the essence of social life as it is experienced by our human selves. 'The difficulty', Baylis and Halpin admit, 'arises as a result of the discrepancy in meaning given to the words "social" and "society" when used in common parlance and when used by entomologists' (1982, pp. 257–8). In view of this discrepancy, entomological accounts of sociality can hardly be expected to throw much light on what, speaking of humans, we normally understand by the term. For while, intuitively, there can be no social life without relationships, an interactive view of sociality appears tantamount to their dissolution, to a reduction of the social below the level at which one can properly speak of 'relationships' at all. Following Bastian, social interaction may be characterized as the 'partial, and usually reciprocal, determination of an indivudal animal's actions by one or more other animals' actions', and as such, every interaction is itself an event of communication, involving the emission and receipt of signals which trigger certain behavioural effects in the recipient (Bastian 1968, p. 576). Yet we commonly interact and communicate, in this sense, with people with whom we do not recognize any relationship. At what point, then, do we deem a relationship to exist between interacting parties?

Hinde (1979, p. 15) admits that the distinction between 'interaction' and 'relationship' cannot be absolutely drawn; nevertheless 'a series of interactions totally independent of each other would not constitute a relationship. An essential character of a relationship is that each interaction is influenced by other interactions in that relationship' (1979, pp. 15–16). If sociality is a quality that inheres in relationships, it must reside not in interaction and communication *per se*, but in an enduring connection among the interacting parties, that is, in an entwining of their respective life histories. Where a relationship exists, every successive interaction between those involved appears as but a moment in its creative unfolding. Rather like a line in a conversation, which seems both to grow out of what was said before and yet to project it forward in directions that could not have been anticipated on the basis of past conditions, every interaction in a relationship is embedded in a previous history of mutual involvement, and will in turn have a bearing on how the participants react to one another in the future. In that sense, whereas every unit of interaction (like every elementary act of speech) is in principle compressible into a single instant, being virtually over in the moment it is begun, a relationship (like a conversation) is a continuous process borne on the current of real time, consubstantial with the lives of the participants. That is why to decompose relationships into their constituent interactions is to eliminate their social content, which cannot be recovered simply through a reverse process of statistical reaggregation, any more than real time—the projection of past into future—can be recovered from a concatenation of present instants.

7.2.4 *The constitution of persons*

A corollary of this view of social relationships is that they exist between persons, and not between individuals; or as Marx and Engels wrote, 'Where there exists a relationship, it exists for me' (1977, p. 51). It is necessary here to make the distinction between the 'person' in its classical sense, corresponding to the Latin *persona*, and its modern sense as a 'category of the self' (Mauss 1979 [1950], p. 87). The former, connoting an artificial role or a part in a performance, was the sense adopted by Radcliffe-Brown in his elaboration of social structure as a systematic arrangement of such parts, making up a regulative order. Yet behind this 'artificial man' of sociological analysis lies 'the real man of our everyday experience', no mere organic individual but a responsible subject endowed with specific intentions and purposes (Dahrendorf 1968, p. 25). Substituting these real people for the parts they play, it is possible to overcome the Durkheimian dichotomy between social life and individual life. By the former we do not understand the life of a pansubjective entity, society, but an intersubjective process, 'the social life of persons'. The life of the particular subject *is* social, because it unfolds in a milieu of relationships with other such subjects; thus what we commonly call *bio*graphy might better be known as *socio*graphy.

Now interactions flow from the association of individuals that arrive on the scene 'ready-made', each teleonomically predisposed to behave in given ways when confronted with other individuals recognizably of the same or of different kinds, or in response to specific signals emitted by such individuals. Persons, however, are *not* ready-made, but undergo a continuous ontogenetic development, as do the relationships between them. It is our relationships with others that make us who we are, and conversely, through those relationships we actively participate in shaping both the identities and the purposes of others around us. Social life, then, is 'concerned with *negotiations*, with negotiations between people rather than with interactions between things' (Shotter 1974, p. 217). It may be regarded as the creative process by which we constitute one another as agents who bear some responsibility for our actions, and who are endowed with what Shotter calls *personal* (in addition to natural) powers (1974, p. 225). For this reason, I refer to social relations in the sense just elaborated as *constitutive* (see Sandel 1982, p. 150), a sense that must be clearly distinguished not only from the *interactive* but also from the *regulative* (Fig. 7.3). These distinctions may be illustrated by means of a simple example.

Consider the relationship between myself and my son. We are, as Schutz put it, 'growing older together', our lives do not merely overlap in time but are bound up to the extent that our respective commitments and anticipations for the future are grounded in the promises of a shared past (Schutz 1962, pp. 16–17). Although classical sociology has it that I am the agent, and my son the patient, of a socialization process, the relationship between us is to some degree mutually constitutive, for as every parent knows, 'children "create parents" as well as parents creating children' (Giddens 1979, p. 129). However, our conduct is regulated by certain customary guidelines, as well as by jurally enforceable laws, and these add up to what a sociologist might call the 'father–son relationship' in our society, being a relation not between real consociates (myself and my son), but between parts in an ideal, symbolically encoded order. Finally, both of us are of course individual organisms (albeit culture-bearing ones), and as such we may be observed to interact with one another. But to regard each interaction as an isolated and disconnected episode involving the emission of parental behaviour, rather than as a moment in the temporal unfolding of a constitutive relationship, is to ignore the experiences of consciousness and selfhood which lift our mutual conduct from the sphere of behaviour into that of social action.

7.2.5 *Sociality and consciousness*

'Sociality', according to Schutz, 'is constituted by communicative acts in which the I turns to the others, apprehending them as persons who turn to him, and both know of this fact' (1970, p. 165). This marvellously succinct characterization of social life as an intersubjective process may be contrasted with the simple definition proposed by Alexander: 'sociality means group-living' (1974, p. 326). Alexander's aim is to construct a general theory of social behaviour on rigidly biological principles, and to this end he begins 'with a description of the selective forces causing and maintaining group-living'. Such an account, however, tells us nothing whatever about the development of the human subject and its orientation towards others, which are the very foundations of sociality in the Schutzian sense.

Consider those species of insects that score most highly in what Wilson (1980, p. 179) calls the 'key properties of social existence', by which he means the scale and complexity of co-operative organization. Their 'group-living' is conducted in the apparent absence of even the most rudimentary forms of awareness (though some would disagree, as does Griffin (1976)), perhaps even *because* of their absence. For in theory the more that consciousness develops, the better equipped is the organism to overcome the determinations of its conspecific environment, and the greater are its powers of autonomous action. And in human history, a scale of co-operation approaching that of the insects has been achieved only through the systematic repression of personal powers, manifested in the functioning of what Mumford has called the 'great labour machine', of the kind that built the pyramids, whose individual components—though human beings—'were reduced to their bare mechanical elements' of bone, nerve, and muscle (Mumford 1967, p. 191).

Clearly, to discover the origins of sociality in its constitutive sense, rather than in the interactive sense adopted by sociobiologists such as Alexander and Wilson, we must seek them not so much in the

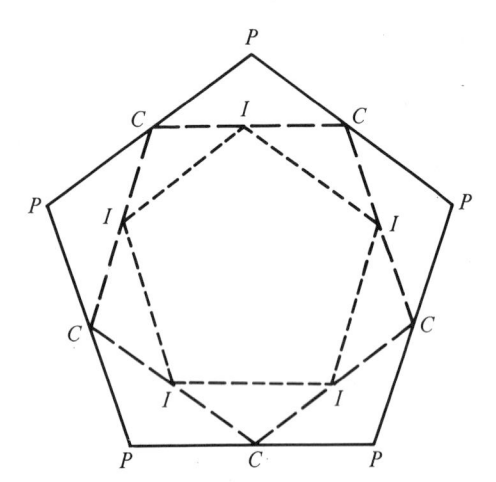

Fig. 7.3 Interactive (I–I), regulative (C–C) and constitutive (P–P) relations (from Ingold 1986b, p. 268). Interactions between individuals, I (inner pentagon) serve to reproduce the relations between symbolic terms, C, of a cultural order (middle pentagon) which serves, in turn, as a vehicle for the conduct of social relations between persons, P (outer pentagon).

adaptive functions of co-operative behaviour as in the evolution of consciousness. Returning to the question of the sociality of non-human animals, we have then to inquire: Do they engage in constitutive relations, or do they merely interact? Drawing again on Shotter's terms, is it true—as he maintains—that animals possess only *natural powers*, bearing no responsibility for what they do, and behaving 'in relation only to their own immediate states of being—as if continually in a "passion"' (Shotter 1984, p. 42)?

Hallowell is undoubtedly correct when he states that the *normative orientation* of conduct, in accordance with a framework of commonly recognized and sanctioned standards, depends upon a 'capacity for self-objectification', that is for the conceptualization of the self as a focus of attention, as something that one can, reflexively, be conscious *of* (Hallowell 1960, p. 346). It may also be agreed that such self-objectification requires a symbolic faculty rooted in the special properties of language (above all, in the use of personal pronouns), and therefore that it is unique to humankind. Human culture and the *concept* of self are thus complementary: each presupposes the other (Lock 1981, p. 19).

This, however, is an argument for reserving for humanity the regulative relations making up the normative framework or moral order to which Hallowell refers, and which is part and parcel of culture. By reference to such an order, human beings may make themselves, or be made, *accountable* for their actions. But accountability is not the same as responsibility; thus action may be consciously *presented* by a subject towards others, even when not selfconsciously *represented* as conforming more or less to norms of conduct or role expectations appropriate to incumbents of positions within an institutionalized, regulative order. In so far as conduct is presented by a responsible agent, we could say that it is directed by *practical* consciousness, whereas the representation of conduct entailed in the rendering of accounts—what Giddens calls its 'reflexive monitoring'—entails the operation of a linguistically-based, *discursive* consciousness (Giddens 1979, pp. 24–5). Non-human animals, we presume, lack discursive consciousness; but this is not to deny them practical consciousness. Just because an animal is unable to render an account of its performance, either beforehand in the form of a plan or retrospectively as a report, we cannot infer that the performance is but an event that happens rather than an action purposively carried out, or that it is an expression of purely natural rather than personal powers. To adopt Searle's useful distinction, conduct that is spontaneous—carried out without previous thought or reflection—may still be informed by *intention in action* although not preceded by a *prior intention*

(1984, p. 65). Much of our own conduct is of this kind, unpremeditated yet not involuntary. We experience it as issuing from ourselves as agents; having thus caused it to happen we are responsible for it, whether or not that responsibility is articulated in the form of an account.

The same must be true of non-human animals, to the extent that they are—like ourselves—both purposive and suffering beings, agents and patients. Their actions, too, may be informed by an intentional component whose source lies in the field of their intersubjective relations. Though Crook may well be right in supposing that 'the development of a capacity for objective self-awareness and description marks the boundary between the animal and the human' (1980, p. 267), intersubjectivity requires that neither *ego* nor *alter* be posited as objects of attention, but only that they join as participants in a shared act of attending, or of purposively doing things together. And whereas the objective self is discursively constituted as an entity within a persistent cultural form, it is through practical immersion in *joint action* that subjects are mutually constituted as vital agents. What I have called constitutive sociality therefore implies subjective awareness, but not necessarily objective self-awareness (on this distinction, see Crook 1980, p. 312).

We can conclude that while regulative sociality is uniquely human, and while interactive sociality can be generalized across the entire animal kingdom, the scope of constitutive sociality is both wider than the former and more narrow than the latter, having emerged in tandem with the evolution of consciousness long before the appearance of the specifically human consciousness of self. It follows from this conclusion that any account of human evolution must be centrally concerned with the ways in which constitutive sociality works its effects upon the physical world of organisms, that is with the mutually transformative interplay between social and ecological systems. This is our central theme in the next part of this chapter; in the final part we turn to consider culture as the mediating term between the two systems, and its alternative mechanisms of transmission and adaptive modification.

7.3 Social and ecological relations

Ecology is the study of the interrelations between organisms and their environments. Human beings are, of course, as much enmeshed within webs of ecological relations as are the individuals of any other species. My objective, in this part of the chapter, is to demarcate the sphere of human ecological relations, so that we can then go on to ask whether, or in what respects, this sphere includes or excludes the sphere of social relations, depending on the different senses of sociality adduced above. We have first to address the difficult question of how to characterize the environment of an organism—a question that admits a rather special kind of answer when posed with regard to objectively self-conscious organisms, namely human beings, whose environment is commonly said to be (at least partially) *artificial*. Given that all organisms physically modify their environments in trifling or substantial ways, in what respects—if at all—does it make sense to say that the human environment is 'man-made', whereas those of the animals are merely 'natural'? I argue that 'making' essentially consists not in the physical modification of objects but in their organization in terms of a framework of concepts. Turning to the social environment, I show that although this naturally includes the conspecifics with which an animal (human or non-human) interacts, its artificiality derives from the imposition upon this interactive network of the institutionalized order of regulative relations otherwise known as 'society'. Forming a part of culture, this order serves to mediate the productive activity of its makers, linked as intentional agents within a field of constitutive relations, upon the natural world. Thus it is in the act of production that the objectives of socially constituted subjects are translated into changes in the state of the ecological system comprising the environmental relations of human organisms. Within the conjunction of social objectives and ecosystemic constraints, I argue, lie the conditions of cultural adaptation.

7.3.1 *The environment of an organism*

An environment is, literally, that which surrounds, and therefore presupposes something to be surrounded. 'Animal' and 'environment' form an inseparable dyad (Gibson 1979, p. 8), or as Lewontin puts it, 'there is no organism without an environment, but there is no environment without an organism' (1982, p. 160). However, we are inclined to think of the environment as a world of nature filled with objects both living and non-living, both mobile and stationary, like a huge room cluttered with furniture and decorations. From this analogy comes the classic ecological concept of the *niche*, signifying a little corner of nature that an organism occupies, and to which it has fitted itself through a process of adaptation. If I remove a vase from an alcove in the wall, a niche remains for a small object that might appropriately fill the vacant space; by analogy it is implied that the ecological niche of an organism is independently specified by the essential properties of the environment, which impose the conditions to which any occupant must conform.

The trouble with this analogy is that it ignores the most fundamental property of living things, that with each there comes into being a particular teleonomic project, which underwrites its subsequent development. The effective environment of an organism consists not of objects as such, but of the opportunities or hindrances they offer for the realization of its project. Gibson (1979, pp. 127–8) expresses this idea nicely by designating the constituents of the niche as a set of *affordances*, but he goes on to argue that each object 'offers what it does because of what it is', or in other words that affordances exist as the invariant properties of environmental objects (1979, p. 139). Though it would appear from this that every niche is constituted 'out there' in the environment, regardless of whether any organism exists to fill it, Gibson's argument is surely fallacious (Shotter 1984, p. 204). For the same essential objects, whether living or non-living, will afford quite different things to different organisms, depending on the nature of their respective behavioural projects: thus a single stone may, at one time or another, serve as shelter for the insect concealed beneath it, as an anvil for the thrush which uses it to break open snail-shells, and as a missile for an angry human to hurl at his adversary. Indeed a catalogue of all the possible affordances of the stone would be as long as that of all the activities of all the organisms that make some use of it.

From this it follows that, far from fitting into a given corner of the world, it is the organism that fits the world to itself, by ascribing functions to the objects it encounters, and thereby integrating them into a coherent system of its own. Hence the environment in which it lives and moves, its so-called niche, is the projection or 'mapping out' of its internal organization on to the world outside its body, or 'nature organized by an organism' (Lewontin 1982, p. 160). Take away the organism, and the niche, in this sense, disappears with it. To be sure, there is still a physical landscape with its array of objects, but these objects have reverted to their primordial status as 'stuff', definable in terms of substance and composition, but not in terms of potential function. What is left, in other words, is a set of essences rather than affordances, constituting a landscape or habitat rather than a niche. As an anvil, the stone is part of the niche of the thrush; as part of the landscape it is merely an object with the essential properties of hardness and durability. As a dwelling or shelter, the cave was part of a niche for ancient humans; as part of the landscape it was merely a semi-enclosed cavity in the rock.

Both stone and cave were there in the landscape long before the arrival of the thrush and the man, but there was no anvil before the thrush began its activity, and no dwelling before man took up residence.

Now for the unselfconscious organism, which lives to execute a received project without reflecting upon its nature or purpose, there can by definition be no such thing as an environment of essences. Hence, what it perceives are not objects as such, but the affordances of objects, already organized (like the body of which they form an extension) for a project coextensive with its own existence. Though there is a sense in which such an organism may be said to 'elect', through its own teleonomic performance, the conditions that make up its effective environment of selection (Waddington 1960, pp. 399–401; Monod 1972, pp. 120–1), this is not a choice deliberately made but a *fait accompli* corresponding to the fact of its own appearance as a living entity. Thus the thrush does not wonder what to do with the stone or the snail, any more than it wonders what to do with its beak. Lewontin therefore goes too far in inferring, from the organisms' *organization* of their environments, that they are 'active *subjects* transforming nature according to its laws' (1982, p. 163).

Such, however, are self-consciously reflective human beings who, to a degree unmatched elsewhere in the animal kingdom, are the *authors* as well as the executors of their projects. This is to say that human conduct, as well as being directed towards certain ends, may at times be motivated by some consciously articulated prior intention to achieve them. As authors, formulating our plans for action, we find ourselves in an environment consisting of objects, substances, or raw materials that have yet to be organized according to the possibilities we may detect in them. Moreover, we perceive an essential disjunction between ourselves as organizers and this inherently resistant raw material. The environment, we feel, is, like our very nature, something that—initially—we *confront*, and that has first to be rendered intelligible within the context of current needs and purposes.

If the environment is to be understood in this sense, as a world of objects apprehended 'in the raw', standing apart from and opposed to the self, it is true to say that for the unselfconscious organism the environment does not, and cannot exist (von Glasersfeld 1976, p. 216). We could of course say that the environment of the thrush includes objects of the kind we call 'stones', but that is not how the bird sees it; what it apprehends are facilities for shell-smashing. Or compare the human hunter–gatherer with the chimpanzee. The environment of the former includes certain species of animals and plants, some of which are regarded as edible, others not. Though we may describe the chimpanzee's environment in the same manner, the chimpanzee neither recognizes nor responds to such verbal taxonomic categories as 'the animal' and 'the plant'. Rather, it discriminates things that afford eating from things that do not (food from non-food), and reacts to their presence accordingly.

7.3.2 *Making an artificial environment*

There is an important conclusion to be drawn from this example: namely that although the hunter–gatherer (almost by definition) may be supposed to inhabit a natural rather than an artificial environment, there is a sense in which that environment is nevertheless man-made. This 'making' consists in the mapping, on to the world of physical things, of a system of ideas or concepts regarding their future use, thereby converting those things—objectively presented to the organizing subject—into use-values. In other words, making here refers to the cultural ordering of nature. But such ordering does not necessarily entail the physical modification of environmental objects; and by the same token, an animal may physically reconstruct parts of its environment without that reconstruction's conforming to a culturally imposed, symbolically encoded plan. This point leads me to draw a key distinction between *constructive modification* and the *self-conscious authorship of design*, a distinction that helps to sort out a great many anomalous instances in comparisons between human works and those of non-human animals.

A classic example concerns the alleged farming activities of certain species of ants. We are in no doubt that the human farmer builds an artificial environment: by converting patches of the landscape into fields, and by planting selected cultigens, he harnesses the ecologically productive potential of the land more efficiently. The ants, apparently, intervene similarly in the reproduction of their food resources, executing operations (albeit in miniature) formally identical to those of cultivation (Reed 1977, pp. 15–17). Should we not, then, admit that they also inhabit an artificial environment? And again, what right have we to claim that the human engineer, in building a dam for the purposes of irrigation, is the constructor of an artificial environment, whereas the beaver, whose feats of environmental modification are equally impressive, is not? The answer is that both farmer and engineer begin work with an idea already formed of the nature of the end-result, and with a knowledge of the procedures required to achieve it. Lacking such knowledge, ant and beaver are constructors of their respective environments, but not designers. Indeed, the designs they embody, and that are 'written out' in the course of their activity, *have no designer*, being the products of an evolutionary process of variation under natural selection.

The same distinction applies if we turn to consider instances where objects are not constructed or modified prior to use. The human hunter–gatherer imposes a symbolic design on the environment, but does not physically transform it like the farmer. Nor does the chimpanzee, though like the ant, and unlike both hunter–gatherer and farming humans, it is not the designer of its project. The harnessing, whether by human hunter–gatherer or chimpanzee, of unmodified environmental objects to a behavioural project may be termed *co-optation*: this is equivalent to what I described earlier as the ascription of functions to objects in the organism's assembly of its niche. The hunter–gatherer's co-optation of the environment, then, is self-conscious, the chimpanzee's is not. Borrowing terms originally coined by Oswalt (1973, p. 14), I propose that the class of objects that are factitious, in the sense that they are harnessed to an intentional design, may be divided into the two subclasses of *artefacts* and *naturefacts*, the former consisting of objects constructed or modified so that they may more efficiently discharge the functions ascribed to them, the latter consisting of objects co-opted without modification. As dwellings, the cave is a naturefact, the igloo an artefact. And the environment of the hunter–gatherer, to the extent that it is not intentionally modified (as through the use of fire to improve plant yields or the construction of facilities for obtaining game), is not so much natural as *natureficial*. As I shall shortly show, this distinction has important implications with regard to our understanding of the concept of production.

7.3.3 *The ecological approach to sociality*

For any individual organism, the environment normally comprises three components: the non-living or abiotic world, the world of other species, and the world of conspecifics. Up to now I have been principally concerned with the former two components; by turning to the last we can link our present discussion to the issues addressed in the first part of this chapter concerning the nature of sociality.

Let me begin with the interactive view of the social, as relations of association between individual conspecifics. An organism's interactions with its conspecific environment fall, just as much as its interactions with predators and prey, or with inanimate objects and substances, within the general field of its ecological relations. From this it follows that, if social relations are understood in exclusively interactive terms, they must form a particular *subset* of ecological relations, and therefore that sociology is a branch of ecology—namely 'socioecology'. Certainly such a subject of study is conceivable; its central focus would be on the nature and evolution of co-operation, that is on what individuals mutually afford to one another (Gibson 1979, p. 135). Social life, from this perspective, involves the co-optation of the capacities of conspecifics to the project of an organism, whose capacities will of course be co-opted by others in their turn. Interactions occur because for each partner, the presence of one or more others provides opportunities for fulfilling a need or disposition; hence the organization of society is a projection, on to the external world of conspecifics, of a project internal to the individual.

All this is perfectly consistent with the Spencerian paradigm of sociality. One basic axiom of the paradigm is that the content of social relations is exclusively instrumental: the interaction is the means, the life of each individual the end. Indeed if we pursue the implications of this axiom to their logical extreme, the line between social interaction and tool-use becomes rather thin. For the human infant, for example, its mother can afford means of access to objects out of immediate reach: Bates *et al.* (1979) refer to the infant's signalling to the mother to retrieve an object as a case of 'social tool use'. Likewise, though a chimpanzee might use a stick to reach food, it could, as Hall (1963, p. 479) notes, equally 'use' another chimpanzee to bring the food to it, whether through positive inducement or threat. In co-operative predation, one individual can 'manipulate' others to bring prey within range for capture. Tool-use and co-operation literally merge in circus acrobatics, where human beings use each others' bodies for a host of purposes, including climbing, swinging, and supporting; likewise weaver ants, by linking their bodies head to abdomen, can form a living chain to bridge a gap between leaves (von Frisch 1974, p. 113). In these instances, as in co-operation generally, the natural powers of individuals are reciprocally augmented through their precise co-ordination, yielding effects greater than the sum of the effects of every individual working independently (Marx 1930, p. 340).

While interactive relations have their parallel in tool-use, a similar correspondence obtains between the functioning of regulative relations and tool-making. By 'making' I refer to the component of self-conscious design rather than constructive modification, and, if my earlier argument is correct—that human beings make their environment through the imposition of a cognitive order upon a world perceived to lie apart from and opposed to the self—it should apply just as well to the conspecific component of the environment as to other components, living and non-living. In tool-making one holds before the mind a picture of the object, or of the projected act, against which one continually monitors actual performance, making appropriate adjustments until the desired end is achieved. The same is true in the normative orientation of social action. In both cases, human beings recursively subject their conduct to rules and standards of which they are themselves the authors, and which precede and govern the subsequent execution. The designs we impose upon the conspecific environment are 'institutions', their integration is 'social structure', and the links between their component parts are regulative relations. Or, as Bidney has put it, where tools are artefacts, institutions are 'socifacts' (1953, p. 130). And a precondition for the construction of these or any other kinds of 'facts' is objective self-awareness: thus a tool-making being, as Hallowell (1960, p. 324) asserted, must also be a moral being, since both capacities attest to the same psychological functions.

7.3.4 *The separation of social and ecological domains*

The analogies between society and artefact, and between social interaction and tool-use, have their limitations; and these are also the limitations of an ecological approach to sociality. For if society is a man-made instrument of co-operation, we must take into account relations between the co-operators as well as the co-operated. In the latter capacity, human beings are but organisms, parts of a multi-bodied 'megamachine' that can be made to perform work. But as fellow operators of the machine they are mutually involved in the exercise of personal rather than natural powers, relating to one another as intentional subjects rather than as interacting objects. This mutual involvement cannot be comprehended within the conventional ecological framework of organism-environment interaction. Adopting a constitutive sense of sociality, we have an essential basis for *separating* the domains of social and ecological relations, instead of treating the former as a subset of the latter. Among human hunter-gatherers, for example, co-operative behaviour in resource-extraction may be comprehended in exclusively ecological terms (Steward 1955, p. 40); but the relations of *sharing* that underwrite the hunter-gatherer's responsibility to his or her group are not ecological but social.

As this example indicates, it would be wrong to draw the interface between social and ecological domains around the boundary of the species, putting all intraspecific relations into the first domain and all interspecific relations into the second. With regard to human beings, intraspecific relations have both a social and an ecological aspect: between persons (intentional subjects) they are social, between individuals (organic objects) they are ecological. The same would hold in the case of non-human animals, to the extent that they are likewise involved in intersubjective relations of a constitutive kind. Moreover, turning from intra-to interspecific relations, an equivalent dualism applies. When a man goes out to hunt, he does so as a responsible agent: his hunting is an exercise of personal powers. And since this responsibility is founded in the relations of sharing, hunting qualifies not just as action but as *social* action. Viewed ecologically, however, hunting reduces to predation, which we define as an interaction between an organism of a certain species (in this case, human) and one or more organisms of another species, resulting in the death of the latter and their conversion into food for consumption by the former. In other words, hunting activity issues from the person, predatory behaviour from the individual. Therefore, hunting is to predation as sharing is to co-operation: in each pair the first term refers to the social aspect, and the second to the ecological aspect, of interspecific and intraspecific relations respectively (Fig. 7.4).

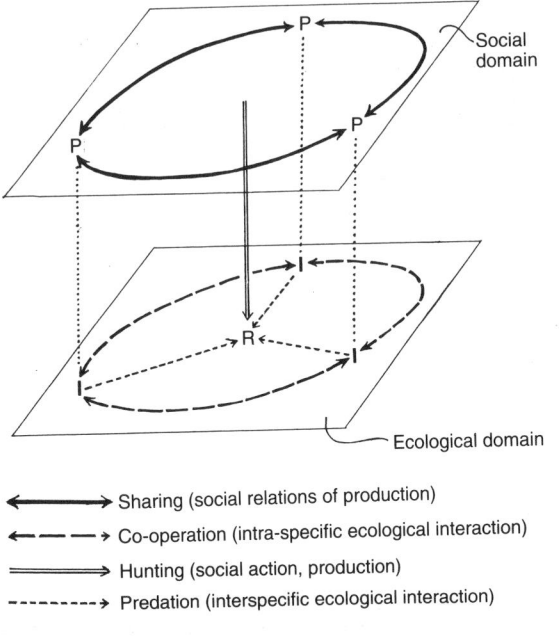

Fig. 7.4 The social and ecological aspects of intra- and interspecific relations: sharing, co-operation, hunting and predation. P, persons; I, individuals; R, food resource (prey). Adapted from Ingold (1988*b*, p. 281).

7.3.5 The essence of production

All human beings are, of necessity, simultaneously caught up in both social and ecological domains of existence, constituting them respectively as bearers of personal and natural powers, as persons and as individuals. Having established a basis for separating the two domains, our next task is to characterize the link between them. This link, as I have argued elsewhere, 'lies in purposive activity that affects the state of the physical world, in other words in social production, work' (Ingold 1983, p. 10). The social domain furnishes the intentional component of production, without which it would fragment into an interminable sequence of behavioural executions. And it is this social purpose that directs and governs the ecological interactions between human individuals and the objective constituents of their environments: other humans, animals, and plants, and inanimate things.

The concept of production is not however free from ambiguity, as is evident from the following definition: 'production designates the process by which the members of a society *appropriate* and *transform* natural resources to satisfy their needs and wants' (Cook 1973, p. 31, my emphases). But appropriation is one thing, transformation quite another; moreover the distinction between them is exactly parallel to the one we made earlier, between intentional design and constructive modification. Over a century ago, Engels pointed to production as the most fundamental criterion of what he saw as a distinctively human 'mastery' of the environment: 'The most that the animal can achieve is to *collect*; man produces.... This makes impossible any unqualified transference of the laws of life in animal society to human society' (1934, p. 308). The essence of production, for Engels, lay in the deliberate planning of activity, by self-conscious organisms, 'towards definite, preconceived ends' (1934, 178, p. 237): this incorporation of environmental objects into a human design, by virtue of which they come to be valorized as resources, is what we call their appropriation.

But when Engels turned to consider concrete examples of human mastery in production, he drew them exclusively from the activities of animal and plant husbandry, involving the physical *transformation* of the environment, or its constructive modification (1934, pp. 34, 178–9). Opposing non-human foraging to human agriculture and pastoralism, Engels ignored the condition of humankind, for the greater part of history, as hunter–gatherers. For this reason the distinction between food-production and food-collection could later be taken up, in virtually identical terms, by the archaeologist V. G. Childe, to demarcate not humans from animals, but 'neolithic' men and their successors from 'palaeolithic' hunters and gatherers (Childe 1942, p. 55). In this form it has become part of the stock-in-trade of modern prehistory.

A result of this shift from appropriation to transformation as the criterion of production was to collapse the original distinction between non-human foragers and human hunter–gatherers, whose activities all came under the single rubric of collection. Supposed to be no more the masters of their environments than foragers, the hunter–gatherers figured as children of nature, living lives wholly encompassed within the ecological domain. This in turn has sustained the widely held assumption that the activities of human hunter–gatherers are more amenable to analysis in purely ecological terms than are those of agriculturalists and pastoralists (Ellen 1979, p. 5). But of course it is precisely in the mastery displayed that hunting and gathering exceed predation and foraging: this mastery is founded in a knowledge of nature in many ways unrivalled even by Western science (Laughlin 1968, pp. 314–15). And it enables hunter–gatherers to live in reasonable comfort in environments that sometimes seem to us scarcely habitable, *without physically modifying them to any significant extent*. For the design they impose upon the world, in mastering it, *does not include the concept of their mastery*, but rather casts them in the role of participant custodians. In short, they use their knowledge to conserve the environment, not to change it. Since their co-optation of nature is self-conscious, they may truly be said to appropriate their resources, even though, by convention, they are said merely to collect from them. For the same reason, as I remarked earlier, their environment is strictly speaking neither natural nor artificial, but 'natureficial'.

Yet if the essence of production lies in its component of intentionality, we have no more cause to assume that it entails appropriation than we have to assume that, to recall Searle's distinction, intention in action presupposes the articulation of prior intentions (corresponding to Engels's 'definite preconceived ends'). It follows that non-human animals can produce too, in so far as their action on the physical world is directed by practical consciousness. Human production is distinctive only to the extent that it is guided by a symbolic projection of the end to be achieved and of the constituent steps towards its realization. The structures of knowledge and taxonomic categories through which hunter–gatherers appropriate the resources of their environments constitute one aspect of the 'forces' of production, defined by Godelier to include 'the material and *intellectual* means that the members of a society implement, within the different

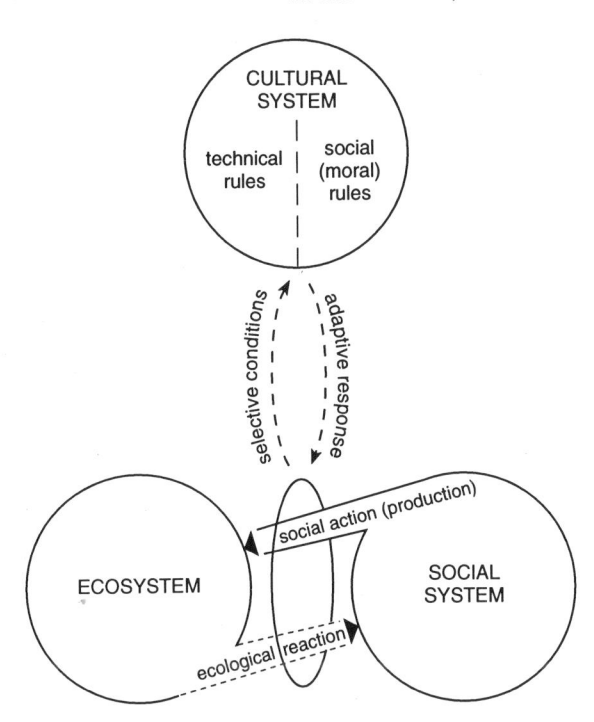

Fig. 7.5 The conditions of cultural adaptation lie in the conjunction of social relations of production (making up the social system) and ecological relations of reproduction (making up the ecosystem). The cultural system includes technical rules, regulating interactions with environmental objects, and social or moral rules, regulating interactions between individuals in society. A schematic representation, adapted from Ingold (1981, p. 128).

"labour" processes, in order to work upon nature' (1978, p. 763). The same applies to the ideal structures of regulative relations, whose locus is likewise in individuals' minds, and which provide a template for practical co-operation. Thus hunting and gathering (or any other human labour process) are productive activities *initiated* by socially constituted subjects, but *mediated* by both 'technical' and 'moral' rules, the former regulating interactions with environmental objects, the latter regulating interactions in society.

Human culture, comprising the totality of such rules, therefore has a basically instrumental function, in that it '*serves to translate a social purpose into practical effectiveness*' (Ingold 1983, p. 14). Indeed it is almost an anthropological cliché to remark that culture is the human means of adaptation. Yet it must be remembered that the conditions of adaptation are never given by the environment alone, but only by the environment as constituted within the act of production itself. The essential properties of an environment do not dictate what should be done in it, but they do impose outer limits on what *can* be done, without jeopardizing the reproduction of both the human population and the resources on which human life depends. Hence, to conclude, the conditions of cultural adaptation must be found in the *conjunction* of social relations of production, constituting the agents and their purposes, and ecological relations of reproduction, within which are constituted populations of both human individuals and other species (Fig. 7.5).

7.4 Culture: its transmission and evolution

Having specified the conditions of cultural adaptation, I turn in this final part of the chapter to consider the mechanisms and processes by which this adaptation occurs. Up to now I have equated culture with human artifice, treating it as a system of symbolically encoded rules and representations. Yet not everything that we are inclined to regard as cultural, by virtue of its non-genetic mode of transmission, can be understood as a product of symbolling. Production need not be (and among non-human animals is not) guided by procedures embodied in a symbolic design; yet to say that procedures not so represented are 'innate' is not to imply that they are necessarily fixed in a set of genetic rather than learned instructions. I aim to show that the mode of transmission specific to 'artificial' culture is not learning but teaching, and that the difference has quite critical implications with regard to the principles of variation and selection which operate on the evolution of cultural forms.

7.4.1 *Reason and tradition*

No concept is more contentious in cultural anthropology than the concept of culture itself. Over thirty years ago, Kroeber and Kluckhohn (1952) compiled a total of 161 different definitions, and since then their number has continued to increase. The root of the definitional problem lies in the original synonymy of culture with the process of civilization, understood as the cultivation of uniquely human faculties of reason and intellect (Tylor 1871, I: 1; see Cassirer 1944, p. 228; Kroeber 1963, pp. 87–8; Stocking 1968, pp. 73–4). A recent advocate of this view is Bidney, who defines 'anthropoculture' as 'the dynamic process of human self-cultivation' (1953, 126). Fashioned by reason, the world of culture is conventionally identified with all that is man-made, or artificial. Non-human animals, all body but devoid of mind, are commonly supposed to remain the automatic slaves of natural instinct: for them there can be neither progress nor history, and by the same token, they perforce lack culture. Though deeply entrenched in Western thought, this verdict on the status of animals has not remained unchallenged. One nineteenth-century dissenter was Lewis Henry Morgan, himself a principal architect of the new discipline of anthropology. On the basis of an intensive study of the behaviour and constructive abilities of the American beaver, Morgan concluded that humans were not uniquely endowed with a 'thinking principle', but rather that all animals could be credited with the capacity to engage in rational deliberation (the beaver being especially gifted in this regard, second only to man). Human pre-eminence, Morgan surmised, lay only in the *degree* to which this capacity had been cultivated in man compared with other animals (Morgan 1868, pp. 276–7; see Fig. 7.6). Today, the same challenge has been taken up by Griffin (1976, 1984), who suspects—just as did Morgan—that animal conduct is guided by 'conscious thinking' to a much greater extent than we are inclined to admit.

Under the dominating influence of Franz Boas, however, the anthropological concept of culture underwent a profound change of significance. Rejecting the classical view of progressive enlightenment, Boas insisted that human behaviour was *not*, for the most part, governed by reason, that it is rather dictated by habits and traditions of which their bearers are only dimly aware, and that the powers of thought are engaged—if at all—only after the event, in the *rationalization* of conduct (Stocking 1968, p. 232). From this view of culture, as a body of received *conventions* rather than as an accumulating stock of progressive *inventions*, it follows that the cultural world is no more artificial or contrived than the world of nature. For the human individual, the elements of culture which he acquires from others, and which he in turn will pass on, constitute a programme which he is destined to follow; but this is not a programme that he or anyone else has *designed*. As Darwin did away with the argument from design in the formation of species, so also did Boas in the formation of cultures. The result was that culture came to be distinguished no longer by its human authorship, but by its mode of transmission, that is, by *learning*. In other words, it was supposed that whereas the components of the animal's project are genetically inherited and in that sense instinctive, those of the human project are 'acquired'—not through a process of active discovery (the sense in which Tylor had used the term) but rather through their passive absorption.

If the 'innate' is constituted by its opposition to the 'artificial', we have then to admit that a great deal (perhaps most) of culture, in so far as it is not the work of reason, must fall in the former category. In short, the boundaries between the innate and the artificial, and between instinct and learning, *do not coincide*. And to complicate matters further, neither boundary is congruent with that separating human from non-human animals. Just as human culture (*pace* Shotter 1984, pp. 41–2) is not entirely man-made, so also a great deal of the behaviour of non-human animals follows instructions that are learned and not genetically inherited. If we were arbitrarily to restrict the cultural to the artificial, we would have to exclude the whole gamut of 'traditional' or 'non-rational' forms, such that science and myth would fall on opposite sides of the boundary, as likewise would mathematics (or similarly artificial symbolic codes) and so-called 'natural' languages. But were we to go to the other extreme, including within culture the sum total of learning-transmitted tradition, we

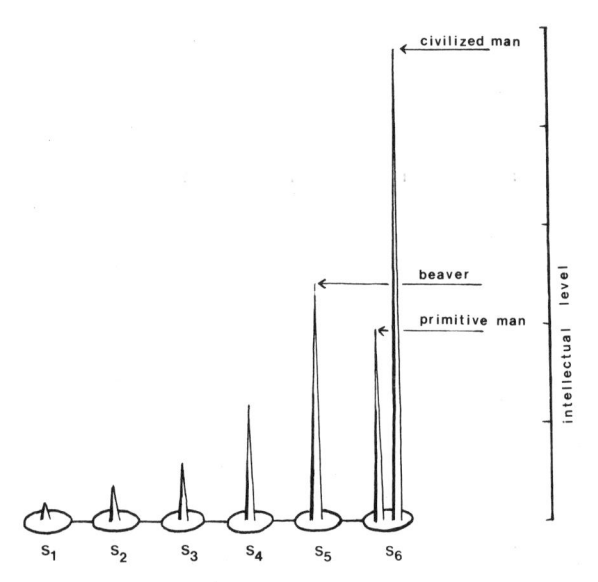

Fig. 7.6 The evolution of consciousness in mankind and other animals: a schematic representation of Morgan's view. S(1–6) represent species ranked on a scale of nature, culminating in man (S6). Morgan thought that not only humans but also all other animal kinds had been endowed by the Creator with a faculty of mind, which had advanced at different rates for different species without involving any change in their bodily forms. While the intellectual powers of civilized humans had far outstripped those of animals, the mind of primitive man was still on a comparable level—indeed the most intelligent of animals, such as beavers, surpassed in their mental powers the most primitive of humans. Adapted from Ingold (1988*a*, p. 88).

would no longer have any grounds at all for reserving culture for humanity. For in many branches of the animal kingdom there is widespread and mounting evidence for traditional transmission, and for the formation of 'behaviour dialects' strictly analogous to the dialectal diversification of human languages (see McGrew and Tutin 1978; Bonner 1980; Beck 1982; and the list of references supplied by Bock 1980, p. 223, n. 28).

Indeed the cliché 'culture is learned behaviour', on which cultural anthropology once staked its claim to disciplinary autonomy, has subsequently proved an acute embarrassment. In order to preserve the study of culture as a branch of the study of humankind, anthropologists have repeatedly sought refuge in definitional circularities to the effect that the traditions of non-human animals are 'not proper culture', ergo proper culture is human. When it comes to the specification of some attribute of humanity underwriting proper culture (or 'euculture'), it is commonly asserted—following Cassirer—that the human being may be defined as an *animal symbolicum*. 'Reason', Cassirer wrote, 'is a very inadequate term with which to comprehend the forms of man's cultural life in all their richness and variety. But all these forms are symbolic forms' (1944, p. 26). The primary reference of the anthropological concept of culture, then, is no longer to non-genetic modes of behavioural transmission, but to the conceptual organization of experience, or 'the imposition of an arbitrary framework of symbolic meaning upon reality' (Geertz 1964, p. 39; see also Holloway 1969, p. 395). The discrepancy between this approach to the definition of culture and its characterization in terms of a learning process has been persistently obscured by the false assumption that whatever behaviour is not symbolically encoded must be crudely instinctive. Sahlins, for example, erects a 'great evolutionary divide' between the instinctive and the symbolic (1972, p. 80), at once consigning to oblivion all traditional practices— whether human or non-human—that do not manifest some kind of symbolic 'blueprint'.

Contemporary biologists, less committed than anthropologists to the premiss of qualitative human uniqueness, have continued to lay primary emphasis on the criterion of transmission. An example is Bonner, who defines culture as 'the transfer of information by behavioural means, most particularly by the process of teaching and learning' (1980, p. 10). The distinction between teaching and learning is, as I shall shortly show, of the utmost importance. But first it is necessary to look rather more closely at what is meant by learning, and in particular to isolate the specific and limited sense in which it may be regarded as a mechanism of transmission.

7.4.2 *Individual and social learning*

There is a certain sense in which *any* organism, maturing in an environment, undergoes a learning process. Its genetic programme does not specify a unique course of development but a continuum of possible trajectories, and the one actually followed will depend upon the particular environmental impacts to which the organism is exposed at successive moments during its life cycle. Thus behaviour, as much as morphology, is in a sense 'acquired' through the environmental steering of development culminating in the mature phenotype. This steering is not simply a matter of conditioning, of the individual's accommodation to a set of pre-existing imperatives; for, as we have already seen, the effective environment of an organism is a function of what it is currently organized to do: hence the environment does not only *evolve* with the organism, but *develops* with it as well (Costall 1985, p. 39). To the extent that learning is equated with the ontogenetic 'acquisition' of behaviour, it is therefore a *two-way* process.

Now this is precisely the view of learning implied in assertions to the effect that culture, being that which the individual acquires from the environment, or that component of behaviour which may be put down to the effects of experience rather than heredity, is *essentially phenotypic* (E. O. Wilson 1980, p. 274). Such assertions apparently result from an unfortunate conflation of the two polarities nature–culture and nature–nurture, each of which is problematic enough in itself (Ellen 1982, p. 15). The homespun concept of nurture corresponds, of course, to the environmental conditions of development, and is opposed to 'nature' understood as a preformed project intrinsic to the original constitution of the organism. What we observe, in the phenotype, is neither nature nor nurture, but an innate potential brought out in a certain way by virtue of environmental experience (Simpson 1958, p. 529). But to regard the environmentally induced component of the variance of phenotypic traits as that part due to *culture* precipitates absurdities. In an ingenious series of experiments, Charles Darwin conclusively demonstrated that earthworms were capable of remarkably flexible behavioural adjustments to diverse environmental conditions (Reed 1982). If culture is simply learned behaviour, a product of nurture, we would have to conclude on the basis of the learning capacities of earthworms that they indeed have culture. Bonner is looking for culture at yet more rudimentary levels of organization, admitting that although 'bacteria are not capable of culture themselves... they do have the basic response system' (1980, p. 56).

Something is obviously wrong, and the mistake lies in the confusion, under the general rubric of learning, of the environmental steering of ontogenetic development with the intergenerational transmission of tradition. If culture is to be distinguished by its mode of transmission, as the transfer of information, and if that mode is 'learning', then clearly the concept of learning has to be understood in the second of the aforementioned senses. For in the first sense no information is transmitted across the generations (save that encoded in the genetic material): each organism learns for itself, and the process of learning is coterminous with its own life-span. In psychological literature, the two sorts of learning are usually distinguished as *individual* and *social* respectively; thus it is said that the possibility of culture rests specifically on social learning (Hull 1982, p. 301). However, this characterization of the distinction in terms of a dichotomy between individual and society is both unfortunate and misleading. For an individual derives the elements of tradition not from society, but from other individuals in the course of their association. Yet the mere fact of association is no guarantee that learning will be 'social' rather than 'individual'. An organism could treat conspecifics just as it treats other components of the animate and inanimate environment, not as sources of instructions for its project, but as affordances and hindrances that call for certain forms of behavioural adjustment and response.

The distinguishing feature of so-called social learning is not, therefore, that it is social, but that it is a process of *replication*. Adopting a term originally proposed by Dawkins (1976), Hull calls the units of replication 'memes', and argues that in the kind of learning basic to culture 'memes encoded in one entity are transmitted to another entity' (1982, p. 301). Now, as replicators, memes are formally equivalent to genes, and both contribute to the constitution of the internal behavioural programme of the organism which is subsequently exposed to external environmental impacts. It follows that *culture is not phenotypic*; rather the relation between cultural rules and manifest behaviour is strictly analogous to that between genotype and phenotype (see Fig. 7.7).

In short, it is through nurture ('individual learning') that a culturally transmitted ('socially learned') programme is brought to fruition in a particular way within the life history of an individual. Since it simultaneously influences the realization of the individual's genetic programme, we must conclude—with Richerson and Boyd—that 'to predict the phenotype of a cultural organism one must know its genotype, its environment and its "culture-type", the cultural message that the organism received from other individuals of the same species' (1978, p. 128). The idea that there exists a cultural analogue of the genotype goes back to an early paper by Gerard, Kluckhohn, and Rapoport, who speak of 'implicit culture'—the inferred cultural structure—as the 'cultural genotype' (Gerard *et al.* 1956, p. 10). Hull refers to the same

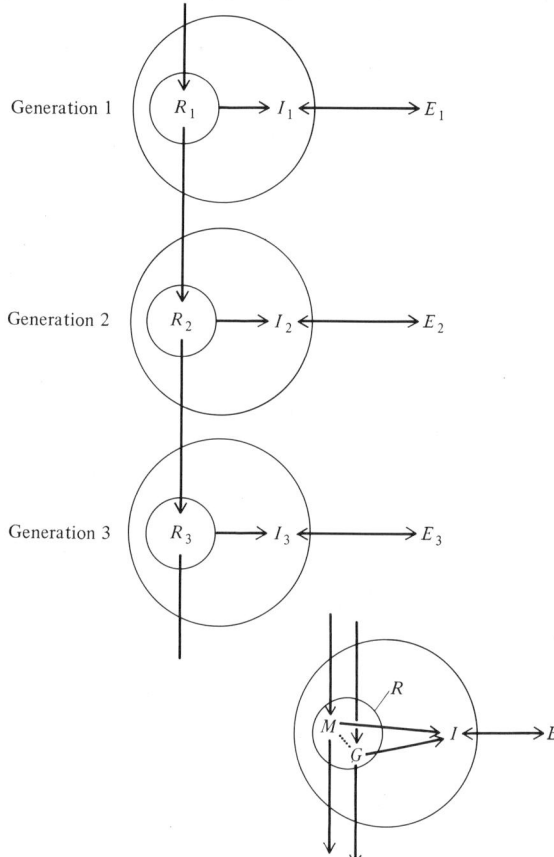

Fig. 7.7 Intergenerational transmission and organism–environment interaction. The diagram shows three individuals, linked in an ancestor-descendant sequence. Replicators (R) are transmitted across the generations through meiosis (in the case of genes) or 'social learning' (in the case of memes). The programme encoded in the replicators is realized in the phenotype, which is the entity that actually interacts with the environment (E). Thus we call the phenotype an *interactor* (I). The conventional distinction between 'individual' and 'social' learning corresponds in the diagram to that between the *horizontal* arrows $E(n) \longleftrightarrow I(n)$ and the *vertical* arrows $R(n) \longrightarrow R(n + 1)$. The lower diagram illustrates the situation where replicators include memes (M) and genes (G). Here, the phenotype (I) is a function of both genes, culture and the environmental conditions of development. Reproduced from Ingold (1986*b*, p. 361).

thing as the 'memotype' (1982: 292; see also Blum 1963), and a number of other competing terms have been proposed in recent years (Lumsden and Wilson 1981, p. 7).

7.4.3 *Learning and teaching*

If culture consists of instructions underlying phenotypic behaviour that are acquired from conspecifics, how are these instructions transmitted? One possible answer is that their transmission is an automatic consequence of the observation and imitation by novices of behaviour emitted by already enculturated individuals (Cloak 1975, pp. 167–8). But although such observational learning depends upon the association of individuals and the overlap of generations, it does *not* require that behavioural instructions should be encoded symbolically in the form of conscious blueprints or models for conduct. For example, the invariant components of bird-song, in so far as they are not genetically transmitted, qualify as memes; but they are certainly not symbols, and for the same reason the bird cannot reflect upon the nature of its song or judge the accuracy of its performance against a conscious ideal standard. Nor can it *teach* its song to other birds. At the very least, teaching entials what Premack (1984) calls *pedagogy*, in which the already trained individual does not merely figure in the novice's environment as a model to imitate, but actively intervenes to bring the behaviour of the novice into conformity with a standard. True teaching, however, goes beyond pedagogy to the extent that it involves instruction not just in particular practices, designed to ensure the accurate replication of an established routine, but also in the abstract rules and principles for the generation of both these and other possible practices. Essential to such instruction is the representation of practices in words, diagrams, or mnemonic or notational symbols. Teaching, as P. J. Wilson puts it, 'is an activity that rests on forms without reality beyond themselves and prepares the learner without requiring him to experience the object taught' (1980, p. 145).

In short, whereas learning involves the transfer of *technique* as a body of practices, teaching involves the transfer of *technology* as an organized body of makers'-cum-users' knowledge. The consequences of this difference are profound. It underlies the distinction that Alexander draws between *unselfconscious* and *self-conscious* design, the former 'learned informally, through imitation and correction', the latter 'taught academically, according to explicit rules' (1964, p. 36). Whereas the unselfconscious executor can only replicate established convention, including any novelty arising from unintended (or 'blind') errors in transmission, the self-conscious designer can use the rules that he has acquired to generate new forms in his imagination, which precede and govern the subsequent execution. That is to say, he can engage in deliberate *invention*.

In evolutionary terms, this priority of rules over execution, technology over technique, or theory over practice, must have been consequent upon the development of the symbolic faculty to some critical threshold, probably marked by the emergence of human language, with its key representational properties of displacement and reflexivity (Kitahara-Frisch 1980, pp. 217–21; Ingold 1983, p. 13). The result was an increase, by several orders of magnitude, in the tempo of cultural change. It is the ability of each generation not merely to act as a vector for the received attributes of predecessors, but to build upon and advance by discovery beyond them, that is responsible for the cumulative or progressive growth of knowledge which is an undeniable and distinctive feature of human history (White 1942, p. 371). And as was long ago recognized by Giambattista Vico (in his *New science* of 1725), and much later by Marx (1930, p. 392, n. 2), the crucial difference between this kind of history and a history of innate forms (whether learned or genetically inherited) is that the former—unlike the latter—consists of works *authored* by humankind.

I have so far outlined three different ways in which practices may be 'acquired': through the environmental steering of ontogenetic development ('individual learning'); the intergenerational transmission of technique ('social learning'); and the communication of symbolic knowledge, in the form of models for conduct, independently of their material realization (teaching). To each of these there corresponds a distinct understanding of the notion of adaptation.

The first is exemplified in Thorpe's classic account of learning in animals. 'The concept of learning', he writes, 'necessarily involves the idea of adaptation ... and accordingly it is defined as *that process which manifests itself by adaptive change in individual behaviour as the result of experience*' (1956, p. 66). This process, of continuous mutual adjustment of organism and environment in the course of a life, corresponds to the kind of adaptation that Toulmin (1981, p. 179) calls 'developmental', and which he contrasts, *inter alia*, to 'populational' and 'calculative' kinds (see also Pittendrigh 1958). The second, the populational sense of adaptation, corresponds to what Darwin originally called 'descent with modification', and to what most contemporary biologists mean by organic evolution. But this kind of adaptation may come about just as well through the differential replication of variant 'memes' in social learning, as through that of variant genes in biological reproduction; or it may be the outcome of a 'co-evolutionary' interdependence of concurrent genetic and memetic replication within the same population of learners (Richerson and Boyd 1978; Durham 1979; Cavalli-Sforza and Feldman 1981; Lumsden and Wilson 1981). Finally, the kind of adaptation that Toulmin calls 'calculative' refers to the rational adjustment of means to ends, or to the *artificial* selection of ideas and concepts in the course of self-conscious design. We suppose this process to be unique to humans, depending as it does upon symbolic intelligence; and it is clearly linked to the distinctive mode of transmission of design through teaching.

7.4.4 *The analogy between organic and cultural evolution*

Even before Darwin adduced his theory of variation under natural selection to account for the evolution of organic forms, something rather similar was being proposed for the development of human knowledge: namely that it involves the systematic testing (or selection) of ideas that originally presented themselves—in the words of William James—'in the shape of random images, fancies, accidental outbirths of spontaneous variation in the functional activity of the excessively unstable human brain' (1898, p. 247). The subsequent development of the analogy between organic evolution and intellectual history has been amply documented by Campbell (1974), and I shall not review it here. Suffice it to say that there persists an equally time-worn analogy between organic and cultural evolution, which has been reasserted on countless occasions in the literature of biology and anthropology (some representative instances are Childe 1951; Huxley 1956; Murdock 1959; Sahlins and Service 1960; and Campbell 1965). The problem at the root of the analogy goes back to our earlier distinction between 'innate' and 'artificial' culture, that is, between tradition passively acquired and executed and the active shaping of the world under human authorship, according to taught rules. Only when understood in the first sense can cultural variation be treated as strictly analogous to genetic variation in organic evolution; yet it is only by regarding culture in the second sense that we can posit an adequate principle of selection. Seeking this principle in rational human deliberation, in *calculative* adaptation, we undermine the foundations of the argument by analogy, which invokes the notion of adaptation in its *populational* sense.

Consider the nature of variation. For cultural evolution to occur, there must be a source of innovations, akin to genetic mutations in organic evolution, which can be combined and recombined in countless ways in the individuals of a population. Cultural innovation is said to come about in two possible ways: by deliberate invention; and by slight, accidental miscopying of elements of tradition during the learning process (Murdock 1971[1956], p. 322; Durham 1976, p. 94). Far from being of minor significance, the distinction between these two kinds of innovation is quite critical. For invention implies the presence of a self-conscious design agent (the invent*or*), having certain intentions and purposes, who brings forms into being as means to their fulfilment. With a Darwinian schema for the interpretation of culture-change, however, the designer disappears along with his purposes, so that all innovation is rendered effectively 'blind' (Steadman 1979, pp. 188–90). People are thus regarded as executors of cultural forms, but not as authors; in other words they are presumed incapable of conceiving solutions to problems of adaptation in advance of their practical implementation. It follows that the distance separating invention and accidental variation also separates self-conscious from unselfconscious design, teaching from learning, the artificial from the innate.

Now in the comparison of organic and cultural evolution it is commonly supposed that the latter is more 'Lamarckian' than 'Darwinian' (Medawar 1960, p. 98; Steadman 1979, pp. 128–9; Gould 1983, pp. 70–1; see also Hull 1982, p. 308). The term 'Lamarckian' has been applied in a wide variety of senses, most of which bear little relation to anything that Lamarck ever wrote. For many it connotes the doctrine of the inheritance of acquired characteristics, although this was neither original with Lamarck nor discounted by Darwin (Zirkle 1946). Accordingly, it is suggested that, while inapplicable to the transmission of genetic characters, the doctrine is valid for cultural transmission. Though superficially true, this view involves a certain mixing of literality and metaphor (Hull 1982, pp. 310–11). If by 'inheritance' is meant *literally genetic* inheritance, then it is not true that traits literally acquired through social learning are inherited, for they are not incorporated into the genome. Only by positing an *analogy* between the intergenerational transmission of genes and memes, or between heredity and heritage, can social learning be regarded as a mechanism of inheritance; yet by the same token, the memes that an individual transmits to the next generation would themselves have been 'inherited' from the previous generation, rather than acquired (through individual learning) during its lifetime.

7.4.5 Variation and invention

Another way of expressing the difference between Darwinian and Lamarckian schemata, of far greater significance for the comparison of organic with cultural evolution, is in terms of the relationship between variations and their conditions of selection. The Darwinian schema, applicable in organic evolution, rests on a principle of 'blind-variation-and-selective-retention' (Campbell 1975, p. 1105), such that whatever the causes of particular innovations (if they are not entirely spontaneous), they bear no relation at all to the current needs and functioning of the individuals in which they occur, or to the nature of their environments. In a Lamarckian schema, to the contrary, the origination of novel traits is attributed to an exercise of will on the part of individuals in actively responding to perceived needs, through the initiation of constructive adaptations subsequently transmissible to offspring.

Following Toulmin (1972, pp. 337–9), Harré argues that the contrasting schemata represent poles on a continuum which may be defined in terms of the *degree of coupling* between what he calls 'M-conditions' and 'S-conditions', where M stands for mutation and S for selection (Harré 1979, pp. 364–6). The Darwinian schema then figures as the special case in which M/S coupling is reduced to zero, whilst at the other end of the spectrum, in a pure Lamarckian schema, M and S are completely coupled, implying that innovations are teleologically motivated by a prior conception of the environmental function they will subsequently fulfil.

Such innovations, as we have seen, are inventions, the products of self-conscious design. Hence it would be wrong to suppose that all culture-change conforms to the Lamarckian schema. Rather it becomes Lamarckian in so far as individuals envisage ends in advance of their realization, and select the means to achieve them in the light of circumstances they expect will occur (Harré 1979, p. 365). Of course, every invention may start its life, as William James described, as one of a myriad of accidental events of spontaneous variation in the human mind, which are subsequently subjected to a selective process of trial and error. But what critically distinguishes invention from the blind variation of the Darwinian schema is that these 'accidents' are *preselected* by an internal representation, in the mind of a selecting agent, of anticipated environmental conditions. That is to say, the inventor already *knows* the adaptive problems he or she sets out to solve, and in this knowledge lie the conditions of their calculative solution. To the extent that his or her projections correspond with actuality, the close coupling of the internally preselected mutation with

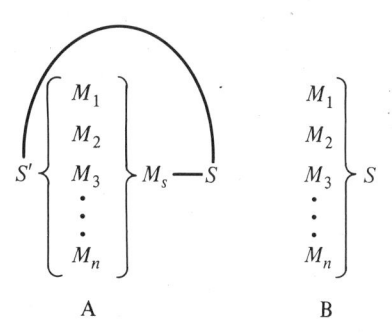

Fig. 7.8 Mutation and selection in the 'Lamarckian' schema (A) and the 'Darwinian' schema (B). In the 'Lamarckian' schema, mutations $M(1 \ldots n)$ are *preselected* by an internal representation (S') of anticipated environmental conditions, yielding the chosen solution $M(s)$. To the extent that S' corresponds to the conditions of selection (S) actually encountered, $M(s)$ and S are coupled. By contrast, in the 'Darwinian' schema, there is no M–S coupling, and variants are selected retroactively by external environmental conditions.

the external conditions of selection is explained (Fig. 7.8). However, complete coupling is probably an impossibility; hence the pure Lamarckian schema represents an extreme case never realized in practice. At the other extreme, a pure Darwinian schema applies in the field of culture-change only if the retroactive selection of novel variants under external environmental conditions is not pre-empted by the operation of the rational intellect in selecting amongst them according to a prior conception of need. Granted that human reason is neither impotent nor omnipotent, cultural evolution probably falls somewhere along the continuum between these two poles; hence adaptation is part 'populational', part 'calculative' (see Table 7.1).

7.4.6 Natural and artificial selection

In the establishment of simple analogies between 'biological' and 'sociocultural' evolution (Campbell 1965; Cavalli-Sforza 1971; Bonner 1980, p. 18), and between 'natural' and 'cultural' selection (Durham 1979), these important distinctions have been overridden and obscured. Darwin, it may be recalled, originally proposed a mechanism of *natural* selection operating on what are now called *genetic* traits. To introduce his ideas, he both compared and contrasted such natural selection with the kind of *artificial* selection exercised by plant- and animal-breeders in the creation of desired strains. Contemporary evolution-

Table 7.1 A summary comparison of 'Darwinian' and 'Lamarckian' schemata of culture change

	'Darwinian'	'Lamarckian'
Coupling of mutation and selection conditions:	Zero	Complete
Innovation:	Accidental variation	Invention
Transmission:	Observational learning	Teaching
Design:	Unselfconscious	Self-conscious
Adaptation:	Populational	Calculative
Selection:	Natural	Artificial

ary biologists are inclined to disregard the contrast, arguing that the artificial selection of domesticates is still a form of natural selection (Hull 1982, p. 317). While this is true in that it involves the adaptive modification of genetically-based characteristics, the fact remains that behind their conditions of selection there lies the scheme of a rational selecting agent. In an earlier section of this chapter (7.3.2) I argued that the dam of the human engineer is an artefact, but not that of the beaver, because although both are physically constructed objects, only the former is governed in its construction by a symbolic blueprint. Likewise it is the deliberate imposition of a programme of selection upon the modification of species that renders it 'artificial'.

We have really to recognize two independent dichotomies: natural vs. artificial selection, of genetically vs. non-genetically transmitted attributes. In combination, they yield four possibilities:

1. natural selection of genetic attributes;

2. artificial selection of genetic attributes;

3. natural selection of memetic attributes; and

4. artificial selection of memetic attributes.

The problem with the inclusive notion of 'cultural selection' is that it is ambiguous with regard to the distinction between (3) and (4). A strict analogy between cultural selection and the Darwinian schema of natural selection implies a situation of the third type. Yet it is difficult, in this situation, to discover any reliable criteria of selection to replace those of differential reproductive success valid only for the selection of genetic attributes. In the absence of suitable criteria, it is tempting to slip from arguing by analogy to arguing by extension, such that as human reason is the product of natural selection, so man is considered the rational selector of his cultural forms. But this is to intimate a situation of type (4), conforming to a Lamarckian schema of intentional design which has no analogue in the natural world. Culture, according to this schema, advances in much the same way as, in the West, scientific knowledge is supposed to advance: institutions 'survive' that are the more effective vehicles of social life, as do hypotheses which have greater explanatory power (compare Tylor 1871, I: 62, and Popper 1972, p. 261). To this, Hull correctly rejoins that 'if science is analogous to anything, it is analogous to artificial selection, not natural selection' (1982, p. 371).

Can the history of Western science be taken as a representative instance of cultural evolution, as the advocates of evolutionary epistemology are inclined to believe? Many anthropologists would object that if culture is analogous to anything, it is to myth, not science. Like myths, they say, cultural projects are largely unauthored by man, existing 'only as they are incarnated in tradition' (Lévi-Strauss 1966, p. 64). Human beings, as Lévi-Strauss famously remarked, do not think in their myths; rather it is the myths that 'think themselves out' through the medium of men's minds and without their knowledge (1966, p. 56). Likewise, in denying the artificiality of culture, Boas had viewed the current of human thought as the revelation of cultural forms rather than the process of their generation (1911, pp. 226–8). But if culture is thus innate, how can we account for all its patterning and coherence? The answer may lie in a mechanism that is neither artificial nor natural, as 'natural selection' is conventionally understood in biology. According to convention, selection is *external*, being a function of the interaction between the organic phenotype and the conditions of its outer environment. But this may be preceded by an *internal* selection in which genetic variants are tested for their compatibility in jointly co-ordinating the development of the individual organism, even before it is significantly exposed to external conditions. Any genotype that fails to establish a viable programme of epigenesis may be automatically eliminated, since the inadequate co-ordination of its mutant components would lead to lethal ontogenetic irregularities. Alternatively, it is hypothetically possible that individual genes may be modified by back-mutation to conform with the necessary conditions for the co-ordination of development, under pressure from the total genetic system of which they are parts. In that case, the 'internal selective environment' of a gene would consist of other genes.

Most biologists remain profoundly sceptical of the notion of internal selection, which was first proposed only twenty years ago, and then by a non-biologist (Whyte 1965). So far as we know, there is no evidence that the genotype as a whole can influence the selection of its individual components, or for what Whyte calls 'genetic reformation'. Yet when we turn to cultural systems, to many anthropologists such influence has long seemed self-evident. Indeed Whyte's views bear a striking resemblance to the Boasian conception, first formulated in 1898, of selection in culture, according to which ideas are 'evolved or accepted' in conformity with 'laws governing the activities of the human mind' (Boas 1974, 155). In this conception, subsequently amplified in the work of Benedict (1935, pp. 33-4), culture is supposed to set up its own conditions of selection, compared with which external environmental conditions are considered relatively inconsequential. Thus the selective environment of a meme would consist, in the first place, of the totality of other memes.

Much more recently, Bateson (1980, pp. 200-1) has drawn a quite explicit parallel between the internal selection of genetic variants in epigenesis and of ideas in the process of learning. Pursuing this parallel, it is possible that many of the problems that attend the extension of the Darwinian schema to the evolution of culture, particularly as regards the specification of adequate selection criteria, would be resolved if the distinction between internal and external mechanisms were duly recognized. For organic evolution the former is problematic, the latter demonstrable; for cultural evolution the reverse obtains. And this leads us to propose the following speculative contrast: organic evolution occurs mainly through the *external* natural selection of genetic variants, whereas in the evolution of culture the adaptation of forms to external conditions is achieved principally through a process of *artificial* selection. Yet these conditions are themselves a projection, on to the environment, of a programme intrinsic to the individual, which owes its coherence to a process of *internal*, natural selection of elements underwritten by the structuring properties of the human mind. In short, to the extent that cultural change is Lamarckian, selection is artificial rather than natural; to the extent that it is Darwinian, it is internal rather than external.

7.5 Conclusion

Human beings are distinguished neither by consciousness nor by their possession of learning-transmitted tradition. They *are* distinguished by the extreme elaboration of certain cognitive specialisms—above all by the faculty of language—which provide a necessary psychological foundation for reflective self-awareness, for society as a moral or regulative order, for tool-making and production as planned activities, for the construction of symbolically-encoded systems of knowledge, and for the transmission of these through teaching.[2] In all these respects it may truthfully be said that the world we inhabit is one that, to an ever greater extent, we have made for ourselves, and that confronts us as the artificial product of human activity. Yet in contrasting the 'man-made' with the 'natural' (Shotter 1984, pp. 40-2), it is important that we do not reduce the latter to a world of unconscious, unfeeling, and inflexible automata. Non-human animals, for all that their conduct is unpremeditated, may nevertheless be considered to be conscious agents responsible for what they do, and, through their immersion in joint action, they may form enduring social relationships. Moreover, consciously directed conduct is guided by instructions that, if untaught, may be equally learned or instinctive. Indeed, though self-consciousness may be a prerequisite for teaching, it is certainly not the case that consciousness is a prerequisite for the transmission of information by learning, and Griffin is quite correct to point out that 'learning is not a reliable criterion of consciousness' (1984, pp. 46-7). But then, neither is thinking an index of consciousness, for there is much that we do consciously, responsibly, sociably, and with feeling, yet unthinkingly. As Whitehead remarked, 'from the moment of birth we are immersed in action, and can only fitfully guide it by taking thought' (1938 [1926], p. 217). If 'taking thought' is the one thing we can do that non-human animals cannot, for most of the time our conduct does not differ all that substantially from the conduct of animals.

I have not, in this chapter, attempted to construct a scenario to account for the evolution of the human capacity for symbolic thought. What I have done is to undertake a job of conceptual clarification, which must be a necessary preliminary to the construction of any such account. We cannot, for example, understand the specific ways in which human beings go about what Goodman (1978) calls 'worldmaking' unless we are first clear about the sense in which any animal, unaided by a symbolic faculty or similarly complex cognitive apparatus, 'organizes' or 'constructs' its environment. Nor can we grasp the implications of symbolic modes of instruction, without first clarifying the nature and limitations of those forms of social learning that do not require the articulation of models 'of', or 'for', conduct (Geertz 1966, pp. 7–8). Likewise, despite the anthropological insistence on the essentially symbolic character of human culture and social relations, it needs to be stressed that neither the historical diversification of tradition, nor the formation of interpersonal relationships, depends upon symbolic representation. Thus even if we accept—with Sahlins—that 'the creation of meaning is the distinguishing and constituting quality of men' (1976, p. 102), it is still important to recognize that the *capacity* to create meaning did not emerge in a vacuum. Once we have characterized those kinds of culture and sociality that do *not* presuppose the operation of a symbolic faculty, we can then go on to ask how this faculty may itself have evolved *within the context of* already existing fields of relationships and bodies of tradition. For in short, though in one sense both 'culture' and 'social life' may be regarded as products of the symbolic imagination, in another they represent the very conditions of its evolutionary emergence.

Notes

1. Author's note: The arguments presented in this chapter are based on ideas elaborated at much greater length in my *Evolution and social life* (1986b, especially Chapters 6 and 7) and *The appropriation of nature* (1987, especially Chapters 1–5). Though I have taken the arguments some way beyond what is presented in these volumes, a certain amount of recapitulation is inevitable. In subsequent publications, however, where I have continued to reflect on the issues dealt with herein, I have come to question some of the epistemological foundations of this previous work (for example, 1989, 1990, 1991, 1992, 1993).

In particular, I now reject the clear-cut dichotomy, drawn in this chapter, between the *person*, as socially constituted subject, and the *individual*, as organic object, and with it the separation of the two domains of social and ecological relations. I would argue instead that personhood, rather than being 'added on' to the human organism, is just one aspect of the agency of the organic being, as it is constituted within the total field of its environmental relations. At the same time, I no longer accept the notion of enculturation as entailing the transmission, across generations, of coded 'information'. What each generation contributes to the next, I would contend, are not rules or instructions for the production of appropriate behaviour, but rather the specific conditions of development under which successors, growing up in a social world, acquire their own embodied skills and dispositions.

2. This parallels Premack's (1985, 1986) conclusion that, from a comparative psychological point of view, the human specializations include quantitative skills, pedagogy, aesthetics, elaborated social attribution, language, and conscious metacognition [eds].

References

Alexander, C. (1964). *Notes on the synthesis of form*. Harvard University Press, Cambridge, Mass.

Alexander, R. D. (1974). The evolution of social behaviour. *Annual Review of Ecology and Systematics*, **5**, 325–83.

Altmann, S. A. (1965). Sociobiology of rhesus monkeys, II: Stochastics of social communication. *Journal of Theoretical Biology*, **8**, 490–522.

Bastian, J. (1968). Psychological perspectives. In *Animal communication* (ed. T. A. Sebeok), pp. 572–91. Indiana University Press, Bloomington.

Bates, E. Benigni, L., Bretherton, I., Camaloni, L., and Voterra, V. (1979). *The emergence of symbols: cognition and communication in infancy*. Academic Press, New York.

Bateson, G. (1980). *Mind and nature*. Fontana, London.

Baylis, J. R. and Halpin, Z. T. (1982). Behavioural antecedents of sociality. In *Learning, development and culture* (ed. H. C. Plotkin), pp. 255–72. Wiley, Chichester.

Beck, B. B. (1982). Chimpocentrism: bias in cognitive ethology. *Journal of Human Evolution*, **11**, 3–17.

Benedict, R. (1935). *Patterns of culture*. Routledge and Kegan Paul, London.

Bidney, D. (1953). *Theoretical anthropology*. Columbia University Press, New York.

Blum, H. F. (1963). On the origin and evolution of human culture. *American Scientist*, **51**, 32–47.

Boas, F. (1911). *The mind of primitive man*. Macmillan, New York.

Boas, F. (1974). *A Franz Boas reader: the shaping of American anthropology 1883–1911* (ed. G. W. Stocking, jun.). University of Chicago Press.

Bock, K. E. (1980). *Human nature and history: a response to sociobiology*. Columbia University Press, New York.

Bonner, J. T. (1980). *The evolution of culture in animals*. Princeton University Press.

Campbell, D. T. (1965). Variation and selective retention in sociocultural evolution. In *Social change in developing areas: a reinterpretation of evolutionary theory* (ed. H. R. Barringer, G. I. Blanksten, and R. W. Mack), pp. 19–49. Schenkman, Cambridge, Mass.

Campbell, D. T. (1974). Unjustified variation and selective retention in scientific discovery. In *Studies in the philosophy of biology* (ed. F. J. Ayala and T. Dobzhansky), pp. 139–61. Macmillan, London.

Campbell, D. T. (1975). On the conflicts between biological and social evolution and between psychology and moral tradition. *American Psychologist*, **30**, 1103–26.

Cassirer, E. (1944). *An essay on man*. Yale University Press, New Haven, Conn.

Cavalli-Sforza, L. L. (1971). Similarities and dissimilarities of sociocultural and biological evolution. In *Mathematics in the archaeological and historical sciences* (ed. F. R. Hodson, D. G. Kendall, and P. Tautu), pp. 535–41. Edinburgh University Press.

Cavalli-Sforza, L. L. and M. W. Feldman (1981). *Cultural transmission and evolution: a quantitative approach*. Princeton University Press.

Childe, V. G. (1942). *What happened in history*. Penguin, Harmondsworth.

Childe, V. G. (1951). *Social evolution*. Fontana, London.

Cloak, F. T. (1975). Is a cultural ethology possible? *Human Ecology*, **3**, 161–82.

Cook, S. (1973). Production, ecology and economic anthropology: notes towards an integrated frame of reference. *Social Science Information*, **12**, 25–52.

Costall, A. (1985). Specious origins? Darwinism and developmental theory. In *Evolution and developmental psychology* (ed. G. Butterworth, J. Rutkowska, and M. Scaife), Harvester Press, Brighton.

Crook, J. H. (1980). *The evolution of human consciousness*. Clarendon Press, Oxford.

Dahrendorf, R. (1968). *Essays in the theory of society*. Routledge and Kegan Paul, London.

Dawkins, R. (1976). *The selfish gene*. Oxford University Press.

Dobzhansky, T. (1962). *Mankind evolving*. Yale University Press, New Haven, Conn.

Durham, W. H. (1976). The adaptive significance of cultural behaviour. *Human Ecology*, **4**, 89–121.

Durham, W. H. (1979). Towards a coevolutionary theory of human biology and culture. In *Evolutionary biology and human social behaviour: an anthropological perspective* (ed. N. A. Chagnon and W. Irons), pp. 39–59. Duxbury Press, North Scituate, Mass.

Durkheim, E. (1982 [1895]). *The rules of sociological method* (trans. W. D. Halls, ed. S. Lukes). Macmillan, London.

Ellen, R. F. (1979). Introduction: anthropology, the environment and ecological systems. In *Social and ecological systems*, ASA monograph 18 (ed. P. C. Burnham and R. F. Ellen), pp. 1–17. Academic Press, London.

Ellen, R. F. (1982). *Environment, subsistence and system: the ecology of small-scale social formations*. Cambridge University Press.

Emerson, A. E. (1958). The evolution of behavior among social insects. In *Behavior and evolution* (ed. A. Roe and G. G. Simpson), pp. 311–35. Yale University Press, New Haven, Conn.

Engels, F. (1934). *Dialectics of nature*. Progress, Moscow.

Fortes, M. (1983). *Rules and the emergence of society*, Royal Anthropological Institute Occasional Paper 39. RAI, London.

Geertz, C. (1964). The transition to humanity. In *Horizons of anthropology* (ed. S. Tax), pp. 37–48. Aldine, Chicago.

Geertz, C. (1966). Religion as a cultural system. In *Anthropological approaches to the study of religion*, ASA monograph 39 (ed. M. Banton), pp. 1–46. Tavistock, London.

Gerard, R. W., Kluckhohn, C., and Rapoport, A. (1956). Biological and cultural evolution: some analogies and explorations. *Behavioural Science*, **1**, 6–34.

Gibson, J. J. (1979). *The ecological approach to visual perception*. Houghton Mifflin, Boston.

Giddens, A. (1979). *Central problems in social theory*. Macmillan, London.

Godelier, M. (1978). Infrastructures, societies and history. *Current Anthropology*, **19**, 763–71.

Goodman, N. (1978). *Ways of worldmaking*. Harvester Press, Brighton.

Gould, S. J. (1983). *The panda's thumb*. Penguin, Harmondsworth.

Griffin, D. R. (1976). *The question of animal awareness: evolutionary continuity of mental experience*. Rockefeller University Press, New York.

Griffin, D. R. (1984). *Animal thinking*. Harvard University Press, Cambridge, Mass.

Hall, K. (1963). Tool-using performances as indicators of behavioural adaptability. *Current Anthropology*, **4**, 479–94.

Hallowell, A. I. (1960). Self, society and culture in phylogenetic perspective. In *Evolution after Darwin, II: The evolution of man* (ed. S. Tax), pp. 309–71. University of Chicago Press.

Harré, R. (1979). *Social being*. Blackwell, Oxford.

Hinde, R. A. (1979). *Towards understanding relationships*. Academic Press, London.

Holloway, R. L. (1969). Culture, a *human* domain. *Current Anthropology*, **10**, 395–412.

Hull, D. L. (1982). The naked meme. In *Learning, development and culture* (ed. H. C. Plotkin), pp. 273–327. John Wiley, Chichester.

Huxley, J. S. (1956). Evolution, cultural and biological. In *Current anthropology* (ed. W. L. Thomas), pp. 3–25. University of Chicago Press.

Ingold, T. (1981). The hunter and his spear: notes on the cultural mediation of social and ecological systems. In *Economic archaeology*, BAR International Series 96 (ed. A. Sheridan and G. Bailey), pp. 119–30. British Archaeological Reports, Oxford.

Ingold, T. (1983). The architect and the bee: reflections on the work of animals and men. *Man* (NS), **18**, 1–20.

Ingold, T. (1986*a*). The sociality of animals: response to Cheater. *Man* (NS), **21**, 347–8.

Ingold, T. (1986*b*). *Evolution and social life*. Cambridge University Press.

Ingold, T. (1987). *The appropriation of nature: essays on human ecology and social relations*. Manchester University Press.

Ingold, T. (1988*a*). The animal in the study of humanity. In *What is an animal?* (ed. T. Ingold), pp. 84–99. Unwin Hyman, London.

Ingold, T. (1988b). Notes on the foraging mode of production. In *Hunters and gatherers, 1: History, evolution and social change* (ed. T. Ingold, D. Riches, and J. Woodburn), pp. 269–85. Berg, Oxford.

Ingold, T. (1989). The social and environmental relations of human beings and other animals. In *Comparative socioecology* (ed. V. Standen and R. Foley). pp. 495–512. Blackwell Scientific, Oxford.

Ingold, T. (1990). An anthropologist looks at biology. *Man* (NS), **25**, 208–29.

Ingold, T. (1991). Becoming persons: consciousness and sociality in human evolution. *Cultural Dynamics*, **4**, 355–78.

Ingold, T. (1992). Culture and the perception of the environment. In *Bush base, forest farm: culture, environment and development* (ed. E. Croll and D. Parkin). pp. 39–56. Routledge, London.

Ingold, T. (1993). Technology, language, intelligence: a reconsideration of basic concepts. In *Tools, language and cognition in human evolution* (ed. K. R. Gibson and T. Ingold), pp. 449–72. Cambridge University Press.

James, W. (1898). *The will to believe, and other essays in popular philosophy*. Longman, New York.

Kitahara-Frisch, J. (1980). Symbolising technology as a key to human evolution. In *Symbol as sense* (ed. M. L. Foster and S. H. Brandes), pp. 211–23. Academic Press, London.

Kluckhohn, C. (1949). *Mirror for man*. McGraw-Hill, New York.

Kroeber, A. L. (1952). *The nature of culture*. University of Chicago Press.

Kroeber, A. L. (1963). *An anthropologist looks at history*. University of California Press, Berkeley.

Kroeber, A. L. and Kluckhohn, C. (1952). *Culture: a critical review of concepts and definitions*, Papers of the Peabody Museum of American Archaeology and Ethnology, Harvard University, Vol. XLVII, no. 1. Cambridge, Mass.

Laughlin, W. S. (1968). Hunting: an integrating biobehavior system and its evolutionary importance. In *Man the hunter* (ed. R. B. Lee and I. DeVore), pp. 304–20. Aldine, Chicago.

Lévi-Strauss, C. (1966). Overture to 'Le cru et le cuit'. *Yale French Studies*, **36/7**, 41–65.

Lewontin, R. C. (1982). Organism and environment. In *Learning, development and culture* (ed. H. C. Plotkin), pp. 151–70. John Wiley, Chichester.

Lock, A. J. (1981). Universals of human conception. In *Indigenous psychologies: the anthropology of the self* (ed. P. L. F. Heelas and A. J. Lock), Academic Press, London.

Lumsden, C. J. and Wilson, E. O. (1981). *Genes, mind and culture*. Harvard University Press, Cambridge, Mass.

McGrew, W. C. and Tutin, C. E. G. (1978). Evidence for a social custom in wild chimpanzees? *Man* (NS), **13**, 234–51.

Marx, K. (1930). *Capital*, Vol. I. Dent, London.

Marx, K. and Engels, F. (1977). *The German ideology* (ed. C. J. Arthur). Lawrence and Wishart, London.

Mauss, M. (1979 [1950]). *Sociology and psychology: essays*. Routledge and Kegan Paul, London.

Medawar, P. B. (1960). *The future of man*. Methuen, London.

Monod, J. (1972). *Chance and necessity*. Collins, London.

Morgan, L. H. (1868). *The American beaver and his works*. Lippincott, Philadelphia.

Mumford, L. (1967). *The myth of the machine: technics and human development*. Secker and Warburg, London.

Murdock, G. P. (1959). Evolution in social organization. In *Evolution and anthropology: a centennial appraisal* (ed. B. J. Meggers), pp. 126–43. The Anthropological Society of Washington, Washington, DC.

Murdock, G. P. (1971 [1956]). How culture changes. In *Man, culture and society* (2nd edn, ed. H. L. Shapiro), pp. 319–32. Oxford University Press.

Murdock, G. P. (1972). Anthropology's mythology. *Proceedings of the Royal Anthropological Institute*, **1971**, 17–24.

Oswalt, W. (1973). *Habitat and technology*. Holt, Rinehart and Winston, New York.

Pittendrigh, C. S. (1958). Adaptation, natural selection and behavior. In *Behavior and evolution* (ed. A. Roe and G. G. Simpson), pp. 390–416. Yale University Press, New Haven, Conn.

Popper, K. R. (1972). *Objective knowledge: an evolutionary approach*. Clarendon Press, Oxford.

Premack, D. (1984). Pedagogy and aesthetics as sources of culture. In *Handbook of cognitive neuroscience* (ed. M. S. Gazzaniga), pp. 15–35, Plenum, New York.

Premack, D. (1985). Gavagai! Or the future history of the animal language controversy. *Cognition*, **19**, 207–96.

Premack, D. (1986). *Gavagai! Or the future history of the animal language controversy*. MIT Press, Cambridge, Mass.

Radcliffe-Brown, A. R. (1947). Evolution, social or cultural? *American Anthropologist*, **49**, 78–83.

Radcliffe-Brown, A. R. (1952). *Structure and function in primitive society*. Cohen and West, London.

Reed, C. A. (1977). The origins of agriculture: prologue. In *The origins of agriculture* (ed. C. A. Reed), pp. 9–21. Mouton, The Hague.

Reed, E. S. (1982). Darwin's earthworms: a case study in evolutionary psychology. *Behaviourism*, **10**, 165–85.

Richerson, P. J. and Boyd, R. (1978). A dual inheritance model of the human evolutionary process, I: Basic postulates and a simple model. *Journal of Social and Biological Structures*, **1**, 127–54.

Sahlins, M. D. (1972). *Stone age economics*. Tavistock, London.

Sahlins, M. D. (1976). *Culture and practical reason*. University of Chicago Press.

Sahlins, M. D. and Service, E. R. (1960). *Evolution and culture*. University of Michigan Press, Ann Arbor.

Sandel, M. (1982). *Liberalism and the limits of justice*. Cambridge University Press.

Schutz, A. (1962). *The problem of social reality*, Collected papers I (ed. M. Natanson). Nijhoff, The Hague.

Schutz, A. (1970). *On phenomenology and social relations* (ed. H. R. Wagner). University of Chicago Press.

Searle, J. R. (1984). *Minds, brains and science.* BBC Publications, London.

Shotter, J. (1974). The development of personal powers. In *The integration of a child into a social world* (ed. M. P. M. Richards), pp. 215–44. Cambridge University Press.

Shotter, J. (1984). *Social accountability and selfhood.* Blackwell, Oxford.

Simpson, G. G. (1958). Behavior and evolution. In *Behavior and evolution* (ed. A. Roe and G. G. Simpson), pp. 507–35. Yale University Press, New Haven, Conn.

Spencer, H. (1876). *The principles of sociology, I.* Williams and Norgate, London.

Steadman, P. (1979). *The evolution of designs: biological analogy in architecture and the applied arts.* Cambridge University Press.

Stern, B. J. (1929). Concerning the distinction between the social and the cultural. *Social Forces*, **8**, 265–71.

Steward, J. H. (1955). *Theory of culture change.* University of Illinois Press, Urbana.

Stocking, G. W. (1968). *Race, culture and evolution.* Free Press, New York.

Thorpe, W. H. (1956). *Learning and instinct in animals.* Methuen, London.

Toulmin, S. (1972). *Human understanding, I.* Clarendon Press, Oxford.

Toulmin, S. (1981). Human adaptation. In *The philosophy of evolution* (ed. U. J. Jensen and R. Harré), pp. 176–95. Harvester Press, Brighton.

Tylor, E. B. (1871). *Primitive culture*, Vol. I. John Murray, London.

Vico, G. (1725). *La scienza nuova e opera sclette* (3rd edn, 1744, trans. T. G. Bergin and M. H. Fisch, 1984). Cornell University Press, Ithaca, NY.

von Frisch, K. (1974). *Animal architecture.* Hutchinson, London.

von Glasersfeld, E. (1976). The development of language as purposive behaviour. In *Origins and evolution of language and speech*, Annals of the New York Academy of Sciences, Vol. 280 (ed. H. B. Steklis, S. R. Harnad, and J. Lancaster), pp. 212–26. Academy of Sciences, New York.

Waddington, C. H. (1960). Evolutionary adaptation. In *Evolution after Darwin, I: The evolution of life* (ed. S. Tax), pp. 381–402. University of Chicago Press.

White, L. A. (1942). On the use of tools by primates. *Journal of Comparative Psychology*, **34**, 369–74.

Whitehead, A. N. (1938 [1926]). *Science and the modern world.* Penguin, Harmondsworth.

Whyte, L. L. (1965). *Internal factors in evolution.* Tavistock, London.

Wilson, E. O. (1978). *On human nature.* Harvard University Press, Cambridge, Mass.

Wilson, E. O. (1980). *Sociobiology: the new synthesis* (abridged edn). Harvard University Press (Belknap), Cambridge, Mass.

Wilson, P. J. (1980). *Man, the promising primate.* Yale University Press, New Haven, Conn.

Zirkle, C. (1946). The early history of the idea of the inheritance of acquired characters and pangenesis. *Transactions of the American Philosophical Society*, **35**, 91–151.

8
Social relations, communication, and cognition

Andrew Lock and Kim Symes

Abstract

The human body can be used to communicate either by itself or as a support for a number of 'props'. Bodily communication is often emotional in nature, relying on underlying physiological and physical responses. What evokes these responses is generally learned, and varies across cultures. Expressive repertoires are used to mark social roles and power. These are also marked by using the body as a 'prop' for various adornments, such as tattooing or clothing. The use of body mutilation is related to the permanence of status assignment of individuals in a society, being generally absent from complex, mobile societies, where temporary adornments such as jewellery or clothing take on a larger burden for marking social roles.

The marking of social roles is also reflected in language forms. Many societies have different forms of male and female speech. Female forms of speech tend to be more conservative than male forms, and reflective of 'prestigious' styles used by high status groups. Status, or social power, is widely marked by particular speech-forms in socially stratified societies. Speakers also shift their speech-forms in predictable ways across different social situations, thus marking a society's construction of social contexts.

A number of indices of a society's complexity or 'differentiation' have been put forward. These indicate a trend for societies to become more complex over time; but as they become more complex their elements become less cohesively integrated. This is not a simple relationship, but one involving factors such as an individual's ability to use the resources provided by more complex social organizations to assay novel activities, and to promote the formation of subcultures. Hunter–gatherer societies tend to have low social complexity, while agricultural ones show increasing degrees of complexity. Family structures tend to be related to food-procurement strategies, being of an 'extended' type mainly in sedentary, agricultural societies. Food-accumulation practices also show a relationship to child-rearing practices, tending to emphasize compliance in agricultural societies versus assertion in hunter–gatherer ones.

Social structure, language, thought, 'individuality', and child-rearing practices are elements of human life that have been claimed as interrelated. A central notion in these interrelationships is 'shared presuppositionality'. The essential argument is that high levels of shared knowledge, such as would hold among members of small-scale, undifferentiated societies and in those where social roles are rigidly marked, reduce the need for highly explicit linguistic coding of information for effective communication. The elements of language, both grammatical and lexical, reflect social needs. Less 'elaborated' linguistic structures provide fewer resources for the handling of complex problem-solving—the view not that language determines thought, but that it provides the equivalent of a problem-solving 'tool-kit', variously facilitating or hindering certain forms of problem formulation and solution. Both language and social structure also mediate child-rearing practices, and hence the process of cultural reproduction. These interrelations have been explored in a number of studies which stem from different theoretical stances, making the comparison of results difficult. But, while at this point in time it is not possible to specify the exact nature of these interrelationships, their broad outlines can be perceived. These suggest that claims as to the availability of complex language in Upper Palaeolithic societies may be unwarranted, as the relatively simple structure of these societies would have promoted high levels of presuppositionality among their members, enabling effective interpersonal communication to be conducted with quite restricted forms of symbolic language.

8.1 Introduction

In this chapter, we review work on communication within human social groups. To begin with, we have divided communication media into non-verbal and linguistic forms. We outline some of the basic findings for both these systems in Sections 8.2 and 8.3. It is apparent that both these forms of communication show patterns of variation that relate to the social structure of the groups they are employed in.

In Section 8.5 we present some of claims that have been put forward to explain differences in the social structures of different societies.

In the remainder of the chapter we summarize and interpret work that claims to elucidate the relations between social structure, communication systems, and cognition. There are a number of almost independent traditions of research and speculation. We believe that it is in the nature of the questions they explore that these traditions will always remain speculative. There are, however, commonalities in the conjectures of these different approaches. Consequently, we present each approach with a view to emphasizing these common elements. In practice this means that Sections 6 to 10 should be read together rather than separately.

8.2 Non-verbal communication in relation to cultural complexity

The body can be used as a vehicle for communication in two ways: by itself, or as the support for a variety of 'props'. In the first category would be placed gesture, facial expression and so on; in the second jewellery, make-up, clothing, and so on.

8.2.1 *The unadorned body*

A pioneering work in this field is Darwin's *The expression of the emotions in man and animals* (1872). Recent work traces itself back to this source, generally via Hall's (1959) *The silent language*, which gave impetus to both an academic and a 'popular' literature on body language. Contemporary reviews include, from an evolutionary perspective, Plutchik and Kellerman (1983); more generally Heslin and Patterson (1982); and, for many of the basic studies especially, Key (1977). The research and descriptive literatures are massive, and here we discuss only the general issues of cultural variation in non-verbal communication, and the emotions involved in its expression.

Given their Darwinian origins, questions of bodily expression have traditionally intertwined a number of aspects of the innate-versus-acquired argument. Firstly, are bodily expressions, especially facial ones, learned or inherited? This question was first brought into focus by Mauss (1935), who put the view that bodily expression was learned and social, rather than innate and biological. Today a crude dichotomy can be drawn between (i) ethological workers (for example, Eibl-Eibesfeldt 1975, pp. 442–88), who point to phylogenetic relations between human and primate facial expressions and the universal features of human expression, and (ii) more socially inclined anthropologists, who follow the tradition of Mauss (1935) and Kluckhohn (for example, 1954, p. 930). Secondly, are human emotions innate or learned? Are all humans alternately happy, sad, surprised, and so on? This leads to a third question concerning the interaction between questions one and two: is the link between emotional states and their expression learned or innate? Do all humans laugh when they are happy, for example?

8.2.1.1 *Innate versus learned repertoires*

There do seem to be a distinguishable set of discrete human facial expressions. These probably rely on inherent patternings of muscle contraction and relaxation sequences (Eibl-Eibesfeldt 1975; cf. Ekman 1972; Scherer and Ekman 1984). Other bodily responses which have a physiological basis likewise appear universal: blushing versus pallor, as blood-flow is directed to or diverted from the skin, is inevitable.

Laughter may have a similar basis, especially as a response by children to being tickled. The remainder of the human gestural repertoire appears to be culturally based, and hence learned: body posture when walking or sitting; nodding for 'yes'; ways of greeting, expressing friendship or hostility; etc. The exception to these is perhaps the pointing gesture for joint orientation of gaze, which may be argued to be learned, though it is so characteristically human in contrast to other primates and social mammals that there is probably some innate component underlying it.

8.2.1.2 *Innate versus learned emotions*

The universality of emotion can be approached in two ways. Firstly, emotional responses are a universal potential: we are all susceptible to happiness and sadness, for example. But at the same time, what makes us respond to an event by becoming happy or sad is culturally determined. This view would seem to be claiming that there is a universally available existential repertoire of characteristically human states.

Alternatively, recent work puts the argument that, in the adult levels of societies at least, emotions are not only culturally relative but also culturally constituted. Love, for example, is not a universal response resulting from a sustained interpersonal relationship, but an existential state resulting from a culturally given framework through which these relationships are construed. Further, the dividing lines between different emotions are delineated in the course of socially learning how to create those states: they are not biologically discrete.

The above approaches result in different ways of conceiving emotional expression. We can distinguish, on the one hand, a view of society as functioning to teach or repress appropriateness of expression via rules of etiquette, thus providing a set of orientations as to how events should be perceived and responded to. On the other hand, there is the view that states are not pre-existing givens to be sorted amongst events. Rather, they are fashioned and articulated out of a cultural patterning and valuing of events that is acquired during socialization and enculturation. Given, for example, that humans conceive of and explain what it is to be human in so many different ways—resulting in very different cross-cultural conceptions as to what a 'self' is—then the English expression 'to feel happy with oneself' cannot be translated across cultures; for someone who conceives of his or her 'self' in such a different way from us cannot experience our sort of happiness with it. To an extent, this dichotomy is just a sophisticated version of the innate–acquired argument. Certainly, it is more a conceptual than an empirical issue.

8.2.1.3 *Empirical research*

1. Within cultures, non-verbal communication functions to distinguish social roles and social power, different social categories often having different repertoires of expression (see, for example, Heslin and Patterson 1982; Plutchik and Kellerman 1983).

2. Between cultures, Mary Douglas (1971) proposed that the 'control imposed' on the non-verbal expression of emotion is related to social structure. That is:

> In its role as an image of society, the body's main scope is to express the relation of the individual to the group. This it does along the dimension from strong to weak control, according to whether the social demands are strong, weak, acceptable or not. From total relaxation to total self-control the body has a wide gamut for expressing this social variable.
>
> What does it mean when one tribe laughs a lot and another tribe rarely? I would argue that it means that the level of social tension has set low or high thresholds for bodily control. In the first case, the full range of the body's power of expression is more readily available to respond fully to a small stimulus. If the general social control settings are slack, the thresholds of tolerance of bodily interruption will be set higher. Comparisons... should take account of the load of social meaning which the body has to carry (p. 390).

8.2.2 *The adorned body*

Social structure is also implicated in the different forms of 'props' which the body provides a 'peg' for. A principal function of body adornment is to provide a visible 'badge' of status. In simple societies there is a tendency for adult social roles to be adopted for life, becoming an integral part of a person's personality structure. The maintenance of social stability is aided by the symbolic use of body adornment and ornament as raw material for the reification of abstract notions such as role, gender, and status.

A cross-cultural examination of adornment styles reveals a striking correlation between *the degree of differentiation and social mobility of a society and the occurrence of indelible forms of decoration*, such as tattooing, scarification, and other forms of mutilation. Such practices are generally absent from complex, socially mobile societies, with the exception of tattooing among minority groups. In all societies, temporary adornments are common: make-up, war-paint, jewellery, etc. Some selected examples are noted below.

8.2.2.1 *Body adornment in simple societies*

a. Tattooing

Tattooing is, or was, common in Melanesian and Polynesian cultures. In New Zealand Maoris the practice was known as *Moko*, which:

like all Polynesian tattooing was a symbol of rank.... Personal individual designs were made on the face, and it was the ambition of all free men to have a finely tattooed face which made them conspicuous in war and attractive to women. A face tattoo was also a kind of personal signature, and the *Maoris believed that their personalities were imprinted on these facial masks.* When the chiefs signed deeds of land sales to Europeans, instead of a signature or a cross, they drew their face patterns without the aid of mirrors and with great ease (Brain 1979, p. 54, emphasis added).

In becoming 'imprinted' in indelible tattoos their personalities became fixed and unchangeable. *Moko* is therefore an example of a technique for maintaining both individual and cultural stability.

b. Scarification

Tattooing pigments do not show up well on dark skin, and amongst many African and Australian Aboriginal groups is replaced by scarification (also known as cicatrization). As with tattooing, the principal function of this technique is to reify social relationships, and thus contribute to their stability: 'scarification, perhaps more than any other body art, tends to indicate social status and social structure, emphasizing the continuity and way of life of a particular tribal group or class (ibid., p. 70).'

Faris (1972) provides a clear example of this in his study of Nuba (Southern Sudan) practices, describing how girls are subjected to a series of scarification rituals as they progress through various life-stages, each associated with a change in status. First, at puberty, a pattern of scars is cut on each side of the abdomen, joining at the navel, and continuing to a point between the breasts. Secondly, at first menstruation, a series of parallel rows under the breasts, continuing round the back and most of the torso, is added. Finally, after the weaning of her first child, she is scarred over her entire back and neck, the back of her arms, the buttocks, and the back of the legs to the knee. Perhaps because of the pain involved, both tattooing and scarification are quite widely associated with the many forms of passage rites employed in many societies.

c. Mutilation

More gross modifications of the body than scarification include piercing the lips, ears, or nasal septum for the insertion of jewellery, elongation of the neck by the use of metal bands, and the binding of the head, waist, or feet to produce unnatural but socially desirable, high-status deformations. Chinese footbinding produced the arched 'lotus foot' which made most forms of work impossible, hence marking its possessor as of high status. It is possible that skull deformations as practised among such cultures as the Mangbetu (Central Africa) and Chinook Indians (North-west America), resulting in an elongated or wedge-shaped skull, were performed for the same reason, making it impossible for the individual to carry heavy baskets, bales, or water-jars.

Brain (1979) outlines how such practices function as a code to demarcate the attainment of successive life-stages in the Suya (Brazil):

Ears are first pierced at adolescence, while the lips are pierced only when the adolescents become fully adult. Becoming adult is associated with ideas about hearing and speaking, and the acquisition of these important "social" faculties is thought to be fully achieved with the wearing of discs. Children with unpierced ears are not expected to "know" or "understand" anything, just as they are not expected to behave in an adult way. At puberty, however, when their ears are pierced, they are expected to listen seriously to their elders and follow their instructions. The bigger boys have their lips pierced when they enter the men's house and leave the world of women; and while they are waiting to marry they learn songs and continue to insert bigger lip-discs (ibid., p. 113).

8.2.2.2 Body adornment in complex societies

In contrast to the simpler small-scale societies discussed above, complex societies are often more fluid, and so provide more opportunities for adults to change their roles and statuses. Westerners, for example, may assume several roles in the course of a lifetime, or even within a single day. This mobility is (i) associated with a rarity of permanent marking, and thus at the same time is (ii) the driving force for the phenomenon of fashion. When their symbols can so easily be assumed, particular groups cannot maintain their positions unless they continually change those symbols. In general, high-status fashions tend towards impractiability, as do the various bodily mutilations described above.

An example of these dynamics is provided by the explosive adoption of wigs in eighteenth-century Europe. The fashion was initiated well before this time by persons of high social rank: Elizabeth I of England and Henry III of France both took to wearing wigs because of the paucity of their own hair. Hence wigs became symbols of status that could be imitated by others. Over the following few centuries wigs became increasingly larger and more elaborate (and because of their expense, the most prestigious wigs could only be afforded by wealthy nobles: hence the origin of the term "bigwig" for persons of high status.) In France, 'one-up-manship' reached the point where there emerged a profession of wig builders capable of fabricating such gigantic, scaffolded creations as that shown in Fig. 8.1.

Given the lack of permanent adornment, other means of inhibiting the social masquerading of un-

Fig. 8.1 Wig-making, France 1788 (from R. Brain 1979, *The decorated body*, p. 119. Harper and Row, New York).

warranted status were necessary. Speech and etiquette systems were one way (see below). Another route was by decree, such as that made by Edward IV of England in the fifteenth century:

No knight under the rank of a Lord ... shall wear any gown, jacket or cloak, that is not long enough when he stands upright, to cover his privities and his buttocks, under the penalty of twenty shillings.... No knight under the rank of a lord ... shall wear any shoes or boots having pikes or points exceeding the length of two inches under the forfeiture of forty pence (cited by Morris 1977, p. 217).

It is clear that clothing and social status were intimately linked.

Currently, such rules do not exist for the general populace (although informal pressures are quite pervasive amongst many occupational groups), but are confined to such groups as prisoners, the military, the police, and some schoolchildren. What these groups have in common is that 'society' desires them to be recognizable as a distinct social category; to have quickly identifiable and unequivocal status; and to encourage the conformity of members to a single ideal rather than individual expression. Uniforms are not something grafted on top of existing status relations, but symbols which help *create* such relations. In complex, mobile societies, then, ritually symbolic body adornment is confined to situations where a rigid role structure pertains.

8.2.3 *Summary*

Non-verbal channels of communication are used in all cultures. The form of the communicative repertoire, and the meanings conveyed, are both variable. There are, however, aspects of this variation which imply relations with social structure. The unadorned body's repertoire has a high emotional content, whereas the adorned body is much more involved in the coding of social status and age-grades, thereby constituting a person's understanding of who he or she is. We find more evidence of the implication of these processes, and their effects, in the use of language.

8.3 Linguistic communication

8.3.1 *Sociolinguistics: origins and perspectives*

Sociolinguistics is that branch of linguistics which studies the relationship between language and society (sociologists have historically shown little interest in the study of language; see Murray 1983, pp. 281–2 on why this has been the case). We can define sociolinguistics as *the study of language in relation to society* (Hudson 1980, p. 1). While there are examples of such work from early in this century (and Hertzler (1953) pointed out the necessity for it in 1952), sociolinguistics as an activity stems from the early 1960s (see Grimshaw 1973). Its recent origins lie in theoretical differences with structuralist linguistics as crystallized in the work of Chomsky (1957).

Structuralist linguistics had shown that sounds were patterned in every language. Research following Bloomfield and Chomsky took the view that any particular language had one phonological system and one grammatical system. In this view, 'language' is an abstracted construct not directly observable; it is an idealization. People possess a linguistic *competence*: every speaker of a language has mastered and internalized a generative grammar that expresses his knowledge of his language (Chomsky 1965, p. 8). Actual speech produced will often depart from the dictates of this grammar, but this variation is of little importance, for structural linguistic theory

is concerned primarily with an ideal speaker–listener in a completely homogenous speech-community, who knows its language perfectly, and is unaffected by such grammatically *irrelevant* conditions as memory limitations, distractions, shifts of attention and interest, errors, random or characteristic, in applying this knowledge of the language in actual behavior (ibid., p. 3, emphasis added).

Performance is therefore treated as an imperfect realization. Empirical variance from the ideal form—abstracted from the linguist's intuitions—was thus dismissed as 'error'.

It is the claim of sociolinguists that much of this so-called error is ordered and socially patterned. Such findings could, in principle, be accommodated within structuralist theories, were it not for their raising deeper issues concerning the relation between language use and language structure. Shared grammatical conventions were recognized by workers such as Hymes (1973, p. 67) as necessary but not sufficient for the production and comprehension of speech, for 'there are rules of use without which rules of syntax are useless'. Grammars can produce grammatical utterances: what they lack is any means of producing relevant utterances.

A child from whom any and all the grammatical sentences of a language might come with equal likelihood would be a social monster. Within the social matrix in which it acquires a system of grammar, a child also acquires a system of use (Hymes 1977, p. 75).

Hymes termed this system of use *communicative competence*, extending the notion of competence introduced by Chomsky. The introduction of competence into theoretical discourse was seen by Hymes as a 'decisive step', changing 'the goal of linguistic description from an object independent of man to a human capacity' (1971, p. 92), but within Chomskian linguistics it was 'an advance more nominal than real, since the "human capacity" was immediately treated as an object' (1970, p. 71). Approaching language from the point of view of communicative competence leads to a very different theoretical system through which to conceptualize language (see, for example, Hudson 1980). Here we will focus on the empirical findings coming from this approach.

8.3.2 *Sex differences in speech*

Haas (1944) reports differences in the speech of men and women in Koasati, an American Indian language of the Muskogean family used in Louisiana. These differences were disappearing at the time the research was carried out. Some examples are given in Table 8.1. In effect, this system is one of speaker-gender marking, in which grammatical form is matched to the sex of the speaker, rather than referent-gender marking (as is the case in many European languages; he, she, etc.). The Yana language of California shows similar differences, in that different verb endings are required depending on the sex of the speaker (Sapir 1929). Differences of this kind have been reported for a number of other American languages.

Other languages show vocabulary differences, rather than grammatical ones, between male and female speech. Thai, for example, has distinct first-person pronouns: *dichan* ('I') for females; *phom* ('I') for males. An extreme example of vocabulary differences occurs in Australian Aborigines. Haviland (1979) observes that

Aboriginal Australians are celebrated for their highly complex social organization, in which people reckon their relationships to one another largely in terms of kinship. Amidst a complicated calculus of social identities that divided everyone into kin or spouse's kin, into friends, neighbors and strangers, or into elders and juniors, many groups of these original Australians observed elaborate etiquette, treating some classes of people with extreme respect and caution and enjoying unrestrained and often ribald relations with

Table 8.1 Male and female forms in Koasati (Akmajian *et al.* 1984)

women's form	men's form	
lakáw	lakáws	'he is lifting it'
lakáwwitak	lakáwwitaks	'let me lift it'
mól	móls	'he is peeling it'
í:p	í:ps	'he is eating it'
tačílw	tacílws	'you are singing'

others. Not surprisingly, this social complexity is mirrored in correspondingly complex speech practices.... Typically, throughout Aboriginal Australia a man was obliged to behave with extreme deference to his wife's mother, or simply to avoid contact with her altogether (1979, pp. 163–4, 211).

When speaking in the presence of his mother-in-law a man would use a special form of language known as *mother-in-law* language. Dixon (1971) describes one such language for Dyirbal in northern Queensland, distinguishing an everyday language variety, Guwal, from Dyalnuy, which had to be spoken in earshot of a mother-in-law. While pronunciation and grammar are similar between the two varieties, vocabulary is entirely distinct, with Dyalnuy having only about a quarter as many words. Similarly, Haviland (1979) describes a *brother-in-law* language in Guugu Yimidhirr spoken in Cooktown, North Queensland, used by a man to speak to his brothers-in-law, father-in-law, and certain other kin. Again, this possesses a restricted vocabulary compared to the everyday variety.

8.3.3 *Sex differences, conservatism, and prestige*

In many languages female speech is more conservative than that of males. Further, the female variety may often be evaluated as 'better', 'more correct' by native speakers. In an early study, Fischer (1958) studied variations in the suffix '-ing/in'' among children in a New England village. He found that '-ing' was used more by girls than by boys. In addition, the higher the socio-economic status of the speaker's parents, the greater the frequency of '-ing'. The form '-ing' may be considered the 'more correct' form of speech for this community, typical of the speech of social groups with high prestige or status. Two other studies (Wolfram, 1969; Trudgill, 1974) are summarized in Fig. 8.2. These deal with the relative occurrence of prestige forms across the spectrum of social classes for males and females in Detroit, USA, and Norwich, UK: the use of post-vocalic 'r' in Detroit; '-ing' in Norwich. In both cases, it is apparent that females use more of the prestigious (or 'correct') form than do males. Greater correctness in female speech, and more slangy, informal male speech have been widely noted (see Thorne and Henley (1975) for a thorough review).

Similar differences in the distribution of prestige forms have been reported which do not fit this male–female dichotomy, and give a clue as to how these differences might be explained. Labov (1972*a*), for example, reports that low-prestige forms of pronunciation were increasing in Martha's Vineyard, USA, among groups who might describe themselves as of high social class or prestige, the tendency being most apparent amongst males. Martha's Vineyard is a relatively isolated island off New England that has recently become a tourist Mecca. Trudgill describes the linguistic changes occurring there as follows:

Natives of the island have come to resent the mass invasion of outsiders and the change and economic exploitation that go with it. So those people who most closely identify with the island way of life have begun to exaggerate the typical island pronunciation, in order to signal their separate social and cultural identity,... This means that the "old-fashioned" pronunciation is in fact most prevalent amongst certain sections of the younger community.... The tendency is most marked amongst young people who have left to work on the mainland *and have come back*—having rejected the mainland way of life. It is least marked amongst those who have ambitions to settle on the mainland (1983, p. 23).

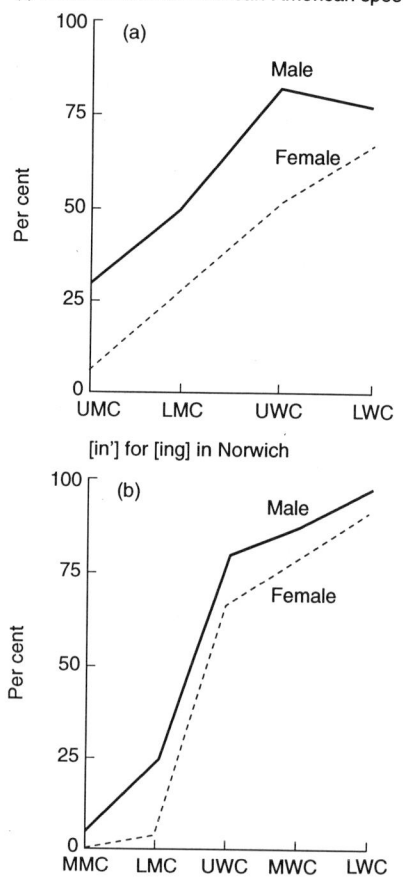

Fig. 8.2 Sex differences in speech pronunciation. (a) after Wolfram 1969; (b) after Trudgill 1974, from Downes 1984). UMC, Upper middle class; LMC, lower middle class; UWC, upper working class; LWC, lower working class.

Thus, it is not so much the male–female distinction that lies at the root of male–female differences in speech as *differential access to social power*. Trudgill concludes that:

> Linguistic sex varieties arise because ... language ... is closely related to social attitudes. Men and women are socially different in that society lays down different social roles for them and expects different behaviour patterns from them. Language simply reflects this social fact. ... What is more, it seems that the larger and more inflexible the difference between the social roles of men and women in a particular community, the larger and more rigid the linguistic differences tend to be. Our English examples have all consisted of *tendencies* for women to use more "correct" forms than men. The examples of *distinct* male and female varieties all come from technologically primitive food-gathering or nomadic communities where sex roles are much more clearly delineated (ibid., p. 88).

Male–female differences may thus be considered an accidental subcategory of more general differences in speech which can be accounted for by reference to social status.

8.3.4 Language styles in socially-stratified language communities

Fischer's (1958) finding on the differential use of '-*ing*/*in* ' is only one of a number of studies that have shown a variation in the use of 'prestige' or 'received pronunciation (RP)' forms within socially-stratified societies. The majority of these stem from Labov's (1966) establishment of an appropriate fieldwork methodology.

8.3.4.1 Rules of address

An Englishman can be variously called *sir, Mr Smith, Smith, Peter, Pete, mate*, or plain *you*. Which of these he is called depends both on the relationship between himself and the speaker, and also on the context in which they converse. This problem of address is complicated in other languages in which different forms of pronouns are available. *Thee* and *thou* are archaic in English, having been replaced by a universal *you*: however, polite and formal second-person singular pronouns are differentiated in many languages: for example, French, *tu, vous*; Spanish, *tu, usted*; Norwegian, *du, ni*; Greek, *esi, esis*; Russian, *ty, vy*. Following the French vocabulary, this distinction is referred to as 'T/V'. (See Brown and Gilman 1960, and Brown and Ford 1961, for the classic investigations of the rules of use for such forms. They argue that the position a speaker perceives his listener to occupy relative to him- or herself on the dimensions of social *power* and *solidarity* determines the choice of T/V pronouns.) In some languages, the shades of meaning to be observed are more complex. In Korean, for example, a speaker has to decide between 6 different verb suffixes depending on his or her relationship to the addressee: intimate, -*na*; familiar, -*e*; plain, -*ta*; polite, -*eyo*; deferential, -*supnita*; and authoritative, -*so*. Similar patterns have been found in other languages (see Murray (1983) for a review of these data). Trudgill (1983, p. 103–5) summarizes the research on the possible historical establishment of these differential uses.

8.3.4.2 Differences between speakers from different social groups

Trudgill (1974) studied the following three speech features in Norwich, England:

(1) the percentage of '-*n*' v. '-*ing*' in verb endings (for instance, 'walkin' ' v. 'walking');
(2) the percentage of *glottal stops* as opposed to *t* (for example, in 'butter'); and
(3) the percentage of *dropped h*'s as opposed to *h* (as in ' 'ammer' v. 'hammer').

His results are shown in Table 8.2.

The pronunciations of these three consonants are quite clearly correlated with different social classes. Comparable results for different phonemes are reported from New York (Labov 1966, 1972*b*), Detroit (Wolfram 1969), and Belfast (Milroy and Milroy 1978), for example.

Differences also occur in speech addressed to listeners in different social groups. While this is generally a result of institutionalized differences in social status and power—for example, Hymes (1972) describes the case of the Abipon of Argentina, who add

Table 8.2 Class differences in the pronunciation of three consonants in Norwich, England. Figures give the percentages of speakers sampled who deviated from British high-social-status 'Received Pronunciation' ('RP') (Trudgill 1983)

Social class	Percentage of speakers deviating from RP for each consonant specified		
	η(ng)	t	h
MMC	31	41	6
LMC	42	62	14
UWC	87	89	40
MWC	95	92	59
LWC	100	94	61

MMC: Middle middle class; LMC: lower middle class; UWC: upper working class; MWC: middle working class; LWC: lower working class

'-in' to the end of every word addressed to a member of the warrior class—some social groups are rather peculiarly drawn. For example, Sapir reports (1915) that the Nootka Indians of Vancouver Island use special word forms when talking to 'children, unusually fat or heavy people, unusually short adults, those suffering from some kind of defect of the eye, hunchbacks, those that are lame, left-handed persons and circumcised males'.

8.3.4.3 *Differences between speakers of different ages*

If linguistic change is occurring in the pronunciation of a variable, then normally it is most pronounced in the youngest speakers. Figure 8.3 (from Chambers and Trudgill 1980) shows this for the pronunciation of *e* in Norwich. Here the higher the index score for each group, the greater is their use of a centralized pronunciation, the change being studied. Note that the rate of change is most apparent in the casual speech of those under thirty, which most closely reveals the basic form of everyday vernacular speech. Hockett has claimed that

the blood-and-bone of many languages is transmitted largely through successive generations of four-to-ten-year-olds: the fires of childhood competition and the twists of childhood prestige do more to shape a given individual's speech patterns, for life, than does any contact with adults (1958, p. 361).

Labov (1972*a*) advances a similar argument in more detail.

Differences also occur in a speaker's production depending on the age of the addressee. It seems likely that *every* language has special forms of talk, called 'motherese', for addressing babies. Ferguson (1971) has suggested that some features of motherese may be universal, such as the absence of inflections and copula verbs where they would normally occur in adult speech. Changes in speech style to older people, except where they are required by explicit social convention, generally result from issues of differential social power, and, doubtless, presuppositions about their deafness or senility.

8.3.4.4 *Differences within an individual's speech depending on social context*

Figure 8.3 further illustrates that a speaker's pronunciation also varies regularly depending on the formality of the situation within which speech is produced. Such findings have led to the concept of *speech registers*, a person having command of a number of varieties of his or her language which are appropriate to different social contexts (Halliday *et al.* 1964; see also Gregory and Carrol 1978). Registers are generally characterized by changes in vocabulary, although it does not seem inappropriate to extend

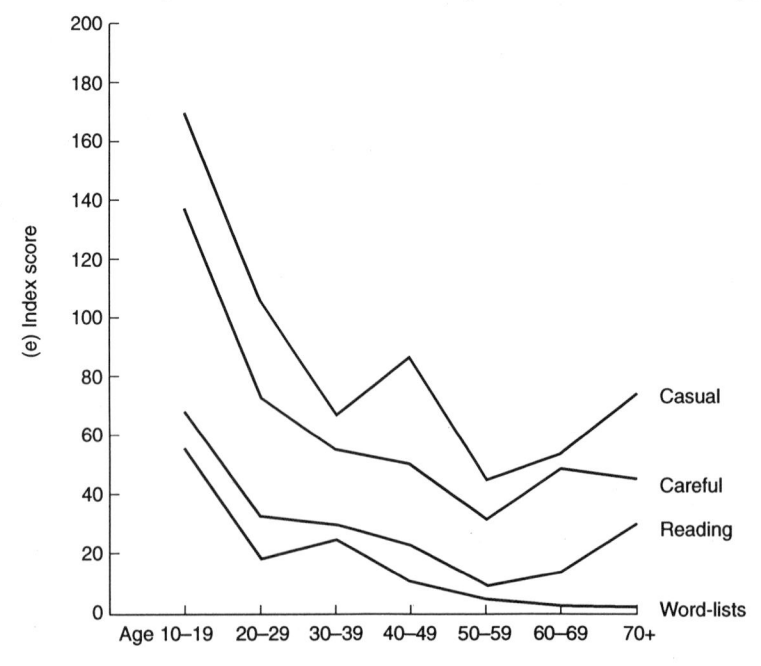

Fig. 8.3 The pronunciation of *e* in Norwich, England, by age and task: casual speech amongst peers, etc.; careful speech to interviewer; reading aloud from prose text; reading word-lists. The higher the index score, the greater the centralization of pronunciation (after Chambers and Trudgill 1980; from Downes 1984).

the term to cover more general differences that can occur.

It is important to note that both in this case and in those discussed above, any particular speaker only switches between the two forms being compared in a statistically predictable pattern. That is, one cannot predict on any one occasion exactly how a speaker will speak, but one can show that if he or she was of a certain age, social class, and sex, then one variant would be used approximately x per cent of the time, on average, in a given situation. This proviso should be born in mind for all the examples, in that often the apparently clear-cut differences noted reflect a continuum of differences rather than distinctly stratified forms of speech.

8.3.4.5 *Institutionalized language contexts*

In contrast to the rather implicit system of rules defining the linguistic selection of pronouns, vocabulary, register, and pronunciation, some linguistic communities define quite explicitly the situations in which different forms are to be used. Ferguson (1959) introduced the term *diglossia* to refer to the occurrence in countries such as Greece and Switzerland of two distinct versions of a language, a 'high' and 'low' variety, respectively *Katharevousa* and *Dhimotiki* in Greece, and *Hochdeutsch* and *Schweizerdeutsch* in Switzerland. Generally, high varieties are used in formal situations, low varieties in informal ones. In Paraguay, Fishman *et al.* (1971) report that the high and low varieties are Spanish and Guarani, an Indian language unrelated to Spanish. Here the same distinctions between formality and informality hold, but in informal situations factors related to social power and solidarity, such as intimacy, also come into play: it is claimed, for example, that courting couples begin in Spanish, later shifting to Guarani as the relationship moves from formal to informal.

8.3.5 *Summary*

In general, four main findings result from the above studies:

1. Differences occur *between* speakers from groups of different social status; *language thus serves to mark social status.*
2. Differences occur *between* speakers of different sexes *within* particular groups (see above); *language thus serves to mark gender.*
3. Differences occur *between* speakers of different ages *within* particular groups; *language thus serves to mark age.* (These three functions of language variation may be subsumed under the more general category of *differential social power*.)
4. Differences occur *within* the speech of different speakers depending on the context in which that speech occurs; *language thus serves to mark social contexts*.[1]

8.4 The significance of linguistic and non-verbal variation

It is quite clear that language, non-verbal communication, and body-adornment systems show empirically-discernible intrasocietal variations that serve to mark different social roles. From an evolutionary perspective, such findings give us a basis for interpreting some aspects of the archaeological record as indicative of marked role-differentiation within societies. But very little of these systems is recoverable, since little in them can leave physical traces. There are, however, aspects of socio-linguistic and allied intra- and cross-cultural work that do suggest the possibility of more substantial evolutionary implications. Here we are thinking of two particular sources and their potential interrelations: work which attempts to link forms of communication to cognition; and work on the communication of cultural knowledge from one generation to the next. This is the work we will be addressing in the next few sections.

We will be dealing with quite separate traditions of research, which to date show almost no cross-citation of each other. It is our claim, however, that there is some fundamental commonality between these traditions. It seems to us that this commonality can be apprehended via the theoretical formulation offered by Basil Bernstein (see Section 8.8). Bernstein's work is an example of 'grand' theorizing. Consequently, it is difficult to summarize, and easy to misrepresent. By contrast, other workers have independently formulated arguments that share features with Bernstein's scheme without having the same sweep to make them so complex. As a result of this, we have often found it necessary in the following to use some studies for a number of purposes simultaneously: for example, to make some substantive point *and* to provide an interpretation that facilitates an outline of Bernstein's position *and* to provide a context to introduce other work that bears on the immediate issues raised. This does not make for the best of chapter structures.

The organization we have opted for is as follows. Our aim is to point out possible relationships between social structure, communicative systems, cognition, and the transgenerational transmission of social structure, communication, and cognition. We begin by outlining some of the general parameters that have been put forward as determinants of social organization and socialization practices. Secondly, we go on to consider proposed relations between communication, social structure, and cognition, and at the same time spell out the logic on which these claims are founded. We do this by focusing on specific studies by Fischer, and use them to introduce further studies of these relationships. This leads us thirdly to an outline of Bernstein's theoretical position. In turn, Bernstein's concern with social relations and the process of their reproduction from one generation to the next, provides a springboard to specific studies of the communicative socialization of cognitive abilities. In addition, at this point we pick up on a separate tradition of research pioneered by Witkin concerning 'psychological differentiation', one that draws work on socialization practices into the arena of the general determinants of social structure itself, thus taking us full circle.

8.5 Ecological determinants of cultural patterns

Our principal aim in what follows is to establish possible relationships between social factors and communicative and cognitive styles. To begin, we review work that has set out to demonstrate the effects of ecological factors on the structural complexity and practices of societies. A first task of such efforts has been in constructing systems of measurement of such factors as 'social complexity' and 'social differentiation' from basic social statistics.

Murdock (1957, 1967) has collated ethnographic information on 862 distinct societies world-wide. This information is most accessible in the 1967 report. These societies have been 'collapsed' to 186 differentiated culture-areas (Murdock and White 1969), for which Murdock and Morrow (1970) and Barry and Paxson (1971) have provided further detailed codings, on subsistence economies and supportive practices, and on infant-and child-rearing practices respectively. From such data a number of workers have investigated the inter-relation of a variety of social variables that bear directly on issues of communication. The importance of two issues in particular will become apparent in subsequent sections: social structure and child-rearing practices. Berry (1976) reviews most of the classic work in this field (see also Naroll 1970). Here, then, we just give a brief summary under three headings.

8.5.1 *Social differentiation and its measurement*

'Differentiation' is a nebulously appealing concept that has a long history in anthropological thought (see Berry 1976, pp. 21–3 for further details), perhaps because of its inclusion in one of the paradigmatic definitions of evolution: 'Evolution is a change from a state of relatively indefinite, incoherent homogeneity to a state of relatively definite, coherent heterogeneity, through continuous differentiations and integrations' (Spencer 1864, p. 216). Thus, with respect to social organization, Naroll (1956, p. 687) states: 'that society is the most evolved which has the highest degree of functional differentiation'. This view has a common-sense appeal, but raises questions concerning how social differentiation might be measured.

Naroll (1956), Freeman and Winch (1957), and Carneiro and Tobias (1963) offer some of the earliest attempts to construct scales of measurement, drawing on the data-base being compiled by Murdock (1967). In general, such scales convert information on political integration, social stratification, religious leadership, and the like into indexical scales. Marsh (1967) offers a ranking of 467 societies from Murdock's listing, along with 114 contemporary national societies.

For 'simple' societies, two indicators of social differentiation are given an ordinal ranking, which is then summed to yield an Index of Differentiation. Modern states are distinguished by the addition of two further indicators: (1) the percentage of males in each society who are in non-agricultural occupations; and (2) gross energy consumption in megawatt hours per caput for one year (see Marsh 1967, pp. 329–36). A comprehensive ranking of all 581 societies in the sample is thus arrived at (ibid., pp. 338–74). Berry (1976, pp. 22–3) concludes that such ordinations show that: 'sociocultural changes have taken place over time, and that the direction of this movement has been toward greater differentiation and complexity.'

However, although there is evidence for a lineal change in differentiation within societies over time, the integration of their elements into a coherent organization has been shown by Lomax and Berkowitz (1972) to be confined to

the central range of this scale (from Amerindian hunters to African agriculturalists); prior to this range, integration of sociocultural elements was relatively high (for example, among African and Australian gatherers), and subsequent to this central range, the integration of elements was relatively low (for example, among classical cultures).... [Hence] the increased diversity of elements is not always accompanied by an increase in their organization or coherence (Berry 1976, p. 23).

The implications of these two trends are not entirely clear. In our view, a number of factors are involved. Firstly, the degree of integration of elements within a culture probably relates to the length of time that culture has possessed its current level of social differentiation. That is, the social structure of a relatively undifferentiated culture may have been stable for a long period of time, allowing the integration of its elements to be accomplished. By contrast, the social structure of a classical culture, as sampled at one particular point, may have had a relatively short history, and may also be transitory. If its social differentiation were to remain stable for a long period, then integration might occur. But in addition to the fact that the integration of items in complex organizations is more difficult than it is in simple ones, we will be pointing later to a second factor of more immediate interest: that as social organizations become more complex they provide their members with a 'cognitive technology' of language symbols that increases their 'individuality'. This individuality is associated with changes in the structure and content of symbolic communication, and hence transforms the entire cultural world within which discourse and thought are situated. One result of this is that it becomes more difficult to integrate elements at a social level. Complex societies not only provide people with more opportunities for novel action, they also enable people to exploit those opportunities better, promoting something more conducive to internal anarchy than integration.

8.5.2 Social organization and 'ecological factors'

Drawing from several indices of role-differentiation, there is consistent evidence that societies of low social complexity tend to be hunter–gatherers, with associated nomadism and low levels of food-accumulation, low population-density, and low energy-consumption. By contrast, societies of high social complexity tend to be agricultural and sedentary, with high levels of food-accumulation and population-density and high energy-consumption. Societies between these two extremes tend to lie at intermediate points on indices of exploitive pattern, settlement pattern, and population-density (see, for example, Naroll 1956; Freeman and Winch 1957; Tatje and Naroll 1970; Carneiro 1970; McNett 1970; and Berry 1976). Table 8.3, from McNett (1970), provides an overview of the relations between social differentiation and settlement pattern. Table 8.4, from Berry (1976) after Nimkoff and Middleton (1960), illustrates the relation between exploitive pattern and social stratification.

Nimkoff and Middleton (1960) have further investigated the relation between family structure and exploitive patterns. They distinguish between 'independent' and 'extended' family structure, the crucial distinction being that, in the independent family, 'the head of a

Table 8.3 Differentiation of socio-cultural elements in relation to settlement patterns (from Berry 1976, p. 52)

Settlement pattern	Sociocultural concomitants Economic	Political	Social
Restricted wandering (wander within owned territory)	Personal property primarily for food-getting. Communal ownership of real estate.	Band of related or friendly families headed by advisory leader.	No status differences.
Central-based wandering (part of year sedentary at central base)	Surpluses, if any, not used exclusively by any group.	Leader is community symbol.	Status based on ability.
Semi-permanent sedentary (move village whenever environment is exhausted)	Family land ownership. Surpluses acquired but redistributed. Some villages show specialization in manufacturing.	Clans or moieties generally basis of organization. Headman is agent of community.	Status based upon surplus distribution.
Simple nuclear centred (self-sufficient village)	Private ownership of real estate. Full-time occupational specialization.	Chief with coercive power in a kinship-based system.	Stratification based on property.
Advanced nuclear centred (permanent administrative center)	Larger surplus controlled by the upper class.	Administrative centres with hierarchy controlled by a king. Law and politics supplant kinship organization.	Hereditary classes.
Supra-nuclear integrated (components integrated into state, typically by conquest)	Commercialism, large-scale circulation of goods, much accumulated wealth, taxes.	Absolute power vested in ruler. Government manipulates population. Professional army.	Large lower class with many slaves.

Source: Condensed from McNett (1970), Table 1, p. 873.

Table 8.4 Relationship between exploitive pattern and general social stratification (from Berry 1976, p. 53)

Exploitive pattern	General social stratification	
	Great	Little
Agriculture dominant or co-dominant	274	90
Animal husbandry dominant or co-dominant with fishing, hunting, or gathering	29	14
Fishing dominant or co-dominant with hunting and gathering	31	20
Hunting and gathering dominant	16	57
	340	181

Source: Nimkoff and Middleton (1960), Table 5, p. 221.

family of procreation is neither subject to the authority of any of his relatives nor economically dependent upon them' (ibid., p. 215).

Berry provides a concise summary of their findings:

With respect to relationships between family type and ecological variables, exploitive pattern and family type were clearly related, as were settlement pattern and family type. For exploitive patterns ... in the 260 societies where agriculture was dominant, 64% had "extended" families while in the 54 societies where hunting or gathering was dominant only 22% had "extended" families; in the 225 societies where there was a mixed exploitive pattern, about half (51%) had extended families. For settlement pattern ... in the 410 societies classed as sedentary, 60% had "extended" families. The demonstration that in sedentary agricultural societies there is a consistent tendency for families to be "extended" to be elements in a stratified social, economic, and familial authority system, may have great import for the techniques of socialization which are employed in such families and societies (1976, p. 54).

8.5.3 Socialization practices and exploitive pattern

Barry *et al.* (1957) devised a method, which they later applied in cross-societal comparison (Barry *et al.* 1959), for assessing socialization practices for their tendency to promote assertive versus compliant outcomes in children who underwent them. Table 8.5 lists their results for two comparisons, between societies at the extremes and in the middle range of food-accumulation practices. The results show a significant relationship between accumulation and child-rearing practices. Whiting (1968) pointed out that this relationship may in fact be attributable to differences between major world regions, which are somewhat confounded, in this sample, with accumulation prac-

Table 8.5 Relationship between 'compliance versus assertion' socialization emphases and degree of food accumulation (from Berry 1976, p. 59)

Compliance versus assertion	Extremes in accumulation		Intermediate in accumulation	
	High (animal husbandry)^n	*Low (hunting fishing)^n*	*High (agriculture only)^n*	*Low (agriculture and hunting)^n*
above median	20	3	14	6
below median	4	19	1	12
$X^2 = 19.6$	df = 1		$X^2 = 9.95$	df = 1
$p < .01$			$p < .01$	

Source: Extracted from Barry *et al.* (1959), Table 2, p. 60.

Table 8.6 Ratings of cultures for sex differences on five variables of childhood socialization pressure (from Barry *et al.* 1957, p. 328)

Variable	Number of cultures	Both judges agree in rating the variable higher in		One judge rates no difference, one rates the variable higher in		Percentage of cultures with evidence of sex difference in direction of		
		Girls	Boys	Girls	Boys	Girls	Boys	Neither
Nurturance	33	17	0	10	0	82%	0%	18%
Obedience	69	6	0	18	2	35%	3%	62%
Responsibility	84	25	2	26	7	61%	11%	28%
Achievement	31	0	17	1	10	3%	87%	10%
Self-reliance	82	0	64	0	6	0%	85%	15%

tices. However, Barry (1969) took 12 matched pairs of closely-related societies and found the relationship still held.

Barry *et al.* (1957) report sex differences in compliance and assertion training across a range of cultures (see Table 8.6). The results confirm a widespread pattern of greater pressure toward compliance in girls (nurturance, obedience, and responsibility) and assertion in boys (self-reliance and achievement-striving); but there is sufficient variation to confirm the cultural rather than biological nature of these differences. With respect to cultural variables, large sex differences in socialization are associated with (1) 'an economy that places a high premium on the superior strength ... [of] the male'; and (2) 'customs that make for a large family group with high cooperative interaction' (Barry *et al.* 1957, p. 330).

Both the above sets of findings can be argued to reflect the broad ecological variables impinging on these different societies. Barry *et al.* (1959, p. 53) put the case thus:

In societies with low accumulation of food resources, adults should tend to be ... assertive and venturesome. By parallel reasoning, adults should tend to be ... compliant and conservative in societies with high accumulation of food resources. ... We may expect the training of children to foreshadow these adaptations ... the emphases in child training will be toward the development of kinds of behavior especially useful for the adult economy.

8.5.4 *Summary*

Role-differentiation, social stratification, and child-rearing practices show interrelationships amongst themselves, and with some of the basic ecological parameters of a society. These relationships are summarized in Table 8.7. We will be exploring in subsequent sections how these factors relate to the communication process.

8.6 Language, thought, and social structure in a cross-cultural perspective

Fischer's (1965, 1966) studies of Truk and Ponape communication provide a paradigm example of work trying to connect cognition and social structure: they show a seemingly large chasm between data and argument. A moment's thought, however, will reveal the problems involved in gathering convincing data. Further, there is at present no equivalent 'taxonomy of ideas' to that which exists for social structures (see last section). Intuitions that there are links in this sphere have led to a theoretical elaboration that looks promising and testable. Thus the field is in an early, somewhat speculative stage. We outline Fischer's work at some length because of its typicality.

8.6.1 *Fischer's work on Truk and Ponape language and social structure*

Both the languages and cultures of Truk, a complex of small islands within a large lagoon, and Ponape, a more remote, larger single land-mass some 400 miles away, are historically related. They have shown a separate development for perhaps two thousand years (Fischer 1966, p. 168). Fischer's claim is that 'the differences which have arisen in the two related languages are produced by certain long-standing differences in general features of the social structure of the two speech communities' (ibid., p. 168). Ponapean social structure is more differentiated than Trukese, possessing a greater variety of significantly different roles. These differences are most pronounced in the spheres of kinship and politics, and are not so obvious in the sphere of technology, perhaps because of the recent influence of Western powers. In sum:

the total effect of the political and kinship systems of the two cultures is to produce one society (Ponape) where individual mobility, the uniqueness of the status of each individual, and his relative ranking compared to others are recognized explicitly and emphasized, as against another society (Truk) where change of status is little affected by personal effort, where unique individual status is of little concern, and where relative rank among individuals of the same age grade is minimized (ibid., p. 176).

8.6.1.1 *Fischer's theoretical perspective*

Fischer (1966; see also 1973) outlines a very clear position for interpreting the relations between language structure, social structure, and, to some extent,

Table 8.7 General overview of elements in ecology and traditional culture components (from Berry 1976, p. 60)

Components and elements	Clustering at poles of dimension	
Ecology		
Exploitive pattern	Hunting and gathering	Agriculture
Food-accumulation	Low	High
Settlement pattern	Nomadic	Sedentary
Mean size	Low	High
Culture		
Role diversity	Low	High
Socio-cultural stratification	Low	High
Socialization emphases	Assertion	Compliance

thought. His views may be said to be founded on the concept of *shared presuppositionality*; how this varies among speakers in different social systems; the effects this has upon language; and the consequence this has for the elaboration of conceptual processes within the social or cultural group.

In the typical simple society, individual personality differences ... tend to be restricted to expressive behavior. In addition, differences in practical roles and practical cultural behavior are much less significant than among ourselves. In matters of everyday life, then, it becomes possible for a member of such a society to share much more of his knowledge with his fellows.... In such a situation abbreviated ways of speech are favoured for communication about frequent shared events, and what are frequent events for anyone in the society are in fact frequent for others as well. Where several words might be required to express an idea to an outsider, only a single word might be required for a member of the in-group.... Words and utterance become concrete in the sense that they express and communicate much detailed information with formal economy (Fischer 1966, pp. 177–8).

An example of this is provided by Lee (1959, p. 90):

If I were to go with a Trobriander to a garden where the taytu, a species of yam, had just been harvested, I would come back and tell you: "There are good taytu there; just the right degree of ripeness, large and perfectly shaped; not a blight to be seen, not one rotten spot; nicely rounded at the tips, with no spiky points; all first-run harvesting, no second gleanings." The Trobriander would come back and say "Taytu"; and he would have said all that I did and more.

This situation can be contrasted with that of more complex societies in which social roles have become more differentiated, where

the realized meaning of words in particular contexts becomes less important than the common or basic meaning. Speakers are forced to assume a greater cognitive gap between themselves and their listeners. At the same time, the basic meaning of the items of the lexicon tends to become more abstract and attenuated, since speakers have less need for words which can express much meaning in compact form to listeners who are conceived of as being much like the self; they have more need, instead, for words which can be used in many different contexts with many different listeners who are conceived as being different from the self and from each other.... Certain common propositions in the complex society may now take longer to utter—their expression may be formally more complex—but at the same time there is a counterbalancing gain in flexibility and explicitness (Fischer 1966, p. 178).

In addition to the effect which social structure has on the process of communication, Fischer argues that there are also consequences for 'general habits of thought' (ibid., p. 176):

the differences in habitual thought patterns between Truk and Ponape may be expressed in terms of several related polar constructs, respectively concrete versus abstract, focal versus lineal, and stimulus-determined versus goal-directed or purposive (ibid.).

This is neither a claim that thought is determined by language and/or social structure, nor that abstract thought is precluded by one system rather than another, but that, in everyday life, 'the average level of abstraction of thought in some cultures is lower than in others, as it is for some individuals within a single society' (ibid.). Attempts to make this general claim more rigorous (for instance Hallpike 1979) are beset with problems (see, for example, Harris and Heelas 1979). How, for example, does one establish an 'average level' without a scale by which to measure 'abstraction'?

Fischer's reasoning is as follows: small, simple, homogeneous societies do not, as we noted above, put a pressure on meaning to be coded in context-independent terms—their characteristics tend to lower the need for the exercise of abstract problem-solving skills. That is, lacking skilled specialists, 'every member of the society ... must have the most useful and time-tested answers to all the most frequent problems of daily living in his own head ...'. Essentially, then, members of such a culture may rely more on traditional knowledge, rule-of-thumb common-sense solutions from which general algorithms need not have been formally abstracted. Further, it is argued that many everyday practices have not been subjected to analysis for why they work, nor does everyday life throw up many situations which require analysis for their solution. An important implication here is that such everyday practices may be acquired by imitation. By contrast, in a more complex society, active problem-solving will be more encouraged, and successful individuals will have mastered cultural skills that can be acquired only through the abstraction of general principles, and not solely through the process of imitation. In this way,

the behavior of the man from the small society is more stimulus-determined ... Purposive, goal-directed behavior is more emphasized among Ponapeans, especially since their political system offers considerable scope for choosing alternate major goals and then choosing between a variety of often devious courses to achieve these goals (Fischer 1966, p. 177).

8.6.1.2 *Fischer's data*

Fischer presents data on the relation between the phonological systems (1965) and the grammatical systems (1966) of Trukese and Ponapean in relation to the social structures of the two cultures that are quite technical, and so details must be sought in the original papers. The main claim is that in Ponapean, both systems show structural characteristics that reflect a greater emphasis on precision, both in speech and in the conveyance of the meanings being spoken. By contrast, Trukese phonology is commensurate with greater fluency, and its grammatical system is less adapted in use to context-independent precision in meaning. In addition, for technical reasons of phrase-construction, a Trukese speaker is more liable to be interrupted; but since within Truk's simple social structure there is a high degree of shared presuppositionality, this creates no great barrier to the communication task. Trukese speech is claimed to be relatively stimulus-directed, being

> treated more as a response to an earlier utterance or to something in the non-linguistic situation than as a purposeful attempt to influence the listener in accordance with a preconceived goal of the speaker. The tighter Ponapean construction offers less encouragement to interruption. It suggests that the speaker has a definite idea in mind which must be communicated in full as a unit to the listener. It suggests further that the speaker assumes the listener to be perhaps quite different from himself and liable to misinterpret fragments of the full proposition (Fischer 1966, p. 179).

This sort of speech Fischer regards as relatively more goal-directed.

Grimshaw (1973, pp. 74–8) provides an accessible source and a critical review of Fischer's work. Consequently, the question we may ask at this point is whether it is possible to establish a general relationship across cultures between linguistic and social structures. This question is not new, and was answered in the affirmative by a number of nineteenth-century philologists, but with arguments that rely more on the smug presuppositions of the 'civilized supremacy' through which they interpreted 'primitive peoples' than any hard evidence. This early work has led to the question's becoming a tainted one, much as theories arguing for the determination of conceptual categories by the semantics of individual natural languages (especially the Whorfian hypothesis: see Lucy 1992 for a full review) and the effect of literacy upon cognition (see Chapter 27) have been given short shrift. As Hymes notes:

> the present temper ... treats mention of differences as grounds for suspicion of prejudice, if not racism, so that poor Whorf, who believed fervently in the universal grounding of language, and extolled the superiority of Hopi, has become, like Machiavelli, a pejorative symbol for unpleasant facts to which he called attention (1973, p. 78).

Such sentiments lead to a paucity of enquiry, for proposals such as those put forward by Jesperson (1894) that there *is* a relationship between the degree of linguistic synthesis and social complexity (see below) have generally been taken as universal statements, and refuted by the judicious choice of counter-examples, rather than seriously tested by appropriate statistical methods. We have, in fact, only come across two such studies.

8.6.2 *Greenberg's analytic–synthetic ordination*

Greenberg (1960) proposed a numerical scale for ranking languages as synthetic or analytic, in which the index of synthesis of a language is given by the number of morphemes in a sample of texts divided by the number of words. The more synthetic a language, the higher its index score, and the less needs to be spoken to convey specific information about objects, processes, etc. Consequently, the simplest *linguistically expressed* concepts of speakers of these languages are necessarily more dense, less 'unpacked', than those of speakers of low-index languages. Kroeber (1960, p. 175 ff.) calculated indices for a very small number of languages that tend to support the hypothesis that synthetic languages occur in simpler societies:

Eskimo	11.33
Sanskrit	7.51
Swahili	7.06
Yakut	7.02
Anglo-Saxon	6.42
English	5.34
Modern Persian	4.88
Annamese	3.13

We need, however, to be cautious in accepting this interpretation, not solely because of the small numbers involved. For example, scientific language would doubtless obtain a high index of syntheticity, but it is difficult to know whether the scientific community should be regarded a simple society. In some senses this would be legitimate, for in a scientific community the individual members share a high degree of presuppositionality about each other (there is massive shared knowledge); and given that one characteristic of empirical work is its replicability by other members of the community who, by definition, possess the same skills, there is actually little role specialization. This interpretation would fit with the general view of language and social structure we are developing here. It does, though, alert us to the problems of moving from linguistic concepts to claims about thought processes, for while there may be some apparent logic in claiming that synthetic language in simply structured societies restricts analytic thought, it is not intuitively obvious that scientists show such a characteristic.

8.6.3 *Colour-term differentiation and social structure*

Ember (1978) has suggested that the elaboration of domains of reference generally is a factor of societal complexity. Armstrong and Katz (1981) pursue this claim for one domain. They note that 'it is possible to talk about the relative elaborateness of systems of reference where the underlying *etic*[2] structure of the referential domain is well worked out' (ibid., p. 334), and that this is the case for colour terminology. They test the hypothesis that colour-term differentiation is related to social complexity for 37 societies (the process by which these societies were selected is given by them in detail, ibid., pp. 335–6). The number of colour terms given for the language of a society by Berlin and Kay (1969) was found to be correlated with the index of societal differentiation calculated according to Marsh's (1967) criteria.

Societal complexity as measured by Marsh's index was found to be a moderate predictor of the number of basic colour terms in a language's lexicon (see Table 8.8). This is congruent with the hypothesis of Berlin and Kay (1969, p. 104) that

color lexicons with few terms tend to occur in association with relatively simple cultures and simple technologies, while color lexicons with many terms tend to occur in association with complex cultures and complex technologies.

Further, if one assumes that the elaboration of other domains of reference is indexed by colour lexicons (and there are important exceptions to this, as many domains of reference are not comparable across cultures, since members of different societies have fundamentally different things to talk about, and this is reflected in their systems of reference (see Boas 1911); for example, Eskimos have a large referential domain associated with snow) then: 'this is also a confirmation of the hypothesis that referential systems become more elaborate as societies become more complex' (Armstrong and Katz 1981, p. 337) [cf. Naroll 1970].

However, Esther Goody reports (personal communication) that this is not the case for a number of West African languages. In these, the number of colour terms relates not to societal complexity, but to the length of a society's trade network. This suggests that it is not societal complexity *per se* which relates to the elaboration of linguistic structures, but, again, *the level of shared presuppositional knowledge between speakers*: one needs be specific with a supplier several hundred miles away. It is quite obvious, though, that levels of shared knowledge between individuals relate to societal complexity in a quite direct way.

Beyond these explicit studies, further indications of links between language, social structure, and thought are somewhat tenuous. One promising area of research has been pioneered by Bloom's (1981) investigation of the relationship between cognition and linguistic structure for Chinese and English speakers.

Table 8.8 A stratified random sample of languages/societies, the number of their basic colour terms (Berlin and Kay 1969), and their Marsh (1967) indices of differentiation (from Armstrong and Katz 1981, p. 338) [cf. Naroll 1970]

language/society	number of basic colour terms	index of differentiation
1. Arawak (Locono)	3	1
2. Pomo	3	2
3. Poto	3	2
4. Toda	3	1
5. Hanunoo	4	0
6. Ibo	4	3
7. Mende	4	5
8. Somali	4	4
9. Arunta	4	1
10. Pukapuka	4	1
11. Batak	5	1
12. Kung	5	1
13. Eskimo (Caribou Eskimo)	5	0
14. Hopi	5	1
15. Navaho	5	1
16. Songhai	5	6
17. Tarasco	5	6
18. Hausa	6	7
19. Masai	6	1
20. Plains Tamil	6	7
21. Mandarin (Taiwan)	6	32.4
22. Bari	7	3
23. Javanese (Sumatra)	7	7
24. Siwi	7	3
25. Dinka	10	2
26. Nandi	8	1
27. Zuni	11	1
28. Arabic (Iraq)	11	26.6
29. Bulgarian	11	23
30. Cantonese (Taiwan)	8	32.4
31. Hungarian	12	36.8
32. Korean (N. and S. Korea)	12	14.7
33. Malay (Malaysia)	10	26.3
34. Tagalog (Philippines)	10	20.9
35. Thai (Thailand)	10	13.7
36. Urdu (Pakistan)	8	16.7
37. Vietnamese (N. and S. Vietnam)	9	16.8

$r = 0.64$, $p < 0.005$

8.6.4 *Chinese and English grammatical structures and problem-solving*

Bloom (1981) argues from experimental investigations that Chinese speakers find some problems more difficult to deal with than English speakers *because the Chinese language lacks explicit grammatical structures that are present in English*. This lack makes it less easy for a Chinese speaker to 'grasp' the problem, and the process of solution is thus hindered. This is perhaps akin to claiming that certain mathematical problems are difficult to solve in the absence of calculus. However, Bloom's claims appear to go further than this in suggesting that, as a result of the lack of these grammatical structures, Chinese speakers conceive of the world in a different way than English ones. In effect, English is more abstract than Chinese conceptualization, which is more rooted in a concrete base-line of reality. Bloom offers arguments from three areas.

8.6.4.1 *Counter-factuality*

Chinese language has no distinct lexical, grammatical, or intonational device to signal entry into the counter-factual realm, to indicate explicitly that the events referred to have definitely not occurred and are being discussed for the purpose only of exploring the might-have-been or the might-be (ibid., p. 16).

This does not mean that Chinese speakers do not think or speak counter-factually in certain situations, for they do. To resolve this apparent contradiction, Bloom proposes that:

a language, by whether it labels or does not label any specific mode of categorizing experience, cannot determine whether its speakers will think that way, but can either encourage or not encourage them to develop a labelled cognitive schema specific to that mode of thought. Even if the English language did not label the notion "bachelor", English speakers could still understand the concept "bachelor" by bringing together in their minds its component elements—unmarried, never previously married, male and adult. But the fact that the English language has a distinct label for bachelor seems to encourage its speakers to develop a schema specifically designed to categorize the world in that way, . . . Analogously, the fact that the English speaker has a distinct label for the counterfactual (i.e. 'had . . . would have'), which the Chinese speaker does not share, cannot by any means be expected to bestow upon the English speaker an exclusive facility for that mode of thought, but it might be expected to encourage him or her, by contrast to his or her Chinese counterpart, to develop a cognitive schema specific to that way of thinking (ibid., pp. 20–1).[3]

8.6.4.2 *Generics*

Chinese does not have any direct means to specify generic concepts, that is, theoretical entities abstracted from the world of actualities. Thus,

perhaps the fact that English has a distinct way of marking the generic concept plays an important role in leading English speakers, by contrast to their Chinese counterparts, to develop schemas specifically designed for creating abstracted

theoretical entities . . . and hence for coming to view and use such entities as supplementary elements of their cognitive worlds (ibid., p. 36).

8.6.4.3 *Entification*

English grammar readily allows its speakers the entification of adjectives and verbs into nouns: from 'red' to 'redness'; 'hard' to 'hardness'; 'to accept' to 'the acceptance of'. This transformation

signals movement from description of the world as it is primarily understood in terms of actions, properties and things, to description of the world in terms of theoretical entities that have been conceptually extracted from the speaker's baseline model of reality and granted, psychologically speaking, a measure of reality of their own (ibid., p. 37).

Chinese lacks any mechanism 'by which to shift to the more theoretically extracted categorization of the world that entification entails' (ibid., p. 38).

8.6.4.4 *Experimental work*

Bloom reports a series of experiments which show how these grammatical differences affect problem-solving between speakers of the two different languages (for the status of these findings, see the discussion between Au (1983, 1984) and Bloom (1984); also Liu (1985) and Takano (1989); for a detailed analysis see Lucy (1992)). An example involving counterfactuals will have to suffice here:

Samples of 173 Taiwanese subjects and 115 American subjects responded in their native language to the following question:
'If all circles were large and this small triangle "∆" were a circle, would it be large?'
Forty-four or 25% of the Taiwanese subjects answered "yes" by contrast to 95 or 83% of the American subjects—a difference significant at $p = 0.00001$ (ibid., p. 31).

From such studies, Bloom goes on to note that

to explain historically why counter-factual and entificational thinking did not develop on a general scale, one would have to look not only to the characteristics of the language but to the social and intellectual determinants of why a perceived need for such thinking did not arise (ibid., p. 59).

but he does not pursue this point. While Chinese social structure had, until this century, perhaps the longest continuous history of any state, and has been remarkably conservative up to that point, and consequently not likely to promote the conditions in which conceptually elaborated or explicit forms of analytic language might flourish (cf. section 7 of this chapter below), it is unlikely that we could pin the responsibility for the lack of counterfactuals and entification on this factor. It is more likely that ideological factors developed within this enduring social stasis contributed to a perceived lack of importance of certain problems, thus not encouraging forms of language that facilitate their handling.

8.7 Self, culture, and communication

There are a number of classical statements on the cultural relativity of human conceptions of being human (for example, Hallowell 1956; Lee 1959). The topic has come in for renewed interest of late (see collections by Marsella *et al.* 1979; Heelas and Lock 1981; Marsella *et al.* 1985; Carrithers *et al.* 1986). Much of this work subscribes to what may be called a *social constructivist* perspective. That is, there is no given way of being a human: both what one is, and how one conceptualizes one's self, is constructed through the behavioural and linguistic resources of the social practices within which one is socialized. Thus, people are not conceived as being everywhere the same, and just adopting different cultural emphases on what aspects of personality are valued or basic, or adopting different vocabularies through which to express a universal commonality of being and experience; rather, what is being referred to through language has itself been created through language and cultural practice (see also Shotter 1984, for a fuller discussion of this view).

In this vein, Harré (1983, p. 87 ff.) argues that the grammatical structure of Eskimo suffix-systems gives us an insight into the sense of personal identity which exists within Eskimo society. Similarly, Lee (1959, p. 134) notes that

a study of the grammatical expression of identity, relationship and otherness, shows that the Wintu [American Indians of Northern California] conceive of the self not as strictly delimited or defined, but as a concentration, at most, which gradually fades and gives place to the other. Most of what is other for us, is for the Wintu completely or partially or upon occasion, identified with the self.

Heelas and Lock (1981, pp. 21–52) suggest that the differing conceptions of self that transpire across human cultures can be usefully located on two orthogonal dimensions: internal–external; and in control–under control. Thus the Dinka of the Sudan conceive their 'selves' as being under the control of agencies located in the external world (Lienhardt 1961); whereas the Maori conceive their 'selves' as in control of them via an internal, inherent agency (Best 1922; Smith 1981). This relativistic, constructivist view of self-conception is itself only now being conceptualized. No studies have been undertaken cross-culturally in an endeavour to relate such concepts to social structure or other ecological factors in an attempt to establish any generalizations.

There are, though, historical studies of the origins of Western self-concepts and emotional concepts that are suggestive of such links. There are, for example, many similarities in the early Indo-European literature, from the *Rig Veda* at its Eastern extremity to the *Ulster Cycle* at its Western one, to that described in the Greek epics by Jaynes (1976a):

there is in general no consciousness in the Iliad . . . There is . . . no concept of will or word for it, the concept developing ominously late in Greek thought. Thus Iliadic men have no will of their own and certainly no notion of free will . . . there is no subjective consciousness, no mind, soul or will in Iliadic men (ibid., pp. 39, 71).[4]

The heroic literature tells of a very different way of being human to ours today. People lacking 'subjective consciousness, mind, soul and will' are effectively not in control of themselves. It is not surprising, then, that there appears to have been much less restraint upon emotional expression. Bloch summarizes these peoples as sharing

the emotionalism of a civilization in which moral or social convention did not yet require well-bred people to repress their tears and their raptures. The despairs, the rages, the impulsive acts, the sudden revulsions of feeling present great difficulties to historians, who are instinctively disposed to reconstruct the past in terms of the rational (1965, p. 73).

But the essential picture of these people that emerges from their literature is that they were not in possession of a sense of self that could be elaborated explicitly in discourse—only implicitly via acts, generally gross, through non-verbal channels, and in the language of boasting. Elias (1978, 194) documents such a form of life as surviving till quite late in Western Europe:

Leaving aside a small élite, rapine, pillage, and murder were standard practice in the warrior society of this time, as is noted by Luchaire, the historian of thirteenth-century French society. There is little evidence that things were different in other countries or in the centuries that followed. Outbursts of cruelty did not exclude one from social life. They were not outlawed. The pleasure in killing and torturing others was great, and it was a socially permitted pleasure.

Elias goes on immediately to make a most important point for our present purposes: 'to a certain extent, the social structure even pushed its members in this direction, making it seem necessary and practically advantageous to behave in this way' (ibid.).

Firstly, this point explicitly alerts us to the role of social structure in fostering this way of life, and prepares us for Elias's subsequent argument that it was changes in the structure of social life that led to a change in the nature of Western being. But secondly, it gives us pause to consider the contexts in which social structures are themselves embedded, so that the changes that we will summarize below as stemming from the pressures that differentiating social roles place upon the linguistic skills required for everyday communication, and the changes in the nature of the self consequent upon them, are also seen to be plausibly motivated by additional mechanisms. That is, in societies with different ecological circumstances to those of early European ones, behavioural restraint may be fostered through other routes. Brutality is viable given a technology of transport and a resourcefull ecosystem: in more sedentary populations at the environmental limits of carrying capacity, restraint may arise by other means than the one outlined here.

Elias's account of the transition from barbarism to civility is complex, detailed, and subtle, and here we can only present a caricature of it (for more detailed summaries, see Dreitzel 1981; Lock 1986). Two quotations convey the essentials of his analysis: 'as the social fabric grows more intricate, the sociogenic apparatus of individual self-control also becomes more differentiated, more all-round and more stable' (1982, p. 234). Elias offers the following explanation for this relationship:

From the earliest period of the history of the Occident to the present, social functions have become more and more differentiated under the pressure of competition. The more differentiated they become, the larger grows the number of functions and thus of people on whom the individual constantly depends in all his actions... As more and more people must attune their conduct to that of others, the web of actions must be organized more and more strictly and accurately, if each action is to fulfil its social function. The individual is compelled to regulate his conduct in an increasingly differentiated, more even and more stable manner (ibid., p. 232).

The factors involved in this transformation are as follows: firstly, the centralization of social power, and the creation of social codes of conduct to be observed through the sanction of punishment; secondly, the specialization of roles within the differentiating society, and the consequent pressure on more explicit forms of communication due to the decline in the number of shared presuppositions; thirdly, the increasing attention an individual must devote to monitoring the effectiveness of his or her communications, and consequently, the attention that must be given to nuance in the presentation of the self, leading to a greater articulation of that self; and fourthly, the increasing number of increasingly context-free problems an individual must attend to, problems removed spatially and temporally from immediate circumstance.

Social differentiation, then, creates pressure for a more articulated mode of communication, a richer vocabulary of linguistic concepts, a reflexive monitoring of self-presentation; effectively, the creation of an articulated self, which has also become open to the socially constructed, but now individually-mediated, emotions of guilt and embarrassment. External control of behaviour by threat of punishment is largely replaced by internalized psychological control by fear of shame or embarrassment. Individuality, 'subjective consciousness, mind, soul and will' have been constituted through the interplay of social and consequent linguistic structures such that the resultant individuality can be expressed explicitly through language, and not be dormant and inarticulate in the non-verbal channels of 'restricted coding' (see discussion of Bernstein's theory below).

8.8 Bernstein's theorizing

The issues of social structure and communication that have been central to the above discussion of cross-cultural research have played a central role in the theorizing of Bernstein (1958, etc.), whose work in the last three decades has largely centred on class-associated variations in linguistic forms within complex modern societies. His ideas are best known for their influence in educational sociology, drawing criticism for the suggestion that a correlation exists between language style, social class, and academic achievement. In concentrating on this aspect of his work, however, Bernstein's critics generally miss the more fundamental and general thesis running—not always explicitly—through it.

At its broadest, Bernstein's theory encompasses the whole field of cultural transmission (of which the educational system is but one channel) via semiotic systems (of which language is the most important—cf. Halliday (1978), who sees the theory as concerning a 'social semiotic' rooted beneath the surface pattern of language, a general theory of the principles and processes of cultural transmission). The theory deals with variation in the content of cultural material and the manner in which it is conveyed between successive generations, and particularly variation in the principles of transmission with respect to social stratification. Recently, Atkinson (1985) has provided an accessible overview to the whole of Bernstein's writings, which we have drawn on here.

In his early papers (for example 1958, 1959), Bernstein was concerned predominantly with the surface features of language, and with the differences in linguistic style of different social classes in the UK. Contrasting styles or 'sociolects' were identified as *formal* and *public*, associated, though not exclusively, with the middle and working classes respectively. Bernstein characterizes these two styles by a listing of their commonly occurring linguistic features

Table 8.9 Characteristics of public and formal languages after Bernstein 1961, p. 169 (from Robinson 1972)

Public language
 1. Short, grammatically simple, often unfinished sentences with a poor syntactical form stressing the active voice.
 2. Simple and repetitive use of conjunctions (so, then, because).
 3. Little use of subordinate clauses to break down the initial categories of the dominant subject.
 4. Inability to hold a formal subject through a speech sequence; thus, a dislocated informational content is facilitated.
 5. Rigid and limited use of adjectives and adverbs.
 6. Infrequent use of impersonal pronouns as subjects of conditional clauses.
 7. Frequent use of statements where the reason and conclusion are confounded to produce a categoric statement.
 8. A large number of statements/phrases which signal a requirement for the previous speech sequence to be reinforced: 'Wouldn't it? You see? You know?', etc. This process is termed 'sympathetic circularity'.
 9. Individual selection from a group of idiomatic phrases or sequences will frequently occur.
10. The individual qualification is implicit in the sentence organization; it is a language of implicit meaning.

Formal language
 1. Accurate grammatical order and syntax regulate what is said.
 2. Logical modifications and stress are mediated through a grammatically complex sentence-construction, especially through the use of a range of conjunctions and subordinate clauses.
 3. Frequent use of prepositions which indicate logical relationships as well as prepositions which indicate temporal and spatial contiguity.
 4. Frequent use of the personal pronoun 'I'.
 5. A discriminative selection from a range of adjectives and adverbs.
 6. Individual qualification is verbally mediated through the structure and relationships within and between sentences.
 7. Expressive symbolism discriminates between meanings within speech sequences, rather than reinforcing dominant words or phrases, or accompanying the sequence in a diffuse, generalized manner.
 8. It is a language-use which points to the possibilities inherent in a complex conceptual hierarchy for the organizing of experience.

(see Table 8.9). The classification of linguistic style into two distinct categories should not be taken as implying that formal and public sociolects are neatly separable and used exclusively by one class—a warning that is common to all sociolinguistic distinctions. They are extreme caricatures—ideal types labelling the poles of a continuously varying scale.

Since language provides the main means of human socialization, the sociolect used in a child's family can have a profound effect on individual development. Those born into an environment of formal language are given access to greater potential for the explicit articulation of subtle ideas and feelings. Moreover, the middle-class child 'is seen and responded to as an individual with his own rights, that is he has a specific social status' (Bernstein 1958, p. 27).

Those born into a community using public language, by contrast, are denied such access, with a number of possible consequences:

Public language use, with its lack of verbal elaboration and explication of meaning and motive is characteristic of a situation where authority is implicit in ascribed roles or positions within the family: here the statuses and roles themselves are, of course, explicit. Such a highly segregated family system will not provide an environment where the sensitive exploration and elaboration of personal intentions is encouraged. Formal language, on the other hand, arises in family settings where complex logical relations are articulated and personal intentions and sensibilities are explored (Atkinson 1985, p. 43).

Public language, therefore, is composed largely of implicit meanings relying on the common background and experience of the interactants. For example, the frequent use of phrases which have acquired an approved, or 'consensus', meaning through repeated use in a limited community leads to the exclusion of outsiders for whom their meaning is not deducible from the words alone.

The ascribed statuses and roles of interactants provide much of the structure for what is implicit in public language. Such categories therefore become reaffirmed and regarded as fixed, concrete entities rather than arbitrary, flexible classifications. Novel expressions which, by definition, do not draw on a pool of established understanding carry no implicit meaning for their recipient. Therefore they are relatively meaningless in public language when compared with its own idiomatic phrases. As a result, creativity in speech and highly individual forms of expression are not facilitated.

In his later publications (after 1962), Bernstein brings in a new terminology of more general application than the public–formal dichotomy. He introduces the concept of *communication code*, identifying two types which he labels *restricted* and *elaborated*. These codes are not in themselves varieties of language, but 'principles of structuration which underpin linguistic and social forms, their variation and their reproduction' (Atkinson 1985, p. 66).

This indicates that the scope of his work goes beyond linguistics, as Atkinson stresses:

> Bernstein's interest lies in expressing not simply issues of language variation, but orientations to means, ends and objects, relationships between objects, the creation and recreation of identities and modes of social control. These are addressed through the medium of language use, but the latter is not the exclusive concern (ibid., p. 40).

Public and formal language styles represent the outward or surface manifestation of restricted and elaborated codes respectively. The codes can therefore be recognized by the same linguistic criteria outlined in Table 8.9. But, in Bernstein's formulation, the codes also control the form of semiotic systems other than language, for example: rule-systems of clothing; food; ritual; and body-adornment. Codes provide the principles of cultural transmission in all channels, and thereby 'regulate the transmission and reproduction of cosmologies and the very social structure itself' (ibid., p. 68).

Even in his early papers, Bernstein was looking below the surface patterns, regarding different styles of verbal language as the outward manifestation of different ways of dealing with and classifying the world:

> It is contended that members of the unskilled and semi-skilled strata, relative to the middle classes, do not merely place different significances upon different classes of objects, but that their perception is of a qualitatively different order (Bernstein 1958, p. 24).

The segregation of codes between different social strata is not random, but related to the requirements of each class. The division of labour and the role-differentiation within large-scale societies has resulted in the specialization of small groups of people into numerous disparate disciplines, each of which has evolved its own specialized vocabulary. Amongst the middle-class controllers of material or symbolic goods, who had to co-ordinate networks of communication between discrete professions, there arose a 'metalanguage' of universalistic terms and abstract notions characteristic of the elaborated code. This emergence of universalistic meanings is claimed to have important psychological side-effects:

> where meanings have this characteristic, then individuals have access to the grounds of their experience and can change the grounds. Where orders of meaning are particularistic, where principles are linguistically implicit, then such meanings are less context-independent and *more* context-bound, that is, tied to a local relationship and to a local social structure (Bernstein 1971a, pp. 175–6).

Thus metalanguage, by transcending the plane of particularistic meanings, allows a bird's-eye view of concrete reality, revealing the arbitrariness of concept-definitions, statuses, and roles. In other words, it facilitates a shift from subjective to objective styles of thought.

One thrust of Bernstein's argument, then, is that social structure is the ground against which meaningful communication is established. In simple societies, that structure does not itself become a topic of the discourse practices it affords. In a manner of speaking, cognition remains embedded in discourse. But in more complex societies, cognition is lifted out of discourse, allowing the perception of the social structure within which it was constituted. Cognition is disembedded from discourse and reconstituted symbolically, then becoming re-embedded within the constraints of the symbol system by which it functions.

Thus elaborated code is seen as emerging out of restricted code as social structure becomes more complex and lowers the shared level of presuppositionality amongst speakers. Lock (1983, p. 255) puts this 'bootstrapping' concept thus:

> there is an historical incorporation of the contexts of time A into the discourse of time B. Discourse at time B is thus conducted in a context that will later be transcended to be incorporated in the later discourse of time C, and so on... That is, contextually-mediated meanings become linguistically-mediated ones.

This transformation occurs in tandem with societal complexification.

Bernstein's work is directly applicable to the explanation of differences we have already noted in body-adornment and non-verbal communication. In addition, it provides a framework for conceptualizing the reproduction of culture. We may consider child development as a process which moves from a simple social structure (the parent–child dyad) to a more complex one (the social networks of adulthood). Language development is effectively a shift from a restricted code of context-dependent gestural communication or one-word speech to an elaborated code of potentially context-independent, grammatically-structured discourse. We may expect to find significant variations in the outcomes of development that is fostered in different intra-and cross-cultural groups.

8.9 The reproduction of culture

8.9.1 *Parent–child interaction and the transmission of cognitive skills*

Bernstein's work provides an explicit model for the process of cultural reproduction (see Bernstein 1981).

There is a large body of literature on parent–child interaction which is broadly compatible with his approach (see Light 1983 for a review). Essentially, the claim is that culture is composed of knowledge that may have an explicit or implicit form of coding; and that the cognitive operations of members of a culture operate upon the symbols available within that culture. Consequently, analytic thought will be facilitated in cultures or subcultures which supply symbols that have resulted from a process of analysis, and hence explicitly code cultural knowledge in 'elaborated chunks'.

Put alternatively in a Vygotskian perspective (see Sinha, this volume, Chapter 17), the argument would be that knowledge is reproduced in interactive situations. Those situations may not only supply explicit symbols, but are also structured in ways that, being informed by an explicit symbolizer, present cultural knowledge to the novitiate in explicit, already analysed 'chunks'. If this is the case, then knowledge will readily make the transition from the implicit, tacit, intermental realm to explicit intramental ones. Conversely, where interactions do not exhibit such a structure, knowledge will remain implicitly coded, and its explicit intramental coding is less likely to be facilitated for those to whom the world is so transacted.

Implicit in these views is the claim that cognitive abilities and their development are inherently social practices, in contradistinction to the Piagetian paradigm in which the knower constructs knowledge on the basis of his or her individual interactions with the world. Hence the notion of 'scaffolding' is of crucial importance. Scaffolding refers to the way an adult structures the problem-solving contexts which provide the bedrock on which cognitive development is founded. Adults may set up problem-solving situations so that infants and children find it more or less difficult to extract the elements of problems and to master the principles that both underly (and constitute) them. These principles are mastered because the data the child requires to formulate a solution to the problem is presented in a way that facilitates the extraction of the essential structure of the problem. Adults who themselves lack practice in explicitly coding such principles will scaffold the interactive teaching situation differently, and make the abstraction of explicit knowledge from the interaction more difficult for their pupils. This is clear in the following excerpt from Hess and Shipman's study (1965), which describes the teaching techniques of three mothers who are showing their children how to solve a sorting problem:

Let us describe the interaction between the mother and child in one of the structured teaching situations. The wide range of individual differences in linguistic and interactional styles of these mothers may be illustrated by excerpts from recordings. The task of the mother is to teach the child how to group or sort a small number of toys.

The first mother outlines the task for the child, gives sufficient help and explanation to permit the child to proceed on her own. She says:
'All right, Susan, this board is the place where we put the little toys; first of all you're supposed to learn how to place them according to color. Can you do that? The things that are all the same color you put in one section; in the second section you put another group of colors, and in the third section you put the last group of colors. Can you do that? Or would you like to see me do it first?'
Child: 'I want to do it.'

This mother has given explicit information about the task and what is expected of the child; she has offered support and help of various kinds; and she has made it clear that she impelled the child to perform.

A second mother's style offers less clarity and precision. She says in introducing the same task:
'Now, I'll take them all off the board; now you put them all back on the board. What are these?'
Child: 'A truck.'
'All right, just put them right here; put the other one right here; all right, put the other one there.'

This mother must rely more on non-verbal communication in her commands; she does not define the task for the child; the child is not provided with ideas or information that she can grasp in attempting to solve the problem; neither is she told what to expect or what the task is, even in general terms.

A third mother is even less explicit. She introduces the task as follows:
'I've got some chairs and cars, do you want to play the games?' Child does not respond. Mother continues: 'O.K. What's this?'
Child: 'A wagon?'
Mother: 'Hm?'
Child: 'A wagon?'
Mother: 'This is not a wagon. What's this?'

The conversation continues with this sort of exchange for several pages. Here again, the child is not provided with the essential information he needs to solve or to understand the problem. There is clearly some impelling on the part of the mother for the child to perform, but the child has not been told what to do (ibid., pp. 881–2).

The overall results of this investigation (see Table 8.10) show a strong correlation between mothers' social class, how they taught a problem, and their child's subsequent ability to solve it and explain the means of solution (that is, not merely to succeed through uninformed imitation). In addition, Hess and Shipman (ibid.) report differences in speech style among the mothers on Bernstein's (1981) 'elaborated–restricted' dimension which underpin these teaching styles, and a theoretically-consistent difference in what they term 'person v. status orientation'. Upper-class mothers legitimize their instructions to their children by appeal to personal responsibility; lower-class mothers to status roles and authority.

Table 8.10 Differences among status groups in children's performance in mastering an object-sorting task taught them by their mothers (figures extracted from Hess and Shipman 1965)

Social status	% of sample placing object correctly	% of those placing object correctly able to verbalize one dimension their decision is based on	% of those placing object correctly able to verbalize both dimensions their decision is based on	Number of subjects in sample
Upper middle	60.0	64.1	45.8	40
Upper lower	48.8	42.1	35.0	41
Lower lower	34.2	26.5	23.1	38
Unemployed receiving Social Security benefits	28.2	20.0	0.0	39

In a similar study, but one where the mother's style of teaching was assessed by having her teach a child other than her own, and the child's mode of learning was assessed by having the experimenter teach a task to the child, Hartmann and Haavind (1981) found a strong relationship between maternal teaching style and child 'educability': not only were children whose mothers' style of teaching presented the task to them in explicit form more likely to learn the task, they also showed a more 'interested' and 'co-operative' approach to being taught. Hartmann and Haavind claim, then, that what is being measured in such experiments is *an enduring style of teaching on the mother's part*, a characteristic mode of action which she exhibits over long periods of time. Further, that *children develop similarly enduring 'traits' in their approaches to learning and problem-solving situations*.

We may conclude that the transmission of analytic skills from a possessor to an initiate requires that the possessor can:

(1) break down the task into its components;
(2) assess the initiate's existing level of competence;
(3) organize the components of the skill into an appropriate hierarchy of subgoals;
(4) explicitly mark the components in the act of communicating them; and
(5) be sensitive to the apparent effectiveness of each communication.

Studies of this transmission process show links between teaching style and the typical language code of the mother. In addition, they show a link between maternal language and social class. That is, effective conveyors of information and principles use explicit, analytic modes of language, and come from high-status social classes. These findings offer some support for the thesis that there are links between social structure, language, and cognition. However, the other important body of literature relevant to this issue, by Witkin and his associates (see below), has been formulated as a different paradigm, and makes no reference to Bernstein's work. As we shall show, this makes it difficult to resolve what, on the surface, appear contradictory findings.

8.9.2 Cognitive style and social practices across cultures

There is a massive literature on cognitive style by Witkin and his associates and co-workers (reviews and details may be found in Witkin 1967; Witkin *et al.* 1962; Witkin and Berry 1975; Berry 1976, 1990; we will not give detailed citations here). The claim is that psychological development progresses from a less- to a more-differentiated state. In perception:

greater differentiation shows itself in the tendency for parts of the field to be experienced as discrete from the field as a whole, rather than as fused with the field, or experienced as global, which is indicative of lesser differentiation (Witkin and Berry 1975, 6).

Individuals show a consistency across a number of different perceptual tasks in their levels of differentiation, this indicating they have a particular 'perceptual style':

Because, at one extreme, perception is dominated by the organization of the field in which it is contained, so that the item cannot easily be disembedded from its context, this mode of perception has been labelled "field dependent". The contrasting mode of perception, in which the organization of the prevailing field is less influential in determining which part of the field is experienced, so that an item can be easily disembedded from surrounding field, has been labelled "field independent" (ibid., p. 8).

These styles reflect tendencies toward modes of perception along a continuum, rather than distinct types.

In addition to showing consistency across perceptual tasks, individuals show a consistency in approach—differentiated vs. undifferentiated, field-dependent vs. field-independent—across many psychological domains, including problem-solving:

relatively field-dependent persons have been found to encounter more difficulty than relatively field-independent ones with that particular class of problems which, to be solved, requires that some critical element be taken out of the context in which it is presented, and the problem material restructured so that the item is used in a different context. With this extension of the picture of self-consistency from the perceptual to the intellectual domains the label of "cognitive styles" becomes more appropriate as a more comprehensive concept than "perceptual styles" (ibid., p. 9).

Work on cognitive style has been carried out quite extensively within one particular culture, the United States, and increasingly in other cultures. What emerges is a cluster of factors that correlate with the different cognitive styles (see Witkin and Berry 1975; Berry 1976 for reviews).

8.9.2.1 *Socialization practices*

Witkin (1967) summarizes the basic claims made concerning the antecedents of cognitive-style differences in socialization practices. Essentially, strict socialization along with restrictions on personal autonomy are predictive of field-dependent styles, the opposite being argued to promote field-independence: 'the weight of evidence suggests that the child-rearing practices which foster the development of greater or more limited differentiation tend to be similar across cultures' (Witkin and Berry 1975, p. 45).

Barry and his associates (for example, Barry *et al.* 1957, 1959; Barry and Paxson 1971) provide evidence which links socialization practices with ecological factors. Socialization for compliance is the general practice in high-food-accumulation societies; for independence in low-food-accumulation societies (see also above, Section 8.5.3).

8.9.2.2 *Social tightness (social conformity)*

This relates to the degree of hierarchical structure in the role-differentiation within a society (cf. Pelto 1968). At the tight end of this dimension societies have, for example, an aristocracy, a distinct clerical class, and labour conscription; at the loose end the bare essentials of political leadership. These different social structures provide different levels of pressure for social conformity in the way they order an individual's scope for self-expression. Witkin and Berry summarize the findings of cross-cultural studies thus:

> The evidence from these studies together suggests that a relatively field-dependent cognitive style, and other characteristics of limited differentiation, are likely to be prevalent in social settings characterized by insistence on adherence to authority both in society and in the family, by the use of strict or even harsh socialization practices to enforce this conformance, and by tight social organization. In contrast,

a relatively field-independent cognitive style and greater differentiation are likely to be prevalent in social settings which are more encouraging of autonomous functioning, which are more lenient in their child-rearing practices, and which are loose in their social organization (1975, p. 48).

8.9.2.3 *Ecological adaptation*

There is an accumulation of evidence which indicates that individuals from hunting-based samples tend to be more field independent on tasks of perceptual differentiation, while those from agriculture-based samples tend to be relatively field dependent (Witkin and Berry 1975, p. 61).

Berry (1966) interprets this as resulting from the ecological demands presented by a hunter–gatherer life-style, which requires the individual to extract particular aspects from the global environment to locate prey, and to combine this information with a precise knowledge of present position in relation to an eventual safe return home. However, such skills could be argued to be essential in agricultural life-styles for the determining of planting seasons. Note that this finding relates to *perceptual* differentiation. There is no evidence supporting *cognitive* style-differences between hunter–gatherers and agriculturalists, bar the assumption of consistency across domains. Attempts at cross-cultural testing of cognitive differences are fraught with difficulty (see Barton and Hamilton this volume, Chapter 27 and also Laboratory of Comparative Human Cognition (1983) for a critical review of this work).

8.9.3 *Are Bernstein's and Witkin's approaches comparable?*

Intuitively, these different approaches, from language codes and psychological style, would appear to share some common ground. Both relate differences in the differentiation of psychological systems to facets of social structure; both point to differences in child-rearing practices that accompany these. They are, though, difficult to integrate because of the central role played by language in one approach, and its complete absence in the other; and because of the different classifications of social structure adopted. Some of these disparities emerge in a study by Cohen (1969), where she used elements of both approaches to investigate cognitive-style differences in school-aged children.

At the time of her study, work drawing on Bernstein's theories was concerned with social-class differences at a quite crude level. Cohen (1969), however, was able to move the analysis away from such gross measures to ones of social-group structure. In addition to language measures as adopted by Hess and Shipman (1965, see above), children's cognitive 'styles' were assessed through procedures standardized by Witkin and his associates. Cohen thus identified two distinct conceptual styles, which she termed *analytic* and *relational*:

The analytic cognitive style is characterized by a formal or analytic mode of abstracting salient information from a stimulus or situation and by a stimulus-centred orientation to reality, and it is parts-specific (i.e. parts or attributes of a given stimulus have meaning in themselves). The relational cognitive style, on the other hand, requires a descriptive mode of abstraction and is self-centred in its orientation to reality; only the global characteristics of a stimulus have meaning to its users, and these only in reference to some total context (1969, pp. 829–30).

With respect to social structure:

relational and analytic conceptual styles were found to be associated with shared-function and formal primary-group participation, respectively, as socialization settings. So intimate were the relationships between primary group styles and conceptual styles that among pupils with experience in both types of groups mixed and conflicting conceptual styles could be observed (ibid., pp. 842).

These findings highlight the problems involved in integrating two different traditions of research. For example, we might expect from a Bernsteinian perspective that a differentiated role-structure would predispose individuals to relate to each other through the functions they are assigned within that social structure, leading to restricted language coding and its accompanying relational, or field-dependent, cognitive strategies. On the other hand, from a Witkinian perspective, we might expect shared-function groups, which we presume to be the characteristic form of hunter–gatherer societies, to be promoting analytic as opposed to relational modes. We will return to this problem below. Here we just draw attention to the fact that, while reliable distinctions between and within cultural groups have been demonstrated, many of the factors put forward as responsible for these differences do not appear, firstly, to be independent of each other, nor, secondly, to be directly comparable from one paradigm to another. Any attempt to draw general conclusions in this area will be speculative at this point in time.

8.9.4 *Cognitive style and brain function*

These strictures apply equally to one final area of research. Paredes and Hepburn (1976) conjecture that the postulated qualitative distinctions in intersocietal, interclass, and interindividual ways of thinking have a basis in brain functioning. Specifically, their claim is that analytic and relational styles (or, more globally, what they contrast as 'scientific' v. 'primitive' thought) rely differentially on the different operative modes of the human cerebral hemispheres. That is, cognitive operations are lateralized, the left hemisphere operating in an analytic mode, the right in a synthetic one. There are a number of similar speculations in the literature. For example:

Ten Houten and Kaplan (1973, pp. 90–108) hypothesize that the modes of thought associated with the English and Hopi languages correspond, albeit roughly, with the propositional and appositional modes of thought of the left and right sides of the brain. To explore this hypothesis, a study of 4th–6th-grade bilingual Hopi children was carried out (Rogers *et al.* 1976). EEG measures were obtained under the experimental conditions of listening to children's stories (with eyes closed) in Hopi and English. Analysis of the data indicated that while listening to the stories in English, alpha rhythm was suppressed in the left-parietal lead, and while listening to the stories in Hopi, the level of alpha suppression was relatively greater in the homotopic right-parietal lead ($n = 16$, $p = 0.025$). This result is consistent with the prediction, but the experiment is in no way definitive, and numerous explanations are not ruled out (Ten Houten 1976, p. 505).

This final warning is appropriate to all work carried out on this topic (see Beaumont *et al.* 1984 for a critical review). Thus, Paredes and Hepburn's conjecture remains just that.

8.10 Overview

Because of the different conceptual schemes involved, the nebulousness of many of the factors invoked, and the consequent lack of precise scales of measurement, any conclusions drawn from this diversity of work on the form and functioning of human communication systems must be provisional. In our view, the problems that lead to this are part and parcel of the field, which by its nature will never be open to rigorous measurement. We believe, however, that some broad generalizations are possible, and these are portrayed in Fig. 8.4.

Firstly, there are factors which are predictive of the degree of differentiation of a society's social structure. Many of these can be located in the ecological domain, although there are doubtless other factors, including those that might be produced by a given level of differentiation itself, that need to be considered. We may term these factors 'pressures' that operate upon social structures, leading to greater or lesser degrees of internal divisions of labour and role-structure. How these pressures have their effect is clearly not simple, and we must resist the temptation to reify them.

Secondly, social structures put varying 'pressures' on the communication systems that sustain them,

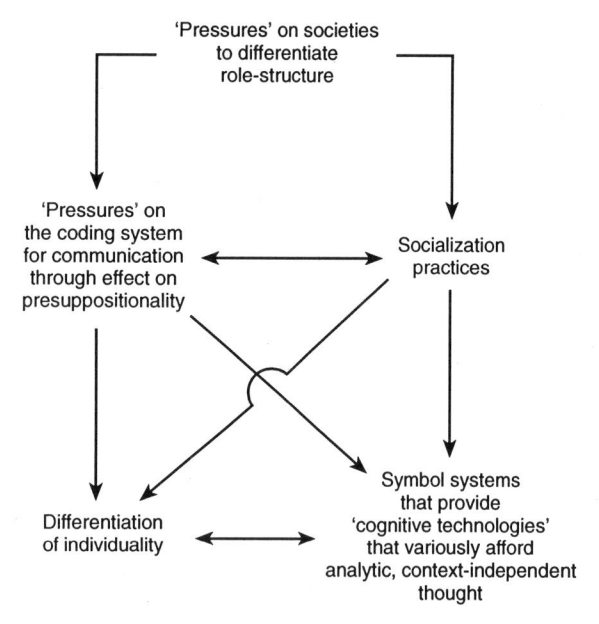

Fig. 8.4 Form and functioning of human communication systems: see text below for elaboration.

through the different levels of presuppositionality each society's members share with each other. With low social differentiation and interchangeable roles, people communicate with each other against a background of common orientation, in a context of interpersonal relations founded on common perceptions, values, interpretative competence, and so on. As social structure differentiates, presuppositionality decreases, so that, to sustain and reproduce themselves, the communicative practices within that society must develop ways of creating common contexts within the medium of communication itself. The contexts which make interpersonal communication possible have to be lifted out of the everyday milieu and created within the symbolic system of the lingua franca. Essentially, contexts will be 'pressured' into symbolic codes, and become conceptualized objects of knowledge rather than processes which sustain knowledge. There is, then, no direct link between the 'pressures' that are responsible for the level of structural differentiation within a society and the form of its symbolic communicative resources; this link is mediated by presuppositionality.

Thirdly, links are apparent between social structure and socialization practices. While it is implicit in this work that socialization practices are somehow acted on by ecological pressures, there is sufficient evidence to implicate the form of the communicative system used in the reproduction of culture from one generation to the next as influential on these practices as well.

All the above combine to give two further outcomes. Firstly, the evidence broadly supports a view of communicative symbol systems as the providers of 'cognitive technologies' or 'tool-kits' that variously afford analytic, context-independent thought. This is not to claim that language determines thought, but that conceptual systems can make some universal mental operations more or less difficult to carry out. As an example of what we mean by this, we draw an analogy with the differences between the Roman and Arabic number systems. The mental operations of multiplication and division are available to users of both systems, but the symbolic 'tool-kits' affect the ease with which these operations can be performed: it is easier to divide 63 by 9 to get 7 than LXIII by IX to get VII. Similarly, it is easier to think about, and even perceive, the structure of society and how communication systems operate within it if one possesses a symbol system which symbolizes these things.

Secondly, the conscious 'individuality' of the members of different social systems is influenced not only by the role-system operating in a society, and its socialization practices, but also by the symbolic resources the communicative system provides for the formulation of individual definition. A system which provides decontextualized symbols will allow an individual to contemplatively formulate plans, etc., from the meanings those symbols allow to be manipulated and explored in individual thought. Anyone can put forward his or her own 'point of view', and develop that point of view into one which is not given solely by the status quo of given meanings and definitions. Concepts can thus be 'coined' by individuals, and fed back into the cultural systems that generated the ability to formulate those concepts in the first place. Individuals can define themselves, and 'find their own voices', rather than speak only with the resources provided them. Finally, we suggest that these last two factors feed back on the entire system, such that the level of differentiation of individuality, for example, can have profound effects on the structural properties of societies.

The linguistic work we have been surveying in this chapter bears on the possible course of elaboration of languages once they have been established. The material concerning non-linguistic forms of human communication provides a context within which more archaic evidence might be interpreted (for example, White 1989a, b, on the interpretation of Upper Palaeolithic personal ornaments from around 30 000–35 000 BP), but extrapolations regarding language-forms to such time-levels are more tenuous. In this respect, we suggest our survey poses questions for the view that many authorities (for example, Marshack 1985; Clark 1981, 1989; Chase and Dibble 1987;

Dibble 1989) appear to agree on: that the richness of Upper Palaeolithic forms of visual art, body ornament, ceremonial burial, specialized hunting strategies, and the like, would not be possible without a complex form of linguistic communication:

> If all these features can be traced to the very early Upper Palaeolithic, then it would seem reasonable to assume that some form of effective, highly structured language had already emerged in at least certain areas of Eurasia by at least 30–35000 B.P. (Mellars 1989, p. 364).

We do not wish to dispute that some form of language was in existence by that time, but would raise two issues. First, this view implies some form of causality; that language provides the conditions for other activities. But this need not be the case. For example, following the role we have argued presupposition plays in determining the nature of a linguistic communication system, a more interactive scenario is quite possible. Thus, rather than body ornamentation's being seen as correlated with, or dependent upon, a differentiated linguistic system, it may well have been a factor in creating linguistic elaboration. That is, if ornamentation is taken as indexing the differentiation of social roles that concomitantly reduces the level of presuppositionality operating within a relatively small group, then role-markers act, not as a reflection of linguistic complexity, but as a factor leading to it, through increasing the need for a more explicit coding of information in the language used.

Our second point is centred on the question of the *relative* complexity of possible languages. The anatomical evidence surveyed by Laitman and his co-workers (this volume, Chapter 3, Editorial Appendix 2) suggests a modern anatomical capacity for *speech* by the Upper Palaeolithic; but speech is not language. Wynn (this volume, Chapter 10) argues conservatively for a much earlier modern intellectual capacity; but capacity need not be exercised. Hewes (this volume, Chapter 21) points to Tasmanian languages as suggesting that phonemically-organized languages were in existence by the end of the Pleistocene. We suggest that two issues we have considered interact with these three points. First, that language symbols act as a cognitive technology, enabling cognitive abilities to function—to provide the forms that enable capacities and potentials to be actualized. And second, the cognitive resources that are embedded in a linguistic system are largely activated by the social practices and problems its users adopt and face.

This mélange leads us to speculate that something akin to a recognizably modern language was in existence in the Upper Palaeolithic, providing the cognitive technology that enabled a pre-existing level of intelligence to be brought to bear on a range of cultural activities, facilitating the elaboration of human life-ways into ones that had essentially modern characteristics (see White, Chapter 9, and Wynn, Chapter 10 this volume). However, the relative internal simplicity of the social groups hypothesized during this period suggests the full range of syntactic, semantic, and pragmatic complexity shown by modern languages may not have been elaborated (see Rolfe, Chapter 26, this volume, for an account of how the complex features of linguistic structure may be elaborated *de novo* from the requirements of highly presuppositional interpersonal communication, and Jaynes 1976*b*, for a similar scenario).

Notes

1. For a more fine-grained analysis of language variation, see Saville-Troike (1989).

2. For an explanation of the terms *emic* and *etic* as used by anthropologists see Note 2 to the Introduction to Part II of the present work, p. 162.

3. An analogous point is made in a developmental context by Dias and Harris (1988) in an investigation of four-and six-year-old children's abilities to reason logically from counterfactual premises, such as 'all cats bark'. Following this premiss, children were introduced to a new member of the category (for example, 'Rex is a cat'), and were asked to evaluate a conclusion (for example, 'Does Rex bark?'). Both groups of children were able to reason consequentially from such premisses provided they heard them in a make-believe context (for example, 'Let's pretend we're on another planet'), and would offer justifications of their answers that referred back to the signalled counterfactuality of the premisses ('"cos cats bark on this planet'). By contrast, children in a 'matter-of-fact' condition gave many more incorrect judgements, and justified their answers by reference to their empirical knowledge ('Rex doesn't bark because I know cats miaow.')

4. Jaynes's conclusions, and the arguments he develops to reach them, are controversial. Interpretations of human life in other times (and other places) are beset with difficulties, but may crudely be characterized as lying somewhere between two poles. On the one hand, humans, everywhere and at all times, have essentially similar experiences of the world and their sense of being in it, but interpret those experiences in ways that differ depending on the cultural resources available for symbolizing those experiences. On the other, humans may have very different experiences of the world and their being in it, to the extent of, analogously, there being different 'psychological

species' at different temporal and spatial locations, since human experiences are constructed and constituted on the basis of different cultural resources.

Jaynes's position is closer to this second pole than that of the majority of classical scholars. It is that the early Greeks possessed

'no conscious minds such as we say we have, and certainly no introspections.... The beginnings of [their] action[s] are not in conscious plans, reasons, and motives; they are in the actions and speeches of gods.... The gods are what we now call hallucinations.... Volition, planning, initiative is organized with no consciousness whatever and then "told" to the individual in his familiar language...' (1976a, pp. 72, 73, 74).

Jaynes's view, then, is not just that this particular culture in question construed the world, or accounted for themselves, in terms that differ from those of cultures existing at other times and places, but that those different terms were an account, an attempt to come to grips with, a very different form of 'self-experience'. Whether the interpretation Jaynes offers holds up for this particular period in Western history does not detract from the questions his position poses for an evolutionary perspective. For while, as Taylor (1989, p. 112) points out, it is the case both that 'we can probably be confident on one level human beings of all times and all places have shared a very similar sense of "me" and "mine"', and that problems of specific forms of self and self-consciousness can be recognized as 'inseparable from existing in a space of moral issues' that are 'historically local' (ibid.), an evolutionary account has to be essayed beyond history and common humanity.

Jaynes's position at least points to this issue: that at some time in the evolutionary continuity stretching back from the present, human experience is rooted in non-human experience. The general point developed in this present section is that 'being a self is inseparable from existing in a space of moral issues, to do with identity and how one ought to be. It is being able to find one's standpoint in this space, being able to occupy, to *be* a perspective in it' (Taylor ibid., p. 112), and that the possibility of this 'space-to-be-occupied'—'being a self'—arises in, and relative to, particular networks of human social relations and the symbol systems they engender.

References

Akmajian, A., Demers, R. A., and Harnish, R. M. (1984). *Linguistics: an introduction to language and communication* (2nd edn). MIT Press, Cambridge, Mass.

Armstrong, D. F. and Katz, S. H. (1981). Brain laterality in signed and spoken language: a synthetic theory of language use. *Sign Language Studies*, 33, 319–50.

Atkinson, P. (1985). *Language, structure and reproduction. An introduction to the sociology of Basil Bernstein.* Methuen, London.

Au, T. K. (1983). Chinese and English counterfactuals: the Sapir–Whorf hypothesis revisited. *Cognition*, 15, 155–87.

Au, T. K. (1984). Counterfactuals: in reply to Bloom. *Cognition*, 17, 289–302.

Barry, H. (1969). Cross-cultural research with matched pairs of societies. *Journal of Social Psychology*, 79, 25–34.

Barry, H. and Paxson, L. M. (1971). Infancy and early childhood: cross-cultural codes 2. *Ethnology*, 19, 466–508.

Barry, H., Bacon, M., and Child, I. (1957). A cross-cultural survey of some sex differences in socialization. *Journal of Abnormal and Social Psychology*, 55, 327–32.

Barry, H., Child, I., and Bacon, M. (1959). Relation of child training to subsistence economy. *American Anthropologist*, 61, 51–63.

Beaumont, J. G., Young, A. W., and McManus, I. C. (1984). Hemisphericity: a critical review. *Cognitive Neuropsychology*, 1, 191–212.

Berlin, B. and Kay, P. (1969). *Basic color terms: their universality and evolution.* University of California Press, Berkeley.

Bernstein, B. (1958). Some sociological determinants of perception. *British Journal of Sociology*, 9, 159–74. (Also in idem, *Class, codes and control*, 1, pp. 23–41.)

Bernstein, B. (1959). A public language: some sociological implications of a linguistic form. *British Journal of Sociology*, 19, 311–26. (Also idem, *Class, codes and control*, 1, pp. 42–60).

Bernstein, B. B. (1961). Social structure, language and learning. *Educational Research*, 3, 163–76.

Bernstein, B. (1971a). Social class, language and socialisation. In *Class, codes and control*, Vol. I: *Theoretical studies towards a sociology of language*, pp. 170–89. Routledge and Kegan Paul, London.

Bernstein, B. (1971b). *Class, codes and control*, Vol. I: *Theoretical studies towards a sociology of language.* Routledge and Kegan Paul, London.

Bernstein, B. (1981). Codes, modalities and the process of cultural reproduction—a model. *Language in Society*, 10, 327–63.

Berry, J. W. (1966). Temne and Eskimo perceptional skills. *International Journal of Psychology*, 1, 207–29.

Berry, J. W. (1976). *Human ecology and cognitive style: comparative studies in cultural and psychological adaptation.* Wiley, Halstead Press Division, New York.

Berry, J. (1990). Cultural variations in cognitive style. In *Bio-psycho-social factors in cognitive style* (ed. S. Wappner), pp. 289–308. Erlbaum, Hillsdale, NJ.

Best, E. (1922). *Spiritual and mental concepts of the Maori.* Dominion Museum, Wellington, NZ.

Bloch, M. (1965). *Feudal society*, Vol. 2. Routledge and Kegan Paul, London.

Bloom, A. H. (1981). *The linguistic shaping of thought: a study in the impact of language on thinking in China and the West.* Erlbaum Associates, Hillsdale, NJ.

Bloom, A. H. (1984). Caution—the words you use may affect what you say: a response to Au. *Cognition*, 17, 275–87.

Boas, F. (1911). Introduction. In *Handbook of American Indian languages*, (ed. F. Boas), pp. 1–83. Smithsonian Institution, Washington.

Brain, R. (1979). *The decorated body*. Harper and Row, New York.

Brown, R. and Ford, M. (1961). Address in American English. *Journal of Abnormal and Social Psychology*, 62, 375–85.

Brown, R. and Gilman, A. (1960). The pronouns of power and solidarity. In *Style in languages* (ed. T. Sebeok), pp. 253–76. MIT Press, Cambridge, Mass.

Carneiro, R. L. (1970). Scale analyses, evolutionary sequences and the rating of cultures. In *A handbook of method in cultural anthropology*, (ed. R. Naroll and R. Cohen), pp. 834–71. Natural History Press, New York.

Carneiro, R. L. and Tobias, S. F. (1963). The application of scale analysis to the study of cultural evolution. *Transactions of the New York Academy of Sciences*, 26, 196–207.

Carrithers, M., Collins, S., and Lukes, S. (1986). *The category of the person: anthropology, philosophy, history*. Cambridge University Press.

Chambers, J. and Trudgill, P. (1980). *Dialectology*. Cambridge University Press.

Chase, P. G. and Dibble, H. L. (1987). Middle Palaeolithic symbolism: a review of current evidence and interpretations. *Journal of Anthropological Archaeology*, 6, 263–93.

Chomsky, N. (1957). *Syntactic structures*. Mouton, The Hague.

Chomsky, N. (1965). *Aspects of the theory of syntax*. Mouton, The Hague.

Clark, J. D. (1981). "New men, strange faces, other minds". An archaeologist's perspective on recent discoveries relating to the origins and spread of modern man. *Proceedings of the British Academy*, 67, 163–92.

Clark, J. D. (1989). The origins and spread of modern humans. A broad perspective on the African evidence. In *The human revolution: behavioural and biological perspectives on the origins of modern humans* (ed. P. Mellars and C. B. Stringer), pp. 565–88. Edinburgh University Press.

Cohen, R. A. (1969). Conceptual styles, cultural conflict, and non-verbal tests of intelligence. *American Anthropologist*, 71, 828–56.

Darwin, C. R. (1872). *The expression of the emotions in man and animals*. John Murray, London.

Dias, M. and Harris, P. L. (1988). The effect of make-believe play on deductive reasoning. *British Journal of Developmental Psychology*, 6, 207–21.

Dibble, H. L. (1989). The implications of stone tool types for the presence of language during the Lower and Middle Palaeolithic. In *The human revolution: behavioural and biological perspectives on the origins of modern humans* (ed. P. Mellars and C. B. Stringer), pp. 415–32. Edinburgh University Press.

Dixon, R. M. W. (1971). A method of semantic description. In *Semantics: an interdisciplinary reader in philosophy, linguistics, and psychology* (ed. D. D. Steinberg and L. A. Jakobovits), pp. 436–71. Cambridge University Press.

Douglas, M. (1971). Do dogs laugh? A cross-cultural approach to body symbolism. *Journal of Psychosomatic Research*, 15, 387–90.

Downes, W. (1984). *Language and Society*. Fontana, London.

Dreitzel, H. P. (1981). The socialization of nature: Western attitudes towards body and emotions. In *Indigenous psychologies: the anthropology of the self* (ed. P. L. F. Heelas and A. J. Lock), pp. 205–24. Academic Press, London.

Eibl-Eibesfeldt, I. (1975). *Ethology* (2nd edn). Holt, Rinehart, and Winston, New York.

Ekman, P. (1972). Universals and cultural differences in facial expressions of emotion. In *Nebraska Symposium on Motivation, 1971* (ed. J. Cole), pp. 207–83. University of Nebraska Press, Lincoln, Nebraska.

Elias, N. (1978). *The civilizing process*, Vol. 1: *The history of manners*. Basil Blackwell, Oxford.

Elias, N. (1982). *The civilizing process*, Vol. 2: *Power and civility*. Basil Blackwell, Oxford.

Ember, M. (1978). Size of color lexicon: interaction of cultural and biological factors. *American Anthropologist*, 80, 364–7.

Faris, J. C. (1972). *Nuba personal art*. University of Toronto Press.

Ferguson, C. A. (1959). Diglossia. *Word*, 15, 325–40.

Ferguson, C. A. (1971). Absence of copula and the notion of simplicity: a study of normal speech, baby talk, foreigner talk and pidgins. In *Pidginization and creolization of language* (ed. D. H. Hymes), pp. 141–50. Cambridge University Press.

Fischer, J. L. (1958). Social influence on the choice of a linguistic variant. *Word*, 14, 47–56.

Fischer, J. L. (1965). The stylistic significance of consonantal sandhi in Trukese and Ponapean. *American Anthropologist*, 67, 1495–1502.

Fischer, J. L. (1966). Syntax and social structure: Truk and Ponape. In *Sociolinguistics: Proceedings of the UCLA Sociolinguistics Conference, 1964* (ed. W. Bright), pp. 168–87. Mouton, The Hague.

Fischer, J. L. (1973). Communication in primitive systems. In *Handbook of communication* (ed. W. Schramm, I. de Sola Pool, N. Maccoby, E. B. Parker and F. W. Frey), pp. 313–36. Rand McNally, Chicago.

Fishman, J. A., Cooper, R. J., and Ma, R. (1971). *Bilingualism in the barrio*. Indiana University Press, Bloomington.

Freeman, L. C. and Winch, R. F. (1957). Societal complexity: an empirical test of a typology of societies. *American Journal of Sociology*, 62, 461–6.

Greenberg, J. (1960). A quantitative approach to the morphological typology of language. *International Journal of American Linguistics*, 26, 178–94.

Gregory, M. and Carrol, S. (1978). *Language and situation: language varieties and their social context*. Routledge and Kegan Paul, London.

Grimshaw, A. D. (1973). Sociolinguistics. In *Handbook of communication* (ed. W. Schramm, I. de Sota Pool,

N. Maccoby, E. B. Parker, and F. W. Frey), pp. 49–92. Rand McNally, Chicago.

Haas, M. R. (1944). Men's and women's speech in Koasati. *Language*, **20**, 142–9.

Hall, E. T. (1959). *The silent language*. Fawcett Publications, Greenwich, Conn.

Halliday, M. A. K. (1978). *Language as social semiotic*. Edward Arnold, London.

Halliday, M. A. K., McIntosh, A., and Stevens, P. (1964). *The linguistic sciences and language teaching*. Longman, London.

Hallowell, A. I. (1956). *Culture and experience*. University of Pennsylvania Press, Philadelphia.

Hallpike, C. R. (1979). *The foundations of primitive thought*. Clarendon Press, Oxford.

Harré, R. (1983). *Personal being*. Basil Blackwell, Oxford.

Harris, P. L. and Heelas, P. L. F. (1979). Cognitive processes and collective representations. *Archives of European Sociology*, **XX**, 211–41.

Hartmann, E. and Haavind, H. (1981). Mothers as teachers and their children as learners: study of the influence of social interaction upon cognitive development. In *Communication in development* (ed. W. P. Robinson), pp. 129–58. Academic Press, London.

Haviland, J. B. (1979). How to talk to your brother-in-law in Guugu Yimidin. In *Languages and their status* (ed. T. Shopen). Winthrop Publishers, Cambridge, Mass. Reprinted University of Pennsylvania Press, Philadelphia, 1987, pp. 160–239.

Heelas, P. L. F. and Lock, A. J. (1981). *Indigenous psychologies: the anthropology of the self*. Academic Press, London.

Hertzler, J. O. (1953). Toward a sociology of language. *Social Forces*, **32**, 109–19.

Heslin, R. and Patterson, M. L. (1982). *Nonverbal behavior and social psychology*. Plenum Press, New York.

Hess, R. and Shipman, V. (1965). Early experience and the socialization of cognitive modes in children. *Child Development*, **36**, 869–86.

Hockett, C. F. (1958). *A course in modern linguistics*. Macmillan, New York.

Hudson, R. A. (1980). *Sociolinguistics*. Cambridge University Press.

Hymes, D. H. (1970). Linguistic method in ethnography. In *Method and theory in linguistics* (ed. P. Garvin), pp. 249–325. Mouton, The Hague.

Hymes, D. H. (1971). *Pidginization and creolization of language*. Cambridge University Press.

Hymes, D. H. (1972). Models of the interaction of language and social life. In *Directions in sociolinguistics: the ethnography of communication* (ed. J. J. Gumperz and D. H. Hymes), pp. 35–71. Holt, Rinehart and Winston, New York.

Hymes, D. H. (1973). On the origins and foundations of inequality among speakers. *Daedalus*, **59**, 86–107.

Hymes, D. H. (1977). *Foundations in sociolinguistics*. Harper and Row, New York.

Jaynes, J. (1976a). *The origin of consciousness in the breakdown of the bicameral mind*. Houghton Mifflin, New York.

Jaynes, J. (1976b). The evolution of language in the Late Pleistocene. In *Origins and evolution of language and speech* (ed. S. R. Harnad, H. D. Steklis, and J. Lancaster), pp. 312–25. *Annals of the New York Academy of Sciences*, Vol. 280.

Jesperson, O. (1894). *Progress in language*. Macmillan, New York.

Key, M. R. (1977). *Nonverbal communication: a research guide and a bibliography*. The Scarecrow Press, Metuchen, NJ.

Kluckhohn, C. (1954). Culture and behavior. In *Handbook of social psychology*, Vol. 2 (ed. G. Lindzey), pp. 921–76. Addison-Wesley, Reading, Mass.

Kroeber, A. (1960). On typological indices I: ranking of languages. *International Journal of American Linguistics*, **26**, 171–7.

Laboratory of Comparative Human Cognition (1983). Culture and cognitive development. In *Handbook of child psychology* (4th edn, ed. P. H. Mussen), Vol. 1 (ed. W. Kessen), pp. 295–356. Wiley, New York.

Labov, W. (1966). *The social stratification of English in New York City*. Centre for Applied Linguistics, Washington DC.

Labov, W. (1972a). *Sociolinguistic patterns*. University of Philadelphia Press, Philadelphia.

Labov, W. (1972b). *Language in the inner city*. University of Pennsylvania Press, Philadelphia.

Lee, D. (1959). *Freedom and culture*. Prentice Hall, New York.

Lienhardt, G. (1961). *Divinity and experience*. Clarendon Press, Oxford.

Light, P. (1983). Social interaction and cognitive development: a review of post-Piagetian research. In *Developing thinking: approaches to children's cognitive development* (ed. S. Meadows), pp. 67–88. Methuen, London.

Liu, L. (1985). Reasoning counterfactually in Chinese: are there any obstacles? *Cognition*, **21**, 239–70.

Lock, A. J. (1986). Metaphenomena and change. In *Mental mirrors: metacognition in social knowledge and communication* (ed. C. Antaki and A. Lewis), pp. 97–116, Sage, London.

Lomax, A. and Berkowitz, N. (1972). The evolutionary taxonomy of culture. *Science*, **177**, 228–39.

Lucy, J. A. (1992) *Linguistic diversity and thought: a reformulation of the linguistic relativity hypothesis*. Cambridge University Press.

McNett, C. W. (1970). A settlement pattern scale of cultural complexity. In *A handbook of method in cultural anthropology* (ed. R. Naroll and R. Cohen), pp. 872–86. Natural History Press, New York.

Marsella, A. J., Tharp, R. G. and Ciborowski, T. J. (1979). *Perspectives on cross-cultural psychology*. Academic Press, New York.

Marsella, A. J., DeVos, G., and Hsu, F. L. K. (1985). *Culture and self: Asian and Western perspectives*. Tavistock, London.

Marsh, R. (1967). *Comparative sociology*. Harcourt Brace & World, New York.

Marshack, A. (1985). *Hierarchical evolution of the human capacity: the palaeolithic evidence*. American Museum of Natural History, New York.

Mauss, M. (1935). Les techniques du corps. *Journal de Psychologie Normale et Pathologique*, **32**, 271–91. English translation in *Economy and Society*, **2**, 70–88 (1973).

Mellars, P. (1989). Major issues in the emergence of modern humans. *Current Anthropology*, **30**, 349–85.

Milroy, J. and Milroy, L. (1978). Belfast: change and variation in an urban vernacular. In *Sociolinguistic patterns in British English* (ed. P. Trudgill), pp. 19–36. Edward Arnold, London.

Morris, D. (1977). *Manwatching: a field guide to human behaviour*. Cape, London.

Murdock, G. P. (1957). World ethnographic sample. *American Anthropologist*, **59**, 664–87.

Murdock, G. P. (1967). Ethnographic Atlas: a summary. *Ethnology*, **6**, 109–236.

Murdock, G. P. and Morrow, D. O. (1970). Subsistence economy and supportive practices: cross-cultural codes 1. *Ethnology*, **9**, 302–30.

Murdock, G. P. and White, D. (1969). Standard cross-cultural sample. *Ethnology*, **8**, 329–69.

Murray, S. O. (1983). *Group formation in social science*. Linguistic Research, Inc., Carbondale and Edmonton.

Naroll, R. (1956). A preliminary index of social development. *American Anthropologist*, **58**, 687–715.

Naroll, R. (1970). Cross-cultural sampling. In *Handbook of method in cultural anthropology* (ed. R. Naroll and R. Cohen), pp. 889–926. Natural History Press, New York.

Nimkoff, M. F. and Middleton, R. (1960). Types of families and types of economy. *American Journal of Sociology*, **66**, 215–25.

Paredes, J. A. and Hepburn, M. J. (1976). The split brain and the culture–cognition paradox. *Current Anthropology*, **17**, 121–7.

Pelto, P. (1968). The difference between tight and loose societies. *Transaction*, **5**, 37–40.

Plutchik, R. and Kellerman, H. (1983). *Emotion: theory, research and experience*, 2 Vols. Academic Press, New York.

Robinson, W. P. (1972). *Language and social behaviour*. Penguin, London.

Rogers, L., Ten Houten, W. D. Kaplan, C. D., and Gardiner, M. (1976). Hemispheric specialization of language: an EEG study of bilingual Hopi Indian children. *International Journal of Neuroscience*, **8**, 1–6.

Sapir, E. (1915). *Abnormal types of speech in Nootka*, Canada Geographical Survey Memoir 62, Anthropological Series 5. Government Printing Bureau, Ottawa. Reprinted in D. G. Mandelbaum (ed.) (1949). *Selected Writings of Edward Sapir in language, culture and personality*. Cambridge University Press.

Sapir, E. (1929). The status of linguistics as a science. *Language*, **5**, 207–14.

Saville-Troike, M. (1989). *The ethnography of communication: an introduction* (2nd edn). Basil Blackwell, Oxford.

Scherer, K. and Ekman, P. (eds) (1984). *Approaches to emotion*. Lawrence Erlbaum, Hillsdale, NJ.

Shotter, J. (1984). *Social accountability and selfhood*. Blackwell, Oxford.

Smith, J. (1981). Self and experience in Maori culture. In *Indigenous psychologies: the anthropology of the self* (ed. P. L. F. Heelas and A. J. Lock), pp. 145–59. Academic Press, London.

Spencer, H. (1864). *First principles*. Appleton, New York.

Takano, Y. (1989). Methodological problems in cross-cultural studies of linguistic relativity. *Cognition*, **31**, 141–62.

Tatje, T. A. and Naroll, R. (1970). Two measures of societal complexity: an empirical cross-cultural comparison. In *A handbook of method in cultural anthropology* (ed. R. Naroll and R. Cohen), pp. 766–833. Natural History Press, New York.

Taylor, C. (1989). *Sources of the self: the making of modern identity*. Harvard University Press, Cambridge, Mass.

Ten Houten, W. D. (1976). More on split-brain research, culture, and cognition. *Current Anthropology*, **17**, 503–6.

Ten Houten, W. D. and Kaplan, C. D. (1973). *Science and its mirror image: a theory of inquiry*. Harper and Row, New York.

Thorne, B. and Henley, N. (1975). *Language and sex*. Newbury, Rowley, MA.

Trudgill, P. (1974). *The social differentiation of English in Norwich*. Cambridge University Press.

Trudgill, P. (1983). *Sociolinguistics* (2nd edn). Penguin, Harmondsworth.

White, R. W. (1989*a*). Visual thinking in the ice-age. *Scientific American*, **261**, 92–9.

White, R. W. (1989*b*). Production complexity and standardization in early Aurignacian bead and pendant manufacture: evolutionary implications. In *The human revolution: behavioural and biological perspectives on the origins of modern humans* (ed. P. Mellars and C. B. Stringer), Vol. 1, pp. 366–90. Edinburgh University Press–Princeton University Press.

Whiting, J. W. M. (1968). Methods and problems in cross-cultural research. In *Handbook of social psychology*, Vol. 2, (ed. G. Lindzey and E. Aronson), pp. 693–128. Addison-Wesley, Reading, Mass.

Witkin, H. A. (1967). A cognitive style approach to cross-cultural research. *International Journal of Psychology*, **2**, 233–50.

Witkin, H. A. and Berry, J. W. (1975). Psychological differentiation in cross-cultural perspective. *Journal of Cross-Cultural Psychology*, **6**, 4–87.

Witkin, H. A., Dyk, R. B., Paterson, H. F., Goodenough, D. R., and Karp, S. A. (1962). *Psychological differentiation*. Wiley, New York.

Wolfram, W. (1969). *A sociolinguistic description of Detroit negro speech*. Center for Applied Linguistics, Washington, DC.

Part II(b)
Palaeoanthropological perspectives

9
On the evolution of human socio-cultural patterns

Randall K. White

Abstract

The reconstruction of the developmental sequence of human socio-cultural evolution is one of the most important and difficult goals of prehistoric archaeology. Until recently, such reconstruction was accomplished speculatively by extrapolating from the present. Archaeologists have now developed a series of methods, many of them unfamiliar to the lay public, for monitoring changes in social patterns and complexity. Applied to the prehistoric record, these methods make possible a general summary of major trends in the evolution of human social patterns. It is clear that prior to 35 000 years ago hominid social patterns were very different from those of the more recent past. Subsequent to 35 000 years ago hominid behaviours reminiscent of those of the present emerged rapidly, and set the stage for the socially complex world of modern times.

9.1 Introduction

The re-emergence of cultural evolutionism in anthropology in the 1960s witnessed a keen interest in the evolution of human social patterns (Sahlins and Service 1960; Service 1962; Fried 1960, 1967). Paradoxically, most of those writing on the subject were anthropologists whose fieldwork had been with contemporary societies. Even as late as the 1960s, then, there was an obvious trend toward seeing contemporary small-scale societies as 'living fossils', a term that matches the contradiction of the conceptual framework. Of course, what was missing was a prehistoric dimension to the analysis of social evolution.

There had been prior attempts in prehistoric archaeology to make contributions. V. Gordon Childe (1936, 1942, 1950) in particular, had demonstrated a very real concern for social evolution while remaining a diffusionist for the most part. Washburn's (1961) symposium on the *Social life of early man* raised the question, for North American anthropologists at least, of the role to be played by archaeologists and physical anthropologists in documenting the evolution of human social behaviour.

At the time, there was a very real contradiction between a desire to understand the nature and causes of human social evolution and the avowedly anti-evolutionary paradigm (Kuhn 1962) in prehistoric archaeology, both within and outside anthropological circles. The ferment of the 'new archaeology' did not reach critical mass until the 1970s, so that, in the 1960s, most emphasis was still upon culture history, and 'explanation' was performed according to the conventions of the 'traditionalist paradigm', that is, that of the migration of people and the diffusion of ideas (L. Binford 1972).

Even if the questions were beginning to be asked, the methods were not there to answer them. In this context, Service (1962), Sahlins (1965), Fried (1960, 1967), and Steward (1955), among others, could project the present into the distant past without even the scantest notion that this was but a hypothetical framework, woven from modern observations and in need of empirical testing. Indeed there seems to have been a certain security in assuming that it *could not* be tested.

Great strides have been made in archaeological epistemology (middle-range theory) in the past decade or so (e.g., L. Binford 1977, 1978, 1979, 1981; Schiffer 1972, 1976). Most would agree with Binford that middle-range theory is the quest for objective means of giving meaning to our observations on the archaeological record. While there *has* been concern with drawing social inferences from the material facts of the archaeological record, most discussion has centred on material aspects of prehistoric human behaviour (subsistence, settlement, technology). As I see it, this is as much a product of an ecological theoretical framework as of the material bias of the archaeological record. That this is so is clearly indicated by the fact that when social behaviour is addressed by archaeologists, it is often squarely within the framework of ecological functionalism (Conkey 1978; Gamble 1983; Jochim 1983). Society is taken to be an adaptive organization within a theoretical framework that views all of culture as 'man's extrasomatic means of adapting to the environment' (L. White 1959). Viewing all aspects of art, religion, social organization, and technology as necessarily and exclusively adaptive represents an unnecessary limitation on our thinking.

9.2 Archaeological methods of social analysis

Perceived theoretical limitations aside, there are a number of aspects of human sociality upon which archaeologists have successfully focused. Four or five of these are introduced below.

9.2.1 *Social demography*

Attempts have been made to ascertain the nature and duration of local groups as expressed archaeologically (Yellen 1977; Monks 1981). Essentially, this is a determination of the position at which a given prehistoric society falls, along two continua—size and mobility. The range is from small mobile social groups to large sedentary ones. Determining this is considerably more difficult than it appears, since: (a) most simple societies exhibit different structural poses, producing annual and spatial variation in the archaeological record; (b) there are few instances in which archaeologists can define the areal extent and/or duration of occupation of a site; and (c) even given a knowledge of the areal extent of occupation, population estimates remain tenuous at best because different cultures use space differently.

9.2.2 *Degree and nature of material exchange*

This procedure, as in the above case, involves location along two continua: (a) a spatial scale from local to planetary networks of material procurement; and (b) an organizational scale from reciprocal to market relations (Sahlins 1965). Here it can be seen that social and economic domains cannot be divorced. Archaeologists work with the assumption that the extent of material exchange is indicative of the level of socio-cultural integration (Steward 1955), since an organizational network is prerequisite to the exchange of goods. It must be recognized, however, that this assumption is limited by the frequent ethnographic observation that some goods can cover distances of a few hundred kilometres without any social interaction occurring between original source and ultimate possessor (Weissner 1977, 1982 for the !Kung San; Sharp 1952 for the Yir Yiront of Australia).

9.2.3 *Degree of internal differentiation*

Archaeologists monitor at least three continua of internal differentiation within prehistoric societies:

(1) degree of differentiation of individuals along a scale from egalitarian, through ranked status differentiation, to stratified status differentiation with implications for individual wealth;

(2) degree of formal differentiation between residence groups within a given society (for example, stylistic differences in ceramics between different clans within the same society, as suggested by Hill 1968 and Deetz 1965); and

(3) degree of differentiation of population units/settlements along a continuum from homogeneous to heterogeneous/hierarchical (for example, towns, villages, cities).

The first is usually indexed by mortuary patterns, either via analysis of grave goods or of skeletal remains (for example, differences in male–female diet) (Tainter 1978; O'Shea 1984). The second is most frequently monitored through the analysis of the spatial distribution of artefact style (see Stanislawski (1977) for a discussion of some of the pitfalls of this approach). The third concerns the establishment of site hierarchies (according to areal extent, for example) and examines the spatial relationships between hierarchical levels (cf. Johnson 1981, 1982).

9.2.4 *Degree of formal boundaries*

This is the degree to which formal boundaries between recognized population units are maintained. The continuum here is from spatially permeable and undifferentiated to rigid boundary-maintenance. Conkey (1978) has argued that this is a critical evolutionary distinction accompanying the emergence of the first anatomically and culturally modern humans. However, there are three epistemological limitations. First, it remains uncertain to what degree regional differences in style or content of *material culture* reflect social differences that were recognized and used as boundary markers by the prehistoric participants in a given social system. Second, as Wobst (1978) has argued, currently recognized boundaries may well be the product of biases in archaeological/ethnographic research as much as they are products of prehistoric human behaviour. Third, the whole notion of boundaries may represent an ethnocentric imposition of twentieth-century conceptions.

9.2.5 *Degree of local integration*

This is really the 'flip side' of boundary formalization, and refers to the degree to which a given social unit of analysis shares a common cosmology, art style, dialect, technology, etc. The continuum here is from homogeneous societies to plural ones.

In many respects, archaeologists do not deal in the same currency as socio-cultural anthropologists, who are most at home with patterns of kinship, residence, marriage, and belief. Archaeologists have been notoriously unsuccessful in addressing these classic domains of anthropological research. In opposition to Harris's (M. Harris 1968) contention that archaeologists are fortunate to have only the material aspects of culture preserved, it is argued here that this is a serious limitation.

9.3 Archaeology's refinement of the social evolutionary framework

The above limitations aside, archaeologists and ethnologists alike have made real progress in contributing to an assessment of the speculative evolutionary frameworks of the 1960s. This has been accomplished in four ways:

1. Through the recognition that contemporary hunter–gatherers and their prehistoric ancestors were socially complex (in many cases sedentary/ranked) and therefore are not representative of some kind of first step across the nature–culture threshold (King 1978). Moreover, in most cases contemporary hunter-gatherers have been in extended contact, even symbiosis, with pastoralists and agriculturalists (Lee and Devore 1968; Wilmsen 1980; Schrire 1984*a*, *b*; Lee and Guenther 1991).

2. By the demonstration that the earliest hominid sites and assemblages, and indeed those up to as recently as 40 000 years ago, do not reflect human social and subsistence organization as we know them from the ethnographic record (L. Binford 1983).

3. By the partial demonstration that, after 40 000 years ago and especially after 12 000 years ago, the process of social stratification and internal differentiation had already begun (see, for example, Conkey, 1978, 1983; White 1982). Thus, some Late Stone Age societies seem not to conform to Service's (1962) notion of egalitarian 'Paleolithic bands' (Soffer 1985).

4. By addressing the issue of the relationship between settled village life, state-formation, and food-production. In the case of this last process, archaeologists have begun to consider that social obligations and internal differentiation have long-term evolutionary consequences for humans (Bender 1978).

9.4 The evolution of human social patterns

The remainder of this work is a general overview of what the archaeological record reveals concerning changes in human social patterns during the past 2 000 000 years. It is not an exhaustive overview; no single scholar has the background for such an endeavour. Moreover, it will be apparent that a consensus does not exist for any of the issues to be discussed. Therefore, an explicit attempt is made to provide the flavour of some selected debates.

9.4.1 *Social life in the Lower Palaeolithic (2 m.y.a. to 125 000 years ago)*[1]

The archaeological record for the Lower Palaeolithic is composed primarily of open-air scatters of simple stone tools, animal bones, and hominid remains. As Binford (L. Binford 1981) has eloquently pointed out, archaeologists have been prone to the assumption that all bones associated with hominid artefacts were directly related to hominid activity. This assumption has begun to crumble as a result of careful work by Behrensmeyer (1975), Behrensmeyer and Dechant-Boaz (1980), Binford (L. Binford 1981, 1984), Brain (1981), Potts and Shipman (1981) and many others. These authors have shown that, in many instances, the association of (1) human activities as implied by stone tools and (2) animal bones, is 'casual' rather than 'causal'. These results have major implications for our understanding of life at early hominid localities, not only in South and East Africa but throughout Eurasia as well.

Throughout the first half of this century anthropologists often concerned themselves with defining humans in relation to the rest of the animal world. Definitions usually had something to do with humans' being culture-bearers (for example, language, tool-using/making, symbolism); and culture was most frequently viewed as an all-or-nothing phenomenon. In other words, if one found stone tools in a hominid site, one inferred that the occupants had been hunter–gatherers possessing much of the cultural baggage that had been observed among ethnographically recorded groups. This perspective prompted Dart (1957, 1960) and Leakey (see M. Leakey 1971, 1979) to generate a picture of early hominids that portrayed them as competent hunters of big game who manufactured stone tools and exhibited social patterns not unlike those of known hunting and gathering peoples.

In the 1970s, Glynn Isaac (1978) proposed a complex model based on (1) primate behaviour, (2) ethnography, and (3) the East African archaeological record. This model provided 'proto-human hominids' with the following set of characteristics:

(1) tool manufacture;
(2) kinship-based social organization;
(3) hunting;
(4) sexual division of labour;
(5) food-sharing; and
(6) home bases.

In the early 1970s we were definitely too willing to impose the ethnographic record on the past. After all, one wonders after examining this list, what was *proto*-human about these early hominid ancestors?

The most widely discussed aspect of Isaac's model is that of home bases, where food-sharing purportedly occurred. These are defined on the basis of (1) high densities of animal bone and stone artefacts and (2) the presence in these sites of materials brought from some distance. This pattern is different from other kinds of sites that have been proposed (Isaac 1971): quarry sites, transitory camps, and kill or butchery sites. Isaac's work clearly indicates that there is internal differentiation in the archaeological record of the Lower Palaeolithic. However, it remains unclear to what extent this reflects internal differentiation in Lower Palaeolithic social units.

At least two lines of evidence have led to a questioning of Isaac's elaborate 1970s framework, especially that part of it that uses modern humans as analogies (see also Isaac (1984), where he discusses the issues):

(1) substantial differences in brain size (and perhaps organization) between early hominids and modern humans; and

(2) evidence that most of the animal bones associated with early hominid tools and fossils were the product of non-human predators, and at most represent carcasses scavenged by early hominids. For example, there are now some examples of animal bones with tool cut-marks overlying carnivore tooth-marks, indicating that the hominids were second in line.

The first point implies, as was pointed out by Lieberman (1975, p. 5) with respect to language evolution, that there may have been intermediary forms of behaviour that are now extinct. The second line of evidence has been clearly drawn for a number of sites by Potts and Shipman (1981), Potts (1983), Hill (1979), Binford (L. Binford 1981, 1984), and Brain (1981), to name only a few. If in all cases the association between artefacts and animal remains is casual rather than causal, then Isaac's model collapses. For additional discussion of the historical development of Isaac's model prior to his untimely death in 1985 see Blumenschine (1991).

The limited amount of evidence available has led to the use of modern primate behaviour as a means of inferring the nature of social life in the Lower Palaeolithic. This has the opposite effect from projecting modern hunter–gatherer behaviour into the distant past: it de-culturalizes our early ancestors. This is truly playing both ends against the middle.

It is becoming evident, then, that life in the Lower Palaeolithic may have been so different from the modern human condition that it is barely recognizable to us. We are presently at a watershed, having cast doubt on previous conceptions but having only begun the process of generating firm inferences concerning early social life.

9.4.2 *Social life and interassemblage variation in the Middle Palaeolithic (125 000–35 000 years ago)*

In 1950, François Bordes (1950) published a pioneering technique for quantifying, visually representing, and classifying Lower and Middle Palaeolithic stone-tool assemblages.[2] This technique formed the basis for a detailed typological comparison of assemblages and, thus, the recognition of patterns in the data generated. The application of this technique to Eurasian assemblages indicated patterns not at all reminiscent of modern hunter–gatherer organization and behaviour. Major differences in the interpretation of this patterning were at the heart of an infamous debate over the significance of Mousterian variability (Bordes 1961, 1972; Bordes and De Sonneville-Bordes 1970; L. Binford 1972, 1973; Binford and Binford 1966, 1969; Mellars 1970; Collins 1969, 1970). This debate is informative with respect to both the organization of life in the Middle Palaeolithic and the processes and problems of archaeological inference.

The following patterns emerged from Bordes's assemblage comparisons:

1. There were four discrete and recognizable assemblage types, which Bordes labelled Charentian, Typical Mousterian, Mousterian of Acheulean Tradition, and Denticulate Mousterian.[3]

2. The above were not chronological variants; that is, they were all present in south-western France throughout the entire 60 000–100 000-year span of the Mousterian. The picture, then, is one of contemporary variation, but virtually no directional change through time.[4]

3. For the most part, these were not geographical variants, although the Charentian group was named after its predominance in that area of south-western France. However, Bordes's Mousterian assemblage types are recognizable as far afield as Central and Eastern Europe and the Near East. The implication here is that they do not conform to our modern conception of spatially delimited cultures.

Bordes's response to these patterns was to interpret them as indicative of the existence of four distinct and long-lasting cultural traditions within the Mousterian. He referred to these as separate and contemporary Mousterian 'tribes'. These were viewed as nearly sedentary, territorial groups having virtually no influence on each others' material culture.

With this interpretative foundation firmly in place, Binford and Binford (1966) presented an alternative analysis and interpretation, all the while using Bordes's type-counts as a common denominator. Rather than using cumulative frequency graphs to monitor interassemblage variation, the Binfords chose factor analysis, assuming from the outset that Bordes's graphic procedure was probably masking a good deal of contemporaneous variation. Given the nature of this analytic technique, the factors that emerged cross-cut Bordes's assemblage types. Interassemblage variation in the contribution of each factor to assemblage composition was interpreted by the Binfords as reflecting differences in activities. This interpretative framework was drawn from analogy with modern hunter–gatherers, who exhibit differences in group size and composition throughout an annual round.

In 1969 the Binfords went so far as to suggest the specific activities that each factor represented. In retrospect, this was premature, given that functional analysis of stone tools was just then being attempted, and especially given that Bordes's tool-types were employed as the analytical units. Bordes never claimed that his morphological types were functional types. The implication, of course, was that the Binfords' units of analysis may have been inappropriate for speaking of function and activity. Nevertheless, interassemblage variation had been recognized and demanded on explanation.

Both Bordes and the Binfords recognized that their interpretative frameworks were based more on a series of modern conventions than on an objective knowledge of the past. This recognition had earlier led Bordes to seek objective understanding of past technology through modern controlled experimentation with regard to the manufacture of stone tools. The same recognition led Binford to make controlled observations of contemporary bone-processing in order to better understand the practical significance of bone-breakage patterns and body-part frequencies in the Mousterian record.

The struggle with the Mousterian problem continues (cf. the contributions to Trinkaus 1983; also Dibble and Montet-White (1988), plus Dibble and Mellars (1992) and Note 4 to this chapter). Nevertheless, there probably exists a consensus to the effect that, even if we do not understand the specifics, the material record for the Mousterian indicates a social system organized very differently from those observable in the present. Technological complexity was quite limited. There was great technological and stylistic homogeneity over vast distances, implying the absence of social or cultural barriers to information-flow such as we find today. Material exchange, if present at all, was geographically very limited. The imposition of symbols on material objects was non-existent. (See Wynn, this volume, Chapter 10). Lifespans seldom if ever exceeded 40 years, perhaps implying a very different make-up of the basic social group; that is, it would contain few old people with their greater accumulated knowledge.[5]

9.4.3 Social life and culture-as-we-know-it: the Upper Palaeolithic/Late Stone Age (35 000–12 000 years ago)

There are a number of developments that characterize the archaeological record for the shift from the Middle to Upper Palaeolithic, which by convention is set at 35 000 years ago (see Mellars 1973; R. White 1982). These contrasts include:

(1) a much greater degree of regional differentiation;

(2) elaboration of the practice of purposeful burial of the dead, which first came about in the late Mousterian;

(3) a much accelerated rate of culture-change;

(4) bodily adornment in the form of pendants and necklaces;

(5) the movement of material goods a few hundred kilometres; and

(6) the emergence of mobiliary and parietal art.

[Also see the editorial material in Table 28.1.]

Conkey (1978) and I (R. White 1982) both suggest a considerable shift in the nature of human social interaction in the vicinity of the Middle–Upper Palaeolithic transition. It can be convincingly argued that the changes occurring across the transition are reflective of the emergence of complex systems of meaning and social action such as we are accustomed to observing ethnographically (Isaac 1976).

It is only this side of 35 000 years ago, and especially after 20 000 years ago, that the continua outlined at the outset become relevant to the patterning evident in the archaeological record. At long last we are dealing with culturally and anatomically modern humans. In the Upper Palaeolithic, for the continua established above, we can see an increase in differentiation within and between cultural systems. For example:

1. A variety of seasonal structural poses that includes large-scale and fairly long-term aggregation of otherwise dispersed groups. I have suggested elsewhere that this is the context within which much of the Upper Palaeolithic *art mobilier* was being produced and used in conjunction with 'intense' forms of social interaction (R. White 1985b).

2. A much wider network of material distribution focusing on precious items, and especially those used in social display. (For the example of body ornaments see R. White 1989, 1993.) This was probably carried out by means of reciprocal relations (Weissner 1977), a pattern observable in the ethnographic record, but virtually absent prior to the Upper Palaeolithic.

3. A much higher degree of internal differentiation, observable in quantitative differences in burial treatment according to sex (S. Binford 1968; Harrold 1980) and also in the investment of large quantities of 'valuables' in the graves of infants and adolescents.

4. A much higher degree of social-boundary maintenance or inter-group distinction. This is evident in the breakdown of the Upper Palaeolithic world into regionally distinct style-zones with respect to both art and technology (assuming these correlate with intergroup boundaries).

In the light of this increasing social complexity, it can be argued that in some instances material shifts (for example, in mode of subsistence) could have been in response to social/organizational requirements. It is also apparent that the world in which hominids lived was more and more a product of their own conception. Hence, responses to environmental stimuli can be hypothesized to be heavily mediated through social systems of belief and cosmology.

9.4.4 *Social life at the end of the Pleistocene (12 000 to c.8000 years ago)*

By the end of the Upper Palaeolithic, conventionally set at $c.12\,000$ years ago, it is apparent that we are dealing with societies of a complexity equal to or greater than that of the hunting and gathering peoples that have been observed ethnographically (see especially Price and Brown 1985; [and also Dodgshon (1987) and Fig. 9.1, which illustrates plausible ecological determinants of social organization for groups practising hunter–gatherer subsistence strategies—eds.]). Economically, the end of the Pleistocene is characterized by intensification. A primary component of this intensification is an increased emphasis on species with higher rates of reproduction (Hayden 1981), such as fish, shellfish, rabbits, hares, birds, land snails, and seed-grasses. These also come in smaller packages, and usually involve mass harvesting rather than individual targeting.

Explanations for intensification have ranged from the passive (settling in) (Braidwood and Reed 1957) to the ardently adaptive (adaptation to reforested environments) (Clark 1980; Cohen 1977) to the primarily demographic (ever-increasing population) (D. Harris 1977); and see also Editorial Appendix.). For present purposes, it is important to point out one recent author who gives emerging social complexity at the end of the Pleistocene a causal role in this shift to intensification. Bender (1978) has proposed the necessity for increased production to meet the material obligations of complex social relations, an ethnographic example being the North-western coast of North America potlach. The potlach and institutions like it in other societies see individuals produce and accumulate large surpluses, which are then given away or destroyed in order to acquire greater prestige in a system of social ranking. Bender's argument is significant in that it allows us to understand the motivation and rationale for the production of surpluses. The surpluses provided by one particular form of intensification, agriculture, formed the material base upon which much more complicated and internally differentiated cultural systems developed (Smith 1972) by 8000 to 5000 years ago.

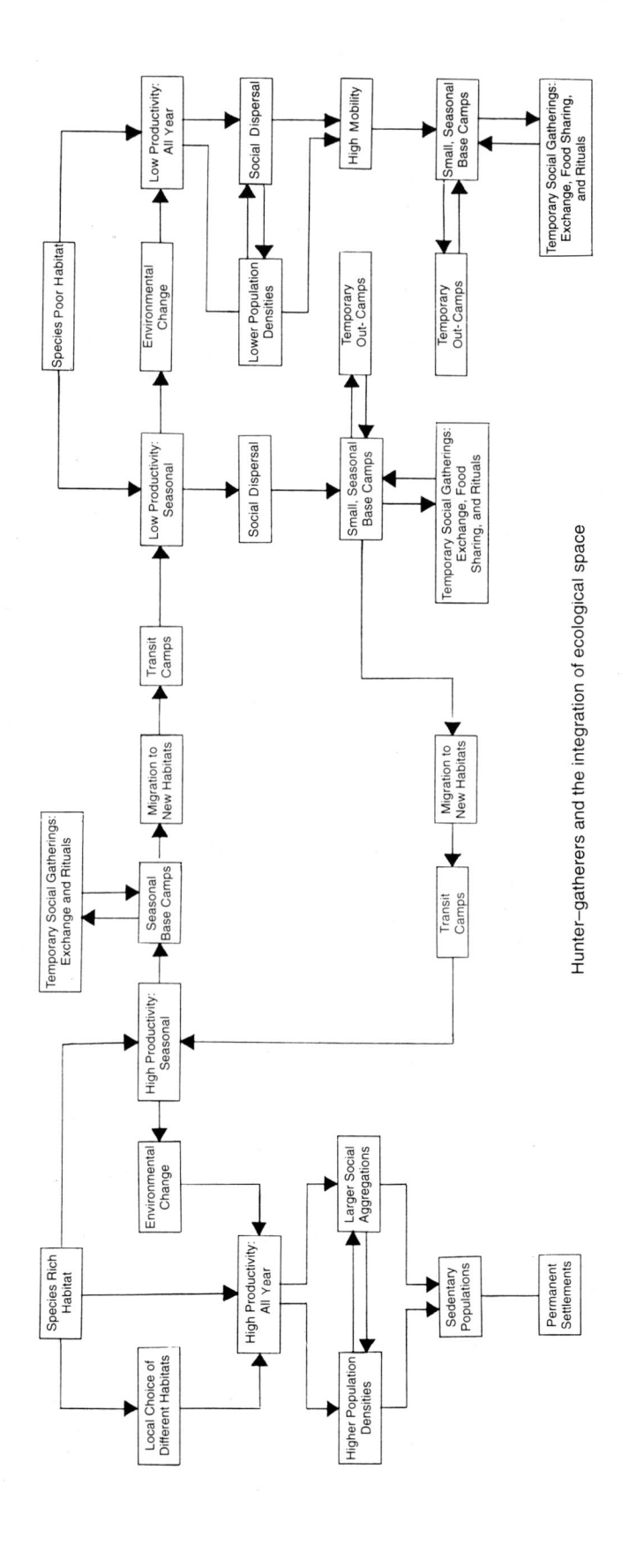

Fig. 9.1 The influence of habitat richness on hunter-gatherer social organization (from Dodgshon (1987) by permission [eds]).

9.4.5 Stratification and the State (c.5000–c.3000 years ago)

The term 'state' is used here to describe societies 'in which there is a centralized and specialized institution of government' (Haas 1982, p. 3). State societies are also characterized by substantial differences in access to wealth, status, and political power. Such societies have emerged in only the last 0.1 per cent of the time since hominids began to make tools.

These days, few would accept surpluses generated through intensification as the *cause* of state-formation in any given area. The reader is referred to Rindos (1984) for a lucid demonstration that agricultural systems do not provide the material stability that is usually assumed. Indeed, there is probably a consensus that it is altered relations of production, not material means of production, that constitute the stuff of state-formation. In other words, the emergence of the state is not to be explained by climatic change or by some major technological invention. For Fried (1978, p. 36).

> ...the state is something more than formally organized society, or even an aggregate of institutions and apparatus of social control at some specified level of complexity. Central to the concept of [the] state [developed by Fried] is an order of stratification, specifically a system whereby different members of a society enjoy invidiously differentiated rights of access to the basic productive necessities of life.

Thus the problem of state origins becomes a socio-political one. From this perspective, at least two poles of opinion can be recognized (see Haas 1982 for a detailed history of thought on the state). One sees the state as a coercive mechanism to maintain the class structure. The other sees the state as basically a positive, integrative mechanism.

However, regardless of one's value judgement of the state, it is generally agreed that in it is seen much greater internal differentiation than in any other form of socio-political organization. At the same time, this differentiation does not amount to 'separate but equal'. There are those who govern and those who are governed; and they share unequally in material wealth and in political power. The archaeological record has very powerfully indicated the extent of internal differentiation in early states. Greater or lesser political power and associated wealth translate directly into massive differences in architecture and burial-treatment. One of the most striking and familiar examples is Pharaonic Egypt, with its massive pyramid burials.

While we are not directly concerned here with the causes of state-formation (but see Editorial Appendix), it is important to provide a sense of how these systems of inequality are thought to have come about; and here, at least at a general level, there is a surprising uniformity of opinion. Specific stimuli vary from decreasing land per capita, in the face of increasing population, resulting in warfare (Carneiro 1978), to external trade (Adams 1966, 1981; Renfrew 1972) for acquiring luxury and/or subsistence goods (see also Wells 1984), to a simple shortage of fertile arable land (Mosely 1975). The common factor in all of this is a strategically controllable commodity that is monopolized or at least manipulated by an entrepreneur (in Wells's terms) who can control access to necessities such as trade goods, well watered land, or military protection. Hallpike (1988) provides additional theoretical discussion as to the role of these various factors in the emergence of the state.

9.5 Some general observations

Over the past two decades our conception of human social evolution has been informed by some very positive trends. The first is the recognition that with respect to the past two million years biology, and perhaps especially neurology (Falk 1980; Holloway 1983), cannot be treated as a given. The term culture when applied to *Homo habilis* means something qualitatively different than it does when applied to biologically modern humans of the late Pleistocene. Leslie White (L. White 1959) was more or less correct when he argued that modern behavioural variation was not somatically, but was culturally based. However, he did not have in mind the full sweep of hominid evolution and the notable long-term behavioural modifications that characterize it. We may well be dealing with very different capacities for culture-as-we-know-it at different evolutionary moments. This possibility is strongly suggested by the observation that there were 'cultural' systems, especially in the Lower Palaeolithic, that do not appear to have changed in any significant way over hundreds of thousands of years.

The second major trend is the emerging recognition that, after 40 000 years ago at least, humans are often responding to a world of their own making (Leroi-Gourhan 1993). Considerable energy is expended not in the food quest, but in manufacturing bodily ornamentation (much of which is very labour-intensive), in more elaborate conventional treatment of the dead, and in a variety of ritual activities that include the exploration, mastery, and organizational appropriation of the remote subterranean world (Leroi-Gourhan 1967; Vialou 1983). It takes little imagination to suppose that many activities, social interactions, and exploitive pursuits were carried out not for their 'practical utility', but because they were considered pleasurable, necessary, or advantageous within a system of conceptions and values created by humans themselves (see Ingold, this volume, Chapter 7, for a full discussion of this point).

Finally, from a methodological perspective, it is most significant that we are coming to realize that the past cannot be understood merely by analogy to some pristine, simple, or primitive present. Human societies, in addition to being the way in which adaptation to the environment is achieved, are also the means by which that environment is conceptualized in the first place (Sahlins 1976). Moreover, the ways in which such conceptualizations are formulated, and indeed the ways in which societies are organized, have much to do with their historical trajectories. It is my clear sense that, more and more, archaeologists are chafing at the bonds of a 'vulgar prehistory' characterized by a high degree of environmental determinism. According a greater consideration to both historical/dialectical and biological factors can only yield a more sophisticated and powerful understanding of the evolution of human social patterns.

Editorial appendix: the development of complex societies

A number of theories to account for the origins of agriculture (generally cited as a prime mover in the instigation of complex human societies), urbanism, and the State have been put forward. Redman (1978a) offers a stage model for the entire spectrum from preagricultural sedentism to urban stratification for the Near East, which provides a context in which these theories may be placed (see Fig. 9.2).

1. *The origins of agriculture*

Stages 1–4 in Fig. 9.2 are concerned with the origins of agriculture. Five main classes of specific theories have been proposed to explain how these stages were motivated:

the *oasis hypothesis* (Childe 1952), Fig. 9.3a;

the *nuclear zone hypothesis* (Braidwood 1970), Fig. 9.3b;

the *neoclimatic change hypothesis* (Wright 1968, 1976), Fig. 9.3c;

the *marginal zone hypothesis* (Binford 1968; Flannery 1969), Fig. 9.3d; and

the *population-pressure hypothesis* (Boserup 1965; Smith and Young 1972), Fig. 9.3e.

In reviewing these theories, Redman points out that three major factors cut across their different foci: those of *environment; culture;* and *social organization*.

Environmental factors

Childe and Wright posit climatic change as a prime mover in the origins of agriculture; Wright and Braidwood emphasize the presence of the necessary wild plants and animals. Wright (for example, 1968, 1977), van Zeist (for example, 1969, 1976), and van Zeist and Wright (1963) have provided the basic impetus to understanding climatic changes in the Near East; but the details of climatic effects on particular sites are still far from certain. Testing environmental hypotheses requires identifying other sites which shared the same environmental features but at which agricul-

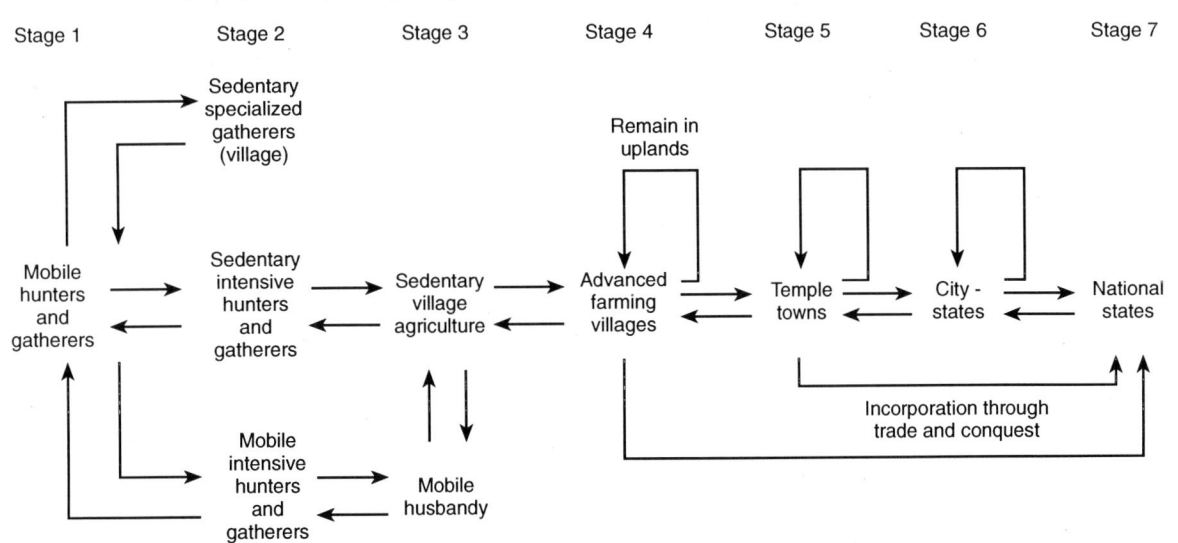

Fig. 9.2 Seven developmental stages of subsistence-settlement development in the Near East (redrawn from Redman 1978a).

ture did not develop; and determining whether different courses of development in different areas show clear relationships with environmental differences.

Cultural factors

Braidwood (for example 1975) suggests that technology and ethno-biological knowledge are essential factors in allowing agriculture to begin. The major question here is whether the invention of techniques came before the practice of agriculture, or occurred in response to the demands of already practising it. Smith and Young (1972) suggest that population density is a factor in promoting cultural innovation, high density leading to innovations to increase productivity. Against this are claims that innovation occurs

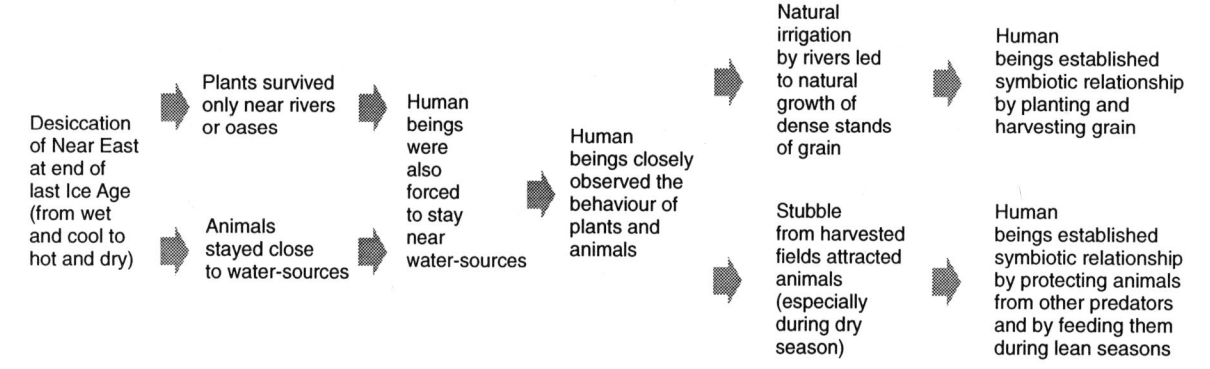

Fig. 9.3a Oasis hypothesis for the origin of agriculture. Agriculture was established when the symbiotic relationships became so strong that plants and animals depended on human beings as much as human beings depended on plants and animals for survival (redrawn from Redman 1978a) (after Childe 1952).

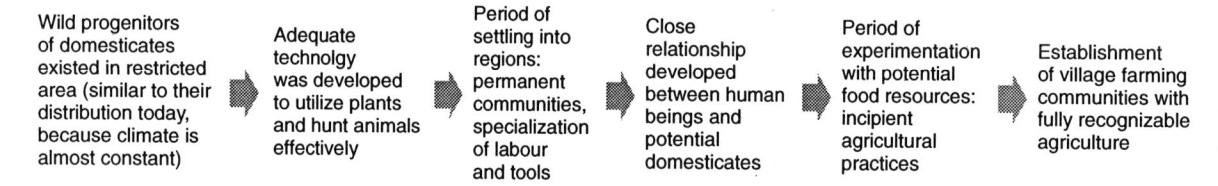

Fig. 9.3b The nuclear zone hypothesis (natural habitat) for the origin of agriculture (redrawn from Redman 1978a) (after Braidwood 1975).

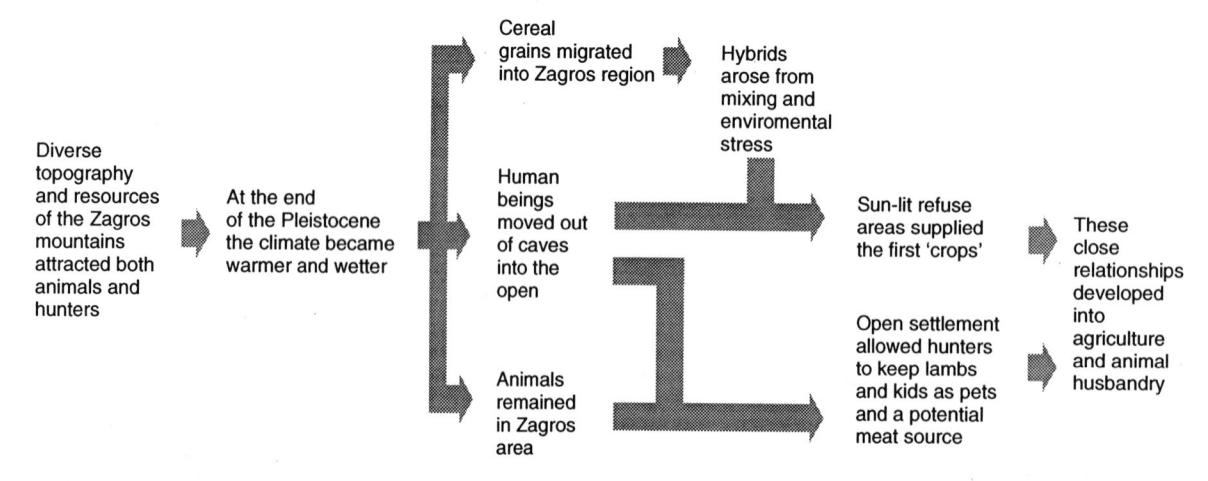

Fig. 9.3c The neoclimatic change hypothesis for the introduction of agriculture (redrawn from Redman 1978a) (after Wright 1968; 1976).

in response to lack of manpower, i.e. *low* population density, and that under conditions of population stress new technologies are not developed, but old ones are intensified. Resolving these issues will require considerably more data.

Social organization

Efficient agriculture requires sedentism, and specialization and division of labour; it allows a greater population density, which leads to concomitant problems of cohesive organization and the scheduling of subsistence activities: all these necessitate major changes in social structure. 'Integrating larger communities, organizing the scheduling of activities, and developing a radically new ethic were amongst the most revolutionary aspects of early agricultural society' (Redman 1978a, p. 105). Whether these developments were a cause or effect of agriculture, however, remains unclear. (Some may have begun earlier with hunter–gatherers, for example increased sedentism—cf. Fig. 9.1.)

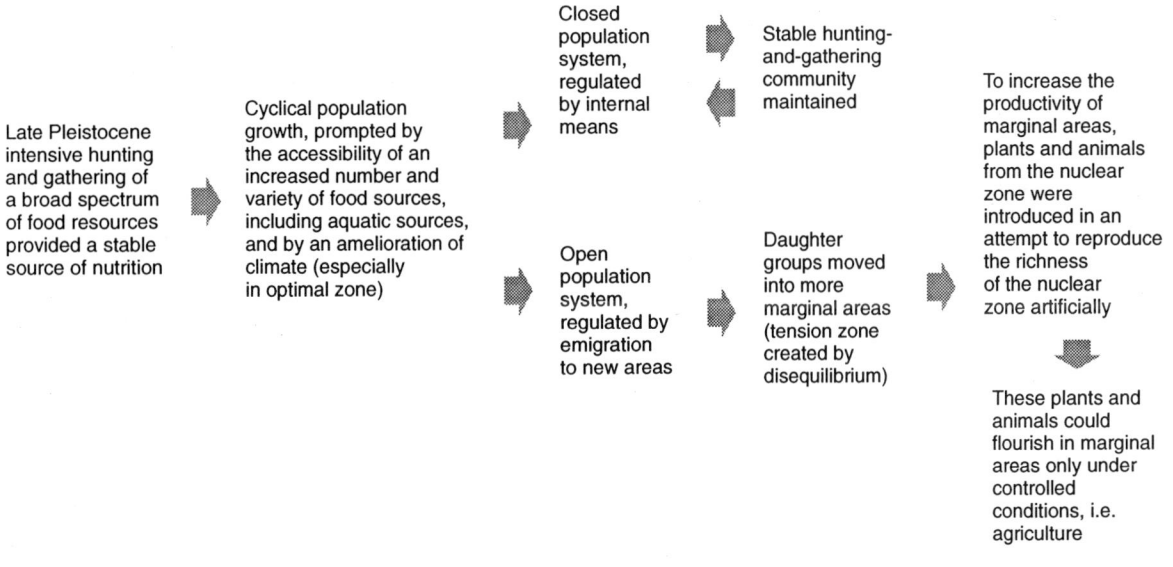

Late Pleistocene intensive hunting and gathering of a broad spectrum of food resources provided a stable source of nutrition ⇒ Cyclical population growth, prompted by the accessibility of an increased number and variety of food sources, including aquatic sources, and by an amelioration of climate (especially in optimal zone) ⇒ Closed population system, regulated by internal means ⇒ Stable hunting-and-gathering community maintained ⇐ To increase the productivity of marginal areas, plants and animals from the nuclear zone were introduced in an attempt to reproduce the richness of the nuclear zone artificially

⇒ Open population system, regulated by emigration to new areas ⇒ Daughter groups moved into more marginal areas (tension zone created by disequilibrium) ⇒

⇓ These plants and animals could flourish in marginal areas only under controlled conditions, i.e. agriculture

Fig. 9.3d The marginal-zone hypothesis for the origin of agriculture (redrawn from Redman 1978a) (after L. Binford 1968; Flannery 1969).

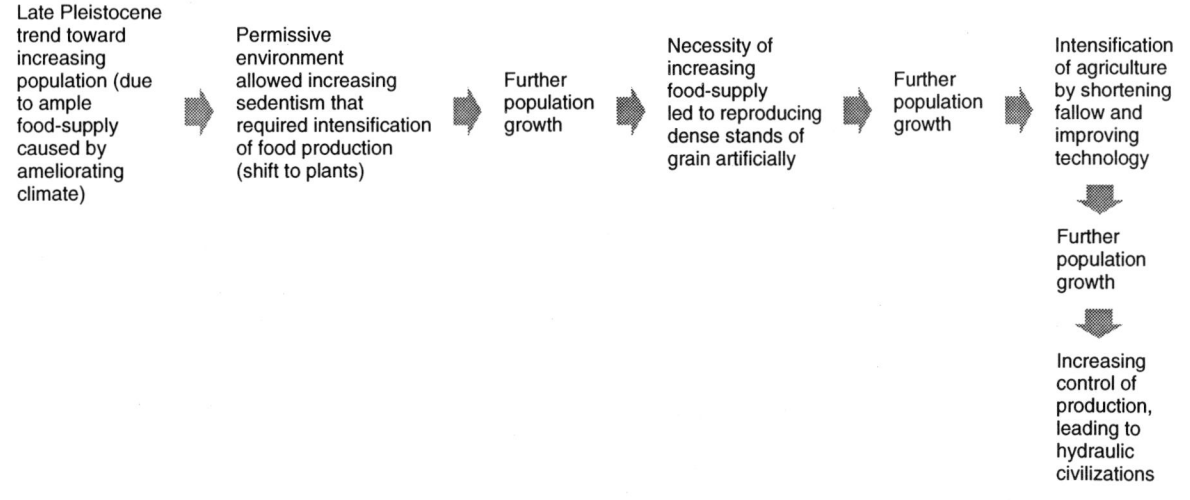

Late Pleistocene trend toward increasing population (due to ample food-supply caused by ameliorating climate) ⇒ Permissive environment allowed increasing sedentism that required intensification of food production (shift to plants) ⇒ Further population growth ⇒ Necessity of increasing food-supply led to reproducing dense stands of grain artificially ⇒ Further population growth ⇒ Intensification of agriculture by shortening fallow and improving technology

⇓ Further population growth

⇓ Increasing control of production, leading to hydraulic civilizations

Fig. 9.3e The population-pressure hypothesis for the origin of agriculture and the rise of early civilizations (redrawn from Redman 1978 *a*) (after Boserup 1965; Smith and Young 1972).

It seems most likely that no single-cause model will be sufficient to account for the origins of agriculture, and it is posited the above factors operated in varying degrees in a positive feedback fashion. The determination of those different degrees requires precise data for the separate and unique pathways by which agriculture was established in different locales:

Although every cultural event was unique and similar events occurring at different sites had distinguishing characteristics, there are regularities to be found in the events that were taking place in the different regions. The most important goals of an inquiry into the origins of agriculture are to discover these regularities and to explain the differences observed (Redman 1978a, p. 106).

2. The origins of urbanism and the State

As with agriculture, several different accounts of urban and state origins have been advanced:

the *hydraulic-managerial hypothesis* (Wittfogel 1957; also Mitchell 1973), Fig. 9.4a;

the *craft-specialization-and-irrigation hypothesis* (Childe 1950), Fig. 9.4b;

the *population-pressure-and-conflict hypotheses* (Carneiro 1961, 1970; Diakonoff 1969; Smith and Young 1972; Gibson 1973), Fig. 9.4c;

the *interregional- and intraregional-exchange hypotheses* (Sanders 1968; Rathje 1971; Flannery 1968), Fig. 9.4d; and

the *multiple-factor-and-organizational hypothesis* (Adams 1966; see also Flannery 1972, 1973), Fig. 9.4e.

Again, each of these hypotheses has its shortcomings: for example, population-pressure hypotheses do not explain why those pressures were initially generated in human groups which previously had kept their numbers relatively stable for long periods of time;

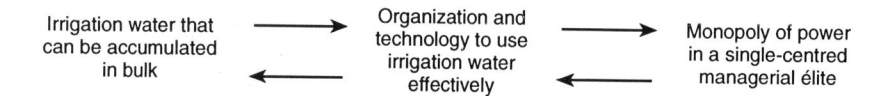

Fig. 9.4a Interrelationships of variables in the hydraulic–managerial hypothesis of the formation of a state (redrawn from Redman 1978a) (after Wittfogel 1957).

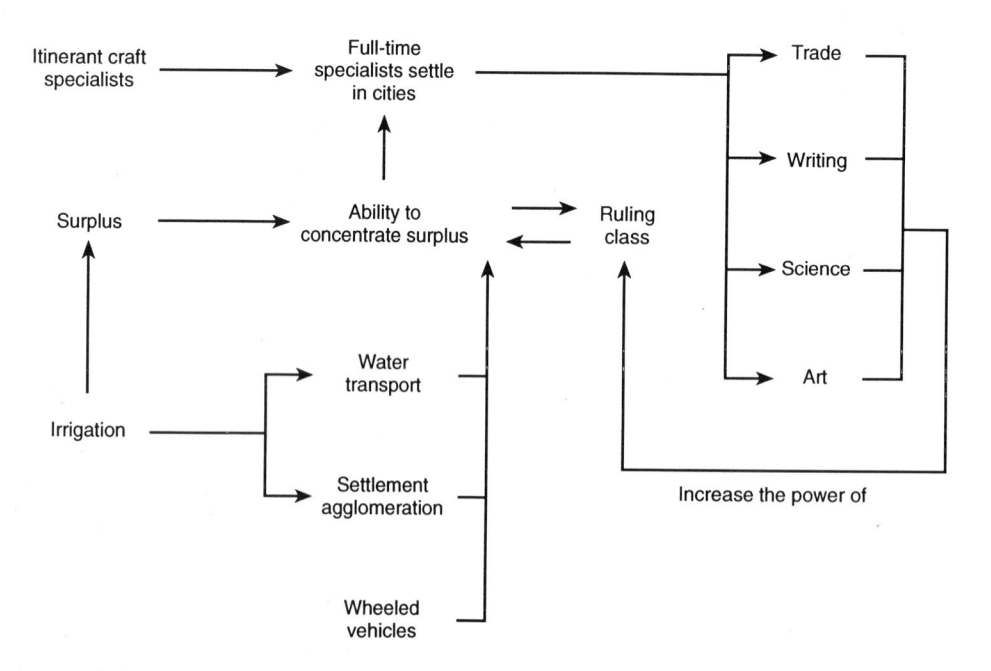

Fig. 9.4b Interrelationships of variables in the craft-specialization- and-irrigation hypothesis for the growth of urbanism (redrawn from Redman 1978a) (after Childe 1950).

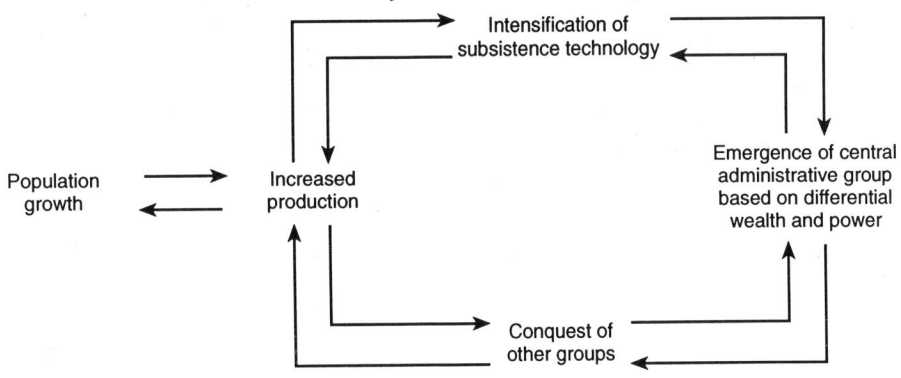

Fig. 9.4c Interrelationships of variables in the population-pressure- and-conflict hypotheses of the formation of a state (redrawn from Redman 1978a) (after Carneiro 1970; Diakonoff 1969; Smith and Young 1972; Gibson 1973).

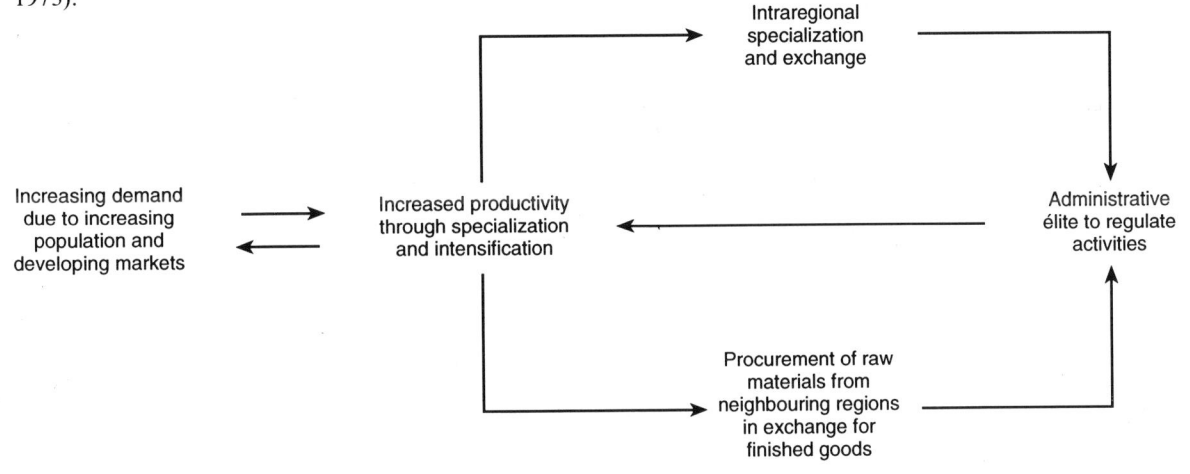

Fig. 9.4d Interrelationships of variables in the interregional- and intraregional-exchange hypotheses of urban development (redrawn from Redman 1978a) (after Sanders 1968; Rathje 1971; Flannery 1968).

exchange hypotheses cannot distinguish between trade as a cause or effect of urbanism; and so on.

Redman (1978a, b) has put forward a systems-ecological model which constitutes an investigatory framework within which urban and state origins can be researched. He describes (1978a, p. 229) the essential features of this model thus:

Urbanization was not a linear arrangement in which one factor caused a change in a second factor, which then caused a change in a third, and so on. Rather, the rise of civilization should be conceptualized as a series of interacting incremental processes that were triggered by favourable ecological and cultural conditions and that continued to develop through mutually reinforcing interactions. The developmental process comprised five positive-feedback interrelationships, three of which [A, B, and C in Fig. 9.5] were prompted by the ecology and gave rise to institutions that characterized early Mesopotamian cities. The fourth and fifth positive-feedback relationships [D and E in Fig. 9.5] were stimulated by early urban developments and helped transform the independent cities into members of... centralized [empires] [referred to as National states in Fig. 9.2] [eds].

Notes

1. Dates given in headings are approximate, and for rough orientation only (eds).

2. Bordes distinguishes 63 separate categories of tools, grouped as varieties of scrapers, backed knives, hand-axes, borers, points, etc. It is the differing proportions of these categories in the tools recovered from a particular geological stratum that provide the bases for the assemblage types. For accessible details, see Bordes (1972, pp. 48–54) and Binford (1983, pp. 87–92).

3. The different assemblages are distinguished on the basis of the proportions of the different tool types

that are represented in distinct excavated strata (see Figs. 9.6 and 9.7). The assemblages are:

Typical Mousterian, defined by most tool types' being present in roughly equal proportions. It differs from

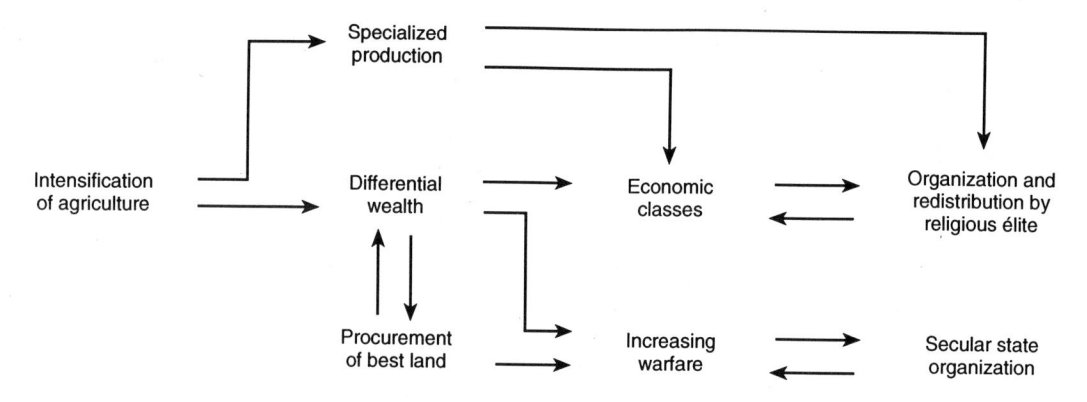

Fig. 9.4e Interrelationships of variables in the multiple-factor-and-organizational hypothesis of the formation of a state (redrawn from Redman 1978a) (after Adams 1966).

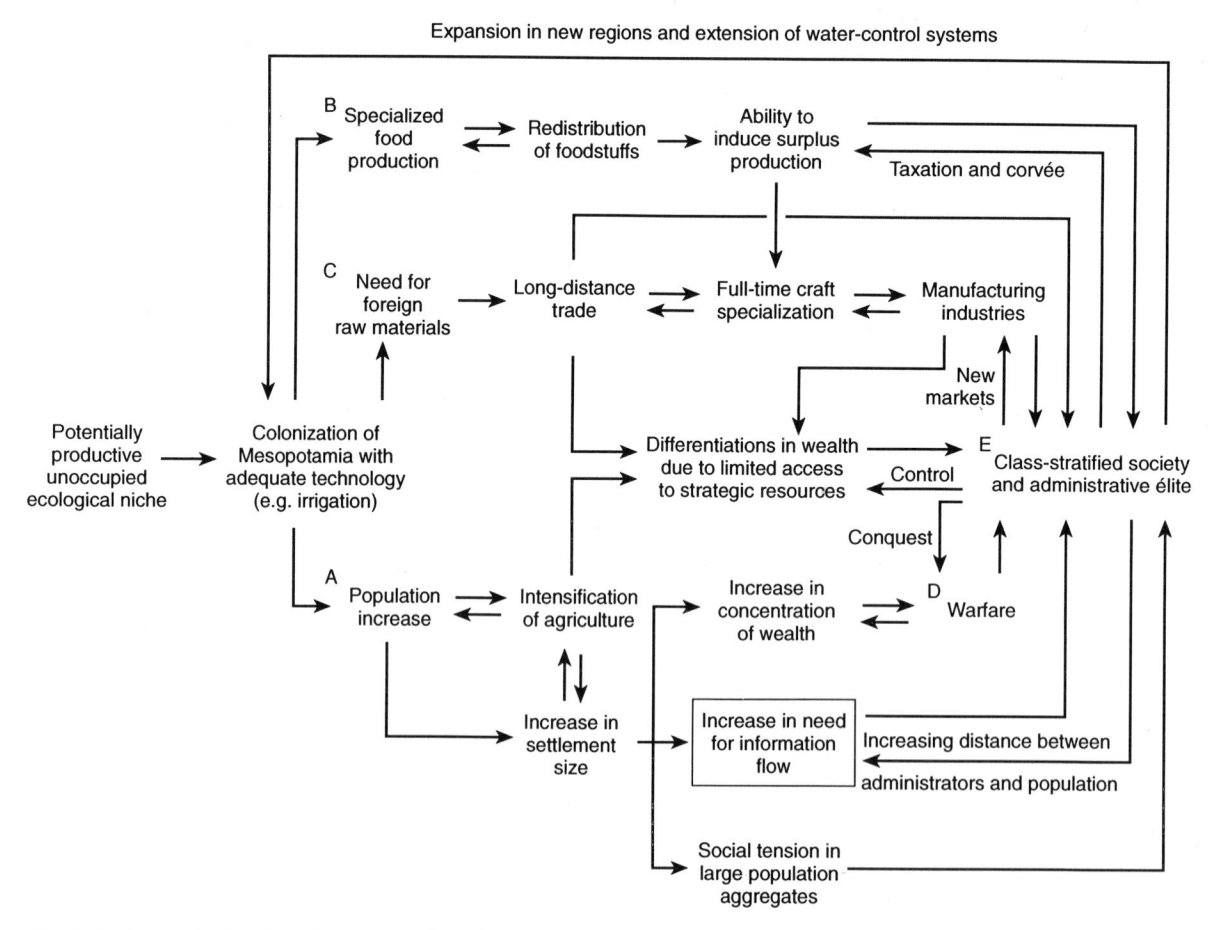

Fig. 9.5 Interrelationships between cultural and environmental variables leading to the increasing stratification of class structure in Mesopotamian society (redrawn from Redman 1978a).

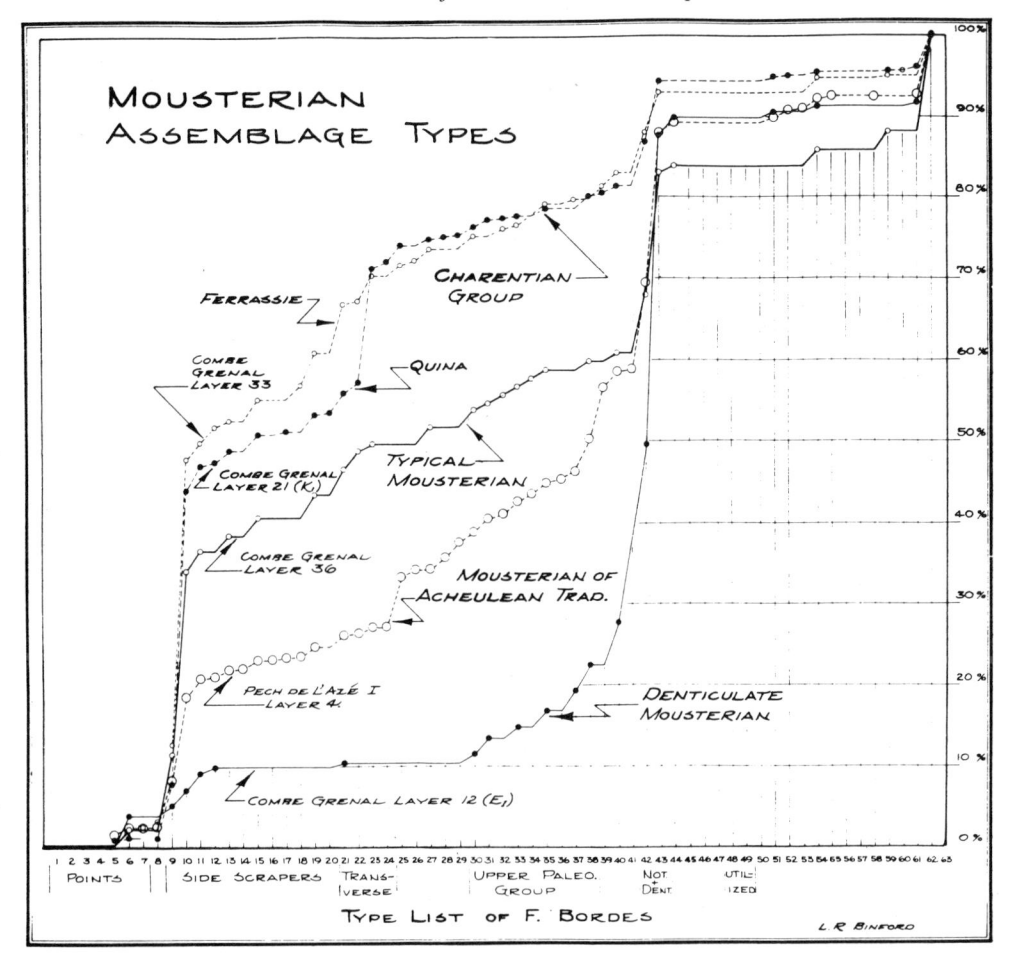

Fig. 9.6 Graphic definition of Bordes's Mousterian tool-assemblage types. Tool types are listed along the horizontal axis. A plot of the cumulative percentages of each tool type within an assemblage generally conforms to one of the four major shape categories (from Binford 1983, by permission of the author).

the Mousterian of Acheulean tradition in the rarity of hand-axes, and lower frequencies of backed knives and other tools analogous to common Upper Palaeolithic types (see Fig. 9.8).

Denticulate Mousterian shows few scrapers and many notches and denticulates; hand-axes are rare to absent; scrapers are poorly made; and backed knives are rare (see Fig. 9.9).

Charentian assemblage is dominated by scrapers; denticulates and notches are infrequent; hand-axes and backed scrapers are rare. Bordes recognized two subtypes of the Charentian: *Quina*, in which scrapers are typically of transverse form and the Levallois (see Wynn, this volume, Chapter 10) technique is rare to absent; and *Ferrassie*, characterized by scrapers manufactured on the sides of flakes, and the frequent presence of the Levallois technique (see Figs. 9.10 and 9.11).

Mousterian of Acheulean Tradition has hand-axes present; moderate numbers of side-scrapers; many denticulates and notches; and a distinctively high frequency of backed knives (see Fig. 9.12). [eds]

4. Mellars (1986a) notes that the traditional interpretations of these assemblages (including those of Bordes) have assumed that the major documented sites, Le Moustier and Combe Grenal, span the same time period, and thus provide parallel records for the greater part of the Mousterian period. Since there was little relation between the sequence of assemblages recovered from the two sites, the idea that these assemblages did not represent a chronological sequence was put forward—these varying manufac-

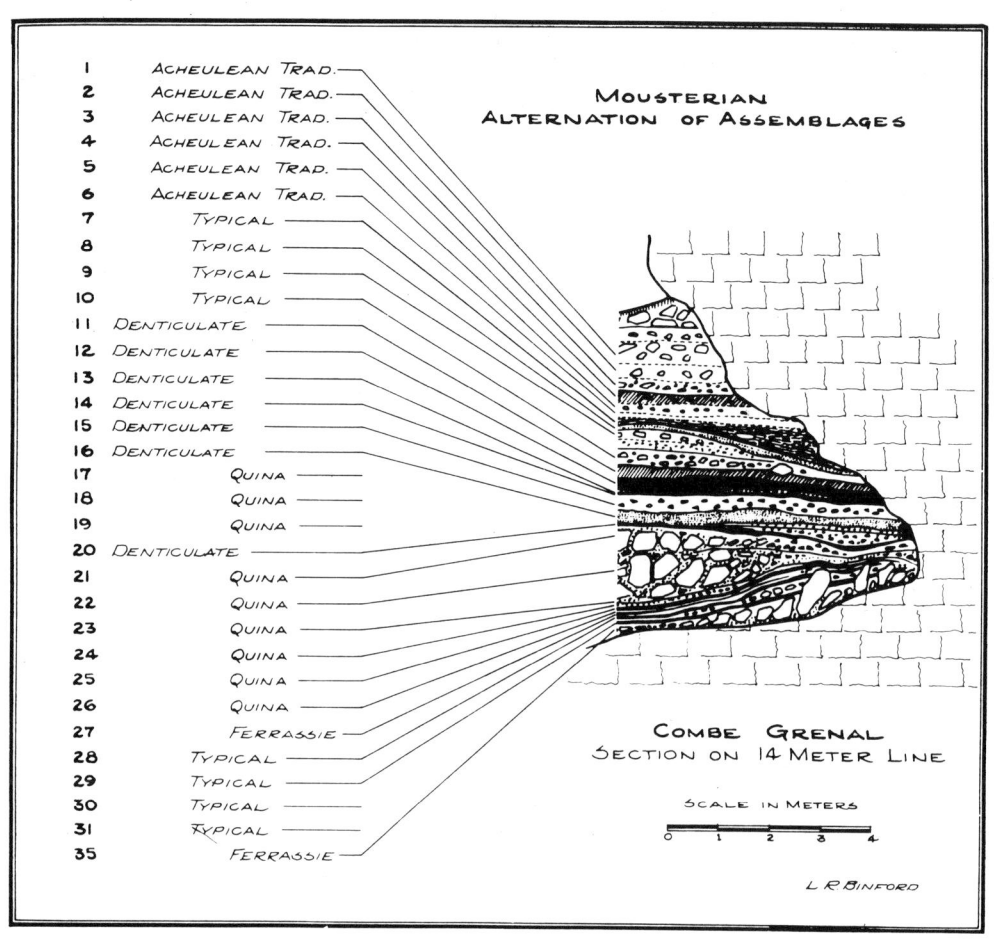

Fig. 9.7 Archeological section from the Mousterian site of Combe Grenal in France, illustrating alternating assemblages (from Binford 1983, by permission of the author).

turing traditions were interpreted as co-existing throughout the Mousterian in south-west France. Additionally, the widespread geographical distribution of these differing assemblages led to Bordes's suggestion that they represent the traditional industries of distinct 'tribes', which had little influence upon each other and tended to settle in a particular locale for a long period before relocating to another locale at which a continuous occupation was re-established.

A problem with interpretations of these assemblages results from the time period (c.40 000 to 115 000 BP) of the Mousterian lying beyond conventional radiocarbon dating techniques. Thus the construction of an internal Mousterian chronology has had to rely on correlations between geological and palaeo-environmental sequences between individual sites—which require a high degree of 'interpretation'; or on equally controversial correlations between the archaeological materials recovered from these sites. However, the technique of thermoluminescence dating of burnt flints has recently been applied to Mousterian deposits (Valladas et al. 1986). Mellars notes that the precise implications of this dating technique for understanding the relationships between various Neanderthal finds and their relation to Cro-Magnon forms remain to be seen. However, in his view:

> The geological, archaeological and therminoluminescence evidence now converges in suggesting that the two sequences at Le Moustier and Combe Grenal represent essentially successive episodes in the total sequence of Mousterian occupation in south-west France, and together provide a largely complete record of human occupation in the area over a period of at least 70 000–80 000 years (1986a, p. 411)

This sequence is illustrated in Fig. 9.13

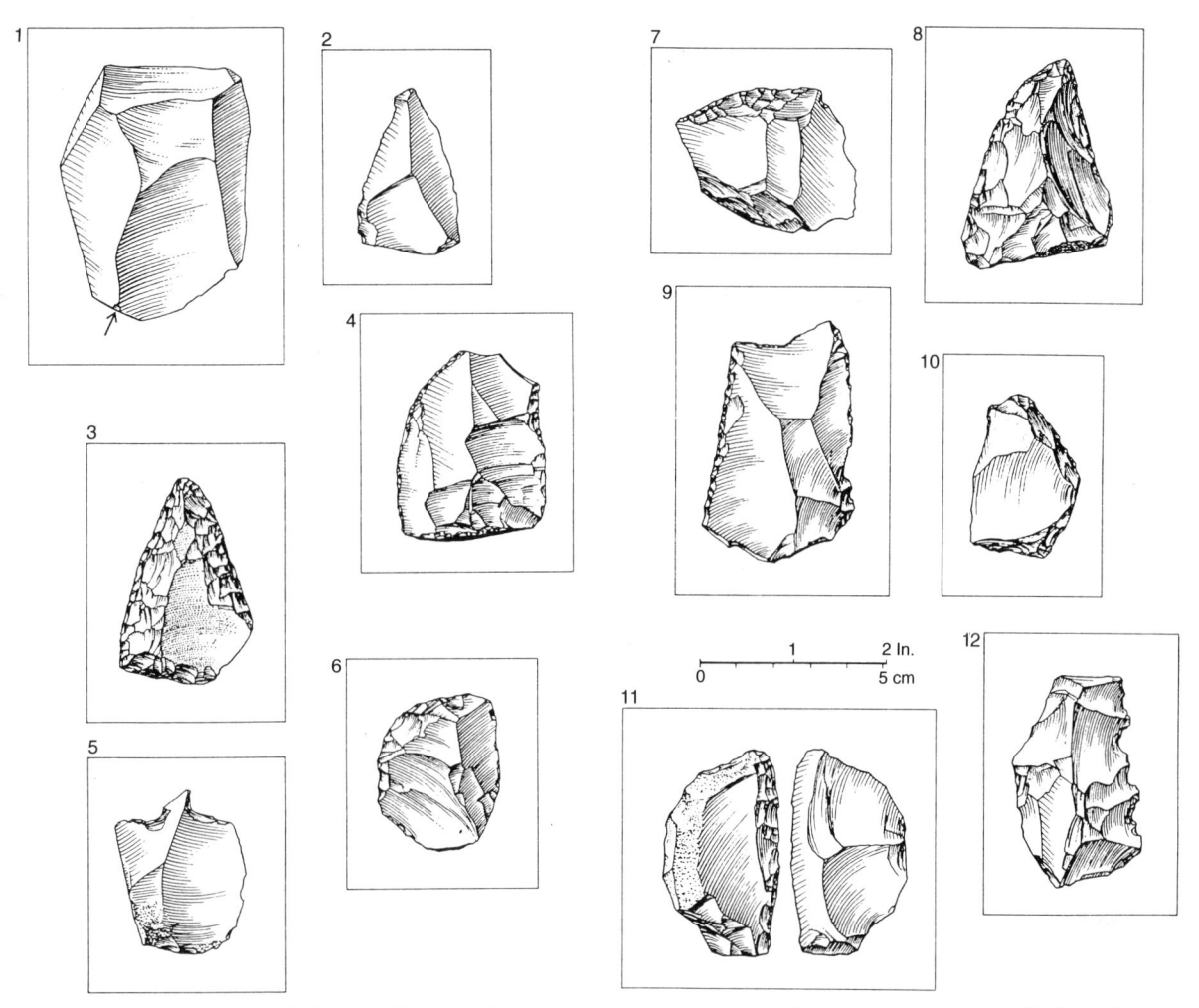

Fig. 9.8 Examples of tools from a Typical Mousterian assemblage, from Combe Grenal, Layer 50 (from Bordes 1972).
1, Levallois flake. 2, Levallois point. 3, Mousterian point. 4, Double side-scraper. 5, Borer. 6, Convex side-scraper. 7, Transverse scraper. 8, Convergent scraper. 9, Straight side-scraper. 10, End-scraper. 11, Side-scraper with a thinned back. 12, Denticulate.

That these sequences may be successive may offer a challenge to Bordes's interpretation (although Mellars's claims are themselves controversial—see Ashton and Cook 1986; Mellars 1986b; also Mellars 1992 for a recent discussion). In addition to these direct implications for the interpretation of the assemblages, there are others, requiring further chains of inference, that bear on the relative and absolute chronology of a number of Neanderthal fossil specimens. Thus, as White observes in this chapter, 'the struggle with the Mousterian problem continues' [eds].

5. Elsewhere I have suggested that this implied different organization may have been linked to the absence of a modern form of language (R. White 1985a), in that the ability to transmit complex ideas across generations was very limited, as was the ability to imagine forms and then achieve them in plastic media such as bone, antler, and wood.

References

Adams, R. McC. (1966). *The evolution of urban society: early Mesopotamia and prehispanic Mexico.* Aldine, Chicago.

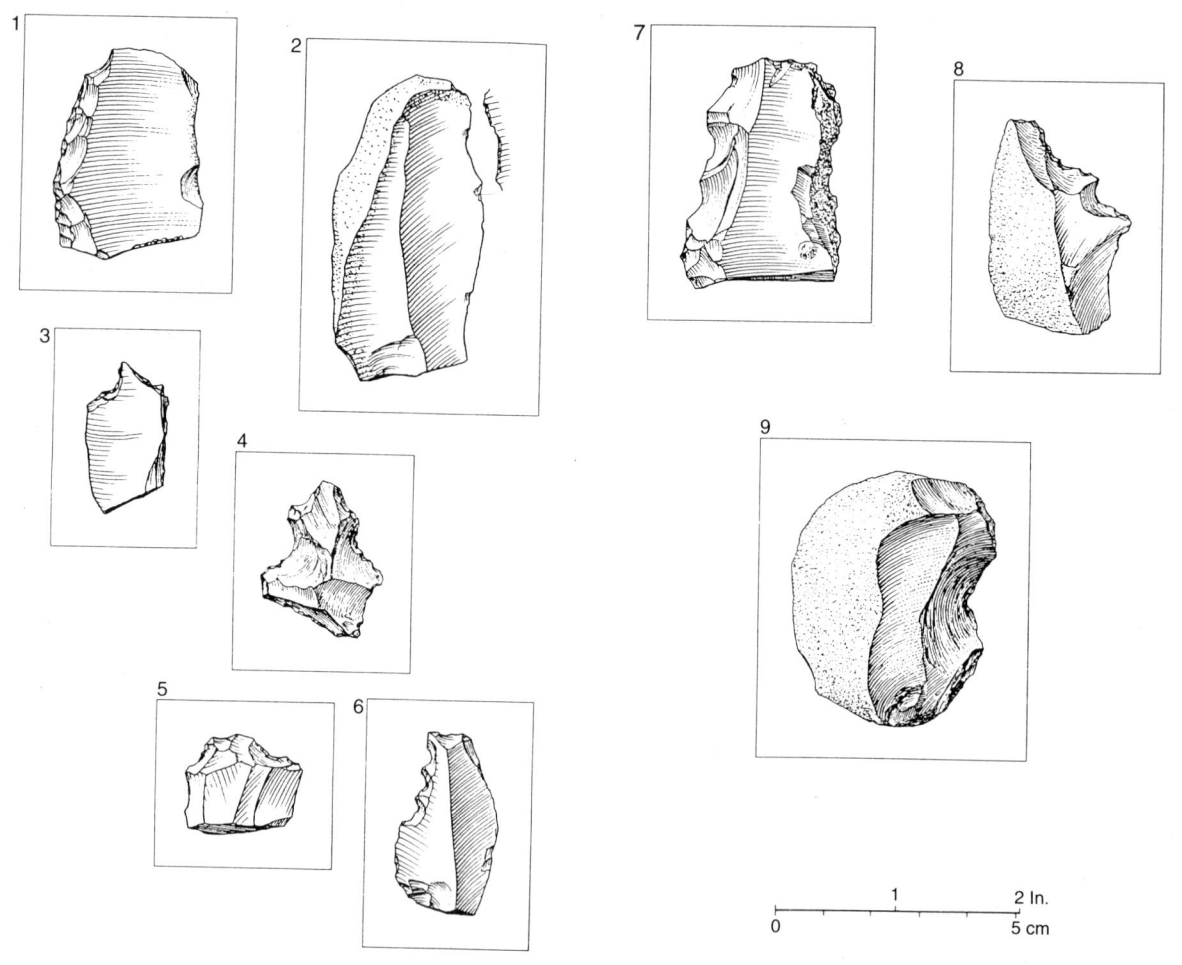

Fig. 9.9 Examples of tools from a Denticulate Mousterian assemblage, from Combe Grenal, Layer 13 (from Bordes 1972).
1, Side-scraper. 2, Natural backed knife. 3, Borer. 4, 6, 7, 8, Denticulates. 5, Atypical end-scraper. 9, Notch.

Adams, R. McC. (1981). *Heartland of cities*. Aldine, Chicago.
Ashton, N. and Cook, J. (1986). Dating and correlating the French Mousterian. *Nature*, **324**, 113.
Behrensmeyer, A. (1975). The taphonomy and paleoecology of Plio-Pleistocene vertebrate assemblages of Lake Rudolf, Kenya. *Bulletin of the Museum of Comparative Zoology*, **146**, 473–578.
Behrensmeyer, A. and Dechant-Boaz, D. (1980). The recent bones of Amboseli Park, Kenya, in relation to East African paleoecology. In *Fossils in the making: vertebrate taphonomy and paleoecology* (ed. A. Behrensmeyer and A. Hill), pp. 72–92. University of Chicago Press.
Bender, B. (1978). Gatherer–hunter to farmer: a social perspective. *World Archaeology*, **10**, 204–22.
Binford, L. R. (1968). Post-Pleistocene adaptations. In *New perspectives in archaeology* (ed. S. R. Binford and L. R. Binford), pp. 313–41. Aldine, Chicago.

Binford, L. (1972). Contemporary model building: paradigms and the current state of Paleolithic research. In *Models in archaeology* (ed. D. Clarke), pp. 109–66. Methuen, London.
Binford, L. (1973). Interassemblage variability: the Mousterian and the 'functional' argument. In *The explanation of culture change* (ed. C. Renfrew), pp. 227–54. Duckworth, London.
Binford, L. (1977). General introduction. In *For theory building in archaeology* (ed. L. Binford), pp. 1–10. Academic Press, New York.
Binford, L. (1978). *Nunamiut ethnoarchaeology*. Academic Press, New York.
Binford, L. (1979). Organization and formation processes: looking at curated technologies. *Journal of Anthropological Research*, **35**, 255–73.
Binford, L. (1981). *Bones: ancient men and modern myths*. Academic Press, New York.

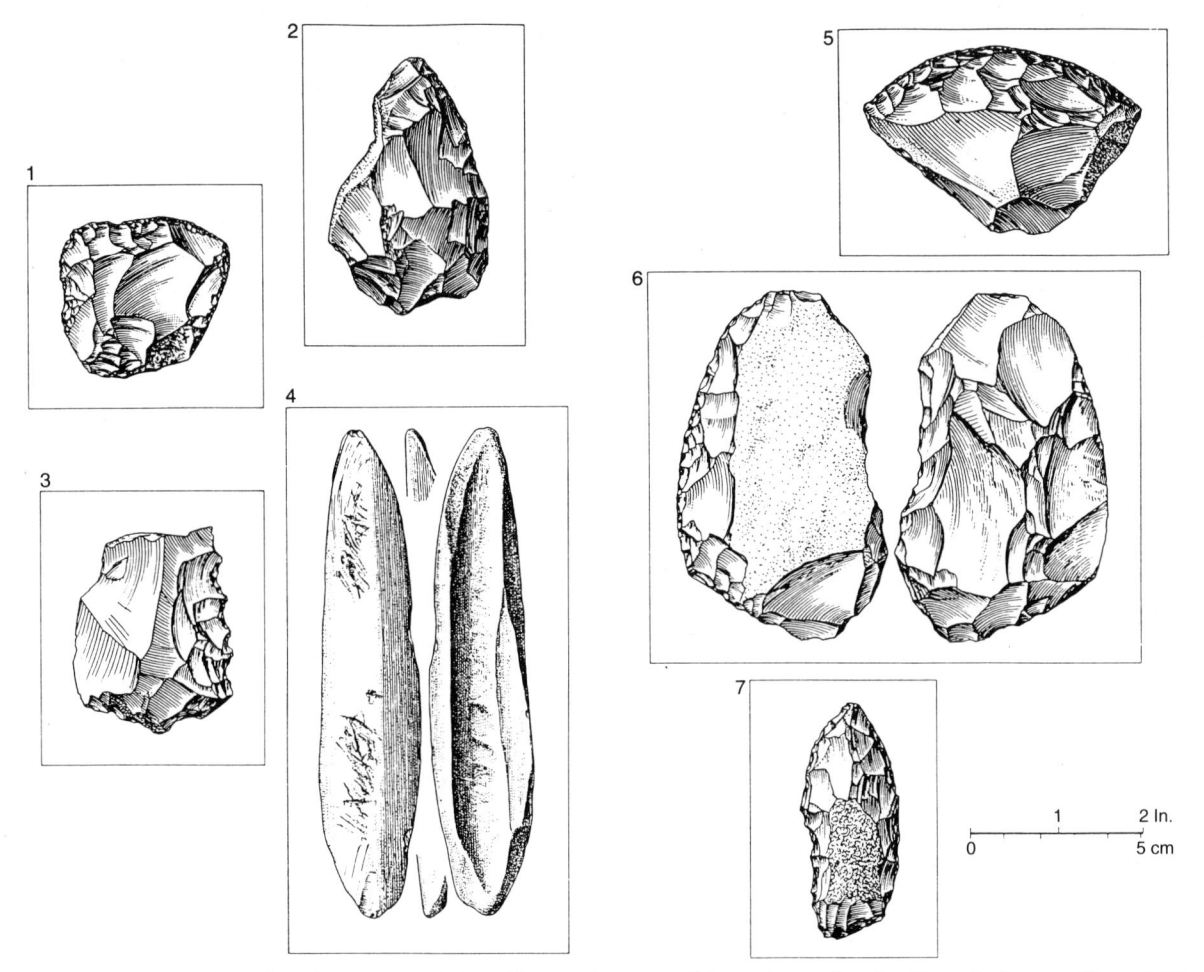

Fig. 9.10 Examples of tools from a Quina Charentian assemblage, from Combe Grenal, Layers 22 and 23 (from Bordes 1972).
1, Déjeté side-scraper. 2, Quina-type convex side-scraper. 3, Denticulate. 4, Utilized bone: the tips are worn smooth; scratches on the side indicate utilization as a flaker. 5, Quina-type transversal scraper. 6, Quina-type bifacial scraper. 7, 'Limace.'

Binford, L. (1983). *In pursuit of the past. Decoding the archaeological record.* Thames and Hudson, New York.

Binford, L. (1984). *Faunal remains from Klasies River Mouth.* Academic Press, New York.

Binford, L. and Binford, S. (1966). A preliminary analysis of functional variability in the Mousterian of Levallois facies. *American Anthropologist,* **68**, 238–95.

Binford, L. and Binford, S. (1969). Stone tools and human behavior. *Scientific American,* **220**, 70–84.

Binford, S. (1968). A structural comparison of disposal of the dead in the Mousterian and the Upper Paleolithic. *Southwestern Journal of Anthropology,* **24**, 139–54.

Blumenschine, R. J. (1991). Breakfast at Olorgesailie: the natural history approach to Early Stone Age archaeology. *Journal of Human Evolution,* **21**, 307–27.

Bordes, F. (1950). Principes d'une méthode d'étude des techniques de débitage et de la typologie du Paléolithique ancien et moyen. *L'Anthropologie,* **54**, 19–34.

Bordes, F. (1961). *Typologie du Paléolithique ancien et moyen.* Mémoires de l'Institut de Préhistoire de l'Université de Bordeaux, No. 1. L'Institut, Bordeaux.

Bordes, F. (1972). *A tale of two caves.* Harper and Row, New York.

Bordes, F. and de Sonneville-Bordes, D. (1970). The significance of variability in Palaeolithic assemblages. *World Archaeology,* **2**, 61–73.

Boserup, E. (1965). *The conditions of agricultural growth.* Aldine, Chicago.

Braidwood, R. J. (1970). Prehistory into history in the Near East. In *Radiocarbon variations and absolute*

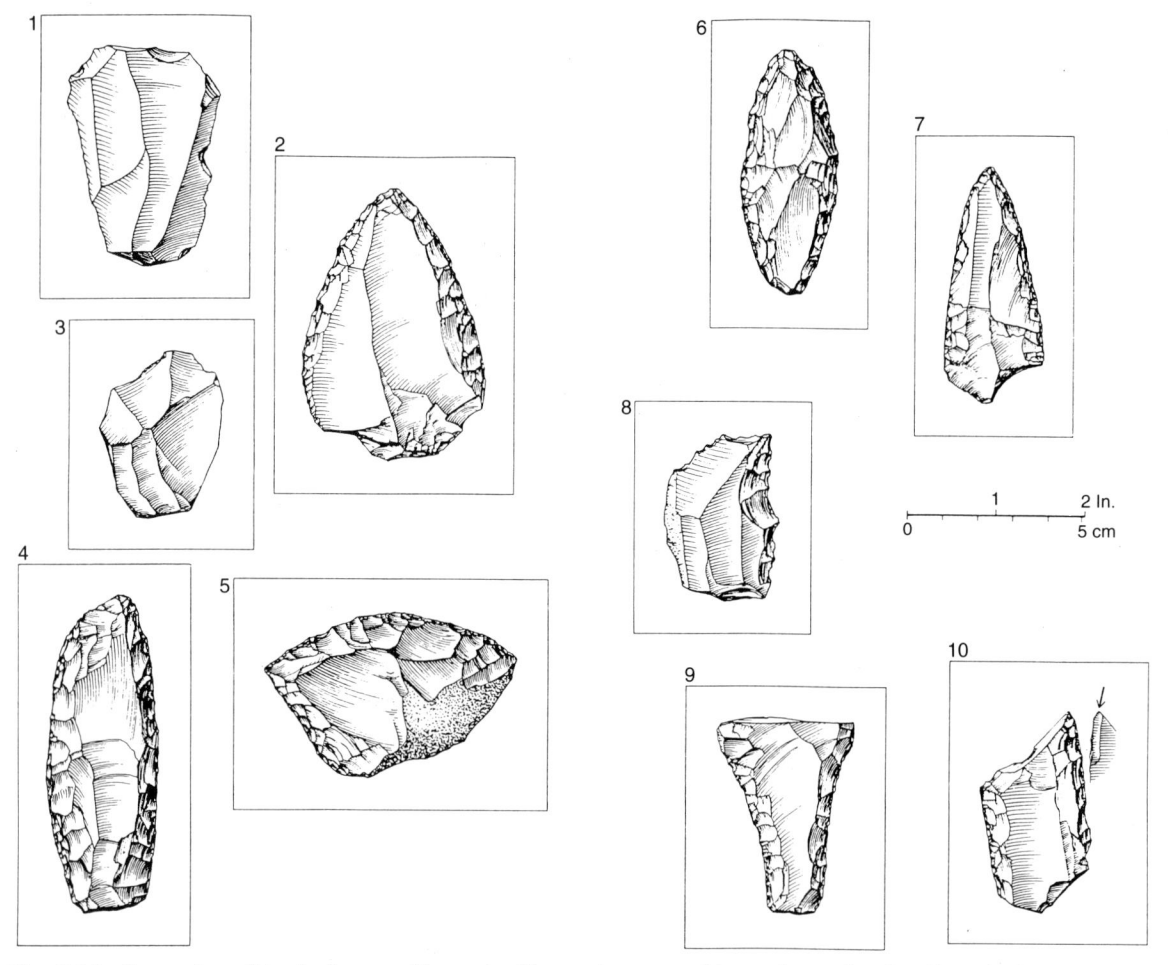

Fig. 9.11 Examples of tools from a Ferrassie Charentian assemblage, from Combe Grenal, Layer 27 (from Bordes 1972).
1, 3, Levallois flakes. 2, Convergent scraper. 4, Double side-scraper. 5, Déjeté scraper. 6, 'Limace.' 7, Mousterian point. 8, Denticulate. 9, Double concave side-scraper. 10, Burin.

chronology (ed. I. U. Olsson), pp. 81–91. Wiley, New York.
Braidwood, R. J. (1975). *Prehistoric men* (8th edn). Scott, Foresman, Glenview, Illinois.
Braidwood, R. and Reed, C. (1957). The achievement and early consequences of food production: a consideration of the archaeological and natural-historical evidence. *Cold Springs Harbor Symposia on Quantitative Biology*, **22**, 19–31.
Brain, C. (1981). *The hunters or the hunted? An introduction to African cave taphonomy*. University of Chicago Press.
Carneiro, R. L. (1961). Slash-and-burn cultivation among the Kuikuru and its implications for cultural development in the Amazon basin. In *The evolution of horticultural systems in native South America, causes and consequences: a symposium* (ed. J. Wilbert). *Anthropologica*, suppl. 2, 47–67. Reprinted in *Man in adaptation: the cultural present* (ed. Y. A. Cohen), pp. 131–45. Aldine, Chicago.
Carneiro, R. L. (1970). A theory of the origin of the state. *Science*, **169**, 733–8.
Carneiro, R. (1978). Political expansion as an expression of the principle of competitive exclusion. In *Origins of the state: the anthropology of political evolution* (ed. R. Cohen and E. Service), pp. 205–23. ISHI, Philadelphia.
Childe, V. G. (1936). *Man makes himself*. New American Library, New York.
Childe, V. G. (1942). *What happened in history*. Penguin, Harmondsworth.
Childe, V. G. (1950). The urban revolution. *Town Planning Review*, **21**, 3–17.
Childe, V. G. (1952). *New light on the most ancient East*. Praeger, New York.

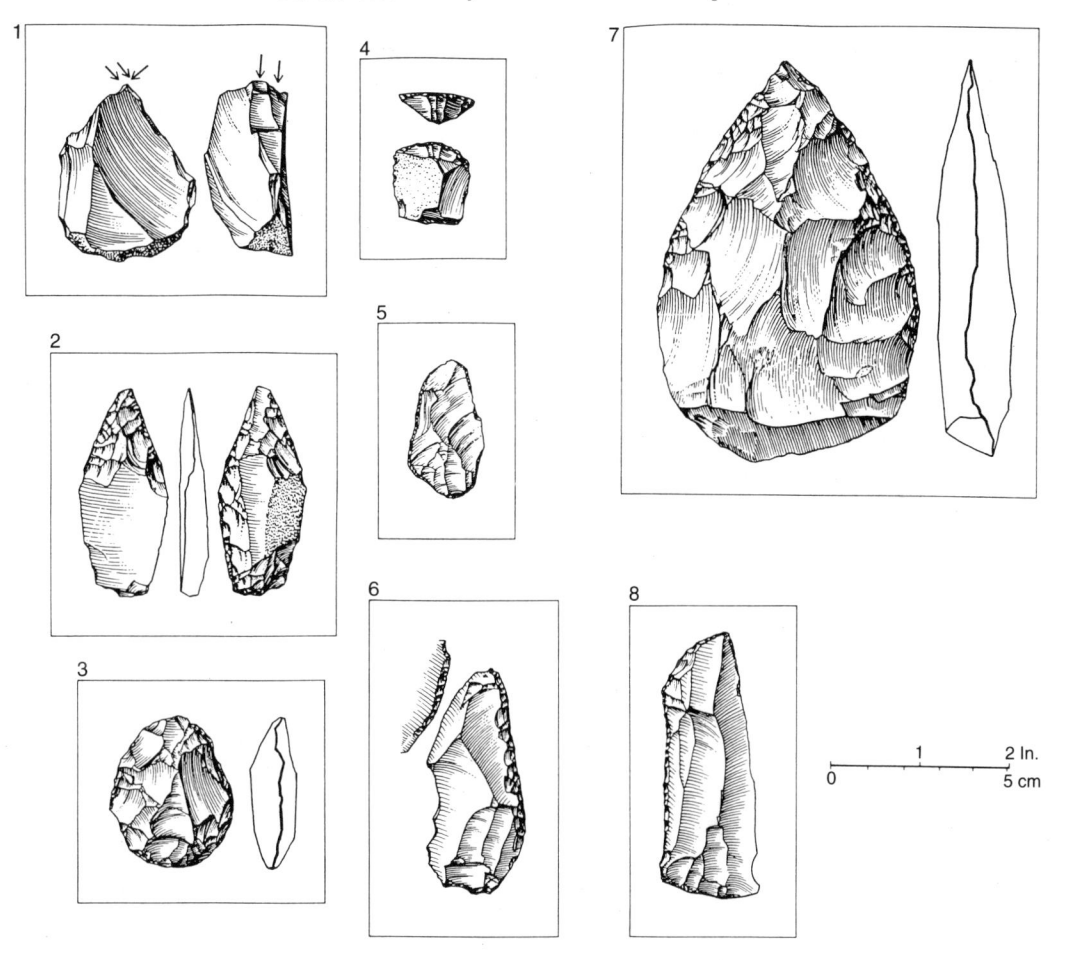

Fig. 9.12 Examples of tools from a Mousterian of the Acheulean tradition assemblage, from Pech de l'Azé, Layer 4 (from Bordes 1972).
1, Burin. 2, Bifacial point. 3, Small discoid hand-axe. 4, Small end-scraper. 5, 'Raclette'. 6, Alternate double side-scraper. 7, Cordiform hand-axe. 8, Backed knife.

Clark, G. (1980). *Mesolithic prelude: the palaeolithic–neolithic transition in Old World prehistory.* Edinburgh University Press.

Cohen, N. (1977). *The food crisis in prehistory.* Yale University Press, New Haven.

Collins, D. (1969). Cultural traditions and environment of early man. *Current Anthropology,* **10**, 267–316.

Collins, D. (1970). Stone artefact analysis and the recognition of cultural traditions. *World Archaeology,* **2**, 17–27.

Conkey, M. (1978). Style and information in cultural evolution: towards a predictive model for the Paleolithic. In *Social archeology: beyond subsistence and dating* (ed. C. Redman et al.), pp. 61–85. Academic Press, New York.

Conkey, M. (1983). On the origins of paleolithic art: a review and some critical thoughts. In *The Mousterian legacy: human biocultural change in the Upper Pleistocene,* BAR International Series 164 (ed. E. Trinkaus), pp. 201–27. BAR, Oxford.

Dart, R. (1957). *The osteodontokeratic culture of Australopithecus Prometheus.* Memoir 10 of the Transvaal Museum, Pretoria.

Dart, R. (1960). The bone tool manufacturing ability of *Australopithecus prometheus. American Anthropologist,* **62**, 134–43.

Deetz, J. (1965). *The dynamics of stylistic change in Arikara ceramics.* University of Illinois Press, Urbana.

Diakonoff, I. M. (1969). The rise of the despotic state in ancient Mesopotamia. In *Ancient Mesopotamia: a socio-economic history* (eds. I. M. Diakonoff), pp. 173–203. 'Nauka' Publishing House, Moscow.

Dibble, H. L. and Mellars, P. (eds.) (1992). *The Middle Paleolithic: adaptation, behavior, and variability.* The University Museum, University of Pennsylvania, Philadelphia.

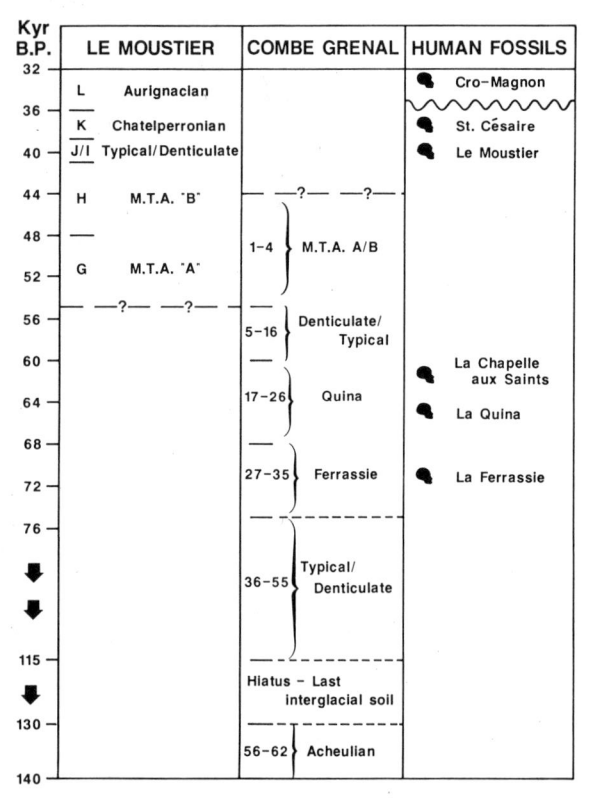

Kyr B.P.	LE MOUSTIER	COMBE GRENAL	HUMAN FOSSILS
32			Cro-Magnon
36	L Aurignacian		
	K Chatelperronian		St. Césaire
40	J/I Typical/Denticulate		Le Moustier
44	H M.T.A. "B"	?—?—	
48		1–4 M.T.A. A/B	
52	G M.T.A. "A"		
56	—?—?—	5–16 Denticulate/Typical	
60			La Chapelle aux Saints
64		17–26 Quina	La Quina
68			
72		27–35 Ferrassie	La Ferrassie
76			
↓		36–55 Typical/Denticulate	
↓			
115			
↓		Hiatus – Last interglacial soil	
130			
		56–62 Acheulian	
140			

Fig. 9.13 Correlation between the archaeological sequences at Le Moustier and Combe Grenal implied by the new thermoluminescence dates for Le Moustier. The inferred chronological distribution of the principal Neanderthal remains from south-west France is shown in the right-hand column. Note that the numbering of the layers in the Combe Grenal succession used in this figure are those given by Bordes, and not those adopted by Binford in Fig. 9.7 of Note 3 to this chapter. Some of the layers, particularly 5, 6, and 31, yield only a very few tools, and their clear diagnosis is perhaps impossible (see Bordes 1972, 112–15). (Figure by courtesy of Paul Mellars.)

Dibble, H. L. and Montet-White, A. (eds) (1988). *Upper Pleistocene prehistory of Western Eurasia*. The University Museum, The University of Pennsylvania, Philadelphia.
Dodgshon, R. A. (1987). *The European past: social evolution and spatial order*. Macmillan, London.
Falk, D. (1980). Language, handedness, and primate brains: Did the Australopithecines sign? *American Anthropologist*, **82**, 72–8.
Flannery, K. V. (1968). The Olmec and the valley of Oaxaca: a model for inter-regional interaction in formative times. In *Dumbarton Oaks Conference on the Olmec, October 28–29, 1967* (ed. E. P. Benson), pp. 79–117. Dumbarton Oaks Research Library and Collection, Trustees for Harvard University, Washington DC.
Flannery, K. V. (1969). Origins and ecological effects of early domestication in Iran and the Near East. In *The domestication and exploitation of plants and animals* (ed. P. J. Ucko and G. W. Dimbleby), pp. 73–100. Aldine, Chicago.
Flannery, K. V. (1972). The cultural evolution of civilizations. *Annual Review of Ecology and Systematics*, **3**, 399–426.
Flannery, K. V. (1973). The origins of agriculture. *Annual Review of Anthropology*, **2**, 271–310.
Fried, M. (1960). On the evolution of social stratification and the state. In *Culture in history: essays in honor of Paul Radin* (ed. S. Diamond), pp. 713–31. Columbia University Press, New York.
Fried, M. (1967). *The evolution of political society: An essay in political anthropology*. Random House, New York.
Fried, M. (1978). The state, the chicken, and the egg, or what came first. In *Origins of the state: the anthropology of political evolution* (ed. R. Cohen and E. Service), pp. 35–48. ISHI, Philadelphia.
Gamble, C. (1983). Culture and society in the Upper Palaeolithic of Europe. In *Hunter–gatherer economy in prehistory: a European perspective* (ed. G. Bailey), pp. 201–11. Cambridge University Press.
Gibson, M. (1973). Population shift and the rise of Mesopotamian civilization. In *The explanation of culture change: models in prehistory* (ed. C. Renfrew), pp. 447–63. Duckworth, London.
Haas, J. (1982). *The evolution of the prehistoric state*. Columbia University Press, New York.
Hallpike, C. R. (1988). *The principles of social evolution*. Clarendon Press, Oxford.
Harris, M. (1968). Comments. In *New perspectives in archeology* (ed. L. and S. Binford), pp. 359–61. Aldine, Chicago.
Harris, D. (1977). Alternative pathways toward agriculture. In *Origins of agriculture* (ed. C. Reed), pp. 179–243. Mouton, The Hague.
Harrold, F. (1980). A comparative analysis of Eurasian Paleolithic burials. *World Archaeology*, **12**, 195–210.
Hayden, B. (1981). Research and development in the Stone Age: technological transitions among hunter-gatherers. *Current Anthropology*, **22**, 519–48.
Hill, J. (1968). Broken K Pueblo: patterns of form and function. In *New perspectives in archaeology* (ed. L. and S. Binford), pp. 103–42. Aldine, Chicago.
Hill, A. (1979). Butchery and natural disarticulation: an investigatory technique. *American Antiquity*, **44**, 739–44.
Holloway, R. (1983). Human brain evolution: a search for units, models and synthesis. *Canadian Journal of Anthropology*, **3**, 215–30.
Isaac, G. (1971). The diet of early man: aspects of archaeological evidence from Lower and Middle Pleistocene sites in Africa. *World Archaeology*, **2**, (3), 278–99.
Isaac, G. (1976). Stages of cultural elaboration in the Pleistocene. Possible archaeological indicators of

the development of language capabilities. *Annals of the New York Academy of Sciences*, **280**, 275–88.
Isaac, G. (1978). The food-sharing behavior of proto-human hominids. *Scientific American*, **238**, 90–108.
Isaac, G. (1984). The archaeology of human origins. *Advances in World Archaeology*, **3**, 1–87.
Jochim, M. (1983). Palaeolithic cave art in ecological perspective. In *Hunter–gatherer economy in prehistory: a European perspective* (ed. G. Bailey), pp. 212–19. Cambridge University Press.
Johnson, G. (1981). Monitoring complex system integration and boundary phenomena with settlement size data. In *Archaeological approaches to the study of complexity* (ed. S. van der Leeuw), pp. 144–88. University of Amsterdam.
Johnson, G. (1982). Organizational structure and scalar stress. In *Theory and explanation in archaeology* (ed. C. Renfrew, M. Rowlands, and B. Segraves), pp. 389–421. Academic Press, New York.
King, T. (1978). Don't that beat the band? Nonegalitarian political organization in prehistoric central California. In *Social archeology. Beyond subsistence and dating* (ed. C. Redman *et al*)., pp. 225–48. Academic Press, New York.
Kuhn, T. (1962). *The structure of scientific revolutions.* University of Chicago Press.
Leakey, M. (1971). *Olduvai Gorge*, Vol. 3: *Excavations in beds I and II, 1960–1963.* Cambridge University Press.
Leakey, M. (1979). *Olduvai Gorge: my search for early man.* Collins, London.
Lee, R. and Devore, I. (1968). Problems in the study of hunters and gatherers. In *Man the hunter* (ed. R. Lee and I. Devore), pp. 3–12. Aldine, Chicago.
Lee, R. and Guenther, M. (1991). Oxen or onions? The search for trade (and truth) in the Kalahari. *Current Anthropology*, **32**, 592–601.
Leroi-Gourhan, A. (1967). *Treasures of prehistoric art.* Abrams, New York.
Leroi-Gourhan, A. (1993). *Gesture and speech.* MIT Press, Cambridge, Mass.
Lieberman, P. (1975). *On the origins of language.* Macmillan, New York.
Mellars, P. (1970). Some comments on the notion of "functional variability" in stone tool assemblages. *World Archaeology*, **2**, 74–89.
Mellars, P. (1973). The character of the Middle/Upper Palaeolithic transition in southwest France. In *The explanation of culture change* (ed. C. Renfrew), pp. 255–76. Duckworth, London.
Mellars, P. (1986a). A new chronology for the French Mousterian period. *Nature*, **322**, 410–11.
Mellars, P. (1986b). [Reply to Ashton and Cook (1986).] *Nature*, **324**, 113–14.
Mellars, P. (1992). Technological change in the Mousterian of southwest France. In *The Middle Paleolithic: adaptation, behavior, and variability* (ed. H. L. Dibble and P. Mellars), pp. 29–43. The University Museum, University of Pennsylvania, Philadelphia.
Mitchell, W. P. (1973) The hydraulic hypothesis: a reappraisal. *Current Anthropology*, **14**, 532–4.

Monks, G. (1981). Seasonality studies. In *Advances in archaeological method and theory*, Vol. 4 (ed. M. Schiffer), pp. 177–240. Academic Press, New York.
Mosely, M. (1975). *The maritime foundations of Andean civilization.* Cummings, Menlo Park, California.
O'Shea, J. (1984). *Mortuary variability.* Academic Press, New York.
Potts, R. (1983). Foraging for faunal resources by early hominids at Olduvai Gorge, Tanzania. In *Animals and archaeology* (ed. J. Clutton-Brock and C. Grigson), pp. 51–62. British Archaeological Reports, Oxford.
Potts, R. and Shipman, P. (1981). Cut marks made by stone tools on bones from Olduvai Gorge, Tanzania. *Nature*, **291**, 577–80.
Price, D. and Brown, J. (eds) (1985). *Prehistoric hunter-gatherers: the emergence of cultural complexity.* Academic Press, New York.
Rathje, W. L. (1971). The origin and development of lowland classic Maya civilization. *American Antiquity*, **36**, 275–85.
Redman, C. L. (1978a). *The rise to civilization: from early farmers to urban society in the ancient Near East.* W. H. Freeman, San Francisco.
Redman, C. L. (1978b). Mesopotamian urban ecology: the systemic context of the emergence of urbanism. In *Social archaeology: beyond subsistence and dating* (ed. C. L. Redman *et al.*), pp. 329–47. Academic Press, New York.
Renfrew, C. (1972). *The emergence of civilization: the Cyclades and the Aegean in the third millennium* BC. Methuen, London.
Rindos, D. (1984). *The origins of agriculture.* Academic Press, New York.
Sahlins, M. (1965). On the sociology of primitive exchange. In *The relevance of models for social anthropology* (ed. M. Banton), pp. 139–236. Association for Social Anthropology, London.
Sahlins, M. (1976). *Culture and practical reason.* University of Chicago Press.
Sahlins, M. and Service, E. (1960). *Evolution and culture.* University of Michigan Press, Ann Arbor.
Sanders, W. T. (1968). Hydraulic agriculture, economic symbiosis, and the evolution of states in central Mexico. In *Anthropological archaeology in the Americas* (ed. B. J. Meggars), pp. 88–107. The Anthropological Society of Washington, Washington DC.
Schiffer, M. (1972). Archaeological context and systemic context. *American Antiquity*, **37**, 156–65.
Schiffer, M. (1976). *Behavioral archeology.* Academic Press, New York.
Schrire, C. (ed.) (1984a). *Past and present in hunter gatherer studies.* Academic Press, New York.
Schrire, C. (1984b). Wild surmises on savage thoughts. In *Past and present in hunter gatherer studies* (ed. C. Schrire), pp. 1–25. Academic Press, New York.
Service, E. (1962). *Primitive social organization.* Random House, New York.
Sharp, L. (1952). Steel axes for Stone Age Australians. In *Human problems in technological change* (ed. E. H. Spicer), pp. 69–90. Wiley, New York.

Smith, P. (1972). *The consequences of food production. Module No. 31, The Philippines*. Addison-Wesley, Reading, Mass.

Smith, P. E. L. and Young, T. C. (1972). The evolution of early agriculture and culture in greater Mesopotamia: a trial model. In *Population growth: anthropological implications* (ed. B. J. Spooner), pp. 1–60. MIT Press, Cambridge, Massachusetts.

Soffer, O. (1985). Patterns of intensification as seen from the Upper Paleolithic of the Central Russian Plain. In *Prehistoric hunter–gatherers. The emergence of cultural complexity* (ed. T. D. Price and J. Brown), pp. 235–70. Academic Press, New York.

Stanislawski, M. (1977). Ethnoarchaeology of Hopi and Hopi-Tiwa pottery making; styles of learning. In *Experimental archaeology* (ed. D. Ingersoll, J. Yellen, and W. Macdonald), pp. 378–408. Columbia University Press, New York.

Steward, J. (1955). *Theory of culture change*. University of Illinois Press, Urbana.

Tainter, R. (1978). Mortuary practices and the study of prehistoric social systems. In *Advances in archaeological method and theory*, Vol. 1 (ed. M. Schiffer), pp. 105–41. Academic Press, New York.

Trinkaus, E. (ed.) (1983). *The Mousterian legacy: human biocultural change in the Upper Pleistocene*. British Archaeological Reports, Oxford.

Valladas, H., Geneste, J. M., Joron, J. L., and Chadelle, J. P. (1986). Thermoluminescence dating of Le Moustier (Dordogne, France). *Nature*, **322**, 452–3.

van Zeist, W. (1969). Reflections on prehistoric environments in the Near East. In *The domestication and exploitation of plants and animals* (ed. P. J. Ucko and G. W. Dimbleby), pp. 35–46. Aldine, Chicago.

van Zeist, W. (1976). On macroscopic traces of food plants in southwestern Asia. *Philosophical Transactions of the Royal Society of London*, B **275**, 27–41.

van Zeist, W. and Wright, H. E., jun. (1963). Preliminary pollen studies at Lake Zeribar, Zagros Mountains, southwestern Iran. *Science*, **140**, 65–9.

Vialou, D. (1983). Art parietal paléolithique ariegois. *L'Anthropologie*, **87**, 83–97.

Washburn, S. (ed.) (1961). *Social life of early man*. Aldine, Chicago.

Weissner, P. (1977). *Hxaro: a regional system of reciprocity for reducing risk among the !Kung San*. University Microfilms, Ann Arbor.

Weissner, P. (1982). Risk, reciprocity and social influences on !Kung San economics. In *Politics and history in band societies* (ed. R. Lee and E. Leacock), pp. 61–84. Cambridge University Press.

Wells, P. (1984). *Farms, villages, and cities: commerce and urban origins in late prehistoric Europe*. Cornell University Press, Ithaca.

White, L. (1959). *The evolution of culture*. McGraw-Hill, New York.

White, R. (1982). Rethinking the Middle/Upper Paleolithic transition. *Current Anthropology*, 23, 169–92.

White, R. (1985a). Thoughts on social relationships and language in hominid evolution. *Journal of Social and Personal Relationships*, **2**, 95–115.

White, R. (1985b). *Upper Paleolithic land use in the Perigord: a topographic approach to subsistence and settlement*. British Archaeological Reports, Oxford.

White, R. (1989). Toward a contextual understanding of the earliest body ornaments. In *The emergence of modern humans* (ed. E. Trinkaus), pp. 211–31. Cambridge University Press.

White, R. (1993). Technological and social dimensions of "Aurignacian-age" body ornaments across Europe. In *Before Lascaux: the complex record of the Early Upper Paleolithic* (ed. H. Knecht, A. Pike-Tay, and R. White), pp. 277–299. CRC Press, Boca Raton.

Wilmsen, E. (1980). *Exchange, interaction and settlement in North Western Botswana: past and present perspective*, African Studies Center Working Paper No. 39. Boston University, Boston.

Wittfogel, K. A. (1957). *Oriental despotism: a comparative study of total power*. Yale University Press, New Haven.

Wobst, M. (1978). The "archaeo-ethnology of hunter-gatherers" or the tyranny of the ethnographic record in archaeology. *American Antiquity*, **43**, 303–9.

Wright, H. E., jun. (1968). Natural environment of early food production north of Mesopotamia. *Science*, **161**, 334–9.

Wright, H. E., jun. (1976). The environmental setting for plant domestication in the Near East. *Science*, **194**, 385–9.

Wright, H. E., jun. (1977). Environmental change and the origin of agriculture in the Old and New Worlds. In *Origins of agriculture* (ed. C. Reed), pp. 281–318. Mouton, The Hague.

Yellen, J. (1977). *Archaeological approaches to the present*. Academic Press, New York.

10
The evolution of tools and symbolic behaviour

Thomas G. Wynn

Abstract

Tools constitute the most abundant evidence of hominid behaviour over the last two million years. While they have undeniably played an important, if not central, role in hominid ecology, they have also played a role in semiotic behaviour. This role probably had its origins in the agonistic use of tools we still see today in non-human primates. When we first encounter extensive use of stone tools, about two million years ago, the ecological context of use is not dramatically different from that of modern apes, and we may assume that the semiotic role of tools was also comparable. By one million years ago tools present patterns well outside the range of anything we know for apes, tempting some scholars to argue for the presence of language. However, given the cognitive and developmental contrasts between tool behaviour and language, such conclusions are unwarranted. At 300 000 BP the ecological context of tool behaviour was much like that of modern hunting and gathering, but the tools present an enigmatic conservatism in style that suggests a semiotic role very different from that of tools in modern culture. And yet the hominids appear to have had an almost modern intelligence. It is not until relatively late in human evolution, certainly by 15 000, that tools present the volatile time and space patterns typical of the indexical role of modern tools.

10.1 Introduction

In modern human behaviour tools operate in the semiotic realm in complex and subtle ways. Tools typically carry information about the user beyond the mere mechanical task performed. A samurai sword is a weapon, an index of social status and cultural affiliation, and a symbol of divine power. Tools are also the most abundant evidence we have of human evolution. Stone tools and pottery sherds number in the millions and millions. The record of tools over the last two million years has, therefore, the potential to inform us about the evolution of semiotic behaviour. The task of deciphering the semiotic role of prehistoric tools is not an easy one, however. Imagine an archaeologist of the future trying to interpret a samurai sword recovered from a site in central North America. He could conclude nothing about specific meanings; but it is possible that by using comparisons to other sites and similar archaeological techniques he could conclude that such meanings must have existed.

In this chapter I will use the shapes of tools and their distribution in space and time to examine four topics of relevance to the evolution of human symbolic behaviour: the ecological utility of tools, the intelligence of the makers, the implications of tool-making for language, and the indexical role of tools. By ecological utility I mean the role tools played more or less directly in the behaviours of feeding and shelter. As we shall see, ecological utility tells us little specific about symbolic behaviour; but it was probably the primary and most common role of tools, and we must understand it in order to isolate other roles. We will examine tools for clues to the evolution of intelligence, because certain symbolic behaviours may require a more powerful intelligence than others. The remaining two topics deal specifically with aspects of symbolic behaviour. The first is language. A curious literature has grown up around the idea that tools can inform us about language. I will present some of the arguments and try to assess their value. The last topic is that of the indexical role of tools. It is in this semiotic realm—not in language—that tools have the most to tell us about the evolution of human symbolic behaviour.

I have organized the chapter chronologically. First, I establish an evolutionary starting point by describing tool behaviour in non-humans, with particular emphasis on non-human primates. I then discuss tools at five successive points in human evolution: 1.8 million, 1 million, 300 000, 85 000, and 15 000 years ago. The dates are arbitrary, but encompass what I believe were the most important steps in the semiotic role of tools. I have attempted to avoid purely archaeological controversies in order to focus on the question in hand. As a consequence the chapter cannot be read as a summary of cultural evolution.

10.2 Non-human tools

Tables 10.1 and 10.2 present a sampling of non-human tool-use. From even this small sample it is clear that the simple fact of tool-using or even tool-making is insufficient to distinguish human behaviour from that of other animals. Indeed, the presence of tool-use has little or no phylogenetic significance. Woodpecker finches modify and use probes to feed on insects—so do chimpanzees. From an evolutionary point of view their tool-use is the result of convergence—opportunistic solutions to specific ecological problems—and has no implications for the relatedness of the taxa. However, if we examine the frequency and manner of tool-use some phylogenetic patterns do emerge that are important in setting the stage for human tool-use, including human symbolic tool-use.

One characteristic of primates—the monkeys and apes at least—is the almost universal use of tools in agonistic situations. Both monkeys and apes brandish, drop, and throw objects in attempts to discourage intruders. Of course other animals do this on occasion; what is remarkable is the near universality of this kind of object-manipulation among the higher primates (see Table 10.2). As van Lawick-Goodall (1970) and Beck (1980) point out, this behaviour owes much to the primate hand, a grasping exploratory organ. A grasping hand is not a sufficient cause, however; and there is something homologous in the behaviour as well. Hall (1963) finds agonistic tool-use in the higher primates so pervasive that he sees it as the source of all other kinds of tool-use. The Great Apes have expanded and refined agonistic use into elaborate displays that are used for general intimidation and, in the case of at least one chimpanzee, the establishment and maintenance of dominance. Humans are hominoids, and we must expect that the use of tools in displays was part of the behaviour of early hominids, as it is in modern humans.

Of the non-human primates, only chimpanzees make significant use of tools in feeding. Termiting, ant dipping, and nutting are seriously practised during certain seasons, and supply a small but significant part of the diet for some groups (Uehara 1982). Interestingly, the frequency and techniques of tool-making and tool-use vary from group to group in cultural fashion (McGrew *et al.* 1979). Of all of the apes, chimpanzees are the most closely related to humans in terms of anatomy and biochemistry. Humans and chimpanzees are also the most similar of the hominoids in terms of the situations in which they use tools. We may conclude that our common ancestor employed tools in both feeding and agonistic displays.

One additional characteristic of chimpanzee tool behaviour is useful in setting the stage for human evolution. The most sophisticated tool behaviours are found in captivity, not in the wild. A few chimpanzees have shown an ability to solve problems requiring the construction of tools from separated parts (rakes and stacks of boxes) (Kohler 1927) and to use make-shift ladders in escape attempts (Menzel 1978). Comparable tools are unknown in the wild. A parallel appears to be true in language. In the wild chimpanzees appear to use a vocal closed call system, but in captivity they can be taught much more complex systems of communication, these being manual or visual rather than vocal (see Ristau, this volume, Chapter 23). In both cases competence exceeds natural performance, at least from a human perspective. Is it possible that tool behaviour and communication are organized by the same mechanisms, and that competence in one informs us about competence in the other? This is the basic assumption of the 'tools and language' argument, and is one to which we will return.

10.3 Hominid tools at 1.8 million years ago

The earliest known hominid tools date to perhaps 2.5 million years ago (Isaac 1984; Harris 1983); but it is not until considerably later that archaeologists have a comparatively large sample. These early tools are known collectively as the *Oldowan industry*, and have been found in many regions in Africa. Oldowan assemblages have been excavated in Ethiopia and Northern Kenya; but the best examples of Oldowan tools come from Olduvai Gorge itself, where they have been excavated from primary contexts dating from 1.8 to 1.6 million years ago (Leakey 1971; Hay

Table 10.1 Sample of tool-use by vertebrates in the wild

	Description of tool	Modifications of tool	Description of tool behaviour	Apparent purpose	Reference
Birds					
Satin bower bird	wad of bark	nibbled	Tool is held at front of bill to control flow of paint solution used to decorate bower.	to attract mate	(van Lawick-Goodall 1970)
Egyptian vulture	stone	none	Stone is held in bill and hurled at or dropped on target (usually an egg) from aloft.	feeding	(Beck 1980)
Woodpecker finch	twig or cactus spine	shortened	Tool is used to probe under bark for insects.	feeding	(Beck 1980)
Non-primate mammals					
Elephants	grass	none	Grass is used to wipe cuts.	hygiene	(Beck 1980)
Horse	branch	none	Branch is held in mouth and used to scratch.	hygiene	(Beck 1980)
Polar bear	chunk of ice	none	Ice is hurled on to heads of walruses (known only anecdotally).	feeding	(Hall 1963; Beck 1980)
Sea otter	stone	none	Stone is balanced on chest and used as an anvil with which to open mussels.	feeding	(Hall 1963)
Monkeys					
Capuchin (*Cebus*)	stone	none	Stone is used to break open shellfish.	feeding	(Beck 1980)
Crab-eating macaque	stone	none	Stone is used to break open shellfish.	feeding	(Beck 1980)
various species	stick	none	Sticks are used to open insect nests.	feeding	(van Lawick-Goodall 1970)
Japanese macaque	stone or stick	none	Female used tool to groom infant.	grooming	(Weinberg and Candland 1981)
Apes					
Chimpanzees	twig or grass stem	stripped of leaves, shortened (including repair while in use)	Used as a probe for extracting termites.	feeding	(Teleki 1974; Sabater Pi 1974; van Lawick-Goodall 1970)
	stone	none	Used to break open nuts.	feeding	(Struhsaker and Hunkeler 1971; Boesch 1978)
	leaves	chewed	Used as a sponge to extract water from inaccessible spots or brains from monkeys.	feeding	(Teleki 1974)

1976). Here they are associated with the refuse of hominid activity; and it is here that archaeologists have the best evidence of this earliest of technologies.

I must enter two important caveats before describing these tools. The first is that archaeologists recover the tools, not the behaviour. The latter must be inferred rather than observed; and this makes many conclusions less certain than those derived from the ethological examples listed earlier. The second caveat is that the sample of early tools is biased toward stone; few other materials have survived the perils of time. Fortunately, stone can be very informative.

Fracturing stones yields sharp edges; this is the basic principle of almost all stone-age technology. Archaeologists usually distinguish between cores and flakes. The former are the stones being fractured, the

Table 10.2 Sample of primate use of tools in display and aggression in captivity and in the wild

	Description of tool	Modifications of tools	Description of tool behaviour	Apparent purpose	Reference
Monkeys					
Capuchin	branch	none	Used to club other monkeys.	aggression	(Beck 1980)
	stick or fruit	none	Used as projectile in aimed throwing or as a club.	discourage intrusion	(Beck 1980)
Colobine	branches and twigs	none	Thrown and dropped.	discourage intrusion	(van Lawick-Goodall 1970)
Patas	stones	none	Thrown at humans.	discourage intrusion	(Beck 1980)
Baboons	stones	none	Thrown at or dropped on humans.	discourage intrusion	(Hamilton et al. 1975; Pettet 1975)
Barbary macaque	roof tiles	none	Thrown at or dropped on humans.	discourage intrusion	(Beck 1980)
Japanese macaque	pine cones	none	Thrown at or dropped on humans.	discourage intrusion	(Beck 1980)
Apes					
Gibbon	branches	none	Shaken or dropped.	discourage intrusion	(van Lawick-Goodall 1970)
Gorilla	vegetation	none	Pulled up and brandished in displays (but no aimed throwing).	display	(Beck 1980)
Orang-utan	branches and stones	none	Dropping, brandishing, and aimed throwing.	display and to discourage intrusion	(Galdikas 1978; 1982)
Chimpanzee	branches	none	Dropping, aimed throwing, and clubbing.	discourage intrusion and to attack one another and baboons	(Beck 1980; van Lawick-Goodall 1970)
	large stones	none	Rolled by males during charging displays.	display	(van Lawick-Goodall 1970)
	stone	none	Aimed throwing.	discourage intrusion	(van Lawick-Goodall 1970)
	paraffin tins	none	Banged together during charging (only one individual known to do this).	increase relative status	(van Lawick-Goodall 1970)

latter the pieces that are removed. A flake has a sharp, thin edge, while a core a sturdier edge. The edges of both flakes and cores can be further modified to make a particular shape, such as a projection or a notch. Stone-knapping is a subtractive task; one removes material to achieve a result. While almost any kind of stone can be knapped, some kinds are easier to work than others, and, consequently, the raw material used does constrain the appearance of stone tools (Jones 1979, 1981).

The Oldowan industry consists of simple core and flake tools, made out of basalt, quartz, and quartzite. There are three basic kinds of chipped stone tools: core tools, modified flakes, and unmodified flakes (Isaac 1976a). Most of the core tools have sinuous, jagged edges produced by the alternate removal of flakes from contiguous faces of a cobble or hunk of stone. If there are one or two such edges the tool is termed a 'chopper' (see Fig. 10.1). Tools with three or more intersecting edges are 'polyhedrons', or, if especially round, 'spheroids'. Many of these core tools have battered edges indicative of use. Unmodified flakes are common, and also have indications of use. At Olduvai there are also modified flakes, most of which have had their edges 'trimmed' by the removal of small chips from one face of the flake. This thickens the edge, and these tools are termed 'scrapers'. There are also a few 'burins', which are flakes with chisel-like edges. Associated with the chipped stone tools are battered stones of various sizes, perhaps

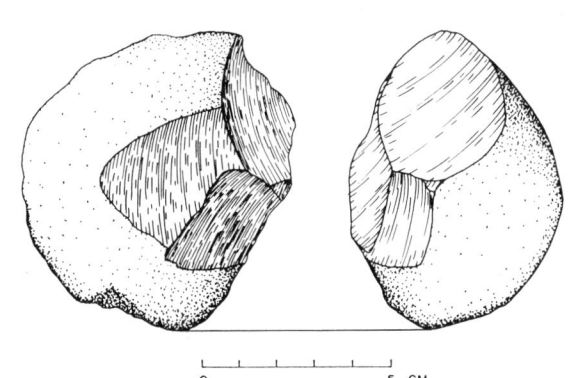

Fig. 10.1 Oldowan chopper, one of the earliest stone tools. This artefact is within the capacities of modern apes (from Wynn 1981).

used as hammers and anvils, and cobbles and nodules of imported raw material that show no modification. The latter are termed manuports. There was some selectivity in raw material. At most Olduvai sites the large tools were made of lava and the small tools of quartz (or chert when available) (Leakey 1971). We should avoid placing too much emphasis on this selectivity; what appear to be comparable kinds of selectivity are practised by West African chimpanzees, who select wooden hammers for one kind of nut and stone hammers for another (Boesch and Boesch 1984).

There is considerable question as to the reality of the tool types in the Oldowan. If by type we mean a well-defined category of tool that existed in the minds of the tool-makers, then we would be unable to argue for their existence. There appear to be only a few distinct divisions within the industry, and these reflect whether the tool was made on a core or on a flake. Within these categories the tool morphology varies more or less continuously (Isaac 1976a). It is doubtful that there were any design criteria whatsoever, beyond perhaps big and little. The continuously varying morphology suggests an almost purely *ad hoc* technology in which the tool-makers chipped a tool for the specific task at hand, a single-edged core tool for one task, a small modified flake for another. The range of sizes and edges would parallel the range of tasks, and would not reflect any pre-existing conceptual designs (Wynn 1978). Isaac, in a similar vein, sees the tools as 'opportunistic least effort solutions to the problem of attaining sharp edges from stone' (Isaac 1981, 184), and not culturally defined categories. Indeed, Oldowan tools are comparable to chimpanzee tools in the degree to which they present arbitrarily imposed shapes (Isaac 1976b). Each termiting probe (Teleki 1974) is an *ad hoc* solution to a single termit-

ing episode, and results from the modification, by removal, of twigs or grass-stems. With chimpanzee tools, as with the Oldowan, size appears to have been the important consideration. Sabater Pi (1974) documents a clear modal size for termiting sticks in one region of West Africa, a size optimal, apparently, for breaking open termite mounds (the technique practised by chimpanzees in Equatorial Guinea).

The ecology of early hominid tool-use appears, however, to have been rather different from the tool-use of chimpanzees. Oldowan hominids employed two behaviours that are rare in chimpanzees: carrying and eating the meat of medium to large prey. At Olduvai and at other East African sites stone tools are associated with the remains of large mammals. While there is no reason to believe that the hominids killed these animals, they were almost certainly butchering portions of carcasses. Other Olduvai sites have quantities of stone tools associated with the broken-up bone of many different individual animals. The animals and animal parts had been carried to the sites (Bunn 1981; Potts 1984), as had the stones for the tools, some of which came from sources as far away as 13 kilometres (Ohel 1984). The animal parts were further processed at the sites (Bunn 1981).

Oldowan tools did not require a particularly sophisticated intelligence. Stone-knapping itself is not difficult, and at least one orang-utan has been taught to flake stone (Wright 1972), though not to manufacture Oldowan-like tools. More specific cognitive prerequisites of Oldowan tools can be assessed using the methods of Piaget (Wynn 1985). Among the many behaviours Piaget used to establish his developmental scheme, he considered children's ability to arrange objects, copy figures, reconstruct scenes and perspectives, draw well-known subjects, and many others. The result is a yardstick of increasingly complex spatial concepts that children develop from infancy through adolescence. It is a yardstick that can be applied to stone tools. The Oldowan tools require only very simple spatial concepts such as proximity and order, concepts that are typical of Piaget's stage of preoperational intelligence (Wynn 1981) (see Sinha's Chapter 17 in the present work for a more detailed look at Piagetian method and theory). The Great Apes fall into this stage of development as well (Parker and Gibson 1979), so there is no reason, based on the tools themselves, to argue for a dramatic difference in intelligence.[1] This assessment concurs with Lancaster's (1968) that the tools do not represent a skilled technology. Of course it is possible that the hominids utilized a more powerful intelligence in realms of behaviour other than tool-making. Indeed, the evidence of fossil brains (see Holloway, this volume, Chapter 4) suggests that significant differences did exist between hominid brains and ape brains at 1.8 m.y.a (Holloway 1983). Parker and Gibson's (1979) argument could be taken to imply that the selective agent for this difference was extractive foraging; and Tobias (1983) maintains that the 'holistic' cultural context of tool-use would have required a more advanced intelligence. However, there is nothing found in the archaeological record itself, including both the tools and the ecological context, that requires an intelligence greater than that of modern apes. If these hominids were more intelligent, and I suspect that they may have been, then selection for intelligence may have been tied to the social demands of 'savannah' living, rather than the technological demands of Oldowan tools.

As Table 10.2 shows, all of the extant Great Apes use tools in agonistic and display behaviour. This kind of use is difficult if not impossible to detect in the archaeological record. Nevertheless there are some possibilities. Leakey (1971) notes the presence of manuports at most of the early Olduvai sites; and there is also the category of spheroids, which have been interpreted as possible missiles (among other things). Use of stone missiles in agonistic encounters is a simple extension of behaviour seen in chimpanzees, and I think we can almost assume its presence, both in interspecific and intraspecific contexts. More provocative would be the use of stone tools in social display behaviour. Much has been written about the social reorganization necessary for 'savannah' living, but little or no attention has been given to the role tools may have played in the maintenance of the social order. Tools can and do play such a role for other hominoids. Given the greater reliance of hominids on tools, it seems likely to me that they also played a greater role in social maintenance, as indices of social dominance for example. Unfortunately, there is nothing about the stone tools themselves that suggests such uses.

It is usual to emphasize the human-like characteristics of culture at 1.8 m.y.a.—stone tools, and meat-eating, carrying, and possibly sharing. It seems just as reasonable to emphasize the ape-like characteristics. Oldowan tools do not represent a qualitative leap in technology. Like chimpanzee tools, they are *ad hoc* in nature, are made of selected materials, lack any apparent design, and have simple geometry. What is different is the evidently greater ecological reliance of Oldowan hominids on these tools; and this is a difference in degree—large perhaps—but not a difference in kind. There is certainly nothing about the tools that suggests anything about language or semiotic ability; and, given the very ape-like nature of these tools, there is no reason to conclude that the symbolic behaviour of these early hominids was much different from that of modern chimpanzees [but see Gibson (1983) for an opposing view—eds].

10.4 Hominid tools at 1 million years ago

Paradoxically, we know less about hominid behaviour at one million BP than we know about behaviour at 1.8 million BP. Sites are fewer and less informative than sites of earlier and later time-periods. This may be a function of bias in the archaeological research or, perhaps, of a change in typical hominid living arrangements. Whatever the reason, archaeologists know less about the ecology of tool-use at one million BP than about the ecology of tool-use in the Oldowan. We do have a number of good assemblages; it is just that they tend not to occur on nice 'living floors' of the variety found in Bed I at Olduvai. Nevertheless we have enough of a sample of the stone tools to contrast them with the earlier Oldowan tools.

The hallmark of this time-period is the biface,[2] a core tool that was extensively modified by bifacial trimming in to an almond-shaped tool with a point and a butt (Fig. 10.2). These tools are quite different from anything found in the Oldowan. One difference is that the knappers imposed an arbitrary shape on the tool. At one million BP this is a two-dimensional shape with a definite long axis and rough bilateral symmetry (Gowlett 1979). A new technique of stone-knapping appears at this time as well; the manufacture of large flakes (>10 cm) (Jones 1981). The knappers used the large flakes as blanks for bifaces. In many

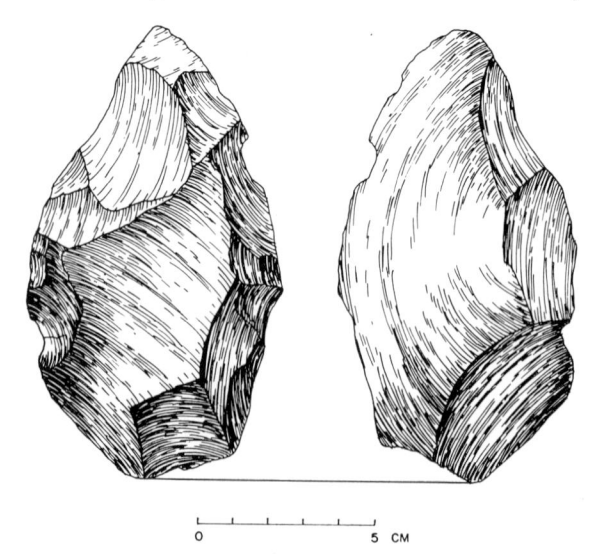

Fig. 10.2 Biface from the early Acheulean site of West Natron. Note the extensive modification and the rough bilateral symmetry (from Wynn 1979).

cases, the knappers trimmed the flake to such an extent that the original flake shape was obliterated, and a new arbitrary shape was imposed. Otherwise the non-biface tools are very similar to the large and small tools of earlier times.

The biface is the first type of tool that is clearly outside the range of an ape technology. As we have seen, Oldowan tools need not have been the result of some pre-existing idea about shape. Each was probably a response to an immediate need. Bifaces, on the other hand, were almost certainly the result of a pre-existing notion about appropriate shape. For artefact after artefact, at site after site, the shape was basically the same. Furthermore the shape was imposed on raw materials with very different flaking properties (Jones 1981). Such redundancy represents standardization. Moreover the shape itself seems mostly arbitrary. Ape tools do not include such standardized and arbitrary shapes, nor does the Oldowan, so we have here the first distinctly hominid kind of technology.

The bilateral 'symmetry' of the earliest bifaces requires further comment, for it presents a knotty problem peculiar to the archaeology of the early Stone Age. Did the makers have a concept of symmetry, or is the concept of symmetry only in the minds of the archaeologists? This problem concerns the filter of modern culture that archaeologists must look through when perceiving prehistoric patterns. One-million-year-old bifaces are, with few exceptions, only vaguely symmetrical, and this only in one plane. Did the knappers possess a concept of midline with mirror-image reversal, a rather sophisticated spatial concept? Given the nature of the blank, usually a large flake, almost any pointed, bifacial tool will approach symmetry, especially if part of the 'idea' was an extensively trimmed edge around the entire circumference. Alternatively, the shape may have been a copy from nature (a leaf), which presents an abundance of symmetrical forms (though few are metrically symmetrical; see Wynn 1989 for a more thorough discussion). Either way, it is possible to produce the early bifaces without a concept of symmetry (see Wynn 1993 for a revision of this point). As I will make clear in the next section, however, later bifaces are more sophisticated.

The apparent symmetry and standardization of bifaces have inspired some interesting arguments concerning the origins of language. It is tempting to use stone tools as informants; language does not preserve, but we have stone tools in abundance. The difficulty lies in drawing a connection between technological and linguistic behaviour. Unfortunately the proposed connections are often based on speculative theoretical position-statements and not on hard evidence. There seem to be two basic approaches. One connects tools and language on a fundamental, even neurological, level, arguing that the mere existence of stone tools informs us about language. A second approach focuses on specific aspects of the stone tools, standardization for example, and searches for parallels or causes in the linguistic realm.

Kitahara-Frisch (1978) presents one of the clearest statements of the first approach:

A closer consideration of the properties of stone tool making further reveals that these imply the exercise of a number of cognitive abilities and that these abilities are the very same found to be operative in other behaviors commonly regarded as specific to man, such as, for instance, language (p. 5).

He argues that making stone tools requires foresight and, more importantly, tools to make tools (hammers, for example). Because language is reflexive and can make metalinguistic statements it '... thereby becomes the most powerful tool to make other tools' (ibid., p. 6). Hence the linguistic ability parallels the technological ability, and the latter can thereby be used as evidence for the former.

Frost (1980) also uses the nature of stone-knapping to argue about language, but his conclusion is very different. His argument revolves around left-hemisphere asymmetry, which is often explained as being a result of language. Frost argues that left-hemisphere asymmetry evolved to accommodate the skilled serial motor behaviour required in stone-knapping, where one hand (fortuitously the right) delivers controlled blows and the other passively holds the stone. Due to a 'competition for neural space' only one hemisphere, the one controlling the skilled hand, would develop the necessary circuitry. By contrast, language and speech, not being asymmetry behaviours like stone-knapping, would not select for brain asymmetry. Therefore stone tools must have preceded speech, which subsequently took advantage of the serial organizing ability of the left hemisphere. From Frost's vantage point the asymmetry of *Homo erectus* brains (Holloway 1981) reflects their technological ability, but not necessarily their linguistic ability.

Hewes (1973) also emphasizes the common sequencing behaviour of tool manufacture and language, and avoids the priority of tools by arguing for a gestural origin for language, which would place both in the same neurological ballpark of motor behaviour.

The above three arguments are similar in that it is the presence of stone tools that is important rather than something specific about tool morphology or the make-up of tool assemblages. An alternative approach emphasizes aspects of the artefacts themselves or of artefact assemblages. An extreme example is Foster (1975), who sees the symmetry and regularity of bifaces as a representation of a principle of opposition, which for her is a key element in all symbolic systems, including language. Isaac (1976b) emphasizes the appearance of rules governing designs—'the degree to which there were distinct target forms in the minds of the craftsmen' (p. 280). He suggests that an increase in the number and definition of designs reflects a more complex rule-system, which in turn has implications for the complexity of other rule-governed behaviour. Isaac is more cautious than Foster, but assumes that 'hominid capacity for conceiving and executing increasingly elaborate material culture designs has been connected with rising capacity for manipulating symbols, naming and speaking' (ibid., p. 276). Unlike Kitahara-Frisch, Isaac sees no reason to argue for language in the Oldowan, but does consider biface industries to represent an increase in complexity, if only by the presence of two or three arbitrary designs.

Rather than deriving tools and language from a common cognitive base, Guilmet (1977) argues that tool patterns reflect the learning patterns typical of certain communication systems. Chimpanzees, for example, learn tool-making by observation and unintentional modelling. The result is a considerable variation in skill and performance. Guilmet suggests that intentional modelling or linguistic instruction would result in less variation in performance, the tools demonstrating greater regularity in form. He is cautious, but his argument implies that standardization, like that seen in bifaces, is an indication of intentional modelling, and, perhaps, of linguistic instruction. Each of the above three interpretations sees evidence for language, at least rudimentary, in one-million-year-old tools.

Most of the language-and-tool arguments assume some parallel between technological and linguistic behaviour, such that the former can be used as evidence for the latter. This assumption may be unwarranted. One of these behaviours, tool-use, may well have been a prior development (cf. Frost 1980). My argument is based on the pre-linguistic behaviour of infants and the technological behaviour of modern adults. Bruner (1983) has studied the way an infant learns to use language, and has observed that the action of pre-linguistic infants '... shows a surprisingly high degree of order and systematicity' (p. 28). By means of conventionalized routines and games, the infant learns such linguistic behaviours as reference and request. These routines and games have a language-like organization: '... from the start, the child becomes readily attuned to "making a lot out of a little" by combination. He typically works on varying a small set of elements to create a large range of possibilities' (ibid., p. 29). Moreover, such systematic routines are early on applied to objects: 'A single act (like banging) is applied successively to a wide range of objects. Everything on which a child can get his hands is banged. Or the child tries out on a single object all the motor routines of which he or she is capable ...' (p. 28). One result of object routines is that children develop skill in object-manipulation before they develop language (Bruner 1972).

The technological behaviour of modern adults is more similar to the systematic routines of pre-linguistic children than to the formal grammars and referents of modern language. Dougherty and Keller's study of blacksmithing (1982) and Gatewood's study of salmon fishing (1985) point to three non-linguistic characteristics. First, technological behaviour is clearly not organized in terms of lexical categories. Indeed, these authors found it difficult to elicit named categories of behaviour from the participants. Rather, the organization was in terms of 'constellations of knowledge', which are tied to a particular task: 'Yet not all of one's actions are represented verbally, outwardly or inwardly. Rather, one experiences visual imagery and muscular tensions appropriate to certain actions, but can only grope for words ...' (Gatewood 1985, p. 206).

Second, the basic organization of action is not equivalent to that of language. 'Each element of a constellation is related to and influences each other in nonhierarchical fashion' (Dougherty and Keller 1982, p. 768). Gatewood terms this a 'string-of-beads' organization, which requires only simple serial memorization: 'First I do job$_1$, then I do job$_2$, then I do job$_3$ ... then I am finished' (Gatewood 1985, p. 206). When a task is mastered, it may be conceived in a more hierarchical, language-like organization; but the initial organization is only serial.

Third, these fundamental organizations do not appear to be shared, but are constructed individually by each novice. Their final organizations are in fact often very different between individuals (Gatewood 1985), a situation unlike that of language. Further ethnographic examples corroborate these findings. Adair (1944) studied Navajo silversmithing, and noted that learning was almost entirely by means of observation. 'Tom [the silversmith] said that the Navajo learn by watching and then doing, following exactly as possible what they have seen their teachers do' (p. 75). Furthermore, when a novice makes an error the teacher insists upon his finishing the piece rather than beginning anew. From Gatewood's viewpoint this would be essential for the novice to construct his own complete 'string-of-beads', or, in Bruner's terms, his own systematic routine. Reichard (1934) also notes that Navajo weaving is learned by observation rather than instruction. The same appears to hold true for stone-knapping. White (White et al. 1977) in a study of a New Guinea group makes the following observation: 'Copying, and trial and error, rather than teaching, are certainly the methods by which young Duna men learn about flaked stone' (p. 381).

When combined with Bruner's study of pre-linguistic children, these studies of adult technological behaviour suggest that tool-use and tool-making are rather different from language, and, in fact, are simpler and developmentally prior. Because no clear connection between the two behaviours can be demonstrated for modern humans, there is no basis for assuming that tools can inform us about prehistoric language. The regularity of earlier bifaces need not have been the result of language-equivalent behaviour, but could rather have been the result of language-like routines.

In sum, the stone tools manufactured by one-million-year-old hominids are different from earlier tools and seem, intuitively, to be more evolved. We must, however, be careful not to over interpret them. While the rough symmetry and standardization of the bifaces is unlike anything known for the apes, this does not mean it is like modern human technology.

10.5 Tools at 300 000 years ago

Three hundred thousand years ago was an intriguing time in human evolution. It was much closer to the present than to the beginnings of culture; and, as one might expect, there were modern elements in hominid behaviour, especially in subsistence patterns. Yet there was also something archaic. There is no evidence of anything resembling modern ideological behaviour—no burial, no ritual, no art. Furthermore there was a strange monotony to the material culture. English bifaces were indistinguishable from those of southern India; and, moreover, the bifaces did not change over time. The absence of ideological behaviour and the monotony of culture may both result from a poverty in the symbolic systems of the time. Indeed, culture may have been very different in the manner in which symbols behaved.

By 300 000 BP technology includes a number of essentially modern elements. Perhaps the most important is fire. While there is some highly problematic evidence for hominid use of fire at 1.4 million BP, which most scholars reject (Isaac 1982), there is some evidence for use of fire and, perhaps, cooking at Zhoukoudian by 700 000 BP (Stringer 1985; Wu Ru-Kang 1980). Fire appears to have been coincident with the movement of hominids out of the tropics, though we must be careful not to assume too much concerning its original uses. While these hominids may not have made fire, they could certainly control it, as evidenced by the hearths at Terra Amata, a site in southern France that dates to between 300 000 and 400 000 BP (de Lumley 1975). Also at Terra Amata are the remains of huts, which consisted of centre and peripheral posts, hearths, and distinct working areas. The largest was an oval structure that measured six by fifteen metres. According to de Lumley as many as eleven huts were super imposed in one location, suggesting that a single group returned several years in succession.[3] The tool-kit at 300 000 BP includes at least one tool for which we have no earlier evidence—the spear. An actual sharpened wooden spear was recovered from the English site of Clacton-on-Sea (Oakley *et al.* 1977); and tips of elephant tusk, presumably used as points, were excavated at Torralba and Ambrona in Spain (Howell and Freeman 1982). These represent the earliest evidence for effective penetrating weapons.

Bifaces were still the mainstay of the tool-kit, in Eurasia and Africa at least. The assemblages present the same range of tool types as the earlier biface assemblages; but there are significant innovations in techniques of manufacture. One of these is the use of 'soft' hammers of bone, antler, or wood for trimming stone tools. Knapping with a soft hammer produces thinner trimming flakes, allowing a finer control of the angle of the working edge and the shape of the tool as a whole. Prepared-core techniques were also common. In these the knapper trims a core in order to predetermine the shape of one or several flakes. The most celebrated prepared-core technique is the Levallois technique, in which the knapper shapes a core and then removes one large, thin flake. Levallois is common in European biface assemblages by 300 000 BP; but other prepared-core techniques appear as early as 700 000 BP in Africa (Clark 1970). These techniques are accorded considerable importance by many prehistorians, perhaps because they require considerable skill. I will return to their cognitive significance later.

The context of tool-use at 300 000 BP resembles in many respects the context of tool-use found with modern hunters and gatherers. As is usual in the archaeological record, the picture of animal exploitation is fairly good, but that for plant exploitation is almost non-existent. Several sites of this age have extensive faunal remains; and there is little question that hominids relied on animal products. There is, however, still some question as to how they were obtaining them. Romance, and some evidence, leads to assertions of big-game hunting, especially at sites such as Hoxne in England (Wymer 1983) and Torralba and Ambrona (Freeman 1975; Howell and Freeman 1982), which have remains of elephant, ox, horse, deer, and rhino. At Olorgesailie in East Africa there are the remains of a number of baboons (Isaac 1977; Shipman *et al.* 1979). But there are other explanations, including scavenging (Binford 1985).

Arguments on this point are technical, covering the mortality patterns of animals, the characteristic modifications of bone produced by human butchery and by carnivorous animals, and body-part analysis. The situation is far from clear, and there is evidence on both sides. However, even the sceptics such as Lewis Binford admit to some evidence for hunting of smaller animals (fallow deer at Hoxne), even if the large animals like elephant were only scavenged (Binford 1985, p. 317). The extensive remains at sites like Torralba and Ambrona and the evidence for some true hunting distinguish this time-period from 1.8 million BP, and suggest a subsistence that was much more like that of modern hunters and gatherers than that of apes.

Like modern hunters, these hunters returned to a particular site for specific activities. Whereas Torralba was a hunting/butchering locality, Montagu Cave (Keller 1973) in South Africa was a manufacturing site. There is no evidence of refuse or habitation, just thousands of unused waste flakes and a few broken bifaces. Terra Amata, on the other hand, appears to have been a habitation site, with many different kinds of domestic activity (de Lumley 1975). This specialization in site activity is reflected in the stone tools. Different sites contain radically different percentages of certain types of tool; and this suggests activity differences, even if the specific tasks are unknown. For example, at some sites bifaces are very common, and occasionally constitute the majority of artefacts; at other sites bifaces are rare, while small tools on flakes are common (Kleindienst 1961; Hansen and Keller 1971; Cole and Kleindienst 1974). The specific differences in activity are harder to determine. Association with refuse provides some indication. Torralba, for example, is an animal-processing site, but has relatively few stone tools; and of these very few are bifaces. One technique for ascertaining the specific function of a tool is the microscopic examination of the wear caused by use. Using this technique on tools from Hoxne, Keeley (1980) has been able to argue that the five bifaces in his sample were in fact used for butchery, while the untrimmed flakes in the assemblage were used for butchery, bone-working, and wood-working.

Hunting and site specialization distinguish the context of tool-use at 300 000 BP from what we know of the context at 1.8 million and 1.0 million BP. But there is still much that is *ad hoc* about the technology. The non-biface tools vary in much the same way that Oldowan tools vary—a large range of unstandardized tools whose shapes appear to have been tied to the particular tasks to hand. At 300 000 BP these tasks were concentrated in specialized sites; but, with the exception of bifaces, there were still no standardized types with specific intended functions. It was the activity that was specialized, not the tools.

The hominids appear to have been intelligent. I base this conclusion on a Piagetian analysis of stone-tool geometry, especially that of the bifaces. One of the concepts achieved late in ontogeny, if at all, is the concept of metrical symmetry, the precise mirror inversion of a figure. While early bifaces have only a rough symmetry, bifaces dating to 300 000 BP have a very fine symmetry. Further, many have fine symmetry in three planes—plan, profile, and cross-section. In the case of a fine biface all of the possible cross-sections are symmetrical: that is, if the artefact were cut through on any plane, the resulting section would be symmetrical (Fig. 10.3). Because most of these

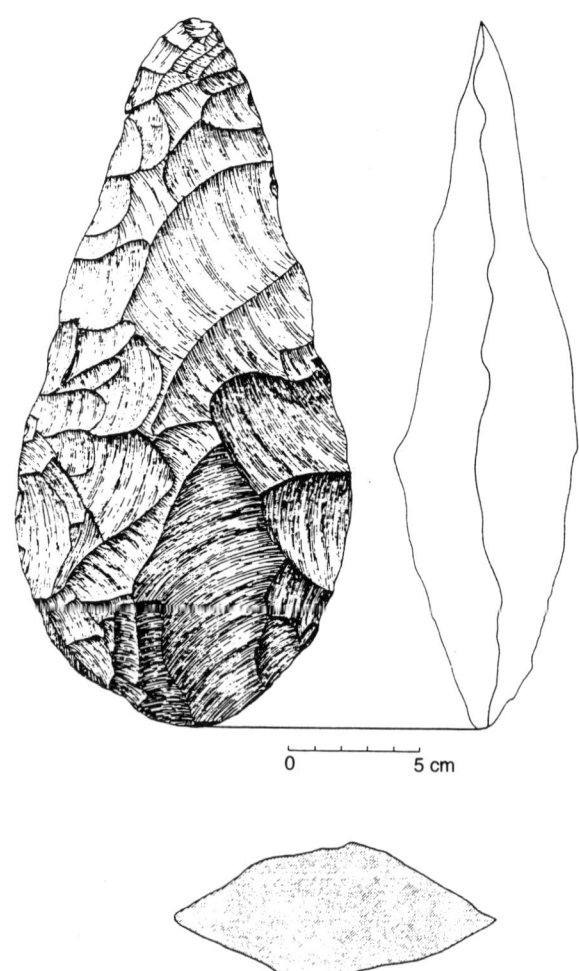

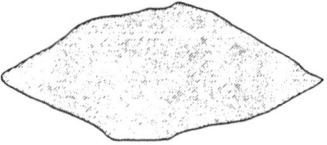

Fig. 10.3 Biface from the late Acheulean site of Isimila. Note the fine symmetry in plan, profile, and cross-section (from Wynn 1979).

hypothetical sections could not have been actually observed by the knapper as he struck with his hammer, he had to have imagined them, and, while trimming the tool, considered them all. This requires an ability Piaget terms 'spatio-temporal substitution' (Piaget and Inhelder 1967), and is characteristic of operational intelligence. Operations are a sophisticated way of organizing thought—one that includes transitivity, for example—and are typical of modern thinking. Thus at 300 000 BP stone-knappers must have been using spatial concepts as sophisticated as those used by modern adults (Wynn 1979). Levallois technique, though demanding considerable skill, requires concepts no more complex than those required for fine bifaces (Wynn 1985). A more difficult question concerns whether or not these people used concepts of equivalent sophistication in other realms of behaviour. Piagetian theory views intelligence as a coherent whole, such that organizational abilities must be equivalent across all behaviours. Atran (1982) argues from a Chomskian perspective that spatial abilities might well be separate from other abilities, and that the sophisticated spatial concepts used at 300 000 BP do not imply equivalent sophistication in other realms. While I favour a Piagetian interpretation, I must add that the matter cannot be resolved in the context of prehistory, but must be resolved in the context of developmental psychology. I must leave that debate to others.

Even though geometrically sophisticated, the distribution of bifaces in time and space is not modern. The shapes of modern tools vary significantly over space and time; Japanese saws differ from Western saws, and Roman ploughs differ from Saxon ploughs. This volatility is characteristic of modern technology, but was not characteristic of biface technology. Indian bifaces are indistinguishable from African, European, and Near Eastern bifaces. Bifaces from early sites like Gadeb in Africa (Clark and Kurashina 1979) are very similar to bifaces from sites hundreds of thousands of years later. This conservatism is one of the great enigmas of prehistory; and, as it may bear on the problem of symbolic behaviour, it requires a somewhat lengthy discussion.

Bifaces are not in fact identical. They differ in size and shape, and include narrow and oval forms and examples with constricted points. All, however, share the same distinctive overall morphology of a biface. What is especially remarkable is that these variants show no consistent correlation with time or space; a particular biface shape would not be out of place in any biface assemblage, of any age, from any part of the world in which bifaces have been found. Some attempts have been made to document such a correlation. Roe (1968), for example, demonstrated a tendency for oval handaxes to become more common in England; and Gilead (1970) and Hours (1975) have made less rigorous demonstrations about Levantine bifaces. However, even in England this tendency is not everywhere true (at Hoxne narrow forms become more common); and it is certainly not a universal trend. There are some geographical differences in the average size of bifaces; but these seem to reflect raw material and the nature of the blank (core or large flake). When these are factored out no identifiable regional traditions appear, only local ones (see Wynn and Tierson 1990 for a revision of this argument) Isaac (1977) has observed that some sites present a considerable range of biface shapes, while others have little variance. When several sites in a region are considered, however, there is no directional tendency, and Isaac argues that shape behaves in a 'random walk' fashion.

In this enigmatic variability we have, I think, the reflections of symbolic behaviour. Archaeologists traditionally explain artefact variance in terms of function or in terms of style. Functional explanations of bifaces have been largely unsatisfactory. While the size and mass of bifaces appears to aid in butchering (Jones 1981), this does not also entail the fine symmetry. O'Brien (1981) has argued that bifaces were projectiles, whose shape was determined by aerodynamics. Most bifaces, however, are not regular and fine. The nice ones are disproportionately represented in figures; but in the museum drawers the unbalanced and cruder ones predominate. If bifaces were intended for aerodynamic purposes, most of them failed. There appears to be no mechanical reason for the biface shape or for the variations. We are left with a stylistic explanation.

Style is a knotty problem in lithic archaeology. Stone is not a malleable medium; and the modifications of stone-knapping are purely subtractive. Furthermore, most stone tools were intended for use; and use obviously has some effect on form. It might seem, then, that stone is an unpromising technology in which to explore stylistic patterns. I hope to show otherwise; and by looking at the styles of bifaces will argue that something was very different about symbolic behaviour 300 000 years ago. My definition of style is based on the work of James Sackett (Sackett 1982) and David Pye (Pye 1964). For Sackett style consists of the choices made within the constraints imposed by the mechanical requirements of the tool. He does not think that the form of an artefact is determined by its function; rather, there is '... a great range of forms that would be more or less equally appropriate' (Sackett 1982, p. 72) for any desired result. The artisan chooses from among these options; and this choice constitutes style. The concept of style closely parallels Pye's concept of design. Like Sackett, Pye considers the desired result to constitute a set of constraints. Even more than Sackett, Pye minimizes the influence of function, and argues that any task could be performed efficiently by a huge number of possible forms.

> The ability of our devices to 'work' and get results depends much less exactly on their shape than we are apt to think. The limitations arise only in small part from the physical nature of the world, but in very large measure from considerations of economy and style. Both are matters purely of choice. All of the works of a man look as they do from his choice, and not from necessity (Pye 1964, p. 11).

It is important to note that Pye is not an archaeologist, and his conception of design is not a convenient definition designed for an archaeological problem. His concept of design is, however, applicable to all media, including stone. Within the constraints imposed by the raw material the knapper has a wide range of choices in technique—prepared-core, soft hammer, etc.—and in overall shape. From the African site of Isimila there are bifaces in such diverse raw materials as granite, quartz, and quartzite (personal observation), all of which have different flaking properties. Yet the knappers imposed the same form on each. It was clearly a matter of artisan choice.

We may thus contrast style as it behaves in modern culture with style as it behaved 300 000 years ago. For modern humans choices about form are determined largely by cultural tradition; choices concerning appropriate form are learned by observation of older, more experienced artisans. In some kinds of technology, pottery for example, style often includes an intended referent—myth, story, whatever—and, from the point of view of semiotics, these would be symbols in the narrow sense. But while symbolic reference is possible with stone tools, it was probably rare. Instead, stone tools act as visual indices. An index signals its referent by physical contiguity (Casson 1981) or direct association. Because choices about appropriate style are learned—culturally transmitted—stone tools can act as indices of the cultural milieu in which an artisan is raised. Because '... chance alone dictates that any single [form] is unlikely to be chosen by two societies which are not ethnically related in some fashion...' style comes to act as a marker of ethnic identity (Sackett 1982, p. 73). Furthermore, there is something about the nature of modern cultural transmission that results in a predictable pattern of change in the stylistic choices made by a particular society. New forms appear, become more or less popular, and then die out. From the semiotic perspective, the signifier changes while the signified remains relatively constant. We see these time–space patterns in modern technology, and we can identify them in the patterns of stylistic variation in stone tools by 25 000 years ago. We do not see them at 300 000 BP.[4]

If biface style acted as an index, it acted in a manner very different from that of modern style. First, the same range of forms, signifiers if you will, were made over three continents. Second, the modern pattern of stylistic change through time was totally lacking. Again, we have only the time–space patterns of signifiers themselves. All that we can conclude with any certainty is that this pattern was different. It suggests that there was something different about the semiotic system or about the nature of cultural transmission, which amounts to the same thing. We can speculate about these differences but our speculations would be very hard to prove.

Nevertheless, I will briefly discuss three possibilities. (1) It is possible that biface style did act as a cultural index, that is, was the result of choices learned by cultural transmission and thereby marked the cultural identity of the maker. In this case the cultural system must have been very different from that of modern culture—one in which there were no social boundaries beyond that of the individual, and one in which an individual in Europe marked his or her identity in precisely the same manner as someone in South Africa. We have no analogues for such a system. As we have seen, modern symbolic systems, such as language, are volatile. The signs and referents change over time and space because they are culturally transmitted and, among other things, are subject to errors in transmission. We would expect the nature of an indexical sign to vary over space and time, especially if the index marked social identity in widely scattered groups. (2) The bifaces may have indexed not social identity but biological identity, sex or maturity, for example. Biological status would have been present world-wide, and would not have changed over time; hence the referent of the index would have been the same everywhere. But this argument is weak because it requires an invariant connection between the index and the referent over vast spans of time and space; and, as I discussed above, this is unlikely in the context of modern cultural transmission. (3) It is possible that bifaces were not indices at all—were not a matter of choice—and hence carried no information about the maker. One could argue that shape was under some kind of genetic control. This need not have been a biface 'gene' but could have been a kind of 'good *Gestalt*', perhaps manifested in other realms of behaviour such as geographic spatial organization, or even kinship 'space'. The advantage of such an argument is that it easily accounts for the time–space distributions. The difficulty is that there are no parallels in the behaviour of modern humans or other primates.

None of the above three speculations is convincing, largely because they are based on so little information. We have only the time and space patterns of the stone tools to go on; we have the indices but not the referents. One thing does seem clear. There was something very different about culture 300 000 years ago; and this archaic character was probably related to the symbolic behaviour of the makers. Whatever it is that makes modern symbolic systems volatile, it was missing 300 000 years ago.

In sum, there are some aspects of 300 000-year-old tools that are modern: in particular, their context of use and the spatial concepts used by their makers. Other aspects are not modern, especially the behaviour of style—the way the stone-knappers made choices about shape. This stylistic difference is not a difference in degree: it is a difference in kind. Apparently the symbolic system was not a simpler form of modern behaviour, but a different kind of behaviour altogether. This assessment is corroborated by the absence of anything that can realistically be interpreted as ritual behaviour.

10.6 Tools at 85 000 years ago

In Europe, 85 000 years ago was the time of the Neanderthals. Because the status of the Neanderthals in human evolution remains uncertain, and because the archaeology of the Neanderthals is riddled with controversy and alternative interpretations—more so than is typical even for archaeology—I am going to focus on South African tools at 85 000. General technological developments were much the same in both areas. Moreover, anatomically modern humans may well have been present in South Africa by 90 000 BP (Rightmire 1979).

By 85 000 BP the monotony of biface culture had broken down, and had been replaced by various regionally specific tool traditions, many of which emphasized the manufacture of tools on flakes. The technique of flake-manufacture is as informative to the archaeologist as the final shape of the tool, which is usually termed its type. While specialized flake-manufacturing techniques appeared by 700 000 BP (see above), it was not until the Middle Stone Age (MSA) that tool types made *on* flakes became characteristic of archaeological assemblages. The MSA in South Africa is characterized by several techniques for manufacturing specific kinds of flakes—one techniques produced long narrow flakes termed 'blades', and another produced triangular flakes (Fig. 10.4). There was even some Levallois. In addition there appear to have been more *types* of tools, by which I mean that the assemblages seem to fall into more discernible categories than the earlier assemblages. We define most of these types on the basis of the shape of the edge and the overall shape of the tool. Types include various types of 'scraper', chisel-like tools termed 'burins', and bifacially trimmed flakes termed 'points'. The names may or may not reflect actual uses; but there does appear to have been a greater variety of tools than there was at 300 000 BP.

These flake-tool assemblages present regional variation. In South Africa there are three such variants: one found in the Transvaal, one in the Orange River Valley, and one on the Cape Coast (Sampson 1974). To a certain extent these divisions reflect areas of research. But there are real differences between the tool assemblages. These include differences in technique and differences in tool type. For example, the Orange River assemblages have a higher frequency of blades and fewer triangular flakes than the Transvaal assemblages (differences in technique), and also have occasional trimmed points, which are lacking in the Transvaal assemblages (a typological difference). The existence of recognizable regional differences is an important contrast to the earlier biface industries, for which no such coherent variation could be seen.

The ecological context of tool-use has few obvious contrasts to that of the biface industries. Indeed they are similar in several respects. Cave 1 at Klasies River Mouth presents one of the longest prehistoric sequences excavated anywhere (Singer and Wymer 1982). Various dating techniques suggest that MSA people used this cave over tens of thousands of years, beginning as early as 120 000 years ago and ending by 70 000 years ago. The deposits contain the remains of large animals such as giant buffalo, eland, Cape buffalo, and blue antelope, as well as smaller animals such as seal and hyrax (Klein 1978, 1983; Binford 1984, 1986). Shells of limpets, periwinkles, and mussels are also common. There is some controversy about what these remains actually indicate about the subsistence of these people. Binford (1984, 1986) argues that the large animals were probably attained by scavenging. Klein (1978, 1983) argues that the animals were probably hunted, but, on the basis of an analysis of the ages of individual animals, suggests that the MSA hunters were not very adept, especially compared with later hunters. A resolution of this controversy is not crucial to my discussion. What is important is that this is the same controversy that we found at 300 000 BP. Not much seems to have changed. Even the presence of shellfish is not particularly remarkable. They could as easily have been gathered by scavengers as by hunters.

If the faunal remains suggest little had changed since 300 000 BP, what about the tools themselves? The MSA variant at Klasies includes bifacially

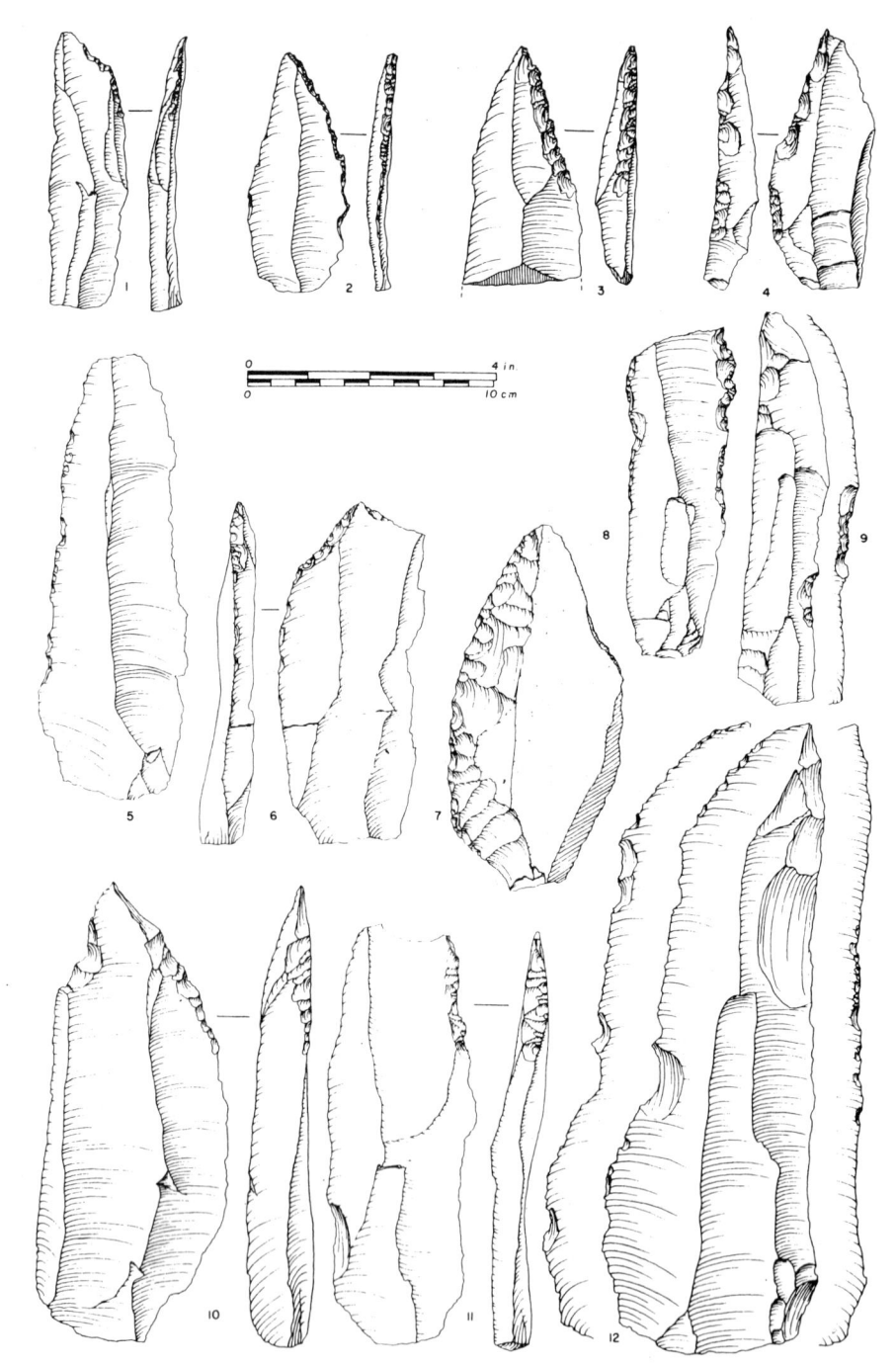

Fig. 10.4 Middle Stone Age tools from the Cave of Hearths in South Africa. Unlike the earlier biface industries, the emphasis here is on the manufacture of many different types of flake tool (from Sampson 1974, by permission).

trimmed 'points', some of which were thinned and notched at the proximal ends. Singer and Wymer argue, on the basis of the shapes of these tools, that they were projectile tips. Binford (1984) sees no evidence in the faunal remains that the animals had to have been killed by projectiles. Scavenging large animals and overpowering small ones are behaviours that could account for all the remains. There is always a problem in assuming function based on shape. As a consequence, we cannot conclude that stone tools were used as projectiles, though it remains a possibility. Furthermore, the MSA has few tool types that could reasonably be interpreted as plant-processing tools. It seems, then, that what we *know* about the context of tool-use during the MSA cannot explain the changes that we see in the tools.

Nothing about the tools suggests a leap in intelligence compared with that of the biface-makers. Any common-sense arguments concerning the planning necessary for prepared cores could also be applied to 700 000-year-old prepared cores. It is not that there was anything new about the shapes and techniques of MSA tools. It is that the emphasis had changed; and it is difficult to argue for a leap in intelligence from a change in emphasis. From a Piagetian point of view, nothing about MSA tools requires a more sophisticated competence than that required for the fine bifaces at 300 000 (Wynn 1985).

Nor can we argue for a more sophisticated language. One could, perhaps, argue that the regional differences in stone-tool assemblages reflect regional differences in dialect or language. Modern differences in material culture occasionally parallel linguistic differences. However, such an argument again assumes a connection between technology and language that, as we have seen, is probably unwarranted. The regionalization is provocative; but it does not require a linguistic explanation.

The indexical role of MSA tools appears to have been neither like that of earlier biface assemblages nor like that of later prehistory and the modern world. Of particular relevance are two patterns documented in the archaeological record—the appearance of regional traditions and the presence of 'interstratified' industries.

I have described the three South African regional MSA traditions briefly earlier. It is important to bear in mind that most of the differences are in terms of relative frequency of tool types and techniques. Sampson (1974), who defined the traditions, considers them to be '... the remains of hunter–gatherer activity in three distinct environments' (p. 157). A correlation certainly exists; but it is hard to imagine precisely how differences in available plant and animal resources would affect the typology and technique of tool manufacture. As we have seen, function *in no sense* determines the form of a tool, and in this case we do not know if there were, in fact, any 'functional' differences between the regions. It seems likely that more or less the same activities would be undertaken regardless of a difference in species. Even if there were major differences in activity, these differences would only set different limits within which choices about tool *shapes* were made. And it seems unlikely that it would have any effect on *technique* of manufacture at all. The latter might be affected by the availability of raw material. Indeed, one of the three traditions, the one from the Orange River area, contains large numbers of unmodified blades and flakes, presumably because of the ready availability of a fine raw material, lydianite (Sampson 1974, 168). But here again the variable does not *determine* the choice of technique; it only sets limits. Something other than environment affected the ultimate choice of shape and technique. Since tool-making was learned, presumably from older artisans, it seems that there must have been differences in culturally transmitted knowledge about appropriate shapes and techniques. However, before we simply accept the presence of three different 'cultures' in South Africa we need to look at the rather knotty problem of interstratified industries.

The cultural stratigraphy at Klasies River Mouth presents an intriguing pattern. The first two stratigraphic levels contain a fairly typical MSA made on flake-blades of local raw material and containing a high frequency of 'scrapers'. Following this is a strikingly different industry (termed 'Howiesonspoort'), made on exotic raw material with high frequencies of crescent-shaped tools made on blades. This is followed by two more stratigraphic units containing typical MSA artefacts on local raw material, with high frequencies of 'scrapers' (Singer and Wymer 1982). This is what is meant by interstratified industries. One industry is replaced by another, only to reappear in a higher level, at the same site, almost unchanged. Such patterning argues against a simple interpretation of the industries as cultural groups, for one would then have to argue that one group departed, only to reappear, virtually unchanged, thousands of years later. There are of course alternative explanations emphasizing environmental changes and changes in activity, all of which require a closer connection between form and function than can reasonably be accepted. The real oddity here is not that one industry replaces another, but that they change so little over time.

What does all of this imply about the indexical role of tools? If the regional differences mark differences in culturally transmitted choices about tools, then, on a very coarse level at least, the tools were indices of groups. But this would only have been a side-effect. Recall that most of the differences are differences in frequency of types. One would have had to observe large numbers of tools before one could determine cultural affiliation. It is therefore unlikely that the tools actually performed an important role as indices. Moreover, the shapes do not change much over time, in marked contrast to what happens in later technology. This suggests to me that, even if there is some indexical quality to the assemblages, it was unintended and irrelevant to the day-to-day behaviour of the people. If tools played an important indexical role, then by analogy one would expect them to change in the volatile fashion of modern tools. Nevertheless, the tools do differ from earlier assemblages, and this difference does carry them in the 'direction' of modern technology.

In sum, tools at 85 000 BP fall between biface assemblages and modern technology. They have lost the conservatism of the former, but do not demonstrate the volatility of the latter. The regionalization, while probably not reflecting an important semiotic role for the tools, is the first evidence of a pattern that does eventually have very important implications for the indexical role of tools. In the light of contemporary evidence for burial and ritual behaviour, this MSA technology perhaps represents the very beginnings of modern forms of culture.

10.7 Tools at 15 000 years ago

From the perspective of the Stone Age, 15 000 years was not very long ago. True, there are immense differences between that time and the modern world—no agriculture, no cities, no market economies, no states. Yet the fundamental organization of culture then appears to have been identical to that of the present. In particular, symbols appear to have operated in the same manner as modern symbols. I base this contention on the tools themselves, and on the way the tools varied in space and time.

If we define complexity as the number of elements or steps involved in a process, then tools at 15 000 BP were more complex than those at 300 000 BP. One simple example is hafting—attaching a stone tool to a handle. There is no evidence of hafting at 300 000 BP, but by 15 000 BP hafting is an essential element of tools in most technologies. In the Later Stone Age of Africa, for example, arrows were made by mounting microlithic stone tools on shafts (Clark 1970). Other kinds of multi-step technologies were also common. In the Solutrean of southern France there were eyed needles made out of antler. The implied sewing technology includes the following steps: making the stone tool or tools; making the needle; and using the needle to make the finished item. More complex still is the use of facilities. 'Facilities are designed to apply human energy indirectly, through attracting, containing, holding, restraining, or redirecting a living mass' (Oswalt 1973, 26), and include such items as hunting blinds, pitfalls, and fish weirs. Evidence for these technologies at 15 000 BP is circumstantial, but nevertheless persuasive. For example, fish are well represented in European Upper Palaeolithic engravings, and hard-to-catch prey such as birds are common at some sites (Dennell 1983). Such resources yield few calories per prey animal and, as a consequence, can be efficiently exploited only by the indirect-energy use of facilities. There is also circumstantial evidence for boats. Humans had arrived in Australia by at least 32 000 years ago; and this required a journey of 50 to 100 kilometres over water (Bowdler 1977).

Much of this increase in technological complexity occurs in raw materials other than stone. Although we have some evidence of bone tools as far back as the Oldowan, they are not common until after 40 000 BP. In addition to bone, artefacts and facilities at 15 000 BP were made of antler, ivory, shell, wood, and fibres. The contemporary stone-tool assemblages include many tools that were used to make and maintain the non-lithic artefacts and facilities (Fig. 10.5). European tool-kits had numerous burins and scrapers for working antler; and Australian and New Guinea assemblages included edge-ground axes, almost certainly woodworking tools, by 23 000 BP (Hayden 1977). This complexity extends to the numbers of tool types in the assemblages as well. These include many different types of extractive tools, and the maintenance tools to make them. One common characteristic of these complex technologies is that they were time-consuming to make and maintain—a factor that must be considered when we look at ecological context.

The ecological context of tool-use at 15 000 BP was still hunting and gathering. However, every region seems to have had a particular emphasis; and this regional specialization is quite different from the more generalized pattern seen at 300 000 BP. Moreover, it is reflected in the technology. For example, in Southeast Asia at 15 000 BP there was a tropical forest and riverine adaptation known as the *Hoabinhian*. Hayden (1977) argues that animal protein constituted a small percentage of the diet, and that plant foods such as cereals were more important. His argument is at least partly based on the technology, which includes mortars and pestles and ground stone knives, the latter presumably for cutting grasses. This Hoabinhian adaptation and technology is very different from that of the European Upper Palaeolithic, where large mammals were the focus of the subsistence system. Bahn (1977) argues that hunters in south-western France followed the reindeer herds, and even practised selective killing of males in the late autumn. Here also the technology was adapted to the resource, and included bone points and spear-throwers. Similar big-game-oriented hunting systems appeared in North America at about this time (Jennings 1974); and adaptations similar to the Hoabinhian appeared in tropical Africa (Miller 1971). In most parts of the world, in other words, people fine-tuned their subsistence systems to the peculiarities of local resources; and one way they did this was through technology. The result was a kaleidoscope of diverse tool-kits. This is in marked contrast to the monotonous technology seen at 300 000 BP. At 15 000 BP both the activities *and* the tools were specialized.

Specialization was not the only factor behind the difference in tools, however. Beginning about 30 000 years ago the shapes of tools in Europe behave in the volatile fashion typical of modern style (Laville *et al.* 1980). New forms appeared, became popular, and then died out. Such stylistic variation was found not only in the mobiliary art, but also in the mundane tools involved in day-to-day activities. Point and harpoon types replace one another over time, as did the types of burins that made them. Distribution in space was also modern in character. Some tool types—Solutrean willow-leaf points, for example—were extremely limited in distribution (Fig. 10.6). Some of the artefacts probably acted as symbols in the narrow sense, that is with specific referents in tradition or mythology. Mobiliary art includes obvious examples; but the very fine Solutrean leaf-shaped points may have been too fragile to use, and so may have played a role in some non-utilitarian realm, perhaps as symbols. It seems likely that the style of most artefacts acted as an index of social identity, and that the time–space patterns of the styles reflected a dynamic in the social environment that had not been present at 300 000 BP. Bahn (1977) observes that these Upper Palaeolithic hunters carried or traded shell and other raw material hundreds of kilometres, and argues that their geographic and, presumably, social range was large. Gamble (1982) argues that the subsistence system of Europe at this time required the exchange of information about far-flung resources and conditions, and that such information could only come from distant kinsmen, real or fictive. Style played a role in communicating information about social identity through its indexical quality. Such a social environment would encourage the imposition of an even greater indexical precision. What is important is that something of social complexity was being communicated, however crudely, through the medium of technology. If this assessment is true, then the semiotic role of tools at 15 000 BP was very different from that at 300 000 BP, and was, indeed, essentially modern.

Earlier in this paper I argued, on the basis of a Piagetian analysis, that hominids had achieved an essentially modern intelligence by 300 000 years ago. Given the greater complexity of technology and the modern patterns of time–space variation at 15 000 BP, this assessment may appear counterintuitive. How can these patterns be explained, if not in terms of conceptual sophistication? I think the most parsimonious explanations are purely cultural ones. The increased complexity of the tools was the result of technological solutions to the ecological fine-tuning discussed earlier—complex technologies solving problems of efficient resource-exploitation. The time–space patterns were, on the other hand, the result of new requirements in the social-communication system, a system in which, as Gamble argues, information was crucial, and tools played a role, however minor, in marking relationships. Neither of these requires greater intelligence.

While the complexity and diversity of the technology did not require a more sophisticated intelligence than that at 300 000 BP, it may have required a more sophisticated language. Such an argument has been advanced by Krantz (1980):

Superior language would have facilitated the teaching of more sophisticated tool making; rapid transmission of new techniques would mean faster technological change as well as easier adaptation to local resources; hafted spear points and other compound tools depend on sophisticated techniques ... (p. 776).

By superior language Krantz means phonemic speech, a principle he argues was discovered and rapidly disseminated about 40 000 years ago. Unlike some of the authors discussed earlier, Krantz does not see a deep cognitive connection between complex tools and language. Rather, like Guilmet, he argues that the communication system would have profound effects on the sophistication of tools. The difficulty here lies in Krantz's common-sense interpretation of technology. Nothing we know about the way technology is learned (see Section 10.4 above) suggests that the relative rapidity of phonemic speech during instruction would have any effect on the complexity of the tools. I do not wish to play down the importance of the phonemic principle in human evolution; I only wish to point out that an argument for its presence cannot be based on tools. Lock (1988) advances a slightly different argument concerning complex tools and language. He observes that '... a grammatical phrase is a composite construction that is treated as a single item by transformational rules' (p. 31). It requires the manipulation of elements into single wholes. A parallel in technology is the construction of self-supporting compound tools such as hafted spears. Such complex tools do not appear until after 300 000 BP. The parallel is real and provocative. But

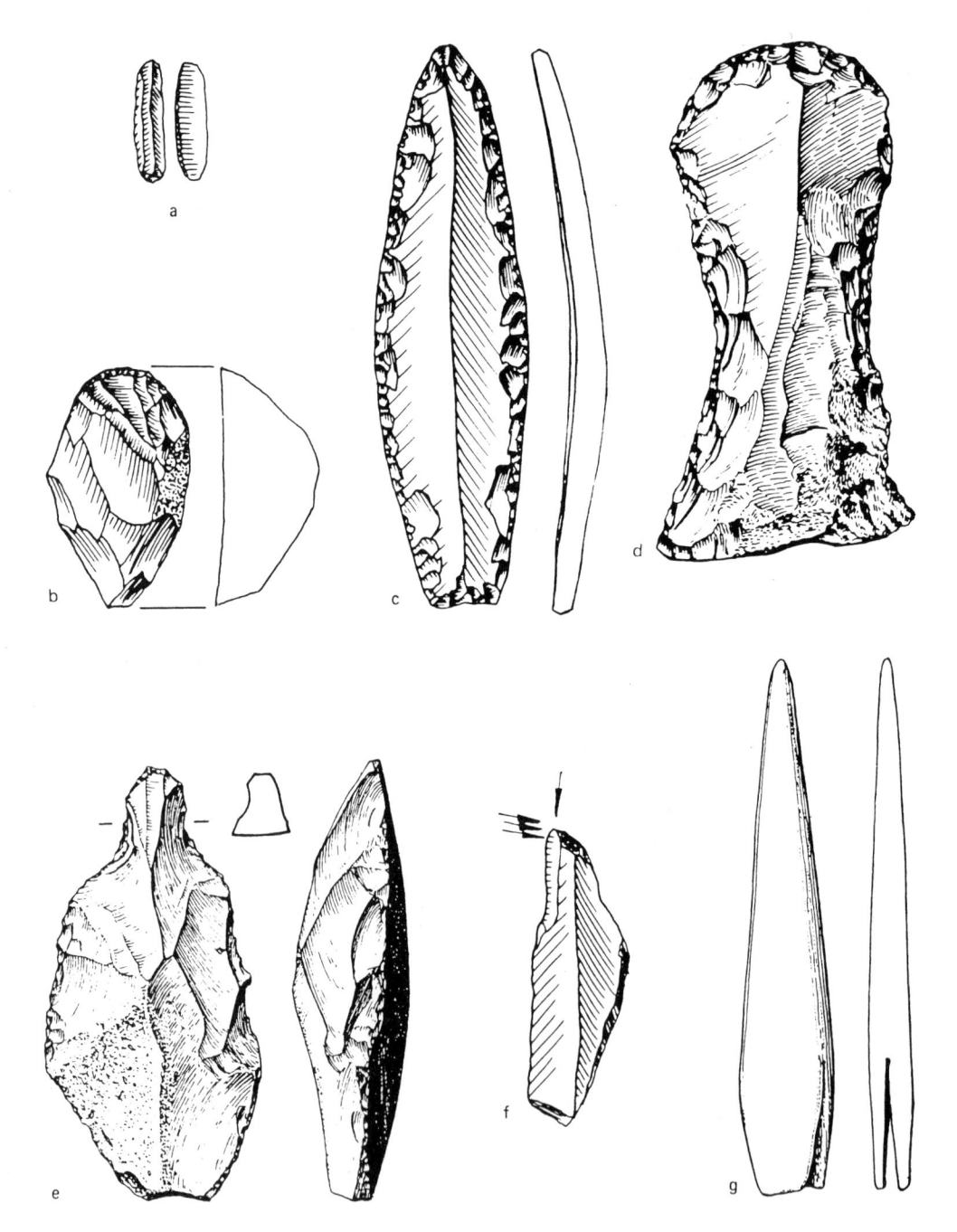

Fig. 10.5 Various Upper Palaeolithic tools from France (from Laville et al. 1980, by permission).

I refer back again to Bruner's pre-linguistic child, whose routines, including routines involving objects, have a script-like quality, and who 'is readily attuned to "making a lot out of a little" by combination ...'. I admit to being very conservative here; but what we know about technological behaviour and learning suggests that Lock's parallel may be coincidence.

In sum, tools at 15 000 BP were very different from tools at 300 000 BP. They were more complex, more

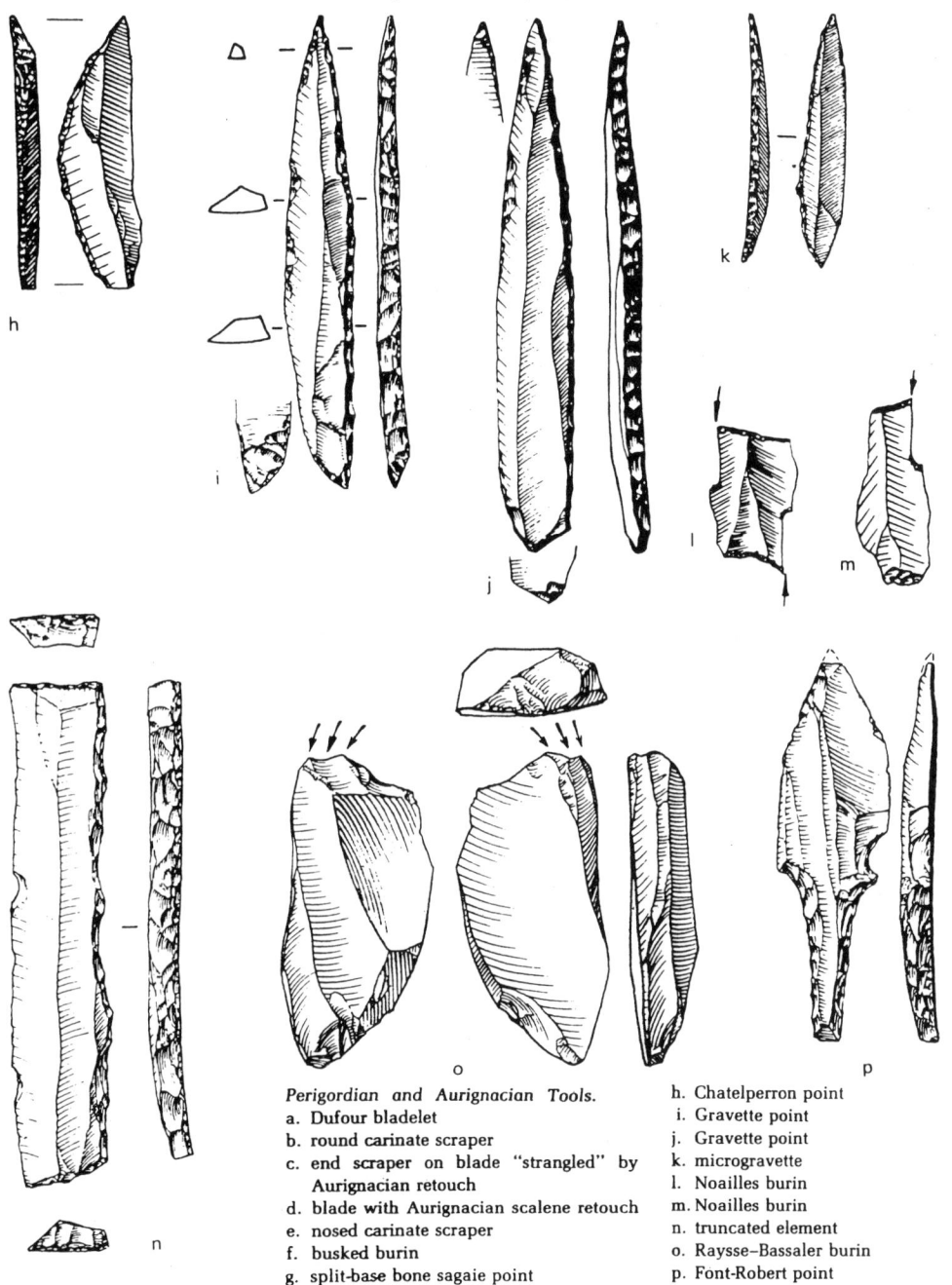

Perigordian and Aurignacian Tools.
a. Dufour bladelet
b. round carinate scraper
c. end scraper on blade "strangled" by Aurignacian retouch
d. blade with Aurignacian scalene retouch
e. nosed carinate scraper
f. busked burin
g. split-base bone sagaie point
h. Chatelperron point
i. Gravette point
j. Gravette point
k. microgravette
l. Noailles burin
m. Noailles burin
n. truncated element
o. Raysse–Bassaler burin
p. Font-Robert point

specialized, and more varied. These developments were undoubtedly related to the increase in general cultural complexity (see White, this volume, Chapter 9), including residence patterns, kinship, and so on. The semiotic role of tools also appears to have changed. Some tools were almost certainly symbols in the narrow sense, including perhaps some of the very fine stone tools. But the more mundane tools also appear to have played an indexical role, so that, in some sense, the entire tool repertory carried information about the social persona of the maker. This is a dramatic change from 300 000, where only a few artefact types may have had semiotic load. It seems that by 15 000 BP human behaviour was infested with semiotic meanings of all sorts. In other words, it was entirely modern.

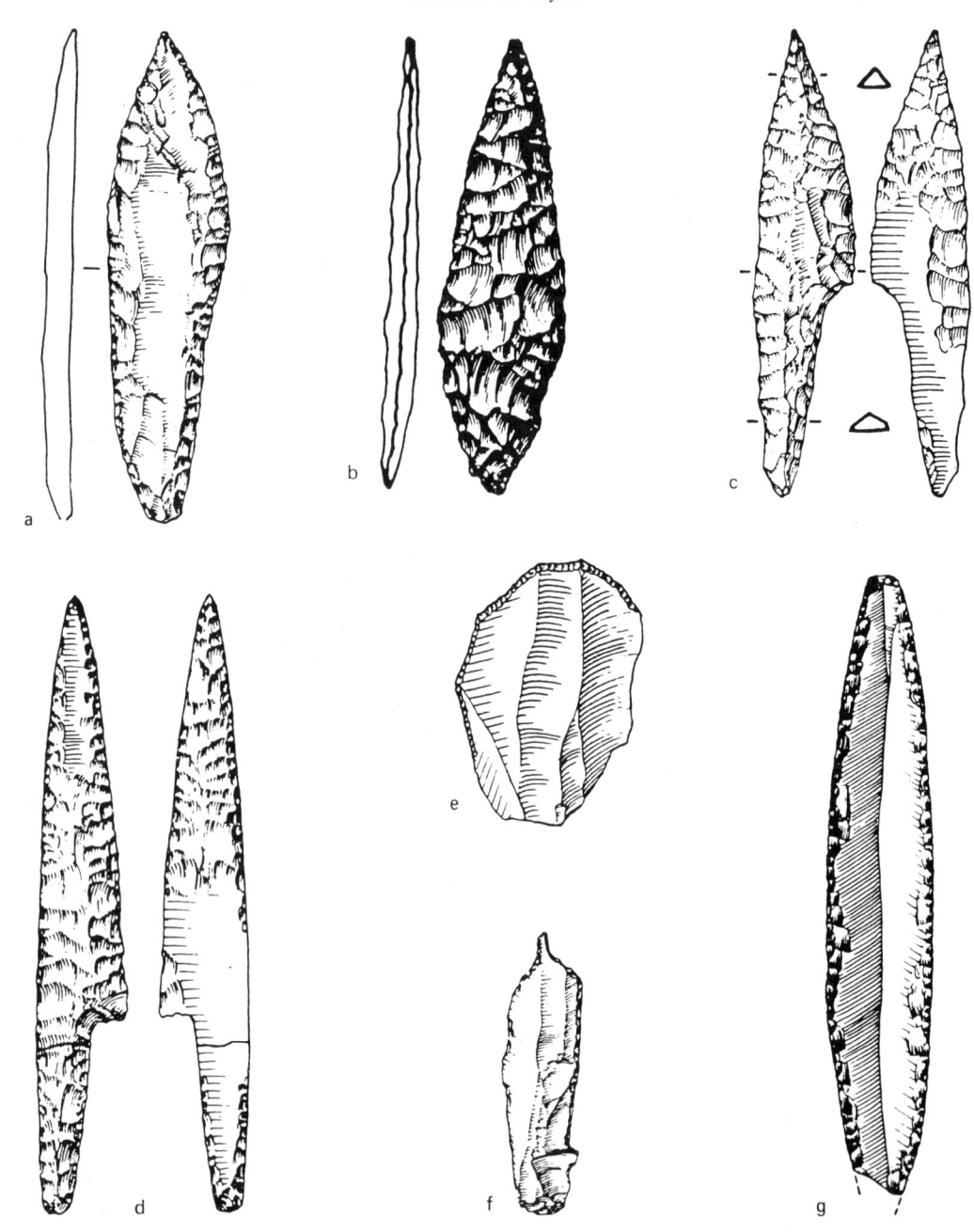

Fig. 10.6 Solutrean and Magdalenian tools. Some of the finest examples of Solutrean blades have a very limited geographical distribution, and may have acted as indices, or even as symbols in the narrow sense (from Laville *et al.* 1980, by permission).

10.8 Recent developments

The basic semiotic role of tools has changed little in the last 15 000 years, despite tremendous changes in human technological sophistication. Tools have become incredibly complex, to the point that many are beyond the understanding of most people. They have also become so effective in 'mediating' the environ-

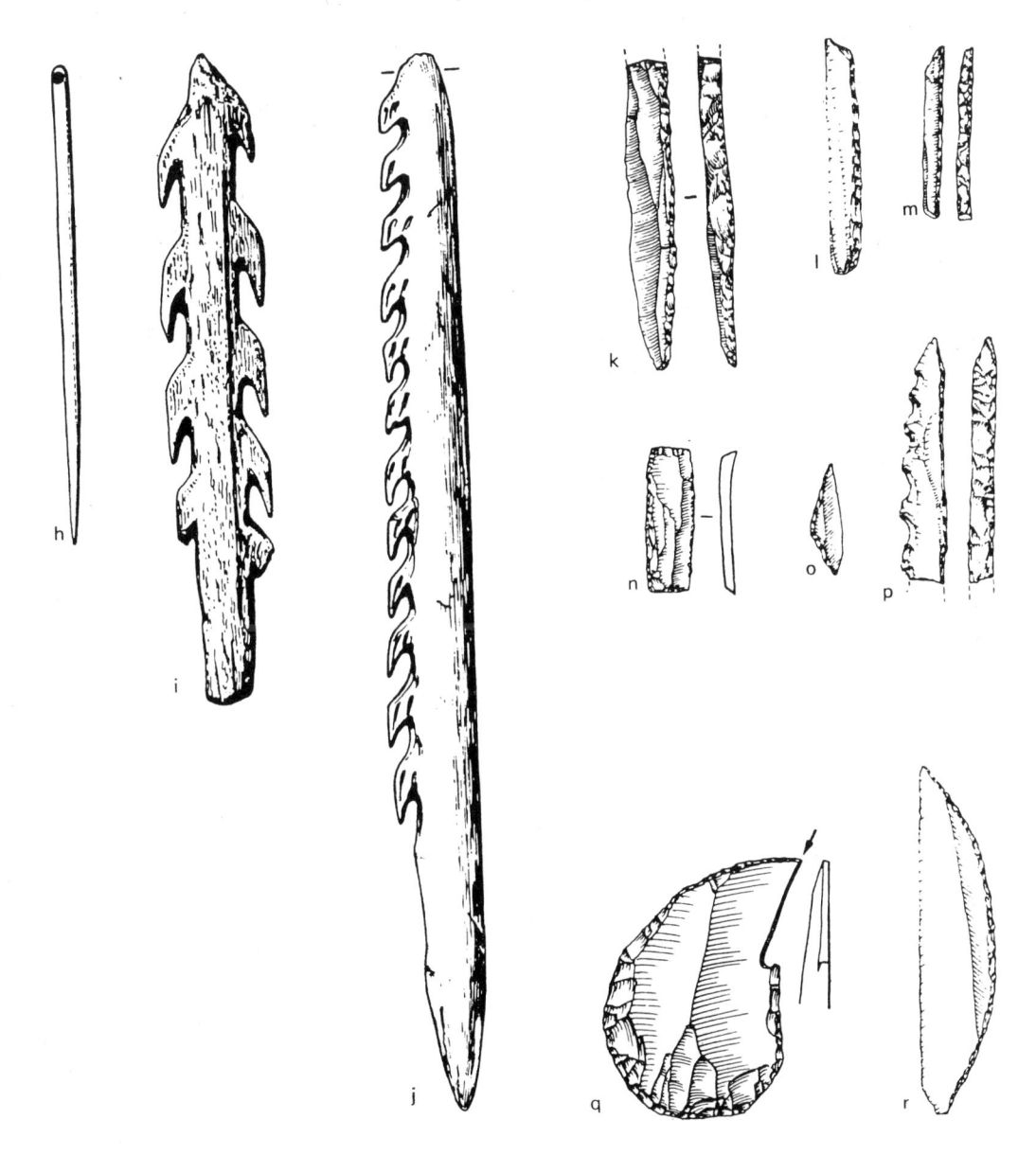

Solutrean and Magdalenian Artifacts.
a. unifacial plane-faced point
b. bifacial laurel leaf point
c. Solutrean shouldered point
d. Solutrean shouldered point
e. raclette
f. perforator
g. Magdalenian I retouched blade
h. eyed needle
i. harpoon with double row of barbs
j. harpoon with single row of barbs
k. backed bladelet
l. backed bladelet
m. backed bladelet
n. rectangular microlith
o. triangular microlith
p. denticulated backed bladelet
q. parrot-beaked truncation burin
r. Azilian point

ment that we must develop technologies to cope with the unintended consequences of other technologies. But their semiotic role has remained relatively unchanged. The time–space patterns of Upper Palaeolithic stone tools are also found in the ceramics, metal tools, and even plastics of later time-periods—the medium has changed but the semiotic role has not. Tools have continued to act as indices of social and cultural status, and, occasionally, as symbols in the narrow sense. Recently, in the last 5 000 years or so, tools have played a part in the development of writing and telecommunication systems. I do not think that these have seriously altered the indexical or symbolic role of tools; but it is possible that they have had an effect on language. I must leave that problem to others.

10.9 Summary and conclusions

In this review I have attempted to cast light on the evolution of human symbolic behaviour by examining the record of human tools. I have used four basic themes: ecological context, intelligence, tools and language, and the indexical role of tools.

For most of human evolution the primary role of tools was in the 'ecological' realm—getting something to eat, keeping warm, and so on. In this the early hominids were not alone. They merely developed to a high degree an ecological speciality found in many species, including the chimpanzees. The near-phylogenetic ubiquity of tool-use argues that ecological utility is in itself unremarkable, and requires no complex symboling behaviour. Otters do not appear to symbol about mussels or stones, even though these are important elements in their feeding. Ecological context is therefore only very indirectly relevant to the question at hand. The ecological role of tools at 1.8 million BP differed from that of chimps only in degree. There is certainly nothing about it that requires symboling behaviour beyond that known for chimpanzees. The ecological role of tools changes over the subsequent two million years primarily in the variety of tasks requiring tools and in the specialization of the tools themselves. The two million years of hunting, scavenging, and gathering are mute in regard to the details of symbolic systems.

Symbolic systems such as language require certain cognitive abilities; and here the tools do give us some information. A Piagetian analysis of the geometry of stone tools suggests that 1.8 million years ago hominids were not much more intelligent than modern apes. This is consistent with the comparability of the ecological contexts. By 300 000 BP, on the other hand, tool geometry required an essentially modern intelligence. While later technological developments are more complex in terms of the numbers of elements involved, this complexity does not *require* a more powerful intelligence than that documented for 300 000 BP. Intelligence does not inform us directly about symbols. Nevertheless, it appears that there were at least no inherent cognitive limits on symbolic behaviour at 300 000 BP.

Tools tell us little about language. True, by one million years ago tools do present redundant patterns, and it is tempting to see this redundancy as somehow akin to language. But technology and language are not equivalent behaviours, and the connection between them is not simple. Modern technological behaviour is neither learned nor performed linguistically; and, even more important, pre-linguistic behaviour includes language-like sequential patterns. These observations suggest that technological redundancy is simpler and probably earlier—phylogenetically as well as ontogenetically. As a consequence, we cannot conclude from the redundant patterns of bifaces at one million years ago, or later for that matter, that any form of language was in use.

While tools tell us little about language, they do inform us about other aspects of semiotic behaviour. The use of tools in display, especially in the social-maintenance displays of chimpanzees, is an indexical use of tools. There are countless examples in modern human behaviour. It seems, then, that there is a phylogenetic continuity in the indexical use of tools; and we can conclude that there must have been phyletic continuity within the human line. But the nature of indexical usage appears to have changed significantly. Tools did not always mark the same kinds of things. We have no reason to conclude that tools at 1.8 million BP were any different from those of chimpanzees in terms of their indexical role. By 300 000 BP bifaces present a remarkable uniformity of shape over a vast space. There is no known functional reason for this monotony, and it seems likely that the shape was an index of something. We do not know what that something was; but it is clear that it has no known analogues in the modern world. In other words, the little that we see of semiotic behavior at 300 000 BP is very different from anything known in non-human primates or modern humans. And this is in the context of an essentially modern intelligence. By 15 000 BP the indexical role of tools was essentially modern. Like modern tools, these tools carried information about the social status of the maker. One consequence was a volatility of tool shapes—a volatility reflecting the nature of the social and communication milieu of the time.

The above conclusions are in a sense general and unsatisfying. Tools are not windows to symbolic behaviour. Rather, they are more like shadows, sometimes giving us a clear outline, but more often supplying only hints. We can combine these hints into tentative assessments; but these assessments lack detail. We know something was very different about symbolic behaviour at 300 000 BP; but the tools do not tell us exactly what. Perhaps if we had more sophisticated theories relating tools to semiotic behaviour we could make more specific assessments. At present such theories do not exist; and common sense is not an adequate substitute.

Notes

1. Gowlett (1979) argues that choppers required 'a complex process involving feed-forward and feed-back'. I am not sure what these mean; and, anyway, the same could be argued for chimpanzee tools.

2. Many are also known as hand-axes, a term with unsubstantiated implications about use.

3. This particular interpretation has been challenged (Villa 1977; 1978), though the evidence for the presence of huts appears sound.

4. This argument has recently received a more comprehensive discussion by Chase (1991).

References

Adair, J. (1944). *The Navajo and Pueblo silversmith*. University of Oklahoma Press, Norman.
Atran, S. (1982). Constraints on a theory of hominid tool-making behavior. *L'Homme*, **22**(2), 35–68.
Bahn, P. G. (1977). Seasonal migration in southwest France during the late glacial period. *Journal of Archaeological Science*, **4**, 245–57.
Beck, B. B. (1980). *Animal tool behavior*. Garland Press, New York.
Binford, L. R. (1984). *Faunal remains from Klasies River Mouth*. Academic Press, Orlando.
Binford, L. (1985). Human ancestors: changing views of their behavior. *Journal of Anthropological Archaeology*, **4**, 292–327.
Binford, L. R. (1986). Reply to Singer and Wymer. *Current Anthropology*, **27**, 57–62.
Boesch, C. (1978). Nouvelles observations sur les chimpanzés de la forêt de Tai (Côte-d'Ivoire). *La Terre et la Vie*, **32**, 195–201.
Boesch, C. and Boesch, H. (1984). Mental map in wild chimpanzees: An analysis of hammer transports for nut cracking. *Primates*, **25**(2), 160–70.
Bowdler, S. (1977). The coastal colonization of Australia. In *Sunda and Sahul* (ed. J. Allen, J. Golson, and R. Jones), pp. 205–46.
Bruner, J. S. (1972). Nature and uses of immaturity. *American Psychologist*, **27**, 687–708.
Bruner, J. S. (1983). *Child's talk: learning to use language*. Norton, New York.
Bunn, H. T. (1981). Archaeological evidence for meat-eating by Plio-Pleistocene hominids from Koobi Fora and Olduvai Gorge. *Nature*, **291**, 574–80.
Casson, E. W. (1981). *Language, culture, and cognition*. Macmillan, New York.
Chase, P. (1991). Symbols and Palaeolithic artefacts: style, standardization and the imposition of arbitrary form. *Journal of Anthropological Archaeology*, **10**, 193–214.
Clark, J. D. (1970). *The prehistory of Africa*. Praeger, New York.
Clark, J. D. and Kurashina, H. (1979). Hominid occupation of the east-central highlands of Ethiopia in the Plio-Pleistocene. *Nature*, **282**, 33–9.
Cole, G. H. and Kleindienst, M. R. (1974). Further reflections on the Isimila Acheulian. *Quaternary Research*, **4**, 346–55.
de Lumley, H. (1975). Cultural evolution in France in its paleoecological setting during the Middle Pleistocene. In *After the australopithecines* (ed. K. Butzer and G. Isaac), pp. 745–808. Mouton, The Hague.
Dennell, R. (1983). *European economic prehistory*. Academic Press, London.
Dougherty, J. W. D. and Keller, C. M. (1982). Taskonomy: a practical approach to knowledge structures. *American Ethnologist*, **5**, 763–74.
Foster, M. L. (1975). Symbolic sets. Paper presented in the symposium *Toward an ideational dimension in archaeology*. Meetings of the Society for American Archaeology, Dallas.
Freeman, L. (1975). Acheulian sites and stratigraphy in Iberia and the Maghreb. In *After the australopithecines* (ed. K. Butzer and G. Isaac), pp. 661–744. Mouton, The Hague.
Frost, G. T. (1980). Tool behavior and the origins of laterality. *Journal of Human Evolution*, **9**, 447–59.
Galdikas, B. (1978). Orang-utans and hominid evolution. In *Spectrum* (ed. S. Udin), pp. 287–309. Dian Rakyat, Jakarta.
Galdikas, B. (1982). Orang-utan tool-use at Tanjung Puting Reserve, Central Indonesian Borneo (Kalimantan Tengah). *Journal of Human Evolution*, **11**(1), 19–33.
Gamble, C. (1982). Interaction and alliance in palaeolithic society. *Man*, **17**, 92–107.
Gatewood, J. (1985). Actions speak louder than words. In *Directions in cognitive anthropology* (ed. J. W. D. Daugherty), pp. 199–220.
Gibson, K. R. (1983). Comparative neuroontogeny and the constructionist approach to the evolution of the brain, object manipulation and language. In *Glossogenetics: the origin and evolution of language* (ed. E. de Grolier), pp. 41–66. Harwood Academic, Paris.
Gilead, D. (1970). Handaxe industries in Israel and the Near East. *World Archaeology*, **2**, 1–11.
Gowlett, J. A. J. (1979). Complexities of cultural evidence in the Lower and Middle Pleistocene. *Nature*, **278**, 14–17.
Guilmet, G. M. (1977). The evolution of tool-using and tool-making behavior. *Man*, **12**, 33–47.
Hall, K. R. L. (1963). Tool-using performances as indicators of behavioral adaptability. *Current Anthropology*, **4**, 479–87.
Hamilton, W. J., Buskirk, R. E., and Buskirk, W. H. (1975). Defensive stoning by baboons. *Nature*, **256**, 488–9.
Hansen, D. L. and Keller, C. M. (1971). Environment and activity patterning at Isimila Korongo, Iringa District, Tanzania: a preliminary report. *American Anthropologist*, **73**(5), 1202–11.
Harris, J. W. K. (1983). Cultural beginnings: Plio-Pleistocene archaeological occurrences from the Afar, Ethiopia. *The African Archaeological Review*, **1**, 3–31.

Hay, R. (1976). *The geology of Olduvai Gorge*. University of California Press, Berkeley.
Hayden, B. (1977). Sticks and stones and ground edge axes: the upper paleolithic in Southeast Asia. In *Sunda and Sahul* (ed. J. Allen, J. Golson, and R. Jones), pp. 73–109.
Hewes, G. W. (1973). An explicit formulation of the relationship between tool-using, tool-making and the emergence of language. *Visible Language*, **7**(2), 101–27.
Holloway, R. (1981). The Indonesian *Homo erectus* brain endocasts revisited. *American Journal of Physical Anthropology*, **55**, 503–21.
Holloway, R. L. (1983). Human brain evolution: a search for units, models and synthesis. *Canadian Journal of Anthropology*, **3**(2), 215–30.
Hours, F. (1975). The Lower Paleolithic of Lebanon and Syria. In *Problems in prehistory: North Africa and the Levant* (ed. F. Wendorf and A. Marks), pp. 249–71. Southern Methodist University Press, Dallas.
Howell, F. C. and Freeman, L. G. (1982). Ambrona: an early stone age site on the Spanish meseta. *The L. S. B. Leakey Foundation News*, **22**, 1–13.
Isaac, G. L. (1976a). Plio-Pleistocene artefact assemblages from East Rudolf, Kenya. In *Earliest man and environments in the Lake Rudolf basin* (ed. Y. Coppens, F. Howell, G. Isaac, and R. Leakey), pp. 552–64. University of Chicago Press.
Isaac, G. L. (1976b). Stages of cultural elaboration in the Pleistocene: possible archaeological indicators of the development of language capabilities. In *Origins and evolution of language and speech* (ed. S. Harnad, H. Steklis, and J. Lancaster), pp. 275–88. Annals of the New York Academy of Sciences, **Vol. 280**.
Isaac, G. L. (1977). *Olorgesailie*. University of Chicago Press.
Isaac, G. L. (1981). Archaeological tests of alternative models of early hominid behavior: excavation and experiments. In *The emergence of man* (ed. J. Young, E. Jope, and K. Oakley), pp. 177–88. The Royal Society and the British Academy, London.
Isaac, G. L. (1982). Early hominids and fire at Chesowanja. *Nature*, **296**, 870.
Isaac, G. L. (1984). The archaeology of human origins. *Advances in World Archaeology*, **3**, 1–87.
Jennings, J. (1974). *Prehistory of North America*. McGraw-Hill, New York.
Jones, P. R. (1979). Effects of raw materials on biface manufacture. *Science*, **204**, 835–6.
Jones, P. R. (1981). Experimental implement manufacture and use: a case study from Olduvai Gorge, Tanzania. In *The emergence of man* (ed. J. Young, E. Jope, and K. Oakley), pp. 189–95. The Royal Society and the British Academy, London.
Keeley, L. H. (1980). *Experimental determination of stone tool uses: a microwear analysis*. Chicago University Press.
Keller, C. M. (1973). *Montagu Cave in prehistory*. University of California Press, Berkeley.
Kitahara-Frisch, J. (1978). Stone tools as indicators of linguistic ability in early man. *Kagaku Kisoron Gakkai Annals*, **5**, 101–9.
Klein, R. G. (1978). Stone age predation on large African bovids. *Journal of Archaeological Science*, **5**, 195–217.
Klein, R. G. (1983). The stone age prehistory of southern Africa. *Annual Review of Anthropology*, **12**, 25–48.
Kleindienst, M. R. (1961). Variability within the late Acheulian assemblage in eastern Africa. *South African Archaeological Bulletin*, **62**, 35–52.
Kohler, W. (1927). *The mentality of apes* (2nd edn). Routledge and Kegan Paul, London.
Krantz, G. (1980). Sapienization and speech. *Current Anthropology*, **21**, 773–92.
Lancaster, J. B. (1968). On the evolution of tool-using behavior. *American Anthropologist*, **70**, 56–66.
Laville, J., Rigaud, J., and Sackett, J. (1980). *Rock shelters of the Perigord*. Academic Press, London.
Leakey, M. (1971). *Olduvai Gorge*, Vol. 3. Cambridge University Press.
Lock, A. (1988). Implicational approaches to the evolution of language. In *The genesis of language: a different judgement of evidence* (ed. M. E. Landsberg), pp. 89–100. Mouton–de Gruyter, Berlin.
McGrew, W. C., Tutin, C. E. G., and Baldwin, P. J. (1979). Chimpanzees, tools, and termites: cross-cultural comparisons of Senegal, Tanzania, and Rio Muni. *Man*, **14**, 185–214.
Menzel, E. W. (1978). Cognitive mapping in chimpanzees. In *Cognitive processes in animal behavior* (ed. S. Hulse, H. Fowler, and W. K. Honig), pp. 375–422. Wiley, New York.
Miller, S. F. (1971). The age of Nachikufan industries in Zambia. *South African Archaeological Bulletin*, **26**, 143–6.
Oakley, K. P., Andrews, P., Keeley, L., and Clark, J. D. (1977). A reappraisal of the Clacton spearpoint. *Proceedings of the Prehistoric Society*, **43**, 13–30.
O'Brien, E. M. (1981). The projectile capabilities of an Acheulian handaxe from Olorgesailie. *Current Anthropology*, **22**, 76–9.
Ohel, M. (1984). Spatial management of hominid groups at Olduvai: a preliminary exercise. *Palaeoecology of Africa*, **14**, 125–46.
Oswalt, W. H. (1973). *Habitat and technology*. Holt, Rinehart and Winston, New York.
Parker, S. and Gibson, K. (1979). A developmental model for the evolution of language and intelligence in early hominids. *The Behavioral and Brain Sciences*, **2**, 367–408.
Pettet, A. (1975). Defensive stoning by baboons. *Nature*, **258**, 549.
Piaget, J. and Inhelder, B. (1967). *The child's conception of space*. (trans. F. Langlon and J. Lunzer). Norton, New York.
Potts, R. (1984). Home base and early hominids. *American Scientist*, **72**, 338–47.
Pye, D. (1964). *The nature of design*. Studio Vista, London.
Reichard, G. A. (1934). *Spider Woman: a story of Navajo weavers and chanters*. Rio Grande Press, Glorietta, NM.
Rightmire, G. P. (1979). Implications of Border Cave skeletal remains for Later Pleistocene human evolution. *Current Anthropology*, **20**, 23–35.

Roe, D. (1968). British Lower and Middle Palaeolithic handaxe groups. *Proceedings of the Prehistoric Society*, **34**, 1–82.

Sabater Pi, J. (1974). An elementary industry of the chimpanzees in the Okorobiko mountains, Rio Muni (Republic of Equatorial Guinea), West Africa. *Primates*, **15**(4), 351–64.

Sackett, J. R. (1982). Approaches to style in lithic archaeology. *Journal of Anthropological Archaeology*, **1**, 59–112.

Sampson, C. G. (1974). *The stone age archaeology of southern Africa*. Academic Press, New York.

Shipman, P., Bosler, W., and Davis, K. L. (1979). Butchering of giant geladas at an Acheulian site. *Current Anthropology*, **22**(3), 257–68.

Singer, R. and Wymer, J. (1982). *The Middle Stone Age at Klasies River mouth in South Africa*. University of Chicago Press.

Stringer, C. B. (1985). On Zhoukoudian. *Current Anthropology*, **26**, 635.

Struhsaker, T. T. and Hunkeler, P. (1971). Evidence of tool-using by chimpanzees in the Ivory Coast. *Folia Primatologica*, **15**, 212–19.

Teleki, G. (1974). Chimpanzee subsistence technology: materials and skills. *Journal of Human Evolution*, **3**, 575–94.

Tobias, P. V. (1983). Hominid evolution in Africa. *Canadian Journal of Anthropology*, **3**(2), 163–85.

Uehara, S. (1982). Seasonal changes in the techniques employed by wild chimpanzees in the Mahale mountains, Tanzania, to feed on termites (*Pseudocanthotermes spiniger*). *Folia Primatologica*, **37**, 44–76.

van Lawick-Goodall, J. (1970). Tool-using in primates and other vertebrates. In *Advances in the study of behavior* (ed. D. Lehrman, R. Hinds, and E. Shaw), Vol. 3, pp. 195–249. Academic Press, New York.

Villa, P. (1977). Sols et niveaux d'habitat du Paléolithique inferieur en Europe et au Proche Orient. *Quaternaria*, **19**, 107–34.

Villa, P. (1978). The stone artifact assemblage from Terra Amata. Unpublished Ph.D. thesis, University of California, Berkeley. University Microfilms, Ann Arbor.

Weinberg, S. M. and Candland, D. K. (1981). Stone-grooming in *Macaca fuscata*. *American Journal of Primatology*, **1**, 465–8.

White, J. P., Modjeska, N., and Hipuya, I. (1977). Group definitions and mental templates. In *Stone tools as cultural markers* (ed. R. V. S. Wright), pp. 380–90. Humanities Press, New Jersey.

Wright, R. (1972). Imitative learning of a flaked stone technology—the case of an orang-utan. *Mankind*, **8**, 296–306.

Wu Ru-Kang (1980). Paleoanthropology in the New China. In *Current argument on early man* (ed. L. Konigsson), pp. 182–206. Pergamon Press, Oxford.

Wymer, J. (1983). The Lower Palaeolithic site at Hoxne. *Proceedings of the Suffolk Institute of Archaeology and History*, **35**, 168–89.

Wynn, T. (1978). Tool-using and tool making. *Man*, **13**, 137–8.

Wynn, T. (1979). The intelligence of later Acheulean hominids. *Man*, **14**, 371–91.

Wynn, T. (1981). The intelligence of Oldowan hominids. *Journal of Human Evolution*, **10**, 529–41.

Wynn, T. (1985). Piaget, stone tools, and the evolution of human intelligence. *World Archaeology*, **17**, 32–43.

Wynn, T. (1989). *The evolution of spatial competence*. University of Illinois Press, Urlsand.

Wynn, T. (1993). Two developments in the mind of early *Homo*. *Journal of Anthropological Archaeology*, **12**, 299–322.

Wynn, T. and Tierson, F. (1990). Regional comparison of the shapes of later Acheulean handaxes *American Anthropologist*, **92**, 73–84.

11
A history of the interpretation of European 'palaeolithic art': magic, mythogram, and metaphors for modernity

Margaret W. Conkey

Abstract

'Palaeolithic art' provides a corpus of evidence that bears on central questions in the study of the evolution of modern human abilities, in that it appears late in the human archaeological record (from around 40 000 years ago), and is almost exclusively associated with *Homo sapiens sapiens* remains. This 'sudden' appearance is germane to debates concerning the evolutionary continuity of human lineages versus the replacement of earlier populations through migration from a single geographical 'homeland' (the 'Eve hypothesis'; see Campbell, and Waddell and Penny, this volume); the Middle to Upper Palaeolithic transition (see White, this volume); and the very sustainability of 'grand narratives' in our understanding of the origins of our symbolic abilities.

How this corpus of 'art' bears on these issues in a question that is only now beginning to be formulated adequately. Our understandings to date have been coloured by our uncritical acceptance of earlier interpretations that have a particular intellectual history. While 'palaeolithic art' has a worldwide distribution, its interpretation has been essayed on the basis of the rich sites of south-western Europe. Interpretations of these sites and their images are embedded in the presuppositions of their period of discovery—the nineteenth-century western *zeitgeist* in which prehistoric 'man' represented a stage in the evolutionary ascent to civilized 'man': a stage associated through analogical metaphors with 'savage', 'primitive', and 'childlike'. Given the sometimes spectacular representation of large animals in this 'classical' corpus at one period, 'palaeolithic art' came to be seen as involved in rituals—particularly as a form of 'hunting magic' (despite the fact that no correlation has been established between the repertoire of animal images and the remains of animals taken as food during this period—the images are a 'bestiary', not a 'menu').

This classic corpus of south-western European imagery is spread over a period of 25 000 years. It was created by a wide variety of techniques. It comprises portable pieces (*art mobilier*) and cave wall decorations (parietal art). It is an open question as to how representative what has survived to the present is of what was created at the time. There is little direct evidence that establishes a clear cut chronological dating of portable or parietal images, either absolutely or relative to each other, with any confidence. Animal images are more frequent than geometric designs or human/anthropomorph images, but the repertoire is quite diverse. It is thus difficult to characterize it adequately, nor can we be confident that any interpretive categories of ours reflect the productive, functional, and symbolic categories of its makers. But neither the diversity of the corpus, nor its problematic status, prevented the historical hegemony of the 'foundation hypothesis' of hunting magic from coming to dominate the interpretive landscape.

The first challenge to the 'foundation hypothesis' was made by Leroi-Gourhan around forty years ago, in the context of 'structuralist thought'. Thus he sought to show that images were not randomly placed on cave walls, but were placed according to the use of regular rules. From this spatial analysis he essayed a semiotic interpretation based on the differential 'maleness' and 'femaleness' of different classes of the imagery, and established a stylistically-based chronological scheme of sequential styles in a continuous evolutionary schema of styles. Irrespective of the sceptical current status of this interpretation, it served to establish the 'modernity' (as opposed to 'primitivity') of Upper Palaeolithic human cultures, and a status for the 'art' as amenable to study by the methods of scientific inquiry.

Subsequent interpretations located the 'art' in adaptive frameworks, hypothesized to have been generated by ecological influences on the interpretive frameworks of the 'artists', often mediated by shifting human social group structures and inter-group relations that the changing climate of the Upper Palaeolithic period 'forced' on human social subsistence practices (and to which they were able to 'react' successfully by such strategies, strategies that had previously been unavailable as possible ways of reacting). Marshack has pursued a related line, attempting to infer the cognitive capabilities—or the 'historical' elaboration of the cognitive technologies that modern cognition uses and is constituted by—from the 'properties' of many of the objects he has studied, particularly with respect to calendrical and incipient mathematical systems. Again, in both cases, the security of the underlying premises from which these accounts are essayed is open to critique.

At present, it has become appreciated that 'palaeolithic art' is polysemic, and no monolithic interpretation is possible. Detailed studies of portions and aspects of the record are being undertaken. On a wider scale, it may prove that the reflexive study of the interpretive frameworks that have been used to characterize 'palaeolithic art' is a useful way of elucidating the principle and processes that generated the 'art' in the first place [eds].

11.1 Introduction

This chapter is about one particular well-documented and well-known corpus of archaeological materials that have been called palaeolithic 'art'.[1] This corpus comprises thousands of images made on bone, stone, antler, ivory, and limestone cave walls, using a variety of techniques and stylistic and artistic conventions. These are not the only images made during the last 25 000 years of the Pleistocene (or Ice Age); but this particular corpus—concentrated in south-western Europe but spread across into Russia—is not only extensive and extraordinarily well-preserved, but has figured prominently in our narratives of human evolution.

The chapter addresses three major topics: (1) what is the palaeolithic material that we call 'art'; (2) how have researchers tried to explain these materials and to make sense of them in this prehistoric context; and (3) where are we now in our study of this 'art'—what are the current and future directions? To begin, however, it is important to emphasize that the study of palaeolithic 'art' is embedded within the prevailing intellectual contexts of how we think about the evolution of modern humans, because this 'art' has always been associated with the early cultural 'success' of anatomically modern humans, and with the establishment of what appears to be a 'fully human' cultural pattern (cf. White, Chapter 9, and Wynn, Chapter 10 of this volume).

11.2 Some intellectual context

There is no doubt that there is a revived and increasing interest in questions about early anatomically modern humans, *Homo sapiens sapiens*: their earliest appearance; their possible interactions with other forms of *Homo*; the development of their expanded symbolic and cultural repertoires; and when and how they spread over the globe. Intimately associated with these questions is the consideration of what humans, or human-ness, are all about, and how we might trace the evolution of what we might call (for lack of a better phrase) 'complex behaviour', which is what we assume to have happened in the hominization process.

Within the archaeological community the discussion on 'complex behaviour' has taken the form of a debate, and an increasingly heated one (see, for example, Foley 1988 vs. Clark 1989; Lewin 1988). In its most simplisitic version the positions in this debate are, on the one hand, that of those who hold that hominids were *different* from all other primates *from the very beginning* (or at least very soon after their divergence from the ancestral hominoid line); and, on the other hand—and this is a more recently elaborated position—that of those who think that the distinctive and 'unique' hominid behavioural traits developed *late*, perhaps only with the appearance of anatomically-modern humans.

The analyses being brought to bear in this debate have centred on aspects of brain development, dentition, and postcranial features (see, for example, Trinkaus 1990*a* for one review); on stone tools (for example, Foley 1988; Sackett 1988; Simek and Price 1990); on faunal expansions and faunal exploitation (for example, Klein 1989; Simek and Snyder 1988); and on reconsiderations of 'key' sites (for example, Zhoukoudian (Binford and Ho 1985); Olduvai (Potts 1988); and Torralba (Shipman and Rose 1983). The two positions in the debate are not unrelated to two of the different 'schools' of thought as to how the evolution of recent humans (especially *Homo sapiens*) took place (see Stringer and Andrews 1988 for one synopsis). One core issue has to do with evolutionary *continuity* within the lineage of *Homo* (see, for example, Clark and Lindly 1989; Wolpoff 1984, 1989) versus the establishment of modern humans across the globe by *replacement* of earlier stocks by one deriving from a single geographical 'homeland' (for example, Stoneking and Cann 1989; Stringer and Andrews 1988).

The combination of exciting and controversial new evidence (for example, Valladas *et al.* 1988) and the re-analysis of old materials (for example, Brooks in press; Simek and Snyder 1988; Gargett 1987) has rekindled interest in the relationship between the hominids of the Middle Palaeolithic and their behaviours and those of the Upper Palaeolithic; that is, rekindled interest in what has been called the Middle-Upper Paleolithic 'transition' (see White, this volume, Chapter 9). By now, common knowledge among Pleistocene prehistorians has come to include the intriguing find of a Neanderthal in association with an early Upper Palaeolithic industry, the Chatelperronian,[2] at St Césaire (France) (Vandermeersch 1984); the suggestive early dates for modern *Homo sapiens* in southern Africa, at some 60 000 (or more) years ago (Deacon 1989); and the so-called 'proto-Cro-Magnon' with Mousterian tools at Qafzeh (Israel), 'dated' to almost 100 000 years ago (for example, Valladas *et al.* 1988). There are now more data and interpretations of the morphology and behaviour of archaic and later *Homo sapiens* in central and eastern Europe (for example, Smith 1982, 1984; Allsworth-Jones 1986); and there are the controversial implications of mitochondrial DNA studies for the kinds of evolutionary replacement models being employed (for example, Stringer and Andrews 1988; and Waddell and Penny, this volume, Chapter 3; for a compendium that considers many of the issues, see Mellars and Stringer 1989).

Speculations on the possible relations between Neanderthals and modern *Homo sapiens* have been vividly detailed and explicitly fictionalized in a burgeoning genre of 'ice age fiction' (for example, Auel 1980; Kurtén 1980, 1986; Thomas 1987). And, at last, the geographical focus that has too long been directed to the archaeological and palaeontological record of south-western Europe is being redirected elsewhere (especially to Africa and the Near East). It is becoming increasingly clear that most of the human biocultural evolution that ends up with the global spread of modern humans probably happened *outside* the so-called 'classic' region for Upper Pleistocene studies—south-western Europe (see Trinkaus 1990*b*).

But the history of prehistory—which 'began' in European contexts (Sackett 1983)—*and* an amazingly well-preserved archaeological record in south-western Europe, packaged mostly in limestone caves and rock-shelters (and including palaeolithic 'art'), have led to a whole range of presuppositions about similarities, differences, and relationships between Neanderthals and succeeding populations of anatomically-modern humans in general, based primarily on this particular regional archaeological record and the attention it has been given for more than a century. The traditional characterizations of European Upper Palaeolithic Cro-Magnons (i.e., early *Homo sapiens sapiens*) have come to 'stand for' a stage in human evolution that is only now beginning to be dislodged.

Intimately bound up with this traditional characterization of European Cro-Magnons (*Homo sapiens sapiens*), and with their purported differences from Neanderthals, is the appearance in the archaeological record of materials that we have labelled 'art'. The presence of a range of materials—from perforated teeth and 'beads' made from stone, to incised stone blocks, ivory and clay figurines, engraved bone and antlers, and traces of 'painting' on cave-wall fragments—has allowed most twentieth-century researchers to assume a certain elevated cultural status for the early Upper Palaeolithic (c.30 000 years ago) makers of such materials, and there is no evidence to suggest that the makers of these assemblages of materials were other than *Homo sapiens sapiens*. But the characterization of the entire European Upper Palaeolithic (c.35 000–40 000 years ago to 10 000 years ago) as a 'cultural bloc', which has yielded evidence of 'art', has created the impression that the 'art'-making might have something to do with the transition from archaic forms of *Homo sapiens* to fully modern ones at about 35 000–40 000 years ago in western Europe.

There are two immediate problems here. *First*, to account for this art as a result of biological change in hominids is a vitalistic account, and as such does not explain how or why the art-making would be the result of a biological change. It is also increasingly the case that there are early specimens of *Homo sapiens sapiens* found in contexts that do *not* include this kind of preserved material culture, so that imagery of this sort was clearly not necessary for the biological transition, nor was it necessarily an immediate result of it (for example, Clark and Lindly 1989; Lindly and Clark 1990). The question 'Why Euro-Russian palaeolithic art?' will have to be addressed on contextual grounds for different regions, and probably for different forms or media (if, that is, we can come to understand how to differentiate, and therefore to classify, materials according to 'medium' or 'form'). This latter point leads to the *second* immediate problem, with the assumption that 'art'-making was causal in, or an immediate product of, the anatomical transition: the 'art', in such scenarios, is usually lumped together as a single phenomenon, as if such things as dynamic cave paintings of animals (see Fig. 11.1)—the usual idea of what palaeolithic 'art' is—and naturalistically carved images in the round (Fig. 11.2) were part of the cultural products of *all* Upper Palaeolithic peoples; which is *not* the case.

There are many (for example, Ucko and Rosenfeld 1967) who have long argued for recognizing the diversity of what we call 'paleolithic art', and for emphasizing the differential historical trajectories and patterning of different media and forms. This emphasis on diversity and on differentials in development should also include cave-painting, which does *not* occur—especially in its most well-known forms of polychrome figures—from the very beginning of the Upper Palaeolithic. It has been argued (for example, by Conkey (1983) and White (1989)) that the celebrated cultural products associated with the Magdalenian period of the later Upper Palaeolithic (c.15 000–10 000 years ago) are not likely to tell us much about the meaning or role of material culture, imagery, and 'art' in the lives of the earliest Upper Palaeolithic representatives of *Homo sapiens sapiens* of over 30 000 years ago, nor about how the cultural products of this earlier period figured in the establishment of fully modern humans in the Euro-Russian regions.

Now most approaches to understanding the origin and dispersal of modern humans, including the most recent (for example, in Mellars and Stringer 1989) still give priority to subsistence (for example, fauna, hunting strategies) and technological factors (for example, stone-tool manufacture and lithic typologies). But there are always *implications* to be drawn from these studies for cognitive and symbolic capacities and behaviours, even if very few scholars have explicitly broached the role of symbolism or 'art' in the transition to modern humans (but cf. Conkey 1983; Marshack 1986, 1990; White 1989). If, however, we are to understand the 'manifestations and evolution of complex human behaviour', there is no doubt that the study of palaeolithic 'art' has much to offer, if only because of its abundance, diversity, and apparent persistence in some form or another over some 25 millennia following the establishment of modern humans in these regions of Europe and Russia.

Furthermore, as this chapter will explore, there are at least two more specific and central questions of contemporary archaeology that palaeolithic 'art' addresses. First, how do we account for something 'new' in the archaeological record? And second, how do we access the 'symbolic' domain of prehistoric life? The following review of palaeolithic 'art' will show that we have barely begun to work at these particular questions, despite more than a century of research. We have allowed certain assumptions about a progressive course of human evolution to 'stand for' answers to these questions, and thus have yet to directly address or to problematize these questions.

Given present understandings of the evolution of *Homo sapiens sapiens*, and given the increasing recognition of the complex and diverse nature of palaeolithic 'art', I am not convinced that we can use the particular inferences from palaeolithic 'art' in south-western Europe to support a grand narrative for the appearance and success of modern humans, although this has implicitly been taken to be the case for most of this century. Rather, I hope to show that the inferences that we *can* draw about palaeolithic 'art' may be used, on the one hand, to understand more fully the relations (during the Upper Palaeolithic of Europe) between particular cultural products (for example, engraved bone and stone), particular meanings, and particular conditions of existence, and, on the other, to explore—in the socio-historical contexts of *some* Pleistocene hunter–gatherers—how these humans can be understood as materialists *and* symbolists. In fact, it is unlikely we could ever adequately address the larger issues that we have previously taken for granted without a good grasp of these more specified and particular ones.

11.3 The materials that we call 'Palaeolithic art'

When Dr François Mayor brought the first two pieces of engraved and decorated antler to light in 1833 in the cave of Veyrier (Haute-Savoie) in eastern France, it was impossible to imagine that not hundreds, but thousands of engraved bones and antlers would be uncovered in archaeological sites over the next hundred and sixty years. These would become part of the very diverse corpus of materials referred to as 'portable art' or *art mobilier*. Included in this category are engravings on stone, bone, and antler, and carvings of ivory. Included are images of humans, often in the form of statuettes; geometrically decorated implements, such as bone points (Fig. 11.4), awls, and harpoons; perforated animal teeth, shells, and bone discs, perhaps originally suspended from or attached to skin clothing (which was being made with distinctive bone needles); and a variety of other figurative objects and images—such as baked clay animal figures (see, for example, Vandiver *et al.* 1989). All these are called 'portable' because they admit the *possibility* of being carried around and being removed from the contexts of production, unlike the painted cave walls.

The story of the 1879 'discovery' of Altamira's painted ceiling has become a classic in the tales of archaeological discoveries (see, for example, Bahn 1988; Fagan 1986); but painted and engraved wall surfaces in caves and rock-shelters are still being discovered in the 1990s despite more than a century of avid spelunking and more or less systematic investigations by prehistorians. These decorated, especially painted, walls became the centrepiece of palaeolithic art studies, and most of the interpretations of the 'art' derive from and refer to the imagery applied in diverse ochres and manganeses to the walls of some two hundred limestone caves, located primarily in the region of what is today south-western France and the northern coast of Spain (Fig. 11.5). This is wall 'art' or 'parietal art', where the context of production (image-making) is also the context of use; people must come to (see) this imagery, although the 'knowledge' of its existence could certainly travel, with people, as ideas, etc.

11.3.1 *Concerns about the evidence: diversity and preservation*

To try to summarize the imagery and media from hundreds of archaeological contexts and prehistorically-visted cave locales that span some 25 000 years is a task far beyond the scope of this or any other single chapter. Few scholars have ever attempted such a catalogue for all media and locales, preferring to emphasize either the cave art (for example, Ministère de la Culture 1984; Ucko and Rosenfeld 1967), or selections from among the portable materials (for example, Barandiaran 1973; Chollot 1964; Sieveking 1987). Further, one should immediately suspect any so-called synthesis: these are imagery and materials spanning two-thirds of human art-history (> 25 000 years) and from archaeological contexts that span the thousands of kilometres from the Iberian peninsula to the Russian plain.

In addition, despite the richness, abundance, and diversity of the materials that *have* been preserved, there is no doubt that this is only a partial repertoire, an incomplete, and not necessarily representative sample of Upper Palaeolithic image-making activities. On the one hand, there are discoveries being made all the time: new decorated caves (for example, Font-Bargeix, Fronsac, France (Carcauzon 1984, 1986)); assemblages of portable materials that change our characterizations of regional 'traditions' (for example, the finds from the Río Nalón, Spain (Fortea *et al.* 1990); and re-'discoveries' of materials that have lain interpretatively dormant in marginal museum collections (for example, White 1986). On the other hand, there is the fact of differential preservation. That painted cave walls are primarily away from the (modern) cave entrances and daylight may say nearly as much about preservation factors as about any deliberate placement of painted images on walls deep in limestone caves by palaeolithic 'artisans'. Although relatively rare, there are occasional finds of 'painted' rock-wall surfaces or plaquettes that have perhaps fallen from the roof or walls of daylit rock-shelters into archaeological deposits, suggesting that other painted surfaces may have disappeared.

In addition, we know that hundreds of portable pieces of engraved bone and antler have *not* been systematically recovered and curated: early investigators left huge 'back dirt' piles in their search for only the finest pieces ('*les belles pièces*'); cave deposits have been mined for their soils, which were then spread out as topsoil, with any artefacts included; pieces were traded among museums (see, for example, Bahn and Cole 1986) and private collections; and entire collections have been misplaced or even—in the case of some apparently undecorated bone and antler implements—buried in an early investigator's back yard! Although most materials come from cave and rock-shelter sites (but cf. Bahn 1985 on the recent discovery of open-air rock engravings), we know that much palaeolithic occupation was open-air, where far less excavation has been undertaken and where bone and antler preservation is practically nil.

Thus the usual archaeological problem of 'negative evidence' must always be considered explicitly in *any* descriptive or interpretative approach to these materials. One must always question how, and to what extent, 'negative evidence' may have figured in drawing inferences from the materials that we do have. For example, just because we have not yet discovered any statuettes of females in the sites of northern Spain, does this mean that the occupants of this area neither made nor had nor carried about such statuettes, which have been found in an extensive distribution across central Europe from France to Russia (Gamble 1982; Delporte 1979)? Most prehistorians believe that what we currently know about Upper Palaeolithic life, derived as it is from an unrepresentative sample of well-preserved, conveniently-located (for the excavators) cave and rock-shelter sites, constitutes no more than a very small and biased peephole into palaeolithic lifeways (cf. for example Rigaud and Simek 1987).

11.3.2 *Concerns about the evidence: chronology*

The chronological observations on palaeolithic 'art' that we can presently offer will not be treated here as a separate topic, although what palaeolithic 'art' might have to offer to the study of the 'origins of art' will be discussed briefly. After some discussion here of aspects of chronology, the next section will present a selective overview of palaeolithic 'art' in terms of techniques, media, and subjects, with suggestions about productive and cultural contexts. As I will suggest in the final section of the chapter, the future of 'Palaeolithic Art' studies will surely lie in our linking the visual images to possible contexts in ways *other than* those dependent upon time and the chronocentric perspective.

The most fundamental tool of traditional archaeological interpretation—chronology—is of relatively little help in 'palaeolithic' art studies, especially when it is the cave wall 'art' that is the subject of inquiry or part of the contextual interpretation of any imagery. Since most interpretations favour wall 'art', this means that most must be subject to serious scrutiny because, as yet, *there is no way to date the imagery that is on cave walls.* Sometimes datable archaeological deposits have covered cave-wall imagery, and some caves have been blocked or filled in since various known periods of palaeolithic occupation (as, for example, at Altamira in Spain or Fontanet in the French Pyrenees); but these 'events' provide only relative dates: the wall 'art' can only be said to have been produced before the accumulation of these cultural or geological deposits. This has not, as we shall see, inhibited the construction of detailed chronological sequences anchored to relative dates, relative stylistic correspondences with relatively-dated objects of portable 'art', and underlying (for example, twentieth-century French) notions as to how the 'stylistic' evolution of an artistic tradition might proceed.

Even many portable 'art' objects lack context and provenience, because they were removed as *belles pièces* without regard for context, or were found opportunistically (as with the female statuette found along the path to the site of Tursac, Dordogne, France). Delporte (1979, pp. 211–13) reports that our confidence in the provenience of most human statuettes, for example, has to be strikingly low (he notes, for example (ibid., p. 212), that we can be confident about the provenience of only 2 out of 20 found in France). His own review of these statuettes focuses on their geographical distributions—where they were found—rather than on any well-founded chronological assignments.

The chronologies that trace the earliest forms to the latest, in even the most general terms, under the label of a single tradition—such as 'palaeolithic art'—falsely invite the reader to derive inferences about the 'progress' of Stone Age artisans, and, by implication, of Stone Age 'cultures'. This urge to look for a chronological sequence with an eye for trends from simple to complex, for example, is rooted in the intellectual contexts of nineteenth-century evolutionist anthropology. Such a perspective also feeds the view of palaeolithic 'art' as the origin and focal point from which we trace the history of art and most 'fully modern' cognition and symbolism. Chronology is thus a serious issue in the interpretation of palaeolithic 'art'.

These rather lengthy comments are an important but often ignored prelude to any discussion of what palaeolithic 'art' comprises, what we can make of the images and their patterns (which is how most archaeological interpretation begins), and what any of the imagery might 'mean' in both a scientific sense and to the original makers and viewers. The *nature* of the evidence in any archaeological inquiry sets certain parameters for its interpretation, and these general observations on factors of preservation, chronology, and the fact that we have tended to collapse some 25 000 years into a single bloc of 'prehistory', are cautionary tales relevant to any past and future interpretations of palaeolithic 'art'.

At minimum, one caveat is not to make too much of the presence or absence of certain forms and images, nor to rely upon questionable chronologies. Although there is much to be said about the images and forms as cultural materials, interpretations of them are even more immediately suspect as 'archaeological' constructions than are those archaeological inferences that appear more securely anchored in the traditional archaeological categories of time and space. Differential preservation—which is always and everywhere an archaeological issue—is a particular difficulty in interpreting the imagery of the Palaeolithic. It must be borne in mind that all interpretations are *archaeo-logical*, in that they are constructed to 'make sense' to the interpreters, and their audiences, in the historically-contingent intellectual and cultural contexts within and for which they are created.

11.3.3 *Techniques, conventions, and media*

Traditionally, palaeolithic 'art' has been presented as a bloc; but even at the grossest levels of differentiation (for example, early vs. late; portable vs. wall) there are, not unexpectedly, variations in media, techniques, subject-matter, and distributional patterns. Portable art is widespread geographically, from Spain to the Russian plain. But the preserved cave paintings and engravings (wall 'art') are strikingly more delimited in geographic extent; they have been found almost exclusively in southern France and northern Spain.

It is important to note, however, that artistic activity or image-making by late Pleistocene anatomically-modern *Homo sapiens sapiens* is *by no means* restricted to this Eurocentric zone (see, for example, Bahn and Vertut 1988). But the images, for example, on cave walls, rock surfaces, or stone plaquettes found in Australia, East Africa, southern Africa, and (possibly) Brazil and New Guinea have not (yet?) been shown to be as abundant in their preservation nor as diverse in their media and subject-matter as the imagery that constitutes the classic corpus of 'palaeolithic art'.

Yet it is never the case that one could safely argue for a superior artistic or aesthetic sense or a richer symbolic life on the part of the classic Upper Palaeolithic peoples as compared to other early *Homo sapiens sapiens* peoples. There are many cultural products and visual images that do not preserve (for example, body decoration, sand painting, hide, and wood, bark); and many rich and evocative symbolic practices and cosmologies that take no material form. Thus, the case can again be made for studies of palaeolithic 'art' to be framed more in terms of its own particular historical contexts than as some sort of 'stage' in, or privileged example of, human artistic or symbolic evolution.

The objects and images of palaeolithic 'art' were made by almost every known technique, using a range of materials that represents a minimum repertoire of raw materials—the materials that have been preserved. Bone, antler, ivory, clay, shell, teeth, stone, and a variety of other minerals were worked with stone tools, or sharp flakes, or small hammerstones. The limestone cave walls, in particular, had applied to them a variety of ochres, manganeses, and a range of other pigments and colouring matters, many of which had to have been ground, mixed, or otherwise processed in order to adhere (see, for example, Clothes *et al.* 1990; Arlette Leroi-Gourhan 1982; Arlette Leroi-Gourhan and Allain 1979; Vandiver 1983).

Other image-making techniques included engraving; carving (often in the round); bas-reliefs and sculptural work (using techniques such as *champleve*, where, for example on the block sculpture of the 'Woman with the Horn' from Laussel (France), the outline of the body is achieved by not just incising but actually hollowing-out (Delporte 1973, p. 125)): painting by various techniques, including blown-on pigments (either chewed and blown directly from the mouth or blown through a bird-bone tube); drawing (in the sense of Rembrandt's crayons, see Vandiver 1983); and a flatwash-type technique. Painting and engraving were sometimes used for the same image, and many engraved bone and antler objects show the use of red ochre *in* the engravings (for example, Buisson *et al.* 1989). Experimental replicative work has shown this to be almost necessary in order to see the imagery as it is being engraved in the fresh bone (for example, Delporte and Mons 1980; Mons 1990). And, in several rare examples, there are clay modelled figures (the famous bison from Le Tuc D'Audoubert; Fig. 11.6), as well as a series of animal figurines made from baked (fired) clay (Dolni Vestonice, Czechoslovakia; Fig 11.7). These latter date to some 29 000 years ago, and suggest that knowledge of clay and its properties and possibilities is much earlier than the Neolithic, when fired clay containers became widespread (Vandiver *et al.* 1989).

Engraving as a technique is there from the very first known images of European palaeolithic 'art', and the general technological strategy of imposing form upon a raw material, or eliciting form from within it by subtracting from it, is expressed in a variety of ways. 'Beads' were made from stone (steatite) or from mammoth ivory (as at Abri Blanchard 34 000–32 000 years ago; see White 1989); an anthropomorphic figure was carved out of a tiny plaquette of ivory (Giessenklösterle, Germany at 32 000–34 000 years ago, see Hahn 1988); and more than fifty limestone blocks dating to over 30 000 years ago from five nearby sites (in the Castelmère Valley of the Vézère in France; see Delluc and Delluc 1978) bear incised motifs of a geometric sort, including a few probable animal or anthropomorphic shapes.

Although bone and antler implements (for example, split-based bone points, double points) appear in the archaeological remains of early Upper Palaeolithic peoples engaged in the manufacture of the so-called Aurignacian types of industry at some 30 000 years ago, it *is* a well-recognized archaeological fact that bone and antler implements are not extensively and regularly decorated—with either geometric and/or figurative motifs—until the later millennia of the Upper Palaeolithic, after 18 000 years ago. But once such implements of bone and antler are regularly incised, they are often found in great numbers (up to 500) at some sites (for example, Le Mas d'Azil (Ariège, France); Laugerie-Basse (Dordogne, France), Fig. 11.8).

While there are a few instances of the use of colour (i.e., applied pigments) on a few stone blocks or plaquettes from both the Castelmère sites in southern France and from an Aurignacian level at Giessenklösterle in south-west Germany (Hahn 1988), there is no doubt that *if* painting existed before 30 000 years ago, it has not been well preserved. On the basis of the *extant* evidence, one would have to say that it is only after 20 000 years ago that there is evidence for the use of pigments/paint to create images in an 'additive' way. By adding colour to cave walls (and often by using the pre-existing wall shapes and features, including even calcite drips or stalactites), Upper Palaeolithic peoples created images—usually of animals—in the limestone caves. One could argue that, conceptually, this may be a very different technique from subtractive engraving or carving to make ('release') images.

Although widely known for its so-called naturalism and realism, not all palaeolithic imagery would fit into these (or other) '-isms' of contemporary art-historical description and classification. There is no doubt that an anatomical and ethological understanding of animals underlies much of the animal imagery that predominates as subject-matter, particularly in wall art (Fig. 11.9). But (as for example, Lorblanchet (1977) has shown) one can identify a range of artistic conventions being employed, even within the same painted cave and within any one so-called 'period' of painting/engraving. Some research has shown that similarities in how animals are rendered may indicate individual artists (Apellaniz 1982), and that renderings of even one species (for example, deer or mammoth), at one site (for example, Lascaux, France, or Gönnersdorf, Germany) do not conform to some stylistic canon (for example, Vialou 1984 and Bosinski 1984, respectively) (Fig. 11.10). Animal imagery may be conveyed by full-body images of specific, species-identifiable forms, complete with careful rendering of the coat and musculature; or they may be conveyed by depicting only a head or a rack of antlers (but seldom by just a tail!).

A number of recent scholars (for example, Lorblanchet 1977; Nodelman 1979–80, 1985) have suggested that we have been somewhat deluded into considering Palaeolithic images to be overwhelmingly 'realistic'. Lorblanchet, for example, points out that whether an animal or a humanoid/anthropomorph is the image, it is usually presented with a realism only for selected details: the mammoths at Pech Merle (Lot, France), for example, are shown with a detailed anus and covering flap but quite devoid of their eyes and their coats; some of the female statuettes have quite detailed and/or exaggerated treatment of buttocks or breasts, but the hands, feet, or even legs and facial features may be minimal or absent (Fig. 11.11).

In suggesting the visual conventions that generated the images are 'profoundly *unlike* those that we have known', Nodelman argues we are unlikely to elucidate the meanings of palaeolithic visual imagery if we focus primarily on what is familiar to us from the traditions of Western art, on what appears 'realistic' or 'iconic'. The focus on realism is chronocentric; we have fixed palaeolithic images in the way of seeing that characterizes our own time. Nodelman (1985) suggests that, if we move away from what we expect and investigate differences, then the repertoire of palaeolithic images is much more diverse than the celebrated animal (and even human) images that predominate in our re-presentation of it.

Yet the visual conventions—how the images are placed in relation to each other, and to the supporting medium—of much of the art have prompted much discussion and interpretative debate, if only because many conventions appear unfamiliar to our Western eyes. For example, the superpositioning of one image over another in all the different media led early researchers to suggest that it was the locale on the cave wall that was sacred and important, not the image itself (Fig. 11.12). And the lack of framing or bounding of images, with infrequent orientational ground lines, has made it almost impossible for researchers to infer whether animals or signs might be in some sort of specific (narrative?) relationship to each other, or if there is any deliberate composition of images, either on walls, antler, or stone (Fig. 11.13). Only a few compositions have been taken as convincing to us as such, for instance, the 'scene' in the well and the so-called swimming deer at Lascaux (Dordogne, France) (Fig. 11.14).

Moreover, there is also quite often a very strong link between the image and the natural setting or raw material. Pre-existing cave-wall bumps and discolorations on walls, for example, appear to have been used in many painted images. The shapes of bone and antler raw materials have frequently been integral to the imagery produced; for example, the edge of a bone may itself provide part of the image (for more discussion on this 'iconic' aspect of much palaeolithic 'art' see Conkey 1980, 1982). It should not be surprising that these palaeolithic peoples often knew well the structures, possibilities, and limitations of the materials that they were working, whether active limestone cave walls, or the durable bone and more pliable antler used in carving.

It is only recently that systematic studies of what appear to us as discrete media have begun to appear: Sieveking's (1987) work on the engraved plaquettes, which may occur as assemblages of more than a thousand at certain sites (for example, Limeuil, La Marche, and Enlène, in France); Cattelain's work (1980, 1986) on the so-called spear-throwers, many of which are elaborately sculpted, but have been shown by him to bear the characteristic use-wear patterns of a functional spear-thrower; Bellier's study (1984) of the animal cut-outs or *decoupées*, which are rather delimited in their temporal and spatial distributions (predominantly found in the Pyrenees during the middle (and late) Magdalenian periods), and are almost exclusively made out of the hyoid (throat) bone of the horse; Baulois's study (1980) of the bone and antler points and the kinds of decorations they bear; and the Dellucs' important study (1978) of the Aurignacian incised blocs from the Castelmère Valley (France) sites; while, although these are not usually taken to be 'art' themselves, there are also the lamps used to light access into caves, some of which were engraved with animals and 'signs', which have been systematically studied by deBeaune (1987). Major treatises on shell-collecting and working (Taborin 1987), and on pigment-processing (Vandiver 1983), and many experiments on the making of ivory statuettes (for instance, Dauvois 1977), engraving (decorating) bone (Delporte and Mons 1980; Mons 1990), and replicating certain cave paintings (for example, Lorblanchet 1980) have all contributed to our understanding of the range and details of specific technologies.

11.3.4 *The 'subject-matter' of the imagery*

Although the 'subject matter' of palaeolithic 'art' is almost always considered to be primarily animals, it is important to recognize that this has been one of the ways in which we have homogenized palaeolithic 'art' into a 'tradition'. This characterization—'an animal art'—holds true only if frequency is the criterion. If the other recognized subjects selected by palaeolithic 'artists' for depiction are the two broad categories of geometric forms or 'signs', and humans/anthropomorphs, then animals, or some recognizable parts of them (for example, antlers), are indeed the most frequent. But, not all animals are depicted in the same ways, in the same locales within caves or regions, or on the same kinds of raw materials, nor with the same techniques. Almost all decorated caves have more than one animal species, but few exhibit all the species in the repertoire of imagery. And, as Leroi-Gourhan (1982), among others (for example, Delporte 1987, p. 117), points out, the list of species depicted is not a menu (animals that were eaten), but a 'bestiary' (animals selected for symbolic reasons) (Fig. 11.15).

A broader approach to the subject-matter has often been to suggest that there are two cross-cutting general categories: on the one hand, there is the portable imagery as against wall imagery, and on the other hand, there are representational and non-representational images. But these are clearly *our* categories, based on criteria that are relevant to twentieth-century Western scholars. In discussing the subject-matter, it is important to keep in mind that the palaeolithic images appear to us as recognizable *or* as enigmatic/ambiguous because of our own visual experiences and expectations. Many of the individual images are recognizable to us as animals—sometimes even as specific species—and as humans. But there are many geometric forms, usually called 'signs', on cave walls or as motifs/design elements on portable bone and antler, that remain enigmatic (Fig. 11.16). And additionally there are a variety of relatively rare anthropomorphic-type images that appear ambiguous: a human form with cervid antlers; a ghost-like face; a bison standing on two hind feet with a curved form (a bow? 'life-breath'?, cf. Smith, n.d.) in front of it (Figs. 11.17, 11.18, 11.19).

There are easily more than 20 000 images on cave walls—there are 1500 engraved images in the cave of Lascaux alone—but it is not really possible to count or to inventory them without entering the domain of interpretation (Clottes 1989). Any inventory will be relative and partial. The only ambitious count in recent years (cf. Reinach 1913 for an early one) is that of cave-wall imagery by Leroi-Gourhan (1965, 1982), which was carried out in the course of laying

Table 11.1 Leroi-Gourhan's percentage figures for the portrayal of different types of animal in Palaeolithic caves

Horse	30%	Principal animals: 60%
Bison and aurochs	30%	
Stag and hind	12%	Other animals: 34.6%
Mammoth	9.6%	
Ibex	8.7%	
Reindeer	4.3%	
Bear	1.5%	Rare animals
Big cats	1.3%	
Rhinoceros	0.7%	
Chamois		Exceptional animals
Wild boar		
Fallow deer		
Giant stag (*Megaceros giganteus*)		
Wolf		
Saiga antelope		
Nocturnal predators		
Salmon		
Musk ox, Elk		Very exceptional or absent animals
Lynx		
Fox		
Wolverine		
Hyena		
Birds		
Large fish		

Source: Leroi-Gourhan (1984).

out a particular interpretative framework to account for palaeolithic 'art' as a single, inclusive 'tradition' (see below). He cites the horse and bison as being the most frequently depicted animals, each comprising 30 per cent of the images he 'counted' in some 120 major caves. He then groups about five different types of animal (mammoth, ibex, and reindeer, and hind and stag) into a single category of 'complementary animals'—those that complement what he considers to be the central figures of horse and bison—that together total another 30 per cent, with the final 10 per cent comprising the 'dangerous animals': bear, felines, rhinos (see Table 11.1).

Whether or not one accepts Leroi-Gourhan's framework and categories (such as 'complementary' animals), it is clear that if one considers *all* cave imagery as a unit, the predominant type of animal depicted is a herbivore, and the two most frequently depicted species are horse and bison. Bison and horse do not, however, necessarily predominate in each individual cave; Lascaux, for example, is characterized by aurochs (wild cow) and horses; Font-de-Gaume (also Dordogne, France), by bison and mammoths.

There are no such systematic inventories of depictions, whether recognizable or enigmatic, on portable 'art', although it appears that there is much more variation and diversity of imagery (see for example, Delporte 1990). In the recognizable animal category, for example, there are many more fish and birds, and even a grasshopper and reptiles (Fig. 11.20). Leroi-Gourhan, however, believes that some of the same fundamental associations he notes for cave 'art' would hold for portable imagery—such as horse (male) with a deer (for example) as a complement—because, to him, the underlying structure or 'rules' for palaeolithic 'art' were expressed in a variety of media.

The so-called 'signs' were originally given names that imply shape, such as 'roof-like' (*tectiforms*) or 'key-like' (*claviforms*), although many were named in more obviously interpretative ways, such as 'wounds', 'traps', or 'huts'. The signs are only painted or engraved; there are none of the recognized forms in bas-relief. They are rarely superimposed on each other in the way that many animal images are, although some may appear *on* animals (an occurrence that helped support the hunting-magic interpretation (see below) of signs as wounds or spears, for example, Fig. 11.21)). Not all caves have signs; only 40 out the 120 studied by Leroi-Gourhan are reported to have them.

The signs have always figured prominently in the

major interpretations that have tried to account for all of palaeolithic 'art'. To Leroi-Gourhan, for example, whose interpretation is discussed in more detail below (Section 11.4.2), all signs originated in the representation, whether abstract or realistic, of male and female genitalia (for example, male as 'thin' and female as 'full') (Fig. 11.22). Furthermore, these signs are just abstract representations, he suggested, of the same fundamental principles—maleness and femaleness—that account for the patterns underlying the depiction and arrangements of the more realistic imagery—the animals. In pursuing the geographical distributions of the signs, Leroi-Gourhan (1978) suggests it is the 'full' (rather than the 'thin') ones that show more distinct regional patternings that imply different social groupings or 'ethnicities'. However, the signs on the portable 'art', he suggests, are different from those on the cave walls; the former are geometricized forms, whereas the latter are of animal origin.

Needless to say, these propositions are far from conclusive or easily supported. Leroi-Gourhan's differentiation of signs into 'full' or 'thin' is itself not always obvious, much less is it obvious that the 'origin' of the shape lies in the male or female genitalia (see Bahn 1986 for one critique of the reliability of interpreting certain signs as 'vulvae'; also Mack, n.d.). There is no obvious answer to what the conceptual relationship, if any, might originally have been between the signs and the animals: are they interdependent (as Leroi-Gourhan suggests for wall 'art'); are they from the same conceptual source or meaning; or are they completely different? (By differentiating between 'representational' and 'non-representational' categories there is an implicit distinction between the signs and the animals.)

Other interpretations for signs have been proposed, such as those who suggest pre-alphabetic systems of graphics for those motifs found on some portable pieces (for example, Chollot 1980; see also Forbes and Crowder 1979); or notational or marking systems of various sorts (for example, Marshack 1970a, 1972a); or as gaming pieces (for example, Dewez 1974). Provocative semiotic analyses have been attempted (for example, Sauvet et al. 1977; Sauvet 1988); and some signs have been suggested to be good examples of entoptic phenomena, i.e., images reported to be 'seen' regularly (and cross-culturally) by people when they are in some sort of a trance or hallucinogenic state (Lewis-Williams and Dowson 1988) (see Fig. 11.31). All these interpretations appear plausible to a certain extent, and it is very likely that perhaps more than one interpretation will have to be entertained.

The other challenging set of imagery that figures prominently in interpretations is the set of anthropomorphs. Although there are still relatively few examples, even in all media combined, this has not inhibited interpreters from inferring such things as an ancient and fundamental Earth Goddess or fertility cult from the existence of only about 60 identifiably female statuettes (out of more than 150 known statuettes of varying shapes, decorations, and forms; see Delporte 1979; Gvozdover 1989). Although the number of known human depictions has increased dramatically with new finds—from about 87 in Reinach's 1913 inventory to well over 500 human images (on plaquettes, on cave walls, in or out of bone, antler, ivory, or stone such as steatite)—the idea that palaeolithic artists rarely depicted humans has persisted. What is interesting is that the kind of depiction (of humans) often differs stylistically from the way in which many animals were rendered, in that there is, for example, more anatomical detail and more attention paid to details, colouring, and shading in the animal imagery (e.g. Figs. 11.23, 11.24). And the predominant kinds of raw materials and locations used for anthropomorphic imagery seem to differ from those used for animals (which are predominant on cave walls), in that there are, for example, more anthropomorphs on portable forms. With such inventories as the La Marche plaquettes (Pales and Tassin de Saint-Péreuse 1976), where there are perhaps at least a hundred human images, it is still correct to say that humans are rare in cave wall 'art', but no longer correct to say that palaeolithic image-makers in all media avoided humans as subjects.

Ucko and Rosenfeld have long held (for instance, 1972) a conservative and appropriately cautious view about the reliability of identifying anthropomorphs and humans. Starting with a corpus of 475 images that, to them, are acceptable as anthropomorphic—in form—in all the media, they would allow only 106 as definitely human. With a relatively low number of such images distributed over such a wide geographic area and through some 20 000 years, it is difficult to show any patterns that might provide clues to significance or meaning. Not surprisingly, few systematic studies of human and anthropomorphic imagery have been carried out (but cf. Delporte 1979; Ucko and Rosenfeld 1972; Pales and Tassin de Saint-Péreuse 1976).

Humans then, like other mammals, are depicted on a wide range of media, not just as the popularized statuettes—bas-reliefs; on sculptured blocks; and engraved on bone and antler, as well as in wall 'art'. There are often just parts of the human body, such as the well-known hands, although these are found in only 20 caves. Although some images *are* clearly identifiable as female or male, many cannot reliably be assigned a sex, though this has not kept researchers from trying, on the basis of such problematic features as long hair or a rounded body (both taken as evidence for a 'female' classification).

With both the humans/anthropomorphs and the animal depictions one could argue for some conceptual similarities, in that often only certain attributes of the animal or human are selected for detailed treatment. In fact, what has led us to think of palaeolithic 'art' as realistic has been this more subtle feature; it may be a realism, but it is a realism of selected details. In the case of many—but not all—female statuettes, it is the apparent realism of such detailed breasts or buttocks that has not only attracted our attention but also contributed to their pervasive, yet unsubstantiated, interpretation as 'fertility figures' (see Rice 1981). Yet this very feature of much palaeolithic imagery—the realism of selected details—is precisely the kind of feature that may provide access to its meanings and significances. These details could be the very 'marked' and 'marking' features that derive from and carry the significances of the imagery. On the other hand, they could also be conventions (some of the many possible) for conveying representation and dynamism to the human observers (only a portion of which we can now appreciate).

11.4 The major interpretations

The time, effort, and debate that was involved in the acceptance of palaeolithic 'art' as being truly Ice Age in date is a well-known story (Bahn and Vertut 1988; Conkey 1981; Ucko and Rosenfeld 1967) that clearly illustrates exactly how preconceptions about the past allow for and select against certain interpretative possibilities. Even though many of the avocational and emergent-professional prehistorians at the turn of the century had come to accept the imagery as being the products of peoples who lived during the late Stone Age, this did not necessarily mean that the imagery was the product of people with cognitive, symbolic, or aesthetic systems that were *comparable* to those systems characteristic of Western civilization. In fact, as I will discuss below, the first primary and persistent interpretation—that the art was in the service of sympathetic hunting magic—was based on the idea that these Ice Age hunter–gatherers had cognitive and cultural systems that were very *unlike* those 'achieved' by nineteenth-century western Europeans.

The early decades of the twentieth century witnessed a variety of accountings for the imagery of palaeolithic 'art', and many of these have been uncritically perpetuated. There is the notion that this imagery was not inspired by abstract or symbolic thought. Instead, it was merely a direct 'reflection of reality [and] intended to serve a specific purpose' (Torbrügge 1968, p. 14), or 'art for art's sake' (see Halverson 1987, especially for references to Verworn 1909 and Van Gennep 1925, and for a review and revival of this idea and the idea that the 'art' is essentially 'thoughtless'). At first, many could only account for the art as a 'natural' result from the leisure time that comes from life in a bountiful environment. And although the idea that the 'art' was made to yield aesthetic and emotional satisfaction and pleasure has never really gone away (cf. for example Halverson 1987; see also Lewis-Williams 1983a for a strong critique of this, with particular reference to rock art of southern Africa), early twentieth-century scholars, in particular, developed and elaborated the notion that there was always a very practical magical purpose for the imagery, involving wish-fulfilment rites for the multiplication, capture, or death of animals. This became the primary interpretation for more than fifty years—what I have described (and discussed at length) elsewhere as 'the foundation interpretation' (Conkey, in press).

11.4.1 *'Art' as hunting magic: the foundation interpretation*

It was the newly 'revealed' ethnographic information of the late nineteenth century (especially Spencer and Gillen 1899), and the theoretical treatises by Tylor (1865), Lubbock (1870), and especially Frazer (1887, 1890) that influenced the first round of interpretations. Indeed, it was the discovery of Australian aboriginal art and the 'harsh' life these peoples were observed to lead that strongly influenced the development of the view that fertility and fecundity were (and *had* to have been!) overriding life concerns that were dealt with in palaeolithic 'art' and its implied associated rites and rituals.

Thus, it was a comparative and palaeoethnographic approach, drawing on notions of 'primitive' totemism and magic, that led to the establishment of the idea that this was magical art, created in a 'primal state of natural aesthetic innocence'. It was Reinach (for example, 1900, 1903), and then the prodigious Abbé Breuil (for example, 1935, 1952), who took these notions of magic—and other broad, supposedly universal generalizations about the conditions and lifestyles of hunter–gatherers, such as their need to struggle against nature—and elaborated upon them in the interpretation of specific palaeolithic 'artforms', *especially* the wall 'art'. Not surprisingly, because of the idea that these peoples were so psychically and practically engaged in hunting, their focus was on the animal depictions. Not surprisingly, the geometric forms ('signs') were interpreted as images that 'made sense' in the context of a hunter's magical art: huts, wounds, traps, arrows. This was a functional interpretation, seeking an account of the imagery in terms of what it might 'do' for the hunting society and its artists. The implication was that, to palaeolithic peoples, animals were 'good to eat'.

Art-as-hunting-magic became a consensus interpretation that allowed for the collection and synthesis of new data—new caves, new finds, new imagery—without much substantive debate over 'meaning'. Combined with Breuil's two-cycle 'chronology' for the 'art', based on notions of how style and technique 'ought' to evolve, there was an agreed-upon framework for organizing and interpreting the images. The use of 'twisted perspective', for example, would be early in the development of 'art', whereas the combined use of engraving and painting would be late.

This is not to suggest that all images and caves—which were rapidly being 'revealed' during the first decades of the twentieth century—were easily accommodated under the basic magic-for-the-hunt interpretation. Rather, the interpretation had to be stretched to accommodate new finds. For example, imagery was found in day-lit rock-shelters (as at Teyjat or La Grèze, both in south-west France) in association with occupation debris, and had the same animal species that were thought to be depicted only in the 'deep' caves that were considered to be sacred 'sanctuaries'. On the one hand, the notion held by Count Bégouën (1929)—that each image was a separate event, a product of a 'special moment' or circumstance—allowed each image to be viewed on its own, as a particularistic phenomenon. Yet, on the other hand, to accommodate many new images and finds the interpreters sometimes had to invoke such marginal but related notions as totemism or even art-for-art's-sake—the very idea the magic-for-the-hunt view had sought to replace.

The portable 'art' was not really dealt with in this foundation interpretation, except where certain animal imagery was concerned, especially that on presumed hunting implements. Breuil long maintained, for no detailed reasons, that the two forms (portable and wall) developed independently, as the products of separate 'lines of thought'. Partly because of the abundance of geometric motifs, which predominate on the engraved bone and antler, the portable 'art' was not so easily accommodated under a single interpretative rubric such as 'hunting magic', although some of these geometrics, then (for example, by Count Bégouën) as well as now (for example, Marshack 1989) could be related, respectively, to hunting as *marques du chasse* (hunter's marks or tallies of kills), or as 'over-marking' on depicted animals to 'ritually kill' the animal (Figs. 11.4, 11.25). Moreover, ever since the turn of the century there have been relatively few attempts (for example, Bégouën 1929; Graziosi 1960) to apply systematically the 'hunting magic' account (or any interpretation, for that matter) to *both* the wall and the portable 'art'. Given these trends, the privileged position of the painting and the wall art (which is not surprising given the privileging of painting over sculpture and other media in modernist thought of the twentieth century—see Burgin 1986), and the pervasiveness of the hunting-magic account, most people came to think that all palaeolithic 'art' could be referred to magical accounts and beliefs.[3]

Because the first serious and systematic challenge to Breuil's work and the hunting-magic interpretation did not emerge until the 1960s, many recent accounts still include portions of the hunting-magic view, which has not been fully dislodged despite the provocative challenges of Leroi-Gourhan (1958, 1965, etc.) and Laming-Emperaire (1962). Although, as will be discussed, there were a variety of reasons to challenge the hunting-magic hypothesis, the most significant problem of such a view is that it tries to account for some 20 000 years (or more) of image-making under a single explanatory rubric, which is, furthermore, based on a variety of assumptions about the cognitive capacities, attitudes towards nature, and intentions of palaeolithic peoples that have not been demonstrated nor shown to be plausible and unproblematic (as we shall see, for example, there is no direct correlation between those animal images depicted and the food remains of the animal species taken as food (for example, Delporte 1987, p. 117)).

11.4.2 *The structuralist 'break-out'*

The inclusive and revolutionary ideas of Leroi-Gourhan, in particular, not only challenged Breuil's work, his chronology for palaeolithic 'art', and the hunting-magic hypothesis, but also established an entire new conceptual field within which palaeolithic imagery could be studied and interpreted. In his most systematic early challenge Leroi-Gourhan brought together what still seem to be two very strange analytical 'bedfellows': quantitative analysis and structuralism. In addition, there were major intellectual and conceptual changes taking place; for example, in how we think about hunter–gatherers, their cognitive systems and socio-ecological relationships (for example, Lee and DeVore 1968; see also Conkey 1984, pp. 255–8). Taken together, we can see how Leroi-Gourhan engendered what I have labelled a 'structuralist break-out', a liberation from the increasingly stretched hunting-magic hypothesis (Conkey 1989). Yet Leroi-Gourhan's structuralist interpretation was also over-inclusive, monolithic, and vulnerable to the charge of having the data stretched or modified to fit it (for early critiques of Leroi-Gourhan's methods see Parkington 1969; Stevens 1975; and of his entire endeavour, Ucko and Rosenfeld 1967).

Leroi-Gourhan assumed that the painted and engraved images were not individual, unique artistic events, nor randomly placed on sacred cave walls, as his predecessors had suggested. Rather, he believed that the images had been deliberately selected by species or type of geometric form, and placed in deliberate locales within any given cave, and in specific relations with respect to each other. That is, he assumed there was an underlying structure or set of structural principles that had generated the imagery and its topographic arrangements: what animal or sign was most likely to be found in the deep recesses of a cave, or on the central panels? The specific images had not been selected for depiction because of some sort of magical powers or needs; rather, they were selected because of the role they could play within a certain 'syntax' that governed the arrangements of images as symbols (Leroi-Gourhan 1966). A well-preserved cave was itself a message (cave-as-text) with elements (frame and figures). The underlying 'formula' for the arrangement of images came to be known as a 'mythogram' (Leroi-Gourhan 1982) (Figs. 11.26, 11.27, 11.28).

The formula was Leroi-Gourhan's way of accounting for images being hierarchically co-ordinated: the A images (horse) and the B images (bison) are the central figures in any composition, comprising 60 per cent of the animal depictions. By formula, they are complemented by the secondary animals, C—stag, hind, mammoth, ibex, or reindeer—that comprise 30 per cent of the depicted creatures. Animals of type D (rhino, bear, feline) compose the remaining 10 per cent, and are remote in relation to the composition as well as within in the cave itself. In their recent book on Lascaux, Arlette, Leroi-Gourhan, and Allain (1979, p. 348) show how the dynamic imagery of the axial alleyway and other specified parts of the cave can be represented with the formulaic code of A, B, C, and D images.

Using an accepted (scientific and empirical) method—quantification (using keypunch techniques of the 1950s and 1960s)—Leroi-Gourhan drew on the developing structuralist ideas of the 1960s, such as searching for underlying generative 'codes' of cultural representations, with a focus on the binary oppositions (such as male–female) as central to the code-making (see Eagleton 1983, 127–50 for a general review; see Conkey 1989 for a review of structural applications to palaeolithic 'art'). In this way, Leroi-Gourhan brought palaeolithic 'art' into the arena of what was culturally and symbolically meaningful. After setting up a typology for animal depictions (by species) and for geometrics (the so-called 'male' or 'thin' signs and the so-called 'female' or 'full' signs), Leroi-Gourhan visited the major caves with wall 'art' to document the frequencies and placements of imagery in each one.

On the one hand, by counting images in over 120 caves and finding repeated (and potentially verifiable) patterns of certain animals in certain types of locales and in certain associations with other animals and geometrics, Leroi-Gourhan 'demystified' the 'art', thus allowing it to become a construct to be classified and analysed, like the object of any other science. Yet, on the other hand, when the 'art' was then subjected to structuralist analysis—what are the fundamental forms, codes, and 'grammars'—the imagery could be considered as a system of signs, in the semiotic sense. Semiotics, which is labelled as the 'science of signs' by its founders and practitioners (for example, C. S. Pierce; see Hawkes 1977 for one summary) refers to the idea that cultural representations are not neutral, but constitute a system of signs, such that the signification—of the signs, the codes, the 'grammars'—can be read. The decorated palaeolithic cave, as a semiotic system, can then be taken as a 'text' and, to Leroi-Gourhan, it could be read as a 'mythological vessel' (Leroi-Gourhan 1966).

The consequence of applying this quantitative method to the imagery of palaeolithic wall 'art' was, paradoxically, 'a move away from the empiricist paradigm to an exploration of anthropological theory' (Lewis-Williams 1983a, p. 5, in reference to a similar development in African rock-art research), to asking 'how do we interpret meaning?' It was an intellectual move that applied to the data of the palaeolithic certain anthropological ideas about structuralism, about semiotics, about art as a system of signs, about asking 'what does it signify?' Animals, as depicted by palaeolithic image-makers, were not merely 'good to eat'; rather, as part of a generative mythogram, animals—in the structuralist view—were 'good to think'.

Leroi-Gourhan himself (1966, 1983) saw his 1960s work as being a study of both the content and chronology of palaeolithic wall 'art'. His stylistically-based chronological scheme of four sequential Styles (I–IV) is a continuous evolutionary schema of styles that involved no change in structure or content. Although Leroi-Gourhan notes trends through time and across space in terms of the four major criteria that determine the characteristic of each Style—form (for example, wall art, plaquettes, decoration of deep sanctuaries, etc.), style (the general trend from abstract to realistic for animal depictions, and from realistic to abstract for the signs), content (what themes and associations or kinds of signs are depicted), and space (certain regional variations in the distributions of certain signs), the chronology itself is based on what few datable cave 'art' sites there are, on dated portable 'art' that can—by stylistic 'associations'—be correlated with certain wall-art depictions, and on a proposed 'evolution' of how human figures and signs are rendered (from realistic to increasingly abstract, cf. Leroi-Gourhan 1968, p. 66).

In essence, this procedure to establish a chronology is the same as that proposed by other prehistorians, with the exception of Leroi-Gourhan's different use of superpositions [since he considers sets of figures to be related according to the principles of the mythogram, rather than as separate images that, by their 'layering' can be used as indices of their having been separate image-making 'events']. The chronology rests on relative archaeological information and on stylistically associational data (Fritz 1969, p. 84).

Thus, Leroi-Gourhan's structuralist analysis took the one class of cultural products (wall 'art') as it appeared to be distributed through time and space, and he argued for a unity because he could infer a single underlying structure across that spatial and temporal spread.

Although there are still no secure means to date the imagery that is on cave walls, the Styles I–IV sequence Leroi-Gourhan developed has been far more favourably received (but cf. Clottes 1990) and more widely adopted than his original generative schema, which proposed certain cultural 'meanings' or referents. This original schema was based on his assumption that there was a male and female valence for different classes of imagery (for example, horse as male, bison as female; 'full' signs as female, 'thin' ones as male). In turn, this postulated male–female symbolism was part of the source for the associations between certain images and between image and cave locale (for instance, horse and bison in central panels), and between certain forms (for example, 'phallic-ended' antler batons with male [horse] imagery on them).

Although the 'reading' of the so-called signs as more-or-less abstract versions of maleness and femaleness has persisted (for example, Leroi-Gourhan 1978), the mythogram itself has come (1982) to centre on just a formula for the hierarchical co-ordination of images that are no longer explicitly or even implicitly associated with/generated by maleness or femaleness. Thus, the 'formula' for palaeolithic cave 'art' has come to comprise first, two major classes of animals, A (horse) and B (bison–aurochs) as the central and predominant images, which are, in turn, seen as complemented by secondary animals, C (stag, hind, mammoth, reindeer, or ibex). Additional types of animal, D (for example, rhino, bear, feline) are remote in relation to the composition, as well as within any given cave, topographically speaking.

By laying out a generative structure based on the probabilities of where any given image would occur in a cave, Leroi-Gourhan could postulate what an 'ideal palaeolithic sanctuary' would look like (Fig. 11.29). Additionally, by laying out this generative structure in the form of a 'formula' for the making and placing of images that Leroi-Gourhan believed was followed by generations of palaeolithic image-makers, he allowed these peoples to be brought into the 1960s views on hunter–gatherers that insisted on their ecological and cognitive sophistication, their well-tuned adaptive cultural systems, and their socio-ecological stability and long-term evolutionary persistence. Palaeolithic 'art', like all the other cultural products taken on by structuralist analysis (for example, the Fang architecture of Africa (Fernandez 1966), North-west Coast art (Holm 1965; Levi-Strauss 1963), all sorts of myths, Tchikrin (Amazonia) body decoration (Turner 1969)) was demystified, and not only came to be understood as a 'model of intelligibility' (Wylie 1982, p. 40), but also secured a position for Upper Palaeolithic anatomically modern *Homo sapiens sapiens* well within the scope of what it means to be fully human (see Conkey 1989, p. 153).

11.5 Alternative interpretative perspectives

Since the middle of the twentieth century, most prehistorians who have worked with the archaeological materials of the European Upper Palaeolithic have tended to marginalize the 'art' as 'too humanistic' (for the prevailing preference for more *scientific* inquiry); without adequate chronological control (and hence unable to contribute to the increasingly preferred reconstructions of palaeolithic 'cultures', which have been based almost exclusively on the identification of relative chronological strata and distinctive tool industries); *or* too dependent upon (palaeo)ethnographic parallels or analogies that were far less acceptable than they had been at the turn of the century (see Sackett 1983 for a historical overview of the practice of palaeolithic archaeology). Breuil and his collaborators had dominated the field, particularly in France, and, as it happened, a major challenge—in theoretical scope, in the systematic assembling of data, and in method—such as Leroi-Gourhan's 1965 work—was necessary to dislodge this privileged and dominant interpretative and analytical position. A 1955 review of the extant literature on the interpretation of palaeolithic 'art' would have revealed that almost everything that was taken at all seriously was some variant or another on the master foundation interpretation of 'hunting magic'.

As a result of the quantitative and structuralist analysis—the 'structuralist break-out' led by Leroi-Gourhan, with help from a number of other colleagues (for instance, Laming-Emperaire 1962)—palaeolithic 'art' came to be seen as an object for serious archaeological study using a variety of alternative interpretative paradigms that were, in fact, simultaneously emerging in certain intellectual circles. For example, within Anglo-American archaeology the 1960s witnessed a major shift in how culture was conceptualized and views as to what the goals of archaeological research and interpretation ought to be: from primarily culture-history to the study of 'culture process' (see Flannery 1967; Redman 1973); from a descriptive archaeology to explanation in archaeology (for example, Watson et al. 1971). This was called the 'new archaeology', which was being invigorated by such theoretical frameworks as cultural ecology, neo-evolutionism, and the culture-as-adaptive-system approach (for instance, Dunnell 1986; Leone 1972). As a result, there was increasing interest in cultural information systems—including 'art' and visual imagery—and how they might be a part of human cultural adaptation, whether in small-scale hunter–gatherer societies or complex states. Palaeolithic art, as 'liberated' from the hunting-magic account, and considered as a system of signs—signs that might have been polysemic and multivalent to the millennia of makers and viewers—could be taken up as a body of cultural materials to be thought of in terms of cultural information systems, in terms of a means of human cultural adaptation (for example, Conkey 1978; Gamble 1982).

Within the practice of cultural anthropology in both Europe and America, the interest in symbolism and structural analysis, as well as an expanded version of cultural ecology that considered ceremonies, rituals, myths, and 'the sacred' to be understandable as part of how humans mediate and regulate their socio-ecological systems (for example, Rappaport 1968, 1971; Vayda and Rappaport 1968) also provided a theoretical and analytical space for such ideas as 'art-as-a-cultural-system', art as adaptive information, art as a vehicle for the playing out and constitution of structural principles and 'grammars' of socio-cultural behaviour (for example, Colby 1975; Faris 1972; Washburn 1983). All these ideas could thus now be applied in alternative interpretations of palaeolithic 'art'.

11.5.1 *Art as adaptive: attempts at context and process*

Many of the new looks at palaeolithic 'art' in the 1970s and early 1980s drew upon these new directions, and an increasing number of analysts from outside France and Spain turned their attention to palaeolithic imagery. In retrospect, we can see that although many of these 'new looks' involved primarily an updated, but still very functional approach—what does the 'art', or some different kinds of 'art', *do* for the makers and users—many interpretations offered something that the structuralist 'reading' did not: a consideration of the contexts—the palaeolithic lifeways—within which the 'art' was made and used. Neither the 'hunting magic' nor 'art-as-mythogram' interpretations made any linkages to the particular (and varying!) features of Upper Palaeolithic life in Europe. Both proposed a single (albeit different) major cultural source or meaning for the 25 000-year time-period; both proposed continuous stylistic evolution and chronology. Subsequent interpretations (for example, Conkey 1978; Gamble 1982; Hammond 1974; Jochim 1983; Pfeiffer 1982; Straus 1987), however, bring in ideas on the socio-cultural contexts in which imagery was made and used, and offer perspectives that are more systematically anthropological, attempting to provide an institutional or even a social analysis of the production of particular artefacts and images.

For example, Gamble (1982) has proposed a socio-ecological context for the making of specific forms, the female statuettes, which are found over a wide geographic area from France across into European Russia, and seem to cluster in a relatively delimited temporal span 26 000–23 000 years ago—although this dating is very tentative and not necessarily accepted (cf. for example Bahn 1989) (Fig. 11.30). First, Gamble hypothesized that certain ecological changes just before this time-period may have made it necessary for the mobile human groups of hunter–gatherers who were occupying this European range to 'realign' their social groups and to reorganize their access to needed food resources. He went on to suggest that as these groups reorganized their regional alliance systems—usually kinship-based networks such as those known among recent hunter–gatherers that 'ally' groups for many social and ecological reasons, such as gaining access to certain resource zones—they would have developed some means for establishing and communicating the reorganized regional alliances. Gamble looked at the widespread distribution of the female statuettes, and took Leroi-Gourhan's (1968) suggestion that these statuettes are quite standardized in the proportions and underlying 'rules' for their manufacture. From this, Gamble suggested that the apparently standardized form of the statuettes was part of the means by which social groups established and communicated the reorganized regional alliances needed for both social and ecological existence. Thus, he links a single class of palaeolithic imagery, its postulated 'limited' chronological occurrence yet its widespread spatial distribution, and its postulated 'standardized' form to a set of very plausible—but still hypothetical—assumptions that begin with environmental and ecological changes that 'trigger' social reorganizations and re-alliances.

For another example, Jochim (1983) suggested that climatic deterioration at the glacial maximum (c.18 000 years ago) reduced—because of the increasing build-up of glacial environments—habitable areas in Germany and central Europe, provoking a wave of 'refugees' into the more hospitable and resource-rich south-western Franco-Hispanic region. He suggested that the florescence of cave-wall 'art' that is attributed to this period, and that is highly concentrated in this area, could be explained as a 'response' to the new demographics and the need for spatial demarcation or other messages about human–land or human–human relations (see also Hammond 1974 or Straus 1987 for variants on the idea that cave 'art' is related to land-use strategies). Perhaps each decorated cave 'marks' the spatial or 'territorial' presence or limits of various groups, or serves as a spatial 'marker' for the social and ritual relationships.

Yet these, and other neo-functional interpretations (for example Conkey (1978), where I suggested that the appearance of paleolithic 'art' as a symbolic system is correlated with the development in the European Upper Palaeolithic of an increasingly complex social geography, and somehow 'served' as a means for storing and sending information about socio-ecological processes) pose problems of verification, and even of warranting assumptions. For example, we have no substantial archaeological or biological evidence for the demographic shift that Jochim suggests. Nor is there systematic, fine-grained ecological and archaeological research to substantiate the early Upper Palaeolithic environmental and resource-zone shifts postulated by Gamble. Both are plausible, *post hoc* hypotheses, but not yet, nor easily, verified, despite the many decades of prehistoric excavation and research in these regions of Europe (see Rigaud and Simek 1987). Further, in most recent interpretative studies of this sort, what remains undeveloped are the theoretical warranting assumptions and inferences that would link cave 'art' to territorial marking, or female figurines to alliances. To say that 'art' and visual imagery are part of adaptational problem-solving is, in large part, to beg the issue.

On the one hand, such functional approaches not only need to address the specific contexts and forms in more detail, but also need to retreat somewhat from the starting position of assuming 'adaptive fit'. Such a position tends to assume that if some cultural product or form exists it must have been 'adaptive' (for critiques of functionalism in archaeology, see, for example, Hodder 1982, 1986). Why and how would cave art be a 'response' to demographic shifts? Why and how would female statuettes be the medium through which social alliances would be maintained? On the other hand, this view of 'art' as adaptive information is one of the factors that has contributed to our current views on the Upper Palaeolithic peoples as being 'just like' ethnographically known human societies. Thus, when ethnographers who study 'art' in the context of recent hunter–gatherer societies show how that art and imagery can be interpreted to be 'about' social regulation, cosmogenic structures, socio-political hierarchies, and the storage of core cultural messages (see for example Anderson 1979), and when interpreters suggest similar sorts of 'functions' for Palaeolithic 'art', it is not surprising that the Upper Palaeolithic societies are considered to be 'just like us: working out strategies for survival', as Pfeiffer (1986) has recently suggested.

Other interpretive suggestions for palaeolithic 'art' have gone even further in basing their assumptions on the behavioural and cultural repertoires of living, modern humans. Although the 'art-as-adaptive' approaches have drawn from general notions on how art 'works' in the cultural systems of modern humans—especially those of ethnographically observed non-literate groups—there are other interpretations that have drawn more directly on contemporary and Western ethnocentric notions about such things as sex-roles and what might be 'universal' in human existence. These interpretations, a few of which are discussed briefly below, have taken quite literally what the popular American news magazine, *Newsweek* recently proclaimed (November 10, 1986) about Upper Palaeolithic peoples in its cover story: [they were] 'The Way We Were'.

For example, drawing on Freud and various other psychoanalytical perspectives, Collins and Onians (1978) propose that the specific contents of the earliest 'art' (for example, a few animal outlines and the 'tactile' engraved signs that they accept as depicting female vulvae (but see Bahn 1986)) were generated by the drive for food and sex-object manipulation by adolescent Cro-Magnon males. In his recent interpretation, Guthrie (1984) has drawn upon certain attributes of the imagery and what is *absent* (i.e., not depicted, such as vegetable resources), and upon the assumptions that it is males who hunt and that females are primarily to be considered as objects of male exchange and male sexual energies, of importance primarily for reproductive 'purposes'. He then proposes that the palaeolithic imagery is the (creative) product of males, who used the visual images as a way to 'talk about' hunting, sex, and fighting. He assumes that these were the central concerns of palaeolithic males.

Faris (1983) offers a very different 'reading' of the 'art', but also draws upon some universalized assumptions about male and female roles (man-the-hunter; woman-the-gatherer), and about the sexual division of labour in hunting and gathering societies. He suggests that the imagery is about the male appropriation of female labour in the increasingly complex societies of Upper Palaeolithic peoples in Europe, where the social relations of production were changing with increases in the size of groups and the ways in which labour had to be organized. Although none of these scenarios had gained a very large following, they all raise quite clearly the problems of uncritical and biased ethnocentric approaches, which are far too complex to discuss here (cf. Conkey and Williams 1991; Mack, n.d.). By extension, they all also raise the important questions as to how one might go about verifying such interpretations.

Eaton also relies upon some questionable assumptions that there are certain universal features to human behaviours. He has tried (1978a, b) to account for all of palaeolithic 'art' in the terms and framework of 'sociobiology'; all works of palaeolithic art, he claims, can be viewed as individual displays or marks of the successful hunter. Eaton argues that palaeolithic art must have played a major role in assuring the evolutionary success of modern *Homo sapiens sapiens* (who had 'art') over the Neanderthals or archaic forms of *Homo sapiens* (who did not have 'art'). Central to this evolutionary success, he suggests, was the evolution of certain social behaviour by males: their abilities to display their hunting prowess would 'earn' them (heterosexual) mates and guarantee their reproductive success. Without even considering the many problems with such a sweeping sociobiological approach, Eaton's theory—which ignores diversity and variation in favour of an inclusive view of the 'art'—has not gained much support despite the fact that it includes features that have characterized all the major interpretations for so long, being both inclusive and very much based on 'the hunt' as a central source of 'meaning' for the 'art'.

There are also those interpretations that take selected attributes of the imagery (as do Collins and Onians 1978 and Guthrie 1984, as cited above) and try to establish linkages to certain kinds of mythology, ritual, or ceremony that is not so explicitly about hunting. For example, Bahn (1978, 1980) has suggested that those caves that *are* decorated—and not all that were known and used by palaeolithic peoples were decorated—are located in particular proximities to various kinds of water-sources (especially 'springs'). Drawing upon widespread mythological notions about springs, he suggests that ' "abnormal water" was probably a major, and hitherto neglected, factor in the beliefs and traditions of which paleolithic art is a manifestation' (Bahn 1978, 132).

In the most developed of the recent hypotheses linking ceremony and cave art, Pfeiffer (1980, 1982) examines such things as the placement and orientation of much imagery within caves, with particular regard for how an observer would approach or 'view' imagery. This is one main line of evidence he uses to suggest that the interiors of caves were perfect 'liminal' or transitional locales; that is, locales where those unfamiliar with them could not easily know their way, and where—through various rituals and/or ceremonies—visitors, or more specifically, potential initiates, could be convinced they were in marginal places and transitional social statuses. That is, in many of the non-literate societies that ethnographers have studied, ceremonies and rituals associated with initiation ceremonies are often conducted in ways so as to create 'liminality'. The images on cave walls, Pfeiffer suggests, not only contribute to feelings of thing's being 'not quite right' by their placements, orientations, or combinations of unusual attributes, but also must be a means to store important social and ecological information. This information, he suggests, is integral to the processes of initiation that he hypothesizes as occuring in many of the caves. Further, he makes a case for these Upper Palaeolithic European societies' having been increasingly complex and increasingly information-rich, thus necessitating the development of new means for transmitting the necessary social and ecological information. As with Gamble's (1982) and Conkey's (1978), Pfeiffer's hypothesis turns on a number of tacit assumptions, not only about the centrality of information-processing, but also to the effect that the 'art' is some sort of response to or way of fulfilling socio-spatial or socio-ecological 'needs'.

Most recently, Lewis-Williams and Dowson (1988) have drawn on the extensive work with the prehistoric rock art of southern Africa that Lewis-Williams (for example, 1982, 1983b) has linked quite securely to trance experiences and trance-related imagery. From this, they suggest that *some* of the European palaeolithic imagery, especially the 'signs', can be accounted for as 'entoptic phenomena', which are said to be cross-culturally, neurologically constant imagery that one 'sees' when in a trance or hallucinogenic state, such as certain zigzag lines. Thus Lewis-Williams and Dowson suggest that some of the palaeolithic cave imagery could have been produced by individuals who either were or had been in a trance state (see Chapter 12, this volume). Although there is little doubt that some of the southern African imagery can best be understood as part of the metaphorical thought deriving from and characterizing trance experiences, the full and credible extension of this localized interpretation to even some of the imagery of palaeolithic 'art' in Europe has yet to be made convincingly (see Fig. 11.31).

11.5.2 Alexander Marshack: from calendars to 'post-structuralist'?

Although not trained within the anthropological paradigms of the 'new archaeology', symbolism, cultural ecology, or structuralisms, Alexander Marshack (for example 1970a, 1972a, b) developed one of the first Anglo-American approaches that offered an alternative, but complementary, direction to the structuralism of Leroi-Gourhan. Although initially inspired, as a scientific journalist, by an investigation into the earliest human calendrical systems, Marshack took a position broadly similar to that underlying Leroi-Gourhan's approach: that the Ice Age image-makers, as *Homo sapiens sapiens*, had conceptual, symbolic, and cognitive systems comparable to those of contemporary humans, and that the images were deliberately structured. In keeping with the reassessments (for example, Lee and DeVore 1968) of hunter–gatherers, Marshack also took as a starting-point the idea that these Ice Age image-makers were keenly aware of ecological and environmental variations and processes, and that they used symbolic and material means to 'mark' such processes (as in 'calendars' that marked the phases of the moon) and to regulate their own adaptations (see, for example, 1970b). (Fig. 11.32)

Much of his early work (for instance, Marshack 1972a) was focused on marshalling evidence from a previously neglected domain of palaeolithic 'art'—the engraved bone and antlers of portable art, especially those with 'geometrics'—to support the anthropological claim for fully modern symbolic and cognitive behaviours. He has continued to provide some very detailed analyses (for instance, Marshack 1969, 1979) of selected Upper Palaeolithic decorated artefacts in support of scenarios for the evolution of the brain and of human cognitive capacities (for instance, Marshack 1976, 1984a, 1985a). For example, Marshack makes inferences as to how certain engraved artefacts were incised, and can link these incising processes, he believes, to the existence of a preferential handedness (for example, 'right-handed') in the engravers, which, in turn, relates to the development and expansion of one cerebral hemisphere.

Although Marshack's initial claims for lunar calendars are not widely acknowledged to have been convincingly *demonstrated*, there are few who would dismiss the plausibility of the claim that ecological 'marking' by means of material culture and imagery might have been a part of some Upper Palaeolithic lifeways, and that such ideas could be one way to account for *some* of the imagery and forms included within palaeolithic 'art'. By establishing an audience with the European scholars from the very beginning (Marshack 1970a; see also 1984b), Marshack has not only contributed to the ideas set out by Leroi-Gourhan concerning the complexity and yet structured nature of palaeolithic imagery (even nicks on bone and antler were not necessarily random, nor only explicable as hunters' tallies), but has also promoted the viability of portable 'art' of the most 'mundane' sort—wall 'art's' neglected step-sister—as evidence in making cultural inferences about the peoples of Ice Age Europe and Russia.

Marshack has continued to use highly developed microphotographic techniques and detailed analyses of selected objects drawn from extensive visits to museums and sites in order to promote ideas concerning the persistence and widespread distribution of what he views as particular 'traditions' of incising and pattern-making (see, for example, Marshack 1977, 1989); as examples of the uses, re-uses, and re-touching of imagery on particular objects and cave walls; and as evidence for symbolic 'marking' that has been neglected in the more dominant debates over the cave painting and what the images there may or may not 'be' or originally 'meant' (Fig. 11.33).

Indeed, as Marshack has argued well, using the benefits of microscopic analyses, many of the images and objects appear to have been retouched and reused, if not also carried around extensively (for example, Marshack 1989; see also Lorblanchet 1980 on repainting) (Fig. 11.34). These images were, in one art historian's interpretation of Marshack's findings, 'the sites and occasions for an on-going process of image generation. Images themselves become the tools for making further images' (Preziosi 1982, p. 325). As research by others has suggested, there is evidence to support this broader 'reading' of what some palaeolithic imagery is all about, even if Marshack himself has not articulated the conceptual framework that Preziosi has attributed to him. For example, many objects of 'palaeolithic art' have been broken, apparently intentionally (see Mons 1986 for a convincing study; see Vandiver *et al.* 1989 for a less convincing one). Some sites have yielded hundreds of fragmentary but incised plaquettes (for example, at Limeuil and Enlène, both in France), which were then often used as elements of 'pavements' that appear to have been 'living floors'. Many of these incised plaquettes have been found with the incised side turned downwards in the 'pavements' (for example, Bégouën et al. 1989).

Preziosi's reading (included in a much wider critique of palaeolithic-art-as-origins research, Preziosi 1982) of Marshack's work is that it appears to be 'post-structural' in the sense that what Marshack shows is the extremely 'open' character of palaeolithic imagery, and that the images 'serve as sites for ongoing palimpsest and modification (both with a single time frame and over time)' (Preziosi 1982, p. 325). Thus, given his own theoretical predispositions—what he wants to see in this early imagery—and a stance that purports to be firmly within semiotics and critical art history (Preziosi 1989), Preziosi suggests some *implications* of Marshack's work that are not, however, self-consciously developed by Marshack himself:

... an important *implication* of Marshack's research is a displacement of the question of meaning or signification from the individual boundaries of particular images to the performative or orchestral context within which artworks come to be invested with (partial) meaning [Preziosi 1982, p. 325, emphasis added].

Although Preziosi very much wants to see in Marshack's detailed analyses the necessary 'post-structural' moves toward consideration of con-texts and processes, rather than a focus on the decontextualized system(s) of meaning preferred by the structuralist accounts (such as those of Leroi-Gourhan), his suggestion that Marshack is primarily responsible for turning the questions to context is easily challenged. Just think how many of the neo-functional and 'adaptive' approaches—such as that of Pfeiffer, for example—might be equally well described by Preziosi's above quote.

On the one hand, Marshack can, I believe, be roundly critiqued for various epistemological weaknesses, such as his continued appeal to intuitive 'meanings' of certain images (for example, ziz-zags as water) as he traces his 'traditions' of incising through time and space (see also the comments after Marshack 1985b and Marshack 1989). On the other hand, *if* Marshack has contributed to this interesting new direction that Preziosi heralds and that I have recently tried to document (Conkey 1987, pp. 422–4), then this is an important contribution. The central idea here is that a very important potential route to understanding the 'meanings' of palaeolithic imagery would be to investigate the productive contexts within which the images and forms came to be, as Preziosi suggests, 'invested with (partial) meaning'. However, this is to move into the interpretative issues of the late 1980s, which derive not only from Leroi-Gourhan and Marshack but from a variety of other approaches that have developed directly out of wider concerns within anthropology, archaeology, and fields that study 'cultural representations' (for example, art history, literary theory, feminist theory).

11.6 Assessing the interpretative terrain: into the 1990s

Contemporary study and inquiry into what we call 'palaeolithic art' has diversified along many intellectual and interpretative planes. One is faced with a multiplicity of possible accounts for different selected sets of imagery, such as for the plaquettes or female statuettes, or for all of cave 'art'. For example, it is argued now that some animals and especially signs on cave walls may be the work of shamans or those who have been in trance (Lewis-Williams and Dowson 1988; Smith, n.d.); much cave 'art' may be part of the dramatic context for initiation ceremonies (Pfeiffer 1982); some of the engraved geometrics on bone/antler, in particular, may be pre-alphabetic graphic signs (Chollot 1980); and, to some, most of the imagery may be semiotic, meaning-bearing imagery (Leroi-Gourhan 1965, 1982; Sauvet 1988); whereas to others (for example, Halverson 1987) 'art-for-art's-sake' is still a viable account! I myself have suggested that some of the engraved bone and antler and other portable 'art' may be part of the ritual uses of material culture in contexts of social gatherings and the concomitant rearrangements of interpersonal relationships (Conkey 1985); and other more or less neofunctional interpretations, as discussed above (section 11.5), have suggested that cave 'art' (for example, Jochim 1983; Straus 1987) and some portable 'art' (for example, Gamble 1982) may be part of the dealing with and marking of socio-spatial arrangements. Delporte (1986) argues that the imagery of palaeolithic 'art' 'must be' about cosmologies and conceptual systems that could broadly be considered 'religious'; whereas another view (Gilman 1983) suggests that the appearance and elaboration of 'art' and abundant material culture during the European Upper Palaeolithic were part of processes embedded in the social relations of production, specifically in the inter-group alliances that led to what he calls the cultural 'revolution' of the Upper Palaeolithic.

Recently, Vialou (for example, 1981, 1983, 1986) has reassessed one of Leroi-Gourhan's central ideas associated with the mythogram—that there is a persistent topographical patterning for cave-wall 'art'—and, as a result, he has challenged the idea that most caves were decorated according to the very same generative principles for selecting and placing images. Rather, Vialou argues, each cave site is best considered a unique and original 'symbolic construction'. There are so many local or individual variations on the basic mythogram, he suggests (for example, by comparing the two closely-situated painted cave sites in the French Pyrenees, Niaux and Fontanet), that it is more the exception (than the rule, as Leroi-Gourhan proposed) to document similarities between any two or several sites. In challenging the pervasiveness and continuity (through time and across space) of Leroi-Gourhan's formula for cave 'art', Vialou introduces his own categories for analysis, or 'compositional axes': themes, thematic liaisons, and 'symbolic constructions' that take into account the volumetrics or 'cave-ness' of each site, going beyond the more linear topographic placements of Leroi-Gourhan's analysis (Vialou 1986; see his 1982 for how this approach is applied to a single cave, in this instance, Niaux). The implications of Vialou's suggestion that each (decorated) palaeolithic site has its own 'irreducible originality' is congruent with the recent research that seeks to understand and document the diversity of imagery and contexts. Vialou's challenge is to the structuralist premiss of Leroi-Gourhan's analysis—that the unity of palaeolithic 'art' derives from a common 'métaphysique' that persisted over the millennia.

Thus, thirty years after the publication (1965) of Leroi-Gourhan's major analysis and challenge to Breuil and the hunting-magic hypothesis, there are now many challenges to the homogeneity that, to a certain extent, both Breuil's and Leroi-Gourhan's inclusive accounts implied. There is questioning of the idea that an underlying epistemology or cosmogenic conceptual schema persisted: the imagery of palaeolithic art is seen as diverse, multifaceted in forms, in contexts, in productive and generative features. Above all, the imagery is taken to be polysemic (i.e., having many meanings/many readings), not merely to us as analysts of the late twentieth century, but as primarily polysemic to those who were engaged with the imagery in the on-goingness of Upper Palaeolithic lives. Now the questions at hand are more complex and more interesting than before. They are certainly more complex than just 'What is left from Leroi-Gourhan's work?' or, 'What, then, are the origins of art?', or 'What new account(s) can be advanced for this imagery?'

I see the study of 'palaeolithic art' to involve now at least two very different but related tasks. On the one hand, we are at last coming to understand the bloc of imagery called 'palaeolithic art' to comprise such extraordinarily diverse forms and cultural products that the concept of *a* 'palaeolithic art' should become obsolete. Furthermore, these diverse images and cultural products were produced, used, and often *re*-used by some—but probably not all—the peoples who lived in a wide range of locales and social collectivities during the last 25 000 years or so of the Pleistocene in Europe and across Russia. Given this, our analytical and interpretative challenges are in trying to understand not just one bloc of 'artistic activity' or one 'art form', but what are very probably many—'perhaps interpenetrating'—sign systems (Davis 1986). Palaeolithic 'art' is not just one cultural output and one record of palaeolithic activity, but surely involves the complex—and, to us, enigmatic and ambiguous—social constructions of things and of meanings. There is no one 'figurative system' of palaeolithic art.

To engage ourselves in this new enterprise does mean taking up the more modest and detailed tasks—more modest than offering an inclusive interpretation for 'the' art, as has been the practice. Among other things, the detailed tasks now include the study of different 'classes' of materials and imagery (for example, of the so-called spear-throwers, the use of shell) and of different sets of imagery and associations that cross-cut materials or 'classes' of decorated objects (for example, how animals are rendered in associations, or what kinds of objects with what kinds of imagery appear to have been reused intentionally or intentionally broken (for example, Mons 1986)). These studies of 'classes' of 'art' or of imagery are not, however, the analytical ends. Rather, these studies of materials and their productive and generative contexts must be taken on in their archaeo-logical contexts. But this is 'contexts' not only in the sense of 'what site', 'what archaeological culture', 'with what other artefacts', and so forth, but hypothetical *contexts* in the sense of social actions and social constructions. These wider conceptual contexts can only be entertained if we do not get drawn into the extant expectations that we can produce a certain semantic fixity, '*a* meaning', even for certain limited sets of objects, such as plaquettes or female statuettes.

The other direction or task of contemporary inquiry into 'palaeolithic art' is that it must be more self-conscious and critical. At one level, many researchers (for example, Preziosi 1982; Ucko 1987; Ucko and Layton 1989) have begun to articulate how we have been confined by the 'strong covert influence(s) of powerful modern ideologies on what constitutes some of the fundamental categories we have defined', such as what constitutes an image, or what 'it' represents (Davis 1985, p. 9). For example, assuming that they even *are* images in our sense of deliberate representations, some analysts have have at last begun to question the idea that certain signs or shapes can be unproblematically equated with specific representations, such as a representation of an ibex, or as a representation of a female vulva (cf. Bahn 1986; see especially Clottes 1989 for discussion) (cf. Fig. 11.35). Given that we still rely upon, and are quite uncritical of, notions about image-making and sign-systems that are derived from our own cultural experiences, there should be break-outs from these particular culture-bound lines of inquiry. We could—as Preziosi (1982) claims Marshack has done—displace the question of meaning from the individual boundaries of particular images to such other aspects as the productive contexts within which the imagery may have come to be invested with meaning (see also Conkey 1987).

But at a more penetrating level, researchers have yet to recognize and comprehend that the objects of our knowledge are not equivalent to the objects we study. The study of palaeolithic 'art' forms and images are not merely about form, image, and meaning. Like so many other sets of archaeological data (i.e., the so-called 'archaeological record' (see Patrik 1985), the palaeolithic images and 'art' forms have been objects of study because they are taken to be manifestations of and evidence for a particular construction of the evolution of 'symbolism' and 'consciousness'. Although we focus on the study of palaeolithic 'art' (our objects of study), it is rarely that anyone further analyses the idea that our inferences are about 'symbolism', 'consciousness', and their 'evolution' (the objects of our knowledge).

Our procedures have usually been to search for certain features: for example, what is there about stone-tool making or the use of perforated shells as necklaces that could be evidence for, respectively, the existence of language or the development of self-awareness? These are features of the archaeological record—stone tools, perforated shells—that we presume to embody the phenomena that are the objects of knowledge: consciousness, symbolism, and their evolution. But, as Lutz has so persuasively argued with regard to one of these concepts, 'consciousness':

As Euro-Americans we have culturally endowed ourselves with a particular kind of consciousness; it entails the notions of rationality, objectivity, control of attentional processes in the interest of solving technical problems, non-emotionality and linear thought, among others.... So construed, the concept of consciousness is fundamentally good and important; it is positively valued and those with less than this are less than fully human (Lutz, 1992).

Furthermore, out of the many senses that one could construct for 'consciousness' or 'symbolism', and that have been selected for elaboration, those that have been constructed are generally used to refer to *self*-awareness, more so than awareness of the self-in-interaction or awareness of relationship. Consequently, both 'consciousness' and 'symbolism' tend to be reified as properties of the mind, rather than conceptualized as emergent properties of social action.

Given such cultural constructions of consciousness or symbolism, it is not surprising that our archaeological analyses have focused on prehistoric technology (as manifestations of rationality and control over problem-solving), on language evolution (as related to the emergence of 'self'), and on what, to us, must be aspects of 'mind' (such as ritual, art, and other 'evidences' for conceptual thought). One classic relevant and recent example of this would be the analysis of Chase and Dibble (1987) who select stone-tool technology, burials, and the presence of 'art' as evidence for and as defining criteria for the evolution of symbolic behaviour and therefore of modern humanness.

Because our focus on technologies is just as much a product of our historical situation and concerns (McGaw 1982), and because our notions of what constitutes consciousness and symbolism (and, therefore, what constitutes the evolution of such phenomena) are historically-contingent constructs—not absolutes or essential phenomena—we must be circumspect and self-critical about our 'uses' of selected features of the archaeological record—such as 'technology' and 'art'—to support an account (of the evolution of consciousness, for example) that is, by definition, predisposed to, if not already defined by, these 'forms' of evidence. The difficulties and challenges in making inferences about the past are there not so much because 'the record' is fragmentary and partial, but because we are so rooted in our own historically-contingent concepts and methods.

11.7 What becomes of the study of origins?

Exactly how the production of palaeolithic imagery fits into the wider picture of human biocultural evolution is still a wide open question for at least three reasons. First, the entire topic of the emergence and spread of *Homo sapiens sapiens* is now being reassessed completely, and with much debate and discussion (see the beginning of the present chapter, also, for example, Mellars and Stringer 1989 and this volume Chapters 2 and 3). Secondly, there are increasing discoveries of what we call imagery and 'art' of late Pleistocene age (from at least 20 000 years ago) in a variety of locations throughout the globe: Australia, Africa, New Guinea (?), and perhaps even South America (for a review see Bahn and Vertut 1988). Thus the Cro-Magnons of south-western

Europe can no longer be seen to have had a monopoly on the production of figurative imagery, on the use of pigment to create imagery, or on the transformation of such materials as stone and bone into image and meaning-bearing forms. Thirdly, until the corpus of Euro-Russian 'palaeolithic art' is better understood, at least in terms of its temporospatial distributions, we can not understand exactly what forms and images, in what contexts of manufacture and use, may be relevant to scenarios of cultural change and elaboration, themselves expected to be variable even within the Euro-Russian geographic zone.

The recent research that now locates early modern humans (*Homo sapiens sapiens*) in such places as southern Africa or the Middle East (for example, in Mellars and Stringer 1989) many millennia before their appearance in south-western Europe lends important and convincing support to the idea that the bulk of what has been preserved in the form of 'palaeolithic art' *clearly post-dates* the establishment of *H. sapiens sapiens* as a biological species. South-western Europe then demands its own contextual and particular historical account. The European Upper Palaeolithic is now liberated from the constraint that it has to be the spatialized period that has 'stood for' *the* emergence of modern humans, of 'humans-as-we-know-them', as a 'segment of ethnography'. But as long as the 'first art' notion is still linked to south-western and central Europe, it will be hard to make the break completely from this kind of view on the 'classic' Upper Palaeolithic. It is the task of the current generation of scholars both within and outside Europe to liberate the corpus of materials called 'Palaeolithic Art' from being merely the beginning of art history and the end of the evolution of human symbolic behaviour.

I have elsewhere (Conkey 1983) reviewed the relationship between the appearance and forms of palaeolithic 'art' and what can be/has been said about its relation to the so-called transition from archaic to modern *H. sapiens*. Although new evidence (fossils, dates, and new 'art')—and entirely new scenarios for the transition—have accumulated since that review, it is still primarily the case that researchers tend to take either of two approaches to palaeolithic 'art', which, in turn, have had implications for understanding its origins and contexts. Increasingly, the case for a unity in the 'art' and all that that implies—homogeneity in conventions, in generative structure, and in stylistic chronologies—is less viable. Increasingly, as this chapter has discussed, the case for diversity and discontinuity can be made, although the meanings and interpretations that follow run quite a gamut, from those rooted in Freudian psychoanalysis to those based on cybernetic information and signalling systems.

Most references to the origins of palaeolithic 'art'— *why* does it appear?—tend to be derived from the meanings attributed by the researchers to the images (for example, if it is magic-for-the-hunt, then the art derived from the intensity of the hunting way of life and the need to 'control' game), or more explicitly from the wider theoretical assumptions that researchers hold about modern (i.e., historic) human cultural systems. For example, if a starting premise is that material culture and imagery are part of the ways in which humans cope, adapt, and store information, then the 'art' can be viewed as coming into existence as part of the ways in which new sociospatial needs were met. There are relatively rare attempts to take on the question of origins directly, even if it is just the issue of 'why does this imagery come into existence at these locales, in these particular forms?'.

For reasons discussed at some length in Conkey (1983), those few attempts that have been made at such contextual accounts for 'origins', such as that by Collins and Onians (1978) (see above, Section 11.5.1) are ethnocentric and not very satisfying. At the same time, there have been recent attempts to take a more sophisticated view on how image-making in any form came into existence (for example, Davis (1986), who inquires into the cognitive and performative processes whereby an 'image' becomes an 'image of'). But this—although drawing upon some of the early Aurignacian images from south-western France—is an inquiry into aspects of more general theories of human cognitive and symbolic evolution. But to suggest, as I am here, that Euro-Russian palaeolithic 'art' is not *the* origins of art does not negate the possibility that such particularistic studies can contribute to how we think about—and *construct*—larger conceptual phenomena, such as symbolism, consciousness, and 'their' evolution.

When we say we want to study palaeolithic 'art' and how it may relate to understanding human symbolic behaviour, we may be able to infer not so much what palaeolithic art 'means' but instead some of the principles by which symbolic evocation 'works', and how it does so in these particular situations. The meanings (both to us and to prehistoric peoples) of symbols come about through use. Meanings may also be derived from such features as the analogies of form, when one item is perceived as somehow being similar to another in a social context. Thus, the uses and re-uses of palaeolithic imagery and of its contexts (such as caves), and the inquiry into the structural principles that may have generated the forms may be the ways in which we will be able to infer some of the principles and processes that generated some of this palaeolithic 'art'.

If we want to inquire into what material symbols 'mean', we must recognize that symbolic knowledge is different from encyclopaedic knowledge (Hodder 1985; Sperber 1974). We cannot go and look up the meaning of a painted pot or of an engraved antler. Material culture items and material symbols are evocative: they evoke conditions, ideas, sentiments, and they 'feed' these evocations right back into the context of social life. They are recursive and reflexive; in social action, humans develop understanding through construction. We could push this idea even further, and suggest that materials can be seen as a way of actualizing, and not merely of reflecting or recording, experience and meanings.

11.8 Some closing thoughts

This chapter has tried to lay out some synopses on the materials and forms that we call 'palaeolithic art' and how researchers have tried to explain these materials and make sense of them within the context of European intellectual history. I have perhaps overemphasized the issues that constrain the study of palaeolithic 'art'. One logical extension of these constraints is that we will never reliably know what the imagery originally meant. The accounts that we do produce will always be archaeo-logical, in that they are our attempts and our constructions in the present to 'make sense' of the past.

Although the 'structuralist break-out' (effected primarily by the work of Leroi-Gourhan) indeed turned palaeolithic 'art' research away, in some important respects, both from the empiricist approach and from thinking there can be a 'direct' reading of the imagery, the necessary theoretical restructuring and re-conceptualization implied has not yet transpired (see comments in for example, Lewis-Williams 1983a, 1984; Preziosi 1982, 1989; Ucko and Layton 1989). And, given that archaeology as a field still has strong preferences for methods that are positivist and empiricist (cf. Hodder 1982, 1986; Wylie 1981), the necessary pluralism and subjectivity that is more obvious when we make inferences about imagery and 'art' are not so easily accommodated by many of our archaeological colleagues (see especially Lewis-Williams 1984).

Even so, we *will* be learning a great deal about the production and uses of palaeolithic imagery in the next decades (see Conkey 1987 for a review of just a few new directions). We will be treated to insights about pigment-processing, engraving techniques, manufacturing sequences, and how caves were used, whether to decorate or to camp in. We are likely to discover some ways to date, at least with more precision, the cave-wall surfaces and, thus, many of the images placed there (Schwarcz, personal communication). We will see many more studies of a comparative and regional, or even cross-regional, sort (for example, Sieveking 1986, 1987; Taborin 1987, Taborin 1992; deBeaune 1987); and, in the challenge to preserve what has remained on cave walls, we will learn much about the techniques and microprocesses of painting and engraving (for example, Aujoulat 1987).

Whoever, among the hundreds of peoples living during the European Upper Palaeolithic, made and used the thousands of images and forms of palaeolithic 'art', indeed left one of the richest and most intriguing sets of cultural materials and forms. The imagery is both familiar and enigmatic; the images are compelling because of their apparent mastery over materials and forms, and because of their suggestive symbolic features. It is both the materialism and the symbolism of these images that have attracted our attention. It is the way in which the makers and users were both materialists and symbolists that has led to their continued celebrated position in our constructions of the past, even if we admit the bias and the subjectivity that have surrounded their emergence and positioning on a centre stage of the human evolutionary play. The challenge of future research is to disengage the materials and their contexts from the (perceived) security of such universal and cross-cultural transformational scenarios, and to conduct simultaneous inquiry into the domain of how meanings may be constructed and enacted—in the contexts of both the Upper Palaeolithic *and* our own lives.

Notes

1. The use of the term *art* is problematic and I use it in quotes ('art') in this chapter to encourage readers to question the term. To call all this imagery 'art' is problematic if only because it presupposes the aesthetic and it classes the imagery in a cultural category of our own, which is a historical and ethnocentric category. By calling it all 'art' we have too easily been led to believe that 'it' is a corpus of symbols and meanings that—perhaps like human culture itself—may be interpreted, and perhaps interpreted definitively.

2. The Chatelperronian is one of many names given to what have been identified as distinctive 'industries' of Palaeolithic materials, usually stone (and bone) tools. These industries have been defined on the basis of certain repetitive patterns in the types of tools

(especially diagnostic ones) and in the relative frequencies with which certain tool types occur in an assemblage. Most industries are named after certain 'type-sites', and many have been subdivided into stages (for example, Magdalenian I to VI). These named industries are almost all based on assemblages from the general Perigord region in south-west France, and the applicability of these labels to assemblages in other, even nearby, regions is always debated (see Fig. 11.3).

The presumed significance and problem of the association between a Neanderthal and the Chatelperronian industry is the idea that the Chatelperronian (albeit somewhat 'transitional' between the Middle Palaeolithic industries and the Upper Palaeolithic ones) had been generally considered closer to an Upper Palaeolithic industry, and therefore, to be implicitly associated with anatomically modern *Homo sapiens*, not with neanderthalers (archaic *H. sapiens*). Does this mean that there was interaction between modern and archaic forms? Did one form learn the others' ways of making tools?

3. The strength and pervasiveness of this hunting-magic interpretation is best exemplified by the way in which Christopher Hawkes (1954) so unquestionably accepts it: 'Paleolithic art *clearly* [*sic*!] has much to do with institutions of hunting magic, and, in the case of the so-called "Venuses", with expressions of desire for human fertility' (C. Hawkes 1954, p. 162).

What is particularly interesting about this kind of acceptance is that this statement of Hawkes's is in a very important article that he wrote, which claims a Comte-ian 'ladder of knowledge' for archaeologists. That is, he argued strongly that there are some aspects of prehistory that are more knowable than others. What is 'fairly easy' to know are aspects of technology, environment, and economics. We can know about social life, he says, with far less certainty, and it is difficult, if possible at all, to know about prehistoric religion and symbolic life. Thus, that he can be so 'certain' about the meanings of palaeolithic art is particularly instructive as to the acceptance of the hunting-magic ideas.

Illustrations

The illustrations for Chapter 11 are to be found on the following pages. Captions by David Phelps [eds].

Prehistoric art 313

Fig. 11.1 An illustration of the 'dynamism' of many of the depictions found in cave-wall paintings: the 'pannel of the leaping cow' from the 'Diverticule axial' at Lascaux, Dordogne, France. The leaping auroch cow is 1.7m long. Probably Magdalenian. Photograph courtesy of Dr. N. Aujoulat, Département d'Art Pariétal, Centre National de Préhistoric Périgueux.

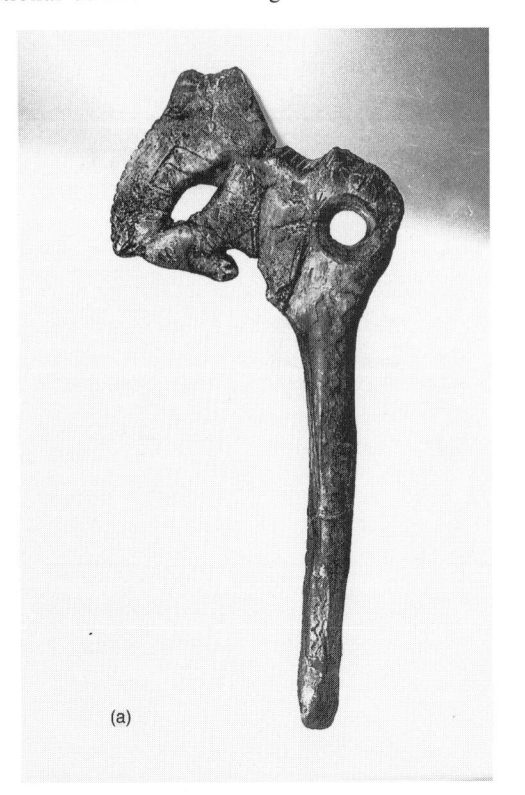

Fig. 11.2 The use of naturalistically carved images in the round in portable art.
(a) The horse baton from Le Mas d'Azil, Ariège, France. Made from perforated antler. The total length of the baton is 20 cm, but that of the head proper, with its raised ears and carefully executed details, is less than 4 cm. The mane is stylized, and the remainder of the shaft is also abstractly decorated. Magdalenian. (b) A carved antler from Le Mas d'Azil, Ariège, France, showing three horseheads, one of which may have been skinned. Possibly a spear-thrower. Magdalenian; length 16.3 cm.
(a) Photograph courtesy of Dr André Alteirac, Le Mas d'Azil, Ariège. (b) Photograph courtesy of the Musée d'Antiquités Nationales, St Germain-en-Laye, Cliché des Musées Nationaux, Paris. © Photo R. M. N.

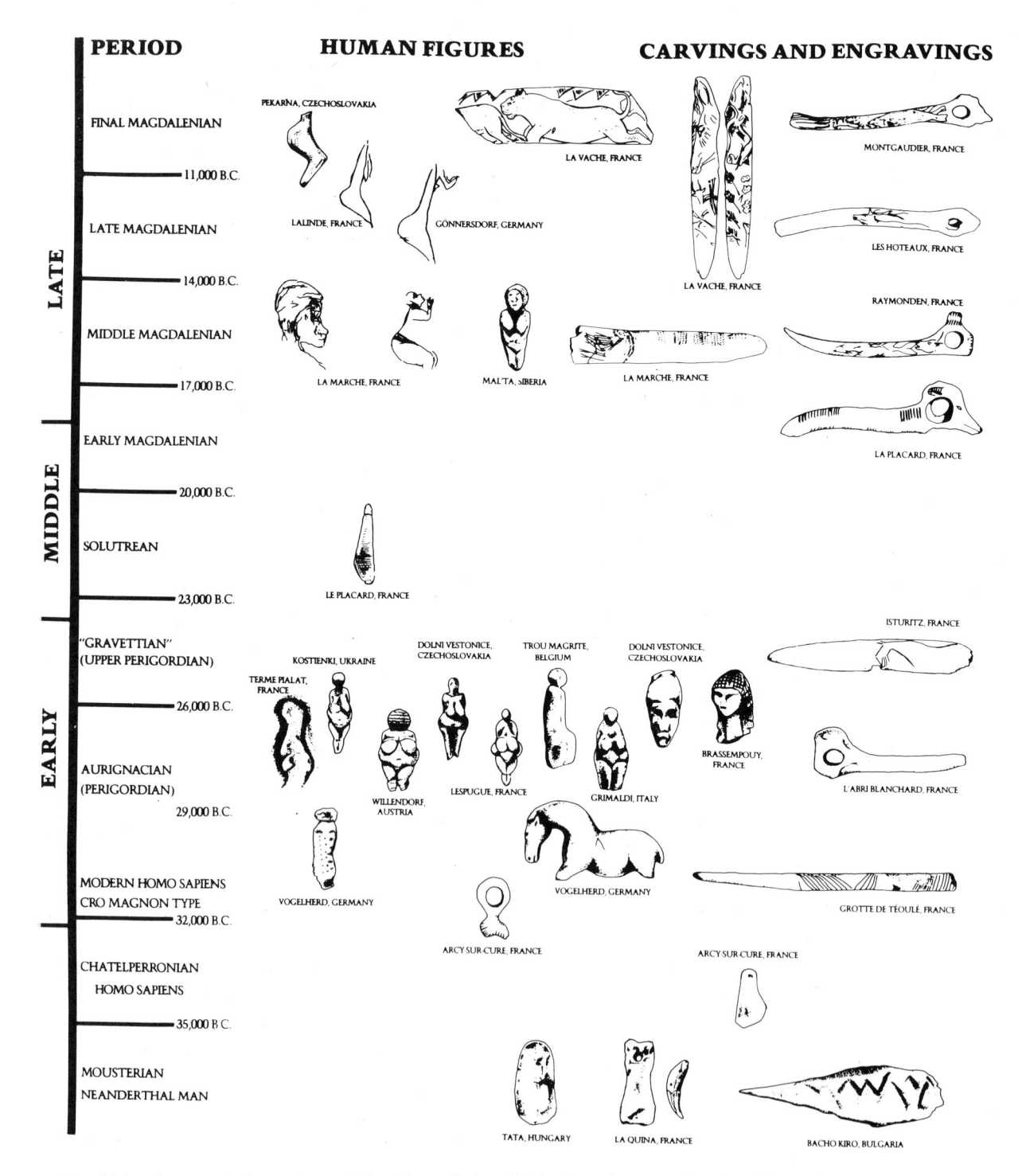

Fig. 11.3 A general chronology of the Upper Palaeolithic. Based on a style chart/chronology prepared by Dr Alexander Marshack for the 1978 *Ice Age Art* exhibition at the American Museum of Natural History. Reproduced courtesy of Dr Marshack.

Prehistoric art

Fig 11.3 (Contd.)

Fig. 11.4 Geometrically engraved or incised spear-points of bone or antler, 9.7 to 14.0 cm long. From the left, the details of the individual spear-points are as follows: (1) from Rivière de Tulle, Lot, France complete; Magdalenian. (2) and (3) from Crozo de Gentillo, Lot, France. Incomplete; Magdalenian. (4) From Jonclas, Lot, France. Incomplete; Solutrean. The marks on No. 1 probably helped to keep the point fixed in the spear-shaft; but Nos. 2–4 have bases with two-sided bevelling, presumably to help fit them into the split end of a spear-shaft, and the marks on them may have served as identifications for individual users or groups of users. All four pieces come from the Logan Museum, Beloit college. Reproduced from White (1986), p. 41, Figs. 37–40, by permission of Professor Randall White.

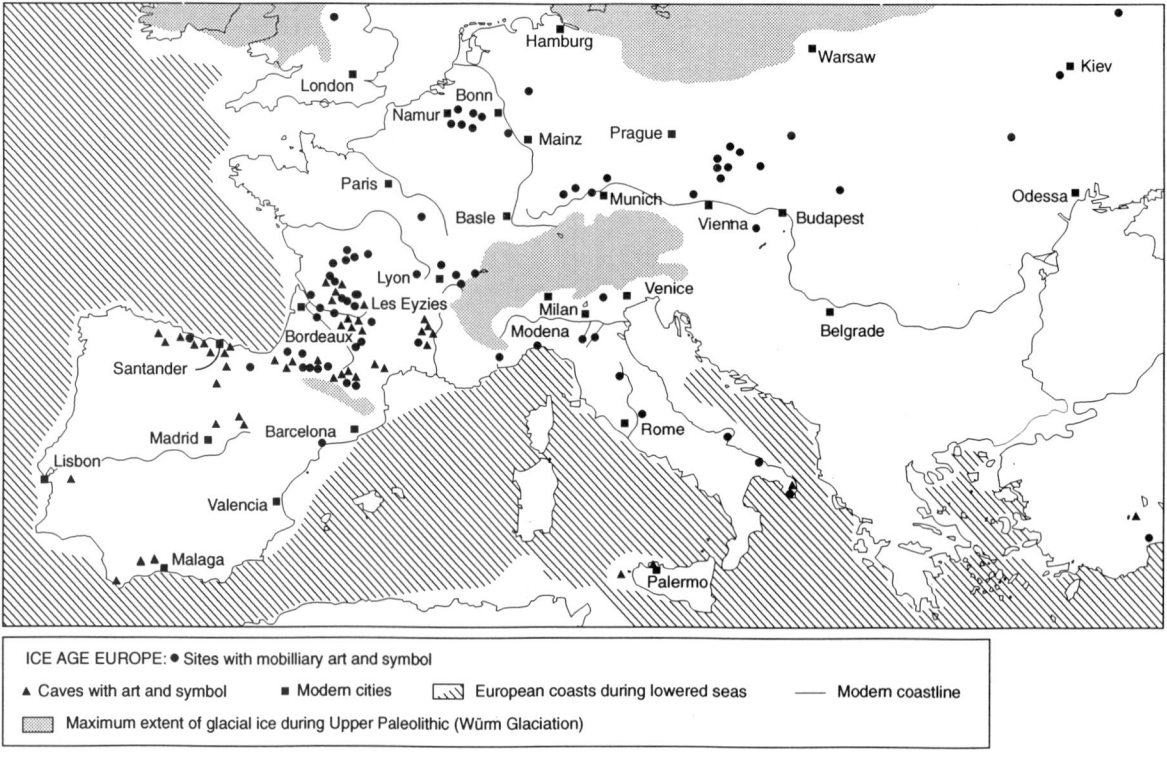

Fig. 11.5 The geographical distribution of Palaeolithic 'art' sites in Europe. After a map prepared by Dr Alexander Marshack for the 1978 *Ice Age Art* exhibition at the American Museum of Natural History. Reproduced courtesy of Dr Marshack.

Fig. 11.6 The Le Tuc bison, from the cave of Le Tuc d'Audoubert, Ariège, France. Modelled in clay. Photograph courtesy of Count Robert Bégouën.

Fig. 11.8 Undecorated Early Aurignacian spear-points. Made from ribs, long bones, or reindeer antler, between 7.0 and 15.8 cm long. Abri Cellier, Dordogne, France. The points have split bases to accommodate the ends of spear-shafts. All four pieces come from the Logan Museum, Beloit College. Reproduced from White (1986), p. 40, figs. 33–6, by permission of Professor Randall White.

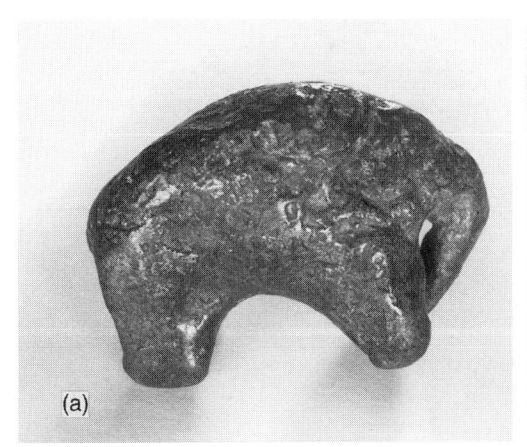

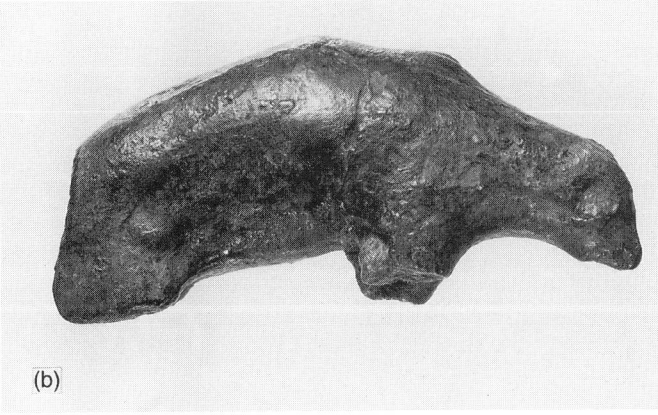

Fig. 11.7 The Dolní Věstonice clay bison. Photographs courtesy of Dr Martin Oliva, Anthropos Institute, Moravské Museum, Brno, Czech Republic.

318   Margaret W. Conkey

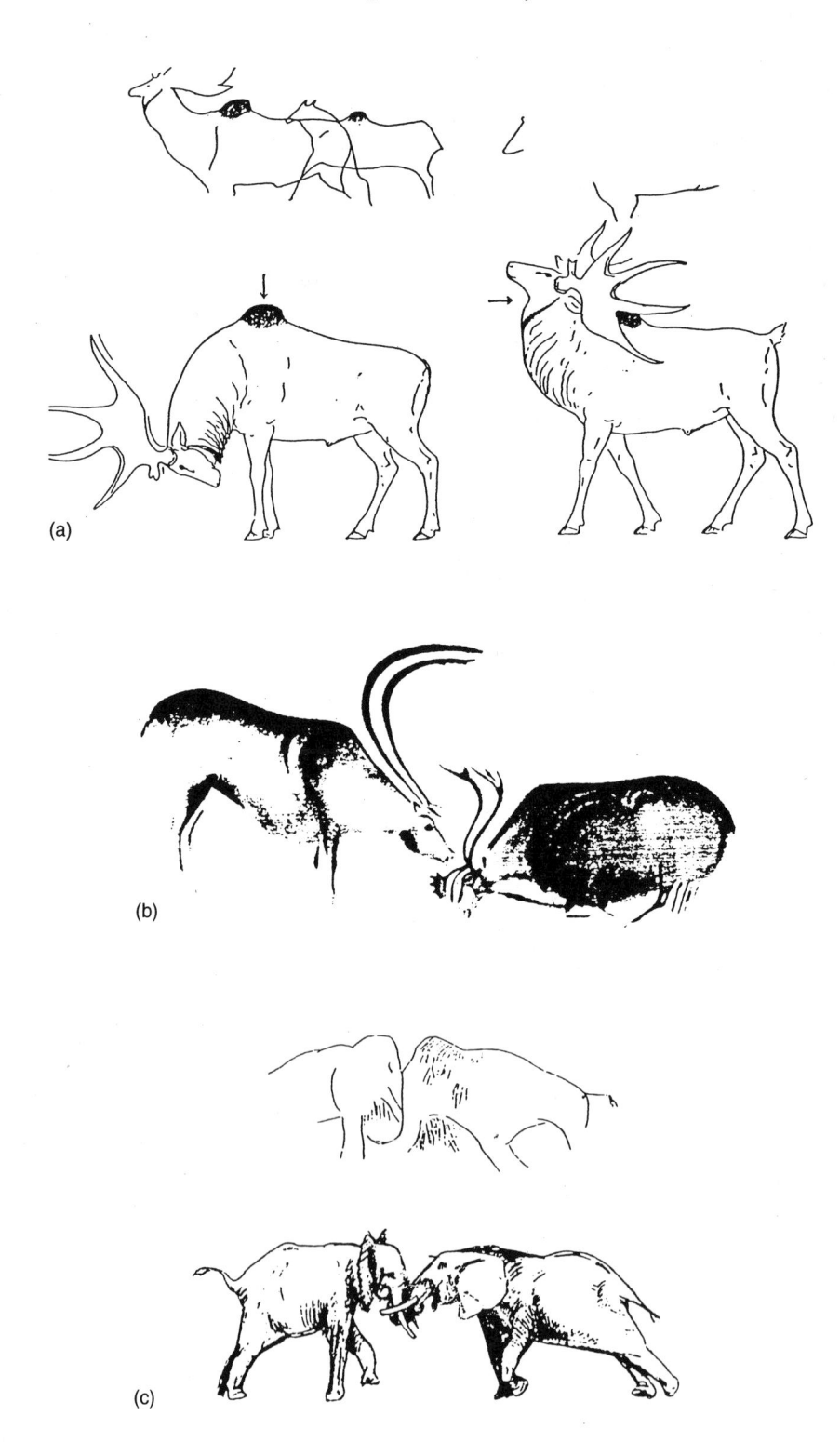

Fig. 11.9 A collection of pieces illustrating the extensive ethnological and anatomical knowledge of the cave artists about the animals they were depicting.

(a) The 'nose-up' agonistic display posture of the giant stag *Megaceros giganteus*. The exaggerated, contrastingly coloured shoulder hump of the male of this species portrayed in the cave drawings indicates an organ used in a display of height or stature, usually associated with a 'head-down' position, as shown in the drawing at the bottom left; whereas the white throat patch outlined in black must have accompanied a 'nose-up' display, as shown at the bottom right. Both these forms of display are known from other ungulates, for many of which the latter is an important threat display, revealing the underside of the anthers by tilting the head upwards with the nose in the air. The unusual palmation of the antlers of *Megaceros* is evidently well adapted to producing the maximum of effect in either of these positions; and the cave-wall drawings from Cognac shown at the top of the figure clearly show the familiarity of the artists with all the principal display features of this species (coloured hump, throat patch, antler palmation), arranged here in what looks distinctly like a preliminary stage of the 'nose-up' display behaviour. Drawings reproduced from Guthrie (1984), Fig. 2, by permission of Editions Universitaires, Fribourg, Fribourg, Switzerland.

(b) Reindeer in confrontation. Copy by Breuil of a polychrome painting at Font-de-Gaume, Dordogne, France. The right-hand animal displays a characteristic brow-ridge, and the smoothness and realism of the modelling of the body masses and the portrayal of the fur is remarkable. Drawing from Rousseau (1984), p. 179, Fig. 2, reproduced by permission of Editions Universitaires Fribourg, Fribourg, Switzerland.

(c) Mammoths in confrontation. A cave-wall drawing from Laugerie-Haute (top), and below it a drawing of modern elephants in confrontation, after Kingdom (1979), p. 60. The parallelism of postures and weight-distributions offers clear evidence of an accurate recollection of first-hand observation. Drawings reproduced from de Spiegeleire (1985); the cave-wall drawing after D. and E. Peyrony (1938), Fig. 18.

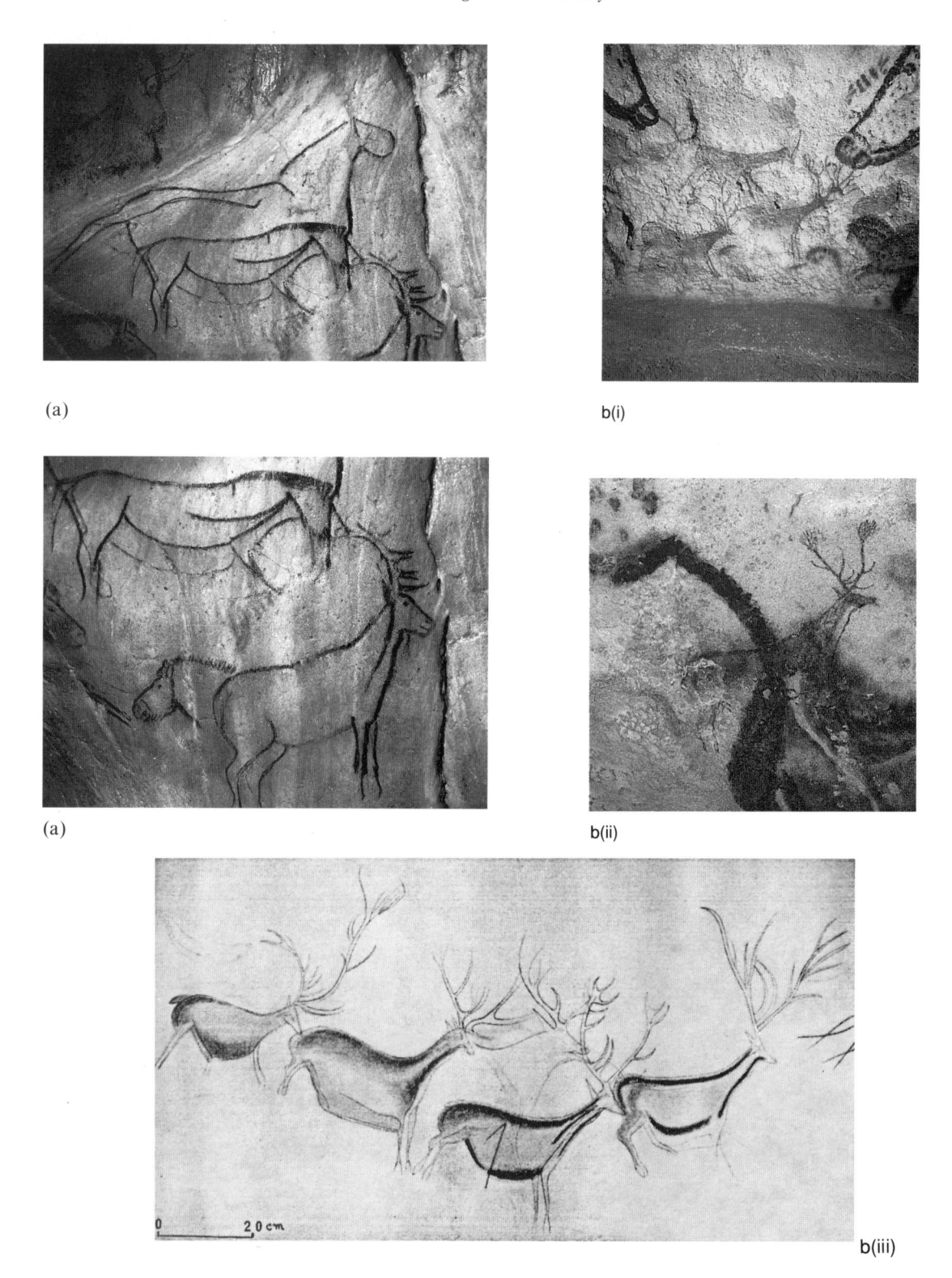

(a)

(a)

b(i)

b(ii)

b(iii)

Fig. 11.10 Illustrations of a considerable degree of variation in the styles employed to produce portraits of different individuals of the same species at a single site.

(a) The stag and neighbouring horses from the Salon Noir at Niaux Ariège, France. Magdalenian. The stag and the horse's head and neck at the same level as the stag are drawn with considerably more elaboration than the horses that are sketched in above them. Photographs courtesy of Dr Jean Clottes.
(b) Three portrayals of deer from Lascaux, Dordogne, France:
 (i) the frieze of the little deer from the Rotunda;
 (ii) the single black deer from the Rotunda; and
 (iii) the frieze of incised and painted deer from the Chamber of the Well Scene (drawing).

A considerable variety can be seen here in the treatment of features such as the antlers and necks: from luxuriant and impressionistic (and outlined rather than solid) in the Well-Scene frieze (iii) to a much more solid and realistic treatment in the single black deer from the Rotunda (ii), with the frieze of the little deer from the Rotunda occupying a somewhat intermediate position between these two, sharing a pattern of pairs of mutually echoing tires (and, to a lesser extent, their long, slender necks) with the deer of the Well-Scene frieze, but with the body outlines and the solid treatment of the antlers closer to the more realistic portrayal of the single black deer from the Rotunda.

Figure 11.10(b)iii is a drawing after A. Glory from Vialou (1984), reproduced by permission of Editions Universitaires Fribourg, Fribourg, Switzerland. Figures 11.10(b)i and 11.10(b)ii are taken from transparencies supplied by CERDAP (the Centre d'Etude et de Recherche Documentaire en Archéologie Préhistorique, Périgueux), courtesy of Dr. N. Aujoulat, Département d'Art Pariétal, Centre National de Préhistoire, Périgueux.

Fig. 11.11 An example of the use of heightened detail in the portrayal of some body parts and not of others: a black outline painting of a mammoth from the cave of Pech-Merle, Lot, France (Length: 60 cm). From a transparency supplied by the Grotte Préhistorique de Pech-Merle. Courtesy of Dr M. Lorblanchet.

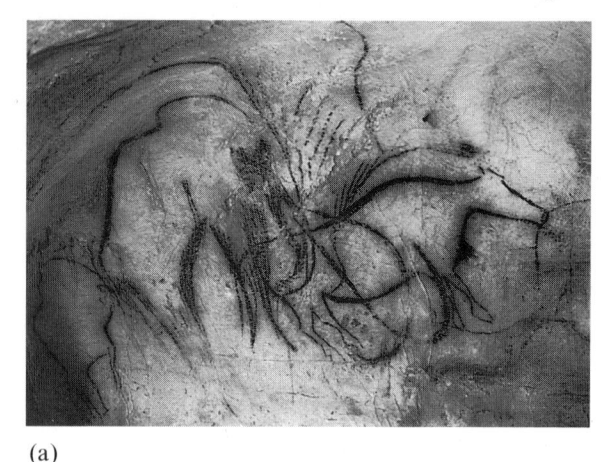

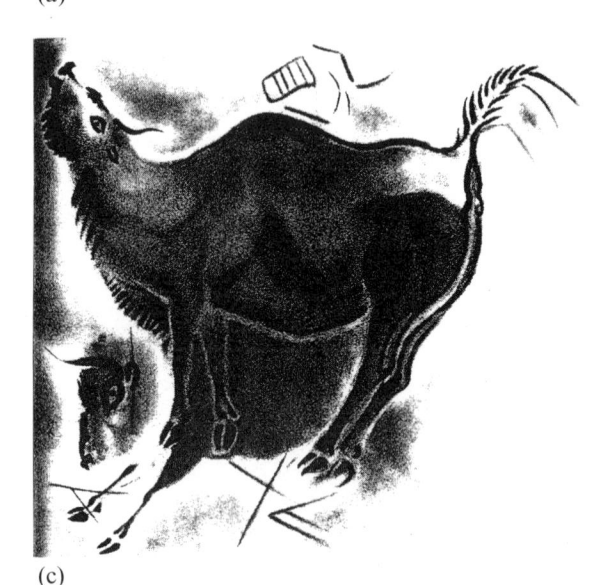

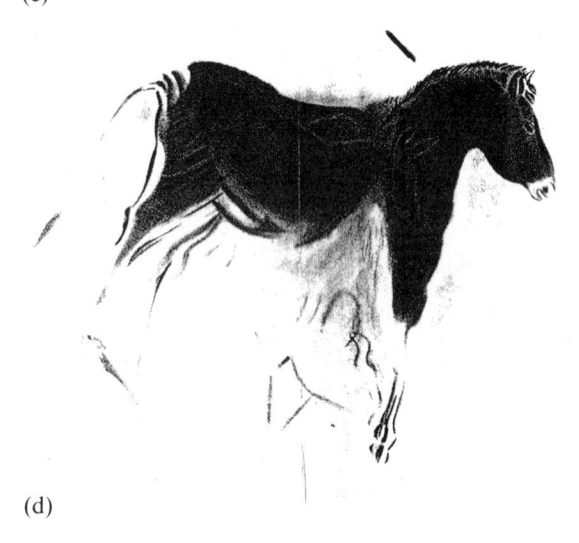

Fig. 11.12 Examples of the superpositioning of one image upon another. See also Fig. 11.10 (a).

(a) An outline figure of a mammoth superimposed on that of an auroch from the black frieze of the cave of Pech-Merle, Lot; France. Solutrean; the mammoth is 1.4 m long. From a transparency supplied by the Grotte Préhistorique de Pech-Merle, courtesy of Dr M. Lorblanchet.

(b) A tracing of some engraved and paraded horses in the Panel of the Handprint at Lascaux, Dordogne, France, with much overdrawing and additional eyes, ears, muzzles, and legs. Probably Magdalenian. The horse covered in arrowlike marks is about a metre long. After Glory, reproduced from Bahn and Vertut (1988) following Leroi-Gourhan (1979), by permission of Paul Bahn and Editions du CNRS.

(c) A bellowing bison, standing, superimposed on another figure of a running bison, with various signs. Altamira, Santander, Spain, after Breuil.

(d) A figure of a red horse, with the superimposed figure of a hind. Altamira, Santander, Spain, after Breuil.

(e) A galloping boar, with a walking boar superimposed: Altamira, Santander, Spain, after Breuil.

(c), (d), and (e) reproduced from Lafuente Ferrari (1981), after Breuil and Obermaier (1935), by permission.

Fig. 11.13 A group of bison from the Great Ceiling at Altamira, Santander, Spain (the group is towards the top right of the ceiling as a whole as shown in the panoramic view of Fig. 11.26 below, the paler of its two central figures being two to the left of the boar at the extreme top right), illustrating the 'lack of framing' of images characteristic of many large-scale Palaeolithic paintings—i.e. the frequent lack of relation between the planes in which adjacent subjects appear to be portrayed, so that there is no evident compositional relationship between the subjects in any straightforward modern pictorial sense, in terms either of naturalistic scene-painting or of obvious decorative symmetry or balance. (This is not to deny that there may be a large measure of compositional balance in a subtler sense and over a wider compass, but merely to point out the absence of such phenomena as for example the simple decorative mutual mirroring of a pair of animals that one would often find on, say, a Greek vase of the Corinthian period.)

None of the four bison of which substantial portions can be seen in this photograph (the centre-right complete dark-red figure with a dark hump; the other, paler, complete figure whose hindquarters lie against the dark-red bison's back and neck; another dark-coloured bison whose tail can be seen at the top left in front of the paler complete bison's forequarters; and another paler bison, whose tail—showing as two roughly parallel curved lines—and whose hindquarters can be seen at bottom left, on a line running roughly from about 10 o'clock to 4 o'clock) appears to stand in a plane that bears any relation to that of any of the others (see also Fig. 11.26). Photograph courtesy of the Museo Arqueologico, Madrid.

Fig. 11.14 An illustration of a (rare) possible 'scene' composition: the so-called 'well' or 'shaft' scene at Lascaux, Dordogne, France. Reproduced from a transparency supplied by CERDAP, courtesy of Dr N. Aujoulat, Département d'Art Pariétal, Centre National de Préhistoire, Périgueux.

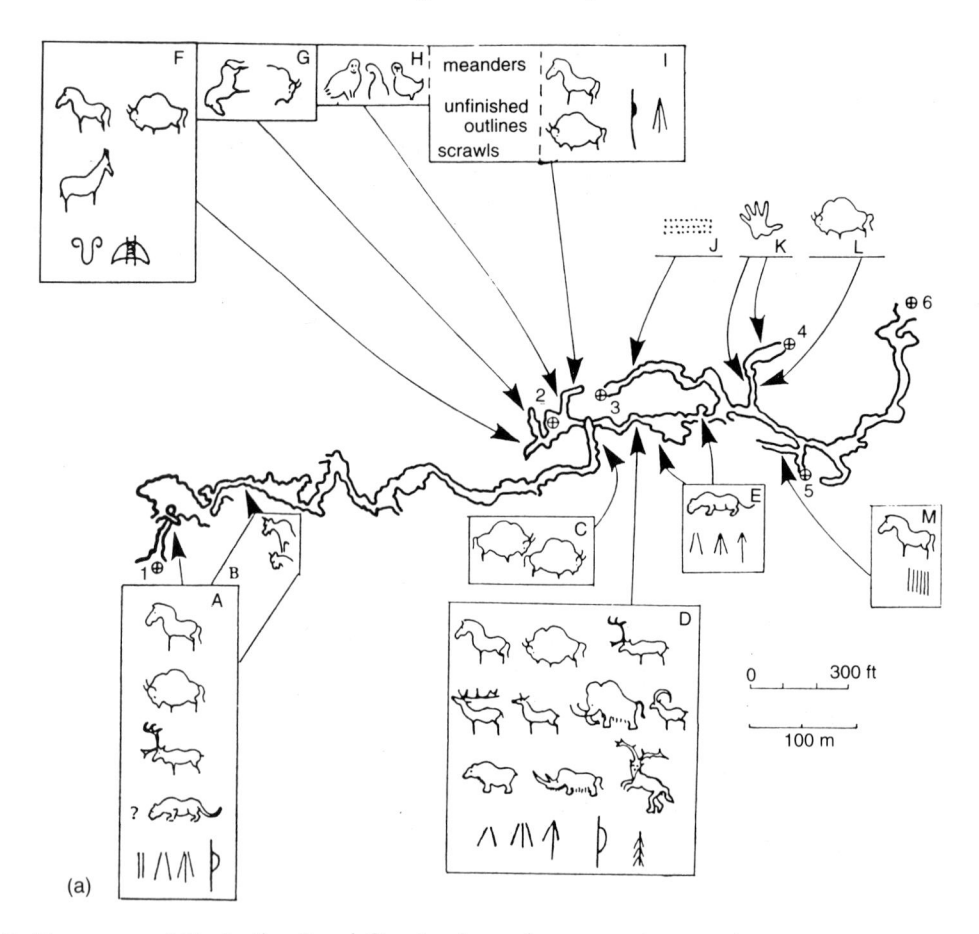

Fig. 11.15 Two ways of illustrating Leroi-Gourhan's work on cave topography.
(a) A schematic map (cave layout) of the subjects portrayed in the cave art of the two associated caves of Les Trois Frères and Le Tuc d'Audoubert, Montesquieu-Avantés, Ariège, France, indicating where each type of portrayal is to be found. Reproduced from Leroi-Gourhan (1965), p. 366, Fig. 155, by permission of Harry N. Abrams, Inc., together with Leroi-Gourhan's original caption to Fig. 155.

	BISON	OX	HORSE	MAMMOTH	HIND	REINDEER	STAG	IBEX	BEAR	FELINE	RHINOCEROS	TOTAL ANIMALS	OVAL	QUADRANGULAR	BRACE-SHAPED	CLAVIFORM	TECTIFORM	WOUND	TOTAL FEMALE SIGNS	MALE FIGURES	BARBED SIGNS	DOTS	STROKES	TOTAL MALE REPRESENTATIONS	HANDS
CENTRAL COMPOSITION	148	46	198	29	3	5	4	3	2	2	4	444	15	17	8	5	6	16	67	3	19	11	36	69	7
SIDE CHAMBER	6	2	5	6	4	2	–	7	3	2	1	38	8	11	12	7	3	2	43	4	1	8	11	24	1
PERIPHERY	–	–	–	10	18	10	13	49	7	1	2	110	3	–	–	1	–	–	4	7	15	8	25	55	–
ENTRANCE	–	1	4	–	5	2	10	3	–	–	–	25	2	1	–	–	3	–	6	–	1	15	13	29	5
PASSAGEWAY	2	–	4	2	3	3	3	4	7	5	1	34	1	–	–	–	–	1	2	6	11	7	29	53	1
BACK OF CAVE	5	1	17	3	3	3	14	9	4	8	1	68	4	4	–	1	1	–	10	12	4	6	14	36	–
TOTAL LOCATIONS	161	50	228	50	36	25	44	75	23	18	9	719	33	33	20	14	13	19	132	32	51	55	128	266	14

PERCENTAGE

	BISON	OX	HORSE	MAMMOTH	HIND	REINDEER	STAG	IBEX	BEAR	FELINE	RHINOCEROS	TOTAL ANIMALS	OVAL	QUADRANGULAR	BRACE-SHAPED	CLAVIFORM	TECTIFORM	WOUND	TOTAL FEMALE SIGNS	MALE FIGURES	BARBED SIGNS	DOTS	STROKES	TOTAL MALE REPRESENTATIONS	HANDS
CENTRAL COMPOSITION	91	92	86	58	8	20	9	4	8	11	44	61	45	51	40	35	46	84	50	9	37	20	28	25	50
SIDE CHAMBER	3	4	2	12	11	8	–	9	13	11	11	5	24	33	60	50	23	10	32	12	2	14	8	9	7
PERIPHERY	–	–	–	20	50	40	29	65	30	5	22	14	9	–	–	7	–	–	3	21	29	14	19	20	–
ENTRANCE	–	2	1	–	13	8	22	4	–	–	–	3	6	3	–	–	23	–	4	–	2	27	10	10	35
PASSAGEWAY	1	–	1	4	8	12	6	5	30	27	11	4	3	–	–	–	–	5	1	18	21	12	22	20	7
BACK OF CAVE	3	2	7	6	8	12	31	12	17	44	11	9	12	12	–	7	7	–	7	37	8	10	10	13	–

(b) General statistics of the frequency of occurrence of various types of subject in cave art, drawn from a study of over 200 caves. Reproduced from Leroi-Gourhan (1965), p. 506, Chart XIX, by Harry N. Abrams, Inc.

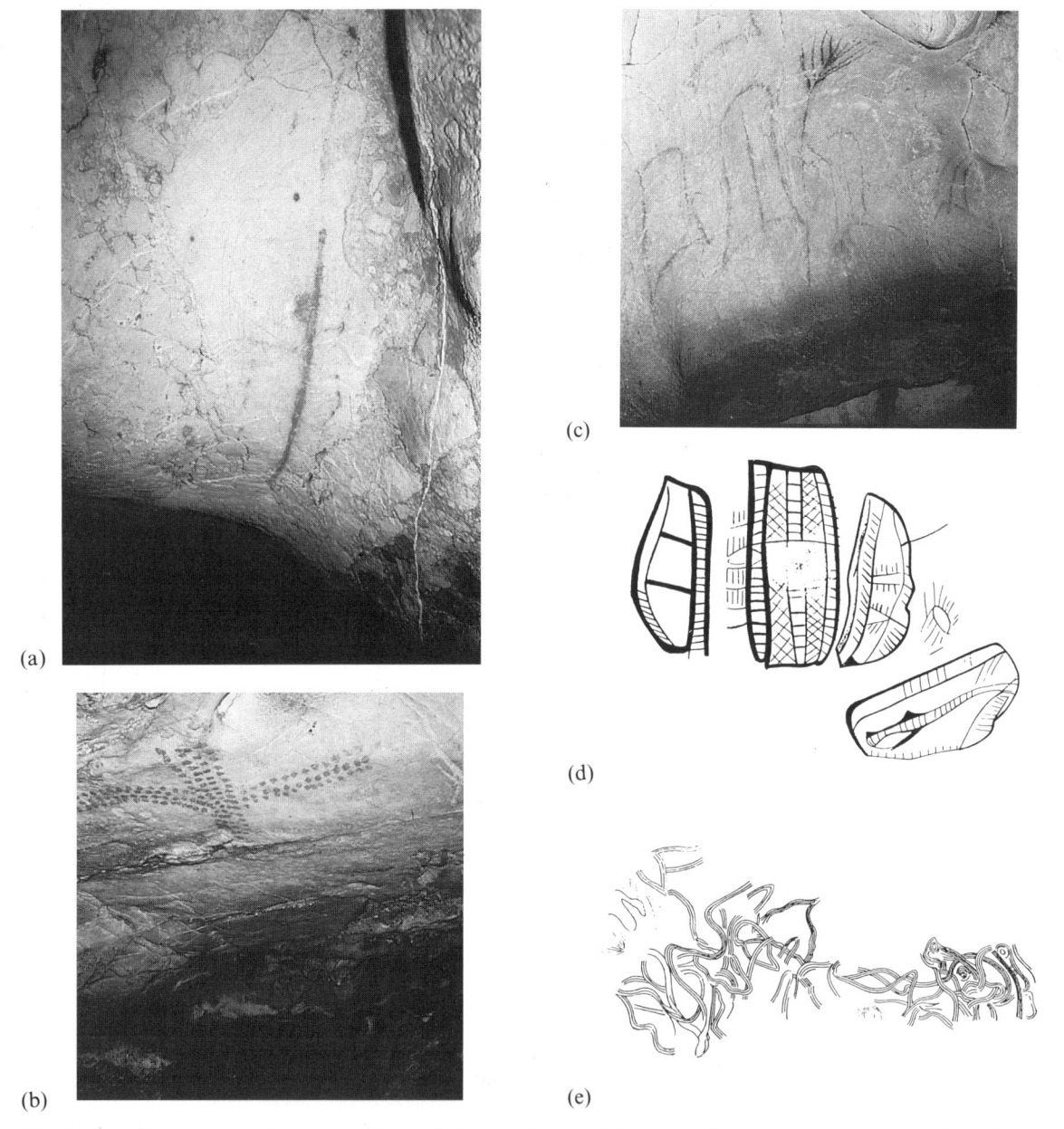

Fig. 11.16 Illustrations of some of the varieties of abstract 'signs', symbols, or patterns found in Palaeolithic cave 'art' along with the animal and human representations.
(a) A red 'claviform' (key-shaped) 'sign' in the cave of Les Trois Frères, Ariège, France, painted on a surface previously prepared by scraping. The mark is 76 cm in length, and probably Magdalenian. Photograph courtesy of Count Robert Bégouên, by permission of the Association Louis Bégouên.
(b) A 'sign' in the form of a pattern of dots from the cave of Castillo, Santander, Spain.
(c) A vulviform 'sign' from the cave of Castillo, Santander, Spain.
(d) 'Tectiform' (roof-like) 'signs' from the terminal corridor of the cave at Altamira, Santander, Spain. Aurigracian.
(e) 'Macaronic' (meander-like) designs from a gallery of the cave at Altamira, Santander, Spain.
(d) and (e) reproduced from Lafuente Ferrari (1981), after Breuil and Obermaier (1935), by permission.

Prehistoric art 327

(a)

(b)

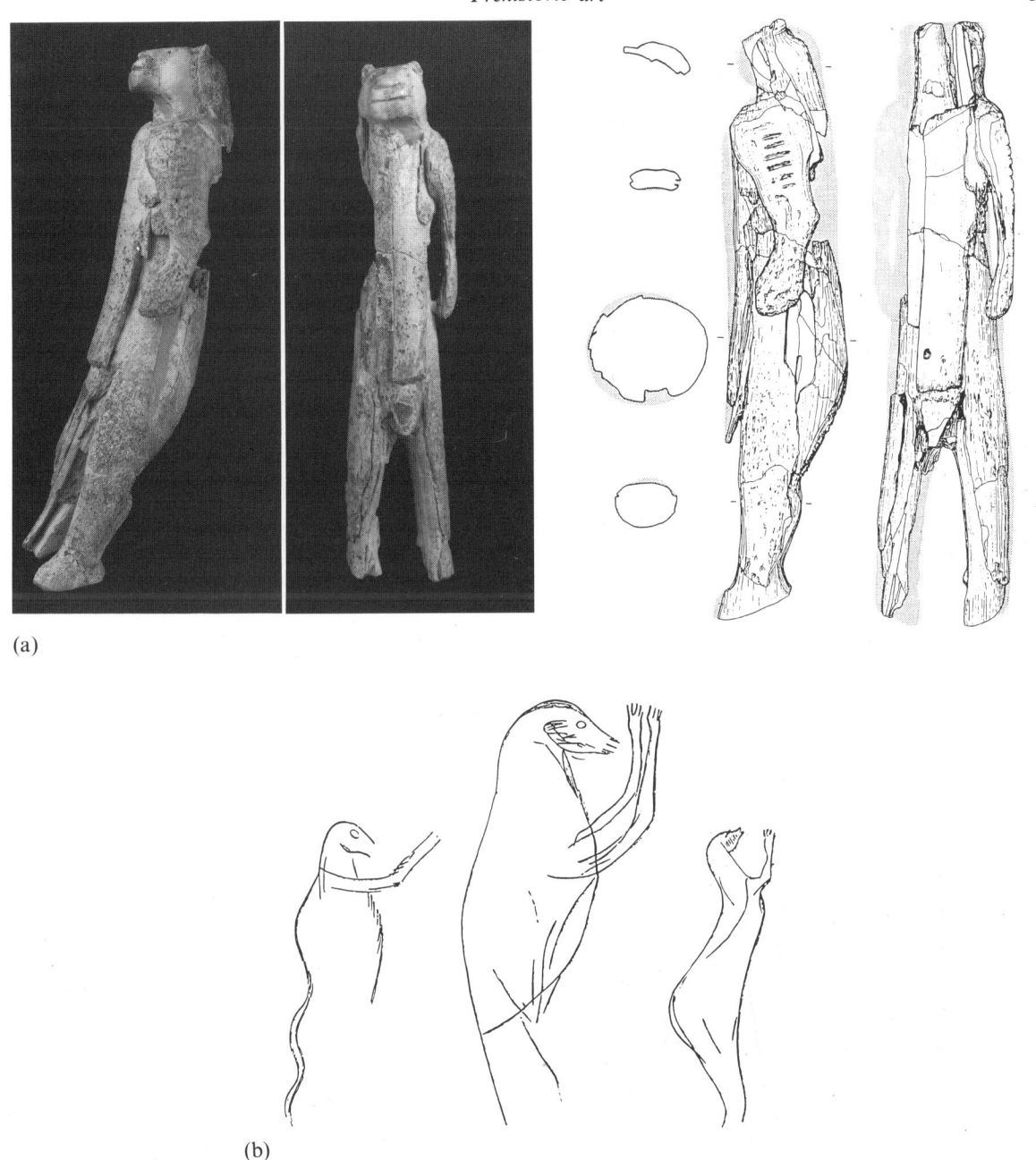

Fig. 11.17 Illustrations of examples of ambiguous or combined (part human, part animal) images of a type sometimes found in Palaeolithic cave art (cf. also Fig. 11.19 below).
(a) The ivory statuette of a human figure with the head of a lioness for Hohlenstein—Stadel, Württemberg, Germany. Aungnacian; 28.1 cm high. Frontal and lateral photographs and drawings courtesy of Dr Kurt Wehrberger, Ulmer Museum. The areas shaded with dots in the drawings indicate the reconstructed dimensions of the figure in its hypothesized original condition.
(b) Anthropomorphic figures (? part human, part seal: 'Silkies'?) engraved on the Great Ceiling of the cave at Altamira, Santander, Spain. Aurignacian. Reproduced from Lafuente Ferrari (1981), after Breuil and Obermaier (1935), by permission.

328 Margaret W. Conkey

Fig. 11.18 Images from various French caves of highly schematic visages and figures of apparently human, animal, aviform, or ambiguous character of the type known as 'ghosts'. Line drawings reproduced from Leroi-Gourhan (1965), p. 512, Chart XXX, by permission of Harry N. Abrams, Inc.

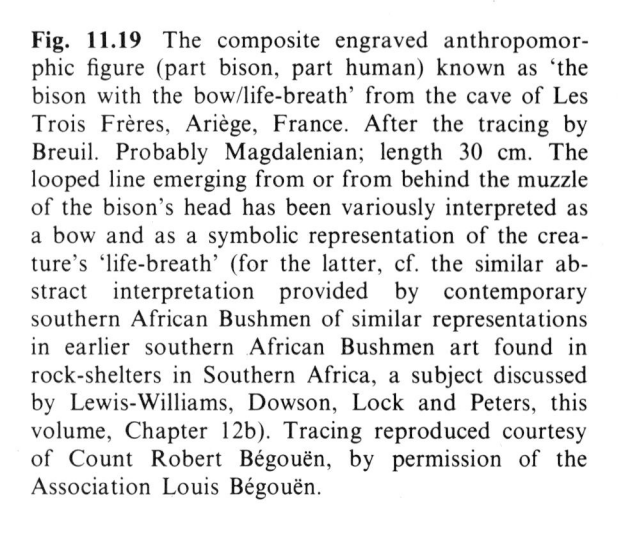

Fig. 11.19 The composite engraved anthropomorphic figure (part bison, part human) known as 'the bison with the bow/life-breath' from the cave of Les Trois Frères, Ariège, France. After the tracing by Breuil. Probably Magdalenian; length 30 cm. The looped line emerging from or from behind the muzzle of the bison's head has been variously interpreted as a bow and as a symbolic representation of the creature's 'life-breath' (for the latter, cf. the similar abstract interpretation provided by contemporary southern African Bushmen of similar representations in earlier southern African Bushmen art found in rock-shelters in Southern Africa, a subject discussed by Lewis-Williams, Dowson, Lock and Peters, this volume, Chapter 12b). Tracing reproduced courtesy of Count Robert Bégouën, by permission of the Association Louis Bégouën.

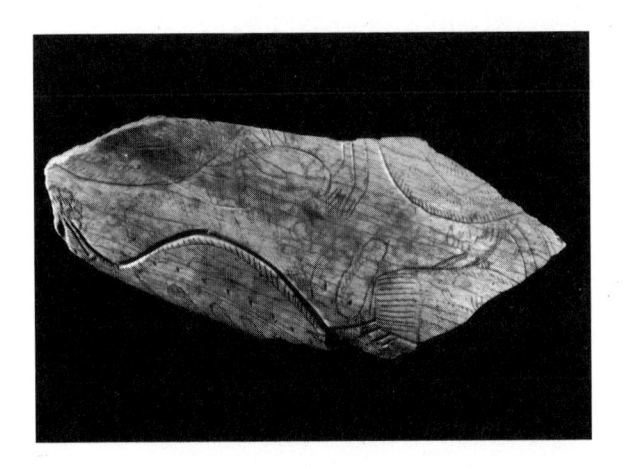

Fig. 11.20 An illustration of the unusual subjects and diverse associations of subjects (when compared with the limited range of large animal, human, anthropomorphic/composite, and abstract subjects commonly to be found in the cave wall-paintings: cf. Figs. 11.15(a), (b), Table 11.1 (animals); Fig. 11.14 (humans); 11.17(b), 11.18, 11.19 (composite/anthropomorphic), and 16 (a), (b), (c), and 11.22 (a), (b) (abstract) that are occasionally to be found portrayed on the portable art-objects found in the caves. Carved representations of birds and a grasshopper on a small plaque found in the cave of Les Trois Frères, Montesquieu—Avantès, Ariège, France (photograph and outline drawing to clarify the positioning of the creatures portrayed.) Photograph courtesy of Count Robert Bégouën; photograph and drawing reproduced by permission of the Association Louis Bégouën. Drawing from Bégouën and Bégouën (1928).

It is also hard to avoid noticing the rarity of the appearances of simple straightforward portrayals of human figures (cf. Fig; 11.14)—as opposed to combined/anthropomorphic portrayals, 'ghosts', etc.—in Palaeolithic parietal art, if this is set in contrast with Mesolithic and Neolithic Saharan rock art, Southern African Bushmen rock art (cf. Lewis-Williams, Dowson, Lock and Peters, this volume, Chapter 12b), or Palaeolithic mobiliary art (cf. Figs. 11.23, 11.30).

It is hard, then, to avoid the conclusion that the parietal art's exceptional concentration on large herbivores in general, and bovines (bison, aurochs) on the one hand and equines on the other in particular (cf. Table 11.1), to the near-total or total exclusion of various other fairly obvious possible types of representation (such as birds, fish, people, or carnivores) must have carried some significance, even if we remain sceptical as to our prospective ability ever to determine exactly what that significance was.

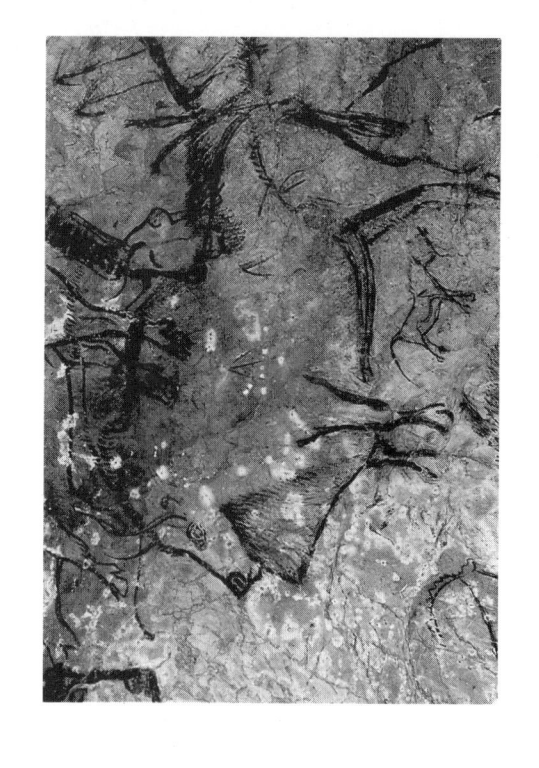

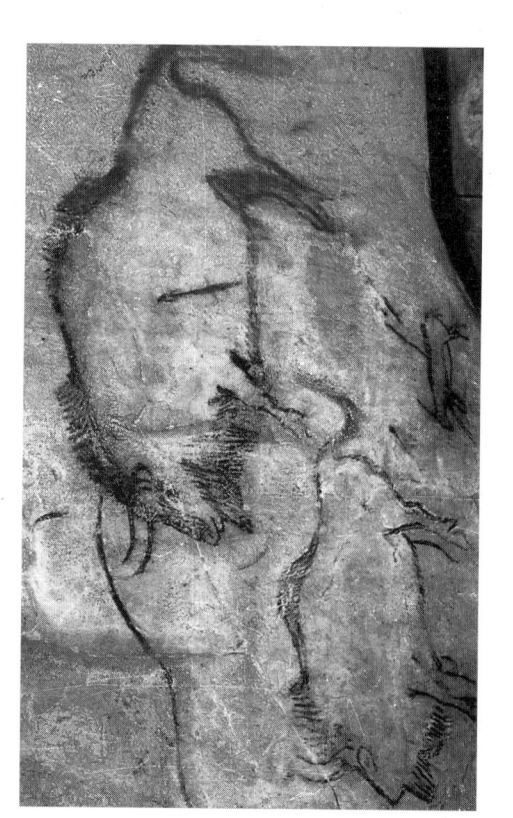

Fig. 11.21 Four specimens of reasonably 'convincing' illustrations of spear-wound marks on representations of living animals. Photographs courtesy of Dr Jean Clottes.

CHART I.
TOPOGRAPHICAL DISTRIBUTION OF MALE AND FEMALE SIGNS

This chart shows that female signs are found almost exclusively in central cave areas, and male signs principally in lateral areas.

		CENTRAL			LATERAL				
		CENTER	SIDE CHAMBER	TOTAL	PERIPHERY	ENTRANCE	PASSAGE	BACK	TOTAL
MALE SIGNS	STROKE	46	11	57	25	13	29	14	81
	BRANCHING SIGN	19	1	20	15	1	11	4	31
	DOT	11	8	19	8	15	7	6	36
	TOTAL	76	20	96	48	29	47	24	148
FEMALE SIGNS	RECTANGLE	17	11	28		1		4	5
	BRACE-SHAPED	8	12	20					
	TECTIFORM	6	3	9		3		1	4
	CLAVIFORM	5	7	12	1			1	2
	OVAL	15	8	23	3	2	1	4	10
	WOUND	16	2	18			1		1
	TOTAL	67	43	110	4	6	2	10	22

CHART II.
TYPOLOGY OF FEMALE SIGNS

A and B. TRIANGULAR SIGNS

In A 3 and B 3 we see "wound" or "arrow" signs. The forms derived from B correspond to the variants of tectiforms in the Les Eyzies region. Other forms are found among the derivations in E.

C. OVAL SIGNS

D. QUADRANGULAR SIGNS
  1. PÉRIGORD-PYRENEES GROUP
  2. CANTABRIAN GROUP

E. CLAVIFORM SIGNS

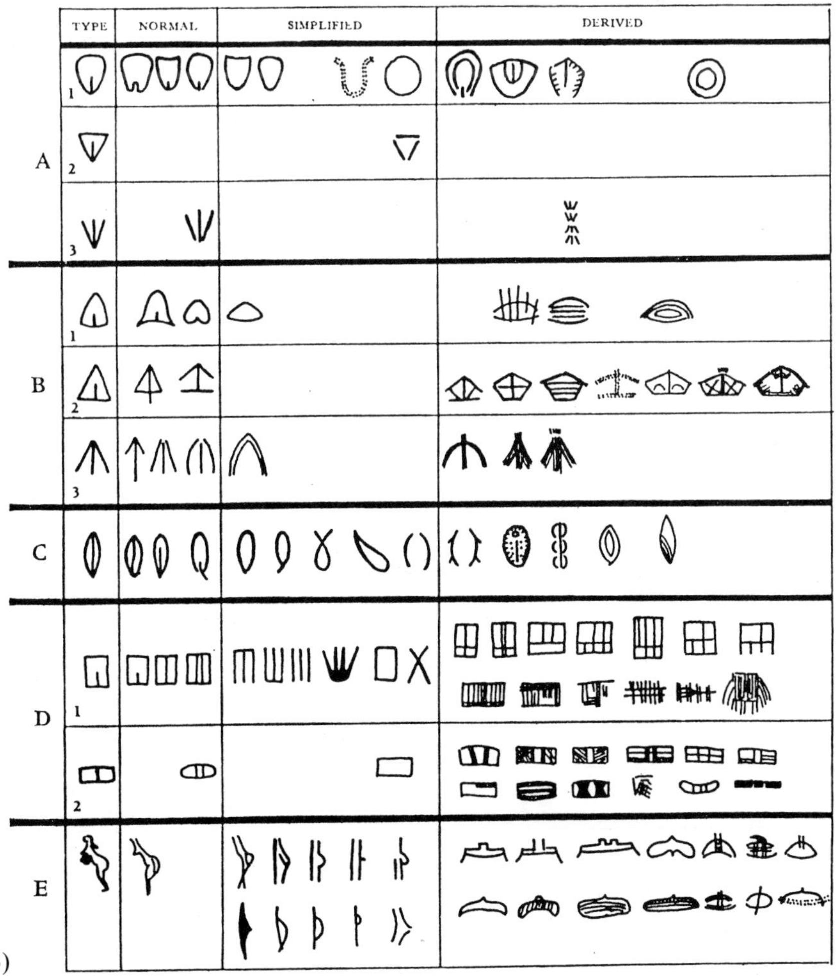

This chart shows typological characteristics only. Obviously, in the case of the derived forms (especially for E), there must have occurred borrowings and cross-cultural influences both between epochs and regions.

Fig. 11.22 (Caption on p. 333)

332   Margaret W. Conkey

CHART III.   TYPOLOGY OF MALE SIGNS

A. HOOKED OR "SPEAR-THROWER" SIGNS.

B. BARBED SIGNS.

C. SINGLE AND DOUBLE STROKES.

D. DOTS AND ROW OF DOTS, SINGLE AND DOUBLE.

CHART IV.

EXAMPLES OF PAIRED SIGNS

A. From left to right: LA PILETA, FONT-DE-GAUME, EBBOU, LA FERRASSIE, PECH MERLE, LA MEAZA.

B. NIAUX, LE GABILLOU, LASCAUX (Chamber of Felines), LASCAUX (Nave), LAS MONEDAS, LA CROZE.

C. FONT-DE-GAUME, EL CASTILLO, LA MOUTHE, BERNIFAL, OULEN.

D. USSAT, ARCY-SUR-CURE, SALLÈLES-CABARDÈS, SAINT-MARCEL, LABASTIDE, LE PORTEL.

E. LES COMBARELLES (group 69), LES COMBARELLES (group 105), USSAT, BERNIFAL, OULEN, OULEN.

F. LASCAUX (Chamber of Felines), LASCAUX (same), LASCAUX (Axial Gallery), LASCAUX (same), LE GABILLOU, LE GABILLOU.

G. LE PORTEL (Gallery 2), ALTAMIRA (terminal gallery), EL CASTILLO, LAS CHIMENEAS, ALTAMIRA, MARSOULAS (half-rounded rods).

H. ALTAMIRA (Painted Ceiling), LAS MONEDAS, NIAUX, LA CULLULVERA, NIAUX, NIAUX.

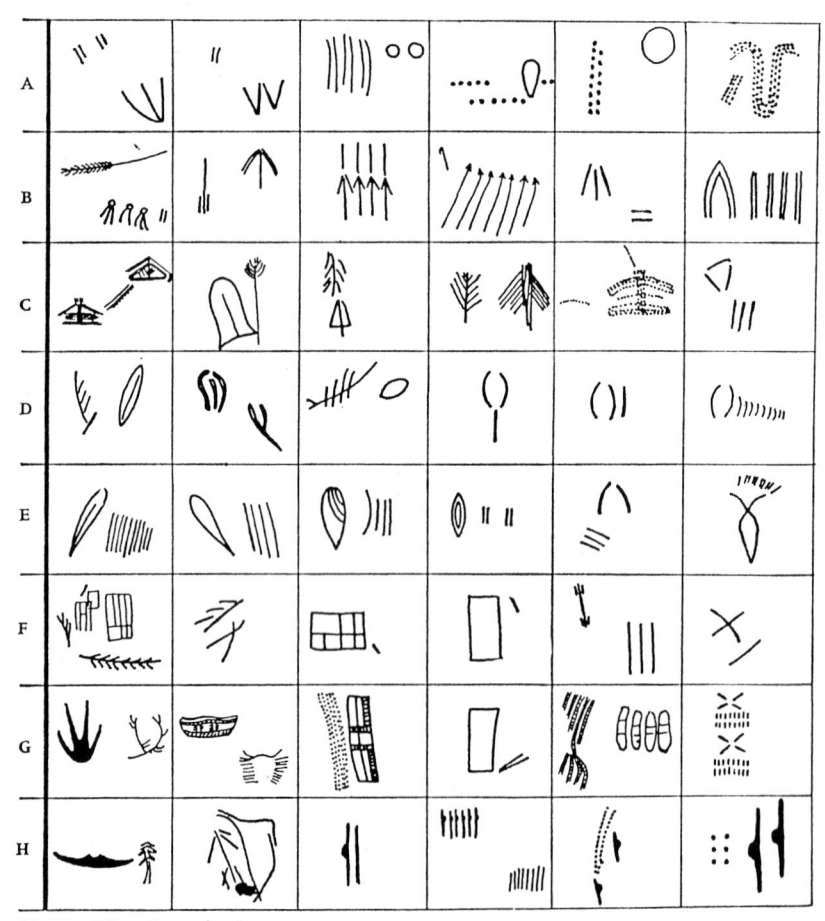

Fig. 11.22 (Continuation)

(A) Realistic depictions	(B) Tectiform, aviform, accoladed markings	(C) Quadrilateral markings	(D) Women and claviform marking
La Ferrassie *Dordogne*	True tectiform marking Font-de-Gaume *Dordogne*	Lascaux *Dordogne*	Pech-Merle *Lot*
The Cellier shelter *Dordogne*	Aviform marking Cougnac *Lot*	La Mouthe *Dordogne*	Les Combarelles *Dordogne*
La Ferrassie *Dordogne*	Accoladed marking Le Portel *Ariège*	Ussat *Ariège*	Lalinde *Dordogne*
Les Combarelles *Dordogne*	Accoladed marking Eniene *Ariège*	Las Chimeneas *Santander*	Labastide *Hautes-Pyrénées*
Pergouset *Lot*	Accoladed marking Oullins *Ardèche*	El Castillo *Santander*	Niaux *Ariège*
El Castillo *Santander*	Accoladed marking La Pasiega *Santander*	La Pasiega *Santander*	Les Trois-Freres *Ariège*
Tito Bustillo *Asturias*	Accoladed marking La Pasiega *Santander*	Altamira *Santander*	Le Portel *Ariège*

(e)

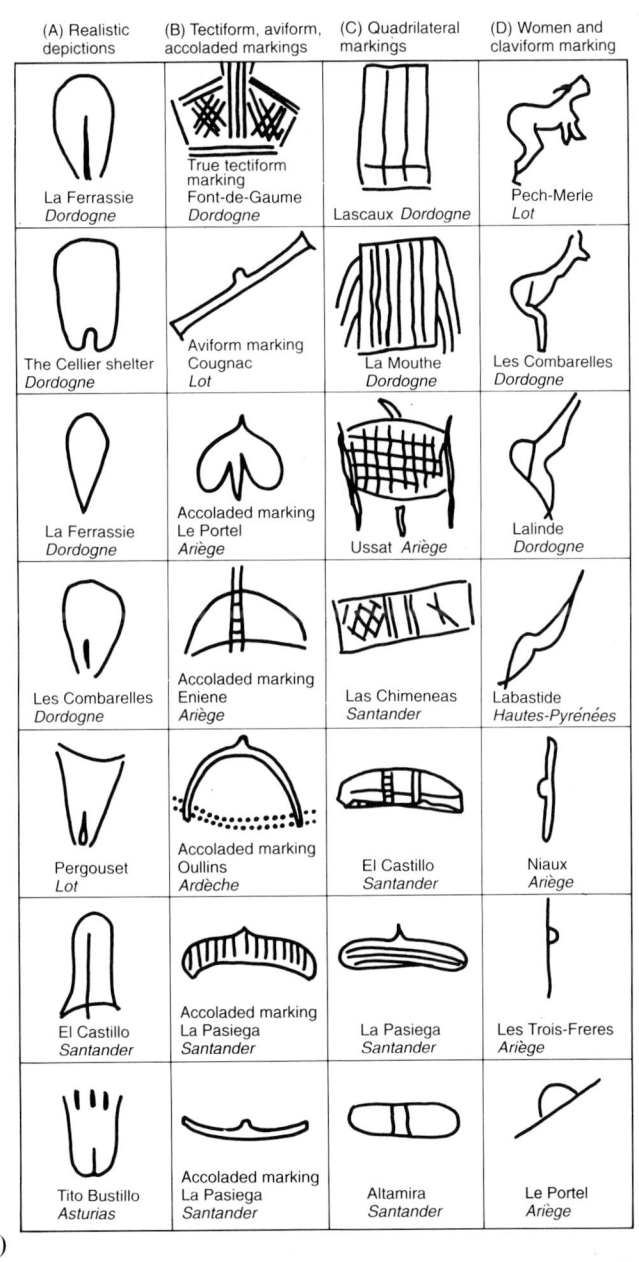

Fig. 11.22 Leroi-Gourhan's distinction between 'male' and 'female' signs; his charts illustrating examples of this distinction.
(a) The topographical distribution of 'male' and 'female' 'signs' within caves: Leroi-Gourhan's Chart I.
(b) A catalogue of the types of sign classified as 'female': Leroi-Gourhan's Chart II.
(c) A catalogue of the types of sign classified as 'male': Leroi-Gourhan's Chart III.
(d) 'Paired signs' ('male' and 'female' signs found in close juxtaposition): Leroi-Gourhan's Chart IV.
(a)–(d) reproduced from Leroi-Gourhan (1965), pp. 513–14, charts I–IV, by permission of Harry N. Abrams, Inc.
(e) The principal types of 'full' signs. Reproduced courtesy of CNRS from Leroi-Gourhan (1978).

Fig. 11.23 Illustration of a style of rendition of human imagery on a plaquette that differs from that normally found in conjunction with the animal images in mural or mobiliary art: a tracing of an engraved plaquette from La Marche, Vienne, France, with the face of a bearded man extracted from the mass of lines. Magdalenian. Reproduced from Airvaux and Pradel (1984).

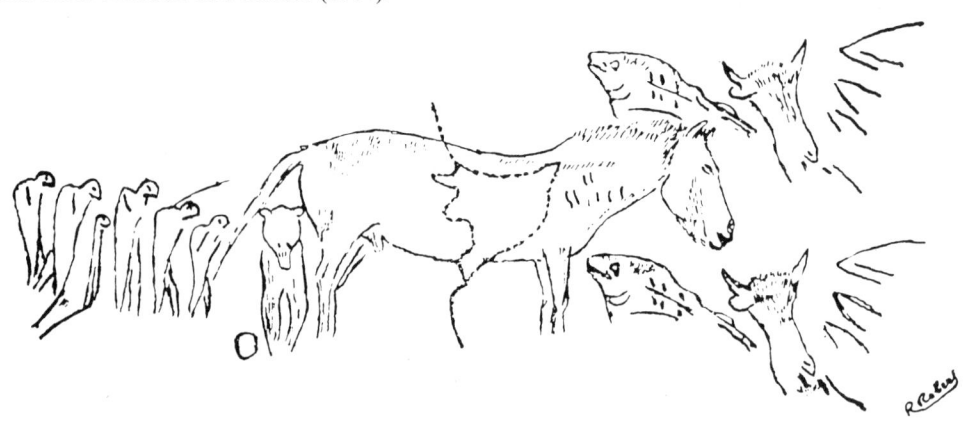

Fig. 11.24 A carving on bone from the cave of La Vache, Ariège, France, illustrating how animals can be depicted with different degrees of naturalistic detail on the same piece. A schematic drawing by Romain Rolland giving a flat projection of the design as a whole. The depiction here has been seen as an initiation scene. Because the design is carved continuously around the surface of the bone it is uncertain in what position some details are intended to be seen. Thus the 'fish' and the bison's head at the far right are both shown in the drawing mapping out the design as a whole as appearing both above and below the head of the great horse. The most logical reading would appear to be to imagine the 'fish' in the lower position and the bison in the upper, one on either side of the head of the great horse. The contrast is striking between the degree of naturalistic detail with which the animals, and particularly the great central horse, are depicted and the stylized and schematic outlines used for sketching in the small anthropomorphic figures at the left (anthropomorhic at least in so far as they are bipedal: but cf. the seal-like 'anthropomorphs' of the Altamira Great Ceiling shown at Fig. 11.17(b) above). Drawing by the late Romain Rolland, reproduced by permission of M. Jean Rolland, Tarascon, Ariège.

Prehistoric art 335

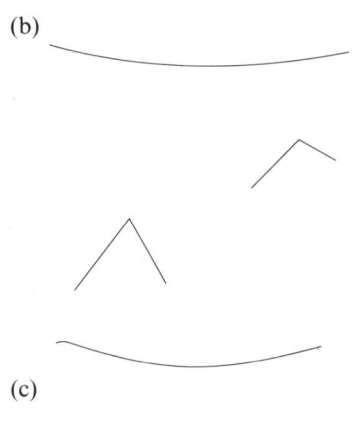

(a)

(b)

(c)

Fig. 11.25 An illustration of what has been seen as an upper Palaeolithic 'tradition' of 'hunters' marks' indicative of an intention (or at least the wish) of killing with weapon-points; the marks, like stylized representations of puncture wounds, are incised into representations of prey or sacrificial animals (cf. the presumed representations of spear-wounds on various animals shown in Fig. 11.21 above).

(a) A mammoth-ivory horse (4.8 cm) from Vogelherd, Lonetal, Germany, with an incised angle in its shoulder. A smaller angle is also incised higher up on the back. Aurignacian, *c.* 32 000 BP.
(b) Close-up of the incised angle in the side of the Vogelherd horse, with an extremely faint second angle incised higher up on the back. These are the only extraneous marks on the horse.
(a) and (b) reproduced from Marshack (1990), by permission of Dr Alexander Marshack; copyright Dr Marshack 1990.

From the stylized nature of the marking on the Vogelherd horse, the relatively superficial nature of the wounds displayed by the animals depicted in Fig. 11.21 above, and the fact that all are depicted upright and presumably still alive, plus the visual absence of portrayal of dead animals as such, one would be led to conclude that these markings represented intentions, designs, or desires of the animals' deaths rather than depictions of the actuality; hence an inclination towards explanations in terms of 'hunters' magic', 'ritual killing', etc. Some sort of votive implication would be another possibility.

Fig. 11.26 Leroi-Gourhan's syntactic formulae/mythograms and their relation to cave art. Reproduced here are (a) Leroi-Gourhan's tabulation of the contents of the whole complex of the journal depictions at Lascaux, broken down into its various 'assemblages' and displayed in coded form, and (b) a panorama of the Great Ceiling at Altamira, perhaps the largest single 'assemblage' of palaeolithic parietal art yet found. Readers may care to contemplate in what ways an encoding of the contents of the Altamira ceiling in a manner similar to that used by Leroi-Gourhan for Lascaux would constitute an aid to the comprehension of its composition, on the one hand, or in what ways the features of the Altamira ceiling that differ from those of Lascaux may call into question the validity of the Lascaux codification on the other. (For example the apparently 'opposed' positions of the two great boars at either top corner of the Altamira ceiling might be brought into higher relief or given an interesting emphasis by such a codification; but also the multiplicity of the orientations of the animals depicted on the Altamira ceiling, if for example one were to represent these by lines drawn from nose to tail-root, might perhaps make one wonder either whether (1) the simple right–left/left–right opposition of orientation normally noted by Leroi-Gourhan at Lascaux can at all usefully be extended from walls to ceilings, or even whether (2) that opposition might itself be a simple artefact of the constraints of normally painting on walls rather than on the freer surface of a ceiling, and therefore perhaps not necessarily very significant.

(a) Leroi-Gourhan's constructional formulae for the various panels of the wall-paintings in the cave at Lascaux, Dordogne, France. The circled numbers indicate the serial number of the composition, the arrows the direction in which the animal or group of animals is facing. The letters conform to the following key: A, horse; B^1, bison; B^2, aurochs; C^{1a}, stag; C^{1b}, doe; C^2, mammoth; C^3, ibex; C^4, reindeer; D^1, bear D^2, feline; D^3, rhinoceros; M, monster. The various species are classified sequentially in terms of their frequency of occurrence in cave wall-paintings in general. Reproduced by permission of CNRS from Leroi-Gourhan (1979), Fig. 352.

(b) Panoramic view of the Great Ceiling at Altamira, Santander, Spain, from the reconstruction at the Museo Arqueológico, Madrid, by permission. Total size roughly 18 m × 9 m; the size of most animals depicted is roughly in the range 1.5–2.25 m. Magdalenian.

(c) The full complex of the design of the Great Ceiling at Altamira, Santander, Spain, as originally discovered in the nineteenth century, after the drawings by Cartaillac and Breuil. Reproduced from Lafuente-Ferrari (1981), after Breuil and Obermaier (1935), by permission.

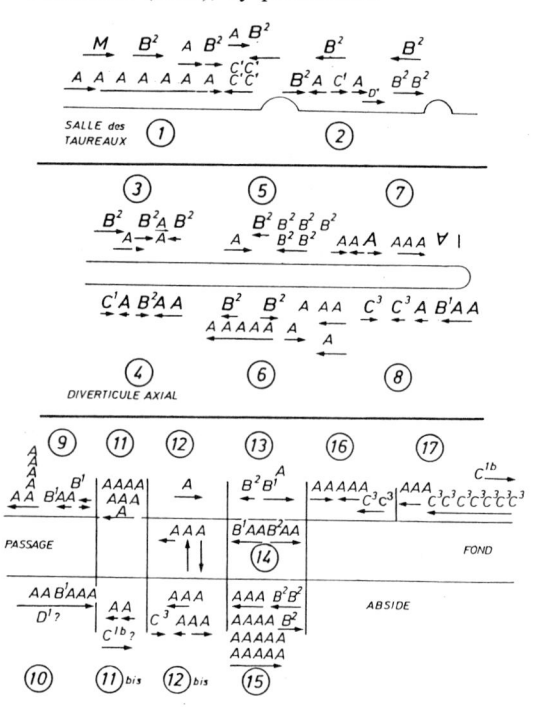

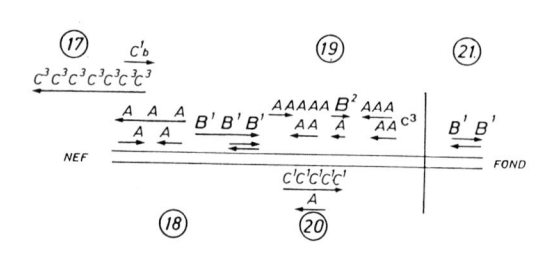

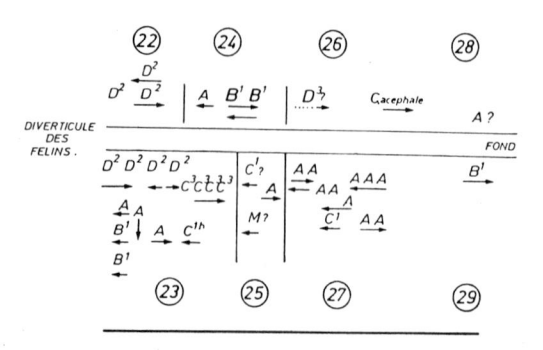

Prehistoric art

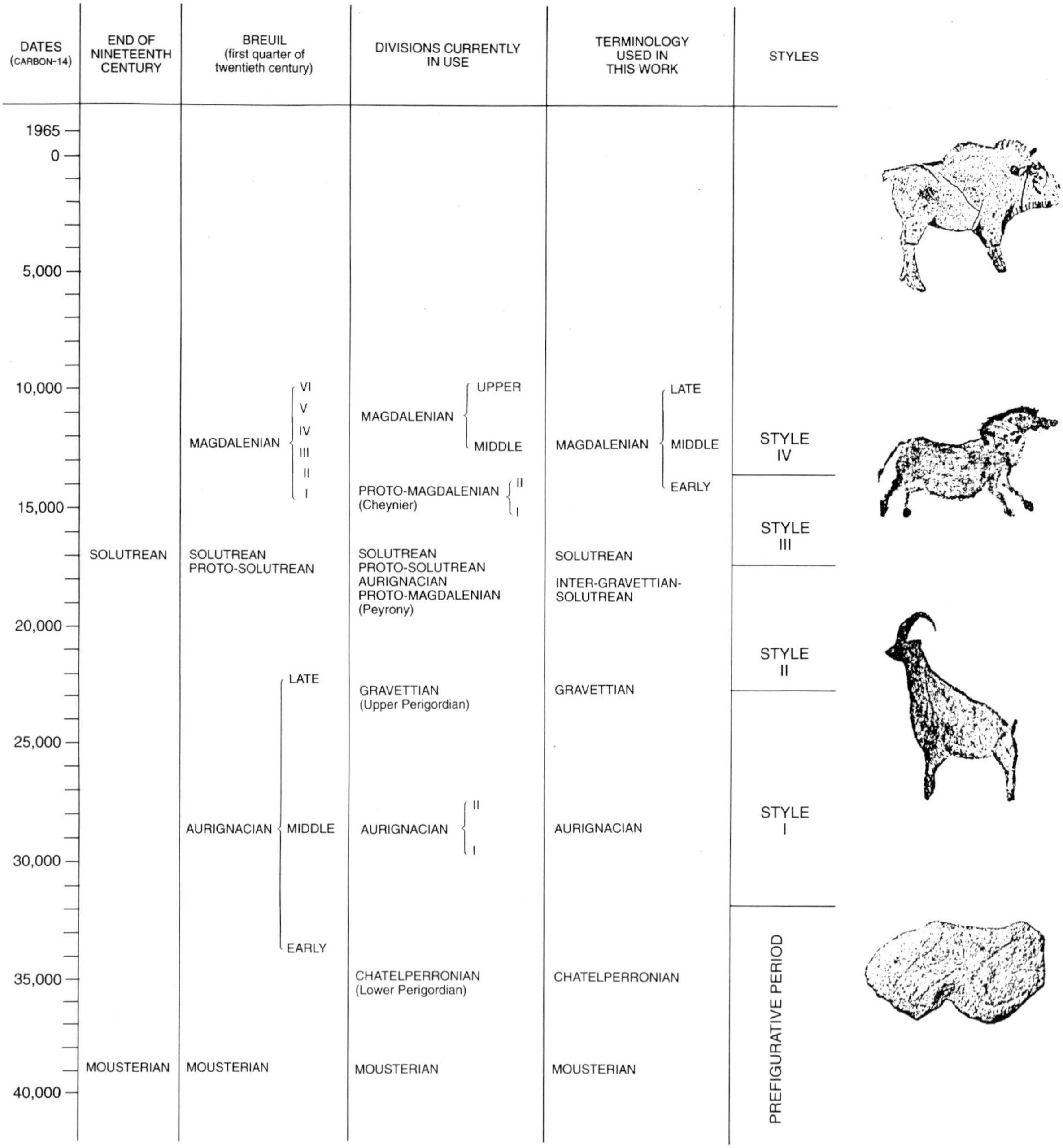

DATES (CARBON-14)	END OF NINETEENTH CENTURY	BREUIL (first quarter of twentieth century)	DIVISIONS CURRENTLY IN USE	TERMINOLOGY USED IN THIS WORK	STYLES
1965 — 0					
5,000					
10,000		MAGDALENIAN { VI, V, IV, III, II, I }	MAGDALENIAN { UPPER, MIDDLE } PROTO-MAGDALENIAN { II, I } (Cheynier)	MAGDALENIAN { LATE, MIDDLE, EARLY }	STYLE IV
15,000	SOLUTREAN	SOLUTREAN PROTO-SOLUTREAN	SOLUTREAN PROTO-SOLUTREAN AURIGNACIAN PROTO-MAGDALENIAN (Peyrony)	SOLUTREAN INTER-GRAVETTIAN-SOLUTREAN	STYLE III
20,000					
25,000		{ LATE	GRAVETTIAN (Upper Perigordian)	GRAVETTIAN	STYLE II
30,000		AURIGNACIAN { MIDDLE	AURIGNACIAN { II, I }	AURIGNACIAN	STYLE I
35,000		EARLY	CHATELPERRONIAN (Lower Perigordian)	CHATELPERRONIAN	PREFIGURATIVE PERIOD
40,000	MOUSTERIAN	MOUSTERIAN	MOUSTERIAN	MOUSTERIAN	

Fig. 11.27 Leroi-Gourhan's own version of his stylistic chronology. Reproduced from Leroi-Gourhan (1965), p. 493, Chart I by permission of Harry N. Abrams, Inc. 'Currently' at the head of col. 4 refers to the date of Leroi-Gourhan's book (1965).

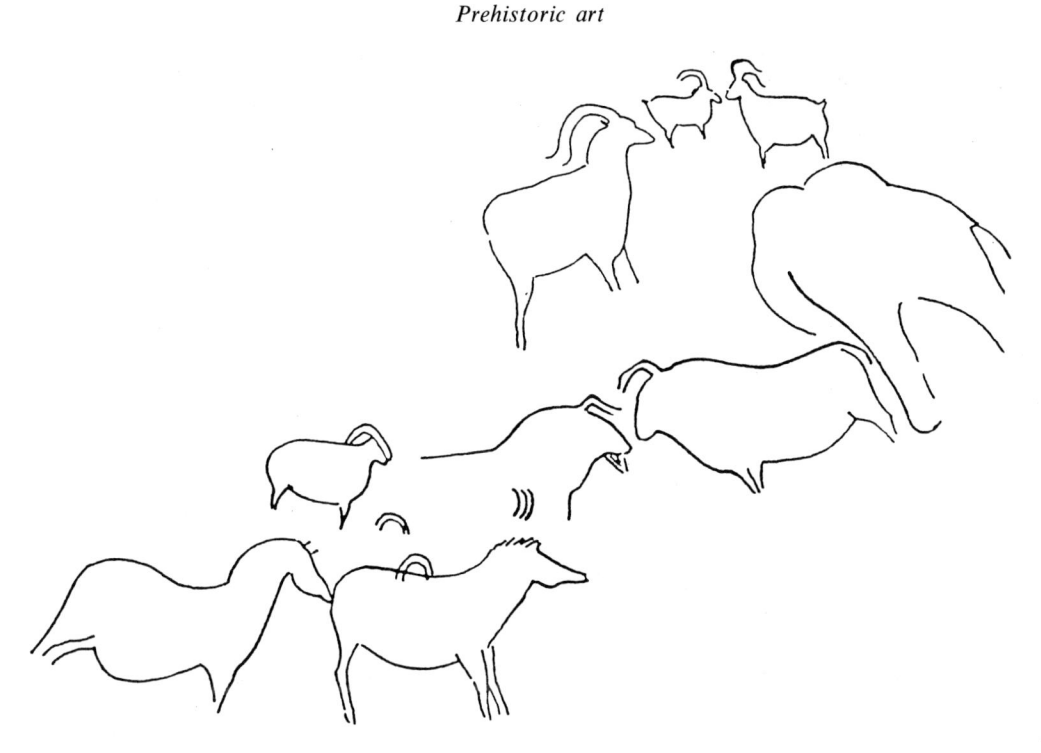

Fig. 11.28 An example of Leroi-Gourhan's ideas of the 'making and placing' of images. Part of Chart XXVI, 'Compositions with two central and two complementary animals', showing 'Bison/horse + ibex and mammoth'. Reproduced from Leroi-Gourhan (1965), p. 510, by permission of Harry N. Abrams, Inc.

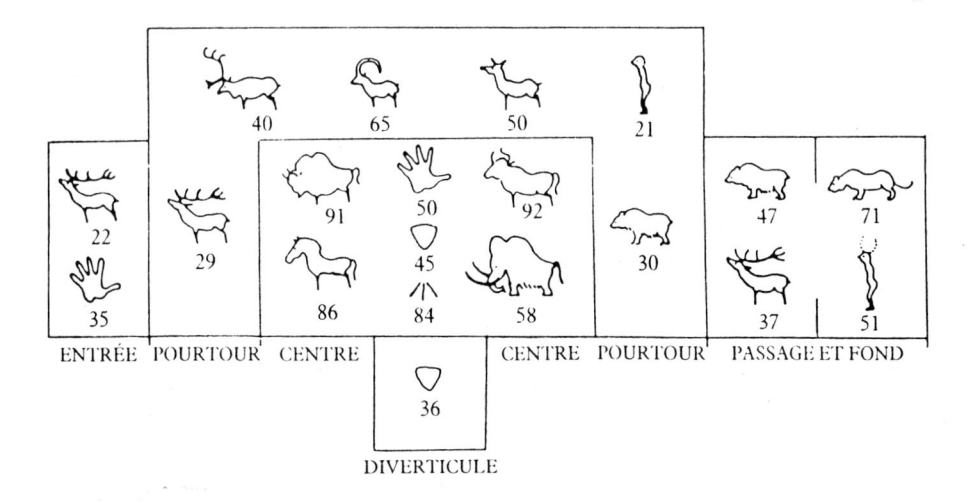

Fig. 11.29 An illustration of Leroi-Gourhan's 'ideal' cave layout. Reproduced from Leroi-Gourhan (1965), p. 501, Chart XV, by permission of Harry N. Abrams, Inc.

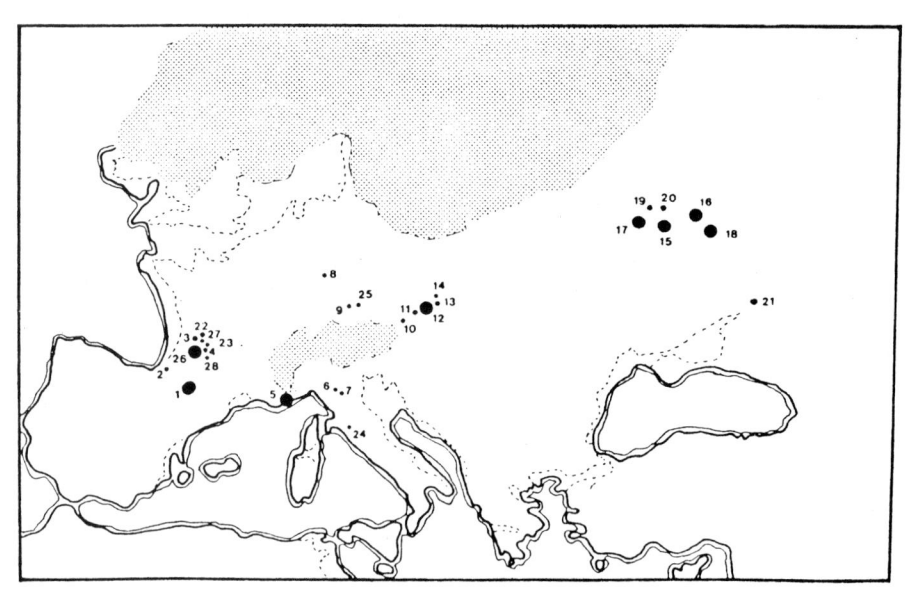

Fig. 11.30 Gamble's map of the distribution of venus figurines c. 25 000–23 000 BP. The numbers refer to the items listed in Gamble's annexed Table 1, and the larger dots indicate a number of figurines from one site. The stippled areas indicate the maximum extension of the ice sheets at c. 18 000 BP. Reproduced from Gamble (1982) by permission of the author. For the original reports of figurine finds cited in the Table (not reproduced here) see Gamble's above paper.

Fig. 11.31 Entoptic phenomena and their relation to cave art: a recombination of the data of Lewis-Williams and Dowson (1988), pp. 206–7, Figs. 1 and 2, reproduced with the permission of Professor J. D. Lewis-Williams. For the original sources from which Dowson and Lewis-Williams's figures were redrawn, see their article pp. 206–7, Figs. 1 and 2 and the references cited therein.

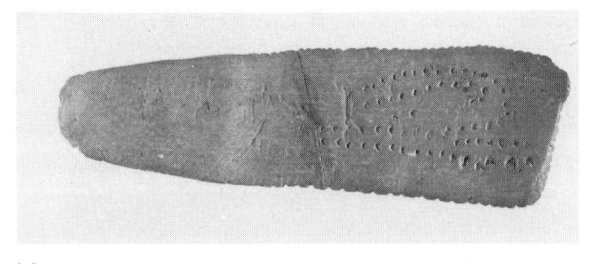

Prehistoric art

(a)

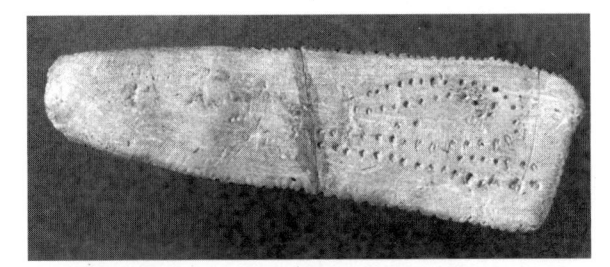

(b)

(c)

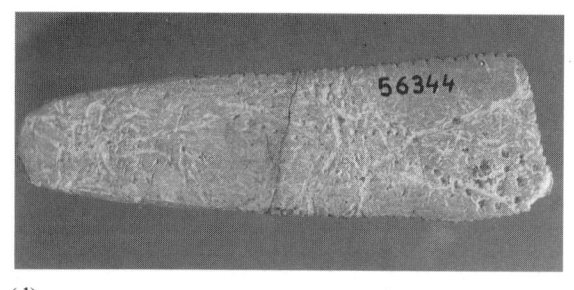

(d)

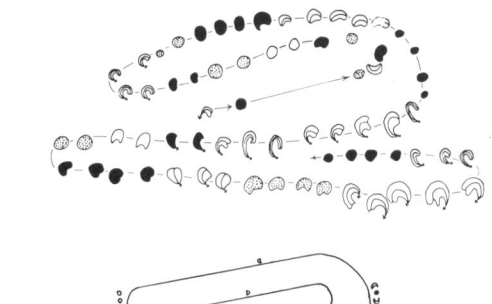

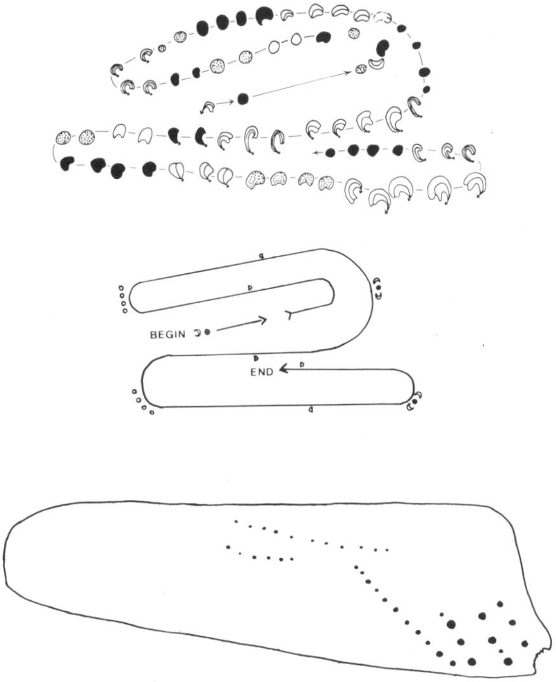

(e)

Fig. 11.32 The Blanchard plaquette and Dr Marshack's interpretation of it.
(a) and (b) Two photographs of the front face of an engraved bone plaquette (11 cm) with punctate markings (edge marks and interior engraving) from Abri Blanchard, Dordogne, France. The front end has been broken back by pressure flaking; the rear is highly polished. Aurignacian.

- (a) Photograph courtesy of the Musée d'Antiquités Nationales, Service Photographique Chantal Dulos.
- (c) Close-up of the interior engraving on the Blanchard plaquette, indicating the sets of marks made by different points and types of marking, some arcing to the left, some to the right, and others being simple gouges. The highly polished edge marks are visible above.
- (d) Photograph of the reverse face of the Blanchard plaquette. The undersurface of the plaquette has deteriorated badly as a result of root action. The markings were determined by microscopic analysis.
- (e) Drawings representing Dr Marshack's interpretation of the Abri Blanchard plaquette. Top: schematic representations of the sequential, serpentine mode of accumulating sets of marks on the plaquette. Assuming that the notation represented an observational lunar marking, the days of the crescents and invisibilities fell to the right and the full moons to the left. Bottom: the subsidiary sets of marks on the reverse face of the plaquette. A comparison of the markings on the two faces indicates that the marking was not intended to be 'decorative'.

Items (b)–(e) reproduced by permission of Dr Alexander Marshack; copyright Dr Marshack 1990.

Fig. 11.33 Mezhwerich, Ukraine. Detail of a portable art engraving of 'ladder' motifs and a schematic representation of a part of it, indicating the containing lines and the lines connecting one set to the next to make a continuous sequence. Included among the 'ladder' motifs are single and double or parallel lines. Reproduced from Marshack (1990) by permission of Dr Alexander Marshack; copyright Dr Marshack 1990.

Fig. 11.34 An illustration of Dr Marshack's notion of the re-using of images: these 'spotty horses' from Pech-Merle, Lot, France had touching-up dots applied to them at various different times with various different pigments. Photograph from a transparency supplied courtesy of Dr M. Lorblanchet, Directeur de Recherche au CNRS.

Fig. 11.35 An image illustrating the cultural nature of image-interpretation. A large limestone block (60 × 60 × 45 cm) from the Abri Cellier, Dordogne, France, deeply engraved with images. Aurignacian. These motifs have often been interpreted as complete and incomplete representations of vulvas. Photograph courtesy of Drs Brigitte and Gilles Delluc.

References

Allsworth-Jones, P. (1986). *The Szeletian and the transition from the Middle to Upper Paleolithic in Central Europe.* Clarendon Press, Oxford.

Anderson, R. L. (1979). *Art in primitive societies.* Prentice-Hall, Englewood Cliffs, NJ.

Appellaniz, J.-M. (1982). *El arte pre-historico del país Vasco y sus vecinos.* Desclee de Brouwer, Bilbao, Spain.

Auel, J. (1980). *The clan of the cave bear.* Crown, New York.

Aujoulat, N. (1987). *Le relevé des œuvres pariétales paléolithiques.* Enregistrement et traitement des données. Documents d'Archéologie Française, 9. Éditions de la Maison des Sciences de l'Homme, Paris.

Bahn, P. G. (1978). Water mythology and the distribution of Paleolithic parietal art. *Proceedings of the Prehistoric Society*, **44**, 125–34.

Bahn, P. G. (1980). "Histoire d'eau": L'art parietal préhistorique des Pyrénées. In *Centenaire de l'Enseignement de la Préhistoire à Toulouse: hommage au Professeur Nougier.* Travaux de l'Institut d'Art Préhistorique, **XXII**, 129–35. Université de Toulouse le Mirail, Toulouse.

Bahn, P. G. (1985). Ice age drawings on open rock faces in the Pyrenees. *Nature*, **313**, 530–1.

Bahn, P. G. (1986). No sex, please, we're Aurignacians. *Rock Art Research*, **3**(2), 99–120.

Bahn, P. G. (1988). Expecting the Spanish Inquisition: Altamira's rejection in its 19th century context. Unpublished paper presented at Conference on Hunting and Gathering Societies (CHAGS), Darwin, Australia.

Bahn, P. G. (1989). Age and the female form. *Nature*, **342**, 345–6.

Bahn, P. G. and Cole, Glen H. (1986). La préhistoire pyrenéenne aux États-Unis. *Bulletin de la Société Préhistorique Ariège-Pyrenées*, **vol. XII**, 95–149.

Bahn, P. G. and Vertut, Jean (1988). *Images of the Ice Age.* Windward, Leicester.

Barandiaran, I. M. (1973). *Arte mueble del Paleolíttico Cantabrico.* Libreria General, Saragossa.

Baulois, A. (1980). Les sagaies decorées du Paléolithique superieur dans la zone Franco-Cantabrique. *Préhistoire Ariègeoise*, **XXXV**, 125–8.

Bégouën, Count H. (1929). The magic origin of prehistoric art. *Antiquity*, Vol. 3.

Bégouën, Count R., Clottes, J., Giraud, J.-P., and Rouzaud, F. (1989). Les foyers de la Caverne d'Enlène (Montesquieu-Avantès, Ariège). In *Actes du Colloque de Nemours, 1987*, pp. 165–79. Mémoires du Musée de Préhistoire de l'Ile de France, 2, Paris.

Bellier, C. (1984). Contribution à l'étude de l'industrie osseuse préhistorique: les contours decoupées du type "Têtes des Herbivores". *Bulletin de la Société Royale Belge d'Anthropologie et de Préhistoire*, **95**, 21–34.

Binford, L. R. and Ho, C. K. (1985). Taphonomy at a distance: Zhoukoudian, "the cave home of Beijing man"? *Current Anthropology*, **26**, 413–42.

Bosinski, G. (1984). The mammoth engravings of the Magdalenian site Gönnersdorf (Rhineland, Germany). In *La contribution de la zoologie et de l'ethologie à l'interpretation de l'art des peuples chasseurs préhistoriques* (ed. H.-G. Bandi, W. Huber, M. Sauter, and B. Sitter), pp. 295–322. Éditions Universitaires, Fribourg, Switzerland.

Breuil, H. (1935). L'évolution de l'art pariétal dans les cavernes et abris ornés de France. In Congrès Préhistorique de France, Compte Rendu de l'onzième session, Perigueux, 1934, pp. 102–18. Bureaux de la société Préhistorique Française, Paris.

Breuil, H. (1952). *Four hundred centuries of cave art.* Centre d'études et de documentation préhistoriqués, Montignac, France.

Breuil, H. and Obermaier, H. (1935). *La cueva de Altamira en Santillana del Mar*, trans. J. P. de Barradas (2nd edn). La Junta de las Cuevas de Altanica, The Hispanic Society of America, and La Academia de la Historia, Madrid.

Brooks, A. L'Aurignacien des niveaux six à quatorze à l'Abri Pataud. In *Les fouilles Movius à l'Abri Pataud.* Éditions du Centre National de Recherche Scientifique, Paris. (In press.)

Buisson, D., Menu, M., Pincon, G., and Walter, P. (1989). Les objets colorés du Paléolithique Supérieur: cas de la grotte de la Vache (Ariège). *Bulletin de la Société Préhistorique Française*, **86**(6), 183–91.

Burgin, Vi. (1986). The absence of presence: conceptualism and postmodernisms. In *The end of art theory: criticism and postmodernity* (ed. V. Burgin), pp. 29–50. Humanities Press International, Atlantic Highlands, NJ.

Carcauzon, C. (1984). Une nouvelle découverte en Dordogne: la grotte préhistorique de Fronsac. *Revue archéologique: Sites. L'Archéologie en France*, **No. 22** (Aout), 7–15.

Carcauzon, C. (1986). La grotte de Font-Bargeix. *Société Historique et Archéologique du Périgord*, **vol. CXIII**, 191–200.

Cattelain, P. (1980). Quelques aspects des propulseurs paléolithiques. *Bulletin de la Société Préhistorique Française*, **77**, 131.

Cattelain, P. (1986). Traces macroscopiques d'utilisation sur les propulseurs paléolithiques. *Helinium*, **XXVI**, 193–205.

Chase, P. G. and Dibble, H. L. (1987). Middle Paleolithic symbolism: a review of current evidence and interpretations. *Journal of Anthropological Archaeology*, **6**, 263–96.

Chollot, M. (1964). *Musée des Antiquités Nationales—Collection Piette* (Art mobilier préhistorique). Musées Nationaux, Paris.

Chollot, M. (1980). *Les origines du graphisme symbolique. Essai d'analyse des écritures primitives en préhistoire.* Singer-Polignac, Paris.

Clark, G. A. (1989). Alternative models of Pleistocene biocultural evolution: a response to Foley. *Antiquity*, **63**, 153–61.

Clark, G. A. and Lindly, J. M. (1989). The case for continuity: observations on the biocultural transition in Europe and West Asia. In *The human revolution: behavioral and biological perspectives on the origins of* modern humans, (ed. P. Mellars and C. Stringer), pp. 626–76. Princeton University Press.

Clottes, J., (1989). The identification of human and animal figures in European Paleolithic art. In *Animals into art* (ed. H. Morphy), pp. 21–56, London. Unwin Hyman.

Clottes, J. (1990). The parietal art of the late Magdalenian. *Antiquity*, 64(22), 524–40.

Clottes, J., Menu, M., and Walter, P. (1990). New light on the Niaux paintings. *Rock Art Research*, 7(2).

Colby, B. (1975). Culture grammars. *Science* 187, 913–18.

Collins, D., and Onians, J. (1978). The origins of art. *Art History*, 1(1), 1–25.

Conkey, M. (1978). Style and information in cultural evolution: toward a predictive model for the Paleolithic. In *Social archaeology* (ed. C. Redman et al.), pp. 61–85. Academic Press, New York.

Conkey, M. (1980). Context, structure, and efficacy in paleolithic art and design. In *Symbol as sense* (ed. M. L. Foster and S. Brandes), pp. 225–48. Academic Press, New York.

Conkey, M. (1981). A century of Paleolithic art. *Archaeology*, 34(4), 20–8.

Conkey, M. (1982). Boundedness in art and society. In *Symbolic and structural archaeology* (ed. I. Hodder), pp. 115–28. Cambridge University Press.

Conkey, M. (1983). On the origins of paleolithic art: a review and some critical thoughts. In *The Mousterian legacy: human biocultural change in the upper Pleistocene*, British Archaeological Reports, International Series no. 164, (ed. E. Trinkhaus), pp. 201–27. BAR, Oxford.

Conkey, M. (1984). To find ourselves: art and social geography of prehistoric hunter–gatherers. In *Past and present in hunter-gatherer studies* (ed. C. Schrire), pp. 253–276. Academic Press, New York.

Conkey, M. (1985). Ritual communication, social elaboration, and the variable trajectories of Paleolithic material culture. In *Prehistoric hunter-gatherers: the emergence of social and cultural complexity*, (ed. T. D. Price and J. A. Brown), pp. 299–323. Academic Press, New York.

Conkey, M. (1987). New approaches in the search for meaning? A review of research in Paleolithic 'art'. *Journal of Field Archaeology*, 14, 413–30.

Conkey, M. (1989). Structural studies of Paleolithic art. In *Archaeological thought in America* (ed. C. C. Lamberg-Karlovsky), pp. 135–54. Cambridge University Press.

Conkey, M. *Paleovisions: interpreting the imagery of Ice Age Europe.* W. H. Freeman, New York, (In press).

Conkey, M., and Williams, S. H. (1991). Original narratives: the political economy of gender in archaeology. In *Gender at the crossroads of knowledge: feminist anthropology in the post-modern era* (ed. M. di Leonardo), pp. 68–89. University of California Press, Berkeley.

Dauvois, Mi. (1977). Travail expérimental de l'ivoire: sculpture d'une statuette féminine. In *Méthodologie appliquée à l'industrie de l'os préhistorique* Colloques Internationaux du CNRS, no. 568, (ed. H. Camps-Fabrer), pp. 270–3. Éditions du Centre National de la Recherche Scientifique, Paris.

Davis, W. (1985). Present and future directions in the study of rock art. *South African Archaeological Bulletin*, 40, 5–10.

Davis, W. (1986). The origins of image-making. *Current Anthropology*, 27, 193–215.

Deacon, H. (1989). Late Pleistocene paleoecology and archaeology in the Southern Cape, South Africa. In *The human revolution: behavioral and biological perspectives on the origins of modern humans* (ed. P. Mellars and C. Stringer), pp. 547–64. Princeton University Press.

deBeaune, S. (1987). *Lampes et godets au Paléolithique*, xxiii/S^e supplement à *Gallia Préhistoire*. Éditions du Centre National de Recherche Scientifique, Paris.

Delluc, B. and Delluc, G. (1978). Les manifestations graphiques Aurignaciennes sur support rocheux des environs des Eyzies (Dordogne). *Gallia Préhistoire*, 21(1–2), 213–438.

Delporte, H. (1973). Les techniques de la gravure Paléolithique. In *Estudios dedicados a Luis Pericot*, Publicaciones Eventuales No. 23, (ed. J. Maluquer de Motes), pp. 119–29. Universidad de Barcelona, Instituto de Arqueología y Prehistoria.

Delporte, H. (1979). *L'image de la femme dans l'art préhistorique.* Picard, Paris.

Delporte, H. (1986). The female image in Upper Paleolithic art. Unpublished paper presented at Life in Ice Age Europe: A Symposium. New York: New York University (October 24 and 25).

Delporte, H. (1987). Piette, pionier de la préhistoire. In *Histoire de l'art primitif* (ed. P. Edouard), pp. 11–182. Picard, Paris.

Delporte, H. (1990). Découverte et classification de l'art mobilier au XIXe siècle. In *L'Art des objets au Paléolithique, tome 1: L'art mobilier et son contexte*, Actes des colloques de la Direction du Patrimoine, sons la direction de Jean Clottes, pp. 9–12. Ministère de la Culture, de la Communication, des Grands Travaux et du Bicentenaire, Paris.

Delporte, H., and Mons, L. (1980). Gravure sur os et argile. *Dossiers de l'Archéologie*, 46, 40–5.

Dewez, M. C. (1974). New hypotheses concerning two engraved bones from La Grotte de Remouchamps, Belgium. *World Archaeology*, 5(3), 337–45.

Dunnell, R. (1986). Five decades of Americna archaeology. In *American archaeology, past and future.* (ed. D. J. Meltzer, D. D. Fowler, and J. A. Sabloff), pp. 23–49. Smithsonian Institution Press, Washington.

Eagleton, T. (1983). *Literary theory: an introduction.* University of Minnesota Press, Minneapolis.

Eaton, R. (1978a). The evolution of trophy hunting. *Carnivore*, I(1) 110–21.

Eaton, R. (1978b). Meditations on the origins of art as trophysim. Manuscript.

Fagan, B. (1986). *The adventure of archaeology.* National Geographic Society, Washington, DC.

Faris, J. (1972). *Nuba personal art.* Duckworth, London.

Faris, J. (1983). From form to content in the structural study of aesthetic systems. In *Structure and cognition in art* (ed. D. Washburn), pp. 90–112. Cambridge University Press.

Fernandez, J. W. (1966). Principles of opposition and vitality in Fang aesthetics. *The Journal of Aesthetics and Art Criticism*, vol. XXV, No. 1 (Fall 1966), 53–64.

Flannery, K. V. (1967). Culture-history vs. culture-process. Review of *An introduction to American archaeology,* Vol. 1: *North and Middle America* by G. R. Willey. *Scientific American*, **217**(2), 119–22.

Foley, R. (1988). Hominid species and stone-tool assemblages. *Antiquity*, **61**, 380.

Forbes, A., jun. and Crowder, T. R. (1979). The problem of Franco-Cantabrian abstract signs: agenda for a new approach. *World Archaeology*, **10**(3), 350–66.

Fortea, J., Corchon, M.-S., Gonzaler-Morales, M., Rodriguez-Asensio, A., Hoyos, M., Laville, H. *et al.* (1990). Travaux récents dans les vallées du Nalón et du Sella (Asturies). In *L'Art des objets au Paléolithique, tome 1: L'art mobilier et son contexte*, Actes des colloques de la Direction du Patrimoine, sous la direction de Jean Clottes, pp. 219–246. Ministère de la Culture, de la Communication, des Grands Travaux et du Bicentenaire, Paris.

Frazer, J. G. (1887). *Totemism*. Adam and Charles Black, Edinburgh.

Frazer, J. G. (1890). *The golden bough*. Republished in 3rd edn (1911–15). Macmillan, London.

Fritz, M. C. (1969). Description and explanation in Paleolithic cave art: a survey of the literature. Master of Arts thesis, Department of Anthropology, University of Chicago.

Gamble, C. (1982). Interaction and alliance in Paleolithic society. *Man* (NS), **17**, 92–107.

Gargett, R. (1987). Grave shortcomings: the evidence for neandertal burial. *Current Anthropology*, **30**(2), 157–90.

Gilman, A. (1983). Explaining the Upper Paleolithic revolution. In *Marxist perspectives in archaeology* (ed. M. Spriggs), pp. 115–26. Cambridge University Press.

Graziosi, P. (1960). *Paleolithic art*. Faber and Faber, London.

Guthrie, R. D. (1984). Ethological observations from Paleolithic art. In *La contribution de la zoologie et de l'ethologie à l'interpretation de l'art des peuples chasseurs préhistoriques* (ed. H.-G. Bandi, W. Huber, M. R. Sauter, and B. Sitter), pp. 35–74. Éditions Universitaires, Fribourg, Switzerland.

Gvozdover, M. D. (1989). The typology of female figurines of the Kostenki Paleolithic culture. In *Female imagery in the Paleolithic, an introduction to the works of M. D. Gvozdover*, special issue (guest ed. O. Soffer-Bobyshev). *Soviet Anthropology and Archaeology*, **27**(4), 32–94.

Hahn, J. (1988). *Das Giessenklösterle, I*. Konrad Theiss Verlag, Stuttgart.

Halverson, J. (1987). Art for art's sake in the Paleolithic. *Current Anthropology*, **28**(1), 63–89.

Hammond, N. (1974). Paleolithic mammalian faunas and parietal art in Cantabria: a comment on Freeman. *American Antiquity*, **39**, 618–19.

Hawkes, C. (1954). Archaeological theory and method: some suggestions from the Old World. *American Anthropologist*, **56**, 155–68.

Hawkes, T. (1977). *Structuralism and semiotics*. University of California Press, Berkeley.

Hodder, I. (1982). Theoretical archaeology: a reactionary view. In *Symbolic and structural archaeology* (ed. I. Hodder), pp. 1–16. Cambridge University Press.

Hodder, I. (1985). Post-processual archaeology. In *Advances in archaeological method and theory*, Vol. 8, (ed. M. B. Schiffer), pp. 1–26.

Hodder, I. (1986). *Reading the past*. Cambridge University Press.

Holm, B. (1965). *Northwest Coast Indian art*. University of Washington Press, Seattle.

Jochim, M. (1983). Paleolithic cave art: some ecological speculations. In *Hunter–gatherer economy in prehistory: a European perspective* (ed. G. Bailey), pp. 212–19. Cambridge University Press.

Klein, R. G. (1989). Biological and behavioural perspectives on modern human origins in Southern Africa. In *The human revolution: behavioral and biological perspectives on the origins of modern humans* (ed. P. Mellars and C. Stringer), pp. 529–46. Princeton University Press.

Kurtén, B. (1980). *Dance of the tiger*. Pantheon, New York.

Kurtén, B. (1986). *Singletusk*. Pantheon, New York.

Lafuente-Ferrari, E. (1981). *El libro de Santillana*. Ediciones de Libreria Estudio, Santander.

Laming-Emperaire, A. (1962). *La signification de l'art rupestre Paléolithique*. Picard, Paris.

Lee, R. B. and De Vore, I. (eds) (1968). *Man the hunter*. Aldine, Chicago.

Leone, M. (1972). Issues in contemporary archaeology. *Contemporary archaeology* (ed. M. Leone), pp. 14–27. Southern Illinois University Press, Carbondale.

Leroi-Gourhan, André (1958). Reparition et groupement des animaux dans l'art pariétal paléolithique. *Bulletin de la Société Préhistorique Française*, **55**, 515–27.

Leroi-Gourhan, André (1965). *Treasures of Paleolithic art*. Abrams, New York.

Leroi-Gourhan, André (1966). La religion des cavernes: magie ou métaphysique? *Sciences et Avenir* **228**, reprinted in Leroi-Gourhan, André (1983). *Le fil du temps. Ethnologie et Préhistoire*, pp. 276–91. Fayard, Paris.

Leroi-Gourhan, André (1968). The evolution of paleolithic art. *Scientific American*, **246**(6), 104–12.

Leroi-Gourhan, André (1978). The mysterious markings in the Paleolithic art of France and Spain. *CNRS Research*, **8**, 26–32.

Leroi-Gourhan, André (1982). *The dawn of European art*. Cambridge University Press.

Leroi-Gourhan, André (1983). *Le fil du temps*. Librairie Arthème Fayard, Paris.

Leroi-Gourhan, André (1984). Le réalisme et le comportement dans l'art paleolithique d'Europe de l'Ouest. In *La contribution de la zoologie et de l'ethologie à interpretation de part des peuples chasseurs préhistoriques* (ed. H.-G. Bandi, W. Huber, M.-R. Santer, and B. Sitter), pp. 75–90. Editions Universitaires

Fribourg, Imprimerie Saint-Paul, Fribourg, Switzerland.

Leroi-Gourhan, Arlette (1982). The archaeology of Lascaux cave. *Scientific American*, **246**, (June) 104–13.

Leroi-Gourhan, Arlette and Allain, J. (1979). *Lascaux inconnu*. Éditions Centre National de Recherche Scientifique, Paris.

Lévi-Strauss, C. (1963). Split representation in the art of Asia and America. In idem, *Structural anthropology*, pp. 245–68. Basic Books, New York.

Lewin, R. (1988). Modern human origins under close scrutiny. *Science*, **239**, 1240–1 (March 11, 1988).

Lewis-Williams, D. (1982). The economic and social context of Southern San rock art. *Current Anthropology*, **23**(4), 429–49.

Lewis-Williams, D. (1983a). Introductory essay: science and rock art. In *New approaches to southern African rock art*, South African Archaeological Society, Johannesberg, Goodwin Series, **4**, pp. 3–13. The Society.

Lewis-Williams, D. (1983b). *Rock art of southern Africa*. Cambridge University Press.

Lewis-Williams, D. (1984). The empiricist impasse in southern African rock art studies. *South African Archaeological Bulletin*, **39**, 58–66.

Lewis-Williams, D. and Dowson, T. (1988). The signs of all times: entoptic phenomena in Upper Paleolithic art. *Current Anthropology*, **29**(2), 201–45.

Lindly, J. M. and Clark, G. A. (1990). Symbolism and modern human origins. *Current Anthropology*, **31**, 233–61.

Lorblanchet, M. (1977). From naturalism to abstraction in European prehistoric rock art. In *Form in indigenous art*, Australian Institute of Aboriginal Studies, Prehistory and Material Culture Series, No. 13, (ed. P. Ucko), pp. 43–56. The Institute, Canberra, Australia.

Lorblanchet, M. (1980). Peinture sur les parois de grottes. *Dossiers de l'Archéologie*, **46**, 33–9.

Lubbock, J. (1870). *The origin of civilization and the primitive condition of man; mental and social condition of savages*. Longman Green, London.

Lutz, C. (1992). Culture and consciousness: a problem in the anthropology of knowing. In *Self and consciousness* (ed. P. Cole, D. Johnson, and F. Kessel), pp. 64–87. Erlbaum, Hillsdale, NJ.

McGaw, J. (1982). Women and the history of American technology. *Signs: Journal of Women in Culture and Society*, **7**(4), 798–828.

Mack, R. (n.d.) The representation of prehistory: archaeology and the female body. Paper on file with the author, Department of the History of Art, University of California, Berkeley.

Marshack, A. (1969). Polesini, a re-examination of the engraved Upper Paleolithic mobiliary materials of Italy by a new methodology. *Rivista di Scienze Prehistoriche*, **24**, 219–81.

Marshack, A. (1970a). *Notation dans les gravures du Paléolithique Supérieur*. Publication de l'Institut de Préhistoire de l'Université de Bordeaux, Memoire 8. Imprimeries Delmas, Bordeaux.

Marshack, A. (1970b). The baton of Montgaudier. *Natural History*, **LXXIX**(3), 57–63.

Marshack, A. (1972a). *The roots of civilization*. McGraw-Hill, New York.

Marshack, A. (1972b). Cognitive aspects of Upper Paleolithic engraving. *Current Anthropology*, **13**, 445–77.

Marshack, A. (1976). Some implications of the Paleolithic symbolic evidence for the origins of language. In *Origins and evolution of language* (ed. S. R. Harnard, H. Stekelis, and J. B. Lancaster), pp. 289–311. New York Academy of Sciences, New York.

Marshack, A. (1977). The meander as a system: the analysis and recognition of iconographic units in Upper Paleolithic compositions. In *Form in indigenous art* (ed. P. Ucko), pp. 285–317. Australian Institute of Aboriginal Studies, Canberra.

Marshack, A. (1979). Upper Paleolithic symbol systems of the Russian Plain. *Current Anthropology*, **20**, 271–311.

Marshack, A. (1984a). The ecology and brain of two-handed bipedalism: an analytic, cognitive, and evolutionary assessment. In *Animal cognition* (ed. H. Terrasce, H. Roitblat and T. G. Bever), pp. 491–511. Erlbaum, Hillsdale, NJ.

Marshack, A. (1984b). Concepts théoriques conduisant à de nouvelles methodes analytiques, de nouveaux procédés de recherche et categories de données. *L'Anthropologie*, **88**, 573–86.

Marshack, A. (1985a). *Hierarchical evolution of the human capacity: the Paleolithic evidence*, Fifty-fourth James Arthur Lecture on the Evolution of the Human Brain, 1984. American Museum of Natural History, New York.

Marshack, A. (1985b). Theoretical concepts that lead to new analytic methods, modes of inquiry and classes of data. *Rock Art Research*, **2**(2), 95–111.

Marshack, A. (198). The Neanderthals and the human capacity for symbolic thought: cognitive and problem-solving aspects of Mousterian symbols. In L'homme neanderthal, tome S: La pensée (co-ordinator O. Bar-Yosef). *Etudes et Recherches Arquéologiques de l'Université de Liège*, **no. 32**, 57–92.

Marshack, A. (1989). Methodology in the analysis and interpretation of Upper Paleolithic image: theory versus contextual analysis. *Rock Art Research*, **6**(1), 17–53.

Marshack, A. (1990). Early hominid symbol and evolution of the human capacity. In *The human revolution: behavioral and biological perspectives on the origins of modern humans*, Vol. 2, (ed. P. Mellars and C. Stringer), pp. 457–98. Princeton University Press.

Mellars, P. and Stringer, C. (eds) (1989). *The human revolution: behavioral and biological perspectives on the origins of modern humans*. Princeton University Press.

Ministère de la Culture (France) (1984). *Atlas des grottes ornées Paléolithiques Françaises*. Ministère de la Culture, Paris.

Mons, L. (1986). Les statuettes animalières en grés de la grotte d'Insuritz (Pyrénées-Atlantiques): observa-

tions et hypothèses de fragmentation volontaire. *L'Anthropologie*, **90**(4), 701–12.
Mons, L. (1990). Les figures animales incompletes dans l'art paléo leurs particularismes techniques et graphiques. In *L'Art des objets au paléolithique, tome 2: Les voies de la recherche*, Actes des Collogues de la direction du Patrimoine, sous la direction de Jean Clottes, pp. 73–8. Ministère de La Culture, de la Communication, des Grands Travaux et du Bicentenaire, Paris.
Nodelman, S. (1979–80). The legacy of the caves. *Portfolio* **I**(5), 48–55.
Nodelman, S. (1985). Theoretical issues in the study of Paleolithic representation. Unpublished paper presented at Symposium: Paleolithic art, the state of the question. Annual Meetings, College Art Association (Los Angeles).
Pales, L. and Tassin de Saint Péreuse, M. (1976). *Les gravures de la Marche. II. Les humains*. Editions Ophyrus, Paris.
Parkington, J. (1969). Symbolism in paleolithic cave art. *South African Archaeological Bulletin*, **24**, 3–13.
Patrik, L. E. (1985) Is there an archaeological record? In *Advances in archaeological method and theory*, 8 (ed. M. B. Schiffer), pp. 27–62. Academic Press, New York.
Pfeiffer, J. (1980). Icons in the shadows. *Science*, **80** 1(4), 72–7.
Pfeiffer, J. (1982). *The creative explosion*. Harper and Row, New York.
Pfeiffer, J. (1986). Cro-Magnon hunters were really us, working out strategies for survival. *Smithsonian*, **7**, (October 17), 74–85.
Potts, R. (1988). *Early hominid activities at Olduvai*. De Gruyter, New York.
Preziosi, D. (1982). Constru(ct)ing the origins of art. *Art Journal*, **1982** (winter), 320–5.
Preziosi, D. (1989). Reckoning with the world. In idem, *Rethinking art history*, pp. 122–55. Yale University Press, New Haven.
Rappaport, R. (1968). *Pigs for the ancestors: ritual in the ecology of a New Guinea people*. Yale University Press, New Haven.
Rappaport, R. (1971). The sacred in human evolution. *Annual Review of Ecology and Systematics*, **2**, 23–42.
Redman, C. L. (ed.) (1973). *Research and theory in current archaeology*. Wiley, New York.
Reinach, S. (1900). Ethnographie. Phénomènes generaux du totémisme animal. *Revue Scientifique* (4e série), 449–57.
Reinach, S. (1903). L'art et la magie. À propos des peintures et des gravures de l'âge du renne, Mémoires originaux. *L'Anthropologie*, **14**, 257–66.
Reinach, S. (1913). *Répertoire de l'art quaternaire*. Éditions Leroux, Paris.
Rice, P. (1981). Prehistoric Venuses: symbols of motherhood or womanhood? *Journal of Anthropological Research*, **37**(4), 402–14.
Rigaud, J.-Ph. and Simek, J. (1987). "Arms too short to box with God": problems and prospects for Paleolithic prehistory in Dordogne, France. In *The Pleistocene old world: regional perspectives* (ed. O. Soffer), pp. 47–62. Plenum, New York.

Sackett, J. (1983). From de Mortillet to Bordes: a century of French Paleolithic research. In *Towards a history of archaeology*, 1, (ed. G. Daniel), pp. 59–112. Thames and Hudson, London.
Sackett, J. (1988). The Mousterian and its aftermath: a view from the Upper Paleolithic. In *Upper Pleistocene prehistory of western Eurasia* (ed. H. Dibble and A. Montet-White), pp. 413–26. The University Museum Press, University of Pennsylvania, Philadelphia.
Sauvet, G. (1988). La communication graphique Paléolithique (De l'analyse quantitative d'un corpus de données à son interpretátion semiologique). *L'Anthropologie*, **92**(1), 3–16.
Sauvet, G., Sauvet, S. and Wlodarczyk, A. (1977). Essai de sémiologie préhistorique (pour une théorie des premiers signes graphiques de l'homme). *Bulletin de la Société Préhistorique Française*, **74**(2), 545–58.
Shipman, P. and Rose, J. (1983). Early hominid hunting, butchering and carcass-processing behaviors: approaches to the fossil record. *Journal of Anthropological Archaeology*, **2**, 57–98.
Sieveking, A. (1986). Les styles régionaux de l'art préhistorique. *La Recherche*, **17**, 104–6.
Sieveking, A. (1987). Engraved Magdalenien art: a regional and stylistic analysis of stone, bone and antler plaquettes from Upper Paleolithic sites in France and Cantabric Spain, British Archaeological Reports, International Series, 396. BAR, Oxford.
Simek, J. F. and Price, H. A. (1990). Chronological change in Perigord lithic assemblage diversity. In *The emergence of modern humans* (ed. P. Mellars), pp. 243–61. Cornell University Press, Ithaca NY.
Simek, J. and Snyder, L. (1988). Changing assemblage diversity in Perigord archaeofaunas. In *The Upper Pleistocene prehistory of western Eurasia* University Museum Symposium Series, Vol. 1, (ed. H. Dibble and A. Montet-White), pp. 321–32. University Museum, University of Pennsylvania, Philadelphia.
Smith, F. (1982). Upper Pleistocene hominid evolution in south central Europe: A review of the evidence and analysis of trends. *Current Anthropology*, **23**, 667–704.
Smith, F. (1984). Fossil hominids from the Upper Pleistocene of Central Europe and the origin of modern humans. In *The origin of modern humans, a world survey of the fossil evidence* (ed. F. Smith and F. Spencer), pp. 137–210. Alan R. Liss, New York.
Smith, N. (n.d.). *A psychology of ice age art*. Manuscript.
Spencer, B. and Gillen, F. J. (1899). *The native tribes of central Australia*. Macmillan, London.
Sperber, D. (1974). *Rethinking symbolism*. Cambridge University Press.
Stevens, A. (1975). Animals in Paleolithic cave art: Leroi-Gourhan's hypothesis. *Antiquity*, **49**, 54–7.
Stoneking, M. and Cann, R. (1989). African origin of human mitochrondial DNA. In *The human revolution: behavioral and biological perspectives on the origins of modern humans* (ed. P. Mellars and C. Stringer), pp. 17–30. Princeton University Press.

Straus, Lawrence G. (1987). The Paleolithic cave art of Vasco-Cantabrian Spain. *Oxford Journal of Archaeology*, **6**(2), 149–63.

Stringer, C. B. and Andrews, P. (1988). Genetic and fossil evidence for the origin of modern humans. *Science*, **239**, (March 11), 1263–8.

Taborin, Y. (1987). Les coquillages dans la parure Paléolithique en France (3 vols). Thèse de doctorat d'État, Université de Paris.

Taborin, Y. (1992). Les espaces d'acheminement de certains coquillages magdaléniens. In *Le peuplement Magdalénien: paleogéographique physique et humaine*, pp. 417–30. Documents Prehistoriques, 2: Actes du Colloque de Chanèclode. Editions du Comit des Travaux Historiques et Scientifiques, Paris.

Thomas, E. M. (1987). *Reindeer moon*. Houghton-Mifflin, Boston.

Törbrugge, W. (1968). *Prehistoric European art*. Abrams, New York.

Trinkaus, E. (1990a). The Upper Pleistocene transition: biocultural patterns and processes. In idem (ed.), *The emergence of modern humans: biocultural adaptations in the Later Pleistocene*. Cambridge University Press.

Trinkaus, E. (1990b). *The emergence of modern humans: biocultural adaptations in the Later Pleistocene*. Cambridge University Press.

Turner, T. (1969). Tchikrin: a Central Brazilian tribe and its symbolic language of bodily adornment. *Natural History*, **78** (October), 50–9, 78.

Tylor, E. G. (1865). *Researches into the early history of mankind and the development of civilization*. J. Murray, London.

Ucko, P. J. (1987). Débuts illusoires dans l'étude de la tradition artistique. *Bulletin de la Société Préhistorique Ariège-Pyrénées*, **42**, 15–81.

Ucko, P. and Layton, R. (1989). La subjectivité et le recensement de l'art Paléolithique, unpublished paper presented at 'L'Art Panétal Paléolithique', Colloque International, Perigueux, France.

Ucko, P. and Rosenfeld, A. (1967). *Paleolithic cave art*. McGraw-Hill, New York.

Ucko, P. and Rosenfeld, A. (1972). Anthropomorphic representations in Paleolithic art. In *Santander symposium*, pp. 149–211. Actos del Symposium Internacional de Arte Prehistorico, Santander.

Valladas, H., Reyss, J. L., Joron, J. L., Valladas, G., Bar-Yosef, O., and Vandermeersch, B. (1988). Thermoluminescence dating of Mousterian "Proto-Cro-Magnon" remains from Israel and the origin of modern man. *Nature*, **331**, 614–16.

Vandermeersch, B. (1984). À propos de la découverte du squelette néandertalien du Saint-Césaire. *Bulletins et Mémoires de la Société Anthropologique de Paris*, **14**(1), 191–6.

Vandiver, P. (1983). Paleolithic pigments and processing. Unpublished Master's thesis. Department of Material Sciences, Massachusetts Institute of Technology.

Vandiver, P., Soffer, O., Klima, B., and Svoboda, Jiri (1989). The origins of ceramic technology at Dolni Vestonice, Czechoslovakia. *Science*, **246**, 1002–8.

Van Gennep, M. (1925). À propos du totémisme préhistorique. In *Actes du Congrès International d'Histoire des Réligions 1923*, pp. 323–38. Campion, Paris.

Vayda, A. and Rappaport, R. (1968). Ecology: cultural and non-cultural. In *Introduction to cultural anthropology* (ed. J. Clifton), pp. 476–98. Houghton-Mifflin, Boston.

Verworn, M. (1909). *Die anfänge der Kunst*. Fischer, Jena.

Vialou, D. (1981). L'Art préhistorique: questions d'interpretations. *Revue des Monuments Historiques*, **118** (November–December), 75–82.

Vialou, D. (1982). Niaux, une construction symbolique Magdalénienne exemplaire. *Ars Praehistorica*, **1**, 19–46.

Vialou, D. (1983). Art pariétal Paléolithique Ariègeois. *L'Anthropologie*, **87**, 83–97.

Vialou, D. (1984). Les cervides de Lascaux. In *La Contribution de la zoologie et de l'ethologie à l'interpretation de l'art des peuples chasseurs préhistoriques* (ed. H. G. Bandi, W. Huber, M.-R. Sauter, and B. Sitter), pp. 199–216. Éditions Universitaires, Fribourg, Switzerland.

Vialou, D. (1986). *L'Art des grottes en Ariège Magdalénienne. XXII Supplement à Gallia Préhistoire*. Éditions du Centre National de la Recherche Scientifique, Paris.

Washburn, D. K. (ed.) (1983). *Structure and cognition in art*. Cambridge University Press.

Watson, P. J., Redman, C. L., and LeBlanc, S. (1971). *Explanation in archaeology*. Columbia University Press, New York.

White, R. (1986). Rediscovering French ice-age art. *Nature*, **320**, 683–4.

White, R. (1989). Production complexity and standardization in early Aurignacian bead and pendant manufacture: evolutionary implications. In *The human revolution: behavioral and biological perspectives on the origins of modern humans* (ed. P. Mellars and C. Stringer), pp. 366–90. Princeton University Press.

Wolpoff, M. H. (1984). Evolution in *Homo erectus*: the question of stasis. *Paleobiology*, **10**, 389–406.

Wolpoff, M. H. (1989). Multiregional evolution: the fossil alternative to Eden. In *The human revolution: behavioral and biological perspectives on the origins of modern humans* (ed. P. Mellars and C. Stringer), pp. 62–108. Princeton University Press.

Wylie, M. A. (1981). Positivism and the new archaeology. Unpublished Ph.D. dissertation, Department of Philosophy, State University of New York, Binghamton.

Wylie, M. A. (1982). Epistemological issues raised by a structural archaeology. In *Symbolic and structural archaeology* (ed. I. Hodder), pp. 39–46. Cambridge University Press.

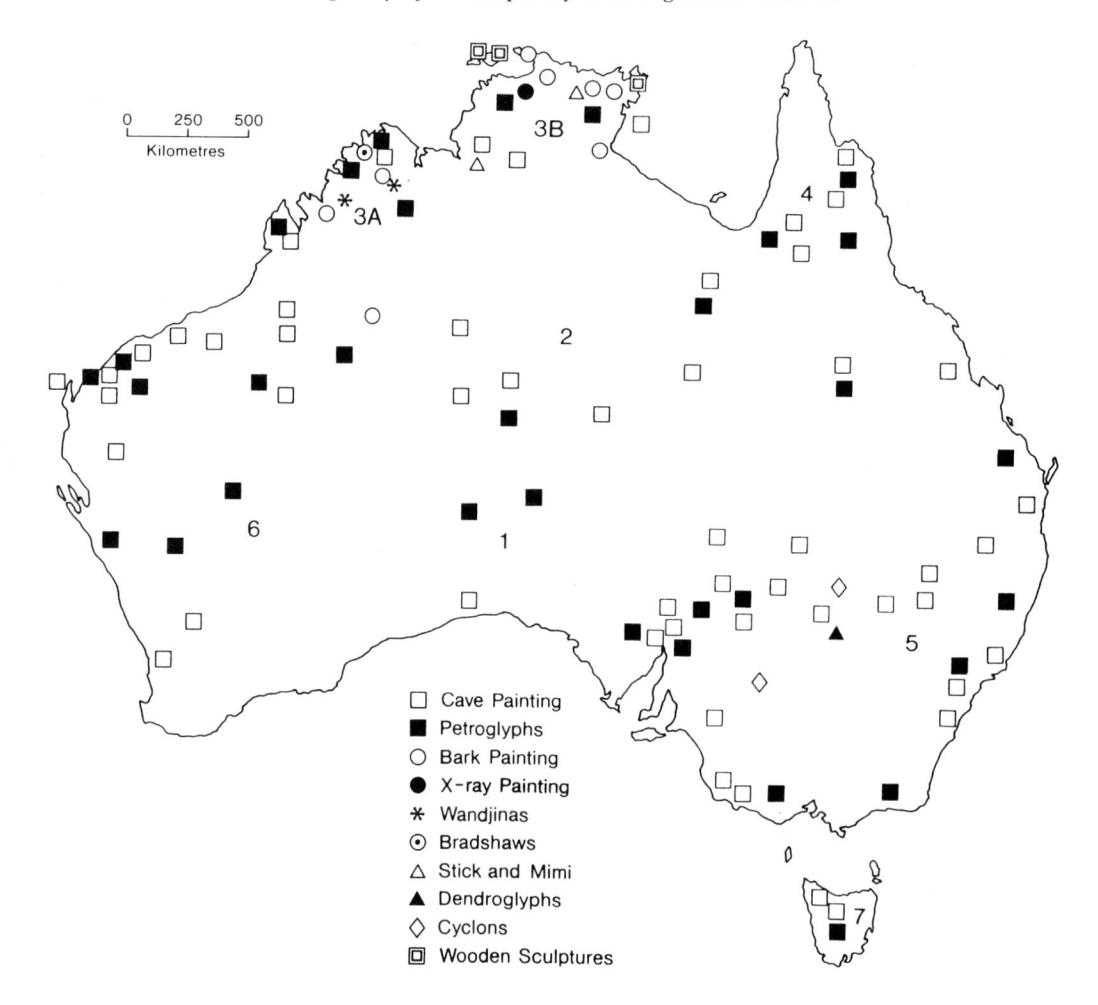

Fig. 12.1 The distribution of Aborginal art.

The size of Australia (*c.*7 700 000 square kilometres) and its climatic and environmental diversity (tropical north, desert hinterland, temperate southern region) defy any definitive attempt to classify art styles or traditions. Many authorities recognize seven distinct regions: 1. *Southern Central*, with an emphasis on weapon decoration, portable ceremonial carved, moulded, and painted constructions of wood, gypsum, and ochre (*toas*), that are complex symbolizations of geography and place, and body-painting. 2. *Central*, containing Uluru (Ayers Rock) and other important sacred spots of Dreamtime significance, marked with rock painting and engraving. The southern part of the central region is particularly rich in rock engravings in the Broken Hill area, and deep cave engravings in the Nullarbor Plain.

3A. *North-West*: the Kimberley region, representing the Wandjina and Bradshaw paintings. 3B. *North*: Arnhem Land, representing Mimi and X-ray-style paintings. Portable art in both these areas includes bark paintings, spears, and figurines. Burial posts and coffins, along with shells, skulls, and body adornments are often embellished with totemic designs (see also Table 12.1). 4. *North-East*: Cape York peninsula; similar to that of the Northern region, though less extensive in occurrence. 5. *South-East*, characterized by rock engraving and rock painting, and two distinct traditions: engraved stone objects (cyclons) of unknown significance; and carved trees (dendroglyphs)—teleteglyphs, marking ceremonial initiation grounds (*Boras*) and taphoglyphs, marking the burial sites of important individuals. 6. *West*: extensive rock paintings and engravings, especially in the area between the Murchison and De Grey rivers, and the Port Headland region. 7. *Tasmania*. (Re-drawn after Stubbs (1974), and Sutton (1988).)

12
Photogallery of contemporary hunter–gatherer rock art

12.1 Australian Aboriginal art

Andrew Lock and Margaret Nobbs

The Aboriginal art of Australia has recently moved from the province of ethno-archaeology into the circles of 'high-art culture' typified by Western art galleries and museums (see Sutton 1988a). Up to the discovery of Australia by Western imperialist global empires, and then more tenuously until the mid-twentieth century, this art was an integral part of the intact cultural practices in which it was naturally situated. Aboriginal cultures are diverse rather than homogeneous, although they are apparently related through a deep mythological tradition, commonly translated into English as 'The Dreaming' (see below), that animates the life of the various linguistic and cultural groups that have elaborated this tradition into its particular extant varieties.

There is no established taxonomy for this body of art, either geographically or historically; but particular regional clusterings can be delineated (see Fig. 12.1). In addition, Chaloupka (for example, 1983, 1984, 1985) has proposed a sequence of styles for Arnhem Land rock art (see Table 12.1 and associated figures). Styles were considered in their stratigraphic context, and the contents of different layers were analysed. A chronological sequence was constructed by the order of superimpositions of different styles, and this relative order was tied to climatological,

Figs. 12.2a and **12.2b** Extinct megafauna.
Fig. 12.2a Depiction of a striped marsupial quadruped, believed to be a thylacine, Arnhem Land (courtesy of G. Chaloupka).

Fig. 12.2b Extinct marsupial fauna from pre-estuarine period of Arnhem Land rock art (cf. Table 12.1): At left a numbat (*Myrmecobius fasciatus*), in the centre a female thylacine (*Thylacinus cynocephallus*) feeding her young, and upper right the long-beaked echidna, *Zaglossus* (from Chaloupka (1984), by permission).

geomorphological, archaeological, and historical data for the region. Additionally, climatic changes during the Pleistocene and Holocene, indexed by changing sea-levels (see Jones and Bowler 1980), brought with them a changing fauna that is represented in the changing repertoire of images, for example, by the painting of extinct Australian megafauna (Figs. 12.2a, 12.2b), followed by the appearance of paintings of estuarine fish such as giant perch and barramundi (Table 12.1), and their subsequent replacement by wetland species such as the magpie goose.

Aboriginals themselves have been reported (for example, by Stubbs 1974) to distinguish three broad categories of art: that created by themselves and used in rituals; that created by the ancestors (see below) when the world was formed (for example, Wandjina and Mimi figures from Arnhem Land, see Figs. 12.3a, 12.3b); and that which they claim no knowledge of, and which plays no part in recorded times in any traditional belief systems (for example Bradshaw figures from Kimberley, and the so-called Panaramitti convention of rock engraving). The range of techniques employed in the preparation of surfaces, materials, and their application, encompasses those used in the European traditions (see Conkey, this volume, Chapter 11), with the addition of prepared tree-bark 'canvasses', sand drawings, and bodypainting.

Aboriginal art has a distinctive and highly stylized visual tradition that depicts images whose sense is embedded in and portrays the body of myths that are collectively termed 'The Dreaming': it is thus a narrative art form, dominated more by conception than perception (see Fig. 12.4). But at the same time as being a body of informative myth, 'The Dreaming' is also the act of perceiving the environment as imbued with the results and continuing power of what was created in the 'Dreamtime', and it embodies the notion of responsibility for the environment:

all life as it is known today, human, animal, bird and fish, is part of one unchanging interconnected system, one vast network of relationships which can be traced back to the great Spirit Ancestors of the Dreamtime, Alcheringa or Tjukurpa (Isaacson 1980, p. 33).

Aboriginal creation myths inform, educate, and narratize Aboriginal consciousness:

The Dreaming is the founding story, the great drama of the creative era, in which the landscape took its present form and the people, animals, plants, and elements of the known world were created. But the Dreaming is also the inner or spiritual dimension of the present. Things contain their own histories. There is no contrast of the natural and the spiritual, and there is no geography without history and meaning. The land is already a narrative—an

Photogallery of contemporary hunter–gatherer rock art 353

Fig. 12.3 Wandjina and Mimi figures
Fig. 12.3a Wandjina figures
As reported by the indigenous Aboriginal peoples of Kimberley (north-west Australia), the Wandjina are creation ancestors, and their images were left at the places where they metamorphosed from their physical form into Spirits, leaving their images on rock. These images are retouched by human hands. They are generally broad-shouldered, cloud-like figures, their heads encircled by curved bands sporting radiating lines; and they conventionally have no mouth. They are both male and female, and have individual names. Their spirits embody the fertility of the land and its human and animal inhabitants. The cyclones and storms that bring the monsoonal rains to the Kimberley region between December and March are caused by their powers. During this period songs are sung to placate them, and their images must be tended and retouched correctly and circumspectly. It is believed that if these representations fall into disrepair that the Wandjina will depart, and the world will become barren. For details of these paintings, see Crawford (1968, 1973, 1977), and Isaacs (1980, 69–75) for the cosmologies within which they are located. Photo: P. Bindon. Reproduced by Courtesy of the University of Western Australia Berndt Museum of Anthropology.

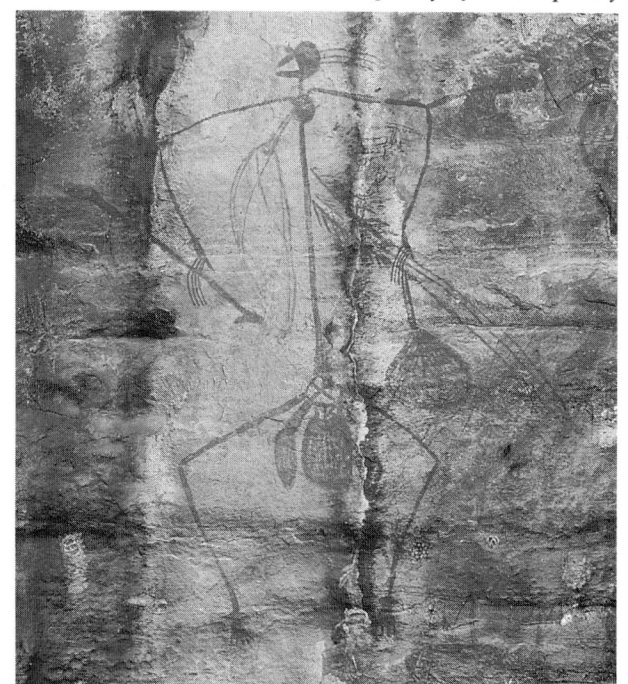

Fig. 12.3b Mimi figures
As reported by the indigenous inhabitants of Arnhem Land (north-eastern Australia), these are representations of very thin spirit hunters who live in cracks in rocks, created by the 'Mimi' themselves. They existed prior to human habitation, and are believed to have taught humans how to spear kangaroos. For details, see Brandl (1973, 1977) and also Edwards (1978). Photos: R. Edwards. Reproduced by courtesy of the University of Western Australia Berndt Museum of Anthropology.

artifact of intellect—before people represent it (Sutton 1988b, p. 1)

These narratives contain a number of elements: some of these relate the narrative to the forms and tempos of life of the environment; some to historical reality; and some to the fictitious exploits of the phenomena chosen to embody form, tempo, and history. Where geographical information is embodied in art, it is symbolically encoded along with the myths that are instantiated in the environment (see Fig. 12.4).

In some cases, the stylistic elements of the visual art have been hypothesized to constitute the oldest continuous human series of conventional representations (for example, Sutton *et al.* (1988, pp. 182–4); but see also Layton 1989, and Maynard 1977, 1979). For example, many of the motifs of modern Central Australian art occur in a sequence of prehistoric examples from the Olary region (South Australia) that yielded cation-ratio dates from 31 400 to 1400 years BP. (Nobbs and Dorn 1988, see Fig. 12.5. In addition, there are a number of prehistoric sites indicating the co-existence of *Homo sapiens sapiens* and a mammalian and aviform fauna that includes very large marsupial species: 3-metre-tall giant kangaroos; giant wombats; *Diprotodon*, a marsupial the size of a rhinoceros; and giant emus, *Genyornis*, the size of the giant moas found in New Zealand in historical times. This fauna was associated with a lush, pre-Ice Age vegetation resulting from a much higher rainfall than has since been the norm for most of the Australian continent.

It is likely that memories of these ancient times are embodied in the extant 'myths' of the 'Dreamtime' that animate Aboriginal art (see Isaacs 1980, and also Figs. 12.2a, 12.2b). Giant animals, and the representation of animals in giant proportions, are a recurring theme, and fossil sites at which their bones are visibly located are used in justification of their historical reality. Such myths not only convey the feelings these giant animals are able to evoke in human consciousness, but perhaps also elaborate other embedded memories of how these animals came to disappear during times of marked climatic change and increasingly sophisticated

Fig. 12.4 The narrative nature of Aboriginal art From Berndt and Berndt (1982), by permission.
Photo: J. E. Stanton. Reproduced by Courtesy of the University of Western Australia Berndt Museum of Anthropology (where the item is held) and the photographer.

This traditional north-eastern example is associated with country and mythology, and is also emblemic. The painting concerns the Wawalag myth, and is localized at Muruwul (or at Mirara-minar) on the mainland near Milingimbi, home of the Yulunggul python (see R. Berndt 1951). From right to left, following each elongated panel (of which there are seven): (1) cliffs at Duridjmanu (*yiridja* m.: all the rest of this country is of the *dua* moiety), with dots representing rocks; (2) tracks of the *gardjambal* kangaroo; (3) this whole panel represents the *gareiga* stringybark tree which Lightning Snake (*wididj*—that is, Yulunggul) has split with his lightning so that it is in pieces (rectangular designs); (4) from top to bottom is spring water running from Riala to Muruwul; the dots are cabbage-palm fruit (or 'nuts') floating in the water; then follow a cabbage palm, further spring water, another palm, and more water; (5) new palm trees, spring water, and 'nuts'; (6) new palm plants, water, and 'nuts'; (7) spring water with floating 'nuts', followed by three pointed stone blades against a black background, the dots in this case representing rocks; then spring water and 'nuts', stone spearheads and spring water and 'nuts'. The stone spear-blades belong to Woial, or Wudal, the mythic Honey Man, associated with the Wawalag Sisters (see C. H. Berndt 1970).

* Painted at Yirrkala, 1946, by Mawulan, Riradjingu *mada*, Wurulul *mala*, *dua* m. Berndt coll. 52 × 60 cm

Fig. 12.5 Continuity of style in Aboriginal petroglyphs in South Australia.

Rock varnish is a thin coating of manganese and iron hydroxides, clay minerals, and trace elements, derived from atmospheric fallout, that forms a stable coating on rock surfaces, especially in arid regions (Dorn and Oberlander 1982). Nobbs and Dorn (1988) took rock varnish samples from 24 petroglyphs inscribed on the outcrop shown in Fig. 12.5a at Karolta, near Olary, South Australia. Only micrograms of material are needed for the determination of varnish cation-ratios, and these can then be compared to ratios obtained from materials of known ages. Twenty-four petroglyphs sampled from this outcrop yielded dates ranging from 1.4k years BP to 31.5k years BP. Ages for closely-spaced petroglyphs were found to vary markedly (see Fig. 12.5b). Animal and bird-track motifs range from 1.4k to 25.8k years BP (Fig. 12.5c). The range of motif categories is constant across this time-period, and all are made by the same technique. This lack of change is similar to the South African situation reported by Lewis-Williams (1984).

Fig. 12.5a General view of inscribed outcrop studied by Nobbs and Dorn (1988), showing placement of motifs numbered 20, 21, 22, and 23.

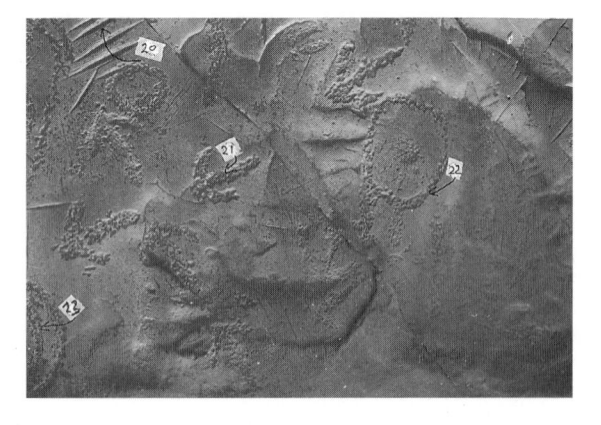

Fig. 12.5b Group of closely spaced motifs of varying ages: 20, 16.1k years BP; 21, 25.7k years BP; 22, 24.7k years BP; 23, 31.6k years BP. 21 is from a line of 6 'bird-tracks', crossed, nearly at right angles, by a line of 7 similar 'bird-tracks'. Both these lines come from a group of dots, suggesting emus leaving nests.

human predation. Simultaneously, it may be that the 'Dreamtime' myths that sustain and support depictive art use such animals, and animals represented in gigantic proportions, as icons from which to symbolically elaborate themes of interpersonal conflict, the temporality of human existence, the importance of water in a dry-climate way of life (or of too much water in the tropics), and other themes. As with the southern African tradition, then, the art of Australian Aboriginals functions polysemically.

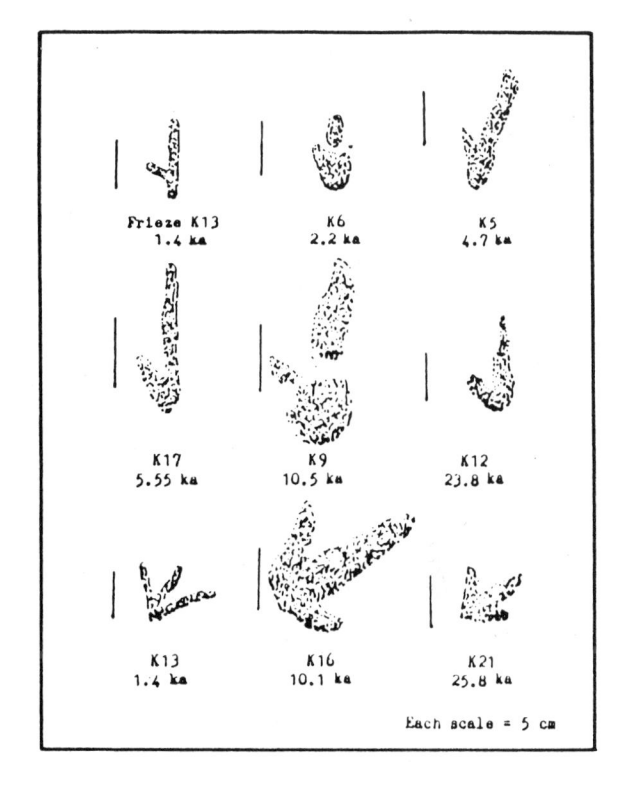

Fig. 12.5c Dating of animal (macropod) and bird-track motifs from outcrop. (Figs. 12.5a–12.5c courtesy of Margaret Nobbs.)

Table 12.1 Proposed chronology of Arnhem Land plateau rock art

Years BP	Period	PHASE	STYLE	MAJOR OR IDENTIFYING SUBJECT
35 000 20 000	Pre-estuarine	Naturalistic	Object imprints Large naturalistic animals and humans	Hand prints, grass prints, thrown objects, *Palorchestes* *Zaglossus*, *Tachyglossus*, thylacine, macropods, rock python, freshwater crocodile (a) (b)
			Dynamic figures	Human beings, animal-headed beings and other anthropomorphs, *Tachyglossus*, macropods, rock python, stencils: hand of 3MF convention,[1] boomerangs, clubs, spears (c) (d)
		Stylization	Post dynamic figures	Mainly human beings, some macropods
		Schematization	Simple figures with boomerangs (SFB)	Mainly human beings, man using fighting pick (e) (f)
		Naturalistic symbolism	Yam figures	Anthropomorphized yam, phytomorphized animals: flying fox, birds, short necked turtle, Rainbow Snake; abstract symbol of segmented circle (g) (h)
9 000 7 000	Estuarine	Intellectual realism	X-ray descriptive	Barramundi, mullet, saltwater crocodile, estuarine catfish, Lightning Man, stone-headed spear (i) (j) (k)

1 000	X-ray decorative	Magpie geese, didjeridu
		Macassan praus, European boats, buffaloes, horses, guns, sorcery paintings
		(m)
		(l)
150	Casual paintings	
Freshwater contact		

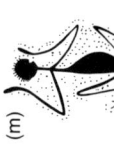

l. 3MF convention refers to the stencilling of the human hand with the three middle fingers held tightly together, and only the thumb and little finger extended, a style unique to this period.

a. Animal tentatively identified as *Palorchestes*, a marsupial tapir, a member of the extinct Australian megafauna. Extinction was at around 18 000 years BP (see Murray and Chaloupka 1984). (From Chaloupka 1983.)

b. *Zaglossus*, a long-beaked echidna extinct in Australia since c.18 000 years BP, but surviving in the highlands of New Guinea (Murray 1978). (From Chaloupka 1983.)

c. Hunter spearing an emu (from Chaloupka 1983). 'Some of the human and anthropoid figures, animals, and even their tracks, are surrounded by dashes, and similar marks emanate from their mouths. These suggest perceptual inductions and inferences, depicting the non-visual aspects of the artists' sensory experience such as sound, smell, force, anxiety, and in the case of the animal tracks perhaps their freshness' (Chaloupka 1983, p. 10).

d. Frontal figures: male in a large head-dress with long hair, boomerangs, and 'skirts'; female carries a digging-stick (from Chaloupka 1984).

e and f. The stylization apparent in the previous style is now further schematized, but head-dresses and 'skirts' continue to be represented. f. may represent a scene of conflict. (From Chaloupka 1984.)

g. A combination of a yam figure in association with a segmented circle and yam strings. The focus of this representational style on yam species of the genus *Dioscora* suggests they have an importance to the artist over and above that found in earlier periods. This style developed during a period of rising, post-glacial sea-levels. Yams require c.120 cm annual rainfall for growth, and may have re-colonized this area at this time, to form an important source of carbohydrate for people who had previously exploited now-submerging grasslands for cereals. The cultivation of yams is associated with rituals in New Guinea and Melanesia, described by Haudricourt (1964) as the 'civilisation de ligname'. Contact between New Guinea and Australian peoples would have been broken during this period, and these paintings may be the expression of an earlier 'yam civilization'.

h. It is during this period that images of the Rainbow Snake first make their appearance. The Rainbow Snake continues to play a major role in present-day stories and ceremonies in this region. It is closely associated with water in all its forms, and it may have gained its significance from the movement of snakes as water levels rose during this period, and thence their theorized metamorphosis (cf. Wandjina figures, Fig. 12.3a) into the serpent-like rainbows that became more common with increasing precipitation. Chaloupka notes (1983, p. 11) that the Rainbow Snake's 'representation in this early rock art style and its portrayal in the subsequent styles documents this as the world's longest continuing religious belief' [more strictly, image, since the continuity of beliefs is conjectural—eds] (from Chaloupka 1983).

i. Giant perch or barramundi (*Lates calcarifer*) in descriptive X-ray style. Tacon (1988) notes that over ten species of fish are identifiable in the contemporary use of this style by their accurately represented internal and external anatomy: 'and hence [their images] can be used to illustrate particular ideas, experiences or myths more effectively when storytelling' (ibid., p. 3) (from Chaloupka 1983).

j. Rock ringtail possum (*Pseudocheirus dahli*) in decorative X-ray style, the subdivision of the animal's body being for decorative purposes, rather than portraying anatomical details of internal organs. Both descriptive and decorative forms have continued down until the most recent times (from Chaloupka 1983).

k. Namarrkon (Namargon), 'The Lightning Man'. Intense electrical storms would have become prevalent during the prolonged wet seasons developing at this period (cf. Rainbow Snake, h above), and these are represented via the image of the Lightning Man, the anthropomorphized creator of thunderstorms (from Chaloupka 1983).

l. Previous styles continue to be used during the period of contact with Europeans and Macassan traders, as in this X-ray depiction of a steamship and its cargo (from Chaloupka 1984).

m. A large number of sorcery figures are associated with this period 'being perhaps a direct product of stresses and sicknesses introduced by contact, such as the documented epidemics of influenza, measles, and leprosy which affected perhaps the majority of the population in the vicinity of Oenpelli (Dashwood 1897), where most of these paintings are found' (Chaloupka 1983, 15). (From Chaloupka 1984.)

12.2 Southern African Bushmen[1] rock art

J. David Lewis-Williams, Thomas Dowson,
Andrew Lock, and Charles Peters

Unlike Australian Aboriginal art, the rock art traditions of the southern African Bushmen came to an end around the turn of this century. There are, however, written accounts of Bushmen world-views surviving from the 1870s, primarily by Wilhelm Bleek, Lucy Lloyd, and Joseph Millerd Orpen (see Lewis-Williams and Dowson 1989) that provide the necessary context for interpreting the copious rock art that survives. In addition, even though present-day Bushmen of the Kalahari Desert in south-western Africa have no rock art tradition of their own—and those who know of paintings from nearby areas claim them to be products of god (Biesele 1979)—recent fieldwork reveals close links between surviving belief-systems and those recorded in the last century. Thus present-day social accounts of the world as Bushmen conceive it can also inform interpretative frameworks within which to understand the art. Here we rely on the recent review by Lewis-Williams and Dowson (1989).

Bushman rock art is widely distributed in southern Africa (see Fig. 12.6). Climatic conditions have prevented the preservation of much parietal art, so that historical sequences such as that reconstructed for Arnhem Land (see Table 12.1) are not attainable. Radiocarbon dates of 500 years BP have been established for one site in the western Cape, working from samples of paint. Portable painted art objects have been dated more thoroughly, largely by radiocarbon dating of the strata from which items have been recovered. The oldest examples currently known, from the Apollo 11 rock-shelter in southern Namibia, are dated at 26 300 years BP (compared with Lascaux cave at 17 000 years BP) (Wendt 1976). Portable engraved art from Wonderwerk Cave near Kimberley has been dated at 10 000 years BP (Thackeray et al. 1981). Along with Australian Aboriginal art, then, the art of southern Africa represents one of 'the longest tradition[s] human beings have produced.... Indeed, some ... depictions ... probably represent concepts that originated deep in the human past' (Lewis-Williams and Dowson 1989, p. 23).

A wide range of techniques were employed in the preparation of materials and their employment. Organic and inorganic pigments were used from untransformed natural sources—ochre, silica, clays, manganese, etc.—and processed ones—ochre, for example, produces a variety of shades when fired. The media in which these pigments were suspended range from antelope blood to urine, egg-white, and plant sap. Paint was sometimes applied by finger; but feathers, quills, and small bones were used for details. Engravings were made in three ways: by pecking, incising, and scraping.

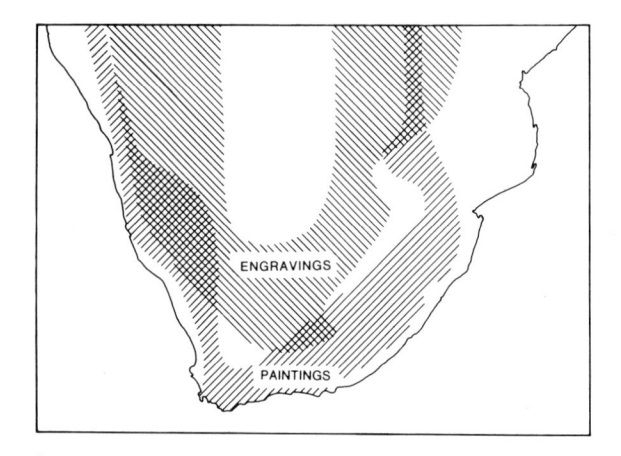

Fig. 12.6 Paintings and engravings are widely found in different regions of southern Africa. Rock paintings are principally found in the mountainous areas that provide the rock-shelters in which they are situated. Engravings are more typical of the interior plateau. Regional divisions of the art in terms of sites, selections of colours, content, and style have been proposed, but presently remain difficult to substantiate.

This art cannot be understood outside the social traditions and practices in which it was embedded. Bushmen draw no clear distinction between secular and sacred. For them, the sacred is an integral part of everyday life, and infuses their perception of the 'real' world. Bushman religion is essentially shamanistic. Their shamans, however, are 'ordinary' people, and not a small, select, privileged class. About half the men and one-third of the women in a present-day Kalahari Bushman camp may be shamans, all of whom will have been through an apprenticeship of several years before they learn to enter and control the trance state that constitutes shamanship. The trance state is one in which a 'supernatural potency' is experienced. This potency imbues many other things in the world, especially large animals. It must be attained and controlled by the shaman, whence it can be used to heal the sick, for the control of animals, for rain-making, and for out-of-the-body travel. A shaman becomes full of potency through a trance-inducing dance, altering his or her state of consciousness through hyperventilation, intense concentration, and sustained rhythmic dancing.

The art tradition of the Bushmen is essentially a pictorial rendition of this trance experience, realized in the context of their conceptions of the supernatural potency of the objects and animals that inhabit their world. Laboratory research, and cross-cultural field studies, on chemically-induced and naturally-attained trances and altered states of consciousness have led to the proposal that 'some trance experiences are human universals' (Lewis-Williams and Dowson 1989, 60; cf. Siegel 1977; Reichel-Dolmatoff 1978), and to the delineation of stages in the attainment of hallucinogenic trance states. Three stages have been proposed.

In Stage 1, while all senses hallucinate, visual imagery appears the most immediate and powerful. This imagery is termed *entoptic*, and comprises luminous geometric shapes, including zigzags, dots, grids, vortices, etc. (see Fig. 11.31, p. 340), that derive from the underlying structure of the human nervous system, and hence are pan-cultural.

In Stage 2, the *construal* stage, people try to make sense of entoptic phenomena by elaborating them into objects with which they are familiar (see Fig. 12.7). This process of making sense is linked to a person's state of mind and expectations. For instance, if a person is thirsty, an ambiguous round (entoptic) shape can be elaborated into a cup of water, or, if he or she is fearful, into an anarchist's bomb (Lewis-Williams and Dowson 1989, p. 63).

Changes in hallucinations occur in Stage 3. Western laboratory subjects report experiencing a vortex or spinning tunnel, and shamans of many cultures talk of 'entering a hole in the ground' (ibid., p. 66). Spontaneous *iconic* hallucinations of people, animals, etc., occur, these forms being culturally variable, as they rely on an individual's memories, and are often associated with powerful emotional experiences, both of which are highly culturally determined. 'For Bushmen, animals are the most emotionally charged things, and it is therefore not surprising that

Fig. 12.7 An image relating to the construal stage of trance-induced visions. U-shapes are common entoptic forms. The honeycombs of wild bees take a nested U-shaped form. Many U-shaped painted images are accompanied by representations of bees, suggesting that shamans interpreted these entoptic images as honeycombs because of bees' being a symbol of potency. The 'buzzing in the ears' that accompanies trance states helps predispose to such an interpretation.

animals feature prominently in their visions' (ibid., p. 66).

Identification with the hallucinated object is a common experience in this third stage of trance—that is, the hallucinating individuals feel that they 'become' the objects. 'Bushman rock art contains numerous depictions of human beings with animal features that depict just such an experience. These depictions are called *therianthropes*' (ibid., p. 60).

Therianthropic images (see Fig. 12.8), then, result from images spontaneously hallucinated in trance states. These images take the forms they do because of the emotional significance certain animals have in the Bushmen concept of potency. They are transformed in the medium of painted representation to meld human and animal forms as a metaphorical depiction of this experience of identification.

Fig. 12.8 Two therianthropic images representing the identification of humans with potent animals. Trance blood pours from one image's nose. The additional digits on one limb of this image are more likely to be a representation of the polymelial hallucinations experienced in trance states than errors of drawing. The other image is inscribed with zigzags and dots. Given the way these 'fall off' the images' legs into heart-shapes similar to those incorporated in the objects wielded by both images, they are more likely to be representative of the entoptic imagery that signals a confusion of reality and hallucination in trance states than to be depictions of actual body-painting.

Simultaneously, there arise somatic hallucinations accompanying the transitions into trance identifications with animals. Katz (1982) was told by one informant: 'In your backbone you feel a pointed something and it works its way up. The base of your spine is tingling, tingling, tingling.'

These feelings are often represented in the art in stylized form. Polymelia, the experience of extra digits and limbs, is also vividly experienced, and is reflected in paintings where the images show polydactyly. A further hallucination is the illusion of changes in body-image. Shamans speak of 'potency in the stomach beginning to rise up the spine until it "explodes" in the head. This experience increases until they experience a sense of attenuation in their limbs and body' (Lewis-Williams and Dowson 1989, p. 77): a man may feel as tall as a tree (see Fig. 12.9).

In addition to direct images arising from the different stages of trance attainment, metaphorical images of the process are also depicted in the rock paintings: 'what we see on the rock is a blend of real and visionary elements . . . the art depicts the shaman's multi-dimensional view of reality' (ibid., p. 51). Death is one powerful metaphor (see Fig. 12.10). First, shamans 'die'—are released from their earthly bodies—when they cross to the spirit world of the trance. Second, there are similarities between the behaviour of the shaman in trance (Fig. 12.11) and that of an animal dying as a result of being shot by a poisoned arrow: staggering, a lowering of the head, bleeding from the nose, sweating, and collapsing. In addition, the hair on a dying eland's back stands on end, and is metaphorically related to the 'hair' that grows on the shaman's back during trance—the 'tingling, tingling, tingling' in the spine. The underwater world is another source of metaphorical representation. Shamans in trance experience difficulty in breathing, sounds in their ears, altered vision, a sense of weightlessness, and finally unconsciousness—all experiential links with near-drowning.

Viewed from this perspective, then, there is clearly a lot more involved in interpreting a painted image of a dying eland than positing it to be part of a representation of a realistic hunt, as the old 'hunting-magic' interpretations have done in the past. Nor may a snake be claimed to be merely a dangerous animal, for snakes often hide underground too, and thus can carry a metaphorical reference to trance-attainment. Snakes are often represented as slithering in and out of the rock faces on which they are painted, and thus allude to the shaman's trance passage from the everyday to the spirit world: the trajectory of a shaman on out-of-the-body travel.[2]

The superpositioning of images one on the other is also a convention in this art tradition (see Fig. 12.12).

Fig. 12.9 'A man may feel as tall as a tree': these images are 67 cm in length. All three images in this group assume trance-dance postures. On the left, the hands are held on the chest—an often represented posture that may be related to the tingling sensation experienced in a person's 'front spine'; the middle figure shows the arms in a typical dancing posture; the right figure shows a backward-facing arm posture that is characteristic of trance-state dancing postures.

It is clearly not the case that a subsequent artist took no account of a predecessor's work. First, there was usually plenty of 'clear' rock available to accommodate a new image, suggesting that the superpositioning is deliberate. Second, a number of images show 'factitious superpositioning'—the addition of a new image in two parts, so that it appears to underlay rather than overlay the original, or even, on occasion, to be 'interleaved' between an original and an overlaid figure. Third, overlays are selective: eland are often overlaid on elands and humans, but seldom are humans overlaid on elands. Complex images built up by successive artists are, therefore, neither 'scenes' nor the result of ignoring previous images, but representations of hallucinations: 'the piling up of images, apparently without regard to any sort of relationship they may have in "real" life, is also part of the trance experience' (ibid., p. 150). But depictions *are* related, 'so that they could interact on what we might call a symbolic level' (ibid., p. 155), within the world view that sustains Bushman socio-cultural practices.

Notes

1. There is no single word used by all Bushmen to designate themselves; each group has its own word. 'Bushmen' is a translation of the Dutch *Boschjesmans*, applied to all indigenous groups by colonists in the seventeenth century. Today the term has acquired racist overtones, and because of this, many scholars refer to them as the *San*. However, within the indigenous Nama KhoiKhoi language and culture from which it is derived this term itself has highly pejorative connotations. It seems inappropriate to apply an indigenous label of opprobrium to these people. There is thus no easy way out of this problem, and here we have retained 'Bushmen'.

2. Solomon (1992) has proposed that, in addition to these interpretations that are centred on the trance dance, many features of the art may be related to other aspects of the Bushmen's culture and worldview, particularly to gender ideology and the complex of notions and practices surrounding feminine gender and female sexuality. She sees the interlinked themes of sexual symbolism, dangerous feminine potency, and rain as probably as important in the art as the shamanistic elements. The gendered forms and sexual idioms that she explores and illustrates certainly do seem to play a vital role in this art.

Fig. 12.10 Metaphorical representations of trance occur because of the difficulty involved in describing the experience without comparing it to more prosaic everyday experiences that the uninitiated can understand. Bushman 'trips' are commonly rendered by metaphors of 'death' or 'dying'. This metaphor underlies rock paintings of dying animals. There are direct similarities between a 'dying' shaman and a dying eland that has been shot by a poisoned arrow (see text). Lewis-Williams and Dowson (1989, p. 51) describe the image in this figure as follows:

A dying eland lowers its head and has exaggerated hair standing up over much of its body. When an animal reaches this stage, its neck muscles relax and its head swings from side to side: here it faces the viewer. The forelegs are giving way under the animals' weight, and the hindlegs are crossed as it stumbles.

With this dying eland are four shamans, three of whom have been transformed so that they share features of the eland. Two have hair standing on end, hoofs, and features of antelope heads. The one holding the eland's tail has his legs crossed in imitation of the eland's crossed legs. Above right, another shaman has an antelope head and fetlocks, and appears to be wearing a kaross [a skin cloak, cf. Fig. 12.12]. The animal characteristics of these figures show that their 'death' is analogous to the death of the eland with which they are painted and that, in this 'death', they become like the eland.

In the centre of the group, a dancer is in the bending-forward posture and has its arms extended [cf. Fig. 12.12]. Its head is raised and a short kaross hangs down from its neck. In contrast to the other figures, it is fairly 'realistic'. Its presence again stresses one of the principles of Bushman rock art: what we see on the rock is a blend of real and visionary elements. In other words, the art depicts the shaman's multidimensional view of reality.

Fig. 12.11 This complex painting includes many references to shamanistic experience of trance-dancing. The bent-forward posture of many of the figures represents the 'boiling' of potency, which is accompanied by a contraction of the stomach muscles into a tight and painful spasm, resulting in this posture. When a shaman is so taken, he will often use one or two dancing-sticks to support his weight (top left). The therianthropic head of this image also alludes to the harnessing of potency in trance states, represented here by a visual identification of the shaman with a potent animal. A number of the dancers hold fly-whisks. These function symbolically to ward off the 'arrows of sickness', which are claimed to be invisible arrows shot by malevolent shamans to make their targets ill, and which are drawn out into the shaman's body during curing rituals, later to be expelled. They may also allude to the 'pins- and-needles' experience that often accompanies trance attainment. Blood pours from the noses of some dancers (left), as it often does from trance dancers. Hands held above the head (top middle) is a typical dancing posture, adopted in the stage prior to attaining trance. One figure (extreme left) clearly represents the 'arms-extended-backwards' position that is often adopted when shamans ask god to imbue them with greater potency. One man in a sitting position wears a skin cloak: shamans feel 'cold inside' when they 'die', and need their cloaks to keep them warm. Many of the figures have dancing-rattles wound around their calves. These may be made from dried cocoons, seed-pods, or antelope ears, and filled with pebbles or pieces of ostrich-egg shells. The rattles are held to possess potency, and are often shaken along a person's back to extract 'arrows of sickness'.

All around the dancers there are white flecks that probably depict potency. At a dance, trancing shamans, but no one else, can see both potency and sickness [thus, these visions may be of entoptic origin]. This scene thus also has an element that suggests it depicts a shaman's, rather than an ordinary person's, view of a dance (Lewis-Williams and Dowson 1989, p. 17).

Fig. 12.12 A complex scene, painted piecemeal over a long period of time by a number of artists. A pooling of spiritual experience was a feature of Bushman shamanism, and it was 'therefore fitting for an artist to add his or her contribution to a developing panel of revelations' (Lewis-Williams and Dowson 1989, 151). Such panels are not 'scenes' in the Western realist tradition, but concatenations of interwoven symbolic images that both reinforce each other and set up internal tensions between themselves. They function polysemically to attain, for members of their culture, an extraordinarily rich and powerful impact. This composition portrays a number of eland, central symbols of potency. The animal at bottom right in the 'head-down' posture is perhaps dying (cf. Fig. 12.10). Immediately above it is an 'arms-back' figure (cf. Fig. 12.11), and to the left a figure with a raised knee, both characteristic dance postures. Neither of these figures is actually part of a dance scene, and hence they are being used here to refer directly to trancing. A cloaked figure (cf. Fig. 12.11) is joined to the dying eland. The eland to the left is wounded, and apparently 'attacked' by an archer, and is positioned above an array of hunting-bags and arrows (bottom left). But these arrows have connotations of trance experiences as well as possibly of real hunting (cf. Fig. 12.11). Middle right there is a snake (see text) with an antelope's head. This apparently chaotic juxtaposition of images is suggested by Lewis-Williams and Dowson (1989, pp. 154–5) to be analogous to the use of images in contemporary Western advertisements: A sword-wielding Samurai warrior, for example, may be associated with a pick-up truck. The combination allows the qualities of the Samurai to interact with the truck on a level far removed from everyday life. In complex panels, the Bushman artists were similarly linking depictions so that they could interact on what we may call a symbolic level.

12.3 Editorial note on the significance of contemporary hunter–gatherer art for understanding the palaeolithic materials

It could be tempting to take aspects of these discussions of contemporary hunter–gatherer art and use them as an interpretative base for understanding palaeolithic art. The two famous therianthropic 'sorcerer' figures from the French cave 'Les trois frères' are shamans indeed, as the blood pouring from the nose of one (see Conkey, Chapter 11, this volume, Fig. 11.19) a second point of parallel, reinforces. These shamans experienced altered states of consciousness, represented by the graphic portrayal of the spine in this latter figure. Cave art in the Palaeolithic has itself an inbuilt hallucinatory property, in that a flickering flame from the torch needed for seeing them has the effect of making the depicted animals seem to move (Bahn and Vertut 1988, p. 110). Different insights are also made possible by taking on board aspects of Australian Aboriginal art. However, it is exactly this tactic that the contemporary materials argue against. Contemporary images are known to be located within unique ideologies and world-views. As for the palaeolithic images, we cannot know the world-views that constituted these images as art, and therefore we cannot definitively interpret them. At this point in scholarship, then, the relevance of recent hunter–gatherer art to an understanding of palaeolithic art is that it is salutary, rather than salvatory; and suggestive rather than inclusive.

References

Bahn, P. and Vertut, J. (1988). *Images of the Ice Age*. Bellew Publishing Company, London.

Berndt, R. M. and Berndt, C. H. (1982). *Australian art: a visual perspective*. Methuen Australia, Sydney.

Biesele, M. (1979). Old K"xau. In *Shamanistic voices: a survey of visionary narratives* (ed. J. Halifax), pp. 54–62. Dutton, New York.

Brandl, E. J. (1973: 2nd Edition, 1982). *Australian Aboriginal paintings in western and cultural Arnhem Land: temporal sequences and elements of Style in Cadell River and Deaf Adder Greek art*. Australian Institute of Aboriginal Studies, Canberra.

Brandl, E. J. (1977). Human stick figures in rock art. In *Form in indigenous art* (ed. P. J. Ucko), pp. 220–42. Australian Institute of Aboriginal Studies, Canberra.

Chaloupka, G. (1983). Kakadu rock art: its cultural, historic and prehistoric significance. In *The rock sites of Kakadu National Park—some preliminary research findings for their conservation and management* (ed. D. Gillèspie), pp. 2–33. Special Publication No. 10, Australian National Parks and Wildlife Service, Canberra.

Chaloupka, G. (1984). *From paleoart to casual paintings*, Monograph Series No. 1, Northern Territory Museum of Arts and Sciences, Darwin.

Chaloupka, G. (1985). Chronological sequence of Arnhem Land plateau rock art. In *Archaeological research in Kakadu National Park*, Special Publication No. 13, (ed. R. Jones), pp. 269–80. Canberra: Australian National Parks and Wildlife Service.

Crawford, I. M. (1968). *The art of the Wandjina*. Oxford University Press, Melbourne.

Crawford, I. M. (1973/1978). Wandjina paintings. In *The Australian Aboriginal heritage: an introduction through the arts* (ed. R. M. Berndt and E. S. Phillips), pp. 108–117. Australian Society for Education through the Arts and Ure Smith, Sydney.

Crawford, I. M. (1977). The relationship of Bradshaw and Wandjina art in north-west Kimberley. In *Form in indigenous art* (ed. P. J. Ucko), pp. 357–69. Australian Institute of Aboriginal Studies, Canberra.

Dashwood, C. J. (1897). *Government Resident report on the Northern Territory*. (Cited in Chaloupka 1983.)

Dorn, R. I. and Oberlander, T. M. (1982). Rock varnish. *Progress in Physical Geography*, 6, 317–67.

Edwards, R. (ed.) (1978). *Aboriginal art in Australia*. Aboriginal Arts Board and Ure Smith, Sydney.

Haudricourt, A. G. (1964). Nature et culture dans la civilisation de l'igname: origins des clones et des clans. *L'Homme*, 4, 93–104.

Isaacs, S. (ed.) (1980). *Australian dreaming: 40,000 years of Aboriginal history*. Lansdowne Press, Sydney.

Jones, R. and Bowler, J. (1980). Struggle for the savanna: northern Australia in ecological and prehistoric perspective. In *Northern Australia: options and implications* (ed. R. Jones), pp. 3–31. Research School of Pacific Studies, Australian National University, Canberra.

Jones, R. and Negerevich, T. (1985). A review of previous archaeological work. In *Archaeological Research in Kakadu National Park*, Special Publication No. 13, (ed. R. Jones), pp. 1–16. Australian National Parks, and Wildlife Service, Canberra.

Katz, R. (1982). *Boiling energy: community healing among the Kalahari ! Kung*. Harvard University Press, Cambridge, Mass.

Layton, R. (1989). The political use of Australian Aboriginal body painting and its archaeological implications. In *The meaning of things: material culture and symbolic expression* (ed. I. Hodder), Chapter 18. Unwin Hyman, London.

Lewis-Williams, J. D. (1984). Ideological continuities in prehistoric southern Africa: the evidence of rock art. In *Past and present in hunter gatherer studies* (ed. C. Schrire), pp. 225–52, Academic Press, New York.

Lewis-Williams, J. D. and Dowson, T. (1989). *Images of power*. Southern Book Publishers, Johannesburg.

Maynard, L. (1977). Classification and terminology in Australian rock art. In *Form in indigenous art* (ed.

P. J. Ucko), pp. 387–402. Australian Institute of Aboriginal Studies, Canberra.

Maynard, L. (1979). The archaeology of Australian Aboriginal art. In *Exploring the visual art of Oceania* (ed. S. M. Mead), pp. 83–110. University of Hawaii Press, Honolulu.

Murray, P. (1978). Late Cenozoic monotreme anteaters. *The Australian Zoologist*, **20**, 29–55.

Murray, P. and Chaloupka, G. (1984). The Dreamtime animals: extinct megafauna in Arnhem Land rock art. *Archaeology in Oceania*, **19**, 105–16.

Nobbs, M. F. and Dorn, R. I. (1988). Age determinations for rock varnish formation within petroglyphs: cation-ratio dating of 24 motifs from the Olary region, South Australia. *Rock Art Research*, **5**, 108–46.

Reichel-Dolmatoff, G. (1978). *Beyond the Milky Way: hallucinatory imagery of the Tukano Indians.* UCLA Latin America Centre, Los Angeles.

Siegel, R. K. (1977). Hallucinations. *Scientific American*, **237**, 132–40.

Solomon, A. (1992). Gender, representation, and power in San ethnography and rock art. *Journal of Anthropological Archaeology*, **11**, 291–329.

Stubbs, D. (1974). *Prehistoric art in Australia.* Macmillan, Melbourne.

Sutton, P. (ed.) (1988*a*). *Dreamings: the art of Aboriginal Australia.* Viking Books, Melbourne.

Sutton, P. (1988*b*). *Dreamings: exhibition notes.* The Asia Society Galleries, New York.

Tacon, P. S. C. (1988). Identifying fish species in the recent rock paintings of Western Arnhem Land. *Rock Art Research*, **5**, 3–15.

Thackeray, A. I., Thackeray, J. F., Beaumont, P. B., and Vogel, J. C. (1981). Dated rock engravings from Wonderwerk Cave, South Africa. *Science*, **214**, 64–7.

Wendt, W. E. (1976). 'Art mobilier' from Apollo 11 Cave, South West Africa: Africa's oldest dated works of art. *South African Archaeological Bulletin*, **31**, 5–11.

Part III
Ontogeny and symbolism

Editorial introduction to Part III: Ontogeny: symbolic development and symbolic evolution

1 Ontogeny and phylogeny

Historically, the juxtaposition of ontogeny and phylogeny has resulted in parallels being proposed between the two, parallels of both form and process. Gould (1977) has provided a detailed history and survey of these issues. Many of the various parallels proposed by biologists have been adopted and elaborated within both psychology and anthropology.[1] But, as Gould demonstrates at length, the relationships that have been proposed to hold between these two processes generally tangle together all kinds of muddled points and thinking, not all of which he clarified. Here we attempt another 'first step' at clearing some of these tangles.

The view we work towards is that the 'evolution' of any behaviour needs to be considered in terms of both the somatic hardware that supports it and the ontogeny of the behaviour itself, and of the relation between the two. With regard to language, for example, the evolution of the *somatic capacity* for language needs to be held separate from the evolution, or elaboration, of language systems. The ontogenetic course of language-learning will probably vary depending on the forms in which the language system being learned is elaborated, and this will feed back into the cognitive systems that are supported by the developing somatic hardware. This feeding-back may be 'responded' to by heterochronic shifts in the control of the elaboration of the somatic hardware that underwrites language. Heterochronic processes that have been proposed as relating ontogeny and phylogeny are appropriate to understanding the evolution of capacities supported by somatic hardware, but not the elaboration of the behavioural systems such as language that they support. Relations between the two domains of change need, in these latter cases, to be established on different grounds, and to take cognizance of the fact that the existence in a developing organism's environment of the system it is acquiring will place demands on the developmental process such that the course of ontogeny cannot be expected to be a direct reflection of the course of phylogeny *in any instance*.

We begin by outlining some of the history of this topic, and the general framework within which these two processes are presently conceived. Our first point is to show the similarity in the conceptual vocabulary of phylogenetic and ontogenetic thought: that both processes are conceived as based on dissociable components that can have independent rates of evolution and development.

The work of the German biologist Haeckel (1834–1919) has had a lasting, and largely negative, effect on this topic. *First*, in a prolific body of writing, Haeckel argued for an all-encompassing 'biogenetic law' that united ontogeny and phylogeny. In his view ontogeny was the short and rapid recapitulation of phylogeny: developing organisms follow a temporally-condensed course that concertinas the evolutionary history of their species. Recapitulation rests on two assumptions that Gould (1977, p. 74) dubs 'the principle of terminal addition', whereby new evolutionary features are successively added to the end of unaltered ancestral ontogenies (so that previously adult stages subsequently become pre-adult ones); and 'the principle of condensation', whereby rates of development must become speeded up so that descendants pass through ancestral ontogeny faster than their ancestors did—something accomplished either by a deletion of previous stages or by a general acceleration of development.

Second, Haeckel unashamedly subsumed history under his monolithic conception as well, arguing that:

the human mind or the psychic life of the whole human race has been gradually evolved from the lower vertebrate soul. . . . the human psyche has proceeded historically from the ape soul (1907, p. 354).

He saw evidence for this process of 'gradual evolution' in the present day, because of his presupposition of a recapitulatory relation between ontogeny and phylogeny:

the wonderful spiritual life of the human race through many thousands of years has been evolved step by step from the lowly psychic life of the lower Vertebrates, and the development of every child-soul is only a brief repetition of that long and complex phylogenetic process (1907, p. 355).

Haeckel's legacy to subsequent studies was twofold. First, he accounted for similarities that are apparent between ontogeny and phylogeny by proposing a mechanism—'recapitulation'—that was shown to be spectacularly wrong (for example, Garstang 1922). Indeed, so spectacularly wrong was it that it had the effect of marginalizing almost all consideration of ontogenetic and phylogenetic relationships for a long period (see Gould 1977). Second, in also subsuming history under biology, he compounded his crime by spawning a lunatic fringe of quasi-racist anthropological writing (again, see Gould 1977, Chapter 5, for a review). It is necessary to note, however, that despite Haeckel's errors, there *must* be relations between ontogeny and phylogeny, even if only at the level that ontogenies have evolutionary histories; and that human history is not to be identified with biology, but is supported by human biological capacities with which it must, *de facto*, interact. (Many recent attempts to use ontogenetic data to inform phylogenetic scenarios of symbol-use are confounded by a similar failure to distinguish biological and socio-cultural historical processes, and thus continue to contribute to the marginalization of this topic.)

1.1 *Mosaic evolution*

'Mosaic evolution' refers to the evolution of different components of a phenotype at highly unequal rates. It was first recognized by Lamarck (1809, p. 58), but stayed outside the mainstream of evolutionary orthodoxy for a long time. In retrospect, it seems strange to modern sensibilities that evolutionary theorists were long held in sway by a vestige of ancient notions of harmony. Historically, many fossils were argued not to be part of an ancestral line if they showed a mixture of early and late-emerging characters, since it was expected that intermediate forms would exhibit a perfect intermediacy between older and newer forms. Thus a fossil such as *Archaeopteryx*, showing a mix of reptilian and avian characteristics, could not be placed as a transitional phase between these two classes, since all its characters are not transitional: evolution was conceived as occurring by a gradual and general transformation of the whole animal. While Cope, Dollo, Lamarck, and Mehnert argued for a phylogenetic dissociation of parts in phylogeny in the nineteenth century (partly as an attempt to rescue recapitulation theory from the recognition that acceleration is not constant across all bodily organs: see Mayr 1982), the legitimacy of 'mosaic' evolution was not fully accepted until de Beer (1954) explicated what he termed 'Watson's Rule'. Watson (1919) had noted that the fossil *Seymouria* showed a mix of amphibian and reptilian features, with little evidence of any intermediate conditions. From this he claimed that:

The curious way in which the structure of *Seymouria* is built up of perfectly well developed amphibian characters and equally decisive reptilian features, those of intermediate type very rare, affords a magnificent example of the way in which the evolution of great groups may have taken place (1919, p. 300).

And hence the concept of mosaic evolution, that different body parts in groups of organisms often evolve in different periods and at different rates. This conception of evolution is important because it links fundamentally to conceptions of development.

1.2 *Embryological mosaics, or dissociations*

The dissociability of different 'parts' of an organism was elucidated earlier in embryology than in evolutionary theory (the basic definition of dissociability was formulated by Needham (1933), although the concept dates back into the nineteenth century). Thus different organ systems were recognized as developing at different, independent rates towards maturity. Recent embryological studies have begun to establish the ways in which these dissociable systems interact with each other during development. These studies provide a starting-point for linking ontogeny and phylogeny. Early developments in a system often provide the basis for later developments by a process termed 'inductive interaction', whereby the expression of a genetically coded character depends on its being 'switched on' by the processes of previous tissue differentiation. Kollar and Fisher (1980), for example, have shown that the embryonic oral epithelium of the chick, which under normal circumstances interacts with oral mesenchyme tissue to produce a beak, can, when grafted with mouse mesenchyme, give rise to complete teeth—an organ last seen in the avian fossil record in the Upper Cretaceous.

Thus, at least one avian genome still retains the genetic information that allows the chick oral epithelium to participate successfully in the sequence of interactions required for tooth morphogenesis and synthesis of the enamel matrix. Evolutionary loss of teeth is therefore the result of a change in the developmental program of avian mesenchyme such that the initial steps of the process fail to occur (Raff and Kaufman 1983, p. 156).

One link, then, between ontogeny and phylogeny is that evolution acts to change developmental pathways. Temporal dissociations occur in ontogenesis between age, growth, and shape that underpin this link and contribute to the evolution of new morphologies. Three ontogenetic processes, growth, development, and maturation, can progress at independent rates (see Gould 1977, p. 234 ff.). Ontogeny thus becomes a possible site for the action of selective evolutionary forces: it is not just that adult forms are selected, but so too are the developmental processes, and the temporal and inductive relationships between them, by which those forms are elaborated. Changes in these temporal relationships form the basis of *heterochronic* processes in evolution, of which Haeckel's recapitulation proposal represents just one possibility. The additional relationships that are possible are noted in the next section.

The essential points so far are as follows: Biologically determined characteristics of adult phenotypes are the result of the ontogenetic integration of quite separate 'organ' systems. The development of these systems is not under strict genetic control, but they interact with each other in the course of their elaboration to yield the final outcome. In addition, genetic instructions can be conserved over very long periods, even though remaining dormant in terms of phenotypic expression. These findings provide the first step in understanding the relation between ontogenetic and phylogenetic events. Evolution can act on separable ontogenetic components that contribute to final phenotypes. These separable components act integrally in determining the ontogenetic course that produces the adult phenotype. The individual's genotype offers a number of possible phenotypes as alternatives subject to selection during this process. These possibilities ensue from the temporal mosaic of interwoven factors in the evolution of ontogeny, and clearly go beyond those captured in the concept of recapitulation. Additional concepts that have been identified are summarized in the next section.

1.3 *Heterochrony and evolution*

While Garstang (1922) and later de Beer (1930) finally laid recapitulation to rest as an all-encompassing explanation, both these theorists *expanded* the range of processes whereby the spheres of development and evolution might be related. They delineated three further processes of *heterochrony*[2] (in addition to recapitulation): progenesis, neoteny, and hypermorphosis.[3]

Paedomorphosis is a phylogenetic change that involves the retention of juvenile characters of ancestral forms in the adult, or in later ontogenetic stages. Paedomorphosis occurs in either of two forms: progenesis or neoteny. *Progenesis* refers to paedomorphosis produced by a precocious sexual maturation of an organism that is still at a morphologically juvenile stage. *Neoteny* refers to paedomorphosis produced by the retardation of somatic development, such that sexual maturity is attained in an organism retaining juvenile characters. *Hypermorphosis* is the phyletic extension of ontogeny beyond its ancestral termination, such that adult ancestral stages become preadult stages of descendants. The gigantic antlers of the extinct Irish Elk are given by Gould (1977) as the paradigm hypermorphotic case.

The four heterochronic possibilities—progenesis, neoteny, hypermorphosis, and recapitulation—establish the major set of genetically-mediated relationships between ontogeny and phylogeny, and are summarized in Table 1. In addition to these heterochronic processes, a further relationship, that of common constraint, has been proposed as explaining parallels between ontogeny and phylogeny—one that does not rest on changes in the timing of developmental components, but more in the design problems involved in going from *any* simple system to a more complex one.

Table 1 Categories of heterochrony*

Timing of: Appearance of somatic feature	Maturation of reproductive organs	Name in de Beer's system	Morphological result
Accelerated	Unchanged	Acceleration	Recapitulation (by acceleration)
Unchanged	Accelerated	Paedogenesis (= progenesis)	Paedomorphosis (by truncation)
Retarded	Unchanged	Neoteny	Paedomorphosis (by retardation)
Unchanged	Retarded	Hypermorphosis	Recapitulation (by prolongation)

* Modified slightly from Gould 1977; used by permission.

The notion of common constraint predates modern biology, and goes back to the German philosophical school of *Naturphilosophie* (cf. Gould 1977). The basic claim was that parallels between ontogeny and phylogeny arise because both are subject to the same universal laws of nature. A century later this notion was reworked by Hertwig (for example, 1901, 1906), who identified the laws of physics and chemistry as the common external constraints.

He argued that the earliest stages of vertebrate development are similar not because they repeat a common ancestry, but because there is no other physical way to develop a many-layered structure from an initial cell (Gould 1977, p. 431).

Thus:

Phylogeny and ontogeny have no direct influence upon each other . . . ; each follows a roughly similar path because it is the only path available. (Mudcracks, basalt pillars, soap bubbles, bee cells, and echinoid plates are all hexagonal because only a few regular forms can fill space completely. The external constraints are identical, but no result has any direct influence upon another) (Gould 1977, p. 144).

This relation has been used in developmental psychology (cf. Gould 1977). Koffka (1928) proposed a 'correspondence' theory, whereby ontogeny appears to parallel phylogeny because the same external constraints apply to both. In Gould's view, this position is also the one adopted by Piaget with respect to the origins of human reasoning. Piaget's apparently recapitulationary view is in fact one that does not imply any link between the ontogenetic and phylogenetic (or historical) growth of knowledge:

The sequences run in parallel, but neither causes the other. They follow similar paths because a common object (the preconscious mind) is pursuing a common history of development (successive assimilations of external reality to produce a sequence in modes of reasoning) (Gould 1977, p. 147).

Apparent recapitulations in both the phylogenetic and ontogenetic spheres may, then, result from constraints to which they are both subject, rather than from genetically-based and coded rearrangements of biological substrate 'organs'.

It should be clear from this brief review that no simple relations can be established between ontogeny and phylogeny, and that slogans such as 'ontogeny recapitulates phylogeny' are over-simplistic. However, the general outlines of how these two domains could be related is emerging. Development is a process whereby separate packages of differentiating 'organs' influence each other to yield a final phenotype. That process of differentiation is determined by a combination of genetically coded information; emerging system interactions; constraints on how the process of moving from simple to complex might be conducted under earthly conditions; and feedback from the external environment. This entire complex of systems that constitute the developmental process is susceptible to evolutionary modification. Evolutionary change can be based in the temporal shifting of developmental events in relation to each other, and the four heterochronic possibilities may account for changes in the somatic hardware, and hence capacities, of evolving organisms. These changes interact with the environmental elaboration of the systems they support, but any relations that might hold between the ontogenetic and phylogenetic elaborations of those systems could be predominantly accountable for under the principle of common constraints upon the construction of complex systems, rather than by biological constraints arising from the heterochronic rearrangements of biological 'hardware' packages. But, at the same time, the products of ontogeny feed back into determining the courses of heterochronic manœuvres.

2 A role for heterochronic processes in human evolution

Heterochronic changes in the timing of developmental events have been postulated as having played a major role in all *biological* evolution. Gould (1977), Montagu (1981), and Groves (1989) place an emphasis on the process of neoteny in human evolution in accounting for the fact that modern humans exhibit a prolonged retention of fetal growth rates, especially where the brain is concerned, and, in general, a delay in maturity. Bolk (1926), Schultz (1926), and Abbie (for example, 1952, 1958) were among the first to assemble the data on which this view is justified. More recently, Gould (1977) has hypothesized that

. . . human beings are 'essentially' neotenous . . . because a general, temporal retardation of development has clearly characterized human evolution. This retardation established a matrix within which all trends in the evolution of human morphology must be assessed (Gould, op. cit., p. 365).

However, more recent analyses cast major doubts over the validity of this claim. Shea (1989, p. 97) concludes from a detailed review that:

Heterochrony in general has undoubtedly played a central role in the evolutionary transformations characterizing our lineage, and neotenic paedomorphism in particular can probably be implicated in several of these shifts. However, a hypothesis of general and pervasive human neoteny is clearly no longer viable.

He outlines a number of reasons for this: (1) that neoteny has been confused with progenesis (and we would agree that terminological confusion is endemic within heterochronic studies); (2) that prolongations of growth periods have been conflated with retardations of shape; (3) that there has been an unwarranted acceptance of superficial and non-homologous similarities between human adults and primate juveniles as primary evidence of underlying neotenic growth processes; and (4) that there has been a failure to appreciate the extent and significance of the many non-paedomorphic features implicated in the evolutionary establishment of modern human morphologies.[4]

McKinney and McNamara (1991) also take issue with the emphasis on neoteny:

(1) there is no single heterochronic process that accounts for all of human evolutionary change, although there is one process that accounts for much of it. However, (2) this process is not neoteny (retardation of growth rate), it is hypermorphosis (prolongation of growth) (op. cit., p. 292).

Their basic arguments echo those put forward by Shea (above).

First, superficial resemblances are a poor guide to underlying processes. Human skull shapes are often equated with those of juvenile primates (hence, neoteny—retention of juvenile features by retardation of growth) when, in fact, the similarities are dependent on our greatly enlarged brains that result from prolonged brain growth (hypermorphosis), neotenous jaws, and prognathic dentition. Again, the claim that we retain such juvenile primate traits as 'curiosity' and 'learning' over longer periods is a result of both our larger brains and a hypermorphic delay in brain maturation, not because we grow 'more slowly' (see point 2 below).

Second, it is a misconception that human development is retarded, or 'slow', across the life-span. Humans do not grow more slowly than other primates, but *grow for a longer time in each phase of growth*.

This is a delay in life history events, not a 'slowing' of them; maturation, like *any single event, is not 'slowed', it is delayed* (late offset, i.e., hypermorphosis). 'Rates' of processes, until that event, are a separate issue. Indeed, late offset of our fetal stage leads to *faster* rates ... the rate of a process is not governed by the time of its offset (1991, p. 292).

And third, not only is hypermorphosis typical of the primates as a group, 'being a major factor in the progressive differentiation of prosimians, monkeys, apes, and finally humans' (ibid., p. 295), but is a characteristic of evolution within the hominid group itself. That is, early hominids were more apelike than humanlike in their ontogeny (see Bromage 1987; Beynon and Dean 1988): 'humans have evolved through a delay in developmental events (hypermorphosis)' (McKinney and Macnamara 1991, pp. 297–8).

However, there is more to human evolution than one single process, be that neoteny, or hypermorphosis, and hence the notion of a matrix within which heterochrony operates is required, in recognition of the fact that developmental features can be acted on both separately and in different ways. The first point, separability, refers to the finding that the relative rates of growth of a suite of features can vary with respect to each other: points of maturation and rates of growth can be subject to independent selection. For example, Schultz (1949, p. 375) notes:

Early in development, when the digits have just separated on the embryonic, plate-like hands and feet, the thumb and the great toe show as yet no sign of rotation in any of the primates. The well-known rotation of the first digit, necessary for opposability, develops gradually during ontogeny and reaches widely different degrees of perfection in the adult hand and feet of the various groups of primates.

Thus a very early stage of foot ontogeny has been frozen in the course of human evolution, the embryonic condition functioning as a preadaptation retained in adulthood as part of the package required for bipedalism (see Fig. 1a). Manley-Buser (1986) considers this a neotenic human feature. By contrast, the growth pattern of the human hand continues beyond the point at which that of the foot has been arrested (see Figs. 1b and 1c).

Secondly—and hence the notion of a general heterochronic matrix within which the evolution of human 'hardware' is embedded—there occur features that develop, not by neotenic retardation, but as a result of other heterochronic processes. There are features that result from acceleration (or terminal addition and condensation, and thus which actually fit the parameters of the discredited concept of recapitulation):

examples include the early fusion of the sternebrae to produce a sternum; the pronounced bending of the spinal column at the lumbo-sacral border; the fusion of the centrale with the naviculare; and several aspects of pelvic shape (Schultz 1949, 1950) (Gould 1977, p. 383).

Harrison *et al.* (1988) note that at birth, the position of the skull relative to the spine is similar in humans and apes. This relation changes in apes, in that the area in front of the occipital condyles grows faster than that behind it, resulting in the joint's being moved backwards. This does not occur in humans, so that the skull retains its *infantile* position with respect to the spine (probably as an adaptation to bipedalism): i.e., this is an example of neoteny. However, as Laitman's work (see this volume, Editorial appendix II, Chapter 4) has demonstrated, additional developmental specializations of the human skull base mean that much of it is more similar to the skull base of *adult* apes: the neotenic feature is thus very localized.

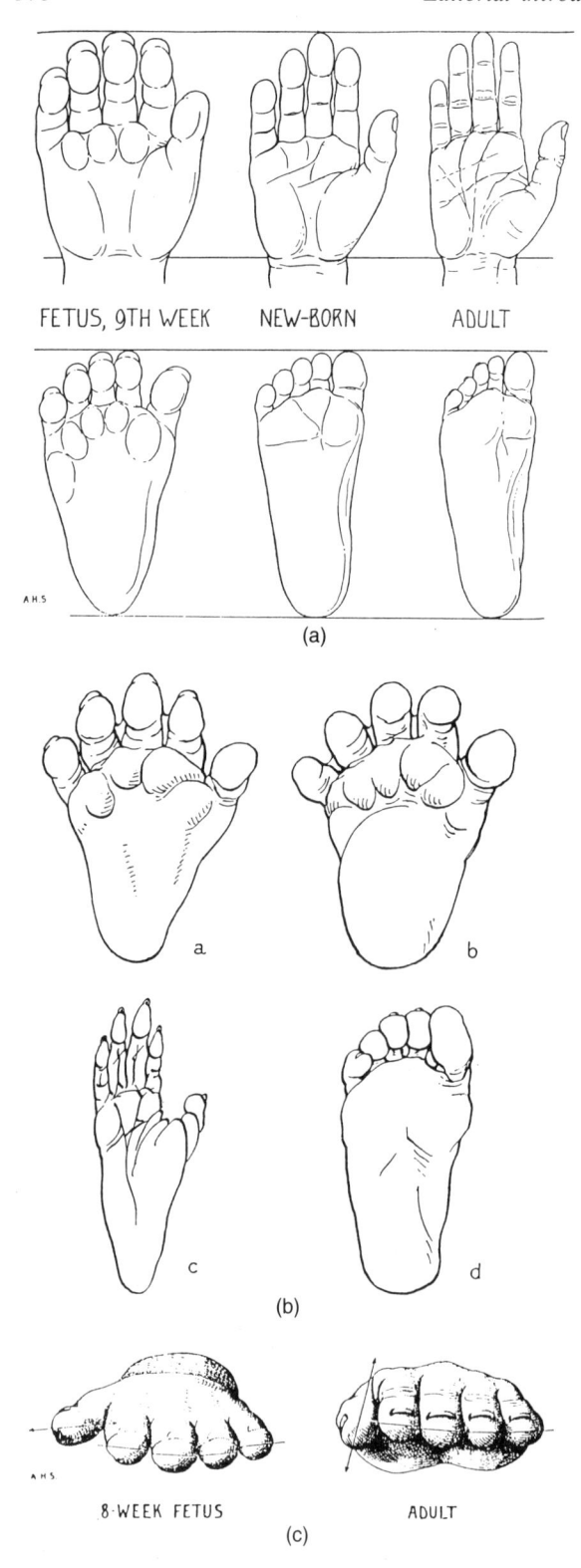

Fig. 1 Superficially, the structure of the human hand and foot retain their embryonic condition from fetal to adult condition (Fig. 1a, from Schultz 1926). That is, the human anatomy appears frozen in the embryonic condition as compared to that of other primates (Fig. 1b, from de Beer 1958, partly after Schultz 1950). However, the opposability of the thumb, attained by a rotation of its plane of alignment with respect to the other digits, is an added development in the case of the human hand (Fig. 1c, from Schultz 1926). (a) Fetal, new-born, and adult human hand and foot morphology. (b) The foot in fetal and adult macaque and human. a: macaque fetus of 23 mm length; b: human fetus of 24 mm length; c. macaque adult; d: human adult. (c) Fetal and adult human hands, viewed from in front, showing the rotation of the thumb (the straight lines run parallel to the transverse axes of the finger-nails).

Further, while modern humans show a general delay in motor development over modern apes (and they in turn over monkeys), and the majority of these are proportionally delayed, Dienske (1986) has noted that the play-face or smile of both apes and humans appears much earlier in ontogeny than it 'should', i.e., it is displaced within the overall developmental motor delay. Dienske (ibid.) suggests that selection has operated so as to favour early smiling in more 'helpless' infants: the slower the rate of motor development, the sooner behaviours that ensure successful mother–infant bonding are needed.

Modern human phenotypes are thus best regarded as the result of a number of heterochronic rearrangements of developmentally independent systems. These rearrangements have resulted in the creation of an ontogeny that provides the hardware to support behavioural adaptations and systems such as language. However, the structural properties of these behavioural systems cannot themselves be the result of these heterochronic rearrangements, for these rearrangements only provide the capacities that underwrite those systems. The elaboration of these systems needs to be accounted for by different processes, as do similarities between the phylogenetic and ontogenetic paths of systems such as language. Of the processes so far reviewed, common constraint is the most likely candidate.

It should be clear, then, that given a dissociation between ontogenetic rates of growth, development, and maturation of organic systems, and the notion of mosaic evolution, *simple* models of the relation between ontogeny and phylogeny become theoretical impossibilities. Somatic forms and structures do not evolve as a result of the direct action of selective forces upon them: rather, selection operates on the reproductive potential of the ontogenetic processes that give rise to those forms and structures. In addition, human ontogeny is embedded in an infant–caregiver relation that is itself embedded in a sociocultural matrix, such that not only does ontogeny evolve, but it probably co-evolves with those infant–caregiver relations. Further, a change in the relative rates of development of just one system in relation to others can markedly effect the overall course of development. For example, changes in rates of growth, development, or maturation will result in the ontogenetic period in which those changes are having their effect being subject to selective pressures of its own. Thus prolonged infancy constitutes a period in which adaptations for infancy itself are likely to be selected for, or against, because the establishment of such adaptations has repercussions further down the developmental line. It is clearly no simple matter to go from human ontogeny to human phylogeny, as the ontogenetic process in no way maps directly on to the phylogenetic one.

3 Relating human behavioural ontogeny and phylogeny: issues

Some progress in relating ontogeny and phylogeny can be made on the basis of recent studies of human cognitive development that conceive the ontogeny of the cognitive capacities, supported by the concurrently emerging somatic systems, as comprising a dissociable suite of elements themselves (see appendix to Butterworth, this volume, Chapter 16). While the links between the hardware and the capacities it enables are unclear, the mosaic nature of these capacities, allied to the fact that they appear to cut across different developmental domains, allows an evolutionary framework to be constructed that is congruent with what we already know of the workings of the evolutionary process.

3.1 *Dissociable systems in cognition and language*

The precise mosaic nature of human cognitive capacities is currently a hotly-debated topic in cognitive science, under the rubric of the *modularity hypothesis* (see Johnson *et al.*, this volume, Chapter 24, and also the introduction to Part IV; Marr 1976, 1982; and Fodor 1983, 1985). The essential claim is that different perceptual and cognitive abilities are served by independent cognitive 'mechanisms'; thus these mechanisms are 'domain-specific'.

There are a number of sources of evidence adduced for these mechanisms. Brain injuries, such as strokes, are one. Here the argument is that some very particular abilities can be affected without the disruption of others: a person may lose the ability to recognize individual faces (a condition termed 'prosopagnosia'), *and nothing else*; or there may be the loss of or impairment in particular linguistic functions that characterize particular types of aphasias, such as the inability to produce grammatical utterances while being able to comprehend them, for instance (for example, Miceli *et al.* 1983). Prodigies and *idiots savants* are another source of evidence, suggesting that some abilities are again quite independent of others. Developmental schedules are a third, as independent parts of 'kit' appear, develop, and reach adult levels of functioning on fixed schedules.

Developmental disorders are a fourth source. Bellugi *et al.* (1990), for example, discuss disruptions occurring in Williams Syndrome, in which linguistic functioning is relatively normal, but spatial cognition grossly deficient, with IQs for their study sample in the range of 41 to 64. The dissociation between language and non-linguistic cognitive capacities is quite dramatic and is consistently found across the children they have studied. Autism is a further area in which domain-specificity of abilities has been hypothesized. Leslie and Thaiss (1992) propose that the results of their structured experiments reveal the existence of a *very* restricted and specialized cognitive 'mechanism' 'which subserves the development of folk psychological notions' (ibid., p. 226); that is, autistic children are autistic because they lack the psychological module that allows reasoning to be carried out when the issues to be thought through are concerned with *beliefs* and other mental states of agents, and nothing else.

This last hypothesis provides a challenge to what has been a widely-held belief in contemporary cognitive psychology, that the rules of inference people use in symbolically-mediated reasoning are content-independent; that is, they generate only true conclusions from true premises, regardless of the content of those premises. Despite the repeated finding that, while this may be true in formal logic, it rarely occurs in humans, who are often content-dependent in their reasoning (see, for example, Johnson-Laird 1982; Pollard 1982), the belief in the content-independence of the rules has been sustained, with the 'content effect' being dealt with in terms of differential experience with particular topics, etc., in a rather *ad hoc* fashion. Recently, however, there has emerged a counter-view that hypothesizes that different cognitive problems are tackled by different domain-specific cognitive processes. And further, that those domains in which reasoning is more accurate are those that were evolutionarily-salient in the human lineage. The essential argument runs thus:

The more important the adaptive problem, the more intensely selection should have specialized and improved the mechanism for solving it (Darwin 1859/1958; Williams 1966). Thus the realization that the human mind evolved to accomplish adaptive ends indicates that natural selection would have produced special-purpose, domain-specific mental algorithms—including rules of inference—for solving important and recurrent adaptive problems (such as learning a language: Chomsky 1975, 1980). It is advantageous to reason adaptively, instead of logically, when this allows one to draw conclusions that are likely to be true, but cannot be inferred by strict adherence to the propositional calculus. Adaptive algorithms would be selected to contain expectations about specific domains that have proven reliable over a species' evolutionary history. These expectations would differ from domain to domain. Consequently, if natural selection had shaped how humans reason, reasoning about different domains would be governed by different, content-dependent, cognitive processes (Cosmides 1985; Cosmides and Tooby 1987, 1989; Rozin 1976; Rozin and Schull 1988; Symons 1987; Tooby 1985) (Cosmides 1989, p. 193).

Cosmides (1989) reports a set of 9 interconnected experiments that establish differences in human abilities to deal with formally identical problems concerned with social cost–benefit and cheating problems as compared to problems lacking those contents, the rationale being that 'evolutionary biology places tight constraints on how humans must process information regarding social exchange' (ibid., p. 196). (For further details, see also contributors to Barkow *et al.* 1992.) Thus, she concludes that 'human reasoning is not unitary and domain general, but instead governed by an array of special purpose mechanisms. . . . These studies . . . provide support for an evolutionary and modular approach outside of psycholinguistics' (Cosmides 1989, p. 260).

None of the above claims are uncontested at this point in time (for example, for autism, see Lock and Colombo, this volume, Chapter 22; for reasoning, see Cheng and Holyoak 1989). If, however, the claims become more substantiated, they pose at least two challenges to the study of human symbolic evolution. First, if complex modern behaviours are based on many separable components, then it becomes necessary to understand the evolution of each of them. Second, it is also necessary to offer an account of their integration. Before the Middle–Upper Palaeolithic transition (see White, Chapter 9, and the Introduction to Part II of this volume), 'art' was lacking, and by 'modern' standards, tool-making and subsistence strategies appear to have been extremely conservative. It may be, then, that the archaeological record is congruent with hominids of this period possessing unintegrated suites of behaviours based on

domain-specific abilities. But after this transition, there appears much more of an integration between technological, social, and economic activities which may have depended on the appearance of a domain-sharing and integrating intelligence. A biological basis for this change cannot be ruled out; but, given the present evidence, this seems unlikely: the behavioural changes involved appear to have resulted from innovations *within* the species *Homo sapiens sapiens*, and are not associated with a transition from one species to another.

The various controversies over the nature of modules can be illustrated with respect to the modular system that has been proposed as underwriting syntax. Two versions are offered in the literature. First, the 'autonomous' view referred to by Johnson *et al.* in Chapter 24 that derives from Chomskyan linguistic theorizing. Here it is more *content* than mechanism that is held to be unique to language, the child being credited with possessing innate knowledge of a specifically linguistic kind. It is this knowledge that enables a child to 'learn' language in the way that it does on the basis of the 'degenerate' input it receives:

the child is, so to speak, "born knowing" certain facts about universal constraints on possible human languages. It is the integration of this innate knowledge with a corpus of "primary linguistic data" (e.g., with the child's observations of utterances produced by adult members of its speech community) that explains the eventual assimilation of mature linguistic capacities (Fodor 1983, p. 4).

A second version emphasizes instead *processes* rather than *contents*, proposing particular mechanisms (or processes) that work solely in some bounded developmental domain, such as syntax, face-recognition, or three-dimensional space. Both horizontal and vertical modules have been postulated. *Horizontal* modules work across domains, handling a variety of inputs; *vertical* ones are domain-specific, and are the variety proposed to be involved in what has been termed the 'language organ'—it is hypothesized that there is a specific syntax module devoted to the acquisition and handling of grammatical rules, for example. In addition, a dichotomy between 'hard-wired modules' and 'constructed modules' has also been drawn.

'Hard-wired modules' are regarded as innately given, and, as Johnson *et al.* (this volume, Chapter 24) point out, they would appear to entail a discontinuity view of evolution. (We say 'appear' because this claim is only proffered by proponents of this view; more strictly, it would seem that this view just makes an evolutionary account more difficult to arrive at. In addition, the claim that modules are innate and hard-wired is often made on the grounds of their apparent localization within the brain. But the argument from localization to innate hard-wiring rests on a very crude model of how brain circuits might be arrived at in ontogeny.) If, however, modules are constructed during development, and result from the exercise and co-ordination of any number of separate, identifiable capacities, then an understanding of their construction provides some clues as to how these capacities might have been recruited in phylogeny to support modern-day behaviours.

Thus studies of the functional elaboration of the brain may yield information as to how functionally specific systems are differentiated in ontogeny from less specific ones (cf. Gibson, this volume, Chapter 14). This must be regarded as a long-term line of investigation, for, impressive though current neuroscience may seem, a sceptic could claim that the only sure facts it has established are that our brains play a crucial role in our behaviours, and that, in the course of our evolution, they got bigger.

We do not, for example, have a full understanding of what Broca's area, which is clearly implicated in language, actually does (it could be a specific site for language-related processing, or a kind of 'telephone exchange' junction of connections between other regions that thereby enables language, or ...), nor exactly how it becomes wired up functionally and structurally with other parts of the brain. Neither do we know exactly what it does for monkeys or apes, nor have we any idea as to what it might have been doing for australopithecines or habilines, given that it was beginning to show greater prominence at this time (cf. Holloway, this volume, Chapter 4). It would seem reasonable, though, that what it was doing, and what it does do in modern ontogeny before language emerges, bears some relation to the way it is eventually implicated in the support of language. We note below two recent lines of study that bear on this issue, thus pointing to the way in which such long-term research programmes could contribute to answering evolutionary questions.

3.1.1 Parallels between action and language: analogy or homology?

Greenfield has recently reported (1990) on work that synthesizes her long-term investigation of parallels between the organizational principles underlying the early structural ontogeny of human action and language. In her early work (for a review, see Greenfield 1978) she and her colleagues emphasized the parallels between the development of the ability to combine objects—particularly blocks and nesting cups—in infancy and the development of word-combinations. A key concept is *hierarchical organization*. The combination of both words and objects occurs in real time, but the structural patterns of these combinations do not arise sequentially on the basis of associations between the contiguous acts, but are generated by some hierarchical ordering that gives them a systematicity.

Her early work established that, first, the underlying structural plans on which both object- and word-combinations were based followed the same developmental pattern of elaboration; and second, that these patterns showed a temporal parallel with each other (see Fig. 2). But this work was not able to distinguish whether this parallel indicated an analogous or homologous relationship between these two developing domains: that is, whether the similar courses they followed rested on parallel structural relationships with no common origin, or ones with a common origin. It is to the resolution of this problem that her recent work has been directed.

One source of data that bears on this question has come from studies of brain-traumatized humans who have been tested directly on tasks derived from Greenfield's original developmental studies. Grossman (1980) thus provides evidence for the existence of a supramodal hierarchical processor in the patterning of the performance of different groups—agrammatic patients with Broca's aphasia,[5] fluent aphasics, non-aphasics with insult lateralized to the right hemisphere, alcoholic Korsakoff patients, and normal controls—in that

the pattern of group differences indicates a *specific deficit* in hierarchical organization associated with lesions in a specific region of the brain, Broca's area in the left hemisphere. Neural specificity is further validated by the fact that this performance was not only associated with Broca's aphasia; it was also *absent* in any other group, pathological or normal (Greenfield 1990).

Cromer (1983) reports similar findings from a group of aphasic children; but here the children were lacking in *all* aspects of language, and not merely hierarchically organized grammar.

By contrast, however, work by Curtiss, Yamada, and Fromkin (Curtiss and Yamada 1981; Curtiss *et al.* 1979; Yamada 1981) on a group of eight mentally retarded individuals aged between six and twenty years provided mixed results: some were competent at Greenfield's manual hierarchical tasks, but poor with grammar; whereas others showed the opposite pattern. These results would argue for separate neural modules for hierarchical organization in each domain. Thus the evidence from neuropsychology relevant to deciding between analogous or homologous bases for hierarchical structures in action and language would appear to be mixed.

Greenfield's proposed resolution of this impasse is to turn to recent work on the possible functions of

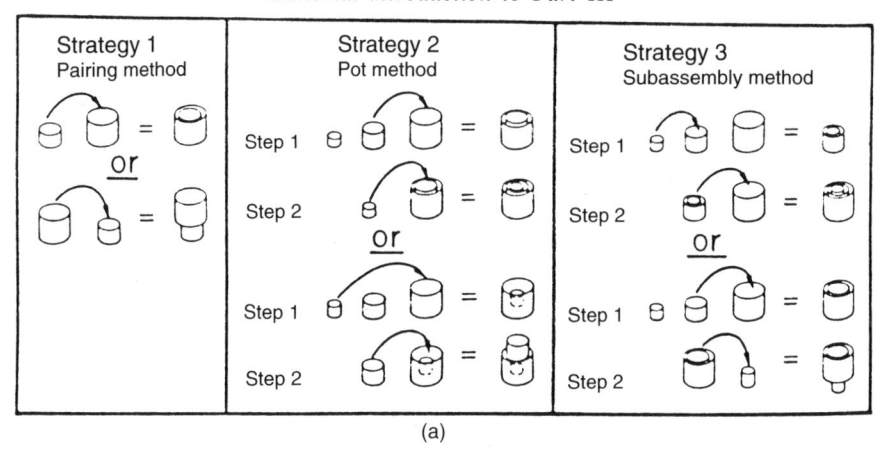

380   *Editorial introduction to Part III*

[Figure (a): Three strategies — Strategy 1 Pairing method; Strategy 2 Pot method; Strategy 3 Subassembly method]

(a)

[Figure (b): Bar chart showing Number of children vs Age in months (11, 12, 16, 20, 24, 28, 32, 36), with Strategy 1, Strategy 2, and Strategy 3 indicated]

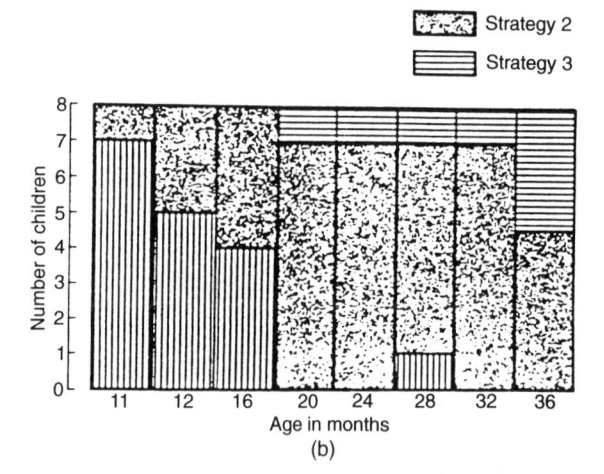

(b)

Broca's area that indicates that it is not lesions to Broca's area alone that lead to agrammatic aphasia, but cortical and subcortical damage to the circuits that allow Broca's area to work in conjunction with more anterior prefrontal areas. For example, research using positron emission tomography (PET scan) by Fox *et al.* (1988) suggests that Broca's area is involved in a number of different cortical circuits with various parts of the motor cortex. Deacon provides data that delineate at least two circuits:

one circuit for the hierarchical organization of manual sequences would include Brodman's Area 9 in the prefrontal cortex (Deacon [1992]). A second circuit for the hierarchical organization of grammar would include an area of the prefrontal cortex just anterior to Broca's area (Deacon 1988). Most likely the first circuit would connect to a superior part of the classical Broca's area, while the second would connect to an inferior part (Deacon, personal communication, 1990) (Greenfield 1990).

This proposal of separate circuits resolves the conflicting results noted above. The topographical proximity of the two allows the possibility of trauma's being sufficiently localized as to allow the occasional sparing of one circuit, thus accounting for the different findings of association and dissociation previously noted. But this proposal does not help in deciding the issue of analogy or homology, since separate pathways might rely on shared or separate processors.

Here, Greenfield turns to developmental data. Developmentally, the circuits connecting the various cortical areas are known *not* to be present from birth, but to be the product of gradual postnatal differentiation (Thatcher *et al.* 1987; cf. appendix to Gibson, this volume, Chapter 14 for details). This yields the possibility that Broca's area might start out, ontogenetically, as an undifferentiated neural region organizing both language and manual action in the same way, and

as the Broca's region developed differentiated circuits or networks with more anterior portions of the prefrontal cortex, the structure of manual action and the structure of language would become more divergent, autonomous, and complex (Greenfield 1990).

From this possibility Greenfield adduces two hypotheses.

First, the structural properties of language and action should be more closely linked in early development than later, when they could be expected to show more autonomy. The evidence that is available on this point is very much in accord with this hypothesis:

from about nine months to two years of age, children pass through parallel stages of hierarchical structure in the domains of word formation and object combination, . . . [and] programs for combining objects become increasingly differentiated from programs for combining words (linguistic grammars), starting around two years of age (ibid.).

Second, the schedule of brain differentiation should correlate with the relative interdependence and autonomy of these two functions. Drawing on a

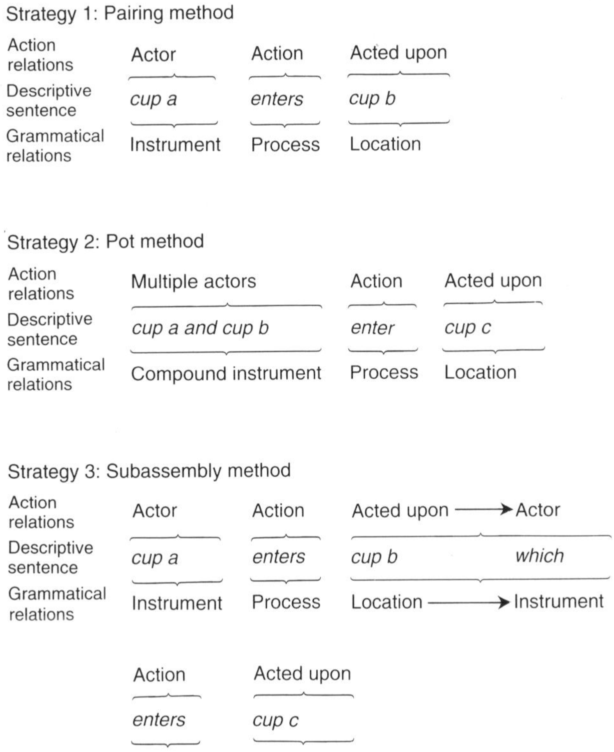

Strategy 1: Pairing method

Action relations	Actor	Action	Acted upon
Descriptive sentence	*cup a*	*enters*	*cup b*
Grammatical relations	Instrument	Process	Location

Strategy 2: Pot method

Action relations	Multiple actors	Action	Acted upon
Descriptive sentence	*cup a and cup b*	*enter*	*cup c*
Grammatical relations	Compound instrument	Process	Location

Strategy 3: Subassembly method

Action relations	Actor	Action	Acted upon → Actor
Descriptive sentence	*cup a*	*enters*	*cup b* *which*
Grammatical relations	Instrument	Process	Location → Instrument

	Action	Acted upon
	enters	*cup c*
(c)	Process	Location

Fig. 2 Part (a) shows the possible strategies available for combining seriated cups, and part (b) the frequency of dominant cup-combining strategies at different ages. Figure 2c illustrates the formal correspondences between manipulative strategies and sentence-types. Here the subassembly strategy is illustrated as parallel to the linguistic one involved in a relative-clause sentence, in which location in the first clause switches roles to become instrument in the second clause. Greenfield *et al.* (1972, 305) note that 'the available evidence suggests parallels between manipulation and grammar not only in form, but also in developmental sequence'. (Figures from Greenfield 1978.)

reanalysis, by Thatcher, of data generated by Thatcher *et al.* (1987) from EEG studies, and neuroanatomical data on the postnatal development of the motor speech area of the brain (Simonds and Scheibel 1989), Greenfield (1990) concludes that 'neural differentiation of higher order programs for language and construction activity occurs in just that period when behavioural differentiation is taking place'.

The final piece in Greenfield's synthesis is to establish, drawing on data from Connolly and Dalgleish (1989) on the course of infants' attempts to master the use of a spoon, which she takes—quite reasonably—as a paradigm case for the development of tool-using abilities, that the early development of tool-use in human infants is a special case of the development of object-combination programs. This step leads to the conclusion that

Broca's area is providing a common neural underpinning for early programs of action in speech and tool use. These programs differentiate from age two on, when Broca's area establishes differentiated circuits with the anterior prefrontal cortex (Greenfield 1990).

From here, Greenfield concludes that the developmental relation of structural combinations in action and language rests on homology, leading her to the view that:

In ontogenesis, the hierarchical programming of language and manual intelligence (including tool use) starts out as a unitary function, but becomes modularized as a result of the developmental process of brain differentiation. *Modularity of language is not present at birth or even [at] the dawn of language; it increases with age and neural differentiation* (ibid.).

She also points out the possible evolutionary implications of her work:

It seems likely that the same progression has occurred in the phylogenetic development of language, tools, and manual intelligence. An undifferentiated Broca's area, evolving to organize sequential action of all kinds, could later in evolutionary history be appropriated to organize sequential language production (ibid.).

3.1.2 Dissociable developmental mechanisms in early language development

A similar line of research has been conducted by Bates and her colleagues (for example, Bates *et al.* 1979; Bates *et al.* 1988), who offer an interlocking set of longitudinal studies of cognitive and linguistic development to argue the view that language development is predicated on a suite of separate component abilities (see the appendix to Butterworth, this volume, Chapter 16). Essentially, their approach is to obtain measures of children's abilities in a number of developmental spheres, and then look for relations, or lack of relations, between these measures both at one particular age *and* between different ages. In general, there emerges a picture of 'what goes with what' in development. For example, it may be found that, at any particular time, children who score highly on measure *a* always score highly on measure *b*: conversely, there may be no relation between these two measures. It is these latter dissociations that lead to the view that what is being measured rests on distinct developmental processes. Patterns of relationships among measures across time indicate the way that separate developmental processes may come to work together to provide for the establishment of a new skill.

In the earlier study (1979) of the period from 9 to 13 months, three measures emerged as developmentally independent of each other (imitation, tool-use, and conventionalized, 'referential' gestural communication), but predictive of later-appearing symbolic language abilities. These measures are taken as indices of different underlying skills. Thus, for example, it is not tool-use *per se* that is a component skill feeding into language, but the analytic ability that informs tool-use, and which is first elaborated in that domain. Later, this ability comes to play a role in the elaboration of language, allowing the child to compare various utterances and abstract underlying patterns amongst them, patterns that establish productive 'rules' for the creation of new utterances from among the child's items of vocabulary.

Their more recent study (1988), in pursuing this line of work out to 28 months of age, establishes patterns of relationships between the measures that point to the existence of three partially dissociable language acquisition mechanisms, which are emphasized to different degrees at different points in development. These are comprehension, rote production, and analysed production. There is no evidence for a split between grammatical and lexical development within or across ages (Bates *et al.* 1988, p. 267).

While these findings make it unlikely there exists an innate grammatical module that is active in language development at this early period, 'it is still possible that some kind of modular division between grammar and semantics will appear further on down the road' (ibid., p. 281). Bates *et al.* note that there are three possibilities:

1. *The maturation hypothesis*. There are innate modules devoted especially to the different subcomponents of language, but they involve more complex aspects of grammar, and they mature later in language development than the period we have covered here [i.e., up to 28 months].

2. *The construction hypothesis*. Domain-specific language modules exist, but they are not innate; instead they must be constructed and exercised over time in order to attain modular status.

3. *Total interactionism*. There are no vertical modules within the language processor; evidence in favor of such modules can be re-interpreted in terms of horizontal architecture (ibid.).

Reviewing the material relevant to these possibilities, they conclude that 'there is enough data on the relationship between grammar and semantics in adult native speakers to support the view that some kind of grammar module does eventually exist', but 'evidence for a late-maturing syntax module is still rather slim' (ibid., p. 283), and while there is some evidence congruent with the total interactionist hypothesis, the current 'best bet' appears to be the construction hypothesis:

Modules are not born; they are made. An independent and ultimately "impenetrable" use of grammar is slowly constructed, piece by piece, practised endlessly until it becomes as effortless and routine as any other fully acquired perceptual–motor skill (ibid., p. 284).

As to the possible evolutionary implications of this work, Bates *et al.* (1979, p. 20) put forward the view that:

Nature builds many new systems out of old parts, and selects for organisms that can carry out the same reconstruction process ontogenetically, jerrybuilding the same new machines from the same old parts in a highly reliable fashion. Human language may be just such a jerrybuilt system, with human infants "discovering" and elaborating their capacity for symbolic communication by a route similar to the one that led our ancestors into language.

The problem is to specify the nature and extent of this similarity, for, on the face of it, the recapitulation thesis entailed by this claim now seems an unlikely one.

3.2 Implications of cognitive ontogeny for phylogeny

The problems involved in establishing relationships between ontogenetic domains pale beside those inherent in establishing them between ontogeny and phylogeny. The above points lead to the conclusion that the dissociable abilities that underpin ontogeny have been heterochronically rearranged during phylogeny to support new outcomes. But any such rearrangement will not be a straightforward one. As Sinha (Chapter 13, this volume) has pointed out, a 'geological strata' model of phylogenesis is not tenable: the outcomes of ontogenetic systems are not grafted on top of a completed phylogeny, but are interactive with, and feed back into, an ongoing phylogeny. Thus the prolonged nature of human infancy constitutes a period that will selectively accrue adaptations to itself, with the result, noted earlier, that the ontogenetic process will not map directly to the phylogenetic one. Some of the factors contributing to this mapping problem are summarized below, and these need to be taken account of in any attempt to essay an evolutionary account informed by ontogenetic data.

3.2.1 Specific ontogenetic adaptations

At birth, the brain of a human baby is much the size that one would expect of that of a new-born monkey or ape of the same body weight (Passingham 1982, p. 112). But humans are born 'too early': Portmann (1941) estimates that we 'would' have a gestation period of 21 months, as opposed to the 9 we do have, if we were to be born at an equivalent stage of development to either chimpanzees or gorillas. Thus, while our gestation is somewhat longer in absolute time than that of any other primate, it has been accelerated quite considerably in relative time, such that a good deal of human 'fetal development' effectively occurs *extra uterum*. If our gestation were to be extended to its 'proper' span, then it seems unlikely that successful parturition would be possible, because the head size at 21 months' gestation would be larger than the female pelvis could deliver; and it is possible that the demands of bipedalism have militated against substantial pelvic modification as a solution to this problem, with the adoption of an earlier termination of pregnancy appearing as a viable alternative. The general drift of many commentators on this point is that this accelerated gestation provides for the subsequent maturation of a number of psychological functions within the socio-cultural environment they will eventually operate in: structural pathways in the brain are laid down in interaction with socio-cultural stimulation, which is advantageous for an organism whose existence is so dependent on acquired, rather than innately-given, forms of life. In addition, this point has been developed with respect to the role of socialization in ensuring the cultural appropriateness of individual human development. This is, in turn, amplified by the continuation for two years or so of the fetal rate of brain growth, which slows down markedly in the chimpanzee, by contrast, just after birth (Passingham 1982, p. 113).

The power of the developmental, socio-cultural environment is evidenced by the effect it can have on the course of development of animals that do not normally experience it (see Ristau, this volume, Chapter 23). However, the difficulty other species experience in 'benefiting' from this environment; the ability of handicapped infants to employ alternative strategies to benefit from it (blind, deaf, and thalidomide infants show some developmental delay, but little other disruption to the elaboration of human skills); and the, at first sight contradictory, evidence that human infants are 'buffered' against a number of perturbations that can occur in their social environment (Messer and Collis, this volume, Chapter 15): all these points indicate that human infants are now strongly adapted to contribute toward their own development via an integrated suite of genetically based abilities. This is *prima facie* evidence that the process of ontogeny, adapted as it is to the tasks of ontogeny, is unlikely to mirror directly the processes whereby ontogeny evolved: nor are these processes likely to show a temporal sequencing equivalent to that of the products of phylogeny.

For example, where in a recapitulationist climate the phenomenon of neonatal reflexive walking might be explained as a vestige of an adaptation for locomotion in a former environment (for example, Jersild 1955), it may well be an adaptation to the particular conditions of ontogeny in which it appears: Prechtl (1984) considers the reflex as one that allows the neonate to turn in the womb, and Ianniruberto (1985) has suggested it plays a role in engaging the neonatal head in the cervix just prior to birth (see also Alberts and Cramer 1988; and Smotherman and Robinson 1988). Such adaptations as may accrue in ontogeny necessarily complicate any reading from one domain (development) to the other (evolution).

3.2.2 Maturational schedules

Changes in the relative timing of developmental events can have consequences for other, apparently unconnected events. Vauclair (1982) notes that the timing of locomotion in ape and human infants leads to the creation of markedly different modes of object-exploration. Apes use all four limbs to locomote towards an object, and then initially prehend it orally via outstretched lips; whereas immobile human infants need to interpose their hands as instruments with which to get objects to their mouths. This hand–mouth co-ordination is maintained once the human infant becomes mobile, rather than being superseded by the ape pattern, resulting in the adoption of quite different object-exploration strategies in the two species at later stages in development, and consequently contributing to markedly different adult outcomes in object skills.

Additional examples are the probable maturational components of the human information-processing system that have been invoked to explain why there is an apparently prolonged single-word period in infant language-development: the adult memory capacity of seven plus or minus two items is one that is not given but develops or matures, and thus may act as a constraint on language production. We should, then, not facilely assume, in the absence of data on the information-processing capacities of adult, pre-modern hominids, that their language capacities would be elaborated through historical epochs of one- and then two-word speech on the grounds of developmental data that are constrained by maturational factors operating in modern infants. Other grounds would need to be adduced in support of such a claim.

Similarly, aimed throwing has a maturational component, and does not become effective in males until around the age of six years (Gesell 1954). Any correlations that might occur between throwing and other ontogenetic events are thus predominantly a result of maturational timing, and are highly unlikely to allow conclusions to be drawn about the evolutionary correlation of any cognitive abilities possessed by pre-modern adults that might have been able to perform aimed throwing. Maturational schedules in ontogeny are thus likely to place the sequencing of various ontogenetic achievements out of kilter with evolutionary events. In addition, these schedules cannot be fully understood purely in the light of individual ontogeny, but need to be located in a framework of the co-evolutionary relationships between infants (as the immediate site of ontogeny) and their care-givers (the immediate context of ontogeny).

3.2.3 *The existence of 'end products' as ontogenetic substrates*

Infants develop within the arena of many sociocultural support systems, and once these are withdrawn, development is hindered. Thus, the possession of the full central nervous system 'wherewithal' for language is no guarantee of the production of language: ontogeny occurs by guided reinvention, a different situation to that of original invention. Cultural practices function to retain and elaborate evolutionary products as substrates for developmental processes. There can be no *a priori* assumption that the order in which the original elaboration of systems occurred is reflected in the ontogenetic learning of them. In addition, obvious short cuts and reworked pathways are used in the ontogenetic mastering of symbol systems compared to the historical processes by which their modern forms were arrived at (for example, in the ontogenetic mastering of alphabetic writing systems; see Barton and Hamilton, this volume, Chapter 27).

Number systems provide two additional clear examples of how cultural systems interact with ontogeny, as argued in the work of Nesher (1988) and Damerow (1988). Nesher's studies concern the early development of children's number concepts, and how this development is affected by the language system of the child's culture. She distinguishes the use of numbers in spoken languages from their use as objects of arithmetic, for example:

> One says in natural language that "five apples *are* too much for me," but in shifting to [arithmetic language] one says that "Five *is* bigger than four" (and not "five *are* bigger than four"). Thus the notion of plurality (of apples) in the first case changes into a notion of singular objects in the latter case (ibid., p. 120).

The essential point she draws with respect to the child's development of a number system is that, in languages where this shift from the predicative function of numbers in spoken language to their substantive function in arithmetic is signalled by the structure of the language, children learn numbers in interaction with others who know the system—distinctions important to their development are socially signalled to them.

In this view, children are not so much 'learning to count' *de novo*, but trying 'to make sense of the adult's counting' (ibid., p. 118). Hence, the child 'given the developed stage of the language used around him, skips many stages in the historical development of the number concept' (ibid., p. 106). In an analogical sense, then, the conservation of historical developments in a cultural system has the effect of shifting the point in ontogeny at which a child comes into contact with what was previously the achievement of an adult: a kind of 'cultural disjunction-heterochrony'. The ontogenetic pathway through which that development is effected 'short-cuts' the original.

Damerow's work (1988) has a similar thrust. He considers numerical sign-systems as preserved on some 3900 clay tablets or 'archaic texts' from the Mesopotamian city of Uruk (*c*.5000 BP). Damerow's study reveals these sign-systems to be *transitional* between the pre-literate clay bullae 'tallying' systems (see Schmandt-Besserat 1982; and Barton and Hamilton, this volume, Chapter 27) that preceded them and abstract mathematical concepts of number that followed them. These sign-systems retain strong relationships to the proto-arithmetical techniques of the bullae medium, but by coding signs *in relation to each other*, rather than in a *one-to-one relation with the objects they signify*, they allow for a quasi-formal manipulation of number, and the direct symbolization of quantities—for instance, a sign with the numerical value of 36 000, which could previously only be apprehended perceptually, and conceptually, as 'lots (and then some)'.

Damerow (1988) considers the historical changes leading towards number concepts and abstract number systems to be contingent on the invention of representational systems:

the... transition from the proto-arithmetical technique of representation through clay tokens [or bullae] to the use of numerical signs in the archaic texts is not... to be understood as an increase in complexity through the extensive use of already available possibilities. Rather, it involves a change of the medium of representation, which as such presupposes neither changed cognitive structures nor an expanded arithmetical technique. Apparently it did, however, have lasting cognitive effects among those who, as specialists in an administrative organization, operated extensively with the new medium (ibid., p. 149).

In sum, the arithmetical techniques that developed alongside the invention of writing result from the possibilities opened up by the systems of representation being used, and 'give evidence that in cultural evolution there is no synchronous appearance of the structural elements of the number concept which Piaget asserts to be universal [in ontogeny]' (ibid., p. 126).

Thus the products of evolutionary processes and historical developments create varying demand characteristics for their acquisition. This is again seen in the case of language acquisition. If, as Johnson *et al.* (this volume, Chapter 24) argue is possible, some form of parameter-setting system were to underlie the mastery of syntactic structures, there is no direct parallel with how those structures were first established, since, in their absence, there are no data by which parameters could be set. Again, the possibility that language-development is a discontinuous process (Johnson *et al.*, this volume, Chapter 24) does not imply that this is the case for its evolution: infants may now be adapted to acquire complex linguistic systems as a result of there being complex linguistic systems previously elaborated to constitute their developmental environment, their phylogenetic elaboration occurring by different means to those now used in ontogeny.

3.2.4 *The 'unit' problem and the co-evolutionary system*

Most attempts to use ontogenetic data in creating evolutionary scenarios assume the individual as their unit of analysis. However, implicit in a number of the above points is that part of what human evolution has accomplished has been an elaboration of the ways in which humans relate to each other. Sinha (this volume, Chapter 13) has outlined how a number of theorists have conceived the importance of these relations for the constitution, maintenance, and transmission of human activities and skills, and Messer and Collis (Chapter 15) review the recent empirical research. Such relations are neglected in individually-focused evolutionary accounts, and it may be that a more appropriate unit around which to essay such an account is a 'co-evolutionary' one, say that between the (usually) closely related adult care-giver and the infant.

The heterochronic nature of human growth and development implies that changes in the nature of this relationship have occurred during human evolution, let alone those that would result from the increasing complexity of the cultural substrates of ontogeny. In addition, older siblings often take the role of care-giver and surrogate parent; patterns of birth-spacing and maturational rates of development will affect the approximation to adult-like care-giving and role-models provided by a sibling, and both patterns of birth-spacing and maturational rates are likely to have changed over the period in which human ontogeny has evolved (and this is not to mention the evolution of grandparenting). Neither the relationships that modern infants establish with care-givers, nor the way in which those relationships are elaborated in the course of ontogeny, are going to have direct counterparts with the relations that pertained between adults in the past, nor the course by which such relationships might have changed during human phylogeny.

3.2.5 *The 'attribution' problem and the socio-cultural system*

If infant–caregiver socio-cultural relationships are highlighted as important factors in human evolution, then a number of previous presuppositions about the conception of what has evolved become open to question, and this raises what we term here the 'attribution' problem. For example, stage theories are often invoked in accounts of cognitive development, and one variety of these, the Piagetian, has often been used in a recapitulatory way to underscore evolutionary scenarios (see, for example, Parker and Gibson 1979; Parker 1985). Piagetian stages are held to reflect cognitive operations that are *possessed* by an individual, and elaborated over the course of ontogeny. These operations are claimed to be independent of linguistic skills. However, the cognitive skills exhibited by language-trained and non-language-trained chimpanzees (see Ristau, Chapter 23, and Lock and Colombo, Chapter 22, this volume) cast a potentially different light on this entire conceptualization of development.

Language-trained chimps exhibit 'enhanced' abilities over other chimps; for example, analogical reasoning and some forms of conservation. Thus, their abilities are, in a sense, not chimpanzee abilities, but consequences of the cognitive technology made available to them via the particular forms of social relationships and cultural patterns their training histories have established between them and humans. Language-trained chimpanzees have in no way evolved the abilities 'they' exhibit. Thus, rather than look for evolutionary stages of human cognition, it may be better to look for evidence of the evolution of the relationships that can establish the resources within which such cognition can be sustained. Hence the attribution problem, for, from this view, Piagetian stages may not be properties of individuals, but 'artefacts' of particular socio-cultural systems that provide and sustain the resources by which cognition is conducted. This point relates back to, and amplifies, the conclusion of Section 3.2.3 above, that the apparent course of modern cognitive ontogeny need bear no straightforward relation to that of evolution, since in this case there is little reason to suppose cognitive development and the elaboration of socio-cultural systems are homologously motivated.

4 Relating human behavioural ontogeny and phylogeny: strategies

The force of these points is that there can be no *direct* reading of human development for evolutionary information. The temporal correlation of events in ontogeny will yield little directly about their possible temporal correlation in phylogeny. That we find relations between specific object-oriented activities and language in early development (Meltzoff 1988) is not a ground for inferring that these skills evolved together: the processes that underlie the two activities may have separate origins and histories, so that phylogenetic evidence for the presence of one skill gives no grounds for assuming the presence of the other. And finally, much of the prehistoric record of human behaviour was produced by the actions of adults, and there is little reason to assume that the products of prehistoric adult activities will be directly comparable to the ontogenetic products of modern infancy and childhood. In sum, with respect to symbol systems, it must be remembered that an ontogenetic problem—for example, how children come to be able to acquire grammatical language—differs from the apparently comparable phylogenetic problem—for instance, how language came to have a grammatical structure for children to acquire.

If, then, there can be no simple reading from development to evolution, what progress can be made in using knowledge of development in pursuit of an evolutionary scenario? This same question can be posed with respect to the use of comparative and cross-cultural psychological data. The following points provide a minimum framework within which to proceed.

1. There is a suite of separable/dissociable abilities contributing to the cognitive skills underlying human symbolic performance.
2. These underlying abilities can be evolved at independent rates.
3. The evolution of these underlying abilities is partly independent of the elaboration of the cognitive/cultural systems they support.
4. The elaboration and cultural conservation of the systems these underlying abilities make possible provides: (a) an environmental substrate that scaffolds and thus enhances the use that can be made of those same underlying abilities; and (b) an environment in which the abilities that have provided for system elaboration can be further acted on by processes of Darwinian selection, i.e. differential reproduction.
5. The rate and direction of the elaboration of these extra-somatic systems is largely prompted by environmental factors. These act to provide the conditions that allow the immanent properties of the systems to be flushed out, to thus 'bootstrap' new 'objects' for both old and newly-required underlying capacities and skills to be elaborated from and operate on.

Ontogenetic information can play a role in developing some of this framework, but only to the extent of hypothesizing possibilities, and thus posing questions that different areas of study could bring evidence to bear on. But even with this limitation, an important advance is made. Ontogeny allows the pieces in the heterochronic jigsaw at least to be discriminated.

4.1 *The suite of separable capacities*

The developmental research that elucidates possible component capacities and their contributions to more complex abilities has only recently begun (see appendix to Johnson *et al.*, this volume, Chapter 24). While it has a clear heuristic value in helping formulate both ontogenetic and phylogenetic accounts of the creation of symbolic systems, a number of questions arise that require further investigation.

4.1.1 *Gestural, referential communication*

Three issues are tangled here: the origins of reference, the origins of gestures, and the origins of communicative motivation. From a phylogenetic perspective, communicative motivation appears to be the most important of these. Great Apes appear to comprehend reference in the wild, as is evidenced by the inferences they appear to be able to draw from the behaviours of their companions (for example, Jolly, this volume, Chapter 6). This comprehension effectively constitutes aspects of those others' behaviours as gesturally referential, and, in human perception at least, quasi-gestural communication is apparent between care-giver and kin, in the way intentional actions become stylized to indicate inter-'personal' goals, for example, 'begging' and raising an arm to initiate being groomed (Plooij 1978).

What appears to be lacking in these animals is the social organization that would foster from these beginnings a motivation to communicate and share information.[6] In fact, many primates live under constraints that actively 'discourage' communication, and it is these that doubtless prompt the increasingly reported instances of their 'deceptive behaviours' (again, see Jolly, this volume, Chapter 6). 'Sharing' information about the world (i.e., referential communication) is potentially costly for an animal that lives in an amoral social world. In that situation, communicating unique knowledge will result in an individual's losing any benefits that might otherwise accrue to itself, and hence motivate it 'not to let on' about some things it knows, an example being Dandy's 'self-control' when he locates hidden food (de Waal, in Jolly, this volume, Chapter 6): this is a clear example of 'necessity being the mother of stasis'. Importantly, though, in some circumstances, such as retrieving plentiful food, dealing with snakes, and escaping from enclosures, communication and co-operation does occur among primates (Menzel, in Lock and Colombo, this volume, Chapter 22).

All this indicates that it may be that we need to seek, phylogenetically, for the origins of a communicative capacity in the social structures that foster the motivation to employ a pre-existing ability. Albeit with additional superstructural or contextual support, the 'ape language' projects (Ristau, this volume, Chapter 23) do indicate communicative capacities in chimpanzees, in social situations that *require* them to communicate, that go beyond any of those presently observed in the wild. (We may expect to find differences in the levels of communicative behaviours in chimpanzee troops that inhabit different environments, in the same way that some groups engage in more tool-using than others, for example, Tai Forest (Boesch 1993) vs. Gombe Stream (Goodall 1964): but the level required of wild chimpanzee communication to put it in the ballpark of what chimpanzees can communicate in the language programmes seems unlikely to be found in their natural situation.) The implication is that the abilities underlying referential, gestural communication have a deep phylogenetic history, and effectively pre-adapted hominids for the elaboration of sophisticated communication systems once social life provided the conditions that motivated or allowed them.

4.1.2 *Imitation*

Imitation and observational learning both occur in the higher primates, especially the Great Apes, but are more constrained than in humans (see Lock and Colombo, this volume, Chapter 22). Thus, there does appear to have been a marked increase in the power of the capacities underlying what Bates *et al.* (1979) have termed the human capacity to perceive contiguity, and to reproduce unanalysed wholes, and 'an increase in the capacity and the motivation to imitate may have been a critical factor in the evolutionary leap to human-like culture, including language' (ibid., p. 337).

Imitation is, though, a poorly understood process, and is in need of an adequate cognitive analysis. In addition, what 'level' of imitation is actually required as a 'threshold', allowing language either to be created *de novo* or learned, has yet to be determined. Neither is it clear what might index the elaboration of this ability in the palaeoanthropological record. The persistence of tool forms may be held to be indicative of imitative abilities, but at this point there is no way of telling what level of capacity this persistence might indicate; nor whether that level has been subsequently increased; nor whether it was sufficient for symbolic purposes; nor how what would have been an adult capacity relates to a present-day infant one; nor the extent to which attaining a symbolic ability might allow a given capacity for imitation to be used in ways that appear to amplify it. There is clearly a lot to be clarified here.

4.1.3 Means–end analysis

Similar problems arise when analytic abilities are considered. Crudely, analytic abilities might reflect sheer 'computing power', in that their basic structure remains constant over time, but that the amount of information and its temporal extension that can be analysed depends upon the 'hardware' of cortical size and the 'software' of representational cognitive technologies. If this is the case, then brain size does give us some index of the increase in this capacity. On the other hand, the achievements and limitations of the ape 'language' projects highlight respectively the important role of socio-cultural practices in scaffolding the attainment of symbolic functioning on the basis of relatively small hominoid brain sizes, and the possible specialization of human analytic abilities over and above a generic means–end capability: for if grammar were simple, a generic system should be capable of handling it.

4.1.4 Comprehension

As we noted above, Bates and her colleagues (1988) have isolated what they term a 'comprehension component' as playing a critical role in language development. Comprehension presupposes perceptual as well as conceptual perception (see also Hirsh-Pasek and Golinkoff 1991). Language-use relies on categorical perception of speech sounds, and this no longer appears, on the empirical evidence (see Ehret 1987, and Lock and Colombo, this volume, Chapter 22), to be a specifically human capability.[7] Savage-Rumbaugh (1987, and continuing research) reports both perceptual and conceptual abilities in the syntactic decoding of speech for bonobos (pygmy chimpanzees; and her preliminary data indicate that the abilities of chimpanzees may previously have been underestimated). At first sight these findings suggest that these prerequisite abilities have a deep phylogeny, and that they play a role in determining the structural properties that are elaborated in human languages.

However, 'comprehension' is much less of a primitive category of cognitive capacity than either imitation or means–end analysis can ever be, and its bases are in need of further elucidation. In addition, while an earlier developing ability to comprehend grammatical language may play a role in the later ontogeny of language production, it is unclear what it might contribute to language creation. Finally, neither bonobos nor chimpanzees could be held to possess an unused but specialized set of cognitive abilities for grammatical language-comprehension that sit around vacantly in their heads waiting for a contemporary Savage-Rumbaugh to activate them: the relation between what can be teased out as independent domains contributing to language ontogeny and what might be independent domains contributing to language creation in phylogeny is by no means clear.

4.1.5 Additional pre-requisites

Lieberman (1984, 1985, 1988) has provided strong grounds for considering the abilities involved in the decoding of speech, plus those involved in speech production, as providing a 'preadaptive basis for human rule-governed syntactic and cognitive ability' (1988, p. 153). This is another variation on the points made above with respect to comprehension, and extends the possible list of pre-requisites further. In addition, changes in memory capacities and attention-span may well be implicated in the achievement of human cognitive abilities. We need to be aware, then, of the possibility that the list of pre-requisites could grow beyond the point at which the general framework becomes less than helpful.

Empirically, replications of the studies by Bates and her colleagues, on which much of this approach rests, need to be carried out. Their claims as to the pathways via which, and the skills upon which, language is ontogenetically elaborated rest on statistical relations holding between a myriad of developmental measures made on a relatively small sample of children. Children's capacities are difficult to measure, and a few percentage points of difference in performance here and there might well have resulted in a different, but still equally plausible, theory of language development.

Notions such as 'imitation' and 'comprehension' need to be analysed, and then have their possible bases established empirically, for few of these proposed dissociable pre-requisites appear to be truly primitive cognitive categories. Conceptually, attention needs to be paid to notions such as 'capacity', 'ability', and 'skill', as these are currently used almost interchangeably in the literature, whereas they have different connotations. 'Memory' and 'attention-span', for example, are of a different nature from 'means–end analysis' and 'the perception of contiguity'; but which of these are best considered as capacities, abilities, or skills has yet to be clarified. Unless problems such as these are confronted and thought through, there is the possibility that a psychology of dissociable 'abilities' will go the way of 'faculty psychology' and, before that, phrenology. That may yet prove to be the case; but currently this approach does look helpful in elucidating phylogenetic questions. For example, the bonobo data (Savage-Rumbaugh 1987) and those from other species at least indicate (if taken at face value) that some of the abilities co-opted to support language and symbolization (for example, those that allow gestural communication) were patent before other abilities.

4.2 *The provision of a scaffolding system*

When we ask about 'the evolution of language' we need to do so in the recognition that there is a distinction between language and the systems that support it. *All* the pre-requisite capacities for modern language could be evolved and thus available to a human organism or population *without* there being any guarantee that an individual or population possessed or used language in anything like its modern sense. Both Hewes (Chapter 21, this volume) and Foster (Chapter 25) raise the possibility of a non-or pre-phonemic language-system. Such a system could have been available to early members of our own species; and to go beyond this system they would not have needed to wait upon the evolution of the necessary somatic hardware to deal with phonemes. Instead, they might only have had to have made the *cultural* 'leap' to such a system, for all the somatic capacities required to support a system of language so organized were presumably already in place. The same could be true for syntactic organization: no evolutionary (genetic) change need be required.

Above we have hypothesized that the skills required for a phonemically organized language could well be in place before such a language-system emerges. This claim entails that cultural systems elaborated at one level from a particular set of abilities act as a substrate on which that same set of abilities can again be applied, resulting in a reorganization of that system, and the creation of a new set of 'objects' for the original package of abilities to work from. Non-contentious examples of this are provided in the cases of the historical development of writing and mathematical systems (see above; also Barton and Hamilton, Chapter 27, this volume, and the Editorial appendices to that chapter). These types of literacy provide a means for externalizing cognition, facilitating new forms of cognition (Gumperz and Levinson 1991; see also the Editorial introduction to Part II of this volume).

An additional implication concerns the creation of syntactic systems in language, in addition to phonemic ones, through a reworking of previously-elaborated systems (see Rolfe, this volume, Chapter 26), and the effect this reworking of a system would have upon the ontogenetic paths via which it was mastered. The claim here would be that the original elaboration of these organizational principles would create a vastly different situation for ontogeny, in that they constitute the goal and medium towards and within which ontogenetic processes operate. The operation of these ontogenetic processes would be allowed 'short cuts' over the original elaboration of these systems, in the sense that learning anything appears easier than inventing it in the first place (given that the skills involved in the original creation are likely to be those involved in its recreation). The notion of 'scaffolding' reviewed by Messer and Collis (this volume, Chapter 15) in the context of the social induction of modern infants into language would thus be just one aspect of the way socially-instantiated systems can provide a superstructure that influences the routes by which they are mastered.

This point may be extended by considering cultural artefacts, as well as practices, as repositories and conservation devices of cognition, in that such objects may act to 'embody' cognition in senses that are currently not well thought through. For example, the calculating skills of an abacus user are to an extent a component of the device itself, 'outside of the user's head'. Note, however, that, for the skilled user, the device can be 'internalized' to such an extent that calculations can after a certain stage be performed mentally without the abacus as accurately, and often faster, than with it. Additionally, such skilled users show very specific enhancements in cognitive ability: digit memory increases to 15-digit strings—forward or backward—while memory for items such as the Roman alphabet, or lists of fruits, remains around the expected 7 ± 2 items (Hatano 1982; Stigler *et al.* 1982).

The point is again implicitly demonstrated in White (1962, p. 113) with respect to compound cranks (see Fig. 3). These appear to have been invented and instantiated as carpenter's braces (drills) in northern Europe around AD 1420. The first indication that the invention was being transferred to Italy comes in a notebook of Giovanni da Fontana dated around 1420–1449. However, da Fontana's sketch is of an unworkable tool. Similarly, his compatriot Mariano di Jacopo Taccola provides from his notebooks the earliest Italian evidence of conjoining a connecting-rod, as a mechanical substitute for a human arm, with a compound crank: but his sketch is again of an unworkable machine, indicating a defective understanding of the motion involved. Such instances provide clues as to how 'cognition' or 'knowledge' may be instantiated in objects rather than 'heads' (although this is clearly an interactive relation rather than an either/or one).

Other examples of scaffolding are embedded in the notion of symbol systems as 'cognitive technologies'

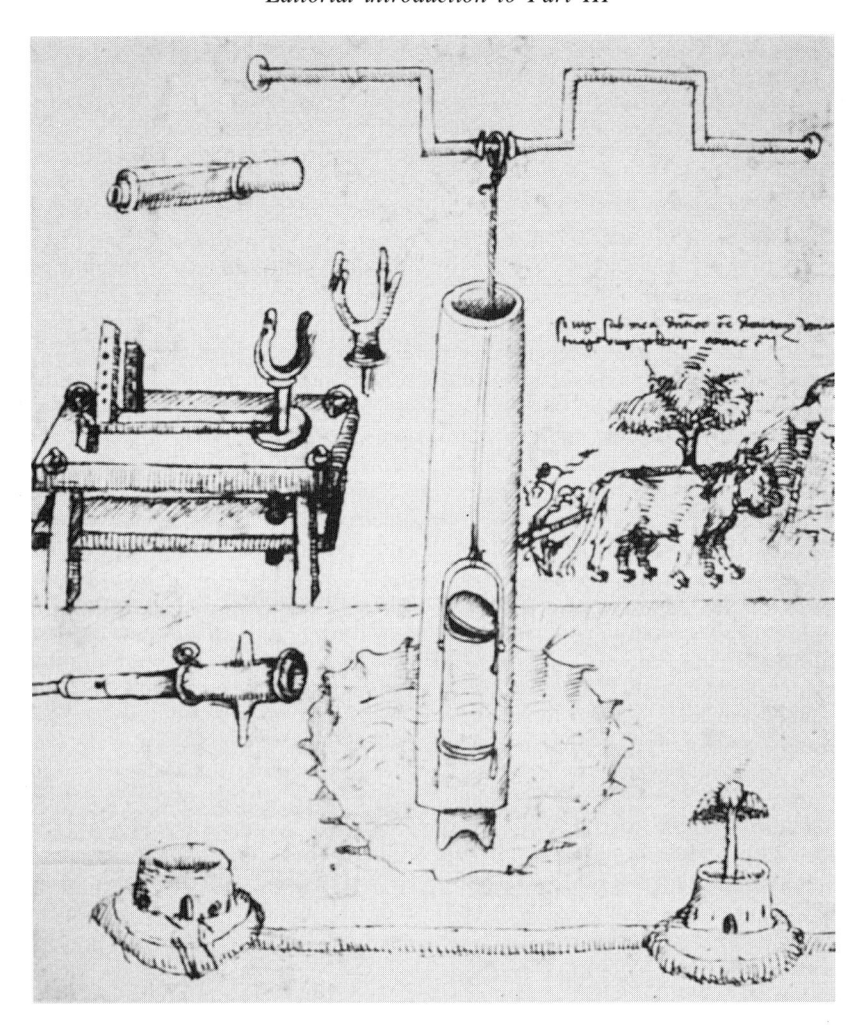

Fig. 3 (a) Giovanni da Fontana's sketch, *c*. 1420–49, of a cranked auger, mechanically misunderstood.

discussed by Lock and Symes (this volume, Chapter 8). The debate over whether Chinese speakers solve some problems more or less easily than English speakers may remain contentious; but it is clearly 'easier' to do long division using the Arabic number system than the Roman, even once one has mastered each system. The relation discussed in that chapter between social structures and conceptual codes argues for another aspect of the role of these symbolic cognitive technologies in scaffolding the outcomes of the same underlying abilities.

A point to note here is that this present view of the bases of cognitive abilities and the way they interact with the media that support them provides a way out of the dilemmas created by oppositions between, for example, 'primitive' and 'scientific' thought, that have racked anthropology in different guises for most of its existence. Claims that there are 'mythical' or 'primitive' modes of thought versus 'scientific' ones, differences in thinking between 'literates' and 'non-literates', or that members of certain societies 'progress' no further than the pre-operational Piagetian stage of thinking as generally shown by Western eight-year-olds, tend to have at their root the Victorian equation of 'evolution' with 'progress'; that is, an inbuilt value judgement that there are 'higher' and 'lower' stages in the elaboration of human mental abilities and social structures.

There are, however, some legacies of the Victorian conception that are useful, and it is a version of 'the doctrine of psychic unity', originating in nineteenth-century social science, that we are implicitly developing here. This notion has assumed a number of guises as paradigms have shifted within anthropology

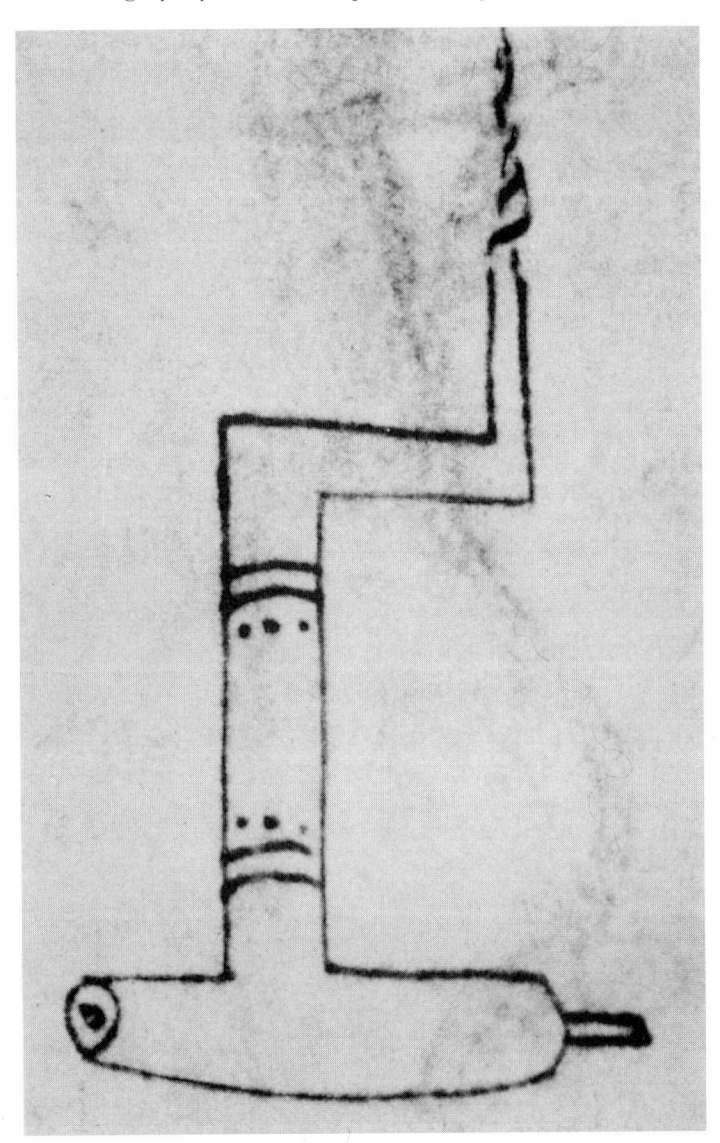

(b) Mariano di Jacopo Taccola's drawing, 1441–58, of a compound crank and connecting rod, mechanically misunderstood. Illustrations taken from White (1962).

(see Harris 1969). In Victorian times the prevailing version was that the thought of 'primitives' was primitive because their cultures were primitive. Primitives were held to be *able* to generalize, for example, as much as Western Europeans; but since their cultures were impoverished in their stock of knowledge, they *did not* generalize very often, nor very correctly. Thus it was cultures that were held to exist at different stages, not individual cognitive abilities.

From Boas's work (1911/1963), the assumption of the unity of a culture—that its level of technology would be on a par with its level of rhetorical skill, for example (and how could they be?)—was upset, but the equivalence of cognitive ability was reaffirmed. Working with a 'primitive' society, the Kwakiutl of the American north-west, Boas convincingly demonstrated the complicated and abstract character of the thought that informed their art and social organization. Thus 'psychic unity' for Boas meant cognitive equivalence, not just in essence, but in actuality: hence, claiming that the 'functions of the human mind are common to the whole of humanity' (1963, p. 135) implied not that they were on the same continuum, but that everywhere they were equally elaborated. According to this view, observed differences are in the area of *content*: the belief-systems and cultural premisses of traditional peoples may differ from those in industrialized societies, but they employ the same logical processes and display the same sort of concern with the relations of cause and effect in practical contexts.

By contrast, much cross-cultural psychology has adopted the opposite emphasis, characterizing cognition by its structural and operational features as proposed by Piaget, and locating these firmly within the individual. These studies come up against the problem of cultural knowledge from this reversed perspective. For example, Dasen (1977) collates data for all cross-cultural studies of liquid conservation conducted up to 1975 (see Fig. 4). These studies are difficult to interpret because 'nothing guarantees the comparability of the studies' (Dasen 1977, p. 169). But, proceeding carefully,

looking only at the general trends indicated by these curves, it remains obvious that the cross-cultural variability in the proportion of individuals attaining this concept of conservation is large.... Of interest are those developmental curves which do not reach 100%, particularly if the study included adolescents or adults. These curves indicate that there is a large proportion of individuals who do not acquire (or at least, display) this particular set of concrete operations. ... What is the significance of this phenomenon? (ibid.).

One significance it does *not* have is made explicit by Dasen:

the curves should not be taken as comparative measures of cognitive capacity in the groups compared. It seems that the following interpretation of such curves may sometimes be made (quoted by Cole and Scribner 1974, 156): 'Tribe X does not mature past the European 11-year stage if 50% of the members of tribe X conserve and 50% do not.' We personally know of no such silly statement in print, but it seems not impossible that it may occur (ibid.).

But, such statements do occur,[8] and serve to emphasize how individually-centred cognitive theories can impale themselves on the other horn of the Victorian dilemma. What, then, as Dasen asks, are we to make of cross-cultural findings that adult members of some

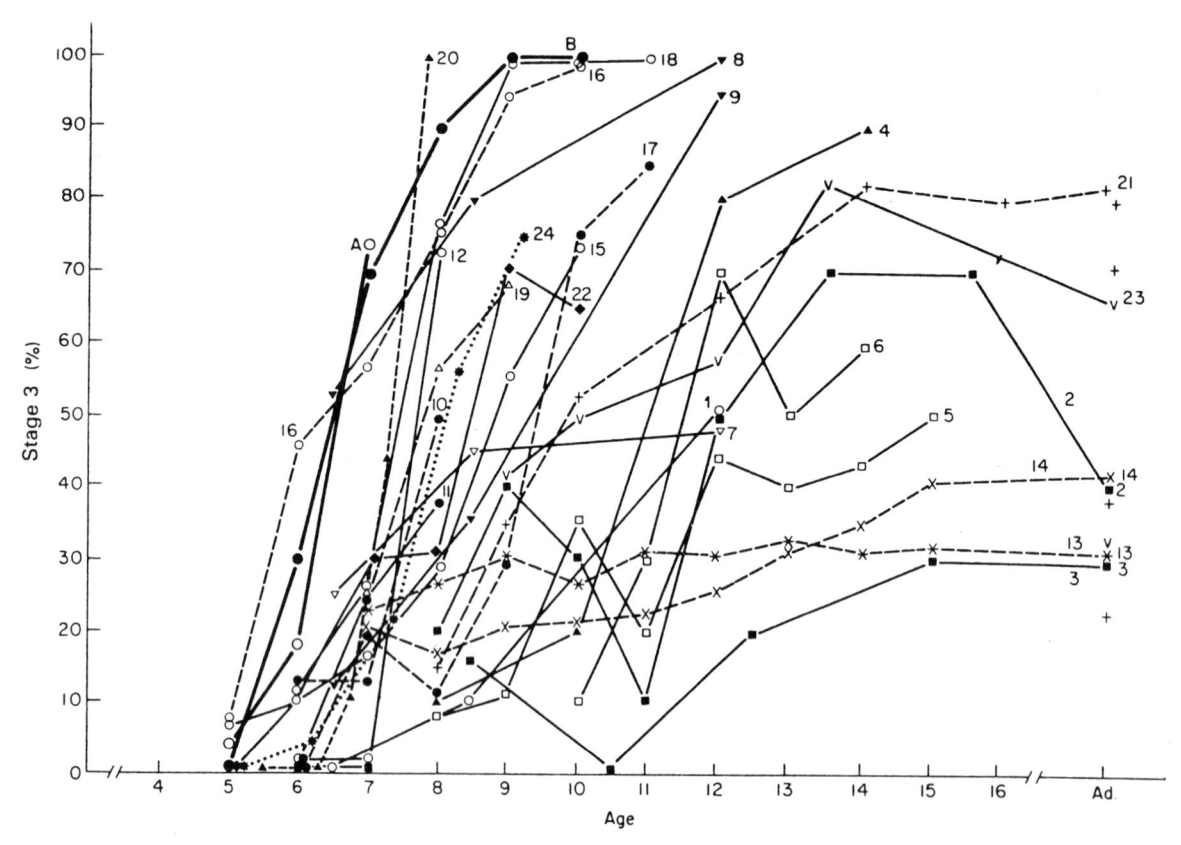

Fig. 4 Conservation of quantity (liquids). Percentage of subjects attaining full conservation. Reprinted from Dasen (1977). Reproduced by permission from *Psychologie Canadienne/Canadian Psychological Review*, University of Calgary, Alberta, Canada.

A	Europeans, Geneva	(Piaget and Inhelder 1963)
B	Europeans, Canberra	(Dasen 1974)
1	Algerians, unschooled	(Bovet 1974)
2	Australian Aborigines (Hermannsburg), medium contact	(Dasen 1974)
3	Australian Aborigines (Areyonga), low contact	(Dasen 1974)
4	Ebrié, Ivory Coast (Adiopodoumé)	(Dasen 1975)
5	Australian Aborigines (Hermannsburg)	(de Lemos 1969)
6	Australian Aborigines (Elcho Island)	(de Lemos 1969)
7	Wolof, Senegal, rural, unschooled	(Greenfield 1966)
8	Wolof, Senegal, rural, schooled	(Greenfield 1966)
9	Wolof, Senegal, Dakar, schooled	(Greenfield 1966)
10	Aden, Arabs	(Hyde 1970)
11	Aden, Indians	(Hyde 1970)
12	Aden, Somali	(Hyde 1970)
13	Papua-New-Guinea, unschooled	(Kelly 1970)
14	Papua-New-Guinea, schooled	(Kelly 1970)
15	Iran, rural	(Mohseni 1966)
16	Iran, urban	(Mohseni 1966)
17	Thailand, rural	(Opper 1971)
18	Thailand, urban	(Opper 1971)
19	Rwanda, rural, schooled and unschooled	(Pinard et al. 1973)
20	Tiv, Nigeria, rural, unschooled	(Price-Williams 1961)
21	Papua-New-Guinea, schooled	(Prince 1969)
22	Oglala Sioux	(Voyat, personal communication)
23	Papua-New-Guinea, schooled	(Waddell, personal communication)
24	Lebanon, Beirut, schooled	(Za'rour 1971)

societies consistently fail to perform on tests as well as adults are believed to in the cultures that originate those tests?

One possibility (often adopted in the so-called 'IQ debate') is to claim that the contents of the tests are inappropriate to the testing of populations other than those of the culture in which they were formulated, since they presuppose culturally specific knowledge. A variant of this can be applied in developmental psychology:

Two year olds are capable of taking the perspective of another (Lempers, Flavell, and Flavell 1977; also see Shatz and Gelman 1973 on perspective taking in four year olds) and detecting categorical structure (Goldberg, Perlmutter, and Myers 1974); as Goldberg et al. note, children can cluster things categorically as soon as they can talk. Three years olds understand many aspects of the concept of causation (e.g. that a cause precedes its consequence—Bullock and Gelman 1979). Four year olds are capable of transitive inference and syllogistic reasoning (Trabasso 1975) and various forms of deductive inference in propositional logic (Macnamara, Baker, and Olson 1976).... The evidence suggests that most mental structures are *available* to the five-year-old child and may well be available earlier, if only we knew how to find them (Shweder 1982, pp. 357–8).

Similar conclusions could be argued from comparative psychology: that, if approached in specific ways, primates can be shown to have a hitherto unsuspected number of mental structures available to them.

Another possibility is the one we would argue for from the ontogenetic points we have been developing thus far: *that the cognitive technologies available to an individual act to scaffold cognition*. In some circumstances it is possible to simulate these technologies by a *contextual scaffolding* of the problem-solving situation. From this view, most mental structures are *not* available to the five-year-old child, as Shweder (1982) claims from his review. Rather, well-versed adult practitioners who do have particular skills can engineer interactions with children so as to enable them to accomplish tasks that on their own they are incapable of completing: the structure of the experimental investigation scaffolds the child's performance.

More generally, Light and Perret-Clermont (1989) adduce the argument that basic Piagetian logical concepts such as amount, area, number, weight, and volume derive from practically-oriented social practices that are associated with the sharing and distribution of commodities. Consequently, they propose that a conceptual ability such as conservation does not entail the possession of a transcendental logical concept, but that conservation concepts are historically elaborated products of practical and social purposes: 'Thus the child's task in mastering conservation concepts is arguably only possible to the extent that he or she is able to share in the purposes and practices to which these concepts relate' (1989, p. 109). (More on this conceptualization of the relation between cognitive ability and social practices can be found in Lave 1990; Rogoff 1990; Saxe 1991; Saxe *et al.* 1987; and Wertsch 1991). In sum, contexts embody properties of the cognitive technologies that create them.

Problems are not solved, then, by the application of particular mental structures, but by the experience the solver has with the symbol systems that constitute the cognitive technology and interactive social contexts applicable to those problems (cf. Scribner and Cole (1981) on the enabling effect of social practices on the inherent possibilities afforded by literacy; and Barton and Hamilton in Chapter 27 of the present volume). Thus, that Australian Aboriginals regularly appear to fail liquid-conservation tests tells us not that they lack a particular mental structure, nor that they are equivalent cognitively to Western six-year-olds, but that in a culture where water is scarce, where the environment is conceived as one to be conserved rather than appropriated, and thus where water is 'stored' where it is found, rather than in some hydraulic technology, the relevant scaffolding cognitive technologies and contexts have not been elaborated to support and report particular problem solutions (see also concluding section to Lock and Colombo, this volume, Chapter 22, and our point above on the 'attribution' problem).

To this it is necessary to add the complexities of the physical instantiation of these abilities within the organic structure of the developing and evolving brain that allows modern humans to be so cognitively human. It is unlikely to be the case, even though we ourselves have in this introduction written as if it were, that brains evolve and develop structures that fully underwrite the functions they subsequently support. It is more likely that the structure of the brain, the way it 'wires itself up', is the result of an interaction between its evolutionary potential, its maturational course, and the structure of experienced reality that it makes possible (see, for example, Deacon 1992). At the same time, it is clearly the case that this experienced reality is not merely 'given', but inherently imbedded in sociality.

The thrust of all this from an evolutionary point of view is that it is necessary to consider many human accomplishments as culturally-constituted ones, based interactively on a biologically-constituted substrate. It is thus necessary to seek both for the 'enabling' conditions that allow not only their constitution but also their creative expression and maintenance, and for the conditions and pre-conditions that 'constrain' their constitution, expression, and maintenance. It is not only that human 'intelligence' has evolved, for example, but that it has also been 'empowered'.

5 Concluding remarks

All the points we have raised require a full working through, clarification, specification, and development to yield specific hypotheses. They are at present imprecise, but represent the first steps in developing a richer view of the contribution ontogenetic material can make to phylogenetic problems than previously discredited ones. Our discussion here in no way comprehensively includes all the issues involved, nor even those raised in numerous contributions to this volume. We have, however, outlined a general framework within which it is possible to locate the evolutionary implications of ontogenetic processes. The day of the slogan 'ontogeny recapitulates phylogeny' has passed: we can replace it with nothing as brief or as simple. Clearly, a more complex formulation is needed.

Notes

1. Borstelmann (1983, 34) economically captures the effect of evolutionary theory on the outlook of Imperial Victorian science:

Darwin's ideas led to the establishment of anthropology, comparative psychology, and to a conception of the child as a natural museum of human phylogeny and history. In the 1800s, parallels between animals and children, primitive societies and the early history of humans were rampant.

The general tenor of this equation, and of its consequences in terms of the views it supported, are easily captured by a few quotations:

During the early stages of human progress, the circumstances under which wandering families and small aggregations of families live furnish experiences comparatively limited in their numbers and kinds; and consequently there can be no considerable exercise of faculties which take cognizance of the *general truths* displayed throughout many special truths (Spencer 1886, p. 521).

The mind of the child and the mind of the savage, when differences due to the presence of manhood and womanhood in the latter, diversity of environment, influence of higher culture, prolonged infancy, social environment, etc., have been taken into consideration, present many interesting parallels of a general sort (Chamberlain 1901, p. 456).

The idol answers to the savage in one province the same purpose that its analogue, the doll, does to the child. It enables him to give a definite existence and a personality to the vague ideas of higher beings, which his mind can hardly grasp without material aid (Tylor 1865, p. 94).

Darwin's conception of evolution as a biological phenomenon was extended by analogy to social and mental phenomena: cultures and minds both evolved; ontogeny recapitulated phylogeny; and savages were living fossils, culturally and mentally. We are now able, with hindsight, to reflect on and offer accounts for why such crass interpretations should have been adopted: for example

Positing another person or another culture as a subject to be understood, rather than an object whose behaviour is to be causally explained (i.e. reduced to external, objectified, circumstances), presupposes a degree of respect, and acceptance of an equality, however relative. Both respect and acceptance of equality were missing... in the era of the "white man's mission" (Bauman 1978, p. 202).

Yet the identification of evolutionary perspectives with such oversimplifications still plays a role—usually that of 'bogey' (see Introduction to Part II, this volume)—in the study of the cross-cultural diversity of human life today, especially in the sphere of 'cross-cultural cognition', and in the drawing of inferences from it, studies of other species, and studies of human ontogeny.

2. Changes in the developmental timing of features relative to their appearance in the ontogeny of an ancestor.

3. The terminology used here follows Lincoln *et al.* (1982) rather than de Beer or Garstang.

4. But, for one hominoid, the pygmy champanzee *Pan paniscus*, the neotenic argument does stack up: 'I have argued previously (Shea 1983, 1984, 1988) that many of the morphological (and perhaps some behavioural) differences between the pygmy chimpanzee..., *Pan paniscus*, and its close relative the common chimpanzee, *Pan troglodytes*, should be viewed as an integrated suite of characters resulting from neoteny' (Shea 1989, p. 90).

5. For a review of the way various aphasias are classified, and a discussion of their neuropsychological implications, see Ellis and Young (1988).

6. The greater propensity for acquiring language that is emerging from Savage-Rumbaugh's and her colleagues' (for example, Greenfield and Savage-Rumbaugh 1990) research on pygmy chimpanzees (or bonobos) (*Pan paniscus*) may be based on differences in the nature of their social life from that of chimpanzees (*Pan troglodytes*):

There does appear to be a distinct possibility that the social life of *Pan paniscus* is considerably more complex than that of *Pan troglodytes*. Male-female relationships are more elaborate and far more frequent in *Pan paniscus*. Infants appear to have contact with a much wider range of adults; the unit group is made up of larger, more stable subgroups; and preliminary observation of presumed attack behavior suggests a remarkable degree of coordination among a large number of individuals of both sexes (Savage-Rumbaugh *et al.* 1989, p. 291).

Alternatively, it may be an 'artefact' of these researchers' particular procedures. The inclusion of Panzee, a female *Pan troglodytes*, into their ongoing research programme may clarify this issue.

7. See also Kojima *et al.* (1989), who report that 'the basic mechanism for the identification of consonants in chimpanzees is similar to that in humans' (1989, p. 403); Kojima and Kiritani (1989), who find similarities underlying vowel perception in the two species; Sinnott and Adams (1987), who report on psychoacoustic sensitivity to speech cues in monkeys, finding them less sensitive than human adults, but perhaps equivalent to human infants; and Owren (1990*a*, *b*) for a more general perspective. By and large, many primates 'apply humanlike processing strategies in decoding [human speech], including use of lateralized neural structures, selective attention for linguistically-relevant acoustic features, and perceptual categorization of physically graded cues' (Owren 1990*a*, p. 20).

8. For example, Hallpike (1979) collapses the distinction between a culture's collective representations and individual thought processes thus: 'The collective representations of a society... reflect... the level of cognitive development of the great majority of the adult members of that society' (ibid., p. 32).

He goes on to advocate a 50 per cent rule that states the mastery of a conceptual notion must be shown by above 50 per cent of the adult members of a society if a notion is to be in the collective representations of a society (ibid., p. 61). Not quite Dasen's point, but very close to it.

References

Abbie, A. A. (1952). A new approach to the problem of human evolution. *Transactions of the Royal Society of South Australia*, **75**, 70–88.

Abbie, A. A. (1958). Timing in human evolution. *Proceedings of the Linnean Society of New South Wales*, **LXXXIII**, 197–214.

Alberts, J. R. and Cramer, C. P. (1988). Ecology and experience: sources of means and meaning of developmental change. In *Handbook of behavioral neurobiology*. Vol. 9. *Developmental psychobiology and behavioral ecology*, (ed. E. M. Blass), pp. 1–40. Plenum Press, New York.

Barkow, J., Cosmides, L., and Tooby, J. (ed.) (1992). *The adapted mind: evolutionary psychology and the generation of culture*. Oxford University Press, New York.

Bates, E., Benigni, L., Bretherton, I., Camaioni, L., and Volterra, V. (1979). *The emergence of symbols:*

cognition and communication in infancy. Academic Press, New York.
Bates, E., Bretherton, B., and Snyder, L. (1988). *From first words to grammar: individual differences and dissociable mechanisms.* Cambridge University Press, New York.
Bauman, Z. (1978). *Hermeneutics in social science.* Hutchinson, London.
Beer, G. de (1930). *Embryology and evolution.* Clarendon Press, Oxford [3rd edn, 1958, *Embryos and ancestors*].
Beer, G. de (1954). *Archaeopteryx lithographica.* British Museum (Natural History), London.
Bellugi, U., Bihrle, A., Jernigan, T., Trauner, D., and Doherty, S. (1990). Neuropsychological, neurological and neuroanatomical profile of Williams Syndrome. *American Journal of Medical Genetics*, **6**, 115–25.
Beynon, A. D. and Dean, M. C. (1988). Distinct dental development patterns in early fossil hominids. *Nature*, **335**, 509–14.
Boas, F. (1911/1963). *The mind of primitive man* (rev. edn). Collier, New York.
Boesch, C. (1993). Aspects of transmission tool use in wild chimpanzees. In *Tools, language and cognition in human evolution*, (ed. K. R. Gibson and T. Ingold), pp. 171–83. Cambridge University Press.
Bolk, L. (1926). On the problem of anthropogenesis. *Proceedings of the Section of Sciences of the Koninklijke Nederlandse Akademie van Wetenschapen, Amsterdam*, **29**, 465–75.
Borstelmann, L. J. (1983). Children before psychology: ideas about children from antiquity to the late 1800s. In *Handbook of child psychology*, Vol. 1, (ed. P. H. Mussen), pp. 1–40. Wiley, New York.
Bovet, M. C. (1974). Cognitive processes among illiterate children and adults. In *Culture and cognition: readings in cross-cultural psychology* (ed. J. W. Berry and P. R. Dasen), pp. 311–34. Methuen, London.
Bromage, T. G. (1987). The biological and chronological maturation of early hominids. *Journal of Human Evolution*, **16**, 257–72.
Bullock, M. and Gelman, R. (1979). Preschool children's assumptions about cause and effect: temporal ordering. *Child Development*, **50**, 89–96 (cited by Shweder 1982).
Chamberlain, F. (1901). *The child: a study in the evolution of man.* Walter Scott, London.
Cheng, P. W. and Holyoak, K. J. (1989). On the natural selection of reasoning theories. *Cognition*, **33**, 285–313.
Chomsky, N. (1975). *Reflections on language.* Random House, New York (cited by Cosmides 1989).
Chomsky, N. (1980). *Rules and representations.* Columbia University Press, New York (cited by Cosmides 1989).
Cole, M. and Scribner, S. (1974). *Culture and thought: a psychological introduction.* Wiley, New York (cited by Dasen 1977).
Connolly, K. and Dalgleish, M. (1989). The emergence of a tool-using skill in infancy. *Developmental Psychology*, **25**, 894–912.
Cosmides, L. (1985). Deduction or Darwinian algorithms?: an explanation of the 'elusive' content effect on the Wason selection task. Doctoral dissertation, Harvard University. University Microfilms 86–02206 (cited by Cosmides 1989).
Cosmides, L. (1989). The logic of social exchange: has natural selection shaped how humans reason? studies with the Wason selection task. *Cognition*, **31**, 187–276.
Cosmides, L. and Tooby, J. (1987). From evolution to behavior: evolutionary psychology as the missing link. In *The latest on the best: essays on evolution and optimality*, (ed. J. Dupré, pp. 227–306). MIT Press, Cambridge, Mass. (cited by Cosmides 1989).
Cosmides, L. and Tooby, J. (1989). Evolutionary psychology and the generation of culture. Part II: Case study: a computational theory of social exchange. *Ethology and Sociobiology*, **10**, 51–97. (cited by Cosmides 1989).
Cromer, R. (1983). Hierarchical planning disability in the drawings and constructions of a special group of severely aphasic children. *Brain and Cognition*, **2**, 144–64.
Curtiss, S. and Yamada, J. (1981). Selectively intact grammatical development in a retarded child. *UCLA Working Papers in Cognitive Linguistics*, **3**, 61–91.
Curtiss, S., Yamada, J., and Fromkin, V. (1979). How independent is language? On the question of formal parallels between grammar and action. *UCLA Working Papers in Cognitive Linguistics*, **1**, 131–57.
Damerow, P. (1988). Individual development and cultural evolution of arithmetical thinking. In *Ontogeny, phylogeny, and historical development* (ed. S. Strauss), pp. 125–52. Ablex, New York.
Darwin, C. (1859/1958). *The origin of species.* New American Library, New York (cited by Cosmides 1989).
Dasen, P. R. (1974). The influence of ecology, culture, and European contact on cognitive development in Australian Aborigines. In *Culture and cognition: readings in cross-cultural psychology* (ed. J. W. Berry and P. R. Dasen), pp. 381–408. Methuen, London.
Dasen, P. R. (1975). Concrete operational development in 3 cultures. *Journal of Cross-cultural Psychology*, **6**, 156–72.
Dasen, P. R. (1977). Are cognitive processes universal? A contribution to cross-cultural Piagetian psychology. In *Studies in cross-cultural psychology*, Vol. 1, (ed. N. Warren), pp. 155–201. Academic Press, London.
Deacon, T. W. (1988). Human brain evolution: 1. Evolution of language circuits. In *Intelligence and evolutionary biology* (ed. H. J. Jerison and I. Jerison), pp. 363–81. Springer-Verlag, Berlin.
Deacon, T. W. (1989). The neural circuitry underlying primate calls and human language. *Human Evolution*, **5**, 367–401.
Deacon, T. W. (1992). Brain–language coevolution. In *The evolution of human languages*, SFI Studies in the Sciences of Complexity. Proceedings, Vol. 10, (ed. J. A. Hawkins and M. Gell-Mann), pp. 49–77. Addison-Wesley, San Francisco.
de Lemos, M. M. (1969). The development of conservation in Aboriginal children. *International Journal of Psychology*, **4**, 255–69.

Dienske, H. (1986). A comparative approach to the question of why human infants develop so slowly. In *Primate ontogeny, cognition, and social behaviour*, (ed. J. G. Else and P. C. Lee), pp. 147–54. Cambridge University Press.

Ehret, G. (1987). Categorical perception of sound signals: facts and hypotheses from animal studies. In *Categorical perception* (ed. S. Harnad), pp. 301–31. Cambridge University Press.

Ellis, A. W. and Young, A. W. (1988). *Human cognitive psychology*. Lawrence Erlbaum, London.

Fodor, J. (1983). *The modularity of mind*. MIT Press, Cambridge, Mass.

Fodor, J. (1985). Précis of *The modularity of mind*. *Behavioral and Brain Sciences*, **8**, 1–42.

Fox, P., Petersen, S., Posner, M., and Raichie, M. (1988). Is Broca's area language-specific? *Neurology*, **38**, Supplement 1, 172.

Garstang, W. (1922). The theory of recapitulation: a critical restatement of the biogenetic law. *Journal of the Linnean Society for Zoology*, **35**, 81–101.

Gesell, A. (ed.) (1954). *The first five years of life: a guide to the study of the pre-school child*. Methuen, London.

Goldberg, S., Perlmutter, M., and Myers, W. (1974). Recall of related and unrelated lists by 2-year-olds. *Journal of Experimental Child Psychology*, **18**, 1–8 (cited by Shweder 1982).

Goodall, J. (1964). Tool using and aimed throwing in a community of free-living chimpanzees. *Nature*, **201**, 1264–6.

Gould, S. J. (1977). *Ontogeny and phylogeny*. Harvard University Press, Cambridge, Mass.

Greenfield, P. M. (1966). On culture and conservation. In *Studies in cognitive growth* (ed. J. S. Bruner, R. R. Olver, and P. M. Greenfield), pp. 225–56. Wiley, New York.

Greenfield, P. M. (1978). Structural parallels between language and action in development. In *Action, gesture and symbol: the emergence of language*, (ed. A. J. Lock), pp. 415–45. Academic Press, London.

Greenfield, P. M. (1990). Language, tools and brain: the development and evolution of hierarchically organized sequential behavior. Unpublished paper prepared for Wenner-Gren Symposium no. 110 'Tools, language and intelligence: evolutionary implications', Cascais, Portugal. (Revised version published as: Greenfield, P. M. (1991). Language, tools and brain: the ontogeny and phylogeny of hierarchically organized sequential behavior. *Behavioral and Brain Sciences*, **14**, 531–95.

Greenfield, P. M. and Savage-Rumbaugh, E. S. (1990). Grammatical combination in *Pan paniscus*. Processes of learning and invention. In *"Language" and intelligence in monkeys and apes: comparative developmental perspectives*, (ed. S. T. Parker and K. R. Gibson), pp. 540–78. Cambridge University Press.

Greenfield, P. M., Nelson, K. and Saltzman, E. (1972). The development of rulebound strategies for manipulating seriated cups: a parallel between action and grammar. *Cognitive Psychology*, **3**, 291–310.

Grossman, M. (1980). A central processor for hierarchically structured material: evidence from Broca's aphasia. *Neuropsychologia*, **18**, 299–308.

Groves, C. P. (1989). *A theory of human and primate evolution*. Clarendon Press, Oxford.

Gumperz, J. J. and Levinson, S. C. (1991). Rethinking linguistic relativity. *Current Anthropology*, **32**, 613–23.

Haeckel, E. (1907). *The evolution of man: a popular scientific study*, Vol. 1: *Human embryology or ontogeny*. Watts and Co, London.

Hallpike, C. R. (1979). *The foundations of primitive thought*. Clarendon Press, Oxford.

Harris, M. (1969). *The rise of anthropological theory: a history of theories of culture*. Routledge and Kegan Paul, London.

Harrison, G., Tanner, J., Pilbeam, D., and Baker, P. (1988). *Human biology*. Oxford University Press.

Hatano, G. (1982). Cognitive consequences of practice in culture specific procedural skills. *Quarterly Newsletter of the Laboratory of Comparative Human Cognition*, **4**, 15–17.

Hertwig, O. (1901). Einleitung und allgemeine Litteraturübersicht. In *Handbuch der vergleichenden und experimentellen Entwickelungslehre der Wirbeltiere*, Vol. 1. (part 1), (ed. O. Hertwig), pp. 1–86. Gustav Fischer, Jena.

Hertwig, O. (1906). Ueber die Stellung der vergleichenden Entwickelungslehre zur vergleichenden anatomie, zur Systematik und Descendenztheorie (Das biogenetische Grundgesetz, Palingenese und Cenogenese). In *Handbuch der vergleichenden und experimentellen Entwickelungslehre der Wirbeltiere*, Vol. 3 (part 3), (ed. O. Hertwig), pp. 149–80. Gustav Fischer, Jena.

Hirsh-Pasek, K. and Golinkoff, R. M. (1991). Language comprehension: a new look at some old themes. In *Biological and behavioral determinants of language development*, (ed. N. A. Krasnegor, D. M. Rumbaugh, R. L. Schiefelbusch, and M. Studdert-Kennedy), pp. 301–20. Erlbaum, Hillsdale, NJ.

Hyde, D. M. G. (1970). *Piaget and conceptual development*. Holt, Rinehart, and Winston, London.

Ianniruberto, A. (1985). Prenatal onset of motor patterns. – Unpublished paper presented to Conference on Motor Skill Acquisition in Children, Nato Advanced Study Institute, Maastricht, Netherlands.

Jersild, A. (1955). *Child psychology* (4th edn). Staples Press, London.

Johnson-Laird, P. N. (1982). Thinking as a skill. *Quarterly Journal of Experimental Psychology*, **34A**, 1–29.

Kelly, M. R. (1971). Some aspects of conservation of quantity and length in Papua and New Guinea in relation to language, sex and years at school. *Territory of Papua and New Guinea Journal of Education*, **7**, 55–60.

Koffka, K. (1928). *The growth of the mind* (2nd edn). Kegan Paul, Trench, Trubner, London.

Kojima, S. and Kiritani, S. (1989). Vocal–auditory functions in the chimpanzee: vowel perception. *International Journal of Primatology*, **10**, 199–213.

Kojima, S., Tatsumi, I. F., Kiritani, S., and Hirose, H. (1989). Vocal–auditory functions of the chimpanzee: consonant perception. *Human Evolution*, **4**, 403–16.

Kollar, E. J. and Fisher, C. (1980). Tooth induction in chick epithelium: expression of quiescent genes for enamel synthesis. *Science*, **207**, 993–5.

Lamarck, J.-B. (1809). *Philosophie zoologique, ou Exposition des considérations relatives à l'histoire naturelle des animaux*. Paris. (English trans. by H. Elliot: *The zoological philosophy*, 1914. Macmillan, London).

Lave, J. (1990). The culture of acquisition and the practice of understanding. In *Cultural psychology: essays on comparative human development*, (ed. J. W. Stigler, R. A. Shweder, and G. Herdt), pp. 309–27. Cambridge University Press.

Lempers, J. D., Flavell, E. R., and Flavell, J. H. (1977). The development in very young children of tacit knowledge concerning visual perception. *Genetic Psychology Monographs*, **95**, 3–54 (cited by Shweder 1982).

Leslie, A. and Thaiss, L. (1992). Domain specificity in conceptual development: neuropsychological evidence from autism. *Cognition*, **43**, 225–51.

Lieberman, P. (1984). *The biology and evolution of language*. Harvard University Press, Cambridge, Mass.

Lieberman, P. (1985). On the evolution of human syntactic ability: its pre-adaptive bases—motor control and speech. *Journal of Human Evolution*, **14**, 657–68.

Lieberman, P. (1988). Language, intelligence, and rule-governed behavior. In *Intelligence and evolutionary biology* (ed. H. J. and I. Jerison), pp. 143–56. Springer-Verlag, Berlin.

Light, P. and Perret-Clermont, A.-N. (1989). Social context effects in learning and testing. In *Cognition and social worlds*, (ed. A. Gellaty, D. Roger, and J. Sloboda), pp. 91–112. Clarendon Press, Oxford.

Lincoln, R. J., Boxshall, G. A., and Clark, P. F. (1982). *A dictionary of ecology, evolution and systematics*. Cambridge University Press.

Macnamara, J., Baker, E., and Olson, C. L. (1976). Four-year-olds' understanding of *pretend*, *forget* and *know*: evidence for propositional operations. *Child Development*, **47**, 62–70 (cited by Shweder 1982).

McKinney, M. L. and McNamara K. J. (1991). *Heterochrony: the evolution of ontogeny*. Plenum Press, New York.

Manley-Buser, K. A. (1986). A heterochronic study of the human foot. *American Journal of Physical Anthropology*, **69**, 235–6.

Marr, D. (1976). Early processing of visual information. *Philosophical Transactions of the Royal Society* (London), **B290**, 199–218.

Marr, D. (1982). *Vision*. W. H. Freeman, San Francisco.

Mayr, E. (1982). *The growth of biological thought: diversity, evolution and inheritance*. Harvard University Press, Cambridge, Mass.

Meltzoff, A. N. (1988). Imitation, objects, tools, and the rudiments of language in humans. *Human Evolution*, **3**, 45–64.

Miceli, G., Mazzucchi, A., Menn, L., and Goodglass, H. (1983). Contrasting cases of Italian agrammatic aphasia without comprehension disorder. *Brain and Language*, **19**, 65–97.

Mohseni, N. (1966). La comparaison des réactions aux épreuves d'intelligence en Iran et en Europe. Unpublished thesis, Université de Paris.

Montagu, M. F. A. (1981). *Growing young*. McGraw-Hill, New York.

Needham, J. (1933). On the dissociability of the fundamental processes in ontogenesis. *Biological Review*, **8**, 180–223.

Nesher, P. (1988). Precursors of number in children: a linguistic perspective. In *Ontogeny, phylogeny, and historical development* (ed. S. Strauss), pp. 106–24. Ablex, New York.

Opper, S. (1971). Intellectual development in Thai children. Unpublished Ph.D. thesis, Cornell University, Ithaca, New York.

Owren, M. J. (1990a). Acoustic classification of alarm calls by vervet monkeys (*Cercopithecus aethiops*) and humans (*Homo sapiens*): I. Natural calls. *Journal of Comparative Psychology*, **104**, 20–8.

Owren, M. J. (1990b). Acoustic classification of alarm calls by vervet monkeys (*Cercopithecus aethiops*) and humans (*Homo sapiens*): II. Synthetic calls. *Journal of Comparative Psychology*, **104**, 29–40.

Parker, S. T. (1985). A social–technological model for the evolution of language. *Current Anthropology*, **26**, 617–39.

Parker, S. T. and Gibson, K. R. (1979). A model of the evolution of language and intelligence in early hominids. *The Behavioral and Brain Sciences*, **2**, 367–408.

Passingham, R. E. (1982). *The human primate*. W. H. Freeman, San Francisco.

Piaget, J. and Inhelder, B. (1963). Les opérations intellectuelles et leur développement. In *Traité de psychologie experimentale* (ed. P. Fraisse and J. Piaget), Vol. 7, *L'Intelligence*. P.U.F., Paris. [English translation: Intellectual operations and their development. In *Experimental psychology: its scope and methods* (ed. P. Fraisse and J. Piaget), Vol. 7, *Intelligence*, pp. 109–55. Routledge and Kegan Paul, London 1969].

Pinard, A., Morin, C., and Lefebvre, M. (1973). Learning of conservation of liquid quantities by Rwandese and French-Canadian children. *International Journal of Psychology*, **8**, 15–23.

Plooij, F. X. (1978). Basic traits of language in wild chimpanzees? In *Action, gesture and symbol: the emergence of language* (ed. A. J. Lock), pp. 111–32. Academic Press, London.

Pollard, P. (1982). Human reasoning: some possible effects of availability. *Cognition*, **10**, 65–96.

Portmann, A. (1941). Die Tragzeiten der Primaten und die Dauer der Schwangerschaft beim Menschen: ein Problem der vergleichen Biologie. *Revue Suisse de Zoologie*, **48**, 511–18.

Prechtl, H. F. R. (1984). Continuity and change in early neural development. In *Continuity of neural function from prenatal to postnatal life* (ed. H. F. R. Prechtl), pp. 1–15. Blackwell, Oxford.

Price-Williams, D. R. (1961). A study concerning concepts of conservation of quantities among primitive children. *Acta Psychologica*, **18**, 297–305.

Prince, J. R. (1969). *Science concepts in a Pacific culture*. Angus and Robertson, Sydney.

Raff, R. A. and Kaufman, T. C. (1983). *Embryos, genes, and evolution: the developmental-genetic basis of evolutionary change*. Macmillan, New York.

Rogoff, B. (1990). *Apprenticeship in thinking: cognitive development in social context*. Oxford University Press, New York.
Rozin, P. (1976). The evolution of intelligence and access to the cognitive unconscious. In *Progress in psychobiology and physiological psychology*, (ed. J. M. Sphuler and A. N. Epstein), pp. 245–80. Academic Press, New York. (cited by Cosmides 1989).
Rozin, P. and Schull, J. (1988). The adaptive–evolutionary point of view in experimental psychology. In *Stevens' handbook of experimental psychology*, (2nd edn, ed. R. C. Atkinson, R. J. Herrnstein, G. Lindzey, and R. D. Luce), pp. 503–46. Wiley, New York. (cited by Cosmides 1989).
Savage-Rumbaugh, S. (1987). A new look at ape language: comprehension of vocal speech and syntax. *Nebraska Symposium on Motivation*, **1987**, 200–55.
Savage-Rumbaugh, S., Romski, M. A., Hopkins, W. D., and Sevcik, R. A. (1989). Symbol acquisition and use by *Pan troglodytes*, *Pan paniscus*, *Homo sapiens*. In *Understanding chimpanzees* (ed. P. G. Heltne and L. A. Marquardt), pp. 266–95. Harvard University Press, Cambridge, Mass.
Saxe, G. B. (1991). *Cultural and cognitive development: studies in mathematical understanding*. Erlbaum, Hillsdale, NJ.
Saxe, G. B., Gruberman, S. R., and Gearhart, M. (1987). Social processes in early number development. *Monographs of the Society for Research in Child Development*, **52**(2), No. 216.
Schmandt-Besserat, D. (1982). The emergence of recording. *American Anthropologist*, **84**, 871–8.
Schultz, A. H. (1926). Fetal growth in man and other primates. *Quarterly Review of Biology*, **1**, 465–521.
Schultz, A. H. (1949). Ontogenetic specializations of man. *Archiv Julius Klaus-Stiftung*, **24**, 197–216.
Schultz, A. H. (1950). The physical distinctions of man. *Proceedings of the American Philosophical Society*, **94**, 428–49 (cited by Gould 1977).
Scribner, S. and Cole, M. (1981). *The psychology of literacy*. Harvard University Press, Cambridge, Mass.
Shatz, M. and Gelman, R. (1973). The development of communication skills: modification in the speech of young children as a function of listener. *Monograph of the Society for Research on Child Development*, **38**, No. 5 (cited by Shweder 1982).
Shea, B. T. (1983). Paedomorphosis and neoteny in the pygmy chimpanzee. *Science*, **222**, 521–2 (cited by Shea 1989).
Shea, B. T. (1984). An allometric perspective on the morphological and evolutionary relationships between pygmy (*Pan paniscus*) and common (*Pan troglodytes*) chimpanzees. In *The pygmy chimpanzee: evolutionary biology and behavior*, (ed. R. L. Sussman), pp. 89–130. Plenum, New York. (cited by Shea 1989).
Shea, B. T. (1988) Heterochrony in primates. In *Heterochrony in evolution*, (ed. M. L. McKinney), pp. 237–66. Plenum, New York. (cited by Shea 1989).
Shea, B. T. (1989). Heterochrony in human evolution: the case for neoteny reconsidered. *Yearbook of Physical Anthropology*, **32**, 69–101.
Shweder, R. (1982). On savages and other children. *American Anthropologist*, **84**, 354–66.
Simonds, R. J. and Scheibel, A. B. (1989). The postnatal development of the motor speech area. A preliminary study. *Brain and Language*, **37**, 42–58.
Sinnott, J. M. and Adams, F. S. (1987). Differences in human and monkey sensitivity to acoustic cues underlying voicing contrasts. *Journal of the Acoustical Society of America*, **82**, 1539–47.
Smotherman, W. P. and Robinson, S. R. (1988). The uterus as environment: the ecology of fetal behavior. In *Handbook of behavioral neurobiology*. Vol. 9. *Developmental psychobiology and behavioral ecology* (ed. E. M. Blass), pp. 149–95. Plenum Press, New York.
Spencer, H. (1886). *The principles of biology* (2 vols). Appleton, New York.
Stigler, J. W., Barclay, C., and Aiello, P. (1982). Motor and mental abacus skills: a preliminary look at an expert. *Quarterly Newsletter of the Laboratory of Comparative Human Cognition*, **4**, 12–14.
Symons, D. (1987). If we're all Darwinians, what's the fuss about? In *Sociobiology and psychology: ideas, issues, applications* (ed. C. Crawford, M. S. Smith, and D. Krebs), pp. 121–46. Erlbaum, Hillsdale, NJ (cited by Cosmides 1989).
Thatcher, R. W., Walker, R. A., and Giudice, S. (1987). Human cerebral hemispheres develop at different rates and ages. *Science*, **236**, 1110–13.
Tooby, J. (1985). The emergence of evolutionary psychology. In *Emerging syntheses in science*, (ed. D. Pines), Santa Fé Institute, Santa Fé, (cited by Cosmides 1989).
Trabasso, T. (1975). Representation, memory, and reasoning: how do we make transitive inferences? In *Minnesota Symposium on Child Psychology*, Vol. 9, (ed. A. Pick), pp. 135–72. University of Minnesota Press, Minneapolis (cited by Shweder 1982).
Tylor, E. B. (1865). *Researches into the early history of mankind and the development of civilization*. John Murray, London.
Vauclair, J. (1982). Sensorimotor intelligence in human and non-human primates. *Journal of Human Evolution*, **11**, 257–64.
Watson, D. M. S. (1919). On *Seymouria*, the most primitive reptile. *Proceedings of the Zoological Society of London*, **1918**, 267–301.
Wertsch, J. V. (1991). *Voices of the mind: a sociocultural approach to mediated action*. Harvard University Press, Cambridge, Mass.
White, L. (1962). *Medieval technology and social change*. Clarendon Press, Oxford.
Williams, G. C. (1966). *Adaptation and natural selection: a critique of some current evolutionary thought*. Princeton University Press, Princeton, NJ.
Yamada, J. (1981). Evidence for the independence of language and cognition: a case study of a 'hyperlinguistic' retarded adolescent. *UCLA Working Papers in Cognitive Linguistics*, **3**, 121–60.
Za'rour, G. I. (1971). Conservation of number and liquid by Lebanese school-children in Beirut. *Journal of Cross-cultural Psychology*, **2**, 165–72.

13

The role of ontogenesis in human evolution and development[1]

Christopher G. Sinha

Abstract

Darwin's theory of evolution caused a revolutionary change in the concept of time. Evolution did not merely extend history backwards, it brought into being an entirely different order of time, in which different time-scales (*durées*) co-existed. Understanding the relations between time-scales—phylogenetic, ontogenetic, historical—was a major preoccupation of both biologists and psychologists. The best-known theory of this type was and is Haeckel's 'biogenetic law' of recapitulation. The metaphor of 'layers' of time—which, because of its association with palaeontology, is christened the 'palaeomorphic metaphor'—was central to the work even of those who, like the Soviet psychologist Vygotsky, rejected recapitulationism. Vygotsky's genetic psychology assumed the 'geological' stratification of 'lower' and 'higher' mental functions, corresponding respectively to biological and socio-cultural stages of evolution.

The mechanism proposed by Vygotsky for the development of 'higher' mental functions in the individual was *internalization*. However, this concept suffers from a logical problem, since it seeks to explain psychological processes in terms which presuppose those very processes. Vygotsky also emphasized the importance of tool-use in both the ontogeny and phylogeny of higher mental processes, drawing an analogy between tool-use and the use of conventional signs, including language. Again, however, this analogy is of limited usefulness, since neither tool-use nor cultural transmission are unique to humans.

An alternative account suggests rather that certain biological features of human infancy were selected, during the stages of human evolution post-dating the invention of tools, for their facilitative value in the process of what Vygotsky's colleague Leontiev called 'appropriation'. Infancy is then seen as a specific niche in which adaptive parameters are set by processes of individual appropriation, in the first instance, of canonical (socially standard) rules governing the use of tools and other artefacts. On this account, the biology of human infancy is a product of the co-evolution of culture and biology. Recent studies of infant cognition and social behaviour lend support to such an account. Infancy, on this account, played a crucial role in the 'socialization' of human biology.

13.1 Adaptation and representation

For any theory of human cognitive and symbolic evolution, as for developmental theories in general, a fundamental problem concerns how time—or rather temporalities in the plural, since more than one time or *durée* is involved—may best be conceptualized.

Darwinian evolutionary theory revolutionized previous European conceptions of time. Darwin's theory went further than simply to valorize the static 'Great Chain of Being' with an evolutionary, temporal dimension. It transformed time itself, splitting and multiplying the scales by which it might be measured. No longer could time be seen as originating at the moment of creation; and, if there was no longer a single First Cause, there was also no longer a single First Event. With the demise of the reciprocal notions of Origin and Finality, it fell to evolutionary theorists to attempt some account of the relations between the time-scales—earth historical, human historical, and psychogenetic—which might also illuminate the course of human evolution itself. One of the earliest and most widely accepted such attempts was Haeckel's recapitulationist theory (Haeckel 1874), which provided a foundation for both Spencerian and Freudian psychobiology; notably in relation to concepts of 'regression' (Sulloway 1979). Haeckel's 'biogenetic law' is long discredited; but I shall suggest that its underlying metaphor for the relationship between the temporalities has continued to dominate most evolutionary thinking.

Although no follower of Haeckel, Piaget too wrote at length on the relations between ontogeny and phylogeny (for example, Piaget 1979), and was a highly original evolutionary thinker, seeking to provide a principled scientific basis for his quasi-Lamarckian views in the concept of epigenesis. Paradoxically, however, in view of his self-designation as a genetic epistemologist, Piaget neglected time and temporality in most of his writings. Time, for Piaget, is reducible to the invariances of sequence which are inbuilt in the relations between the logico-mathematical structures characterizing the developmental stages of the cognitive or epistemic subject.

It is by now a common criticism of Piaget's developmental psychobiology that it neglects the social, interpersonal, and communicative dimensions of human development. Piaget's relative neglect of the social was manifested also in his failure to address the scale of temporality specific to society: that is, of human history. The same cannot be said of Vygotsky (for example Vygotsky 1962, 1978), whose work hinged on the notion of the 'internalization' of the cultural and the historical by the individual. Vygotsky's remarks on this subject were, however, as I shall suggest below, limited by his acceptance of the same 'palaeomorphic' model of the temporalities that inspired Haeckel's recapitulationist theory.

The problem of time also has a somewhat contradictory status in twentieth-century linguistic and social theory. In one sense, it is central to the structuralist project founded in the early years of the century by Saussure (see Saussure 1966); the time-referring concepts of synchrony and diachrony underlie the *langue–parôle* distinction which he took to constitute the very possibility of a science of language. On the other hand, as writers from the Prague Linguistic Circle of the 1930s (see Jakobson and Tynjanov 1985) to the present day (Bailey and Harris 1985) have noted, the structuralist focus on synchrony, *langue*, and abstract universal competence, has tended to obscure the importance both of linguistic change and development, and of the particularities of socio-cultural process.

The structuralist tradition, following in the footsteps of the Enlightenment philosophers, has viewed language as that which sets humanity off from other species, enabling the development of culture and rationality. Inevitably, such an approach tends, first, to treat 'language' and 'culture' as universal terms, and, second, to view the problem of ontogenesis in terms of a transition of the infant from an animal/biological to a human/social state. This basic notion of 'socialization' is common to such diverse thinkers as Freud, Piaget, and Vygotsky, whatever their other differences. For all of these theorists, language/culture/symbolization stands as the *sine qua non* of the human condition itself—and thus of human history; and history/society/culture as the sine-qua-non of the socialization of the individual.

If the opposition between 'nature' and 'culture' can be seen as having its roots in the work of the eighteenth-century *philosophers*, the idea that time itself is 'layered' is of more recent origin; being principally inspired by Darwin's revolutionary reassessment of the implications of the geological and palaeontological investigations of Lyell and others. I would suggest, in fact, that the importance of palaeontology in establishing Darwinian evolutionary theory led to the appropriation by many evolutionists of palaeontology as a fundamental metaphor, not just for changes *within* a *durée*, but for relations *between durées*. What underlies this 'palaeomorphic' model, or metaphor, is a hierarchy of temporalities: evolutionary-biological time (itself embedded within geological time) is presupposed by social-historical time, which is presupposed by ontogenetic-developmental time. Within the terms of the model, one may postulate parallelisms of sequence or mechanism, as did Freud and Piaget with the respective concepts of recapitulation and epigenesis; and one can argue over the relative influences of more 'fundamental' levels over more 'superficial' ones, as in Vygotsky's insistence upon the equivalent importance of the social and the biological in individual development. What remains unchallenged, however, is the basic assumption that time itself is stratified, each 'layer'—geological, evolutionary, historical, individual—being sedimented upon the previous one (see Fig. 13.1).

The 'palaeomorphic metaphor', as I shall term it, was explicitly stated in relationship to behavioural development by Vygotsky, in the following terms:

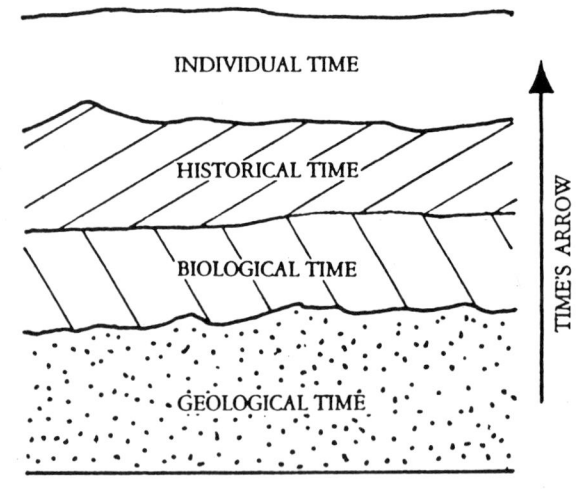

Fig. 13.1 Time in the palaeomorphic metaphor.

One of the most fruitful theoretical ideas genetic psychology has adopted is that the structure of behavioural development to some degree resembles the geological structure of the earth's core. Research has established the presence of genetically differentiated layers in human behaviour. In this sense the geology of human behaviour is undoubtedly a reflection of 'geological' descent and brain development. If we turn to the history of brain development, we see what Kretschmer calls the law of stratification in the history of development... lower centres are retained as subordinated structures in the development of higher ones and... brain development proceeds in accordance with the laws of stratification, or construction of new levels on old ones... Instinct is not destroyed, but 'copied' in conditioned reflexes, as a function of the ancient brain which is now to be found in the new one. Similarly, the conditioned reflex is 'copied' in intellectual action... the behaviour of the modern, cultural, adult human can be understood only 'geologically', since various genetic layers, which reflect all the stages through which humans have travelled in their psychological development, are reflected in it (Vygotsky 1981, pp. 155–6).

Although, in strict terms, Vygotsky's proposals are not identical to those of Haeckel, they clearly have much in common with the latter's 'biogenetic law'. The ideas expressed by Vygotsky continue to be influential in some areas of neuroscientific research, having much in common, for example, with the hypotheses advanced by MacLean (1972) regarding the development of the emotions.

In the same article, Vygotsky proposes that the development of 'higher mental functions' be viewed as representing a 'completely new level of development', in which 'any higher mental function was [once] external, because it was social, at some point before becoming an internal, truly mental function. It was first a social relation between two people.' Vygotsky presents this as a general genetic law of cultural development: any function in the child's cultural development appears twice, or on two planes. First it appears on the social plane, and then on the psychological plane... higher functions are not developed in biology [phylogenesis]. Rather, the very mechanism underlying higher mental functions is a copy from social interaction; all higher mental functions are internalized social relationships.

Whatever virtues Vygotsky's approach may have as a general orientation towards the fundamentally social nature of developmental and cognitive processes, the specific mechanism of 'internalization' which he postulated suffers from a serious logical problem. If the individual cognitive subject is seen as being an internalized product of social life and organization, and not a product of biology, then what is the nature of the subject (or proto-subject) which is initially responsible for the act(s) of internalization? To say that this is itself biological is simply to push the problem down a level, for the capacity to become 'fully human' is also a uniquely human characteristic. An equivalent dilemma is faced by theories which, in a Vygotskian fashion, attempt to solve the problem by appeal to the intersubjective structuration of early actions and interactions (Trevarthen and Hubley 1978; Lock 1980). Granted the importance of attributions of intentionality to infants by adults, and of adult 'scaffolding' (Bruner 1983) of early interactions, these again presuppose a subject capable of 'copying' the interpersonal attributions and transactions on to the 'inner-psychological' plane.

I shall return below to the role of interpersonal transactions in human evolution and development. For now, I want to note that, despite its interactionist and dialectical impulses, the Vygotskian theory of internalization reproduces in its internal logic the very divisions between the natural and the cultural, and the individual and the social, which it strives to overcome. Further, the palaeomorphic model of layering and sedimentation of neurological psychological function, corresponding to the layering of temporalities (ontogenetic upon socio-cultural upon phylogenetic), reproduces the classical division between 'instinctual' and 'learned' behaviour; leaving the Vygotskian account open to the criticism that it does not really break with the behaviourist and empiricist paradigm (Fodor 1972).

This is particularly unfortunate at the present time, since much current work in cognitive science is being conducted within a theoretical framework which I have described elsewhere (Sinha 1984) as 'neo-rationalist'; that is, in which innate 'mental faculties', of a computational nature, are held responsible for what Vygotsky called the 'higher mental functions'. The model of mind held by these theorists (for example, Chomsky 1980; Fodor 1976) is one which is modular in nature: rather than Vygotsky's palaeomorphic, vertical sedimentation model of mental function, a horizontally-specified set of processing modules is proposed, each of which operates with respect to a different cognitive domain. Such a model can be criticized for its lack of attention to evolutionary and biological processes, and its (idealistic) over-reliance on the computational metaphor. In many ways, the Vygotskian cultural-historical approach, and certain related neo-Piagetian approaches, are the only serious contenders, since the demise of behaviourism, against the neo-rationalist account. If, as I suggest, the Vygotskian account is flawed by its problematic reliance on the palaeomorphic metaphor, then it is important to develop an alternative.

To begin with, there is an obvious truth in the palaeomorphic model. Both individual development and social life do indeed depend upon an evolutionary biological process which for the greatest part pre-dates the emergence of our species, and whose *durée* is of an entirely different order from that of even the longest historical span. We may, for example, suppose that if a new-born Neolithic infant were somehow transposed with one from a present-day delivery room, the subsequent physical, psychological, and social developments of the changeling would be commensurable with those of its new-found peers. Thus, *both* what is constant in human development *and* what is variable across cultures are equally supported by a genetically transmitted biological 'core'. This much is unquestionable, as is the relevance of genetic transmission for variation between individuals within cultures.

What *is* open to question, in my view, is the assumption that the biological 'core' of individual development—the organismic 'support' necessary for the growth of human subjectivity—is itself a product *solely* of biological evolution. This proposition, stated negatively—'biology is not a product solely of biology'—has a paradoxical ring to it, and so is perhaps better stated in the following terms, as a positive hypothesis: 'the biology of human development is a product of the interaction of biological and cultural evolution at the specific site of ontogenesis'.

What I am proposing, then, is that rather than seeing cultural evolution as 'taking off' from a terminal point of biological evolution, we should rather see evolutionary biological processes as having been, as it were, 'captured' by an emergent cultural process, with ontogenetic processes—especially those involving representation, symbolization, and communication—as a crucial catalyst and product of the co-evolution of culture and biology. As Vygotsky, and many others including his colleagues Luria and Leontiev have emphasized, the emergence of human symbolic and representational capacities, in both ontogenesis and phylogenesis, is intimately connected with the emergence of cultural transmission and tool-use. However, merely to state this is, in a way, to state the obvious; and, since neither tool-use nor cultural transmission are unique to the human species, without further specification such a theory lacks explanatory power.

The theoretical position that I wish to advocate begins with a reassessment of the notion of representation, with particular emphasis on the materiality of representation, and the representational environment. The human environment, for any concrete individual, has always-and-already been intentionally shaped by previous generations of human agents into a material culture. Of course, biology—and in particular neurobiology—is an indispensable precondition or support for symbolic representation. However, a crucial aspect of the human environment is that it involves representational systems which extend *beyond the boundaries of the individual organism*. Representation, on this account, is not simply a distinctively human cognitive and cultural capacity, but a constitutive property of the material surround into which the human infant is born, and which, like all environments, supports and constrains the activities of the developing organism.

As Leontiev (Leontiev 1981 [1931–1960], p. 134) put it: 'Before the individual entering upon life is not Heidegger's "nothing", but the objective world transformed by the activity of generations'. To this we might add that neither are the 'objects' in the world encountered by the child *simply* Newtonian particles obeying universal physical laws, nor are the actions in which they are embedded *simply* exemplifications of abstract logico-mathematical operations, as a reading of Piaget might suggest.

In the first place, as many authors have noted (Trevarthen and Hubley 1978), the infant early encounters 'social' or 'personal' objects—simply, other people—whose qualities are different from those of impersonal objects. In the second place, however, *impersonal* objects encountered by the infant are *also* social, in a different sense, in that they are encountered within a context of particular social practices and social relations. Furthermore, in the case of artefacts in general, as for tools in particular, the context-of-use of the object commonly achieves a representational status in the structure of the object itself. It is in this highly specific sense that we may speak of the materiality of representation.

Most artefacts are designed to fulfil a certain purpose, which we may designate the *canonical* (or socially standard) function of the object, and this canonical function (containment, cutting, etc.) in part determines the *form* of the object (possession of a cavity or blade, etc.). Such relations between functions and forms may be termed, in general, design rules, and any artefact may be seen as a material representation of a subset of the design rules current in a culture. Such rules are not strictly deterministic, since although they are constrained by physical laws, such constraint is week. A particular function may be subserved by many different forms: for example, the function 'key' may be subserved by a hand- or machine-engineered implement, a mechanical combination-lock sequence, or a sequence of characters entered on an electronic keyboard. In general, we may say that the range of available instantiations of a specified function will increase with increasing knowledge and technique, as will the range of actually available functions. Furthermore, certain instantiations may become obsolete, so that design rules are subject to historical change.

Nevertheless, although design rules are underdetermined by function, they are not arbitrary, since they embody constraints of nature and of social practice, of habit, and sometimes of aesthetic value. Further, a certain core of design rules, governing the use of artefacts common to all cultures—implements for containing, supporting, cutting, pounding, tying, etc.—remain relatively speaking invariant across space and time, and it is this subset of invariant design rules, correlating canonical (that is, socially standard) functions and canonical forms, which may be designated as *canonical* rules. My proposal is that such canonical rules act as a material, representational core to object-usage, and possess a privileged status in both ontogenetic and phylogenetic cognitive and symbolic development.

Although this chapter is not intended as an empirical review, it is to be noted that there is evidence that the ontogenetically earliest conceptual representations of objects appear to implicate a rudimentary grasp of canonical rules (Freeman *et al.* 1980), and that canonicality is a central aspect of the early organization of actions and interpretative strategies for language (Sinha 1982, 1983). Thus I shall suggest that the child's early accommodations to, and assimilations of, objects are partially predetermined by the social and intentional shaping of these objects to represent, in their structural properties, canonical rules. In this respect, the object provides an external epigenetic pathway (Waddington 1977) for what Leontiev (1981) calls the 'appropriation' of knowledge and culture by the developing human organism.

The canalization of cognitive growth by the object is reinforced by the tuition provided by adult co-participants in joint and routine interpersonal and communicative activities. To quote Leontiev (1981, p. 135) again, 'the individual, the child, is not simply thrown into the world; it is introduced into this world by the people around it, and they guide it in that world'. The extraordinarily high degree of sensitivity displayed by the human infant and young child to the cues provided, directly through interpersonal communication channels, and indirectly through context-setting, by adult interlocutors has been documented by many developmental psychologists in recent years, as has the dyadic structure of early tutorial and pedagogical relationships (Wood *et al.* 1976).

Representational and symbolic development do not, however, end with the consolidation of knowledge of canonical rules: rather, such rules form the basis for subsequent developments in which the child systematically violates canonical rules, in symbolic and substitutional pretend-play routines, and in constructional goal-oriented activities. The ability to decouple canonical function–form relations is, of course, crucial to language development, particularly the acquisition of syntax; and in general to the generation of novel means–ends relations. This fundamentally creative or innovative capacity is as much, or indeed more, a specifically human cognitive capacity than the ability to produce enduring tools and artefacts; furthermore, the ability to go 'beyond the information given' (Bruner 1974) is combined, in the human species, with the emergence of constructional activities which involve the co-ordination of different but complementary roles in the social organization of the task (Reynolds 1982).

Thus, if the grasp or appropriation of canonical rules may be seen as a fundamental mechanism for the social transmission of the core invariants of culture, symbolic play and co-operative action provide ontogenetic mechanisms for the generation of complex substitutional and combinatorial systems, and ultimately for the appropriation of the generative and combinatorial rules (language, kinship systems, etc.) which are specific to particular cultures. What implications do these ontogenetic processes hold for possible mechanisms of the phylogenetic evolution of cognition and symbolization?

My proposal is that the emergence of culture and symbolization, in its human form, was made possible by the evolution of infancy itself as a specific niche, whose adaptive parameters were set by the prior emergence of hominid artefact manufacture and use. Ontogenesis, then, was the specific evolutionary adaptation permitting the intergenerational transmission through appropriation of the earliest cultural artefacts, those instantiating canonical rules. According to this hypothesis, it was not tool-use *per se* which was responsible for the emergence of human cognitive and symbolic capacities, but the enduring products of such tool-use in the form of artefacts (including tools themselves).

As Leontiev has emphasized, the process of appropriation is not simply one of adaptation to a given reality, but of the active mastery of the transformative procedures which tools and artefacts represent and potentiate; furthermore, the process of appropriation is fundamentally social, in two distinct but related senses. First, in that the procedures mastered and reproduced through appropriation are inseparable from the social order as a whole, and second in that the guidance of adults is an intrinsic part of the developmental process. Therefore, it is not the case that infancy is adapted to tool-and artefact-use *per se*, and I am not arguing that tool-use is 'innate' in the conventional sense.

Rather, the biological adaptations necessary for human evolution were selected by the requirements of appropriation as an ontogenetic mechanism. Thus, the evolution of infancy was the biological mechanism through which the potential for intergenerational cultural transmission created by tool-use was optimized, and appropriation was the specific social and ontogenetic process in response to which the niche of infancy evolved. In this respect, the evolution of infancy was not so much a terminal point of biological evolution, but the crucial inaugural moment in the socialization of biology (Riley 1978).

Such an evolutionary process, I suggest, would lead to a rapidly expanding endogenous spiral of change and adaptation. Those aspects of infant psychology facilitating appropriation processes, such as social-contextual sensitivity, representational and symbolic control of sensorimotor processes, and 'dedicated' perceptuomotor mechanisms for processing speech and gesture, are also those which are most clearly implicated in the crucial features of a specifically human form of life: symbolic and representational systems, technological innovation, complex classificatory systems, and so forth.

The hypothesis that I have presented brings ontogenetic processes once again to the fore in theories of human symbolic, cognitive, and cultural evolution, while departing significantly from the assumptions of the 'palaeomorphic metaphor'. Specifically, this hypothesis suggests that certain features of the biology of human infancy and early childhood—those associated with appropriation procedures—are evolutionary adaptations which postdate the emergence of artefact use, but potentiate the emergence of complex categorization and social-technological capacities. This theoretical account also leaves room for the evolution of certain 'modular' human capacities, perhaps specifically those involved with processing language and gesture, as part of the adaptive response to the niche of infancy. However, this is not the same as saying that 'language' is innate, any more than to suggest that innate pre-adaptation to the appropriation of canonical rules is the same as saying that these rules themselves are innate. The innateness of the rules (of either grammar or artefact-use) cannot be excluded by, but nor is it implied by, the theory I have presented.

Note

1. The material in this chapter is presented more fully in my book *Language and representation: a socionaturalistic approach to human development* (The Harvester Press, Brighton, 1988).

References

Bailey, C. J. and Harris, R. (1985). *Developmental mechanisms of language*. Pergamon, London.
Bruner, J. (1974). *Beyond the information given: studies in the psychology of knowing*. Allen & Unwin, London.
Bruner, J. (1983). *Child's talk*. Oxford University Press.
Chomsky, N. (1980). Rules and representations. *Behavioural and Brain Sciences*, **3**, 1–15.
Fodor, J. (1972). Some reflections on L. S. Vygotsky's 'Thought and Language'. *Cognition*, **1**, 83–96.
Fodor, J. (1976). *The language of thought*. Harvester, Hassocks.
Freeman, N., Lloyd, S. and Sinha, C. (1980). Infant search tasks reveal early concepts of containment and canonical usage of objects. *Cognition*, **8**, 243–62.
Haeckel, E. (1874). *The evolution of man: a popular exposition of the principal points of human ontogeny and phylogeny*. International Science Library, New York.
Jakobson, R. and Tynjanov, J. (1985). Problems in the study of language and literature. In *Roman Jakobson: Verbal art, verbal sign, verbal time* (ed. K. Pomorska and S. Rudy), pp. 25–7. Basil Blackwell, Oxford.
Leontiev, A. (1981). *Problems of the development of the mind*. Progress Publishers, Moscow.
Lock, A. (1980). *The guided reinvention of language*. Academic Press, London.
MacLean, P. (1972). Cerebral evolution and emotional processes. *Annals of the New York Academy of Sciences*, **193**, 137–49.
Piaget, J. (1979). *Behaviour and evolution*. Routledge & Kegan Paul, London.
Reynolds, P. (1982). The primate constructional system: the theory and description of instrumental object use in humans and chimpanzees. In *The analysis of action* (ed. M. von Cranach and R. Harré), pp. 343–83. Cambridge University Press.
Riley, D. (1978). Developmental psychology, biology and marxism. *Ideology and Consciousness*, **4**, 73–91.

Saussure, F. de (1966). *Cours de linguistique générale*. McGraw-Hill, New York.

Sinha, C. (1982). Representational development and the structure of action. In *Social cognition: studies in the development of understanding* (ed. G. Butterworth and P. Light), pp. 137–62. Harvester, Brighton.

Sinha, C. (1983). Background knowledge, presupposition and canonicality. In *Concept development and the development of word meaning* (ed. T. Seiler and W. Wannenmacher), pp. 269–96. Springer-Verlag, Berlin.

Sinha, C. (1984). A socio-naturalistic approach to human development. In *Beyond neo-Darwinism: an introduction to the new evolutionary paradigm* (ed. M.-W. Ho and P. Saunders), pp. 331–62. Academic Press, London.

Sulloway, F. (1979). *Freud: biologist of the mind*. Basic Books, New York.

Trevarthen, C. and Hubley, P. (1978). Secondary intersubjectivity: confidence, confiding and acts of meaning in the first year. In *Action, gesture and symbol: the emergence of language* (ed. A. Lock), pp. 183–229. Academic Press, London.

Vygotsky, L. S. (1962). *Thought and language*. MIT Press, Cambridge, Mass.

Vygotsky, L. S. (1978). *Mind in society: the development of higher mental processes*. Harvard University Press, Cambridge, Mass.

Vygotsky, L. S. (1981). The genesis of higher mental functions. In *The concept of activity in Soviet psychology* (ed. J. Wertsch), pp. 144–88. M. E. Sharpe, New York.

Waddington, C. (1977). *Tools for thought*. Paladin, St Albans.

Wood, D., Bruner, J. and Ross, G. (1976). The role of tutoring in problem solving. *Journal of Child Psychology and Psychiatry*, **17**, 89–100.

14

The ontogeny and evolution of the brain, cognition, and language

Kathleen Gibson

Abstract

Brain maturation data provide no factual support for the common view that the human brain is a neotenous organ. Nor can the human brain be considered unusually altricial at birth. Limited recapitulation, however, does characterize human neural maturation. Specifically, the neocortical association areas are late to mature and have demonstrated the greatest phylogenetic expansion. This parallel provides a basis for attempts to unravel the evolution of language phylogeny by examining its ontogeny. In particular, the enlargement of the neocortical association areas has most probably provided the quantitatively based, hierarchical mental constructional skills which Case suggests underlie the maturation of human intelligence and which appear to distinguish human and ape tool-using and communication behaviours. These considerations suggest that Case's quantitative developmental framework provides a logical base for theories of language evolution. In Case's framework, the development of intelligence is not only quantitative and hierarchical in nature, it is also interactive. Object-manipulation skills, social behaviour, and language mature in a synchronous and mutually facilitatory fashion. This implies that not only brain size, but also cultural remains such as tools, shelters, or evidence of feeding techniques can provide important clues to the evolution of language. On the basis of these considerations, language, like tool-use and brain size, is postulated to have evolved slowly over several million years. With each cognitive and linguistic advance, new foraging and social interactive skills would have arisen. It is suggested that, as in ontogeny, language evolution began in mother–child dyads with the communication of simple needs and desires by one- to two-'word' utterances. As grammatical skills increased, hominids became capable of discussing co-operative endeavours and/or absent 'rendezvous' or other sites. At still later stages, linguistic skills permitted the discussion of animal and plant ecology and, finally, the prediction of seasonal events.

14.1 Neoteny and recapitulation

Several hundred years before the birth of Christ, Aristotle depicted human mental development as a blossoming of three successive stages of the soul: nutritive, sensitive, and rational, and likened the nutritive and sensitive stages to the souls of plants and animals. In so doing, he planted seeds that were to be reaped millennia later in the great theme—ontogeny recapitulates phylogeny—or—as organisms mature they pass through stages reminiscent of ancestral adults (Gould 1977).

Initially, recapitulation was a derived hypothesis based on the observations of living forms. In the writings of Ernst Haeckel, however, recapitulation was elevated into the biogenetic law, a theoretical formulation which postulated recapitulation to be the expected result of all phylogenetic processes (Haeckel 1874). Though now often maligned, the biogenetic law followed logically from the theories of the times which considered evolution to be a progressive event that proceeded largely through the interactions of two universal processes. New traits were added to the end of ancestral ontogenies by means of terminal addition, while, at the same time, the ontogenies, themselves, were truncated (Gould 1977). Thus, elevated into law, the concept of recapitulation held great promise. For, it implied that if the ontogeny of an organism could be unravelled, its phylogeny would fall neatly into place. As such, it stimulated much embryological research.

With the advent of modern genetic theory and the realization that new traits could appear at any point in ontogeny, however, the biogenetic law met its demise. In fact, serious doubts arose as to whether recapitulation ever occurred. Instead, the theory of another nineteenth-century biologist, Von Baer, was resurrected and enshrined as a biological axiom—as organisms mature they proceed through stages reminiscent of ancestral embryos, rather than through stages reminiscent of ancestral adults (Von Baer 1828).

So strongly was the law of Haeckel eclipsed by that of Von Baer that recapitulation suffered a fate nearly akin to that of Lamarckism. It is, today, a 'taboo' or 'bogey' subject among many theorists (Brainerd 1979; Ebbesson 1984; Gould 1977; Lock 1985; Northcutt 1984). Nevertheless, concepts of ontogenetic and phylogenetic parallels have remained popular among those who would chart the evolution of human language and intelligence (Baldwin 1894; Piaget 1969, 1978). Indeed ontogenetic data are routinely examined by theorists in this field, and several models of linguistic and cognitive evolution are frankly based on our knowledge of language development in children (Gibson 1981, 1983, 1988; Lamendella 1976; Lock 1983; Parisi 1983; Parker 1985; Parker and Gibson 1979).

This continuing popularity of recapitulative models reflects, in part, the fact that language does not fossilize. Hence, an analysis of modern language development is one of the few potential means of gaining clues to the linguistic capacities of earlier humans. In addition, as language seems to have only one means of development, from the simple to the complex, it has been argued that common constraints dictate similarities between linguistic ontogenetic and phylogenetic processes (Lock 1983, 1985; Piaget 1969, 1978).

Intuition and convenience, however, are not sufficient to render a scientific method valid, especially in the face of serious scholarly opposition. Moreover, the extent to which concepts of common constraint can be applied within an evolutionary time-frame remain unclear. If, for instance, the various anatomical structures mediating language evolved rapidly then, in tandem, fully elaborated language could have appeared quite suddenly in evolutionary terms despite ontogenetic constraints on language acquisition. Alternately, a gradual evolution of the anatomical foundations of language could have produced gradual phyletic increases in linguistic competence. Consequently, if concepts of recapitulation are to be used for the reconstruction of language evolution, a credible theory of the biological foundations of recapitulation must be elaborated and its application to the hominid archaeological record must be demonstrated.

A recent theoretical formulation by Gould (1977) has again placed recapitulation frameworks firmly within the fold of scientific respectability. According to Gould, the demise of the biogenetic law as 'law' did not automatically discount the possibility that genuine recapitulation might sometimes occur. Rather, under certain conditions recapitulation is almost inevitable (1977, p. 269). In fact, many probable recapitulations were described by nineteenth-century anatomists, and such recapitulations continue to be described up to the present day, especially among students of the evolution of a critical linguistic organ—the brain (Ebbesson 1984; Jacobson 1978).

In Gould's framework, evolution can occur either by the introduction of new, genetically controlled traits at any point in the life cycle or by alterations in the rates of growth and length of the growth period, a process he terms *heterochrony*. That is, heterochrony is the process of changes in developmental timing leading to phylogenetic change. In the presence of this process, pronounced changes in form may occur despite little genetic change. In particular, changes in developmental timing can lead both to paedomorphism, the resemblance of descendent adult to its ancestral juvenile form, and to recapitulation, the resemblance of the young of descendent species to ancestral adults.

While giving legitimacy to certain recapitulatory frameworks, in his examination of human development Gould chose to ignore a century or more of neuroanatomical research indicating parallels between neural ontogeny and phylogeny, and concluded, instead, that the human brain is a neotenous organ. This conclusion of human neoteny has been eagerly accepted by many developmental psychologists. In particular, the immaturity of the brain at birth and the unusually protracted postnatal developmental cycle are considered to provide an optimum environment for childhood learning, while the juvenilization of the adult is considered fundamental to the human ability to learn throughout the life-span (Lerner 1984; Rader 1985; Rutkowska 1985). To these scholars, human neoteny is perhaps the most fundamental parameter in the evolution of language and the brain.

Curiously, then, although neoteny and recapitulation can be seen as contradictory processes, both are currently evoked as 'explanations' of increased brain size and of the evolution of language by leading theorists in the field. Few have even stopped to question whether it is really possible for the brain, language, and cognition to be simultaneously neotenous—i.e. like that possessed by the ancestral juvenile—and recapitulative—i.e. advanced beyond the stage achieved by the ancestral adult. Nor have many questioned the possible contradictions between the neoteny and recapitulative models and a viable third alternative that human neural morphology may reflect a major structural reorganization of the brain and/or the addition of new structures or pathways.

Many scholars, however, have proposed that human linguistic capacity reflects just such neural changes. If this is the case, then changes in developmental timing may be largely irrelevant to the linguistic endeavour and to our attempts to reconstruct the evolution of language. For, if the human brain contains structures not present in the brains of other primates or is organized in a radically different fashion, then the human brain cannot be neotenous, i.e. like the brain of an ancestral juvenile. Nor can it be recapitulative, unless the newly emergent functional entities or reorganizational properties are added at the end of the ontogenetic sequence. On the other hand, if there is nothing really new about the human brain, there is just more of it or more of certain parts of it, then changes in developmental rates may, in fact, have been the critical factor in neural and linguistic evolution. Neoteny and recapitulation then remain viable hypotheses.

14.2 Quantity versus reorganization in the evolution of the brain

At an average of 1300 grams, the human brain is approximately three times as large as the average ape brain. Of all of the features which distinguish human from ape, it is perhaps this difference in brain size which is the most impressive. For despite a century or more of avid search, no functionally unique structures are known to exist in the human brain. Even Broca's area, the inferior parietal association area, the hippocampus minor, and the uncinate fasciculus, which were all at one time considered uniquely human, are now known to occur in the brains of other primates.[1] Nor has the often stated concept that the human brain alone displays anatomical and functional lateralization been substantiated.[2] For this reason, over the decades, many scholars have considered that it is just this first-mentioned aspect of the brain—its size—which is also the most important in the determination of specifically human behaviours, including language.[3]

This view has been strongly contested, however, by Holloway (1966, 1968, 1979; Holloway and Post 1982) and his followers, who have argued that increased brain size alone cannot account for the emergence of uniquely human behavioural traits such as language. Rather, they argue that neural reorganization has been the critical ingredient in the evolution of specifically human behaviours (see also Holloway, this volume, Chapter 4).

For, as Holloway points out, as evolution has occurred, certain neural structures have increased in size more than others. As a result, the human brain is proportioned differently from that of the apes. In particular, the neocortex, the neocortical association areas, the cerebellum, and certain thalamic and limbic nuclei are relatively larger (Armstrong 1982, 1986; Passingham 1973). Increased cortical fissurization, increased complexity of dendritic branching, and an increased ratio of cortical glial cells to cortical neurons have accompanied this increase in neocortical size. The human brain, therefore, is a more cortically dominated brain than that of the ape, and the human neocortex is more heavily fissured and exhibits more complex interconnectivity.

When enumerated in detail, these changes seem to be of major magnitude, and readily lead to the conclusion that the human brain is a very different brain indeed from that of the ape. Other scholars, however, have noted that many of these differences may simply reflect an underlying primary change in brain size (Jerison 1982). Studies by Passingham, for instance, have indicated that among all primates, an increase in brain size is regularly accompanied by disproportionately greater increases in the size of the neocortex than of other brain regions (Passingham 1973, 1975; Passingham and Ettlinger 1975). Moreover, with neocortical enlargement, the association areas exhibit greater increases in size than the primary sensory and motor areas. According to Passingham's data, neither the human neocortex as a whole, nor the human neocortical association areas, are any greater in size than would be predicted for an ape with the same-sized brain.

Similarly, Jerison has demonstrated that increasing cortical fissurization regularly accompanies increased cortical size throughout the mammalian class (Jerison 1979, 1982). Hence, rather than reflecting the addition of new functional areas, the increasing fissurization of the human brain reflects the expansion of pre-existing functional entities. Likewise, according to Jerison's data, increasing ratios of glial cells and dendritic branches to neurons correlate with increasing cortical size (also see Hofman 1983).

Hence, the data suggest that while Holloway may be correct in his interpretation that it is organization, not size *per se*, that has been critical in the evolution of the human brain, his point may still be moot: much, if not all, of the reorganization that has occurred in the human brain is of a quantitative nature.

The implications of these findings for the evolution of language are profound. For they imply that linguistic capacity may be a predictable result of an expansion of neural areas already present in the ape. Moreover, the recognition that reorganization of the human brain is primarily a quantitative phenomenon paves the way for a more detailed focus on the meaning of brain size, and opens the discussion to the implications of changes in developmental timing.

14.3 Comparative brain maturation patterns

The brain of laboratory rats, rhesus monkeys, and humans all continue to exhibit maturational changes throughout the period of body maturation, suggesting that this is a general mammalian pattern.[4] Monkeys, depending on environmental conditions, may reach menarche between two and three years of age, while apes may do so between eight and eleven years. By comparison, average age at menarche in human populations varies from about twelve to sixteen years of age. Gould's conclusion that the human brain is a heterochronic organ was based primarily on this comparatively lengthy human maturational period and the prolonged period of neural growth which accompanies it.

In Gould's framework, however, protracted maturation can lead to neoteny and/or hypermorphosis (Gould 1977, pp. 254–7, 376). The distinction rests upon the nature of the developmental delay and the ultimate form attained. If fetal growth processes are prolonged into postnatal life *and* result in the retention of fetal or youthful proportions, the result is neoteny. Alternately, protracted growth may reflect the allometric elaboration of late-, rather than early-, developing structures (McGuire 1966; Rensch 1959). In that case, an organ or species may exhibit positive allometry and successively pass through stages in which it exhibits proportions similar to those possessed by smaller ancestors. The eventual result is a recapitulation of proportions. Classical examples of hypermorphotic organs include the unusually enlarged antlers of various elk and deer.

Although in his elaboration of the hypermorphotic process, Rensch cited the enlargement of the vertebrate brain as a possible example, Gould concludes that the human brain is a genuinely neotenous rather

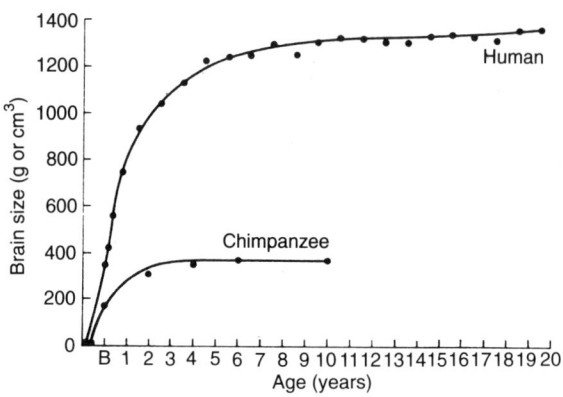

Fig. 14.1 Growth of the brain in human and chimpanzee. The figures for chimpanzee are for cranial capacity cm^3, and those for human for brain weight (grams) (from Passingham 1982).

than a recapitulative organ. Gould's conclusion was based on two grounds. First, the neonatal human brain is approximately 25 per cent as large as the adult organ. By contrast, some studies suggest that the neonatal rhesus monkey brain has already achieved 65 per cent of its final adult size (Schultz 1941) (see also Fig. 14.1). Secondly, human brain-size–body size growth ratios maintain a positive allometry for a much longer postnatal developmental time-frame than is the case for other primates (Count 1947; Holt *et al.* 1975; Weidenreich 1941).

In Gould's view, these facts imply that the infantile human brain is growing at fetal rates, and hence maintaining fetal proportions. There are several problems with this conclusion, however. For one, the limited comparison of human to rhesus monkey is highly misleading. Although it is true that Schultz estimated that the rhesus brain attains 65 per cent of its eventual size by birth, others have estimated that the neonatal rhesus brain has achieved as little as 48 per cent of its final size (Kerr *et al.* 1969). This latter estimate appears to be more in keeping with estimates of rhesus neonatal brain maturity based on myelination data (Gibson 1986, 1991). Moreover, even if the 65 per cent estimate should prove correct, this level of maturity at birth cannot be considered typical of mammals as whole. Sacher and Staffeldt (1974), for instance, compiled neonatal brain maturation data for over a hundred mammalian species: only four exceeded the 65 per cent estimate often quoted for the rhesus monkey. The neonates of many species, including most carnivores and insectivores and many rodents, have acquired 10 per cent or less of adult brain size. Hence, by no means can the neonatal human brain be considered unusually immature by comparison to those of mammals as a whole.

Moreover, the rhesus monkey may be unusually precocial in both its skeletal and neural development, even by comparison to other primates (Gibson 1970, 1986, Gibson, in press *a*; Newell-Morris and Fahrenbruch 1985; Watts 1986). Even such closely related primate taxa as baboons appear to be much less neurologically mature than the rhesus at birth. In fact, one estimate suggests that the neonatal *Hamadryas* baboon has achieved only 29 per cent of its total adult brain size (Kennard and Willner 1941). With respect to human evolution, the most critical comparisons are those of human and ape. Here, too, the situation is not as clear as Gould implies. Estimates of chimpanzee neonatal brain maturity vary from 35 per cent to 46 per cent; while those for humans vary from 23 per cent to 29 per cent (Sacher and Staffeldt 1974). Moreover, the human data are somewhat difficult to interpret, as they derive from infants who died at birth and, in most cases, the potential impact of the processes leading to neonatal death on brain size is not known. Hence, while the human brain at birth is probably less mature than that of other primates in terms of the percentage of adult brain size achieved, the primate–human divergence is not as great as Gould implies by his use of the 65 per cent estimate of completed brain size in the neonatal rhesus as the general primate standard.

Moreover, in his focus on the twin factors of neonatal brain size and brain–body growth ratios, Gould ignored the fact that neither one of these factors reflects the actual anatomy of the brain itself. After all, the human brain is composed of numerous functionally diverse subcomponents. The critical issue for language development is which of these diverse regions has actually enlarged. The linguistic and behavioural results of brain enlargement would, for instance, be quite different if the massive expansion of the brain had primarily involved the enlargement of brain-stem, rather than cortical, areas. Had the brain stem expanded, however, the superficial anatomical results of lengthy postnatal growth periods and positive brain–body allometry could well be the same. Distinguishing between the various hypotheses—neoteny, hypermorphotic recapitulation, or reorganization independent of gross developments in timing—demands an understanding of what it is that is actually growing in the postnatal time-frame.

When neural development is examined from this perspective, no support can be found for the neoteny hypothesis. All mammalian brains experience similar maturational patterns in that each proceeds through three general stages: embryogenesis, neurogenesis, and histogenesis (Gabriel 1980). During the first of these stages, which is prenatal in all placental mammals, the major portions of the nervous system are formed. During the second, neurons are formed and migrate to their ultimate destination in the brain. During the third, neurons achieve their mature function, by making synaptic contact with other neurons, by acquiring complex dendritic branching patterns, and by developing speed and specificity of neuronal transmission through the acquisition of the myelin sheath. During this histogenetic period the blood-supply to the nervous system also expands, and glial cells proliferate.

The timing of these second and third stages differs among species. In general, the degree of neuronal maturity at birth correlates with the percentage of adult brain size that is achieved. In the altricial laboratory rat, for instance, considerable neurogenesis occurs postnatally, and neither myelination nor cortical synaptogenesis have begun at birth (Altman 1969; Bloom and Aghajanian 1968; Jacobsen 1963). The brains of cats and dogs also exhibit very little myelination neonatally (Fox 1971; Langworthy 1929*a*, *b*). By contrast, in the praecocial guinea-pig, neurogenesis and myelination are nearly complete at birth, while cortical synaptogenesis is a perinatal rather than a postnatal event. Both the rhesus monkey and the human are relatively mature with respect to each of these parameters at birth. Neurogenesis is essentially a prenatal event in both forms (Rakic 1985); synaptogenesis begins months prior to birth (Huttenlocher 1979; Rakic *et al.* 1986); and considerable myelin exists in the neonatal brain stem and cerebral white matter (Flechsig 1920; Gibson 1970, 1991). Hence, while the human brain may grow at high fetal rates until long after birth, this growth does not reflect the retention of early fetal growth processes, but rather a quantitative elaboration of developmental events which are of a postnatal nature in many mammals.

As mammalian brains develop, they also experience predictable alterations in external and internal brain structure. During early fetal life, the brain is anatomically dominated by the brain stem; the cerebrum is small. Later neurogenetic and histogenetic stages result in an elaboration of the cerebrum, which encompasses an ever greater percentage of total brain size as development progresses. In addition, as the cerebrum enlarges it changes from a smooth to a highly fissured structure, and from a structure with high neuronal density to one in which neuronal density is sparse by comparison with the density of dendritic branches and glial cells.

From this perspective, it is clear that paedomorphism is not the final result of the expanded human growth period. The adult human brain in no way resembles the brains of the young of any animal in either its external or its internal form. Were it a paedomorphic organ, the brain stem would be relatively large by comparison to the cortex, the cortex would be smooth, the cortical neuronal density would be high, and the dendrites would be relatively unbranched. None of these factors prevail. Quite the opposite: the human brain displays an extreme cortical dominance, the greatest degree of cortical fissurization of any primate brain, the lowest neuronal density, and the most extreme dendritic branching. Hence it can be concluded that the human brain is not a neotenous organ. Whatever it is that determines the emergence of human language and the expansion of human intelligence, it is not the retention of an immature brain structure.

In many respects, however, the developmental changes described above are quite similar to the changes which have accrued during the evolution of the human brain, particularly during the expansion of the human neocortex. Similar ontogenetic–phylogenetic parallels can be seen in the internal maturation of the brain. Specifically, neural tracts characteristic of the adults of lower vertebrates are now known to form in fetal mammalian brains, only to be resorbed at a later date (Ebbesson 1984). In addition, general parallels between the myelination of neural areas and tracts have been reported (Gibson 1970, 1991; Jacobson 1978).

Despite these parallels, the brain cannot be considered a recapitulative organ in all aspects of its development. Data collected by Holloway (1979) and Armstrong (1982, 1986), for instance, indicate that the brain is not merely an allometrically enlarged ape brain, as certain regions of the brain have enlarged more than would be expected on the basis of allometric equations, while others have enlarged less. Hence it cannot be considered a hypermorphotic organ in Rensch's sense. That is, it is not a brain that passes through successive stages of development each manifesting the neuroanatomical shape and proportions characteristic of ancestral primate adults.

Moreover, among all vertebrates the nervous system myelinates in the same general order: peripheral and cranial nerves, brain stem and spinal cord, thalamus, basal ganglia, limbic system, and, lastly, the neocortex. In addition, the three mammalian species in which cortical myelination has been studied in detail—rat, rhesus monkey, and human—all display the same pattern. Within the cortex, primary sensory and motor areas myelinate early; association areas, last. A number of brain-stem, thalamic, and limbic areas, as well as certain primary cortical areas, are known to have enlarged in human evolution. They mature, however, in tandem with other earlier evolving constituents of the same neural regions. These data suggest that the primary determinant of maturational sequence may not be the point in phylogenetic time at which a structure evolved or enlarged, but where in the nervous system it is located. This pattern may reflect inborn constraints on the system. For, if peripheral structures are ablated, their central representations may fail to form at all. Similarly, the ablation of brain-stem areas results in failure of formation of their cortical projections (Bates and Killackey 1985; Killackey 1979).

Hence, these data provide definitive, but limited, grounds for the reconstruction of linguistic evolution based on maturational parameters. Specifically, they suggest that to the extent that language is based on the functioning of the late-maturing and late-evolving cortical association areas, ontogenetic–phylogenetic parallels are to be expected. To the extent that it is based on the functioning of primary sensory and motor areas and/or on the functioning of subcortical regions, ontogeny might not parallel phylogeny.

14.4 Cognitive maturation and neocortical development

Within a few weeks of postnatal life, human infants can distinguish phonetic sounds (Eimas 1985), recognize objects by means of cross-modal sensory perceptions (Meltzoff and Borton 1979), turn their eyes and heads in the direction of auditory stimuli (Alegria and Noirot 1978; Butterworth and Castillo 1976; Wertheimer 1961), reach and grasp for objects presented in auditory or visual modes (Bower *et al.* 1970; Bower 1972, 1974; Wishart *et al.* 1978), imitate simple facial expressions (Meltzoff and Moore 1977), and distinguish between a looming object headed directly toward the body and one on a near-miss path (Ball and Vurpillot 1981). Researchers have often expressed surprise at the scope of infantile behaviours, as the neocortex is minimally functional during the first weeks of postnatal life. Many animals with minimal neocortex, however, can reach and grasp, learn, orient towards the source of a sound, and engage in other complex activities. If brain-stem processing mechanisms can mediate such behaviours in other species, there is no reason why they should not be able to do so in humans.

That some of these skills are likely to be brain-stem-mediated is also evident from the fact that their organizational form is often quite different from that manifested by adult forms of the same behaviour, and rather similar to that shown by smaller-brained animals (Gibson 1981). The infantile reach-and-grasp, for instance, is extremely stereotyped. Infants first reach and then grasp. All five fingers work in unison. They do not first grasp a near object and then reach with it. They do not use single digits at a time, and they do not touch and feel objects without grasping them (Bower 1974). In other words, infants cannot break the reach-and-grasp down into discrete motor actions which can then be combined and rearranged into new and more versatile patterns. A similar holistic approach is manifested by many neonatal behaviours and perceptions, and sets them apart from the very varied behaviours of adults.

By about three to six months postnatally, many of the holistic and stereotyped neonatal behaviours disappear, and infantile perceptions and actions become more highly differentiated (Bower 1974; Mounoud and Vinter 1981). Before the end of the first year, discrete actions and perceptions are combined to produce varied, but co-ordinated sensorimotor schemata (Piaget 1952). During the second year, the infant learns to use tools and words. By the end of that year, his communication skills are genuinely symbolic.

These successive achievements correlate rather nicely with the myelination of the neocortex (Gibson 1977, 1981). For instance, the differentiation of perceptions and actions into component parts coincides in time with the beginning myelination of sensorimotor areas (Conel 1939–1967; Gibson 1977). These areas are known to function in the recognition of discrete sensations and in the mediation of fine motor movements (Hubel and Wiesel 1968; Kuypers 1962; Mountcastle 1978). Similarly, the initiation of perceptual and motor combinatory actions occurs in synchrony with the initial myelination of secondary sensory and motor regions. These regions possess more holistic and synthetic sensorimotor functions (Duffy and Buschfiel 1971; Gross *et al.* 1972; Luria 1966). Finally, the emergence of tool-use and symbolization during the second year of life coincides in time with the myelination of the neocortical association areas (Table 14.1).

Cognitive development is, of course, still incomplete at two years of age. So, too, the neocortex is still far from mature. While all neocortical areas contain some myelin and appear to be functioning in at least a rudimentary fashion, all areas also continue to mature during the early childhood years (Conel 1939–1967). The cortical association areas continue to do so until well into adolescence or later (Yakovlev and Lecours 1967). Consequently, cognitive maturation during childhood and adolescence depends primarily on the continuing maturation of cortical areas already functional in late infancy. Further, the cortical association areas would appear to be in a position to play the greatest role in the progressive development of linguistic and cognitive skills during the later developmental stages. Hence, whatever the functions of the neocortex as a whole, and of the neocortical association areas in particular, they must be of a sufficiently general nature to account not only for elementary tool-use and symbolization, but also for the abstract intellectual capacities of the adult.

In the works of Piaget, two prime mental processes are postulated to underlie all cognitive development: *differentiation*, or the breakdown of holistic sensory and motor perceptions into component parts; and *construction*—the assembly of differentiated schemata into higher-order mental constructs in such a manner that the parts are subordinated to the whole (Piaget 1952, 1954, 1955, 1969). It has been suggested on the basis of a variety of neurological data that these two processes are prime neocortical functions (Gibson 1977, 1981, 1983, 1985, 1988, 1990). The association areas, in particular, are critical for the assembly of diverse cross-modal sensorimotor skills into higher-order cognitive constructs such as object-naming, grammar, tool-use, object-construction, and drawing (Geschwind 1965; Luria 1966); while the primary sensorimotor areas are critical in the perception and construction of precisely differentiated sensorimotor acts. In this mental-constructionist view, the post-infancy, progressive ontogenetic development of higher cognition depends upon the continuing differentiation and integration of previously developed schemata and newly acquired knowledge, and not on newly acquired neurological capacities.

Although the differentiational–constructional framework was derived, in part, from the works of Piaget, it is at variance with one basic tenet of classical Piagetian psychology. According to Piaget cognitive development involves a series of discrete stages, each delineated by a major qualitative leap in the organization of intelligence and by the emergence of new types of concepts or abilities. This 'stage' concept would seem to demand the mediation of entirely new cortical processing mechanisms at subsequent stages of cognitive development. No support can be found for this concept in terms of neuroanatomical development subsequent to two years of age in the human child (Gibson 1978).[5]

Recent neo-Piagetian analyses of cognition by Case (1985), however, are more compatible with the neurological data. Case analyses the cognitive jumps that characterize the Piagetian stages in quantitative terms. While recognizing the importance of differentiational capacities, Case emphasizes maturational increases in mental constructional capacity. In his view, as children progress from one stage to the next, two or more distinct cognitive achievements are integrated in a hierarchical fashion—that is one scheme or concept is subordinated to another in the pursuit of a common goal. These newly integrated structures appear to be qualitatively different from those that

Table 14.1 Human intellectual development: major stages and milestones

Stages (from Case 1985)	Neural parameters	Major intellectual achievements by end of stage	Major linguistic achievements by end of stage	Pragmatic linguistic skills by end of stage
Sensorimotor (birth to eighteen months)	myelination of brain stem and of sensorimotor cortex synaptogenesis	simple tool-use, object concepts; imitation of facial expressions and vocalizations; imitation of mother's actions on objects	the first words	demarcation of objects by words; instrumental use of words to obtain objects or aid use or words or other vocalizations to maintain maternal interest; 'what' questions—e.g. 'What name?'
Relational (eighteen months to five years)	continuing myelination of sensorimotor regions; myelination of sensory association areas; beginning myelination of association regions; dendritic arborization	the understanding of relationships between objects, words, people; simple drawings; role-taking during play; comprehension of simple rules	sentences six to eight words in length; negation; phrases; word-inflections; references to future and past	ability to follow simple verbal instructions and to give simple commands; play characterized by collective monologues as each child describes what he or she is doing; play characterized by 'communications of action': 'I'll do this.' 'You do that.'; arguments involving simple assertions and clashes of opinion; discussions of present events or of events known to the collective memory of both participants; 'why' and 'when' questions
Dimensional (five to eleven years)	acquisition of some myelin by all cortical association areas and layers; continuing myelination in earlier-myelinating areas; continuing dendritic arborization	the understanding of quantitative concepts, e.g. weight, size, length; the understanding of quantitative relationships, e.g. arithmetic; the comprehension of complex social rules; the comprehension of physical causality	the ability to describe precisely absent or distant events without using gestures; the ability to recount precisely stories or events in proper sequential order; the understanding and use of advanced grammatical structures; the ability to follow a sequence of instructions: e.g. 'Sit down, take out a pencil and paper, and write your name.'	factual discussions of events actually experienced by only one of the participants; arguments based on factual knowledge and logic; discussions of physical causality; discussions about social and moral rules; collaborations in acquisition of factual knowledge based on concrete events
Abstract (eleven to eighteen years)	essentially adult nervous system	The understanding of vectorial relationships	verbal analogies	Discussions of abstract concepts of morality,

Stages (from Case 1985)	Neural parameters	Major intellectual achievements by end of stage	Major linguistic achievements by end of stage	Pragmatic linguistic skills by end of stage
	but continuing myelination and dendritic arborization perhaps into old age influence on brain function of hormonal changes of puberty	between dimensions: e.g., algebra, physics The development of abstract concepts of morality Attempts to hypothesize and predict events The development of reasoning based on abstract premisses	metaphor understanding of literary and poetic symbolism	social rules, and physical casuality (some people) ability to compose mature literary works (some people) collaboration in acquisition of factual knowledge based on prediction, hypothesis, and accumulated cultural knowledge

The data summarized in this table are derived primarily from Case (1985), Conel (1939–1967), and Piaget (1952, 1954, 1955).

preceded them; but their underlying base is a quantitative increase in the numbers of schemes that can be focused on simultaneously and co-ordinated together. The result is not the emergence of new types of abilities, as Piaget had claimed, but rather the emergence of higher levels of pre-existing abilities. Ultimate cognitive levels are achieved when the greatest capacities for hierarchical and quantitative integration are reached, in late adolescence. This increase in the numbers of schemes that can be subordinated to one overall goal results from increases in the operational efficiency of short-term storage space (i.e. memory), rather than from the addition of totally new mental processes.

As an example, Case discusses the properties that are needed for a child to speak his first words. These include a variety of pre-requisite skills, each of which has its own complex developmental history: the ability to use complex behavioural structures in social interaction (for example, motor actions and gestures); the ability to vocalize and to substitute patterns of vocalization for other behaviours that gain social attention; the ability to understand the vocalizations of others; the ability of the infant to modify his own vocalizations on the basis of those of others in the community; and an interest in and understanding of objects' properties. At about twelve to eighteen months these varied, and previously somewhat independent, skills are integrated into a new superordinate concept with major linguistic and cognitive import—words (Table 14.2).

From that point on, words serve as fundamental building-blocks for new hierarchical linguistic interactions. Between the ages of eighteen months and two years, children begin to combine two words to express one relationship, 'Daddy hit' or 'hit ball'. Between two years and three and a half years, grammatical modifications are introduced—'the ball' or 'sleeping'—and words are combined into sentences expressing two relationships, 'Daddy hit ball.' At the end of this period, expressions of two relationships and grammatical modifications are merged into single, somewhat longer, sentences, 'Daddy hitted the ball.' Later, children begin to use even longer sentences, containing elaborations of one or more of the sentence components—'Daddy hit the red ball hard.' Later still come the abilities to construct complex sentences and to combine sentences into coherent stories and/or reconstructions of sequential events (Tables 14.1 and 14.3).

Each of these successive achievements depends on the ability to integrate hierarchically ever-increasing quantities of precisely differentiated linguistic information with ever-expanding understandings of the physical and social environment. Except for progressive expansions in this integrational capacity, no new neurological processing mechanisms are needed once the basic elements of vocalization, object-manipulation, and non-verbal social communication have been acquired in infancy. In Case's framework, this capacity for hierarchical integration extends across cognitive domains, and continually expands throughout the maturational period. It reflects itself in the development of a variety of achievements, including spatial constructs, social cognition, and causal, verbal, and mathematical reasoning abilities.

Table 14.2 The sensorimotor period and the achievement of the first words as described by Case (1985)*

Intellectual substage	General cognitive style	Linguistically relevant behaviours
Substage 0 (one to four months) **Operational consolidation**	Extension and consolidation of individual sensorimotor schemata	sucking, grasping, crying, cooing visual following of moving objects visual scanning of stationary objects stereotyped reaching and grasping for objects turning head toward source of sound eye contact with mother accompanied by cooing or smiling smiling, kicking, and cooing while observing mother's response imitation of simple facial expressions already in the infant's repertoire
Substage 1 (four to eight months) **Operational co-ordination**	Behavioural structures assembled from two qualitatively distinct sensorimotor precursors one behaviour subordinated to another: one structure is used as a means towards an end specified by the other Greater differentiation of preceding sensorimotor structures	babbling babble to gain attention stretch out arms to be picked up while babbling or cooing 'play', such as pulling mother's hair and laughing and cooing at the same time shared visual regard: i.e. looking at object mother is looking at 'turn-taking': i.e. making sounds and waiting for mother to make sounds
Substage 2 (eight to twelve months) **Bifocal co-ordination**	Focus on two objects at once while executing a voluntary action	bang spoon or rattle while looking at mother and smiling babble at adult while pointing to object look at object pointed out by mother and listen to name two-syllable utterances imitation of sounds already in infant's repertoire
Substage 3 (twelve to eighteen months) **Elaborated co-ordination**	Similar to Substage 2, but tasks are more complicated and take more features into account	imitate mother's actions on objects invite mother to imitate infant's actions actively present objects to mother imitate novel adult actions and vocalizations use of spoons and other simple tools obtain objects by pulling on supports or using simple rakes babble at object and listen to mother name it name objects obtain object by stating its name and pointing

* Case derives many of his data from other sources including the cognitive developmental studies of Piaget (1952) and the studies of language-development of Brown (1973) and Bates et al. (1979).

Table 14.3 Relational thought in early childhood, based on Case (1985), Piaget (1955), and Brown (1973)

Intellectual substage	General cognitive style	Linguistically relevant behaviours
Substage 0 (twelve to eighteen months) **Operational consolidation**	Consolidation of achievements of sensorimotor period Comprehension and manipulation of a single relationship between two objects e.g., if one end of a balance beam moves up, the other will move down	same as in Substage 3 of sensorimotor period

Intellectual substage	General cognitive style	Linguistically relevant behaviours
Substage 1 (eighteen to twenty-four months) **Operational co-ordination**	The comprehension of two different kinds of relationship or the utilization of one relationship to effect a change in another: e.g., in order to move one end of a balance beam up, a block must be removed from the opposite end so that it can be moved down	meaningful combination of two different kinds of words actor–action, action–object, agent–object possession negation engage in complementary object actions i.e. co-operative actions involving a division and differentiation of labour: e.g. 'rake' lawn that has been cut by parent attempt to 'play' games involving both object and social manipulation: e.g. hit ball with racket towards parent draw a line
Substage 2 (two to three and a half years) **Bifocal co-ordination**	Focus on two systemic interrelations rather than one: e.g. move balance beam down by removing supports and placing a weight on one side of beam	combine words to express two or more relationships: e.g. agent–action–object, agent–action–modifier grammatical sentences generalization of grammatical rules, e.g. 'I runned.' relative clauses inflection of words 'why', 'when' questions prepositions demands, commands count objects arranged in a line focus on two social-object interrelationships at once: e.g. retrieve a ball bounced by person on left and, in turn, bounce it to person on right; or parent cuts grass, sibling rakes it, child places it in basket draw a cross in a circle
Substage 3 (three and a half to five years) **Elaborated co-ordination**	As in stage 2, but taking into account one additional feature	draw a circle with lines coming down from it copy numerals and letters count objects arranged randomly or count only those matching a specific criterion such as colour play one role while simultaneously monitoring role of another child play with two dolls having complementary functions, such as doctor–nurse sentences of six to eight words clauses beginning use of subjunctive follow simple verbal instructions, e.g. 'Sit down.' talk about what he/she is doing or about to do talk about what he/she wants others to do engage in play involving each child's engaging in monologue describing his/her own actions play at games requiring co-operative actions and 'I'll do this, you do that.' statements commentaries on ongoing events

While Case's analysis is of too recent a vintage to have been rigorously tested by other investigators, his basic postulate of a common quantitative underpinning to all cognitive processes is compatible with many other cognitive and developmental theories. Previous researchers have suggested that similar mechanisms underlie linguistic and cognitive development, and the development of memory, language, morality, object-manipulation skills, and social intelligence have all been described as outcomes of mental constructional processes.[6] Moreover, recent research demonstrates that the primary characteristic separating skilled chess-players, speakers of a second language, and athletes from their less skilled colleagues is the quantity of information that can be focused on at once and organized into complex hierarchical schemes (Dreyfus and Dreyfus 1986; Glaser 1984).

From an evolutionary perspective, three points are critical in Case's analysis. First, it would seem that if human language and other cognitive skills can mature primarily by means of hierarchical integration and differentiation of pre-existing elements, then they could have evolved by a similar mechanism. Hence, modern linguistic skills could be a result of quantitative changes in information-processing capacities, rather than of the addition of qualitatively different neural functions. Secondly, if all cognitive processes are subject to similar developmental constraints, then they may also have been subject to similar constraints evolutionarily. Finally, if the maturation of object concepts and of object and social manipulation skills is essential to the development of language, then such skills may also have been essential to its evolution. In other words, tool-use and tool-manufacture, social skills, language, art, and other seemingly distinctive human behaviours may have evolved in tandem both because they are subject to similar cognitive constraints and because they interact developmentally. Why, for instance, would hominids have invented the most basic three-word sentence—agent, action, object—if they were not aware of and interested in object manipulation?

14.5 Can the neo-Piagetian ontogenetic framework be placed in an evolutionary perspective?

If the brain contains few, if any, unique structures, attempts to identify distinctive behavioural gaps between man and apes, and endless arguments about whether apes do or do not possess language are probably counterproductive. Rather, emphasis should be placed on evolutionary continuities, and on the search for the behavioural foundations of human skills in other animals. The task, then, becomes one of developing a viable model of how quantitative changes in neural structures could result in the expansion of apelike capacities into modern linguistic skills (Gibson 1988).

Not so long ago such a reconstructional task seemed impossible, as the behavioural gaps between man and ape appeared monumental and unbridgeable. With the expansion of serious research in the field of animal behaviour, however, these gaps have progressively narrowed, particularly with respect to linguistically relevant skills and to behaviours often thought to have served as prime foci of early linguistic discourse (Ettlinger 1984). In particular, apes are now known to use gestural symbols in captivity, to make and use tools, to hunt co-operatively upon occasion, and to 'share' their food in the wild.[7] By contrast to previous assertions, it is also clear that primate vocalizations exhibit rudimentary 'syntax' and 'phonemic' differentiation; occasionally refer to environmental objects, at least in vervet monkeys; are subject to conditioning and learning; and may be under cortical control.[8]

The realization that the rudiments of human skills are present in the apes suggests that major evolutionary advances in cognition and language may have resulted from mental differentiational and hierarchical constructional processes similar to those which are operative in the maturation of the human child (Gibson 1988, 1990). Whereas in the child these capacities result from increasing functional maturation of neocortical areas, in evolution they may have been mediated by the increasing size of the same areas.

Analyses of neocortical function suggest that, in fact, expanded differentiational and mental constructional capacities are predictable results of an enlarged brain. For instance, increases in size of the motor and somatosensory representations of particular anatomical structures are regularly accompanied by increases in the sensory acuity and precision control of those organs.[9] Although apes can utilize rudimentary symbolic gestures, and they possess well-developed powers of long-call vocalization, their gestures are crudely formed by comparison to human gestures (Armstrong 1983, 1985), and they lack the vocal precision to produce human speech-sounds. Expansion of the primary sensorimotor areas would be expected to provide the increased precision control of vocal and manual organs necessary, but not sufficient, to mimic human communication and tool-use. By contrast, size increases in the motor and somatosensory association areas subserving the hands and the oral cavity would provide increased bimanual co-ordinations and increased ability to combine separate movements of the various speech organs into new simultaneous and sequential patterns: for example, some consonants and words.

Although the neural areas which provide the sensorimotor foundations for tool-use and language are somewhat expanded in man as compared to apes, it is the association areas which exhibit the greatest enlargement (Passingham 1973, 1975). Hence, just as in the later stages of ontogeny the greatest intellectual advances reside in hierarchical integrational capacities, so, too, the major cognitive differences between ape and man should reside in the domain of hierarchical mental integration or mental construction (Gibson 1983, 1988, 1990; Reynolds 1981, 1982, 1983). Behaviourally, an expansion of these latter areas

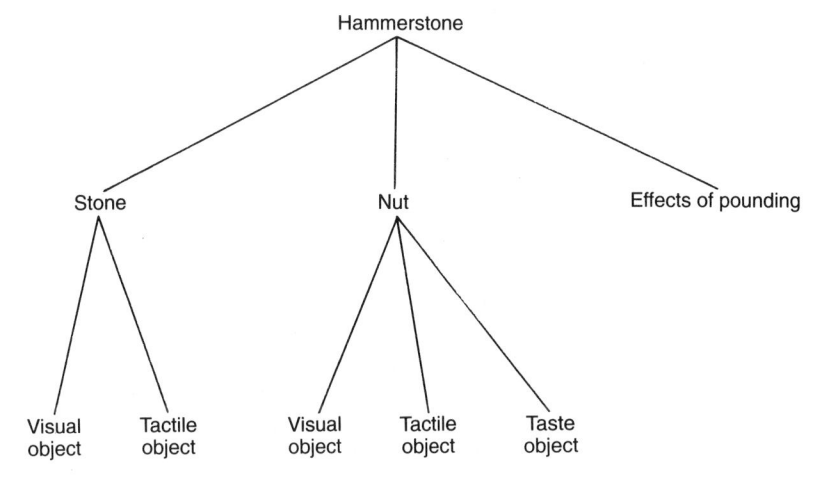

Fig. 14.2 The hierarchical construction of the concept of a hammerstone. A number of primates can master hierarchical tasks at this level.

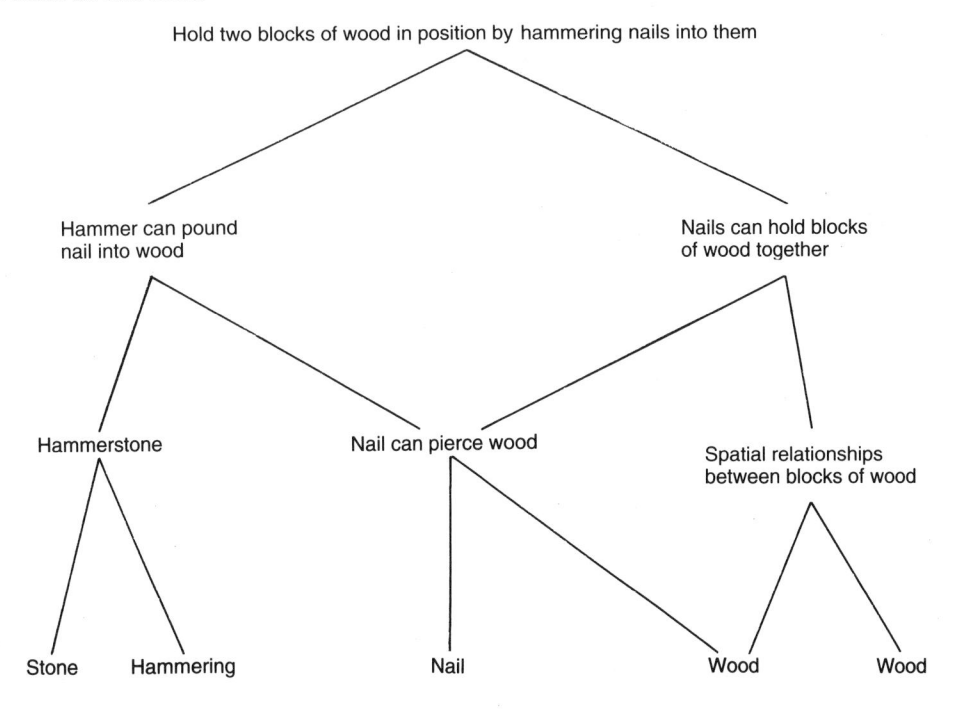

Fig. 14.3 The hierarchical construction of a more complex tool-using task which is beyond the capacities spontaneously exhibited by wild apes.

would have resulted in the ability to take action patterns already present in the apes and combine and recombine them in a hierarchical fashion to produce the novel and elaborate behaviours of modern humans.

Apes, for instance, can use simple hammers, probes, digging-sticks, and containers. Each tool, however, tends to be used alone. The mental integrational capacity necessary for the combined use of two of these tools at once or in sequence to meet a single long-term goal appears to be lacking. Hence, apes have never been known to engage in what may be one of the most basic of all human foraging patterns—digging tubers or roots from the ground, and then transporting them, perhaps along with other items, in a container, to be pounded into a pulp at some later time. Similarly, although apes will, in captivity, hammer pegs into holes if presented with children's peg-boards, and can stack blocks (Gibson, personal observations), they have never been observed combining these diverse skills into a single compound action, such as using a hammer to pound a probe-like object (nail) into two blocks of wood to provide a permanent construction. Figures 14.2 and 14.3 illustrate the hierarchical nature of these hammering tasks.

Apes can also make tools by subtractive methods. For instance, they can make probes by removing side-branches and leaves from twigs (van Lawick-Goodall 1968, 1970). They do not, however, manufacture tools by constructive methods: that is, by joining two disparate objects together to make a more efficient tool. Hence, although wild chimpanzees use both sticks and stones as tools, they apparently do not possess the mental constructional capacity to join the two together to make a more efficient hammer or a spear. Nor do they make tools which require the mental capacity to envision the end-result of numerous sequential steps, as in making a tool that is then used to make another tool, or in making a tool to match a preconceived geometric or symmetrical form.

Similarly, although apes cannot speak, captive apes can use isolated symbolic gestures and combine gestures into two-, or occasionally, three-'word' utterances (Fouts 1973; Gardner and Gardner 1969, 1980; Miles 1983). Their ability to combine manual gestures into grammatical sequences appears to be quite limited, however (Terrace 1979, 1980), perhaps because they lack the necessary hierarchical integrational capacity (Figs. 14.4 and 14.5). Ape gestures are used to request or name objects in the immediate environment or objects with which they are familiar: toys, food, pets. They may also request aid in reaching a desired immediate goal—to be tickled, to go for a ride, etc. In this respect, ape language-usage is very similar to that of small children.

Like small children, apes lack the ability to recount stories in proper sequential order, to discuss objects or events that are not part of the common experience, and to give complicated, but precise, linguistic instructions (for example, on how to perform a difficult procedure). In children the development of these skills

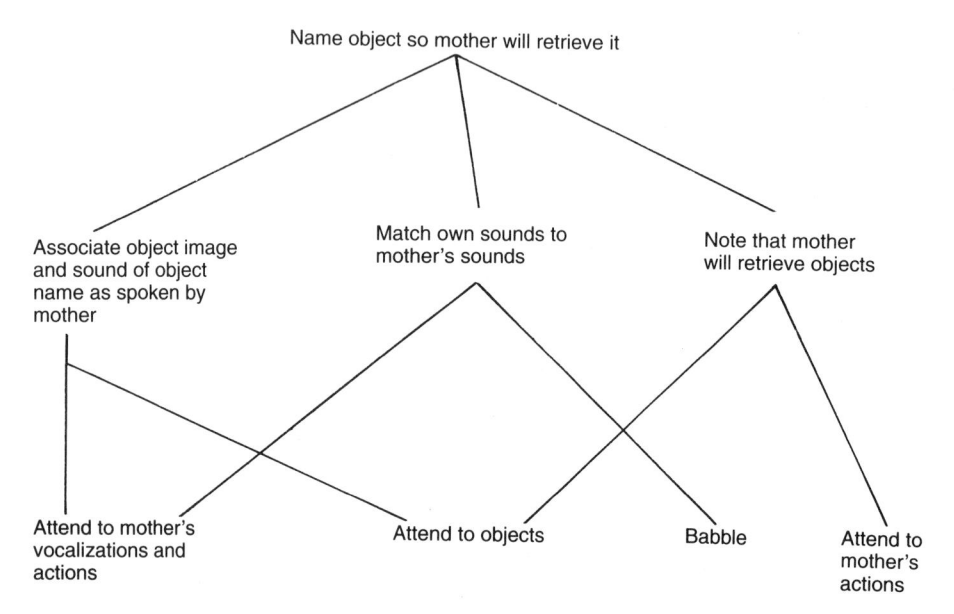

Fig. 14.4 The hierarchical construction of object-names by human infants. Although no ape can do this vocally, they reach a similar constructional level with gestural names.

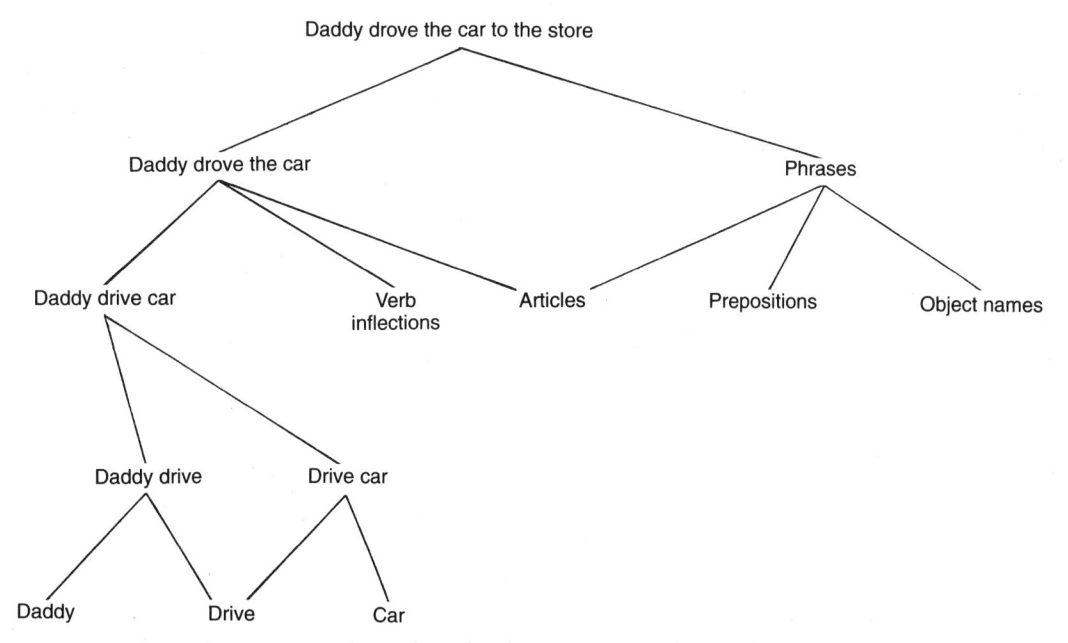

The ontogeny and evolution of the brain, cognition, and language 421

Fig. 14.5 The hierarchical construction of a simple sentence. This requires greater mental constructional capacity than is generally exhibited by apes.

is contingent upon advances in mental constructional capacity. First, they must be able to keep numerous concepts in mind at once. Second, they must possess advanced grammatical and sentence-constructional skills. The ability, in fact, to communicate new information accurately and to give directions in the absence of manual demonstration and gesture requires such complex mental skills that children do not develop them until about seven years of age (Karmiloff-Smith 1979; Piaget 1955). According to the thesis of this paper, apes never develop these skills because they possess only rudimentary mental constructional capacities by comparison with older human children and adults.

Similar analyses could be applied to social interaction. Apes have very complex social skills. Yet highly differentiated and hierarchically co-ordinated, symbolic and rule-based political interactions and economic divisions of labour do not exist in ape societies (De Waal 1983; Goodall 1986).

14.6 What can be 'reconstructed' phylogenetically?

Hence, contrary to the views of some, the expansion of the human brain was among the most fundamental of all evolutionary advances. By itself, it may have accounted for much of the evolution of cognition and language. In particular, it is likely that evolutionary advances in brain size correlated very strongly with advances in the potential for precision control of the speech and manipulative organs, with mental constructional ability, and with all the behaviours that depend upon constructional skills—symbolization, grammar, complex tool-use and tool-making, drawing, etc. Specifically, just as ontogeny results in the ability to hold ever-greater numbers of concepts in mind simultaneously, so, too, brain-size advancements should have been accompanied by ever-increasing *abilities* to organize large quantities of information into complex linguistic and cognitive constructions.

An examination of the fossil record indicates that by two to three million years ago the australopithecines possessed brains that were probably somewhat enlarged relative to body size in comparison to Great Ape brains (Falk 1985; Holloway and Post 1982; Tobias 1971). From that time on, the hominid brain gradually expanded until about 100 000 years ago, at which time it reached its modern size in archaic forms of *Homo sapiens* such as the neanderthalers. Although no brain expansion occurred subsequent to neanderthaler times (see Holloway, this volume, Chapter 4), obvious changes did occur subsequently in the form of the cranium and in the external configuration of the brain. In particular, the occipital region of the neanderthaler brain was more protuberant than in modern brains, and some authors consider the frontal lobes to have been relatively smaller (Kochetkova

1978; Trinkaus and LeMay 1982; although see Holloway 1985 for an opposing view). These points suggest to me that, over a three-million-year period, hominids experienced gradual increases in the fundamental mental constructional skills that could be applied to language and other behaviours, and that modern neurological capacities may not have been reached until the appearance of *Homo sapiens sapiens*.

The fact that mental constructional skills are applied across cognitive domains suggests that further clues to the potential linguistic abilities of early hominids may be gained by examining their cultural remains—tools, shelters, feeding-sites, etc.—for clues to the level of mental constructional capacity present (Falk 1980; Holloway 1969, 1972; *contra* Wynn, this volume, Chapter 10; Lewin 1986). In this respect, the archaeological record parallels the brain-size record. By two million years ago, the australopithecine-like *Homo habilis*, if not all the australopithecines, was making stone tools. This behaviour involves the use of a tool to make a tool. No ape has been observed spontaneously engaging in similar manufacturing tasks in the wild.

By 700 000 to 300 000 years ago *Homo erectus* or early *Homo sapiens* was manufacturing finely formed, bilaterally symmetrical 'hand-axes'. According to Wynn (1979), the manufacture of these stone artefacts requires the level of operational intelligence exhibited by at least a modern Western seven-year-old child. *Homo erectus*, however, was not yet applying modern adult mental constructional skills to his tool-making endeavours. His tools were limited to a few forms, and their manufacture required a small number of steps and a minimal amount of advanced planning (Kochetkova 1978; Prideaux 1973). In addition, we have no evidence that *Homo erectus* constructed tools of two or more separate elements. It is not until the time of the neanderthalers, approximately 100 000 years ago, that this critical advance apparently occurred, in the form of stone points which some investigators think may have been hafted to wooden spear-shafts to make spears.

With the advent of skeletally modern humans tools became more varied in type, and were manufactured from bone, ivory, and other materials which can only be effectively used with advanced preparation that may require days or even weeks of planning and work. Compound tools composed of separate but interacting parts, such as bows and arrows, were invented, and artistic skills flowered (Binford 1982; Kotchekova 1978; Prideaux 1973). Whether or not the changes in neural form which occurred in Late Stone Age humans were of sufficient functional significance to account for this expansion in cultural remains is as yet unclear, and will probably remain so for some time, as the various scholars who have examined fossil *Homo sapiens sapiens* endocasts and/or theorized about the possible meaning of changes in brain form have come to divergent conclusions on this matter (Gibson 1985; Holloway 1985, Holloway, this volume, Chapter 4; Kochetkova 1978; Trinkaus and LeMay 1982).

On the basis of the slow increments in brain size and tool complexity, it is plausible that language should also have emerged gradually over a three-million-year period, and that the cognitive skills necessary for complex modern languages were finally achieved some time between 100 000 and 35 000 years ago. While, however, brain size and the archaeological record can provide critical clues to the pace of language evolution, by themselves they can tell us little about the nature of the communication skills possessed by early hominids. It is in this respect that an understanding of ontogenetic–phylogenetic relationships becomes critical. In particular, as both ontogeny and phylogeny are characterized by expanded functioning of the neocortical association areas, an examination of the ontogeny of the behaviours controlled by these areas can provide a more detailed view of possible parallel evolutionary pathways.

Several authors have attempted to reconstruct language evolution based on ontogenetic data. Each has focused on divergent aspects of language development. Lock (1983) has emphasized the logistics of language maturation and evolution, beginning with the non-communicative action patterns of infants and apes, and advancing to gestures, symbols, and more complex linguistic skills. Parisi (1983), by contrast, has focused on the evolution and development of lexical and syntactic systems. Peters (1974) and Parker and Gibson, both conjointly and separately, have discussed the pragmatics and adaptive significance of emerging linguistic systems (Gibson 1983, 1988; Parker 1985; Parker and Gibson 1979).

Each of these models is compatible with the basic tenets of this paper, in that each views language as a system emergent from the communicative skills of the apes, and each proposes a phylogenetic sequence that is in accord with the neo-Piagetian view of gradually developing mental constructional skills.

A brief synthesis of these works would suggest the following general evolutionary pattern. The transition from ape-like communication systems to language began gradually as manual *and* vocal action patterns were 'ritualized' into communicative gestures (Lock 1983). This would require no advance in neurological capacities beyond what is possessed by the apes (Lock 1983). In human children gestures develop when mothers interpret and reinforce the infant's action patterns as if they were meaningful. The prehistoric ecological transition to extractive foraging on foods that were both difficult to obtain and process would have resulted in mandatory parental provisioning of post-weanling children (Parker and Gibson 1979). Abortive attempts by children to open tough nuts, dig deep tubers from the ground, and engage in other complex activities would have resulted in need for parental aid. Many parents would have anticipated their children's difficulties in accomplishing these tasks, and would have come to their aid as soon as interest was evidenced by the child by pointing, vocalizing, reaching, etc. The probable result would have been that certain vocal or manual gestures would have acquired specific meaning within individual mother–infant pairs. One can visualize such 'ritualized' meanings then being used by the mother in her interactions with each of her subsequent children, and eventually being passed on through her children to her grandchildren. The result would have been a kin-group-specific protolanguage composed of simple vocal or manual gestures. Depending on the degree of interactions between kin-groups, such a protolanguage could have been passed to other members of the population.

Later, as overall brain size and intelligence increased, the mental capability of hierarchically organizing manual and vocal gestural systems into symbolic units would also have increased, as it does in modern human children who have attained a sufficient cognitive level. At this early stage, symbolic units would probably have been organized in a lexical manner (Parisi 1983). In Parisi's framework, these lexical systems would have been of a combinatory nature. Signals would have been divided into parts, with each subpart contributing only a portion of the meaning of the total signal. Although lexical systems represent a major advance over typical animal systems, they have major limitations. In particular, according to Parisi, purely lexical communications are very context-dependent. It is impossible in a purely lexical system, for instance, to distinguish the meaning of 'house behind tree' from 'tree behind house' by linguistic means alone. Pointing, other context-dependent gestures, or pantomime are needed to clarify the linguistic meaning. For precise linguistic description of complex, non-context-dependent, object relationships, syntactical systems in which meaning is derived from word order or other grammatical markers is essential (Parisi 1983).

It is unclear just when syntactic structures would have become an essential feature of hominid communication systems. At a simple level, agent–action–object sentences, accompanied by pointing and pantomine, would have demanded little in the way of cognitive advance over that present in the apes, and could have been quite useful as soon as co-operative actions or adult–adult food-sharing arose. Perhaps, as suggested by Peters (1974), rudimentary syntax arose rather early in phylogeny, as human groups developed habits of splitting into small foraging groups during the day and congregating at various rendezvous sites in the evenings. Syntax could have been quite helpful for such groups in terms of the advanced delineation of each evening's rendezvous.

On the basis of ontogenetic data, although rudimentary syntax accompanied by pointing and pantomime may have developed early, the transition to fully developed modern syntactic capacity would have been gradual. While three-year-old children clearly possess rudimentary syntax, it is not until seven or eight years of age that their syntactic capacity is sufficient to permit accurate communication by linguistic means alone about distant events and/or events known to only one of the participants (Karmiloff-Smith 1979; Piaget 1955).

Clues to the actual communicative content of early lexical or rudimentary syntactic systems can be gained from a knowledge of what children who possess such systems can talk about (Tables 14.1, 14.2, and 14.3). According to Piaget, the discourse of very small children can be characterized as the communication of desires (needs, wants). Four- to six-year-old children engage in much 'on-the-spot' communication of action: for instance, 'I'll do this, this way'; 'You do that.' These communications are of necessity accompanied by much gesturing and demonstration. By seven years of age, syntactic and cognitive structures are sufficiently developed that children become capable of discussing and debating factual knowledge about the environment and about physical and behavioural causality. As adolescence approaches, hypothesis and propositional logic enter linguistic discourse.

It has been suggested that hominid evolution exhibited similar linguistic phases (Gibson 1988). The earliest language usage corresponding to the beginning of parental food-provisioning could be termed a 'communication of desire'. As hominids began to engage in on-the-spot co-operative activities, such as co-operative hunting, food-collecting, or shelter-building, their communication skills would also have expanded, and perhaps reached a stage that could be called a 'communication of action'. That is they would have engaged in on-the-spot co-operative activities demarcated by simple linguistic or 'proto-linguistic' statements—'I'll do this; you do that,'—accompanied by pointing and gesture.

Despite the academic emphasis on co-operative hunting, other aspects of the food-procurement techniques of modern hunters and gatherers are far more unique. Specifically, humans develop a broad base of symbolically organized knowledge pertaining to various aspects of animal ecology: i.e. animal habits, plant growth cycles, weather conditions, terrain, seasonal change, etc. This permits a social division of labour in the procurement of food, and enhances the effectiveness with which food can be obtained, especially in difficult environments. Such knowledge, however, cannot be obtained by individuals acting alone, or even by small groups who always travel and act together. Rather it is a group knowledge that is acquired and collected by many individuals over many generations. Its acquisition demands the ability to debate factual knowledge. Hence it is unlikely that hominids could have successfully inhabited difficult environments until they reached the linguistic stage of at least a seven- to eight-year-old child. In accordance with Piaget, this stage could be termed the 'communication of knowledge'. Eventually, with the development of oral or artistic record-keeping skills (Marshack 1972), the communication of knowledge became a 'communication of hypothesis and prediction'. By this point, reliance on migratory herds, salmon runs, and seasonal plant harvests became a reality.

As the complexity of food-procurement techniques increased, co-operative actions, collective knowledge, and symbolic planning became necessary adjuncts of the food-procurement process. Food-sharing and economic divisions of labour developed. As result, the complexity of social interactions also increased. Systems of rules were elaborated to assure optimum social functioning. This placed further demands on linguistic systems, which have been elaborated by Parker (1985).

While the general frameworks proposed by Lock, Parker, Parisi, Peters, and Gibson are basically compatible in terms of ontogenetic sequences, they do differ considerably in the proposed time-frame—even among papers by the same authors. These divergent opinions result from a focus on differing aspects of the fossil and archaeological literature. A few years ago, many investigators considered learned vocalizations to be beyond the capacity of non-human primates. Moreover, it was thought that the ability to pronounce modern human vowels was not acquired until about 35 000 years ago (Lieberman 1983). These considerations seemed to suggest a very late emergence of vocal morphophonemic systems. More recent analyses, however, have suggested an earlier emergence of a modern laryngeal tract (Laitman 1983 and this volume, Editorial Appendix, Chapter 4) and have indicated a much greater vocal capacity in non-human primates than was previously supposed (Seyfarth and Cheney 1982). Hence it no longer seems necessary to postulate a late emergence of vocal languages and vowel-triangle sounds.

Palaeoneurological data can be interpreted as meaning that advances in brain size and in the degree of neural lateralization had already occurred by two million years ago (Falk 1980); while archaeological data suggest that, in terms of stone-tool manufacture, by 500 000 years ago *Homo erectus* was only relying on a mental capacity similar to that available to a seven-year-old modern child (Wynn 1979).[10] These data suggest the possibility of an early emergence of linguistic capability. Ultimately, however, the setting of a proper time-frame for linguistic evolution will demand improved methods of interpreting the fossil and archaeological records. The framework presented in this paper suggests that further knowledge of the ontogenetic interrelationships between tool-use, object-constructional skills, and linguistic development may be among the most important lines of research in terms of eventually unravelling the threads of language phylogeny. In particular, a detailed analysis of the maturation of human object-manipulation and linguistic skills, using the quantitative framework pioneered by Case, holds much promise in this regard, particularly if combined with a detailed analysis of stone tools from the same quantitative and hierarchical perspective. For if, as much modern work suggests, object-manipulation and language depend on the same cognitive substrates and mature in parallel, then, given sufficient understanding of the ontogenetic levels required for specific manipulative and linguistic tasks, I believe it should be possible to determine linguistic capabilities from prehistoric artefacts.

Admittedly, this view of the potential implications of stone tools for linguistic assessment is contrary to that expressed by Wynn (this volume, Chapter 10). These contrasting assessments result from the divergent approaches to tool-use utilized by Wynn and myself. Wynn does not utilize the ontogenetic and brain-size data presented here. Instead, his opinion basically hinges on a comparison of the learning of the highly specialized and complex tool-using traditions of blacksmithing, salmon fishing, and silversmithing with the protolinguistic behaviour of modern human infants. In the view of this author, these are 'apples to oranges' comparisons on several grounds. First, they focus on tool-using/-making traditions that involve metalworking and/or the accurate prediction of seasonal fish migrations. None of these traditions could possibly have been practised until the advent of the modern human brain, and they are therefore of questionable relevance to early hominid tool traditions and/or communicative skills.[11]

Secondly, Wynn's analysis ignores the earlier culturally-based trial-and-error activities that must have preceded the perfection of these tool-using, tool-making traditions. Instead, Wynn focuses on how individuals who presumably have already learned certain rudiments of tool-using, such as how to bang, cut, probe, etc., learn to expand and apply their skills in the replication of highly complex technological routines that have previously been perfected by others. Realistically, there can be only a few correct ways to perform such complex tasks; hence, imitation and the precise following of demonstrations in a 'string-of-beads' fashion become, of necessity, the method of choice. If instead, the analysis had focused on how these complex routines were originally invented and developed, the conclusions about the role of imitation, memorization, and 'string-of-beads' organization in tool-use would undoubtedly have been quite different.

In my view the proper linguistic analogy to the learning of complex, but pre-existing tool-using and tool-making routines such as blacksmithing is not the learning of an entire language by infants, as is assumed by Wynn, but the learning of an already-perfected linguistic routine such as the lines of a play, poem, or prayer by a child who already possesses a basic linguistic repertoire. Alternatively, the learning of an entire culturally-based tool-using repertoire should be compared to the learning of language: for example, the learning of the use of spoons, forks, knives, levers, scissors, hammers, etc. to the learning of words and grammar. This is, in fact, what Piaget, Case, and other cognitive developmental psychologists have tried to do. On the basis of these comparisons, tool-use and language do indeed seem to develop through similar pathways and involve similar cognitive skills.

In any event, as Wynn also points out, the unravelling of the pathways of language evolution will eventually require the elaboration of appropriate theoretical frameworks (Parker and Gibson 1982). It is the belief of this author that the neo-Piagetian developmental framework is one appropriate theoretical approach. Hopefully, this paper will encourage others to develop this line of thought.

Notes

1. Bailey *et al.* 1950; Bonin 1952; Bonin and Bailey 1947; Ettlinger 1984; Galaburda and Pandya 1982; Jones and Powell 1970; Krieg 1954; Passingham 1973.

2. Cain and Wadja 1979; Dewson and Burlingame 1975; Diamond *et al.* 1981; Falk 1978; Hamilton and Vermier 1982; LeMay and Geschwind 1975; LeMay *et al.* 1982; Nottebohm 1977.

3. Gibson 1988, 1990; Jerison 1973; Keith 1948; Krantz 1961; Passingham 1982.

4. Conel 1939–1967; Epstein 1974; Gibson 1970, 1991; Jacobsen 1963; Krogman 1972; Schultz 1949; Yakovlev and Lecours 1967.

5. There is some electroencephalographic, rather than neuroanatomical, evidence that is claimed to show a degree of relation to Piagetian stages. Matousek and Petersen (1973) report a large-sample, cross-sectional study of human electroencephalographic (EEG) relative power. Subsequent analyses of these values (Thatcher 1980; Hudspeth 1985; Epstein 1986) revealed a continuous growth function containing abrupt and significant increments in relative power in specific cortical areas; they also indicated that the ages at which these increments occurred overlapped Piagetian stages of development.

Subsequently, Thatcher *et al.* (1987) have reported data on EEG coherence and phase from a large sample aged from two months to early adulthood. These measures allow more specific measures of developing cortical functioning to be resolved. Thatcher *et al.* (1987, p. 1110) report on the basis of their data that:

The left and right hemispheres developed at different rates and with different post-natal onset times with the timing of growth spurts overlapping the timing of the major developmental stages described by Piaget.

They caution that the data do not lend themselves to a simple model that seeks to relate the development of psychological functioning to developmental differences in the maturation of the two cerebral hemispheres that might be simplistically related to apparently different modes of their functioning in adults. They do conclude, however, that:

the strength and specificity of the patterns in the data strongly favour the ontogenetic hypothesis of human cortical development in which there is a genetically programmed unfolding of specific corticocortical connections at relatively specific postnatal ages (ibid., p. 1113),

and that these 'overlap quite well with the timing for the Piaget theory of human cognitive development' (ibid.) [eds].

6. Bates 1976; Bates *et al.* 1979; Beilen 1975; Greenfield 1978; Greenfield *et al.* 1972; Kohlberg 1984; Lock 1983; Sinclair 1971; Slobin 1973.

7. Boesch and Boesch 1981, 1983, 1984*a, b*; Fouts 1973; Gardner and Gardner 1969, 1980; Jones and Sabater-Pi 1969; Kortlandt 1962, 1967, 1972; Kortlandt and Kooij 1963; McBeath and McGrew 1982; McGrew 1974; McGrew and Rodgers 1983; Miles 1983; Nishida and Hiraiwa 1982; Nishida and Uehara 1980; Savage-Rumbaugh *et al.* 1978; Silk 1978;

Telecki 1973, 1974, 1975; Van Lawick-Goodall 1968, 1970.

8. Seyfarth and Cheney 1982; Snowden 1982; Steklis 1985; Steklis and Raleigh 1979; Sutton 1979.

9. Campos and Welker 1976; Pimential-Sousa et al. 1980; Pubols and Pubols 1972; Welker and Carlson 1976; Welker and Seidenstein 1959; Welker et al. 1964, 1976.

10. Wynn, however, does not accept that he has ever made any comparison of the *mental capacity* or *intelligence* of *Homo erectus* with that of a modern seven-year-old child; such a comparison he would regard as methodologically unacceptable, preferring to speak in terms of the probable nature and organization of the skills apparently evidenced by the artefacts, and the possible comparability of these to those manifested at various developmental stages by modern infants [eds].

11. Wynn, however, would argue that if complex tool behaviours such as blacksmithing do not resemble language in their organization then presumably simpler tool behaviours do not do so either, pointing out that he does in fact also cite the simpler technologies of weaving and stone-knapping as further examples. He adds that even if tool behaviour and language did both use some of the same cognitive processes (long-term memory, for example) one could not therefore argue that evidence for the one is also evidence for the other: one could only argue for the presence of the general cognitive process as such, which may be a necessary, but is certainly not obviously a sufficient, condition for language development [eds].

References

Alegria, J. and Noirot, E. (1978). Neonate orientation behavior towards human voice. *International Journal of Behavioral Development*, **1**(4), 291–312.

Altman, J. (1969). Histological studies of postnatal neurogenesis 4. Cell proliferation and migration in the anterior forebrain. *Journal of Comparative Neurology*, **137**, 433–58.

Armstrong, D. F. (1983). Iconicity, arbitrariness, and duality of patterning in signed and spoken languages: perspectives on language evolution. *Sign Language Studies*, **38**, 51–69.

Armstrong, D. F. (1985). Commentary. *Current Anthropology*, **26**, 626–7.

Armstrong, E. (1982). Mosaic evolution in the primate brain: differences and similarities in the hominid thalamus. In *Primate brain evolution* (ed. E. Armstrong and D. Falk), pp. 131–62. Plenum Press, New York.

Armstrong, E. (1986). Enlarged limbic structures in the human brain: the anterior thalamus and medial mammillary body. *Brain Research*, **362**, 394–7.

Bailey, P., Bonin, G. von, and McCulloch, W. S. (1950). *The isocortex of the chimpanzee*. University of Illinois Press, Urbana, Illinois.

Baldwin, J. M. (1894). *The development of the child and of the race*. Macmillan, New York.

Ball, W. and Vurpillot, E. (1981). Action and perception of displacement in infancy. In *Infancy and epistemology* (ed. G. Butterworth), pp. 115–36. Harvester Press, Brighton, England.

Bates, C. and Killackey, H. (1985). The organization of the neonatal rat's brainstem trigeminal complex and its role in the formation of central trigeminal patterns. *Journal of Comparative Neurology*, **240**, 265–87.

Bates, E. (1976). *Language and context*. Academic Press, New York.

Bates, E., Benigni, L., Bretherton, I., Camaioni, L., and Volterra, V. (1979). *The emergence of symbols: communication and cognition in infancy*. Academic Press, New York.

Beilin, H. (1975). *Studies on the cognitive basis of language development*. Academic Press, New York.

Binford, L. R. (1982). Commentary on R. White, 'Rethinking the Middle/Upper Paleolithic transition'. *Current Anthropology*, **23**, 177–81.

Bloom, F. E. and Aghajanian, G. K. (1968). Fine structural and cytochemical analysis of the staining of synaptic junctions with phosphotungstic acid. *Journal of Ultrastructural Research*, **22**, 361–75.

Boesch, C. and Boesch, H. (1981). Sex differences in the use of natural hammers by wild chimpanzees: a preliminary report. *Journal of Human Evolution*, **10**, 585–93.

Boesch, C. and Boesch, H. (1983). Optimisation of nut cracking with natural hammers in chimpanzees. *Behavior*, **83**, 265–86.

Boesch, C. and Boesch, H. (1984a). Possible causes of sex differences in the use of natural hammers by wild chimpanzees. *Journal of Human Evolution*, **13**, 415–40.

Boesch, C. and Boesch, H. (1984b). Mental map in wild chimpanzees: an analysis of hammer transports for nut cracking. *Primates*, **25**, 160–70.

Bonin, G. von (1952). Notes on cortical evolution. *Archives of Neurology and Psychiatry* (London), **67**, 135–43.

Bonin, G. von and Bailey, P. (1947). *The neocortex of Macaca mulatta*. University of Illinois Press, Urbana, Illinois.

Bower, T. G. R. (1972). Object perception in infancy. *Perception*, **1**, 15–30.

Bower, T. G. R. (1974). *Development in infancy*. W. H. Freeman, New York.

Bower, T. G. R., Broughton, J. M., and Moore, M. K. (1970). Demonstration of intention in the reaching behavior of neonate humans. *Nature*, **228**, 679–81.

Brainerd, C. J. (1979). Recapitulationism, Piaget, and the evolution of intelligence: déjà vu. *The Behavioral and Brain Sciences*, **2**, 381–2.

Brown, R. (1973). *A first language: the early stages*. Harvard University Press Cambridge, Mass.

Butterworth, G. and Castillo, M. (1976). Coordination of auditory and visual space in new-born human infants. *Perception*, **5**, 155–60.

Cain, D. P. and Wadja, J. A. (1979). An anatomical asymmetry in the baboon brain. *Brain, Behavior and Evolution*, **15**, 222–6.

Campos, G. B. and Welker, W. I. (1976). Comparisons between brains of a large and a small hystricomorph rodent: capybara, *Hydrochoerus* and guinea pig, *Cavia*. *Brain, Behavior and Evolution*, **13**, 243–6.

Case, R. (1985). *Intellectual development: birth to adulthood*. Academic Press, London.

Conel, J. L. (1939–1967). *The postnatal development of the human cerebral cortex*, Vols. 1–8. Harvard University Press, Cambridge, Mass.

Count, E. W. (1947). Brain and body weight in man: their antecedents in growth and evolution. *Annals of the New York Academy of Sciences*, **46**, 993–1122.

De Waal, F. (1983). *Chimpanzee politics*. Harper and Row, New York.

Dewson, J. H. and Burlingame, A. C. (1975). Auditory discrimination and recall in monkeys. *Science*, **187**, 267–8.

Diamond, M. C., Dowling, G. A., and Johnson, R. E. (1981). Morphological cerebral cortical asymmetry in male and female rats. *Experimental Neurology*, **71**, 261–8.

Dreyfus, H. L. and Dreyfus, S. E. (1986). *The power of human intuition and expertise in the era of the computer*. Macmillan, New York.

Duffy, F. H. and Buschfiel, J. L. (1971). Somatosensory system: organizational hierarchy from single units in monkey Area 5. *Science*, **172**, 273–5.

Ebbesson, S. O. E. (1984). Evolution and ontogeny of neural circuits. *The Behavioral and Brain Sciences*, **7**, 321–66.

Eimas, P. (1985). The perception of speech in early infancy. *Scientific American*, **252**, 46–53.

Epstein, H. T. (1974). Phrenoblysis: special brain and mind growth periods. I. Human brain and skull development. *Developmental Psychobiology*, **7**, 207–16.

Epstein, H. T. (1986). Stages in human brain development. *Developments in Brain Research*, **30**, 114–19.

Ettlinger, G. (1984). Humans, apes and monkeys: the changing neuropsychological viewpoint. *Neuropsychologia*, **22**, 685–96.

Falk, D. (1978). Cerebral asymmetry in monkeys. *Acta Anatomica*, **101**, 195–9.

Falk, D. (1980). Language, handedness. and primate brains: did the Australopithecines sign? *American Anthropologist*, **82**, 71–8.

Falk, D. (1985). Hadar AL 162–28 endocast as evidence that brain enlargement preceded cortical reorganization in hominid evolution. *Nature*, **313**, 45–7.

Flechsig, P. (1920). *Anatomie des menschlichen Gehirns und Rueckenmarks auf myelogenetishcher Grundlage*. George Thomas, Liepzig.

Fouts, R. S. (1973). Acquisition and testing of gestural signs in four young chimpanzees. *Science*, **180**, 973–80.

Fox, M. W. (1971). *Integrative development of brain and behavior in the dog*. University of Chicago Press.

Gabriel, R. S. (1980). Malformations of the central nervous system. In *Textbook of child neurology* (ed. J. H. Menkes), pp. 161–237. Lea and Febiger, Philadelphia.

Galaburda, A. M. and Pandya, D. N. (1982). Role of architectonics and connections in primate brain evolution. In *Primate brain evolution* (ed. E. Armstrong and D. Falk), pp. 203–16. Plenum Press, New York.

Gardner, R. A. and Gardner, B. T. (1969). Teaching sign language to a chimpanzee. *Science*, **165**, 664–72.

Gardner, R. A. and Gardner, B. T. (1980). Comparative psychology and language acquisition. In *Speaking of apes* (ed. T. A. Sebeok and J. Umiker-Sebeok), pp. 287–330. Plenum Press, New York.

Geschwind, N. (1965). Disconnection syndromes in animals and man. *Brain*, **88**, 237–94.

Gibson, K. R. (1970). Sequence of myelinization in the brain of *Macaca mulatta*. Ph.D. dissertation, University of California, Berkeley.

Gibson, K. R. (1977). Brain structure and intelligence in macaques and human infants from a Piagetian perspective. In *Primate biosocial development* (ed. S. Chevalier-Skolnikoff and F. Poirer), pp. 113–57. Garland Press, New York.

Gibson, K. R. (1978). Cortical maturation: an antecedent of Piaget's behavioral stages. *The Behavioral and Brain Sciences*, **1**, 188.

Gibson, K. R. (1981). Comparative neuroontogeny; its implications for the development of human intelligence. In *Infancy and epistemology* (ed. G. Butterworth), pp. 52–83. Harvester Press, Brighton, England.

Gibson, K. R. (1983). Comparative neuroontogeny and the constructionist approach to the evolution of the brain, object manipulation and language. In *Glossogenetics: the origin and evolution of language* (ed. E. de Grolier), pp. 37–62. Harwood Academic Publishers, Paris.

Gibson, K. R. (1985). Has the evolution of intelligence stagnated since Neanderthal Man? In *Evolution and developmental psychology* (ed. G. Butterworth, J. Rutkowska, and M. Scaife), pp. 102–14. The Harvester Press, Brighton, England.

Gibson, K. R. (1986). Comparative brain myelination in relationship to concepts of primate ontogenetic and cognitive evolution. *Primate Report*, **14**, 181.

Gibson, K. R. (1988). Brain size and the evolution of language. In *The genesis of language: a different judgement of evidence* (ed. M. Landsberg), pp. 149–72. Mouton Press, The Hague.

Gibson, K. R. (1990). New perspectives on instinct and intelligence: brain size and the emergence of hierarchical constructional skills. In *"Language" and intelligence in monkeys and apes: comparative developmental perspectives* (ed. S. T. Parker and K. R. Gibson), pp. 97–128. Cambridge University Press.

Gibson, K. R. (1991). Myelination and behavioral development: a comparative perspective on questions of neoteny, altriciality and intelligence. In *Brain maturation and cognitive development: comparative and cross-cultural perspectives* (ed. K. R. Gibson and A. C. Petersen), pp. 29–64. Aldine de Gruyter, Hawthorne, NY.

Glaser, R. (1984). Education and thinking: the role of knowledge. *American Psychologist*, **39**, 93–104.

Goodall, J. (1986). *The chimpanzees of Gombe*. Harvard University Press, Cambridge, Mass.

Gould, S. J. (1977). *Ontogeny and phylogeny*. Harvard University Press, Cambridge, Mass.

Greenfield, P. (1978). Structural parallels between language and action in development. In *Action, gesture, and symbol* (ed. A. Lock), pp. 415–45. Academic Press, New York.

Greenfield, P. M., Nelson, K., and Salzman, E. (1972). The development of rule-bound strategies for manipulating seriated cups: a parallel between action and grammar. *Cognitive Psychology*, **3**, 291–310.

Gross, C. G., Roche-Miranda, C. E., and Bender, D. B. (1972). Visual properties of neurons in the inferotemporal cortex of the macaque. *Journal of Neurophysiology*, **35**, 96–111.

Haeckel, E. (1874). *Anthropogenie: Keimes-und Stammes-Geschichte des Menschen*. W. Engelmann, Leipzig.

Hamilton, C. R. and Vermier, B. A. (1982). Hemispheric differences in split-brain monkeys learning sequential comparisons. *Neuropsychologia*, **20**, 691–8.

Hofman, M. A. (1983). Encephalization in hominids: evidence for the model of punctuationalism. *Brain, Behavior and Evolution*, **22**, 102–17.

Holloway, R. L. (1966). Cranial capacity, neural reorganization, and hominid evolution: a search for more suitable parameters. *American Anthropologist*, **68**, 103–21.

Holloway, R. L. (1968). The evolution of the primate brain: some aspects of quantitative relationships. *Brain Research*, **7**, 121–72.

Holloway, R. L. (1969). Culture, a human domain. *Current Anthropology*, **10**, 395–412.

Holloway, R. L. (1972). Australopithecine endocasts, brain evolution in the Hominoidea, and a model of hominid evolution. In *The functional and evolutionary biology of primates* (ed. R. Tuttle), pp. 185–204. Aldine Press, Chicago.

Holloway, R. L. (1979). Brain size, allometry and reorganization: a synthesis. In *Development and evolution of brain size* (ed. M. E. Hahn, B. C. Dudek, and C. Jensen), pp. 59–88. Academic Press, New York.

Holloway, R. L. (1985). The poor brain of *Homo sapiens neanderthalensis*: see what you please. In *Ancestors: the hard evidence* (ed. E. Delson), pp. 319–24. Alan Liss, New York.

Holloway, R. L. and Post, D. (1982). The relativity of relative brain size measures and hominid evolution. In *Primate brain evolution, methods and concepts* (ed. E. Armstrong and D. Falk), pp. 57–76. Plenum Publishing Corporation, New York.

Holt, A. B., Cheek, D. B., Mellets, E. D., and Hill, D. E. (1975). Brain size and the relation of the primate to the nonprimate. In *Fetal and postnatal cellular growth: hormones and nutrition* (ed. D. B. Cheek), pp. 23–44. John Wiley, New York.

Hubel, D. H. and Wiesel, T. N. (1968). Receptive fields and functional architecture of monkey striate cortex. *Journal of Physiology*, **195**, 215–43.

Hudspeth, W. J. (1985). Developmental neuropsychology: functional implications of quantitative EEG maturation. *Journal of Clinical and Experimental Neuropsychology*, **7**, 606.

Huttenlocher, P. R. (1979). Synaptic density in human frontal cortex—developmental changes and effects of aging. *Brain Research*, **163**, 195–205.

Jacobsen, S. (1963). Sequence of myelinization in the brain of the albino rat: A: cerebral cortex, thalamus and related structures. *Journal of Comparative Neurology*, **121**, 5–29.

Jacobsen, M. (1978). *Developmental neurobiology*. Plenum Press, New York.

Jerison, H. J. (1973). *Evolution of the brain and intelligence*. Academic Press, New York.

Jerison, H. J. (1979). The evolution of diversity in brain size. In *Development and evolution of brain size: behavioral implications* (ed. M. Hahn, C. Jensen, and B. Dudek), pp. 29–57. Academic Press, New York.

Jerison, H. J. (1982). Allometry, brain size, cortical surface, and convolutedness. In *Primate brain evolution* (ed. E. Armstrong and D. Falk), pp. 77–84. Plenum Press, New York.

Jones, C. and Sabater Pi, J. (1969). Sticks used by chimpanzees in Rio Muni, West Africa. *Nature*, **223**, 100–1.

Jones, E. G. and Powell, T. P. S. (1970). An anatomical study of converging sensory pathways within the cerebral cortex of the monkey. *Brain*, **93**, 793–820.

Karmiloff-Smith, A. (1979). *A functional approach to child language*. Cambridge University Press.

Keith, A. (1948). *A new theory of human evolution*. Watts, London.

Kennard, M. A. and Willner, M. D. (1941). Weights of brains and organs of 132 new and old world monkeys. *Endocrinology*, **28**, 937–84.

Kerr, G. R., Kennan, H. A., Waisman, H. A., and Allen, J. R. (1969). Growth and development of the fetal rhesus monkey. I. Physical growth. *Growth*, **33**, 201–15.

Killackey, H. P. (1979). Peripheral influences on connectivity in the developing rat trigeminal system. In *The developmental neurobiology of vision* (ed. R. Freeman and W. Singer), pp. 381–90. Plenum Press, New York.

Kochetkova, V. I. (1978). *Paleoneurology*. V. H. Winston, Washington, DC.

Kohlberg, L. (1984). *The moral development of the child:* Vol. II *the psychology of moral development*. Harper and Row, San Francisco.

Kortlandt, A. (1962). Chimpanzees in the wild. *Scientific American*, **206**, 128–38.

Kortlandt, A. (1967). Experimentation with chimpanzees in the wild. In *Neue Ergebnisse der Primatologie* (ed. D. Starck, R. Schneider, and H. J. Kuhn), pp. 208–24. Gustav Fischer, Stuttgart.

Kortlandt, A. (1972). *New perspectives on ape and human evolution*. Stichting vor Psychobiologie, Amsterdam.

Kortlandt, A. and Kooij, M. (1963). Protohominid behavior in primates. *The Primates—Symposium of the Zoological Society of London*, **10**, 61–88.

Krantz, G. (1961). Pithecanthropine brain size and its cultural consequences. *Man*, **103**, 85–7.

Krieg, W. J. S. (1954). *Connections of the frontal lobe of the monkey*. Thomas, Springfield, Illinois.

Krogman, W. M. (1972). *Child growth*. University of Michigan Press, Ann Arbor, Michigan.

Kuypers, H. G. J. (1962). Corticospinal connections: postnatal development in the Rhesus monkey. *Science*, **138**, 678–80.

Laitman, J. T. (1983). The evolution of the hominid upper respiratory system and implications for the origins of speech. In *Glossogenetics: the origin and evolution of language* (ed. E. de Grolier), pp. 63–90. Harwood Academic Publishers, Paris.

Lamendella, J. T. (1976). Relations between the ontogeny and phylogeny of language: a neorecapitulationist view. In *Annals of the New York Academy of Sciences* (ed. S. R. Harnard, H. D. Steklis, and J. Lancaster), **280**, 396–412.

Langworthy, O. L. (1929a). A correlated study of the development of reflex activity in fetal and young kittens and the myelinization of tracts in the nervous system. *Carnegie Institute of Washington, Contributions to Embryology*, **20**, 127–72.

Langworthy, O. L. (1929b). Histological development of cerebral motor areas in young kittens correlated with their physiological reaction to electrical stimulation. *Carnegie Institute of Washington, Contributions to Embryology*, **20**, 37–52

LeMay, M. and Geschwind, N. (1975). hemispheric differences in the brains of great apes. *Brain, Behavior and Evolution*, **11**, 48–52.

LeMay, M., Billig, M. S., and Geschwind, N. (1982). Asymmetries of the brains and skulls of nonhuman primates. In *Primate brain evolution: methods and concepts* (ed. E. Armstrong and D. Falk), pp. 263–78. Plenum Press, New York.

Lerner, R. M. (1984). *On the nature of human plasticity*. Cambridge University Press, New York.

Lewin, R. (1986). Anthropologist argues that language cannot be read in stones. *Science*, **233**, 23–4.

Lieberman, P. (1983). On the nature and evolution of the biological foundations of human language. *Glossogenetics the origin and evolution of language* (ed. E. de Grolier) pp. 91–114. Harwood Academic Publishers, New York.

Lock, A. (1983). "Recapitulation" in the ontogeny and phylogeny of language. In *Glossogenetics: the origin and evolution of language* (ed. E. de Grolier), pp. 255–73. Harwood Academic Publishers, Paris.

Lock, A. (1985). Commentary. *Current Anthropology*, **26**, 628–9.

Luria, A. (1966). *Higher cortical functions in man*. Basic Books, New York.

McBeath, N. M. and McGrew, W. C. (1982). Tools used by wild chimpanzees to obtain termites at Mt. Assirik, Senegal: the influence of habitat. *Journal of Human Evolution*, **11**, 65–72.

McGrew, W. C. (1974). Tool use by wild chimpanzees in feeding upon driver ants. *Journal of Human Evolution*, **3**, 501–8.

McGrew, W. C. and Rodgers, M. E. (1983). Chimpanzees, tools and termites: new record from Gabon. *American Journal of Primatology*, **5**, 171–4.

McGuire, O. S. (1966). Population studies of the ostracod genus *Polytylites* from the Chester series. *Journal of Paleontology*, **40**, 883–910.

Marshack, A. (1972). Cognitive aspects of Upper Paleolithic engraving. *Current Anthropology*, **13**, 445–77.

Matousek, M. and Petersen, I. (1973). Frequency analysis of the EEG in normal children and adolescents. In *Automation of electroencephalography* (ed. P. Kellaway and I. Petersen), pp. 75–102. Raven Press, New York.

Meltzoff, A. N. and Borton, R. W. (1979). Intermodal matching by human neonates. *Nature*, **282**, 403–4.

Meltzoff, A. N. and Moore, M. K. (1977). Imitation of facial and manual gestures by human infants. *Science*, **205**, 217–19.

Miles, H. L. (1983). Two-way communication with apes and the evolution of language. In *Glossogenetics: the origin and evolution of language* (ed. E. de Grolier), pp. 201–10. Harwood Academic Publishers, Paris.

Mounoud, P. and Vinter, A. (1981). Representation and sensorimotor development. In *Infancy and epistemology* (ed. G. Butterworth), pp. 200–35. The Harvester Press, Brighton, England.

Mountcastle, V. (1978). An organizing principle for cerebral function: the unit module and the distributed system. In *The mindful brain* (ed. G. M. Edelmann and V. B. Mountcastle), pp. 7–50. MIT Press, Cambridge, Mass.

Newell-Morris, L. and Fahrenbruch, C. E. (1985). Practical and evolutionary considerations for use of the nonhuman primate model in prenatal research. In *Nonhuman primate models for human growth and development* (ed. E. S. Watts), pp. 9–40. Alan Liss, New York.

Nishida, T. and Hiraiwa, M. (1982). Natural history of a tool-using behavior by wild chimpanzees in feeding upon wood-boring ants. *Journal of Human Evolution*, **11**, 73–99.

Nishida, T. and Uehara, S. (1980). Chimpanzees, tools and termites: another example from Tanzania. *Current Anthropology*, **21**, 671–2.

Northcutt, R. G. (1984). Parcellation: the resurrection of Hartsoeker and Haeckel. *The Behavioral and Brain Sciences*, **7**, 345.

Nottebohm, F. (1977). Asymmetries in neural vocal control of the canary. In *Lateralization in the nervous system* (ed. S. Harnard, R. W. Doty, L. Goldstein, J. Jaynes, and G. Krauthamer), pp. 23–44. Academic Press, New York.

Parisi, D. (1983). A three-stage model of language evolution: from pantomime to syntax. In *Glossogenetics: the origin and evolution of language* (ed. E. de Grolier), pp. 419–34. Harwood Academic Publishers, Paris.

Parker, S. T. (1985). A social–technological model for the evolution of language. *Current Anthropology*, **26**, 617–39.

Parker, S. T. and Gibson, K. R. (1979). A model of the evolution of language and intelligence in early hominids. *The Behavioral and Brain Sciences*, **2**, 367–407.

Parker, S. T. and Gibson, K. R. (1982). The importance of theory for reconstructing the evolution of language and intelligence in early hominids. *Advanced views in primate biology* (ed. A. B. Chiarelli and R. S. Corruccini), pp. 42–62. Springer-Verlag, Berlin–New York.

Passingham, R. E. (1973). Anatomical differences between the neocortex of man and other primates. *Brain, Behavior, and Evolution*, 7, 337–59.

Passingham, R. E. (1975). Changes in the size and organization of the brain in man and his ancestors. *Brain, Behavior and Evolution*, 11, 73–90.

Passingham, R. E. (1982). *The human primate*. W. H. Freeman, Oxford.

Passingham, R. E. and Ettlinger, G. (1975). A comparison of cortical function in man and other primates. *International Review of Neurobiology*, 16, 233–301.

Peters, C. (1974). On the possible contribution of ambiguity of expression to the development of protolinguistic performance. In *Language origins* (ed. R. W. Wescott) pp. 83–102. Linstok Press, Silver Springs, MD.

Piaget, J. (1952). *The origins of intelligence in children*. W. W. Norton, New York.

Piaget, J. (1954). *The construction of reality in the child*. Basic Books, New York.

Piaget, J. (1955). *The language and thought of the child*. World Publishing Company, Cleveland, New York.

Piaget, J. (1969). Genetic epistemology. *Columbia Forum*, 12, 4–11.

Piaget, J. (1978). *Behavior and evolution*. Pantheon, New York.

Pimential-Souza, F., Cosanza, G., Campos, G. B., and Johnson, J. J. (1980). Somatic sensory cortical regions of the agouti. *Brain, Behavior and Evolution*, 17, 218–40.

Prideaux, T. (1973). *CroMagnon man*. Time–Life Books, New York.

Pubols, B. H. and Pubols, L. M. (1972). Neural organization of somatic sensory representation in the spider monkey. *Brain, Behavior and Evolution*, 5, 342–66.

Rader, N. (1985). Change and variation: on the importance of heterochrony for development. In *Evolution and developmental psychology* (ed. G. Butterworth, J. Rutkowska, and M. Scaife), pp. 22–9. The Harvester Press, Brighton, England.

Rakic, P. (1985). Limits of neurogenesis in primates. *Science*, 227, 1054–6.

Rakic, P., Bourgeois, J. P., Eckenhoff, M. F., Zecevic, N., and Goldman-Rakic, P. (1986). Concurrent overproduction of synapses in diverse regions of the primate cerebral cortex. *Science*, 232, 232–5.

Rensch, B. (1959). *Evolution above the species level*. Columbia University Press, New York.

Reynolds, P. C. (1981). *On the evolution of human behavior*. University of California Press, Berkeley and Los Angeles.

Reynolds, P. C. (1982). The primate constructional system: the theory and description of instrumental object use in humans and chimpanzees. In *The analysis of action* (ed. M. Von Cranach and R. Harré), pp. 343–83. Cambridge University Press.

Reynolds, P. C. (1983). Ape constructional ability and the origin of linguistic structure. In *Glossogenetics: the origin and evolution of language* (ed. E. de Grolier), pp. 185–200. Harwood Academic Publishers, Paris.

Rutkowska, J. (1985). Does the phylogeny of conceptual development increase our understanding of concepts or of development? In *Evolution and developmental psychology* (ed. G. Butterworth, J. Rutkowska, and M. Scaife), pp. 115–29. The Harvester Press, Brighton, England.

Sacher, G. A. and Staffeldt, E. F. (1974). Relation of gestation time to brain weight for placental mammals: implications for the theory of vertebrate growth. *American Naturalist*, 108, 593–615.

Savage-Rumbaugh, S., Rumbaugh, D., and Boysen, S. (1978). Symbolic communication between two chimpanzees (*Pan troglydytes*). *Science*, 201, 641–4.

Schultz, A. H. (1941). The relative size of the cranial capacity in primates. *American Journal of Physical Anthropology*, 28, 273–87.

Schultz, A. H. (1949). Ontogenetic specializations of man. *Archiv Julius Klaus-Stiftung*, 24, 197–216.

Seyfarth, R. M. and Cheney, D. (1982). How monkeys see the world: a review of recent research on East African vervet monkeys. In *Primate communication* (ed. C. T. Snowden, C. H. Brown, and M. R. Petersen), pp. 239–52. Cambridge University Press.

Silk, J. (1978). Patterns of food-sharing among mother and infant chimpanzees of the Gombe Stream National Park in Tanzania. *Folia Primatologica*, 29, 129–41.

Sinclair, H. (1971). Sensorimotor action patterns as a condition for the acquisition of syntax. In *Language acquisition: models and methods* (ed. R. Huxley and E. Ingram), pp. 121–35. Oxford University Press, New York.

Slobin, D. I. (1973). Cognitive prerequisites for the development of grammer. In *Studies of child language development* (ed. C. A. Ferguson and D. I. Slobin), pp. 175–276. Holt, New York.

Snowden, C. T. (1982). Linguistic and psycholinguistic approaches to primate communication. In *Primate communication* (ed. C. T. Snowden, C. H. Brown, and M. R. Peterson), pp. 212–38. Academic Press, New York.

Steklis, H. D. (1985). Primate communication, comparative neurology and the origin of language reexamined. *Journal of Human Evolution*, 14, 157–73.

Steklis, H. D. and Raleigh, M. (1979). *Neurobiology of social communication in primates*. Academic Press, New York.

Sutton, D. (1979). Mechanisms underlying vocal control in nonhuman primates. In *Neurobiology of social communication in primates* (ed. H. Steklis and M. Raleigh), pp. 45–68. Academic Press, New York.

Telecki, G. (1973). *The predatory behavior of wild chimpanzees*. Bucknell University Press, Lewisburg, Pennsylvania.

Telecki, G. (1974). Chimpanzee subsistance technology: materials and skills. *Journal of Human Evolution*, 3, 575–94.

Telecki, G. (1975). Primate subsistance patterns: collector–predators and gatherer–hunters. *Journal of Human Evolution*, **4**, 125–84.

Terrace, H. S. (1979). *Nim: a chimpanzee who learned sign language*. Knopf, New York.

Terrace, H. S. (1980). Is problem-solving language? In *Speaking of apes* (ed. T. A. Sebeok and J. Umiker-Sebeok), pp. 385–405. Plenum Press, New York.

Thatcher, R. W. (1980). Neurolinguistics: theoretical and evolutionary perspectives. *Brain and Language*, **11**, 235–60.

Thatcher, R. W., Walker, R. A., and Giudice, S. (1987). Human cerebral hemispheres develop at different rates and ages. *Science*, **236**, 1110–13.

Tobias, P. V. (1971). *The brain in hominid evolution*. Columbia University Press, New York.

Trinkaus, E. and LeMay, M. (1982). Occipital bunning among later Pleistocene hominids. *American Journal of Physical Anthropology*, **57**, 27–36.

van Lawick-Goodall, J. (1968). The behaviour of free-ranging chimpanzees in the Gombe Stream Reserve. *Animal Behaviour Monographs*, **1**, 161–311.

van Lawick-Goodall, J. (1970). Tool-using in primates and other vertebrates. In *Advances in the study of behavior* (ed. D. S. Lehrman, R. A. Hinde, and E. Shaw), Vol. 3, pp. 195–248. Academic Press, New York.

Von Baer, K. E. (1828). *Entwicklungsgeschichte der Thiere: Beobachtung und Reflexion*. Borntraeger, Königsberg.

Watts, E. S. (1986). Evolution of the human growth curve. In *Human growth*, Vol. 1, (ed. F. Falkner and J. M. Tanner), pp. 153–66. Plenum Press, New York.

Weidenreich, F. (1941). The brain and its role in the phylogenetic transformation of the human skull. *Transactions of the American Philosophical Society*, **31**, 321–42.

Welker, W. I. and Carlson, M. (1976). Somatic sensory cortex of hyrax (*Procavia*). *Brain, Behavior and Evolution*, **13**, 294–301.

Welker, W. I. and Seidenstein, S. (1959). Somatic sensory representation in the cerebral cortex of the raccoon (*Procyon lotor*). *Journal of Comparative Neurology*, **111**, 469–501.

Welker, W. I., Johnson, J. I., and Pubols, B. H. (1964). Some morphological and physiological characteristics of the somatic sensory system in raccoons. *American Zoologist*, **4**, 75–94.

Welker, W. I., Adrian, H. O., Lifschutz, W., Kaulen, R., Caviedes, E., and Gutman, W. (1976). Somatic sensory cortex of llama (*Lama glama*). *Brain, Behavior, and Evolution*, **13**, 184–293.

Wertheimer, M. (1961). Psychomotor coordination of auditory and visual space at birth. *Science*, **134**, 1692.

Wishart, J. G., Bower, T. G. R., and Dunkeld, J. (1978). Reaching in the dark. *Perception*, **7**, 507–12.

Wynn, T. G. (1979). The intelligence of later Acheulean hominids. *Man*, **14**, 371–91.

Yakovlev, P. I. and Lecours, A. R. (1967). The myelenogenetic cycles of regional maturation of the brain. In *Regional development of the brain in early life* (ed. A. Minkowski), pp. 3–70. Blackwell, Oxford.

15
Early interaction and cognitive skills: implications for the acquisition of culture

David Messer and Glyn Collis

Abstract

The precise origin of infants' social powers is the subject of conflicting views. However, there is general agreement that infants are attracted to the physical and behavioural characteristics of people, and that such capacities are likely to be the product of evolutionary processes. In these terms infants appear to have a basic social disposition which is part of our evolutionary heritage. However, this social disposition does not appear to extend to the way that infants are able to contribute to the structure of social activities in which they engage. As illustrated by studies of gaze and vocalization, infants are not full partners in the interactive process; rather interaction is structured by Western adults to appear as a co-ordinated interpersonal process.

We adopt the view that social interaction between young infants and primary caregivers provides the basis for the formation of relationships, particularly attachments. There may be considerable variability in the way these relationships are formed, given the diversity of child-rearing patterns across different cultures; but it seems likely that all relationships involve infants having quite sophisticated representations of their caregivers. As infants become older the continuing social interaction with their caregivers increases infants' social skills, so that communication with other members of the culture becomes more efficient and effective.

Social interaction with caregivers makes available, in addition, various forms of information which can be utilized to assist cognitive development. Adult behaviour provides a model for infant activities, and by the end of the first year infants are capable of imitating a range of activities. Social interaction also contains a variety of forms of information which co-ordinate the interests of infants and adults. Such procedures effectively highlight culturally appropriate objects and events, thereby promoting a shared understanding between adult and child. Cross-cultural studies have done much to call into question the idea that certain forms of social interaction common in Western societies provide an essential basis for language acquisition; and it is usually assumed that social interaction does not have a direct relevance to the acquisition of syntactic abilities. However, cross-cultural studies also strongly point to the way that the pattern of early social interaction is influenced by the characteristics of a culture. As a result infants, before they speak, are able to tune in to the values, procedures, and assumptions that are present in their culture. Furthermore, one should not forget that for older children social interaction increases the power of interpersonal activities to promote cognitive skills. Two important procedures that occur in such circumstances are the way that discussion can promote cognitive change, and the way that effective communication with adults appears to facilitate general cognitive development.

In this chapter we argue that participation in social interaction provides a basis for relationships with care-givers; these relationships in turn facilitate the acquisition of social skills necessary for interaction, and help to develop a culturally based perception in infants of salient aspects of their environment. It is in such social circumstances that symbols and language are employed; and this provides a crucial step towards children becoming full members of their culture. This is because the use of symbols and language provides the means for children to move beyond their first relationships to communicate about complex issues with members of their culture who are less familiar to them.

15.1 Introduction

Two hallmarks of the human species are the degree to which use is made of symbols when employing sophisticated communicative and cognitive skills, and the way that culture has emerged, presumably as a consequence of these skills. Human language, intellect, and culture are not without parallels; but it is clear that humans have achieved a degree of control and knowledge of their environment which is not matched by other species.

Current ideas about evolutionary pressures which could have provoked these achievements focus primarily on the increasing complexity of social organization that is possible with sophisticated communicative and cognitive processes (for example, Humphrey 1976), and the selective advantages of complex and flexible social organization. It is unlikely that a single selective pressure has been responsible for these developments. Take the example of attachment, an emotional bond between care-givers and children (see Section 15.3). In this case, theory has emphasized the importance of attachment in ensuring the safety of the young. However, the advantages of developing this special kind of relationship are likely to go beyond ensuring the safety of the child. As we will argue, this relationship is most effective when it is supported by sophisticated socio-cognitive processes, which in turn may be utilized to provide the basis for further increases in the complexity of communication. Thus, certain selective pressures may have provoked the initial development of skills beyond the level of related species. However, once these skills had emerged they were probably put to further uses, being refined and differentiated by new pressures which were quite unrelated to the original ones.

Our main task in this chapter is to examine the ontogeny rather than the phylogeny of communicative and cognitive processes. None the less, both ontogeny and phylogeny are pieces of a single, larger story, and our understanding of each can be enhanced by a consideration of the other. Despite being an attractive idea, it is now well understood that ontogeny need not, and often does not, recapitulate phylogeny. Indeed, evolutionary changes in the *scheduling* of development, which would disturb any simple relationship between an ontogenetic and phylogenetic sequence, can be seen as important aspects of evolutionary change and adaptation. An example is the idea that humans have an extended period of immaturity, and that it is this that has made cognitive sophistication possible (see Bates *et al.* 1979; Bruner 1972; and Gibson, this volume, Chapter 14, for a discussion of these ideas).

There are traps awaiting us when we seek to understand the adaptive value of cognitive and behavioural processes that occur in infancy and early childhood. Are we looking at adaptations which have the concurrent effect of enhancing the survival of children, or are we looking at processes whose primary adaptive value is developmental in nature, such that effects on fitness will not become apparent until later in life? To illustrate the two possibilities, one could argue that the language of the two-or three-year-old conveys various selective advantages to children at that age. Alternatively, one could argue that the advantage of language only comes later, but it needs to start at an early age because of the time it takes to develop to the point of being adaptively useful: after all, social relationships in infancy and childhood are relatively simple and affective in nature, and do not seem to require the full subtlety of linguistic communication to work effectively.

There is also the conceptual trap that, even if we can establish that some attribute or process is the product of biological evolution, presumably having involved genetic change, it does not follow that the precise expression of the attribute is directly governed by genetic material. For example, we may accept it as adaptive for a duckling to be able to distinguish its mother from other adult female ducks, and members of its own species from those of another species; yet experience plays an important role in the development of these capacities, through the mechanism known as imprinting. Thus it is perfectly possible for evolution to utilize non-genetic processes in the development of a favoured outcome, provided that these processes work with a high probability of success.

There is a further subtlety in the general point: evolution involves developmental processes that are neither genetic not environmental, nor some simple interaction between the two. Instead, functional constraints imposed by the logical, mathematical, or physical properties of the developmental task can be important determinants of development. Bates and her colleagues (Bates *et al.* 1979; Bates and McWhinney 1982) discuss this point at some length in the context of a constructivist view of psychological development. They see, for example, the emergence of grammar as an inevitable solution to the task of encoding and decoding information under the linear-order constraint imposed by the acoustic–articulatory channel of communication. Genetic processes may be primarily reponsible for humans' adopting this channel of communication; but they are not responsible for the fact that a temporal coding system that is hierachical in nature is the most efficient method of transmitting complex information over this channel. It is this latter principle that guides language development down the pathway to grammar. Bates *et al.* (1979) do not have in mind a trial-and-error learning process as a means of finding an optimum mode of communicating with sound, any more than a soap bubble uses trial-and-error to find that a sphere is the optimal shape for attaining maximal volume with a minimal surface area. The key is the nature of the task constraint. Bates *et al* (ibid.) offer the principle that, in the evolution of new capacities such as language and complex cognitive processes, 'old genotypes, confronted with new but heavily constrained tasks, need very little new genetic control to fall into the most efficient solutions to those tasks' (ibid., p. 20).

A final issue concerning the relationship between ontogeny and phylogeny concerns the difficulty of disentangling the consequences of biological (genetic) and cultural evolution. There is hope for progress in this area based on recent theoretical work which aims to incorporate both types of process within a single framework, so that their interaction can be explored (Boyd and Richerson 1986; Petrovich and Gewirtz 1985).

Our main task in this chapter is to examine the ontogeny of social and cognitive skills in young children, and the acquisition of culture. Here the focus is on the way that social interaction plays a part in the emergence of cognitive skills. When considering this topic, our definition of cognitive skills is a broad and somewhat loose one; it includes specific competences such as imitation and language, as well more general capacities such as those identified by intelligence tests. An important component underlying these competencies is the ability of children to process and manipulate symbols.

An extremely simple-minded view of the role of enculturation in the early years would be to conceive it as no more than a matter of elements of culture being acquired directly by the child during this time, a process which would continue in much the same way as the child grew older. Of course, this is not a false conception; it is just woefully incomplete. Part of the motivation for studying early interaction is that the skills and knowledge children acquire during this time will in turn serve as a means for acquiring more. In short, we favour a bootstrapping hypothesis: what is being acquired by the child in the early years are the skills of interaction, communication, and understanding that compose the basis through which culture is subsequently acquired and the child becomes socialized. Thus, our objective is to examine the way that the socialization of children into their culture facilitates *and* is facilitated by the acquisition of cognitive skills.

The chapter starts with an examination of theories about the origins of infants' social dispositions. The next section considers the nature of early social interaction, and the contributions of infants and caregivers to this process. The third section involves an examination of the way that social interaction is associated with the development of attachment relationships between infants and care-givers. This in turn leads on to an outline of the ways in which social interaction between infants and individuals to whom they are attached can lead on to further cognitive advances. The fifth section draws attention to cross-cultural studies. The sixth section outlines the relevance of social interaction to cognitive development in later childhood.

15.2 Developmental theory and the origins of social behaviour

Table 15.1 provides an outline of the social abilities of infants during the first fifteen months of their life. It can be seen that infants begin life with a number of capacities which are important for social responsiveness. Towards the end of the first year more sophisticated social capacities emerge. A lively topic of debate has been the origin of these social dispositions; and in this first section we consider some of the theoretical stances that have been adopted. This is followed by a section that discusses the form and pattern of early social interaction.

15.2.1 *Trevarthen*

Trevarthen has taken the position that children from an early age have the capacity for 'intersubjectivity'. He explains the idea as follows:

Table 15.1 The social abilities of infants

Age	Hearing and vocalization	Face and gestures		Attachment
New-born	Distinguish mother's voice from another[1] More responsive to speech than non-speech[2]	Components of adult facial expression present[21] Imitation of adult facial expression.[22]	New-born	Preference for smell of mother after 6 days[31] Discrimination of mother's face from stranger's.[32]
3 wks		3–4 wks: social smile[23] sadness expression[24] 6 wks: smiles randomly distributed in interaction.[25]		
2 m.	2–3m. Laughing and cooing	Anger expressions[24]	2 m.	
3 m.	Systematic manipulation of sounds[3]	Distinquish different emotional expressions on same face[26]	3 m.	3–4m: responsiveness to familiar people develops. Discrimination of mother and father from stranger[33]
4 m.	Ability to distinguish phonemes[4]	13–26 wks: clustering of smiling during interaction[25]	4 m.	
5 m.	Similarities with the sounds of parents' language[5]		5 m.	
5–6m.	Consonants used		5 m.	
6 m.	6–10m: canonical babbling		6 m.	6–8m: attachment to particular individuals; following and distress at separation[34]
7 m.		Fear expressions[24]	7 m.	7–9 m: wariness of strangers shown by a high proportion of children.[35]
9 m.	Comprehension of words[6]	Follow simple pointing	9 m.	9–12 m: social referencing; child will look at adult when uncertain about a situation.[36]
10 m.	10–13 m: use of single words[7] Frequent imitation of words[8] Loss of ability to discriminate some phonemes[9]			
12 m.	12–18m: Illocutionary communication[10]	Follow eye gaze of another[29]	12 m.	10–14 m. children will stay with babysister after 10–12 hrs previous contact but distressed if briefer contact[33]
13 m.	overextension of words		13 m.	13–20 m: Peak distress when left regularly at daycare[37]
14 m.		Follow most pointing[27,28]	13 m.	
15 m.	Average age of vocabulary of 10 words[11]			80–90% infants will stay with babysitter.[38]

References are listed at the top of p. 438.

1. De Casper and Fifer 1980.
2. Hutt et al. 1968.
3. Uzgiris 1976.
4. Eimas et al. 1971.
5. Boysson-Bardies et al. 1984; Trevarthen 1977.
6. Benedict 1979; Bates et al. 1987.
7. Dale 1976.
8. Bates et al. 1979.
9. Werker et al. 1981.
10. Bates et al. 1975, 1979; Harding and Golinkoff 1979.
11. Nelson 1973.
21. Oster 1978.
22. Field et al. 1982, 1984.
23. Wolff 1963.
24. Izard and Malatesta 1987.
25. Kaye 1982.
26. Young-Browne et al. 1977.
27. Murphy and Messer 1977.
28. Lempers et al. 1977.
29. Scaife and Bruner 1975; Collis 1977.
31. MacFarlane 1975.
32. Bushnell et al. 1989.
33. Maccoby 1980.
34. Ainsworth 1973.
35. Batter and Davidson 1979.
36. Campos and Stenberg 1981.
37. Kagan 1979.
38. Smith and Noble 1987.

---

for infants to share mental control with other persons they must have two skills. First, they must be able to exhibit to others at least the rudiments of individual consciousness and intentionality. This attribute of acting as agents I call *subjectivity*. In order to communicate, infants must also be able to fit this subjective control to the subjectivity of others: they must also demonstrate *intersubjectivity* (Trevarthen 1979a, p. 322).

Trevarthen sees even neonates as exhibiting latent sociability and intentionality. He claims that, a few days after birth, the infant plays an active role in interaction, co-operating with the mother and anticipating her signals in situations such as feeding and holding (Trevarthen 1979b). From two months onward, infants go through a phase he calls *primary intersubjectivity*. During this phase, gaze and facial expressions are selectively directed towards persons, as are the movements of mouth and limbs that Trevarthen identifies as primitive forms of articulatory movements and gestures. The nature of such signals is taken to indicate their source in a person, and thus to be indicative of subjectivity in Trevarthen's sense. There is a further claim that two subjectivities, the adult's and the infant's, are coupled in reciprocal and shared control of the ongoing interaction (hence intersubjectivity).

Trevarthen sees social and cognitive development not as proceeding through a relationship with a more competent adult, but as the maturational unfolding of innate capacities. The qualitative changes in social functioning during infancy are believed to be a consequence of changes in the balance of control between two modes of functioning. From birth, infants are believed to have distinct modes of operation for interacting with persons (the communicative mode) and with objects (the praxic mode). Actions typifying the operation of the praxic mode are reaching and grasping. The idea that neonates can distinguish inanimate and animate objects seems to be based on an assumption that something akin to direct perception of human movement is in operation (see Collis 1981),

a point that has been questioned in experimental studies (Sylvester-Bradley 1985).

The characteristics of the period of primary intersubjectivity are seen as a consequence of communicative functioning dominating the praxic mode for a period. The onset of well-controlled reaching and grasping signals the beginning of a phase when the communicative mode is dominated by the praxic—hence the cooling of interest in persons, and the avid interest in objects and the inanimate environment that is commonly described from four months onwards. During this period, adults often have to make use of the child's interest in objects as a vehicle for social interaction (Collis 1977). Trevarthen (1979b) calls this phase the *epoch of games*.

A further developmental phase is ushered in when the two modes of functioning regain a new balance and become capable of integration, so that objects can be shared with people, usually at around nine months. Trevarthen calls this phase *secondary intersubjectivity* (Trevarthen and Hubley 1978). Trevarthen concludes that the many new achievements of the nine-month-old could not be mastered unless they were dependent on the maturation of an innate endowment for sociability. Such a theory allows little scope for the differentiation of divergent cultures in the early months, as other people are seen to play a minor role in the epigenetic process. Indeed, it is as if psychological (as opposed to neurological) processes have a minimal role in development.

15.2.2 *Piagetian approaches*

The Piagetian position is in stark contrast to Trevarthen's in two respects. Firstly, for Piaget there is little that is special about *social* cognition, and there are no cognitive structures serving specifically linguistic functions. Hence, social and communicative developments are believed to come about as a result of developments in general cognitive capacity. Secondly, whereas Trevarthen stresses the maturational unfolding of abilities through infancy, the Piagetian view is that cognitive structures are *constructed* as a consequence of action and interaction with the world. In this connection, Piaget seems to emphasize the importance of infants' physically acting on the world. However, the development of spatial cognition in infants with motor disorders (Lewis 1987) and of language development in children with disorders of the speech apparatus (Bishop 1988) seriously call into question this description.

According to Piaget, towards the end of the first year infants achieve stage 5 of sensorimotor development, where they are able to use new means to familiar ends. Bates *et al.* (1975) argue that at around ten to eleven months infants employ this capacity to use objects as devices or props for obtaining an adult's attention, and use adults to obtain their needs (Bates *et al.* are clearly describing the same behavioural phenomena that Trevarthen describes as 'secondary intersubjectivity'). Later on, the general capacity to use symbols is followed by word-use. In other words, cognitive advances result in infants' directing their communication to adults.

A particularly interesting account of the general constructivist position is given by Bates *et al.* (1979). Bates *et al.* dismiss the learning-theory approach to cognitive development, and also play down the role of genetic determination. Instead, their emphasis is on development being guided by the influence of task constraints—the restricted set of possible solutions to a developmental problem. For Bates *et al.* the essence of the constructivist process is that 'the child's solutions to a set of similar tasks become internalized, i.e. embodied in the form of rules and principles that can be applied to new tasks without going through the whole construction process all over again' (p. 19). The role of genetic factors is simply to provide predispositions that 'have their effect by guaranteeing that the animal will enter into situations with particular task demands which are likely to produce certain high-probability solutions' (p. 24).

15.2.3 *Vygotsky*

Vygotsky's approach is founded on the belief that understanding has to occur in a social context *before* it can be incorporated into a person's cognitive structures. This gives prominence to the role of social behaviour in developing skills. Vygotsky's work is considered in more detail elsewhere in this volume (see Sinha, this volume, Chapters 13, 17).

Vygotsky's ideas about development have been extended to help explain the transition between social interaction, intentional communication, and language-acquisition (Clark 1978; Lock 1978, 1980). This viewpoint assumes that, initially, infant actions do not involve any intent to communicate with others. However, adults respond to certain infant actions (crying, reaching, etc.) *as if* they are intentional communications. The result is that infants come to recognize that their actions can influence others and that they have communicative powers. The breakthrough to speech is thought to occur through adults' indicating objects and labelling them. This leads infants to imitate words, even though at this stage they have little or no understanding of their meaning. Understanding is supposed to develop through interactions with adults which involve the 'original word game'—adults' asking infants for the names of objects and asking them to locate named objects. Thus, theories in the Vygotskian tradition propose that social capacities relevant to a culture will develop out of interaction with adults, and that the social responses to the production of actions and words lead to developments in communication.

15.2.4 *Kaye*

The position adopted by Kaye (1982) is in many ways similar to that of Vygotsky. His ideas will be given a detailed coverage because of their relevance to the theme of our chapter. Kaye holds that social interaction is responsible for many of the cognitive developments of infancy. He accepts that infants have the inherent potential to develop the cognitive characteristics of the mature human, but he emphasizes that the realization of this potential depends crucially on the special fit between adult and infant behaviour, a fit that is the product of evolutionary processes.

A longitudinal study by Kaye was originally expected to show that parameters from a fine-grained analysis of early adult–infant interaction were related to later functioning. However, few links were found once continuities in parental behaviour were partialled out. This apparent lack of predictability led Kaye to the view that infants have little direct influence on the structure of early social interactions; in a particular sense of the term used by Kaye, young infants cannot be considered to be part of a *social system* (or what anthropologists would call a sociocultural system) because they have no experience or knowledge of the nature of social activities.

Following from this longitudinal study, the central question for Kaye concerned the means by which the cognitive skills involved in communication develop in an organism that lacks a mind capable of purposive social control. He rejected the hypothesis that infants engage in social relations only after certain cognitive skills have developed; and he also rejected the notion that infants are innately endowed with certain social skills. Instead, he suggested that infants develop social and cognitive skills in social interaction with adults by a process which he likens to an apprenticeship. An important concept here is the *parental framing of infant behaviour*. Framing refers to a process whereby adults provide support or context for an infant's activities; where infants lack the skill to provide some feature of interaction, that function is performed by adults. Thus skills develop through the provision of assistance by parents so that children can complete their plans and not become enmeshed in the details of subgoals.

Kaye identifies a sequence of social skills that parents are concerned with developing in their child. In the first three months, Kaye suggests that shared rhythms emerge as parents build a semblance of dialogue (for example, turn-taking) by fitting in with the temporal regularities in the infant's behaviour. At this early stage, it is the parent who provides the interaction with an appearance of intentionality. Between two and eight months, parents recognize intentions in their infant's behaviour, and, by so doing, foster the emergence of *shared intentions*. For much of this time, it is adults who use their memory of past events to enhance social activities and thereby help to build up a repertoire of shared activities and knowledge. In time the infant takes on a role in this domain too, so that *shared memories* emerge to form the basis of the comprehension of conventional signs. Shared memory also makes possible the development of self-consciousness and an understanding of the social roles, rules, and conventions that characterize human social life, together with the acquisition of shared language. Kaye sees language and socialization as being intimately related: 'a little language changes the nature of thought a little, advances the socialisation process a little, which enables language to advance to a more sophisticated level, and so forth' (Kaye 1982, p. 238).

One general difficulty with Kaye's account of cognitive development is that it ignores the possibility that non-social experience or maturation may lead to changes in functioning. Kaye provides a compelling model of the way social interaction leads to cognitive development; but he does not establish that it does. Kaye's ideas may be too closely related to Western culture. A review by Ochs and Schieffelin (1984) suggests that other cultures do not provide the intense social interaction characteristic of our own. This detracts from Kaye's approach as a satisfactory explanation of the development of social and cognitive skills that are consistent across cultures; but his apprenticeship idea remains attractive as an explanation of the way people, objects, and activities become significant to children within their cultural setting.

15.2.5 *Summary*

Each of these major theories has given prominence to particular processes which contribute to development, maturational or experiential, social or non-social; but there is a lack of attention to the other contributions. This is not to say that the theorists have denied a role to the contributions they choose not to highlight; rather these other contributions have been largely ignored. We have seen that the approaches of Trevarthen and Kaye provide a stark contrast. Trevarthen emphasizes maturational processes and hardly mentions the possibility that the social interactions experienced by infants have implications for the course of development. He explicitly dismisses not only learning-theory-based environmental approaches, but also constructivist approaches. On the other hand, Kaye scarcely mentions the implications of maturational changes for changes in psychological functioning. An embarrassment for Kaye's position is that it is very closely tied to child-rearing practices in modern Western culture. Cultures which do not provide the same level of intense social interactions (Ochs and Schieffelin 1983) would seem to require a different developmental route. Trevarthen's approach does not suffer from this drawback; but this is achieved by virtually ruling out social processes as playing a role in infant development. Instead, the emphasis is on neurological development, which does not receive much attention from other theorists, though Bates *et al.* (1988) are prepared to speculate about the role of neurological variation as a substrate of individual differences in early language development (see also Bates *et al.* 1993).

The views of Piaget and Vygotsky also provide a contrast. For Piaget the issue of the way infants acquire social powers is a minor one. Instead he concentrates on the way development occurs by virtue of children interacting with their environment and resolving conflicts between their model of the world and their experience of how the world behaves. Social cognition is seen as a simple extension of cognitive concerns to an animate environment. In contrast, Vygotsky and other theorists see the social events as fundamental to development: without social experiences significant advances would not occur. Current thinking about the evolution of human intelligence favours the Vygotskian principle that intelligence is primarily social in nature over the Piagetian view that social cognition is just one part of a general cognitive system which is primarily an adaptation for interaction with the non-social environment.

What does an evolutionary perspective tell us about these various theoretical positions? Evolution does not necessarily favour the 'hard-wired' types of developmental process of the kind implied by Trevarthen's maturational theory. A necessary and sufficient condition for an ontogenetic process to evolve is a high probability that a favoured outcome will result. If the cognitive and social powers achieved in the first few years could develop by neurological maturation operating under genetic control, then Trevarthen's position would pass this test. But the constructivist approach could pass the test too. Bates *et al.* (1979) would argue that a construction process is more likely, because, in principle, it is more economical in terms of the genetic changes that would be required to evolve a cognitively sophisticated organism from a less well-endowed one. The Vygotskyian view could also pass the 'high probability of outcome test', provided that one of two conditions is met: either the crucial aspects of the role of social beings are invariant, or there is a range of alternative social processes that could serve the same developmental role. Kaye argues that parental behaviour has been selected so as to fit the needs of the developing child. However, it is not really convincing that his model would work, given the degree of variation that is commonly observed in various aspects of parent–infant interaction.

A compromise model of social development would involve a construction process based on task constraints, where the task constraints and some hints to their solution are presented to the developing child through the medium of interaction with other persons. In such a model it would be hypothesized that the critical aspects of parental behaviour would be held constant by the task constraints. A corollary of this is that variable aspects of parental behaviour are unlikely to have much predictive value for the child's development. In these terms parental social behaviour would be seen as essential to development, but not necessarily to result in individual differences in cognitive abilities.

15.3 The beginnings of cultural understanding: infant capacities and social interaction

This section considers the basis and development of the infants' early interactions with adults. The newborn infant appears to be tuned to the types of stimuli that adults provide. However, although they have the dispositions to attend to and to be entertained by social interaction, it would appear that these abilities do not extend to synchronizing their behaviour with the pattern of social interaction provided by adults. Such flexibility on the infants' part permits the diversity that is present in the social behaviour of different cultures.

The following review illustrates the way that infants are enmeshed in social behaviour from the beginning of their life. However, it also suggests that some of the studies in the 1970s overestimated the capacity of young infants to take part in social interaction. Our conclusion is that adult–infant social interaction has three important consequences for the child: first, relationships are established with the adults with whom infants are frequently in contact; second, within these relationships children develop a broader range of social skills that later can be used with a more diverse range of people; and third, culturally-based understandings of salient aspects of the environment are established. As a result, infants move from rudimentary social activities into co-ordinated interactions which form the basis for emotional bonds, and then into shared understanding of culturally-appropriate objects, events, and conventions. These social activities form the basis for the child to be able to use symbols and language.

15.3.1 *Social responsiveness and infant capacities*

In the first few months after birth, infants have neither the physical capacities (nor the cognitive capacities) to survive by themselves (Schaffer 1984). Instead they have a repertoire of activities and capacities which enable them to survive in a social environment. For example, crying and smiling are two powerful classes of signal by which infants influence adults. Adults find the cries of infants disturbing (Frodi 1985), and they are likely to respond by trying to alleviate whatever factors they believe might be causing distress. There is evidence that different sources of distress—pain, hunger, etc.—elicit perceptually different types of cry (Wasz-Hockert *et al.* 1968). Smiling is also a compelling signal. From about 2–3 months infants are likely to smile when presented with the characteristics of a human face. Such smiles are rewarding for parents, and later can provide indications of the activities which infants enjoy. Infants also exhibit a variety of other facial expressions which adults may interpret as expressive. A detailed analysis of the expressions of very young infants by Oster (1978) revealed a complexity that is done scant justice with descriptions such as 'smiling' or 'angry' face.

Other actions which, at first sight, may not seem to be intrinsically communicative in nature, may none the less have important social implications. For example, although sucking may be primarily an act of feeding, it inevitably brings people into contact with the baby. Unsatisfactory feeding caused, for instance, by the baby's failing to place the whole nipple into his or her mouth, or by the breast's occluding the nostrils, will upset not only the baby but also the mother (Gunther 1961).

The perceptual characteristics of people attract the interest of infants. Neonates are short-sighted, yet, as Stern (1977) has suggested, this may direct their attention to the adults with whom they come into close proximity (but see Banks and Salapatek (1983) for a discussion of the complexities of visual accommodation in infants). Newborn infants respond preferentially to face-like stimuli (Goren et al. 1975; Maurer and Young 1983) and to voices (Eisenberg 1976), and very rapidly learn to recognize familiar faces (Bushnell et al. 1989) and familiar voices (DeCasper and Fifer 1980). Newborns can also discriminate canonical and non-canonical syllables, which are believed to be fundamental units in the processing of speech sounds (Moon et al. 1992), and appear to like the special characteristics of the manner of speaking that adults use with young children, such as higher pitch and simple, pronounced, and repetitive intonational contours (Fernald 1984). They can also recognize movement patterns characteristic of human movement (Bertenthal et al. 1985).

Recently, there has been a return to the debate about whether the attractiveness of faces is because they contain a special arrangement of features or whether it is because they simply have a general class of perceptual properties which are especially attractive to infants. Kleiner (1987) has claimed that infants are attracted to stimuli with a high amplitude (i.e. a high contrast between light and dark areas), and visual preferences will be based on this characteristic. However, Johnson and Morton (1991) dispute this, and maintain that newborn infants show preferences for diagrammatic faces with normally-organized internal features (eyes, nose, mouth, etc.) over the same diagrammatic faces with their internal features randomly rearranged. In fact, when a wider body of work is reviewed, both the stimulus energy *and* the arrangement of features seem important, so that, as Johnson and Morton propose, the presence of face-like features seems to be important when two stimuli have similar stimulus energy; but when there are large differences in visual energy, then it is the high-energy stimulus that will attract infant attention.

15.3.2 *Social interaction and infant capacities*

What sorts of social encounters do infants experience? What role do infants play in determining the structure and content of these encounters? Do the social experiences and activities of such interactions have consequences for the development of further skills? The 1970s saw a flood of studies aimed at providing answers to these and related questions. Many such studies aimed to separate the roles of infants and adults in determining the structure and course of the interaction. To the extent that infants could be said to influence the course of these interactions, they could be conceived as having some elements of social, communicative, or interactive skill.

One early and dramatic capacity attributed to infants was the ability to synchronize their movement with the speech they hear. Condon and Sander (1974) reported that the initiation and changes in direction of infant movements were synchronized with the phonemic elements of adult speech. Subsequent studies have failed to replicate these findings, and the methods of data analysis used in the study have been questioned (Dowd and Tronick 1986; Pack 1983; Rosenfield 1981).

An important dimension of social activity between infants and parents which has long been thought to have special significance is gaze (Robson 1967). Gaze at a person can be used to signify interest and attention; gaze away from another can be used to signal a lack of interest and a wish to terminate interaction. Initial studies suggested that, from an early age, infants and mothers responded to one anothers' gaze behaviour. Stern (1974) reported that, even at three months, infants are more likely to start gazing at their mother when she is already gazing at them than when she is not. Similarly, he reported that infant gaze at the mother is less likely to be terminated when the mother is looking at her infant than when she is not. From this it appeared that even very young infants are altering their social gaze according to the activities of their partner.

Hayes and Elliott (1979) questioned these findings. They reported that the conditional probability of infants' social gaze was similar to the probabilities obtained from the same data when they were randomly shuffled. The random shuffling disrupted the temporal relationship between the two individuals' gaze patterns, while preserving their overall frequency and duration. Consequently, Hayes and Elliott argued that the pattern of gaze reported by Stern was simply the result of the overall duration and frequency of maternal and infant gaze, rather than of the sensitivity of the infant to the mother and of the mother to the infant. However, the picture may be even more complex. Analyses conducted by Messer and Vietze

(1988) suggest that, although the overall frequency and duration of gaze during social interaction may be the largest influence on infant social gaze, there appears to be some deviation from what would be expected by chance, and this might be attributable to the infants' sensitivity to the behaviour of their partners.

A number of investigations have been made into vocal interaction between infants and their mothers, most often motivated by an interest in the possibility of a degree of continuity with turn-taking in adult conversation. The notion of turn-taking implies that conversation has two properties: participants speak alternately rather than simultaneously, and successive turns have coherence of meaning. It is alternation rather than coherence of meaning that has formed the focus of investigations into preverbal interactions. One can observe adults talking, cooing, and gurgling to infants in such a way that interruptions seem to be avoided, and the participants seem to be answering one another. Three questions arise from such observations. Firstly, can the impression of vocal alternation (a lack of simultaneous vocalizing) be substantiated? Secondly, what does the infant contribute to controlling the interpersonal regularities in vocal behaviour? And thirdly, what is the connection with later linguistic exchanges?

With respect to the first question, in one of the earliest investigations of this topic Stern et al. (1975) reported that, far from alternating, many of the mother–infant pairs they studied produced more simultaneous vocalizing than would be expected in a simple probabilistic model (but see comments by Elias et al. 1984). Stern et al. postulated the existence of two distinct modes of social interaction, alternating and co-actional, and suggested that co-actional (simultaneous) vocalizing was associated with moments of high arousal. Some support for the latter idea has been provided by Schaffer et al. (1977) and Slee (1983). Collis (1985) has taken a more functional perspective, arguing that co-actional vocalizing is most likely to occur when the content of vocalizing is not particularly important to the interaction, the alternating mode being used when adults act 'as if' they are conveying meaning to the infant via speech. The overall conclusion is that episodes where vocal interaction involves precise alternation or precise co-actional synchrony are relatively brief and isolated occurrences in the stream of interaction. Qualitative analyses of selected episodes (for example Trevarthen 1977) are difficult to interpret, as are quantitative analyses of the prevalance of simultaneous or alternating vocalizing over any substantial period of time, as in Stern et al. (1975).

In attempting to answer the second question, whether the infant plays a role in determining the temporal structure of early interactions, studies have suffered from difficulties and ambiguities similar to those discussed in connection with gaze. Trevarthen (1977) argued that very young infants play a role by entraining with and influencing an interactive rhythm. This seems to suggest that infants synchronize their sometimes very brief vocalizations with parental speech at a time-scale of a fraction of a second. Such a process would require very precise rhythms with very brief cycle duration, for which there appears to be no good evidence (see Collis 1985). Arguments by Mayer and Tronick (1985) that very young infants have marked skills for regulating alternation do not seem to be supported by the data that the authors themselves present. Analysing vocal sequences using a random-shuffling technique, Elias et al. (1986) found evidence for maternal control of interpersonal sequencing. From an analysis of interruptions of children by mothers, and of mothers by children, Rutter and Durkin (1987) conclude that it is not until the third year that children begin to play a significant role in controlling interpersonal regularities in vocal behaviour. The implication is that, with younger children, it is the adult who adapts to the child's pattern of vocalizing, so that the interaction has an alternating pattern. Data on peer interaction (de Maio 1982) seems to be consistent with this view.

Regarding the third question, concerning the continuity with later linguistic exchanges, Kaye (1977, 1982) has hypothesized that alternating structures in prelinguistic interaction foster an expectation that human communicative interaction is based on alternating turns by participants. Kaye's arguments are not based solely on vocal interaction but extend to other kinds of behaviour, an example being when bouts of sucking by an infant alternate with periods when the mother jiggles the baby. An attractive feature of his hypothesis is that, arguably, the task of learning to talk might be facilitated if infants already knew some of the ground rules of how to manage conversations, one ground rule being—take turns.

In his discussion, however, Kaye did not separate the influence of interpersonal contingencies in general, which are surely important in any kind of social interaction, from the specific temporal relationship that can be described as alternation. Moreover, much prelinguistic interaction, including that which could be described as proto-referential, is characterized by simultaneous or overlapping actions by the participants. As Collis (1985) pointed out, adults do not wait for the cessation of a child's visual fixation before commenting on the object of attention; in these instances an alternating pattern is neither necessary nor desirable. Concerning vocalization, alternation suits some circumstances, but there are many contexts where it is not particularly appropriate—for example, choruses of cooing, laughing, singing, etc. The appropriateness or otherwise of alternation depends on the nature of the communicative act and its context. Considerations such as these led Collis to argue that infants probably could not learn anything useful about the rules for taking turns in conversation before they could process the content of language. In this view, alternation is a superficial aspect of pre-verbal interaction, which adults employ, neglect, or play with at will, and has little significance until it comes together with coherence of meaning to constitute turn-taking as an important component of conversational interchange.

Even if infants do not exert direct control on the fine timing of social interaction it is quite possible that infant characteristics influence interaction in less direct ways, which yet still have implications for development. Babies with various types of disorder, for example Down syndrome, appear to cause difficulties in social interaction with adults (Berger and Cunningham 1983). It has also been suggested that genetic similarity between children and parents results in more harmonious interactions than between unrelated children and adults, and that this influences the development of intelligence (Bergeman and Plomin 1988).

In summary, infants come into the world with an interest in the types of stimuli that people produce. This prepares them to take part in conversation-like social exchanges. However, recent analyses of movement, social gaze, and vocal turn-taking have suggested that during the first few months infants are not equipped to be full partners in these exchanges, but that, in our culture at least, interaction is structured by the adult so that it appears to have the pattern of a social exchange. The findings indicate that infants are often put in the position of appearing to function as competent members of our culture before they actually achieve this. The result is that infants become enmeshed in social exchanges from an early age. This, of course, fits the Vygotskian view outlined earlier, in that such exchanges may provide a basis for the later development of symbolic and other cognitive capacities; however, before moving on to this topic it is necessary to consider the impact that social interaction has on the relationship of the participants.

15.4 Social interaction and the development of relationships

We have seen that interaction enmeshes infants in social activities with particular adults. Our argument is that this has consequences for the relationship between infants and adults, and that the interactions within the dyad and the security afforded by relationships provides a basis for the further development of cognitive skills. If infants were not exposed to the social world in this way, they would miss experiencing much of the complexity of human functioning. Thus, early social relationships set the groundwork for the development of later cognitive skills within a culture.

The way in which relationships are conceptualized, particularly in connection with the social development of infants and young children, has been greatly influenced by the work of Hinde (1976, 1979). Our discussion of adult–infant interactions showed that there is not a simple one-to-one mapping between the properties of interactions (such as turn-taking) and the capabilities of the individuals who participate in those interactions (so, for example, infants do not regulate turn-taking). In a similar manner, relationships have properties and implications for development over and above the properties of the interactions which, in some senses, are their component parts. In particular, expectancies will emerge in infants as to how specific individuals will behave towards them. Moreover, the infants will have expectancies about the feelings that they themselves are likely to experience as a consequence of the anticipated responses of those individuals.

15.4.1 *Attachment*

By far the greatest concern with relationships in developmental psychology has to do with the concept of attachment. The concept has its origins in psychoanalysis and ethology, and was firmly established at the core of developmental psychology by Bowlby (1969) and by Ainsworth and her co-workers (for example, Ainsworth and Wittig 1969; Ainsworth 1973; Ainsworth *et al.* 1974; Ainsworth *et al.* (1978).

At a descriptive level, Bowlly's view of attachment was that, initially, young infants discriminate poorly or not at all among the various persons they encounter. Subsequently, infants discriminate and increasingly organize their behaviour with reference to a single person, the *discriminated figure* or *attachment figure*, or to a small number of such figures. This organization of the infant's behaviour functions as a tie or bond to the attachment figure. Most usually, the primary attachment figure is the mother. Bowlby (1969) sketched out four phases in the development of attachment behaviour:

Phase 1 (birth until 8 to 12 weeks). Orientation and signalling toward people without discrimination. During this phase, the infant shows behaviour directed towards other persons. An effect of this behaviour on adults is to increase the amount of time the infant has the attention of companions. Signalling behaviour continues to develop, but, according to Bowlby, abilities to discriminate between one person and another are limited, and probably confined to auditory stimuli. However, recent research has shown that even new-borns can recognize their mothers' voices (De Casper and Fifer 1980), and young infants can probably discriminate their mothers on the bases of sight (Bushnell *et al.* 1989) and smell (Cernoch and Porter 1985).

Phase 2 (8 to 12 weeks to 6 months). Orientation and signalling towards one or more discriminated figures. Friendly social behaviour continues, and becomes a little more marked toward the mother-figure than toward others.

Phase 3 (6 months until into the third year). Maintenance of proximity to a discriminated figure by means of locomotion as well as signals. The attachment figure is recognized as having an independent existence of his or her own, moving in time and space. The repertoire of behaviour expands to include following the attachment figure and greeting him or her upon return. The attachment figure is now used as a base from which to explore the environment, and behaviour toward him or her becomes goal-directed. Sociability towards other persons wanes, and strangers become treated with alarm and withdrawal.

Phase 4 (three years onwards). Formation of a goal-directed partnership. The child gains insight into the feelings and motives of the attachment figure, and the fact that he or she has goals of his or her own which might conflict with those of the child.

Beyond this descriptive sketch, two key theoretical principles underpinned Bowlby's view of the nature of attachment. Firstly, child-to-parent attachment behaviour was postulated to be controlled by a biologically-based *motivational system*. Secondly, on the basis of the first attachment, the young child begins to organize experiences and expectations of social interactions into *internal working models* or cognitive representations of the self, the attachment figure, and the attachment relationship. A third key concept, *felt security/insecurity* provides an important link between these two mechanisms.

The concept of a motivational system was taken from ethological theory (Hinde 1970, 1983), and refers to the complex of causal factors controlling a set of activities that serve a common function, where function is to be understood in terms of survival and evolutionary fitness. A key feature of the attachment system was that presence of the attachment figure (typically the mother) enhanced felt security; absence of the mother, in a broad variety of circumstances, resulted in feelings of insecurity and the production of attachment behaviour. This behaviour was variable according to the age of the child and the context, but had the goal of re-establishing proximity.

Such a system was seen as the product of evolutionary pressures, and as having a primary function of ensuring the protection and safety of the child, without the child's having the cognitive capability of comprehending the nature of danger, except in the most rudimentary way through feeling insecurity. Secondary consequences were that attachment focused many social experiences and opportunities for learning on a single person, and that the security offered by the attachment figure facilitated the child's exploration of other features of the social and inanimate environment. Postulating a biological system provided Bowlby with an alternative to views that early relationships merely reflect infants' learning about which adult comforts or feeds them, and the functional role ascribed to feelings of insecurity provided an alternative to the view that childhood fears were irrational products of deep unconscious urges.

The concept of *Internal working models* comes from cognitive psychology. An internal working model is a system of cognitive representations that can work as a knowledge base about the self and others in relationships. For Bowlby, this conceptual framework provided the link between early attachments and subsequent relationships through to adulthood. Working models formed on the basis of the earliest relationships were postulated to influence working models of relationships in general. It is possible to identify different views as to how much or how little the early internal working model of the parent–child attachment relationship influences later working models for other relationships. The influence might be relatively minor, with early models being subject to extensive modification by subsequent experiences (Bretherton 1985) or the influence might be rather profound, particularly in the way that feelings of security and insecurity feature in the model and influence social behaviour and relationships of various types in adulthood (Main *et al.* 1985).

Bowlby originally proposed that it was important for infants to establish relationships with one particular individual, usually the biological mother. However, this notion of 'monotropy' has been criticized from various viewpoints (Tavecchio and van IJzendoorn 1987; Nash and Hay 1993). For example, Tronick *et al.* (1985) discuss the multiple care-giving practised by the Efé in Zaïre. In this tribe, the infant is usually suckled by another lactating mother until the biological mother no longer produces colostrum; later on care-giving is often shared by several individuals. Tronick *et al.* argue that such multiple care-taking is adaptive in its own culture, and may have certain selective advantages, such as more support for the mother, and the presence of care-takers who can take over the mother's functions if she is unwell or absent. Thus the care-giving system in which attachment is embedded may vary from culture to culture, and may not be restricted to the state of having only one functional mother.

Within Western cultures, the notion that attachment is a monotropic relationship with a single other person rests uneasily with a more balanced understanding of the range of different kinds of social relationships that infants can engage in, including relationships with fathers (Lamb 1986; Lewis and O'Brien 1987), siblings (Dunn 1983, 1993), grandmothers (Myers et al. 1987) and professional caregivers (Sagi et al. 1985). It might be possible to rescue the idea of monotropy if it could be shown that these other kinds of relationships were not based on the same motivational system as classical attachment, and it is worth noting that Harlow and Zimmerman (1959) postulated a number of quite distinct affectional systems. However, it has been possible to assess these different kinds of relationships as though they were attachments, by focusing on aspects of security (Nash and Hay 1993). If this genuinely captures the variety of social relationships entered into by infants (which it might not—Dunn 1993) then the principle of monotropy founders completely.

A key feature of Bowlby's approach was that he used the concept of attachment as a label for a motivational system organized to achieve a certain biological goal rather than for a set of specific behaviours. A variety of behavioural acts can serve the function of ensuring the physical safety of the child through the protection afforded by the attachment figure. Bowlby used the parlance of control theory in expounding his idea; and this approach was taken further by others (Bischof 1975; Sroufe 1977; Waters 1981). Other workers have conceptualized attachment in quite different ways. Some saw it as a characteristic which could be expected to have the properties of stability and predictive validity expected in trait psychology; the disappointing outcome of psychometric analysis (Masters and Wellman 1974) has been held by some to support Bowlby's view that attachment is a motivational system rather than a trait. Others conceived attachment as something more than a within-individual process, and sought to identify it with inter-individual interaction (Cairns 1972; Rosenthal 1973; Gewirtz 1978). Recently, the concept of a relationship has become more sharply distinguished from the concept of interaction. Clearly, both concepts refer to inter-individual processes, but at two different levels of analysis, particularly in terms of time and space (Hinde 1976, 1979). Interactions involve the communication of both individuals in moment-to-moment, minute-to-minute, or possibly hour-to-hour processes: relationships persist over days, months, or years and through periods of separation and non-communication. The concept of internal working models—the cognitive component of attachment theory—clearly goes beyond here-and-now phenomena; and for this reason attachment is more appropriately considered as a relationship than as an interaction.

Broadly, there are two possible views of attachment: a process within the child, or a relationship between the child and one or more adults—though there is a degree of overlap between these views. Bowlby's original view, developed by Ainsworth, was of attachment as a motivational control system within the infant or young child, which organized the balance between seeking security by maintaining proximity or contact with the attachment figure and exploiting the environment when it was safe to do so. Much empirical work on this motivational system view has been concerned with individual differences in the organization of attachment behaviour, presumed to be a consequence of individual differences in the operation of the motivational system. Ainsworth et al. (1971, 1972) postulated a threefold typology of patterns of attachment based on the reactions of children when they were reunited with their mothers after a few minutes' brief separation. The type B pattern, presumed to be optimal in normal conditions, was for the children to greet the mother and/or approach her. In contrast, the type A pattern was for the children to avoid the mother, and the type C pattern was angry or resistant behaviour interspersed with contact-seeking.

Various associations have been found between this typology and the behaviour of the adult to whom the child is attached. Theoretically the most compelling of these associations are to do with the sensitivity of the attachment figure to cues from the child. Mothers of group B infants were reported as being most likely to respond appropriately and promptly to their infant's crying in the first month of life; there was more tender holding of the babies, and these mothers' behaviour in feeding and face-to-face interaction was more contingent upon the child's behaviour. Mothers of group A infants frequently rejected their infant's initiatives, especially bids for physical contact. Group C mothers were not very consistent in responding to their infants (Ainsworth and Bell 1969; Blehar et al. 1977; J. E. Bates et al. 1985). Other studies have shown that classifications of infants based on this typology are stable over a period of months (Waters 1978; Main and Weston 1981). However, stressful events may lead to changes in the type of attachment shown by an infant (Vaughn et al. 1979), and attachments to mother and to father may be of a different type (Main and Weston 1981).

Although many investigations have provided support for Ainsworth's approach, it should be noted that questions have been raised about the methodology used to investigate the origins of the attachment types (for example, the observers were not blind to the typology, and the coding was based on written transcripts of the sessions) and about some of the theoretical interpretations (whether the typology reflects attachment or some other infant characteristic such as arousal—see Campos et al. 1983).

The type of attachment between the infant and mother appears to be related to later cognitive and motivational functioning of the infant. Matas et al. (1978) have reported that infants who were identified as securely attached at twelve or eighteen months, in comparison to insecurely attached infants, were more enthusiastic, more co-operative, and less likely to become angry when attempting a tool-use problem with their mother at two years. A similar pattern of findings has been reported by Main et al. (1979). Waters et al. (1979) have found a relationship between the security of attachment in infancy and later characteristics at pre-school. Securely attached infants, in comparison to insecurely attached infants, were more likely at three and a half years to be social leaders and popular with other children.

The bridge between an individual-differences approach to attachment, with an emphasis on processes internal to the child, and a relationships approach, is to be found in the hypothesis that the attachment system involves the construction and operation of working models of the self and attachment figures (Bowlby 1969; Bretherton 1985; Main et al. 1985). These working models develop as the child experiences interactions with an attachment figure, so that, in a sense, what is built is a model of a relationship. Such a view further implies that the attachment figure also has a model of the relationship, though, as Bretherton (1985) points out, this aspect has been almost entirely ignored in empirical research. The point is important, as the adult's view of the relationship will very probably be detected by the child and influence his or her own view of the relationship.

The development of a cognitive model of the attachment figure implies that the child's perception of the world will be influenced by the adult's perception of the world. The child is most likely to attend to those parameters in the environment, especially the social environment, that the adult indicates as being salient, a process that has come to be known as social referencing (Feinman and Lewis 1983; Klinnert et al. 1986; Zarbatany and Lamb 1985; Feinman 1992). Furthermore, in conditions of uncertainty children of nine months and older tend to look at their caregivers to monitor their reactions (Campos and Stenberg 1981).

The possible persistence of cognitive working models of early attachment relationships has led to an increasing interest in the possibility that some adult relationships are essentially attachments. Hazan and Shaver (1987) attempted a formulation of romantic love between adults on this basis. Using questionnaire data, they provide evidence of three types of romantic love, corresponding to Ainsworth's A, B, and C types of attachment in the early years. This idea has been extended further by Bartholomew and Horowitz (1991). However, there is clearly a danger of attachment's being extended so broadly among adult relationships as to have very little explanatory power.

Ainsworth (1989) has discussed with great care which relationships may usefully be termed attachments. Ainsworth suggests that a subset of adult human relationships are properly regarded as *affectional bonds* if based on a long-enduring tie in which the partner is important as a unique individual. The key psychological process underlying the tie is an internal working model of the relationship built up during the history of the relationship. The tie may be maintained during absences, but there is a desire to come together and pleasure on so doing, so there are affectional aspects to the tie. Typically, separation will cause distress, and loss will cause grief. Some, but not all, affectional bonds are attachments; the criterial feature of an attachment is the experience of security and comfort in the presence of the partner (which is a consequence of the activation of the attachment motivational system). Ainsworth quite explicitly indicates that a mother's bond to her child is not properly called an attachment because this feature is lacking. Any particular affectional bond may have components from different affectional systems. For example marital relationships typically involve sexual attraction, caring for offspring, caring for the partner, and experiencing comfort and security from the presence of the partner. Only if the last component is present would such a relationship properly be called an attachment. Without a narrow conceptualization of adult attachments in a scheme something like this, it seems likely that the concept will become so broad and cover so many long-standing close human relationships that it will lose all explanatory power.

The possibility of very long-term consequences of cognitive models of early relationships raises the issue of the transmission of attachment patterns across generations—whether it might be the case that the relationship a child constructed with his/her own parents influences the subsequent behaviour of that child towards his/her own children (Ricks 1985). The retrospective nature of most data on intergenerational effects makes evaluation of the evidence difficult, as it is uncertain whether an adult's recall of attachment relationships experienced as child is accurate, or whether it is simply a reconstruction based on current cognitions, including an awareness of the view, often accepted as 'established' in Western developed societies, that childhood experiences are potent in shaping later personality.

15.5 Relationships and social cognition in non-human primates

In view of the evolutionary background of Bowlby's original formulation of attachment, it is natural to look for parallels among non-human primates. Patterns of infant care-taking vary widely among different species of primate. In some species (vervets, *Cercopithecus aethiops*, langurs, *Presbytis* spp. and patas monkeys, *Erythrocebus patas*), individuals other than the mother may participate in care-taking, while in other species, especially *Macaca* spp., other individuals may exhibit hostility towards the infant (Lee 1983; Hrdy 1976). Even within a species, the optimal pattern of care will depend on circumstances. For example, Altmann (1980) suggests that dominant female yellow baboons (*Papio cynocephalus*) can adopt a relatively *laissez faire* style of mothering because their rank reduces the danger to their offspring. A *laissez faire* style will produce infants who are relatively independent, and thus more likely to survive if they are orphaned. On the other hand, mothers of lower rank cannot afford to adopt a *laissez faire* style, because their rank offers less protection from harassment from conspecifics, even though the more restrictive mothering style means the infants are less independent and more vulnerable should they become orphaned.

Clearly, evolutionary thinking offers no easy answers as to what pattern of infant care is optimal. It is worth noting that at least one study of human attachment in a Western European sample (Grossman *et al.* 1981) indicates a predominance of the avoidant category, a pattern that has been regarded as sub-optimal, largely on the ground that it was a minority pattern in North America (Ainsworth *et al.* 1978).

The disruptive effects for infant monkeys of a lack of experience with a mother over relatively lengthy period are well documented in the experimental work of Harlow and his co-workers (for example, Suomi and Harlow 1977). The consequences of severe social isolation were thought to be due to interference with social learning rather than with cognitive development. Work on the less devastating effects of short-term separations, particularly by Hinde and his co-workers (for example, Hinde and McGinnis 1977; Hinde *et al.* 1977) have brought to prominence the role of social relationships external to the mother-infant dyad. For instance, where primate mothers are separated from the infant and from the social group, the infants are more severely affected on their reunion than in the case where the infants were removed and the mothers stayed in the social group. In the former case, the mothers spent a lot of time re-establishing their own social relationships with other members of the group, and appeared to have little time for their infants.

Field studies of free-living primates have revealed the commonplace employment of cognitive skills that are quite different from the cognitive skills usually used in the laboratory, and these newly-recognized skills are primarily to do with the management of social relationships and social structure (Cheney *et al.* 1986). For example, there are many indications that chimpanzees and various species of monkey can use other individuals as 'social tools'. Indeed this appears to be more widespread than tool-use in the more usual sense, where inanimate instruments are used to achieve some end. Of course, using the phrase 'tool-use' need not necessarily imply common cognitive processes in the social and non-social domains, since that is still an open question—see Bates *et al.* (1979) for a discussion of possible relationships between tool-use in the usual sense and language as the use of symbols as 'social tools'.

There is considerable evidence that chimpanzees and monkeys can use some kind of understanding of the relationships between third parties, i.e. relationships in which they themselves are not directly involved. For example, female vervet monkeys respond to a scream from a juvenile by looking not at the juvenile but at its mother (Cheney and Seyfarth 1980). Male baboons have been shown to be less likely to challenge another male if the latter is with a female that strongly prefers him (Smuts 1985; see also Jolly, this volume, Chapter 6). These examples suggest the ability to comprehend the existence of a relationship with a third party, and the strength of that relationship. Moreover, the finding that many species form transitive dominance hierarchies—hierarchies with a relative paucity of 'circular' relationships such that A dominates B, B dominates C, and C dominates A—suggest an ability to comprehend patterns among several relationships so as to avoid inconsistency.

The current evidence on relationships among non-human primates (Hinde 1983; Cheney *et al.* 1986; Byrne and Whiten 1988) thus suggests that it is in the area of social relationships that their most sophisticated cognitive abilities are to be found. While it is important not to fall into the trap of believing on *a priori* grounds that these species must use cognitive processes which are essentially similar to those employed by humans in similar situations, and which must develop in the same way, it is a reasonable line of enquiry to look for evidence of parallels in development. If primates manage social relationships by using internal working models of relationships, then these models must have a developmental history. As in the human case, it is possible to hypothesize that the first relationship(s) developed by an infant non-human primate with a care-taker may influence subsequent relationships. A number of kinds of influence are possible. The first relationship could form a prototype of subsequent relationships, providing a default style that subsequent relationships would only diverge from under specific pressures. Alternatively, social skills learned during the course of developing the first relationship might be used in the negotiation of subsequent relationships. Third, having established a first relationship, knowledge gained specifically from the partner might provide an important focus in the content of subsequent relationships.

15.6 Social interaction and the development of cognitive skills

The formation of attachments is accompanied by changes in the range and complexity of social behaviour. In this section we concentrate on three skills which develop during the first two years: imitation, the co-ordination of attention, and language.

15.6.1 *Imitation*

One powerful mechanism which enables infants to develop cognitive and other skills is imitation. If infants can imitate the actions of another person then they have a powerful device for acquiring new skills. If children can use imitation then they do not need to be involved in a lengthy discovery process or be capable of understanding complex verbal instructions in order to carry out everyday tasks, which can range from filling a drinking cup to planting seeds. Instead they can attempt to reproduce the actions of another person, and in this way master quite complex activities.

Piaget's writings on the subject were, for a long time, widely treated as a definitive statement about the development of imitation. For Piaget, imitation had an important place in infant cognitive development because of its position as a precursor to symbolic representation. He suggested that in the first months there is limited imitation of the activities of others. In these instances infants do not distinguish between their own acts and those of a model; rather they are simply assimilating and then reproducing an activity that they experience. In addition, he supposed that infants only imitate models which involve actions familiar to them. For example, they may cry because they hear the crying of others. Piaget claimed that from about eight months infants develop the capacity to imitate activities which are not in their repertoire and actions which they cannot see themselves doing. Then at about eighteen months the development of mental representation allows infants to be able to imitate complex new models and to be able to use deferred imitation (the imitation of actions that have occurred some time previously).

The validity of Piaget's ideas about imitation has been seriously called into question by Meltzoff and Moore's experiments (1977, 1983*a*). They reported that new-born infants can imitate the actions of others even when they cannot see their own imitative acts, as in the case of tongue-protrusion and mouth-opening (see Fig. 15.1). This raises the issue of the way infants represent these movements, and whether these representations have a relation to later symbolic capacities. Meltzoff and Moore suggest that imitation is possible because at birth infants can translate what they see into actions.

Other studies also have reported successful imitation by young infants; but some authors have offered different interpretations as to its mechanism. A study conducted by Jacobsen (1979) revealed that tongue-protrusion and hand-opening were elicited not only by the appropriate modelling behaviour of an adult, but also by moving a pen backwards and forwards, or a ring up and down. She interpreted her findings as indicating that infants' imitation is caused by a releasing mechanism which can be triggered by general classes of activities, and is not the copying of specific behaviours.

Field *et al.* (1982) presented two-day-olds with happy, sad, and surprised expressions. Observers were reliably able to identify which expression had been presented by viewing only the expression on the infants' faces (see Fig. 15.2). Field *et al.* follow Meltzoff and Moore in suggesting that imitative acts are a result of infants' being able to integrate what they see with their actions.

Abravanel and Sigafoos (1984) conducted a longitudinal study to examine the occurrence of imitation between 4 and 21 weeks. The only evidence they found of imitation was for partial imitation of tongue-protrusion at 4–6 weeks. Abravanel and Sigafoos interpret their findings as indicating that

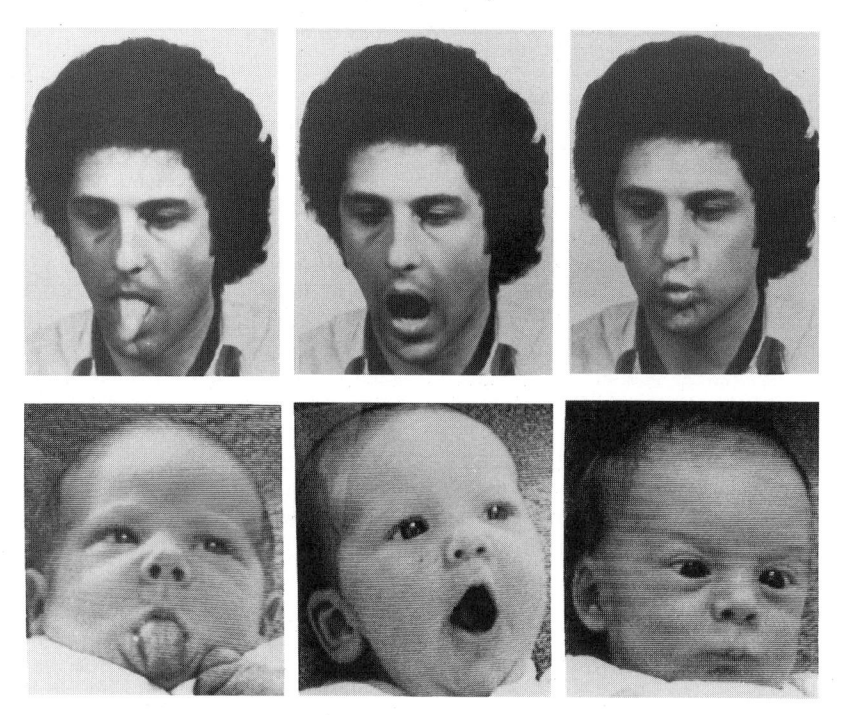

Fig. 15.1 Sample photographs from videotaped recordings of 2-to 3-week-old infants imitating tongue protrusion, mouth opening, and lip protrusion (from Meltzoff, A. N. and Moore, M. K. (1977). Imitation of facial and manual gestures by human neonates. *Science*, **198**, 75–8).

imitation may occur in the first few weeks after birth, but that the ability to imitate declines, only to re-emerge towards the end of the first year. They speculate that this effect may be due to early imitation's being an automatic process similar to a fixed action pattern, whereas later imitation is a voluntary process. However, more recently Meltzoff and Moore (1992) have demonstrated imitation over a wider range of ages.

A number of experiments initially called into question infants' ability to imitate (for example Hayes and Watson 1981; McKenzie and Over 1983a; Koepke *et al.* 1983), and this led to rebuttals by Meltzoff and Moore (1983b) and counter-arguments by McKenzie and Over (1983b). Further studies by Meltzoff and his colleagues have led to an increasing acceptance of his findings. More recently, Meltzoff and Gopnik (1993) have suggested that imitation may be due to infants' detecting that people are 'like me', and that this ability is coupled with the capacity to translate actions they see into movements of their own. This is a radical claim, somewhat similar to the position taken by Trevarthen. The claim about a 'like me' response suggests that infants, from very soon after birth, are able to distinguish people from other objects and possibly other animals. However, Meltzoff and Gopnik do not specify precisely what characteristics elicit a 'like me' response, so it is unclear whether imitation would be more likely and stronger if, for instance, other babies provided a model for behaviour, and whether the response would be extended to other living creatures. From our point of view the importance of this claim is that it suggests that very young infants begin life tuned into the characteristics of conspecifics, and are therefore more likely to attend to and follow the behaviour of these individuals.

The role that imitation plays in the development of older children has attracted limited interest. One exception to this is the proposal by Bates *et al.* (1979) that children will start to use symbols when the capacities of imitation, tool-use, and communicative intent all reach a certain threshold level of competence. This proposal is partly based on the findings of correlations between these three capacities and the age at which symbols start to be used. Bates *et al.* (1979, Chapter 7) also believe that these three capacities evolved before language, and were critical to the emergence of linguistic capacities. She and her colleagues point out that imitation is used much more by human infants than by most other species, with imitation being chiefly confined to primates and song-

Kaye (1982) has highlighted the social basis of later imitation. He suggests that imitation is not an isolated cognitive skill but an interpersonal process, in which the skill of the model may be as important as the capacity of the imitator. He supposes that there are various forms of imitation, and that the capacity to imitate will depend on the infant's level of functioning; this in turn develops as a consequence of social interaction. Kaye, like Piaget, regards imitation as an important capacity because of its links with the use of signs. Kaye states that representation by the means of symbols is 'only possible because of the ways adults organize the infant's world and selectively present imitable models to him' (p. 159). Thus for Kaye imitation is closely related to both the development of general skills and the capacity to use signs.

At present there is still uncertainty about whether or not imitation is a capacity of the new-born, and over the mechanism that produces these responses. What is more certain is that imitation does reliably occur towards the end of the first year. Here we see that there appears to be the emergence of a cognitive capacity which can utilize information present in social interaction to further develop the infants communicative and social skills. Imitation provides a powerful mechanism for infants to incorporate adult actions into their repertoire.

It should not be forgotten that adults do a lot of imitating of the behaviour of infants (Pawlby 1977). In this case, not only is there more temporal contingency between adult and infant behaviours for the infant to detect, but there is also a match between the type of behaviour shared by the two partners. As Newson (1979) has suggested, this could highlight for a young infant the fact that it is another socially communicative organism with which he or she is interacting. The implication of a matching of actions takes on even more significance for establishing shared meanings when the actions are less arbitrary than poking out a tongue or gurgling a syllable. Collis and Schaffer (1975) describe how mothers 'imitate' the gaze-direction of their infants, so establishing shared visual attention toward a common focus of interest.

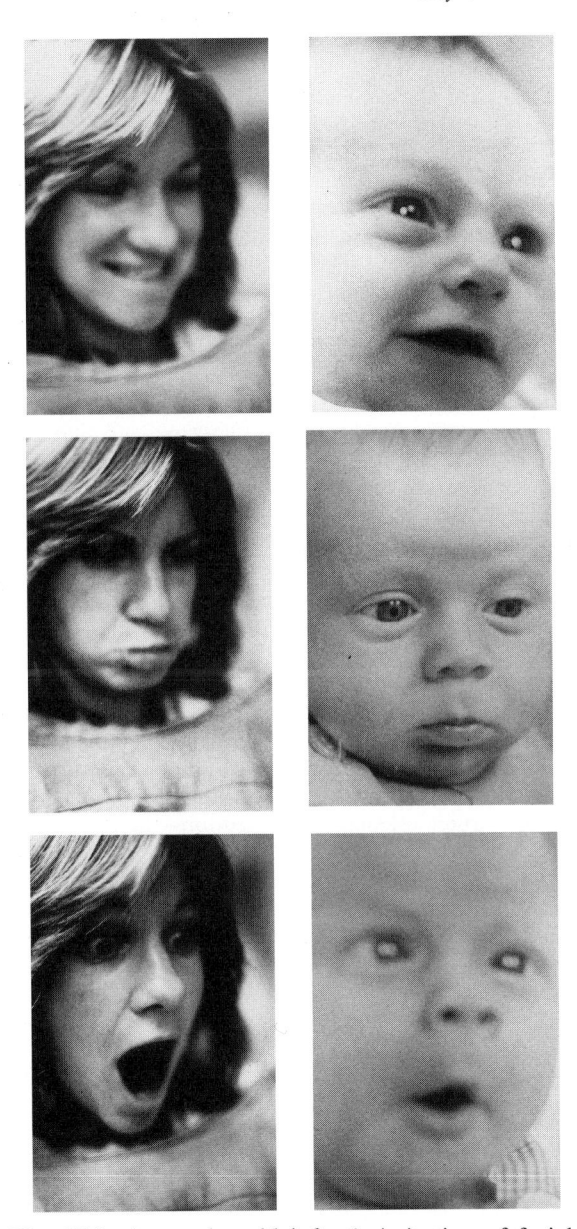

Fig. 15.2 A two-day-old infant's imitation of facial expressions (happy, sad, and surprised) (from Field, T. M., Woodsom, R., Greenberg, R., and Cohen, D. (1982). Discrimination and imitation of facial expressions by neonates. *Science*, **218**, 179–82. Copyright 1982 by the AAAS.

birds. Furthermore, primates tend to be worse at imitation than other comparable sensorimotor skills. Bates argues for imitation's having played a crucial role in the emergence of human culture, including language.

15.6.2 *The co-ordination of attention*

Various social behaviours such as gaze, pointing, and manipulation, can be employed to co-ordinate the interest of infants and a familiar adult. The development of co-ordinated attention is important because it enables the establishment of topics of common interest around which social interaction can take place. It seems likely that co-ordination of attention helps infants develop a shared understanding of culturally appropriate topics of conversation. Both these processes are necessary if infants are to come to use symbols.

One way that co-ordination is achieved is by adults timing their behaviour so that it is integrated with the interests of infants. Collis (1977) followed up the Collis and Schaffer (1975) observations, and described how mothers tend to follow the direction of infants' gaze, and are likely to talk about the object of the infants' attention. Consequently, the careful timing of maternal behaviour appears to allow the infant to have the initiative in establishing subjects of shared interest. In a similar way, Messer (1978) reported that maternal speech often refers to objects that are being manipulated by either the infant or the mother. As a result, maternal speech concerns objects in which the child has a manipulative interest.

Towards the end of their first year infants actively direct the attention of adults towards objects with pointing gestures. Trevarthen (1977) and Fogel and Hannan (1985) have reported that infants a few months after birth will employ manual activities which involve the extension of the index finger and curling of the others. Fogel and Hannan also report that this pointing is associated with mouthing activity, and interpret such pointing as fullfiling affective rather than referential functions. The relation of these early social activities to later referential pointing remains unclear. It is generally accepted that the referential use of pointing becomes an important activity for children older than seven months (Murphy and Messer 1977; Leung and Rheingold 1981). Adults very frequently respond to infant points, and Murphy (1978) has reported that the points of fourteen-month infants are accompanied by maternal naming of the relevant object more often than one would expect by chance. The points of older children tend to be accompanied by mothers' asking questions, a change which may reflect the mothers' sensitivity to the increasing verbal capacity of their children. Indeed, the emergence of the ability to point is widely considered to be of profound significance for referential and personal aspects of language (Bates *et al.* 1979). Thus it is apparent that adults integrate their activities with the interests of the infant, and that by the end of the first year infants can use gestures to direct adults' interest. However, these processes are not unidirectional. By the end of the first year infants are able to follow the direction of an adult's head-turn (Lempers 1976; Scaife and Bruner 1975; Butterworth and Cochran 1980; Churcher and Scaife 1982). And, from eight months onwards, pointing can be used by adults with increasing effectiveness to direct children's attention to objects (Lempers 1976; Murphy and Messer 1977; Churcher and Scaife 1982).

It is clear that one should not regard interactions merely as a process by which one individual directs the attention of the other by the use of explicit behaviours. Instead, it is possible to view the organization of interaction as providing a structure which helps both participants locate a topic of shared interest. For example, it would appear that much of maternal speech consists of sequences of utterances which refer to the same object (Messer 1980). What is of particular interest is that the beginnings of the sequences tend to occur after the infant or mother have manipulated a new object, and after a longer than normal pause between utterances; and that the first utterance of an episode is more likely to contain the name of the referent. Thus, it would appear that a complex arrangement of verbal and non-verbal behaviours helps to mark out the topic of conversation.

Another important process of structuring which may allow shared interests to develop is the use of behaviour rituals. Bruner (1975, 1983) has argued that taking part in the same activity over a number of months allows infants to come to understand the demands of these situations and the appropriate forms of communication that are required. For example, during book-reading (Ninio and Bruner 1978) maternal utterances appeared to occur in a specific order: a demand for attention—'look'; a question about the referent—'what's that'; and a label—'it's a teddy'. Bruner suggests that the repeated exposure to rituals such as these enables children to come to understand their role in supplying conversation within an established format. The ritualized nature of these formats means that children have less difficulty in understanding the pragmatics of the situation—they already know how and when to respond. Instead, they can concentrate on working out the intricacies of the rules of language and how it maps on to the situation. In addition, such experiences are supposed to allow the infant to come to share with the adult a view of what is important, relevant, and of interest.

The use of all these procedures is likely to emphasize cultural values. For example, the value that our culture places on certain cognitive skills which are related to common topics of social interaction can be seen by examining the problems and pictures present in developmental tests which are clearly culturally relative (for example, tasks involving brick-building, naming referents, and puzzles). Furthermore, the presence of culturally favoured topics of interaction allows any member of the culture to use them to facilitate interaction and thereby build on the child's existing skills.

The issue of infant attention has recently become an important topic in discussions about the development of social understanding and in related discussions about the nature and development of autism. Two very different explanations have been put forward about essentially the same kind of behaviour. Baron-Cohen and Ring (1994) postulate a cognitive module which they call the Eye Direction Detector (EDD), which detects whether another person is gazing directly at oneself and, if not, where the gaze is directed. They also postulate a second module, the Shared Attention Mechanism (SAM), which comes into play between 6 and 9 months, and which is sensitive to, and controls, behaviour relevant to attention jointly engaged in by the infant and by another person. Baron-Cohen believes that it is the SAM which provides the basis for understanding the mental states of others; this is possible because other mental states (such as interest) can be substituted for gaze. He contends that, in autistic children, the EDD is intact but the SAM is deficient.

A rather different theoretical account is provided by Hobson (1994), who suggests that at about 9 months infants are able to deal with 'triadic relationships' which involve themselves, another person, and another entity (for example an object). Such relationships occur when infants follow another person's gaze, or direct the interest of the other person. Hobson believes that these attention-based triadic relationships provide a basis for infants' understanding of a perspective different from their own, which occurs simply because infants notice that others do not always have the same reactions to events as themselves. In this way there develops an understanding that there can be more than one representation of an event or thing. This in turn provides a basis for symbolic play and metarepresentational abilities. Hobson uses these principles as a model to explain observations that some blind children, who are denied some of the earliest and most basic experiences of triadic attention, show symptoms, such as echolalia and pronoun reversal, which are also characteristic of autistic children.

The co-ordination of attention is of fundamental relevance to the acquisition of words and the expansion of children's vocabularies. At some level, behaviours involved in the co-ordination of attention must play a part in these processes, but what degree of exposure is necessary before a word is associated with a class of objects or events? A study on shared reference by Tomasello and Todd (1983) has indicated that the amount of time mothers and infants jointly attend to the same focus is related to the vocabulary size at eighteen months, while the frequency of attempts by mothers to direct their infant's attention is negatively related to vocabulary size. This latter finding was interpreted as indicating that it is more effective to follow an infant's interest than to attempt to direct it. A later study by Tomasello and Farrar (1986) revealed that reference to objects during joint attention, and particularly when it followed the child's interest, was positively related to concurrent vocabulary size. In addition, a training study suggested that speech which followed a child's attention, rather than directing it, was more effective in promoting comprehension of words.

It would also seem that relatively young infants can make surprisingly good use of subtle forms of information to help them identify the referents of adult speech. For example, in a study conducted by Baldwin (1991) the experimenters named an object while looking at it themselves, but while the children were looking elsewhere. If the 16-month-old children were simply linking a word with the object they were looking at, then we would expect them to make predictable errors when later tested. In fact the majority of children were able to infer the correct referent of the adult utterance, as was shown in a later post-test. Studies by Tomasello with 2-year-olds have also revealed similarly high levels of sophistication in using contextual information to help understand the pragmatics of the message, and from this to discover the meaning of unfamiliar words (Tomasello and Barton 1994). In one study the experimenter announced he/she was looking for an object that was referred to by a nonsense word; the experimenter then proceeded to search through and handle a series of objects, making suitable reactions to his/her lack of success; finally, on finding the 'appropriate' object an exclamation of pleasure was uttered. When tested later the children showed that they understood the novel word as referring to the appropriate object, thus showing that they did not simply assume that the word referred to the next object that was touched.

A different perspective on the acquisition of words, which seems to minimize the role of social and cultural factors, comes from suggestions about constraints on the way that children interpret the relation between words, referents, and context. It has been proposed that children have a set of assumptions that help them to make sense of circumstances when they hear a new word. For instance, it has been proposed that there is a 'whole object assumption', so that children will assume that a new word refers to a whole object rather than to an attribute or property of the object. In this way much of the uncertainty about reference can be eliminated. There is also a suggestion in some of this writing that these constraints are innate predispositions (see Gelman and Byrnes 1991).

These rather mechanistic ideas have been criticized by Carey (1993) and by Tomasello and Barton (1994). Tomasello and Barton use their own data to argue that children are not functioning in a mindless way, but rather are using a problem-solving approach to make the best use of the limited information that is available. In the same way we are arguing in this chapter that the experience of social interaction in the first years of life gives children a rich knowledge of cultural assumptions about what is important, interesting, and relevant, a process that Zukow-Goldring (1993) has termed 'educating children's attention'. Further support for this general position comes from crosslinguistic evidence assembled by Bowerman (1985, 1989) that the boundaries of spatial concepts that are present in early speech are profoundly influenced by the structure of the language that is being acquired.

Thus there is considerable evidence that, even before the onset of speech, social interaction between adults and infants who have formed a relationship allows the establishment of procedures which enable each of them to attend to and direct the attention of the other. These procedures are likely to be necessary before experience within a social setting can have an impact on the infant's cognitive skills, for without ways to regulate mutual attention, the task of acquiring knowledge and symbols becomes much more difficult. Because these skills are implicated in the acquisition of vocabulary, it is likely that they are necessary for further cognitive advances.

15.6.3 Social interaction and the acquisition of language

The most investigated topic concerning the relation between social interaction and the acquisition of cognitive skills is that of language development. With language we have a system which involves rules concerned with the production of symbols. Three sets of relationships between social interaction and language development have attracted interest: the relevance of simplified speech for development; the relationship between speech and non-verbal activities; and the importance of pre-linguistic forms of communication.

15.6.3.1 Simplified speech to children

The recognition that child-directed speech (also termed motherese, baby talk, etc.) is generally simpler than adult-to-adult speech, and has various distinctive characteristics (Snow and Ferguson 1977; Hoff-Ginsberg and Shatz 1982) led to a number of studies describing the characteristics of this speech pattern. In so far as the earlier view was that the language experienced by the child was ill-formed, replete with ungrammatical constructions, false starts, and hesitations, and generally unsuitable as a basis for infants to work out the linguistic code, this evidence seemed to shift the burden of explanation of language acquisition away from innate language-acquisition devices to environmentally driven processes.

However, enthusiasm for the idea that simplified input eases the problem of understanding language acquisition may be misplaced, the existence of a causal link between simplified input and language acquisition is controversial. The obvious research strategy of correlating input with later language abilities is fraught with methodological problems, and the data are inconsistent (Newport *et al.* 1977; Furrow *et al.* 1986; Gleitman *et al.* 1979). More fundamental is the point made by Gleitman *et al.* (1984) that simplified input will have fewer clues for the child to use in working out the syntax of the language, and, in this respect, would be less helpful than if more complex instructions were available (although it is conceded that shorter utterances might ease pressures on a young child's attentional processes). In this view, the simplification of speech directed toward children is simply an adaptation to the need to interact and communicate with children in the here and now; the simplified register is not motivated by a tutorial purpose, nor does it serve such a function, except perhaps for the limited case of vocabulary teaching (for example, Ninio and Bruner 1978).

This still leaves the problem of how the child's experience of language contributes to the acquisition process in general. The learnability theorists (for example, Pinker 1979, 1984; Wexler and Culicover 1980; Wexler 1982) agree with the intuition that simply processing many examples of correct syntactic constructions is not enough. One potentially useful addition would be feedback from adults about the correctness of the child's early attempts at the production of syntax.

For some time Brown and Hanlon's (1970) report about the absence of explicit feedback has been taken as indicating an absence of any feedback about whether utterances are syntactically acceptable. Brown and Hanlon discussed the issue in terms of the absence of explicit indications about errors in children's syntax. However, recent investigations have suggested that some subtle forms of feedback are available to children (Hirsh-Pasek *et al.* 1984; Demetras *et al.* 1986; Penner 1987; Bohannon and Stanowicz 1988). This has led to a dispute between those that maintain that the adult responses are sufficient to enable children to identify grammatically ill-formed utterances (for example Bohannon et al. 1990), and those who dispute this claim (Morgan and Travis 1989; Pinker 1989; Marcus 1993). Central to these arguments has been the issue of whether or not children can utilize probabilistic information to help them identify grammatically ill-formed utterances, and whether or not these patterns are universal. In respect to the latter point it does seem that some individuals can develop language (for example by learning to control the output of a computer word processor) even though they cannot produce speech because of motor problems (Bishop 1988).

These discussions are further complicated by theory and evidence suggesting that some individual differences (largely non-syntactic) in language performance in children result from their adopting different strategies and/or following different developmental routes to adult language (Nelson 1981; Furrow and Nelson 1986; Wells 1986; Goldfield and Snow 1985; Bates et al. 1988, 1993). The most important difference between children appears to be in terms of whether or not they employ a referential or an expressive style. The referential style involves the initial vocabulary's containing a very high proportion of object-names, earlier language development, and greater consistency in their use of linguistic rules. Expressive children tend to use speech for social rather than declarative purposes, and contrast with referential children on the dimensions already described.

15.6.3.2 *Speech and non-linguistic context*

Another source of information that is widely thought to be crucial in enabling the child to crack the linguistic code is the presumed ability of the child to work out the meaning of speech from situations and contexts. This idea was brought into prominence by Macnamara (1972), and plays a crucial role in recent theorizing (Pinker 1984). However, while there is clear evidence of redundancy between linguistic and non-linguistic input to children (for example, Collis 1977; Messer 1978, 1983; Harris et al. 1983), as well as more general support for communicative exchanges from the structure of interaction (Bruner 1983), attempts to demonstrate *precise* mapping between non-linguistic and linguistic signals have largely failed (see, for example, Shatz 1982; Schaffer et al. 1983).

An additional source of clues for discovering the rules of language are interactions between third parties witnessed by the child (Forrester 1988, 1992). Work in the last twenty years has emphasized the importance of the child's *participation* in social interaction. The implications of the child's presence in a bystander role have been virtually ignored; but the time seems ripe for an exploration of the consequences of such experiences for communicative development. Perceiving the connections between successive utterances of interacting third parties seems a likely additional source of information for the child on meaning–syntax relations, though little attention has been paid to this as yet.

15.6.3.3 *Pre-linguistic and linguistic communication*

Studies which have attempted to relate patterns of pre-linguistic social interaction to language development have not met with much success. A review by Bates et al. (1982) of studies which have examined the relation between patterns of social interaction and later linguistic competence revealed that most investigations have failed to discover significant correlations. Such findings indicate that caution should be exercised before assuming that early patterns of social interaction are directly related to language and other associated cognitive abilities. It is of course possible that the lack of significant findings was due to the measures of social and cognitive skills being either inappropriate or not sufficiently sensitive. Another explanation for these results is the presence of a threshold effect; that is, infants typically receive sufficient experience of interaction for the development of cognitive skills such as language, and as a result differences beyond this threshold in pre-verbal communication are not related to later competence.

In considering the lack of statistical relationships between social input to the infant and later linguistic skills it should be remembered that there has been considerable difficulty in identifying capacities of infants which are related to later non-linguistic cognitive skills. Most of the measures of infant competence have been found to provide a poor prediction of later IQ in non-handicapped samples (McCall 1983). Similarly, important predictors of children's and adults' capacities such as SES (socio-economic status) fail to predict IQ in infancy (Golden and Birns 1983). Such findings suggest that differences between infants in their competence and environment do not result in associated differences in ability until two to three years. This makes it a little less surprising that other measures taken in infancy fail to predict later cognitive skills, including language. A major exception to this seems to be the relation between measures of infant attention and competence in childhood (see Bornstein and Sigman 1986). The habituation paradigm used in these experiments could be seen as assessing infants' efficiency at processing information about the similarity and differences between stimuli, a process which has some similarity to concept-formation.

15.6.3.4 *Social interaction and linguistic innovation*

It is apparent that there has been difficulty in providing convincing evidence that social interaction is clearly related to the development of formal linguistic abilities. This has strengthened the position of those who argue the transition from pre-linguistic to linguistic communication is the result of some innate capacity. A prominent version of this argument is Chomsky's theory of parameter-setting. This suggests that infants innately possess hypotheses about the possible linguistic forms that any language can contain. Exposure to the appropriate syntax will set a parameter, which will direct language-acquisition toward a restricted subset of possible languages; and as a result the environment is seen as providing only a limited contribution to the acquisition of seemingly diverse languages.

More recently discussions have concerned the precise way in which parameters are set. It has been suggested that there are default settings (termed unmarked parameters), and that in some circumstances children will adopt the unmarked parameter when producing utterances. For instance, Hyams (1986; 1992*a*) has proposed that the telegraphic speech of children acquiring English is a result of their adoption of an unmarked parameter. Originally, Hyams (1986) suggested that this unmarked parameter resulted in speech being similar to that in languages which do not require the subject to be present at the beginning of the main clause (for example, the equivalent of 'am going to the shop' would be acceptable in some Italian utterances): such an explanation helps to account for the lack of grammatical subjects in telegraphic speech. Hyams (1992*b*) has developed these ideas to take account of criticisms (for example Pizzuto and Caselli 1992); but the proposal still remains controversial, as there are questions about the equivalence of such utterances in Italian and English and about other issues (see Valian 1990; Weissenborn *et al*. 1992).

An issue that faces any parameter-setting account is how to explain the fact that children produce ungrammatical utterances even though they hear the grammatical utterances of adults. If parameter-setting were operating without any errors this should not occur. Furthermore, it is also necessary to account for the developmental change in children's speech: why does language develop if parameters have been set? Two types of explanation of this phenomenon have been put forward. One, termed the continuity hypothesis, supposes that children's ability to process adult speech changes with age (Clahsen 1992). In this way it can still be assumed that children have access to the principles of universal grammar and to examples of appropriate language, but fail to produce grammatical speech from an early age. Thus, the limitation in language development is located in the capacity of children to identify and process grammatical structures in the speech that they hear.

An alternative view has been termed the maturational hypothesis. In a 'strong form' of this approach, Felix (1987; 1992) has argued that that the various principles of universal grammar bcome operational at different ages. The evidence Felix uses to support this argument comes from two-word and telegraphic speech, with the examples consisting of children's utterances which have grammatical characteristics unlike those that are present in naturally occurring adult languages. A problem with this type of evidence is that some authorities argue that two-word and telegraphic speech cannot be considered language (Radford 1990). In addition, Clahsen (1992) has questioned whether the examples provided by Felix violate the principles of universal grammar.

Bates (1993) has criticized the parameter-setting approach on a number of grounds. A focus for these criticisms is the idea that language processing involves cognitive modules that are specialized for this function and consequently are separable from other, more general cognitive processes. Such claims seek to separate language from more general aspects of human abilities. Bates *et al*. reject this argument; instead, Bates and her colleagues argue that there are local homologies between cognition and language at certain ages, so that both seem to develop as a result of a single underlying change. For example, Bates *et al*. (1987) have suggested that the change from one-to two-word speech may be a product of an increasing capacity to process more units of information. Some support for this prediction has come from the changes which occur at about 18 months in non-delayed children (Bates *et al*. 1987), and from similar changes which occur in children with Down's syndrome at about the same mental age (Messer and Hasan 1990). Bates (1993) also joins with those that believe that the mechanism of language-acquisition is similar to that occurring in connectionist networks, and consequently is a general rather than a specific cognitive process.

There are difficulties in evolutionary terms in explaining the evolution of a parameter-setting mechanism (Bates *et al*. 1991): how could the mechanism be acquired in a relatively short historical time-scale; what was the phylogenetic sequence between the development of speech and the development of parameter-setting; and why in genetic terms was it necessary to have a complex process whereby not one but many languages can be acquired from a variety of inputs (see Lieberman 1989)?

Although current perspectives tend to minimize the relevance of social interaction for language-acquisition, there are some isolated findings which suggest it is of importance. It would appear that mere exposure to linguistic input which is not tied to social interaction is not sufficient for language-acquisition. Ervin-Tripp (1973) has reported that Dutch children who are exposed to German television do not acquire the German language. Similarly, Moscowitz (1978) reports that hearing children of deaf parents do not acquire spoken language when they are only exposed to radio and television. In contrast, some studies have identified certain characteristics of child-rearing environments which are related to later linguistic and cognitive gains. One of the most important characteristics appears to be the responsiveness of parents to infant communication (Clarke-Stewart et al. 1979; Bee et al. 1982; Olson et al. 1986).

In addition, it would appear that language innovation occurs in circumstances where there has been a history of communication between individuals. For example, there are claims that some forms of language can develop without a child's being exposed to a linguistic model. Such a case is the development of idioglossia. This phenomena has occurred with twins who have developed their own language (Gorney 1979; Howes 1977), which did not appear to be closely related to that provided by adult models. Furthermore, it is claimed that deaf children can develop their own very simple syntax of sign-communication independently of adult models (Feldman et al. 1978; Goldin-Meadow and Mylander 1984). However, some investigators working with children in similar circumstances have failed to find such constructions (Gregory and Mogford 1981). Additionally, Bickerton (1977) has suggested that the transition from simple pidgin to more complex creole languages occurs in conversations between children whose parents use a pidgin language. The interesting implication from these findings is that social interaction rather than examples of language is sufficient for the development of linguistic forms. However, it is worth noting that Bickerton (1977) interprets his data as suggesting that new forms of language are the result of children's utilizing the rules contained in unmarked parameters to formulate new forms of syntax based on the pidgin language.

15.6.3.5 *Summary*

The role of social interaction in language-acquisition is still far from clear. We know that most children succeed in learning the language to which they are exposed, even though there must be wide variation in the nature of their linguistic experiences. In that sense, language-acquisition is clearly robust and relatively undemanding of special types of social input. This point is emphasized by cross-cultural studies, which are considered in the next section. However, it would appear that experience of linguistic input in the context of social interaction is needed for speech to be acquired; furthermore, it would seem that given a basic pattern of social interaction children can produce linguistic innovations without being exposed to model utterances or even language itself.

15.7 The influence of social processes

15.7.1 *Cross-cultural comparisons of social interaction and language-acquisition*

It should be recognized that definitions of intelligence and cognitive skills reflect cultural assumptions. Luntz and Levine (1983) have examined several cultures in order to show that there are differences in the assumptions made about infants' capacities and the way that these develop. Thus, one should not forget that different cultures will view the capacities of infants and the eventual expression of cognitive skills in different ways.

Most of the research on adult–infant social interaction has been conducted on white, middle-class, Western mothers and infants. Without counter-evidence it has been easy to assume that behaviours observed in these samples reflect universal patterns of the way that children are introduced into their culture. These assumptions have been challenged by a number of reports about child-rearing practices in other cultures. In reviewing this material Ochs and Schieffelin (1984) have suggested that social interaction between infants and adults reflects cultural attitudes and beliefs about children. They suggest that Western culture adapts behaviour and objects to the capacities of infants, and infants are treated as if they have the capacity to engage in communication. By way of contrast they draw attention to quite different child-rearing practices in Samoa and in Papua New Guinea.

One of these divergent patterns of child-rearing described by Ochs and Schieffelin is that of the Kaluli of Papua New Guinea. Their society assigns an important role to conversation. Conversations take place against a complex pattern of relationships based on obligation and reciprocity. Ochs and Schieffelin state that acquiring language and independence is an important goal of Kaluli child-rearing. However, the Kaluli do not talk about the feelings or thoughts of others, a norm which has consequences for child-rearing practices.

The Kaluli infant's mother is the primary caregiver, and the mothers are responsive and attentive to their children. Mothers almost never leave their infants, but often engage in activities with others. However, infants are not treated as conversational partners, because of the belief that infants are not capable of understanding verbal exchanges. As a result, few utterances are addressed to infants; those that are usually involve calling the infant's name or the use of expressive vocalizations. Although the mothers do not talk *to* their infants, the mothers will talk *for* their infants. For example, if another child addresses an infant, the mother of the infant will reply in a high-pitched, nasalized voice. Furthermore, the mother's speech will be appropriate for the older child, not similar to the baby speech used in Western homes. Unlike the attempts at conversation in Western families, the speech of the Kaluli mother appears unrelated to the activities of the infant. Ochs and Schieffelin suggest that these exchanges are designed to foster social relationships rather than to teach language. Another feature of these early interactions is that Kaluli mothers do not engage in mutual gaze; instead they usually put their infants in a position where they can observe and be observed by others.

When the infants are between six and twelve months old adults start to speak to the infants using imperatives. These are usually short utterances, and although the utterances may contain questions these are usually rhetorical, with no verbal response from the infant being required; rather, the infant is expected to comply with the imperative. These limited verbal interactions continue until children start to use the words for mother and breast, at which point it is considered that children can be taught to speak. Teaching involves providing a model utterance and instructing the child to repeat it. Such utterances are usually concerned with the social uses of language rather than object-labelling. These model utterances are not modified according to the child's linguistic ability. In fact, mistakes by the child are corrected by adults, in an apparent effort to make the child a more competent speaker.

Some caution should be exercised before concluding on the basis of these cross-cultural studies that motherese-type modifications of adult speech to infants are not important for language-acquisition. It should be remembered that modification of speech to infants appears to occur in a wide variety of cultures (Ferguson 1964); that these modifications are even made by children of three or four years of age (Shatz and Gelman 1973); and that the modifications appear finely tuned to infants' perceptual capacities (Fernald 1984).

Given these studies of social interaction, what can we conclude about the role of culture in developing cognitive skills? To begin with, it would appear that by two years of age infants have not only become sensitive to verbal and non-verbal communication, but that this sensitivity also reflects the pattern of communication that occurs in the culture to which the child belongs. This latter process occurs because the communication of adults to children reflects the wider assumptions of the culture. It would also appear that the acquisition of one of the most important cognitive skills, that of language, is not necessarily through adaptation to the child's needs, but through an emphasis on certain patterns of culturally-appropriate exchanges. Different cultures emphasize different functions to which language can be put, and these functions are those emphasized in speech to children.

15.7.2 Social interaction and cognitive development

Tronick *et al.* (1985) have speculated about the relation between culture, social interaction, and the development of skills. They follow Sander (1977) in suggesting that infants become entrained to the stimulation that they receive, be this social interaction or diurnal rhythms. Infants adapt to the entraining stimuli, and later become able to function without them. As Tronick *et al.* point out, infants in different cultures will become entrained to the patterns of interaction in their own cultures. The implication from this position is that the development of skills depends not on the absolute sensory experience of the infant, but on the sensory experience of the infant *in relation* to the experiences which are characteristic of the culture. Thus, infants may fail to progress when their experience departs from that of their culture, rather than when it falls below some absolute level. This interpretation helps to explain why child-rearing practices which, from a Western perspective, might be thought to have devastating consequences on development do not appear to adversely affect the child. A classic example of this is Kagan and Klein's (1973) finding that although infants of a tribe in Guatemala are confined to a darkened hut for their first year of life, this does not appear to hinder their later development.

Similarly, a number of investigations of infants who have experienced severely depriving conditions have reported remarkable recoveries. One of the most famous is a study by Dennis (1973) which involved children in a Lebanese orphanage who received minimal care and social interaction during their first years of life. The infants were left in their cots most of the day, there was a poor ratio of staff to children, the children received little individual attention, and there were few toys or provisions for stimulation. At twelve months these children had a mean score of only 50 on standardized tests, where the average score would be 100. However, children who were adopted before two years showed a recovery of performance to average levels, and even children who remained in the institution, but were later given suitable schooling, made sufficient gains for their IQ to be considered within the range for non-retarded children.

A more recent case-report has documented how a child who suffered both the lack of social care and persistent ill health during his first eighteen months but was later adopted into a North American family, made recovery to above-average performance on tests of intelligence (Thompson 1986). Thus it is apparent that sustained social interaction during the first two years of life is not necessary for the development of cognitive and symbolic skills. Children who experience severe social and physical deprivation may at the time have impaired performance; but quite dramatic improvements can occur if they later experience more normal child-rearing conditions.

It should also be remembered that there is an important tradition which emphasizes that children's cognitive development is in part due to their own motivation to develop competence and to master problems. White (1959) drew attention to the pleasure that children derived from mastering problems, and suggested that there is an instrinsic motivation for competence. Hunt (1965) suggested that children have a motivation to resolve incongruity, and that such a motivation will lead to cognitive development. Harter (1978, 1983) has added a social dimension to these theories by proposing that individual differences emerge because of different patterns of socialization. According to Harter's model the way that adults react to successful and unsuccessful attempts to solve problems will lead to an internalization of the adult's standards by the child. In this way children may be encouraged either to make their own independent attempts or to become dependent on the advice and evaluation of others.

More recently, Messer (1993) has speculated that the form of social interaction occurring in the first half-year of life might have important consequences for the development of sustained attention and habituation. Parents who are able to maintain the interest of their infant by providing appropriate changes in interaction when infant attention starts to diminish may be developing the capacity of their infant to engage in sustained and persistent activities. Furthermore, changing the social stimuli when infants are becoming inattentive may promote the growth of the information-processing capacities of infants.

There have been difficulties in measuring infant motivation and competence. However, studies are beginning to suggest that there may be a relation between infants' attention and persistence in tasks at six and twelve months and their later performance at thirty months on an intelligence test (Messer *et al.* 1986; Yarrow *et al.* 1983). In these studies the relations were found to be stronger for girls than boys, and stronger at six months than at twelve months. It will be of interest to discover whether early differences in attention and persistence can be related to differences in social experience. There is already a suggestion that parental sensory stimulation may be associated with persistence at problem-solving (Yarrow *et al.* 1984).

Thus, although we have emphasized the role of social interaction in the development of cognitive skills, we also acknowledge that there must be reservations about its influence. For example, the supposition that pre-linguistic interaction is a precursor to early forms of syntactic structure is probably incorrect. In addition, it would appear that individuals who have been deprived of infant social experiences can recover and function within their culture, as the studies of Dennis and of Kagan and Klein suggest. Thus very early social interaction may not be necessary for later social and cognitive skills, since such skills can be largely recovered at 2–3 years or more of age. Findings such as these raise issues about the way that influences on infant development should be conceptualized. It has been extremely difficult to establish that particular experiences are necessary for the development of any particular psychological capacity. A reason for this may be that normally human psychological development is so well buffered that the absence of certain experiences can be compensated by the presence of other experiences, or even by the presence of similar experiences at a somewhat later age (Snow and Gilbreath 1983). If such buffering is common it would mean that even if social processes are usually involved in the development of cognitive skills, it would be difficult empirically to isolate the contribution of any one particular form of social experience.

15.8 Communication with language and the extension of cognitive skills

Social interaction appears to put children in a position to gain information which could help cognitive advancement. In industrial societies, probably the most important cognitive skill acquired during the pre-school years is language. As has already been discussed, the prevailing view is that social interaction is not implicated in the acquisition of syntax beyond providing examples of language which can supply the information necessary to work out the rules and regularities present in the adults' language. However, it is clear that the emergence of language changes the range and complexity of social interaction. This section considers the influence of interaction using language on the development of cognitive skills and abilities.

The examination of the effects of social interaction on cognitive development of school-aged children has received increasing attention (for reviews see Light 1983; Tudge and Rogoff 1989). A number of studies have investigated whether, as Piaget predicted, conflict amongst peers forces children to decentre and take account of the views of another, thereby initiating cognitive advancement. Investigations have found that children will make cognitive advances after working in groups (Light 1983), although such effects are not always present (Messer et al. 1993; Tudge 1992). Similarly, children have been found to make cognitive advances when exposed to conditions where the experimenter discusses with children the reasons for their attempts, rather than simply demonstrating the solution (Heber 1981). Thus school-age children when communicating in groups sometimes appear to derive benefit from social interaction with individuals of similar ability, and in circumstances where discussion occurs.

Other research has examined adult–child social interaction and children's competence. The relation between effective interpersonal communication and cognitive development has been highlighted by findings from Dickson et al. (1979). Mothers and their four-year-old children took part in a referential task where the mothers could only use speech to instruct their children to select one of four pictures. The accuracy with which children selected the correct picture was found to be significantly related to the children's later IQ and other cognitive scores which were obtained at six years. This relation remained when the influence of maternal education, SES, and children's competence were partialled out, suggesting that the correlation was not simply one of cognitive ability. Robinson and Robinson (1981) examined the relation between early naturally occurring conversations and later performance on a referential task. They found that in families where mothers told their six-year-old child when they could not understand his or her utterance, the child performed better on referential communication tasks. Robinson and Robinson (1982) have also been able to demonstrate experimentally that giving explicit feedback about ambiguity in children's referential messages improves their performance on a task.

The relation between maternal communication and children's competence has also been investigated by Hartmann and Haavind (1981). They assessed mothers' teaching strategies with their own and another six-year-old child, while the children's competence was assessed from a competitive game with the experimenter. It was found that mothers who were sensitive to children's misunderstandings when giving instructions or advice were the mothers whose children performed better on a task with an experimenter. Thus, with these older children, the mothers' ability to engage in effective communication with their children appears to be related to various measures of children's cognitive skill. A series of studies by Siegel (1986) suggest that the level of cognitive demands placed on three-to six-year-old children by their parents is related to concurrent scores on standardized tests and to representational abilities. Siegel has interpreted his findings as supporting the notion that the use of optimal demands promotes cognitive growth.

Another line of research suggesting that maternal responsiveness is important for successful performance in social activities concerns scaffolding activities. This term has been used by Wood and Middleton (1975) to refer to the way that adults structure interaction to support children's problem-solving activities. Wood and Middleton (1975) found that four-year-old children's independent performance on a problem was related to the way their mothers had previously aided them on the same problem. Mothers who were flexible and sensitive to the needs of their children tended to have children who subsequently performed better. An experimental investigation of this effect confirmed these findings (Wood et al. 1978).

As has already been mentioned, one of the best predictors of children's IQ is the SES of the parents (Lock and Symes, this volume, Chapter 8). A number of studies have suggested that the effects of SES may be related to communication between parents and children. Hess and Shipman (1965) reported differences between mothers in their teaching style with children corresponding to SES, and these differences were related to the children's performance. Other investigators have speculated that differences in maternal style may be responsible for class-related differences in cognitive performance (Brophy 1970).

Currently, there is increasing interest being shown in the relation between social interaction and the development of specific cognitive skills. Durkin et al. (1986) have called attention to the role that conversations about number may have in the acquisition of number concepts. Edwards and Middleton (1988) have argued that the investigation of the development of children's capacity to use memory should be firmly placed within an interactive context. Their view is that the process of remembering should be seen as a capacity which develops between people, rather than in isolation.

Language allows children to extend the range and complexity of the topics of social interaction. Such conversations usually take place in the context of culturally agreed conventions and activities. The material reviewed in this section indicates that interaction amongst peers facilitates cognitive advances, while effective communication from parents is associated with higher ability levels. These findings illustrate the continuing importance of social interaction for the cognitive development of children, and also suggest the importance of early experiences in laying the foundations for later skills.

15.9 Summary

New-borns appear to have a basic set of inherited responses which ensure that they can engage in social interaction with adults. Infants are not passive respondents in this process. They have also inherited a repertoire of social activities and biases about the types of stimuli to which they attend and respond. In a similar way it is likely that adults have inherited dispositions which influence the way that they respond to infant signals and the way they establish relationships with their offspring. Analysis of social interaction between mothers and young infants suggests that much of its structure is derived from the mothers', rather than the infants', contributions to its synchrony and integration.

The behaviour occurring during social interaction is likely to stimulate and foster cognitive development. Even if it were true that, during the first two years, maturational changes were responsible for most of the advances in cognitive skills, it would be important to remember that these advances do not occur in a vacuum, but in the context of the infants' social environment. The physical and social worlds of infants are largely the result of caregivers' making decisions about what is appropriate and suitable for their child. These decisions in turn are influenced by cultural assumptions about child-rearing.

Social interaction provides the first exposure of infants to their culture; but this exposure is filtered by the social interaction they experience with particular caregivers. These early interactions may be important in providing the basis for infants to develop a working model of the individuals to whom they are attached, and may also influence the type of attachment pattern that occurs. It would be surprising if some type of attachment pattern is not part of our species-typical behaviour. The lack of attachment only appears to occur in exceptional and pathological circumstances; otherwise attachment will develop between infants and their care-givers. Within these attachment relationships, infants can 'utilize' adult knowledge and memories to provide a scaffold for shared experiences. The presence of these shared experiences provides the basis for children to interact with other members of their culture, and to establish a culturally-based perception of salient aspects of their environment.

As children's social relationships diversify so their acquisition of culture will become less dependent on the filter provided by the primary care-givers; the important primary relationships which were formed during the attachment period in infancy remain, but these are supplemented with a more widespread set of relationships with adults and other children. What is unsurprising, but often unnoticed, is that at the same time that children are beginning to use symbols and language they are also beginning to move into a wider set of relationships that permit them to make fuller use of these skills. The ability to use symbols and language allows them to understand and acquire knowledge from a wider circle of persons in their culture.

In considering the influence of social interaction on cognition it is important to make the basic contrast between the raw machinery of cognition (such as the information-processing capacity, the memory capacity, and the ability to use symbols) and the content of cognition (such as the memories, strategies, thought-processes, and types of symbols). The role of social interaction in developing the raw machinery of cognition is usually dismissed, because it is commonly believed that such developments are under genetic control; but if we extrapolate from animal models, it is not unreasonable to suppose that the level of stimulation in the infant's social environment could make a contribution to the physiological processes underlying the development of cognitive capacity (Kolata 1984). Furthermore, the studies of children in impoverished environments illustrate the way in which early cognitive functioning can not only be impaired but can also recover if children are exposed to improved conditions. Here it is relevant to note that the easiest way to provide a richer environment for an infant is to provide people rather than inanimate objects or events.

Turning to the content of cognition, the memories, strategies, thought-processes, and types of symbol are, to a large extent, influenced by the culture the child experiences. The shared memories which develop between care-giver and child may provide an entry into the one-word stage; the use of shared topics establishes objects and events which are of interest to caregiver and child; and imitation can be used to acquire new gestures and actions.

We have also argued that the development of infants is influenced by three overlapping and interrelated dimensions: those of social interaction, cognitive skills, and cultural practices. The use of symbols marks an important transition between the period of the infant's early social interaction and cognition, which is largely based on experiences within a limited but secure social environment, and the period when the use of language takes children into a wider commerce with the members of their culture. Basic to our conception of the relationship between social interaction and cognition is that these dimensions influence one another in a transactional manner, much as Sameroff and Chandler (1975) have described the effects of children and their environment on one another. Changes in cognition allow there to be changes in the form of social interaction, while the process of and information contained in, social interaction influence cognition and can result in cognitive advances. Each influence of one dimension on the other results in a change of the affected dimension, which in turn promotes a reciprocal change in the dimension which originally produced the effect.

Although we believe that a transactional model provides a useful perspective to view the interplay between social interaction and cognitive skills, we also believe that for some advances there need to be particular characteristics in the child's social environment and cognitive apparatus. A prime candidate as an example of this is the acquisition of language. There are indications from a limited literature that children *will not* develop language in cases where there is exposure to speech without social interaction, yet *will* develop a form of language where there is no exposure to speech or sign, but there is exposure to social interaction.

Our view is that the use of symbols provides a crucial step between the child's first social world, where interaction occurs in relation to a limited set of caregivers with whom there are usually strong attachments, and the wider social interactions with other diverse members of the culture. These latter interactions no longer draw on the shared memories which have been built up by caregiver and infant. Instead, such social interactions take place in the context of the shared assumptions of the culture. The use of symbols makes possible the use of language, and, with it, the ability to construct and interpret new meanings. Thus, cognition and interaction provide the basis for the development of symbols. In turn the use of symbols allows for communication to be separated from the here and now, and for explanations and knowledge to be represented. This, in turn, allows conventions, rituals, laws, and organizations to be passed from one generation to another.

The ability to use symbols marked a profound advance in human evolution. How such a change came remains a mystery. However, ontogenetic development suggests that both social and cognitive capacities would underline such an advance. Bates *et al.* (1979) have suggested that, phylogenetically, symbol use was the result of different skills' coming together, such as tool-use, imitation, and means–end relations. A similar ontogenetic claim about the way language emerges has been put forward by Bloom (1983) and Shatz (1983), who both see language as occurring because of the integration of different domains. An alternative viewpoint suggested in Bates's (Bates *et al.* 1987) later writings is that ontogenetic changes in the information-processing capacity of infants might enable both cognitive and communicative advances to occur.

We have seen that symbol development can be regarded as an important facet of children's cognitive development, located in a web of interpersonal and cultural influences. In turn, the ability to use symbols is crucial to the development of culture. Our argument is that, throughout early development, there is a complex interplay between the development of cognitive structures, which permit new levels of social functioning, and the processes of social interaction, which support and give content to these cognitive structures. In our view it would be a mistake to try to identify a single causal or essential factor in the phylogenetic development of symbol use, because of the obvious complexity of the whole process. The evolutionary influences on symbol-formation could range from the specific—in terms of developing the cognitive underpinnings for symbol use—to the very general—in terms of social adaptations of infants and social dispositions of adults to infants. Thus, the study of ontogeny shows that there are a number of complex factors (social, cognitive, cultural, physiological) which contribute to the development of human capacities.

References

Abravanel, E. and Sigafoos, A. O. (1984). Exploring the presence of imitation during early infancy. *Child Development*, **55**, 381–92.

Ainsworth, M. D. S. (1973). The development of infant–mother attachment. In *Review of child development research*, Vol. 3 (ed. B. M. Caldwell and H. N. Ricciuti), pp. 1–94. University of Chicago Press.

Ainsworth, M. D. S. (1989). Attachment beyond infancy. *American Psychologist*, **44**, 709–16.

Ainsworth, M. D. S. and Bell, S. M. (1969). Some contemporary patterns of mother–infant interaction in the feeding situation. *Stimulation in early infancy* (ed. J. A. Ambrose), pp. 133–70. Academic Press, London.

Ainsworth, M. D. S. and Wittig, B. A. (1969). Attachment and exploratory behaviour of one-year-olds in a strange situation. In *Determinants of infant behaviour*, Vol. 4, (ed. B. M. Foss), pp. 111–36. Methuen, London.

Ainsworth, M. D. S., Bell, S. M., and Stayton, D. J. (1971). Individual differences in strange situation behaviour of one-year-olds. In *The origins of human social relations* (ed. H. R. Schaffer), pp. 17–52. Academic Press, London.

Ainsworth, M. D. S., Bell, S. M., and Stayton, D. J. (1972). Individual differences in the development of some attachment behaviours. *Merrill–Palmer Quarterly*, 18, 123–43.

Ainsworth, M. D. S., Bell, S. M. and Stayton, D. J. (1974). Infant–mother attachment and social development. In *The integration of a child into a social world* (ed. M. P. M. Richards), pp. 99–135. Cambridge University Press, London.

Ainsworth, M. D. S., Blehar, M. C., Waters, E., and Wall, S. (1978). *Patterns of Attachment*. Erlbaum, Hillsdale, N. J.

Altmann, J. (1980). *Baboon mothers and infants*. Harvard University Press, Cambridge, Mass.

Baldwin, D. A. (1991). Infants' contribution to the achievement of joint reference. *Child Development*, 62, 875–90.

Banks, M. S. and Salapatek, P. (1983). Infant visual perception. In *Handbook of child psychology*, Vol. 2, *Infancy and development psychobiology* (ed. M. M. Haith and J. J. Campos), pp. 436–571. Wiley, New York.

Baron-Cohen, S. and Ring, H. (1994). A model of the mindreading system: neuropsychological and neurobiological perspectives. In *Origins of an understanding of mind* (ed. C. Lewis and P. Mitchell), 183–207. Erlbaum, Hillsdale, NJ.

Bartholomew, K. and Horowitz, L. M. (1991). Attachment styles among adults: a test of a four-category model. *Journal of Personality and Social Psychology*, 61, 226–44.

Bates, E. (1993). *Modularity, domain specificity and the development of language, Project in Cognitive Neuroscience Technical Report 9305*. Center for Research in Language, University of California, San Diego.

Bates, E. and McWinney, B. (1982). Functionalist approaches to grammar. In *Language acquisition: the state of the art*, (ed. E. Warner and C. R. Gleitman), pp. 173–218. Cambridge University Press.

Bates, E., Camaioni, L., and Volterra, V. (1975). The acquisition of performatives prior to speech. *Merrill–Palmer Quarterly*, 21, 205–26.

Bates, E., Benigni, L., Bretherton, I., Camaioni, L., and Volterra, V. (1979). *The emergence of symbols: cognition and communication in infancy*. Academic Press, New York.

Bates, E., Bretherton, I., Beeghly-Smith, M., and McNew, S. (1982). Social bases of language development: a reassessment. In *Advances in child development and behavior* (ed. H. W. Reese and L. P. Lipsitt), pp. 173–218. Academic Press, New York.

Bates, E., O'Connell, B., and Shore, C. (1987). Language and communication in infancy. In *Handbook of infant development* 2nd edn, (ed. J. Osofsky), pp. 149–203. Wiley, New York.

Bates, E., Bretherton, I., and Snyder, L. (1988). *From first words to grammar*. Cambridge University Press.

Bates, E., Thal, D. L., and Marchman, V. (1991). Symbols and syntax: a Darwinian approach to language development. In *Biological and behavioural determinants of language development* (ed. N. Krasnegor, D. Rumbaugh, R. Schiefelbusch, and M. Studdert-Kennedy), pp. 29–65. Erlbaum, Hillsdale, NJ.

Bates, E., Dale, P. S., and Thal, D. (1993). Individual differences and their implications for theories of language development. In *Handbook of child language* (ed. P. Fletcher and B. MacWhinney). Basil Blackwell, Oxford.

Bates, J. E., Maslin, C. A., and Frankel, K. A. (1985). Attachment security, mother–infant interaction, and temperament as predictors of behaviour-problem ratings at three years. *Monographs of the Society for Research in Child Development*, 50, Serial No. 209, 167–93.

Batter, B. S. and Davidson, C. V. (1979). Wariness of strangers: reality or artifact? *Journal of Child Psychology and Psychiatry*, 20, 93–110.

Bee, H. L., Barnard, K. E., Eyres, S. J., Gray, C. A., Hammond, M. A., Spietz, A. L., et al. (1982). Prediction of I.Q. and language skill from perinatal status, child performance, family characteristics, and mother–infant interaction. *Child Development*, 53, 1134–56.

Benedict, H. (1979). Early lexical development: comprehension and production. *Journal of Child Language*, 6, 183–200.

Bergeman, C. S. and Plomin, R. (1988). Parental mediators of the genetic relationship between home environment and infant mental development. *British Journal of Developmental Psychology*, 6, 11–21.

Berger, J. and Cunningham, C. C. (1983). Development of early vocal behaviour and interactions in Down's syndrome and non-handicapped infant–mother pairs. *Developmental Psychology*, 19, 322–31.

Bertenthal, B. I., Proffitt, D. R., Spetner, N. B., and Thomas, M. A. (1985). The development of infant sensitivity to biomechanical motions. *Child Development*, 56, 531–43.

Bickerton, D. (1977). Pidginization and creolization: language acquisition and language universals. In *Pidgin and creole linguistics* (ed. A. Valdman), pp. 49–70. Indiana University Press, Bloomington.

Bischof, N. (1975). A systems approach towards the functional connections of attachment and fear. *Child Development*, 46, 801–17.

Bishop, D. (1988). Language development in children with abnormal structure or function of the speech apparatus. In *Language development in exceptional circumstances* (ed. D. Bishop and K. Mogford), pp. 203–19. Churchill Livingstone, Edinburgh.

Blehar, M. C., Lieberman, A. F., and Ainsworth, M. D. S. (1977). Early face-to-face interaction and its

relation to later mother–infant attachment. *Child Development*, **48**, 182–94.
Bloom, L. (1983). Of continuity and discontinuity, and the magic of language development. In *The transition from prelinguistic to linguistic communication* (ed. R. M. Golinkoff), pp. 79–92. Erlbaum, Hillsdale, NJ.
Bohannon, J. N. and Stanowicz, L. (1988). The issue of negative evidence: Adult responses to children's language errors. *Developmental Psychology*, **24**, 684–689.
Bohannon, J. N., MacWhinney, B., and Snow, C. (1990). No negative evidence revisited: beyond learnability, or who has to prove what to whom. *Developmental Psychology*, **26**, 221–6.
Bornstein, M. H. and Sigman, M. D. (1986). Continuity in mental development from infancy. *Child Development*, **57**, 251–74.
Bowerman, M. (1989). Learning a semantic system: what role do cognitive predispositions play? In *The teachability of language* (ed. M. L. Rice and R. L. Schiefelbusch), pp. 133–70. Brooks, Baltimore.
Bowlby, J. (1969). *Attachment and loss*. The Hogarth Press, London.
Boyd, R. and Richerson, P. J. (1986). *Culture and the evolutionary process*. Chicago University Press.
Boysson-Bardies, B., Sagart, L., and Durand, C. (1984). Discernible differences in the babbling of infants according to target language. *Journal of Child Language*, **11**, 1–15.
Bretherton, I. (1985). Attachment theory: retrospect and prospect. *Monographs of the Society for Research in Child Development*, **50**, 30–5.
Brophy, J. E. (1970). Mothers as teachers of their own preschool children. *Child Development*, **41**, 79–94.
Brown, R. and Hanlon, C. (1970). Derivational complexity and order of acquisition in child speech. In *Cognition and the development of language* (ed. J. R. Hayes), Wiley, New York.
Bruner, J. S. (1972). The nature and uses of immaturity. *American Psychologist*, **27**, 1–22.
Bruner, J. S. (1975). From communication to language—a psychological perspective. *Cognition*, **3**, 255–87.
Bruner, J. (1983). The acquisition of pragmatic commitments. In *The transition from pre-linguistic to linguistic communication* (ed. R. M. Golinkoff), pp. 27–42. Erlbaum, Hillsdale, New Jersey.
Bushnell, I. W. R. (1982). Discrimination of faces by young infants, *Journal of Experimental Child Psychology*, **33**, 298–308.
Bushnell I. W. R., Sai, F., and Mullin, J. T. (1989). Neonatal recognition of the mother's face. *British Journal of Developmental Psychology*, **7**, 3–16.
Butterworth, G. E. and Cochran, E. (1980). Towards a mechanism of joint visual attention in human infancy. *International Journal of Behavioural Development*, **3**, 253–72.
Byrne, R. W. and Whiten, A. (1988). *Machiavellian intelligence*. Oxford University Press.
Cairns, R. B. (1972). Attachment and dependency: a psycho-biological and social-learning synthesis. In *Attachment and dependency*, (ed. J. L. Gewirtz), pp. 29–80, Winston, New York.

Campos, J., and Stenberg, C. (1981). Perception, appraisal and motion: the onset of social referencing. In *Infant social cognition* (ed. M. Lamb and L. Sherrod), pp. 273–314. Erlbaum, Hillsdale, NJ.
Campos, J. J., Barrett, K. C., Lamb, M. E., Goldsmith, H. H., and Stenberg, C. (1983). Socioemotional development. In *Handbook of child psychology, Vol. 2: Infancy and developmental psychobiology* (ed. M. M. Haith and J. J. Campos), pp. 783–915. Wiley, New York.
Carey, S. (1993). Ontology and meaning. In *Language and cognition: a developmental perpsective* (ed. E. Dromi), pp. 88–103, Ablex, Norwood, NJ.
Cernoch, J. M. and Porter, R. H. (1985). Recognition of maternal axillary odors by infants. *Child Development*, **56**, 1593–8.
Cheney, D. and Seyfarth, R. (1980). Vocal recognition in free-ranging vervet monkeys. *Animal Behaviour*, **28**, 362–7.
Cheney, D., Seyfarth, R., and Smuts, B. (1986). Social relationships and social cognition in nonhuman primates. *Science*, **234**, 1361–6.
Churcher, J. and Scaife, M. (1982). How infants see the point. In *Social cognition* (ed. G. Butterworth and P. Light), pp. 110–36. Harvester, Brighton.
Clahsen, H. (1992). Learnability theory and the problem of development in language acquisition. In *Theoretical issues in language acquisition* (ed. J. Weissenborn, H. Goodluck, and T. Roeper), pp. 53–76. Erlbaum, Hillsdale, NJ.
Clark, R. A. (1978). The transition from action to gesture. In *Action, gesture and symbol: the emergence of language*, (ed. A. J. Lock), pp. 231–57. Academic Press, London.
Clarke-Stewart, K. A., Vanderstoep, L. P., and Killian, G. A. (1979). Analysis and replication of mother–child relations at two years of age. *Child Development*, **50**, 777–93.
Collis, G. M. (1977). Visual co-orientation and maternal speech. In *Studies in mother–infant interaction* (ed. H. R. Schaffer), pp. 355–78. Academic Press, London.
Collis, G. M. (1981). Social interaction with objects: a perspective on human infancy. In *Behavioural development* (ed. K. Immelmann, G. W. Barlow, L. Petrinovich, and M. Main), pp. 603–20. Cambridge University Press.
Collis, G. M. (1985). On the origins of turn-taking: alternation and meaning. In *Children's single-word speech* (ed. M. D. Barrett), pp. 217–30. Wiley, Chichester.
Collis, G. M. and Schaffer, H. R. (1975). Synchronization of visual attention in mother–infant pairs, *Journal of Child Psychology and Psychiatry*, **16**, 315–20.
Condon, W. S. and Sander, L. W. (1974). Neonate movement is synchronized with adult speech, intentional participation in language acquisition. *Science*, **183**, 99–101.
Dale, P. S. (1976). *Language development*. Holt, Rinehart, and Winston, New York.
De Casper, A. J., and Fifer, W. P., (1980). Of human bonding: newborns prefer their mothers' voices. *Science*, **208**, 1174–6.

de Maio, L. J. (1982). Conversational turn-taking: a salient dimension of children's language learning. In *Speech and language: advances in basic research and practice,* Vol. 8. (ed. N. J. Lass) pp. 159–90. Academic Press, New York.

Demetras, M. J., Post, K. N., and Snow, C. E. (1986). Feedback to first language learners: the role of repetitions and clarification questions. *Journal of Child Language*, **13**, 275–92.

Dennis, W. (1973). *Children of the creche.* Appleton Century Crofts, New York.

Dickson, W., Hess, R., Mikabe, N., and Azum, H. (1979). Referential communication accuracy between mother and child as a predictor of cognitive development in the United States and Japan. *Child Development*, **50**, 53–9.

Dowd, J. M. C. and Tronick, E. Z. (1986). Temporal co-ordination of arm movements in early infancy. *Child Development*, **57**, 762–76.

Dunn, J. (1983). Sibling relationships in early childhood. *Child Development*, **54**, 787–811.

Dunn, J. (1993). *Young children's close relationships: beyond attachment.* Sage, London.

Durkin, K., Shire, B., Riem, R., Crowther, R. D., and Rutter, D. R. (1986). The social and linguistic context of early number word use. *British Journal of Developmental Psychology*, **4**, 269–89.

Edwards, D. and Middleton, D. (1988). Conversational remembering and family relationships: how children learn to remember. *Journal of Social and Personal Relationships*, **5**, 109–14.

Eimas, P., Siqueland, E. R., Jusczylke, P., and Vigorito, J. (1971). Speech perception in infants, *Science*, **171**, 303–6.

Eisenberg, R. (1976). *Auditory competence in early life: the roots of communicative behavior.* University Park Press, Baltimore.

Elias, G., Broerse, J., Hayes, A., and Jackson, K. (1984). Comments on the use of conversational features in studies of the vocalisation behaviours of mothers and infants. *International Journal of Behavioural Development*, **6**, 177–91.

Elias, G., Hayes, A., and Broerse, J. (1986). Maternal control of covocalisation and inter-speaker silences in mother–infant vocal engagements. *Journal of Child Psychology and Psychiatry*, **27**, 409–15.

Ervin-Tripp, S. M. (1973). Some strategies for the first two years. In *Cognitive development and the acquisition of language,* (ed. T. E. Moore), pp. 261–86. Academic Press, New York.

Feinman, S. (1992). *Social referencing and the social construction of reality in infancy.* Plenum, New York.

Feinman, S. and Lewis, M. (1983). Social referencing at ten months: a second order effect on infant's responses to strangers. *Child Development*, **54**, 878–87.

Feldman, H. S., Goldin-Meadow, S., and Gleitman, L. (1978). Beyond Herodotus: the creation of language by linguistically deprived deaf children. In *Action, symbol and gesture: the emergence of language,* (ed. A. Lock), pp. 351–414. Academic Press, New York.

Felix, S. (1987). *Cognition and language growth.* Foris, Dordrecht.

Felix, S. (1992). Language acquisition as a maturational process. In *Theoretical issues in language acquisition* (ed. J. Weissenborn, H. Goodluck, and T. Roeper), pp. 25–52. Erlbaum, Hillsdale, NJ.

Ferguson, C. A. (1964). Baby talk in six languages. *American Anthropologist*, **66**, 103–14.

Fernald, A. (1984). The perceptual and affective salience of mothers' speech to infants. In *The origins and growth of communication,* (ed. L. Feagans, C. Garvey, and R. Golinkoff), pp. 5–29. Ablex, Norwood.

Field, T. M., Woodsom, R., Greenberg, R., and Cohen, D. (1982). Discrimination and imitation of facial expressions by neonates. *Science*, **218**, 179–82.

Field, T., Cohen, D., Garcia, R., and Greenberg, R. (1984). Mother-stranger face discrimination by the ewborn. *Infant Behavior and Development*, **7**, 19–26.

Fogel, A. and Hannan, T. E. (1985). Manual actions of nine-to-fifteen-week-old human infants during face-to-face interaction with their mothers. *Child Development*, **56**, 1271–9.

Forrester, M. A. (1988). Young children's polyadic conversation-monitoring skills. *First Language*, **8**, 201–26.

Forrester, M. (1992). *The development of young children's social cognitive skills.* Erlbaum, Hove.

Forrester, M. (1993). Affording social-cognitive skills in young children: the overhearing context. In *Critical influences on child language acquisition and development* (ed. D. Messer and G. Turner), pp. 40–64. Macmillan, London.

Frodi, A. (1985). Variation in parental and non-parental response to early infant communication. In *The psychobiology of attachment and separation* (ed. M. Reite and T. Field), pp. 263–77. Academic Press, London.

Furrow, D. and Nelson, K. (1986). A further look at the motherese hypothesis: a reply to Gleitman, Newport and Gleitman. *Journal of Child Language*, **13**, 163–76.

Furrow, D., Nelson, K., and Benedict, H. (1979). Mothers' speech to children and syntactic development: some simple relationships. *Journal of Child Language*, **6**, 423–42.

Gelman, S. A. and Byrnes, J. P. (1991). *Perspectives on language and thought.* Cambridge, University Press.

Gewirtz, J. L. (1961). A learning analysis of the effects of normal stimulation, privation, and deprivation on the acquisition of social motivation and attachment. In *Determinants of infant behaviour,* (ed. B. M. Foss), pp. 213–29. Methuen, London.

Gewirtz, J. L. (1978). Social learning in early human development. In *Handbook of applied behaviour analysis,* (ed. A. C. Catamia and T. A. Brigham), pp. Irvington, New York.

Gleitman, L., Newport, E., and Gleitman, H. (1979). The current status of the motherese hypothesis. *Journal of Child Language*, **6**, 423–42.

Gleitman, L., Newport, E., and Gleitman, H. (1984). The current status of the motherese hypothesis. *Journal of Child Language*, **11**, 43–79.

Golden, M. and Birns, B. (1983). Social class and infant intelligence. In *Origins of intelligence* (2nd edn, ed. M. Lewis), pp. 347–98. Plenum, New York.

Goldfield, B. A. and Snow, C. E. (1985). Individual differences in language acquisition. In *The development of language* (ed. J. Berko Gleason), pp. 307–30. Merrill, Columbus, Ohio.

Goldin-Meadow, S. and Mylander, C. (1984). Gestural communication in deaf children: The effects and non-effects of parental input on early language development. *Monographs of the Society for Research in Child Development*, **49**, (serial No. 207). The society, Chicago.

Goren, C. C., Sarty, M., and Wu, P. Y. K. (1975). Visual following and pattern discrimination of face-like stimuli by newborn infants. *Paediatrics*, **50**, 544–9.

Gorney, C. (1979). Twins with a language of their own. *The Boston Globe*, 23 July, p. 21.

Gregory, S. and Mogford, K. (1981). Early language development in the deaf. In *Perspectives on British Sign Language and deafness* (ed. B. Woll, J. Kyle, and M. Deuchar), pp. 218–37. Croom Helm, London.

Grossman, K. E., Grossman, K., Huber, F., and Wartner, U. (1981). German children's behaviour toward their mothers at 12 months and their fathers at 18 months in Ainsworth's strange situation. *International Journal of Behavioural Development*, **4**, 157–81.

Gunther, M. (1961). Infant behaviour at the breast. In *Determinants of infant behaviour*, Vol. 1, (ed. B. M. Foss), Methuen, London.

Harding, C. G. and Golinkoff, R. M. (1979). The origins of intentional vocalizations in prelinguistic infants. *Child Development*, **50**, 33–40.

Harlow, H. F. and Zimmerman, R. R. (1959). Affectional responses in the infant monkey. *Science*, **130**, 421–32.

Harris, M., Jones, D., and Grant, J. (1983). The non-verbal context of mothers' speech to infants. *First Language*, **4**, 21–30.

Harter, S. (1978). Effectance motivation reconsidered: toward a developmental model. *Human Development*, **1**, 34–64.

Harter, S. (1983). A model of intrinsic mastery motivation in children. In *Minnesota Symposium on Child Psychology*, **14**, (ed. W. A. Collins), pp. 241–73. Erlbaum, Hillsdale, NJ.

Hartmann, E. and Haavind, H. (1981). Mothers as teachers and their children as learners: a study of the influence of social interaction upon cognitive development. In *Communication in development* (ed. W. P. Robinson), pp. 129–58. Academic Press, London.

Hayes, A. and Elliott, T. (1979). *Gaze and vocalization in mother–infant dyads: conversation or coincidence?* Unpublished paper presented at the biennial meeting of the Society for Research in Child Development, San Francisco.

Hayes, L. A. and Watson, J. S. (1981). Neonatal imitation: fact or artifact? *Developmental Psychology*, **17**, 655–60.

Hazan, C. and Shaver, P. (1987), Romantic love conceptualised as an attachment process. *Journal of Personality and Social Psychology*, **52**, 511–24.

Heber, M. (1981). Instruction versus conversation as opportunities for learning. In *Communication in development* (ed. W. P. Robinson), pp. 183–202. Academic Press, London.

Hess, R. D. and Shipman, V. (1965). Early experience and the socialization of cognitive modes in children. *Child Development*, **36**, 869–86.

Hinde, R. A. (1970). *Animal behaviour*. McGraw-Hill, New York.

Hinde, R. A. (1976). Interactions, relationships and social structure. *Man*, **11**, 1–17.

Hinde, R. A. (1979). *Towards understanding relationships*. Academic Press, London.

Hinde, R. A. (1983). *Primate social relationships: an integrated approach*. Blackwell Scientific Publications, Oxford.

Hinde, R. A. and McGinnis, L. (1977). Some factors influencing the effects of temporary mother–infant separation—some experiments with rhesus monkeys. *Psychological Medicine*, **7**, 197–212.

Hinde, R. A., Leighton-Shapiro, M. E., and McGinnis, L. (1977). Effects of various types of separation experience on rhesus monkeys 5 months later. *Journal of Child Psychology and Psychiatry*, **19**, 199–211.

Hirsh-Pasek, K., Treiman, R., and Schneiderman, M. (1984). Brown and Hanlon revistied: mothers' sensitivity to ungrammatical forms. *Journal of Child Language*, **11**, 81–8.

Hobson, R. P. (1994). Perceiving attitudes, conceiving minds. In *Origins of an understanding of mind* (ed. C. Lewis and P. Mitchell), pp. 71–93. Erlbaum, Hillsdale, NJ.

Hoff-Ginsberg, E. and Shatz, M. (1982). Linguistic input and the child's acquisition of language. *Psychological Bulletin*, **92**, 3–26.

Howes, E. R. (1977). Twin speech: a language of their own. *The New York Times*, Sept. 11th, p. 54.

Hardy, S. B. (1976). Care and exploitation of non-human primate infants by conspecifics other than the mother. In *Advances in the study of behaviour*, Vol. 6, (ed. J. S. Rosenblatt, R. A. Hinde, E. Shaw, and C. Beer, pp. 101–58. Academic Press, London.

Humphrey, N. K. (1976). The social function of intellect. In *Growing points in ethology* (ed. P. P. G. Bateson and R. A. Hinde), pp. 303–17. Cambridge University Press.

Hunt, J. McV. (1965). Intrinsic motivation and its role in psychological development. In *Nebraska Symposium on Motivation*, Vol. 13, (ed. D. Levine), pp. 189–270. University of Nebraska Press, Lincoln., Nebraska.

Hutt, S. J., Hutt, C., Lenard, H. G., Bernuth, H. V., and Muntjewerff, W. J. (1968). Auditory responsivity in the human neonate. *Nature* **218**, 888–90.

Hyams, N. (1986). *Language acquisition and the theory of parameters*. Reidel, Dordrecht.

Hyams N. (1992a). A reanalysis of null Subjects in child language. In *Theoretical issues in language acquisition* (ed. J. Weissenborn, H. Goodluck, and T. Roeper), pp. 249–68. Erlbaum, Hillsdale, NJ.

Hyams, N. (1992b). Morphosyntactic development in Italian and its relevance to parameter setting models: comments on the paper by Pizzuto and Casselli. *Journal of Child Language*, **19**, 695–709.

Izard, C. E. and Malatesta, C. Z. (1987). Perspectives on emotional development 1: Differential emotions theory of early emotional development. In *Handbook of infant development* (ed. J. Osofsky), pp. 494–554. Wiley, New York.

Jacobsen, S. (1979). Matching behavior in the young infant. *Child Development*, **50**, 425–30.

Johnson, M. H. and Morton, J. (1991). *Biology and cognitive development: the case of face recognition.* Blackwell, Oxford.

Kagan, J. (1979). *The growth of the child.* Methuen, London.

Kagan, J. and Klein, R. E. (1973). Cross-cultural perspectives on early development. *American Psychologist*, **28**, 947–61.

Kaye, K. (1977). Toward the origin of dialogue. In *Studies in mother–infant interaction* (ed. H. R. Schaffer), pp. 89–118. Academic Press, London.

Kaye, K. (1982). *The mental and social life of babies.* Harvester Press, Brighton.

Kleiner, K. A. (1987). Amplitude and phase spectra as indices of infants' pattern preferences. *Infant Behavior and Development*, **10**, 49–59.

Klinnert, M. D., Emde, R. N., Butterfield, P., and Campos, J. J. (1986). Social referencing: the infant's use of emotional signals from a friendly adult with mother present. *Developmental Psychology*, **22**, 427–32.

Koepke, J. E., Hamm, M., and Legerstee, M. (1983). Neonatal imitation: two failures to replicate. *Infant Behavior and Development*, **6**, 97–102.

Kolata, G. (1984). New neurons form in adulthood. *Science*, **224**, 1325–6.

Lamb, M. E. (1986). *The father's role, applied perspectives.* Wiley, Chichester.

Lee, P. C. (1983). Caretaking of infants and mother–infant relationships. In *Primate social relationships* (ed. R. A. Hinde), pp. 146–51. Blackwell Scientific Publications, Oxford.

Lempers, J. D. (1976). Production of pointing, comprehension of pointing and understanding of looking behavior in young children. Unpublished Ph.D. dissertation, University of Minnesota.

Lempers, J. D., Flavell, E. R., and Flavell, J. H. (1977). The development in very young children of tacit knowledge concerning visual perception. *Genetic Psychology Monographs*, **95**, 3–53.

Leung, E. H. L. and Rheingold, H. (1981). Development of pointing as a social gesture. *Developmental Psychology*, **17**, 215–20.

Lewis, V. (1987). *Development and handicap.* Blackwell, Oxford.

Lewis, C. and O'Brien, M. (1987). *Reassessing fatherhood: new observations on fathers and the modern family.* Sage, London.

Lieberman, P. (1989). Some biological constraints on universal grammar and learnability. In *The teachability of language* (ed. M. L. Rice and R. L. Schiefelbusch), pp. 199–266. Brooks, Baltimore.

Light, P. (1983). Social interaction and cognitive development: a review of post-Piagetian research. In *Developing thinking* (ed. S. Meadows), pp. 67–88. Methuen, London.

Lock, A. J. (1978). The emergence of language. In *Action, Gesture and symbol: the emergence of language* (ed. A. J. Lock), pp. 3–18. Academic Press, London.

Lock, A. J. (1980). *The guided reinvention of language.* Academic Press, London.

Luntz, C. and Levine, R. A. (1983). Culture and intelligence in infancy: An ethnopsychological view. In *Origins of intelligence*, (2nd edn, ed. M. Lewis), pp. 327–45. Plenum, New York.

McCall, R. B. (1983). A conceptual approach to baby mental development. In *Origins of intelligence* (ed. M. Lewis), pp. 107–34. Plenum, New York.

Maccoby, E. E. (1980). *Social development.* Harcourt Brace Jovanovich, New York.

MacFarlane, A. (1975). Olfaction in the development of social preferences in the human neonate. In *Parent-infant interaction*, Ciba Foundation Symposium, 33. Elsevier, Amsterdam.

McKenzie, B. and Over, R. (1983a). Young infants fail to replicate facial and manual gestures. *Infant Behavior and Development*, **6**, 85–9.

McKenzie, B. and Over, R. (1983b). Do neonatal infants imitate?: a reply to Meltzoff and Moore. *Infant Behavior and Development*, **6**, 109–11.

Macnamara, J. (1972). Cognitive basis of language learning in infants. *Psychological Review*, **79**, 1–11.

Main, M. and Weston, D. (1981). The quality of the toddler's relationship to mother and to father: related to conflict behaviour and the readiness to establish new relationships. *Child Development*, **52**, 932–40.

Main, M., Tomasini, L. and Tolam, W. (1979). Differences among mothers of infants judged to differ in security. *Developmental Psychology*, **15**, 472–3.

Main, M., Kaplan, N., and Cassidy, J. (1985). Security in infancy, childhood, and adulthood: a move to a level of representation. In *Growing points of attachment theory and research*, Monographs of the Society for Research in Child Development, 50 (ed. I. Bretherton and F. Waters).

Marcus, G. F. (1993). Negative evidence in language acquisition. *Cognition*, **46**, 53–85.

Masters, J. and Wellman, H. (1974). Human infant attachment: a procedural critique. *Psychological Bulletin*, **81**, 218–37.

Matas, L., Arend, R. A., and Sroufe, L. A. (1978). Continuity of adaptation in the second year: the relationship between quality of attachment and later competence. *Child Development*, **49**, 547–56.

Maurer, D. and Young, R. (1983). Newborns' following of natural and distorted arrangements of facial features. *Infant Behavior and Development*, **6**, 127–31.

Mayer, N. K. and Tronick, E. Z. (1985). Mothers' turn-giving signals and infant turn-taking in mother–infant interaction. In *Social perception in infants* (ed. T. M. Field and N. A. Fox), pp. 199–216. Ablex, Norwood, NJ.

Meltzoff, A. and Gopnik, A. (1993). The role of imitation in understanding persons and developing a theory of mind. In *Understanding other minds—perspectives from autism* (ed. S. Baron-Cohen, H. Tager-Flusberg, and D. Cohen), pp. 335–66. Oxford University Press.

Meltzoff, A. N. and Moore, K. M. (1977). Imitation of facial and manual gestures by human neonates. *Science*, **198**, 75–8.

Meltzoff, A. N. and Moore, K. M. (1983*a*). Newborn infants imitate adult facial gestures. *Child Development*, **31**, 78–84.

Meltzoff, A. N. and Moore, K. M. (1983*b*). Methodological issues in studies of imitation: comments on McKenzie and Over and Koepke, *et al. Infant Behavior and Development*, **6**, 103–8.

Meltzoff, A. N. and Moore, M. K. (1992). Early imitation within a functional framework: the importance of person identity, movement, and development. *Infant Behavior and Development*, **15**, 479–505.

Messer, D. J. (1978). The integration of mothers' referential speech with joint play. *Child Development*, **49**, 781–7.

Messer, D. J. (1980). The episodic structure of maternal speech to young children. *Journal of Child Language*, **7**, 29–40.

Messer, D. J. (1983). Adult speech and non-verbal interaction: a contribution to acquisition? In *The transition from prelinguistic to linguistic communication* (ed. R. M. Golinkoff), Erlbaum, Hillsdale, New Jersey.

Messer, D. J. (1993). Mastery, attention, I.Q. and parent-infant social interaction. In *Mastery motivation in early childhood* (ed. D. Messer), pp. 19–35. Routledge, London.

Messer, D. and Hasan, P. (1990). The relationship between cognition, speech comprehension and speech production in children with Down's syndrome. Paper presented at the Vth IASCL Congress, Budapest.

Messer, D. J. and Vietze, P. M. (1988). Does mutual influence occur during mother-infant social gaze? *Infant Behavior and Development*, **11**, 97–111.

Messer, D. J., McCarthy, M. E., McQuiston, S., MacTurk, R. H., Yarrow, L. J. and Vietze, P. M. (1986). Relationship between mastery behavior in infancy and competence in early childhood. *Developmental Psychology*, **22**, 366–72.

Messer, D. J., Joiner, R., Loveridge, N., Light, P., and Littleton, K. (1993). Influences on the effectiveness of peer interaction: children's level of cognitive development and the relative ability of partners. *Social Development*, **2**, 279–94.

Moon, C., Bever, T. G., and Fifer, W. P. (1992). Canonical and non-canonical syllable discrimination by two-day-old infants. *Journal of Child Language*, **19**, 1–17.

Morgan, J. L. and Travis, L. L. (1989). Limits on negative information in language input. *Journal of Child Language*, **16**, 531–52.

Moscowitz, B. A. (1978). The acquisition of language. *Scientific American*, **239**, (November) 92–108.

Murphy, C. M. (1978). Pointing in the context of a shared activity. *Child Development*, **49**, 371–80.

Murphy, C. M. and Messer, D. J. (1977). Mothers, infants and pointing: a study of a gesture. In *Studies in mother-infant interaction* (ed. H. R. Schaffer), pp. 325–54. Academic Press, London.

Myers, R. J., Jarvis, P. A., and Creasey, G. L. (1987). Infants' behavior with their mothers and grandmothers. *Infant Behavior and Development*, **10**, 245–59.

Nash, A. and Hay, D. F. (1993). Relationships in infancy as precursors and causes of later relationships and psychopathology. In *Precursors and causes in development and psychopathology* (ed. D. F. Hay and A. Angold), pp. 199–232. Wiley, Chichester.

Nelson, K. (1973). *Structure and strategy in learning to talk*, Monographs of the Society for Research in Child Development, **38**, (Serial No. 149).

Nelson, K. (1981). Individual differences in language development: implications for development and language. *Developmental Psychology*, **17**, 170–87.

Newport, E. L., Gleitman, H., and Gleitman, I. R. (1977). Mother, I'd rather do it myself: some effects and non-effects of maternal speech style. In *Talking to children* (ed. C. E. Snow and C. A. Ferguson), pp. 109–49. Cambridge, University Press.

Newson, J. (1979). The growth of shared understanding between infant and caregiver. In *Before speech* (ed. M. Bullowa), pp. 207–22. Cambridge University Press.

Ninio, A. and Bruner, J. (1978). The achievement and antecedents of labelling. *Journal of Child Language*, **5**, 1–15.

Ochs, E. and Schieffelin, B. B. (1984). Language acquisition and socialization. In *Culture theory* (ed. R. A. Shweder and R. A. Levine) pp. 276–322. Cambridge University Press.

Olson, S. L., Bayles, K., and Bates, J. C. (1986). Mother-infant interaction and children's speech progress. *Merrill-Palmer Quarterly*, **32**, 1–20.

Oster, H. (1978). Facial expression and affect development. In *The development of affect* (ed. M. Lewis and L. A. Rosenblum), pp. 43–75. Plenum, New York.

Pack, C. (1983). Interactional synchrony as an artifact of microanalysis. *Dissertation Abstracts International*, **43**, 3423A.

Pawlby, S. J. (1977). Imitative interaction. In *Studies in mother-infant interaction* (ed. H. R. Schaffer), pp. 203–26. Academic Press, London.

Penner, S. (1987). Parental responses to grammatical and ungrammatical child utterances. *Child Development*, **58**, 376–84.

Petrovich, S. B. and Gewirtz, J. L. (1985). The attachment and learning process and its relation to cultural and biological evolution. In *The psychobiology of attachment and separation* (ed. M. Reite and T. Field), pp. 259–92. Academic Press, Orlando.

Pinker, S. (1979). Formal models of language learning. *Cognition*, **10**, 243–8.

Pinker, S. (1984). *Language learnability and language development*, Harvard University Press, Cambridge, Mass.

Pinker, S. (1989). *Learnability and cognition: the acquisition of argument structure*. MIT Press, Cambridge, Mass.

Pizzuto, E. and Casselli, M. C. (1992). The acquisition of Italian morphology: implications for models of language development. *Journal of Child Language*, **19**, 491–557.

Radford, A. (1990). *Syntactic theory and the acquisition of English syntax: the nature of early child grammars in English*. Blackwell, Oxford.

Ricks, M. H. (1985). The social transmission of parental behaviour: attachment across generations. *Monographs of the Society for Research in Child Development*, 50, 211–27.

Robinson, E. J. and Robinson, W. P. (1981). Ways of reacting to communication failure in relation to the development of the child's understanding about verbal communication. *European Journal of Social Psychology*, II, 189–208.

Robinson, E. J. and Robinson, W. P. (1982). The advancement of children's verbal referential communication skills: the role of metacognitive guidance. *International Journal of Behavioural Development*, 5, 329–55.

Robson, K. S. (1967). The role of eye-to-eye contact in maternal–infant attachment. *Journal of Child Psychology and Psychiatry*, 9, 13–25.

Rosenfield, H. M. (1981). Whither interactional synchrony? In *Prospective issues in infancy research* (ed. K. Bloom), pp. 71–97. Erlbaum, Hillsdale, NJ.

Rosenthal, M. K. (1973). Attachment and mother–infant interaction: some research impasses and a suggested change in orientation. *Journal of Child Psychology and Psychiatry*, 14, 201–7.

Rutter, D. and Durkin, K. (1987). Turn-taking in mother–infant interaction: an examination of vocalization and gaze. *Development Psychology*, 23, 54–61.

Sagi, A., Lamb., M. E., Lewkowicz, K. S., Shoham, R., Dvir, R. and Estes, D. (1985). Security of infant–mother, –father and –metapelet attachments among kibbutz-reared Israeli children. In *Growing points in attachment theory and research*, Monographs of the Society for Research in Child Development, serial no. 209 (ed. I. Bretheron and E. Waters), pp. 257–75.

Sameroff, A. and Chandler, M. (1975). Reproductive risk and the continuum of caretaking causality. In *Review of child development research* (ed. F. Horowitz), pp. 187–244. University of Chicago Press.

Sander, L. W. (1977). The regulation of exchange in the infant–caretaker system and some aspects of the context–content relationship. In *Interaction, conversation and the development of language* (ed. M. Lewis and L. A. Rosenblum), pp. 133–156. Wiley, New York.

Scaife, M. and Bruner, J. S. (1975). The capacity for joint visual attention in the infant. *Nature*, 253, 265.

Schaffer, H. R. (1971). *The growth of sociability*. Penguin, Harmondsworth.

Schaffer, H. R. (1984). *The child's entry into a social world*. Academic Press, London.

Schaffer, H. R. and Emerson, P. E. (1964). The development of social attachments in infancy. *Monographs of the Society for Research in Child Development*, 29, 1–77.

Schaffer, H. R., Collis, G. M., and Parsons, G. (1977). Vocal interchange and visual regard in verbal and pre-verbal children. In *Studies in mother–infant interaction* (ed. H. R. Schaffer), pp. 291–328. Academic Press, London.

Schaffer, H. R., Hepburn, A., and Collis, G. M. (1983). Verbal and nonverbal aspects of mothers' directives. *Journal of Child Language*, 10, 337–55.

Shatz, M. (1982). On mechanisms of language acquisition: can features of the communicative environment account for development? In *Language acquisition: the state of the art* (ed. E. Wanner and L. Gleitman), pp. 102–27. Cambridge University Press.

Shatz, M. (1983). On transition, continuity, and coupling: an alternative approach to communicative development. In *The transition from prelinguistic to linguistic communication* (ed. R. M. Golinkoff), pp. 337–55. Erlbaum, Hillsdale, NJ.

Shatz, M. and Gelman, R. (1973). The development of communication skills: modification in the speech of young children as a function of the listener. *Monographs of the Society for Research in Child Development*, 38.

Siegel, I. (1986). Early social experience and the development of representational competence. In *Early experience and the development of competence* (ed. W. Fowler), pp. 49–66. Jossey Bass, San Francisco.

Slee, P. T. (1983). Mother–infant vocal interaction as a function of emotional expression. *Early Child Development and Care*, 11, 33–44.

Smith, P. K. and Noble, R. (1987). Factors affecting the development of caregiver–infant relationships. In *Attachment in social networks* (ed. L. W. C. Tavecchio and M. H. IJzendoorn), pp. 93–134. North-Holland, Amsterdam.

Smuts, B. (1985). *Sex and friendship in baboons*. Aldine, New York.

Snow, C. E. and Ferguson, C. A. (1977). *Talking to children*. Cambridge University Press.

Snow, C. and Gilbreath, B. J. (1983). Explaining transitions. In *The transition from pre-linguistic to linguistic communication* (ed. R. M. Golinkoff), pp. 281–96. Erlbaum, Hillsdale, NJ.

Sroufe, L. A. (1977). Wariness of strangers and the study of infant development. *Child Development*, 48, 731–46.

Stern, D. N. (1974). Mother and infant at play: the dyadic interaction involving facial, vocal, and gaze behaviors. In *The effects of the infant on its caregiver* (ed. M. Lewis and L. A. Rosenblum), pp. 187–214. Wiley, New York.

Stern, D. (1977). *The first relationship*. Fontana, London.

Stern, D. N., Jaffe, J., Beebe, B., and Bennett, S. L. (1975). Vocalising in unison and in alternation: two modes of communication within the mother–infant dyad. *Annals of the New York Academy of Sciences*, 19, (263), 89–100.

Suomi, S. J. and Harlow, H. F. (1977). Early separation and behavioural maturation. In *Genetics, environment and intelligence* (ed. A. Oliviero), pp. 197–214. Elsevier, Amsterdam.

Sylvester-Bradley, B. (1985). Failure to distinguish between people and things in early infancy. *British Journal of Developmental Psychology*, 3, 281–92.

Tavecchio, L. W. C. and van IJzendoorn, M. H. (eds) (1987). *Attachment in social networks*. North Holland, Amsterdam.

Thompson, A. M. (1986). Adam—a severely-deprived Colombian orphan: a case report. *Journal Child Psychology and Psychiatry*, **27**, 689–97.
Tomasello, M. and Barton, M. (1994). Learning words in non-ostensive contexts. *Developmental Psychology*, **30**, 639–50.
Tomasello, M. and Farrar, M. J. (1986). Joint attention and early language. *Child Development*, **57**, 1454–63.
Tomasello, M. and Todd, J. (1983). Joint attention and lexical acquisition style. *First Language*, **4**, 197–212.
Trevarthen, C. (1977). Descriptive analyses of infant communicative behavior. In *Studies in mother–infant interaction* (ed. H. R. Schaffer), pp. 227–70. Academic Press, London.
Trevarthen, C. (1979*a*). Communication and co-operation in early infancy: a description of primary intersubjectivity. In *Before speech* (ed. M. Bullova), pp. 321–47. Cambridge University Press.
Trevarthen, C. (1979*b*). Instincts for human understanding are for cultural co-operation: their development in infancy. In *Human ethology* (ed. M. von Cranach, K. Foppa, W. Lepenies and D. Ploog), pp. 530–71. Cambridge University Press.
Trevarthen, C. B. and Hubley, P. (1978). Secondary intersubjectivity: confidence, confiding and acts of meaning in the first year. In *Action, gesture and symbol: the emergence of language* (ed. A. J. Lock), pp. 183–229. Academic Press, London.
Tronick, E. Z., Winn, S., and Morelli, G. A. (1985). Multiple caretaking in the context of human evolution: Why don't the Efé know the Western prescription for child care? In *The psychobiology of attachment and separation* (ed. M. Reite and T. Field), pp. 293–322. Academic Press, Orlando.
Tudge, J. (1992). Processes and consequences of peer collaboration: a Vygotskian analysis. *Child Development*, **63**, 1364–79.
Tudge, J. and Rogoff, B. (1989). Peer influences on cognitive development: Piagetian and Vygotskian perspectives. In *Interaction in human development* (ed. M. H. Bornstein and J. S. Bruner), pp. 17–40. Erlbaum, Hillsdale, NJ.
Uzgiris, I. C. (1976). Organization of sensorimotor intelligence. In *Origins of intelligence* (ed. M. Lewis), pp. 123–63. Plenum, New York.
Valian, V. (1990). Null subjects: a problem for parameter-setting models of language acquisition. *Cognition*, **35**, 105–22.
Vaughn, B., Egeland, B., Sroufe, L. A., and Waters, E. (1979). Individual differences in infant–mother attachment at twelve and eighteen months, stability and change in families under stress. *Child Development*, **50**, 971–5.
Wasz-Hockert, O., Lind, J., Vuorenkoski, T., and Valanne, E. (1968). *The infant cry*. Heinemann, London.
Waters, E. (1978). The reliability and stability of individual differences in infant–mother attachment. *Child Development*, **49**, 483–94.
Waters, E. (1981). Traits, behavioural systems, and relationships: three models of infant–adult attachment.

In *Behavioural development* (ed. K. Immelmann, G. W. Barlow, L. Petrinovich, and M. Main), pp. 621–50. Cambridge University Press.
Waters, E., Wippman, J., and Sroufe, L. A. (1979). Attachment, positive affect, and competence in the peer group: two studies in construct validation. *Child Development*, **50**, 821–9.
Weissenborn, J., Goodluck, H., and Roeper, T. (1992). Introduction: old and new problems in the study of language acquisition. In *Theoretical issues in language acquisition* (ed. J. Weissenborn, H. Goodluck, and T. Roeper), pp. 1–24. Erlbaum, Hillsdale, NJ.
Wells, G. (1986). *The meaning makers: children learning language and using language to learn*. Hodder and Stoughton, London.
Werker, J., Gilbert, J., Humphrey, K., and Tees, R. (1981). Developmental aspects of cross-language speech perception. *Child Development*, **52**, 349–55.
Wexler, K. (1982). A principle for language acquisition. In *Language acquisition: the state of the art* (ed. E. Wanner and L. Gleitman), pp. 288–318. Cambridge. University Press.
Wexler, K. and Culicover, P. W. (1980). *Formal principles of language acquisition*. MIT Press, Cambridge, Mass.
White, R. W. (1959). Motivation reconsidered: the concept of competence. *Psychological Review*, **66**, 297–333.
Wolff, P. H. (1963). Observations on the early development of smiling. In *Determinants of infant behaviour*, Vol. 2, (ed. B. M. Foss), pp. 113–38. Methuen, London.
Wood, D. and Middleton, W. (1975). A study of assisted problem-solving. *British Journal of Psychology*, **66**, 181–91.
Wood, D., Wood, H., and Middleton, D. (1978). An experimental evaluation of four face-to-face teaching strategies. *International Journal of Behavioural Development*, **1**, 131–47.
Yarrow, L. J., McQuiston, S., MacTurk, R. H., McCarthy, M. E., Klein, R. P. and Vietze, P. M. (1983). Assessment of mastery motivation during the first year of life: contemporaneous and cross-age relationships. *Developmental Psychology*, **19**, 159–71.
Yarrow, L. J., MacTurk, R. H., Vietze, P. M., MacCarthy, M. E., Klein, R. P. and McQuiston, S. (1984). Developmental course of parental stimulation and its relationship to mastery motivation during infancy. *Developmental Psychology*, **20**, 492–503.
Young-Browne, G., Rosenfeld, H. M., and Horowitz, F. D. (1977). Infant discrimination of facial expressions. *Child Development*, **48**, 555–62.
Zarbatany, L. and Lamb, M. E. (1985). Social referencing as a function of information source: mothers vs. strangers. *Infant Behaviour and Development*, **8**, 25–33.
Zukow-Golding, P. (1993). When gestures speak louder than words: an ecological approach to achieving consensus. Paper presented at the XI ZASCL Conference, Trieste.

16
The origins of language and thought in early childhood

George Butterworth

Abstract

The classical theories of the relation between language and thought in developmental psychology are those of Piaget and Vygotsky. Piaget's claim is that language depends on thought for its development, and is based on four sources of evidence: the period of infancy, in which fundamental principles of thought are exhibited well before language; the simultaneous emergence of language, deferred imitation, symbolic play, evocative memory, and mental imagery, suggesting language is but one outcome of more fundamental changes in cognitive abilities; the lack of effect of language upon reasoning abilities in middle childhood; and the nature of speech in early childhood, the claim being that the communicative function of speech results from cognitive developments. By contrast Vygotsky, while seeing thought and language as initially separate systems, considers the two merge at around two years of age, producing verbal thought. Mental operations are regarded as embodied in the structure of language, and hence cognitive development results from an internalization of language.

Current research on infancy has elucidated the perceptual and social sophistication of the neonate, and points to developments occurring from this base during the course of adult–infant social interchanges. Preverbal gestural communication is established between six and nine months, and by twelve months is under intentional control. The shift to referential communication is again mediated by social interaction, particularly the development of routines to bring about the joint attention of adult and infant upon the same object—especially the production and mutual comprehension of manual pointing. In addition, underlying changes in the infant's abilities to relate 'parts' to 'wholes' and to construct relations between means and ends appear to inform the elaboration of the simultaneously emerging cognitive abilities noted by Piaget. This suggests that the entire symbolic function is a separate cognitive domain to which wider cognitive abilities may be applied.

For older children the influence of language on thought has proved difficult to investigate conclusively. Evidence for the Whorfian hypothesis is scarce, and is incomplete for the claim that language plays a major role in the developing self-regulation of the child's behaviour.

Previously, this material has been used somewhat uncritically to inform phylogenetic speculation on the role of language in the evolution of human cognitive abilities. Recapitulatory theories of 'terminal addition' have overlooked the possibility that behavioural development may not occur in stages, and that such stages may not be additive; 'neotenous' theories do not deal satisfactorily with how a rearrangement of the timing of abilities can lead to 'qualitative' changes in 'behavioural capacities'. Recent work explains parallels in ontogeny and phylogeny by appeal to common constraints on information-processing that reflect the demands of changing levels of the structure of knowledge as it interacts with more basic perceptual competencies [eds].

16.1 Introduction

Developmental psychology is only one among many disciplines that have contributed to our understanding of the relation between language and thought. It is unique, however, in making development both the object and the method of enquiry; by examining the precursors of language and thought from the ontogenetic perspective it is possible to arrive at a theory based on empirical evidence. In recent years, extensive studies of the perceptual and cognitive abilities of human infants have offered the possibility of mapping the emergence of thought and language in development, and this in turn provides a basis for more informed comparison with the phylogenetic, comparative perspective. Chomsky's (1980) proposal that there is a genetically based initial state of mind which, through epigenetic transformation, more or less universally culminates in the acquisition of language provides the rationale for such a programme. The main result has been to bring our concepts of language and cognition more closely together. Language may be regarded as one aspect of cognition, and language development as one component of a developing cognitive system. From this point of view we may suppose not only that thought will have its influence on language, but also that the acquisition of language will have an influence on thought. Detailed studies may allow us to observe cognitive precursors to language and the linguistic influences on thought as they develop.

Rather than enter into controversies about the definition of language, a simple-minded approach will be adopted here. It will be assumed that the general contrast between, on the one hand, language as communication by means of spoken symbols and, on the other hand, thought as reflective mental activity will suffice to carry the preliminary distinctions necessary.

16.2 Classical theories of the relation between language and thought in young children: Piaget and Vygotsky

The classical accounts within developmental psychology of the relation between language and thought are due to Jean Piaget (1926) and Lev Vygotsky (1962 [1934]). These are often presented as mutually contradictory theories, the former founded on an individualistic epistemology and the latter on a collectivist approach. Both theorists agree that communication through speech depends upon intellectual development during the first year of life, but they disagree over the status of verbal thought in early childhood. Piaget's and Vygotsky's theories will first be outlined as background, before recent evidence is reviewed (see also Sinha, this volume, Chapter 17, Section 2).

16.2.1 *Piaget's theory*

Piaget (1926) asserted that thinking has its developmental roots in the sensorimotor activities of babies, and is not founded upon language. He argued that there arises in infant development a 'logic of action' which forms the necessary foundation for mental operations which appear in the second year of life. Four major sources of evidence led him to conclude that mental operations precede language in development and so give structure to it (Sinclair (1982) provides a detailed exposition and defence of Piaget's position):

1. The crux of Piaget's argument rests on his studies of infant development (Piaget 1951, 1953, 1954). He argued that the fundamental precursors of thought can be observed in the infant's behaviour long before the child utters his or her first words. Thus, language cannot give rise to thought; on the contrary, certain elementary forms of thought are a necessary condition for language-acquisition. Piaget describes the concept of object permanence as the fulcrum of the cognitive system; and as such this must be the most basic cognitive precursor of language. The object concept includes understanding that objects have their own unique identity—a type of knowledge that would seem intimately linked to naming. He suggests that knowing that objects have names may be seen as a natural outgrowth of more elementary forms of knowledge about objects. The general view, then, is that cognitive development during infancy is a necessary (but not sufficient) condition for language-acquisition.

2. A second source of evidence against the identity of language and thought comes from Piaget's theory of symbolic development (Piaget 1951). Spoken language emerges almost simultaneously with four other behaviour patterns at the end of infancy. These are: deferred imitation, symbolic play, evocative memory, and mental imagery. Piaget therefore argues that the important changes in cognitive processes that give rise to language are much more extensive than is apparent if the focus is restricted to speech. The first words are just one consequence of development of the 'semiotic function', best understood as the ability to reason in terms of symbols. According to Piaget spoken language emerges at the end of the sensorimotor period because it depends upon mental representation, which in turn derives from 'internalized' imitation of objects

and events. It therefore has many qualities in common with the other activities, such as play, in which objects may be made to stand as symbols for other objects. These similarities are taken as additional evidence that language has its roots in more general cognitive precursors.

3. Further evidence concerning the relation between language and thought comes from studies of reasoning in young children. Having speech does not prevent the child from making errors typical of his or her stage of intellectual development. The pre-school child uses words which, to the adult mind, incorporate mental operations such as quantifiers (i.e. words referring to number, volume, weight, or any aspect of quantity), and comparatives (such as 'more', 'less'), yet this does not prevent errors in conservation tasks which require the child to assert the invariance of the basic properties of matter (mass, number, volume), despite appearances to the contrary. Other difficulties with logical reasoning problems were extensively demonstrated by the Piagetian school even into adolescence, when language is obviously well established. Hence Piaget argued that the words constituting the language are themselves being defined and re-defined in terms of the child's level of mental development; in this theory language is always subordinate to the mental operations of thought.

Equivalent evidence, in terms of its importance for the debate on language and thought, comes from research showing that children born deaf nevertheless pass through broadly the same stages in cognitive development as those who can hear and speak. A classic study of thinking in the deaf is by Furth (1963).

4. One of Piaget's main sources of evidence on the relation between language and thought in young children came from his studies of 'egocentric speech' (Piaget 1926). Egocentric speech has a number of characteristics. It is defined by several aspects of the verbal behaviour of pre-school children:

- monologue: where the child talks with no apparent audience;
- collective monologue: where the child addresses speech aloud to herself while in the presence of other people;
- apparently aimless repetition, or playing with words.

Piaget characterized roughly 50 per cent of the speech of pre-school children as egocentric, on the grounds that it did not result in any effective communication with others. It is perhaps as well to point out that the pre-school child in Piaget's view clearly *intends* to communicate with others, but often fails because he or she lacks the intellectual operations necessary to reciprocate appropriately in verbal exchanges. Egocentrism, according to Piaget, is the natural consequence of inability to take the position of another person. The mental operations that underlie reciprocity and facilitate the exchange of information are not yet available. This is a very clear assertion that the communicative function of speech depends on mental development.

In summary, Piaget insists that language depends upon intellectual operations for its development. Language is a 'figurative' content of thought, a kind of 'substance' upon which the mind can operate.

16.2.2 *Vygotsky's theory*

Vygotsky (1962 [1934]) is widely considered to have been the most influential Soviet developmental psychologist. He agreed with Piaget that thought and speech spring from different developmental roots. Phylogenetic comparison enables the separate influences to be observed. A pre-linguistic phase can be observed in the evolution of thought, (since chimpanzees clearly can solve problems requiring thought, yet they do not speak) and a pre-intellectual phase can be observed in the evolution of speech (since parrots can imitate speech, yet they have only limited intellectual abilities by comparison with humans and the primates).

Man's particular achievement is to combine intellectual and imitative capacities in such a way as to give rise to spoken language. Ontogenetic investigation also reveals the separate roots of thought and speech in infancy. His evidence for the separate origins of language is entirely anecdotal, and comes from its presumed precursors such as infants' babbling, the very early reactions of the baby to the human voice, early communicative gestures, and social responsiveness before there is any evidence of thought.

From the age of two the separate developmental streams underlying thought and speech are said by Vygotsky to merge. The child discovers the symbolic function of words with the realization that every object has its own name. Speech begins to serve the intellect—thoughts begin to be spoken, and language becomes a particularly human vehicle for thought. For Vygotsky, verbal thought marks a point where language and thought unite. Verbal thought is not the only type of thought available to humankind; however, through its analysis, and especially through an understanding of how word-meanings are acquired by children, the specifically verbal aspects of thought can be studied.

In summary, Vygotsky, unlike Piaget, considers verbal thought as one area of cognition where language is not subordinate to the development of more general logical operations.

Vygotsky differs crucially from Piaget in his analysis of egocentric speech. From the Vygotskian perspective mental operations are already embodied within the structure of language, and hence the internalization of language (inner speech) may contribute to the development of thought. According to Vygotsky, verbal thought is the outcome of a prolonged developmental process in which the child transfers initially overt speech, which serves a social function and is of social origin, inwards to the mental plane. On their way 'inwards' thought and language pass through an egocentric stage, in which thoughts are spoken aloud (or, one might equally say, speech is thought aloud). Vygotsky emphatically denies that egocentric speech is not yet social. On the contrary, he asserts that since language is rooted in social relations the direction of development is from social to private speech (rather than from private to socialized, as Piaget argued). A universal developmental transition is postulated, from overt speech, to speech for oneself, to inner speech, and this transition is complete by about eight years of age. To illustrate the social, communicative function of egocentric speech Vygotsky carried out a famous series of experiments, which included observing the young child among deaf-mute children, or in a situation where the child could not hear his or her own voice because loud music was playing. In each case, the amount of egocentric speech dropped virtually to zero. Vygotsky considered that the feeling of being understood is essential to the child displaying egocentric speech; that communication through language is the child's intention.

The public nature of language explains why the first words the child acquires are global and undifferentiated. They carry an excess of meaning because they represent to the child not only the specific referent intended by the adult, but also incidental aspects of the context in which the utterance is made and which from the child's point of view are an integral aspect of the communication. Ambiguity of reference gives rise to the syncretic, overinclusive characteristics typical of thought in young children. The implication of Vygotsky's analysis is that Piaget overestimates the extent to which the pre-school child's thought is illogical. When circumstances are favourable, the child will demonstrate an aptitude for logical reasoning that is necessarily undetected by Piaget in his rather abstract tests of the child's cognition and language. Vygotsky's theory anticipates the contemporary distinction between performance (which may be deficient on a variety of reasoning tasks) and the underlying competence (which may be observed under specially controlled circumstances). Vygotsky also anticipates contemporary research which has shown that the social context in which tests of mental abilities are made may actually contribute to mistakes by children. Where communication is ambiguous the child may have difficulty comprehending the referent of the adult's speech, and this in turn may lead to errors in reasoning (Donaldson 1982). The child's logic may be perfectly adequate, but may not yet be encoded in linguistic form.

16.3 Perceptual, intellectual, and social precursors of thought and language in young children

A great deal of research has been carried out in the various branches of developmental psychology that are relevant to the origins of language and thinking, and it has proved necessary to be highly selective here in the material to be discussed. The aim is to illuminate the extent to which language and thinking may be considered as mutually interacting influences in development.

16.3.1 *Intellectual development and the pre-requisites for linguistic communication*

A number of workers have put forward claims that thought provides a basis for language. From a developmental point of view, such theories inevitably require further elaboration in order to establish what may be the precursors of thought and communication in ontogeny.

The origins of language have often been traced to systems of pre-verbal communication; Goodwin (1980), for example, has suggested that the following constitute the minimum set of cognitive and behavioural pre-requisites:

1. Intersubjectivity: infants must recognize others as conspecifics with whom they may interact and communicate.
2. There must be some means of co-ordinating activity between the infant and adult participants, and there must be some means of regulating 'turn-taking' in the communication process.
3. There must exist an adequate repertoire of communicative acts at the non-verbal level to control the joint activity of the participants.
4. There must be co-ordinated attention to objects in the environment if infant and care-taker are to

agree on the topic which is at the focus of their joint activity.

In fact, the perceptual and social sophistication of the neonate make it quite possible that the minimum conditions Goodwin describes may actually be satisfied in the innate repertoire. For example, new-born babies are particularly sensitive to the rhythmic properties of speech (Condon and Sander 1974), and will move in synchrony with patterned sound. Methods have also been developed which demonstrate that the auditory system is particularly attuned to the phonetic structure of speech from the earliest days of infancy (Kuhl 1985). New-born babies will imitate mouth and tongue movements and hand movements, and this may have important implications for the acquisition of speech and gesture (Meltzoff 1982; Vinter 1983). There appears to be rudimentary turn-taking and joint attention in various social activities; this has important implications for reciprocity in pre-verbal communication (Trevarthen 1982; Scaife and Bruner 1975). The origins of turn-taking and many of the reciprocal aspects of communication through language may have precursors in the various aspects of the pre-attuned interaction between parents and babies (see, for example, Bullowa (1979); Golinkoff (1983)).

Great emphasis has also been placed on the skills of adults in regulating their social interaction with infants in ways that make comprehension easier for the child, even to the point of 'baby talk' which accentuates the critical perceptual dimensions of speech for the infant (see for example Schaffer (1984); Gottlieb (1985)). Furthermore, it should not be forgotten that, however sophisticated or rudimentary the infant's abilities, they are supplemented and supported by adults; the adult's cognitive abilities ensure the success of early communication. Vygotsky emphasized the importance of the tuition provided by the adult in creating the 'zone of proximal development' (ZPD). The ZPD is the developmental stage that the child is about to attain. It is always a critical distance ahead of the present level of functioning, and arises through instruction by parents and peers or through play. Bruner (1983) has taken this Vygotskian point to heart, and given the adult's role in 'scaffolding' the child's language development the acronym 'LASS' ('language acquisition support structure'). He did this to emphasize the complementarity of the social structure with any innate abilities that may comprise Chomsky's 'LAD' ('language acquisition device').

It is necessary to remember that, although various abilities may be general pre-requisites for communication (and some of them may therefore apply to species other than humans), the problem with respect to the relation between language and thought is to determine the extent to which changes in the child's intellectual abilities during infancy give rise to communication by means of spoken language. Although knowledge of the innate repertoire is useful in showing where development originates, it does not really help in pinpointing the cognitive changes that may support the emergence of speech. Early communication, for example by smiling or other non-verbal signals, may serve to maintain social interaction; but any communication that occurs is entirely in the context of the immediate social interaction. There is no need to dismiss such communication as merely reflexive or inadvertent; such 'protocommunication' may nevertheless involve shared meanings, even though it does not yet constitute a language.

Golinkoff (1983) offers a useful distinction between an early kind of communicative competence that can be observed in the first six months of life and a more sophisticated form of instrumental communication that can be observed in babies toward the end of the first year. Instrumental communication may be one of the cognitive precursors of language that arise in the course of the child's own cognitive development. It can be observed once infants begin to use pre-verbal communication intentionally to influence others to carry out their wishes, as for instance when the fourteen-month-old baby vocalizes to catch mother's attention, and then points to a desired object. Even though the infant may not immediately succeed in conveying what is meant to the mother, the baby will persevere until the message is understood. In fact, the child's failed messages are particularly revealing of communicative intent, since the child will often reject the mother's interpretation. Something in the intellectual development of the infant gives rise to perseverance in the face of message-failure. The child insists on communicating; the communication cannot be simply dismissed as the adult's attributing communicative intent to the child. According to Piaget's theory, cognitive development in the one-year-old led to the necessary understanding of causality, and gave rise to the extended spatio-temporal framework that enables intentional communication beyond the immediate 'here and now', even though speech is absent. The emphasis is on general changes in the child's abilities that give rise to intentional control of manual and articulatory gestures, and may also underpin the instrumental use of language.

This example of an intellectual precursor to spoken language (i.e. the intentional use of signals) also raises the centrally important topic of reference and its role in language-acquisition. Put at its simplest, reference concerns how one individual may indicate to another an object for their joint attention. How an infant understands reference is important in unravelling the relation between thought and language in young children, because reference enables a connection to be established between words and things. In the example above, given by Golinkoff, the infant points at the desired object; and even though not immediately successful in fulfilling the intention of obtaining the object, she does eventually succeed. Could there be a relation between such pre-linguistic forms of referring and the redirection of attention by linguistic means?

Bruner (1983) discusses the emergence of reference in detail. The earliest form that can be observed consists in joint visual attention to an interesting object; when the mother turns to look at an object that catches her attention, her infant, even at as young as two months, will turn and look in the same direction, as if aware of a potentially interesting sight at the terminus of their joint direction of gaze. Although evidence from other species which never acquire speech shows that comunication based on joint visual attention does not depend on a particularly linguistic structure, it is necessary to be aware that gaze may nevertheless serve a linguistic function for humans.

Detailed studies of joint visual attention between infants and their mothers have revealed that 'looking where someone else is looking' is governed by three progressively more sophisticated cognitive mechanisms, which emerge sequentially in the first eighteen months of life (Butterworth and Cochran 1980; Butterworth and Jarrett 1991). The first mechanism, which governs joint visual attention up to about nine months, consists in the infant looking in the same direction as the mother and selecting an object for attention which is intrinsically attractive. That is, the communication between mother and infant depends to some extent on the fact that the infant is distractible, both by the mother's change of attention and by the intrinsic properties of objects in the world. This mechanism is superseded, somewhere between nine and twelve months, by a much more precise 'geometric' process, whereby the infant can determine which of two objects the mother is looking at even when the objects themselves are identical. This process has been called 'geometric', because it seems to involve extrapolating an invisible line along the mother's line of gaze to the target, using the angular displacement of the mother's head and eyes as the cue. Acquiring this ability coincides with the ability to comprehend the mother's manual pointing. In a third phase of development, between twelve and eighteen months, the infant extends her comprehension of looking and pointing to areas outside her own immediate visual field. Now, if the mother looks or points behind the baby, the infant will understand the reference to a hidden object, and search for the object singled out by the mother. By the time the infant has entered the second year of life she is well able to comprehend the meaning of indexicals, words such as 'where' or 'what', which the mother will use extensively when presenting objects to the baby in teaching the names of things (see Bruner 1983).

Clearly, this aspect of language-acquisition does not only have its roots in innate forms of intersubjectivity, such as joint visual attention, but also develops by incorporating the primitive form into later-appearing control systems. Cognitive development extends the infant's comprehension of looking, and seems also to give rise to comprehension of manual pointing. There is an underlying change in competence. Intellectual development also enters into the intentional production of communicative gestures such as manual pointing to redirect the attention of another, which can be observed toward the end of the first year, before speech has been acquired.

Other examples of cognitive development in infancy that seem intimately related to the production (as well as the comprehension) of language have been described by Langer (1983, 1986) and Sugarman (1983). Langer has published an extensive series of analyses of changes in the cognitive abilities of babies that occur between six and eighteen months. He argues that conceptual development in infancy takes two forms: (i) changes in the ways in which the baby can relate 'parts' and 'wholes' and (ii) changes in the ways babies construct relations between means and ends. The former series of developments ensure that by eighteen months the infant has the ability to unite as many as four single objects into composite structures, which themselves can be reunited into recompositions, involving substitution, exchange, or replacement of components. Parallel developments occur in means–ends relations, so that by eighteen months there is systematic variation of means and ends in the child's exploration of causality. Langer shows that the infant has already developed an extensive repertoire of logical abilities that can now be applied to language, most obviously in the mastery of grammatical distinctions such as actor, agent, and object, and in the substitution of words within a basic syntactic framework.

The implication of Langer's analysis is that the symbolic function of which Piaget speaks constitutes a separate branch of the cognitive system, to which intellectual operations may be applied. To take one classic example (from Stern 1924), some names may be acquired in infancy by a kind of part–whole analysis. Names such as 'woof-woof' for a dog may commonly occur because the child will use a part of the object (in this case the auditory 'part') to stand for the whole. It has also been argued that symbols only gradually develop, and that early words such as these are not yet symbols. Kaye (1982) makes a distinction between a gesture (for example, pointing which can be observed around one year); an index (for example, 'woof-woof' to indicate a dog); a sign (for example, an involuntary cry of pain); and a symbol (a conventional and intentional sign). On this strict definition, a symbol enables meaning to be shared through intentional signification because the members of a community have arbitrarily assigned the symbol a particular meaning.

The logical abilities which Langer describes are not all that is required for meaningful speech to emerge. Harris (1982) has argued that what determines the timetable for language acquisition, i.e. when certain concepts will be expressed verbally, may be the ease with which the child's ideas can be mapped into the particular language of the community which the child hears. Language-acquisition not only depends on development of thought, it also depends on the particular language being acquired, and on contact with the culture (see also editorial appendix).

16.3.2 Linguistic aspects of thought in young children

There are at least two areas in which the influence of language on thought may be demonstrated in young children. The first concerns linguistic categorization and its effects on thought; the second concerns the child's ability to use language in the self-regulation of behaviour. This is a topic that is particularly important in Soviet psychology (Luria and Yudovitch 1956) although it is becoming accepted in Western psychology, under the general title of 'metacognition'.

16.3.2.1 Effects of language on categorization: the Whorfian hypothesis

Benjamin Lee Whorf (1897-1941) suggested that conceptual categorizations of reality may, at least in part, be determined by the structure of the native language (Whorf 1956). Developmental evidence for the Whorfian hypothesis is scarce. Sugarman (1983), using the techniques developed by Langer, has suggested that children by two and a half years may occasionally demonstrate that language influences the ways in which objects are categorized. She gives an example in which a child groups a square with a triangular block because 'it would make a good house' to suggest a reciprocal influence of language on thought. Thus not only does classification allow entry into the language, but language itself may lead to particular ways of classifying things.

Another way to discover the effect of language on thought is through the study of bilingual children, who offer a natural control for the effects of cognition upon language, since whatever their stage of intellectual development in the Piagetian sense, this is presumably a factor common to both languages. There are examples (for example Slobin 1973) of bilingual children who make errors on Piagetian tasks when they are questioned in one language, and yet do not make errors on the same task when questioned in their other language. These phenomena are observed where relational judgements such as 'more than' or 'less than' are involved, and have been found for bilingual Turkish-English children, and in comparisons between Spanish and English children who are at the same Piagetian stage of cognitive development. The child will make errors in English but not in the other language in tasks such as the 'class-inclusion' problem, which require judgement of the numerical relation of parts to the whole. Performance seems to depend upon differences in the ways in which words in particular languages express particular meanings, on contextual cues, and on aspects of references derived from social interaction with the adult (Donaldson and Balfour 1968; Donaldson 1982).

Perhaps the most parsimonious explanation is that some languages express relational concepts in a more complicated way than others, i.e. it is the structure of the language itself that influences the child's thought. Where languages encode reality in idiosyncratic fashion, at least as judged from the standpoint of one's own language, an influence on thought can be observed.

16.3.2.2 The verbal regulation of behaviour

Vygotsky argued that between three and seven years of age the child's egocentric speech differentiates from social speech. Vocalization drops out (although it will return if the child is faced with unexpected difficulties), and the child begins to 'think in words' without pronouncing them. Since it is speech for the self, it possesses its own peculiar syntax: it is condensed; it omits the subject while preserving the predicate of the sentence; and it uses many fewer words.

The main tests of Vygotsky's theory were carried out by his student A. R. Luria (Luria and Yudovitch 1956). Luria proposed that there are three main stages in the acquisition of verbal control over behaviour. He carried out simple experiments in which the child had to make a manual response of squeezing a rubber bulb whenever a signal light came on. He concluded that between eighteen months and three years verbal control resides in social relations with adults; their commands can initiate actions in the child, but fail to inhibit them. Between three and five years verbal control still lies mainly with adults, but the child's actions may now be inhibited by their commands. This stage is a transitional one, and Luria shows that children can activate motor acts for themselves if they verbalize aloud. However, action is simply energized by the self-initiated command; the child cannot inhibit behaviour through her own speech. Transfer from control by the impulsive effect of language to control based on word-meaning is said to occur at around four and a half years, and is shown through the child's ability to use language to inhibit or activate behaviour in a variety of simple tasks. In summary, verbal control occurs first in overt form (egocentric speech), and eventually becomes covert (verbal thought) at about seven years.

There are a number of difficulties with this theory. First, Luria's experiments are controversial, and there have been a number of failures to replicate his findings by American researchers (for example Bloor 1977). Second, it is often argued that there are logical problems with the theory of verbal regulation of behaviour. In order to affect the listener, speech must be understood in terms of its meaning. Since meaning depends on knowledge which the listener already possesses, it is difficult to see how inner speech can serve to tell the listener something that isn't already known. Third, if language regulates behaviour, what regulates the behaviour called language?

Wozniak (1975) and Bloor (1977) rescue the theory from this infinite regress by suggesting that speech for the self acts as a feedback system at phonetic, syntactic, and semantic levels. Considered as a feedback process, speech can embody information about the success (or failure) of acts that themselves arise from complex intercoordinations of systems, and in that way it can contribute to self-regulation. Karmiloff-Smith (1984) has suggested another metacognitive function for speech (and verbal thought). It may serve as a superordinate means of re-representing other types of stored experience, for example spatial or kinaesthetic memory, so that the single, common code provided by language may aid in the intercoordination of aspects of experience. Contemporary psychology therefore offers some support for Vygotsky's theory of the verbal regulation of behaviour. The theory has become easier to understand as cybernetic systems have become more common, and his work is currently enjoying a revival of influence (for example Wertsch 1985).

In summary, although evidence on the effects of language on thought is hard to come by (cf. Lock and Symes, this volume, Chapter 8), developmental studies do suggest that the different systems of categorization imposed by different languages may influence children's thought. This seems especially to be the case in tasks requiring relational judgements. At the more general, metacognitive level, it has also been suggested that speech and verbal thought may have a general self-regulatory and integrating function. The effect of the self-regulatory functions of speech on experience and behaviour can be observed in the child from the middle of the fourth year of life.

16.4 Ontogeny and phylogeny: stages of development and the recapitulation hypothesis

Evolutionary biology has profoundly influenced theories of the development of language, and, in recent years, data from child development has been used to sketch possible theories of the evolution of language in hominid ancestry. One powerful source of evidence on the relation between ontogeny and phylogeny comes from the comparative literature on teaching sign languages to apes. It is sufficient to say that this literature (see Ristau, this volume, Chapter 23) seems to demonstrate that non-human primates can acquire symbol systems with special training; but extensive communication by means of symbols is not typical of primates in nature (although some manual signing may occur in the wild). As Bates (1979) puts it, the higher primates are at some brink in tool-use, imitation, and symbolism which children rush across with far less assistance. Given that man and chimpanzee are more than 99 per cent genetically identical, an explanation is required that can account for the great morphological and intellectual differences that nevertheless exist between them.

In the contemporary literature, modern forms of recapitulationary theory have been most influential in explaining the relation between phylogeny and ontogeny. The theory of recapitulation was originally propounded by Haeckel (1874). He argued that the individual passes through a series of stages in embryo in which the the adult forms of ancestors is repeated in embryological development, i.e.: fish, amphibian, mammal, primate, in the serial order of phylogeny. Hence, ontogeny was said to recapitulate phylogeny. Although Haeckel's strict biogenetic principle has been refuted by modern research, it nevertheless remains influential in the form of more subtle recapitulationary accounts (see Gibson, this volume, Chapter 14). Such theories stress constraints on form and function of thought and speech that may give rise to similarities between the development of language in the child and the forms that language may have taken in phylogeny.

16.4.1 'Terminal addition'

The first approach to be discussed is called 'recapitulation with terminal addition', and the suggestion is that distinct stages have been added to the sequence that can be observed in primate development. This theory has been most strongly advocated by Parker (1985) and Gibson (1982). It is a modified recapitulationary view which makes no attempt to assert a literal recapitulation of ancestry.

The argument is that species differences reflect the addition of later stages of intelligence to the end of the developmental cycle. Early intellectual development in all primate species is similar because shared fundamental processes are pre-requisites for the appearance of primate intelligence. The differences between man and the higher primates are to be explained by the addition of intellectual abilities in evolution that begin to become apparent in children at around four years of age (Gibson 1982, p. 74). This account therefore explains the limited extent to which gestural languages are possible in chimpanzees (because they depend on intellectual abilities shared by all higher primates), but also emphasizes the important contribution made by novel intellectual abilities acquired in the late pre-operational period to the further elaboration of language.

Parker (1985) offers an extensive review of the evidence on early language-acquisition to support a theory of the evolution of language in relation to tool-use and subsistence. She suggests that man's earliest ancestor *Australopithecus* (four million years BP) communicated by iconic–mimetic signs allowing reference to and requests for distant or hidden foods. The shift to vocal encoding occurred with the evolution of *Homo erectus* (two million years BP). The acquisition of the upright posture and the shift in the position of the larynx required for speech coincided with the freeing of the hands for use of tools, butchery, and co-operative hunting. Parker suggests, on the basis of evidence from tool production, shelter-building, and other aspects of social organization that the level of intelligence of this human ancestor corresponded roughly to that of the five- to six-year-old human child entering Piaget's concrete operational period. She suggests that vocal communication would have expanded to include syntactic relations such as actor, agent, object, and vocal encoding of prototypical events. With *Homo sapiens* (40 000 years BP) she suggests that further 'terminal additions' of intelligence occurred, corresponding roughly to Piaget's formal operational stage of early adolescence. These new intellectual abilities enabled both machine technologies and the propositional uses of language, embedded syntax, and the concepts of possibility, necessity, and inference.

Such reconstruction of the intellectual and linguistic abilities of our ancestors on the basis of stages of intellectual development of modern children is fraught with problems. For example, developmental stages may not be additive, and hence there need be no systematic relationship between the ordering of stages in ontogeny and in phylogeny. Furthermore, the psychological processes involved may be quite different, even though aspects of behaviour may be structurally isomorphic in phylogeny and ontogeny. For example, it is almost certainly the case that tool construction as revealed by the fossil record and elementary tool-use in infancy may not reveal equivalent levels of cognitive development. The elementary understanding of principles of leverage, of cause and effect, etc., evidenced in the spontaneous actions of babies with ready-made implements may not exhaust the mental capacities required for the effective construction of tools. Tool construction is a highly skilled performance, well beyond the capacity of the young child, and one which may require an extensive knowledge-base of the properties of materials and their suitability for particular functions. The mental development of the adult ancestor may, therefore, not be very precisely revealed by the cognitive level of the contemporary child. Today's child is not yesterday's ancestral form, as the strong version of the recapitulation doctrine would have it. The doctrine is useful if the modern child is simply taken as a yardstick for the general level of intellectual development that might apply to our ancestors, but runs into difficulties if a literal recapitulation of ancestry is implied (see Gibson 1985 and Rutkowska 1985 for critiques of Parker's theory). Nevertheless, it is an approach founded on the assumption that language is based on cognition, and that there exists a systematic relationship between the two.

16.4.2 Neoteny

The second recapitulationary approach, 'neoteny', suggests that the timing of developmental processes has changed over evolution, so that the juvenile form of the common ancestor of man and of the higher primates is retained in humans. Furthermore, changes during evolution in the relative developmental rates of different primate abilities (heterochrony) may give rise to speech as a consequence of the reordering of the developmental timetable governing the constituent abilities. Vocal communication may come to dominate gestural communication in humans as a simple consequence of changes in the relative rate of development compared to primate ancestors.

On a strict interpretation of this hypothesis there is no need to postulate capacities added to general abilities of higher primates to account for the emergence of language. Rather, reorganization of these abilities and retention of juvenile form (and function) are sufficient to explain the acquisition of speech. Certainly, this hypothesis receives impressive support in terms of morphology (Gould 1977); but it is more difficult to pin down the general, pre-linguistic capacities that are the pre-requisites for speech and language and are held in common with higher primates.

Bates (1979), for example, suggests that intentional communication through conventional signals can be understood as a particular manifestation of tool-use. Symbolic communication as tool-use emerges from a capacity shared with other primates for imitation, social motivation, and shared reference to objects. However, Bates also makes the point that humans are actually better at imitation and tool-use than chimpanzees. To explain this additional capacity may require that neoteny and terminal addition theories of recapitulation should be recombined if an evolutionary account is to capture both the added capacity of the cognitive system and the shared precursors that have allowed language to evolve.

Lock (1985) suggests that language has its evolutionary origins in gestures which became functionally elaborated both at the vocal and non-vocal level to carry holistic meanings. This contextually and socially dependent repertoire is eventually transcended by truly symbolic, propositional language and thought. He argues that the capacity for speech was probably available to *Homo erectus c.*400 000 BP. On his view elaborated, spoken language evolved under the pressure of tool-use, perhaps not simply because the upright posture freed the hands and the mouth, but also because it becomes difficult to communicate by means of gesture with full hands. Tool construction also reveals repeatedly recursive means–ends thinking at least as long ago as 300 000 BP. Nevertheless, a fully propositional language may not have been fully exploited much before 40 000 BP, when social and cultural conditions favoured it. On Lock's view, resemblences between ontogeny and phylogeny arise because there is a necessary sequence in the emergence of language from a holistically-constituted precursor; stages reflect systems-constraints in the emergence of explicitly defined forms of communication from their implicit, holistic roots.

16.4.3 *Common constraint*

There is another, more naïve aspect of recapitulationary thinking in popular speculation on the origins of language. This concerns the possibility that the child may acquire words in an order revealing of their etymology (or that etymology somehow sets constraints on the order of development of words). Contemporary approaches to cognitive development, however, can explain how the resemblance arises between the order of acquisition of words and the etymology of language. For example, perhaps the clearest recent account is by Keil (1986), who points out that strict-stage models of cognitive development (upon which strict recapitulationary theories are based) are undermined by recent research demonstrating the perceptual and cognitive competence of the young child. In Keil's view, stage transitions in development arise because the innate perceptual competence of the child exerts constraints on cognitive development that shape the knowledge-base. He describes a typical sequence in language-acquisition whereby the child first defines a concept by a set of highly characteristic features (for example, 'uncle': someone who bought you toys on your birthday, who watched sport with your father, etc., etc.), moving on later to a set of definitive features (i.e.: your father's brother). This transition from a word-meaning based on characteristic relations to one based on abstract relations is said to occur as a result of the interaction of locally defined knowledge structures—for instance, 'uncle' is first understood in the abstract in relation to other locally defined kinship terms, such as 'aunt', 'grandmother'. Keil's 'characteristic to defining shift' seems like a stage phenomenon; but the transition occurs on a time-scale defined by the knowledge-structure itself, rather than in terms of biological processes of maturation or the acquisition of new forms of logic. Keil's account is similar to that of Lock (1985), except that he lays greater stress on the constraints on knowing imposed by the structure of knowledge already acquired.

The general implication of such an approach to language-acquisition is that any resemblance that does exist between the etymology of language and the order of acquisition of words in childhood is not strictly speaking a recapitulationary phenomenon. There is no literal repetition of ancestry; resemblance between ontogeny and phylogeny reflects initial constraints on information-processing that that have been the common basis of perceptual competence for generations past. Differences and similarities between the structure of domains of knowledge are said to account for differences in their rate of acquisition in ontogeny (rather than the acquisition of new logical forms of thought) and for commonalities in the representation of knowledge over longer time-scales. Of course, it is precisely in terms of such common constraints on development that resemblances in embryology and evolution are explained today, (Butterworth *et al.* 1985) rather than in the literal recapitulation of Haeckel (1874).

16.5 Conclusion

This chapter has progressed through a very large number of issues with the aid of a simple-minded dichotomy between language and thought. It should be clear by now that a better characterization of their relations is to be found with Chomsky's insight that language is one aspect of cognition, and, as such, that there is a reciprocal influence between language and thought which can be observed in development. The importance of the ancient debate on the relation between language and thought lies in its centrality to explaining the link between human biology and human culture. Speech enables the transmission of culture through the symbolic medium of language. Recent research on the origins of speech and language in early childhood has helped us to understand something of their interrelation, and to speculate more adequately on their evolution.

Editorial appendix: Recent studies of the relation between cognition and language development

1 *Temporal relations between domains*

Recent investigations of the relation between cognition and language development represent, by and large, refinements of Piaget's original claim that symbolism arises out of the attainment of representational intelligence at the end of the sensorimotor period (Stage 6), and, as such, is relatively modality-free, with language being just one of the media in which it is exhibited. This very *general* claim appears to hold up: objects are used symbolically in much the same way as early vocalizations (for example Musatti and Mayer 1987). By contrast, Piaget's more *specific* claim that achieving Stage 6 levels of intelligence is a pre-requisite for the emergence of language has gained little support. What has emerged, however, is a more detailed picture of the interrelations between specific abilities in the two domains at this period.

Gopnik and Meltzoff (1986) offer what they term a 'specificity hypothesis'. This argues that, rather than Stage 6 intelligence as a whole being a precursor to the emergence of language, there will be very specific links between cognition and language. Thus they predicted and confirmed a close coupling between the developmental mastery of object permanence and the acquisition of disappearance words; and between insightful tool use to reach a goal and the acquisition of success/failure words. At the same time, no cross-coupling between abilities in the two domains was detected. A second study (1987) obtained a similar relation between the ability to categorize objects (that is, to spontaneously sort a presented pile into similar groupings) and the 'naming explosion' (see Johnson *et al.*, this volume, Chapter 24).

Bates *et al.* (1979, 1988) have put forward a view of the relations between cognition and early language from data gathered from three cohorts of children, two in America and one in Italy, between the ages of 9 and 28 months (a total of 52 children being involved). Both studies employ a correlational approach, abstracting out from a plethora of measures what goes with what in development at any particular point in time, and what measures at one time predict the appearance of abilities at a later time. The original study covered the period from nine to thirteen months, and established three separable domains of developing capacities that they argue to be pre-requisites for the emergence of language symbols. The second study covered the range thirteen to twenty-eight months, in an attempt to follow the contribution of these capacities in the move from linguistic symbolling to linguistic patterning.

The three domains isolated, and claimed to be prerequisites for symbolic development, in the sense that each was both independent of the others and required to pass some (unspecified) threshold of elaboration before providing the foundation for linguistic symbols to be constructed, were: (1) conventional gesturing that implicates objects beyond the infant's body (as opposed to ritualized 'showing off'); (2) means–end analysis (as indexed by tool-use); and (3) imitation (both of actions *per se* and actions upon objects. Means–end analysis and imitation abilities are indexed through the child's performance on standard developmental tests, and do not yield as precisely specifiable a relationship with language as do the measures used by Gopnik and Meltzoff (1986), but, none the less, provide evidence for what Bates *et al.* (1979, p. 128 ff.) term a 'local homology' or 'skill-specific' model of development:

a package of related structures, capacities that are implicated in the development of linguistic and nonlinguistic symbols. There is also supporting evidence from abnormal language development, and from comparisons across species, suggesting that the same capacities that are present when language emerges are absent when language fails to emerge (ibid., p. 316).

In addition, there are indications in their data that the relationships that hold do so only at certain 'sensitive' periods. One way of interpreting this is that, for example:

tool use at 9 months involves a capacity that feeds into the 13-month discovery that things have names. Beyond that point, however, "more" tool use capacity may be irrelevant to "more" naming (Bates *et al.* 1979, pp. 367–8).

This interpretation leads to their presenting a 'threshold model', hypothetically applicable to both the phylogenetic and ontogenetic emergence of symbols:

once certain critical threshold levels were reached in each of these three domains [communicative intent, tool-use, and imitation], it was possible for the same three capacities to join in the service of a new function, the symbolic capacity (ibid.).

In their more recent study (1988), data was not available regarding the early establishment of communicative intent, but the remaining underlying capacities of analysis and rote reproduction were found to continue to play a constructive role in language development, with the addition of an unspecified language 'comprehension' factor. Again, these capacities were found to make varying contributions at different points in development. In addition, no evidence emerged for a split between lexical and grammatical development, meaning that the same processes operating in the elaboration of vocabulary are also operative in the establishment of early grammar: thus, if there is any specifically linguistic knowledge brought to bear in language development—the autonomous model, see Johnson *et al.*, this volume, Chapter 24—it only plays a role after this point.

2 The emergence of combinatorial skills

Combinatorial abilities appear and develop between around twelve to thirty months of age, and do so in most areas of the child's life: language, from one-, to two-, to multiword utterances; motor imitation; pretence play; peer social interaction; and problem-solving. A number of studies have documented parallel developments during the second year of life in numerous of these domains: between language development, combinatorial abilities, and decentration in symbolic play; problem-solving tasks such as obeying directions, building towers, and replacing puzzle pieces; self-awareness, problem-solving, and representational skills; and peer social interaction, imitation, and language (see Brownell 1986, 1988, for reviews of the literature). It is thus *possible* that these convergent developments derive from some common source: perhaps the maturation of a general combinatorial ability; something related to the possession of symbolic representation; the attainment of predication; or to changes in the memory and information-processing capacities of infants. (Note, however, that the crucial empirical experiments have yet to be conducted: that is, if these parallel developments are dependent on some common underlying structure, then stimulation in one area should lead to development in other areas as a result of induced changes in the underlying structure.)

With respect to structural (grammatical) properties of language, Goodnow and Levine (1973) pointed out that the act of copying a design could be described in terms of grammatical rules, suggesting direct parallels between the organization of language and action (see also Greenfield 1978). However, a large number of activities are so describable, and it may just be that complex activities in general tend to be hierarchically organized, and thus that there is only an analogical relationship between their structure and that of language (for example, Cleveland and Snowdon (1982) argue the sequencing of calls in cotton-top tamarins requires a simple phrase-structure grammar to account for it, and Peters (1972) has written a simple grammar for rhesus monkey copulation).

This issue of analogy v. homology is of fundamental importance. For example, the argument could be made that object manipulation provides the basis for later language, in that operating on an object held in one's hands yields the basic language structure agent–action–patient. To be useful in an evolutionary scenario it needs to be shown that this structure is *homologous* to that involved in language construction. Evolutionary data on this issue are difficult to come by, but evolutionary relationships between domains may be inferred from ontogenetic studies that address it (see Greenfield 1990, discussed in the Introduction to Part III of this volume, and 1991).

References

Bates, E. (1979). The emergence of symbols: ontogeny and phylogeny. In *Children's language and communication: The Minnesota Symposium on Child Psychology*, Vol. 12, (ed. W. A. Collins), pp. 121–56. Erlbaum, New Jersey.

Bates, E., Benigni, L., Bretherton, I., Camaioni, L., and Volterra, V. (1979). *The emergence of symbols: cognition and communication in infancy*. Academic Press, New York.

Bates, E., Bretherton, B., and Snyder, L. (1988). *From first words to grammar: individual differences and dissociable mechanisms*. Cambridge University Press, New York.

Bloor, D. (1977). The regulatory function of language. An analysis and contribution to the current controversy over Soviet theory. In *Psycholinguistics, Series I: Developmental and Pathological* (ed. J. Morton and J. C. Marshall), pp. 73–97. Paul Elek, London.

Brownell, C. A. (1986). Convergent developments: cognitive–developmental correlates of growth in infant/toddler peer skills. *Child Development*, **57**, 275–86.

Brownell, C. A. (1988). Combinatorial skills: converging developments over the second year. *Child Development*, **59**, 675–85.

Bruner, J. S. (1983). *Child's talk: learning to use language*. Oxford University Press.

Bullowa, M. (1979). *Before speech: the beginnings of interpersonal communication*. Cambridge University Press.

Butterworth, G. E. and Cochran, E. (1980). Towards a mechanism of joint visual attention in human infancy. *International Journal of Behavioral Development*, **3**, 253–72.

Butterworth, G. E. and Jarrett, N. L. M. (1991). What minds have in common is space: spatial mechanisms serving joint visual attention in infancy. *British Journal of Developmental Psychology*, **9**, 55–72.

Butterworth, G. E., Rutkowska, J. and Scaife, M. (1985). *Evolution and developmental psychology*. Harvester, Brighton.

Chomsky, N. (1980). *Rules and representations*. Blackwell, Oxford.

Cleveland, J. and Snowdon, C. T. (1982). The complex vocal repertoire of the adult cotton-top tamarin (*Saquinus oedipus oedipus*). *Zeitschrift für Tierpsychologie*, **58**, 231–70.

Condon, W. S. and Sander, L. W. (1974). Neonate movement is synchronized with adult speech: interactional participation and language acquisition. *Science*, **183**, 99–101.

Donaldson, M. (1982). Conservation, what is the question? *British Journal of Psychology*, **73**, 199–207.

Donaldson, M. and Balfour, G. (1968). Less is more: a study of language comprehension in children. *British Journal of Psychology*, **59**, 461–71.

Furth, H. G. (1963). *Thinking without language*. Free Press, New York.

Gibson, K. (1982). Comparative neuro-ontogeny: its implications for the development of human intelligence. In *Infancy and epistemology: an evaluation of Piaget's theory* (ed. G. E. Butterworth), pp. 52–84. Harvester, Brighton.

Gibson, K. (1985). Has the evolution of intelligence stagnated since neanderthal man? In *Evolution and developmental psychology* (ed. G. E. Butterworth, J. Rutkowska, and M. Scaife, pp. 102–14. Harvester, Brighton.

Golinkoff, R. M. (ed.) (1983). *The transition from prelinguistic to linguistic communication*. Erlbaum, Hillsdale, NJ.

Goodnow, J. and Levine, R. (1973). 'The grammar of action': sequence and syntax in children's copying. *Cognitive Psychology*, **4**, 82–98.

Goodwin, R. (1980). Two decades of research into early language acquisition. In *Developmental psychology and society* (ed. J. Sants), pp. 169–217. Macmillan, London.

Gopnik, A. and Meltzoff, A. N. (1986). Relations between semantic and cognitive development in the one-word stage: the specificity hypothesis. *Child Development*, **57**, 1040–53.

Gopnik, A. and Meltzoff, A. N. (1987). The development of categorization in the second year and its relation to other cognitive and linguistic developments. *Child Development*, **58**, 1523–31.

Gottlieb, G. (1985). On discovering significant acoustic dimensions of auditory stimulation for infants. In *Measurement of audition and vision in the first year of life. A methodological overview* (ed. G. Gottlieb and N. A. Krasnegor), pp. 3–37. Ablex, Norwood, NJ.

Gould, S. J. (1977). *Ontogeny and phylogeny*. Harvard University Press, Cambridge, Mass.

Greenfield, P. M. (1978). Structural parallels between language and action in development. In *Action, gesture and symbol: the emergence of language* (ed. A. J. Lock), pp. 415–45. Academic Press, London.

Greenfield, P. M. (1990). Language, tools and brain: the development and evolution of hierarchically organized sequential behavior. Paper prepared for Wenner-Gren Symposium no. 110 'Tools, Language and Intelligence: Evolutionary Implications', Cascais, Portugal.

Greenfield, P. M. (1991). Language, tools and the brain: the ontogeny and phylogeny of hierarchically organized sequential behavior. *Behavioral and Brain Sciences*, **14**, 531–95.

Haeckel, E. (1874). *The evolution of man* (English translation: Watts, London, 1906).

Harris, P. L. (1982). Cognitive prerequisites to language? *British Journal of Psychology*, **73**, 187–95.

Karmiloff-Smith, A. (1984). Children's problem solving. In *Advances in developmental psychology*, Vol. 3, (ed. M. E. Lamb, A. L. Brown, and B. Rogoff), pp. 39–90. Erlbaum, Hillsdale, NJ.

Kaye, K. (1982). *The mental and social life of babies: how parents create persons*. Harvester, Brighton.

Keil, F. C. (1986). On the structure-dependent nature of stages of cognitive development. In *Stage and structure: reopening the debate* (ed. I. Levin), pp. 144–63. Ablex, Norwood, New Jersey.

Kuhl, P. (1985). Methods in the study of infant speech perception. In *Measurement of audition and vision in the first year of life. A methodological overview* (ed. G. Gottlieb and N. A. Krasnegor), pp. 223–49. Ablex, Norwood, New Jersey.

Langer, J. (1983). Concept and symbol formation by infants. In *Towards a holistic developmental psychology* (ed. S. Wapner and B. Kaplan), pp. 221–34. Erlbaum, Hillsdale, NJ.

Langer, J. (1986). *The origins of logic: one to two years*. Academic Press, New York.

Lock, A. (1985). Processes of change and the elaboration of language. In *The comparative development of adaptive skills: evolutionary implications* (ed. E. S. Gollin), pp. 239–70. Erlbaum, Hillsdale, NJ.

Luria, A. R. and Yudovitch, F. (1956). *Speech and the development of mental processes in the child*. Penguin, London.

Meltzoff, A. N. (1982). Imitation, intermodal co-ordination and representation in early infancy. In

Infancy and epistemology (ed. G. E. Butterworth), pp. 85–114. Harvester, Brighton.
Musatti, T. and Mayer, S. (1987). Object substitution: its nature and function in early pretend play. *Human Development*, **30**, 225–35.
Parker, S. (1985). A social-technological model for the evolution of language. *Current Anthropology*, **26**, 617–39.
Peters, C. R. (1972). Evolution of the capacity for language: a new start on an old problem. *Man*, NS, **7**, 33–49.
Piaget, J. (1926). *The language and thought of the child.* (paperback edn, Routledge and Kegan Paul, London, 1960).
Piaget, J. (1951). *Play, dreams and imitation in childhood.* Routledge and Kegan Paul, London.
Piaget, J. (1953). *The origins of intelligence in the child.* Routledge and Kegan Paul, London.
Piaget, J. (1954). *The construction of reality in the child.* Routledge and Kegan Paul, London.
Rutkowska, J. (1985). Does the phylogeny of conceptual development increase our understanding of concepts or of development? In *Evolution and developmental psychology* (ed. G. E. Butterworth, J. Rutkowska, and M. Scaife), pp. 115–32. Harvester, Brighton.
Scaife, M. and Bruner, J. S. (1975). The capacity for joint visual attention in the infant. *Nature*, **253**, 265.
Schaffer, H. R. (1984). *The child's entry into a social world.* Academic Press, New York.
Sinclair, H. (1982). Piaget on language: a perspective. In *Jean Piaget: consensus and controversy* (ed. S. Modgil and C. Modgil), pp. 167–77. Holt, Rinehart and Winston, London.
Slobin, D. (1973). Cognitive pre-requisites for the development of grammar. In *Studies of child language development* (ed. C. A. Ferguson and D. Slobin), pp. 175–208. Holt, Rinehart and Winston, New York.
Stern, W. (1924). *Psychology of early childhood up to the sixth year of age.* George Allen and Unwin, London.
Sugarman, S. (1983). *Children's early thought: developments in classification.* Cambridge University Press.
Trevarthen, C. (1982). The primary motives for cooperative understanding. In *Social cognition: studies of the development of understanding* (ed. G. E. Butterworth and P. H. Light), pp. 77–109. Harvester, Brighton.
Vinter, A. (1983). Imitation, representation, et mouvement dans les premiers mois de la vie. Unpublished thesis, Docteur en Psychologie, University of Geneva.
Vygotsky, L. S. (1962). *Thought and language.* MIT Press, Cambridge, Mass. (First published 1934.)
Wertsch, J. V. (1985). *Culture, communication and cognition: Vygotskian perspectives.* Cambridge University Press.
Whorf, B. L. (1956). *Language, thought and reality.* MIT Press, Cambridge, Mass.
Wozniak, R. (1975). Speech for self as a multiply reafferent human action system. In *The developing individual in a changing world* (ed. K. Riegel and J. Markham), Vol. 1, pp. 151–60. Mouton, The Hague.

17
Theories of symbolization and development[1]

Christopher G. Sinha

Abstract

An interdisciplinary approach to human symbolic evolution draws upon psychological, semiotic, and social theory, as well as upon evolutionary biology. This chapter provides a historical and theoretical overview of main trends in the theory of signs, and examines the fusion of semiotic, psychological, and biological themes in the classic works of genetic psychology.

Contemporary semiotic theory, as well as some key issues in the philosophy of language, have their origins in the work of three major figures at the end of the nineteenth and the beginning of the twentieth century. Charles Sanders Peirce attempted to found a general theory of knowledge upon his analysis of the nature and function of signs in cognition and communication; he introduced the term 'semiotics', and can also be considered as the founder of pragmatics. The logician Gottlob Frege introduced the distinction between *sense* and *reference* which was one of the major foundations of analytic philosophy. Ferdinand de Saussure, the founder of structural linguistics, analysed language as a system of 'signifying differences', in which the value of an element is dependent upon its relations with other elements.

More recently, analyses of language as a communicative vehicle—as pragmatic instrument, that is—have been coupled with criticisms of the way in which traditional linguistic theory treats language as an abstract object independently of use and of context. The theoretical contributions of Mead, Bakhtin, and Barthes provide important insights into the role of language in the creation and maintenance of social life.

Freud, Piaget, and Vygotsky—the three principal figures in the foundation of 'genetic psychology'—were all concerned, in elaborating their theories, to understand processes which can be conceived at one and the same time as semiotic, cognitive, and biological. These three psychologists employed concepts from Darwinian (and, in the cases of Freud and Piaget, Lamarckian) evolutionary theory to fashion their observations of child development (and, in the case of Freud, of clinical symptoms) into integrated theories in which human nature, culture and society, and semiotic and cognitive processes are treated as evolutionary and developmental phenomena.

17.1 Sign and object

17.1.1 *Signification, representation, and language*

Regardless of whether we regard thought as the prisoner of language, or language as the highest expression of creative rationality, the indissolubility of this pair continues to inspire, or, depending on one's point of view, bedevil the efforts of workers in a variety of disciplines. The multifaceted character of natural language, its ubiquity in our social, intellectual, and emotional lives, and its unrivalled flexibility and complexity amongst communicative and symbolic systems, impart to it a natural pre-eminence in the study of human symbolization. By the same token, however, it is important to bear in mind that the acquisition of language does not occur in a vacuum, but as a part of the totality of human developmental processes, including those involving other symbolic and representational systems.

The notions of 'representation', 'symbolization', and 'signification' (and others such as 'denotation', 'designation', 'reference', etc.) can only satisfactorily be defined in a recursive (self-referential) fashion; they 'stand for' the relation which they name, that of 'standing for'. The notion, like the relation, is deceptive in its apparent simplicity. Even the most apparently transparent representation, such as a photograph, depends for its interpretation upon a complex and culturally specified repertoire of norms and beliefs, themselves inscribed within a regulative apparatus for the reproduction of a social order (Berger 1980; Sontag 1982). There is no such thing as 'absolute' likeness, and no ultimate standard of verisimilitude against which representations may be judged, in the manner in which nineteenth-century scientists judged length by reference to absolutes embodied in bars of metal at Greenwich or in Paris. Yet this does not mean that there are *no* canons or rules governing the structure of representations, nor that such canons are entirely independent of the nature of the world and of our own species-being within it. An even greater degree of opacity attaches to symbolic entities, such as linguistic signs, whose relation to what they stand for, in the world or in the mind of the language-user, appears to be grounded not at all in resemblance or likeness, but solely in formal rules and arbitrary conventions.

All representation presupposes some medium within, upon, or by means of which the message, intended or unintended, is carried. The material substrate of representation, however, is not restricted to the medium as such, whether this is conceived of as paper and ink, computer hardware, or the neural architecture supporting mental and linguistic processes. It is also to be sought, at least in the case of human symbolic processes, in the objective, systemic properties of languages and sign systems, as manifested in spoken utterances, written texts, rituals and so forth; in the social practices and relations which are realized and sustained by linguistic communication; and in the tools, artefacts, and enduring symbols which represent the technical–practical capacities of the society, the results of past labour, and the values of the culture.

All evolutionary theories seek their explanatory principles in the dynamic relationships between the life cycle of the organism and the (changing) environment. In the case of the evolution of human symbolic activities, the relevant material environment includes as a decisive factor the representational environment, which is the product of symbolic and constructional practices. An adequate theory must, therefore, amongst other issues, address that of the nature of signs, sign systems, and symbolic behaviour.

17.1.2 *Theories of the sign: Peirce, Frege, Saussure*

Semiotics, or the science of signs and sign-systems, has its origins in the classical and scholastic study of the arts of Logic, Rhetoric, and Poetics. In the late nineteenth and early twentieth centuries, however, the study of language diverged into different disciplinary pathways, resulting in a division of labour which, despite the rise of new 'interdisciplines' such as cognitive science, persists to this day. Philosophers, in search of new and more powerful logics, have explored and evaluated natural-language expressions against the background of their preoccupation with formal, propositional truth. Linguists, at least until recently, eschewed such problems in favour of an essentially descriptive enterprise, seeking to systematize the facts of language structure independently of its referential or communicative functions. Psychologists have looked to linguistic processes for exemplification of the general principles underlying learning, perception, and cognition. Sociologists and anthropologists have viewed language as a cultural form reflecting or implementing the ideological and 'superstructural' reproduction of social relations (see Williams (1977) for a critical discussion).

In spite of these distinctions, however, 'modern' linguistic and semiotic science, as essentially adhered to by all these currents of thought, consists largely in the elaboration and critique of the crucial insights of three figures, two philosophers and a linguist, whose ideas—more or less contemporaneously developed around the turn of the century—have profoundly influenced our understanding of the sign and signification.

17.1.2.1 *Charles Sanders Peirce (1839–1914)*

Peirce was by no means the first philosopher to treat of language and reasoning. His distinction consists principally in that he constructed a theory and systematics of signs which embraced linguistic and non-linguistic semiotic processes within a unified framework. Further, problems of the nature and ground of knowledge—epistemological questions—were approached by Peirce within the context of the theory of action, and in particular semiotically mediated action.

Peirce's theory consists in essence of the juxtaposition of two sets of threefold, or triadic, distinctions; one concerning the *mechanism* of the sign, and its internal relations; the other concerning the *nature* of different types of sign, according to the manner in which the internal relations operate. All signs share a common structure, in that they depend upon and are defined in terms of the relations between three elements. The first element, for which Peirce himself frequently used the simple term 'sign', consists in the perceptible unit (for example, a group of phonemes, a printed word, a traffic light) which conveys the meaning of the sign. McNeill (1979), whose account I shall largely follow, therefore calls this element the 'sign vehicle' (see also Bates *et al*. 1979). In order to standardize terminology within this chapter, I shall refer to this element as the *signifier*. The second element of the sign is the *object* which the signifier can be interpreted as 'standing for'—which need not be either a physical or a real object, and may indeed be another sign. The third element of the sign is the *interpretant* of the sign, by virtue of which the signifier signifies the object for a subject.

The interpretant is at once: (1) the *cause* of the signifying relation between the signifier and the object, in that the relation between the interpretant and the object duplicates the relation between the signifier and the object; and (2) the *result* of the act of signification, in that the interpretant is the 'significate outcome' (in the mind of the subject) of the interpretation of the signifier as signifying the object. According to Peirce, the interpretant is therefore *also* a signifier, albeit of a mental or cognitive nature.

Peirce's most distinctive contribution to semiotics was to examine the different forms that the relation between signifier and object can take in different types of signs. He proposed that relations between signifier and object (and hence also between interpretant and object) may be of three kinds. In the case of an *indexical sign*, the signifier is related to the object by virtue of a real relationship (such as physical causation, or part–whole relations) obtaining between signifier and object, or a relationship such that the signifier directs attention by 'blind compulsion' to the object. Thus, the symptoms of a disease, the smoke of a fire, the tip of an iceberg, and a pointing finger are all examples of indexical signifiers. In the case of an *iconic sign*, the signifier is related to the object by virtue of its resemblance to the object, as is the case with pictures, images, or diagrams. In the case of *symbolic signs*, the signifier is related to the object by virtue of a rule or convention, as is the case with a word such as 'dog', which signifies by virtue of a complex of linguistic rules and conventions, rather than by virtue either of necessity (indexicality) or similarity (iconicity).

Much has been written about the problems associated with Peirce's typology of signs, and I shall restrict myself here to three comments. First, it is frequently the case that different signifying relations may be combined in a single sign. For example, as Bates *et al*. (1979) point out, the footprint of an animal is iconically related to the shape of the animal's foot, and is simultaneously an indexical signifier of the presence of the animal at this place at a previous time. Equally, although language is an essentially symbolic signifying system, it also presents aspects of iconicity (as in the temporal ordering of narrative sequences) and indexicality (as in the contextual determination of the reference of deictic terms, such as 'this', 'I', and 'here'). Second, the nature of symbolic signs as being governed by convention is often taken to imply that all conventions in symbolic systems such as language are arbitrary, an assumption which is largely grounded in the work of Saussure, which is discussed below. Not only have the iconic aspects of language, as McNeill (1979) points out, frequently been underemphasized, but the *motivation* of certain conventions by functional considerations has also frequently been underplayed (Paprotté and Sinha 1987).

The third issue is more complex, and relates to the extent to which Peirce's theory applies only to the internal workings of signs, or can also be seen to solve the problem of how signs are used to refer to the real world. I have presented here a largely 'internalist' account, emphasizing that the process by which symbolic signs are interpreted is one involving further signifying relations, between the subject's concept of an object, and the object itself (bear in mind that not only is the 'interpretant' also a signifier, but the 'object' may itself also be a sign). Peirce himself did frequently use the terms 'refer' and 'reference', but he lacked a theory of reference, and in this respect his most important contribution may be seen precisely in producing a theoretical apparatus capable of comprehending the recursive and multiple nature of semiotic processes.

17.1.2.2 *Gottlob Frege (1848–1925)*

Frege was a logician and mathematician, as well as a philosopher of language, and his approach to linguistic signification was motivated by an attempt to understand how the substitution of one referring expression for another preserves or does not preserve the truth of a proposition. Traditional logics distinguished between propositions expressing 'analytic' truths—those which, in modern terms, are necessarily true in all possible worlds, such as 'all humans are mortal'; and those expressing 'empirical' truths, true contingently within our knowledge and experience, such as 'Mount Everest is the highest mountain on earth'.

The notion of analyticity remains problematic (see Kripke 1980; Putnam 1981), but Frege (1892) started his analysis by observing that the same individual entity may appropriately be referred to on different occasions using different expressions. To use his own example, the expressions 'the morning star' and 'the evening star', while unrelated by analytical logical entailments, share a common referent. This example is in some respects a special case; but the phenomenon of the multiplicity of possible referring expressions for a given entity is a general one (Brown 1958). Thus, Frege distinguished between the *sense* (*Sinn*) of a referring expression and its *reference* (*Bedeutung*): two expressions, then, may have quite different senses (for example 'my neighbour' and 'the doctor'), allowing quite unrelated inferences (for example 'lives next door', 'is medically qualified'), while sharing the same referent. This phenomenon also, although this was not part of Frege's analysis, underlies the *informative potential* of language, by permitting multiple predications of the same referent: for example, 'my neighbour is a doctor'.

On the other hand, while the same expression (for example 'the doctor') may be used to refer to different *individuals*, it is also the case that expressions which are analytically identical in their sense (for example 'doctor' and 'medically qualified person'—NB the expression 'doctor' is used here *in the sense of* 'medical doctor') have identical *scope* of reference; they refer to the same *class* of individuals. Thus, a general asymmetry exists in the relations between sense and reference: while referring expressions of different sense may refer to the same individual, a referring expression may not be used to refer to an individual if that individual cannot be referred to by an expression with identical sense. Therefore, according to Frege, it is the sense of a term which determines its reference; and, further, the sense of a complex expression (and hence its reference) is determined by the senses of the constituent expressions into which it may be analysed.

Our understanding of language, then, depends upon our grasp of sense, or meaning, since it is meaning which determines and delimits appropriate reference, and not reference which determines meaning. If this be the case, what do 'senses' consist of? Frege's answer was that 'sense' is an intrinsic and objective property of language, which is grasped or apprehended by the cognitive activity of the subject. Senses, then, and their role in cognition, may be regarded in much the same way as Peirce's interpretants: as signs, but, in this case, signs which are proper to language and which underlie the mechanisms of linguistic reference.

Versions of Peircean typologies of the sign have been used by theorists from Piaget onwards to analyse the development of communication and cognition. The recent upsurge of studies of gestural communication (Bruner 1975; Bates *et al.* 1979; Lock 1980), as well as studies of the later development of discourse processes (Karmiloff-Smith 1979), has brought questions of reference and indexicality to the fore in developmental psycholinguistics, and both Bates (1979) and McNeill (1979) have advanced 'neo-Piagetian' theories of cognitive development based upon Peirce's typology of the sign.

In contrast, the more formal approach to meaning in Frege's work has led indirectly to contemporary logical approaches to model-theoretic semantics, and more generally to that discipline known as 'linguistic' or 'analytic' philosophy. Until recently, this work has tended to fall outside the central concerns of developmental psycholinguists (but see Macnamara 1982). However, the interdisciplinary nature of modern cognitive science, and the contribution to it of the philosophy of language, suggest that the Fregean and analytic tradition will ultimately offer much to developmental theory too.

17.1.2.3 *Ferdinand de Saussure (1857–1913)*

Saussure is frequently cited as the founder of modern linguistic science. His work, published posthumously from the collated lecture notes of his students as the *Cours de linguistique générale* (Saussure 1966 [1915])[2] has had a tremendous influence, not merely within linguistics, but upon the human sciences as a whole, since in this text he lays out and exemplifies the general principles of the method and philosophy which has come to be known as 'structuralism'.

Before Saussure, linguistics had been dominated by historical and philological concerns, and by the study of written texts as opposed to spoken language. Saussure, on the contrary, insisted that the foundations of linguistic science consisted in the study of the *synchronic* structure of a particular language at a given time, and that the *diachronic* study of language-change had necessarily to be secondary to this. Thus, rather than viewing linguistic signs in terms of their origins and historical transmutations, he sought to define the rules governing their lawful combinations and interrelations within a language. Saussure distinguished between *la parôle* (speech), consisting of the actual speech acts of language-users, and *la langue* (language), the linguistic system adhered to by a community of language-users and regulating their speech conduct (*parôle*). The object of linguistic science is the *langue*, the facts of language structure, and its goal is their systematic description for a given language.

According to Saussure, a linguistic sign consists of a coupling of a *signifier* with a *signified*. The signifier is the acoustic or graphic 'mark', and the signified is the meaning or concept which it conveys—the Saussurean 'signified' may thus be thought of as equivalent to Frege's 'sense' and Peirce's 'interpretant'. Saussure, however, was concerned not so much with the inner workings of the sign, or with its referential function, as with its functioning within the sign-system as a totality. At the level of the individual sign, Saussure argued, there is no necessary connection between the signifier and the signified. For example, the English word 'tree' and the French word *arbre* signify in the different languages according to different conventions. Further, the realm of the signified—or sense—may be 'cut' differently by the signifiers of different languages. For example, the French word *mouton* signifies a domain which may, according to context, be signified by either of the English words 'mutton' or 'sheep'. For this reason Saussure insisted that the connection between signifier and signified is 'arbitrary'.

A linguistic sign, he suggested, may therefore adequately be defined only in terms of the *relations* which it 'contracts' with other signs: linguistic units possess a purely relational identity. Such relations may be of two sorts. *Paradigmatic* relations are contracted between elements which may be substituted for each other in an expression, and *syntagmatic* relations are contracted between elements which may be combined with each other in an expression. These relations are in turn governed by the sets of *contrasts* or *oppositions* which constitute the rule system of a language (*langue*). Such contrast sets exist at different levels. Thus, the sounds *p* and *b* contrast with each other inasmuch as they may be substituted for one another in some (but not all) phonological contexts, and the permissible and impermissible contexts of substitution are equally governed by the totality of contrasts in the sound-system of a language. By the same token, the word *dog* contrasts with the word *cat*, and also with the word *bitch*, and the totality of such meaning contrasts constitutes the semantic system of the language.

Thus, by applying Saussure's structural and contrastive method, we may conclude that the *sense* of a term consists in the meaning-relations it contracts with other terms; and, in essence, this remains the basis of much contemporary semantic theory, although contrast in the strict sense of antonymy or 'oppositeness' is not the only relevant meaning-relation (Lyons 1977). Like Peirce, Saussure conceived of linguistics as one branch of a general science of signs, semiotics, or semiology, and believed his structuralist method to be appropriate for the analysis of all semiotic systems. The structuralist method has been applied by anthropologists to the study of kinship systems, culinary practices, and so forth, and the general proposition that social practices may best be seen as signifying and symbolic practices has become central to many areas of twentieth-century social theory.

Many theorists within the structuralist tradition have tended to emphasize the dependence of thought or cognition upon language and its categories or meanings. At the extreme, this has involved the hypothesis of 'linguistic determinism', also known as the Whorf–Sapir hypothesis, which asserts that human thought-processes are determined by the syntactic and lexical forms of the particular language into which the individual is born. This hypothesis is now generally discredited, at least in this strong form (Rosch 1977).

Other varieties of structuralism, such as Lévi-Strauss's theory of myth, and Piaget's genetic epistemology (described in Section 17.2), emphasize rather the universal properties of the human mind, as does Chomsky's generative linguistics (for example Chomsky 1957), which, despite its departures from Saussurean linguistics, shares its synchronic orientation.[3]

The work of Peirce, Frege, and Saussure laid the ground for the subsequent study of language in the disciplines of linguistics, philosophy, and psychology, as well as establishing the foundations of the science of semiotics. In the next subsection, I shall briefly characterize the ideas of three further thinkers, whose influence has perhaps not been so uniformly pervasive, but whose work may be considered as representative of the kind of concerns which have guided recent approaches to the role of symbolization in social life.

17.1.3 *Social life as semiosis: Mead, Bakhtin, Barthes*

17.1.3.1 *George Herbert Mead (1863–1931)*

The American psychologist G. H. Mead stands in direct line of descent from Peirce, as an adherent of the 'pragmatist' philosophy also espoused by James and Dewey. His influence has been widespread in social psychology and microsociology, but of particular interest in the present context is that Mead was centrally interested in the problem of how children come to participate in the world of symbolically mediated social action and interaction. For Mead, language (and symbolization) is not merely a reflection of the external world, but is constitutive both of the social world and, crucially, of the self in that world. 'Meaning', for Mead, resides in the social process as a totality, and it is by virtue of symbolization that intelligent human action, and the 'selves' that originate intelligent action, are possible. Symbolization and the human social order are thus mutually interdependent, since meanings both organize experience and depend upon the social process for their signification (Mead 1934, p. 89):

> a universe of discourse is always implied as the context within which significant gestures or symbols do in fact have significance. This universe of discourse is constituted by a group of individuals carrying on and participating in a common social process of experience and behaviour... a universe of discourse is simply a system of common or social meanings.

Meaning, according to Mead, is intimately connected with action, inasmuch as actions are governed or 'controlled' by meanings which are inherent in the properties of the social environment. This environment is social both in that it includes the responses of others to the acts of the self, and in that the objects towards which acts are directed are meaningful within the matrix of the social process. In the beginning, the control of acts by meanings is effected by the actual responses of the other (Mead 1934, pp. 76–7), within the social and communicative interaction between infant and care-taker adult. Thus meaning is invested in an act by virtue of the place of the act within the organization and sequencing of a socio-communicative interaction, beginning with mother–infant interactions (see also Messer and Collis, this volume, Chapter 15).

Mead, like many of his contemporaries, was profoundly influenced by evolutionary theory, and his account of symbolization attempted to integrate developmental and evolutionary considerations into semiotic theory (in this respect, he was also influenced by the work of J. M. Baldwin: see Russell 1978; Sinha 1984). The critical evolutionary and developmental achievement of human beings, according to Mead, lies in the use of 'significant symbols' as vehicles both for social exchange and psychological control processes. The earliest manifestation of the significant symbol he held to be the 'vocal gesture', which signifies aspects of the 'implicit' meanings of social situations in such a way as to 'lift' them into consciousness, thus constituting the self and reflective intelligence. Many commentators have noted the similarity between this view of Mead's, and the role assigned by Vygotsky (see Section 17.3 below) to the internalization of speech in the ontogenesis of socialized reasoning (Lock 1978).

17.1.3.2 *Mikhail Bakhtin (1895–1975)*

Mead's social-pragmatic account of the nature of symbolization, and its role and origin in social life, is comparable in many respects to that offered by the Russian semiotician M. M. Bakhtin.[4] Bakhtin's approach to linguistic and semiotic theory was informed by dialectical and historical materialist philosophy; but he was perhaps the first Marxist to acknowledge the lack of an adequate theory of the subject in orthodox Marxism, and explicitly to locate the material substrate of consciousness in the sign. Like Mead, Bakhtin held social interaction and process to both constitute, and be governed by, the use and exchange of signs. Further, the sign was seen by Bakhtin, as it was by Mead, to be the vehicle of subjective consciousness and experience:

> signs can arise only on *interindividual territory*. It is territory that cannot be called 'natural' in the direct sense of the word ... It is essential that the two individuals be *organised socially*, that they compose a group (a social unit); only then can the medium of signs take shape between them ... individual consciousness is not the architect of the ideological superstructure, but only a tenant lodging in the social edifice of ideological signs ... the reality of the inner psyche is the same reality as that of the sign. Outside the material of signs there is no psyche ... the subjective psyche is to be localised somewhere between the organism and the outside world, on the borderline separating these spheres of reality ... but the encounter is not a physical one: the organism and the outside world meet here in the sign (Volosinov 1973 [1929]: pp. 12, 13, 26).

Unlike Mead, however, Bakhtin was also familiar with the principles of structuralist linguistics, as developed by Saussure, and with the related work of the Russian Formalist school (Ehrlich 1965). Whilst recognizing the scientific advance constituted by the structuralist method, Bakhtin took issue with its dualistic distinction between *langue* and *parôle*, and with the priority assigned by it to the formal systemic analysis of the former. The actual and dynamic process of linguistic exchange, asserted Bakhtin, can only adequately be understood by recognizing the inherently *dialogic* character of the sign and signification. Bakhtin's linguistic analysis thus focused not upon the form and structure of language, as revealed by the study of sentences, but upon the negotiation and implementation of speech-acts by interlocutors, within extended discourse. From this analysis, Bakhtin attempted to derive a theory of speech and behavioural 'genres', and of the articulation of the speaking subject within discourse structures characteristic of such genres. Such discursive genres, he argued, must be understood with reference to 'the objective conditions of social–verbal interaction'; that is to say, overall social relations as crystallized in 'behavioural ideologies'.

Paradoxically, given Bakhtin's indictment of the secondary role assigned by structuralism to 'parôle', and his insistence on the primacy of speech, his concrete research addressed the problem of the production of speech and behavioural genres through the analysis of written texts. Bakhtin's work here anticipated many of the present-day concerns of text-linguistic and semiotic theory (Beaugrande and Dressler 1981; Eco 1976, 1979), and 'post-structuralist' theories of the subject (Henriques *et al.* 1984).

17.1.3.3 *Roland Barthes 1915–1980*

The work of the French semiotician and literary and cultural critic Roland Barthes, a key figure in 'post-structuralist' analyses, bears many affinities to that of Bakhtin, although the latter was not a direct influence. Barthe's early work (for example *Mythologies*, Barthes 1973 [1957]) situated itself firmly within the Saussurean tradition, in attempting to analyse non-linguistic sign systems, such as photography, or fashion, in terms of internal structural oppositions understood with reference to underlying systems of 'ideology as myth'. None the less, Barthes maintained that such signifying systems themselves depend crucially upon language, inasmuch as language both mediates their representational functioning, and apparently offers an analytical discourse—a 'metalanguage'—capable of effecting the 'unmasking' of this (ideological) function.

> As for collections of objects (clothes, food), they enjoy the status of systems only in so far as they pass through the relay of language ... we are, much more than in former times, and in spite of the spread of pictorial illustration, a civilization of the written word (Barthes 1967 [1964], p. 10).

From the outset, then, Barthes's project was one of 'de-naturalizing' representation, demonstrating that the most apparently spontaneous and 'realistic' image or text is 'read' through a grid of linguistic signifiers: signs refer to signs, and not to a pregiven (pre-signified) 'untexted nature' or object (Silverman and Torode 1980). Barthes's most crucial move, marking a decisive break with the Saussurean tradition, came however when he extended the 'dialogic' premiss of Bakhtin's analysis of utterances to the understanding of texts themselves. Texts do not arise *de novo*, fully-armed from the brow of the singular author; nor do they merely chart an unproblematic reality, for the 'semblance of reality'—the *vraisemblable*—is, as we have seen, itself produced as an effect of signifying processes. Rather, texts can only be understood in terms of their *intertextuality*.

Thus, other texts are not merely the disambiguating 'context' of the text, but the invoked and evoked, actual and virtual, alterities and *différances* (Derrida 1973) which are set in motion through the 'play' of signifiers; through dialogue, parody, contradiction, and contestation. Barthes's dialectical method therefore shifts the ground of semiotic analysis away from its initial preoccupation with 'signification'—that is, signifier–signified relations—towards 'significance'—that is, the productive work of signifiers in the textual and discursive practices within which 'readings' (and their subjects) are positioned and produced as multiple, and possibly contradictory, sites. From this perspective, the monologic and univocal 'referring expression' of linguistic analytic philosophy in the tradition of Frege; the 'unity of the sign' of Saussurean linguistics and semiotics; and the individual subject, the *cogito* of Descartes and the *competence* of Chomsky, are all equally problematized by Barthes, and by other post-structuralist critiques.

Although it would be satisfying to conclude the first part of this chapter by indicating an outline of a comprehensive theory of the sign (and language), the prospects for such a synthetic account are remote. All of the approaches referred to continue to be influential in varying degrees in the study of the genesis of symbolization, as well as in wider fields of linguistic, psychological, social, and literary studies. Rather than attempt an evaluation, then, it is more appropriate to mention some of the important issues arising from the discussion, and their relation to problems of genesis.

There are perhaps four crucial (and perennial) problems which face all theories of the sign (including general metatheories of language). The first is that of the relative priority of environmental and innate factors in determining the form and the mode of acquisition of language and symbolization. The second is the relative importance accorded to the referential function of language, and the extent to which the 'reality' which is referred to is independent of and 'outside' language and discourse. The third is the extent to which the structural properties of sign-systems—particularly language—are governed by functional and communicative constraints, and the extent to which they represent formal and universal constraints on the products of the human mind. The fourth is the extent to which symbolization and language (and specific systems thereof) are *dependent* upon general cognition, *determinant* of general cognition, or *autonomous* from other cognitive processes. I shall hazard no definitive answers to these questions, but they may usefully be borne in mind when considering the theories outlined in the next section.

17.2 Language, thought, and symbol in human development: Freud, Piaget, Vygotsky

17.2.1 *Darwin and the mind of the child*

Probably the single most important influence in establishing child psychology as a field of scientific study was that of evolutionary theory. Indeed, one of the earliest 'modern' texts of child psychology was Darwin's own study of his son's early development, *A biographical sketch of an infant* (Darwin 1877), which covered such areas as motor, emotional, communicative, intellectual, and moral development. Darwin's most powerful and enduring influence on the emerging discipline of child psychology, however, was indirect, through such disciples as his friend and colleague George Romanes, to whom Darwin entrusted his unpublished psychological manuscripts, James Mark Baldwin, James Sully, Karl Groos, and G. Stanley Hall (see Gruber 1974). Whatever their other differences, all these psychologists shared a common commitment to, first, a comparative and evolutionary approach to human psychology, and, second, the 'biogenetic law' of 'recapitulation' formulated by Haeckel (1874). Ontogeny was seen as the key to unlock the secrets of 'mental evolution', including the evolution of language and symbols (See also Gibson, this volume, Chapter 14, Section 1). This Darwinian and evolutionary legacy was, in different ways, central to the work of Freud, Piaget, and Vygotsky discussed in this section.

17.2.2 *Freud: Ursprache and the unconscious*

Mention of 'Freudian symbolism' usually conjures images of trains roaring into tunnels, and other tokens of sexual imagery. In fact, the scientific concern of Sigmund Freud (1856–1939) with symbolic processes both pre-dated his mature theory of psychosexual development, and encompassed a general preoccupation with what is now known as 'cognitive psychology', and the relations between biology and mind. As Sulloway (1979) makes clear in his biography *Freud: biologist of the mind*,

many, if not most, of Freud's fundamental conceptions were biological by *inspiration* as well as by implication ... Freud stands squarely within an intellectual heritage where he is, at once, a principal scientific heir of Charles Darwin and other evolutionary thinkers in the nineteenth century and a major forerunner of the ethologists and sociobiologists of the twentieth century (p. 5).

Sulloway argues that the 'Freud' of the Freudian myth is a 'crypto-biologist', inasmuch as the myth—of which Freud himself was the principal author as well as player—has sought to play down or disguise the true extent of his intellectual debt to the psychobiological milieu within which his theories took shape. The expression 'crypto-biologist' is a fortunate one, also, because, intertwined with the biological themes which inform Freud's work, are themes which, from our historical vantage point, can be recognized as 'semiological': Freud was indeed a cryptologist, a decipherer and interpreter of the processes by which psychic symptoms arise as representations/representatives of desires and memories censored from consciousness.

17.2.2.1 *Freud as evolutionist*

As Sulloway (1979) has convincingly demonstrated, evolutionary biological theory was centrally implicated in the Freudian project, in the form of the 'biogenetic-Lamarckian' synthesis. Neither Haeckel's 'biogenetic law' nor Lamarckian 'inheritance of acquired characteristics' are any longer seriously entertained in evolutionary biology (at least in their original meanings—see Ho and Saunders 1984), the latter in particular having been proscribed by the twentieth-century 'neo-Darwinian synthesis'. Yet, during the period of both Freud's and Piaget's intellectual formation, these theories were not merely current, but orthodox: Darwin himself, for example, who did not live to see the rediscovery of Mendel's genetic experiments, accepted Lamarckian inheritance as a supplement to inheritance by natural and sexual selection.

The fusion of Lamarckian and biogenetic (recapitulationist) assumptions provided Freud with the rationale for his attempt, which became increasingly explicit in his writings, to adduce the phylogenetic history of culture and symbolization through the study of the neuroses, and other manifestations of unconscious processes, such as jokes, slips of the tongue, and, above all, dreams. Thus, Freudian 'cryptobiology' was simultaneously Freudian 'phyloculturalism'. References to phylogenesis recur, for example, throughout Freud's *Introductory lectures* (Freud 1922):

The era to which the dream-world takes us back is 'primitive' in a two-fold sense: in the first place, it means the early days of the *individual*—his childhood—and, secondly, in so far as each individual repeats in some abbreviated fashion during the course of childhood the whole course of the development of the human race, the reference is *phylogenetic* ... It seems to me quite possible that all that today is narrated in the form of phantasy—seduction in childhood, stimulation of sexual excitement upon observation of parental coitus, the threat of castration— ... was in prehistoric periods of the human family a reality; and that the child in its phantasy simply fills out the gaps in its true individual experiences with true prehistoric experiences (pp. 168, p. 311).

Freud's phyloculturalist preoccupations came, with the passage of time, to weigh as heavily as clinical considerations in the development of his later, 'metapsychological' theories—that is, theories of the overall (evolutionary–cultural) matrix within which psychoanalysis was to be situated. In *Totem and taboo*

(Freud 1953–74, vol. 13 [1912–13]) for example, Freud developed the hypothesis that the universality of the incest-taboo, the establishment of the super-ego, and the centrality of the Oedipus complex, could all be traced to a 'primal scene'—itself based upon Darwin's speculations in *The descent of man* (Darwin 1871), wherein the ejected sons killed the Father/Ancestor.

Freud's propensity to weave myth out of myth is, of course, all too open to scientific criticism—though we would do well to remember that, at the time, the theoretical underpinning of the metapsychology was plausible enough. Still, if phyloculturalist speculation were *all* to Freudian theory, we would be justified in regarding it as yet another outgrowth of the rich foliage of nineteenth-century social biology. Aside from the clinical aspects of psychoanalysis, however, which do not directly concern us here, what compels continuing attention to the Freudian synthesis is its specific combination of methods, findings, and hypotheses regarding the ontogenesis of symbolic and representational processes, their emotional investments, and their continuing consequences in adulthood.

17.2.2.2 *Freud as semiotician*

The preceding quotation from Freud emphasized the correspondences which Freud postulated between the 'primal' scenes of, respectively, the infancy of the individual and the 'infancy' of the species. A complementary, and methodologically more fundamental, correspondence between (ontogenetic) infant experience and adult manifestations of unconscious (or primary) processes was succinctly expressed in *The interpretation of dreams* (Freud 1954 [1900], p. 546): 'a dream might be described as *a substitute for an infantile scene modified by being transferred on to a recent experience*'.

The hypothesis of the origin of dreams in the unconscious was the cornerstone of Freud's analysis, and the basis for his declaration that 'The interpretation of dreams is the royal road to a knowledge of the unconscious activities of the mind' (1954, 608). Freud's application of psychoanalytic principles to the analysis of dreams also marked the extension, after he had abandoned earlier theories in which *actual* childhood seductions played a central role in the genesis of neuroses, of the theory to apply to 'normal' as well as pathological psychic phenomena. From the point of view of the ontogeny of symbolization, perhaps the most important aspect of the theory is the way in which it conceives dream-content and other psychic 'representatives', such as symptoms, as being the result of the operation of opposing dual principles.

In the first place, the wishes generated in the unconscious, which is in effect the repository of repressed infantile memories and desires, 'cathect' or invest 'trains of thought' initiated in daytime consciousness; in the second place, secondary processes of censorship transform the dream-thoughts in such a way that the manifest dream content does not provoke the anxiety which would result from the direct expression in consciousness of the inadmissible wishes of what Freud would later call the 'Id'. The 'semiotic' aspect of Freudian theory thus consists in the specific analysis of the mechanisms of transformation effected by the secondary (pre-conscious) processes upon the psychic representatives of primary process (unconscious) material, in order to render it admissible to consciousness. Thus, Freud wrote:

The dream-thoughts and the dream-content are presented to us like two different versions of the same subject-matter in two different languages. Or, more properly, the dream-content seems like a transcript of the dream-thoughts into another mode of expression, whose characters and syntactic laws it is our business to discover by comparing the original and the translation (1954, p. 277).

The task of psychoanalysis, as Freud saw it, is simultaneously to elucidate the nature of the 'dream work' and to relate it systematically to the theory of the unconscious and to unconscious processes, such as repression, identification, denial, and so on. The two main processes constituting the dream work were termed by Freud *condensation* and *displacement*. Freud observed that very frequently a single element of dream-content may be interpreted as simultaneously 'standing for' a variety of elements in the chains of different 'dream thoughts' representing different unconscious wishes. This 'overdetermination' of the unconscious meaning of elements of dream content lends to dreams their quality of 'condensation', as well as their apparently illogical character:

Not only are the elements of a dream determined by the dream thoughts many times over, but the individual dream-thoughts are represented in the dream by several elements (Freud 1954, p. 284).

The process of condensation, resulting in the overdetermination of dream content, is frequently accomplished linguistically by punning, alliteration, and other devices to which Freud drew attention in *The psychopathology of everyday life* (Freud 1953–1974, vol. 6 [1901]); and the principal technique for investigating such processes psychoanalytically is the free association of verbal material. The work of condensation, however, is not the only means by which censorship is achieved:

in the dream work a psychical force is operating which on the one hand strips the elements which have a high psychical value of their intensity, and on the other hand, *by means of overdetermination*, creates from elements of low psychical value new values, which afterwards find their way into the dream content. If that is so, a *transference and displacement of psychical intensities* occurs in the process of dream-formation, and it is as a result of these that the difference between the text of the dream content and that of the dream thoughts comes about (Freud 1954, p. 308).

17.2.2.3 *Freud as cognitive theorist*

Although Freud stresses the significatory nature of condensation, and the energetic nature of displacement, the two aspects in fact coexist in both processes: the energetic aspect providing the motivation, and the semiotic or significatory the mechanism, for the dream work and for symptom-formation in general. The applicability of linguistic–semiotic analyses of metaphoric (similarity) and metonymic (contiguity) relations in signifying systems (a distinction roughly equivalent to that between paradigmatic and syntagmatic relations: cf. Section 17.1) to the concepts of condensation and displacement has been noted by a number of writers, amongst them Jakobson (1956) and Lacan (1966), who maintains that 'the psychoanalysable symptom... is supported by a structure identical to that of language' (1966, p. 21). Space does not permit an extensive discussion of these issues, but it can be seen that the general notions of primary and secondary process are as much 'informational' as they are 'energetic' (see also Pribram and Gill 1976).

In Freud's early work, a persistent theme was in fact the attempt to understand the relationship between motivational–affective and neuro-cognitive aspects of the development of symbolization and the self; an attempt which, according to Sulloway, was subsequently abandoned as unattainable, and replaced by an evolutionary–cultural paradigm in Freud's later work. Nevertheless, as many authors have noted (see for example Wolff 1960; Greenspan 1979; Ingleby 1983), a definite and dualistic theory of the relationship between affect and cognition underlies the entire Freudian approach to the ontogenesis (and phylogenesis) of symbolization, in which the tension between primary-process 'irrealism' or phantasy in the wish for gratification and the reality principle resulting from the psychic structures engendered by secondary processes remains perpetually and inevitably unresolved, while none the less providing a 'motor' for psychological development. In this respect, Freudian theory contrasts with Piaget's conception of symbolic and representational development, in which both cognition and affect are governed by the same processes of structural growth.

17.2.3 *Piaget: from solipsism to structure*

Sigmund Freud and Jean Piaget (1896–1980) have probably been responsible more than any other individuals for shaping the discipline and practice of child and developmental psychology. Amongst other similarities, they were both architects of grand synthetic theories which resist easy summary, and they were both profoundly influenced by evolutionary theories which, while they were current during their intellectual formations, now appear distinctly heterodox. Piaget, in particular, and even more explicitly than Freud, was a lifelong adherent and exponent of a neo-Lamarckian approach to evolution; and, while rejecting the simplicities of Haeckel's recapitulationism, he maintained a continuous dialogue in his developmental psychological theorizing with evolutionary biology. Piaget never in fact characterized himself as a psychologist, preferring the term *genetic epistemology* to designate the scientific discipline which he sought if not to found (the phrase was apparently coined by J. M. Baldwin), at least to establish on a secure empirical footing. The lasting impact of Piaget upon our understanding of the development of mind is, no less than Freud's, undiminished by the problematic light cast upon his specific theoretical formulations by more recent evidence.

As well as their common commitment to a genetic psychobiology embracing both ontogenetic and phylogenetic development, Freud and Piaget shared a common assumption regarding the psychic status of the new-born infant. This assumption may be summarized as a hypothesis of *original a-dualism*; that is, a lack of psychic differentiation between infant and environment, internal state and sensory surround. Freud called this state 'primary narcissism', a term emphasizing the affective connotations of the hypothesis; Piaget dwelt rather upon the intellectual 'egocentrism' of the infant and the young child. For both theorists, however, 'His Majesty the Baby' is only gradually displaced from centre stage, in a process involving cognitive and affective conflict.

Despite these similarities, however, the Freudian and Piagetian theories of the actual developmental process are quite different. For Piaget, the growth of subjectivity proceeds in an essentially harmonious manner, via the incorporation of *disequilibrated* conflictual states into higher-order, *equilibrated* '*structures d'ensembles*'; whereas for Freud, the process is *inherently* contradictory, involving the gradual, and never wholly completed, subordination of primary to secondary process. Again, for Piaget, original a-dualism is a logical initial postulate in a monistic, constructivist theory in which subject and object are co-constituted through action; in contrast, Freud's notion of original a-dualism is one of a primal condition which is interrupted and split by the emergent dualism of psychic processes, leading eventually to the establishment of the ego and super-ego. Consequently, the 'subjects' of Piagetian and Freudian theories are radically different entities. The Piagetian 'epistemic subject', situated within a progressive dialectical spiral of the growth of knowledge, is on a one-way track to higher, more equilibrated structures; though progress may be halted or retarded, it is irreversible. The 'progress' of the Freudian psychoanalytic subject, on the other hand, is altogether more precarious—it is continually prone to regression and dis-integration; earlier infantile and primordial states and experiences, though repressed, threaten to return and haunt the present.

As Ingleby (1983) has noted, however, Piaget and Freud share a further presupposition, closely bound in both cases with the postulate of original a-dualism: that is, of the *a-sociality* of the infant. In Freud's case, consistently with the point of view I have outlined, this a-sociality is *dis*-placed in adulthood, but persists in the Id; whereas, in the Piagetian scheme, the egocentrism of childhood is *re*-placed by socialized cognition. Yet, and this is perhaps the most crucial aspect of Piagetian theory, the growth of egocentric thinking into de-centred, socialized thinking is accomplished *without* any significant role being accorded to social interaction or social structure *per se*. (The apparently a-social nature of Piagetian cognitive-developmental theory has increasingly been criticized in recent developmental psychology.)

Piaget's is a stage theory, in which the basic characteristic of any given stage of cognitive development is the manner in which actions (actual or internalized) are co-ordinated with respect to the objects, and relations between objects, to which the actions are directed. Action is a more fundamental category than either subject or object, for through successive stages of the co-ordination of action, the object is *constructed* by the organism, in terms of categories such as causality, space, time, and motion. Since these categories, as Kant argued, form the necessary basis for logical reasoning, the growth of understanding on the part of the epistemic (or knowing) subject is to be construed as the growth of logic. The logical (or logico-mathematical) structures constituting a given developmental stage are also to be understood as being *constructed*: logic, for Piaget, is both the measure and the product of the organism's developing interaction with the environment. This, in the baldest possible terms, is Piaget's theory, elaborated throughout a lifetime of empirical and theoretical work devoted to tracing the growth of logico-mathematical, operational thinking from the earliest stages of infant pre-intellectual action.

Piagetian theory conceives of human cognitive development as passing through three general stages or periods: the *sensorimotor*, lasting from birth until about 18–24 months, and itself divided into six substages; the *pre-operational*, from the end of the sensorimotor period until the emergence of 'concrete operations' at around 6–8 years; and the *operational*, divided into the concrete and the formal operational stages, the latter emerging in late pre-adolescence or early adolescence.

The central achievement of cognitive development at the end of the sensorimotor period consists in the *internalization* of actions as *representations*. Piaget is here his own best summarist (Piaget 1952, in Gruber and Voneche 1977, p. 456–7):

Sensori-motor intelligence is not yet ... operational in character, as the child's actions have not yet been internalised in the form of representations (thought). But in practice even this limited type of intelligence shows a tendency towards reversibility, which is already evidence of the construction of certain invariants.

The most important of these invariants is that involved in the construction of the permanent object. An object can be said to attain a permanent character when it is recognized as continuing to exist beyond the limits of the perceptual field, when it is no longer felt, seen or heard, etc. At first, objects are never thought of as permanent; the infant gives up any attempt to find them as soon as they are hidden under or behind a screen ... Only towards the end of the first year does the object become permanent in its surrounding spatial field.

The object's permanent character results from the organization of the spatial field, which is brought about by the child's movements. These co-ordinations presuppose that the child is able to return to his starting point (reversibility), and to change the direction of his movements (associativity), and hence they tend to take on the form of a 'group'. The construction of this first invariant is thus a resultant of reversibility in its initial phase. Sensori-motor space, in its development, attains an equilibrium by becoming organised by such a 'group of displacements' ... The permanent object is then an invariant constructed by means of such a group; and thus even at the sensori-motor stage one observes the dual tendency of intelligence towards reversibility and conservation.

This quotation exemplifies many of the central themes of Piaget's work: the roots of knowledge in movement and action; the construction of reality by co-ordination of these actions; the subordination of perception to cognition; and the formalization of cognitive processes in terms of mathematics and logic (in particular, the logico-mathematical theory of sets, groups, and transformations). To continue, however: the stages of sensorimotor development may, at one level, be defined by the stages in the development of the object concept; however, another fundamental aspect of sensorimotor development is *the co-ordination of means and ends* in the organization of action schemata, and the co-ordination of schemata through reciprocal assimilation (Piaget 1953 [1936]).

Whereas the *structural* aspect of Piaget's developmental theory is characterized in terms of logico-mathematical operations (or pre-logical 'operative actions'), ordered according to their degree of formal completion and flexible closure, the dynamic, *functional* aspect of development is characterized in terms of the dialectical interplay of *assimilation* and *accommodation*, of which the former is the more fundamental. Cognitive structures, at every stage of development, function as *assimilatory schemata*, analogous to the behavioural and physiological systems which enable the organism to assimilate food. All intelligence presupposes and depends upon assimilation, by which alone the structure of reality can be apprehended, and upon which perception itself depends. *Accommodation*, however, is also implied by the very process of assimilation: every assimilatory act must be precisely 'fitted' (accommodated) to the particular object, event, or logico-mathematical structure to which it is directed. Adaptation, for Piaget, consists of a process of *equilibration* between assimilation and accommodation.

The gropings of the infant towards means to re-enact pleasurable or interesting events (the event and the causative action being as yet undifferentiated) give rise to *circular reactions* in which elementary schemata are eventually assimilated to each other in parallel with the assimilation of object to schema. Eventually, this leads to the accommodatory ability to invent *new* means to achieve the same ends, through insight, which Piaget interprets as indicating the ability to co-ordinate schemata mentally (the internalization of action). Thus, the acquisition of a mature object concept, indexed by the ability to infer the position of a hidden object subjected to invisible displacements; the acquisition of elementary knowledge of spatial displacements independent of the actions of the subject; and the differentiation of means from ends are all fundamental aspects of the emergence of *representation* at the end of the sensorimotor period.

Piaget (Piaget 1954 [1937]; Piaget 1962 [1945]) defined representation—also termed the 'symbolic function'—in two (not always consistent) ways: first, in terms of the ability mentally to evoke absent realities, as implied by the notion of object permanence; and second, in terms of his own reading of semiotic theory. In this latter sense, Piaget distinguished between 'indices' or 'signals', which are already present in the early stages of sensorimotor development, and which directly 'trigger' assimilatory schemata; and 'symbols', which *represent* schemata, in virtue of their having a 'figurative' content different from the objects which they 'signify' via the schemata. Thus, 'symbolic signs' (including conventional linguistic signs) are closely related for Piaget to accommodation, inasmuch as the differentiation of the figurative signifier from the signified schema involves the internal differentiation of the schema itself (including means from ends).

Representational thinking provides the basis for developments in the *pre-operational* stage, including those involving language, deferred imitation, and symbolic play; all of which necessitate accommodatory adaptations to 'figurative' aspects of the environment.[5] However, Piaget is quite clear that figurative knowledge depends upon, and is secondary to, operative (and ultimately logico-mathematical) knowledge. In other words, symbolization, for Piaget, is a *consequence* and not a *cause* of cognitive development. In particular, Piaget insists upon the primacy of action, and the cognitions deriving from the co-ordination of action, over linguistic knowledge: 'there exists a logic of co-ordination of actions. This logic is more profound than the logic attached to language and appears well before the logic of propositions in the strict sense' (Piaget 1962, in Furth 1969, p. 122).

Piaget's denial of a formative developmental role to language and symbolization is of a part with his general belief that the figurative aspects of the environment in general are of only secondary importance: the fundamental basis of cognition being related to the internalization in operational thinking of the logico-mathematical structures which Piaget believed to be the necessary foundations of all knowledge, and thus the basis for genetic epistemology. For this reason, Piaget viewed all forms of linguistic determinism as incompatible with his constructivist approach.

Furthermore, the emergence of symbolization does not in itself bring about the development of reversible logical operations which Piaget takes to be the hallmark of concrete operational thinking; as evidenced, for example, in conservation, perspective-taking, and class-inclusion tasks. The attainment of representation and the concept of object permanence are only a first step in the passage from egocentric to de-centred thinking; and although the emergence of representation permits the development of language, the mastery of reversible thinking is not dependent upon, for example, the acquisition of certain lexical items, nor can it be hastened by training in verbal expression and comprehension. The same applies in principle to the transition from concrete to formal operations (co-ordinations of co-ordinations), although Piaget conceded that mastery of certain symbolic capacities may be a pre-requisite for this final stage of cognitive development.

Two particular aspects of Piagetian theory are worthy of note, which relate both to recent critical discussions and to the concerns of this volume and chapter (see also Bates *et al.* 1979). The first concerns the relation, during sensorimotor development, between (a) the differentiation and co-ordination of means and ends; and (b) the construction of object permanence. Piaget considered these to be concurrent processes, both of which are necessary for the development of language and representational thinking. However, he acknowledged the possibility of a partial decoupling of the two during development, a phenomenon generally referred to in Piagetian theory as *décalage*. The crucial question may then be posed as to *which* of these particular aspects of sensorimotor development is relevant and necessary to *which* particular aspects of language development.

Further, it may be noted that the significatory relations involved in both of these aspects of development, including their early stages, are of a *syntagmatic* or *metonymic* character, involving part–whole or contiguity relations (Saussure 1966; Jakobson 1956). Those aspects of signification involving paradigmatic, metaphoric *substitution* relations are under-emphasized by Piaget; particularly inasmuch as they involve relations of iconic *similarity*. This is of course consistent with Piaget's general lack of emphasis on the 'figurative'.

The neglect of the figurative is symptomatic of the more general emphasis upon assimilation as more fundamental than accommodation, which underlies the whole of Piaget's theory, and in particular its 'internalist', a-social depiction of cognitive developmental processes. Within this scheme, no *productive* role can be accorded to significatory processes, in the manner conceived by Barthes. Nor, since the ultimate source of all signification is to be sought in action schemata, is there any room either for a theory of *reference*, as conceived by Frege, or for a theory of symbol-usage as a process of social *exchange*, as conceived by Saussure, involving specific universes and modes of discourse, as conceived by Mead and Bakhtin. Because Piaget attempts to account for all cognitive processes in terms of the intra-individual co-ordination of actions, there can be no formative role for signs themselves in cognition. An alternative perspective, which sees signs as the vehicle of both inter-individual and intra-individual cognitive and communicative processes, is outlined in the next subsection.

17.2.4 Vygotsky: language and the sociogenesis of reasoning

It has been said that the latter half of the nineteenth century witnessed three great dethronements of Man and Reason, in the theories of Darwin, Marx, and Freud. Darwin not only evolutionized the 'Great Chain of Being', but, more radically, suggested that rationality itself was an adaptive *product* of nature, not a God-given faculty for the better understanding thereof. Freud pointed to the still-extant psychobiological inheritance of prehistoric dramas, and to the unconscious source in this prehistory of neurotic illnesses and other discontents of civilization. Marx, reversing the priority accorded since Descartes to the individual consciousness as the seat of Reason, stated that 'the human essence is no abstraction inherent in each single individual. In its reality it is the ensemble of the social relations' (*Theses on Feuerbach*: Marx 1975 [1845] p. 423). Furthermore, social relations change, their transformations driven by the dialectic of class struggle and contradiction; thus, a materialist understanding of consciousness must also be of a historical and dialectical nature. Historical materialist analysis was therefore to provide the foundations both for a scientific analysis of class society, and for revolutionary socialist practice.

Lev Semenovich Vygotsky (1896–1934) was no less influenced by evolutionary theory than Freud and Piaget; but the particular forms of evolutionary thought which he took as his point of departure reflected the social, intellectual, and ideological climate of post-revolutionary Russia in the 1920s, when he carried out most of his psychological work. That is to say, Vygotsky's approach to psychology (often referred to as the 'historical–cultural approach'), like Bakhtin's approach to linguistics, to which it bears many affinities, was deeply marked by Marxist historical materialist theory. Vygotskian psychology is both *genetic*, or evolutionary–developmental, and *social*, emphasizing the historical and cultural determination of forms of thought and symbolization.

This does not mean that Darwinian–evolutionary theories were unimportant in Vygotsky's thought. On the contrary, like most contemporary Marxists, he held evolutionary and historical materialisms to be complementary and mutually-reinforcing doctrines. Moreover, Vygotsky strove to achieve a concrete synthesis of the two in his analysis of symbolization as tool-use, which provided the keystone for his psychology.

In the early 1920s, the materialist reflexology of Sechenov (for example 1935 [1863] and Pavlov (1927)) was consecrated as official Soviet psychology, in a repudiation of the hitherto-dominant introspectionist psychology. While Vygotsky welcomed the objective and scientific method underlying reflexological psychology, he criticized its neglect of consciousness and subjectivity. What distinguished Vygotsky's position from, for example, that of the *Gestalt* psychologists, was his insistence that the material substrate of consciousness was to be sought partly in the socially and historically determinate conditions of its acquisition and development:

> Within a general process of development, two qualitatively different lines of development, differing in origin, can be distinguished: the elementary processes, which are of biological origin, on the one hand, and the higher psychological functions, of sociocultural origin, on the other. *The history of child behaviour is born from the interweaving of these two lines.* The history of the development of the higher psychological functions is impossible without a study of their prehistory, their biological roots, and their organic disposition. The developmental roots of two fundamental, cultural forms of behaviour arise during infancy: the use of *tools* and human *speech* (Vygotsky 1978 [1930–1934], p. 46).

Speech and symbolization—sign-usage—stand at the centre of Vygotsky's theory of the sociogenesis of reasoning. Pavlov had already recognized the importance of speech as a 'second signal system', a *mediator* between stimulus and response underlying neo-cortical functioning. Vygotsky drew on this analysis of mediating function in comparing (but not identifying) sign and tool-use, and in analysing their co-functioning:

> The basic analogy between sign and tool rests on the mediating function that characterises each of them.... However a most essential difference between sign and tool, and the basis for the real divergence of the two lines [of their development], is the different ways that they orient human behaviour. The tool's function is to serve as the conductor of human influence on the object of activity; it is *externally* oriented; it must lead to changes in objects. It is a means by which human activity is aimed at mastering ... nature. The sign, on the other hand, changes nothing in the object of a psychological operation. It is a means of internal activity aimed at mastering oneself; the sign is *internally* oriented ... The mastering of nature and the mastering of behaviour are mutually linked, just as Man's alteration of nature alters Man's own nature. In phylogenesis we can reconstruct this link through fragmentary ... evidence, while in ontogenesis we can trace it experimentally ... The use of artificial means, the transition to mediated activity, fundamentally changes all psychological operations, just as the use of tools limitlessly broadens the range of activities within which the new psychological functions may operate. In this context, we may use the term *higher* psychological function, or *higher behaviour*, as referring to the combination of tool and sign in psychological activity (Vygotsky 1978, pp. 54–5).

The joint mediation of activity, by sign and tool, enables the child to gain increasing control over his or her own actions, through, first, the use of an external sign as a 'tool' of consciousness (Vygotsky uses the example of tying a knot in one's handkerchief as an *aide-mémoire*); and, second, through the later *internalization* of such sign-usage in the form of 'higher functions' such as voluntarily-directed memory and attention. The notion of 'internalization', together with that of mediation, is fundamental to Vygotsky's historical–cultural approach:

> The internalization of cultural forms of behaviour involves the reconstruction of psychological activity on the basis of sign operations. Psychological processes as they appear in animals actually cease to exist; they are incorporated into this system of behaviour and are culturally reconstituted and developed to form a new psychological entity ... The internalization of socially rooted and historically developed activities is the distinguishing feature of human psychology (Vygotsky 1978, p. 57).

Vygotsky also laid considerable stress upon the role of actual social interactions in the internalization of higher functions, an emphasis which, as already noted, has led many recent writers directly to compare his position with that of Mead. A much-quoted dictum of Vygotsky would certainly support such a comparison:

> Every function in the child's cultural development appears twice: first, on the social level, and later, on the individual level; first, *between* people (interpsychological), and then *inside* the child (intrapsychological) ... All the higher functions originate as actual relations between human individuals (Vygotsky 1978, p. 57).

As Holzman (1985) has noted, however, despite the similarities between the two theorists, while Mead took communication and interaction to be constitutive of 'social process', for Vygotsky the historical process rather determined the available repertoire of signifying and other psychological activities, and thus, ultimately, the actual course of development. The historical materialist basis of Vygotsky's developmental psychological theory has subsequently been developed by Soviet psychologists such as Luria and Leontiev, and I shall discuss the role of history in ontogenesis more fully in the conclusion to this chapter. Meanwhile, no summary of Vygotsky's theory is complete without reference to his notion of 'inner speech', and the critique he offered of Piaget's conception of infantile egocentrism.

Vygotsky noted that 'the development of thought is, to Piaget, a story of the gradual socialization of deeply private, personal, autistic mental states. Even social speech is represented as following, not preceding, egocentric speech' (Vygotsky 1962 [1934], p. 18). Piaget had introduced the notion of 'egocentric speech' in his first book, *The language and thought of the child* (Piaget 1926 [1923]), in which he observed that many of the utterances of pre-school children appeared to be self-rather than other-directed. This he took to be symptomatic of the general intellectual egocentrism of the child; thus, in an intepretative style which was later to become characteristic, he assigned a structural–development, or 'stage', significance to his observations.

Vygotsky, on the other hand, both emphasized the *functional* significance of 'egocentric' speech, and interpreted its developmental status quite differently, in terms of his concept of internalization. This is how Vygotsky summarized his viewpoint:

> The primary function of speech, in both children and adults, is communication, social contact. The earliest speech of the child is therefore essentially social. At first it is global and multifunctional: later its functions become differentiated. At a certain age the social speech of the child is quite sharply divided into egocentric and communicative speech... Egocentric speech emerges when the child transfers social, collaborative forms of behaviour to the sphere of inner-personal [intrapsychological] functions... Egocentric speech as a separate linguistic form is the highly important genetic link in the transition from vocal to inner speech, an intermediate stage between the differentiation of the functions of vocal speech and the final transformation of one part of vocal speech into inner speech... Thus our schema of development—first social, then egocentric, then inner speech—contrasts both with the traditional behaviourist schema—vocal speech, whisper, inner speech—and with Piaget's sequence—from non-verbal 'autistic' thought through egocentric thought and speech to socialized speech and logical thinking. In our conception, the true direction of the development of thinking is not from the individual to the socialized, but from the social to the individual (Vygotsky 1962, pp. 19–20).

According to Vygotsky, inner speech represents the fusion of the hitherto-separate developmental paths of 'pre-linguistic thought' and 'pre-intellectual speech', giving rise to consciously directed thought processes and, ultimately, logical and 'scientific' thinking. Thus, the notion of inner speech was closely tied to another Vygotskian theme, that of the *directive* function of speech and language in cognitive activity. This aspect of Vygotsky's was taken up by his co-workers (notably A. R. Luria) after Vygotsky's own early death from tuberculosis at the age of thirty-eight. It also reflected a general emphasis in Vygotsky's work on the *interdependence* of language and thought, a position further distinguishing Vygotskian from Piagetian developmental theory. For Piaget, as we have seen, the 'figurative' aspects of symbol acquisition and use, including language, were subordinate to operative development. For Vygotsky, on the contrary, the development of language itself permitted the differentiation and elaboration of thought: 'Thought is not merely expressed in words; it comes into existence through them' (Vygotsky 1962, pp. 119–20).

17.3 Conclusion

Although Piaget and Vygotsky differed considerably in their developmental theories—over the roles which they assigned, for example, to social interaction and language in cognitive development—there was also much which they shared. Both theorists stressed the lack of 'logical' or 'scientific' concepts in young children; both stressed the differences, rather than similarities, between the representational and symbolic capacities of children and adults; and both proposed a form of constructivism as an alternative to both empiricism and nativism.

More generally, Piaget, Vygotsky, and Freud all drew inspiration, in their attempts to understand the development and functioning of the human mind, from the findings of evolutionary biology. There are undoubtedly aspects of their theories which, from our present vantage point, appear both scientifically naïve and socio-politically questionable. The discovery of the evolution of species in the nineteenth century, as is well known, was frequently coupled with the propagation of 'Social Darwinism', in which the laws of nature were inappropriately transposed into the realm of human society. Furthermore, the idea of a 'parallelism' between phylogenesis, ontogenesis, and the evolution of cultures was deeply embedded in the pioneering works of developmental child psychology. The 'parallelism' notion was not restricted to Haeckel's law of recapitulation, for it embraced many other comparisons between the 'Childhood of the Individual' and the 'Childhood of Man' which we would now consider not only unconvincing, but also invidious and ethnocentric.

The 'discovery', for example, of the 'Savage Mind' by cultural anthropologists occured at much the same time as the 'discovery' of the 'Childish Mind'—and it was but a short step to equate the manifestations of 'primitive' thinking in children with those in 'primitive' societies, and in mentally disordered individuals. Scribner (1985) has recently undertaken a detailed analysis of some of Vygotsky's most important texts, in order to establish the extent to which his concept of 'general' or 'cultural' history may be understood as implying a theory of parallelism, and the extent to which it marks a break from such theories. She concludes that Vygotsky's struggle to reformulate the relationships between mind, nature, and society, while subject to many ambiguities, still represents an important point of departure not only for developmental psychology, but for psychological theories in general. In Chapter 13 of this volume, I have examined certain other aspects of Vygotsky's approach, notably the concept of 'internalization', in an attempt further to clarify the extent to which his highly original theories yet bore the imprint of evolutionary metaphors which should, perhaps, now be considered as dead; and to recast his theories in the light of more recent research.

Recent research in language and cognitive development (reviewed by George Butterworth in this volume) does not, in any case, present a uniform picture which can be used unequivocally to criticize or to support the theories of either Vygotsky or Piaget. On the one hand, the depiction by Piaget and others of the infant as cognitively limited, and of early childhood as being characterized by an entirely different 'mentality' from that of adulthood, can no longer be sustained in the light of accumulating evidence. On the other hand, development in thought and language does indeed possess both a high degree of intrinsic regularity and organizational structure attesting to the importance of biological factors, and an equally striking dependence on rich micro-environmental factors. The interactionist and constructivist approaches of Piaget and Vygotsky, while by no means commanding universal assent amongst developmental psychologists, continue to provide crucial reference points for research and theory construction.

Freudian psychoanalytic theory, too, though frequently criticized by developmental psychologists for its haphazard and cavalier approach to empirical matters, has recently enjoyed a considerable resurgence in relation to semiotic and discourse-theoretic approaches to the constitution of subjectivity, particularly with regard to questions of language, gender, and power-relations (see for example Urwin 1985). It seems reasonable to hazard a guess that a major focus of research in the next few years will be the attempt to reintegrate accounts of social, affective, and cognitive development within a framework informed by recent work in semiotic and pragmatic theory. These new approaches, it is to be hoped, will go beyond the increasingly sterile oppositions between biological and environmental determinisms, and seek to examine the process by which biology both contributes to, and is constructed by, the process of human psychosocial development in its linguistic and communicative context.

Notes

1. The material in this chapter is presented more fully in my book, *Language and representation: a socio-naturalistic approach to human development* (Harvester–Wheatsheaf, London, 1988).

2. Here and elsewhere in the text the date of original publication of a work, in square brackets, follows the citation of the work in translation, collection, and/or later or standard edition.

3. Chomskyan linguistics is based upon a distinction between underlying competence and actual performance which, despite its apparent similarities with Saussure's *langue–parôle* distinction, is very different in kind. Competence, as understood by Chomsky, is intended as a description of the actual (though tacit) knowledge of the speaking subject; it is thus a psychological—and, ultimately, biological—notion, rather than one pertaining to societies and communities, as was Saussure's *langue*. In this respect, Chomskyan theory falls squarely within the tradition of Cartesian rationalism, in which questions of knowledge are cast in terms of innate capacities. Further, such capacities are defined 'computationally', in terms of strictly formal properties; in Chomskyan theory, issues of function and content are subordinate to those of form, an emphasis which has led in the past to an almost exclusive emphasis, in language-acquisition studies, upon syntax (see Sinha (1984) for further discussion).

4. The importance of Bakhtin and his 'circle' has only recently been recognized (Clark and Holquist 1984). A particular problem here has been the attribution of texts written by Bakhtin to other authors, particularly V. N. Volosinov, another member of the Bakhtin circle. It is clear, at any rate, that although collective work was a normal practice for Bakhtin and his associates, and in that sense the name 'Bakhtin' is a metonymic reduction of 'Bakhtin/ Volosinov/Medvedev/etc.', Bakhtin himself was the undisputed leader of the circle. No doubt Roland Barthes would have relished such intertextual ambiguities.

5. The term 'figurative' in Piaget's terminology refers to 'those aspects of thought that are related more directly to the states of objects than to their transformations' (Gruber and Voneche 1977, p. 645) [eds].

References

Barthes, R. (1967). *Elements of semiology*. Jonathan Cape, London.
Barthes, R. (1973). *Mythologies*. Paladin, St Albans.
Bates, E., Benigni, L., Bretherton, I., Camaioni, L., Volterra, V., et al. (1979). *The emergence of symbols: cognition and communication in infancy*. Academic Press, New York.
Beaugrande, R. de, and Dressler, W. (1981). *Introduction to text linguistics*. Longman, London.
Berger, J. (1980). *About looking*. Writers and Readers, London.
Brown, R. (1958). How shall a thing be called? *Psychological Review*, **65**, 14–21.
Bruner, J. S. (1975). From communication to language: a psychological perspective. *Cognition*, **3**, 225–87.
Chomsky, N. (1957). *Syntactic structures*. Mouton, The Hague.
Clark, K. and Holquist, M. (1984). *Mikhail Bakhtin*. Harvard University Press, Cambridge, Mass.
Darwin, C. (1871). *The descent of man, and selection in relation to sex*. John Murray, London.
Darwin, C. (1877). A biographical sketch of an infant. *Mind*, **2**, 285–94.
Derrida, J. (1973). *Speech and phenomena*. Northwestern University Press, Evanston.
Eco, U. (1976). *A theory of semiotics*. Macmillan, London.

Eco, U. (1979). *The role of the reader: explorations in the semiotics of texts.* Hutchinson, London.

Ehrlich, V. (1965). *Russian Formalism: history doctrine.* Mouton, The Hague.

Frege, G. (1892). *Über Sinn und Bedeutung—Zeitschrift für Philosophie und Philosophische Kritik,* **100**, 25–50.

Freud, S. (1922). *Introductory lectures on psychoanalysis.* Allen and Unwin and the International Psychoanalytical Institute, London.

Freud, S. (1953–1974). *The standard edition of the complete psychological works of Sigmund Freud.* Hogarth Press and the Institute of Psychoanalysis, London.

Freud, S. (1954). *The interpretation of dreams.* Allen and Unwin, London.

Furth, H. (1969). *Piaget and knowledge.* Prentice-Hall, Englewood Cliffs, NJ.

Greenspan, S. I. (1979). *Intelligence and adaptation: an integration of psychoanalytic and Piagetian developmental psychology.* International Universities Press, New York.

Gruber, H. E. (1974). *Darwin on man: a psychological study of scientific creativity.* Wildwood House, London.

Gruber, H. and Voneche, J. (eds) (1977). *The essential Piaget: an interpretive reference and guide.* Routledge and Kegan Paul, London.

Haeckel, E. (1874). *The evolution of man: a popular exposition of the principal points of human ontogeny and phylogeny.* International Science Library, New York.

Henriques, J., Hollway, W., Urwin, C., Venn, C., and Walkerdine, V. (1984). *Changing the subject: psychology, social regulation and subjectivity.* Methuen, London.

Ho, M.-W. and Saunders, P. T. (eds) (1984). *Beyond neo-Darwinism: an introduction to the new evolutionary paradigm.* Academic Press, London.

Holzman, L. H. (1985). Pragmatism and dialectical materialism in language development. In *Children's language,* Vol. 5, (ed. K. Nelson), pp. 345–68. Erlbaum, Hillsdale, NJ.

Ingleby, D. (1983). Freud and Piaget: the phoney war. *New Ideas in Psychology,* **1**, 123–44.

Jakobson, R. (1956). Two aspects of language and two types of aphasic disturbance. In *Fundamentals of Language,* (ed. R. Jakobson and M. Halle), pp. 53–82. Mouton, The Hague.

Karmiloff-Smith, A. (1979). *A functional approach to child language: a study of determiners and reference.* Cambridge University Press.

Kripke, S. (1980). *Naming and necessity.* Basil Blackwell, Oxford.

Lacan, J. (1966). *Écrits.* Editions du Seuil, Paris.

Lock, A. (ed.) (1978). *Action, gesture and symbol: the emergence of language.* Academic Press, London.

Lock, A. (1980). *The guided reinvention of language.* Academic Press, London.

Lyons, J. (1977). *Semantics.* Cambridge University Press.

Macnamara, J. (1982). *Names for things: a study of human learning.* The MIT Press, Cambridge, Mass.

McNeill, D. (1979). *The conceptual basis of language.* John Wiley, New York.

Marx, K. (1975). *Early writings.* Penguin and New Left Review, Harmondsworth and London.

Mead, G. H. (1934). *Mind, self and society.* University of Chicago Press.

Paprotté, W. and Sinha, C. (1987). A functional approach to early language development. In *Social and functional approaches to language and thought* (ed. M. Hickmann), pp. 203–22. Academic Press, Orlando, Florida.

Pavlov, I. P. (1927). *Conditioned reflexes: an investigation of the physiological activity of the cerebral cortex.* Oxford University Press.

Piaget, J. (1926 [1923]). *The language and thought of the child.* Kegan Paul, London.

Piaget, J. (1952). *Logic and psychology.* Manchester University Press.

Piaget, J. (1953). *The origins of intelligence in the child.* Routledge and Kegan Paul, London.

Piaget, J. (1954). *The construction of reality in the child.* Basic Books, New York.

Piaget, J. (1962). *Play, dreams and imitation.* Routledge and Kegan Paul, London.

Pribram, K. and Gill, M. (1976). *Freud's project' reassessed.* Hutchinson, London.

Putnam, H. (1981). *Reason, truth and history.* Cambridge University Press.

Rosch, E. (1977). Classification of real-world objects: origins and representations in cognition. In *Thinking* (ed. P. Johnson-Laird and P. Wason), pp. 501–19. Cambridge University Press.

Russell, J. (1978). *The acquisition of knowledge.* Macmillan, London.

Saussure, F. de (1966). *Cours de Linguistique Générale.* McGraw-Hill, New York.

Scribner, S. (1985). Vygotsky's uses of history. In *Culture, communication and cognition: Vygotskian approaches* (ed. J. Wertsch), pp. 119–45. Cambridge University Press.

Sechenov, I. (1935 [1863]). *Selected works.* State Publishing House for Biological and Medical Literature, Moscow–Leningrad.

Silverman, B. and Torode, B. (1980). *The material word: some theories of language and its limits.* Routledge and Kegan Paul, London.

Sinha, C. (1984). A socio-naturalistic approach to human development. In *Beyond neo-Darwinism: an introduction to the new evolutionary paradigm* (ed. M.-W. Ho and P. Saunders), pp. 331–64. Academic Press, London.

Sontag, S. (1982). *A Susan Sontag reader.* Farrar, Straus, and Giroux, New York.

Sulloway, F. (1979). *Freud: biologist of the mind.* Basic Books, New York.

Urwin, C. (1985). Power relations in the emergence of language. In *Changing the subject* (ed. J. Henriques *et al.*), pp. 164–202. Methuen, London.

Volosinov, V. N. (1973). *Marxism and the philosophy of language.* Seminar Press, New York.

Vygotsky, L. S. (1962). *Thought and language.* MIT Press, Cambridge, Mass.

Vygotsky, L. S. (1978). *Mind in society: the development of higher mental processes.* Harvard University Press, Cambridge, Mass.

Williams, R. (1977). *Marxism and literature.* Oxford University Press.

Wolff, P. H. (1960). *The developmental psychologies of Jean Piaget and psychoanalysis.* International Universities Press, New York.

18
Children's drawings and the evolution of art

J. Gavin Bremner

Abstract

Early psychological studies of children's art aimed to gain access to children's cognitive development through investigating the way their drawings developed, the assumption being that peculiarities or shortcomings in their drawings reflected immature cognition. Recent work on children's drawing has cast doubt on this assumption for two main reasons. Firstly, children's drawings are unlikely to be a direct reflection of how they understand reality; production problems are bound to intervene, ranging from simple motor-skill limitations, to relatively high-level graphic planning problems. For this reason, recent work has studied drawing as a specific skill rather than as a phenomenon reflecting general principles of cognitive development. Secondly, implicit in much early work was the view that children's drawing was simply a poorly developed version of the adult form. Recent work, however, suggests that children have different aims when they draw. At the extreme, there is the suggestion that early drawing is aesthetic rather than representational, and that its gradual development towards representational art arises because adults push children in that direction. And even in theories that view children's art as representational, there is growing recognition that children aim to represent different things, for instance how objects are arranged relative to each other rather than how they appear from a single viewpoint.

A number of early workers held the view that parallels could be drawn between the development of child art and the evolution of art through historical time. This is currently a controversial view, however. On the one hand Gablik (1976) claims that there are clear parallels between the developmental sequence seen in children's cognitive development and the sequence seen in the evolution of art, and goes on to argue that the historical development of art was a direct function of cognitive evolution. On the other hand, Gombrich (1960) dismisses such a connection, claiming that in art-history there is no discernible developmental sequence from primitive to sophisticated art, and suggests instead that changes in artistic style arise as the intentions of the artist change to suit the culture within which s/he is working.

Current accounts of the development of children's art have something to contribute to this controversy. Firstly, the view that children's drawings are not a direct reflection of their general cognitive level should lead us to ask whether Gablik is safe in drawing a parallel between cognitive evolution and art-history. If there was a clear link of this sort would we not also expect to see it in ontogeny? There is more in the developmental literature in support of Gombrich, since many of the phenomena and developmental changes appear to relate more to the child's intentions than to limitations in his or her cognitive structures. However, most developmentalists would see Gombrich's account as too extreme in dismissing any developmental aspect in the history of art. Although changes in style may relate to changes in the artist's intentions, Gombrich recognizes that these intentions relate to the demands of the culture in which the artist lives. There are strong arguments to support the hypothesis that cultural evolution is closely tied to cognitive evolution of the individuals within it. Assuming that this is true, there may still be an important sense in which the evolution of art occurs on a developmental sequence, not because it reflects directly the developing cognitions of the artist, but because it reflects an adaptation to the developing demands of cultures that are evolving new ways of thinking about the world.

18.1 Introduction

Drawings and paintings are rich sources of information in a number of ways. The survival of examples of graphic art from prehistoric times to the present day provides not only a vital record on which art-historians can base their accounts, but also provides data on which anthropologists and biologists have developed hypotheses about the lifestyle and intellectual abilities of early humanity. The situation is similar in the case of artistic development during childhood. Again, the fact that children are prolific drawers has provided a mass of data on the development of graphic art, but, in this area in particular, developmental psychologists have used these data as evidence bearing directly on the development of intellect. At the root of much of this work has been the assumption that children's drawings are external representations of the child's reality; and the notion was that the strange features of the drawings reflected the limitations of the child's internal representations of the world. In addition, these two areas—evolution and development—have not always been studied in isolation from one another. There are numerous references to child art in the literature on the evolution of art, and some workers have identified very close parallels between the historical evolution of graphic art and its development in childhood. And by extension they have identified parallels between the evolution of intelligence from prehistory onwards and the development of intelligence in childhood. In particular, Gablik (1976) has argued that the history of art is best understood by assuming that the evolution of intelligence followed a similar sequence to the one Piaget described in the case of the individual child. In contrast, however, Gombrich (1960) argues against the notion that a consistent evolution can be identified in art, claiming instead that the emergence of particular styles related more to the changing use to which art was put, which itself related to the requirements of the viewer. In this view, artistic style more or less follows the whim of culture.

The main aim of this chapter is to review evidence and theory on the development of drawing in childhood, and to discuss the arguments of art-historians in the light of this evidence and in terms of recent changes in the form of developmental theorizing. I will suggest that the 'ecological' approach, currently gaining support within developmental psychology, is more applicable to problems in evolution than Piagetian theory is, and that the emergence of this approach has created considerable scope for drawing parallels between evolutionary and developmental processes.

18.2 Children's spontaneous drawings

18.2.1 *Eng (1931)*

Many of the well-known studies of children's drawing consisted of the systematic collection and categorization of spontaneous drawings produced in the child's day-to-day life. Much of the early work is of this sort. Eng (1931) made an extensive study of the drawings produced by one child (her niece) from her second to her eighth year, and her data are valuable since they give a detailed history of the development of drawing by an individual child over a long period. Of course, such a method carries with it the disadvantage that the generality of the developmental sequence cannot be directly assessed. However, Eng refers extensively to other contemporary case-studies, pointing to where the sequences seemed to be similar and where they seemed different. She found that early attempts at drawing consisted of wavy scribbles that gradually became more circular. Then, during the latter part of the second year, compositions consisting of a variety of scribbles appeared (for instance, circles, crosses, waves, etc.); and it was around this time that drawings were named for the first time. Initially, the name was applied after the drawing had been produced; but within the same month naming was occurring before the drawing was made (this very rapid transition to naming before drawing is at variance with the commonly held belief that children go through a lengthy period during which drawing is aimless, and naming results purely from the recognition of a resemblance between the drawing and something in real life).

Shortly after this, the first of what Eng calls 'formula drawings' appear. These consist of the systematic placement of two or more elements, for instance a straight line with a circle at one end as a first attempt at drawing the human figure. Eng sees these formulae as dominating the child's subsequent efforts. Children develop formulae for drawing familiar objects, and attempt to modify them in the case of unfamiliar ones (it is worth noting that this is just what, according to Gombrich (1972), can be seen happening in the history of art). Also, their reliance on formulae means that they very rarely pay attention to a model when drawing it. The formula is either run off automatically when the subject-matter is known, or, if the model is novel, an existing formula is at least tried out.

It is probably Eng's observations of very early drawing attempts that are of most value today, since subsequent work has done much to clarify the later developments in drawing. Eng dismisses Buhler's (1933) claim that early scribbles are purely the results of motor exercise, pointing out that the visual result is important for the child. In general, in the work of this period there was a clear tendency towards rich interpretation of early scribbles. For instance, Eng interprets some of her data as examples of what Lukens (1896) called 'local arrangement', in which the elements of the drawings are scribbles but can be seen to represent an object through maintaining the correct spatial relationships between elements.

18.2.2 *Luquet (1927)*

Eng draws heavily on the writings of others for her theorizing. Of these, the worker whose claims have received the greatest attention since is Luquet (1927). Part of the reason for a continuing interest in his theory is undoubtedly the fact that Piaget and Inhelder (1956 [1948]) incorporated his account with little modification within general Piagetian stage theory. However, Luquet's theory has an intuitive appeal of its own, and would have continued to attract attention without the seal of approval of Piagetian theory. According to Luquet, there are three stages in the development of drawing. Firstly, there is a stage of 'synthetic incapacity', during which the child's drawings are composed of disorganized or poorly organized elements. Eyes, nose, and mouth may be contained within the outline of a face, but they may be wrongly arranged relative to one another. Once problems concerning the arrangements of elements are surmounted, the child enters a stage of 'intellectual realism', during which s/he is said to draw what s/he knows rather than what s/he sees. Rather than drawing from a particular perspective, the child draws 'everything that is there' (Luquet 1927). Finally, children enter a stage of 'visual realism', in which objects are drawn as they are viewed, and no invisible elements are included.

Although subsequent research has shown that young children do not draw *all* that is there, there is plenty of evidence that they often include features that are not visible to them at the time of drawing, whether this be achieved by transparency drawings or by drawings in which some parts are turned round so that they can be shown, or by running off a formula in the face of a model that lacks some of the features contained in the formula. It has been suggested (Freeman and Janikoun 1972) that the child includes key features that serve to define an object. Again, there is a parallel argument in art-history here. Gombrich (1972), dismissing the notion that Egyptian artists saw reality as they painted it, says, 'They merely followed a rule which allowed them to include everything in the human form that they considered important.'

18.2.3 *Kellogg (1970)*

The analyses discussed up to this point have focused on children's drawings as representations of reality, and have said nothing about their aesthetic qualities. This reflects a research bias that has continued to the present day; but a notable exception is the work of Kellogg (1970). She also based her developmental account on the spontaneous drawing and painting of children, but her data-base is very much larger than Eng's, being composed of around a million drawings produced by large numbers of children of varying age. Although, through this method, she loses out on the detailed longitudinal component to her data, she gains by having data that can be assumed to be a representative sample of the spontaneous work of children.

Kellogg finds it possible to categorize even the most early graphic products, identifying twenty different types of basic scribbles. But these early scribbles are not just a passing phase, since she argues that they are the basic components of all the drawings that follow. Although it is possible in principle for these scribbles to be made without visual guidance, Kellogg argues that another aspect of early drawing, the pattern of placement on the page, has to be visually monitored. These placement patterns, made up of circumscribed scribbles or combinations of scribbles, form the forerunners for the six basic diagrams (rectangle/square, oval/circle, triangle, Greek cross, diagonal cross, and irregular shape). What makes the diagrams stand out from earlier attempts is the fact that they are fairly precise and generally composed of single lines. Later, diagrams are produced in pairs

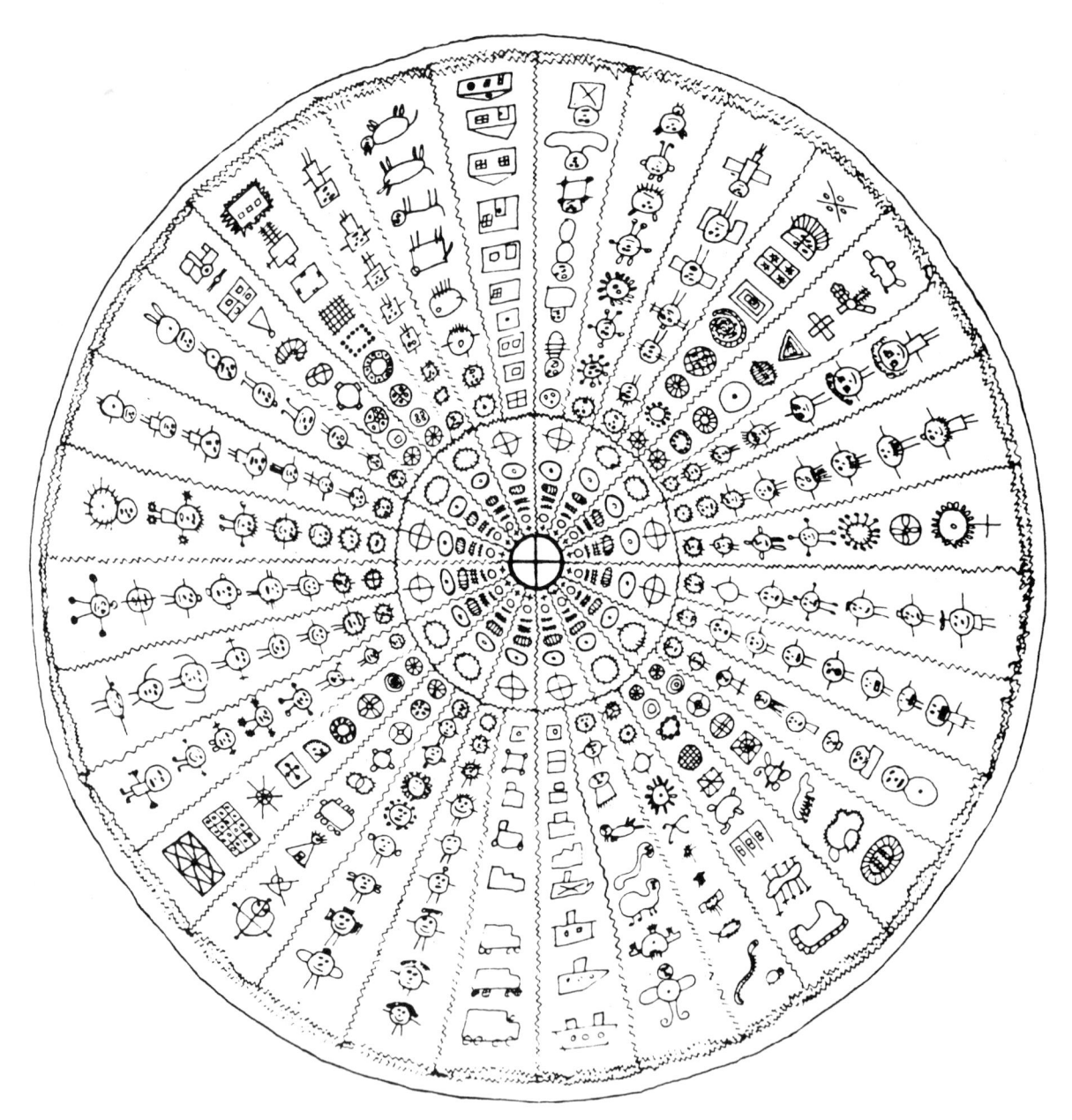

Fig. 18.1 Kellogg's 'mandalas'. Her original caption to this figure, from p. 273 of *Analyzing children's art* (Mayfield, Palo Alto, 1970) runs as follows: 'Scheme of the evolution of pictorial work from earlier drawings, beginning with the structures around the center and extending out. The scheme shows common sequences in the classifications of child art, but many other sequences are possible (author's sketch).' Reproduced by permission of the publisher.

(combines) or in larger groupings (aggregates), but particular significance is attached to concentric groupings (for instance, a square or circle divided into quarters by a cross). Kellogg names these 'mandalas', and claims that they are a basic aesthetic form, perceived by children from at least as early as two years, and produced with more and more precision as their graphic skills develop.

This is where it becomes clear that Kellogg's account has departed from the conventional path. She goes on to argue that early stereotyped drawings, for instance, of houses or people, are not intended by the child to represent reality in any way. They are simply aesthetically pleasing patterns in the child's eye (or, in the terms of *Gestalt* psychology, 'good forms'). They are produced as patterns in their own right, and any representational interpretation is imposed by the well-meaning but (in Kellogg's view) misguided adult.

There are reasons to be worried about this analysis, since there seems to be no external standard used to determine what should be or should not be aesthetically pleasing. We could appeal to *Gestalt* principles, but, in noting the rare occurrence of the diamond in children's early drawings, Kellogg claims that it cannot be a good form in the eyes of the child, despite the fact that it is a good form in *Gestalt* terms. This seems a rather blatant case of writing one's own rules, and could easily lead to any oddity or orderly form in drawing being explained as holding particular aesthetic appeal. Another problem with this analysis is the very scant attention to the role of construction skill in the drawing process. For instance, diamonds may not occur because they are, for one reason or another, difficult to produce. Admittedly, Kellogg does mention the issue of production; not as a limiting factor on the forms produced, though, but as an extension of the aesthetics argument, certain activities being pleasing in the same way as the graphic patterns that they produce.

Despite these problems, Kellogg's analysis provides an important counterbalance for the predominantly representational analysis that exists in the literature on child art, which also dominates certain notable analyses of the history of art. Kellogg's argument is that child art, rather than being a poor approximation to adult art, is a rich and sophisticated art form in itself. Children have a strong aesthetic sense which they apply more or less naturally in their drawings. They are concerned to produce works containing line balance, proportion, and shaping, while adults intentionally or unintentionally push them towards conventional adult art based on the visual appearance of objects. If nothing else, such an argument should alert us to the dangers of interpreting children's drawings solely in terms of their approximation to adult artistic conventions.

It is worth noting two points of contact here with work on chimpanzee 'art'. Firstly, like Kellogg, early investigators in this field (for example, Morris 1962; Schiller 1951) tended to interpret their subjects' graphic products in aesthetic terms, apparently finding evidence for introduction of symmetry or overall balance to pictures and *Gestalt*-like completion of partial figures.

However, subsequent work (Boysen *et al.* 1987) in which more tight experimental control was applied yielded more modest conclusions. Sure enough, chimpanzees' scribbles are not random: they tend towards the centre and bottom of the page, and there is also a tendency to mark existing patterns. This would at least indicate that chimpanzees, like children, are concerned about the visual outcome of their scribbling. However, Boysen *et al.* obtained no evidence that products were influenced by aesthetic qualities such as symmetry or completeness. This, of course, indicates nothing about a possible aesthetic basis for children's very early drawings; but it raises the possibility that controlled experimental study might similarly fail to yield evidence of aesthetic principles in the art of the young child.

Secondly, one point emerging from the literature on chimpanzee art is that there is little or no evidence that this art form ever becomes representational. Whether or not we accept Kellogg's aesthetic analysis of early child art, her account may provide a good reason for the chimpanzee's failure to adopt representational art. If it is true that representational qualities only emerge in children's drawings because parents apply representational interpretations, we would not expect the same progression in chimpanzees, who either have natural parents who make no such interpretations, or who are unlikely to be able to pick up any interpretation imposed by a human care-taker.

18.3 The development of drawing within the Piagetian framework

Children's drawing did not seem to be a central concern for Piaget. Rather, the topic seems to have been just another phenomenon that had to be incorporated within the general theoretical framework. In *The child's conception of space* (Piaget and Inhelder 1956 [1948]) one chapter is devoted to the issue, and a large part of that is concerned with children's copies of geometrical figures. Piaget and Inhelder are content to incorporate Luquet's theory within their general theory of the development of spatial cognition. Luquet's stage of synthetic incapacity corresponds to their first stage, during which the child is governed by topological spatial principles. Elements of patterns are appropriately arranged in terms of proximity, enclosure, etc., but cannot be systematically arranged in order along vertical or horizontal axes. The stage of intellectual realism corresponds to a transitional period, during which Euclidean and projective geometry is beginning to emerge, but is applied in an incoherent manner. Hence, an object is drawn as if viewed from multiple irreconcilable perspectives, and may include impossible perspectives (for example, transparency drawings). The child concentrates on the correct arrangement of elements relative to each other, but has not reached the point of producing an arrangement as seen from the fixed perspective of the viewer. The stage of visual realism, emerging around the age of eight or nine years, marks the development of Euclidean and projective space to the point when the principles are applied in a coherent manner, resulting in the accurate construction of perspective views.

This apparent convergence of theories might be taken as encouragement that thought is proceeding along the right lines. However, Luquet's formulations are vague, and Piagetian theory is extremely adaptable; so we should not be over-impressed. Indeed, the adaptability of Piagetian theory is emphasized by the paradox that the same theoretical principles are taken to explain, on the one hand, the fact that when asked to identify another's view of an array, children persistently pick or construct their own view, and on the other hand, that when asked to draw, they do everything else but show their own view.

At the root of the Piagetian arguments on drawing is the same assumption that was less explicitly made by earlier writers like Eng and Luquet: that is, that children's drawings provide, in quite a strong sense, a window on their cognitive state. Piaget and Inhelder make the link without apology; and Eng points to similar stages in the development of other abilities (for instance, language) as indication that the drawing phenomena are rooted in a general developmental process. However, just as the performance–competence distinction became a concern in the study of language development, workers studying drawing became more concerned with the performance factors involved. If children produced poor graphical representations, was this because their internal representations were limited, or was it because they had difficulty in putting them on paper? Thus, for most workers, drawing has become at best a very murky window on children's intellect.

This, of course, does not mean that we should discard the Piagetian analysis of children's drawings, just that we must be cautious in the generality of the conclusions that we reach. There is in fact evidence that changes in children's human figure drawing relate to development of an explicit dimensional reference system as an organizer of graphic space (Teufel and Lange-Kuettner 1993). There is, however, little research on relationships between drawing and spatial reference systems used by children in real space. Elsewhere, I have argued that the development of dimensional reference systems is fairly domain-specific, relating to classroom activities such as mapping, graphing, and drawing (Bremner et al. 1993), and my suspicion is that any relationships that exist between drawing and other spatial abilities will lie within this relatively focused domain.

18.4 Some recent experimental approaches to children's drawings

18.4.1 *Drawing as problem-solving*

Goodnow (1977) analyses children's drawings in terms of problem-solving. The flavour of her approach can be seen by taking the example of a human figure drawn without arms. Whereas Kellogg would argue that this arises because addition of arms would reduce the aesthetic appeal of the drawing, Goodnow suggests that arms are often omitted because they are, for one reason or another, difficult to add. For instance, arms and hands are difficult to incorporate in a human figure produced as a single outline, or the space where arms should go may already be occupied by long hair, or the arms, being the last to be added, may simply be forgotten. From Goodnow's point of view, part of the child's problem is the development of graphical 'equivalents' for real objects (in this sense, an equivalent is something that stands for something else, but retains only some of its properties). For example, the child learns that a circle is an equivalent of a human face, or a ball, or the sun. But other important problems crop up when more complex equivalents are developed, such as whole human figures, or houses. Here, children seem to adopt a rule that says that two elements must not occupy the same space on paper (sensible enough if they do not do so in reality), and will go to considerable lengths to maintain this rule when two elements compete for the same space. Goodnow presents experimental evidence on this. For instance, she shows how children will add arms to a human figure in odd places (for example, on the lower half of the torso, or attached to the legs), because their normal space is already partially taken up by long hair. Also, if asked to add wheels to a train which offers no more space underneath, they do not add wheels which overlap others (although this could be a reasonable perspective solution), but add them to the sides or top of the coach.

Another aspect of drawing skill is sequencing. Children who adopt a systematic sequence are less likely to omit features, but some of the idiosyncrasies in drawing are also, according to Goodnow, due to the sequencing of the drawing. For instance, transparency drawings such as people with see-through skirts occur because the skirt equivalent (a triangular outline) is added last. The problem the child fails to tackle here

Fig. 18.2 An example of transparency drawing through lack of forward planning: the legs are drawn first, and show through the skirt once it is added—'Mum', by Andrew Bremner, aged 7.

is one of forward planning. If they are to show only what is visible, they must start by drawing the major visible parts.

These considerations apply to equivalents that children may have developed to quite a large extent on their own; and the question arises as to how children come to adopt orthodox equivalents, that is, artistic conventions. Goodnow's answer is that they do this with great conservatism, progressing to new equivalents by minimal modification of existing ones. For instance, to show a person running, they may simply lengthen the legs or change their angles. Or if asked to draw their school from an unusual perspective (from up in the air), children initially only make it smaller. Then the roof may be focused on in a number of ways: for instance, by showing it alone but as viewed from the side, or by showing it as viewed from above, but placed atop a building viewed from the side. All these examples, according to Goodnow, show the child exercising problem-solving skills that have application beyond the sphere of drawing. Hence for her, drawing is a window on specific aspects of cognition, but in a way that is much less fraught with problems than were earlier formulations of the notion.

18.4.2 *Drawing as a spatial skill*

Freeman (1980) has probably done more than anyone else to bring analytic experimental methods to bear on the study of drawing. The analytic style of his approach is shown in the way he tackles an odd proportion effect in human-figure drawings, that is, the fact that children often draw the head far too large in proportion to the body. Previous workers blithely claimed that the head was drawn large because it was more important than the rest of the body. For the purposes of drawing parallels with art-history, this would be a nice simple conclusion to reach, since it is suggested that size was used to indicate importance in Egyptian art (Gombrich 1972); however, Freeman is far more cautious. Firstly, he points out that since the head is usually drawn first, it does not have to compete for space with other parts. When later parts are added, they may have to be out of proportion simply because there is less space on the page. Another possibility is that the head outline is drawn bigger because it has to contain more features (such as eyes, nose, mouth) than the trunk, which often contains none.

If the conventional tendency to draw the head first leads the body to be represented as smaller in proportion owing to space constraints, one would assume that the effect would be reduced or even reversed if children could be persuaded to draw the body before the head. Freeman found that children were reluctant to do this, often refusing. However, Thomas and Tsalimi (1988) had more success, finding that if 5- to 8-year-olds were asked to draw a human figure starting with the trunk they produced drawings in which the head–body ratio was much closer to the actual proportions in the human figure. This seems like clear evidence that this sort of planning problem is responsible for the effect. However, could it be that the correct proportions are arrived at fortuitously, through the children's not leaving enough room for a head which they would otherwise draw larger? This appears not to be the case, since Thomas and Tsalimi found that children arrived at the more or less correct ratio by enlarging the body and not by reducing the size of the head.

It turns out, however, that this is not the whole story. Henderson and Thomas (1990) investigated the possiblity that inclusion of contained features (such as eyes, nose, and mouth) increased the size of the head. They found that the request to include details of a jacket increased the size of the trunk outline and that the request to draw a person from behind (and hence with no facial features) let to a reduction in the size of the head. Thus it appears that at least two factors influence the normal overestimation of the head relative to the body. And it is important to bear in mind that both the tendency to draw the head first and to include internal features may stem from the importance accorded to the head in the human figure. Thus, this research does not allow us to rule out any

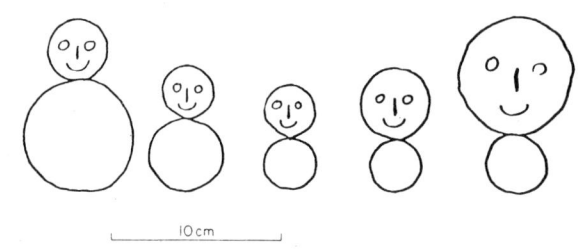

Fig. 18.3 Incomplete drawings with head–trunk ratio scaled according to method of constant stimuli. Reproduced from Fig. 2 in Freeman (1975). Reprinted by permission from *Nature*, Vol. 254, pp. 416–17. Copyright (c) 1975 Macmillan Magazines Ltd.

one account, but it does allow us to understand very much better how different factors interact in determining the form of the product.

The real merits of Freeman's approach become evident in the case of his analysis of tadpole drawings of human figures (those in which arms and legs are attached to a single circle containing the facial features). Freeman starts off by pointing out that such a figure could arise for a multitude of different reasons, and for our purposes here it is sufficient to pick just a few. Firstly, it may be that the circle is meant to represent the whole body including the head; in which case the error, apart from lack of differentiation, consists of failing to keep the facial features to the top of the outline. Another possibility is that the tadpole figure is preferred because it is radially symmetrical, and so approximates more to a mandala form (Kellogg). Another is that the circle represents the head, the trunk is omitted, and the arms and legs are wrongly attached to the head.

To distinguish between these alternatives and others, Freeman (1975) used a completion task in which children were invited to add arms and legs to partial figures comprising two circles to represent head and body. If the circle produced by 'tadpole drawers' represents head and body, then they should position the arms on the trunk when both trunk and head are provided; whereas if they are guided by the principle that the arms should be attached to the head, they should persist in this despite the provision of a trunk. However, if they are guided by radial symmetry, the best placement for the arms in this case is at the junction between the two circles (leg placement seems less likely to be a sensitive indicator here, owing to the likelihood that legs will be attached below, whatever the figure). To simplify Freeman's results considerably, children whose normal drawings included a trunk usually attached the arms appropriately in these completion studies. However, 'tadpole drawers' turned out to be sensitive to what Freeman named the body-proportion effect. If the head was larger than the body, they attached the arms to the head; whereas if the body was larger than the head, they did the opposite. None of them attached the arms in such a way as to produce radial symmetry. Thus, the only account to come close to explaining the data is the one based on the assumption that children believe that the arms should be attached to the head; and even this one fails to fit if the trunk is made larger than the head. It looks more as if children are using the largest element as some sort of anchor point.

On the basis of these data as well as the results of other studies, Freeman explains human-figure drawing as a problem in serial positioning. End-points of a series are more easily remembered than middle ones, and a three-term series should be more difficult to construct than a two-term one. Hence, children may use one end-point as an anchor on which to base the rest of the construction, which is composed of other end-points (legs and arms) and no middle points. In spontaneous drawing, the head provides the anchor; but in completion tasks, when both head and trunk are provided, the larger element provides the anchor.

Whether or not other contemporary workers have agreed with Freeman's analysis of the drawing process, his is probably the best worked-out example of the general style of analysis of children's drawing that is prevalent today. The assumption is still that children's drawings reveal things about children's cognition, but not in the way earlier workers seemed to think. Their drawings do not reveal how they see or understand the real world, but they do show how good or bad they are at translating relationships in the real world into relationships on the page. In other words, drawings tell us important things about the fairly specific spatial skills involved in the process itself. This should become clear in the next section when we discuss portrayal of depth. It is clear that children whose drawings appear flat do not perceive the world as flat. Their problem (if it is a problem) is that they have not developed the drawing methods that permit portrayal of the third dimension.

18.4.3 *Representation of the third dimension in drawings*

Freeman *et al.* (1977) asked children ranging in age from five to ten years to draw from imagination an apple, and then to draw another one 'behind it'. The question, of course, was how they would portray the depth relationship on paper. The main developmental effect was a shift with age from segregation of elements to partial occlusion of the far element by the

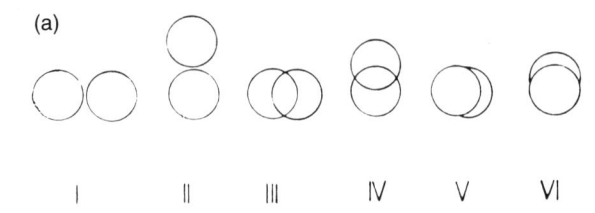

Fig. 18.4a Methods of representing one apple behind another. I and II are segregated, III and IV show interposition, and V and VI show hidden-line elimination. Will these be picked up as stages of development in an experiment? (Reproduced from Fig. 1 in Freeman *et al.* (1977), by permission of Academic Press.)

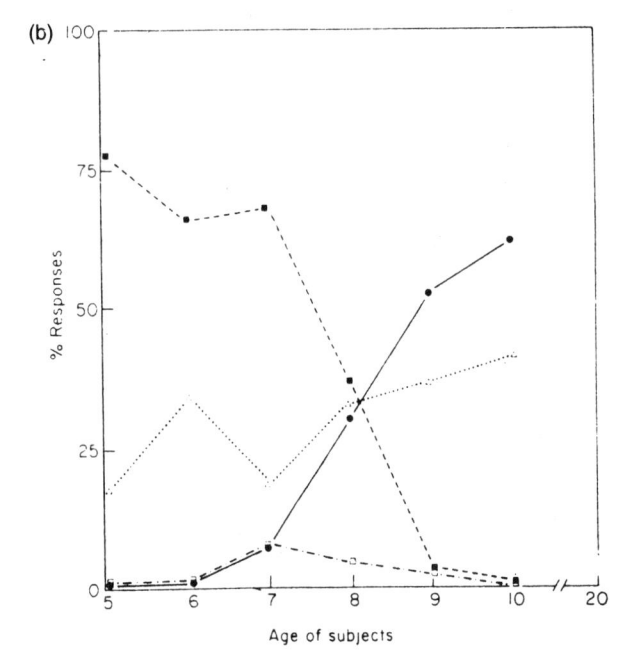

Fig. 18.4b Drawing devices used in representing one apple behind another over a wide age-range. One drawing device, enclosure, was found which is not contained in Fig. 18.4a.
■------■ segregation; □–·–□ interposition;
●——● hidden-line elimination; ············ enclosure.
(Reproduced from Fig. 2 in Freeman *et al.* (1977) by permission of Academic Press.)

intermediate stage of interposition drawings (partial occlusion drawings in which hidden lines had not been omitted); but there was some evidence of transparency drawings—around a quarter of the children of all ages showed the outline of the further apple contained within the outline of the near one.

At first sight, the segregation drawings characteristic of early attempts would seem to be totally uninformative about the depth relationship; however, within this drawing type there was a developmental shift from showing the apples side by side to showing them one above the other on the page. This shift to a consistent vertical arrangement appears to be the first stage of depth representation through a fairly crude use of height in picture to indicate distance, something which only begins to emerge at about seven years. Other data seem to confirm the notion that young children do not portray depth. For instance, Cox (1981) found that all five-year-olds produced horizontal segregated drawings, even though the children were drawing two real objects arranged one behind the other. Also, Light and Simmons (1983) found that 84 per cent of five- and six-year-olds produced horizontal drawings of two objects arranged in depth.

All this fits the notion of an orderly development in the portrayal of depth relationships. Initial segregation drawings do not indicate the third dimension at all, while later ones do through simple use of height-in-picture, and this crude indication of depth is then gradually replaced by the more graphically complex device of partial occlusion. However, recent studies indicate that this is not a stage sequence in a strong sense. Cox (1978) found that if the relationship between the objects was stressed by asking children to point to the object 'in front/behind', the majority of five- and six-year-olds adopted vertical arrangement in a subsequent drawing. Also, Bremner (1985) found that five-year-old children who had drawn a near and a far object side by side, shifted to drawing them one above the other when an extra object was placed to one side of the near object. The argument was that the pairing of a depth relationship with a real side-by-side relationship led children to abandon the left–right arrangement for near and far objects, and to adopt an appropriate vertical arrangement. It was also found that children who had responded in this way to the three-object array continued to use vertical portrayal when faced with only two objects one behind the other—so they seemed to stick with the strategy provoked by this particular array.

Similar findings have emerged from studies of the use of occlusion as a depth cue. For instance, Light and Simmons (1983) found that although seven- to eight-year-old children normally used segregation drawings of two objects arranged in depth, they would

near one. The predominant response up to the age of seven years involved drawing the two apples quite separately on the page, but after this age, the segregation response fell off, to be replaced by drawings involving partial occlusion. Surprisingly, from Freeman's point of view, there was no evidence of an adopt occlusion if they were asked to make a drawing that would tell another child where they had sat when they made it. In addition, Radkey and Enns (1987) found that constraining children's viewpoint by having them view the array through an aperture led to a higher incidence of total and partial occlusion drawings by five- and six-year-olds respectively.

Cox (1981) found a high incidence of partial occlusion drawings by six-year-olds when the task was set in the meaningful context of a robber trying unsuccessfully to hide behind a wall. Her interpretation is that the context elicits such drawings because occlusion is a crucial part of the hiding activity—another type of salience effect. However, Light and Foot (1986) found similar results when no meaningful context was supplied, and they suggest that two separate factors interact to lead to partial occlusion when drawing an object behind a wall. First, children have a strong predisposition to draw objects from particular angles: in the case of a wall a side view is preferred over an end view. Second, children strive to give an accurate portrayal of the relationship *between* objects (what Light and his colleagues call the *array relationship*). The outcome is that children first draw the wall in side view, and are then forced to use partial occlusion to show the correct relationship between it and the further object. If they had started with an end view of the wall, occlusion would not have been necessary, since the array relationship would have been accurately represented through horizontal portrayal. There are probably other explanations possible; but, whatever the final conclusion on this issue, the main point is again that young children are not rigidly tied to one method of portrayal: under appropriate circumstances they will use fairly complex methods such as partial occlusion to depict the third dimension.

Light and Humphreys (1981) suggest that even horizontal portrayals should not be viewed as representational failures. Children may be more concerned to show the relationship within the array than the relationship from their point of view. But, if the task demands are changed, they show that they are quite capable of portraying their point of view when it seems appropriate. This is an important point that I think runs through most of the contemporary thinking about children's drawing. Although young children clearly lack the sophisticated graphic skills of the adult artist, we run the risk of severely underestimating their capabilities if we analyse their attempts in conventional terms. Children may have quite different aims in their everyday drawings; but these simple experimental manipulations show they are quite able to adapt their efforts in the direction of convention if the requirements are made explicit. Additionally, it turns out that even at an age when it is not clear that children are portraying their viewpoint, their drawings are still systematically related to viewpoint, so that an array of objects will be drawn differently depending on the viewpoint adopted (Bremner and Batten 1991). Thus there is evidence for a sensitivity to viewpoint before the stage at which viewpoint begins to be accurately portrayed, and this is in keeping with the finding that viewpoint can be portrayed if the child understands that this is an important requirement.

Study of older children's drawings again indicated a stage-like sequence. Willats (1977), asking subjects to draw a table, recorded an orderly development in the use of artistic projection systems in children between the ages of five and seventeen years. Orthographic projection was the first system to emerge (mean age 9.7 years). Then oblique projection developed at around twelve years, followed finally by true perspective at about fourteen. Again, however, subsequent work has shown that this sequence is by no means rigid. While Lee more or less replicated Willats's findings in the case of drawing a table from observation (Lee and Bremner 1987), she obtained quite different results when the task was drawing from imagination (Lee 1987). In this case, very few children or adults used any version of linear perspective, adopting oblique projection instead. The few perspective drawings that did occur appeared at around the age when children would be receiving training in perspective in art classes; and their numbers fell off again thereafter. So again to a large extent the task-demands determine the system used, and there is little evidence of a rigid invariant developmental sequence.

18.4.4 *Intellectual realism re-assessed*

(a) **Conceptual issues**. One of the main problems with the intellectual realism concept is that it is poorly defined. The notion that the child draws what s/he knows leaves us little the wiser unless we have independent access to what the child knows. What is needed are more precise formulations of what it is about objects that children are at pains to show in their drawings. Light's notion that children are normally concerned to show the spatial relationships within an array rather than the relationships between array and viewer could be treated as a better-specified version of the intellectual realism account. At least in this case predictions can be made about what children will include in their drawings. However, one of the key notions in Luquet's account was that young children could not even begin to draw a representation of their viewpoint, and we now know that this is not true. On the other hand, the phenomena that prompted Luquet's account can be obtained in an experimental setting quite readily. For instance, Freeman and Janikoun (1972) found that when children were asked to draw a cup with a flower pattern in view but with the handle out of sight, those under seven included the handle but generally omitted the flower. In contrast, older children were much less likely to include the handle, and were more likely to include the flower. Their interpretation is that young children include the defining features of an object even if they are out of view, whereas older children include only the visible features of an object even if they are non-defining. Again, this account is more detailed, homing in more specifically on the sorts of object properties that are likely to appear in children's drawings.

However, Barrett and Light (1976) identified a further ambiguity in the intellectual realism concept. They pointed out that it was not clear whether young children were drawing from knowledge of the particular object before them, or from more general knowledge of the class to which the object belonged. Thus, they distinguished between *intellectual realism*—knowledge of the particular object—and *symbolism*—knowledge of the general properties of the class to which the object belongs. To establish which sort of knowledge was guiding children's drawings, they asked children to draw a cup that lacked a handle and a house that lacked a door. Despite the fact that the absence of these features was stressed, five- to seven-year-old children showed a strong tendency to include them in their drawings. In contrast, older children were much more responsive to the properties of the object before them. Barrett and Light took this as evidence for symbolism in young children's drawings, and put forward the suggestion that there is a developmental sequence from symbolism, through intellectual realism, to visual realism.

As already noted, though, Light's later work on portrayal of the third dimension shows that if such a sequence exists it is not so rigid that young children will never produce visually realistic pictures. Also, this tendency to include hidden features is not totally consistent. Taylor and Bacharach (1982) found that young children rarely included a handle in drawings of a mug that either had the handle broken off or out of sight. Such a replication failure detracts from the generality of the phenomenon, but leaves us in doubt as to why different results were obtained in the two cases. Taylor and Bacharach suggest that their different results arose because they did not name the objects being drawn, whereas Barrett and Light did. If they are correct (and I shall shortly present evidence that they are) this would again point to considerable flexibility in young children's drawing strategies.

Davis (1983) has produced a more striking demonstration of flexibility. She found that children would either include or omit a hidden handle on a mug, depending on the form of the drawing task. If a single mug was presented with its handle hidden, they included the handle in their drawings, but if a mug with handle hidden was placed alongside a mug with handle in view, they omitted the hidden handle. Apparently, children feel that it is important to mark the visual contrast between the differently oriented mugs, and they adopt view-specific drawing to achieve this. So, if the child sees the need, s/he will be visually realistic. But Davis has also demonstrated that contrast effects can act in the opposite direction too. Children who had just omitted the hidden handle in the paired-cup task went on to reintroduce it when the cup with handle hidden was paired with a cup without a handle. In this case it appears that children are at pains to show the physical contrast between two objects although they look the same from the drawer's viewpoint.

(b) **Interpretations**. *All these studies show that children are not rigidly committed to one way of portrayal at any one point in development*, and that quite subtle experimental manipulations can provoke them into adopting methods that seem more advanced from the viewpoint of conventional art. The fact remains, however, that *if left more to their own devices, young children do not portray their viewpoint*, either by failing to show the self-referent relationship between objects in an array, or through showing invisible features. There is also good evidence that older children do much more to show their view in a drawing. The conventional interpretation of this would be that younger children are, for a number of reasons, poorer artists than older ones. Yet the degree of sensitivity they show to task demands should warn us to be cautious of such an interpretation. Here it is worth recalling Kellogg's claim that child art is a rich aesthetic medium, quite distinct from adult art. Just as there has been little attempt in the recent experimental work to interpret drawings in aesthetic terms, there has probably been too little attention to the possibility that children have different aims from adults in their representational drawings. Children may simply want to represent different properties of an object or array of objects, properties that are not easily shown by a view-specific drawing. After all, engineers draw mechanical components as if seen from three separate points of view in order to impart all the information about them, and even an untrained 'technical draughtsman' would be likely to draw a component from the angle that imparted most information about its form. In this light, *the developments towards view-specific drawing seen in middle childhood may be better interpreted as conformity to* the constraints of adult convention than as a real cognitive advance.

Even so, it might still be considered a little surprising that young children show so little tendency to show their own view. At first sight it might seem to be the easiest thing for them to do, since they have continual information about that view while they make their drawing. One answer would be that despite the presence of this perceptual input, they lack the skills to transfer their perspective to the page. Certainly, young children do not seem to pay very much attention to the model during the drawing process. But another answer arises from reconsideration of the perceptual process itself. The conventional view of visual perception treats static retinal images as the basic units; and, in this framework, part of perceptual development involves the integration of separate images obtained from different viewpoints. From this we might expect portrayal of one's own viewpoint to be simpler than anything else, because it involves a relatively direct transformation from static perceptual image to the picture on the page. In contrast to this is the diametrically opposed view of J. J. Gibson (1979), who claims that perception is a dynamic process which is continuous over time. According to Gibson, static views, rather than being the building-blocks of perception, are exceptional, and have to be extracted from the temporally continuous perceptual flux. In this account, there is no reason to be surprised that children fail to portray their viewpoint or a single viewpoint, and it is only convention that leads adults to do so (the same convention that leads us to think of perception as a series of static snapshots).

Again, then, the intellectual realism–visual realism distinction takes a further knock. Children at all ages could be drawing what they see. Younger ones may be drawing what they have seen of the object over time, rather than what they see at a given point. There is some evidence in keeping with this interpretation (Bremner and Moore 1984). When asked to draw an object with a hidden feature, those children who had inspected the object from all angles before drawing were much more likely to include the feature in their drawing than those who had not had this opportunity, and this applied even if the object was familiar (a cup). So perceptual experience is important, and in a way that is in keeping with the notion that young children draw what they have seen over time. However, this is only one of the determining factors. Although very few children drew a handle on a mug in a situation in which they were never allowed to see the handle, almost all did under the same conditions if the mug was named before drawing. This was not simply a recognition problem, since most children in the 'no-naming' condition were able to name the mug after making their drawing. So while perceptual factors are important determinants of what is drawn, this strong effect of object-naming suggests that cognitive factors are also important. It seems likely that the name calls up the category, so that the child then draws a canonical exemplar of the category (cf. Barrett and Light's 'symbolism'), rather than relying on perception of the specific object.

18.5 Summary

There are two main trends in current work on children's drawing. Firstly, there is the move towards studying the development of drawing as a more or less specific spatial skill. Few workers are now willing to draw conclusions about the general cognitive state of children from the form of their drawings. There clearly must be connections between drawing skills and other cognitive activities, but we are only just beginning to find out about these links. Secondly, there is now plenty of evidence against the notion of rigid stages in the development of drawing. Quite young children are extremely sensitive to experimental manipulations, often producing drawings that are more advanced by the criterion of approximation to adult conventions. In fact, we need to question whether this criterion is appropriate at all. As Hagen (1985) points out, most adults, let alone children, never develop to a level at which they can use artistic conventions such as linear perspective. But here I think Hagen carries the argument too far, by suggesting that there is no development in drawing. There are quite obvious changes; and our failing so far is that we have not found the best way of explaining them. Current evidence suggests that many of the changes relate to the intent behind the drawing. In many cases younger children's drawings may differ from those of older children because the different age-groups are simply trying to show different things, and important theoretical progress may result from asking why these intentions change during development.

18.6 Two accounts of the evolution of art

Early writers on child art were quick to draw parallels with 'primitive art'. Eng devotes a chapter to such comparisons, and concludes:

The unquestioned parallels, which can be shown between the drawing of primitive man and that of children—formalism, transparency, turning-over, spacelessness, want of synthesis—are based on features common to the psyche of the child and primitive man, on the want of firm voluntary attention, of penetrating analysis and higher synthesis, on the weakness of the power of abstraction and of logical and realistic thinking. When we find these features of children's drawings repeated again in primitive art, we may conclude that their executants were similar in their psychical make-up.

This is a strong claim, considering current thinking about the relation between drawing and cognition in children, and one feels happier with some of the specific parallels drawn in early work than with the deeper conclusions. For instance, Luquet (1923) points to a way in which the study of child art may help us to interpret prehistoric art. Referring to an example of the latter in which 'animal' figures are shown in upright postures, he rejects the need for any sort of mystical interpretation, suggesting instead that this is just a case of the prehistoric artist failing to escape from the much used animal form when attempting to draw the human figure. This interpretation springs from examples of the same conservative tendency in children's drawing, although in this case, since the human figure is much more commonly drawn than the animal figure, it is the human form that is retained when drawing an animal.

This general notion seems plausible, particularly since examples of animals with 'human-like' heads can be identified much later in the history of art (Gombrich 1972). In addition, Kellogg's (1970) account should be borne in mind by interpreters of early art. For example, she claims that the mandala has aesthetic significance first and foremost, and only gained religious significance through cultural imposition. If this is true, we should be cautious in assuming religious significance for some of the forms found in early art.

But useful as these pointers might be, they are little more than surface parallels (often not at all obvious), and to feel confident of a real link between child art and the art of the past we need to show that the phenomena in both areas can be accounted for by a similar developmental mechanism. If there is a parallel between evolution and child development, it should be possible not only to see similar phenomena but to see similar sequences of development in the history of art as appear in the developing art of the individual child. And we need to produce a theory that can plausibly be applied to both domains.

18.6.1 *Gablik (1976)*

Probably the most thorough attempt to show a parallel between child development and art-history has been made by Gablik (1976), whose contention is that Piagetian stage theory can be applied to the evolution of intelligence just as it is applied to cognitive development in the child. The assumption is that human mentality has evolved through the same stages, from prehistoric times to the present day, as are represented by the Piagetian stages of cognitive development in individuals from the age of two years to adulthood. Seen in this light, art reflects the mental sophistication of the artist, and Gablik identifies three artistic periods which she ties to three Piagetian stages of intelligence. In doing this, however, she makes little reference to child art itself, being content to construct links between general Piagetian theory and the evolution of art.

These historic periods are arrived at through analysis of the spatial principles to be found in the artistic products.

1. The first period includes ancient and medieval art, which is characterized by Gablik as follows: 'Distance between objects is based on their proximity to one another on a two-dimensional plane which only takes height and breadth into account. Absence of depth, no unified global space which conserves size and distance' (p. 43). She identifies this with Piaget's Pre-operational Stage, in which the child's spatial organization is primarily based on topological principles such as proximity, separation, neighbourhood, etc., the main point being that the pre-operational child has no grasp of a Euclidean or projective spatial framework in which objects can be coherently arranged in three dimensions.

2. The second period covers the Renaissance, in which pictures are organized in terms of Euclidean relationships based on a single static viewpoint. This period is identified with Piaget's Concrete Operational Stage, in which the child develops a grasp of Euclidean spatial geometry, but can only apply it to the concrete reality of perceived space.

3. The final period corresponds to modern art, including late impressionism, cubism, and formalism. Here, her description of the spatial characteristics of art becomes more obscure: 'Indeterminate Atmospheric Space... Space as an all-over extension in which all points are of equal status and are relative to each other' (p. 43). This period is related to Piaget's Formal Operational Stage, in which the spatial logic developed in the Concrete Operational Stage can now be applied beyond perceived reality. In other words, cognition becomes hypothetical, and in art we see the development of systems based on abstract logic with no basis in direct experience. For instance, Cubism abandons the single point of view to give an amalgam of viewpoints in the same picture, and formalism is based on formal geometry, with no attention to physical objects.

Fig. 18.5 Examples of art from different periods of history.
(a) Period I: Ancient and medieval art. *Hunting the hippopotami in the marshes* (Egyptian, 4th–5th dynasty). Copyright British Musuem.

According to Gablik, art could have developed in no other way, since the evolution of art depended on the evolution of cognition, which followed the same fixed sequence of stages as seen in the developing child.

18.6.2 *Gombrich (1960)*

A contrasting view is presented by Gombrich (1960), in which he rejects the notion that changes in artistic style can be related to the evolution of cognitive structure. For instance, in discussing similarities between child art and the art of two painters of the fourteenth and fifteenth centuries, he claims 'One thing we can be sure of: neither Duccio nor Sassetta had a childish, undeveloped mentality' (p. 248). Gombrich lays far more emphasis on the motivation of the artist, and ties this in turn to the demands of the culture of the time. He claims that the earliest aim in art was the *making* of things in their own right, and that only later on came the preoccupation with *matching* the product with reality as viewed from a set perspective.

Seen in this light, early paintings should not be seen as less developed because they were poorer approximations to a perspective view; their form simply reflected the different intentions of the artist. The artist painted according to his/her particular 'mental set', and according to the demands of the audience, which in turn were determined by their mental set. Gombrich's meaning is

Fig. 18.5 (b) Period II: the Renaissance. *The Marriage of Mary* (1504) (alterpiece) by Raphael Sanzio of Urbino (1483–1520) Pinacoteca di Brera, Milan/Bridgman Art Library, London.

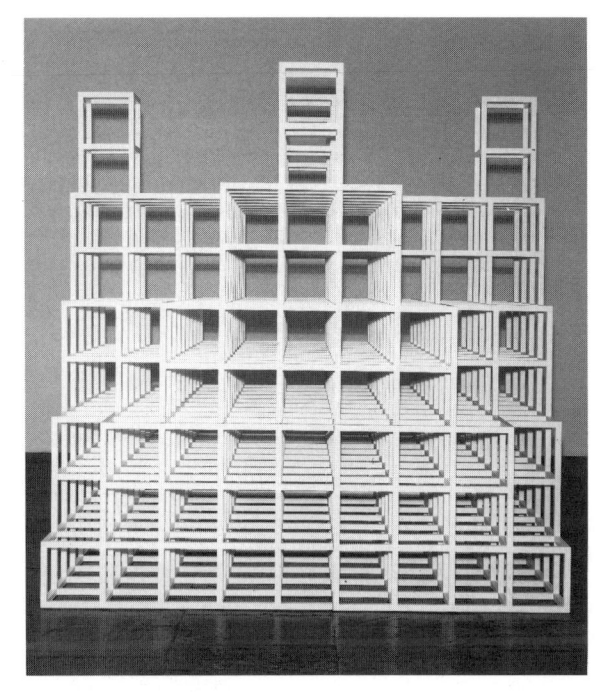

Fig. 18.5 (c) Period III: Modern Art, Sol LeWitt: *Untitled* (1966). Copyright 1995: Whitney Museum of Art.

best seen in his discussion of the nonchalant brushwork (*sprezzatura*) of Renaissance art, of which he says:

> It is an art in which the painter's skill in suggesting must be matched by the public's skill in taking hints. The literal minded Philistine ... does not understand the magic of *sprezzatura* because he has not learned to use his own imagination to project. He lacks the appropriate mental set to recognize in the loose brushstrokes of a 'careless work' the images intended by the artist ... (p. 165).

The point here is that the limitation is a matter of the constraint of set rather than a cognitive limitation. People of that period had the mental capacity to interpret new methods, but were constrained more or less by inertia to do otherwise. When change occurs, it is due to the changing demands of the culture. So, for instance, when society became preoccupied with scientific thinking, this both crept into the thinking of the artist and the demands of the audience. It is only to this extent that Gablik's analysis of art in terms of scientific spatial principles becomes even partially appropriate; partially so, since the emergence of Euclidean principles comes from cultural vogue rather than internal mental structure.

The differing views put forward by Gablik and Gombrich can be traced back to deep philosophical differences. Gombrich sees the 'matching' component of art as a progressive discovery of the illusionist capacities of art to appear like objective reality. In contrast, Gablik believes that art only has this property because both reality and artistic representations of it are constructed according to the same principles; there is no reality independent of our construction of it. Earlier peoples saw the world differently because they were less mentally sophisticated; so there is a real sense in which their art reflected a different perception of the world. In propounding this view, Gablik accepts Piaget's theory as the way in which this construction proceeds, although there is reason to suspect that her version of the theory is a more extreme form of constructionism than Piaget intended. She argues that the developments taking place in the transition from Pre-operational to Concrete Operational thought reorganize the way the world is perceived; whereas Piaget was talking about our mental manipulations of a reality that had become objective at an experiential level during the Sensorimotor Stage (a period that Gablik omits from her analysis).

But I believe that there is an important sense in which Gablik's philosophy is mistaken. There *is* a structure to the environment, one that permits consistent outcomes of the organism's actions on it. This is empirical fact, and should not be confused with our struggles to construct mental accounts of reality to explain the consistencies. Hence we can pick up and act appropriately to the structure of the environment without being able to produce a coherent mental account of the whole thing. This view is developed by J. J. Gibson in his claim that the structure of the environment can be directly perceived, without the need for the construction of cognitive mediating systems. The organism is adapted to the environment, and operates as one component of an interactive system. This is not to deny the reality of cognitive systems, but simply to put them in their proper place. We may need cognition to explain how we interact with the environment; but not in order to interact in the first place. In addition, cognitive systems, through imposing a formalism on our acts, may in turn modify them to make them more efficient.

This analysis may also be applied to the process of artistic portrayal by the suggestion that, for example, artists produced approximations to perspective before knowing the formal geometry of perspective, but that they improved their perspective techniques once a coherent descriptive system was developed to explain why perspective was such a good approximation to reality. Artists discovered approximations to perspective because they looked good, and discovered the rules of perspective later.

But there is also something missing from Gombrich's account. Although it seems very plausible that the demands and prevalent ways of thinking of a culture should have a determining influence on the forms of art, we are left with the question of where these ways of thinking came from. Are we to rule out cognitive development altogether in the development of culture? To be fair, Gombrich may not have done this, simply feeling that he had relegated the role of cognitive development to a sufficient distance from art for it to fall outside the scope of his argument. But it may be relevant to bring it back into focus, since this may take a large part of the randomness out of his account of artistic change.

18.6.3 *Pointers from developmental psychology*

This is an appropriate point at which to go back to see whether developmental psychology can offer any pointers to a modified account of the evolution of art. We can do this both by reference to evidence on the development of art in children, and by reference to recent progress in general developmental theory. If cognitive development is reflected in art-history, it should also be reflected in the developing art of the child. Despite this apparently fundamental point, Gablik makes little use of data on children's art, preferring to relate the history of artistic style to Piaget's theory as it applies to representation of space. This is a shortcoming in two ways. First, at a theoretical level, she treats problems of pictorial portrayal and spatial representation as synonymous when they are not. Mental representation involves imposing an organization on reality; whereas pictorial portrayal involves transformation of reality. Although organization may facilitate transformation, the two processes are at least one step removed from each other. Secondly, she assumes that these cognitive–geometric considerations are at the root of what is developing in art. Although she admits to neglecting certain aspects of art, she seems to assume that these other aspects of the problem are sufficiently minor to leave her model intact. Both of these assumptions become problematic when we recall that one of the main things to emerge from recent studies of child art is that the evidence cannot be adequately explained in terms of the development of general cognitive skills. The progression now emerging in child art seems to relate primarily to two factors: (i) growth of specific mental skills related to drawing, and (ii) the changing intentions of the artist. If this is true of the development of art, is it not at least as likely to be true of the evolution of art?

At a more specific level, the form of young children's drawings may lead us to a rethink of Gablik's account. One property often noted in the drawings of children around the four to six age-range is their 'multi-perspective' quality, as if a single drawing contains multiple perspectives on the same object. This is what Cubism is supposed to do—in Gablik's view, a product of formal operational thought. Although we should be wary of the way such similarities can mislead, this should at least give us pause for thought. Maybe development in art is more to do with the suppression of conceptual knowledge, as many (for instance, Fry 1934; Ruskin 1857) have suggested in the past. However, whereas the conventional account has it that conceptual knowledge is suppressed in order to gain access to the retinal image, the present version is that it is suppressed in order to escape the concept of perception as a series of static images. The young child, unhampered by such a conception, may be happy to do what it has taken the artist a long time to return to. This is not to deny that it took considerable mental sophistication to achieve this return. But it entirely changes the role of cognition in the process of art.

Another problem in Gablik's account is the apparent wholesale acceptance of Piaget's account as if it were the accepted model of cognitive development. Over the past twenty years or so, developmental psychology has been very much preoccupied not only with theoretical criticisms of Piagetian theory, but with the growing body of data that his theory is ill equipped to explain. The main data here consist of evidence of objective perception in infants and logical capabilities in young children long before they should appear according to the theory. In a milder way, the data on children's drawings fit this picture, with young children producing relatively sophisticated products when the experimental context is right. But the main problem for Piagetian theory here is that the drawing data do not seem to fit terribly well at all. For instance, children make drawings that do not focus on their own view just at the same time as Piagetian theory says that they are unable to consider any perspective other than their own (Piaget and Inhelder 1956 [1948]).[1]

On the other hand, Gombrich's account generates more parallels with developmental phenomena, despite its less developmental flavour. The idea that early art was concerned more with making than with matching is very much in keeping with many accounts of the Wrst steps in children's drawing, which in one way or another stress the idea that the child is initially only preoccupied with making patterns on paper, with matching to reality only coming later (Eng 1931; Kellogg 1970). Also, the notion that artists are conservative when faced with new art objects, modifying old schemas to represent the new object, seems very similar to the conservatism identified by Eng, in which children attempt to apply old formulae to new objects. But such similarities are not necessarily underpinned by the same general developmental mechanism, and we need to look for links with developmental theory. Here it is worth returning to J. J. Gibson's theory, and the way he applies it to pictorial representation. The basic assumption of the theory is that the organism extracts invariants from the incoming perceptual flux. These invariants are ready specified, and serve to indicate the three-dimensional properties of the world (the theory is hard to follow in many places, and there is not space to expand on it here; hence the reader is referred to the original work: Gibson 1979). In the child's early scribbles there exist pictorial invariants, which the child picks up. The next step is the identification of the link between pictorial invariants and environmental ones; and once this link is made, the child can begin to replicate reality on paper. This analysis comes very close in some respects to Gombrich's transition from making to matching. Initially drawings are made for their own sake, and later they are identified with and progressively matched to reality.

It may be that we can pull many of the strings together on the issue of how art developed if we apply a recent form of developmental theorizing that springs directly from Gibson's ecological theory of perception. Ecological theory emphasizes the interaction between environment and individual, and insists that we should study the organism–environment system rather than one or the other in isolation. The notion of an inseparable system of this sort is implicit in some of Piaget's writings, but was never fully elaborated. In Piagetian theory, the motive force behind the development of cognitive structure is situated in the organism, and the form of intellectual structure is determined by the fixed ways in which environmental information is sampled and assimilated to existing structure. From this point of view, an environment is needed if development is to proceed, but it does not determine the structure of intelligence. In contrast, ecological theory gives more weight in the equation to environmental structure. The Gibsonian version marks the extreme form of ecological theorizing by dismissing the need for internal representation for most forms of interaction with the environment. The structure is in the environment, and the organism needs simply to pick it up. However, there are problems with this extreme, since the organism must be suitably adapted to pick this information up, and so there must be some form of internal structure, whether or not we want to call it representation. A better alternative may be to view development as the product of an interplay between internal and external structure, in which each play a complementary part. An organism in a given state of development encounters particular environmental structures through engaging in the particular activities allowed by that state of development. These environmental 'problems' are solved through adaptation of internal structure in ways quite specific to the structure of the external problem, and so the environment does not simply permit development; it leaves its stamp on it at every point.

In the light of this sort of developmental model, both Gablik and Gombrich can be seen to be at least partly correct. We may assume that knowledge about artistic technique, rather than just the technique itself, does develop. But we should be wary in assuming that this corresponds in any close way to general levels of mental ability. Change in artistic technique occurs when the demands made of the artist by society change (an example given by Gombrich is the notion that infusion of life into hitherto rigid human figures took place in Greek art because it had to be placed in '. . . narrative contexts that demanded a convincing recreation of a situation' (Gombrich 1960, p. 113). So here specific problems thrown up by cultural changes lead to artistic changes as solutions. But this simple model is unrealistic unless viewed as a specific part of a more general transaction between individual and environment (society takes a privileged position here, since it is on both sides of the interaction at once). A society possessing a particular level of knowledge will explore problems thrown up by that level of knowledge, which will in turn determine the form that the next level of knowledge takes. Thus societies may have gone through cognitive stages rather like Piaget's, but in a much less predetermined way than his theory suggests. A different problem facing a society at a given time might have altered the course of subsequent development considerably. But the point to be made is that the relationship between general knowledge structures and art is indirect. For instance, the Greek narrative style may have been a product of more sophisticated thinking (hypothetical—about events temporally displaced, etc.), and this in turn may have led to the infusion of life into art. Also, more advanced knowledge may have permitted artists to analyse their art more rigorously. However, it remains a question whether this analysis led to changes in their art. My guess is that the sort of geometric knowledge discussed by Gablik has been of more use to those studying art than to those practising it; but I have to be cautious here, since I cannot really claim to be doing either.

Note

1. It is worth noting a contemporary and apparently entirely independent analysis by Gowans (1979) in which broadly similar conclusions are reached about the relation between Piagetian cognitive development and evolution of art. Although Gowans makes direct comparisons between art-history and child art, the literature referred to is somewhat limited and outdated, and again the acceptance of Piagetian theory is wholesale.

References

Barrett, M. D. and Light, P. H. (1976). Symbolism and intellectual realism in children's drawings. *British Journal of Educational Psychology*, **46**, 198–202.

Boysen, S. T., Berntson, G. G., and Prentice, J. (1987). Simian scribbles: a reappraisal of drawing in the chimpanzee (*Pan troglodytes*). *Journal of Comparative Psychology*, **101**, 82–9.

Bremner, J. G. (1985). Provoked use of height in picture as a depth cue in young children's drawings. *British Journal of Developmental Psychology*, **3**, 95–8.

Bremner, J. G. and Batten, A. (1991). Sensitivity to viewpoint in children's drawings of objects and relations between object. *Journal of Experimental Child Psychology*, **52**, 375–94.

Bremner, J. G. and Moore, S. (1984). Prior visual inspection and object naming: two factors that enhance hidden feature inclusion in young children's drawings. *British Journal of Developmental Psychology*, **2**, 371–6.

Bremner, J. G., Andreasen, G., Kendall, G., and Adams, L. (1993). Conditions for success in coordination of spatial dimensions by four-year-old children. *Journal of Experimental Child Psychology*, **56**, 149–72.

Buhler, K. (1933). *The mental development of the child*. Routledge and Kegan Paul, London.

Cox, M. V. (1978). Spatial depth relationships in young children's drawings. *Journal of Experimental Child Psychology*, **26**, 551–4.

Cox, M. V. (1981). One thing behind another: problems of representation in children's drawings. *Educational Psychology*, **1**, 275–87.

Davis, A. M. (1983). Contextual sensitivity in young children's drawings. *Journal of Experimental Child Psychology*, **35**, 478–86.

Eng, H. (1931). *The psychology of children's drawings*. Routledge and Kegan Paul, London.

Freeman, N. H. (1975). Do children draw men with arms coming out of the head? *Nature*, **254**, 416–17.

Freeman, N. H. (1980). *Strategies of representation in young children: analysis of spatial skills and drawing processes*. Academic Press, London.

Freeman, N. H. and Janikoun, R. (1972). Intellectual realism in children's drawings of a familiar object with distinct features. *Child Development*, **43**, 1116–21.

Freeman, N. H., Eiser, C., and Sayers, J. (1977). Children's strategies in producing three-dimensional relationships on a two-dimensional surface. *Journal of Experimental Child Psychology*, **23**, 305–14.

Fry, R. (1934). *Reflections on British painting*. Faber and Faber, London.

Gablik, S. (1976). *Progress in art*. Thames and Hudson, London.

Gibson, J. J. (1979). *The ecological approach to visual perception*. Houghton Mifflin, Boston.

Gombrich, E. H. (1960). *Art and illusion: a study in the psychology of pictorial representation*. Phaidon, London.

Gombrich, E. H. (1972). *The story of art* (12th edn). Phaidon, London.

Goodnow, J. (1977). *Children's drawing*. Open Books, London.

Gowans, A. (1979). Child art as an instrument for studying history. *Art History*, **2**, 247–74.

Hagen, M. A. (1985). There is no development in art. In *Visual order: the nature and development of pictorial representation* (ed. N. H. Freeman and M. V. Cox), pp. 59–77. Cambridge University Press.

Henderson, J. A. and Thomas, G. V. (1990). Looking ahead: planning for the inclusion of detail affects relative sizes of head and trunk in children's human figure drawings. *British Journal of Developmental Psychology*, **8**, 383–91.

Kellogg, R. (1970). *Analyzing children's art*. Mayfield, Palo Alto.

Lee, M. M. (1987). Task dependency and development of table drawing. Unpublished paper presented to the Annual Conference of the Developmental Psychology Section of the British Psychological Society, York.

Lee, M. and Bremner, G. (1987). The representation of depth in children's drawings of a table. *Quarterly Journal of Experimental Psychology*, **39A**, 479–96.

Light, P. and Foot, T. (1986). Partial occlusion in young children's drawings. *Journal of Experimental Child Psychology*, **41**, 38–48.

Light, P. H. and Humphreys, J. (1981). Internal spatial relationships in young children's drawings. *Journal of Experimental Child Psychology*, **31**, 521–30.

Light, P. H. and Simmons, B. (1983). The effects of a communication task upon the representation of depth relationships in young children's drawings. *Journal of Experimental Child Psychology*, **35**, 81–92.

Lukens, H. T. (1896). A study of children's drawings in the early years. *Pedagogical Seminary*, **4**, 79–109.

Luquet, G. H. (1923). Le réalisme dans l'art paléolithique. *L'anthropologie*, **33**, 347–55.

Luquet, G. H. (1927). *Le dessin enfantin*. Alcan, Paris.

Morris, D. (1962). *The biology of art*. Methuen, London.

Piaget, J. and Inhelder, B. (1956 [1948]). *The child's conception of space* (trans. F. J. Langdon and J. L. Lunzer). Routledge and Kegan Paul, London. (First published in French, 1948.)

Radkey, A. L. and Enns, J. T. (1987). Da Vinci's window facilitates drawings of total and partial occlusion in young children. *Journal of Experimental Child Psychology*, **44**, 222–35.

Ruskin, J. (1857). *The elements of drawing*. Smith, Elder and Co., London.

Schiller, P. (1951). Figural preferences in the drawings of a chimpanzee. *Journal of Comparative and Physiological Psychology*, **44**, 101–11.

Taylor, M. and Bacharach, V. R. (1982). Constraints on the visual accuracy of drawings produced by young children. *Journal of Experimental Child Psychology*, **34**, 311–29.

Teufel, S. S. and Lange-Küttner, C. (1993). What has human figure drawing to do with the graphic space concept? Poster presented to the VIth European Conference on Developmental Psychology, Bonn, September 1993.

Thomas, G. V. and Tsalimi, A. (1988). Effects of order of drawing head and trunk on their relative sizes in children's human figure drawings. *British Journal of Developmental Psychology*, **6**, 191–203.

Willats, J. (1977). How children learn to draw realistic pictures. *Quarterly Journal of Experimental Psychology*, **29**, 367–82.

Part IV
Language systems

Editorial introduction to Part IV: Language systems in an evolutionary perspective

No theme in linguistic science is more often and more voluminously treated than this, and by scholars of every grade and tendency; nor any, it may be added, with less profitable result in proportion to the labor extended.

So wrote W. D. Whitney in his 1870 review in the *Transactions of the American Philological Association*, 'On the present state of the question as to the origin of language' (cited by Marshall (1985, p. 686), and reprinted in Whitney's (1973) *Oriental and linguistic studies*. Scribner, New York). Over 125 years later it is clear definitive statements as to *the* origin of language will never be possible. Yet the range of possible accounts is becoming narrower.

To formulate any account of the evolution of language, it must be recognized, from the outset, that three different spheres of change are operating simultaneously, and that feedback loops hold between them. Thus:

1. The abilities that make language possible have to be assembled to allow a language system to be formulated.
2. A language system is unlikely to be formulated fully-fledged initially, but will be elaborated over time within cultural practices. This elaboration may involve the recruitment of additional skills over and above those required for its original impetus.
3. The cultural practices that sustain language will themselves be changed by the possibilities opened by the possession of language, feeding back to the way in which the language system is further elaborated, and thus, in turn, feeding back to the skills required for the acquisition of what might be far removed from the originally created system. Additionally, reversing this perspective, the skills that provide for the establishment of earlier language forms will themselves act as constraints on if and how the task of subsequently elaborating those forms is carried out, as will the cultural practices that any language both supports and is embedded in at any particular time.

The first point in this list has been considered in detail in the introduction to Part III, and material relevant to it in this Part will be found in the two comparative chapters, Chapter 23 by Ristau and Chapter 22 by Lock and Colombo.

Points 2 and 3 have received less attention in the past than they warrant. The non-evolutionary perspective of much modern linguistics, noted by Johnson *et al.* (this volume, Chapter 24) as implying a 'discontinuity view' of language evolution, appears the major reason for this neglect. Four chapters in this section, by Hewes, Ristau, Foster, and Rolfe, directly bear on point 2: point 3 lies at the root of the material reviewed by Barton and Hamilton, and the appendices to that chapter. At this time, neither of these points, nor their implications, have been fully worked through from an evolutionary perspective.

1 Construction vs. discontinuity: modularity and grammar again

Bickerton (1988, 1990) proposes a two-stage model of human spoken language that synthesizes a generally-held position. He distinguishes two sets of phenomena. First, what he terms *Dubiously language or Protolanguage:* (a) Children's utterances between 1 and 2 years; (b) Early-stage pidgin languages; (c) The output of linguistically-deprived children; and (d) The output of linguistically-trained apes. Second, *Indisputably language, or Language:* (a) Children's utterances between ages 2 and 4 years; and (b) Creole languages. He bases his distinction on four points.

First, where a speaker in the first category places words in order on the basis of semantic or pragmatic considerations, in the second category these ordering decisions fall under the formal constraints of syntax. Second, the nature of omitting sentence constituents is different for those speaking language versus protolanguage, in that in the former omissions are systematically related to some overt constituent elsewhere in the sentence. Third, protolanguages lack virtually all devices for the expansion of sentences. And finally, they also lack 'the devices which glue language together ... the determiners which indicate the referential scope of nouns, the verbal auxiliaries or inflections ... which link subject and predicate by governing the former while attaching themselves to the latter' (ibid., 93).

Bickerton places his distinction alongside that put forward by Chomsky (1980; and see also Johnson *et al.*, this volume, Chapter 24), that language can be divided into a conceptual and a computational component, such that:

"Genuine language" is produced by the conceptual and computational components working in tandem; protolanguage is produced by the conceptual component working in isolation, or at most supplemented by very general (not language-specific) processing capacities (1988, p. 94).

From here he goes on to three proposals, each of which he hypothesizes as corresponding to evolutionary stages:

(a) the conceptual component (a general primate characteristic); (b) the conceptual component plus communicative use of that component (a characteristic of ?); (c) the conceptual component plus the computational component plus the communicative use of the conjunction of the two components (a characteristic unique to *H. s. sapiens*) (ibid., p. 95).

Bickerton also develops an evolutionary scenario that we will not discuss here (see Bickerton 1990 for details and Mufwene 1991 for criticisms).

This tripartite division receives general support from the material reviewed in this volume. First, the primate work is indicative of the early existence of the bases for a cognitive component. The comparative primate data generated by Rumbaugh and his associates on reversal learning and transfer-of-training (see Lock and Colombo, this volume, Chapter 22) indicate that increasing encephalization within the primate order is associated with a shift from associative learning to relational, mediated learning. It is clear that the Great Apes have sufficient cognitive powers to support a protolanguage system when those powers are scaffolded by their relations with humans in the language-learning projects.

Second, the dissociative ontogenetic models reviewed here do suggest that, with the creation of a communicative motivation, early hominids that were more encephalized than extant Great Apes would have been able, in theory, to support some form of simple symbolic communication system. What is unclear is *to what extent, and in what form, such a system may have been elaborated*, and under what conditions a 'communicative motivation' might have arisen. Additionally, the dissociative evidence, along with the ontogenetic and comparative findings in general, indicates the possibility of there being functional symbolic communication systems based on an organizational structure that did *not* involve either syntactic or phonological principles. Again, then, the second stage Bickerton acknowledges, like the first, is in line with the hypotheses that present data suggest.

Bickerton's third stage is also congruent with the hypotheses that can be generated. But it is at this point that the 'modularity claim' again arises with respect to the hypothesizing of possible evolutionary courses. The issue is one of continuity versus discontinuity; that "the gulf between protolanguage and syntactic language looks too sharp and too deep to have ever been gradually bridged" (1988, p. 100). Bickerton puts it in focus thus:

The more closely one examines language, the less plausible it seems that syntax could have developed by any kind of piecemeal accretion of more complex and yet more complex structures, as the account of Bickerton (1981 [see Deuchar, this volume, Chapter 20]) tacitly assumed. Such a view [i.e., 'piecemeal accretion'] would have been more plausible a decade or so ago, when generative grammars consisted of "just one damn rule after another", and the discovery of a novel linguistic phenomenon was handled by simply tacking another rule onto the grammar. Since then, a new generation of restrictive theories of grammar has been developed [see Johnson *et al.*, this volume, Chapter 24], theories which generate a wide range of complex phenomena from the interaction of a small number of (in themselves) quite simple principles. These interactions are often so subtle that it is hard to imagine how language could have developed if only one or two of those principles had been in place at the beginning, and if the others had had to be added serially and gradually over time. Such principles seem more like aspects of a single complex schematism which must have sprung into place more or less ready-made (ibid., p. 99).

Bickerton considers that a pre-adapted neurological base could not have been 'co-opted' from some original purpose, 'since the types of constraint that produce syntactic organization seem to have nothing comparable to them in the rest of human behavior' (ibid., p. 99), even though such a possibility would make more evolutionary sense than a single saltation.

We have discussed the possible neurological and psychological bases for supporting language in detail in the introduction to Part III. That discussion indicates, in conjunction with the ontogenetic histories of the underlying components allowing language, that a position relying on a 'hard', preadaptive modularity is difficult to sustain: it is naïve in its conception of neurobiology. Apparent modern modules are probably the product of systems interacting with the particularly constructed languages used in real language communities.

In this section, two chapters, by Rolfe and Foster, implicitly offer an evolutionary focus on the 'modularity' question from another angle. That is, even if we accept a 'sudden' appearance for the ability to handle grammatically organized, modern spoken language, we are still faced with accounting for how grammatically organized, modern spoken language came into being as a spoken system used for communication. It is most unlikely that one or two humans 'suddenly' began to talk a language as-we-presently-know-it. The situation would seem to be more akin to that involved in the development of writing, where it appears that all normal modern humans can master reading and writing, even though, until recently, *everyone* used to be illiterate. Thus, having the ability to do something may be a necessary condition for being able to do it, but is by no means a sufficient one.

From very different perspectives, both Foster and Rolfe develop accounts of possible courses for the elaboration of modern language systems. Foster, through the methodology of reconstructive linguistics, argues that early forms of language were very different from modern ones, in that they were not phonemically organized. Rather, the sounds that now form the phonemes of language systems were originally meaningful in themselves—Foster terms these 'phememes'—and their meanings involve analogies between the shaping of the vocal tract during their production and features perceived in the referred-to world. She hypothesizes that the attainment of morphological grammars, and subsequently syntactic grammars, which act as organizing principles in language, occurred in only the last 10 000 years.

Rolfe approaches the elaboration of linguistic systems from the view that grammar acts to handle the expression of communicative intentions, and that these intentions are elaborated over time, as different discourse practices are demanded to sustain particular forms of social organization and interpersonal relation. Consequently, communicative intentions motivate grammatical form, and Rolfe's view is that newly-emerged forms come to 'dominate' older ones, so that where, for example, a topic–comment structure might still be workable for many aspects of everyday speech—subject–predicate structures only being required for the handling of specific communicative intentions generated in specific social settings—topic–comment structures have been 'imperialistically' taken over by the newer subject–predicate forms. The process by which this occurs he terms 'reworking'. Again like Foster, though on different grounds, Rolfe hypothesizes modern language forms to be of recent origin.

Ristau's review of the ape 'language' studies serves to emphasize the importance of cultural practices in the phylogenesis and maintenance of human languages. It looks like hard work to get chimpanzees to use symbols for communicative purposes. Once this is done, then it does not seem so difficult for them to begin combining a few symbols on their own. From there on, it looks almost impossible to get them up to scratch in grammar. These facts speak unequivocally for the importance of social practices in language maintenance, since without the social practices they engage in with their trainers, potential ape abilities would never have been realized. With respect to phylogeny, that chimpanzees use their hands to realize these abilities, and do not progress to grammar, nor at the same time enter into social practices that would require grammatical language for their sustenance: all these points confirm the importance of social practices and their role in fostering the evolutionary elaboration of grammatical language. The scenario Hewes adduces as a possible evolutionary route for language ultimately serves to confirm these points, even though his specific line of argument again differs in emphasis and detail from that of Foster and Rolfe. In sum, the chapters in this section are sufficiently infused with an evolutionary perspective to call for a repeal of the historical ban reproduced at the beginning of this volume on the topic of language evolution.

2 Language reconstruction

The Map Gallery which is Chapter 19 of this *Handbook* provides one recent view of the world's language families. It represents the culmination of linguistic research over the past two-hundred-plus years, and is itself controversial in a number of its aspects (Ruhlen 1987; also see Greenberg 1992, and Ruhlen 1992; for a critical perspective see Ringe 1992). It is the hierarchy of relations among these language families that remains the most methodologically difficult dimension, and therefore the most controversial.

As Ruhlen (1992) notes, it is customary to recognize as the starting-point of scientific historical and comparative linguistics the 1786 lecture of Sir William Jones, in which he asserted that Sanskrit, Greek, and Latin bear

a stronger affinity, both in the roots of verbs and in the forms of grammar, than could have been produced by accident; so strong that no philologer could examine Sanskrit, Greek, and Latin, without believing them to have sprung from some common source, which, perhaps, no longer exists. There is a similar reason, though not quite so forcible, for supposing that both the Gothic and the Celtic had the same origin with the Sanskrit (quoted by Ruhlen 1992, p. 159).[1]

Foster outlines, in her chapter, many of the ways in which *correspondences* between different languages have been traced to yield reconstructed 'proto'-languages. The general criteria for establishing genetic

Table 1 Criteria for establishing genetic relationship. [After A. B. Dolgopol'skij (1967). Ot Saxary do Kamčatki jazyki iščut rodstvennikov, *Znanie–sila*, **42**(1), 43–6.]

Agreement in different parts of language	Can this result *accidentally*?	Can this result easily from *borrowing*?	Can this be *inherited*?	Evidence or proof of *relationship*?
Agreement in the principles of syntax, morphology, and sound system	Yes	Yes	Yes	No
Agreement in descriptive and onomatopoeic vocabulary	Yes	Yes	Yes	No
Agreement in easily borrowable vocabulary	No	Yes	Yes	No
Multiple agreement in the basic and rather unborrowable vocabulary with *sound correspondences*	No	No	Yes	Yes
Considerable and frequent agreement in grammatical formants (endings, prefixes, auxiliaries) and *sound correspondences*	No	No	Yes	Yes

relationships between languages are summarized in Table 1, (see also Foster, this volume, Chapter 25, Section 4) and Foster outlines the criteria for remote reconstructions in Chapter 25, Section 7.

There are numerous controversies in what is a controversial field. Here we note just three. *First*, as Foster points out, Swadesh (for example 1971) is one of the few workers committed to a monogenetic

Table 2 Basic 100-word vocabulary for measuring glottochronological divergence

1. I	26. root	51. breasts	76. rain
2. you	27. bark	52. heart	77. stone
3. we	28. skin	53. liver	78. sand
4. this	29. flesh	54. drink	79. earth
5. that	30. blood	55. eat	80. cloud
6. who	31. bone	56. bite	81. smoke
7. what	32. grease	57. see	82. fire
8. not	33. egg	58. hear	83. ash
9. all	34. horn	59. know	84. burn
10. many	35. tail	60. sleep	85. path
11. one	36. feather	61. die	86. mountain
12. two	37. hair	62. kill	87. red
13. big	38. head	63. swim	88. green
14. long	39. ear	64. fly	89. yellow
15. small	40. eye	65. walk	90. white
16. woman	41. nose	66. come	91. black
17. man	42. mouth	67. lie	92. night
18. person	43. tooth	68. sit	93. hot
19. fish	44. tongue	69. stand	94. cold
20. bird	45. claw	70. give	95. full
21. dog	46. foot	71. say	96. new
22. louse	47. knee	72. sun	97. good
23. tree	48. hand	73. moon	98. round
24. seed	49. belly	74. star	99. dry
25. leaf	50. neck	75. water	100. name

position. His position has been developed, along with the work of others (for example Embleton 1983, 1986; Lees 1953; Rea 1973; Sankoff 1973; and see also Bergsland and Vogt 1962) into one termed *glottochronology*. The root notion is that the greater the amount of time from the point at which two languages separated from a common ancestor, then the greater will be the differences between them. As an index of divergence Swadesh proposed a basic, 'culture-free' vocabulary (either of 200, or later 100, words, see Table 2) that would not be subject to changes brought about by migration or social or technological change.

These words can then be compared for any two languages, and on the basis of known laws of sound-change the number of cognate words can be calculated. These are assumed to be retentions from a common ancestor, while non-cognate words are assumed to differ because either one or both languages have lost the original form. The preliminary study (Lees 1953) used pairs of languages that are documented for over 1000 years to obtain an index of the rate of change between the earlier and later vocabularies of the 'same' language. This rate of change (or retention) is surprisingly similar amongst the different languages studied, words being lost, added, and replaced at around either 14 per cent for the 100-word list or 19 per cent for the 200-word list per millennium.

From this, the date at which any two related languages split can be calculated, using the formula $t = \log c / 2 \log r$, where c is the percentage of cognates, and r the assumed percentage of cognates to be expected after 1000 years of separation (see Swadesh 1955 for detail; Sankoff 1973 and Embleton 1991 for refinements). The claim is that dates so resulting are valid until one reaches the level of chance similarity, determined as around 8 per cent, giving a time-depth for these calculations of around 12 000 years. There are a number of criticisms that have lead to a widespread rejection of this entire approach;[2] but, as Renfrew (1987, 116) notes: 'the astonishing thing is that the answers from such calculations are in some cases so close to a date of differentiation which can be established on independent grounds.' For example, Embleton (1991) has calculated the divergence of Germanic languages by two methods—the 'standard' glottochronic formula, and one in which this is modified to take account of language contact, etc.—and compares the calculated dates with historically available material that bears on these. The reconstructed dates appear remarkably accurate (see Fig. 1 and Table 3). Essentially, then, glottochronology seems more or less to work, but there is little rational basis for assuming it should, nor for explaining why it does.

Second, since Foster's chapter was drafted, Bichakjian has published (1988) a position statement, previously outlined in earlier publications (for example 1986, 1987) that language evolution is an inherently neotenous process. Now, given that 'language' is not an organism, any application of the term 'neoteny' with relation to it is necessarily metaphorical. Bichakjian's argument is that there is a directionality inherent in language-change over time toward shorter forms, smaller paradigms, fewer irregularities, and more transparent combinations, and that these changes can be accounted for by the elimination from languages of forms that are known to be acquired late in ontogeny. Aside from the problem of how, if neoteny is a guiding principle, languages became complex in the first place (an objection that *could* be resolved under Rolfe's proposal (this volume, Chapter 26) of the pragmatic construction of language functions, and the subsequent reworking of these functions into coherent structural forms), and reservations some linguists (for example Klein 1990) have over the accuracy of the data on which Bichakjian's arguments are based, an explanatory account based on a single principle would appear less satisfactory in this context than one proposing multi-determinations of present language structures (such as that offered by Slobin (see Deuchar, this volume, Chapter 20).

Both glottochronology and the neoteny argument treat language as an object that can be studied in its own right, and focus on the *form* of language alone. That languages are *used* tends to be overlooked (although see Lock and Symes, this volume, Chapter 8 for some of the sociolinguistic models of language-change); yet attempting to take this into account in reconstructive work brings with it a whole new set of methodological problems. For example, a *third* controversial issue concerns the use of historical linguistic work to infer something of the physical surroundings, social organization, and culture of the users of reconstructed languages, an approach sometimes termed 'linguistic palaeontology'.[3]

The basic principle here is that if members of a particular cultural group, for example proto-Indo-Europeans (see for instance Lehmann 1973; Friedrich 1970; Thieme 1964) or proto-Algonquians (Siebert 1967; Walker 1975), can be shown via linguistic reconstruction to have had a name for something, then they can be assumed to have had knowledge of it. Unfortunately, most 'linguistic palaeontology' also takes negative cases on board, the absence of a word being taken to indicate a lack of knowledge of something. Thus:

Proceeding to the everyday life of the Indo-European community we find terms for "herd, cow, sheep, goat, pig, dog, horse, wolf, bear, goose, duck, bee, oak, beech, willow,

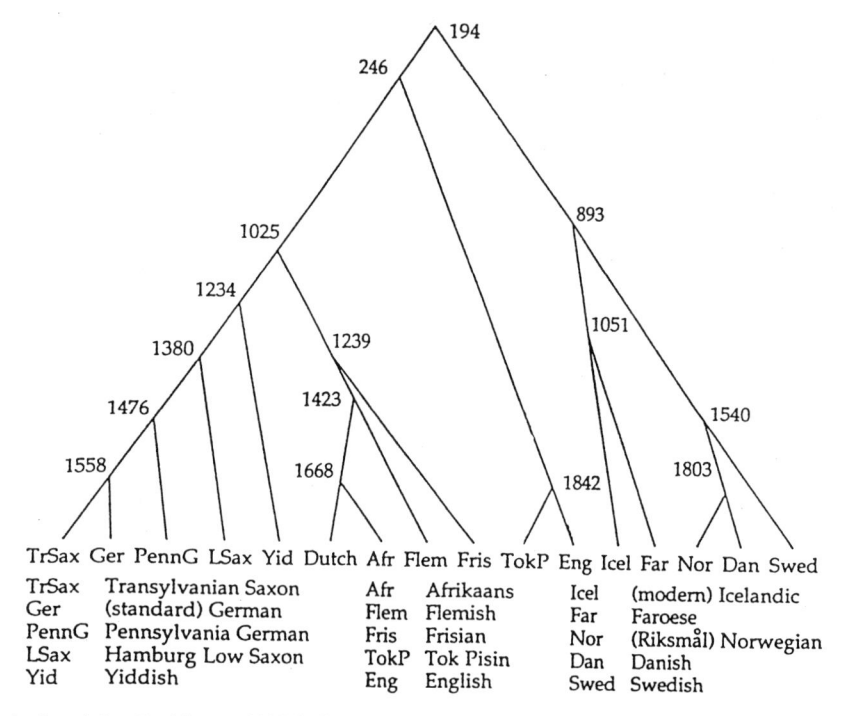

Fig. 1 Dates calculated by Embleton (1991) for the separation of modern and historically attested languages in the Germanic family after attempting a correction for the rates at which anticipated changes in the 200-word Swadesh list may have been affected by borrowings between geographically contiguous languages (adapted from Embleton 1991).

grain". The lack of specific terms for grains or vegetables indicates a heavy reliance on animals for food (Lehmann 1973, p. 232).

Negative data are always problematic, and it is difficult to know where to draw the line in this case. Arguing a societal artefact or practice on purely linguistic grounds is even more problematic.

Whatever, not only has a way of life, nomadic in the above case, thus been adduced for a culture, so also has its chronological and geographical location:

When we attempt to reconstruct words for metals, we can ascribe to the Indo-European vocabulary no words even for "silver" or "gold", let alone "iron" and scarcely even a general term for "metal, bronze, copper"... On the basis of such vocabulary we characterize the Indo-European community as late neolithic (Lehmann 1973, p. 233).

Similarly, on the basis of reconstructed terms for 53 plants, mammals, birds, and fish in proto-Algonquian, Siebert (1967) has been able to isolate 3— woodland caribou, lake trout, and harbour seal—as indicating the likely homeland of this linguistic community, since they must have lived in an area where all three species were present:

between Lake Huron and Georgian Bay and the middle course of the Ottawa River, bounded on the north by Lake Nipissing and the Mattewa River and on the south by the northern shore of Lake Ontario, the headwaters of the Grand River, and the Saugeen River.

The problems inherent in this mode of drawing inferences have been stated satirically by Pulgram (1958, pp. 146–7):

If we reconstructed Latin on the evidence of the Romanic languages alone, ignoring and neglecting the existence of Greek, Keltic [*sic*], Germanic and other ancient Indo-European dialects, and if thereupon we derived from the state of the common Romanic vocabulary conclusions on the culture of the speakers of Latin . . . we might well arrive at the following results: Proto-Romanic *regem* and *imperatorem* show us that the Latins lived in a monarchy under kings or emperors (but what shall we make of *rem publicam* which could presuppose a Latin republic?); since all Romanic languages contain words cognate with French *prêtre* and *évêque*, "priest" and "bishop", the Latins were Christians; also words cognate with French *bière, tabac, café* are common Romanic, evoking a picture of Caesar's soldiers guzzling beer and smoking cigars in sidewalk cafés; and since all Romanic languages name a certain animal *cheval, caballo, cal*, etc., and have words for "war" like *guerre, guerra*, the Latins called the horse *caballum* and war *guerram* and were no doubt warlike people with a strong cavalry (cited by Renfrew 1987, p. 85).

Table 3 Known historical information indicative of dates of language splitting in the Germanic family. Abbreviations for languages are given in Fig. 1.

Divergence	Reconstructed	Historical information
Nor–Dan	1803	Peace of Kiel, 1814
TokP–Eng	1842	Early to mid-nineteenth century (from migration evidence)
Dutch–Afr	1668	First European settlement, 1652; first comment on divergence, 1685
Ger–TrSax	1558	Waves of settlers from 1211 to 1846
Dan–Swed	1540	Accession of the Vasas, 1523
Ger–PennG	1476	Settlement, 1863, but admixture from other dialects and divergence from German before emigration
Dutch–Flem	1423	Both developed from thirteenth-century Low Franconian standard, but considered separate by sixteenth century
Ger–LSax	1380*	Low Saxon texts from 770, more numerous from ninth century
Dutch–Fris	1239*	First written records, fourteenth century
Ger–Yid	1234	End of the Crusade period in late thirteenth century
Icel–Far	1051*	Settled from Norway and Iceland after Norse occupation of Iceland
Ger–Dutch	1025*	Dutch 'forged' out of various dialects, 1200–1500
Icel–Dan	893	Settlement of Iceland, 874 until 930
Eng–Ger	246	Traditional date for settlement of England is 449, but divergence already in Continental homeland
West–North	194	Second century AD

Dates marked * show poor agreement between reconstructed date and historically-indicated date, but may result from influences not accounted for in the borrowing 'correction factor'. Thus language 'prestige' may have influenced the lack of divergence apparent between standard German and Low Saxon through motivating a retention of a higher proportion of German words than otherwise expected. Given the almost unspecifiable number of factors that could be expected to influence rates of historical change, the fit between the two sets of data is remarkably good.

There are clearly major difficulties inherent in reconstructing social and cultural forms from reconstructed languages, just as there are from material artefacts. It is difficult, however, for a non-specialist in this area to winnow his or her way through the proposals in the literature, and the topic demands an accessible critical review. We have appended a representative list of sources here, and merely point out that this is a sub-area of the study of human symbolic evolution in which the advocacy of a position currently tends to dominate over critical assessment.

Despite their shortcomings, attempts to reconstruct the pathways by which language could have been elaborated are important. They offer the promise of a logical framework that provokes empirical questions. This promise should be kept in sight when assessing present efforts in the field.

Notes

1. Ruhlen (1987) provides a brief introduction to the methodological history of the discovery of and subsequent research on the Indo-European language family.

2. The method assumes an absolute family-tree model of language-change, whereas in reality this is unlikely

to be the whole case; loan-words adopted by both compared languages are bound to increase the number of cognates; if synonyms exist for 'basic' vocabulary items, of which one is cognate and the other not, there are no criteria for choosing which to select; there is no *a priori* reason for assuming languages would suffer word-loss at a constant rate—in fact, sociolinguistic theories would predict the opposite (see Swadesh 1971, 280–2; Renfrew 1987, Chapter 5; and Embleton 1991 for details).

3. For full reviews of 'glottochronology' (and 'lexicostatistics', of which it is a sub-branch), and the historical development of 'linguistic palaeontology' in general, see Gudschinsky (1964) and Hymes (1960*a* and *b*).

References

Bergsland, K. and Vogt, H. (1962). On the validity of glottochronology. *Current Anthropology*, **3**, 115–53.

Bichakjian, B. H. (1986). When do lengthened vowels become long? Evidence from Latin and French, and a paedomorphic explanation. In *Studies in compensatory lengthening* (ed. W. L. Wetzels and E. Sezer), pp. 11–36. Foris, Dordrecht.

Bichakjian, B. H. (1987). The evolution of word order: a paedomorphic explanation. In *Papers from the VIIth International Conference on Historical Linguistics, Pavia, Italy, 1985* (ed. A. G. Ramat, O. Carruba, and G. Bernini), pp. 87–108. Benjamins, Amsterdam.

Bichakjian, B. H. (1988). *Evolution in language*. Karoma, Ann Arbor.

Bickerton, D. (1988). A two-stage model of the human language faculty. In *Ontogeny, phylogeny, and historical development* (ed. S. Strauss), pp. 86–105. Ablex, Norwood, NJ.

Bickerton, D. (1990). *Language and species*. University of Chicago Press.

Chomsky, N. (1980). *Rules and representations*. Columbia University Press, New York.

Embleton, S. M. (1983). Incorporating borrowing rates in lexicostatistical tree reconstruction. In *Historical linguistics* (ed. B. Brainerd), pp. 1–24 Brockmeyer, Bochum.

Embleton S. M.(1986). *Statistics in historical linguistics*. Brockmeyer, Bochum.

Embleton S. M.(1991) Mathematical methods in genetic classification. In *Sprung from some common source. Investigations into the prehistory of language* (ed. S. M. Lamb and E. D. Mitchell), pp. 365–88. Stanford University Press.

Friedrich, P. (1970). *Proto-Indo-European trees*. Chicago University Press.

Greenberg, J. (1992). Preliminaries to a systematic comparison between biological and linguistic evolution. In *The evolution of human languages, SFI studies in the sciences of complexity, Proc. Vol. XI* (ed. J. A. Hawkins and M. Gell-Mann), pp. 139–58. Addison-Wesley, Redwood City, California.

Gudschinsky, S. C. (1964). The A.B.C.'s of lexicostatistics (glottochronology). In *Language in culture and society* (ed. D. Hymes), pp. 612–23. Harper and Row, New York.

Hymes, D. (1960*a*). Lexicostatistics so far. *Current Anthropology*, **1**, 3–44.

Hymes, D. (1960*b*). More on lexicostatistics. *Current Anthropology*, **1**, 338–45.

Jones, Sir W. (1786). Third anniversary discourse: 'On the Hindus'. In *The collected works of Sir William Jones*, Vol. 3 (1807), pp. 24–46. John Stockdale, London.

Klein, J. (1990). Review of *Evolution in language* by B. H. Bichakjian. *Journal of Pidgin and Creole Languages*, **5**, 321–6.

Lees, R. B. (1953). The basis of glottochronology. *Language*, **29**, 113–25.

Lehmann, W. P. (1973). *Historical linguistics, an introduction*. Holt, Rinehart, and Winston, New York.

Marshall, J. C. (1985). Speechless in Java. *Nature*, **316**, 685–6.

Mufwene, S. S. (1991). Language genesis and human evolution. *Diachronica*, **8**, 239–54.

Pulgram, E. (1958). *The tongues of Italy*. Harvard University Press, Cambridge, Mass.

Rea, J. A. (1973). The Romance data of the pilot studies for glottochronology. In *Current trends in linguistics II: Diachronic and typological linguistics* (ed. T. A. Sebeok), pp. 355–67. Mouton, The Hague.

Renfrew, C. (1987). *Archaeology and language: the puzzle of Indo-European origins*. Jonathan Cape, London.

Ringe, D. A. (1992). On calculating the factor of chance in language comparison. *Transactions of the American Philosophical Society*, V. 82, pt 1. American Philosophical Society, Philadelphia.

Ruhlen, M. (1987). *A guide to the world's languages*. Stanford University Press.

Ruhlen, M. (1992). An overview of genetic classification. In *The evolution of human languages, SFI studies in the sciences of complexity, Proc. Vol. XI* (ed. J. A. Hawkins and M. Gell Mann), pp.159–89. Addison-Wesley, Redwood City, California.

Sankoff, D. (1973). Mathematical developments in lexicostatistic theory. In *Current trends in liguistics II: Diachronic, areal and typographical linguistics*. (ed. T. A. Sebeok), pp. 93–113. Mouton, The Hague.

Siebert, F. S. (1967). *The original home of the Proto-Algonquian people* Anthropological Series, Bulletin No. 214. National Museum of Canada, Ottawa.

Swadesh, M. (1955). Towards greater accuracy in lexicostatistic dating. *International Journal of American Linguistics*, **21**, 121–37.

Swadesh, M. (1971). *The origin and diversification of language*. Aldine-Atherton, Chicago.

Thieme, P. (1964). The comparative method for reconstruction in linguistics. In *Language in culture and society* (ed. D. H. Hymes), pp. 585–98. Harper Row, New York.

Walker, W. (1975). The Proto-Algonkians. In *Linguistics and anthropology: In honor of C. F. Voegelin* (ed. M. D. Kincaid, K. L. Hale, and O. Werner), pp. 633–47. P. de Ridder Press, Lisse.

Whitney, W. D. (1973). *Oriental and linguistic studies.* Scribner, New York.

Sources for historical reconstructions undertaken on the basis of linguistic evidence

Antilla, A. (1972). *An introduction to historical and comparative linguistics.* Macmillan, New York.
Benveniste, E. (1973). *Indo-European language and society.* University of Miami Press, Coral Gables.
Bomhard, A. (1984). *Toward Proto-Nostratic: a new beginning.* Benjamins, Amsterdam.
Buck, C. D. (1949). *A dictionary of selected synonyms in the principal Indo-European languages.* University of Chicago Press.
Bynon, T. (1983). *Historical linguistics* (2nd edn). Cambridge University Press.
Cardona, C., Hoenigswald, H. M. and Senn, A. (eds) (1970). *Indo-European and Indo-Europeans.* University of Pennsylvania Press, Philadelphia.
Gamkrelidze, T. V. and Ivanov, V. V. (1985). The ancient Near East and the IE question: temporal and territorial characteristics of Proto-Indo-European based on linguistic and historico-cultural data; *and* the migration of tribes speaking Indo-European dialects from their original homeland in the Near East to their historical habitations in Eurasia. *Journal of Indo-European Studies*, **13**, 3–91.
Goyvaerts, D. L. (1975). *Present-day historical and comparative linguistics: an introductory guide to theory and method.* Story-Scientia, Ghent.
Hickerson, N. P. (1980). *Linguistic anthropology.* Holt, Rinehart, and Winston, New York.
Hock, H. H. (1986). *Principles of historical linguistics.* De Gruyter, Berlin.
Lamb, S. M. and Mitchell, E. D. (eds.) (1991). *Sprung from some common source: investigations into the prehistory of languages.* Stanford University Press.
Lehmann, W. P. (1973). *Historical linguistics: an introduction.* Holt, New York.
Meillet, A. (1970). *The comparative method in historical linguistics,* trans. G. B. Ford. Champion, Paris.
Puhvel. J. (1987). *Comparative mythology.* Johns Hopkins University Press, Baltimore.
Renfrew, C. (1987). *Archeology and language: the puzzle of Indo-European origins.* Jonathan Cape, London.

19
Map gallery of the distribution and classification of extant human languages

The maps, captions and table for this map gallery of the distribution and classification of extant human languages are from Ruhlen (1981) *A guide to the world's languages*, Vol. 1 (Stanford University Press). Although aspects of the classification remain controversial, the maps are testimony to the diversity of human languages that have evolved in the past few thousand years.

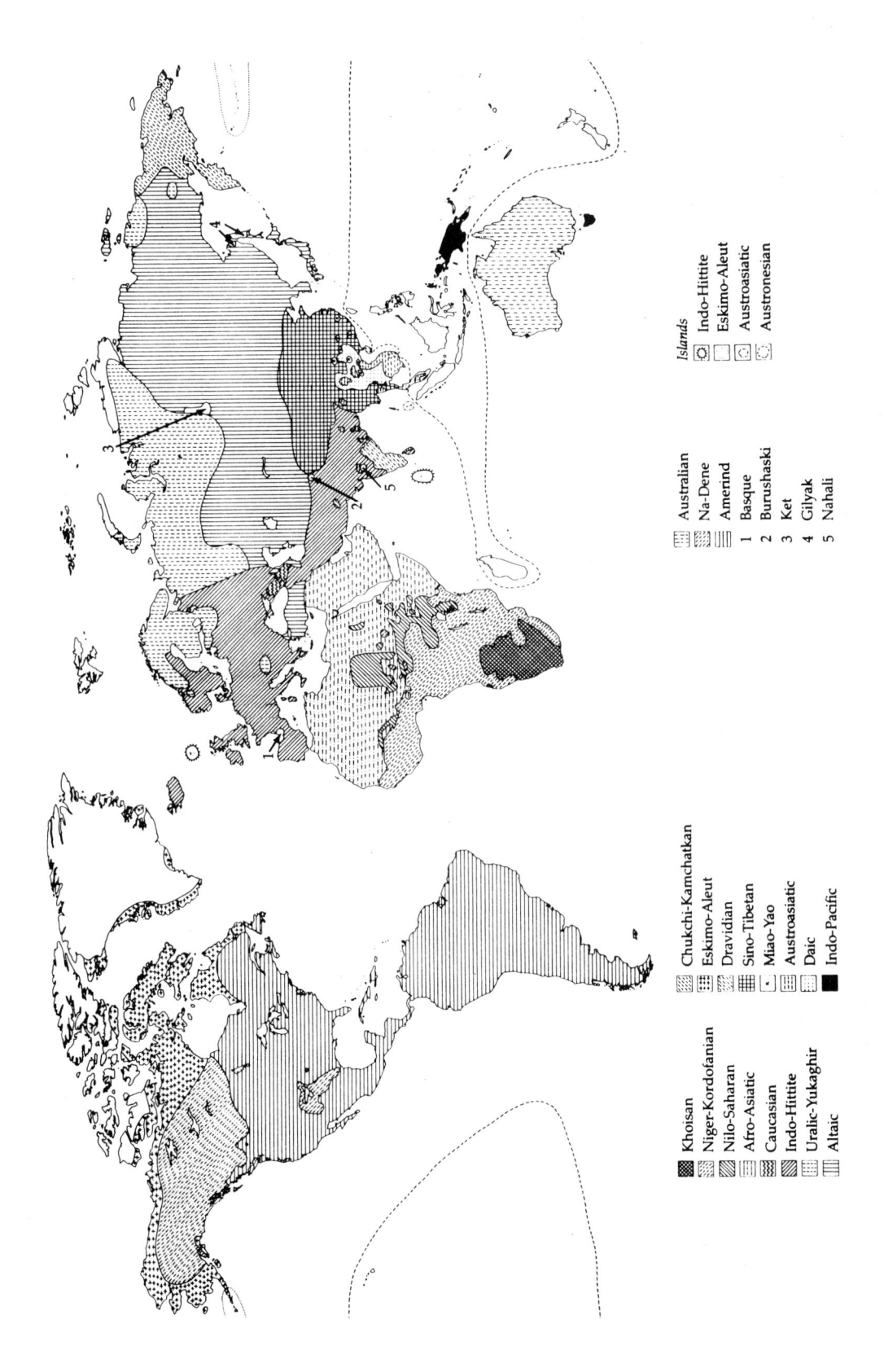

Fig. 19.1 The world's language families

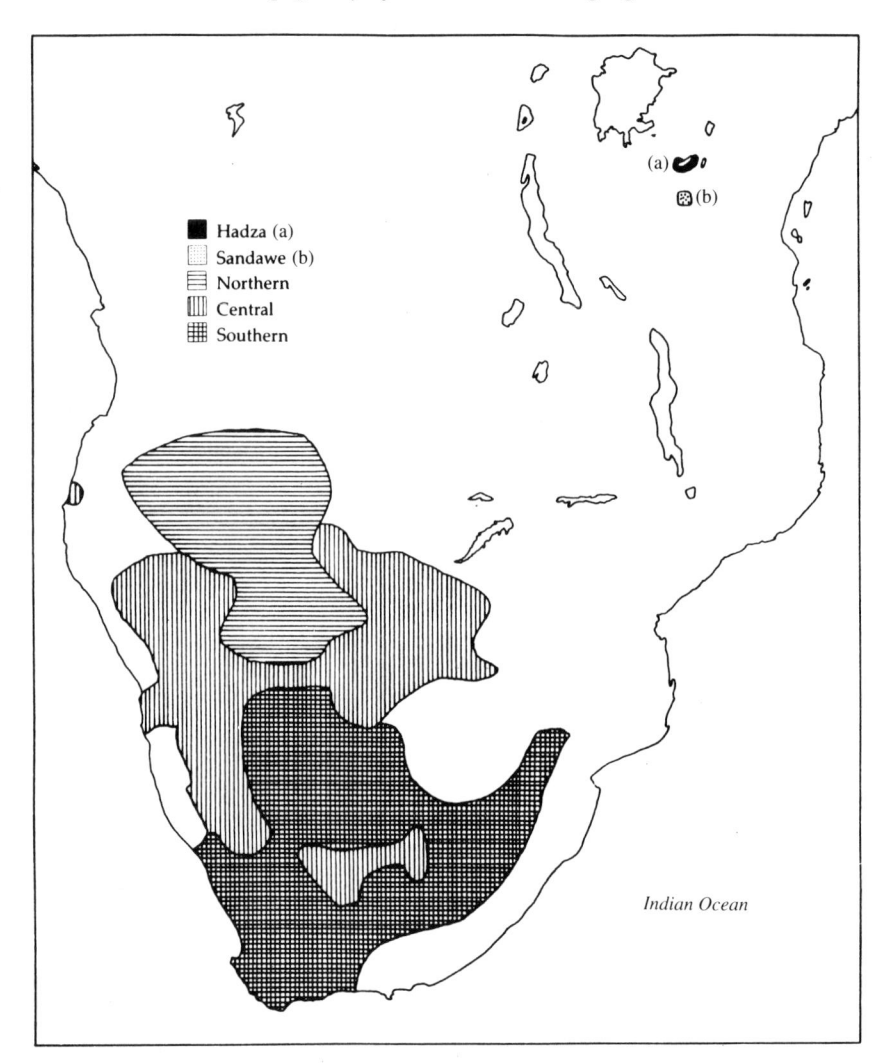

Fig. 19.2 The Khoisan family

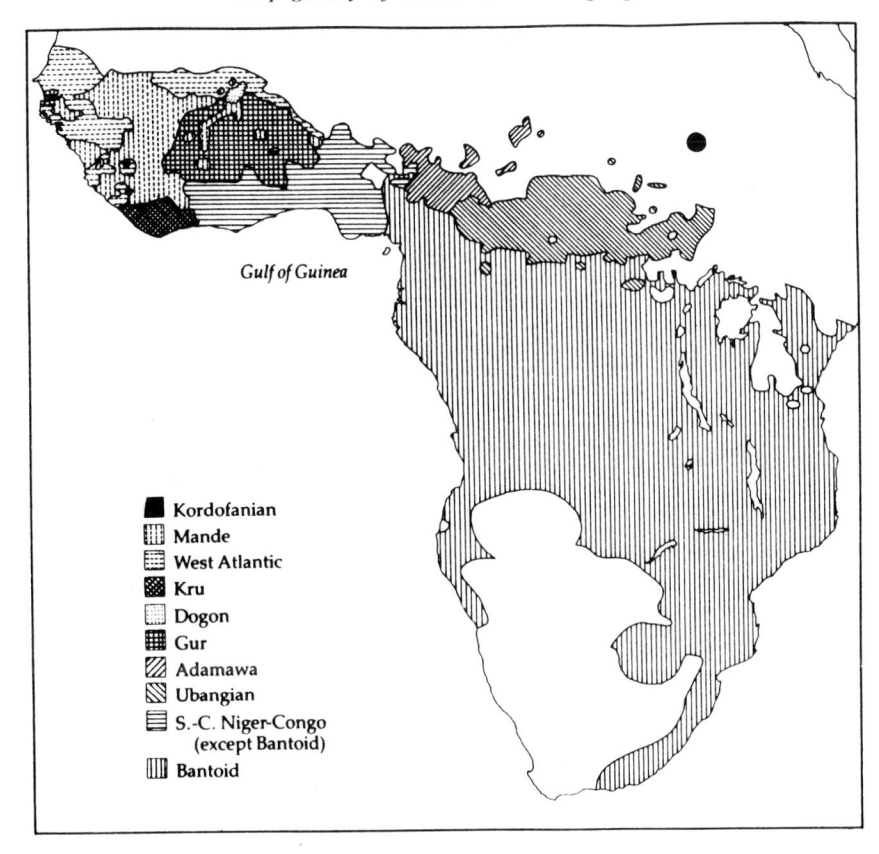

Fig. 19.3 The Niger–Kordofanian family

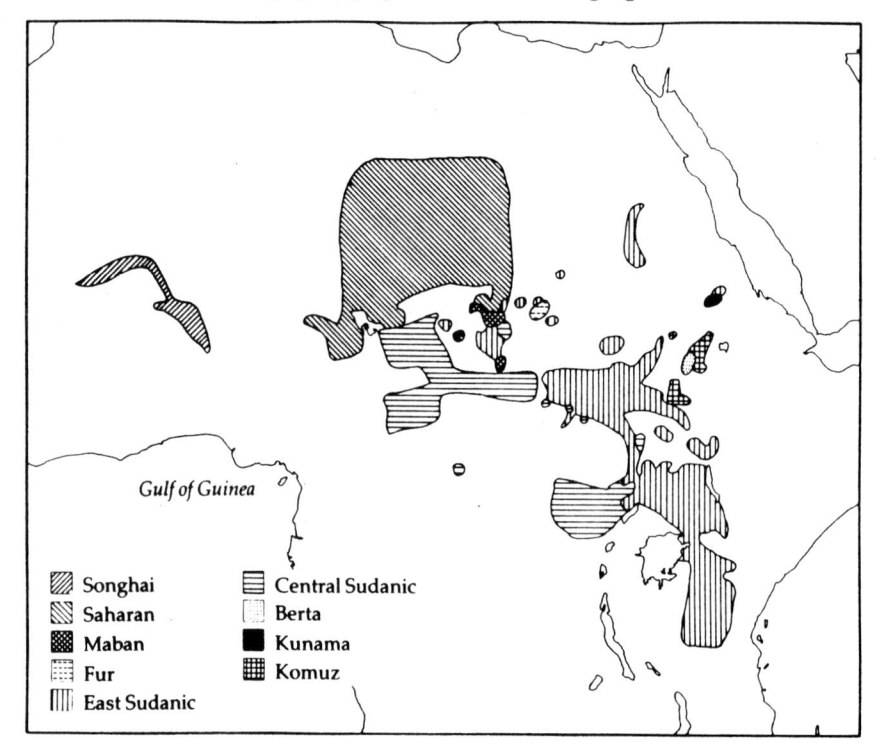

Fig. 19.4 The Nilo–Saharan family

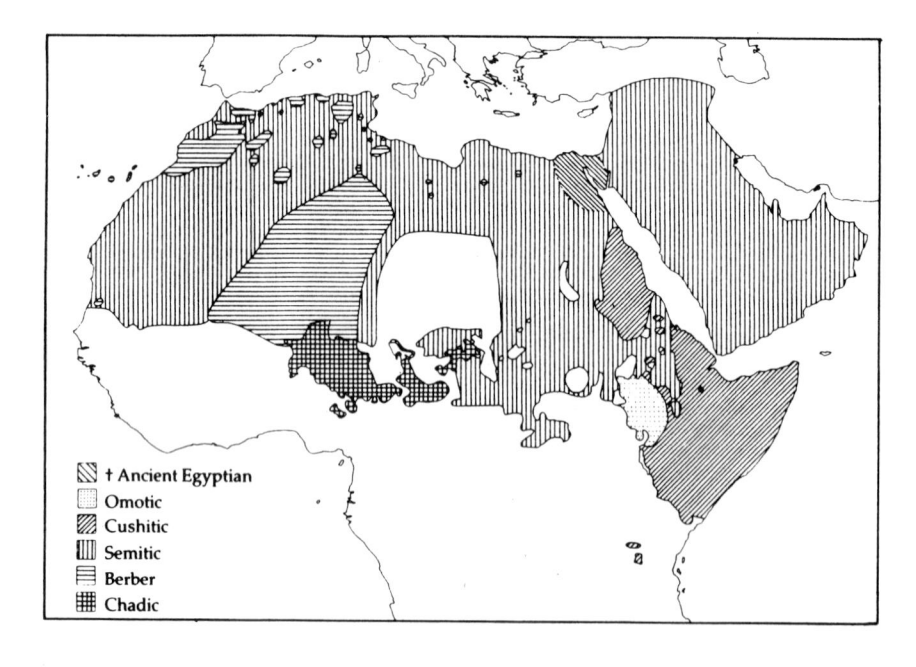

Fig. 19.5 The Afro-Asiatic family. Arabic is now spoken in the area where Ancient Egyptian once prevailed.

Map gallery of extant human languages 537

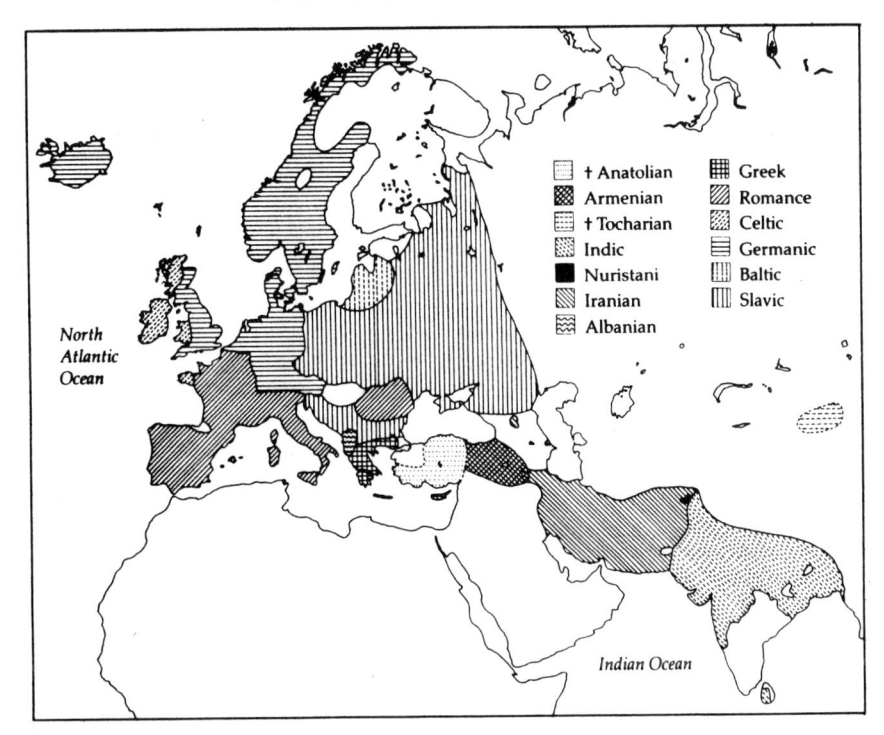

Fig. 19.6 The Indo-Hittite family. Turkish is now spoken where the Anatolian languages were spoken; several Turkic languages have supplanted Tocharian

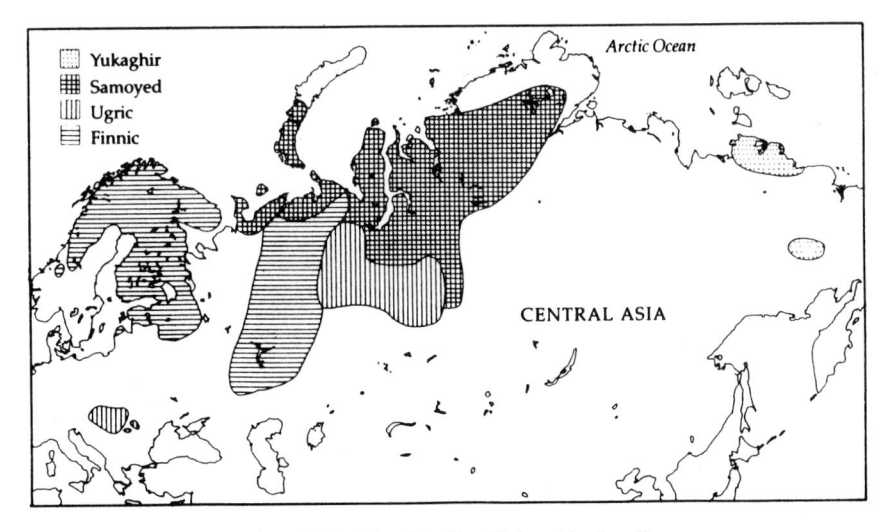

Fig. 19.7 The Uralic–Yukaghir family

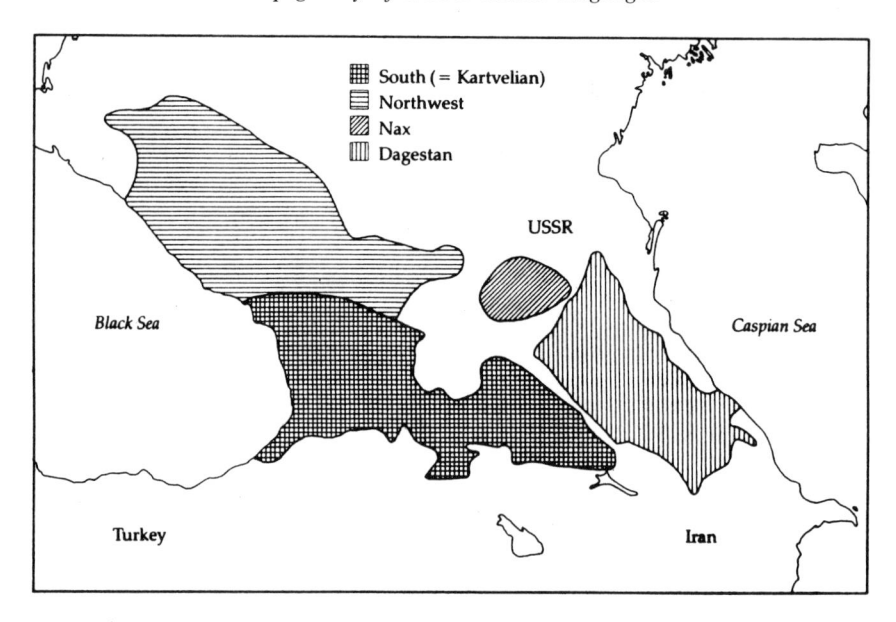

Fig. 19.8 The Caucasian family

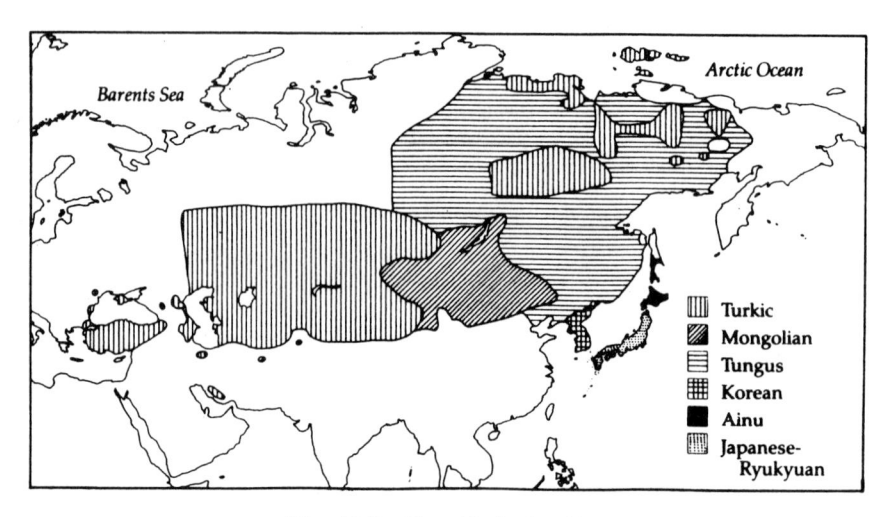

Fig. 19.9 The Altaic family

Map gallery of extant human languages 539

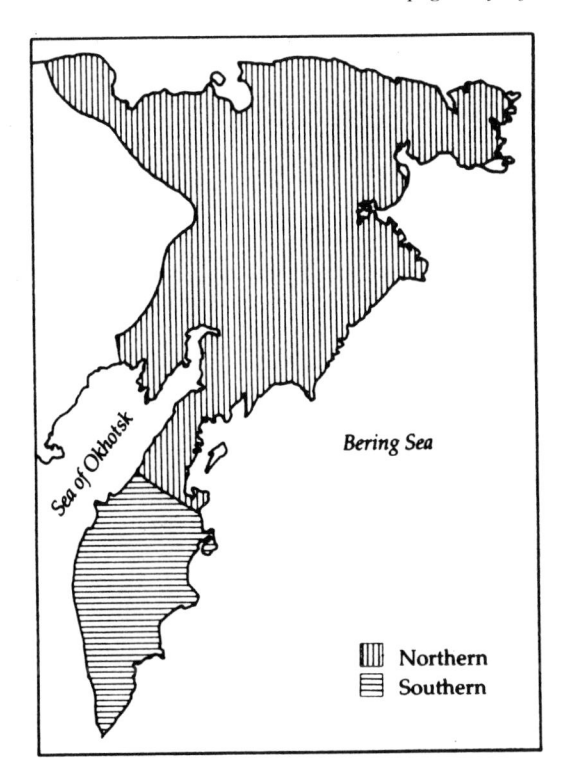

Fig. 19.10 The Chukchi–Kamchatkan family

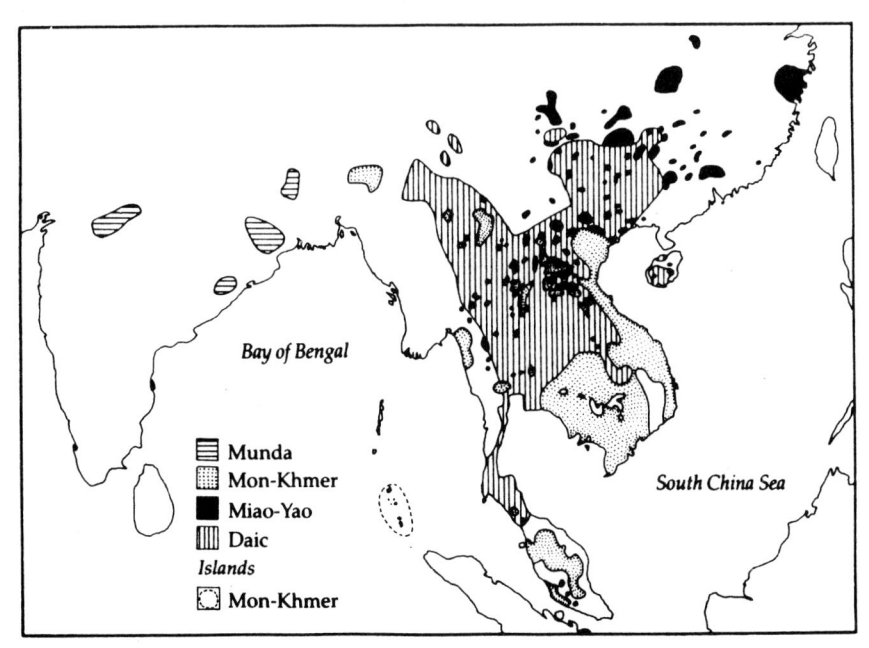

Fig. 19.11 The Miao–Yao, Austrosiatic, and Daic families

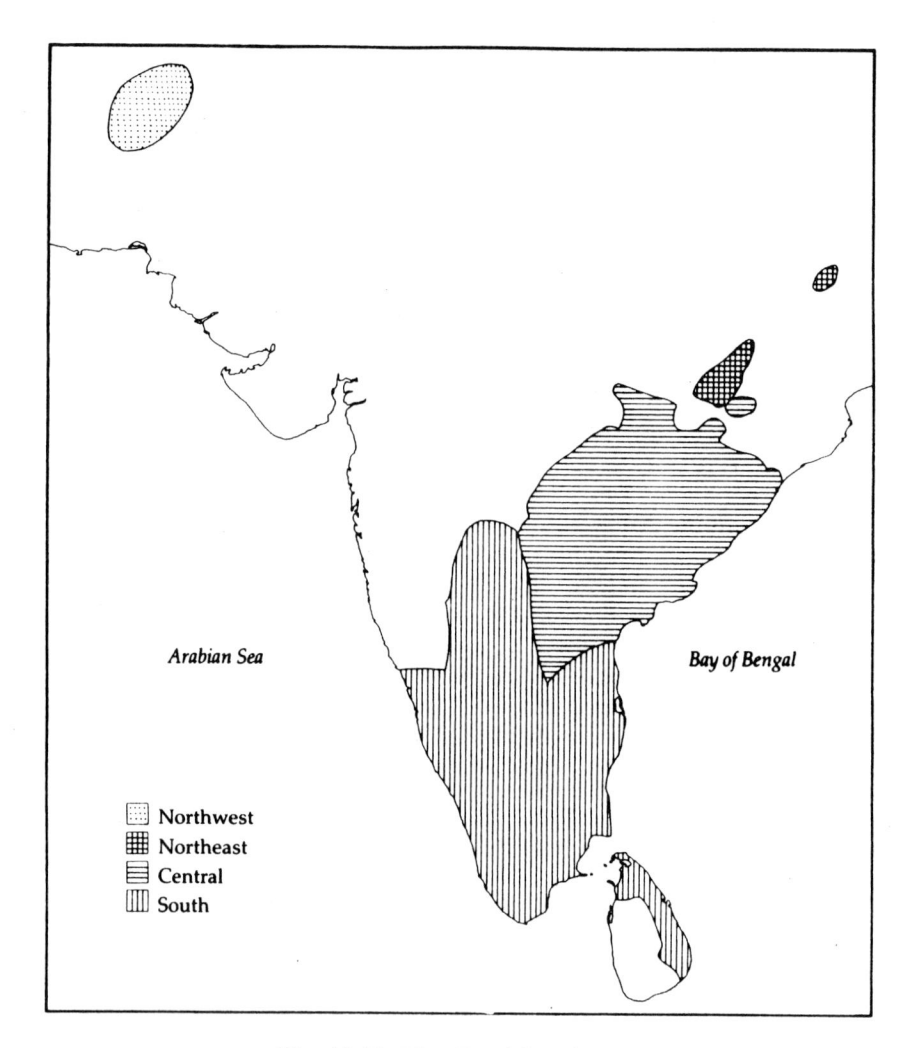

Fig. 19.12 The Dravidian family

Map gallery of extant human languages

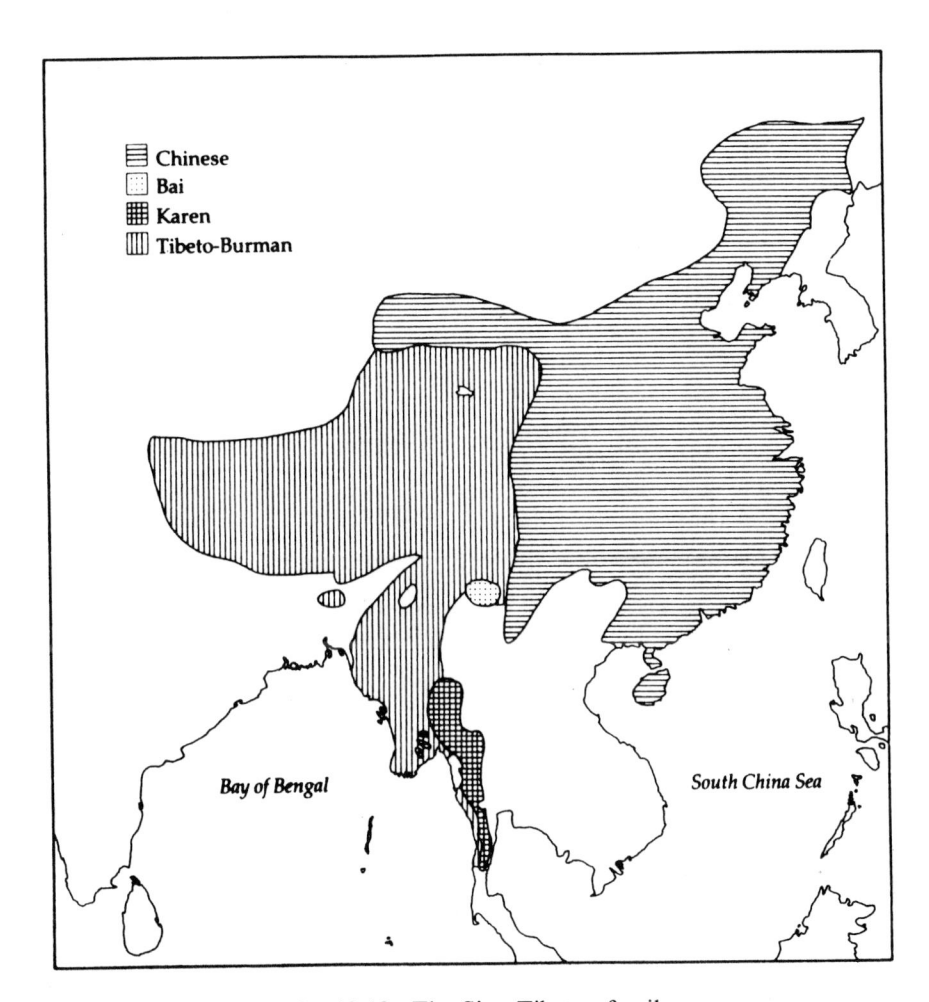

Fig. 19.13 The Sino-Tibetan family

542 *Map gallery of extant human languages*

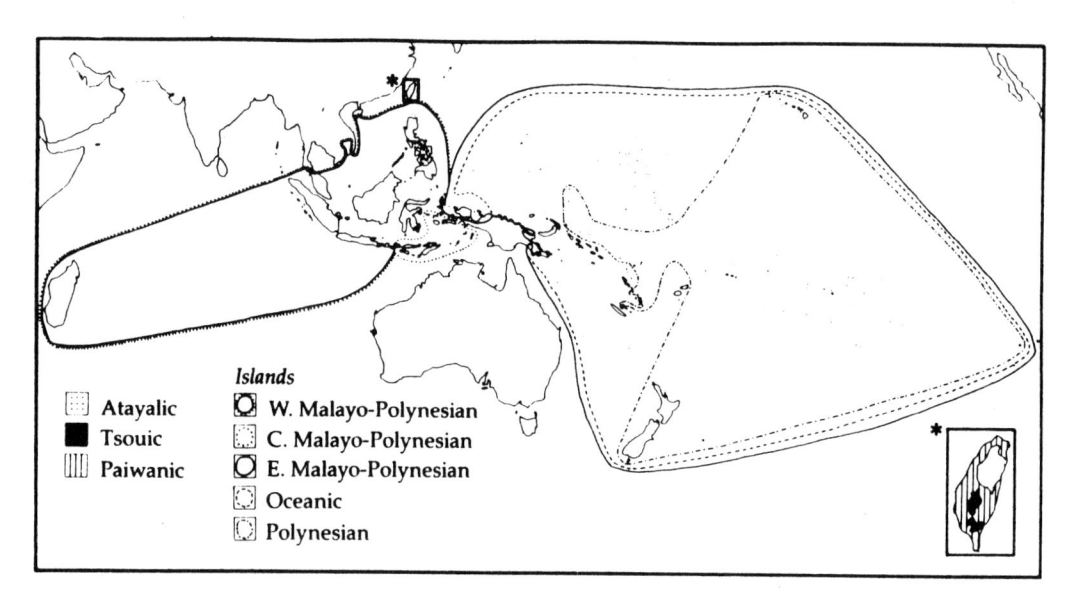

Fig. 19.14 The Austronesian family

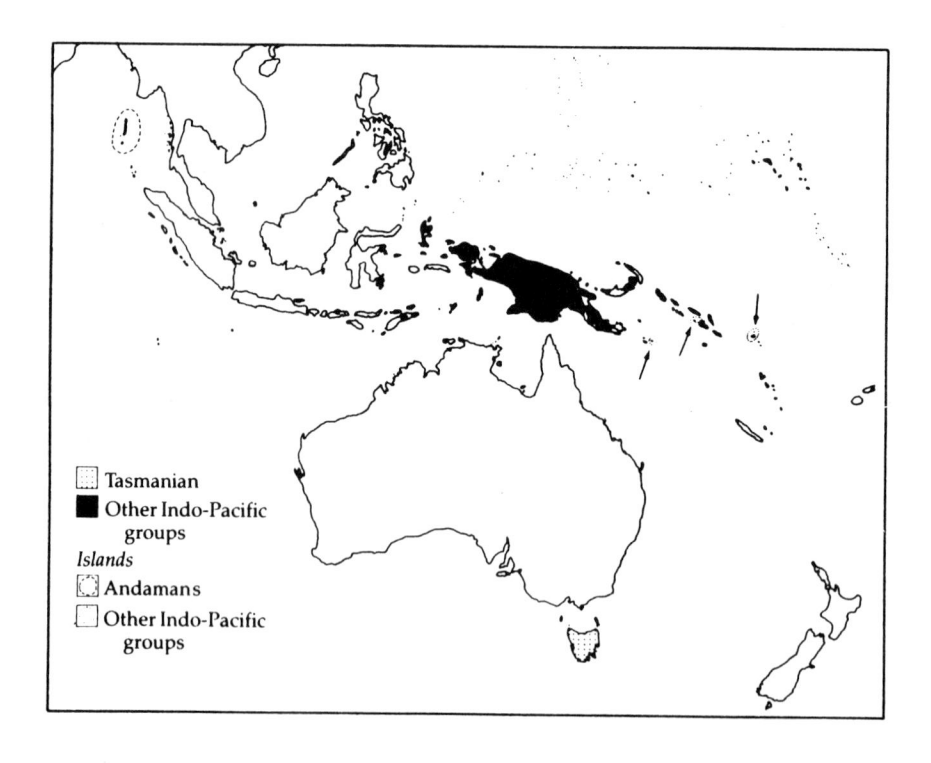

Legend:
- Atayalic
- Tsouic
- Paiwanic

Islands
- W. Malayo-Polynesian
- C. Malayo-Polynesian
- E. Malayo-Polynesian
- Oceanic
- Polynesian

Fig. 19.15 The Indo-Pacific family

Legend:
- Tasmanian
- Other Indo-Pacific groups

Islands
- Andamans
- Other Indo-Pacific groups

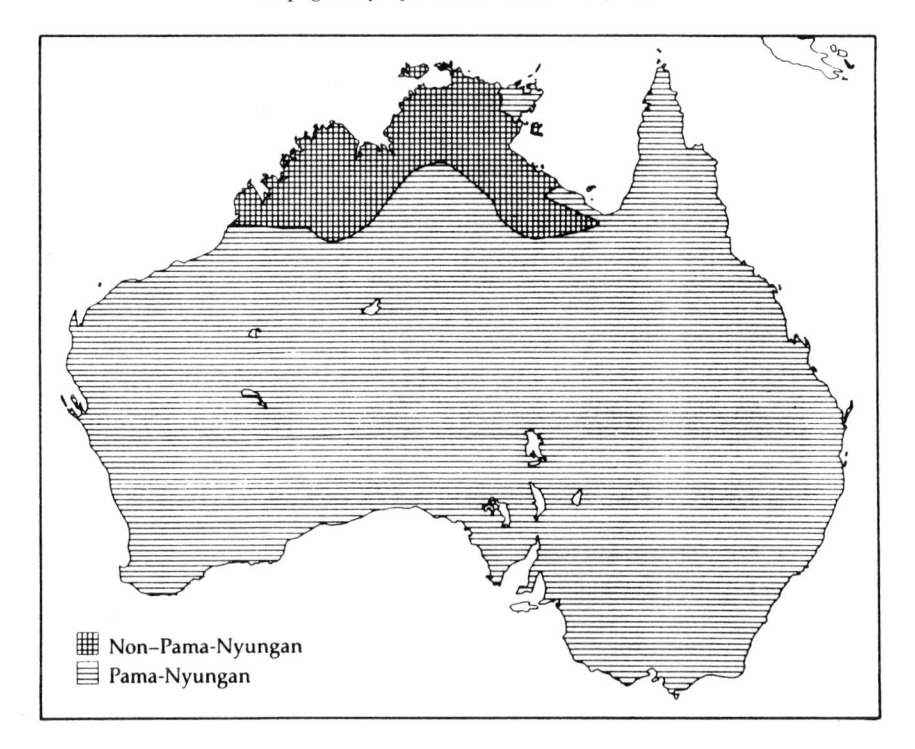

Fig. 19.16 The Australian family

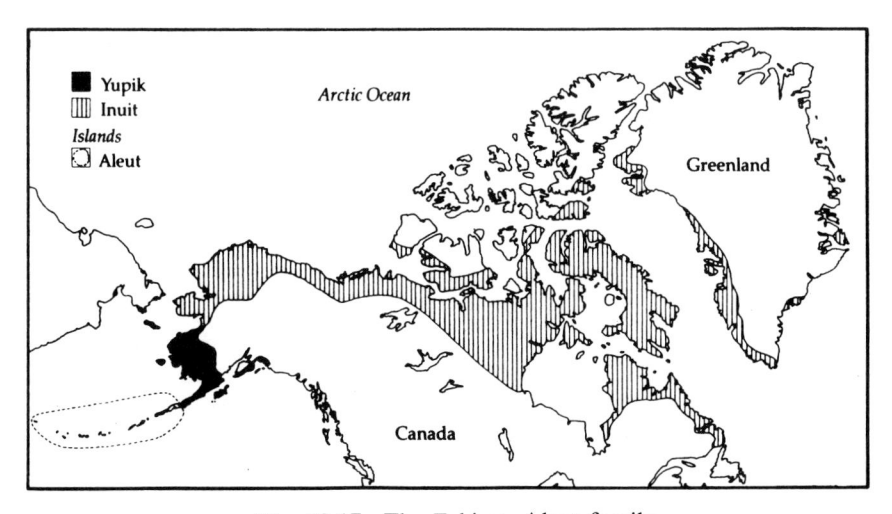

Fig. 19.17 The Eskimo–Aleut family

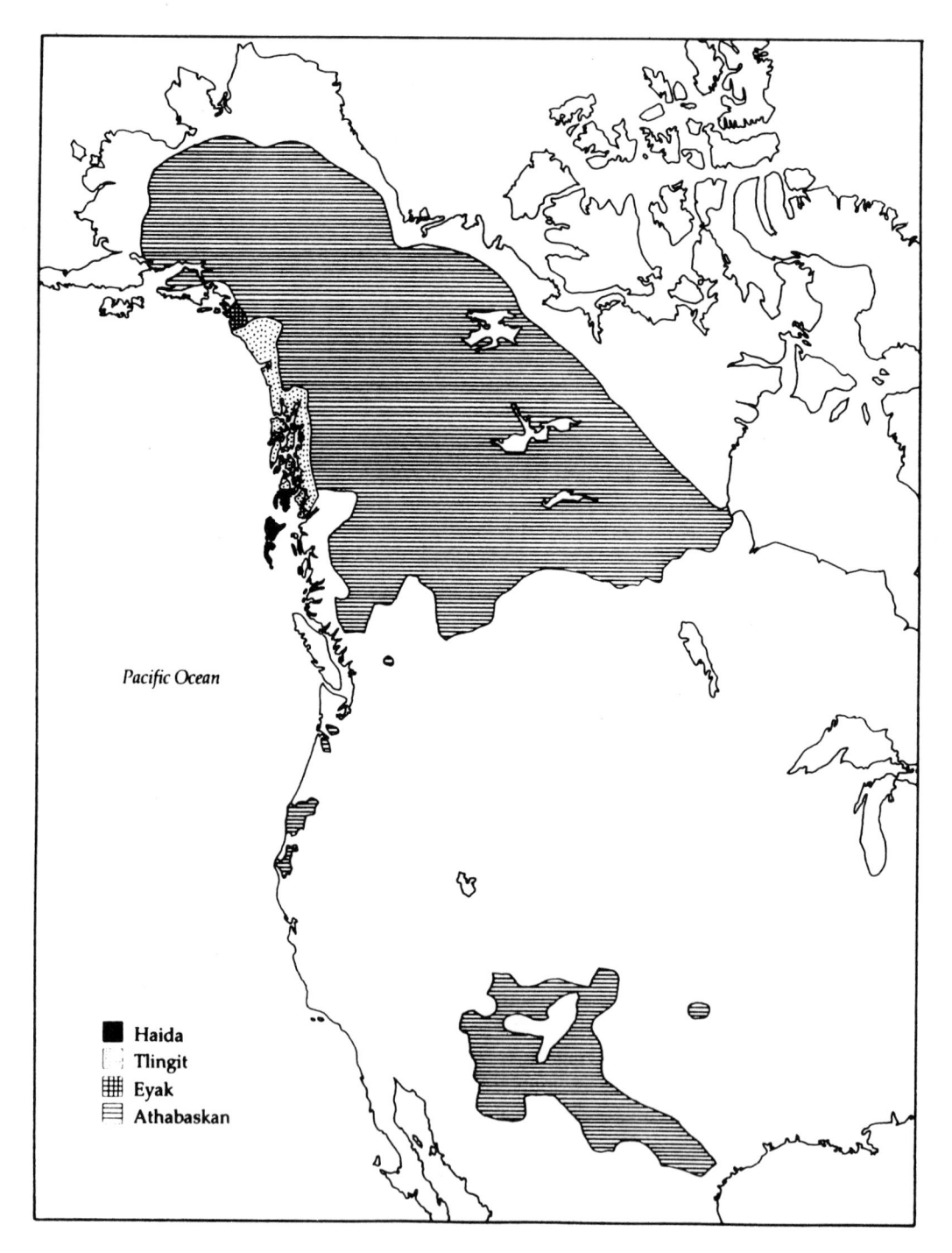

Fig. 19.18 The Na-Dene family

Amerind
Eskimo-Aleut
Na-Dene

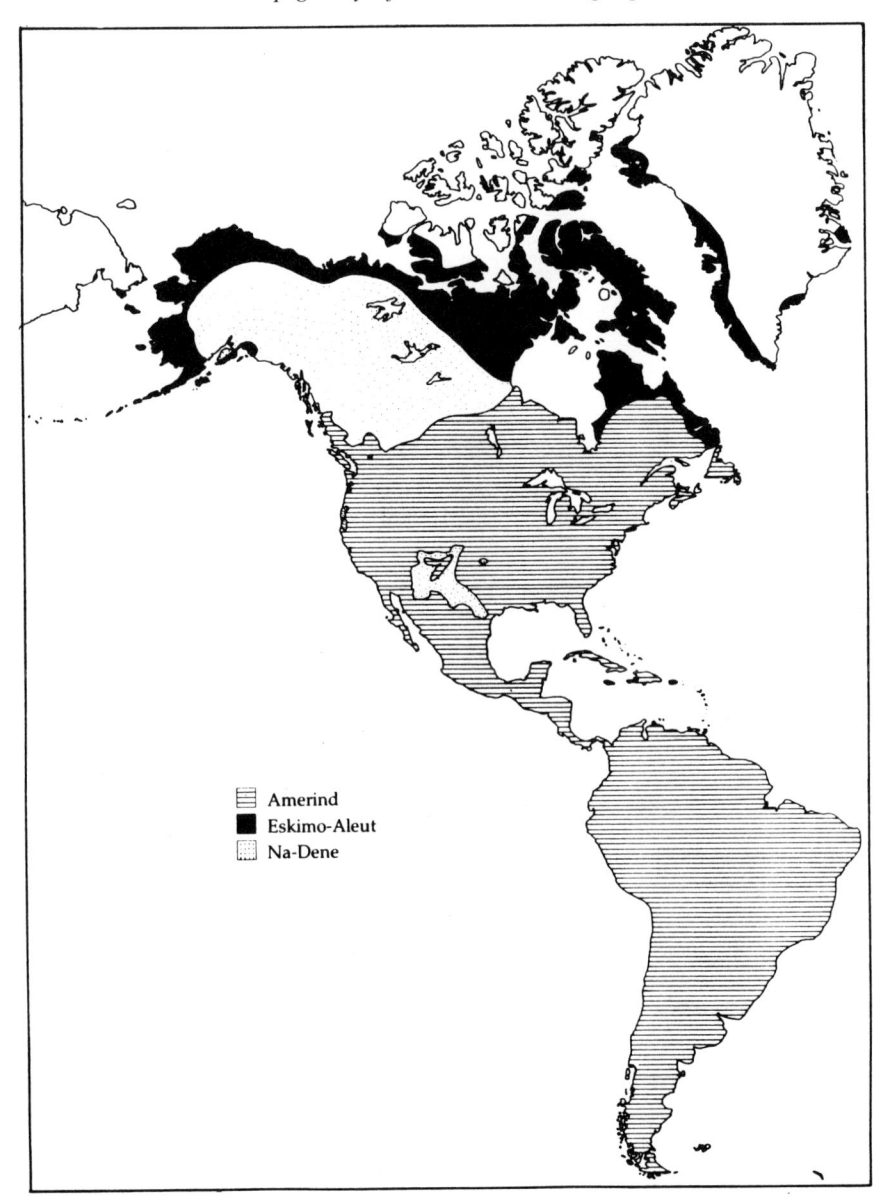

Fig. 19.19 The Amerind family

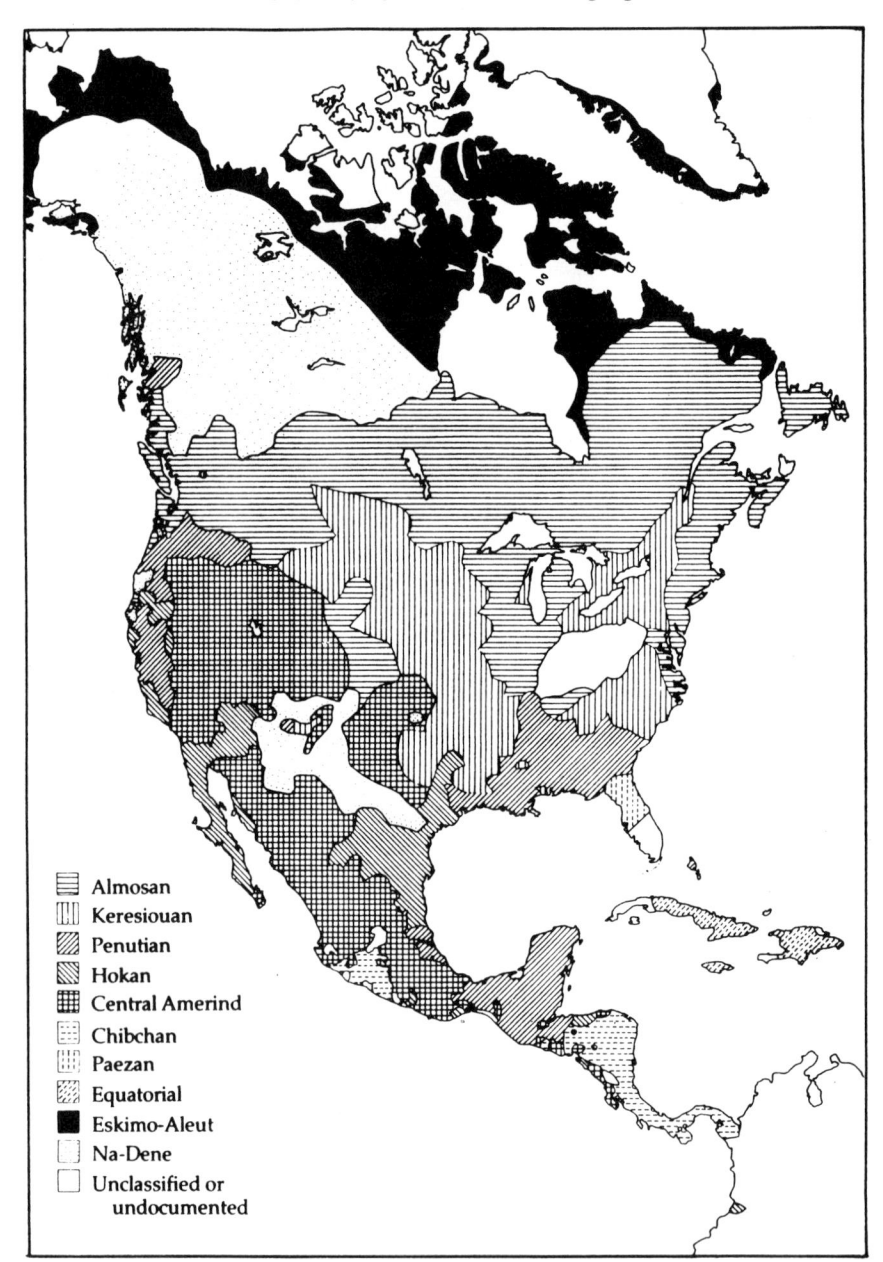

Fig. 19.20 The Amerind family, North and Central America

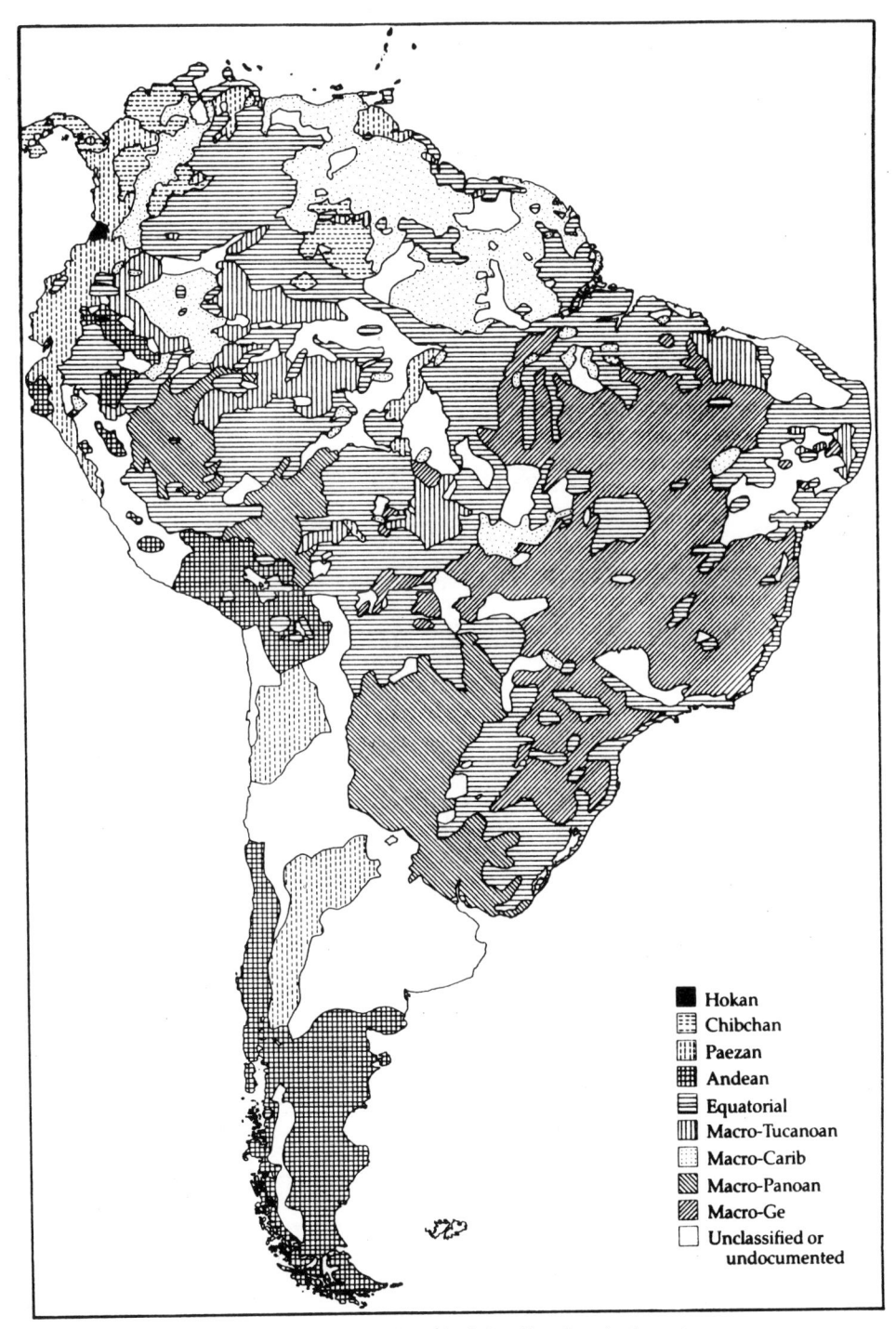

Fig. 19.21 The Amerind family, South America

Table 19.1 Overview of language phyla

Phylum	Number of extant languages	Number of speakers	Location or country	Better-known languages
KHOISAN:	31	120 000	South Africa, Namibia, S Angola, Botswana, N Tanzania	Nama (= Hottentot)
NIGER-KORDOFANIAN: Fula (= Fulani), Mandinka, Yoruba, Igbo	1064	181 000 000	Central and S Africa	Fula (=Fulani, Mandinka, Yoruba, Igbo, Swohini, Kongo, Luganda, Rwanda, Shona, Tswana, Xhosa, Zulu
NILO-SAHARAN:	138	11 000 000	Central Africa	Kanuri, Luo, Nubian, Maasai, Songhai
AFRO-ASIATIC:	241	175 000 000	N Africa, Near East	
†Ancient Egyptian	–	–	Egypt	†Ancient Egyptian
Berber	30	11 000 000	Algeria, Morocco, Tunisia, Libya, Mauritania, Senegal	Shilha, Kabyle, Riff, Tuareg, Tamazight
Chadic	123	30 000 000	Chad, Niger, Ghana, Nigeria, Cameroon, Central African Republic, Togo, Benin	Hausa
Omotic	34	1 000 000	W Ethiopia, N Kenya	Ometo
Cushitic	35	12 000 000	Somalia, Ethiopia, Sudan, Kenya, Tanzania	Somali, Oromo
Semitic	19	121 000 000	N Africa, Near East	Arabic, Hebrew, Aramaic, Tigrinya
CAUCASIAN:	38	5 000 000	Caucasus (USSR)	Georgian
INDO-HITTITE:	144	2 billion	Europe, SW Asia, India, Americas, Australia, South Africa, New Zealand	
†Anatolian	–	–	Turkey	†Hittite
Armenian	1	5 000 000	USSR	Armenian
†Tocharian	–	–	W China	†Tocharian
Indo-Iranian	93	700 000 000	Iran, Afghanistan, Pakistan, India	Romany, Farsi (= Persian), Kurdish, Pashto, Punjabi, Gujarati, Hindi-Urdu, Marathi, Bengali
Albanian	1	4 000 000	Albania	Albanian
Greek	2	10 000 000	Greece	Greek
Italic	16	500 000 000	Romania, Italy, France, Spain, Portugal, Central and South America	†Latin, Romanian, Italian, French, Provençal, Catalan, Spanish, Portuguese
Celtic	4	2 500 000	Ireland, Wales, N France	Irish, Welsh, Breton
Germanic	12	450 000 000	Germany, Holland, Scandinavia, Great Britain, North America, Australia, New Zealand, South Africa	German, Yiddish, Dutch, Afrikaans, English, Damish, Swedish, Norwegian, Icelandic

Table 19.1 (contd.)

Balto-Slavic	15	290 000 000	USSR, Poland, Czechoslovakia, Yugoslavia, Bulgaria	Lithuanian, Latvian, Russian, Ukrainian, Byelorussian, Polish, Czech, Slovak, Serbo-Croatian, Bulgarian
URALIC-YUKAGHIR:	24	22 000 000	Finland, Estonia, Hungary, USSR	Hungarian, Finnish, Saami (= Lapp), Estonian
ALTAIC:	63	250 000 000	Asia	
Turkic	31	80 000 000	Turkey, USSR, Iran	Turkish, Uzbek, Uighur, Azerbaijani, Turkmen, Tatar, Kazakh, Kirghiz, Chuvash, Bashkir
Mongolian	12	3 000 000	Mongolia, China, USSR	Khalkha (= Mongolian)
Tungus	16	80 000	E USSR, China	Manchu, Evenki
Korean	1	55 000 000	Korea	Korean
Japanese–Ryukyuan	2	115 000 000	Japan	Japanese
Ainu	1	few	N Japan, S Sakhalin Island (USSR)	Ainu
CHUKCHI-KAMCHATKAN:	5	23 000	NE Siberia (USSR)	Chukchi
ESKIMO-ALEUT:	9	85 000	Alaska, N Canada, Greenland, NE USSR	Eskimo, Aleut
ELAMO-DRAVIDIAN:	28	145 000 000	S and E India, S Pakistan	Telugu, Kannada, Tamil, Malayalam
SINO-TIBETAN:	258	1 billion	China, Tibet, Nepal, India, Burma, Thailand, Laos	Mandarin, Wu, Yue (= Cantonese), Tibetan, Burmese, Karen
AUSTRIC:	1175	293 000 000	SE Asia, Oceania	
Miao–Yao	4	7 000 000	S China, N Vietnam, N Laos, N Thailand	Miao, Mien (= Yao)
Austroasiatic:	155	56 000 000	NE India, SE Asia	
Munda	17	6 000 000	NE India	Santali, Mundari
Mon–Khmer	138	50 000 000	NE India, SE Asia, Nicobar Islands	Vietnamese, Mon, Khmer (= Cambodian)
Daic	57	50 000 000	S China, SE Asia	Thai (= Siamese), Lao (= Laotian), Li, Shan, Zhuang, Kam
Austronesian:	959	180 000 000	Oceania, S Vietnam, Madagascar	
Western	533	179 000 000	Madagascar, Formosa, Indonesia, Philippines, S Vietnam, Kampuchea	Malagasy, Javanese, Sundanese, Malay, Tagalog, Cebuano, Ilokano, Hiligaynon
Eastern (= Oceanic)	426	1 500 000	Melanesia, Micronesia, Polynesia	Fijian, Samoan, Tahitian, Hawaiian
INDO-PACIFIC:	731	2 735 000	New Guinea, Timor, Alor, Pantar, Halmahera, New Britain, New Ireland, Bougainville, Solomons, Reef Islands, Santa Cruz, Andaman Islands, Tasmania	†Tasmanian, Enga, Wantoat, Telefol, Iatmul, Asmat

Table 19.1 (contd.)

Phylum	Number of extant languages	Number of speakers	Location or country	Better-known languages
AUSTRALIAN:	170	30 000	Australia	Western Desert Language
NA-DENE:	34	202 000	Alaska, W Canada, Oregon, California, Arizona, New Mexico	Navajo, Apache
AMERIND:	583	18 000 000	North, Central, and South America	
Kutenai	1	200 (in 1977)	Montana, Idaho	Kutenai
Algic:	16	91 000	Canada, USA	
Ritwan	1	10 (in 1980)	N California	Yurok
Algonquian	15	91 000	Central and E Canada, Central and E USA	Ojibwa, Cree, Blackfoot, Cheyenne
Mosan:	27	9 500	NW USA, SW Canada	
Chimakuan	1	10 (in 1977)	NW Washington	Quileute
Wakashan	6	2 700	SW Canada	Nootka, Kwakwala (= Kwakiutl)
Salish	20	6 800	SW Canada, NW USA	Shuswap, Kalispel, Squamish
Keresan	2	7 000	New Mexico	Keres
Siouan-Yuchi	11	21 000	Central USA	Dakota, Crow
Caddoan	4	1 000	Central USA	Wichita, Pawnee
Iroquoian	7	15 000	E USA	Cherokee, Mohawk
Penutian	68	3 200 000	W Canada, W and SE USA, S Mexico	Chinook, Zuni, Muskogee (= Creek), Quiche, Cakchiquel, Kekchi, Mam, Yucatec
Hokan	28	55 000	California, Arizona, Texas, Mexico, Colombia	Mohave, Yuma, Tlapanec, Tequistlatec
Tanoan	7	7 400	New Mexico, Oklahoma	Kiowa, Tewa
Uto-Aztecan	25	1 100 000	W USA, Mexico	Comanche, Hopi, Nahuatl (= Aztec)
Oto-Manguean	17	1 700 000	S Mexico	Otomi, Mixtec, Zapotec
Chibchan-Paezan	43	200 000	Florida, S Mexico, Central America, W South America	Tarascan, Yanomami, Guaymi, Cuna, Paez, Warao, Embera, Cayapa
Andean	18	8 500 000	W South America	Quechua, Aymara, Mapudungu (= Mapuche)
Macro-Tucanoan	47	35 000	NW and E South America	Ticuna, Tucano, Nambikuara, Puinave
Equatorial	145	3 000 000	South America, Caribbean	Guarani, Tupi, Goajiro, Arawak
Macro-Carib	47	50 000	N South America	Galibi (= Carib), Witoto
Macro-Panoan	49	50 000	W South America	Toba, Tacana
Macro-Ge	21	10 000	E South America	Bororo, Chavante

† Languages no longer extant.

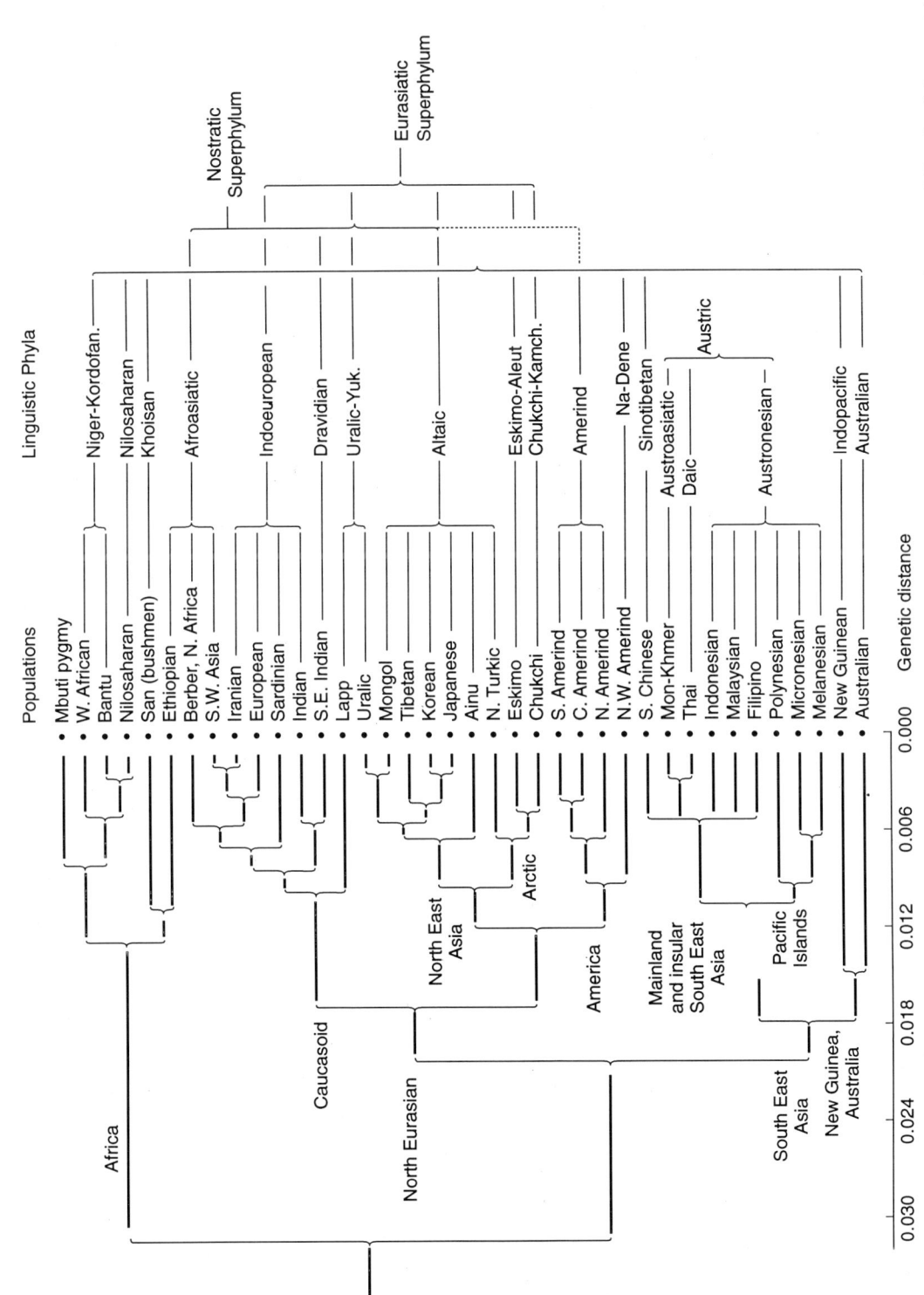

Fig. 19.22 Comparison of a genetic tree (based on enzyme allele frequencies) representing 42 populations of the world's aborigines, with linguistic phyla (based on Ruhlen 1987) (modified from Cavalli-Sforza et al. 1988). 'Genetic distance' is a measure that gives an indication of how distinct one aboriginal population is from another, and provides a basis for reconstructing the phylogeny of contemporary populations. Average genetic distances between these populations are claimed to be proportional to archaeologically-attested separation times. Linguistic phyla closely correspond to groups of populations with few, and potentially explainable (on the basis of contact, etc.) exceptions, and thus their origins may be given a time-frame. See Cavalli-Sforza et al. (1988) for details; Bateman et al. (1990) for criticisms and a reply by Cavalli-Sforza et al.; and Ruhlen (1992) for a fuller review.

References and further reading

Bateman, R., Goddard, I., O'Grady, R., Funk, V. A., Mooi, R., Kress, W. J., and Cannell, P. (1990). Speaking of forked tongues: the feasibility of reconciling human phylogeny and the history of language. *Current Anthropology*, **31**, 1-24, 177-83 and 313-16.

Cavalli-Sforza, L. L., Piazza, A., Menozzi, P., and Mountain, J. (1988). Reconstruction of human evolution: bringing together genetic, archaeological, and linguistic data. *Proceedings of the National Academy of Sciences, USA*, **85**, 6002-6.

Greenberg, J. H. (1992). Preliminaries to a systematic comparison between biological and linguistic evolution. In *The evolution of human languages* (ed. J. A. Hawkins and M. Gell-Mann), pp. 139-58. Addison-Wesley, Reading, Mass.

Moore, J. H. (1994). Putting anthropology back together again: the ethnogenetic critique of cladistic theory. *American Anthropologist*, **96**, 925-48.

Ruhlen, M. (1987). *A guide to the world's languages*, Vol. 1. Stanford University Press.

Ruhlen, M. (1992). An overview of genetic classification. In *The evolution of human languages* (ed. J. A. Hawkins and M. Gell-Man), pp. 159-89. Addison-Wesley, Reading, Mass.

20
Spoken language and sign language

Margaret Deuchar

Abstract

'Language' has been defined in three ways: by listing its 'design features'; its structural properties, particularly its 'rule-governed creativity'; and its uses or functions.

In terms of design features, spoken and signed languages differ trivially in terms of the latter's not using the vocal–auditory channel; but have been claimed to differ more significantly in the extent to which spoken language is composed of arbitrary signs, while sign languages are based on more iconic signs. This has led to an erroneous demotion of the status of sign languages. A more careful analysis shows that both types of language are comparable on this dimension in their contemporary forms, although there is some evidence that languages of both media have become more arbitrary over time. Similarly, structural analyses of both systems reveal they show a similar degree of 'duality of patterning', both below the structure of the word or sign, and above it at the level of grammar.

Children go through very similar stages in acquiring either system when it is the 'natural' language of their early environment, although at different rates. Initially sign language is learned earlier; but later this advantage diminishes. However, most deaf children learn sign language under unusual conditions, since most ($c.90$ per cent) do not have native signing parents. Such children are comparable to those 'learning' spoken creoles on the basis of pidgin inputs: features of their signing system are creations of the 'learners' themselves. This provides some support for Chomsky's contention that language acquisition is not heavily dependent on the nature of the linguistic input.

Creoles are languages which have developed out of pidgins into native languages. Structural similarities between spoken creoles and sign languages may reflect language universals. This has been claimed as supporting the existence of an innate language faculty that strongly constrains the properties of individual languages. An alternative argument is that language structure is more constrained by functional demands, structural characteristics resulting as compromises between the need for a usable language to be clear, cognitively processable, 'quick and easy', and expressive. These two theories may well complement each other: the initial stages of language (both developmentally and historically) might be accounted for in terms of what is biologically given, whereas later language may change more in response to its expanding uses [eds].

20.1 Introduction

This chapter will take a look at the properties of language, both spoken and signed, with the aim of determining to what extent these are dependent on the medium in which they are produced. Having determined what structural properties all languages have in common, I shall then discuss how languages evolve in two different ways, both in children acquiring them, and as new linguistic systems. I shall note the special conditions under which sign languages are acquired, and how this is related to their status as newly emergent languages. A comparison between sign languages and spoken creoles will lead to a discussion of the possible explanations for the development and evolution of language.

20.2 Spoken and signed languages

In this chapter we shall be considering both spoken and signed languages in order to understand as much as we can about language in general. By 'spoken languages' I mean languages that are primarily manifested in the vocal–auditory medium, which means most of the world's languages as we know them today. This includes familiar languages such as English and French as well as less familiar ones (in our culture) such as Hopi (an American Indian language) or Yoruba (a language spoken in West Africa). By 'signed languages' I mean sign languages of the deaf: visual–gestural systems which are used by deaf people as first languages. I shall use the term 'sign language' to refer specifically to these languages of the deaf. In this chapter, the term will not also have the broader reference, found elsewhere, to systems as diverse as sign languages as used by aborigines or Indians (cf. for example Umiker-Sebeok and Sebeok 1978), individual gestures used to accompany speech (cf. for example Morris *et al.* 1979), and the rather unsystematic gestures which we may use when we do not share a spoken language with someone, for example on a holiday abroad. What these usages have in common is that they refer to situations where the users have a primary spoken language, but where gestures are being used as a substitute for, or as an accompaniment to, speech. Sign languages of the deaf, on the other hand, are used by people who do not have a primary spoken language (because of their deafness); and they are quite independent of spoken languages. Although special sign systems have been developed by educators for the purpose of teaching spoken languages to deaf people, these are quite different from the sign languages evolved by the deaf themselves.

20.2.1 *Spoken languages*

There is a much longer research tradition for investigating spoken languages than for sign languages. Estimates of the number of spoken languages in the world vary, but Hudson (1980) suggests that there are four or five thousand (see also Foster, this volume, Chapter 25, and the Map gallery, Chapter 19). The problem with any estimate is that it is by no means clear exactly how dialects should be grouped into languages. It is somewhat clearer, however, that languages can generally be assigned to language families, based on their structural characteristics and the historical relations between them. The language family on which the greatest amount of research has been done is probably the Indo-European, which includes languages such as French, English, and Russian. Figure 20.1 gives some idea of the family relationships between the Indo-European languages.

20.2.2 *Signed languages*

While research on spoken languages extends back to the time of the Greeks, sign languages of the deaf have only received widespread attention from linguists for the past thirty years or so. Although there was an attempt (Stokoe 1973) to assign sign languages to families in a way similar to the family trees for spoken languages illustrated in Fig. 20.1, it is not clear that historical relationships of this kind are well motivated, given that sign languages must generally have developed in isolation from one another. A possible exception to this is the relation between French Sign Language and American Sign Language (ASL). It used to be taken for granted that American Sign Language was descended from French Sign Language because of contacts between American and French educators in the nineteenth century. However, as Woodward (1978) points out, creolization (see sections 20.6 and 20.7) is likely to have been involved in the development of ASL, and a direct historical relationship seems most unlikely. ASL is the sign language which to date has received the most attention from researchers; but considerable research has also been done on British Sign Language and Swedish Sign Language, for example. Table 20.1 contains a list of some of the sign languages on which research has begun, together with illustrative references. It is important to realize, as Table 20.1 shows, that sign languages throughout the world differ from one another, contrary to popular belief. The fact that British and American Sign Languages are different from one another, despite the fact that they are both used in English-speaking countries, is also proof of the point made earlier, that sign languages are independent of spoken languages.

20.3 Definition of language

Considerable efforts have been directed by linguists towards formulating a definition of language. Most such definitions do not take sign language into account; but a consideration of the extent to which they do apply to sign language is useful in determining

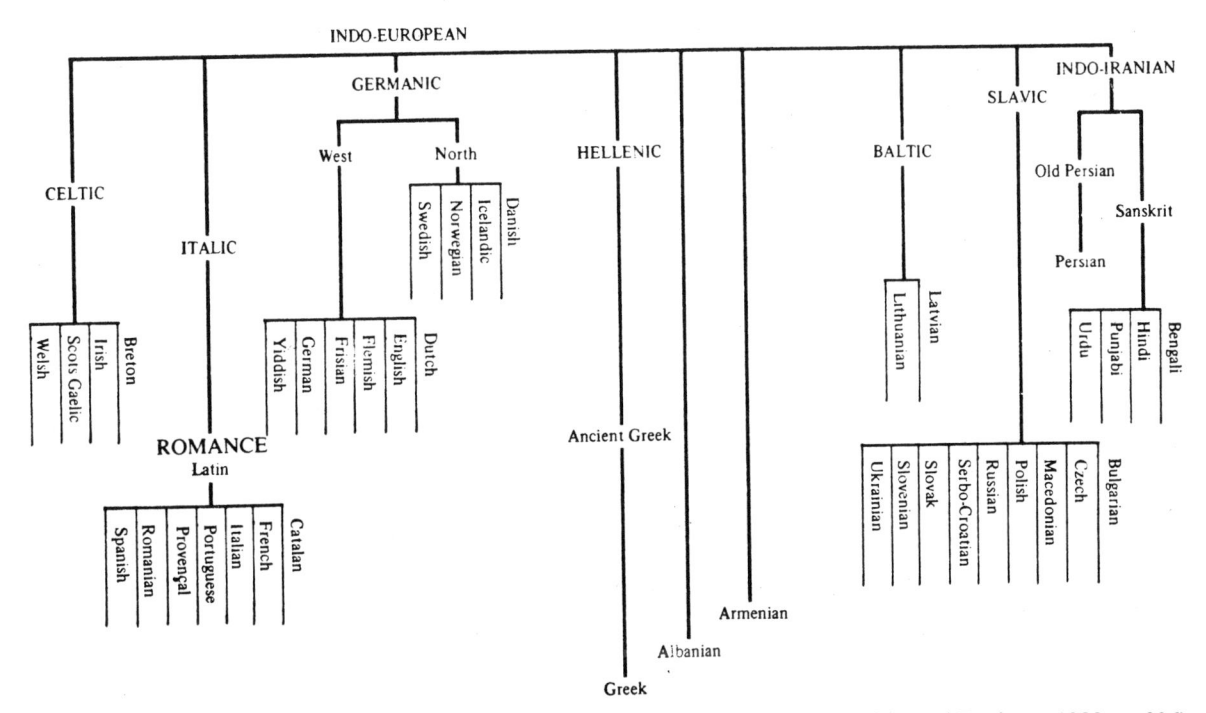

Fig. 20.1 Family relationships among the Indo-European languages (from Fromkin and Rodman 1983, p. 306).

what the essential properties of language are. Lyons (1981) lists a selection of twentieth-century definitions, which are given in Table 20.2.

It is notable that the first three definitions given in Table 20.2 assert or imply that language is a peculiarly human system. Since language has often been considered a uniquely human attribute, at least since the time of Rousseau, many scholars' motivation in defining language has been to determine how it differs from animal communication. The approach represented in these definitions focuses on what languages have in common rather than on how they differ, and a particular contribution to this question this century has been made by the structural linguist Charles

Table 20.1 Some sign languages that have been studied up to 1991

Sign language	Illustrative references
American Sign Language	Baker and Cokely (1980)
	Lane and Grosjean (1980)
Brazilian Sign Language	Hoemann et al. (1981, 1985)
British Sign Language	Deuchar (1984)
	Kyle and Woll (1983)
Danish Sign Language	Albertsen (1985)
	Engberg-Pedersen and Pedersen (1985)
Dutch Sign Language	Schermer (1985)
	Stroombergen and Schermer (1985)
French Sign Language	Moody (1983)
Italian Sign Language	Corazza et al. (1985)
Japanese Sign Language	Peng (1985)
Norwegian Sign Language	Vogt-Svendsen (1983)
Russian Sign Language	Zaitseva (1983)
Swedish Sign Language	Bergman and Wallin (1985)

Table 20.2 Definitions of language cited in Lyons (1981)

(i) 'Language is a purely human and non-instinctive method of communicating ideas, emotions and desires by means of voluntarily produced symbols' (Sapir 1921, 8).

(ii) 'A language is a system of arbitrary vocal symbols by means of which a social group co-operates' (Bloch and Trager 1942, 5).

(iii) Language is 'the institution whereby humans communicate and interact with each other by means of habitually used oral–auditory arbitrary symbols' (Hall 1968, 158).

(iv) Languages are 'symbol systems ... almost wholly based on pure or arbitrary convention' (Robins 1980, 13).

(v) A language is 'a set (finite or infinite) of sentences, each finite in length and constructed out of a finite set of elements' (Chomsky 1957, 13).

Hockett. Hockett (1960a) uses a set of thirteen design features to try and determine the differences between human and animal communication. He believes that this approach might throw some light on the question of the origin of language, on the assumption that human language evolved from some more basic, animal-like system. Whether or not we share this particular assumption, his work has led to some interesting conclusions as to what all languages have in common. Hockett's thirteen 'design features' are given in Hockett (1960a), and are a mixture of structural and functional properties. Hockett claims that while all thirteen features are found in human languages, animal communication systems have fewer than the total set. The thirteen features are listed in Table 20.3, together with a rough explanation of what each means.

Hockett compares a number of communication systems with regard to these features, and finds that only language possesses them all. Bee dancing, for example, lacks the vocal–auditory channel, arbitrariness, discreteness, and duality of patterning, and it is doubtful whether the song of the western meadowlark bird possesses interchangeability and semanticity. Hockett does not include human sign languages in his investigation; but we can see immediately that they would be ruled out as human languages because of their lack of a vocal–auditory channel. They do, however, have broadcast transmission and directional reception, at least if we allow the signals to be received visually rather than heard. Rapid fading is clearly found in sign language, in that the signals disappear shortly after being produced. Interchangeability is found, in that senders can be receivers; total feedback obtains in that speakers get visual and kinaesthetic feedback; and specialization and semanticity are found. Arbitrariness is found at least partially, but since it is one of the most debatable design features for sign languages, it will be further discussed in section 20.4. Discreteness is related to duality of patterning, and they will both be demonstrated for sign language in section 20.5. Displacement is found, in that signers can refer to things not actually present at the time of signing; and productivity is found, in that signers

Table 20.3 Hockett's design features

1. Vocal–auditory channel: produced by voice and received by ear.
2. Rapid fading: the signal fades soon after its production, leaving the channel free.
3. Broadcast transmission and directional reception: signals can be heard by anyone within range, and the source located.
4. Interchangeability: senders can be receivers and vice versa.
5. Total feedback: speakers can hear their own signals as they are produced.
6. Specialization: signals are purely communicative, having no other function.
7. Semanticity: there are fixed associations between elements of signals and what they refer to.
8. Arbitrariness: the connections between signals and their meanings are arbitrary.
9. Discreteness: the units of signals are distinct from one another.
10. Displacement: things remote in space or time can be referred to.
11. Productivity: novel combinations of signs are possible.
12. Traditional transmission: the system is transmitted by learning and teaching.
13. Duality of patterning: there is independent structure at at least two levels.

produce novel combinations of signs. Finally, traditional transmission or language learning will be discussed later, in Section 20.6, but generally proceeds in a way parallel to that in spoken language.

The only design feature which is totally absent in sign languages is that of a vocal–auditory channel, which, being a purely physical feature, may be considered fairly superficial. Lyons (1972) argues that language can actually be considered independently of the medium in which it is manifest, and suggests that a non-vocal, visual communication system might indeed be a language, though he does not appear to have sign language in mind. The features arbitrariness and duality of patterning will be discussed further below; but meanwhile we may briefly look at approaches to the definition of language which do not focus mainly on the distinction between language and non-language. Such approaches may be roughly divided into two categories, those which focus on the structure of language, and those which focus on its function.

20.3.1 *Focus on structure*

The standard approach to language by linguists involves a focus on structure rather than function, and is particularly well represented in its current form by the work of Chomsky (for example 1965, 1972), whose definition of language as it appeared in his influential *Syntactic structures* (1957) is quoted as no. (v) in Table 20.2. Chomsky's main focus in his definition, as indeed in much of his work, is on the infinite use of finite means, or the way in which an infinite number of combinations can be produced from the finite building-blocks of language, or sounds. This reminds us of Hockett's feature of productivity, which he explains in the following way: 'Language is open, or 'productive', in the sense that one can coin new utterances by putting together pieces familiar from old utterances, assembling them by patterns of arrangement also familiar in old utterances' (Hockett 1960a, p. 90).

Although Hockett's article appeared after Chomsky's book, Chomsky's focus represents the way in which he has clearly selected from the structuralist tradition which he inherited, of which Hockett was one of the main exponents. Chomsky's focus has come to be described as 'creativity', or more specifically, 'rule-governed creativity' (cf. Lyons 1981, p. 23), since Chomsky's particular contribution has been in formulating grammatical rules. This 'creative' or 'productive' characteristic of language is clearly related to Hockett's 'duality of patterning', since there are rules about the way in which the finite set of elements at one level are combined in infinite ways to form the next level up. We shall explore how this works for both spoken and sign languages in Section 20.4.

20.3.2 *Focus on function*

Functional approaches to language tend to concentrate on producing taxonomies of language functions as an end in themselves, rather than on using such functions as criteria for the definition of language. However, if one were to seek a definition of language in functional terms, one might well look at one of the best-known classifications of language functions, that produced by Jakobson (1960). Jakobson isolates six factors which he considers to be constitutive of any speech event, and then identifies six functions, each of which involves a focus on one of the six factors. Figure 20.2 is a conflation of two figures from Jakobson (1960): the six factors are indicated in uppercase letters, and their associated functions in lowercase letters.

The referential function, which is often taken by linguists to be the central function of language, and which is sometimes also described by terms such as 'cognitive' or 'descriptive', involves the conveying of information from the addresser to the addressee about a particular referent (person or thing) in the context. The narration of events in one's life, or the material in academic textbooks, would be an example of the use of language to fulfil the referential function. The emotive function involves the use of language to express the feelings of the addresser, as in interjections like 'Ouch' to express pain, or statements like 'that hurts', which may fulfil the same function. The conative function involves the use of language by the addresser to get the addressee to do something, as in orders like 'Close the door', or requests like 'Could you help me move this furniture?' The poetic function

```
                    CONTEXT
                    referential

                    MESSAGE
                    poetic
ADDRESSER                                      ADDRESSEE
emotive             CONTACT                    conative
                    phatic

                    CODE
                    metalingual
```

Fig. 20.2 Six factors constitutive of speech events, with the language functions characterized by focusing on each of them (adapted from Jakobson 1960, pp. 353 and 357).

involves the use of language for aesthetic purposes, as in poetry or literature, where Jakobson says that the focus is on the message for its own sake. The phatic function of language is fulfilled when language is used merely to establish contact with the addressee, without the content of what is being said being important. This is true of comments about the weather in British culture, where the main purpose of an expression like 'It's a nice day today' is to make contact, rather than to convey information about an obvious state of affairs. Finally, the metalingual function of language is fulfilled when there is focus on the code, and language is used to talk about itself, as in this chapter or in everyday language where people may ask one another, for example, about the meaning of unfamiliar words.

Since we shall not discuss the functions of sign languages later to the same extent that we shall discuss their structure, it is probably worth pointing out at this stage that just as English, or French, or Swahili is used for all six of Jakobson's functions, this is also true of the sign languages of the deaf. To take an example from the British deaf community, British Sign Language (BSL) is used for the following purposes (Jakobsonian functions are listed in brackets after each example): narrating events (referential); expressing anger (emotive); telling a child what to do (conative); composing poetry (by at least one British sign poet) (poetic); exchanging greetings (phatic); and discussing differences between signs used by signers who attended different schools for the deaf (metalingual).

20.4 Arbitrariness

I suggested above that arbitrariness was one of the most debatable design features as far as sign languages are concerned, so this section will be devoted to a discussion of what arbitrariness actually involves, for both spoken and signed languages. Arbitrariness in language is usually explained as applying to the connection between the form of words in a language and their meaning. Although there has been debate from the time of the Greeks onwards about the exact nature of the relation between language and what it represents, arbitrariness has been considered a particularly important property of language in the twentieth century as a result of the influential work of the Swiss linguist, Ferdinand de Saussure. Saussure is particularly well known for his principle of *l'arbitraire du signe* (see Sinha, this volume, Chapter 17, Section 12.3).

Saussure saw a linguistic sign as being made up of two aspects, the sequence of sounds (what he called the *signifiant*) and what it represented (the *signifié*). He argued that the relation between these two aspects was totally arbitrary, so that the idea of 'sister', for example, has no intrinsic relation to the sequence of sounds making up the French word *soeur*.

It is doubtless because of Saussure's emphasis on the importance of arbitrariness as a property of language that arbitrariness has achieved the status of being almost a hallmark and criterion of language. The effect of this has been that non-arbitrary aspects of language have been seen to detract from the linguistic status of communication systems, particularly when that very linguistic status is in question, as in the case of sign languages. However, arbitrariness may have both advantages and disadvantages, as Hewes (1976) and Hockett (1960b, 1978) point out. Hewes suggests that iconicity, which is commonly considered to be the converse of arbitrariness, may be helpful in decoding, and Hockett (1960b) that it may be useful in the early stages of the construction of a communication system. Hockett (1978) even suggests that arbitrariness might be considered a limitation of spoken languages, and that 'it would be stupid not to resort to picturing, pantomiming, or pointing whenever convenient' (Hockett 1978, p. 55).

20.4.1 *Iconicity versus arbitrariness*

So what is meant by iconicity? Hockett (1960b) says that 'A symbol means what it does iconically to the extent that it resembles its meaning in physical contours, or to the extent that the whole repertory of symbols in the system shows a geometrical similarity to the whole repertory of meanings. To the extent that a symbol or system is not iconic, it is arbitrary' (Hockett 1960b, p. 410). As Hockett's definition points out, iconicity and arbitrariness are not mutually exclusive, but may occur in varying degrees. Nevertheless, there has been a tendency to view the two properties as categorical opposites, and, in particular, to assume that spoken language is totally arbitrary, with no iconic properties. This may be largely due to Saussure's emphasis on the importance of arbitrariness as a property of language.

20.4.2 *Iconicity in spoken language*

If we allow a slightly broader definition of iconicity to allow for similarity between the sound of a symbol and its meaning, we may see that there are non-arbitrary or iconic properties of spoken language which have been largely played down. These include the existence of onomatopoeic words where there is some relation between the sounds and what they represent: English 'splash' or French *cocorico* (for the noise made by a cock) would be examples. We may also note the possibility of iconic mapping between the intensity of a form and its meaning in both spoken and signed language. The English word 'big', for example, pronounced slowly and loudly, can be interpreted as describing a larger size than the same word said normally. This is also true of the BSL sign BIG: the larger the signing space covered in making this sign, the larger is the size referred to. Although attention in discussions of arbitrariness tends to be confined to the level of the word only, Bolinger (1980) has suggested that there may be even more iconicity above the level of the word, saying that 'The smaller the unit, the greater the arbitrariness, as a rule' (Bolinger 1980, p. 192). He points out, for example, that events in English are usually narrated in the sequence in which they occurred, so that we find a sentence like 'John came in and sat down' rather than 'John sat down and came in' (Bolinger 1980, p. 20). Bolinger is clearly using the notion of iconicity in an even broader sense here, since he allows for temporal mapping of media (whether visual–spatial or auditory). Nevertheless his argument serves to illustrate the fact that the notion of iconicity is more complex than might at first sight appear, and that, if it is defined more broadly, it must be seen as a much more pervasive phenomenon in both spoken and signed languages. (For further discussion see Armstrong 1983.)

20.4.3 Iconicity versus arbitrariness in sign languages

Just as spoken languages are not completely arbitrary, so signed languages are not completely iconic. It is not actually all that surprising if sign languages do have a greater degree of iconicity than spoken languages, at least at the word or sign level, because many more entities in the world have characteristics that are visually salient than that are auditorily salient. (This fact would also explain the partial iconicity of Chinese logograms (cf. Barton and Hamilton, this volume, Chapter 27).) Also, as Hockett (1978) points out, sign languages have more possibility of iconicity, since they have more dimensions (three of space and one of time) available to them. Nevertheless, we may find both arbitrary and iconic aspects in sign languages. To take some concrete examples, the BSL signs HOUSE, COME, and NEAR could all be considered iconic to some extent, and they are illustrated in Fig. 20.3.

There is a clear relationship between the outline traced by the hands and the shape of a house, between the movement of the index finger and the movement of a person approaching, and between the actual distance between the index fingers in NEAR and what is meant by this sign. However, other signs show no obvious relation between a sign and its meaning. The examples in Fig. 20.4, which are also from BSL, show that both concrete entities and abstract ideas may have arbitrary signs.

To emphasize the arbitrary nature of these signs, we can point out that the sign for 'good' in BSL means 'male' in Japanese Sign Language, and that the BSL sign for 'true' means 'stop' in American Sign Language.

A further indication that we should not over-emphasize iconicity in sign languages comes from psycholinguistics. As we shall see in Section 20.4, linguists have analysed signs not in iconic terms, but in terms of abstract parameters, so that for each sign, a location, movement, and hand configuration can be identified. Bellugi and Klima (1976) tried to determine to what extent these abstract parameters had psychological reality in testing what signers' errors in recalling signs presented to them revealed about the way that the signs were being coded in memory. They found that the most frequent errors tended to get just

HOUSE: hands outline roof and walls

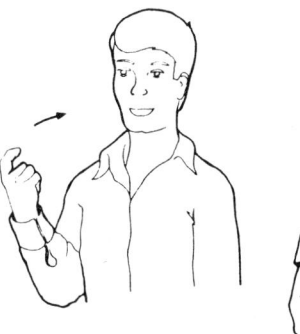

COME: index finger moves towards the signer

NEAR: one index finger is held upright near the other

Fig. 20.3 Iconic features of the BSL signs HOUSE, COME, and NEAR (from Deuchar 1984, p. 12).

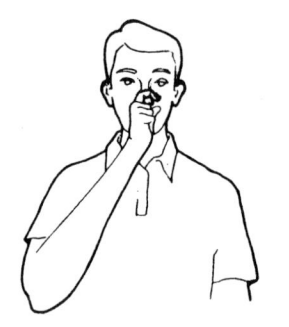

SISTER: bent index finger touching nose

GOOD: fist held with thumb upwards

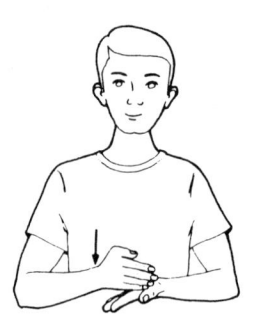

TRUE: right hand, flat palm, contacts left hand, flat palm upwards, so that right hand is above left and the hands at right angles and perpendicular to one another

Fig. 20.4 BSL signs without any obvious iconic relations to their meanings (from Deuchar 1984, pp. 13–14).

one structural parameter wrong, and thus indicated that signers were coding the signs in abstract rather than more iconic ways. For example, the sign NEWSPAPER, which was frequently remembered for BIRD, shares handshape and movement with BIRD, but differs from it in location. Thus iconicity does not appear to be significant in the mental representation of signs (but see the claims of Feldman *et al.* about children, reported below in Section 20.4.4).

Bellugi and Klima are also responsible for drawing an important distinction between two notions relating to iconicity: 'transparency' and 'translucency'. By transparency they mean the 'guessability' of a sign, or the chances that someone totally unacquainted with a sign language might be able to guess what a particular sign meant. Translucency is the extent to which a relation between a sign and its meaning may be discerned if the meaning of the sign is already known. In testing the transparency and translucency of 90 signs from ASL (American Sign Language) for people with no previous knowledge of ASL, they found that the degree of translucency was much higher than that of transparency. Of 90 signs, 81 were not guessed by any of the ten subjects; whereas, once the meaning was known, the subjects managed to agree on the basis for the relation between the signs and their meaning for over half the signs presented.

This distinction between transparency and translucency helps to explain why the signs for a similar thing in different sign languages may all be considered iconic, and yet are different. Bellugi and Klima (1976) illustrated this with the signs for tree in American, Danish, and Chinese Sign Languages. Their illustrations are reproduced in Fig. 20.5.

These examples serve also to illustrate that iconicity need not exclude conventionality, which is normally strongly associated with arbitrariness. In the case of arbitrary signs, it is generally argued that their relationship is governed by convention; but we can see that this is none the less true of iconic signs, particularly since there are different conventions in different countries about how to represent the entity 'tree'. The concepts of iconicity and arbitrariness both focus on the relation between the physical form of a sign or word and its meaning; but, as we have seen, all signs

Spoken language and sign language 561

Fig. 20.5 Three different, yet equally iconic, signs for 'tree': American, Danish, and Chinese Sign Languages (Copyright by U. Bellugi and E. Klima: Two faces of sign: iconic and abstract. *Annals of the New York Academy of Sciences*, **280**, 514–38, p. 523.)

and words, whether arbitrary or iconic, must have the property of conventionality, that is, they must be shared by a particular community. This, however, is a feature of language having to do with its social function rather than its structure, and as we have seen, linguists have been primarily concerned with structure. However, Baron (1981), who advocates a functional approach to language, argues that conventionality is more important than arbitrariness. She says:

... is the issue of the arbitrariness of the sign really comparable to the issue of whether signs are shared between language users? Not at all. In the first instance, we might decide to abandon the notion of arbitrariness altogether as a design feature of human language without changing our conception of language itself or of the fundamental ways in which we analyse language. However, in the case of the sign, to acknowledge that pairings of *signifié* and *signifiant* are not shared is to deny the possibility of studying language as a social activity (ibid., p. 10).

20.4.4 *Arbitrariness and evolution*

The idea that languages may have evolved from more iconic to more arbitrary systems (see for example Hockett 1978; Armstrong 1983) gains possible support from the little we know about the evolution of sign languages themselves. In a historical study of ASL, Frishberg (1975) found that American signs appear to have become more arbitrary and less iconic over time. To give an example, the sign for HOME used to be a compound consisting of the signs for EAT (bunched hand touching mouth) and for SLEEP (flat hand on cheek). Now, however, the sign for HOME involves just two touches on the cheek, both involving the bunched ('O') handshape. Frishberg's study is interesting to compare with that of Tervoort (1961), who studied the way in which a sign system developed among deaf children in Dutch and American schools. He found that signs were highly iconic when they were first invented or 'coined', and relied for their comprehension on their resemblance to what they represented. However, after a time they were recognized because of an established convention as to what they represented, and not because of what they looked like. Tervoort's findings may be compared with those of Feldman *et al.* (1978), who found that young deaf children coining signs in the absence of any linguistic input tended to use actions as the iconic base for signs. They interpreted this as evidence that the children had action-like substructures in their mental representations. They proposed, however, in a way analogous to Frishberg (1975), that the 'action-iconic' psychological status of such signs might disappear over time.

20.4.5 *Summary*

In this section on arbitrariness in language we have seen that it is not an absolute concept, in total opposition to iconicity. We have seen that both arbitrariness and iconicity may be found to varying degrees in spoken and signed languages. Iconicity, moreover, is not an undifferentiated concept, and does not preclude conventionality, which may be a more important feature of language than arbitrariness from a functional point of view. Finally, we noted that there is a little evidence to support the view that languages (including sign languages) may become more arbitrary over time.

20.5 Structure in spoken and signed language

The purpose of this section is to show how structure can be identified in spoken and signed languages. In so far as it is possible to discern structure at more than one level of the language, it will have satisfied Hockett's (1960*a*) criterion of duality of patterning. In this section we shall look at structure at two levels: below the level of the word or sign, and above that level.

20.5.1 *Structure below the word/sign level*

In spoken languages, the words are made up of elements of sound, which structural linguistics has identified as phonemes. Although there is considerable controversy about the status of the phoneme, and many alternative approaches to phonology have dispensed with it (see, for example, the review by Sommerstein 1977), the phoneme remains a useful construct which is still widely used in informal phonological descriptions. I shall use it here, partly because the 'classical' approach to phonology is so well known, and partly because the approach is most comparable with the work so far done on phonology in sign languages. In classical phonology, the sounds of a given language are assigned to separate phonemes if they contrast with one another in the same position in words. Whether or not sounds do in fact contrast can be determined by the 'minimal pair' test. An example of a minimal pair in English is the pair 'pit' and 'bit' where the difference lies in the contrast between /p/ and /b/, which occur in the same position in the word. Thus /p/ and /b/ are assigned to separate phonemes in English. The phoneme /p/ in English actually consists of at least two sounds or phones, 'aspirated' or 'unaspirated' [p]. Aspirated [p] (with a small puff of air following the release of the [p]) is found at the beginning of words such as 'pin', but after an /s/ (another phoneme of English) we find an unaspirated [p]. However, because the aspirated and unaspirated p's do not contrast, that is, do not occur in the same position in different words, they are not considered to be separate phonemes of English. In another language, Thai, however, aspirated [p] and unaspirated [p] are separate phonemes, because they serve to distinguish two separate words, that for 'forest', which is *paa* with unaspirated [p], and that for 'to split', which is *paa* with aspirated [p].

In spoken languages phonemes are viewed as occurring in linear sequence, so that the English word 'pin' is made up of three phonemes in linear sequence, /p/, /i/, and /n/. In sign languages we can also identify components like phonemes that occur below the level of the individual sign; but the difference is that they do not occur sequentially, but simultaneously. Linguists working on sign languages generally agree that they are made up of at least three components, 'tab', or location, 'dez', or handshape, and 'sig' or move-

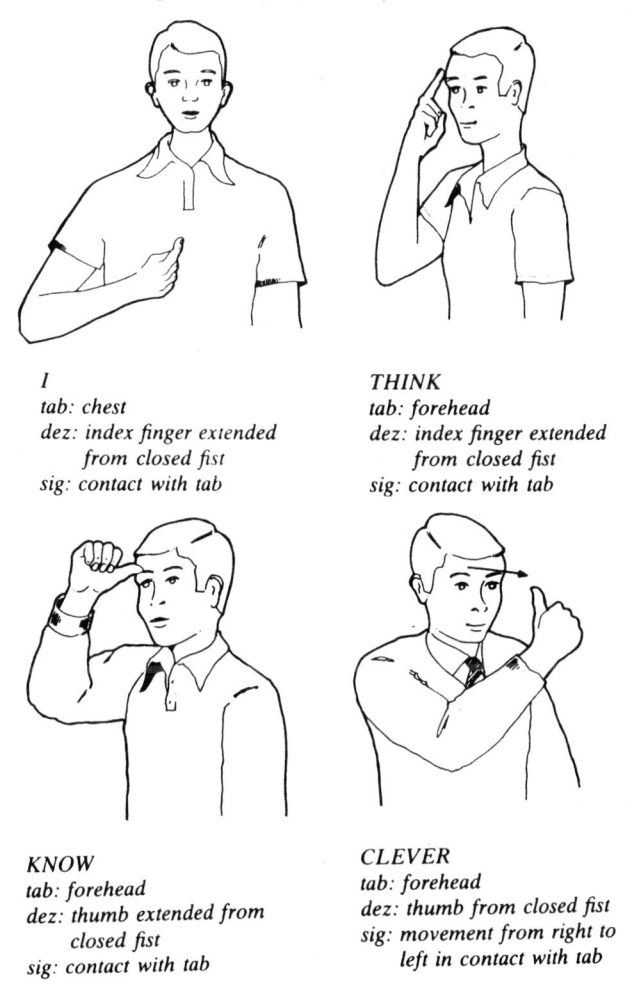

I
tab: chest
dez: index finger extended
   from closed fist
sig: contact with tab

THINK
tab: forehead
dez: index finger extended
   from closed fist
sig: contact with tab

KNOW
tab: forehead
dez: thumb extended from
   closed fist
sig: contact with tab

CLEVER
tab: forehead
dez: thumb from closed fist
sig: movement from right to
   left in contact with tab

Fig. 20.6 Minimally contrasting pairs of signs in BSL: I and THINK, and KNOW and CLEVER (from Deuchar 1984, pp. 54–5).

ment. The illustrations in Fig. 20.6 show how the BSL signs I, THINK, KNOW, and CLEVER can be analysed according to tab, dez, and sig.

The examples in Fig. 20.6 show not only how signs can be analysed into tab, dez, and sig, but also how these components can be seen to contrast with one another in minimal pairs. The signs I and THINK can be considered a minimal pair because they only contrast in tab; THINK and KNOW are a minimal pair contrasting in dez, and KNOW and CLEVER contrast in sig. Using the minimal-pair test in this way, it is possible to determine the 'phonological' components of sign languages in a way similar to the determination of the contrasting features that create phonemes in spoken languages.

We saw that spoken languages could differ according to their phonemes, as in the case of English and Thai; and this is also true of signed languages. For example, as far as tabs or location are concerned, the cheek and ear are considered distinct tabs in BSL, but not in ASL. This is because the cheek and ear appear to contrast with one another in BSL in the minimal pair CHEEKY and LUCKY (identified by Brennan *et al.* 1980, p. 65). In these signs the thumb and index finger hold the cheek or ear (respectively), shaking it to and fro. In ASL, however, the cheek and ear do not contrast, and indeed the sign DEAF, which originally involved a contact on the ear, now makes that contact on the cheek (see Friedman 1977, p. 39). Not only do locations, handshapes, and movements differ according to whether or not they contrast in different signed languages; they also differ according to whether or not they occur at all. For example, BSL has a dez or handshape consisting of the middle finger alone extended from the first, whereas this handshape does not occur at all in ASL, possibly because of its obscene connotations in American society. Klima and Bellugi (1979) show that an underarm tab is found in Chinese Sign Language, whereas this is not possible in ASL, or indeed BSL. Examples of this kind are similar to finding in a spoken language a sound that simply does not occur in some other language: this is true of the velar fricative ('ch' or /x/) in German, which does not occur in French or standard English, although it does occur in Scottish English, for example in the word 'loch'.

20.5.2 *Grammar*

We shall now look at structure at the level above the word or sign in both spoken and signed languages, or what can be termed grammatical as opposed to phonological structure. Since an exhaustive account of the grammar of spoken and signed languages is not possible here, we shall choose just a few areas: case marking, time marking, and negation. In spoken languages case is commonly indicated by word order or special case markers. In an English sentence like 'The woman saw the man' we know who saw whom because of the standard subject–verb–object word order in English, and not because of any case markers. In German, however, the verb occurs at the end of the sentence under certain circumstances, and it is only through case markers that we can be fully clear which is the subject and which is the object. Thus, both 'die Frau hat den Mann gesehen' and 'den Mann hat die Frau gesehen' mean the same as the English sentence: we identify 'den Mann' as the object because of the accusative case marking on the definite article 'den'.

In research on sign languages there has been considerable controversy as to whether sign order is at all significant to indicate case (cf. Wilbur 1979, for discussion); but it seems that one kind of case marking is found in the form of inflections on certain verbs. So in BSL, for example, direction of movement in a small set of verbs indicates whether the first person is subject or object: movement away from the signer indicates first-person subject, while movement towards indicates first-person object. Thus the sign GIVE, for example, as shown in Fig. 20.7, with movement away from the signer, indicates first-person subject, as in 'I give you'.

Movement towards the signer, however, indicates first-person object, as in 'You give me'. (For a more detailed discussion of case marking in BSL see Deuchar 1984.) This phenomenon is also found in ASL and FSL, where the verbs GIVE in both sign languages also show case marking. Although case marking on certain verbs is found in all three sign languages, it does not affect the same verbs in all three. For example, case marking verbs in BSL include GIVE, EXPLAIN, ASK, SAY; in ASL, GIVE, MOCK, TEACH, TELL; in FSL, GIVE, ASK, SHOW, LOOK-AT (see Edge and Herrmann 1977 and Moody 1983 for more details of ASL and FSL).

Fig. 20.7 Case-marking in verbs in BSL: the sign GIVE with movement away from the signer, indicating first-person subject, as in 'I give you' (from Deuchar 1984, p. 56).

Time is marked in languages in two main ways: by inflections on the verbs or verb auxiliaries (usually referred to as 'tense'), and by time adverbials. In English, the main means is by tense, as in the past '-ed' inflection added to verbs, for example 'walk' v. 'walked', or in the use of a future auxiliary 'will', as in 'I will go'. French uses tense, in the form of inflections on the verb, for example *il comprit* ('he understood') versus *il comprendra* ('he will understand'); and constructions with auxiliaries, for example, *j'ai travaillé* ('I worked') versus *je vais travailler* (I shall work). If we now compare English and French with British and French Sign Languages, we may note that the sign languages are more similar to one another in the way they mark time than they are to the spoken languages spoken in the same country. Both British and French Sign Language mark future tense by the use of an auxiliary placed next to the verb (glossed WILL and VA-VA respectively), and both mark past time by adverbials rather than inflections on the verbs. This is illustrated by the following examples:

BSL: WILL ASK ('I will ask.')

FSL: MOI S'OCCUPER VA-VA ('I'll take care of it.')

BSL: WASHINGTON BEFORE NAME WASHINGTON SEE ('I saw the place called Washington.')

FSL: HIER MOI TRAVAILLE MOI ('I worked yesterday.')

(BSL examples are from my own data, FSL examples from Moody 1983.)

To end this section we shall take a look at negation in spoken and signed languages. In English the rule is to place the word 'not' after the first auxiliary verb, so that for example 'I will come' becomes 'I will not come'. If there is no auxiliary verb, as in a sentence like 'I come', the verb 'do' is added before the main verb, and the negator 'not' placed after this, to give 'I do not come'. The exact form of negation varies from language to language; and instead of having a separate negation marker, it may be done by an inflection on the verb, as in Japanese, where verb forms ending in '-masu' are modified to end in '-masen', and forms ending in '-u' are modified to end in '-anai'. In sign languages, the most common form of negation seems to involve a shaking of the head during an otherwise affirmative utterance. So in BSL, shaking one's head while signing GO, for example, can mean 'I'm not going', as opposed to signing GO without the headshaking, to mean 'I am going'. A very similar means of negation appears to be used in some other sign languages, including ASL, FSL, and Swedish Sign Language. All these sign languages also have a kind of negative inflection which is somewhat akin to the Japanese form of negation, but is restricted to a small set of verbs. The form of the inflection is not identical in all sign languages, but in general involves an opening and outward movement of the hand during the formation of the verb sign. The signs HAVE, KNOW, LIKE and WANT all allow negative inflection (also called negative incorporation, cf. Woodward and DeSantis 1977) in BSL, ASL, and FSL; and GOOD (which functions as a verb) allows it in BSL and ASL. Negative inflection is also found in Swedish Sign Language and Danish Sign Language (Bergman and Hansen, personal communication), again in a restricted set of signs.

20.5.3 Summary

In this section I have tried to show how structure can be discerned both below and above the level of the word or sign in spoken and signed languages, and thus how both types of language have Hockett's design feature 'duality of patterning'. The kinds of structure we have examined look quite similar for spoken and signed languages, most of the differences, particularly in phonology, being attributable to the difference of medium.

20.6 Language acquisition

Having seen that there is linguistic structure of a comparable kind in both spoken and signed languages, it is interesting to determine whether acquisition of the two kinds of language proceeds in a parallel way. In addition, comparing signed language and spoken language acquisition can help us to resolve a particular theoretical issue relating to language acquisition, and that is the role of linguistic input. While some investigators (see for example Snow and Ferguson 1977) argue that the course of language acquisition is significantly affected by the nature of parents' speech to children, the Chomskyan position is that linguistic input plays only a small role in language acquisition. As we shall see below, spoken and signed language acquisition prove very similar under comparable conditions of input; and even where input is severely lacking, signed language acquisition nevertheless proceeds.

Research on the acquisition of spoken languages has shown that children go through very similar stages, even though they may pass through these stages at very different ages and speeds. After producing single words which generally refer to objects or people in their environment, they go on to produce two-word utterances which appear to express a variety of grammatical relations, such as 'action–object' (for example 'put book': see Dale 1976). The length of their utterances gradually increases, and they also start to produce inflections. In the early stages of learning inflections children tend to 'overgeneralize', and apply them more widely than is correct in adult language: for example, children learning English commonly apply the past inflection '-ed' to an irregular verb like 'bring', producing 'bringed' instead of 'brought'. Like their grammar, their phonology also develops gradually from a simple system with few contrasting items towards the full adult system. By the age of five children have usually acquired something close to the full adult system, though there are a few remaining constructions waiting to be learned (see Johnson et al., this volume, Chapter 24).

Deaf children learning sign language on the basis of signing produced by their deaf parents proceed in a very similar way to children learning spoken languages, passing through similar stages (see Schlesinger and Meadow 1972). They initially use one sign alone, then two signs in combination, then more. In acquiring inflections they show the same phenomenon of overgeneralization, and master a similar range of semantic relations to that of children learning spoken languages. Handshapes, like speech sounds, are also acquired in a generally predictable order (see for example McIntire 1977). It is interesting to note that the first sign is on average learned earlier than the first word; but this advantage seems to diminish during the later stages of language acquisition (see Bonvillian 1983). Bonvillian suggests that the difference may be due to the motor system's developing more rapidly than the speech system.

So far we have compared language acquisition of spoken language (mostly English) and sign language (ASL) under similar input conditions, that is, where the children's parents were using the language which they were acquiring.

However, most deaf children do not learn sign language under these conditions, since most of them (about 90 per cent) do not have deaf parents. Such children usually begin to learn sign language only when they go to a school for the deaf, and meet other deaf children. Only a minority of those deaf children have deaf parents, and thus already know sign language; and teachers do not generally use sign language in the classroom. So there develops a school signing system, which appears to be at least partially invented by the children themselves. Since the children are creating a system on the basis of very little input, their situation may well be compared to that of children acquiring a creole language on the basis of pidgin input. (A 'pidgin' language may be defined as a limited language used for restricted communication between two people or groups not sharing one another's language, while a 'creole' is a fully fledged language with a particular history: it has been developed as their native language by children exposed to a pidgin language as input. For further information see for example Hymes 1971.) When children leave school and join the adult deaf community, which is based in Britain on a network of deaf clubs, they adapt their system to the adult sign language, apparently with very little difficulty. This may be because the adult system is fairly similar to the children's system, and some evidence for this was found by James (1985) in a pilot study of the signing in a British school for the deaf. This similarity is perhaps not surprising, since the adult sign language will have been developed in a similar way by the previous generation, and indeed may also be creole-like. In the next section we shall consider evidence for the idea that sign languages may in fact be creoles on structural grounds, and thus may be considered to be newly emerging languages.

20.6.1 *Summary*

In this section we have seen that signed and spoken languages are acquired in a similar way when the conditions of input are similar. However, when the input is limited, it is interesting to note that language development nevertheless proceeds. Although we do not yet have sufficient detailed studies of the kind of development under these conditions, the fact that it occurs provides some support for Chomsky's contention that language acquisition is not heavily dependent on the nature of the linguistic input. Nevertheless, we may note that a triggering social environment, with conditions conducive to communication, may well be necessary for language development to proceed (cf. the study by Curtiss 1977, where this was not the case).

20.7 Sign languages and creoles

We saw in the previous section that sign languages appear to be acquired even under conditions of limited linguistic input. We noted that they could be considered similar to creoles in this respect, creoles being languages which have developed as native languages from a more limited type of language known as a pidgin. Creoles which have recently developed out of pidgins, and sign languages developed by children with limited input, are of interest to students of human symbolic evolution in that they may be considered examples of language in its early stages. In this section we shall note some structural similarities between sign languages and early creoles, and I shall suggest that early languages of this kind may reflect universal constraints on language.

In Section 20.5.2 we examined some of the grammatical characteristics of sign languages. Some of these were quite similar across sign languages: for example, case marking and the marking of time and negation. Similarities between languages are sometimes accounted for on the grounds of common historical descent, but this is unlikely for sign languages, given that they developed separately in fairly isolated communities (see Section 20.2.2).

Another way of accounting for similarities is to assume that they reflect language universals, which, as Chomsky has suggested, may result from the properties of the human mind. Since the similarities we have noted are specific to sign languages, we would have to claim that they reflect language universals in a way specific to the visual medium. Before we pursue this idea any further, we may note that a development from Chomsky's notion of language universals has been Bickerton's (1981) idea of a bioprogram that may affect the structure of languages differentially, and have a particularly strong effect on languages in early stages of development, such as creoles developing from pidgins with very little other linguistic input. I suggested in the previous section that signed languages appear to be acquired in rather a similar way to creoles; and now we may note that structurally, despite their difference of medium, they appear to share some characteristics with spoken creoles. Bickerton (1981) surveys the areas of grammar which he finds to be similar in many early creoles, and we may pick out several which also appear to be shared by many sign languages. (Fischer (1978) and Ladd and Edwards (1982) have also suggested that ASL and BSL respectively may be creoles.) The following are the features from Bickerton's list that we shall look at in relation to sign languages: (1) Aspect-marking is closest to verb, then modality, then time; (2) existential and possessive share the same lexical item; (3) the copula is not found; (4) adjectives function as verbs; and (5) there is no passive. (Note that while some of these features are found in non-creoles, it is the *co-occurrence* of several features in which I am interested.)

1. We saw in Section 3 that sign languages tend to mark time by adverbials such as BEFORE that do not have to be placed next to the verb. On the other hand, FSL and BSL were seen to have an auxiliary, WILL, that does have to be placed next to the verb. BSL, ASL, and FSL all have other auxiliaries such as MUST and CAN which, equally, have to be placed next to the verb. These are clearly examples of modality markers (marking uncertainty or possibility), and we could also include WILL in that category, since futurity has associated with it the uncertainty or possibility generally included in the category of modality. As we have seen, these modality markers are closer to the verb than the adverbial time markers, so far conforming with Bickerton's pattern. Finally, Bickerton finds that aspect marking (indicating something about, for example, the duration of the activity) is closest to the verb; and indeed, we find aspectual inflections on verbs in BSL, ASL, FSL, and Russian Sign Language. These inflections mostly take the form of a repetition, either fast or slow, of the basic form of the sign. For example, slow repetition indicates iterative (i.e. repeated) action for punctual verbs (verbs whose action takes place at one point in time), such as FIND in ASL; while it indicates durative (long-lasting) action for non-punctual verbs such as DRINK (see Fischer 1973). In BSL, slow repetition indicates iterative aspect for both punctual and non-punctual verbs (for further details, see Deuchar 1984). Thus sign languages seem to conform to the general pattern found in creoles for the way in which time, modality, and aspect are marked.

2. The existential and possessive share the same lexical item in at least BSL, ASL, and FSL. For example, the sign glossed as HAVE in BSL, and made with a closed fist held in front of the body, can indicate possession, as in for example HAVE SISTER ('I have a sister'), and also existence, as in HAVE HOUSE THERE ('There is a building there'). The same pattern is found in ASL and FSL.

3. Copulas seem in general to be absent from sign languages as from creoles, as the following examples will illustrate: BSL: HE NO-GOOD ('He is no good'); ASL: LYNN SINGLE ('Lynn was single'); FSL: LOUP LUI ('he is a wolf'). (The ASL example is taken from Edge and Herrmann 1977, and the FSL example from Moody 1983.)

4. Related to the absence of a copula is the fact that adjectives function as verbs. Evidence for this in sign languages is found in the fact that in BSL, the sign glossed as GOOD takes negative incorporation (see Section 20.4), like several verbs; in ASL the sign glossed as SICK takes durative aspect marking, like several verbs (see Klima and Bellugi 1979); and in FSL the adjective GRAVE can be negated, for example, by the addition of the negative sign PAS, as in GRAVE PAS ('It's not serious') (see Moody 1983).

5. Finally, the passive is absent from sign languages, as it is from creoles. I checked the following sign languages, and found it in none of them: BSL, ASL, FSL, Danish Sign Language, Swedish Sign Language, Russian Sign Language.

Thus we have seen that there are structural similarities between two types of language which differ in medium, but which have in common that they are both newly emerging languages. One way of accounting for the similarities would be in terms of Bickerton's suggestion that there is a bioprogram or language faculty which strongly constrains the properties of individual languages. Bickerton follows Chomsky in this view (for example Bickerton, 1984, 178), but makes the specific suggestion that the influence of the language faculty will be particularly strong in the case of newly emergent languages, where other factors (he mentions those of culture and processing) have not yet been brought to bear.

20.8 Language use and language evolution

Whereas Bickerton, following Chomsky, emphasizes the influence of biology on language, other scholars give more weight to the influence of language use on its structure.

For example, Slobin (1977) argues that language structure is constrained by its use, in particular, by four 'charges' to language, as follows: (1) 'Be clear'; (2) 'Be humanly processable in ongoing time'; (3) 'Be quick and easy'; and (4) 'Be expressive'. (Slobin 1977, p. 186). He argues that whereas the first two charges are particularly influential in child language, pidgins, and the early stages of creoles, the third and fourth charges affect languages at a later stage of development. Slobin's charges might lead to some different interpretations of some of Bickerton's findings. The charge to be clear, for example, might explain the way in which time, aspect, and modality are marked in clearly separate ways in sign languages and creoles. It would not, however, easily account for the similarity in form of the existential and the possessive noted in the last section, or the lack of a copula. These could be explained, however, in Bickerton's terms, by assuming that language has some 'given' properties which are not determined by its use.

Slobin's 'charges' may be useful in accounting for how language changes over time, whether in a child becoming an adult, or in a developing creole. He argues that the creole Tok Pisin developed relative clauses in order to satisfy the fourth charge, just as an adult English speaker needs relative clauses in order to be rhetorically expressive. Bickerton's and Slobin's theories might complement one another well in a theory of language evolution. The initial stages of language, particularly those developed in the absence of linguistic input, might be accounted for partially in terms of what is biologically given, and partially by the way in which language is used. As language develops, we might then expect it to change in accordance with its expanding uses.

As far as sign language evolution is concerned, we have seen that its early stages may perhaps be accounted for in terms of Bickerton's bioprogram and Slobin's first two charges to language. But what of its later stages? Can they be accounted for by the other 'charges', in the same way as Slobin claims for spoken language? This is hard to determine for British Sign Language, which, for reasons explained in Section 20.6, remains in the state of an early creole. However, there is some evidence that 'second generation' users of ASL (i.e. those who learned ASL in infancy from parents who learned it later in life) have more complex morphology than their parents. Reporting these findings, Newport (1982, p. 481) concludes that 'complex internal morphological analysis is performed by second-generation deaf on an input that does not itself contain this morphology'. (For further details see also Newport (1981).) It would seem possible to account for this change in terms of Slobin's 'charge' to be quick and easy, for while complex morphology may replace a one-to-one relationship between form and meaning with a one-to-many relationship, it is quicker and easier (at least in terms of overall effort) from the production point of view.

20.9 Spoken and signed languages: a summary

In this chapter I have not addressed at all the familiar question of whether language originated in the gestural or the vocal medium, (cf. Hewes, this volume, Chapter 21); but I have compared currently existing languages in those two media from various points of view. In Sections 20.2–20.5 I tried to show that spoken languages and signed languages are closely parallel in their structure and function, and that the differences attributable to their different media are relatively minor. Those who are interested in the medium used by the earliest form of language will be able to base discussion on the current existence of a fully viable language in the visual medium, and, hopefully, will no longer make the mistake of assuming that visual languages are totally iconic, vocal languages totally arbitrary. In Section 20.6 I provided further support for the idea that signed and spoken languages are fairly similar in structure and function by pointing to the similarities in their acquisition under similar input conditions. In addition, I suggested that sign language as it is normally acquired (i.e. with limited input) provides a useful test case for the role of the environment in language acquisition. I argued that the evidence indicates that Chomsky was right to insist on the relative unimportance of the linguistic input. In Section 20.7 I discussed the similarities between sign languages and creoles, and the possibility that they may reflect universal biological constraints on language. In Section 20.8 I discussed an alternative approach to the explanation of language structure that is based on the way language is used, and suggested that it might be particularly useful in accounting for the evolution of language beyond the initial stages.

References

Albertsen, K. (1985). Verbal morphology in Danish Sign Language. In *Proceedings of the III International Symposium on Sign Language Research* (ed. W. Stokoe and

V. Volterra), pp. 210–16. Linstok Press, Silver Spring, Maryland.
Armstrong, D. (1983). Iconicity, arbitrariness, and duality of patterning in signed and spoken languages: perspectives on language evolution. *Sign Language Studies*, **38**, 51–69.
Baker, C. and Cokely, D. (1980). *American Sign Language: a teacher's resource text on grammar and culture*. T. J. Publishers, Silver Spring, Maryland.
Baron, N. S. (1981). *Speech, writing and sign. A functional view of linguistic representation*. Indiana University Press, Bloomington.
Bellugi, U. and Klima, E. (1976). Two faces of sign: iconic and abstract. *Annals of the New York Academy of Sciences*, **280**, 514–38.
Bergman, B. and Wallin, L. (1985). Sentence structure in Swedish Sign Language. In *Proceedings of the III International Symposium on Sign Language Research* (ed. W. Stokoe and V. Volterra), pp. 217–25. Linstok Press, Silver Spring, Maryland.
Bickerton, D. (1981). *Roots of language*. Karoma Publishers, Ann Arbor.
Bickerton, D. (1984). The language bioprogram hypothesis. *Behavioral and Brain Sciences*, **7**, 173–221.
Bloch, B. and Trager, G. (1942). *Outline of linguistic analysis*. Linguistic Society of America/Waverly Press, Baltimore.
Bolinger, D. L. (1980). *Language: the loaded weapon*. Longman, London and New York.
Bonvillian, J. (1983). Early sign language acquisition and its relation to cognitive and motor development. In *Language in sign: an international perspective on sign language* (ed. J. Kyle and B. Woll), pp. 116–25. Croom Helm, London.
Brennan, M., Colville, M. D., and Lawson, L. (1980). *Words in hand*. British Sign Language Research Project, Moray House College of Education, Edinburgh.
Chomsky, N. (1957). *Syntactic structures*. Mouton, The Hague.
Chomsky, N. (1965). *Aspects of the theory of syntax*. The MIT Press, Cambridge, Massachusetts.
Chomsky, N. (1972). *Language and mind* (enlarged edn). Harcourt Brace Jovanovich, New York.
Corazza, S., Radutzky, E., Santarelli, B., Verdirosi, M., Volterra, V., and Zingarini, A. (1985). Italian Sign Language: general summary of research. In *Proceedings of the III International Symposium on Sign Language Research* (ed. W. Stokoe and V. Volterra), pp. 289–98. Linstok Press, Silver Spring, Maryland.
Curtiss, S. (1977). *Genie. A Psycholinguistic Study of a modern-day 'wild child'*. Academic Press, New York.
Dale, P. S. (1976). *Language development. Structure and function* (2nd edn). Holt, Rinehart and Winston, New York.
Deuchar, M. (1984). *British Sign Language*. Routledge and Kegan Paul, London.
Edge, V., and Herrmann, L. (1977). Verbs and the determination of subject in ASL. In *On the otherhand. New perspectives on American Sign Language* (ed. L. A. Friedman), pp. 137–79. Academic Press, New York.

Engberg-Pedersen, E., and Pedersen A. (1985). Proforms in Danish Sign Language: their use in figurative signing. In *Proceedings of the III International Symposium on Sign Language Research* (ed. W. Stokoe and V. Volterra), pp. 202–9. Linstok Press, Silver Spring, Maryland.
Feldman, H., Goldin-Meadow, S., and Gleitman, L. (1978). Beyond Herodotus: the creation of language by linguistically deprived deaf children. In *Action, gesture and symbol. The emergence of language* (ed. A. Lock), pp. 351–414. Academic Press, London.
Fischer, S. D. (1973). Two processes of reduplication in the American Sign Language. *Foundations of Language*, **9**, 469–80.
Fischer, S. D. (1978). Sign language and creoles. In *Understanding language through sign language research* (ed. P. Siple), pp. 309–31. Academic Press, New York.
Friedman, L. A. (ed.). (1977). *On the other hand. New perspectives on American Sign Language*. Academic Press, New York.
Frishberg, N. (1975). Arbitrariness and iconicity in American Sign Language. *Language*, **51**, 696–719.
Fromkin, V., and Rodman, R. (1983). *An introduction to language* (3rd edn). Holt, Rinehart, and Winston, New York.
Hall, R. A. (1968). *An essay on language*. Chilton Books, Philadelphia and New York.
Hewes, G. (1976). The current status of the gestural theory of language origin. *Annals of the New York Academy of Sciences*, **280**, 482–504.
Hockett, C. F. (1960*a*). The origin of speech. *Scientific American*, **203**, 88–96.
Hockett, C. F. (1960*b*). Logical considerations in the study of animal communication. In *Animal sounds and communication*, Symposium Series no. 7, (ed. W. E. Lanyon and W. N. Tavolga), pp. 392–430. American Institute of Biological Sciences, Washington, DC.
Hockett, C. F. (1978). In search of Jove's Brow. *American Speech*, **53**, 243–313.
Hoemann, H., Oates, E., and Hoemann, S. (eds) (1981). *The sign language of Brazil*. Mill Neck Foundation, New York.
Hoemann, H., Hoemann, S., and Rehfeldt, G. (1985). Major features of Brazilian Sign Language. In *Proceedings of the III International Symposium on Sign Language Research* (ed. W. Stokoe and V. Volterra), pp. 274–80. Linstok Press, Silver Spring, Maryland.
Hudson, R. A. (1980). *Sociolinguistics*. Cambridge University Press.
Hymes, D. (1971) (ed.). *Pidginization and creolization of languages*. Cambridge University Press.
Jakobson, R. (1960). Closing statement: linguistics and poetics. In *Style in Language* (ed. T. A. Sebeok), pp. 350–77. Wiley, New York.
James, H. (1985). Pidgin sign English in the classroom? In *Proceedings of the III International Symposium on Sign Language Research* (ed. W. Stokoe and V. Volterra), pp. 351–5. Linstok Press, Silver Spring, Maryland.
Klima, E. S., and Bellugi, U. (1979). *The signs of language*. Harvard University Press, Cambridge, Massachusetts.

Kyle, J. and Woll, B. (1983). *Sign language: the study of deaf people and their language.* Cambridge University Press.

Ladd, P. and Edwards, V. (1982). British Sign Language and West Indian Creole. *Sign Language Studies*, **35**, 101–26.

Lane, H. and Grosjean, F. (eds) (1980). *Recent perspectives on American Sign Language.* Erlbaum, Hillsdale, New Jersey.

Lyons, J. (1972). Human language. In *Non-verbal communication* (ed. R. A. Hinde), pp. 49–85. Cambridge University Press.

Lyons, J. (1981). *Language and linguistics.* Cambridge University Press.

McIntire, M. (1977). The acquisition of ASL hand configurations. *Sign Language Studies*, **16**, 247–66.

Moody, B. (1983). *La langue des signes.* International Visual Theatre, Vincennes.

Morris, D., Collett, P., Marsh, P., and O'Shaughnessy, M. (1979). *Gestures, their origins and distribution.* Jonathan Cape, London.

Newport, E. L. (1981). Constraints on structure: evidence from American Sign Language and language learning. In *Aspects of the development of competence. The Minnesota Symposium on Child Psychology,* Vol. 14, (ed. W. A. Collins), pp. 93–124. Erlbaum, Hillsdale, New Jersey.

Newport, E. L. (1982). Task specificity in language learning? Evidence from speech perception and American Sign Language. In *Language acquisition: the state of the art* (ed. E. Wanner and L. R. Gleitman), pp. 450–86. Cambridge University Press.

Peng, F. (1985). Some morphological considerations of Japanese Sign Language. In *Proceedings of the III International Symposium on Sign Language Research* (ed. W. Stokoe and V. Volterra), pp. 226–38. Linstok Press, Silver Spring, Maryland.

Robins, R. (1980). *General linguistics. An introductory survey.* Longman, London.

Sapir, E. (1921). *Language.* Harcourt Brace, New York.

Schermer, T. (1985). Analysis of natural discourse of deaf adults in the Netherlands: observations on Dutch Sign Language. In *Proceedings of the III International Symposium on Sign Language Research* (ed. W. Stokoe and V. Volterra), pp. 281–8. Linstok Press, Silver Spring, Maryland.

Schlesinger, H. and Meadow, K. (1972). *Sound and sign: childhood deafness and mental health.* University of California Press, Berkeley.

Slobin, D. (1977). Language change in childhood and in history. In *Language learning and thought* (ed. J. Macnamara), pp. 185–214. Academic Press, New York.

Snow, C. and Ferguson, C. (eds) (1977). *Talking to children. Language input and acquisition.* Cambridge University Press.

Sommerstein, A. H. (1977). *Modern phonology.* Arnold, London.

Stokoe, W. (1973). Classification and description of sign languages. In *Current trends in linguistics*, Vol. 12, (ed. T. Sebeok), pp. 345–71. Mouton, The Hague.

Stroombergen, M. and Schermer, T. (1985). The Dutch sign dictionary project: a progress report. In *Proceedings of the III International Symposium on Sign Language Research* (ed. W. Stokoe and V. Volterra), pp. 269–73. Linstok Press, Silver Spring, Maryland.

Tervoort, B. (1961). Esoteric symbolism in the communicative behaviour of young deaf children. *American Annals of the Deaf,* **106**, 436–80.

Umiker-Sebeok, J. and Sebeok, T. (eds) (1978). *Aboriginal sign languages of the Americas and Australia,* (2 vols). Plenum Press, New York.

Vogt-Svendsen, M. (1983). Lip movements in Norwegian Sign Language. In *Language in Sign: an international perspective on sign language* (ed. J. Kyle and B. Woll, pp. 85–96. Croom Helm, London.

Wilbur, R. B. (1979). *American sign language and sign systems.* University Park Press, Baltimore.

Woodward, J. (1978). Historical bases of American Sign Language. In *Understanding language through sign language research* (ed. P. Siple), pp. 333–48. Academic Press, New York.

Woodward, J. and DeSantis, S. (1977). Negative incorporation in French and American Sign Languages. *Language in Society,* **6**, 379–88.

Zaitseva, G. L. (1983). The sign language of the deaf as a colloquial system. In *Language in Sign: an international perspective on sign language* (ed. J. Kyle and B. Woll), pp. 77–84. Croom Helm, London.

21

A history of the study of language origins and the gestural primacy hypothesis

Gordon W. Hewes

Abstract

Speculative writings on language origins seem mainly to be confined, until very recently, to the Classical and Judaeo-Christian West. Until the Enlightenment nearly everything that was said about language origins in the West proceeded from the assumption that language began with Adam in the Garden of Eden.

In the eighteenth century a gestural origin for language was proposed by several writers, and some thought that apes might have a capacity for language. Renewed interest in these ideas developed in the mid-twentieth century, with systematic studies of human sign-languages, and then experiments to teach Great Apes visual languages.

The modern argument for gestural primacy in language origins draws on several lines of evidence, including the following. Sound is of questionable suitability as the original basis for language, given the greater creative capacity and open-endedness of higher primate manual and digital operations. Regular tool-using in hominids probably evolved before vocal language, and the human brain's left-lateralization for speech could have been tacked on to a previous specialization for predominantly right-handed gestural language and precise sequences of manual manipulations. In relatively simple contexts gestural communication has the distinct advantage of greater transparency and ease of communication.

It is speculated that with increasing manual preoccupations proto-speech developed out of mouth-gestures patterned after hand gestures and combined with vocalizations. Now the development of more abstract conceptual thinking was possible, given that gestural language suffers from proneness to commit the logical fallacy of misplaced concreteness. More recently the invention and diffusion of phonemicized speech acted as the principal stimulus for what we recognize as the cultural revolution of the Upper Palaeolithic. The basic advantage of a small set of phonemic units lies in their cognitive indexical function, and their facilitation of the storage and rapid retrieval of information from an increasingly larger mental lexicon [eds].

21.1 Introduction

Among the theories or models for the origin and evolution of language, the claims (Jóhannesson 1952; Hewes 1973a; Stokoe 1974; Hockett 1978; Kendon 1983) that language began in gesture rather than as a vocal system have, in my view, considerable plausibility. I shall offer a scenario[1] based on the primacy of gesture, fully aware that any such account must also show how gesture came to be superseded by speech to be at all credible.

Definitions

a. By *gesture* is meant any of a wide range of visible movements or positionings of parts of the body, functioning as signs in social communication (Morris *et al.* 1979; Kendon 1982). They may include movements of the facial muscles (Zajonc 1985), directions of glance, extensions of the tongue, and changes in the posture of the body as a whole. In particular, gestures which might have served as primordial language or proto-language are those of the fingers and hands and arms, and contacts these may make with other parts of the body, such as the ears, nose, mouth, and so on. I exclude vocal 'gestures', since their sign-values are mainly conveyed acoustically, even though some aspects of human speech are not only externally visible, but can be decoded, as in lip-reading, by specially trained persons.

b. *Origins* refers to the initial conditions, factors, and environment, including the then existing biological capabilities, which might have been associated causally with the emergence of language in the early hominids.

c. *Theories* is used here loosely, usually in the sense of a conjectural account, rather than in the narrow sense preferred by some philosophers, i.e. statements which can be tested by the canons of science or for which explicit directions for their falsification can be supplied.

d. *Language* is the most difficult part of the topic of this problem to define to everyone's satisfaction (Hockett 1959; Premack 1986). At a minimum it means here a communication system capable of stating propositions over a varied field of encodable items, and which differs from the social communication systems of most, if not all, non-human animals in that it is open-ended rather than closed. The waggle-dance of the honey-bee (von Frisch 1967) or the call-system of certain monkeys (Seyfarth *et al.* 1980; Gouzoules *et al.* 1984) can specify which of a small set of entities or attributes of a situation frequently encountered is part of a message. However, such messages are contextually limited: to the direction, distance, and relative amount of potential food (for the bee); or to which of three possible kinds of dangerous predator is present (for vervet monkeys, *Cercopithecus aethiops*); or to variations in signals used to recruit assistance during agonistic encounters with conspecifics (for the rhesus macaque, *Macaca mulatta*). I prefer to exclude, as non-language, communication systems with such limited flexibility, however suggestive they may be for the eventual emergence of language among the hominids. One of the essences of language is that a relatively large number of different lexical units can be inserted or substituted for others in utterances, so as to differentiate, locate, or describe a variety of phenomena, and to elicit a variety of actions from others (von Glasersfeld 1977).

The purpose of this chapter is to present the view that language, as minimally defined above, arose in the hominid family in the gestural mode at first; and to suggest how it might have later given way, in large part, to spoken language.

21.2 A brief history of speculations about language origins

21.2.1 *Classical and pre-Renaissance times*

Although some language-origin myths, etc., are found in the traditions of other civilizations, an explicit concern with language origins, successively religious, philosophical, and scientific, arose mainly in Western civilization. Its roots lie partly in the origin myth of the ancient Middle East, presented in the first book of the Bible. In it, as in other such glottogonic myths, language was spoken from the start, and appeared fully fashioned, by divine inspiration or creative fiat. Speculative writings on language origins seem to be confined, until very recently, to the Classical and Judaeo-Christian West. I have presented a more detailed history of glottogonic thought elsewhere (Hewes 1977); but it is useful to review some of that background here. Tables 21.1–21.3 provide a summary. Although the Renaissance showed an increase in writing on the subject, nearly everything that was said about language origins proceeded on the assumption that language began with Adam in the Garden of Eden, a few millennia ago.

21.2.2 *The seventeenth to mid-nineteenth centuries*

Following the Renaissance, the burgeoning of scientific enquiry and geographical exploration began to accumulate a larger body of relevant information bearing on the study of language. By the eighteenth century a few writers were sufficiently free of religious constraints to dare to speculate about how language might have arisen other than in the fashion set forth in the scriptural account. Several Enlightenment writers, including Rousseau, suggested that language must have emerged from natural cries and gestures (Kendon 1982). A few, more daring, said that apes might be taught to speak or use sign-language, or even more bold, like Lord Monboddo, that apes had once had language, but had abandoned it. The gestural origin of language was proposed by several writers in the eighteenth century, including some of the contestants in the 1769 prize essay competition sponsored by the Royal Prussian Academy of Sciences, although the winner, J. G. Herder, saw the origin in a much vaguer communicational propensity. The eighteenth century

Table 21.1 The development of glottogonic thought: the earliest sources to 1500 AD

Religious and mythological accounts	Language natural for mankind	Language by agreement	Gesture and signing	Other and general works
Bible—Gen. **2**, 19–20 (Adam) Gen. **11**, 1–19 (Babel) Aeschylus *Prometheus bound* ll. 442–641 Language a gift of the gods Sophocles d. 406 BC *Antigone* ll. 332–56 *India*—Sanskrit the language of the Cosmos *China*—writing given to mankind by a sage-emperor	Herodotus d. 400 BC Psammetichus, King of Egypt, experiments to determine the oldest language Zeno of Citium 4th century BC Epicurus d. 270 BC Lucretius d. 55 BC *De rerum natura* Dionysius of Halicarnassus d. 7 BC	(Conventionalist theory) Plato d. 348 BC *Cratylus* Vitruvius *fl.* 1st century BC Diodorus Siculus	Xenophon d. 431 BC—signing encounter in *Anabasis* Chrysippus d. 206 BC—on rhetorical gesture	*Yāska Nirukta*—Sanskrit linguistic treatise c.5th century BC *Pānini—Astādyāyī*—Sanskrit grammar 4th century BC? *Dong Zhong Shi*—essay on language 104 BC Dionysius Thrax—Greek grammarian d. 100 BC Horace d. 8 BC—general views on language

BC

AD Origen d. 254—language a divine gift Eunomius d. 360—language a divine gift Theodore of Mopsuestia d. 428—language a divine gift *India*: Mimansa school; c.500 language cosmic in origin	Gregory of Nyssa d. 394 *Contra Eunomium*—language natural to mankind	*India*: Nyāya School	Onasander d. 58—use of gestures in military communications Pliny the Elder d. 79—finger gestures Quintilian d. 90—rhetorical gesture Lucian of Samosata d. 192 mentions sign language Cyril of Alexandria d. 444—on finger computation	Xu Shen d. 121 *Shuo-wen jia-zi* on pictorial origin of Chinese writing Apollonius Dyscolus *fl.* 2nd century—on syntax Aelian *fl.* 200—on animal communication Donatus *fl.* 450—Latin grammarian Priscian *fl.* 500—Latin grammarian

500

Midrash—compiled 5th–6th century Hebrew the first language, Adam the first speaker *Holy Qūr'ān*—early 7th century. Arabic the first language, Adam the first speaker al-Mas'ūdī d. 957 — Syriac the first language			Isidore of Seville d. 636 *Etymologiae* Remarks on sign language Bede d. 735—on finger computation (also in 8th century on finger computation: Albinus Artabasda Nigellus Ermoldus)	Isidore of Seville—on Latin word origins *Ælfric of Eynsham* c.1000—comparative Latin-English grammar

900

| **Hebrew the 1st language** Rashi d. 1105 Andrew of St. Victor d. 1125 John of Salisbury d. 1180 Jakob van Maerlant d. 1300 Dante d. 1321 Johannes Historiographus d. 1350 John Capgrave d. 1464 Nicholas of Cusa d. 1464 Johann von Hagen d. 1475 | **Language other than Hebrew original, or unknown language** Michael of Syria d. 1199—Original language Syriac Gregory (Abul Faraj) d. 1286—Original language Syriac Abul Feda d. 1331—Original language now lost | **Language invented by Adam (or not a direct gift of God)** Peter Abelard d. 1142 Thomas Aquinas d. 1274 Petrus Hispanus d. c.1277 (later Pope John XXI) Konrad von Megenburg d. 1374—Language not an inborn ability | **Animal communication** William of Shyreswood d. 1267—parallel between animal and human cries Albertus Magnus d. 1280 on language of 'apes' | **Gesture and signing** Giraldus Cambrensis d. 1146 on monastic sign-language Leonardo Fibonacci d. c.1240 on finger computation Bartolus de Saxoferrato d. 1357 on sign-language of a deaf person Rudolfus Agricola d. 1485 on the education of the deaf | Peter Helias c.1150— grammarian *de modis significandi* 'universal grammars' concept—with Latin as model underlying all languages **Modistae:** Michel de Marbais 13th century Martin of Dacia *fl. c.*1250 Robert Kilwardby c.1250 John of Dacia *fl. c.*1250 Roger Bacon d. 1294 Siger de Courtrai d. 1341 Thomas of Erfurt *fl. c.*1350 |

1500

Table 21.2 The development of glottogonic thought: 1500–1700

Voyages and explorations greatly expand knowledge of the world's languages

Hebrew the 1st language	Language other than Hebrew original	Language 'by nature'	Language 'by convention'	Gesture and signing	Other and general works
1500					
Aventinus, Johann Turmair d. 1534 Juan Luis Vives d. 1540 Sebastian Franck d. 1542 Theodorus Bibliander fl. 1548	Joachim Perion d. 1559–61—French the first language Philipp Melancthon d. 1560	**Doctrine of natural signatures** Giovanni Pontano 1538 Juan Luis Vives d. 1540 (cf. col.1) Pietro Bembo d. 1547—Nature, not God, gave man language Paracelsus (Theophrastus von Hohenheim) d. 1541	François Rabelais d. 1553 Gargantua and Pantagruel—languages are conventional in origin	Juan Andres Mossen fl. 1515—finger computation Johann Thurmair = Aventinus, d. 1537—finger computation Rabelais—on gestures Thomas Murner d. 1537—manual alphabet Robert Recorde d. 1558—finger computation	Paracelsus (Theophrastus von Hohenheim) d. 1541—man used to understand language of birds Theodorus Bibliander fl. 1548 universal grammar Pierre de la Ramée (Ramus), d. 1594—notion of deep structure
Guillaume Postel d. 1581	Johann Becanus van Gorp d. 1572 (Goropius)—First language was Flemish	Michel de Montaigne d. 1592—isolated child would invent language	Francisco Sanchez (Sanctius) 1587—speech consists of conventional signs; interjections not a part of discourse	Juan Perez de Moya 1562—finger computation Girolamo de Cardano d. 1576—deaf can express abstract ideas Michel de Montaigne d. 1592—gestures understood by animals and children	Sieur de Fresnes Canaye fl. 1589 universal grammar Andrew Battell—1st report in Europe of chimpanzee and gorilla, 1590
1600			Animals and language		
Claude Duret d. 1611—God taught Adam in Hebrew Christian Avianus fl. 1620 Thomas van Erpe d. 1624 Johann Heinrich Alsted d. 1638 Cornelis Jansen the younger d. 1660 Georg Calixt d. 1665 Samuel Bouchart d. 1667 Andreas Sennert 1681 Olaus Borch d. 1690 Franz Mercurius von Helmont d. 1699 Hebrew letters = mouth forms Christoph Crinesius d. 1699	Böhme—Adamic language lost at Babel Inigo Jones d. 1651—Chinese the first language Wilkins—first humans immediately understood voice of God in Garden of Eden Matthew Hale d. 1676—an original 'radical language' (of roots)	Fausto Soccino d. 1604—Adam needed no Divine authority for language Jacob Böhme d. 1624—an unformed primordial language: natural 'signatures' John Webster fl. 1654—Adam understood natural signatures Jan Komenský (Comenius) d. 1670—Human language is man's work John Wilkins d. 1672 invented a philosophical language, 'a real character' Johann Erich fl. 1697—Adam's language instinctive; 1st word 'Oh!' of astonishment	René Descartes d. 1650—Animals lack all reason; only mankind can have language Samuel Pepys 1661—suggests that an ape might 'speak or make signs' Pierre Gassendi d. 1655—opposed Descartes's views Edward Tyson 1699 dissects chimpanzee: could chimps speak?	Antonius Deusingius c.1600 on sign-language of deaf Giovanni Bonifaccio 1616-sign-language as universal language Juan Bonet d. 1629—education of deaf with manual alphabet Francis Bacon d. 1626—on gestures Ludovicus Cresollius 1620—gesture language prior to vocal language John Bulwer fl. 1650 Chirologia—on manual sign-language George Dalgarno d. 1687—sign language of deaf	Johann-Heinrich Alsted d. 1638—general v. special grammar Bassett Jones 1659—'reason is the same in all countries' Arnauld and Lancelot 1660 Grammaire générale et raisonée **The Port-Royal grammar** Juan Caramuel y Lobkowitz 1682—speculative grammar
1700					

Table 21.3 The development of glottogonic thought: 1700–1800

Hebrew the 1st language	Language other than Hebrew original	Language 'by nature'	Animals and language	Gesture and signing	Other and general works
1700					
A. Berndt 1700	P. Y. Pezron—Celtic the language of Gomer 1703	John Wallis d. 1703 Onomatopoeia	E. A. von Pernau—on birdsong 1702	John Wallis d. 1703—teaching deaf, lip-reading	John Locke d. 1709—words are arbitrary conventions
J. Ungar 1716	William Wotton d. 1727—original language was Semitic	G. W. von Leibniz d. 1716—onomatopoeia origin of language	G. H. Bougéant 1739 language of animals	J. C. Amman d. 1724—teacher of deaf	
G. M. Ayroli d. 1721	M. de Larramendi 1729—Basque one of the languages formed at Babel			G. W. von Leibniz d. 1716—on gesture language, Trappist sign language	B. Mandeville d. 1733—on language origin
C. Vitringa d. 1737	J. A. Egenolff d. 1729—some 'oriental language'			E. Condillac 1754	G. Vico d. 1744—complex, multiple origins of language
J. Hutchinson d. 1737	G. Sale d. 1736—Adam's language later lost				C. Wolff d. 1754—language originates in society
J. M. Budde (Buddeus) d. 1744	S. Shuckford d. 1754 original language either Hebrew or Chinese		J. d'O. de la Mettrie d. 1751 'L'homme machine'—ape could be taught language	P.-L. M. de Maupertuis d. 1759—cries and gestures, also convention	C. C. du Marsais d. 1756—general grammar, deep structure
J. A. Bengel 1752	G. da Venezia d. 1764—lost at Babel			M. V. Lomonosov d. 1765—opposed gestural origin theory	J. G. Wachter fl. 1752—alphabet forms represent speech sounds in mouth
B. J. Feijóo y Montenegro d. 1764	W. Shaw—original language Gaelic 1778	Berlin Academy Prize for language-origin essay 1769 won by J. G. Herder (d. 1803)	F. Vicq d'Azyr 1779 compares human and animal 'voices'	C. A. Helvetius d. 1771—gestural origin	
J. C. Gottsched d. 1766	J. P. Süssmilch d. 1767—language 'from God'	1770—language outcome of human social nature		J.-J. Rousseau d. 1778—language begins with natural cries, gestures	W. Jones 1786—on Sanskrit
M. Sarmiento d. 1771		E. Swedenborg d. 1772—original language 'silent, of respiration' or signs	G.-L. de Buffon d. 1788 monkeys could speak if they had suitable ideas	W. Warburton d. 1779 gestural origin	
G. Sharpe d. 1771		F. M. A. de Voltaire d. 1778 Original language monosyllabic 'like Chinese'		E. B. de Condillac d. 1780—gestural origin	
J. Belli 1788		J.-J. Rousseau d. 1778—natural cries	P. Camper d. 1789—impossibility of ape language	C. M. Epée 1776—education of deaf	
J. W. Meiner d. 1789	G. Plouquet d. 1790—God made man able to speak	Moses Mendelssohn d. 1786—natural origin			Lord Monboddo d. 1799—general work on language origins, 1773
N. J. Bergier 1790	P. F. Suhm d. 1798—original language not Hebrew, not Chinese	G. Plouquet d. 1790 made man able to speak	Lord Monboddo (J. Burnett) d. 1799 apes once spoke		
1800					

was a period of very active, if still very weakly supported, glottogonic speculation (see Table 21.3).

21.2.3 The mid-nineteenth century and the Darwinian revolution

Without great elongation of geological time, or the rejection of the view that humankind was the outcome of a separate creation, there could not be a real advance in moving to a more scientific model for language origins. Table 21.4 lists some of the many names and intellectual landmarks of this period in their relation to glottogonic theorizing.

The publication of Darwin's *On the origin of species* in 1859, and *The descent of man* in 1871, reopened many lines of scientific thought, some nearly dormant since the French Revolution and the Napoleonic era. Alfred Russel Wallace (1895), who had arrived at the theory of evolution through natural selection at the same time as Darwin, and Edward B. Tylor (1904), and other major scientists, wrote about language origins, and on gestural origins in particular. Darwin himself contributed a pioneer study of the expression of the emotions in man and animals expressly based on his evolutionary views (1872), although he did not commit himself to the gestural origin of language.

Max Müller, the famous Oxford philologist, coined pejorative labels for a number of theories: for the onomatopoeic model, 'the bow-bow theory'; for the work-chant model, 'the yo-he-ho theory'; and for the emotional cry or expletive model, the 'pooh-pooh theory' (1891). His own notion that speech arose from 'resonances' in nature was called the 'ding-dong theory'. Linguists pointed out that onomatopoeic words form a very tiny part of the lexicon of any spoken language (hardly a devastating criticism), and that there was little evidence for the other suggestions. Recent research (see, for example, Foster, this volume, Chapter 25) indicates that there may be more to sound-symbolism than the nineteenth-century sceptics realized, but perhaps it played a later rather than a creative role in the evolution of spoken language. This period (1860–1890) also saw a debate on the status of signed languages, both of the deaf or of various tribal peoples, such as the North American Plains Indians (Mallery 1881). The ideas were either that signed languages consisted of 'universal' gestures, or of signs comparable to spoken words without self-evident meanings.

21.2.4 The early twentieth century

Towards the end of the nineteenth century new aspects of the problem of language origins were recognized. Work in neurology was revealing that language capabilities are usually localized in the left cerebral hemisphere, much as is right-handedness, though not in the same part of the left cortex. Animal communication began to be seriously studied, starting with bird-song. There was a primitive attempt to study the vocal call systems of monkeys and apes by R. L. Garner, utilizing the newly invented wax-cylinder phonograph in the West African forest (1900). In Philadelphia, W. H. Furness managed to teach an orang-utan to produce a few very poorly enunciated 'words' (1916). The last such effort, by K. J. Hayes and C. Hayes (1951), prompted the Gardners (Gardner and Gardner 1969) to employ a gestural rather than a visual language-code for working with their chimpanzee Washoe (see below, and also Ristau, this volume, Chapter 23). During the First World War Wolfgang Köhler, a German *Gestalt* psychologist, aroused great interest with his studies of the 'mentality of apes' in a captive chimpanzee colony in the Canary Islands. By the 1920s, Köhler's work (1925) was one of the stimuli leading to the founding by R. M. Yerkes of the first American research institute devoted to anthropoid ape behaviour. Also during the first decades of this century the famous experimental psychologist Wilhelm Wundt had become quite interested in language and language origins, and he espoused a gestural origin model in his magnum opus, *Voelkerpsychologie* (1912–22). An overview of all the work undertaken in fields relevant to glottogenesis during the first half of the twentieth century can be obtained from Table 21.5.

21.2.5 The 1920s to the 1950s

A few finds of fossil man had been made before the First World War. These included several remains of essentially modern humankind such as Cro-Magnon, still fewer specimens of neanderthalers, and, of early hominids, only a single very incomplete specimen of what is now known as *Homo erectus* (Java Man, 1891–4), and the isolated Heidelberg lower jaw (1908). There was also the large-brained 'Piltdown Man', from southern England, announced 1911–12, but revealed to be an ingenious hoax in the 1950s. Even so, these few specimens prompted the question of their owners' possible language capabilities, and of whether they might have had a spoken language in particular.

The Rhodesian skull found at Broken Hill in Zambia in 1925 did not seem at the time to shed much light on these matters; but R. Dart's report on the first australopithecine skull, issued that year, turned out to be the first on a series of early small-brained hominids. It was not until the late 1930s and

Table 21.4 The development of glottogonic thought: 1800–1900

	Divine origin	Natural origin	Children and language	Sound symbolism	Gesture	Brain studies	Animals	Philological studies	Evolution
1800	James Beattie d. 1803	J. Herder d. 1803	J. Itard 1801—'The Wild Child' (Victor of Aveyron)		J. M. Degérando 1800 on gestures D. Tiedemann d. 1803			F. V. Schlegel 1803	W. Paley—'natural theology' 1802 J.-B. de Lamarck 1801, 1801, 1815—inheritance of acquired characteristics
1825	Joseph de Maistre d. 1820 F. Willner 1831—'enspiritualization of natural sounds' L. Bonald d. 1840	A. Murray 1823 Jeremy Bentham (d. 1813) W. von Humboldt d. 1835		S. Henshaw— 1807	**The ideologues** B. St Pierre d. 1814 S. H. Long 1823—Plains sign language J. M. Degérando 1827 on deaf education	C. Bell 1811—brain anatomy F. Magendie 1822 J. Bouillaud 1825 C. Bell 1832—on organs of voice	J. Brandt 1816—on vocal anatomy in animals T. S. Traill 1821—orang speech deficiency is mental, not anatomical W. Youatt 1835—'why can't chimpanzees speak?'	Sanskrit studies Franz Bopp 1816 W. v. Humboldt d. 1835	C. Lyell 1830-3 Principles of geology Bridgewater Treatises 1833 C. Darwin 1831-6—Voyage of HMS Beagle R. Chambers 1844—Vestiges of natural history of creation
1850	*General works* H. Steinthal 1851, 1855 E. Renan 1858—transcendental origin of language	*Interjections/emotional origins* H. Steinthal 1871 F. Engels—language begins in work 1876			J. M. Degérando d. 1842 J. B. Barrois 1850 Hyde Clarke 1852—mouth gesture F. W. Farrar 1860—opposes gesture theory L. Geiger 1868 J. Rae 1862—mouth gesture theory		R. S. Savage and J. Wyman on chimpanzees' tool-use and on newly-found gorilla	E. Burnouf d. 1852	First Neanderthal find 1856 C. Darwin 1859—Origin of species A. Schleicher 1863—Darwinism and language Cro-Magnon man 1868
1870	*General work on language* F. Max Müller 1864-1882 critical views on language origins A. Marty 1875 O. Caspari 1877	L. Noiré—work chants 1877, 1880	E. B. Tylor 1863—on feral children M. Müller 1867—feral children M. Hale 1868 H. A. Taine 1870, 1876 Baudouin de Courtenay 1870 H. Hirt 1883—child language and evolution	A. F. Pott 1865 W. Bleek 1868 H. Wedgwood 1866, 1872 J. Lubbock 1875 M. Müller— opposes onomatopoeia 1877	G. Jäger 1867 E. B. Tylor 1871, 1880	P. P. Broca 1861, 1878—lateralization of language functions J. Hughlings Jackson 1873, 1893 aphasia C. Wernicke 1874	A. Bastian—1868 'Apes cannot speak' J. Ward 1883 apes generally silent G. U. Romanes 1883—mental evolution in animals	Jacob Grimm d. 1863 **Paris linguistic society bans language-origin papers** 1866 c.1875 **The Neogrammarians** K. Brugmann, et al.	C. Darwin 1871—Descent of man; 1872 Expression of the emotions
		P. Reynaud 1888 R. Collignon 1890—mechanism of interjection		E. R. Hun 1886—private language in a child H. Winkler 1884 D. Brinton 1894—on sound symbolism	C. Mallery 1884—on sign language A. R. Wallace 1881—mouth gesture J. R. Rambosson 1880, 1881 R. Kleinpaul 1888 F. H. Cushing 1892 T. Ziehen 1895	P. P. Broca d. 1880 J. Marigue 1886 C. Beevor and V. Horsley 1888—monkey brain studies S. Freud 1891 on aphasia M. Marchand 1893 F. Sano 1897—cortical centres for language	J. Lubbock 1884—teaching animals to converse R. L. Garner 1896—'languages of apes and monkeys'	W. D. Whitney d. 1894	**Altamira cave art** found 1879 G. de Mortillet 1883—Chellean man lacked speech A. Gaudry 1890 **Java man**—1891-4 E. DuBois
1900	J. Donovan 1893—festal origins of language		C. Franke—1899						

Table 21.5 The development of glottogonic thought: 1900–1950

General works	Children and language	Sound symbolism	Gesture	Brain studies	Animals	Evolution
A. Trombetti 1905—monogenesis of language J. v. Ginneken 1907—linguistic psychology	A. Lemaitre 1902 J. Sully 1903 C. Stern and W. Stern 1907 (English trans. 1920)	M. Grammont 1901—onomatopoeia A. Timmermans 1903 W. Wundt 1907			H. Coupin 1901 birdsong H. S. Jennings 1906	E. Mehnert 1901—larynx and bipedalism Heidelberg jaw 1907 La Chapelle neanderthaler 1908 C. Franke 1911—reconstruction of prehistoric language 'Piltdown man' (fraud) 1912 R. Anthony 1913—La Quina Neanderthal—did it have language?
W. Wundt (d. 1920) 1912–22—*Völkerpsychologie* F. de Saussure 1915—'Arbitrariness of linguistic sign'	E. Clarparède 1911 V. Bekhterev 1913	A. von Vetics 1909	W. Jerusalem 1907 K. Skraup 1908 W. Wundt 1912, 1916 H. Höffding 1912 C. Town 1912 E. T. Seton 1915 on 'Sign-talk'	K. Goldstein 1906 C. Sherrington 1906 'The integrative action of the nervous system' W. J. Sollas 1908—Neanderthal brain M. Boule and R. Anthony 1911—brain of La Chapelle (Neanderthal man)	L. Boutan 1913 on gibbon calls W. Furness 1916—attempt to teach orang to speak J. Némai 1920—vocal organs of primates K. v. Frisch 1923—language of bees	
Prague circle—N. S. Trubetskoy d. 1938 E. Cassirer 1923 O. Jespersen 1924 on origin theories (d. 1943) G. de Laguna 1927—sketch of language origins	H. Rogier 1923	R. Blanchard 1917 B. Liebich 1923 K. Lang 1925	H. G. Wells 1920 F. Allport 1924 M. Nice 1925 M. Jousse 1925, 1936 W. Schmidt 1926 J. Morlaas 1928 R. Paget 1927, 1928	A. Gray 1912—evolution of speech centres J. Dejerine 1914 P. Marie and C. Foix 1917 O. Hauger 1921 B. Naunyn 1925 H. Head 1926 G. E. Smith 1927	G. Hackenberg 1924 R. M. Yerkes 1927—suggests apes could learn sign language G. Schwidetzsky 1932 W. Kellogg 1931 *Humanizing the ape* H. C. Raven 1932—chimpanzee gesture	Rhodesian man 1925 Australopithecus I 1925 Peking man 1927–31 F. Tilney 1928, 1933—Java man could speak W. Oppenheimer 1932 (orang tongue)
Behaviourist approaches in linguistics E. H. Sturtevant 1922—sketch of language origins N. Y. Marr 1925—Marxist origin theory of language **Structural linguistics** L. Bloomfield 1933—Language-gesture is secondary to vocal language	P. Guillaume 1927 J. Piaget 1926—*The language and thought of the child*	K. Bühler 1933 M. Grammont 1939—sound symbolism R. Stopa 1935—clicks and sound symbolism H. Müller 1935 G. W. Allport 1935	A. Flach 1928 M. Krout 1931 N. Y. Marr 1933, 1936 R. Paget 1936 G. Allport and P. Vernon 1935 P. Eisenberg 1937 M. Critchley 1939 J. v. Ginneken 1939	F. Tilney 1928 K. S. Lashley 1929 F. Tilney 1933 K. Kleist 1934 K. Isserlin 1936 K. Goldstein and H. Nissen 1939—Prelinguistic behaviour in chimpanzees	L. Bloomfield 1933—rejects animal–human continuity in language J. Bierens de Haan 1934—human and animal language E. C. Tolman 1932 R. M. Yerkes and H. Nissen 1939—Prelinguistic behaviour in chimpanzees	Further australopithecine finds 1936 D. Black and Teilhard de Chardin 1937—did Peking man speak?
F. Ameghino 1935—polygenesis of language L. Thorndike 1943—Babble-luck origin theory A. Martinet 1949—duality	I. Latif 1934 M. Lewis 1936 S. Liljegren 1938 R. Zingg 1940—feral children A. Gesell et al. 1940—*Language development* A. Gesell 1941—*Wolf child* W. F. Leopold 1947, 1956 O. C. Irwin 1948 H. Werner 1948 R. Jakobson 1949 L. Stein 1949	M. Macdermott 1940	D. Efron 1941 A. Jóhannesson 1944, 1948 H. Werner 1948	J. Marmor 1938 J. Nielsen 1939, 1941 D. O. Hebb 1940 H. Hecaen 1944–71 L. Lotmar 1949	B. Grzimek 1940 on chimpanzee 'language' G. Schwidetzky 1948 on gorilla language	E. Cassirer 1944—distinction between language and animal cries P. Hirschlen 1942 R. Broom and G. W. H. Schepers 1946 A. Keith 1948 V. Negus 1949—on the larynx

1900

1950

later that more australopithecines were found, and not until the late 1940s that physical anthropologists began to realize their critical importance for hominid evolution and for hominid behavioural development in particular (Dart 1959).

Excavation of a set of well-preserved remains of what is now called *Homo erectus* near Peking in 1927-9 challenged the validity of the Piltdown specimens as genuinely ancient. Franz Weidenreich's painstaking reconstructions of the Peking materials made it clear that if the 'Piltdown chimera' were left aside, there was a reasonable progression in hominid brain size, which, with the australopithecines as Stage I, demonstrated a tripling in cranial capacity by the time of the neanderthalers and modern *Homo sapiens*. To be sure, it required the finding of additional australopithecine specimens, including pelvic and limb-bone fragments, to warrant their identification as legitimate hominids. The definitive proof that the controversial 'Piltdown Man' was a fraud did not come until 1953-4.

It is true that the enlargement of the hominid fossil record did not bear directly on language origins except for further confirmation of the tripling of brain size from the end of the Pliocene to the Middle Palaeolithic. The australopithecines were fully evolved for bipedal locomotion. There did not appear to be any way to tell whether they might have possessed the capacity for articulate spoken language, nor was it certain that they made or used stone tools.

In the years after the Second World War a new model of hominid evolution emerged, not without serious disagreements among the specialists. It was now certain that there had been early hominids, walking about as habitual bipeds, but who probably made no tools of durable stone. They probably did not hunt large mammals—recent evidence tends to indicate that the meat part of their diet came from scavenged carcasses killed by big carnivores, or from small live game; and the main part of their diet came from the collection of wild plant foods (Isaac 1983; Lucas and Corlett 1985). But little could be said about their cognitive capacities. An overview of developments in this and other fields relevant to glottogenesis between 1950 and 1986 can be obtained from Table 21.6.

21.2.5.1 *Primate field studies*

The early 1950s witnessed the beginning of field research on wild primate behaviour on a much larger scale than before. Whereas primatology in Japan arose from a more general interest in animal ecology, and initially with the Japanese macaque, American and British investigators were partly impelled by the hope that such studies might shed light on palaeoanthropological theory, as is shown by the participation of S. Washburn, an anthropologist, on the first American team to study baboon troupes in east and southern Africa, and the fact that J. Goodall's continuing study of wild chimpanzees was directly suggested to her by L. S. B. Leakey, then deeply engaged in early hominid research at Olduvai Gorge (Goodall 1963, 1972). Significantly, it was Goodall who reported the first observations of chimpanzee tool-using in the wild (1964). To be sure, the Japanese primatologists also recognized the bearing of some of their findings regarding Japanese macaques on human evolutionary problems, notably in connection with the observed transmission of learned behaviours (Kawai 1965; Itani 1958). Primatologists at a field station in Puerto Rico and in Japan meanwhile discovered that macaques regularly avoided mother–son sexual relations, weakening the previous anthropological dogma that incest avoidance was strictly a matter of cultural learning, restricted to humans and based upon language. Investigators of primates now collected detailed data on call-systems (Lanyon and Tavolga 1960; Green and Marler 1979), using greatly improved recording devices.

21.2.5.2 *The Chomskyan revolution in linguistics*

The 1950s also witnessed a major revolution in linguistics, led by N. Chomsky, which undermined the paradigm generally accepted since the time of de Saussure (Chomsky 1957, 1959). The pre-Chomskyan linguistics establishment adhered to a rigorously behaviourist paradigm, allowing practically no role for biologically built-in factors, and insisting that language phenomena were therefore entirely arbitrary (or conventional), and hence endlessly variable, within the limits of only the most general human capacities for learning. Instead, Chomsky argued for significant universal language features, especially of syntax, which he held to be present from birth, and which constituted what he called a 'language-acquisition device', unique to our species. Chomsky was not alone in pursuing an interest in language universals, nor in a willingness to entertain the possibility of deep biological roots for the seemingly species-specific phenomena of language; but his unique approach soon became influential. He did not attempt to explain how humankind's peculiar language-acquisition propensity came into being, other than to suppose that it must have arisen from a singular mutation. Nevertheless, Chomsky's revolutionary linguistic views, though focused rather narrowly on syntax, in contrast to the previous obsession with phonemic analysis and simplistic learning theory, helped revive interest in neurological as well as vocal-tract morphological factors bearing on human language capabilities. Not inappropriately, Chomsky

Table 21.6 The development of glottogonic thought: 1950–1986

	General works	Children and language	Sound symbolism	Gesture	Brain studies	Animals	Evolution
1950	O. Assirelli 1950, 1951—origin theories J. Stalin 1951—repudiates N. Y. Marr's theories of language R. Jakobson and M. Halle 1956—*Fundamentals of language* N. Chomsky 1959—Review of B. F. Skinner, *Verbal behavior* A. S. Diamond 1960—*History and origin of language* S. K. Langer 1962—on origins of speech L. Vygotsky 1962 N. Chomsky 1966—*Cartesian linguistics* W. Orr and S. Cappanari 1967—evolution of language R. Wescott 1967—reopening 'closed subject of language origins' M. Swadesh 1971—origin and diversification of language (d. 1967)	C. Tagliavini 1950—indirect language ideas M. Swadesh 1951—monogenesis D. A. McCarthy 1954 (Bibliography) on child language) J. Berko 1958 A. R. Luria 1959 S. Ervin and W. Miller 1963 R. Brown and U. Bellugi 1964 J. Bruner 1964—*The course of cognitive growth* G. A. Miller and F. Smith 1965 R. Jakobson 1968 R. Brown 1970 (in *Psycholinguistics*)		A. Borst 1957–63—*The tower of Babel. Survey of history of language-origin theories* G. Revesz 1959 *Origin and prehistory of language* (English trans. 1970) **Generative (transformational) Grammar** N. Chomsky et al. *Semiotica* (journal) Vol. 1, 1964 **Expansion of semiotic studies** **Toronto**—symposium on language origins 1972 (ed. R. Wescott 1974)	G. Smithers 1954 M. Chastaing 1958 M. Wertheimer 1958 L. Solomon 1959 O. S. Akhmanova 1960 I. Fonagy 1961 I. K. Taylor 1963–7 S. Tsuru and H. Fries 1962	A. Johannesson, 1950, 1952 B. Tervoort 1953 on deaf children and signing E. H. Sturtevant 1956 R. Stopa 1954—clicks and gestures E. Froeschels—mouth gesture 1952, 1959 A. Leroi-Gourhan 1958–64 L. West 1960 A. S. Diamond 1960 M. Critchley 1960, 1963 A. Burgess 1964 U. Bellugi 1965 A. Kortlandt 1967—gesture of chimpanzees R. A. Gardner and B. Gardner 1969—**Washoe experiment** Trân Duc Thao 1970	R. W. Sperry 1952 J. C. Eccles 1953 C. J. Herrick 1956 R. Dart 1956—brain size and 'human status' R. Jakobson 1955 (on aphasia) A. R. Luria 1958 K. Pribram 1958 W. Penfield and L. Roberts 1959 W. R. Brain 1961, 1965 H. Jerison 1963 T. Alajouanine and F. Lhermitte 1964 J. Lilly 1963 N. Geschwind 1965 onwards R. L. Holloway 1966, 1968 E. H. Lenneberg 1967 *Biological foundations of language* R. Jakobson 1968	K. and C. Hayes 1951—attempt to teach chimpanzee to speak N. Kohts-Ladygina 1955—chimpanzee thought differs from human K. Lorenz 1958 D. Morris 1962—*Biology of art* P. Marler 1961 J. Lilly 1965 (dolphin) S. A. Altman 1965 R. A. Hinde 1966 C. A. Greenwalt 1968 (birds) M. Adler 1967—rejects animal-human continuity in mind P. A. Sebeok 1968, 1970 E. Menzel 1969 P. Marler 1969 D. Premack 1970—chimpanzee language study	A. Kortlandt 1953—signals in wild chimpanzees *Primates* (Journal, Vol. 1, 1958) N. Tinbergen 1962 C. F. Hockett 1960, 1964—design features of language M. Kawai 1963, 1965—cultural behaviour in macaques R. A. Gardner and B. Gardner 1969—**Washoe experiment** S. K. Langer 1971 *The great shift* (to language from animals)	G. Heberer 1956—critical of endocast data G. v. Koenigswald 1956—speech in fossil man? E. DuBrul 1958, 1966—evolution of speech apparatus R. Dart 1959—evolution of language V. Kochetkova 1960 M. Critchley 1960 B. Kraus 1964 A. Marshack 1964—palaeolithic markings R. Holloway 1966—and later L. Leakey 1967, 1968—Olduvai Gorge *Journal of Human Evolution* v. 1 V. Kochetkova 1970 J. Wind 1970—evolution of larynx A. Marshack 1972

Table 21.6 (*Contd.*)

General works	Children and language	Sound symbolism	Gesture	Brain studies	Animals	Evolution
New York—Conference on Origins and Evolution of Language, New York Academy of Sciences 1975 (eds S. Harnad, H. Steklis, and J. Lancaster, 1976)	J. Macnamara 1972 *Journal of Child Language* v. 1 1975	C. Tanz 1971 M. L. Foster 1975, 1978	*Sign Language Studies* (journal) v. 1, 1972 G. Hewes 1973 W. Stokoe 1972, 1978	A. R. Luria 1970 K. Przibram 1971 N. Geschwind 1972 J. W. Brown 1972 *Aphasias... Brain and Language* (journal, Vol. 1, 1974)	J. Bronowski and U. Bellugi 1970 A. Healy 1971 L. Miles 1976 F. Marler 1976, 1977	P. Lieberman and E. Crelin 1972 D. Morris 1974 *Neanderthal speech*
	J. Bruner 1975		A. Kendon 1975		D. Griffin 1977	A. Jolly 1972—evolution of primate behaviour
G. Hewes 1975—Language origins bibliography	R. Brown 1978—*A first language*		M. L. Foster 1975—mouth gesture		D. Rumbaugh 1977—Lana project	T. Wynn 1979
W. O. Dingwall 1979—*Evolution of human communication systems*	A. Lock (ed.) 1978—*Action, gesture, symbol* E. Bates *et al.* 1979	A. Mandelker 1982	D. Morris *et al. Gestures* 1979 C. F. Hockett 1979—*In search of Jove's brow*	J. Levy 1976 J. Jaynes 1976—laterality and consciousness *Behavioral and Brain Sciences* (journal, Vol. 1, 1977)	H. Terrace *et al.* 1979 R. Seyfarth *et al.* 1980 Fouts, R. *et al.* 1984	D. Johanson and M. Edey 1981 G. Krantz 1981 T. Wynn 1981 R. Passingham 1982—*The human primate*
	K. E. Nelson (ed.) 1980 *Children's language*		D. Kimura 1978 W. Raffler-Engel 1983	S. Blumstein 1981	D. A. Hamburg and E. McCown (eds) 1979—*The Great Apes* T. A. Sebeok 1981	J. Laitman 1983—basicranium
C. Bickerton 1981—language-origin parallels with modern creoles	A. Lock 1980, 1985	M. L. Foster Chapter 25, this volume and 1983	A. Kendon 1983 E. Pulleyblank 1983	H. Jerison 1983—brain evolution	F. Patterson and E. Linden *The education of Koko* C. T. Snowdon *et al.* 1982	C. Lumsden 1983 J. Pfeiffer 1982
P. Lieberman 1984, 1985—origin and evolution of vocal language	E. Wanner and L. Gleitman (eds) 1983 E. Bates 1984 R. Brown 1983 S. Romaine 1985		M. L. Foster 1984 D. Armstrong 1983 H. Steklis 1985—doubts gestural origin J. Kyle, B. Wohl *et al.* 1985 Sign language	D. Falk 1984 D. Bickerton 1984 R. Holloway 1983	D. J. Gilian and D. Premack 1981 S. Savage-Rumbaugh 1984, 1985 R. Schusterman and K. Krieger 1984 D. Premack 1983, 1986—*Gavagai*	S. Parker 1985—social technological model for evolution of language
Oxford—Language Origins Society Conference 1986					S. Gouzoules *et al.* 1984 D. Rumbaugh 1985	

1986

contributed an appendix to E. Lenneberg's *Biological foundations of language* (1967).

Not all linguists involved in this paradigm-change of the 1950s and 1960s accepted Chomsky's programme of 'Cartesian linguistics' (Chomsky 1966; cf. Kronenfeld 1978/79). Other prominent linguists equally devoted to possible biological aspects of language and social communication in animals generally included C. F. Hockett, who examined the design features of language from a comparative zoological perspective (1959, 1961), and T. A. Sebeok, who promoted work in zoosemiotics, the study of sign systems used by non-human animals (1968). These biologically-orientated investigations have made more positive contributions to language-origin theory than Chomsky's critique of Saussurean and behaviourist linguistic doctrines.

21.2.6 *Renewed interest in gestural languages*

Partly aroused by the turmoil in linguistics instigated by Chomsky, but also by the widened interest in semiotics, there was a revival of research on gestural language systems, both as used by the deaf and by other special groups. The established linguistic view was that all such non-vocal systems were merely transformations of or derivations from spoken language, and of little importance. On the contrary, investigators such as W. C. Stokoe, jun., a linguist at the Gallaudet College for the deaf in Washington, DC, demonstrated that such signed languages were independent languages in their own right (Stokoe 1960), even though such languages contain borrowings from both spoken and written languages. The choice of ASL (American Sign Language of the Deaf) as the vehicle for the experiments of the Gardners for their chimpanzee Washoe was not accidental (Gardner and Gardner 1969).

21.2.7 *1970 to the present*

Although there had been some expressions of curiosity regarding the evolutionary consequences of language in human evolution in connection with the new australopithecine and primate behaviour data in the 1950s and 1960s (for example, Dart 1959; Bunak 1959); the convergent Chomskyan shift in linguistics; the interest of semioticians in gestural sign-systems; and newly-gained neurological evidence, as exemplified in Lenneberg's book (1967), these were not the only factors making glottogonic speculation more reasonable or more respectable by the early 1970s.

The systematic study of child language-acquisition, previously a marginal concern, now began to show that simplistic learning theory could not account for the cross-cultural, cross-linguistic similarities of the process now being reported. Piaget's work (1955) showed that simple reinforcement theory was insufficient to explain the cognitive development of the child. Piagetian models have since been used in studies of the mental growth of monkeys and apes (Chevalier-Skolnikoff and Poirier 1977; cf. Lock 1978; Bates *et al.* 1977). There are deep neurolinguistic factors, presumably in part genetically based, involved in the acquisition of language by young human children.

The reports of the ape 'language' researchers not only generated a bitter and continuing controversy (cf. the many attacks on their results, by Sebeok and Umiker-Sebeok 1980; Terrace 1984; Seidenberg 1984), but prompted some anthropologists, including Wescott and this writer, to reopen the old problem of language origins (Wescott 1967; Hewes 1969, 1973*b*). In 1972 there was enough interest expressed to justify a symposium on language origins at an anthropological meeting in Toronto, which led to a volume in 1974 (Wescott 1974). In 1975 a larger interdisciplinary conference was sponsored by the New York Academy of Sciences (Harnad *et al.* 1976), which covered a wide range of approaches, including the gestural origins theory. In 1981 there was a smaller conference in Paris, sponsored by UNESCO (de Grolier 1983), and another in Vancouver, Canada, at the International Congress of Anthropological and Ethnological Sciences, in 1983 (Landsberg 1988). An international Language Origins Society was founded following that meeting, which sponsored a conference at Wadham College, Oxford, in the autumn of 1986. The same period also witnessed a marked increase in journal articles on this topic, particularly in *Current Anthropology*, *Man*, and the *Journal of Human Evolution*. This renewed concern with language origins has also been reflected in recent textbooks in anthropology, and in books on human evolution for the general reader.

21.3 Specific evidence for the gestural origin of language

There are only two major models for the emergence of language aside from myths of its divine creation, or the notion that it 'just arose' as part of the high cognitive competence attained in our species. The first major hypothesis holds that language originated in the vocal–auditory channel as a refinement of the call-systems already present for social communication in the primates and other mammals. The other is that the earliest medium for language lay in gesture, with a later shift to the vocal medium. The higher vertebrates use both vision and hearing for social communication, along with touch and olfaction. It is the present predominance of speech which makes it so easy to suppose that language is only an elaboration of vocal calls or cries (but cf. Steklis 1985).[2]

If the human cultural record, including its linguistic records, had ended before the invention of phonographic and other sound-recording techniques, an analogous argument might lead to the erroneous conclusion that human beings had always communicated mainly in writing—on the ground that the principal surviving evidence for language consisted of graphic signs.

21.3.1 *The questionable suitability of sound as the original basis for language*

Sounds convey much useful information to primates, but, on the whole, less than is carried on the visual channel. The alarm calls produced by primates are generally evoked by the sightings of predators (Seyfarth *et al.* 1980); the most effective predators on land—mammals—hunt as silently as possible, to minimize acoustic cues indicating their proximity (cf. Valsiner and Allik 1982).

The sound signals produced vocally by primates consist of twenty-odd genetically-programmed calls, subject to little or no modification through learning (Gouzoules *et al.* 1984), although it seems possible to train some monkeys to produce such calls in response to other than specific natural stimuli. The facial expressions, patterned body movements, and postures ordinarily used in primate social communication appear to be under similar genetically-programmed control—evocable by specific kinds of stimuli, and generally resistant to modification (Zoloth *et al.* 1979). By contrast, primate manipulatory movements appear to be both more open-ended in their diversity of functions and their application to novel tasks, especially having to do with the food-quest (Parker and Gibson 1979, 1982), and, among the higher primates, are subject to modification by learning, including social learning through visual observation.

This greater open-endedness of manual and digital operations arises from the omnivorous character of primates (like *Cebus*, the cercopithecines, and the Great Apes), who must inspect, handle, peel, crack open, or otherwise process a very large range of possibly edible natural objects of different sizes, shapes, textures, and hardness. Much of the feeding behaviour of these primates requires considerable social learning on the part of young animals (Itani 1958; Goodall 1963, 1972), as it does in the more specialized carnivores.

The development of the capacity to produce new and voluntary alarm calls, simple as it seems for us, apparently involved cortical restructuring in our ancestors. On the basis of present evidence, it seems highly unlikely that any monkeys of the species now known to emit alarm calls relating differentially to predators could be simply trained to produce quite different vocalizations in response to other kinds of threatening events. Such changed capabilities may have to await accumulations of mutations affecting neural linkages of a fairly complex kind (but cf. Walker 1981). As it is, the chief evidence we now have bearing on this is that between the earliest australopithecine hominids and *Homo erectus* brain size more than doubled, and since then has increased by a further third (Jerison 1983; Lumsden 1983; Dingwall 1988; Holloway, this volume, Chapter 4). This striking increase could have arisen in part from natural selective pressures related to the advantages of voluntary signalling modifiable through social learning—and of course not only in situations of danger from roaming predators, or conspecific antagonists.

21.3.2 *Weaknesses of some alternative models for the emergence of language*

As noted, in the nineteenth century Müller bestowed trivial labels on language-origin theories: 'bow-wow', 'pooh-pooh', etc. (1891). The 'pooh-pooh' theory asserted that language originated in emotional cries—an idea advanced in the eighteenth century by Rousseau among others. The 'yo-he-ho' theory saw the basis for language in co-operative work-songs or chants, a view advocated by Noiré (1917). Müller himself preferred the 'ding-dong' theory, which claimed that a kind of psychic resonance existed between certain combinations of sounds and natural objects. The 'mouth gesture' theory was advanced by Rae, and elaborated by Wallace (1895); it derived articulate speech from approximations by the tongue and other parts of the vocal tract to replicating movements of the hands and fingers in gestural communication. Only this last (and by no means convincing!) model derived spoken language from an already existing non-vocal language. For a modern and much more carefully developed version of the mouth-gesture model, see the work of Foster (this volume, Chapter 25).

The most telling objection to the emotional-cry pathway to speech relates to the human cortex. Emotional cries, both in man and in other primates, and mammals generally, have been shown to be mediated, from their ultimate limbic triggers, through cortical areas not only bilaterally represented, but also far from the specific cortical areas which are involved in speech and other language functions (Lenneberg 1967; Lieberman 1984; cf. Section 21.3.3).

The work-chant theory (Donovan 1893; Noiré 1917) has no plausible basis in the reconstructions of the daily life of early hominids, nor in fact of that of any human groups until activities had reached the level of building elaborate structures of heavy logs or stones. Synchronized muscular effort of the sort seen among rowing crews, sailors hauling in sails, or the Volga boatmen may appear in advanced pre-literate societies in connection with launching large canoes, lifting roof-beams, or dragging in walrus or dugong carcasses, but is not a frequent or necessary part of the daily life of many existing or recent hunting and gathering peoples. It seems even less likely that strenuous group work was frequent enough to bring about patterned vocalizations leading to language under far simpler prehistoric social conditions.

21.3.3 *Manual motor control and the cerebral lateralization of language*

The importance of the left cerebral hemisphere to language abilities has been widely acknowledged since the nineteenth century (Bryden 1982; Needham 1982; cf. Holloway, this volume, Chapter 4). Moreover, both in modern humans and well back into the hominid past, as indicated by stone artefacts made by right-handed flint-chippers, and to fit the hands of right-handed users, hominids have been left-hemisphere dominant for precise manual manipulations. Apes and monkeys exhibit individual hand-preferences; but, on the whole, left- and right-handedness, and intermediate ambidextrality, are about equally distributed in non-human primates (Warren 1953). Marked lateral asymmetry occurs in the limbs of some otherwise bilaterally symmetrical animals, such as fiddler crabs, and songbirds are brain-lateralized with respect to learned song-patterns (Nottebohm 1978). Eye preferences are usually correlated with handedness in humans, and, not unexpectedly, so are skills involving the use of the feet for kicking, etc. Explanations for this state of affairs have varied widely; but, whatever the original causes of prevalent hominid dextrality, it seems to be both ancient and genetically based.

If the earliest language or proto-language had been gestural rather than vocal, it could well be that its subsequent striking left-lateralization in the brain resulted from language's having been a function tacked on to the kinds of precise manual manipulations involved in the predominantly right-handed making and using of tools (Hewes 1973*b*). Preferred right-handedness, and its concomitant cortical representation in the left hemisphere, could thus have been a prerequisite of hominid evolution for the sinistral localization of language functions (Kinsbourne 1978). Recent experiments show that observational learning of a complex manual skill is far easier when the demonstrator and the learner share the same preferred handedness. It is more difficult for a right-handed individual to learn how to tie knots (and presumably, how to fashion a serviceable flint cutting tool) from a left-handed teacher than from another right-hander. No such problem arises in speech-acquisition where teacher and learner differ with respect to handedness, but I suggest that for optimal acquisition of a manual sign-language teacher and learner should share the same preferred handedness (unless, of course, all the signs are performed simultaneously and identically with both hands). I have discussed further implications of the possible relationships between tool-making, tool-using, and language elsewhere (Hewes 1973*b*). The cause for the striking predominance of right-handedness in our species (and thus one of our species-specific attributes, compared to the other primates) may lie in the selective advantage for regular hand-tool and weapon users to share the same handedness, so as to optimize the social transmission of tool-making and tool-using skills. This does not explain the initial impetus for dextral rather than sinistral choice.

Given the rudimentary tool-making and tool-using ability present in modern wild chimpanzees, and the great antiquity of the earliest datable stone tools (Isaac 1983), I propose that implemental behaviour is a more ancient attribute of the hominids than language. This argument is perhaps logically weak, but not as weak as the contention that language arose before tool-making. However, assuming that a gestural (rather than vocal) language arose after the earliest regular tool-using in the hominids, such a language had a disadvantage—minor at the start, but perhaps eventually disadvantageous enough to favour the utilization of the vocal tract for language communication. While the hands are engaged in manipulating tools or weapons, using the hands for signing would create interference—an interference recognized by modern teachers of the deaf. To be sure, human beings can usually organize their activities to minimize the effects of such competing behaviours; few of us would attempt to write a letter while rowing a boat, or to deliver a lecture while using a pneumatic drill.

21.3.4 *Some advantages of gestural communication*

Gestural communication has some advantages over speech (see Bellugi 1988 for the critical difference between 'gesture' and sign language—[eds]). Although spoken languages are culturally universal, it is well known that resort to gesture and miming is more convenient in the short run than acquiring another spoken language in situations where members of different speech communities must interact (see below). Isolated deaf individuals or communities have created viable signed languages (Goldin-Meadow 1982; Goldin-Meadow and Feldman 1977); but aside from incompletely documented cases we do not have good evidence for invention of *ad hoc* spoken languages. Lingua francas which later develop into creoles are all derivable from pre-existing spoken languages, though Bickerton (1981) believes that creoles present syntactical features which may resemble primordial spoken language. What the data on sign-languages versus spoken languages suggest is that the former are somewhat easier to create, and teach to others, and certainly far more effective to use in situations where the participants do not share a common spoken language, and have had no prior contact with developed sign-languages (Hewes 1974).

Gestural and vocal communication begins to manifest itself earlier during infancy than speech (Lock 1978), although in infants with normal hearing both are soon superseded by rudimentary speech. Speech lost in adults with brain damage is often partly replaced by gestural communication, although perhaps at a lowered level of cognitive complexity. Gestures are routinely employed for communication with severely retarded patients in institutions. The Great Apes, as we have noted in the historical survey above, can be taught some elements of sign language, but have uniformly failed to acquire speech under conditions of intensive training (Furness 1916; Hayes and Hayes 1951). The pygmy chimpanzee, *Pan paniscus*, uses several hand and arm gestures without human instruction, chiefly for sexual communication (Savage-Rumbaugh 1984; Savage-Rumbaugh *et al.* 1977; Savage Rumbaugh and Wilkerson 1978; Savage-Rumbaugh *et al.* 1985).

Even some non-primates can acquire limb gestures useful for communicating with human beings, of which the canine signal for summoning someone to open a door is perhaps the most familiar example. There are birds which can learn to imitate human speech sounds, and even entire sentences, but their instrumental use of this ability is usually not as impressive as the canine signal to open the door. There has been a surprising lack of serious research, however, on the extent to which many common domestic animals, and zoo animals as well, learn to respond appropriately to human spoken and gestural commands. So far, aside from primates, the main experiments have been carried out on dolphins and other cetaceans (Lilly 1967).

Delivery of gesture language (for further details on sign languages, see Deuchar, this volume, Chapter 20), as against its reception, requires less precision in movement of body parts than the production of articulate speech, because the spatial area of articulation for speech is much smaller. Gestural signs can be decoded even when the sign is made a centimetre or more from its standard position, whereas no such tolerance is permissible if speech sounds are to be decoded, for articulation errors of a few millimetres may render a common spoken word almost unintelligible. Timing may also be less crucial for gestural signing, as compared, for example, to the precision involved in the temporal onset of voicing in speech (Calvin 1983).

At an early stage of language evolution, gestural-sign lexicons may well have been as capable of providing sufficient numbers of labels for things and concepts as spoken words came to be. But at some undetermined point, it may have become easier for the human brain to cope with thousands of terms coded and filed phonetically than with similar numbers of terms coded in terms of finger and hand positions. Although readers can cope with great numbers of visual signs, especially in the case of 'logographic' scripts such as Chinese (cf. Barton and Hamilton, this volume, Chapter 27), such Chinese written characters are already linked to spoken words, and thus, ultimately, represented phonetically. Traces of these linkages occur in the majority of the written graphs (Newnham 1971).

21.3.5 *Distribution of gestural communication*

The deaf and those who regularly interact with them have commonly resorted to gesture, and many sign languages have developed (Critchley 1939; Paget 1963). Sign languages have also been generated among hearing groups, for specialized communication—among North American Indian tribes (Mallery 1881; West 1960; Kendon 1988), in certain cloistered religious orders (Hutt 1968), and for use in unusually noisy working environments.

There is abundant evidence in accounts of European voyages and travels from the fifteenth century onward that *ad hoc* sign communication not only was attempted, but usually succeeded, in obtaining information (Hewes 1974). Linguists and others who have denied that such signing could overcome cultural and spoken-language barriers are simply ignorant of the relevant literature. Such signals are still essential in the operations of small infantry units, even in modern warfare; but they have other familiar functions.

Shortly after birth, human infants begin to make finger and hand movements that are later incorporated in gestural signs, although they are apparently unintentional. Neonates are also able to imitate rudimentary hand and tongue-thrust movements. Later in infancy, around 9–12 months, pointing gestures appear, though it has not been determined whether such pointing arises from infant imitation of mother's pointing, or is endogenous (Trân Duc Thao 1984; Hewes 1981; Lock 1978; Dobrich and Hollis 1984; see also Messer and Collis, this volume, Chapter 15).

The Great Apes readily learn to point and respond to pointing by humans, but do not seem to employ pointing or many manual gestures in the wild. The best-documented manual gesture in apes is the 'begging gesture', with palm upward, used to solicit food (Goodall 1971). Kortlandt reports a sole-of-the-foot gesture used on the trail by wild chimpanzees (1967). As noted, greater use of gestures, including pointing, in apes not specifically trained by humans is reported for the captive pygmy chimpanzee. So far, observers in the natural forest habitat of the pygmy chimpanzee have reported only a few such gestures (Mori 1984).[3]

The differences in the proportions of the digits of man and the pongids do not seem to be a serious barrier to gestural understanding, according to the Gardners, Fouts, Patterson, and others, all of whom have worked with ASL sign-using apes. In any case, the australopithecines, some of which are our putative ancestors, had hands of more human proportions (for details, see Marzke, this volume, Chapter 5).

Both Old World macaques and New World capuchin monkeys have been trained to imitate a variety of human manual behaviours, traditionally for street entertainment. In some countries, such as Japan, macaques have been trained to perform as actors in complex dramatic episodes, involving even more intricate imitations of human activities (*Dai Hyakka Jiten* 1932). For primates, upper limb morphology *per se* would not have been a significant impediment to the development of signed language at any time during the past several million years, although it continues to prevent some of the other highly intelligent mammals, such as dolphins, killer whales, dogs, and elephants from acquiring the ability to imitate human hand gestures.

21.3.6 *Volar depigmentation as evidence for the primacy of gestural communication*

The ridged skin of the palms and soles (the 'volar skin'), as well as the nailbeds on fingers and toes, are relatively depigmented, often light pinkish-brown even in otherwise very dark-skinned human populations. In a comparative study of volar and nail pigmentation covering many living non-human primates, I found (Hewes 1983a) uniformly dark brown to blackish nails, and dark volar pigmentation in adults, although lighter volar skin was not uncommon as an infant or juvenile feature. This led to the suggestion that the lighter volar skin and pale nailbeds present in all human groups may be a very ancient hominid trait, with the selective advantage of enhancing the visibility of the palms and fingertips both in tool-making and tool-using, since these skills are normally transmitted by visual observation. Modern interpreters for the deaf deliberately emphasize the visibility of their fingers and hands by wearing dark upper garments.

21.3.7 *Phonemes and language*

All spoken languages, including those known indirectly from ancient scripts, where phonetic values were sometimes inadequately represented, appear to be based on the phonemic principle. Duyzentkunst (1978) raises some disturbing questions regarding the very existence of these constituents of spoken language. Without going into a very complicated matter, phonemes are defined as the psycholinguistically minimal units of speech sounds, even though acoustic analysis can resolve speech sounds into still smaller components. Phonemes usually lack statable meanings, and they usually occur in patterned and contrasting sets within the sound-system of a given language. However, it is not difficult to conceive of a non-phonemic spoken language in which each such minimal sound unit would possess referential meaning. Whether this is true of any existing spoken language, fully comprehensible language codes exist in which not only phonemes, but phonetics are absent, whatever may have been their original derivation from speech or a written version of speech. An example would be the telegraphic code used to transmit Chinese characters, which consists of sets of digits which have no direct relationship to Chinese phonology.

But this is precisely the point being made. Without a phonemic system in addition to a code based on small set of vocal sounds, speech would, I suggest, remain exceedingly inefficient, except for short and highly predictable messages. Decoding long messages would then resemble the task of decoding communications in 'dictionary codes', which require resort to a codebook (where, of course, the system has been deliberately contrived to prevent easy decipherment). Pre-phonemic mental lexicons were perhaps organized mainly by semantic categories, much as sets of high-frequency words are grouped into semantic sets in travellers' phrase-books, with words for food in one list, and words for clothing in another.

The indexing efficiency provided by phonemicization must have transformed speech from a tedious and error-prone process into a rapid and dependable vehicle for increasingly complex communication (Hewes 1983b). Eventually, writing systems based on phonemicized speech would raise the speed of decoding scripts far beyond the rates attainable for speech.

Some linguists working with gestural sign languages such as ASL have contended that they can also be regarded as having 'phonemes'—more properly, visible sign-units approximating to phonemes in function (Stokoe 1960; Armstrong 1984; cf. Deuchar, this volume, Chapter 20).[4] If this is the case, it could be an expression of neurolinguistic properties by now genetically programmed in all human beings, deaf or hearing. I shall argue below (Section 21.4.7), in agreement with Krantz (1980), that the phonemic 'principle' came into spoken language at a comparatively recent date, and for functions other than lexicon-enlargement, and that the use of phonemes diffused as a culture-trait rather than by genetic processes (Hewes 1983b).

21.3.8 *Fetal eavesdropping*

A plausible theory of the primacy of gestural language over speech must, as has already been noted, account for its general replacement by spoken language. Spoken-language advantages may include the fact that fetuses *in utero* overhear, probably in a muffled way, vocalizations produced by their mothers, along with other internal and external noises. This has been tested in fetal sheep, who have been shown to exhibit some learning of repeated extra-uterine sounds. Human neonates attend more closely to the patterns of their own mothers' voices than to the voices of other adult females, as measured by instruments attached to nipped devices, indicating similar prenatal learning (DeCasper and Fifer 1980; von Raffler-Engel 1983; Kolata 1984).

Had there been gestural language prior to the development of speech, fetuses obviously could not have acquired much prenatal familiarity with visible signing, unless such gestures were made with considerable force, so as to actually jar the fetus. One of the selective advantages of speech may have lain in the possibility of some learned familiarity with certain aspects of spoken language even before birth. Had there been a lengthy period of combined gestural and vocal language, fetal familiarity with the latter might have favoured the shift in the direction of speech. Alternatively, fetal eavesdropping could be interpreted as an argument for vocal primacy.

21.3.9 *Possible cognitive and other advantages of speech*

At the risk of alienating some colleagues (for example, Lieberman 1985), who insist on the full equivalency of gestural sign languages and spoken languages, I contend that there are some very likely, if presently untestable, cognitive advantages to speech. These could lie in the realm of memory storage of the components of spoken language, and their rapid access, which I have linked to phonemicization (see above, Section 21.3.8), but which may inhere in other aspects of speech as well.

It is conceivable that the visual explicitness of gestural language might inhibit the development of more abstract conceptual thinking by being over-concrete. In modern languages with a long civilized past we employ a great many words and phrases which we may not ever concretize in the sense of rendering them in visual imagery.[5] Although most common words come from visualizable or kinaesthetic roots, it may interfere with our easy use of them to be reminded of their basis in tangible things or visible actions. 'Understanding' is a convenient English word, but I am not sure that we would find it so useful were we always reminded by it of something standing beneath something else. Perhaps gestural languages, more than spoken ones, suffer from proneness to the logical fallacy of misplaced concreteness.

Differences in 'dialect' are well known in the non-human world as mechanisms to facilitate dispersion and boundary-keeping, notably in the case of songbirds. Similarly, spoken languages appear to serve their users' group-marking needs, possibly more effectively than gestural language could.

21.4 A gestural-origin model for language

Here I shall offer a model for the gestural origin of language, with its necessary corollary, a model for the replacement of gestural language by speech.

21.4.1 *Biological and cultural pressures for continuing language evolution*

Most linguists have been reluctant to face the possibilities of drastic differences between languages of the past few millennia and those of tens or hundreds of thousands of years ago. Indeed, this reluctance has frozen into a dogma, which in effect denies that language has undergone any fundamental transformations since the very first language came into use.

In contrast to this steady-state model, I shall argue here that profound changes probably occurred in the course of language evolution, involving a shift from a gestural to vocal basis, from a non-phonemic vocal language condition to phonemicized languages, and from gestural languages with a low capacity for handling complex concepts to the specialized mathematized and logical languages which underlie the achievements of recent science and technology.

This viewpoint has a minor corollary: different language systems, far from being equally suitable for advanced cognitive operations, have been significantly unequal for at least several millennia. This does not mean, to be sure, that a language with such deficiencies cannot be modified to overcome them. Hebrew, which survived into modern times chiefly as a religious language, needed considerable and deliberate updating to become a suitable linguistic vehicle for modern science and technology. In that case, the modernization was chiefly lexical; but it is conceivable that for some languages, syntactic reorganization might also be required, (cf. Rolfe, this volume, Chapter 26).

Going back to the postulated early gestural proto-language, it is unreasonable to suppose that it appeared as a consequence of changes in the hominid genome. In instances where non-human species have embarked on new behavioural adventures, such as the tomtits who acquired the new habit of plucking caps off milk bottles on British front doorsteps, we have not had to postulate genetic changes in the birds who participated in this novel addition to their subsistence pattern. If tomtit interaction with milk-bottle caps were to persist for hundreds of thousands or millions of years, it would indeed be likely that selective modifications would appear in their bills, etc., to optimize bottle-top removal. Our scenario for the origin and subsequent development and transformations of language conforms to the established evolutionary view that new functions are built upon or out of already-existing ones, and do not appear *ex nihilo*.

21.4.2 Cerebral lateralization

Among the factors probably entering the mix which fostered the emergence of language was, as I have suggested, the rise of right-handedness in connection with regular tool-making and tool-using. I have offered an explanation for why it would have been advantageous for tool-using hominids to become more homogeneous in respect to handedness (see Section 21.3.3), but the basis for the right-hand advantage remains obscure, aside from the notion that the right arm and hand have a slight circulatory edge over the left. This topic is sufficiently important to merit a separate chapter, but I shall leave it at this point (cf. Kinsbourne 1978; Needham 1982).

Right-handed precision manipulation of objects by close to 95 per cent of modern human individuals, and according to some stone-tool evidence, of similarly high percentages of prehistoric right-handers (for example, Toth 1985), indicates that dextrality is a deeply-rooted and species-specific hominid trait. It is something which may have developed along with the regular making and use of tools, which is to say, at the latest some time before two million years ago. Right-handedness and the comparable localization of controls for the production of a gestural proto-language on the same left cerebral hemispheres might thus have developed in tandem. The social learning involved in the transmission of tool-making and tool-using behaviours is comparable to that involved in the transmission of gestural signalling, and it is reasonable to suppose that both activities tended to focus on the right hand, and for their visual monitoring, on the left hemisphere projection areas served by the usually dominant right eye. Modern pongids readily learn to imitate manual behaviours providing obvious subsistence benefits; it seems logical to credit the earliest hominids with comparable capabilities.

21.4.3 *Deixis as a starting-point for the early hominids*

For early hominids inhabiting the savannah–woodland mosaic a few rudimentary gestures such as pointing would serve admirably, once instituted. Trân Duc Thao (1984) emphasized the productive elaborations possible for a sign language starting from simple deixis. Pointing and its variants can serve command or prescriptive functions, as well as deictic ones (see Rolfe, this volume, Chapter 26). Such gestures are still of everyday importance, and include the common command for silence (hand or finger against the mouth), come-hither commands or requests, indicators of sensory inputs by means of touching the ear, nose, tongue, or eye, the 'halt' gesture, the 'follow me' gesture, the 'go away' gesture, and so on. Although none of these specific signs may have been used by early humans, I believe that most of them are now universally understandable within our species. The few cultural groups in which finger-pointing is tabooed, for example the Navajo of North America, do not appear to me to constitute genuine exceptions to the universality of deixis, but only specialized rules of etiquette which would be unnecessary if the gesture were in fact totally unknown.

21.4.4 *Hypothetical onset of gestural communication*

I suppose that some simple gestures began to be used habitually by early hominids before there was much, if any, hunting of big game, and probably before stone tools were being made on a regular basis, although tools of perishable wood, etc., were probably already in use. This state of affairs could have lasted for many millennia before the propensities for either tool-using or gesture were enhanced by genetic factors, and not just through the social transmission of learning of the sort seen in sweet-potato washing by Japanese macaques.

The feedback from a half million to a million years or more of sustained employment of tools and gestures beyond what we can see in today's chimpanzees (which I take as a rough model for pre-bipedal, proto-hominid behaviour) may have gradually led to an enlargement of the cerebral hemispheres such as can be seen in the cranium of KNM-ER 1470 (see Campbell, and also Holloway, this volume, Chapters 2 and 4). Such changes in the skull almost certainly imply significant genetic changes arising from selective pressures arising in behaviour. Hominids with average brain volumes of 700 cc may well have been able to sustain a proto-sign-language, i.e., a system going beyond the production of isolated gestural signals. This point is based on the fact that modern microcephalics of comparable cranial size are often capable of limited spoken language. Microcephalic humans today tend to be seriously retarded quite apart from language; whereas early hominids with 700 cc cranial capacity were obviously fully capable of surviving on their own (cf. Lenneberg 1967). Incidentally, there has been little scientific interest in microcephalics, who are usually placed in institutions at an early age, and are rarely considered worth studying (but see Seckel 1960), and certainly not from the standpoint of hominid evolution.

21.4.5 *Beginnings of vocal language, hypothetically as an accompaniment of gestural communication*

Some time between the level of KNM-ER 1470 and company (about 2 m.y.a.) and about 100 000 years ago, at a rough guess, vocally based language came to coexist with gesture language. The later date has been chosen because by then a marked upswing in technological productivity was under way, as exemplified by the Levallois flake technique. Moreover, some of the surviving skulls of the hominids of this period (500 000–100 000 years ago), such as the Broken Hill ('Rhodesian Man') specimen from Zambia, retain many archaic facial features, such as very heavy brow-ridges and prognathism, but have essentially modern vocal-tract morphology, to the extent that it can be reconstructed from the basicranium (Lieberman 1984; Laitman, this volume, Chapter 3). These considerations would tend to place the advent of vocal language some time in the later part of the span from 1.5 million years ago to 100 000 years before the present.

The ideas about the movement from a postulated pre-speech language to a rudimentary spoken one are admittedly the weakest part of my model. Partly by default, I can only insert at this point the mouth-gesture hypothesis, which goes back to Wallace (1895) and to Rae (1862), who first rather casually suggested it. The notion was elaborated by Jóhannesson (1952) and by Paget (1963).

The basic idea, in brief, holds that the various positions and movements of the mouth parts employed in speech are traces or approximations, by the voluntary muscles of the upper vocal tract, of actions of the arms, hands, fingers, etc., in dealing with the external world, and hence able to serve as signs of actions (cf. Foster, this volume, Chapter 25). There is a kind of evolutionary recapitulation loop in this view, in that the forelimbs in many higher vertebrates took over some of the food-handling actions originally limited to the mouth, but now, recursively, the mouth in turn assumed some of the communicative functions previously performed by the forelimbs. The mouth-gesture theory asserts that mouth movements, patterned after hand gestures, were combined with vocalizations so as to produce articulations: distinctive vowel and consonant sounds. Perhaps the very specialized click consonants, now limited to a few languages in Africa, played some role in this process (Stopa 1968). All of this is highly conjectural in the extreme, and I only offer it for lack of anything better.

The banal advantages of speech over gesture probably also played a part. Speech is handier to use in the dark, and can overcome visual barriers, such as vegetation, which might interfere with gestural communication. Finally, speech can be carried on while performing various manual tasks so long as such tasks are not too noisy. Eventually, such advantages *might* help to tilt the balance between gestural and vocal language in favour of the latter. Gestural communication did not disappear, and it continues to play a very important role in human language, though one generally subordinate or auxiliary to speech.

There are other conceivable pathways to speech, which have often been advanced as the direct source of speech sounds and the 'words' we construct from them. These include onomatopoeia or imitation of natural sounds. There is evidence that at least small parts of most spoken languages consist of sound imitations, notably in the names for birds, and more widely as sound-symbolism. Imitation of the sounds made by other species, whether of prey animals, or of birds, does play a role in the hunting methods of modern hunting people, and would have been advantageous from the start. Hunters even in modern civilized cultures employ imitations of animal calls, although the most reliable ones are produced by special artefacts, such as whistles, which can attract game, from ducks to elk. Imitations of natural birdcalls are also sometimes used for signalling to one's fellow soldiers in clandestine military situations, or to fellow poachers, where it is important to conceal the fact that human beings are communicating with one another. Unfortunately, attainment of the ability to produce such imitations or quasi-imitations of naturally occurring noises demands the very troublesome achievement of fully voluntary (pre-phonemic) control of the organs of the vocal tract.

To summarize this somewhat unsatisfactory part of my argument, I am assuming a phase of mixed gestural and spoken language, the latter quite unlike anything now existing, and perhaps resting on the principle of mouth-gesture.

21.4.6 Attainment of habitual but still non-phonemic vocal language

The phonological systems of all existent spoken languages, and of all languages known from the past via writing, seem to be based on phonemic principles (Pulleyblank 1983). Yet it is unnecessary to insist that the earliest spoken languages had this elegant design feature. Instead, it is reasonable to suppose that regular use of vocal communication of a kind very different from any existing spoken language had appeared by 100 000 years ago, if not well before that. Evidence for this is admittedly weak. Mainly, it comes down to the evidence some fossil specimens from roughly that period had human-like vocal-tract morphology.

21.4.7 World-wide phonemicization of speech

Krantz (1980) originally hypothesized that the remodelling of the human face and jaws into their characteristic form in *Homo sapiens* was due to what he called the 'phonemicization of speech'. He proposed that the cause of the transformation lay in the very rapid diffusion of a new kind of vocal language, utilizing phonemes. Building on Krantz's ideas, I examined (Hewes 1983b) the advantages of phonemicized speech, and how phonemes could be considered cultural/linguistic inventions, capable of diffusion, not necessarily at first as a ready-made set of orderly replacements for earlier phonological units, but as de-semanticized phonetic bits, as in modern loanwords. I therefore modify Krantz's model by suggesting that the spread of phonemicized language entailed a dramatic change in overall human cognitive processing, ultimately affecting, I believe, almost everything in everyday life, from subsistence techniques and kinship relations to supernatural beliefs. Thus, if we assume the spread of phonemicized speech as the diffusion of a set of linguistic inventions, we could further speculate that it did not simply coincide with the pan-human cultural revolution which we call the Upper Palaeolithic, but was its principal stimulus. The final subservience of gestural language in providing no more than the suprasegmental markers which accompany modern speech, or being employed in specialized situations, such as communication across spoken-language boundaries, may have occurred by the beginning of the Upper Palaeolithic.

Most previous explanations of the utility of phonemes focus on the fact that, with a small set of phonemic units, very large lexicons can be built. While this is undoubtedly true, the more basic advantage seems to me to lie in the indexical function of phonemes, and the facilitation of storage and retrieval from the mental lexicon. My claim is that phonemes constitute a largely subconscious equivalent for speakers of what alphabetization provides for readers and writers, enabling users to find particular lexical items with greater efficiency than if lexemes were arranged in brain storage according to semantic categories, or according to other principles such as rhyme. It took many centuries for people in literate societies to hit upon alphabetization (Isidore of Seville, d. AD 636, was an early user of the principle in his encyclopedic *Etymologiae*) for word-listing purposes. Analogies to alphabetical filing have developed in cultures with non-alphabetic scripts, such as Chinese, although they remain somewhat more cumbersome, and one of the Chinese word-finding systems still rests on semantic domain-markers (Newnham 1971).

By 25 000 to 15 000 years ago, the phonemic principle may have been adopted in all or almost all spoken languages. The now extinct Tasmanian languages, cut off from contact with the Australian continent by sea-level changes at the end of the Pleistocene c.14 000–12 000, provide the evidence for a *terminus ad quem* for the pan-human diffusion of phonemicized speech. World-wide examples of sound symbolism may indicate that the process was not uniformly successful in eliminating semantic linkages to individual phonological units.

Most modern linguists reject the notion that spoken language has undergone any progressive evolution since its earliest establishment. Where anything of the sort is grudgingly admitted, it is usually reserved for the growth of lexicons, since languages spoken in complex civilized societies generally possess much larger numbers of words than languages of small, isolated social groups. While basic syntactical systems may not have changed greatly, examination of the specialized languages of law, philosophy, and science suggests that some syntactic development has also occurred, especially in written languages, to overcome the ambiguities which in ordinary speech are usually resolved by the immediate context of discourse (see also Rolfe, this volume, Chapter 26).

21.5 Conclusions

The foregoing ideas are built on data and speculations with a wide range of plausibility. We are still very far from a scientific model for the origin and evolution of language which can be supported at most points by convincing evidence, logical argument, and the canon of falsifiability. That there may have been a very long period in the prehistoric past when language was mainly gestural rather than vocal is still debatable. Yet even if the gestural model can be shown to be wholly invalid, there remain all sorts of problems with its rivals—especially with the notion that language was vocal from the beginning.

Notes

1. Scientific scenarios are complex hypotheses expressed in semi-narrative form. Guidelines for their adequate development have yet to be fully agreed upon, but following J. A. Harris Van Couvering (1983) one can advance a possible working format to bring rigour to scenario-weaving.
This would include:
   1. a statement of hypotheses;
   2. acknowledgement of background paradigms and cultural biases;
   3. detailed reference to the data-base, original or cited;
   4. examination of hypotheses using these data; and
   5. iteration of the scenario.

This type of explicit format has yet to be adopted by the theoreticians working on the evolutionary origins of language [eds].

2. Steklis (for example, Steklis and Raleigh 1979; Steklis 1985) is the main proponent of the counter-argument to gestural primacy theory. On the basis of a comparative review of the cognitive, productive, and perceptual aspects of language, and their anatomical correlates, Steklis and Raleigh (1979, pp. 301–2) conclude:

First, currently there is no evidence of a qualitative distinction between humans and African apes in language-related cognitive abilities (e.g., cross-modal perception, propositional symbol use). African apes possess (at least) the rudiments of these abilities, which humans have significantly elaborated and facilitated by brain expansion and language development. Furthermore, because these complex abilities are manifested by two ape genera and by humans, it is likely that they are homologous among these hominoids and therefore may be inferred to have been present in the earliest Hominidae. Second, vocal tract differences between humans and other primates are not likely to represent critical limitations to speech development. Despite differences in peripheral anatomy, several non-human primates are able to produce speech sounds (i.e., formants and vowels) not predicted from their peripheral anatomy. Third, all primates appear to have some neocortical control of the vocal apparatus; however, in humans, additional cortical association areas in the frontal, temporal, and parietal lobes support speech motor mechanisms. In addition, the control mechanisms of speech production, unlike those mediating vocalization in other primates, are hemispherically lateralized at cortical and subcortical (i.e., thalamic) levels. This speech motor system has in humans become linked closely to neural mechanisms, which perhaps in all higher primates govern sequential oral-facial and arm movements. Fourth, the categorical nature of speech perception does not appear to be unique to speech sounds or the auditory modality. Categorical decoding of speech may be found in information processing capacities evolved by all (at least) higher primates. Furthermore, in two primate species this vocal decoding system appears to involve hemispheric specialization, which therefore occurred independently of and perhaps prior to the development of human language.

The implications of this review are claimed (ibid., pp. 304–5) to be that the

primordial speech system emerged in the earlier phases of human evolution and was unlikely to have been preceded by a gestural language system.

Later, Steklis (1985, p. 169) concluded in a similar vein:

"gestural" models of language origin (e.g. Hewes 1973a; Steklis and Harnad 1976) now appear less compatible than "vocalist" models with the present comparative data on primate communicative abilities. [eds]

3. Plooij (1978, 1984), however, reports remarkable parallels (once different ecological demands are taken into account) between infant feral chimpanzees and

human infants in their early preverbal development of a gestural repertoire [eds].

4. Sublexical structure consists of simultaneous contrasting features without phoneme-like units. Contrasts of handshape, movement, and place of articulation are used to produce morphemes directly (see Bellugi 1988) [eds].

5. I suspect that it would be difficult to devise tests for this, given the fact that as users of languages with vocabularies long 'contaminated' by literary, philosophical, and legal metaphors, etc., we could not respond naïvely to tests couched in the very terms to which we have reference. With respect to tests bearing on the role of visual imagery accompanying verbal language, the obvious difficulty is that, except for subjects with unusual drawing ability, the only way to elicit information about visual imagery is by means of further verbalizations.

References

Armstrong, D. F. (1984). Iconicity, arbitrariness, and duality of patterning in signed and spoken languages: perspectives on language evolution. *Sign Language Studies*, **38**, 31–83.

Bates, E., Benigni, L., Bretherton, I., Camaioni, L., and Volterra, V. (1977). From gesture to the first word: on cognitive and social prerequisites. In *Origins of behavior: communication and language* (ed. M. Lewis and L. Rosenblum), pp. 247–307. Wiley, New York.

Bellugi, U. (1988). The acquisition of a spatial language. In *The development of language and language researchers: essays in honor of Roger Brown* (ed. F. Kessell), Erlbaum, Hillsdale, NJ.

Bickerton, D. (1981). *Roots of language*. Karoma Publishers, Ann Arbor.

Bryden, M. P. (1982). *Laterality: a functional asymmetry in the intact brain*. Academic Press, New York.

Bunak, V. V. (1959). Present state of the problem of the origin of speech and the early stages of its evolution. *Journal of World History*, **5**, 310–24.

Calvin, W. H. (1983). *The throwing madonna*. McGraw-Hill, New York.

Chevalier-Skolnikoff, S. and Poirier, F. E. (1977). *Primate bio-social development: social and ecological determinants*. Garland, New York.

Chomsky, N. (1957). *Syntactic structures*. Mouton, The Hague.

Chomsky, N. (1959). Review of B. F. Skinner, *Verbal behavior*, 1957. *Language*, **35**, 26–58.

Chomsky, N. (1966). *Cartesian linguistics: a chapter in the history of rationalist thought*. Harper and Row, New York.

Critchley, M. (1939). *The language of gesture*. Edward Arnold, London.

Dai Hyakka Jiten [encyclopaedia] (1932). Vol. 10, p. 625 s.v. Sarushibai. Heibonsha, Tokyo.

Dart, R. A. (1959). On the evolution of language and articulate speech. *Homo*, **10**, 154–65.

Darwin, C. (1859). *The origin of species*. J. Murray, London.

Darwin, C. (1871). *The descent of man and selection in relation to sex*. D. Appleton and Company, New York.

Darwin, C. (1872). *The expression of the emotions in man and animals*. John Murray, London.

DeCasper, A. J. and Fifer, W. P. (1980). Of human bonding—newborns prefer their mothers' voices. *Science*, **208**, 174–6.

de Grolier, E. (ed.) (1983). *Glossogenetics: the origin and evolution of language*. Harwood Academic Publishers, Chur.

Dingwall, W. O. (1988). The evolution of human communicative behaviour. In *Linguistics: the Cambridge Survey*, Vol. 4, (ed. F. J. Newman), pp. 274–313. Cambridge University Press, New York.

Dobrich, W., and Hollis, S. S. (1984). Form and function in early communication: language and pointing gestures. *Journal of Experimental Child Psychology*, **38**, 475–90.

Donovan, J. (1893). The festal origin of speech. *Mind*, **16**, 498–506; **17**, 325–39.

Duyzenkunst, F. Balk-Smit (1978). Phoneme and alphabet. In *Linguistics in the Netherlands* (ed. Wim Zonneveld), pp. 1–6. The Peter D. Ridder Press, Lisse.

Furness, W. H. (1916). Observations on the intelligence of chimpanzees and orang-utans. *Proceedings of the American Philosophical Society*, **55**, 281–90.

Gardner, R. A. and Gardner, B. T. (1969). Teaching sign-language to a chimpanzee. *Science*, **165**, 664–72.

Garner, R. L. (1900). *Apes and monkeys: their life and language*. Ginn and Company, Boston.

Goldin-Meadow, S. (1982). The resilience of recursion: a study of a communicative system developed without a conventional language model. In *Language acquisition: the state of the art* (ed. L. R. Gleitman and E. Wenner), pp. 51–77. Cambridge University Press, New York.

Goldin-Meadow, S. and Feldman, H. (1977). The development of language-like communication without a language model. *Science*, **197**, 401–3.

Goodall, J. (1963). My life among wild chimpanzees. *National Geographic Magazine*, **124**, 272–308.

Goodall, J. (1964). Tool-using and aimed throwing among wild chimpanzees. *Nature*, **201**, 1264–6.

Goodall, J. (1971). *In the shadow of man*. Houghton-Mifflin, Boston.

Goodall, J. (1972). Observational learning among the Gombe Stream chimpanzees. *Abstracts, 4th International Congress of Primatology*, Portland, Oregon.

Gouzoules, S., Gouzoules, H., and Marler, P. (1984). Rhesus monkey (*Macaca mulatta*) screams: representational signalling in the recruitment of agonistic aid. *Animal Behaviour*, **32**, 182–93.

Green, S. and Marler, P. (1979). Analysis of animal communication. In *Social behavior and communication: handbook of behavioral neurobiology*, Vol. 3, (ed. J.

Vandenbergh and P. Marler), pp. 73–158. Plenum Press, New York.

Harnad, S. R., Steklis, H. D., and Lancaster, J. (eds) (1976). *Origins and evolution of language and speech. Annals of the New York Academy of Sciences*, **Vol. 280**.

Harris Van Couvering, J. A. (1983). Human origins, scenarios, and ethics. *American Journal of Physical Anthropology*, **60**, 204 [Abstract].

Hayes, K. J., and Hayes, C. (1951). The intellectual development of a home-raised chimp. *Proceedings of the American Philosophical Society*, **95**, 105–9.

Hewes, Gordon W. (1969). Human language origins and the Washoe sign language experiment. *Colorado–Wyoming Academy of Sciences: Proceedings, 40th Annual Meeting, Abstracts*, p. 41.

Hewes, G. W. (1973a). Primate communication and the gestural origin of language. *Current Anthropology*, **14**, 5–24.

Hewes, G. W. (1973b). An explicit formulation of the relation between tool-using and early human language emergence. *Visible Language*, **7**, 101–27.

Hewes, G. W. (1974). Gesture language in culture contact. *Sign Language Studies*, **1**, 1–34.

Hewes, G. W. (1977). Language origin theories. In *Language learning by a chimpanzee. The LANA project* (ed. D. M. Rumbaugh), pp. 5–53. Academic Press, New York.

Hewes, G. W. (1981). Pointing and language. In *The cognitive representation of speech* (ed. T. Myers, J. Laver, and J. Anderson), pp. 263–9. North Holland, Amsterdam.

Hewes, G. W. (1983a). The communicative function of palmar pigmentation in man. *Journal of Human Evolution*, **12**, 297–303.

Hewes, G. W. (1983b). The invention of phonemically-based language. In *Glossogenetics: the origin and evolution of language* (ed. E. de Grolier), pp. 143–62. Harwood Academic Publishers, Chur.

Hockett, C. F. (1959). Animal languages and human language. *Human Biology*, **31**, 32–9.

Hockett, C. F. (1961). Logical considerations in the study of animal communication. In *Animal sounds and communication*, American Institute of Biological Science, Publication 7, (ed. W. E. Lanyon and W. N. Tavolga), pp. 392–430. The Institute, Washington DC.

Hockett, C. F. (1978). In search of Jove's brow. *American Speech*, **53**, 243–319.

Hutt, C. (1968). Étude d'un corpus: dictionnaire du langage gestuel chez les Trappistes. *Langages*, **10**, 107–18.

Isaac, G. L. (1983). Aspects of human evolution. In *Evolution from molecules to men* (ed. D. S. Bendell), pp. 509–43. Cambridge University Press.

Isidore of Seville (d. AD 636). *Etymologiae*. Spanish trans.: L. Cortés y Gongora, *Etimológias*. Biblioteca de Autores Cristianos, Madrid, 1951.

Itani, J. (1958). On the acquisition and propagation of a new food habit in the troop of Japanese monkeys at Takasakiyama. *Primates*, **1**, 84–98 [in Japanese].

Jerison, H. J. (1983). Evolutionary neurobiology and the origin of language as a cognitive adaptation. Unpublished paper read at the International Congress of Anthropological and Ethnological Sciences, Vancouver, British Columbia.

Jóhannesson, A. (1952). *Gestural origin of language*. H. F. Leiftur, Reykjavik, and B. H. Blackwell, Oxford.

Kawai, M. (1965). Newly acquired pre-cultural behaviour in the natural troop of Japanese monkeys on Koshima Islet. *Primates*, **6**, 1–30.

Kendon, A. (1982). The study of gestures: some observations on its history. *Recherches Sémiotiques/ Semiotic Inquiry: Journal of the Canadian Semiotic Association*, **March 1982**, p. 27.

Kendon, A. (1983). *Some considerations for a theory of language origins*, Litteris XI: Language. Connecticut College, New London.

Kendon, A. (1988). *Sign languages of Aboriginal Australia*. Cambridge University Press.

Kinsbourne, M. (ed.) (1978). *Asymmetrical function of the brain*. Cambridge University Press.

Köhler, W. (1925). *The mentality of apes* (repr. 1959). Vintage Books, New York.

Kortlandt, A. (1967). Handgebrauch bei freilebenden Schimpansen. In *Handgebrauch und Verstaendigung bei Affen und Fruehmenschen* (ed. B. Rensch), Hueber, Berne.

Kolata, G. (1984). Studying learning in the womb. *Science*, **225**, 302–3.

Krantz, G. S. (1980). Sapienization and speech. *Current Anthropology*, **21**, 773–92.

Kronenfeld, D. B. (1978/79). Innate language. *Language Sciences*, **1**, 209–39.

Landsberg, M. E. (ed.) (1988). *The genesis of language: a different judgement of evidence*. Mouton de Gruyter, Berlin.

Lanyon, W. E. and Tavolga, W. N. (eds) (1960). *Animal sounds and communication*. American Institute of Biological Sciences, Washington DC.

Lenneberg, E. H. (1967). *Biological foundations of language*. Wiley, New York.

Lieberman, P. (1984). *The biology and evolution of language*. Harvard University Press, Cambridge, Mass.

Lieberman, P. (1985). On the evolution of human syntactic ability: its preadaptive bases—motor control and speech. *Journal of Human Evolution*, **14**, 657–68.

Lilly, J. C. (1967). *The mind of a dolphin*. Doubleday, New York.

Lock, A. (ed.) (1978). *Action, gesture and symbol: the emergence of language*. Academic Press, London.

Lucas, P. W. and Corlett, R. T. (1985). PlioPleistocene hominid diets: an approach combining masticatory and ecological analysis. *Journal of Human Evolution*, **14**, 187–202.

Lumsden, C. J. (1983). Neuronal group selection and the evolution of hominid cranial capacity. *Journal of Human Evolution*, **12**, 169–84.

Mallery, G. (1881). Sign language among the North American Indians compared with that among other peoples and deaf-mutes. *Bureau of American Ethnology, Annual Report*, **1**, 263–552.

Mori, A. (1984). An extended study of pygmy chimpanzees in Wamba, Zaire: a comparison with chimpanzees. *Primates*, **25**, 255–78.

Morris, D., Collett, P., Marsh, P., and O'Shaughnessy, M. (1979). *Gestures: their origins and distribution*. Stein and Day, New York.

Müller, F. M. (1891). *The science of language*, 2 vols, (1st edn, 1877). Longman, Green and Company, London.

Needham, C. W. (1982). *The principles of cerebral dominance*. Charles C. Thomas, Springfield, Illinois.

Newnham, R. (1971). *About Chinese*. Penguin Books, Harmondsworth.

Noiré, L. (1917). *The origin and philosophy of language* (2nd rev. edn [1st edn, 1877]). The Open Court Publishing Company, Chicago.

Nottebohm, F. (1978). Origins and mechanisms in the establishment of cerebral dominance. In *Handbook of behavioral neurology* (ed. M. S. Gazzaniga), Vol. 2, pp. 295–344. Plenum Press, New York.

Paget, R. A. S. (1963). *Human speech: some observations, experiments and conclusions as to the nature, origin, purpose and possible improvement of human speech*. Routledge and Kegan Paul, London.

Parker, S. T., and Gibson, K. R. (1979). A developmental model for the evolution of language and intelligence in early hominids. *The Behavioural and Brain Sciences*, **2**, 367–408.

Parker, S. T. and Gibson, K. R. (1982). The importance of theory for reconstructing the evolution of language and intelligence in hominids. In *Advanced views in primate biology* (ed. A. B. Chiarelli and R. S. Coruccini), pp. 42–65. Springer-Verlag, Berlin.

Piaget, J. (1955). *The language and thought of the child*. Meridian, New York and Routledge and Kegan Paul, London.

Plooij, F. X. (1978). Some basic traits of language in wild chimpanzees? In *Action, gesture and symbol: the emergence of language* (ed. A. J. Lock), pp. 111–31. Academic Press, London.

Plooij, F. X. (1984). *The behavioural development of free-living chimpanzee babies and infants*. Ablex, Norwood, NJ.

Premack, D. (1986). *Gavagai, or the future history of the animal language controversy*. The MIT Press, Cambridge, Mass.

Pulleyblank, E. G. (1983). The beginnings of duality of patterning in language. In *Glossogenetics: the origin and evolution of language* (ed. E. de Grolier), pp. 369–410. Harwood Academic Publishers, Chur.

Rae, J. (1862). *The Polynesian* (repr. as an appendix to Paget (1963), q.v.).

Savage-Rumbaugh, E. S. (1984). *Pan paniscus* and *Pan troglodytes*: contrasts in preverbal communicative competence. In *The pygmy chimpanzee: evolutionary biology and behaviour* (ed. R. L. Susman), pp. 395–413. Plenum Press, New York.

Savage-Rumbaugh, S. and Wilkerson, B. J. (1978). Socio-sexual behaviour in *Pan paniscus* and *Pan troglodytes*: a comparative study. *Journal of Human Evolution*, **7**, 327–44.

Savage-Rumbaugh, S., Wilkerson, B. J., and Bakeman, R. (1977). Spontaneous gestural communication among conspecifics in the pygmy chimpanzee (*Pan paniscus*) In *Progress in ape research* (ed. G. H. Bourne), pp. 97–116. Academic Press, New York.

Savage-Rumbaugh, S., Sevic, R. A., Rumbaugh, D. M., and Rubert, E. (1985). The capacity of animals to acquire language: do species differences have anything to say to us? *Royal Society of London, Philosophical Transactions*, **B 308**, 177–85.

Sebeok, T. A. (ed.) (1968). *Animal communication: techniques of study and results of research*. Indiana University Press, Bloomington.

Sebeok, T. A., and Umiker-Sebeok, D. J. (eds) (1980). *Speaking of apes: a critical anthology of two-way communication with man*. Plenum Press, New York.

Seckel, H. P. G. (1960). *Bird-headed dwarfs: studies in developmental anthropology including human proportions*. Karger, Basle.

Seidenberg, M. S. (1984). Aping language. *Semiotica*, **44**(1/2), 177–94.

Seyfarth, R. M., Cheney, D. L., and Marler, P. (1980). Vervet monkey alarm calls: semantic communication in a free-ranging primate. *Animal Behaviour*, **28**, 1070–94.

Steklis, H. D. (1985). Primate communication, comparative neurology, and the origin of language re-examined. *Journal of Human Evolution*, **14**, 157–73.

Steklis, H. D. and Harnad, S. R. (1976). Some critical stages in the evolution of language. In *Origins and evolution of language and speech* (ed. S. R. Harnad, H. D. Steklis, and J. Lancaster). *Annals of the New York Academy of Sciences*, **Vol. 280**.

Steklis, H. D. and Raleigh, M. J. (1979). Requisites for language: interspecific and evolutionary aspects. In *Neurobiology of social communication in primates: an evolutionary perspective* (ed. H. D. Steklis and M. J. Raleigh), Academic Press, New York.

Stokoe, W. C., jun. (1960). *Sign language structure: an outline of the visual communication systems of the American deaf*, University of Buffalo Studies in Linguistics, Occasional Paper, 8. University of Buffalo.

Stokoe, W. C., jun. (1974). Motor signs as the first form of language. In *Language origins* (ed. R. Wescott), pp. 35–49. Linstok Press, Silver Spring, Maryland.

Stokoe, W. C., jun. (1978). Sign language versus spoken language. *Sign Language Studies*, **18**, 69–90.

Stopa, R. (1968). Kann man eine Bruecke schlagen zwischen der Kommunikation der Primaten und derjenigen der Urmensche? *Homo, Zeitschrift für die vergleichende Forschung am Menschen*, **19**, 129–51.

Terrace, H. (1984). Language in apes. In *The meaning of primate signals* (ed. R. Harré and V. Reynolds), pp. 179–207. Cambridge University Press.

Toth, N. (1985). Archaeological evidence of preferential right-handedness in the Lower and Middle Pleistocene, and its possible implications. *Journal of Human Evolution*, **14**, 607–14.

Trân Duc Thao (1984). *Investigations into the origin of language and consciousness* (trans. D. J. Herman and R. L. Armstrong). Reidel, Dordrecht.

Tylor, E. B. (1904). *Anthropology: an introduction to the study of man and civilization.* J. A. Hill and Company, New York.

Valsiner, J. and Allik, J. (1982). General semantic capabilities of the higher primates: some hypotheses on communication and cognition in the evolution of human semiotic systems. In *Nonverbal communication today: current research* (ed. Mary Ritchie Key), pp. 245–57. Mouton, Berlin.

von Frisch, K. (1967). *The dance language and orientation of bees.* Harvard University Press, Cambridge, Mass.

von Glasersfeld, E. C. (1977). Linguistic communication: theory and definition. In *Language learning by a chimpanzee: the Lana project* (ed. D. M. Rumbaugh), Academic Press, New York.

von Raffler-Engel, W. (1983). On the synchronous development of gesticulation and vocalization in man's early communicative behaviour. In *Glossogenetics: the origin and evolution of language* (ed. E. de Grolier), pp. 295–311. Harwood Academic Publishers, Chur.

Walker, L. C. (1981). The ontogeny of the neural substrate for language. *Journal of Human Evolution*, **10**, 429–41.

Wallace, A. R. (1895). Expressiveness of speech; or, mouth gestures as a factor in the origin of language. *Fortnightly Review*, **64**, NS 58, 528–43.

Warren, J. M. (1953). Handedness in the rhesus monkey. *Science*, **118**, 622–3.

Wescott, R. W. (1967). The evolution of language: re-opening a closed subject. *Studies in Linguistics*, **19**, 67–82.

Wescott, R. W. (ed.) (1974). *Language origins.* Linstok Press, Silver Spring, Maryland.

Wescott, R. W. (1984). Semogenesis and paleosemiotics. *Semiotica*, **48**, 181–5.

West, L., jun. (1960). The sign language: an analysis. Unpublished Ph. D. dissertation, Indiana University.

Wundt, W. (1912). *Voelkerpsychologie; eine Untersuchung der Entwicklungsgesetze von Sprache, Mythos und Sitte*, Part I, 2 vols, (3rd edn). W. Engelmann, Leipzig.

Zajonc, B. (1985). Emotion and facial expression: a theory reclaimed. *Science*, **228**, 15–21.

Zoloth, S. R., Petersen, M. R., Beecher, M. D., Green, S., Marler, P., Moody, D. B., *et al.* (1979). Species-specific perceptual processing of vocal sounds by monkeys. *Science*, **204**, 870–2.

22
Cognitive abilities in a comparative perspective

Andrew Lock and Michael Colombo

Abstract

Recent studies of non-human animals indicate that cognitive processes mediate many areas of their behaviour. A number of human cognitive systems, and their properties—for example memorial processes, categorical auditory and visual perception—appear to have quite deep phylogenetic roots. It is not yet possible, however, to provide a precise evolutionary classification of these systems, for ecological factors play as large a role in the elaboration of an animal's cognitive abilities as does its phylogenetic status. Thus, many 'indices' that have been proposed as differentiating phylogenetic groups in terms of their 'learning abilities' or 'intelligence' have not been substantiated: initially promising proposals have been confounded by animals from 'lower' taxa showing 'unexpected' levels of ability in sensory domains relevant to their ecological niches. A restricted focus on primates, however, does tend to show an improvement in levels of performance from prosimians to Great Apes on tasks such as reversal learning; and it seems likely that these changes are based in differences in the underlying cognitive abilities and strategies these species employ.

The ability to form concepts has been shown for a number of non-human species. Most of these concepts have physical instantiations. Conflicting claims are made regarding the possession of the concepts of 'same' and 'different'. These appear to be absent in pigeons and goldfish; their status in monkeys is subject to dispute; they are quite well elaborated in the Great Apes, especially in chimpanzees, where most of the experimental effort has been focused. In this last-mentioned species, 'same–different' judgements extend into areas of analogy and transitive inference, which are perhaps closer to reasoning abilities than merely conceptual ones.

Observational learning (or imitation) has been divided into the categories of social facilitation, stimulus enhancement, and imitative copying. The last of these is largely confined to the Great Apes; whereas the other two occur across all the primate groups. Imitative copying is less developed in the Great Apes than in humans, and 'teaching' plays little or no role in the transfer of skills. Chimpanzees are capable of adaptive novel responses in a problem-solving context; but these are more possible for them in some situations than in others. Chimpanzees and orang-utans show self-recognition when confronted with a mirror; but all other primates tested tend to react to mirrors socially, as if they were confronting another conspecific.

Monkeys and apes occasionally act as though they recognize that other individuals have beliefs, but even the most compelling naturalistic and experimentally-induced observations can usually be explained in terms of learned behavioural contingencies, without invoking a higher-order 'understanding' of intentionality. What little evidence there is at this time does point towards chimpanzees (and possibly the other Great Apes') having some 'theory of other minds', although its precise nature is not yet clear.

There are some Piagetian-inspired investigations of comparative cognitive abilities. The Great Apes appear to reach sensorimotor Stage 6 in object permanence, spatial concepts, imitation, and the understanding of causality. Piagetian investigations of possible representational intelligence have proved disappointing to date.

Two oversimplified but none-the-less useful generalizations would characterize monkeys as possessing the ability to form conceptual representations, and apes as able to manipulate representations; and monkeys as more dominated by immediate perceptual experience than apes. In both cases, abilities are often restricted to particular domains of action. The elaboration of cross-modal and cross-situational abilities appears a major factor in the evolution of primate cognitive skills towards those possessed by modern humans.

22.1 Animal cognition

The view that animals have minds, and hence cognitions, was a dominant one in Victorian times (for example, Romanes 1883; and reviews by Demarest 1983 and Wasserman 1984). It became unfashionable with the ascent of the behaviourist paradigm. Griffin (1976), in his book *The question of animal awareness*, re-opened inquiries into whether non-human animals have mental experiences; and there has since been a revival of interest in the issue from a number of different fields, for example comparative neuropsychology (for example MacPhail 1982); the information-processing paradigm (Hulse et al. 1978; Roitblat et al. 1984; Roitblat 1987); and from students of animal communication (see Jolly, Chapter 6, and Ristau, Chapter 23, in this volume). In these approaches the concern is with elucidating the psychological processes, *internal* to the animal, that are held to underlie and mediate its behavioural capacities. Such studies of animal cognition imply neither consciousness nor mentalism. Rather, they imply a model positing some form of representation of the world, to which an animal can refer, so to speak, in the pursuit of adaptive behaviour (Terrace 1984).

The counter-claim has a long history, and is subtly embedded in the way evolutionary theory has entered into Western culture's presuppositions; that evolution is progressive, and humans are placed at its apex, and differ quite markedly from animals. Seneca (54 BC–AD 39) is credited with the view that:

Animals are unable to recall their past experience, their memory capacity being limited to recognition. Thus the horse may recognize a road over which it has travelled before, but remembers nothing of it when in the stable afterwards (cited by Davey 1981).

The classical S–R framework takes recognitory memory as axiomatic: animals must retain some memory of a learned response for it to be subsequently elicited on presentation of the appropriate stimuli. More recently, evidence has accrued that *representational memory* is possessed and used by a number of organisms: that is, the ability 'to respond appropriately in a situation when controlling stimuli are *not* present or when there has been no opportunity to form an S–R bond' (Davey 1981, p. 280).

For example, D'Amato and Worsham (1974) developed a delayed conditional matching task and used it with capuchin monkeys (*Cebus apella*). Here, a monkey is first shown either a red disk or a vertical line; after a delay it is presented with two comparison stimuli, an inverted triangle and a small circle. If the red disk had appeared previously, the correct response is the inverted triangle; if the vertical line, then the small circle. Thus,

the stimuli at the time of choice provide no information whatever regarding the identity of the standard stimulus. Or in somewhat different terms, the comparison stimuli serve no differential retrieval function (D'Amato 1973, p. 254).

The experimental monkeys were successful in performing this task, indicating they have a representational memory of some kind. Since this report, there have been a great number of studies of animal cognition. The first topic reviewed here concerns memory.

22.2 Memory

22.2.1 *Spatial memory*

Work by Olton and Samuelson (1976) and Olton (1978, 1979) showed that rats in a radial maze, in which each arm was baited with food, rarely re-entered an arm where they had previously eaten in a particular session, unless they had not consumed all the food available there at their first visit. It has been shown that rats rely on visual cues in the room surrounding the maze to guide their behaviour (Mazmanian and Roberts 1983; Olton and Samuelson 1976). This plus the fact that rats show impressive retention of arms visited despite delay periods of up to four hours (Beatty and Shavalia 1980) suggests that the rats have formed a 'spatial map' of their surroundings by encoding spatial information in the form of extra-maze cues into memory. In addition it has been shown that rats employ efficient memory strategies in solving radial-arm mazes, remembering the number of arms which places fewest demands on memory. Thus in a 12-arm maze, rats will tend to retrospectively code the number of arms visited up to six arms, at which point they switch to a prospective coding strategy and remember the number of arms that still need to be visited, thereby always maintaining no more than six items in memory (Cook et al. 1985).

Similarly, Menzel's (1978) work on chimpanzee search-behaviour showed, among other things, that the animals had something like a 'cognitive map' of their enclosures, in that, having been shown where food could be found, they would use the most economical route to locate it from wherever they were subsequently released. Figure 22.1 shows the routes used by four individuals in this experiment, and all of them show some evidence of a 'least distance' principle, in that their search is efficient rather than random.

22.2.2 *Delayed matching to sample paradigm (DMTS)*

The DMTS paradigm is outlined in Fig. 22.2. An animal is presented with a sample stimulus for a brief period, and then after a delay is presented with two comparison stimuli, one of which matches the sample stimulus. The animal's task is to 'choose' the comparison stimulus which matches the sample stimulus. While it may be claimed that this paradigm taps recognition rather than representational memory, many workers in the field now subscribe to the view that, since there is evidence from other experiments that the species involved possess *cognitive representations*, those representations are probably drawn on by these animals in this present case (cf. note 1). Early studies of short-term memory using this paradigm on pigeons (Grant and Roberts 1973; Shimp and Moffit 1974; Zentall 1973), rats (Roberts 1972, 1974), and monkeys (D'Amato 1973; Jarvik et al. 1969; Moise 1970) have provided data showing that a number of species can successfully master this task. Additionally, work in this paradigm raises general issues, discussed below, concerning the processes of memory as they occur across different species.

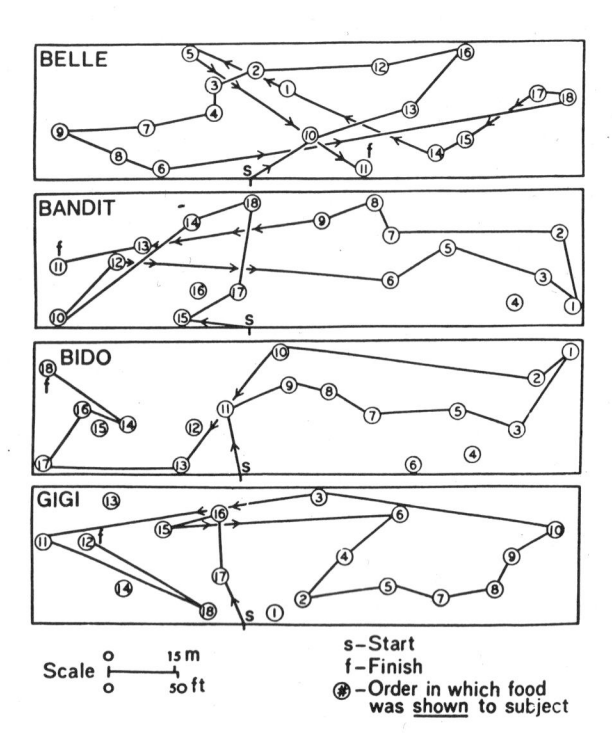

Fig. 22.1 Maps showing the exact order in which various places were searched, with the connecting lines giving an approximation of the animal's travel routes, for the trial (out of four) on which the largest number of hidden foods were retrieved after the animal had previously been shown their location (numbers in circle denote the order in which locations were shown). Comparisons with baseline data for the mobility of control animals within the enclosure, and models that generate all possible itineraries, strongly indicate the use of cognitive maps to guide searching and retrieval. (From Menzel (1973). Copyright 1973 by the American Association for the Advancement of Science. Reproduced by permission of AAAS and E. W. Menzel.)

22.2.3 *Serial position and clustering effects in recall*

'Serial position effects' is a term denoting the well established finding in human memory research that, when presented with a sequential list of twenty or so items, early (primacy effect) and late (recency effect)

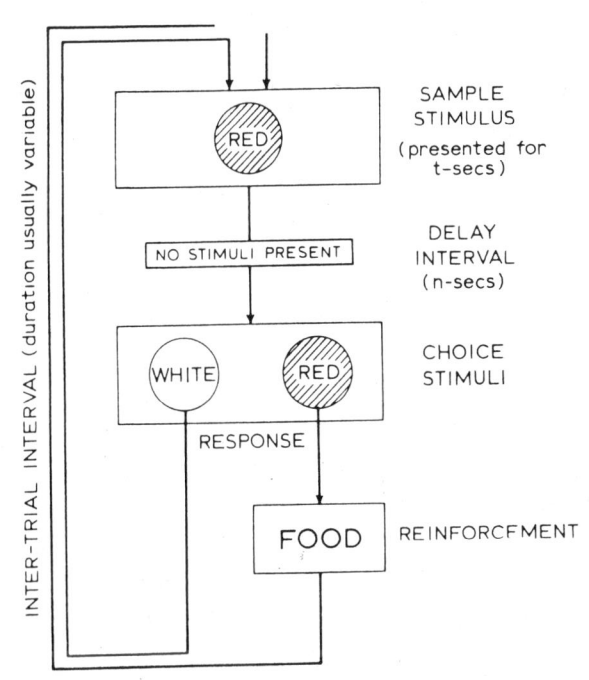

Fig. 22.2 Schematic representation of the delayed matching to sample (DMTS) paradigm. (From Davey (1981). Reproduced by permission.)

items in the list are more often recalled than items in the middle (for example Murdock 1962; see Crowder 1976, for a review). 'Category clustering' refers to the finding that if adult humans are given a sequential list of words that can be grouped into particular taxonomic classes, the order of recall of items will tend to be grouped taxonomically, irrespective of the serial ordering of the items in the list presented (for example Bousfield *et al*. 1958).

Serial-position effects have been studied in two ways, through serial probe recognition (SPR) techniques and through free recall. The SPR was introduced by Wickelgren and Norman (1966) for human subjects. Lists of items, generally words, digits, or pictures, are presented. Then, following some delay, a probe item is presented, which is either from the list (Same) or not (Different). The subject has to determine whether this probe item was in the original list, responding 'same' or 'different'. This procedure has been adapted for use with non-humans. Full reviews of this topic can be found in Sands *et al*. (1984) and Wright *et al*. (1984).

A number of studies with non-human primates have reported quite low levels of performance (for example Devine and Jones 1975; Gaffan 1977*a*, for rhesus monkeys), compared to humans (cf. Wickelgren and Norman 1966); but these findings may result from the methodologies employed rather than from disabilities within the animals. More success has been obtained with dolphins (Thompson and Herman 1977) using a tone list. For up to six items, dolphins show a significant recency, but no primacy effect. However, the dolphin's serial position recall curve was within the range of those reported by Wickelgren and Norman (1966; see also Crowder 1976) for humans; in addition, a number of studies have reported serial-position curves which show a recency but no primacy effect for humans (Hines 1975; Potter and Levy 1969). Sands and Wright (1980) demonstrated both primacy and recency effects in rhesus monkeys and a human subject under the same conditions, and although, again, the monkeys' primacy effect was weaker than their recency effect, they justifiably claim that their results 'encourage the view of similar mechanisms of memory in monkey and man' (1980, p. 386). With somewhat different techniques, serial-position effects have also been found for rats (Kesner and Novak 1982; Roberts and Smythe 1979) and rabbits (Wagner and Pfautz 1978), suggesting commonalities of memory function across a wide spectrum of mammals.

Closer parallels to human memory emerge with chimpanzees trained to use symbols. For example, Buchanan *et al*. (1981) studied the free-recall capabilities of Lana, a chimpanzee trained in 'Yerkish' (see Ristau, this volume, Chapter 23 for further details). Lana was presented with lists of up to nine symbols drawn from three taxonomic categories (food/drink; objects; and colours). Lana's task was to then 'recall' those symbols presented by tapping them out on a keyboard. Lana's performance showed both serial and category-clustering effects, and overall her performance was comparable to that of a human child of about five years of age. Clustering effects have also been reported by Premack (1976) for the chimpanzee Sarah, using a recognition rather than a recall task.

22.3 Other cognitive systems

In addition to there being an apparent commonality in the nature of mammalian recall systems, common functioning is present elsewhere. While the psychophysical parameters may vary in different species and families, the basic functioning of the visual system is taken to be similar amongst the mammals (indeed, most medical, pharmacological, and physiological work is predicated on this fact). Some have even gone so far as to draw analogies between primate and avian visual systems, implying a commonality of visual function between these species as well (Karten and Shimuzu 1989).

Auditory perception again has different psychophysical profiles in different species, but is taken to have a common basis. Like humans, monkeys show a right-ear advantage (left hemispheric dominance) when processing 'species-specific' sounds (Petersen *et al*. 1978). Furthermore, damage to the left auditory cortex of the monkey brain impairs discrimination of species-specific sounds, whereas right lesions have little or no effect, indicating that monkeys may possess an analogue to Wernicke's area in the human brain (Heffner and Heffner 1984).

Across these different modalities there is currently emerging the view that some quite basic processes may be held in common. For example, the phenomenon of *categorical perception* in human speech perception has been shown to be a property of other species responses to human speech (see Kuhl 1987 for a review), and has also been demonstrated for several kinds of non-linguistic and visual stimuli (Aslin *et al*. 1983). It would seem, then, that many aspects of the human cognitive system have quite deep phylogenetic roots, and were established well before language became part of our cognitive apparatus (which is not to deny that extant humans may possess cognitive capacities or systems that are specifically tailored to language).

22.4 Some caveats

It is important to note, however, that these basic similarities of systems do not mean animal and human cognition necessarily function in the same way. With humans, for example, in the delayed-matching-to-sample paradigm, the longer the sample to be matched is presented, the better the recall of it. It has been suggested that longer presentations allow the detection and coding of more features of the stimulus, and it is these features which are hypothesized as establishing the 'representational code' via which memory operates. D'Amato and Worsham (1972), however, found this was not the case with capuchin monkeys. Using presentations of 0.075, 0.1, and 0.15 seconds resulted in equally high performance, hence equally good recall, hence equally good encoding. Why?

One reason is that animals *are not* humans. Thus, even when we are looking for *similarities* we must expect to find *differences*. For example, Hulse *et al*. (1984) report that while starlings, like humans, can discriminate a tune across different frequency ranges, unlike humans they cannot generalize this ability beyond an octave. A similar finding was reported by D'Amato and Colombo (1988*a*) in monkeys. In addition, D'Amato and Colombo (1989) found that the ability of two cebus monkeys to execute a delayed matching-to-sample task was impaired when the comparison stimuli, which had been appearing on the bottom two projectors of a five-projector array, were allowed to appear on any of the five projectors.

Such apparently minor task transformations have been known for some time to disrupt monkey but not human performance. For example, Weinstein (1941) trained two monkeys and two three-year-old children with three-dimensional objects and then tested for transfer of the matching concept to a number of different stimuli and task situations. Both monkeys and children transferred the matching concept to novel three-dimensional objects. When the transfer stimuli were changed to two-dimensional forms and colours, however, the children began to outperform the monkeys. The performance differences increased further when the two-dimensional stimuli were affixed to lids on boxes, so that the instrumental response changed from pushing aside the stimuli to lifting the lid of a box. The results from these and other transfer tests showed that the children were able to apply the matching concept to a wider range of circumstance than the monkeys. Similarly, Rumbaugh (1986, p. 35) comments that

the so-called complex learning-set skills of apes and monkeys are remarkably brittle when the nature of the test apparatus is changed; something one would not expect if there were really a knowledge base of a human type underlying the mastery manifested by the animals. But that is exactly the point. Animals are not humans. Their cognitions, if extant, are surely both different and more circumscribed than in humans.

This is a reminder that we must beware of the trap of considering the abilities of modern animals as indexical of phylogenetic history. Essentially common abilities quite probably show differing adaptive manifestations, and may also reflect completely different internal mechanisms (see Section 22.8 for further examples).

A second reason, advanced by D'Amato (1973), is that the particular experimental capuchins were heavily overtrained, and hence their familiarity with the stimuli and the task may have produced a ceiling effect well before sample durations were manipulated. The general point here is that much of the work to be summarized below involves either highly-trained or otherwise 'test-wise' animals. The status of the abilities reported thus needs some thought, as, while the bogey of 'Clever Hans' does not arise directly here, in that these abilities have been established in a paradigm that incorporates adequate controls, it is not certain that one is always dealing with 'genuine' (in the senses of feral, naïve, or pristine) animals.

Third (and again raised by D'Amato (1973)), is the issue of language and the verbal coding of information. It may be that presentation duration is only relevant when the 'subject' can form verbally-mediated representations. This is an important point, since many of the more challenging reports of animal cognitive abilities (see below) involve apes which have been through some 'linguistic' training. Another point here is that while verbal mediation may play an important role in human memory, its presence or absence is not the only factor influencing recall abilities. For example, in the present context pigeons are not good candidates for possessing verbal abilities; but, unlike capuchin monkeys, their DMTS recall is facilitated by longer presentation durations (Roberts 1972; Roberts and Grant 1974). Memory, and doubtless cognitive processes in general, cannot be assumed to be invariant in their nature across *all* species; indeed, there are commentators whose caution would prevent them going along with the similarity thesis advanced here:

verbal organisms, such as man, may verbally encode attributes of the stimulus whereas more "lowly" organisms, such as the pigeon, may have to rely on a rapidly decaying sensory trace.... Non-human primates ... may have entirely different strategies; these may involve neither simple trace decay nor symbolic encoding of attributes, or alternatively may involve even a combination of both. Given differential storage strategies a variable such as sample duration is unlikely to have a constant effect across species, or even across different recall tasks (Davey 1981, p. 288).

It is clear, then, that there are a number of difficulties in looking at the evolution of cognition. For example, as is implicit in the above, no simple-minded distinction between what might be termed the *substrate* and the *operations* of a cognitive system can be maintained. That is, what is being encoded interacts with how it is encoded. In addition, ecological factors interact over time with cognitive systems, leading to specializations, enhancements, and suppressions of various bits of 'kit' in animals, obscuring any general patterns that might be expected. And finally, there is a lack of extensive empirical data on many species, and, given the point on ecology, it is difficult to posit clear generalizations about 'monkey' or 'ape' cognition- analytic categories that have traditionally been adopted in evolutionary speculation.

Fourth, in addition to this lack of detailed data, there is also a lack of conceptual rigour in the terminology used to distinguish apparently different cognitive abilities amongst animals. This point will become clearer below, where the distinction between such categories as 'concepts' and 'stimulus generalization', and 'concepts' and 'reasoning', often appears quite blurred. This lack of rigour is, in fact, instantiated in the above discussion of memory, where terms such as 'storage strategies', 'knowledge base... underlying the mastery manifested by animals' and 'mechanisms of memory' have been used by different authors in apparently similar contexts, generally without explicit definition. Do such terms mean the same thing, or do they reflect different authors' differing presuppositions as to the nature of the processes they are studying? Finally, there is the comparative psychologist's version of Heisenberg's Uncertainty Principle to be contended with. In subatomic physics this principle establishes that the more precisely one measures the speed of a particle, the less certain one can be about where it is, and vice versa. The analogous problem here is that the more precisely one tests an animal, the more it becomes trained and 'test-wise', leading one to wonder whether the precise ability is located in the animal, the experimenter, or the experimental situation.

Given these caveats, the approach here conceptually borrows from a statistical model. Hence, a number of 'factors' are admitted as contributing to the 'variance' between animal and human cognition, but a main effect of 'similarity' is claimed. The central claim thus far is that many basic cognitive operations are commonly instantiated among vertebrates: that is, much of the human information-processing system is *not* unique to humans, but has a long phylogenetic history.

22.5 The uses of cognition: what can animals do or 'learn'?

Students of animal learning tend to divide the field into comparatively simple tasks (such as conditioning and discrimination learning) and complex or higher-order problems (such as tool-mastery and insightful problem-solving). From a cognitive viewpoint it is not at present clear, when reviewing the literature, whether this divide is held to reflect quantitative or qualitative differences in the way in which animals, especially primates, perform on these tasks. For example, it is possible that simple tasks may be accomplished by stimulus–response association, and that more complex ones require some cognitive mediational processes for their accomplishment; hence a qualitative difference. Alternatively, King and Fobes (1982, p. 327) introduce their version of the distinction between simplex and complex problems or tasks with the claim that complex tasks 'typically require the learner to master more subtle and intricate concepts than those required for [simplex tasks]': hence a quantitative difference. Additionally, Meador *et al.* (1987, p. 19) note that an important trend in recent work is

Table 22.1 Summary review of investigations of primate learning abilities. See text for further details.

Sub-Order	Prosimii		Anthropoidea			Hominoidea	
Super family							
Family	Tupaiidae	Other families	Ceboidea	Cercopithecoidea		Hylobatidae	Pongidae and Panidae
Simple conditioning and discrimination learning	Nothing distinctive in comparison to other vertebrates. Fobes and King 1978.	Ehrlich et al. 1976. Good discrimination on brightness, pattern, and object. Size and colour not known.	Similar to prosimians. Interspecies differences probably relate to ecological factors: Fragaszy 1981; Devine 1970.	Similar to prosimians: Warren 1965b Social rewards effective: Swartz and Rosenblum 1980. Well-developed visual pattern discrimination abilities: Gaffan 1977; Rosenfeld and Van Hoesen 1979.		Insufficient data.	Simple discriminations learned rapidly: Fobes and King 1982. *Pongo* lowest performance, and more easily disruptible than *Pan* and *Gorilla*: Essock 1975; Rumbaugh et al. 1973.
Complex discrimination learning	Can develop object-discrimination learning sets, but lower performance than other prosimians: Fobes and King 1978. Learning sets formed in reversal shift experiments: Fobes and King 1978; Riddell et al. 1974.	Develop object-discrimination learning sets: Ehrlich et al. 1976; Wilkerson and Rumbaugh 1979. Similar level of performance to *Callithrix* and *Saimiri* monkeys: Ehrlich et al. 1976. Reversal performance poor: Rumbaugh and Pate 1984.	Develop object-discrimination learning sets: Shell and Riopelle 1958. Mediational strategies at very moderate level: Rumbaugh and Pate 1984.	Better performance than prosimians and most New World monkeys. General rule process of learning: Fobes and King 1982. Reversal performance similar to pongids.		Performance on object discriminations typically poor, but can be elevated to comparable levels to pongids: Rumbaugh et al. 1973. Possibly transitory between associative and mediational learning: Rumbaugh and Pate 1984.	Inter-species differences, but higher levels of performance than other groups: Rumbaugh and McCormack, 1967. Strong evidence for mediational processes: Essock-Vitale 1978.
Time-based learning	Poor: Ehrlich et al. 1976; Fobes and King 1978, 1982; Wilkerson and Rumbaugh 1979.		Below Great Apes, but similar to Old World monkeys. See Fig. 22.8.	See Fig. 22.8		Poor. See Fig. 22.8.	See Fig. 22.8. Good, and similar for all species. Harlow et al. 1932. Maier and Schneirla 1935. Rumbaugh 1970. Insufficient data.
Response-patterned learning	Interspecific differences in performance related to size of prefrontal brain system: Masterton and Skeen 1972.		Insufficient data.	Most data for rhesus monkey (*Macaca mulatta*). Performance often poor (Livesey 1965, 1969), but some evidence for generalization and rule-mediation.		Poor: Schusterman and Bernstein 1962.	

Table 22.1 (contd.)

Match-to-sample learning	Insufficient data.	Insufficient data.	*Macaca mulatta* shows similar performance to chimpanzee, requiring 1000–1500 trials to learn matching: Riopelle and Hill 1973.	Insufficient data.	*Pan* learns more rapidly than rhesus monkey: Riopelle and Hill 1973; but slower than humans: Nissen et al. 1949. Performance suggests use of higher-order mediating processes.
Probability learning	Data not available.	Data not available.	Clearly established for *Macaca mulatta*: Fobes and King 1982.	Data not available.	*Pan* and human children show same patterns: Fobes and King 1982; Schusterman 1963.
Cross-modal recognition	Data not available.	Small literature. Appear similar to Old World monkeys and *Pan* for recognition from touch to vision: Elliott 1977; Hewett and Ettlinger 1978, 1979; Jarvis and Ettlinger 1975, 1978.	Large literature, e.g.: Bolster 1978; Malone et al. 1980; Norris and Ettlinger 1978; Tolan et al. 1981; Weiskrantz and Cowey 1975. Comparable abilities in macaques and *Pan*; Malone et al. 1980; Tolan et al. 1981.	Data not available.	Large literature demonstrating this capacity, e.g.: Davenport and Rogers 1970, 1971; Davenport et al. 1973, 1975.
Cross-modal transfer of learning	Tree shrews (*T. glis*) transfer from vision to audition (Ettlinger 1977), but not from audition to vision (Ward et al. 1970).	Bush-babies (*Galago senegalensis*) transfer from vision to audition, and vice versa: Ehrlich et al. 1976; Ettlinger 1977; Ward et al. 1970.	Negative results for *Cebus albifrons griseus* and *Saimiri sciureus*: Blakeslee and Gunter 1966; Wilson and Zieler 1976. One positive result for one *Cebus* subject: Blakeslee and Gunter 1966. See also Ettlinger 1977.	Mostly negative results (Ettlinger 1977), though some indication of vision to touch and vision to audition (Ettlinger 1977; Wilson 1964, 1966).	Results unclear. Present in language-trained apes for vision to audition (Fouts 1974), but not demonstrable for vision and touch in others (Ettlinger 1977).

The phylogenetic status of the Tupaiidae has been a matter of some dispute see, for example Campbell 1974; 1967), the issue being whether they are insectivores of prosimians. Present thought has them as insectivores, and thus they do not appear in current classification of primates such as, for instance Campbell, this volume, Chapter 2). Whatever the conclusion, modern species in this group represent the vestiges of the transition from more primitive mammals to primates, and are included here for reference only.

an increasing recognition of the possible influence of higher mental processes on learning controlled by lower brain functions.... Briefly, by the adoption of Skinner's functional analysis of verbal behaviour (Skinner 1957), a significant difference between pigeon behaviour (Epstein *et al.* 1980) and chimpanzee behaviour (Savage-Rumbaugh *et al.* 1978a, b) is demonstrable. The higher mental processes of chimpanzees produce behaviour that more closely approximates human verbal behaviour.

Thus, the underlying mechanisms responsible for producing behaviours presently classed as the same, for instance, as 'conditioned responses', may be quite different in different species.[1]

The field is only just beginning to give attention to such theoretical issues. King and Fobes (1982, p. 357) conclude their review of complex learning by noting:

most of the research is simply a demonstration that some primate species can learn some complex problems. Little research has been directed toward the variables affecting complex learning and even less toward theoretical explanations. It is ironic that the aspect of primate learning that is most similar to complex mental processing has been the most neglected.

This 'state of the art' means that any evolutionary points to be drawn from the comparative field must at present be given a provisional status. Information concerning primate learning under a number of headings is tabulated in Table 22.1, with some further details and additional topics being discussed in the following subsections.

22.5.1 Simple conditioning and discrimination learning

Primates have not been popular subjects in either the classical or the instrumental conditioning paradigms. This is due to the fact that primates can accomplish much more complex learning, and cataloguing variations in performance between species at these basic levels has therefore been eschewed in favour of investigating their 'higher' learning abilities. Conditioning studies of primates have thus been held to be of little comparative significance (for example, Warren 1965a, b, 1973, 1974). However, their possible significance has recently been re-evaluated (see Meador *et al.* 1987). First, as noted above, performance on conditioning tasks appears to be related to the complexity of an animal's cognitions. Second, the comparative significance of these studies can now be recognized as having an ecological dimension: for example, animals appear to be differently 'prepared' for conditioning in different domains (Breland and Breland 1961), so that social rewards, for example, may be more effective than food rewards in some species compared to others, and that visual stimuli may be more easily discriminated than auditory, and so on. These differences are likely to be consequences of an animal's ecological situation, and an understanding of them from an ecological perspective simultaneously refines our understanding the role ecological factors play in the evolution of behavioural abilities.

22.5.2 Complex discrimination learning

22.5.2.1 Discrimination learning set (DLS)

The basic DLS paradigm was formulated by Harlow (1944), who noted that 'once appropriate reaction sets have been formed in monkeys, these sets may be *transferred* from one pair of discrimination objects to another' (p. 11). Essentially, two objects are presented to a monkey for six trials. On the first trial, one object is arbitrarily designated 'correct', and if that is the one chosen by the monkey, it gets a reward. Obviously, there is a 50 per cent chance of the animal's being successful at this point. Over the six trials the correct object is held constant: thus the six trials constitute one particular problem. Over these trials, trial-1 contains all the information the animal needs to be successful: i.e., if it is rewarded for its choice on trial-1 it will continue to obtain rewards if it keeps choosing that object; if it is not rewarded for its trial-1 choice, it will be if it chooses the other object for the subsequent trials. Once the animal has completed the trials for this problem, it is presented with the next problem, comprising two new objects, one arbitrarily designated 'correct', for six more trials. In Harlow's original work, each animal was tested over 344 problems.

Figure 22.3 illustrates his experimental results. Trial 2 performance is taken as a measure of 'learning to learn', the development by the animal of a *discrimination learning set*. By the final 55 problems the monkeys were 96 per cent correct in their choices on trial 2; that is, information from previous trials was being transferred to inform their choices. Data for prosimians and New World monkeys tested in this paradigm are given in Fig. 22.4a; for Old World monkeys, Great Apes, and humans in Fig. 22.4b. Within the primates, DLS performance may provide a reasonable measure of learning ability. In terms of evolutionary status there is a clear increase, apparent in Figs. 22.4a and 22.4b, in the ease with which learning sets are formed.

There are, however, clear effects of stimulus dimension upon a given species' performance—dolphins, for example, are poor on these tasks in the visual modality, but substantially superior in the auditory one (Herman and Arbeit 1973)—so that no precise phylogenetic correlations can be established across

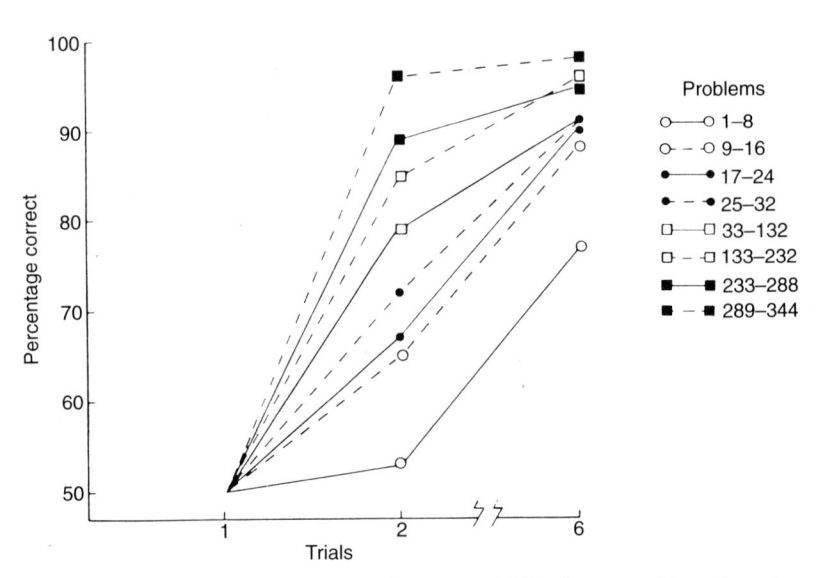

Fig. 22.3 The development of discrimination learning set (DLS); inter-problem learning is evidenced by increases in the percentages correct on trial-2 performance (connected points on graph). Intra-problem learning improvement is indicated by performance on trials 1–6. (From Fobes and King (1981), after H. Harlow (1951). By permission of Academic Press and the authors.)

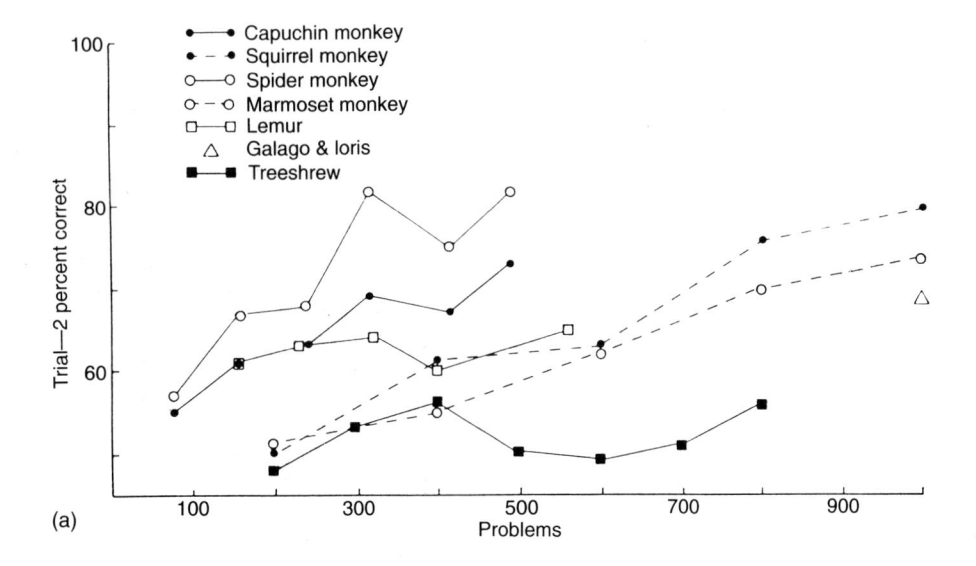

Fig. 22.4a Object discrimination learning set (DLS) for prosimians: ring-tailed lemurs (*Lemur catta*) (Stevens 1965), thick-tailed galagos (*Galago crassicaudatus*) and slow lorises (*Nycticebus coucang*) (Musicant 1975), and common treeshrews (Leonard *et al.* 1966); and New World monkeys: squirrel-monkeys (Miles 1957b; Shell and Riopelle 1958), black-handed spider monkeys (Shell and Riopelle 1958), common marmosets (*Callithrix jacchus*) (Miles 1957b), and white-fronted capuchin monkeys (Shell and Riopelle 1958). These results are typically based on testing with 6-trial problems, although data are included for lemurs and squirrel-monkeys (Shell and Riopelle 1958), spider monkeys, and capuchin monkeys tested to a criterion. Stimuli consisted of junk objects. (From Fobes and King (1981). By permission of Academic Press and the authors.)

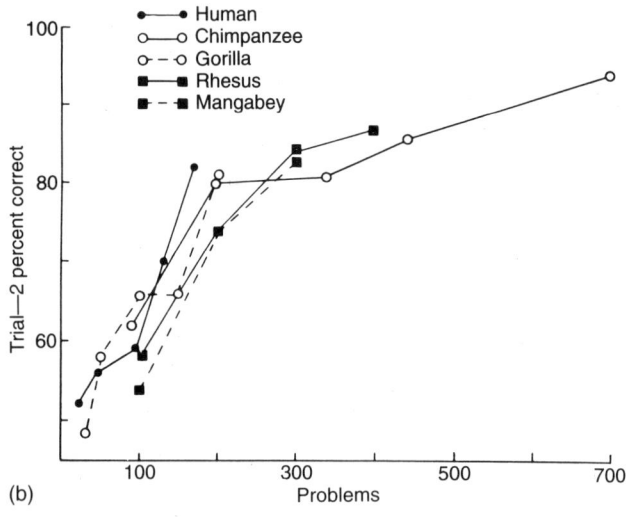

Fig. 22.4b Object discrimination learning set (DLS) for Old World monkeys: rhesus (Behar 1962; Miles 1957*b*) and white-collared mangabey monkeys (*Cercocebus torquatus*) (Behar 1962); Great Apes: chimpanzees (Hayes *et al*. 1953; Davenport *et al*. 1969) and western lowland gorillas (Fischer 1962); and human children (Koch and Meyer 1959). These results represent testing with junk objects for all but gorillas (pictures) and children (colours). However, training procedures differed in terms of the number of trials per problem. Rhesus and mangabeys were tested for 10-, 7-, and 4-trial problems (Behar 1962), and rhesus were also tested with 6-trial problems (Miles 1957*b*). Chimps were given 10- and 2-trial problems and Criterion training (Hayes *et al*. 1953) as well as 10- and 4-trial problems (Davenport *et al*. 1969). Gorillas were given both 50- and 6-trial problems, and children were tested on 6-trial problems. (From Fobes and King (1981). By permission of Academic Press and the authors.)

taxa. This is clear in Fig. 22.5, where the performance of chickens, blue-jays, and mink is anomalous with their taxonomic standing *vis-à-vis* the other vertebrates measured. Such 'poor fits' are probably due to the ecologically-related factors referred to under simple conditioning: high performances on visual discriminations are found among animals in which vision is the dominant sensory modality (Warren 1974), and primates are unusual among mammals since their visual systems are particularly well developed (birds and mink probably have good acuity and also colour vision; rats, by contrast, have a higher DLS performance with olfactory as opposed to visual stimuli (Slotnik and Katz 1974).

22.5.2.2 *Reversal learning*

In reversal learning, an animal learns a particular discrimination; for example, that choosing the brighter of two stimuli gains a reward. After reaching a particular criterion of performance on this, the reward structure is reversed, i.e. the previously correct stimulus becomes incorrect, and vice versa. Once this reversal is learnt, the process is continued with successive reversals. Bitterman (1965) proposed that this paradigm could be used to provide a means of ordering species' abilities by taxa, but this has not proved to be the case. Within the primates, however, a general improvement in performance from prosimians to Great Apes is discernible (see Fig. 22.6).

Rumbaugh (1971) developed what he termed a modified discrimination reversal task, which, he argues, allows the demonstration of different learning strategies being used by different primate species. The procedure was to train an animal on a multiple discrimination task (i.e. stimulus A+ is designated correct and B− incorrect) to a criterion of 9 out of 10 responses. Then, one standard reversal trial is given (i.e., B+A−), followed by 10 trials under one of three conditions: (1) the standard reversal (B+A−) is continued, and termed the control condition; (2) a new positive condition, the correct object being replaced with a new one (C+A−), is used; or (3) a new negative condition, the incorrect object being replaced with a new one (B+C−), is introduced. Rumbaugh (1971) argues that these three conditions allow the separation of three strategies that animals might employ in dealing with these problems.

First, if an animal is strictly following the rules of stimulus–response association, its performance on the control condition will contain more errors than in

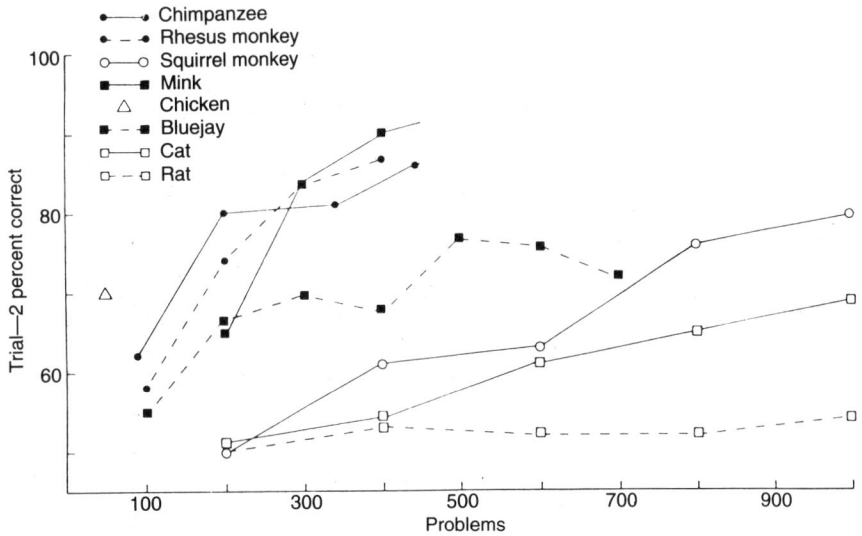

Fig. 22.5 Discrimination learning set (DLS) as a measure of animal intelligence. The good news: trial-2 performance systematically separates the abilities of chimpanzees (Davenport *et al.* 1969; Hayes *et al.* 1953), rhesus monkeys (Behar 1962; Miles 1957*b*), squirrel-monkeys (Miles 1957*b*; Shell and Riopelle 1958), cats (Meyers *et al.* 1962), and rats (Warren 1965*b*). The bad news: White Plymouth Rock chickens (Plotnik and Tallarico 1966), Mink (Doty *et al.* 1967), and northern bluejays (Hunter III and Kamil 1971) evidence an ability strikingly disproportionate to their taxonomic standing. These results represent a wide range of test procedures. In addition to primate testing features mentioned in Figs. 22.4a and 22.4b, cats were tested on 6-trial problems with junk objects; rats for 6-trial problems with unspecified stimuli; chickens to a criterion with junk objects; mink for 6-trial problems with three-dimensional wooden blocks differing along various dimensions; and bluejays for 25-, 15-, 10-, and 6-trial problems with junk objects. (From Fobes and King (1981). By permission of Academic Press and the authors.)

the other two conditions. The argument is that in the control condition the animal has to learn two new cue patterns, whereas in the other two conditions it only has to learn one new relationship (either A or B). Learning two new cue values should be more difficult than learning one if an animal is employing a stimulus-response strategy.

A second possible strategy is to respond to novelty, and respond to the new stimulus (C) irrespective of its value. An animal following this strategy should thus make most errors in the new negative condition (B+C−), the least errors in the new positive condition C+A−), and an intermediate number in the control condition (A−B+).

Finally, a third possible strategy would be to abstract the general rule that applies to correct performance on discrimination and reversal tasks. That is, if an object was rewarded on the last trial, stay with it; if it was not, shift to the other object on the next trial (a strategy termed '*win-stay; lose-shift*' by Levine (1959)). An animal following this strategy will show a similar performance on all three reversal conditions. The results from using this procedure with different species are shown in Fig. 22.7.

The results for three species (talapoin monkeys, *Miopithecus talapoin*, gibbons, *Hylobates lar*, and gorillas, *Gorilla gorilla*), which all learned the pre-reversal phase with equal facility, indicate from their subsequent performance that they were employing qualitatively different strategies in dealing with the reversal phases. Talapoins demonstrate a stimulus–response associational pattern; gibbons a novelty strategy; and gorillas a hypothetical or general rule strategy. These findings have proved robust in further studies (Gill and Rumbaugh 1974; Essock-Vitale 1978; see Rumbaugh and Pate 1984 for a review). In Essock-Vitale's study an interesting finding was that as rhesus and stump-tailed monkeys gained experience of the tasks, they shifted from an initial associational strategy to the hypothesis-learning one.

22.5.3 Time-based, response-patterned, and match-to-sample learning

These three paradigms all represent attempts to differentiate among taxonomic groups in terms of their 'higher-order' learning capacities. The original

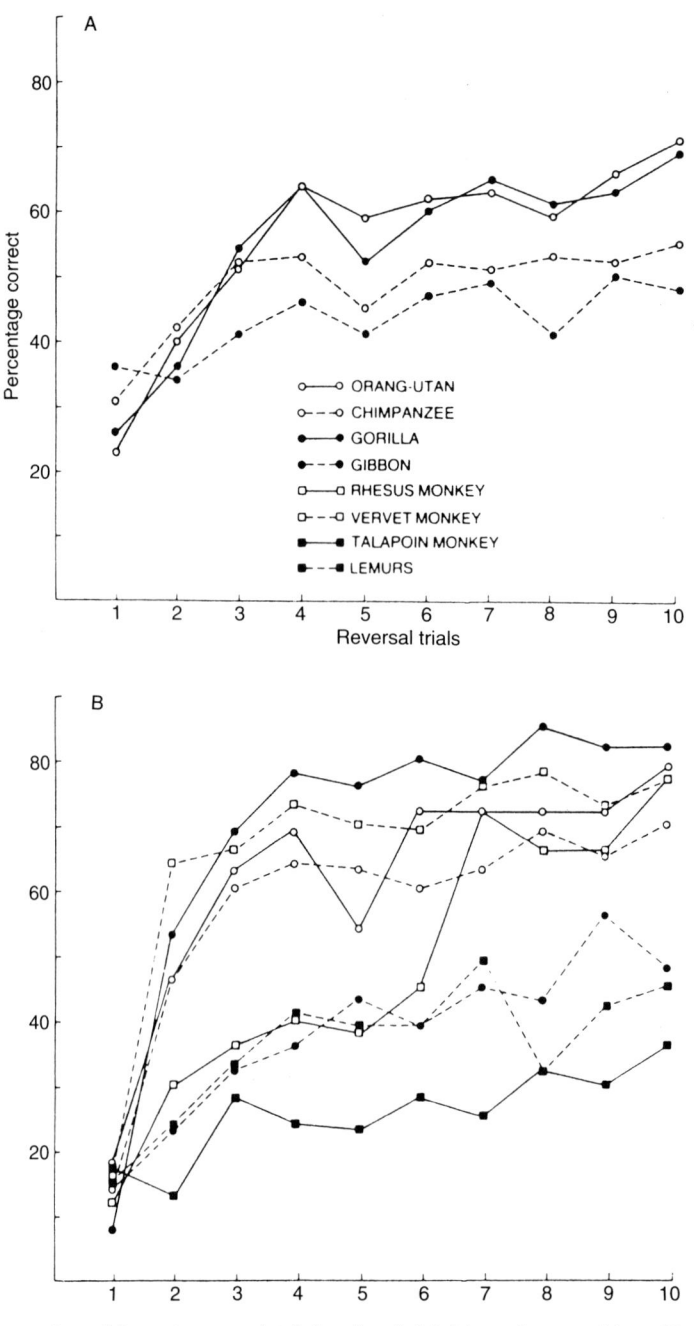

Fig. 22.6 Percentage correct on 10 post-reversal trials after initial learning to either 67 per cent (A) or 84 per cent (B) correct. Data are shown for Great Apes, Old World monkeys, and prosimians, from Rumbaugh and Gill (1973) and Wilkerson and Rumbaugh (1979). (Modified from Fobes and J. King (1981). By permission of Academic Press, the authors, and D. M. Rumbaugh.)

time-based paradigm is the delayed response procedure introduced by Hunter (1913). Here, food is hidden, and the animal is delayed from searching for it. Since an animal that subsequently searches would be responding when the stimulus was no longer visible to it, its performance, if successful, was claimed to be

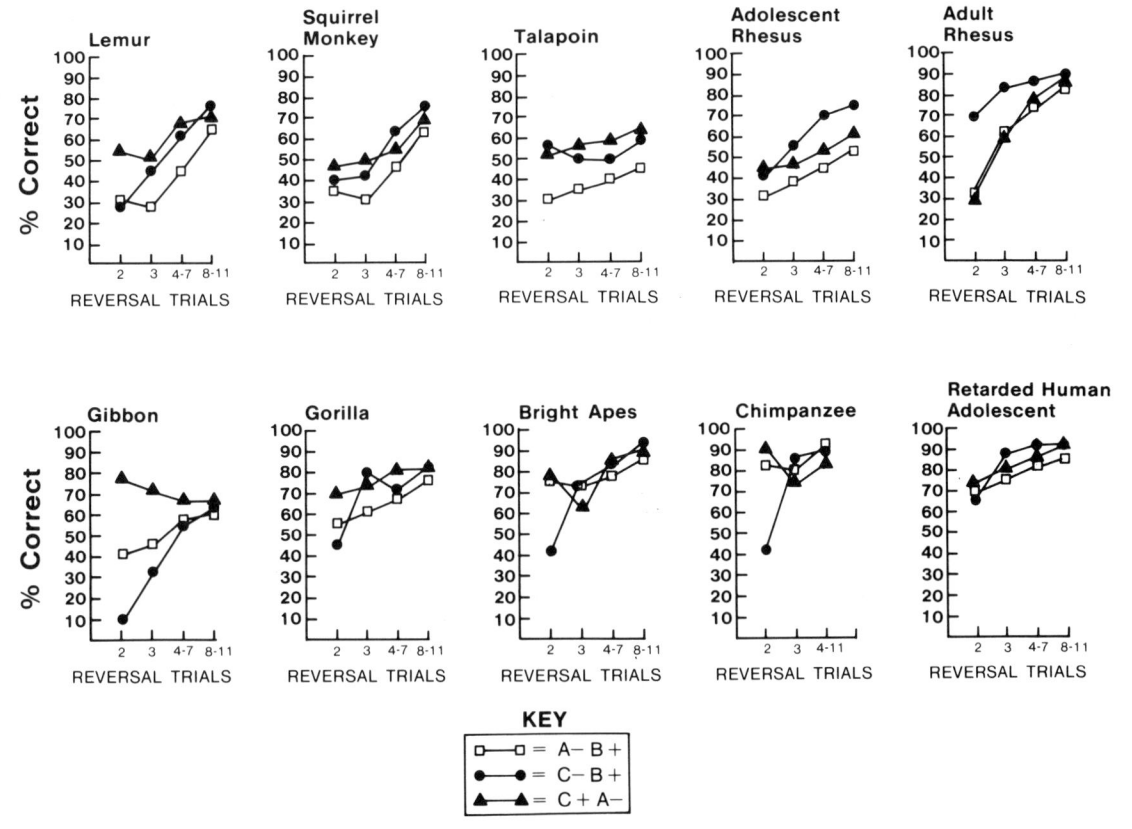

Fig. 22.7 Reversal performances for each of the conditions in the associative–mediational assessment paradigm (see text for expansion). (Figure reproduced by permission of D. M. Rumbaugh, incorporating data for adult rhesus monkeys from Essock-Vitale 1978 and for retarded humans from Meador and Rumbaugh 1981.)

mediated by 'symbolic' or 'representational' processes. Great variability in a species' performance has been found by different experimenters, suggesting that the various testing situations employed have a substantial effect on the way an animal performs in such tests (Fletcher 1965). This measure has not turned out to be of comparative significance (see Fig. 22.8).

In response-patterned learning the dominant experimental design, the double-alternation problem, was again introduced by Hunter (1920). Here, two identical objects may be presented, and the animal has to make four spatially defined responses on the pattern right-right-left-left. While initially this procedure failed to discriminate amongst various vertebrates in a comprehensible way (racoons being good (Johnson 1961); rabbits and cats less good (Livesey 1965); rats and rhesus monkeys the worst (Livesey 1965, 1969)), further studies (see Fobes and King 1982 for a review) have revealed that different species use different strategies in this task. When animals trained on this 4-trial task are presented with 8-trial double-alternation tasks, rhesus monkeys generalize rapidly, indicating some understanding of the double-alternation rule, whereas cats and racoons are unable to generalize. This problem situation thus provides additional evidence for the existence of mediational abilities in primates.

22.5.4 Probability learning

In probability learning an animal is faced with two stimuli. Rewards are not given on an all-or-none basis, but a percentage of responses to each of the stimuli is rewarded, but at a different rate, for example 8:2 or 7:3. Animals show a number of response strategies in this situation, including: 'maximizing', indicated by responses' being directed almost exclusively to the more frequently rewarded stimulus alternative; and 'matching', in which the proportion of responses directed to each stimulus is closely equivalent to the probability of reinforcement.

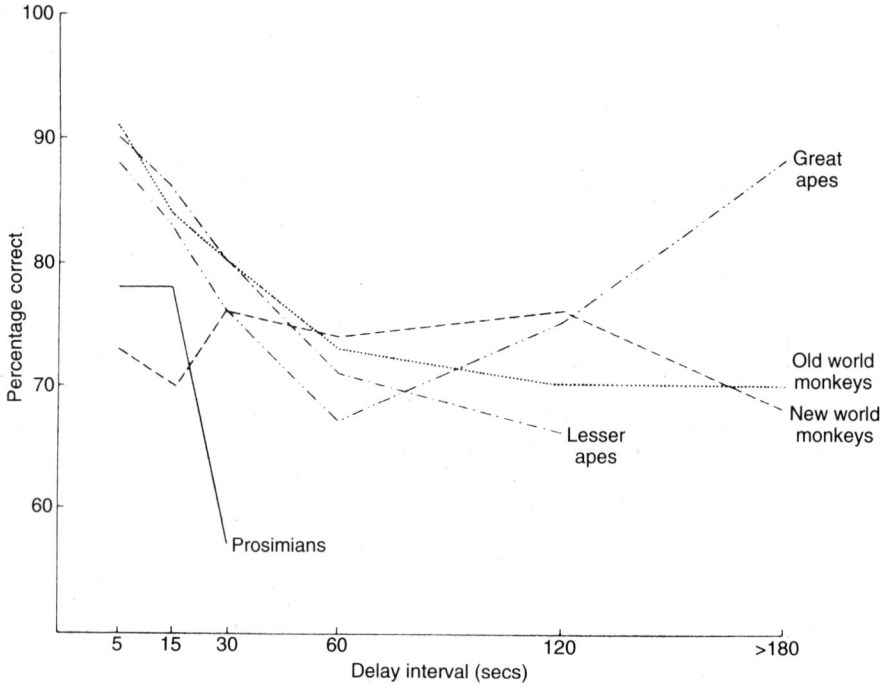

Fig. 22.8 Percentage correct after various delay intervals for prosimians: common treeshrews (*Tupaiaglis*), ruffed and mongoose lemurs (*Lemur variegatus* and *mongoz*); New World monkeys: marmosets (sp.), white-throated, white-fronted, and black-capped capuchins (*Cebus capucinus, albifrons*, and *apella*), Squirrel-monkeys (*Saimiri sciureus*), black-handed and sp. spider monkeys (*Ateles geoffroyi* and sp.), and Humboldt's woolly monkeys (*Lagothrix lagothricha*); Old World monkeys: green, mona, and grivet guenons (*Cercopithecus sabaeus, mona*, and *aethiops*), drills and mandrills (*Mandrillus leucophaeus* and *sphinx*), chacma, yellow, sacred, and Guinea baboons (*Papio ursinus, cynocephalus, hamadryas*, and *papio*), pig-tailed, crab-eating, and Barbary ape macaques (*Macaca nemestrina, fascicularis*, and *sylvana*); lesser apes: white-handed gibbon (*Hylobates lar*); and Great Apes: orang-utans, chimpanzees, and gorillas (*Pongo pygmaeus, Pan troglodytes*, and *Gorilla gorilla*). Data were obtained by the direct method of testing, and are from Harlow and Bromer (1939), Harlow *et al.* (1932), King *et al.* (1968), Leonard *et al.* (1966), Maslow and Harlow (1932), and Miles (1957a). (From Fobes and King (1981). By permission of Academic Press and the authors.)

Early work on matching led Bitterman (1965) to propose that response strategies served to differentiate animals, the more highly evolved mammals showing maximizing and matching, with less evolved animals, such as birds, fish, and reptiles responding randomly. This proposal has not been borne out by the data: all species tested have been shown to maximize under certain conditions (W. A. Wilson *et al.* 1964; Warren 1973).

22.5.5 Cross-modal recognition and transfer of learning

A similar role has been proposed for cross-modal transfer: that its presence in Great Apes and Old World macaques demonstrates a discontinuity in animal learning capacities. However, there is now some evidence of cross-modal abilities for all the major primate groups (see Ettlinger and Ettlinger and Wilson 1990 for reviews of theory and research). Meador *et al.* (1987, p. 42) comment:

With demonstrations of cross-modal abilities in preverbal [human] infants (Bryant *et al.* 1972; Wagner *et al.* 1981) the possibility of cognitive, non-language-based processes of cross-modal performance in non-human primates is being considered (Ettlinger 1981).

Cross-modal abilities clearly implicate representational processes of some form or another.

22.5.6 Representation: categorization and concepts

The findings just discussed raise the question of what sorts of representations animal cognitive systems use.

The underlying basis for categorization by animals is at present unclear (although it is no more clear in the human case). There is, however, wide agreement that animals are able to form *concepts* (see Roitblat 1987 for a recent review). The most prominent work to date is that of Herrnstein and his students with pigeons. In successful experiments, the animals are concurrently taught several discriminations, and then the common feature of these discriminations (i.e. the concept) is immediately transferred to new stimuli, with the level of performance attained in discriminating the learning set being maintained with the new set, i.e., almost complete transfer of discrimination to the novel material.

The concepts investigated include humans in general (Herrnstein and Loveland 1964; Mallot and Siddal 1972; Siegel and Honig 1970), a particular human individual, trees, bodies of water (Herrnstein 1979; Herrnstein *et al.* 1976), fish (Herrnstein and de Villiers 1980), other pigeons (Poole and Lander 1971), and oak leaves (Cerella 1979). Similar work, using a variety of experimental techniques, points to the formation of equivalent conceptual classes, for example, letters, humans, and other conspecifics, both individually and generally, in some species of monkeys and non-primates: for example squirrel-monkeys (Burdyn and Thomas 1984) stumptailed monkeys (Schrier *et al.* 1984); rhesus monkeys (Schrier and Brady 1987); cebus monkeys (D'Amato and van Sant 1988); and dolphins (Herman and Forestell 1985).

Are some concepts easier for some species to learn than others? For example, it would not have been surprising to find that pigeons were very good at discriminating between the leaves of different trees, yet poor when it came to recognizing fish: but it turns out that they are good at this task as well. Could it be that animals which are less adept at generalizing their abilities from one class of phenomena to another are just those we naïvely regard as less adaptable? At present information clarifying the relationship between ecological and cognitive factors is not available. The effect of its absence in determining our naïvety is brought into relief when experimental results bring us face-to-face with our naïve presuppositions. For example, Schrier *et al.* (1984) set up a very elaborate experiment to record monkey eye-movements, in order to investigate what information these animals might be using in learning concepts, following procedures that had been developed successfully with pigeons. Schrier and Brady (1987, p. 136) ruefully noted later that

it didn't occur to these investigators that they might have trouble demonstrating the basic outcome, because they assumed that anything pigeons could do, monkeys could do better or at least as well.

However, those original investigators, Schrier among them, certainly found their intentions more difficult to pursue than they had expected: their naïve expectations as to the relative competence of pigeons and monkeys proved to be false. Again, their surprise at this reinforces the point that extant animals are not behavioural fossils.

22.6 Same/different judgements by animals

What cognitive *operations* can animals perform? One area of work bearing on this question claims some animals have concepts of 'same' and 'different'. If an animal can form a concept of, say, a particular letter, it is implicitly making a judgement in the experimental situation that a particular stimulus is either the same as, or different from, the category it is acting upon. There have been numerous techniques deployed in investigating these judgements, making it difficult to resolve all the findings coherently (see, for example, Premack 1983*a*). In addition, the topic is rather confused conceptually.

For example, the term 'concept' is used with a number of meanings in this work. These stretch all the way from possibly non-conceptually mediated performance on the basis of stimulus generalization to rule-mediated performance which verges on reasoning. Some of the questions that have not been clearly demarcated include: does the fact that an animal may be taught to discriminate different stimuli as equivalent, say as letter As, implying it sees these stimuli as the 'same', mean it possesses the concept that they are the same? Again, an animal may learn 'if red light, peck green key': does this mean it has learnt anything of a conceptual nature? The answer to this depends upon how it transfers this ability to new test situations. If it shows high transfer to orange/turquoise but not circle/square, this may only indicate stimulus generalization, rather than the abstraction of the notion of similarity, or sameness, of the relationship. But, unless the appropriate controls are performed, one cannot rule out in this event that it has not in fact abstracted a colour-domain-specific rule that red/green, orange/turquoise are related in the same way, and cannot extend this rule to other domains. Further,

A true same/different judgement is more complex than may meet the eye. When an individual calls two apples the same, two elephants the same, two toads the same, etc., he is judging that the relation between the apples (same1) is the same as that between the elephants (same2) and that both these relations are the same as that between the toads (same3). Judging relations between relations is more complex than discriminating between items that have and have not been experienced before or reacting to items on the basis of their physical similarity (Premack 1983b, p. 355).

This conceptual confusion invades experimental techniques, irrespective of whether adequate control procedures have been adopted. For example, the experimental literature often tends to conflate the *successive* testing of 'similarity'—the same stimulus on different occasions—with the *simultaneous*—two stimuli of one type at the same time. Given that animals perform these tests with remarkably different levels of success, it may be that the former technique is in fact tapping the animals judgement of familiar/unfamiliar rather than same/different. In sum, Premack (1983b, p. 354) offers a salient caution:

The study of identity—the conditions under which different species react differentially to items that humans consider to be same or different—has been heating up in recent years. As a new field, it is somewhat casual in its distinctions and runs the risk of bogging down in confusion.

While the field has progressed since Premack wrote, his is still a point to bear in mind.

22.6.1 Non-primates

Despite a large literature investigating the matching concept (the identity or same/different concept) in species such as goldfish (Zerbolio and Royalty 1983) and in particular pigeons (Lombardi *et al.* 1984;

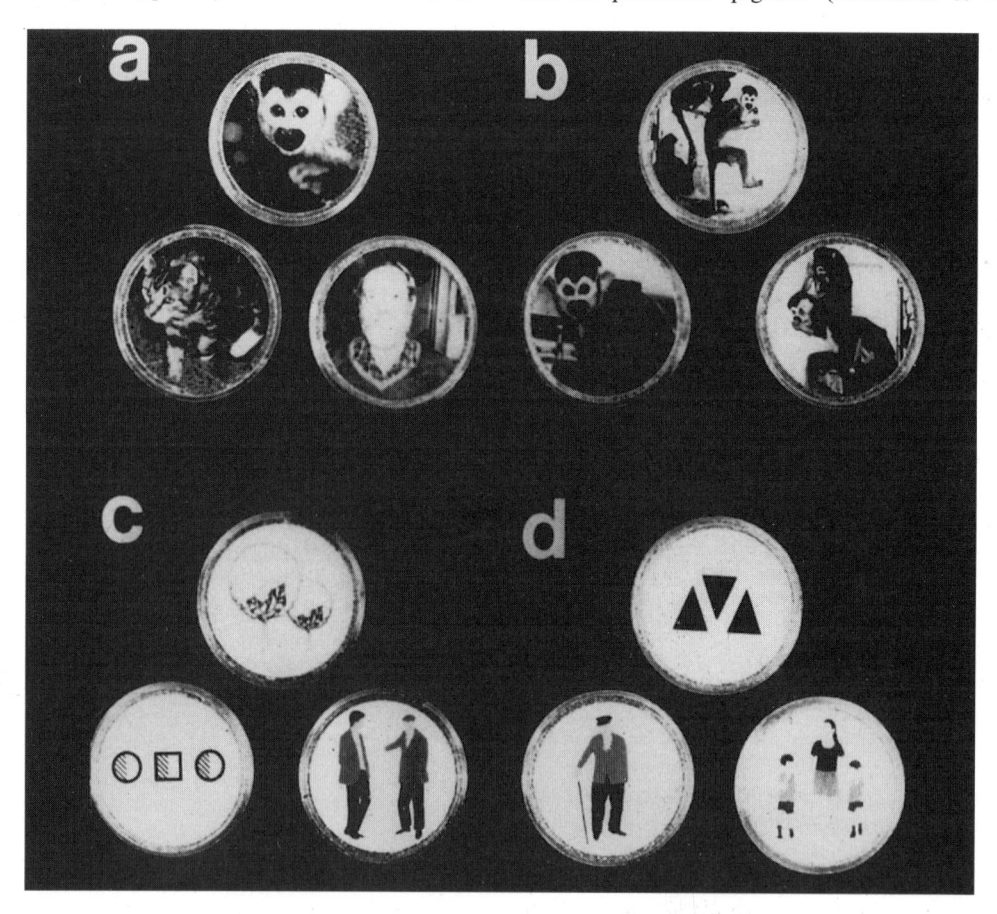

Fig. 22.9 Representative stimuli used by Irle and Markowitsch (1987), for conceptual testing of monkey and human subjects, and results obtained for different conceptual classes. (From Irle and Markowitsch (1987). Copyright (1987) by the American Psychological Society. Reprinted by permission.)
(a) Representative trials for testing the conceptual classes primates versus non-primates (Panel a) and numerousness concepts (Panels b–d). (For an easier orientation to the reader the sample stimuli are represented at the top, and the correct ['new'] response stimuli are always arranged at the bottom left. All correct ['new'] and wrong response stimuli have *equal* numbers of differences from their sample stimuli with respect to general visual properties.)

Zentall and Hogan 1974, 1976, 1978; Zentall *et al.* 1981; Zentall *et al.* 1984), D'Amato *et al.* (1985) (cf. Wilson *et al.* 1985*a, b*; Zerbolio 1985) argue that claimed demonstrations of the matching concept in pigeons and goldfish could be accounted for by simpler (non-conceptual) processes such as stimulus generalization and differential transfer of specific response predispositions. On the basis of further work, D'Amato *et al.* (1986, p. 372) conclude:

> An appreciation for the abstract quality of 'sameness' seems to be either completely lacking or very poorly developed in pigeons... although other avian species (corvids) may prove more talented in this regard (Wilson *et al.* 1985*b*).

Pigeons may have concepts (see Section 22.5.6), but they do not appear to have relational ones.

22.6.2 *Monkeys*

There is more evidence for a same/different concept in monkeys. Within the visual mode, D'Amato *et al.* (1985*a*) have demonstrated transfer to new stimuli with training on as few as two exemplars. Irle and Markowitsch (1987) were successful in showing same/different judgements in squirrel-monkeys for a variety of conceptual classes (see Fig. 22.9) after only one presentation of an exemplar. Interestingly, they tested the same procedure with three human subjects, and found that:

> The overall performance in the 3 conceptual trials did not differ significantly between monkeys and human subjects...

The monkey with the best performance outperformed the human subject with the poorest performance (ibid., p. 37–9).

This does not mean, however, that monkeys have a well-articulated concept of sameness. For example, although the monkeys in the D'Amato *et al.* (1985*a*) study showed high levels of transfer to novel geometric stimuli, they failed to transfer the matching concept to stimuli that varied along a dynamic dimension (flashing vs. steady green disk). Furthermore, there is no evidence that once they have formed a concept in the visual modality, monkeys can extend the matching concept to the auditory modality (D'Amato 1973; Salmon 1984), or touch (Milner 1973), despite the fact that experiments have shown that monkeys are capable of forming a matching concept in the auditory modality (D'Amato and Colombo 1985*a*; Wright *et al.* 1990).

It seems as if monkeys must develop the matching concept anew each time they meet a new class of 'referents': they may have a conceptual facility, but not be able to deploy it in a generalizable sense,[2] as might be the case for humans (cf. Weinstein 1941). By contrast, both Old and New World monkeys show a greater degree of generalization in their performance on the opposite of the matching problem, i.e. the oddity problem. Investigations of this problem use different techniques to those investigating sameness, and probably involve quite different underlying skills (for reviews see Fobes and King 1982; and D'Amato and Colombo 1989).

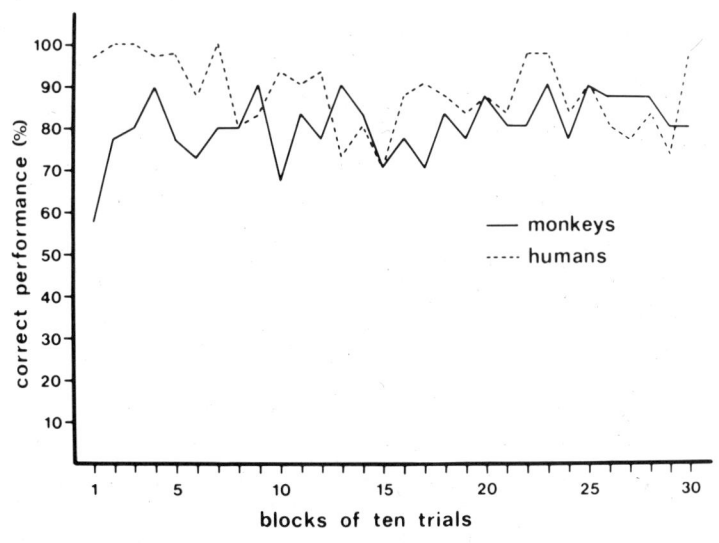

(b) Percentage of correct responses of monkeys and humans during the time course of training. (Ten trials were tested during one session for the monkeys [resulting in 30 training sessions], and 100 trials were tested per session for the humans [resulting in 3 training sessions]. The 300 trials were definitely the same, and applied in the same order, for both monkeys and humans.)

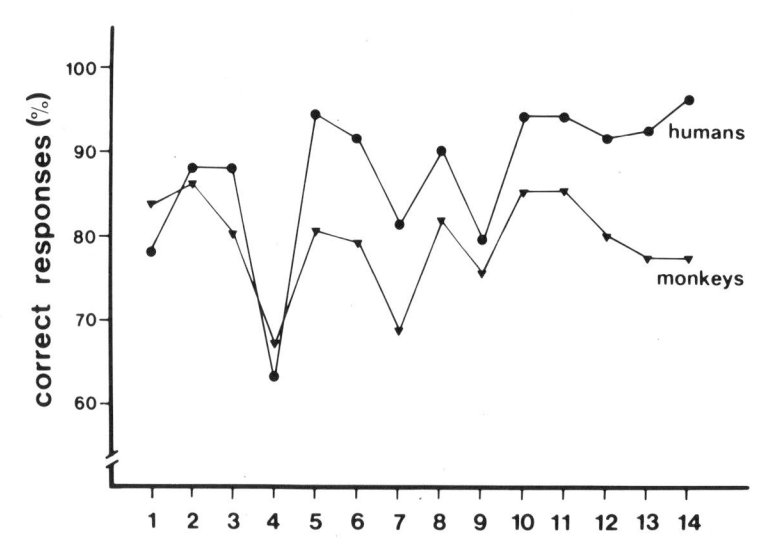

(c) Percentage of correct responses of monkeys and humans for individual conceptual classes tested. (Note that the diverse conceptual classes were trained according to random sequences so as to guarantee that each conceptual class was tested with similar frequency during all training sessions. Significant performance differences [$p < .01$] between monkeys and human subjects [Mann–Whitney U tests] is indicated by asterisk. 1 = monkeys vs. non-primates [46 trials]; 2 = humans vs. non-primates [28 trials]; 3 = monkeys vs. humans [20 trials]; 4 = primates vs. non-primates [10 trials]; 5 = non-primates vs. other non-primates [10 trials]; 6 = animal [incl. human] species vs. fruits [31 trials]; 7 = apples vs. bananas [15 trials]; 8 = animal species vs. trees [21 trials]; 9 = numerousness concepts [28 trials]; 10 = sameness concepts [45 trials]; 11 = arrows vs. geometrical forms [17 trials]; 12 = *E*s vs. geometrical forms [11 trials]; 13 = *O*s vs. geometrical forms [7 trials]; 14 = telephones vs. geometrical forms [11 trials].)

22.6.3 *Apes*

Apes were the earliest subjects for sameness/difference experiments (Robinson 1955, 1960). Chimpanzees showed great ability on these tasks, and were able to generalize to new stimuli. Subsequent work by King (1973) confirmed this finding, and extended it to the orang-utan, yielding no significant interspecies differences. It is, however, from within the 'language' training programmes that most of the information on anthropoid conceptual abilities has been generated. These programmes unequivocally demonstrate that chimpanzees, gorillas, and orang-utans are able to form and use symbolic concepts of quite a wide variety (see Ristau, this volume, Chapter 23). The facility these species show in the formation of concepts has proved to be such that little of the 'basic' type of research discussed above has been carried out. Rather, most investigations have focused on the use these animals can make of symbolic concepts: thus, studies of apes can be best thought of as 'meta-conceptual'.

That apes can perform better than monkeys on 'same/different' tests is indicated not only by what direct experimental work there is (see Premack 1976, 1983*a, b*), but also by their performance on tasks which blur the distinction between operating at a conceptual level and *reasoning*. Figure 22.10 gives an example of the sorts of problems used by Gillan *et al*. (1981). Here, the chimpanzee Sarah is being tested as to whether she can classify *relationships* between objects as 'same/different', and not just 'similarities' or 'equivalences' between or among objects themselves. The results of this investigation, taken as a whole,

strongly support the hypothesis that Sarah can reason analogically in a variety of circumstances.
The finding that a chimpanzee can reason analogically invalidates the extreme claims that reasoning is an exclusively human cognitive ability (e.g., Huxley 1897; James 1890; Morgan 1894; Thorndike 1898). . . . [But] it is difficult to assess the phylogenetic generality of reasoning since these experiments studied only one chimpanzee. Sarah has a unique experimental history which may have contributed to her reasoning abilities.
Although the present experiments do not allow broad claims about analogical reasoning in chimpanzees other than Sarah, previous experiments suggest that other chimpanzees may be able to reason analogically (Gillan *et al*. 1981, pp. 1–11).

22.7 Symmetry and transitivity of conditional relations

22.7.1 *Non-primates and monkeys*

D'Amato's work on symmetry and transitivity of conditional relations represents another line of research in this general area of same/different investigations, and also another attempt in a long line to establish features of interspecies cognitive differences (D'Amato *et al.* 1985*b*). In what is termed a conditional or symbolic matching task, an animal associates a pair of standard stimuli, say A and B, with a pair of comparison stimuli, say X and Y, such that if presented with A it responds by choosing X, and if B then Y. Asch and Ebenholtz (1962) argued that for humans a backward relationship is simultaneously formed with the conditioned forward relationship—what they termed the 'principle of associative symmetry'—such that if, after training, subjects were presented with X they would respond A, and if Y then B. While the claim for symmetry seems not to have been established, strong backward associations are often reported for human subjects (for example Houston 1981).

In contrast to the case for humans, and despite quite intensive training, D'Amato *et al.* (1985*b*) failed to find any evidence for monkeys (*Cebus apella*) or pigeons forming backward associations:

our results, in conjunction with those of Sidman *et al.* (1982) and the data from relevant pigeon studies, suggest that a sharp difference exists between humans and other species (monkeys, baboons, and pigeons) in the degree to which associations occur bidirectionally (D'Amato *et al.* 1985*b*, p. 41).

Sharp differences also exist between humans and other species with respect to inferential transitivity, although whether the apes can show this is still debatable (see below). A subject demonstrates inferential transitivity if after having learned that A > B and B > C it concludes that A > C. Although little work has been done with monkeys (McGonigle and Chalmers 1977; Menzel and Draper 1965), the evidence does not favour the view that monkeys are capable of inferential transitivity of the type exhibited by children (Bryant and Trabasso 1971). Recently there have been suggestions that both pigeons (von Fersen *et al.* 1991) and rats (Davis 1992) are capable of inferential transitivity. In the case of the pigeon study (and probably also the rat study) simpler interpretations that do not invoke the notion of inferential transitivity can fully account for the data (Couvillon and Bitterman 1992; Markovits and Dumas 1992).

Although monkeys seem not to show inferential transitivity, they are able to demonstrate 'associative

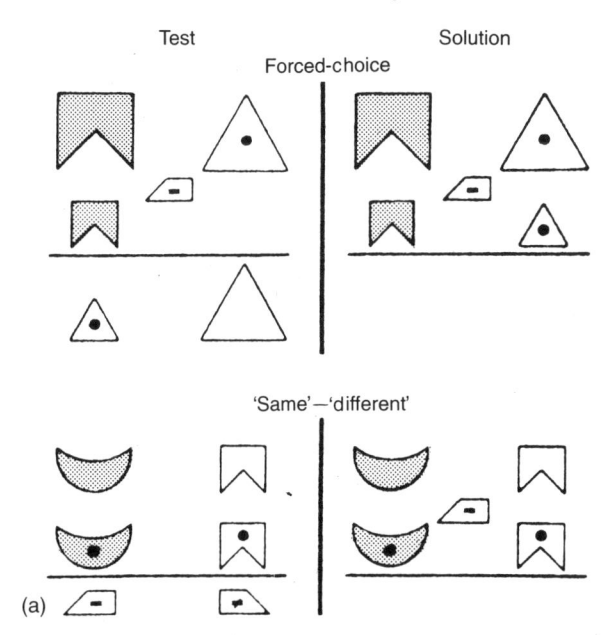

Fig. 22.10 (a) Example of analogy problems: a forced-choice, figural analogy problem (Experiments 1A and 1B) and solution, and a same–different analogy problem (Experiment 2) and solution.

(b) Example of the conceptual analogy problems used in Experiment 3A. (Problem A shows a closed lock [A], a key [A′], and a closed, painted can [B] in the problem display, and a can-opener [B′] and paintbrush [C] as the alternatives. Problem B shows marked paper [A], a pencil [A′], and a closed, painted can [B] in the problem display and a paintbrush [B′] and can-opener [C] as the alternatives.)
(From D. Gillan *et al.* (1981). Copyright (1981) by the American Psychological Society. Reprinted by permission.)

transitivity'. Here, an animal is trained to associate A and B with X and Y as before, and then X and Y with a new pair J and K. Associative transitivity is shown if when tested with A and B as stimuli and J and K as possible responses an animal selects J for A and K for B. They note

such an outcome ... is by no means inevitable and may be indicative of a cognitive capacity not equally represented in all animals. For example, one plausible interpretation of associative transitivity is that, during the test trials, the sample elicits a representation of the original correct comparison stimulus, which then serves as the covert or surrogate sample. Clearly, a cognitive capacity of this sort goes beyond the learning mechanisms of conditional matching tasks (1985, pp. 21–2).

No evidence for associative transitivity was found for pigeons, suggesting a difference in the ability of monkeys and pigeons to utilize implicit or surrogate stimuli, or in the language of cognitive models, representations.

22.7.2 Apes

Further evidence for representational reasoning comes from Gillan's (1981) study of transitive inferences in chimpanzees. Here, non-language-trained chimps were tested on a Piagetian problem: if they 'know', for example, that A is bigger than B, and that B is bigger than C, can they infer that A is bigger than C? The results provide evidence that chimpanzees can perform the reasoning required, although, again

it is not clear whether transitive inference is an ability in a wide range of primates. However, previous research suggests that another primate species, the squirrel monkey, can also make transitive inferences (McGonigle and Chalmers 1977). ... In addition, [the experimental] data indicate that language is not a necessary condition for transitive inferences (Gillan 1981, pp. 161–2).

As with all results in this field, there is dispute over their status. D'Amato and Salmon (1984) argue on methodological grounds that 'the presently available data do not seem conclusively to demonstrate inferential transitivity in nonhuman primates, not even in apes (D'Amato *et al.* 1985, p. 44). (But it is not clear when developmental psychologists might agree on the ontogenetic, as opposed to phylogenetic appearance of this ability either: see, for example, Breslow (1981).)

22.8 Representations of lists in pigeons and monkeys

Species; in particular pigeons and monkeys, not only differ in the representation of the matching concept but also in their representation of lists of items. This capacity is tested using a serial-order task, originally developed for use with pigeons (Straub and Terrace 1981) and subsequently adapted for use with monkeys (D'Amato and Colombo 1988b). The procedure is relatively simple. The animals are trained to respond to five simultaneously presented stimuli, termed here A, B, C, D, and E for the sake of exposition, in the following order to obtain a reward: A→B→C→D→E. Any deviation from this sequence, for example A→B→D (a forward error) or A→B→B (a backward error) is incorrect. Both pigeons and monkeys can master this task, and it is perhaps not surprising that monkeys solve such a problem faster than pigeons. The interesting differences are revealed in what is termed a 'pairwise' test, in which the animals are presented with all possible pairs that can be constructed from the 5 items (10 in all: AB, AC, AD, AE, BC, BD, BE, CD, CE, and DE). Monkeys perform at high levels on all 10 pairs, responding to the items in a pair in the order in which they appeared in the original sequence. Thus when presented with pair BD, for example, monkeys will respond to item B first, followed by item D. In contrast, pigeons respond at high levels only on those pairs beginning with item A (AB, AC, AD, and AE) or ending with item E (BE, CE, and DE). Their performance on the internal pairs (BC, BD, and CD) is often at chance levels. It is clear that despite the fact that both pigeons and moneys can acquire a five-item series, even to similar criterial levels, how each species represents the serial list is fundamentally different.

This difference is even more apparent when one observes the differences in latency to respond to the first and second item of each list. In monkeys, latency to respond to the first item increases monotonically as a function of the position of the first item in the series (latency to A is determined by averaging across pairs AB, AC, AD, and AE; B across pairs BC, BD and BE; C across pairs CD and CE; and latency to D is derived from pair DE only). In addition, latency to respond to the second item also increases monotonically as a function of the number of missing items between the first and second items of the pair (0: AB, BC, CD, and DE; 1: AC, BD, CE; 2: AD and BE; 3: AE). In contrast, pigeons show no 'first item' or 'missing item' latency effects, that is, their latency functions are flat across all conditions (Terrace 1993).

These findings strongly argue that in learning the original five-item series, the monkeys form a linear representation of the five items. Thus when presented with two stimuli from the original list, the monkey appears to begin with item A and progress through the list in an orderly fashion until an item on the display matches the one currently accessed in memory, in much the same way that we might determine whether the letter H comes before or after the letter K in the alphabet. Pigeons, on the other hand, do not seem to form such a linear representation, relying instead on simple rules such as 'always press A first' and 'always press E last'. As the position was stated succinctly by Terrace (1993, p. 168):

> Differences in performance by pigeons and monkeys on the [pairwise] test provide a clear exception to MacPhail's 'null hypothesis' that cognitive processes in non-human animals differ in degree but not kind (MacPhail 1985). According to MacPhail, qualitative differences in the performance of different species on cognitive tasks can be attributed to non-cognitive factors such as differences in sensory capacities, specific test procedures, and so on. As applied to the present results, one obvious flaw in that argument is that, for both species, the [pairwise] test was derived from similar lists of the same length. Further, the pigeons and the monkeys could not be distinguished with respect to accuracy of responding, distributions of forward and backward errors, serial position effects, and other details of their performance on 5-item list. It seems reasonable to conclude, therefore, that the results of the [pairwise] test reflect real differences in a pigeon's and a monkey's representation of the original list and that, in general, the success of MacPhail's null hypothesis is an artifact of insensitive test of cognitive differences between species.

22.8.1 Interim conclusion

The Cartesian notion of 'animals as robots' is clearly untenable. Most mammals do appear to have cognitive representations of the world, which function in quite similar ways to support, for example, memorial capacities and, perhaps only in the higher primates, something akin to conceptual thought. There does appear to be some evidence that the likelihood of demonstrating such capacities increases amongst primates in relation to the recency of the evolutionary emergence of the superfamily to which a particular species belongs.

22.9 Observational learning

Beck (1975) has differentiated three categories of observational learning:

1. *Social facilitation* refers to the synchronization of actions between two or more animals, such that the performance of a particular activity by an individual is more likely if it has observed the actions being performed by other individuals. Here observation is obviously involved, but learning need not be.

2. *Stimulus enhancement* occurs when the activities of one individual serve to increase the likelihood of another's orienting to some aspect of the environment. As with social facilitation, an observational component is necessarily implicated, and while learning one need not be (for example, in escaping from predators which have been observed only as a consequence of observing another animal—which in a natural setting would be difficult to distinguish from social facilitation), it often is, in the sense of modifying what the individual would otherwise pay attention to in the environment (for example, what an individual selects as 'food').

3. *Imitative copying* refers to the common-sense meaning of observational learning. Beck's definition is rather stringent (1975, p. 437), requiring an 'exact' duplication of the actions of one individual by another, indicating that there is again a range of activities which come under this concept. For example, we might naïvely conclude that one individual learns to open a box with a key by imitating another when in fact it actually apprehends the means–end relationship between the key (means) and opening (end), or vice versa. Learning is clearly involved in this category of activities, but the form of learning, from copying to facilitated problem-solving, shades across a wide spectrum of abilities.

The empirical research on observational learning in primates is summarized in Table 22.2. Social facilitation and stimulus enhancement occur across the entire order. Imitative copying, however, is predominantly an ability of the apes, there being only a few unequivocal cases reported in monkeys. But even in apes there are more constraints on this ability than with humans.

First, factors of social organization often affect who observes and copies whom. In some senses this is applicable to human 'fashions', so that the diffusion of activities and ideas is constrained by differential social power; but it is a more immediate constraint among the other primates. Second, Jolly (1972) notes that 'teaching' plays no role in non-human primate imitative skill transfer.[3] Individuals do not modify their activities for pedagogic ends, by giving a good model or demonstration or by 'correcting' the learner. Third, it has only been with the recent upsurge of primate research that imitation has been 'discovered' as part of a primate's 'natural' behavioural repertoire, suggesting it has a low frequency of occurrence.

22.10 Piagetian studies of animal cognition

Piaget's concern with the evolutionary, as opposed to ontogenetic, foundations of knowledge only became explicit in his later work (for example, 1970, 1978,

Table 22.2 Summary review of investigations of primate observational learning, categorized by Beck's (1975) scheme

Sub-Order	Prosimii	Anthropoidea			
Super Family		Ceboidea	Cercopithecoidea	Hominoidea	
Family				Hylobatidae	Pongidae and Panidae
Social facilitation Stimulus enhancement	Learning of paths in home range: Pariente 1979. Juvenile lemurs learn from watching mothers, but not vice versa: Feldman and Klopfer 1972; see also Hladik 1979.	Demonstrated for marmosets: Menzel and Juno 1982. Wild howler monkeys learn food habits from mothers: Hall 1963. Marmosets: Menzel and Juno 1982. Squirrel-monkeys: Mahan and Rumbaugh 1963.	Demonstrated for patas monkeys: Hall and Goswell 1964. Rhesus monkey: Darby and Riopelle 1959. Food-washing in Japanese macaque: Kawai 1965. Chacma baboon food habits: Hall 1963. Tool-use learning in Guinea baboons: Beck 1973.	Data not available. Data not available.	Demonstrated for chimpanzees: Goodall 1968. Chimpanzee: Goodall 1968. Gorilla: Fossey 1979. Orangutan: Rijksen 1978.
Imitative copying	Not demonstrated.	Not demonstrated.	Demonstrated for 1 individual on 1 occasion for each of: pig-tail macaque: Beck 1978; Rhesus monkey: Kinnamon 1902. (see Hall 1963).	Data not available.	Termite-fishing: Goodall 1968. Ant-dipping: McGrew, 1977. Water-avoiding: McGrew 1977. Nest-construction: Bernstein 1962. (All for chimpanzee.)

1980). The first suggestion that it could provide a valuable perspective for the comparative study of animal cognition and development was made by Jolly (1972) in the first edition of her book *The evolution of primate behaviour*. The first Piagetian-inspired study of animal behaviour was of the development of object-permanence in the cat, by Gruber *et al.* (1971). Recently, Doré and Dumas (1987) have provided a comprehensive review of the body of work that has amassed since then. Here work on primates will be briefly discussed.

22.10.1 Sensorimotor intelligence

Data are available for several primate species—chimpanzees (*Pan troglodytes*): Bergeron (1979); Chevalier-Skolnikoff (1977, 1982); Hallock and Worobey (1984); Mathieu and Bergeron (1981); gorillas (*Gorilla gorilla*): Chevalier-Skolnikoff (1977, 1982); Redshaw (1978); orang-utans (*Pongo pygmaeus*): Chevalier-Skolnikoff (1982); Chevalier-Skolnikoff *et al.* (1982); Laidler (1978); stump-tailed macaques (*Macaca arctoides*): Chevalier-Skolnikoff (1977, 1982); Parker (1977); hanuman langurs (*Presbytis entellus*): Chevalier-Skolnikoff (1982); and Capuchin monkeys (*Cebus apella*): Fragaszy *et al.* (1990). Reviewing this material, Doré and Dumas (1987, p. 222; see also the contributors to Antinucci 1989) conclude, with respect to the Great Apes:

> Chimpanzee, gorilla and human infant go through similar stages and in the same order [see Fig. 22.11]. However, important differences between great apes and humans emerge. Although the developmental sequence is similar, the early appearance of locomotion in great apes gives them a temporary advantage over human infants. On the other hand, the role of interactions with physical objects is more limited in great apes than in humans, and circular reactions (primary, secondary, and tertiary) are displayed in fewer modalities and contexts. Indexes indicate that the beginning of mental representation, which characterizes Stage 6 in humans, is also present in great apes by the end of the second year.

They add two qualifications to these points. First, despite the similarities, there are structural differences, in that some of the sensorimotor structures described for humans by Piaget appear to have sometimes greater and sometimes lesser prominence in animal development. Second, these studies confirm Vauclair (1982) and Vauclair and Bard's (1983) assertion that whereas object manipulation is an important determinant of human development, social interactions are much more influential in primates (Doré and Dumas 1987, p. 223).

22.10.2 Object permanence

Longitudinal data is available only for stump-tailed macaques (Parker 1977) and chimpanzees (Bergeron 1979). Primates studied in experimental situations

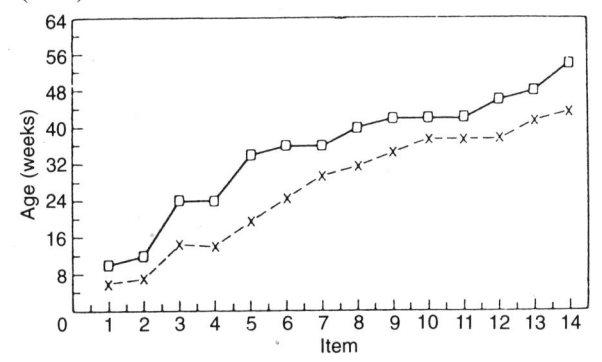

Scale

1. Follows a slowly moving object through 180° arc smoothly
2. Gaze lingers at the point of disappearance of a slowly moving object
3. Finds an object which is partially hidden
4. Gaze returns to starting point on disappearance of a slowly moving object
5. Finds an object which is completely covered by one screen
6. Finds an object which is hidden under one of two screens
7. Finds an object hidden under one of three screens
8. Finds an object hidden under a number of superimposed screens
9. Finds an object following a visible displacement under one of the three screens
10. Finds an object following an invisible displacement under a single screen
11. Finds an object following an invisible displacement under one of two screens
12. Finds an object following one invisible displacement with two screens alternated
13. Finds an object following one invisible displacement under one of three screens
14. Finds an object following a series of invisible displacements under three screens

Fig. 22.11 Developmental profiles of human and gorilla infants for visual pursuit and object permanence, as tested on the developmental scales of I. Uzgiris and J. McV. Hunt (1975). Human—□—. While gorillas are precocious over human infants, there is clear evidence of fundamental similarity in the courses of development. This similarity is not as substantial for spatial knowledge, operational causality, and means–end behaviour scales. (From M. Redshaw (1978). Copyright (1978) by Academic Press. Used by permission.)

that are not totally confounded by methodological problems are squirrel-monkeys (*Saimiri sciureus*) (Vaughter *et al.* 1972); rhesus monkeys (*Macaca mulatta*) (Wise *et al.* 1974); *Macaca fuscata* (Natale *et al.* 1986); woolly monkeys (*Lagothrica flavicauda*) and white-coated capuchins (*Cebus capucinus*) (Mathieu *et al.* 1976); chimpanzees (Mathieu *et al.* 1976; Wood *et al.* 1980); and gorillas (Natale *et al.* 1986; Redshaw 1978).

There are numerous methodological problems in studying this topic, and very often conclusions drawn about primate performance come from situations very different from those under which human infants are tested. Present claims for animals beyond the establishment of Stage 4 behaviours should thus presently be taken as provisional: they are that

some monkeys (stump-tailed macaques, Japanese macaques, and probably rhesus and wooly monkeys) reach State 5 of cognitive development, other monkeys (white-throated capuchins) and great apes (chimpanzees and gorillas) reach Stage 6 of object permanence (Doré and Dumas 1987, p. 225).

(But see also Natale and Antinucci 1989.)

Recently, Dumas (1992) has shown that cats, who were believed capable only of Stage 5 behaviours (Gagnon and Doré 1992), were able to display Stage 6a behaviours when tested on an ecologically-relevant version of the invisible displacement task. These findings emphasize the point raised above regarding the provisional nature of the data on the comparative capacities of primates with respect to object permanence, especially for those animals that have not had the benefit of testing under ecologically-relevant conditions.

The case for object permanence in non-mammals is much less clear. Although object permanence has been shown in a number of species of psittacine birds (Pepperberg and Kozak 1986; Pepperberg and Funk 1990), Gagnon and Doré (1992) have argued that the solution of these tasks may be explained by invoking 'simpler' mechanisms.

22.10.3 *Other sensorimotor acquisitions*

There are a few studies of spatial concepts, understanding causality, and imitation in primates (see Doré and Dumas 1987, p. 226 for details). There is some evidence, again, that Great Apes reach Stage 6 levels of behaviour.

22.10.4 *Representational intelligence*

Given that apes, at least, may attain Stage 6 Sensorimotor abilities, it is also possible they possess some of the rudiments of pre-operational intelligence. A

Fig. 22.12 Conservation test for liquid quantity. A While Sarah watches, trainer pours liquid from one container into another of different proportions. Trainer then gives 'same' and 'different' symbols to Sarah and leaves room. B: Sarah opens dish and removes a symbol. C: Sarah places 'same' symbol in circle between the two containers. She then calls trainer back into the room by ringing bell. (From Woodruff *et al.* (1978). Copyright (1978) by the AAAS. Used by permission.)

second reason for this possibility is the ability of apes to use symbols in the language training experiments discussed by Ristau (this volume, Chapter 23), since, within the Piagetian framework, the representational abilities which characterize the pre-operational stage are necessary pre-requisites for symbol use. Third, the strategies employed by many primates in their social lives (see Jolly, this volume, Chapter 6) imply that social knowledge is mentally represented and manipulated in the conduct of everyday life. Finally, such representative abilities, and even true operations as opposed to pre-operations, are implicated in the level of 'same/different' judgements reported by Woodruff and his colleagues (see above, Section 22.6).

To date, however, Piagetian-inspired investigations have been thin on the ground and disappointing in their results, largely owing to the methodological problems inherent in testing Piagetian notions by methods far removed from their original conversational settings. There are some indications that chimpanzees can display the basic elements of pre-operational intelligence (Woodruff et al. 1978; Muncer 1983 with respect to conservation; and Perrault 1982 with respect to seriation), and thus that they have at least a rudimentary form of representational (symbolic) intelligence (see Fig. 22.12). Squirrel-monkeys (Czerny and Thomas 1975) and macaques (Pasnak 1979) show some of the skills required for conservation tests, but only after extensive training. Classification skills have not yet been demonstrated in primates, even for non-linguistic chimpanzees.

The difference in the conceptualizing of these issues in the Piagetian framework as compared to the more traditional experimental one that underlies the work discussed previously in Section 22.6 makes a comparison of these two sets of data quite difficult. 'Same/different' judgements could be argued to rely on, or constitute, classification skills, so why can they not be substantiated by Piagetian-inspired techniques? Very grossly, the temptation is to conclude that, while many mammals appear to possess representational abilities (from the experimental tradition), they give little evidence of 'knowing how to use them' (from the Piagetian tradition). There is, however, another set of data that bears on this issue.

22.11 Tool-use, insight, self-recognition, and mirror-use in primates

22.11.1 Tool-use and insight

It is not easy to distinguish reliably imitative behaviours from creative problem-solving, yet this distinction raises again the question of representation and cognitive operation. Is an animal performing some 'trick', or is it 'thinking'? If the former, then cognition need not be invoked; if the latter, then cognition is generally invoked. The topics in this section all bear directly on this cognitive, representational issue. If it is true that, with enough monkeys, time, and typewriters, the works of Shakespeare could be created, then it is possible that chimpanzees came across 'termiting', for example, by pure fortune, and maintain it across generations by 'blind' imitation. Yet it is also reasonable to speculate these apes have some 'comprehension' of what they are about.

Köhler's classic work (1973 [1925]) on problem-solving in chimpanzees is directly relevant here. Köhler's claim was that animals, especially primates, were capable of solving a variety of problems if all the components necessary for a solution were available to them in their perceptual field (see Fig. 22.13). Trial-and-error strategies could thus be bypassed, a correct solution being obtained by some sudden reorganization of the elements of the perceptual field, a process he termed 'insight'. Similar abilities to solve spatial problems have also been demonstrated for gorillas (Riesen et al. 1953) and gibbons (Yerkes and Yerkes 1929).

As with all the topics being considered in this review, Lloyd Morgan's canon can be invoked, and other explanations put forward which do not imply any cognitive 'processing'. For example, Birch (1945) and Menzel et al. (1970) have shown that prior familiarity with the elements of a problem has a marked effect on an individual chimpanzee's ability to achieve a solution. Schiller (1952, 1957) has taken this factor further, and concluded that a reinforcement explanation, whereby the component motor skills required for solutions are developed spontaneously in free play with the objects involved in particular problems, could account for the abilities in question.

While the existence of the motor components required for solutions may play a role in these activities, it does not, however, guarantee successful problem-solving. For example, while the chimpanzees studied by Menzel (1972, 1973; see also Fig. 22.14) had prior experience of using poles, their spontaneous use of them as ladders required them to go beyond problems of motor co-ordination to solving the perceptual problem of the placement of the pole's top and bottom to use it as a ladder. There is also social co-operation involved in the solving of this problem, one chimp holding the 'ladder' in place for another to climb: the entire activity clearly represents the emergence of a novel strategy.

Yet, on other occasions, chimpanzees can appear remarkably perverse. This is especially the case when the problem involved requires the animal to obtain out-of-reach yet visible food (see, for example, Birch 1945). Reviewing this work, King and Fobes (1982, pp. 332-3) conclude:

> chimpanzees and, no doubt, other nonhuman primates are fascinating and perplexing mixtures of adaptive intelligence and stupidity. Contrary to the arguments of Schiller, the chimpanzee is not a mindless creature whose successful problem solving is solely determined by contingent reinforcement of fortuitously correct behaviors. Chimpanzees are clearly capable of adaptive novel responses in a problem-solving context. Yet, as Birch's food-in-pipe problem shows, chimpanzees can also be remarkably obtuse in other circumstances. This seeming contradiction is confusing only if chimpanzees' problem-solving activities are viewed from a narrow human perspective. The difficulty of a problem from a human viewpoint is usually an unreliable guide to prediction of an animal's success, even one as closely related to human beings as the chimpanzee. Understanding and ultimately proper theoretical treatment of primate problem solving and tool using will depend upon data showing: (a) the variables that affect particular types of problems and (b) the extent to which already learned behaviors will generalize to related versions of problem solving and tool use. To date, in spite of promising pioneer efforts (Kluver 1933; Kohler 1925; Yerkes 1916) as well as more recent data . . . few data addressed to the preceding questions exist.

For recent reviews, see contributors to Chavaillon and Berthelet (in press) and Parker and Gibson (1990).

22.11.2 *Mirror-use and self-recognition*

Both chimpanzees and orang-utans will quickly use a mirror to inspect normally unseen parts of their bodies, such as their ears or the insides of their mouths (see Fig. 22.15). They also show a capacity for self-recognition, in that they will use their reflection to guide their hands and fingers to unfamiliar marks placed on their faces (Gallup 1970; Lethmate and Ducker 1973; Suarez and Gallup 1981). To date, all other primates tested by this method have shown no evidence of self-recognition: gibbons (*Hylobates*); gorillas (*Gorilla*); Old World macaques (*Macaca*), mandrills (*Mandrillus*), and baboons (*Papio*); New World spider monkeys (*Ateles*), capuchin monkeys (*Cebus*), and squirrel-monkeys (*Saimiri*) all tend to react to a mirror image *socially*, as if they were seeing *another* individual (Anderson 1983, 1984*a*, *b*; Bayart and Anderson 1985; Benhar *et al.* 1975; Gallup 1977*a*, 1987; Gallup *et al.* 1980). Indeed, this social response can be quite robust, persisting in two macque

Fig. 22.13 (a) A chimpanzee having successfully stacked a tower of boxes in order to obtain bananas hanging from the ceiling of the room.

(b) The 'fitting-sticks' problem being attempted by one of Köhler's chimpanzees.
From *The Mentality of Apes* by Wolfgang Köhler (1973), copyright © 1973 by Routledge and Kegan Paul Ltd. Reprinted by permission.

Fig. 22.14 Three chimpanzees collaborate in setting a pole ladder against a tree to avoid electrified wires. Note how one animal holds the base of the ladder while another climbs. (From Menzel (1972). Copyright (1972) by S. Karger AG, Basel. Used by permission.)

monkeys despite a lifetime (13 years) of exposure to mirrors. Although such prolonged exposure has been shown to result in a decrease in social responding over time (Suarez and Gallup 1986) there is no indi-

Fig. 22.14 (*Contd*)

cation that this decline indicates that the monkeys recognize their reflection. The reason is that moving the mirror to a different position in the cage, or leaving the mirror in the same position but turning it away from the cage for a period of 5 days and then turning it back again, will often temporarilly reinstate social responding (Suarez and Gallup 1986; Gallup and Suarez 1991). In contrast, chimpanzees will show normal self-recognition upon being re-exposed to a mirror that had been removed for one-year period (Calhoun and Thompson 1988). The failure to demonstrate self-recognition in gorillas is particularly puzzling (but see Law and Lock, 1994).

Along with the ability to recognize themselves, chimpanzees are also capable of using mirror or televised images 'to guide their hands or limbs to otherwise unlocatable targets' (Menzel *et al.* 1985, p. 211; see Fig. 22.16), and can do so even when, through manipulation of the image, there is no direct relation between the image they see and the real world (for example, with inverted or reversed images) (ibid.). Anderson (1986) has provided evidence that macaque monkeys (*M. tonkeana* and *M. fascicularis*) can locate food that is only visible in a mirror; but it is not clear whether this ability involves mirror-guided reaching, or the more simple associative learning that is shown by monkeys and human infants that learn to turn and look at an object first seen in a mirror (see Anderson 1984*a*). (For recent reviews and discussion see Mitchell 1993*a*, *b*.)

22.11.3 *'Theory of mind': understanding others*

In 1978, Premack and Woodruff posed the question 'Does a chimpanzee have a theory of mind?', that is, do they attribute beliefs, knowledge, intentions, and emotional states to others (and ultimately, to their 'selves')? That adult humans are adept at attributing mental states to others is self-evident, and

Fig. 22.15 A chimpanzee, Austin, using a video-monitor to inspect his mouth as he swirls water around his lower lip. (From E. Sue Savage-Rumbaugh (1986). (Copyright (1986) Columbia University Press, New York. Used by permission.)

prompts the question of what are the ontogenetic and phylogenetic origins of these abilities? At present, the investigation of these questions represents one of the most interdisciplinary enterprises of the behavioural sciences, bringing together contributions from philosophers, computer scientists, primatologists, and developmental psychologists (see, for example, Mitchell 1986; Astington *et al.* 1988; Byrne and Whiten 1988;

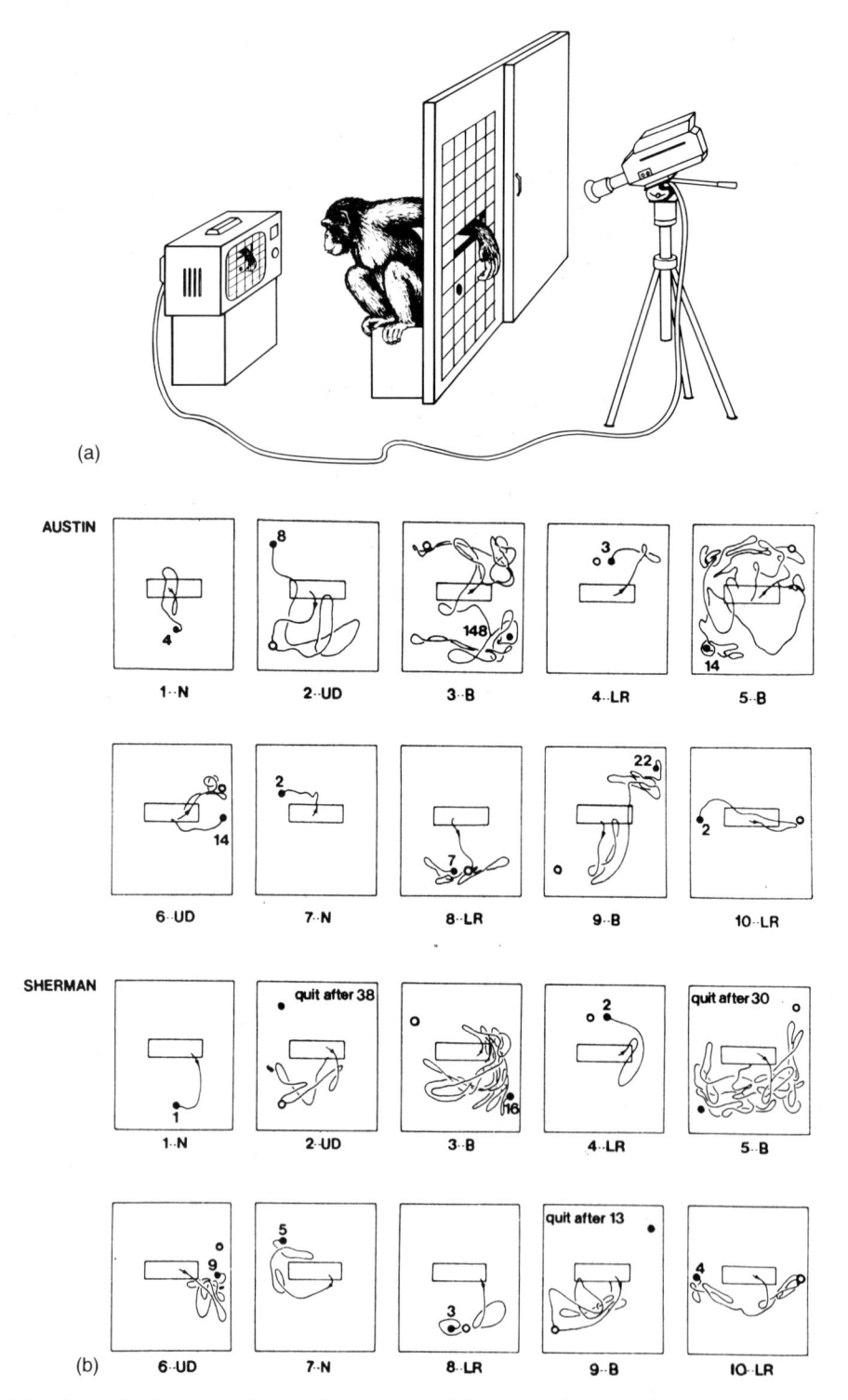

Fig. 22.16 (a) A schematic diagram of one of the tests used by Menzel *et al.* (1985) to test chimpanzee abilities to search for objects only visible on a video-monitor.

Harris 1989; Cheney and Seyfarth 1990; Ristau 1990; Wellman 1990; Perner 1991; Whiten 1991).

22.11.3.1 *Phylogenetic origins*

Determining the extent to which species other than humans possess a theory of mind has proved difficult for two reasons. First, there are many different taxonomies of attributional states, making it difficult to choose which should be used to determine if a particular species satisfies the criteria for possessing a 'theory of mind'. Second, much of mental-state attribution revolves around linguistic competence, thereby making it particularly difficult to examine such an issue in species other than humans. Despite these problems, a number of researchers have examined the ability of different animals, mostly primates, to attribute mental states to others. In general, the results fall in line with the many other examples discussed in this chapter, in that this ability is highly represented in humans, less so in apes, and virtually absent in monkeys.

Perhaps the most elegant series of experiments have been conducted by Povinelli and colleagues, who compared two forms of mental attribution capacities in chimpanzees and monkeys (Povinelli *et al.* 1990; Povinelli *et al.* 1991; Povinelli 1993). The first experiment examined a form of social attribution known as role-reversal. A diagram of the apparatus used in this experiment is shown in Fig. 22.17(A) and (B). On the operator side are handles which when pulled result in food trays being delivered to both operator and informant. Opaque shields cover the food trays, making the food visible only to the informant, whose task it is to signal to the operator which handle should be pulled to obtain the food. The chimpanzee and monkey subjects were paired with adult humans in such a way that half the animals in each group learned the task as either informant or operator. Evidence for social attribution comes not from simply learning to perform the task as either operator or informant, but rather from performance on a transfer test in which subjects are required to switch roles, that is, subjects serving as operator are now required to perform in the capacity of informant, and vice versa. As was stated by Povinelli (1993, p. 499):

Organisms that see the task solely from a behaviorist standpoint ought to perform quite poorly on role reversal, despite having learned to carry out their own original role with near-perfect accuracy. On the other hand, organisms that not only utilize cues from their partner successfully but also comprehend the meaning of such cues within the context of the task, ought to understand what to do when the roles are reversed.

Three of the four chimpanzees showed immediate transfer on role reversal. By contrast, although all four monkeys were able to perform their original role, in contradistinction to the chimpanzees none was able to solve the task once the roles were reversed.

A second experiment examined a more complex form of mental-state attribution, the ability to determine between guessing and knowing. This ability underlies the attribution of false belief in children (Wimmer and Perner 1983), which is considered by many to be an acid test of whether an organism possesses a 'theory of mind'. A diagram of the apparatus is shown in Fig. 22.17(c). At the start of a trial the monkey or chimpanzee subject witnessed one of two human participants, termed the 'guesser', leave the room. The other participant, termed the 'knower', then proceeded to hide food under one of the opaque cups situated over the food wells. The subject was able to see the 'knower' hide the food, but did not see where the food was hidden. The 'guesser' then returned into the room, and both human participants simultaneously pointed to different cups, the 'knower' naturally pointing to the correct cup, and the 'guesser' to a predetermined incorrect cup. To obtain the food the subject should have pulled the handle of the food tray indicated by the 'knower' rather than the 'guesser'. Again three of the four chimpanzees were able to distinguish 'knower' from 'guesser'. In contrast, none of the four rhesus monkeys showed any evidence for this discrimination, despite in some cases extensive training on the task.

The inability of monkeys to attribute mental states to others is by no means restricted to the above examples. Cheney and Seyfarth (1991), for instance, have shown that macaque mothers are not able to attribute the conditions of knowledge and ignorance to their offspring. The experiment was conducted in the following manner. In the 'knowledgeable' condition both mother and offspring witnessed a human place apple slices in a food bin. In the 'ignorant' condition, only the mother witnessed this event. The offspring were then released in the direction of the food bin, and the number of food calls (macaques are

---

(b) Performances of each of the two chimpanzees on their first 10 trials of a session involving video images that were 'normal' (N), reversed 180° left to right (LR), inverted 180° (UD), or both reversed and inverted (B). (Each panel shows a rough map of the movements of the chimpanzee's reaching hand on a single trial and latency [in seconds] to first contact the target spot. Solid circles show the actual location of the target; arrows, the direction of the chimpanzee's initial move.)
(From Menzel *et al.* 1985. Copyright (1985) by the American Psychological Association. Used by permission.)

known to elicit a distinctive food call for preferred food items) elicited by the mother were noted. The assumption was that if the mothers were sensitive to the knowledge state of their offspring they would utter more calls when the offspring was ignorant rather than knowledgeable. The same procedure was followed in a second experiment, where the discriminative stimulus was a human approaching with a net (which elicits a distinctive alarm call). In neither case, however, did the mothers alter their behaviour as a function of whether the offspring was knowledgeable or ignorant.

Gallup (1982, 1985) has proposed that mirror self-recognition can serve as an index for the ability to attribute mental states to others. Thus animals that are able to use mirrors to guide their attention to parts of their body should be in a better position to impart mental states to others than animals that show no evidence for this ability. To a certain extent this does appear to be the case. For example, in line with the evidence that chimpanzees but not monkeys are sensitive to the mental states of others are the findings that chimpanzees but not monkeys are capable of mirror self-recognition (for example, Section 22.11.2, above; for reviews, see Gallup 1991). In further support of Gallup's position is the finding that the ability to attribute mental states to others emerges along with self-recognition in children (Lewis *et al.* 1989).

Although the evidence favours the view that attributional states are poorly represented in monkeys, the issue is far from resolved. The reason is that there are many different classification schemes of attributional skills, and whereas monkeys may fail on role reversal and false belief-like tests, they may yet succeed on more lower-order form of mental attribution. Finally, Cheney and Seyfarth (1991, pp. 253–4) raise a relevant caution when they state that what is being tapped in these instances of mental-state attribution may be nothing more than 'learned behavioral contingencies' rather than 'higher-order intentionality'. Chimpanzees may fare better than rhesus monkeys on the knower–guesser task not because they can impart mental states to the 'knower' and 'guesser', but because they may be faster at learning the behavioural cues which distinguish between 'knower' and 'guesser', the most prominent cue being that the 'guesser' leaves

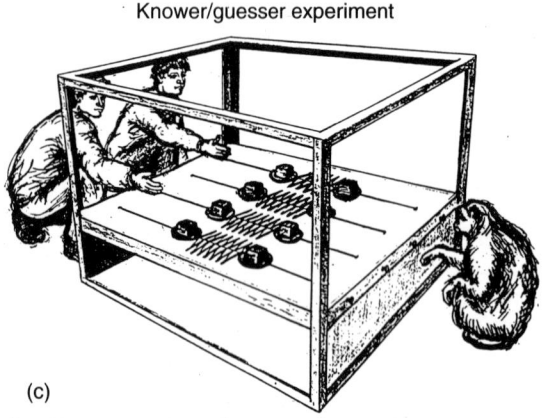

Fig. 22.17 In the role-reversal experiment the monkey and chimpanzee subjects either learned the task in the role of operator (a) or informant (b). The informant must indicate which tray, and hence which handle, should be pulled by the operator to obtain the food. Once the task was learned the ability of the monkey and chimpanzee subjects to serve in the opposite role was assessed. In the knower/guesser experiment (c) the monkey or chimpanzee subjects must choose the tray pointed to by the 'knower', rather than the 'guesser', to obtain food. For further details see text. © P. Penno.

the room (cf. Povinelli 1993). And the control for this behavioural interpretation, which was to have the guesser stay in the room and place a bag over his head, may reduce, but by no means eliminates, the behavioural cues that could be used, and thus by no means does it invalidate the behaviouristic interpretation of these data.

22.11.3.2 *Ontogenetic origins*

In contrast to the above, an intentional stance is pervasive in the ontogenetic literature. For example, Wellman (see for instance 1990), following Davidson (1980), characterizes the everyday, mentalistic 'folk psychology' of (occidental) humans as a belief–desire psychology: 'the fundamental, obvious idea is that people engage in actions because they *believe* those actions will satisfy certain *desires*' (Wellman 1991, p. 20). The human ontogenetic literature indicates that it is around 3 years of age that children begin to use such a system in understanding others; earlier than that, at 2 years, a more simple desire-only psychology is employed (see for example, Wellman and Woolley 1990), and before that again, at 1 year, prelinguistic infants can monitor another's gaze direction, giving some evidence that they are going beyond just attending to another's behaviour, and have some rudimentary understanding of a mentalistic nature. Further, only from about 4 years of age can *false* beliefs be dealt with—a capacity which really does evidence a mentalistic stance to others. An understanding of *intention*, as mediating between *want* and *action*, is a still later achievement, at 5 years (Shultz and Shamash 1981; Astington 1992).

The ontogenetic literature is bedevilled by a lack of an accepted terminology. Thus, Whiten and Perner (1991, pp. 14–15) note that the situation described as 'Sarah knows her trainer thinks that she wants to eat bananas' would be a third-order mental state in Dennett's scheme; that Perner and Wimmer (1985) would describe this as Sarah *attributing* a second-order mental state to her trainer (he thinks she wants); and Whiten and Byrne (1988) would regard the situation as a third-order representation (Sarah represents her trainer's representation of her representation).[4] Further, conceptual differences between investigators are indicative of the field's still being in a pre-paradigmatic state (see, for example, the footnotes to Harris 1991). Despite these differences, however, the overall recursive nature of these systems is agreed on at a fundamental level, and it is this agreement that signals a potential bringing-together of a number of diverse aspects of animal and human behaviours under some coherent scheme (Leslie 1987, for example, provides a theoretical account that is a first attempt in this direction).

It is Leslie's claim (see, for example, 1987, 1988) that despite their independence, both pretence and 'mindreading', for example, share a common underlying cognitive basis in modern humans: an ability to *represent representations*. The developmental assessment of a correlation in the ontogenetic establishment of these two abilities is supportive of their being related, but by no means conclusive as yet. The most compelling evidence comes from autistic children (see, for example, Baron-Cohen 1987; Frith 1989, for reviews), in whom both abilities show simultaneous deficits, suggesting an underlying link between them. Leslie and Thaiss (1992) present evidence that there are sub-domains within this meta-representational ability itself, in that autistic children understand pictures, sentences, and diagrams as representations, but cannot deal with false beliefs, i.e., false-belief and representation tasks do not tap the same cognitive mechanisms. Their evidence leads them to claim the existence of 'a specialized cognitive mechanism, which subserves the development of folk psychological notions, and which is dissociably damaged in autism' (ibid., p. 226), and that the elements of the characteristic triad of impairments—in social competence, in communicative skills, and in pretence—that characterizes autism can all be tied to an impairment of this single cognitive mechanism. Thus, normal humans not only have 'a theory of mind', but 'a theory-of-mind mechanism (or module)'.

This last conjecture adds complexity to the possible phylogenetic implications of current ontogenetic research.[5] Without it, we could be looking for a general increase in the ability to represent representations on which other abilities were similarly founded. Imitation would be one instance. Bruner (1972), for example, has put the view that the ability to imitate requires that the mimic can put him-or herself in the mental position of the target, a feat which requires representing the knowledge and goals of another. Again, there are other candidate abilities that could be brought into this framework. The interpretation of mirror or video images so as either to recognize oneself as oneself (rather than other) or use the representation of one's hand to guide one's real hand (Section 22.11.2 above) probably involve second-order representation. Tool-use, or the preparation of a stick as a probe, similarly implicates the ability simultaneously to represent a stick as a stick while also metarepresenting it as a probe. Insightful problem-solving, teaching, informing, and empathizing with another can also be brought into this *mélange*. All these are abilities that have yet to be given any convincing demonstration for monkeys, show some appearance in the apes (and perhaps, intriguingly, at different levels of elaboration, with imitation and pretence (see below, Section 22.12) perhaps being the least developed (for example, Tomasello et al. 1987), and yet are characteristically interrelated in normal human development (see, for example, Bates et al. 1979 and the editorial introduction to Part III of this volume). If there is, though, a single and very specific cognitive 'module' underpinning only the ability to meta-represent beliefs about others, then a more complex phylogenetic account is indicated. That is, rather than having to deal with the evolution of one overarching ability, we have to account for the evolution of a potential myriad of independent abilities.

Finally, with respect to developing an evolutionary account, we need to note a topic of current concern, and that is the nature and status of much of the evidence that can be marshalled. With respect to many of these particular abilities, much of the primate evidence is still anecdotal. The value of such evidence is currently being re-evaluated (see, for example, Lock 1992 and the contributors to Mitchell and Thompson 1993 and Davis and Balfour 1992). It may be, however, that low-frequency behaviours in primates actually represent something of the variation from which natural selection has fashioned the modern suite of human abilities. On this point, Whiten and Byrne (1991, p. 277–8) implicitly indicate the problems to be faced in their insightful observation in respect of chimpanzees:

that mindreading is manifest in natural contexts, whereas pretend play is only apparent in non-natural environments, suggests that metarepresentation may have evolved initially to support the first of these. If second-order capacities have been selected for in chimpanzees, their natural function appears to be to underwrite these animals' acknowledged Machiavellian expertise (de Waal 1982; Byrne and Whiten 1988; Whiten and Byrne 1989). Pretend play may emerge as a product of such abilities only in the rarefied environment of rearing by humans—although it must be acknowledged that methodological difficulties may have prevented pretend play being so clearly recorded in the wild.

That is, the problem of the apparent frequency of a behaviour and the quality of observation from which that apparent frequency has been concluded aside, the selective environment to which natural selection has responded has been that of the social world of primates themselves, not the pressures of their 'natural' environment. We pick up on this point further in our concluding comments.

22.12 Imitation, problem-solving, and cognition

Imitation and (creative) problem-solving are related in the sense that they are both goal-oriented techniques which imply that the organism employing them is able to represent cognitively a state of affairs other than that which perceptually confronts it, and then to pursue some strategy aimed at transforming that state of affairs into the one represented cognitively. Given this reliance of both processes on representation, it is clear why imitation and creative problem-solving are best evidenced by the higher primates, since it is with them that the most robust forms of representation are found.

In evolutionary terms, it might be said that monkeys have evolved brains that produce cognitive representations, whereas apes have evolved brains that can manipulate these phylogenetically-older cognitive representations. It is clear, though, that an ape's representational manipulations are limited in the extent to which they can be divorced from the dominating effect that immediate perception exerts upon the older-established, basic primate cognition. In monkeys, cognition appears to be used to inform immediate action. Past experience can mediate present activity in the immediately perceived world. Apes, by contrast, appear to be able to escape, to some extent, their immediate world, by both representing to themselves a future world, based on a perception of the present mediated by representations stemming from past experience, and also manipulating these representations prior to selecting present actions. They are, however, not always able to accomplish this: it is as if, on occasion, the immediate world exerts too much 'pull' on them. Thus, the perversity chimpanzees show with Birch's 'food-in-pipe' problem (1945), and Schiller's findings (1952, 1957) that the likelihood of chimpanzees' 'learning' to fit two sticks together to make a long rake was lowered if food was visible both suggest that the ability to manipulate representations is affected by the 'strength' of immediate perceptions.

While at this level imitation and problem-solving show similarities, they are probably quite different in terms of the cognitive abilities they involve (see, for example, Bates et al. 1979, Chapters 2 and 7). Imitation can be employed without 'understanding': by contrast, problem-solving requires an ability to break down means–end relations into their component parts, establish the relationships between these parts, find substitutes for missing parts, and then assemble a whole 'package' of actions. Imitation requires cross-modal transfer of observed packages of activity into reproduced action. The transfer involved in problem-solving is more one across perceptual contexts: an item useful in one context has to play a new role in a different context.

In terms of their evolution, both imitation and problem-solving could be thought of as arising from a combination of separate 'modules' of previously established abilities. Imitation requires, at least, an input from 'associative learning'—a perception of contiguity, that action *a* is followed by consequence *b*; from cross-modal transfer abilities—such that what is seen can be reproduced in action; and from memory capacities—otherwise what was observed cannot be reproduced. Problem-solving requires input from the abilities that underlie similarity judgements, memory, and representation, and an ability to attend to representations at the expense of immediate perceptual presentations: the represented world must come to have an equality in its motivation of activities comparable to that of the presented world.

22.13 Evolutionary implications of comparative studies of 'cognition'

Shifting from the level of analysis appropriate for outlining the comparative data to one which considers its evolutionary implications for the foundation and elaboration of human abilities requires a change in the 'grain' of our perception. Every experimental result is hedged with qualification and reservation; but, when viewed as a whole, the cumulative body of findings takes on a different status, since 'trends' do appear to emerge. But the reservations themselves reflect on the way these trends can be drawn out. It is not possible, for example, to draw up a list of clear evolutionary stages from the data, as was the practice of Victorian biology, to yield some kind of *scala naturae* of cognitive abilities possessed by individual representatives of the Mammalia, for reasons that are discussed below. General statements need to take account of the ways in which any framework within which those statements are located is drawn up. In the evolutionary case, there appear three main issues that influence the construction of a framework: the ecology of cognition; the testing of cognition; and the attribution of cognition.

22.13.1 *The ecology of cognition*

Mammalian cognitive abilities appear to be quite 'context-dependent'. In some areas of its life an animal might show abilities that have no formal analogue in other areas. For example, an animal might appear to be able to solve quite complex social tasks, and apparently be able to 'use' another animal as if it were a 'tool' that was instrumental in its attaining some goal; and yet show no ability to transfer the formal structure of this to its use of physical tools. As King and Fobes (1982) observed (above, Section 22.9.1), chimpanzees and other primates are a per-plexing mix of intelligence and stupidity. Thus, rather than trying to assign a 'level of ability' to a particular species, it is necessary to take account of the range of circumstances in which it appears to show those abilities.

The economics of evolution dictate that a minimal amount of energy will be invested in the support of any ability an animal possesses, and that any 'short cuts' to establishing something new will be taken. Thus, skills developed in one facet of life can be expected to be economically transferred to a different one by the elaboration of cross-modal and cross-situational abilities, rather than needing to be constructed *de novo*. Consequently, the comparative investigation of these abilities demands a great deal of future prominence, for they would seem ideal candidates for increasing our understanding of the evolutionary history of primate abilities. This prescription implies that, to elucidate the origins of human abilities, it is necessary to place in the forefront an understanding of any specimen's ecological situation, and to posit changes in the range of application of its abilities as being ecologically motivated.

If this re-focusing of research occurs, it needs to be borne in mind that an animal's conspecifics can be a major component of its ecology, on top of the physical world it inhabits. For example, the social skills apparent in chimpanzees, such as their ability for self-restraint and deception, could be construed as the foundations for some human symbolic skills, only if it is recognized that major changes in intra-group ecology are required to motivate these abilities towards a communication system. It would seem that there is a pressure in chimpanzee societies *not* to communicate what one knows, and thus, without a change in intra-species relations, the abilities that underpin deceit would be unlikely to be pushed towards supporting 'telling the truth' (cf. Jolly, this volume, Chapter 6).

22.13.2 *The testing of cognition*

There appears to be a gulf between what an animal does 'in the wild' and what it can be got to do 'in the laboratory'. This gulf cuts in two ways. First, if ecology plays such an important role in determining an animal's abilities, then these abilities are more likely to be apprehended when it is in its natural habitat: thus, it may intuitively appear 'more intelligent' in that situation than in a laboratory setting that excludes many of its natural dimensions of motivation and perception. However, this leads to the problem of anecdotal evidence and its value (cf. Ristau, this volume, Chapter 23).

But second, animals can be shown to have abilities in the laboratory that there is no hint of their having in the wild. There is, for example, no evidence of symbolic communication systems being used by feral apes. This raises the issue of 'if they can do it, why don't they?' This issue concerns the 'potentiality' and 'actuality' of abilities, and what these terms are taken to mean. These terms are, presently, genuinely problematic. For example, two abilities no animal clearly demonstrates in its natural habitat are to perceive and respond to straight lines, since nature provides straight lines in short supply. But these abilities are so easily demonstrated in the laboratory that we assume them to be *actual* abilities possessed by an animal: we seem to assume them as a natural part of the animal's repertoire, perhaps because our own environment is so full of straight lines.

What, in fact, we are faced with is an animal that is being tested with stimuli that 'evolved' *after* the animal itself evolved: stimuli that have thus exerted no selective pressures in establishing the animal's abilities. Until a precise vocabulary for dealing with such subtleties of the meaning of potentiality is elaborated, one that goes beyond the blanket concept of *preadaptation*, the exact status of laboratory findings is hard to gauge with respect to the role they could play in an evolutionary paradigm. Thus both laboratory data and 'wild' anecdotes are problematic, though in different ways.

22.13.3 *The attribution of cognition*

One way out of the above problem has been developed by Premack (for example, 1980) with his concept of 'an upgraded mind'. His view is that language-trained apes are no longer 'really' apes: 'as a result of engaging in these [activities] the ape's mind is significantly upgraded' (1980, p. 66). In line with the above point, however, another interpretation of this 'effect' would be not to regard the ape's mind as upgraded, but its activities as 'scaffolded' to such an extent that they are supported by a minimum of cognition on the ape's part. Somewhere between these extremes could be essayed the view that large areas we take to be human skills are themselves supported by a minimal elaboration of cognitive skill above that of a chimpanzee, but by a massive elaboration in the support-scaffolding of culture. Rather than asking 'How have human cognitive abilities evolved?', this perspective more naturally poses the question 'How has the cognitive support system of culture been elaborated?'

22.13.4 *Accounting for human cognition from a comparative perspective*

The possibilities for developing an evolutionary account of human cognitive abilities become clearer if the above points are given a concrete context. Here they are explored with respect to the work reported by Menzel *et al.* (1985; see Section 22.9.2 and Fig. 22.15). This experimental investigation explores a chimpanzee's ability to deal with a technologically presented view of the world that no feral chimpanzee would ever experience. It is rich in its implications. First, as a technique, it represents a way of exploring the limits of a chimpanzee's spatial cognition. But second, it does so in a way that shows that cognition to have properties that transcend the normal range of its application—to be preadapted to support new levels of behaviour. What the technique shows is that a chimpanzee has the ability to act on a perception of the world that is not its own *as if it were its own.*

Symbolic language is founded on exactly this ability. Extant language systems are composed of symbols that represent the world from the standpoint termed by Mead (1934) that of the 'significant other' (see Sinha, this volume, Chapter 17). However, in the same way that chimpanzees do not take spatial perspectives other than their own in the wild, neither do they use such symbol systems, even though it appears they have the capacity to do so. At this point, the necessity for a conceptual overhaul of the way in which the evolution of cognition is accounted for becomes clear. Menzel's work does not yield information about 'real' chimpanzee abilities, as they might be employed in the 'natural' life of chimpanzees, but delineates 'possible' worlds that chimpanzees are able to deal with (this is the attribution problem of section 22.11.3 again). And those possible worlds are actual worlds for human experimenters—worlds that chimpanzees are able to deal with to some extent because they are worlds that are built on extrapolations of earlier real abilities.

This line of analysis has repercussions on the way the evolution of human abilities is interpreted. Our investigative techniques do not tell us anything about relative degrees of animal intelligence, as Victorian science thought, but about the extent to which those 'abilities' can operate in, and with the resources provided by, a world that is structured on principles extrapolated through an evolutionary elaboration and instantiation of the animal's possibilities for action. Comparative experiments are, by their nature, evolutionary experiments founded on deep and subtle temporal transformations of an animal's situation.

Menzel's work is thus a study of how a chimpanzee deals with spatial displacements via a methodology based on temporal displacements that confront a chimpanzee with a reality literally beyond itself. These issues are ones that need attention if a fuller understanding of the evolutionary significance of comparative work is to be attained.

Despite this, there are evolutionary trends apparent in the comparative literature. Whatever support it is given, a monkey appears less able to act in a context-independent fashion than an ape; so does an ape in comparison to a human. The basic properties of the cognitive support required for context-independent abilities, such as memorial abilities, appear to be more widely available among the vertebrates. Most measures that tap the components of context-independence—transfer of learning; reversal learning; analogical reasoning; cross-modal transfer; etc.— show their greatest levels of elaboration in the Great Apes. The evidence on ape capabilities when exposed to humanly-supported cognition points to there having been an evolutionary rearrangement of skills relative to each other, and a generalization of their spheres of employment from one particular domain to another, during the course of human evolution. The ways in which the products of these rearrangements have become instantiated in an animal's experienced environment, and thus become selective feedback pressures influencing the process of rearrangement, appears a more profitable area for investigation than does solely asking quantitative questions about relative animal intelligence. Thus, the role that cross-species studies can play in elucidating evolutionary pathways is not one that can be settled at this point, nor one that would appear capable of being settled by more data alone; but rather one that is also in need of conceptual analysis.

Notes

1. This issue arose earlier in a different guise when noting that the DMTS task appears, on the face of it, to tap recognition rather than representational memory, whereas it has been claimed to be a task involving representation, on the ground that since many of the animals studied are claimed from other investigations to possess representational abilities, these are probably involved in the way they perform DMTS tasks. While this may be the case, it is necessary to be wary of the argument.

2. Other writers, however, draw different conclusions. For example, Premack (1983*a*, p. 135) is of the view that:

The reaction to similarity, to the likeness between the appearances of things, is found in both primates and non-primates and does not depend on special experience. Moreover, in the primate (and probably the non-primate as well) the reaction is of an abstract kind: an animal trained to respond to the similarity between, say, colors transfers to the similarity between (say) shapes. Not only cross-dimensional but also cross-modal transfer can be found in the animal's reaction to the similarity between properties and objects.

Irle and Markowitsch (1987, p. 310) are likewise of the view that this is a quite general vertebrate ability:

The concept of sameness is frequently trained using the match-to-sample procedure or pair-wise *same–different* comparisons (Burdyn and Thomas 1984; Premack 1983*b*; Sands *et al.* 1982). Otherwise, an oddity training or, for apes, an analogical reasoning procedure [see below] has been applied (Gillan *et al.* 1981; King 1973; Thomas and Frost 1983). The concept of sameness may be unlearned and may also, in fact, be the most common concept among vertebrates (Weiskrantz 1985).

Whether the 'state of the art' is sufficiently developed to warrant these larger claims is not clear. Certainly there is evidence for such concepts across a number of vertebrates, but there are also differences between species in the extent to which the many facets of these concepts are attained and employed. D'Amato *et al.*'s (1986) conservative view is probably the safest to adopt at the moment; for further discussion of this point with respect to monkeys, see also D'Amato and Colombo 1989).

3. But, as with all such categorical statements regarding what primates do not do, further fieldwork is likely to undermine it. Indeed, Boesch (1993) reports a number of observations of chimpanzees at Tai National Park, Ivory Coast, where tool-use is a regular occurrence as a result of the prominence of hard-shelled nuts in their diet, which provide some evidence of rudimentary demonstration and error-correction by mothers of their infants' efforts at wielding stones upon nuts.

4. This confusion is compounded by the fact that a number of developmental psychologists have questioned whether the concept of representation needs to be invoked in the case of very young children: these children might know others want something, but do not understand the representational nature of wanting, etc. (e.g. Astington and Gopnik 1991; Harris 1991).

5. Note that here we are not looking for congruence between the developmental elaboration of an understanding of others and its phylogenetic elaboration as evidence of recapitulation, 'for any correspondencies [*sic*] which are found between ontogenetic and phylogenetic elaboration may simply reflect general principles in the biological construction of mind reading systems' (Whiten 1991, p. 319).

References

Anderson, J. R. (1983). Responses to mirror-image stimulation, and assessment of self-recognition in mirror- and peer-reared stumptail macaques. *Quarterly Journal of Experimental Psychology*, **35B**, 201–22.

Anderson, J. R. (1984a). The development of self-recognition: a review. *Developmental Psychology*, **17**, 35–49.

Anderson, J. R. (1984b). Monkeys with mirrors: some questions for primate psychology. *International Journal of Primatology*, **5**, 81–98.

Anderson, J. R. (1986). Mirror-mediated finding of hidden food by monkeys (*Macaca tonkeana* and *M. fascicularis*). *Journal of Comparative Psychology*, **100**, 237–42.

Antinucci, F. (ed.) (1989). *Cognitive structure and development in nonhuman primates*. Erlbaum, Hillsdale, NJ.

Asch, S. E. and Ebenholtz, S. M. (1962). The principle of associative symmetry. *Proceedings of the American Philosophical Society*, **106**, 135–63.

Aslin, R., Pisoni, D., and Jusczyk, P. (1983). Auditory development and speech perception in infancy. In *Handbook of child psychology: infant development* (ed. M. Haith and J. Campos), pp. 573–687. Wiley, New York.

Astington, J. W. (1992). Intention in the child's theory of mind. In *Children's theories of mind* (ed. C. Moore and D. Frye), pp. 157–72. Erlbaum, Hillsdale, NJ.

Astington, J. W. and Gopnik, A. (1991). Developing understanding of desire and intention. In *Natural theories of mind: evolution, development, and simulation of everyday mindreading* (ed. A. Whiten), pp. 39–50. Blackwell, Oxford.

Astington, J. W., Harris, P. L., and Olson, D. (eds) (1988). *Developing theories of mind*. Cambridge University Press.

Baron-Cohen, S. (1987). Autism and symbolic play. *British Journal of Developmental Psychology*, **5**, 139–48.

Bates, E., Benigni, L., Bretherton, I., Camaioni, L., and Volterra, V. (1979). *The emergence of symbols: cognition and communication in infancy*. Academic Press, New York.

Bayart, F. and Anderson, J. R. (1985). Mirror-image reactions in a tool-using, adult male *Macaca tonkeana*. *Behavioural Processes*, **10**, 219–27.

Beatty, W. W., and Shavalia, D. A. (1980). Rat spatial memory: resistance to retroactive interference at long retention intervals. *Animal Learning and Behavior*, **8**, 550–2.

Beck, B. (1973). Observational learning of tool-use by captive Guinea baboons (*Papio papio*). *American Journal of Physical Anthropology*, **38**, 579–82.

Beck, B. (1975). Primate tool behaviour. In *Socioecology and psychology of primates* (ed. R. H. Tuttle), pp. 413–47. Mouton, The Hague.

Beck, B. (1978). Ontogeny of tool use by non-human animals. In *The development of behavior: comparative and evolutionary aspects* (ed. G. M. Burghardt and M. Bekoff), pp. 405–19. Garland STPM Press, New York.

Behar, I. (1962). Evaluation of the significance of positive and negative cue in discrimination learning. *Journal of Comparative and Physiological Psychology*, **55**, 502–4.

Benhar, E., Carlton, P. L., and Samuel, D. A. (1975). A search for mirror-image reinforcement and self-recognition in the baboon. In *Contemporary primatology* (ed. S. Kondo, M. Kawai, and A. Ehara), pp. 202–8. Japan Science Press, Tokyo.

Bergeron, G. (1979). Développement sensori-moteur du jeune chimpanzé (*Pan troglodytes*) en liberté restreinte [Sensorimotor development of chimpanzee infants in semi-captivity]. Unpublished Master's dissertation, Université de Montréal, Québec, Canada.

Bernstein, I. S. (1962). Response to nesting materials of wild-born and captive chimpanzees. *Animal Behavior*, **10**, 1–6.

Birch, H. G. (1945). The relation of previous experience to insightful problem solving. *Journal of Comparative Psychology*, **38**, 367–83.

Bitterman, M. E. (1965). Phyletic differences in learning. *American Psychologist*, **2**, 396–409.

Blakeslee, P. and Gunter, R. (1966). Cross-modal transfer of discrimination learning in *Cebus* monkeys. *Behaviour*, **26**, 76–90.

Boesch, C. (1993). Aspects of transmission of tool-use in wild chimpanzees. In *Tools, language and cognition in human evolution* (ed. K. R. Gibson and T. Ingold), pp. 171–83. Cambridge University Press.

Bolster, R. B. (1978). Cross-modal matching in the monkey (*Macaca fascicularis*). *Neuropsychologia*, **16**, 407–16.

Bousfield, W. A., Cohen, B. H., and Whitmarsh, G. A. (1958). Associative clustering in the recall of words of different taxonomic frequency of occurrence. *Psychological Reports*, **4**, 39–44.

Breland, K. and Breland, M. (1961). The misbehavior of organisms. *American Psychologist*, **16**, 681–4.

Breslow, L. (1981). Re-evaluation of the literature on the development of transitive inferences. *Psychological Bulletin*, **89**, 325–51.

Bruner, J. S. (1972). Nature and uses of immaturity. *American Psychologist*, **7**, 687–702.

Bryant, P. E., Jones, P., Claxton, C. C., and Perkins, G. M. (1972). Recognition of shapes across modalities by infants. *Nature*, **240**, 303–4. (Cited by Meador *et al*. 1987.)

Bryant, P. E., and Trabasso, T. (1971). Transitive inference and memory in young children. *Nature*, **232**, 456–8.

Buchanan, J. P., Gill, T. V., and Braggio, J. T. (1981). Serial position and clustering effects in a chimpanzee's "free recall". *Memory and Cognition*, **9**, 651–60.

Burdyn, L. E. and Thomas, R. K. (1984). Conditional discrimination with conceptual simultaneous and successive cues in the squirrel monkey (*Saimiri sciureus*). *Journal of Comparative Psychology*, **98**, 405–13.

Byrne, R. W. and Whiten, A. (eds) (1988). *Machiavellian intelligence: social expertise and the evolution of intellect in monkeys, apes, and humans*. Oxford University Press.

Calhoun, S. and Thompson, R. L. (1988). Long-term retention of self-recognition by chimpanzees. *American Journal of Primatology*, **15**, 361–5.

Campbell, C. B. (1974). On the phyletic relationships of the tree shrews. *Mammal Review*, **4**, 125–43.

Cerella, J. (1979). Visual classes and natural categories in the pigeon. *Journal of Experimental Psychology: Human Perception and Performance*, **5**, 68–77.

Chavaillon, J. and Berthelet, A. *The use of tools in human and non-human primates*. Oxford University Press. (In press.)

Cheney, D. L. and Seyfarth, R. M. (1991). Reading minds or reading behaviour? Tests for a theory of mind in monkeys. In *Natural theories of mind* (ed. A. Whiten), pp. 175–94. Blackwell, Oxford.

Chevalier-Skolnikoff, S. (1977). A Piagetian model for describing and comparing socialization in monkey, ape and human infants. In *Primate biosocial development: biological, social and ecological determinants* (ed. S. Chevalier-Skolnikoff and F. E. Poirier), pp. 159–88. Garland, New York.

Chevalier-Skolnikoff, S. (1982). A cognitive analysis of facial behaviour in old world monkeys, apes and human beings. In *Primate communication* (ed. C. T. Snowdon, C. H. Brown, and M. R. Petersen), pp. 303–68. Cambridge University Press.

Chevalier-Skolnikoff, S., Galdikas, B. M. F., and Skolnikoff, A. Z. (1982). The adaptive significance of higher intelligence in wild orang-utans: a preliminary report. *Journal of Human Evolution*, **11**, 639–52.

Cook, R. G., Brown, M. F., and Riley, D. A. (1985). Flexible memory processing by rats: use of prospective and retrospective information in the radial maze. *Journal of Experimental Psychology: Animal Behaviour Processes*, **11**, 453–69.

Couvillon, P. A. and Bitterman, M. E. (1992). A conventional conditioning analysis of "transitive inference" in pigeons. *Journal of Experimental Psychology: Animal Behaviour Processes*, **18**, 308–10.

Crowder, R. G. (1976). *Principles of learning and memory*. Erlbaum, Hillsdale, NJ.

Czerny, P. and Thomas, R. K. (1975). Sameness–difference judgements in *Saimiri sciureus* based on volumetric cues. *Animal Learning and Behaviour*, **3**, 375–9.

D'Amato, M. R. (1973). Delayed matching and short-term memory in monkeys. In *The psychology of learning and motivation: advances in theory and research*, Vol. 7, (ed. G. H. Bower), pp. 227–69. Academic Press, New York.

D'Amato, M. R. and Colombo, M. (1985). Auditory matching-to-sample in monkeys (*Cebus apella*). *Animal Learning and Behaviour*, **13**, 375–82.

D'Amato, M. R. and Colombo, M. (1988a). On tonal pattern perception in monkeys (*Cebus apella*). *Animal Learning and Behaviour*, **16**, 417–24.

D'Amato, M. R. and Colombo, M. (1988b). Representation of serial order in monkeys (*Cebus apella*). *Journal of Experimental Psychology: Animal Behaviour Processes*, **14**, 131–9.

D'Amato, M. R. and Colombo, M. (1989). On the limits of the matching concept in monkeys (*Cebus apella*). *Journal of the Experimental Analysis of Behavior*, **52**, 225–36.

D'Amato, M. R. and Salmon, D. P. (1984). Cognitive processes in cebus monkeys. In *Animal cognition* (ed. H. L. Roitblat, T. G. Bever, and H. S. Terrace), pp. 149–68. Erlbaum, Hillsdale, NJ.

D'Amato, M. R. and Van Sant, P. (1988). The person concept in monkeys (*Cebus apella*). *Journal of Experimental Psychology: Animal Behaviour Processes*, **14**, 43–55.

D'Amato, M. R. and Worsham, R. W. (1972). Delayed matching in the capuchin monkey with brief sample durations. *Learning and Motivation*, **3**, 304–12.

D'Amato, M. R. and Worsham, R. W. (1974). Retrieval cues and short-term memory in capuchin monkeys. *Journal of Comparative and Physiological Psychology*, **86**, 274–82.

D'Amato, M. R., Salmon, D. P., and Colombo, M. (1985a). Extent and limits of the matching concept in monkeys (*Cebus apella*). *Journal of Experimental Psychology: Animal Behaviour Processes*, **11**, 35–51.

D'Amato, M. R., Salmon, D. P., Loukas, E., and Tomie, A. (1985b). Symmetry and transitivity of conditional relations in monkeys (*Cebus apella*) and pigeons (*Columba livia*). *Journal of the Experimental Analysis of Behavior*, **44**, 35–47.

D'Amato, M. R., Salmon, D. P., Loukas, E., and Tomie, A. (1986). Processing of identity and conditional relations in monkeys (*Cebus apella*) and pigeons (*Columba livia*). *Animal Learning and Behavior*, **14**, 365–73.

Darby, C. L. and Riopelle, A. J. (1959). Observational learning in the rhesus monkey. *Journal of Comparative and Physiological Psychology*, **52**, 94–8.

Davenport, R. K. and Rogers, C. M. (1970). Intermodal equivalence of stimuli in apes. *Science*, **168**, 279–80.

Davenport, R. K. and Rogers, C. M. (1971). Perception of photographs by apes. *Behavior*, **39**, 318–20.

Davenport, R. K., Rogers, C. M., and Menzel, E. W. (1969). Intellectual performance of differentially reared chimpanzees: II. *American Journal of Mental Deficiency*, **73**, 963–9.

Davenport, R. K., Rogers, C. M., and Russell, I. S. (1973). Cross-modal perception in apes. *Neuropsychologia*, **11**, 21–8.

Davenport, R. K., Rogers, C. M., and Russell, I. S. (1975). Cross-modal perception in apes. Altered visual cues and delay. *Neuropsychologia*, **13**, 229–35.

Davey, G. (1981). *Animal learning and conditioning*. Macmillan, London.

Davidson, D. (1980). Psychology as philosophy. In *Essays on actions and events* (ed. D. Davidson), pp. 229–44. Oxford University Press.

Davis, H. (1992). Transitive inference in rats (*Rettus norvegicus*). *Journal of Comparative Psychology*, **106**, 342–9.

Davis, H. and Balfour, A. D. (eds) (1992). *The inevitable bond*. Cambridge University Press.

Demarest, J. (1983). The ideas of change, progress and continuity in the comparative psychology of learning. In *Comparing behavior: studying man studying animals* (ed. D. W. Rajecki), pp. 143–79. Erlbaum, Hillsdale, NJ.

Devine, J. V. (1970). Stimulus attributes and training procedures in learning set formation of rhesus and cebus monkeys. *Journal of Comparative and Physiological Psychology*, **73**, 62–7.

Devine, J. V. and Jones, L. C. (1975). Matching-to-successive samples: a multiple-unit memory task with rhesus monkeys. *Behaviour Research Methods and Instrumentation*, **7**, 438–40.

de Waal, F. (1982). *Chimpanzee politics*. Harper and Row, New York. (Cited by Whiten and Byrne, 1991.)

Doré, F. Y. and Dumas, C. (1987). Psychology of animal cognition: Piagetian studies. *Psychological Bulletin*, **102**, 219–33.

Doty, B. A., Jones, C. N., and Doty, L. A. (1967). Learning-set formation by mink, ferrets, skunks, and cats. *Science*, **155**, 1579–80.

Dumas, C. (1992). Object permanence in cats (*Felix catus*): an ecological approach to the study of invisible displacements. *Journal of Comparative Psychology*, **106**, 404–10.

Ehrlich, A., Fobes, J. L., and King, J. E. (1976). Prosimian learning capacities. *Journal of Human Evolution*, **5**, 599–617.

Elliott, R. C. (1977). Cross-modal recognition in three primates. *Neuropsychologia*, **15**, 183–6.

Epstein, R., Lanza, R. P., and Skinner, B. F. (1980). Symbolic communication between two pigeons (*Columba livia domestica*). *Science*, **207**, 543–5. (Cited by Meador *et al.* 1987.)

Essock, S. M. (1975). Disruption of performance of a lab-born orang-utan after introduction of irrelevant foreground cues. *Perceptual and Motor Skills*, **40**, 645–6.

Essock-Vitale, S. M. (1978). Comparison of ape and monkey modes of problem solution. *Journal of Comparative and Physiological Psychology*, **92**, 942–57.

Ettlinger, G. (1977). Interactions between sensory modalities in non-human primates. In *Behavioural primatology: advances in research and theory*, Vol. 1, (ed. A. M. Schrier), pp. 71–104. Erlbaum, Hillsdale, NJ.

Ettlinger, G. (1981). The relationship between metaphorical cross-modal abilities: failure to demonstrate metaphorical recognition in chimpanzees capable of cross-modal recognition. *Neuropsychologia*, **19**, 583–6. (Cited by Meador *et al.* 1987.)

Ettlinger, G. and Wilson, W. A. (1990). Cross-modal performance: behavioral processes, phylogenetic considerations and neural mechanisms. *Behavioral Brain Research*, **40**, 169–92.

Feldman, D. W. and Klopfer, P. H. (1972). A study of observational learning in lemurs. *Zeitschrift für Tierpsychologie*, **30**, 297–304.

Fischer, G. J. (1962). The formation of learning set in young gorillas. *Journal of Comparative and Physiological Psychology*, **55**, 924–5.

Fletcher, H. J. (1965). The delayed-response problem. In *Behavior of non-human primates*, Vol. 1, (ed. A. Schrier, H. Harlow, and F. Stollnitz), pp. 129–66. Academic Press, New York.

Fobes, J. L. and King, J. E. (1978). Learning capacities of Tupaiidae—the transitional Insectivora–Primates. *Journal of Human Evolution*, **7**, 609–18.

Fobes, J. L. and King, J. E. (1982). Measuring primate learning abilities. In *Primate behavior*, (ed. idem), pp. 289–326. Academic Press, New York.

Fossey, D. (1979). Development of the mountain gorilla (*Gorilla gorilla gorilla*): the first 36 months. In *The Great Apes* (ed. D. A. Hamburg and E. R. McCown), pp. 139–84. Benjamin–Cummings, Menlo Park, CA.

Fouts, R. S. (1974). Language: origins, definitions and chimpanzees. *Journal of Human Evolution*, **3**, 475–82.

Fragaszy, D. M. (1981). Comparative performance in discrimination learning tasks in two New World primates *Saimiri sciureus* and *Callicebus moloch*. *Animal Learning and Behavior*, **9**, 127–34.

Fragaszy, D., Visalberghi, E., and Robinson, J. (eds) (1990). Special issue on *Cebus. Folia Primatologica*, **54**, (3–4).

Frith, U. (1989). *Autism: explaining the enigma*. Blackwell, Oxford.

Gaffan, D. (1977). Discrimination of word-like compound visual stimuli by monkeys. *Quarterly Journal of Experimental Psychology*, **29**, 589–96.

Gagnon, S. and Doré, F. Y. (1992). Search behavior in various breeds of adult dogs (*Canis familiaris*): object permanence and olfactory cues. *Journal of Comparative Psychology*, **106**, 58–68.

Gallup, G. G. (1970). Chimpanzees: self-recognition. *Science*, **167**, 86–7.

Gallup, G. G. (1977a). Absence of self-recognition in a monkey (*Macaca fascicularis*) following prolonged exposure to a mirror. *Developmental Psychobiology*, **10**, 281–4.

Gallup, G. G. (1977b). Self-recognition in primates: a comparative approach to the bidirectional properties of consciousness. *American Psychologist*, **32**, 329–38.

Gallup, G. G. (1982). Self-awareness and the emergence of mind in primates. *American Journal of Primatology*, **2**, 237–48.

Gallup, G. G. (1985). Do minds exist in species other than our own? *Neuroscience and Biobehavioral Reviews*, **9**, 631–41.

Gallup, G. G. (1987). Self-awareness. In *Comparative primate biology*, Vol. 2B, *Behavior, cognition, and motivation*, (ed. G. Mitchell and J. Erwin), pp. 3–16. Alan R. Liss, Atlanta.

Gallup, G. G. (1991). Toward a comparative psychology of self-awareness: species limitations and cognitive consequences. In *The self: an interdisciplinary approach* (ed. G. R. Goethals and J. Strauss), pp. 121–35. Springer-Verlag, New York.

Gallup, G. G. and Suarez, S. D. (1991). Social responding to mirrors in rhesus monkeys (*Macaca mulatta*): effects of temporary mirror removal. *Journal of Comparative Psychology*, **105**, 376–9.

Gallup, G. G., Wallnau, L. B., and Suarez, S. D. (1980). Failure to find self-recognition in mother–infant and infant–infant rhesus monkey pairs. *Folia Primatologica*, **33**, 210–19.

Gill, T. V. and Rumbaugh, D. M. (1974). Learning processes of bright and dull apes. *American Journal of Mental Deficiency*, **78**, 683–7.

Gillan, D. J. (1981). Reasoning in the chimpanzee. II. Transitive inference. *Journal of Experimental Psychology: Animal Behaviour Processes*, **7**, 150–64.

Gillan, D. J., Premack, D., and Woodruff, G. (1981). Reasoning in the chimpanzee. I. Analogical reasoning. *Journal of Experimental Psychology: Animal Behaviour Processes*, **7**, 1–17.

Goodall, J. (1968). The behaviour of free-living chimpanzees in the Gombe Stream area. *Animal Behaviour Monographs*, **1**, 161–311.

Grant, D. S. and Roberts, W. A. (1973). Trace interaction in pigeon short-term memory. *Journal of Experimental Psychology*, **101**, 21–9.

Griffin, D. R. (1976). *The question of animal awareness: evolutionary continuity of mental experience*. Rockefeller University Press, New York.

Gruber, H. E., Girgus, J. S., and Banuazizi, A. (1971). The development of object permanence in the cat. *Developmental Psychology*, **4**, 9–15.

Hall, K. R. L. (1963). Observational learning in monkeys and apes. *British Journal of Psychology*, **54**, 201–26.

Hall, K. R. L. and Goswell, M. J. (1964). Aspects of social learning in captive patas monkeys. *Primates*, **5**, 59–70.

Hallock, M. B. and Worobey, J. (1984). Cognitive development in chimpanzee infants (*Pan troglodytes*). *Journal of Human Evolution*, **13**, 441–7.

Harlow, H. F. (1944). Studies in discrimination learning by monkeys: I. The learning of discrimination series and the reversal of discrimination series. *Journal of General Psychology*, **30**, 3–12.

Harlow, H. F. (1951). Primate learning. In *Comparative psychology* (3rd edn, ed. C. Stone), pp. 183–238. Prentice-Hall, New York.

Harlow, H. F. and Bromer, J. A. (1939). Comparative behavior of primates. VIII. The capacity of platyrrhine monkeys to solve delayed reaction tests. *Journal of Comparative Psychology*, **28**, 299–304.

Harlow, H. F., Uehling, H., and Maslow, A. H. (1932). Comparative behaviour of primates. I. Delayed reaction tests of primates from the lemur to the orangutan. *Journal of Comparative and Physiological Psychology*, **13**, 313–43.

Harris, P. L. (1989). *Children and emotion: the development of psychological understanding*. Blackwell, Oxford.

Harris, P. L. (1991). The work of the imagination. In *Natural theories of mind: evolution, development and simulation of everyday mindreading* (ed. A. Whiten), pp. 283–304. Blackwell, Oxford.

Hayes, K. J., Thompson, R., and Hayes, C. (1953). Discrimination learning sets in chimpanzees. *Journal of Comparative Psychology*, **46**, 99–104.

Heffner, H. E. and Heffner, R. S. (1984). Temporal lobe lesions and perception of species-specific vocalization by macaques. *Science*, **226**, 75–6.

Herman, L. M. and Arbeit, W. R. (1973). Stimulus control and auditory discrimination learning sets in the bottlenose dolphin. *Journal of the Experimental Analysis of Behavior*, **19**, 379–94.

Herman, L. M. and Forestell, P. H. (1985). Reporting presence or absence of named objects by a language-trained dolphin. *Neuroscience and Biobehavioral Review*, **9**, 667–81.

Herrnstein, R. J. (1979). Acquisition, generalization, and discrimination reversal of a natural concept. *Journal of Experimental Psychology: Animal Behaviour Processes*, **5**, 116–29.

Herrnstein, R. J. and de Villiers, P. A. (1980). Fish as a natural category for people and pigeons. In *The psychology of learning and motivation*, Vol. 14, (ed. G. H. Bower), pp. 59–95. Academic Press, New York.

Herrnstein, R. J. and Loveland, D. H. (1964). Complex visual concept in the pigeon. *Science*, **146**, 549–51.

Herrnstein, R. J., Loveland, D. H., and Cable, C. (1976). Natural concepts in pigeons. *Journal of Experimental Psychology: Animal Behaviour Processes*, **2**, 285–311.

Hewett, T. D. and Ettlinger, G. (1978). Cross-modal performance: the absence of transfer in nonhuman primates capable of recognition. *Neuropsychologia*, **16**, 361–6.

Hewett, T. D. and Ettlinger, G. (1979). Cross-modal performance: the nature of the failure at "transfer" in non-human primates capable of "recognition". *Neuropsychologia*, **17**, 511–4.

Hines, D. (1975). Immediate and delayed recognition of sequentially presented random shapes. *Journal of Experimental Psychology: Human Learning and Memory*, **1**, 634–9.

Hladik, C. M. (1979). Diet and ecology of prosimians. In *The study of prosimian behavior*, (ed. G. E. Doyle and R. D. Martin), pp. 307–57. Academic Press, New York.

Houston, J. P. (1981). *Fundamentals of learning and memory* (2nd edn). Academic Press, New York.

Hulse, S. H., Fowler, H., and Honig, W. K. (eds) (1978). *Cognitive processes in animal behavior*. Erlbaum, Hillsdale, NJ.

Hulse, S. H., Cynz, J., and Humpal, J. (1984). Cognitive processing of pitch and rhythm structures by birds. In *Animal cognition* (ed. H. L. Roitblat, T. G. Bever, and H. S. Terrace), pp. 183–98. Erlbaum, Hillsdale, NJ.

Hunter, W. S. (1913). The delayed reaction in animals and children. *Behavior Monographs*, **2**, 1–86.

Hunter, W. S. (1920). The temporal maze and kinaesthetic sensory processes in the white rat. *Psychobiology*, **2**, 1–17.

Hunter, M. W. III, and Kamil, A. C. (1971). Object-discrimination learning set and hypothesis behavior in the northern bluejay (*Cyanocitta cristata*). *Psychonomic Science*, **22**, 271–4.

Huxley, T. H. (1897). On the hypothesis that animals are autonoma, and its history. In *idem*, *Methods and results: essays*, pp. 199–250. Appleton, New York. (Cited by Gillan *et al.* 1981.)

Irle, E. and Markowitsch, H. J. (1987). Conceptualization without specific training in squirrel monkeys (*Saimiri sciureus*): a test using the non-match-to-sample procedure. *Journal of Comparative Psychology*, **101**, 305–11.

James, W. (1890). *The principles of psychology*, Vol. I. Holt, New York. (Cited by Gillan *et al.* 1981.)

Jarvik, M. E., Goldfarb, T. L., and Carley, J. L. (1969). Influence of interference on delayed matching in monkeys. *Journal of Experimental Psychology*, **81**, 1–6.

Jarvis, M. J. and Ettlinger, G. (1975). Transfer of spatial alternation between responding in the light and in the dark. *Neuropsychologia*, **13**, 115–6.

Jarvis, M. J. and Ettlinger, G. (1978). Cross-modal performance in monkeys and apes: is there a substantial difference? In *Recent advances in primatology 1* (ed. D. J. Chivers and J. Herbert), pp. 953–6. Academic Press, London.

Johnson, J. I. (1961). Double alternation by raccoons. *Journal of Comparative and Physiological Psychology*, **54**, 248–51.

Jolly, A. (1972). *The evolution of primate behavior*. Macmillan, New York.

Karten, H. J. and Shimuzu, T. (1989). The origins of neocortex: connections and lamination as distinct events in evolution. *Journal of Cognitive Neuroscience*, **1**, 291–301.

Kawai, M. (1965). Newly acquired pre-cultural behavior of a natural group of Japanese monkeys on Koshima Island. *Primates*, **6**, 1–30.

Kesner, R. P. and Novak, J. M. (1982). Serial position curve in rats: role of the dorsal hippocampus. *Science*, **218**, 173–5.

King, J. E. (1973). Learning and generalization of a two-dimensional sameness–difference concept by chimpanzees and orang-utans. *Journal of Comparative and Physiological Psychology*, **84**, 140–8.

King, J. E. and Fobes, J. L. (1982). Complex learning by primates. In *Primate behavior* (ed. J. L. Fobes and J. E. King), pp. 327–60. Academic Press, New York.

King, J. E., Flaningam, M. R., and Rees, W. W. (1968). Delayed response with different delay conditions by squirrel monkeys and fox squirrels. *Animal Behavior*, **16**, 271–7.

Kinnamon, A. J. (1902). Mental life of two *Macaca rhesus* monkeys in captivity. *American Journal of Psychology*, **13**, 98–148, 173–218.

Kluver, H. (1933). *Behavior mechanisms in monkeys*. Chicago University Press.

Koch, M. B. and Meyer, D. K. (1959). A relationship of mental age to learning-set formation in the pre-school child. *Journal of Comparative and Physiological Psychology*, **52**, 387–9.

Köhler (1973 [1925]). *The mentality of apes*. Routledge and Kegan Paul, London. (Originally published Harcourt Brace, New York, 1925.)

Kuhl, P. (1987). The special-mechanisms debate in speech: contribution of tests on animals (and the relations of these tests to studies using non-speech signals). In *Categorical perception* (ed. S. Harnad), pp. 355–86. Cambridge University Press.

Laidler, K. (1978). Aspects of physical and cognitive development in the infant orang-utan (*Pongo pygmaeus*) during the first 15 months of life. Unpublished Ph. D. dissertation, Department of Anthropology, University of Durham.

Law, L. E. and Lock, A. J. (1994). Do gorillas recognize themselves on television?. In *Self-awareness in animals and humans: developmental perspectives* (ed. S. T. Parker, R. W. Mitchell, and M. L. Boccia), pp. 308–12. Cambridge University Press.

Leonard, C., Schneider, G. E., and Gross, C. G. (1966). Performance on learning set and delayed response tasks by tree shrews (*Tupaia glis*). *Journal of Comparative and Physiological Psychology*, **62**, 501–4.

Leslie, A. M. (1987). Pretence and representation in infancy: the origins of 'theory of mind'. *Psychological Review*, **94**, 84–106.

Leslie, A. M. (1988). Some implications of pretence for mechanisms underlying the child's theory of mind. In *Developing theories of mind* (ed. J. W. Astington, P. L. Harris, and D. Olson), pp. 19–46. Cambridge University Press.

Leslie, A. M. and Thaiss, L. (1992). Domain specificity in conceptual development: neuropsychological evidence from autism. *Cognition*, **43**, 225–51.

Lethmate, J. and Ducker, G. (1973). Untersuchungen zum Selbsterkennen im Spiegel bei Orang-utans und einigen anderen Affenarten. *Zeitschrift für Tierpsychologie*, **33**, 248–69.

Levine, M. (1959). A model of hypothesis behavior in discrimination learning set. *Psychological Review*, **66**, 353–66.

Lewis, M., Sullivan, M. W., Stanger, C., and Weiss, M. (1989). Self-development and self-conscious emotions. *Child Development*, **60**, 146–56.

Livesey, P. J. (1965). Comparison of double alternation performance of white rats, rabbits and cats. *Journal of Comparative and Physiological Psychology*, **59**, 155–8.

Livesey, P. J. (1969). Double and single alternation learning by rhesus monkeys. *Journal of Comparative and Physiological Psychology*, **67**, 526–30.

Lock, A. J. (1992). Anecdotal occurrences may be really important events. *Journal of Human Movement Science*, **11**, 497–504.

Lombardi, C. M., Fachinelli, C. C., and Delius, J. D. (1984). Oddity of visual patterns conceptualized by pigeons. *Animal Learning and Behavior*, **12**, 2–6.

McGonigle, B. O. and Chalmers, M. (1977). Are monkeys logical? *Nature*, **267**, 694–6.

McGrew, W. (1977). Socialization and object manipulation of wild chimpanzees. In *Primate biosocial development: biological, social and ecological determinants* (ed. S. Chevalier-Skolnikoff and F. Poirier), pp. 261–88. Garland, New York.

MacPhail, E. M. (1982). *Brain and intelligence in vertebrates*. Oxford University Press.

MacPhail,985). Vertebrate intelligence: the null hypothesis. In *Animal intelligence* (ed. L. Weiskrantz), Oxford University Press, New York. (Cited by Terrace 1993.)

Mahan, J. L. and Rumbaugh, D. M. (1963). Observational learning in the squirrel monkey. *Perceptual and Motor Skills*, **17**, 686.

Maier, N. R. and Schneirla, T. C. (1935). *Principles of animal psychology*. McGraw-Hill, New York.

Mallot, R. and Siddal, J. W. (1972). Acquisition of the people concept in pigeons. *Psychological Reports*, **31**, 3–13.

Malone, D. R., Tolan, J. C., and Rogers, C. M. (1980). Cross-modal matching of objects and photographs in the monkey. *Neuropsychologia*, **18**, 693–7.

Markovits, H. and Dumas, C. (1992). Can pigeons really make transitive inferences? *Journal of Experimental Psychology: Animal Behaviour Processes*, **18**, 311–12.

Maslow, A. H. and Harlow, H. F. (1932). Comparative behavior of primates. II. Delayed reaction tests on primates at Bronx Park Zoo. *Journal of Comparative Psychology*, **14**, 97–108.

Masterton, B. and Skeen, L. C. (1972). Origins of anthropoid intelligence: prefrontal system and delayed alternation in hedgehog, tree shrew and bush baby. *Journal of Comparative and Physiological Psychology*, **81**, 423–33.

Mathieu, M. and Bergeron, G. (1981). Piagetian assessment of cognitive development in chimpanzees (*Pan troglodytes*). In *Primate behavior and sociobiology* (ed. A. B. Chiarelli and R. S. Corruccini), pp. 142–7. Springer-Verlag, Berlin.

Mathieu, M., Bouchard, M. A., Granger, L., and Herscovitch, J. (1976). Piagetian object-permanence in *Cebus capucinus, Lagothrica flavicauda* and *Pan troglodytes*. *Animal Behavior*, **24**, 585–8.

Mazmanian, D. S. and Roberts, W. A. (1983). Spatial memory in rats under restricted viewing conditions. *Learning and Motivation*, **14**, 123–39.

Mead, G. H. (1934). *Mind, self, and society*. Chicago University Press.

Meador, D. M. and Rumbaugh, D. M. (1981). Quality of learning in severely retarded adolescents. *American Journal of Mental Deficiency*, **85**, 404–9.

Meador, D. M., Rumbaugh, D. M., Pate, J. L., and Bard, K. A. (1987). Learning, problem-solving, cognition, and intelligence. In *Comparative primate biology*, Vol. 2B: *Cognition and motivation* (ed. G. Mitchell and J. Erwin), pp. 17–83. Alan R. Liss, Atlanta.

Menzel, E. W. (1972). Spontaneous invention of ladders in a group of young chimpanzees. *Folia Primatologica*, **17**, 87–106.

Menzel, E. W. (1973). Chimpanzee spatial memory organization. *Science*, **182**, 943–5.

Menzel, E. W. (1978). Cognitive mapping in chimpanzees. In *Cognitive processes in animal behavior* (ed. S. H. Hulse, H. Fowler, and W. K. Honig), pp. 375–422. Erlbaum, Hillsdale, NJ.

Menzel, E. W. and Draper, W. A. (1965). Primate selection of food by size: visible versus invisible rewards. *Journal of Comparative and Physiological Psychology*, **59**, 231–9.

Menzel, E. W. and Juno, C. (1982). Marmosets (*Saguinus fuscicollis*): are learning sets learned? *Science*, **217**, 750–2.

Menzel, E. W., Davenport, R. K., and Rogers, C. M. (1970). The development of tool using in wild-born and restriction-reared chimpanzees. *Folia Primatologica*, **12**, 273–83.

Menzel, E. W., Savage-Rumbaugh, E. S., and Lawson, J. (1985). Chimpanzee (*Pan troglodytes*) spatial problem-solving with the use of mirrors and televised equivalents of mirrors. *Journal of Comparative Psychology*, **99**, 211–17.

Meyers, W. J., McQuiston, M. D., and Miles, R. C. (1962). Delayed-response and learning-set performance of cats. *Journal of Comparative and Physiological Psychology*, **55**, 515–7.

Miles, R. C. (1957a). Delayed-response learning in the marmoset and the macaque. *Journal of Comparative and Physiological Psychology*, **50**, 352–5.

Miles, R. C. (1957b). Learning-set formation in the squirrel monkey. *Journal of Comparative and Physiological Psychology*, **50**, 356–7.

Milner, A. D. (1973). Matching within and between sense modalities in the monkey (*Macaca mulatta*). *Journal of Comparative and Physiological Psychology*, **83**, 278–84.

Mitchell, R. W. (1986). A framework for discussing deception. In *Deception: perspectives on human and non-human deceit* (ed. R. W. Mitchell and N. S. Thompson), pp. 3–40. State University of New York Press, Albany.

Mitchell, R. W. (1993a). Mental models of mirror-self-recognition: two theories. *New Ideas in Psychology*, **11**, 295–325.

Mitchell, R. W. (1993b). Recognizing one's self in a mirror. *New Ideas in Psychology*, **11**, 351–77.

Mitchell, R. W. and Thompson, N. S. (eds) (1993). *Anthropomorphism, anecdotes and animals: the Emperor's new clothes?* University of Nebraska Press, Lincoln.

Moise, S. L. (1970). Short-term retention in *Macaca speciosa* following interpolated activity during matching from sample. *Journal of Comparative and Physiological Psychology*, **73**, 506–14.

Morgan, C. L. (1894). *Introduction to comparative psychology*. Walter Scott, London. (Cited by Gillan *et al.* 1981.)

Muncer, S. J. (1983). "Conservations" with a chimpanzee. *Developmental Psychobiology*, **16**, 1–11.

Murdock, B. B. (1962). The serial position effect in free recall. *Journal of Experimental Psychology*, **64**, 482–8.

Musicant, A. (1975). Object discrimination learning set in prosimian primates. Unpublished paper presented at the meeting of the Western Psychological Association, Sacramento, 1975.

Natale, F. and Antinucci, F. (1989). Stage 6 object-concept and representation. In *Cognitive structure and development in nonhuman primates* (ed. F. Antinucci), pp. 97–112. Erlbaum, Hillsdale, NJ.

Natale, F., Antinucci, F., Spinozzi, G., and Poti, P. (1986). Stage 6 object concept in nonhuman primate cognition: a comparison between gorilla (*Gorilla gorilla gorilla*) and Japanese macaque (*Macaca fuscata*). *Journal of Comparative Psychology*, **100**, 335–9.

Nissen, H. W., Blum, J. S., and Blum, R. A. (1949). Conditional matching behaviour in chimpanzee: implications for the comparative study of intelligence. *Journal of Comparative and Physiological Psychology*, **42**, 239–356.

Norris, E. and Ettlinger, G. (1978). Cross-modal performance in the monkey: non-matching to sample with edible shapes. *Neuropsychologia*, **16**, 99–102.

Olton, D. S. (1978). Characteristics of spatial memory. In *Cognitive processes in animal behavior* (ed. S. H. Hulse, H. Fowler, and W. K. Honig), pp. 341–73. Erlbaum, Hillsdale, NJ.

Olton, D. S. (1979). Mazes, maps and memory. *American Psychologist*, **34**, 583–96.

Olton, D. S. and Samuelson, R. J. (1976). Remembrance of places past: spatial memory in rats. *Journal of Experimental Psychology: Animal Behaviour Processes*, **2**, 97–116.

Pariente, G. (1979). The role of vision in prosimian behavior. In *The study of prosimian behavior* (ed. G. E. Doyle and R. D. Martin), pp. 411–59. Academic Press, New York.

Parker, S. T. (1977). Piaget's sensorimotor series in an infant macaque: a model for comparing unstereotyped behavior and intelligence in human and non-human primates. In *Primate bio-social development: biological, social and ecological determinants* (ed. S. Chevalier-Skolnikoff and F. E. Poirier), pp. 43–112. Garland, New York.

Parker, S. and Gibson, K. (eds) (1990). *"Language" and intelligence in monkeys and apes*. Cambridge University Press.

Pasnak, R. (1979). Acquisition of prerequisites to conservation by macaques. *Journal of Experimental Psychology: Animal Behaviour Processes*, **5**, 194–210.

Pepperberg, I. M. and Funk, M. S. (1990). Object permanence in four species of psittacine birds: an African grey parrot (*Psittacus erithacus*), an illiger mini macaw (*Ara maracana*), a parakeet (*Melopsittacus undulatus*), and a cockatiel (*Nymphicus hollandicus*). *Animal Learning and Behavior*, **18**, 97–108.

Pepperberg, I. M. and Kozak, F. A. (1986). Object permanence in the African grey parrot (*Psittacus erithacus*). *Animal Learning and Behavior*, **14**, 322–30.

Perner, J. (1991). *Towards understanding representation and mind*. Bradford Books, MIT Press, Cambridge, Mass.

Perner, J. and Wimmer, H. (1985). 'John thinks that Mary thinks that...': attribution of second-order beliefs by 5- to 10-year-old children. *Journal of Experimental Child Psychology*, **39**, 437–71.

Perrault, J. (1982). Étude comparée des conduites de sériation chez le chimpanzé et le jeune enfant. Unpublished Master's dissertation, University of Montreal.

Petersen, M. R., Beecher, M. D., Zoloth, S. R., Moody, D. B., and Stebbins, W. C. (1978). Neural lateralization of species-specific vocalizations by Japanese macaques (*Macaca fuscata*). *Science*, **202**, 324–7.

Piaget, J. (1970[1967]). *Biology and knowledge*. Edinburgh University Press. (French original, 1967.)

Piaget, J. (1978[1976]). *Behaviour and evolution*. Pantheon Books, New York. (French original, 1976.)

Piaget, J. (1980[1974]). *Adaptation and intelligence: organic selection and phenocopy*. Chicago University Press. (French original, 1974.)

Plotnik, R. J. and Tallarico, R. B. (1966). Object-quality learning-set formation in the young chicken. *Psychonomic Science*, **5**, 195–6.

Poole, J. and Lander, D. G. (1971). The pigeon's concept of pigeon. *Psychonomic Science*, **25**, 157–8.

Potter, M. C. and Levy, E. I. (1969). Recognition memory for a rapid sequence of pictures. *Journal of Experimental Psychology*, **81**, 10–15.

Povinelli, D. J. (1993). Reconstructing the evolution of mind. *American Psychologist*, **48**, 493–509.

Povinelli, D. J., Nelson, K. E., and Boysen, S. T. (1990). Inferences about guessing and knowing by chimpanzees (*Pan troglodytes*). *Journal of Comparative Psychology*, **104**, 203–10.

Povinelli, D. J., Parks, K. A., and Novak, M. A. (1991). Do rhesus monkeys (*Macaca mulatta*) attribute knowledge and ignorance to others? *Journal of Comparative Psychology*, **105**, 318–25.

Premack, D. (1976). *Intelligence in ape and man*. Erlbaum, Hillsdale, NJ.

Premack D. (1980). Characteristics of an upgraded mind. In *Bericht über den 32. Kongress der Deutschen Gesellschaft für Psychologie in Zürich* (ed. W. Michaelis), Vol. 1, pp. 49–70.

Premack, D. (1983a). The codes of man and beasts. *The Behavioral and Brain Sciences*, **6**, 125–67.

Premack, D. (1983b). Animal cognition. *Annual Review of Psychology*, **34**, 351–62.

Premack, D. and Woodruff, G. (1978). Does a chimpanzee have a theory of mind? *The Behavioral and Brain Sciences*, **4**, 515–26.

Redshaw, M. (1978). Cognitive development in human and gorilla infants. *Journal of Human Evolution*, **7**, 133–41.

Riddell, N., Cori, K., Bennett, V., and Reimers, O. (1974). Discrimination learning differences and similarities as a function of brain index. *Physiology and Behaviour*, **13**, 401–5.

Riesen, A. H., Greenberg, B., Granston, A. S., and Fantz, R. L. (1953). Solutions of patterned string problems by young gorillas. *Journal of Comparative and Physiological Psychology*, **46**, 19–22.

Rijksen, H. D. (1978). A field study on Sumatran orang-utans (*Pongo pygmaeus abelii* Lesson, 1827): Ecology, behaviour and conservation. *Mededelingen Landbouhogeschool, Wageningen (Nederland)*, **78-2**, 1–420.

Riopelle, A. J. and Hill, C. W. (1973). Complex processes. In *Comparative psychology: a modern survey* (ed. D. A. Dewsbury and D. A. Rethlingshafer), pp. 510–48. McGraw-Hill, New York.

Ristau, C. A. (ed.) (1990). *Cognitive ethology: the minds of other animals*. Erlbaum, Hillsdale, NJ.

Roberts, W. A. (1972). Spatial separation and visual differentiation of cues of factors influencing short-term memory in the rat. *Journal of Comparative and Physiological Psychology*, **78**, 284–91.

Roberts, W. A. (1974). Spaced repetition facilitates short-term retention in the rat. *Journal of Comparative and Physiological Psychology*, **86**, 164–71.

Roberts, W. A. and Grant, D. S. (1976). Studies of short-term memory in the pigeon using the delayed matching to sample procedure. In *Processes of animal memory* (ed. D. L. Medin, W. A. Roberts, and R. T. Davis), pp. 79–112. Lawrence Erlbaum, Hillsdale, NJ.

Roberts, W. A. and Smythe, W. E. (1979). Memory for lists of spatial events in the rat. *Learning and Motivation*, **10**, 313–6.

Robinson, J. S. (1955). The sameness–difference discrimination problem in chimpanzees. *Journal of Comparative and Physiological Psychology*, **48**, 195–213.

Robinson, J. S. (1960). The conceptual basis of chimpanzees' performance on the sameness–difference discrimination problem. *Journal of Comparative and Physiological Psychology*, **53**, 368–70.

Roitblat, H. L. (1987). *Introduction to comparative cognition.* Freeman, New York.

Roitblat, H. L., Bever, T. G., and Terrace, H. S. (eds) (1984). *Animal cognition.* Erlbaum, Hillsdale, NJ.

Romanes, G. J. (1883). *Mental evolution in animals.* John Murray, London.

Romer, A. S. (1967). Major steps in vertebrate evolution. *Science*, **158**, 1629–37.

Rosenfeld, S. A. and Van Hoesen, G. W. (1979). Face recognition in the rhesus monkey. *Neuropsychologia*, **17**, 503–9.

Rumbaugh, D. M. (1970). Learning skills in anthropoids. In *Primate behavior: developments in field and laboratory research* (ed. L. A. Rosenblum), pp. 1–70. Academic Press, New York.

Rumbaugh, D. M. (1971). Evidence of qualitative differences in learning processes among primates. *Journal of Comparative and Physiological Psychology*, **76**, 250–5.

Rumbaugh, D. M. (1986). Animal thinking—by stimulation or simulation. In *Primate ontogeny, cognition and social behaviour*, Vol. 3, (ed. J. G. Else and P. C. Lee), pp. 31–40. Cambridge University Press.

Rumbaugh, D. M. and Gill, T. V. (1973). The learning skills of great apes. *Journal of Human Evolution*, **2**, 171–9.

Rumbaugh, D. M. and McCormack, C. (1967). The learning skills of primates: a comparative study of apes and monkeys. In *Progress in primatology* (ed. D. Starck, R. Schneider, and H. Kuhn), pp. 289–306. Fischer, Stuttgart.

Rumbaugh, D. M. and Pate, J. L. (1984). The evolution of cognition in primates: a comparative perspective. In *Animal cognition* (ed. H. L. Roitblat, T. G. Bever, and A. S. Terrace), pp. 569–87. Erlbaum, Hillsdale, NJ.

Rumbaugh, D. M., Gill, T. V., and Wright, S. C. (1973). Readiness to attend to visual foreground cues. *Journal of Human Evolution*, **2**, 181–8.

Salmon, D. P. (1984). An investigation of modality specificity in the cognitive processes of monkeys (*Cebus apella*). *Dissertation Abstracts International*, **45**, 706B.

Sands, S. F. and Wright, A. A. (1980). Primate memory: retention of serial list items by a rhesus monkey. *Science*, **209**, 938–40.

Sands, S. F., Lincoln, C. E., and Wright, A. A. (1982). Pictorial similarity judgements and the organization of visual memory in the rhesus monkey. *Journal of Experimental Psychology: General*, **111**, 369–89.

Sands, S. F., Urcuioli, P. J., Wright, A. A., and Santiago, H. C. (1984). Serial position effects and rehearsal in primate visual memory. In *Animal cognition* (ed. H. L. Roitblat, T. G. Bever, and H. S. Terrace), pp. 375–88. Erlbaum, Hillsdale, NJ.

Savage-Rumbaugh, E. S. (1986). *Ape language: from conditioned response to symbol.* Columbia University Press, New York.

Savage-Rumbaugh, E. S., Rumbaugh, D. M., and Boysen, S. (1978a). Linguistically mediated tool use and exchange by chimpanzees (*Pan troglodytes*). *Behavioural and Brain Sciences*, **4**, 539–54. (Cited by Meador *et al.* 1987.)

Savage-Rumbaugh, E. S., Rumbaugh, D. M., and Boysen, S. (1978b). Symbolic communication between two chimpanzees (*Pan troglodytes*). *Science*, **201**, 641–4. (Cited by Meador *et al.* 1987.)

Schiller, P. H. (1952). Innate constituents of complex responses in primates. *Psychological Review*, **59**, 177–91.

Schiller, P. H. (1957). Innate motor action as a basis of learning. In *Instinctive behavior* (ed. P. H. Schiller), pp. 264–87. International Universities Press, New York.

Schrier, A. M. and Brady, P. M. (1987). Categorization of natural stimuli by monkeys (*Macaca mulatta*): effects of stimulus set size and modification of exemplars. *Journal of Experimental Psychology: Animal Behaviour Processes*, **13**, 136–43.

Schrier, A. M., Angarella, R., and Povar, M. L. (1984). Studies of concept formation by stumptailed monkeys: concepts *humans, monkeys,* and letter *A. Journal of Experimental Psychology: Animal Behaviour Processes*, **10**, 564–84.

Schusterman, R. J. (1963). The use of strategies in two-choice behavior of children and chimpanzees. *Journal of Comparative and Physiological Psychology*, **56**, 96–100.

Schusterman, R. J. and Bernstein, I. S. (1962). Response tendencies of gibbons in single and double alternation tasks. *Psychological Reports*, **11**, 521.

Shell, W. F. and Riopelle, A. J. (1958). Progressive discrimination learning in platyrrhine monkeys. *Journal of Comparative and Physiological Psychology*, **51**, 467–70.

Shimp, C. P., and Moffit, M. (1974). Short-term memory in the pigeon: stimulus–response associations. *Journal of the Experimental Analysis of Behavior*, **12**, 745–57.

Shultz, T. R. and Shamash, F. (1981). The child's conception of intending act and consequence. *Canadian Journal of Behavioural Science*, **13**, 368–72.

Sidman, M., Rauzin, R., Lazar, R., Cunningham, S., Tailby, W., and Carrigan, P. (1982). A search for symmetry in the conditional discriminations of rhesus monkeys, baboons, and children. *Journal of the Experimental Analysis of Behavior*, **37**, 23–44.

Siegel, R. K. and Honig, W. K. (1970). Pigeon concept formation: successive and simultaneous acquisition. *Journal of the Experimental Analysis of Behavior*, **13**, 385–90.

Skinner, B. F. (1957). *Verbal behavior*. Appleton–Century–Crofts, New York. (Cited by Meador et al. 1987.)

Slotnick, B. M. and Katz, H. (1974). Olfactory learning-set formation in rats. *Science*, **185**, 796–8.

Stevens, D. A. (1965). A comparison of learning in rhesus monkeys, cebus monkeys, lemurs, and Burmese cats. Unpublished Doctoral dissertation, University of Oregon.

Straub, R. O. and Terrace, H. S. (1981). Generalization of serial learning in the pigeon. *Animal Learning and Behavior*, **9**, 454–68.

Suarez, S. D. and Gallup, G. G. (1981). Self-recognition in chimpanzees and orangutans, but not gorillas. *Journal of Human Evolution*, **10**, 175–88.

Swartz, K. B. and Rosenblum, L. A. (1980). Operant responding by bonnet macaques for color videotape recordings of social stimuli. *Animal Learning and Behavior*, **8**, 31–321.

Terrace, H. S. (1984). Animal cognition. In *Animal cognition* (ed. H. L. Roitblat, T. G. Bever, and H. S. Terrace), pp. 7–28. Erlbaum, Hillsdale, NJ.

Terrace, H. S. (1993). The phylogeny and ontogeny of serial memory: list learning by pigeons and monkeys. *Psychological Science*, **4**, 162–9.

Thomas, R. K. and Frost, T. (1983). Oddity and dimension-abstracted oddity (DAO) in squirrel monkeys. *American Journal of Psychology*, **96**, 51–64.

Thompson, R. K. R. and Herman, L. M. (1977). Memory for lists of sounds by the bottle-nosed dolphin: convergence of memory processes with humans? *Science*, **195**, 501–3.

Thorndike, E. L. (1898). Animal intelligence: an experimental study of the associative processes of animals. *Psychological Monographs*, **2**, (4, whole of No. 8). (Cited by Gillan et al. 1981.)

Tolan, J. C., Rogers, C. M., and Malone, D. R. (1981). Cross-modal matching in monkeys: altered visual cues and delay. *Neuropsychologia*, **19**, 289–300.

Tomasello, M., Davis-Dasilva, M., Camak, L., and Dard, K. (1987). Observational learning of tool-use by young chimpanzees. *Human Evolution*, **2**, 175–83.

Uzgiris, I. and Hunt, J. McV. (1975). *Assessment in infancy: ordinal scales of psychological development*. University of Illinois Press, Urbana.

Vauclair, J. (1982). Sensorimotor intelligence in human and non-human primates. *Journal of Human Evolution*, **11**, 257–64. (Cited by Doré and Dumas 1987.)

Vauclair, J., and Bard, K. A. (1983). Development of manipulation with objects in ape and human infants. *Journal of Human Evolution*, **12**, 631–45. (Cited by Doré and Dumas 1987.)

Vaughter, R. M., Smotherman, W., and Ordy, J. M. (1972). Development of object permanence in the infant squirrel monkey. *Developmental Psychology*, **7**, 34–8.

von Fersen, L., Wynne, C. D. L., Delius, J. D., and Staddon, J. E. R. (1991). Transitive inference formation in pigeons. *Journal of Experimental Psychology: Animal Behaviour Processes*, **17**, 334–41.

Wagner, A. R. and Pfautz, P. L. (1978). A bowed serial-position function in habituation of sequential stimuli. *Animal Learning and Behavior*, **6**, 395–400.

Wagner, S., Winner, E., Ciccheti, D., and Gardner, H. (1981). "Metaphorical" mapping in human infants. *Child Development*, **52**, 728–31.

Ward, J. P., Yehle, A. L., and Doerflein, R. S. (1970). Cross-modal transfer of a specific discrimination in the bushbaby (*Galago senegalensis*). *Journal of Comparative and Physiological Psychology*, **73**, 74–7.

Warren, J. M. (1965*a*). Primate learning in comparative perspective. In *Behavior of nonhuman primates* Vol. 1, (ed. A. M. Schrier, H. F. Harlow, and F. Stollnitz), pp. 249–82. Academic Press, New York.

Warren, J. M. (1965*b*). The comparative psychology of learning. *Annual Review of Psychology*, **16**, 95–118.

Warren, J. M. (1973). Learning in vertebrates. In *Comparative psychology: a modern survey* (ed. D. A. Dewsbury and D. A. Rethlingshafer), pp. 471–509. McGraw-Hill, New York.

Warren, J. M. (1974). Possibly unique characteristics of learning by primates. *Journal of Human Evolution*, **3**, 445–54.

Wasserman, E. A. (1984). Animal intelligence: understanding the minds of animals through their behavioral "ambassadors". In *Animal cognition* (ed. H. L. Roitblat, T. G. Bever, and H. S. Terrace), pp. 45–60. Erlbaum, Hillsdale, NJ.

Wasserman, E. A. (1993). Comparative cognition: beginning the second century of the study of animal intelligence. *Psychological Bulletin*, **113**, 211–28.

Weinstein, B. (1941). Matching-from-sample by rhesus monkeys and by children. *Journal of Comparative Psychology*, **31**, 195–213.

Weiskrantz, L. (1985). Categorization, cleverness and consciousness. In *Animal intelligence* (ed. L. Weiskrantz), pp. 3–19. Clarendon Press, Oxford.

Weiskrantz, L. and Cowey, A. (1975). Cross-modal matching in the rhesus monkey using a single pair of stimuli. *Neuropsychologia*, **13**, 257–61.

Wellman, H. M. (1990). *The child's theory of mind*. Bradford Books, MIT Press, Cambridge, Mass.

Wellman, H. M. (1991). From desires to beliefs: acquisition of a theory of mind. In *Natural theories of mind: evolution, development and simulation of everyday mindreading* (ed. A. Whiten), pp. 19–38. Blackwell, Oxford.

Wellman, H. and Wooley, J. D. (1990). From simple desires to ordinary beliefs: the early development of everyday psychology. *Cognition*, **35**, 245–75.

Whiten, A. (ed.) (1991). *Natural theories of mind: evolution, development and simulation of everyday mindreading*. Blackwell, Oxford.

Whiten, A. and Byrne, R. W. (1988). Tactical deception in primates. *Behavioural and Brain Sciences*, **11**, 233–73.

Whiten, A. and Byrne, R. W. (1989). Machiavellian monkeys: cognitive evolution and the social world of primates. In *Cognition and social worlds* (ed. A. R. H. Gellaty, D. R. Rogers, and J. A. Sloboda). Oxford University Press. (Cited by Whiten and Byrne 1991.)

Whiten, A. and Byrne, R. W. (1991). The emergence of metarepresentation in human ontogeny and primate phylogeny. In *Natural theories of mind: evolution, development and simulation of everyday mindreading* (ed. A. Whiten), pp. 267–82. Blackwell, Oxford.

Whiten, A. and Perner, J. (1991). Fundamental issues in the multidisciplinary study of mindreading. In *Natural theories of mind: evolution, development and simulation of everyday mindreading* (ed. A. Whiten), pp. 1–18. Blackwell, Oxford.

Wickelgren, W. A. and Norman, D. A. (1966). Strength models and serial position in short-term recognition memory. *Journal of Mathematical Psychology*, **3**, 316–47.

Wilkerson, B. J. and Rumbaugh, D. M. (1979). Learning and intelligence in prosimians. In *The study of prosimian behaviour* (ed. G. E. Dayle and R. D. Martin), pp. 207–46. Academic Press, New York.

Wilson, B., Mackintosh, N. J., and Boakes, R. A. (1985a). Matching and oddity learning in the pigeon: transfer effects and the absence of relational learning. *Quarterly Journal of Experimental Psychology*, **37B**, 295–311.

Wilson, B., Mackintosh, N. J., and Boakes, R. A. (1985b). Transfer of relational rules in matching and oddity learning by pigeons and corvids. *Quarterly Journal of Experimental Psychology*, **37B**, 313–32.

Wilson, M. (1964). Further analysis of intersensory facilitation of learning set by monkeys. *Perceptual and Motor Skills*, **18**, 917–20.

Wilson, M. (1966). Strategies and cross-modal transfer in monkeys. *Psychonomic Science*, **4**, 321–2.

Wilson, W. A. and Zieler, S. (1976). Some tests of intermodality transfer of intensity in squirrel monkeys. *Neuropsychologia*, **14**, 237–41.

Wilson, W. A., Oscar, M., and Bitterman, M. E. (1964). Visual probability learning in the monkey. *Psychonomic Science*, **1**, 71–2.

Wimmer, H. and Perner, J. (1983). Beliefs about beliefs: representation and constraining function of wrong beliefs in young children's understanding of deception. *Cognition*, **13**, 1103–28.

Wise, K. L., Wise, L. A., and Zimmerman, R. R. (1974). Piagetian object permanence in the infant rhesus monkey. *Developmental Psychology*, **10**, 429–37.

Wood, S., Moriarty, K. M., Gardiner, B. T., and Gardiner, R. A. (1980). Object permanence in child and chimpanzee. *Animal Learning and Behavior*, **8**, 3–9.

Woodruff, G., Premack, D., and Kennel, K. (1978). Conservation of liquid and solid quantity. *Science*, **202**, 991–4.

Wright, A. A., Santiago, H. C., Sands, S. F., and Urcuioli, P. J. (1984). Pigeon and monkey serial probe recognition: acquisition, strategies and serial position effects. In *Animal cognition* (ed. H. L. Roitblat, T. G. Bever, and H. S. Terrace), pp. 353–73. Erlbaum, Hillsdale, NJ.

Wright, A. A., Shyan, M. R., and Jitsumori, M. (1990). Auditory same/different concept learning by monkeys. *Animal Learning and Behavior*, **18**, 287–94.

Yerkes, R. M. (1916). The mental life of monkeys and apes: a study of ideational behavior. *Behavior Monographs*, **3**, No. 1.

Yerkes, R. and Yerkes, A. (1929). *The Great Apes*. Yale University Press, New Haven.

Zentall, T. R. (1973). Memory in the pigeon: retroactive inhibition in a delayed matching task. *Bulletin of the Psychonomic Society*, **1**, 126–8.

Zentall, T. and Hogan, D. (1974). Abstract concept learning in the pigeon. *Journal of Experimental Psychology*, **102**, 393–8.

Zentall, T. R. and Hogan, D. E. (1976). Pigeons can learn identity or difference, or both. *Science*, **191**, 408–9.

Zentall, T. R. and Hogan, D. E. (1978). Same/different concept learning in the pigeon: the effect of negative instances and prior adaptation to the transfer stimuli. *Journal of the Experimental Analysis of Behavior*, **30**, 177–86.

Zentall, T. R., Edwards, C. A., Moore, B. S., and Hogan, D. E. (1981). Identity: the basis for matching and oddity learning in pigeons. *Journal of Experimental Psychology: Animal Behaviour Processes*, **7**, 70–86.

Zentall, T. R., Hogan, D. E., and Edwards, C. A. (1984). Cognitive factors in conditional learning by pigeons. In *Animal cognition* (ed. H. L. Roitblat, T. G. Bever, and H. S. Terrace), pp. 389–405. Erlbaum, Hillsdale, NJ.

Zerbolio, D. J. (1985). Categorical color coding in goldfish. *Animal Learning and Behavior*, **13**, 269–73.

Zerbolio, D. J. and Royalty, J. L. (1983). Matching and oddity condition discrimination in the goldfish as avoidance responses: evidence for conceptual avoidance learning. *Animal Learning and Behavior*, **11**, 341–8.

23
Animal language and cognition projects

Carolyn A. Ristau

Abstract

Interpreting the data on the cognitive and linguistic abilities of non-human species needs to be done with care. The results of particular studies need to be understood in the light of (1) the ontogeny of the ability in question and the specific training procedures involved in demonstrating it; (2) the settings within which abilities are observed—laboratory versus field; (3) a precise description of and the limits of the abilities claimed to have been demonstrated; (4) the number of individual animals that are claimed to show the abilities in question; and (5) possible problems in experimental designs. Given all these constraints, summary statements must be somewhat provisional.

Some of the Great Apes, in some situations, have achieved the use of rudimentary symbols. This ability can be used to support symbolic forms of communication, especially requests, between apes and their human 'carers'. This symbolic communication ability does not show the multiplicity of functions that humans employ. Particularly in the past, the training methods used to establish these abilities did not bear much resemblance to the procedures by which human children develop language skills; more recent methods do (Savage-Rumbaugh *et al.* 1993; Boysen 1993*b*). There is little evidence that most apes use grammar in their communications, though one bonobo has attained at least a rudimentary grammar. Furthermore, in some highly specific situations, apes, sea-lions (*Zalophus californianus*), and dolphins (*Tursiops truncatus*) are able to comprehend the order of lexical items, and some apes and pigeons can reproduce certain simple series reliably.

The relation between comprehension and the production of symbols in non-human species is unclear, though the two systems appear more independent of each other than in humans. With specific training to do so, highly 'language'-trained chimpanzees (*Pan troglodytes*) can transfer symbols learned in one mode to the other, and then generalize this ability to new symbols. The bonobo (*Pan paniscus*) seems able to make such transfers far more readily. Some enhanced quantitative and reasoning abilities can be demonstrated for 'language'-trained apes as compared to apes not so trained. In some cases these findings depend on the performance of one particularly apt ape; and it is not yet clear that it is only 'language' training that is responsible for the differences found: this research should be extended.

The demonstration of symbol use and category formation in non-primate species implies that it is not the unique organization of the primate brain, nor any special property of their social and physical environments alone, that is responsible for the possession of some symbolic ability. Furthermore, since the apes are our contemporaries and not our ancestors, the question remains open as to the evolutionary significance of those abilities for human abilities. It is clear, however, both that apes can acquire symbols and that humans have a far greater facility for acquiring and using symbolic systems than any other species.

23.1 Introduction

Are apes capable of language? If so, is man's unique status threatened? It did manage to remain intact when other species were found to make and use tools and to engage in co-operative hunting (Beck 1980). And presumably, humans will still remain unique whatever sort of language science and the press decide that the apes have or do not have. At least it is clear as we enter the fourth decade of modern ape-language studies that ape linguistic abilities, or whatever term we wish to use, are not the same as adult humans'; and, indeed, no scientist has ever made this claim. Thus let us dispense with the controversial aura and dispassionately attempt to understand the nature and limits of apes' possible linguistic and cognitive abilities as compared to those of humans and other species. And, so far as possible, let us examine the evolutionary implications. Which of the ape's abilities, if any, are likely to be precursors to humans' abilities? (See also earlier critical reviews of these issues in Ristau and Robbins 1979, 1982a, b.)

23.2 A brief history of the ape-language projects

Probably because humans communicate primarily by the spoken word, scientists first attempted to teach apes a vocal language. Near the beginning of the century, a female orang-utan learned to produce vocally 'papa', 'cup', and 'Th' after eleven months of instruction (Furness 1916). The home-raised chimpanzee Vicki was taught after six years of prodigious training to produce four raspy utterances, 'mama', 'papa', 'up', and 'cup' (Hayes 1951; Hayes and Hayes 1951). Another home-raised chimpanzee, Gua, despite intensive efforts, could not produce a single spoken word (Kellogg and Kellogg 1933). Using operant conditioning techniques modelled after those used successfully with autistic children, Laidler (1978) trained an infant male orang-utan to produce four sounds and to use them with presumed meanings of 'beverage in a mug', 'contact comfort', 'food', and 'more brushing' (i.e. grooming). The source of the difficulties in these approaches is controversial, and is presumed to lie either in the structure of the vocal apparatus, or in the peripheral or central nervous control of those structures, or in a lack of the necessary cortical association areas.

Yet apes use manual gestures in their natural communication systems, and show great manual dexterity. Perhaps, therefore, Yerkes reasoned some fifty years ago, apes could learn a manual sign language (Yerkes and Yerkes 1929). It was such a project that the Gardners undertook with one chimpanzee (Washoe) and then several others. The language they and subsequent researchers have used is a simplified or pidgin version of American Sign Language (ASL) (see Fig. 23.1). With it Washoe learned about 132 signs after about four years of training, and used them in interactions with humans Gardner *et al.* (1989). But can chimpanzees communicate to each other with hand signs? Whether they can and whether a chimpanzee mother can teach ASL to an infant chimpanzee are among the concerns of Fouts' work (Fouts *et al.* 1982). Other subjects for sign-language studies include gorillas (Patterson 1978a,b) and an orang-utan (Miles 1983); the gorillas have been exposed to spoken English as well. The chimpanzee Nim Chimpsky, named for the linguist, has likewise been taught hand-signing. Extensive use of video tapes has permitted detailed analysis of the experimenter-ape interactions (Terrace 1979a; Terrace *et al.* 1980).

To permit more precise data-collection (hand signs can sometimes be difficult to distinguish from each other), visual lexicons were adopted in some projects, namely coloured plastic shapes (Premack 1976) and coloured geometric figures (Rumbaugh 1977) (see Fig. 23.2). The latter were available on a computer keyboard. Such a device theoretically permitted the recording of all 'linguistic' interactions between chimpanzee and experimenter, as well as any chimpanzee monologues. Selected portions of the record have been analysed (Savage-Rumbaugh 1986). The Lana project enabled the chimpanzee to gain considerable control over her environment by sending lexigram commands to the computer. The ape could, for instance, gain access to food, drink, movie segments, a window opening, etc. Premack's study emphasized cognitive abilities of the apes, investigating such matters as the use of prepositions, same-different judgments, and metalinguistic capacities such as 'name of' and 'colour of'.

Public and scientific interest in these controversial projects was high. But the scientists', and even much of the public's, attitude reversed as a result of several factors. Initial media hype (and sometimes scientists' announcements) resulted in overblown claims. Then Terrace, also initially hopeful, published his findings about the paucity of Nim Chimpsky's language skills. These received wide public attention, as did assertions by the Sebeoks that the positive findings of the ape-language research were spurious, and due, in fact, to witting and unwitting experimenter bias and subtle cueing by experimenters—the 'Clever Hans' phenomenon (see the discussion later in this chapter).

More recent developments are simultaneously more basic and more speculative. Premack and colleagues have investigated the nature of the mental states of their primate subjects, such as intentions, beliefs, beliefs about beliefs, and deception, as well as reasoning and quantitative abilities (Premack and Woodruff 1978a, b; Woodruff and Premack 1979). The nature of the 'word' has been investigated by both Premack and Savage-Rumbaugh et al. (1978a,b). The ability of chimpanzees to interact via an artificial lexicon has been investigated in tool-use and exchange experiments in which chimpanzees obtain and share food through mutual co-operation. Most recently quantitative concepts have also been investigated by Matsuzawa (1985a), Boysen (1992, 1993a), Boysen and Berntson (1989), Boysen et al. (1993), and Rumbaugh et al. (1987). Comparative studies between the common chimpanzee (*Pan troglodytes*), heretofore the only chimpanzee species used as a subject in the ape-language and cognition studies, and the bonobo or pygmy chimpanzee (*Pan paniscus*) and a human infant have recently become of considerable interest, since the bonobo is apparently a far more apt student 'linguistically' than the common chimpanzee, and has abilities in some senses comparable to those of a two-year-old human. These abilities include comprehension of simple spoken English sentences.

The ape-language research has likewise had its impact on research with humans. Prior to the sign-language studies with chimpanzees, relatively little was known about the acquisition of human signing languages; but research in that area is now flourishing. Previously, the method of 'rich interpretation' of human children's vocal linguistic performance was widely used (Bloom 1973; Bloom et al. 1976), with little concern about the precise deductions to be made from the methods. Typically this caused little problem, for the human children did usually grow to be proficient language-users. When the same methods were applied to the 'linguistic' apes, the deficiencies in the theoretical underpinnings became very obvious.

Overall, the ape-language projects initially began with attempts at cross-fostered chimpanzees' being raised in human homes, often with human child 'peers'. The emphasis was on amassing vocabulary. Interest shifted to a concern with word order and the possible grammatical capabilities of the apes. When these appeared to be negligible, a concern with the meaning of the 'words' learned by apes became prominent, as did a renewed interest in their various cognitive and mental capacities, such as spatial maps, quantitative and measuring abilities, and notions of 'self'. Yet more recently, interest in the social interactive use of language has been emphasized (Greenfield and Savage-Rumbaugh 1990; Savage-Rumbaugh et al. 1993). Simultaneously, comparable 'linguistic' abilities were being sought in other species—sometimes so termed, and sometimes carefully framed as examples of various classical learning skills. The species investigated include bottle-nosed dolphins (Herman et al. 1984; Herman and Forestell 1985), sea lions (Schusterman and Krieger 1984), and a parrot (Pepperberg 1981). A tabular overview of recent ape language projects and non-primate language projects is given in Tables 23.1–23.3 at the end of this chapter (pp. 672–81).

23.3 Provisos when interpreting the results of the ape-language and cognition studies

The above brief history alludes to some of the controversies this field of research has generated. Because of them, I think it is worth listing here, before considering the research in more detail, a number of points that serve to contextualize the studies. First, a most critical issue is the kinds of evidence needed to support claims made. Such evidence should include information about the *ontogeny* of the behaviour studied. For example, chimpanzees are reported to share food with each other (Savage-Rumbaugh 1986); but this occurs only after extensive training to do so in a specific structured situation. The point is that the nature of the behaviour, e.g. sharing, is important; but, perhaps *as* important, is how that behaviour developed.

Second, the kinds of evidence typically used to support claims derive from the laboratory or captive situation. But most interesting too would be information about social interactions obtained from natural groups in the field or from semi-natural social groups in large enclosures. Such information aids in understanding how an ability exhibited in the lab is used in the apes' natural life. Alternatively, the ability may apparently not be used naturally. It is possible that we, by Socratic methods, can assist in the emergence of abilities which typically are not part of the animal's repertoire. The reverse is also very often true. Primates, in their social behaviour, can exhibit far more advanced abilities than we are able to demonstrate clearly in our restricted laboratory situations; the chimpanzee's complex deceptive acts are such an example (de Waal 1986; see also the review of 'anecdotes' from fieldwork compiled by Whiten and Byrne (1988), and their discussion of the status of such observations in establishing the status of primate capacities).

Third, the relevant abilities of other species such as monkeys and humans are of particular interest. (Recall that some natural animal communication systems apparently convey semantic information. For example, the vocalizations made by vervet monkeys

(Seyfarth et al. 1980) and by rhesus monkeys (Gouzoules et al. 1984, 1985) are, respectively, most reasonably interpreted as providing information about type of predator on the one hand and kinship and status of aggressor on the other—see Jolly, this volume, Chapter 6). If species of very different lineage, such as birds, can be shown to exhibit some of the same advanced abilities as apes, then we cannot argue that there is something unique about the neural organization of the apes' brains or about the social circumstances of the ape that has given rise to these abilities. At the least, we shall have to argue that various brains and circumstances can produce similar capacities.

Fourth, interspecific comparisons highlight the need to investigate the *limits* and the extent of the linguistic and cognitive abilities claimed for apes. It is important to *characterize* the nature of the ability, rather than to 'name' it as a human ability, such as 'reading' or 'writing'. We must also recognize that there are likely to be Great Ape 'geniuses', or at least very talented individuals, just as there are human ones.

Last, as a result of differing individual talents, but also, even more importantly, of the specific nature of the training procedures, the ability exhibited by one chimpanzee in one training situation is *not* necessarily the same as the *seemingly* similar ability of another chimpanzee undergoing another training technique. In short, if, after training, the chimpanzees Sherman and Austin exhibit some characteristics of symbolic use of a lexical item, this does not imply that all chimpanzees, whatever their training history, have a similar ability. Likewise, demonstrating Sherman and Austin's rudimentary symbolic ability, for example in classifying particular exemplars (apple, M&M candies) as belonging to the category 'food', does not imply that the chimpanzees also have the ability to so classify members of other categories (for example, animal) for which they have not been specifically trained.

23.4 Theoretical issues

23.4.1 *Definition of language: the meaning of the word and the utterance*

In addition to the above points, this field of research brings with it a host of conceptual questions and issues for which there are no clear empirical answers. What, for example, is a 'word'?. There is little agreement in the ontogenetic literature as to when a human child might be said to actually be using 'words' *as* words (see Johnson *et al.*, this volume, Chapter 24, Section 24.5.4).

One characteristic of the human use of words is humans' ability to refer to events remote in time and space. Some researchers, however, argue about the degree of requisite remoteness, given young children's difficulty in comprehending how long it is to Christmas, or how much longer a journey will take before they 'arrive at Granny's'. On the contrary, one can note that 'Granny' is not present here and now, so any use of her name constitutes displacement, a view closer, I think, to the essential property of displacement. Such definitions encounter similar difficulties in the comparative domain.

Further, the *meaning* of the utterance is of great importance. Do chimps mean what they say? Do they have intentions? A (non-intentional) behaviouristic interpretation of chimpanzee actions in terms of stimulus–response sequences has not proved adequate as an explanation. Can the chimps express intentions and understand those of others; i.e. can they attribute intentions? These issues are, again, not settled with regard to human ontogeny, not just because of a lack of empirical studies, but also because of their philosophical complexities.

No one has, as yet, an adequate definition of language. At one extreme are the views of the linguist Noam Chomsky: for example, he writes,

To determine whether music, or mathematics, or the communication system of bees, or the system of ape calls, is a *language*, we must first be told what it is to count as a *language*. If by *language* is meant *human language* the answer will be trivially negative in all these cases. If by *language* we mean *symbolic system*, or *system of communication*, then all these examples will be languages, as will numerous other systems—e.g. style of walking which is in some respects a conventional culturally determined system used to communicate attitude, etc. If something else is intended, it must be clarified before inquiry can begin (Chomsky 1980, p. 430).

To a linguist of this persuasion, human language is essentially *human*. The genetically-based human linguistic system began developing long after the evolutionary separation of the ancestors of man and apes. According to this viewpoint, it is therefore foolishness to look for linguistic continuity between human and ape. Chomsky further considers that many processes, such as intentionality, informativeness, symbolic representation, word order, and various iterative procedures (as discussed in Granier-Deferr and Kodratoff 1986), often claimed to be definitional for language, cannot be, for they are shared by other communication systems or general cognitive capacities.

In Chomsky's view, a necessary characteristic of language is the existence of *grammar*. The grammar, or system of rules, is biologically based in humans and unique to them: it is a human universal. Through grammar an infinity of expressions can be generated.

Principles for constructing a hierarchy of phrases are a crucial feature of grammar, recursive embedding being one such rule. Except for a small segment of Premack's work with Sarah, the ape-language projects have not dealt with phrase hierarchies, for most of the ape's utterances are simply too short.

From the Chomskyans' viewpoint, could any aspects of language be shared by apes and humans? The first stage of the two-or three-stage model of language acquisition held by most linguists could be similar, in that it involves general cognitive development. Even young children to about age five, Chomsky asserts, use a primitive communication system completely independent developmentally and functionally from the adult human language system.

To some, all this emphasis on grammatical structure and cognition misses the recognition that language is an essential part of human life. To describe language merely in terms of structural rules is to lose its essence. This has been termed 'languages as forms of life' (Wittgenstein 1953).

The most extreme view on the other end of this spectrum of positions is probably represented by B. F. Skinner's (1957) analysis of human language as verbal behaviour that is reinforced. Just as predictions can be made for non-verbal responses, so they could for verbal responses or operants. There are several kinds of verbal operants essential to verbal development: the tact, the mand, and echoic response.[1] The community (parents in particular) reinforces the verbal operants until the child has a large repertoire. Eventually, the child begins to build up sentences which are composed of intraverbal associations, i.e. combinations predictable by linear associations as opposed to hierarchical relations.

The Chomskyan approach is opposed both to Skinner's and to the delineation of language as a list of features such as that offered by Hockett (1958; Hockett and Altmann 1968; see also Deuchar, this volume, Chapter 20). Such lists are considered very weak approaches by Chomskyan linguists, for they claim that language is an integrated system, deriving from genetically based modules or capacities that are intertwined with each other (see Johnson *et al.*, this volume, Chapter 24). The modules promote and constrain learning; and are only possessed by humans. Therefore, the argument runs, animals cannot possibly learn anything which can be construed as language.

However, each of Hockett's design features, with the exception of reflectiveness, is considered by some researchers to be a characteristic of at least some system of animal communication, and also of the ape artificial languages. In fact, the application would be yet broader if one took the position that such features should not be applied in an all-or-none fashion to these communication systems, but that one should rather search for precursors and levels of the features.

None of these approaches would satisfy some linguists, for demonstrating that animals have the capacity for any list of enumerated features will merely reveal superficial similarities to language; the underlying integration will be missing (Hoban 1986; Chomsky 1959, 1980). Yet this seems to leave us in an impossible situation. For, by this argument, no matter how thoroughly we characterize language, no animal will ever possess it, because there will always be some undisclosed integrative ability that the animal does not, cannot possess.

23.4.2 *Definitions of language and the nature of American Sign Language (ASL)*

From the Chomskyan viewpoint it is absurd to search for linguistic continuity between humans and other animals, because human language is uniquely human, having developed (he asserts) after human and ape evolutionary lines diverged. Thus, the nature of ASL becomes of particular import (1) with respect to evidence that it qualifies as a *bona fide* language and (2) in terms of the questions whether the signing systems taught the apes were ASL, a pidgin language, or not a language at all. For, in order to evaluate the possible linguistic capacities of the apes, it is necessary to understand thoroughly the characteristics of the 'language' they were taught in the various projects.

ASL is considered an actual language with the grammatical complexity and expressive capability of any spoken language, although it took the analyses by Stokoe (1960) and others to achieve such status (cf. Deuchar, Chapter 20, and Hewes, Chapter 21, this volume). Briefly, ASL has syntax, but much of this is incorporated into modulations of the signs, and word order is of far less importance than in a spoken language such as English for conveying grammatical information. Thus a change in the movement patterning of a sign corresponds to a change in its meaning or grammatical function. At least the following parameters affect meaning (Stokoe 1960): hand configuration, place of articulation, movement, and possibly orientation (this latter being proposed by Klima and Bellugi 1979). This is illustrated by the example of the ASL word 'sick' (Bellugi and Klima 1979; also described in Hoban 1986). By modulating only the movement parameter, the sign may be changed to indicate meanings such as when the state began, how long it lasted, and the frequency of its recurrence and strength—i.e. 'intensely ill', 'repeatedly sick', etc.

Were the signing apes taught ASL? For the most part they were not, because the instructors were typically novices to ASL, though every study had one or more fluent signers. However, in the later work of the Gardners with four young chimpanzees, the researchers were careful to hire only trainers experienced in ASL. For most of the work of the Gardners, Fouts, Patterson, and Terrace, the apes can best be described as being exposed to 'pidgin signed English', that is, a very simplified version of ASL which made little use of its complex inflectional characteristics, thereby losing most of the grammatical capacity of ASL. Typically an English or English-like word order was applied to the signs, so that word order once again became important. This was particularly necessary in those studies, especially Patterson's, when the trainers spoke English words along with the signing. But then neither was exact word order always recorded.

Thus, from the point of view of some linguists, the apes were typically never exposed to a language *per se*. Therefore, it could be argued, one cannot hope to discover grammar in the ape's productions, for it was not there in the system being taught. Further, since such linguists' definition of language relies on a complex grammatical structure, the signing apes cannot be described as having a language. But such linguists seem to want it both ways. ASL is described as being too rich and too complex for us to dare to expect apes to achieve it, while at the same time pidgin ASL is considered too impoverished to count as a language.

However, the use of English word order by 'teachers' at least offers some hope for rule learning in word use by the signing apes so taught. For the apes exposed to the complex inflectional grammar of ASL, as the later Gardner chimps were, there is interest in their possible modulation abilities. These capacities are being analysed through the use of diary records and videotapes). For example, van Cantfort and Rimpau (1982), students of the Gardners, describe how the chimpanzees contrast the use of 'that' to indicate statements or questions. In the latter usage, the sign is held longer and eye contact is made and maintained.

Detailed analysis by Rimpau *et al.* (1989) shows several types of modulations in sign usage, each type used by each of the five chimpanzees studied. Among the most interesting were the ape's molding the human's hands into the sign just made by the ape (for example 'Out') after the human had ignored the ape's own sign to go out. Making the sign on or near the referent was another modulation which is of especial interest because objects, not just signs or words, are being incorporated into the communicative act to convey meaning: for example 'Open' made near the door or container to be opened. One chimp, Dar, when the human Tony stopped brushing him, used the sign 'brush' three times sequentially: 'brush' on Dar's shoulder, again as a citation or simple sign, and again on the human's arm—all of which seems appropriate to the probable interpretation that the chimp wanted Tony to continue brushing. These are, plausibly, simple expansions of meaning, similar to young children's, though not as extensive.

In sum, the reader must be aware that the signing apes have, by and large, not been exposed to ASL, but appear to have been exposed to some word-order information, thereby permitting the potential acquisition of at least simple rules. Whether a linguist would accept these as true grammatical rules is another matter of discussion.

23.4.3 *Production versus comprehension*

Human children learning language use the productive and comprehensive modes interchangeably, although comprehension typically precedes production (although these processes appear to be more independent of each other in at least some language-retarded humans (Romski *et al.* 1984, 1985, 1994)). Possession of a language, then, could be claimed to require ability in both modalities. This claim creates immediate difficulties in the comparative literature, for the relation between production and comprehension appears to vary across species.

Some mammals, for example dolphins (Herman 1987; Herman *et al.* 1984; Langbauer 1982) and sea lions (Schusterman 1988; Schusterman and Gisiner 1988; Schusterman and Krieger 1984, 1986), exhibit good receptive skills (i.e. comprehension), but poor productive abilities in an artificial language. A dolphin taught an artificial whistle language can comprehend whistle commands, but can *use* the whistle in a productive mode only in a very limited fashion (five 'words' are used).[2]

Related work has been done with other species (baboons, monkeys, and pigeons) and with children; the procedures are described as investigations of the symmetry of conditional relations. Young children were given a conditional matching task in which they were required to match correctly each of a pair of 'comparison' stimuli to each of a pair of 'sample' stimuli. The children could then do the reverse task, that is, match the previous 'sample' to the previous 'comparison'. Unlike the children, who exhibited 'backward association' to the point of symmetry (Sidman *et al.* 1982), baboons, monkeys, and pigeons (Sidman *et al.* 1982; D'Amato *et al.* 1985) showed at best only weak and transient backward associations (see also Lock and Colombo, this volume, Chapter 22).

There is similar evidence for the separation of these systems in *Pan troglodytes* deriving from work with Lana (Rumbaugh 1977; Rumbaugh and Gill 1977), and experiments with two chimpanzees studied by Premack (1986). Lana, using computer-based lexigrams of various geometric line figures and colours, had used the expression, 'Please machine give M&M [a small piece of chocolate candy]' hundreds of times, after which she had received M&Ms from the dispenser. But when given a new task, namely to push the lexigram for M&M when the experimenter asked (via a string of lexigrams), 'What name of this?', an additional 1600 trials were needed for her to be able to label two items 'banana' and 'M&M'. In short, she knew *how* to obtain a reward by using a string of lexigrams, but apparently did not understand that the lexigrams were labels.

An obvious criticism of the interpretation that these are two different systems is that when asked 'What name of this?' Lana might simply not have understood the nature of the task; she had, in effect, been overtrained on a different task. But lack of comprehending the task turns out not to be the complete explanation of the difficulties, as is indicated by Premack's work testing transfer between production and comprehension. Each of two chimpanzees was taught ten or more words in each mode, i.e. one set of words was learned productively and a different set was learned receptively. Thus each chimpanzee was familiar with the demands of each kind of experimental situation. Then transfer tests were conducted; i.e. words learned in the productive mode were tested in the receptive mode, and vice versa. The apes failed miserably, performing at about chance level over 200 trials. When the apes finally did learn, success at switching modalities for only five or six words was sufficient to obtain total transfer of subsequent items (Premack 1986, pp. 62–6). Production and comprehension *can* be interwoven as systems in the common chimpanzee, although they seem not to be that way initially.

The pygmy chimpanzee or bonobo (*Pan paniscus*), appears to present a different case. While an adult female bonobo was being taught lexigrams (not very successfully) over a period of about two years, her infant Kanzi usually lingered near her or was taken to the side and played with by humans. When the mother was removed for breeding purposes, the infant chimpanzee revealed at the lexigram keyboard that he knew a considerable number of lexigrams, both productively and receptively, including those his mother had never mastered. Within the first six months after his mother's absence, Kanzi was using about half the 60 lexigrams assigned to the keyboard appropriately and spontaneously (Savage-Rumbaugh 1984, 1986; Savage-Rumbaugh *et al.* 1986). During continuing training of Kanzi and another bonobo, both are continuing to exhibit observational learning and spontaneous use of lexigrams in both receptive and productive modes without specific training to do so.

There are a number of issues here that will be developed subsequently in this chapter. What I want to emphasize from the material already noted is the care needed in drawing conclusions from comparative work. It is not sufficient to ask a question such as 'Can chimpanzees learn language?', of the kind we often ask undergraduate students to write essays on. We need to take account of which particular animal we are dealing with, the way it has been taught, and how its performance has been assessed. We need also take care in the way we conceptually phrase the questions we seek to answer from the empirical studies. For example, does the work just outlined indicate there *are* separate production and comprehension systems for language, and that they have different relationships to each other in different species? Or is it that their apparent separation depends on whether the particular animal studied has, or has not, apprehended the notion of 'naming' (cf. Terrace, 1985*a*). Either there is a different relationship between these two modalities in the two species of chimps, or pygmy chimps are 'smarter' than standard chimps; or they have been taught differently; or . . .

23.5 The findings with regard to symbolic abilities

23.5.1 *Linguistic-like capacities*

23.5.1.1 *Grammar*

A capacity for generative grammar is essential for any language. It permits the understanding and production of novel statements, i.e. it is an 'open' system. For human languages, the structure-dependent rules are hierarchical as opposed to linear, and include recursive rules (see Deuchar, this volume, Chapter 20). Humans possess a tacit knowledge of linguistic rules: an individual's use of language, and understanding of that of others, must conform to those rules, while he or she need not be able to verbalize or specify precisely what rules are being used. Thus, from the point of view of researchers interested in probing the continuity of cognitive abilities in humans and other animals, the apes' following of any rule, however simple, is of consequence.[3]

Most ape utterances are too short or too highly repetitive for any ability to use the complex rule structures of human languages to be directly detected.[4] But the use of simpler rules has been claimed. McNeill (1974) has analysed Washoe's signed *productions* and found that a recurring order, namely 'addressee–addresser', represents the predominant ordering of her signs. However, in view of the fact that the observers in this project did not stress accurate signing order when they recorded the chimpanzee's productions, this rule may more readily reflect the human observer's rule than the chimpanzee's. If there is a body of data for which signing order is correctly preserved, it would be most interesting to determine if this and a few other rules McNeill (1974) abstracted do apply.

By contrast, Terrace and colleagues (1979) have analysed a large sample of Nim's utterances over an eighteen-month period—about 19 000 utterances. They found no recurring use of word order except in two sign utterances. Again for lack of sufficient numbers of long utterances, the highest strings analysed were composed of only three or four signs. Nim did produce some longer strings, but they were highly repetitious, as noted in this sixteen-sign utterance produced by Nim when his mean length of utterance (MLU) was 1.6, 'Give orange me give eat orange me eat orange give me eat orange give me you.' Thompson and Church (1980) analysed a corpus of about 14 000 of Lana's contributions. They concluded that these could be accounted for without recourse to grammatical processes, by conditional discrimination and paired-associate learning. Greenfield and Savage-Rumbaugh (1984, 206) conclude for the multilexigram symbols of Sherman and Austin that 'no evidence was found that combinations are, in any way, related to syntactic rules'. If the chimpanzees are not using English or pidgin signed English word order in their signing, what is it they are doing? Certainly they are repeating words of considerable interest to them, as the previous quotation from Nim shows. This is not unlike children who excitedly shout 'cookie, cookie, me cookie, cookie, cookie', or the chimpanzees who make repeated pant-hoots and other excited vocalizations. At the least, we are seeing emphasis through repeated use of certain signs.

In other domains of chimpanzee performance, however, a sensitivity to order is apparent. Premack set Sarah, and two other language-trained apes, the task of putting in proper sequence objects which represented a causal relationship. For example, the act of cutting was represented by the sequential ordering 'apple, knife, slices of apple'. Sarah could arrange these and novel objects appropriately, and all the language-trained apes tested could substitute the missing item when a space was left in the sequence.[5] Such orders are, though, not arbitrary. They represent a real order of objects and events. To be capable of any sort of rudimentary grammar, in which events in the real world are represented by a conjunction of arbitrary symbols, it would seem that a very simple communication system would require only a sensitivity to some natural order. This natural order would then have to be assigned to symbols. Anything slightly more complex requires sensitivity to, and an ability to learn, an *arbitrary* order.

Indications of a sensitivity to the role of the arbitrary ordering of materials comes from subjects in several of the animal-language research studies, for example dolphins (Herman), sea lions (Schusterman), and chimpanzees (Premack) and the Lana project (Rumbaugh) (see, Tables 23.2 and 23.3). In both these chimpanzee projects, correct order was required of the subjects in order to complete a problem successfully (Premack's research) or receive rewards (Lana project). Neither used hand signs for communication. Both Lana and Premack's chimpanzees could also *produce* correct order of the lexical items.[6]

Interpretation of the sensitivity of sea lions and dolphins to the order information has produced an active debate between Schusterman and Herman, the respective principal investigators. Using gestural signs in procedures modelled after those of Herman, and achieving results for sea lions similar in a number of respects to the dolphins, Schusterman claims that the sea lions' accomplishments can be described in traditional learning and operant conditioning terms, such as 'paired-associate learning' and 'conditional discriminations about the relationships between sequentially presented signs' (Schusterman and Gisiner 1989, p. 3). Herman uses linguistic terminology, i.e. 'words' and 'grammatical rules', to describe the dolphins' capabilities, and considers that the dolphin has furthermore achieved 'a level of representation lacking or weak for the sea lion' (Herman 1989, p. 19). This touches on a pervasive debate in animal language research (see for example Petito and Seidenberg 1987), and highlights the difficulties involved both in defining and establishing criteria for 'grammar', 'words', and 'representations' and in the interpretation of behaviour.

Recently, a study of a bonobo chimpanzee Kanzi has provided much stronger evidence of understanding a simple grammar, generalized over a wide variety of novel sentences and comparing favourably to the abilities of a two-year-old human child who was tested on the same *spoken* English sentences (Savage-Rumbaugh *et al.* 1993; Linden 1993). Note, before any further description, this is *not* to say that the chimpanzee's linguistic abilities *are* the *same* as the child's. Their brains are very different; the child quickly outpaces the chimpanzee. But a fairly rich, socially interactive experience was sufficient for both to achieve a similar level of ability... at least by these particular tests. It is always possible that other measures or observations would reveal interesting differences that lead to the later very large difference between the two species. Perhaps, too, the experience critically necessary is different; a human child might be able to 'make do' with less. Nevertheless, Kanzi's abilities, and, in particular, the careful testing used to assess them, reveal, in a very convincing manner, an ability we are hard put to term anything but a simple grammar. It is not that previous studies did not claim grammatical abilities about equivalent to those of a human 1½- to 2½-year-old for some apes (for example the Gardners' common chimpanzee Washoe, Patterson's gorilla Koko, and Lyn Miles's orang-utan); but in this later research, (1) the tests can withstand any methodological criticisms to date; (2) they were identical for child and chimpanzee; (3) the chimpanzee learned through observation and engagement in social routines and chores, that is, in manners similar to a human child's; and (4) the sentences were spoken English, thus being simultaneously a more dramatic show of abilities and making it easier for human English-speakers to grasp the chimpanzee's achievements. Furthermore, a publicly available video, 'Can Chimps Speak?' provides convincing examples of some of Kanzi's interactions and response to English sentences (Ikeo 1994).

In selecting routines in which Kanzi was exposed to language, Sue Savage-Rumbaugh chose activities to approximate to those engaged in by the chimp in its natural habitat. Choosing and carrying items needed on an outing, meandering along paths through the 55-acre Georgia woods, walking to get food in various locations, preparing/eating food inside in a kitchen or at an outdoor campfire, grooming and otherwise playing with each other... these are among the ways the staff interacted with Kanzi. Spoken English was always used; often lexigrams on a portable board as well. Using what are best (but admittedly vaguely) described as normal human parenting and teaching skills, the staff encouraged, hugged, and occasionally chided Kanzi (and other younger chimps also under study); the mother of the human child was likewise a care-taker for Kanzi.

In blind testing of Kanzi's linguistic skills, there were carefully controlled procedures. A one-way mirror prevented Kanzi and the child Alia from seeing the experimenter giving them directions, while the observers/human interactants wore earphones with loud music to prevent their hearing the requests. Seven different sentence types were tested: putting an object in/on another; giving or showing object(s) to a live creature or a toy; performing an action on another creature or toy, sometimes using an object in the process (for example 'knife the sweet potato'); taking objects to or from places (for example 'get the carrot that's in the microwave'), and having toys interact with each other. Some sentences announced information—where a goodie/object was located, or that some tickle or chase game was about to begin with them—these were designed to provoke a reaction from Alia or Kanzi. There were as well several more complex sentences not so readily categorized, including, for example, multiple actions.

Kanzi's performance was 74 per cent correct on blind test trials, and Alia's, 65 per cent; the total for all trials was about the same. A particularly interesting subset consisted of 'those sentences in which the order of the key words remained constant but the nature of the appropriate response did not ("Take the rock outdoors" / "Go get the rock that's outdoors") ' (Savage-Rumbaugh et al. 1993, p. 92). On these Kanzi's performance was 79 per cent correct, and 57 per cent of the pairs were correct (8/14 pairs); while Alia's performance was 67 per cent correct, with 38 per cent (5/13 pairs) of pairs correct. Obviously the child's scores depend on the age of testing; Alia was between 1.5 and 2.0 years old; Kanzi was 8.0 years old.

Granted these sentences do not show recursion; neither is the passive voice or the subjunctive mood used. It can be noted, as various critiques have concerning Kanzi's lexigram usage (for example Seidenberg and Petito 1987; Zuckerman 1991; Cenami Spada 1994) that the linguistic order information is implicitly revealed in the nature of the performance necessary to service the instrumental requests. This topic is better discussed in the section on meaning. Suffice it to say that such was necessary in the testing context, because a behavioural indication of Kanzi's and Alia's understanding was needed. Perhaps future tests could use a child and a chimpanzee indicating 'True' and 'False' to statements, and answering 'Where?' questions... though this is easy for a commentator to say, and much work for an experimenter and a subject to accomplish.

Nevertheless, we can state that Kanzi apprehends order information, and does so in a wide array of contexts. He understands simple English, and understands it more quickly than he learns lexigrams. He can't speak it, but then he can't produce the consonant sounds that we are capable of making, and that are critical to speaking any human language. (The experimenters' attempts to create a 'compromise' spoken language that could be produced by both chimpanzee and human, i.e. one consisting only of vowels, were incomprehensible to the humans.) He communicates spontaneously and in many varied contexts... but that leads us to other concerns: meaning, which we discuss next.

In further investigations of chimpanzee understanding of causality and associated categories, Premack attempted to teach Sarah to place markers for 'agent of action', 'object of action', and 'instrument of action' on a video screen where an action was depicted. On simple tests, Sarah was very good. However, her performance deteriorated badly when the test was made more complex, so that there were two potential instruments of action and two possible objects, with only one action actually occurring. She was most successful with the agent marker (85 per cent correct overall). When pressed—i.e. when the procedure was changed so as to correct Sarah's errors—she simply stopped playing the game. She put all the markers in a blank portion of the screen and did nothing more (Premack 1986, pp. 128–30).[7]

In short, sensitivity to simple, arbitrary ordering rules has been established. None of the non-human species studied, however, simply 'pick up' such rules spontaneously (but see note 4). Rather it takes very precise, repetitious, highly-constrained training to have the subjects, be they ape, dolphin, sea lion, or pigeon, become sensitive to such order information. Except for limited skills in some apes (for example Lana from Rumbaugh's research and some of the apes in Premack's), there is also no evidence for *use* of a grammar *per se*, but simply for *receptive* capacity in a highly constrained learning paradigm.[8]

23.5.1.2 *Meaning*

The 'meaning' of an utterance can refer both to the conceptual complexity of the individual items in the lexicon, for example the meaning of words, and to the meaning of the phrase or proposition. The animal-language researchers have concentrated their studies primarily upon word, not utterance, meaning, and therefore this discussion will focus on the word. Typically, the ape-language researchers, who are, for the most part, experimental psychologists by training, test for 'meaning' by so-called 'labelling' paradigms. In these designs, the ape must, when shown an object, produce or select from an array, the appropriate lexical item (productive capacity) and vice versa (receptive capacity). Such procedures are not far different from an associative learning paradigm, with the exception that exemplars in the ape-language research can be quite varied (for example many instances of 'ball' are possible). Premack and Savage-Rumbaugh have ventured furthest in this domain.

The meaning of an individual word involves reference both to the representation of the word (by which I do not mean a particular neurophysiological substrate, but rather, information the organism has concerning the word) and to whether the word functions as a linguistic referent. For example, in the various artificial animal languages we might ask, 'Are the object names understood by the animal as symbolic representations of the objects? That is, do they really function as words in natural language?' (Hoban 1986, p. 134). To be able to refer to an object, as opposed to its merely being a paired associate with some object or event, to be a symbol rather than merely a sign (see Sinha, this volume, Chapter 17), it is generally agreed that the user of a word must exhibit displacement: i.e. must be able to refer to objects or events remote in space and time. Recall that displacement is one of the design features of human language initially enumerated by Hockett (1958; Hockett and Altmann 1968; see also Deuchar, this volume Chapter 20). It is typically required as well that a word should be used in a variety of contexts, to further ascertain that the inferred meaning is not due to contextual cues. In addition, a symbol can not only refer, it can denote; and it can also convey mood.

It was only after emphasizing vocabulary-acquisition and possible grammatical use of lexical items that many researchers turned their full attention to the meaning of those items. Certain aspects of meaning have been investigated: (1) elucidating procedures which establish and do not establish reference (for example Savage-Rumbaugh 1986; Savage-Rumbaugh *et al.* 1978a, 1978b, 1980); (2) the nature of the internal image or representation (for example. Premack 1976, 1986); and (3) tests to reveal whether a particular word is functioning as a linguistic referent.

23.5.1.2.1 *Procedures to establish reference*

Savage-Rumbaugh, Rumbaugh, and their colleagues (Savage-Rumbaugh *et al.* 1978a, 1978b) have suggested that labelling paradigms and embedding a word in a string are two methods which, when used in the ape-language projects, do *not* lead to the use of symbols in a word-like manner. Training the meaning for an object's name through functional use of the object is said to be far more effective. *Labelling experiments* are, however, quite common. Typically the experimenter holds up an object or food and requires the ape (in the Savage-Rumbaugh paradigm) to press a lighted lexigram key. If this is done, the ape receives a food reward. When this was tried, their apes failed to learn even after 3000 trials.

In another variant, an experimenter held up a piece of food and the ape was expected to push the relevant lexigram key and then receive *that* food from the experimenter. This labelling procedure seems more likely to be effective in having the apes attend to the food and associate it with the lexigram, because the ape both sees the experimenter with the food and gets to have it—but the apes failed again. Lana, and initially Sherman and Austin as well, were trained to produce a string of lexigrams which we might gloss as 'Give orange' or 'Pour coke' or 'Please machine give M&M', after which the subject received the requested item. However, tests of individual meaning and the spontaneous use of strange combinations by the apes suggested the apes had not learned much about the semantic content of the words. For example, if the ape knew significant amounts about the semantic content of 'give' and 'pour' he should not have used both verbs in one sentence for the same substance; but he did so 16 per cent of the time (Savage-Rumbaugh and Rumbaugh 1978).

The functional learning paradigm was initially used by Savage-Rumbaugh and colleagues (1986) to teach the names of tools. The ape first saw the human use a tool to obtain food, and secondly used the tool to get food itself. Then it saw the human label the tool with the lexigram, for example 'This wrench', and then had to *use* the correct lexigram to request the tool from the human. To be certain that the label 'wrench' was associated with the object *wrench*, and not with the whole activity and context of using the wrench, a later phase was introduced. After naming the tool displayed by the experimenter, chimpanzees did not use it to obtain food, but merely received social praise or a token to be later used in a vending machine.

The chimpanzees readily learned the tool names by this functional learning paradigm. These methods were contrasted with a *labelling paradigm* using objects familiar to the chimpanzees. In this paradigm, the humans labelled an item, allowed the ape to play with it, and rewarded him with food or social praise if he correctly named the item when asked. After hundreds of trials, neither ape learned the label to a criterion of nine consecutive correct responses. It is extremely difficult to evaluate these experiments or to understand precisely which aspects of the training procedure produced the clear results of an ease in learning tool names by the 'functional procedure' and a difficulty in learning object names by the 'labelling procedure'. There are simply too many differences in the two situations (for a more complete discussion see Ristau and Robbins (1982*b*)).

The clear differences in the results obtained from the two paradigms, however, suggests an interpretation similar to those offered by Piaget and Bruner (Piaget and Inhelder 1969; Bruner 1966) to explain early concepts/word learning by a young child.

Piaget argues that a child, in the sensory motor stage of development, incorporates his bodily actions, i.e. his sensory motor involvements into the schema or, loosely interpreted, 'mental image' of the concept. Thus the meaning of 'a hole' is 'to dig'. Many early childhood curricula are based upon such a 'hands on' philosophy, in which children are specifically encouraged to work with their hands and bodies in order better to understand concepts of volume, number, and even objects (Ristau and Robbins 1982*b*, p. 204).

Apropos of this, the Gardners and others (Gardner and Gardner 1979; Nelson 1973) have noted that frequent exposure to a referent does not guarantee learning to name the item. Referents commonly named were 'changeable, moveable, and manipulable' (Gardner and Gardner 1979, p. 343), such as the electric lights children or apes switch on and off.

The differences between the 'labelling' and the 'functional' paradigms are probably most important in the initial stages of acquiring a sign, i.e. in the *development* of word acquisition. As Savage-Rumbaugh (1986) notes, there came a time when the very experienced apes Sherman and Austin could not only learn new labels merely by seeing a new lexigram when presented with a new object, but when Sherman was described as himself spontaneously *assigning* a new lexigram to an object in this way. This suggests that a critical difference between the 'labelling' and 'functional' paradigms for a less experienced ape may lie in the difficulty of the ape's understanding or attending to the relevant functions of the lexigram, as intended by the experimenter.

In another procedure termed '*categorical sorting*', Savage-Rumbaugh *et al.* (1980) investigated whether lexigrams used by the chimpanzees were functioning at a symbolic level or whether the lexigrams were merely paired associates with the items they 'labelled'. The experiments are more properly viewed, I think, as demonstration of a procedure to *establish* category names rather than as a test of whether *any* label used by the chimpanzees can function as a category name. In the categorical sorting procedure, indeed an ingenious one, three apes served as subjects, Lana, Sherman, and Austin, all well experienced in the 'ape-language' experiments. They were required to sort into one bin three edibles (orange, bread, bean cake) and into another bin, three inedibles, the so-called 'tools' (sticks, key, and money). When the apes were sorting 90 per cent correctly, lexigrams were introduced for 'food' and 'tool'; the apes had not previously been taught these. The chimps then had to sort each food and tool into its proper bin and then select the appropriate lexigram for the category 'food' or 'tool'. The other apes did well, but Lana did very poorly at this point, Lana being the ape who had a history of learning lexigrams by 'labelling', i.e. learning 'specific paired associative responses required by her training' (Savage-Rumbaugh et al. 1980, p. 923). In short she could sort the actual items, but could not associate the appropriate class lexigrams, namely 'food' and 'tool', with either the items or the sorting procedure.

Next objects, with their photographs attached, were sorted and labelled as 'food' or 'tool'; then merely photographs were so used. The most critical phase of this study was labelling lexigrams. Lexigrams for the foods and tools used in training were taped to photos of the items, sorted as before, and finally, just the lexigrams were so sorted. In the final phase, the chimpanzees were tested on lexigrams for items most of which had not been encountered in this particular experiment. One ape had perfect scores on first-trial presentations, while the other erred only once, classifying the sponge from which he occasionally sucked water as a 'food' rather than a 'tool'. No other project has used this detailed training procedure. The apes are clearly exhibiting an ability which seems understandable as categorization, namely, they can classify members of the category 'food' with that category-name.[9]

Note that this work does not indicate that all the apes, nor indeed that Sherman and Austin, actually do use lexigrams in a categorical way in circumstances other than the one just described, which entails the elaborate training procedure used. But these cautions apart, this 'categorization' study has clearly demonstrated that chimpanzees, at least language-trained ones, under certain training procedures have at least a rudimentary capacity to use symbols.

23.5.1.2.2 *The nature of the internal representation*

It is of great interest to attempt to understand what the apes may be understanding in their use of lexical items. It is of course never clear exactly how the ape is 'glossing', i.e. interpreting, a particular sign or other lexical item, especially in comparison to our word or interpretation of the sign. There is a difficulty inherent in determining the chimpanzee's meaning for a sign that we may gloss quite differently with respect to referent and semantic class, i.e. whether the sign functions something like a verb, noun, preposition, etc., to the ape (discussed in Ristau and Robbins 1982b, p. 193). Just about all the ape-language researchers have issued warnings on this problem, though some more adamantly than others.

McDermott (1981) terms the tendency to overinterpret specific and limited capacities '*wishful mnemonics.*' This is a very general problem of determining the abilities underlying specific behaviours of an organism, or, as in his field, the particular aspects of a computer program attempting to model artificial intelligence. Instead, one should label a capacity neutrally, for example call it 6PX4, and then, by virtue of the actual behaviours of the organism, slowly build a notion of what 6PX4 might entail. The point is well taken (more fully discussed in Ristau and Robbins 1982b, p. 191).

Any discussion of representation is necessarily involved with a discussion of displacement, though I shall discuss separately specific experimental tests for displacement. As I, and certainly many others, have noted (for example Ristau and Robbins 1982a,b; Griffin 1976, 1984, 1986; Premack 1986), it is quite conceivable that many species besides humans can have representations of objects or events which they carry with them well outside the immediate environs of that object or event. Premack suggests: 'It is not language *per se*, but a far more basic property of mind that makes displacement possible' (Premack 1986, p. 65). In Ristau and Robbins (1982b) we speculated that the ape or other creature might well profit by having a concept or representation of something like 'red, ripe berry' to guide its foraging. Again, Griffin (1976, p. 84) has argued that the anticipation of future enjoyment of food and mating or fear of injury could certainly be adaptive by leading to behaviour that increases the likelihood of positive reinforcement and decreases the probability of pain or injury. There are, then, good biological arguments for expecting some form of representation to be present in the apes at least, in addition to the experimental and observational work that points in this direction (see Lock and Colombo, this volume, Chapter 22, for a more detailed treatment).

Something of the nature of the representation, or at least the difference between the apes' understanding and ours for a particular 'word', may be determined by noting *errors* made in response to identification questions by an experimenter, in generalizations when the ape uses the lexical item, and in novel uses of that item, i.e. extensions of generalization (see also Johnson *et al.*, this volume, Chapter 24 with regard to error in human language-development).

For example, Washoe most readily remembered the category of the correct response, if erring on the specific: for example, she was likely to make the sign for another colour, rather than for size or kind of object, when erring on the specific colour (Gardner and Gardner 1986). In a study of the learning of colour names by three-year-old human children (S. Carey and E. J. Bartlett, as reported in Miller and Gildea 1987), it was noted that the children assigned a new word to a semantic category (for example, colour) before they could use the word correctly either productively or receptively. In an example concerning generalization of new signs:

Washoe was introduced to the sign for flower by being presented with a real flower. However she apparently interpreted this sign to refer to smell, so she generalized it to pipe tobacco and kitchen fumes (Gardner and Gardner 1969; Gardner and Gardner 1975). Since the sign for flower entails holding the finger tips below the nose and breathing in, an act similar to smelling the fingers, generalizing the sign to odour is not unlikely.... yet the fact remains, there are many features of this situation that could have been selected as outstanding to the ape and used when employing the new sign and new context; the ape selected odour or the act of smelling. (Ristau and Robbins 1982b, p. 193).

Novel word combinations can reveal meaning as they are applied to new situations. Almost all the artificial-language researchers have published examples of such instances. For example, the gorilla Koko called a stale cake 'cookie rock', and a face-mask 'eye hat' (Patterson 1978a,b, 1979; Patterson and Linden 1981). Moja, a chimpanzee studied by the Gardners (1979) termed, among other things, alka seltzer 'listen-drink' and a cigarette holder 'metal hot'. The first bite of a radish provoked the chimpanzee Lucy to sign 'cry hurt food', and she continued to use that sign afterwards, though she had, before tasting it, called a radish 'that Lucy fruit' (Fouts 1975; Linden 1975, pp. 105–9).

Probably the most important caveat here is that one needs a statistical analysis of word combinations and error frequencies drawn from an appropriate, unbiased sample of the ape's productions in order to evaluate the significance of 'creative naming' (Desmond 1979; Petitto and Seidenberg 1979; Ristau 1980, Ristau and Robbins 1982b; Terrace *et al.* 1979, 1980). If the majority of word combinations are nonsensical, then the few meaningful ones that are reported are inappropriate samples. Apes have indeed been described as random sign-generators (Petitto and Seidenberg 1979), although mistakes in word combinations are rarely reported in the literature.

Savage-Rumbaugh, however, has published (1986) a corpus of utterances which includes all lexigram combinations produced outside of training tasks over a three-month period by Sherman and Austin, and also a detailed description of a session between a teacher and two apes. In the analysis of this corpus, single-word utterances were deleted. Immediately after it occurred, each utterance was scored as correct or incorrect given the context. The listing also includes an indication of whether each production was an imitation of the teacher's first or second prior utterance, a partial imitation of those, or a complete imitation. Of Sherman's total utterances, 95.5 per cent were judged appropriate to the context; of Austin's, 99 per cent were so judged. For each ape, approximately one-third of the utterances were novel (for that time-period) and non-imitative, while the rest were either partial or complete imitations of the teacher. Novelty is clearly not a rare event, and future analyses of this phenomenon are likely to be of great value.

Premack has also conducted a number of experiments investigating the meaning of the lexical item for the chimpanzee; recall that Premack used coloured plastic chips as 'words'. In one experiment, he tested the kinds of features the ape associated with an object and with the plastic chip for that object (Fig. 23.4).

A match-to-sample procedure was used. In the first test, the sample was an actual apple, and Sarah chose from pairs of alternatives that did and did not instance some feature of apple. All her choices were acknowledged with approving tones, and none rewarded in preference to others. She consistently chose red over green, circle over square, and square with stem over a plain square, and tended to choose roundness more frequently than square with stem. The test was repeated, except that the object apple was replaced with the 'word' for apple, namely a small blue triangular piece of plastic; her choices were essentially the same.

This paradigm could be criticized, in that the ape could simply be repeating with the plastic chip the choices she made for the actual apple. Premack, however, did the following experiment, showing that this objection is not substantiated. He reversed the order of the task: i.e. presented choices for the 'word' for the object first and *then* the choices for the object itself second. In this test Sarah was shown the 'word' for a caramel and then given four pairs of alternatives intended to instance the shape, colour, size, and texture of a caramel (which is a brown cube of candy wrapped in cellophane). Again the features she chose generally accorded with human analysis and were approximately the same for the word and for the object.

Premack also sought to ask questions which he phrased as

How effective *is* ape mental representation? How preserving of information? Do ape representations retain *all* the information contained in perception or consistently lose some proportion? Do representations vary with domain? (Premack 1986, p. 65).

The pertinent experiments utilized fruit—clearly objects of great interest to the apes. Eight kinds of fruit in their diet were divided into features and components: wedge, stem, peel, seed, white outlines of shape, colour patches, and one non-visual cue, taste. Again the apes were given match-to-sample tests requiring them to match one feature of a fruit with another feature. 'For instance, we gave them a taste of peach and then required them to chose between a red and a yellow patch... or presented a peach pit and required them to choose between a peel of peach and of banana, etc.' (Premack 1986, p. 66).

The four apes tested did well, and the star pupil Sarah used every cue correctly. Once again, information in the errors made by the apes was illuminating. As one might expect, the whole fruit was the most informative, but colour and peel were very effective as well. Taste was less informative, and shape, wedge, and stem even less so; seed, the least. 'Chimpanzees ... can store detailed representations of items such as fruit' (Premack and Premack 1983). Therefore, reasons Premack, the chimpanzee should be able to think about objects not present, and plastic chips may aid this function by reliably evoking the image of the fruit.

There were suggestions in his, and indeed other projects, that this could be done. In many projects, the apes would request preferred foods not present via their lexical items. The typical criticism of these occurrences is that we have no independent way of verifying that the ape intended to ask for that specific food. Alternatively, the context could evoke a particular lexical unit, or the ape might simply randomly choose some well-rewarded one (see further discussion in the section below on 'displacement testing').

Another kind of evidence that lexical items can evoke a mental representation is found in the instance in which Sarah was successfully taught the word 'brown' by being instructed 'brown colour of chocolate' without chocolate or any brown object's being present (Premack 1976, p. 202). More formal tests of this were done by using the previously described match-to-sample test with features of fruit as the sample, but plastic *names* of the fruit as alternatives. Sarah was completely successful in this test, though the other apes again made interesting errors. Colour was the most effective cue; plastic words were more effective than any physical part of the fruit, and were about as effective as the actual fruit. As Premack summarizes the results, the apes' mental representation of the fruit is sufficiently detailed to be effective in these kinds of tests, and the plastic words are very effective devices for retrieving the information. I have no reason to disagree.[10]

23.5.1.2.3 *Displacement testing*
In theory, much of the above testing involves displacement: that is, the ape demonstrates that it is able to use a word to refer to the item in the absence of that item. As noted previously, the criticism usually advanced is that the ape is not using the lexical item in a referential way, but rather is exhibiting mere paired associate learning. Typically, something in the context is said to evoke the particular lexical item used by the ape.

The traditional learning paradigm which most closely approximates a stimulus evoking an object in the absence of the object is that of backward association after training in a 0-second-delay conditional ('symbolic') matching procedure. Translated from jargonese, this means that if subjects are shown sample stimulus A, they are to respond to the comparison stimulus X (having a choice of X and Y); similarly, when shown B, they are to respond to Y. To test backward association, subjects are shown X, and must choose A with A and B before them; analogously, when they are shown Y they must choose B. Both pigeons and monkeys (*Cebus apella*) showed only very weak evidence of backward association (results indicate trends in the data, but not statistical significance) (D'Amato *et al.* 1985), while most of the children tested at six to seven-and-a-half years old showed backward association to the point of symmetry (Sidman *et al.* 1982, as discussed in D'Amato *et al.* 1985) (see also Lock and Colombo, this volume, Chapter 22). Completely analogous experiments have not been conducted with chimpanzees, though the ape-language tasks are often even more demanding. For example, teaching a chimp the colour brown by the phrase 'brown colour of chocolate' (Premack 1976) seems to me to be evidence for the referential function of brown. I am unable to construct any reasonable arguments which account for this solely by paired associate learning.

A series of 'displacement' tests have also been done with a dolphin (Herman 1987), in which the dolphin was given two-word commands with varying delays before the object to be acted upon was made available. A command might be 'Ball under', glossed as 'swim under the ball'. After the delay, various objects were thrown into the tank simultaneously, and the dolphin's accuracy of response was scored. With no delay, each of the two dolphins (Ake and Phoenix) responded about 80 per cent correctly to the 43 commands. With delays, one dolphin (Ake) was tested and responded 90 per cent correctly to 75 commands. Delays were 0, 7.5, 15, and 30 seconds, as well as a control condition in which objects were introduced prior to the signing. Five of the seven errors were at the 30-second delay (67 per cent correct). *Sometimes*, but not always, the dolphin exhibited behaviours which appeared to be 'rehearsals' of the required action, for example, holding the pectoral fin as she does for a 'Ball pec-touch' command. In summary, a dolphin *can* remember signed commands entailing an action and an object for delays as long as 30 seconds.[11]

23.5.1.3 *Other instances of possible symbolism by human-reared apes, signing apes, and apes in the wild*

What may be symbolic play has been observed in at least a few apes raised by humans and in the wild. With close, frequent interactions with humans, sometimes with children, these human-reared apes have a rich social life. A widely reported case is that of Viki and her imaginary pull-toy. She often played with it in the bathroom, where she spent long times being 'trained'. The toy seemed to have a long imaginary string which occasionally got caught on knobs and such, whereupon Viki would untangle it. The toy continued to amuse Viki for several weeks. But once the string got so tangled that Viki seemed to request Cathy's help in untangling. Viki yanked at the imaginary string to no avail. Then she stared at Cathy's face in the mirror, and breathily vocalized 'Mama, Mama'—one of her four 'words'. Cathy, surprised at this scene, untangled the string, and Viki ran off, pulling her 'toy'.

Days later, Cathy decided to invent her own imaginary pull-toy, which even clattered. Viki watched intently, then ran to stare at the floor *where the toy would have been*. The next day, this imaginary performance terrified Viki, who whimpered and leapt into 'mother' Cathy's arms. Neither played with the imaginary pull-toy again (Hayes 1951, pp. 80–5; see also Jolly, this volume, Chapter 6, Section 6.5).

The apes in the contemporary language projects also perform apparent symbolic play. Sarah, Premack's star pupil, after wearing a hat, placed a triangular piece of paper atop the picture of a face (Premack 1975). Savage-Rumbaugh (1986) describes a complicated scene with Sherman and Austin and a frightening imaginary object in a cage—all of this occurring after they'd watched the film 'King Kong'.

This is all extremely difficult to evaluate. Certainly even kittens pounce at what seem to be imaginary, non-existent items. Yet the examples from the apes, certainly Viki's imaginary pull-toy, are far more elaborate than anything ever described for kittens. A further particular difficulty in evaluating the symbolic content of these activities is that even young human children have strong tendencies to mimic actions with little comprehension of the function of the action. For example, a child two or three years old 'swept' with his own small broom as Mummy was doing with hers, but without regard to the location of the litter being 'swept' (pers. obs.).

From observations in the wild, Goodall (1986, pp. 590–1) reports the following:

Once four-year-old Wunda watched intently from a safe distance as her mother, using a long stick, fished for fierce driver ants from a branch overhanging the nest. Presently Wunda picked a tiny twig, perched herself on a low branch of a sapling in the same attitude as her mother, and poked her little tool down—into an imaginary nest?

Or is Wunda merely moving her stick as her mother does, as unaware of imaginary ants and a nest as the human sweeper was of real litter? Jolly (1988) provides a more complete discussion of symbolic play in the apes. What is not clear from this literature is the extent to which the apes are engaging in symbolic activity and the extent to which human observers are engaging in 'wishful mnemonics'. Yet an over-critical attitude may make us blind to apes' symbolic abilities.

Apes have also been given crayons and permitted to 'draw'. Typically they produce what we humans would call a 'scribble', though the scribbles can be quite different from one another, and have some concentrated area. It is reported that the apes often indicate that they are 'finished' with their drawing, while at other times they are not, and will try to grab the drawing back if it is removed. If asked of their drawings, 'What that?', they have often supplied 'titles'. And to our human imagination, sometimes those titles seem appropriate. Intriguing, not to be dismissed: but enormously difficult to evaluate.

In a brief summary, Beach (1985) reports observations of drawings self-initiated by chimpanzees and other drawings made in response to requests to draw specific items. She indicates evidence for representation and use of consistent schemata by the apes. Beach's work bears further consideration (see also Bremner, this volume, Chapter 18). The use of 'language'-trained apes offers some hope out of the conundrum of a *human*'s attempting to determine whether a given scribble represents a bug or some other item to an *ape*, even though it may not to the human (or vice versa). Of course, the essential problem then still remains of determining what the 'word' 'bag' means to the human and to the ape.

23.5.2 *Cognitive and symbolic capacities*

23.5.2.1 *Quantitative abilities*

The ability to count and a more primitive ability to 'subitize' (see below) have also been investigated. Let us consider the abilities which seem to be necessary for counting. In the terms in which the subject was first discussed by Piaget (see also Gelman and Gallistel 1978; Davis and Perusse 1988) an organism which is said to count must also: (1) know a *sequence* of numbers (i.e. labels—what Piaget called 'ordination'. Thus one knows that the order of the sequence is 'one, two, three, four, etc.'); (2) be able to do 1:1 matching with this sequence (i.e. the chain of cardinal numbers must be capable of being applied in 1:1 correspondence to a set of objects, so that the subject can 'tag' the objects); and (3) know that one more than the last in a sequence is the next in the sequence. Though this is not specifically stated by Piaget, an implicit assumption is that these abilities can be demonstrated in both receptive and productive tasks, and can be generalized across a wide variety of circumstances. Few of the experiments with chimpanzees (or other species for that matter) exhibit all these characteristics, so the researchers often conservatively term their abilities 'numerical competence' rather than 'counting'.

Quantitative concepts that have been studied range from a knowledge of 'greater than' to the ability to place items on an ordinal scale or at least a sequential one, as well as receptive and productive counting abilities (or, more conservatively, 'numerical competence') and rudimentary summation. There again appear to be differences manifested between language-trained and non-language-trained chimps in relational knowledge (Premack 1983a,b; 1986).

All the findings of quantitative prowess must be tempered by the fact that other species also have been shown to possess some rudimentary skills, such as numerical competence (cf. Barton and Hamilton, this volume, Chapter 27). In particular, various bird species apparently do rudimentary counting (for example Köhler's studies of jackdaws reported in Bateson 1979; Pepperberg's (1987a, 1994) studies of a grey parrot). Racoons, rats, and rhesus monkeys also have primitive numerical abilities (Davis and Memmott 1982). In this specific area, as well as in the entire field, both the difficulty and the fascination lie in specifying precisely the nature of the skill that is demonstrated by a particular experiment, and how it relates to the skills of other species. I will review the research about quantitative abilities, beginning with those that appear to be most basic (and see also appendix to Barton and Hamilton, this volume, Chapter 27).

A fairly simple ability is learning to place several items reliably in the same sequence (discussed in Section 23.5.1.1). This skill is a pre-requisite to learning any kind of grammar and to understanding temporal order, the concepts of past, present, and future, any cause–effect relationships, and the ability to count. Any ape, such as Lana, that has to produce a correct sequencing of lexigrams or plastic chips has mastered a sequential skill. A similar, yet simpler, skill has been mastered by pigeons (Straub *et al.* 1979). Likewise, some apes, dolphins, and sea lions have exhibited the analogous receptive ability—correctly responding to information conveyed by the *order* of lexical items.

Premack and Woodruff have investigated the apes' ability to match proportions (Premack 1983a,b; Woodruff and Premack 1981). Both Sarah and four non-language-trained juveniles were given match-to-sample procedures in which a glass cylinder was filled with tinted water either completely or to the ¼, ½, or ¾ level. Fruits (apple, grapefruit, or potato) were similarly either left intact or cut into quarters, halves, or three-quarter pieces. In the first set of tasks, the animal was presented with a sample fruit and then given as alternatives, the same kind of fruit, cut either identically or to a different size. Then the same task was presented using a glass cylinder filled to different levels instead of the fruit. All the animals passed these tests. However, when the glass cylinder was presented as a sample, with the pieces of fruit cut to different proportions as the alternatives, the juveniles flunked and Sarah was 100 per cent correct on first trials. At one point, Premack interpreted these findings as revealing the difference between an imaginal code used by the non-language-trained juveniles and an abstract code possessed by the language-trained chimpanzee Sarah (Premack 1983a). He later expressed more uncertainty as to the nature of the codes (Premack 1983b).

The ability to count and a more primitive ability to 'subitize' (see below) have also been investigated. Working with a female chimpanzee, Matsuzawa (1985a,b; Matsuzawa *et al.* 1986) has shown that she can correctly label 1–6 items in a productive task (choosing the correct arabic numeral when presented with a set of objects). In his experiments, there were 300 different training sets composed of 5 different three-dimensional objects of 5 different colours, presented 1–6 at a time, which the ape described in terms of name of object, colour, and quantity. A long training procedure was needed. To get 98.5 per cent accuracy with the numbers 1–5, required 95 sessions (almost 29 000 trials). And an additional 3000 trials were needed to learn the label 'six' to 90 per cent accuracy.

As McGonigle (1985) notes, these abilities may indicate a rudimentary counting mechanism, 'subitizing'—'the simultaneous apprehension of up to four elements through direct pattern recognition" (1985, p. 17). In human children, a disjunction in reaction-times separates that process of simultaneous number perception from the more costly process of counting. The break may not be exactly the same in apes; there is a particularly large number of errors made in discriminating between five and six items, which suggests a disjunction there. However, just because a display is of four or fewer items does not necessarily imply that the process involved in apprehending them is subitizing; as McGonigle also notes, 'true' counting begins with small numbers.

By means of a productive task, a chimpanzee, Lana, has demonstrated an ability to produce the number of items needed when presented with arabic numerals (Rumbaugh et al. 1987; Rumbaugh and Washburn 1993). At this time the work is still in progress, and only numerals 2, 3, and recently 4 have been trained. The procedure is a most interesting one. Computer-generated training begins with a task that probably relies on perceptual grouping; but later tasks may not be so simply explained.

In early training, an Arabic numeral, say 2, appears in the appropriate position (second) in a row of 'feedback' boxes at the top of a computer screen. A variable number of geometric shapes appears below. (These shapes often had a dot in the centre, a feature presumably reminiscent of other tracking 'games' Lana had played.) Lana must touch a cursor to the computer screen to make the geometric shapes disappear one at a time. Each disappearance is accompanied by a tone, and initially by coloration of the feedback boxes and an Arabic numeral in the place of the missing shape. Gradually the feedback boxes and feedback numerals are phased out of the procedure. The ape signals to the computer when she has finished; i.e. Lana must *remember* the number of shapes which have disappeared (or the number of tones heard). The initial set of shapes varies in number from trial to trial. It is the total number of *missing* shapes that corresponds to the Arabic numeral. An impressive feat, although, like most of the other numerical training tasks, it took thousands of trials to master.

In subsequent work with both Lana and other language-experienced apes (Sherman and Austin), numerical competence to 4 or 5 was demonstrated by their capacity to place either Arabic numerals or black dots on boxes on a computer screen (Rumbaugh and Washburn 1993). However, the chimpanzees have not been tested for these abilities in a wide array of circumstances. And, as Rumbaugh and Washburn note, 'they have not yet given us reason to believe they have any comprehension of the principle that one can always count to a next higher value' (p. 100).

Work with another chimpanzee, Sheba, also provides evidence of numerical competence with numbers 1–3, and, in some experiments, with 0 and 4 as well (Boysen 1992, 1993a; Boysen and Berntson 1989). The human experimenter apparently served as the model in this research, exhibiting the behaviours she wanted the chimp to learn, but then simply provided the opportunity for the chimpanzee to do the task. Sheba learned one–one correspondence between disks with 1–3 small metal disks attached and those numbers of small food items, and then between Arabic numerals and the corresponding bits of food. To train/test for generalization, the items were either homogeneous food, heterogeneous food (gumdrops, M&Ms, or grapes), and then variously sized inedible objects unfamiliar to Sheba (for example flashlight batteries or spools).

The same chimpanzee has also exhibited an ability Rumbaugh terms 'summation', which may better be described as an ability to judge which of two groupings is 'greater than' the other. The chimpanzee can do this even with substantial spatial separation of the group items. 'Perceptual grouping' may suffice as an explanation for Rumbaugh's 'summation' experiments, and perhaps for Matsuzawa's work with chimps applying numerical labels to groups of objects; but we should beware of using it as a catch-all term without exploring the possible cognitive aspects of the process.

Earlier experiments with a chimpanzee have indicated similar abilities, in tasks requiring her to choose the middle food-well from a linear array of wells (Rohles and Devine 1966, 1967). With training, the ape performed well with sets composed of as many as seventeen wells. The wells were often irregularly spaced, so that the chimpanzee could not simply choose the spatial middle of the array. An ability to judge 'same amount' is the simplest explanation of these results. Alternatively, the ape may have memorized specific correct spatial arrangements—the first-trial data with new configurations would refute this interpretation. A useful experimental manipulation for both this work and Rumbaugh's 'summation' studies is to require a judgement to be made of equal numbers of differently sized objects, so that a perceptual judgement based on total area or volume would be incorrect.

In other work, another chimpanzee, Sheba, demonstrated that she could sum arrays of 0–4 oranges placed in 2 of 3 possible sites in the laboratory (Boysen and Berntson 1989). For training, during the first three trials, the experimenter walked to all the sites with Sheba, noting the presence of the hidden oranges. Upon return to the start position, Sheba, not the experimenter, chose the Arabic numeral that represented the sum. She did the task immediately. Her performance was significantly above chance from the first sessions, and she was correct at levels of 84 per cent (non-blind) and 74 per cent (blind control; chance = 25 per cent) by the end of training. Importantly, for this ape in this highly interactive procedure with this experimenter, training was only about 200 trials, in contrast to the thousands required to train the chimpanzees Lana, Sherman, and Austin in a computer-based procedure.

In the next experiment Arabic numerals replaced the hidden oranges, and Sheba, again, performed the task immediately, recognizing what was to be done. In only 21 trials, she was 76 per cent correct, and 85 per cent correct in the next 20 blind trials.

Subsequent experiments investigated abilities in understanding ordinality and transitivity in three chimpanzees including Sheba, utilizing 5 differently coloured boxes (replicating Gillan 1981) and then the Arabic numerals 1–5 painted on a card (Boysen *et al.* 1993). Sheba already had a good working knowledge of the numbers 1–5 and was receiving training on 6 and 7 in other tasks during the course of the experiments. She outperformed the other chimps, although they, too, were able to do most of the tasks. Their successes included novel probes which could not possibly be explained by prior reinforcement history; thus when required to select the large number, they correctly chose 1 over 0, though the selection of 1 had never been reinforced throughout the 2½ years of the entire study.

In summary, though we may still quibble as to whether all the characteristics of counting have been met, apes have been shown to display, often spontaneously, and at least very rapidly, many of Piaget's original criteria. In some cases, the early stages of correctly labelling a set of objects with the correct Arabic numeral have required hundreds, even thousands of trials. In other cases, such as Boysen's studies, progress has been much more rapid. Although the apes involved may differ in intellect, another possible explanation lies in the intense, interpersonal interactions and 'gamelike' atmosphere that seem to characterize her training procedures. Again, the importance of social interaction in the apes' learning weaves through the examples.

23.5.2.2 *Reasoning*

The classic work is Köhler's (1925) book, *The mentality of apes*. I will, however, concentrate on contemporary research which bears directly upon the ape-language studies. This narrows the field primarily to Premack (1983*a*, *b*, 1984, 1986) and his colleagues (Gillan 1981; Gillan *et al.* 1981). Premack's problem-solving experiments were conducted with three or four non-language-trained chimpanzees and usually with three language-trained animals, as well as young human children about 2½ years old in some experiments and 3½ in others. In the experiments, there are two kinds of tasks: those that are and those that are not enhanced by language training.

1. *Judging proportions*

These are discussed more fully in Section 23.5.2.1 on quantitative abilities. Essentially, both non-language-trained and language-trained chimpanzees can do problems involving simple matching of proportions that are soluble on a *sensory* basis: but only language-trained apes can immediately match the same proportions of different objects. Three-and-a-half-year-old children pass as well; younger children have not been tested.

2. *Action or casuality*

A simple physical action might be cutting an apple or moving it. Such an action can be represented by three-element sequences: an object in its initial state (the uncut apple), an instrument capable of changing the object (a knife), and the object in its terminal state (the cut apple). Experiments were conducted requiring an understanding of the necessary sequential order of these events or the ability to select the appropriate item when one element was missing (see the discussion in Section 23.5.1.1.). The language-trained apes passed all the tests, but the non-language-trained could not, despite repeated training (Premack 1984).

Earlier work on transitive-inference-type problems (Gillan 1981) is relevant as well. A non-language-trained chimpanzee (Sadie) could solve such problems, but failed in tests that seemed to require an abstract coding of the items rather than an imaginal one (Gillan 1981; Premack 1983*a*, *b*). The language-trained chimpanzee (Sarah) could do related more difficult tasks, including those apparently requiring abstract coding (Premack 1983*a*, *b*).

3. *Analogies*

Only a language-trained ape (Sarah) could solve problems dealing with analogies; the four non-language-trained chimpanzees tested could not (Gillan *et al.* 1981). Sarah's tests required either choosing an item to complete an analogy or indicating whether the sample relation was one depicting 'same' or 'different'. For example, the first kind of test could be represented as '*a* is to *a'* the same as *b* is to ?', with *b'* and *c* as alternatives. '*a"* might be a small blue triangle, '*a'* a large blue triangle, '*b* a small yellow circle with a dot in it, '*b"* a large yellow circle with a dot in it, and '*c* a large blue circle. Note that the animal was not explicitly trained to perform these tests. She was about 85 per cent correct on both conceptual and perceptual material. The conceptual problems might be 'key is to door the same as can-opener is to can' (not words but objects or pictures of objects were used). Since the non-language-trained animals did not have the words 'same' and 'different' they were offered XX as sample with YY and CD as alternatives. Sarah, the four non-language-trained animals, and children from 3½ to 6 years were tested on these formats. The first series consisted of 12 trials, 6 of the XXYY variety (same-same), and 6 of the XYCD variety (different-different). Common items, toys, hardware, and such, were the stimuli. Sarah was correct on all 12, while the non-language-trained animals responded at a chance level, showing no progress over 15 additional sessions. Premack does not report the children's results. Note that Sarah was the only language-trained animal to be used in these particular tests.

4. Natural reasoning

In these tasks both language-trained and non-language-trained animals succeeded (Premack 1984). Reasoning tasks for animals usually have two characteristics: information is given to the animal perceptually instead of verbally; and the problem tends to be spatial, a distance or barrier to be overcome. Premack added an additional feature: that the perceptual information given the animal was not a sufficient basis for solving the problem. The animal had to add to it in some sense.

Examples of these problems are the following. Two widely separated containers are pointed out to the subject, and it is shown that an apple is put into one container and a banana into the other. The subject is then removed from the field, and, when brought back, finds a second individual standing between the two containers eating either an apple or a banana. After the second individual leaves, the subject is released to choose one container or the other. The apes either do so correctly during the first trial, or learn to do so; children 4½ years old do as well as the chimpanzees, but many younger ones fail.

In some of the variants of these procedures, the ape does not see the actual placement (or alternatively the retrieval) of the fruit by a human, but must infer that such a thing happened by observing the human walking towards the containers or returning from them. The non-language-trained apes were successful at these tasks, as were 3½- to 5-year-old children for the conditions in which they were tested; the language-trained Sarah, requiring caging, was not tested with these latter procedures.

Premack distinguishes between the intelligence manifested in the two processes as between what he calls 'conceptual problem-solving versus natural reasoning'. He wonders why language-training should enable the chimpanzee to 'deal with relations between relations'. The answer may lie in the truth claim. 'A truth claim ... depends on the ability to map sentences onto conditions, or vice-versa.' He notes that 'different exemplars of the same *relation* do instantiate the same invariant', one that can be given a *physical* description, however complex. But the agreement between sentences and conditions is not of this kind. 'The invariance we use in describing sentences will not reappear in the conditions that are described by the sentences' (Premack 1984, p. 280).

5. Comparing results of experiments that differentiate between language-trained and non-language-trained apes

A major concern about these experiments is the number of animals tested; often the language-trained animal is only the very apt subject, Sarah; so it is sometimes difficult to differentiate the effects of language-training *per se* from those of an extremely astute chimpanzee. One needs some less astute language-trained apes and more apt apes with no language training: Sadie, another of Premack's subjects, may be such an apt ape. (Unfortunately, Premack's ape projected has been discontinued. Boysen's project now has Sarah and several non-language-trained chimpanzees.)

23.6 Methodological problems

23.6.1 *The 'Clever Hans' phenomenon*

'Clever Hans' has become an unfortunate 'buzz phrase' which some scholars blithely use to dismiss the entirety of the ape-language/cognition research. Clever Hans was a horse, and reputed to be a mathematically gifted creature at the turn of the century. He was touted about for stage demonstrations of his abilities, which apparently included counting, adding, multiplying, and other wonders. Not only his trainer, but even audience members, could elicit such talents from him. Hans indicated the answer by tapping his hoof the appropriate number of times. But a slight mystery; if the questioner did not know the correct answer to a problem (usually because it was so difficult), neither could Hans solve it. The psychologist, Otto Pfungst (1911), demonstrated that Hans was responding to subtle relaxations and postural changes of his human questioner when the horse tapped out the *correct* number. The experimenters' cueing in these cases seemed inadvertent.

Some critics, Umiker-Sebeok and Sebeok (1980) in particular, have invoked the Clever Hans phenomenon as a means of discrediting the ape-language/cognition research. But to invoke the phenomenon is not to prove its relevance. Inadvertent cueing through non-verbal signals could reasonably explain inhibition of hoof-tapping by a horse; but it is difficult to see how a body posture or eye-glance could reveal to a signing ape which particular hand configuration to use (see also Ristau and Robbins 1982b, pp. 237–9). Eye-glances might direct the attention of an ape, dolphin, or other creature to the correct one of several objects to be chosen by them. And indeed, in some of the early research, there were experimental paradigms in which this was at least a possibility. [However, in some recent work the number of alternatives an animal might be choosing among can be as many as 100–200. In these cases the ways cueing could occur are vastly different and more limited than in the go–no go situation of the original 'Clever Hans' (Savage-Rumbaugh, in litt.) (eds).]

But experimental techniques have improved, and inadvertent cueing is no longer a difficulty. Perhaps that is the major, and not insignificant, contribution of the Sebeoks' raising of the Clever Hans spectre. As Premack (1986, pp. 12 ff.) notes, many experimenters adopt elaborate measures to preclude social cueing, which, none the less, had never been conclusively demonstrated anyway. He wonders as well why a horse given an insoluble problem is a good model for an ape given an arguably soluble one. Furthermore, some species are likely to have an intrinsic disposition to solve problems and like to do so, and even to understand the situation as involving a problem. As some examples, Premack notes that Sultan the chimpanzee (Köhler 1925) often solved problems for his inept peers, leaving them the banana which was the supposed 'reward' reachable when the task was done. Rhesus monkeys stopped responding, though given food rewards, to insoluble versions of problems they had previously solved (Pasnak 1979). Not by bread alone, but by solving problems, are these primate kin of ours motivated.

There is, however, one further issue here that complicates the picture. Any symbolically based system of communication both requires, and is embedded in, a set of shared presuppositions between the communicants. There is, by nature, a level of intersubjectivity involved in agreeing upon the meaning of a symbol, and in the shared perception of the contexts within which those symbols are used. In some sense, then, a meaningful world shared by two individuals could be said to 'cue' each participant's activities, in that as it unfolds any interaction continually creates the conditions which provide for its future elaboration or termination. We enter a minefield of confusion where it is difficult to disentangle what constitutes a cue and what an understanding. To an extent, these terms lie on a continuum, and the questions involved over and above the simple 'Clever Hans' phenomenon cannot be dismissed simply. And this lead us to the next problem.

23.6.2 *Strict experimental control or casual interaction between ape and communicator—which is more appropriate?*

When the experimental situation is strictly controlled, with specified trials and readily gathered data (for example Premack, some testing procedures used by the Gardners, and Rumbaugh and Savage-Rumbaugh), the criticism can be raised, as indeed Terrace (1979b) has raised it, that the ape has merely engaged in problem-solving; the behaviour has nothing to do with language-use. Further, a certain rigidity of interaction has been criticized in the work with Nim by Terrace and his associates. Nim has often been repeatedly questioned by trainers asking, 'What this? What this?'. An ape may balk at such testing, and often does, as, indeed, human children do. This pressed questioning contrasts with casual signing episodes, which more closely parallel the more normal situation for humans.

The nature of the training and testing procedures is of more than passing interest, because it influences characteristics of the ape's abilities important in comparisons with human language-use, for example the possibility of spontaneity, casual comments on surroundings, creativity, turn-taking, and conversational aspects of interactions. All these are largely antithetical to a controlled experimental situation. In fact, the impact of testing procedures has arisen in a comparison between the abilities of Nim when tested by Terrace *et al.* (1979; Sanders 1980, 1985) and his abilities when, having left the Terrace project, he was tested in more relaxed outdoor situations by Yeager *et al.* (1981). Under the latter conditions, his rate of imitation was sharply decreased from 44 per cent in Terrace *et al.* (1979) to 6 per cent; the majority of his signs were spontaneous or novel (90 per cent); and he exhibited more turn-taking.

Methodological differences in analyses may account for *some*, but not all, of the discrepancies. In Terrace *et al.* (1979), *only* adjacent utterances, namely those which immediately followed a teacher's utterance, were analysed for their relationship to the preceding utterance. Both Sanders (1980, 1985) and Yeager *et al.* (1981) examined a larger segment of the prior context for its relationship to Nim's signing, and included Nim's spontaneous utterances in the analysis. The latter would decrease the level of imitation calculated. The major causes of the reduced imitation, however, are interpreted by Yeager *et al.* (1981, p. 4) as due to the casual social setting and to the established social relationship between Nim and the experimenters. This seems a reasonable interpretation, especially since Yeager *et al.* had the same high rate of adjacent utterances in their mock 'training' session as did Terrace *et al.* in their data. In short, how we test determines the results we get.

This fact raises the comparability of the results obtained from different projects, since training and testing techniques differ between experimenters. However, this is becoming a less intractable problem as more apes of various species have been studied, often producing convergent results despite the procedural differences.

23.6.3 The significance of the novel or few-time event

This is a difficult issue, permeating to the core of our scientific methodology. If we hope to observe examples of intelligence and creativity, they are to be found in an organism's novel solution of a problem not encountered before; they are not as readily found in oft-repeated, learned abilities of the organism (see also Dennett 1983, 1987). Careful observation by trained observers is crucial to an understanding of humans and other animals. But the reporting of 'anecdotes' is considered anathema. Especially on the subject of animal mental states, we must contend with numerous anecdotes reported by Romanes (1884). He seems all too charitable in his acceptance of reports by various and sundry observers. Experimenters, to date wisely for their reputation as scientists, omit from published scientific reports those one-time or few-time events which are so illuminating, and often provoke future experimentation. Yet Piaget based an important body of research on careful observations of his children, often constructing experimental paradigms from them, as others have continued to do since his day. And those observations were essential to the progress of developmental psychology. With luck a video camera is sometimes available to record an especially interesting event; but indeed this is rare, for one requires not only the camera, but adequate lighting, a correct angle of view, etc. Usually we must rely on more conventional recording of information.

In the ape-language and cognition field, a counter-culture of information seems to exist, that published in more popular presses such as *National Geographic* and heard in casual conversation with the investigators. It is a great loss that their carefully gathered observations are missing from the body of information which is available and professionally acceptable to other researchers. Many observations are insufficient evidence in themselves, and *should* suggest experiments to further test the suspected ability. But the observations are definitely not to be ignored. There may be a growing trend to report such observations carefully. I think it is a necessary trend if we are to deal intelligently with possibly intelligent behaviour by animals.[12]

23.6.4 The importance of ethological indices of behaviour

As has previously been suggested (Ristau and Robbins 1982a), ethological indices can support the cognitive interpretations applied to the chimpanzees' actions. It is likely that these very indices are an important part of the basis for our 'intuitive' sense of when an act is 'cognitive' and when rote.

One example is the sense that an organism is 'thinking about' the problem at hand. A classic case was that described by Köhler (1925) for the chimpanzee Sultan just before he solved the problem of reaching inaccessible food by putting two sticks together so as to make a longer pole. The ape is said to have looked back and forth between the sticks and the food for a while, and then simply walked over to the sticks, put them together, and used them to reach the bananas out of reach outside the cage.

Consider, also, the chimpanzees in Woodruff and Premack's (1979) studies of deception, requiring discrimination between 'good guy' and 'bad guy' trainers, who, respectively, shared and did not share hidden food with the apes. The chimpanzees, who knew which container held the food, progressed through interesting stages of, first, impulsively showing excitement when either trainer was near the container with hidden food; next to generalized agitation and extraneous behaviours that functioned to make it less obvious which container held food; then to gradual inhibition of, first, involuntary body-turning, and then even of eye-gaze toward the correct container when the 'bad guy' trainer was present; and finally, to the use of a voluntary movement, pointing, towards the correct container (the one with food) to the 'good guy' and the container without food to the 'bad guy'. (See further discussion in Ristau and Robbins 1982a.) This sequence is probably a good starting-point for examining the ontogeny of human children, as well as for comparative studies of deceptive abilities.

In the comparative studies of mirror recognition (see Lock and Colombo, this volume, Chapter 22), it is important not merely to note whether an animal has touched a mark on its head while looking in the mirror, but also the kind of touch: a brief swipe, or a slower examination, perhaps sniffing the digit that has touched the mark as well.

In trying to determine whether the numerical competence or 'counting' exhibited by primates is indeed that or instead some rote ability or subitizing, we note suggestions of possible 'thinking'. Lana, the chimpanzee in the Rumbaugh project, is described as pausing on 21 per cent of 1442 videotaped trials of her training programme for 'counting'. She then either continued moving the cursor on its previous path or changed its route, attaining accuracies of 71 per cent and 61 per cent respectively, to be contrasted to an overall 81 per cent correct (Rumbaugh and Washburn 1993).

In a study of the quantitative abilities 'larger/smaller than' the chimpanzee Sheba had to select the smaller of two piles of candy in order finally to be given the larger pile to eat while another chimp received the dish she selected; she was unable to do so (Boysen 1993a, b). Noticing her apparent 'frustration' and 'irritability' just after she made her incorrect choice, the experimenter Boysen replaced the piles with Arabic numerals; the ape was immediately correct. Switching back to actual candy always resulted in an incorrect response. The impulsive nature of Sheba's desire for the candy overshadowed her cognitive capacity to make the necessary discrimination. This kind of lack of inhibition, also apparent in Woodruff and Premack's chimps in the deception experiment, is probably a characteristic widespread in children and in animal species, and even in adult humans at times.

In another quantitative task, Sheba was photographed 'tagging' the items she was to be counting. This behaviour emerged when, after training with the numerals 1–3, the number 4 was introduced, and her performance deteriorated somewhat: '... she was observed to begin to touch, point to, or move items in the array before making her final decision' (Boysen and Berntson 1989, p. 24). During the preceding 18 months of number training, her teacher had consistently tagged items. In very young human children, the tagging and partitioning behaviour appears in early stages of learning to count and, presumably, helps them keep track of items counted and yet to be counted (Gelman and Gallistel 1978). Although the functional significance of the behaviour has not been determined for Sheba the chimpanzee, it is intriguing to have noted it.

The chimpanzee language projects and studies in the wild of chimpanzees have many examples of eye-gaze directed to the experimenter, seeming to signify a request, and toward an event or object, seeming to be a direction of another's attention. By compiling these examples, and better still, by more detailed study in the wild of natural chimpanzee communication, we are likely to gain better insight into their other capacities.

23.7 The changing interpretations of linguistic development and of the ape language projects

Philosophers have sometimes been drawn to consider the animal language/cognition projects, in particular the research conducted with common chimpanzees and bonobos by Sue Savage-Rumbaugh and her colleagues. I presume a major part of the interest in that project has to do both with the use of 'language games' that are more like those of human parents and children than the usual more rigid experimental settings, and with the extraordinary accomplishments of the apes, particularly the bonobo Kanzi, in that project.

Interpretations of the apes' accomplishments vary enormously. Gauker, for example, reviews the studies done in the early and mid-eighties of the common chimpanzees Sherman and Austin (research with the bonobo Kanzi was not published until later). For Gauker, language is conceived of 'as a tool—not a tool for representing the world or expressing one's thoughts but a tool for effecting changes in one's environment'. The '... chimpanzees learn[ed] to use symbols by means of a cause–effect analysis...' (Gauker 1990, p. 31). In short, the apes learned how to get rewards via the lexigrams; in his view, humans do the same. He does not consider the notion of reference to be useful in analysis of language, because of the deep philosophical problems reference entails. Although Gauker's ideas may initially seem to be a kind of Skinnerian behaviourism, they are not so easily dismissed, in the light of the generalizations or transfers the chimpanzees can make to similar kinds of problems in which the actual stimuli involved are completely different—a capacity not handled by a Skinnerian interpretation.

Quite differently, Shanker considers that Savage-Rumbaugh has made a paradigm revolution. 'The paradigm revolution which Savage-Rumbaugh is doing so much to forge is one which situates and makes sense of an agent's actions—what an agent is doing or trying to do—in the context of social practices; which looks at language learning in terms of an agent's mastering the rules of these social practices rather than as a private inductive process...' (Shanker 1994, p. 36). As with Gauker, the concept of 'reference' does not play a significant explanatory role in the interpretation. Shanker further explains 'The operative picture here is not *making sense*; it is *interacting*' (p. 35). The chimpanzee or the child has desires and expectations and tries to satisfy them by learning how to gain and direct attention, using sounds and gestures, and then words 'in order to change its situation, obtain an object which is not present, or simply to play' (p. 35).

Linguists, such as Greenfield, who has co-authored papers with Sue Savage-Rumbaugh (Greenfield and Savage-Rumbaugh 1990), would, I think, agree with Shanker. To sounds and gestures in the young child's (and chimp's) developing communicative skills she would add the use of objects. (Recall how the Gardners' chimps incorporated objects into their discourse, both as the place on or near which to make a sign, thus modulating the meaning of the sign, and as part of the interaction.) Such emphasis on language as a communicative act, importantly incorporating objects, gestures, sounds, and gaze to convey meaning, runs counter to a Chomskyan analysis, with its reliance on grammatical definitions of language.

A further remarkable development in the field of neurophysiology is likely to have significant impact on our interpretations of the language-continuity issue in humans and other animals, and probably on the nature of the Chomskyan 'language organ' itself, if such a thing exists. This is work by Paula Tallal and her colleagues that shows left-hemisphere specialization/advantage in rapid temporal processing (Tallal et al. 1993a). The left hemisphere can process changes in tens of milliseconds (as needed to perceive consonants), while the right hemisphere deals with processing of steady-state or slowly changing information (such as vowels), regardless of whether stimuli are verbal or non-verbal. Data derive from brain-scanning techniques and behavioural measures, not only in humans, but in animals that have been tested, specifically rats (Fitch et al. 1993). Left-hemisphere dysfunction that results in temporal processing disorders now links not only speech-perception, but speech-production as well, and even includes defects in rapid fine motor skills and visual processing; specific disorders such as dysphasia and dyslexia (reading disorders) (Tallal et al. 1993b). It is not clear what the impact on human and animal language/cognition research will be. Some intriguing relationships might be the following. The ability to comprehend grammar in spoken language normally requires the ability to process consonants: is that the primary characteristic of the 'language organ'? Do children who have difficulties with spoken language also show difficulties with signing? If not, that's a remedial technique to improve their language skills. If so, would merely slowing down signing improve their condition? If so, do these children using 'slow signing' have the same grammatical capacities as normally signing and normally speaking children? If so, that would place a 'language organ' somewhere other than in the rapid temporal processing abilities of the left hemisphere.

Furthermore, different parts of the brain have been implicated in different developmental phases of human language use, in particular the learning of language as distinct from the maintenance and use of fluent language in a mature adult (discussed in Bates 1993, p. 239). This is interpreted as providing 'little evidence for a circumscribed language organ. The whole brain participates in language...' (ibid., p. 239). (Strictly speaking, it is not necessary that a language organ, if such there be, is centred in only one portion of the brain.) Bates further considers that we should abandon a 'vanilla' kind of general cognition. The brain, however, is not organized around content-specific faculties (for example, for language, music, face perception); rather, neural systems have different computational properties, some of which *indirectly* support language learning and use in humans. Chimpanzees have some different neural systems, resulting in 'quantitative and qualitative differences in language ability' (ibid.).

From my own vantage point, it is possible that the 'universal grammar' presumably found in different human languages (though that point is now also being questioned) reflects the functional roles of any language. A language needs at least to describe attributes of the world, including space and time, the needs of oneself and others and the interaction of both, and reference. Thus, intrinsic to any grammar are subject, action, object, spatial relations ('on', 'under'), ownership ('of'), reference ('about'). The objects and events have characteristics: adjectives, adverbs, and embedded clauses accomplish that description. Events were, are, will be, or might be: hence tenses and the subjective.

One would expect that a *very* primitive language might not have all these attributes, concentrating, perhaps, on instrumental requests/commands and spatial locations. Interestingly enough, the primitive so-called 'bee language' indicates spatial location; the chimpanzees' initial errors in a number of situations, in particular the 'functional training' for meaning, were to associate the lexigrams for tools, not with tools or their uses, but with specific spatial locations where the tools were encountered (Savage-Rumbaugh and Rumbaugh 1978, as discussed in Ristau and Robbins 1982a, pp. 198–205). A common criticism of the ape 'language' is that the productions are instrumental. Unfortunately, there is no extant primitive language nor any records of such. Human children raised without any adult linguistic input would provide relevant data. Such unfortunate situations occasionally arise in wartime, but the absence of adult linguistic input cannot be verified.

Finally a point about the nature of evidence in science. There comes a time when we must recognize that, after millions of years of evolution, we are very powerful analytic devices, prone though we be as well to certain kinds of biases and simplifying heuristics. We are fairly good at understanding each other's social interactions; better than any gadget yet devised. We are probably especially good at understanding our 'next-of-kin', the other Great Apes, in contrast to other species. If our present, verbal, analytic concepts and tools don't 'jibe' with the understanding that we have when we take part in or see ape social interactions and communications, alive or on video, that should give us great pause. It may be necessary to revamp our scientific tools, rather than to toss away our millennia-old social comprehension skills.

23.8 Evolutionary significance

What is the evolutionary significance of the ape-language studies for the evolution of human symbolism? This question is often understood as pitting two extreme points of view against each other: 'human language evolved from animals' natural communication systems' versus 'human language is an innate capacity which is the unique domain of humans' (cf. for example Chomsky passim, Cenami Spada 1994). Those who hold the innate view often cite Hewes's (1977) point that the neural control of the natural or so-called 'survival' communication systems is at a different location from that of human language. Animal and human screams of pain can be evoked by stimulating the limbic system. Hence it is said that if human language evolved from those call systems, it should have replaced them naturally, or at least should have been located nearby (see also Hoban 1986). Yet this neural argument is not really so simple. It is not clear what is meant by 'cries of pain' being 'located' in the limbic system. If stimulating the limbic system produces such cries, it may mean only that the stimulation produced pain, and that in turn the organism made the typically involuntary call associated with the sudden feeling of pain. Even to produce the sound requires a connection to the motor-control parts of the vocal system. So there is indeed interconnection with other neural areas for even so simple a call.

There are, of course, reasonable positions between the two extreme views of language origin. It is quite plausible that humans and animals share similar cognitive functions (see, for example, Ristau 1991), many of which form the foundation for language-use by humans. The same functions may also be accessed by other species to solve everyday problems, and perhaps while using their natural communication systems. A capacity for displacement may be such an example. Indeed, the ape-language work has been progressing towards more research about cognition, with a growing understanding of the important role of cognition in assessing apes' language capabilities.

Although a discussion of natural animal communication systems is outside the scope of this paper, certain points must be established because of their relevance to interpretations of the ape language and cognition research. Animal communication is often considered to be involuntary and emotive, i.e. expressing an emotional state of the animal. It may also be used to advertise a general mood, receptivity, or behavioural tendency of the organism. An example, is the song of birds in the springtime, which apparently communicates a male bird's readiness to receive a female into a newly established territory, and simultaneously warns conspecific males that this is indeed an established territory: 'keep away; or the territory-holder's behaviour is likely to be aggressive'. Animal communication is, however, far more complicated than this (Smith 1977; 1991). There is evidence that at least some animal communication conveys semantic information (Seyfarth *et al.* 1980; Gouzoules *et al.* 1984, 1985) (see the discussions in Section 23.3 of this chapter, Jolly, this volume, Chapter 6, and Ristau 1988), and is sensitive to the presence of an audience and its type (research with chickens, Gyger *et al.* 1986; Marler *et al.* 1986, 1991). Another basic fact that should inhibit tendencies to make broad evolutionary statements about the origin of language from research with contemporary apes is that indeed the apes are contemporary; they are not our forebears. We and they are but mere twigs on the evolutionary bush (Gould 1977).

23.9 Conclusions

One essential question is which of an ape's abilities are evolutionary precursors to human abilities, particularly in the realm of symbolism. Language abilities are of interest, because they are human's most prevalent and potent expression of symbolic content. At least some of the apes have, in some situations, achieved the use of rudimentary symbols. Though their use of symbols is not fairly characterized as dealing only with requests, they do not show the multiplicity of functions of language which humans do. It should also be noted that most of the training methods little resemble the procedures by which human children are believed to 'pick up' language; though an exception is the recent work with bonobo chimpanzees, in particular, Kanzi (Savage-Rumbaugh *et al.* 1993), who did 'pick up' the meanings of lexigrams and of spoken English words and simple sentences. Only a few years ago, I would have concluded that the apes do not show evidence for the use of grammar, except that in some highly specific situations apes (Premack 1976), sea lions (Schusterman and Krieger 1984; Schusterman and Gisiner 1988, and dolphins (Herman *et al.* 1984; Herman 1986, 1987) have exhibited an ability to *comprehend* the order of lexical items, and *some* apes (Rumbaugh 1977; Premack 1976) and pigeons (Straub *et al.* 1979), at least, can *produce* certain simple series reliably. Yet the ability to comprehend and use grammar certainly increases the scope of symbolic expression.

Notwithstanding the paucity of grammar, the symbolic abilities exhibited by apes could suggest an evolutionary precursor to human abilities. Experiments with other species, in particular work with the parrot (Pepperberg 1986) that explored the potential meaning of lexical items, reveal abilities to form categories and rudimentary symbols. Such results imply that it is not the unique organization of the primate brain nor the special circumstances of the social or other environment of the primate that are critical for the appearance of some symbolic ability. What is critical may simply be that, under divergent social situations and divergent environments, there are recurring needs for cognitive abilities.

Notes

1. An example of a tact is a child naming an object; in Skinner's terms the discriminative stimulus for that tact is the object. A mand is a demand; the discriminative stimulus is a deprivation condition, such as hunger or thirst, to which a child might say *cookie* or *juice*. Echoic responses mimic sounds that are heard.

2. They are capable of producing somewhat similar whistles as part of their *natural* communication system. The issue has recently generated further interest, because though separation between the two processes seems obvious in the common chimpanzee, current work with the pygmy chimpanzee (Savage-Rumbaugh 1984a, 1986) suggests close intertwining of the two processes somewhat comparable to that which occurs in humans.

3. To expand upon the nature of rules as understood by linguists: Examples of structure-dependent rules in which grammatical relationships are conveyed by means other than mere word order are the following: 'Bob hates liver. The liver is hated by Bob. It is Bob who hates liver.' These complexities are no where entertained in the artificial ape languages.

A recursive rule is frequently helpful in language. Recursion is the possibility of applying the same rule indefinitely. Occasionally, its use can strain even our intelligence. A familiar example is: 'This is the house that Jack built ... And this is the dog that chased the cat that ate the mouse that lived in the house that Jack built.' Again, none of the artificial animal languages have such complexities.

4. However, according to an analysis by Granier-Deferre and Kodratoff (1986), chimpanzees doing certain problem-solving tasks can, in those situations, use recursive rules, at least to a limited extent; macaque monkeys could not. The experiments analysed were those by Tinklepaugh (1932) and by Menzel (1973, 1978) and Menzel and Halperin (1975). Using Artificial Intelligence (AI) concepts, Granier-Deferre and Kodratoff (1986) suggest that the minimal cognitive abilities required for the chimpanzee's successes require the same skills needed to manage context-free and context-independent grammars. Thus, state the authors, the ape's behaviour, if not its linguistic expression to date, indicates the presence of important prerequisites for the understanding of structural properties of language. Their analysis suggests the fruitfulness of encouraging scholars from the field of AI to consider these and other problem-solving experiments and their relevance to linguistic and cognitive faculties.

5. It is interesting to note that in these studies of causality, all three language-trained animals passed both versions of the test in the first session consisting of 12 trials, whereas all four non-language trained animals failed both versions, and continued to fail even though given 8 replications of the same 12 object–implement pairs.

6. It is necessary to be careful in interpreting these findings. For example, Straub et al. (1979) have shown that pigeons, when tested with coloured lights, are also able to produce a specific order. The pigeons were required to peck in correct sequence up to five differently coloured buttons which were arranged in different orders on different trials; the pigeons could do this. Straub and colleagues then attempted to draw an analogy to the ape-language research, noting that the experimenter could have assigned a different 'meaning' to each coloured button, perhaps, 'Please machine give me grain'. The pigeon would have been judged as capable of using those items in correct order, i.e. of producing a 'sentence'. Of course such a test does not reveal such an ability; but it does alert us to dangers of over-interpretation. Neither is this case a good analogy to the ape experiments, for the pigeon has only five potential 'lexical items' and one prescribed order. The apes can have to deal with over a hundred units, and various prescribed orders depending on context and function. Similarly, Epstein et al. (1980) demonstrated, via operant conditioning techniques, sequential key-pecking of several keys by and between two pigeons. This, too, was deemed by the researchers to be analogous to the ape-language capabilities, but faces at least the same criticisms just noted.

These provisos serve to draw attention to the need for the precise nature of apparent rule-learning to be investigated. As an illustration of this, pigeons were trained to peck four colours in a specific sequence, symbolized as ABCD, and then tested with three element sequences. They did best on ABD and worst on BCD; none of the results were up to the levels for the ABCD sequence (Straub and Terrace 1981). The authors' interpret the pigeons' behaviour as accounted for by rules 'start at A' and 'respond to D last', that is, far simpler rules than 'learn that A precedes B precedes C precedes D' or the like. Similar tests have not been conducted on monkeys or apes, but the accomplishments of some of Premack's chimpanzees, the dolphins, and the sea lions in the use of lexical order make it most unlikely that such simple rules suffice.

7. Note that when four-year-old children (Dolgin 1981) were given the same tasks, they, too, could not pass the transfer tests. With, however, an interesting exception: they could be told the meaning of 'agent', 'object', and 'instrument', and then they passed the transfer tests at a high level of accuracy. College sophomores do not even require verbal definitions.

8. Realize, though, that grammar conveyed by word order is the device used by English, not by all human languages. Inflection of key words to convey grammatical relationships is used by many human languages, and could be a mechanism utilized by non-human species (Griffin 1986). That possibility has not yet been investigated, though the Gardners are reviewing their accumulated data for the significance of inflections in the chimpanzees' signing.

9. A few niceties: it would have been preferable to test the apes with yet a third classification to ascertain that the divisions as understood by the ape were indeed 'food' versus 'tool' rather than 'edible' versus 'inedible', or 'edible' versus 'anything else in the world'.

10. In general, it has become more common to speak in terms of diverse species of animals having mental 'representations' which serve them in various learning and cognitive tasks (see Lock and Colombo, this volume, Chapter 22).

11. Critics (for example, Hoban 1986, p. 136) argue that the dolphin is simply withholding a response for a brief period, which is different from achieving displacement. Several comments are in order. The dolphin is doing more than withholding a response, because, until the goal object is present, it must keep a representation present in memory. Other species, including the pigeon, can perform simpler delayed match-to-sample tasks, which often require maintaining the memory of a stimulus. Maximum delays tolerated vary with species and sometimes with tasks, but, except for the Great Apes, are less than a minute and, for the pigeon, are typically of the order of 10–15 seconds. Performance has often been explained in terms of fading stimulus–response traces; whether this is appropriate is another controversial issue.

It must be agreed, however, that the dolphin has not exhibited the more advanced notion of displacement which would be exemplified by a test Hoban suggests. Humans, upon hearing of a traffic jam, can alter their route. Could a dolphin, given information of a change in its surroundings, do the same? For example, the dolphin may have put a ball in a basket. Surreptitiously remove the ball from the basket and place it elsewhere. Then tell the dolphin the ball's new location and ask the dolphin to retrieve it. Does the dolphin look for the ball where she left it, as she normally does, or can she use the trainer's information to alter her behaviour? It seems that a critical difference between Herman's experiment and the one suggested by Hoban lies in the manipulations upon a memory or cognitive map required in Hoban's experiment as opposed to merely carrying a notion of an absent object in Herman's. Another facet of Hoban's description is that it entails variety in the kinds of use of the lexical item—i.e. the lexical item is not merely used in requests, but can function simply to give information. As previously noted, evidence that a lexical item is used as a referent should include both displacement and the ability to use that lexical item in a variety of contexts or function. Herman is currently developing experimental paradigms in which more varied functions of lexical items will be possible.

12. See also Whiten and Byrne (1988), and commentary and authors' reply, for a longer discussion of the status of 'anecdotal' evidence.

Table 23.1 An overview of recent ape language cognition projects: the signing apes

Principal investigator(s)	Gardner and Gardner I	Gardner and Gardner II	Terrace	Fouts	Patterson	Miles
Chief interests of principal investigator(s) regarding the project	Two-way communication between human and apes	Develop and describe 'verbal' communication from birth on to determine limits and extent of chimpanzee linguistic capacities	Initially to amass corpus of 'utterances' to evaluate linguistically. What features of a natural (human?) language can chimpanzees master?	Communication between chimpanzees using signs. Can apes learn a language? Cultural transmission of signing between apes.	Investigation of linguistic capacities in the gorilla. Comparative cognitive abilities among apes and humans	To investigate communicative and cognitive skills involved in the acquisition of signing language. Particular interest in meaning of signs, use of deception.
Species	Chimpanzee (*Pan troglodytes*)	Chimpanzee (*Pan troglodytes*)	Chimpanzee (*Pan troglodytes*)	Chimpanzee (*Pan troglodytes*)	*Gorilla gorilla*	Orang-utan (*Pongo pygmaeus*)
Sex, age when acquired, and age when training began of principal subject(s)	Washoe (♀); acquired and training begun at 8–14 months old.	Moja (♀), Tatu (♀), and Peli (♀) (died before second birthday), and Dar (♀); all acquired (and exposure to ASL began) at 1–4 days old	Nim (♀); acquired, and exposure to ASL began, at 2 weeks old; formal classroom training from age of 10 months.	Booee (♂), 3-year-old born in captivity, with partial split brain*; Ally (♂), 3 years old, born in captivity; Bruno (♂), 32 months old, born in captivity; Lucy (♀), 7 years old; Cindy (♀), 45–51 months old, African-born; Thelma (♀), 33–39 months old, African-born. (Training for all begun at ages indicated.) 1970: acquired Washoe from Gardners. 1979–80: acquired Moja, Tatu, and Dar from Gardners. Louis (♂), 10 months old, born 1978 in captivity, adopted by Washoe for cultural transmission study.	Koko (♀), born in captivity; began at approx. 1 year of age. Michael ♂; began at approx. 3.5 years old.	Chantek ♂, born in captivity in 1978; acquired (and language training began) at nine months of age (Oct. 1978).
Living environment	Housetrailer, with access to inside, and fenced-in yard	Home-reared	Home-reared; then partial home environment	Combinations of home-reared and partial home environment and caged, initially at Inst. for Primate Studies, Univ. of Oklahoma, then at Central Washing University in 4 connecting cages; now in new, bigger facility.	Zoo nursery, then five-room trailer: limited access to outside and some rooms; total access to kitchen and living-room.	House-trailer with access to yard and play gym; now at Language Center, Ga.
Training environment	Living environment	Living environment	Living environment and small classroom (8 ft square) at Columbia University	Living environment	Living environment	Living environment, including outdoors
Language medium	Manual signing	Manual signing	Manual signing	Manual signing (Ally also heard spoken English).	Manual signing and spoken English ('simultaneous communication')	Manual signing in 'Pidgin Sign English'

Trainers: number	A few principal trainers and student volunteers	A few project research assistants (at least six on any one day); total not specified	Sixty teachers, most of whom were 'occasional playmates'. Core group of eight teachers.	Some principal trainers and others of brief durations	One primary trainer and 14 assistants	'Small staff'; 24-hour companionship for first six months
Trainers: qualifications	Most not fluent in ASL	Many deaf and fluent signers, research assistants, and others fluent in ASL	Many student volunteers; most not fluent in ASL; core group has 1+ years' ASL training.	Initially most not fluent in ASL. Then more training.	Many deaf or native signers	Unspecified
Major training methods	Moulding, modelling, prompting, and observation of signing; no specific training for utterance size	Moulding, prompting, and observation of signing; used fluent native signers as human models.	Moulding, modelling, and prompting	Moulding, modelling, and prompting	Moulding, modelling, and prompting. Is in presence of signing companions 5 hrs/day when in zoo; 8-12 hrs/day in the project's living environment.	Casual use of signing by staff in course of tending and playing with Chantek
Rewards	Social interactions; food object or event being taught was often 'delivered'.	Same as Gardner and Gardner I	Social, food	Social, food (a raisin was often given).	Social; object or event being taught or topic of conversation often 'delivered'.	Not specifically contingent upon signing; but social interactions, goodies
Major testing methods	Vocabulary tests: 'Wh-' questions	Formal Vocabulary tests:	'Wh-' questions	Formal vocabulary tests	'Wh-' questions; signing alone; spoken English alone; infant intelligence tests	
Other data-collection methods	Tape-recorded 'comprehensive' samples; video tape; written observations of sign and context	Diary records, videotapes, inventory of phrases, samples of 'verbal' input and output; method of 'obligatory contexts'	Videotape; teacher reports; written observations of sign and context	Videotape, inc. some 'remotely', without human present; recording of outputs and contexts	Videotape; daily diary; tape-recorded sampling of signing and context; some timed samples of Koko's signed output	Initially full contextual information gathered for all signed communications by ape. From Nov. 1979 only different signs, combinations, and contexts recorded. Videotape 1-hour monthly sample of relaxed sign exchanges
Reliability tests	Double-blind test of sign knowledge for sample of objects; use of naïve (with regard to Washoe) native signers; inter-judge reliability.	Double-blind; independent data-recorder and trainer for samples of interchanges; inter-judge reliability.	Comparison of teacher reports; 77% agreement before teachers consulted one another, 94% after.	Double-blind tests (83–100%, $\bar{x}$ = 94% agreement)	Double-blind tests; Koko often 'avoided' task; overall performance = 60% correct.	Not indicated
Criteria for the acquisition of a vocabulary item	Must occur at least once 'spontaneously' and appropriately on	Identified by at least three of the fluent human companions,	Three independent observers note spontaneous	Same as Gardner and Gardner I	Must be recorded by two independent observers and be	Spontaneous and appropriate usage on half the days

Table 23.1 (*Contd.*)

Principal investigator(s)	Gardner and Gardner I	Gardner and Gardner II	Terrace	Fouts	Patterson	Miles
	each of 15 consecutive days; 'spontaneous' = occurring without prompt *or* in response to 'What this?'	followed by 15 consecutive days of spontaneous appropriate usage	occurrence; then sign must occur spontaneously on 5 successive days; 'spontaneous' = signed in an appropriate context without aid.		used spontaneously and appropriately on at least half the days of a given month.	of a given month
Size of expressive vocabulary	132 signs after 51 months of training	First signs appear at 3 months of age; accepted 'immature' variants of signs; over 50 signs within 2 years.	125 signs after approximately 45 months of training	Approximately 40–70+ signs after 2–4 years of training	Koko had 264 signs after 66 months of training (approximately 185 signs by the Gardners' criteria). She has more now, and somewhat more than signing chimpanzees have generally.	Over 100 signs by age 4½ years
Abilities and major results claimed by researchers	Spontaneous appropriate use of combinations and phrases; signs to self; overgeneralization to novel objects and situations; word-creation.	Abstract use of 'no'; first 50-sign vocabularies not different from first 50-word vocabularies of approximately 2-year-old humans; new word-combinations (all Moja), e.g., alka seltzer = *listen-drink*	Multisign utterances; MLU remains at 1.6 signs/utterance last 1.5 years; increase in length did not add information, e.g., *Play me to Play me Nim;* 'discourse' analysis of videotape transcripts showed signs often imitative of teacher's prior sign.	Performance in vocabulary tests ranged from 26% to 90% correct; communication by signs and cultural transmissions of signs between chimps; novel combinations: e.g., Washoe: *water-bird* for swan	Engages in deceit; signs to self; over-generalization to novel objects and situations; word-creation; novel combinations—e.g., *cookie rock* for stale roll. Claimed to rhyme signs and understand spoken English.	Highly spontaneous and non-imitative use of signed communications, including novel sequences; referential sign usage; deceptive skills of several levels
Future research plans	See Gardner and Gardner II.	Analysing past accumulated data on videotape, daily diaries, and written observations. Especial emphasis on subtleties of apes' use of signing, such as modifying the form of the sign to vary meaning. Detailed analysis of errors to indicate apes' understanding of signs.	Collaborative work with Savage-Rumbaugh, Rumbaugh, and Menzel on spatial knowledge and/or cognitive tasks in the chimpanzee. Cognitive studies in the pigeon (*Columba livia*), including serial learning.	Continuing ongoing studies. Analysing videotape for contextual details. Simultaneous natural communication occurring with signing, and sign modulations by the apes.	To continue studies with gorillas Koko and Michael, including spelling and reading.	To continue studies of Chantek, now housed at the Language Center in Georgia; collaborative work with Savage-Rumbaugh, Rumbaugh, and colleagues.

* Booee's cerebral hemispheres and spine were surgically split except for the medulla.

Table 23.2 An overview of recent ape language cognition projects: artificial lexicons

	Premack I	Premack II	Rumbaugh, T. Gill, E. von Glasersfeld (linguist)	Savage-Rumbaugh and Rumbaugh I	Savage-Rumbaugh III	Matsuzawa	Boysen
Principal investigator(s)							
Other major investigators associated with project		D. J. Gillan, G. Woodruff, Cognition: match-to-sample problems, reasoning, quantitative abilities, map-reading, conservation, deception, attribution of intention and other mental states	T. Gill, E. von Glasersfeld (linguist)	S. Boysen			
Chief interests of principal investigator(s) regarding the project	Language training; components of intelligence		Linguistic ability of apes; the nature of language; methods and tactics for language training of mentally retarded children	Development of a 'word' in a chimpanzee, communication between chimpanzees	Comparative cognitive and linguistic studies in a different *Pan* species; comparison of bonobo's and human child's ability to comprehend novel spoken English sentences.	Language-like skills used as tool to study cognitive abilities of chimpanzees; comparison of such abilities to humans'	Comparative cognitive skills, including numerical, in chimpanzees and other species
Species	Chimpanzee (*Pan troglodytes*)	Chimpanzee (*Pan troglodytes*)	Chimpanzee (*Pan troglodytes*)	Chimpanzee (*Pan troglodytes*)	Bonobo or 'pygmy' Chimpanzee (*Pan paniscus*).	Chimpanzee (*Pan troglodytes*) Human child.	Chimpanzee (*Pan troglodytes*)
Sex, age when acquired, and age when training began of principal subject(s)	Sarah ♀, Peony ♀, Elizabeth ♀; acquired at <1 year old, African-born; training begun at 5–6 years old for Sarah, approx. 2.0–2.5 years for others.	The 3 language-trained apes and Bert ♂, Sadie ♀, Luvie ♀, Jessie ♂, these all acquired at 1.0–1.5 years old, African-born; training begun at approximately 2.0–2.5 years old (exc. Sarah)	Lana ♀; acquired at 1.5 years old, born at Yerkes Primate Center; training begun at 2.25 years old.	Sherman ♀, Austin ♀; acquired at 2 and 3 years old, born at Yerkes Primate Center; training begun at 3.5 and 4.5 years old.	Kanzi (♂), Matata (♀), Mulika (♂, died). Matata acquired at 5–7 years old, African-born; others her offspring, born in captivity. Matata began language training as an adult; Kanzi exposed to her training at age 6 months, to her own training at 2 1/2 years; Mulika exposed as an infant. Human child Alia (♀) studied between 1.5 and 2.0 years old.	Ai (♀): acquired at 1 year old, African-born; language training began at 2 years old. 9 other chimps available.	Sheeba (main subject) (♀), Sarah (♀), Kermit (♂), Darrell (♂). Captive-born. Sheeba human cross-fostered from 4 months until came to project at 2.5 years. Sarah from Premack research. Males peer-raised in laboratory nursery until 3.0 and 3.5 years, when joined project.
Living environment	Standard caged/laboratory environment; reared in close contact with humans.	Standard caged/laboratory environment; reared in close contact with humans.	Large room with computer keyboard	Similar to Gardner I and to Terrace, but apes not separated.	During the present study, Matata in large, caged living area; Kanzi has access to 5 large indoor living areas and 55 acres of natural Southern US forest. Home and laboratory for human child.	Individual cages, outdoor playground with other chimpanzees	Males have shared home cage, including an outdoor area. Sheeba has individual cage.
Training environment	Living environment	Living environment	Living environment	Living environment	Living environment for apes. Home and laboratory for human child.	Laboratory	Living environment; Sheeba, smaller and younger, tested unrestrained in open area.

Table 23.2 (*Contd.*)

Principal investigator(s)	Premack I	Premack II	Rumbaugh, T. Gill, E. von Glasersfeld (linguist)	Savage-Rumbaugh and Rumbaugh I	Savage-Rumbaugh III	Matsuzawa	Boysen
Language medium	Plastic chips of various colours and shapes arranged vertically on magnetic board; each chip was 'word-like' in function.	Photographs, videotape, human actors, small toys and objects	Computer lexigrams, colour-coded with regard to gross semantic classification; Yerkish grammar	Computer lexigrams (not colour-coded); Yerkish grammar.	Lexigrams on stationary and portable computer and board (grammar approximates English). Subjects hear simultaneous spoken English and English without lexigrams. Child also exposed to both media.	Computer lexigrams (different symbols from Savage-Rumbaugh)	No language testing done on apes.
Trainers: number	Three principal trainers, and at least six who stayed briefly	Some principal trainers, and others of brief duration	A few principal trainers plus others	Some principal trainers, and others	Unspecified. Kanzi and child share a common care-giver, the child's mother.	Unspecified	Principal investigation and others. Total unspecified.
Trainers: qualifications	Unspecified	Unspecified	Unspecified	Early phase of project: had taught 'language' to apes or intellectually impaired children. At present: training programme for new personel in lab.	Unspecified; some are Ph.D. candidates.	Unspecified	Unspecified
Major testing methods	Purely observational failed; non-correction training for 'component skills'; training gradually increased utterance length.	Match-to-sample procedure requiring comprehension of problems faced by other individuals; learning how to acquire hidden foods in co-operative or competitive contexts	'Labelling' by associating lexigram and object it denoted; 'whole phrase' training.	'Functional' training for lexigram meaning; tool-use and social co-operation; categorization training	Matata by discrete trial symbol training tasks similar to Sherman and Austin; Kanzi and Mulika by observational learning and social interactions, including daily routines. Attempt to foster spontaneous, casual use of lexigrams on portable board. Child is reared normally, with additional lexigram exposure similar to Kanzi's.	Symbolic match-to-sample	Experiments with intensive human–ape interaction. Most 5 days/week.

Rewards	If correct: food, verbal hand-patting ('Good girl'); if incorrect: verbal ('No, you dummy'), withdrawing chips from board, trainer requesting ape to give up chips.	If correct: food; if incorrect: no food and removal from room.	Food, social praise, social interactions, movies, and various events (when correct on ape generates 'grammatically' correct 'sentences')	Food, social praise, tickling games, objects, exploring out-of-doors, etc.; reward-sharing	Access to food, desirable objects, activities and social interactions by request using lexigrams. Less emphasis on rewards for Kanzi and Mulika. Social rewards of praise, hugs, and pats. Child rewarded 'normally'.	Different buzzers for correct and incorrect responses. Raisin or apple piece delivered after consecutive correct responses.	Social (praise) and food reward for correct answers. Apparently correction procedure used.
Major testing methods	Multiple-choice tests; homogenous problem sets	Forced choice photograph tests; production and comprehension tests	Multiple-choice tests; relatively small problem sets; analysis of 'conversations'	Inter-animal communication in co-operative tool-use paradigm; problem-solving tasks, often automated	Emphasis on conversational use of lexigrams in daily social interactions with humans, including meals, play, walks in woods. Ongoing record of much lexigram usage; notes on context and accuracy.	Same as late-stage training methods. *Only* productive abilities, not receptive, tested.	Same as late-stage training methods. Blind tests also used, with no cueing possible.
Other data-collection methods	Some descriptions of her behaviour	Videotape of a subset of trials	Computer has all productions	Computer has all productions; videotape sample of non-'linguistic' productions.	Computer has most indoor productions, and outdoor utterances when the portable computer is used. Some videorecord. Blind tests use one-way mirror to hide the human speaking; two observers wear earphones with loud music to prevent their hearing directions given to Kanzi or the child.	Computer has all productions.	No computerized procedures. Some video record.
Reliability tests	'Dumb' trainer method illustrates reliability and evaluates role of 'Clever Hans' cueing.	Human observers not in same room or different trainers involved in baiting and testing.	Humans not necessarily involved in many interactions with computer; however, human often present during 'novel productions' Performance on multiple-choice tests	'Blind' tests with 'dumb' trainer who cannot see ape give response	Many usages on computer; no reliability tests for contextual information and categorization of utterance type. Tests to control for possible cueing: to select lexigrams and photographs in response to spoken English words; to select photo when shown lexigram.	Not applicable; responses on computer.	Probably unnecessary since apes' responses are typically limited and discrete.
Criteria for the acquisition of a vocabulary item	Performance on multiple-choice tests	Not relevant	Performance on multiple-choice tests	Performance on multiple-choice and tool-use tests; 95% correct on monthly tests of all vocabulary items in all semantic functions (receptive, labelling, statement, and request)	Spontaneous occurrence in 9 out of 10 consecutive 'appropriate' occasions (must be followed by 'behavioural concordance' in each).	Performance on computer console with lexigram keyboard—in effect, a multiple-choice test	Not applicable

Table 23.2 (*Contd.*)

Principal investigator(s)	Premack I	Premack II	Rumbaugh	Savage-Rumbaugh and Rumbaugh I	Savage-Rumbaugh II	Matsuzawa	Boysen
Size of expressive vocabulary	Approximately 130 items	Not relevant	Approximately 75 items	Over 50 items	Matata: 6 food symbols after 1.5 years of training; Kanzi: 50 from age 30 to 46 months.	31	Not applicable
Abilities and major *results* claimed by researchers	Hierarchally organized sentences; conceptual classes, 'colour of', 'name of', 'shape of', 'size of', predicates, some quantifiers, logical connectives, negative article, 'wh-?'	Cognitive capacities, including some proportions, analogical reasoning, causal and transitive inference. Language-trained chimps could solve classes of problems (those requiring abstract code?) the non-language-trained could not. Ape understood actor's intention; some chimpanzees able to convey and utilize accurate and misleading information; capacity for intentional communication.	'Name of', relational concepts, numerical concepts, novel use of stock sentences, colour identification, cross-modal matching	Symbolic and communicative use of lexigrams as well as social co-operation (food-sharing); comprehending numerosity, summation, behavioural cause-effect relationships.	Via observational learning begun in infancy, Kanzi and Mulika (1) exhibit spontaneous use of symbols for intentional communication; (2) comprehend many spoken English words and simple English sentences; (3) understand simple grammar; and (4) Kanzi invented his own grammatical rules.	Appropriate use of 11 colour names and numbers 1–6; labelling 14 objects; and appropriate combinations of above	Some apes: numerical competence to number 6; summing, fractions, ordinality, transitivity. Evidence for same-different concept, cross-modal recognition, recognition of conspecifics, drawing skills. Preliminary studies show better transfer from gestural to vocal communication for pigs than dogs.
Future research plans	See Premack II.	Unspecified	See Savage-Rumbaugh and Rumbaugh I.	To train in spelling, spatial map comprehension (with E. Menzel and H. Terrace), and other cognitive tasks.	Comparison of language-acquisition by *Pan paniscus* and *P. troglodytes* reared simultaneously from infancy, and by human child, investigating other cognitive skills	To continue study of cognitive abilities, including reading and spelling.	To continue study of cognitive abilities, expanding to other species such as dogs and pigs. Vocal and gestural symbolic communication with dogs (esp. border collie) and pigs.

Table 23.3 An overview of recent non-primate language cognition projects

	Herman	Schusterman	Pepperberg
Principal investigator(s)			
Chief interests of principal investigator(s) regarding the project	Emphasis on comprehension, e.g. 'Can a dolphin understand a sentence?'	Receptive abilities of sea lions in symbolic communication experiments (modelled after Herman's studies with the dolphins).	Initially stated as the study of language and cognitive acquisition. Now the stated goal is to train a non-human, non-mammalian species to communicate using the sounds of English speech, and to use this code to examine its cognitive abilities.
Species sex, age when acquired, and age when training began principal subject(s)	Bottlenosed dolphins (*Tursiops truncatus*) Phoenix (♀) Akeakamai (Ake) (♀)	California sea lion (*Zalophus californianus*) Rocky ♀, born in wild. Training began at approx. 4 years old. Bucky ♂, born in captivity. Training began at approx. 1 year old; not now in research programme. Gertie ♀, born in captivity. Training began at approx. 2 years old.	African Grey parrot (*Psittacus erithacus*) Alex ♂; acquired from pet store; training began at 13 months old.
Living environment	Housed in 2 interconnected outdoor tanks, each 15.2 m diameter and 1.5 m deep.	Rocky: originally in tank 3.5 m × 11.1 m, water 1.2 m deep. Bucky: originally in tank of 7.6 m diameter, water 3 m deep. Rocky and Gertie now housed communally in tank 7.6 m diameter, water 3m deep.	Standard parrot cage during sleeping hours. Access to lab. while trainers are present (about 8 hours per day). Toys and food available at intervals.
Training environment	Living environment. Tower by each tank permitted unobtrusive observations.	Living environment	Laboratory room
Language medium	Two different media. Phoenix: acoustic signals (whistle-like) produced by computer and broadcast into tank by underwater speaker. Used linear grammar. Ake: gestural signals (arms and hands) invented for the dolphins. Used inverse grammar.	Rocky and Bucky: Gestural signing devised for sea lions. Use inverse grammar. Gertie: Object stands for another object; gestural signing for actions.	Spoken English words
Trainers: number	Numerous, but unspecified	Unspecified, but termed 'many'	Mainly the principal investigator; several students over the years (12 over the original 26 months of the project)
Trainers: qualifications	Volunteers: undergraduate and graduate student apprentices with a minimum of approx. 2 months on-job training	Unspecified. Many are undergraduate volunteers, and some graduate students.	Undergraduates
Major training methods	Rewarded. Paired associate learning; prompting via indicative gestures (e.g. pointing); physically guiding animal to target object; 'stimulus substitution' (e.g. by fading techniques)	Symbolic matching paradigm and conditional discrimination procedures (similar to Herman's). 2 sessions/day, approx. 5 days/week, 50-100 trials/session.	Various, depending on the experiment. Two-trainer model/rival method, whereby humans demonstrate interactive behaviour to be acquired by the parrot (i.e. the role to be imitated by the parrot). In other sessions, one trainer shapes the correct pronunciation.
Rewards	When correct, hand-clapping and/or an acoustic string glossed as 'Yes Ake (or Phoenix) Fish'. Then dolphin receives social reward (physical contact and vocal praise) and food (a fish). When incorrect, acoustic signal for its name.	Pieces of fish for correct responses and trainer blows whistle. When incorrect, trainer says 'No'.	If correct, praise, petting, and acquisition of an object. Withholding the object and an emphatic 'No' if incorrect. Bonding between parrot and trainers seems extremely important.

Table 23.3 (*Contd.*)

Principal investigator(s)	Herman	Schusterman	Pepperberg
Major testing methods	Computer produces whistles or human signs combinations to which dolphin responds by performing named action to named object or by constructing a relationship between two objects (e.g. 'inside', 'on top of'). On all test trials, observer is blind to instructions given to dolphin.	Human, wearing opaque goggles, signs a combination of lexical units to which sea lion responds, either by bringing one specified object (transport item—TI) to another (goal item—GI), or by operating directly upon an object. One or two observers record data on paper. All training and testing now on video.	Principal investigator sits in same room but does not look at bird; her presence encourages bird's participation. Examiner administers test; PI repeats bird's response. Then rewards or 'No' by the examiner follow as appropriate.
Other data-collection methods	Video tape used during new training procedures and during all testing.	Daily log with comments. Noting sea lion's head-orientation and search patterns.	Initially complete daily written record of vocalizations; now samples. Occasional videotaped sessions and overnight audiotapings.
Reliability tests	Videotape available.	Occasional blind tests in which trainer is blind and deaf. No blind tests for responses, but videotape is available.	Initially vocalizations had to be *identified* by trainers at least 80% of the time. Now 98% inter-observer agreement on the identity of taped vocalizations.
Criteria for acquisition of a vocabulary item	The action and the object manipulated first (also called Direct Object (DO) by Herman and Transport Item (TI) by Schusterman) is rarely mistaken (85–95% correct). Responses were wholly correct to lexically novel sentences 67% (Phoenix) and 60% (Ake) of time.	At least 9–10 correct labels for object.	(1) For acquisition of correct pronunciation, see item no. 12. (2) 80% accuracy on labelling tasks (receptive abilities).
Size of expressive vocabulary	Expressive: Ake seven: 5 objects and 2 functional words (use of 'present–not present' also interpreted as 'Yes–No'). Receptive: approx. 35–40 lexical units, including spatial modifiers 'left', 'right', 'bottom', and 'surface'. Over 1000 2–6-unit sentences.	Expressive: 2 functional words (reporting 'Yes'–'No' by pushing appropriate paddles). Receptive: Rocky (after 6 years of training) 23 signs: 13 objects, 4 modifiers designating object properties ('large', 'small', 'black', 'grey'; no longer used—'white'); 6 actions. Bucky (after 20 months training) 16 signs: 8 objects, 2 modifiers, 5 actions. Gertie 13 signs: 8 objects, 5 actions. Over 7000 different sequences possible for Rocky, over 300 for Gertie; Bucky no longer used.	Over 70.

Abilities and major results claimed by researchers	*Receptive skills*: understanding sentences, including novel ones up to 4 units in length for Ake, 5–6 for Phoenix. Ake understands reference to objects not immediately present, can respond to imperative or interrogative form about presence or absence of an object. *Productive (expressive) skills*: Ake: vocal mimicry of 5 whistle sounds, uses sounds to label 5 different objects.	Schusterman uses standard learning and operant conditioning terms rather than linguistic ones to interpret results, e.g. 'Conditional discrimination' procedures define relations between stimuli (gestural signs and their referents) and have been used to study language comprehension. Sea lion understands sequences up to 7 units in length of the forms: (1) 'modifier–object–action', glossed as 'go to the object and perform the designated action on it'—this procedure was used to demonstrate both absolute and relational properties of modifiers such as 'large' and 'small'; and (2) 'modifier(s)–object–modifier(s)–object', glossed as 'bring the second object to the first'. The abilities are considered to be cognitive precursors to language. Evidence for a symmetrical (reflexive) relationship between gestural signs and their referents. The perceptual world of the sea lion is to be investigated by the 'Yes-No' procedure. Possible interest in keyboard system to study production. Further study of equivalence relations.	Labels for approx. 32, objects; 6 colours; quantity for sets of up to 6 items; 5 shapes (called '2-', '3-', etc. 'cornered' objects). Use of 'no'. Simple requests such as 'come here', 'I want X' (object), 'I wanna go Y' (locations). Concepts such as 'same–different', 'shape of', 'colour of'. That Alex passes object-permanence tests.
Future research plans	Production of symbols and symbol sequences via keyboard for both Phoenix and Ake in order to control environment, refer to objects. Attempt to establish 2-way communication between dolphins, in particular when one dolphin has information needed by the other (as in the Savage-Rumbaugh tool-use and exchange paradigm).		Recognition of two-dimensional objects. Acquisition of more cognitive concepts, such as 'none' (non-existence); labelling of ordinal numerals (e.g. 'third'). Replicating and extending research with another recently acquired young parrot.

References

Bates, E. (1993). Commentary: comprehension and production in early language development. In *Language comprehension in ape and child*, Monographs of the Society for Research in Child Development, (ed. E. S. Savage-Rumbaugh, J. Murphy, R. A. Sevcik, K. E. Brakke, S. L. Williams, and D. M. Rumbaugh), Serial 233, 58.

Bateson, G. (1979). *Mind and nature: a necessary unity*. Wildwood House, London.

Beach, K. H. (1985). Evidence for representational drawings in chimpanzees. Abstracts of Papers Presented at the 38th Annual Northwest Anthropological Conference, 18–20 April 1985, Ellensburg, Washington. *Northwest Anthropological Research Notes*, **19**, p. 43.

Beck, B. B. (1980). *Animal tool behavior*. Garland, New York.

Bellugi, U. and Klima, E. S. (1979). The structured use of space and movement: morphological processes. In *The signs of language* (ed. E. S. Klima and U. Bellugi), pp. 272–315. Harvard University Press, Cambridge, Mass.

Bloom, L. M. (1973). *One word at a time: the use of single-word utterances before syntax*. Mouton, The Hague.

Bloom, L. M., Rocissano, L., and Hood, L. (1976). Adult–child discourse. *Cognitive Psychology*, **3**, 521–52.

Boysen, S. T. (1992). Counting as the chimpanzee views it. In *Complex extended stimuli in animals* (ed. W. K. Honig and J. G. Fetterman), pp. 367–83. Erlbaum, Hillsdale, NJ.

Boysen, S. T. (1993a). Counting in chimpanzees: nonhuman principles and emergent properties of number. In *The development of numerical competence: animal and human models* (ed. S. T. Boysen and E. J. Capaldi), pp. 39–59. Erlbaum Hillsdale, NJ.

Boysen, S. T. (1993b). Paper presented at XXIII International Ethological Congress. Torremolinos, Spain.

Boysen, S. T. and Berntson, G. G. (1989). Numerical competence in a chimpanzee (*Pan troglodytes*). *Journal of Comparative Psychology*, **103**, 23–31.

Boysen, S. T., Berntson, G. G., Shreyer, T. A., and Quigley, K. S. (1993). Processing of ordinality and transitivity by chimpanzees (*Pan troglodytes*). *Journal of Comparative Psychology*, **107**, 208–15.

Bruner, J. S. (1966). On cognitive growth, I. and II. In *Studies in cognitive growth* (ed. J. S. Bruner, R. R. Olver, and P. M. Greenfield), pp. 1–67. Wiley, New York.

Cenami Spada, E. (1994 in press). Animal mind–human mind: the continuity of mental experience with or without language. *International Journal of Comparative Psychology*, **7**(4), 159–93.

Chomsky, N. (1959). Review of Skinner's *Verbal behavior*. *Language*, **35**, 26–58. Reprinted (1964) with comments as The structure of language. In *The philosophy of language* (ed. J. A. Fodor and J. J. Katz), pp. 547–78. Prentice-Hall, Englewood Cliffs, New Jersey.

Chomsky, N. (1980[1979]). Human language and other semiotic systems. *Semiotics*, **25**, 31–44. Reprinted (1980) In *Speaking of apes: a critical anthology of two-way communication with man* (ed. T. A. Sebeok and J. Umiker-Sebeok), pp. 429–40. Plenum, New York.

D'Amato, M. R., Salmon, D. P., Loukas, E., and Tomie, A. (1985). Symmetry and transitivity of conditional relations in monkeys (*Cebus apella*) and pigeons (*Columba livia*). *Journal of the Experimental Analysis of Behavior*, **44**, 35–47.

Davis, H. and Memmott, J. (1982). Counting behavior in animals: a critical evaluation. *Psychological Bulletin*, **92**, 547–71.

Davis, H. and Perusse, R. (1988). Numerical competence in animals: definitional issues, current evidence and a new research agenda. *Behavioral and Brain Sciences*, **11**, 561–615.

Dennett, D. C. (1983). Intentional systems in cognitive ethology: the "Panglossian" paradigm defended. *Behavioral and Brain Sciences*, **6**, 343–90.

Dennett, D. C. (1987). *The intentional stance*. Bradford Books, MIT Press, Cambridge, Mass.

Desmond, A. J. (1979). *The ape's reflexion*. Dial, New York.

de Waal, F. (1986). Deception in the natural communication of chimpanzees. In *Deception: perspectives on human and non-human deceit* (ed. R. W. Mitchell and N. S. Thompson), pp. 221–4. State University of New York Press, Binghampton.

Dolgin, K. G. (1981). A developmental study of cognitive predisposition: a study of the relative salience of form and function in adult and four-year-old subjects. Unpublished Ph.D. dissertation, University of Pennsylvania (as reported in Premack 1986).

Epstein, R., Lanza, R. P., and Skinner, B. F. (1980). Symbolic communication between two pigeons (*Columba livia domestica*). *Science*, **107**, 543–5.

Fouts, R. S. (1975). Capacities for language in the great apes. In *Socioecology and psychology of primates* (ed. R. H. Tuttle), pp. 371–90. Mouton, The Hague–Paris.

Fouts, R. S., Hirsch, A., and Fouts, D. H. (1982). Cultural transmission of a human language in a chimpanzee mother/infant relationship. In *Psychobiological perspectives: Child nurturance series Vol. III* (ed. H. E. Fitzgerald, J. A. Mullins, and P. Page), pp. 159–93. Plenum Press, New York.

Furness, W. (1916). Observations on the mentality of chimpanzees and orangutans. *Proceedings of the American Philosophical Society*, **65**, 281–90.

Gardner, R. A. and Gardner, B. T. (1969). Teaching sign language to a chimpanzee. *Science*, **165**, 664–72.

Gardner, B. T. and Gardner, R. A. (1975). Evidence for sentence constituents in the early utterances of child and chimpanzee. *Journal of Experimental Psychology: General*, **104**, 244–67.

Gardner, B. T. and Gardner, R. A. (1979). Two comparative psychologists look at language acquisition. In *Children's language*, Vol. 2, (ed. K. E. Nelson), pp. 309–69. Halsted, New York.

Gardner, R. A. and Gardner, B. T. (1986). Discovering and understanding the meaning of primate signals.

British Journal for the Philosophy of Science, **37**, 477–95.
Gardner, B. T., Gardner, R. A., and Nichols, S. G. (1989). The shapes and uses of signs in a cross-fostering laboratory. In *Teaching sign language to chimpanzees* (ed. R. A. Gardner, B. T. Gardner, and T. E. Van Cantfort), pp. 55–180. SUNY Press, Albany, NY.
Gauker, C. (1990). How to learn a language like a chimpanzee. *Philosophical Psychology*, **3**, 31–53.
Gelman, R. and Gallistel, R. (1978). *The child's understanding of number*. Harvard University Press, Cambridge, Mass.
Gillan, D. J. (1981). Reasoning in the chimpanzee: II. Transitive inference. *Journal of Experimental Psychology*, **7**, 150–64.
Gillan, D. J., Premack, D., and Woodruff, G. (1981). Reasoning in the chimpanzee: I. Analogical reasoning. *Journal of Experimental Psychology*, **7**, 1–17.
Goodall, J. (1986). *The chimpanzees of Gombe*. The Bellknap Press, Cambridge, Mass.
Gould, S. J. (1977). *Ever since Darwin*. Norton, New York.
Gouzoules, H., Gouzoules, S., and Marler, P. (1984). Rhesus monkey (*Macaca mulatta*) screams: representational signalling in the recruitment of agonistic aid. *Animal Behaviour*, **32**, 182–93.
Gouzoules, H., Gouzoules, S., and Marler, P. (1985). External reference in mammalian vocal communication. In *The development of expressive behavior: biology–environment interactions* (ed. G. Zivin), pp. 77–101 (Chap. 4). Academic Press, New York.
Granier-Deferre, C. and Kodratoff, Y. (1986). Iterative and recursive behaviors in chimpanzees during problem solving: a new descriptive model inspired from the artificial intelligence approach. *Cahiers de Psychologie Cognitive*, **6**, 483–500.
Greenfield, P. M. and Savage-Rumbaugh, E. S. (1984). Perceived variability and symbol use: a common language–cognition interface in children and chimpanzees (*Pan troglodytes*). *Journal of Comparative Psychology*, **98**, 201–18.
Greenfield, P. M. and Savage-Rumbaugh, E. S. (1990). Grammatical combination in *Pan paniscus*: processes of learning and invention in the evolution and development of language. In *'Language' and intelligence in monkeys and apes: comparative developmental perspectives* (ed. S. T. Parker and K. R. Gibson), pp. 540–79. Cambridge University Press, New York.
Griffin, D. R. (1976). *The question of animal awareness: evolutionary continuity of mental experience* (2nd edn, 1981). The Rockefeller University Press, New York.
Griffin, D. R. (1984). *Animal thinking*. Harvard University Press, Cambridge, Mass.
Griffin, D. R. (1986). Foreword. In *Dolphin cognition and behavior: a comparative approach*. (ed. R. J. Schusterman, J. A. Thomas, and F. G. Wood). Erlbaum, Hillsdale, NJ.
Gyger, M., Karakashian, S. and Marler, P. (1986). Avian alarm calling: is there an audience effect? *Animal Behaviour*, **34**, 1570–2.
Hayes, C. (1951). *The ape in our house*. Harper, New York.

Hayes, K., and Hayes, C. (1951). The intellectual development of a home-raised chimpanzee. *Proceedings of the American Philosophical Society*, **95**, 105–9.
Herman, L. M. (1986). Cognition and language competencies of bottlenosed dolphins. In *Dolphin cognition and behavior: a comparative approach* (ed. R. Schusterman, J. Thomas, and F. Wood), pp. 221–52. Erlbaum, Hillsdale, NJ.
Herman, L. M. (1987). Receptive competencies of language-trained animals. In *Advances in the study of behavior* (ed. J. S. Rosenblatt, C. Beer, and M.-C. Busnel), pp. 1–60. Academic Press, New York.
Herman, L. M. (1989). In which procrustean bed does the sea lion sleep tonight? *The Psychological Record*, **39**, 19–50.
Herman, L. M., and Forestell, P. H. (1985). Reporting on the presence or absence of named objects by a language-trained dolphin. *Neuroscience and Biobehavioral Reviews*, **9**, 667–81.
Herman, L. M., Richards, D. B., and Wolz, J. P. (1984). Comprehension of sentences by bottlenosed dolphins. *Cognition*, **16**, 129–219.
Hewes, G. W. (1977). Language origin theories. In *Language learning by a chimpanzee: the Lana project* (ed. D. M. Rumbaugh), pp. 3–53. Academic Press, New York.
Hoban, E. (1986). The promise of animal language research. Unpublished D.Phil. thesis, University of Hawaii.
Hockett, C. F. (1958). *A course in modern linguistics*. Macmillan, New York.
Hockett, C. F. and Altmann, S. A. (1968). A note on design features. In *Animal communication* (ed. T. A. Sebeok), pp. 574–5. Indiana University Press, Bloomington.
Ikeo, M. (director) (1994). *Can chimps talk?* NOVA films.
Jolly, A. (1988). The evolution of purpose in Machiavellian intelligence. In *Social expertise and the evolution of intellect: experience from monkeys, apes and humans* (ed. R. W. Byrne and A. Whiten), pp. 363–78. Oxford University Press.
Kellogg, W. N. and Kellogg, L. A. (1933). *The ape and the child: a study of environmental influence upon early behavior*. Whittlesey House, New York. (Reprinted (1967) by Haffner, New York.)
Klima, E. S. and Bellugi, U. (eds) (1979). *The signs of language*. Harvard University Press, Cambridge, Mass.
Köhler, W. (1925). *The mentality of apes*. Harcourt, New York.
Laidler, K. (1978). Language in the orang-utan. In *Action, gesture and symbol: the emergence of language* (A. Lock, ed.), pp. 133–155. Academic Press, New York.
Langbauer, W. (1982). Prerequisites for language in a porpoise? Unpublished Ph.D. dissertation, Boston University.
Linden, E. (1975). *Apes, men, and language*. Saturday Review Press–Dutton, New York.
Linden, E. (1993). Can animals think? *Time*, 22 March 1993, 56–61.
McDermott, D. (1981). Artificial intelligence meets natural stupidity. In *Mind design* (ed. J. Haugeland), pp. 143–60. Bradford Books, Montgomery, Vermont.

McGonigle, B. (1985). Can apes learn to count? *Nature*, **315**, 16–17.

McNeill, D. (1974). Sentence structure in chimpanzee communication. In *The growth of competence* (ed. K. Connolly, and J. Bruner), pp. 75–94. Academic Press, New York.

Marler, P., Dufty, A. and Pickert, R. (1986). Vocal communication in the domestic chicken: II. Is a sender sensitive to the presence and nature of a receiver? *Animal Behaviour*, **34**, 194–8.

Marler, P., Karakashian, S., and Gyger, M. (1991). Do animals have the option of withholding signals when communication is inappropriate? In *Cognitive ethology: the minds of other animals: essays in honor of Donald R. Griffin* (ed. C. A. Ristau), pp. 187–208. Erlbaum, Hillsdale, NJ.

Matsuzawa, T. (1985a). Use of numbers by a chimpanzee. *Nature*, **315**, 57–9.

Matsuzawa, T. (1985b). Color naming and classification in a chimpanzee (*Pan troglodytes*). *Journal of Human Evolution*, **14**, 283–9.

Matsuzawa, T., Asano, T., Kubota, K., and Murofushi, K. (1986). Acquisition and generalization of numerical labeling by a chimpanzee. In *Current perspectives in primate social dynamics* (ed. D. Taub and F. King), pp. 416–30. Van Nostrand Reinhold, New York.

Menzel, E. W. (1973). Chimpanzee spatial memory organization. *Science*, **182**, 943–5.

Menzel, E. W. (1978). Cognitive mapping in chimpanzees. In *Cognitive processes in animal behavior* (ed. S. H. Hulse, H. Fowler, and W. K. Honig), pp. 375–422. Erlbaum, Hillsdale NJ.

Menzel, E. W. and Malperin, S. (1975). Purposive behavior as a basis for objective communication between chimpanzees. *Science*, **189**, 652–4.

Miles, H. L. (1983). Apes and language: the search for communicative competence. In *Language in primates: perspectives and implications* (ed. J. de Luce and H. T. Wilder), pp. 43–61. Springer-Verlag, New York.

Miller, G. A. and Gildea, P. M. (1987). How children learn words. *Scientific American*, **257**(3), 94–9.

Nelson, K. (1973). Structure and strategy in learning to talk. *Monographs of the Society for Research in Child Development*, **38**, 1–137.

Pasnak, R. (1979). Acquisition of prerequisites to conservation by macaques. *Journal of Experimental Psychology: Animal Behavior Processes*, **5**, 194–210.

Patterson, F. G. (1978a). The gestures of a gorilla: Language acquisition in another pongid. *Brain and Language*, **5**, 72–97.

Patterson, F. G. (1978b). Linguistic capabilities of a young lowland gorilla. In *Sign language and language acquisition in man and ape: new dimensions in comparative pedolinguistics* (ed. F. C. Peng), pp. 161–201. Westview, Boulder, Colorado.

Patterson, F. G. (1979). Innovative uses of language by a gorilla: a case study. In *Children's language*, Vol. 2, (ed. K. E. Nelson), pp. 497–561. Gardner, New York.

Patterson, F., and Linden, E. (1981). *The education of Koko*. Holt, New York.

Pepperberg, I. M. (1981). Functional vocalizations by an African grey parrot (*Psittacus erithacus*). *Zeitschriff für Tierpsychologie*, **5**, 139–60.

Pepperberg, I. M. (1986). Acquisition of anomalous communicatory systems: implications of studies on interspecies communication. In *Dolphin behavior and cognition: comparative and ethological aspects* (ed. R. Buhr, R. Schusterman, J. Thomas, and F. Wood), pp. 289–302. Erlbaum, Hillsdale, NJ.

Pepperberg, I. M. (1987a). Evidence for conceptual quantitative abilities in the African Grey parrot: labeling of cardinal sets. *Ethology*, **75**, 37–61.

Petitto, L. A. and Seidenberg, M. S. (1979). On the evidence for linguistic abilities in signing apes. *Brain and Language*, **8**, 162–83.

Pfungst, O. (1911). *Clever Hans, the horse of Otto von Osten*. (trans. R. L. Rahn). Holt, Rinehart and Winston, New York. (With preface of J. R. Angell and introduction by C. Stumpf.) (Reprinted in 1965 in the English translation by C. L. Rahn, with an introduction by R. Rosenthal.)

Piaget, J. and Inhelder, B. (1969). *The psychology of the child*. Basic Books, New York.

Premack, D. (1975). Putting a face together. *Science*, **172**, 808–22.

Premack, D. (1976). *Intelligence in ape and man*. Erlbaum, Hilsdale, NJ.

Premack, D. (1983a). The codes of man and beasts. *The Behavioral and Brain Sciences*, **6**, 125–67.

Premack, D. (1983b). Author's response. *The Behavioral and Brain Sciences*, **6**, 125–67.

Premack, D. (1984). Possible general effects of language training on the chimpanzee. *Human Development*, **27**, 268–81.

Premack, D. (1986). *Gavagai! Or the future history of the animal language controversy*, Bradford Books, MIT Press, Cambridge, Mass.

Premack, D. and Premack, A. J. (1983). *The mind of an ape*. Norton, New York.

Premack, D. and Woodruff, G. (1978a). Does the chimpanzee have a theory of mind? *The Behavioral and Brain Sciences*, **4**, 515–26.

Premack, D. and Woodruff, G. (1978b). Cognition and consciousness in nonhuman species: authors' responses. *The Behavioral and Brain Sciences*, **4**, 616–29.

Rimpau, J. B., Gardner, R. A. and Gardner, B. T. (1989). Expression of person, place and instrument in ASL utterances of children and chimpanzees. In *Teaching sign language to chimpanzees* (ed. R. A. Gardner, B. T. Gardner, and T. E. van Cantfort), pp. 240–68. SUNY Press, Albany, NY.

Ristau, C. A. (1980). Ape language, cognition and culture, review of Desmond, Adrian, *The ape's reflexion*, Dell, New York, 1979. *Nature* (London), **284**, 684–5.

Ristau, C. A. (ed.) (1991). *Cognitive ethology: the minds of other animals. Essays in honor of Donald R. Griffin*. Erlbaum, Hillsdale NJ.

Ristau, C. A. and Robbins, D. (1979). A threat to man's uniqueness? Language and communication in the chimpanzee. *Journal of Psycholinguistic Research*, **8**, 267–300.

Ristau, C. A. and Robbins, D. (1982*a*). Cognitive aspects of ape language experiments. In *Animal mind–human mind* (ed. D. R. Griffin), pp. 299–331. Springer-Verlag, Berlin and New York.

Ristau, C. A. and Robbins, D. (1982*b*). Language in the Great Apes: a critical review. In *Advances in the study of behavior*, Vol. 12 (ed. J. S. Rosenblat, R. A. Hinde, C. Beer, and M.-C. Busnel), pp. 142–255. Academic Press, New York.

Rohles, F. H. and Devine, J. V. (1966). Chimpanzee performance on a problem involving the concept of middleness. *Animal Behaviour*, **14**, 159–62.

Rohles, F. H. and Devine, J. V. (1967). Further studies of the middleness concept with the chimpanzee. *Animal Behaviour*, **15**, 107–12.

Romanes, G. J. (1884). *Mental evolution in animals*. Reprint 1969, with posthumous essay on instinct by Charles Darwin. ABS Press, New York.

Romski, M. A., Sevcik, R. A., and Joyner, S. E. (1984). Nonspeech communication systems: implications for mentally retarded children. *Topics in Language Disorders*, **5**, 66–81.

Romski, M. A., Sevcik, R. A., and Rumbaugh, D. M. (1985). Retention of symbolic communication skills in five severely retarded persons. *American Journal of Mental Deficiency*, **89**, 441–4.

Rumbaugh, D. M. (ed.) (1977). *Language learning by a chimpanzee: the Lana project*. Academic Press, New York.

Rumbaugh, D. M., and Gill, T. V. (1977). Lana's acquisition of language skills. In *Language learning by a chimpanzee: the Lana project* (ed. D. M. Rumbaugh), pp. 165–92. Academic Press, New York.

Rumbaugh, D. M. and Washburn, D. A. (1993). Counting by chimpanzees and ordinality by macaques in video-formated tasks. In *The development of numerical competence*, (ed. S. T. Boysen and E. J. Capaldi), pp. 87–106. Erlbaum, Hillsdale, NJ.

Rumbaugh, D. M., Savage-Rumbaugh, S., and Hegel, M. T. (1987). Summation in the chimpanzee (*Pan troglodytes*). *Journal of Experimental Psychology: Animal Behavior Processes*, **13**, 107–15.

Sanders, R. J. (1980). The influence of verbal and nonverbal context on the sign language conversations of a chimpanzee. unpublished Ph.D. dissertation. Columbia University.

Sanders, R. J. (1985). Teaching apes to ape language: explaining the imitative and nonimitative signing of a chimpanzee (*Pan troglodytes*). *Journal of Comparative Psychology*, **99**, 197–210.

Savage-Rumbaugh, E. S. (1984*a*). *Pan paniscus* and *Pan troglodytes*. Contrasts in preverbal communicative competence. In *The pygmy chimpanzee: evolutionary biology and behavior* (ed. R. S. Sussman), pp. 395–414. Plenum Press, New York.

Savage-Rumbaugh, E. S. (1984*b*). Verbal behavior at a procedural level in the chimpanzee. *Journal of the Experimental Analysis of Behavior*, **41**, 223–50.

Savage-Rumbaugh, E. S. (1986). *Ape language: from conditioned response to symbol*. Columbia University Press, New York.

Savage-Rumbaugh, E. S., and Rumbaugh, D. M. (1978). Symbolization, language and chimpanzees: a theoretical reevaluation based on initial language acquisition processes in four young *Pan troglodytes*. *Brain and Language*, **6**, 265–300.

Savage-Rumbaugh, E. S., Rumbaugh, D. M., and Boysen, S. (1978*a*). Symbolic communication between two chimpanzees (*Pan troglodytes*). *Science*, **201**, 641–4.

Savage-Rumbaugh, E. S., Rumbaugh, D. M., and Boysen, S. (1978*b*). Linguistically mediated tool use and exchange by chimpanzees (*Pan troglodytes*). *The Behavioral and Brain Sciences*, **4**, 539–4.

Savage-Rumbaugh, E. S., Rumbaugh, D. M., Smith, S. T., and Lawson, J. (1980). Reference: the linguistic essential. *Science*, **210**, 922–5.

Savage-Rumbaugh, E. S., McDonald, K., Sevcik, R., Hopkins, W. D., and Rupert, E. (1986). Spontaneous symbol acquisition and communicative use by pygmy chimpanzee (*Pan paniscus*). *Journal of Experimental Psychology: General*, **115**, 211–35.

Savage-Rumbaugh, E. S., Murphy, J., Sevcik, R. A., Brakke, K. E., Williams, S. L., and Rumbaugh, D. M. (1993). *Language comprehension in ape and child*, Monograph of the Society for Research in Child Development, Serial 233, 58.

Schusterman, R. J. (1988). Artificial language comprehension in dolphins and sea lions: the essential cognitive skills. *The Psychological Record*, **38**, 311–48.

Schusterman, R. J. and Gisiner, R. (1988). Animal language research: marine mammals re-enter the controversy. In *Intelligence and evolutionary biology* (ed. H. J. Jerison and I. Jerison), pp. 319–50. Springer-Verlag, Berlin.

Schusterman, R. J. and Gisiner, R. C. (1989). Please parse the sentence: animal cognition in the procrustean bed of linguistics. *The Psychological Record*, **39**, 3–18.

Schusterman, R. J., and Krieger, K. (1984). California sea lions are capable of semantic comprehension. *The Psychological Record*, **34**, 3–23.

Schusterman, R. J. and Krieger, K. (1986). Artificial language comprehension and size transportation by a California Sea Lion (*Zalophus californianus*). *Journal of Comparative Psychology*, **100**, 348–55.

Seidenberg, M. S. and Petitto, L. A. (1987). Communication, symbolic communication, and language: comment on Savage-Rumbaugh, McDonald, Sevcik, Hopkins and Rupert (1986). *Journal of Experimental Psychology: General*, **116**, No. 3, 279–87.

Seyfarth, R. M., Cheney, D. L., and Marler, P. (1980). Monkey responses to three different alarm calls: evidence of predator classification and semantic communication. *Science*, **210**, 801–3.

Shanker, S. G. (1994). Ape language in a new light. In *Language and communication*, Vol. 14, No. 1, (ed. B. King).

Sidman, M., Rauzin, R., Lazar, R., Cunningham, S., Tailby, W., and Carrigan, P. (1982). A search for symmetry in the conditional discriminations of rhesus monkeys, baboons, and children. *Journal of the Experimental Analysis of Behavior*, **37**, 23–44.

Skinner, B. F. (1957). *Verbal behavior*. Appleton, New York.
Smith, W. J. (1977). *The behavior of communicating*. Harvard University Press, Cambridge, Mass.
Smith, W. J. (1991). Animal communication and the study of cognition. In *Cognitive ethology: the minds of other animals. Essays in honor of Donald R. Griffin* (ed. C. A. Ristau), pp. 209–30. Erlbaum, Hillsdale NJ.
Stokoe, W. J., Jn. (1960). *Sign language structure: an outline of the visual communication systems of the American deaf*, Studies in Linguistics, Occasional Papers 8. University of Buffalo Press, Buffalo, New York.
Straub, R. O. and Terrace, H. S. (1981). Generalization of serial learning in the pigeon. *Animal Learning and Behavior*, **9**, 454–68.
Straub, R. O., Seidenberg, M. S., Terrace, H. S., and Bever, T. G. (1979). Serial learning in the pigeon. *Journal of the Experimental Analysis of Behavior*, **32**, 137–48.
Tallal, P., Miller, S., and Fitch, R. H. (1993a). Neurobiological basis of speech: a case for the preeminence of temporal processing. In *Temporal information processing in the nervous system: special reference to dyslexia and dysphasia*, (ed. P. Tallal, A. R. Galaburda, R. R. Linas, and C. von Euler). *Annals of the New York Academy of Sciences*, **682**. New York Academy of Sciences, New York.
Terrace, H. S. (1979a). *Nim*. Knopf, New York.
Terrace, H. S. (1979b). Is problem-solving language? *Journal of the Experimental Analysis of Behavior*, **31**, 161–75.
Terrace, H. S. (1984). Language in apes. In *The meaning of primate signals* (ed. R. Harré and V. Reynolds), pp. 179–207. Cambridge University Press.
Terrace, H. S. (1985a). In the beginning was the "name". *Journal of the American Psychological Association*, **40**, 1011–28.
Terrace, H. S., Pettito, L. A., Sanders, R. J., and Bever, T. G. (1979). Can an ape create a sentence? *Science*, **206**, 891–902.
Terrace, H. S., Petitto, L. A., Sanders, R. J., and Bever, T. G. (1980). On the grammatical capacity of apes. In *Children's language*, Vol. 2, (ed. K. Nelson), pp. 371–495. Gardner, New York.
Thompson, C. R. and Church, R. M. (1980). An explanation of the language of a chimpanzee. *Science*, **206**, 313–14.
Tinklepaugh, O. L. (1932). Multiple delayed-reaction with chimpanzees and monkeys. *Journal of Comparative and Physiological Psychology*, **13**, 207–43.
Umiker-Sebeok, J. and Sebeok, T. A. (1980). Introduction: questioning apes. In *Speaking of apes: a critical anthology of two-way communication with man* (ed. T. A. Sebeok and J. Umiker-Sebeok), pp. 1–59. Plenum, New York.
van Cantfort, T. E. and Rimpau, J. B. (1982). Sign language studies with children and chimpanzees. *Sign Language Studies*, **34**, 15–72.
Whiten, A. and Byrne, R. W. (1988). Tactical deception in primates. *The Behavioral and Brain Sciences*, **11**, 233–73.
Wittengenstein, L. (1953). *Philosophical investigations*. Macmillan, New York.
Woodruff, G. and Premack, D. (1979). Intentional communication in the chimpanzee: the development of deception. *Cognition*, **7**, 333–62.
Woodruff, G. and Premack, D. (1981). Primitive mathematical concepts in the chimpanzee: proportionality and numerosity. *Nature* (London), **293**, 568–70.
Yeager, C., O'Sullivan, C., and Autry, D. (1981). Communicative competence in *Pan troglodytes*: rising (or falling) to the occasion. Paper presented at meeting of the American Psychological Association, Los Angeles, August 1981.
Yerkes, R. M. and Yerkes, A. (1929). *The Great Apes: a study of anthropoid life*. Yale University Press, New Haven, Connecticut.
Zuckerman, L. (1991). Apes r not us. *The New York Review*, 30, May 1991, 43–9.

Further reading

Fouts, R. S., Chown, W., and Goodin, L. T. (1976). Transfer of signed responses in American Sign Language from vocal English stimuli to physical object stimuli by a chimpanzee (*Pan*). *Learning and Motivation*, **7**, 458–75.
Gallup, G. G. (jun.) (1977). Self-recognition in primates: a comparative approach to the bidirectional properties of consciousness. *American Psychologist*, **32**, 329–38.
Gardner, R. A., Gardner, B. T., and van Cantfort, T. E. (eds) (1989). *Teaching sign language to chimpanzees*. State University of New York Press, Allany, New York.
Gibbons, A. (1991). Déjà vu all over again: chimp-language wars. *Science*, **251**, 1561–2.
Gisiner, R. and Schusterman, R. J. (1992). Sequence, syntax and semantics: responses of a language-trained sea lion (*Zalophus californianus*) to novel sign combinations. *Journal of Comparative Psychology*, **106**, No. 1, 78–91.
Hanggi, E. B. and Schusterman, R. J. (1990). Kin recognition in captive California sea lions (*Zalophus californianus*). *Journal of Comparative Psychology*, **104**, No. 4, 368–72.
Hediger, H. (1968). *The psychology and behavior of animals in zoos and circuses*. Dover, New York.
King, B. J. (1994). Evolutionism, essentialism and an evolutionary perspective on language: moving beyond a human standard. *Language and Communication*, **14**, No. 1, 1–13.
Lieberman, P. (1985). *The biology and evolution of language*, MIT Press, Cambridge, Mass.
Menzel, E. W. (1974). A group of young chimpanzees in a one acre field. In *Behavior of non-human primates*, Vol. 4, (ed. A. M. Schrier and F. Stollnitz), pp. 83–153. Academic Press, New York.
Mignault, C. (1985). Transition between sensorimotor and symbolic activities in nursery-reared chimpanzees (*Pan troglodytes*). *Journal of Human Evolution*, **14**, 747–58.

Miles, H. L. (1986). How can I tell a lie? Apes, language and the problem of deception. In *Deception: perspectives on human and nonhuman deceit*, (ed. R. W. Mitchell and N. S. Thompson), pp. 245–66. State University of New York Press, Albany.

Parker, S. and Gibson, K. R. (1994). *Language and intelligence in animals: developmental perspectives*. Cambridge University Press, London.

Pepperberg, I. M. (1987b). Acquisition of the same different concept by an African Grey Parrot (*Psittacus erithacus*): learning with respect to categories of color, shape and material. *Animal Learning and Behavior*, **15**, 423–32.

Pepperberg, I. M. (1994). Numerical competence in an African Gray Parrot (*Psittacus erithacus*). *Journal of Comparative Psychology*, **108**, No. 1, 36–44.

Ristau, C. A. (1983). Symbols and indication in apes and other species? Comment on Savage-Rumbaugh *et al. Journal of Experimental Psychology: General*, **112**, 498–507.

Ristau, C. A. (1988). Thinking, communicating and deceiving: means to master the social environment. In *Evolution of social behavior: integrative levels* (ed. G. Greenberg and E. Tobach), pp. 213–40. Erlbaum, Hillsdale, NJ.

Romski, M. A., Sevcik, R. A., and Wilkinson, K. M. (1994). Peer-directed communicative interactions of augmented language learners with mental retardation. *American Journal on Mental Retardation*, **98**, No. 4, 527–38.

Savage-Rumbaugh, E. S. (1988). A new look at ape language: comprehension of vocal speech and syntax. In *Comparative perspectives in modern psychology* (ed. D. Leger), Nebraska Symposium on Motivation, Vol. 35, pp. 201–55. University of Nebraska Press, Lincoln, NE.

Savage-Rumbaugh, E. S. (1993). Language learnability in man, ape and dolphin. In *Language and communication* (ed. H. Roitblat, L. Herman, and P. Nachtigall), pp. 457–84. Erlbaum, Hillsdale, NJ.

Savage-Rumbaugh, E. S. and Brakke, K. (1990). Animal language: methodological and interpretive issues. In *Interpretation and explanation in the study of animal behavior*: Vol. 1. *Interpretation, intentionality and communication* (ed. M. Bekoff and D. Jamieson), pp. 313–43. Westview, Boulder, Co.

Savage-Rumbaugh, E. S. and Lewin, R. (1994). *Kanzi: the ape at the brink of the human mind*. Wiley/Doubleday, New York.

Savage-Rumbaugh, E. S. and McDonald, K. (1988). Deception and social manipulation in symbol using. In *Machiavellian intelligence: social expertise and the evolution of intellect in monkeys, apes, and humans* (ed. R. W. Byrne and A. Whiten), pp. 224–37. Oxford University Press.

Savage-Rumbaugh, E. S. and Rumbaugh, D. M. (1979). Chimpanzee problem comprehension: insufficient evidence. *Science*, **206**, 1201–2.

Savage-Rumbaugh, E. S. and Sevcik, R. (1984). Levels of communicative competency in the chimpanzee: representational and representational. In *Behavioral evolution and integrative levels* (ed. G. Greenberg and F. Tobach), pp. 197–219. Erlbaum, Hillsdale, NJ.

Savage-Rumbaugh, E. S., Romski, M. A., Hopkins, W. D., and Sevcik, R. A. (1989). Symbol acquisition and use by *Pan troglodytes*, *Pan paniscus* and *Homo sapiens*. In *Understanding chimpanzees* (ed. P. G. Heltne and L. A. Marquardt), 266–95. Harvard University Press, Cambridge, Mass.

Seidenberg, M. S. and Pettito, L. A. (1987). Communication, symbolic communication and language: Comment on Savage-Rumbaugh, McDonald, Sevcik, Hopkins, and Rupert (1986). *Journal of Experimental Psychology: General*, **116**, 279–87.

Schusterman, R. J. and Kastak, D. (1993). A California sea lion (*Zalophus californianus*) is capable of forming equivalence relations. *The Psychological Record*, **43**, 823–39.

Schusterman, R. J., Gisiner, R., Grimm, B. G., and Hanggi, E. B. (1993). Behavior control by exclusion and attempts at establishing semanticity in marine mammals using matching-to-sample paradigms. In *Language and communications: a comparative perspective* (ed. H. Roitblat, L. Herman, and P. Nachtigall), pp. 249–75. Erlbaum, Hillsdale, NJ.

Stokoe, W. C. (1989). Comparative and development sign language studies: a review of recent advances. In *Teaching sign language to chimpanzees* (ed. R. A. Gardner, B. T. Gardner, and T. E. van Cantfort), pp. 308–16. SUNY Press, Albany, NY.

Terrace, H. S. (1985b). Animal cognition: thinking without language. *Philosophical Transactions of the Royal Society*, (London), **B**, **308**, 113–28.

24
Symbols and structures in language-acquisition

Carolyn Johnson, Henry Davis, and Marlys Macken

Abstract

Children learning natural languages need to master a symbol system with both a constituent and a combinatorial structure. There are at present three main schools of theorizing as to how this mastery is accomplished: the 'interactive', which locates language-learning within its context of use; the 'cognitive', which locates language-learning as part of symbolic development in general; and the 'autonomous', which emphasizes the independence of the study of grammatical competence from both language use and general principles of cognitive development. Additionally, the task of language-learning is generally investigated under a number of independent headings: phonological acquisition; lexical acquisition; morphological and syntactic acquisition; and pragmatic acquisition.

The phonological (sound) system of a language is rule-based. Very young human infants can discriminate the majority of the features which comprise the adult sound system, and can do so across complex physical dimensions. At a minimum, infants' abilities provide an initial grid for segmenting, sorting, and classifying phonetic categories, and for mapping these to the higher levels of phonological systems. For production, there is continuity from the babbling period into early language. The early productive sound systems appear to be based on words rather than smaller units, words not being initially analysed into their component phonemes. This analysis begins between the ages of two and four years, although the child's system still remains simplified compared to that of the adult. The process of phonological development follows no invariant sequence, and can show regressions at the same time as the system becomes more complex. Some form of 'cognitive' theory at present provides the best explanatory framework.

Verbal communication is preceded by gestural communication, beginning at about nine months of age. Accounts have been offered in which gestural communication is claimed to be necessary for later language development, and continuous with it; but the evidence on both these points is equivocal. Meaningful 'words' are used from about one year of age. There is a spurt in vocabulary development from around eighteen months. Errors in word use provide the major data source for theories of semantic development. These data have formed the basis of a number of theoretical accounts; but none of these are yet comprehensive in their explanatory powers. Crucially lacking is an account of how the child learns to relate words to concepts, and clear criteria for determining when a child's 'words' are truly symbolic.

The acquisition of syntax can be divided into three stages: presyntactic; syntactic; and postsyntactic. Early presyntactic word-combinations are of three types: the combination of words that reflect grammatically relevant real-world relations; a word plus an intonationally integrated but meaningless extra syllable; and formulaic or rote-learned sequences. It is not clear whether there is any continuity between the combinatorial structures of this stage and those of the next, syntactic, stage, which lasts from about two to five years of age. First-language learners make very few errors in constructing complex grammars—a fact supporting the claim that there are innate constraints, whether specifically linguistic or not, on a learner's 'hypothesis space'. Analysis of errors in learning suggests that morphological learning may occur via a probabilistic mechanism, whereas syntactic learning may be based on different, possibly innate, principles. The final postsyntactic stage represents the integration of the newly-emerged syntactic system with the real-world knowledge the child has accumulated.

Early functions of language are of three types: solicitation of action, social interaction, and joint attention; expression of affect; and participation in games. With the advent of naming, children can also label and request labels. Later developments, such as threatening, promising, and deceiving, have not been systematically studied.

The best-supported theories of language-development are of the 'autonomous' school. It appears that language-learning skills are domain-specific, and there are few parallels with non-linguistic domains, or precursors to the formal systems. Thus, the autonomous view presents difficulties for both onto- and phylogenetic accounts, in that it is a discontinuity view. However, the 'interactive' and 'cognitive' schools at present offer less plausible accounts of the acquisition of linguistic abilities than 'autonomous' ones, even though they offer more continuous views of development, and hence might be more attractive to an evolutionary scenario at first sight [eds].

24.1 Introduction

This chapter reviews the theories and data of language-acquisition, particularly as they pertain to the ontogeny and phylogeny of the human symbolic capacity. For each of three language systems—phonology, lexical semantics, and syntax—we will address the learner's task, the data base, and principal theories of acquisition.[1] We first present our view of language as a symbol system, and then turn to a brief historical overview of the study and theories of child language. The main body of the chapter deals with the acquisition of the separate systems of phonology, semantics, and syntax. We conclude with brief considerations of pragmatic development and the ontogeny of language as it might relate to evolutionary concerns.

24.2 Language as a symbol system

Let us take as a starting-point the notion that language is indeed symbolic, in that the forms of language represent or convey meaning (see Deuchar, this volume, Chapter 20, and Sinha, this volume, Chapter 17). There is a degree of direct connectedness between linguistic forms and the objects they represent, where the symbolism is in some measure iconic: what can count as a phoneme in a spoken language is partly determined by the physics of sound waves; many lexical categories have a structure determined by the structure of natural kinds in the physical world (for example, the class *mammals*); and some syntactic categories can be defined in terms of semantic concepts such as 'agent' or 'action'. More typically, however, the 'stand for' relation in language is 'by convention'. But even where there is a principled connectedness between linguistic forms and their meanings, linguistic forms do not reduce to real-world referents.

When we turn to language as a system, we might expect it to resemble other symbol systems devised by humans, such as the formal systems of mathematics and symbolic logic. These systems consist of conventionally named objects—for example, numbers and rules for combining them—that have nothing to do with the actual content of the formal objects. For example, the rule for combining 'two' and 'seven' is *never* affected by whether what is being combined is apples or pecans. The formal languages of symbolic logic are languages for which there is a grammar that characterizes the set of strings on a particular finite alphabet. This grammar gives the vocabulary of symbols, the syntactic rules for combining these symbols into well-formed (i.e. grammatical) strings or formulae, and the semantic rules for assigning meanings or interpretations to such formulae. There is a close relation between natural language seen in this way and the formal languages of symbolic logic—a similarity that is the foundation of the generative approach to language description and analysis (see, for example, Chomsky 1965; Chomsky and Halle 1968).

The fact, then, that natural language is in some ways tied directly to real-world objects, *and* defined by convention, *and* yet also like a formal calculus, make it quite unlike other human symbol systems. On the one hand, the relation between the symbols of language and their referents is much like that seen in other, non-linguistic, symbol–referent pairs. But on the other, the symbols of natural language have a formal dimension that is not found in other human social–cultural symbol systems (and probably not found outside mathematical systems). At the very least, it is only in natural language (and in formal logic languages) that we find a system with both a constituent structure and a combinatorial structure. The symbols of language 'break down' into parts (or constituents) that can be recombined to make *new* symbols. Moreover, the symbols combine in specific ways to make *sentences*, something for which there is no real analogue in other symbol systems. And to a significant degree, these constituent and combinatorial structures are independent of the physical world to which the symbols of language refer. Normal human children master language—actually a system of such symbol systems—as a matter of routine, most probably using those symbols of the system that are most directly related to the real world as an 'opening wedge' (Slobin 1985).

24.3 Child-language study in a historical perspective

The mystery of children's language-acquisition has intrigued scholars for centuries. In the fifth century BC, Herodotus (*The Histories*, II 2) reported a child-language 'experiment' that addressed a phylogenetic question. The Egyptian king Psammetichus I sent two infants into the wilderness with a shepherd care-giver who was forbidden to talk to them. The objective was to determine the world's original language, and, thus, the most ancient human race. The infants' first word was *bekos*, the Phrygian for 'bread', forcing Psammetichus to concede that the first people were Phrygians. [In addition to such experiments are reports of the language abilities of 'feral' children (see Table 24.1). These 'natural experiments' have been used in support of the claim that there is a 'critical period' involved in language-acquisition (for example, Lenneberg 1967), and that unless language-learning begins within it, language will not be developed. A problem with using these 'data' in support of this claim is the lack of certainty as to the initial 'normality' of the children involved (eds).] Since this early experiment, interest in language-acquisition has sporadically found its way into print; for example, in his *Confessions* (Book I), St. Augustine 'remembered' how he learned his first words. Language-acquisition was a central focus of the great Rationalist–Empiricist debate of the seventeenth, eighteenth, and early nineteenth centuries (see Chomsky 1966; Hewes, this volume, Chapter 21). Opposing views are represented by Locke's (1690) strongly empiricist description of word-learning and von Humboldt's (1792) claims about the innateness of the language capacity; both of these echo in current theoretical debates. Beginning with Tiedemann (1787), nineteenth- and early twentieth-century parent–scholars from disciplines as diverse as philosophy, physiology, medicine, psychology, education, and linguistics published diaries or observations of infant behaviour and development. The diaries are typically general developmental accounts, providing somewhat scanty but valuable cross-linguistic information about early words and sounds—information that cumulatively is a rich data-source for current hypotheses about phonological and lexical development (see, for example, Clark 1973; Ingram 1978a; Jakobson 1968).[2]

From the early part of this century, reaching their peak in the heyday of behaviourism, large-scale studies of children's language predominated. These studies, carried out mainly by educators and psychologists, provided normative information about children's speech (for example, Templin 1957; Wellman *et al.* 1931; some of the basic normative data for the early stages of language development are summarized in Figs. 24.1–24.3 and Table 24.2 [eds]), but are not very interesting to linguists or cognitive psychologists.[3] McCarthy (1954) reviews the research of this era.

Child-language study as we know it today began in the mid-1950s, partly in response to Skinner's new slant on behaviourist learning theory (1957); but the greatest impetus to the field was the 'Chomskyan revolution' in linguistics associated with the publication of *Syntactic structures* (Chomsky 1957). Chomskyan linguistics charted a clear break with earlier behaviourist and American structuralist theories by advancing the view that no general learning theory can account for language-acquisition, and that language is an abstract, formal system, a separate faculty of mind (for example, Chomsky 1959, 1968). Consequently, syntax was the focus of child-language research in the 1960s. Psycholinguists—who came into existence at this time—wrote 'grammars' to account for the distributional characteristics of children's sentences (for example, Braine 1963; Miller and Ervin 1964); non-adult-type grammars, which focused on the child as a rule-learner—in contrast to the earlier view of a more passive child, whose grammar was deficient when compared to an adult standard.

By the end of that decade, investigators were disenchanted with the prevailing research programme because it failed to account for much of what even two-year-olds know about their language; semantics became the new research focus, and cognitive

Table 24.1 Feral and attic/dungeon children[a]

Acquired some language	No language	Language ability unknown
Feral children	*Feral children*	*Feral children*
1344 Wetteravian wolf-boy 12	1344 Hesse wolf-boy 7	1767 Bear-girl of Hungary 18
16?? Denmark bear-boy 14	16?? Jean de Liège 21	185? Bankipur child 10
1694 2nd Lithuanian bear-boy 10[b]	1661 1st Lithuanian bear-boy 12	185? Captain Egerton's child
1731 Wild girl of Châlons	1672 Irish sheep-boy 16	185? Overdyke wolf-child
1767 Mlle LeBlanc, Songi girl 10	1680 Bamberger calf-boy	1858 Shahjahanpur child 6
1767 Karpfen bear-girl 18	1694 3rd Lithuanian bear-boy 12	1872 2nd Sikandra wolf-child 10
1767 Tomko of Zips, Hungary[c]	1717 Pyrenees man 30	1876 2nd Lucknow child
180? Clemens, Overdyke pig-boy	1717 Pyrenees girl 16	1891 Skiron, the Pindus boy
1860 Sultanpur child 4	1784 Wild man of Kronstadt 24	1893 Batzipur boy 14
1920 Kamala of Mednipur 8[d]	182? Salzburg swine-woman 22	189? Justedal snow-hen 12
1933 Tarzancito of El Salvador 6[e]	1849 2nd Sultanpur wolf-boy 9	1910 1st leopard-boy
	1851 1st Lucknow child 10[n]	1916 Satna wolf-boy
	1858 Shahjahanpur wolf-man 2	1920 Indian panther-child
Attic/dungeon children	1867 Dina Sanichar, Sikandra wolf-boy 6	1923 Jackal-girl
1828 Kaspar Hauser 17	1887 Jalpaiguri bear-girl 3	1933 Jhansi wolf-child
1937 Anna of Pennsylvania 5	1920 Amala of Mednipur 2[p]	193? Indian wolf-child
1938 Isabelle of Ohio 6	1927 Maiwana wolf-boy 10	193? Casamance child
1970 Genie 13	1946 Syrian gazelle child 12	193? Assicia of Liberia
		193? 2nd leopard child 8
Marginal language	**Clear fakes**[q]	1954 Ramu, the Lucknow wolf-boy 12
Feral children	*Feral children*	1960 Mauritanian gazelle-child 6
1717 Kranenburg girl 18[f]	1903 Lucas, the baboon-boy 13	1961 Two wild children in Lebanon?
1724 Peter of Hanover 13[g]	1959 Parasram, the Agra wolf-boy 6	1970 Tehran ape-child 14
1799 Victor of Aveyron 11[h]	1974 John, the Burundi monkey-boy	
1843 Hasanpur wolf-boy 12[j]		*Attic/dungeon children*
1943 1st Sultanpur wolf-boy 12[g,k]		19? Patrick, the chicken-boy 7
1847 Chandour wolf-boy 9[g]		1963 Yves Cheneau
1937 Mt Olympus bear-girl 9[g,m]		

[a] Each entry gives date of discovery, name, and estimated age at discovery, where reported. This table was compiled from: Armen 1971; Briffault 1927; Curtiss 1977; Lane and Pillard 1978; Maclean 1978; Malson 1972; Mandelbaum 1943; Ogburn 1959; Singh and Zingg 1942; Zingg 1940. [b] Speech described as being in a hoarse and inhuman tone. [c] Became marginally bilingual—spoke Slovak, understood German, but persisted in using invented language of his own. [d] Had learned 50 words before death at age 17. [e] Completely recovered from wild behaviour in three years. [f] Unintelligible speech. [g] Learned to imitate gestures. [h] Learned sign language. [j] Learned to read and write but not to speak. [k] Never spoke until a few minutes before death, when he said 'it ached', and asked for a drink of water. [m] Understood 'come' and 'go'. [n] Spoke one word—'Aboodeea'—the name of a girl who had treated him kindly. [p] Died within a year of return to civilization. [q] Many or most cases of feral children have been questioned at some time. (From P. Reich (1986), by permission.)

questions were added to linguistic ones. Piaget's work (for example, 1952) was acknowledged and built upon, and general cognitive developments were recruited to explain language development (in, for example, Cromer 1974; Macnamara 1972; Schlesinger 1974; Sinclair-de Zwart 1969; Slobin 1973). By the mid-1970s research interests had expanded to include functional and social-interactional concerns; 'communicative competence' (Hymes 1971) largely replaced linguistic competence as the object of enquiry. In the 1980s there has been a resurgence of interest in the acquisition of autonomous linguistic systems. Investigations of phonological acquisition have been a less dominant theme through these decades, but have proliferated recently as the field of speech/language disorders has become more linguistic in orientation.

24.4 Theories of language-acquisition

In the 1980s child-language investigators were many, and represented several disciplines—particularly developmental cognitive psychology and linguistics of various sorts (psycho-, socio-, anthropological, formal). Research activity was also vigorous in the applied areas of speech/language pathology and second-language teaching. Research interests and

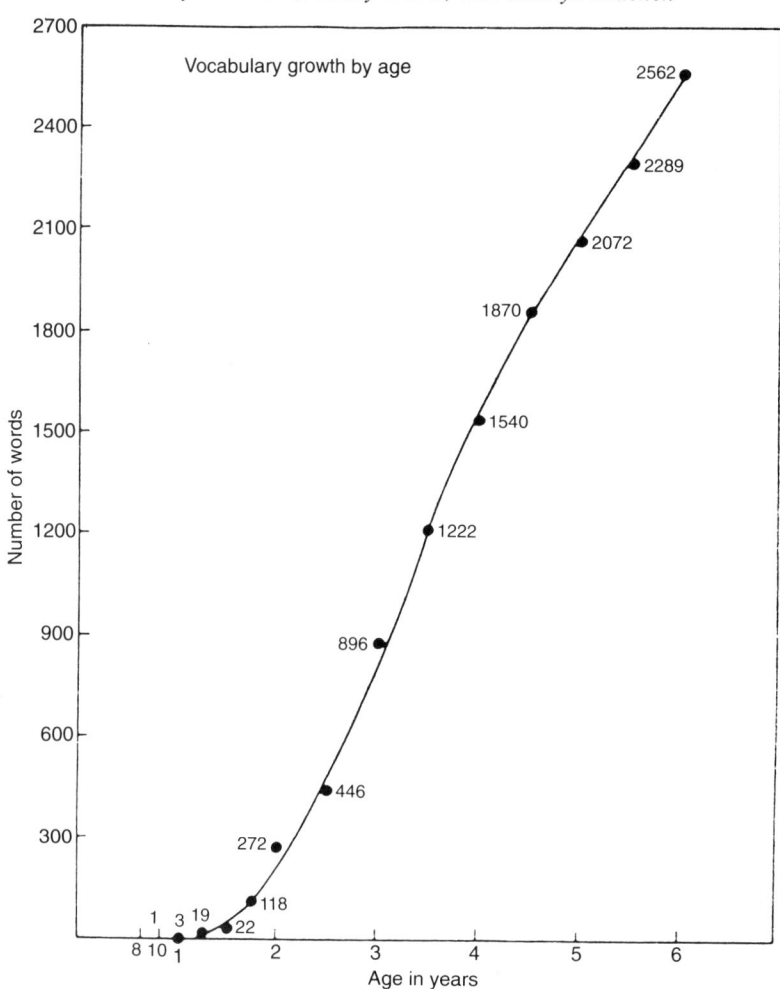

Fig. 24.1 Vocabulary growth by age. Based on M. E. Smith (1926), who elicited productive vocabulary by using objects, pictures, and questions. At any point in language-acquisition a child's comprehension vocabulary is greater. For example, at six years a child's comprehension vocabulary has been estimated at 8000 words (Templin 1957). Peter A. Reich, *Language development*, © 1986, p. 43. Reprinted by permission of Prentice-Hall Inc., Englewood Cliffs, New Jersey.

methods were correspondingly diverse (see Bennett-Kastor, 1985/6, for a discussion). Acquisition theories and specific research questions reflected the (sub)discipline of origin and depended on the coherence of theories in the parent discipline. The demands on language-acquisition theories are particularly stringent, in that a comprehensive theory must ultimately be able to account for linguistic, communicative, cognitive, biological, and developmental facts. To date, productive acquisition theories—ones that are coherent, make testable predictions, and have provoked a considerable amount of research—do not meet this requirement; they tend to be relevant to narrow domains of enquiry and focus on either developmental processes or knowledge states at different points in time. Broader theories are more likely to lead to polemical debate than advancement of knowledge. One such debate was the Chomsky versus Skinner nature–nurture controversy (Chomsky 1959).

The relation of language to general cognition has been a long-standing controversy within psycholinguistics—one that divides the field—and theories—along disciplinary lines. Following Chomsky, most linguists today hold the 'autonomy' view. The independence of language from other cognitive symbolic capacities is, in more recent terminology, the 'lan-

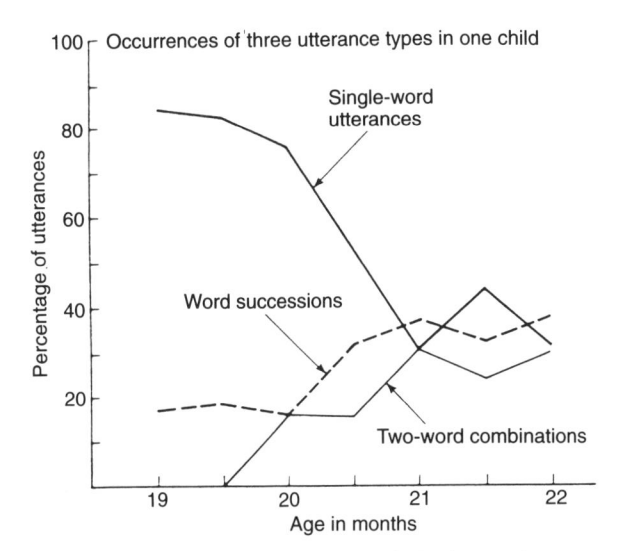

Fig. 24.2 Relative proportions of single-word utterances, word successions, and two-word combinations in one child. (From Garman 1979, p. 187, based on Branigan 1976.) Peter A. Reich, *Language development*, © 1986, p. 72. Reprinted by permission of Prentice-Hall, Inc., Englewood Cliffs, New Jersey.

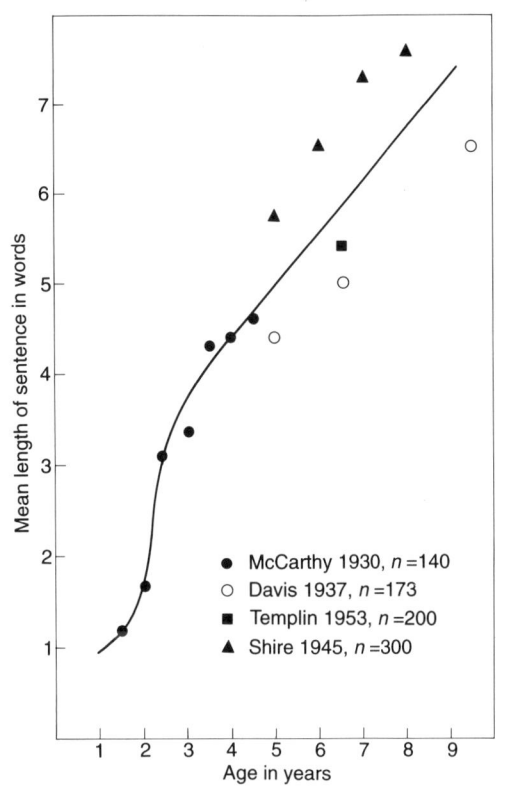

Fig. 24.3 Growth of sentence length (in words) with age. Peter A. Reich, *Language development*, © 1986, p. 73. Reprinted by permission of Prentice-Hall, Inc., Englewood Cliffs, New Jersey.

guage module' thesis: there is a special and autonomous language faculty unique to humans and unique among human symbolic capacities, a faculty that is domain-specific, characterized by specialized neural architecture, and independent of other cognitive systems and processes (see Fodor 1983; Chomsky 1980). In contrast, most psychologists have argued, with Piaget, that language is just one component of a general human symbolic capacity. This is the 'unitary mind' view of the relation between language and cognition.

In this section we compare and contrast three different perspectives on the acquisition of language. The first, which we shall refer to as 'interactive', stresses language-learning as an aspect of acquiring communicative competence, on the premise that grammatical systems are best understood in the context of language use. The second, 'cognitive' approach sees the acquisition of language as one part of an overall picture of cognitive development, contingent on and connected with other forms of symbolic development. In contrast, the third, 'autonomous', approach emphasizes the independence of the study of grammatical competence from the study of language-use, arguing that grammatical principles constitute one of a set of separate components involved in communication and cognition whose relationship cannot be reduced either to superordinate communicative principles or to general principles of cognitive development. In theory, these three approaches should lead to different predictions about the acquisition of language; for example, the interactive and cognitive approaches define 'language-acquisition' in a much broader way than the autonomous approach, which concerns itself almost exclusively with syntactic (and sometimes phonological) acquisition. However, it is sometimes very difficult to convert these theoretical differences into testable hypotheses, particularly with respect to the early stages of acquisition.

24.4.1 The interactive approach

The interactive approach, with its emphasis on communication, sees language as emerging from a matrix of social interaction, mediated by gesture, gaze, and conversational timing, and crucially dependent on the child's relationships with primary care-givers, who are responsible for facilitating the infant's transition into a world of meaningful communication. In the field of

Table 24.2 Composite table showing age in months at which selected language items are reported in eight major studies of infant development

	Strictly longitudinal		Principally cross-sectional					
	Bayley (1933)	Shirley (1933)	C. Bühler (1930)	C. Bühler and Hetzer (1935)	Gesell, Thompson, and Amatruda (1938)	Gesell and Thompson (1934)	Gesell (1925)	Cattell (1940)
1 Vocal grunt		0.25						
2 Differential cries for discomfort, pain, and hunger						1		
3 Vocalizes small throaty noises					1.3			
4 Vocalizations	1.5							
5 Makes several different vocalizations						2		
6 Makes several vocalizations							4	
7 One syllable		2						
8 Vocalizes *ah*, *uh*, *eh*					1.3			
See Items 26–33.								
9 Attends readily to speaking voice						2		
10 Reacts positively to human voice				2				
11 Responds to voice	1.3							
12 Turns head on sound of voice					4			
13 Voice, attends (supine)								2
14 Voice, turns to (sitting)								4
15 Cooing			2	3				
16 Coos					3		4	
17 Babbles or coos								2
18 Returning glance with smiling or cooing				3				
19 Coos to music							6	
See Item 22.								
20 Two syllables		3						
21 Gives vocal expression to feelings of pleasure						3		
22 Actively vocalizes pleasure with crowing or cooing						6		
23 Vocalizes pleasure	5.9							
See Items 15–19, 36–37, 43–44								
24 Vocalizes to social stimulus	3.1							
25 Responds vocally when socially stimulated						4		
See Items 38, 60.								
26 Vocalizes in self-initiated sound play						4		
27 Articulates many syllables in spontaneous vocalizations							6	
28 Vocalizes several well-defined syllables						6		
29 Says several syllables	6.3							
30 Vocalizes *ma* or *mu*					6.5			
31 Vocalizes *da*					7			
32 Two syllables—2d repetition of 1st—*mama* or *dada*	8				7			

	Strictly longitudinal		Principally cross-sectional					
	Bayley (1933)	Shirley (1933)	C. Bühler (1930)	C. Bühler and Hetzer (1935)	Gesell, Thompson, and Amatruda (1938)	Gesell and Thompson (1934)	Gesell (1925)	Cattell (1940)
33 Says *da-da* or equivalent	8.5					9		
34 Gives vocal expression of eagerness						5		
35 Vocalizes eagerness	5.6							
36 Vocalizes displeasure on withdrawal of coveted object						5		
37 Vocalizes displeasure	5.9							
38 'Talks' to a person See Items 25, 60.		6						
39 Distinguishes between friendly and angry talking			6					
40 Imitating sounds *re-re-re*—immediate or delayed response			6					
41 Imitates sounds								9
42 Incipient or rudimentary imitation of sounds See Items 65, 66, 68.						10		
43 Vocalizes satisfaction	6.5							
44 Vocalizes satisfaction in attaining an object See Item 21–23.						7		
45 Singing tones		7.3						
46 Vocalizes recognition	7.4							
47 Gives vocal expression to recognition						8		
48 Single consonants See Items 30, 31.		8						
49 Adjusts to words See items 55, 62				8				9
50 Vocalizes in interjectional manner						8		
51 Vocal interjection	8.1							
52 Listens to familiar words See Item 61.	8.5							
53 Listens with selective interest to familiar words See Item 62.						9		
54 Understands gestures			9					
55 Responds to *bye-bye*				9				
56 Can wave *bye-bye* and often can say it							12	
57 Expressive sounds		9						
58 Expressive jargon	13.5							
59 Uses expressive jargon						15		
60 Uses jargon conversationally						18		

Table 24.2 (*Contd.*)

	Strictly longitudinal		Principally cross-sectional					
	Bayley (1933)	Shirley (1933)	C. Bühler (1930)	C. Bühler and Hetzer (1935)	Gesell, Thompson, and Amatruda (1938)	Gesell and Thompson (1934)	Gesell (1925)	Cattell (1940)
61 Differentiates words See Item 52.	9.8							
62 Makes conditioned adjustment to certain words See Items 69–77.						10		
63 Vocalizes in cup–spoon situation					10			
64 Vocalizes in 2-cube situation					10			
65 Imitating syllables, *mama, papa, dada*			11					
66 Imitates words See Items 40–42, 68.	11.7							
67 One word		14						11
68 First imitative word (*bow-wow*, etc.) See Items 40–42, 65, 66.		15						
69 Adjusts to commands					10			
70 Inhibits on command	11.5							
71 Adjusts to simple commands						12		
72 Places cube in or over cup on command						12		
73 Comprehends simple verbal commissions							12	
74 Understanding simple commands				13–15				
75 Understanding a demand ('Give me that' with gesture)			15–17					
76 Understanding a command ('Sit down' or 'lie down' or 'stand up' with gesture)			21–23					
77 Putting watch to ear on command See Items 62, 95.			21–23					
78 Responds to inhibitory words					12			
79 Understanding a prohibition				16–18				
80 Understanding a forbidding			18–20					
81 Says 2 words	12.9					12		12
82 Says 2 words or more					12			
83 Says 2 words besides *mama* and *dada*							12	
84 Vocalizes when looking in mirror					12			
85 Says 3 words or more					13			13–14
86 Says 4 words or more					13	15		
87 Words, 5						18	18	15–16
88 Names 1 object (ball, pencil, cup, watch, scissors)	17.4							
89 Names picture in book (dog)		19						

Table 24.2 (*Contd.*)

	Strictly longitudinal		Principally cross-sectional					
	Bayley (1933)	Shirley (1933)	C. Bühler (1930)	C. Bühler and Hetzer (1935)	Gesell, Thompson, and Amatruda (1938)	Gesell and Thompson (1934)	Gesell (1925)	Cattell (1940)
90 Naming 1 object or more			19–24					
91 Names 1 picture	18.7					21		
92 Names picture in book (baby)		22.5						
93 Asks with words See Items 101–103.								17–18
94 Says 'Hello', 'Thank you', or equivalent							18	
95 Points to nose, eyes, or hair						18	18	
96 Comprehends simple questions See Items 69–77.							18	
97 Names Gesell watch on fifth picture See Item 113.	19.4							
98 Names 2 objects	19.6							
99 Repeats things said						21		
100 Repeats 4 syllables (2 words)				30				
101 Joins 2 words in speech						21		
102 Words, combines								21–22
103 Uses words in combination See Item 93.						24		
104 Names 3 pictures	21.2							
105 Picture vocabulary 3								23–24
106 Names 3 objects	21.5							23–24
107 Names 3 objects in picture							36	
108 Identifies 4 objects by name								23–24
109 Names 3 of 5 objects See Items 104–107.						24		
110 Names familiar objects like key, penny, watch							24	
111 Points to 5 objects on card						24		
112 Names 5 pictures	24.4					30		
113 Names Gesell watch and picture See Item 97.	24.5							
114 Points to 7 of 10 simple pictures							24	
115 Points to 7 pictures	25.1					30		28–30
116 Picture vocabulary 7 (1937 Stanford–Binet)								25–27
117 Names 7 pictures	32.9							25–27
118 Pictures, points to 6								25–27
119 First pronoun		23						
120 Uses pronouns, past, and plural							36	

Table 24.2 (*Contd.*)

	Strictly longitudinal		Principally cross-sectional					
	Bayley (1933)	Shirley (1933)	C. Bühler (1930)	C. Bühler and Hetzer (1935)	Gesell, Thompson, and Amatruda (1938)	Gesell and Thompson (1934)	Gesell (1925)	Cattell (1940)
121 First phrase See Items 124–126.		23						
122 First sentence		23						
123 Uses simple sentences and phrases								24
124 Distinguishes *in* and *under*								24
125 Understands 2 prepositions	25							
126 Understands 3 prepositions	28							

From McCarthy (1954), by permission.

language-acquisition, proponents of this view have been heavily influenced by the work of Vygotsky (1962) as well as by speech-act theory (Searle 1969) and the theory of conversational implicature first developed by Grice (1967); both the latter theories in turn owe a heavy intellectual debt to the seminal work of Wittgenstein (1953).

The interactive approach to language-acquisition predicts that new linguistic forms will develop to express new communicative functions. In the study of language development, this has led to a number of attempts to code early child-language in terms of communicative functions: see in particular the work of Dore (1973, 1979) and Halliday (1975). Of course, in the adult language, there is no straightforward relationship between the syntactic form of an utterance and its intended meaning: the sentence *You'll go shopping*, for example, can be construed as a statement, a question, a command, or a request, depending on linguistic and extralinguistic context and on clues such as intonation pattern.

Thus the investigator of child language committed to a functionalist perspective must adopt one of two positions with regard to the relation between child and adult language: either the two are based on different principles (i.e., communicative versus formal/grammatical) or both are based on communicative principles. The former hypothesis leads to the prediction that there will be a radical discontinuity in language-acquisition, since somehow the child must effect a transition from a communicatively-based to a grammatically-based system (this has been referred to as the 'developmental shift hypothesis' by Bates and MacWhinney (1982)); the latter hypothesis, on the other hand, starts out by assuming that higher-order communicative principles can provide an explanatory synchronic theory of adult grammar.

24.4.2 *The cognitive approach*

The cognitive approach to language-acquisition is based on the hypothesis that linguistic competence can be reduced to or explained by general cognitive principles also manifested in other forms of symbolic behaviour; this is the 'unitary mind' view mentioned above. This approach is clearly inspired by and indebted to the work of Piaget, sometimes directly (for example Sinclair-de Zwart 1969; Ingram 1978*b*; Corrigan 1978), sometimes more indirectly (Bloom 1973, 1974; Karmiloff-Smith 1979; O'Grady 1987). Psycholinguists who have adopted the cognitive approach have focused on two aspects of the relation between linguistic and cognitive development: first, correlations between the acquisition of non-linguistic cognitive abilities and language-acquisition; and second, structural parallels between grammatical systems and non-linguistic cognitive systems.

24.4.2.1 *Correlations between linguistic and cognitive developments*

Much research has been devoted to correlations between linguistic and other forms of cognitive development, and, while results are often inconclusive, it can be stated with confidence that the predicted correlations do indeed occur in at least some areas. For example, some types of tense-system that require a sophisticated cognitive apparatus, including at least three temporal reference-points (speech time, event time, and reference time) are acquired relatively late (C. Smith 1980; Weist 1986). The 'here-and-now' quality of early language can be straightforwardly related to the young child's inability to 'decentre' the discourse from his or her immediate spatio-temporal environment, and the concept of 'reference time', which involves a temporal perspective different from either that of the speech-event itself or that of the event to which the speech refers, seems particularly difficult to acquire.[4]

While there is undoubtedly a relation between conceptual and linguistic development in the acquisition of syntactically expressed temporal systems, once again the existence of a correlation does not provide evidence that the cognitive structures underlying temporal reference are in any way constitutive of the syntactic structures which express them. In other words, though the child's discourse is constrained by many factors, obviously including the limits of the child's cognitive abilities, none of the studies demonstrating correlations between cognitive and linguistic development provide evidence for a common structure underlying both.

24.4.2.2 General cognitive mechanisms

The second main focus of cognitively-based approaches to language-acquisition is the attempt to discover cognitive mechanisms specific enough to account for the acquisition of natural language, yet general enough to account for other aspects of cognitive growth as well. Not surprisingly, perhaps, there has been considerably less attention paid to this theoretically crucial aspect of the cognitive approach. The most recent, and most sophisticated attempt is that of O'Grady (1987), who proposes a theory of 'general nativism' (as opposed to the 'specific nativism' of autonomous approaches to language-acquisition) to account for both linguistic and non-linguistic development. His claim is that children can develop grammars without any specifically grammatical equipment, using notions of adjacency and dependency to map a 'conceptual structure' (Jackendoff 1983) on to a syntactic representation. Whether such an approach can ultimately explain the acquisition of syntax is an open question; to be successful, the relevant conceptual structures would have to be both specific enough to have explanatory power as grammatical representations and general enough to apply to other aspects of cognitive organization, such as vision. This is an ambitious undertaking, whose overall success or failure cannot be judged in advance, particularly given our comparatively primitive understanding of the interaction of different cognitive systems. There is, however, no doubt that the mixture of specificity and universality that characterizes many syntactic constraints poses a formidable challenge to any attempt to reduce at least syntactic principles to more general cognitive or conceptual schemata.[5]

24.4.3 The autonomous approach

The autonomous approach views cognition as 'modular' (Fodor 1983)—the 'language module' thesis mentioned in Section 24.4. According to this view, there is no general system of communication, but rather a set of self-contained though interacting subsystems each organized according to independent principles. Grammatical knowledge constitutes one such subsystem. Linguists and psycholinguists working within this research paradigm take the view that the child is relatively 'hard-wired'—that is, that many specifically grammatical principles are not, and cannot be, learned by general cognitive mechanisms under the normal circumstances of language development. This has led to the hypothesis that language-'learning' is rather minimal; at a certain level of abstraction, the child is not so much 'acquiring' a grammar as choosing between a set of predetermined grammatical possibilities.

The most recent approach within the general paradigm of autonomous acquisition has been the 'principles-and-parameters theory' (Hyams 1986a; Roeper and Williams 1987). According to this theory, children are innately equipped with a set of invariant specifically grammatical principles and a set of semi-invariant parameters where a value must be set, such as the presence or absence of auxiliary verbs. These parameters are set on the basis of (rather minimal) exposure to data from a target language. The actual 'triggering evidence' which sets the parameter one way or the other need not even be related to the construction to be acquired; it is entirely possible (though somewhat outlandish) to imagine a system in which a grammatical 'flag' (for example, overt movement of an interrogative pronoun to the beginning of the sentence) sets a completely unrelated parameter (for example, a zero-pronoun option, where the subject pronoun does not occur in some sentences).

This view of language-acquisition makes a number of interesting predictions. To start with, since the relationship between the child and the primary linguistic data of the target language is indirect, the theory predicts that linguistic data need not be presented to the child learner in any particular way. Next, the theory makes predictions about what kinds of mistake a child might make in acquiring language: since the child only has available a finite number of grammars to select from, we should find that errors in acquisition correspond to other possible (but unrealized) grammars. Finally, in conjunction with the theory of markedness developed within generative grammar in the last few years, the principles-and-parameters approach predicts a particular sequence of 'stages' in acquisition. This is because less marked grammars in the sense of 'more frequent cross-linguistically' are meant to correspond to less marked grammars in the sense of 'hypothesized earlier in acquisition'. In the most extreme interpretation of this correlation, the child runs through a sequence of grammars from least to most marked, stopping when no counter-evidence to the currently hypothesized grammar is encountered; thus, we should find a series of stages in acquisition, corresponding to successively greater degrees of markedness (see Hyams 1986a; Wexler and Manzini 1987; Roeper and Williams 1987).

Specific theories within these three general approaches are reviewed in Sections 24.5, 24.6, and 24.7, which deal with language subsystems. We now turn to a review of phonological acquisition.

24.5 Acquisition of the phonological system

In this section we consider how the child learns to perceive and produce the sounds of a language in a systematic, conventional, and meaningful way.

24.5.1 *The learner's task*

Phonology, the sound system of a language, encompasses the sound patterns across the words of the lexicon and some kinds of sentential and discourse phenomena, such as sentence intonation and changes to sounds when words are combined in a sentence. Given that the lexicon is just a memorized list of words, are these words holistic acoustic entries, phonologically unanalysed and memorized individually, or are they are broken down into smaller units that enter into a productive system of phonological rules? What is the nature of the 'system' that must be learned?

Consider English plurals. These are formed quite uniformly for newly coined words. For, say, *frobish* ('an object smaller than a breadbin/breadbox') speakers of English will systematically give [frobɪʃɪz]. This creativity is typical of many kinds of phonological behaviour. Clearly, some kind of rule governs this result. As for the unit 'represented' in the rule, two possibilities are the alphabetic-type segment and something smaller than the segment, a 'feature'. For words like *boxes*, *sashes*, and *churches*, our rule could be R1: if a noun ends in /s, z, ʃ, ʒ, tʃ, dʒ/, add /ɪz/ or R1': if a noun ends in [+ coronal + strident] (i.e. the feature that the set of segments listed in R1 have in common), add /ɪz/. For words like *cats, coughs*, and *paths*, we need R2: if a noun ends with /p, t, k, f, θ/, add /s/ or R2': if a noun ends with [−voice, −strident], add /s/. Finally, for all the other cases (such as *bags, cars*, etc.), we need R3: ... otherwise add /z/. Now, what is the plural of *Bach*, a 'new' noun ending with a novel, non-English consonant [x] ([−voice, +velar, +strident])? If speakers use R1 and R2 that list segments, they must use the default R3 to produce [baxz]. However, if speakers have rules like R1' and R2' that employ features, they will use R2' to produce [baxs]. And, it is indeed the latter that speakers of English give as the plural of [bax], thus indicating that words are decomposable into features.

A wide range of evidence shows that both features and segments are units in the lexical representation of words, and that other units, such as syllables, are as well (see Deuchar, this volume, Chapter 20). For example, one of the reasons we know that words are broken down into syllables (or minimal combinations of consonants and vowels) is that people, when searching for a word they want to use but can't remember fully, can often recall the number of syllables even when they can't remember anything more of the word. Rhyming games also show that syllables have at least a psychological reality, and hence qualify as units of the phonology. Another kind of evidence showing that speakers have analysed or learned abstract rules about the sound-shapes of words comes from speaker judgements about possible words of English. Consider the following nonsense syllables (from Halle 1985, p. 104):

flib slin vlim smid fnit vrig plit trit brid blim tnig bnin.

Speakers generally have clear intuitions about which of these nonsense syllables could be words of English, namely *flib, slin, smid, plit, trit, brid*, and *blim*, whereas they will rule out *vlim, fnit, vrig, tnig, bnin*, though each of the latter group is a possible word in other languages. This example suggests that speakers follow general principles that hold over the words of their language, i.e. that the phonology is a system of rules and not a memorized list. The evidence we present below will show that children learn rules covering phonological generalizations, and that this rule learning is the key mechanism underlying phonological acquisition.

The phonological component, then, is a system that is 'symbolic,' though the latter term is not part of the specialized vocabulary of phonology. By 'symbolic' here, we mean two things. First, phonological categories are collections of discrete, often quite different 'objects' that none the less count as the same, where, for each category, one object is privileged, i.e. names the set. For example, the phoneme /p/ includes the acoustically different phones [p], [p⁻] and [pʰ], the first of which names the category. Second, phonology is a formal system, and, like syntax, similar in particular ways to the languages of symbolic logic. At a simple level, a phonology is a grammar that characterizes the set of strings on a particular finite alphabet. The grammar gives the vocabulary of symbols, the syntactic rules for combining these symbols into well-formed (i.e. grammatical or "possible") strings or formulae, and the semantic rules for assigning meanings or interpretations to such formulae. But the phonology is strongly linked to its phonetic base. Wherever this link is direct, the symbolic status is minimized, and less learning (of the kind described in Section 24.5.3.5) is required. Wherever the phonology is independent of its interpretative system of phonetics it is symbolic, and, in addition, requires a learning mechanism or innate learning schema sufficiently powerful to account for generalization or abstraction over the relevant input.[6]

24.5.2 Three developmental issues

Given that there is a phonological component—along with the semantic and syntactic components—to be mastered by the child, we shall address three issues:

1. The first issue is the neurobiological status of the initial state: how much 'knowledge' of the sounds of language is innate or early-maturing? Here, we will consider whether the abilities of the very young infant are due to general properties of mammalian auditory systems or whether they are the reflection of special structures dedicated to language. This is the 'unitary mind' versus 'language module' debate introduced in Section 24.4. The initial-state issue is relevant to the child's first ten to twelve months of life.

2. The second issue concerns the child's transition at eight to twelve months from babbling to the use of words: how does the earlier 'pre-phonological' knowledge relate to the 'learning' of the phonological system? A key question here concerns the status of the so-called 'pre-linguistic' system and of the 'first' phonological system, which is characteristic of infants of ten or twelve months up to eighteen or twenty-four months.

3. The third issue is the acquisition of the more complex phonological system, when the child is two to four years old. Discussion of this period proceeds to evaluation of the leading theories of phonological acquisition in Section 24.5.3.

Discussion of issue (1) will focus on the infant's auditory perception[7]; the perceptual domain during infancy has been studied more extensively than the production domain. Discussion of issues (2) and (3), however, will focus on production, the domain that is best understood for these stages.[8]

24.5.2.1 The initial state

Nearly two decades of research have shown that infants can make fine distinctions with each of the acoustic parameters of frequency, intensity, and duration, distinctions that underlie speech-sound categories. At issue is whether this early ability validates a 'language module/special mechanisms' or a 'unitary mind/general capacity' theory of language-acquisition. This implicitly includes the question of whether there is a truly 'pre-linguistic' stage. This issue centrally involves the perception task, which concerns the relation of phonetic representation to the physical waveform. One of the remarkable discoveries made possible by the invention of the sound spectrograph was that, although the brain hears a string of segment-sized units (phones or speech sounds), the physical signal is nearly continuous. Since then, much has been learned about the acoustic cues underlying speech. For example, for each phonetic feature that distinguishes two phones, types, a *set* of acoustic cues can be shown to be involved in the perception of a phone containing that phonetic feature. However, these acoustic cues do not explain perception: the same acoustic cue can trigger the perception of different phones, and quite different acoustic cues can separately trigger the same phone.[9] This means there is no one-to-one relation between acoustic events and phonetic perception. These properties of language are called the 'lack of linearity' and the 'lack of invariance' (see, for example, Liberman 1970):

Lack of linearity: the speech stream is not temporally segmented in a linear, left-to-right string of phonemes or segments, but rather is a continuous, time-varying acoustic waveform, not easily segmented by eye or machine using physical criteria alone.

Lack of invariance: phonetic segments are multiply cued and the cues vary greatly by context, so greatly that, for example, the transition cue for [d] in [di] bears no resemblance to the cue for [d] in [du].

This complexity of acoustic-phonetic mapping was the first of three findings to suggest that mechanisms specific to language are needed to account for speech perception. The second finding was the discovery of categorical perception: for a continuum of sounds created along an acoustic dimension altered in small but equal steps, say twelve steps, people 'hear' only two discrete categories. In other words, people hear the supposed continuum in a highly discontinuous way (Liberman *et al.* 1967). The categories that are discriminated are closely associated with the phonological categories available in the individual's language, on the evidence that discrimination and labelling experimental tasks yield the same categories; hence experimenters concluded that the phenomenon of categorical perception was a specifically linguistic phenomenon.

The last finding suggesting mechanisms specific to language was the work of Eimas, Siqueland, Jusczyk, and Vigorito (Eimas *et al.* 1971) which showed that infants of one and four months of age could discriminate voiced/voiceless stop-consonant stimuli [ba]–[pa] and, further, that they did so in a way that strikingly resembles categorical perception in adults.[10] Since 1971 more than thirty published studies involving over fifty different speech contrasts have demonstrated the remarkable abilities of very young infants (for reviews, see Kuhl 1985; Eimas *et al.* 1987). The original work was interpreted to demonstrate a biological predisposition for human infants to acquire language, and an innate or early-maturing set of phonetic-feature detectors especially sensitive to the acoustic features underlying speech. This work, with the other two findings, seemed to argue strongly for a special language module.

Over the next decade, however, research shifted the field toward the 'unitary mind' view. First, studies done after 1971 showed that the categorical effect could be shown with non-speech signals (for example Pisoni 1977). Second, and more surprisingly, non-human mammals demonstrated categorical perception over a wide range of acoustic stimuli in a way that resembles what human adults and infants do. For example, Kuhl and Miller (1975) found that chinchillas divided the [ba–pa] continuum into two categories, as do human adults and infants, and the chinchillas' categories were nearly identical to the humans'. This led to the view that linguistic phonetic categories are a function of general properties of the mammalian auditory tract, and this view held until around 1980.

However, the work of Kuhl and her students shows that infants not only achieve a categorical discrimination of acoustic stimuli, but also categorize stimuli across complex physical variations: infants can use trading relations in the speech-signal to extract categories, they can make context adjustments (adjusting for speaking rate, which notoriously changes acoustic cues for some contrasts, and, most impressively, adjusting for consonant identification across vowel contexts, where cues for the 'same' consonant can be radically different); further, they can categorize stimuli across differences (for example, in pitch) of speaker voices (Hillenbrand 1984; Kuhl 1985). In addition, Meltzoff and Kuhl (1986) have demonstrated cross-modal perception in infants. That is, infants can use cues that link visual and auditory dimensions for speech categorization. Meltzoff and Kuhl argue that these results are best explained by a phonetic model that is specifically linguistic in nature. These kinds of data and arguments are also behind the resurgence of language-module theories for general speech perception (see Liberman and Mattingly 1985).

Without question, infants' speech-perception abilities are remarkable: infants can discriminate, for example, consonants differing in place or manner of articulation; they can discriminate a number of vowels (even when the formant-relation patterns of the contrasted vowels are similar); they discriminate pitch differences; and they can categorize in a linguistically relevant manner across complex physical dimensions. The early age at which human infants show these categorization abilities for speech, as well as in other sensory domains such as vision, suggests a biological basis for these abilities. But while there is now little disagreement over the remarkable capabilities of infants, there is controversy over the nature of these abilities, both over what they tell us about the language faculty and over the form of the representation supporting them.

Corresponding to the 'language module' thesis and the 'unitary mind' view discussed in section 24.4, there are two views of speech perception and the infant's initial state. Under the language-module view, infants are biologically endowed with mechanisms that have uniquely evolved to analyse the complex acoustic wave-form of speech into the categories of language (see, for example, Liberman and Mattingly 1985). Thus infants essentially begin processing input into linguistic categories by virtue of the operation of a pre-established phonetic system. In contrast, the 'unitary mind' view holds that infants do not begin with specialized mechanisms, but rather their initial apparent categorization is due to general cognitive and auditory abilities (Pisoni 1977; Aslin and Pisoni 1980). Consequently, infants do not start out processing 'speech'; rather, they acquire a phonetic representation of speech over the first year or so of life. In either case, it looks as if there is no truly 'pre-linguistic' stage. At a minimum, infants' abilities provide an initial grid for segmenting, sorting, and classifying phonetic categories, and for mapping to the higher levels of the phonology (Macken 1987).

There remains a key question concerning the representational structure of the infant's categories. Do infants use acoustic cues to perceive phones fairly directly? If that were the case, it would be puzzling if the child's earliest phonological unit were not the segment but a much larger one, most often described as a 'word'-sized unit (see Section 24.5.2.2). Yet in a recent review of both infant and adult speech-perception data, Eimas et al. (1987) suggest that the categorical representations of the infant have a syllabic structure, and that these structures are modified over the first year by linguistic experience, providing a basis for the construction of the phonological and lexical systems during the second year and onward. This view is more compatible with the findings of the later period; but the critical representational question is far from answered.

24.5.2.2 *The transition from babbling to language*

The speech-perception abilities of young infants are, then, remarkably well-attuned to the task of learning language; infants can probably discriminate, in minimally simple syllables and under maximally simple conditions, most if not all the phonetic features basic to the phonological contrasts found in spoken languages. In a similar way, the stages in speech-sound production that infants go through before arriving at the onset of speech, in a sense, 'prepare' the infant for speaking: the early stages of crying and cooing are followed by the babbling stage (from six to eight months on), where infants first utter strings of similar consonant + vowel strings (for example [ba ba ba]) and then strings of differentiated consonant + vowel strings (for example [ba ba de]). The sound-inventory during the first part of the babbling period is fairly constrained, consisting mainly of stop, nasal, and glide consonants, and lax vowels. Toward the end of the babbling period the sound-inventory increases, but the same stops, nasals, and glides predominate, with the newly added fricatives, affricates, and liquids accounting for a much smaller proportion of most children's repertoires (Stark 1986; Locke 1983). In general the sounds of these early babbling stages are similar across languages, as is the timing of the stages (though individual infants can be strikingly different from other infants of the same language environment, and indeed, some infants do not produce much of the canonical and variegated consonant + vowel babbling just described). In addition, though ten- to twelve-month-old infants may use no recognizable words, they communicate and have learned many conversationally relevant skills such as turn-taking (see Messer and Collis, this volume, Chapter 15).

What are the language phenomena of this transition period? In addition to the various types of babble, this period is characterized by sound-play, phrasal jargon, and 'protowords' (i.e. phonetically fairly stable, meaningful, wordlike units that are variously child inventions, highly emotive rather than referential, and severely limited in meaning and/or context-bound), as well as some first adult-lexicon-based words.[11] In all these utterance types, each child will use the same sounds and sound structures: the continuity between the so-called non-language of babbling and language is quite strong, *contra* the predictions about the autonomy of language from babbling put forth by Jakobson (1968) (Elbers 1982; Vihman et al. 1985).

During the later part of this period, from about 14 to 18 or 20 months, children typically have a small vocabulary of twenty to fifty words, but the nature of their phonological system during this period is widely considered to differ from the end-state phonological system of the adult. Ferguson and Farwell (1975) were among the earliest to suggest that the first phonological 'system' is based not on contrastive segment-sized units, but on the phonological word (see also Macken 1979; Peters 1983). That is, at this stage whole words function as contrastive units in the child's small lexicon; words are not as yet analysed into their component phonemes. The child recognizes and pronounces words holistically, on the basis of canonical shape plus certain 'free-floating' phonetic characteristics. For example, *pen* might be remembered simply as a monosyllable with nasality. Evidence for this stage includes children's highly variable pronunciation of particular words, preference for or avoidance of certain word-shapes, and 'progressive phonological idioms' (words pronounced almost perfectly in terms of the adult model and far in advance of other words). In general, there is little evidence of children's awareness of phoneme-sized contrastive units or, indeed, of many cross-lexical-item phonological generalizations at all during this first stage of learning the phonological system proper.

24.5.2.3 *Learning the phonological system*

Between the ages of two and four years, children learn much of the phonological system of their language. In this section we point out central facts about children's production and perception during this period. Other properties of this important period are discussed in conjunction with theories of phonological acquisition in Section 24.5.3.

Figure 24.4 illustrates the gradualness and temporal extent of children's progress toward the adult inventory of sounds in the system of a particular language, in this case English. It is important to note that such 'norms' can only be stated with respect to a particular language, since what has been tested is children's ability to match adult phonemes. These norms are limited in their representativeness and usefulness for a number of reasons. Most norms are based on only one production of each phoneme in each possible word-position per child. Thus, they do not reveal variability across productions—which is frequent for young children—or variations caused by phonetic environment (for example, assimilation to a contiguous phoneme). Norms are often different from study to study because researchers collected and recorded children's productions in different ways, used more or less stringent criteria for judging a child's production 'correct', and used different criteria for defining 'acquired'. A serious problem is that normative studies typically make assumptions about what acquisition means, but do not take a theoretical stance. Norms also mask many of the most interesting facts of acquisition—often those that individual theories address, some of which we identify in the next section.

During the period from two to four years children's productions are strikingly simplified compared to adults' productions of the same words, with systematic omission or neutralization of many of the sounds easily discriminated several months earlier (at the infant stage).[12] For example, English-speaking children frequently pronounce both *bad* and *bat* as [bat] during the early stages. Secondly, children's identifiication of many of these words containing neutralized contrasts is better when the input is the adult form than if it is their own form of the word. In many cases, children can discriminate minimally contrasting words, like *bad* versus *bat* in decontextualized settings (N. Smith 1973; Barton 1980). Thus, for a large percentage of the words involved, it can be shown that two-year-olds can perceive quite a bit more than they can produce. On the other hand, there is also evidence that at least some of children's pronunciation deviations are due to perceptually based errors encoded in the child's lexical representation (Macken 1980). Consonant with this finding are results of perception tasks showing that two-year-olds cannot discriminate some early discriminated segments if they are contrasted in complex phonological environments (Barton 1980). Clearly, an adequate theory of phonological acquisition must explain a number of facts which, taken together, are puzzling.

24.5.3 *Theories of phonological acquisition*

What accounts for the apparent discrepancy between the infant's extraordinary discrimination abilities and the young child's perceptual and articulatory limitations? That is, between the initial state, the interim states of the young child, and the ultimate end-state phonology of the adult, what is changing and why? In this section, we briefly review the major theories of phonology-acquisition, and then turn to the evidence for one of these theories, namely what has been called the cognitive theory of acquisition.

Theories of acquisition are distinguished by both the theory of phonology and the theory of the learner that they assume. The task of acquiring a phonology

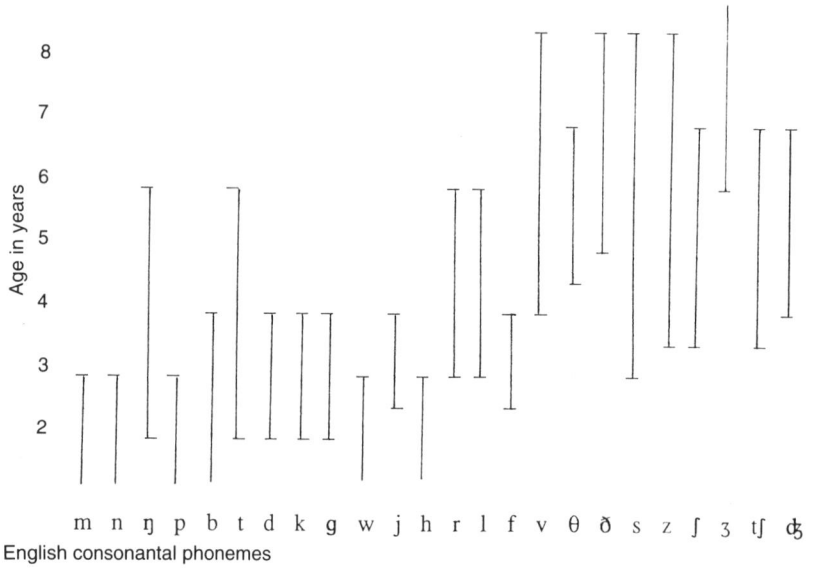

Fig. 24.4 Age-range between children's customary production[a] and acquisition[b] of English consonantal phonemes. Adapted from Sander (1972), by permission. Data from large group studies by Templin (1957) and Wellman (1931).

[a] 'customary production': median age when 50 per cent of children tested produced phoneme correctly in all word positions.

[b] 'acquisition': age when 90 per cent of children tested produced phoneme correctly in all word positions.

may be characterized as learning a set of rules that map between a more abstract underlying representation and a more concrete surface representation. The two levels require two subtasks: learning the relationships among words (phonological rules) and learning how to pronounce individual segments and pieces of words (phonetic rules). We argue that the learner is an active sorter and classifier, a kind of scaled-down hypothesis-tester. Other theories define the domain and the learner differently.

Up to the 1960s in North America the prevailing acquisition theory for phonology was the behaviourist theory of Mowrer (1960) and Olmsted (1966, 1971), which was based on the American structuralist theory of phonology. According to this theory, the child's babbling is shaped and directed towards the adult pronunciation of words through selective reinforcement of sounds progressively closer to the target language. The child is a passive learner, the task is phonetic (i.e., the learning of the articulations necessary to produce the desired speech-sounds), and there are no abstract phonological rules in the sense envisaged in this chapter. This theory is explicitly anti-universalist. In contrast, nearly all the theories of the past twenty years have focused on the universal character of phonology. We turn now to the major current theories, each of which still has support within the field.[13]

24.5.3.1 *Jakobson's structuralist theory*

The most influential universalist theory is that of Jakobson (1968), which assumes that there is a universal hierarchy of structural laws that determine the inventory of phonemic systems and the relative frequency, combinatorial distribution, and assimilatory power of particular phonemes.[14] Jakobson's acquisition theory is that children proceed by learning *phonological* oppositions, or contrasts, in a universal order determined by the structural laws of phonology. According to Jakobson, the autonomy of phonology from the phonetic roots of babbling is at the heart of the discontinuity between the initial state of the infant and the child learning language. Here, we have a phonological task, and development unfolding in a rigid sequence.

24.5.3.2 *Stampe's natural phonology*

In Stampe's theory of natural phonology, there is a universal system of processes (mental operations) governed directly by the possibilities and limitations of human articulation and perception. The phonological system of a particular language is based on retaining some of these processes and suppressing others. During acquisition, those processes that do not apply to the language being learned must be constrained by suppression, limitation, or ordering (Stampe 1979; Donegan and Stampe 1979). Acquisition is the same across children, and the learner is relatively passive. Menn's (1983) 'output routines' theory, which deals with the early stages of acquisition, shares with Stampe a focus on the phonetic aspects of acquisition, but employs a different kind of 'template' structure and, unlike Stampe, recognizes widespread individual differences (and an accordingly more flexible learner).

24.5.3.3 *Generative phonology*

The nativist theory of generative phonology concurs with Stampean and Jakobsonian theories on the issue of universals, but its model of phonology is different. The Chomsky and Halle (1968) theory provided a system of abstract rules that operate on segment-sized units composed of distinctive phonological features. Current generative theories of autosegmental and metrical phonology, known collectively as non-linear phonology, have displaced the earlier version of generative phonology (see, for example, van der Hulst and Smith 1982). The newer theories are characterized by less abstract rules and representations, and are more closely tied to phonetic facts. Acquisition studies that assume various versions of the generative phonology framework propose that the child is acquiring phonological generalizations across the lexicon, as well as lower-level pronunciation, or output, skills (Kiparsky and Menn 1977; Macken and Ferguson 1983; Smith 1973; and Spencer 1985).

24.5.3.4 *Prosodic theory*

The Firthian prosodic theory of Waterson (1971) posits representational structures larger than the segment, like those of non-linear phonology (for example, Spencer 1985). A general, major difference between the two theories is that the American non-linear theory, but not the British prosodic theory, claims that there is a universal structure to phonology and language. Waterson holds that individual differences are the norm. Waterson's (1971) theory is unique among acquisition theories in emphasizing the role of perceptual errors in accounting for the structures children use.

24.5.3.5 *Cognitive theory*

In common with Jakobsonian (Section 24.5.3.1) and generative (Section 24.5.3.3) theories, the cognitive theory must explain the central problem of how children arrive at the phonology of their language; for the phonology is a fairly abstract system of rules, or generalizations across the lexicon. How are such abstractions and generalizations learned? What distinguishes the cognitive theory from Jakobsonian and generative acquisition theories is a focus on the *process* of acquisition, a focus that reveals how the child gradually progresses toward the more complex adult phonological system. At the same time, this focus also clarifies the phonological nature of the structures being acquired.[15] The shift to this cognitive framework was initially triggered by data that could not easily be handled within the other, universalist, linguistic theories, namely, significant widespread individual differences between children acquiring the same language. The newer data, moreover, show that the acquisition process is not a linear progression of unfolding abilities, as earlier theories predict. Rather, acquisition is driven in large part by an active learner, who sorts information, forms hypotheses, makes some kinds of choices, 'tests' and revises earlier hypotheses (or rules), and constructs progressively more complex systems of rules. Individual variation in development is seen to be a function of the active learner, the diversity of local input, and the variety of solutions to particular phonological problems possible within general universal constraints on phonologies. With respect to the latter, substantial cross-linguistic research has shown strong statistical patterns favouring some phonological solutions over others, and consistent limits to variation; thus the cognitive theory assumes that there is an innate phonological module that, in effect, constrains the 'problem space' of the phonology to be learned. In this way and others the cognitive theory is classically linguistic and only weakly cognitive: the claim is that there is some resemblance between the generalization process seen in phonology (and syntax) and the generalization process seen in cognitive, non-linguistic domains. The attempt is *not* to reduce phonology (or syntax) learning to general cognitive processes.

What is the evidence for the cognitive theory? Primary are the aforementioned variations, or individual differences, in acquisition. *Contra* Jakobson, there is no invariant order of acquisition of oppositions, nor any invariant substitution or assimilatory hierarchy. *Contra* Stampe and other rigid-template theorists, some children go through a stage of experimentation, trying out different rules for the same data; and they can be quite accurate for a while, before losing that accuracy as they learn a new rule—a phenomenon of regression (Macken and Ferguson 1983). The classic example of the latter is Hildegard Leopold's *pretty*, her first permanent word at the age of 10 months, which she produced with near perfect accuracy for nearly a year, but subsequently changed to [pɪti] and, later, [bɪdi] as rules for consonant-cluster reduction and then voicing appeared in her system. Leopold reports that nine words during the first years followed a similar development from early accuracy to late reduction (Leopold 1939–49, Vol. 2, 1947, pp. 164–5), and the phenomenon has been documented by others as well (for example, N. Smith 1973; Macken and Ferguson 1983).

In the cognitive view, children discover rules. Early words are assumed to be largely unanalysed, as in the example from Hildegard Leopold. As children learn to break down whole word-shapes into segments and other phonological constituents, they discover or invent rules that systematize the relations between these constituents in different contexts. The general set of stages is:

(1) early piecemeal, unintegrated data, with occasional isolated accuracy on a few words;

(2) experimentation, as the child tries to find acceptable pronunciations for particular words or sounds;

(3) regularization, as the child settles on a rule, producing several words in the same way for the first time; and

(4) a period of overgeneralization, i.e. extension of a rule to inappropriate instances, sometimes resulting in loss of accuracy in previously correct forms (see Macken and Ferguson 1983).

In addition, many of children's invented rules are novel, though rule-governed in principled phonological ways. Examples:

1. The child reported by Macken (1979) favoured /bVdV/ word-shapes, using metathesis, for example *sopa* 'soup': [bota]; and assimilation, for example *tenedor* 'fork': [bede], as well as other processes, to achieve this target.

2. The child reported by Priestley (1977) produced all adult disyllabic words as [CVjVC], for example *sucker*: [fajak], *cupboard*: [kajat], *panda*: [pajan].

3. The child reported by N. Smith (1973) invented a rule converting all unstressed syllables to [ri-], for example *attack*: [ritæk], *guitar*: [ritar].

4. Children learning English discover different ways of (or rules for) handling final voicing, such as 'produce an extra syllable,' for example *pig*: [pɪgə] (Macken and Ferguson 1983); or 'produce a nasally released segment', *ride*: [raidn] (Fey and Gandour 1982); or 'contrastively lengthen the vowels', for example *ride*: [rai:t] versus *write*: [rait] (Hooper 1977).

These examples show the basic flexibility or *creativity* in the system, an aspect to which *selectiveness* also attests: children have been found to systematically avoid words with sounds difficult for them or to prefer words of a particular type that is easy for them (see, for example, Ferguson and Farwell 1975).

The parallelism between the process of rule-formation above and stages in the acquisition of non-linguistic knowledge (Karmiloff-Smith and Inhelder 1974/5) suggests that both types of learning invoke the same mechanism, yielding similar stages:

(1) single-item match;
(2) gradual recognition of a pattern;
(3) period of experimentation or exploration;
(4) construction of a theory/rule; followed by
(5) overgeneralization and apparent loss of ability;
(6) gradual recognition of regularity of other data/counter-examples;
(7) construction of a new theory/rule, distinct from the first; and
(8) gradual development of a single unified theory, in this case the end-state phonological system.

While this progression of stages has been developed to partially explain acquisition of the phonological system, there is increasing evidence that the same process of rule-formation applies to acquisition of the lexicon and the morphological and syntactic systems (for example Bowerman 1978, 1982, 1983; C. E. Johnson 1981; Peters 1983).

24.6 Acquisition of the lexical semantic system

This section focuses on how the child learns the meaning of what are usually called general nominals, or words for things (Brown 1958). The naming ability, or the capacity for symbolic use of words, is usually interpreted to be the key to language, for individuals and for species. Ontogenetically, attainment of this ability is correlated with a rapid increase in vocabulary, and the beginning of phonological and propositional language at about 18 to 24 months of age (Lock 1980; Macken and Ferguson 1987). Phylogenetically, symbolic reference and development of a lexical system are seen as prior to, and giving rise to, the more formal phonological and syntactic systems (Lock 1985; Parker 1985). True referential use of linguistic symbols is an ability our closest phylogenetic cousins do not yet share (Petitto 1987).

24.6.1 *The learner's task*

Using names to refer to things is the paradigm case for linguistic symbolic ability. Put simply, a general nominal is transparently a symbol, in that it stands for a set of entities in the physical world. Children typically develop the ability to refer with linguistic symbols sometime in their second year, as the culmination of several separate strands of development (see our discussion of phonology, above, and Butterworth, this volume, Chapter 16). In learning a new name, a child must pair a phonological representation of the word with a mental representation of its meaning, which must be, in turn, connected to a concept of the word's referents.[16] Over time, the learner will discover that the word may have not one, but several meanings, some of which may be culturally—as opposed to 'naturally'—determined. In a given instance of use the meaning required may be selected by another word in the sentence, as in the example 'Harry is a good father to his stepson, whose real father seems to have disappeared' (example based on Miller 1978). Before the learner is finished with the task, the lexical term will form part of a highly structured semantic domain, and the 'dictionary entry' for the word will contain information about permitted derivational and inflectional morphology (father*hood*, father*s*, etc.) and pronunciation, as well as thematic and grammatical roles the word may fill (agent, noun, subject, etc.). Words in the learner's lexicon will refer not only to objects and people, but will include other 'open class' words of a more relational nature (such as *give* and *tall*), and a smaller set of 'closed class' words, the meaning of which is primarily relational (such as *and*, *the*, and *some*).[17]

Templin (1957) reports that children typically understand 14 000 such words (including derived and inflected forms; 8 000 root words) by the time they are six years old. Using the larger figure, Carey (1978) calculates that the average child, from years two to six, learns one new word during each waking hour, or approximately nine words every day. The child's prodigious feat is a triumph of the human mind that defies simple explanation—or brief exposition. In this section we will limit our discourse to a short overview of early lexical development, a review of theories of the acquisition of word-meaning, and discussion of issues particularly relevant to the developing semantic system *vis-à-vis* the ontogeny and phylogeny of symbols more generally.

24.6.2 *Early lexical development*

Children are most commonly judged to enter their language community when they utter their 'first words' at about one year of age. Over the next few months they add new words to their vocabularies at an average rate of nine words per month (Benedict 1979). New vocabulary consists primarily of general nominals used to name things that move and things a person can act on, but will also include proper names, action words, modifiers, and personal–social words (such as *hi* and *uhoh*) (Benedict 1979; Nelson 1973; Rescorla 1980). By the time children say ten words, they already understand more than fifty, of which a high proportion are action words (Benedict 1979). In the latter half of the second year, rate of vocabulary-growth increases suddenly and dramatically, in what investigators have called the 'word spurt' or 'vocabulary explosion'; 'vertical constructions', of semantically linked single words, and then two-word constructions appear (Scollon 1976; see Section 24.7.2.1).

This account, a highly oversimplified amalgam of mean ages, rates, and word-types from several reports of diary and observational studies, glosses over many complexities and individual differences among children. The age and rate at which particular children acquire early words varies widely, as does the onset and rate of production versus comprehension of vocabulary (Ingram 1989). The vocabulary explosion occurs at different points relative to vocabulary size for different children, ranging at least between 50 and 120 words. The content of children's vocabularies is determined to some extent by phonological constraints (see Section 24.5.3.5). In addition, some children are highly 'referential', with a particularly high proportion of common nouns in their vocabularies, while others fall toward the 'expressive' end of a functional continuum, using a higher than average proportion of action, modifier, or personal–social words (Nelson 1973). Nelson (1985) suggests that this difference in functions of language may reflect differences in when children begin to talk relative to their conceptual development (but see Bowerman 1979, 1980).[18]

24.6.2.1 Children's errors in word-use

Children's 'errors' in word-use during the second year constitute the major data addressed by the best-developed theories of semantic acquisition. Children's extension of a word—the range of things to which the word refers—does not always match adults' extension of the same word. A child's mismatch can be one of several types. The most widely reported, probably because it is easiest to observe, is 'overextension', when the child applies a word to a wider range of referents than the language allows. An example—one of many from nineteenth- and early-twentieth-century diary studies—is one child's use of the word *moon* to refer first to the moon, subsequently to cakes, then round marks on a frosty window, writing on windows and in books, tooling on leather book covers, round postmarks, and the letter 'O' (Chamberlain and Chamberlain 1904, cited in Clark 1973). More difficult to observe are cases of 'underextension', when application of a word is limited to a subset of the adult extensional domain. An example comes from Bowerman's daughter, Eva, who over a period of several months used the word *off* to refer only to the removal of objects from the body; referents, in chronological order, included sleepshades, shoes, her car safety harness, glasses, a pinned-on pacifier, a diaper, and her bib (Bowerman 1976, 1980). A third type is the 'associative complex' (Vygotsky 1962), in which new referents of a word share one or more attributes with the original exemplar, but do not share the same attributes with each other. Eva provides an example of this too, with her word *kick*, first used when she kicked a floor fan with her foot. Subsequently she said *kick* while looking at a picture of a kitten with a ball near its paw, while watching a row of cartoon turtles on television doing the can-can, just before throwing something, as she made a ball roll by bumping it with her kiddicar wheel, and as she pushed her chest against a sink. Bowerman (1980) enumerates relevant attributes of the original referent as (a) a waving limb, (b) sudden sharp contact between a body part and another object, and (c) an object propelled. Each ensuing use of *kick* included one or more of these attributes, but it is difficult to see what all the instances have in common. Less frequently discussed are errors of 'overlap', which involve both overextension and underextension, illustrated by one child who said *muffin* to refer to blueberry muffins and blueberries, but not other kinds of muffins (Dale 1976; see also Anglin 1977; Schlesinger 1974); and errors of 'non-overlap', when the child's use of a word is completely outside the adult domain of extension, as when a child interprets *old* as if it meant 'big' or 'tall' (Kuczaj and Lederberg 1977). In addition to these 'errors', there is the—at least logical—possibility that children's word-use may sometimes coincide exactly with adult usage.

There is little agreement among scholars about the relative frequency of these error types or how they relate to each other. One claim is that words are first underextended; then the range of use increases, sometimes going as far as overextension (Anglin 1977; Griffiths 1986). Underextensions are reported for children as young as 8 months and as old as 2½ years (Griffiths 1985; Reich 1976). Overextensions are most often linked to the age-range one to 2½ years (Clark 1983; but Bowerman discusses overextensions of much older children; see below). Bloom (1973) argues that complexive uses of words (as in Vygotsky's 'associative complexes', defined above) predate non-complexive uses, and indicate that a child has not yet achieved object permanence. Bowerman (1980), on the other hand, observes that her daughters made some non-complexive errors from the onset of word production, and then went through a period during which both complexive and non-complexive word uses were observed. Specific theories of meaning acquisition were developed to account for one or—more recently—several of these error types.

24.6.3 Theories of the acquisition of word-meaning

Speculations about how children learn the meanings of words come to us across many centuries. St Augustine reminisced about how he was taught his first words. His account has a modern analogue in associationist theories, which assume the basic tenets of the behaviourist classical conditioning model (see Brown 1958, pp. 82–108, for discussion). John Locke had clear ideas on this topic:

If we will observe how children learn languages, we shall find that, to make them understand what the names of simple ideas or substances stand for, people ordinarily show them the thing whereof they would have them have the idea; and then repeat to them the name that stands for it, as 'white', 'sweet', 'milk', 'sugar', 'cat', 'dog'. (Locke 1690, cited in Landau and Gleitman, 1985, p. 1)

His empiricist view shares assumptions about the nature of concepts and word-meanings with current componential theories of semantic development; these theories fall within what is now called the 'classical' view of concepts and semantics. In more recent years various interpretations of component-by-component lexical learning have been proposed (Anglin 1977; McNeill 1970; see Clark 1973 for critical discussion). The period since 1970 has been one of vigorous theory-construction and research activity specifically related to early lexical semantic development.

Current theories make competing claims and predictions about meaning-acquisition, but all (with the possible exception of some versions of prototype theory, discussed below) share the classical view that concepts and word-meanings can be stated in terms of hierarchically ordered constituent features, or attributes, or components (see Miller and Johnson-Laird 1976; Smith and Medin 1981 for discussion). They assume that first words are mapped on to existing concepts. They also use observations of children's 'extensions' of a word (the referents to which a word can apply) to infer the word's 'intensions' (the word's meaning, which may have an internal structure), on the assumption that a word's intension provides information necessary and sufficient to determine its extension (Carey 1982). The theories share a common focus on referential word-use during the early period of lexical acquisition (summarized in Section 24.6.2), although specific theories extend their explanations to other word types, later stages of semantic development, and even aspects of syntactic development. Theories sharing these characteristics include the semantic feature hypothesis, the functional core hypothesis, several variants of prototype theory, and—most recently—the lexical contrast theory, or contrastive analysis hypothesis.

24.6.3.1 The semantic feature hypothesis

Clark's (1973) semantic feature hypothesis provided researchers with coherent, empirically testable proposals about the intensional structure of words, the relation of child words to adult words, and the developmental path from child to adult meanings. Consequently, this framework led to a considerable amount of research, much of which refuted Clark's strongest claims and led to modifications of the original statement (Clark 1975, 1977). Clark's theory is based on the tenet that the meaning of a word is best characterized in terms of criterial features, i.e. smaller units of meaning that determine whether a real-world entity is or is not a proper referent of the word. The component binary [+/−] features can be seen as symbols representing conditions the referent must fulfil. To apply a referential word, the speaker must test perceptual attributes of the 'thing', specifically size, shape, texture, sound, movement, or activity, against relevant meaning features. To take a well-worn example, a speaker may call a thing a *dog* after ascertaining that it is an [animal] within a certain [smallish] size-range, with [four legs], a [bark], and a [wagging tail]. If the thing lacks a characteristic corresponding to a criterial feature, (if, say, it doesn't bark), it cannot be a true referent of the word.

According to the semantic feature hypothesis, children's word-meanings differ from adult meanings by being incomplete. The child starts with a word-meaning specified in terms of only one or two of the more general semantic features constituting the adult meaning, and, consequently, overextends the word to improper referents because the immature semantic domain is too broad. To continue the *dog* example, a child who does not include the feature [+bark] as a criterion for using the word may incorrectly say *dog* while pointing at a cat. (Clark cautions that this does not mean the child cannot distinguish between cats and dogs; rather, the child does not yet know which characteristics are encoded in the word-meaning.) As the child's meaning of a word develops, more features are associated with the lexical entry, with the later added features more specific than the early ones. The word's extension is narrowed, and new words enter the lexicon to cover the abandoned semantic space. Causal direction is not clear here; observations of children's word-use indicate that overextensions disappear at about the same time new words for wrongly included referents appear (Clark 1983). Errors other than overextension are not predicted by the semantic feature hypothesis—one of the empirical failures of this theory.

Clark proposed the semantic feature hypothesis to explain not only early use of referential words, but also older children's knowledge about polar relational terms such as *before* and *after*, *more* and *less*, *big* and *small* (or *wee*). The principles are the same: componential meanings, incomplete lexical entries, and addition of defining features in the order from general to specific. Experimental work has shown that the semantic feature hypothesis does not explain the acquisition of these terms, largely because syntactic features and non-linguistic strategies interact with semantic complexity.[19] Overall, Clark's theory has been empirically refuted (predictably so, because it makes strong, empirically testable statements) and criticized on the basis of its componential analysis of word-meaning (see discussion below). The theory has been modified several times to accommodate new evidence and arguments, to the extent that it is now a different theory with a different name—lexical contrast theory, described in Section 24.6.3.3 below.

24.6.3.2 *The functional core hypothesis*

Nelson (1974, 1977, 1985) developed the functional core hypothesis as an attempt to explain how children break into the language system in the first place. The theory is built on a more Piagetian cognitive (as opposed to linguistic) view of the 'mapping problem'; the focus is on infants' active involvement with their environments, and developing concepts that reflect this involvement. The theory predicts that the sensorimotor child will develop hierarchically structured concepts with functional characteristics at their core (see Sinha, Chapter 17, and Butterworth, Chapter 16, this volume). Thus, the concept 'ball' has a core of the type 'something that rolls, bounces'. The concept, which is at first holistic, includes non-core components, such as who rolled the ball, where the action took place, attributes of the ball, and the final location of the ball. Over developmental time, the child must learn to sort out and eliminate non-core components of the concept. In the case of 'ball' the characteristics of specific balls, the contexts in which the child first formed the concept, the particular rollers or throwers of those balls, will not ultimately be part of the 'ball' concept. The perceptual attributes of balls [roundness] will remain in the concept (though not at its core), serving as a basis for recognizing new instances of 'ball'. Thus, the functional core hypothesis provides not only for initial concept-formation centred on functional qualities, but also—as a second step—generalization to new exemplars on the basis of perceptual similarities. The child begins to name after noticing that a word is consistently paired with an established concept. Nelson does not say how this might happen, except that children may be 'specifically tuned' to recognize word–concept pairs (1974, p. 279), and she gives some credence to abstraction theory (which takes us back to John Locke).

Most criticisms of Nelson's theory result from comparing it directly with the semantic feature hypothesis, on the (mistaken) assumption that it predicts function-based overextensions. Analysis of children's word-use reveals that overextensions are based on perceptual attributes, particularly shape, much more frequently than on function—when these factors can be teased apart—so Nelson's theory loses points (Barrett 1978; Bowerman 1976; Clark 1983). In fact, Nelson's theory predicts early underextensions (if names are used before holistic concepts are analysed), followed by perception-based overextensions. The seeming incompatibility of the two theories diminishes when we note that Nelson and Clark are actually concerned with different aspects of the mapping problem: Nelson deals with children in an earlier phase of development, and her primary interest is in concept structure as a basis for word-meaning, while Clark's focus is on semantic features included in initial word-meanings, and the way these influence word-use. An aspect of Nelson's theory that is not often mentioned is its provision for the child's future word-combinations. Nelson speculates that, as children eliminate non-core relations from their holistic concepts, the core concept and non-core relations can be expressed independently; words fulfilling the roles of agent, recipient or goals, location, etc., can then be used in combination with the concept name. Barrett (1978) and Clark (1983) note that Nelson's theory does not make predictions about the relative difficulty of learning different words or acquiring different semantic domains.

Both the semantic feature hypothesis and the functional core hypothesis derive from the classical view of concept structure. This approach has been heuristically very useful, because it provides the warrant for inferring meaning intensions from extensions, and allows us to account for meaning similarities within semantic domains (Miller and Johnson-Laird 1976). It is also satisfying to refer to primitive features, such as [round], that are perceptually 'true', and to think of children gradually building up complex concepts out of an increasingly available collection of irreducible parts. But this view has recently come under attack—independent of child language—from linguists, philosophers, and psychologists.[20] In the area of semantic development, theorists have grappled with the problem of specifying which—and how many—features are necessary and sufficient to define a word and/or a concept, hence allowing the child to extend the word to appropriate new referents, but not inappropriate ones (and the contrastive hypothesis proposes an answer; see Section 24.6.3.3 below). After considering problems like whether a dog is still a *dog* if it has only three legs or doesn't bark, or a ball is still a *ball* if it is deflated, we are forced to conclude that meaning is not equivalent to a list of criterial features. Carey (1982) systematically examines the classical theory's equating of definitional with computational and with developmental features, concluding that this equivalence does not hold up against empirical evidence. (For example, words that express complex concepts do not take as much time to compute in experimental tasks as phrases that atomistically express the same meaning, for example *bachelor* versus *unmarried adult male*). One such problem is that definitional features are too coarse-grained, abstract, and culturally codified to account for early word-meanings, which are tied to a narrow range of contexts. How, she asks, is a child to abstract the meaning feature common to *thick* orange-peels and *thick* boards?

24.6.3.3 *The contrastive hypothesis*

One recent refinement of feature-based theory, motivated in part by the 'which features?' question, is the contrastive hypothesis (Barrett 1978). This theory, and Clark's (1983) lexical contrast theory—originally motivated by broader and slightly different concerns—share the claim that children approach word-learning with the assumption that 'The conventional meanings of every pair of words . . . contrast' (Clark's Principle of Contrast, Clark 1983, p. 820). This principle, a basic tenet of European structural linguistics (see, for example, de Saussure 1959 [1916]), is actually implied in the 1973 version of the semantic feature hypothesis (Gathercole 1987; Merriman 1986; Nelson 1979). Barrett built on the principle, originally claiming that the child needed both positive and negative evidence about a word's referents to acquire its meaning (1978; he has since then retreated from this claim, cf. for example 1982, 1986). The contrastive hypothesis predicts that overextensions will disappear as children learn new words within a particular semantic domain, and that young children do not tolerate synonyms. Barrett and Clark both recruit a substantial amount of evidence in support of these predictions; but there is also counter-evidence (some of it from the same data-sources, including Barrett's own; see Gathercole 1987). A closely related hypothesis is that children organize words in structured semantic domains from the time of their earliest vocabulary development. While there is isolated confirming evidence (for example Griffiths's 1986 report of a child who said *tea* after several unsuccessful attempts to pronounce *coffee*), majority opinion in the field is against it. Also, 'semantic domain', as invoked by the contrastive hypothesis, is beset with a number of theoretical problems, such as what might count as evidence that a child has set up a semantic domain when she or he learns a new contrast (Gathercole 1987; Nelson 1979). Clark's version of the theory additionally postulates that children's lexical innovations fill (only) lexical gaps, and give way to conventional words when these are learned, and that the same principles of contrast and conventionality motivate the child's retreat from morphological and syntactic overgeneralizations (1983, 1987). A major issue contrast theory must address is the degree to which children's lexical contrasts are determined by the prior establishment of contrasts at the general cognitive (non-linguistic) level and/or the presence of contrasts in the adult linguistic input that serve as models for the child (for example 'That's not a dog, it's a horse'). Contrast theory needs a considerable amount of revision and elaboration, but it is too recent to be dismissed out of hand (Gathercole 1987; Merriman 1986). Its ultimate weakness may be that it is still a classical decompositional theory.

24.6.3.4 *Prototype theories*

Prototype theories of early lexical acquisition were proposed to account for complexive word-extension errors, and to solve some of the general problems associated with criterial features (raised at the end of Section 24.6.3.2). While there are several substantially different prototype theories, they make common appeal to Wittgenstein's (1953) qualities. Although all family members do not possess all of these qualities, each member possesses a sufficient number of them to be recognizable as belonging to that family. Translated to the domain of meaning, we get Putnam's (1962) 'cluster concepts', or Rosch's (1973) 'family-resemblance concepts'. These concepts have structural components but—in contrast to the classical view—the components are probabalistic, not criterial. At the heart of the theory is the notion that children erect concepts on the basis of central (prototypical) instances, as opposed to building them up component by component.

The several prototype theories differ on details about the resulting structure of concepts, and how children assign novel instances to a conceptual category. Anglin (1977) proposes that a child determines category membership by comparing a new instance in a non-analytic way with a central example of the concept. The concept itself is a complex of perceptual and functional attributes, everything the child knows about the concept, based on first experiences with a 'nuclear object'. A match will be determined by degree of overlap—implicitly, number of shared attributes—although Anglin maintains that children do not use concept intension to determine category extension. Peripheral members will not be included in the category, category boundaries will be vague, and the ategory centre may even change as the child reorganizes the concept in response to new knowledge. Bowerman's (1976, 1978) prototypes are similar, but she argues that unanalysed wholes cannot really be compared, and some attributes are more central to a given concept than others (also see Palermo 1982). Thus the child builds a concept on the basis of a prototypical example, abstracts non-criterial characteristic features, and determines category extension by fit in terms of central features. Mulford (1979) proposes a more analytic origin for prototypes. In her view, the child starts by abstracting *correlated sets* of attributes associated with several characteristic instances of a category, and then synthesizes a holistic concept. The concept structure initially includes hierarchically organized iconic representations of specific exemplars. Greenberg and Kuczaj (1982) and Griffiths (1986) agree with Mulford's view of a concept composed of iconic representations—Griffiths refers to 'something like videotape recordings of one or more [object]-involving episodes' from the child's life (p. 298)—but see the origin of prototypes as holistic first instances as opposed to syntheses of features into an abstract whole. Greenberg and Kuczaj claim that attributes that characterize concepts are descriptive qualities of the full-blown adult concepts that have no reality during development: this issue was also raised by Carey (see above).

Prototype theories are appealing; they are compatible with what we know about the sensorimotor infant (see Butterworth, this volume, Chapter 16), and can account for all attested types of children's early word-extension errors. They also allow us to capture the fuzziness that characterizes category boundaries, and the fact that the boundaries—and even the category centres—change on the basis of cultural and contextual information.[21] Prototype theory can also be extended to account for some syntactic facts, for example that boundaries are fuzzy between some grammatical categories (for example, adjective and verb) and some nouns and verbs are better nouns and verbs than others in that they participate in more rules, are learned earlier, and are the locus for other grammatical learning (de Villiers 1980; Ross 1972). Despite these advantages, prototype theories present us with (perhaps not unsolvable) problems. First, there is experimental evidence that adults are happy to see fuzzy prototype-based categories, even where the categories are, in fact, clearly categorical. For example, Armstrong *et al.* (1983) found that 7 is the *best* odd number, more 'odd' than, say, *23*. This may mean that, in many cases, prototype characterizations are in the adult analyst's head—not the child's. However, this research programme is yet young, and particular theories are still amorphous. Palermo points out that the main problem with abandoning a 'feature theoretical approach to semantic development lies in the lack of a comparable heuristically simple conceptual structure for dealing with the synthetic nature of meaning' (1982, p. 335). Carey is an even harsher judge. She notes the 'the cluster view of concepts cannot do the epistemological work' of the classical view, but is in other ways similar to it, in that it ultimately has definitional primitives and uses intensions (in a more complex way) to determine extensions (1982, p. 352; see also Atkinson 1982).[22]

24.6.3.5 *The state of the art*

In the midst of all of this theory-building, it is difficult to find information specifically relevant to how infants solve the mapping problem, how they know to attach words to things, and to which (parts of) things to attach names. (How does the child know *rabbit* means the whole animal, not rabbit ears, or rabbit whiskers? How does the child find the unit 'word' in the first place? See Gleitman and Wanner 1982; Landau and Gleitman 1985; Slobin 1985, for discussion.) It is likely that human infants are genetically biased to auditorily perceive linguistically relevant units (Section 24.5.2.1, above), and to attend to and form concepts relevant to visual wholes, and that their predisposition to map words on to concepts is a fundamental part of the species-specific language-capacity. These biological endowments aid the child in a graduated history of learning to refer (Atkinson 1974; Lock 1980; Ninio and Bruner 1978). It is clear that the infant *is* able to solve the mapping problem; the rapid rate of word-learning is convincing evidence that children learn from simply hearing (or seeing, in the case of signed languages) new words in conjunction with examples of their proper referents, and from ostensive definitions. Current theories are narrowly focused—the ones we discussed relate primarily to early referential word-learning—and leave large areas of semantic knowledge development unaccounted for. In the next section we address unresolved theoretical and empirical issues particularly relevant to general symbolic development.

24.6.4 The lexicon and human symbolic capacity: developmental questions

24.6.4.1 When is a word a symbol?

The state of the art of language development studies is such that we can confidently say that baby's first 'word' at about 12 months is somehow not a real word, but is better referred to by a different term, such as 'sensorimotor morpheme' or 'protoword' (see Section 24.5.2.2). Some investigators of infant communication go to great lengths to provide operationally viable definitions of 'word'; and here there is some consensus, but details vary (see Ferguson 1978). To illustrate, Dore (1983, p. 175) only counts as words those forms that are phonemically stable and largely conventional; have intensions that at least partially overlap the adult's; are used detached from the immediate context and displaced in time and space; are well-established in the child's memory; and participate at least minimally in semantic contrasts. This is a stringent set of criteria; but it clearly identifies a point on a developmental scale when we can say unequivocally that the word is a true symbol fulfilling definitional characteristics of arbitrariness and displacement, and able to take its place in a structured symbol system. The issue of when infants' vocalizations should be counted as words is germane to our discussion because we want to ask what an adequate theory of lexical acquisition is meant to account for, and whether the same set of principles must be able to account for developments before and after this point (leaving aside very general principles, such as contrast and conventionality, that are definitional to language itself). In relation to phylogenetic questions, we need to identify what might be 'evolutionary hurdles' that other species have yet to jump. For example, Petitto and Seidenberg (1979) concluded, after careful and detailed analysis, that chimpanzees taught to communicate with arbitrary gesture and plastic 'symbols' did not use them displaced in time, space, or context from very specific well-defined interactions.

While it is easy to claim that there is a qualitative change in children's knowledge about and use of words sometime before their second birthday, it is difficult to pinpoint the transition itself, because development appears to be gradual and continuous. It is possible to chart very small increments of change as a child uses 'words' in more and more decentred and decontextualized ways, at some moment understanding the general principle that things have names (cf. for example Bates *et al.* 1979; Lock 1978, 1980, 1985). McShane (1980) has called this the 'naming insight', something that must be explained. McShane's own explanation is that this insight, 'a relatively sudden realization of some previously unseen structural relationship', is a psychological transformation facilitated by the child's participation in ritual 'naming' interactions with their care-givers (p. 49; also see Ninio and Bruner 1978). The naming insight can be documented only in terms of its consequence, the rapid increase in rate of word-learning mentioned earlier (see also Halliday 1975), although some parents may observe the moment itself, sometimes described as the cartoonlike 'light-bulb phenomenon'. (Not all children show accelerated vocabulary growth at this time; McShane (1980) and Nelson (1985) propose that these children are social-functional learners who have not participated seriously in the 'naming game'.) Other correlated behaviours are a sudden interest in categorizing and naming things, a relational use of referential words, the appearance of relational words and constructed single-word sequences, and the beginnings of a true phonological system and true dialogue, with word-combinations following shortly.

This difficult-to-identify naming insight is the key to language, in that the child must somehow come to know just how to relate language units and structures to concepts and intentions. Remarkable is the fact that *not one* of the theories we have reviewed explicitly confronts this issue (though Nelson does try to relate the beginning of naming to concept-differentiation).[23] Theories of lexical acquisition, especially the semantic feature hypothesis and the contrastive analysis hypothesis, implicitly assume that the learner already possesses this fundamental concept, but their data derive—at least in part—from a pre-insight period. This must influence our evaluation of theories and central acquisition issues, such as the apparent comprehension–production gap. To take this 'gap' as an example, various explanations of the developmental priority of lexical comprehension have been proposed, primarily appealing to processing and memory factors: to understand, a child must recognize a word and search memory for a concept to match the usual referents of the word, while to produce a word the child must have a concept in mind, and then make an active search of memory to retrieve a known word (see Clark 1983; Nelson *et al.* 1978, for discussion). Can this interpretation hold up if children who show an asymmetry in comprehension and production are in a pre-naming insight state? The children in Benedict's study—the study most cited for the temporal relation of the two language-performance modes—were between nine and ten months old at the beginning of the study, and no more than two years at the end (with an additional observation at two years and three months). The children were most likely not symbolic word-users in the first phase of the study, which ended when they were one year and five months, and had not yet produced more than fifty words. The criterion for comprehension was the child's 'clear, immediate and correct response' to a word; the responses reported were pointing or touching present objects (Benedict 1979, p. 186). We must ask on what basis the children were 'comprehending' in this early associative-referential period, and whether the basis for comprehension (and thus, perhaps, also the relation between comprehension and production) changes when children are using 'real' words. This sample issue brings us to consideration of a related question: is lexical acquisition directly attributable to the developing general symbolic capacity?

24.6.4.2 Is the developing lexical system part of a general symbolic capacity?

The Piagetian answer to this question is 'yes' (see Butterworth, Chapter 16, and Sinha, Chapter 17, this volume, for discussion). This view is what Bates (Bates et al. 1979) calls a 'direct causal model', in which cognitive prerequisites are counted as necessary and sufficient conditions for language—here lexical—acquisition. This is but one version of the 'unitary mind' view (Section 24.4). It predicts both strict sequencing and positive correlations between cognitive and language domains. Evidence from sequencing studies has largely disconfirmed this strong position; predicted antecedent cognitive accomplishments sometimes postdate their language equivalents, particularly when stated in general terms (see Corrigan 1979, for a review of this research). There are strong positive correlations between the acquisition of some very specific words and the solutions to related specific cognitive problems; but the sequential order within each pair may vary. For example, the (approximately) 18-month-old child will use the word *gone* and solve the Piagetian object-permanence task at the serial invisible displacement level at about the same time, but not always in the same order. To take one more example, the naming explosion is closely temporally related to spontaneous sorting of objects (Gopnik and Meltzoff 1986).

The psychologist's most typical retreat from the direct causal model has been to a position where the development of a general symbolic capacity is said to underlie and explain correlations between language and cognition. An example is Bates's (Bates *et al.* 1979) 'homology through shared structure' model. She argues that the human symbolic capacity arises from the combination of pre-existing cognitive and communicative skills that originally served other functions.[24] Bates supports her proposal with data she and her colleagues assembled in the search for pre-requisites: development of non-linguistic and linguistic symbols, and correlations between cognitive and communicative developments, and between these together and mother–child social developments. The correlational studies were designed to isolate 'old parts', components that combine in a way (determined by heterochrony) that leads to symbol developments both ontogenetically and phylogenetically. Arrival at the symbolic level depends on criterial levels of component skills which human infants attain. Bates pinpoints the period between nine and thirteen months as 'the critical moment for the emergence of symbols', and the idea that things have names. Note that this is much earlier than the 18-month to 2-year-old period we have identified. Bates counts as symbolic referential acts that are quite context-bound, for example a child's use of *wufwuf* in the presence of a dog or upon hearing a dog bark. While she refers to a slow process of decontextualization, she does not consider displacement to be developmentally prior to the moment of discontinuity that marks the beginning of language proper (see Lock 1980, 1985, for an opposing view). The homology model is not the only retreat from the direct causal model. Gopnik and Meltzoff (1986) propose that separate linguistic and cognitive domains interact importantly from the earliest stages. Children acquire words for just those cognitive problems they are working on at the time, and the words may even help solve the problems. This is not dissimilar to Vygotsky's view, and fits our theory of an active, experimental learner (Section 24.5.3.5; see also Dromi 1988; Lock 1980).

Although we have no definitive answer to the question we posed at the beginning of this section, we raise two issues to keep in mind when reflecting on the language–thought connection in the early language period: (1) the relation between gesture and words and (2) the relation between concepts and words.

24.6.4.2.1 Gesture and words

For normal human infants, gestural communication precedes verbal communication and—during the period of single-word utterances—supplements it. Psychologists who see language emerging from a general cognitive–communicative base call attention to the temporal precedence of gesture and its continuity with language as demonstrated by similarities in the developmental history and function of each communication mode, and the apparent 'substitution' of words for gestures in equivalent messages once the child has become verbal. In particular, communicative pointing is interpreted to be the foundation for reference.[25] Lock (1980, 1985) traces gestural communication back to its origins in the infant's actions, which are interpreted as 'communicative' by the mother. Toward the end of the first year, the infant uses gestures singly and then in combination (for example, crying plus pointing) in intentional but still underspecified communication. The infant's first 'words' are interpreted to be holistic verbal gestures, constituting functional differentiation of gestures, which allows greater communicative specificity and 'produces the bedrock from which symbolic communication of an analytic nature may arise' (Lock 1985, p. 258). Lock follows the gesture–verbalization continuity as far as two-word utterances, in which he finds gestural origins of each word (which is transformed from, rather than substituted for the gesture) and equivalent structural relations between the two elements in both gestured and spoken communications. An example is the functional and structural equivalence of the gestured [cry + arm raising] with the more specific verbal *mama down* (Lock 1980, p. 30). By this point in developmental time, children are in possession of separate elements of knowledge that are necessary and sufficient—by new combination—to inevitably catapult them into a new cognitive arena where propositional language and analytic thought are possible.

This is a very appealing developmental tale, from both the ontogenetic and the phylogenetic vantage points, but we must at least query the *necessity* of prior gestural communication and the *apparent continuity* of development between gestural and verbal systems. Information relevant to this query comes from reports of blind children, who are not observed to spontaneously use natural (manual) gestures such as pointing and open-handed reaching communicatively (for example Urwin 1978). In a current longitudinal study of identical twins, one of whom is blind, C. Johnson found that the twins were on the same developmental timetable for their early language development. First 'words' appeared at virtually the same time, as did word combinations, although the blind twin did not gesture communicatively. Of course, the mother responded to both twins' actions and crying 'gestures' as if they were communicative, and it may be significant that the blind child used what we interpreted to be a vocative *mama* more than a month earlier than his twin did; this would presumably count as a 'verbal gesture'. But even if we use Lock's expanded definition of 'gesture', there is a huge gap in this child's gestural communication system where there should be particular gestures that are transformed into words. This raises the possibility that gesture is not the only way into verbal language, i.e. that language is by now so robust in humans that it is multidetermined.

Deaf infants learning a signed language provide an additional testing ground for the hypothesized continuity between pre-linguistic gestural communication and symbolic (propositional) linguistic communication. If language truly builds on a foundation of communicative natural gestures, signed language should be developmentally continuous with, and facilitated by, pre-linguistic gestures. A specific test case is congenitally deaf children's acquisition of the personal pronoun system in American Sign Language (ASL), in which the form of the pronouns *I, you*, etc., is deictic pointing. Hence the sign is non-arbitrary and the meaning should be transparent. Petitto's (1987) observational and experimental data show that two deaf children learned personal pronouns at about the same time and in the same order as hearing children (beginning at 17 to 20 months). This result demonstrates that pre-linguistic deictic pointing did not facilitate acquisition of this linguistic system, *contra* predictions of the claim that words develop as direct transformations of gestures. But Petitto's study provides even stronger evidence against the continuity between deictic pointing—used extensively by these children—and linguistic (pronoun) pointing: just prior to learning first- and second-person pronouns, the children avoided pointing at another person meaning *you*, using the person's name instead. Shortly thereafter, the children went through a brief period of pronoun-reversal errors. For example, one child used the sign YOU (point at other) to indicate ME (herself), as if this gesture were her name—an error that some hearing children also make, though in the spoken mode—and this error was resistant to correction. Petitto interprets this error to be a problem of lexical learning (i.e. not general cognitive), because it occurred when the child's deictic pointings to objects had begun to decline and her vocabulary was increasing rapidly. The child ignored apparently obvious gestural information to interpret the pointing as an arbitrary symbol. (See Petitto 1987 for further evidence and argument that pre-linguistic gesture is radically discontinuous with language.)

24.6.4.2.2 *Words and concepts*
The relation of content-words to concepts and real-world referents is much more direct and transparent than it is for other linguistic symbols such as, for example, phonemes or complementizers. The lexical semantic system is very different in nature from the phonological and syntactic systems of language, which were somehow designed to do an enormous amount of creative work with very limited inventories of symbols and rules. While the lexical semantic system is rule-governed (including intensional and formational rules, and orderly relations between words at the same and superordinate levels), it contains an exponentially greater number of symbols, and the 'open class' lexicon is just that—open to admit new

members. The lexicon is the language component most closely tied to general cognition. Jackendoff (1983) sees semantics as a link between language and thought, inasmuch as semantic theory is constrained by both grammatical and cognitive facts, though the 'grammatical constraint' has been neglected in most philosophical and psychological theories of meaning. We will address both grammatical and cognitive facts in our consideration of the relation between the lexical and cognitive domains.

In a discussion of the *relation* between two domains it is important to keep the constructs of each domain separate; but in most accounts of children's early word-meanings, the assumption that meanings are isomorphic with concepts prevails. Although Clark (for example 1983) points out the importance of distinguishing concept and meaning, the distinction is blurred in most theoretical statements (including Clark's early ones), and the main problem—precisely how a child relates a word's intension to a structured concept—is not explicitly addressed. An illustration of this problem is use of the term 'intension' itself, which refers variously to the concept the word expresses (for example Carey 1982; Greenberg and Kuczaj 1982) or the meaning-structure of the word (for example Clark 1973, 1983). This apparent confusion is probably due, at least in part, to the lack of consensus on the separate issues of the nature and structure of concepts—independent of language—and the meaning-structure (intension) and extension of words. A stance on each of these issues must be adopted before the child's solution to the mapping problem can be investigated. But the conflation is also linked to the 'unitary mind' theoretical emphasis on cognitive development as foundation, pace-setter, and constraint on language development.

The cognitivist views the child's early language-learning task as 'mapping or translating from one representational system (cognition) into another (language)' (Bowerman 1980, p. 289), where encoded concepts are formed before their semantic analogues (words) as the infant interacts with the social and physical environment. Many linguistic semantic categories, especially ones relevant to one-year-olds' concerns, are highly constrained by biologically determined perceptual and/or cognitive factors, and children's nonlinguistic understanding of the world and its 'natural categories' clearly guides them in their attempts to fathom meanings of new words. But word-meanings are not completely determined by non-linguistic categories; the concept–word connection is not a direct one, but is rather mediated by the organization of the tightly structured semantic system of a language. Semantic categories in different languages honour some, but not other 'natural' categories, so the young child's success in recruiting universal categorizational principles to decode word-meaning will depend on the language itself. The learner's hypothesis in a particular word-learning situation may be correct or incorrect, or possibly marginally correct by metaphorical extension (Bowerman 1983). This is illustrated by an example from one of Bowerman's daughters, who said *night night* to refer to 'normally vertical objects now horizontal'. Bowerman goes on to explain that, although English does not routinely encode this change in spatial orientation, there are some American Indian languages in which almost every sentence obligatorily codes the standing, sitting, or lying position of both animate and inanimate objects (1983, p. 297). This example is not taken to support the Sapir–Whorf hypothesis that language determines thought. Rather, it shows that the child learner, prepared for entry into the language of the community with a set of interpretative biases for categorization, tests cognitively motivated hypotheses to solve word-meaning problems set by the language. It demonstrates that the relation between semantic and conceptual categories is indirect, that the fit between these domains is loose enough to allow cross-linguistic mapping differences—and this must be so for children to learn the different languages of the world. Slobin (1985) refers to a flexible, undifferentiated semantic space as an unbiased 'opening wedge' to the acquisition of language.[26] The child's ability to perceive and categorize objects, events, and states in a (constrained) variety of ways provides the basis for an, at first, language-neutral semantic space. As the learner deals with the input from the language at hand, he or she structures, and then restructures, that space. The lexical semantic system—to a much greater degree than the other language subsystems—continues to evolve throughout an individual's life-span, both by the addition of lexical entries and by the reorganization of old ones as new knowledge is assimilated.

Other problems with assuming too close a relation between words and referents emerge as we look at the act of referring. First, referents can be called by different names. To take an adult example from Brown and Yule (1983, p. 54), a given individual (doing away here with classes of things, for simplicity) might be talked about as Ellen, Ellen Blair, my friend Ellen Blair, the former chairman Ellen Blair, or a nurse in the ward called Ellen Blair, depending on the relationship of the speaker to the individual and the conversational partner(s), the participants' shared presuppositional pool, and the speaker's purpose in bringing the name into the conversation. Second, some referents undergo a change of state between the beginning and end of a unit of discourse. This should imply a corresponding conceptual change—but the name is the same.

Another example from Brown and Yule (p. 175) in the form of a recipe illustrates this point: 'Slice [one medium] onion finely, brown in the butter... Crush the browned onion with a spoon and add it to the chicken.' The 'thing' changes from a whole raw onion, to raw onion slices (only elliptically referred to), to butter-browned onion slices, to crushed bits of butter-browned onion, each state conjuring up a different picture in the mind; but in its final transformation it is still called 'the onion'. A third aspect of referential words is polysemy; children must gradually learn multiple meanings for single words as they grow up.[27] At first glance it may seem that such problems of reference are outside the realm of the 18-month-old; but children are presented with these learning problems quite early. To create one imaginary parallel, consider the family pet, Stormy, who on different occasions might be named *Stormy*, a nice *doggie*, a cute *puppy*, or a *nuisance*.

A final example of the complexity of the child's mapping problem is illustrated by words that encode components that are not transparently conceptual. One category of this type is the personal pronoun system. First, children must learn that, while personal pronouns have a symbolic meaning as part of their structure ('person'), part of the meaning changes with speaker–hearer participant roles, and so reference 'shifts' (Jespersen 1964), or is indexical as well as symbolic. While children sometimes make reversal errors, as illustrated by Petitto's example (1987; see Section 24.6.4.2.1), they typically acquire the deictic–symbolic combination in *I* and *you* with ease, by the time they are two. Tomasello *et al.* (1984/5) report that two- to three-year-old children know pronouns do not refer as explicitly as nouns, because they maintain or add deictic pointings more frequently with repeated pronominal reference than with repeated nominal reference.[28] Several years later, children learn that pronouns can 'point' back to reference in the discourse, not just present referents. Pronoun-learning is further complicated by components in the lexical entry—person, number, and (often arbitrary) gender distinctions—which figure importantly in morphosyntactic agreement rules in many languages.[29] When the conceptual basis of the component is unavailable, the child must map syntactic, not conceptual, distinctions on to semantic categories. Carey (1982) argues that gender is not a good candidate for semantic marking, because the young child does not understand natural gender, but makes distinctions based on superficial parameters (for example, clothing); she then goes on to cite cross-linguistic evidence that children master grammatical gender-marking before they learn the conceptual-semantic basis for this marking. English-learning children make gender errors in using third-person pronouns at an age when Hebrew and French learners are able to mark all nouns for gender. With pronouns, even the 'obvious' conceptual distinctions are not consistently marked in individual languages. Bowerman (1985) takes the example of *we*, which in different languages can mean simply first person plural (as in English); or 'we-including' or 'we-excluding' the listener (for example, in Tamil); or 'we-two' as opposed to 'we-more than two' (for example, in West Greenland Eskimo); or the last two combined for a four-way split (for example, in Hawaiian); or a further breakdown of number into two, three, and more-than-three distinctions, combined with inclusive versus exclusive for a six-way split (for example, in Nogogu, a Melanesian language).

The *we* example takes us back to our point about the 'naturalness' of semantic categories; but we have one more syntactic point to raise. The flip side of the syntactic–semantic mapping in pronouns is children's use of syntactic information to discover words and their meanings. Experimental studies demonstrate that children as young as 18 months can use morpho-syntactic cues to isolate words and discover whether they refer to actions, or countable or uncountable things, or individuals or classes, and so on.[30] Although these discussions differ on details about how children exploit syntactic information, taken together they indicate that here is one more deviation from a simple, direct word-on-concept mapping.

In this section we have raised issues we consider relevant to charting the relation between cognitive and semantic space. While cognitive development (with social interaction) surely lays the foundation for linguistic communication, and non-linguistic categorizational predispositions are essential if children are to crack the code at all, the precise relation between the domains remains an open question. The evidence we have presented here cautions that we must avoid assuming over-simplistic connections. There are important discontinuities between gestured communication and language, as well as between early use of 'words' and the beginning of propositional language; and the mapping between words and concepts is flexible, and influenced by factors unrelated to general cognition. No research to date has answered the fundamental question: how does the mapping from one cognitive domain to another happen? Our only retreat at this time is to reiterate that the child's cognitive biases and abilities interact importantly with the organization of the input language from the earliest period of verbal communication.

The scope of our discussion has been very narrow indeed, a scope that cannot be maintained without losing sight of what meaning—and language in general—is really like. We have ignored the acquisition of relational words and the relational use and pragmatic functions of referential words in the early word period, as well as aspects of later semantic development that clearly illustrate the high level of organization and the repeated extension and restructuring of the semantic system and would further elucidate our consideration of the mind–language issue.[31] But our focus was not a random choice. Following Brown (1958), who states that theories of meaning begin with the abilities to name instances of referents and to respond to names as symbols of referents, we have at least made a beginning.

24.7 Acquisition of the syntactic system

In this section we will examine the ontogenesis of syntax and morphology. Since it is impossible in such a short space to give a comprehensive as well as a detailed picture of morphosyntactic acquisition, we will confine ourselves here to the following more limited aims: (1) a brief statement of the learner's task, and (2) a broad overview of the sequence of acquisition, highlighting areas of particular theoretical interest.

24.7.1 *The learner's task*

The acquisition of syntax is an unparalleled intellectual feat, generally accomplished in a remarkably short time (from five to ten years), and under conditions which are arguably far from optimal. How children accomplish this feat is the subject-matter not only of developmental psycholinguistics (which concerns itself directly with children during the process of language-acquisition), but also of the theory of grammar itself (see, for example, Chomsky 1965). Both disciplines are concerned with the central question of how a child, given the evidence available, comes to choose the correct grammar for a natural language.

This question is anything but easy to answer. Let us illustrate this with an example, that of 'structure-dependency'. Structure-dependency refers to the hierarchical constituent structures that underlie strings of words in the sentences of natural languages. Thus, the words in an English sentence such as *The cat chased a rat* are organized into superordinate phrasal units on the basis of various distributional regularities; for example *the cat* and *a rat* are noun phrases, *chased a rat* is a verb phrase, and the whole string is a sentence. Each phrasal unit has its own characteristics. A noun phrase, for example, is the only type of constituent that may be introduced by determiners such as *the* and *a*, may normally appear in subject position of a sentence, and may be replaced by a pronoun.

Syntactic rules are 'structure-dependent' in the sense that they crucially refer to constituent structures, as opposed to simple strings of words. This can be illustrated by the rule of subject–auxiliary inversion, which operates in questions and various other non-declarative environments in English. This rule moves a tensed auxiliary verb in front of the subject noun phrase, as shown in the examples in (a) below, where the auxiliary is italicized:

(a) *Did* the president fall asleep again?
What *are* we going to do?
Never again *will* we put our trust in such a man!

Subject–auxiliary inversion could, at first sight, be formulated as:

R1: Move the leftmost verb in a string to the front of a main clause.

This is a partially structure-independent rule, in that it refers to the linear order of constituents (as in 'leftmost verb') rather than to the hierarchical relations between them. Alternatively, we could formulate the rule as:

R1′: Move an auxiliary verb to the front of the main clause that immediately dominates it.

Here we have replaced reference to linear order with the notion of 'immediate dominance', which refers not to strings, but to structures; R1′ is thus a structure-dependent operation. Now we are in a position to ask which of these two possible formulations of subject–auxiliary inversion most accurately characterizes the behaviour of native speakers of English. The examples in (a) provide no evidence one way or the other. On the other hand, the application of the inversion rule to examples such as (b) does provide a relevant test:

(b) The man who is tall is in the room.

Sentence (b) contains a relative clause embedded in a subject noun phrase. If we were to apply rule R1 to this structure, we would move the leftmost verb (the *is* immediately following *who*) to the front of the main clause, to give us (c) as output:

(c) *Is the man who tall is in the room?

This sentence is ungrammatical, as is indicated by the asterisk. The structure-dependent rule R1′, on the other hand, will correctly adjoin the main-clause auxiliary verb to the front of the sentence that immediately dominates it (i.e. the main clause itself), leading to a grammatical output:

(d) Is the man who is tall in the room?

Thus more complex sentences show us that native speakers employ structure-dependent syntactic rules, even where in simpler cases structure-independent rules appear to characterize the data adequately.

This conclusion has important implications for first-language acquisition. The child learning syntax must be able to choose the structure-dependent rule over the structure-independent rule, in order to avoid generating sentences such as (c). Of course, it might be claimed that the learner proceeds in exactly the same way as the linguist, systematically noting the contrast between grammatical examples such as (d) and ungrammatical examples such as (c). However, there is strong evidence that the data available to first-language learners is qualitatively different from, and much more restricted than, that available to syntacticians. In particular, it appears that children neither systematically receive nor employ so-called 'negative evidence', that is, evidence based on the *un*-grammaticality of examples such as (c) (Brown and Hanlon 1970). It might be argued that, even without negative evidence, it is still possible for the child to formulate the correct structure-dependent rule of subject–auxiliary inversion on the basis of 'positive evidence' provided by sentences such as (d), which could not be generated by a rule such as R1. However, this leads to another implausible assumption about the data base available to children learning the syntax of their first language: that complex structures containing relative clauses, as in (d), are readily available to, and can be processed by, young children learning the rule of subject–auxiliary inversion. In fact, relative clauses are mastered comparatively late in the acquisition of English (see Section 24.7.2 below)—certainly well after subject–auxiliary inversion. It appears unlikely, then, that sentences such as (d) are useful to young children at the stage in question.

One other possibility is that children initially formulate a structure-independent rule of subject–auxiliary inversion, and then, on the basis of more complex data, switch to a structure-dependent version. This view of learning makes the prediction that, prior to encountering data such as the sentence in (d), children will have a very different type of inversion rule from that which characterizes the syntactic knowledge of adult speakers. In other words, there will be a radical discontinuity in the acquisition of inversion, and indeed in many other syntactic processes which share similar properties. However, evidence to date gives very little support to the claim that children ever have a structure-independent rule of subject–auxiliary inversion; both a complete absence in the child-language literature of recorded errors of the type in (c) and recent experimental work (Crain and Nakayama 1987) seem to show that, though children do make characteristic errors in the acquisition of inversion (see Section 24.7.2.2 below), they never adopt structure-independent inversion rules.

To summarize, the task of the child learning syntax involves the acquisition of a set of rules characterized by complex properties (such as structure-dependency) that are by no means readily accessible from the restricted data available to the learner. In fact, on a more formal level, it has been known since the pioneering work of Gold (1967), who established a machine analogue for human language-learning known as 'identification in the limit', that certain types of language are unlearnable under certain conditions. In particular, in Gold's paradigm, only those languages which constitute finite sets of sentences are learnable without the use of both grammatical and ungrammatical examples (appropriately labelled) in the learner's input data. Now, it is well-known that all natural languages consist of *in*finite sets of sentences; and as we have pointed out above, children apparently make no use of ungrammatical sentences in acquiring their native syntax. This leads to the curious conclusion that, under the observed conditions of human language-acquisition, human languages are unlearnable.

One way out of this paradox is to present the data to the language-learner in an ordered fashion, from least to most complicated; it has been claimed, particularly by proponents of the so-called 'motherese hypothesis', that primary care-givers do indeed order the data in this fashion. Another way is to allow the learner to calculate the probability of occurrence of a given sentence, and to infer the ungrammaticality of a missing sentence from its absence. In this way, the learner can gain access to information about both grammatical and ungrammatical sentences, and, thus, using what is often termed 'indirect negative evidence', successfully overcome the learnability problem. A third way out of the paradox, adopted by some proponents of the 'principles-and-parameters' approach to the problem of language-acquisition (see Section 24.4.3 above), is to claim that only a finite number of grammars (in fact, possibly even a single grammar with a limited number of variants) are *a priori* (i.e. innately) possible for the child; in this case, 'learning' a grammar consists not in identifying one of an infinite number of possible grammars, but in choosing one out of a finite set.

These various solutions to the problem of language-learnability have very different implications for the study of language-acquisition. The 'ordering solution' just described predicts that the environment necessary for a child to acquire a natural language must be highly structured, since the care-giver will impart the language to the child in a highly systematic way. This is indeed the position adopted by many 'interactive' approaches to the acquisition of syntax (see Section 24.4.1 above). On the other hand, both the 'indirect negative evidence solution' and the 'principles-and-parameters solution' give greater weight to the abilities of the child than to abilities of the care-giver, though in different ways: the first of these two solutions sees the child as a sophisticated calculating machine, sensitive to the frequency of strings in its input; whereas the second sees linguistic input as serving a more limited 'triggering' function, in the sense that certain crucial types of data are necessary to set a given parameter (see Section 24.4.3). In other words, the probabilistic language-learner is relatively data-dependent in comparison with the principles-and-parameters learner; we might expect this difference to show up in studies of actual language-acquisition. Thus tentative answers to fundamental questions of learnability lead to rather different programmes for the study of language-acquisition, as we shall see.

24.7.2 Overview of syntactic acquisition

The acquisition of syntax (and of morphology) may be said to follow three stages, which we will label 'presyntactic', 'syntactic', and 'postsyntactic'. The presyntactic stage covers the period from the emergence of the first words up to the appearance of early word combinations. The syntactic stage lasts from the onset of complex syntax (marked by the acquisition of inflections and the 'closed class' syntactic categories, i.e. those such as determiners, auxiliary verbs, and complementizers, which consist of a small set of grammatically important items) up to the point where the child has a superficially adultlike system (usually around the beginning of the school age). Postsyntactic acquisition deals with those areas of syntactic competence with which older children are known to have difficulty. The transitions between these stages are not abrupt; some morphological paradigms (for example the past participle form of irregular English verbs, such as *given* and *written*), for example, are learned at a comparatively late stage in acquisition, and are sometimes not even completely mastered by adults, while others are acquired much earlier and more successfully.

24.7.2.1 *The presyntactic stage*

We will take the position here that there is very little to be learned about specifically *syntactic* acquisition from the one-word or two-word stage of linguistic development, though there is a great deal of semantic and conceptual information to be extracted from this period. Presyntactic development has received a great deal of attention from investigators of child language.[32] It is relatively easier to give a descriptive account of this period than it is to provide an explanatory framework.

Acquisition proceeds roughly as follows. During an initial period, beginning usually in the second year of life and lasting up to several months, only single-word utterances are produced. These generally lack any morphological marking. At roughly eighteen months of age—though chronological age is a notoriously bad indicator of syntactic development—the learner begins to 'chain' single-word utterances (Bloom and Lahey 1978; Branigan 1979; Scollon 1976). These 'successive single-word utterances' are intonationally separated but thematically connected words, for example *throw* followed by the separate utterance *ball* while the child is throwing a ball, as opposed to *throw ball* under a single intonation contour. Such constructions of single words often occupy a transitional phase between single-word and multi-word utterances, which begin to appear shortly thereafter. There is some evidence that the onset of early word combinations is correlated with the well-known 'word-explosion' described in Section 24.6.2 above.

There appear to be three types of early word combinations. The most well-known characteristically consists of content words (i.e., nouns, verbs, and adjectives), the 'open class' defined above. Morphological differentiation is typically absent, and determiners, auxiliary verbs, and other 'functional' categories are missing—hence the tradition of calling this stage 'telegraphic'. The following examples, from Brown (1973), illustrate such 'telegraphic speech':

Bambi go	(agent + action)
See sock	(action + object/theme)
Want more	(action + object/theme)
Pretty boat	(object + attribute)
Baby highchair	(entity + location)

Beside each utterance is its description in terms of 'thematic relations' (Gruber 1967). This type of description grew out of observations that early multi-word utterances tend to reflect a restricted set of grammatically relevant real-world relations (see in particular Bowerman 1973 and Brown 1973). Word order is generally adultlike, though Braine (1976) points out the existence of a transient 'groping pattern', characterized by thematically linked but randomly ordered word combinations, such as one child's production of *juice Daddy* and *Daddy juice*, both with the possessive meaning, 'Daddy's juice'. For the 'groping' child, these two utterances are synonymous; no meaning-difference is attached to word order. This suggests that, at least for some children, word order lags behind what might be called 'thematic syntax'.

Recent research has pointed out that there are two further types of early multi-word utterance. The first consists of a single-word utterance supplemented by intonationally appropriate but meaningless extra syllables (Braine 1976; Ramer 1976; Peters 1977; Macken 1979). The second consists of unanalysed or partially analysed rote-learned sequences, which the child employs as formulae and only subsequently segments into separate syntactic units (C. E. Johnson 1981; Peters 1983). It appears that most children—in all languages from which relevant information is available—produce various combinations of these types of utterance, though a rough division can be made between children who employ mainly 'bottom-up' strategies (i.e., concatenate single units) and those who employ 'top-down' strategies (i.e., segment formulaic multi-unit utterances).

Several major methodological problems confront investigators of the presyntactic stage. The first of these is the issue of productivity. As has been pointed out, it cannot safely be assumed that children's early word combinations represent the output of productive syntactic rules—a considerable proportion of those utterances may be rote-learned formulae. Several investigators have begun to develop procedures for assessing productivity, which promise to clarify this important issue (Braine, 1976; Ingram, 1981).

A second and perhaps more fundamentally problematic issue concerns what we shall call 'underdetermination'. Given corpora of two- or three-word utterances, it is possible to write very simple grammars—certainly far less complex than those required to account for adult utterances. Moreover, on the basis of child-language data alone, it often seems preferable (i.e., more 'natural') to construct grammars whose primitive categories differ qualitatively from those in adult grammars; for example, it has been argued by Bowerman (1973) and Brown (1973) that thematically-based grammars are more appropriate for generating early word combinations than grammars based on conventional syntactic categories such as noun, verb, and the like. This approach raises the 'continuity problem' discussed in Section 24.7.1 once again: how do children switch from early, simple, possibly thematically-based grammars to far more complex, syntactically-based adult grammars?

In the absence of any empirical evidence for a radical discontinuity in syntactic acquisition, the usual answer to this question is to deny that early grammars differ qualitatively from later ones; according to this view, the observed differences are epiphenomenal, and caused by independent factors. For example, Pye (1983) claims that the apparently 'thematic' profile of many children's early word combinations is actually phonological in origin: it so happens that in English (and in many other languages) content-words bear focal stress, and thus are phonologically salient, while function-words typically bear little or no stress, and are thus more difficult to acquire. Pye gives evidence from the acquisition of Quiché Mayan, which has a stress-pattern independent of the content-word–function-word distinction, that 'telegraphic' speech is phonologically reduced rather than thematically based. For example, one Quiché child at this early stage reduced the verb form *k-0-aw-il-oh* ('you see it') to *loh*. The child's form corresponds to the last syllable of the verb form, which is stressed, but does not respect morpheme boundaries (marked by hyphens in the adult form) or represent lexical meaning; it consists of the last consonant of the verb root and a 'terminating' grammatical morpheme that does not carry meaning. To take an English example for contrast, the child utterance *baby highchair* cited on p. 718 is a reduced form of a hypothetical adult utterance *The baby is in the highchair*, retaining just those words that carry stress and—in addition—carry the essential meaning of the sentence.

Another solution to the continuity problem has become known as the 'semantic bootstrapping hypothesis' (Grimshaw 1981; Pinker 1984). According to this view, the apparent thematic basis of early grammars is caused by an acquisition mechanism which exploits a syntax–semantics parallel between thematic and syntactic categories in order to lay the foundations for a purely syntactic system. At the stage in question, it is hypothesized that the child exploits certain canonical correspondences between syntactic and semantic categories in order to construct primitive grammatical representations; thus nouns are typically associated with things and people, verbs with actions, adjectives with attributes, and so forth, and the child can use these limited correspondences to establish a set of 'core categories'. Such correspondences, however, are positively misleading during the later stages of acquisition, when syntactic categories must be established on the basis of distributional regularities alone; at this stage, the semantic scaffolding which the child has used to construct core categories is kicked away, leaving a purely syntactic system.

Yet another view, that advocated by Maratsos and Chalkley (1980) and Maratsos (1982), denies any role whatsoever to thematic categories. According to this view, syntactic categories such as 'noun', 'subject', and 'verb phrase' are established from the very beginning on distributional evidence alone: the child acts as a 'little linguist', establishing syntactic regularities on the basis of syntactic evidence. The 'thematic' flavour of early word combinations is simply a reflection of what young children tend to talk about (i.e., the objects, people, actions, and relations in their environment), and provides no evidence for thematically-based grammars.

In any case, it should be noted that, however interesting children's early word combinations might be from the point of view of theories of the role of conceptually and semantically based structure in language development, the presyntactic period throws little light on the acquisition of specifically syntactic structures, since such structures are conspicuously absent until the onset of the 'syntactic' stage of development, to which we now turn.

24.7.2.2 *The syntactic stage*

For the investigator of morphosyntactic acquisition, perhaps the most crucial period of language development takes place during the period which lasts from the appearance of the first characteristically grammatical elements (for example, inflectional morphemes, such as past-tense or plural-number markers, depending on the language) until the acquisition of the full range of syntactic constructions in the target language (including co-ordination, complementation, relativization, and various processes of ellipsis and pronominalization, again depending on the language). This stage lasts roughly from the ages of two to five years, though acquisition is by no means complete at the end of this period.

The syntactic stage is also the stage of typological differentiation. At the presyntactic stage, children learning diverse languages proceed in a very similar fashion (see Gentner 1982, for a survey). However, with the onset of the syntactic stage and at subsequent points in development, language-specific factors come into play far more sharply. One important divergence takes place between what might be termed 'morphologically biased systems' (those which express the major syntactic relations mainly through inflectional morphology) and 'configurationally biased systems' (those which express these relations mainly through word order). It appears that early in the syntactic stage, children focus either on inflectional paradigms (in morphologically biased systems) or on word order (in configurationally biased systems); thus Slobin (1985) reports that children learning Turkish, a morphologically biased system, learn inflectional morphology much more rapidly and easily than those learning English or Italian, both of which are strongly configurationally biased; on the other hand, Turkish children experience unexpected difficulties with configurational structures such as relative clauses, which English and Italian children appear to acquire with greater ease.[33]

Perhaps the most remarkable aspect of the syntactic stage is how rapidly and successfully most children pass through it. In general, first-language learners appear to make very few errors in constructing complex grammars on the basis of a limited range of data, an observation which appears to support the contention that the learner's 'hypothesis space' is severely limited by innate constraints on possible grammars, whether such constraints are specifically linguistic or not.

However, it is not the case that syntactic acquisition is errorless: certain highly consistent error-patterns characterize several aspects of syntactic and morphological development. These error-patterns have proved to be of great interest for the investigation of the relation between syntactic theory and language-acquisition, since they represent cases where the child 'misprojects' from the available data to an incorrect grammar. As such, they provide evidence bearing on three related issues: (1) the question of the relation between misprojections (i.e. intermediate grammars) and possible adult grammars; (2) the question of the scope of children's syntactic generalizations; and (3) the question of 'retreat mechanisms' from overgeneralized grammars. Let us briefly discuss each in turn, before turning to some specific examples.

First, the relation between misprojections in acquisition and the notion of 'possible grammar'. While theories that involve a discontinuity in syntactic acquisition make no strong predictions about the relation between 'child' and 'adult' grammars, theories that adopt the 'continuity hypothesis' (see Section 24.7.1 above) predict that any intermediate grammar must, *ceteris paribus*, correspond to a possible adult grammar (White 1981). The next question concerns the scope of syntactic generalizations in acquisition: do children make categorical generalizations? If so, we might expect misprojections to involve entire syntactic categories (verbs, for example). On the other hand, if first-language learners are more cautious, we might expect misprojection errors based on subcategories (verbs of existence, for example) or possibly on individual lexical items (the verb *to be*, for example).

Finally, there is the question of 'retreat'. Recall the generally accepted observation that children acquiring natural languages make no use of negative evidence (i.e., overt corrections by the care-giver; see Section 24.7.1 above). Now, suppose a child misprojects a syntactic rule and ends up with an 'overgeneral' grammar—that is, one which generates all the grammatical sentences of the target adult language as well as ungrammatical sentences that are *not* in the target adult language. In that case, nothing short of negative evidence appears to be capable of telling the child that the grammar s/he has generated is over-general;

hence, the problem of retreat. In theory, this situation should never occur in language-acquisition—and several authors, following Berwick (1983), have proposed principles which specifically preclude it. However, as we shall see, children quite regularly do overgeneralize both morphological and syntactic rules; this has led to a vast and burgeoning literature on putative retreat mechanisms.[34]

Let us, then, turn to some specific cases of misprojections. The first and perhaps most familiar type is exemplified by the acquisition of inflectional morphology. Take, for example, the well-known and much-investigated case of the English past-tense morpheme.[35] Typically, acquisition of this morpheme takes place in three stages: (1) specific, generally irregular forms (including 'strong' verbs which form their past tense by a change in the stem vowel, for example, *see/saw, eat/ate, break/broke*, as well as complete alternations such as *go/went*) are acquired in an item-by-item fashion, with little or no generalization; (2) then there is a period of overgeneralization, where regular forms often either supplant or supplement irregular inflections (thus, we get forms such as *goed, seed, breaked* as well as—less frequently—*broked* and *sawed*); and (3) finally the adult system is acquired. This pattern is very typical of morphological acquisition in general, as well as of phonological, lexical, and syntactic acquisition (see Section 24.5.3.5). It has been named the 'U-shaped learning curve', after the typical graphic representation of an apparent drop in performance (due to incorrectly overgeneralized forms) followed by a subsequent improvement as the adult system is gradually acquired.

Morphological misprojections provide evidence on at least two of the questions we raised above. As far as the scope of generalization is concerned, it is clear from the very existence of overgeneralizations that, at least during the over-regularization stage just described, children are not acquiring inflectional alternations one by one. On the other hand, the relevant overgeneralizations are not uniform across the whole set of past-tense forms; that is, irregular forms (for example *broke*), over-regularized forms (for example *breaked*), and over-regularized irregular forms (for example *broked*) appear to compete during the misprojection stage. This variability is of two types. The first, which we will refer to as 'type variability', characterizes variable application of a rule to different lexical items of the same syntactic category. For example, a child who consistently over-regularized the past-tense form of some irregular verbs (producing from the verbs *break, see*, and *fall* forms such as *broked, sawed*, and *felled*) but not others (producing correct forms such as *went, knew*, and *ran*) would be applying the rule of past-tense formation in a type-variable way. The other kind of variability, which we will call 'token variability', refers to production of two or more forms of the same lexical item (for example *breaked* and *broked*) during the same period by the same child. Both type- and token-variability are characteristic of the acquisition of the English past-tense morpheme, and of the acquisition of morphology and even of language in general. This poses problems for a theory of morphosyntactic learning based on generalization across whole grammatical categories, since under this interpretation we would expect all irregular verb forms, for example, to be regularized at the same time. However, token-variability in particular also poses problems for a more 'cautious' theory of the language-learner, where the application of a rule is learned separately for each individual lexical item; this theory predicts no overgeneralization at all.

It appears that neither of these models is by itself sufficiently complex to account for the actual pattern of morphological acquisition. At least two other factors must be taken into account. First, learners are sensitive to input frequency, both 'type frequency' (i.e., the number of lexical items which take a given morphological affix) and 'token frequency' (i.e., the relative frequency in discourse of a given lexical item); second, children appear to employ a partially stochastic procedure (based on guesswork) to acquire morphological rules; this accounts for the apparent coexistence of several different analyses at the same time, for example, irregular, over-regularized, and over-regularized irregular past-tense forms of the same verb, as discussed above. In recent work, computational linguists working within the Parallel Distributed Processing (PDP) model of cognitive organization have introduced both of these factors into a computer simulation of the acquisition of English past-tense morphology, with some surprisingly accurate results (see Rumelhart and McClelland 1987).

A pattern very similar to the pattern of variable morphological overgeneralization seems to characterize certain types of inflectionally linked syntactic misprojection. It is proposed that a widely accepted series of 'stages' in the acquisition of the rule of subject–auxiliary inversion in English (Bellugi 1971) involve morphologically linked misprojections; in particular, the claim concerns the 'failure-to-invert' stage, exemplified by utterances such as *What you did?, Where that man went?*, and *Why I can't go?* Davis (1987) argues that such interrogative sentences, in which the subject and auxiliary are not appropriately inverted, are the result of the child's misanalysis of [WH-phrase + inverted auxiliary] sequences as [WH-phrase + agreement-clitic] sequences. Close examination of 'stages' in the acquisition of subject–auxiliary inversion shows a developmental pattern with properties very similar to that shown by morphological acquisition. In particular, subject–auxiliary inversion misprojections are type- and token-variable, apparently sensitive to the relative frequency of relevant structures in the child's data base, and partially mediated by guesswork.

It should be emphasized that the most striking aspect of grammatical development is not the occurrence of overgeneralization, but rather its absence. This certainly appears to be true for the acquisition of major grammatical structures, and, in particular, for the specifically syntactic principles that apply not to specific lexical items, but to complex syntactic structures that the first-language learner is unlikely to encounter frequently, if at all; nevertheless, the evidence available (for example, Otsu 1981) suggests that children apply these principles correctly as soon as they can process the relevant structures.

Thus we appear to be faced with two rather different types of learning: on the one hand, morphologically linked learning, with attendant variability and misprojection; on the other, rapid, successful acquisition of major syntactic structures and principles. A number of theorists are now offering hybrid explanations that propose a combination of language-acquisition components (Macken 1987; Roeper and Williams 1987). For example, Davis (1987) suggests that the best way to reconcile these two types of acquisition is through a 'two-tiered' model. On the first tier, we find a very general type of learning, possibly of the type advocated in the principles-and-parameters approach (Section 24.4.3), in which the fundamental parameters of the target language are set on the basis of rather minimal 'triggering' evidence. On the second tier, we find specific learning, linked to particular morphological paradigms, and characterized by a variable, probabalistic learning mechanism of the type proposed above.

24.7.2.3 *The postsyntactic stage*

Ever since the pioneering work of Carol Chomsky (1969), it has been known that, though superficially adultlike, the grammatical competence of children in the 'postsyntactic' period (which lasts roughly from the age of five until adulthood) differs in some significant ways from that of adults. Such differences have been discovered mainly in the following areas:

1. Children up to the age of ten consistently misinterpret the 'understood subject' in complement clauses (C. Chomsky 1969; Cromer 1972). For example, in the sentence *I promised him to go*, an adult understands that *I* is the subject of the verb *go*, but a child understands that the *him* is the one who will go. Likewise, in the sentence *He is easy to draw*, the child understands that *he* is the one who is drawing, as if the sentence were actually *He finds it easy to draw* (see Figs. 24.5 and 24.6—[eds]).

2. Children older than five often misunderstand sentences whose order fails to match that of the order of events which they describe. Thus they will correctly understand *The girl washed the boy before the boy went upstairs* but incorrectly interpret *Before the girl washed the boy, the boy went upstairs* as if it meant *The girl washed the boy; then the boy went upstairs* (Ferreiro 1971; Weil and Stenning 1978; Trosborg 1982).

3. Children in this age-group often depend more on their knowledge of the world than their knowledge of syntax to interpret passive sentences like *The dog was chased by the cat*. Since dogs usually chase cats, and not vice versa, the more 'natural' interpretation of this passive sentence is to ignore the passive morphology and treat it as an active sentence equivalent to *The dog chased the cat* (see Fig. 24.7 for an illustration of the early phase of active and passive comprehension—[eds]). Indeed, Bever (1970) and Maratsos (1974) show that older children's performance on such sentences actually declines before improving again.[36]

4. Older children often systematically misunderstand certain sentences with relative clauses, in particular those like *The pig that the dog bites chases the cat*, where the noun modified by the relative clause (here *pig*) is the subject of the main clause, but the object of the relative clause (that is, *the dog bites the*

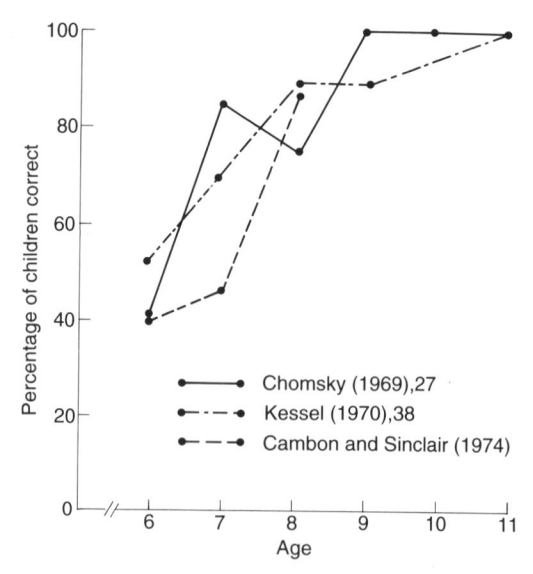

Fig. 24.5 Acquisition of *easy to X* in three studies. Peter A. Reich, *Language development*, © 1986, p. 134. Reprinted by permission of Prentice-Hall, Inc., Englewood Cliffs, New Jersey.

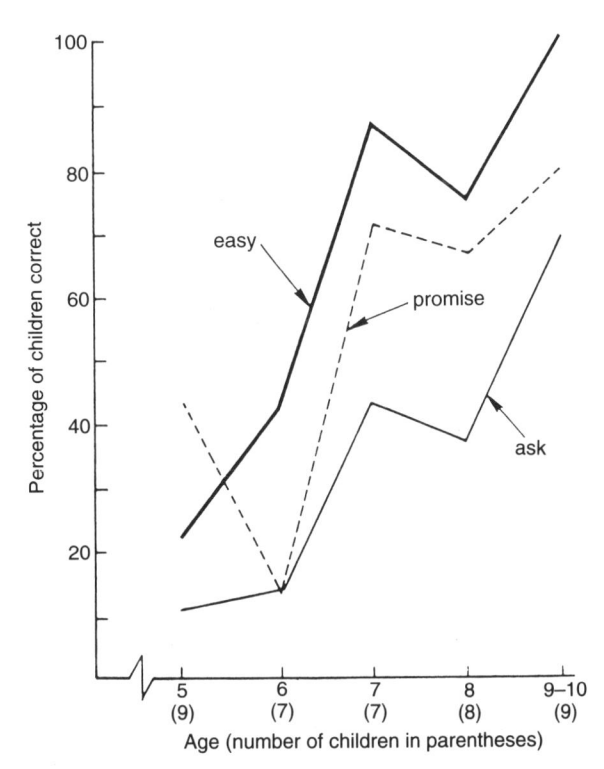

Fig. 24.6 Acquisition of complement constructions by 40 children (based on C. Chomsky 1969). Peter A. Reich, *Language development*, © 1986, p. 135. Reprinted by permission of Prentice-Hall, Inc., Englewood Cliffs, New Jersey.

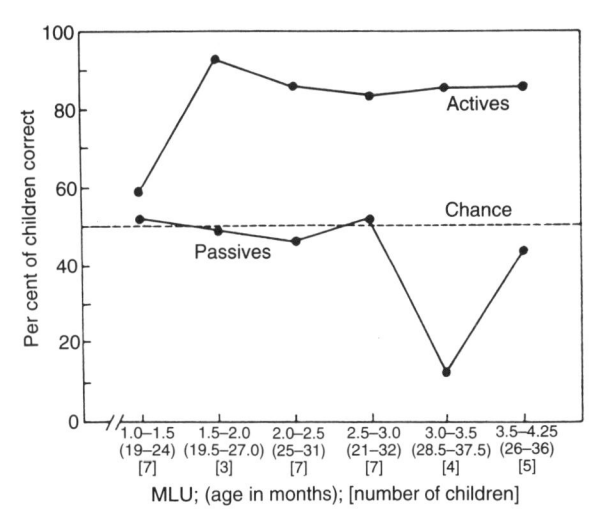

Fig. 24.7 Development of comprehension of reversible actives and passives: the early stages. Peter A. Reich, *Language development*, © 1986, p. 127. Reprinted by permission of Prentice-Hall, Inc., Englewood Cliffs, New Jersey.

[*pig*]). On the other hand, they have no difficulty in correctly understanding sentences like *The pig that chases the cat bites the dog*, in which *pig* is the subject of both main and relative clauses (de Villiers and de Villiers 1985).

5. Another area where older children still appear to have great difficulty is that of discourse-related phenomena. Karmiloff-Smith (1980), for example, found that while sentence-internal pronominalization had been acquired more or less successfully by the age of five, pronominalization in discourse contexts was used extralinguistically (in a deictic manner) rather than intralinguistically (in an anaphoric manner) up to the age of eight or nine. This can be seen in the following story told by a four-year-old in Karmiloff-Smith's study: 'There's a little boy in red. He's walking along and he sees a ballon man and he gives him a green one and he walks off home and it flies away into the sky so he cries.'

Clearly, the business of acquiring a native language is by no means over at the age of five. However, it is questionable to what extent the 'postsyntactic' learning which takes place later in linguistic development is linked to fundamental principles of grammar, as opposed to the interaction of grammatical and real-world knowledge, as in the example of older children's problems with the interpretation of passive sentences presented above. Here, it is significant that performance actually deteriorates; this suggests that early, grammatically-based competence is temporarily overridden by later-acquired pragmatic principles (for example, the real-world knowledge which leads to the equation of agents with animate subjects). Even such apparently purely linguistic problems as the discovery of complement-clause subjects discussed in (1) above are not necessarily grammar-based; these may arise from difficulties with pragmatic disambiguation (Davis 1987).

If this hypothesis is correct (at present, it must be regarded as tentative at best), then our initial division of syntactic acquisition into presyntactic, syntactic, and postsyntactic stages actually bears some theoretical weight: the presyntactic stage lays the cognitive and conceptual framework within which language is situated; the syntactic stage contains the specifically linguistic aspects of grammatical acquisition, in which innate principles are brought to bear upon the fundamental syntactic structures and processes of the target language; and the postsyntactic stage represents the integration of this newly emerged syntactic system with the real-world knowledge which the older child has accumulated.

24.8 Acquisition of speech acts

In discussing the acquisition of phonology, the lexicon, and morphosyntax as separate, coherent subsystems of language, we do not mean to lose sight of the child as a social being who learns language in the context of social interaction and largely for social-interactional purposes, and for whom language plays a major socializing role. In order to become a competent communicator, a child must—in addition to acquiring the symbols and rules of language we have already discussed—learn how to use these symbols and rules to encode a variety of communicative intentions appropriately. This 'communicative competence' includes 'competence as to when to speak, when not, and as to what to talk about with whom, when, where, in what manner' (Hymes 1972, p. 277). Considering the more linguistic aspects of this competence, we briefly broaden our scope to consider the development of pragmatics, which we define to mean principles underlying the use of language.[37]

Within the domain of pragmatics, we use the framework of speech-act theory to focus on the development of communicative functions as it relates to language development.[38] This topic has been most studied by investigators working within the interactive view of language development (See Section 24.4.1). Studies tend to fit into three main categories: (1) detailed and comprehensive descriptions of an individual child or a small number of children between roughly 9 and 18 months old; (2) taxonomic descriptions of the range of communicative functions expressed by pre-school or young school-age children; and (3) detailed analysis of the development of a single communicative function, most commonly 'requests', in terms of speaker- and listener-variables as they relate to variations in linguistic form. For the purposes of our discussion, we emphasize that communication development and language development do not mean the same thing: the abilities involved are different in each domain, and the development of one is not explained by development of the other (for discussion, see Shatz 1983, pp. 844–7). A child may have normal language ability but disordered communicative ability, or disordered language but normal communicative ability in terms of number and type of communicative functions conveyed (Johnston 1985).

Speech-act theory is the most complete and explicit formal means available for relating linguistic forms and communicative functions of utterances. Its basic premiss is that

> speaking a language is performing speech acts, acts such as making statements, giving commands, asking questions, making promises, and so on; and more abstractly, acts such as referring and predicating; and secondly, that these acts are in general made possible by and are performed in accordance with certain rules for the use of linguistic elements (Searle 1969, p. 16).

At present there is little agreement on how many types of speech acts there are, or precisely how these should be defined. Searle (1975) defines five basic speech-act types, or 'illocutionary acts'—representatives, directives, commissives, expressives, declarations—on the primary parameters of 'illocutionary point' (i.e. purpose), 'direction of fit', and 'expressed sincerity condition'.[39] To explicate these terms by illustration, we classify a 'request' as a 'directive', because the illocutionary point is to attempt to get the hearer to do something; the direction of fit is world-to-word, i.e. in using directives speakers try to get the world to match the words; and the expressed sincerity condition is 'want, wish', i.e. the speaker expresses a desire that the hearer perform an act. Each act-type is paired with a characteristic sentence-type that grammaticalizes the speaker's attitude, for example statement–declarative and request–interrogative; but there are a number of alternatives to this canonical pairing, as in the following request–declarative: *It would be nice if someone could take out the garbage.* Thus the child's task in learning these pairings is not trivial. Interrelations in the development of forms and functions have been most thoroughly studied for a few infants in the transition to linguistic communication.

24.8.1 *Primitive speech acts*

A key issue in this first group of speech act/communicative function studies of infants up to about 18 months old is attribution of intentionality to an infant whose apparent communicative behaviour is being described. We distinguish infants' early 'informative' behaviour from true 'communicative' behaviour, which definitionally requires that the communicator not only intend to send a particular message (i.e. have a particular goal in mind), but also be aware of the recipient of the message and formulate the message with both goal and recipient in mind. This contrasts with 'informative' signals, which must simply have meaning for the receiver, independent of the sender's intention or ability (Lyons 1977; Shatz 1983).[40] Developmental psychologists attribute communicative ability to an infant on the basis of the infant's attainment of a pre-requisite level of cognitive development, namely Piaget's sensorimotor stage 5, when the infant demonstrates tool-use, particularly the ability to use an adult in an instrumental way to obtain an object (Bates *et al.* 1979). Another statement of this criterion is 'person–object co-ordination', i.e. co-ordination of a goal-oriented action involving an object (such as reaching) with social signalling (such as eye contact) in a single action unit (Sugarman 1983). Other behaviours taken to indicate intentional communication are eye contact + vocalization sequences (Harding and Golinkoff 1979); persistence in attempts to achieve goals (Bates 1976; Bates et al. 1979; Greenfield 1980); alteration of the signal in response to the adult's change of behaviour; and shortening or ritualization of the signal (signifying conventionalization of the communication) (Bates et al. 1979). To date, the pre-requisite cognitive developmental level that supports these behaviours has been demonstrated to be necessary but not sufficient to explain the transition to linguistic communication. For example, institutionalized infants attain co-ordinated person–object orientation on the same timetable as non-institutionalized children, but the onset of language is significantly later for them (Sugarman 1983). In any case, attribution of intentionality is closely tied to a middle-class Euro-American view of 'communicating' infants. Western Samoans, to take a contradictory example, would judge such behaviours to be physiological reflexive actions devoid of intentionality (Ochs 1982). Strictly speaking, whether pre-linguistic communication is intentional is an unresolved empirical issue.

Studies of 'primitive speech acts' are based on the assumption that infants communicate intentionally by the age of 9 or 10 months. If we accept this assumption for heuristic purposes, questions that arise are (1) whether and how this period is continuous with later, linguistic communication, (2) which communicative functions are first to emerge, (3) how many communicative functions the infant expresses, and (4) the extent to which early functions are tied to emerging linguistic forms. Making a claim for continuity, Bates (1976) describes the development from unintentional to fully linguistic communication in terms of three periods: (1) the 'perlocutionary period', when infants' behaviour is unintentionally informative, as defined above; (2) the 'illocutionary period'—that period with which we are concerned here—of intentional but not linguistic communication; and (3) the 'locutionary period', when the communicator's intentions are expressed linguistically. Although Bates' terminology is a misappropriation of speech-act terms,[41] it does point out the essential intentionality of communication during the period of transition to language (while ignoring the lack of conventionality dependent on knowing the code). The extent to which communicative development actually is continuous over these periods is still at issue; the answer probably varies according to which specific aspect of communication is in question, with continuity more likely in the case of communicative functions than in the case of linguistic devices (Haslett 1987).

During the 'illocutionary' period, the first illocutionary acts to appear are reported to be the 'protoimperative', with which the infant 'requests' an object or an action of an adult, and the complementary 'protodeclarative', in which the child uses an object to engage an adult's attention.[42] These acts are first accomplished completely non-linguistically, with a combination of gesture, eye gaze, and non-linguistic vocalization, and then, within a few months, a combination of a word with a gesture, typically an indicative pointing or an open-handed reaching out (Bates 1976; Greenfield and Smith 1976).

By the time an infant communicates with protowords or single words, it is clear that additional communicative intents are being expressed, but the number and labels for these vary according to researcher. Dore (1973), for example, provides the following inventory of primitive speech-acts for two one-year-olds: labelling, repeating, answering, requesting action, requesting information, calling, greeting, protesting, and practising.[43] Outside the speech-act framework, Halliday (1975) attributes 'instrumental' and 'regulatory' intentional vocal communications (both requesting-type functions) in addition to 'interactional' and 'personal' communications (both of which are affective–social) to one child between 9 and 12 months old. When this child was 16 to 18 months old, his intentional vocal communications included the 'heuristic', which has to do with obtaining or reflecting information about the world, and the 'imaginative', which includes various types of play. A distillation of categories defined by various researchers—in most cases for only one or two children—yields three main categories of communicative intent expressed by protowords: (1) solicitation of action, social interaction, and joint attention; (2) expression of affect, such as pleasure, surprise, and rejection or dislike; and (3) participation in games and routines, such as peek-a-boo. By the time a variety of single words appear—predominantly names for things (see Section 24.6.2)—the child can label objects, and request labels. A significant development at this point is that the child's speech begins to be contingent on other speech; the child's labels can be responses to routine questions, and the child may verbally acknowledge what another speaker says (Chapman 1981). A little later (22 months for the child Halliday studied) children are able to use words to communicate new information to others. Because this typically involves talking about objects or events that are not part of the immediate context it is strong evidence of the child's use of words in a symbolic way.

There is some disagreement among researchers about how closely vocal forms and communicative functions are tied during this early period. Carter (1978), Halliday (1975), and Greenfield and Smith (1976) argue for an initial one-to-one mapping between form and function, where form includes intonation and/or gesture as well as vocal syllable; and, in fact, this 'packaging' of form with intent and context is definitional to what we have called protowords (see Section 24.5.2.2). But Dore (1973) argues that, at least pre-linguistically—through Bates's 'illocutionary period'—linguistic form and communicative intent are two separate strands of development which proceed on their own timetables; co-ordinating these strands during the one-word stage constitutes learning to perform primitive speech-acts. This claim predicts individual differences among children; for example, a child whose communicative development is outpacing linguistic-form development might appear to be 'expressive', whereas a child advancing linguistic form at the expense of communicative functions would be more 'referential' (see Section 24.6.2). By the time a child has an inventory of object names, there is no question that a given single word can be used for a variety of communicative functions. A hypothetical twenty-month-old might say *cookie* to name, or request, or comment on the unexpected absence of, or herald the appearance of, the named object.

24.8.2 *Communicative functions in early childhood*

Studies of young children's communicative functions are largely taxonomies of communicative intents, and virtually all these studies have been done with children learning to speak English. Although the taxonomies are much more elaborate than taxonomies of younger children's communications, the elaboration reflects the increasing complexity of what children talk about, the linguistic forms they use, and the relation of children's utterances to other speakers' utterances rather than new functions *per se*. For example, Dore (1979) adds the category 'acknowledgements' when coding pre-schoolers' communicative functions; these approve, disagree with, or neutrally recognize another speaker's utterance, or return the floor to the other speaker. Tough's (1977) categories, defined for the age-range 3–7 ½ years, underline cognitive rather than communicative developments; for example, her 'other-directing' category includes subcategories of 'demonstrating', 'instructing', 'forward planning', and 'anticipating collaborative action'. Starting in the later pre-school years, children begin to use speech for new purposes—to threaten, warn, and promise: but the origins and development of these speech-acts have not been systematically studied.

24.8.3 *The relation between communicative functions and linguistic forms in childhood*

A major development of communication in early childhood concerns the interrelation of communicative intent and linguistic form. Most of our information about this development comes from studies of children's directive utterances, particularly 'requests for action', and interrogative sentences. From at least as early as the age of two, children respond successfully to a wide variety of linguistic encodings of requests. For example, a two-year-old will perform the appropriate action whether an adult says 'Close the door', 'Will you please close the door?' or 'Would you mind closing the door?' (Shatz 1975; see also Bates 1976). These results are interpreted to mean that the child assumes the message will be relevant, recognizes at least one word in the message (here 'door'), and will perform a relevant action if one is possible. They also reinforce the observation that children hear a variety of function–form pairings from a very early age (Shatz 1982).

Children's own directive speech-acts appear to develop continuously (from a functional point of view) from the pre-linguistic period, when they are typically expressed by a grunt plus an open-handed reaching out (Carter 1978). When the child says single words, the grunt may be replaced by a general 'want' word or a label for the desired object or action; but the gesture remains. The first loosening of the connection between function and form occurs at this stage, as the child encodes expectations about outcomes, for example using rising intonation or a whining tone if the request is tentative or likely to be denied. Early linguistically encoded directives are 'direct' in that the desired object or action is explicitly mentioned, even after directives can be expressed by a variety of sentence-types (Ervin-Tripp 1977).

Two-year-olds are already aware of social variables such as power, familiarity, rights, and obligations, and encode them in their directive utterances. For example, one two-year-old reported by Gordon and Ervin-Tripp said 'Mommy, I want milk' to her mother, but 'You want milk, Daddy? Milk in there, Daddy? Daddy, I want some, please? Please Daddy, huh?' to her father for the same purpose. In general, children address more commands to mothers than fathers, tell siblings what to do (for example 'stop sucking your fingers'), and speak politely to less familiar adults (Gordon and Ervin-Tripp, 1984, p. 298).[44] Bates (1976) describes a shift at the age of two from efficiency to politeness in communication (Gordon and Ervin-Tripp do not agree). New options acquired in the syntactic period facilitate this shift, but early expressions of politeness also include lexical items such as *please* and *thank you*, rote-learned forms that would otherwise be beyond the child's linguistic ability, for example *would you*, and other ways of 'softening' the request, for example by asking for 'just a little' (Bates 1976; Ervin-Tripp 1977). Garvey (1975) argues that pre-schoolers already demonstrate conventional linguistic encoding of speech-act appropriateness conditions such as querying the listener's ability (*Can you . . .*) or willingness (*Will you . . .*) to perform an act, or stating the speaker's sincere desire (*I want . . .*) or need (*I need you to . . .*) to have an act performed.

Further developments in directives, as the available example of speech-acts in general, largely involve refining syntactic and semantic means of expression. The formal aspects of this development are just one aspect of the intersection of syntactic rules and pragmatic factors in the 'postsyntactic' stage (Section 24.7.2.3). The particulars of the syntactic refinement are language-specific. For example, learning to make polite requests in English involves learning to be indirect, thereby giving the listener the option of non-compliance and the speaker the possibility of saving face (for example 'Don't you think it's a little cold in here?' meaning 'Close the door'; see Ervin-Tripp 1976). In Italian, learning to be polite involves learning the syntax of conditional forms (Bates 1976). Becker (1982) discusses the development of directives from an ethnographic point of view. She argues that learning different ways to express directives is adaptive for children just as non-linguistic signalling is for other species; it allows them to signal a proper relationship to other members of the society. For example, an indirect request is not only 'polite', but also conveys a deferential attitude, acknowledging a submissive or inferior status (see also Gordon and Ervin-Tripp 1984).

The flip side of the development of the directive function is the development of functions served by interrogative sentences—the sentence-type canonically paired with requests. There is a massive amount of cross-linguistic evidence (for middle-class Euro-American culture) that children's interrogatives begin as highly constrained routines (for example *What's that?* and *Where's this go?*) as unanalysed holistic phrases used in very specific interactions (C. E. Johnson 1981). Concomitant with the development of productive interrogative syntax, the child will begin to use this sentence-type to perform more and more communicative and discourse functions, for example to solicit joint attention, to begin conversations, to request actions and objects, to request information, to solicit approval, to change the topic under discussion, and to communicate new information (*Did you know I got a dog?*).

Language forms and communicative functions become increasingly complex during childhood, both separately and in their interaction. In addition, single utterances come to convey more than one communicative or discourse function at a time. For example, *Did you know I got a dog?* might invite a friend to engage in conversation, establish the topic for discussion, announce new information, and elicit a verbal response (see Halliday 1975, 1978). Fine tuning across systems of language and communication continues over a life-span.

24.9 Language-acquisition and the evolution of language: some closing thoughts

Questions about the evolution of language motivated some of the earliest recorded observations of children's language development (see Section 24.3). Although these questions are outside mainstream linguistic and language-acquisition enquiry, they continue to tickle the imagination of a few language scholars. But extending ontogenetic arguments and observations concerning language development in individuals to language development in species is not a simple process. It brings together issues related to theories of evolution, theories of language, theories of development, and theories of language-acquisition—not to mention theories of how all these now disparate theories relate to each other. Taking a stance on these issues (inevitable, whether or not explicitly done) leads scholars to select a line of argumentation and a corpus of relevant ontogenetic and phylogenetic data.

Given the enormous complexity of the task, it is predictable that resulting speculations will be various and at odds, restricted to a narrow range of data, and, for the most part, over-simplistic. To take an example, language scholars may or may not refer to a specific theory of evolution. Among those who do, Lieberman (1984) takes a Darwinian approach, and argues that data concerning the evolution of speech argue strongly against the theory of neoteny; Bates (Bates *et al.* 1979), on the other hand, propounds Gould's theory of evolution, and goes to considerable trouble to argue for neoteny. Working from a strong interest in speech and speech perception, Lieberman focuses on the development of auditory and articulatory abilities specific to humans. Working in a cognitive–functionalist framework, Bates concentrates on a general symbolic ability that arises from the convergence of pre-existing cognitive and communicative skills which originally served other functions; speech is not of central interest to her, and is considered only inasmuch as it postdates gesture both ontogenetically and phylogenetically. In fact, it may not do so. If, as Lieberman argues, language depends on the development of neural mechanisms that allow automatization of rule-governed behaviour—which includes subroutines necessary for both speech and syntax—spoken language may actually predate *language* in the manual mode; this does not negate the temporal priority of gestured *communication*. Such an interpretation would be consistent with our observations of the discontinuity between deaf children's pre-linguistic gesture and language and blind children's successful language-learning despite the absence of communicative gesture (Section 24.6.4.2.1).

24.9.1 *Theories of language-acquisition and symbolic evolution*

Let us focus for a moment on how theories of language-acquisition might fare in relation to theories of symbolic evolution. Considering the ontogeny of language, particularly phonology and syntax, there is little evidence for the 'interactive' approach (Section 24.4.1). Most of the work done in this framework focuses on the presyntactic stage, and provides very little evidence for any particular theory of phonology or syntax. Moreover, while there is some evidence that child-directed speech differs from adult-directed speech, such evidence is not nearly sophisticated or detailed enough to account for the specifics of language-acquisition. At first sight this is disappointing, because the interactive view entails a relatively transparent relations between the acquisition of language by children and language-evolution; according to this view human social interaction constitutes both the ontogenetic and the phylogenetic basis of language. The infant's induction into the world of discourse appears to parallel the 'discovery' of language by prehistoric societies. Indeed, this relationship seems to be one of the main attractions of the interactionist viewpoint, and probably the reason why it has attracted so many investigators in spite of its theoretical problems and lack of empirical success. Nevertheless, it can be argued than even from the perspective of symbolic evolution, the interactionist viewpoint is not as attractive as it seems, since the putative ontogeny–phylogeny correspondence it invokes is based on an oversimplified interpretation of evolutionary theory. Put crudely, this view involves the claim that, since language evolved to meet communicative needs, language (or at least its ontogeny) should be described in communicative terms. However, this view is at odds with the central principles of Darwinian natural selection, which hold that the environment (communicative, in this case) determines the direction of genetic change only indirectly, via the phenotype. In that case, we might expect human linguistic systems to meet communicative needs, but we cannot hope to explain their structure on the basis of those needs, anymore than we can account for the colours on a peacock's tail by appealing to its function in sexual selection. Thus we conclude that, though it might tell us something about the development of communicative competence, the interactive perspective has little real to offer as an explanation of the development of purely linguistic competence.

While we cannot—and have not—ruled out an approach to language-acquisition based on general cognitive principles (Section 24.4.2), it seems fair to say that there is not yet a coherent proposal that accounts for specifically linguistic development and at the same time for communicative and other aspects of cognitive development (once again, with the possible exception of O'Grady 1987). It is certainly possible that organizing principles of this type might emerge as cognitive science matures; at present, the prospect appears remote, and, as far as productive work in both theoretical linguistics and language-acquisition is concerned, it seems more sensible at this point to work with the specifically linguistic principles that appear to provide the most explanatory theory of human linguistic competence. Turning to the relation between ontogeny and phylogeny, the cognitive approach—at least on a metatheoretical level—does appear more attractive than the autonomous approach (Section 24.4.3). Instead of having to explain the appearance of specifically linguistic principles, the cognitive theorist of human symbolic evolution can appeal to the evolution of a general symbolic function, which would also include principles of logical reasoning, mathematics, and so forth. However, in the absence of a cognitive theory linking these aspects of human symbolic competence, theories of evolution of a general symbolic function appear to be somewhat premature.

It should be clear by now that our theoretical bias is towards an autonomous approach to at least the 'core' aspects of the acquisition of language. Our focus on the formal dimension of language is due to our conviction that it is critical to an understanding both of language and of how it is acquired. This approach provides an elegant solution to the learnability paradox, accounts for the rapidity, robustness, and comparatively errorless nature of language-acquisition, and—supplemented with an appropriate theory of the learner—accords well with empirical data on first-language development. However, at first sight, the autonomous approach leads to rather implausible conclusions about human linguistic evolution, because it is a discontinuity view of both language development and evolution (the punctualist or cataclysmic theory). This is because language is domain-specific. There are few parallels with non-linguistic domains (except those concerning the nature of the learner), and there are few precursors to the formal systems.[45] But the autonomous linguist's position is that we know more about grammatical competence than about prehistory or evolutionary theory, and that therefore an implausible evolutionary scenario is preferable to an implausible theory about language-acquisition. This implies that there is a certain logical progression in the study of human language, beginning with a purely synchronic examination of language systems, and only then moving on to ontogenetic and phylogenetic considerations.

24.10 Conclusion

In this chapter we have reviewed facts and theories of language-acquisition. The research contributing to the current knowledge-base represents a very broad field of enquiry; but individual contributions are, of necessity, narrowly focused and relevant to specific theories of language and language-acquisition. It is tempting to compile a profile of an idealized language-learning child by abstracting relevant bits of information from studies of acquisition of the various systems of language. For example, it would be satisfying to say that our hypothetical child, at the age of 18 months, understands 200 words, says 50 words, has a rudimentary phonological system based on phoneme-sized units, and is combining words in vertical constructions which are used to demand actions and verbal interactions. But this is a dangerous undertaking. As we have shown in this chapter, conclusions about the acquisition of, or the timetable for acquiring, a specific aspect of language depend on the theoretical framework of the research; and there is no generally accepted or generally relevant (in the sense of being equally applicable or equally tested across language domains) theory of language-acquisition. The 'facts' of language-acquisition in our current data base, in consequence, are not equivalent in size, scope, or quality—and they are sometimes even contradictory; so it is difficult and misleading simply to add them together to get a bigger picture.

A further consequence of the narrow focus of language-acquisition studies is a lack of knowledge about how different aspects of language relate to each other. Theories of language take a stance on this issue, to some extent; but development adds a new dimension. In this chapter we have reviewed the separate language systems of phonology, the lexicon, syntax, and pragmatics. These systems are autonomous in the sense that each is coherent and described by a set of rules and terms relevant to that system alone. Autonomy in development means that the developmental timetable for each system may proceed, to some extent, on its own schedule. This leads to individual differences among children; so, for example, one child may have a 50-word vocabulary when she or he starts to combine words, while another may have a 300-word vocabulary before reaching the same syntactic milestone. An extreme mismatch of level of development across language systems is characteristic of some childhood language disorders.

But autonomy does not entail complete separateness, and there are orderly interactions between the systems that constitute language. For example, some syntactic rules apply only to verbs belonging to a specific semantic category. Interaction between language systems in development means that the form or nature of a new rule in one language system can be affected by the developmental level of a different language system. For example, a surge in vocabulary development may provide the impetus for a child to change from holistic word-based phonological rules to phonological segment-based phonological rules (Macken and Ferguson 1987). Because the systems are autonomous, they do not always interact at synchronized developmental points. Because the systems are developing, the interactions are dynamic, with the results of interaction dependent on the current state of each system involved.[46] At present there is virtually no information about the interaction of language systems during development. This has important consequences for theories of language-evolution, particularly homology theories based on the differentiation and co-ordination of existing elements in both the child and the species. These theories depend on heavily constrained 'next steps' (see for example Bates et al. 1979; Lock 1985). If dynamic interaction between systems means that the 'next steps' in ontogeny are constrained in a more complex way, and perhaps more various, the analogy between ontogeny and phylogeny becomes more difficult to draw. A clear direction for future research, then, is investigation of the relationship between separate systems of language in flux.

Editorial appendix: Gestures and words: the early stages of communicative development

The early stages of infant communication have come to be more widely researched in the last fifteen years or so. Before that, they were effectively ignored, largely because in the prevailing linguistic paradigms language was seen as a unique system, especially in its grammatical organization, and distinct from other forms of communication. Pre-verbal systems were not seen as being of relevance to language as so defined. It is not clear from any of the subsequent research that there is anything particularly linguistic about these systems (see, for example Atkinson 1985, and the earlier sections of this chapter).

However, aside from the issue of continuity, there are two particular developments that occur in these early systems that are of prime importance: the development of reference and the development of predication. Given that there is currently almost no evidence for either of these in non-human communication systems, they, along with grammar and duality of patterning, represent defining characteristics of human language systems, and present a real challenge to evolutionary theories. Modern ontogeny provides one of only two places (the other being the ape-language projects, see Ristau, this volume, Chapter 23) in which their establishment can be observed; and, while this does not equate with the evolutionary situation, how child development currently occurs provides some clues as to the possible pathways for beginning a symbolic communication system.

In what follows the early stages of the ontogenetic elaboration of human communication are outlined as being part of a continuous process. This continuity should not be taken as implying one underlying set of processes or mechanisms. Nor should the use of the notion of 'stages' be misunderstood. There is, for example, a period in which children predominantly use single words. It is *convenient* to refer to this as 'the one-word stage'. It is not, however, a stage with a definite beginning or end, but merges with other productions before and after it; nor is it a homogenous stage, for there are changes going on within it; and it is probably not founded on a single and unchanging underlying set of 'mechanisms'. It is a stage for expositional convenience only, as are all the stages given here. Further details can be found in collections of papers in Golinkoff (1983), Barrett (1985), von Tetzchner *et al.* (1989), and Volterra and Erting (1990).

Stage 1: Pre-communication (0–9 months)

The pre-communication stage is that period in which human infants convey information to others by default. In the terminology of Vygotskian psychology, their abilities are inter-mental: that is, constituted in the social relationships between them and their cultural milieu. This is 'communication-in-the-eye-of-the-beholder' only. Infants in this period give no evidence of intending to communicate, but only communicate by virtue of the fact of their being within a socio-cultural human group, in which it is impossible *not* to communicate. Thus an infant might yell out of frustration, yet be treated as if he or she had communicated 'Help me'.

In this stage the infant is inducted into a set of meaningful social episodes with adults that provide a structure within which the infant can discover the meanings its activities are being accorded (see, for example, Lock 1980). Inchoate motives are structured with respect to particular practices, so that they can be pursued in a goal-oriented manner. Thus, 'discomfort', for example, is structured into particular 'discomfort removing strategies', yielding a culturally-patterned set of refined motivations or goals for action.

These developments may be described as providing the child with the resources to uncover and control the implications of its 'being-as-a-child'. Thus early discomfort could be said to have an expressed value for the infant 'whatever this bodily state is, I do not want it'. As the child is relieved of discomforts, it is given information that allows it to uncover and structure the implication of this, 'I want something else', as 'I want that'. The altricial nature of human 'being-as-a-child' is such that 'I want that' has the further implication 'you do something to get me that'. But, in this first stage, infants only accomplish a structuring of their motives, and do not come to control this additional implication—thus their actions remain pre-communicative.

Stage 2: Pre-symbolic intentional communication (9–15 months)

At around 9 months of age, infants begin to give evidence of controlling the implication in their actions of 'you do something to get me that': that is, they *address* messages to other actors (see Fig 24.8). They begin to use a number of communicative gestures, these often being actions that are 'lifted' from their original direct manipulation of the world (see Table 24.3). In addition to this direct origin, it is likely that some of an infant's repertoire is developed by imitation of parental models. There are marked individual differences in the adoption of these gestures, both in relation to their order and frequency of appearance, and in whether and when they are accompanied by vocalizations (see, for example, Zinober and Martlew 1985). The ontogenetic origins of pointing are less clear than for the other gestural categories (Lock *et al.* 1990); but it appears to be a gesture with a universally similar form, essentially the same as that used by adults right from the outset.

In addition to these non-vocal gestures, the infant develops vocal counterparts to them (see examples in Table 24.4). Various categorization systems have been

Fig. 24.8 Pre-symbolic intentional communication in the gestural mode by a 12-month-old infant (Stage 2). Failing to reach a desired object (1), the infant vocally attracts (2) and obtains (3) the care-taker's attention, and then directs it (4). Infants in Stage 1 do not appeal to adults for assistance when they fail in some action: if they cry (and thereby attain assistance), that cry appears more a result of frustration than an intended communication (from Lock 1980).

Table 24.3 A possible classification scheme for pre-symbolic gestural communication (from Lock *et al.* 1988, based on data from Zinober and Martlew 1985)

	Function	*Origin*	*Variation*	*Vocalization*
Expressive E.g. arm-waving, smiling, clapping, banging feet	To convey emotional states, ±	Stylization of early rhythmic movements	Depends on infant's 'temperament'?	Accompanies from 10 months
Instrumental E.g. pick-up, headshake/nod, open palm	To control the behaviour of another person	From the act of doing the thing itself	Cultural variation in form, perhaps, but universal?	Accompanies from 10 months—conveys same meaning
Enactive E.g. symbolic play—pantomime	To represent actions and attributes of people and objects	From the technical act itself	Variable at level of dyad	13–21 months
Deictic E.g. pointing, reaching	To isolate an object from its general context	?	Probably universal	Accompanies from 10 months

Table 24.4 Early pre-symbolic vocal counterparts to manual gestures for one child (from Dore 1985)

Forms	Applications
	Part 1: Object-related signs
	Animal-related
bee	—flying, crawling, or stationary insects; birds; aeroplanes; specks of dirt; breadcrumbs; bleeps on a TV screen; when a table is slapped or hands are clapped once
/kiki/	—family cat ('Kiki'); any small (pictures of) animals or toys
	Food-related
/ku/	—cookies, crackers, or candies; bags of other sweets or dried fruits;
/kəkə/	jars; when pantry doors are opened or closed
/baba/	—bottles and jars of liquid; glasses, cups, containers
	People-related
mama	—fixed associations of car colours to person 'labels'; excited actions
dada	like 'requesting' something
nana	
baby	—children; small dolls
/ejə/	—to call out to his sister, Angela
	Plaything-related
/bɔ/	—small, round balls; anything rolling, to initiate games
/ki/	—keys; key chains; screwdrivers; twirling of fingers; car ignition; locks; tinkling noises
	Vehicle-related
/bo/	—boats; bulky objects on water; toy trains; blocks of wood; lettered cubes
	Part 2: Personal/social interactionally-related signs
	Onomatopoetics
brrrr	—all engine noises
grrrr	—mimicking 'monsters'
tsk-tsk	—when inappropriate action occurs (falling, spilling, etc.)
shhhh	—when seeing or hearing flowing water
boom	—after hearing loud noises, falling down, or banging something
	Expressionals
/ow::/	—when excitedly pointing to some object, or otherwise orienting
/u::/	himself to an interesting activity
/au/	—when he is hurt

proposed for these vocalizations (see Nelson and Lucariello 1985, 61–3, for a brief review). Common to these systems is the view that these vocalizations are not truly symbolic, are closely tied to specific contexts, and are not phonetically structured. The complementarity between these vocalizations and the gestures that are used alongside them is apparent in the empirical findings of Bates *et al.* (1979), who conclude these systems are equivalent:

At this point in development, the only difference between the two domains is in the modality of expression.... we see no evidence to suggest that a 13-month-old is in any way biased toward the development of vocal language as opposed to gestural language (ibid., p. 177).

Grieve and Hoogenraad (1979) have characterized these early forms as a means of sharing *experiences* rather than *meanings*: that is, they are not yet truly referential. The shift to the referential, symbolic domain is accomplished in the next stage.

Stage 2 is also open to the sort of implicational analysis introduced earlier. It would seem that what is occurring is not a shift *beyond* the level of implication the child has already come to control, but an elaboration *at* the level already attained. That is, motives and situations are being more clearly discriminated by the infant; and this is reflected in the specificity of the means used to communicate them. A general ability to convey, say, 'you do something to get me that' is being differentiated into 'you pick me up', or 'you give me X', where X can be specified by a specific gesture or vocalization or by pointing, while still having linked with it the entire message; that is, communicative 'items' have holistic meanings at this point.

Stage 3: Symbolic referential communication (15–24 months)

A: Reference and predication

The transition to symbolic, referential communication is poorly understood. It involves the establishment of *names*. Vocalizations become less tied to contexts, and, apparently, more to objects. A 'naming explosion' has often been reported, and is taken as evidence that a child has made the insight that, as a general principle, things have names. A number of interactive situations have been described that doubtless feed into the establishment of 'names' as 'names' (see Goodwin 1985). Many of these will be specific to particular cultures; none of them can be posited as necessary pre-requisites in their own right. The common factor amongst them is that they all establish ritual formats which serve to increase the saliency of the small number of elements varying across them. In 'naming games' the format is generally for the adult to highlight an object, and the child to provide a response: different objects require different responses, but within ritualized formats. Objects similar in adult eyes can be substituted for each other while requiring the same response, so that sounds are not linked to, say, just one picture of a dog, but many different forms of dog; and can be substituted in other game formats, a straw, say, being used to function as a comb. Gestures or vocalizations used as demands also support aspects of referential symbolization, in that what is being demanded can be temporally displaced from the present.

Productive vocabulary grows slowly during the early part of this stage (see Table 24.5). Fifty words is often taken as the point at which the 'naming explosion' may be expected (Nelson 1973; Kagan 1981), meaning that the infant can have a fairly large repertoire of functional vocalizations prior to the establishment of a 'linguistic' referential system. Linguists such as Bloom (1973) and Dore (1985) credit the first signs of linguistic organization to late in the one-word period, when the child is able to produce words that relate syntagmatically to another element in the context; that is, in using a word communicatively the child 'presupposes shared knowledge of an aspect of context (which could have been made explicit), but ... focuses on another' (Dore ibid., 36).

There are marked individual differences in the rate of vocabulary development. Nice (1926/7) tabulated 47 published vocabularies of 24-month occidental infants, the average being 328 words, with a range from 5 to 1212 words. In a more recent study of 30 similar children Bates *et al.* (1988) report a mean vocabulary at 20 months of 142 words, with a range from 1 to 404 words. There is little evidence of words' being phonologically organized along the lines of the adult system during this period. Rather, they are organized holistically in contrast to each other, and not yet analysed into their component phonemes (Ferguson and Farwell 1975).

Along with the attainment of reference at this age come some profound changes in what the child apparently refers to. Nelson and Lucariello (1986, p. 80) summarize these:

during the first half of the second year the event representation remains unanalyzed in terms of specific concepts of objects, actions and actors. It is only during the second half of the second year, in general, that discrete concepts are differentiated from the whole.

Table 24.5 Early vocabulary growth (based on Smith 1926; see also Fig. 24.1 to this chapter)

Age (months)	Number of children	Number of words	Gain
8	13	0	
10	17	1	1
12	52	3	2
15	19	19	16
18	14	22	3
21	14	118	96
24	25	272	154

Thus, once children have abstracted the general principle of word-meaning from the various contexts in which it is embeddedly presented to them, they re-embed words into diverse contexts that allow them to abstractively create new meanings for them, and progress to predication.

In terms of the implicational framework already introduced, early words are being proposed as having holistic meanings, such that when a child says 'down', for example, as it puts an object down, it means all of 'I am putting that object down', and is not coding just one item of this meaning. The separate components coded in the adult language need to be constructed: actions, in English, need to be constructed as separate from agents and patients: where 'kick' might originally mean 'I am kicking the ball', it has to come to just represent the act of kicking; similarly for 'ball', uttered in the same circumstances. Once these differentiations are made, then the child is moving beyond the realm of reference to that of predication; not just identifying something as a ball, but identifying a ball and saying something about it (see Lock 1980: 148–76, for a detailed account of these developmental changes).

The point to be drawn here is that this stage of language development is one in which some quite fundamental aspects of the language system are being constructed. To call it a 'one-word stage', as is often done in the literature, obscures this. The child enters this stage with a communicative repertoire that is slowly re-worked into a linguistic system based upon reference and predication, and in this re-working some major conceptual achievements on which language can be further elaborated are developed. At the same time, however, it should be recognized that this symbolic referential and predicative system, while pre-grammatical, is effective in specifying a range of intentions and meanings. Embedded in the adults' socio-cultural system (an important constraint for a possible evolutionary account), the infant is able to convey implied meanings of a far greater range and power than it is able to render or control explicitly.

B: First combinations

There are two further reasons why, for present purposes, the notion of a 'one-word stage' is inappropriate. First, many early 'words' are in fact produced in combination with gestures: for example, naming is often accompanied by pointing; and requests are often produced with their gestural equivalents—'up' with outstretched arms, for example. Zinober and Martlew (1986) note that, for what they term 'instrumental gestures' at least (see Table 24.3), there is a developmental progression in the form of word-gesture combination. Prior to 21 months their subjects used gestures in a supportive role to words, in that word and gesture appeared to have the same meaning. From 21 months, however, they were used in complementary roles. For example, a child says 'Book' while placing mother's hand on a pile of books:

the utterance "book" gives the name of the object being negotiated. The gesture indicates who shall perform the action (the mother) and what action is required (picking up the book). Together [word] and gesture communicate "you pick the book up, mummy" (ibid., p. 200).

Gestures thus continue to accompany words across the emergence of predication, and, indeed, Michael and Willis (1968) have catalogued 12 gestures that commonly signal messages such as 'go away' and 'come here' among four- to seven-year-olds.

Second, some word-combinations do occur quite early in many children's productions. On the one hand there appear 'rote' productions. These are not in any sense based on the application of a rule system, but are more 'formula-like', and do not serve as productive templates for new combinations. On the other hand, as the conceptual distinctions noted above come to be made, they allow the construction of utterances in which these distinctions are represented: 'Daddy pipe' (possessor), 'more book' (recurrence), 'all gone stick' (non-existence) appear to be pre-propositional combinations, as opposed to the later appearing post-propositional ones such as 'Daddy sit' (agent), 'baby cry' (experiencer) (examples from Leonard 1976). Thus, before some uses of words emerge as single-word utterances, other uses are being combined together, obscuring further the separation of a one- from a two-word stage.

None of the above combinational categories appear to be based on any knowledge of syntax, and if there is a 'grammar module' involved in the adult human linguistic system, there is no evidence that it plays a role in establishing early grammatical abilities (see the introduction to Part III, this volume). It seems more likely that these early combinations are ordered on the basis of perceptual, semantic, and pragmatic considerations (see Lempert 1984, for a concise review); but exactly what is going on at this time remains unclear.

Notes

1. For more general discussion of communicative development, particularly in the first two years of life, the reader is referred to Anisfeld (1984), Bates (1976), Bates *et al.* (1979), Golinkoff (1983), Greenfield and Smith (1976), Halliday (1975), Lock (1980), and Shatz

(1983). Also see Butterworth (Chapter 16) and Sinha (Chapter 17) in this volume. Although our discussion is limited to language in the vocal–auditory mode, it concerns language-acquisition generally, independent of mode. Thus, with substitution of reference to a visual–manual mode in the 'phonology' section—specifically features of handshape, movement, location, and orientation—the main points of our chapter are equally applicable to the acquisition of the signed languages of the deaf.

2. References and short excerpts from a number of these diaries can be found in Bar-Adon and Leopold (1971).

3. Normative speech-sound information is presented in Fig. 24.4 in Section 24.5.2.3. Other quantitative information collected during this era concerns, for example, amount and rate of speech, and relation of vocabulary size to IQ and sex.

4. See Cromer (1968, 1974), Antinucci and Miller (1976), and the extensive literature on the acquisition of the temporal conjunctions *before* and *after* that arose from the work of Clark (1971).

5. For example, the 'fixed subject constraint' prevents the extraction of a subject noun phrase from a subordinate clause whenever that clause is headed by the complementizer *that*. It is this constraint which accounts for the contrast between the grammatical sentence *Who did you say that Bill left?* and the ungrammatical *Who did you say that left?* (Note that the grammaticality judgement here does not depend on the possible who/whom alternation, but rather on the movement of *who* out of the clause introduced by *that*.)

6. It is beyond the scope of the present chapter to explain or motivate each of the constructs below. For readers unfamiliar with phonology, good introductory sources are Hyman (1975) on phonology, Kenstowicz and Kisseberth (1975) on generative phonology, Ladefoged (1982) on phonetics, van der Hulst and Smith (1982) on autosegmental and metrical phonology, Clements and Keyser (1983) on syllable structure, and Ohala (1986) on the psychological reality of phonology.

7. For a review of the data on infant production during this period, the reader should consult Stark (1986).

8. For a review of the data on perception during these stages, see Barton (1980) and the other articles in Yeni-Komshian *et al.* (1980), Vol. 2.

9. To illustrate this point, the phonetic feature 'voicing' distinguishes the phones [d] and [t]. The primary acoustic cue, voice onset time (VOT) co-varies with characteristics of the stop-release burst, intensity of aspiration, pitch perturbations in the following vowel, and several other cues. None of these cues is uniquely or necessarily present in all or present in only tokens of [d]:[t] contrasts. A variety of cues, on different instances of use or in different phonetic contexts, may all lead to the perception of [d].

10. Although the series of stop-consonant and vowel syllables was generated by using different values of voice onset time (VOT, the time between the burst of the stop and the onset of voicing) stepped over the single VOT continuum in small but equal steps, the infants' perception of the stop-consonants did not go in continuous steps, but rather resulted in only two (as opposed to thirteen) categories. Further, the infants discriminated the stimuli into the two categories that correspond with adult [b] and [p].

11. See, for example, Dore *et al.* (1976), Ferguson (1978), Halliday (1975), Macken and Ferguson (1987), Menn (1976), and Stoel-Gammon and Cooper (1984).

12. For a summary of the general characteristics and common phonological processes of this period, see Macken and Ferguson (1983), Menn (1983), and the phonology chapters in Fletcher and Garman (1986).

13. See Ferguson and Garnica (1975) for a more detailed review.

14. This view of phonology is a cornerstone of the European structuralist school known as the Prague School of Linguistics; despite the structuralist name, the Prague School of Jakobson and Trubetskoy has little philosophy or theory in common with the American structuralist school of Bloomfield.

15. For more thorough presentation of cognitive theories of phonological acquisition, see Kiparsky and Menn (1977) and Macken and Ferguson (1983).

16. No part of this seemingly straightforward description is uncontroversial. Debates rage about the nature of mental representation and the architecture of the relationship between word and referent. See Block (1981), Lyons (1977), and Sinha, this volume, Chapter 17 for several points of view on each controversy.

17. 'Open class' words are essentially 'contentives'—nouns, verbs, and adjectives. This class of words is large and freely admits new members. The 'closed class' is a small class of words that does not easily admit new members. Its members, also referred to as 'functors', do not make reference, but rather modulate and mark grammatical meaning. Closed-class words include articles, conjunctions, prepositions, auxiliary verbs, and complementizers.

18. For more thorough discussion of the early period of word-acquisition, the reader is referred to Bowerman (1976, 1979, 1982), Dromi (1988), and Griffiths (1986).

19. See, for example, Amidon and Carey (1972), Coker (1978), Ehri and Galanis (1980), French and Brown (1977), H. L. Johnson (1975), Kavanaugh (1979), and Trosborg (1982) on *before* and *after*.

20. For example, one criticism comes from Putnam (1975), who argues that the intensions of natural-kind terms such as *tiger* and *electron* do not determine the terms' extensions, because the meaning components of a lexical item (which comprise its intension and are a set of descriptions) are characteristics (or 'essence') of natural kinds (a matter for science). See Carey (1982), Katz (1977), Kripke (1972), Putnam (1975), and Smith and Medin (1981) for discussion and alternatives.

21. For example, Andersen (1975) demonstrates that children's notion of 'cup' changes over developmental time, partly as a result of experience with fast-food restaurants. Palermo (1982) shows that the centre of the concept 'furniture' changes, depending on whether we are talking about the living-room or the bedroom.

22. See Lakoff (1987) for further critical discussion of prototype theory.

23. Some psychologists bypass this problem by appealing to a strict cognitive pre-requisites model of language-acquisition. This framework allows them to infer the symbolic status of words from prior or correlated symbolic behaviours in other cognitive domains.

24. Here is the basis for Bates's phylogenetic claim, which she relates to Gould's (1977) version of evolutionary theory.

25. See, for example, Bates (1976), Bates and MacWhinney (1982), Burner (1975), Greenfield and Smith (1976), Lock (1980), Werner and Kaplan (1963).

26. See Bowerman (1985) for further discussion, particularly as relevant to categories encoded in grammars. She argues that unbiased wedges are not always possible.

27. For further discussion of theories and problems of naming, see Barwise and Perry (1983), Frege (1960), Macnamara (1982), and Miller (1978). 'Reference', or the act of referring, belongs in the pragmatic domain.

28. The authors' main claim is that the child's *separate* gesture and language systems begin to be co-ordinated by age two years.

29. For example, in Spanish the article and noun agree in both grammatical gender and number, as in the word 'cat', which is classified as masculine due to word- final -o, not the cat's actual sex: *el gato* but *los gatos* (plural).

30. The reader is referred to Carey (1982), Gleitman and Wanner (1982), Landau and Gleitman (1985), Macnamara (1982), and Maratsos (1982) for elaboration.

31. Some of the later developments include hierarchical structure within particular semantic domains (for example Carey 1978; Macnamara 1982); restructuring of the established lexicon due to the discovery of common meaning-components in words first learned more holistically, or of 'hidden' meanings (Bowerman 1978, 1982, 1983); learning words that express logical relationships (French and Nelson 1985); figurative use of language (Gardner *et al.* 1978); lexical innovation (Clark 1982); and metalinguistic awareness (Hakes 1980).

32. See, among others, Bloom (1970, 1973), Bowerman (1973), Brown (1973), Braine (1976), Scollon (1976), Ingram (1978*b*), Peters (1983), and Hill (1983).

33. See Hyams (1986*b*) for very similar conclusions.

34. See Wexler and Culicover (1980), Mazurkewich and White (1984), Clark (1987), and Randall (1987), among others.

35. See MacWhinney (1978), Bybee and Slobin (1982), and Rumelhart and McClelland (1987), among many others.

36. For other studies of passive acquisition, see Horgan (1978) and Maratsos *et al.* (1979).

37. The term 'pragmatics' is notoriously difficult to define. A key source of this problem is the issue of whether pragmatics is taken to mean language in use or principles (i.e. rules) of use. Failure to come to agreement on this issue is one cause of the rejection of pragmatics as a legitimate domain of linguistic enquiry by some linguists. In addition, pragmatics is frequently called preparadigmatic, pretheoretical, or plagued with 'terminological profusion and underlying vagueness' (Levinson 1983, p. x). To some extent, this confusion results from the diversity of approaches to the study of pragmatic phenomena and from researchers' use of underspecified or partial theories. See Haslett (1987), Levinson (1983), and Lyons (1977) for definitions and discussion.

38. Other topics generally included in the domain of pragmatics are presupposition, conversational implicature, inference, foregrounding, and various aspects of discourse structure. Most of these topics are of equal interest in the fields of discourse analysis (Brown and Yule 1983), linguistic and philosophical pragmatics (Levinson 1983; Sadock 1978; Searle 1969, 1975), functional grammar (Givón 1979*a*, *b*; Halliday 1978), and sociolinguistics (Ervin-Tripp 1976; Lock and Symes, this volume, Chapter 8), and the lines that divide these disciplines are fuzzy indeed. The usual distinction is one of focus. For example, utterances classified as 'requests' interest a number of scholars:

the language philosopher wants to know about appropriateness conditions (whether the speaker truly desires the act to be done, whether the listener is willing and able to do the act, etc.); the sociolinguist wants to know what social variables influence the form of the utterance (interlocutor status relationships, territoriality, unusualness of the act requested, etc.); the discourse analyst wants to know how the speaker uses conversational implicature to formulate the request, how the listener makes inferences from the utterance, and how request and response are sequenced; the functional grammarian wants to know how the request function is grammaticalized. The transformational grammarian declines to be interested, considering these matters to be in the domain of performance theories or theories of social interaction.

39. Searle provides a series of additional dimensions relevant to classification, including degree of strength of illocutionary point, status relations of speaker and hearer, etc.

40. This requirement is built into speech-act theory, in that illocutionary acts must, by definition, have an illocutionary point, express a sincerity condition, and be conventionally formulated in accordance with a set of listener variables. Our distinction does not deny the importance of a growing literature on the social interactivity of young infants and their care-givers, and the communicative effect of the infants' actions and vocalizations (for example Bateson 1975; Bullowa 1979; Stern et al. 1975; Trevarthen 1977). These studies are based on the assumption that there is continuity between pre-linguistic and linguistic communication.

41. Bates's terminology is a misappropriation, in that a speech-act accomplishes these three acts simultaneously; each utterance is (1) a locutionary act in so far as words are spoken; (2) an illocutionary act, in that it conventionally conveys a particular speaker-intention; and (3) a perlocutionary act, in that it has an effect on the listener, which may or may not be intentional (Austin; 1962; Searle 1969). In speech-act research it is particularly difficult to define 'perlocutionary act', and the hearer's 'uptake' of the illocutionary act is hard to distinguish from perlocutionary effect. In fact, a number of investigators have used perlocutionary effect as evidence of illocutionary uptake; although this inappropriately confounds constructs of speech-act theory, it is otherwise difficult to measure illocutionary uptake.

42. There is no consistent evidence that one of these primitive speech-acts is developmentally prior to the other; and it is possible that limitation to these two speech-acts reflects researcher grounding in Piagetian stages more than it accounts for what children intentionally communicate during this developmental period. Greetings and systematic expressions of pleasure or dislike do not fit into this description, for example (but see Greenfield and Smith 1976).

43. Note that these labels are not all defined on the speech-act parameters outlined at the beginning of 24.8: some categories are defined on the basis of discourse role, as opposed to speaker's intent (for example, repeating, answering).

44. For a more general discussion of how social variables and linguistic forms interact, see Ervin-Tripp (1976).

45. But the issue is far from settled. But theories that posit continuity in ontogeny and phylogeny focus on what we believe are lower-level issues, for example phonetic issues in the phonological theories of Lindblom (1986) and Stevens (1972). These theories do not contain mechanisms that might 'scale up' to the level of rule-learning needed for more abstract phonological and morphophonemic structures. Similar is the connectionist modelling of language done by Rumelhart and McClelland (1987) and others, which shows little promise of explaining either syntax or phonology.

46. For a similar argument with reference to evolution, see Gollin (1985).

References

Amidon, A. and Carey, P. (1972). Why five-year-olds can not understand *before* and *after*. *Journal of Verbal Learning and Verbal Behavior*, **11**, 417–23.

Andersen, E. (1975). Cups and glasses: learning that boundaries are vague. *Journal of Child Language*, **2**, 79–103.

Anglin, J. (1977). *Word, object and conceptual development*. Norton, New York.

Anisfeld, M. (1984). *Language development from birth to three*. Erlbaum, Hillsdale, NJ.

Antinucci, F. and Miller, R. (1976). How children talk about what happened. *Journal of Child Language*, **3**, 167–89.

Armen, J. C. (1971). *L'Enfant sauvage de Grand Désert*. Delachaux et Niestlé, Neuchâtel, Switzerland. Translation: *Gazelle-boy: a child brought up by gazelles in the Sahara Desert* (trans. S. Hardman). The Bodley Head, London, 1974.

Armstrong, S., Gleitman, L., and Gleitman, H. (1983). What some concepts might not be. *Cognition*, **13**, 263–308.

Aslin, R. and Pisoni, D. (1980). Some developmental processes in speech perception. In *Child phonology*, Vol. 2: *Perception* (ed. G. Yeni-Komshian, J. Kavanagh, and C. Ferguson), pp. 67–96. Academic Press, New York.

Atkinson, M. (1974). Prerequisites for reference. In *Developmental pragmatics* (ed. E. Ochs and B. Schieffelin), pp. 229–49. Academic Press, New York.

Atkinson, M. (1982). *Explanations in the study of child language development*. Cambridge University Press.

Atkinson, M. (1985). How linguistic is the one-word stage? In *Children's single-word speech* (ed. M. Barrett), pp. 289–312. Wiley, Chichester.

Austin, J. (1962). *How to do things with words*. Harvard University Press, Cambridge, Mass.

Bar-Adon, A. and Leopold, W. (eds) (1971). *Child language: a book of readings*. Prentice Hall, Englewood Cliffs, NJ.

Barrett, M. (1978). Lexical development and overextension in child language. *Journal of Child Language*, 5, 205–19.

Barrett, M. (1982). Distinguishing between prototypes: the early acquisition of the meaning of object names. In *Language Development*, Vol. 1: *Syntax and semantics* (ed. S. Kuczaj II), pp. 313–34. Erlbaum, Hillsdale, NJ.

Barrett, M. (ed.) (1985). *Children's single-word speech*. Wiley, Chichester.

Barrett, M. (1986). Early semantic representations and early word-usage. In *The development of word meaning* (ed. M. Barrett and S. Kuczaj II), pp. 39–67. Springer-Verlag, New York.

Barton, D. (1980). Phonemic perception. In *Child Phonology*, Vol. 2: *Perception* (ed. G. Yeni-Komishian, J. Kavanagh, and C. Ferguson), pp. 97–116. Academic Press, New York.

Barwise, J. and Perry, J. (1983). *Situations and attitudes*. MIT Press, Cambridge, Mass.

Bates, E. (1976). *Language and context: the acquisition of pragmatics*. Academic Press, New York.

Bates, E. and MacWhinney, B. (1982). Functionalist approaches to grammar. In *Language acquisition: the state of the art* (ed. E. Wanner and L. Gleitman), pp. 173–229. Cambridge University Press, New York.

Bates, E., Benigni, L., Bretherton, I., Camaioni, L., and Volterra, V. (1979). *The emergence of symbols: cognition and communication in infancy*. Academic Press, New York.

Bates, E., Bretherton, I., and Snyder, L. (1988). *From first words to grammar: individual differences and dissociable mechanisms*. Cambridge University Press, New York.

Bateson, M. (1975). Mother–infant exchanges: the epigenesis of conversational interaction. In *Developmental psycholinguistics and communication disorders*. Annals of the New York Academy of Sciences, Vol. 263.

Bayley, N. (1933). Mental growth during the first three years. *Genetic Psychology Monographs*, 14(1), 92.

Becker, J. (1982). Children's strategic use of requests to mark and manipulate social status. In *Language development*, Vol. 2: *Language, thought and culture*, (ed. S. Kuczaj II), pp. 1–17. Erlbaum, Hillsdale, NJ.

Bellugi, U. (1971). Simplification in children's language. In *Language acquisition: models and methods* (ed. R. Huxley and E. Ingram), pp. 95–117. Academic Press, New York.

Benedict, H. (1979). Early lexical development: comprehension and production. *Journal of Child Language*, 6, 183–200.

Bennett-Kastor, T. (1985/6). The two fields of child language research. *First Language*, 6, 161–74.

Berwick, R. (1983). *The grammatical Basis of linguistic performance: language use and acquisition*. MIT Press, Cambridge, Mass.

Bever, T. (1970). The cognitive basis for linguistic structures. In *Cognition and the development of language* (ed. J. R. Hayes), pp. 279–362. Wiley, New York.

Block, N. (1981). *Readings in philosophy of psychology*, Vol. 2. Harvard University Press, Cambridge, Mass.

Bloom, L. (1970). *Language development: form and function in emerging grammars*. MIT Press, Cambridge, Mass.

Bloom, L. (1973). *One word at a time: the use of single word utterances before syntax*. Mouton, The Hague.

Bloom, L. (1974). Talking, understanding and thinking. In *Language and perspectives: acquisition, retardation, intervention* (ed. R. L. Schiefelbusch and L. L. Lloyd), pp. 285–311. University Park Press, Baltimore, Md.

Bloom, L. and Lahey, M. (1978). *Language development and language disorders*. Wiley, New York.

Bowerman, M. (1973). *Early syntactic development: a cross-linguistic study with special reference to Finnish*. Cambridge University Press.

Bowerman, M. (1976). Semantic factors in the acquisition of rules for word use and sentence construction. In *Normal and deficient child language*, (ed. D. Morehead and A. Morehead), pp. 99–179. University Park Press, Baltimore, Md.

Bowerman, M. (1978). Systematizing semantic knowledge: changes over time in the child's organization of word meaning. *Child Development*, 49, 977–87.

Bowerman, M. (1979). Words and sentences: uniformity, individual variation, and shifts over time in patterns of acquisition. In *Communicative and cognitive abilities: early behavioral assessment* (ed. F. Minifie and L. Lloyd), pp. 349–96. University Park Press, Baltimore, Md.

Bowerman, M. (1980). The structure and origin of semantic categories in the language-learning child. In *Symbol as sense: new approaches to the analysis of meaning* (ed. M. Foster and S. Brandes), pp. 277–99. Academic Press, New York.

Bowerman, M. (1982). Reorganizational processes in lexical and syntactic development. In *Language acquisition: the state of the art* (ed. E. Wanner and L. Gleitman), pp. 319–46. Cambridge University Press, New York.

Bowerman, M. (1983). Hidden meanings: the role of covert conceptual structures in children's development of language. In *Acquisition of symbolic skills*, (ed. D. Rogers and J. Sloboda), pp. 445–70. Plenum, New York.

Bowerman, M. (1985). What shapes children's grammars? In *The crosslinguistic study of language acquisi-*

tion, Vol. 2: *Theoretical issues* (ed. D. Slobin), pp. 1257–319. Erlbaum, Hillsdale, NJ.
Braine, M. (1963). The ontogeny of English phrase structure: the first phase. *Language*, **39**, 1–14.
Braine, M. (1976). Children's first word combinations. *Monographs of the Society for Research in Child Development*, 41, Serial No. 164.
Branigan, G. (1976). Sequences of single words as structural units. *Papers and Reports on Child Language Development*, **11**, 60–70.
Branigan, G. (1979). Some reasons why successive single word utterances are not. *Journal of Child Language*, **6**, 411–21.
Briffault, R. (1927). *The mothers*. Macmillan, New York.
Brown, R. (1958). *Words and things*. Free Press, Glencoe, Ill.
Brown, R. (1973). *A first language: the early stages*. Harvard University Press, Cambridge, Mass.
Brown, R. and Hanlon, C. (1970). Derivational complexity and order of acquisition in child speech. In *Cognition and the development of langauge* (ed. J. R. Hayes), pp. 11–53. Wiley, New York.
Brown, G. and Yule, G. (1983). *Discourse analysis*. Cambridge University Press.
Bruner, J. (1975). The ontogenesis of speech acts. *Journal of Child Language*, **2**, 1–19.
Buhler, C. (1930). *The first year of life*. Day, New York.
Buhler, C. and Hetzer, H. (1935). *Testing children's development from birth to school age*. Farrar and Rinehart, New York.
Bullowa, M. (ed.) (1979). *Before speech*. Cambridge University Press.
Bybee, J. and Slobin, D. (1982). Rules and schemes in the development and use of the English past. *Language*, **58**, 265–89.
Carey, S. (1978). The child as word learner. In *Linguistic theory and psychological reality* (ed. M. Halle, J. Bresnan, and G. Miller), pp. 264–93. MIT Press, Cambridge, Mass.
Carey, S. (1982). Semantic development: the state of the art. In *Language acquisition: the state of the art* (ed. E. Wanner and L. Gleitman), pp. 347–89. Cambridge University Press, New York.
Carter, A. (1978). The development of systematic vocalizations prior to words: a case study. In *The development of communication* (ed. N. Waterson and C. Snow), pp. 127–138. Wiley, New York.
Cattell, P. (1940). *The measurement of intelligence of infants and young children*. The Psychological Corporation, New York; Science Press, Lancaster, Pa.
Chamberlain, A. and Chamberlain, J. (1904). Studies of a child. *Pedagogical Seminary*, **11**, 264–91.
Chapman, R. (1981). Exploring children's communicative intents. In *Assessing language production in children* (ed. J. Miller), pp. 11–36. University Park Press, Baltimore.
Chomsky, C. (1969). *The acquisition of syntax in children from 5 to 10*. MIT Press, Cambridge, Mass.
Chomsky, N. (1957). *Syntactic structures*. Mouton, The Hague.
Chomsky, N. (1959). Reveiew of B. F. Skinner's *Verbal behaviour*. *Language*, **35**, 26–58.
Chomsky, N. (1965). *Aspects of the theory of syntax*. MIT Press, Cambridge, Mass.
Chomsky, N. (1966). *Cartesian linguistics*. Harper & Row, New York.
Chomsky, N. (1968). *Language and mind*. Harcourt, Brace & World Inc., New York.
Chomsky, N. (1980). *Rules and representation*. Columbia University Press, New York.
Chomsky, N. and Halle, M. (1968). *The sound pattern of English*. Harper & Row, New York.
Clark, E. (1971). On the acquisition of the meaning of before and after. *Journal of Verbal Learning and Verbal Behavior*, **10**, 266–75.
Clark, E. (1973). What's in a word? On the child's acquisition of semantics in his first language. In *Cognitive development and the acquisition of language* (ed. T. E. Moore), pp. 65–110. Academic Press, New York.
Clark, E. (1975). Knowledge, context and strategy in the acquisition of meaning. In *Georgetown University Round Table on Languages and Linguistics* (ed. D. Dato), pp. 77–98. Georgetown University Press, Washington DC.
Clark, E. (1977). Strategies and the mapping problem in first language acquisition. In *Language learning and thought* (ed. J. Macnamara), pp. 147–68. Academic Press, New York.
Clark, E. (1982). The young word maker: a case study of innovation in the child's lexicon. In *Language acquisition: the state of the art* (ed. E. Wanner and L. Gleitman), pp. 390–425. Cambridge University Press, New York.
Clark, E. (1983). Meaning and concepts. In *Handbook of child psychology*, Vol. III: *Cognitive development* (ed. J. Flavell and E. Markman), pp. 787–840. Wiley, New York.
Clark, E. (1987). The principle of contrast: a constraint on language acquisition. In *Mechanisms of language acquisition* (ed. B. MacWhinney), pp. 1–33. Erlbaum, Hillsdale, NJ.
Clements, G. and Keyser, S. (1983). *CV phonology: a Generative theory of the syllable*. MIT Press, Cambridge, Mass.
Coker, P. (1978). Syntactic and semantic factors in the acquisition of before and after. *Journal of Child Language*, **5**, 261–77.
Corrigan, R. (1978). Language development as related to stage 6 object permanence development. *Journal of Child Language*, **5**, 173–89.
Corrigan, R. (1979). Cognitive correlates of language: differential criteria yield differential results. *Child Development*, **50**, 617–31.
Crain, S. and Nakayama, M. (1987). Structure dependence in grammar formation. *Language*, **63**, 522–43.
Cromer, R. (1968). The development of temporal reference during the acquisition of language. Unpublished Ph.D. dissertation, Harvard University.
Cromer, R. (1972). The learning of surface structure clues to deep structure by a puppet show technique.

Quarterly Journal of Experimental Psychology, **24**, 66–76.

Cromer, R. (1974). The development of language and cognition: the cognitive hypothesis. In *New perspectives in child development* (ed. B. Foss), pp. 184–252. Penguin Books, Harmondsworth, Middlesex.

Curtiss, S. (1977). *Genie: a psycholinguistic study of a modern-day "Wild Child"*. Academic Press, New York.

Dale, P. (1976). *Language development: structure and function*. Holt, Rinehart, & Winston, New York.

Davis, D. M. (1937). *The development of linguistic skills in twins, singletons, and only children from age five to ten years*, Institute of Child Welfare Monograph Series, 14. University of Minnesota Press, Minneapolis.

Davis, H. (1987). The acquisition of the English auxiliary system and its relation to linguistic theory. Unpublished Ph.D. dissertation, University of British Columbia.

de Saussure, F. (1959 [1916]). *Course in general linguistics*. McGraw-Hill, New York.

de Villiers, J. (1980). The process of rule learning in child speech: a new look. In *Children's language*, Vol. 2, (ed. K. E. Nelson), pp. 1–44. Gardner Press, New York.

de Villiers, J. and de Villiers, P. (1985). The acquisition of English. In *The crosslinguistic study of language acquisition*, Vol. 1: *The data* (ed. D. Slobin), pp. 27–139. Erlbaum, Hillsdale, NJ.

Donegan, P. and Stampe, D. (1979). The study of natural phonology. In *Current approaches to phonological theory* (ed. D. A. Dinnsen), pp. 126–73. Indiana University Press, Bloomington, Indiana.

Dore, J. (1973). The development of speech acts. Unpublished Ph.D. dissertation, The City University of New York.

Dore, J. (1979). Conversational acts and the acquisition of language. In *Developmental pragmatics* (ed. E. Ochs and B. Schieffelin), pp. 339–61. Academic Press, New York.

Dore, J. (1983). Feeling, form and intention in the baby's transition to language. In *The transition from prelinguistic to linguistic communication* (ed. R. Golinkoff), pp. 167–90. Erlbaum, Hillsdale, NJ.

Dore, J. (1985). Holophrases revisited: their "logical" development from dialogue. In Children's single-word speech (ed. M. Barrett), pp. 23–58. Wiley, Chichester.

Dore, J., Franklin, M., Miller, R., and Ramer, A. (1976). Transitional phenomena in early language acquisition. *Journal of Child Language*, **3**, 13–28.

Dromi, E. (1988). *Early lexical development*. Cambridge University Press.

Ehri, L. and Galanis, H. (1980). Teaching children to comprehend propositions conjoined by before and after. *Journal of Experimental Child Psychology*, **30**, 308–24.

Eimas, P., Siqueland, E., Jusczyk, P., and Vigorito, J. (1971). Speech perception in infants. *Science*, **171**, 303–6.

Eimas, P., Miller, J., and Jusczyk, P. (1987). Infant speech perception. In *Categorical perception* (ed. S. Harnad), pp. 161–95. Cambridge University Press, New York.

Elbers, L. (1982). Operating principles in repetititve babbling: a cognitive continuity approach. *Cognition*, **12**, 45–63.

Ervin-Tripp, S. (1976). Is Sybil there? The structure of American English directives. *Language in Society*, **5**, 25–66.

Ervin-Tripp, S. (1977). Wait for me, roller-skate! In *Child discourse* (ed. S. Ervin-Tripp and C. Mitchell-Kernan), pp. 165–88. Academic Press, New York.

Ferguson, C. (1978). Learning to pronounce: the earliest stages of phonological development in the child. In *Communicative and cognitive abilities: early behavioral assessment* (ed. F. D. Minifie and L. L. Lloyd), pp. 273–97. University Park Press, Baltimore, Md.

Ferguson, C. and Farwell, C. (1975). Words and sounds in early language acquisition. *Language*, **31**, 419–39.

Ferguson, C. A. and Garnica, O. (1975). Theories of phonological development. In *Foundations of language development*, Vol. 1 (ed. E. H. and E. Lenneberg), pp. 153–80. Academic Press, New York.

Ferreiro, E. (1971). *Les relations temporelles dans le langage de l'enfant*. Droz, Paris.

Fey, M. and Gandour, J. (1982). Rule discovery in phonological acquisition. *Journal of Child Language*, **9**, 71–88.

Fletcher, P. and Garman, M. (eds) (1986). *Language acquisition*, (2nd edn). Cambridge University Press.

Fodor, J. A. (1983). *The modularity of mind*. MIT Press, Cambridge, Mass.

Frege, G. (1960 [1892]). On sense and reference. In *Translations from the philosophical writings of G. Frege* (ed. P. Geach and M. Black) Blackwell, Oxford.

French, L. and Brown, A. (1977). Comprehension of *before* and *after* in logical and arbitrary sequences. *Journal of Child Language*, **4**, 247–56.

French, L. and Nelson, K. (1985). *Children's acquisition of relational terms: some ifs, ors and buts*. Springer-Verlag, New York.

Gardner, H., Winner, E., Bechhofer, R., and Wolf, D. (1978). The development of figurative language. In *Children's language*, Vol. 1, (ed. K. E. Nelson), pp. 1–38. Gardner Press, New York.

Garman, M. (1979). Early grammatical development. In *Language acquisition: studies in first language development* (ed. P. Fletcher and M. Garman), pp. 177–208. Cambridge University Press.

Garvey, C. (1975). Request and responses in children's speech. *Journal of Child Language*, **2**, 41–60.

Gathercole, V. (1987). The contrastive hypothesis for the acquisition of word meaning: a reconsideration of the theory. *Journal of Child Language*, **14**, 493–531.

Gentner, D. (1982). Why nouns are learned before verbs: linguistic relativity versus natural partitioning. In *Language development*. Vol. 2: *Language, thought, and culture* (ed. S. Kuczaj II), pp. 301–34. Erlbaum, Hillsdale, NJ.

Gesell, A. (1925). *The mental growth of the preschool child: a psychological outline of normal development*

from birth to six years, including a system of developmental diagnosis. Macmillan, New York.
Gesell, A. and Thompson, H. (1934). *Infant behavior: its genesis and growth.* McGraw-Hill, New York.
Gesell, A., Thompson, H., and Amatruda, C. S. (1938). *The psychology of early growth.* Macmillan, New York.
Givón, T. (1979a). From discourse to syntax: grammar as a processing strategy. In *Syntax and semantics, Vol. 12: Discourse and syntax* (ed. T. Givón), pp. 81–112. Academic Press, New York.
Givón, T. (1979b). *On understanding grammar.* Academic Press, New York.
Gleitman, L. and Wanner, E. (1982). The state of the state of the art. In *Language acquisition: the state of the art* (ed. E. Wanner and L. Gleitman), pp. 3–46. Cambridge University Press, New York.
Gold, E. M. (1967). Language identification in the limit. *Information and Control,* 10, 447–74.
Golinkoff, R. M. (ed.) (1983). *The transition from prelinguistic to linguistic communication.* Erlbaum, Hillsdale NJ.
Gollin, E. (ed.) (1985). *The comparative development of adaptive skills: evolutionary implications.* Erlbaum, Hillsdale NJ.
Goodwin, R. (1985). A word in edgeways? The development of conversation in the single-word period. In *Children's single-word speech* (ed. M. Barrett), pp. 113–47. Wiley, Chichester.
Gopnik, A. and Meltzoff, A. (1986). Words, plans, things and locations: interactions between semantic and cognitive development in the one-word stage. In *The development of word meaning* (ed. M. Barrett and S. Kuczaj II), pp. 199–223. Springer-Verlag, New York.
Gordon, D. and Ervin-Tripp, S. (1984). The structure of children's requests. In *The acquisition of communicative competence* (ed. R. Schiefelbusch and J. Pickar), pp. 295–321. University Park Press, Baltimore.
Gould, S. (1977). *Ontogeny and phylogeny.* Harvard University Press, Cambridge, Mass.
Greenberg, J. and Kuczaj, S., II (1982). Towards a theory of substantive word-meaning acquisition. In *Language development. Vol. 1: Syntax and semantics* (ed. S. Kuczaj II), pp. 275–313. Erlbaum, Hillsdale, NJ.
Greenfield, P. (1980). Towards an operational and logical analysis of intentionality: the use of discourse in early child language. In *The social foundations of language and thought* (ed. D. Olson), pp. 254–79. Norton, New York.
Greenfield, P. and Smith J. (1976). *The structure of communication in early language development.* Academic Press, New York.
Grice, H. P. (1967). William James Lectures, Harvard, published in part in 1975 as *Logic and conversation.* In *Syntax and semantics,* Vol. III: *Speech acts* (ed. P. Cole and J. L. Morgan), pp. 41–58. Academic Press, New York.
Grieve, R. and Hoogenraad, R. (1979). First words. In *Language acquisition* (ed. P. Fletcher and M. Garman), pp. 93–104. Cambridge University Press.

Griffiths, P. (1986). Early vocabulary. In *Language acquisition,* (2nd edn, ed. P. Fletcher and M. Garman), pp. 279–306. Cambridge University Press.
Grimshaw, J. (1981). Form, function, and the language acquisition device. In *The logical problem of language acquisition* (ed. C. L. Baker and J. J. McCarthy), pp. 163–87. MIT Press, Cambridge, Mass.
Gruber, J. (1967). Topicalization in child language. *Foundations of Language,* 3, 37–65.
Hakes, D. (1980). *The development of metalinguistic abilities in children.* Springer-Verlag, New York.
Halle, M. (1985). The representation of words in memory. In *Phonetic linguistics* (ed. V. Fromkin), pp. 101–14. Academic Press, New York.
Halliday, M. A. K. (1975). *Learning how to mean: explorations in the development of language.* Arnold, London.
Halliday, M. A. K. (1978). *Language as a social semiotic.* Arnold, London.
Harding, C. and Golinkoff, R. (1979). The origins of intentional vocalizations in prelinguistic infants. *Child Development,* 50, 33–40.
Haslett, B. (1987). *Communication: strategic action in context.* Erlbaum, Hillsdale, NJ.
Hill, J. A. C. (1983). *A computational model of language acquisition in the two-year-old.* Indiana University Linguistics Club, Bloomington, Indiana.
Hillenbrand, J. (1984). Speech perception by infants: categorization based on nasal consonant place of articulation. *Journal of the Acoustical Society of America,* 75, 1613–22.
Hooper, J. B. (1977). Substantive evidence for linearity: vowel length and nasality in English. In *Papers from the 13th Regional Meeting of the Chicago Linguistic Society,* pp. 152–64. University of Chicago, Chicago.
Horgan, D. (1978). The development of the full passive. *Journal of Child Language,* 5, 63–80.
Hyams, N. (1986a). *The acquisition of parameterized grammars.* Reidel, Dordrecht.
Hyams, N. (1986b). Core and peripheral grammar and the acquisition of inflection. Unpublished paper presented at the 11th Annual Boston University Conference on Language Development.
Hyman, L. (1975). *Phonology: theory and analysis.* Holt, Rinehart, & Winston, New York.
Hymes, D. (1971). Competence and performance in linguistic theory. In *Acquisition of language: models and methods* (ed. R. Huxley and E. Ingram), pp. 3–28. Tavistock, London.
Hymes, D. (1972). On communicative competence. In *Sociolinguistics: selected readings* (ed. J. Pride), pp. 269–93. Penguin, Baltimore.
Ingram, D. (1978a). The production of word-initial fricatives and affricates by normal and linguistically deviant children. In *Language acquisition and language breakdown* (ed. A. Caramazza and E. Zurif), pp. 63–85. The Johns Hopkins University Press, Baltimore.
Ingram, D. (1978b). Sensori-motor intelligence and language development. In *Action, gesture, and symbol: the*

Ingram, D. (1981). The transition from early symbols to syntax. In *Early language: acquisition and intervention* (ed. R. Schiefelbusch and D. Bricker), pp. 259–86. University Park Press, Baltimore.

Ingram, D. (1989). *Child language: method, description and explanation*. Cambridge University Press.

Jackendoff, R. (1983). *Semantics and cognition*. MIT Press, Cambridge, Mass.

Jakobson, R. (1968 [1941]). *Child language, aphasia and phonological universals*. Mouton, The Hague. (German publication 1941; English translation 1968.)

Jespersen, D. (1964 [1924]). *Language: its nature, development and origin*. Norton, New York.

Johnson, C. E. (1981). Children's questions and the discovery of interrogative syntax. Unpublished Ph.D. dissertation, Stanford University.

Johnson, H. L. (1975). The meaning of *before* and *after* for preschool children. *Journal of Experimental Child Psychology*, 19, 88–99.

Johnston, J. (1985). The discourse symptoms of developmental disorders. In *Handbook of discourse Analysis*, Vol. 3: *Discourse and dialogue* (ed. T. van Dijk), pp. 79–93. Academic Press, New York.

Kagan, J. (1981). *The second year of life: the emergence of self-awareness*. Harvard University Press, Cambridge, Mass.

Karmiloff-Smith, A. (1979). *A functional approach to child language: a study of determiners and reference*. Cambridge University Press.

Karmiloff-Smith, A. (1980). Psychological processes underlying pronominalization and non-pronominalization in children's connected discourse. In *Papers from the Parasession on Pronouns and Anaphora* (ed. J. Kreiman and E. Ojedo), pp. 231–50. Chicago Linguistic Society, Chicago.

Karmiloff-Smith, A. and Inhelder, B. (1974/5). If you want to get ahead, get a theory. *Cognition*, 3, 195–212.

Katz, J. (1977). A proper theory of names. *Philosophical Studies*, 31, 1–80.

Kavanaugh, R. (1979). Observations on the role of logically constrained sentences in the comprehension of before and after. *Journal of Child Language*, 6, 353–7.

Kenstowicz, M. and Kisseberth, C. (1975). *Generative phonology*. Academic Press, New York.

Kiparsky, P. and Menn, L. (1977). On the acquisition of phonology. In *Language learning and thought* (ed. J. Macnamara), pp. 47–78. Academic Press, New York.

Kripke, S. (1972). Naming and necessity. In *Semantics of natural language* (ed. D. Davidson and G. Harmon), pp. 253–355. Reidel, Dordrech.

Kuczaj, S., II and Lederberg, A. (1977). Height, age and function: differing influences on children's comprehension of 'younger' and 'older'. *Journal of Child Language*, 4, 395–416.

Kuhl, P. (1985). Categorization of speech by infants. In *Neonate cognition* (ed. J. Mehler and R. Fox), pp. 231–62. Erlbaum, Hillsdale, NJ.

Kuhl, P. and Miller, J. (1975). Speech perception by the chinchilla. *Science*, 190, 69–72.

Ladefoged, P. (1982). *A course in phonetics*, (2nd edn). Harcourt Brace Jovanovich, New York.

Lakoff, G. (1987). *Women, fire, and dangerous things*. The University of Chicago Press.

Landau, B. and Gleitman, L. (1985). *Language and experience: evidence from the blind child*. Harvard University Press, Cambridge, Mass.

Lane, H. and Pillard, R. (1978). *The wild boy of Burundi: a study of an outcast child*. Random House, New York.

Lempert, H. (1984). Topic as starting point for syntax. *Monographs of the Society for Research in Child Development*, Serial No. 208, Vol. 49.

Lenneberg, E. H. (1967). *The biological foundations of language*. Wiley, New York.

Leonard, L. B. (1976). *Meaning in child language*. Grune & Stratton, New York.

Leopold. W. (1939–49) [1939, 1947, 1949]. *Speech development of a bilingual child*, Vol. I–IV. AMS Press, New York, and Northwestern University Press.

Levinson, S. (1983). *Pragmatics*. Cambridge University Press.

Liberman, A. M. (1970). The grammars of speech and language. *Cognitive Psychology*, 1, 301–23.

Liberman, A. M. and Mattingly, I. (1985). Motor theory of speech perception revised. *Cognition*, 21, 1–36.

Liberman, A. M., Cooper, F., Shankweiler, D., and Studdert-Kennedy, M. (1967). Perception of the speech code. *Psychological Review*, 74, 431–61.

Lieberman. P. (1984). *The biology and evolution of language*. Harvard University Press, Cambridge, Mass.

Lindblom, B. (1986). Phonetic universals in vowel systems. In *Experimental phonology* (ed. J. Ohala and J. Jaeger), pp. 13–44. Academic Press, Orlando, FL.

Lock, A. J. (1978). The emergence of language. In *Action, gesture and symbol: the emergence of language* (ed. A. Lock), pp. 3–18. Academic Press, London.

Lock, A. J. (1980). *The guided reinvention of language*. Academic Press, London.

Lock, A. J. (1985). Processes of change and elaboration of language. In *The comparative development of adaptive skills: evolutionary implications* (ed. E. Gollin), pp. 239–70. Erlbaum, Hillsdale, NJ.

Lock, A. J., Young, A. W., Service, V., and Chandler, P. (1990). Some observations on the origins of the pointing gesture. In *From gesture to language in hearing and deaf children* (ed. V. Volterra and C. Erting), pp. 42–55. Springer-Verlag, Berlin and New York.

Locke, J. (1690). *An essay concerning human understanding* (ed. A. D. Woozley). Meridian Books, Cleveland, 1964.

Locke, John L. (1983). *Phonological acquisition and change*. Academic Press, New York.

Lyons, J. (1977). *Semantics*. Cambridge University Press.

McCarthy, D. (1930). *The Language development of the preschool child*, Institute of Child Welfare Monograph Series, 4. University of Minnesota Press, Minneapolis.

McCarthy, D. (1954). Language development in children. In *Manual of child psychology* (2nd edn, ed. L. Carmichael), pp. 492–630. Wiley, New York.

Macken, M. (1979). Developmental reorganization of phonology: a hierarchy of basic units of acquisition. *Lingua*, **49**, 11–49.

Macken, M. (1980). The child's lexical representation: the puzzle–puddle–pickle evidence. *Journal of Linguistics*, **16**, 1–17.

Macken, M. (1987). Representation, rules and overgeneralization in phonology. In *Mechanisms of language acquisition* (ed. B. MacWhinney), pp. 367–97. Erlbaum, Hillsdale, NJ.

Macken, M. and Ferguson, C. (1983). Cognitive aspects of phonological development: model, evidence and issues. In *Children's language*, Vol. 4 (ed. K. E. Nelson), pp. 256–82. Erlbaum, Hillsdale, NJ.

Macken, M. and Ferguson, C. (1987). Protowords: linguistic structure and function. Unpublished paper presented at the 12th Annual Conference on Language Development, Boston University.

Maclean, C. (1978). *The wolf children*. Allen Lane, London.

Macnamara, J. (1972). Cognitive basis of language learning in infants. *Psychological Review*, **79**, 1–13.

Macnamara, J. (1982). *Names for things: a study of human learning*. MIT Press, Cambridge, Mass.

McNeill, D. (1970). *The acquisition of language: the study of development psycholinguistics*. Harper & Row, New York.

McShane, J. (1980). *Learning to talk*. Cambridge University Press.

MacWhinney, B. (1978). The acquisition of morphophonology. *Monographs of the Society for Research in Child Development*, 43, Serial No. 174.

Malson, L. (1972). *Wolf children and the problem of human nature*. Monthly Review Press, New York.

Mandelbaum, D. G. (1943). Wolf-child histories from India. *Journal of Social Psychology*, **17**, 25–44.

Maratsos, M. (1974). Children who get worse at understanding the passive: a replication of Bever. *Journal of Psycholinguistic Research*, **3**, 65–74.

Maratsos, M. (1982). The child's construction of grammatical categories. In *Language acquisition: the state of the art* (ed. E. Wanner and L. Gleitman), pp. 240–66. Cambridge University Press, New York.

Maratsos, M. and Chalkley, M. A. (1980). The internal language of children's syntax: the ontogenesis and representation of children's syntactic categories. In *Children's language*, Vol. 2 (ed. K. E. Nelson), pp. 127–214. Gardner Press, New York.

Maratsos, M., Chalkley, M. A., Fox, D. E., and Kuczaj, S. II, (1979). Some empirical findings in the acquisition of transformational relations. In *Minnesota Symposia on Child Psychology*, Vol. 12, (ed. W. A. Collins), pp. 1–45. Erlbaum, Hillsdale, NJ.

Mazurkewich, I. and White, L. (1984). The acquisition of the dative alternation: unlearning overgeneralizations. *Cognition*, **16**, 589–600.

Meltzoff, A. and Kuhl, P. (1986). Infants' perception of faces and speech sounds: challenges to theory. In *Challenges to developmental paradigms* (ed. P. Zelazo), pp. 67–91. Erlbaum, Hillsdale, NJ.

Menn, L. (1976). Pattern, control and contrast in beginning speech. Unpublished Ph.D. dissertation, University of Illinois at Urbana-Champaign.

Menn, L. (1983). Development of articulatory, phonetic, and phonological capabilities. In *Language Production*, Vol. 2, (ed. B. Butterworth), pp. 1–50. Academic Press, London.

Merriman, W. (1986). How children learn the reference of concrete nouns: a critique of current hypotheses. In *The Development of word meaning* (ed. M. Barrett and S. Kuczaj II), pp. 1–38. Springer-Verlag, New York.

Michael, G. and Willis, F. (1968). Development of gestures as a function of social class, education and sex. *Psychological Record*, **18**, 515–19.

Miller, G. (1978). The acquisition of word meaning. *Child Development*, **49**, 999–1004.

Miller, G. and Johnson-Laird, P. (1976). *Language and perception*. Harvard University Press, Cambridge, Mass.

Miller, W. and Ervin, S. (1964). The development of grammar in child language. In *The acquisition of language* (ed. U. Bellugi and R. Brown), pp. 9–34. *Monographs of the Society for Research in Child Development*, 29; University of Chicago Press.

Mowrer, O. (1960). *Learning theory and symbolic processes*. Wiley, New York.

Mulford, R. (1979). Prototypicality and the development of categorization. *Papers and Reports on Child Language Development*, **16**, pp. 13–25. Stanford University.

Nelson, K. (1973). Structure and strategy in learning to talk. *Monographs of the Society for Research in Child Development*, 38, Serial No. 1249.

Nelson, K. (1974). Concept, word, and sentence: interrelations in acquisition and development. *Psychological Review*, **81**, 267–85.

Nelson, K. (1977). The conceptual basis for naming. In *Language learning and thought* (ed. J. MacNamara), pp. 117–36. Academic Press, New York.

Nelson, K. (1979). Features, contrasts and the FCH: some comments on Barrett's lexical development hypothesis. *Journal of Child Language*, **6**, 139–46.

Nelson, K. (1985). *Making sense: the acquisition of shared meaning*. Academic Press, New York.

Nelson, K. and Lucariello, J. (1985). The development of meaning in first words. In *Children's single-word speech* (ed. M. Barrett), pp. 59–86. Wiley, Chichester.

Nelson, K., Rescorla, L., Gruendel, J., and Benedict, H. (1978). Early lexicons: what do they mean? *Child Development*, **9**, 960–8.

Nice, M. M. (1926/7). On the size of vocabularies. *American Journal of Speech*, **2**, 1–7.

Ninio, A. and Bruner, J. (1978). The achievements and antecedents of labelling. *Journal of Child Language*, **7**, 565–73.

Ochs, E. (1982). Talking to children in Western Samoa. *Language in Society*, **11**, 77–104.

Ogburn, W. F. (1959). The wild boy of Agra. *American Journal of Sociology*, **64**, 449–54.
O'Grady, W. (1987). *Principles of grammar and learning.* University of Chicago Press.
Ohala, J. (guest ed.) (1986). The validation of phonological theory. Special issue of *Phonology Yearbook*.
Olmsted, D. L. (1966). A theory of the child's learning of phonology. *Language*, **42**, 531–5.
Olmsted, D. L. (1971). *Out of the mouth of babes.* Mouton, The Hague.
Otsu, Y. (1981). Universal grammar and syntactic development in children. Unpublished Ph.D. dissertation, Massachusetts Institute of Technology.
Palermo, D. (1982). Theoretical issues in semantic development. In *Language development*. Vol. 1: *Syntax and semantics* (ed. S. Kuczaj II), pp. 335–64. Erlbaum, Hillsdale, NJ.
Parker, S. (1985). A social-technological model for the evolution of language. *Current Anthropology*, **26**, 617–39.
Peters, A. (1977). Language learning strategies: does the whole equal the sum of the parts? *Language*, **53**, 560–73.
Peters, A. (1983). *The units of language acquisition.* Cambridge University Press.
Petitto, L. (1987). "Language" in the pre-linguistic child. In *The development of language and language researchers* (ed. F. Kessel), pp. 187–221. Erlbaum, Hillsdale, NJ.
Petitto, L. and Seidenberg, M. (1979). On the evidence for linguistic abilities in signing apes. *Brain and Language*, **8**, 72–88.
Piaget, J. (1952). *The origins of intelligence in children* (1st edn, 1936). International Universities Press, New York.
Pinker, S. (1984). *Language learnability and language development.* Harvard University Press, Cambridge, Mass.
Pisoni, D. (1977). Identification and discrimination of the relative onset time of two component tones. *Journal of the Acoustical Society of America*, **61**, 1352–61.
Priestley, T. (1977). One idiosyncratic strategy in the acquisition of phonology. *Journal of Child Language*, **4**, 45–66.
Putnam, H. (1962). The analytic and the synthetic. In *Minnesota Studies in the Philosophy of Science*, Vol. 3, pp. 358–97. University of Minnesota Press, Minneapolis.
Putnam, H. (1975). *Mind, language and reality.* Cambridge University Press.
Pye, C. (1983). Mayan telegraphese: intonational determinants of inflectional development in Quiché Mayan. *Language*, **59**, 583–604.
Ramer, A. (1976). Syntactic styles in emerging language. *Journal of Child Language*, **3**, 49–62.
Randall, J. (1987). *Indirect positive evidence: overturning overgeneralizations in language acquisition.* Indiana University Linguistics Club, Bloomington, Ind.
Reich, P. (1976). The early acquisition of word meaning. *Journal of Child Language*, **3**, 117–24.
Reich, P. (1986). *Language development.* Prentice-Hall, Englewood Cliffs, NIJ.

Rescorla, L. (1980). Overextension in early language development. *Journal of Child Language*, **7**, 321–35.
Roeper, T. and Williams, E. (eds) (1987). *Parameters and linguistic theory.* Reidel, Dordrecht.
Rosch, E. (1973). On the internal structure of perceptual and semantic categories. In *Cognitive development and the acquisition of language* (ed. T. E. Moore), pp. 111–44. Academic Press, New York.
Ross, J. R. (1972). Act. In *Semantics of natural languages* (ed. D. Davidson and G. Harman), pp. 70–126. Reidel, Dordrecht.
Rumelhart, D. and McClelland, J. (1987). Learning the past tenses of English verbs: implicit rules or parallel distributed processing. *Monographs of the Society for Research in Child Development*, **43**, Serial No. 174, 195–248.
Sadock, J. (1978). *Toward a linguistic theory of speech acts.* Academic Press, New York.
Sander, E. (1972). When are speech sounds learned? *Journal of Speech and Hearing Disorders*, **37**, 55–63.
Schlesinger, I. (1974). Relational concepts underlying language. In *Language perspectives: acquisition retardation and intervention* (ed. R. Schiefelbusch and L. Lloyd), pp. 129–51. University Park Press, Baltimore.
Scollon, R. (1976). *Conversations with a one-year-old.* University of Hawaii Press, Honolulu, Hawaii.
Searle, J. (1969). *Speech acts.* Cambridge University Press.
Searle, J. (1975). A taxonomy of illocutionary acts. In *Language, mind and knowledge* (ed. K. Gunderson). University of Minnesota Press, Minneapolis. Reprinted in Searle, J. (1979). *Expression and meaning*, pp. 1–29. Cambridge University Press.
Shatz, M. (1975). How young children respond to language: procedures for answering. *Papers and Reports on Child Language Development*, **10**, 97–110.
Shatz, M. (1982). On mechanisms of language acquisition: can features of the communicative environment account for development? In *Language acquisition: the state of the art* (ed. E. Wanner and L. Gleitman), pp. 102–27. Cambridge University Press, New York.
Shatz, M. (1983). Communication. In *Handbook of child psychology, Vol. III: Cognitive development* (ed. J. Flavell and E. Markman), pp. 841–89. Wiley, New York.
Shire, M. L. (1945). The relation of certain linguistic factors to reading achievement in first-grade children. Unpublished Ph.D. dissertation, Fordham University.
Shirley, M. (1933). *The first two years: a study of twenty-five babies.* University of Minnesota Press, Minneapolis.
Sinclair-de Zwart, H. (1969). Developmental psycholinguistics. In *Studies in cognitive development: essays in honour of Jean Piaget* (ed. D. Elkind and J. H. Flavell), pp. 315–36. Oxford University Press, London.
Singh, J. A. L. and Zingg, R. M. (1942). *Wolf children and feral man.* Harper and Brothers, New York.
Skinner, B. F. (1957). *Verbal behaviour.* Prentice-Hall, Englewood Cliffs, NJ.
Slobin, D. (1973). Cognitive prerequisites for the development of grammar. In *Studies of child language development* (ed. C. Ferguson and D. Slobin) pp. 175–208. Holt, Rinehart, & Winston, New York.

Slobin, D. (1985). *The crosslinguistic study of language acquisition*. Erlbaum, Hillsdale, NJ.

Smith, C. (1980). The acquisition of time talk: relations between child and adult grammars. *Journal of Child Language*, 7, 263–78.

Smith, E. and Medin, D. (1981). *Categories and concepts*. Harvard University Press, Cambridge, Mass.

Smith, M. E. (1926). An investigation of the development of the sentence and the extent of the vocabulary in young children. *University of Iowa Studies in Child Welfare*, 3, No.5.

Smith, N. (1973). *The acquisition of phonology*. Cambridge University Press.

Spencer, A. (1985). Toward a theory of phonological development. *Lingua*, 68, 3–38.

Stampe, D. (1979). *A dissertation on natural phonology* (1973 dissertation). Garland Press, New York.

Stark, R. (1986). Prespeech segmental feature development. In *Child Phonology. Vol. 1: Production* (ed. G. Yeni-Komshian, J. Kavanagh, and C. Ferguson), pp. 149–73. Academic Press, New York.

Stern, D., Jaffe, J., Beebe, B., and Bennett, S. (1975). Vocalizing in unison and in alternation: two modes of communication within the mother–infant dyad. In *Development psycholinguistics and communication disorders*, Annals of the New York Academy of Sciences, Vol. 263.

Stevens, K. (1972). The quantal nature of speech: evidence from articulatory-acoustic data. In *Human communication: a unified view* (ed. E. David jun. and P. Denes), pp. 51–65. McGraw-Hill, New York.

Stoel-Gammon, C. and Cooper, J. (1984). Patterns of early lexical and phonological development. *Journal of Child Language*, 11, 247–71.

Sugarman, S. (1983). Empirical versus logical issues in the transition from prelinguistic to linguistic communication. In *The transition from prelinguistic to linguistic communication* (ed. R. Golinkoff), pp. 133–45. Erlbaum, Hillsdale, NJ.

Templin, M. (1953). *The development and interrelations of language skills in children*, Institute of Child Welfare Monograph Series. University of Minnesota Press, Minneapolis.

Templin, M. (1957). *Certain language skills in children: their development and interrelationships*. University of Minnesota Institute of Child Welfare, Monograph 26. University of Minnesota Press, Minneapolis.

Tiedemann, D. (1787). *Über die Entwickelung der Seelenfähigkeiten bei Kindern*, Hessische Beiträge zur Gelehrsamkeit und Kunst. Trans. Murchison, C. and Langer, S. (1927). Tiedemann's 'observations on the development of the mental faculties of children', *The Pedagogical Seminary and Journal of Genetic Psychology*, 34, 205–30.

Tomasello, M., Anselmi, D., and Farrar, M. J. (1984/5). Young children's coordination of gestural and linguistic reference. *First Language*, 5, 199–210.

Tough, J. (1977). *The development of meaning*. Wiley, New York.

Trevarthen, C. (1977). Descriptive analyses of infant communicative behavior. In *Studies in mother–infant interaction* (ed. H. Schaffer), pp. 227–70. Academic Press, London.

Trosborg, A. (1982). Children's comprehension of *before* and *after* reinvestigated. *Journal of Child Language*, 9, 281–402.

Urwin, C. (1978). The development of communication between blind infants and their parents. In *Action, gesture and symbol: the emergence of language* (ed. A. Lock), pp. 79–108. Academic Press, London.

van der Hulst, H. and Smith, N. (1982). An overview of autosegmental and metrical phonology. In *The structure of phonological representations*, Part I (ed. H. van der Hulst and N. Smith), pp. 1–45. Foris, Dordrecht.

Vihman, M., Macken, M., Miller, R., Simmons, H., and Miller, J. (1985). From babbling to speech: a reassessment of the continuity issue. *Language*, 61, 397–445.

Volterra, V. and Erting, C. (eds.) (1990). *From gesture to language in hearing and deaf children*. Springer-Verlag, Berlin and New York.

von Humboldt, W. (1792). *Ideen zu einem Versuch die Grenzen der Wirksamkeit des Staats zu bestimmen*. Deutsche Bibliothek, Berlin. Trans. as *The limits of state action* (ed. J. Burrow). Cambridge University Press, London, 1960.

von Tetzchner, S., Siegel, L. S., and Smith, S. (eds) (1989). *The social and cognitive aspects of normal and atypical language development*. Springer-Verlag, New York.

Vygotsky, L. (1962). *Thought and language*. MIT Press, Cambridge, Mass.

Waterson, N. (1971). Child phonology: a prosodic view. *Journal of Linguistics*, 7, 179–211.

Weil, J. and Stenning, K. (1978). A comparison of young children's comprehension and memory for statements of temporal relations. In *Recent advances in the psychology of language: language development and Mother–Child interaction*, Vol. 4A (ed, R. Campbell and P. T. Smith), pp. 395–410. Plenum, New York and London.

Weist, R. M. (1986). Tense and aspect. In *Language acquisition*, (2nd edn, ed. P. Fletcher and M. Garman), pp. 356–74. Cambridge University Press.

Wellman, B., Case, I., Megert, I., and Bradbury, D. (1931). Speech sounds of young children, University of Iowa Studies in Child Welfare, 5. University of Iowa, Iowa City.

Werner, H. and Kaplan, B. (1963). *Symbol formation*. Wiley, New York.

Wexler, K. and Culicover, P. (1980). *Formal principles of language acquisition*. MIT Press, Cambridge, Mass.

Wexler, K. and Manzini, M. R. (1987). Parameters and learnability. In *Parameter setting* (ed. T. Roeper and E. Williams), pp. 41–76. Reidel, Dordrecht.

White, L. (1981). The responsibility of grammatical theory to language acquisitional data. In *Explanations in linguistics: the logical problem of language acquisition* (ed. N. Hornstein and D. Lightfoot) pp. 241–71. Longman, New York and London.

Wittgenstein, L. (1953). *Philosophical investigations*. Basil Blackwell, Oxford.

Yeni-Komshian, G., Kavanagh, J. and Ferguson C. (eds.) (1980). *Child phonology*, Vol. 1: *Production*; Vol. 2: *Perception*. Academic Press, New York.

Zingg, R. M. (1940). Feral man and extreme cases of isolation. *American Journal of Psychology*, **52**, 487–517.

Zinober, B. and Martlew, M. (1985). The development of communicative gestures. In *Children's single-word speech* (ed. M. Barret), pp. 183–215. Wiley, Chichester.

25
The reconstruction of the evolution of human spoken language

Mary LeCron Foster

Abstract

Language is an analogical system for classification on multiple levels. Language systems build upon semantic analogies and analogies in phonological, morphological, and syntactic distributions (positional analogies). New meanings are created through the process of metaphorical extension. The direction of language change is determined in large part by this process and by analogical systematization—hierarchical congruence of classes.

The regularities of sound-change reconstructed by the comparative method provide the most reliable diagnoses of remote linguistic relations; but these are limited to 'families', or, in a few cases, 'stocks' made up of interrelated families. Broader groupings, 'phyla' or 'super-stocks', are suggested on the basis of typological relations, rather than on firmly established sound-correspondences. The basis for going even further and attempting to reconstruct a single prototype for all the world's spoken languages is not agreed upon; but the reconstruction should reflect systematic correspondences in sound and meaning throughout, whether insights were intially gained from typological studies of phonology and/or from internal reconstructions. Hypotheses must show system. While individual meanings underlying reconstructed forms need not be identical, differences should be minimized. Once correspondences are firmly established, culturally influenced semantic variations are useful in assessing degrees of interrelationship among languages.

Pursuing the monogenetic reconstruction through this bare-bones phonemic approach, refined by a series of simplifications, leads to the startling hypothesis that the sounds of which the VC and CVC roots are composed were originally themselves meaning-bearers. These *phememes*, as they are termed, were minimal units of sound whose meaning derived from the shaping and movement of the articulatory tract. In other words, the phonemes of language, as well as the combinations into which they unite within the word, were originally not arbitrary signs, but abstract, highly motivated analogical symbols.

In the earliest stage of primordial language, single *phememes* expressed notions of space and motion. Across the evolution of the genus *Homo* these were differentiated and new *phememes* created, hypothetically in stages, until the *phememic* inventory was completed during the Upper Palaeolithic. In the Neolithic period, it is hypothesized, syllabic concatenation with morphophonemic merging increasingly obscured the analogical significance of *phememes*, which gradually became what we now know as phonemes. The linguistic comparative method applied globally reveals a single primordial structure of *phememically* constructed roots reflected distinctively in each historical language.

25.1 Introduction

The evolutionary model for reconstruction that I will describe is an analogical model. Basically it relies on linguistic reconstruction by means of the comparative method, coupled with a strong reliance on the systematization of sound and meaning. Since all languages have system, and system is based on the operation of analogy, it will be necessary to keep in mind system, and the directions in which systems have the possibility to change, as we attempt to move backward in time through the many millennia of language use.

In trying to understand how language came into being and how it developed into language-as-we-know-it, we must look at the underlying universal properties of language and hypothesize the steps that must have been taken in their development.

Unless the nature and complexity of language is understood, it is impossible to understand mechanisms of change, possibilities for change, or the systematic nature of change. Without such understanding it is not possible to go beyond very elementary types of reconstruction. By stressing the fundamental principles of analogy and showing its operation in the construction of linguistic systems, I am setting the stage for discussion of the changes in whole systems that any single, enduring change brings in its wake. A language is a well-integrated whole; but as it is lived and spoken, holes appear in its fabric that require not just mending or patching, but systematic reorganization or reweaving. Yet each reweaving reveals irregularities in the system, and further readjustments must be made. This is a dynamic process that shows both the human drive toward constitution of order and the inevitable destruction of particular orders over time.

If it is to be biologically reinforced by examination of the behaviour of other sentient creatures, the focus should be on the degree to which such creatures are able to classify their experience and extend their classifications by means of analogical inventiveness. If it is to be reinforced through archaeological examination of the remains of early man, these should be explored for evidence of the analogical organization of experience reflected in material culture. If it is to be reinforced by study of children's language learning, a major focus should be the child's progressive exploration of classification possibilities.

25.2 Language as classification by analogy

Language is the major vehicle for the conveyance of symbolic meaning. One can detect the rudiments of a symbolic way of thinking in the behaviour of certain other sentient creatures than humans, but it is hard to imagine any degree of symbolic complexity without language. Body movements can send signals ('danger!', 'all clear!', 'I am willing!' and the like), or signs ('There is no danger,' 'I am in a bad/good mood,' etc.), but symbolism requires deliberate representation of something by something else of a different order. No sign or symbol by itself conveys an idea of the nature of symbolization, which is an exceedingly complex phenomenon, however simple it may have been in its earliest beginnings.

Since the publication of Saussure's (1959) lectures early in this century, linguists have accepted unquestioningly the arbitrary nature of spoken language. This postulate of arbitrariness is based on the fact that, except for a certain amount of onomatopoeia in all spoken languages, there is no necessary relationship between the sounds in any word and that word's meaning. For Saussure, language was not a system of symbols based on analogy, but a system of signs. Symbolism is variously defined; but I would argue for analogy as an important criterion, and for language as an analogical system.

This is the case because language is a system of classification, and classification is an analogical method of organizing experience so that it may be comprehended and dealt with effectively. In methods of classification, including hierarchical ordering, the human animal has far outstripped other species; and much of this is directly attributable to language as the primary classification tool.

There is nothing random about language. In its classificational methods it is far from arbitrary. Items (words, parts of words, sounds, sentences, parts of sentences, methods of discourse) are ordered in classes because they are in some degree *like*. The likeness does not have to be based on likeness between sound and meaning; there are many other methods of establishing likeness. One method is semantic. Many nouns represent things; others are treated syntactically as if they were things. Likeness is principally conveyed through placement in the spoken or written chain. Order is never random, but is carefully organized in multiple ways.

Language is easily learned by children, not because they have a genetically programmed aptitude for language, but because they have a genetically programmed aptitude for classification on multiple levels. The whole of culture and the learning of culture is similarly an exercise in classification—hence it is symbolic, not arbitrary. Typical of symbolism is its capacity to proliferate meaning. The notion of 'one linguistic sign: one meaning' is palpably false, as a brief look at any dictionary will show. And dictionaries only supply meanings that are culturally widely shared. They give no account of meanings-in-the-making, or the capacity of any word to generate new meanings through metaphorical extension in order to cope adequately with new experiences. Thus, there are always more denotational meanings than one, and the potential always exists for extending denotation through connotational processes, which always go back ultimately to intentionality—or the capacity of the individual to create meanings which may come to have common currency.

Language, like all cultural systems, is thus an active process, and its users are continually experimenting with new ways to use it to make experience more meaningful, and hence less banal—for banality sets in with overuse. Thus intentionality accounts for linguistic change; but the direction that change will take is also determined by processes of analogical extension—expanding old classes to make room for new members—and analogical systematization—the re-imposition of hierarchical congruence on classes.

25.3 Analogy and language as system

Language systems build upon semantic analogies and analogies in phonological, morphological, and syntactic distributions (positional analogies). The sounds of any given language are organized systematically in accordance with both position and manner of articulation. Sounds that are *like* in either position or manner of articulation tend to operate in accordance with like rules (paradigms). This results in a kind of symmetry that is never absolute, but is characteristic of all phonological systems (see Figs. 25.1–25.5).

While phonemes *per se* are meaningless, but signal differences of meaning through their contrastiveness, morphemes are the minimal meaning-bearers in every language. They constitute paradigms because they can be linked with other morphemes in accordance with some semantic/distributional *similarity* between the morphemes with which linkage is possible. Thus, in English, the form class (or paradigm) that we call nouns is singular in isolation but plural when a phonological

Manner of articulation		Activity of vocal chords	Position (point) of articulation						Point in tongue
			Bilabial	Labio-Dental	Apical	Frontal	Dorsal	Faucal	
					Dental Alveolar	Palatal	Velar	Glottal	Point in 'mouth'
stop (closure)		voiceless voiced	p b		t d		k g	q = ʔ	
affricate		voiceless voiced	pΦ	pf	ts = c dz	tš = č dž = ǰ	(kx)		
spirant = fricative	slit	voiceless voiced	Φ β	f v	θ = þ ð	ç y	x γ	(h) (ɦ)	
	groove	voiceless voiced			s z	š ž			
nasal		voiced	m		n	ñ = ɲ	ŋ		
liquid	lateral	voiced			l	λ = l'	L, ł = π		
	trill	vibrant voiced			r		R		
flap		voiced			D = ř				
glide		voiced	w			y	w		

Fig. 25.1 Basic consonant chart defined by three articulatory parameters. A row for glides is added to emphasize the frequent patterning of [y] and [w]. Note that both occur twice in the diagram. Note that [q] is used for the glottal stop and not for a back velar. (From Anttila 1972, p. 8.)

	Lip position		Tongue advancement					
			Front		Central		Back	
			U	R	U	R	U	R
Tongue height	High	H	i	ü = y	ɨ		ï = ɯ	u
		L	ɪ	Ü	ɨ			U
	Mid	H	e	ö = ø	ə			o
		L	ɛ	ö = œ			ʌ	ɔ
	Low	H	æ = ä					
		L	a		ɑ		ɒ	

Fig. 25.2 Basic vowel chart defined by three articulatory parameters. (From Anttila 1972, p. 7.)

```
p                              k        ʔ
m        n        l
w                                       h
```

Fig. 25.3 Hawaiian consonantal phonemes (adapted from Pukui and Elbert 1957).

```
p     t     c     ṭ     č     k
pʔ    tʔ    cʔ    ṭʔ    čʔ    kʔ        ʔ
b     d     z     ḍ     ž     g
            s     ṣ           x     h
m     n     l                 ŋ
mʔ    nʔ    lʔ                ŋʔ
w                       y
wʔ                      yʔ
```

Fig. 25.4 Yokuts consonantal phonemes (adapted from Newman 1944).

Point of articulation:	Labial	Alveolar	Velar
Manner of articulation:			
Voiceless	p	t	k
Voiced:	b	d	g

Fig. 25.5 English stop consonants.

variant of the plural element (a bound morpheme in English) is added. Syntactic paradigms are formed from *similarities* between word-linkage possibilities. Thus, in English, they can be as short as those formed of articles or demonstratives plus nouns. Analogical groupings are not the same from language to language, but there are underlying similarities in form-class possibilities. For example, whether grammatically marked or not, all languages have syntactically selective lexical categories. It is impossible to construct rules, particular or comprehensive, for the operation of language (speech or writing), or to learn to use a language, without prior discovery of its analogical possibilities.

When structuralism was the dominant paradigm (in the Kuhnian sense) for the construction of grammars, the paradigm (in the grammatical sense) was of paramount importance. Generative-transformational grammar since Chomsky has obscured the importance of paradigms, although generative-transformational rules could not be written without their implicit recognition. However, in that tradition, these virtual (i.e. never actualized in speech) groupings are never specifically isolated. I would argue that comparative linguistics, the reconstruction of earlier language stages, is greatly hampered by the Chomskyan exclusion of these abstract systems from consciousness—or at least visibility—*qua* systems in which analogy is the motivating principle.

For nearly thirty years linguists have been influenced by the generative tradition into virtually ignoring the central role of analogy in language. Although Anttila (1977) provided a useful discussion of its importance and of its eclipse after the Chomskyan revolution, this has been largely ignored.

25.4 Reconstructive methodology

During the nineteenth century European philologists developed the comparative method for linguistic reconstruction. Details of the method were worked out during a concerted effort to understand the history of those languages of Europe and Asia that came to be known as Indo-European (IE). Lexical inspection had indicated a genetic interconnection, which scholars set out to explore. Some Indo-European prehistory was already known because of texts in, and continued religious use of, certain ancient forms of the IE daughter languages, such as Latin, Ancient Greek, Gothic, Old Church Slavonic, and Sanskrit. During the course of the reconstruction it became apparent that, in general, sound-changes followed regular patterns; for example, a sound that was /p/ in Latin, Greek, and Sanskrit was realized as /f/ in the Germanic languages. Thus we had Sanskrit pad-, Latin ped-, Greek pod-, Gothic fōt- as various versions of the root underlying instances of the words for 'foot'. Vowel differences indicate an e/o vowel alternation, as well as differential vowel length, phenomena common to IE roots. Agreement of consonants in the IE reflexes, except for Germanic, indicates that the originals were *p and *d, and that Germanic changed these to /f/ and /t/ respectively.[1] Comparison of words for 'father' in the same languages show Skt. pitar-, Lat. pater-, Gk. patēr, Goth. fader, with similar consonantal regularity, but an original /t/ rather than /d/ this time.

It is on this kind of regularity that the comparative method was founded. If the regular reflex (expected phonological shape associated with a given gloss (meaning)) did not occur in a reconstructible form, some rule to account for the disruption had to be discovered if the reconstruction were to be considered reliable. Sound-change regularity is the rock upon which the comparative method is founded. It is what provides the verifiability of remote linguistic relationships.

It is generally assumed (see, for example, Kroeber 1964) that lexical and grammatical attrition beyond a time-span of perhaps 8 000 years makes reconstruction impossible. Such an assumption overlooks the fact the virtually all the roots, stems, and grammatical morphemes held in common by IE languages have been reconstructed and are attested by the comparative method (Buck 1949; Pokorny 1959). A corollary assumption is that the comparative method is effective only if the reconstruction is carried out stage by stage, backward through time, thoroughly establishing each stage before earlier stages can be apprehended (Haas 1969; Campbell and Mithun 1979). Haas (1969, pp. 71–77) is optimistic that such a method can work if only enough dictionaries are compiled and grammars analysed, and the efforts of scores of trained scholars are then mobilized to write comparative grammars and dictionaries for each stage. She feels that if this were done for the languages of North America it would be possible to achieve a good knowledge of the various protolanguages in use around 2000–5000 years ago, and then to determine whether application of the comparative method to those protolanguages might carry us back another two to five millennia. However, this would take us no farther back for the New World than to the period when writing systems probably first began in the Old World.

By co-ordinating various methodologies and pointing to seminal work by a variety of scholars, Birnbaum (1977) provides an insightful survey of the prospects for more remote linguistic reconstruction. He examines works of typology and internal reconstruction and uses of the comparative method, focusing on the systematicity of change and the prospects for understanding change provided by application of these methodologies to phonology, morphology, syntax, and semantics.

25.4.1 *The comparative method*

As we have seen, it is the assumed regularity of sound-change that has made the comparative method a reliable means of recovering language history.

	PIE				Germanic		
	labial	dental	velar		label	dental	velar
voiceless:	p	t	k	> fricative:	f	Θ	x
voiced:	b	d	g	> voiceless:	p	t	k
voiced aspirated:	bh	dh	gh	> voiced:	b	d	g

Fig. 25.6 Grimm's Law.

Despite its long history, the use of this method remains virtually unchanged today. It traditionally works best where inspection reveals similarities of both sound and meaning, and where regular shifts in sound-reflexes for obviously related languages can be worked out. Under these circumstances, it is usually possible to discover grammatical sharing as well, and to account similarly for sound-shifts in grammatical elements. Both meanings and sounds often change, and for reconstruction variant meanings are considered acceptable as long as they maintain a rough synonymity.

For example, the reconstructed PIE (Proto-Indo-European) root *wed-/(or *awed-) (reconstructions are asterisked) meant something like 'wet', or 'flow', and underlies a variety of forms suggesting liquid. In English we have, of course, 'wet' as well as 'water', both showing the regular Germanic reflex of */t/ for PIE *d. In Greek, as in Sanskrit, a reduced form eliminates the medial vowel and changes the *w to its vocalic allophone, *[u]: Gk. üdōr, Skt. udán-'water'.

25.4.2 *Systems in sound-change*

The most famous, and early, discovery of system in sound-change is known as Grimm's law, after the discoverer. Grimm's law (Grimm 1852) illustrates the fact that sound-change tends to operate systematically across phonemic sets that were originally (as well as after the change) symmetrical. The law shows regular changes from the PIE (Proto-Indo-European) to the Germanic system. It is illustrated in Fig. 25.6.

Contemporaries of Grimm found exceptions to his 'law', and, according to the comparative method, rules (or laws) of sound-change must not have unexplained exceptions. Other famous 'laws' resolved certain of these problems. For example, Verner (1875) postulated a law to account for the failure of PIE voiceless stops to be reflected as Germanic /d/ in certain cases. Thus, Latin (which preserves PIE voiceless stops) has pater ('father') while Gothic has fadar. However Latin frāter ('brother') is Gothic brōðar ('brother'). Instead of the voiced dental stop /d/, in Gothic (and English) we find a voiced dental fricative /ð/. There are other similar examples involving other voiced stops becoming fricatives at the same point of articulation. Verner discovered that Sanskrit has preserved ancient stress patterns, lost in some other languages, such that a stress before the voiceless, medial stop in Sanskrit shows a fricative in Gothic, whereas, if the stress is after the stop, the stop follows Grimm's law and is still a stop, but voiced. Thus Sanskrit bhrắta 'brother' and Gothic brōðar, but Sanskrit pitá 'father' and Gothic fadar. It was discoveries of this kind that led ultimately to the decision that sound laws, instead of allowing no exceptions, could brook no *unexplained* exceptions.

Use of the comparative method in particular cases is greatly strengthened if systematic change of this kind can be shown to pertain, and if isoglosses between languages can be generalized to several languages. Isoglosses are lines of separation between genetically related languages that show where innovation and geographic distance have caused dissimilarities between branches of what was once a unified language. The line that separates the Germanic dialect family from other IE dialect families is marked by isoglosses such as the consonant shifts seen in Grimm's law.

25.4.3 *Family trees and wave theory*

Reconstruction by means of the comparative method is predicated on a 'family tree' model, which assumes that shared isoglosses indicate splits from former unity, and that these quite possibly took place in serial rather than simultaneous order. Thus, relative dating of language separation is indicated by the degree of isogloss-sharing at any point on the 'tree'. Such a tree diagram can be devised for the Indo-European family to show isoglossic splitting into 10 branches, each with a variety of sub-branches (see, for example, Figs. 25.7a and b).

However, comparative linguists were also aware that isoglosses were not necessarily as clear-cut as Grimm's law would indicate. Languages separating from one another were not necessarily completely isolated, and considerable borrowing back and forth among them could take place after the separation. In 1872 a substitute theory was proposed by Johannes Schmidt (Schmidt 1872), to show a more accurate

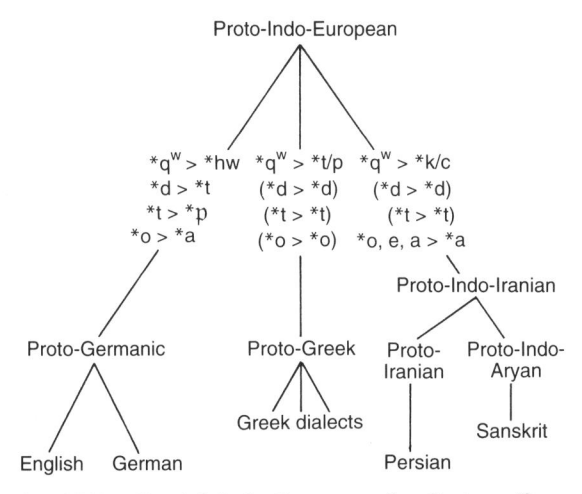

Fig. 25.7a Partial Indo-European family tree (from Bynon 1977, p. 99).

depiction of the relationships between families within a linguistic stock. This was called a 'wave theory', because certain innovations seemed to spread from one language to another like ripples or waves caused by a dropped stone. This can also be illustrated diagrammatically for Indo-European (Fig. 25.8).

25.4.4 Linguistic groupings

Terms for linguistic groupings are quite loosely applied. In popular usage, *dialect* refers to a localized variation of a language, distinguished only by relatively insignificant isoglosses of pronunciation and grammatical or syntactic usage. Sometimes it is used disparagingly to indicate difference from the standardized version of the language. In point of fact, we all speak a dialect; none of us speak a language. If dialects are similar enough, phonological and grammatical rules can be devised in such a way that two or more dialects can be described as a single language. Dialect geography is the study of dialect isoglosses.

The term *language*, then, is usually used to cover a group of dialects among which there is mutual intelligibility. However, sometimes the term *language* is reserved for the dialect of that group which has achieved political and social prestige. The dialect of English that is described by English grammars is essentially a written language, spoken in all of its 'purity' by relatively few. A standard language always begins to deviate from its 'pure' form, causing scholars (or English teachers) to deplore what has often already become common usage.

When a proto-form of a language, such as PIE, is reconstructed, it becomes commonplace for linguists to speak of its as a language, and the isoglossically differentiated branches as 'dialects'. Actually, variants termed dialects of a given language are not always mutually intelligible. Less well standardized languages than, for example, Italian, French, Spanish, and Portuguese, if not legitimated politically, would under this usage be designated as dialects of some larger grouping—perhaps Romance, which is the name of the family to which they belong.

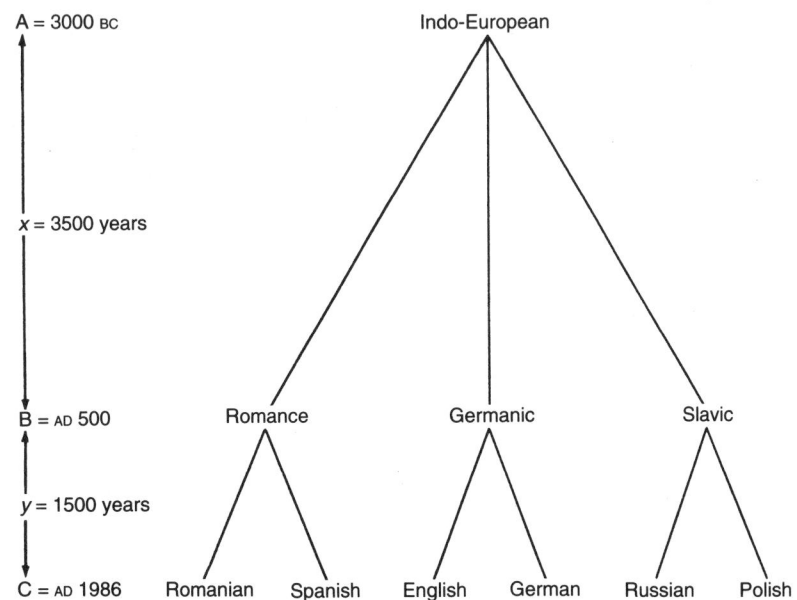

Fig. 25.7b Differentiation of the Indo-European family (partial) (from Ruhlen 1987, Fig. 1.8) [eds].

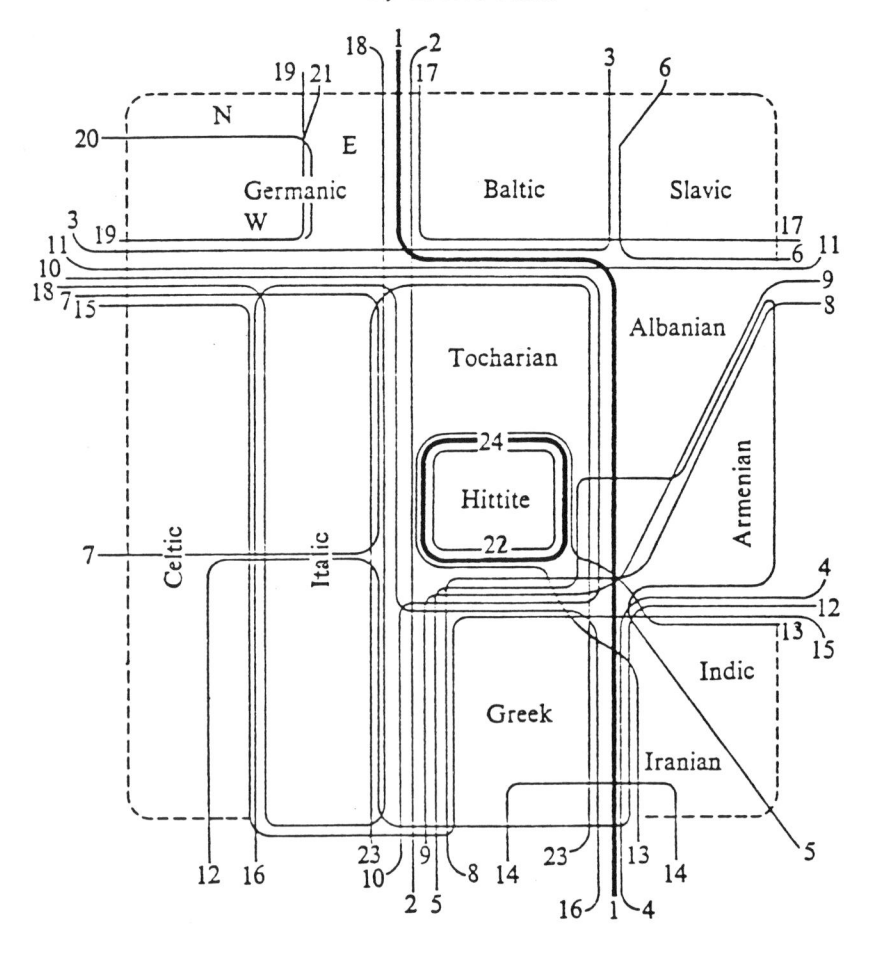

1. centum | satem [right]
2. -ss- | -st-, -tt- [right]
3. aoə | a, āō | ō [inside]
4. eao | a [inside]
5. s | h [inside]
6. CVRC | CRVC [inside]
7. k^w | p [inside]
8. e- | ø 'past' [left, outside]
9. -osyo 'genitive' [right, inside]
10. -r | -i 'present' [right, outside]
11. -m- | -bh- 'case marker' [below]
12. -to- | -mo- 'ordinal' [below]
13. -u 'imperative' [inside]
14. proti | poti 'preposition' [inside]
15. secondary endings (without no. 10) [below]
16. feminine nouns with masculine endings [inside]
17. -ad 'ablative' | 'genitive' [inside]
18. new tense system from perfect [inside]
19. umlaut [inside]
20. -ww-, -jj- | stop + w, j [outside]
21. -ggj- | -ddj [right]
22. laryngeals as h's [inside]
23. uncontracted reflexes of sequence *yH [inside]
24. unit pronouns | particles + enclitic pronouns [inside]

Fig. 25.8 A dialect map of the Indo-European languages (from Anttila 1972, p. 305).

25.4.5 Typology

Most of the groupings that can be called phyla are established on the basis of typological similarities rather than (or without being supported by) phonological regularities such as those noted by Grimm's law. Typologies can be: phonological—similarities of phoneme types and articulatory organization; morphological—similarities of affixation and the semantics of affixation; or syntactic—similarities of lexical ordering and meanings that are expressed through such ordering. Often typological similarities are found in geographical areas, between contiguous languages that are not obviously genetically related.

Levin's (1971) equation of IE and Semitic structure is a very complex typological study. Edward Sapir (1929) used typological criteria rather than comparative evidence in order to reduce to six super-stocks (phyla) a whole host of North American and some Central American Indian languages. Haas (1969, 105-6), however, argues that classificatory (typological) linguistics is a 'millstone' that should be removed from comparative classification if real progress is to be made in establishing prehistoric family relationships. Typological resemblances should be admissible only if supported by comparative phonological criteria.

A major use of typological studies lies in establishing language universals and universals of implication. Begun by Jakobson (1958), this type of analysis has been continued especially by Greenberg (1963, 1966a, 1978), with very fruitful results. One typological implication that Jakobson (1958, 528-9) suggested is that 'no language adds to the pair /t/–/d/ a voiced aspirate /d^h/ without having its voiceless counterpart /t^h/, while /t/, /d/, and /t^h/ frequently occur without the comparatively rare /d^h/'. Such typological discoveries have important implications for historical linguistics. Anttila (1972, p. 317) puts it thus:

> Largely through Greenberg's efforts, we now have a doctrine of dynamic comparison between language types, where the objects of comparison are the processes themselves. It is often possible to establish the relative origin of types and we might call this method *seriation of types*. There are two classes of languages, those that have nasal vowels and those that do not. There is the further implicational hierarchy that languages with nasal vowels also have oral vowels. Now, a hypothesis can be asserted about the relative origin of nasal vowels: they result from the loss of a nasal consonant, apart from borrowings and analogical creations. A typical course of events is VN > ṼN > Ṽ, but the nasal of course can also precede the vowel. This is pure *relative* chronology, because nasal vowels need not arise; when they do, however, they originate this way. There are further hierarchical scales about nasal vowels; for example, their number is always the same or less than that of oral vowels, never the reverse, and mergers within nasal vowels occur more often than among oral vowels.

The above correlates with Greenberg's (1966b) theory of *marking*. A marked category has a characteristic that an unmarked category lacks (in this case nasalization). Thus, a marked category always implies an unmarked category, but not vice versa. It also implies that chronologically an unmarked category preceded the marked one. To quote Anttila (p. 318) again:

> The ultimate task of typology is to determine what structures are possible, and why they are possible to the exclusion of others. In this way typology is hierarchically superior to genetic linguistics. It permits us to understand the general laws of change and the possibilities of change within a given language type.

25.4.6 Internal reconstruction

Close comparison of forms within an actual language, or within a reconstructed (and thus by definition hypothetical) language, will reveal many structural characteristics of still earlier stages in that language or that language-family. This kind of analysis is known as *internal reconstruction*. In conjunction with the comparative method it can yield far-reaching results.

Like revisions of sound-laws in the comparative method, such as that resolved by Verner's discovery of the role of ancient stress, internal reconstruction builds on discrepancies or puzzles, to discover older, underlying patterns that have been ironed out or partially obscured through change.

The most famous (and provocative) example of internal reconstruction was presented as a hypothesis by Saussure (1879) in an attempt to solve the problem of alternation in the root vowels (vowel ablaut) in Indo-European languages.

The typical 'Indo-European root is of a C[V]C pattern, that is, it consists either of consonant–vowel–consonant or of a pair of consonants, of which the latter must come from a range of sonants that can be more or less independently sounded—examples for Greek would be lip- 'leave' for the first type and dr̥- 'see' for the second. In Saussure's reconstruction of the alternation of vowel infixes in these basic forms of the root, each root will then be capable of generating a range of three distinct forms by the infixing in the basic root of either *e or *o or nothing, with the resultant forms then becoming available to fulfil various differentiated syntactic or semantic roles connected with the basic meaning of the root.

Thus we find in Greek from the basic root form dr̥- 'see' not only ĕdrakon 'I saw', which contains the unmodified basic or zero- infix form of the root, but also dérkomai 'I see', which contains the infixed *e form of the root, and *dédorka 'I have seen', which contains the infixed *o form. And from the root lip- 'I leave' we can similarly generate not only élipon 'I left', with the basic form without any additional vowel infix, but also léipō, 'I leave', with the *e infix (here coalescing with the /i/ of the root to form the Greek diphthong /ei/), and léloipa 'I have left', with the *o infix (here coalescing with the /i/ of the root to form the Greek diphthong /oi/). A similar root-alternation occurs in all branches of IE.

Alternations between *a or *e/*o/*ə (a reduced vowel) also occur, with vowel lengthening, as in PIE *bha-, *bho-, *bhə- 'to shine', and *eg-, *og-, *əg- 'to speak, say'. Saussure (1879) speculated that there were two original sounds (which he did not specify, but called 'coefficients sonantiques') which had originally occurred either before the vocalic nucleus, or after in the case of vowel-lengthening, which coloured the vowel *e with an [a] sound or an [o] sound. After the discovery of Hittite, Kurylowicz (1927) observed that in some cases Hittite had /h/ in the position in which the lost consonants were supposed to have occurred, but not in others. He therefore postulated four lost 'laryngeals', $e^1 = *e$, $e^2 = *o$, $e^3 = *a$, $e^4 = *a^h$ (but this last without the h- in Hittite). Because of Hittite it was assumed that the lost consonants had a laryngeal quality, and the hypothesis of the lost vowel-colouring consonants came to be known as 'the laryngeal hypothesis'. The 'laryngeals' (anywhere from three to six, depending on the investigator) have been written either with the letter e plus a number, following Kurylowicz; with H plus a number (the system used below, as it makes for easier reading); or as E, O, A, placed either before or after a vowel to indicate the vowel coloration that it induces.

Martinet (1953), noticing that certain roots, such as *dō- 'give', tended to develop a *w* before a following vowel, as in Sanskrit davane, postulated a [w] offglide for the o-coloured laryngeal, writing it as A^w. Diver (1959) provided a similar explanation for H_1, reinterpreting it as H^y.

The laryngeal theory has always been controversial, and has been unacceptable to some scholars. Others have supplied still more 'laryngeals' to account for further unsolved PIE problems. Excellent accounts of the laryngeal hypothesis appear in Winter (1965) and Lehmann (1955). Implications for PIE laryngeal theory derived from more remote constructions will be discussed in Section 25.8.3 below.

25.5 Reconstructed languages

The languages of the world are usually said to number around 3000. But if we take into account the fact that many so-called 'dialects' are in fact unintelligible to one another, and that the term 'language' is used loosely, the number far exceeds that figure [Ruhlen 1987, gives a figure of 'roughly 5000 languages'—eds].

Some of these languages are known to be related, either because relationship is apparent through inspection alone, or because it has been established through careful application of the comparative method. Proven relationships constitute 'families', or, in a few cases, 'stocks' made up of interrelated families. Terms such as 'phyla', or 'super-stock' are generally assumed from typological resemblances rather than on the basis of firmly established sound-correspondences. Until sound-correspondence is adequately demonstrated, phyla must remain tentative groupings (see the Map gallery of language maps, this volume, Chapter 19).

An early standard reference for global relationships is Meillet and Cohen (1952). This work is a conservative global survey, and includes short grammatical sketches of the structure of many languages. More recent global discussions of language classification are those of Voegelin and Voegelin (1977) and Ruhlen (1987). Other sources are briefly discussed in Comrie (1992), and more fully in articles on specific languages in Bright (1992), as Comrie mentions. Investigation of Old World languages has a longer history than that of New World languages, and is in many ways more advanced. New World languages are considered to derive ultimately from those of the Old World, as the New World is assumed to have been populated later through waves of migration across the Bering Straits. To date no solid proof of trans-oceanic relationships has been published.

The most thoroughly investigated genetic interrelationship is that between Indo-European (IE) languages, stretching geographically across Europe and into Asia (see language map 19.6 in Chapter 19). All European languages except Basque (an isolate) and the Finno-Ugric languages (such as Finnish, Hungarian, and Estonian) are members of the IE stock. Relationships between families have been depicted through both family-tree and wave (or network) models. IE is the most exhaustively studied stock and, while not without remaining problems, has been provided with etymological dictionaries (Pokorny 1959; Buck 1949; Watkins 1985) that list virtually every root and stem, and with thoroughgoing grammatical reconstruction (Meillet 1964; Kurylowicz and Mayrhofer 1968–86). Hittite has been seen to be closely related to IE; but, because recently discovered, it has not yet been fully integrated into PIE reconstruction.

Semitic, spoken in North Africa, and earlier in Mesopotamia, is another well-studied stock. Akkadian, long extinct, was one of the earliest written languages, utilizing the cuneiform writing of the Sumerians. Akkadian belongs to the group of Northeast Semitic languages which also includes Babylonian and Assyrian. These are all 'dead' languages, with the earliest evidence for Old Akkadian dating to between *c.*2500 and 2000 BC. North-west Semitic forms can be traced back to around the second millennium BC, and this grouping includes Canaanite (which in turn includes Hebrew, Phoenician, Punic, and Moabite), Aramaic, and probably Ugaritic. South-west Semitic includes North Arabic, South Arabic, and Ethiopic. Proto-Semitic is a stock comparable to PIE, and as with PIE Semitic includes ancient languages that can be used in its reconstruction. Major comparative studies of Semitic are Moscati *et al.* (1969), Gray (1971), and the partially completed root lexicon of Cohen (1970, 1976).

The Ancient Sumerian of Mesopotamia is still considered to be an isolate.

It has long been recognized that ancient Egyptian (with its now extinct offspring, Coptic) was related to the Semitic languages. Greenberg (1966*b*) gives evidence of its linkage to other languages of Africa in a grouping called Hamitic. Others that have discussed this grouping are Bynon and Bynon (1975) and Diakonoff (1974). Hamito-Semitic is increasingly termed Afroasiatic (see language map 19.1 and 19.5, Chapter 19).

Greenberg's (1966*b*) grouping of African languages into seven major families and a number of other smaller groups, while not bolstered by thorough application of the comparative method, is nevertheless convincing because it is supported by groups of lexical or morphological sets, similar in phonological structure and meaning.

Collinder's (1955, 1960, 1965) work on the Uralic languages makes this another of the best-studied groupings (these languages are spoken on either side of the Ural mountains). Anttila (1972, p. 301) provides a family tree for that family that shows the relative dates for branchings, supported by isoglosses (Fig. 25.9).

While Poppe (1965) applied the comparative method successfully to the Altaic languages, Turkic, Mongol, and Tungus, and Martin (1966) used comparative evidence to link Japanese and Korean genetically, the Altaic stock as a whole is still controversial.

The languages of Polynesia are well established as a family (Elbert 1953; Krupa 1982), and have been firmly connected to Maori (Tregear 1969) as well as to Fijian (Hockett 1976). Thus Austronesian (Malayo-Polynesian) as a grouping has not been in question since Dempfwolff's (1934–8) pioneering comparative study, and has been discussed extensively by Dahl (1973) and others; but the status of, or inclusion of, the eastern languages within such a group can be questioned (Grace 1959).

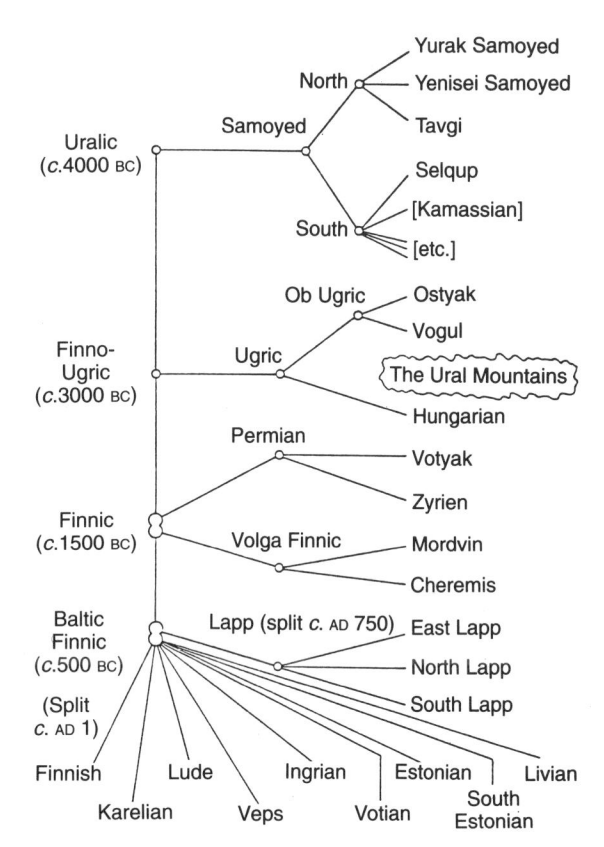

Fig. 25.9 Family tree of the Uralic languages (from Anttila 1972, p. 301).

The bulk of the Asian languages is only tentatively classifiable. Whether they form a larger group, called Sino-Tibetan, is controversial. Voegelin and Voegelin (1964) organize these languages into nine groups which have been reconstructed (or are reconstructable) into nine different parent languages. The nine are: (1) Han Chinese; (2) Miao–Yao; (3) Kam–Thai; (4) Burmese–Lolo; (5) Karen; (6) Bodo–Naga$_1$–Kachin; (7) Naga$_2$–Chin; (8) Gyarung–Mishmi; and (9) Tibetan. The Voegelins (1965) group all languages that are non-Caucasian, non-Indo-European, or non-Sino-Tibetan into an areal group which they call Boreo-Oriental: Uralic, Altaic, Japanese–Korean, Palaeosiberian, and Ainu.

Other reconstructed or reconstructable northern groups are Eskimo–Aleut and Kartvelian (South Caucasian).

Dravidian of southern India is etymologically well attested (Burrow and Emereau 1961), although reconstructed forms are not provided.

In North America many of the 58 American Indian families established by Powell (1891) on the basis of strong phonological and semantic resemblance have since been established by rigorous application of the comparative method. Much of the work of applying this method has been carried out by Edward Sapir and his students, especially Mary Haas and linguists trained by her. Despite various efforts to reduce the number of North American families (most notably the major typological effort of Sapir (1929) mentioned above), most Americanists are at present cautious about 'lumping' (Campbell and Mithun 1979, 37). The Campbell and Mithun volume gives a good overview of well-established North and Central American linguistic relationships. Haas (1969) discusses comparative problems. Sebeok (1977) provides a comprehensive survey of accomplishments in the complex task of sorting out and reconstructing linguistic relationships in the Americas. According to Longacre (1977, p. 99):

> In Mesoamerica comparative reconstruction is reaching a very mature stage, especially in Otomanguean and Mayan stocks and in the Mixe-Zoque family. Proto-Otomanguean, embracing some thirty languages grouped into seven component families, is currently on the drawing boards.... Etymological dictionaries—sizable bodies of cognate sets—will soon be available for component families and the stock as a whole. Current work in the Mayan stock is also very promising with a proven link of Mayan to one South American language. Mixe-Zoque is shaping up well. Good work has been recently done in Utoaztecan, but we still lack an etymological dictionary of adequate size for that stock. Less advanced are available comparative studies in Totonac–Tepehua (not to mention the vexed question of 'Mexican Penutian': Mayan, Totonac-Tepehua, Mixe-Zoquean) and in Yuman (Hokan). In South America and in the Caribbean, comparative projects are at early stages.

According to Landar (1977, 408), 'Very little work of distinction has been done in the classification of South American languages.' The major reason for this is lack of adequate lexical material on which to base reconstructions. The monumental nature of the task can easily be seen by looking at a tribal map of only one part of South America, as represented in Fig. 25.10.

25.6 Hypotheses suggesting further groupings

With a few notable exceptions, the comparative method has not been applied to groupings that are more remote in time than those that, in the last section, I referred to as 'families'—a term that I have stretched to some extent for this purpose. Some of the most comprehensive suggestions have only been substantiated by phono-semantic 'look-alikes'.

1. One of the most far-reaching of these is Lahovary's (1963) study, which presents similarities among languages from a wide geographic area reaching across Europe, North Africa, and Asia, and includes Dravidian, Semitic, Hamitic, Basque, Sumerian, Caucasian, Elamite, and Etruscan. Of these, McAlpin (1981) has since reconstructed ancient Elamite with Dravidian, and Hamitic and Semitic have long been known to be related.

2. Key's (1978, 1979, 1980–1, 1981) efforts to link South American languages to one another and, finally, to languages of North America have provided sets of look-alikes with suggestions for comparative regularities, but without reconstruction. More recently (1984) she has, by the same methods, linked American Indian languages with those of Polynesia.

3. Brunner (1982) has, at the same time, linked Malayo-Polynesian, Semitic, and Indo-European roots, using examples of regularity of sound correspondences, but without actual reconstruction.

4. Following suggestions of Poppe (1965), Miller (1971) has used typological and limited phonological evidence in order to include Japanese in the Altaic grouping.

5. Radin (1919) suggested that all North American Indian languages were related; and Swadesh (1954) used statistical analysis of numbers of look-alikes to suggest that this could be proved without the discovery of sound regularities. Greenberg (1987) increases the number of look-alikes in postulating Proto-Amerind.

6. Making no attempt at reconstruction, Gostony (1975) has shown resemblances between Sumerian and other languages of Europe, Asia, and Africa, giving 938 examples. His best case is made for Hungarian (he has not made use of Uralic reconstructions).

Without the comparative method, the cognacy of such scattered look-alike sets is hard to assess (see, for example, Greenberg 1960). Even with attempts at reconstruction, assessment is not necessarily easy. For example, Matteson (1972) provides 974 sets in which resemblances between a wide variety of North and South American languages are shown, and even reconstructed. However, no language family is consistently represented, and, in order to show regularities, it has been necessary to set up a phonemic system for Proto-Amerindian which has forty consonants and six vowels (see below, Fig. 25.11). Matteson provides a chart (pp. 26–7) showing numbers of forms reconstructed across families and the percentage of cognates found between languages from particular families. The highest percentage is 37.2 between Proto-Tucanoan and Proto-Arawakan, and the lowest is 1.0 between Witoto and Bora. The average is around 10.

Fig. 25.10 NW South American tribal areas (from Landar 1977, p. 403).

p	t	ṭ	ts	k̂	k	ḳ
pʔ	tʔ	ṭʔ	tsʔ	k̂ʔ	kʔ	ḳʔ
b	d	ḍ			g	ʔ
		č		č̂		
		čʔ		č̂ʔ		
f	s	š		š̂	x	h
	z			ǰ		
m	n			ñ	ŋ	
	r					
	l			lʸ		
w				y		
				i	ɨ	u
				e		o
				a		

Fig. 25.11 Proto-Amerind consonants (from Matteson 1972).

The comparative method has been used by Bomhard (1984) to differentiate Hittite (or more comprehensively, Anatolian) from the other IE dialects, and then to relate PIE to Afroasiatic (Hamito-Semitic). Because of extensive use of vowel gradation (ablaut) in all of the languages, he has not attempted to reconstruct vowels, but instead has focused on CVC or CVCC roots or stems as the major meaning-bearers. He has shown regularity of consonant reflexes in 290 sets.

In calling his study 'an approach to Proto-Nostratic' Bomhard is following the lead of a group of Soviet scientists who have set up a 'Nostratic' macrofamily that comprises IE, Afroasiatic, Kartvelian (South Caucasian), Uralic, Altaic, and Dravidian. Japanese, Korean, and Eskimo–Aleut are tentatively included. Nostratic was first proposed by Pedersen (1903) as a language group consisting of IE, Hamito-Semitic, Uralic, and possibly the sub-Saharic languages Bantu and Sudanese.

Soviet Nostratic studies were started by the late Illič-Svityč and have been continued by a group of his colleagues and students. Three hundred and fifty-three extensively annotated sets reconstructed by Illič-Svityč (1971, 1976, 1984) have been edited by B. A. Dybo, his successor as head of this Soviet group. Although the commentary is in Russian, the reconstructions and the cognates are presented in the Roman alphabet, with the standard linguistic symbols to show phonemic contrasts. None of the Nostratic material has been translated (but see Shevoroshkin and Ramer, 1991 [eds]). Other leading names associated with the Nostratic theory are A. B. Dolgopol'skij, Y. G. Testelec, E. A. Helimski, and S. A. Starostin. Each of these scholars is a specialist in a particular linguistic group, which assures that the reconstructive scholarship is grounded in the structural reality of each language family with which it is concerned.

Like Matteson, Nostratic scholars postulate a complex (and to some extent similar) phonological system (Fig. 25.12).

All these studies are provocative. Whether or not the groupings postulated can really be shown to share isoglosses pointing to branchings from some protosystem, it is apparent that, as more languages are documented and analysed, more resemblances between them are observed and more scholars become interested in finding means to demonstrate as yet unattested relationships. But, on the whole, the studies fail to take into account the full range of remote reconstructive methodology available to them. Guidelines for methodology will be discussed in Section 25.7.

Meanwhile, despite the many and overlapping groupings that have been proposed, only two linguists have hypothesized that all the world's spoken languages had a single origin—developed from a single prototype—and that means are available for demonstrating this. Swadesh's (1971) proposal for monogenesis relies principally on typological and statistical criteria for substantiation. Foster's (1978, 1980, 1983, 1990a, 1992) relies on the comparative method coupled with typological implication, internal reconstruction, and use of whatever insights are available

p	t	c	č	ć	k	q		
ṗ	ṭ	c̣	č̣	ć̣	k̇	q̇		ʔ
b	d	z	ž	ź	g	G		
m	n		ñ	ń	(ŋ)			
w	r			(ŕ)				
	l			(ĺ)				
		s	š	ś				
		(z)				x	h	

(A dot under a stop = glottalization.)

Fig. 25.12 Proto-Nostratic (adapted from Illic-Svityč).

from the pre-proto enigmas remaining in reconstructions of particular language families.

25.7 Guidelines for remote reconstruction

I would like to propose a set of guidelines for remote linguistic reconstruction, that seem to me realistic and conducive to achieve satisfactory results.

1. Believable reconstruction must reflect coincidence of sound and meaning, with 'coincidence' depending upon definable structural criteria which unmistakably show the presence of system.

2. Reconstruction should reflect system: system in the reconstructed prototype, system in the organization of each of the daughter languages, and system in processes of change, both shared at particular levels and unique at others.

3. Similarities of form and meaning are only useful if they can be used as the bases for hypothesizing and testing regularities of sound-correspondence. They can be demonstrated for any two of the world's languages, so that by themselves, or reconstructed in insufficient quantity, they have little meaning.

4. Insights derived from the reconstruction of particular languages should be tested for relevance cross-linguistically.

5. Internal reconstruction of both particular languages and of reconstructed prototypes should be carried out. Possibilities for this will reveal themselves as the reconstruction progresses.

6. Typological implications are a function of system. Those relating to phonology are particularly important in making and testing hypotheses having to do with phonological system.

7. Any insights derived from linguistic theory can be helpful. Diachronic linguistics should synthesize these and make use of them in reconstruction wherever possible.

Birnbaum (1977) does more to integrate historical and comparative theory than any other source that I am acquainted with. Anttila (1972) uses a similar approach and provides a great many examples. I find his focus very much in agreement with my own experience in his emphasis on system and the role of analogy. Anttila (1977) is also a useful source for this point of view.

Campbell and Mithun (1979) also posit a set of cautionary guidelines for remote linguistic reconstructions.

1. Only purely linguistic considerations count (cultural information is irrelevant).
2. Only resemblances involving both sound and meaning are relevant.
3. Both lexical and grammatical evidence should be included.
4. Hypotheses should show system.
5. The reconstructed form must be a complete word.
6. Reconstructed forms should be from basic (i.e., non-cultural) vocabulary.
7. Proposed cognates should be semantically equivalent.
8. Length of cognates is important.
9. The number of corresponding segments is important.
10. Longer forms should match in their entirety.

This list is counterproductive for reasons outlined below. Adherence to these requirements would have prevented reconstruction of Proto-Indo-European, and would allow for no remote reconstructions beyond the most obvious. Only point number 4 is completely acceptable. Focus on systematicity in every aspect of change provides strength to the reconstructive hypothesis.

Problems with the other guidelines are as follows:

1. A linguistic reconstruction should obviously be linguistic. However, meaning is always culturally determined, and language is the most basic cultural system. All definitions shift their semantic ground over time. Shifts must be taken into account in determining cognacy, and both phonological and semantic changes are crucial for the establishment of family branching. The more elaborate and idiosyncratic the semantic system revealed, the less chance there is of parallel development, and thus the more plausible the reconstruction. This is consistent with point 4, that hypotheses must show system.

2. If this means that the only relevant data are sound–meaning correspondences, I cannot agree. Other relevant data stem from internal reconstruction and from typographical implications, which are related to (4), i.e. they reflect system. If what is meant is that both criteria of sound and meaning must be taken into account if sets presumed to be cognate are to be included, (2) is indeed important.

3. Grammatical evidence is only relevant at fairly shallow levels of reconstruction. Grammar changes more drastically over time than do root morphemes or stems, and because it is the outer lexical layer and can be added, substituted for, or subtracted easily, it is more apt to be permanently subtracted or supplanted.

5. The claim that the reconstructed form must be a complete word is untenable, both for the reason given in response to (3)—that outer lexical layers are more variable and roots more durable—but also because roots reappear in more than one word and are relatively indestructible. Even at fairly shallow time-depths complete words are relatively rarely reconstructed, as an examination of the Pokorny (1959) PIE dictionary will testify. If this rule were adhered to, very little would get reconstructed, and certainly almost nothing at remote depths.

6. Since all vocabulary is cultural, earliest attempts at reconstruction should establish sound regularity by finding cross-language sequences in which both sound and meaning are similar. After regularity is attested by frequencies of examples of these, examples with greater variation can be explored and tested. Regularity and system should be the criteria against which possibilities are tested. Lexical items which might be considered to be 'basic' are often very disparate, even in closely related languages, as examination of PIE reconstruction can easily show.

7. This depends completely upon one's definition of semantic equivalence. Obviously, complete semantic equivalence is desirable, but it is rarely achieved even at shallow time-depths, as a look at Buck's (1949) dictionary of IE synonyms will quickly reveal. Meanings have considerable flexibility. Few words have single fixed meanings. Metaphor is the rule rather than the exception. The more exact the semantic equivalence the closer the relationship, so that adherence to (3) can be used to show closeness of relationship and isoglossic points of departure for particular branches. This is important in establishing relative time-levels.

8. 'Important' is a pretty vague word. Length of cognates is only important if reconstructed roots or primary stems can be shown to be composed of a number of phonemes. It is more important to demonstrate that reconstructed segments reflect system than to discover cognate length, except at shallow time-depths.

9. Numbers of corresponding segments are of limited relevance for reasons stated under (5) and (8). Again, segment numbers indicate depth or recency of branching.

10. Of course length always has significance, but as was argued in response to (5), (8), and (9), long cognates may be unattainable at remote time-depths.

25.8 Towards primordial language: the monogenetic hypothesis

25.8.1 Background

Since I am the only linguist who to my knowledge has published on the subject of reconstruction based on an assumption of monogenesis, I will in this section rely solely on my own experience of thirty years of work in this area. I first became convinced of the unity of all American Indian languages in the early 1960s because of my analysis of two Mexican Indian languages (Foster and Foster 1948; Foster 1969), comparative work on the family (Mixe-Zoque) of the language described in the Sierra Popoluca grammar, and efforts to understand directions of borrowing in the New World, which led me to review a great deal of lexical data from many languages. I soon began to discover that the striking resemblances that I uncovered in the New World were also shared with languages of the Old World that were geographically very remote; thus my conviction of a single origin of the world's languages grew. Trained by Mary Haas at Berkeley, I was imbued with the decisive value of the comparative method, and convinced that if languages were all interrelated their joint heritage could be recovered by that means.

25.8.2 *The phememic system*

As I formed and tested hypotheses about sound-correspondences in PL (primordial language), I found my primordial phonemic inventory steadily expanding, much as have those of Matteson (Fig. 25.11 above) and Illič-Svityč and his colleagues (Fig. 25.12 above). It was probably my conviction of monogenesis and the simpler beginnings that this seemed to entail that made me explore the possibilities of a simpler system. The means of simplification are virtually identical to typological implication in phonology: marked categories imply unmarked categories, hence the unmarked must have been prior. Therefore, I tested an almost barebones system which essentially retained only unmarked phonemes. Later I realized that it was faulty to retain palatalized and labialized stops, as I had at the start of simplification (Foster 1978, 1980), as these might well be derived from simple voiceless stops (Foster 1983). Also, internal examination of Egyptian convinced me that at least one laryngeal was essential, and on the Egyptian model I added *ˀ (The symbol represents a glottal stop).

Root-length words are often uncommon. The Egyptian lexicon includes many of these as well as longer sequences using root extensions. The three-consonant Arabic root canon reflects such extensions. These can be compared easily with IE CVC roots and extensions (see Benveniste 1962) as well as with the canons of other, more remote, languages. Such cross-language comparisons convinced me that the phoneme /x/ in Afroasiatic was originally a laryngeal spirant *h, that this *h in combination with semivowels had served to create two additional Afroasiatic laryngeal phonemes, /h/ < *h^w and /ḥ/ < *h^y. Comparison with IE and other languages convinced me that it was these palatalized and labialized sounds in fusion with the non-phememic, epenthetic vowel, [a], that were responsible for ablaut (vowel gradations) in IE and many other languages. While this is totally lacking in Arabic, traces of it exist in Egyptian and some Semitic languages.

The laryngeal hypothesis from PIE reconstruction has always influenced my thinking about primordial vowels (see, for example, Lehmann 1955; Winter 1965). Insights from Afroasiatic, and especially Egyptian, internal reconstruction began to suggest a solution to the PIE laryngeal hypothesis.

Very early I had conjectured that the original 'laryngeals' were primordial spirants (Foster 1978), all of which had been lost in common IE, but /h/ was retained before a vowel in Hittite. In early attempts at reconstruction it had seemed that other spirants than *h would be needed because of some unexplained consonants in various languages, either spirants or stops at the established points of articulation (Foster 1978). There was also a problem of voicing that must be explained. It appeared that this must involve either spirants or laryngeals.

Semitic aleph (glottal stop: /ʔ/) seemed sometimes to be voiced, and reconstruction suggested that in IE the central vowel *schwa: /ə/ was traceable to a lost laryngeal in some environments. Aleph (ʔ) seemed the best candidate for voicing, and responsible for voicing of stops. The voiced aspirated set in PIE seemed then to require stop fusion with the sequence *ʔh. Arabic internal reconstruction reinforced this hypothesis. I eliminated *r as a phoneme. Internal reconstruction showed it to have resulted from *ʔl fusion.

As I broke down the system into the VC and CVC roots that are characteristic of so many languages and that seemed to be fundamental to the earlier system as well, comparison of these 'roots' with one another startled me by revealing that *the sounds of which they were composed were themselves meaning-bearers*,[2] and that combination of these in sequences modified but did not basically change the underlying meanings that I had discovered.

The bare-bones phonemic approach was crucial to the discovery of these units, which I termed *phememes*, as minimal units of sound and meaning.

The fronted velar series in PIE (e.g. *k̄), realized in the east as dental spirants and in the west as velar stops, I had originally (Foster 1978, 1980) thought to be derived from velar palatization (i.e. *k^y), similar to velar labialization in the IE labialized velars (*e.g. /*k^w, *ʔk^w, *ʔhk^w). Examination of Afroasiatic and other languages convinced me of the need to substitute an alveopalatal sequence: *c, *ʔc, *ʔhc. This proved valid cross-linguistically.

The meanings of the sounds derive from the shaping and movement of the articulatory tract (lips, tongue, upper mouth surface, larynx). The meanings are very abstract, as will be seen in Fig. 25.13. Some meanings are easier to determine than others, and these latter may well be somewhat inexact; but the general design seems essentially accurate.

The most striking linguistic fact to stem from this discovery is that the phonemes of language, as well as the combinations into which they unite within the word, were originally not arbitrary signs, as Saussure (1959) found them to be, but highly motivated analogical symbols, or *phememes*.

Point of Articulation

	Labial	Dental	Alveopalatal	Velar	Laryngeal
Resonants:	m bilateral	n internal	l unconstrained		
Stops:	p projective	t introjective	c extrinsic	k divergent	ʔ discontinuous
Glides	w curvate		y linear		h continuous

Fig. 25.13 The consonantal phememic system.

25.8.3 Comparison, internal reconstruction, and system for phememes

Rather than giving illustrations from many languages, in this chapter system is emphasized, and PIE, Hittite (H), and Egyptian (E) are primarily compared. Semivocalic Japanese (J) is included in some cases to provide a suggestion of range.

Internal reconstruction of Egyptian convinced me that the responsibility for the vowel-gradation ubiquitous in the world's languages lay not with laryngeals but with glides. PL semivocalic glides can occur between any consonant and a following vocalic nucleus, or become vowels themselves where such a nucleus is absent. In many languages they also fuse with a preceding or following consonant, the fusion resulting in a marked phoneme.

In Egyptian and the Semitic languages, the proliferation of velar and laryngeal spirants is a function of fusion of *h with a following semivowel, or with a preceding *ʔ. Examination of Coptic reflexes of Egyptian sequences, where these are available, provides insight into Egyptian vowel qualities. Comparison of Egyptian sequences with similar meanings shows the fusion process.

Velar/laryngeal spirant values in Afroasiatic (AA) are these:

PL *h gives rise to AA *x, with [a] vowel-colouring (PIE H_3);

PL *h^y gives rise to AA *ḫ, with [e] vowel-colouring (PIE H_1); and

PL h^w gives rise to AA *h, with [o] vowel-colouring (PIE H_2)

PL *ʔh gives rise to AA *ʕ, which can have any of the three vowel values, depending on the original presence or absence of a vowel-colouring glide. In other words, there is only one voiced phoneme to represent all three situations. Only the following vowel indicates which is intended. In many languages, *h has been lost in all or some of these positions.

Hittite does not differentiate between PIE *a and *o—both are written as /a/. *hw and *hy are reflected as H hu- and hi- respectively, and *h^{w}a as ha, *h^{y}a as he. Egyptian /y/ and /w/ are usually transcribed as /i/ and /u/ respectively, but semivowels rather show syllabification than syllable nucleii. Vowels are not written in heiroglyphic script, but the provision of vowel nucleii syllables in Roman transcription was conventionalized with the letter *e* between consonants. I have omitted these as unnecessarily confusing. When *h becomes voiced by a preceding *ʔ, Hittite is similar to common PIE in loss of the consonant when it occurs initially in the word. Japanese retention of *h as /h/ (which otherwise derives from *p), and loss of *ʔh is interestingly similar to these reflexes in Hittite. Examples (vowels are not provided in the Egyptian writing, but are sometimes available from its daughter language, Coptic) are provided in Table 25.1.

Tables 25.1 and 25.2 show the primordial voicing effect produced by fusion of *ʔ with a following voiceless phememe. Fusions of laryngeals with following stops occurred early in language evolution. Also, differential fusions of phememes with following semi-

Table 25.1 Examples of simple velar and laryngeal spirant reflexes

Examples	Primordial language	Egyptian	Hittite	Proto-Indoeuropean	Japanese
1	*ḥm(ham), HOLD, ENCLOSE, JOIN	xmꜥ, 'to grasp, hold, possess'	hamank-, hamenk-, 'bind, promise (a girl) in marriage'	*am(m)a 'mother' *ambhi 'clear around, on both sides'	hamaru, 'get into, fit into, fall into' hameru 'put/pull on, set into, entrap'
2	*ḥym *(hʸam)	ḥmꜣg, 'to grasp, clasp tightly, sack bag' ḥm-t 'woman, wife' (C hime, 'uterus, matrix, vulva')*	henk-, 'fix, fasten, devote'	*em- 'hold, hold together, join'	himitsu 'secrecy, privacy' himo 'string, cord, fillet, band'
3	*hwm (hʷam)	hmhm 'to enter into' hmw 'to but/gore with the horns'	hūmanza 'whole, all, every'	*om(e)so-s, 'shoulder'	humiyápat 'hold, grasp' hōmō 'the clutches (of the law)' hōmuru 'bury, entomb'
4	*ʔm (ʔam, ʔʸam, ʔʷam)	ʔm, ʔmw 'seize, grasp'	in(in)eya 'mix, become involved'	*em 'to take'	ami 'net, casting net' amu 'knit, braid' emono 'catch, captive, take' umeru 'bury, inter, inlay, fill'
5	*ʔhm (ʔh(a)m, ʔhʸ(a)m, ʔhʷ(a)m)	ꜥm 'to grasp, fist' ꜥmꜥ 'to nurse'	(as 4)	(as 1, 2, 3)	(as 4)

*C denotes Coptic.

Table 25.2 Examples of simple stops and *ʔ plus stop reflexes

Examples	Primordial language	Egyptian	Hittite	Proto-Indoeuropean	Japanese
1	*t(a)ht(w)y, TO PENETRATE, BREAK THROUGH	thy 'to cross over, transgress, infringe, violate a frontier, break into, trespass, invade, attack'		*tei- 'sharp point' *teig- 'to stick into, sharp'	taiha 'serious damage, ruin decay' taihai 'crushing defeat' taiho 'a gun, cannon'
2	*tyʔl, DESTROY, SCRAPE, RUB	tr 'to destroy, wipe out, efface' trf 'to write'	teripp- 'to plough'	*ter-, *tēr-, *teru-, *teri 'to rub, bore through'	taira 'flatness, evenness' tairageru 'subdue, subjugate' čiri 'dust, trash' čiru 'fall, scatter'
3	*ʔaty, *ʔt(y)a TAKE/MOVE FROM	ʔt 'loss, diminution, not' dg 'to fail, hide'	(see 5, below)	*at- 'go' *ati- 'away from' *deg- English 'take'	ate 'an aim, a goal' ta 'others, besides'
4	*ʔt(y)ʔl, *ʔt(w)ʔl REMOVE FROM, DESTROY	dr 'to drive out, expel, erase, expunge, conquer, destroy'	arha tarupp- 'remove'	*der- 'split, peel, skin'	toru, 'take' tarasu 'drop, let drop' tareru 'evacuate'
5	*ʔ(a)tw, *ʔ(a)ty GIVE, TAKE	d, dw, dʔ 'to give'	dā 'take, take from'	dō-, dō-u-, dau-, du 'to give' pres. dido-mi	ataeru 'to give' atai 'value, price'
6	*ʔcʸm, *ɕm SHARP POINT	smʔ 'knife, sword' smʔ 'to slay, kill'	gimras military campaign	*kem- or *k̂am 'sharp point, horn'	kama 'sickle' kami 'the top, the head'
7	*ʔcam, *ʔacm, *ʔcym BREAK OPEN, BITE, CRUSH	zmʔx-t 'to eat' zmʔm 'to crush' zmyw 'devourers'		*ĝembh-, *ĝmbh- 'to bite'	kamu 'bite, gnaw' akima 'a vacant room' kimei 'inscription, engraving'
8	*ʔac CUT, OPEN, SHARP	ʔsbw 'to crush' ʔsx 'to sickle'		*ak̂- 'sharp'	aky 'open, be opened' aki 'a gap, an opening'
9	*cyh, *cyhw (USE) SHARP POINT	shm 'to pound, crush, grind' shq 'to cut, hack in pieces'		*k̂e(i)- 'to sharpen, whet' *k̂eipo 'sharp point'	kiho 'point' keikaku, 'angle, corner'

vowels produced either secondary consonants or secondary vowel nucleii—an early evolutionary isoglossic split.

For various reasons, but largely because vowel-coloration was provided by a vowel-preceding consonant, I became convinced that consonant voicing and stop aspiration in many languages, including PIE, were caused by fusion with a stop-preceding phememe. At first I assigned *f the voicing job (Foster 1978, in some examples), assuming laryngeals to be spirants. As *h was clearly the aspiration feature, and *s only occurs in PIE (but not Hittite) before a stop, but becomes fused in many other languages (for example, Egyptian *č derives from *st and *sk), *f seemed a good candidate. Internal reconstruction of Egyptian led me to recognize *ʔ as the correct voicing factor. PIE and Egyptian simple voiced stops derive from *ʔ plus stop, and PIE voiced aspirated stops from a combination of *ʔ and *h (which may be labialized or palatalized) plus a stop. In contrast, Egyptian and Japanese stops so derived are still only voiced, not aspirated.

It is interesting that Gamkrelidze (see Bomhard 1984, 26) interpreted PIE voiced stops as glottalized stops primarily on the basis of typological criteria. This is the system used in Soviet Nostratic, as well as by Bomhard (1984) for stops that are voiced (but not aspirated) in PIE (written by Nostratic scholars as p, t, k, but understood as glottalized).

Hittite shows a medial distinction between voicing and lack of voicing by writing voiced consonants with a single letter and voiceless with a double letter. Thus resonants are found to be voiceless as well as voiced medially, voicelessness reflecting an earlier *h plus resonant. Spirants are also subject to medial, but not initial, voicing in Hittite.

Juxtaposition of consonants is reflected variously. When two consonants come together without an intervening vowel nucleus, either fusion occurs and a new phoneme is formed, or else one or the other of the two consonants (sometimes both) is lost. Both are lost in PIE and in Hittite, where voicing of *h (>*∅) results from fusion of a preceding *ʔ with *h.

*h meaning 'trajectory' or 'passage' (see Fig. 25.13 above) is given a wide variety of interpretations. The primary meaning seems to derive from the movement of the sun across the heavens, arching upwards to the zenith from the east, and downwards toward the horizon in the west. Thus, we find *hʷ in a preponderance of words for rising, lighting, burning, going forward, etc. and *hʸ in words for withdrawal, going backwards, going down, etc. This *phememe* proliferated especially in words for daylight and burning. Primary categories of meaning are light/burn, time/day, and go/move. Since Egyptian shows contrast between *h, *hʸ, and *hʷ (as do the Semitic languages)

it is diagnostic. Some examples of uses of *h in Egyptian (E) and Coptic (C) are:

TIME: E: ḥʔ, ḥʔw 'day, time, season' (C: how); hrw 'day'; hrhr 'to prolong' (C. helhol), 'before, a second time, at once, immediately'; ḫʔ-t 'a second of time'

GO/MOVE: E: hh 'to go, to march' (C: hua); hwʔ 'to throw, drive, growth', etc.; ḥʔ 'to go back, retreat'; hwnw 'to escape from'; ḥmm 'to retreat, withdraw'; xb 'to dance, do gymnastic feats'; xr 'to fall'; ʕbʕ bw 'to dance'.

LIGHT/HEAT: E: x-t 'fire, flame, heat, to burn up'; xw 'fire, flame'; xy 'heaven, sky'; hʔ 'to burn, break into flame, heat, fire, warmth'; ḥʔy-t 'sky'; ḥʔmy 'to shine'; ʔxw 'splendour, light, brightness'; ʕn 'to turn a glance toward something, to be beautiful, beauty'; ʕnw-t 'ray of light, beam'; ʕnj 'light, radiance, splendour'; āḫ 'moon'.

As can be seen, possibilities for affixation are multiple, and sometimes, but not always, are found in other languages as well. These examples, except for ʔh fusion (E), are all suffixed. It would seem that the root *phememe* always occurred first, or only preceded by *ʔ.

Egyptian has preserved shorter sequences than are found in many other languages. However, this is usually deceptive, since it masks considerable consonant fusion. Sumerian has also preserved short sequences, but is less useful for comparison than Egyptian and other Afroasiatic languages because it has fewer contrastive phonemes.

Because of the virtually limitless possibilities for affixation (or compounding, as it should probably be considered in the earliest stages), not every stem will be found in every language; but all languages have a fair range, making it possible to isolate cognate sets and determine regularities.

25.8.4 *Toward a theory of the PL stem*

Having established that the PL 'root' is a single *phememe*, we may now use the internal evidence that we have amassed in order to establish stem canons—or whatever constitutes the 'root' in most latter grammars, such as PIE. Following Benveniste (1962), it would appear that the vast majority of PIE root forms, as far as they can be reconstructed, have the pattern C[V]C. Benveniste (ibid., pp. 147–73) analysed the PIE root as consisting of a sequence CVC. In order to do this he built from the laryngeal hypothesis the thesis that one of the missing laryngeals constituted the first or the last consonant where one or the other was missing (for example, *er- 'rise, begin', and dō-'give' had missing first and last consonants respectively). Movement of stress from the first to the second syllable of the word occurred when a suffixed consonant (called *élargissement*) occurred after the root (CVC segment). Often, in such a case, an initial laryngeal would be lost (for example, *areg- (Gk. órégō, Lat. reg-)). He postulated that a verbal theme may have both a suffix and an *élargissement*, but not more than one of each. If the root were full degree (i.e. with a root vowel), the suffix must be zero degree, and no *élargissement* could occur. Thus, it is possible to have per-k and pr-ek-s, but not pr-k-s. Suffixes are distinguished from *élargissements* by the fact that they may have alternative forms, as -et-/-t, while an *élargissement* can only have -t. This was an elegant solution to syllabification in PIE, and has considerable bearing on syllable canons as well as on the loss of contrasts in various languages.

Starting at the earliest stage of incipient language, single *phememes* expressed notions of space and motion, which were very abstract. As time wore on, *phememes* became concatenated. At first this was undoubtedly a slow process, and silences, however short, occurred between segments. Later, as the number of segments in each utterance expanded, the silent spaces diminished, and words that were originally spaced as units came to be made up of more than one segment. Early words contained only one nucleus (i.e. vowel), and could only be as long as production of consonants with a single nucleus could be maintained. As sequences stretched, consonants became either fused or lost. We thus see that many of what Benveniste and other Indo-Europeanists considered to be single stop consonants were actually composed of from one to four segments, only one of them a stop.

In order to ascertain the structure of the early stem it is necessary to discover the *phememe* that expressed the basic meaning and then determine the range of possibility for prefixation and suffixation. The Egyptian *h examples above illustrate this process. After a number of *phememes* have been so analysed the ranges can be compared for similarities, and compared again across languages to discover the common ground.

As sequences expand, it is possible to trace the shifts in form and meaning associated with each successive accretion. For example, Egyptian ʔhy 'to go, to march' (from *ʔahʸay) expanded with *-m to become ʔhm 'to advance'; and again with *-r, but now losing the initial ʔh-, as mʔr 'travellers'. That this loss did not occur in all languages is apparent from Karok (of California), in which we find *ihmara* 'to run', with metathesis (exchange of position) of *h and *y.

More research on problems of affixation and stress in Egyptian and other languages should refine the theory of the permissibility of adding phonemes.

Benveniste's suffixes and *élargissements* were drawn from the pre-existing stock of *phememes*, just as was, ultimately, outer-layer, or inflectional grammar. Because inflection as we know it is both outer and looser, it would seem that more deeply buried *phemic* levels were the original grammar, which gradually became supplanted as older forms became more and more frozen or obscured. It should be possible to recover these inner and early grammatical forms, and to begin to understand the IE difference between a suffix and an *élargissement*.

25.8.5 *Sorting the isoglosses*

Certain important isoglosses are implicit in this presentation of the monogenetic theory. Some languages have what is reconstructable as a dual-stop system (Egyptian, Hittite, Japanese); others have a three-way stop series (reconstructed IE); and others a four-way stop series (Sanskrit). Dravidian and Uralic are examples of the single-stop type, but medially both language families show vestiges of greater complexity, reflecting pre-glottalization and pre-aspiration.

Japanese, Hittite, and Tarascan of Mexico are like Egyptian in their two-stop systematization (although Japanese voiced stops only reflect *ʔ h plus stop), to give just a few examples, as well as like Egyptian (i.e. Coptic) in their vowel development, while Uralic is quite different. Dravidian has similar vowel development, while that of Sumerian is also similar. In Sumerian, the voiced and voiceless stop sets are reversed (i.e. glottalization produced voiceless stops in Sumerian, and stops that are voiced or *lenis* (as Tarascan) in most other languages are voiceless in Sumerian). Basque is like Sumerian in this regard. Reconstructed Malayo-Polynesian, like Egyptian, Japanese, Hittite, and Tarascan, has a two-stop system.

Because most languages seem to show at least traces of greater stop complexity, it must be assumed that stop-fusion took place at a very early stage. It is not necessarily clear that languages with simple systems lost their complexity through fusion; it may never have developed in cases where spirants and glides vanished rather than becoming fused. This can only be determined if significant numbers of isoglosses are found associated with one another in particular languages or language groups.

These isoglossic suggestions must be validated by many more isoglosses if they are to hold as indicative of genetic family-tree branchings; but many are obviously very old. However, it would be premature to construct family trees.

Establishment of isoglosses is the most important task ahead. Assumption of monogenesis yields no insight as to degree of relationship unless the relative time of shared changes can be demonstrated. This is a time-consuming job; but it can be done, and its importance cannot be overestimated.

At this point, too little has been done to make it certain that either Nostratic or Matteson's (1972) Amerind are valid sub-groupings. I think it is doubtful that they are, for the above reasons. On the other hand, I think that Brunner (1982), Bomhard (1984), and Levin (1971) were on the right track in grouping Afroasiatic and Indo-European together, and Brunner in including Malayo-Polynesian in this package.

Since all languages can be assumed to be related, linguists can find and explore the relationships between the languages with which they are most familiar. Thus I have been interested in how Tarascan (Foster 1969) and the Mixe-Zoque languages (Foster and Foster 1948) fit into this picture, and always look for their closest relationships. Those of Tarascan are unmistakably of the Afroasiatic, IE, and Japanese package. Mixe-Zoque shows close affiliation with Egyptian.

In searching for isoglosses, semantic closeness of long reconstructed sequences is an excellent source. Such identity of prefixation and suffixation cannot be found if connections between languages are very remote. As examples 4 and 5 in Table 25.1, and examples 6, 7, and 9 in Table 25.2 indicate, the semantics of PIE and J sequences are often very similar. Verbs, and especially CVC verbal stems (usually thought of as roots) are more apt to reconstruct than are nouns; while reconstruction of grammar is even less likely at remote time-depths.

25.9 Wider implications of monogenetic reconstruction

25.9.1 *The role of analogy*

The most striking implication to emerge from this study is the crucial role of analogy of articulation and meaning in language origin. What this means in human evolution is that the potential of classification (i.e. finding and using similarities productively) increases markedly as we move up the evolutionary ladder. Thus, primate behaviour, as well as the behaviour of early hominids that can be deduced from their artificial remains, should be carefully explored with this in mind. Many animals have long memories and easily make associations. The Great Apes have demonstrated an interesting capacity for memorization and simple association in their 'language' training.

Evidence of this kind is indirect confirmation of the analogical beginnings of language. I believe that it will be found that both early markings of the kind discussed by Marshack (1972), and the abstract markings on Palaeolithic cave walls, are associated with early linguistic analogies (see Foster 1990). It seems very likely to me that the meanders described by Marshack (op. cit.), which are a fundamental part of early cave markings, represented the *phememe* *h 'transitory' or 'continuity', for example, and that the turns and bends, ups and downs, have an association with the very prevalent *hy and *hw sequences found in PL reconstructions of words for water and fire. Examination of early alphabets and their similarities to palaeolithic abstract signs and the underlying analogical base of the Chinese writing system are other avenues to be explored.

25.9.2 *Internal and external analogy*

J. Sapir's (1977) discussion of metaphor as a form of analogy makes distinctions that are useful in considering PL reconstruction and its meaning for human evolution. Sapir speaks of two kinds of analogy: internal and external. Internal analogy or metaphor (expressing analogies in implicit form through language) puts two terms into juxtaposition in such a way that their samenesses are revealed through their differences, neither of which are made explicit. Thus, if I say that Jane is a social butterfly, the implicit meaning stems from some characteristic that Jane and the insect share: they move (shift ground, flit) from one thing (person, flower) to another, not remaining long with any. This kind of metaphoric usage becomes total transposition in mythology, where statements are made about fantasized characters whose traits and actions coincide in some unstated way with real human situations or problems to which the myth through analogical indirection applies solutions.

In Primordial Language, from the very beginning, different activities were treated as if they were the same. Common ground was discovered which made it possible to encompass a variety of meanings within a single term. A *phememe* was a form of metaphor, or internal analogy, implicitly linking two terms which were different but the same in some semantic or phonological features.

The second kind of analogy (or metaphor) discussed by Sapir interconnects not just two terms, but four: a is to b as x is to y or, formulaically, a:b::x:y. In terms of oral articulation: p:m::t:n. Or, expressed in corollary semantic terms in accordance with the *phememes*: projecting: enclosing::penetrating: enclosed. Sapir calls this external analogy (or metaphor). It entered into oral communication the minute the first four *phememes* were produced.

In terms of language evolution, the first *allophemic* split (i.e., creation of two *phememes* from a single *phememe* with two *allophemes*) moved language from a schema of internal analogy to one of external analogy. This would seem to be the single most important cognitive development in hominid history. According to J. Sapir (1977, 28) 'Once an external metaphor is set out as an opposition, further metaphors can be developed from the original simply by adding to or slightly altering the criteria of opposition.' And further (ibid., 31), 'The more remote the terms are from each other... the greater are the possibilities of making a variety of non-arbitrary connections.'

This is surely the cognitive Rubicon whose crossing gave rise to what may be truly called 'symbolic culture'. It set in motion a powerful chain-reaction that changed the world.

I believe the first *allophemic* split to have been the conversion of *N (any nasal) into the two *phememes* *m and *n, and the conversion of *P (any lip or tongue protrusion) into the two *phememes* *p and *t. This allowed two external analogies to arise: *p:m::*t:n, and *p:t::*m:n. In terms of the semantics of the *phememe* chart (Fig. 25.13 above) this can be restated as: externalization is to projection as internalization is to introjection, and, projection is to introjection as externalization is to internalization. As new *phememes* were added and in their turn underwent *allophemic* splitting, the possibilities for external analogizing increased steadily, and human cognitive horizons expanded.

Phememes in closest proximity on the chart (see Fig. 25.13 above) share the greatest number of both semantic and phonological features. As splitting occurred and new *phememes* began to occupy spaces in the back as well as the front of the mouth, new semantic possibilities became available. A great deal of polysemy was possible, and often made for semantic overlap between *phememes*. Inevitably semantic change also sometimes makes it difficult to determine with any certainty the original *phememic* semantics. For this reason I have modified my charts from time to time (Foster 1978, 1980, 1983), and believe the one presented here more closely mirrors the original schema, which was in place during the Upper Palaeolithic with the advent of *Homo sapiens sapiens*.

25.9.3 *The complexity of modern 'roots'*

I have been struck with the fact that unmarked phonemes are much less subject to change than marked phonemes. In most languages, stops and most resonant *phememes* are still recognizable, but a great deal of work is needed on the problems of *phememe* fusion which created marked phonemes.

I have also been struck with the fact that, despite the obvious simplicity of the early *phememic* system, the search for means of making communication less abstract has led to construction of underlying sequences that are much more complex than what have been assumed to be the roots of each language. Thus, Campbell and Mithun's (1979) eighth guideline holds true after all, that length of reconstruction is important, and can be achieved even at very early levels.

25.9.4 *Advantages of inclusion of dissimilar languages*

As I have worked with the theory, I have constantly found possible refinements for it and new avenues for exploration. Each language that I have worked with has revealed some conservative aspects and some drastically changed features. Thus, each language is helpful from a different standpoint. As I have found languages that share isoglosses I have worked to include languages in the study that are quite obviously different. While one can see almost at a glance, on the basis of previous experience, that any given language will turn out to fit the model, it often takes a great deal of work to determine in just what way this is so. This is because lexemes of similar sound and meaning can turn out to take many different forms, as can be seen in the examples that I have provided with underlying *h. It is never easy to sort out this kind of grouping, but earlier experience with possibilities facilitates the matching.

It will take many years of effort and many trained workers before the sorting of the world's languages into sub-groupings can be brought to any kind of conclusion. But it is an exciting task, full of constant 'aha' rewards—one of the world's most intriguing puzzles.

25.10 Systematic correlations in human evolution

25.10.1 *The beginning*

Sometime before three million years ago, hominoids began walking upright. The earliest ancestral hominid forms have been found in Africa and given the name *Australopithecus*. Early australopithecines apparently left no stone tools, but it can be assumed that their mimetic skills and fabrication of simple vegetal tools surpassed those of the Great Apes, and perhaps passed from the *ad hoc* to involve some notable degree of planning.

Bipedalism appears to have preceded crude stone-tool manufacture. *Homo habilis*, with whose remains the stone tools are temporally associated, initiated the genus *Homo*. Remnants of *Homo habilis*' tool-making have been found in East Africa. From this evidence Isaac (1978*a*, *b*) postulated a social model for *Homo habilis* that included a home base to which stones for tool-manufacture were brought and to which butchered pieces of large mammals were brought back for sharing. This model carries the implication of wide-ranging, mammal-butchering males and less mobile females who gathered vegetal foods and cared for infants in and around the home base, and all the co-operation that such a division of labour implies (see White, this volume, Chapter 9, for criticism of this model).

Division of labour on gender lines was the first evidence of the conceptual binarism that underlies all phonetic systems and must be assumed to have underlain the start of the conventionalization of meaning associated with vocal sound-production. Just as tools were at first crude—choppers and flakes made from striking a rock with a hammerstone—so must have been the earlist beginnings of spoken language: visibly apprehended movements made toward the front of the mouth: sharp sounds, [p] and [t] (probably with protruded lips and tongue respectively), and more internally produced soft, or nasal, sounds, [m] and [n]. The sharp sounds mimicked striking, outward motion, and thrust. The soft sounds mimicked internally satisfying events such as eating, sexual gratification, and embracing. Jakobson and Halle (1962) and Jakobson (1962) laid the groundwork for such a theory of early sounds, both in ontogeny and phylogeny, and Foster (1981) correlated it with Isaac's now controversial home-base model.

Theoretically, in the beginning, [m] and [n] were *allophemes* of the same *phememe* with the abstract meaning of 'internal, inward, enclosure' which also encompassed such semantic variations as 'vagina, mother, mouth, holding, food, food extraction'. By the same token [p] and [t] were *allophemes* of a *phememe* meaning 'outside, outward, thrust' and encompassing such variations as 'striking, giving, penis erection and intromission, ejaculation, stone tool, bone'.

This period of human development seems to have lasted for at least half a million years with no apparent change before a second stage is visible in the archaeological record. Changes came slowly at first and gradually accelerated. Each stage in technological advancement finds correlates with increased cognition and associated body changes, until *c*.35 000 years BP, after which hominid physical development seemed to stabilize while technological and social complexity continued to grow. The seven stages outlined below carry the evolution of humankind into the dawn of history. Each consists of assumptions about language development that can be postulated from monogenetic reconstruction.

25.10.2 *Hypothetical evolutionary stages*[3]

I. Palaeolithic period

C_1 (communication): body mimesis, imitative learning; *Australopithecus*.

C_2: Rudimentary visual–vocal communicative mimesis using lips, tongue, and open-mouth vocalization; *Homo*.

Phememes:
*M = [m] [n]: 'inward, containing, enclosing, rounded, female'
*T = [p] [t]: 'outward, thrusting, projecting, linear, male'
Further analogical semantic correlations:
*M with 'home base', mothering, gestation, eating, feeding, fullness, sexual satisfaction, pleasure, receiving.
*T with distance, fathers, male erection and introjection, striking, butchering, tool-manufacture, bringing.

C_3: Increase in communicative need over distance leading to *allophemic* splitting and creation of new *phememes* symmetrically organized, addition of vocalic differentiation through lip-shaping and tongue-raising and lowering. *Pa* structure (*P* = phememe, *a* = undifferentiated vocalism)

Phememes:

p	t
from	toward
m	n
bounding	bounded

Beginning of external metaphor:
*p is to *t as *m is to *n (with a variety of semantic interpretations carried through analogy).

Analogical correlations:
p: going, giving, penis erection, pointing (arm/hand), killing, stone as weapon, fire
t: butchering, tool manufacture, flaked tool for butchering, copulation, entering
m: mouth, vagina, eating, suckling, grasping/holding (arm/hand), shelter, receptivity, nurture, mothering
n: pregnancy, gestation, food contained, hearth, nurtured, mothered, blood.

C_4: Beginning of CVC or CVCV concatenation (theoretically equated with hafting).

Phememes:
Consonantal

p	t
projection	introjection
m	n
externalization	internalization

Semi-vocalic

w(u)	y(i)
3-dimensional	2-dimensional

h(a)
1-dimensional
(= beginning, ending, potentiality)

External metaphor: extended through equation of beginning and ending as revealed by burial, and other, perhaps incipient, reversal transformations.

Analogical correlations: creation of new *phememes* necessitates limiting the semantic area occupied by the old. Thus, *w takes over some of the semantic territory previously occupied by *m and *n, and *y some of that previously occupied by *p and *t. Older forms are retained with same meanings as before, while new are added to provide greater precision. This makes for a degree of semantic overlap.

C_5: Expansion of *phememic* inventory to include alveolars, velars, and fricatives, increasing concatenation of *phememes* to create more precise meanings. *Phememic* inventory completed (see Fig. 25.13) during the Upper Palaeolithic.

External metaphor correlations: Increasing ability to perceive and act upon samenesses and differences correlates with increase in cultural productivity, inventiveness, and subtle differentiation; maximum potential for analogical *phememic* development correlates with transfer of analogizing potential to creating of morphological classes and hierarchies through concatenation rules; social practices increasingly rule-dominated.

II. Neolithic period, *c*.8000–3000 BC

C_6: Syllabic concatenation with morphophonemic merging increasingly obscuring analogical significance of original *phememes*, which gradually became what we now know as 'phonemes' (no longer analogically motivated). Birth of grammar through morphological synthesis and development of voiced/voiceless stop differentiation or other kinds of marking through phonemic synthesis; increasing dialectal differentiation leading to loss of mutual intelligibility.

Analogical correlations: This was a period of consolidation coupled with experimentation with new uses of metaphor. Language, like other cultural systems, became both more abstract in the sense of the obscuring of its systemic foundations and more concrete in its ability to express finer semantic distinctions. Social hierarchy was extended to include deities.

III. From the Bronze Age: the beginning of history *c*.3 000 BC

C_7: Language as we know it from early documents was similar in structure to modern languages. It must, therefore, be assumed that structural consolidation of hierarchical systems in all their modern complexity was completed before the dawn of history.

At that time mythology showed an advanced use of metaphor, as did astrology. These were religious manifestations used to validate the social status quo with its hierarchy of privilege, at least in major centres of power. Deities were invented as a metaphorical hierarchical extension of social classes.

Syntax has, in languages associated with social hierarchy, increasingly replaced morphology as a grammatical device. With the growth of technology, greater overt emphasis has been placed on science and the overt logic of cause and effect rather than on the covert logic of metaphorical equation. Metaphor continues to guide and inspire practice as it always has, but this is hidden because devalued. Thus, cognitive activities grounded in languages have become more lineal. This equates with increased male social dominance in the older, metaphoric, sense of a male equation with lineality, a female with circularity.

25.11 Language in culture

The point at which human behaviour began to be guided by external metaphor marked the true beginning of culture. It is these external metaphors and analogies that constitute system in both language and culture. This is not a conscious activity, but is the stuff of everyday life. Although it is far beneath the surface of cognition, it profoundly influences both the reception of experience and the individual's reaction to that experience.

Because of the exploitation of external metaphor and the kinds of transformations that it encourages, culture both came into being and has constantly expanded and changed. Since the connections made through external analogy are non-arbitrary, it is apparent that language and culture are non-arbitrary (i.e. symbolic) systems.

Throughout the Palaeolithic, as external metaphor was exploited in novel ways, culture, including language, grew and expanded. It would seem that technology has changed more and grown more than has language during the past 5 000 years at least. It is not clear what this will mean for humankind if our current runaway technology allows for survival of life on the planet. Visual communicative technologies have been expanding and exploiting symbolic potential in new ways. These may supplant spoken language and make for new methods of adaptation. While this would seem a severe loss from the point of view of dedicated language-users of today, it may have the potential to become a more powerful symbolic tool than spoken language has proved to be, thus releasing new springs of cognitive development.

Notes

1. An asterisk indicates a reconstructed form. / / indicates a phonemic construction in a particular language.

2. That discussion of the mimetic origin of phonemes or roots has a history, albeit intermittent, is revealed in discussions by Paget (1951) and Whorf (Carroll, ed., 1956, pp. 74–6). Paget discovered the earliest such suggestion in the 1792 work of M. De Kempelen, *Le méchanisme de la Parôle*. Whorf was impressed by a very similar hypothesis, however differently derived, in the 1815–16 work of Antoine Fabre d'Olivet, *La langue hébraïque restituée*. He believed this work to show great psychological penetration, and insight far in advance of its time. De Kempelen derived his hypothesis of the original meaningfulness of linguistic sounds primarily from the lip-reading abilities of the deaf, while Fabre d'Olivet drew his from internal reconstruction of Hebrew, which, according to Whorf gave it a strictly linguistic (i.e. formal and systematic) validity. Both works antedated the comparative method, so comparative material was limited. Carroll, in his introduction (pp. 8–9, 11–13, 16, 74–6), expands on the influence of Fabre d'Olivet on Whorf's thinking. Whorf called a type of language structure in which all, or nearly all of the vocabulary stems from a very small number of significant elements combining sound and meaning 'oligosynthesis'. For him, Nahuatl (Ancient Aztec) exhibited such a structure. He considered this structure to be ancestral to the historic language, and possibly to be indicative of an original common language (pp. 12–13).

3. For an earlier hypothesis of stages of language evolution see Foster 1990*b*, pp. 30–4.

References

Anttila, R. (1972). *An introduction to historical and comparative linguistics*. Macmillan, New York.

Anttila, R. (1977). *Analogy*, Trends in linguistics state-of-the-art reports, 10. Mouton, The Hague.

Benveniste, E. (1962). *Origines de la formation des mots en Indo-Européen*. Adrien-Maisonneuve, Paris.

Birnbaum, H. (1977). *Linguistic reconstruction: its potentials and limitations in new perspective*, Journal of Indo-European Studies, Monograph No. 2. Institute for the Study of Man, Washington DC.

Bomhard, A. R. (1984). *Toward Proto-Nostratic: a new approach to the comparison of Proto-Indo-European and Proto-Afroasiatic*, Amsterdam Studies in the Theory and History of Linguistic Science, 27. John Benjamins, Amsterdam and Philadelphia.

Bright, W. (ed.) (1992). *International encyclopedia of linguistics*, 4 vols. Oxford University Press.

Brunner, L. (1982). *Malayo-Polynesian vocabulary with Semitic and Indo-European roots*, The Epigraphic Society, Occasional Publications 10, Part 2. The Society, San Diego.

Buck, C. D. (1949). *A dictionary of selected synonyms in the principal Indo-European languages*. University of Chicago Press.

Burrow, T. and Emeneau, M. B. (1961). *Dravidian etymological dictionary*. Oxford University Press.

Bynon, J. and Bynon, T. (eds) (1975). *Hamito-Semitica*. Mouton, The Hague.

Bynon, T. (1977). *Historical linguistics*. Cambridge University Press.

Campbell, L. and Mithun, M. (eds) (1979). *The languages of Native America: historical and comparative assessment*. University of Texas Press, Austin.

Carroll, J. B. (ed.) (1956). *Language, thought, and reality: selected writings of Benjamin Lee Whorf*. MIT Press, Cambridge, Mass.

Cohen, D. (1970, 1976). *Dictionnaire des racines sémitiques. Fascicule 1: /H–TN; Fascicule 2: TN–GLGL*. Mouton, Paris.

Collinder, B. (1955). *Fenno-Ugric vocabulary: an etymological dictionary of the Uralic language*. Almquist and Wiksell, Stockholm.

Collinder, B. (1960). *Comparative grammar of the Uralic languages*. Almquist and Wiksell, Stockholm.

Collinder, B. (1965). *An introduction to the Uralic languages*. University of California Press, Berkeley and Los Angeles.

Comrie, B. (1992). Languages of the world. In *International encyclopedia of linguistics* (ed. W. Bright), Vol. 2, pp. 305–10. Oxford University Press.

Dahl, O. C. (1973). *Proto-Austronesian*, Scandinavian Institute of Asian Studies Monograph Series, 15. Studentlitteratur, Lund.

Dempwolff, O. (1934–8). Vergleichende Lautlehre des Austronesischen Wortschatzes, 3 vols. *Beihefte zur Zeitschrift für Eingeborenen-Sprachen*, **XV, XVII, XIX**. Berlin.

Diakonoff, I. M. (1974). *Semito-Hamitic languages*. Nauka, Moscow.

Diver, W. (1959). Palatal quality and vocalic length in Indo-European. *Word*, **15**, 185–93.

Elbert, S. (1953). Internal relationships of Polynesian languages and dialects. *Southwestern Journal of Anthropology*, **9**, 147–73.

Foster, M. L. (1969). *The Tarascan language*, University of California Publications in Linguistics 56. University of California Press, Berkeley.

Foster, M. L. (1978). The symbolic structure of primordial language. In *Human evolution*, Vol. IV: *Biosocial perspectives I: Perspectives on human evolution* (ed. S. L. Washburn and E. R. McCown). Benjamin/Cummings, Menlo Park, California.

Foster, M. L. (1980). The growth of symbolism in culture. In *Symbol as sense* (ed. Mary LeCron Foster and Stanley H. Brandes), pp. 371–97. Academic Press, New York.

Foster, M. L. (1981). Culture as metaphor: A new look at language and culture in the Pleistocene. *Kroeber Anthropological Society Papers*, **59** and **60**, 1–12.

Foster, M. L. (1983). Solving the insoluble: language genetics today. In *Glossogenetics, the origin and evolution of language* (ed. Eric de Grolier), pp. 455–80. Harwood Academic Press, Paris.

Foster, M. L. (1990a). The birth and life of signs. In *The life of symbols* (ed. Mary LeCron Foster and Lucy Jane Botscharow). Westview Press, Boulder, Colorado.

Foster, M. L. (1990b). Symbolic origins and transitions in the Palaeolithic. In *The emergence of modern humans* (ed. Paul Mellors). Edinburgh University Press.

Foster, M. L. (1992). Body process in the evolution of language. In *Giving the body its due* (ed. Maxine Sheets-Johnstone). SUNY Press, Albany, New York.

Foster, M. L., and Foster, G. M. (1948). *Sierra Populuca speech*, Smithsonian Institution of Social Anthropology Publication, No. 8. US Government Printing Office, Washington, DC.

Gostony, C. G. (1975). *Dictionnaire d'etymologie Sumerienne et grammaire comparée*. Editions E. de Boccard, Paris.

Grace, G. W. (1959). *The position of the Polynesian languages within the Austronesian (Malayo-Polynesian) language family*, Indiana University publications in anthropology and linguistics, Memoir 16 of the International Journal of American Linguistics. Indiana University Press, Bloomington, Indiana.

Gray, L. H. (1971). *Introduction to Semitic comparative linguistics*. Philo Press, Amsterdam.

Greenberg, J. H. (1960). The general classification of Central and South American languages. In, pp. 791–4.

Greenberg, J. H. (ed.) (1963). *Universals of language*. MIT Press, Cambridge, Massachusetts.

Greenberg, J. H. (1966a). Language universals. In *Current trends in linguistics, 3: Theoretical foundations* (ed. Thomas A. Sebeok). Mouton, The Hague and Paris.

Greenberg, J. H. (1966b). *The languages of Africa*. Indiana University Press, Bloomington, Indiana; Mouton, The Hague.

Greenberg, J. H. (ed.) (1978). *Universals of human language*, 4 vols. Stanford University Press.

Greenberg, J. H. (1987). *Language in the Americas*. Stanford University Press.

Grimm, J. L. C. (1852). Über den Ursprung der Sprache. *Abhandlungen der Königlichen Akademie der Wissenschaften*, Berlin, vom Jahr 1851.

Haas, M. (1969). *The prehistory of languages*. Mouton, The Hague and Paris.

Hockett, C. F. (1976). The reconstruction of Proto Central Pacific. *Anthropological linguistics*, **18**(5), 187–235.

Illič-Svityč, V. M. (1971). *Opyt sravnenija nostraticeskix jazykov (semitoxamitskij, kartvelskij, indoevropejskij, ural'skij, dravidiskij, altajskij): Sravnitel'nyj slovar'* (b–k). Nauka, Moscow.

Illič-Svityč, V. M. (1976). *Opyt sravnenija nostraticeskix jazykov (semitoxamitskij, kartvelskij, indoevropejskij, ural'skij, dravidiskij, altajskij): Sravnitel'nyj slovar'* (1–ž). Nauka, Moscow.

Illič-Svityč, V. M. (1984). *Opyt sravnenija nostraticeskix jazykov (semitoxamitskij, kartvelskij, indoevropejskij, ural'skij, dravidiskij, altajskij): Sravnitel'nyj slovar'* (p–q). Nauka, Moscow.

Isaac, G. Ll. (1978a). Food sharing and human evolution: Archaeological evidence from the Plio-Pleistocene of East Africa. *Journal of Anthropological Research*, **34**, 311–25.

Isaac, G. Ll. (1978b). The food-sharing behavior of proto-human hominids. *Scientific American*, **238**(4), 90–108.

Jakobson, R. (1958). Typological studies and their contribution to historical linguistics. In *Proceedings of the Eighth International Congress of Linguists* (ed. Eva Sivertsen), pp. 17–35. University Press, Oslo.

Jakobson, R. (1962). *Selected writings*. Mouton, 's-Gravenshage.

Jakobson, R. and Halle, M. (1962). Phonology and phonetics. In *Selected writings*. Mouton, 's-Gravenshage.

Key, M. R. (1978). Araucanian genetic relationships. *International Journal of American Linguistics*, **44**(4), 280–93.

Key, M. R. (1979). *The grouping of South American Indian languages*. Gunter Narr, Tübingen.

Key, M. R. (1980–1). South American relationships with North American Indian languages. In *Homenaje a Ambrosio Rabanales*, Boletín de Filologia 31, pp. 331–50. Universidad de Chile, Santiago, Chile.

Key, M. R. (1981). *Intercontinental linguistic connections*. University of California, Irvine, California.

Key, M. R. (1984). *Polynesians and American linguistic connections*. Jupiter Press, Lake Bluff, Illinois.

Kroeber, A. L. (1964). On taxonomy of languages and cultures. In *Language in culture and society* (ed. Dell Hymes), pp. 654–63. Harper and Row, New York, Evanston, London.

Krupa, V. (1982). *The Polynesian languages: a guide*, Languages of Asia and Africa, Vol. 4. Routledge and Kegan Paul, London.

Kurylowicz, J. (1927). Indo-européen et Hittite. In *Symbolae grammaticae in honorem Ioannis Rozwadowski*, Vol. 1, pp. 95–104. Cracow.

Kurylowicz, J. and Mayrhofer, M. (eds) (1968–86). *Indogermanische Grammatik*, 3 vols. Winter, Heidelberg.

Lahovary, N. (1963). *Dravidian origins and the west*. Orient Longman, Bombay.

Landar, H. (1977). South and central American Indian languages. In *Native languages of the Americas* (ed. Thomas A. Sebeok), pp. 401–527. Plenum, New York.

Lehmann, W. P. (1955). *Proto-Indo-European phonology*. University of Texas Press, Austin.

Levin, S. (1971). *The Indo-European and Semitic languages*. State University of New York Press, Albany, New York.

Longacre, R. E. (1977). Comparative reconstruction of indigenous languages. In *Native languages of the Americas* (ed. Thomas A. Sebeok), pp. 99–139. Plenum, New York.

McAlpin, D. W. (1981). *Proto-Elamo-Dravidian: the evidence and its implications*, Transactions of the American Philosophical Society, 71, part 3. The American Philosophical Society, Philadelphia.

Marshack, A. (1972). Cognitive aspects of Upper Paleolithic engraving. *Current Anthropology*, **13**, 445–77.

Martin, S. E. (1966). Lexical evidence relating Korean to Japanese. *Language*, **42**, 185–251.

Martinet, A. (1953). Non-apophonic o-vocalism in Indo-European. *Word*, **9**, 253–7.

Matteson, E. (1972). Toward Proto Amerindian. In *Comparative studies in Amerindian languages* (ed. Esther Matteson et al.), Janua Linguarum, Series Practica, 127, pp. 21–89. Mouton, The Hague and Paris.

Meillet, A. (1964). *Introduction a l'étude comparative des langues Indo-Européennes*. University of Alabama Press, University, Alabama.

Meillet, A. and Cohen, M. (1952). *Les langues du monde*. Centre National de la Recherche Scientifique, Paris.

Miller, R. A. (1971). *Japanese and other Altaic languages*. University of Chicago Press.

Moscati, S., Spitaler, A., Ullendorff, E., and von Soden, W. (1969). *An introduction to the comparative grammar of the Semitic languages: phonology and morphology*. Otto Harrassowitz, Wiesbaden.

Newman, S. (1944). *Yokuts language of California*, Viking Fund Publications in Anthropology, No. 2. Viking Fund, New York.

Olmsted, D. L. (1966). *Achumawi dictionary*, University of California Publications in Linguistics 45. University of California Press, Berkeley.

Paget, R. A. S. (1951). *The origin of language*, Science News 20. Penguin, Harmondsworth.

Pedersen, H. (1903). Türkische Lautgesetze. *Zeitschrift der Deutschen Morgenlandischen Gesellschaft*, **57**, 535–61.

Pokorny, J. (1959). *Indogermanisches etymologisches Wörterbuch*, 2 vols. Francke Verlag, Bern and Munich.

Poppe, N. (1965). *Introduction to Altaic languages*, Ural Altaische Bibliothek. Otto Harrassowitz, Wiesbaden.

Pukui, M. K. and Elbert, S. H. (1957). *Hawaiian–English dictionary*. University of Hawaii Press, Honolulu.

Powell, J. W. (1891). *Linguistic families of North America north of Mexico*. Bureau of American Ethnology Reports, 7, 1–142.

Radin, P. (1919). The genetic relationship of the North American Indian languages. *Publications in American Archaeology and Ethnology*, **14**(5), 489–505. University of California Press, Berkeley.

Ruhlen, M. (1987). *A guide to the world's languages*. Vol. 1, *Classification*. Stanford University Press.

Sapir, E. (1929). Central and North American languages. *Encyclopedia Britannica*, 14th edn, **5**, 138–41.

Sapir, J. D. (1977). The anatomy of metaphor. In *The social use of metaphor: essays on the anthropology of rhetoric* (ed. J. David Sapir and J. Christopher Crocker), pp. 3–32. University of Pennsylvania Press, Philadelphia.

Saussure, F. de (1879). *Mémoire sur le système primitif des voyelles dans les langues indo- européennes*. Teubner, Leipzig.

Saussure, F. de (1959). *Course in general linguistics*. Philosophical Library, New York.

Schmidt, J. (1872). *Die Verwandtschaftsverhältnisse der Indogermanischen Sprachen*. Bohlau, Weimar.

Sebeok, T. A. (ed.) (1977). *Native languages of the Americas*. Plenum, New York.

Shevoroshkin, V. and Ramer, A. M. (1991). Some recent work on the remote relations of language. In *Sprung from some common source: investigations into the prehistory of languages* (ed. S. M. Lamb and E.D. Mitchell), pp. 178–99. Stanford University Press.

Swadesh, M. (1954). Perspectives and problems of Amerindian comparative linguistics. *Word*, **10**, 306–32.

Swadesh, M. (1971). *The origin and diversification of language*. Aldine–Atherton, Chicago.

Tregear, E. (1969). *The Maori–Polynesian comparative dictionary*. Anthropological Publications, Oosterhout, The Netherlands.

Verner, K. (1875). Eine Ausnahme der ersten Lautverschiebung. *Zeitschrift für vergleichende Sprachforschung auf dem Gebiete der indogermanishchen Sprachen*, **23**, 97–130.

Voegelin, C. F. and Voegelin, F. M. (1964). Languages of the world: Sino-Tibetan, Fascicle One. *Anthropological linguistics*, **6**(3) 1–109. University of Indiana Press, Bloomington, Indiana.

Voegelin, C. F. and Voegelin, F. M. (1965). Languages of the world: Boreo-Oriental, Fascicle One. *Anthropological linguistics*, **7**(1), 1–143. University of Indiana Press, Bloomington, Indiana.

Voegelin, C.F. and Voegelin, F. M. (1977). *Classification and index of the world's languages*. Elsevier, New York.

Watkins, C. (1985). *The American Heritage Dictionary of Indo-European roofs* (2nd edn). Houghton Miffin, Boston.

Winter, Werner (ed.) (1965). *Evidence for laryngeals*. Mouton, The Hague.

26
Theoretical stages in the prehistory of grammar

Leonard Rolfe

Abstract

This chapter conjectures that the hierarchical structure of present-day grammars might be the result of an evolutionary process. Grammar is taken to be a communicative device patterned to cater for various communicative intentions such as asking questions, making statements, and expressing comments. These intentions are themselves elaborated in the course of an evolving dialogic system. Each communicative intention might be thought of as having a corresponding pattern for its expression. If so, this would lead to a non-integrated grammatical system. Hence the notion of 'recency dominance' is introduced here, whereby a newly-emerged pattern becomes dominant and 'reworks' older patterns into conformity with it. Eight stages of elaboration are proposed. These are probably not discrete; rather, language evolution should be viewed as proceeding in a more mosaic pattern.

The first stage—necessarily without any prior pattern to build on—concerns getting in touch with another, the intention to 'address'. Address is motivated by the seeking of some obligement: termed here 'solicitation'. Granting these obligements constitutes 'compliance'. Finally, a 'close-out' indicates the exchange is finished. This sequence constitutes a 'frame of dialogue' that represents the source of later grammatical functions, which become necessary to handle communicative functions as they become more elaborated—vocative from address; imperative and interrogative moods of the verb from solicitation; affirmative (and negative) from compliance (and refusal); and various markers for turn-taking. This frame also contributes to the provision of first and second persons of the verb when they are later grammaticalized.

A minor complexification of these abilities yields the next stage, ostension—pointing at a visible item with one's index finger. Ostension has three important facets: it is for another (and is hence situated in the earlier dialogic frame); it implies the addressee understands what is being pointed at; and it is oriented on the speaker—that, is, it is 'deictic'. Ostension primarily concerns visible items, and distinguishes between those within and beyond reach, but can be extended to indicate non-visible phenomena. In this instance, however, ostension can no longer be contextually supported, but needs a new form, 'identification', which is secured by 'naming'. This builds on ostention, but extends it to constitute a new stage.

Identification can be secured by gestural imitation of shape or activity. Vocal units may have taken the first step towards being words by reformulating some of this gestural inventory, or by iconic representation. The notion of referential symbolism via words may not have been fully perceived at their creation, and they may well have arisen and been worked into the communicative system in an *ad hoc* manner. One method of securing referential symbolism is 'thematization' and this is proposed as the next stage.

'Thematic roots' are an organizational principle exploited in some language systems whereby a set of phonological alternances around a vowel allow the expression of particular aspects of the notion contained in the root. Verbs and nouns are not distinguished, but are implicit in the particular semantic features expressed by members of a theme. Thematic clusters constitute a semantic route towards grammar, in that their semantic features are sorted out in modern languages into the basic noun/verb distinction and grammaticalized into syntax.

Topic–comment structures are proposed as the first stage in the transition toward syntax. Initial topic–comment structures are asyntactic, since their relations are purely pragmatically based. These motivate the possibility of expressing third-person action, and the realization of this enables a speaker to depict scenes, which itself leads to a forefronting of implicit case relations, and moves topic/comment structures toward the realm of 'narrative' which requires considerable grammatical support for its effective handling. Narrativity begins to shift the relating of events from the intersubjective realm of the dialogic participants towards the objectification of events, yielding an 'epistemic' patterning of discourse, and motivating syntactic devices that handle the hitherto implicit features of narrative. This last, epistemic, stage may be relatively recent, and characteristic of a level of social organization that produces the state [eds].

26.1 Introduction

The origin of human language has been considered a worthy subject for enquiry since at least the days of the Classical philosophers. Modern theory on the subject takes its impetus from the new scientism of Descartes and Vico at the turn of the seventeenth century (Stam 1976). For Descartes, language was not developmental, but an innate characteristic of being human. By contrast, Vico envisaged a prehistorical development from some putative point of origin. These—and ensuing—formulations lacked the underpinning of a scientific apparatus; but in the nineteenth century comparative philology became established. The new science made the reconstruction of 'parent' languages possible; for the first time language development had to some extent a prehistory.

Modern reconstructional research, seeking a common parent for all languages, gets no farther back than about 6000–7000 BC (see Section 26.5 below). One element of these reconstructions is the notion of sound symbolism, i.e. that the sounds of a word evoke its meaning. This notion has been exploited in various ways (see for example Fónagy 1983; Key 1985) to reach back further, on reductionist principles, to a possible primordial form of language in which the speech sounds, rather than evoking objects and events, were psychologically associated with their qualities, such as 'hard' or 'quick'. In this way objects and events could be identified by association with their readily observable qualities (see Section 26.4 below).

The nineteenth century also produced the notion of biological evolution, as propounded by Darwin. According to this notion all modern forms must be developments from and adaptations of previous forms; and 'development' implies a progress from the primordial and simple to the extant, elaborated form. This notion is a major underlying assumption in this chapter as indeed, I take it, it is of this entire *Handbook of human symbolic evolution*. Given this perspective, nineteenth-century enquirers into language origins proposed various sources for language (for example Müller's (1871) 'bow-wow' theory) that might be deemed to arise naturally out of pre-existing ecological and biological motivations.

A difficulty for these theories was how to proceed developmentally from such motivations to modern language, particularly as the Darwinian model required gradualism, i.e. very slight, often imperceptible modifications having a cumulative effect over a long time-scale. How could gradualism operate? Language today is so complex and yet so integrated in all its parts that should one apparently minor element of it be taken away, the fabric collapses. Despite the implication of Darwinism that there is no special creationist niche for the human race, both the complexity and the integrality of modern language do plausibly suggest a 'one-off' creation, particularly as some of its design features (Hockett 1960) are unique as compared with other animal communicative systems.

Perhaps because of these considerations current linguistic thinking (for example Chomsky 1972) tends to adopt an 'innateness' explanation for the origin of language, such that at some point a gross mutational process suddenly produced modern humans out of an anthropoid setting. This 'punctuation' endows modern humans with the complete neural capacity for modern language. That neural capacity dictates the form that the grammars of all languages must take (i.e. 'universal grammar'). Their encoding systems are innately understood. Only the surface form in which the common code is clothed is conventional, according to the forms of expression of a given speech community. In dealing in this way with the evolutionary aspect of language it is interesting that Chomsky (1966, for example) makes appeal to Cartesian rationalism.

Yet Darwinism remains. May not language, too, be a matter of development and adaptation that marches hand in hand with biological evolution? One glimpses such ideas behind the views of Bolinger (1975, Chap. 10) and Givón (1979), and, more strongly, in the writings of Bickerton (1981). Meanwhile some biologists (for example Stanley 1985) have reconsidered the notion of gradualism. According to this thinking there is indeed gradualism; but once the accumulation reaches a critical point on a 'plateau' rapid changes are triggered. Further, there are quite a number of such plateaux and spurts in the evolutionary descent of an organism. Such thinking is adopted in this chapter, for it accords *grosso modo* with palaeontological findings as to hominoid species in their evolution to human status. The Chomskyan view subscribes to the notion of a spurt; but it has to be disproved that there was not one, but a number of spurts for language, and these I shall call 'stages' (noting that Popper (1966) and Pulleyblank (1967) regard language phylogeny as a matter of discrete stages). The notion of stages has been found by many linguists (for example Dale (1976) among many others) to apply to the development of the child's acquisition of language, although ontogeny does not necessarily correspond to the phylogeny of language.

The exercise of parsing, i.e. sentence analysis, reveals that grammar is hierarchically structured. Might not that structuring correspond to an evolutionary progress from the simplex to the hierarchically complex? A further insight of modern linguistics is that there is a functional basis for discourse, such that in some way the form of discourse is dictated by the speaker's intentions. Perhaps these intentions may be associated with the stages mentioned above, and even be regarded as motivations that trigger those stages. Perhaps also, features of modern grammar may be examined that correspond functionally with those intentions, and the insights of 'universal grammar' as to hierarchical ordering—and indeed other structural features of modern grammar—may be considered according to reductionist principles as above, to see whether their dependencies can be ordered on a developmental sequence. In this way an attempt may be made to synthesize the various insights that have been discussed.

26.2 Patterns of grammar

Language is generally regarded as comprising a number of components. The lexicon is a store of the items that provide the semantic component. The syntactic component serves to organize these items according to the grammatical structure of language, so that sentences may be formed. The phonological component concerns the structure of language in sound. The pragmatic component is the sum total of functions which language serves within the context of speaker-hearer dialogue: the presuppositions and communicative intentions held or adopted by the interlocutors, and what Rommetveit (1968) calls the 'deictic anchorage' of discourse: the spatial, temporal, and directional orientation to the 'here and now' of the speaker (occasionally, the addressee).

If phonological considerations are left out of account, we may choose to give priority to grammar as a formal structure—an approach which has dominated recent linguistic theorizing. Under this procedure grammar is to be studied as a context-free system separable from the universe of discourse in which it functions. The system embodies 'mental representations'; items and processes of the pragmatic component are applied to the system after it has been elaborated. By contrast, priority may be given to the pragmatic component, in which case grammar becomes a communicative device, one that is patterned to cater for various communicative intentions. On this basis, syntactic structures and semantic arrangements exist not so much in their own right, but as the servants of various communicative intentions in discourse. The latter view, taken among others by Halliday (1978), Dik (1978), and Foley and Van Valin (1984), is the one adopted here. Table 26.1 outlines the two approaches.

Examples of the 'intentions' that have just been posited are: asking questions, making statements, and expressing comments. As they are 'communicative' these intentions also include pragmatic elements such as phatic communion (expressions that establish empathy between the interlocutors) and making sure that the message will be understood within its context of utterance. Correspondingly, various levels of discourse are found; a simple imperative like 'Go!' is of a different order, at least structurally, from an utterance like 'I say, look here!', and even more so from a three-page sentence to be found in James Joyce's *Ulysses*. The notion of structural difference suggests that for each communicative intention there is a corresponding pattern for its expression. These discrete patterns are to be found in modern language, provided we take 'patterns' to include not only grammatical structures (for example sentence structure) but

Table 26.1 Two possible ways of assigning grammatical function

-A-
1. The speaker has MENTAL REPRESENTATIONS (assumed to be logical in nature?).
2. These are interpreted as semantic structures whose logical form is expressed syntactically,
3. and items from the lexicon provide the expression of their semantic content.
4. Pragmatic functions are then included (e.g. 'speech act' functions where deemed appropriate),
5. and the resultant structure is then processed by the phonological 'component' so as to produce the speech utterance.

-B-
1. The speaker has INTENTIONS, e.g.
   (a) wants to agree, deny, question, command, give an opinion, offer information, tell a story, announce a basic truth, etc.; and
   (b) wants to express his/her viewpoint, or attitude, about what is being said.
2. The relevant patterns for expressing elements from (a) and (b) are incorporated into a coherent structure by means of a system of grammar,
3. and items from the lexicon provide the expression of the relevant semantic content.
4. The resultant structure is then processed by the phonological 'component' so as to produce the speech utterance.

---

also arrangements that express discourse functions (for example the formulas to be observed when dialogue is entered into).

Such an analysis offers the picture of a loose amalgam of patterns that is likely to be an inefficient means of communication because of its lack of uniformity of patterning. We would need to explain how discrete patterns come to be welded together to form the coherent overall grammatical patterns of extant languages. The explanation would seem to lie in postulating 'recency dominance': in effect, the later pattern incorporates earlier patterns by reworking them and making them conform, grammatically, to the newest and hence dominant pattern. Currently we would consider the dominant pattern is subject–predicate, for most utterances in language today consist of sentences cast in the subject–predicate mould, whatever may be their underlying communicative intention.[1] But we might envisage an evolutionary stage wherein, say, topic–comment was the dominant pattern to which earlier patterns would be made to conform, and which subsequently may be an underlying structure forced into the subject–predicate mould. For the process of being made to conform to a dominant pattern the expression 'rework' is used in this chapter. Reworking implies the increasing grammaticalization of communicative intentions and of the conceptual notions that are woven into them.

Bearing in mind the notion of reworking the various patterns may be arranged sequentially. This is based on the assumption that a given pattern requires a prior grammatical system or systems on which it can build or, on occasion, rework. The sequence can be envisaged as revealing a number of cumulative stages.

26.3 The frame of dialogue

The first stage for 'evolving' any language—of necessity without any prior pattern to build on—concerns getting in touch with fellow-members of one's species, a motivation presumably common not only to our ancestors but also, to a limited degree, to the higher mammals as well. Now, getting in touch requires making the correct approach or 'address'. To address another is a privilege, particularly since it automatically calls forth 'attending' on the part of the addressee. Even today, the importance of address may be seen if, say, someone walks into an office and announces: 'Water is made of hydrogen and oxygen.' The hearers attend unwillingly, for there was no address.

Address in motivated by some sort of obligement that is sought; this may be called 'solicitation'. The solicitation may be for fellow-feeling, or for an action to be performed, or for the provision of information, or for attending while the speaker furnishes comments, opinions, or information. The granting of these obligements constitutes 'compliance', though at times the solicitation may be met with a refusal. Compliance is a favour given, so suitable acknowledgement is called for; but as the giving of a favour upsets the 'balance of power' between the interlocutors, it frequently happens that compliance is accompanied by a relief of some sort (Goffman 1971). Finally, when the sequence has run its course, there is 'close-out' to indicate that the exchange is completed. The sequence may be thought of as 'the frame of dialogue'.

The frame of dialogue is not so much a pattern of grammar but of discourse function. Yet it contains the source of what is later to be grammaticalized as the vocative (from the address element); as the imperative mood of the verb together with the various forms of modified imperative (as 'Go!', simple imperative, vs. 'Would you kindly pass me the salt?' (modified imperative)) and the interrogative, both of these from the solicitation element; as the affirmative and the negative, these from the compliance (or alternatively, refusal) element; and as various markers for turn-taking, for example 'Really? (...)' or 'Yes indeed! (...)'. This frame contributes to the provision of the first and second persons of the verb when later they are grammaticalized, for these notional persons represent the two participants to the dialogue.

An entire message may be non-verbal, as when one addresses someone with a look of appeal and points to the mustard, hoping that it will be duly passed along. This is communication by gesture. Whatever medium gestures might be performed in, these dialogic captures of attention can be claimed to be among the most primitive.

26.4 Ostension and deixis

Making the previous abilities only a little more complex, the next stage would be ostension (cf. Gans 1981). Ostension is rooted in gesture, but begins to widen the range of communicative possibilities. Ostension, simply, is pointing at a visible item with one's index finger. In modern communicative behaviour ostension is infrequent: to point at things or persons is often regarded as childish, or naïve, or in some way inappropriate. The devices of modern language are preferred: we use markers, for example 'Look!' or 'Why there's (...)', to indicate that we are pointing something out; we name the item, and we use an ostensive, for example 'over there' (plus a directed gaze) to indicate where the item is in relation to us as the speaker. The communicative elements in the act of pointing are turned into a grammatically structured sentence—but they can be accomplished much more simply.

Three aspects of ostension may be noted. First, it is for someone else's benefit. Without the presence of an addressee or addressees whose attention can be drawn there is no motivation for ostension. Thus ostension requires a partner, it is within the situation of dialogue. Further, it communicates in the sense that it indicates.

Secondly, ostension implies that the addressee gathers just what is being pointed at. This understanding is aided by the situation of dialogue, which helps the addressee to infer just which item or phenomenon has been selected by the speaker. This second aspect leads on ultimately to noun-phrase determination, and will contribute towards the development of noun phrases after many steps, which will be discussed as they arise (see Sections 26.7 and 26.11 below).

Thirdly, ostension is orientated on the speaker. That is to say, the addressee knows that what is being pointed at is being seen from where the speaker *is*, he/she is the 'datum point' of reference as it were. The notion of the speaker's being taken as the point of reference is called 'deixis'. Deixis may be spatial and, by extension, temporal; and it may be directional, by indicating motion to or from the speaker (an example of deictic employments is: 'He came here after I got home yesterday'). Speaker deixis—the indication of something near at hand to him/her—together with ostension—the pointing at something a little farther away—enables a basic opposition of near vs. far, with 'near' being in the speaker's proximity and 'far' being something beyond his or her reach. Deixis on the speaker together with the near–far opposition also applies to the situation of dialogue itself, and helps to 'fix' the notion of the first two persons of the verb. While ostensives do not express directionality as it is fully realized in modern language (that is, language as it is today) they imply deictic directionality, i.e. motion to and from the speaker. These elements may, I suggest, be derived from ostension.

These three aspects provide such powerful communicative aids that ostension may be regarded as a candidate for a clear-cut stage of development. If so, the stage corresponds evolutionarily to the development of the ability to point to something. Ostension is its own intention: the desire or intention to point out things.

Ostension indicates phenomena in an indeterminate way. When we say: 'Look at that!', 'that' does not need to be a thing, but refers indifferently to the event and any participants associated with it. The addressee relies on inferences drawn from the situation of dialogue and the perceived salience of an event (or alternatively a thing) within the field of purview.

Ostension primarily concerns visible items, and distinguishes between those within and out of reach. But doubtless as the emerging human species acquired increased memory capacity ostension could be extended to indicate non-visible phenomena. These are phenomena registered in our mental store which the speaker can invite the addressee to recall. Ostension now extends 'far' to the opposition 'near distant' vs. 'far distant' that corresponds to visible vs. non-visible. However, far-distant indication can no longer be supported by inferences drawn from the situation of dialogue (for far distant is outside that situation) or by the visible salience of events. The process of ostension has to be reworked to produce a form of symbolic reference that may be called 'identification'; and identification is secured by a process of labelling or 'naming'. The development of naming builds on ostension, but so further extends it as to produce a further stage with its patterns.

26.5 Naming and words

The previous stage, dominated by pointing, might be considered as being expressed entirely or at least mainly in the gestural mode. The identification of touchable items is total, it is one-to-one; but non-touchable or 'far' items lack such precision. Identification can be secured in the gestural mode by the imitation of a given shape or an activity (and indeed this is how some gesture proceeds today). Such imitation, however, can only apply to things by association, the thing (or person) being the most salient feature of the activity. This labelling by association with an activity is the basis of the notion 'a thing is what it does'.

These remarks suggest the possibility of a gestural mode for human communication that was prior to the vocal mode (Hewes 1976, 1983), the latter arising initially as amplifying or redundant accompanying vocalizations to the gestures. If so, not only the first two stages but also the first steps in the formulation of words are within the range of gesture. We can only be sure that at some point the vocal mode emerged. If it emerged after the gestural mode, then from being the lesser accompaniment of imitation and its ensuing repertoire of conventional gestural signs the vocal mode became predominant—though it has not completely taken over all the ground covered by gesture, as may be seen if we watch a television speaker with the sound turned off, or a mathematician going through elementary Euclidean proofs at the blackboard, where gesture is reworked in a technologically augmented form.

As to the path taken by the vocal units the first step may have been to reformulate some of the inventory of gestural signs so as to produce words. By 'words' I mean integral units of vocalization that have referential content, that refer to phenomena, or at least to aspects of phenomena such as size or directionality. Words in that sense are to be distinguished from items in the vocabulary which, though counting as lexical items and thereby entering into sentence structures, yet do not have referential content, for example general alarm calls, affective expressions, iconic vocalizations such as 'miaow' which may just mimic rather than refer, markers of refusal and consent, and markers for the formulas of the frame of dialogue. These all carry over into modern language, incorporated to a greater or lesser degree into the current dominant pattern. For example, when the negative in English is a marker of refusal it is either free-standing or not incorporated into the sentence, for example 'Stand up!' 'No [, I won't]!' On the other hand when the marker is reworked as a word for the logical negative it is incorporated, as in 'No children are unkind' (exclusion from set membership).

Instead of reformulating some of the inventory of gestural signs (or, in addition to doing so) words may have started as vocal gestures that offered iconic representations (cf. Pulleyblank 1983). These vocal gestures would then be organized into a system of contrasts that could properly be described as sound symbolism: some evidence for this process is offered in the next section (see also Foster, this volume, Chapter 25). Ultimately the system becomes so conventionalized that as an end-result the allotment of speech-sounds to meanings becomes arbitrary, in contrast to any iconic or sound-symbolism-based allotment. Thus, for example, the morpheme *ink* means 'ink' for no apparent reason.

Foster (this volume, Chapter 25,) suggests that the primordial symbolism was basically analytical, that is, the symbolism did not refer to concrete items but, rather, to basic physical attributes of the world such as 'in' (or, 'penetration') and 'around' and 'open'. The first words (whatever these may have been) could also have been holistic expressions that indicated a given phenomenon as a complete experience. In modern language, holistic expressions appear with the beginnings of language-acquisition by the child. At this early stage they consist of single units of expression such as 'allgone' (Peters 1983). At the adult stage they are more likely to be in the form of formulae such as in English 'What's the matter?' or 'It's raining.' In form these are sentences; but in substance they are unanalysed units (Bloomfield 1915, 1917). Bloomfield regards the holistic form of expression as being very widespread. Peters considers that not only is the holistic form the most prevalent for the onset of language, it also remains the preferred form in adult life. Peters is not concerned with prehistory; but Bloomfield discerns an evolutionary sequence for language from the holistic to the analytic, i.e. the form where logical relations are specified. We cannot claim that ontogeny (in this case the child's acquisition of language) always recapitulates phylogeny (in this case the development of grammar). The principle of recapitulation is useful, however, in corroborating other evidence.

Yet whichever analysis is adopted the problem remains: how does language get words to have referential content? Clearly words are not the referents themselves, but (as in the case of 'ink', usually highly arbitrary) symbols for the referent. Then in some way there must be an association between referent and symbol which is processed by the mind in language apprehension. One essential feature of that association is identification: reference is to one given item to the exclusion of all others. Words are able to refer to events and things and states and processes, in general to perceived phenomena of the world partly (and importantly) because they identify them. This 'power to identify' is vouchsafed in the long run by indication. Indication is ultimately a feature of ostension. The act of identification or 'naming' that provides the stock of words with referential content may be regarded as a stage that follows on from ostension, as an intention to identify far or non-visible phenomena that can no longer be indicated by simple ostension.

Some writers (for example Diamond 1959) envisage that identification was initially on acts that were sought via urgent solicitation, giving rise to imperatives such as 'Cut!' This may well have been the case; but earnestly desired acts are often readily to be inferred from the situation of dialogue, whereas the specification of visible but non-touchable items is not so readily to be inferred. Perhaps act solicitation and ostension were combined in the first instance.

The notion of referential symbolism via words may not have been fully perceived at their creation. One method of securing referential symbolism is the thematization of 'root forms', as discussed in the next section. In addition to thematization, modern language employs several grammatical expedients to coin new words, such as affixing, compounding, verbalization, nominalization, metaphor, and shifts of meaning values. There is in general an extensive recategorization of functions (for example verbs are reanalysed as prepositions, as in Chinese). Whole sentences are taken as words; and at the other extreme stopgap items are employed ('that', 'it') when identification fails. None of this suggests a sudden and complete invention of the notion of 'words', but rather an *ad hoc* process, a 'groping towards' that utilizes resources as and when in turn they are developed. The rise of words counts as a stage in the development of grammar, but one which has undergone many subsequent adaptations.

26.6 Thematization—the phonemic principle

So far internal evidence from modern language has been relied upon, and appeal has been made to various psychological and pragmatic considerations as to what is implied in the situation of dialogue itself. This section discusses some reconstructions of comparative philology. These cannot trace the form of language back very far. The most completely reconstructed language, Indo-European, is generally thought to date from around 4500 BC at the earliest, meaning that 'reliable' reconstructional methods can take us back around 6500 years at most. It is perhaps significant that the reconstructions agree on a number of features. These I now present briefly.

Grammatical and syntactic morphemes are scarcely discernible. This may be due to such markers' being purely vocalic utterances, whereas the 'roots' that will be discussed typically include consonants in their structure, and etymologically, consonants are more stable than vowels. For Indo-European, for example, Benveniste (1935) mentions only two suffixes that were outside the morphemic system that will be outlined, that is, that were applied independently of its morphological rules. One is the agglutinative *-AI, which has a hortative value. The other is *-0, which, added to a root, offered a kind of protogenitive. Instead, the vocabulary consists in the first instance of roots. These express some basic notion: for example, the Indo-European root *WEL ('want', 'wish', 'good'). Roots have a uniform morphology: for example, in language X all roots must be, say, biconsonantal syllables. This uniformity allows for phonological alternances (whether as to front, median, or back articulation; whether voiced or unvoiced consonant; etc.) to be exploited, so that alternances of the root phonemes give clusters, each member of which expresses a particular aspect of the notion expressed by the root.

Indo-European reconstructional evidence stops at the point where roots are identified. Yet there does appear to be toponymic evidence concerning a race of speakers prior to the advent of the Indo-Europeans in Europe. Rostaing (1945, p. 24) points out that 'in paleolithic times Western Europe was inhabited by a population whose language is unknown but who must have given names to the places where they lived, above all mountains and rivers; these names have survived to our day, for there is a toponymic residue that escapes any explanation by a known language'. According to Fouché (1939) pre-Indo-European (PIE), placed by him between Indo-European and Semitic, had a morphological system related to these two language groups. There is a consonant+vowel+consonant triliteral base, the first consonant generally being an occlusive. The quality of the median vowel may vary; and in the same way the consonants may differ, both in the manner and in the point of articulation (for example, voiced/ unvoiced, etc.; and front, median, back). A voiced consonant may alternate with a sonant, an *-L with an *-R, so that even without considering vowel alternations the same radical may adopt the forms *KAL, *GAL, *KAR, *GAR; and, with apophony, *AL and *AR, or even simply the voiced *L or *R. The same root may have its full form *KAL; or a zero vowel, to give *KL, *GL, *KR, and *GR. From the variants of this root derive, in French, according to Fouché, topographical terms such as *causse* ('limestone area'), *garrigue* ('rock-strewn area'), *calanque* ('steep-sided inlet') and place and river names such as Chelles, Cala, Carry, Cher, les Alpes and many others. All these items are associated with the notion 'stone', and presumably the variants of the root form that are provided by the alternances expand the basic notion to its various aspects and qualities.

Swadesh (1960, p. 901) sees a similar play of alternances in his reconstructions for proto-Ancient-American. For nominal themes the contrasts of basic vs. small or large are effected, first, by vowel interchange: mid-vowels for the basic grade, high vowels for diminutive, and the low vowel *a* for augmentative. Consonant change is according to Swadesh both positional and articulatory. A three-way set comprising, say, variants of the sibilant+n/l/r+variants of the dental gives normal, diminutive, and augmentative. In this complex application of sound symbolism, in addition to the alternances, all the phonemes may be changed in the same direction.

A further development can expand these simple phoneme units to give an even wider range of meanings. An illustration of this process is to be found in Korean of a word cluster having a basic notion (in this case the activity of 'knocking against'); the semantic alternations, which in this illustration deal with various kinds of sounds evoked by that activity, are symbolized by the phonological alternances:

water movement: chalang 'a small amount'; cholong 'as a bathful'; chulong 'as on the seashore';

objects hitting each other, their dimensions: calang 'small, less width'; c'alang 'small, greater width'; colong 'large, less width'; c'olong 'large, greater width' (Lee 1982).

The notion common to all members of a cluster may be called a 'theme' and the aspect of meaning unique to one member may be called a 'semantic feature'. Then we may compare thematic processes with some of the syntactic functions of modern language. For themes that realize essentially verblike notions the semantic features realize what in modern language becomes aspect or manner, that is, how, or why, or in what way the action indicated by the verb is performed; such aspect is usually expressed in modern language by inflection (as, 'hunt' vs. 'hunt*ed*') or by lexical means ('she cried *continuously*'). Where a theme realizes essentially a nounlike notion—say 'man'—then the different semantic features realize qualities that in modern language are usually expressed as adjectives or relative clauses.

However, it is not clear that the verb–noun distinction is available for themes, and—in the indeterminate manner noted for ostension in Section 26.4 above—their semantic features express aspects of phenomena rather than of things or events specifically (cf. Benveniste (1935) for the putative emergence of specifically noun themes in PIE). It is the context of utterance, and the salience of being a thing, or an event, that counts. We see this in modern language for items such as in English 'bump', whose verb vs. noun saliency may vary. In a similar fashion the semantic features of themes may offer verblike or nounlike aspect for the same item; and we see this, again, in modern language: for example 'a *strong* hero' and 'a *yesterday's* hero' offer noun attribution and verb aspect respectively. The emergence of the verb–noun contrast is an important element in the development of later stages, as in the next two sections; but in considering themes another contrast is to be noted: themes offer semantic features, whereas modern language sorts them out according to the basic verb–noun distinction and grammaticalizes them in syntax. On that basis it may be said that the system of thematic clusters is as it were a semantic route to grammar, but one without syntax.

Each unit in the thematic system is a package, it is a holistic expression. One might call them 'micro-sentences' (Diffloth 1976): they are conflated topic+ comments; the thematic notion running through the whole cluster would be the topic, and the specific semantic feature the comment about it; but the elements 'topic' and 'comment' have not been polarized out. A similar arrangement is frequent in modern language, as in 'What a pretty dress!' Here 'what (a)' is a pragmatic expression of approval, and 'pretty dress' is a reference to an item in the immediate purview of the interlocutors, not any token of pretty dresses. Moreover, we might, as a linguistic exercise, analyse 'pretty dress' as: '[concerning the topic] (this present) dress, [my comment is] (it's) pretty'; but we are unpacking a holistic expression in which topic+ comment are conflated.

Not only are theme units holistic, they are independent of one another. There is no syntax to combine them. They may be associated—we must surmise that they often were; but then the second thematic utterance merely takes up and extends the general purport of the first. This arrangement is to be seen, in modern language, in book titles, for example *Jack and the Beanstalk*.

The system outlined above exploits the alternances inherent in speech-sounds in an ingenious fashion, but only for the provision of 'words' grouped in semantic matrices. As noted above, the attempts at reconstructions generally reveal few or no syntactic markers for the thematic system for languages, nor is there any apparent formal or marked deixis, i.e. items that allude to the speaker or sender or to the context of utterance. Instead, a speaker would have had to rely on inference drawn from the situation of dialogue, on intonation perhaps, and on conventionalized gestures.

In support of these reconstructions the survival of basically thematic arrangements in modern language have been mentioned. Some neurological evidence, offered by the researches of Bechtereva et al. (1979), may also be noted. They find that portions of the thalamus transcribe auditory nerve signals as words, but these are grouped according to semantic fields. Since the grouping is according to a neural lattice, a 'generic' item—one that summarizes the parts of the lattice most commonly activated—is made available. The words are then distributed by the thalamus to the appropriate areas of the neocortex, presumably for their sequence to be processed syntactically. On this basis the mental process calls for the identification of words, according to a species of thematic grouping, before the syntax is dealt with; might this order reflect the evolution of grammar?

Thematization, then, appears to be the culmination of a morphophonological development that could go no further; it does not produce syntax in the modern sense. The beginnings of syntax constitute a further stage.

26.7 Topic–comment and nominals

Table 26.2 gives examples of the topic–comment pattern. These all have a common structure: they are binomials, of which the one constituent is the topic and the other, the comment. The 'topic' column shows that the one syntactic requirement for this pattern is that the topic is to be a nominal of some sort; no such restriction applies to comments. The labels 'topic' and 'comment' that are given to the two constituents indicate that a pragmatic consideration operates: an item has been selected by the speaker as an appropriate 'subject for discussion' as it were; and then the speaker makes his/her personal comment about it. The binomial structure and the requirement for nominal topics reflect this pragmatic consideration. The comment is speaker oriented, there being freedom of choice as to how he/she sees the matter. On the other hand, there are restrictions on the topic: it must be something already under discussion, or known to both interlocutors, or generally known to the speech community. Topics have to be identified information that is shared.

The fact that comments are speaker-oriented allows a further development. Orientation also provides perspective, that is, the speaker's attitude to the comment. This perspective allows for the switching of comments to be topics, thereby allowing the discussion to expand. For example, from 'Cars sell well here' we may proceed, say, to 'Their sales arouse great interest.'

The presupposition on the part of the speaker that the topic is known information, part of the common stock shared by the speech community, enables a further step in the identifying process. From ostension there is a progress to the indication of non-visible items, as discussed in 'Naming and words' above. The indexing gets to be not merely on immediate

Table 26.2 Some types of topic–comment

type	example	
	topic	comment
(a) asyntactic; pragmatically based:		
subjective assessment	John('s)	a great guy.
subjective association	That('s)	your problem!
(b) syntactic; there is a semantic relation:		
ascription	Cars	sell well there.
experiencer–experience	I	feel fine
actor–act	John	works at Woolworth's
agent–agentive act	John	is cutting the cake
causative agent	John	grows tomatoes

phenomena but on identified information that has been mentally inventoried; it is ostension at a further remove. The availability of words (items with referential content) allows the inventorying to take place, and the expression of this inventory is the lexicon of a language.

'Ostension at a remove' may be seen in the determination of noun phrases in modern language.[2] In English, for instance, noun-phrase determination includes the definite article 'the'. This determiner derives from a former demonstrative adjective with the value 'that'. The adjective in turn derives probably from the corresponding demonstrative pronoun or ostensive expression. Moreover the indefinite pronoun 'it', which includes a demonstrative element, derives from an Old English demonstrative which is cognate with the Gothic 'hita'='this'. In effect the category of noun phrases[2] reflects a pragmatic presupposition: they are all known, inventoried information or, alternatively, information which is being alluded to in the context of discourse.

Thus the speaker's intention is to make comments; but speakers are restricted to selecting topics that are known, referable information. In this restriction is to be found the agency for emerging nominals.

To express a topic as a nominal means that it has been reified, it is regarded as being metaphorically a 'thing'. This process of reification is productive in modern language of nominalizations such as 'the homecoming'='X comes home'. It also contributes to the process of 'identifying by association' to form actor nominals, for example, 'a player'='one who plays'.

The speaker's intention to make comments provides the motivation for this further stage in language development that produces binomials. Binomials are a first step toward modern syntax, which has the overall binomiality of subject–predicate structure. Further, where topic–comment survives as a structure in modern language 'it is applied in a variety of ways in the service of a variety of motives' (Bates and MacWhinney 1982) such as perspective and salience; this variety suggests a long history of reworking and incorporation.

The simple binomial structure that we have been considering is asyntactic, as in the 'asyntactic' examples offered in Table 26.2(a). In these examples the only relation between the two constituents is that of topic vs. comment itself. A further stage is needed to produce syntactically structured binomials.

26.8 Topic–comment and case-relations

The asyntactic topic–comments discussed in the previous section consist of purely subjective assessments or associations, as in the examples of Table 26.2(a). In these examples no perceived, inherent relation holds between the topic and its comment; the relation is indeterminate because the topic–comment relation is pragmatically based. But in the examples of Table 26.2(b) it is seen that the comments in each instance stand in a specific, determinate relation to the topic. These are perceived or 'logical' relations, inherent in the phenomena presented. Thus, the examples of Table 26.2(b) all offer a syntactic structure. (Syntax provides a logical form which embodies the semantic relations holding between items. One may distinguish semantic reference, provided by lexical items, and semantic relations, provided by syntax. The logical form may be encoded by inflections or prepositions or even formal word order.)

The examples of 26.2(b) compared with those of 26.2(a) show a development towards grammatical structure. The sequence of examples in 26.2(b) further suggests the route taken in this development. Ascription is rather close to subjective association; but now a 'factive' element of 'inherently so' has been added. 'Experiential' topic–comments have the validity of the speaker's self-perception. The remaining three examples extend the range of relations that may hold between the topic and its comment.

'Experiential' topic–comments seem to have a psychological motivation. Two very likely topics are the first and second persons of the verb—and these, developing from ostension, derive from the situation of dialogue itself. Again for social reasons what people *do* is the most likely candidate for comment-making. Thus the most likely form of comment (as remains true for modern language) would utilize a verbal theme as expressing some kind of activity. On this basis the first and second persons of the verb would have been developed, whether the expression of these persons was by free-standing pronouns or by some form of inflection of the verb theme.

However, the requirement arises to express what other persons ('third persons') are doing, how they are involved in the action. Moreover, particularly if there are a number of them, the precise relation of the various participants in the event needs to be clarified. This 'spelling out', as it were, of the various relations to an event is graphic: it enables the speaker to depict scenes. And the 'spelling out' is performed, grammatically, by case relations.

Case-relations (later formalized into case systems for various languages) are syntactic means of expressing one set of semantic relations, namely those holding between a number of participants to a given event. In modern language these systems vary in the range of relations they express. Crosslinguistically we may discern—among others—the following relations to a given event: actor/agent/causative agent; affected object or person; recipient; beneficiary; instrument; bystander; interested or concerned party; and—tacked on to the case system—direction and location adopted by a participant. It is apparent that these relations are all graphic: they tell us who does what to whom; who is moving in or out of the picture; and who is 'on the edge of' the action. It seems likely that the motivation for the development of this pattern of grammar is an intention to depict scenes. Indeed in Fillmore's words (1977, p. 59) 'meanings [of the case relations] are relativized to scenes'.

The depiction of scenes appears to be a fairly modern development in vocal language. Its developmental steps may be traced, at least in regard to the Indo-European case system (see Benveniste (1935) for the rise of the animate vs. neuter distinction that allows for animate topic subject; Kuryłowicz (1964) for a reworking of locative and directional expressions (for example 'at', 'towards') to derive a case system). It is interesting that an element of this development was contributed by the active vs. passive distinction, which enabled affected-object comments to become, as passives, topics in their turn: a shift in the perspective from which a scene is viewed. For example 'The Romans have annihilated the Greeks' may be contrasted with 'The Greeks have been annihilated.'

Thus a developmental stage is envisaged here as making syntactic topic–comments available. The continuing process of language evolution builds on the availability of nominals. The subjective assessments and associations as in Table 26.2(a) are reworked as genuine or 'factive' ascriptions. This is a cognitive advance, revealing a greater perception of external realities. At the same time there arises not only the 'self-ascription' of subjective experiences but an ascription of external acts to actors: a person is associated with an event; he/she is the topic; and the comment is, what he/she is doing. As soon as there is a clearly defined actor associated with an act, there is a scene that is being depicted. Scene depiction allows the mention of other participants. The part those participants play is expressed in visual terms, for scenes are what is being looked at. The visual terms are encoded into the grammatical pattern by reworking verblike themes of motion and arrival as grammatical markers for directionality and 'place where'.

As they stand, the last three examples of Table 26.2(b) are topic–comments. But the scenes that these examples evoke ('works at Woolworth's', 'is cutting the cake', etc.) are virtually narrative; the distinction is finely drawn. It is the distinction between 'He goes to work by car' (topic–comment) and 'he went to work by car last year and then he switched to Amtrak' (narrative). In effect scenes can be sequenced by adding a time dimension ('went', 'switched') and by repeating the same participant ('he'). The sequencing of scenes into narrative, and the considerable developments in grammar involved, constitute the next proposed stage.

26.9 Narrative

Sustained narrative is, grammatically, a highly complex matter that calls for the integration and coherence of a large number of grammatical processes whose scope has great variation, whether over one lexical item, a phrase, a clause, a sentence, or a complete narration. These grammatical processes include case-relation patterns; a complete pattern for tense, aspect, and voice indication; a pattern to express the notion of 'result'; hierarchic organization of clauses within sentences and of sentences within sustained discourse; patterns for foregrounding and backgrounding; markers for sequentiality, both intra- and extrasentential; and proformas that provide anaphora, or the reference back to a previously mentioned item. The following specimen illustrates some of these patterns:

After [a sequential marker] that [anaphora on the preceding paragraph],* he [reference back to the main actor in the narrative] wrote [past tense, telic[3] aspect] a letter*, **while [marker of subordinate clause] Mary cooked [past tense, continuous aspect implied] the meal.**

* = foregrounded material; ** = backgrounded material.

I shall attempt to hypothesize the main steps in the development of narrative grammar, noting that the development was motivated by the intention to narrate, to tell stories—whether myths, sagas, or the exploits of interesting people.

An early step, as in the previous section, would be the establishment of case relations. This allows the depicting of scenes. But narrative is a recital; it stresses events, whereas the participants are usually inferable from the context of the ongoing discourse. This focusing on events leads to a great augmentation of the scope of the verb. A very wide range of aspect and mood—how the event happens and how the speaker reacts to it—clusters round the verb, to produce inflectional paradigms.[4] The extent of these paradigms can be seen in the conjugations of, say, Shona, an African Bantu language (see Hannan 1974, xii–xvii) or Navajo, a North American Indian language (see Young and Morgan 1980, 146–52). They provide for various tense aspects, each with various implications such as 'I have done so before now', 'I have not done so up till now'; there is an extensive array of modals (cf. English 'would,' 'could', etc.) and a great number of 'manner' aspectives to deal with notions such as incipient, progressive, continuous, habitual, and the like. Such a system was characteristic of early Indo-European (Kuryłowicz 1964); it was an inflectional system, as also was its case relation system. Narration is facilitated by an inflectional system; the conjugations and declensions serve to organize the presentation of both the events narrated and the participants in those events.

An important feature in the development of verb conjugation is the third person of the verb. The reference of the first two persons is specific; but the third person indicates any, unspecified, participant. The Italian 'canta' for example means 'he/she/it sings', and precise identification is afforded either by the context of discourse or by the anaphora that is available in narrative, as in the above specimen. In narrative, main participants are introduced by presentation (Givón 1976), as in 'Once upon a time there was a [...]', and thereafter their mention is discontinued except by anaphora that refers back to the presented participant, as in '... and then he [...]'. 'Reference back' allows for the linkage of scenes into a narrative sequence, with the main participants being carried forward into the next series of events. Anaphora is extended to other proformas such as 'who' and 'there'.

But for narrative sequencing to be complete a linear dimension for time is necessarily required. Modern languages generally provide for this with a sequence of tenses, past, present, and future. This sequence, as it stands, expresses aspects of an event, i.e. whether it is completed, or ongoing, or not as yet realized. If we can supply a notional fixed point (say, 'now') in relation to which events occur then there is linearity; one event is deemed to happen before or after another, so that there is anteriority. Modern tense systems, that include anteriority, have been built on 'tense aspect'. For example, Moulton (1906) notes that New Testament Greek had tense aspect but not as yet, anteriority; and Gilman (1970) discusses the lack of 'formal' (as distinct from 'subjective') anteriority in medieval Spanish. Therefore, anteriority may be considered a latecomer, and, owing to the mosaic progress of language development, belongs at least partly to the later epistemic stage yet to be discussed. An intermediate step towards anteriority would appear to be the linking of narrative units by time-words (for example 'after', 'next') that are extensions of words for spatial relations. Such a form of linkage is very common in modern languages, and is evident for European languages (cf. German 'nachher' and French 'après' for 'afterwards').

Anteriority also provides for 'subordinate anteriority' in the manner of 'after he had [...], he then [...]'. In this way events outside the mainstream of the narrative may be introduced. A further process of clause subordination would rework simple linkages ('John [...], and he [...]') as relative clauses (John, who [...]'). Similarly, subordinate clauses would, and do, add further background information to main events; entire clauses are reworked by the process of nominalization to become words, so that for example 'after he arrived' becomes 'his arrival'. Finally anteriority allows for constructions in the manner of 'As/Because it rained, we stayed at home' that express result.

From the foregoing it will be noted that the highly structured narrative pattern is basically hierarchical: events and viewpoints, as well as participants, are ranked in importance, and there is foregrounding and backgrounding. These features allow for stylistic elements in the relation of narrative: focus, choice of vocabulary, coining of epithets, and the like.

The above paragraphs have outlined only the barest details of the narrative pattern. There still remains a further development to be considered, one which concerns focusing and harks back to topic–comment. The previous section discussed case relations as arising from the basis of topic–comment. But as a number of participants get introduced into narrative sentences, the notion of topic–coment is obscured. Deixis on the speaker (for a comment is his/her comment, it is how they see it) is transferred to the event itself, so that a kind of objectivity operates which will contribute towards the later epistemic pattern. This focusing on the event, inherent in narrative, is grammaticalized in the narrative pattern wherein the verb is central and becomes the only required element of the sentence; noun phrases (that express participants) are satellite to it.

Such a pattern is proposed here for PIE (cf. Lehmann 1974). PIE had a complex set of personal endings on verbs, and the function of these endings was to mark the person of the subject. But because of the relatively free word-order of PIE certain items, whatever their case, could be fronted, and may have been accorded a certain salience which could be associated with topichood. Owing to human egocentricity, however, it was commonly those items having the semantic case of 'actor' or 'agent' that came to be selected preferentially as topics, for usually these are not merely animates, but humans too. In this way there arose a 'pecking order' for the selection of topic subjects (Givón 1976).

26.10 The epistemic pattern: objectivity in discourse

Some discourse is 'epistemic', in that it informs, it increases the addressee's knowledge of the world, and this knowledge is considered to be public knowledge; it purports to be factual and truthful, and it is objective, that is, it eliminates as far as possible the speaker's subjective stance. Epistemic discourse is in demand for science and philosophy and historical narrative. It reckons to generalize from the particular, and thus is useful for matters of law, government, and all kinds of rule-giving.

These cultural activities are modern to the extent that they characterize a level of social organization that produces states. (It could be that simpler forms of society also employed epistemic patterns in their languages, but they would seem to provide less motivation for such patterns.) I propose that systematic objective discourse arose, not from any abstract philosophical enquiry that our forebears decided to undertake, but, in the first instance, from the requirements of complex larger-scale societies. It is also proposed here that the development of epistemic patterns of grammar is to some extent due to the adoption and widespread use in a society of writing (Goody 1977), a new mode that competes with the vocal. Face-to-face oral discourse includes the powerful communicative aids of posture, intonation, and even more powerfully, inference drawn from the general suppositions built into the situation of dialogue. None of these aids are available in writing: in this medium discourse has to be more carefully articulated to make good their absence, and a variety of new opportunities for style emerge.

Table 26.3, column 1 lists some of the grammatical features of the epistemic pattern, and column 3 gives the corresponding epistemic functions that the grammatical features express. From the examples given in column 2 it will be seen that these grammatical features are mostly reworkings of features that exist side by side with them in modern language. As a result, a number of expressions have two kinds of grammatical featuring, according to whether they are used with epistemic function or not. 'There' for example retains its original value as a deictic. The generalizing quantifiers, for example 'all', 'some', are matched by the particularizing quantifiers, so that against 'all men' we find 'all the men.' Partee (1978) notes the dual function of some terms, for example 'or', according to whether they are used in logic or in 'ordinary' language; similarly 'and' may express conjunction, linkage, or arithmetical summation, another manual symbol system (Weinreich 1966, 170). In addition there is a range of conceptual abstract nouns that have been coined to express abstract, generalized notions.

These grammatical features are incorporated into the formal structure comprising (grammatical) subject + predicate[5] as distinct from topic + comment. In the previous section it was noted that narrative patterns produce a topic subject. However, that topic may be in any case, for example dative, as in the earlier English 'Methinks [. . .]'. But after extensive reworking an arbitrary arrangement was established whereby any noun phrase that is focused and hence selected as subject acquires nominative or 'subject' case, so that, for English for example, 'Methinks' has switched to 'I think'. In modern language the subject requires the verb's concord (if the language expresses concord); also, all the rest of the sentence is to be the 'predicate', or, 'what is said about the subject'. Once there is a (grammatical) subject + predicate the reworking is complete: there is topic–comment in a new, but now objective, guise.

In order to achieve objectivity the epistemic pattern revalues pragmatically-based patterns; and anteriority in narrative is adapted for historical narrative by assigning a given date as a fixed point. The 'result' narrative pattern becomes a causative construction (for example 'John grew tomatoes') that allows the depiction of phenomena as being due to an effective causative agency.

Table 26.3 Some features of the epistemic pattern

feature	example	expresses
gnomic present	Sugar *dissolves* in water	omnitemporality
the copula	Hydrogen *is* an element	equation
connectives	*either* John *or* Mary will come	inference
'There is/are (...)'	*There are* angels in heaven	existentiality
quantifiers	*All/no* children are wise	logical inclusion/exclusion
generic 'the'	*The* lark is a passerine bird	set membership
If (...)	*If* that is true, then I am right	conditionality
conceptual abstract nouns	electronics; musicianship; grace	encapsulated concepts

Similarly the epistemic pattern extends the lexical category of nouns. The previous sections of this chapter have traced the development of nounlike forms out of nominals; the final development is the formation of a special class of abstract nouns, the conceptual abstract nouns illustrated in Table 26.3. This development arises, firstly, from an awareness of physical phenomena through pointing at them, and thereafter their identification by means of thematic units (see Sections 26.4, 26.5, and 26.6 above). Then there is the distinction between animates and neuters, a universal distinction in modern language; but neuters are still concrete items. Next abstract nouns are formed, yet these still directly refer to immediate experience; the results of activities are regarded as things, so that for instance 'a pit' means what results from digging. (According to Benveniste (1935) PIE utilized a marker of passive/ perfective aspect for this purpose; the value of this affix was something like 'has undergone change, has been completed'.) Then abstract nouns are formed, for example 'living', that cannot be identified with any observable unitary phenomenon but represent features abstracted from an entire class of phenomena. All these formations are validated by the shared knowledge and experience of the speech community. But the epistemic pattern moves on to coinages like 'education'. Such abstract nouns express concepts, they are proposals or analyses or hypotheses drawn *from* knowledge or experience; and these nouns are able to encapsulate a very wide range of phenomena. They are validated in discourse by the presupposition on the part of the addressee that the speaker is offering these projections not as opinions but as possible or even probable truths. The development whose progress has been conjectured above is a lexical and grammatical one; but it reflects a cognitive development, from an awareness limited to immediately visible items (indicated by ostension) to an ultimate awareness of the most abstract qualities to be discerned in phenomena, such qualities being not visible, but rather to be deduced from our experience. Such an awareness may be called conceptuality.

Objectivity and conceptuality, the achievements of the epistemic pattern, mark a powerful development of human symbolism.

26.11 Conclusion

The programme of cumulative stages and their corresponding patterns is summarized in Table 26.4. The table is notional and the stages are conjectural. In particular the attempt at claiming the discreteness of the stages must remain suspect. Rather (and perhaps the remarks in the foregoing sections imply this) language development is to be viewed as proceeding mosaic-fashion, with a feature being acquired at one stage but only fully exploited in a later one. This mosaic progress tends to mask the presence of discrete stages. But on the other hand, it allows the survival of earlier patterns side by side with later ones, so that we may appeal to examples of these patterns surviving in modern language as I have attempted to do throughout this chapter. Some of the earlier patterns, particularly those that apply directly to the situation of dialogue (for example sentence adverbs such as 'Incidentally, [. . .]') are scarcely incorporated, grammatically, into the dominant sentence-structure of modern language. On the other hand a pattern such as topic–comment may be reworked so as to incorporate it into the epistemic pattern.

It would be encouraging if the sequence of stages for the development of grammar, as envisaged in this chapter, could be made to correspond to some extent with evolutionary development in a biological sense. It is not apparent that developments in grammar have any bearing on biological development except for those earlier stages ('Naming and words', and possibly 'Thematization') where language is closely linked to sound-symbolism, for then it is phonologically-based, and may be envisaged as progressively exploiting speech mechanisms and their neural control as it came to be perfected. Rather, the development of grammar seems to be concomitant with cognitive advances, and in its most modern developments (sustained narrative, the epistemic pattern) to be motivated by social and cultural developments.

Yet I have tried to show that there is a continuity in the elaboration of language over time, not any fresh invention *ex nihilo* but a building on and a readaptation of previous existing patterns. It is only in these terms that we may venture to entertain the notion of a prehistory of grammar. With that caveat in mind perhaps Table 26.4 might serve as a summary of one possible approach to the prehistory of grammar and its significance for human symbolic evolution.

Notes

1. For example, context-free parsing (for example, for computer translation) would render the sentence: 'Waiter, there's a fly in my soup' as a sentence adjunct ('waiter') and an existential sentence. But this subject–predicate format is a reworking of the gist comprising three elements: 'O waiter!' (vocative); 'Lo, a fly!' (ostensive); and '[concerning] my soup, [my comment is] a fly "ins"' (topic–comment). This analysis is defended by noting that the example sentence

Table 26.4 Theoretical cumulative stages in the development of grammar

Subsystem or stage formation	Type of pattern	Features generated	Are the source of
the frame of dialogue	no grammatical structure; pragmatically based	the sequence: address, attending; solicitation, compliance (or refusal); acknowledgement, close-out	the vocative; 1st/2nd persons of verb (partly); imperative, interrogative; negative, affirmative; markers for turn-taking; affective expressions, interjections, and other items that don't get incorporated grammatically into the sentences of modern language.
ostension	no grammatical structure; based on indication	deixis; demonstration; the opposition near–far	semantic polarities (?); noun-phrase determination (ultimately); 1st/2nd persons of the verb (partly); spatial and temporal relations; directionality.
words	pre-grammatical; based on identification	integral ideational units	words as integral units having referential content.
thematization	grammatically holistic; phonologically based	word units (associable paratactically?) in thematic clusters, each unit offering different features of the theme	(after extensive reworking) aspect and manner for verb phrases, modifiers for noun phrases.
topic–comment, asyntactic	grammatical, but only as having binominal structure; pragmatically based.	the structure, topic–comment; the topic is a nominal (i.e. it is indexed or 'determined')	the binominal structure (i.e. subject–predicate) of modern language; contributes to the lexical category 'noun'.
topic–comment, syntactic	grammatical, relational	events, participants, and their relative participation are identified	case relations.
narrative	grammatical, relational, sequential, relative grounding	sentences are organized hierarchically via subordination and are sequenced in sustained discourse; temporality is provided	resultative and causative constructions; 3rd person of the verb; entry of noun phrases into sentence structure as being 'in' the 3rd person.
the epistemic pattern	grammatical, relational, sequential, explicitly employs logical operators; based on the notion 'objective'	abstract nouns with non-experiential reference; noun-phrase determination encodes genericness, set membership; omnitemporal verb aspect; co-ordination expresses notions of set functions, e.g. conditionality, logical inclusion/exclusion, equation, existentiality; topic subjects are reworked as grammatical subjects	(?) predicate logic

demands a context of utterance, and the utterance when made in that context will reveal the three elements by its intonation pattern.

2. Lyons (1977, 425) points out that 'whereas nouns have denotation, nominals have (or may have) reference' and 'definite referring noun-phrases... always contain a deictic element' (ibid., 654) (see also McCawley (1971) for the indexation of the arguments of propositions). The referential indexation may be overt, as determiners, specifiers, or classifiers, etc.; but is conflated in proper names, and may, for the grammars of some languages, be null, as is seen by comparing the two sentences 'The elephants trumpeted' vs. Elephants are mammals'.

3. Of a conjunction or clause: expressing end of purpose [eds].

4. The so-called 'analytic' languages, for example Chinese, employ auxiliary verbs and adverb phrases instead, so that there is no longer a formal paradigm or extended conjugation. European languages combine inflection with these other resources.

5. At least, in some extant languages. According to Li and Thompson (1976) languages vary on a range between 'subject [= grammatical subject in the terms of this chapter]-prominent' and 'topic-prominent'.

References

Bates, E. and MacWhinney, B. (1982). Functionalist approaches to grammar. In *Language acquisition: the state of the art* (ed. Eric Wanner and Lila R. Gleitman), pp. 173–218. Cambridge University Press.

Benveniste, E. (1935). *Origines de la formation des noms en Indo-Européen*. Librairie Adrien-Maisonneuve, Paris.

Bechtevera, N. P., Bundzen, P. V., Gogolitsin, Y. L., Malyshev, V. N. and Perepelkin, P. D. (1979). Neurophysiological codes of words in subcortical structures of the human brain. In *Brain and Language*, 7, 145–63.

Bickerton, D. (1981). *Roots of language*. Karoma, Ann Arbor.

Bloomfield, L. (1915). Sentence and word. *Transactions of the American Philological Association*, 45, 65–75.

Bloomfield, L. (1917). Subject and predicate. *Transactions of the American Philological Association*, 47, 13–22.

Bolinger, D. (1975). *Aspects of language* (2nd edn). Harcourt Brace Jovanovich, New York.

Chomsky, N. (1966). *Cartesian linguistics*. Harper and Row, New York.

Chomsky, N. (1972). *Language and mind* (enlarged edn). Harcourt Brace Jovanovich, New York.

Dale, P. S. (1976). *Language development: structure and function*. Holt, Rinehart and Winston, New York.

Diamond, A. S. (1959). *The history and origins of language*. Methuen, London.

Diffloth, G. (1976). Expressives in Semai. In *Austroasiatic studies, Part 1*, Oceanic Linguistics Special Publication No. 13, (ed. Philip N. Jenner, Laurence C. Thompson, and Stanley Starosta), pp. 249–64. University Press of Hawaii, Honolulu.

Dik, S. C. (1978). *Functional grammar*. North-Holland, Amsterdam.

Fillmore, C. J. (1977). The case for case reopened. In *Syntax and semantics*, Vol. 8: *Grammatical relations* (ed. Peter Cole and J. M. Sadok), pp. 59–81. Academic Press, New York.

Foley, W. A. and Van Valin, R. D., jun. (1984). *Functional syntax and universal grammar*. Cambridge University Press.

Fónagy, I. (1983). Preconceptual thinking in language (an essay in linguistic paleontology). In *Glossogenetics: the origin and evolution of language* (ed. Eric de Grolier), pp. 329–53. Harwood Academic, Chur.

Fouché, P. (1939). Quelques considérations sur la 'base toponymique', à propos du pré-IndoEuropéen *KAL—'Pierre'*. *Revue des Langues Romanes*, 68, 295–326.

Gans, E. (1981). *The origin of language*. University of California Press, Berkeley and Los Angeles.

Gilman, S. (1970). Time and tense in Spanish epic poetry. In *Explorations in communication* (ed. Edmund Carpenter and Marshall McLuhan), pp. 24–35. Jonathan Cape, London.

Givón, T. (1976). Topic, pronoun and grammatical agreement. In *Subject and topic* (ed. Charles N. Li), pp. 151–86. Academic Press, New York.

Givón, T. (1979). *On understanding grammar*. Academic Press, New York.

Goffman, I. (1971). *Relations in public*. Allen Lane/The Penguin Press, London.

Goody, J. (1977). *The domestication of the savage mind*. Cambridge University Press.

Halliday, M. A. K. (1978). *Language as a social semiotic*. Edward Arnold, London.

Hannan, M. (1974). *Standard Shona dictionary*. Rhodesia Literature Bureau, Harare, Zimbabwe.

Hewes, G. W. (1976). The current status of the gestural theory of language origin. In *Origins and evolution of language and speech* (ed. Stevan R. Harnad, Horst D. Steklis, and Jane Lancaster), pp. 482–584. The New York Academy of Sciences.

Hewes, G. W. (1983). The invention of phonemically-based language. In *Glossogenetics: the origin and evolution of language* (ed. Eric de Grolier), pp. 143–62. Harwood Academic, Chur.

Hockett, C. F. (1960). The origin of speech. *Scientific American*, **September**, 88–96.

Joyce, J. (1922). *Ulysses*. Shakespeare and Company, Paris.

Key, M. R. (1985). An approach to semantics through comparative linguistics. In *Studia linguistica diachronica et synchronica* (ed. Ursula Pieper and Gerhad Stickel), pp. 467–74. Mouton-de Gruyter, Berlin.

Kuryłowicz, J. (1964). *The inflectional categories of Indo-European*. Carl Winter Universitätsverlag, Heidelberg.

Lee, C. (1982). 'Honorific Expressions in Korean', paper presented to the Conference on Formulaicity, 50th Linguistic Institute of the Linguistic Society of America, University of Maryland (unpublished).

Lehmann, W. P. (1974). *Proto-Indo-European syntax*. University of Texas Press, Austin.

Li, C. N. and Thompson, S. A. (1976). Subject and topic: a new typology of language. In *Subject and topic* (ed. Charles N. Li), pp. 457–89. Academic Press, New York.

Lyons, J. (1977). *Semantics* (2 vols). Cambridge University Press.

McCawley, J. D. (1971). Where do noun phrases come from? In *Semantics: an interdisciplinary reader in philosophy, linguistics and psychology* (ed. Danny D. Steinberg and Leon A. Jakobovits, pp. 217–31. Cambridge University Press. (An extended version of a paper in R. A. Jacobs and P. S. Rosenbaum (eds), *Readings in*

English transformational grammar, Blaisdell, Waltham, Mass., 1970.)

Moulton, J. H. (1906). *A grammar of New Testament Greek,* Vol. I. T. & T. Clark, Edinburgh.

Müller, F. M. (1871). *The science of language: founded on lectures delivered at the Royal Institution in 1861 and 1863* (6th edn). Longmans Green, London.

Partee, B. H. (1978). *Fundamentals of mathematics for linguistics.* Reidel, Dordrecht and London, and Greylock, Stamford, Conn.

Peters, A. M. (1983). *The units of language acquisition.* Cambridge University Press.

Popper, K. R. (1966). *Of clouds and clocks*: being the Arthur Holly Compton Memorial Lecture presented at Washington University, April 21, 1965. Washington University, St Louis, Missouri.

Pulleyblank, E. G. (1967). The evolution of language: reopening a closed subject. *Studies in Linguistics,* **19**, 67–81.

Pulleyblank, E. G. (1983). The beginnings of duality of patterning in language. In *Glossogenetics: the origin and evolution of language* (ed. Eric de Grolier), pp. 369–410. Harwood Academic, Chur.

Rommetveit, R. (1968). *Words, meanings and messages.* Academic Press, New York.

Rostaing, C. (1945). *Les noms des lieux.* Presses Universitaires de France, Paris.

Stam, J. H. (1976). *Inquiries into the origin of language.* Harper and Row, New York.

Stanley, S. M. (1985). *The new evolutionary timetable.* Harper and Row, New York.

Swadesh, M. (1960). On interhemisphere linguistic connections. In *Culture and history: essays in honor of Paul Radin* (ed. Stanley A. Diamond), pp. 894–924. Columbia University Press, New York.

Weinreich, U. (1966). On the semantic structure of language. In *Universals of language* (ed. Joseph H. Greenberg), pp. 142–216. MIT Press, Cambridge, Mass.

Young, R. W. and Morgan, W. (1980). *The Navajo language: a grammar and colloquial dictionary.* University of New Mexico Press, Albuquerque.

27

Social and cognitive factors in the historical elaboration of writing

David Barton and Mary Hamilton

Abstract

Writing originated separately in Mesopotamia and Egypt, China, pre-Columbian America, and, possibly, the Indus Valley. The earliest evidence of writing is cuneiform script from Mesopotamia at *c.*3500 BC. Six earlier classes of visual representation contributed to the development of writing-systems: the expressive and ritualistic markings found in cave art; tallying devices; property markings and totems; tokens; mnemonic devices; and pictographic/ideographic narrative forms. Early writing-systems were used for political and economic, religious, and historical–literary functions. There is no single order of functional development that applies to all cultures.

Writing-systems are classified into three types: logographic systems, which represent morphemes; syllabic systems, which represent syllables; and alphabetic systems, which represent units more closely related to phonemes than to syllables. Writing-systems tend to develop from the logographic to the syllabic, though this is not always the case. As syllabic systems interact with the structure of the spoken language they are trying to capture they adapt themselves through a variety of devices. This historical elaboration is not well served by considering it to be an evolutionary sequence, as has often been claimed.

Strong claims that literacy *per se* qualitatively affects cognitive abilities are not well supported by evidence. Literacy is better seen as a communicative technology involved in the production and reproduction of shared meaning or knowledge. It is the social practices sustaining these meanings that determine the consequent skills associated with literacy. Arguments that credit literacy as a prime causal factor underlying social change are thus oversimplistic. Rather, literacy is just one factor in a nexus which includes social and political institutions.

Printing led to a restructuring of literate activities through the incorporation of a technical invention into the social organization and production of knowledge. Printing assists in the cultural diffusion of ideas, and in the standardization of knowledge and linguistic forms. However, social factors again play a role in determining access to literacy, and thus the extent to which printing can act as an agent in the diffusion of literacy [eds].

27.1 What is writing?

27.1.1 Introduction

Our intention is to examine the historical development of writing as a symbolic system and to provide a way to begin to think about how such systems might have developed and how change in such systems might be described. A symbolic system such as writing mediates between individual cognition and social phenomena. Stating that writing is an individual system means that it has a psychological basis and that any piece of writing is an external representation or outcome of internal cognitive processes. At the same time, writing is 'out there'; it exists along with other social artefacts of culture, and forms part of a broader social context. Most studies of writing have placed it either in its individual *or* in its social context. However, a fuller description needs to take account of both perspectives, and much of this article is concerned with clarifying the relationship between individual and social phenomena.

We do not provide a history of dates and facts associated with writing; where others have attempted this, it has been done as adequately as current knowledge permits, and we will refer to these works. Rather, we devote considerable space to outlining a richer description of what we mean by writing. This in itself is a major undertaking if we are to avoid over-simplistic dichotomies between language and thought or between the individual and society. We also wish to be wary of trying to impose simplistic evolutionary mechanisms on the development of this symbolic system. We hold that it is not possible to plot development and to posit mechanisms until there exists an adequate understanding of the phenomenon being described.

Like others we see literacy as a technology; but, again, in characterizing literacy like this we need to avoid simply viewing technologies as being independent of the cognition that produced them and the social context in which they exist. Our analysis of writing in this respect serves as a particular example of the discussion of more general issues.

We begin by examining how writing has been defined in relation to other forms of visual representation. This enables us to discuss its origins in these other forms. This is followed, in Section 27.1.3, by a brief overview of the functions of writing in early complex societies and the materials and techniques they employed. Section 27.2 covers the well-documented area of the development of writing-systems, and whether it is possible to evaluate one writing-system as 'better' than another. We then move on from the structure of writing to consider more broadly the functions of writing.

Sections 27.3 and 27.4 can be seen as the beginnings of a functional history of writing; this appears to be a novel undertaking. Existing histories concentrate on the technical features of scripts, and deal unevenly with aspects such as the degree of penetration of literacy skills in a given society (i.e. who reads and writes, and for what purposes) and the relationship of literacy activities to the economic, political, religious, and legal systems of the culture of which they form a part. It is difficult to piece together existing information into a coherent account of the development or diversification of functions of reading and writing—the contexts and social practices of literacy—but this is essential if we are to fully account for the phenomenon. Section 27.3 examines hypotheses about the functions of literacy at the individual level—the effects that literacy has been said to have on cognition. Section 27.4 investigates critically some of the features of the relationship between literacy and society and the claims made about the social effects of transitions from oral-based cultures to those based on literate traditions. In this way, we begin to break down and analyse the complex notion of 'function' as it applies to literacy, and to approach it on a number of different levels.

Our account of the development of written language draws on three distinct sources of information. The obvious first source of data is the evidence which can be gleaned from the historical remains of early literate societies. To this we can add ethnographic and other information which can be obtained from comparisons between contemporary literate societies. Thirdly we will include data from oral cultures, both the language of oral cultures and the other means of communication used; this is a source of information which is often not considered, but which is essential for a full picture of the nature of literacy.

Tracing the prehistory and development of established writing-systems is hampered by the disintegration of crucial evidence; this is a restriction on historical reconstruction that we should bear in mind. As Gaur points out (1984, p. 35)

Many ancient systems of writing, like those used in Egypt or India, seem to have appeared more or less fully fashioned, simply because we encounter them first on imperishable material, mostly on stone. But as we know from later examples, writing on imperishable material is nearly always preceded (and accompanied) by writing on perishable material.

27.1.2 *Defining writing*

The two most widely quoted authorities on the development of writing are two works which were first published over four decades ago: they are Diringer (1968, orig. 1948) and Gelb (1963, orig. 1952). Both are rich in sources of data and anecdote, both being presented within a framework of gradual development toward our modern Western alphabet. More recent overviews are presented by Goody (1981) and Gaur (1984). These four authors are the sources of the data in the remainder of this chapter, unless otherwise indicated.

There is, in human terms, a huge gap between the earliest cave paintings and the emergence of writing. Evidence exists from many parts of the world of marks incised on rock faces by humans some 30 000 years ago. The 'explosion' of visual representation at

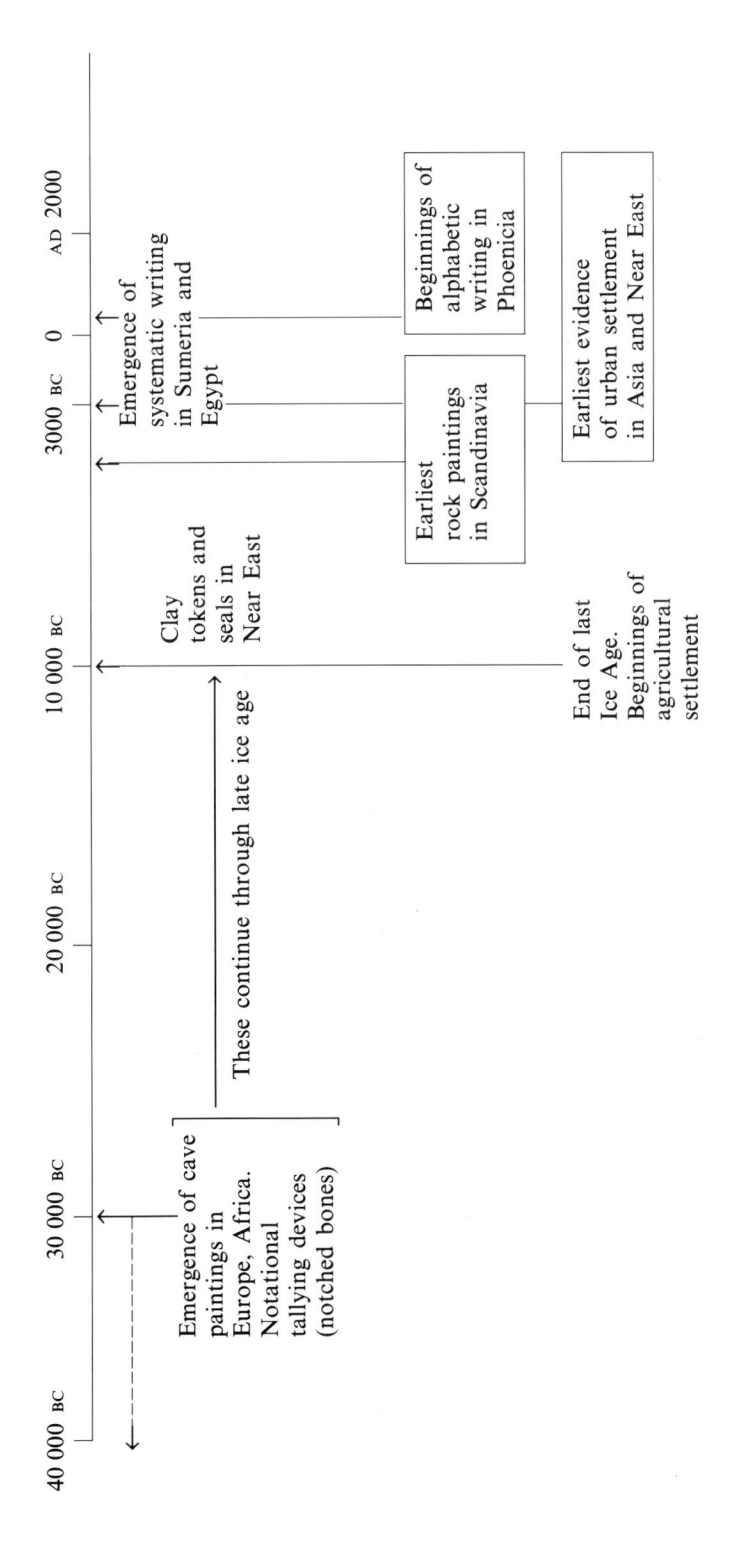

Fig. 27.1 The emergence of visual symbolic representation.

this time is in fact used as one marker of the qualitatively different achievements of our species (see Conkey, this volume, Chapter 11). Many and varied examples of cave paintings and other marks continue throughout history. However, the emergence of what is regarded as a true writing-system dates back less than 6000 years from the present.

Between cave paintings and computer graphics lie many varieties of visual symbolic representation that any study of writing must situate itself within. Activities that we may wish to call 'writing' extend far beyond the use of an alphabetic script written with a pen on paper: typing symbols into a video display unit, carving Mayan hieroglyphs on a stone sarcophagus, incising cuneiform characters on a clay tablet, stamping Chinese logograms on silk fabric are all writing activities. Some may even want to claim etching a mythic 'map' on birch bark to transmit a narrative within a tribal culture as a form of writing; and our own alphabetic writing-system is frequently mixed with other symbols in particular layouts, for example when we read advertisements or statistical reports. We also 'read' minds, maps, road signs, and tea-leaves. It is not clear what the boundaries are and where the metaphors begin. Which of these activities involve literacy? Which are writing-systems?

27.1.2.1 *Definitions of writing-systems*

Definitions of writing vary mainly in how inclusive they are. There is little disagreement about the phenomena associated with literate activity, only about the range of phenomena that should be designated 'true' writing. The dilemma centres around the connections between the system of visual representation and spoken language, which may be more or less tenuous for a given system. Thus Gelb (1963, p. 12) suggests a broad definition which encompasses, but does not restrict writing to, systems based on linguistic forms: 'Writing is a system of human intercommunication by means of conventional visible marks.' Diringer (1968, p. 8) however, offers a more restricted definition that emphasizes the correspondence of a writing system with spoken language: 'Writing is the graphic counterpart of speech. Each element, symbol, letter, hieroglyph, written word, in the system of writing corresponds to a specific element, sound or group of sounds such as a syllable or spoken word in the primary system.'

The importance of the issue of whether writing should be seen as visible language, or whether it is more usefully characterized as a means of reproducing thoughts, or storing information, can be seen by considering writing in relation to other forms of visual and symbolic representation. A basic distinction is between writing and purely expressive forms of visual representation, such as pictures, carvings, and sculptures. In contrast to these forms, writing is for purposes of communicating from one human being to another (or others). The marks of a writing-system, therefore, are to some degree stylized and conventional, and not freely varying. Crucially, the writing-system is composed of a *sequence* of conventional marks. It is distinguishable from the static representation of a single picture: it aims to convey 'not the pure representation of the event, but a narrative of the event' (Diringer 1968, p. 10). Gibson and Levin (1975, p. 166) further propose a set of 'universal' design features distinguishing writing from drawing or painting (such as directionality and size). These describe the graphic features of a display rather than its relationship with the spoken language.

A second distinction is between writing as a carrier of symbolic meanings and other vehicles for such meanings. These range from the use of objects or sets of objects, for example notched sticks and knotted ropes, to convey messages (see Gelb 1963, pp. 5–6; Diringer 1968, p. 9) to other extremely complex and abstract systems of notation. These latter include mathematical and logical notation, chemical or alchemical formulae, musical notation, spatial representation in maps and diagrams, and codes of various kinds. Although these systems are formally distinguishable from writing in that they have separate sets of symbols to communicate distinct kinds of information and relationships, in practice their use and development is closely tied in with writing itself. More recently, Gelb (1980) has drawn attention to these. He terms them 'para-graphic' systems, and proposes that a major task for understanding writing-systems is to document and integrate these aspects of visual representation and their significance for human symbolic thought.

The distinction between writing proper and other notational systems would seem to centre around the type of information being communicated. It is possible to make the distinction clear by restricting 'full' writing to those systems which are based on some correspondence with the spoken language. 'Writing' then becomes one restricted variety of notation (for conveying linguistic information) comparable to other notational systems. Ideographic systems without a linguistic link are thereby relegated to a category of 'proto-writing' or 'forerunners' of writing. This is, in fact, the solution favoured by Gelb and others (see also Haas 1976). Such a solution partly reflects a central preoccupation of these authors with spoken language as a starting-point for investigating writing. If one's starting-point is the development and variety of visual manifestations of human symbolic thought, however, such divisions are less pivotal. Gaur (1984) has pointed this out, challenging traditional approaches to the history of writing by looking at writing primarily as one particular means of information-storage that facilitates human communication through space and time.

In order to understand the historical development of writing, we need to adopt an integrated approach which addresses all the varieties of representation mentioned above. This includes functional questions—examining the characteristics of reading and writing in terms of their uses in societies and for individuals, and how these uses have changed and influenced writing-systems themselves. Such an approach can retain essential distinctions between writing and other types of visual and symbolic representation, while enabling activities which on the surface seem very diverse to be considered on the same continuum.

27.1.3 *Writing and earlier forms of symbolic representation*

Writing originated in at least three separate cultural areas: in Mesopotamia and Egypt, in China, and in pre-Columbian America (Hammond 1986). A further possible source, the Harappa civilization of the Indus valley (now Pakistan) has yielded a script of uncertain origins that is as yet undeciphered (Mahadevan 1982). These details are summarized in Fig. 27.2. Of these systems, only the Chinese has maintained an unbroken tradition to the present. Use of the pre-Columbian scripts and the Indus valley script ended with the demise of the civilizations of which they were a part. The Mesopotamian and Egyptian scripts went through a chain of transformations from pictographic beginnings to the development of alphabetic forms spreading through different cultures and languages, so that most current writing-systems can be traced back to them (see Appendix 1). The earliest evidence of writing occurs in Mesopotamia around 3500 BC with the cuneiform writing of the Sumerian people. This script was originally pictographic, but exact details of where, when, and by whom it was invented are unknown.

We identify six classes of visual representation which can be considered to give us an understanding of the roots of writing and of the impetus toward developing writing-systems (summarized in Fig. 27.1).

1. Expressive and ritualistic markings of cave paintings and carvings. As Gelb puts it (1963, p. 27) 'At the basis of all writing stands the picture.' Paintings and etchings on rock-surfaces began to appear during the Upper Palaeolithic period around 32 000 years ago (Marshack 1972). Such markings are found worldwide, and consist mainly of representational pictures of human and animal figures and geometrical markings (combinations of lines, circles, dots, etc.). In many instances, human and animal figures appear in conventionalized and abstract forms. However, these markings are distinguishable from writing in that there is no sequence to the pictures or symbols. They appear to be static and self-contained impressions, rather than being narrative in intent.

We can only speculate about the reasons why these early markings were made and their symbolic meanings, if any. Many of them may have been expressive and aesthetic rather than deliberately communicating (or recording) meanings, or they may have been of ritualistic or magical significance (as is suggested by the fact that many of the markings appear in areas of caves that must have been quite separate from living quarters—see Paturi 1979).

2. Tallying. It has been proposed (see Marshack 1972) that the geometrical markings were tallying devices for counting and reckoning time, acting as rudimentary lunar calendars. Tallying devices appear frequently in cultures without systematic writing-systems. The development of tallying systems for reckoning time and objects seems to have preceded other aspects of writing by thousands of years (see Flegg 1983, pp. 40–3). Indeed it may have been a major impetus toward developing more complex systems of symbolic notation, both mathematical and linguistic (see Appendix 2).

3. Property markings and totems. Markings which are symbolic of ownership or personal identity are extremely common in both oral and literate societies. Such marking ranges from branding of livestock, through craftsmen's hallmarks and tattoos and totems indicating community of kinship identity, to the symbols of heraldry, seals, badges, and flags. Historical examples (from Diringer 1968) include Scandinavian runes which were used to identity personal weapons and Egyptian ceramic craftsmen's marks. As Gaur points out (1984, p. 23):

Property marks are in many ways already a utilitarian form of writing—they can act as 'signatures', establishing authority, indicating ownership. They are closely connected with elements congenial to the development of systematic writing: a growing awareness of the importance of personal property, a realization that in a differentiated society property can bestow status; a desire to protect and/or exchange such property and the realization that property must be administratively identifiable.

4. Tokens. For record-keeping and exchange, used to unambiguously identify and reckon commodities in a commercial context. The connection of these with rudimentary tallying devices and also with the wider class of property marks is obvious. Recent investigations have suggested that the original Sumerian cuneiform

Fertile crescent[1]

Sumerian 3100 BC – AD 75

Egyptian 3100 BC – AD 200

Other distinct writing systems:
Proto-Elamite 3000 BC – 2200 BC
Cretan 2000 BC – 12th Century BC
Hittite and Luwian 1500 BC – 700 BC

These writing systems are the roots of all current alphabetic systems, including Indian and SE Asian, via Phoenician (c.1500 BC) and Greek, and developed from pictographic beginnings.

Pre-columbian America[2]

Mayan AD 300 – AD 1500

Toltec and Aztec scripts later derived from Mayan

The Mayan writing system died with the Spanish conquest of the indigenous civilizations. The Mayan system was hieroglyphic, and included phonetic elements.

Indus valley[3]

Proto-Indic, (c.2400 BC – 1700 BC)

Undeciphered seals. The writing system died with the culture, and later Indian writing systems originated elsewhere

1500 BC to present

Chinese writing is a stable, ideographic system. It is the longest unbroken tradition of writing, and is the root of Japanese, and early Vietnamese

4000 BC 3000 BC 2000 BC 1000 BC 0 AD 1000 AD 2000

Fig. 27.2 The origins of writing.

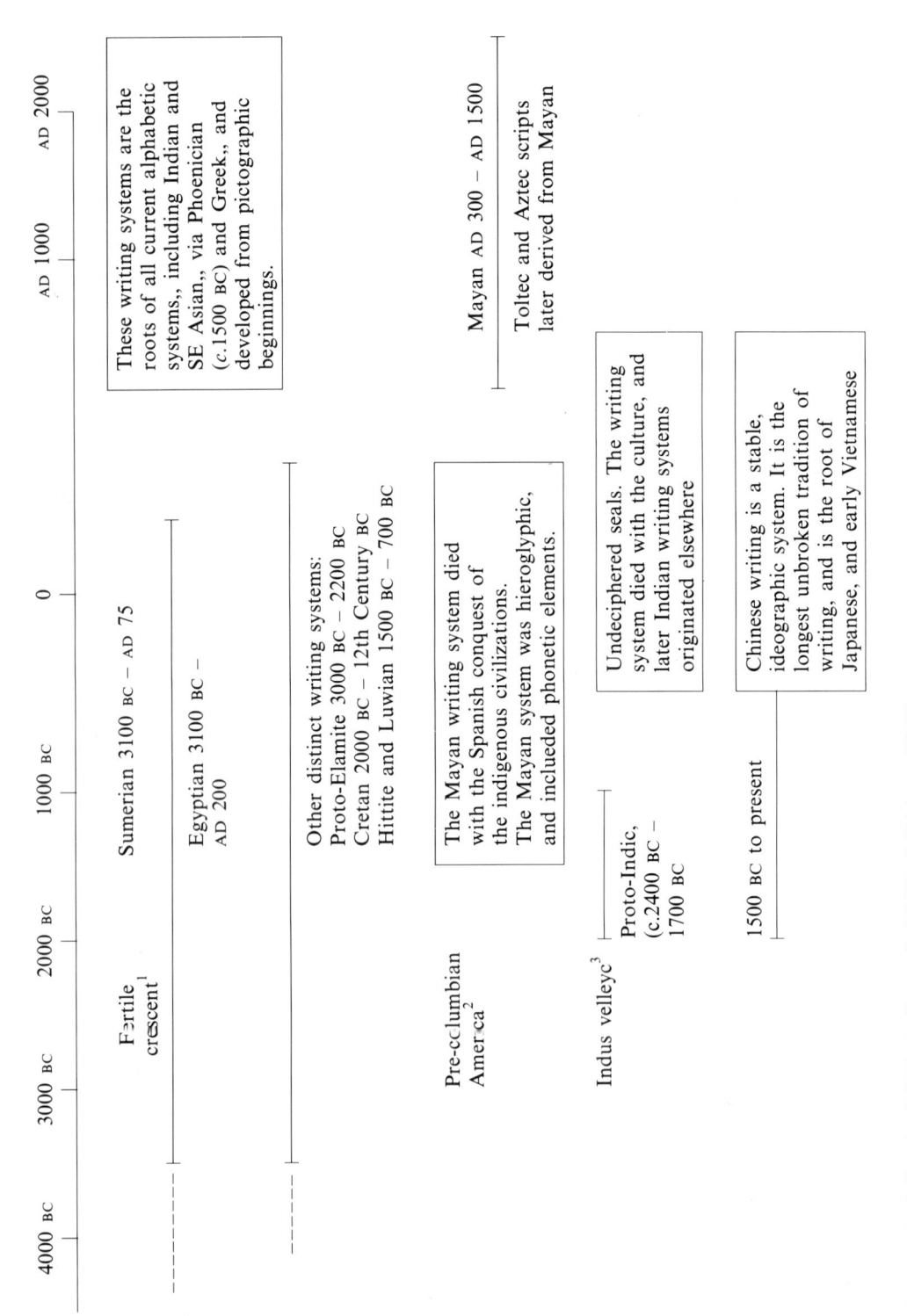

Source: 1, 4 Goody (1981); 2 Cotterell (1980); 3 Diringer (1968).

writing-system may have developed directly from the use of such tokens, some of which were in use as early as 10 000 years ago (Schmandt-Besserat 1978).[1]

5. Mnemonic devices. The use of visual memory-aids has been widespread in past and contemporary societies, and likewise occurs in both oral and literate societies. A variety of devices have been used, and they are often extremely complex, ranging from the use of notched sticks for calendars and reckoning, to the wampum-beaded belts of the North American Iroquois, with diverse functions including narrating stories and carrying messages. These devices continue in literate societies, as in the Catholic rosary, and illustrate a widespread function of symbolic representation as an amplification of human memory.

6. Pictographic or purely ideographic narrative forms consisting of *sequences* of pictures which may be more or less stylized and systematized and have the clear intent of communicating a message. Examples can be found among Eskimo–Alaskan peoples, North American Indians, and in Australian aboriginal 'churingas', which are narratives of a mystic dream time expressed via abstract symbols on wooden tablets. As the pictographic signs become more and more abstract and systematic, the structure verges toward a system like Chinese writing, and can provide a bridge between these early forms and actual writing-systems.

27.1.4 *Apparent functions and processes in early societies*

As already noted, early societies have had a variety of uses for graphic notation: even cave paintings and early pictographs included abstract and conventionalized characters, which it has been suggested were used for ritual purposes and record-keeping; possible token and calendrical systems are also in evidence. The mythic 'maps' of the Ojibwa Indians and the 'churingas' of Australian aboriginals are examples of spatial and narrative representations that can be extended in complex ways (see Goody 1981).

27.1.4.1 *Functions of writing*

If we look at the different civilizations that have engendered writing, we again see a variety of early uses here too. It is notoriously difficult to classify functions. Here we will keep to Goody's (1981) broad distinctions of the *political and economic*, the *religious* (in the sense of mystical, divinatory, and ritual uses), and the *historical–literary* functions. In ancient Mesopotamia the original impetus for developing writing appears to have been trade and commerce, and the earliest users were merchants and accountants. In Egypt, the considerations were also economic, but involved the development of a calendar for predicting the floods of the Nile, and writing was practised by priests and administrators (Baines 1983). In these societies, therefore, the direction of development of writing was from the economic into the historical and literary spheres.

In China, the first written 'documents' that we know of include examples of divinatory uses and administrative uses. The Chinese early developed extensive historical and literary records, maps, and astronomy, but the development of writing does not seem to have been primarily due to commercial needs (see Ho Peng Yoke 1980). In the Indus valley, the surviving (but largely undeciphered) records seem to be of seals or tokens, indicating commercial and administrative uses. In later Indian civilizations, writing was first used for commerce and administrative purposes, and gradually established itself as a carrier of religious tradition after initial resistance. Historical records were not highly developed, reflecting, Gough (1968) suggests, a different attitude toward the significance of chronological time. In pre-Columbian America, the commercial and administrative uses of writing appear to have been less important than the religious and divinatory uses, involving the development of a calendar and astronomical science. Since these societies were theocratic, however, religious uses overlapped with administrative ones. As Hammond (1986, p. 106) put it: 'The most important functions served by ... [literacy and numeracy] included depicting the gods and showing how the relation between gods and human beings underlay the succession of rulers in the Maya community.' Looking at each of these societies in turn, the picture suggests that we can identify a pattern of development in any given society (as Baines (1983) has done for Egypt), but that there is no single or necessary order that applies to all societies.

27.1.4.2 *Materials and techniques for writing*

The *surfaces* used for writing vary in their perishability and the amount of preparation and manufacture they require. In terms of natural materials, stone, clay, wood, plant leaves and barks, animal skins, bones, and shells have all been used and adapted for writing. The earliest known writing, cuneiform, was written on clay tablets. Indian writing has largely been on highly perishable palm leaves. The earliest known Chinese writing is engraved on bones, used as a divinatory oracle (Ho Peng Yoke 1980).

Prepared surfaces range from fabrics such as silk and linen, to metals, to vellum and parchments derived from animal skins. Papyrus, invented and marketed by the ancient Egyptians, was the most widely favoured in the Near East for almost 4000 years, until the advent of paper overtook it.

Most societies have used a variety of materials for different writing purposes. 'Permanent' surfaces such as stone or metal were used for inscriptions that were intended to endure; 'soft' re-usable surfaces, such as the Greek and Roman wax tablets, for note-taking and drafts; paper-like or parchment and bark surfaces for text that needs to be reasonably durable and portable (see Bowman and Thomas 1974).

Paper was invented by the Chinese around the beginning of the second century AD, although it did not reach Europe for a further thousand years. During this time the indigenous people of pre-Columbian America were using their own version of paper—'amatl' paper, made from plant fibres and varnished. In China, paper was probably developed as a result of deliberate experiments with writing materials such as silk and other woven fabrics.

A wide range of *techniques* have always been used for writing. Written characters may be incised, painted, or stamped on to surfaces, and all three techniques have been liberally used historically. Incision involves the use of cutting tools appropriate to the medium—wood and stone chisels. Ink and varnish have been used since the earliest times, pre-dating writing and extending back to palaeolithic cave paintings; their suitability has to be seen in terms of the available surfaces and the instruments used, along with the durability of the mixtures.

This brief review suggests that writing has been used in early societies and civilizations for many of the *broad* functions that we can identify today, although in detail and complexity there are still huge variations. From the beginning many materials and techniques were utilized as they were available, and there has been a continual search for greater refinement and efficiency. It should also be clear from this section that there is a great deal of continuity from other forms of visual representation to actual writing in terms of function and processes. In later sections we will return to discussing how these developments have continued in society; but as a first step we now turn to the structure of different writing-systems.

27.2 The development of writing

27.2.1 *Types of writing-system*

Before investigating how writing-systems have developed and discussing possible advantages of particular systems, we will describe the standard classification of types of writing-systems. This identifies three types: *logographic* writing-systems, *syllabic* ones, and *alphabetic* ones. These differ from each other crucially in the units they are composed of; other differences follow on from this. In reality, describing a particular writing-system involves much more than specifying the units that are its building-blocks, and includes punctuation, layout, and script. However, the units represent the principles underlying the construction. Inevitably, the description which follows is an oversimplification. Fuller descriptions can be found in several contemporary sources, such as Gibson and Levin (1975), Taylor (1981), and Sampson (1985).

Chinese is usually given as a modern example of a *logographic* writing-system. The basic unit of the Chinese writing-system is the character. The character is a unit of *meaning*, so that the simplest characters represent individual morphemes (the minimal units of meaning in a language). Japanese is an oft-quoted example of a *syllabic* writing-system. Japanese writing is in fact mixed. It contains characters, as in Chinese, but in addition it uses two syllabic systems. Here the basic unit is the syllable, a unit of sound rather than of meaning. The third type of writing-system, the *alphabetic*, is represented by the familiar example of English. The unit is again based on sound; but here the unit chosen is more closely related to phonemes than to syllables.

27.2.1.1 *Logographic script*

Returning to the Chinese characters, there is a traditional classification into six distinct types of character in terms of how they are constructed (as described by Martin 1972). Some characters are visual abstractions of the meaning of the word. For example, the character for *tree* is an abstraction of a picture of a tree. However, most characters are more complex than this, and contain some clue to the pronunciation of the word as well as a clue to its meaning. The most common type of character is made up of two parts, a *radical*, which contains a 'clue' to the meaning, and a *phonetic*, which contains a 'clue' to the pronunciation. There is a relatively small fixed number of radicals (around two hundred), and they are used as components of a large number of characters. Such characters are known as 'phonetic compounds', and it is estimated that around 90 per cent of characters are composed in this way (Martin 1972). Thus the great majority of Chinese characters have an internal structure (the image of the Chinese having to learn by memory 40 000 unrelated squiggles has a fine intellectual history in the West (see for example Halle 1969, p. 18; Ong 1982, p. 87), but is totally incorrect).

The existence of the *phonetic* in most Chinese characters means that they do generally contain some 'clue' to their pronunciation. However, it is only a clue: historical sound-change and the existence of many dialects of Chinese ensure that it can only ever be approximate. Further, in contrast to the situation with syllabaries and alphabets, in cases of uncertainty reference can never be made to a relatively brief but complete list of fixed and canonic forms of all the symbols that are in use.

27.2.1.2 *Syllabic script*

Japanese is straightforward, in that the spoken language is fairly simple and regular in its syllable structure, and only a small number of syllable symbols (around forty) are needed to write it. However, in practice the situation is far more complex: in addition to Chinese characters, Japanese uses two distinct syllabaries to write the language—words of Japanese origin and grammatical morphemes are written in *hiragana*, while words of foreign origin are written in *katakana*. Up to a third of all words are written using Chinese characters. These three systems coexist in any one sentence, and the reader of a text is in fact contending with three different systems at the same time.

27.2.1.3 *Alphabetic script*

English uses an alphabet, where 26 symbols are used singly and in combination to represent the forty or so distinct sounds of the language. The relationship between sound and symbol is complex, one sound being represented by several different symbols and one symbol representing several different sounds. The English spelling system is notoriously complex (others such as Spanish and Finnish being less so); but there is no alphabet of any language which has a completely simple relationship between sound and symbol. Further complications arise when we accept that the English spelling system provides information about word structure, as well as being a guide to potential pronunciations (see below).

27.2.2 *Development from one system to another*

How could these different writing-systems have developed, and to what extent could there have been evolution from one type to another? We will begin with the logographic systems. Goody (1981, p. 111) identifies seven logographic systems used in the ancient world, the earliest being the Sumerian. However, we will keep to our example of Chinese, for which there is evidence only from 1500 BC onwards, since, being in use today, it is better documented, and we can check any claims made about the language with living speakers of it.

Some Chinese logographs, the ones which developed from pictorial representations (such as the character for 'tree'), were gradually made simpler and more stylized. Problems arise in needing more and more characters and in representing words which are not as easily picturable as 'tree'. There are two obvious directions which the expansion of characters can take: they can pursue similarities of meaning—such as having similar characters for 'river', 'water', 'drink', and 'thirsty'—or they can pursue similarities of pronunciation. The crucial step here is in the looseness of the tie between the meaning of a character and the pronunciation of that character. This opens up the possibility of characters' being used to stand for words similar in pronunciation but unrelated in meaning. As has been pointed out by Martin (1972), cognitively this is no different from the processes involved in punning, and it is used today in Chinese, for example when creating characters to write foreign surnames. This ability is essential for the development of writing-systems; but in terms of people's mental abilities it rests upon common forms of abstraction. It was also a very obvious feature of other writing-systems as far apart as Mayan logographs and Egyptian hieroglyphs with their *rebus* writing—for example using a picture of an eye to represent the word *I* in English. Crucially, examples of this are found in the symbol systems of oral cultures. Gaur (1984) gives an example from the Yoruba of Nigeria, where cowrie shells are used for messages: eight shells means *I agree*, since *ejo*, the word for 'eight', is homophonous with the word for 'agreed'. We will return to these two dimensions, meaning and sound, in the next section.

Before looking at how other writing-systems could develop we need to make two important points. Firstly, spoken languages differ widely in the ways they are constructed, and we cannot discuss any writing-system in isolation from the language it is being applied to. The Chinese writing-system, for instance, suits the Chinese language in certain ways because of particular facts about the way in which Chinese is constructed. Secondly, the most common way for languages to get a writing-system is through contact with another language which already has one: a writing-system is borrowed and adapted to new circumstances. It is by means of this process of adapting to the demands of a new language that most of the significant developments in writing-systems have taken place. Each development is small but significant, and is explicable in terms of the context in which it occurs (see Goody 1981 for examples). It is not necessary to stare wide-eyed at the 'invention' of the alphabet as if some individual genius sat down with a crisp piece of blank paper and at a stroke made a great leap for humankind. The alphabet was developed gradually and awkwardly as writing-systems were adapted to suit different languages. So it probably does not make sense to try to pinpoint the actual emergence of a distinct alphabet.

However, we are running ahead. We should first return to Chinese, and the question of how a syllabic system might develop from a logographic one. Chinese has very few inflections to add to the ends of words; it also has a large number of homophones. These two different facts about Chinese contribute to a logographic system suiting that language fairly well. That Chinese also has a simple syllable structure and that a large proportion of its words are monosyllables means that the next 'step', of regarding each character as representing a single syllable, can easily be taken. Thus, the seeds of a syllabic writing-system are comprehensively contained within the Chinese system.

This was the development which took place when Chinese characters were adapted to the needs of the Japanese language. Japanese also has a simple syllabic structure; but it is more highly inflected than Chinese (meaning that it has a large number of 'endings'), and fewer of its words are monosyllabic. Because of these characteristics Japanese is well suited to having individual symbols which represent the limited number of syllables in the language, but themselves do not carry meaning. We can thus understand how Chinese characters were adapted in this way and simplified to form a list of the syllables of Japanese, and also how Japanese came to have a syllabic system as part of its writing-system.

This is how Japanese syllabaries developed; but we are not implying that a syllabary has to have a logographic system as a forerunner. It is noteworthy that when indigenous peoples have invented writing-systems after exposure to the mere idea of writing (as in examples in West Africa and North America) they have tended to invent syllabaries (see, for example, Walker 1981).

The steps from a syllabic system to an alphabetic one can also be traced by realizing that languages have particular properties of their own and that these different properties will make different demands on a writing-system. The historical details of the development of the alphabet, as far as they are known, are clearly laid out by Goody 1981 (see also Goody 1983). Here we want to abstract out the principles underlying this development. Several languages of the Near East have phonological systems in which words are constructed in such a way that vowels are relatively predictable and words can be written unambiguously with only the consonants. We can therefore understand how a language might borrow the syllabic units of another written language and use them to represent consonants alone. However, it may be the case, as with Hebrew and Arabic, that while one can usually write only the consonants, vowels are sometimes needed to make writing comprehensible. Vowel symbols can then be added, attached to the consonant symbols as diacritics.

Whether at this point one has a syllabary or an alphabetic system seems largely a question of definition (and one that has vexed those who wish to attribute all things good—including the invention of the alphabet—to the ancient Greeks). One can view this situation as consonant symbols and vowel symbols joined together; or one can regard this as being a 'symmetrical' syllabary where the consonant elements and the vowel elements take a common form in different syllable symbols. The answer might depend on whether the vowel symbols can occur independently or whether they only exist joined up to consonants. Posing these questions serves to demonstrate how blurred the distinction between syllabic and alphabetic systems can become. (See the discussion in French 1976; Haas 1983; and Taylor's discussion (1980) on Korean as an alphabet, a syllabary, and a logography.)

If one has a system where vowels are optional appendages and one attempts to apply it to a language like Greek where, because of the structure of the spoken language, the vowels in a word are not predictable, then a great deal of ambiguity (or more likely, incomprehensibility) would result if the vowels were not always marked. It is then a short step to having separate symbols for each of the vowels, and thus to having arrived at the forerunner of our modern alphabet.

Having given a general outline of how an alphabet might develop from a syllabic system, we can now investigate whether this development has been an 'evolution'.

27.2.3 *Advantages and disadvantages of particular types of writing-system*

What has been the nature of this development from one writing-system to another? Has it been some unbridled march of 'progress', some natural evolution towards the 'best' system? This is certainly the position taken by Gelb (1963) and Diringer (1968), the two most widely quoted documenters of the writing-systems of the world. Gelb's book begins with a family tree which has our own English alphabet right in the middle at the bottom, as if all changes were an inexorable development towards this point. Diringer proclaims that 'alphabetic writing is now universally employed by civilised peoples' (1968, p. 13), and turns up the most extraordinary explanations as to why users of lesser writing systems have failed to adopt our alphabet (p. 127). These views on the alphabet seem to have originated as part of a post-Second World War ethnocentricism, so entrenched that in 1958 Berry was able to claim that 'it is generally accepted on all grounds an alphabetic system is best' (p. 753). It is depressing that a recent popularization of 'the technologizing of the word' (Ong 1982) should be cast in this same mould.

The first qualifying point is that, as we have already explained, different writing-systems may suit different languages, so that a syllabic system may be totally appropriate for a language such as Japanese, which has a simple and regular syllabic structure, while such a system would be very cumbersome in another language such as English. We can go further than this, and point out that different languages themselves vary in terms of how easy they are to write down: spoken English, for example, has certain properties, such as its ubiquitous reduced vowels, which the writing-system has never adequately captured, that make it a fairly 'messy' language to write down.

To say that certain writing-systems suit certain languages is not to suggest that writing-systems are adopted for a language solely on this basis. Writing-systems are adopted for political reasons; maps of the spread of writing-systems and maps charting the extent of Chinese-based writing, or Arabic-based writing, or English-based writing are maps of economic and religious domination. The choice of a writing-system is firstly a political decision; and within that constraint further decisions are then made (see, for example, Wellisch 1978).

The second point is that there is not one simple dimension on which to base this evaluation. Writing-systems serve many different purposes, and these will impose different and sometimes contradictory demands (see Berry 1958, 1977; Venezky 1970). We will examine some of these demands to see if alphabetic systems are automatically superior in all, or any, of these situations.

Two demands to begin with are those made by someone learning to read and those made by a fluent reader: do writing-systems differ in how easy they are to learn, and do writing-systems differ in how quickly or efficiently they can be read by the mature reader? Neither of these demands is a straightforward matter. They are distinct demands, as there is no *a priori* reason to expect that the characteristics aiding the mature reader should be the same as those which make it easy for the learner. Further complexity is added when we consider that the learner can be an adult or a child, and the language can be their first or their second. Most focus has been on the child learner; and we will restrict ourselves to this situation.

Although fluent reading need not be acquired in relation to the spoken language, learning to read typically starts from there. Different writing-systems require the learner to attend to different aspects of the language and the meaning of sounds at different levels of detail, syllables or segments. A logographic system is probably the easiest to master at the very beginning stages of learning to read, and 'whole word' methods in English (Gibson and Levin 1975) effectively treat English words as logographs. Syllable-based systems demand less detailed analysis of the spoken language, and it is claimed that with such systems children learn to read more quickly and do not experience reading problems such as dyslexia (Makita 1968: but see Paradis *et al.* 1985). It seems that at the very beginning stages alphabetic systems such as English are over-complex. Significantly, one study of English-speaking children experiencing reading problems claimed that such children had no difficulties when beginning to read with a method that involved teaching them to read English using Chinese characters (Rozin *et al.* 1971). On the other hand pure alphabetic systems have the advantage that once one has 'cracked the code' one can read any word, and the case is similar with a syllabic system: there are more symbols to learn, but once one has mastered them any word is readable. With logographs the learning continues into adulthood, with parts of each new character being learned separately.

27.2.3.1 *The efficiency of scripts*

There is now considerable interest in what is referred to as the 'efficiency' of different scripts (see, for example, Taylor 1981; Henderson 1982). It is certainly clear at the surface level of dealing with the visual information that the processes involved in reading a text will differ with different writing-systems. The question is posed whether a skilled reader can read more efficiently, that is more quickly and accurately, in one system than another. Most of the research on reading processes has concentrated on English (but see the work of Tzeng and colleagues on other scripts, such as Tzeng *et al.* 1978). While comparisons between scripts are difficult to make without bringing in a host of other factors, probably all researchers would agree that there are not global differences in efficiency attributable solely to the type of writing-system. There is no support for the idea of alphabetic scripts being more efficient in this sense than other scripts (see Martin 1972, for example, on Japanese texts being shorter than comparable English ones, and thus probably quicker to read).

The lists of demands that can be made on a writing-system can easily be extended beyond ease of learning to read and efficiency for the mature reader. One could discuss ease of learning to write and efficiency for the skilled adult writer. Writing involves a whole set of different skills from reading, and places different demands on a writing-system. Another demand is the ease with which it can be incorporated into forms of technology. Japanese and Chinese have proved cumbersome for printing and typing. However, computing presents a new configuration of factors, and with this new technology Chinese characters present no problems of storage, retrieval, or printing (see Stallings 1975; Becker 1984). (At first the alphabet was thought to be particularly suited to the computer; but it may be that the computer design was geared to the alphabet, rather than the other way round.) One could easily continue with this list of demands, adding, subdividing, and reorganizing the criteria. Here, all that needs to be demonstrated is that the alphabet is not superior in all respects; rather each writing-system has advantages and disadvantages, and these have to be discussed in relation to the specific language which they are expressing.

27.2.3.2 *Two principles of writing-systems*

Let us return to our original classification of writing-systems. Rather than try to fit them uncomfortably into the straightjacket of three distinct 'types' that have evolved one from another, let us take a fresh look at them. There are two principles upon which writing-systems can be based. These two principles reflect the commonly stated fact that language can be analysed on two levels; it can be analysed in terms of *sounds* or in terms of *meanings*. Either of these principles can be used to express language in a written form. (Differences between syllable-based systems and segment-based systems are differences in the size of unit chosen; they are not differences in the principle underlying construction.) In practice writing-systems make use of both principles.

Chinese makes use of both principles, in that the most common type of character contains both a radical, a clue to meaning, and a phonetic, a clue to pronunciation. Japanese clearly uses both principles, in that it contains both logographic and syllabic components. That English utilizes both principles is also true, although it may be less obvious: firstly, written English contains many logographs, such as &, 6, etc., and secondly, many idiosyncratic spellings, such as *right*, *write*, *rite*, act as logographs. As in Japanese and Chinese, they serve to keep potential homophones apart in writing, and have to be learnt individually. Similarly, the morphophonemic information in our spelling system (the relatedness of *sign* and *signature*, for example) often provides a clue to meaning.

Describing writing-systems in this way should clarify some potential problems with the different writing systems. With English we should realize that the aim is not solely to provide a phonetic representation, and that English is not necessarily inadequate where it fails to have regular sign–symbol correspondences. When we accept that two distinct principles are being used, then the Japanese writing-system makes much more sense to the outsider; there may in fact be advantages in an explicit mixture of the two systems. This could also be true of Chinese. China has been promoting the *pinyin* alphabet for the written language for several decades along with other reforms of the language (see Lehmann 1975). Slowness to adopt this seeming advantage may have its source in the other advantages inherent in using logographs for representing Chinese (such as to distinguish homophones and to make the written language accessible across diverse dialects). Currently *pinyin* is used primarily as a supplementary system, for example in providing children learning to read with a phonetic representation of new characters and when introducing a new character into the adult language. This leads us to surmise that in fact Chinese will never abandon characters: rather, like Japanese, Chinese will develop a mixed writing-system.

27.3 Literacy and cognition

27.3.1 *Cognitive 'effects'*

As societies acquire written language, are there any implications for the way people think? That is, do literate people think in different ways from people who are not literate, and can any such differences be attributed to literacy? Anthropologists and psychologists have addressed these questions and studied literate and non-literate people in an attempt to provide some answers. That there is a *great divide* between non-literate and literate societies, and that there exist *cognitive effects* of literacy are two significant and pervasive views that have provided a framework for much of the research on the topic. It is these assumptions that have influenced psychologists such as Greenfield (1972) and Olson (for example Olson and Nickerson 1978 and see also the contributors to Olson *et al.* 1985) who have carried out experimental studies, and it has also been the position that others such as Scribner and Cole (1981) and Hamilton and Barton (1983) have started out from. In this section we will examine these views critically.

An influential paper which can act as a landmark is one by Goody and Watt (1963, reprinted 1968). The core of Goody and Watt's argument is that oral and literate societies differ in how the cultural tradition is conveyed (Goody 1968, p. 67):

> In oral societies the cultural tradition is transmitted almost entirely by face-to-face communication; and changes in its content are accompanied by the homeostatic process of forgetting or transforming those parts of the tradition that cease to be either necessary or relevant. Literate societies, on the other hand, cannot discard, absorb or transmute the past in the same way. Instead, their members are faced with permanently recorded versions of the past and its beliefs; and because the past is thus set apart from the present, historical enquiry becomes possible.

They continue with how this can lead to different forms of thought and specifically to new forms of logical reasoning (ibid., pp. 67–8; emphasis added):

> This in turn encourages scepticism.... From here the next step is to see how to build up and to test alternative explanations.... The kinds of analysis involved in the syllogism, and in other forms of logical procedure, are *clearly dependent on writing*, indeed upon a form of writing sufficiently simple and cursive to make possible widespread and habitual recourse both to the recording of verbal statements and then to the dissecting of them.

A similar position on the relation of the development of writing and the development of logic is taken by the psychologist David Olson, who argues (1977) that a transition from spoken utterance to written text has implications for how we think. Changes can be described in terms of increasing explicitness; written language is able to stand as an unambiguous representation of meaning. Olson argues that, with writing, meaning can stand on its own, divorced from the shared meanings of speaker and hearer; with speech, the meaning resides partly in the context, whereas writing gives the possibility of decontextualized language and the creation of explicit autonomous text: the reader searches for *the* meaning (i.e. the one meaning) in the text, whereas the listener derives meaning from the intentions of the speaker (see Sinha, this volume, Chapter 17, for a fuller discussion).

What empirical support is there for these views? As societies gain literacy, if they cross a great divide and if profound effects are wrought on people's thinking processes, then we might expect to find clear evidence of these changes in contemporary societies where people are becoming literate. Although caution needs to be exercised in comparing historical developments with contemporary situations, it is instructive to examine evidence from cross-cultural studies. Such studies are notoriously difficult to carry out and interpret, and certainly some studies, such as Greenfield's (1972) of concept-formation in Wolof children, have been heavily criticized for using Western-style experimental situations inappropriately (see, for example, Street 1984.)

The most thorough and carefully executed study is undoubtedly Scribner and Cole's (1981) investigation of Vai literacy. While we cannot report on this study in detail here, it is worth looking at their overall findings. They took advantage of a fascinating linguistic situation that exists among the Vai of northwest Liberia, where three different literacies which are acquired in different manners co-exist side by side. Some people are literate in English, taught in formal schools; some are literate in Arabic, taught via the Koran; and some people are literate in Vai, an indigenous syllable-based script passed on informally from person to person. Many people there are not literate in any language, and some are literate in more than one of the languages. With a full panoply of tests and surveys and the collecting of background information, Scribner and Cole examined more than a thousand people over a period of four years, making detailed comparisons between the different groups in order to tease apart the possible effects of literacy. The skills investigated included memory, logical abilities, metalinguistic awareness, and classification and communicative abilities. For our purposes here it is worth while quoting their own evaluation of their search for cognitive effects of literacy (1981, p. 251): 'Our results are in direct conflict with persistent claims that "deep psychological differences" divide literate and non-literate populations... On no task ... did we find all non-literates performing at lower levels than all literates.' Similarly, in a much smaller study comparing literate adults with people with low levels of literacy in the United States, we set out to find cognitive effects of being literate and were unable to identify qualitative shifts in the way literates think (Hamilton and Barton 1983; Barton 1985).

27.3.2 *Speech as 'deficit'*

Since detailed empirical investigations of adults fail to pinpoint qualitative differences in thought between literates and non-literates, how can we explain the claim made by Goody and Olson? Two points can be made, one in this section concerning differences between written and spoken language and the other in the next section on the way we characterize literacy.

Olson compares written and spoken language and identifies specific phenomena in written language not present in spoken language, a crucial example being the explicitness that is associated with 'decontextualized' language. These properties of written language are then linked with specific reasoning abilities which are explicit and decontextualized, and the connection is made that the reasoning abilities have their source in learning to read and write.

The problem with this argument is the oversimple way in which differences between written and spoken language are described. Spoken language is represented by free conversation, while written language is represented by expository texts. However, there exist many forms of spoken language and many forms of written language: love letters, political debates, notes to the milkman, invoices, religious incantations, radio phone-ins, etc. Free conversation is one of the more context-bound forms of language, while written expository texts are more formal and explicit. However, to pick out conversation to represent spoken language and expository texts to represent written language is misleading. In any culture a whole range of forms of language exists; any person will have available a broad repertoire of 'ways of speaking', and one way in which these will vary will be in terms of the extent to which they rely on context.

To assume that, since expository texts are explicit, if in fact they are so, they are the only source of explicitness in a culture, and that cultures without expository texts lack a certain form of reasoning, is a false argument. It is a form of the 'deficit' argument: its reasoning is as follows. We (i.e. literate cultures) have something—literacy—which leads to something else—explicitness. Another culture lacks the first attribute, *therefore* it lacks the second. This is the falseness of the argument: it is totally possible, and in fact true, that there exist other possible sources or causes of explicitness. There is no reason to assume a simple unitary causation. This error in arguing comes from examining only our own language or culture in detail and then identifying certain aspects which other cultures *lack*. This is done without investigating these other cultures in detail, and, crucially, without examining how they might use different means to achieve the same ends.

The language of oral cultures has not been studied to the extent that the language of written cultures has been. However, it is essential to dispel the myth that people in oral cultures sit on their haunches all day in the sun chatting to each other, and that idle conversation is their main form of speaking. Oral cultures can have schools, legal systems, and political structures (see, for example, Vansina 1965; Bloch 1975); and these situations demand particular forms of language. Those who have examined ways of speaking in oral cultures have demonstrated that people use other means to achieve the effects attributed to writing in literate cultures. Several examples of this can be given.

Firstly, one of Olson's claims is that writing enables the originators of a message to distance themselves from the content of the message. This may be true, but it is incorrect to imply that oral cultures do not have ways for speakers to do this it may be achieved by form of words, by tone of voice, by stance or by costume; but it is certainly a feature of the discourse of oral cultures.

Secondly, Olson's suggestion (1982) that oral cultures lack metalinguistic terms for discussing language is easily countered by several of the contributors to Bloch (1975) and by Heeschen (1978). Each of these studies of very different oral cultures demonstrates that non-literates can talk about talking in very sophisticated ways. Further, oral cultures can have forms of speech which are as distinct from everyday conversation as expository text is in our culture. Akinnaso's analysis of Yoruba ritual communication demonstrates this point (Akinnaso 1982), as do studies of oral poetry and oral literature (Finnegan 1977). (For more discussion and examples, see Horton and Finnegan 1973; Street 1984.)

27.4 Social influences

27.4.1 *Characterizing literacy*

In discussing the cognitive effects of literacy, problems arise in the very way we characterize literacy. (Problems also arise in how we characterize thinking, but we will not pursue this here.) Literacy is seen as an independent variable, a set of technical kills, which can be separated out from other influences such as the social context and the educational situation. This 'autonomous' view of literacy (Street 1984) permeates the work on cognitive effects, but it cannot account for the results of studies such as Scribner and Cole's. We need new definitions of literacy in order to understand its implications for human cognition.

What might a new definition of literacy be like? Firstly, many authors consider literacy as a technology (although in several different senses). We believe it is useful to retain some notion of literacy as technology, but to look critically at what this implies about the role of literacy and of technologies generally in society and in human cognitive activities. The view of literacy as technology has frequently treated 'literacy', 'the individual', and 'society' as if they are independent entities that meet at some points. This does not allow for the dynamic and interactive nature of these relationships. Further, technologies are by no means neutral or autonomous (see Winner 1977), and literacy is a good case in point.

It is useful to see literacy as a *communications technology* concerned with the production and reproduction of shared meaning or knowledge. This is the perspective taken by Williams (1981). Since communication is a form of social relationship, communication systems have to be seen as social institutions and as part of a wider process of social organization and production.

Literacy, in this view, can only be understood in the context of other social institutions and the role which written communication has within these institutions. The uses of literacy need to be explored in detail: to know that a society has or does not have writing of itself tells us very little.

A parallel point has been developed in more detail by Street, who rejects the 'autonomous' model of literacy in favour of an 'ideological' model that assumes that 'the meaning of literacy depends on the social institutions in which it is embedded' and that 'the particular practices of reading and writing that are taught in any context depend upon such aspects of social structure as stratification... and the role of educational institutions' (1984, p. 8). A similar position is developed by Heath (1983). Scribner and Cole (1981) edge towards an alternative notion with their 'practice account' of literacy, arguing that literacy can only be understood in the context of the social practices in which it is acquired and used. They conclude their study of Vai literacy thus (1981, p. 236; see also Cole and Griffin 1980):

Instead of focussing exclusively on the technology of a writing system and its reputed consequences... we approach literacy as a set of socially organised practices which make use of a symbol system and a technology for producing and disseminating it. Literacy is not simply knowing how to read and write a particular script but applying this knowledge for specific purposes in specific contexts of use. The nature of these practices, including, of course, their technological aspects, will determine the kinds of skills ('consequences') associated with literacy.

These newer conceptions of literacy are being articulated by sociologists, anthropologists, and cognitive psychologists converging on the same problem. They are also crucial for understanding how literacy developed historically. They argue for us to connect the development of literacy with that of other communications technologies, and to integrate knowledge of social context and the uses of literacy with the search for cognitive consequences. In the next section, we will elaborate on some of the suggestions that have been made as to how literacy is embedded in social and economic institutions and activities, and what the consequences of this might have been. We will begin by looking at the role of literacy in Greek society, to continue our examination of the 'great divide' proposal.

27.4.2 *Greek literacy*

An important aspect of the 'great divide' proposal is the notion that modern literate societies are fundamentally different in many aspects of social organization from earlier, simpler societies, and that these differences are ultimately attributable to literacy. Aspects of modern societies that are said to hinge upon the existence of literacy include the development of democracy, certain forms of political organization, and the possibility for technological advance.

All these factors are identified from the sixth century BC onwards in Greek society, the foundation stone upon which Western civilization has apparently been built. Goody and Watt see the Greek example as crucial: 'The rise of Greek civilisation, then, is the prime historical example of the transition to a really literate society' and 'offers not only the first instance of this change, but also the essential one for any attempt to isolate the cultural consequences of alphabetic literacy' (1963, p. 42). Although accepting that any claims are extremely tentative because of lack of evidence, Goody and Watt continue (p. 43):

... the fact that the essential basis both of the writing systems and of many characteristic cultural institutions of the Western tradition as a whole are derived from Greece, and that they both arose there simultaneously, would seem to justify the present attempt to outline the possible relationships between the writing system and those cultural innovations of early Greece which are common to all alphabetically literate societies.

The source of these arguments appears to be Havelock (1963; see also 1976), who sets out at length his case that alphabetic literacy was the source of much Greek cultural achievement. However, having examined his work in detail, the only evidence we can find is that these two phenomena occurred one after the other in early Greek society. Close temporal sequence merges into causality. In addition to the crude causality attributed to these two supposed facts—that the Greeks 'invented' the alphabet and that they 'invented' new forms of logic which form the key to Western thought—they suffer from the additional liability that neither fact appears to be supported by the historical evidence. Supporters of the Greek thesis have always accepted that the invention of the alphabet was somewhat earlier than the Greek renaissance, and it is argued that the Greeks somehow perfected a budding alphabet. However, as Goody has recently pointed out, discoveries which are still being made have pushed the existence of alphabets to earlier times: he draws attention to evidence of a Proto-Canaanite alphabet 'invented' around 1500 BC (Goody 1983). The second part of the argument, concerning the invention of new forms of thought, also comes under attack in Goody's more recent article; he accepts that some of the forms of proof and argument attributed to the Greeks existed far earlier in Babylon. As Goody concludes (1983, p. 83): 'We need to modify not only the idea of the uniqueness of the Greek alphabet but also the uniqueness of their intellectual achievement. Both rest firmly on earlier developments in the Near East.'

A particularly bizarre aspect of the 'Greek' argument is the notion that the alphabet itself, rather than writing (or even the development of printing), is directly responsible for certain aspects of abstract thought. This position is again attributable to Havelock (see also McLuhan 1962). It has been developed and spread by Olson (1977) and others, and it is in danger of becoming accepted fact (for example with Cole and Griffin (1983) surmising that perhaps 'the ability to send astronauts into space owes a lot to the analytic power of the alphabet'). To some extent the claim rests on the alphabet's being constructed on more 'abstract' units than other writing-systems. However, as we have shown earlier, this is an oversimplification. (In addition, if abstractness were the most significant criterion then we would look to Korean, with its writing-system based on distinctive features, rather than to the alphabet.)

Once again, beyond the claim in Havelock's work, we are unable to find any more direct historical evidence that the alphabet, rather than writing itself, is crucial. As we have tried to demonstrate earlier, differences between writing-systems are not as clearcut as they might at first appear, and they need to be discussed in the context of specific languages. Pinning everything on the Greek alphabet seems to be more an argument about the significance of Greek civilization than about the purported effects of literacy. It appears to be difficult for some of those who study Greek culture to accept that writing (and civilization) existed before 500 BC; to confirm their prejudices it is necessary for them to depreciate any forms of writing that existed before the Greeks (see also Gaur 1984, p. 136; and for a parallel point about the existence of painting prior to the classical period, also, Paturi 1979, p. 36). We believe, therefore, that the crucialness of the alphabet *per se* has never been demonstrated; rather it arises as part of a more general adulation of the achievements of the Greeks that still blinkers modern scholarship.

27.4.3 *Social and economic correlates of literacy*

The Greek example is a particular instance of a more general question: how do variations in literacy levels and practices correlate with other important indicators of the social and economic organization of a society? In the Nineteenth- and twentieth-century mind this has been seen as a very simple question with an equally simple answer: literacy is viewed almost entirely in positive terms. Literacy is equated with progress, and via literacy, benefits accrue to nations and to individuals. As a result, levels of literacy in a society are hypothesized to correlate positively with any and all indicators of social and economic progress. As Graff (1981) says 'the uses of literacy are still debated; its basic value is not'.

Notice that again this hypothesis is couched in terms of correlations; but it frequently implies a strong, causative link between literacy and the condition of society. It is suggested that literacy, via the attitudes and cognitive skills it fosters in individuals, promotes economic development and prosperity. Literacy therefore becomes essential in the attempt to develop 'underdeveloped' countries and in the elimination of poverty, disease, and traditional but unproductive forms of social organization. Hence literacy education has been one of the main planks of UNESCO's modernization policies since the 1950s, with the figure of a 40 per cent literacy rate being quoted as essential for economic take-off (Anderson 1965). This is still taken up as a rationale for literacy campaigns, despite the very dubious validity of such a simplistic generalization.

It is important to evaluate these claims, both in terms of whether the claimed correlations actually exist, and secondly in terms of what such correlations could tell us about causality. On the first of these points, the question assumes a single, easily identified index of 'literacy-level' which is very problematic. 'Functional' literacy levels vary a great deal from society to society, as do the actual practices of literacy in which an individual must show competence. It is not easy to reduce these complexities to a quantitative expression which can be used to compare different societies. A number of recent careful ethnographic studies of literacy practices illustrate this point—especially where very different types of literacy exist side by side within the same society (for example, Scribner and Cole's work (1981) among the Vai and Street's work (1984) in Iran).

On the second point, we should be cautious in interpreting causality from correlation. Although the prevailing wisdom is that literacy contributes to social and economic change, it is equally plausible that the reverse is true—that social and economic changes foster literacy. Gaur (1984) holds the view that literacy is not used by a society—although its existence may be recognized—until that society is ready to exploit its possibilities. She quotes several examples in support of this: literacy was introduced into India in the sixth century BC but not used widely for several centuries; the Chinese had techniques of printing long before they used them for the widespread dissemination of texts.

Before we go on to examine specific proposals about the social and economic correlates of literacy, it is worth pointing out that attitudes to literacy have varied in other cultures and times, and it has not always been seen to have unmitigated positive value. In India, writing was viewed with great hostility by the carriers of the traditional Vedic culture, and although we can see in this hostility the threat that writing posed to the power interests of an élite group, bent on preserving their monopoly of knowledge, there seems also to have been a genuine feeling of the superiority of oral modes for the absorption and use of valued knowledge, and a concern about the debilitating effects of literacy. Such fears were also expressed by the Greeks, as Havelock (1963) reports. In addition, Clanchy (1979), in his study of the spread of literacy in the Middle Ages in England, describes the gradual acceptance of literate conventions in the legal sphere of titles and charters; there was resistance to the use of written documents because of the widespread danger of fraud, and preference for relying on the oral testimony of 'good men and true'. It is necessary to bear these reservations about literacy in mind, because they are no less valid than the positive claims that are made for literacy. Significantly, they recur in relation to new technologies such as television and computers. Witness, for example, the reluctance to rely on tape-recorded evidence or photographs in court because of the possibility of fraud (Brand *et al.* 1985). We will now turn to some of the main claims that have been made about the social and economic correlates of literacy.

27.4.3.1 *Franchise and political participation*

This has been a central claim for literacy among widely different sources, from Goody and Watt (1963) in relation to Greek democracy and the mass-access possibilities of the alphabet, to UNESCO in its rationale for literacy in developing countries. The weight of evidence, however, seems to contradict this claim. As Gough (1968) states:

> ...it is hard to generalise directly from literacy to political structure. Pre-modern states with substantial literacy have included aristocratic, oligopolistic [*sic*; *sc.* 'oligarchic'], and democratic city-states, feudal regimes, and bureaucratic despotisms of the 'Oriental' type. Modern mass society includes both fascism and parliamentary democracy, as well as military regimes with varying popularity, in the capitalist bloc, and both highly bureaucratic centralism and more decentralised forms of popular participation in the communist world.

Freire (1972) makes a distinction between the 'domesticating' and 'liberating' possibilities for literacy, depending on the ideology and purposes of literacy instruction and the social conditions in which literacy is acquired. Contemporary literacy campaigns, particularly on the 'liberating' Freirian model, have frequently been perceived as a threat by non-democratic governments (as in Brazil in the 1950s). However, *there appears to be no necessary relationship between literacy and democratic political participation.*

27.4.3.2 *Employment and economic development*

It is frequently claimed that literacy ensures the more skilled and productive labour force essential for economic development, and this has been a major foundation of international 'functional literacy' programmes during the 1950s and 1960s. However, the UNESCO experimental world literacy scheme (see UNESCO 1976) produced disappointing results in its attempts to inject doses of literacy into societies striving for economic take-off. In contrast, there have been successful mass literacy campaigns carried out in countries such as Cuba, where both educational and economic development occurred in the wake of a revolutionary situation in the society as a whole (see Kozol 1980). If anything, this implies that social structure and economic development affect literacy rate, rather than the other way around. In addition, well-documented examples exist both of societies with high rates of literacy but little or no economic development (such as Sweden, where historically religion established literacy but the country remained poor for many years—see Johansson 1981), and of highly centralized and technologically sophisticated societies without literacy (for example the Inca civilization in Peru achieved large-scale, precision building construction without the benefit of literacy to make plans).

Several writers have argued for a more discriminating analysis of both literacy practices and economic activity in order to evaluate the claims made about their relationship. Scribner and Cole (1981) found that occupational categories differed for each of the three types of literacy found in Vai society, English, Koranic, and Vai, suggesting that different literacies have different economic significance. Graff (1981) distinguishes between *commercial* development and *industrial* development, each of which has a different relationship with literacy, and makes different demands on a workforce; and Goody (1968) makes further distinctions between different traditions of commerce which make use of literacy in different ways (see also Gough 1968; Street 1984).

A related claim posits a correlation at the individual level between economic and social well-being and literacy level. Graff (1979, 1981) has been a consistent critic of this claim, which he terms 'the literacy myth', and shows via detailed evidence of occupational mobility data and literacy levels in nineteenth-century Canada that correlations are far smaller than this claim would lead us to expect, and can in any case equally plausibly be explained by other characteristics of people's social situation. (Graff sees the claim for individual mobility to be a rationalization given by employers and the state to encourage working people to participate in schooling, which is of more benefit to industry through the moral discipline and acculturation it imposes than through the development of cognitive and technical skills *per se*).

27.4.3.3 *Modernity*

The concept of 'modernity' has been developed to deal with a complex of factors and attitudes that are associated with the transition from traditional to 'developed' cultures (see Lerner 1958; Inkeles and Smith 1974). Among these factors are such things as: urban residence (versus rural residence); extent of geographical mobility; contact with mass media; and attitudes toward religion, morality, political participation, birth control, and education.

This constellation of factors correlates quite strongly with literacy acquired through formal Western-style schooling, but, as Scribner and Cole (1981) and others (Inkeles 1973) have argued, this begs the question as to whether the causative factor is literacy itself, or rather exposure to and absorption of the Western schooling experience in general. One suspects the latter to be the more plausible explanation, particularly in view of the breadth of characteristics covered by 'modernity' and by their coincidence with Western values themselves. Scribner and Cole indeed provide some evidence on this point, since they were able to dissociate 'schooling' from 'literacy' in their study of Vai culture. Even their non-literate informants had reasonably high 'modernity' scores; but there were variations associated with literacy. They summarize their findings thus (1981, p. 95):

> Our results tend to confirm findings in earlier studies ... that English schooling is an important determinant of modern attitudes, and that urban experience is related to knowledge of the larger world and to open-mindedness about religion. They give a mixed message to the more general hypothesis that literacy *per se* functions as a modernising influence. Learning associated with mastery of the Qu'ran and achievement of Arabic literacy appears on the whole to be a retarding influence, while Vai script literacy exerts a modest positive influence on modern attitudes... Tradition as tapped in our questionnaire and modernity as measured in this internationally used scale do not emerge as opposites, an outcome with intriguing implications for theories of both social and psychological change.

Scribner and Cole's findings suggest that the different practices associated with literacy, and the social context and ideology in which these practices are embedded, are the key factors to understanding the relationship of literacy to social and economic development. The weight of evidence in this section again tends away from the view that 'literacy' is an autonomous set of skills that can be applied to any society in order to produce a regular, predictable set of consequences. The reality is much more complex, and involves a feedback effect from the initial social and economic context into which literacy is introduced.

27.4.4 *The technology of printing*

As we have already seen, the processes of reading and writing have varied considerably from society to society and in different historical periods, depending on the materials available for writing and the methods of manufacture and dissemination. The manufacture of paper and the development of printing and of the electronic media are the most visible technological milestones in the spread of literacy. In this section we concentrate on the technology of printing as a powerful example of the restructuring of literate activity through technical invention incorporated into the social organization and production of knowledge. We draw primarily on the ideas of Eisenstein (1979, 1981).

Techniques of stamping can be traced back to the ancient use of individual stamping seals, and on through block printing, which was in use in China by 600 AD, to movable-type printing, which was widespread in China by 980 AD and invented in Europe in the mid-fifteenth century, giving the possibility of mass copying of texts. Developments in Europe were apparently independent of those in the Far East, although Chinese block-printing techniques were known (see Febvre and Martin 1984).

In Europe, printing accompanied the explosion of knowledge which occurred during the Renaissance. Many hypotheses have been advanced about the impact of printing via the wide dissemination of texts, both in terms of breadth of ideas suddenly available to people and how printing may have shaped those ideas. Eisenstein integrates a great deal of scholarly material to suggest that printing had complex and contradictory effects in terms of productive knowledge and the secularization of thought. She also suggests that many of the effects claimed for 'literacy' may be due to printing and the wider circulation and preservation of texts that mass copying made possible. Scribal culture had many of the same limitations as oral culture: individual copyists produced texts with idiosyncratic formats and conventions, and were liable to make mistakes. The existence of only a few copies of a given manuscript makes for restricted accessibility and a high incidence of the loss or destruction of texts. Scribal culture was highly selective, since it could not sustain advances in knowledge on many fronts at once.

Eisenstein discusses two main processes that printing assisted—*cultural diffusion* and *standardization*. In terms of cultural diffusion the availability of multiple copies of the same text both enables many people to read the same texts, and one person to read and compare many different texts, thus encouraging the cross-fertilization of new ideas. However, Eisenstein suggests that while printing facilitated the wide circulation of new knowledge, and in this way contributed to the breaking down of old religious traditions and authorities, it could also have some opposite effects: not only were secular and 'enlightened' writings circulated widely, but also medieval religious, occult, and mystical writings that reinforced past traditions. Of itself, then, printing is not selective of the kind of ideas it exposes readers to.

In another sense, however, Eisenstein believes printing to be highly selective, contributing to the channelling of certain messages and emphasizing particular stereotypes. She describes this amplification and reinforcement as follows (1981, p. 67):

> It is produced by an unwitting collaboration between countless authors of new books and articles. For 500 years, authors have jointly transmitted certain old messages with augmented frequency even while separately reporting on new events or spinning out new ideas. ... As printed materials proliferate this effect becomes more pronounced ... The more wide-ranging the reader at present, the more frequent will be the encounter with the identical version and the deeper the impression it will leave.

Those amplified messages are channelled by fixed linguistic frontiers, so that different 'stereotypes' develop in different vernacular literatures, though other frontiers, for example religious ones, cut across these. The reverse side of this amplification is, of course, the effective suppression of other ideas and messages via neglect, or their deliberate putting aside when things fail to reach print or go out of print. Such elimination is all the more effective because of the weight of truth and authority accorded the printed word.

The process of standardization has perhaps been the most widely recognized effect of printing: the elimination of scribal errors and idiosyncracies, in repeated copies, the fixing of linguistic maps and conventions, and the mass reproduction of texts that are particularly difficult to copy accurately by hand, such as pictures, maps, and diagrams. Eisenstein emphasizes the powerful role that early print-shops played in the systematization of literate knowledge. They were 'laboratories of erudition' in their own right, compiling and producing reference manuals and guides and indexes, devising notation systems, and arranging contents, running heads, and footnotes, and so vastly changing the shape of the 'book' and guiding the thoughts of readers via these new built-in aids.

The advent of print, then, appears to have played a very dynamic role in the way people think about and read texts, a role that is not simply the effect of mass circulation of new ideas, but more complex and contradictory than this. Printing certainly fostered a 'systematic' approach to learning, via the opportunities it presented to scrutinize a variety of ideas side by side and to compare historically different versions of the same text for inconsistencies and developments, and also through the efforts of printers themselves in classifying and cross-referencing existing texts. As a result of the increased preservation and accumulation of texts and fixed records, Eisenstein suggests that the pursuit of truth itself took a new direction: it became the discovery of *new* knowledge rather than the constant effort to recover and preserve traditional knowledge. This in turn gave rise to the idea of the steady advance of knowledge with fixed records to mark its stages of development. If this is true, it is potentially a very radical change in cognitive orientation made possible by the extension of collective cultural 'memory' via mass literacy.

27.4.5 *Restricted access to written language*

An understanding of the role and development of literacy is not complete without consideration of the ways in which literacy skills and activities are distributed within a given society—what Eisenstein (1981) refers to as the 'social penetration of literacy' (p. 68). Although the practice of literacy skills appears to be an individual affair, access to those skills and distribution of literate activities is a *social* phenomenon. We again approach literacy as a technology which is shaped by the existing social relations in a culture and co-varies with other social institutions. From this viewpoint, the structuring of access to literacy is seen as an essential characteristic of any communications system, and exploring the patterns in different societies can tell us much about the social practices and meanings of literacy.

Goody identifies this issue with his notion of 'restricted' literacy, but does not follow through all its implications. Goody is concerned to understand why the potentialities of literacy (specifically alphabetic literacy) are not fully realized in many societies. Why, for example, have other societies not developed democratic forms of participation and the 'rationality' of scientific thought and method that the coming of alphabetic literacy supposedly brought to the Greeks? Goody looks for explanation to such factors as religious 'restrictive practices' and mismatches between the language of literacy and vernacular languages.

It seems to us that Goody has identified an important area in his notion of 'restricted' literacy, but that it should not be regarded as a peripheral topic. Rather, it is central to any characterization of literacy in a society. Our premise is that all societies control access to the written word in some way, because literacy involves information and idea-transmission, and is practised in a context where its uses may both maintain and challenge existing social institutions.

Although issues of access are not reliably documented in historical studies of writing, we can identify five distinct types of restriction (as in Table 27.1).

The picture that has often been painted is one of fewer and fewer restrictions being placed on the spread of literacy. However, tempting though this evolutionary picture of a tendency toward unrestricted literacy is, it is hard to substantiate. First, even Goody admits it is difficult to determine exactly how widespread literacy was in ancient societies. In looking more closely into the mass literacy of ancient Greece the obvious point can be made that exclusion of non-citizens (women, foreigners, and slaves) lowers the percentage of literate people considerably. Gough argues (1968, p. 70) that the figures are in fact probably comparable with the literacy rates of China and India at the time. Secondly, even in contemporary advanced industrial societies, the notion of 'universal literacy' may still be more of an ideal than an actuality, because there are very unequal patterns in the acquisition of literacy skills and their practice, particularly writing skills.

In addition the level of literacy skills required to function effectively in a complex urban society changes, so that, for example, although the ability to sign one's name was once enough to be designated literate, someone who was able to do no more than this in modern-day Europe, Japan, or America would not be considered to be literate in any real sense. In mass-media based cultures, too, the ability to read and write may be of less significance on its own than access to the media, not only for receiving information, but for transmitting it too. In this respect, all contemporary societies are still extremely hierarchical and restrictive, and as Williams points out (1981, p. 228) questions of access become more important as technology becomes more complex.

Table 27.1 Restricted access to literacy

1. Overt political restriction:	explicit restrictions imposed by physical force or via the legal/religious system	e.g.	—book-burning; legal controls of the radical press (see Goody 1968; Gaur 1984)
2. Structural restrictions:	implicit restrictions patterned by the social structure	e.g.	—only certain classes literate in ancient societies (see Goody 1968) —a class of scribes in Europe during the Middle Ages (see Clanchy 1979)
3. Restriction by function:	literacy developed only for a narrow range of purposes	e.g.	—via use of specific languages and scripts for certain contexts (see Scribner and Cole 1981) —UNESCO literacy programmes for 'key' workers (see UNESCO 1976)
4. Restriction by skill:	only certain literacy skills acquired or practised	e.g.	—Fijians taught to read but not to write or use printing presses (see Clammer 1976) —women as 'the writers' in Appalachian working-class communities (see Heath 1983)
5. Restriction by language:	mass literacy establishes 'official' languages and marginalises vernacular languages and dialects	e.g.	—standardization of Spanish with advent of printing (see Illich 1981; other examples in Fishman et al. 1968; Fishman 1977; Leith 1983)

27.5 Conclusions

We have traced in outline the historical development of several components of the complex phenomenon of literacy. At the same time we have been sketching out what we regard as a more adequate characterization of the nature of writing. This is a pre-requisite for explaining the mechanisms and principles involved in the development of a symbolic system.

Our overview has covered developments in structural aspects of writing systems and in the functions of writing; we have traced some purported effects of literacy on individual cognition and on society; we have discussed literacy as technology; and we have examined the significance of restrictions on access to literacy in society. Common themes have recurred in these diverse areas: that many views of literacy are simplistic in treating it as a single isolable variable; that many views of literacy are ethnocentric in evaluating all developments against our Western alphabetic society; that writing is intimately entwined with other visual symbolic systems; and that writing is simultaneously a psychological phenomenon and a sociological phenomenon.

We have made some beginnings. One of our first landmarks was Gelb's pioneering work. We agree with his more recent assessment (1980, p. 22):

It is necessary to extend our horizons both vertically and horizontally: vertically back in time, as far back as the Upper Paleolithic, some 30 000 years ago, even farther back to the Acheulian some 150 000 to 300 000 years ago and horizontally, to all kinds of para- and meta-aspects of writing which occur within or in addition to writing proper in modern times and earlier.

We would add two further dimensions: firstly, the need for a functional history of writing, and, secondly, the need to provide a theory that integrates individual cognition and social context. Bringing these strands together promises to be an exciting and fruitful area of enquiry.

Editorial appendices

Editorial appendix I: Script evolution

With very few exceptions, extant script systems derive from either Chinese or Near Eastern forms. Fig. 27.3 (from Gaur 1984) traces the developments that have diffused from the Near Eastern proto-Semitic script of between 1800 and 1300 BC. (The origin of proto-Semitic script is obscure: it may have been an independent invention, or have adopted elements from already existing Egyptian, cuneiform, and Hittite scripts.) Proto-Semitic appears to be the first script solely based on symbol–sound correspondence, with no elements of picto-or ideograms. It is neither alphabetic nor syllabic, but a consonantal script (see Gaur 1984, p. 88). Its diffusion to non-Semitic peoples

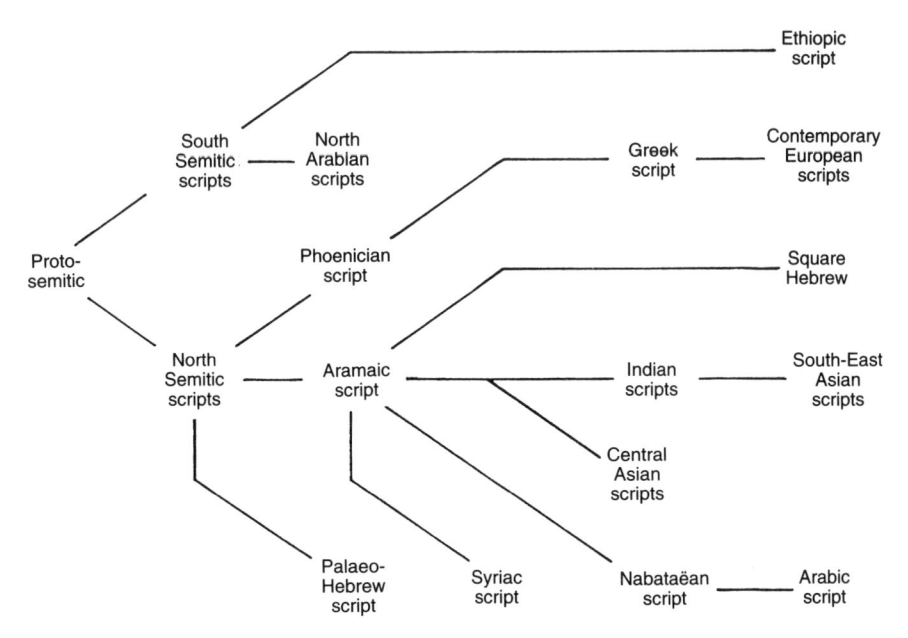

Fig. 27.3 Development of scripts from Proto-Semitic scripts of 1800–1300 BC (from Gaur 1984).

led to the development of the alphabet in Europe and syllabic script forms in South Asia.

The stylistic variation of the present-day scripts is large, and that they have a common origin is only apparent from the historical record, where transitional stages are recoverable. These variations have arisen by means noted by Barton and Hamilton: to deal better with the demands of specific languages—additional consonants, vowel disambiguation, etc.; political factors; and so on.

Additionally, some stylistic changes may involve the factors put forward by Slobin (1978; see Deuchar, this volume, Chapter 20) as general principles underlying language change, especially the 'charge' to language to be 'quick and easy'. Thus, in the case of cuneiform script systems, as they are adopted by more and more individuals, a stylization taking them away from their early and fairly transparent iconic base towards a more arbitrary and opaque form occurs (see Fig. 27.4). In this particular case, the instruments and medium of writing—styli and wet clay—themselves push towards a simplification for practical reasons. We should not forget, either, that there may well arise a political motivation for making scripts less transparent (see also Goody 1981).

Editorial appendix II: Ehlich's developmental account of writing

Ehlich (1983) provides an account of the creation of writing that is developmental in its essentials, yet is sympathetic in its handling of the points Barton and Hamilton argue as militating against a straightforward 'progressive' interpretation of the history of writing systems. There are three strands to Ehlich's account. Firstly, the social practices of a human group are seen as generating problems which 'need' to be solved to maintain the society: one such set of problems could be solved by a writing system. Secondly, the nature of the solution to any particular problem is constrained by the inherent limits of previous solution techniques: some systems cannot be transferred successfully to new problems they did not originally arise to solve. Thirdly, in the general problem area of writing down speech, spoken languages differ in the demands they place on a writing system.

Ehlich is thus attempting to deal with the reality behind the historical diffusion of scripts from one culture to another. Presented with a technique developed in another culture for solving particular problems, what use can a new culture, with different needs and a different language, make of it; how might the technique be modified; what are its new possibilities and limitations? Ehlich develops his views using the Near East as an example; but he takes it as a paradigm case. The main motivation he sees for the creation of writing is that of overcoming the evanescence of spoken language. This can be done by the creation of *oral 'texts'*, using spoken language to organize memory, perhaps by rote learning, to preserve essential information.

This solution is adequate for speech-acts which are integrated into repetitive speech situations or rituals, but is inapplicable for occasional ones, which are often repeated as 'types', but on each occasion with a different 'content':

I am speaking of simple economic transactions such as became necessary under the more highly developed social and economic conditions of the centralized high river cultures which had developed the division of labour to a relatively great extent [see White, this volume, Chapter 9] (Ehlich 1983, p. 101).

Schmandt-Besserat (1978) discusses the early evidence of 'book-keeping' by the use of 'count stones' (9000–2000 BC). Individual count stones were miniature models of the objects they represented. In addition, large earthenware vessels ('bullae') were used as repositories for collections of tokens. These were sealed so that their contents could not be tampered with. But then one would not know what was in the vessel. To solve this, the sealing-clay was stamped with impressions of the count stones it contained.

This is an important step, for where the count stones are symbols for objects, impressions are symbols for symbols:

The bulla markings were obviously not invented to take the place of the token system of record keeping. Nevertheless, this is what happened. One can visualize the process. At first the innovation flourished because of its convenience: anyone could read what tokens [count stones] a bulla contained without destroying the envelope and its seal impressions. What then happened was virtually inevitable, and the substitution of two-dimensional portrayals of the tokens for the tokens themselves would seem to have been the crucial link between the archaic recording system and writing (ibid., p. 47).

While this scenario is essentially hypothetical, it gains some credence from the similarities between the form of the three-dimensional tokens and the two-dimensional written symbols of early Sumerian writing (see Fig 27.5).

At the same time, a co-ordination is set up between the written sign, the object it represents, and the word in the language for it, such that a link is made at the level of written symbol standing for linguistic symbol.

Pictograms are well suited to represent objects which (1) have a visible form, (2) have a lasting duration, and (3) can be reproduced in drawing. But

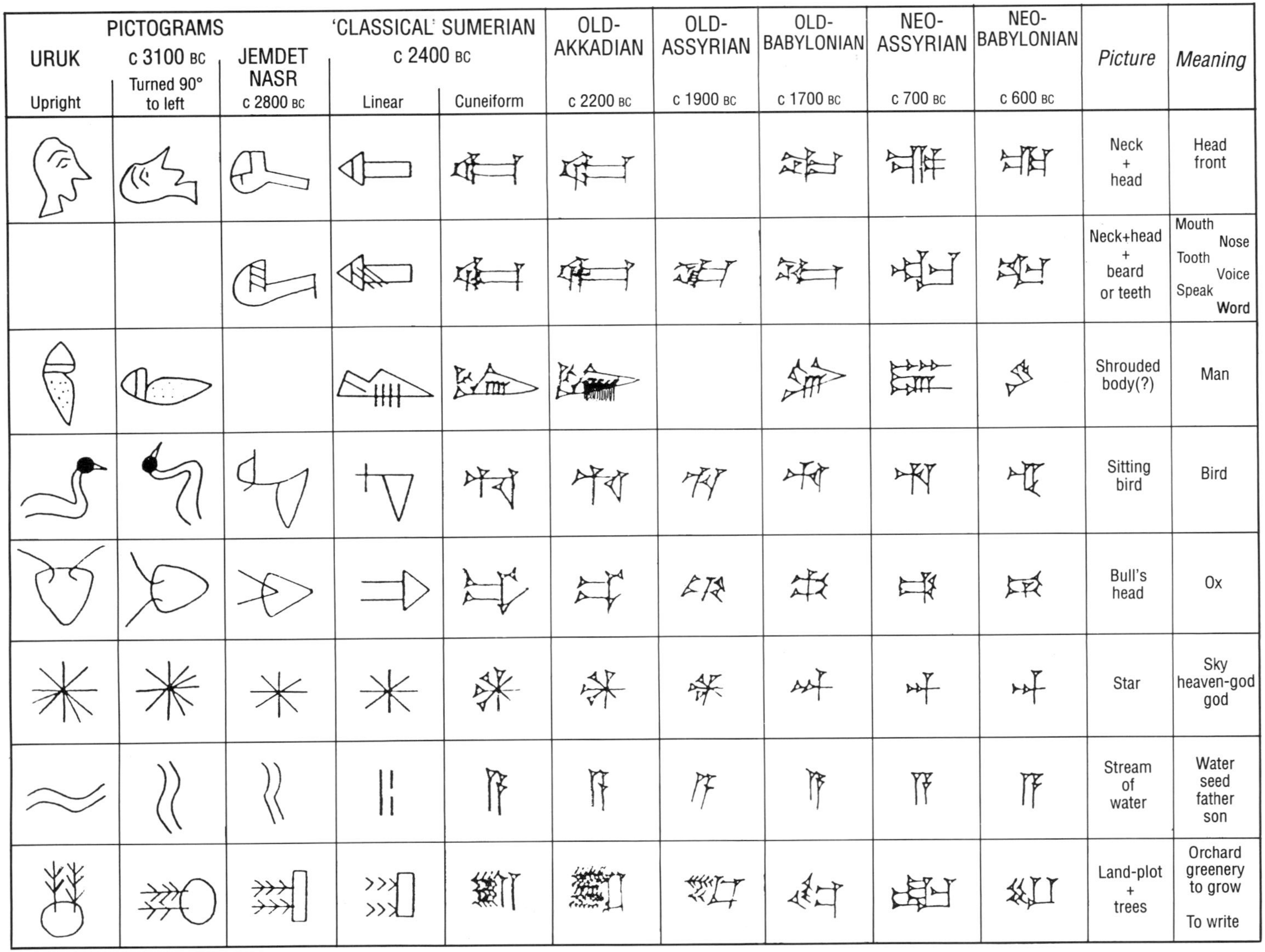

Fig. 27.4 The evolution of cuneiform script from fairly transparent pictograms to more arbitrary and opaque forms (from Gaur 1984).

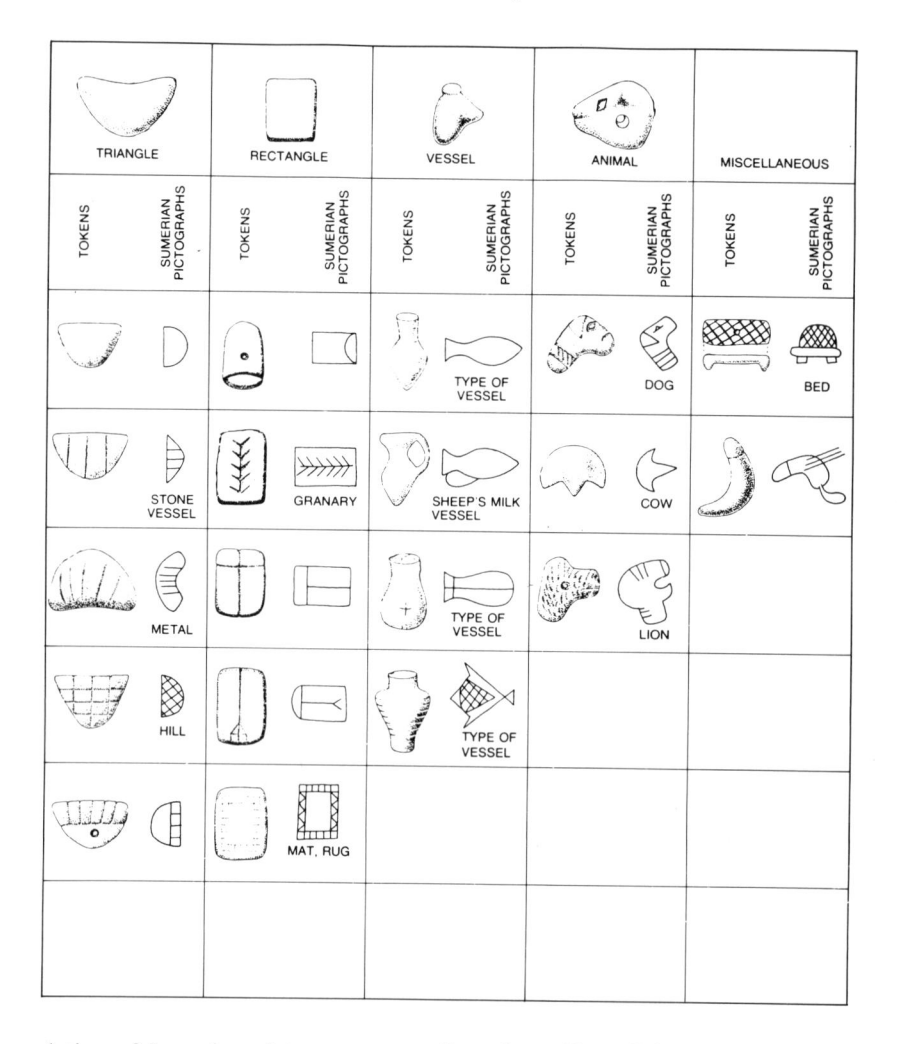

Fig. 27.5 The relation of Sumerian pictograms to earlier tokens (from Schmandt-Besserat 1978).

language provides a means for communicating further aspects of reality: (a) qualities; (b) abstract ideas and positions; (c) movements and conditions; (d) complex facts and relations; and (e) 'operative units' (prefixes, suffixes, articles, demonstratives, etc.). Thus, a new problem situation is set up, to be dealt with only when social conditions raise the problem of representing and conveying information that speech can already encode beyond the practical sphere of accounting.

When this became necessary, ideographic systems reacted in general through using certain *analogizations* which 'abstract and transfer features of the primary writing form, i.e. the ideograms' themselves (Ehlich 1983, p. 111):

Thus the problem solution consisted in systematically guiding the reader's association (with a certain degree of arbitrariness): Guiding the reader's associating aims towards standardized identification patterns which enable the reader (a) to recognize the pictogram or ideogram does not stand for what it directly represents and (b) to choose, in an appropriate way, a related alternative (ibid., p. 112).

Figure 27.6 shows how this process operated for Egyptian: 'cool' is represented by water flowing out of a pitcher; 'south' by a lily, the symbol for Upper Egypt; 'flying' by a bird in the process of taking off; 'ruling' by a staff; etc. In Sumerian (Fig. 27.7), combinations of two different symbols are often found (for example 'mouth' + 'bread' = 'eating').

This strategy does not work for abstract class (e) above. One solution is not to try to represent these units in written text. When this is no longer adequate, the *phonic* dimension in language, as opposed to the

to rule   to lead   the south   to find   old age   cool

to strike   to fly   to eat   to go

to fight   to row   to step   to cry

Fig. 27.6 Egyptian hieroglyphics (see text for discussion). (From Jensen 1958).

Fig. 27.7 Compound signs that combine meanings to represent new concepts in Sumerian (From Jensen 1958).

hitherto focused-on semantic dimension, needs to be extracted from the complexity of the linguistic sign for the further production of written language. The first attempts at accomplishing this led to problems, in that written symbols could now be read as direct representations, associated ideas, or phonic indicators: text became highly ambiguous.

The first solutions to these new problems were an expansion of *phonic writing* and the introduction of *determinatives*. Phonic specification worked as follows:

the symbol can be read either as '*an*' ('heaven') or as '*dingir*' ('god') .... In order to show which one of the possible readings is meant, a second cuneiform symbol, e.g. that of '*na*', is added in the first case. The ambiguity is thus resolved for the reader, who knows he should read '*an*', that is, the word which ends with the first consonant of the following symbol (Jensen 1970, p. 92).... If, on the other hand, the reader finds the syllable symbol for '*ri*', then he, in the same way, knows that he should read '*dingir*' (god) (Ehlich 1983, p. 115).

Determinatives are symbols which indicate the semantic class membership of the preceding or following word. Thus '*dingir*' stands as a determinative next to the names of gods, '*lu*' ('man') next to the names of tribes, and so on:

The determinatives serve purely to assist in the better recognition and easier classification of the symbols. They thus give a sort of 'reading instruction' which makes clear that they perform a mere auxiliary function that is supposed to help solve the problems of the previous problem solution (ibid.).

The above provides the essential basis for a full syllabic system of writing. It is by no means a 'perfect' system, since a high degree of ambiguity is inherent in any text, and *ad hoc* solutions to combat this serve to make the writing system more and more cumbersome and opaque. The early solution to these problems was not accomplished by changes in the writing system itself, but in the growth of a scribal trade (for example, in Sumer–Akkad and Egypt) with a long apprenticeship beginning at an early age: writing and reading became a professional skill.

The Sumerian–Akkadian and Egyptian writing types retained their complex forms throughout their history. A further concentration on the phonic side of writing, leaving the semantic aspect of representation completely aside, occurred on the borders of these areas, in regions of Palestine and the Sinai peninsula using the northern Semitic languages, perhaps because of the particular characteristics of these languages' structures:

The known peculiarities of Semitic already demand of its speakers/hearers a comparatively analytic approach to the production and comprehension of utterances. The identification of the consonants is absolutely necessary, especially in order to recognize the meaning of the word or root semantics. Use could be made of this in the further development of the phonic–syllabic problem solution type. The consequence is the construction of *consonant alphabets* (Ehlich 1983, p. 123).

Initially, these early alphabets retained a cuneiform script (see Fig. 27.8). Thus, there is a continuity of script form conserved over the change in the structural principles of writing. These symbols were soon abandoned for other forms, which led directly to modern forms: Ethiopic, Aramaic, Arabic, and Modern European via Phoenician (see above, Editorial Appendix 1). Figure 27.9 gives a clear representation of the development of European script.

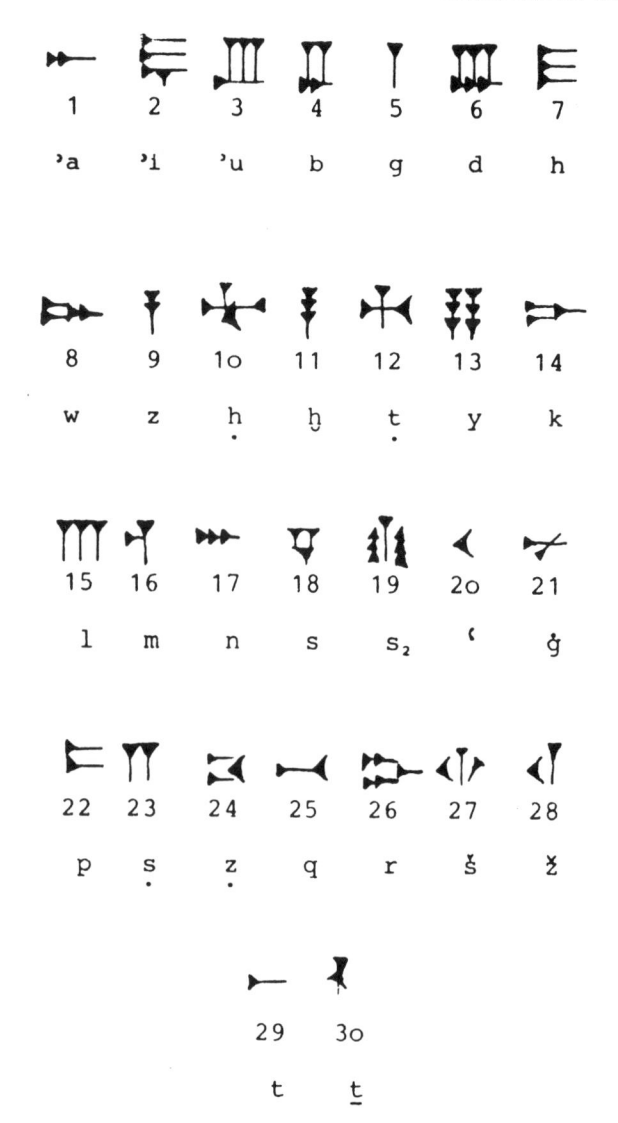

Fig. 27.8 Use of cuneiform symbols to code sound: the Ugaritic alphabet (from Ehlich 1983).

Editorial appendix III: Numbers and mathematics

At first sight, number systems may appear a parallel case of graphical representation to writing. However, number systems are at the same time *conceptual systems*. Both historically and ontogenetically this leads to a marked difference between them and writing systems: crudely, numerical concepts do not seem to have existed prior to their being written down, in the way that speech did. It seems more the case that what is represented graphically interacts with what can be represented conceptually. Thus, a parallel claim to the one critically reviewed in the case of literacy by Barton and Hamilton—that literacy affects cognition—is advanced more strongly in the case of numerical representation. That is, number systems are indices of conceptual systems that change in interaction with their representational systems. For example:

> The tremendous spurt in mathematical analysis during the seventeenth century was not, it seems fair to say, merely a concomitant of the general cultural advances being made in Europe at the time.... For on analyzing the mathematical progress of that time, one is struck by how much it actually consisted of the invention of a new and powerful symbolic apparatus (Wilder 1968, p. 165).

This trade between representational and conceptual systems is captured in Heinreich Hertz's remark that:

> One cannot escape the feeling that these mathematical formulas have an independent existence and an intelligence of their own, that they are wiser than we are, wiser even than their discoverers, that we get more out of them than was originally put into them (quoted by Bell 1937, p. 16).

The 'wiseness' of mathematics is clear when one considers that it is possible to use the numbers one to ten, quite effectively and for many purposes, without knowing they have properties beyond those that were 'originally put into them': for example, that there is an odd and even set; that some (1, 3, 5, and 7) are primes; that 6 is a perfect number (the sum of its factors, 1 + 2 + 3 = 6), etc.

This 'bootstrapping' of mathematical concepts suggests that there is a hierarchical order in the elaboration of mathematical knowledge that results from a process of *abstraction*, setting up a representational system which contains implicit further properties which can then subsequently be uncovered by a further act of abstraction:

> Comparison with the number science of their predecessors reveals ... that the Greeks introduced new components into the process of mathematical evolution, especially those of *abstraction* and *generalization*. To be sure, this is a matter of degree—these did not *start* with the Greeks, any more than an individual person started the counting process. The development of primitive concepts of number and length (as well as standards of measurement) in the Babylonian and Egyptian cultures had already stimulated some abstraction and generalization of an elementary kind. And, indeed, the same can be stated of the symbolization that underlay these achievements. But it is essentially with the Greek development of mathematics that one finds the peculiarly mathematical form of abstraction and generalization that is familiar to every modern mathematician. The pre-Hellenic types of abstraction and generalization were somewhat analogous to those employed by the modern engineer when he sets up suitable mathematical models to fit a given 'real life' situation. The latter might be called a 'first-order' type of abstraction and generalization, whereas the Greeks introduced a 'second-order' type, in that it built upon the 'first-order' elements

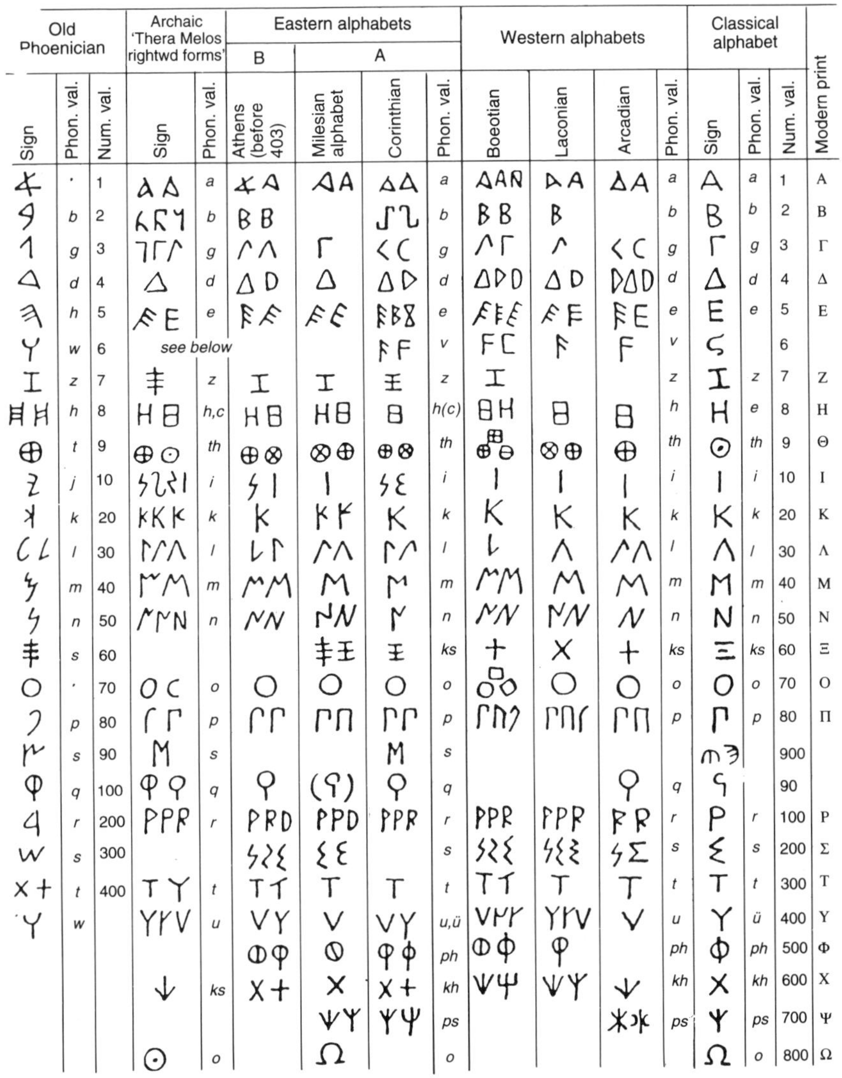

Fig. 27.9 The developmental transition from Old Phoenician to the Classical Greek alphabet (from Jensen 1958).

already existent in number science and rules of mensuration (Wilder 1968, p. 167)

At this level of analysis, Hughes's conclusion that 'the evolution of numerals ... followed a very similar pattern to that seen in the representations of quantity made by contemporary [Western] children' (1986, p. 88) is indicative of the constraints that operate in the construction of mathematical conceptual systems, an epistemological rather than biological recapitulation. Because there are so many wierd and wonderful twists that any emerging numerical system can take, there can be no direct correspondence between the historical and developmental courses of elaboration. But some general parallels are apparent. (For recent reviews, see Klein and Starkey 1987, and Damerow 1988.)

1 The direct perception of numerical quantity

It is clear that humans share with many animals[2] an ability to to recognize small numbers 'at a glance' (a process often termed *subitizing*—see for example Mandler and Shebo 1982). For numbers larger than four this process is unreliable, and a system of *counting* is necessary. There is some dispute as to the status of subitizing. Mandler and Shebo (ibid) are of the view that it involves a holistic recognition of canonical perceptual patterns that takes no account of the ordinal relations between numbers. By contrast, van Loosbroek and Smitsman (1990), Gallistel (1990), and Wynn (1992) present evidence that suggests that subitization does encode ordinal information.

Using a habituation paradigm, Starkey and Cooper (1980) were the first to suggest that this ability is present in infancy. Subsequent work, for example, Strauss and Curtis (1981), Antell and Keating (1983), and Starkey *et al.* (1990), has confirmed this. However, work by Wynn (1992) (see also commentary by Bryant 1992) opens up the possibility that infants can go beyond perceptually-based abilities, and may possess true numerical concepts, in that she showed 5-month-old infants to be capable of simple arithmetical operations on small numbers of items. This suggests 'that subitization is a process that encodes ordinal information, not a pattern-recognition process yielding non-numerical concepts' (Wynn 1992, p. 749).

Her findings may necessitate a revision to the following account of the elaboration of number systems. On the other hand, it is also worth putting her results in the context of Nesher's remarks (1988; see this volume, Editorial Introduction to Part III) on the interaction of culture and the development of counting skills, and those of Lock and Colombo (this volume, Chapter 22) concerning the problem of attributing experimental results to the skills of the individual being tested or the skills of the experimenter, before accepting at face value the concomitant claim that 'humans are innately endowed with arithmetical abilities' (Wynn, op. cit., p. 749).

Historical accounts of low-technology societies in this century point to many cultures' not developing numerical concepts beyond the range encompassed by subitization:

Members of the Aranda tribe in Australia had only two basic number words: *ninta* (one) and *tara* (two). For three and four they said *tara-ma-ninta* (one-and-two) and *tara-ma-tara* (two-and-two). Beyond *tara-ma-tara* they used a word meaning 'many'.

Islanders of the Torres Strait... had only these number words: *netat* (one), *neis* (two), *neis-netat* (three, literally two-one) and *neis-neis* (four, literally two-two); beyond that, they used a word meaning something like 'a multitude'. Among other examples of the same kind, we can mention the Indians of Tierra del Fuego, the Abipones in Paraguay, the Bushmen and Pygmies in Africa, and the Botocoudos of Brazil (when the Botocoudos said their word for 'many', they pointed to their hair, as if to say, 'Beyond four, things are as countless as the hairs on my head') (Ifrah 1981, pp. 6–7).

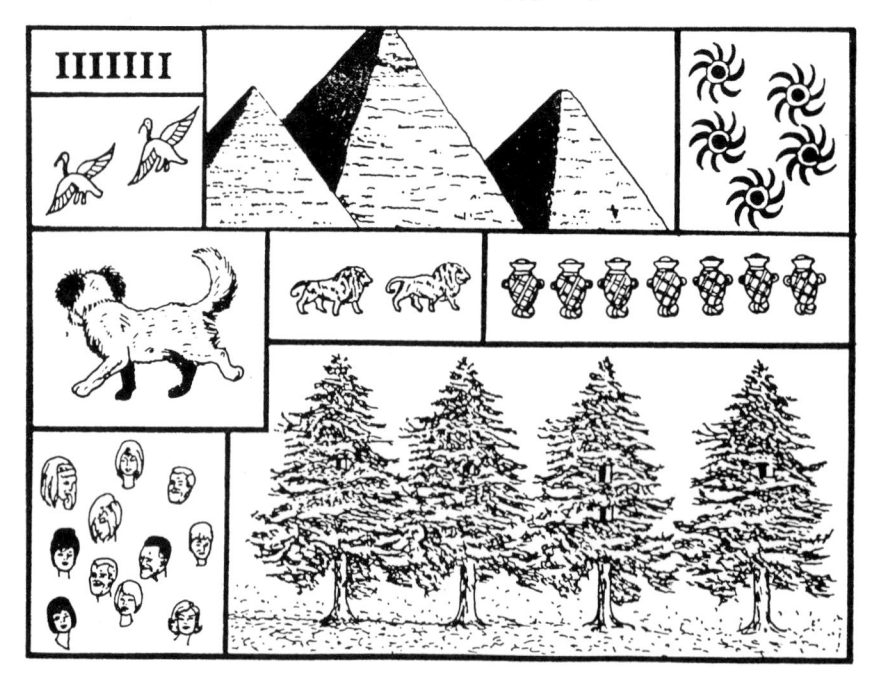

Fig. 27.10 Subitizing: three to four elements can be perceived directly; beyond that (although five may be a transitional case) number must be discriminated by counting (from Ifrah 1981).

Lévy-Bruhl (1926) reports that some Oceanic tribes declined and conjugated verbs in the singular, dual, trial, quadruple, and finally the plural. Roman numerals retain a vestige of much earlier systems (Ifrah 1981, Chapter 9); and similarly the vestiges of a declension system comparable to those of Oceania are apparent in the Latin language:

Only the first four number words are declined; beginning with five, they have neither declension nor gender. Similarly, only the first four months of the Roman year have real names: Martius, Aprilis, etc. Beginning with the fifth, Quinctilis, they are only 'serial numbers', up to the last month of the original year, December (Gerschel 1960). (Ifrah 1981, pp. 7–8).

To go beyond four requires the ability to count, but a lack of this ability does not mean that numerical procedures are barred from human exploitation, for 'one-to-one' correspondence can be conducted as an efficient substitute for counting; hence Ifrah's claim (1981, p. 9) that 'there are excellent reasons ... to support the conjecture that *for centuries people used many numbers without ... [conceiving of] them abstractly*'.

2 Concrete counting

If one cannot count, then one can have no understanding of what 'the number 63' means. But using material objects, such as pebbles, body parts, bones, notched wood, etc., it is possible to keep track of 63 objects: if one has 63 sheep, to make a notch each time one passes on the way out to the fields, and to cross each notch off when they return; and then, without knowing one has 63 sheep, one will know whether they have all returned safely or not. Similar systems can be used for keeping track of time without requiring an abstract calendar. Ifrah (1981, pp. 11–14) gives an example of combined 'visual counting' and 'tallying'. In the nineteenth century Torres Islanders counted by association with parts of the body (see Fig. 27.11).

Numbers are not conceived abstractly, but quantity is associated with the movements of touching various parts of the body in a fixed order. Now, suppose a special occasion is scheduled for what we would call 'the tenth day of the seventh moon'. Lacking such abstract numbers, the date can be expressed via the bodily system of Fig. 27.11 as 'Moon, right elbow; day, left shoulder'.

So as not to forget this important date, the chief of the tribe uses some sort of durable colouring substance to mark his own right elbow and left shoulder; he may, for example, draw a line on his left shoulder to indicate the day of the ceremony, and a circle on his right elbow for the 'rank' of the corresponding moon. He then tells some of his subordinates to do the same in order to take the message to the other tribes.

To make sure they will recognize the date when it arrives, the tribe members combine observation of the next new moon with one of the ingenious procedures handed down to them by tradition and originally developed by their ancestors after generations of trial and error.

On the first day of the next moon after the meeting of the council, the chief of the tribe takes one of the bones with thirty notches in it that he uses whenever he needs to consider the days of a single moon in their order of succession. Then he ties a string around the first notch of this 'lunar calendar'. The next day, he ties another string around the second notch, and so on till the end of the lunar month [Fig. 27.12]. When he has tied the thirtieth string, he unties it and all the others, and to show that one moon has passed he draws a circle on the little finger of his right hand.

At the beginning of the next moon he again ties a string around the first notch of the bone, and continues in this way till the end of the moon, when he draws a circle on the third finger of his right hand. This time, however, he stops at the twenty-ninth notch, since his ancestors observed long ago that the lunar month alternates between twenty-nine and thirty days.

He continues this procedure, alternating between twenty-nine and thirty days, until he marks his right elbow with a circle. He now knows that there are no more moons to count. He also knows that the tribe must soon leave for the place of the ceremony because there are only ten more days—or rather, from his viewpoint, the time to reach his left shoulder [Fig. 27.11]—before the day that has been agreed on (Ifrah 1981, pp. 12–13).

This plausible reconstruction (several elements of which are found among the Australian aborigines, for example, and in Papua New Guinea societies; see Saxe (1982) for a developmental analysis) shows that such techniques make it possible to reach relatively large numbers when parts of the body, touched in a fixed order, are associated with other objects: knotted strings, sticks, pebbles, notched bones, etc.

Such abilities seem to be transitional between the earlier concreteness of number as an inseparable part of the objects to which it perceptually relates and number as an abstract quality. Lévy-Bruhl (1926) captures this transitional nature in noting

such numeration may unconsciously become half-abstract and half-concrete, as the names (especially the first five) gradually bring before the mind a fainter representation of the parts of the body and a stronger idea of a certain number which tends to separate itself and become applicable to any object whatsoever.

3 Abstract counting

The principle of one-to-one correspondence underlies counting via body or hand parts, tallying, and the child's early attempts to represent number (Hughes

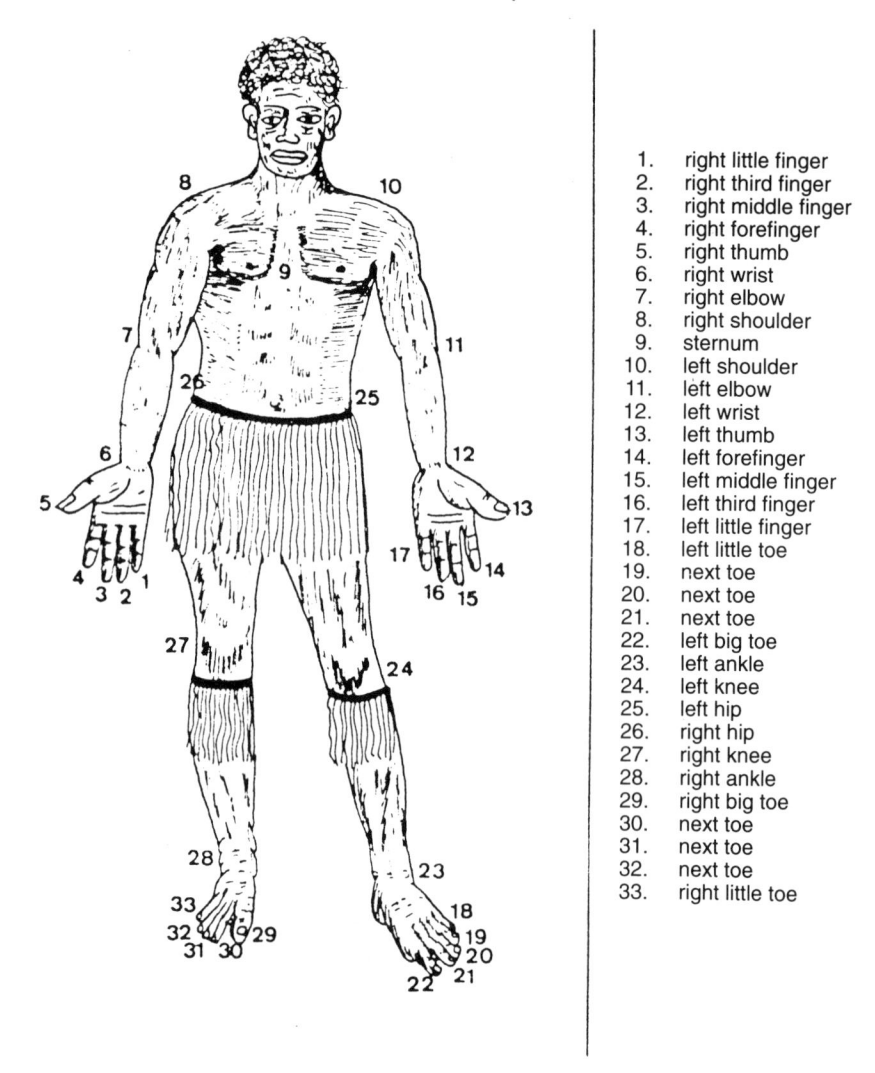

1.	right little finger
2.	right third finger
3.	right middle finger
4.	right forefinger
5.	right thumb
6.	right wrist
7.	right elbow
8.	right shoulder
9.	sternum
10.	left shoulder
11.	left elbow
12.	left wrist
13.	left thumb
14.	left forefinger
15.	left middle finger
16.	left third finger
17.	left little finger
18.	left little toe
19.	next toe
20.	next toe
21.	next toe
22.	left big toe
23.	left ankle
24.	left knee
25.	left hip
26.	right hip
27.	right knee
28.	right ankle
29.	right big toe
30.	next toe
31.	next toe
32.	next toe
33.	right little toe

Fig. 27.11 Counting method used by Torres Strait Islanders in the nineteenth century (from Ifrah 1981). Such body systems are still employed in New Guinea—see for example Saxe (1982).

1986, p. 85). It also forms the basis for early numerical systems such as the Roman, Babylonian, etc. These, however, are predicated on a greater abstraction of the concept of number than the above examples. Thus, the Egyptian hieroglyphic number system represented the numbers 1 to 9 by the appropriate number of vertical strokes. This system is easy to understand, but cumbersome. Such early systems were replaced by *cipherized* ones, in which each number below the base of the system has its own unique symbol (for example 0, 1, 2, 3, 4 . . . 9). These allow greater efficiency, but are opaque to the uninitiated. It is only after the Middle Ages that more abstract numbers—complex, negative, real, transfinite, etc.—

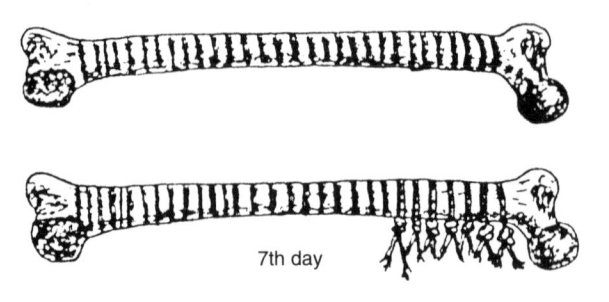

Fig. 27.12 Use of notched bones to count the days of a lunar month (see text for discussion) (from Ifrah 1981).

appear. Further details of these developments can be found in the sources listed.

4 The 'forces' of mathematical evolution

Wilder (1968) lists eleven sources (see below) which promulgate developmental change in mathematics, which in concert lead to a picture similar to the 'practice account' of literacy put forward by Scribner and Cole (see the main text of Chapter 27), and the role of social differentiation, presupposition, and communicative codes discussed by Lock and Symes (this volume, Chapter 8) (see also Damerow 1988; Wilder 1981). That is, mathematical symbol systems have their own intrinsic possibilities and potentialities, but these need to be 'discovered' by their users; and these discoveries appear to be facilitated by the social practices in which mathematics is used.

1. *Environmental stress*: it is only in 'simple' societies that quantity does not become truly numerical (see above, on subitizing and languages without an extended numerical vocabulary).

Only when the stresses evoked by the growing complexity of a culture become great enough to induce it does the more refined method of *counting* emerge. One is reminded here of the situation in some primitive tribes relative to color words. That certain tribes have only one word for green and blue was originally considered to mean that they were unable to distinguish between the two colors; now it is recognized that they could distinguish amongst them visually, but that the cultural need for such a distinction was not strong enough to compel a verbal equivalent for the differentiation (Wilder 1968, p. 33).

Similarly, the problems created by a culture in other spheres of its action can produce stresses that lead to the development of new mathematical techniques; physics and mechanics can create problems that motivate new methods.

2. *Hereditary stress*: here Wilder's meaning would be better conveyed by the term *self-constituted* stress, for he is concerned with pointing to the problems that are created intrinsically by a mathematical system itself (cf. above on odd–even, perfect, prime). Greek mathematics created stresses for itself by allowing the discovery of incommensurability and the space–time paradoxes of Zeno, which in turn were

a major factor in the introduction of both the axiomatic method and the development of geometry as a tool for investigating number theory, as well as for the study of special forms such as the triangle and the circle (ibid., p. 164).

Similarly, higher-degree equations led to a tension that required an extension of the notion of number to include $\sqrt{-1}$ as a number, in addition to the earlier extension to include negative numbers themselves.

3. *Symbolization*: 'So long as one was tied to the natural language, the language of common discourse, or even special *words* created expressly for mathematical purposes, mathematical advance was hampered' (ibid., p. 164). Without a specifically mathematical notational system, both calculation and conception are made more difficult: quadratic equations cannot be easily done in English; a description of the normal curve is virtually impossible to communicate with ordinary-language symbols alone.

4. *Diffusion*: diffusion means culture contact. This does not seem to be a necessary component of mathematical development, although

without new cultural contacts, arithmetic and geometry of the Babylonian–Egyptian type could well have remained in the virtually static state that Chinese mathematics found itself.... The initial impetus for the great advance made by the Greek culture in mathematics was given by *diffusion*, whereby Babylonian and Egyptian mathematics met Greek philosophy and produced a fusion that was a new and entirely different kind of mathematics (ibid., p. 166).

Presumably, diffusion could be compensated for by processes internal to a particular society, although synergistic or syncretic processes may provide unique results.

5. *Abstraction* and 6. *Generalization*: these have already been discussed above.

7. *Consolidation*: the process of bringing together diverse and scattered mathematical systems to be encompassed by one new system.

In cultures that employed varying types of numerals for different categories of objects, the ultimate transition to a single numeral for all categories may have been a case of consolidation involving cultural stress or an elementary form of abstraction (ibid., p. 168).

There are many specific examples of consolidation that could be cited in the case of modern mathematics.

8. *Diversification*: counting diversifies, to provide the basis for the operational and measuring aspects of mathematics:

Eventually the operational and measuring aspects led to fractions and, ultimately, to the real numbers; the primitive counting aspect was extended to the transfinite cardinal numbers and the ordering aspect to the transfinite ordinals (ibid., p. 170).

9. *Cultural lag* and 10. *Cultural resistance*: these are aspects of cultural conservatism that presumably provide the basis for the role of cultural diffusion, as well as for the possibility that diffusion will not occur.

11. *Selection*: 'With the passing of time one frequently observes that various symbolic devices (or merely special symbols) emerge for the expression or handling

of a concept, with the eventual survival of only one' (ibid., p. 171). This is an elementary example of 'selection'. It is a process that does not entail 'survival of the fittest' or the retention of the most efficient way of accomplishing something; power groups within the mathematical sub-culture may determine 'fashions'.

5 Numerical representations

There have been a variety of number-writing systems invented during historical time (see Ifrah 1981, especially Part III). As with language scripts (see Appendix 1 above), the diffusion of many Indo-European present-day forms can be traced out, though this time to an Indian rather than Semitic origin (see Fig. 27.13, next page). In the case of these number-representations, however, changes in them have been fundamentally stylistic, and all retain the basic decimal base of the original. Unlike written language systems, where the nature of the language to be represented affects the nature of the writing system settled on (see, for example, Barton and Hamilton on the functional motivation for adopting syllabic versus alphabetic scripts Section 27.2.3), the nature of the number system is invariant.

Further reading

Cajori, F. (1928). *A history of mathematical notations*, Vol. 1: *Notations in elementary mathematics*. Open Court, La Salle, Ill.

Dantzig, T. (1930). *Number: the language of science*. Macmillan, New York.

Flegg, G. (1984). *Numbers: their history and meaning*. Penguin Books, Harmondsworth.

Editorial appendix IV: Music

'Evolutionary' accounts of music have often been put forward, but are now regarded as contentious.[3] In general, their claims have been that music has become more complex over time. Thus rhythm is regarded as having been exhibited from very early periods, increasingly accompanied by the use of simple song patterns. These led to a slow elaboration of the laws of tonality, the arrangement of sounds of definite pitch so that they have meaning in relation to one another.

And from this eventually came an understanding of the laws of harmony, the use of several notes sounded at the same time.... It is amazing to see how few cultures of the world, even today, have developed a true polyphonic music, a system of music that employs more than one melodic line. . . . Melody and harmony imply the development of a musical scale (Alberti 1974, p. 16).

The basis of Greek music was the tetrachord, a system of only four notes. The pentatonic scale was an early elaboration, found in Chinese, Celtic, and African music. The hexachord was a six-note system used in medieval European music; whereas the modern diatonic scale made its appearance in the Renaissance, around the time of polyphonic harmony.

This order of musical elaboration was first given an explicitly evolutionary interpretation by Rowbotham (1885–7) in a history of music that borrowed from Spencerian philosophy. More pervasive in its influence was Parry's *The art of music*, originally published in 1893, which argued that music possessed an inherent logical form which had been elaborated, and hence by analogy had evolved, in the order that is apparent historically and under the constraint of the laws of logic. Perhaps one of the most thoroughgoing accounts in this vein is Glyn's (1934) *Theory of musical evolution*, which contains an appendix chart showing an evolutionary series classifying, in modern musical notation, a sequence of examples from an Australian Aboriginal monotone to the folk and art music of the Renaissance.

Psychological accounts were also in vogue for a while. Wallace (1908), for example, considered music to be a separate mental faculty, undergoing an evolutionary development of its own independently of other abilities, as is witnessed by his noting the time when humans developed the ability to harmonize (put two different notes together): 'ludicrously simple as this appears, it is not much older than Westminster Abbey' (ibid., p. 65);[4] and, more explicitly:

If we contrast the highest musical achievement even of a hundred years ago with the music which we have today, we see an advance in thought and imagination which is almost incredible, and it is only to be explained on the supposition that a new faculty has come into existence in recent years (ibid., p. 19).

Such speculation was accompanied by such recapitulationary claims as those of Isaacson (1928):

The baby of today reproduces in rapid order the gradual evolution of the savage's musical possession. His tone-making sang with the music of the cat ... Then he began clapping his hands ... and he danced with hand-clappings ... [etc.] (Chap. 10).

As far as we can ascertain, such inanities have resulted in evolutionary theory becoming a *bête noire* among musicologists, and there are no recent accounts of the historical elaboration of musical systems or notations, and the possible interaction between the two, that are comparable to the sorts of accounts put forward for writing and mathematical developments.

Social and cognitive factors in the historical elaboration of writing 825

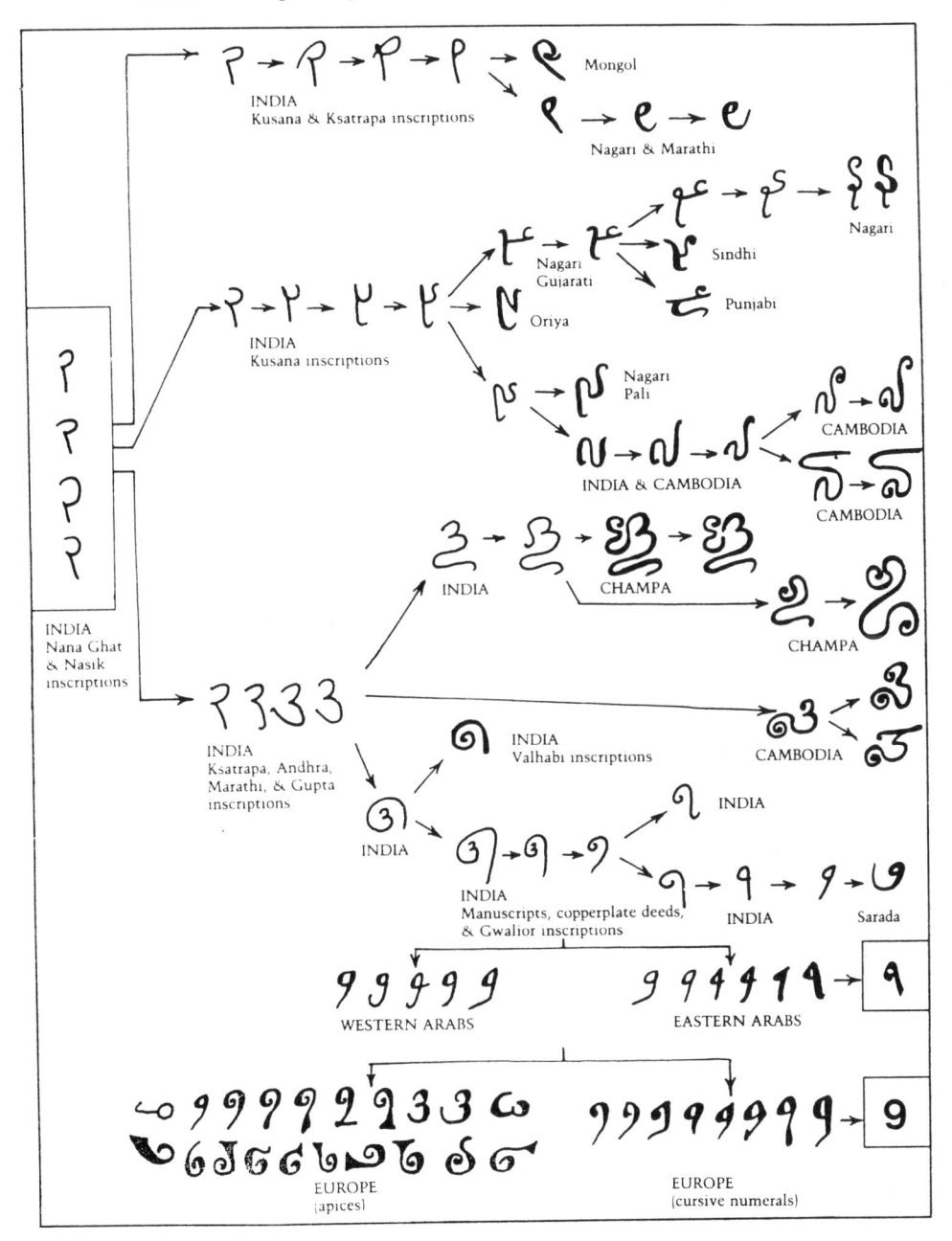

Fig. 27.13 Origin and historical development of the numeral '9' (from Ifrah 1981).

However, accessible and sober introductions to the substantive topic areas can be found in *The New Grove Dictionary of music and musicians* (for example Bent *et al.* 1980; Porter 1980). For a recent psychological account of musical development in children, see Hargreaves (1986); for the cognitive psychology of music, see Sloboda (1985), who also provides an overview of the relations between music and language, and suggests music evolved alongside language as a social communication system. Little musical ability is shown by animals. Reviewing the literature, D'Amato (1988, p. 474) notes:

frequency contour detection of non-species-specific acoustic stimuli is a poorly developed capacity in cebus monkeys and rats. This cognitive limitation is very likely not restricted to these two species, as there are strong indications of it in other animals.

In fact, it is possible in D'Amato's view that the only animal that might show such abilities is the pygmy chimpanzee (*Pan paniscus*); but unambiguous data is not presently available. In less technical terms, D'Amato's conclusion is that 'monkeys can't hum a tune ... because they don't hear them' (ibid., p. 478).

Editorial appendix V: Dance and choreography

Dance and its notation (choreography) have suffered a similar fate to music, being badly served by the evolutionary enthusiasms of the turn of the century. Sachs (1937), for example, places his treatment of 'primitive' and 'Stone Age' dance in the context of Menghin's *Weltgeschichte der Eisezeit (World history of the glacial ages)* (1931), where it is claimed that the cultures of 'diminutive races' found in central Africa, Asia, the Andaman Islands, central New Guinea, etc., are a survival of Middle Palaeolithic cultures, and a fascist-inspired anthropology is proposed. From this mish-mash, Sachs derives the evolution of the dance.

A more reasonable historical survey is given by Kraus and Chapman (1981), who make it apparent that while there is a degree of evidence about the early history of dance from early Sumerian times, little is known of its origins, beyond speculations that take 'primitives' as cultural fossils. The written representation or choreography of dance dates from the 1500s, and Guest (1977) provides the most accessible source on the origins and development of this form of written symbolism. For the anthropology of dance, see Hanna (1978), and for a symbolic communication model of dance in a cross-cultural perspective, see Hanna (1979 *a* and *b*).

Editorial appendix VI: Cartography

Only a few maps exist from before the time of Ptolemy (AD 87–150). These range from a clay map of Northern Mesopotamia at *c*.3800 BC and a Babylonian world map of *c*.500 BC to some second-century BC maps from China. An historical overview of cartographic history is shown in Timecharts 1–4 (on pp. 828–33 from Raisz 1938). From early times maps appear to have had three broad functions:

(1) a record of geographical features;
(2) a guide for travellers; and
(3) a vehicle to express figuratively abstract, hypothetical, or religious concepts.

Thematic maps are much more recent, beginning only in the West in the late 1600s. It is possible that some early maps may have been of resources, for example geological resources, rather than topological. Contemporary maps created in pre-literate societies tend to be used more as guides than as geographical representations. Stick-and-shell maps from the Pacific, and Eskimo carved maps are remarkably accurate by modern standards (see Figs. 27.14 and 27.15).

Developmental studies of children's understanding of spatial relations and their construction and use of maps are becoming more common. Two useful sources from which to enter into this literature are Hart and Berzok (1982) and Downs (1985). Parallels between the cultural elaboration of maps and the history of art (for example Rees 1980) have been drawn, and have led to some speculations on the relation between the culturally-changing representations of space in maps and the ontogenesis of spatial cognition in children (see for example Wood 1977). Such parallels face the same problems as those proposed by Gablik (1976) between cognitive development and the history of pictorial representation (see discussion in Bremner, this volume, Chapter 18).

Further reading

Bagrow, L. and Skelton, R. A. (1964). *History of cartography*. Watts, London.
Carpenter, E. S. (1955). Space concepts of the Aivilik Eskimos. *Explorations*, **5**, 131–45.
Davenport, W. (1960*a*). Marshall Island navigation charts. *Imago Mundi*, **15**, 19–26.
Davenport, W. (1960*b*). *Marshall Islands navigation charts*. Bobbs-Merill, Indianapolis.
De Hurtorowicz, H. (1911). Maps of primitive peoples. *Bulletin of the American Geographical Society*, **43**, 669–79.
Harley, J. B. and Woodward, D. (eds) (1987). *History of cartography*. Vol. 1: *Cartography in prehistoric, ancient and mediaeval Europe and the Mediterranean*. University of Chicago Press.
Harvey, P. D. A. (1980). *The history of topographical maps: symbols, pictures and surveys*. Thames and Hudson, London.
Lyons, H. (1928). The sailing charts of the Marshall Islanders. *Geographical Journal*, **72**, 325–8.
Robinson, A. H. (1982). *Early thematic mapping in the history of cartography*. University of Chicago Press.
Thrower, N. J. (1972). *Maps and man: an examination of cartography in relation to culture and civilization*. Prentice-Hall, Englewood Cliffs, NJ.
Turnbull, D. (1991). *Mapping the world in the mind: an investigation of the unwrtten knowledge of the Micronesian navigators*. Deakin University Press Geelong, Victoria.

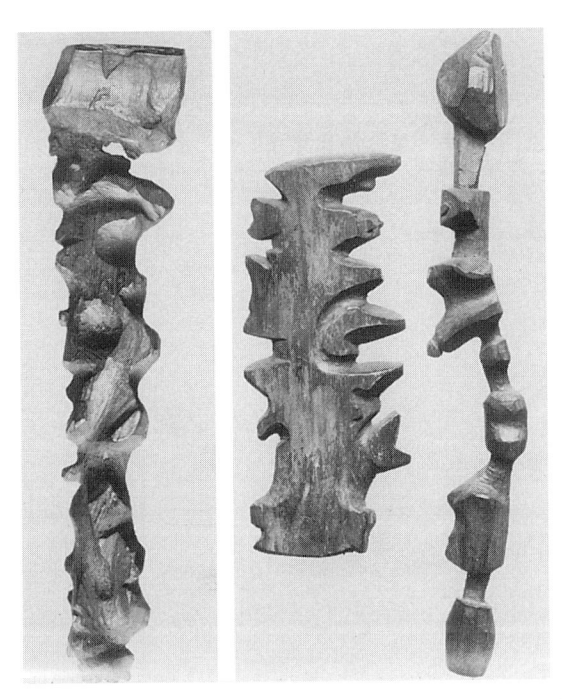

Fig. 27.14 Eskimo coastline maps. Bays and headlands are modelled with a high degree of accuracy (from L. Bagrow (1951) *Geschichte der Kartographie*, (2nd edn). Safari-Verlag, Berlin).

Ungar, E. (1935). Ancient Babylonian maps and plans. *Antiquity*, **9**, 311–22.
Ungar, E. (1937). From cosmos pictures to world maps. *Imago Mundi*, **2**, 1–7.

Fig. 27.15 Marshall Island stick charts. The square grid represents the open sea, over an area of several hundred square miles, and functions as a framework for the map elements proper. Shells are tied on to show the relative position of islands. The curved sticks show the prevailing wave fronts. Navigation was performed by lying in the bottom of a canoe so as to detect the rhythmical wave patterns in the sea at that point, as they swell from varying directions. The map can then be 'read' to determine the location of the canoe with respect to particular islands. (British Museum, Department of Ethnography, No. 2289.)

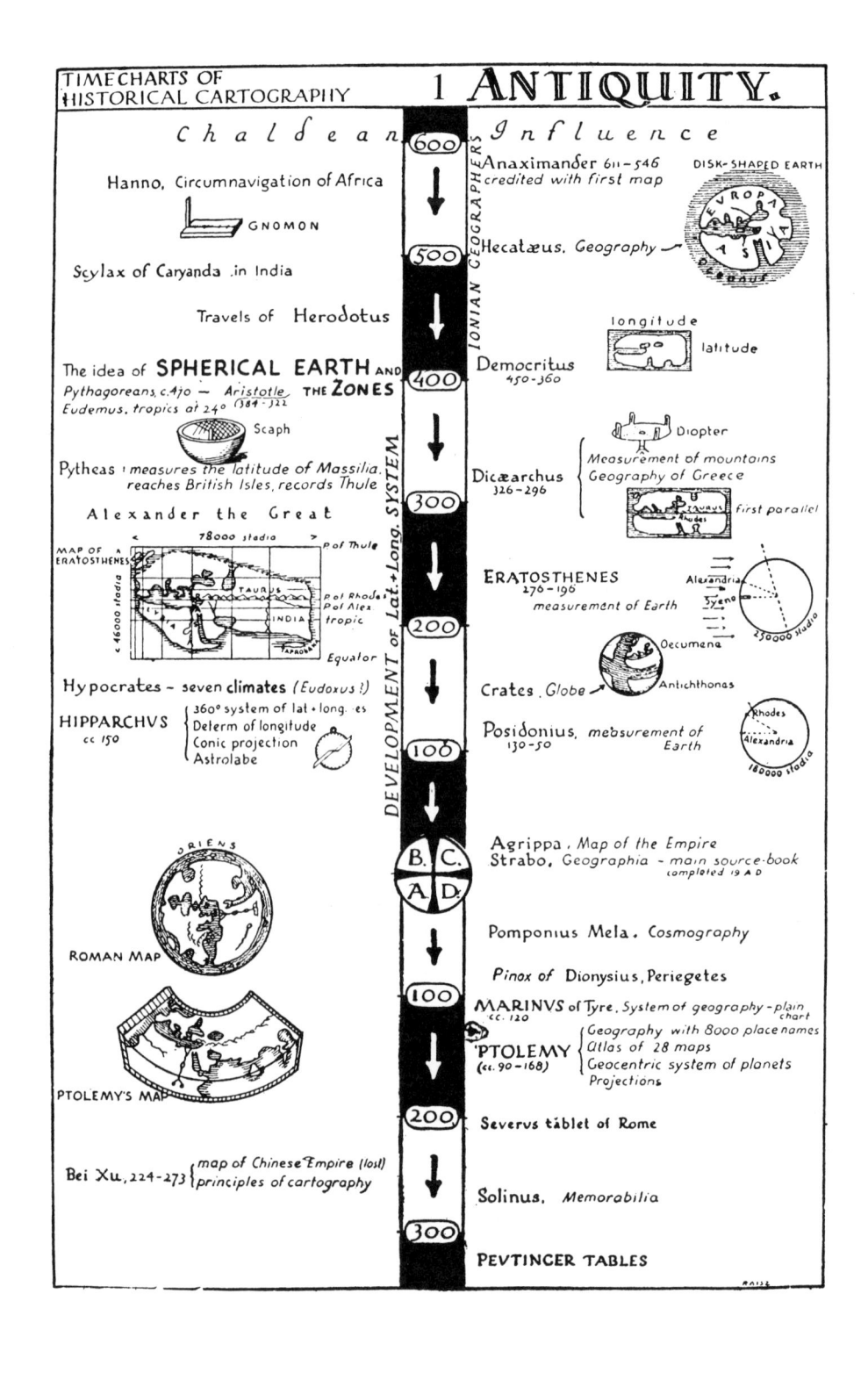

TIMECHARTS OF HISTORICAL CARTOGRAPHY — 1 ANTIQUITY

Chaldean Influence

600 BC
- Hanno, Circumnavigation of Africa
- GNOMON
- Anaximander 611–546, credited with first map — DISK-SHAPED EARTH

500 BC
- Scylax of Caryanda in India
- Hecataeus, Geography

- Travels of Herodotus
- Democritus 450–360 (longitude, latitude)

400 BC
- The idea of **SPHERICAL EARTH** and **THE ZONES**
- Pythagoreans, c.470 — Aristotle (384–322)
- Eudemus, tropics at 24°
- Scaph
- Pytheas measures the latitude of Massilia, reaches British Isles, records Thule
- Dicaearchus 326–296 — Diopter, Measurement of mountains, Geography of Greece, first parallel

300 BC
- Alexander the Great
- MAP OF ERATOSTHENES (78000 stadia, 46000 stadia, P. of Thule, Taurus, P. of Rhodes, P. of Alex, tropic, India, Taprob., Equator)
- ERATOSTHENES 276–196, measurement of Earth (Alexandria, Syene, 250000 stadia)

200 BC
- Hypocrates — seven climates (Eudoxus?)
- Crates, Globe (Oecumena, Antichthonas)
- HIPPARCHUS cc 150
  - 360° system of lat + long.
  - Determ of longitude
  - Conic projection
  - Astrolabe
- Posidonius, measurement of Earth 130–50 (Rhodes, Alexandria, 180000 stadia)

100 BC

B.C. / A.D.
- Agrippa, Map of the Empire
- Strabo, Geographia — main source-book completed 19 AD

- ROMAN MAP (ORIENS)
- Pomponius Mela, Cosmography

100 AD
- Pinox of Dionysius, Periegetes
- MARINUS of Tyre, System of geography — plain chart cc. 120
- PTOLEMY (cc. 90–168)
  - Geography with 8000 place names
  - Atlas of 28 maps
  - Geocentric system of planets
  - Projections
- PTOLEMY'S MAP

200 AD
- Severus tablet of Rome

- Bei Xu, 224–273 — map of Chinese Empire (lost), principles of cartography
- Solinus, Memorabilia

300 AD
- PEUTINGER TABLES

DEVELOPMENT of Lat + Long. SYSTEM
IONIAN GEOGRAPHERS

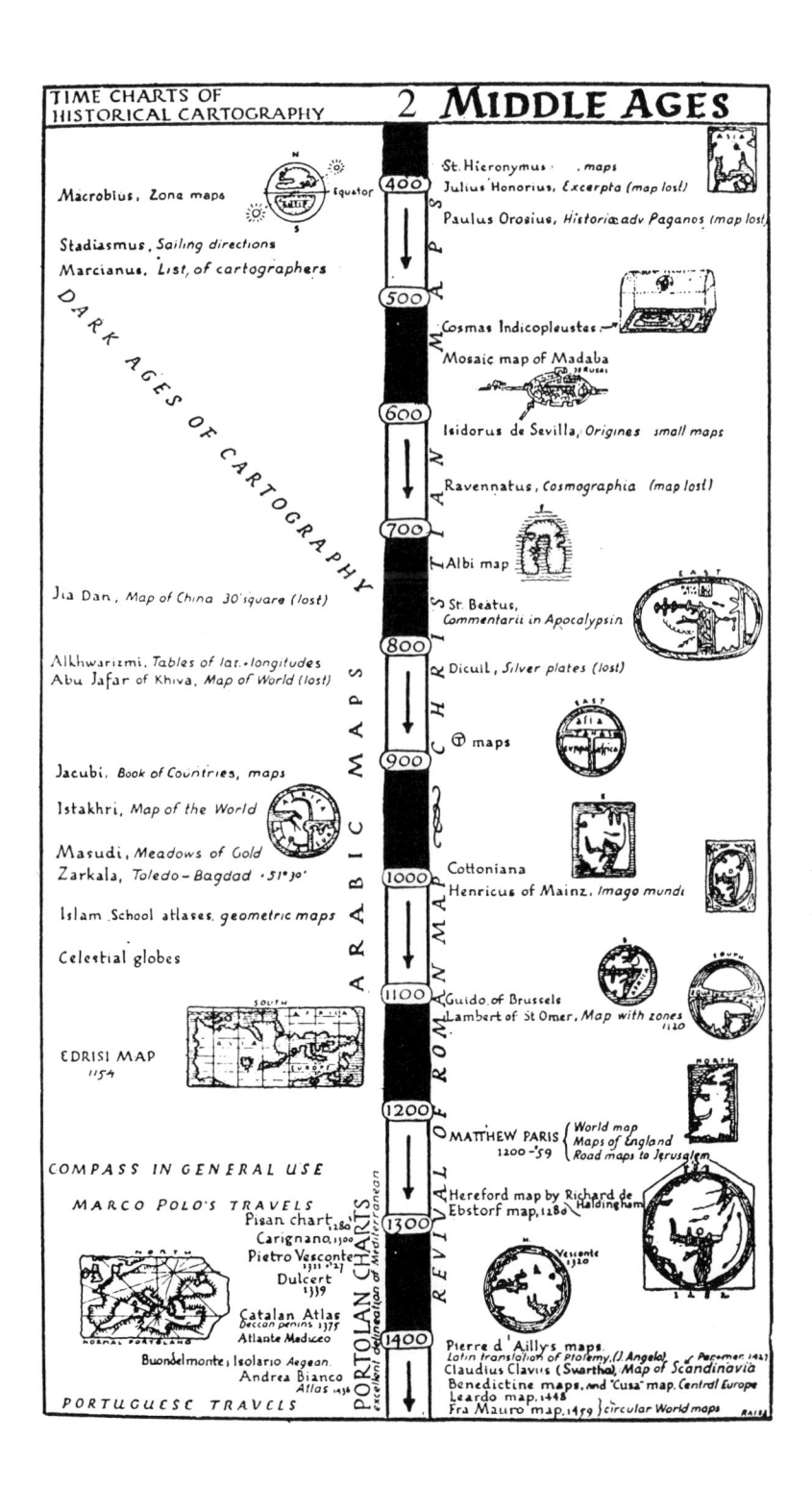

Social and cognitive factors in the historical elaboration of writing

TIME CHARTS OF HISTORICAL CARTOGRAPHY — 2 MIDDLE AGES

DARK AGES OF CARTOGRAPHY

- Macrobius, Zone maps
- Stadiasmus, Sailing directions
- Marcianus, List of cartographers

400
- St. Hieronymus, maps
- Julius Honorius, Excerpta (map lost)
- Paulus Orosius, Historiæ adv Paganos (map lost)

500
- Cosmas Indicopleustes
- Mosaic map of Madaba

600
- Isidorus de Sevilla, Origines, small maps

700
- Ravennatus, Cosmographia (map lost)
- Albi map

- Jia Dan, Map of China 30' square (lost)

- St. Beatus, Commentarii in Apocalypsin

800
- Alkhwarizmi, Tables of lat.+longitudes
- Abu Jafar of Khiva, Map of World (lost)
- Dicuil, Silver plates (lost)

900
- ⊕ maps

ARABIC MAPS

- Jacubi, Book of Countries, maps
- Istakhri, Map of the World
- Masudi, Meadows of Gold
- Zarkala, Toledo–Bagdad 51°30'

1000
- Cottoniana
- Henricus of Mainz, Imago mundi

- Islam School atlases, geometric maps
- Celestial globes

1100
- Guido of Brussels
- Lambert of St Omer, Map with zones 1120

- EDRISI MAP 1154

REVIVAL OF ROMAN MAP

1200
- MATTHEW PARIS 1200–'59 { World map / Maps of England / Road maps to Jerusalem }

- COMPASS IN GENERAL USE
- MARCO POLO'S TRAVELS

- Hereford map by Richard de Haldingham
- Ebstorf map, 1284

- Pisan chart 1280
- Carignano 1300
- Pietro Vesconte 1311–'27
- Dulcert 1339

1300
- Vesconte 1320

PORTOLAN CHARTS (excellent delineation of Mediterranean)

- Catalan Atlas (Deccan penins. 1375)
- Atlante Mediceo
- Buondelmonte, Isolario Aegean
- Andrea Bianco Atlas 1436

1400
- Pierre d'Ailly's maps
- Latin translation of Ptolemy (J. Angelo)
- Claudius Clavus (Swartha) Map of Scandinavia
- Benedictine maps, and "Cusa" map, Central Europe
- Leardo map 1448
- Fra Mauro map 1459 } circular World maps
- Pesemer 1423

PORTUGUESE TRAVELS

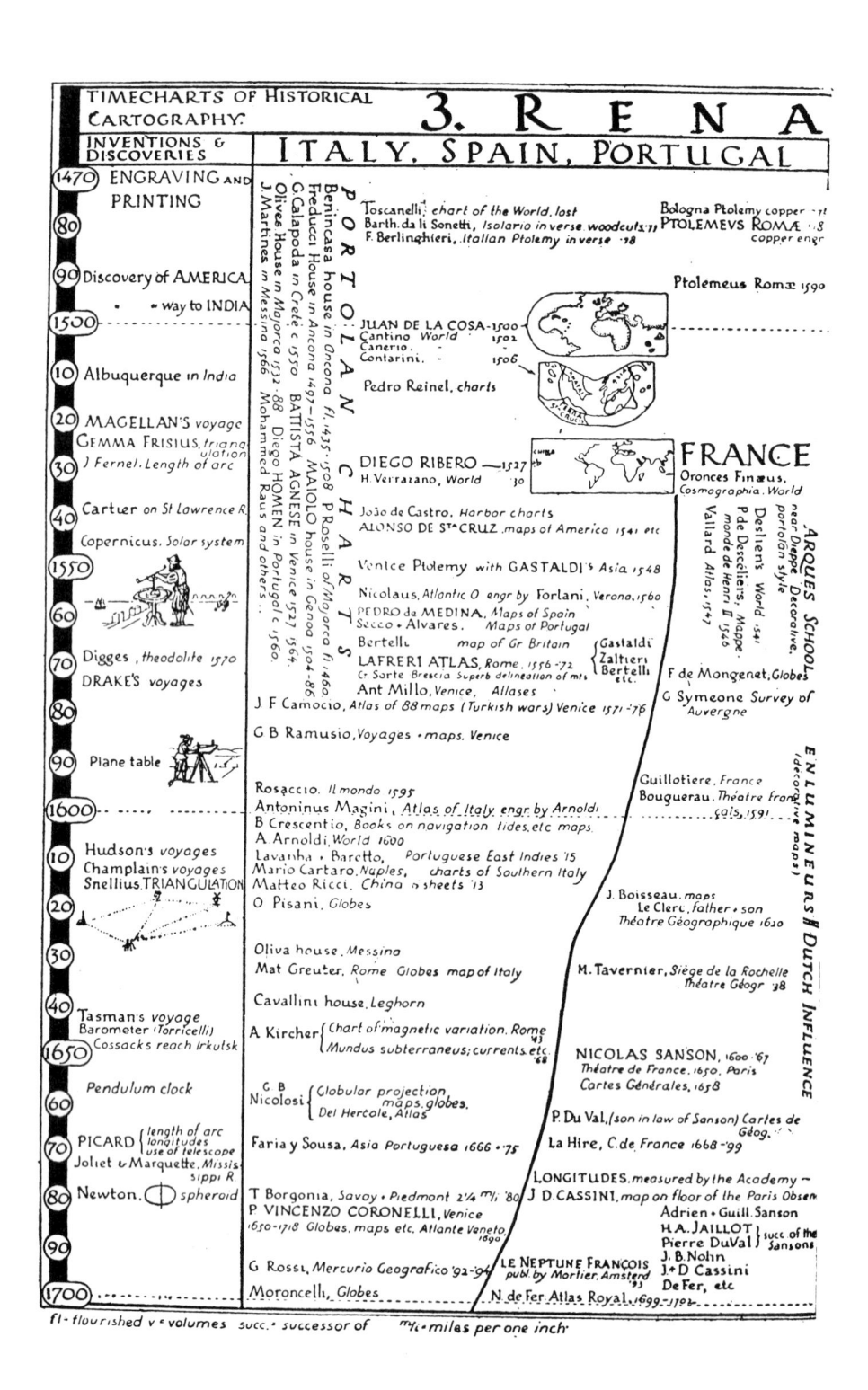

TIMECHARTS OF HISTORICAL CARTOGRAPHY:

3. RENA — ITALY, SPAIN, PORTUGAL

INVENTIONS & DISCOVERIES

- **1470** — ENGRAVING AND PRINTING
- **1480**
- **1490** — Discovery of AMERICA
  - " way to INDIA
- **1500**
- **1510** — Albuquerque in India
- **1520** — MAGELLAN's voyage; GEMMA FRISIUS, triangulation
- **1530** — J Fernel. Length of arc
- **1540** — Cartier on St Lawrence R.
  - Copernicus. Solar system
- **1550**
- **1560**
- **1570** — Digges, theodolite 1570; DRAKE's voyages
- **1580**
- **1590** — Plane table
- **1600**
- **1610** — Hudson's voyages; Champlain's voyages; Snellius TRIANGULATION
- **1620**
- **1630**
- **1640** — Tasman's voyage; Barometer (Torricelli); Cossacks reach Irkutsk
- **1650**
- **1660** — Pendulum clock
- **1670** — PICARD {length of arc, longitudes, use of telescope}; Joliet & Marquette. Mississippi R
- **1680** — Newton. ○ spheroid
- **1690**
- **1700**

PORTOLAN CHARTS

Benincasa house in Oncona fl. 1435-1508 P Roselli of Majorca fl. 1460
Freducci house in Ancona 1497-1556 MAIOLO house in Venice 1527-1564
G Calapoda in Crete c 1550 BATTISTA AGNESE in Venice 1527-1564
Olives house in Majorca 1532-88 Diego HOMEN in Portugal c 1560
J Martines in Messina 1566 Mohammed Raus and others...

Toscanelli, chart of the World, lost
Barth. da li Sonetti, Isolario in verse, woodcuts '71
F. Berlinghieri, Italian Ptolemy in verse '78

Bologna Ptolemy copper '71
PTOLEMEVS ROMÆ '8 copper engr

Ptolemeus Romæ 1590

JUAN DE LA COSA — 1500
Cantino World   1502
Canerio.
Contarini.   1506

Pedro Reinel, charts

DIEGO RIBERO — 1527
H. Verrazano, World '30

João de Castro. Harbor charts
ALONSO DE Sᵗᵃ CRUZ, maps of America 1541 etc

Venice Ptolemy with GASTALDI's Asia 1548
Nicolaus, Atlantic O engr by Forlani, Verona, 1560
PEDRO de MEDINA. Maps of Spain
Secco & Alvares. Maps of Portugal
Bertelli   map of Gr Britain   {Gastaldi
LAFRERI ATLAS, Rome, 1556-72   {Zaltieri
C. Sorte Brescia Superb delineation of mts {Bertelli etc
Ant Millo, Venice, Atlases
J F Camocio, Atlas of 88 maps (Turkish wars) Venice 1571-76

G B Ramusio, Voyages • maps. Venice

Rosaccio. Il mondo 1595
Antoninus Magini, Atlas of Italy engr by Arnoldi
B Crescentio, Books on navigation tides, etc maps
A. Arnoldi, World 1600
Lavanha • Baretto,   Portuguese East Indies '15
Mario Cartaro, Naples,   charts of Southern Italy
Matteo Ricci, China 6 sheets '13
O Pisani, Globes

Oliva house, Messina
Mat Greuter, Rome Globes map of Italy

Cavallini house, Leghorn
A Kircher { Chart of magnetic variation, Rome '43
          { Mundus subterraneus; currents etc. '68

G B
Nicolosi { Globular projection, maps. globes.
         { Del Hercole, Atlas

Faria y Sousa, Asia Portuguesa 1666 • 75

T Borgonia, Savoy • Piedmont 2¼ m/i '80
P. VINCENZO CORONELLI, Venice
1650-1718 Globes. maps etc. Atlante Veneto. 1690

G Rossi, Mercurio Geografico '92-'94
Moroncelli, Globes

FRANCE

Oronces Finæus, Cosmographia. World

ARQUES SCHOOL near Dieppe. Decorative, portolan style
Desliens World 1541
P.de Descellers, Mappe monde de Henri II 1546
Vallard Atlas, 1547

F de Mongenet, Globes
G Symeone Survey of Auvergne

Guillotiere. France
Bouguerau. Théâtre français, 1591

J. Boisseau. maps
Le Clerc, father & son
Théâtre Géographique 1620

M. Tavernier, Siège de la Rochelle, Théâtre Géogr '38

NICOLAS SANSON, 1600-'67
Théâtre de France, 1650. Paris
Cartes Générales, 1658

P. Du Val, (son in law of Sanson) Cartes de Géog.
La Hire, C.de France 1668-'99
LONGITUDES, measured by the Academy
J D CASSINI, map on floor of the Paris Obsᵛ
Adrien • Guill. Sanson
H.A. JAILLOT } succ of the Sansons
Pierre DuVal
J. B. Nohn
LE NEPTUNE FRANÇOIS J • D Cassini
publ by Mortier, Amsterd De Fer, etc
N de Fer Atlas Royal 1699-1702

ENLUMINEURS / DUTCH INFLUENCE (decorative maps)

---

fl - flourished v - volumes succ.- successor of m/i - miles per one inch

ISSANCE

GERMANY, NETHERLANDS | OTHERS

1470

Nicolaus Germanus *introduces trapezoidal (Donis) projection and adds modern maps to his Ptolemy ms. (Italy) 1466–'82*

'80

Ulm Ptolemy, N. Germanus Woodcuts, 1482–'86
Eichstätt and Martellus version of 'Cusa' map of Central Europe, 1490

'90

Martellus Germanus. World showing Portuguese discoveries
BEHAIM GLOBE, *first detailed terrestrial globe based on Ptolemy*
Etzlaub map of Germany, showing roads, 1492·1501, etc.

Conrad Türst, Helvetia

1500

Stabius–Werner, projection
J. Ruysch, World 1508 — 1507 WORLD MAP
WALDSEEMÜLLER { Wall map of Europe 1511
San Dié 1470–1527 { STRASBOURG PTOLEMY 1513. 20 modern maps
{ Carta marina 1516. 12 sheets

Piri Reis. Chart of Atlantic 1513

10

J. Schöner, Globes of 1515 + 1529 showing Terra Australis

Nic. Claudianus Bohemia '18

20

Petrus APIANUS. Cosmographia 1524

B. Wapowski, Poland
Ziegler, Scandinavia 1532

30

Jacob of Deventer, Netherlands 1536–'39
Gerardus MERCATOR, World 1536
1512–94

Tschudi, Helvetia 1538
OLAUS MAGNUS, Scand. '39
Anton Wied, Muscovy '42

40

Sebastian Münster Cosmographia, 1544
Joh. Honter, Rudimenta cosmogr.

Herberstein, Russia '49

1550

Caspar Vopel, Cologne, Globes, etc.

Lily, England 1546

Indian Ocean, Arab chart '54

Mercator, Europe 1554

60

Diego Gutierrez, America, Antwerp 1562
Philip Apianus, Bayerische Landtafeln, 1:144 000
Mercator projection. World map, 1569
A. ORTELIUS, THEATRUM ORBIS TERRARUM 1570
Hogenberg–Höfnagl, Civitates Orbis Terrarum
Gerard de Jode Speculum Orbis Terrarum
J. Metellus, Itinerarium Orbis Christiani, 1579
WAGHENÆR Spieghel der Zeevaerdt

Lazius, Austria & Hungary '61
Jenkinson, Russia 1562
J. Le Moyne, Florida, 1565
Fabricius, Moravia '69

Lhuyd, England '69

70

Chr. SAXTON County maps 74–79
c. 1542–1611

80

John White Virginia, manusc. '85

E. Molyneux, Globes table
J. Norden, Estate surveys

Oeder, Saxonia 1:144 000
1586–1607

90

Plancius. World 1592 Globes
W. Barentszon. Mediterr. Charts, 1595
Mercator's Atlas 1595
Linschoten Itineraries 1596
Jodocus } HONDIUS, succ. of Mercator
Hendric } World 1601·08 etc.

Edward WRIGHT, World
in Mercator proj. 1600

A. Bureus Sweden

1600

Bertius. Atlases

Timothy Pont Scotl'd '08
J. Speed. Theatre of Great Britaine, 1610

Simancas (Spain) map. N. Am. '10
Smith Powell Virginia 1612,
CHAMPLAIN { New England, 1616
{ Acadia 1607
{ St Lawrence 1612
{ New France 1632

10

20

J. A. RAUCH.
Landtafeln
(hachuring)

W. Janszoon BLAEU. Atlas Novus 1634+45
1596–1673 Theatrum 3–6 v. '35+'54
J. JANSZOON. Nieuwe Atlas, 3 v. 1638+53, etc.

Schickhart Württemberg
(triangulated) 1:30 000 1624'5)

30

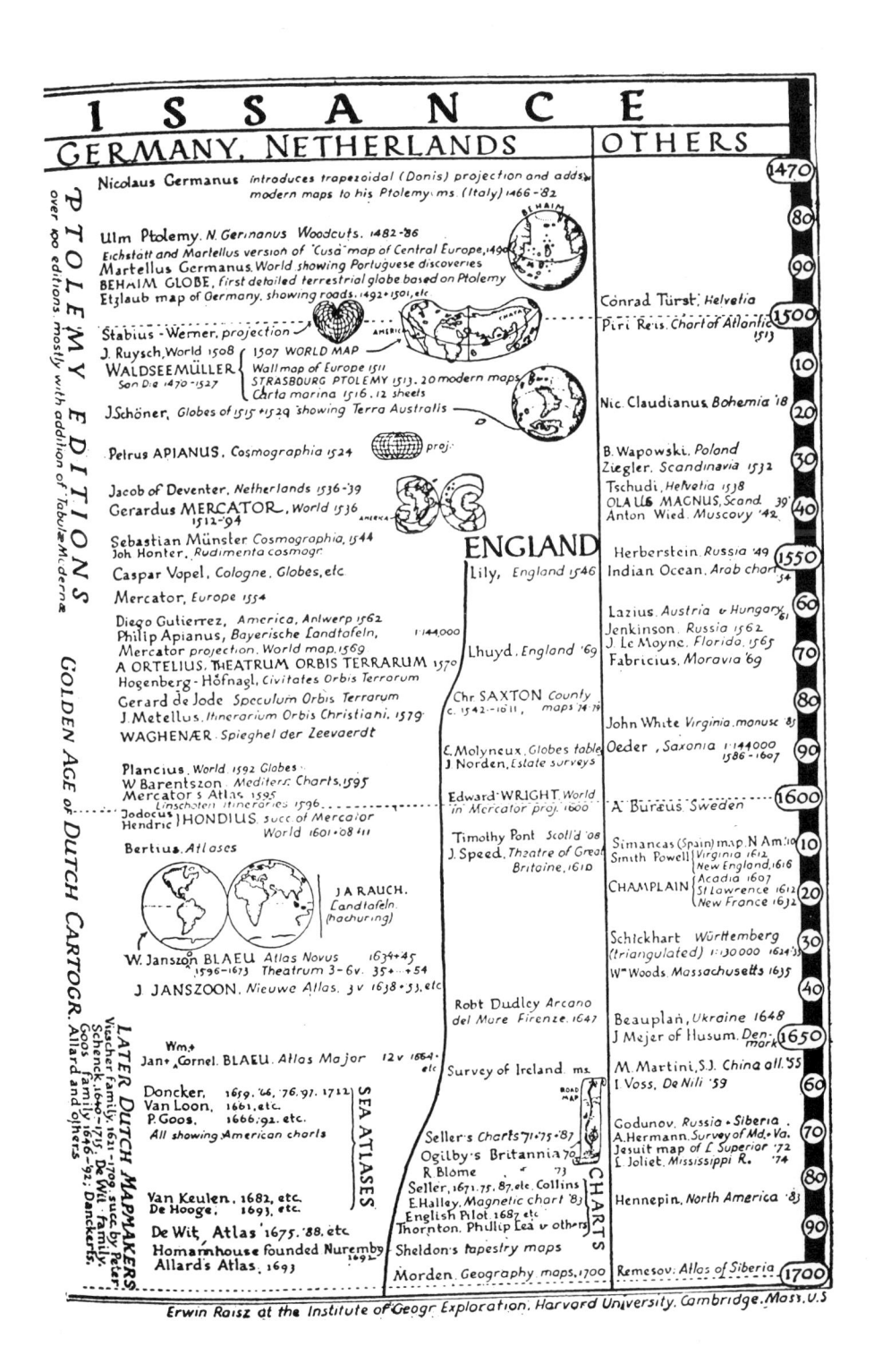

Wm Woods. Massachusetts 1635

40

Robt. Dudley Arcano
del Mare Firenze, 1647

Beauplan, Ukraine 1648
J. Mejer of Husum. Denmark

1650

Wm + Jan + Cornel. BLAEU. Atlas Major 12 v. 1664, etc.

Survey of Ireland. ms.

M. Martini, S.J. China atl. '55
I. Voss, De Nili '59

60

Doncker. 1659,'66,'76,'91, 1712,
Van Loon. 1661, etc.
P. Goos. 1666,'92, etc.
All showing American charts

70

Seller's Charts '71·'75·'87
Ogilby's Britannia '70
R. Blome   "   '73

Godunov, Russia + Siberia
A. Hermann, Survey of Md.+ Va.
Jesuit map of L. Superior '72
L. Joliet, Mississippi R. '74

80

Van Keulen. 1682, etc.
De Hooge. 1693, etc.

Seller, 1671,'75,'87, etc. Collins
E. Halley. Magnetic chart '83
English Pilot 1687 etc.
Thornton, Philip Lea & others

Hennepin, North America '83

90

De Wit, Atlas 1675,'88, etc.
Homannhouse founded Nuremberg 1692
Allard's Atlas, 1693

Sheldon's tapestry maps
Morden, Geography maps, 1700

Remesov: Atlas of Siberia

1700

Left margin: PTOLEMY EDITIONS over 100 editions mostly with addition of Tabulæ Modernæ

GOLDEN AGE of DUTCH CARTOGR.

LATER DUTCH MAPMAKERS: Visscher Family, 1621–1709, succ. by Peter Schenck, 1645–1715, De Wit family, Goos family, 1645–92; Danckerts, Allard and others

Right column labels: SEA ATLASES · CHARTS · ROAD MAP · ENGLAND

Erwin Raisz at the Institute of Geogr. Exploration. Harvard University. Cambridge. Mass. U.S.

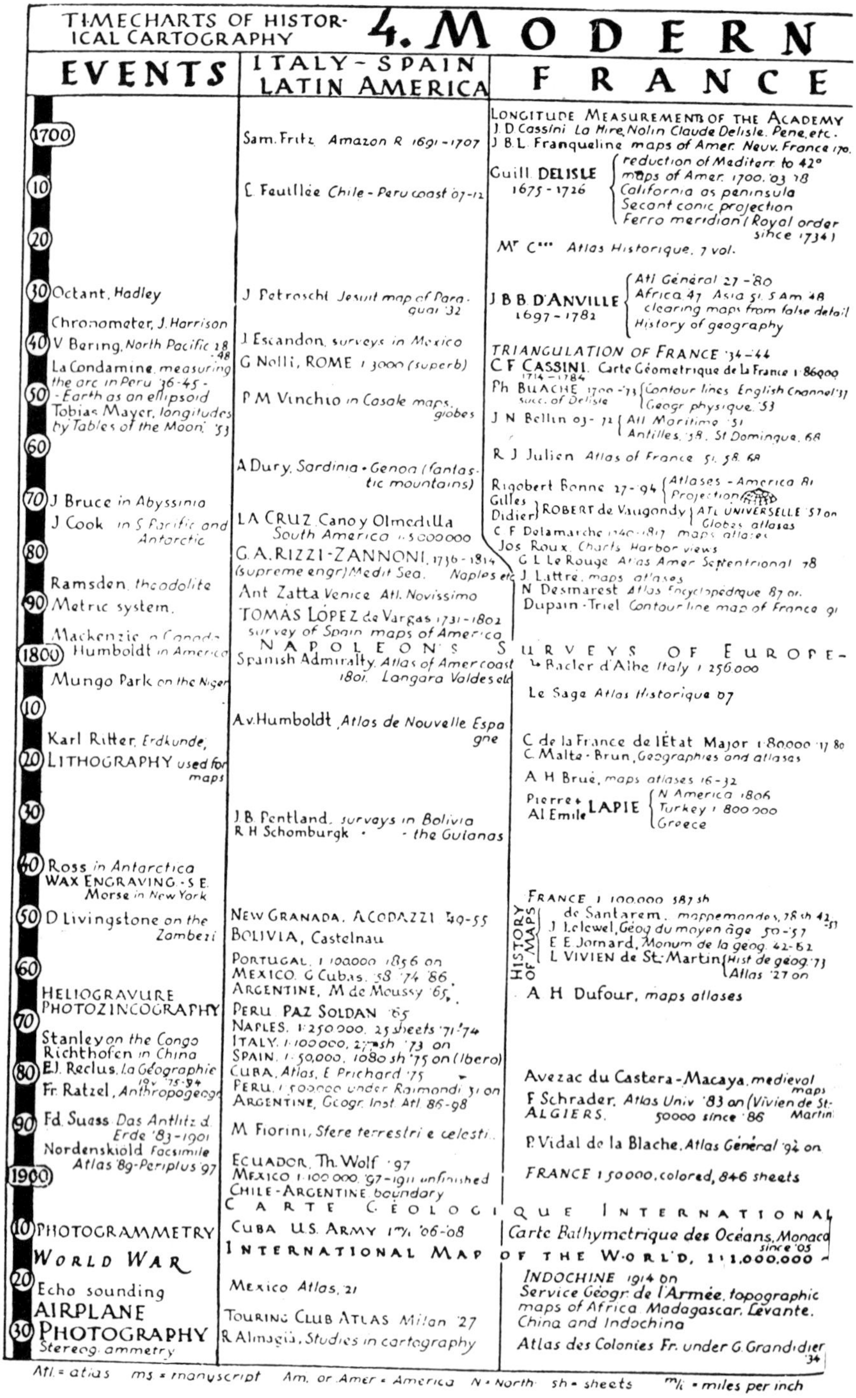

MAPS — Age of National Surveys, 1700–present

NETHERLANDS / GERMANY – AUSTRIA	GR. BRITAIN	OTHERS	
Homann house in Nuremberg; Seütter – Augsburg prolific mapmakers. Dutch style. HESSEN-KASSEL 1:540000 Schleusteinlen. WÜRTTEMBERG, triangulation (?) 1710 by Johann Mayer	English Pilot. 1685 – 1792. Herman Moll (from Amsterdam) fl. 98-32. John Senex d'49 maps, geographies, atlases, globes		1700
LATE DUTCH MAPMAKERS: R.+J. Ottens, Covens + Mortier, J. Loots, v.d. Aa, G.L. Valk, C. Allard, F. de Wit, J.v. Keulen, Isaac Tirrion, P. Schenck, etc.		J.J. Scheuchzer, Switzerl. RUSSIA, Coast survey of Peter the Great	10
		J. Chr. Muller Bohemia '27	20
First contourlines on maps – Merwede River bottom by Cruquius 1728 '30	Henry Popple, America in 20 sheets. '33	Kirilov, Atlas of Russia '34	30
J.G. Doppelmayr, Atl. Coelestis '42		de Marsigli, The Danube 31 sh. '41. Russian Acad. Atl. 19 sh '45	40
	John Mitchell, N.Am '55. SCOTLAND. Watson surv. J. Rocque d'62 Atl of forts '63	Lewis Evans, Middle British Colonies, Philadelph. '55	50
T.C. Lotter, Augsburg, maps.		TYROL: Anich + Huber '60– (Bauernkarte) 1:104000	60
		Kanter, Poland '70 Königsbg. de Lacy, Hungary.	70
BRANDENBURG 1:50000, 270 sh. v. Schmettau 67–87 ms. Projections { J.H. LAMBERT / Leonhard Euler } SILESIA, SAXONY, MORAVIA 1:100000 Geusau Tobias Mayer Mappa critica '80 '80 ms. HANNOVER 1:21333, 185 sh. '64–'86. MECKLENBURG 1:33900 '80+'88 Schmettau J.W.A. Jaeger, Gr. Atlas d'Allemagne '89. J.G. Lehman system of hachuring '99. Tranchot – Moreau etc. 1:100000	J.F.W. Desbarres, Atlant. Neptune '74 – N.Am Pilot '79. J. Rennel, Bengal Atlas '81. TRIANGULATION of Engld. ORDNANCE SURVEY '91. HYDROGRAPHIC OFFICE '95. SURVEY OF INDIA 1800. *LONDON MAPMAKERS and PUBLISHERS: John + Thomas Bowles, Emmanuel + Thomas Bowen fl. '20. Thomas Kitchin '38-'84, John Rocque fl.'34-'62, Thomas JEFFERYS fl.'31-'71 succ.by Sayer+Bennet– Wm. Faden, Whittle+Laurie fl.'98-1818, John CARY 1:69-1896, Aaron Arrowsmith fl. '90-'23/successors up to '58/J. Imray, charts A. Dalrymple charts*	BELGIUM, de Ferraris '71-'77 275 sh. 1:11520 (triangulated) DENMARK, Acad. survey SWITZERL. J.H. Weiss '76-'18 Schimeck, Turkish War Atl.'78 La Perouse, North Pacific '85 NETHERLANDS, Krayenhoff '02-'14 SWEDEN, Hermelin '97-'18. TURKEY in Arabic, 22 sh. EGYPT, 55 sh. Napoleon HUNGARY, Görög 4 m. '96-'04 de Lipszky 12 sh. '06	80 / 90 / 1800
Projections, Mollweide '05, Albers '05. Karl Ritter, Physical map of Europe '06. Weimar Inst. Germany, 1:177000, 254 sh. Justus Perthes, Gotha, maps since 1786 '07		Takahashi Inō, Japan, surveys GREECE 10 m. Fr. Müller, Vienna	10
		John Melish, U.S.A.	20
A.v. Humboldt, Isotherm map '17. C.F. Gauss, projections '22	Wm. Smith, Geological maps '24. IRELAND survey 6'/m.	RUSSIA 1:420000, 22 sh. '21-'39. H.S. TANNER in Philadelphia. A. de KRUSENSTERN Atlas of the Pacific O. Petersburg '27 SWEDEN 1:500000, altitude	30
	A.K. Johnston, Edinburgh since '25 '30. ROYAL GEOGRAPHICAL SOC. '30	Sam. Aug. Mitchell, U.S.A. tints	40
WALL MAPS, Emil Sydow. ALTITUDE TINTS { v. Hauslab '42 / E. Sydow '42 } WÜRTTEMBERG 1:50000 '21-'24. BAVARIA 1:50000 '12-'68. PRUSSIA 1:100000 '40 on.	Ch.+A. Black, London since '40. 6'/m Ordnance S maps '46 on.	GREECE 1:200000, 20 sh. '52-'80. SWITZERLD. DUFOUR map oblique hachuring '42-'65. RUSSIA 1:126000, 845 sh. '57– CAUCASIA 1:210000, – '63-'85. HOLLAND 1:25000, 776 sh. '66.	50 / 60
Petermanns Mitteilungen '55 on	John Bartholomew, Edinburgh. Edw Stanford, London	BELGIUM 1:20000, 527 sh. – NORWAY 1:100000, 331 sh. '69– BALKANS 1:200000, 36 (Russian) 1:200000, (Austrian)	70
AUSTRIA-HUNGARY 1:75000, 165 sh. CENTRAL EUROPE 1:200000, 192 sh. GERMANY 1:100000, 675 sh.	Ordnance Survey 1'/m 696 sh. since '72. E.G. Ravenstein, E. Equatorial Africa '82 Bartholomew, J.G. Physical atlases George Philip + Son, Liverpool	SWEDEN 1:100000, 234 sh. JAPAN 1:100000 '87 on. CHINA Kordt, Facsimile Atlas of Russia, 1690-1931.	80 / 90
K. Kretschmer, Atl. discovery of Amer. Vogel, Karte d. Deutschen Reiches. K. Peucker, Plastic colorshading, Vienna DE EUROPE – Beirich '96 Konrad Miller, studies of early maps. Max Eckert, projections – proposed by A. Penck in 1891 – Committee London 1909. – about 300 sheets Geogr. Section of General Staff maps of Balkans, Near East, Sudan. Asia 1:4000000, Africa 1:2000000 etc.	R.G.S. maps of explorations in Geog.J. The Times Atlas, 1900 on. Hauchecorne 1900	FINLAND, Atlas '99 +1911 1:1000000, '03 on Beyshlag '04 – 1913 Kümmerly + Frey, Berne – (20 %) published in 1935 CANADA Atlas '06 + '15 SIBERIA '14 NORWAY '22 EGYPT '28 CZECHOSLOVAKIA '28	1900 / 10 / 20 / 30
Joseph Fisher S.J. studies of Ptolemy. F.C. Wieder, Monumenta Cartogr. Leyden '25 on			NATIONAL ATLASES

Prepared by Erwin Raisz at the Inst of Geogr. Exploration, Harvard University

Editorial appendix VII: Printing

Barton and Hamilton have pointed out that generalizations about the general issues of literacy—the media used and their historical elaboration, how these interact with cognition, and so on, about which there have been strong claims put forward in the literature—rest on shaky empirical foundations. This is also the case for the mechanization of written symbolic communication as 'printing and publishing': claims often refer to printing as a 'revolution', according it an importance similar to the effects claimed for the alphabet on Greek civilization. For example:

> The invention and development of printing with movable type brought about the most radical transformation in the conditions of intellectual life in the history of western civilization. It opened new horizons in education and in the communication of ideas. Its effects were sooner or later felt in every department of human activity (Gilmore 1952, p. 186).

Such claims are difficult to test. Even straightforward questions regarding, say, the number of books published cannot be answered exactly.

For example, statistical sources relating to publishing are available (see below). That there are more books published today than in the past would seem a reasonable statement, and one borne out by the data. However, this statement needs to be approached with care. What, for example, constitutes 'a book'? Prior to the adoption of (and it is by no means clear when all countries party to this adoption put it into practice) 'A Recommendation concerning the International Standardization of Statistics relating to Book Production and Periodicals' by the General Conference of UNESCO on 19 November 1964—that a book is 'a non-periodical publication of at least 49 pages, exclusive of the cover pages'—quite various and disparate criteria were adopted (see, for example Barker 1956), and these can colour the statistics. Additionally, what constitutes 'more'—more titles or more copies? And further, what constitutes a title—a first edition, a re-edition, a reprint, a translation?

Again, what do the bald statistics of production reveal about the *use* of material? It is not certain that all books sold by publishers are read (even leaving aside those that are 'recycled' or 'pulped'); nor is it the case that all books are read in the same way—a novel may be read from cover to cover, and never referred to again; and the Bible may be read from cover to cover, but referred to again; whereas logarithm tables may never be read from cover to cover, but only referred to 'again and again'. Such uncertainties necessarily constrain the way in which the basic statistics of the printing trade's *production* (summarized below) can be interpreted as an index of its *consumption* (the information crucial to evaluating claims of the effects of printing).

Escarpit (1966) (see also Febvre and Martin 1976, pp. 217–22) estimates that the average print-run for a book in the middle of the sixteenth century was c.1000 copies, and between 2000 and 3000 until the end of the eighteenth century (for Western European publishers). Between 1800 and 1820 the technical innovations of the metal press, the foot-operated cylinder press, and the mechanical steam press 'revolutionized' the mode of publishing production—'before the end of Napoleon's reign, more sheets could be printed in an hour than had been possible in a day fifteen years earlier' (Escarpit 1966, p. 23). Average print-runs increased slightly (see below), but the 'best seller' emerged, so that individual books can have a disproportinate influence on the 'average figure': for example, Byron's *The Corsair* (1814) sold 10 000 copies on its day of publication; later in the nineteenth century *Uncle Tom's Cabin* (1852) sold 1 500 000 copies within a year of its publication (Escarpit 1966).

From the late nineteenth century more representative, but by no means accurate, figures for numbers of books published are available from sources such as *Publishers' Weekly*, *Bibliographie de la France*, *Borsenblatt*, and *Publisher's Circular*. For 1881 these give total titles published of 2991, 12 261, 15 191, and 5404 for the USA, France, Germany, and Britain respectively. Basic data for the United States have been compiled from *Publishers' Weekly* by Tebbel (1975, 1978) for the period 1865–1945 (see, for example, Table 27.2 for the period 1915–45), and for Britain from *The English Catalogue of Books* and *The Bookseller* by Norrie (1982) for the period 1900–81 (see Table 27.3 and Table 27.4 for subject areas for the period 1937–81). Both series of figures reflect the influence of major social events—world wars and economic depressions—as they impinge on the economics of publishing.

Global data for the modern period are reported yearly in *The United Nations Statistical Yearbook*. The last major review based on these data was compiled by Escarpit (1966). Table 27.5 shows world production of books by number of titles published for 1952 and 1962 (or nearest year to these for which data are available, given in brackets where applicable). Six countries—the USSR, mainland China, the UK, the Federal Republic of Germany, Japan, and the USA—produced more than 20 000 titles in 1962, and six others, France, India, Spain, Italy, the Netherlands, and Czechoslovakia, around 7500 or more titles. These twelve countries thus account for three-quarters of world production. The Soviet bloc countries account for approximately 36 per cent of world production, as does the 'Western bloc' of English, German, Spanish, and French producers. The remaining forty nations included in Table 27.5 that are not encompassed in either of the above groups (leaving aside China, for which figures are 'unreliable')—nine European and thirty-one Afro-Asian countries—contribute 28 per cent of world production by title.

Figures for total numbers of books produced, and average numbers of copies of each title are more difficult to establish, but certainly represent an increase on the figures for the sixteenth to the eighteenth centuries given above. Reviewing the evidence, Escarpit (op. cit.: 64–5) concludes that average edition size for the world was in the region of 10 000 copies in 1952 and 13 000 in 1962. These figures put world production of numbers of copies at approximately

Table 27.2 US book title production, 1915–1945

Year	Total titles*	Percent change from previous year	New books	New editions	Percent of new editions
1915	8,202	−20.73	8,349†	1,385†	14.22†
1916	8,504	+ 3.68	9,160†	1,285†	12.30†
1917	8,009	− 5.82	8,849†	1,211†	12.03†
1918	6,861	−14.33	8,085†	1,152†	12.47†
1919	5,741	−16.32	7,625†	969†	11.27†
1920	6,187	+ 7.76	5,101	1,086	17.55
1921	6,446	+ 4.18	5,438	1,008	15.63
1922	6,863	+ 6.46	5,998	865	12.60
1923	7,188	+ 4.73	6,267	921	12.81
1924	7,538	+ 4.86	6,380	1,158	15.36
1925	8,173	+ 8.42	6,680	1,493	18.26
1926	8,359	+ 2.27	6,832	1,527	18.26
1927	8,899	+ 6.46	7,450	1,449	16.28
1928	9,176	+ 3.11	7,614	1,562	17.02
1929	10,187	+11.01	8,342	1,845	18.11
1930	10,027	− 1.57	8,134	1,893	18.87
1931	10,307	+ 2.79	8,506	1,801	17.47
1932	9,035	−12.34	7,556	1,479	16.36
1933	8,092	−10.43	6,813	1,279	15.80
1934	8,198	+ 1.30	6,788	1,410	17.19
1935	8,766	+ 6.92	6,914	1,852	21.12
1936	10,436	+19.05	8,584	1,852	17.74
1937	10,912	+ 4.56	9,273	1,639	15.02
1938	11,067	+ 1.42	9,464	1,603	14.48
1939	10,640	− 3.85	9,015	1,625	15.27
1940	11,328	+ 6.46	9,515	1,813	16.00
1941	11,112	− 1.90	9,337	1,775	15.97
1942	9,525	−14.28	7,786	1,739	18.25
1943	8,325	−12.59	6,794	1,561	18.75
1944	6,970	−16.27	5,807	1,163	16.68
1945	6,548	− 6.05	5,386	1,162	17.74

Source: *Publishers' Weekly*
* Overall total: 267 621 titles (annual average: 8 633), books only
† "New books" and "New editions" figures include pamphlets for the years 1915–1919 inclusive; for that reason, they do not add up to the corresponding figures in the "Total titles" column, which includes books only. The "Percent of new editions" figures for the years 1915–1919 inclusive are for books and pamphlets combined.

Table 27.3 Total number of UK titles published

Year	New books	Total, including new editions
1900	5670	7149
1905	6817	8252
1910	8468	10804
1915	8499	10665
1918	6750	7716 (lowest war year total)
1920	8738	11004
1925	9977	13202
1930	11856	15494
1935	11410	16678
1937	11327	17137 (pre-war record year)
1940	7523	11053
1943	5504	6705 (lowest war year total)
1945	5826	6747
1950	11638	17072
1955	14192	19962
1960	17794	23783
1965	21045	26358
1970	23512	33489
1975	27247	35608
1976	26207	34434
1977	27684	36322
1978	29530	38766
1979	32854	41940
1980	37382	48158
1981	33651	43083

Source: J. Whitaker & Sons Ltd. (Figures pre-1928 taken from *The English Catalogue of Books*, thereafter from *The Bookseller*.)

2500 million and 4500 million for 1952 and 1962 respectively.

The above figures can also be considered in terms of *per caput* production. Table 27.6 compares figures from Barker (1956) and Escarpit (1966) for the years 1952 and 1962, ranked in descending order for 1962. The main European nations, including the USSR and various socialist countries, all lie in the intermediate range (between Portugal and Italy). The smaller, highly developed and educated European countries clearly have a very high demand for published material, as does the Israeli reading population. The United States's position is in marked contrast to the apparent size of its publishing industry (Table 27.5), which, while one of the largest in the world at this time, had one of the smallest *per caput* productions of titles (but see Escarpit 1966, p. 67, who notes that magazine production in the USA was ten times greater than that of Europe at this time).

Table 27.7 categorizes countries as high, medium, or small translators in terms of the number of translated editions produced in relation to national production. These indicate that the high figures in Table 27.6 for the smaller European countries reflect a public demand that cannot be satisfied solely by national production, and that the reading practices of individuals are quite highly employed. The case of Israel perhaps represents a variant on this, the population having such a heterogenous background, both geographically and intellectually. The UK, by contrast, appears insular in the extreme in the material its population reads, though national production is high, and thus, presumably, caters for the gamut of UK demand.

Table 27.8 lists works classed under Class 8 (Literature) of the Dewey decimal classification. These figures are subject to a number of vagaries (and vagueries!). In the same sense that it is difficult to define a 'book' (above), it is by no means always clear where the dividing lines between literature, travel books, history, philosophy, etc., are to be drawn. Again, Western countries generally produce a higher percentage of literature (around 30 per cent of total titles) than do other countries. Production in most socialist countries of eastern Europe is around or below 20 per cent (although the absolute number for the USSR is the highest for any country). Literary production in economically developing countries is at an even lower level. Escarpit (1966, p. 76) notes with respect to these figures:

In the last resort, the popularity of literary books in the various countries... depends, on the one hand, on a given country's political and social institutions and its inhabitants' educational level and free time and, on the other, on its demographic situation—i.e., the existence of a population capable of providing simultaneously an adequate number of writers and an adequate number of readers.

One main point can be be drawn from this brief survey. The data reflect a marked increase in the production of written symbols, and thus index an increased penetration of them into modern life as compared with what occurred during previous periods in history. Pre-literate indications of the mobility of human 'ideas' can be inferred from the geographical distribution of exotic materials at Upper Palaeolithic sites (for example, marine shells at Mezin (Ukraine) (Gladkih *et al.* 1984), Pa Sangir (Iran), (Smith 1986), and Perigord (France) (Mellars 1973), and widely distributed 'art' objects, such as Venus figures (Gamble 1980, 1982). The distances involved appear to be around a maximum of 1000 kilometres; in addition, the 'informational' content of such objects is of a diVerent order to that encapsulated in written material. Written works are clearly more widely distributed, and transmit more explicit information, acting to amplify the number of 'points of view' that any individual may (vicariously) have available to him or her.

Claims as to the 'revolutionary effects' of print media for 'human consciousness' cannot be tested with any rigour on the basis of the available statistics. Printing, from an evolutionary perspective, along with the increased dissemination of its products, represents an increase in the temporal and spatial 'binding' of symbolic media. It more efficiently makes available a heterogeneity of ideas and interpretations of experience, while simultaneously providing for a standardization of forms of language and permitted interpretation. In this sense, an evolutionary account of printing as one form of literacy is possible; but in the 'progressive' sense in which the invention has often been invoked in such contexts, evolutionary accounts are, as Barton and Hamilton have been at pains to point out for literacy in general, 'difficult to substantiate'.

Table 27.4 Subject areas of titles published

These tables, reprinted by permission of J. Whitaker & Sons, Ltd, show the books recorded in *The Bookseller*

Classification	1937 (Pre-war record year) Total	Reprints and new editions.	Trans.	Edns. deluxe	Classification	1981 Total	Reprints and new editions.	Trans.	Ltd editions
Aeronautics	50	9	—	—	Aeronautics	237	35	—	—
Annuals and Serials	123	101	—	—	Agriculture & Forestry	451	79	4	1
Anthropology	46	5	2	—	Architecture	347	69	9	1
Archaeology	60	6	3	—	Art	1383	238	92	7
Art and Architecture	230	46	6	9	Astronomy	120	35	1	—
Astronomy	43	3	1	—	Bibliography and Library Economy	788	138	2	—
Banking and Finance	42	13	—	—	Biography	1243	302	48	4
Bibliography	98	44	3	3	Chemistry and Physics	682	115	19	—
Biography	789	188	43	4	Children's Books	2934	496	97	1
Botany and Agriculture	178	39	1	1	Commerce	1213	312	4	1
Calendars	67	31	1	—	Customs, Costume, Folklore	158	37	6	—
Chemistry and Physics	133	27	4	—	Domestic Science	695	181	13	2
Children's Books	1597	552	6	1	Education	1040	194	6	—
Classics and Translations	87	57	40	3	Engineering	1488	239	29	1
Dictionaries	62	15	—	—	Entertainment	630	117	12	—
Directories	166	92	—	—	Fiction	4747	1837	118	5
Domestic Economy	79	12	—	—	General	557	96	5	—
Educational	1337	223	20	—	Geography and Archaeology	476	102	9	—
Engineering	155	53	3	—	Geology and Meteorology	340	41	5	—
Essays	462	96	16	1	History	1432	347	50	2
Facetiae	61	9	—	—	Humour	171	24	1	—
Fiction	5097	2944	81	6	Industry	492	96	1	—
Geology	56	10	—	—	Language	657	136	10	—
History	458	87	22	1	Law and Public Administration	1399	304	9	—
Illustrated Gift Books	234	31	2	21	Literature	1151	190	54	5
Law and Parliamentary	248	97	2	—	Mathematics	726	138	8	—
Maps and Atlases	30	4	1	—	Medical Science	2838	497	27	—
Mathematics	38	4	1	—	Military Science	113	28	1	—
Medical and Surgical	543	155	9	1	Music	365	97	12	1
Music	83	13	3	—	Natural Sciences	1234	190	14	1
Natural History	186	34	1	1	Occultism	251	59	19	—
Nautical	99	18	—	1	Philosophy	431	111	49	—
Naval and Military	62	6	—	—	Photography	237	27	—	1

Table 27.4 (*Contd.*)

Classification	1937 (Pre-war record year) Total	Reprints and new editions.	Trans.	Edns. deluxe	Classification	1981 Total	Reprints and new editions.	Trans.	Ltd editions
Occultism	58	13	2	—	Plays	256	102	51	2
Oriental	169	25	11	2	Poetry	620	70	61	38
Philately	13	6	—	—	Political Science and Economy	3764	868	79	—
Philosophy and Science	164	36	18	—	Psychology	725	121	14	—
Poetry and Drama	569	211	20	9	Religion and Theology	1363	274	138	3
Politics	633	66	23	2	School Textbooks	1991	261	16	—
Psychology	59	1	—	—	Science, General	55	12	1	—
Religion	927	135	60	3	Sociology	1031	149	19	—
Sociology	264	26	2	—	Sports and Outdoor Games	511	87	15	—
Sports and Pastimes	260	34	5	2	Stockbreeding	264	69	2	—
Technical Handbooks	322	80	2	—	Trade	536	141	3	1
Topography	139	18	—	—	Travel and Guidebooks	677	279	8	1
Trade	81	22	—	—	Wireless and Television	264	47	3	—
Travel and Adventure	411	107	19	—					
Veterinary and Stock keeping	38	11	1	—					
Wireless	19	5	—	—					
Totals	17137	5810	434	17	Totals	43083	9387	1144	78

Table 27.5 World production by titles from 1952 to 1962

Country	1881	1952	1962
Afghanistan		—	60(63)
Albania		98	571
Argentina		4,257	3,323
Australia		627	1,793
Austria		3,179	3,557
Belgium		4,610	3,465
Brazil		3,208	3,911(61)
Bulgaria		2,031	3,716(61) or 3,767(63)
Burma		82(51)	330(60)
Cambodia		392(53)	159
Cameroons		—	18
Canada		684	3,600
Ceylon		268	1,969
Chile		—	1,040
China (Taiwan)		427	2,625
China (Mainland)		2,507	26,414(58)
Costa Rica		—	164
Cuba		615(53)	736
Czechoslovakia		5,837	8,703
Denmark		2,186	4,157
Dominican Republic		115(49)	71(63)
El Salvador		—	139
Ethiopia		—	178(61)

Table 27.5 (*Contd.*)

Country	1881	1952	1962
Finland		1,748	2,646
France*	12,261	11,954	13,282
Germany (Federal Republic)	15,191	13,913	21,481
Germany (Eastern)		4,310(53)	6,540
Ghana		—	269(63)
Greece		1,016	1,277
Guatemala		70(53)	500
Guinea		—	4(63)
Honduras		70(53)	189
Hungary		3,195	5,256
Iceland		420	665(59)
India		18,252	11,086
Indonesia		778	869(61)
Iran		391(54)	569(61)
Iraq		248(53)	143(59)
Ireland		149	217
Israel		822(50)	2,532(61)
Italy		8,949	7,401(61)
Japan		17,306	22,010
Jordan		—	162(63)
Kenya		—	98
Korea (Republic of)		1,393	3,720
Kuwait		—	161(63)
Lebanon		396(50)	402
Liberia		—	4(60)
Libya		—	5(60)
Luxembourg		420	134(61)
Malaysia		—	338
Mexico		—	3,760
Monaco		104	38
Morocco		100	161(60)
Netherlands		6,728	9,674
New Zealand		327	1,212
Nicaragua		122(47)	—
Nigeria		—	262(63)
Norway		2,704	3,119
Pakistan		—	1,787
Panama		22	—
Peru		702	791
Philippines		195(53)	595(63)
Poland		6,632	7,162
Portugal		4,153	4,461
Rhodesia		—	369
Romania		5,381(53)	7,359
Rwanda		—	23
Saudi Arabia		—	321
Senegal		—	67
Sierra Leone		—	48
Singapore		—	237
South Africa		834	1,289(63)
Spain		3,445	9,556
Sudan		—	83(63)

Table 27.5 (*Contd.*)

Country	1881	1952	1962
Sweden		3,286	5,472
Switzerland		3,245	5,633
Thailand		3,953	1,397(61)
Tunisia		56(53)	—
Turkey		2,447	4,842
Uganda		—	46(63)
Union of Soviet Socialist Republics		43,135	79,140
United Arab Republic		654(53)	3,294
United Kingdom of Great Britain	5,404	18,741	25,079
United States	2,991	11,840	21,904
Uruguay		—	217
Venezuela		—	338(61) or 743(63)
Viet-Nam (Republic of)		936	1,515
Yugoslavia		5,184	5,637
Zanzibar		—	75

* The figures for 1952 and 1962 include the total production of books (locally produced works, translations, publications in foreign languages) represented by the 'duty copies' deposited and listed in the *Bibliographie de la France* and the *Annuaire Statistique de la France*. (Source: Unesco.)

Table 27.6 Production (by titles) per million of population

Country	1952	1962
Israel	750(50)	1,150
Switzerland	645	995
Denmark	504	893
Norway	812	857
Netherlands	673	820
Sweden	469	724
Czechoslovakia	455	628
Finland	427	587
Portugal	461	500
Austria	558	499
Hungary	341(50)	495
United Kingdom	375	469
Romania	158	394
Federal Republic of Germany	290	392
Belgium	512	376
Spain	119	310
Yugoslavia	305	299
France	242	270
Poland	265	237
Japan	199	231
Union of Soviet Socialist Republics*	188	195
Canada	47	193
Turkey	111	166
Italy	206	162

Social and cognitive factors in the historical elaboration of writing 841

Table 27.6 (*Contd.*)

Country	1952	1962
Argentina	237	155
United States	74	117
Mexico	114	101
China (Mainland)	5(approx.)	38(58)
India	47	25

(Sources: R. E. Barker, *Books for all*. United Nations Statistical Yearbook. Unesco statistics.)
* Commercially distributed books only.

Table 27.7 The proportion of translations: percentage of translations in relation to national production (1960)

Large translators		Average translators		Small translators	
Country	%	Country	%	Country	%
Israel	34.0	Italy	19.7	Romania	9.4
Albania	25.5	Denmark	18.9	Brazil	9.4
Finland	24.8	Iceland	18.9	Mexico	8.8
Belgium	23.9	Sweden	18.5	United States	8.6
Norway	23.3	Iran	17.9	Hungary	7.6
Spain	23.3	Czechoslovakia	17.0	China (Taiwan)	6.3
		Bulgaria	16.3	Ceylon	6.2
		Netherlands	16.1	India	5.8
		Yugoslavia	15.8	Viet-Nam (Republic of)	5.0
		Korea (Republic of)	14.4	Pakistan	4.8
		Switzerland	13.8	Indonesia	4.5
		UAR	13.3	Japan	4.1
		Greece	12.2	USSR†	4.0
		Portugal	12.0	Austria	3.0
		Turkey	11.6	Chile	2.7
		France	11.2	South Africa	2.0
		Poland	11.0	Canada	1.9
		Germany*	10.9	United Kingdom	1.7
		Burma	10.6		
		Argentina	10.4		

* Federal Republic of Germany and Eastern Germany.
† Not including works translated from Russian into another Soviet language.
Source: Escarpit (1966).

Table 27.8 World production (by titles) of works in Class 8 (Literature): evolution from 1952 to 1962

	1952		1962	
	Production	Percentage of total	Production	Percentage of total
Afghanistan	—	—	0	0
Albania	—	—	137	24

Table 27.8 (*Contd.*)

	1952		1962	
	Production	Percentage of total	Production	Percentage of total
Argentina	3,258	76	1,891	57
Australia	159	25	211	12
Austria	733	23	741	21
Belgium	1,126	24	1,294	37
Brazil	870	27	716	18
Bulgaria	324	16	608(61) or 790(63)	16(61) or 21(63)
Burma	16	20	32(60)	10
Cambodia	97(53)	25	21	13
Cameroons	—	—	3	17
Canada	200	29	654	18
Ceylon	38	14	522	28
Chile	—	—	203	20
China (Taiwan)	—	—	1,438	55
China (Mainland)	511	20	2,851	20
Costa Rica	—	—	16	10
Cuba	70(53)	11	156	21
Czechoslovakia	1,014	17	1,617	19
Denmark	588	27	767	18
Dominican Republic	38(49)	33	17(63)	24
El Salvador	—	—	20	14
Ethiopia	—	—	2	1
Finland	501	29	841	32
France	4,063	36	4,440	33
Germany (Federal Republic)	3,535	25	4,957	23
Germany (Eastern)	899(53)	21	1,737	25
Ghana	—	—	1	0
Greece	468	31	443	35
Guatemala	25(53)	36	37	7
Guinea	—	—	0	0
Honduras	4(53)	6	7	4
Hungary	415(53)	14	1,031	20
Iceland	133	32	194(59)	29
India	2,467	14	3,534	32
Indonesia	100	13	97(61)	11
Iran	202(54)	52	211(61)	37
Iraq	23(53)	9	30(59)	21
Ireland	49	33	49	23
Israel	287(50)	35	751(61)	30
Italy	2,979	33	2,574(61)	35
Japan	5,650	33	5,063	23
Jordan	—	—	10(63)	6
Kenya	—	—	4(63)	4
Korea (Republic of)	537	39	540	15
Kuwait	—	—	6(63)	4
Lebanon	95(50)	24	158	39

Table 27.8 (Contd.)

	1952 Production	1952 Percentage of total	1962 Production	1962 Percentage of total
Liberia	—	—	2(60)	50
Libya	—	—	2(60)	40
Luxembourg	34	8	13(61)	10
Malaysia	—	—	9	3
Mexico	—	—	690	18
Monaco	72	70	20	53
Morocco	18	18	13(60)	8
Netherlands	1,557	23	2,721	28
New Zealand	29	9	71	6
Nicaragua	35(47)	29	—	—
Nigeria	—	—	12(63)	5
Norway	752	28	898	29
Pakistan	—	—	483(63)	21
Panama	2	9	—	—
Peru	54(50)	7	57	7
Philippines	43(53)	22	103(63)	17
Poland	1,280(55)	18	1,332	19
Portugal	314	8	913	20
Romania	561(53)	10	988	13
Rwanda	—	—	0	0
Saudi Arabia	—	—	55	17
Senegal	—	—	0	0
Sierra Leone	—	—	2	4
Singapore	11(55)	23	43	18
South Africa	209	25	340(63)	26
Spain	1,547	45	3,738	39
Sudan	—	—	10(63)	12
Sweden	1,179	36	1,772	32
Switzerland	753	23	1,263	22
Thailand	571	14	405(61)	29
Tunisia	9(53)	16	—	—
Turkey	409	17	805	17
Uganda	—	—	2(63)	3
Union of Soviet Socialist Republics	5,858(54)	12	8,083	10
United Arab Republic	122(53)	19	465	14
United Kingdom	6,533	35	8,077	32
United States	4,423	37	7,259	33
Uruguay	16(55)	25	68(61) or 50(63)	26(61) or 36(63)
Venezuela	106(55)	20	51(61) or 107(63)	15(61) or 14(63)
Viet-Nam (Republic of)	213	23	171	11
Yugoslavia	1,209	23	1,659	29
Zanzibar	—	—	0	0

(Sources: *United Nations Statistical Yearbook*; R. E. Barker, *Books for all*, Unesco.)

Table 27.9 Epigenesis of writing, cognitive technology, and socio-cultural evolution: a largely Western perspective

The entries in the following table have been selected from a number of sources (see below). They allow a sight of the interplay between social complexity, technology, and symbol systems; the extent to which the pace of change in human technology has accelerated, especially over the last two centuries; and the geographical localization of the bulk of that change over the last thousand years in Occidental cultures.

Sources

Berry, W. T. and Poole, H. E. (1966). *Annals of printing: a chronological encyclopaedia from the earliest times to 1950*. Blandford, London.
Carter, T. F. (1925). *The invention of printing in China and its spread westward*. Columbia University Press, New York.
Dawson, L. H. (ed.) (1935). *The march of man*. Encyclopaedia Britannica, London.
Grun, B. (1982). *The timetables of history: a horizontal linkage of people and events* (based on W. Stein's *Kulturjahrplan*). Simon and Schuster, New York.
McEvedy, C. and Jones, R. (1978). *Atlas of world population history*. Penguin, London.
Westwood, J. (1987). *The atlas of mysterious places*. Weidenfeld and Nicolson, London.
Paxton, J. and Fairfield, S. (1980). *Calendar of creative man*. Macmillan, London.

	Writing and literature	Notations and great symbols	Technology and science	Social complexity
	Circulation figures of British daily newspapers: *Express* 3.4 million, *Mail* 2.4 million, *Mirror* 5 million, *Telegraph* 1.3 million, *Times* 250 000 (1965). *Silent spring* by Rachel Carson Published 1962. 1768 USA newspapers publish 59 million copies daily (1954)	'Information superhighway'	Apollo 11 places humans on moon. DNA synthesized, USA. First human orbit of planet, USSR. First weather satellite, USA. First artificial satellite, USSR. DNA decoded, UK. Hydrogen bomb. Electric power station for commercial use, USA	World human population over 5 billion (1987). World oceanic trade over 1 billion metric tons per annum (1983). Transatlantic cable telephone inaugurated. 71 cities with population over 1 million as against 16 in 1914.
1950	Xerography invented by Carlson.		Colour TV service, USA. Transistor invented, USA. Atomic bomb, Los Alamos. Magnetic recording tape invented. First electronic computer, USA. Fermi splits atom.	END OF SECOND WORLD WAR Creation of United Nations SECOND WORLD WAR Scheduled flight service between Europe and North America.
	Biro invents ball-point pen.		Whittle: first jet engine.	BBC inaugurates first television service, England
			Fleming discovers penicillin. First talking film. Goddard: first liquid-fuel rocket. Baird transmits first TV picture.	

Social and cognitive factors in the historical elaboration of writing 845

Table 27.9 (*Contd.*)

	Writing and literature	Notations and great symbols	Technology and science	Social complexity
1920	Sanger: first book on birth control.	$E = mc^2$	Einstein: General theory of relativity. Bakelite, first plastic, manufactured. Einstein: Special theory of relativity. First powered flight, Wright brothers. Fessenden transmits voice by radio. Planck formulates quantum theory.	First public broadcasting station in England. BBC founded. FIRST WORLD WAR
AD 1900				
			First magnetic sound recording.	Trans-Siberian railway begun Canadian Pacific railroad completed. First run of Orient Express.
			Aspirin invented.	Australian frozen meat on sale in UK.
			Lumière: motion-picture camera	Zulus defeat British forces. Frozen meat shipped from Argentina to Europe
			Beginnings of wireless telegraphy—Marconi. Daimler and Benz produce first motor car. Hertz identifies radio waves Kodak 'box-camera' Goodwin invents celluloid film Swan produces synthetic fibre. First skyscraper, Chicago, ten storeys.	
			Edison and Swan produce practical electric lights. Edison invents phonograph. Bell invents telephone. Mendeleyev: periodic table of elements. Mendel's Laws of Heredity.	Canned meat and fruit appear in stores. Electric street lighting, London and New York. World human population $c.$ 1.2 million
	Remington begins to produce typewriters.		Pasteur: germ theory of disease and 'Pasteurization'.	First European settlers arrive in New Zealand.

Table 27.9 (*Contd.*)

Writing and literature	Notations and great symbols	Technology and science	Social complexity
		Darwin: Natural Selection theory of evolution.	
		Goebel: first electric light bulb.	
		Fizeau measures speed of light.	Beginning of Irish Potato Famine and emigration of 1.6 million to USA.
Daily News, first cheap daily newspaper, England.	Poisson publishes his *Recherches sur la probabilité des jugements*.		Slave population in Cuba *c.* 436 000
Wood-pulp paper invented by Keller.	Morse code.	Mayer and Joule found thermodynamics.	Penny post established in England.
	Pitman devises shorthand.		Official birth registration introduced, England.
	Braille invents reading notation for blind.		First passenger railway, England.
	Faraday proposes pictorial representation of magnetism.	Ohm's law.	Working day limited to 12 hrs for juveniles in England.
London *Times* printed by Steam Press.		Faraday: first electric motor.	
		Babbage constructs calculating machine.	First steamship crossing of Atlantic.
		Stephenson invents first steam locomotive.	
		Appert invents food canning.	
		Dalton's atomic theory.	
	Jacquard invents punch card for controlling loom.	Volta produces first battery.	British cotton industry employs 90 000 factory workers and 184 000 handloom weavers.

AD 1800
___

Senefelder invents lithography.	Rosetta Stone discovered.		
	Metric system adopted in France.		
		École Polytechnique in Paris—first technical college.	
		Jussieu proposes modern classification of plants.	First telegraph—Paris to Lille.
	Lavoisier: first table of chemical elements.	First steam-driven cotton mill, Manchester, UK	**FRENCH REVOLUTION.**
		Steam power begins to supersede water power in English manufacturing.	
		First mechanically driven boat.	
		First cast-iron bridge, England.	

Table 27.9 (*Contd.*)

Writing and literature	Notations and great symbols	Technology and science	Social complexity
		BEGINNINGS OF INDUSTRIAL REVOLUTION IN ENGLAND. Arkwright invents spinning mule. Watt perfects steam engine. Cavendish, Rutherford, and Priestley discover composition of air.	
Encyclopaedia Britannica begins as a weekly publication. Voltaire: *Philosophical dictionary*.	Sussmulch initiates statistics.	Watt invents steam condenser.	World human population *c.* 720 million Numbering of houses, in London.
	Pereire invents sign language for deaf mutes.	Celsius proposes centigrade scale of temperature. Hadley invents navigational sextant.	
First marriage announcement in newspaper, in Manchester. Jablonski publishes first short encyclopaedia in Danzig. First copyright act, in England. Jakob Christoph Le Blon invents three-colour printing.		Fahrenheit constructs mercury thermometer.	

AD 1700

	Halley draws first meteorological map.		First directory of addresses published, in Paris. First modern attempts at street lighting, in London.
		Finite velocity of light discovered by Olaus Romer.	
	First minute hands in watches. Newton and Leibniz independently invent calculus.	Newton constructs reflecting telescope. Newton develops theory of gravitation. Robert Boyle defines chemical elements.	First modern census, in Quebec.
James Howell publishes English–French–Italian–Spanish dictionary. First fountain pens produced in Paris. Division of publisher and printer begins in book trade.			First bank note issued, by Swedish Bank.
		Pascal and Fermat formulate theory of probability.	
			World population estimated at 500 million.
		Torricelli invents barometer. Descartes develops analytical geometry.	
Académie Française compiles French dictionary.			

Table 27.9 (*Contd.*)

	Writing and literature	Notations and great symbols	Technology and science	Social complexity
		Oughtred proposes × as symbol for multiplication.		Beginning of public advertising in France.
		Gerard adopts brackets and other symbols in mathematics.		
		Taj Mahal built at Agra.		
		First systematic use of decimal point.		
	John Cowell—first law dictionary.	John Napier invents logarithms.	Galileo investigates gravitation.	
			Invention of telescope.	
	Sevus Calvisius: first history of music.		William Gilbert writes of magnetism and electricity.	
AD 1600				
	First English–Italian dictionary.			
			Galileo's Golden Rule.	First water-closets in Queen's Palace, Richmond, England.
	Dafne by Jacopo Peri: first opera.			
		François Viète proposes using letters for algebraic quantities.		
			William Lee invents first knitting machine.	Forks used at French court for first time.
		Vatican Library in Rome rejuvenated: oldest public library in Europe	Galileo experiments with pendulum.	
			Bernadino Telesio: *De rerum natura* foreshadows empirical method in science	
		Michelangelo begins work on dome of St Peters, Rome.	Michael Servetus discovers pulmonary circulation of blood.	Pope Paul III institutes Inquisition in Rome.
	First complete edition of Aristotle published by Erasmus.		Petrus Apianus: *Cosmographia*, first textbook of theoretical geography.	
			Scipione de Ferro solves cubic equations.	
	First book printed in Arabic type.	Luther's 95 theses, Wittenberg.	Copernicus proposes heliocentric solar system.	
			Da Vinci designs forerunner of water turbine.	Beginning of slave trade to New World.
	Spanish Polyglot Bible (Greek, Latin, Hebrew) published 1517.		First attempts to restrict practice of medicine to qualified doctors.	Vienna–Brussels post service extended to Madrid.
	Swift development of printing: since 1445, 1000+ printing offices opened producing c.35 000 books with c.10 million copies.		First watch: Peter Henlein of Nuremberg.	Pocket handkerchief comes into use.
			First black-lead pencils used in England.	First regular postal service, Vienna–Brussels.
AD 1500				

Table 27.9 (*Contd.*)

Writing and literature	Notations and great symbols	Technology and science	Social complexity
	The symbols +/− adopted in mathematics.		World human population *c.* 450 million Beginnings of ballet at Italian courts.
	First terrestrial globe constructed in Nuremberg by Martin Behaim. First printed music.	Caxton prints first book in English. First printing press in Paris. 42-line bible printed at Mainz. Metal plates used for printing.	Institution of the Spanish Inquisition 1478.
Gutenberg prints 'Constance Mass Book'.	Counterpoint in music introduced by John Dunstable. Compilation of the *Yong-Le da-dian*, a 22 973-chapter encyclopaedia. Only three copies made.		Bethlehem Hospital, London (Bedlam) becomes institution for insane.

AD 1400

Langland—*Piers Plowman*. Boccaccio—*Decameron*.	Mechanical clock at Strasburg cathedral.	First scientific weather forecasts by William Merlee of Oxford. Invention of sawmill.	Black Death in Europe kills *c.*75 million people.
Dante—*Divina commedia*.		Gunpowder reinvented in Germany.	Philip IV builds indoor tennis court in Paris.
Block printing in Ravenna.		Invention of spectacles. Invention of glass mirror.	Marco Polo in service of Kublai Khan.
Lohengrin—German epic poem. *Tristan and Isolde*—written version. *Parsifal*—German epic poem.			Fourth Lateran council forbids trial by ordeal.

AD 1200

The *Nibelungenlied* written down	Western façade of Chartres Cathedral built. Beginnings of Easter Island monuments, Polynesia. Angkor Wat, Cambodia.	First water-driven mechanical clock constructed in Peking.	Pueblo Bonito, Chaco Canyon, New Mexico.
	Time values given in musical notation. Polyphonic singing replaces Gregorian chant.		

Table 27.9 (*Contd.*)

	Writing and literature	Notations and great symbols	Technology and science	Social complexity
AD 1000	*Beowulf* written in Old English.	'Sight singing' introduced at Pamposa Monastery, Ravenna. Indian mathematician Sridhara recognizes importance of zero.	Chinese perfect gunpowder.	World human population *c.*265 million.
AD 800	Chinese encyclopaedia of 1000 chapters (978–84). Book printing in China.	Development of systematic musical notation. Arabic numerals brought to Europe. Earliest known attempts at polyphonic music.	Astrolabe perfected by Arabs.	Public street lighting in Ummayad Cordoba. Great Zimbabwe, South Central Africa. Calibrated candles used for first time in England to measure time.
AD 600	State-owned paper mills established in Baghdad. Pictorial book printing in Japan. Arabs learn technique of paper-making in Samarkand from Chinese. Arabs discover library at Alexandria holds 300 000 scrolls.	Borobudur, Java. Arabic numerals derived from Indian. Great Mosque, Mecca. Greek becomes official language of Eastern Roman Empire instead of Latin. Dome of the Rock, Jerusalem. First year of Muslim calendar.	Jabir founds chemistry as distinct from alchemy. Mill-wheels driven by water in Europe. 'Greek fire' invented and used in siege of Constantinople. Cotton introduced into Arab countries.	Sugar planted in Egypt. First organized news system introduced by Caliphs. Civil Service entrance exams introduced in China.
AD 400	St Augustine *The City of God*.	Decimal system in India. Church bells introduced in France. St Sophia Basilica, Constantinople. Boethius introduces Greek musical letter notation to West.	First paddle-wheel boats. Founding of Constantinople University. Beginnings of alchemical science.	End of plague which halved population of Europe (542–94). Beginning of chess in India. End of Western Roman Empire. Vandals sack Rome.
	Books begin to replace scrolls.	Basilican Church of St Peter, Rome.	First book on algebra— Diophantus of Alexandria.	Hymn-singing introduced. Gladiatorial combat forbidden. Partition of Roman Empire into East and West.

Table 27.9 (Contd.)

	Writing and literature	Notations and great symbols	Technology and science	Social complexity
				Roman citizenship extended to all freeborn subjects of Empire.
AD 200	First paper made in China	Oldest Mayan monuments. Earliest known Sanskrit inscriptions from India. Pyramid of the Sun, Teotihuacan, Mexico	Ptolemy draws maps of 26 countries.	World human population $c.$170 million.
0	Library of Ptolemy I in Alexandria destroyed in fire.	Adoption of Julian calendar of 365.25 days and leap year. White Horse, Uffington, Oxfordshire		First recorded wrestling match in Japan. Petra, Jordan.
100 BC	Xu Shen produces Chinese dictionary of 10 000 characters.	Rosetta Stone.	Heron founds first College of Technology at Alexandria. Use of gears leads to invention of ox-driven water-wheel for irrigation.	Romans destroy Carthage, killing 450 000 of 500 000 population, selling rest as slaves. Earliest known paved streets appear in Rome.
200 BC		Great Wall of China—215.	Eratosthenes estimates earth's circumference. Introduction of oil lamps to Greece. Archimedes.	Hannibal crossses Alps. First public gladiatorial combat in Rome—264. Full equality between plebeians and patricians in Rome—287.
300 BC	Periclean Age: Anaxagoras, Protagoras, Socrates (Philosophy), Aeschylus, Sophocles and Euripides (Tragedy), Aristophanes, Cratinus (Comedy), Herodotus, Thucydides (History). Beginning of Greek historical writing.	Mime as a dramatic form— Sophron of Syracuse. Beginnings of Nazca line markings, Peru.	Hippocrates, 'Father of Medicine'. Development of technology and agriculture in China.	Carrier pigeons used in Greece. Athenian soldiers and judges receive regular salary. Ball games in Greece. Coinage becomes common as legal tender. World human population $c.$100 million
500 BC				

Table 27.9 (*Contd.*)

	Writing and literature	Notations and great symbols	Technology and science	Social complexity
		Buddha preaches first sermon.	Veins, arteries, and nerves distinguished by Alcmaeon of Croton.	
			Modern calendar devised in Babylon.	First Persian coin with picture of ruler.
		Building of theatre at Delphi.	Thales of Miletus. First occidental prediction of solar eclipse and oldest occidental mathematical theorem.	Theodorus of Samos invents lock and key. Cyrus —Persian empire founded 559.
	Aesop's fables.		Sundial (gnomon) in Greece and China.	Solon's laws in Athens.
	Introduction of papyrus to Greece.	Pythagoras introduces octave to music.	First circumnavigation of Africa.	The prophet Zoroaster, Persia
600 BC				
				Nebuchadnezzar Babylon (605–562)
				Egyptians begin canal between Nile and Red Sea.
	Indian Vedas (other than Rig Veda) completed: Upanishad tradition.		Water-clocks in Assyria.	'Graffiti' by Greek soldiers in Nubia suggest good basic education.
	King Assurbanipal's library contains 22 000 clay tablets on history, science, etc.			First written Athenian laws—by Draco. First mention of 'alpine sports'. Sennacherib, King of Nineveh.
700 BC				
	Collection of 'Sayings of Solomon'.		First iron utensils.	Late period Pharaohs in Egypt (712–525BC).
		Earliest recorded (cuneiform) music—Sumeria.	Etruscans use hand-cranks.	First recorded Olympic games (776 BC).
	First Chinese poems.			Battlefield surgery reported by Homer.
800 BC				
	Iliad and *Odyssey*—Homer.	Beginning of verified Chinese historical chronology (841 BC).		Carthage founded as trading centre with Tyre. Cumaeon Sibyl and Delphic oracle
900 BC				
		New calendar in Assyria, verified at solar eclipse, 15 June 763 BC.	Water-filled cubes for measuring time, weight, and length, Chaldea.	Rise of Assyrian Empire (c.900–612 BC)
	Beginning of Hebrew literature—'Song of Songs'. Hebrew alphabet developed.			
		Indian lunar year has 360 days, adjusted at random to coincide with solar year.	Subterranean water-supply tunnels at Jerusalem.	
	Chinese script fully developed.			World human population c.50 million
1000 BC				

Table 27.9 (*Contd.*)

Writing and literature	Notations and great symbols	Technology and science	Social complexity
	Mathematical permutations and 'magic square' known in Chinese mathematics.	Kikleuli of Mitanni writes the first treatise on horse breeding and training.	Prohibition decreed in China.
	Possible earliest use of labyrinth maze symbol, Sardinia	Height of Sun in relation to incline of polar axis measured in China.	Labour strike in Thebes, Egypt.
Gilgamesh Epic is recorded. Library at Hittite capital has tablets in eight languages. Primitive Greek syllabary at Cnossus, Pylos, Mycenae. 'Hymns of the Rig Veda in India.	Properties of Pythagorean triangle theory also known in China. Moses receives the Ten Commandments on Mt. Sinai. Ikhnaten (Amenhotep IV) of Egypt sets up short-lived monotheistic religion: wife, Nefertiti; successor, Tutankhamen.	Beginning of Iron Age in Syria and Palestine. Intricate clock, measuring flow of water, found in tomb of Amenhotep III. Obelisks in Egypt serve as sundials: e.g. 'Cleopatra's Needle', London. Silk fabrics in China.	Regulations concerning the sale of beer in Egypt. Extensive export and import trade in Egypt. Abolition of monarchy in Athens: Medon becomes first archon. Phoenicians become predominant trading power in Mediterranean area: they import tin from mines in England.

1500 BC

	Egyptians use knotted rope triangle with 'Pythagorean' numbers to construct right angles.	Mercury used in Egypt.	18th Dynasty brings Egypt to height of its pre-Hellenistic power and achievements. New Kingdom of Egypt (1550–712 BC).
First Hittite cuneiform inscriptions. Beginning of Semitic alphabet. The *Story of Simuke*, the oldest form of novel, written in Egypt.	Beginnings of Stonehenge, England. Geometry used as a basis for astronomic measurements.	Four basic elements known in India: earth, air, fire, and water. Palace at Cnossus, Crete, has light and air shafts, bathrooms with water-supply.	Decline of Babylonian Empire under Hammurabi's son. First Babylonian Empire (1850 BC–1700 BC).
Egyptians use an alphabet of 24 signs.	Code of Hammurabi, Babylon.	Horses used to draw vehicles.	Shang dynasty in China (1760 BC–1122 BC)

2000 BC

In Egyptian literature, lamentations and scepticism about meaning of life. Epic poetry in Babylonia celebrates re-creation of the world.	Ziggurat of Ur. Map of Babylonia.	Oldest pictorial representation of skiing: carving on a rock, South Norway.	Middle Kingdom of Egypt (2100–1700 BC). Indus cities of Harappa and Mohenjo-Daro. African Pygmies appear at Egyptian court.
First libraries in Egypt.	The Great Sphinx of Gizeh. Avebury Circle, England.	Egyptians use papyrus.	Height of Sumerian – Akkadian civilization
Sumerian cuneiform writing reduces pictographs, still in use to c.550. Sumerian poetry, lamenting death of Dumuzi the shepherd god (Tammuz).	The Great Pyramid, Gizeh. Construction of Silbury Hill Sumerian numerical system based on multiples of 6 and 12.	Metal mirrors used in Egypt. Oil-burning lamps, Sumeria. Beginning of systematic astronomical observations in Egypt, Babylonia, India, and China.	Old Kingdom of Egypt (2800–2300 BC).

3000 BC

Table 27.9 (*Contd.*)

	Writing and literature	Notations and great symbols	Technology and science	Social complexity
		Earliest known numerals in Egypt. Newgrange tomb, Ireland	Wheeled vehicles in use in Sumeria.	Continuation of Neolithic in W. Europe (to 1700 BC).
			Copper alloys used by Sumerians and Egyptians.	Mesopotamian influence predominant in Mediterranean regions of Asia (to 1600 BC).
	Sumerian writing, done on clay tablets, shows *c*.2000 pictographic signs.	Carnac megalith, France. Egyptian calendar, regulated by sun and moon: 360 days, 12 months of 30 days each.		Earliest cities in Mesopotamia.
5000 BC				
		5600–7500 BC complex religious shrines in the neolithic village of Çatal Hüyük, Anatolia		
8000 BC				
			Joman Pottery, Japan.	
12000 BC		Altamira cave paintings.		

Notes

1. In this context, Ellis (1985, 130) notes with respect to written language:

Since the first known use of written language seems to have surrounded property transactions... it might be hypothesized that a major reproductive advantage underlying the evolution of human writing is that it has facilitated ownership.

For further details of early token systems and their relation to the possible origins of writing systems, see Schmandt-Besserat (1980, 1981, 1982, 1986). For a critical perspective on this approach, see Gelb (1980) [eds].

2. There are at present a number of investigations of numerical skills in animals in progress. As with comparative language programmes, evidence is emerging that higher primates can demonstrate only limited counting and arithmetical skills compared with humans, and those only with respect to small (i.e., less than 10) numbers. See Davis and Perusse (1988), and the ensuing peer commentary, and Boysen and Berntson (1989) for representative studies and discussions.

3. See Allen (1962)—for example: 'The doctrine of development found favor with scholars who believed that nineteenth-century European culture represented the highest stage of development in time and space' (p. 230)—and Nettl (1983) (especially (Chapters 11–17) for fuller accounts of the difficulties inherent in evolutionary approaches to music; for an introduction to the cross-cultural variety of music, see May (1980) and Dowling and Harwood (1986).

4. Forsyth (1986) offers the suggestion that the harmony characteristic of formal Western music may have originated unintentionally as a result of the unplanned overlap of resonant echoes within such large stone edifices.

References

Akinnaso, F. N. (1982). The literate writes and the non-literate chants: written language and ritual communication in sociolinguistic perspective. In *Linguistics and literacy* (ed. W. Frawley, pp. 7–36. Plenum, New York.

Alberti, L. (1974). *Music through the ages.* Cassell, London.

Allen, W. D. (1962). *Philosophies of music history: a study of general histories of music, 1600–1960.* Dover Books, New York.

Anderson, C. A. (1965). Literacy and schooling on the development threshold: some historical cases. In *Education and economic development* (ed. C. A. Anderson and M. J. Bowman), pp. 347–62. Aldine, Chicago.

Antell, S. and Keating, D. (1983). Perception of numerical invariance in neonates. *Child Development* 54, 695–701.

Baines, J. (1983). Literacy and Ancient Egyptian society. *Man*, **18**, 572–99.

Barker, R. E. (1956). *Books for all*. UNESCO, Paris.

Barton, D. (1985). Awareness of language units in adults and children. In *The psychology of language*, Vol. 1, (ed. A. Ellis). Erlbaum, London.

Becker, J. D. (1984). Multilingual word processing. *Scientific American*, **251**, 82–93.

Bell, E. T. (1937). *Men of mathematics*. Simon and Schuster, New York.

Bent, I. D., Hiley, D., Bent, M., and Chew, G. (1980). Notation. In *The New Grove Dictionary of music and musicians* (ed. S. Sadie), pp. 333–420. Vol. 13, Macmillan, London.

Berry, J. (1958). The making of alphabets. In *Proceedings of the VIII International Congress of Linguistics* (ed. E. Silversten), pp. 752–64. Oslo University Press.

Berry, J. (1977). The making of alphabets revisited. In *Advances in the creation and revision of writing systems* (ed. J. A. Fishman), pp. 3–16. Mouton, The Hague.

Bloch, M. (1975). *Political language and oratory in traditional society*. Academic Press, New York.

Bowman, A. K. and Thomas, J. D. (1974). *The Vindolanda writing tablets*. Frank Graham, Newcastle upon Tyne.

Boysen, S. T. and Berntson, G. G. (1989). Numerical competence in a chimpanzee (*Pan troglodytes*). *Journal of Comparative Psychology*, **103**, 23–31.

Brand, S., Kelly, K., and Kinney, J. (1985). Digital retouching; the end of photography as evidence of anything. *Whole Earth Review*, **47**, 42–9.

Bryant, P. E. (1992) Arithmetic in the cradle. *Nature* **358**, 712–13.

Clammer, J. P. (1976). *Literacy and social change*. Brill, Leiden.

Clanchy, M. T. (1979). *From memory to written record, England 1066–1307*. Arnold, London.

Cole, M. and Griffin, P. (1980). Cultural amplifiers reconsidered. In *The social foundations of language and thought* (ed. D. R. Olson), pp. 343–64. Norton, New York.

Cole, M. and Griffin, P. (1983). A socio-cultural approach to remediation. In *Quarterly Newsletter of the Laboratory for Human Comparative Cognition*, **5**, 69–71.

D'Amato, M. (1988). A search for tonal pattern perception in *Cebus* monkeys: why monkeys can't hum a tune. *Music Perception*, **5**, 453–80.

Damerow, P. (1988). Individual thinking and cultural evolution of arithmetical thinking. In *Ontogeny, phylogeny, and historical development* (ed. S. Strauss), pp. 125–52. Ablex, New York.

Davis, H. and Perusse, R. (1988). Numerical competence in animals: definitional issues, current evidence, and a new research agenda. *Behavioral and Brain Sciences*, **11**, 561–615.

Diringer, D. (1968). *The alphabet: a key to the history of mankind*, 2 vols, (3rd edn.) (orig. 1948). Hutchinson, London.

Dowling, W. J. and Harwood, D. L. (1986). *Music cognition*. Academic Press, New York.

Downs, R. M. (1985). The representation of space: its development in children and cartography. In *The development of spatial cognition* (ed. R. Cohen), pp. 323–45. Erlbaum, Hillsdale, NJ.

Ehlich, K. (1983). Development of writing as social problem solving. In *Writing in focus* (ed. F. Coulmas and K. Ehlich), pp. 99–130. Mouton, Berlin.

Eisenstein, E. (1979). *The printing press as an agent of change*, 2 vols. Cambridge University Press.

Eisenstein, E. (1981). Some conjectures about the impact of printing on Western society and thought: a preliminary report. In *Literacy and social development in the West* (ed. H. Graff), pp. 53–68.

Ellis, L. (1985). On the rudiments of possessions and property. *Social Science Information*, **24**, 113–43.

Escarpit, R. (1966). *The book revolution*. Harrap, London and UNESCO, Paris.

Febvre, L. and Martin, H. J. (1984). *The coming of the book: the impact of printing 1450–1800*. (First published 1958.) Verso–New Left Books, London.

Finnegan, R. (1977). *Oral poetry: its nature, significance and social context*. Cambridge University Press.

Fishman, J. A. (ed.) (1977). *Advances in the creation and revision of writing systems*. Mouton, The Hague.

Fishman, J. A., Ferguson, C. A., and Das Gupta, J. (eds) (1968). *Language problems of developing nations*. Wiley, New York.

Flegg, G. (1983). *Numbers, their history and meaning*. André Deutsch, London.

Forsyth, M. (1986). *Buildings for music: the architect, the musician, and the listener from the seventeenth century to the present day*. MIT Press, Cambridge, Mass.

Freire, P. (1972). *The pedagogy of the oppressed*. Penguin, Harmondsworth.

French, M. A. (1976). Observations on the Chinese script and the classifications of the writing system. In *Writing without letters* (ed. W. Haas), pp. 101–29. Manchester University Press.

Gablik, S. (1976). *Progress in art*. Thames and Hudson, London.

Gallistel, C. R. (1990) *The organization of learning*. MIT Press, Cambridge, MA.

Gamble, C. S. (1980). Information exchange in the Palaeolithic. *Nature*, **283**, 522–33.

Gamble, C. S. (1982). Interaction and alliance in palaeolithic society. *Man*, NS, **17**, 92–107.

Gaur, A. (1984). *A history of writing*. The British Library, London.

Gelb, I. J. (1963). *A study of writing* (2nd edn; first published 1952). University of Chicago Press.

Gelb, I. J. (1980). Principles of writing systems within the frame of visual communication. In *Processing of visible language*, 2, (ed. P. Kolers, M. Wrolstad, and H. Bonma), pp. 7–24. Plenum, New York.

Gerschel, L. (1960). Comment compatient les anciens Romains? In *Hommages à Leon Herrmann*. Latomus, Brussels. (as cited by Ifrah 1986).

Gibson, E. J. and Levin, H. (1975). *The psychology of reading*. MIT Press, Cambridge, MA.

Gilmore, M. P. (1952). *The world of humanism 1453–1517*. Harper and Row, New York.

Gladkih, M. I., Kornietz, N. L., and Soffer, O. (1984). Mammoth bone dwellings on the Russian plain. *Scientific American*, **251**, 134–43.

Glyn, M. H. (1934). *Theory of musical evolution*. Dent, London.

Goody, J. (ed.) (1968). *Literacy in traditional societies*. Cambridge University Press.

Goody, J. (1981). Alphabets and writing. In *Contact: human communication and its history* (ed. R. Williams), pp. 105–26. Thames & Hudson, London.

Goody, J. (1983). Literacy and achievement in the ancient world. In *Writing in focus* (ed. F. Coulmas and K. Ehlich), pp. 83–97. Mouton, The Hague.

Goody, J. and Watt, I. P. (1963). The consequences of literacy. *Comparative Studies in History and Society*, **5**, 304–45. (Reprinted in J. Goody (ed.) (1968): *Literacy in traditional societies*, pp. 27–68. Cambridge University Press.)

Gough, K. (1968). Implications of literacy in traditional China and India. In *Literacy in traditional societies* (ed. J. Goody), pp. 69–84. Cambridge University Press.

Graff, H. (1979). *The literacy myth: literacy and social structure in the 19th century city*. Academic Press, New York.

Graff, H. (ed.) (1981). *Literacy and social development in the West*. Cambridge University Press.

Greenfield, P. M. (1972). Oral or written language: the consequences for cognitive development in Africa, the United States and England. *Language and Speech*, **15**, 169–78.

Guest, A. H. (1977). Choreography and dance notation. *Encyclopaedia Britannica*, 15th edn, Macropaedia, Vol. 4. Encyclopaedia Britannica, Chicago.

Haas, W. (ed.) (1976). *Writing without letters*. Manchester University Press.

Haas, W. (1983). Determining the level of a script. In *Writing in focus* (ed. F. Coulmas and K. Ehlich), pp. 15–30. Mouton, Berlin.

Halle, M. (1969). Some thoughts on spelling. In *Psycholinguistics and the teaching of reading* (ed. K. S. Goodman and J. T. Fleming), pp. 17–24. International Reading Association, Newark, NJ.

Hamilton, M. and Barton, D. (1983). Adults' definitions of 'word': the effects of literacy and development. *Journal of Pragmatics*, **7**, 581–94.

Hammond, N. (1986). The emergence of Maya civilization. *Scientific American*, **255**(2), 98–107.

Hanna, J. L. (1978). *The anthropology of dance: a selected bibliography* (Rev. edn). Hanna, College Park, Md.

Hanna, J. L. (1979a). Movements towards understanding humans through the anthropological study of dance. *Current Anthropology*, **20**, 313–39.

Hanna, J.L. (1979b). *To dance is human: a theory of non-verbal communication*. University of Texas Press, Austin.

Hargreaves, D. J. (1986). *The developmental psychology of music*. Cambridge University Press.

Hart, R. and Berzok, M. (1982). Children's strategies for mapping the geographic-scale environment. In *Spatial abilities: development and physiological foundations* (ed. M. Potegal), pp. 147–69. Academic Press, New York.

Havelock, E. (1963). *Preface to Plato*. Harvard University Press, Cambridge, MA.

Havelock, E. (1976). *Origins of Western literacy*. Ontario Institute for Studies in Education, Toronto.

Heath, S. B. (1983). *Ways with words*. Cambridge University Press.

Heeschen, V. (1978). The metalinguistic vocabulary of a speech community in the highlands of Irian Jaya (West New Guinea). In *The child's conception of language* (ed. A Sinclair et al.), pp. 155–87. Springer-Verlag, New York.

Henderson, L. (1982). *Orthography and word recognition in reading*. Academic Press, New York.

Ho, Peng Yoke (1980). Early science and technology in China. In *The encyclopedia of ancient civilizations* (ed. A. Cotterell), pp. 304–10. Windward, London.

Horton, R. and Finnegan, R. (eds) (1973). *Modes of thought: essays on thinking in Western and non-Western societies*. Faber, London.

Hughes, M. (1986). *Children and number: difficulties in learning mathematics*. Blackwell, Oxford.

Ifrah, G. (1981). *Histoire universelle des chiffres*. Editions Seghers, Paris. English translation *From one to zero: a universal history of numbers*: Viking–Penguin, New York, 1985 and Penguin, Harmondsworth, 1987.

Illich, I. (1981). Vernacular values. In I. Illich, *Shadow work*, pp. 27–52. Boyars, London.

Inkeles, A. (1973). The school as a context for modernisation. *International Journal of Comparative Sociology*, **40**, 163–78.

Inkeles, A. and Smith, D. H. (1974). *Becoming modern: individual change in six developing countries*. Heinemann, Portsmouth, NH.

Isaacson, C. D. (1928). *The simple story of music*. Macy-Masius, New York.

Jensen, H. (1970). *Sign, symbol and script*. Allen and Unwin, London. (cited by Ehlich (1983)).

Johansson, E. (1981). The history of literacy in Sweden. In *Literacy and social development in the West* (ed. H. Graff), pp. 151–82. Cambridge University Press.

Klein, A. and Starkey, P. (1987). The origins and development of numerical cognition: a comparative analysis. In *Cognitive processes in mathematics* (ed. J. A. Sloboda and D. Rogers), pp. 1–25. Clarendon Press, Oxford.

Kolers, P., Wrolstad, M., and Bouma, H. (eds) (1980). *Processing of visible language*, **2**, Plenum, New York.

Kozol, J. (1980). *Children of the revolution*. Delta, New York.

Kraus, R. and Chapman, S. A. (1981). *History of the dance in art and education* (2nd edn). Prentice-Hall, Englewood Cliffs, NJ.

Lehmann, W. P. (1975). *Language and linguistics in the People's Republic of China*. University of Texas Press, Austin.

Leith, D. (1983). *A social history of English*. Routledge and Kegan Paul, London.

Lerner, D. (1958). *The passing of traditional society*. Glencoe Free Press, New York.

Lévy-Bruhl, L. (1926). *How natives think*. Knopf, New York.

McLuhan, M. (1962). *The Gutenberg galaxy*. Routledge and Kegan Paul, London.

Mahadevan J. I. (1982). Terminal ideograms in the Indus script. In *Harappan civilization* (ed. G. L. Possehl), pp. 311–17. Aris & Phillips, Warminster.

Makita, K. (1968). The rarity of reading disability in Japanese children. *American Journal of Orthopsychiatry*, **38**, 599–614.

Mandler, G. and Shebo, B. J. (1982). Subitizing: an analysis of its component processes. *Journal of Experimental Psychology: General*, **111**, 1–22.

Marshack, A. (1972). Upper Paleolithic notation and symbol. *Science*, **178**, 817–28.

Martin, S. E. (1972). Non alphabetic writing systems: some observations. In *Language by ear and by eye*. (ed. J. F. Kavanagh and I. G. Mattingly), pp. 81–102. MIT Press, Cambridge, MA.

May, E. (1980). *Music of many cultures: an introduction*. University of California Press, San Francisco.

Mellars, P. A. (1973). The character of the middle–upper Palaeolithic transition in South West France. In *The explanation of Culture Change: models in prehistory* (ed. C. Renfrew), pp. 255–76. Duckworth, London.

Nesher, P. (1988). Precursors of number in children: a linguistic perspective. In *Ontogeny, phylogeny, and historical development* (ed. S. Strauss), pp. 106–24. Ablex, New York.

Nettl, B. (1983). *The study of ethnomusicology: twenty-nine issues and concepts*. University of Illinois Press, Urbana, Ill.

Norrie, I. (1982). *Mumby's publishing and bookselling in the twentieth century* (6th edn). Bell and Hyman, London.

Olson, D. R. (1977). From utterance to text: the bias of language in speech and writing. *Harvard Educational Review*, **47**, 257–81.

Olson, D. R. (1982). What is said and what is meant in speech and writing. *Visible Language*, **XVI**, 151–61.

Olson, D. R. and Nickerson, N. (1978). Language development through the school years: learning to confine interpretation to the information in the text. In *Children's language*, I, (ed. K. E. Nelson), pp. 117–69. Gardner Press, New York.

Olson, D. R., Torrance, N., and Hildyard, A. (eds) (1985). *Literacy, language and learning: the nature and consequences of reading and writing*. Cambridge University Press.

Ong, W. J. (1982). *Orality and literacy: the technologizing of the word*. Methuen, London.

Paradis, M., Hagiwara, H. and Hilderbrandt, N. (1985). *Neurolinguistic aspects of the Japanese writing system*. Academic Press, New York.

Parry, C. H. H. (1893). *The art of music*. Kegan Paul, London.

Paturi, F. R. (1979). *Pre-historic heritage*. Scribner, New York.

Porter, J. (1980). Europe, prehistory. In *The New Grove Dictionary of music and musicians* (ed. S. Sadie), Vol. 6, pp. 312–15. Macmillan, London.

Raisz, E. (1938). *General cartography*. McGraw-Hill, New York.

Rees, R. (1980). Historical links between cartography and art. *The Geographical Review*, **70**, 60–78.

Rowbotham, J. F. (1885–7). *A history of music to the time of the troubadors*, 3 Vols. Trubner, London.

Rozin, P., Poritsky, S., and Sotsky, R. (1971). American children with reading problems can easily learn to read English represented by Chinese characters. *Science*, **171**, 1264–7.

Sachs, C. (1937). *World history of the dance*. Norton, New York.

Sampson, G. R. (1985). *Writing systems*. Blackwell, Oxford.

Saxe, G. B. (1982). Culture and the development of numerical cognition: studies among the Oksapmin of Papua New Guinea. In *Children's logical and mathematical cognition: progress in cognitive developmental research* (ed. C. J. Brainerd), pp. 157–76. Springer-Verlag, New York.

Schmandt-Besserat, D. (1978). The earliest precursor of writing. *Scientific American*, **238**(6), 38–47.

Schmandt-Besserat, D. (1980). The envelopes that bear the first writing. *Technology and Culture*, **21**, 357–85.

Schmandt-Besserat, D. (1981). From tokens to tablets: a re-evaluation of the so-called numerical tablets. *Visible Language*, **15**, 321–44.

Schmandt-Besserat, D. (1982). The emergence of recording. *American Anthropologist*, **84**, 871–8.

Schmandt-Besserat, D. (1986). An ancient token system: the precursor to numerals and writing. *Archaeology*, **39**, 32–9.

Scribner, S. and Cole, M. (1981). *The psychology of literacy*. Harvard University Press, Cambridge, MA.

Slobin, D. I. (1978). Language change in childhood and history. In *Language learning and thought* (ed. J. Macnamara), pp. 185–214. Academic Press, New York.

Sloboda, J. A. (1985). *The musical mind: the cognitive psychology of music*. Clarendon Press, Oxford.

Smith, P. E. L. (1986). *The Palaeolithic archaeology of Iran*. American Institute of Iranian Studies, University of Pennsylvania Museum, Philadelphia.

Stallings, W. (1975). The morphology of Chinese characters: a survey of models and application. *Computers and the Humanities*, **9**, 13–24.

Starkey, P. and Cooper, R. G. (1980). Perception of numbers by human infants. *Science*, **210**, 1033–5.

Starkey, P., Spelke, E.S., and Gelman, R. (1990). Numerical abstraction by human infants. *Cognition* **36**, 97–127.

Strauss, M. S. and Curtis, L.E. (1981). Infant perception of numerosity. *Child Development*, **52**, 1146–42.

Street, B. V. (1984). *Literacy in theory and practice*. Cambridge University Press.

Taylor, I. (1980). The Korean writing system: an alphabet? a syllabary? a logography? In *Processing of visual language*, 2 (ed. P. Kolers, M. Wrolstad, and H. Bouma), pp. 67–82. Plenum, New York.

Taylor, I. (1981). Writing systems and reading. In *Reading research: advances in theory and practice* (ed. T. G.

Walker and G. E. MacKinnon), pp. 1–51. Vol. II, Academic Press, New York.

Tebbel, J. (1975). *A history of book publishing in the United States*, Vol. II: *The expansion of an industry, 1865–1919*. Bowker, New York.

Tebbel, J. (1978). *A history of book publishing in the United States*. Vol. III: *The golden age between two wars, 1920–1940*. Bowker, New York.

Tzeng, O. J. L., Hung, D. L. and Garro, L. (1978). Reading the Chinese characters: an information processing view. *Journal of Chinese Linguistics*, **6**, 287–305.

UNESCO (United Nations Educational, Scientific, and Cultural Organization) (1976). *The experimental world literacy programme: a critical assessment*. The Unesco Press, Paris.

van Loosbroek, E. and Smitsman, A. W. (1990). Visual perception of numerosity in infancy. *Developmental Psychology* **26**, 916–22.

Vansina, J. (1973). *Oral tradition*. Penguin, Harmondsworth.

Venezky, R. L. (1970). Principles for the design of practical writing systems. *Anthropological Linguistics*, **12**, 256–70. (Reprinted in J. A. Fishman (ed.) 1977): *Advances in the creation and revision of writing systems*, Mouton, The Hague.)

Walker, W. (1981). Native American writing systems. In *Language in the U.S.A.* (ed. C. A. Ferguson and S. B. Heath), pp. 145–74. Cambridge University Press.

Wallace, W. (1908). *The threshold of music: an inquiry into the development of the musical sense*. Macmillan, London.

Wellisch, H. H. (1978). *The conversion of scripts: its nature, history and utilization*. Wiley, New York.

Wilder, R. L. (1968). *Evolution of mathematical concepts: an elementary study*. Wiley, London.

Wilder, R. L. (1981). *Mathematics as a cultural system*. Pergamon Press, Oxford

Williams, R. (1981). *Contact: human communication*. Thames and Hudson, London.

Winner, L. (1977). *Autonomous technology*. MIT Press, Cambridge, MA.

Wood, D. (1977). Now and then: comparisons of ordinary Americans' symbol conventions with those of past cartographers. *Prologue*, **8**, 151–61.

Wynn, K. (1992). Addition and subtraction by human infants. *Nature*, **358**, 749–50.

Part V
Epilogue

28

Tempo and mode of change in the evolution of symbolism (a partial overview)

Charles R. Peters

28.1 Introduction

The original purpose of this concluding chapter was to present key points from the preceding sections about human symbolic evolution in a readily perused chart form, and to provide some introductory notes on the issues and possible theories that might help account for the major transformations on evolutionary to protohistorical time-scales. It falls short of that. Instead, it is a partial overview. This overview draws freely on the *Handbook*'s chapters, providing references only when points are made that have not appeared previously in the relevant sections. It assumes that the reader is already familiar with the chapters, if not their master. It is thus not a substitute for the contributions that are the *Handbook*. Much of the ontogeny material, in particular, came too late to be conceptually integrated here. A full synthesis of these materials, if that is possible, must await a resolution of issues and future theoretical developments that may only emerge in the next few decades. This is not a long time by palaeoanthropological standards, where for example the time-lag between new discoveries and a consensus about their significance may take a generation or more.

The time charts (pp. 878-85) are four in number, and in the style of Universal History they cover the time-frames of: (1) the last four million years; (2) the last one hundred thousand years; (3) the last forty thousand years; and (4) the last ten thousand years. They are internally arranged in three layers. The information displayed across the bottom of each chart is of an observational or factual type—realizing the common meanings of these terms are being relied upon here. The apparent truth-status of a given fact can of course change on the intergenerational human time-scale, owing both to the pursuit of 'truth' and, unfortunately, to vigorous advocacy. The user of this *Handbook* will find it useful to review the status of the facts set forth here in the seasoned view of another decade, if not sooner. The general advice of a senior scientist and scholar to the young apprentice may be relevant here: in so far as you can, you should always check for yourself the facts that you build your own work upon. A similar heuristic would apply to the other facts and products of the pursuit of knowledge about human origins, as it relates to symbolism.

The layer that runs across the middle of the time-charts, the second layer as we move up from the bottom, presents encapsulated empirical generalities or inductive inferences—for example, that the brains and vocal apparatus of different populations of prehistoric hominids were probably not all of the same kind. A slight example, to be sure. The capsules themselves are not trivial, however. They represent in most cases the cumulative work of two or more generations of researchers interested in the various specific facts of this work. Some of the capsules are hybrids of fact and of these more tentative inferences. Their respective shares are indicated by differences in the type that they are set in. Fortunately, most cases dealt with here appear straightforward enough to make this kind of naïve empirical categorization practical.

The top layer of the charts presents theory or explicit hypothetical constructs. This layer seems very scantly provided. It is impoverished both because of its relatively undeveloped nature with regard to the newly refined empirical inputs that mark this latter half of the twentieth century's contribution to understanding the origins of human symbolism, and because much previous theorizing was founded on empirically unsound premises or inherently weak forms of pre-modern argumentation (for example, wishing it were so makes it so), untestable in principle. Systematic application of theory and a sounder empirical base offer the next generation new scope for theorizing and model-building, even in those areas, such as the emergence and development of complex societies founded upon agriculture, which are already conceptually relatively well developed (see, for example, Hodder 1987, 1989).

This chapter is roughly organized around the four time-charts, but it goes on to recognize briefly some of the fashionable and long-lasting issues that characterize inquiry into our evolutionary and historical development. It is far from an exhaustive synthesis. It suffers from lack of expertise. Do not let your consumer expectations run wild. It is a hurried overview and commentary that had to await the completion of numerous tardy chapters, and felt the full pressure of that seemingly elusive 'press deadline'.

28.2 Chart I

The left-hand part of this chart is an account of the so-called 'gracile' australopithecines. These bipedal but ape-like earliest hominids were (by analogy with the more socially complex higher primates) probably readily capable of communicating their identity, location, emotional state, and probable intentions (Passingham 1982). It is hypothesized that the communication skill they may have possessed beyond that of extant apes was primarily gestural in modality. This hypothesis is supported by their hands being freed from locomotor needs by habitual bipedality, and their somewhat more human-like hand morphology. The contrasting hypothesis of vocal primacy is called into question by their lack of incipient basicranial specialization, indicating an absence of change toward a human-like vocal apparatus. (Evidence for their use or manufacture of stone tools and changes in the brain are equivocal.) This then is the baseline or primordial hominid state.[1]

Theoretically, ostension (i.e. pointing to things, events and phenomena, context-dependent in near-time and near-space) plus the use of iconic names was the dominant grammatical pattern for the gracile australopithecines and the australopithecine-like forms of earliest *Homo*, until the incipient changes inferred for the vocal apparatus emerge in early *Homo erectus* (if not *H. habilis*, the unknown case). Further discussion of the australopithecine grammatical pattern must await future theoretical work. The test implications of even the simple pattern postulated above are unclear, and the ecological/social motivations for such a system have yet to be systematically developed. Lexicon was almost certainly very small, and the grammatical pattern would have lacked grammatically marked syntax (that work being done by 'frame of dialogue'; see Rolfe, this volume, Chapter 26).

Early *Homo* is the hominid that fills the middle of the Chart.[2] We would like to focus here on the sometimes australopithecine-like polymorphic *Homo habilis*, but unfortunately the fossilized skulls are too incomplete to provide the needed evidence as to whether the basicranium displayed an early manifestation of specializations related to development of the human vocal apparatus (see this volume, Chapter 4, Editorial appendix II). There is palaeoneurological evidence for human-like specialization of the brain of *H. habilis* (see Holloway, this volume, Chapter 4), and the Olduvai hand (see Marzke, this volume, Chapter 5) and the earliest stone artefacts are attributed by many researchers to *H. habilis* (not the robust australopithecines contemporaneous with them); but without corroborative evidence from the basicranium we can go no further.

It is true that the identification of an enlarged third inferior frontal convolution of the left hemisphere on the brain endocast of ER 1470 suggests a voluntary control of vocalization more complex than that displayed by the Great Apes, but it is impossible to evaluate this apparent specialization independently of evidence from the basicranium; especially if one can hypothesize (cf. Lieberman 1988) that the early stage of development of Broca's area could have been related to enhancement of precise, sequential right-hand motor control.[3] The bottom line is that one can no longer assume that endocranial evidence for 'reorganization' of the regions including Broca's cap is *prima facie* evidence for the emergence of specialized speech areas. (For Wernicke's area, see Walker's 1993 comments.) At best, enhancement of manual specializations is suggested. The strong pattern of cerebral hemispheric asymmetry in ER 1470 is also equivocal evidence for increased capacity for vocalization, since left-hemisphere dominance has recently been shown to be as characteristic of human production of signed language as of spoken language (Poizner *et al.* 1987). The left hemisphere is dominant for sign-language, even though processing sign-language involves processing spatial relations at all linguistic levels. Furthermore, the right hemisphere shows complementary specialization for visuospatial non-language functions, even when these appear within the signed sentences themselves as the topographic representation of actual spatial relations (Bellugi 1988). As noted in Chapter 4, Frost (1980) has argued that in hominid evolution lateralized representation was an evolutionary consequence of the requirement for asymmetric employment of the forelimbs in the making and using of tools. The colateralization of language mechanisms was held to be a consequence of the coupling of these to the motoric mechanisms already lateralized at an earlier point in hominid evolution.

Fortunately the even more human-like pre-Acheulean *Homo erectus* is in evidence by 1.7 m.y.a. in Africa, and its manifestations of a human-like basicranium indicate that an enlarged vocal tract was present even then. If this is substantiated (see also Duchin 1990 and the subsequent criticisms by Lieberman *et al.* 1992) then the theoretical implications of this fact for human symbolism should be examined in detail. I can suggest two theoretical lines of inquiry that could prove fruitful for conceptualizing this stage of the evolution of human vocal communication. One can be derived from the hypothesis that incipient enlargement of the vocal tract in early *Homo erectus* provided for a greater range of sounds and more complex intonation patterns. Here I avoid the phrase 'speech sounds', because parsimony requires an unequivocally advanced argument for the attribution of protospeech, although some researchers have posited this state without apparent linguistic justification. The hypothesis of a greater range of sounds and more complex intonation patterns is more cautious. It implies a capability to mimic more faithfully the animal and inanimate sounds in nature, for example, without implying sound symbolism or the phonemic principle. Some singing or melody in voice would be possible, for example, but protospeech would not be implied. Full resolution of these issues must, of course, await their thorough theoretical treatment and discussion. See Deacon (1992, pp. 73-4) for some discussion of the possibility of a differential capacity for the production of consonantal vs. vocalic-like sounds.

A second theoretical line of inquiry into the more human-like vocalization of *H. erectus* that I can suggest is derived from the hypothesis that obligate terrestriality in *Homo erectus* necessitated an amplified coevolution of the caregiver–infant communication system. Unlike *Australopithecus* and the australopithecine-like *H. habilis* specimens from Olduvai Gorge (especially OH 62), the skeleton of *H. erectus* does not suggest arboreal capabilities. There is apparently nothing about the limb proportions or individual skeletal elements that suggests arboreality. If the australopithecines can be described as scansorial and bipedal (i.e. including both some arboreal activities and habitual terrestrial bipedality in their day-to-day activities), *H. erectus* appears to have been more or less exclusively terrestrial.

What, then, are the life-history implications of obligate or complete terrestriality for the immature phase of the life cycle and the caregiver–infant interactions for *H. erectus*? Lack of notable arboreal retentions and modern limb proportions in adult *H. erectus* suggests, by analogy with modern humans, that there may have been a crawling or quadrupedal infantile stage, rather than relative locomotor inactivity, preceding the young childhood transition to efficient bipedality. In either case, independent terrestrial movement of the infant and slow-running bipedal child, without effective access to arboreal means of escape, implies high risk of terrestrial predation from even the medium-sized carnivores. Given the low reproductive rates of the Great Apes and hunting–gathering humans, by demographic analogy this translates into a serious demographic risk. It is doubtful that early *Homo* could have occupied open-country habitats (thornbush and savannah) without some more effective means of predation-avoidance than that exhibited by chimpanzees, for example. Theoretically then, this is a mandatory, not optional, requirement for *H. erectus*. With increased vocalization an apparently habitual attribute of *H. erectus* (see this volume, Chapter 4, Editorial appendix II) some amplified or refined child–caregiver predation-avoidance specializations can be theoretically motivated. Attention to, discrimination of, and communication about more details in the social and ambient environment are called for. For example, with vocal as opposed to gestural communication *H. erectus* caregivers and infants could continuously monitor each other's immediate locations and conditions of safety. Caregivers could more successfully direct or control the movements of children unable to defend themselves. Anticipatory communicative exchanges between adults directing each other's attention to potential risks would also seem mandatory if the slow-speed movements of hominid bipedality (compared to the speed of carnivore movements) could not be compensated for by an essentially instantaneous and extremely refined selective use of defensive weaponry. Given the rapid movements of smaller carnivores (for example, the jackal) it seems that something more than distraction and diversionary displays is called for. Perhaps this exaggerates the risks and therefore the need for compensatory performance. A more knowledgeable assessment of risk is required, as well as theoretical discussion of alternative means of survival. But the hypothesis, even at this stage of its development, is open to falsification in the sense that palaeoanthropological finds of *H. erectus* infant fossils with evidence for significant arboreal capabilities would render the approach a non-starter. This possibility gives the hypothesizing at least some rudimentary scientific status.

These comments have focused on the vocal mode; but there can be little theoretical doubt that the gestural mode was more developed than in *Australopithecus*, in part because the gestural mode offers pre-grammatical solutions to critical communication problems. How can scenes be depicted, for example?

The syntactic topic–comment pattern seems to be required as a first evolutionary step in the vocal mode. Mime and gesture, intelligently directed, offer a paratactic means for the communication of simple scenes without syntax, however, relying only on 'frame of dialogue' and the analogy of temporal order. Such a device might be necessary to ensure co-operative social organization in a hominid with daily dispersal and regrouping of the labour force, for example, since important social actions routinely unobserved by absent members could repeatedly disturb the fabric of expected social relations in a highly intradependent social system. The capability for differentiation and co-ordination of social labour beyond that seen in higher primates and social carnivores requires increased capacity for, not just heightened disturbance of, social integration (see Jolly, this volume, Chapter 6). Sexual division of labour and food-sharing at rendezvous points or base camps is the level of social organization that has received the most speculative attention; but archaeological investigations have yet to provide an unequivocal index of this system in the materials associated geologically with early *Homo*. This will no doubt remain an issue for the next generation of investigations to sort out, if they are successful in developing some fine-grained observational techniques and telling interpretations of the palimpsest-like prehistoric record.

Another issue deserves at least a passing mention. Some archaeologists assert that well-made symmetrical hand-axes (in evidence by as early as *c.*900 000 years ago in Africa: M. P. Meneses, pers. comm.) required language for their manufacture. Outside archaeology such claims may be viewed as mere conjecture and linguistically uninformed. Clearly this issue is one that would appear to be amenable to scientific investigation, albeit it is restricted to living subjects.

Contrasting with this view of protolanguage as a pre-requisite to early developments in technology is the idea that refinement in technologically expressed motor skills provided repeated neurological preadaptations for language development (see for example Calvin 1983, 1991). But the apparent lack of evolutionary correspondences between the first appearances of hominid taxa and the major stages of prehistoric technology do not seem to fit into this speculation. There is pre-Acheulean *H. erectus* in Africa (for example ER 3733), early archaic *H. sapiens* in Europe with an Acheulean technology (for example Swanscomb) and early anatomically modern *H. sapiens* in the Near East with a Middle Palaeolithic technology (for example Jebel Qafzeh: see Section 28.3). It seems that major changes in technology occur after, not concurrently with, the appearance of new taxa of *Homo*. Recognizably new species develop new technologies as a part of the unfolding of their potential.

This brings us to the penultimate phase of Chart I, the emergence of *Homo sapiens*. Consilience of evidence indicates that *Homo sapiens* was present in the latter fourth of the Pleistocene, between 400 000 and 100 000 years ago (see Campbell, this volume, Chapter 2). It is a remarkable finding of palaeoanthropology that the basicranium reflecting the structure of the vocal apparatus in early *Homo sapiens* was modern, not archaic, in form. Their otherwise largely archaic-looking skulls (see the photogallery of fossil skulls, this volume, Chapter 1) are a mute contrast to this finding (but see Lieberman (1984) for a more conservative position on capacity and performance).

The correlated theoretical assertion for symbolic evolution is a hypothesis of the emergence of human vocalization in the form of sound symbolism in early *Homo sapiens*. (See Jakobson and Waugh (1979) for an introduction to sound symbolism in modern languages.) Theoretically these were vocal gestures that offered iconic representation, and which were organized into a system of contrasts. For Foster (this volume, Chapter 25) these were analogical symbols ('phememes') that were minimal units of sound and meaning derived iconically from the shape and movement of the articulatory tract. One theoretical issue is whether there are other types of sound symbolism that might be appropriate for characterizing the Palaeolithic period and the primordial languages of *Homo sapiens*.

Lastly the inference that anatomically modern populations (i.e. *H. sapiens sapiens*) probably arose over 100 000 years ago is also a notable development in our view of prehistory. I should point out that the implications for language origins of an African point-of-origin for *Homo sapiens sapiens* have yet to be examined. What theoretical difference does it make that our ancestral tongue was African? Obviously much remains to be explored and discussed. Perhaps the next decade will not be time enough.

28.3 Charts II and III

A number of changes took place in the palaeolithic cultures of the Late Pleistocene between *c.*100 000 and 12 000 years ago. This is relatively well documented only for parts of Europe and the Near East. Several apparent differences in the socio-cultural systems of the Middle Palaeolithic (specifically the Mousterian of Western Europe and the Near East) compared with the Upper Palaeolithic are summarized in Table 28.1.

Although *Homo sapiens sapiens* appears to have evolved elsewhere at least 60 000 years before the advent of the European Upper Palaeolithic and cave art, these people do not appear in western Europe until quite late (*c*.35 000 years ago), and the contrast between their culture and that of the neandertalers of western Europe is quite striking. The contrast is useful in the sense that it heightens our awareness of the nature of the Middle to Upper Palaeolithic transition, but the peculiarities of Neandertal skeletal anatomy and their geographic semi-isolation in what amounts to a small western cul-de-sac of Eurasia cautions us that they may not be broadly representative of archaic *Homo sapiens*. Still, we realize that *H. sapiens sapiens* was probably associated with Middle Palaeolithic technologies throughout most of its (i.e. our) prehistory. For example, the Near Eastern Skhūl and Qafzeh skeletons, although classified as robust but anatomically modern (i.e. *H. sapiens sapiens*), are Middle Palaeolithic burials in the Mousterian tradition.[4] It is unfortunate that as yet we know very little about the *in situ* evolution that was occurring in Africa. This makes it impossible to tell anything exact about the tempo of evolution, except to say that it had no doubt accelerated markedly in comparison to the period covered by Time Chart I.

There is an apparent contrast in both mode and tempo of evolution for energy vs. information technology in the Middle vs Upper Palaeolithic. Middle Palaeolithic energy technology was expressed in both simple energy constructs (see Campbell, this volume, Chapter 2; including hafted tools and constructed shelters, plus, presumably, containers of various sorts) and some control of fire (cooking with unconfined hearths on the ground surface). In contrast to these apparent energy system advances, we have no systematic evidence for cognitive technology, no systematic evidence for symbolically embodied externalized cognition in the Middle Palaeolithic.

As for the essential nature of the Middle to Upper Palaeolithic transition, whatever the obscure sociocultural system was really like before, after the transition in the Late Pleistocene we recognize a cultural way of life that was in essence apparently similar to that of hunting and gathering people in historical times, including the rich cultural systems developed by coastal fishing societies. To put it another way, if animal behavioural traditions are quasi-cultural, and the Lower (Early Stone Age) and Middle Palaeolithic are protocultural, then only with the advent of the Upper Palaeolithic and Ice Age Art do we have clear evidence for a human type of culture, volatile, capable of being impregnated with style (see Wynn, this volume, Chapter 10; see also Noble and Davidson 1991, the discussion on fallacies in Chase 1991, and the commentaries in Lindly and Clark 1990). This may be an exaggerated view, but the apparent prolonged stability characteristic of the Lower Palaeolithic and Middle Palaeolithic traditions suggests that although they were tool-assisted or tool-dependent in nature, they were not culture-based biobehavioural systems. This conceptualization contrasts with the growing tendency until quite recently to de-emphasize the differences between ourselves and our remote ancestors. If *Homo sapiens sapiens* emerged at least 100 000 years ago, then culture can be seen as a recent development for even anatomically modern *Homo sapiens*. This requires a conceptualization of culture separate from the social and ecological systems (see Ingold, this volume, Chapter 7). It also requires understanding how externalized cognition (the heart of human culture) can promote various forms of practical, speculative, and discursive consciousness (both awareness and judgement, in addition to tacit knowledge), which themselves may be socially partitioned and reunited under different social conditions to produce new forms of culture (Zilsel's 1942 thesis provides a recent historical example), and which seem to take on and express an evolutionary life of their own. Suffice it to say here that Upper Palaeolithic 'arts' (the artefacts of personal adornment, the figurines, the notational devices, and then the cave paintings) represent our first systematic evidence for symbolically embodied externalized cognition. The shift in paleoanthropological interpretation is from emphasizing the indirect evidence for a change across the Middle to Upper Palaeolithic in how individuals *perceived* or cognized the forms and process of social reproduction to that of recognizing the emergence of new patterns of socially distributed externalized cognition.

Chart II attempts to generalize some of these points on a broad scale, and add the linguistic perspective. Foster (this volume, Chapter 25) is logically able to posit a number of stages to the elaboration or expansion of the *phememic* inventory or lexicon. She sees this as not completed, or expanded to its maximum, until the Upper Palaeolithic (the age of the early cave-art). The Neolithic is, in her view, the period during which the phonemic principle was developed. Other hypotheses are possible. The most apparent one is the hypothesis that the phonemic principle was developed during the Upper Palaeolithic, as well as the grammatical pattern of topic–comment. The phonemic principle (Rolfe's 'thematization') provides thematic units or groupings whose internal variants are phonologically realized (for example, the ablaut, sing/sang/ sung/song) by the exploitation of alternances inherent in human speech-sounds. Not only do the items with their referential content (i.e. the words) allow mental inventorying to take place

Table 28.1 Apparent differences in Middle Palaeolithic vs. Upper Palaeolithic socio-cultural systems (designed to be read from the bottom up)*

	Communication and common symbols (Perigord, SW France) [Based on White 1982 and others]	Burial practices (Europe and Near East) [Based on Harrold 1980]	Technology (Europe) [Based on Dennell 1983, White 1989, and Anderson-Gerfaud 1990, plus Mellars 1989]	Society and mode of production (SW France and Europe) [Based on White 1982, Binford's 1982 commentary, and others]
Upper Palaeolithic	Hypothesis that the ability to anticipate events (e.g. herd movements) and conditions not yet experienced became one of the strengths of palaeolithic peoples at the time of the appearance of clear evidence of symboling, e.g. personal ornaments and graphic art (Binford's 1982 comments on White 1982). Communication augmented with visual arts, personal adornment, and music (Dennell 1983). Formal variation in stone artefacts through time seems to appear first in the Upper Palaeolithic along with clear regional (stylistic) differences in artefact morphology. Presence of exotic materials from distant sources is frequent (also the case in European Russia) and implies long-distance contacts.	Ochre in graves common ($n = 35/67$). The common body position in burial was extended limbs and trunk, occurring far more often than the strongly flexed position ($n = 24$ to 13). Upper Palaeolithic burials tend to have a greater diversity of grave goods (c.50% with art/decorative items vs. none in the Middle Palaeolithic) and a greater quantity of grave goods (e.g. more tools and animal bones). The inference is that Upper Palaeolithic societies were more complex, with greater symbolic marking of status distinctions in burials. Female burials still less common than males ($n = 21$ to 42), but females typically buried with grave goods. The inference from a generally similar range of grave goods in male and female burials is that at least some females have social statuses as complex and richly marked as those of males.	Hypothesis of transport with snowshoes and dog-traction. Weapons included spear-thrower and ivory/bone-tipped projectiles. Cooking in cobble-lined hearths in the ground. Inference: artefacts with more than three components: ladders; multi-pieced clothing; necklaces; nets. As below, plus: pressure-flaking of stone; grooving/splintering and drilling of bone, antler, and ivory; carving and engraving of stone; kneading of clay; grinding/polishing of stone and bone; compression straightening of ivory; and twining (?). Technological shift to increased stone-tool processing of hide, bone, antler, ivory, and shell products, as opposed to wood. A substantial increase in the scale of blade production, with tighter degrees of 'standardization' and more sharply defined *separation* between discrete morphological types of stone tools.	Hypothesis of logistically organized hunters and collectors. Broadening of the subsistence base to include fish and birds. Apparent greater maximum size and internal variance in settlements. Some occupation of northern regions of Europe under more extreme climatic conditions (Gamble 1982), although the early Magdalenian saw total depopulation of parts of NW Europe (Meiklejohn 1982). First peopling of the Pyrenees in the Magdalenian (Straus 1982). Appearance of regional stylistic differences marks the point at which local population size passes the minimal level necessary to be a locally viable mating system. Area the size of south-western France too small to support more than one biologically effective breeding population (c.500–1000 people) during the Upper Palaeolithic or Mesolithic (Meiklejohn 1982). Relatively independent mating systems could have operated in areas as close to France as Germany, England, and Belgium (my inference from White).

Table 28.1 (*Contd.*)

	Communication and common symbols (Perigord, SW France) [Based on White 1982 and others]	*Burial practices (Europe and Near East)* [Based on Harrold 1980]	*Technology (Europe)* [Based on Dennell 1983, White 1989, and Anderson-Gerfaud 1990]	*Society and mode of production (SW France and Europe)* [Based on White 1982, Binford's 1982 commentary, and others]
Middle Palaeolithic	Vocal communication with no apparent material-culture augmentation (Dennell 1983). Presence of exotic materials, such as marine shells and flint from distant sources, is rare.	Ochred graves very rare (cf. Harrold 1980 and Wreschner 1985). The common body position in burial was with strongly flexed limbs, as if the corpse were crowded into a grave of inadequate size. Females buried without grave goods ($n = 0/7$). Males ($n = 8/10$) often buried with grave goods consisting of unmodified stones, stone tools, and animal bones. (Inference: a greater variety of social information and/or a more complex *persona* is expressed in male burials compared with female burials.)	Cooking with unconfined hearths on the ground surface. Inference: at most two components in the reconstructed artefacts (e.g. stone-tipped spears, lashed clothing); evidence for hafting. Inference: six basic techniques (percussion, chopping, whittling, scraping–planing, sawing, and cutting) used to work a few raw materials (bone, stone, skin, hide and especially wood).	Hypothesis of low-density nomadic scavenger/hunters and foragers with spatially extensive open-ended mating networks.

* Based on White (1982) and commentaries plus White's replies, and Dennell (1983), Harrold (1980), and Wreschner (1985), with additional information from White (1989), Anderson-Gerfaud (1990), Mellars (1989), Bar-Yosef and Meignen (1992), and Knecht *et al.* (1993).

systematically, the phonemes themselves theoretically allow the development of a new cognitive indexing system (see Hewes, this volume, Chapter 21), as by analogy does the more recent alphabet system in writing. One could also hypothesize that early manifestations of the syntactic topic–comment and the grammaticalized narrative pattern appeared in the vocal repertoire of the Upper Palaeolithic, but the equivocal evidence for the depiction of scenes in the art from the Late Pleistocene suggests that neither case-relations nor narrative were the dominant cognitive patterns. Another issue in need of systematic theoretical attention is the relationship between this developing primordial vocal language and the gestural language system, which one can assume was also continuing to develop. The future contributions of theoretical linguistics to these topics are potentially the most interesting for prehistorians. Developing their test implications relative to the prehistoric record will no doubt take us well into the twenty-first century.

The parallel socio-cultural theory that would articulate with this also remains largely underdeveloped. From an evolutionary perspective what have been partially developed are system-state models, i.e. the biological hominid-type is fixed, and the model deals with (1) socio-cultural amplification, or (2) a change in the mode of production. The Price and Brown (1985) diagram that appears here as Fig. 28.1 is an example of the former. Increasing complexity results in social status differences, increased ritual activity, and new institutions such as sodalities. Redman's (1978) graphic models of the origins of agriculture, urbanism, and the political state (see the appendix to White's Chapter 9, this volume) provide examples of the latter (see also the discussion below on Time Chart IV). Models that develop socially determinist explanations probably have greater potential than those that are merely environmentally determinist, at least in so far as the forcing functions that drive language and complex symbolism are concerned. But what of the models that depict transformations associated with the change from archaic to anatomically modern *H. sapiens*, i.e. a change in hominid-type? These are virtually undeveloped, and there is little hope that the system-state models will provide the explanatory principles for this kind of biological transformation.

Lastly, I should point out the 'missing link' with regard to what were probably important changes in the caregiver–child interaction and communication system. Some new and perhaps unexpected behaviours are in evidence for the Upper Palaeolithic, such as elaborate burials of children, and small footprints in deep caves suggesting psychological confirmation of aspects of a supernatural world-view through a sharing of dramatic supernatural-like experiences (Pfeiffer 1982).[5] Moreover, we can hypothesize on the basis of interpretations of anatomical studies of the skeletons of Neandertalers that the life-history pattern or ontogeny of archaic *H. sapiens*, relative to that of *H. sapiens sapiens*, was characterized by: (1) precocious neonates (Trinkaus 1984; Dean *et al.* 1986); (2) full brain growth completed earlier in ontogeny (Dean *et al.* 1986); and (3) extremely rare if any survival past middle age or the reproductive years (40–45 years, Trinkaus and Thompson 1987). Physically precocious, powerfully built children, and a society without grandparenting suggest less organizational complexity for the social group, in comparison with *H. sapiens sapiens* (see the various viewpoints expressed in Fisher 1988). In contrast, altricial offspring that show a marked delay in the attainment of independent self-maintenance were apparently a hallmark of anatomically modern *H. sapiens*. Without elaborated and more specialized socio-cultural support systems this reproductive pattern would have been unable to outcompete that of archaic *H. sapiens*. Contrary to previous interpretation, the altricial, anatomically-modern infant may have evolved before the appearance of Upper Palaeolithic culture; but it was not until the Middle to Upper Palaeolithic transition was complete that this reproductive pattern was able to overcome its relative handicap and fully exhibit the competitive advantages inherent in delayed maturation and extended ontogenetic development. One of the future values of the ontogeny sections of the *Handbook* lies in motivating richer models of language development that recognize the possibility of ontogenetic systems' differing profoundly across the different phases of human evolution.

28.4 Chart IV

Moving from the late Pleistocene into the early Holocene we encounter societies of increasing socio-cultural complexity, although it is perhaps surprising that the initial tempo of cultural take-off is relatively slow, given the assumption that the phonemic principle was developed in the Neolithic, if not sooner. This reminds us that there are at least three faces of evolution that we must understand and explain: (1) directional change, the unfolding; (2) non-directional change and variation; and (3) unchanging features or persisting patterns.

With the emergence of early states (and then empires), beginning in the mid-Holocene, most smaller societies are ultimately destroyed and/or incorporated into political systems orders of magnitude more

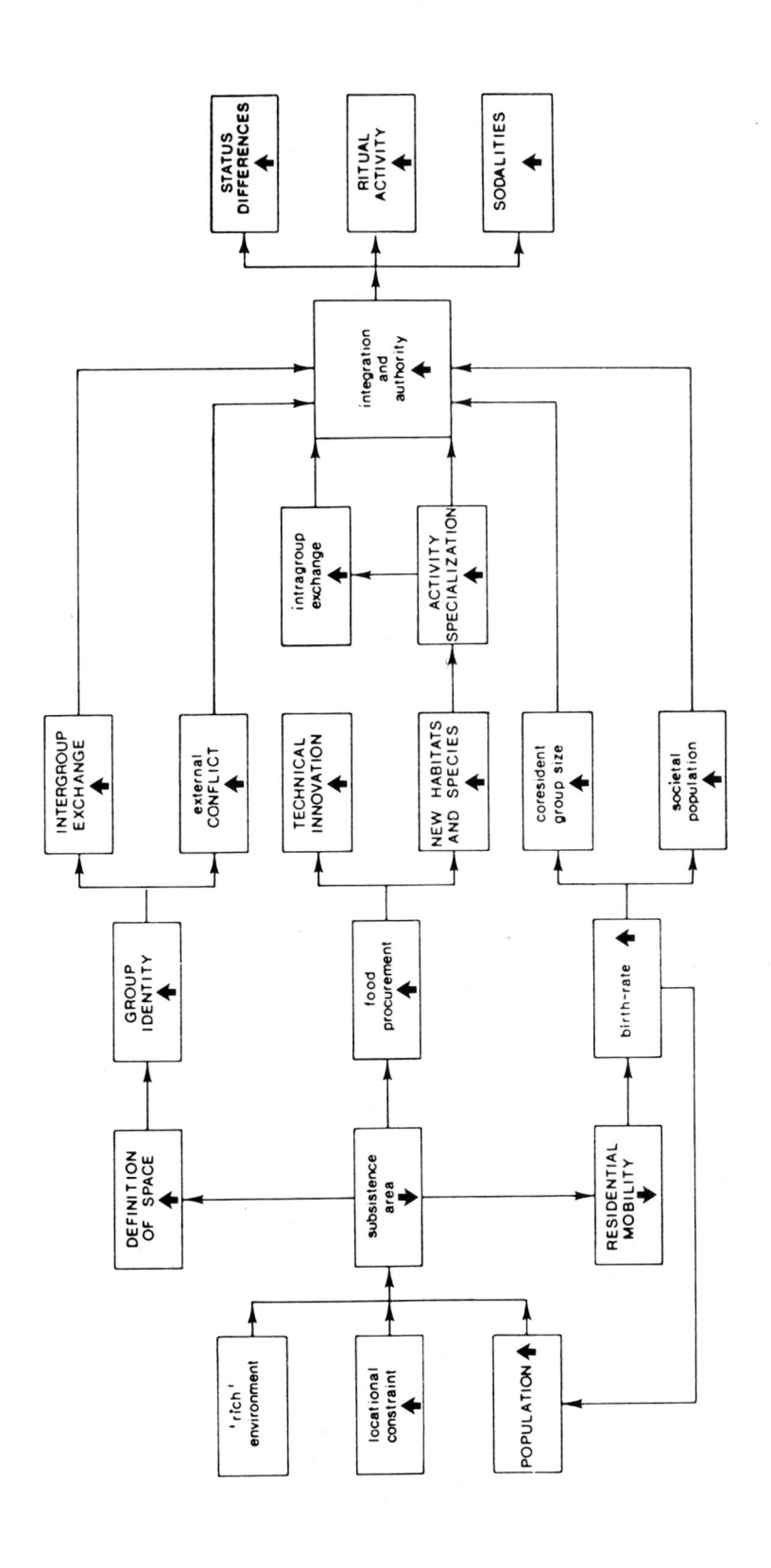

Fig. 28.1 Conditions, causes, and consequences of the phenomenon of increasing complexity among hunter–gatherers. This heuristic device portrays some of the important components of and possible relations of this process. The diagram is not a model of the real-world situation, nor does it provide a sequence of events involved in increasing complexity. The items shown in upper-case letters are regarded as more accessible in the archaeological record. (Figure and caption from Price and Brown (1985), Fig. 1.1.)

complex, i.e. modern states and the world system(s). Human socio-cultural evolution in the Holocene, if not before, is in large part the endogenous type: new forms are the result of an unfolding and re-working of the implicit developmental potential or latent properties of earlier cultural institutions and beliefs.

Complexification of symbolism in the Holocene can be thought of as involving at least three processes: (1) the amplification of earlier principles of symbolism; (2) the realization of latent properties previously neglected; and (3) the creation of new principles. These may not be easily distinguished. As an example of the latter from the New World, Knight (1989) compares animal-symbolism in the representational art from burials of the pre-Columbian Eastern Woodland USA (Hopewellian period) with that of the Mississippian period that followed. In the pre-Mississippian animal-symbolism the animals portrayed are real species, not inventions or fabrications. With few exceptions there are no monsters, no animal composites. These tribal peoples seem to have relied upon real natural distinctions to symbolize and clarify social ones (see also Vinnicombe 1976). Knight refers to this as an *allegorical* use of symbolism. All that is required to understand the metaphors is common knowledge of the animals and people. The animal symbolism of Mississippian chiefdoms, a few centuries more recent, represents a very different symbolic mode. We now have synthetic images of pure imagination, for example, antlered rattlesnakes with wings and human-headed reptilian panthers (see Fig. 28.2).

These are imaginative unnatural symbols, at times seemingly surrealistic. Knight refers to this as *arcane* symbolism, and notes that most Mississippian commoners may have been no better off than we are in understanding their esoteric significance. He sees their invention as a particular instance of the invention of power. In this case they are part of the development of early south-eastern chiefdoms, a level of political integration and social control that did not previously exist in eastern North America. In Knight's words, these élite ritual symbols played a part in this political innovation as leaders struggled to associate themselves with supernatural power by creating and monopolizing an esoteric lore, a world of fantastic symbols and obscure meanings whose interpretation rested secure with the privileged. Mystification was the objective, the élite alone being capable of tapping the latent power in these symbols.

Chart IV is largely a partial recounting of highlights from historical evolution in the Old World (based primarily on Grun 1982, see Table 27.9, pp. 844–54, this volume). The notes suggest a late emergence of something more than new record-keeping technologies and mental aids in the service of developing political economies, although the broad time-blocks employed here mask a good deal of diversity in the form and function of these early civilizations (see also Larsen 1988). But seen in this way the poverty of our theoretical notions is clear. For a later example, the Nuremberg globe is a landmark symbolizing not only the earth but also a new stage in the externalization of our knowledge and awareness of the earth as a global system.

The study of socio-cultural evolution in the New World reminds us that although we currently have only incomplete notions as to the why of symbolic evolution over the past few thousand years, the tempo has been accelerating, if uneven, with initially long 'delays' in cultural take-off typical after the presumed development of the phonemic principle in speech during the early Neolithic. The patient reader will note that it has taken us a long time to come to true

Fig. 28.2 Examples of arcane symbolism from the pre-Columbian Mississippian period. (Phillips and Brown 1978 (with plate 80) and 1984 (plate 223)).

Table 28.2 Historical modes of symbolic evolution*

Reworking of expressive devices to serve new epistemological levels
Increasing degrees of abstraction
Expanding generalization
Extending hierarchies of analogy

Increasing the behaviour that is symbolically marked
Differentiation of specialized vocabularies

Working out inherent stresses introduced by the creation of a new system
Diversification of depiction and notation
Iconic representations in notation increasingly replaced by symbols
Multiplexing
Consolidating and standardizing symbols
Differential conservation
Stylistic selection

Diffusion plus syncretic to synergistic transformations

Increasing investment in apprenticeships

Paper-and-pencil working-out of logical or syntactic principles and constraints
Development of new devices for the externalization of cognition.**

* See especially the appendices to Barton and Hamilton, this volume, Chapter 27.
** Cf. Leroi-Gourhan 1993 and Gumperz and Levinson, in press.

symbols—more than two million years. For this final period of development our theoretical understandings are not obviously superior to those characterizing earlier periods. A number of questions have yet to be adequately formulated, much less addressed. Are the processes operating in the historic period essentially similar to those that operated in the Upper Palaeolithic and the Neolithic, only recently serving new levels of complexity? Is scalar stress (or hierarchical information-flow and control) really the key to all of this? Does symbolism itself become a driving force (not merely a facilitator) for change in socio-cultural complexity? Even if it is clear that new principles are involved, these last two (now traditional) questions do not seem to capture the fabric or web of processes contributing to the historical transformations. (See Wright 1986 for related comments on the need for new theoretical constructs.) Some of the historical modes of symbolic evolution suggested by various sections of the *Handbook* are listed in Table 28.2. A number of issues could be developed around these processes and the nature of their realization in historical time. It is not clear, for example, to what extent the paper and pencil factor has been responsible for the creative explosion in mathematics seen over the past few hundred years. Moreover, a deeper understanding of the creative socio-cultural setting is required, including the ontogenetic reworking of language systems, since this is the means both for the learning and for the recreating of these systems across the generations.

With the evolution of states and empires we can see many of the earlier-mentioned processes contributing to the increasing complexity of symbolism. To continue with examples from the New World, Conrad and Demarest (1984) provide a detailed analysis of how traditional beliefs and practices were reworked and converted into policies of imperial expansion by the Aztec and the Inca. The leaders of these developing states created a sense of divine mission by a manipulation of fundamental religious concepts and symbols, which helped to create pressures for territorial expansion. These leadership skills were further encouraged in the sons of the élite by formal programmes of training: in addition to learning military skills they were thoroughly indoctrinated in the state religion and the authorized version of their history (see also Berdan 1982, for the Aztecs). Some of the differences in the ideologies of these two state systems are particularly fascinating; but only their evolutionary parallels will be summarized here.

To borrow freely from the analysis of Conrad and Demarest (pp. 180–5), the ideological adaptations of the Aztec and Inca leadership were of two main types: manipulations of the upper religious pantheons and reworkings of ancient, basic institutions. Many of the reforms were consciously directed. Both polities inherited long-standing traditions of multi-aspected deities based on the movements and transformations of astronomical phenomena. Manipulations of the upper pantheons began in pre-imperial times with the crystallization of patron deities: in the Aztec case through the fusion of a tribal deity/culture hero with the divine complex of Central Mexican civilization; in the Inca case through the manipulation of existing elements of the traditional Southern Highland sky-god. Both peoples came to place their greatest emphasis on the solar associations of the ancient divine complexes, and eventually to isolate one solarized aspect as a national symbol and dynastic ancestor. Agricultural intensification centred upon the all-important staple crop, maize, and a great elaboration of ritual related to corn and the sun ensued in both imperial religions.

The ideological reforms were also revolutionary in the sense that the new state religions were able to integrate economic, social, political, and religious factors into unified cults of imperial expansion. Imperialism via conquest became a sacred duty.

By institutionalizing virtually fanatical drives for conquest, the reformed state religions transformed Mexican and Inca society into the most dynamic and ferocious war machines in New World prehistory. The all-pervasive militarism of their cultures gave the Mexican and the Inca the decisive advantage over their neighbors and propelled them outward as irresistible forces. In turn, their expansions were favored by environmental selective forces. Access to the products of varied ecological zones, to land and labor for agricultural intensification, made both societies stronger and richer than any potential competitor. Conquest poured wealth into the hands of Mexican and Inca leaders. As the rulers channeled some of their newly won riches downward through their patronage networks, a wide spectrum of Mexican and Inca society shared in the economic, social, and political benefits. Both peoples flourished because of their supremely successful ideological adaptations to their natural and social environments. . . .

However, the successes of the Mexican and Inca ideological adaptations were short-lived. No one can predict what would have happened if European invaders had not truncated the independent evolution of Mesoamerican and Central Andean cultures, but in their final decades the Aztec and Inca Empires were obviously deeply troubled (Conrad and Demarest 1984, p. 183).[6]

By late imperial times both empires desperately needed major internal reorganizations, major reforms that would promote stabilization in a shift from expansion to consolidation. But as history reminds us, they never got the chance. Because of their fatal encounter with the Spanish Empire, we will never know if they could have created the new principles of secular symbolism, ritual, and cosmology apparently required to make the transition from early into more mature forms of complex statehood and nationalism (see, for example, Kehoe 1990, Linenthal 1990), which in turn are embedded in an expanded world system with its own international and global symbolisms.

28.5 Modernity's complexity

With the episodic rise of increasingly complex cultures and societies across the past few hundred years we see multilineal trajectories culminating in the webs of layered reworkings that characterize symbolism in modern pluralistic nation states. To say that surviving 'traditional societies' (if not encapsulated) have become incorporated into these regional and world systems is not to say that they have become homogenized or capitalist in their entirety. They continue as aspects of complex multiple histories, for example, as part of societies with both pagan and Christian pantheons, subsistence and production-of-value economic duplexities, domestic and public languages, ethnic and nationalistic appearances. The myth of socio-political postmodernism is equifinality and homogenization. In most countries of the world, formal federally controlled education does not have the resources to create an ideological homogenization of the populace. As Gonzalez Chavez (1988, p. 340) has argued, the development of educational centralization, through which the modern state gains a monopoly of formal schooling, can be regarded as a contradictory historical process. In its only partially successful attempt to unify and homogenize the nation, the state increases social differentiation and heterogeneity. Moreover, continuing his argument, this process acts to limit the range and influence of the state in society, thereby creating space for other forms of action and social institutions. These may be relatively disorganized and govern more limited realms of power, but they are autonomously controlled and define important processes of social change. From the perspective of our *Handbook*, this can be seen to contribute significantly to the complexity of modern symbolism.

Overprinted on this, at a broader global level, a new simplification has developed in the creation of transnational functions and their accompanying symbolisms. Now central-sphere élites create esoteric knowledge as part of a monopoly on the natural world, the effects sometimes appearing to commoners (myself included) as supernatural.

However, even in the high-energy core areas of the modern world system the ancient symbolisms have not disappeared altogether. They are preserved (manifest) in reworkings whose complexity and pattern have yet to attract systematic investigation. In the mood, style, and status advertisements of high-tech

Fig. 28.3 An example of modern critical humour, showing the reworking and layering of symbolism. (Rough cuts by J. A. Reid, from *In these times*, 29 March–4 April, 1989, courtesy of the author. © 1989, J. A. Reid.).

society one can juxtapose the icons of nature and the working class with signatures of wealth and aristocratic leisure; in the mediascape of high modernity one can simulate being wealthy by way of playfully mimicking the working class, appropriating critiques of class privilege while relegitimizing it (Kroker *et al.* 1989, 132–3). Symbolic pluralism and syncretism now operate differentially on multiplex scales (for example see Fig. 28.3). Cognitive technology amplifies the diversity (see for example Halio 1990). These are hallmarks of the modern world, or, some might argue, of post-modernity.

Notes

1. As for the robust australopithecines (*Paranthropus* spp., *Australopithecus boisei*), to my knowledge there has been no systematic discussion of the implications for vocalization of the unique anatomy inferred by Aiello and Dean (1990) for their vocal tract.

2. In the past few years the hypothesis has been advanced that some of the early specimens of *Homo* found in East Africa previously placed into *Homo habilis* Leakey, Tobias, and Napier 1964 and *Homo erectus* Dubois 1892 should be recognized as new species (Groves 1989; Wood 1992).

Within *Homo habilis sensu lato* the Olduvai specimens originally assigned to that taxon remain valid members in both the Groves (1989) and the Wood (1992) treatments. Both authors are also in agreement in hypothesizing that the specimens from East Turkana represented by ER 1470 are a different species, one that would bear the name *Homo rudolfensis* (Alexeev) Groves. The specimens represented by ER 1813 are assigned by Groves (1989) to *Homo ergaster* Groves and Mazak, while Wood (1992) assigns them to *Homo habilis* Leakey et al.

Within *Homo erectus sensu lato* both Groves (1989) and Wood (1992) are in agreement in hypothesizing that the specimens from East Turkana represented by ER 3733 are a different species. For Groves (1989) the species remains unnamed, but Wood (1992) assigns the name of *Homo ergaster* Groves and Mazak to this taxon.

One can see from this that a certain amount of confusion might occur in informal discourse about these species. It may be some time before a consensus emerges. The problem revolves around assignment of the type specimen of *H. ergaster*, mandible ER 992. This mandible was originally thought to belong to the small-brained *H. habilis*-like form (for example ER 1813: Leakey *et al.* 1978; Groves 1989), but recent analysis on the bases of the Nariokotome skeleton (WT 15000) reassigns ER 992 to the earliest '*H. erectus*-grade' form, including ER 3733 and ER 3883 (Walker 1993). Wood's (1992) taxonomic assignments are strengthened.

3. In humans the third inferior frontal cortical convolution of the left hemisphere is the motor programming 'association' area ('adjacent' to the part of the brain that controls motor activity) that Broca announced in 1861 was the seat of articulate speech. Linguistic deficits in speech-production associated with brain damage in Broca's 'area' include lack of fluent articulation, phonetic and phonemic errors, and grammatical errors. In addition to deficits in speech motor programming and syntax, comprehension of complex sentences is also impaired, as is simple speech-perception. Performances can be impaired on virtually all linguistic, perceptual, and abstract cognitive tasks. (From Lieberman's 1984 summary. See also related notes in the Introduction to Part III: Ontogeny and Symbolism.) Furthermore, Begun and Walker (1993) cite the studies of Petersen *et al.* (1988, 1989) and their co-workers showing that

speech production (repeating words aloud) activates brain regions near to (together surrounding) Broca's area that are also activated by simple movements of the mouth and tongue, as well as actual and imagined simple hand movements. They, in turn, cite the review by Mohr *et al.* (1978) showing that small lesions confined to the classically defined Broca's area most frequently cause effortful articulation and oral dyspraxia without specific language involvement, rather than the full-blown Broca's aphasia (including protracted mutism, verbal stereotypes, and agrammatism) that is produced by much larger lesions in that area. They conclude that these regions are a general motor programming area, rather than being speech- or language-specific.

4. See McDermott *et al.* (1993) for absolute dating of these deposits, which range from *c.*40 000 to *c.*100 000 years ago.

5. Pfeiffer (1982) theorizes that the amount of information to be transmitted from generation to generation in the most complex early Upper Palaeolithic societies had risen to such levels that indoctrination involving the dramatic presentation of materials was essential to insure accurate uptake and further propagation. (Cf. Isaac's 1983 review of Pfeiffer's work.) Dramatic rituals not only helped to establish the authority of those who controlled them, but also encoded parts of the programmes in unforgettable ways (op. cit.). As Isaac points out, what is new in this is the idea that the information load to be transmitted had become so complex that encoding and dramatic mnemonic systems became developed for the first time. The hypothesis helps to explain deep-cave art by its focus on drama and situation as memory-imparting devices. The implications for socialization, enculturation, and integrative ontogenetic transition through puberty into young and even later adulthood would also be of theoretical interest.

6. 'The dwindling cost–benefit ratio of imperialism was aggravated by shifting demographic patterns... The results were localized imbalances between population and resources, especially in the capital districts, with their rising concentrations of hereditary nobles and non-food-producing specialists... attempts to correct these imbalances were counterproductive. Continued expansionism served only to intensify existing strains. In established provinces, increasing tax burdens and ethnic tensions between rulers and ruled combined to create smoldering resentments that regularly ignited into rebellions. Such uprisings were beaten down in campaigns that further decreased local agricultural work forces, making future tribute or tax payments even harder to meet. Likewise, agricultural intensification in the capital districts and established provinces eventually became a self-defeating measure. As the demands for surplus production grew, increasing reliance on marginal land actually served to decrease the security of the imperial subsistence economies by heightening the risk of crop failure. This problem is evidenced by Ahuitzotl's disastrous aqueduct extension [Aztec] and Polo's testimony that in many regions of Tawantinsuyu crops failed three years out of five [Inca]' (Conrad and Demarest 1984, p. 184).

References

Asfaw, B., Beyene, Y., Suwa, G., Walter, R. C., White T. D., WoldeGabriel, G., and Yemane, T. (1992). The earliest Acheulean from Konso-Gardula. *Nature*, **360**, 732–5.

Aiello, L. and Dean, C. (1990). *An introduction to human evolutionary anatomy*. Academic Press, London.

Anderson-Gerfaud, P, (1990). Aspects of behaviour in the Middle Palaeolithic: functional analysis of stone tools from southwest France. In *The emergence of modern humans: an archaeological perspective* (ed. P. Mellars), pp. 389–418. Cornell University Press, Ithiaca.

Bahn, P. G. and Vertut, J. (1988). *Images of the Ice Age*. Facts on File, New York.

Bar-Yosef, O. and Meignen, L. (1992). Insights into Levantine Middle Paleolithic cultural variability. In *The Middle Paleolithic: adaptation, behavior and variability* (ed. H. L. Dibble and P. Mellars), pp. 163–82. The University Museum, The University of Pennsylvania, Philadelphia.

Begun, D. and Walker, A. (1993). The endocast of the Nariokotome hominid. In *The Nariokotome* Homo erectus *skeleton* (ed. A. Walker and R. E. Leakey), pp. 326–58. Harvard University Press, Cambridge, Mass.

Bellugi, U. (1988). The acquisition of a spatial language. In *The development of language and language researchers: essays in honor of Roger Brown* (ed. F. Kessell), pp. 153–85. Erlbaum, Hillsdale, NJ.

Berdan, F. F. (1982). *The Aztecs of central Mexico: an imperial society*. Holt, Rinehart, and Winston, New York.

Binford, L. R. (1982). Comments on R. White's 'Rethinking the Middle/Upper Paleolithic transition'. *Current Anthropology*, **23**, 177–81.

Calvin, W. H. (1983). A stone's throw and its launch window: timing precision and its implications for language and hominid brains. *Journal of Theoretical Biology*, **104**, 121–35.

Calvin, W. H. (1991). *The ascent of mind: Ice Age climates and the evolution of intelligence*. Bantam Books, New York.

Chase, P. G. (1991). Symbols and paleolithic artifacts: style, standardization, and the imposition of arbitrary form. *Journal of Anthropological Archaeology*, **10**, 193–214.

Conrad, G. W., and Demarest, A. A. (1984). *Religion and empire: the dynamics of Aztec and Inca expansionism.* Cambridge University Press.

Davis, W. (1986). The origins of image making. *Current Anthropology*, **27**, 193–215, 371, 515–16.

Deacon, T. W. (1992). Brain–language co-evolution. In *The evolution of human languages* (ed. J. A. Hawkins and M. Gell-Mann), pp. 49–83. Addison-Wesley, Reading, Mass.

Dean, M. C., Stringer, C. B., and Bromage, T. G. (1986). Age at death in the Neanderthal child from Devil's Tower, Gibraltar, and the implications for studies of general growth and development in Neanderthals. *American Journal of Physical Anthropology*, **70**, 301–9.

Dennell, R. (1983). *European economic prehistory.* Academic Press, London.

Duchin, L. E. (1990). The evolution of articulate speech: comparative anatomy of the oral cavity in *Pan* and *Homo. Journal of Human Evolution*, **19**, 687–97.

Fisher, A. (1988). On the emergence of humanness. *Mosaic*, **19** (1), 34–45.

Frost, G. T. (1980). Tool behavior and the origins of laterality. *Journal of Human Evolution*, **9**, 447–59.

Gamble, C. (1982). Comments on R. White's 'Rethinking the Middle/Upper Paleolithic transition'. *Current Anthropology*, **23**, 183.

Gonzalez Chavez, H. (1988). The centralization of education in Mexico: subordination and autonomy. In *State and society: the emergence and development of social hierarchy and political centralization,* (ed. J. Gledhill, B. Bender, and M. T. Larsen), pp. 320–43. Unwin Hyman, London.

Groves, C. P. (1989). *A theory of human and primate evolution.* Clarendon Press, Oxford.

Grun, B. (1982). *The timetables of history: a horizontal linkage of people and events* (based on W. Stein's *Kulturfahrplan*). Simon and Schuster, New York.

Gumperz, J. J. and Levinson, S. C. (eds) (in press). *Rethinking linguistic relativity.* Cambridge Univerisity Press.

Halio, M. P. (1990). Student writing: can the machine maim the message? *Academic Computing*, **4**(4), 16–19, 45, 52–3.

Harrold, F. B. (1980). A comparative analysis of Eurasian Paleolithic burials. *World Archaeology*, **12**, 195–211.

Hodder, I. (ed.) (1987). *The archaeology of contextual meanings.* Cambridge University Press.

Hodder, I. (ed.) (1989). *The meanings of things: material culture and symbolic expression.* Unwin Hyman, London.

Howell, F. C., Haèsaerts, P., and de Heinzelin, J. (1987). Depositional environments, archaeological occurrences and hominids from Member E and F of the Shungura Formation (Omo basin, Ethiopia). *Journal of Human Evolution*, **16**, 665–700.

Isaac, G. (1983). Art of indoctrination: book review of *The creative explosion—an inquiry into the origin of art and religion*, by J. E. Pfeiffer, Harper and Row, New York, 1982. *Nature*, **302**, 764–5.

Isaac, G. (1984). The archaeology of human origins: studies of the Lower Pleistocene in east Africa 1971–1981. *Advances in World Archaeology*, **3**, 1–87.

Jakobson, R. and Waugh, L. (1979). *The sound shape of language.* Indiana University Press, Bloomington.

Kay, P. (1977). Language evolution and speech style. In *Sociocultural dimensions of language change*, (ed. B. G. Blount and M. Sanches), pp. 21–33. Academic Press, London.

Kehoe, A. B. (1990). 'In fourteen hundred and ninety-two, Columbus sailed . . .': the primacy of the national myth in US schools. In *The excluded past—archaeology in education* (ed. P. Stone and R. MacKenzie), pp. 201–16. Unwin Hyman, London.

Kibunjia, M., Roche, H., Brown, F. H., and Leakey, R. E. (1992). Pliocene and pleistocene archeological sites west of Lake Turkana, Kenya. *Journal of Human Evolution*, **23**, 431–8.

Knecht, H., Pike-Tay, A., and White, R. (eds) (1993). *Before Lascaux: the complex record of the Early Upper Paleolithic.* CRC Press, Boca Raton, Florida.

Knight, V. J. (1989). Some speculations on Mississippian monsters. In *The southeastern ceremonial complex: artifacts and analysis* (ed. P. Galloway), pp. 205–10. University of Nebraska Press, Lincoln, Nebraska.

Kolb, E. (1989). When women finally got the word. *The New York Times Book Review*, **9 July**, pp. 28–9.

Kroker, A., Kroker, M., and Cook, D. (1989). *Panic encyclopedia: the definitive guide to the postmodern scene.* St Martin's Press, New York.

Larsen, M. T. (1988). Introduction: literacy and social complexity. In *State and society: the emergence and development of social hierarchy and political centralization* (ed. J. Gledhill, B. Bender, and M. T. Larsen), pp. 173–91. Unwin Hyman, London.

Leakey, R. E., Leakey, M. G., and Behrensmeyer, A. K. (1978). The hominid catalogue. In *The fossil hominids and an introduction to their context: Koobi Fora research project* Vol. 1 (ed. M. G. Leakey and R. E. Leakey), pp. 86–90. Clarendon Press, Oxford.

Leroi-Gourhan, A. (1993) *Gesture and speech.* (Translated from the 1964 edition of *Le Geste et la parôle* by A. Bostock Berger.) MIT Press, Cambridge, Mass.

Lieberman, P. (1984). *The biology and evolution of language.* Harvard University Press, Cambridge, Mass.

Lieberman, P. (1988). Some of the brain mechanisms for human language and thinking: neuroanatomy, behavior and evolution. *Program and abstracts of the Fourth Meeting of the Language Origins Society*, Cortona, Italy, July 8–22, 1988, pp. 23–6. (abstract.)

Lieberman, P., Laitman, J. T., Reidenberg, J. S., and Gannon, P. J. (1992). The anatomy, physiology, acoustics and perception of speech: essential elements in analysis of the evolution of human speech. *Journal of Human Evolution*, **23**, 447–67.

Lindly, J. M. and Clark, G. A. (1990). Symbolism and modern human origins. *Current Anthropology*, **31**, 233–61.

Linenthal, E. T. (1990). Symbolic warfare on America's battlefields. *Cultural Resources Management Bulletin (US National Park Service)*, **13**(3), 7–8.

McDermott, F., Grün, R., Stringer, C. B., and Hawkesworth, C. J. (1993). Mass-spectrometric U-series dates for Israeli Neanderthal/early modern hominid sites. *Nature*, **363**, 252–5.

Meiklejohn, C. (1982). Comments on R. White's 'Rethinking the Middle/Upper Paleolithic transition'. *Current Anthropology*, **23**, 183–4.

Mellars, P. (1989). Technological changes across the Middle–Upper Palaeolithic transition: economic, social and cognitive perspectives. In *The human revolution: behavioural and biological perspectives on the origins of modern humans* (ed. P. Mellars and C. Stringer), pp. 339–65. Princeton University Press.

Mohr, J. P., Pessin, M. S., Finkelstein, S., Funkenstein, H. H., Duncan, G. W., and Davis, K. R. (1978). Broca aphasia: pathologic and clinical. *Neurology*, **28**, 311–24.

Noble, W. and Davidson, I. (1991). The evolutionary emergence of modern human behavior: language and its archaeology. *Man* (NS), **26**, 223–53.

Passingham, R. E. (1982). *The human primate*. Freeman, San Francisco.

Petersen, S. E., Fox, P. T., Posner, M. I., Mintum, M., and Raichle, M. E. (1988). Positron emission tomographic studies of the cortical anatomy of single word processing. *Nature*, **331**, 585–9.

Petersen, S. E., Fox, P. T., Posner, M. I., Mintum, M., and Raichle, M. E. (1989). Positron emission tomographic studies of the processing of single words. *Journal of Cognitive Neuroscience*, **1**, 153–70.

Pfeiffer, J. E. (1982). *The creative explosion—an inquiry into the origin of art and religion*. Harper and Row, New York.

Phillips, P. and Brown, J. A. (1978 and 1984). *Pre-Columbian shell engravings: From the Craig Mound at Spiro, Oklahoma*, Vols 3 and 5. Peabody Museum Press, Cambridge, Mass.

Poizner, H., Klima, E. S., and Bellugi, U. (1987). *What the hands reveal about the brain*. MIT Press, Cambridge, Mass.

Price, T. D. and Brown J. A. (1985). Aspects of hunter–gatherer complexity. In *Prehistoric hunter–gatherers: the emergence of cultural complexity* (ed. T. D. Price and J. A. Brown), pp. 3–20. Academic Press, London.

Straus, L. G. (1982). Comments on R. White's 'Rethinking the Middle/Upper Paleolithic transition'. *Current Anthropology*, **23**, 185–6.

Toth, N. (1985). The Oldowan reassessed: a close look at early stone artifacts. *Journal of Archaeological Science*, **12**, 101–20.

Trinkaus, E. (1984). Neandertal pubic morphology and gestation length. *Current Anthropology*, **25**, 509–14.

Trinkaus, E. and Thompson, D. D. (1987). Femoral diaphyseal histomorphometric age determinations for the Shanidar 3, 4, 5, and 6 Neandertals and Neandertal longevity. *American Journal of Physical Anthropology*, **72**, 123–9.

Vinnicombe, P. (1976). *People of the eland: rock paintings of the Drakensberg Bushmen as a reflection of their life and thought*. University of Natal Press, Pietermaritzburg.

Walker, A. (1993). Perspectives on the Nariokotome discovery. In *The Nariokotome* Homo erectus *skeleton* (ed. A. Walker and R. E. Leakey), pp. 411–30. Harvard University Press, Cambridge, Mass.

White, R. (1982). Rethinking the Middle/Upper Paleolithic transition: plus Comments by N. Arts, etc., and replies by White. *Current Anthropology*, **23**, 169–92; 238–40; 355–9.

White, R. (1989). Production complexity and standardization in early Aurignacian bead and pendant manufacture: evolutionary implications. In *The human revolution: behavioural and biological perspectives on the origins of modern humans* (ed. P. Mellars and C. Stringer), pp. 366–90. Princeton University Press.

Wood, B. (1992). Origin and evolution of the genus *Homo*. *Nature*, **355**, 783–90.

Wreschner, E. E. (1985). Evidence and interpretation of red ochre in the early prehistoric sequences. In *Hominid evolution: past, present and future* (ed. P. V. Tobias), pp. 387–94. Alan R. Liss, New York.

Wright, H. T. (1986). The evolution of civilizations. In *American archaeology past and future* (ed. D. J. Meltzer, D. D. Fowler, and J. A. Sobloff), pp. 323–65. Smithsonian Institution Press, Washington, DC.

Zilsel, E. (1942). The sociological roots of science. *The American Journal of Sociology*, **47**, 544–62.

Time charts

TIME CHART I

Hypotheses of Gestural Primacy

Australopithecines small brained bi-pedal early hominids

Australopithecines' basicranium *(and inferred vocal tract)* like that of an ape

Australopithecine hands (Hadar): mosaic morphology capable of strong hook-like grasp (but no evidence of knuckle-walking) combined with clear departures from the extant ape pattern approaching the modern human morphology facilitating a greater variety and more firm and controlled grips of objects by the thumb, index, and third fingers

Oldest Confirmed Stone Artefacts: flaked lava cobbles and smashed quartz pebbles in East Africa (Isaac 1984; Howell *et al.* 1987 Kibunjia *et al.* 1992)

4   3

Millions of years ago

Time Chart I

Hypothesis of Incipient Enlargement of the Vocal Tract in Early *Homo erectus* Providing for a Greater Range of Sounds and More Complex Intonation Patterns

Hypothesis of Emergence of Human Vocalization in Sound Symbolism

Hypothesis of Obligate Terrestriality in Early *Homo erectus* Neccessitating Amplified Co-evolution of the Caregiver-Infant Communication System

Habitual Carrying: consilience of evidence from morphology of the hand and bipedality plus transport and simple curation of stone artefacts implies that ordinarily carrying was part of the behavioral repertoire of at least some early hominids

DNA Research Indicates an African Origin for Modern Humans

Oldowan Industrial Complex: marginally curated pre-Acheulean stone-tool technology; expedient stone-tools produced by ad hoc smashing of pebbles and flaking of cobbles; *absence of stylistic norms (Toth, 1985)*

Stone artefact analyses suggests early tool-making hominids were preferentially right-handed

Homo habilis specimen ER1470 displays the first clear increase in hominid brain size and a strong pattern of cerebral hemispheric asymmetry, plus a somewhat more modern-human-like frontal region containing Broca's area

Early *Homo erectus* specimen ER 3733 shows some shift has occurred toward a modern-human-like morphology of the basicranium, *and by inference the earliest example of a more human-like vocal apparatus*

Fully modern Basicranium (*and inferred vocal apparatus*) in early *Homo sapiens* before the emergence of *H. sapiens sapiens*

Olduvai hand: retains capacity for strong flexion but notably more modern-like than Hadar hands in the structure of the thumb and fingertips

Archaic *Homo sapiens* brain size and hemispheric asymmentry well within modern range of variation

Acheulean Industrial Complex Begins: hallmark = large flakes retouched to make large bifacial forms, esp. hand axes and cleavers; some later assemblages include highly standardized and symmetrical forms (Isaac 1984; Asfaw et al. 1992)

2		1		0

Millions of years ago

TIME CHART II

Before Forty Thousand Years Ago the Social and Subsistence Organization (i.e. the Mode of Production) Was Not Modern

Consilience of evidence from stone tools & burial indicates possible beginnings of modern forms of material-culture-symbolism in the later half of MSA Times

First substantiated human burials and the use of red ochre: Mousterian cultures of W. Europe and Near East. (Wreschner 1985)

The Middle Stone Age shows greater regional variety in stone-flake shapes but these lithic tools do not show the morphological volatility typical of Late Stone Age stone-tool high cultures

|100|90|80|70|60|

Thousands of years ago

Time Chart II

Socio-Cultural Systems Emerge
in the Upper Paleolithic
Similar in Many Ways to Those of Historic
Hunting and Gathering People

Hypothesis of
Maximum Phememic
Inventory
Created During the
Upper Paleolithic

Hypothesis of
Development of the Phonemic Principle
During the
Upper Paleolithic

Tanzania, Africa:
Rock painting
tradition may begin
this early
(Anati in
Davis 1986)

Aurignacian Images

Development of
agriculture,
urbanism
and the
political state

Portable
stone fragments
with painted
animals,
Namibia, Africa

Early manifestations
of modern-human-type
stylistic variation
in stone tools

|50|40|30|20|10|0|

Thousands of years ago

TIME CHART III

Larger Population Groups Symbolically Differentiated on a Regional to Subregional Scale, with Logistically Organized Hunting and Collecting Modes of Production; Increased Information Density, Including Anticipatory Scheduling of Subsistence

Hypotheses of
Pre-writing, Pre-arithmetic
Forms of Notation,
Metonymy and Metaphor
In the Ice Age Art and Ornament
of the Upper Palaeolithic

Adornment:
perforated animal-tooth
pendants

Presence of exotic materials from distant sources is frequent
and implies long-distance contacts

Dolni Vestonice,
Czech Republic

Vogelherd, Germany

Willendorf,
Austria

Grimaldi, Italy

Brassempouy,
France

Aurignacian Images:
ivory figurines
of animals and humans

Gravettian Images:
small terracotta figurines
of animals and humans

40 — 30

Thousand years ago

Time Chart III

Hypothesis of
Innovation of
Dramatic Ritual Mnemonic Devices
(Esp. Deep-Cave Art)
to Establish Authority and
Encode More Complex Cultural Programmes
in the Upper Palaeolithic

Hypothesis of
Emergence of the
Phonemic Principle
During the
Neolithic

Increased Need
for Information Flow
Results in Reworkings of the
Dominant Grammatical Patterns
to Reduce the Need
for Inference
in Communication

Lascaux, France

Altamira, Spain

Development of writing, mathematics, and the
commincation technologies

Lascaux
cave paintings

Altamira
cave paintings

Early manifestations
of modern-human-type
stylistic variation
in stone tools

Natufian
proto-neolithic

---------------- Metal Ages ----------------

---------------- Neolithic ----------------

20   10   0
     End of   Holocene
     Pleistocene

Thousand years ago

TIME CHART IV

Socio-Cultural Selection Pressures,
Weak Throughout Most
of the Pleistocene,
Begin to Take Off
In the Increasingly
Complex Societies of the Holocene

Coping with and Creating
Higher Population Densities
and Greater Pluralism
(Scalar Stress)
by Increasing Symbolically
Marked Behaviour,
Particularly Ritual Behaviour,
and Material Culture

Functional
Diversification of Speech
Community Requiring
Development of More
Autonomous Speech Forms
(More Precise and
Explicit Forms)
(Kay 1977)

Specialists produce finer
lexical distinctions and
they also produce more
abstract terms with which
they communicate their
specialized knowledge

Height of
Sumerian civilization.
Sumerian writing,
done on clay tablets,
shows c. 2000
pictographic signs

Earliest cities
in Mesopotamia

Egyptian calendar,
regulated by sun and moon:
360 days, 12 months of
30 days each

Earliest known numerals
in Egypt

Early village life
plus
domestication of plants
and animals

|10|9|8|7|6|

Thousand years before present (BP)

Time Chart IV

Societies of Intermediate Complexity
(Early States and Other Tributary
Modes of Production)
Emphasize Kin Selection,
Based Upon Status Differences
(incl. Despotism),
Ritualization of High Status,
and
Afterlife Preparations

Complexity of Modern Symbolism the Result of the
Development of New Principles
Plus the Reworking and Multilayering of Previous Systems

Coinage
becomes common
as legal tender

Music
is a part of
daily life

Indo-European
begins to
differentiate

'Graffiti' by Greek soldiers
in Nubia *suggest good
basic education*

Development of the
written vernacular
in medieval Japan and Europe
(Kolb 1989)

Swift development
of printing:
Europe, AD 1450–1550
1000+ printing offices opened
producing *c.*35 000
books with *c.*10 million
copies

Shang Dynasty:
first of seven periods of
Chinese literature.
Height of sun
in relation to incline of
polar axis
measured in China

Chinese script
fully developed.
Chinese textbook of
mathematics includes
planimetry, proportions,
'rule of 3' arithmetic,
root multiplication,
geometry, and equations with
one and more unknown
quantities

Chinese encyclopaedia
of 1000 chapters.
Civil Service entrance exams
introduced in China.
Book printing in China

Sumerian cuneiform writing
reduces pictographs
still in use to *c.* 550.
Sumerian poetry, lamenting
death of Dumuzi,
the shepherd god.
Sumerian numerical system
based upon multiples
of 6 and 12

First terrestrial globe
constructed in Nuremberg
by Martin Behaim

18th Dynasty brings Egypt
to the height
of its pre-Hellenistic
power and achievements

Hebrew alphabet developed.
Beginning of Hebrew
literature,
Song of Deborah

First book on algebra —
Diophantus of Alexandria.
Ptolemy draws maps
of 26 countries

Invention of logarithms,
development of
analytical geometry
and probability theory,
adoption of common symbols
in mathematics and
systematic use of the
decimal point

First libraries in Egypt.
In Egyptian literature,
lamentations and scepticism
about meaning of life

Oldest musical notation,
in cuneiform
(Babylon).
Code of Hammurabi
includes guidelines for
medical practices
and permissible fees

King Assurbanipal's library
contains 22 000 clay
tablets: covers history,
medicine, astronomy,
and astrology

Arabs discover library
at Alexandria
with *c.*300 000 scrolls.
Jabir founds chemistry
as distinct from alchemy.
Astrolabe
perfected by Arabs

First regular postal service

First human sets foot
on the moon

Epic poetry in Babylonia
celebrates re-creation
of the world.
Map of Babylonia

Beginning of Semitic
alphabet

Egyptians begin canal
between Nile and Red Sea.
Eratosthenes estimates
earth's circumference.
Heron, the mathematician,
founds first College of
Technology at Alexandria.
Rosetta Stone

Decimal system in India.
Arabic numerals derived
from Indian.
Arabic numerals brought
to Europe

Second Edition
Oxford English Dictionary
defines 616 500
words and terms,
using nearly 60 million
words

Beginning of systematic
astronomical observations
in Egypt, Babylonia, India,
and China

Library at Hittite capital
has tablets in 8 languages

Kikleuli of Mitanni writes first
treatise on
horse-breeding and training

Development of systematic
musical notation

|5|4|3|2|1|0|

Thousand years before present (BP)

Index

Note: *passim* denotes where the subject matter is not in one continuous piece. *v.* denotes versus

Aborigines, Australian, *see* Australian region
Abri Cellier 343
abstract counting 821–3
abstraction in mathematics 818
abstract thought/concepts
  alphabet being responsible for 808
  in development 414–15
academic achievement, Bernstein's theory concerning social class/language and 223–5
accommodation and assimilation, interplay between 494
Acheulean
  bifaces 268–75
  hand-axes 864
acoustic perception, *see* hearing
action
  interpersonal, *see* interaction
  meaning and 488
  means and end in organization of, *see* means and end
  structure/organization of language and, parallels between 379–82, 379
  understanding by ape of 661
actives in language, acquisition 722, 724
Adam, language invented by 573
adaptation in ontogenesis 400–5
addressing a person
  grammar and 779–80
  rules 211
adornments/ornaments, body as prop for 204, 230–1
  intrasocietal variations 213–14
a-dualism, original 492–3
adverbials, time 564
Aegyptopithecus sp. 33
affectional bonds 445
affixation 767, 768
affordances, constituents of niche as set of 186
Africa, human origin in (mitochondrial Eve; out-of-Africa hypothesis) 3, 43, 49, 60–6, 70
  when and where of 64–5
  who and how of 66
African languages, classification/reconstruction 757
Afro-Asiatic (AA) language family 533, 536

velar/laryngeal spirant values 763
agonist use of tools, primates 264, 266, 268
agriculture, *see* farming
airways, upper (vocal tract), evolution 116–25, 477
Akkadian writing 817
alarm calls 583
Algonquian, proto- 527, 528
alleles, phylogenetic relationships and frequency of 66–7
allometric studies of behaviour and anatomic variables (brain size etc.) 80–1
allophemes 769–70, 771
alphabetic scripts/systems 800, 801, 817
  Greek 807–8
  pinyin (Chinese) 804
  reading 803
  transition from syllabic to 802
  Ugaritic 817
Altaic language family
  map 533, 538
  statistics/countries/members 549
Altamira, cave art 292, 322, 323, 326
  part human/part animal 327
  syntactic formulae/mythograms and 336
American Sign Language 648–9
  French and, relationship between 554
  pronoun system, acquisition 713
Americas/New World
  migration of humans to 66
  monkeys, taxonomy 169
  symbolism 870
  see also Central America; Latin America; North America; Pre-Columbian America; South America; United States
Amerind language family
  map 533, 545–7
  North/Central America 546, 550
  proto-language 758, 760, 783
  reconstruction 758
  South America, *see* South American language family
  statistics/countries/members 550
analogy
  apes dealing with problems of 615, 661–2

homology *v.*, in parallels between action and organization of language 379–82
language and 769–70
  Palaeolithic 771, 772
  as system 749–51
language as classification by 748–9
analytic cognitive style 229
analytic language 219, 791
analytic skills, transmission 227
analytic truths, notions expressing 485–6
animals/fauna
  behavioural foundation of human skills found in 418–21
  brain maturation patterns, comparative 410–12
  cognition 596–685
    evolutionary implications of studies 631–3
    historical views 597
    learning 195–6, 601–11
    provisos in interpreting study results 646–7
    same/different judgments 611–15, 633
    uses 601–11
  communication/vocalization/ language, *see* communication; vocalization
  hunting of, *see* hunting
  images/paintings of, contemporary (rock art) 351, 354
  images/paintings of, Palaeolithic 291, 295, 296, 298
    anatomical and ethological understanding 295, 318–19
    examples 313–43 *passim*
    Leroi-Gourhan/structuralist analysis 297, 298, 301
    part human/part animal (anthropomorphism) 296, 298–9, 327, 328, 329
    spear wounds 331, 335
    techniques/conventions/media 294
  memory 597–9
  society 179–85, 572
    human *v.* 180
  tool use 264, 265, 266, 284, 621–2
    as behavioural foundation of human skills 418
  see also individual genera/species

ant(s), farming 187
anteriority 787
Anthropoidea, learning in 602–3
  observational 618
anthropomorphism in Palaeolithic art 296, 298–9, 327, 328, 329
antler, implements made from 295, 317
apes
  behavioural foundation of human skills found in 418–21
  cognition 596
    learning abilities 602–3, 604–11 *passim*
    numerical skills, see numbers
    pre-operational intelligence 620–1
    provisos in interpreting study results 646–7
    same/different concepts 614, 621
    symmetry and transitivity of conditional relations 616
    theory of mind 169, 624–30
  communication 386, 644–85
    gestural 387, 420, 586, 645
    linguistic/vocal, see subheading below
    referential 387
  deception 173–4
  drawings/art 505, 658
  evolution 31–2, 32–6
    evolutionary trees 53–73
  hands 128–30
    compared with humans 132
  language (and vocal communication) 386, 420, 576, 582, 644–85
    changing interpretations of study results 665–6
    historical studies 645–6
    methodological problems 662–5
    provisos in interpreting study results 646–7
    theoretical issues 647–50
    upgraded mind in language-trained apes 632
  symbolic abilities/behaviour 174–5, 650–62, 667
  taxonomy/classification 31–2, 32–6, 169–70
  tool use 264, 265, 267–8, 419–20
  see also primates *and genera/species*
appropriation of natural resources 190
arcane symbolism 870
archaeological evidence
  evolutionary tree dating 65–6
  of sociality 239–62
    methods used 240–1
arithmetic, see mathematics; numbers/numerical systems
Arnhem Land rock art 352
  chronology 358
art (drawings/paintings/sculptures etc.), contemporary
  ape 505, 658
  children's, see children
  hunter–gatherer rock 350–68

and palaeolithic art, significance of former in understanding the latter 363–7
art (drawings/paintings/sculptures etc.), evolution (in general), child art and 512–18
art (drawings/paintings/sculptures etc.), Palaeolithic (cave paintings; figures etc.) 288–349, 797
  as adaptive 303–6
  biocultural evolution and studies of 309–11
  chronology 293–4, 314–15, 338, 358
  concerns about evidence 292–4
  defining 311
  diversity 292–3
  in intellectual context 289–92
  interpretations 299–309
    alternative 302–7
    assessment 307–9
    contemporary hunter–gatherer art (drawings/paintings/sculptures etc.) aiding 363–7
    cultural nature 309, 343
    major 299–302
  preservation 292–3
  retouched/re-used 306–7, 343
  subject matter 296–9
  techniques/conventions/media 294–6
  what we call 292–9
artefacts
  canonical rules governing use 403–4
  harnessed to intentional design 188
articulation and meaning, analogy of, see analogy
artificial environment, making and 187–8
artificial selection (in social evolution) 197–9
a-sociality of infant 493
aspiration, stop 767
assertion *v.* compliance socialization 216–17
assimilation and accommodation, interplay between 494
associations, primate understanding of 615, 649–50, 653
attachment (to figure) 435, 442–6
  primate parallels 446
attention, co-ordination of 449–52
attic children 689
attribution
  of cognition to primates 632
  of mental states to others by primates 169, 624–30
audition, see hearing
Aurignacian tools 280, 281
auroch 313, 322, 336–7
Austin (chimp)
  language studies 654, 655, 656, 658, 673–6
  self-recognition 625
Australian language family
  map 533, 543

statistics/countries/members 550
Australian region
  Aborigines
    art 351–7, 799
    vocabulary differences 209–10
  colonization 65–6
Australopithecus spp. 31, 36–41, 770, 862
  basicrania 120, 122
  bipedality 41, 138–9
  cerebral hemisphere asymmetries 95
  tools 42, 770
  vocal (upper respiratory) tract anatomy 118, 120
Australopithecus aethiopicus, cranial volume 112
Australopithecus afarensis 37–8
  A. africanus compared with 39
  brain size/volume 87, 88, 95, 108, 112, 113
  Hadar, see Hadar *A. afarensis*
  hands and locomotion 138–9
Australopithecus africanus 38–40
  A. robustus compared with 40
  brain
    organization 91
    size/volume 88, 95, 108–9, 112, 113
  hands and locomotion 139
  skull 4–5, 38
    base (basicrania) 122
Australopithecus boisei 40
  brain volume 88, 109, 112, 113
  skull 6–8
    base (basicrania) 122
Australopithecus boisei skull, *H. erectus* and, comparisons 16
Australopithecus robustus 40
  brain volume 88, 109, 112, 113
Austroasiatic language family
  map 533, 539
  statistics/countries/members 549
Austronesian language family
  map 533, 542
  statistics/countries/members 549
autism 377
  representational abilities 629
autonomy, language acquisition and 687, 689–90, 697–8, 729
Aztecs 871–2

babbling, transition to words from 699, 701
babies, see infants; neonates
backward associations/relationships 615, 649–50, 657
Bakhtin M, linguistic and semiotic theory 488–9, 498
barbarism 223
Barthes R, semiotic theory 489
basicranium of fossil hominids, upper respiratory tract anatomy and 119–23, 124
Bates E, on language acquisition 382, 712, 725, 727

beavers, dam building 187
behaviour
 animal, as behavioural foundation
  of human skills 418–21
 brain and
  sex differences 99–100, 101–2
  size of 80
 chimpanzee, ethological indices of
  664–5
 insect 181, 184
 interspecies differences in 79–80
 intraspecies differences in 79
 learned 193
  culture as 192
 learning/development, palaeomorphic
  metaphor 401–3
 locomotor, *see* locomotion
 ontogeny and phylogeny of 377–94
  relating 386–94
 products of, resulting from past
  hominid activity 87
 social, *see* social behaviour; social
  systems/socio-cultural systems
 symbolic, evolution 263–87, 371–99,
  667, 728–9
 verbal, language as reinforced 648
 verbal regulation of 475–6
behaviourist theory of language
 acquisition 703
behaviour rituals, children's exposure
 to 450
beliefs 377, 629
Bernstein B, on theory of
 language/social class/academic
 achievement 223–5, 228–9
Bible, glottogonic thought in 572, 573
bifacial core tools
 1.0m years ago 268–71
 300000 years ago 271, 272
biogenetic law of Haeckel 407–8, 476
biology and history, distinction 159
bipedality 126–54
 fossil hominoids/hominids 135–41,
  144–5, 771
 Australopithecus 41, 138–9
 tool use and 134–5
birds, tool use 265
bison 339
 Altamira 322, 323
 Dolni Vestonice 317
 Les Trois Frères, part human and
  328
 Le Tuc 294, 317
Blanchard plaquette 341
boar, Altamira 322, 336
body
 adorned, *see* adornments
 size/weight, brain size and 44–5, 80–1
 unadorned 205–6
bones
 carving on
  La Vache 334
  notching, counting system 306,
   341, 822

human, hand 129
 implements made from 295, 316, 317
 orang-utan, hand 128
bonobo, *see* pygmy chimpanzee
book(s)
 printing and publishing, *see* printing
  and publishing
 translations 836, 841
book-keeping, early 814
bootstrapping, semantic (hypothesis)
 719
boundaries, formal, degree of 240–1
Boysen ST, ape-language/numeracy
 studies 660, 660–1, 673–6
brain 74–125, 407–31
 asymmetries/laterality, *see* cerebral
  hemispheres
 evolution of 74–125, 407–31, 862
  lines of evidence regarding 85–7
  quantity v. reorganization in
   409–10
 function, cognitive style and 229
 language and the 666, 699
  brain injury studies 377, 379,
   874–5
  evolution of 577–8, 580–1
 maps 78
 maturation patterns, comparative
  410–12
 ontogeny/development 377–82,
  407–31
  disorders 377
  genetic control 79
 size/volume (cranial
  volume/capacity) 77–8, 84–5,
  87–90, 95, 96
  absolute 87–9
  behaviour and 80
  H. erectus 44–5, 88, 109–10, 112,
   113, 583
  see also palaeoneurology
 structure/organization 76–85, 90–7
  changes in development 411
  changes in evolution 96, 409–10
  hierarchical 77
  sex differences 98–100, 101–2
  tool-use and the 135
brain-stem and cognitive maturation
 412–17
branches/branching points of
 evolutionary trees 57–8, 62
breathing, larynx and 118
British Sign Language 558
Broca's area
 hominid endocast studies 91–3,
  862
 lesions 380, 873–4
 role 379, 380, 873–4
Brodmann's areas 82, 83
Broken Hill man, skull 21
Bronze age language 772
brother-in-law language 210
brutality 223
Bühler C, infant language 692–6

burial practices, Middle/Upper
 Palaeolithic compared 866–7
Bushmen rock art, Southern Africa
 357–63, 364–6, 367

calcarine fissure 83
Callitricidae, taxonomy 34, 168, 169
canonical rules governing use of
 artefacts 403–4
capuchin monkey, delayed conditional
 matching task 597
carnivores, social skills 175
carpals of fossils
 Miocene hominoids 136
 Pliocene/Pleistocene hominids 137
cartography 826–33
carvings 313, 797
 chronology 314
 La Vache 334
case-marking in verbs 564
Case R, on brain maturation 413–18
case-relations and topic–comment 785–6
Castillo 326
categorical perception 600
categorization
 in animals 610–11
  ape-language studies and 654–5
 in humans, language and its effects
  on 475
Caucasian language family, map 533,
 538
causal interaction between ape and
 communicator 663
causality, chimpanzee understanding
 653, 661
causal model of cognition in language
 acquisition, direct 712
cave art 288–349, 797
 chronology 315, 338
  problems 293
 dynamism 313
 entoptic phenomena and 340
 Leroi-Gourhan's ideal layout 339
 see also art
Cebidae, taxonomy 34, 168, 169
Ceboidea
 learning in 602–3
  observational 618
 taxonomy 34, 168
Cebus apella, delayed conditional
 matching task 597
Central America (Mesoamerica),
 language family 546, 550
 reconstruction 758
central nervous system, *see* nervous
 system *and specific components*
central sulcus 83, 85
Cercopithecidae, taxonomy 34, 168
Cercopithecoidea
 learning in 602–3
  observational 618
 taxonomy 34, 168
cerebral cortex
 cognitive maturation and 412–18

cerebral cortex (*Contd.*)
  frontal lobe as most favoured part of 82
  proportion of brain (relative volume) 81–2
  structural/functional subdivisions 77, 78, 83, 96
  tool-use and 135
  see also specific areas
cerebral hemispheres,
  asymmetries/laterality 93–5, 862
  language and 584, 588, 666, 862
  non-human 666
Chantek (orang-utan) 670–2
Chatelperronian 311–12
children
  attic/dungeon 689
  cognitive development 383, 393–4, 412–13, 472–6
    parent–child interactions and, *see* parent
  counting/number concepts 384
  drawings/art 501–18
    evolution of art and, parallels drawn between 512–18
    flexible strategies 511
    in Piagetian framework 505–6
    recent experimental approaches to 506–12
  feral 688, 689
  language acquisition, *see* language
  mind of, *see* mind
  numerical ability 384, 820
  parent/caregiver, *see* parent
  rearing practices, food accumulation and 216–17
  social behaviour, *see* social behaviour
  socialization pressures, sex difference 216, 217
  speech of/speech heard by, *see* speech
  thought in, origins, *see* thought
  very young, *see* infants; neonates
  vocal tract 117–18
chimpanzee, common (*Pan troglodytes*) (and its evolutionary relationship to humans) 53, 59, 70
  'art' 505
  basicranial line, ontogenetic development 121
  brain 83, 84
    comparative neuroanatomy 86, 93
    growth rate 410
    size/volume 113, 114
  cognition 596
    discrimination learning 606
    memory 598, 599
    mirror-use/self-recognition 174, 622
    observational learning 618
    problem-solving 621–2, 623
    representational reasoning 616
    same/different concepts 613
    self-recognition 624
    sensorimotor intelligence 619
    social 446

  theory of mind 624–30
  communication 172
    gestural/referential 387
    linguistic/vocal, *see* subheading below
    deception 173
    divergence dates 57, 60, 65
    ethological indices of behaviour 664–5
    hand 129
      compared to human 132
    language (and vocal communication) 174, 386, 395, 579, 644–85 *passim*
      methodological problems 661–5
      relationships in 446
    tool use 264
chimpanzee, pygmy, *see* pygmy chimpanzee
Chinese language
  problem-solving and English language *v.* 220–2, 390
  writing 799, 804
    function 799
    function/processes 799
    logographic 796, 800–1, 801–2
    printing 810
    syllabic 802
Chomskyan theory
  linguistics 498, 557, 566–7, 579–82, 688
  ape 647
  parameter-setting 454
chopper, Oldowan 266, 267
choreography 826
chronology of art
  Arnhem Land rock art 358
  Palaeolithic 293–4, 314–15, 338, 358
Chukchi–Kamchatkan language family
  map 533, 539
  statistics/countries/members 549
cipherized number systems 822
civilization, 'superorganic' as synonymous with 181
Clark's semantic feature hypothesis 707–8
classification of primates/hominids, *see* taxonomy
claviform sign 326
'Clever Hans' phenomenon 662–3
climate, *see* neoclimatic change hypothesis; temperate climate
clothing and social status 208
clubbing 145
clustering effects in memory experiments 598–9
cognition/intellect 225–9, 377–82, 432–68, 596–685
  brain injury studies 377, 379, 874–5
  comparative perspectives (animals and humans) 596–685
    evolutionary implications 631–3
    provisos in interpreting study results 646–7
    tests used 631–2

  development/ontogeny 377–86, 407–31, 412–18, 432–68, 472–6
  children, *see* children
  Freud and theories of 492
  language acquisition and 472–5, 477, 479–80, 690–1, 696–7, 712, 728–9
  major stages 414–15
  Piagetian analysis, *see* Piagetian analysis
  scaffolding 226, 389–94
  social interaction and 456–7
  dissociable systems 377–82
  evolution/phylogeny 383–6, 421–5, 627–8, 631–3
  externalized 157, 865, 871
  literacy and its effects on 794, 804–6
  Palaeolithic art in context of 289–92
cognitive model/view
  of attachment figure 445
  of phonological acquisition 703
cognitive organization, parallel distributed processing model of 721
cognitive representation, *see* mental representation
cognitive style 227–8, 229
  analytic 229
  brain function and 229
  cross-cultural social practices and 227–8
  relational 229
Colobidae, taxonomy 34
colour, images using 295
colour-term differentiation 219–20
combinatorial skills in development 379–82, 480
  with words 414, 415, 420, 421, 691, 718–20, 734
  apes 656
comment, *see* topic–comment
communication 204–35, 432–82
  animals
    non-primate/in general, historical ideas 573, 574–5, 577–8, 580–1, 645–6
    primate, *see* primates
  competence, language acquisition and 209, 691
  of desires, young children 423
  function in early childhood 726
  infant stage before 730
  of knowledge, stage in development termed 424
  language evolution and its use for 524
  see also language
  middle/upper Palaeolithic compared 866–7
  non-verbal/linguistic, *see* non-verbal communication; pre-linguistic communication
  ontogeny 432–68, 469–82
  self and culture and 222–4

social structures putting pressure on
systems of 229–30
see also interaction; relationships
communication code 224–5
communication technology, writing as
806–7
complement clauses, acquisition 722,
723
compliance
assertion *v.*, in assessing
socialization practices 216–17
dialogue eliciting 779–80
comprehension in language
development 388
production *v.* 649–50
computational component in language
evolution 524
concepts (ability to form)
animals 610–15
same/different 611–15, 633
symbolic 614
human
infant, words and 713–16
language evolution and 524
concrete counting 821
concrete operational stage of
development (Piagetian) 494
Renaissance art and, parallels
between 513
condensation (dreams) 491
conditional relations
symmetry 615–16, 649
transitivity of 615–16
conditioning studies, primates 602,
604
consciousness
Aboriginal, art and its relevance to
352–4
evolution 192
sociality and 184–5
conservatism in speech, sex differences
210
consolidation in mathematical
evolution 823
consonant(s)
chart 749
juxtaposition 767
lost vowel-colouring, laryngeal
quality of 756, 763, 767–8
Proto-Amerind 758, 760
social group differences in
pronunciation 211
stop, *see* stop consonants
voicing 767
consonantal phonemes 764
English, acquisition and customary
production 702
Hawaiian 750
Yokuts 750
consonant–[vowel]–consonant pattern
767–8
constitution of persons 183–4
constitutive social relations 184,
185

construal stage of trance-induced
visions 360–1
construction
of artificial environment 187–8
of cognitive structures 437
in language 523–5
in acquisition 382
contrastive hypothesis (contrast
theory) 709
convolutions (of brain surface) 82–4
third inferior frontal 91–3, 862, 873
co-optation 188
Coptic language 767
copulas in sign language 567
copying, imitative, *see* imitation
core tools
1.8m years ago 265–6
bifacial, *see* bifacial core tools
corpus callosum, sexual dimorphism
98–100
cortex of brain, *see* cerebral cortex
counter-factuality and English/Chinese
languages 221
counting, *see* numbers/numerical
systems
craft-specialization-and-conflict
hypotheses (growth of
urbanism) 250
cranium, *see* skull
cranks, compound, mechanics
misunderstood 389
creation myths, Aboriginal 352–4
creoles 553, 566–7
definition 566
Cro-Magnons
art 290–1
skull 26–8, 30
cross-modal recognition, primates 603,
610
cross-modal transfer of learning in
primates 603
crying infant 439
crypto-biologist, Freud as 490
cubism and formal operational stage
of development 513
cueing in ape-language studies,
inadvertent 663
culture 178–203
acquisition in tiny unrelated
snippets 159
agriculture and, origins of 248–9
art/imagery and 303, 303–6
interpretation of 309, 343
child's acquisition of 432–68
cognitive styles and social practices
related to 227–8
complexity, non-verbal
communication and 205–8
definition 191–2
ecological determinants of patterns
214–17
evolution 179, 191–9, 239–52
analogy with organic evolution 196
existence of, mode 181

infant evolution allowing emergence
of 400, 404–5
language and 217, 455–6, 772–3
origins of 587–8
literacy and attitudes of different 809
in mathematical evolution
contact between 823
lag/resistance 823
reproduction 225–9
self and communication and 222–3
social systems and 217, 455–6
society *v.* 181–2
'superorganic' as synonymous with
181
thought and 217
transmission of, mode 181, 191–9
see also social systems/socio-cultural
systems
cuneiform script 799, 817
evolution 814, 815

Daic language family 533, 539
dancing
evolution 826
trance 364–5
Darwin C, and mind of child 489–90
Darwinian ideas
cultural change 196, 197, 198
language origins 575
deaf, sign language, *see* sign language
death, metaphorical images of 362, 364
décalage in Piagetian theory 495
deception, *see* lies
deer
antler, implements made from 295,
317
art/imagery 295, 318–21
deixis 588, 731, 780–1
delayed matching to sample (DTMS)
in animals 598, 633
monkeys 597
deliberation, rational, engaging in 192
demography, social 240
dentition, *see* teeth
depth (dimensional) relationship in
children's drawings 508–10, 511
de Saussure F, linguistic science of
486–7
design features of language 553,
556–7, 648
desires
belief that action will satisfy 629
communication of, in young
children 423
determinatives 817
development, *see* ontogeny
dexterity, hominid, *see* manipulative
facility; tools
diachronic study of language-change
486
dialect 753
map, Indo-European languages 754
dialogue, frame of 779–81, 790
diencephalon 77

differentiation
  in cognitive development 413
  within society 214–15, 229
    colour-term 219–20
    internal, degree of 240, 244
    in upper Palaeolithic 244
digging 144
digits
  of hand, see fingers
  ontogeny 375, 376
dimensions (as stage in child development) 414
  in drawings, representation 508–10, 511
ding-dong theory of language origin 576, 583
direct causal model of cognition in language acquisition 712
directive speech-acts 726–7
discourse 780
  epistemic pattern 787–9, 790
  objectivity in 787–9
  post-syntactic stage in language development and 723
discreteness in language 556
discrimination, by infant with persons encountered 442–3
discrimination learning 604–6
  by primates
    complex 602, 604
    simple 602, 604
displacement, in dreams 491, 492
displacement tests 657, 669
  dolphins 657, 669
display and tool use, primate 264, 266, 268
dissipative social systems 161
dissociability/dissociable systems
  in cognition/language 377–82
  in embryology 372–3
divergence dates for hominids/hominoids 56–60, 68, 69
  total error on 59–60
diversification in counting 823
D-loop sequences 64–5
DNA (evolutionary studies) 48–9, 53–73
  fossil, sequence data 69
  hybridization 59
  mitochondrial, see mitochondrial DNA
  nuclear 66–9
Dolni Vestonice 294, 317
dolphins
  displacement tests 657, 669
  language abilities 651, 657, 677–9
    comprehension v. production 649
  learning abilities 604
Dravidian/Elamo-Dravidian language family
  map 533, 540
  statistics/countries/members 549
drawing, see art; cartography

dreaming
  Freud on 491–2
  The (aboriginal art) 352–4
dryomorphs, hand bones and locomotion 135, 136
dungeon children 689
dying, metaphorical images of 362, 364

e (letter), social group differences in pronunciation 212
echoic responses 648, 668
ecology (human) 179, 185–91
  art and 303–4, 306
  cultural patterns and 214–17
  hunter-gathering and 228
  Maring-speaking people 159–61
  social relations and 185–91
  sociology as branch of, see socioecology
  tool use by hominids 284
    15000 years ago 278–9
    85000 years ago 275
    300000 years ago 271–2
    1.8m years ago 267
ecology (non-human animals), cognition and 631
economic correlates of literacy 808–9
economic elements related to settlement patterns 215
edge lengths, evolution and 60
education, see teaching
egocentrism in children 492
  speech 471, 472, 496, 497
Egyptian language (historic aspects) 757
  reconstruction 757, 763, 764–7, 768
  writing 817
    function 799
    hieroglyphic 816, 817, 822
    origins 797
Ehlich K, developmental account writing 814
Elamo-Dravidian language family, see Dravidian
élargissment 768
electroencephalographic data correlating with Piagetian studies 425
embryological mosaics 372–3
emic factors 159, 161, 162
emotion(s), bodily expression, see expression of emotions
emotional cry model of language origin 576, 583–4
employment and literacy 809–10
  see also work
emu, hunter spearing 259
enaction in pre-symbolic gestural communication 731
encephalization quotients 89–90
endocast studies, see palaeoneurology

energy technology, evolution in hominids 865
  H. erectus 45–6
Eng H, on children's drawings 502–3
English language
  alphabet 801
  consonantal phonemes, acquisition and customary production 702
  plurals, acquisition 698
  problem solving compared with Chinese language 220–2, 390
  stop consonants 750
  written 804
engravings 294–5, 296
  chronology 314
  spear-point 316
entification, English/Chinese languages 221
entoptic phenomena
  Palaeolithic cave art 340
  Southern Africa Bushmen rock art 360–1
environment (organism's) 186–7
  agriculture and the, origins of 247–8
  artificial, making 187–8
  H. erectus ability to modify 45–6
  interactions with organism, see ecology
  see also natural resources
epistemic pattern in discourse 787–9, 790
epistemology, genetic 492
errors in word use/syntax 706–7, 720
ER specimens, see KNM-ER
Eskimo—Aleut language
  map of 533, 543
  statistics/countries/members 549
  suffix-systems 222
Eskimo coastline maps 829
ethological indices of chimpanzee behaviour 664–5
ethological understanding of animals, cave artists 295, 318–19
etic operations 161, 162
etymology of language and its acquisition of words 478–9
Eurasian hominid species 49
Europe (maps)
  art sites (Palaeolithic) 316
  Neanderthals 44
  see also Indo-Hittite/Indo-European language family
Europeans, modern, origin/relationship to other populations 68
Eve, mitochondrial, see Africa
evolution (biological [phylogeny] and sociocultural)
  bipedality, see bipedality
  brain, see brain; cognition
  cartography 826–33
  consciousness 192
  culture, see culture
  dancing 826
  deceptive behaviour 167

definition/meaning vii (preface)
energy technology, *see* energy technology
hand 126–54
heterochronic processes in, *see* heterochronic processes
human 31–154
hypotheses 49
  out-of-Africa, *see* Africa
information technology in Palaeolithic 865
language, *see* language
mathematics 818–24
molecular genetic studies 48–9, 53–73
mosaic 372
music 824–6, 854
ontogeny and, *see* ontogeny
outline 31–52
Palaeolithic art studies and its place in 309–11
social/socio-cultural systems 158, 239–62, 861–76
symbolism/symbolic behaviour 263–87, 371–99, 667, 728–9, 861–76
  tempo and mode of change in 861–76
theory of mind 624–30
tools, *see* tools
vocal apparatus 116–25, 477
writing 793, 794, 797–804, 813–33, 844–55
evolutionary trees 53–73
  dating 56–8
    with archaeological evidence 65–6
  ingroup 65–6, 67–9
  obtaining, basic steps 54–6
  types 56
existential in sign language 567
experiences, sharing of, infant communication and 732
expletive (emotional cry) model of language origin 576, 583–4
exploitative pattern and social stratification 215–16
expression of emotions (bodily) 205–6
  facial, *see* face
  by infant 731
expressive markings, Palaeolithic 797
Eye Direction Detector 451

face
  attractiveness to infants 440
  expressions 205–6
    adult's, infant imitation of 447–9
    infant's 435
false beliefs 629
'family tree' in language reconstruction 752–3
  Uralic language 757
family type and exploitative patterns 215–16

farming/agriculture
  ants 187
  humans, origins 247–8
  see also food
fauna, *see* animals
features (in phonology) 698
  lexical semantic systems and 707–8, 709
feeding, *see* food
feelings, bodily expression, *see* expression
female
  in art 298–9
    genitalia 298
    role depiction 304–5
    signs 331–3
    statuettes 303–4
  differences from male, *see* sex differences
feral children 688, 689
fertility figures 299
fetal eavesdropping 587
finger(s)
  communication using 586
  grip, *see* grip
  index, pointing with, *see* ostension
  ontogeny 375, 376
fire, *H. erectus* ability to control 46
first person verbs, topic–comments and 785
Fischer JL, on Truk and Ponape communication 217–19
fish, Southern Africa Bushmen art 359, 360
flake tools
  1.8m years ago 265–6, 267
  85000 years ago 275, 276
Font-de-Gaume 319
food (and feeding)
  accumulation/procurement of food
    compliance *v.* assertion socialization and 216–17
    linguistic evolution and 423–4
  chimpanzee tool use 264
  sources of food, intensification at end of Pleistocene 244–5
  see also farming; hunting
food-calls, primate 167, 172
  see also carnivores
food-sharing, proto-human hominid 242
foot, ontogeny 375, 376
formal boundaries, degree of 240–1
formal language styles 223–5
formal operational stage of development (Piagetian) 494
  modern art, parallels between 513
formal situations, linguistic differences in informal and 213
fossil(s) (in evolutionary tree studies) 68, 68–9
  in divergence date estimation 59–60, 69
  DNA sequence from 69
fossil hominids and hominoids, *see* hominids; hominoids

Fouts RS, ape-language studies 670–2
France, cartography 830, 832
Freeman NH on children's drawings 507, 508
Frege G, sign theory 485–6
French and American Sign Language, relationship between 554
Freud S
  Ursprache and the unconscious 490–2
  as evolutionist 490–1
frontal convolution, third inferior 91–3, 862, 873
frontal lobe 91–3
  as cerebral cortex's most favoured part 82
function, language, *see* language
functional core hypothesis 708–9
functional learning paradigm, ape-language studies 654

Gablik S, on child development and art-history 513–14, 516–18
games, epoch of 436
  see also play
Gardner RA and BT, ape-language studies 654, 655–6, 670–2
gaze, infant social 440–1
generalization in mathematics 818
  see also overgeneralization
general nativism, theory of 697
generations, culture transmitted through 194
generative grammar, mastering and internalizing 209
generative phonology, nativist theory 703
generic concepts, English/Chinese languages 221
genetic control of brain development 79
  see also biogenetic law; hereditary stress
genetic epistemology 492
genetic relationship between languages 525–7, 551, 756–8
  see also monogenetic reconstruction
genetics, molecular, human ancestry data via 48–9, 53–73
genitalia in art 298
geographical distribution, *see* map
Germanic language family
  divergences 527, 528, 529
  statistics/countries/members 548
Gestalt psychology applied to child art 505
gestation of humans, 'ideal' period of 21 months 383
gestures/gestural communication 571–95, 712–13
  advantages of using 584–5
  definition 571–2

gestures/gestural communication (*Contd.*)
  distribution of use 585–6
  human adult, infant abilities with 435
  in human evolution (human ancestors) 478, 580–91, 781
    historical ideas 573–5, 577–8, 580–1
    language origin based on 580–91
    vocal language accompanying 589–90
  human infant 450, 585, 586, 686, 712–13
    words and 712–13, 729–34
  primate 586
    ape 387, 420, 586, 645
'ghost' images 328
gibbon
  cranial volume 113, 114
  reversal learning 607
glacial periods, *see* ice age
glides 764
globin gene, evolutionary trees 58–9
glottochronology 526, 527, 530
  see also language
gluteus maximus and bipedality 135
Gombrich EH, on child development and art-history 514–15, 517–18
gorilla (*Gorilla gorilla*)
  evolutionary relationship to humans 59
    basicranial line, ontogenetic development 121
    divergence dates 57
    hand 129, 131, 132
    social structure 170
  language skills 670–2
  learning by
    discrimination 606
    reversal 607
  tool use 266
Gothic 752
government, *see* state
grammar 480, 564–5, 650–3, 666, 777–92
  ape use, evidence 647–8, 650–3, 666, 667, 668
  Chinese *v.* English, problem-solving 220–2
  in development 697, 720
    mastering and internalizing generative grammar 209
  patterns 778–9
  prehistory, theoretical stages 777–92
  reconstruction methods employing 762
  semantics and, modular division between 382
  see also structure
graphic notational systems
  Palaeolithic 298
  uses 799
  writing and other, distinction between 796
  see also visual representation
Great Apes, *see* Apes

Greek
  ancient
    cartography 827
    consciousness 222, 231–2
    literacy 807–8
    mathematics 818–19
  relationship to other languages 525, 755–6
Greenberg's theory of marking 755
Grimm's law 752
grips
  A. afarensis (Hadar) 138
  African apes 129
  humans 131–4, 146
Guarani and Spanish in Paraguayans 213
guessing and knowing, chimpanzee ability to determine between 627, 628–9
gyri of brain 83

habitat richness and hunter–gather social organization 245
Hadar (*A. afarensis*)
  hand 138–9, 143
  locomotion 138–9
Haeckel E, biogenetic law and recapitulation hypothesis 407–9, 476–9
hallucination-related imagery
  Bushmen of Southern Africa 360
  Palaeolithic 305–6
hand 126–54
  communication using 586
  gestural, *see* gestures
  development 375, 376
  evolution 126–54
  in symbolic behaviour 145
  see also fingers; manual motor control
handedness in hominoids 135
  language and 584, 588
Haplorhini, classification 168, 169
Hawaiian consonantal phonemes 750
hearing/auditory/acoustic perception 124
  animals 600
  infants 435, 699–701
Hebrew language 573–4
hemispheres, cerebral, *see* cerebral hemispheres
Herman LM, non-primate language cognition projects 677–9
heterochronic processes in evolution 373, 373–7, 408
  of humans 374–7
  terminological confusions 374–5
hiding activity
  children's drawings 510, 511
  primates 173
hierarchical brain organization 77
hierarchical mental capability (integration or construction) 419–21

fossil hominids and 422, 423
hieroglyphics
  Egyptian 816, 817, 822
  Mayan 796
history
  all human life as matter of 159
  of animal cognition studies 597
  biology and, distinction 159
  of language development studies in children 577–8, 580–1, 688–9
  of language origin studies 571–95, 645–6, 777–8
  prehistory and, distinction 159
Hittite, reconstruction/relationships 763, 764–7
  see also Indo-Hittite
Hockett C on design features of language 556, 648
Hohlenstein–Stadel statuette 327
Holocene, early/mid-, sociocultural complexity 868–70
  see also Neolithic
home bases, proto-human hominid 242, 771
hominids, fossil/past 31–154
  activity, products of behaviour resulting from 87
  cognitive evolution 383–6, 421–5
  gestural communication, *see* gestures
  hand 137–42, 143–5
  language evolution, *see* language
  lineage/evolution 31–154
  locomotion 137–42, 144–5
  proto-human, set of characteristics 242
  skulls of
    base 119–23
    brain, *see* brain
    photogallery 3–30
    taxonomy 49–50
  tools, *see* tools
  vocal (upper respiratory) tract
    anatomy 118
    palaeoanthropological inferences about 119–23
  see also palaeoanthropological perspectives *and individual genera/species*
hominoid(s)
  fossil
    hands 135–42, 143
    locomotion 135–42
  living, *see* Hominoidea
Hominoidea (living hominoids)
  cranial volumes 113, 114
  evolutionary trees 53–73
  learning in 602–3
    observational 618
  taxonomy 34, 168
Homo erectus 43–6, 120–3, 863
  in evolutionary tree 60, 61, 65, 69
  hemispheric asymmetries 95
  linguistic capacity 422, 477, 478, 576, 863

skull 14–15, 44
  base (basicrania) 120–3
  capacity/brain volume 44–5, 88, 109–10, 112, 113, 583
  comparisons with *A. boisei* 16–17
  vocal (upper respiratory) tract anatomy 118, 120–3
Homo ergaster 112, 873
Homo habilis 31, 41–3, 771, 862, 873
  brain 862
    size 95–7, 109, 112, 113
    KNM-ER 1470, *see* KNM-ER 1470
  language beginning with 93, 101, 589, 862
  skull 10–13
    base (basicrania) 122
  social model 771
homology *v.* analogy in parallels between action and organization of language 379–82, 379
Homo rudolfensis, 112, 873
Homo sapiens (archaic) 31, 46–9, 864
  skulls 18–30, 46–7
    base 122, 123
    brain volume 88, 110–11, 112
    transition to anatomically modern man 29
  vocal (upper respiratory) tract anatomy 118
  see also human
Homo sapiens neanderthalensis, *see* Neanderthals
Homo sapiens sapiens (anatomically modern humans) 31, 864, 865, 868
  African origin, *see* Africa
  art, *see* art
  cranial volume 111, 112
  evolution, *see* evolution
  hand 130–5
    bones 129
    compared with Great Apes 132
    muscles 130
    tool-use 131–5
  hunting, *see* hunting
  populations, phylogenetic relationship of, estimating 67
  skull 26–8, 31
    base (basicrania) 122
  society 179–85
  see also proto-human hominids
Hopi language 229
horses, portraits 320, 321, 322, 339
  touching-up 343
  see also Clever Hans
human (conceptions of being), cultural relativity 222
  see also body; individual; person
human (form in art)
  children's art 506–7, 507–8
  palaeolithic art 298–9
    chronology 314
    examples 329, 334

part animal and (anthropomorphism) 296, 298–9, 327, 328, 329
human (species), *see Homo sapiens sapiens*
hunting 189
  in art
    in contemporary rock art, *see* art, contemporary
    in Palaeolithic art, marks of killing 335
hunting and gathering people 188
  art
    contemporary 350–68
    Palaeolithic 303
  ecological demands 228
  male/female roles 304–5
  social life and its complexity 241, 244, 245, 869
  end of Pleistocene 244, 245
  tool use
    15000 years ago 278–9
    85000 years ago 275
    300000 years ago 271–2
    1.8m years ago 267
hunting magic, art as 299–300, 312, 335
hydraulic-managerial hypothesis (formation of state) 250
Hylobates lar, cranial volume 113, 114
Hylobatidae, learning in 602–3
  observational 618
hypermorphosis 373, 375, 410

ibex 339
ice ages/glacial periods 47
  cave art 304
iconicity in language 556, 558–62
  arbitrariness *v.* 558
  in sign language 559–61
  in spoken language 558–9
iconic sign 485
Id 491
ideas, primate 167–77
identification, *see* naming
identity (sameness)/difference
  judgments by animals 611–15, 621, 633
ideographic narrative forms 799
illocutionary period 725, 726, 737
image-making, *see* art
imitation 387, 447–9
  by adults 449
  by animals 617, 618, 630–1
  by infants 387, 447–9
implicational framework in communicative development 730, 732–3, 734
Incas 871–2
Index of Differentiation (in society) 214
index finger, pointing with, *see* ostension

indexical sign 485
'indisputably' language 523
individual
  relationships between, *see* interaction; relationships
  speech differences depending on social context 212
  as unit of analysis in evolutionary scenarios 385
  see also person
individual learning 193–5
Indo–Hittite/Indo-European (IE) language family
  map 533, 537
  dialect 754
  proto-language (PIE) 752, 753, 763, 765, 787
  relationships/reconstruction 555, 751–6 *passim*, 782–3
  statistics/countries/members 548–9
Indo–Pacific language family
  map 533, 542
  statistics/countries/members 549
infants
  born 'too early' (ideal gestation of 21 months) 383
  brain, growth 410
  cognitive/intellectual development 412–13, 472–6
  culture and socialization and the role of 400, 404–5
  gestural communication 450, 585, 586, 686, 712–13
  intersubjectivity, *see* intersubjectivity
  linguistic learning 270, 414, 415, 473–4, 686–746
    stages 692–6
  newborn, *see* neonates
  numerical ability 820
  pre-communication stage 730
  social abilities 435–6
    a-sociality 493
  thought 469–82
  vocal communication, *see* vocalization
  vocal tract 117
  see also children
inferential transitivity/transitive inferences in animals 615
  chimpanzees 616, 661
informal *v.* formal situations, linguistic differences in 213
information, coding/storing
  verbal 601
  written 796
information-processing systems in humans, maturational components 384
information technology, Palaeolithic evolution 865
ingroup dating of evolutionary trees 65–6, 67–9
innate emotions 206
innate expressions 205–6

innovations
  cultural 196, 197
  linguistic, social interaction and 454–5
  see also technology
insects (social/co-operative behaviour) 181, 184
  see also ants
insight
  human, naming 711
  primate 621–2
institutionalized language contexts 213
instrumental communication, infants 473
intelligence/intellect
  hominid tool use and 284
    85000 years ago 277
    300000 years ago 272–3
  language development and 472–5, 477, 479–80, 690–1, 696–7, 728–9
  primate 168–70
    discrimination learning set as measure of 607
    representational intelligence 611, 620–1
    sensorimotor intelligence 619–20
  thought development and importance of 472–5, 477
  see also realism, intellectual
intensification (food sources), shift, at end of Pleistocene 244–5
intentionality 436, 478, 629
  as component in production 190–1
  development/ontogeny 629, 726–7, 730–3
  grammatical function and 778, 779
interaction(s) (and interactive social systems/relations) 161, 182–3, 184, 188, 402, 432–68, 496
  parent–child, see parent
  relationship and, distinction between 183
interactionism/interactive approach in language acquisition 382, 691–6
internalization, Vygotskian theory 402, 497
internal working model in cognitive psychology 43
interpersonal transaction, see interaction
interregional-and-intraregional-exchange hypotheses (of urban development) 250, 251
interrogative sentences 727
intersubjectivity (infant) 434–6, 472
  primary 436
  secondary 436
intraparietal sulcus 83
invariance (speech), lack of 699
invention, deliberate, in cultural innovation 196, 197
isoglosses, sorting 768–9

Italic language family, statistics/countries/members 548

Jakobsen R
  structuralist theory of phonological acquisition 703
  typological similarities between languages 755
Japanese language
  relationships/reconstruction 765
  syllabaries in 801
Japanese macaques
  communication 171
  tool use 265, 266
joints, hand
  human 132–3
    structures stabilizing 131, 133
  orang-utan 129

Kalahari Bushmen rock art 357, 360
Kaluli child-rearing 455–6
Kanzi (chimp), language skills 650, 651–2, 673–6
Kaye K, on social interaction 437–8, 438, 439, 441
Kellogg R, drawings 503–4
key-shaped sign 326
Khoisan language family
  map 533, 534
  statistics/countries/members 548
Klaisies River Mouth, tools 275–6
knapping of stone 268–9
KNM-ER 406 (*A. boisei*) skull 16
KNM-ER 1470 (*H. habilis*) 74
  brain organization 91, 92, 93
  language 589
  skull 10–13
KNM-ER 3733 (*H. erectus*) skull 14–16
knowledge, communication of, stage in development termed 424
Koasati, sex differences 209
Koko (gorilla), language skills 656, 670–2
Koobi Fora, hand bones/locomotion 140–1
Kung 63, 64

labelling experiments, ape-language projects 653–4
laboratory tests with animals
  cognitive 631–2
  linguistic
    methodological problems 662–5
    precaution in interpreting results 662
labour, see work; work-chant
La Chapelle-aux-Saints, Neanderthal skull 22–3
ladder motifs 342
La Ferassie, Neanderthal skull 24–5
Lake Turkana, hand and locomotion 141
La Marche plaquettes 298, 334

Lamarckian schema of cultural change 196, 197, 198
Lana (chimp)
  language skills 645, 654–5, 673–6
  number skills 660, 664
language/linguistic communication (predominantly spoken language/verbal communication) 123–4, 208–14, 217–25, 230–1, 407–31, 469–82, 470–2, 489–97, 521–858
  action and, parallels between 379–82, 480
  analogy and, see analogy
  analytic 219, 791
  animals (in general), see animals
  brain and, see brain
  categorization and effects of 475
  culture and, see culture
  dead 756–7
  definition 554, 554–8, 572, 647–9, 753
  design features 553, 556–7, 648
  diachronic study of change 486
  evolution/origins/emergence (glottogonic thought) 123–4, 421–5, 477, 523–31, 667, 727–9, 747–858
    arbitrariness and 562
    gestural primacy hypothesis 580–91
    H. erectus and 422, 477, 478, 577, 863
    H. habilis/KNM-ER 1470 and 93, 101, 589, 862
    history of studies of 571–95, 645–6, 777–8
    influence of language use on 568
    reconstruction, see reconstruction
  function 557–8, 726–7
    early 687, 726
  genetic relationships in 525–7, 551, 756–8
  groupings 753
  learning/development/acquisition 270, 377–82, 384–5, 395, 401, 407–31, 469–82, 489–97, 523, 553, 565–6, 686–746
    etymology and 478–9
    historical ideas 577–8, 580–1, 688–9
    intellect/cognition and 472–5, 477, 479–80, 690–1, 696–7, 712, 728–9
    mental construction capacity in 420–1
    perceptual/social precursors 472–6
    production v. comprehension 649–50
    stages 414–15
    symbols and 489–96, 687–8, 687–746
    theories 689–98, 728–9
  map gallery of distribution/classification of extant forms 532–52

numbers in, *see* numbers
primate, *see* primate
primordial 762–9
proto-, *see* proto-language
public nature 472
relative complexity of possible languages 231
representation and 483–4, 714–15
signification and, *see* sign; signing/signification
society and, *see* society
structure in, *see* structuralist analysis; structure
synchronic structure 486
synthetic 219
systems in 521–858
  dissociable 377–82
  evolutionary perspective 523–31
  supporting (=scaffolding) 389–94
  syntactic, *see* syntax
thought and 470–2, 489–97, 714, 796–7, 808
  cross-cultural perspective on 217
  theories of relation between 470–2, 487, 714
tools and 270, 284, 424–5, 426, 584, 588
  15000 years ago 279
  85000 years ago 277
  1.0m years ago 270
written, *see* writing
see also speech; spoken language; verbal thought; vocabulary; words
language acquisition support structure (LASS) 473
laryngeal quality of lost vowel-colouring consonants (laryngeal hypothesis) 756, 763, 767–8
larynx and evolution of vocal apparatus 116–18, 120, 121, 122
Lascaux cave art 295, 296, 297, 301
  auroch (leaping) 313
  deer 321
  horse 322
  syntactic formulae/mythograms and 336
  well/shaft scene 296, 323
laterality of brain, *see* cerebral hemispheres
Latin, reconstruction/relationships to other languages 525, 528, 752
Latin America, cartography 832
Laugerie-Haute cave wall drawing 319
La Vache 334
learned emotions 206
learned expressions 205–6
learning 193–6, 601–11
  animal 601–11
  culture acquired via 192–3
  individual 193–5
  language, *see* language
  social 193–5

see also teaching
Le Mas d'Azil 313
Lemuroidea, taxonomy 168
Leroi-Gourhan A, on Palaeolithic art 296–9 *passim*, 300–2, 324–5, 336–9
Les Trois Frères 236
  anthropomorphism 328, 329
Le Tuc bison 294, 317
lexical category of nouns, epistemic pattern extending 789
lexical semantic system, acquisition 705–16
  theories 707–10
lexicons 711–12
  artificial, language cognition projects using 673–6
  development, early 705–6
  see also vocabulary; words
lies/deceptive behaviour 172–4
  chimpanzee 664, 665
  evolution 167
life-history patterns, Neanderthals vs. *Homo Sapiens Sapiens*
ligaments, hand, orang-utan 128
limbs in gestural communication 585, 586
  see also bipedality
linearity (speech), lack of 699
linguistic communication, *see* language
lioness, human with head of 327
lists, representations of, pigeons/monkeys 616–17
literacy 793, 804–12
  cognition and effects of 794, 804–6
  definition 806
  education campaigns 808, 809
  functions 794, 804–6
  restricted 811–12
  society and 806–12
literature
  1950–present day 841, 844
  12000BC–1950AD 844–54
locomotion/movement
  of fossil hominoids/human ancestors 135–42
    evidence for 142–3
  of infant synchronized with speech heard 440
locutionary period 725
logic, writing and development of 805
logographic scripts 800–1
  Chinese 796, 800–1, 801–2
  reading 803
  transition to syllabic from 801–2
Lorisoidea, taxonomy 168
Lucy (chimp), language skills 656
lunate sulcus 83, 85
  position 90–1
Luquet GH, on children's drawings 502, 505
Luria AR, on verbal regulation of behaviour 475–6

lying, *see* lies
macaques
  communication 171
  mirror-use 622–4
  tool use 265, 266
  see also rhesus monkey
macaronic sign 326
Magdalenian tools 282, 283
magic, hunting, art as 299–300, 312, 335
making of things, aim of art as 514
male
  in art 298–9, 304–5
    genitalia 298
    role depiction 304–5
    signs 331–3
  art by 304–5
  differences from female, *see* sex differences
mammals (non-primate), tool use 265
  see also specific genera/species
mammoths 339
  Laugerie-Haute 319
  Pech-Merle 295, 321, 322
mand 648, 668
mandalas, Kellogg's 504
manipulative facility/dexterity, hominid 135, 143
  source 143
  see also tools
manual gestures, *see* gestures
manual motor control/manual operation, language and, cerebral lateralization of 584
Maoris, tattooing 206–7
map (of brain) 78
map (of world)
  art (Palaeolithic) 316
    venus figurines 340
  drawing 826–33
  hominids (fossil) 43, 44
  language distribution and classification 532–52
mapping/translating from one representational system to another 714–15
marginal zone hypothesis (origins of agriculture) 247, 249
Maring-speaking people 159–61
marking
  expressive and ritualistic (cave art) 797
  property 797
  in spoken language, theory of 755
Marshack A, on Palaeolithic art 306–7, 341, 343
Marshall Island stick charts 829
marsupials, rock art images 351, 352
matching component/concept
  animals 612–13
    language experiments 656
    learning/cognition experiments 603, 607–9, 659

matching component/concept
    animals (*Contd.*)
        in memory experiments 597, 598
        human art 514, 515
material exchange, degree of 240
mathematics, evolution 818–24
Matsuzawa T, ape-language studies 659, 673–6
maturation hypothesis in cognitive/language development 382, 383–4, 454
Mayan writing 798
    hieroglyphs 796
maze experiments, rats 597–8
Mead GH, on language/symbolization/social life 487–8, 496
meander-like sign 326
meaning 487–8; *see also* semantics
    analogy of articulation and, *see* analogy
    analysis of language in terms of 804
    discovery by infant that its activities are being accorded 730
    minimal units of, *see* phenemes
    of utterance by ape 647–8, 653–7
    of words for things, *see* lexical semantics
    writing as carrier of 796
means and end (co-ordination) 493–4
    language development and 388
median evolutionary tree 62–3
Megaceros giganteus 318, 319
memes/memetic attributes 194
    natural and artificial selection of 198
memory 597–9
    representational 597
    spatial 597–8
men, *see* male; sex differences
mental capability, hierarchical, *see* hierarchical mental capability
mental representation (cognitive representation), in animals 598, 657
mental state attributed to others (theory of mind), apes 169, 624–30
mesencephalon 77
Mesoamerica, *see* Central America
Mesopotamia
    numerical sign systems 384–5
    writing 797
message, writing enabling originators of message to distance themselves from content of 806
metacarpals of fossils
    Miocene hominoids 137
    Pliocene/Pleistocene hominids 137
metaphor
    external 769
        Palaeolithic 772
    as form of analogy 769
    internal 769
metaphorical images arising from trance states 362, 364
Mezhwerich ladder motifs 342

Miao–Yao language family
    map 533, 539
    statistics/countries/members 549
middle-range theory 239–40
Miles HL, ape-language studies 670–2
Mimi figures 352, 354
mind (child) 497
    Darwin and 489–90
    unitary, language/cognition and the 691, 700
mind (primate) 175
    theory of (in apes) 169, 624–30
    upgraded, in language-trained apes 632
Miocene fossil hominoid hand and locomotion 135–7, 143
mirror-use in primates 622–4, 628, 664
misprojections in syntax 720
missiles, stone 268
Mississippian period, symbolism 870
mitochondrial DNA (human ancestry data from) 48–9, 60–6, 70
    out-of-Africa hypothesis and, *see* Africa
mnemonic devices 799
mnemonic(s), wishful 655
modernity 810
    complexity 872–3
modularity in cognitive/language development 377–82, 388, 523–4, 629, 690–1, 697–8
Moja (chimp), language skills 656, 670–2
Moko 206–7
molecular clock hypothesis 56–7
molecular genetics, human ancestry data via 48–9, 53–73
monkeys
    cognition
        auditory perception 600
        imitation 618, 630
        learning, *see* subheading below
        memory 597
        mental attribution capacities 627
        minor task transformation and 600
        mirror-use and its significance 622–4, 628
        representation of lists 616–17
        same/different concepts 613
        social 446
        symmetry and transitivity of conditional relations 615–16
    communication/vocalization 171–2, 576
    learning by 604
        discrimination 606
        reversal 607, 608
        relationships in 446
    taxonomy 34, 168, 169
    tool use 264, 265
    see also individual genera/species
monogenetic reconstruction of spoken language 747, 762–70
    wider implications 769–70
monotropy 443–4

morphemes 749, 782
mosaic, embryological 372–3
mosaic evolution 372
mother, *see* parent
mother-in-law language 210
motivation 457
    attachment behaviour and 443, 444, 445
motor control
    development 376, 414, 493–4
        first words and 414, 416
    manual, *see* manual motor control
    primates, problem-solving and 621
    see also sensorimotor abilities
Mousterian 243
    of Acheulean tradition 253, 259
    Charentian 243, 253, 257
    Denticulate 253, 256
    tool types 252–4, 255, 256, 259
    Typical 252–3, 255
mouth-gesture theory 583
movement, *see* locomotion
multiple-factor-and-organizational hypothesis (of formation of state) 250, 252
multi-regional hypothesis of human evolution 60–1
muscles, hand
    African apes 129
    orang-utan 128
music, evolution 824–6, 854
'mutation' (M) conditions and social evolution 197
mutilation 207
myelination 412
    neocortex 413
myth(s)
    creation, Aboriginal 352–4
    language-origin 572, 573
mythograms 301, 307–8, 336–7

Na-Dene language family
    map 533, 544
    statistics/countries/members 550
naming/identification
    in language acquisition 694–5, 705–16
        insight 711
    in language evolution 781–2
narrative 786–7, 790
    in art 799
        Australian rock art 352–4, 355, 799
nativism, general, theory of 697
natural characteristics, *see also* entries *under* innate
naturalism in art
    Arnhem Land rock art 358
    Palaeolithic 291, 295, 313, 358
natural resources, appropriation and transformation 190
    see also environment
natural selection (in social evolution) 197–9
naturefacts harnessed to intentional design 188

Neanderthals (*H. sapiens neanderthalensis*) 47–8
  disappearance 48
  distribution 44
  in evolutionary tree 68
  hand 141–2, 143
  locomotion 141–2
  skull/cranium 22–5
    base (basicrania) 122, 123
    volume/size 111, 421
  tool-use 143
Near East, subsistence-settlement patterns, developmental stages 247
negation in language 565
negotiation 183–4
Nelson K, functional core hypothesis 708–9
neoclimatic change hypothesis (origins of agriculture) 247, 248
neocortex
  and behavioural foundation of human skills found in animals 418
  cognitive maturation and 412–18
Neolithic, language evolution 772, 865
  see also Holocene
neonates/newborn babies
  born 'too early' (ideal gestation of 21 months) 383
  brain, growth 410
  hand/tongue movements 586
  mother's voice, attention to 587
  social abilities 435, 472
  see also infants
neo-Piagetian analyses of cognition 413–21
neoteny 373, 375, 395, 407–9, 477–8
  language evolution and 527
nervous system (primarily central nervous system)
  anatomy, comparative 86–7
  development 410–12
  language studies 699
    human and animal comparative studies 666
  variability and behaviour 79
  see also brain; palaeoneurology
neurobiology, *see* brain; nervous system
neurones, formation/development 411
New World, *see* Americas; Amerind language family
Niaux Ariège 321
niche 186
Niger-Kordofanian language family
  map 533, 535
  statistics/countries/members 548
Nilo-Saharan language family
  map 533, 536
  statistics/countries/members 548
Nim Chimpsky, language skills 645, 651, 670–2
  methodological problems 663
nine (number), origin and historical development 825

non-verbal/linguistic communication 205–8, 230
  intrasocietal variation in 213–14
  see also pre-linguistic communication *and specific forms of communication*
North American language family 546, 550
  reconstruction 758
  see also United States
Nostratic macrofamily 760
notational systems
  dance 826
  evolution alongside writing/literature/science/technology/society 844–55
  graphic, *see* graphic notational systems
'not noticing' in primates 173
nouns, epistemic pattern extending lexical category of 789
nuclear zone hypothesis (origins of agriculture) 247, 248
numbers/numerical systems and counting 384–5, 818–24
  human concepts/perception 819–24
    infant/children's 384, 820
  primate (ape)
    concepts/perception/skills 658–9, 854
    methodological problems 662, 664

oasis hypothesis (origins of agriculture) 247
object
  permanence (primates) 619–20
  sign and 483–9, 732
objectivity in discourse 787–9
observational learning in animals 596, 617, 618
  common-sense meaning, *see* imitation
occlusion in children's drawings 510
  see also hiding activity
OH5 *A. boisei* skull 6–8
OH24 *H. habilis* skull 10–13
Oldowan tools 266–7, 269
  bifacial tools compared to 268
Olduvai Gorge
  hand bones 139–40, 143
  locomotion 139–40
  tools 266–7
Old World monkeys, taxonomy 169
onomatopoeics in communication
  infants 732
  language origin models and 576, 583–4
ontogeny/development 369–519
  basicranial line 121
  brain, *see* brain
  cognition, *see* cognition
  intentionality in 629, 726–7
  language, *see* language
  phylogeny/evolution and 371–4, 377–94, 400–6, 868

behavioural, *see* behaviour
  role of ontogenesis in evolution and development 400–6
  representation in 400–5, 494, 700–1
  symbolization and/in 371–99, 483–500
  see also specific skills
operation(s), of cognitive systems 601
operational stage of development (Piagetian) 494
oral cultures, language 806
orang-utan (*Pongo pygmaeus*)
  cognition 596
    mirror-use/self-recognition 596
  cranial volume 113, 114
  evolutionary origin 60
  hand 128–9
    human and other ape hands compared with 132
  language skills 670–2
  social structures 170
    communication 172
  tool use 266
ordering solution in child language acquisition 717–18
ordination/ordinality, ape understanding 658
organization, social, *see* social systems
original a-dualism 492–3
ornaments, *see* adornments
ostension 776, 780–1, 790
  at a remove 785
out-of-Africa hypothesis, *see* Africa
overdetermination in dreams 491, 492
overgeneralization in grammar 720–1

pad-to-side grip 131, 133, 134
paediatric perspectives, *see* children
paedomorphosis 373
painting, *see* art
palaeoanthropological perspectives
  social/socio-cultural systems 158, 237–62
  upper respiratory tract anatomy 119–23
Palaeolithic (in general)
  art, *see* art
  language evolution 771–2
  see also Pleistocene
Palaeolithic, Lower (2m–125000 years ago)
  social life 241–2
  tools 271–5
Palaeolithic, Middle (125000–35000 years ago)
  social life/socio-cultural systems 864–8
    compared with upper Palaeolithic 864–8
  and interassemblage variation 242–3
  tool use 275–8

Palaeolithic, Upper (35000–12000 years ago)
  culture 243–4
  H. sapiens skulls 29
  linguistic communication 230–1
  social life/socio-cultural systems 243–4
    compared with middle Palaeolithic 864–8
  tools 278–81
palaeomorphic metaphor for behavioural development 401–3
palaeoneurology (endocast studies) 74, 85–6, 87–97, 108–16, 862
  brain size determinations, *see* brain
  linguistic capacity and 424
Pan, *see* chimpanzee
Panidae, learning abilities 602–3
paper 800
papyrus 800
paradigmatic relations 487
Paraguayans, Spanish and Guarani 213
parallel distributed processing model of cognitive organization 721
parameter-setting, Chomsky's theory of 454
parent (mother/other caregiver)
  food accumulation and child-rearing practices 216–17
  interactions with child 432–68
    attachment of child, *see* attachment
    cognitive development and 225–7, 385–6, 432–68
    cross-cultural comparisons 455–6
    early evolution of communication 863
    joint visual attention 474
    signs and 732
    vocal 441
  neonate paying attention to voice of 587
parietal lobe association cortex, relative increase 90–1
parrot, language cognition skills 677–9
parsimony and evolutionary trees 62
passive in language
  acquisition 722, 724
  sign language 567
Patterson FG, ape-language studies 670–2
Pech-Merle
  horse 343
  mammoth 295, 321, 322
Peirce CS, sign theory 484–5
pelvis of chimpanzee/*Australopithecus*/humans 41
Pepperberg IM, non-primate language cognition projects 677–9
perception
  auditory, *see* hearing
  categorical 600
  numerical, *see* numbers
perceptual characteristics of people, infants attracted to 440

perceptual precursors in thought and language development 472–6
Perigordian tools 280, 281
perlocutionary period 725
person (as category of self), constitution of 183–4
  see also individual
Petralona *H. sapiens* skull 18–19
petroglyphs, Aboriginal 356
phalanges
  African apes 129
  Miocene hominoids 136–7
  Pliocene/Pleistocene hominids 137
pharynx, supralaryngeal portion of 117, 118
phememes 747, 763–8, 769–70
  creation of two from one (=allophemic split) 769–70, 771
  palaeolithic period 771, 772, 865
  systems 764–7, 770
philological studies of language origins 577
phone(s), perception 699, 735
phonemes 562–4, 586–7, 590–1, 772, 782–4
  in sign language, components like 562–4, 587
  in speech 389, 562, 586–7, 773, 782–4
    children's ability to match those of adults 701–2
    consonantal, *see* consonantal phonemes
    in evolution 279, 590–1, 772, 773, 865
    the phonemic principle 590, 782, 865, 868, 870
phonetics in logographic script 800, 801
phonic writing 817
phonology in language, acquisition 686, 698–705, 735
  theories 702–5
phyloculturalism, Freudian 490–1
phylogeny, *see* evolution
Piagetian analysis (of cognition/intelligence etc.) 386, 401, 492–5, 497–8
  animal studies 617–21
  drawings of children 505–6
    and evolution of art 513, 515
  EEG data correlating with 425
  language and thought development, relationship between 470–1, 497–8
  social interactions and 436–7, 438–9, 447
  stone tool geometry 284
    15000 years ago 279
    300000 years ago 272–3
pictograms 814–16
  cuneiform script 814, 815
  narrative forms 799
  Sumerian 814, 816
pidgin language, definition 566

pigeons
  concepts in 613
  representation of lists in 616–17
  sequential skills 668
pinyin alphabet 804
play, symbolic 174–5
  apes 658
  see also games
Pleistocene hominid
  Early/Lower, hand 137–41, 143
  Late/Upper
    hand 141–2
    social life 244–5
  locomotion 137–42
  Middle, hand 137, 141
  see also Palaeolithic
Pliocene hominids
  hand 137–41, 143
  locomotion 137–41
pointing 588
  with index finger, *see* ostension
  infant 450, 588, 730
polar relational terms, learning 708
politics
  of literacy 809
  of its restriction 812
  of settlement patterns 215
polymorphism, molecular 59, 66–7
  divergence dates and 59, 60
Polynesian languages, reconstruction 757
Ponapean communication 217–19
Pongidae/pongids
  hand
    A. afarensis (Hadar) hand compared with 128, 138
    human hand compared with 142
  learning 602–3, 618
  mirror-use/self-recognition 622
  symbolic abilities 670–2
pooh-pooh theory of language origin 576, 583
population(s)
  genetic difference between two, measurement 67
  human
    formal boundaries between, degree of 240–1
    phylogenetic relationships of, estimating 67
    polymorphisms in 59
population-pressure-and-conflict hypotheses (formation of state) 250, 251
population-pressure hypothesis (origins of agriculture) 247, 249
possessive in sign language 567
post-structuralist analysis
  linguistics and semiotics 489
  Palaeolithic art 306–7
post-syntactic stage in language development 687, 722–3
posture, bipedal, *see* bipedality
pounding 145

Index

pragmatics in language, development 724, 736–7
precision grip 133–4
Pre-Columbian America
  Mississippian period, symbolism 870
  writing 799
pre-communication stage 730
predation and hunting, distinction 189
predication 733–4
prefixation 768
prefrontal cortex, amount 82
prehistory
  anthropological studies, see palaeoanthropological perspectives
  of grammar, theoretical stages 777–92
  history and, distinction 159
pre-intellectual speech 497
pre-linguistic communication 453, 686, 699
pre-linguistic thought 497
Premack D *et al.*, ape-language studies 651, 656, 661, 664, 673–6
pre-operational stage of development (Piagetian) 494
  ape rudiments of 620–1
  art-history and, parallels between 513
prestige forms in speech, sex differences 210
presyntactic stage in language 686, 718–20
  methodological problems in investigation of 719
primates
  behavioural foundation of human skills found in 418–21
  brain
    maturation patterns, comparative 410–12
    size (cranial capacities) 45
  classification 33, 34–6
  cognition
    auditory perception 600
    historical studies 597
    imitation 618, 630–1
    insight 621–2
    language, see subheading below
    learning abilities 602–3, 604–11 *passim*
    memory 598, 599
    minor task transformation and 600
    mirror-use 622–4, 628, 664
    numerical skills, see numbers
    object permanence 619–20
    provisos in interpreting study results 646–7
    representational abilities 611, 620–1
    same/different judgments 610–15 *passim*
    self-recognition 622–4, 628, 664
    sensorimotor intelligence 619–20
    theory of mind 169, 624–30

tool use 264, 265, 266, 419–20, 621–2
communication 167–77, 387, 644–85
  gestural, see gestures
  linguistic/vocal, see subheading below
evolutionary trees 53–73
hands 127–30
  comparative morphology, as evolutionary evidence 142–3
  shared morphological characteristics 127–8
intelligence, see intelligence
language (and vocal communication) 171–2, 386, 395, 420, 525, 576, 579, 583, 644–85
  changing interpretations of study results 665–6
  provisos in interpreting study results 646–7
  theoretical issues 647–50
  upgraded mind in language-trained apes 632
social structure 168–70
  relationships 446–7
see also apes; monkeys *and genera/species*
primordial language 762–9
principles-and-parameters theory 697
printing and publishing (of books etc.) 810–11, 834–43
  statistics on book publication 834–43
    subject area 837–8
    world per capita comparisons 836, 840–1
    world production in class 8 (literature) 836, 841–3
    world production by titles 834–5, 838–40
    world translation comparisons 836, 841
privacy in primates 170–2
probability learning 609–10
  primates 603
probing 144
problem-solving 393–4
  development of skill 394
  and the transmission of cognitive skills 226–7
  drawing (of children) as 506–7
  English and Chinese language and 220–2, 390
  primates 621–2, 668
Proconsul spp. 33
  hand bones and locomotion 135–6
production
  evolving modes of, Middle/Upper Palaeolithic compared 866–7
  of language (in learning), comprehension *v.* 649–50
  social relations of 190–1
  see also construction
progenesis 373

pronoun system in American Sign Language, acquisition 713
pronunciation, differences in
  sexes 210
  social groups 211–12
property marking 797
Propliopithecus spp. 33
proportions, judging, ape abilities 659, 661
prosimians, learning by 602–3
  discrimination 605
  observational 618
  reversal 608
  time-based 610
prosodic theory of phonological acquisition 703
proto-declarative acts 725
  Semitic 813–14
proto-human hominids, set of characteristics 242
proto-imperative acts 725
proto-language (dubiously language) 525, 584, 751
  Amerind 758, 760, 783
  Indo-European (PIE) 752, 753, 763, 765, 787
  v. indisputable language 523
  written 796
  see also primordial language
Proto-Nostratic 760
prototype theories of lexical acquisition 709–10
Psittacus erithacus, language cognition skills 677–9
psychoanalysis, Freud on 491
psycholinguistics 688
  sign languages and 559–60
psychology, Gestalt, applied to child art 505
public language 223–4
  characteristics 223–5
public nature of language 472
publishing, see printing and publishing
pygmy chimpanzee (bonobo; *P. paniscus*)
  brain size/volume 113, 114
  language skills 395, 646, 650, 651–2
  study conditions 673–6
  neoteny 395
  social structures 170

quantitative abilities, apes 658–61, 664–5
  see also numbers
radicals in logographic script 800
Rainbow Snake, images of 359, 360
ramamorphs, hand bones and locomotion 135, 136
rat, maze experiments 597–8
raw materials for art
  Bushmen of Southern Africa 357
  Palaeolithic 294, 296
reading of writing systems 803

realism, intellectual
  Arnhem Land rock art 358
  children's drawings 510–12
  v. symbolism 511
  Palaeolithic art 295
reasoning
  in apes 614, 661, 662
    natural 662
    representational 616
  in humans 192
    development of 471, 495–7
recapitulation 373, 374, 383, 407–9, 476–9
  with terminal addition 477
recognition in primates
  cross-modal 603, 610
  of self 622–4, 628, 664
reconstruction of language 525–30, 747–75, 782–4
  methods 751–6, 777
    comparative 751–2, 764–7
    guidelines 761–2
    internal 755–6, 764–7
    monogenetic, *see* monogenetic reconstruction
recursion rule in language 668
reference and referential communication 714–15, 733–4
  apes 387
    procedures establishing 653–4
  grammatical prehistory and 782
  infant/child 473–4, 707–8, 714–15, 733–4
  theories of 485–6, 495
reflection, primate ability to recognize 622–4, 628, 664
regulative social systems/relations 161, 182, 184, 185
  tool-making and 188–9
reindeer
  antler, spear-points made from 317
  art/imagery 318, 319
reinforced verbal behaviour, language as 648
relation(s), conditional, *see* conditional relations
relational cognitive style 229
relational thought in development 414, 416–17
relationships (between animals)
  animal–human 182–3
    interactions and, distinction between 183
    interspecific v. intraspecific 189
  human, development 442–7
    social interaction and 442–7
    triadic relationship 450
  primates 446–7
relationships (between any subjects/objects), words expressing 414, 415
relationships (between signs) 487
religion accounts of language origin 572, 573
Renaissance times

art of, child development and, parallels between 513
language origins and the ideas of 572
replication, social learning as process of 194
representation (and representational abilities) 483–4, 655–7
  in animals 610–11, 616, 620–1, 630, 655–7
  cognitive/mental representation, *see* mental representation
  internal 655–7
  lists of items 616–17
  in humans
    in development 400–5, 494, 700–1, 714–15
    language and 483–4, 714–15
    numerical 824
    signification and 483–4
    visual, *see* visual representation
representational memory 597
representational reasoning in apes 616
respiratory tract, upper (vocal tract), evolution 116–25, 477
response-based learning in animals 607–9
  primates 602, 609
retreat mechanisms in language acquisition 720–1
reversal learning, animals 606–7, 608, 609
reworking, the process 776, 779, 781, 785, 789, 871
rhesus monkeys
  communication 171–2
  learning by, discrimination 606
right-handedness in hominids 135
  language and 584, 588
ritual(s), behaviour, children's exposure to 450
ritualistic markings 797
ritual regulation in Maring-speaking people 160
rock art, contemporary hunter–gatherer, *see* art, contemporary
Rolandic (central) sulcus 83, 85
role reversal in chimpanzees, transfer on 627
roof-like sign 326
rule-based acquisition of phonology 686, 703–5
Rumbaugh DM, ape-language studies 645, 653, 664, 673–6

Sadie (chimp), transitive-inference-type problem 661
same/different judgments by animals 611–15, 621, 633
San (Bushmen) rock art, Southern Africa 357–63, 364–6, 367
Sanskrit, reconstruction/affinity with other languages 525, 752

Sapir–Whorf hypothesis on language and thought 475, 714
Sarah (chimp)
  classifying relationships 614
  language skills 651, 657, 673–6
  transitive-inference-type problem 661
Saussure, F de, linguistic science 486–7
Savage-Rumbaugh ES, ape-language studies 645, 650–8 *passim*, 665, 673–6
scaffolding in cognitive development 226, 389–94
scarification 207
scenes, depiction in vocal language 786
Schusterman RJ, non-primate language cognition projects 677–9
science, writing/society/technology evolving with 844–55
scripts (writing-systems) 793, 800–1
  advantages and disadvantages of different 802–4
  development/evolution 813–14
    from one type to another 801–2
  efficiency 803–4
  see also specific scripts
sculpture, Palaeolithic 294
  see also statuettes
seal/human hybrid figure 327
sea lion, language cognition skills 651, 677–9
search behaviour, chimpanzee 598
  self-recognition and 624, 626
second person verbs, topic–comments and 785
segments, alphabetic-type 698
selection (in social evolution)
  conditions of 197
  natural and artificial 197–9
self
  culture and communication and 222–3
  recognition in primates 622–4, 628, 664
semantic bootstrapping hypothesis 719
semantic feature hypothesis 707–8, 709
semantics/semantic communication
  acquisition of skill 705–16
  theories 707–10
  grammar and, modular division between 382
  vocalization of monkeys and 171–2
semiosis, social life as 487–9
semiotician, Freud as 491
semiotics/semiotic systems 223, 484
  art and 301
  tools and (last 15000 years) 282–4, 284
  see also sign; signing/signification
Semitic languages
  proto- 813–14
  relationships to others 756
  writing 817
sensorimotor abilities

humans, development 414, 416, 493–4
  primates 619–20
sentences
  construction 414, 415, 420, 421, 691, 722–3
  interrogative 727
sequencing
  apes 658–9
  children's drawings 506–7
  narrative 787
  pigeons 668
  writing 796
serial position effects in memory experiments 598–9
seriation of types 755
settlement patterns, differentiation of socio-cultural elements related to 215
  see also subsistence-settlement patterns
sex differences
  brain 98–100, 101–2
  childhood socialization pressures 216
  labour (e.g. hunter–gatherers) 304–5, 771, 864
  speech 204, 209–11
  see also female; male
sexual organs in art 298
shamanism and rock art 357–62, 364–6
Shared Attention Mechanism (SAS) 451, 458
Sheba (chimp) 673–6
  numerical skills 660, 660–1, 664–5
  study conditions 673–6
Sherman (chimp), language studies 654, 655, 656, 658, 673–6
siamang, cranial volume 113, 114
sign(s) (and symbols) 483–9
  making, see signing/signification
  numerical, Mesopotamia 384–5
  object and 483–9, 732
  in Palaeolithic art 296, 297–8, 302, 326
    chronology 315
    male/female 331–3
    personal-related 732
    social interactions and 732
    in child development 460
  theories of 484–7
  tools and 496
  see also semiotics
signing/signification
  coupling between signifier and signified 487
  historical ideas 573–5
  representation and language and 483–4
sign language (deaf persons) 553–70, 648–9
  apes 645, 648–9, 670–2
    specific studies 670–2
    symbolism and 658
  definition 554

design features 556–7
evolution 568
iconicity v. arbitrariness 559–61
learning by children 565–6, 713
structure in 562–5
various types 555
sign language (hearing persons) 585–6
similarity testing in animals 612
  see also same/different judgments
Sino-Tibetan language family 533
Sivapithecus spp. 57–8, 70
  hand bones and locomotion 136
skills in development
  combinatorial, see combinatorial skills
  specific 479
skin, depigmented volar, gestural communication and 586
skull/cranium
  development 375
  fossil hominid 3–30
    Australopithecus 4–8, 38
    base (basicranium), upper respiratory tract anatomy and 119–23, 124
    capacity, see brain
    H. erectus, see *Homo erectus*
    H. habilis, capacity 42
    H. sapiens, see *Homo sapiens*
    photogallery 3–30
    see also palaeoneurology
smiling infant 439–40
Snake, Rainbow, images of 359, 360
social behaviour of children 434–57
  origins
    developmental theory and 434–9
    neonates 435, 472
    thought and language development and importance of 472–6
social classes/groups/strata
  agriculture and, origins of 249–50
  ancient and prehistoric 241
  the state and 246
  language style differences 211–12
social context, individuals speech differences depending on 212
social demography 240
social facilitation, animals 617, 618
sociality
  consciousness and 184–5
  ecological approach to 188–9
  see also a-sociality
socialization
  compliance v. assertion 216–17
  infancy as a crucial role in 400, 404–5
  practices 228
    social structure and, links between 230
social learning 193–5
social systems/socio-cultural systems 155–368, 385–6
  art and 303–6
  attribution problem and 385–6

children and, see children
comparative perspectives 167–235
evolution/origins 158, 239–62, 861–76
infant–caregiver 385–6, 863, 868
interactive, see interaction
language and, see society
organization 215–16
  see also social class
palaeoanthropological perspectives 158, 237–62
primates, see primates
semiosis and 487–9
social tightness/conformity 228
society 179–85
  animal, see animals
  complex
    body adornment in 207
    development 247–55
  complexity 204, 861–76
    Early Holocene 868–70
    hunter–gathers 241, 244, 245, 869
    late Pleistocene 244, 868
    modern times 872–3
    writing/science/technology evolving alongside 844–55
  culture and
    cognitive style and 227–8
    distinction 181–2
  differentiation within, see differentiation
  language and 208–14, 217–22, 455–6
    intrasocietal variations in language 213–14
    thought and, cross-cultural perspective 217–22
    written (literacy) 806–12
  simple, body adornment in 206–7
  writing and its function in early 799–800
socioecology 188
  art and 303–4
sociolinguistics, origins and perspectives 208–9
solicitation 779, 780
Solutrean tools 282, 283
sound
  cuneiform symbols coding 818
  language analysed in terms of 804
  language origins based on 583
  minimal units of, see phenemes
  see also phonology
sound-change, language systems in 752
sound symbolism
  early evolution 771, 863, 864
  historical ideas 577–8, 580–1
South Africa/Southern Africa
  Bushmen rock art 357–63, 364–6, 367
  hominid tool use 85000 years ago 275–8
South American language family 547, 550
  tribal map 758, 759
Spain, cartography 830, 832

Spanish and Guarani in Paraguayans 213
spatial skill
  children's drawings as 507–8
  memory 597–8
speaking, *see* language; speech; spoken language
spearing
  in contemporary rock art 354, 359
  in Palaeolithic art, wound representation 330, 335
spear-points 317
  engraved 316
species
  behavioural differences between 79–80
  behavioural differences within 79
  relationships between 189
  relationships within 189
specificity hypothesis in cognitive and linguistic development 479
speech 209–11, 496–7
  acoustic cues underlying 699
  of children 496–7
    egocentric 471, 472, 496, 497
    inner 497
    meaningful, emergence 475
    pre-intellectual 497
  to children/infants 452–3
    ability to work out meaning from situations and context 453
    infant synchronizing movements with speech heard 440
    modification in various cultures 456
    neonatal responses 473
    simplified 452–3
  as 'deficit' 805–6
  perception, *see* hearing
  phonemic, *see* phonemes
  production, evolving ability 124
  sex differences 204, 209–11
  vocal tract evolution required for 116–25, 477
  see also language; vocal tract; words
speech acts 724–7, 814
  acquisition 724–7, 726, 737
  primitive 724–5
speech registers 212–13
spirants 763
  values, velar/laryngeal 764, 766–7
splenial portion of corpus callosum, sexual dimorphism 99
spoken/vocal language 553–70
  evolution 123–4, 589–91
  reconstruction 747–92
  iconicity *v.* arbitrariness 558–9
  structure in 562–5
  written and, differences 805–6
  see also language
stag 318, 319, 320, 321
state (the)
  origins 250–1, 871
  social stratification and 246
statuettes

female 303–4
part animal/part human 327
see also sculpture
Steinheim skull 20, 47
Sterkfontein *A. africanus*
  hands and locomotion 139
  skull 38–40
stick
  clubbing 145
  digging and probing 144
stimulus–response (S–R) framework 597
stimulus enhancement, animals 617, 618
stone ages, middle/old, *see* Neolithic; Palaeolithic
stone tools 265–81
  language and 424
  see also tools
stone weapons 268, 273
stop aspiration 767
stop consonants 768
  English 750
storytelling, *see* narrative
Strepsirhini, classification 168–9
structuralist analysis
  language development 401, 487, 703
  Palaeolithic art 300–2, 303, 311
  see also post-structuralist analysis
structuralist linguistics 209
structure (organization) in language 480, 557, 562–5, 686–746
  action and, parallels between 379–82, 480
  ape language and 648
  grammatical, *see* grammar
  influence of language use on 568
  phonological 562–4
  see also structuralist analysis
subitizing 820
  primates 658, 659
subsistence-settlement patterns, developmental stages in Near East 247
substrate of cognitive systems 601
suffixation 768
sulci of brain 83
Sultan (chimp), language studies 663
Sumerian writing 814, 816, 817
summation by chimpanzee 660
  see also numbers
superorganic, notion of 180–1
superpositioning of images in Southern Africa Bushmen rock art 362, 366
supralaryngeal portion of pharynx 117, 118
surface convolutions of brain, *see* convolutions
suspensory behaviour, ancestral 142–3
swallowing, larynx and 118
Swartkrans, hand bones and locomotion 140
syllabic scripts 800, 801
  reading 803

  transition from, to alphabetic 802
  transition from logographic to 801–2
syllables
  concatenation in Neolithic 772
  nonsense 698
  stress moved from first to second 768
symbol, definition/meaning vii (preface)
  see also signs and symbols
symbolic sign 485
symbolization and development 371–99, 483–500
symmetry of conditional relations 615–16, 649
Symphalangus syndactylus, cranial volume 113, 114
synchronic structure of language 486
syntactic formulae/mythogram in Palaeolithic art 301, 307–8, 336–7
syntactic stage in language development 686, 720–2
syntactic topic–comment 785–6, 790
syntagmatic relations 487, 495
syntax in language
  acquisition in children 686–7, 716–24
  evolution of systems 389, 423
synthetic language 219
systems theories 161

tact 648, 668
talapoins, reversal learning 607
tallying 797
Tasmanian language 590–1
tattooing 206–7
taxonomy/classification 49–50
  hominids 49–50
  primates 33, 34–6
teaching/education 195–6
  literacy 808, 809
  by parent of children, and transmission of cognitive skills 226–7
technology
  evolution of
    in Palaeolithic evolution 865, 866–7
    writing evolving alongside 844–55
  writing as 806–7
  see also innovations
tectiform sign 326
teeth/dentition
  Australopithecus 37, 38, 39, 40
  H. habilis 42
temperate climates/regions
  H. erectus in 46, 47
  ice ages 47
temporal events, *see* time
terminal addition, recapitulation with 477
Terrace HS, ape-language studies 670–2
thematization/thematic categories 777, 782, 790, 865
  utterances in terms of 718–20
theory of mind, apes 169, 624–30

therianthropic images, trance-related 361, 362, 364–5
thinking, *see* thought
third dimension represented in children's drawings 508–10, 511
third person verbs, topic–comments and 785
thought/thinking
　abstract, *see* abstract thought
　chimpanzee 664
　development (children) 414–15, 416–17, 469–82, 497
　　perceptual/intellectual/social precursors 472–6
　　symbols and 489–96
　dimensional, *see* dimensions
　language and, *see* language
　relational 414, 416–17
　social structure and, in a cross-cultural perspective 217
three-dimensional representation in children's drawings 508–10, 511
three-jaw-chuck thumb/finger grip 131–3, 146
throwing
　bipedalism and 134, 135, 144–5
　maturational component 384
thumb
　African ape 129
　orang-utan 129
thumb/finger grip, three-jaw-chuck 131–3, 146
time
　learning tasks based on (in animals) 607–9
　　primates 602, 609
　left cerebral hemisphere processing capacity relating to 666
　marking in language 564–5, 567
toes, ontogeny 375, 376
tokens 797–9
tools (making and use) 131–5, 144–5, 145–6, 188–9
　canonical rules governing use 403–4
　human 131–5
　　bipedalism contributing to 134–5
　　brain and 135
　　intentional communication and 478
　　language and, *see* language
　human ancestors (hominids) 143, 251–9, 263–87, 264–81
　　15000 years ago 278–81
　　85000 years ago 275–8
　　300000 years ago 271–5
　　1.0m years ago 268–71
　　1.8m years ago 264–8
　　Australopithecus 42, 770
　　bilateral symmetry 269–70
　　H. habilis 42
　　mental development and 477
　　role in evolution of hand and bipedality 144–5
　non-human, *see* animals
　signs and 496
　social aspects 188–9
tools (teaching name of), apes 654
topic–comment 777, 784–6
　asyntactic 785, 790
　case-relations and 785–6
　syntactic 785–6, 790
Torres Strait Islanders, number skills 820, 821, 822
total interactionism in language acquisition 382
totems 797
trance-related imagery
　Bushmen of Southern Africa 357–62, 364–6
　Palaeolithic 305–6
transfer
　of learning in primates, cross-modal 603
　on role reversal, chimpanzees 627
transference in dreams 492
transformation of environment/natural resources 190
transitivity 661
　of conditional relations 615–16
translation of books 836, 841
translucency in sign language 560
transparency
　in children's drawings 506–7
　in sign language 560
Trevarthen C, on infant social behaviour 434–6, 438, 439
triadic relationships 450
Trukese communication 217–19
Tupaiidae, learning abilities 602–3
turn-taking, vocal, infant–mother 441
Tursiops trunctatus, *see* dolphins
typological differentiation stage in development 720
typological similarities between languages 755

Ugaritic alphabet 817
unconscious, ursprache and the 490–2
undetermination 719
UNESCO, literacy education 808, 809
uniforms 208
unitary mind, language/cognition and the 691, 700
Urak (Mesopotamia), numerical sign systems 384–5
Uralic–Yukaghir language family
　family tree 757
　map 533, 537
　statistics/countries/members 549
urbanism, origins 250–1
ursprache and the unconscious 490–2
utterance, *see* vocalization

Vai literacy 804–5, 810
variation
　cultural evolution and 196, 197
　intrasocietal, in language/non-verbal communication/body adornment systems 213–14

velar spirant values 764, 766–7
venus figurines 340
verb(s)
　case-marking 564
　first/second/third person, topic–comments and 785
verbal behaviour, reinforced, language as 648
verbal coding of information 601
verbal communication, *see* language
verbal regulation of behaviour 475–6
verbal thought, language and thought uniting with 471–2
viewpoint in art
　child portraying own 510, 511–12
　in Renaissance art 513
Viki (ape), symbolic play 658
visions, trance-related, *see* trance-related imagery
visual attention of mother–child, joint 474
visual images, sex differences in responses to 99
visual representation
　emergence 795
　writing in relation to other forms of 794
　see also graphic notational systems
visual striate cortex, primary, relative amount 90, 90–1
vocabulary (child's), expansion
　age-related 690, 705–6, 733
　co-ordination of attention and 451
　see also lexicon; words
vocal interactions, infant–mother 441
vocalization/vocal communication/utterance
　infant 435, 584, 730–2
　　counted as words 711
　primate 171–2, 576
　　meaning 647–8, 653–7
　see also language; spoken language
vocal tract, evolution 116–25, 477
voicing, consonant 767
volar depigmentation and gestural communication 586
vowel(s)
　chart 750
　root, Indo-European languages 755
vowel-colouring consonants, lost, laryngeal quality of 756, 763, 767–8
Vygotsky LS (on cognitive/behavioural development etc.) 401–3, 471–2, 495–7, 497–8
　on social behaviour/interactions 437, 438–9, 496
　thought and language development and their relationship 471–2, 495–7, 497–8
　tests of theory about 475–6

walking, *see* bipedality
Wandjina figures 352, 353

Washoe (chimp), language skills 651, 655–6, 670–2
Watson's rule and mosaic evolution 372
wave theory with languages 752–3
weapon, stone as 268, 273
  see also spear; spear-point
Wernicke's area 93
West Lake Turkana, hand and locomotion 141
whistle language, dolphin 649
Whorf–Sapir hypothesis on language and thought 475, 714
wigs 207–8
Williams syndrome 377
wishful mnemonics 655
women, *see* female; sex differences
words (and their acquisition) 712–16
  in ape language
    meaning 647–8
    novel combinations 656
  closed-class 705, 735
  combining (in infants) 414, 415, 420, 421, 691, 718–20, 734
  concepts and 713–16
  co-ordination of attention and 451
  errors in use 706–7, 720
  etymology of language and order of acquisition of 478–9
  evolution of language systems and 526, 527
  first and single 414, 415, 416, 691, 730
    transition from babbling to 699, 701
  gestures and 712–13, 729–34
  grammar development and 790
  learning about meaning of words for things, *see* lexical semantics
  open-class 705, 735
  as symbols 711–12
work, sex/gender differences 304–5, 771, 864
  see also employment
work-chant theory of language origin 576, 584
writing (and written language) 793–858
  definition 793–4, 795–800
  function 799
  materials/techniques 799–800
  origin/development/evolution 793, 794, 797–804, 813–33, 844–55
  restricted access 811–12
  social influences 806–12
  speech and, differences between 805–6
writing-systems 796–7
  definitions 796–7
  principles 804
  types, *see* scripts
Wunda (chimpanzee), language skills 658

yam figure 360
yo-he-yo theory of language origin 576, 583
Yokuts consonantal phonemes 750

Zalophus californicus (Californian sea lion), language cognition skills 651, 677–9
zone of proximal development (Vygotsky's) 473